PHYSICS FOR SCIENTISTS AND ENGINEERS

Prefixes for Powers of 10*

Multiple	Prefix	Abbreviation
10^{24}	yotta	Y
10^{21}	zetta	Z
10^{18}	exa	E
10^{15}	peta	P
10^{12}	tera	T
10^{9}	giga	G
10^{6}	mega	M
10^{3}	kilo	k
10^{2}	hecto	h
10^{1}	deka	da
10^{-1}	deci	d
10^{-2}	centi	c
10^{-3}	milli	m
10^{-6}	micro	μ
10^{-9}	nano	n
10^{-12}	pico	p
10^{-15}	femto	f
10^{-18}	atto	a
10^{-21}	zepto	z
10^{-24}	yocto	y

*Commonly used prefixes are in bold. All prefixes are pronounced with the accent on the first syllable.

Terrestrial and Astronomical Data*

Acceleration of gravity at Earth's surface	g	$9.81\ \text{m/s}^2 = 32.2\ \text{ft/s}^2$
Radius of Earth R_E	R_E	6371 km = 3959 mi
Mass of Earth	M_E	5.97×10^{24} kg
Mass of the Sun		1.99×10^{30} kg
Mass of the moon		7.35×10^{22} kg
Escape speed at Earth's surface		11.2 km/s = 6.95 mi/s
Standard temperature and pressure (STP)		0°C = 273.15 K 1 atm = 101.3 kPa
Earth–moon distance†		3.84×10^{8} m = 2.39×10^{5} mi
Earth–Sun distance (mean)†		1.50×10^{11} m = 9.30×10^{7} mi
Speed of sound in dry air (at STP)		331 m/s
Speed of sound in dry air (20°C, 1 atm)		343 m/s
Density of dry air (STP)		$1.29\ \text{kg/m}^3$
Density of dry air (20°C, 1 atm)		$1.20\ \text{kg/m}^3$
Density of water (4°C, 1 atm)		$1000\ \text{kg/m}^3$
Heat of fusion of water (0°C, 1 atm)	L_f	333.5 kJ/kg
Heat of vaporization of water (100°C, 1 atm)	L_v	2.257 MJ/kg

*Additional data on the solar system can be found in Appendix B and at http://nssdc.gsfc.nasa.gov/planetary/planetfact.html.

†Center to center.

The Greek Alphabet

Alpha	A	α	Nu	N	ν
Beta	B	β	Xi	Ξ	ξ
Gamma	Γ	γ	Omicron	O	o
Delta	Δ	δ	Pi	Π	π
Epsilon	E	ϵ, ε	Rho	P	ρ
Zeta	Z	ζ	Sigma	Σ	σ
Eta	H	η	Tau	T	τ
Theta	Θ	θ	Upsilon	Υ	υ
Iota	I	ι	Phi	Φ	ϕ
Kappa	K	κ	Chi	X	χ
Lambda	Λ	λ	Psi	Ψ	ψ
Mu	M	μ	Omega	Ω	ω

Mathematical Symbols

$=$	is equal to
$\equiv$	is defined by
$\neq$	is not equal to
$\approx$	is approximately equal to
$\sim$	is of the order of
$\propto$	is proportional to
$>$	is greater than
$\geq$	is greater than or equal to
$\gg$	is much greater than
$<$	is less than
$\leq$	is less than or equal to
$\ll$	is much less than
Δx	change in x
dx	differential change in x
$\lvert x \rvert$	absolute value of x
$\lvert \vec{v} \rvert$	magnitude of $\vec{v}$
$n!$	$n(n-1)(n-2)\ldots 1$
Σ	sum
lim	limit
$\Delta t \to 0$	Δt approaches zero
$\dfrac{dx}{dt}$	derivative of x with respect to t
$\dfrac{\partial x}{\partial t}$	partial derivative of x with respect to t
$\int_{x_1}^{x_2} f(x)dx$	definite integral $= F(x)\Big\vert_{x_1}^{x_2} = F(x_2) - F(x_1)$

Abbreviations for Units

A	ampere	H	henry	nm	nanometer (10^{-9} m)
Å	angstrom (10^{-10} m)	h	hour	pt	pint
atm	atmosphere	Hz	hertz	qt	quart
Btu	British thermal unit	in	inch	rev	revolution
Bq	becquerel	J	joule	R	roentgen
C	coulomb	K	kelvin	Sv	seivert
°C	degree Celsius	kg	kilogram	s	second
cal	calorie	km	kilometer	T	tesla
Ci	curie	keV	kilo-electron volt	u	unified mass unit
cm	centimeter	lb	pound	V	volt
dyn	dyne	L	liter	W	watt
eV	electron volt	m	meter	Wb	weber
°F	degree Fahrenheit	MeV	mega-electron volt	y	year
fm	femtometer, fermi (10^{-15} m)	Mm	megameter (10^6 m)	yd	yard
ft	foot	mi	mile	μm	micrometer (10^{-6} m)
Gm	gigameter (10^9 m)	min	minute	μs	microsecond
G	gauss	mm	millimeter	μC	microcoulomb
Gy	gray	ms	millisecond	Ω	ohm
g	gram	N	newton		

Some Conversion Factors

Length

1 m = 39.37 in = 3.281 ft = 1.094 yd

1 m = 10^{15} fm = 10^{10} Å = 10^9 nm

1 km = 0.6214 mi

1 mi = 5280 ft = 1.609 km

1 lightyear = 1 $c \cdot$ y = 9.461×10^{15} m

1 in = 2.540 cm

Volume

1 L = 10^3 cm^3 = 10^{-3} m^3 = 1.057 qt

Time

1 h = 3600 s = 3.6 ks

1 y = 365.24 d = 3.156×10^7 s

Speed

1 km/h = 0.278 m/s = 0.6214 mi/h

1 ft/s = 0.3048 m/s = 0.6818 mi/h

Angle–angular speed

1 rev = 2π rad = 360°

1 rad = 57.30°

1 rev/min = 0.1047 rad/s

Force–pressure

1 N = 10^5 dyn = 0.2248 lb

1 lb = 4.448 N

1 atm = 101.3 kPa = 1.013 bar = 76.00 cmHg = 14.70 lb/in^2

Mass

1 u = $[(10^{-3}\ \text{mol}^{-1})/N_A]$ kg = 1.661×10^{-27} kg

1 tonne = 10^3 kg = 1 Mg

1 slug = 14.59 kg

1 kg weighs about 2.205 lb

Energy–power

1 J = 10^7 erg = 0.7376 ft · lb = 9.869×10^{-3} L · atm

1 kW · h = 3.6 MJ

1 cal = 4.184 J = 4.129×10^{-2} L · atm

1 L · atm = 101.325 J = 24.22 cal

1 eV = 1.602×10^{-19} J

1 Btu = 778 ft · lb = 252 cal = 1054 J

1 horsepower = 550 ft · lb/s = 746 W

Thermal conductivity

1 W/(m · K) = 6.938 Btu · in/(h · ft^2 · °F)

Magnetic field

1 T = 10^4 G

Viscosity

1 Pa · s = 10 poise

SIXTH EDITION

PHYSICS FOR SCIENTISTS AND ENGINEERS

WITH MODERN PHYSICS

Paul A. Tipler
Gene Mosca

macmillan learning
Austin • Boston • New York • Plymouth

PT: For Claudia

GM: For Vivian

Senior Vice President:	Daryl Fox
Program Director, Physical Sciences:	Andrew Dunaway
Program Manager:	Lori Stover
Marketing Manager:	Nancy Bradshaw
Senior Developmental Editor:	Kharissia Pettus
Media Editor:	Jeanette Picerno
Editorial Assistants:	Janie Chan, Kathryn Treadway
Photo Editor:	Ted Szczepanski
Photo Researcher:	Dena Digilio Betz
Cover Designer:	Blake Logan
Text Designer:	Marsha Cohen/Parallelogram Graphics
Senior Project Editor:	Georgia Lee Hadler
Copy Editors:	Connie Parks, Trumbull Rogers
Illustrations:	Network Graphics
Illustration Coordinator:	Bill Page
Senior Workflow Project Supervisor:	Susan Wein
Production Supervisor:	Lawrence Guerra
Composition:	Lumina Datamatics, Inc.
Printing and Binding:	King Printing Co., Inc.

Library of Congress Control Number: 2007010418

ISBN-13: 978-1-319-36581-3
ISBN-10: 1-319-36581-7

Printed in the United States of America

4 5 6 28 27 26 25

Macmillan Learning
One New York Plaza
Suite 4600
New York, NY 10004-1562
www.macmillanlearning.com

w.h.freeman
Macmillan Learning

In 1946, William Freeman founded W. H. Freeman and Company and published Linus Pauling's *General Chemistry*, which revolutionized the chemistry curriculum and established the prototype for a Freeman text. W. H. Freeman quickly became a publishing house where leading researchers can make significant contributions to mathematics and science. In 1996, W. H. Freeman joined Macmillan and we have since proudly continued the legacy of providing revolutionary, quality educational tools for teaching and learning in STEM.

Contents in Brief

Thinkstock/Alamy

PART IV ELECTRICITY AND MAGNETISM

PART V LIGHT

PART VI MODERN PHYSICS: QUANTUM MECHANICS, RELATIVITY, AND THE STRUCTURE OF MATTER

APPENDICES

Contents

Extended

Courtesy of Rossignol Ski Company

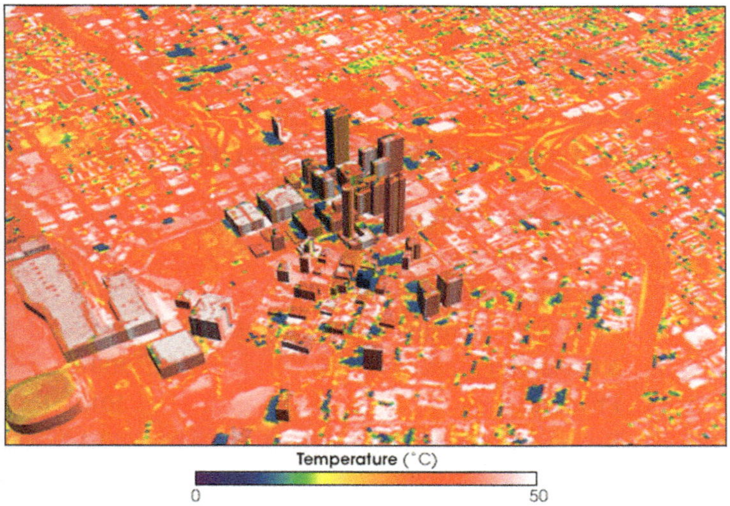

NASA/Goddard Space Flight Center Scientific Visualization Studio

Atlas Photo Bank/Photo Researchers, Inc.

NASA

The text you trust, with support like never before

For years, instructors have trusted Tipler's *Physics for Scientists and Engineers* to prepare physics, science, and engineering majors for future coursework. The text has long been respected for its rigorous emphasis on foundational skills, depth of coverage, and integration of calculus.

The new *Physics for Scientists and Engineers* Media Update makes it easier than ever for instructors to create lecture materials, engage students during class time, and assign homework exercises. These new resources can be easily edited, assigned, and integrated with your campus LMS to save time and effort for instructors and students alike. The text and its supplements are supported by Macmillan's personal customer service to fit your course needs and objectives.

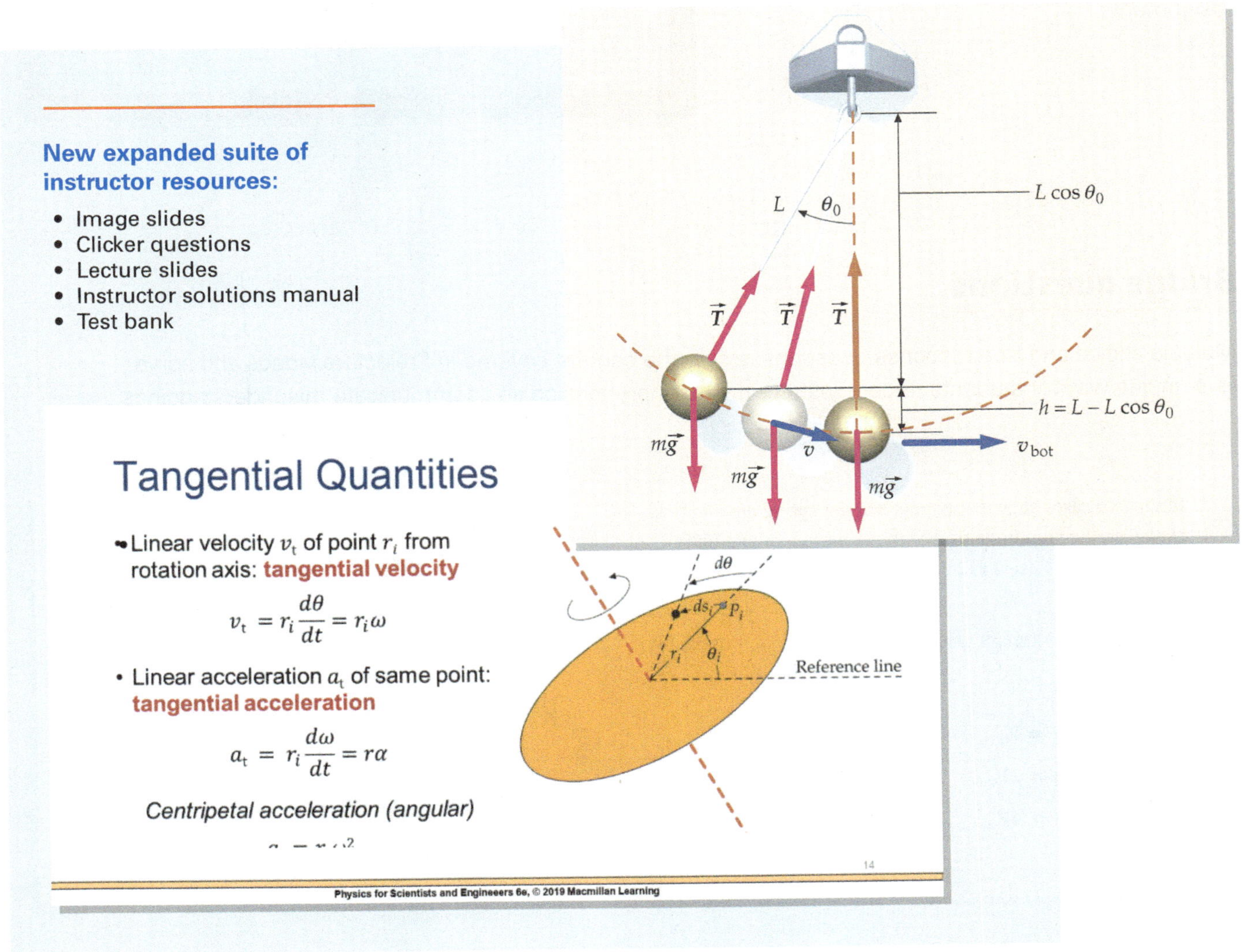

Prepare and Practice

Prelecture videos

Animated, narrated videos introduce core physics topics, laying the groundwork for conceptual understanding before lecture. Every chapter in the standard edition contains a series of 8-10 short videos; the activity can be assigned in its entirety or divided into smaller, more focused assignments.

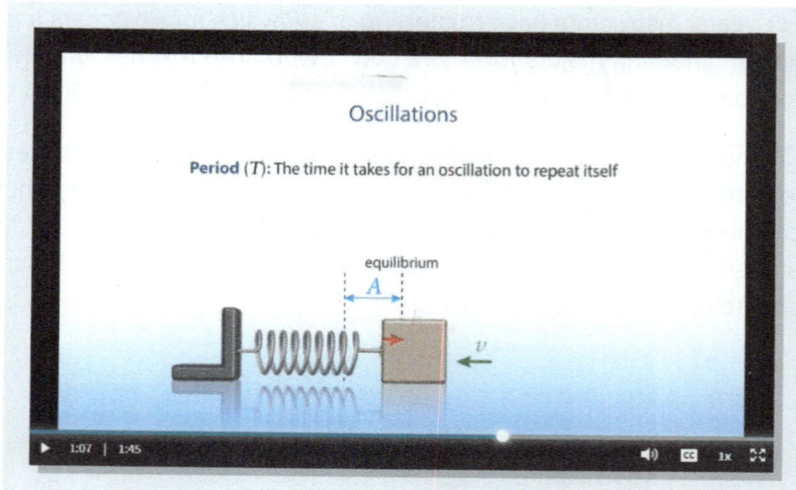

Bridge questions

Multiple choice and free response questions review the content covered in Prelecture videos and serve as a unique way for students to demonstrate their comprehension and communicate misunderstandings.

In the two cases shown, the mass and the spring are identical. However, the amplitude of the simple harmonic motion is twice as big in Case 2 as in Case 1.

How are the maximum velocities $V_{max,1}$ and $V_{max,2}$ in the two cases related?

- ○ $V_{max,2} = V_{max,1}$
- ○ $V_{max,2} = 2V_{max,1}$
- ○ $V_{max,2} = 4V_{max,1}$

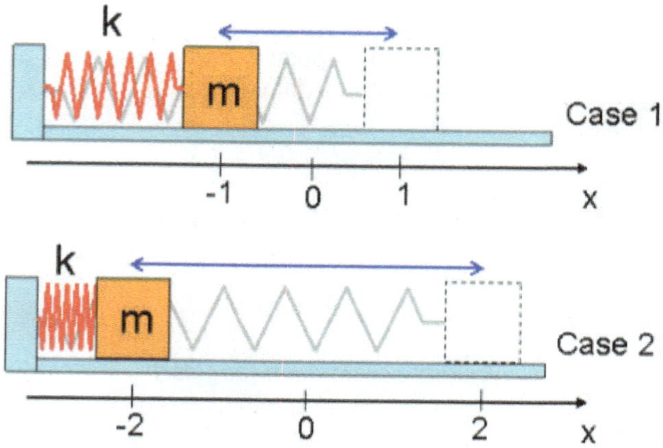

Every Problem Counts with SaplingPlus

Homework questions

Every homework question is built to test core competencies while coaching students through errors and common misconceptions.

Suppose Imani is on an amusement park ride that consists of a wide drum of radius R that rotates at ω rad/s and slowly increases its angle θ with the ground. Assume that Imani's back and heels are in contact with the wall of the drum.

In the configuration shown, which forces contribute to the centripetal force on Imani?

- [x] the normal force applied by the drum wall
- [ ] the normal force applied by the drum floor
- [x] the gravitational force
- [ ] the centrifugal force
- [ ] the frictional force

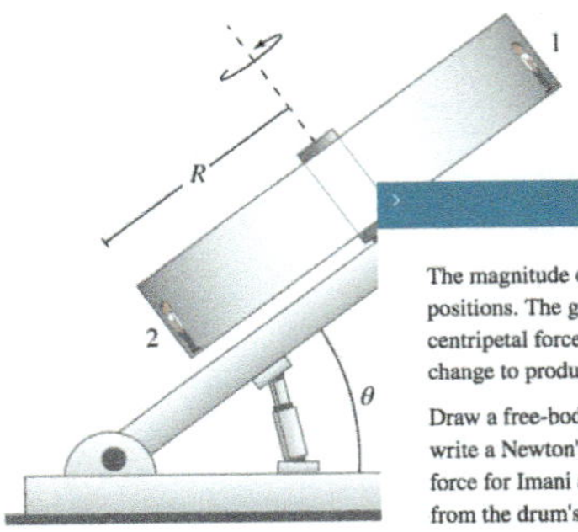

In which position is the force of the drum's wall on Imani the greatest?

- ○ 1
- ● 2
- ○ neither

Hint

A centripetal force is a *net* force that generates uniform circular motion.

Use free-body diagrams and Newton's second law to compare the forces contributing to the centripetal force at each position.

Hints encourage critical thinking by providing suggestions without giving away the answer.

Feedback

The magnitude of the force of the drum's wall is not the same at both positions. The gravitational force component does not contribute to the centripetal force in the same way at both positions, so the normal force must change to produce the same centripetal force.

Draw a free-body diagram for Imani at both position 1 and position 2. Then, write a Newton's second law equation along the direction of the centripetal force for Imani at each position. Compare the magnitude of the normal force from the drum's wall in each position.

Wrong-answer-specific feedback targets and addresses students' misconceptions.

To determine in which position the normal force $\vec{F}_{N,wall}$ from the wall has the greatest magnitude, draw a free-body diagram and solve for $F_{N,wall}$ in each position using Newton's second law. Use a coordinate system in which the x-axis is parallel to the floor of the drum and the y-axis coincides with the drum's rotation axis.

In position 1, the normal force from the floor points in the positive y-direction, the normal force from the wall points in the negative x-direction, and the x-component $\vec{F}_{g,x}$ of the gravitational force points in the negative x-direction.

Write Newton's second law in the x-direction for Imani in position 1, then solve for $F_{N,wall}$.

$$-F_{N,wall} - F_{g,x} = -ma_c$$

$$F_{N,wall} = ma_c - F_{g,x}$$

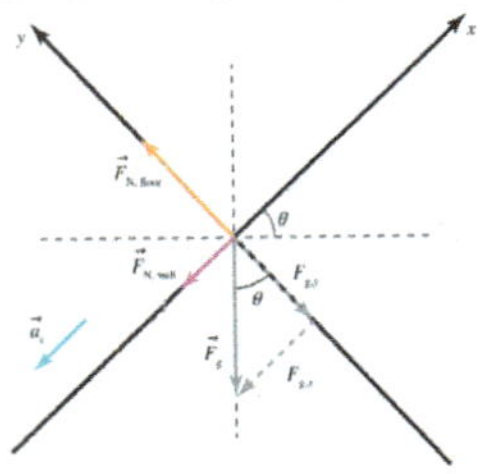

Detailed solutions provide an in-product study guide for every problem in SaplingPlus.

Custom Solutions and Ordering Options from Macmillan Learning Physics

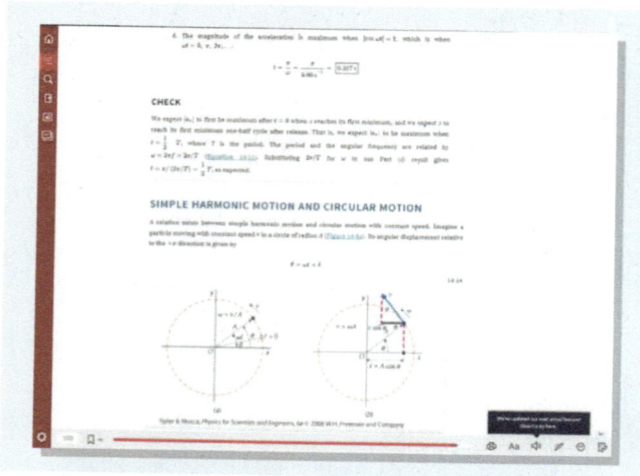

The new, enhanced e-book for Tipler's *Physics for Scientists and Engineers* is Macmillan's most affordable option for students. In partnership with VitalSource, the e-book contains highlighting and note-taking functionality, as well as offline reading and an accessible screen reader.

Custom texts and isolated volumes are easy to make in Freeman Select. Choose only the chapters you need and add your own content for your own custom text. Priced per page, it's an affordable solution for classes and students looking for a print solution. Try it for yourself at **macmillanlearning.com/freemanselect.**

Ordering options for *Physics for Scientists and Engineers 6e*:

For more information and ordering options, including rental options, visit **macmillanlearning.com/physics**.

E-book: 978-1-319-32171-0

E-book + SaplingPlus: 978-1-319-32169-7 (multi-term access)

Looseleaf: 978-1-319-36314-7

Looseleaf + SaplingPlus: 978-1-319-36875-3 (multi-term access)

Hardcover + SaplingPlus: 978-1-319-36828-9 (multi-term access)

Purposeful Classroom Engagement

iClicker

iClicker's market-leading student engagement solutions allow students to participate using laptops, mobile devices, or iClicker remotes—all in the same classroom. To learn more, visit **iclicker.com**.

Clicker questions for Tipler's *Physics for Scientists and Engineers* are available in SaplingPlus or on the Macmillan online catalog.

The iClicker app can be packaged with any book or digital version of *Physics for Scientists and Engineers* for an additional $5! Contact your local rep to learn more.

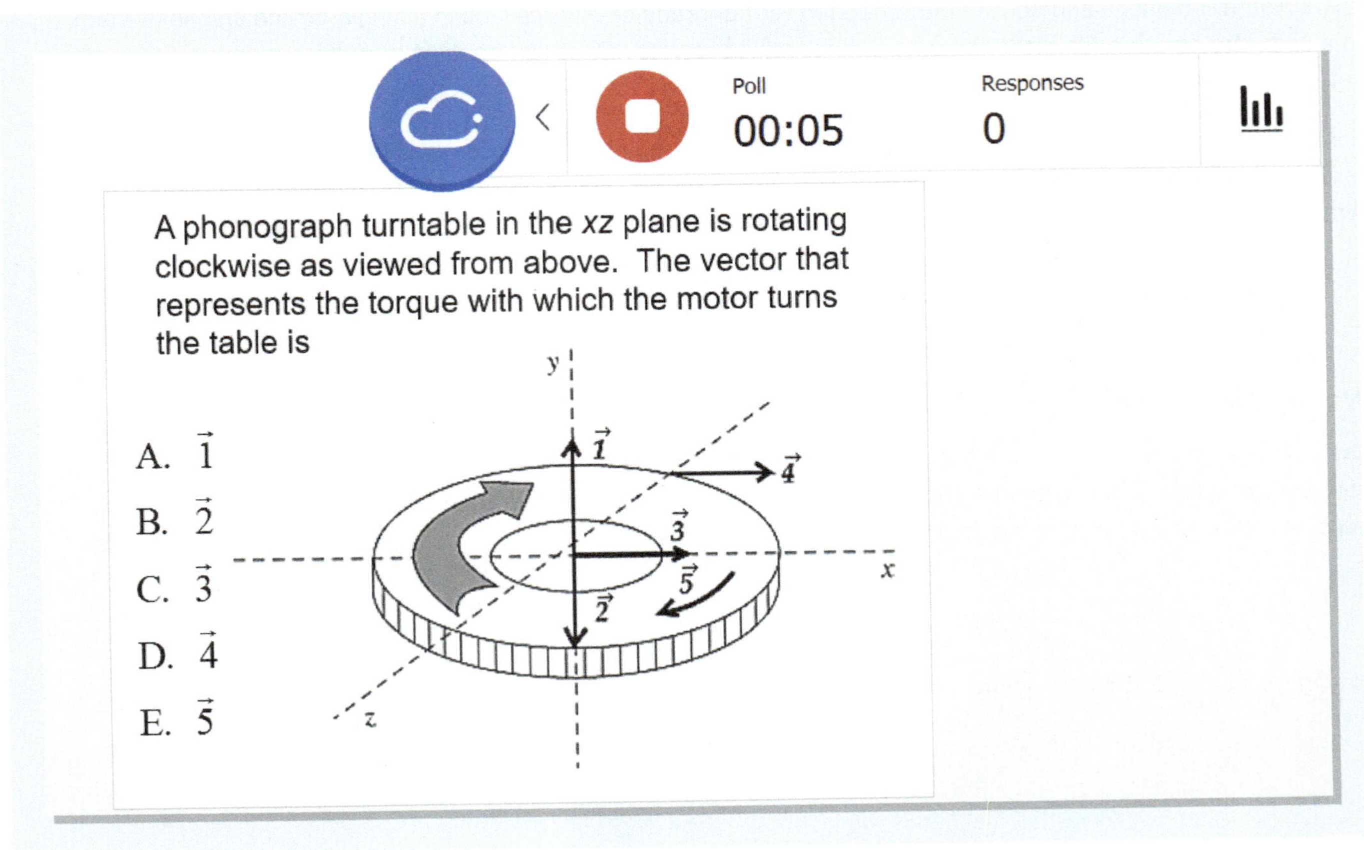

Lab Solutions From Macmillan Physics

Set Your Physics Lab Free

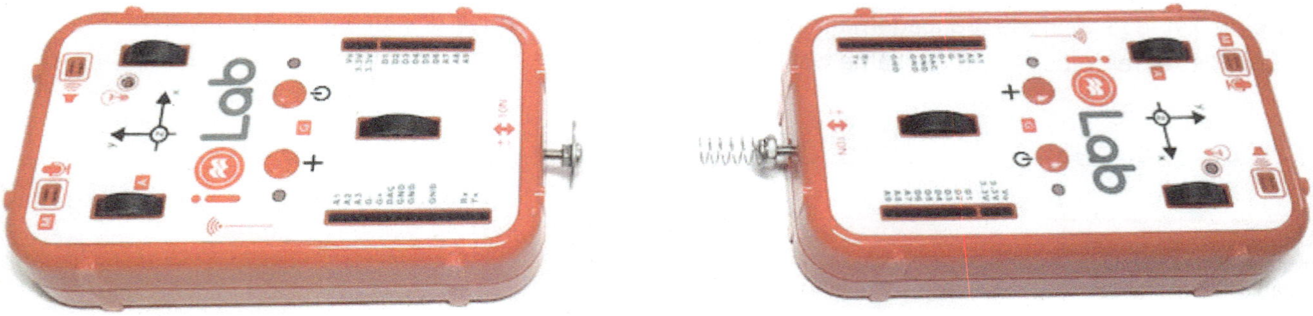

iOLab is a revolutionary new hardware and software solution for your physics lab. It combines all the measurement devices and components needed for hundreds of physics labs in a single device and links them to a software solution for gathering data and recording results. iOLab can replace or augment your current lab set up and is perfect for online, hybrid, or traditional classes. For more information, visit **iolabworks.com**.

For print or digital lab manuals, Macmillan's custom publishing group, **Hayden-McNeil** works with you to build lab resources that reflect your unique teaching objectives and are affordable for students. Faculty can write their own labs or borrow content from our Hayden-McNeil Content Collection to get started.

We help you curate your lab materials so that you can run your lab course, your way.

Our print and digital lab solutions are affordable for students. Each project is custom built with you to meet your budget goals. The Hayden-McNeil team will help you save time, save money, and save resources while building a quality lab solution for your course. For more information on lab manuals, visit **haydenmcneil.com**.

Acknowledgments

We are grateful to the many instructors, students, colleagues, and friends who have contributed to this edition and to earlier editions.

Anthony J. Buffa, professor emeritus at California Polytechnic State University in California, wrote many new end-of-chapter problems and edited the end-of-chapter problems sections. Laura Runkle wrote the Physics Spotlights. Richard Mickey revised the Math Review of the fifth edition, which is now the Math Tutorial of the sixth edition. David Mills, professor emeritus at the College of the Redwoods in California, extensively revised the Solutions Manual. We received invaluable help in creating text and checking the accuracy of text and problems from the following professors:

Thomas Foster
Southern Illinois University

Karamjeet Arya
San Jose State University

Mirley Bala
Texas A&M University—Corpus Christi

Michael Crivello
San Diego Mesa College

Carlos Delgado
Community College of Southern Nevada

David Faust
Mt. Hood Community College

Robin Jordan
Florida Atlantic University

Jerome Licini
Lehigh University

Dan Lucas
University of Wisconsin

Laura McCullough
University of Wisconsin, Stout

Jeannette Myers
Francis Marion University

Marian Peters
Appalachian State University

Todd K. Pedlar
Luther College

Paul Quinn
Kutztown University

Peter Sheldon
Randolph-Macon Woman's College

Michael G. Strauss
University of Oklahoma

Brad Trees
Ohio Wesleyan University

George Zober
Yough Senior High School

Patricia Zober
Ringgold High School

Many instructors and students have provided extensive and helpful reviews of one or more chapters of this edition. They have each made a fundamental contribution to the quality of this revision, and deserve our gratitude. We would like to thank the following reviewers:

Ahmad H. Abdelhadi
James Madison University

Edward Adelson
Ohio State University

Royal Albridge
Vanderbilt University

J. Robert Anderson
University of Maryland, College Park

Toby S. Anderson
Tennessee State University

Wickram Ariyasinghe
Baylor University

Yildirim Aktas
University of North Carolina, Charlotte

Eric Ayars
California State University

James Battat
Harvard University

Eugene W. Beier
University of Pennsylvania

Peter Beyersdorf
San Jose State University

Richard Bone
Florida International University

Juliet W. Brosing
Pacific University

Ronald Brown
California Polytechnic State University

Richard L. Cardenas
St. Mary's University

Troy Carter
University of California, Los Angeles

Alice D. Churukian
Concordia College

N. John DiNardo
Drexel University

Jianjun Dong
Auburn University

Fivos R Drymiotis
Clemson University

Mark A. Edwards
Hofstra University

James Evans
Broken Arrow Senior High

Nicola Fameli
University of British Columbia

N. G. Fazleev
University of Texas at Arlington

Thomas Furtak
Colorado School of Mines

Richard Gelderman
Western Kentucky University

Yuri Gershtein
Florida State University

Paolo Gondolo
University of Utah

Benjamin Grinstein
University of California, San Diego

Parameswar Hari
University of Tulsa

Joseph Harrison
University of Alabama—Birmingham

Patrick C. Hecking
Thiel College

Kristi R. G. Hendrickson
University of Puget Sound

Linnea Hess
Olympic College

Mark Hollabaugh
Normandale Community College

Daniel Holland
Illinois State University

Richard D. Holland II
Southern Illinois University

Eric Hudson
Massachusetts Institute of Technology

David C. Ingram
Ohio University

Colin Inglefield
Weber State University

Nathan Israeloff
Northeastern University

Donald J. Jacobs
California State University, Northridge

Erik L. Jensen
Chemeketa Community College

Colin P Jessop
University of Notre Dame

Ed Kearns
Boston University

Alice K. Kolakowska
Mississippi State University

Douglas Kurtze
Saint Joseph's University

Eric T. Lane
University of Tennessee at Chattanooga

Christie L. Larochelle
Franklin & Marshall College

Mary Lu Larsen
Towson University

Clifford L. Laurence
Colorado Technical University

Bruce W. Liby
Manhattan College

Ramon E. Lopez
Florida Institute of Technology

Ntungwa Maasha
Coastal Georgia Community Collegee and University Center

Jane H MacGibbon
University of North Florida

A. James Mallmann
Milwaukee School of Engineering

Rahul Mehta
University of Central Arkansas

R. A. McCorkle
University of Rhode Island

Linda McDonald
North Park University

Kenneth McLaughlin
Loras College

Eric R. Murray
Georgia Institute of Technology

Jeffrey S. Olafsen
University of Kansas

Richard P. Olenick
University of Dallas

Halina Opyrchal
New Jersey Institute of Technology

Russell L. Palma
Minnesota State University—Mankato

Todd K. Pedlar
Luther College

Daniel Phillips
Ohio University

Edward Pollack
University of Connecticut

Michael Politano
Marquette University

Robert L. Pompi
SUNY Binghamton

Damon A. Resnick
Montana State University

Richard Robinett
Pennsylvania State University

John Rollino
Rutgers University

Daniel V. Schroeder
Weber State University

Douglas Sherman
San Jose State University

Christopher Sirola
Marquette University

Larry K. Smith
Snow College

George Smoot
University of California at Berkeley

Zbigniew M. Stadnik
University of Ottawa

Kenny Stephens
Hardin-Simmons University

Daniel Stump
Michigan State University

Jorge Talamantes
California State University, Bakersfield

Charles G. Torre
Utah State University

Brad Trees
Ohio Wesleyan University

John K. Vassiliou
Villanova University

Theodore D. Violett
Western State College

Hai-Sheng Wu
Minnesota State University—Mankato

Anthony C. Zable
Portland Community College

Ulrich Zurcher
Cleveland State University

We also remain indebted to the reviewers of past editions. We would therefore like to thank the following reviewers, who provided immeasurable support as we developed the fourth and fifth editions:

Edward Adelson
The Ohio State University

Michael Arnett
Kirkwood Community College

Todd Averett
The College of William and Mary

Yildirim M. Aktas
University of North Carolina at Charlotte

Karamjeet Arya
San Jose State University

Alison Baski
Virginia Commonwealth University

William Bassichis
Texas A&M University

Joel C. Berlinghieri
The Citadel

Gary Stephen Blanpied
University of South Carolina

Frank Blatt
Michigan State University

Ronald Brown
California Polytechnic State University

Anthony J. Buffa
California Polytechnic State University

John E. Byrne
Gonzaga University

Wayne Carr
Stevens Institute of Technology

George Cassidy
University of Utah

Lay Nam Chang
Virginia Polytechnic Institute

I. V. Chivets
Trinity College, University of Dublin

Harry T. Chu
University of Akron

Alan Cresswell
Shippensburg University

Robert Coakley
University of Southern Maine

Robert Coleman
Emory University

Brent A. Corbin
UCLA

Andrew Cornelius
University of Nevada at Las Vegas

Mark W. Coffey
Colorado School of Mines

Peter P. Crooker
University of Hawaii

Jeff Culbert
London, Ontario

Paul Debevec
University of Illinois

Ricardo S. Decca
Indiana University-Purdue University

Robert W. Detenbeck
University of Vermont

N. John DiNardo
Drexel University

Bruce Doak
Arizona State University

Michael Dubson
University of Colorado at Boulder

John Elliott
University of Manchester, England

William Ellis
University of Technology — Sydney

Colonel Rolf Enger
U.S. Air Force Academy

John W. Farley
University of Nevada at Las Vegas

David Faust
Mount Hood Community College

Mirela S. Fetea
University of Richmond

David Flammer
Colorado School of Mines

Philip Fraundorf
University of Missouri, Saint Louis

Tom Furtak
Colorado School of Mines

James Garland
Retired

James Garner
University of North Florida

Ian Gatland
Georgia Institute of Technology

Ron Gautreau
New Jersey Institute of Technology

David Gavenda
University of Texas at Austin

Patrick C. Gibbons
Washington University

David Gordon Wilson
Massachusetts Institute of Technology

Christopher Gould
University of Southern California

Newton Greenberg
SUNY Binghamton

John B. Gruber
San Jose State University

Huidong Guo
Columbia University

Phuoc Ha
Creighton University

Richard Haracz
Drexel University

Clint Harper
Moorpark College

Michael Harris
University of Washington

Randy Harris
University of California at Davis

Tina Harriott
Mount Saint Vincent, Canada

Dieter Hartmann
Clemson University

Theresa Peggy Hartsell
Clark College

Kristi R.G. Hendrickson
University of Puget Sound

Michael Hildreth
University of Notre Dame

Robert Hollebeek
University of Pennsylvania

David Ingram
Ohio University

Shawn Jackson
The University of Tulsa

Madya Jalil
University of Malaya

Monwhea Jeng
University of California — Santa Barbara

James W. Johnson
Tallahassee Community College

Edwin R. Jones
University of South Carolina

Ilon Joseph
Columbia University

David Kaplan
University of California — Santa Barbara

William C. Kerr
Wake Forest University

John Kidder
Dartmouth College

Roger King
City College of San Francisco

James J. Kolata
University of Notre Dame

Boris Korsunsky
Northfield Mt. Hermon School

Thomas O. Krause
Towson University

Eric Lane
University of Tennessee, Chattanooga

Andrew Lang (graduate student)
University of Missouri

David Lange
University of California — Santa Barbara

Donald C. Larson
Drexel University

Paul L. Lee
California State University, Northridge

Peter M. Levy
New York University

Jerome Licini
Lehigh University

Isaac Leichter
Jerusalem College of Technology

William Lichten
Yale University

Robert Lieberman
Cornell University

Fred Lipschultz
University of Connecticut

Graeme Luke
Columbia University

Dan MacIsaac
Northern Arizona University

Edward McCliment
University of Iowa

Robert R. Marchini
The University of Memphis

Peter E. C. Markowitz
Florida International University

Daniel Marlow
Princeton University

Fernando Medina
Florida Atlantic University

Howard McAllister
University of Hawaii

John A. McClelland
University of Richmond

Laura McCullough
University of Wisconsin at Stout

M. Howard Miles
Washington State University

Matthew Moelter
University of Puget Sound

Eugene Mosca
U.S. Naval Academy

Carl Mungan
U.S. Naval Academy

Taha Mzoughi
Mississippi State University

Charles Niederriter
Gustavus Adolphus College

John W. Norbury
University of Wisconsin at Milwaukee

Aileen O'Donughue
St. Lawrence University

Jack Ord
University of Waterloo

Jeffry S. Olafsen
University of Kansas

Melvyn Jay Oremland
Pace University

Richard Packard
University of California

Antonio Pagnamenta
University of Illinois at Chicago

George W. Parker
North Carolina State University

John Parsons
Columbia University

Dinko Pocanic
University of Virginia

Edward Pollack
University of Connecticut

Robert Pompi
The State University of New York at Binghamton

Bernard G. Pope
Michigan State University

John M. Pratte
Clayton College and State University

Brooke Pridmore
Clayton State College

Yong-Zhong Qian
University of Minnesota

David Roberts
Brandeis University

Lyle D. Roelofs
Haverford College

R. J. Rollefson
Wesleyan University

Larry Rowan
University of North Carolina at Chapel Hill

Ajit S. Rupaal
Western Washington University

Todd G. Ruskell
Colorado School of Mines

Lewis H. Ryder
University of Kent, Canterbury

Andrew Scherbakov
Georgia Institute of Technology

Bruce A. Schumm
University of California, Santa Cruz

Cindy Schwarz
Vassar College

Mesgun Sebhatu
Winthrop University

Bernd Schuttler
University of Georgia

Murray Scureman
Amdahl Corporation

Marllin L. Simon
Auburn University

Scott Sinawi
Columbia University

Dave Smith
University of the Virgin Islands

Wesley H. Smith
University of Wisconsin

Kevork Spartalian
University of Vermont

Zbigniew M. Stadnik
University of Ottawa

G. R. Stewart
University of Florida

Michael G. Strauss
University of Oklahoma

Kaare Stegavik
University of Trondheim, Norway

Jay D. Strieb
Villanova University

Dan Styer
Oberlin College

Chun Fu Su
Mississippi State University

Jeffrey Sundquist
Palm Beach Community College — South

Cyrus Taylor
Case Western Reserve University

Martin Tiersten
City College of New York

Chin-Che Tin
Auburn University

Oscar Vilches
University of Washington

D. J. Wagner
Grove City College
Columbia University

George Watson
University of Delaware

Fred Watts
College of Charleston

David Winter

John A. Underwood
Austin Community College

John Weinstein
University of Mississippi

Stephen Weppner
Eckerd College

Suzanne E. Willis
Northern Illinois University

Frank L. H. Wolfe
University of Rochester

Frank Wolfs
University of Rochester

Roy C. Wood
New Mexico State University

Ron Zammit
California Polytechnic State University

Yuriy Zhestkov
Columbia University

Dean Zollman
Kansas State University

Fulin Zuo
University of Miami

Finally, we would like to thank our friends at W. H. Freeman and Company for their help and encouragement. Susan Brennan, Clancy Marshall, Kharissia Pettus, Georgia Lee Hadler, Susan Wein, Trumbull Rogers, Connie Parks, John Smith, Dena Digilio Betz, Ted Szczepanski, and Liz Geller were extremely generous with their creativity and hard work at every stage of the process.

We are also grateful for the contributions and help of our colleagues Larry Tankersley, John Ertel, Steve Montgomery, and Don Treacy.

About the Authors

Paul Tipler was born in the small farming town of Antigo, Wisconsin, in 1933. He graduated from high school in Oshkosh, Wisconsin, where his father was superintendent of the public schools. He received his B.S. from Purdue University in 1955 and his Ph.D. at the University of Illinois in 1962, where he studied the structure of nuclei. He taught for one year at Wesleyan University in Connecticut while writing his thesis, then moved to Oakland University in Michigan, where he was one of the original members of the physics department, playing a major role in developing the physics curriculum. During the next 20 years, he taught nearly all the physics courses and wrote the first and second editions of his widely used textbooks *Modern Physics* (1969, 1978) and *Physics* (1976, 1982). In 1982, he moved to Berkeley, California, where he now resides, and where he wrote *College Physics* (1987) and the third edition of *Physics* (1991). In addition to physics, his interests include music, hiking, and camping, and he is an accomplished jazz pianist and poker player.

Gene Mosca was born in New York City and grew up on Shelter Island, New York. He studied at Villanova University, the University of Michigan, and the University of Vermont, where he received his Ph.D. in physics. Gene recently retired from his teaching position at the U.S. Naval Academy, where as coordinator of the core physics course he instituted numerous enhancements to both the laboratory and classroom. Proclaimed by Paul Tipler "the best reviewer I ever had," Mosca became his coauthor beginning with the fifth edition of this book.

CHAPTER 1

THE NUMBER OF GRAINS OF SAND ON A BEACH MAY BE TOO GREAT TO COUNT, BUT WE CAN ESTIMATE THE NUMBER BY USING REASONABLE ASSUMPTIONS AND SIMPLE CALCULATIONS. *(Corbis.)*

? How many grains of sand are on your favorite beach? (See Example 1-7.)

Measurement and Vectors

We have always been curious about the world around us. Since the beginnings of recorded thought, we have sought to understand the bewildering diversity of events that we observe—the color of the sky, the change in sound of a passing car, the swaying of a tree in the wind, the rising and setting of the Sun, the flight of a bird or plane. This search for understanding has taken a variety of forms: one is religion, one is art, and one is science. Although the word *science* comes from the Latin verb meaning "to know," science has come to mean not merely knowledge but specifically knowledge of the natural world. Physics attempts to describe the fundamental nature of the universe and how it works. It is the science of matter and energy, space and time.

Like all science, physics is a body of knowledge organized in a specific and rational way. Physicists build, test, and connect models in an effort to describe, explain, and predict reality. This process involves hypotheses, repeatable experiments and observations, and new hypotheses. The end result is a set of fundamental principles and laws that describe the phenomena of the world around us.

These laws and principles apply both to the exotic—such as black holes, dark energy, and particles with names like leptoquarks and bosons—and to everyday life. As you will see, countless questions about our world can be answered with a basic knowledge of physics: Why is the sky blue? How do astronauts float in space? How do CD players work? Why does an oboe sound different from a flute? Why must a helicopter have two rotors? Why do metal objects feel colder than wood objects at the same temperature? How do moving clocks run slow?

In this book, you will learn how to apply the principles of physics to answer these, and many other questions. You will encounter the standard topics of physics, including mechanics, sound, light, heat, electricity, magnetism, atomic physics, and nuclear physics. You will also learn some useful techniques for solving physics problems. In the process, we hope you gain a greater awareness, appreciation, and understanding of the beauty of physics.

In this chapter, we'll begin by addressing some preliminary concepts that you will need throughout your study of physics. We'll briefly examine the nature of physics, establish some basic definitions, introduce systems of units and how to use them, and present an introduction to vector mathematics. We'll also look at the accuracy of measurements, significant figures, and estimations.

1-1 THE NATURE OF PHYSICS

The word physics comes from the Greek word meaning the knowledge of the natural world. It should come as no surprise, therefore, that the earliest recorded efforts to systematically assemble knowledge concerning motion came from ancient Greece. In Aristotle's (384–322 B.C.) system of natural philosophy, explanations of physical phenomena were deduced from assumptions about the world, rather than derived from experimentation. For example, it was a fundamental assumption that every substance had a "natural place" in the universe. Motion was thought to be the result of a substance trying to reach its natural place. Because of the agreement between the deductions of Aristotelian physics and the motions observed throughout the physical universe and the lack of experimentation that could overturn the ancient physical ideas, the Greek view was accepted for nearly two thousand years. It was the Italian scientist Galileo Galilei (1564–1642) whose brilliant experiments on motion established the absolute necessity of experimentation in physics. Within a hundred years, Isaac Newton had generalized the results of Galileo's experiments into his three spectacularly successful laws of motion, and the reign of the natural philosophy of Aristotle was over.

Experimentation during the next two hundred years brought a flood of discoveries—and raised a flood of new questions. Some of these discoveries involved electrical and thermal phenomena, and some involved the expansion and compression of gases. These discoveries and questions inspired the development of new models to explain them. By the end of the nineteenth century, Newton's laws for the motions of mechanical systems had been joined by equally impressive laws from James Maxwell, James Joule, Sadi Carnot, and others to describe electromagnetism and thermodynamics. The subjects that occupied physical scientists through the end of the nineteenth century—mechanics, light, heat, sound, electricity and magnetism—are usually referred to as *classical physics*. Because classical physics is what we need to understand the macroscopic world we live in, it dominates Parts I through V of this text.

The remarkable success of classical physics led many scientists to believe that the description of the physical universe was complete. However, the discovery of X rays by Wilhelm Röntgen in 1895 and of radioactivity by Antoine Becquerel and

Marie and Pierre Curie a few years later seemed to be outside the framework of classical physics. The theory of special relativity proposed by Albert Einstein in 1905 expanded the classical ideas of space and time promoted by Galileo and Newton. In the same year, Einstein suggested that light energy is quantized; that is, that light comes in discrete packets rather than being wavelike and continuous as was thought in classical physics. The generalization of this insight to the quantization of all types of energy is a central idea of quantum mechanics, one that has many amazing and important consequences. The application of special relativity, and particularly quantum theory, to extremely small systems such as atoms, molecules, and nuclei, has led to a detailed understanding of solids, liquids, and gases. This application is often referred to as *modern physics*. Modern physics is the subject of Part VI of this text.

While classical physics is the main subject of this book, from time to time in the earlier parts of the text we will note the relationship between classical and modern physics. For example, when we discuss velocity in Chapter 2, we will take a moment to consider velocities near the speed of light and briefly cross over to the relativistic universe first imagined by Einstein. After discussing the conservation of energy in Chapter 7, we will discuss the quantization of energy and Einstein's famous relation between mass and energy, $E = mc^2$. Four chapters later, in Chapter R, we will study the nature of space and time as revealed by Einstein in 1903.

1-2 UNITS

The laws of physics express relationships among physical quantities. **Physical quantities** are numbers that are obtained by measuring physical phenomena. For example, the length of this book is a physical quantity, as is the amount of time it takes for you to read this sentence and the temperature of the air in your classroom.

Measurement of any physical quantity involves comparing that quantity to some precisely defined standard, or **unit,** of that quantity. For example, to measure the distance between two points, we need a standard unit of distance, such as an inch, a meter, or a kilometer. The statement that a certain distance is 25 meters means that it is 25 times the length of the unit meter. It is important to include the unit, in this case meters, along with the number, 25, when expressing this distance because different units can be used to measure distance. To say that a distance is 25 is meaningless.

! When you use a number to describe a physical quantity, the number must always be accompanied by a unit.

Some of the most basic physical quantities—time, length, and mass—are defined by the processes of measuring them. The length of a pole, for example, is defined to be the number of some unit of length that is required to equal the length of the pole. A physical quantity is often defined using an **operational definition,** a statement that defines a physical quantity by the operation or procedure that should be carried out to measure the physical quantity. Other physical quantities are defined by describing how to calculate them from these fundamental quantities. The speed of an object, for example, is equal to a length divide by a time. Many of the quantities that you will be studying, such as velocity, force, momentum, work, energy, and power, can be expressed in terms of time, length, and mass. Thus, a small number of basic units are sufficient to express all physical quantities. These basic units are called base units, and the choice of base units determines a system of units.

Water clock used to measure time intervals in the thirteenth century. *(The Granger Collection.)*

THE INTERNATIONAL SYSTEM OF UNITS

In physics, it is important to use a consistent set of units. In 1960, an international committee established a set of standards for the scientific community called SI (for *Système International*). There are seven base quantities in the SI system. They are length, mass, time, electric current, thermodynamic temperature, amount of substance, and luminous intensity, and each base quantity has a base unit. The base SI unit of time is the second, the base unit of length is the meter, and the base unit of mass is the kilogram. Later, when you study thermodynamics and electricity, you will need to use the base SI units for temperature (the kelvin, K), for the amount of a substance (the mole, mol), and one for electrical current (the ampere, A). The seventh base SI unit, the candela (cd) for luminous intensity, we shall have no occasion to use in this book. Complete definitions of the SI units are given in Appendix A, along with commonly used units derived from these units.

Cesium fountain clock with developers Steve Jefferts and Dawn Meekhof. *(© 1999 Geoffrey Wheeler.)*

Time The unit of time, the **second** (s), was historically defined in terms of the rotation of Earth and was equal to $(1/60)(1/60)(1/24)$ of the mean solar day. However, scientists have observed that the rate of rotation of Earth is gradually slowing down. The second is now defined in terms of a characteristic frequency associated with the cesium atom. All atoms, after absorbing energy, emit light with frequencies and wavelengths characteristic of the particular element. There is a set of frequencies and wavelengths for each element, with a particular frequency and wavelength associated with each energy transition within the atom. As far as we know, these frequencies remain constant. The second is now defined so that the frequency of the light from a certain transition in cesium is exactly 9 192 631 770 cycles per second.

Length The **meter** (m) is the SI unit of length. Historically, this length was defined as one ten-millionth of the distance between the equator and the North Pole along the meridian through Paris (Figure 1-1). This distance proved to be difficult to measure accurately. So in 1889, the distance between two scratches on a bar made of platinum-iridium alloy held at a specified temperature was adopted as the new standard. In time, the precision of this standard also proved inadequate and other standards were created for the meter. Currently, the meter is determined using the speed of light through empty space, which is defined to be exactly 299 792 458 m/s. The meter, then, is the distance light travels through empty space in 1/(299 729 458) second. By using these definitions, the units of time and length are accessible to laboratories throughout the world.

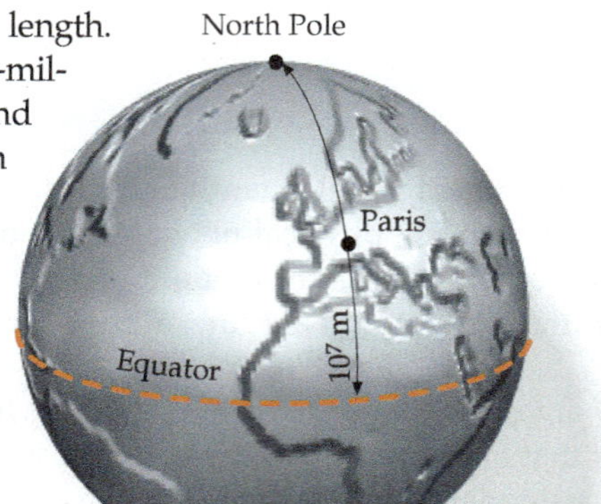

FIGURE 1-1 The meter was originally chosen so that the distance from the equator to the North Pole along the meridian through Paris would be 10 million meters (10 thousand kilometers).

Mass The SI unit of mass, the **kilogram** (kg) was once defined as the mass of one liter of water at 4°C. (A volume of one liter is equal to the volume of a cube 10 cm on an edge.) Like the standards for time and length, the kilogram standard has changed over time. The kilogram is now defined to be the mass of a specific platinum-iridium alloy cylinder. This cylinder, called the *standard body*, is kept at the International Bureau of Weights and Measures in Sèvres, France. A duplicate of the standard body is kept at the National Institute of Standards and Technology (NIST) in Gaithersburg, Maryland. We shall discuss the concept of mass in detail in Chapter 4, where we will see that the weight of an object at a given location is proportional to its mass. Thus, by comparing the weights of different objects of ordinary size with the weight of the standard body, the masses of the objects can be compared with each other.

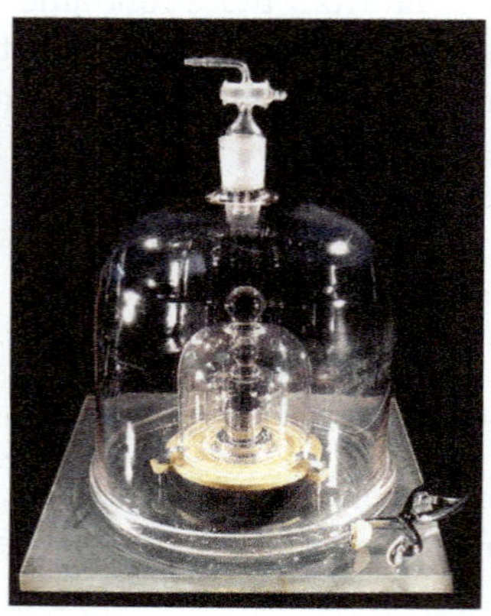

The standard body is the mass of a specific platinum-iridium alloy cylinder that is kept at the International Bureau of Weights and Measures in Sèvres, France. *(© BIPM; www.bipm.org.)*

Table 1-1 Prefixes for Powers of 10*

Multiple	Prefix	Abbreviation
10^{18}	exa	E
10^{15}	peta	P
10^{12}	tera	T
10^{9}	giga	G
10^{6}	mega	M
10^{3}	kilo	k
10^{2}	hecto	h
10^{1}	deka	da
10^{-1}	deci	d
10^{-2}	centi	c
10^{-3}	milli	m
10^{-6}	micro	μ
10^{-9}	nano	n
10^{-12}	pico	p
10^{-15}	femto	f
10^{-18}	atto	a

* The prefixes hecto (h), deka (da) and deci (d) are not multiples of 10^3 or 10^{-3} and are rarely used. The other prefix that is not a multiple of 10^3 or 10^{-3} is centi (c). The prefixes frequently used in this book are printed in red. Note that all prefix abbreviations for multiples 10^6 and higher are uppercase letters, all others are lowercase letters.

UNIT PREFIXES

Sometimes it is necessary to work with measurements that are much smaller or much larger than the standard SI units. In these situations, we can use other units that are related to the standard SI units by a multiple of ten. Prefixes are used to denote the different powers of ten. For example, the prefix "kilo" means 1000, or 10^3, while the prefix "micro" means 0.000 001, or 10^{-6}. Table 1-1 lists prefixes for common multiples of SI units. These prefixes can be applied to any SI unit; for example, 0.001 second is 1 millisecond (ms) and 1 000 000 watts is 1 megawatt (MW).

PRACTICE PROBLEM 1-1

Use prefixes to describe the following: (*a*) the delay caused by scrambling a cable television broadcast, which is about 0.000 000 3 second and (*b*) the circumference of Earth, which is about 40 000 000 meters.

OTHER SYSTEMS OF UNITS

In addition to SI, other systems of units are sometimes used. One such system is the *cgs system*. The fundamental units of the cgs system are the centimeter for length, the gram for mass, and the second for time. Other cgs units include the dyne (force) and the erg (work or energy).

The system of units with which you are probably most familiar is the U.S. customary system. In this system, the base unit of length is the foot and the base unit of time is the second. Also, a unit of force (the pound-force) rather than mass is considered a base unit. You will see in Chapter 4 that mass is a better choice for a fundamental unit than force, because mass is an intrinsic property of an object, independent of its location. The base U.S. customary units are now defined in terms of the base SI units.

1-3 CONVERSION OF UNITS

Because different systems of units are in use, it is important to know how to covert from one unit to another unit. When physical quantities are added, subtracted, multiplied, or divided in an algebraic equation, the unit can be treated like any other algebraic quantity. For example, suppose you want to find the distance traveled in 3 hours (h) by a car moving at a constant rate of 80 kilometers per hour (km/h). The distance is the product of the speed v and the time t:

$$x = vt = \frac{80\ \text{km}}{\cancel{\text{h}}} \times 3\ \cancel{\text{h}} = 240\ \text{km}$$

We cancel the unit of time, the hours, just as we would any algebraic quantity to obtain the distance in the proper unit of length, the kilometer. This method of treating units makes it easy to convert from one unit of distance to another. Now, suppose we want to convert the units in our answer from kilometers (km) to miles (mi). First, we need to find the relationship between kilometers and miles, which is 1 mi = 1.609 km (see either the front pages or Appendix A). Then, we divide each side of this equality by 1.609 km to obtain

$$\frac{1\ \text{mi}}{1.609\ \text{km}} = 1$$

Notice that the relationship is a ratio equal to 1. A ratio such as (1 mi)/(1.609 km) is called a **conversion factor,** which is a ratio equal to 1 and expresses a quantity expressed in some unit or units divided by its equal expressed some different unit or units. Because any quantity can be multiplied by 1 without changing its value, we can multiply the original quantity by the conversion factor to convert the units:

> ! If the units of the quantity and the conversion factor do not combine to give the desired final units, the conversion has not been properly carried out.

$$240\ \text{km} = 240\ \cancel{\text{km}} \times \frac{1\ \text{mi}}{1.609\ \cancel{\text{km}}} = 149\ \text{mi}$$

By writing out the units explicitly and canceling them, you do not need to think about whether you multiply by 1.609 or divide by 1.609 to change kilometers to miles, because the units tell you whether you have chosen the correct or incorrect factor.

(a)

(b)

(*a*) Laser beam from the Macdonald Observatory used to measure the distance to the moon. The distance can be measured within a few centimeters by measuring the time required for the beam to go to the moon and back after reflecting off a mirror (*b*) placed on the moon by the Apollo 14 astronauts. *(a, McDonald Observatory; b, Bruce Coleman).*

Example 1-1 Using Conversion Factors

Your employer sends you on a trip to a foreign country where the road signs give distances in kilometers and the automobile speedometers are calibrated in kilometers per hour. If you drive 90 km/h, how fast are you going in meters per second and in miles per hour?

PICTURE First we have to find the appropriate conversion factors for hours to seconds and kilometers to meters. We can use the facts that 1000 m = 1 km and 1 h = 60 min = 3600 s. The quantity 90 km/h is multiplied by the conversion factors, so the unwanted units cancel. (Each conversion factor has the value 1, so the value of the speed is not changed.) To convert to miles per hour, we use the conversion factor 1 mi/1.609 km.

SOLVE

1. Multiply 90 km/h by the conversion factors 1 h/3600 s and 1000 m/1 km to convert km to m and h to s:

$$\frac{90\ \cancel{\text{km}}}{\text{h}} \times \frac{1\ \cancel{\text{h}}}{3600\ \text{s}} \times \frac{1000\ \text{m}}{1\ \cancel{\text{km}}} = \boxed{25\ \text{m/s}}$$

2. Multiply 90 km/h by 1 mi/1.609 km:

$$\frac{90\ \cancel{\text{km}}}{\text{h}} \times \frac{1\ \text{mi}}{1.609\ \cancel{\text{km}}} = \boxed{56\ \text{mi/h}}$$

CHECK Notice that the final units in each step are correct. If you had not set up the conversion factors correctly, for example if you multiplied by 1 km/1000 m instead of 1000 m/1 km, the final units would not be correct.

TAKING IT FURTHER Step 1 can be shortened by writing 1 h/3600 s as 1 h/3.6 ks and canceling the prefixes in ks and km. That is,

$$\frac{90\ \text{km}}{\cancel{\text{h}}} \times \frac{1\ \cancel{\text{h}}}{3.6\ \text{ks}} = \boxed{25\ \text{m/s}}$$

Canceling these prefixes is equivalent to dividing the numerator and the denominator by 1000.

You may find it helpful to memorize the conversion results in Example 1-1. These results are

$$25\ \text{m/s} = 90\ \text{km/h} \approx (60\ \text{mi/h})$$

Knowing these values can provide you with a quick way to convert speeds to units you are more familiar with.

(Eunice Harris/Photo Researchers.)

1-4 DIMENSIONS OF PHYSICAL QUANTITIES

Recall that a physical quantity includes both a number and a unit. The unit tells the standard that is used for the measurement and the number gives the comparison of the quantity to the standard. To tell *what* you are measuring, however, you need to state the *dimension* of the physical quantity. Length, time, and mass are all dimensions. The distance d between two objects has dimensions of length. We express this relation as $[d] = \text{L}$, where $[d]$ represents the dimension of the distance d and L represents the dimension of length. All dimensions are represented by upper-case roman (nonitalic) letters. The letters T and M represent the dimensions of time and mass, respectively. The dimensions of a number of quantities can be written in terms of these fundamental dimensions. For example, the area A of a surface is found by multiplying one length by another. Because area is the product of two lengths, it is said to have the dimensions of length multiplied by length, or length squared, written $[A] = \text{L}^2$. In this equation, $[A]$ represents the dimension of the quantity A and L represents the dimension of length. Speed has the dimensions of length divided by time, or L/T. The dimensions of other quantities such as force or energy are written in terms of the fundamental quantities of length, time, and mass. Adding or subtracting two physical quantities makes sense only if the quantities have the same dimensions. For example, we cannot add an area to a speed to obtain a meaningful sum. For the equation

$$A = B + C$$

the quantities A, B, and C must all have the same dimensions. The addition of B and C also requires that these quantities be in the same units. For example, if B is an area of 500 in.^{2} and C is 4 ft^2, we must either convert B into square feet or C into square inches in order to find the sum of the two areas.

You can often find mistakes in a calculation by checking the dimensions or units of the quantities in your result. Suppose, for example, that you mistakenly use the formula $A = 2\pi r$ for the area of a circle. You can see immediately that this cannot be correct because $2\pi r$ has dimensions of length whereas area must have dimensions of length squared.

! Evaluating the dimensions of an expression will tell you only if the dimensions are correct, not whether the entire expression is correct. While expressing the area of a circle, for instance, dimensional analysis will not tell you the correct expression is πr^2 or $2\pi r^2$. (The correct expression is πr^2.)

Example 1-2 Dimensions of Pressure

The pressure P in a fluid in motion depends on its density ρ and its speed v. Find a simple combination of density and speed that gives the correct dimensions of pressure.

PICTURE Using Table 1-2, we can see that pressure has the dimensions M/(LT^2), density is M/L^3, and speed is L/T. In addition, both the dimensions of pressure and density have mass in the numerator, whereas the dimensions of speed do not contain mass. Therefore, the expression must involve multiplying or dividing dimensions of density and dimensions of speed to obtain the unit of mass in the dimensions of pressure. To find out the exact relationship, we can start by dividing the dimensions of pressure by those of density, and then evaluate the result with respect to the dimensions for speed.

SOLVE

1. Divide the dimensions of pressure by those of density to obtain an expression with no M in it:

$$\frac{[P]}{[\rho]} = \frac{\text{M/LT}^2}{\text{M/L}^3} = \frac{\text{L}^2}{\text{T}^2}$$

2. By inspection, we note that the result has dimensions of v^2. The dimensions of pressure are thus the same as the dimensions of density multiplied by speed squared:

$$[P] = [\rho][v^2] = \frac{\text{M}}{\text{L}^3} \times \left(\frac{\text{L}}{\text{T}}\right)^2 = \frac{\text{M}}{\text{L}^3} \times \frac{\text{L}^2}{\text{T}^2} = \boxed{\frac{\text{M}}{\text{LT}^2}}$$

CHECK Divide the dimensions of pressure by the dimensions of speed squared and the result is the dimensions of density $[P]/[v^2] = (\text{M/LT}^2)/(\text{L}^2/\text{T}^2) = \text{M/L}^3 = [\rho]$.

1-5 SIGNIFICANT FIGURES AND ORDER OF MAGNITUDE

Many of the numbers in science are the result of measurement and are therefore known only to within a degree of experimental uncertainty. The magnitude of the uncertainty, which depends on both the skill of the experimenter and the apparatus used, often can only be estimated. A rough indication of the uncertainty in a measurement is inferred by the number of digits used. For example, if a tag on a table in a furniture store states that a table is 2.50 m long, it is saying that its length is close to, but not exactly, 2.50 m. The rightmost digit, the 0, is uncertain. If we use a tape measure with millimeter markings and measured the table length carefully, we might estimate that we could measure the length to ±0.6 mm of its true length. We would indicate this precision when giving the length by using four digits, such as 2.503 m. A reliably known digit (other than a zero used to locate the decimal point) is called a **significant figure.** The

Table 1-2 Dimensions of Physical Quantities

Quantity	Symbol	Dimension
Area	A	L^2
Volume	V	L^3
Speed	v	L/T
Acceleration	a	L/T^2
Force	F	ML/T^2
Pressure (F/A)	p	M/LT^2
Density (M/V)	ρ	M/L^3
Energy	E	ML^2/T^2
Power (E/T)	P	ML^2/T^3

number 2.50 has three significant figures; 2.503 m has four. The number 0.00130 has three significant figures. (The first three zeroes are not significant figures but are merely markers to locate the decimal point.) The number 2300. has four significant figures, but the number 2300 (the same as 2300. but without the decimal point) could have as few as two or as many as four significant figures. The number of significant figures in numbers with trailing zeros and no decimal point is ambiguous.

Suppose, for example, that you measure the area of a circular playing field by pacing off the radius and using the formula for the area of a circle, $A = \pi r^2$. If you estimate the radius to be 8 m and use a 10-digit calculator to compute the area, you obtain $\pi(8\text{ m})^2 = 201.0619298\text{ m}^2$. The digits after the decimal point give a false indication of the accuracy with which you know the area. To determine the appropriate numbers of significant figures for calculations involving multiplication and division, you can follow this general rule:

> When multiplying or dividing quantities, the number of significant figures in the final answer is no greater than that in the quantity with the fewest significant figures.

In the previous example, the radius is known to only one significant figure, so the area is also known only to one significant figure, or 200 m^2. This number indicates that the area is likely somewhere between 150 m^2 and 250 m^2.

The accuracy of the sum or difference of measurements is only as good as the accuracy of the *least* accurate of the measurements. A general rule is

> When adding or subtracting quantities, the number of decimal places in the answer should match that of the term with the smallest number of decimal places.

CONCEPT CHECK 1-1

How many significant figures does the number 0.010457 have?

! Exact values have an unlimited number of significant figures. For example, a value determined by counting, such as 2 tables, has no uncertainty and is an exact value. In addition, the conversion factor 1 m/100 cm is an exact value because 1 m *is* exactly equal to 100 cm.

! When you work with numbers that have uncertainties, you should be careful not to include more digits than the certainty of measurement warrants.

Example 1-3 Significant Figures

Subtract 1.040 from 1.21342.

PICTURE The first number, 1.040, has only three significant figures beyond the decimal point, whereas the second, 1.21342, has five. According to the rule stated for the addition and subtraction of numbers, the difference can have only three figures beyond the decimal point.

SOLVE

Sum the numbers, keeping only three digits beyond the decimal point: $1.21342 - 1.040 = 0.173\not{4}\not{2} = \boxed{0.173}$

CHECK The answer cannot be more accurate than the least accurate number or 1.040. The answer has the same number of significant figures beyond the decimal point as 1.040.

TAKING IT FURTHER In this example, the given numbers have four and six significant figures, but the difference has only three significant figures. Most examples and exercises in this book will be done with data to two, three, or occasionally four significant figures.

PRACTICE PROBLEM 1-2 Apply the appropriate rule for significant figures to calculate the following: (*a*) 1.58×0.03, (*b*) $1.4 + 2.53$, (*c*) $2.456 - 2.453$

SCIENTIFIC NOTATION

When we work with very large or very small numbers we can show significant figures more easily by using scientific notation. In this notation, the number is written as a product of a number between 1 and 10 and a power of 10, such as 10^2 (= 100) or 10^3 (= 1000). For example, the number 12 000 000 is written 1.2×10^7; the distance from Earth to the Sun, which is about 150 000 000 000 m, is written 1.5×10^{11} m. We have assumed that none of the trailing zeros in this number are significant. If two of the trailing zeros were significant we could express this unambiguously by writing the number as 1.500×10^{11} m. The number 11 in 10^{11} is called the exponent. For numbers smaller than 1, the exponent is negative. For example, $0.1 = 10^{-1}$, and $0.0001 = 10^{-4}$. The diameter of a virus, which is about 0.000 000 01 m, is written 1×10^{-8} m. Notice that by writing numbers in this form, you can easily identify the number of significant figures. For example, 1.5×10^{11} m contains two significant figures (1 and 5).

See **Math Tutorial** *for more information on* **Exponents**

PRACTICE PROBLEM 1-3

Apply the appropriate rule for significant figures to calculate $2.34 \times 10^2 + 4.93$.

Use the following Problem-Solving Strategy to do calculations with numbers in scientific notation.

PROBLEM-SOLVING STRATEGY

Scientific Notation

PICTURE If the numbers involved in a calculation are very large or very small, you may want to rewrite these numbers in scientific notation. This notation often makes it easier for you to determine the number of significant figures that a number has and makes it easier for you to perform calculations.

SOLVE Use these items to solve problems that involve scientific notation.

1. When numbers in scientific notation are multiplied, the exponents are added; when numbers in scientific notation are divided, the exponents are subtracted.

   Example: $10^2 \times 10^3 = 100 \times 1{,}000 = 100\,000 = 10^5$

   Example: $\dfrac{10^2}{10^3} = \dfrac{100}{1000} = \dfrac{1}{10} = 10^{-1}$

2. In scientific notation, 10^0 is defined to be 1. To see why, suppose we divide 1000 by 1000.

   Example: $\dfrac{1000}{1000} = \dfrac{10^3}{10^3} = 10^{3-3} = 10^0 = 1$

3. Be careful when adding or subtracting numbers written in scientific notation when their exponents do not match.

   Example: $(1.200 \times 10^2) + (8 \times 10^{-1}) = 120.0 + 0.8 = 120.8$

4. To find the sum without converting both numbers into ordinary decimal form, rewrite either of the numbers so that its power of 10 is the same as that of the other.

   Example: $(1200 \times 10^{-1}) + (8 \times 10^{-1}) = 1208 \times 10^{-1} = 120.8$

5. When raising a power to another power, the exponents are multiplied.

   Example: $(10^2)^4 = 10^2 \times 10^2 \times 10^2 \times 10^2 = 10^8$

CHECK Make sure that when you convert numbers smaller than one into scientific notation, the exponent is negative. You should also remember when exponents are added, subtracted, or multiplied, because performing the wrong operation can cause your answer to be inaccurate by powers of 10.

TAKING IT FURTHER During a calculation, avoid entering intermediate results via keyboard entry. Instead store these results in the calculator memory. If you must enter intermediate results via the keyboard, key in one or two additional (non significant) digits, called *guard digits*. This practice serves to minimize round-off errors.

! All exponents are dimensionless and have no units.

Example 1-4 How Much Water?

A liter (L) is the volume of a cube that is 10 cm by 10 cm by 10 cm. If you drink 1 L (exact) of water, how much volume in cubic centimeters and in cubic meters would it occupy in your stomach?

PICTURE The volume V of a cube of edge length ℓ is ℓ^3. The volume in cubic centimeters is found directly from $\ell = 10$ cm. To find the volume in cubic meters, convert cm^3 to m^3 using the conversion factor 1 cm $= 10^{-2}$ m.

SOLVE

1. Calculate the volume in cm^3:

$$V = \ell^3 = (10\ \text{cm})^3 = 1000\ \text{cm}^3 = \boxed{10^3\ \text{cm}^3}$$

2. Convert to m^3:

$$10^3\ \text{cm}^3 = 10^3\ \text{cm}^3 \times \left(\frac{10^{-2}\ \text{m}}{1\ \text{cm}}\right)^3$$

Notice that the conversion factor (which equals 1) can be raised to the third power without changing its value, enabling us to cancel units.

$$= 10^3\ \text{cm}^3 \times \frac{10^{-6}\ \text{m}^3}{1\ \text{cm}^3} = \boxed{10^{-3}\ \text{m}^3}$$

CHECK Notice that the answers are in cubic centimeters and cubic meters. These answers are consistent with volume having dimensions of length cubed. Also note that the physical quantity 10^3 is greater than the physical quantity 10^{-3}, which is consistent with a meter being larger than a centimeter.

Example 1-5 Counting Atoms

In 12.0 g of carbon, there are $N_A = 6.02 \times 10^{23}$ carbon atoms (Avogadro's number). If you could count 1 atom per second, how long would it take to count the atoms in 1.00 g of carbon? Express your answer in years.

PICTURE We need to find the total number of atoms to be counted, N, and then use the fact that the number counted equals the counting rate R multiplied by the time t.

SOLVE

1. The time is related to the total number of atoms N, and the rate of counting $R = 1$ atom/s:

$$N = Rt$$

2. Find the number of carbon atoms in 1.00 g:

$$N = \frac{6.02 \times 10^{23}\ \text{atoms}}{12.0\ \text{g}} = 5.02 \times 10^{22}\ \text{atoms}$$

3. Calculate the number of seconds it takes to count these at 1 per second:

$$t = \frac{N}{R} = \frac{5.02 \times 10^{22}\ \text{atoms}}{1\ \text{atom/s}} = 5.02 \times 10^{22}\ \text{s}$$

4. Calculate the number n of seconds in a year:

$$n = \frac{365\ \cancel{\text{d}}}{1.00\ \text{y}} \times \frac{24\ \cancel{\text{h}}}{1\ \cancel{\text{d}}} \times \frac{3600\ \text{s}}{1\ \cancel{\text{h}}} = 3.15 \times 10^7\ \text{s/y}$$

5. Use the conversion factor 3.15×10^7 s/y (a handy quantity to remember) to convert the answer in step 3 to years:

$$t = 5.02 \times 10^{22}\ \text{s} \times \frac{1.00\ \text{y}}{3.15 \times 10^7\ \text{s}}$$

$$= \frac{5.02}{3.15} \times 10^{22-7}\ \text{y} = \boxed{1.59 \times 10^{15}\ \text{y}}$$

CHECK The answer can be checked by estimation. If you need approximately 10^{22} seconds to count the number of atoms in a gram of carbon and there are approximately 10^7 seconds in a year, then you would need $10^{22}/10^7 = 10^{15}$ y.

TAKING IT FURTHER The time required is about 100000 times the current known value for the age of the universe.

PRACTICE PROBLEM 1-4 How long would it take for 5 billion (5×10^9) people to count the atoms in 1 g of carbon?

Table 1-3 The Universe by Orders of Magnitude

Size or Distance	(m)	Mass	(kg)	Time Interval	(s)
Proton	10^{-15}	Electron	10^{-30}	Time for light to cross nucleus	10^{-23}
Atom	10^{-10}	Proton	10^{-27}	Period of visible light radiation	10^{-15}
Virus	10^{-7}	Amino acid	10^{-25}	Period of microwaves	10^{-10}
Giant amoeba	10^{-4}	Hemoglobin	10^{-22}	Half-life of muon	10^{-6}
Walnut	10^{-2}	Flu virus	10^{-19}	Period of highest audible sound	10^{-4}
Human being	10^{0}	Giant amoeba	10^{-8}	Period of human heartbeat	10^{0}
Highest mountain	10^{4}	Raindrop	10^{-6}	Half-life of free neutron	10^{3}
Earth	10^{7}	Ant	10^{-4}	Period of Earth's rotation	10^{3}
Sun	10^{9}	Human being	10^{2}	Period of Earth's revolution around the Sun	10^{7}
Distance from Earth to the Sun	10^{11}	Saturn V rocket	10^{6}		
		Pyramid	10^{10}	Lifetime of human being	10^{9}
Solar system	10^{13}	Earth	10^{24}	Half-life of plutonium-239	10^{12}
Distance to nearest star	10^{16}	Sun	10^{30}	Lifetime of mountain range	10^{15}
Milky Way galaxy	10^{21}	Milky Way galaxy	10^{41}	Age of Earth	10^{17}
Visible universe	10^{26}	Universe	10^{52}	Age of universe	10^{18}

ORDER OF MAGNITUDE

In doing rough calculations, we sometimes round off a number to the nearest power of 10. Such a number is called an **order of magnitude.** For example, the height of an ant might be 8×10^{-4} m or approximately 10^{-3} m. We would say that the order of magnitude of an ant's height is 10^{-3} m. Similarly, though the typical height h of most people is about 2 m, we might round that off further and say that $h \sim 10^{0}$ m, where the symbol $\sim$ means "is the order of magnitude of." By saying $h \sim 10^{0}$ m we do not mean that a typical height is really 1 m but that it is closer to 1 m than to 10 m or to 10^{-1} m. We might say that a human being is three orders of magnitude taller than an ant, meaning that the ratio of heights is about 1000 to 1. An order of magnitude does not provide any digits that are reliably known, so it has no significant figures. Table 1-3 gives some order-of-magnitude values for a variety of sizes, masses, and time intervals encountered in physics.

In many cases, the order of magnitude of a quantity can be estimated using plausible assumptions and simple calculations. The physicist Enrico Fermi was a master at using order-of-magnitude estimations to generate answers for questions that seemed impossible to calculate because of lack of information. Problems like these are often called **Fermi questions.** The following examples are Fermi questions.

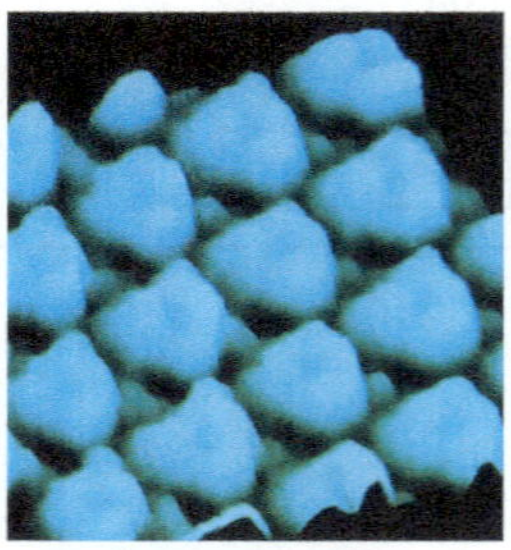

Benzene molecules of the order of 10^{-10} m in diameter as seen in a scanning electron microscope. *(IBM Research, Almaden Research Center.)*

The diameter of the Andromeda galaxy is of the order of 10^{21} m. *(Smithsonian Institution.)*

Distances familiar in our everyday world. The height of the woman is of the order of 10^{0} m and that of the mountain is of the order of 10^{4} m. *(Kent and Donnan Dannon/Photo Researchers.)*

Example 1-6 Burning Rubber

What thickness of rubber tread is worn off the tire of your automobile as it travels 1 km (0.6 mi)?

PICTURE Let's assume the tread thickness of a new tire is 1 cm. This estimation may be off by a factor of two or so, but 1 mm is certainly too small and 10 cm is too large. Because tires have to be replaced after about 60000 km (about 37000 mi), we will also assume that the tread is completely worn off after 60000 km. In other words, the rate of wear is 1 cm of tire per 60000 km of travel.

(Corbis.)

SOLVE

Use 1 cm wear per 60000 km travel to compute the thickness worn after 1 km of travel:

$$\frac{1\text{ cm wear}}{60000\text{ km travel}} = \frac{1.7 \times 10^{-5}\text{ cm wear}}{1\text{ km travel}}$$

$$\approx \boxed{2 \times 10^{-7}\text{ m wear per km of travel}}$$

CHECK If you multiply 1.7×10^{-5} cm/km by 60000 km, you will get approximately 1 cm, which is the thickness of tread on a new tire.

TAKING IT FURTHER Atoms have diameters equal to about 2×10^{-10} m. Thus, the thickness worn off for each kilometer of travel is about 1000 atomic diameters thick.

Example 1-7 How Many Grains

Context-Rich

You have been placed on detention for falling asleep in class. Your teacher says you can get off detention early by estimating the number of grains of sand on a beach. You decide to give it your best shot.

PICTURE First, you make some assumptions about the size of the beach and the size of each grain of sand. Let's assume the beach is about 500 m long, 100 m wide, and 3 m deep. Searching the Internet, you learn that the diameter of a grain varies from 0.04 mm to 2 mm. You assume that each grain is a 1-mm-diameter sphere. Let's also assume that the grains are so tightly packed that the volume of the space between the grains is negligible compared to the volume of the sand itself.

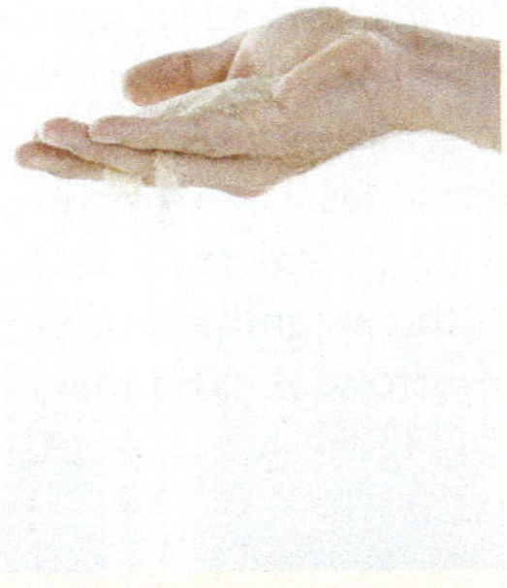

(Corbis.)

SOLVE

1. The volume V_B of the beach is equal to the number N of grains times the volume V_G of a single grain:

$$V_B = NV_G$$

2. Using the formula for the volume of a sphere, find the volume of a single grain of sand:

$$V_G = \frac{4}{3}\pi R^3$$

3. Solve for the number of grains. The numbers in our assumptions are specified to only one significant figure, so the answer will be expressed with one significant figure:

$$V_B = NV_G = N\frac{4}{3}\pi R^3$$

so

$$N = \frac{3V_B}{4\pi R^3} = \frac{3(500\text{ m})(100\text{ m})(3\text{ m})}{4\pi(0.5 \times 10^{-3}\text{ m})^3} = 2.9 \times 10^{14} \approx \boxed{3 \times 10^{14}}$$

CHECK To check the answer, divide the volume of the beach by the number of grains the beach holds. The result is $1.5 \times 10^5\text{ m}^3/3 \times 10^{14}$ grains $= 5 \times 10^{-10}\text{ m}^3/\text{grain}$. This result is the estimated volume of one grain of sand or $4/[3\pi(5 \times 10^{-4}\text{ m})^3]$.

TAKING IT FURTHER The volume of the space between grains can be found by first filling a one-liter container with dry sand, and then slowly pouring water into the container until the sand is saturated with water. If we assume that one-tenth of a liter of water is needed to fully saturate the sand in the container, the actual volume of the sand in the one-liter container is only nine-tenths of a liter. Our estimate of the number of grains on the beach is too high. Taking into account that the sand actually occupies only, say, 90 percent of the volume of its container, the number of grains on the beach would be only 90 percent of the value obtained in step 3 of our solution.

PRACTICE PROBLEM 1-5 How many grains of sand are on a 2-km stretch of beach that is 500 m wide? *Hint: Assume that the sand is* 3.00 m *deep and the diameter of one grain of sand is* 1.00 mm.

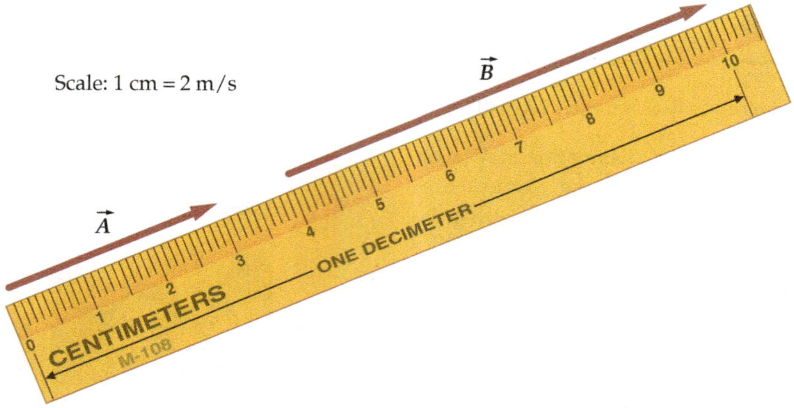

FIGURE 1-2 Velocity vectors $\vec{A}$ and $\vec{B}$ have magnitudes of 6 m/s and 12 m/s, respectively. The arrows representing them are drawn using the scale 1 cm = 2 m/s, so the arrows are drawn 3 and 6 cm long.

1-6 VECTORS

If an object moves in a straight line, we can describe its motion by describing how far or how fast it moves, and whether it moves to the left or right of the origin. But when we look at the motion of an object that is moving in two or three dimensions, we need more than just plus and minus signs to indicate direction. Quantities that have magnitude and direction, such as velocity, acceleration, and force, are called **vectors.** Quantities with magnitude but no associated direction, such as speed, mass, volume, and time, are called **scalars.**

We represent a vector graphically using an arrow. The length of the arrow, drawn to scale, indicates the magnitude of the vector quantity. The direction of the arrow indicates the direction of the vector quantity. Figure 1-2, for example, shows a graphical representation of two velocity vectors. One velocity vector has twice the magnitude of the other. We denote vectors by italic letters with an overhead arrow, $\vec{A}$. The magnitude of $\vec{A}$ is written $|\vec{A}|$, $\|\vec{A}\|$, or simply A. For the vectors in Figure 1-2, $A = |\vec{A}| = 6$ m/s and $B = |\vec{B}| = 12$ m/s.

> ! While working with vectors, you should always include an arrow over the letter to indicate a vector quantity. The letter without the arrow represents only the magnitude of the vector quantity. Note that the magnitude of a vector is never negative.

1-7 GENERAL PROPERTIES OF VECTORS

Like scalar quantities, vector quantities can be added, subtracted, and multiplied. However, manipulating vectors algebraically requires taking into account their direction. In this section, we will examine some of the general properties of vectors and how to work with them (multiplication of two vectors will be discussed in Chapters 6 and 9). Throughout most of the discussion, we will consider displacement vectors—vectors that represent change of position—because they are the most basic of vectors. However, keep in mind that the properties apply to all vectors, not just displacement vectors.

BASIC DEFINITIONS

If an object moves from location A to location B, we can represent its **displacement** with an arrow pointing from A to B, as shown in Figure 1-3*a*. The length of the arrow represents the distance, or magnitude, between the two locations. The direction of the arrow represents the direction from A to B. A displacement vector is a directed straight-line segment from the initial location to the final location that represents the change in position of an object. It does not necessarily represent the

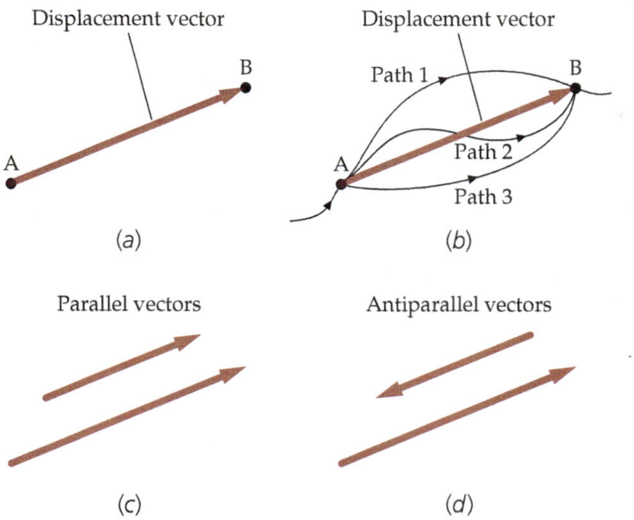

FIGURE 1-3 (*a*) shows a displacement vector from a point A to a point B; (*b*) shows the same displacement vector with three different paths between the two points; (*c*) shows the same displacement vector next to a second displacement vector that is parallel but a different length; (*d*) shows the same displacement vector next to a vector that is antiparallel (the head and tail are reversed) and a different length.

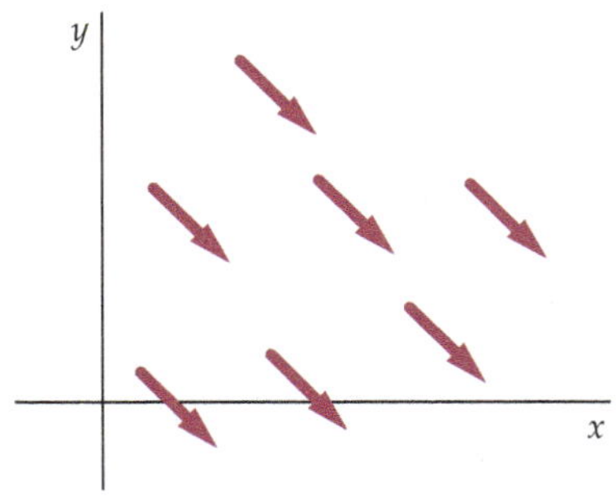

FIGURE 1-4 Vectors are equal if their magnitudes and directions are the same. All vectors in this figure are equal.

actual path the object takes. For example, in Figure 1-3*b*, the same displacement vector corresponds to all three paths between points A and B.

If two displacement vectors have the same direction, as shown in Figure 1-3*c*, they are **parallel.** If they have opposite directions, as shown in Figure 1-3*d*, they are **antiparallel.** If two vectors have *both* the same magnitude and the same direction, they are said to be equal. Graphically, this means that they have the same length and are parallel to each other. A vector can be drawn at different locations as long as it is drawn with the correct magnitude (length) and in the correct direction. Thus, all the vectors in Figure 1-4 are equal. In addition, vectors do not depend on the *coordinate system* used to represent them (except for position vectors, which are introduced in Chapter 3). Two or three mutually perpendicular coordinate axes form a coordinate system.

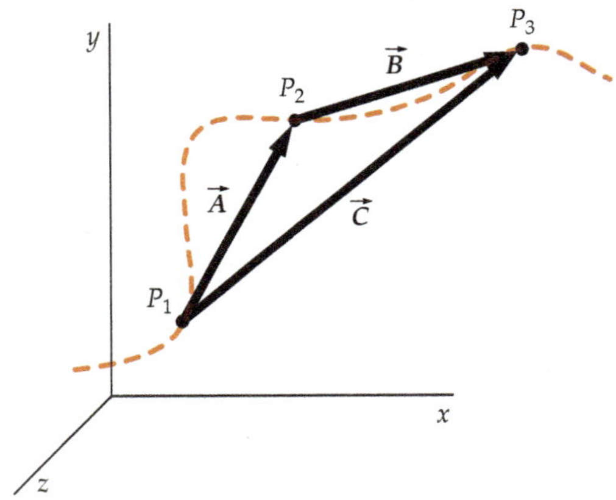

FIGURE 1-5

ADDITION AND SUBTRACTION OF VECTORS

Suppose you decide to take a hike along a trail through a forest. Figure 1-5 shows your path as you move from point P_1 to a second point P_2 and then to a third point P_3. The vector $\vec{A}$ represents your displacement from P_1 to P_2, while $\vec{B}$ represents your displacement from P_2 to P_3. Note that these displacement vectors depend only on the endpoints and not on the actual path taken. Your *net* displacement from P_1 to P_3, is a new vector, labeled $\vec{C}$ in the figure, and is called the sum of the two successive displacements $\vec{A}$ and $\vec{B}$:

$$\vec{C} = \vec{A} + \vec{B} \qquad \text{1-1}$$

The sum of two vectors is called the sum, **vector sum,** or **resultant.**

The plus sign in Equation 1-1 refers to a process called *vector addition.* We find the sum using a geometric process that takes into account both the magnitudes and the directions of the quantities. To add two displacement vectors graphically, we draw the second vector $\vec{B}$ with its tail $\vec{B}$ at the head of the first vector $\vec{A}$ (Figure 1-6). The resultant vector is then drawn from the tail of the first to the head of the second. This method of adding vectors is called the **head-to-tail method.**

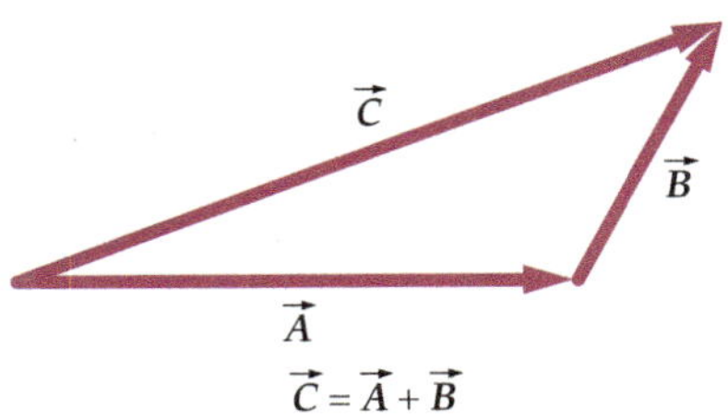

FIGURE 1-6 Head-to-tail method of vector addition.

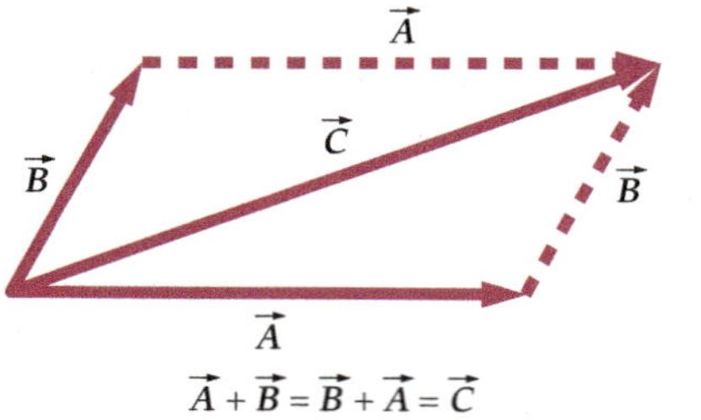

FIGURE 1-7 Parallelogram method of vector addition.

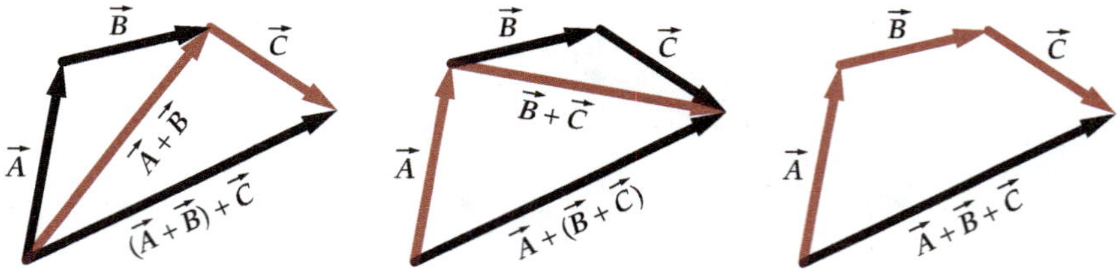

FIGURE 1-8 Vector addition is associative. That is, $(\vec{A} + \vec{B}) + \vec{C} = \vec{A} + (\vec{B} + \vec{C})$.

An equivalent way of adding vectors, called the **parallelogram method,** involves drawing $\vec{B}$ so that it is tail-to-tail with $\vec{A}$ (Figure 1-7). A diagonal of the parallelogram formed by $\vec{A}$ and $\vec{B}$ then equals $\vec{C}$ as shown (Figure 1-7). As you can see in the figure, it makes no difference in which order we add two vectors; that is, $\vec{A} + \vec{B} = \vec{B} + \vec{A}$. Therefore, vector addition obeys the commutative law.

To add more than two vectors—for example, $\vec{A}$, $\vec{B}$, and $\vec{C}$—we first add two vectors (Figure 1-8), and then add the third vector to the vector sum of the first two. The order in which the vectors are grouped before adding does not matter; that is $(\vec{A} + \vec{B}) + \vec{C} = \vec{A} + (\vec{B} + \vec{C})$. This reveals that like the addition of ordinary numbers, vector addition is associative.

> ! C does not equal $A + B$ unless $\vec{A}$ and $\vec{B}$ are in the same direction. That is, $\vec{C} = \vec{A} + \vec{B}$ does not imply that $C = A + B$.

If vectors $\vec{A}$ and $\vec{B}$ are equal in magnitude and opposite in direction, then the vector $\vec{C} = \vec{A} + \vec{B}$ is a vector with a magnitude of zero. This can be shown by using the head-to-tail method of vector addition to graphically construct the sum $\vec{A} + \vec{B}$. Any vector with a magnitude of zero is called the zero vector $\vec{0}$. The direction of a vector with zero magnitude has no meaning, so in this book we will not use vector notation for the zero vector. That is, we will use 0 rather than $\vec{0}$ to denote the zero vector. If $\vec{A} + \vec{B} = 0$, then $\vec{B}$ is said to be the negative of $\vec{A}$ and vice versa. Note that $\vec{B}$ is the negative of $\vec{A}$ if $\vec{B}$ has the same magnitude as $\vec{A}$ but is in the opposite direction. The negative of $\vec{A}$ is written $-\vec{A}$, so if $\vec{A} + \vec{B} = 0$, then $\vec{B} = -\vec{A}$ (Figure 1-9).

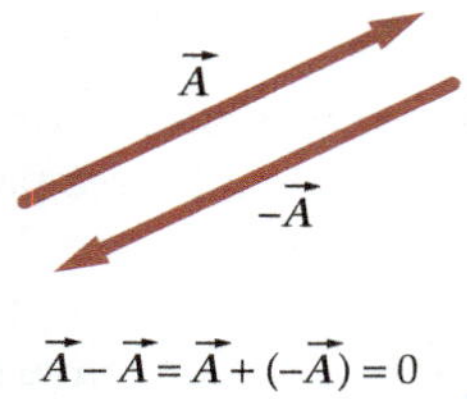

FIGURE 1-9

To subtract vector $\vec{B}$ from vector $\vec{A}$, add the negative of $\vec{B}$ to $\vec{A}$. The result is $\vec{C} = \vec{A} - \vec{B} = \vec{A} + (-\vec{B})$ (Figure 1-10*a*). An alternative method of subtracting $\vec{B}$ from $\vec{A}$ is to add $\vec{B}$ to both sides of the equation $\vec{C} = \vec{A} + (-\vec{B})$ to obtain $\vec{B} + \vec{C} = \vec{A}$, and then graphically add $\vec{B}$ and $\vec{C}$ to get $\vec{A}$ using the head-to-tail method. This is accomplished by first drawing $\vec{A}$ and $\vec{B}$ tail-to-tail (Figure 1-10*b*), and then drawing $\vec{C}$ from the head of $\vec{B}$ to the head of $\vec{A}$.

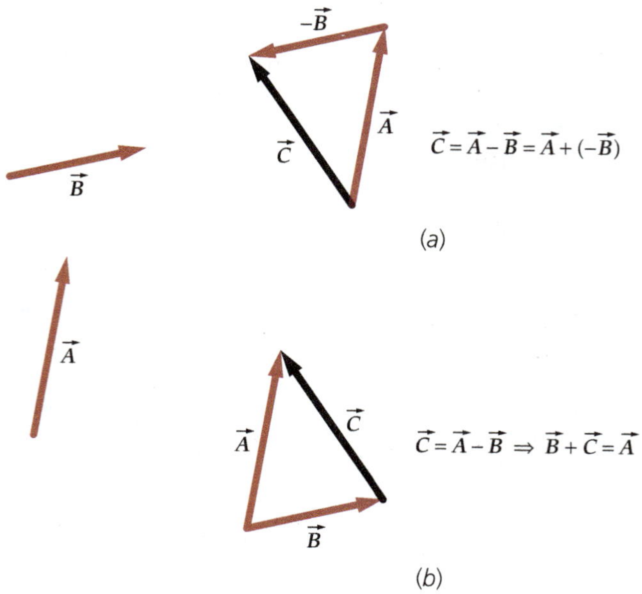

FIGURE 1-10 Alternative ways of subtracting vectors. Let $\vec{C} = \vec{A} - \vec{B}$. (*a*) To obtain $\vec{C}$, we add $-\vec{B}$ to $\vec{A}$. (*b*) To obtain $\vec{C}$, we first draw $\vec{A}$ and $\vec{B}$ with their tails together. Then, $\vec{C}$ is the vector we add to $\vec{B}$ to get $\vec{A}$.

Example 1-8 Your Displacement — Conceptual

You walk 3.00 km due east and then 4.00 km due north. Determine your resultant displacement by adding these two displacement vectors graphically.

PICTURE Your displacement is the vector from your initial position to your final position. You can add the two individual displacement vectors graphically to find the resultant displacement. To accurately draw the resultant, you must use a scale such as 1 cm on the drawing = 1 km on the ground.

SOLVE

1. Let $\vec{A}$ and $\vec{B}$ represent displacements of 3.00 km due east and 4.00 km due north, respectively, and let $\vec{C} = \vec{A} + \vec{B}$. Draw $\vec{A}$ and $\vec{B}$ with the tail of $\vec{B}$ at the head of $\vec{A}$, and with $\vec{C}$ drawn from the tail of $\vec{A}$ to the head of $\vec{B}$ (Figure 1-11). Use the scale 1 cm = 1 km. Include axes indicating the directions north and east.
2. Determine the magnitude and direction of $\vec{C}$ using your diagram, the scale 1 cm = 1 km, and a protractor.

The arrow representing $\vec{C}$ has a length of 5.00 cm, so the magnitude of $\vec{C}$ is 5.00 km. The direction of $\vec{C}$ is approximately 53° north of east.

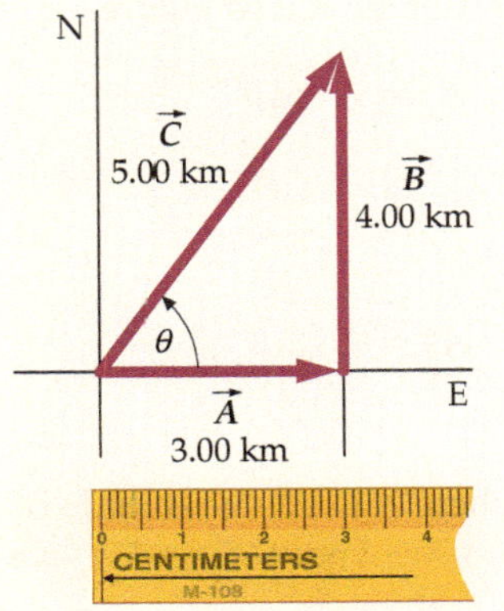

FIGURE 1-11

CHECK The distance traveled is 3.00 km + 4.00 km = 7.00 km and the magnitude of the net displacement is 5 km. This is consistent with the adage "the shortest distance between two points is a straight line." Also, if you go 3 km east and 4 km north, you should expect to be somewhat more than 45° north of east from your starting point.

TAKING IT FURTHER A vector is described by its magnitude and its direction. Your resultant displacement is therefore a vector of length 5.00 km in a direction approximately 53° north of east.

MULTIPLYING A VECTOR BY A SCALAR

The expression $3\vec{A}$, where $\vec{A}$ is an arbitrary vector, represents the sum $\vec{A} + \vec{A} + \vec{A}$. That is, $\vec{A} + \vec{A} + \vec{A} = 3\vec{A}$. (In like manner, $(-\vec{A}) + (-\vec{A}) + (-\vec{A}) = 3(-\vec{A}) = -3\vec{A}$.) More generally, the vector $\vec{A}$ multiplied by a scalar s is the vector $\vec{B} = s\vec{A}$, where $\vec{B}$ has magnitude $|s|A$. $\vec{B}$ is in the same direction as $\vec{A}$ if s is positive and is in the opposite direction if s is negative. The dimensions of $s\vec{A}$ are those of s multiplied by those of A. (In addition, to divide $\vec{A}$ by a scalar s, you multiply $\vec{A}$ by $1/s$.)

COMPONENTS OF VECTORS

We can add or subtract vectors algebraically by first breaking down the vectors into their components. The **component** of a vector in a given direction is the projection of the vector onto an axis in that direction. We can find the components of a vector by drawing perpendicular lines from the ends of the vector to the axis, as shown in Figure 1-12. The process of finding the x, y, and z components of a

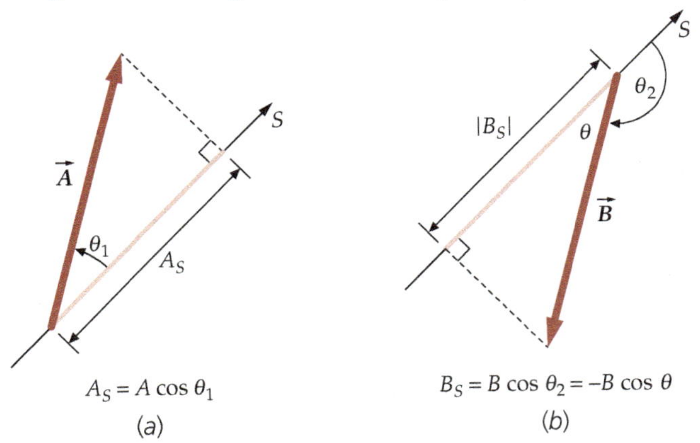

FIGURE 1-12 The component of a vector in a specified direction is equal to the magnitude of the vector times the cosine of the angle between the direction of the vector and the specified direction. The component of the vector $\vec{A}$ in the $+S$ direction is A_S, and A_S is positive. The component of the vector $\vec{B}$ in the $+S$ direction is B_S, and B_S is negative.

vector is called **resolving the vector** into its components. The components of a vector along the x, y, and z directions, illustrated in Figure 1-13 for a vector in the xy plane, are called the rectangular (or Cartesian) components. Note that the components of a vector *do* depend on the coordinate system used, although the vector itself does not.

We can use right-triangle geometry to find the rectangular components of a vector. If θ is the angle measured counterclockwise* from the $+x$ direction to the direction of $\vec{A}$ (see Figure 1-13), then

$$A_x = A\cos\theta \qquad \text{1-2}$$

x COMPONENT OF A VECTOR

and

$$A_y = A\sin\theta \qquad \text{1-3}$$

y COMPONENT OF A VECTOR

where A is the magnitude of $\vec{A}$.

If we know A_x and A_y we can find the angle θ from

$$\tan\theta = \frac{A_y}{A_x} \qquad \theta = \tan^{-1}\frac{A_y}{A_x} \qquad \text{1-4}$$

and the magnitude A from the Pythagorean theorem:

$$A = \sqrt{A_x^2 + A_y^2} \qquad \text{1-5}a$$

In three dimensions,

$$A = \sqrt{A_x^2 + A_y^2 + A_z^2} \qquad \text{1-5}b$$

Components can be positive or negative. The x component of a vector is positive if the x coordinate of an ant as it walks from the tail to the head of the vector increases. Thus, if $\vec{A}$ points in the positive x direction, then A_x is positive, and if $\vec{A}$ points in the negative x direction, then A_x is negative.

It is important to note that in Equation 1-4, the inverse tangent function is multiple valued. This issue is clarified in Example 1-9.

PRACTICE PROBLEM 1-6

A car travels 20.0 km in a direction 30.0° north of west. Let east be the $+x$ direction and north be the $+y$ direction, as in Figure 1-14. Find the x and y components of the displacement vector of the car.

Once we have resolved a vector into its components, we can manipulate the individual components. Consider two vectors $\vec{A}$ and $\vec{B}$ that lie in the xy plane. The rectangular components of each vector and those of the sum $\vec{C} = \vec{A} + \vec{B}$ are shown in Figure 1-15. We see that the rectangular components of each vector and those of the sum $\vec{C} = \vec{A} + \vec{B}$ are equivalent to the two component equations

$$C_x = A_x + B_x \qquad \text{1-6}a$$

and

$$C_y = A_y + B_y \qquad \text{1-6}b$$

In other words, the sum of the x components equals the x component of the resultant, and the sum of the y components equals the y component of the resultant. The angle and magnitude of the resultant vector can be found using Equations 1-4 and 1-5a, respectively.

* This assumes the $+y$ direction is 90° counterclockwise from the $+x$ direction.

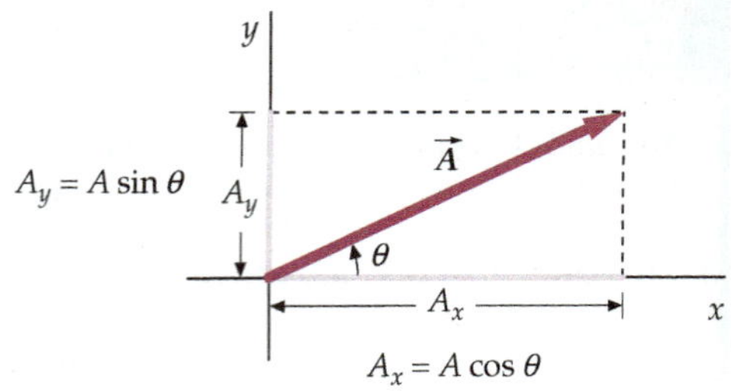

FIGURE 1-13 The rectangular components of a vector. θ is the angle between the direction of the vector and the $+x$ direction. The angle is positive if it is measured counterclockwise from the $+x$ direction, as shown.

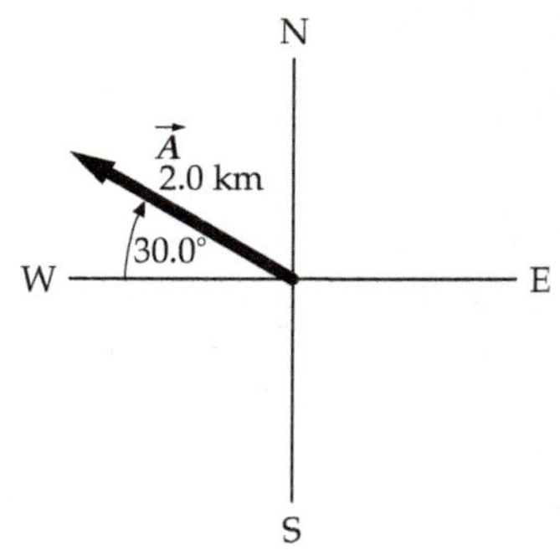

FIGURE 1-14

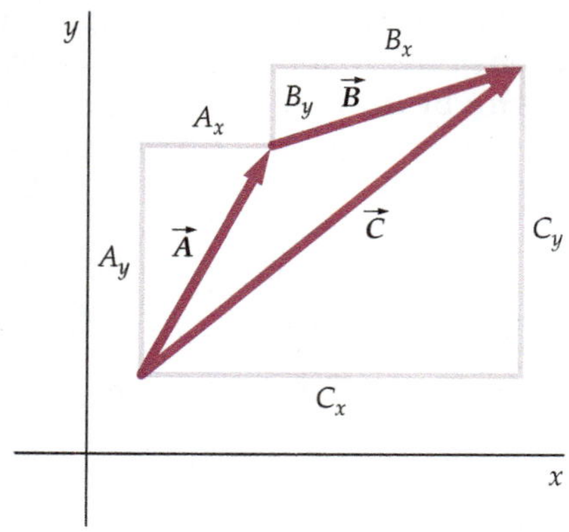

FIGURE 1-15

Example 1-9 A Treasure Map — Context-Rich

You are working at a tropical resort, and are setting up a treasure hunt activity for the guests. You've been given a map and told to follow its directions in order to bury a "treasure" at a specific location. You don't want to waste time walking around the island, because you want to finish early and go surfing. The directions are to walk 3.00 km headed 60.0° north of due east and then 4.00 km headed 40.0° north of due west. Where should you head and how far must you walk to get the job done quickly? Find your answer (*a*) graphically and (*b*) using components.

PICTURE In both cases you need to find your resultant displacement. In Part (*a*), use the head-to-tail method of vector addition and solve for the resultant vector graphically. You can do this by drawing each of the displacements to scale and then measuring the resultant displacement directly from your sketch. For Part (*b*), you will need to resolve the vectors into their individual components and then use the components to find the resultant displacement.

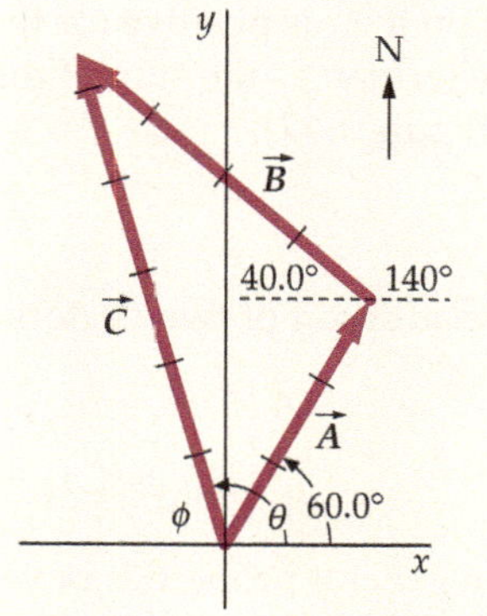

FIGURE 1-16

SOLVE

(*a*) 1. Draw a vector-addition diagram to scale (Figure 1-16). First draw coordinate axes, with the $+x$ direction toward the east and the $+y$ direction toward the north. Next, starting at the origin draw the first displacement vector $\vec{A}$ 3.00 cm long at 60.0° north of east. Beginning at the head of $\vec{A}$, draw the second vector $\vec{B}$ 4.00 cm long at 40.0° north of west. (You will need a protractor to measure the angles.) Then draw the resultant vector $\vec{C}$ from the tail of $\vec{A}$ to the head of $\vec{B}$:

2. Measure the length of $\vec{C}$. Using a protractor, measure the angle between the direction of $\vec{C}$ and the $+x$ direction:

$\vec{C}$ is about 5.40 cm long. Thus, the magnitude of the resultant displacement is $\boxed{5.40\text{ km}}$. The angle ϕ made between $\vec{C}$ and due west is about 73.2°. Therefore, you should walk 5.40 km headed $\boxed{73.2°\text{ north of west}}$.

(*b*) 1. To solve using components, let $\vec{A}$ denote the first displacement and choose the $+x$ direction toward the east and the $+y$ direction toward the north. Compute A_x and A_y from Equations 1-2 and 1-3:

$$A_x = (3.00\text{ km})\cos 60° = 1.50\text{ km}$$
$$A_y = (3.00\text{ km})\sin 60° = 2.60\text{ km}$$

2. Similarly, compute the components of the second displacement $\vec{B}$. The angle between the direction of $\vec{B}$ and the $+x$ direction is $180.0° - 40.0° = 140°$:

$$B_x = (4.00\text{ km})\cos 140° = -3.06\text{ km}$$
$$B_y = (4.00\text{ km})\sin 140° = +2.57\text{ km}$$

3. The components of the resultant displacement $\vec{C} = \vec{A} + \vec{B}$ are found by addition:

$$C_x = A_x + B_x = 1.50\text{ km} - 3.06\text{ km} = -1.56\text{ km}$$
$$C_y = A_y + B_y = 2.60\text{ km} + 2.57\text{ km} = 5.17\text{ km}$$

4. The Pythagorean theorem gives the magnitude of $\vec{C}$:

$$C^2 = C_x^2 + C_y^2 = (-1.56\text{ km})^2 + (5.17\text{ km})^2 = 29.2\text{ km}^2$$
$$C = \sqrt{29.2\text{ km}^2} = \boxed{5.40\text{ km}}$$

5. The ratio of C_y to C_x equals the tangent of the angle θ between $\vec{C}$ and the positive x direction. Be careful, the value you are seeking may be 180° larger than the value returned by your calculator for the inverse tangent:

$$\tan\theta = \frac{C_y}{C_x} \quad \text{so}$$
$$\theta = \tan^{-1}\frac{5.17\text{ km}}{-1.56\text{ km}} = \tan^{-1}(-3.31)$$
$$= \text{either } -73.2° \quad \text{or} \quad (-73.2° + 180°)$$
$$= \text{either } -73.2° \quad \text{or} \quad +107°$$

6. Because C_y is positive and C_x is negative we know to select the value for θ in the second quadrant:

$$\theta = \boxed{107°\text{ counterclockwise from east}}$$
$$\phi = \boxed{73.2°\text{ north of west}}$$

CHECK Step 4 of Part (*b*) gives the magnitude as 5.40 km and step 6 gives the direction as 73.2° north of west. This agrees with the results in Part (*a*) within the accuracy of our measurement.

TAKING IT FURTHER To specify a vector, you need to specify either the magnitude *and* direction, or both components. In this example, the magnitude and direction was specifically asked for.

UNIT VECTORS

A **unit vector** is a *dimensionless* vector with magnitude exactly equal to 1. The vector $\hat{A} = \vec{A}/A$ is an example of a unit vector that points in the direction of $\vec{A}$. The circumflex, or hat, denotes that it is a unit vector. Unit vectors that point in the positive x, y, and z directions are convenient for expressing vectors in terms of their rectangular components. These unit vectors are usually written $\hat{i}$, $\hat{j}$, and $\hat{k}$, respectively. For example, the vector $A_x\hat{i}$ has magnitude $|A_x|$ and points in the $+x$ direction if A_x is positive (or the $-x$ direction if A_x is negative). A general vector $\vec{A}$ can be written as the sum of three vectors, each of which is parallel to a coordinate axis (Figure 1-17):

$$\vec{A} = A_x\hat{i} + A_y\hat{j} + A_z\hat{k} \tag{1-7}$$

The addition of two vectors $\vec{A}$ and $\vec{B}$ can be written in terms of unit vectors as

$$\begin{aligned}\vec{A} + \vec{B} &= (A_x\hat{i} + A_y\hat{j} + A_z\hat{k}) + (B_x\hat{i} + B_y\hat{j} + B_z\hat{k}) \\ &= (A_x + B_x)\hat{i} + (A_y + B_y)\hat{j} + (A_z + B_z)\hat{k}\end{aligned} \tag{1-8}$$

The general properties of vectors are summarized in Table 1-4.

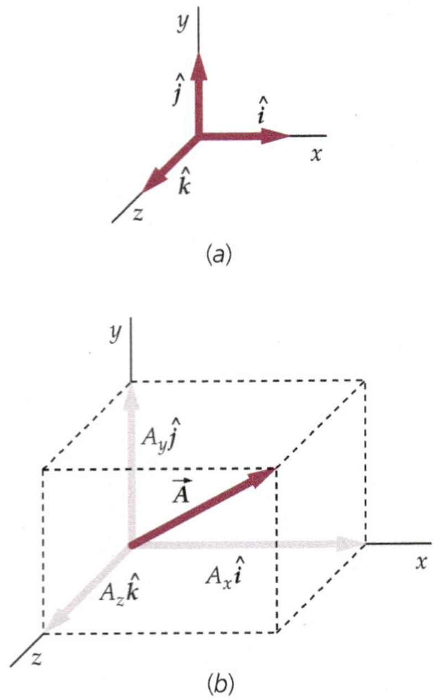

FIGURE 1-17 (*a*) The unit vectors $\hat{i}$, $\hat{j}$, and $\hat{k}$ in a rectangular coordinate system. (*b*) The vector $\vec{A}$ in terms of the unit vectors: $\vec{A} = A_x\hat{i} + A_y\hat{j} + A_z\hat{k}$.

PRACTICE PROBLEM 1-7

Given vectors $\vec{A} = (4.00\text{ m})\hat{i} + (3.00\text{ m})\hat{j}$ and $\vec{B} = (2.00\text{ m})\hat{i} - (3.00\text{ m})\hat{j}$, find (*a*) A, (*b*) B, (*c*) $\vec{A} + \vec{B}$, and (*d*) $\vec{A} - \vec{B}$.

Table 1-4 Properties of Vectors

Property	Explanation	Figure	Component Representation
Equality	$\vec{A} = \vec{B}$ if $\lvert\vec{A}\rvert = \lvert\vec{B}\rvert$ and their directions are the same		$A_x = B_x$ $A_y = B_y$ $A_z = B_z$
Addition	$\vec{C} = \vec{A} + \vec{B}$	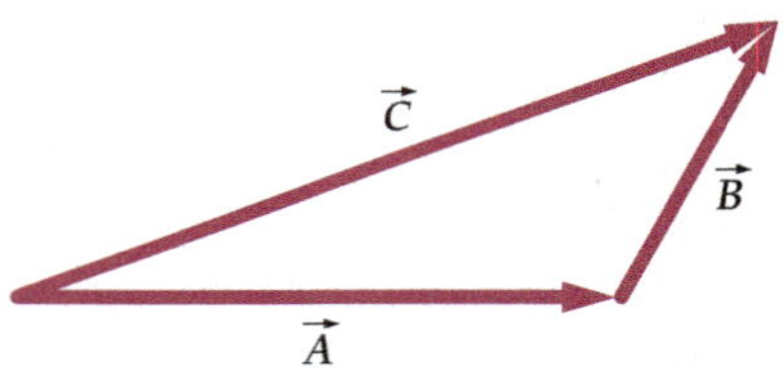	$C_x = A_x + B_x$ $C_y = A_y + B_y$ $C_z = A_z + B_z$
Negative of a vector	$\vec{A} = -\vec{B}$ if $\lvert\vec{B}\rvert = \lvert\vec{A}\rvert$ and their directions are opposite		$A_x = -B_x$ $A_y = -B_y$ $A_z = -B_z$
Subtraction	$\vec{C} = \vec{A} - \vec{B}$	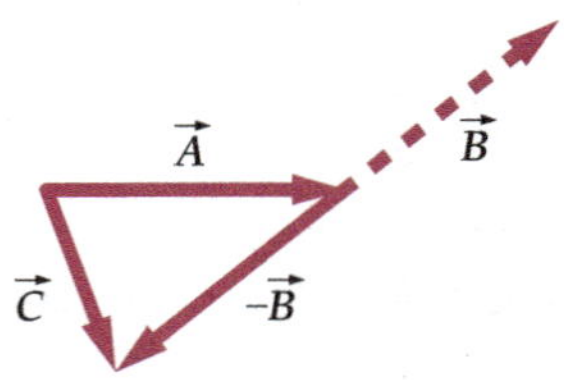	$C_x = A_x - B_x$ $C_y = A_y - B_y$ $C_z = A_z - B_z$
Multiplication by a scalar	$\vec{B} = s\vec{A}$ has magnitude $\lvert\vec{B}\rvert = \lvert s\rvert\lvert\vec{A}\rvert$ and has the same direction as $\vec{A}$ if s is positive or $-\vec{A}$ if s is negative		$B_x = sA_x$ $B_y = sA_y$ $B_z = sA_z$

Physics Spotlight

The 2005 Leap Second

The calendar year 2005 was longer—by exactly one second, known officially as a "leap second." This adjustment was necessary to synchronize two systems of keeping time, one based on Earth's rotation and the other based on a select group of atomic clocks.

Throughout history, timekeeping has been related to the position of the Sun in the sky, a factor determined by Earth's rotation on its axis and around the Sun. This astronomical time, now called Universal Time (UT1), assumed that the rate of Earth's rotation was uniform. But as more accurate methods of measurement were developed, it became evident that there were slight irregularities in the rotation rate of Earth. This meant that there would also be some variability in the scientific standard unit for time, the second, as long as its definition—$(1/60)(1/60)(1/24)$ of a mean solar day—depended on astronomical time.

In 1955 the National Physical Laboratory in Britain developed the first cesium atomic clock, a device of far greater accuracy than any clock formerly in existence. Timekeeping could now be independent of astronomical observations, and a much more precise definition of the second could be given based on the frequency of radiation emitted in the transition between two energy levels of the cesium-133 atom. However, the more familiar UT1 continues to be of importance for systems such as navigation and astronomy. Thus it is important that atomic time and UT1 be synchronized.

According to the National Physical Laboratory, UK, "The solution adopted [for synchronization] was to construct an atomic time scale called Coordinated Universal Time … (UTC) as the basis of international timekeeping. It combines all the regularity of atomic time with most of the convenience of UT1, and many countries have adopted it as the legal basis for time."* The International Bureau of Weights and Measures in Sèvres, France, takes data from select time laboratories around the world, including the U.S. Naval Observatory in Washington, DC, to provide the international standard UTC.

When slight differences accrue between UTC and UT1 because Earth's rotation varies slightly (usually slowing) over time, a leap second is added to close the gap. The concept is similar to the way that leap years are used to correct the calendar. A year is not exactly 365 days, but rather 365.242 days. To account for this, an extra day is added to the calendar every four years and designated February 29.

Since 1972 when the world shifted to atomic timekeeping, 23 leap seconds have been added to UTC. By international agreement, a leap second is added whenever the difference between UT1 and UTC approaches 0.9 seconds. The International Earth Rotation and Reference Systems (IERS) Service, through its center at the Paris Observatory, announces the need for a leap second months in advance.

In a year without any leap second, the last second of the year would be 23:59:59 UTC on December 31, while the first second of the new year would be 00:00:00 UTC on January 1 of the new year. But for 2005 a leap second was added at 23:59:59 UTC on December 31, so that atomic clocks read 23:59:60 UTC before changing to all zeros.

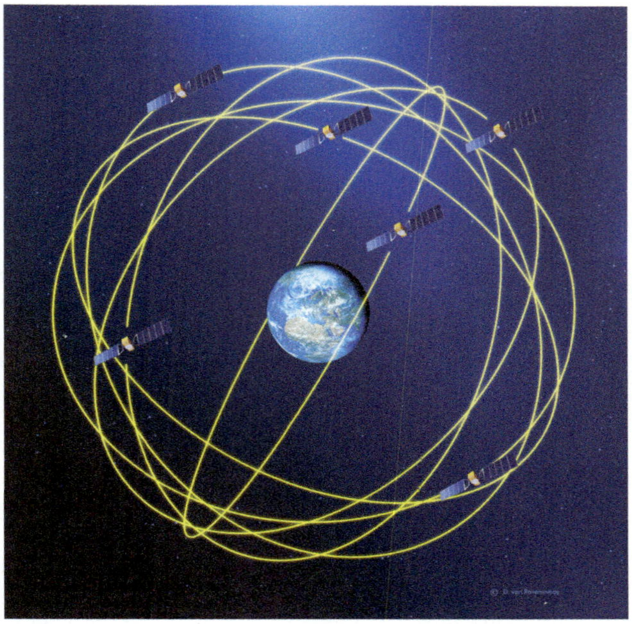

The global positioning system (GPS) requires that there be 24 satellites in primary service at least 70 percent of the time. Each primary satellite has an orbital period of 1/2 a sidereal day (1 sidereal day = ~23 h 56 min) and an orbital radius about 4 times the radius of Earth. There are 6 equally spaced orbital planes, each of which is inclined 55° with respect to the equatorial plane of Earth, and each of these planes contains 4 primary satellites. In addition, there are several other GPS satellites that serve as in-orbit spares in the event that one or more of the primary satellites fails. At the time of this writing (May 2006) there are 29 operational satellites in orbit. *(Detlev Van Ravenswaay/Photo Researchers.)*

* http://www.npl.co.uk/time/leap_second.html

Summary

TOPIC	RELEVANT EQUATIONS AND REMARKS	
1. Units	**Physical quantities** are numbers that are obtained by taking measurements of physical objects. Operational definitions specify operations or procedures that, if followed, define physical quantities. The magnitude of a physical quantity is expressed as a number times a unit.	
2. Base Units	The base units in the SI system are the meter (m), the second (s), the kilogram (kg), the kelvin (K), the ampere (A), the mole (mol), and the candela (cd). The unit(s) of every physical quantity can be expressed in terms of these base units.	
3. Units in Equations	Units in equations are treated just like any other algebraic quantity.	
4. Conversion	**Conversion factors,** which are always equal to 1, provide a convenient method for converting from one kind of unit to another.	
5. Dimensions	The terms of an equation must have the same dimensions.	
6. Scientific Notation	For convenience, very small and very large numbers are generally written as a number between 1 and 10 times a power of 10.	
7. Exponents		
Multiplication	When multiplying two numbers, the exponents are added.	
Division	When dividing two numbers, the exponents are subtracted.	
Raising to a power	When a number containing an exponent is itself raised to a power, the exponents are multiplied.	
8. Significant Figures		
Multiplication and division	The number of significant figures in the result of multiplication or division is *no greater than* the least number of significant figures in any of the numbers.	
Addition and subtraction	The result of addition or subtraction of two numbers has no significant figures beyond the last decimal place where both of the numbers being added or subracted have significant figures.	
9. Order of Magnitude	A number rounded to the nearest power of 10 is called an order of magnitude. The order of magnitude of a quantity can often be estimated using plausible assumptions and simple calculations.	
10. Vectors		
Definition	Vectors are quantities that have both magnitude and direction. Vectors add like displacements.	
Components	The component of a vector in a direction in space is the projection of the vector on an axis in that direction. If $\vec{A}$ makes an angle θ with the positive x direction, its x and y components are	
	$A_x = A\cos\theta$	1-2
	$A_y = A\sin\theta$	1-3
Magnitude	$A = \sqrt{A_x^2 + A_y^2}$	1-5*a*
Adding vectors graphically	Two vectors may be added graphically by drawing them with the tail of the second arrow at the head of the first arrow. The arrow representing the resultant vector is drawn from the tail of the first vector to the head of the second.	
Adding vectors using components	If $\vec{C} = \vec{A} + \vec{B}$, then	
	$C_x = A_x + B_x$	1-6*a*
	and	
	$C_y = A_y + B_y$	1-6*b*
Unit vectors	A vector $\vec{A}$ can be written in terms of unit vectors $\hat{i}$, $\hat{j}$, and $\hat{k}$, which are dimensionless, have unit magnitude and lie along the x, y, and z axes, respectively	
	$\vec{A} = A_x\hat{i} + A_y\hat{j} + A_z\hat{k}$	1-7

Answers to Concept Checks

1-1 5

Answers to Practice Problems

1-1 (*a*) 300 ns; (*b*) 40 Mm

1-2 (*a*) 0.05, (*b*) 3.9, (*c*) 0.003

1-3 2.39×10^2

1-4 3.2×10^5 y

1-5 $\approx 6 \times 10^{15}$

1-6 $A_x = -17.3$ km, $A_y = 10.0$ km

1-7 (*a*) $A = 5.00$ m, (*b*) $B = 3.61$ m, (*c*) $\vec{A} + \vec{B} = (6.00\text{ m})\hat{i}$, (*d*) $\vec{A} - \vec{B} = (2.00\text{ m})\hat{i} + (6.00\text{ m})\hat{j}$

Problems

In a few problems, you are given more data than you actually need; in a few other problems, you are required to supply data from your general knowledge, outside sources, or informed estimate.

Interpret as significant all digits in numerical values that have trailing zeroes and no decimal points.

- • Single-concept, single-step, relatively easy
- •• Intermediate-level, may require synthesis of concepts
- ••• Challenging, for advanced students
- SSM Solution is in the *Student Solutions Manual*
- Consecutive problems that are shaded are paired problems.

CONCEPTUAL PROBLEMS

1 • Which of the following *is not* one of the base quantities in the SI system? (*a*) mass, (*b*) length, (*c*) energy, (*d*) time, (*e*) All of the above are base quantities. SSM

2 • In doing a calculation, you end up with m/s in the numerator and m/s^2 in the denominator. What are your final units? (*a*) m^2/s^3, (*b*) 1/s, (*c*) s^3/m^2, (*d*) s, (*e*) m/s

3 • The prefix giga means (*a*) 10^3, (*b*) 10^6, (*c*) 10^9, (*d*) 10^{12}, (*e*) 10^{15}.

4 • The prefix mega means (*a*) 10^{-9}, (*b*) 10^{-6}, (*c*) 10^{-3}, (*d*) 10^6, (*e*) 10^9.

5 • Show that there are 30.48 cm per foot. How many centimeters are there in one mile? SSM

6 • The number 0.000 513 0 has ______ significant figures. (*a*) one, (*b*) three, (*c*) four, (*d*) seven, (*e*) eight

7 • The number 23.0040 has ______ significant figures. (*a*) two, (*b*) three, (*c*) four, (*d*) five, (*e*) six

8 • Force has dimensions of mass times acceleration. Acceleration has dimensions of speed divided by time. Pressure is defined as force divided by area. What are the dimensions of pressure? Express pressure in terms of the SI base units kilogram, meter, and second.

9 • True or false: Two quantities must have the same dimensions in order to be multiplied.

10 • A vector has a negative x component and a positive y component. Its angle measured counterclockwise from the positive x axis is (*a*) between zero and 90 degrees, (*b*) between 90 and 180 degrees, (*c*) more than 180 degrees.

11 • A vector $\vec{A}$ points in the $+x$ direction. Show graphically at least three choices for a vector $\vec{B}$ such that $\vec{B} + \vec{A}$ points in the $+y$ direction. SSM

12 • A vector $\vec{A}$ points in the $+y$ direction. Show graphically at least three choices for a vector $\vec{B}$ such that $\vec{B} - \vec{A}$ points in the $+x$ direction.

13 • Is it possible for three equal-magnitude vectors to add to zero? If so, sketch a graphical answer. If not, explain why not. SSM

ESTIMATION AND APPROXIMATION

14 • The angle subtended by the moon's diameter at a point on Earth is about 0.524° (Figure 1-18). Use this information and the fact that the moon is about 384 Mm away to find the diameter of the moon. *Hint: The angle can be determined from the diameter of the moon and the distance to the moon.*

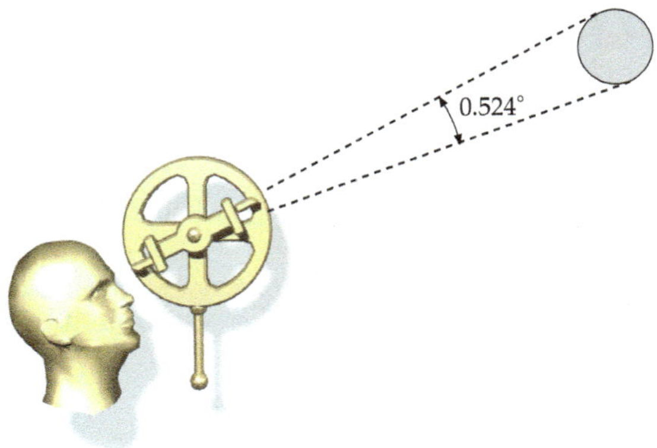

FIGURE 1-18 Problem 14

15 • **BIOLOGICAL APPLICATION** Some good estimates about the human body can be made if it is assumed that we are made mostly of water. The mass of a water molecule is 29.9×10^{-27} kg. If the mass of a person is 60 kg, estimate the number of water molecules in that person. SSM

16 •• **ENGINEERING APPLICATION** In 1989, IBM scientists moved atoms with a scanning tunneling microscope (STM). One of the first STM images seen by the general public was of the letters IBM spelled with xenon atoms on a nickel surface. The letters IBM were 15 xenon atoms across. If the space between the centers of adjacent xenon atoms is 5 nm (5×10^{-9} m), estimate how many times "IBM" could be written across this 8.5-inch page.

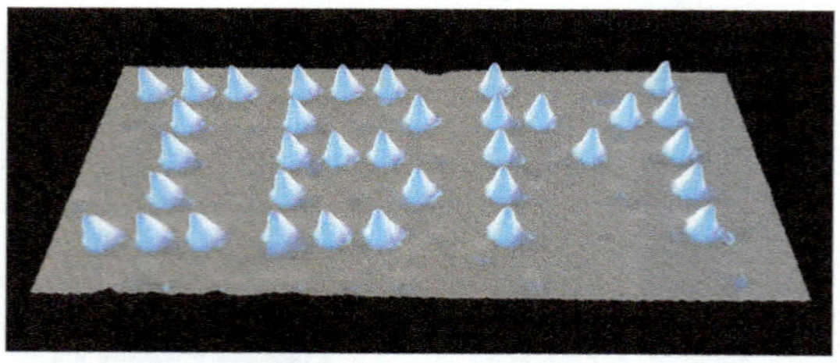

(By permission of IBM Reasearch, Almaden Research Center.)

17 •• There is an environmental debate over the use of cloth versus disposable diapers. (*a*) If we assume that between birth and 2.5 y of age, a child uses 3 diapers per day, estimate the total number of disposable diapers used in the United States per year. (*b*) Estimate the total landfill volume due to these diapers, assuming that 1000 kg of waste fills about 1 m^3 of landfill volume. (*c*) How many square miles of landfill area at an average height of 10 m is needed for the disposal of diapers each year?

18 •• (*a*) Estimate the number of gallons of gasoline used per day by automobiles in the United States and the total amount of money spent on it. (*b*) If 19.4 gal of gasoline can be made from one barrel of crude oil, estimate the total number of barrels of oil imported into the United States per year to make gasoline. How many barrels per day is this?

19 •• **ENGINEERING APPLICATION** A megabyte (MB) is a unit of computer memory storage. A CD has a storage capacity of 700 MB and can store approximately 70 min of high-quality music. (*a*) If a typical song is 5 min long, how many megabytes are required for each song? (*b*) If a page of printed text takes approximately 5 kilobytes, estimate the number of novels that could be saved on a CD. SSM

UNITS

20 • Express the following quantities using the prefixes listed in Table 1-1 and the unit abbreviations listed in the table Abbreviations for Units. For example, 10 000 meters = 10 km. (*a*) 1 000 000 watts, (*b*) 0.002 gram, (*c*) 3×10^{-6} meter, (*d*) 30 000 seconds.

21 • Write each of the following without using prefixes: (*a*) 40 μW, (*b*) 4 ns, (*c*) 3 MW, (*d*) 25 km.

22 • Write the following (which are not SI units) using prefixes (but not their abbreviations). For example, 10^3 meters = 1 kilometer: (*a*) 10^{-12} boo, (*b*) 10^9 low, (*c*) 10^{-6} phone, (*d*) 10^{-18} boy, (*e*) 10^6 phone, (*f*) 10^{-9} goat, (*g*) 10^{12} bull.

23 •• In the following equations, the distance x is in meters, the time t is in seconds, and the velocity v is in meters per second. What are the SI units of the constants C_1 and C_2? (*a*) $x = C_1 + C_2 t$, (*b*) $x = \frac{1}{2}C_1 t^2$, (*c*) $v^2 = 2C_1 x$, (*d*) $x = C_1 \cos C_2 t$, (*e*) $v^2 = 2C_1 v - (C_2 x)^2$ SSM

24 •• If x is in feet, t is in milliseconds, and v is in feet per second, what are the units of the constants C_1 and C_2 in each part of Problem 23?

CONVERSION OF UNITS

25 • **MULTISTEP** From the original definition of the meter in terms of the distance along a meridian from the equator to the North Pole, find in meters (*a*) the circumference of Earth and (*b*) the radius of Earth. (*c*) Convert your answers for (*a*) and (*b*) from meters into miles.

26 • The speed of sound in air is 343 m/s. What is the speed of a supersonic plane that travels at twice the speed of sound? Give your answer in kilometers per hour and miles per hour.

27 • A basketball player is 6 ft $10\frac{1}{2}$ in tall. What is his height in centimeters?

28 • Complete the following: (*a*) 100 km/h = ______ mi/h, (*b*) 60 cm = ______ in., (*c*) 100 yd = ______ m.

29 • The main span of the Golden Gate Bridge is 4200 ft. Express this distance in kilometers.

30 • Find the conversion factor to convert from miles per hour into kilometers per hour.

31 • Complete the following: (*a*) 1.296×10^5 km/h^2 = ______ km/(h · s), (*b*) 1.296×10^5 km/h^2 = ______ m/s^2, (*c*) 60 mi/h = ______ ft/s, (*d*) 60 mi/h = ______ m/s.

32 • There are 640 acres in a square mile. How many square meters are there in one acre?

33 •• **CONTEXT-RICH** You are a delivery person for the Fresh Aqua Spring Water Company. Your truck carries 4 pallets. Each pallet carries 60 cases of water. Each case of water has 24 one-liter bottles. The dolly you use to carry the water into the stores has a weight limit of 250 lb. (*a*) If a milliliter of water has a mass of 1 g, and a kilogram has a weight of 2.2 lb, what is the weight, in pounds, of all the water in your truck? (*b*) How many full cases of water can you carry on the cart? SSM

34 •• A right circular cylinder has a diameter of 6.8 in. and a height of 2 ft. What is the volume of the cylinder in (*a*) cubic feet, (*b*) cubic meters, (*c*) liters?

35 •• In the following, x is in meters, t is in seconds, v is in meters per second, and the acceleration a is in meters per second squared. Find the SI units of each combination: (*a*) v^2/x, (*b*) $\sqrt{x/a}$, (*c*) $\frac{1}{2}at^2$. SSM

DIMENSIONS OF PHYSICAL QUANTITIES

36 • What are the dimensions of the constants in each part of Problem 23?

37 • The law of radioactive decay is $N(t) = N_0 e^{-\lambda t}$, where N_0 is the number of radioactive nuclei at $t = 0$, $N(t)$ is the number remaining at time t, and λ is a quantity known as the decay constant. What is the dimension of λ?

38 •• The SI unit of force, the kilogram-meter per second squared (kg · m/s^2) is called the newton (N). Find the dimensions and the SI units of the constant G in Newton's law of gravitation $F = Gm_1m_2/r^2$.

39 •• The magnitude of the force (F) that a spring exerts when it is stretched a distance x from its unstressed length is governed by Hooke's law, $F = kx$. (*a*) What are the dimensions of the *force constant*, k? (*b*) What are the dimensions and SI units of the quantity kx^2?

40 •• Show that the product of mass, acceleration, and speed has the dimensions of power.

41 •• The momentum of an object is the product of its velocity and mass. Show that momentum has the dimensions of force multiplied by time. SSM

42 •• What combination of force and one other physical quantity has the dimensions of power?

43 •• When an object falls through air, there is a drag force that depends on the product of the cross sectional area of the object and the square of its velocity, that is, $F_{air} = CAv^2$, where C is a constant. Determine the dimensions of C. SSM

44 •• Kepler's third law relates the period of a planet to its orbital radius r, the constant G in Newton's law of gravitation ($F = Gm_1m_2/r^2$), and the mass of the Sun M_s. What combination of these factors gives the correct dimensions for the period of a planet?

SCIENTIFIC NOTATION AND SIGNIFICANT FIGURES

45 • Express as a decimal number without using powers of 10 notation: (*a*) 3×10^4, (*b*) 6.2×10^{-3}, (*c*) 4×10^{-6}, (*d*) 2.17×10^5. SSM

46 • Write the following in scientific notation: (*a*) 1 345 100 m = ______ km, (*b*) 12 340 kW = ______ MW, (*c*) 54.32 ps = ______ s, (*d*) 3.0 m = ______ mm.

47 • Calculate the following, round off to the correct number of significant figures, and express your result in scientific notation: (*a*) $(1.14)(9.99 \times 10^4)$, (*b*) $(2.78 \times 10^{-8}) - (5.31 \times 10^{-9})$, (*c*) $12\pi/(4.56 \times 10^{-3})$, (*d*) $27.6 + (5.99 \times 10^2)$. SSM

48 • Calculate the following, round off to the correct number of significant figures, and express your result in scientific notation: (*a*) $(200.9)(569.3)$, (*b*) $(0.000\,000\,513)(62.3 \times 10^7)$, (*c*) $28\,401 + (5.78 \times 10^4)$, (*d*) $63.25/(4.17 \times 10^{-3})$.

49 • **BIOLOGICAL APPLICATION** A cell membrane has a thickness of 7.0 nm. How many cell membranes would it take to make a stack 1.0 in. high? SSM

50 •• **ENGINEERING APPLICATION** A circular hole of radius 8.470×10^{-1} cm must be cut into the front panel of a display unit. The *tolerance* is 1.0×10^{-3} cm, which means the actual hole cannot differ by more than this quantity from the desired radius. If the actual hole is larger than the desired radius by the allowed tolerance, what is the difference between the actual area and the desired area of the hole?

51 •• **ENGINEERING APPLICATION** A square peg must be made to fit through a square hole. If you have a square peg that has an edge length of 42.9 mm, and the square hole has an edge length of 43.2 mm, (*a*) what is the area of the space available when the peg is in the hole? (*b*) If the peg is made rectangular by removing 0.10 mm of material from one side, what is the area available now? SSM

VECTORS AND THEIR PROPERTIES

52 • **MULTISTEP** A vector that is 7.0 units long and a vector that is 5.5 units long are added. Their sum is a vector 10.0 units long. (*a*) Show graphically at least one way that the vectors can be added. (*b*) Using your sketch in Part (*a*), determine the angle between the original two vectors.

53 • Determine the x and y components of the following three vectors in the xy plane. (*a*) A 10-m displacement vector that makes an angle of 30° clockwise from the $+y$ direction. (*b*) A 25-m/s velocity vector that makes an angle of 40° counterclockwise from the $-x$ direction. (*c*) A 40-lb force vector that makes an angle of 120° counterclockwise from the $-y$ direction. SSM

54 • Rewrite the following vectors in terms of their magnitude and angle (counterclockwise from the $+x$ direction). (*a*) A displacement vector with an x component of +8.5 m and a y component of −5.5 m. (*b*) A velocity vector with an x component of −75 m/s and a y component of +35 m/s. (*c*) A force vector with a magnitude of 50 lb that is in the third quadrant with an x component whose magnitude is 40 lb.

55 • **CONCEPTUAL** You walk 100 m in a straight line on a horizontal plane. If this walk took you 50 m east, what are your possible north or south movements? What are the possible angles that your walk made with respect to due east?

56 • **ESTIMATION** The final destination of your journey is 300 m due east of your starting point. The first leg of this journey is the walk described in Problem 55, and the second leg in also a walk along a single straight-line path. Estimate graphically the length and heading for the second leg of your journey.

57 •• Given the following vectors: $\vec{A} = 3.4\hat{i} + 4.7\hat{j}$, $\vec{B} = (-7.7)\hat{i} + 3.2\hat{j}$, and $\vec{C} = 5.4\hat{i} + (-9.1)\hat{j}$. (*a*) Find the vector $\vec{D}$, in unit vector notation, such that $\vec{D} + 2\vec{A} - 3\vec{C} + 4\vec{B} = 0$. (*b*) Express your answer in Part (*a*) in terms of magnitude and angle with the $+x$ direction.

58 •• Given the following force vectors: $\vec{A}$ is 25 lb at an angle of 30° clockwise from the $+x$ axis, and $\vec{B}$ is 42 lb at an angle of 50° clockwise from the $+y$ axis. (*a*) Make a sketch and visually estimate the magnitude and angle of the vector $\vec{C}$ such that $2\vec{A} + \vec{C} - \vec{B}$ results in a vector with a magnitude of 35 lb pointing in the $+x$ direction. (*b*) Repeat the calculation in Part (*a*) using the method of components and compare your result to the estimate in (*a*).

59 •• Calculate the unit vector (in terms of $\hat{i}$ and $\hat{j}$) in the direction opposite to the direction of each of vectors $\vec{A}$, $\vec{B}$ and $\vec{C}$ in Problem 57. SSM

60 •• Unit vectors $\hat{i}$ and $\hat{j}$ are directed east and north, respectively. Calculate the unit vector (in terms of $\hat{i}$ and $\hat{j}$) in the following directions. (*a*) northeast, (*b*) 70° clockwise from the $-y$ axis, (*c*) southwest

GENERAL PROBLEMS

61 • The Apollo trips to the moon in the 1960s and 1970s typically took 3 days to travel the Earth–moon distance once they left Earth orbit. Estimate the spacecraft's average speed in kilometers per hour, miles per hour, and meters per second. SSM

62 • On many of the roads in Canada the speed limit is 100 km/h. What is this speed limit in miles per hour?

63 • If you could count $1.00 per second, how many years would it take to count 1.00 billion dollars?

64 • (*a*) The speed of light in vacuum is 186 000 mi/s = 3.00×10^8 m/s. Use this fact to find the number of kilometers in a mile. (*b*) The weight of 1.00 ft^3 of water is 62.4 lb, and 1.00 ft = 30.5 cm. Use this information and the fact that 1.00 cm^3 of water has a mass of 1.00 g to find the weight in pounds of a 1.00-kg mass.

65 • The mass of one uranium atom is 4.0×10^{-26} kg. How many uranium atoms are there in 8.0 g of pure uranium?

66 •• During a thunderstorm, a total of 1.4 in. of rain falls. How much water falls on one acre of land? (1 mi^2 = 640 acres.) Express your answer in (*a*) cubic inches, (*b*) cubic feet, (*c*) cubic meters, and (*d*) kilograms. Note that the density of water is 1000 kg/m^3.

67 •• An iron nucleus has a radius of 5.4×10^{-15} m and a mass of 9.3×10^{-26} kg. (*a*) What is its mass per unit volume in kg/m^3? (*b*) If Earth had the same mass per unit volume, what would be its radius? (The mass of Earth is 5.98×10^{24} kg.)

68 •• **ENGINEERING APPLICATION** The Canadian Norman Wells Oil Pipeline extends from Norman Wells, Northwest Territories, to Zama, Alberta. The 8.68×10^5-m-long pipeline has an inside diameter of 12 in. and can be supplied with oil at 35 L/s. (*a*) What is the volume of oil in the pipeline if it is full at some instant in time? (*b*) How long would it take to fill the pipeline with oil if it is initially empty?

69 •• The astronomical unit (AU) is defined as the mean center-to-center distance from Earth to the Sun, namely 1.496×10^{11} m. The parsec is the radius of a circle for which a central angle of 1 s intercepts an arc of length 1 AU. The light-year is the distance that light travels in 1 y. (*a*) How many parsecs are there in one astronomical unit? (*b*) How many meters are in a parsec? (*c*) How many meters in a light-year? (*d*) How many astronomical units in a light-year? (*e*) How many light-years in a parsec?

70 •• If the average density of the universe is at least 6×10^{-27} kg/m^3, then the universe will eventually stop expanding and begin contracting. (*a*) How many electrons are needed in each cubic meter to produce the critical density? (*b*) How many protons per cubic meter would produce the critical density? ($m_e = 9.11 \times 10^{-31}$ kg; $m_p = 1.67 \times 10^{-27}$ kg.)

71 ••• **CONTEXT-RICH, ENGINEERING APPLICATION, SPREADSHEET** You are an astronaut doing physics experiments on the moon. You are interested in the experimental relationship between distance fallen, y, and time elapsed, t, of falling objects dropped from rest. You have taken some data for a falling penny, which is represented in the table below.

(*a*) y (m)	10	20	30	40	50
(*b*) t (s)	3.5	5.2	6.0	7.3	7.9

You expect that a general relationship between distance y and time t is $y = Bt^C$, where B and C are constants to be determined experimentally. To accomplish this, create a log-log plot of the data: (*a*) graph $\log(y)$ vs. $\log(t)$, with $\log(y)$ the ordinate variable and $\log(t)$ the abscissa variable. (*b*) Show that if you take the log of each side of your expected relationship, you get $\log(y) = \log(B) + C\log(t)$. (*c*) By comparing this linear relationship to the graph of the data, estimate the values of B and C. (*d*) If you drop a penny, how long should it take to fall 1.0 m? (*e*) In the next chapter, we will show that the expected relationship between y and t is $y = \frac{1}{2}at^2$, where a is the acceleration of the object. What is the acceleration of objects dropped on the moon? SSM

72 ••• **SPREADSHEET** A particular company's stock prices vary with the market and with the company's type of business, and can be very unpredictable, but people often try to look for mathematical patterns where they may not belong. Corning is a materials-engineering company located in upstate New York. Below is a table of the price of Corning stock on August 3, for every 5 years from 1981 to 2001. Assume that the price follows a power law: price (in \$) $= Bt^C$ where t is expressed in years. (*a*) Evaluate the constants B and C (see methods suggested for the previous problem). (*b*) According to the power law, what should the price of Corning stock have been on August 3, 2000? (It was actually \$82.83!)

(*a*) Price (dollars)	2.10	4.19	9.14	10.82	16.85
(*b*) Years since 1980	1	6	11	16	21

73 ••• **ENGINEERING APPLICATION** The Super-Kamiokande neutrino detector in Japan is a large transparent cylinder filled with ultra-pure water. The height of the cylinder is 41.4 m and the diameter is 39.3 m. Calculate the mass of the water in the cylinder. Does this match the claim posted on the official Super-K Web site that the detector uses 50,000 tons of water? SSM

74 ••• **CONTEXT-RICH** You and a friend are out hiking across a large flat plain and decide to determine the height of a distant mountain peak, and also the horizontal distance from you to the peak (Figure 1-19). In order to do this, you stand in one spot and determine that the sightline to the top of the peak is inclined at 7.5° above the horizontal. You also make note of the heading to the peak at that point: 13° east of north. You stand at the original position, and your friend hikes due west for 1.5 km. He then sights the peak and determines that its sightline has a heading of 15° east of north. How far is the mountain from your position, and how high is its summit above your position?

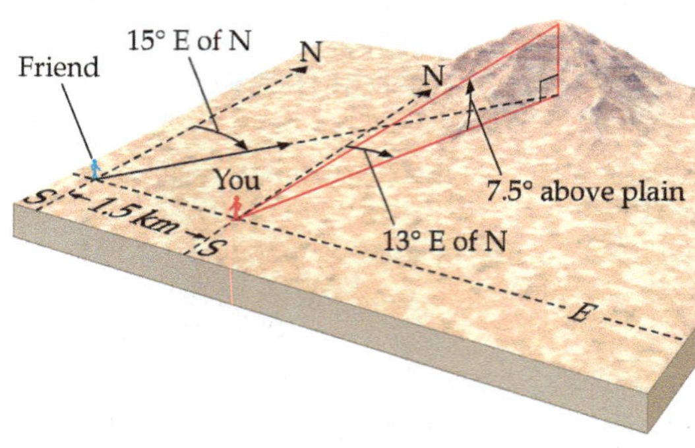

FIGURE 1-19 Problem 74

75 ••• **SPREADSHEET** The table below gives the periods T and orbit radii r for the motions of four satellites orbiting a dense, heavy asteroid. (*a*) These data can be fitted by the formula $T = Cr^n$. Find the values of the constants C and n. (*b*) A fifth satellite is discovered to have a period of 6.20 y. Find the radius for the orbit of this satellite, which fits the same formula.

(*a*) Period T, y	0.44	1.61	3.88	7.89
(*b*) Radius r, Gm	0.088	0.208	0.374	0.600

76 ••• **MULTISTEP** The period T of a simple pendulum depends on the length L of the pendulum and the acceleration of gravity g (dimensions L/T^2). (*a*) Find a simple combination of L and g that has the dimensions of time. (*b*) Check the dependence of the period T on the length L by measuring the period (time for a complete swing back and forth) of a pendulum for two different values of L. (*c*) The correct formula relating T to L and g involves a constant that is a multiple of π, and cannot be obtained by the dimensional analysis of Part (*a*). It can be found by experiment as in Part (*b*) if g is known. Using the value $g = 9.81$ m/s^2 and your experimental results from Part (*b*), find the formula relating T to L and g.

77 ••• A sled at rest is suddenly pulled in three horizontal directions at the same time but it goes nowhere. Paul pulls to the northeast with a force of 50 lb. Johnny pulls at an angle of 35° south of due west with a force of 65 lb. Connie pulls with a force to be determined. (*a*) Express the boys' two forces in terms of the usual unit vectors. (b) Determine the third force (from Connie), expressing it first in component form and then as a magnitude and angle (direction).

78 ••• You spot a plane that is 1.50 km north, 2.5 km east, and at an altitude 5.0 km above your position. (*a*) How far from you is the plane? (*b*) At what angle from due north (in the horizontal plane) are you looking? (*c*) Determine the plane's position vector (from your location) in terms of the unit vectors, letting $\hat{i}$ be toward the east direction, $\hat{j}$ be toward the north direction, and $\hat{k}$ be in vertically upward. (*d*) At what elevation angle (above the horizontal plane of Earth) is the airplane?

PART I MECHANICS

CHAPTER 2

MOTION IN ONE DIMENSION IS MOTION ALONG A STRAIGHT LINE LIKE THAT OF A CAR ON A STRAIGHT ROAD. THIS DRIVER ENCOUNTERS STOPLIGHTS AND DIFFERENT SPEED LIMITS ON HER COMMUTE ALONG A STRAIGHT HIGHWAY TO SCHOOL. *(Medio Images/Getty Images.)*

? How can she estimate her arrival time? (See Example 2-3.)

Motion in One Dimension

2-1 Displacement, Velocity, and Speed

2-2 Acceleration

2-3 Motion with Constant Acceleration

2-4 Integration

Imagine a car speeding down a highway. There are a number of ways in which you could describe the car's motion to someone else. For example, you could describe the change in the car's position as it travels from one point to another, how fast the car is moving and the direction in which it travels, and whether the car is speeding up or slowing down as it moves. These basic descriptors of motion—known as displacement, velocity, and acceleration—are an essential part of physics. In fact, the attempt to describe the motion of objects gave birth to physics more than 400 years ago.

The study of motion, and the related concepts of force and mass, is called **mechanics.** We begin our investigation into motion by examining **kinematics,** the branch of mechanics that deals with the characteristics of motion. You will need to understand kinematics to understand the rest of this book. Motion permeates all of physics, and an understanding of kinematics is needed to understand how force and mass effect motion. Starting in Chapter 4, we look at **dynamics,** which relates motion, force, and mass.

We study the simplest case of kinematics in this chapter — motion along a straight line. We will develop the models and tools you will need to describe motion in one dimension, and introduce the precise definitions of words commonly used to describe motion, such as displacement, speed, velocity, and acceleration. We will also look at the special case of straight-line motion when acceleration is constant. Finally, we consider the ways in which integration can be used to describe motion. In this chapter, moving objects are restricted to motion along a straight line. To describe such motion, it is not necessary to use the full vector notation developed in Chapter 1. A + or − sign are all that is needed to specify direction along a straight line.

2-1 DISPLACEMENT, VELOCITY, AND SPEED

In a horse race, the winner is the horse whose nose first crosses the finish line. One could argue that all that really matters during the race is the motion of that single point on the horse, and that the size, shape, and motion of the rest of the horse is unimportant. In physics, this type of simplification turns out to be useful for examining the motion of other objects as well. We can often describe the motion of an object by describing the motion of a single point of the object. For example, as a car moves in a straight line along a road, you could describe the motion of the car by examining the motion of a single point on the side of the car. An object that can be represented in this idealized manner is called a **particle.** In kinematics, any object can be considered a particle as long as we are not interested in its size, shape, or internal motion. For example, we can consider cars, trains, and rockets particles. Earth and other planets can also be thought of as particles as they move around the Sun. Even people and galaxies can be treated as particles.

(*Bettmann/Corbis.*)

POSITION AND DISPLACEMENT

To describe the motion of a particle, we need to be able to describe the **position** of the particle and how that position changes as the particle moves. For one-dimensional motion, we often choose the x axis as the line along which the motion takes place. For example, Figure 2-1 shows a student on a bicycle at position x_i at time t_i. At a later time, t_f, the student is at position x_f. The change in the student's position, $x_f - x_i$, is called a **displacement.** We use the Greek letter Δ (uppercase delta) to indicate the change in a quantity; thus, the change in x can be written as

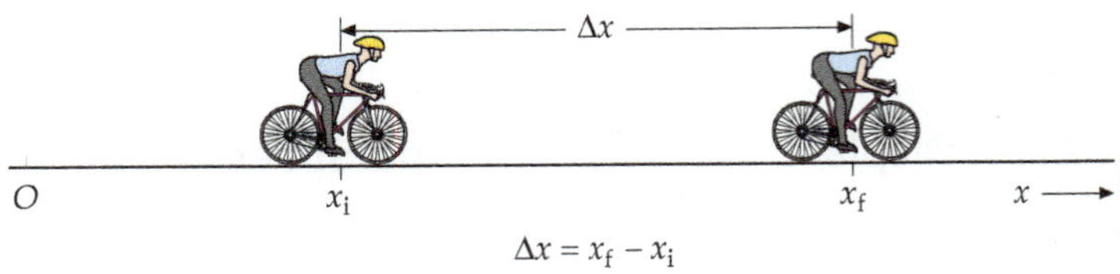

FIGURE 2-1 A student on a bicycle is moving in a straight line. A coordinate axis consists of a line along the path of the bicycle. A point on this line is chosen to be the origin O. Other points on it are assigned a number x, the value of x being proportional to its distance from O. The numbers assigned to points to the right of O are positive as shown, and those assigned to points to the left of O are negative. When the bicycle travels from point x_i to point x_f, its displacement is $\Delta x = x_f - x_i$.

$$\Delta x = x_f - x_i \qquad 2\text{-}1$$

DEFINITION—DISPLACEMENT

It is important to recognize the difference between displacement and distance traveled. The distance traveled by a particle is the length of the path a particle takes from its initial position to its final position. Distance is a scalar quantity and is always indicated by a positive number. Displacement is the *change in position* of the particle. It is positive if the change in position is in the direction of increasing x (the $+x$ direction), and negative if it is in the $-x$ direction. Displacement can be represented by vectors, as shown in Chapter 1. We will use the full vector notation developed in Chapter 1 when we study motion in two and three dimensions in Chapter 3.

! The notation Δx (read "delta x") stands for a single quantity that is the change in x. Δx is not a product of Δ and x any more than $\cos\theta$ is a product of cos and θ. By convention, the change in a quantity is always its final value minus its initial value.

Example 2-1 Distance and Displacement of a Dog

You are playing a game of catch with a dog. The dog is initially standing near your feet. Then he jogs 20 feet in a straight line to retrieve a stick, and carries the stick 15 feet back toward you before lying on the ground to chew on the stick. (*a*) What is the total distance the dog travels? (*b*) What is the net displacement of the dog? (*c*) Show that the net displacement for the trip is the sum of the sequential displacements that make up the trip.

PICTURE The total distance, s, is determined by summing the individual distances the dog travels. The displacement is the dog's final position minus the dog's initial position. The dog leaves your side at time 0, gets the stick at time 1, and lies down to chew it at time 2.

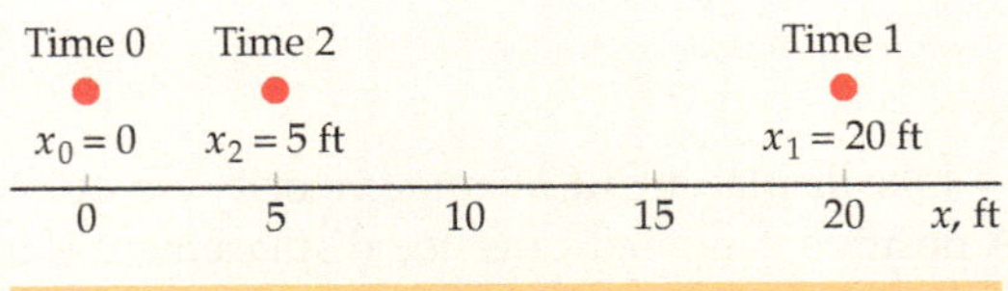

FIGURE 2-2 The red dots represent the dog's position at different times.

SOLVE

(*a*) 1. Make a diagram of the motion (Figure 2-2). Include a coordinate axis:

2. Calculate the total distance traveled:

$$s_{02} = s_{01} + s_{12} = (20\text{ ft}) + (15\text{ ft}) = \boxed{35}\text{ ft}$$

(The subscripts indicate the time intervals, where s_{01} is the distance traveled during the interval from time 0 to time 1, and so forth.)

(*b*) The net displacement is found from its definition, $\Delta x = x_f - x_i$, where $x_i = x_0 = 0$ is the dog's initial position. Five feet from the initial position or $x_f = x_2 = 5$ ft is the dog's final position:

$$\Delta x_{02} = x_2 - x_0 = 5\text{ ft} - 0\text{ ft} = \boxed{5\text{ ft}}$$

where Δx_{02} is the displacement during the interval from time 0 to time 2.

(*c*) The net displacement is also found by adding the displacement for the first leg to the displacement for the second leg.

$$\Delta x_{01} = x_1 - x_0 = 20\text{ ft} - 0\text{ ft} = 20\text{ ft}$$

$$\Delta x_{12} = x_2 - x_1 = 5\text{ ft} - 20\text{ ft} = -15\text{ ft}$$

adding, we obtain

$$\Delta x_{01} + \Delta x_{12} = (x_1 - x_0) + (x_2 - x_1) = x_2 - x_0 = \Delta x_{02}$$

so

$$\Delta x_{02} = \Delta x_{01} + \Delta x_{12} = 20\text{ ft} - 15\text{ ft} = \boxed{5\text{ ft}}$$

CHECK The magnitude of the displacement for any part of the trip is never greater than the total distance traveled for that part. The magnitude of the Part-(*b*) result (5 ft) is less than the Part-(*a*) result (35 ft), so the Part-(*b*) result is plausible.

TAKING IT FURTHER The total distance traveled for a trip is always equal to the sum of the distances traveled for the individual legs of the trip. The total or net displacement for a trip is always equal to the sum of the displacements for the individual legs of the trip.

AVERAGE VELOCITY AND SPEED

We often are interested in the speed something is moving. The **average speed** of a particle is the total distance traveled by the particle divided by the total time from start to finish:

$$\text{Average speed} = \frac{\text{total distance}}{\text{total time}} = \frac{s}{\Delta t} \qquad 2\text{-}2$$

DEFINITION—AVERAGE SPEED

Because the total distance and total time are both always positive, the average speed is always positive.

Although speed is a useful idea, it does not reveal anything about the direction of motion because neither the total distance nor the total time has an associated direction. A more useful quantity is one that describes both how fast and in what direction an object moves. The term used to describe this quantity is *velocity*. The **average velocity,** $v_{\text{av}\,x}$, of a particle is defined as the ratio of the displacement Δx to the time interval Δt:

$$v_{\text{av}\,x} = \frac{\Delta x}{\Delta t} = \frac{x_{\text{f}} - x_{\text{i}}}{t_{\text{f}} - t_{\text{i}}} \qquad (\text{so } \Delta x = v_{\text{av}\,x}\Delta t) \qquad 2\text{-}3$$

DEFINITION—AVERAGE VELOCITY

Like displacement, average velocity is a quantity that may be positive or negative. A positive value indicates the displacement is in the $+x$ direction. A negative value indicates the displacement is in the $-x$ direction. The dimensions of velocity are L/T and the SI unit of velocity is meters per second (m/s). Other common units include kilometers per hour (km/h), feet per second (ft/s), and miles per hour (mi/h).

Figure 2-3 is a graph of a particle's position as a function of time. Each point represents the position x of a particle at a particular time t. A straight line connects points P_1 and P_2 and forms the hypotenuse of the triangle having sides $\Delta x = x_2 - x_1$ and $\Delta t = t_2 - t_1$. Notice that the ratio $\Delta x/\Delta t$ is the line's **slope,** which gives us a geometric interpretation of average velocity:

The average velocity for the interval between $t = t_1$ and $t = t_2$ is the slope of the straight line connecting the points (t_1, x_1) and (t_2, x_2) on an x versus t graph.

GEOMETRIC INTERPRETATION OF AVERAGE VELOCITY

See **Math Tutorial** *for more information on* **Linear Equations**

Notice that the average velocity depends on the time interval on which it is based. In Figure 2-3, for example, the smaller time interval indicated by t_2' and P_2' gives a larger average velocity, as shown by the greater steepness of the line connecting points P_1 and P_2'.

! The definitions of average speed and average velocity are the most basic of the kinematic parameters. You will need to know these definitions and the definitions that appear later in this chapter to effectively solve kinematics problems.

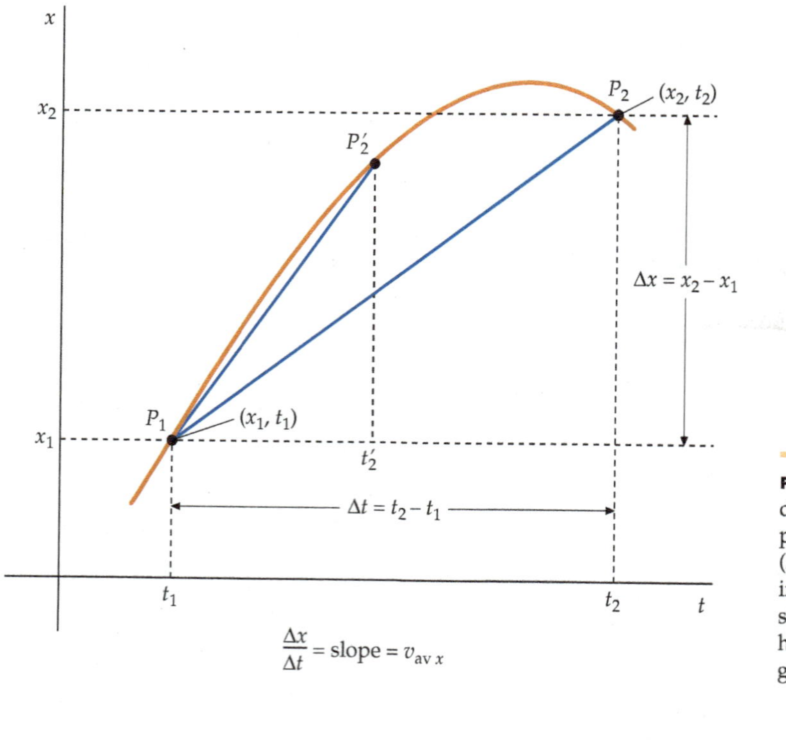

FIGURE 2-3 Graph of x versus t for a particle moving in one dimension. Each point on the curve represents the position x at a particular time t. We have drawn a straight line through points (x_1, t_1) and (x_2, t_2). The displacement $\Delta x = x_2 - x_1$ and the time interval $\Delta t = t_2 - t_1$ between these points are indicated. The straight line between P_1 and P_2 is the hypotenuse of the triangle having sides Δx and Δt, and the ratio $\Delta x/\Delta t$ is its slope. In geometric terms, the slope is a measure of the line's steepness.

Example 2-2 Average Speed and Velocity of the Dog

The dog that you were playing catch with in Example 2-1 jogged 20.0 ft away from you in 1.0 s to retrieve the stick and ambled back 15.0 ft in 1.5 s (Figure 2-4). Calculate (*a*) the dog's average speed, and (*b*) the dog's average velocity for the total trip.

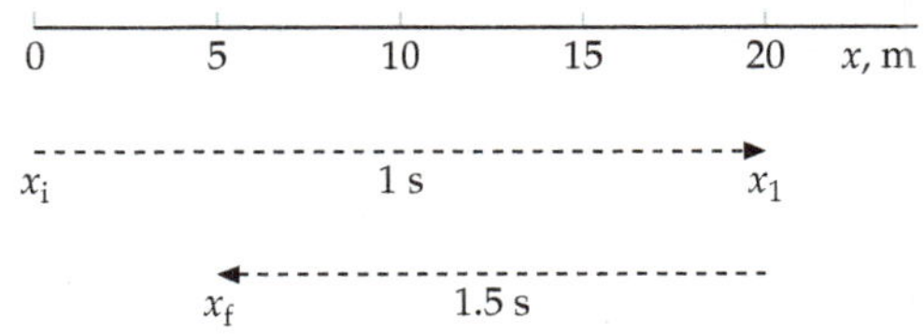

FIGURE 2-4

PICTURE We can solve this problem using the definitions of average speed and average velocity, noting that average *speed* is the total *distance* divided by the total time Δt, whereas the average *velocity* is the net *displacement* divided by Δt:

SOLVE

(*a*) 1. The dog's average speed equals the total distance divided by the total time:

$$\text{Average speed} = \frac{s}{\Delta t}$$

2. Calculate the total distance traveled and the total time:

$$s = s_1 + s_2 = 20.0\text{ ft} + 15.0\text{ ft} = 35.0\text{ ft}$$

$$\Delta t = (t_1 - t_i) + (t_f - t_2) = 1.0\text{ s} + 1.5\text{ s} = 2.5\text{ s}$$

3. Use s and Δt to find the dog's average speed:

$$\text{Average speed} = \frac{35.0\text{ ft}}{2.5\text{ s}} = \boxed{14\text{ ft/s}}$$

(*b*) 1. The dog's average velocity is the ratio of the net displacement Δx to the time interval Δt:

$$v_{\text{av}\,x} = \frac{\Delta x}{\Delta t}$$

2. The dog's net displacement is $x_f - x_i$, where $x_i = 0.0$ ft is the initial position of the dog and $x_f = 5.0$ ft is the dog's final position:

$$\Delta x = x_f - x_i = 5.0\text{ ft} - 0.0\text{ ft} = 5.0\text{ ft}$$

3. Use Δx and Δt to find the dog's average velocity:

$$v_{\text{av}\,x} = \frac{\Delta x}{\Delta t} = \frac{5.0\text{ ft}}{2.5\text{ s}} = \boxed{2.0\text{ ft/s}}$$

CHECK An Internet search reveals a greyhound can have an average speed of approximately 66 ft/s (45 mi/h), so our dog should easily be able to jog 14 ft/s (9.5 mi/h). A Part-(*a*) result greater than 66 ft/s would not be plausible.

TAKING IT FURTHER Note that the dog's speed is greater than the dog's average velocity because the total distance traveled is greater than the magnitude of the total displacement. Also, note that the total displacement is the sum of the individual displacements. That is, $\Delta x = \Delta x_1 + \Delta x_2 = (20.0\text{ ft}) + (-15.0\text{ ft}) = 5.0\text{ ft}$, which is the Part (*b*), step 2 result.

Example 2-3 Driving to School

It normally takes you 10 min to travel 5.0 mi to school along a straight road. You leave home 15 min before class begins. Delays caused by a broken traffic light slow down traffic to 20 mi/h for the first 2.0 mi of the trip. Will you be late for class?

PICTURE You need to find the total time that it will take you to travel to class. To do so, you must find the time $\Delta t_{2\text{ mi}}$ that you will be driving at 20 mi/h, and the time $\Delta t_{3\text{ mi}}$ for the remainder of the trip, during which you are driving at your usual velocity.

SOLVE

1. The total time equals the time to travel the first 2.0 mi plus the time to travel the remaining 3.0 mi:

$$\Delta t_{\text{tot}} = \Delta t_{2\text{ mi}} + \Delta t_{3\text{ mi}}$$

2. Using $\Delta x = v_{\text{av}\,x}\,\Delta t$, solve for the time taken to travel 2.0 mi at 20 mi/h:

$$\Delta t_{2\text{ mi}} = \frac{\Delta x_1}{v_{\text{av}\,x}} = \frac{2.0\text{ mi}}{20\text{ mi/h}} = 0.10\text{ h} = 6.0\text{ min}$$

3. Using $\Delta x = v_{\text{av}\,x}\,\Delta t$, relate the time taken to travel 3 mi at the usual velocity:

$$\Delta t_{3\text{ mi}} = \frac{\Delta x_2}{v_{\text{av}\,x}} = \frac{3.0\text{ mi}}{v_{\text{usual}\,x}}$$

4. Using $\Delta x = v_{\text{av}\,x}\,\Delta t$, solve for $v_{\text{usual}\,x}$, the velocity needed for you to travel the 5.0 mi in 10 min:

$$v_{\text{usual}\,x} = \frac{\Delta x_{\text{tot}}}{\Delta t_{\text{usual}}} = \frac{5.0\text{ mi}}{10\text{ min}} = 0.50\text{ mi/min}$$

5. Using the results from steps 3 and 4, solve for $\Delta t_{3\text{ mi}}$:

$$\Delta t_{3\text{ mi}} = \frac{\Delta x_2}{v_{\text{usual }x}} = \frac{3.0\text{ mi}}{0.50\text{ mi/min}} = 6.0\text{ min}$$

6. Solve for the total time:

$$\Delta t_{\text{tot}} = \Delta t_{2\text{ mi}} + \Delta t_{3\text{ mi}} = 12\text{ min}$$

7. The trip takes 12 min with the delay, compared to the usual 10 min. Because you wisely allowed yourself 15 min for the trip, *you will not be late for class*.

CHECK Note that 20 mi/h = 20 mi/60 min = 1.0 mi/3.0 min. Traveling the entire 5.0 miles at one mile every three minutes, it would take 15 minutes for the trip to school. You allowed yourself 15 minutes for the trip, so you would get there on time even if you traveled at the slow speed of 20 mi/h for the entire 5.0 miles.

Example 2-4 A Train-Hopping Bird

Two trains 60 km apart approach each other on parallel tracks, each moving at 15 km/h. A bird flies back and forth between the trains at 20 km/h until the trains pass each other. How far does the bird fly?

PICTURE In this problem, you are asked to find the total distance flown by the bird. You are given the bird's speed, the trains' speeds, and the initial distance between the trains. At first glance, it might seem like you should find and sum the distances flown by the bird each time it moves from one train to the other. However, a much simpler approach is to use the facts that the time t the bird is flying is the time taken for the trains to meet. The total distance flown is the bird's speed multiplied by the time the bird is flying. Therefore, we can approach this problem by first writing an equation for the quantity to be found, the total distance s flown by the bird.

SOLVE

1. The total distance s_{bird} traveled by the bird equals its speed times the time of flight:

$$s_{\text{bird}} = (\text{average speed})_{\text{bird}} \times t = (\text{speed})_{\text{av bird}} \times t$$

2. The time t that the bird is in the air is the time taken for one of the trains to travel half the initial distance D separating the trains. (Because the trains are traveling at the same speed, each train travels half of the 60 km, which is 30 km, before they meet.):

$$\tfrac{1}{2}D = (\text{speed})_{\text{av train}} \times t$$

so

$$t = \frac{D}{2(\text{speed})_{\text{av train}}}$$

3. Substitute the step-2 result for the time into the step-1 result. The initial separation of the two trains is $D = 60$ km. The total distance traveled by the bird is therefore:

$$s_{\text{bird}} = (\text{speed})_{\text{av bird}}\, t = (\text{speed})_{\text{av bird}} \frac{D}{2(\text{speed})_{\text{av train}}}$$

$$= 20\text{ km/h}\,\frac{60\text{ km}}{2\,(15\text{ km/h})} = \boxed{40\text{ km}}$$

CHECK The speed of each train is three-fourths the speed of the bird, so the distance traveled by one of the trains will be three-fourths the distance the bird travels. Each train travels 30 km. Because 30 km is three-fourths of 40 km, our result of 40 km for the distance the bird travels is very plausible.

INSTANTANEOUS VELOCITY AND SPEED

Suppose your average velocity for a long trip was 60 km/h. Because this value is an average, it does not convey any information about how your velocity changed during the trip. For example, there may have been some parts of the journey where you were stopped at a traffic light and other parts where you went faster to make up time. To learn more about the details of your motion, we have to look at the velocity at any given instant during the trip. On first consideration, defining the velocity of a particle at a single instant might seem impossible. At a given instant, a particle is at a single point. If it is at a single point, how can it be moving? If it is not moving, how can it have a velocity? This age-old paradox is resolved when we realize that observing and defining motion requires that we look at the position of the object at more than one instant of time. For example, consider the graph of position versus time in Figure 2-5. As we consider successively shorter time intervals beginning at t_P, the average velocity for the interval approaches the slope of the tangent at t_P. We define the slope of this tangent as the **instantaneous velocity,** $v_x(t)$, at t_P. This tangent is the limit of the ratio $\Delta x/\Delta t$ as Δt, and therefore Δx, approaches zero. So we can say:

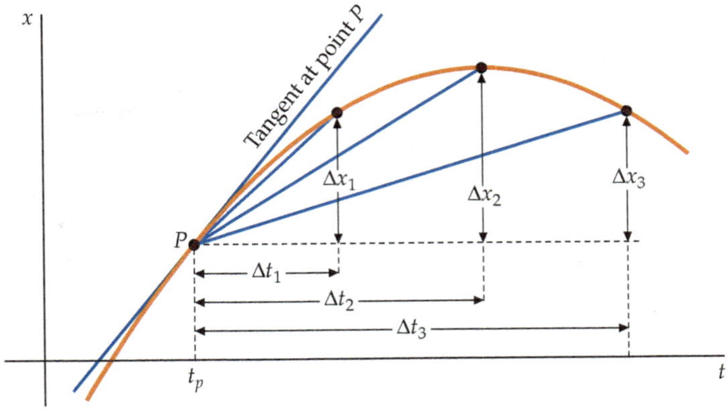

FIGURE 2-5 Graph of x versus t. Note the sequence of successively smaller time intervals, $\Delta t_3, \Delta t_2, \Delta t_1, \ldots$. The average velocity of each interval is the slope of the straight line for that interval. As the time intervals become smaller, these slopes approach the slope of the tangent to the curve at point t_P. The slope of this tangent line is defined as the instantaneous velocity at time t_P.

The instantaneous velocity v_x is the limit of the ratio $\Delta x/\Delta t$ as Δt approaches zero.

$$v_x(t) = \lim_{\Delta t \to 0} \frac{\Delta x}{\Delta t} = \text{slope of the line tangent to the } x\text{-versus-}t \text{ curve} \qquad 2\text{-}4$$

DEFINITION—INSTANTANEOUS VELOCITY

In calculus, this limit is called the **derivative** of x with respect to t and is written dx/dt. Using this notation, Equation 2-4 becomes:

$$v_x(t) = \lim_{\Delta t \to 0} \frac{\Delta x}{\Delta t} = \frac{dx}{dt} \qquad 2\text{-}5$$

See **Math Tutorial** *for more information on* **Differential Calculus**

A line's slope may be positive, negative, or zero; consequently, instantaneous velocity (in one-dimensional motion) may be positive (x increasing), negative (x decreasing), or zero (no motion). For an object moving with constant velocity, the object's instantaneous velocity is equal to its average velocity. The position versus time graph of this motion (Figure 2-6) will be a straight line whose slope equals the velocity.

The instantaneous velocity is a vector, and the magnitude of the instantaneous velocity is the **instantaneous speed.** Throughout the rest of the text, we shall use "velocity" in place of "instantaneous velocity" and "speed" in place of "instantaneous speed," except when emphasis or clarity is better served by the use of the adjective "instantaneous."

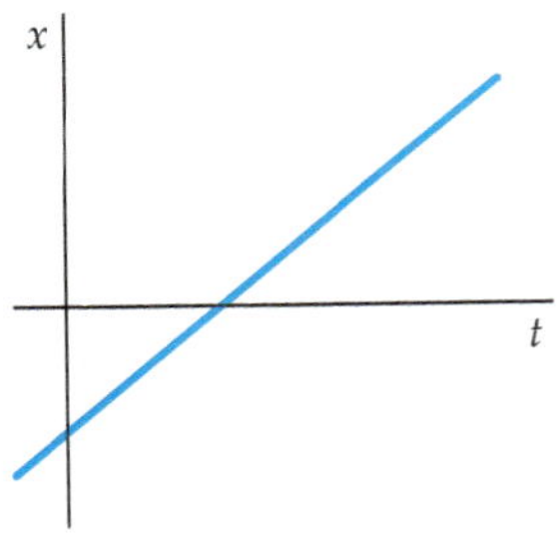

FIGURE 2-6 The position-versus-time graph for a particle moving at constant velocity.

Example 2-5 Position of a Particle as a Function of Time

Try It Yourself

The position of a particle as a function of time is given by the curve shown in Figure 2-7. Find the instantaneous velocity at time $t = 1.8$ s. When is the velocity greatest? When is it zero? Is it ever negative?

PICTURE In Figure 2-7, we have sketched the line tangent to the curve at $t = 1.8$ s. The tangent line's slope is the instantaneous velocity of the particle at the given time. You can measure the slope of the tangent line directly off this figure.

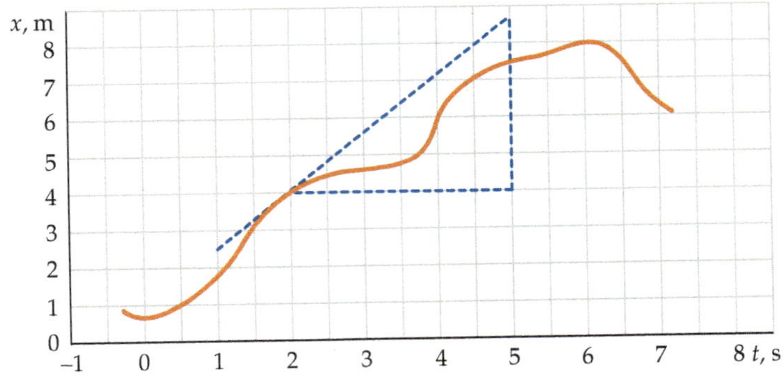

FIGURE 2-7

SOLVE

Cover the column to the right and try these on your own before looking at the answers.

Steps	Answers
1. Find the values x_1 and x_2 for the points on the tangent line at times $t_1 = 2.0$ s and $t_2 = 5.0$ s.	$x_1 \approx 4.0$ m, $x_2 \approx 8.5$ m
2. Compute the slope of the tangent line from these values. This slope equals the instantaneous velocity at $t = 2.0$ s.	$v_x = \text{slope} \approx \dfrac{8.5\text{ m} - 4.0\text{ m}}{5.0\text{ s} - 2.0\text{ s}} = \boxed{1.5\text{ m/s}}$

3. From the figure, the tangent line is steepest, and, therefore, the slope is greatest at about $t = 4.0$ s. The velocity is $\boxed{\text{greatest at } t \approx 4.0\text{ s}}$. The slope and the velocity both are $\boxed{\text{zero at } t = 0.0 \text{ and } t = 6.0\text{ s}}$ and are $\boxed{\text{negative for } t < 0.0 \text{ and } t > 6.0\text{ s}}$.

CHECK The position of the particle changes from about 1.8 m at 1.0 s to 4.0 m at 2.0 s, so the average velocity for the interval from 1.0 s to 2.0 s is 2.2 m/s. This is the same order of magnitude as the value for the instantaneous velocity at 1.8 s, so the step-2 result is plausible.

PRACTICE PROBLEM 2-1 Estimate the average velocity of this particle between $t = 2.0$ s and $t = 5.0$ s.

Example 2-6 A Stone Dropped from a Cliff

The position of a stone dropped from a cliff is described approximately by $x = 5t^2$, where x is in meters and t is in seconds. The $+x$ direction is downwards and the origin is at the top of the cliff. Find the velocity of the stone during its fall as a function of time t.

PICTURE We can compute the velocity at some time t by computing the derivative dx/dt directly from the definition in Equation 2-4. The corresponding curve giving x versus t is shown in Figure 2-8. Tangent lines are drawn at times t_1, t_2, and t_3. The slopes of these tangent lines increase steadily with increasing time, indicating that the instantaneous velocity increases steadily with time.

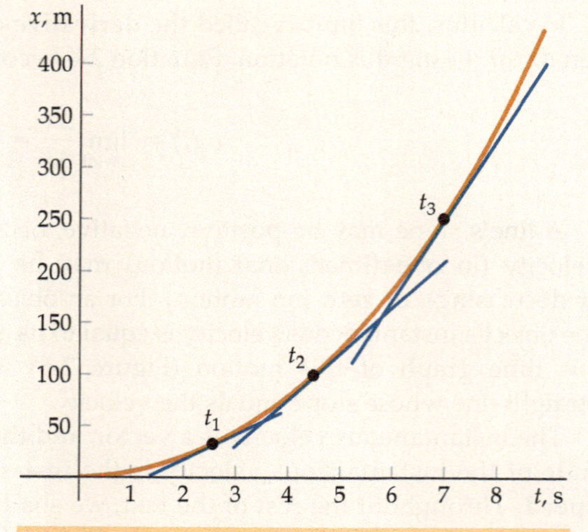

FIGURE 2-8

SOLVE

1. By definition the instantaneous velocity is

$$v_x(t) = \lim_{\Delta t \to 0} \frac{\Delta x}{\Delta t} = \lim_{\Delta t \to 0} \frac{x(t + \Delta t) - x(t)}{\Delta t}$$

2. We compute the displacement Δx from the position function $x(t)$:

$$x(t) = 5t^2$$

3. At a later time $t + \Delta t$, the position is $x(t + \Delta t)$, given by:

$$x(t + \Delta t) = 5(t + \Delta t)^2 = 5[t^2 + 2t\,\Delta t + (\Delta t)^2] = 5t^2 + 10t\,\Delta t + 5(\Delta t)^2$$

4. The displacement for this time interval is thus:

$$\Delta x = x(t + \Delta t) - x(t) = [5t^2 + 10t\,\Delta t + 5(\Delta t)^2] - 5t^2 = 10t\,\Delta t + 5(\Delta t)^2$$

5. Divide Δx by Δt to find the average velocity for this time interval:

$$v_{\text{av}\,x} = \frac{\Delta x}{\Delta t} = \frac{10t\,\Delta t + 5(\Delta t)^2}{\Delta t} = 10t + 5\,\Delta t$$

6. As we consider shorter and shorter time intervals, Δt approaches zero and the second term $5\Delta t$ approaches zero, though the first term, $10t$, remains unchanged:

$$v_x(t) = \lim_{\Delta t \to 0} \frac{\Delta x}{\Delta t} = \lim_{\Delta t \to 0}(10t + 5\,\Delta t) = \boxed{10t}$$

where v_x is in m/s and t is in s.

CHECK The stone starts at rest and goes faster and faster as it moves in the positive direction. Our result for the velocity, $v_x = 10t$, is zero at $t = 0$ and gets larger as t increases. Thus, $v_x = 10t$ is a plausible result.

TAKING IT FURTHER If we had set $\Delta t = 0$ in steps 4 and 5, the displacement would be $\Delta x = 0$, in which case the ratio $\Delta x/\Delta t$ would be undefined. Instead, we leave Δt as a variable until the final step, when the limit $\Delta t \to 0$ is well defined.

To find derivatives quickly, we use rules based on the limiting process above (see Appendix Table A-4). A particularly useful rule is

$$\text{If} \quad x = Ct^n, \quad \text{then} \quad \frac{dx}{dt} = Cnt^{n-1} \qquad \text{2-6}$$

where C and n are any constants. Using this rule in Example 2-6, we have $x = 5t^2$ and $v_x = dx/dt = 10t$, in agreement with our previous results.

2-2 ACCELERATION

When you step on your car's gas pedal or brake, you expect your velocity to change. An object whose velocity changes is said to be accelerating. **Acceleration** is the rate of change of velocity with respect to time. The **average acceleration,** $a_{\text{av}\,x}$, for a particular time interval Δt is defined as the change in velocity, Δv, divided by that time interval:

$$a_{\text{av}\,x} = \frac{\Delta v_x}{\Delta t} = \frac{v_{\text{f}x} - v_{\text{i}x}}{t_\text{f} - t_\text{i}} \qquad (\text{so } \Delta v_x = a_{\text{av}\,x}\,\Delta t) \qquad \text{2-7}$$

DEFINITION—AVERAGE ACCELERATION

Notice that acceleration has the dimensions of velocity (L/T) divided by time (T), which is the same as length divided by time squared (L/T^2). The SI unit is meters per second squared, m/s^2. Furthermore, like displacement and velocity, acceleration is a vector quantity. For one-dimensional motion, we can use + and − to indicate the direction of the acceleration. Equation 2-7 tells us that for $a_{\text{av}\,x}$ to be positive, Δv_x must be positive, and for $a_{\text{av}\,x}$ to be negative, Δv_x must be negative.

Instantaneous acceleration is the limit of the ratio $\Delta v_x/\Delta t$ as Δt approaches zero. On a plot of velocity versus time, the instantaneous acceleration at time t is the slope of the line tangent to the curve at that time:

$$a_x = \lim_{\Delta t \to 0} \frac{\Delta v_x}{\Delta t}$$

$$= \text{slope of the line tangent to the } v\text{-versus-}t \text{ curve} \qquad \text{2-8}$$

DEFINITION—INSTANTANEOUS ACCELERATION

Thus, acceleration is the derivative of velocity v_x with respect to time, dv_x/dt. Because velocity is the derivative of the position x with respect to t, acceleration is the second derivative of x with respect to t, d^2x/dt^2. We can see the reason for this notation when we write the acceleration as dv_x/dt and replace v_x with dx/dt:

$$a_x = \frac{dv_x}{dt} = \frac{d(dx/dt)}{dt} = \frac{d^2x}{dt^2} \qquad \text{2-9}$$

Notice that when the time interval becomes extremely small, the average acceleration and the instantaneous acceleration become equal. Therefore, we will use the word acceleration to mean "instantaneous acceleration."

It is important to note that the sign of an object's acceleration does not tell you whether the object is speeding up or slowing down. To determine this, you need to compare the signs of both the velocity and the acceleration of the object. If v_x and a_x are both positive, v_x is positive and becoming more positive so the speed is increasing. If v_x and a_x are both negative, v_x is negative and becoming more negative so the speed is again increasing. When v_x and a_x have opposite signs, the object is slowing down. If v_x is positive and a_x is negative, v_x is positive but is becoming less positive

! Deceleration does not mean the acceleration is negative. Deceleration does mean that v_x and a_x have opposite signs.

CONCEPT CHECK 2-1

You are following the car in front of you at high speed when the driver of the car in front of you brakes hard, bringing his car to a stop to avoid running over a huge pothole. Three tenths of a second after you see the brake lights on the lead car flash, you too brake hard. Assume that the two cars are initially traveling at the same speed, and that once both cars are braking hard, they lose speed at the same rate. Does the distance between the two cars remain constant while the two cars are both braking hard?

so the speed is decreasing. If v_x is negative and a_x is positive, v_x is negative but is becoming less negative so the speed is again decreasing. In summary, if v_x and a_x have the same sign, the speed is increasing; if v_x and a_x have opposite signs, the speed is decreasing. When an object is slowing down, we sometimes say it is decelerating.

If acceleration remains zero, there is no change in velocity over time—velocity is constant. In this case, the plot of x versus t is a straight line. If acceleration is nonzero and constant, as in Example 2-13 , then velocity varies linearly with time and x varies quadratically with time.

Example 2-7 A Fast Cat

A cheetah can accelerate from 0 to 96 km/h (60 mi/h) in 2.0 s, whereas a Corvette requires 4.5 s. Compute the average accelerations for the cheetah and Corvette and compare them with the free-fall acceleration, $g = 9.81\ \text{m/s}^2$.

(*Gunther Ziesler/Peter Arnold.*)

PICTURE Because we are given the initial and final velocities, as well as the change in time for both the cat and the car, we can simply use Equation 2-7 to find the acceleration for each object.

SOLVE

1. Convert 96 km/h into a velocity of m/s:

$$96\ \frac{\text{km}}{\text{h}}\left(\frac{1\ \text{h}}{3600\ \text{s}}\right)\left(\frac{1000\ \text{m}}{1\ \text{km}}\right) = 26.7\ \text{m/s}$$

2. Find the average acceleration from the information given:

$$\text{cat}\quad a_{\text{av}\,x} = \frac{\Delta v_x}{\Delta t} = \frac{26.7\ \text{m/s} - 0}{2.0\ \text{s}} = 13.3\ \text{m/s}^2 = \boxed{13\ \text{m/s}^2}$$

$$\text{car}\quad a_{\text{av}\,x} = \frac{\Delta v_x}{\Delta t} = \frac{26.7\ \text{m/s} - 0}{4.5\ \text{s}} = 5.93\ \text{m/s}^2 = \boxed{5.9\ \text{m/s}^2}$$

3. To compare the result with the acceleration due to gravity, multiply each by the conversion factor $1g/9.81\ \text{m/s}^2$:

$$\text{cat}\quad 13.3\ \text{m/s}^2 \times \frac{1g}{9.81\ \text{m/s}^2} = 1.36g = \boxed{1.4g}$$

$$\text{car}\quad 5.93\ \text{m/s}^2 \times \frac{1g}{9.81\ \text{m/s}^2} = 0.604g = \boxed{0.60g}$$

CHECK Because the car takes slightly more than twice as long as the cheetah to accelerate to the same velocity, it makes sense that the car's acceleration is slightly less than half that of the cat's.

TAKING IT FURTHER To reduce round-off errors, calculations are carried out using values with at least three digits even though the answers are reported using only two significant digits. These extra digits used in the calculations are called *guard digits.*

PRACTICE PROBLEM 2-2 A car is traveling at 45 km/h at time $t = 0$. It accelerates at a constant rate of 10 km/(h·s). (*a*) How fast is it traveling at $t = 2.0$ s? (*b*) At what time is the car traveling at 70 km/h?

Example 2-8 Velocity and Acceleration as Functions of Time

The position of a particle is given by $x = Ct^3$, where C is a constant. Find the dimensions of C. In addition, find both the velocity and the acceleration as functions of time.

PICTURE We can find the velocity by applying $dx/dt = Cnt^{n-1}$ (Equation 2-6) to the position of the particle, where n in this case equals 3. Then, we repeat the process to find the time derivative of velocity.

SOLVE

1. The dimensions of x and t are L and T, respectively:

$$C = \frac{x}{t^3} \Rightarrow [C] = \frac{[x]}{[t]^3} = \boxed{\frac{\text{L}}{\text{T}^3}}$$

2. We find the velocity by applying $dx/dt = Cnt^{n-1}$ (Equation 2-6):

$$x = Ct^n = Ct^3$$

$$v_x = \frac{dx}{dt} = Cnt^{n-1} = C3t^{3-1} = \boxed{3Ct^2}$$

3. The time derivative of velocity gives the acceleration: $a = \frac{dv_x}{dt} = Cnt^{n-1} = 3C(2)(t^{2-1}) = \boxed{6Ct}$

CHECK We can check the dimensions of our answers. For velocity, $[v_x] = [C][t^2] = (L/T^3)(T^2) = L/T$. For acceleration, $[a_x] = [C][t] = (L/T^3)(T) = L/T^2$.

PRACTICE PROBLEM 2-3 If a car starts from rest at $x = 0$ with constant acceleration a_x, its velocity v_x depends on a_x and the distance traveled x. Which of the following equations has the correct dimensions and therefore could possibly be an equation relating x, a_x, and v_x?

(*a*) $v_x = 2a_x x$ (*b*) $v_x^2 = 2a_x/x$ (*c*) $v_x = 2a_x x^2$ (*d*) $v_x^2 = 2a_x x$

MOTION DIAGRAMS

In studying physics, you will often wish to estimate the direction of the acceleration vector from a description of the motion. Motion diagrams can help. In a motion diagram the moving object is drawn at a sequence of equally spaced time intervals. For example, suppose you are on a trampoline and, following a high bounce, you are falling back toward the trampoline. As you descend, you keep going faster and faster. A motion diagram of this motion is shown in Figure 2-9*a*. The dots represent your position at equally spaced time intervals, so the space between successive dots increases as your speed increases. The numbers placed next to the dots are there to indicate the progression of time and an arrow representing your velocity is drawn next to each dot. The direction of each arrow represents the direction of your velocity at that instant, and the length of the arrow represents how fast you are going. Your acceleration vector* is in the direction that your velocity vector is changing—which is downward. In general, if the velocity arrows get longer as time progresses, then the acceleration is in the same direction of the velocity. On the other hand, if the velocity arrows get shorter as time progresses (Figure 2-9*b*), the acceleration is in the direction opposite to that of the velocity. Figure 2-9*b* is a motion diagram of your motion as you rise toward the ceiling following a bounce on the trampoline.

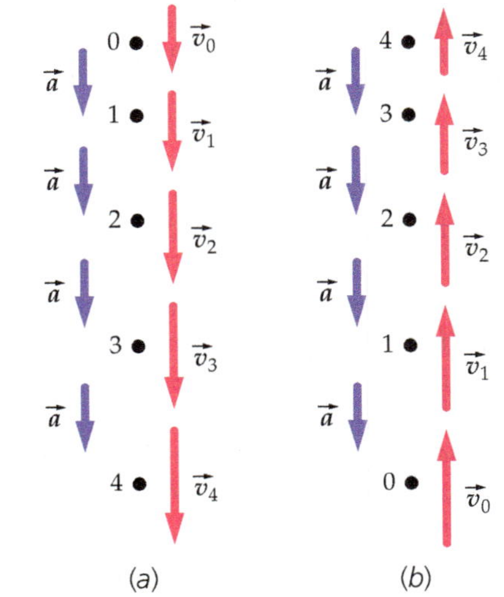

FIGURE 2-9 Motion diagrams. The time intervals between successive dots are identical. (*a*) The velocity vector is increasing, so the acceleration is in the direction of the velocity vector. (*b*) The velocity vector is decreasing, so the acceleration is in the direction opposite to that of the velocity vector.

2-3 MOTION WITH CONSTANT ACCELERATION

The motion of a particle that has nearly constant acceleration is found in nature. For example, near Earth's surface all unsupported objects fall vertically with nearly constant acceleration (provided air resistance is negligible). Other examples of near constant acceleration might include a plane accelerating along a runway for takeoff, and the motion of a car braking for a red light or speeding up at a green light. For a moving particle, the final velocity v_x equals the initial velocity plus the change in velocity, and the change in velocity equals the average acceleration multiplied by the time. That is,

$$v_x = v_{0x} + \Delta v = v_{0x} + a_{\text{av}\,x}\,\Delta t \qquad \text{2-10}$$

If a particle has constant acceleration a_x, it follows that the instantaneous acceleration and the average acceleration are equal. That is,

$$a_x = a_{\text{av}\,x} \quad (a_x \text{ is constant}) \qquad \text{2-11}$$

Because situations involving nearly constant acceleration are common, we can use the equations for acceleration and velocity to derive a special set of **kinematic equations** for problems involving one-dimensional motion at constant acceleration.

* The velocity vector and the acceleration vector were introduced in Chapter 1 and are further developed in Chapter 3.

DERIVING THE CONSTANT-ACCELERATION KINEMATIC EQUATIONS

Suppose a particle moving with constant acceleration a_x has a velocity v_{0x} at time $t_0 = 0$, and v_x at some later time t. Combining Equations 2-10 and 2-11, we have

$$v_x = v_{0x} + a_x t \qquad (a_x \text{ is constant}) \tag{2-12}$$

CONSTANT ACCELERATION: $v_x(t)$

A v_x-versus-t plot (Figure 2-10) of this equation is a straight line. The line's slope is the acceleration a_x.

To obtain an equation for the position x as a function of time, we first look at the special case of motion with a constant velocity $v_x = v_{0x}$ (Figure 2-11). The change in position Δx during an interval of time Δt is

$$\Delta x = v_{0x}\,\Delta t \qquad (a_x = 0)$$

The area of the shaded rectangle under the v_x-versus-t curve (Figure 2-11*a*) is its height v_{0x} times its width Δt, so the area under the curve is the displacement Δx. If v_{0x} is negative (Figure 2-11*b*), both the displacement Δx and the area under the curve are negative. We normally think of area as a quantity that cannot be negative. However, in this context that is not the case. If v_{0x} is negative, the "height" of the curve is negative and the "area under the curve" is the negative quantity $v_{0x}\,\Delta t$.

The geometric interpretation of the displacement as the area under the v_x-versus-t curve is true not only for constant velocity, but it is true in general, as illustrated in Figure 2-12. To show that this statement is true, we first divide the time interval into numerous small intervals, Δt_1, Δt_2, and so on. Then, we draw a set of rectangles as shown. The area of the rectangle corresponding to the ith time interval Δt_i (shaded in the figure) is $v_i\,\Delta t_i$, which is approximately equal

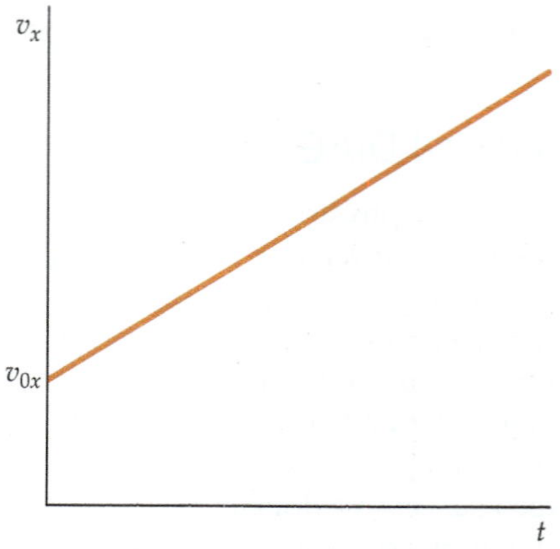

FIGURE 2-10 Graph of velocity versus time for constant acceleration.

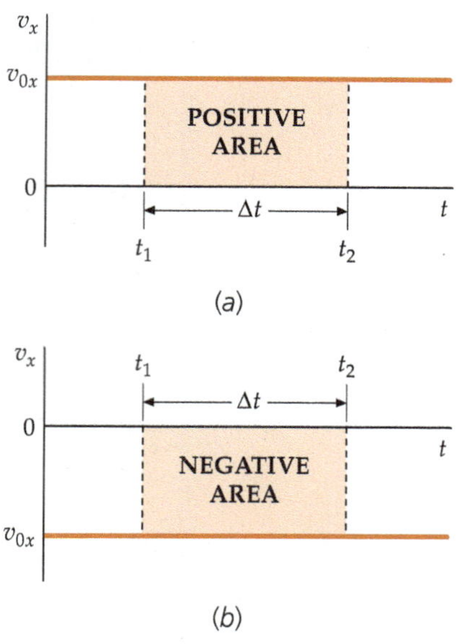

FIGURE 2-11 Motion with constant velocity.

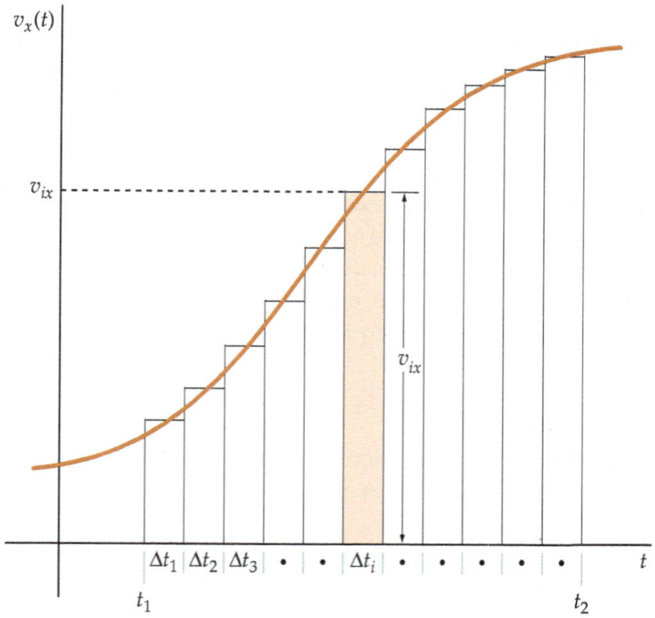

FIGURE 2-12 Graph of a general $v_x(t)$-versus-t curve. The total displacement from t_1 to t_2 is the area under the curve for this interval, which can be approximated by summing the areas of the rectangles.

to the displacement Δx_i during the interval Δt_i. The sum of the rectangular areas is therefore approximately the sum of the displacements during the time intervals and is approximately equal to the total displacement from time t_1 to t_2. We can make the approximation as accurate as we wish by putting enough rectangles under the curve, each rectangle having a sufficiently small value for Δt. For the limit of smaller and smaller time intervals (and more and more rectangles), the resulting sum approaches the area under the curve, which in turn equals the displacement. The displacement Δx is thus the area under the v_x-versus-t curve.

For motion with constant acceleration (Figure 2-13*a*), Δx is equal to the area of the shaded region. This region is divided into a rectangle and a triangle of areas $v_{1x}\,\Delta t$ and $\frac{1}{2}a_x(\Delta t)^2$, respectively, where $\Delta t = t_2 - t_1$. It follows that

$$\Delta x = v_{1x}\,\Delta t + \tfrac{1}{2}a_x(\Delta t)^2 \qquad \text{2-13}$$

If we set $t_1 = 0$ and $t_2 = t$, then Equation 2-13 becomes

$$x - x_0 = v_{0x}t + \tfrac{1}{2}a_x t^2 \qquad \text{2-14}$$

CONSTANT ACCELERATION: $x(t)$

where x_0 and v_{0x} are the position and velocity at time $t = 0$, and $x = x(t)$ is the position at time t. The first term on the right, $v_{0x}t$, is the displacement that would occur if a_x were zero, and the second term, $\frac{1}{2}a_x t^2$, is the additional displacement due to the constant acceleration.

We next use Equations 2-12 and 2-14 to obtain two additional kinematic equations for constant acceleration. Solving Equation 2-12 for t, and substituting for t, in Equation 2-14 gives

$$\Delta x = v_{0x}\frac{v_x - v_{0x}}{a_x} + \frac{1}{2}a_x\left(\frac{v_x - v_{0x}}{a_x}\right)^2$$

Multiplying both sides by $2a_x$ we obtain

$$2a_x\,\Delta x = 2v_{0x}(v_x - v_{0x}) + (v_x - v_{0x})^2$$

Simplifying and rearranging terms gives

$$v_x^2 = v_{0x}^2 + 2a_x\,\Delta x \qquad \text{2-15}$$

CONSTANT ACCELERATION: $v_x(x)$

The definition of average velocity (Equation 2-3) is:

$$\Delta x = v_{\text{av}\,x}\,\Delta t$$

where $v_{\text{av}\,x}\,\Delta t$ is the area under the horizontal line at height $v_{\text{av}\,x}$ in Figure 2-13*b* and Δx is the area under the v_x versus t curve in Figure 2-13*a*. We can see that if $v_{\text{av}\,x} = \frac{1}{2}(v_{1x} + v_{2x})$, the area under the line at height v_{av} in Figure 2-13*a* and the area under the v_x versus t curve in Figure 2-13*b* will be equal. Thus,

$$v_{\text{av}\,x} = \tfrac{1}{2}(v_{1x} + v_{2x}) \qquad \text{2-16}$$

CONSTANT ACCELERATION: $v_{\text{av}\,x}$ AND v_x

For motion with constant acceleration, the average velocity is the mean of the initial and final velocities.

For an example of an instance where Equation 2-16 is not applicable, consider the motion of a runner during a 10.0-km run that takes 40.0 min to complete. The

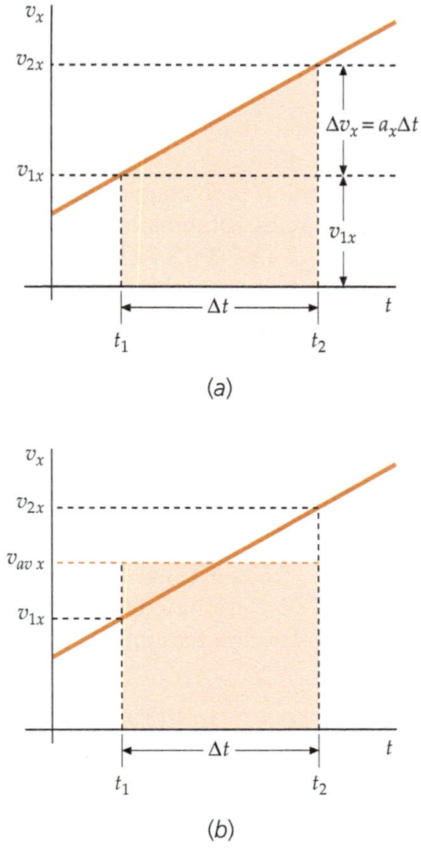

FIGURE 2-13 Motion with constant acceleration.

"It goes from zero to 60 in about 3 seconds." (*© Sydney Harris.*)

! Equation 2-16 is applicable only for time intervals during which the acceleration remains constant.

average velocity for the run is 0.250 km/min, which we compute using the definition of average velocity ($v_{\text{av }x} = \Delta x/\Delta t$). The runner starts from rest ($v_{1x} = 0$), and during the first one or two seconds his velocity increases rapidly, reaching a constant value v_{2x} that is sustained for the remainder of the run. The value of v_{2x} is just slightly greater than 0.250 km/min, so Equation 2-16 gives a value of about 0.125 km/min for the average velocity, a value almost 50% below the value given by the definition of average velocity. Equation 2-16 is not applicable because the acceleration does not remain constant for the entire run.

Equations 2-12, 2-14, 2-15, and 2-16 can be used to solve kinematics problems involving one-dimensional motion with constant acceleration. The choice of which equation or equations to use for a particular problem depends on what information you are given in the problem and what you are asked to find. Equation 2-15 is useful, for example, if we want to find the final velocity of a ball dropped from rest at some height x and we are not interested in the time the fall takes.

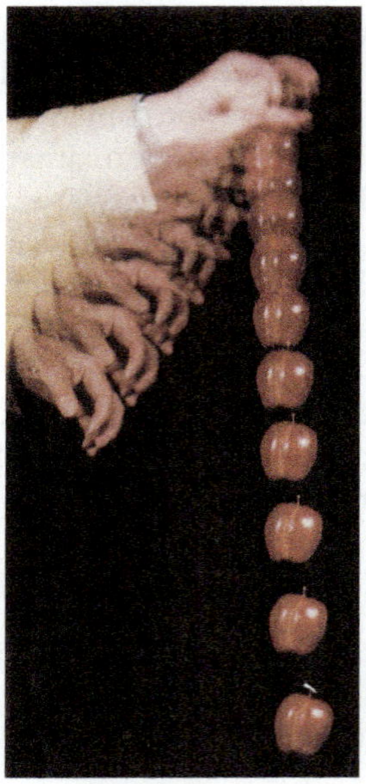

A falling apple captured by strobe photography at 60 flashes per second. The acceleration of the apple is indicated by the widening of the spaces between the images. *(Estate of Harold E. Edgerton/Palm Press.)*

USING THE CONSTANT-ACCELERATION KINEMATIC EQUATIONS

Review the Problem-Solving Strategy for solving problems using kinematic equations. Then, examine the examples involving one-dimensional motion with constant acceleration that follow.

PROBLEM-SOLVING STRATEGY

1-D Motion with Constant Acceleration

PICTURE Determine if a problem is asking you to find the time, distance, velocity, or acceleration for an object.

SOLVE Use the following steps to solve problems that involve one-dimensional motion and constant acceleration.

1. Draw a figure showing the particle in its initial and final positions. Include a coordinate axis and label the initial and final coordinates of the position. Indicate the + and − directions for the axis. Label the initial and final velocities, and label the acceleration.
2. Select one of the constant-acceleration kinematic equations (Equations 2-12, 2-14, 2-15, and 2-16). Substitute the given values into the selected equation and, if possible, solve for the desired value.
3. If necessary, select another of the constant-acceleration kinematic equations, substitute the given values into it, and solve for the desired value.

CHECK You should make sure that your answers are dimensionally consistent and the units of the answers are correct. In addition, check to make sure the magnitudes and signs of your answers agree with your expectations.

Problems with one object We will begin by looking at several examples that involve the motion of a single object.

Example 2-9 A Car's Stopping Distance

On a highway at night you see a stalled vehicle and brake your car to a stop. As you brake, the velocity of your car decreases at a constant rate of (5.0 m/s)/s. What is the car's stopping distance if your initial velocity is (*a*) 15 m/s (about 34 mi/h) or (*b*) 30 m/s?

PICTURE Use the Problem-Solving Strategy that precedes this example. The car is drawn as a dot to indicate a particle. We choose the direction of motion as $+x$ direction, and we choose the initial position $x_0 = 0$. The initial velocity is $v_{0x} = +15\text{ m/s}$ and the final velocity $v_x = 0$. Because the velocity is decreasing, the acceleration is negative. It is $a_x = -5.0\text{ m/s}^2$. We seek the distance traveled, which is the magnitude of the displacement Δx. We are neither given nor asked for the time, so $v_x^2 = v_{0x}^2 + 2a_x\Delta x$ (Equation 2-15) will provide a one-step solution.

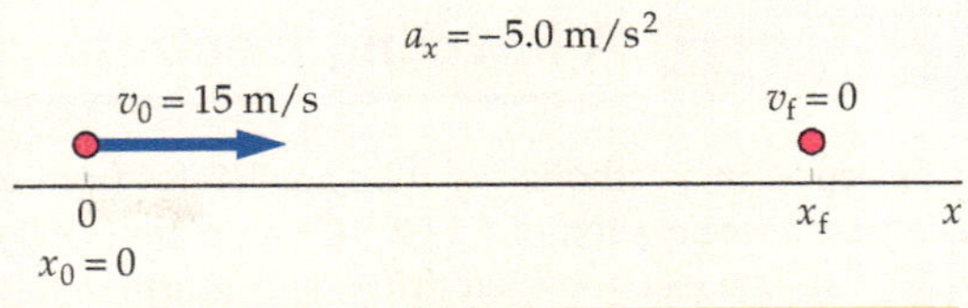

FIGURE 2-14

SOLVE

(*a*) 1. Draw the car (as a small dot) in its initial and final positions (Figure 2-14). Include a coordinate axis and label the drawing with the kinematic parameters.

2. Using Equation 2-15, calculate the displacement Δx:

$$v_x^2 = v_{0x}^2 + 2a_x\Delta x$$
$$0 = (15\text{ m/s})^2 + 2(-5.0\text{ m/s}^2)\Delta x$$
$$\Delta x = 22.5\text{ m} = \boxed{23\text{ m}}$$

(*b*) Substitute an initial speed of 30 m/s into the expression for Δx obtained in Part (*a*) (see Figure 2-14):

$$v_x^2 = v_{0x}^2 + 2a_x\Delta x$$
$$0 = (30\text{ m/s})^2 + 2(-5.0\text{ m/s}^2)\Delta x$$
$$\Delta x = \boxed{90\text{ m}}$$

CHECK The car's velocity decreases by 5.0 m/s each second. If its initial velocity is 15 m/s, it would take 3.0 s for it to come to rest. During the 3.0 s, it has an average velocity of half 15 m/s, so it would travel $\frac{1}{2}(15\text{ m/s})(3.0\text{ s}) = 23\text{ m}$. This confirms our Part-(*a*) result. Our Part-(*b*) result can be confirmed in the same manner.

Example 2-10 Stopping Distance

Try It Yourself

In the situation described in Example 2-9, (*a*) how much time does it take for the car to stop if its initial velocity is 30 m/s, and (*b*) how far does the car travel in the last second?

PICTURE Use the Problem-Solving Strategy that precedes Example 2-9. (*a*) In this part of the problem, you are asked to find the time it takes the car to stop. You are given the initial velocity $v_{0x} = 30\text{ m/s}$. From Example 2-9, you know the car has an acceleration $a_x = -5.0\text{ m/s}^2$. A relationship between time, velocity, and acceleration is given by Equation 2-12. (*b*) Because the car's velocity decreases by 5.0 m/s each second, the velocity 1.0 s before the car stops must be 5.0 m/s. Find the average velocity during the last second and use that to find the distance traveled.

SOLVE

Cover the column to the right and try these on your own before looking at the answers.

Steps	Answers
(*a*) 1. Draw the car (as a small dot) in its initial and final positions (Figure 2-15). Include a coordinate axis and label the drawing with the kinematic parameters.	
2. Use Equation 2-12 to find the total stopping time Δt.	$\Delta t = \boxed{6.0\text{ s}}$
(*b*) 1. Draw the car (as a small dot) in its initial and final positions (Figure 2-16). Include a coordinate axis.	
2. Find the average velocity during the last second from $v_{\text{av}\,x} = \frac{1}{2}(v_{\text{i}\,x} + v_{\text{f}\,x})$.	$v_{\text{av}\,x} = 2.5\text{ m/s}$
3. Compute the distance traveled from $\Delta x = v_{\text{av}\,x}\,\Delta t$.	$\Delta x = v_{\text{av}\,x}\,\Delta t = \boxed{2.5\text{ m}}$

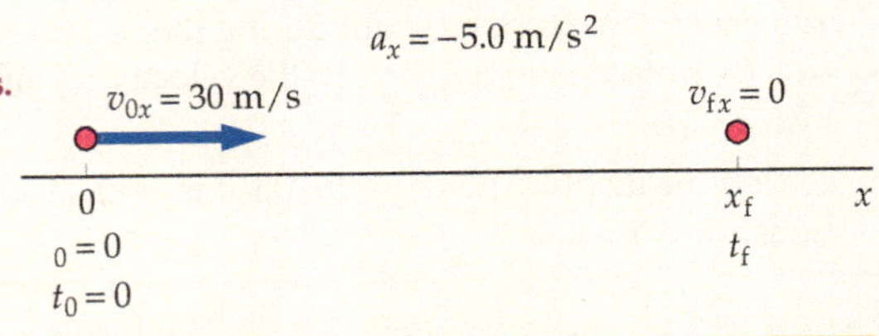

FIGURE 2-15

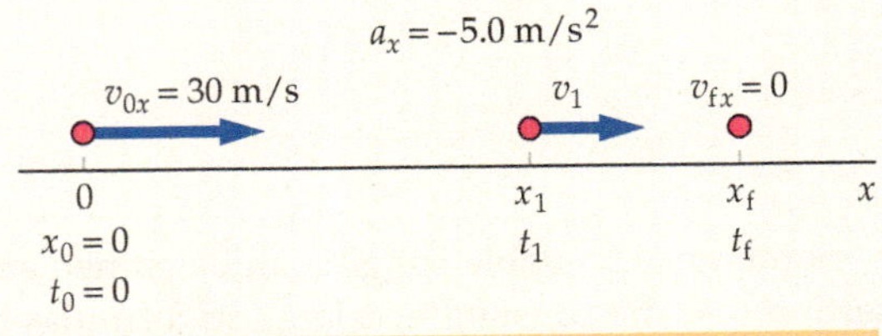

FIGURE 2-16

CHECK We would not expect the car to travel very far during the last second because it is moving relatively slowly. The Part-(*b*) result of 2.5 m is a plausible result.

Example 2-11 A Traveling Electron *Try It Yourself*

An electron in a cathode-ray tube accelerates from rest with a constant acceleration of $5.33 \times 10^{12}\ \mathrm{m/s^2}$ for $0.150\ \mu s$ ($1\ \mu s = 10^{-6}$ s). The electron then drifts with constant velocity for $0.200\ \mu s$. Finally, it slows to a stop with an acceleration of $-2.67 \times 10^{13}\ \mathrm{m/s^2}$. How far does the electron travel?

The two-mile-long linear accelerator at Stanford University used to accelerate electrons and positrons in a straight line to nearly the speed of light. Cross section of the accelerator's electron beam as shown on a video monitor. *(Stanford Limear Accelerator, U.S. Department of Energy.)*

PICTURE The equations for constant acceleration do not apply to the full duration of the electron's motion because the acceleration changes twice during that time. However, we can divide the electron's motion into three intervals, each with a different constant acceleration, and use the final position and velocity for the first interval as the initial conditions for the second interval, and the final position and velocity for the second interval as the initial conditions for the third. Apply the Problem-Solving Strategy that precedes Example 2-9 to each of the three constant-acceleration intervals. We will choose the origin to be at the electron's starting position, and the $+x$ direction to be the direction of motion.

SOLVE

Cover the column to the right and try these on your own before looking at the answers.

Steps

1. Draw the electron in its initial and final positions for each constant-acceleration interval (Figure 2-17). Include a coordinate axis and label the drawing with the kinematic parameters.

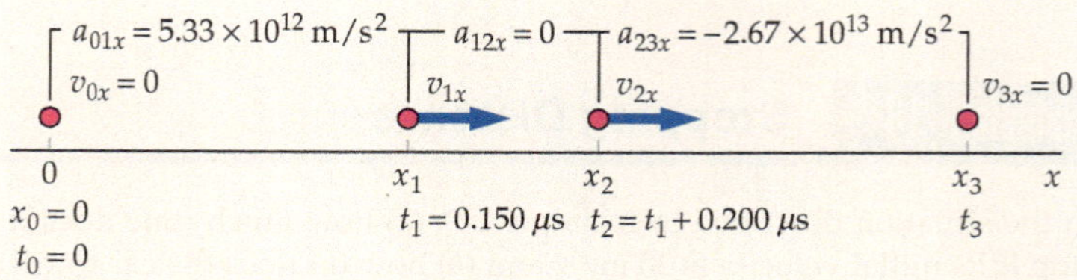

FIGURE 2-17

Answers

2. Set $v_{0x} = 0$ (because the electron starts from rest), use Equations 2-12 and 2-14 to find position x_1 and velocity v_{1x} at the end of the first 0.150-μs interval.

   $x_1 = 6.00$ cm, $v_{1x} = 8.00 \times 10^5$ m/s

3. The acceleration is zero during the second interval, so the velocity remains constant.

   $v_{2x} = v_{1x} = 8.00 \times 10^5$ m/s

4. The velocity remains constant during the second interval, so the displacement Δx_{12} equals the velocity v_{1x} multiplied by $0.200\ \mu s$.

   $\Delta x_{12} = 16.0$ cm, so $x_2 = 22.0$ cm

5. To find the displacement for the third interval, use Equation 2-15 with $v_{3x} = 0$.

   $\Delta x_{23} = 1.20$ cm, so $x_3 = \boxed{23.2\ \text{cm}}$

CHECK The average velocities are large, but the time intervals are small. Thus, the distances traveled are modest as we might expect.

Sometimes valuable insight can be gained about the motion of an object by asserting that the constant-acceleration formulas still apply even when the acceleration is not constant. The results are then estimates and not exact calculations. This is the case in the following example.

Example 2-12 The Crash Test

Context-Rich

In a crash test that you are performing, a car traveling 100 km/h (about 62 mi/h) hits an immovable concrete wall. What is the acceleration of the car during the crash?

PICTURE In this example, different parts of the vehicle will have different accelerations as the car crumples to a halt. The front bumper stops virtually instantly while the rear bumper stops some time later. We will solve for the acceleration of a part of the car that is in the passenger compartment and out of the crumple zone. A bolt holding the driver's seat belt to the floor is such a part. We do not really expect the acceleration of this bolt to be constant. We need additional information to solve this problem–either the stopping distance or the stopping time. We can estimate the stopping distance using common sense. Upon impact, the center of the car will certainly move forward less than half the length of the car. We will choose 0.75 m as a reasonable estimate of the distance the center of the car will move during the crash. Because the problem neither asks for nor provides the time, we will use the equation $v_x^2 = v_{0x}^2 + 2a_x\,\Delta x$.

SOLVE

1. Draw the bolt (as a small circle) at the center of the car at its initial and final positions (Figure 2-18). Include a coordinate axis and label the drawing with the kinematic parameters.

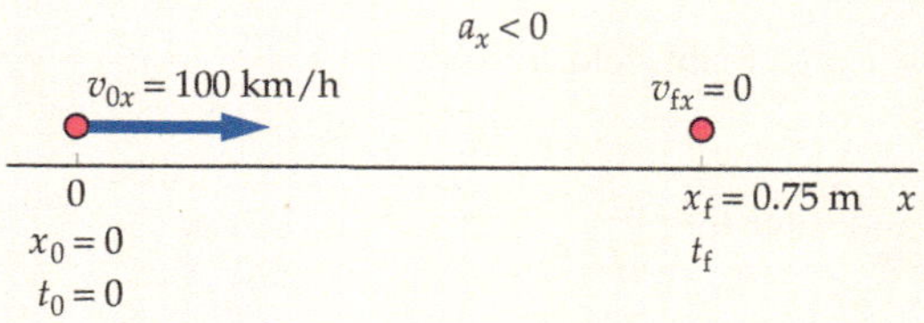

FIGURE 2-18

2. Convert the velocity from km/h to m/s.

$$(100\ \text{km/h}) \times \left(\frac{1\ \text{h}}{60\ \text{min}}\right) \times \left(\frac{1\ \text{min}}{60\ \text{s}}\right) \times \left(\frac{1000\ \text{m}}{1\ \text{km}}\right) = 27.8\ \text{m/s}$$

3. Using $v_x^2 = v_{0x}^2 + 2a_x\,\Delta x$, solve for the acceleration:

$$v_x^2 = v_{0x}^2 + 2a_x\,\Delta x$$

so

$$a_x = \frac{v_x^2 - v_{0x}^2}{2\Delta x} = \frac{0^2 - (27.8\ \text{m/s})^2}{2(0.75\ \text{m})}$$

4. Complete the calculation of the acceleration:

$$a_x = -\frac{(27.8\ \text{m/s})^2}{1.5\ \text{m}} = -514\ \text{m/s}^2 \approx \boxed{-500\ \text{m/s}^2}$$

CHECK The magnitude of the acceleration is about 50 times greater than the acceleration caused by the car breaking hard on a dry concrete road. The result is plausible because a large acceleration is expected for a high-speed head-on crash into an immovable object.

PRACTICE PROBLEM 2-4 Estimate the stopping time of the car.

Free-Fall Many practical problems deal with objects in free-fall, that is, objects that fall freely under the influence of gravity only. All objects in free-fall with the same initial velocity move identically. As shown in Figure 2-19, an apple and a feather, simultaneously released from rest in a large vacuum chamber, fall with identical motions. Thus, we know that the apple and the feather fall with the same acceleration. The magnitude of this acceleration, designated by g, has the approximate value $a = g \approx 9.81\ \text{m/s}^2 = 32.2\ \text{ft/s}^2$. If downward is designated as the $+y$ direction, then $a_y = +g$; if upward is designated as the $+y$ direction, then $a_y = -g$.

FIGURE 2-19 In a vacuum the apple and the feather, released simultaneously from rest, fall with identical motion. *(James Sugar/Black Star.)*

> ! Because g is the *magnitude* of the acceleration, g is *always* positive.

Example 2-13 The Flying Cap

Upon graduation, a joyful physics student throws her cap straight upward with an initial speed of 14.7 m/s. Given that its acceleration has a magnitude of 9.81 m/s² and is directed downward (we neglect air resistance), (*a*) how long does it take for the cap to reach its highest point? (*b*) What is the distance to the highest point? (*c*) Assuming the cap is caught at the same height from which it was released, what is the total time the cap is in flight?

PICTURE When the cap is at its highest point, its instantaneous velocity is zero. (When a problem specifies that an object is "at its highest point," translate this condition into the mathematical condition $v_y = 0$.)

SOLVE

(*a*) 1. Make a sketch of the cap in its initial position and again at its highest point (Figure 2-20). Include a coordinate axis and label the origin and the two specified positions of the cap.

2. The time is related to the velocity and acceleration: $v_y = v_{0y} + a_y t$

3. Set $v_y = 0$ and solve for t: $t = \dfrac{0 - v_{0y}}{a_y} = \dfrac{-14.7\text{ m/s}}{-9.81\text{ m/s}^2} = \boxed{1.50\text{ s}}$

(*b*) We can find the displacement from the time t and the average velocity:

$$\Delta y = v_{\text{av }y}t = \tfrac{1}{2}(v_{0y} + v_y)\Delta t = \tfrac{1}{2}(14.7\text{ m/s} + 0)(1.50\text{ s}) = \boxed{11.0\text{ m}}$$

(*c*) 1. Set $y = y_0$ in Equation 2-14 and solve for t:

$$\Delta y = v_{0y}t + \tfrac{1}{2}a_y t^2$$
$$0 = (v_{0y} + \tfrac{1}{2}a_y t)t$$

2. There are two solutions for t when $y = y_0$. The first corresponds to the time at which the cap is released, the second to the time at which the cap is caught:

$t = 0$ (first solution)

$t = -\dfrac{2v_{0y}}{a_y} = -\dfrac{2(14.7\text{ m/s})}{-9.81\text{ m/s}^2} = \boxed{3.00\text{ s}}$ (second solution)

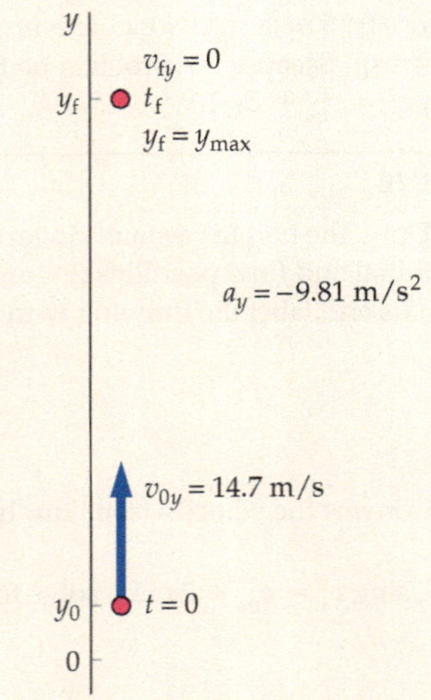

FIGURE 2-20

CHECK On the way up, the cap loses speed at the rate of 9.81 m/s each second. Because its initial speed is 14.7 m/s, we expect it rise for more than 1.00 s, but less than 2.00 s. Thus, a rise time of 1.50 s is quite plausible.

TAKING IT FURTHER On the plot of velocity versus time (Figure 2-21*b*) note that the slope is the same at all times, including the instant that $v_y = 0$. The slope is equal to the instantaneous acceleration, which is a constant -9.81 m/s^2. On the plot of height versus time (Figure 2-21*a*), note that the rise time equals the fall time. In reality, the cap will not have a constant acceleration because air resistance has a significant effect on a light object like a cap. If air resistance is not negligible, the fall time will exceed the rise time.

PRACTICE PROBLEM 2-5 Find $y_{max} - y_0$ using Equation 2-15. Find the velocity of the cap when it returns to its starting point.

PRACTICE PROBLEM 2-6 What is the velocity of the cap at the following points in time? (*a*) 0.100 s before it reaches its highest point; (*b*) 0.100 s after it reaches its highest point. (*c*) Compute $\Delta v_y/\Delta t$ for this 0.200-s-long time interval.

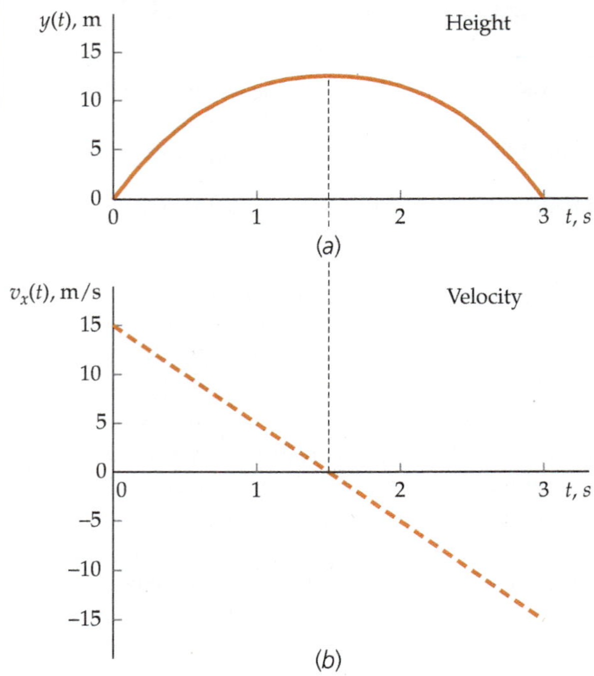

FIGURE 2-21 The height and velocity graphs are drawn one above the other so that both the height and the velocity can be observed at each instant of time.

Problems with two objects We now give some examples of problems involving two objects moving with constant acceleration.

Example 2-14 Catching a Speeding Car

A car is speeding at a constant 25 m/s (= 90 km/h ≈ 56 mi/h) in a school zone. A police car starts from rest just as the speeder passes by it and accelerates at a constant rate of 5.0 m/s^2. (*a*) When does the police car catch the speeding car? (*b*) How fast is the police car traveling when it catches up with the speeder?

PICTURE To determine when the two cars will be at the same position, we write the positions of the speeder x_S and of the police car x_P as functions of time and solve for the time t_c when $x_S = x_P$. Once we determine when the police car will catch up to the speeder, we can determine the velocity of the police car when it catches up to the speeder using the equation $v_x = a_x t$.

SOLVE

(*a*) 1. Draw the two cars at their initial positions (at $t = 0$) and again at their final positions (at $t = t_c$) (Figure 2-22). Include a coordinate axis and label the drawing with the kinematic parameters.

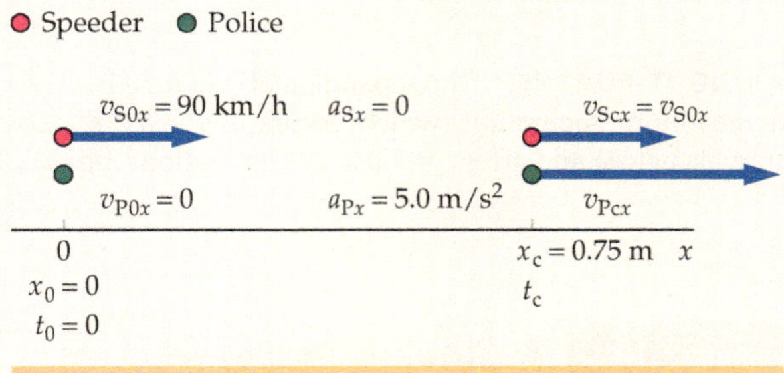

FIGURE 2-22 The speeder and the police car have the same position at $t = 0$ and again at $t = t_c$.

2. Write the position functions for the speeder and the police car:

$$x_S = v_{Sx}t \text{ and } x_P = \tfrac{1}{2}a_{Px}t^2$$

3. Set $x_S = x_P$ and solve for the time t_c, for $t_c > 0$:

$$v_{Sx}t_c = \tfrac{1}{2}a_{Px}t_c^2 \Rightarrow v_{Sx} = \tfrac{1}{2}a_{Px}t_c \qquad t_c \neq 0$$

$$t_c = \frac{2v_{Sx}}{a_{Px}} = \frac{2(25\text{ m/s})}{5.0\text{ m/s}^2} = \boxed{10\text{ s}}$$

(*b*) The velocity of the police car is given by $v_x = v_{0x} + a_x t$, with $v_{0x} = 0$:

$$v_{Px} = a_{Px}t_c = (5.0\text{ m/s}^2)(10\text{ s}) = \boxed{50\text{ m/s}}$$

CHECK Notice that the final velocity of the police car in (*b*) is exactly twice that of the speeder. Because the two cars covered the same distance in the same time, they must have had the same average velocity. The speeder's average velocity, of course, is 25 m/s. For the police car to start from rest, maintain a constant acceleration, and have an average velocity of 25 m/s, it must reach a final velocity of 50 m/s.

PRACTICE PROBLEM 2-7 How far have the cars traveled when the police car catches the speeder?

Example 2-15 The Police Car — *Try It Yourself*

How fast is the police car in Example 2-14 traveling when it is 25 m behind the speeding car?

PICTURE The speed is given by $v_P = a_x t_1$, where t_1 is the time at which $x_S - x_P = 25$ m.

SOLVE

Cover the column to the right and try these on your own before looking at the answers.

Steps

1. Sketch an x-versus-t graph showing the positions of the two cars (Figure 2-23). On this graph identify the distance $D = x_S - x_P$ between the two cars at a given instant.

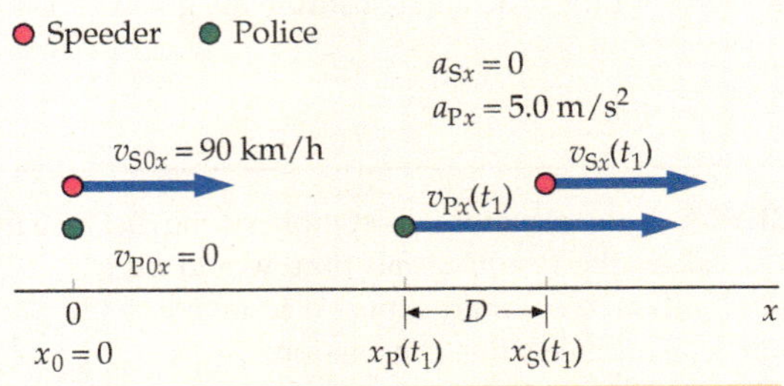

FIGURE 2-23

Answers

2. Using the equations for x_P and x_S from Example 2-14, solve for t_1 when $x_S - x_P = 25$ m. We expect two solutions, one shortly after the start time and one shortly before the speeder is caught. — $t_1 = (5 \pm \sqrt{15})$ s
3. Use $v_{P1} = a_{Px}t_1$ to compute the speed of the police car when $x_S - x_P = 25$ m. — $v_{P1} = \boxed{5.64 \text{ m/s}}$ and $\boxed{44.4 \text{ m/s}}$

CHECK We see from Figure 2-24 that the distance between the cars starts at zero, increases to a maximum value, and then decreases. We would expect two speeds for a given separation distance.

TAKING IT FURTHER The separation at any time is $D = x_S - x_P = v_{Sx}t - \frac{1}{2}a_{Px}t^2$. At maximum separation, which occurs at $t = 5.0$ s, $dD/dt = 0$. At equal time intervals before and after $t = 5.0$ s, the separations are equal.

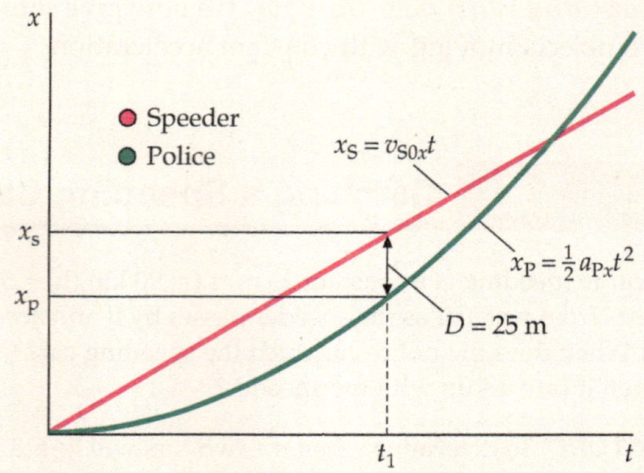

FIGURE 2-24

Example 2-16 A Moving Elevator

While standing in an elevator, you see a screw fall from the ceiling. The ceiling is 3.0 m above the floor. How long does it take the screw to hit the floor if the elevator is moving upward and gaining speed at a constant rate of 4.0 m/s² at the instant the screw leaves the ceiling?

PICTURE When the screw hits the floor, the positions of the screw and the floor are equal. Equate the these positions and solve for the time.

SOLVE

1. Draw a diagram showing the initial and final positions of the screw and the elevator floor (Figure 2-25). Include a coordinate axis and label the drawing with the kinematic parameters. The screw and the floor have the same initial velocity, but different accelerations. Choose the origin to be the initial position of the floor, and designate "upward" as the positive y direction. The screw hits the floor at time t_f:

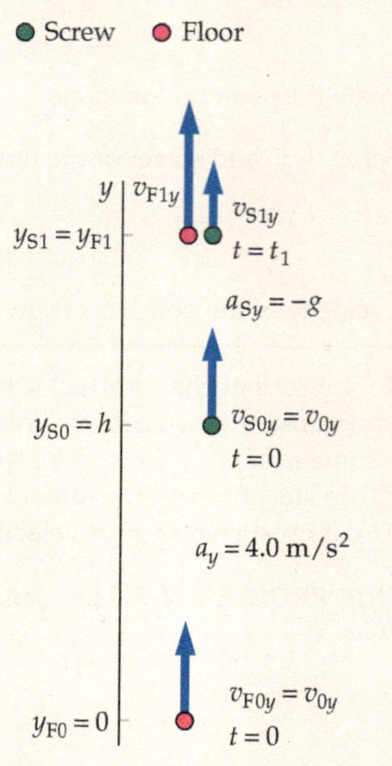

FIGURE 2-25 The y axis is fixed to the building.

2. Write equations specifying the *position* y_F of the elevator floor and the *position* y_S of the screw as functions of time. The screw and the elevator have the same initial velocity v_{0y}:

$$y_F - y_{F0} = v_{F0y}t + \tfrac{1}{2}a_{Fy}t^2$$
$$y_F - 0 = v_{0y}t + \tfrac{1}{2}a_{Fy}t^2$$
$$y_S - y_{S0} = v_{S0y}t + \tfrac{1}{2}a_{Sy}t^2$$
$$y_S - h = v_{0y}t + \tfrac{1}{2}(-g)t^2$$

3. Equate the expressions for y_S and y_F at $t = t_f$ and simplify:

$$y_S = y_F$$
$$h + v_{0y}t_f - \tfrac{1}{2}gt_f^2 = v_{0y}t_f + \tfrac{1}{2}a_{Fy}t_f^2$$
$$h - \tfrac{1}{2}gt_f^2 = \tfrac{1}{2}a_{Fy}t_f^2$$

4. Solve for the time and substitute the given values:

$$h = \tfrac{1}{2}(a_F + g)t_f^2 \quad \text{so}$$
$$t_f = \sqrt{\frac{2h}{a_F + g}} = \sqrt{\frac{2(3.0 \text{ m})}{4.0 \text{ m/s}^2 + 9.81 \text{ m/s}^2}} = 0.659 \text{ s} = \boxed{0.66 \text{ s}}$$

CHECK If the elevator was stationary, the distance the screw falls is given by $h = \frac{1}{2}gt_f^2$. With $h = 3.0$ m, the resulting fall time would be $t_f = 0.78$ s. Because of the elevator's upward acceleration, we would expect it to take less than 0.78 s for the screw to hit the floor. Our 0.66-s result meets this expectation.

Example 2-17 The Moving Elevator *Try It Yourself*

Consider the elevator and screw in Example 2-16. Assume the velocity of the elevator is 16 m/s upward when the screw separates from the ceiling. (*a*) How far does the elevator rise while the screw is in freefall? What is the displacement of the screw during freefall? (*b*) What are the velocity of the screw and the velocity of the elevator at impact?

PICTURE The time of flight of the screw is obtained in the solution of Example 2-16. Use this time to solve Parts (*a*) and (*b*).

SOLVE

Cover the column to the right and try these on your own before looking at the answers.

Steps	Answers
(*a*) 1. Using Equation 2-13, find the distance the floor rises between $t = 0$ and $t = t_f$, where t_f is calculated in step 4 of Example 2-16.	$\Delta y_F = v_{Fi}t_f + \frac{1}{2}a_F t_f^2 = \boxed{11.4\text{ m}}$
2. Between $t = 0$ and $t = t_f$, the displacement of the screw is less then that of the floor by 3.0 m.	$\Delta y_S = \boxed{+8.4\text{ m}}$
(*b*) Using $v_y = v_{0y} + a_y t$ (Equation 2-12), find the velocities of the screw and of the floor at impact.	$v_{Sy} = v_{Siy} - gt_f = \boxed{9.5\text{ m/s}}$ $v_{Fy} = v_{Fiy} + a_{Fy}t_f = \boxed{19\text{ m/s}}$

CHECK The Part-(*b*) results (velocity of the screw and the velocity of the floor at impact) are both positive indicating both velocities are directed upward. For impact to occur, the floor must be moving upward faster then the screw so it can catch up with the screw. This result is consistent with our Part-(*b*) results.

TAKING IT FURTHER The screw strikes the floor 8.4 m above its position when it leaves the ceiling. At impact, the velocity of the screw relative to the building is positive (upward). Relative to the building, the screw is still rising when it and the floor come in contact.

2-4 INTEGRATION

In this section, we use integral calculus to derive the equations of motion. A concise treatment of calculus can be found in the Math Tutorial.

To find the velocity from a given acceleration, we note that the velocity is the function $v_x(t)$ whose time derivative is the acceleration $a_x(t)$:

$$\frac{dv_x(t)}{dt} = a_x(t)$$

If the acceleration is constant, the velocity is that function of time which, when differentiated, equals this constant. One such function is

$$v_x = a_x t \qquad a_x \text{ is constant}$$

More generally, we can add any constant to $a_x t$ without changing the time derivative. Calling this constant c, we have

$$v_x = a_x t + c$$

When $t = 0$, $v_x = c$. Thus, c is the velocity v_{0x} at time $t = 0$.

Similarly, the position function $x(t)$ is that function whose derivative is the velocity:

$$\frac{dx}{dt} = v_x = v_{0x} + a_x t$$

We can treat each term separately. The function whose derivative is the constant v_{0x} is $v_{0x}t$ plus any constant. The function whose derivative is $a_x t$ is $\frac{1}{2}at^2$ plus any constant. Writing x_0 for the combined arbitrary constants, we have

$$x = x_0 + v_{0x}t + \tfrac{1}{2}at_x^2$$

When $t = 0$, $x = x_0$. Thus, x_0 is the position at time $t = 0$.

Whenever we find a function from its derivative, we must include an arbitrary constant in the general form of the function. Because we go through the integration process twice to find $x(t)$ from the acceleration, two constants arise. These constants are usually determined from the velocity and position at some given time, which is usually chosen to be $t = 0$. They are therefore called the **initial conditions.** A common problem, called the **initial-value problem,** takes the form "given $a_x(t)$ and the initial values of x and v_x, find $x(t)$." This problem is particularly important in physics because the acceleration of a particle is determined by the forces acting on it. Thus, if we know the forces acting on a particle and the position and velocity of the particle at some particular time, we can find its position and velocity at all other times.

A function $F(t)$ whose derivative (with respect to t) equals the function $f(t)$ is called the **antiderivative** of $f(t)$. (Because $v_x = dx/dt$ and $a_x = dv_x/dt$, x is the antiderivative of v_x and v_x is the antiderivative of a_x.) Finding the antiderivative of a function is related to the problem of finding the area under a curve.

In deriving Equation 2-14 it was shown that the change in position Δx is equal to the area under the velocity-versus-time curve. To show this (see Figure 2-12), we first divided the time interval into numerous small intervals, Δt_1, Δt_2, and so on. Then, we drew a set of rectangles as shown. The area of the rectangle corresponding to the ith time interval Δt_i (shaded in the figure) is $v_{ix}\,\Delta t_i$, which is approximately equal to the displacement Δx_i during the interval Δt_i. The sum of the rectangular areas is therefore approximately the sum of the displacements during the time intervals and is approximately equal to the total displacement from time t_1 to t_2. Mathematically, we write this as

$$\Delta x \approx \sum_i v_{ix}\,\Delta t_i$$

For the limit of smaller and smaller time intervals (and more and more rectangles), the resulting sum approaches the area under the curve, which in turn equals the displacement. The limit of the sum as Δt approaches zero (and the number of rectangles approaches infinity) is called an **integral** and is written

$$\Delta x = x(t_2) - x(t_1) = \lim_{\Delta t \to 0}\left(\sum_i v_{ix}\,\Delta t_i\right) = \int_{t_1}^{t_2} v_x\,dt \qquad \text{2-17}$$

See **Math Tutorial** *for more information on* **Integrals**

It is helpful to think of the integral sign $\int$ as an elongated S indicating a sum. The limits t_1 and t_2 indicate the initial and final values of the integration variable t.

The process of computing an integral is called **integration.** In Equation 2-17, v_x is the derivative of x, and x is the antiderivative of v_x. This is an example of the fundamental theorem of calculus, whose formulation in the seventeenth century greatly accelerated the mathematical development of physics. If

$$f(t) = \frac{dF(t)}{dt}, \quad \text{then} \quad F(t_2) - F(t_1) = \int_{t_1}^{t_2} f(t)\,dt \qquad \text{2-18}$$

FUNDAMENTAL THEOREM OF CALCULUS

The antiderivative of a function is also called the indefinite integral of the function and is written without limits on the integral sign, as in

$$x = \int v_x\, dt$$

Finding the function x from its derivative v_x (that is, finding the antiderivative) is also called integration. For example, if $v_x = v_{0x}$, a constant, then

$$x = \int v_{0x}\, dt = v_{0x}t + x_0$$

where x_0 is the arbitrary constant of integration. We can find a general rule for the integration of a power of t from Equation 2-6, which gives the general rule for the derivative of a power. The result is

$$\int t^n\, dt = \frac{t^{n+1}}{n+1} + C,\ n \neq -1 \qquad \text{2-19}$$

where C is an arbitrary constant. This equation can be checked by differentiating the right side using the rule of Equation 2-6. (For the special case $n = -1$, $\int t^{-1}dt = \ln t + C$, where $\ln t$ is the natural logarithm of t.)

Because $a_x = dv_x/dt$, the change in velocity for some time interval can similarly be interpreted as the area under the a_x-versus-t curve for that interval. This change is written

$$\Delta v_x = \lim_{\Delta t \to 0}\left(\sum_i a_{ix}\, \Delta t_i\right) = \int_{t_1}^{t_2} a_x\, dt \qquad \text{2-20}$$

We can now derive the constant-acceleration equations by computing the indefinite integrals of the acceleration and velocity. If a_x is constant, we have

$$v_x = \int a_x dt = a_x \int dt = v_{0x} + a_x t \qquad \text{2-21}$$

where we have expressed the product of a_x and the constant of integration as v_{0x}. Integrating again, and writing x_0 for the constant of integration, gives

$$x = \int (v_{0x} + a_x t)dt = x_0 + v_{0x}t + \tfrac{1}{2}a_x t^2 \qquad \text{2-22}$$

It is instructive to derive Equations 2-21 and 2-22 using definite integrals instead of indefinite ones. For constant acceleration, Equation 2-20, with $t_1 = 0$, gives

$$v_x(t_2) - v_x(0) = a_x \int_0^{t_2} dt = a_x(t_2 - 0)$$

where the time t_2 is arbitrary. Because it is arbitrary, we can set $t_2 = t$ to obtain

$$v_x = v_{0x} + a_x t$$

where $v_x = v_x(t)$ and $v_{0x} = v_x(0)$. To derive Equation 2-22, we substitute $v_{0x} + a_x t$ for v_x in Equation 2-17 with $t_1 = 0$. This gives

$$x(t_2) - x(0) = \int_0^{t_2} (v_{0x} + a_x t)dt$$

This integral is equal to the area under the v_x-versus-t curve (Figure 2-26). Evaluating the integral and solving for x gives

$$x(t_2) - x(0) = \int_0^{t_2} (v_{0x} + a_x t)dt = v_{0x}t + \tfrac{1}{2}a_x t^2\Big|_0^{t_2} = v_{0x}t_2 + \tfrac{1}{2}a_x t_2^2$$

where t_2 is arbitrary. Setting $t_2 = t$, we obtain

$$x = x_0 + v_{0x}t + \tfrac{1}{2}a_x t^2$$

where $x = x(t)$ and $x_0 = x(0)$.

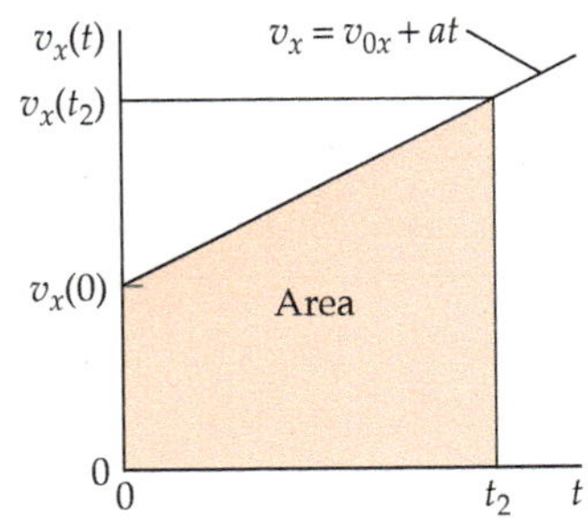

FIGURE 2-26 The area under the v_x-versus-t curve equals the displacement $\Delta x = x(t_2) - x(0)$.

The definition of average velocity is $\Delta x = v_{\text{av}\,x}\,\Delta t$ (Equation 2-3). In addition, $\Delta x = \int_{t_1}^{t_2} v_x\,dt$ (Equation 2-17). Equating the right sides of these equations and solving for $v_{\text{av}\,x}$ gives

$$v_{\text{av}\,x} = \frac{1}{\Delta t}\int_{t_1}^{t_2} v_x\,dt \qquad \text{2-23}$$

ALTERNATIVE DEFINITION OF AVERAGE VELOCITY

where $\Delta t = t_2 - t_1$. Equation 2-23 is mathematically equivalent to the definition of average velocity, so either equation can serve as a definition of average velocity.

Example 2-18 A Coasting Boat

A Shelter Island ferryboat moves with constant velocity $v_{0x} = 8.0\text{ m/s}$ for 60 s. It then shuts off its engines and coasts. Its coasting velocity is given by $v_x = v_{0x}t_1^2/t^2$, where $t_1 = 60$ s. What is the displacement of the boat for the interval $0 < t < \infty$?

(*Gene Mosca.*)

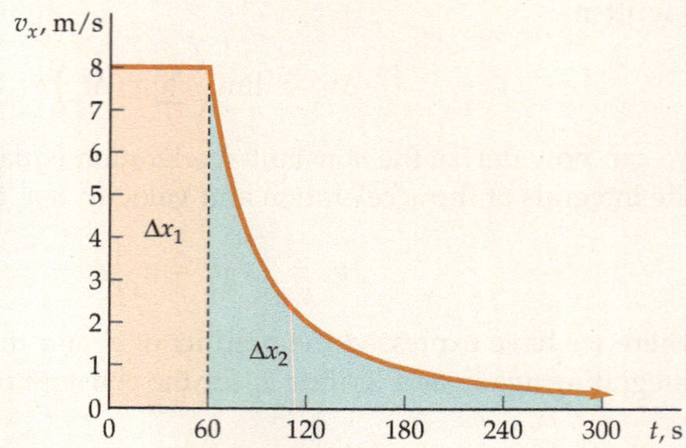

FIGURE 2-27

PICTURE The velocity function for the boat is shown in Figure 2-27. The total displacement is calculated as the sum of the displacement Δx_1 during the interval $0 < t < t_1 = 60$ s and the displacement Δx_2 during the interval $t_1 < t < \infty$.

SOLVE

1. The velocity of the boat is constant during the first 60 s; thus the displacement is simply the velocity times the elapsed time:

$$\Delta x_1 = v_{0x}\,\Delta t = v_{0x}t_1 = (8.0\text{ m/s})(60\text{ s}) = 480\text{ m}$$

2. The remaining displacement is given by the integral of the velocity from $t = t_1$ to $t = \infty$. We use Equation 2-17 to calculate the integral:

$$\Delta x_2 = \int_{t_1}^{\infty} v_x\,dt = \int_{t_1}^{\infty}\frac{v_{0x}t_1^2}{t^2}\,dt = v_{0x}t_1^2\int_{t_1}^{\infty} t^{-2}\,dt$$

$$= v_{0x}t_1^2\left.\frac{t^{-1}}{-1}\right|_{t_1}^{\infty} = -v_{0x}t_1^2\left(\frac{1}{\infty} - \frac{1}{t_1}\right)$$

$$= -(0 - v_0t_1) = (8\text{ m/s})(60\text{ s}) = 480\text{ m}$$

3. The total displacement is the sum of the displacements found above:

$$\Delta x = \Delta x_1 + \Delta x_2 = 480\text{ m} + 480\text{ m} = \boxed{960\text{ m}}$$

CHECK The expressions obtained for the displacements in both steps 1 and 2 are velocity multiplied by time, so they are both dimensionally correct.

TAKING IT FURTHER Note that the area under the v_x-versus-t curve (Figure 2-27) is finite. Thus, even though the boat never stops moving, it travels only a finite distance. A better representation of the velocity of a coasting boat might be the exponentially decreasing function $v_x = v_{0x}e^{-b(t-t_1)}$, where b is a positive constant. In that case, the boat would also coast a finite distance in the interval $t_1 \le t \le \infty$.

Physics Spotlight

Linear Accelerators

Linear accelerators are instruments that accelerate electrically charged particles to high speeds along a long, straight track to collide with a target. Large accelerators can impart very high kinetic energies (on the order of billions of electron volts) to charged particles that serve as probes for studying the fundamental particles of matter and the forces that hold them together. (The energy required to remove an electron from an atom is on the order of one electron volt.) In the two-mile-long linear accelerator at Stanford University, electromagnetic waves boost the speed of electrons or positrons as they move through an evacuated copper pipe. When the high-speed particles collide with a target, several different kinds of subatomic particles are produced along with X rays and gamma rays. These particles then pass into devices called particle detectors.

The beige cylinder in the background is the linear accelerator at the heart of the Naval Academy Tandem Accelerator Laboratory. A beam of high-speed protons travels from the accelerator to the target area in the foreground. *(Gene Mosca.)*

Through experiments with such accelerators, physicists have determined that protons and neutrons, once thought to be the ultimate particles of the nucleus, are themselves composed of more fundamental particles called quarks. Another group of particles known as leptons, which include electrons, neutrinos, and a few other particles, have also been identified. Most large accelerator research centers such as the Fermi National Accelerator Laboratory in Batavia, Illinois, use a series of linear and circular accelerators to achieve higher particle speeds. As the speed of a particle approaches the speed of light, the energy required to accelerate it to that speed approaches infinity.

Although the big accelerators may have a high profile, thousands of linear accelerators are used worldwide for a host of practical applications. One of the most common applications is the cathode ray tube (CRT) of a television set or computer monitor. In a CRT, electrons from the cathode (a heated filament) are accelerated in a vacuum toward a positively charged anode. Electromagnets control the direction of the electron beam onto the inside of a screen coated with a phosphor, a material that emits light when struck by electrons. The kinetic energy of electrons in a CRT ranges to a maximum of about 30,000 electron volts. The speed of an electron that has this kinetic energy is about one third of the speed of light.

In the field of medicine, linear accelerators about a thousand times more powerful than a CRT are used for radiation treatment of cancer. "The linear accelerator uses microwave technology (similar to that used for radar) to accelerate electrons in a part of the accelerator called the 'wave guide,' then allows these electrons to collide with a heavy metal target. As a result of the collisions, high-energy x-rays are scattered from the target. A portion of these x-rays is collected and then shaped to form a beam that matches the patient's tumor."*

Other applications of accelerators include the production of radioisotopes for tracers in medicine and biology, sterilization of surgical tools, and analysis of materials to determine their composition. For example, in a technique called particle-induced X-ray emission (PIXE), an ion beam, often consisting of protons, causes target atoms to emit X rays that identify the type of atoms present. This technique has been applied to the study of archeological materials and variety of other types of samples.

* The American College of Radiology and the Radiological Society of North America, http//www.radiologyinfo.org/content/therapy/linear_accelerator.htm.

SUMMARY

Displacement, velocity, and acceleration are important *defined* kinematic quantities.

TOPIC	RELEVANT EQUATIONS AND REMARKS	
1. Displacement	$\Delta x = x_2 - x_1$	2-1
Graphical interpretation	Displacement is the area under the v_x-versus-t curve.	
2. Velocity		
Average velocity	$v_{\text{av}\,x} = \dfrac{\Delta x}{\Delta t}$ or $v_{\text{av}\,x} = \dfrac{1}{\Delta t}\displaystyle\int_{t_1}^{t_2} v_x\,dt$	2-3, 2-23
Instantaneous velocity	$v_x(t) = \lim\limits_{\Delta t \to 0} \dfrac{\Delta x}{\Delta t} = \dfrac{dx}{dt}$	2-5
Graphical interpretation	The instantaneous velocity is the slope of the x-versus-t curve.	
3. Speed		
Average speed	$\text{average speed} = \dfrac{\text{total distance}}{\text{total time}} = \dfrac{s}{t}$	2-2
Instantaneous speed	Instantaneous speed is the magnitude of the instantaneous velocity $\text{speed} = \lvert v_x \rvert$	
4. Acceleration		
Average acceleration	$a_{\text{av}\,x} = \dfrac{\Delta v_x}{\Delta t}$	2-7
Instantaneous acceleration	$a_x = \dfrac{dv_x}{dt} = \dfrac{d^2x}{dt^2}$	2-9
Graphical interpretation	The instantaneous acceleration is the slope of the v_x-versus-t curve.	
Acceleration due to gravity	The acceleration of an object near the surface of Earth in free-fall under the influence of gravity alone is directed downward and has magnitude $g = 9.81\ \text{m/s}^2 = 32.2\ \text{ft/s}^2$	
5. Kinematic equations for constant acceleration		
Velocity	$v_x = v_{0x} + a_x t$	2-12
Average velocity	$v_{\text{av}\,x} = \frac{1}{2}(v_{0x} + v_x)$	2-16
Displacement in terms of $v_{\text{av}\,x}$	$\Delta x = x - x_0 = v_{\text{av}\,x} t = \frac{1}{2}(v_{0x} + v_x)t$	
Displacement as a function of time	$\Delta x = x - x_0 = v_{0x} t + \frac{1}{2} a_x t^2$	2-14
v_x^2 as a function of Δx	$v_x^2 = v_{0x}^2 + 2a_x\,\Delta x$	2-15
6. Displacement and velocity as integrals	Displacement is represented graphically as the area under the v_x-versus-t curve. This area is the integral of v_x over time from some initial time t_1 to some final time t_2 and is written $\Delta x = \lim\limits_{\Delta t \to 0} \sum_i v_{i\,x}\,\Delta t_i = \displaystyle\int_{t_1}^{t_2} v_x\,dt$ (2-17) Similarly, change in velocity is represented graphically as the area under the a_x-versus-t curve: $\Delta v_x = \lim\limits_{\Delta t \to 0} \sum_i a_{i\,x}\,\Delta t_i = \displaystyle\int_{t_1}^{t_2} a_x\,dt$ (2-20)	2-17, 2-20

Answers to Concept Checks

2-1 No. The distance between cars will not remain constant. Instead, the distance will continuously decrease. When you first begin breaking, the speed of your car is greater than the speed of the car in front. That is because the car in front began breaking 0.3 s earlier. Because the cars lose speed at the same rate, the speed of your car will remain greater than the speed of the car in front throughout the braking period.

Answers to Practice Problems

2-1 1.2 m/s

2-2 (*a*) 65 km/h (*b*) 2.5 s

2-3 Only (*d*) has the same dimensions on both sides of the equation. Although we cannot obtain the exact equation from dimensional analysis, we can often obtain the functional dependence.

2-4 54 ms

2-5 (a) and (*b*) $y_{max} - y_0 = 11.0$ m (*c*) -14.7 m/s; notice that the final speed is the same as the initial speed

2-6 (*a*) $+0.981$ m/s (*b*) -0.981 m/s
(*c*) $[(-0.981\text{ m/s}) - (+0.981\text{ m/s}]/(0.200\text{ s}) = -9.81\text{ m/s}^2$

2-7 250 m

PROBLEMS

In a few problems, you are given more data than you actually need; in a few other problems, you are required to supply data from your general knowledge, outside sources, or informed estimates.

Interpret as significant all digits in numerical values that have trailing zeroes and no decimal points.

For all problems, use $g = 9.81$ m/s² for the free-fall acceleration due to gravity and neglect friction and air resistance unless instructed to do otherwise.

- • Single-concept, single-step, relatively easy
- •• Intermediate-level, may require synthesis of concepts
- ••• Challenging
- SSM Solution is in the Student Solutions Manual
- Consecutive problems that are shaded are paired problems.

CONCEPTUAL PROBLEMS

1 • What is the average velocity over the "round trip" of an object that is launched straight up from the ground and falls straight back down to the ground?

2 • An object thrown straight up falls back and is caught at the same place it is launched from. Its time of flight is T; its maximum height is H. Neglect air resistance. The correct expression for its average speed for the entire flight is (*a*) H/T, (*b*) 0, (*c*) $H/(2T)$, (*d*) $2H/T$.

3 • Using the information in the previous question, what is its average speed just for the first half of the trip? What is its average velocity for the second half of the trip? (Answer in terms of H and T.)

4 • Give an everyday example of one-dimensional motion where (*a*) the velocity is westward and the acceleration is eastward, and (*b*) the velocity is northward and the acceleration is northward.

5 • Stand in the center of a large room. Call the direction to your right "positive," and the direction to your left "negative." Walk across the room along a straight line, using a constant acceleration to quickly reach a steady speed along a straight line in the negative direction. After reaching this steady speed, keep your velocity negative but make your acceleration positive. (*a*) Describe how your speed varied as you walked. (*b*) Sketch a graph of x versus t for your motion. Assume you started at $x = 0$. (*c*) Directly under the graph of Part (*b*), sketch a graph of v_x versus t. SSM

6 • True/false: The displacement *always* equals the product of the average velocity and the time interval. Explain your choice.

7 • Is the statement "for an object's velocity to remain constant, its acceleration *must* remain zero" true or false? Explain your choice.

8 •• **MULTISTEP** Draw careful graphs of the position and velocity and acceleration over the time period $0 \leq t \leq 30$ s for a cart that, in succession, has the following motion. The cart is moving at the constant speed of 5.0 m/s in the $+x$ direction. It passes by the origin at $t = 0.0$ s. It continues on at 5.0 m/s for 5.0 s, after which it gains speed at the constant rate of 0.50 m/s each second for 10.0 s. After gaining speed for 10.0 s, the cart loses speed at the constant rate of 0.50 m/s for the next 15.0 s.

9 • True/false: Average velocity *always* equals one-half the sum of the initial and final velocities. Explain your choice.

10 • Identical twin brothers standing on a horizontal bridge each throw a rock straight down into the water below. They throw rocks at exactly the same time, but one hits the water before the other. How can this be? Explain what they did differently. Ignore any effects due to air resistance.

11 •• Dr. Josiah S. Carberry stands at the top of the Sears Tower in Chicago. Wanting to emulate Galileo, and ignoring the safety of the pedestrians below, he drops a bowling ball from the top of the tower. One second later, he drops a second bowling ball. While the balls are in the air, does their separation (*a*) increase over time, (*b*) decrease, (*c*) stay the same? Ignore any effects due to air resistance. SSM

12 •• Which of the position-versus-time curves in Figure 2-28 best shows the motion of an object (*a*) with positive acceleration, (*b*) with constant positive velocity, (*c*) that is always at rest, and

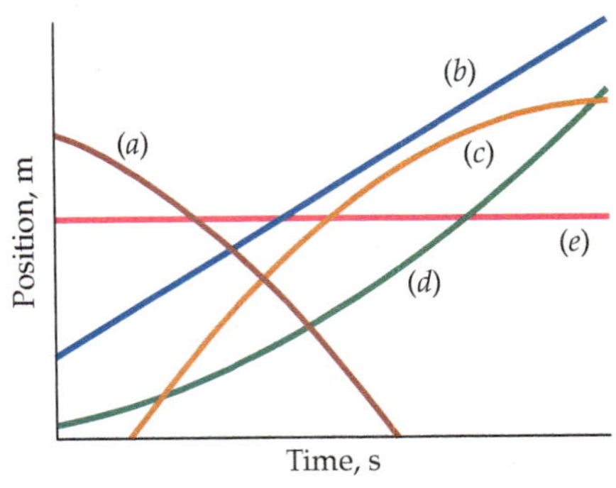

FIGURE 2-28 Problem 12

(*d*) with negative acceleration? (There may be more than one correct answer for each part of the problem.)

13 •• Which of the velocity-versus-time curves in Figure 2-29 best describes the motion of an object (*a*) with constant positive acceleration, (*b*) with positive acceleration that is decreasing with time, (*c*) with positive acceleration that is increasing with time, and (*d*) with no acceleration? (There may be more than one correct answer for each part of the problem.) SSM

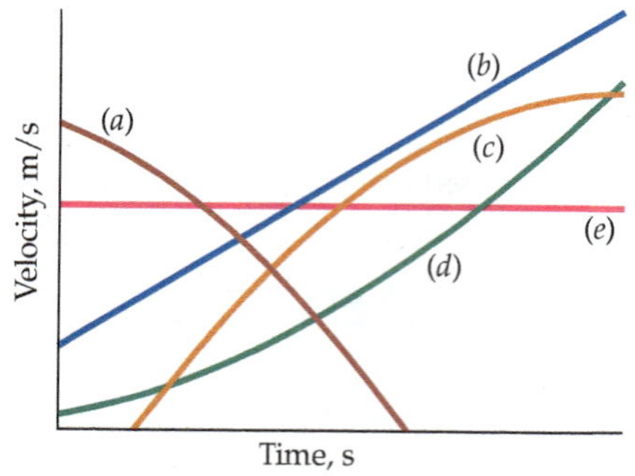

FIGURE 2-29 Problem 13

14 •• The diagram in Figure 2-30 tracks the location of an object moving in a straight line along the x axis. Assume that the object is at the origin at $t = 0$. Of the five times shown, which time (or times) represents when the object is (*a*) farthest from the origin, (*b*) at rest for an instant, (*c*) in the midst of being at rest for a while, and (*d*) moving away from the origin?

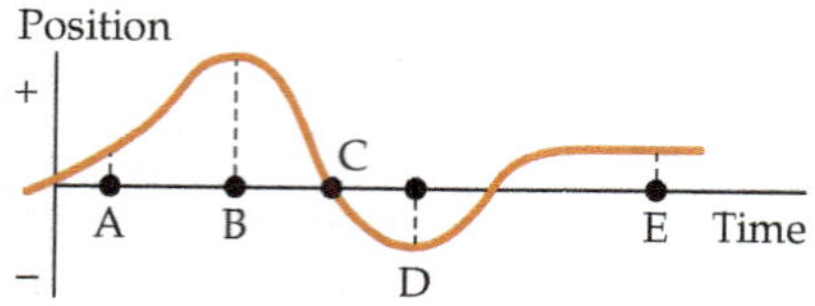

FIGURE 2-30 Problems 14 and 15

15 •• An object moves along a straight line. Its position-versus-time graph is shown in Figure 2-30. At which time or times is its (*a*) speed at a minimum, (*b*) acceleration positive, and (*c*) velocity negative? SSM

16 •• For each of the four graphs of x versus t in Figure 2-31 answer the following questions. (*a*) Is the velocity at time t_2 greater than, less than, or equal to the velocity at time t_1? (*b*) Is the speed at time t_2 greater than, less than, or equal to the speed at time t_1?

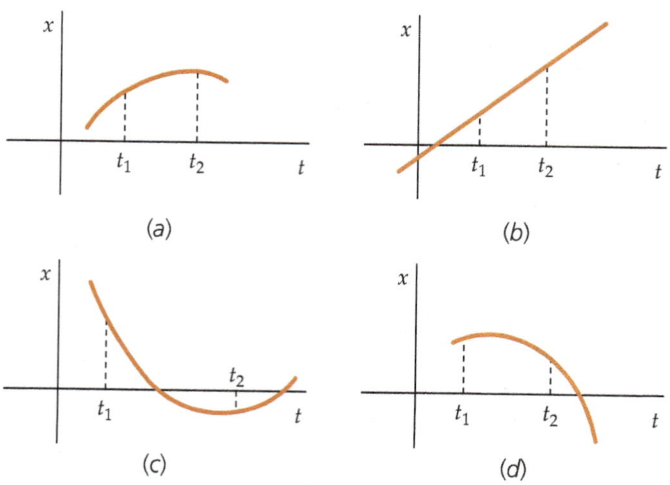

FIGURE 2-31 Problem 16

17 •• True/false:

(*a*) If the acceleration of an object is always zero, then it cannot be moving.
(*b*) If the acceleration of an object is always zero, then its x-versus-t curve must be a straight line.
(*c*) If the acceleration of an object is nonzero at an instant, it may be momentarily at rest at that instant.

Explain your reasoning for each answer. If an answer is *true*, give an example.

18 •• A hard-thrown tennis ball is moving horizontally when it bangs into a vertical concrete wall at perpendicular incidence. The ball rebounds straight back off the wall. Neglect any effects due to gravity for the small time interval described here. Assume that toward the wall is the $+x$ direction. What are the directions of its velocity and acceleration (*a*) just before hitting the wall, (*b*) at maximum impact, and (*c*) just after leaving the wall?

19 •• A ball is thrown straight up. Neglect any effects due to air resistance. (*a*) What is the velocity of the ball at the top of its flight? (*b*) What is its acceleration at that point? (*c*) What is different about the velocity and acceleration at the top of the flight if instead the ball impacts a horizontal ceiling very hard and then returns. SSM

20 •• An object that is launched straight up from the ground, reaches a maximum height H, and falls straight back down to the ground, hitting it T seconds after launch. Neglect any effects due to air resistance. (*a*) Express the average speed for the entire trip as a function of H and T. (*b*) Express the average speed for the same interval of time as a function of the initial launch speed v_0.

21 •• A small lead ball is thrown directly upward. True or false: (Neglect any effects due to air resistance.) (*a*) The magnitude of its acceleration decreases on the way up. (*b*) The direction of its acceleration on its way down is opposite to the direction of its acceleration on its way up. (*c*) The direction of its velocity on its way down is opposite to the direction of its velocity on its way up.

22 •• At $t = 0$, object A is dropped from the roof of a building. At the same instant, object B is dropped from a window 10 m below the roof. Air resistance is negligible. During the descent of B to the ground, the distance between the two objects (*a*) is proportional to t, (*b*) is proportional to t^2, (*c*) decreases, (*d*) remains 10 m throughout.

23 •• **CONTEXT-RICH** You are driving a Porsche that accelerates uniformly from 80.5 km/h (50 mi/h) at $t = 0.00$ s to 113 km/h (70 mi/h) at $t = 9.00$ s. (*a*) Which graph in Figure 2-32 best describes the velocity of your car? (*b*) Sketch a position-versus-time graph showing the location of your car during these nine seconds, assuming we let its position x be zero at $t = 0$.

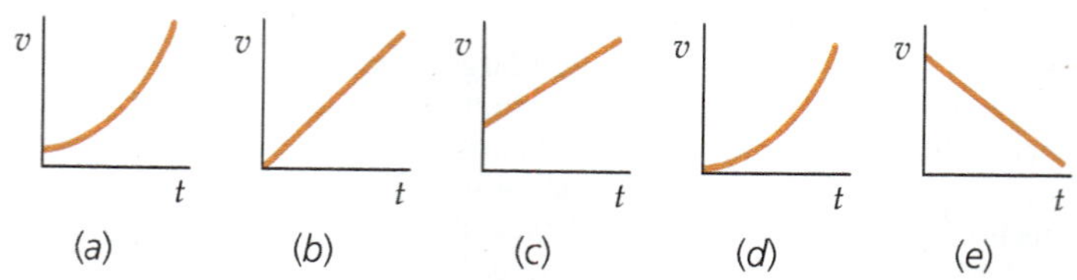

FIGURE 2-32 Problem 23

24 •• A small heavy object is dropped from rest and falls a distance D in a time T. After it has fallen for a time $2T$, what will be its (*a*) fall distance from its initial location, (*b*) its speed, and (*c*) its acceleration? (Neglect air resistance.)

25 •• In a race, at an instant when two horses are running right next to each other and in the same direction (the $+x$ direction), Horse A's instantaneous velocity and acceleration are $+10$ m/s and $+2.0$ m/s^2, respectively, and horse B's instantaneous velocity and acceleration are $+12$ m/s and -1.0 m/s^2, respectively. Which horse is passing the other at this instant? Explain.

26 •• True or false: (*a*) The equation $x - x_0 = v_{0x}t + \frac{1}{2}a_xt^2$ is always valid for particle motion in one dimension. (*b*) If the velocity at a given instant is zero, the acceleration at that instant must also be zero. (*c*) The equation $\Delta x = v_{av}\, \Delta t$ holds for all particle motion in one dimension.

27 •• If an object is moving in a straight line at constant acceleration, its instantaneous velocity halfway through any time interval is (*a*) greater than its average velocity, (*b*) less than its average velocity, (*c*) equal to its average velocity, (*d*) half its average velocity, (*e*) twice its average velocity.

28 •• A turtle, seeing his owner put some fresh lettuce on the opposite side of his terrarium, begins to accelerate (at a constant rate) from rest at time $t = 0$, heading directly toward the food. Let t_1 be the time at which the turtle has covered half the distance to his lunch. Derive an expression for the ratio of t_2 to t_1, where t_2 is the time at which the turtle reaches the lettuce.

29 •• The positions of two cars in parallel lanes of a straight stretch of highway are plotted as functions of time in the Figure 2-33. Take positive values of x as being to the right of the origin. Qualitatively answer the following: (*a*) Are the two cars ever side by side? If so, indicate that time (those times) on the axis. (*b*) Are they always traveling in the same direction, or are they moving in opposite directions for some of the time? If so, when? (*c*) Are they ever traveling at the same velocity? If so, when? (*d*) When are the two cars the farthest apart? (*e*) Sketch (no numbers) the velocity versus time curve for each car. SSM

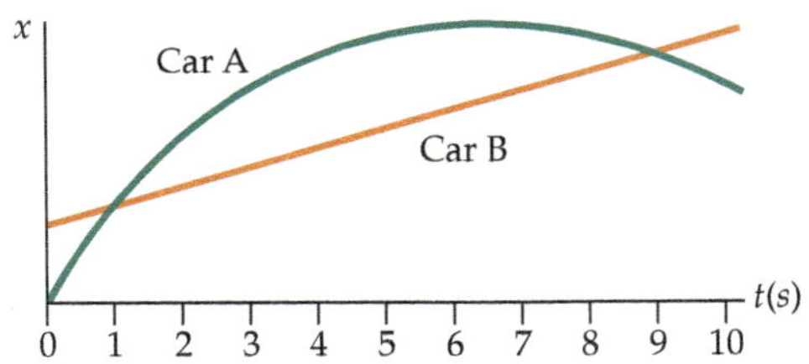

FIGURE 2-33 Problem 29

30 •• A car driving at constant velocity passes the origin at time $t = 0$. At that instant, a truck, at rest at the origin, begins to accelerate uniformly from rest. Figure 2-34 shows a qualitative plot of the velocities of truck and car as functions of time. Compare their displacements (from the origin), velocities, and accelerations at the instant that their curves intersect.

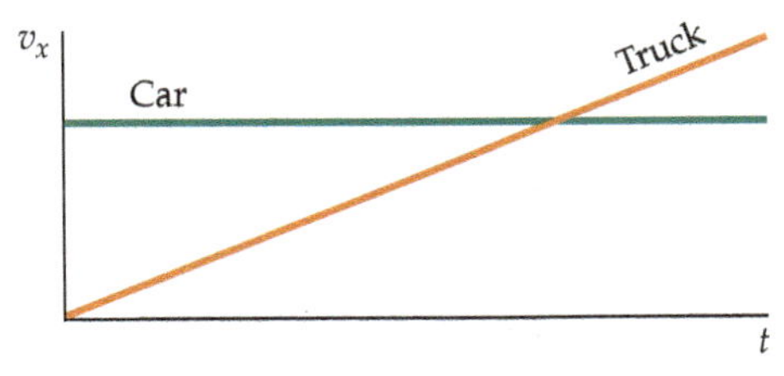

FIGURE 2-34 Problem 30

31 •• Reginald is out for a morning jog, and during the course of his run on a straight track, he has a velocity that depends upon time as shown in Figure 2-35. That is, he begins at rest, and ends at rest, peaking at a maximum velocity v_{max} at an arbitrary time t_{max}. A second runner, Josie, runs throughout the time interval $t = 0$ to $t = t_f$ at a constant speed v_J, so that each has the same displacement during the time interval. (*Note:* t_f is NOT twice t_{max}, but represents an arbitrary time.) What is the relation between v_J *and* v_{max}?

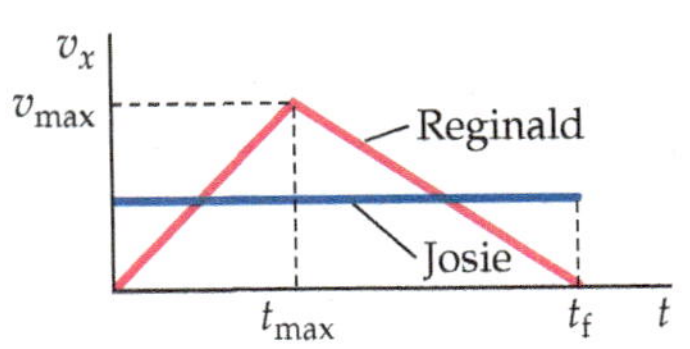

FIGURE 2-35 Problem 31

32 •• Which graph (or graphs), if any, of v_x versus t in Figure 2-36 best describes the motion of a particle with (*a*) positive velocity and increasing speed, (*b*) positive velocity and zero acceleration, (*c*) constant nonzero acceleration, and (*d*) a speed decrease?

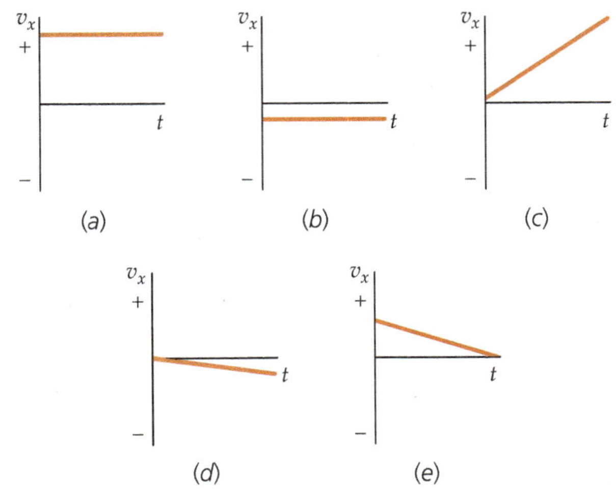

FIGURE 2-36 Problems 32 and 33

33 •• Which graph (or graphs), if any, of v_x versus t in Figure 2-36 best describes the motion of a particle with (*a*) negative velocity and increasing speed, (*b*) negative velocity and zero acceleration, (*c*) variable acceleration, and (*d*) increasing speed?

34 •• Sketch a *v*-versus-*t* curve for each of the following conditions: (*a*) Acceleration is zero and constant while velocity is not zero. (*b*) Acceleration is constant but not zero. (*c*) Velocity and acceleration are both positive. (*d*) Velocity and acceleration are both negative. (*e*) Velocity is positive and acceleration is negative. (*f*) Velocity is negative and acceleration is positive. (*g*) Velocity is momentarily zero but the acceleration is not zero.

35 •• Figure 2-37 shows nine graphs of position, velocity, and acceleration for objects in motion along a straight line. Indicate the graphs that meet the following conditions: (*a*) Velocity is constant,

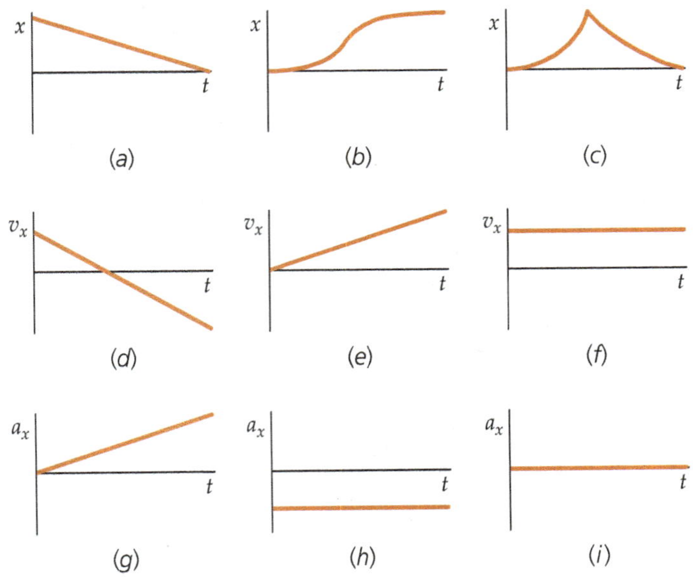

FIGURE 2-37 Problem 35

(*b*) velocity reverses its direction, (*c*) acceleration is constant, and (*d*) acceleration is not constant. (*e*) Which graphs of position, velocity, and acceleration are mutually consistent?

ESTIMATION AND APPROXIMATION

36 • **CONTEXT-RICH** While engrossed in thought about the scintillating lecture just delivered by your physics professor you mistakenly walk directly into the wall (rather than through the open lecture hall door). Estimate the magnitude of your average acceleration as you rapidly come to a halt.

37 • **BIOLOGICAL APPLICATION** Occasionally, people can survive falling large distances if the surface they land on is soft enough. During a traverse of the Eiger's infamous Nordvand, mountaineer Carlos Ragone's rock anchor gave way and he plummeted 500 feet to land in snow. Amazingly, he suffered only a few bruises and a wrenched shoulder. Assuming that his impact left a hole in the snow 4.0 ft deep, estimate his average acceleration as he slowed to a stop (that is, while he was impacting the snow). SSM

38 •• When we solve free-fall problems near Earth, it's important to remember that air resistance may play a significant role. If its effects are significant, we may get answers that are wrong by orders of magnitude if we ignore it. How can we tell when it is valid to ignore the effects of air resistance? One way is to realize that air resistance increases with increasing speed. Thus, as an object falls and its speed *increases*, its downward acceleration *decreases*. Under these circumstances, the object's speed will approach, as a limit, a value called its *terminal speed*. This terminal speed depends upon such things as the mass and cross-sectional area of the body. Upon reaching its terminal speed, its acceleration is zero. For a "typical" skydiver falling through the air, a typical terminal speed is about 50 m/s (roughly 120 mph). At half its terminal speed, the skydiver's acceleration will be about $\frac{3}{4}g$. Let us take half the terminal speed as a reasonable "upper bound" beyond which we should not use our constant acceleration free-fall relationships. Assuming the skydiver started from rest, (*a*) estimate how far, and for how long, the skydiver falls before we can no longer neglect air resistance. (*b*) Repeat the analysis for a Ping-Pong ball, which has a terminal speed of about 5.0 m/s. (*c*) What can you conclude by comparing your answers for Parts (*a*) and (*b*)?

39 •• **BIOLOGICAL APPLICATION** On June 14, 2005, Asafa Powell of Jamaica set a world's record for the 100-m dash with a time $t = 9.77$ s. Assuming he reached his maximum speed in 3.00 s, and then maintained that speed until the finish, estimate his acceleration during the first 3.00 s.

40 •• The photograph in Figure 2-38 is a short-time exposure (1/30 s) of a juggler with two tennis balls in the air. (*a*) The tennis ball near the top of its trajectory is less blurred than the lower one. Why is that? (*b*) Estimate the speed of the ball that he is just releasing from his right hand. (*c*) Determine how high the ball should have gone above the launch point and compare it to an estimate from the picture. *Hint: You have a built-in distance scale if you assume some reasonable value for the height of the juggler.*

FIGURE 2-38
Problem 40
(Courtesy of Chuck Adler.)

41 •• A rough rule of thumb for determining the distance between you and a lightning strike is to start counting the seconds that elapse ("one-Mississippi, two-Mississippi, . . .") until you hear the thunder (sound emitted by the lightning as it rapidly heats the air around it). Assuming the speed of sound is about 750 mi/h, (*a*) estimate how far away is a lightning strike if you counted about 5 s until you heard the thunder. (*b*) Estimate the uncertainty in the distance to the strike in Part (*a*). Be sure to explain your assumptions and reasoning. *Hint: The speed of sound depends on the air temperature, and your counting is far from exact!*

SPEED, DISPLACEMENT, AND VELOCITY

42 • **ENGINEERING APPLICATION** (*a*) An electron in a television tube travels the 16-cm distance from the grid to the screen at an average speed of 4.0×10^7 m/s. How long does the trip take? (*b*) An electron in a current-carrying wire travels at an average speed of 4.0×10^{-5} m/s. How long does it take to travel 16 cm?

43 • A runner runs 2.5 km, in a straight line, in 9.0 min and then takes 30 min to walk back to the starting point. (*a*) What is the runner's average velocity for the first 9.0 min? (*b*) What is the average velocity for the time spent walking? (*c*) What is the average velocity for the whole trip? (*d*) What is the average speed for the whole trip? SSM

44 • A car travels in a straight line with an average velocity of 80 km/h for 2.5 h and then with an average velocity of 40 km/h for 1.5 h. (*a*) What is the total displacement for the 4.0-h trip? (*b*) What is the average velocity for the total trip?

45 • One busy air route across the Atlantic Ocean is about 5500 km. The now-retired Concord, a supersonic jet capable of flying at twice the speed of sound, was used for traveling such routes. (*a*) Roughly how long did it take for a one-way flight? (Use 343 m/s for the speed of sound.) (*b*) Compare this time to the time taken by a subsonic jet flying at 0.90 times the speed of sound.

46 • The speed of light, designated by the universally recognized symbol *c*, has a value, to two significant figures, of 3.0×10^8 m/s. (*a*) How long does it take for light to travel from the Sun to Earth, a distance of 1.5×10^{11} m? (*b*) How long does it take light to travel from the moon to Earth, a distance of 3.8×10^8 m?

47 • Proxima Centauri, the closest star to us besides our own Sun, is 4.1×10^{13} km from Earth. From Zorg, a planet orbiting this star, a Gregor places an order at Tony's Pizza in Hoboken, New Jersey, communicating by light signals. Tony's fastest delivery craft travels at $1.00 \times 10^{-4}c$ (see Problem 46). (*a*) How long does it take Gregor's order to reach Tony's Pizza? (*b*) How long does Gregor wait between sending the signal and receiving the pizza? If Tony's has a "1000-years-or-it's-free" delivery policy, does Gregor have to pay for the pizza? SSM

48 • A car making a 100-km journey travels 40 km/h for the first 50 km. How fast must it go during the second 50 km to average 50 km/h?

49 •• **CONTEXT-RICH** Late in ice hockey games, the team that is losing sometimes "pulls" their goalkeeper off the ice to add an additional offensive player and increase their chances of scoring. In such cases, the goalie on the opposing team might have an opportunity to score into the unguarded net 55.0 m away. Suppose you are the goaltender for your university team and are in just such a situation. You launch a shot (in hopes of getting your first career goal) on the frictionless ice. You eventually hear a disappointing "clang" as the puck strikes a goalpost (instead of going in!) exactly 2.50 s later. In this case, how fast did the puck travel? You should assume 343 m/s for the speed of sound.

50 •• Cosmonaut Andrei, your co-worker at the International Space Station, tosses a banana at you at a speed of 15 m/s. At exactly the same instant, you fling a scoop of ice cream at Andrei along exactly the same path. The collision between banana and ice cream produces a banana split 7.2 m from your location 1.2 s after the banana and ice cream were launched. (*a*) How fast did you toss the ice cream? (*b*) How far were you from Andrei when you tossed the ice cream? (Neglect any effects due to gravity.)

51 •• Figure 2-39 shows the position of a particle as a function of time. Find the average velocities for the time intervals *a*, *b*, *c*, and *d* indicated in the figure.

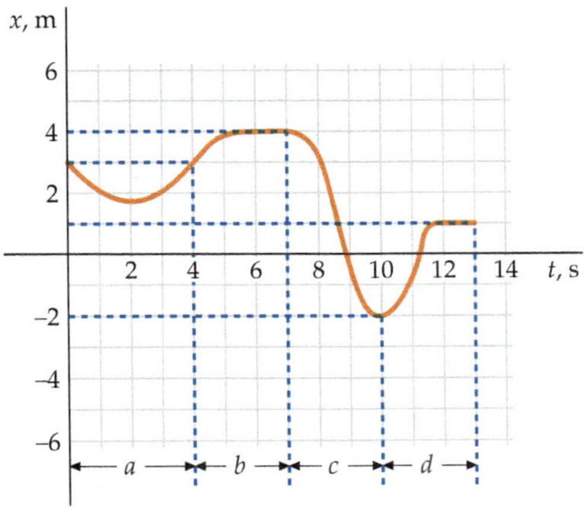

FIGURE 2-39 Problem 51

52 •• **ENGINEERING APPLICATION** It has been found that, on average, galaxies are moving away from Earth at a speed that is proportional to their distance from Earth. This discovery is known as Hubble's law, named for its discoverer, astrophysicist Sir Edwin Hubble. He found that the recessional speed v of a galaxy a distance r from Earth is given by $v = Hr$, where $H = 1.58 \times 10^{-18}\ s^{-1}$ is called the Hubble constant. What are the expected recessional speeds of galaxies (*a*) 5.00×10^{22} m from Earth, and (*b*) 2.00×10^{25} m from Earth? (*c*) If the galaxies at each of these distances had traveled at their expected recessional speeds, how long ago would they have been at our location?

53 •• The cheetah can run as fast as 113 km/h, the falcon can fly as fast as 161 km/h, and the sailfish can swim as fast as 105 km/h. The three of them run a relay with each covering a distance L at maximum speed. What is the average speed of this relay team for the entire relay? Compare this average speed with the numerical average of the three individual speeds. Explain carefully why the average speed of the relay team is *not* equal to the numerical average of the three individual speeds. SSM

54 •• Two cars are traveling along a straight road. Car A maintains a constant speed of 80 km/h and car B maintains a constant speed of 110 km/h. At $t = 0$, car B is 45 km behind car A. (*a*) How much farther will car A travel before car B overtakes it? (*b*) How much ahead of A will B be 30 s after it overtakes A?

55 •• **MULTISTEP** A car traveling at a constant speed of 20 m/s passes an intersection at time $t = 0$. A second car traveling at a constant speed of 30 m/s in the same direction passes the same intersection 5.0 s later. (*a*) Sketch the position functions $x_1(t)$ and $x_2(t)$ for the two cars for the interval $0 \le t \le 20$ s. (*b*) Determine when the second car will overtake the first. (*c*) How far from the intersection will the two cars be when they pull even? (*d*) Where is the first car when the second car passes the intersection? SSM

56 •• **BIOLOGICAL APPLICATION** Bats use echolocation to determine their distance from objects they cannot easily see in the dark. The time between the emission of a high-frequency sound pulse (a click) and the detection of its echo is used to determine such distances. A bat, flying at a constant speed of 19.5 m/s in a straight line toward a vertical cave wall, makes a single clicking noise and hears the echo 0.15 s later. Assuming that she continued flying at her original speed, how close was she to the wall when she received the echo? Assume a speed of 343 m/s for the speed of sound.

57 ••• **ENGINEERING APPLICATION** A submarine can use *sonar* (sound traveling through water) to determine its distance from other objects. The time between the emission of a sound pulse (a "ping") and the detection of its echo can be used to determine such distances. Alternatively, by measuring the time between *successive* echo receptions of a *regularly timed set* of pings, the submarine's *speed* may be determined by comparing the time between echoes to the time between pings. Assume you are the sonar operator in a submarine traveling at a constant velocity underwater. Your boat is in the eastern Mediterranean Sea, where the speed of sound is known to be 1522 m/s. If you send out pings every 2.00 s, and your apparatus receives echoes reflected from an undersea cliff every 1.98 s, how fast is your submarine traveling?

ACCELERATION

58 • A sports car accelerates in third gear from 48.3 km/h (about 30 mi/h) to 80.5 km/h (about 50 mi/h) in 3.70 s. (*a*) What is the average acceleration of this car in m/s²? (*b*) If the car maintained this acceleration, how fast would it be moving one second later?

59 • An object is moving along the x axis. At $t = 5.0$ s, the object is at $x = +3.0$ m and has a velocity of $+5.0$ m/s. At $t = 8.0$ s, it is at $x = +9.0$ m and its velocity is -1.0 m/s. Find its average acceleration during the time interval $5.0\ s < t < 8.0\ s$. SSM

60 •• A particle moves along the x axis with velocity $v_x = (8.0\ m/s^2)t - 7.0\ m/s$. (*a*) Find the average acceleration for two different one-second intervals, one beginning at $t = 3.0$ s and the other beginning at $t = 4.0$ s. (*b*) Sketch v_x versus t over the interval $0 < t < 10$ s. (*c*) How do the instantaneous accelerations at the middle of each of the two time intervals specified in Part (*a*) compare to the average accelerations found in Part (*a*)? Explain.

61 •• **MULTISTEP** The position of a certain particle depends on time according to the equation $x(t) = t^2 - 5.0t + 1.0$, where x is in meters if t is in seconds. (*a*) Find the displacement and average velocity for the interval $3.0\ s \le t \le 4.0\ s$. (*b*) Find the general formula for the displacement for the time interval from t to $t + \Delta t$. (*c*) Use the limiting process to obtain the instantaneous velocity for any time t. SSM

62 •• The position of an object as a function of time is given by $x = At^2 - Bt + C$, where $A = 8.0\ m/s^2$, $B = 6.0$ m/s, and $C = 4.0$ m. Find the instantaneous velocity and acceleration as functions of time.

63 ••• The one-dimensional motion of a particle is plotted in Figure 2-40. (*a*) What is the average acceleration in each of the intervals *AB*, *BC*, and *CE*? (*b*) How far is the particle from its starting point after 10 s? (*c*) Sketch the displacement of the particle as a function of time; label the instants *A*, *B*, *C*, *D*, and *E* on your graph. (*d*) At what time is the particle traveling most slowly?

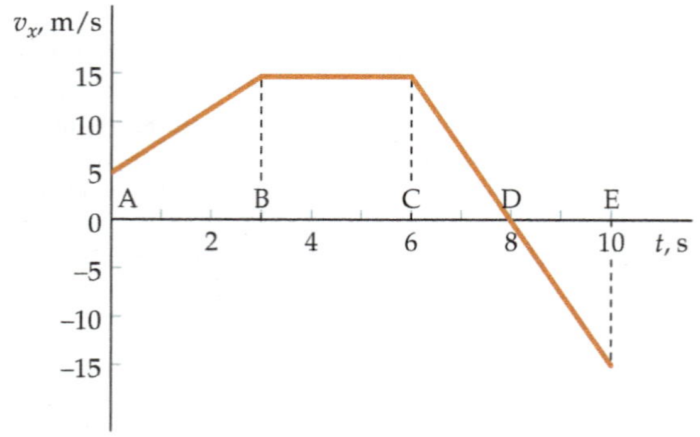

FIGURE 2-40 Problem 63

CONSTANT ACCELERATION AND FREE-FALL

64 • An object projected vertically upward with initial speed v_0 attains a maximum height h above its launch point. Another object projected up with initial speed $2v_0$ from the same height will attain a maximum height of (*a*) $4h$, (*b*) $3h$, (*c*) $2h$, (*d*) h. (Air resistance is negligible.)

65 • A car traveling along the x axis starts from rest at $x = 50$ m and accelerates at a constant rate of 8.0 m/s². (*a*) How fast is it going after 10 s? (*b*) How far has it gone after 10 s? (*c*) What is its average velocity for the interval $0 \leq t \leq 10$ s?

66 • An object traveling along the x axis with an initial velocity of +5.0 m/s has a constant acceleration of +2.0 m/s². When its speed is 15 m/s, how far has it traveled?

67 • An object traveling along the x axis at constant acceleration has a velocity of +10 m/s when it is at $x = 6.0$ m and of +15 m/s when it is at $x = 10.0$ m. What is its acceleration? SSM

68 • The speed of an object traveling along the x axis increases at the constant rate of +4.0 m/s each second. At $t = 0.0$ s, its velocity is +1.0 m/s and its position is +7.0 m. How fast is it moving when its position is +8.0 m, and how much time has elapsed from the start at $t = 0.0$ s?

69 •• A ball is launched directly upward from ground level with an initial speed of 20 m/s. (Air resistance is negligible.) (*a*) How long is the ball in the air? (*b*) What is the greatest height reached by the ball? (*c*) How many seconds after launch is the ball 15 m above the release point?

70 •• In the Blackhawk landslide in California, a mass of rock and mud fell 460 m down a mountain and then traveled 8.00 km across a level plain. It has been theorized that the rock and mud moved on a cushion of water vapor. Assume that the mass dropped with the free-fall acceleration and then slid horizontally, losing speed at a constant rate. (*a*) How long did the mud take to drop the 460 m? (*b*) How fast was it traveling when it reached the bottom? (*c*) How long did the mud take to slide the 8.00 km horizontally?

71 •• A load of bricks is lifted by a crane at a steady velocity of 5.0 m/s when one brick falls off 6.0 m above the ground. (*a*) Sketch the position of the brick $y(t)$ versus time, from the moment it leaves the pallet until it hits the ground. (*b*) What is the greatest height the brick reaches above the ground? (*c*) How long does it take to reach the ground? (*d*) What is its speed just before it hits the ground? SSM

72 •• A bolt comes loose from underneath an elevator that is moving upward at a constant speed of 6.0 m/s. The bolt reaches the bottom of the elevator shaft in 3.0 s. (*a*) How high above the bottom of the shaft was the elevator when the bolt came loose? (*b*) What is the speed of the bolt when it hits the bottom of the shaft?

73 •• An object is dropped from rest at a height of 120 m. Find the distance it falls during its final second in the air.

74 •• An object is released from rest at a height h. During the final second of its fall, it traverses a distance of 38 m. Determine h.

75 •• A stone is thrown vertically downward from the top of a 200-m cliff. During the last half second of its flight, the stone travels a distance of 45 m. Find the initial speed of the stone. SSM

76 •• An object is released from rest at a height h. It travels $0.4h$ during the first second of its descent. Determine the average velocity of the object during its entire descent.

77 •• A bus accelerates from rest at 1.5 m/s² for 12 s. It then travels at constant velocity for 25 s, after which it slows to a stop with an acceleration of magnitude 1.5 m/s². (*a*) What is the total distance that the bus travels? (*b*) What is its average velocity?

78 •• Al and Bert are jogging side-by-side on a trail in the woods at a speed of 0.75 m/s. Suddenly Al sees the end of the trail 35 m ahead and decides to speed up to reach it. He accelerates at a constant rate of 0.50 m/s² while Bert continues on at a constant speed. (*a*) How long does it take Al to reach the end of the trail? (*b*) Once he reaches the end of the trail, he immediately turns around and heads back along the trail with a constant speed of 0.85 m/s. How long does it take him to meet up with Bert? (*c*) How far are they from the end of the trail when they meet?

79 •• You have designed a rocket to be used to sample the local atmosphere for pollution. It is fired vertically with a constant upward acceleration of 20 m/s². After 25 s, the engine shuts off and the rocket continues rising (in freefall) for a while. (Air resistance is negligible.) The rocket eventually stops rising and then falls back to the ground. You want to get a sample of air that is 20 km above the ground. (*a*) Did you reach your height goal? If not, what would you change so that the rocket reaches 20 km? (*b*) Determine the total time the rocket is in the air. (*c*) Find the speed of the rocket just before it hits the ground.

80 •• A flowerpot falls from a windowsill of an apartment that is on the tenth floor of an apartment building. A person in an apartment below, coincidentally in possession of a high-speed high-precision timing system, notices that it takes 0.20 s for the pot to fall past his window, which is 4.0-m from top to bottom. How far above the top of the window is the windowsill from which the pot fell? (Neglect any effects due to air resistance.)

81 •• In a classroom demonstration, a glider moves along an inclined track with constant acceleration. It is projected from the low end of the track with an initial velocity. After 8.00 s have elapsed, it is 100 cm from the low end and is moving along the track at a velocity of −15 cm/s. Find the initial velocity and the acceleration. SSM

82 •• A rock dropped from a cliff covers one-third of its total distance to the ground in the last second of its fall. Air resistance is negligible. How high is the cliff?

83 •• A typical automobile under hard braking loses speed at a rate of about 7.0 m/s^2; the typical reaction time to engage the brakes is 0.50 s. A local school board sets the speed limit in a school zone such that all cars should be able to stop in 4.0 m. (*a*) What maximum speed does this imply for an automobile in this zone? (*b*) What fraction of the 4.0 m is due to the reaction time? SSM

84 •• Two trains face each other on adjacent tracks. They are initially at rest, and their front ends are 40 m apart. The train on the left accelerates rightward at 1.0 m/s^2. The train on the right accelerates leftward at 1.3 m/s^2. (*a*) How far does the train on the left travel before the front ends of the trains pass? (*b*) If the trains are each 150 m in length, how long after the start are they completely past one another, assuming their accelerations are constant?

85 •• Two stones are dropped from the edge of a 60-m cliff, the second stone 1.6 s after the first. How far below the top of the cliff is the second stone when the separation between the two stones is 36 m?

86 •• A motorcycle officer hidden at an intersection observes a car driven by an oblivious driver who ignores a stop sign and continues through the intersection at constant speed. The police officer takes off in pursuit 2.0 s after the car has passed the stop sign. She accelerates at 4.2 m/s^2 until her speed is 110 km/h, and then continues at this speed until she catches the car. At that instant, the car is 1.4 km from the intersection. (a) How long did it take for the officer to catch up to the car? (*b*) How fast was the car traveling?

87 •• At $t = 0$, a stone is dropped from the top of a cliff above a lake. Another stone is thrown downward 1.6 s later from the same point with an initial speed of 32 m/s. Both stones hit the water at the same instant. Find the height of the cliff.

88 •• A passenger train is traveling at 29 m/s when the engineer sees a freight train 360 m ahead of his train traveling in the same direction on the same track. The freight train is moving at a speed of 6.0 m/s. (*a*) If the reaction time of the engineer is 0.40 s, what is the minimum (constant) rate at which the passenger train must lose speed if a collision is to be avoided? (*b*) If the engineer's reaction time is 0.80 s and the train loses speed at the minimum rate described in Part (*a*), at what rate is the passenger train approaching the freight train when the two collide? (*c*) For both reaction times, how far will the passenger train have traveled in the time between the sighting of the freight train and the collision?

89 •• **BIOLOGICAL APPLICATION** The click beetle can project itself vertically with an acceleration of about 400*g* (an order of magnitude more than a human could survive!). The beetle jumps by "unfolding" its 0.60-cm-long legs. (*a*) How high can the click beetle jump? (*b*) How long is the beetle in the air? (Assume constant acceleration while in contact with the ground and neglect air resistance.)

90 •• An automobile accelerates from rest at 2.0 m/s^2 for 20 s. The speed is then held constant for 20 s, after which there is an acceleration of -3.0 m/s^2 until the automobile stops. What is the total distance traveled?

91 •• Consider measuring the free-fall motion of a particle (neglect air resistance). Before the advent of computer-driven data-logging software, these experiments typically employed a wax-coated tape placed vertically next to the path of a dropped electrically conductive object. A spark generator would cause an arc to jump between two vertical wires through the falling object and through the tape, thereby marking the tape at fixed time intervals Δt. Show that the change in height during successive time intervals for an object falling from rest follows *Galileo's Rule of Odd Numbers*: $\Delta y_{21} = 3\Delta y_{10}$, $\Delta y_{32} = 5\Delta y_{10}, \ldots$, where Δy_{10} is the change in y during the first interval of duration Δt, Δy_{21} is the change in y during the second interval of duration Δt, etc. SSM

92 •• Starting from rest, a particle travels along the x axis with a constant acceleration of $+3.0$ m/s^2. At a time 4.0 s following its start, it is at $x = -100$ m. At a time 6.0 s later it has a velocity of $+15$ m/s. Find its position at this later time.

93 •• If it were possible for a spacecraft to maintain a constant acceleration indefinitely, trips to the planets of the Solar System could be undertaken in days or weeks, while voyages to the nearer stars would only take a few years. (*a*) Using data from the tables at the back of the book, find the time it would take for a one-way trip from Earth to Mars (at Mars' closest approach to Earth). Assume that the spacecraft starts from rest, travels along a straight line, accelerates halfway at 1*g*, flips around, and decelerates at 1*g* for the rest of the trip. (*b*) Repeat the calculation for a 4.1×10^{13}-km trip to Proxima Centauri, our nearest stellar neighbor outside of the Sun. (See Problem 47.) SSM

94 •• The Stratosphere Tower in Las Vegas is 1137 ft high. It takes 1 min, 20 s to ascend from the ground floor to the top of the tower using the high-speed elevator. The elevator starts and ends at rest. Assume that it maintains a constant upward acceleration until it reaches its maximum speed, and then maintains a constant acceleration of equal magnitude until it comes to a stop. Find the magnitude of the acceleration of the elevator. Express this acceleration magnitude as a multiple of g (the acceleration due to gravity).

95 •• A train pulls away from a station with a constant acceleration of 0.40 m/s^2. A passenger arrives at a point next to the track 6.0 s after the end of the train has passed the very same point. What is the slowest constant speed at which she can run and still catch the train? On a single graph, plot the position versus time curves for both the train and the passenger.

96 ••• Ball A is dropped from the top of a building of height h at the same instant that ball B is thrown vertically upward from the ground. When the balls collide, they are moving in opposite directions, and the speed of A is twice the speed of B. At what height does the collision occur?

97 ••• Solve Problem 96 if the collision occurs when the balls are moving in the same direction and the speed of A is 4 times that of B.

98 ••• Starting at one station, a subway train accelerates from rest at a constant rate of 1.00 m/s^2 for half the distance to the next station, then slows down at the same rate for the second half of the journey. The total distance between stations is 900 m. (*a*) Sketch a graph of the velocity v_x as a function of time over the full journey. (*b*) Sketch a graph of the position as a function of time over the full journey. Place appropriate numerical values on both axes.

99 ••• A speeder traveling at a constant speed of 125 km/h races past a billboard. A patrol car pursues from rest with constant acceleration of (8.0 km/h)/s until it reaches its maximum speed of 190 km/h, which it maintains until it catches up with the speeder. (*a*) How long does it take the patrol car to catch the speeder if it starts moving just as the speeder passes? (*b*) How far does each car travel? (*c*) Sketch $x(t)$ for each car. SSM

100 ••• When the patrol car in Problem 99 (traveling at 190 km/h) is 100 m behind the speeder (traveling at 125 km/h), the speeder sees the police car and slams on his brakes, locking the wheels. (*a*) Assuming that each car can brake at 6.0 m/s^2 and that the driver of the police car brakes instantly as she sees the brake lights of the speeder (reaction time = 0.0 s), show that the cars collide. (*b*) At what time after the speeder applies his brakes do the two cars collide? (*c*) Discuss how reaction time would affect this problem.

101 ••• Leadfoot Lou enters the "Rest-to-Rest" auto competition, in which each contestant's car begins and ends at rest, covering a fixed distance L in as short a time as possible. The intention is to demonstrate driving skills, and to find which car is the best at the *total combination* of speeding up and slowing down. The course is designed so that maximum speeds of the cars are never reached. (*a*) If Lou's car maintains an acceleration (magnitude) of a during speedup, and maintains a deceleration (magnitude) of $2a$ during braking, at what fraction of L should Lou move his foot from the gas pedal to the brake? (*b*) What fraction of the total time for the trip has elapsed at that point? (*c*) What is the fastest speed Lou's car ever reaches? Neglect Lou's reaction time, and answer in terms of a and L.

102 ••• A physics professor, equipped with a rocket backpack, steps out of a helicopter at an altitude of 575 m with zero initial velocity. (Neglect air resistance.) For 8.0 s, she falls freely. At that time, she fires her rockets and slows her rate of descent at 15 m/s^2 until her rate of descent reaches 5.0 m/s. At this point, she adjusts her rocket engine controls to maintain that rate of descent until she reaches the ground. (*a*) On a single graph, sketch her acceleration and velocity as functions of time. (Take upward to be positive.) (*b*) What is her speed at the end of the first 8.0 s? (*c*) What is the duration of her slowing-down period? (*d*) How far does she travel while slowing down? (*e*) How much time is required for the entire trip from the helicopter to the ground? (*f*) What is her average velocity for the entire trip?

INTEGRATION OF THE EQUATIONS OF MOTION

103 • The velocity of a particle is given by $v_x(t) = (6.0\ \text{m/s}^2)t + (3.0\ \text{m/s})$. (*a*) Sketch v versus t and find the area under the curve for the interval $t = 0$ to $t = 5.0$ s. (*b*) Find the position function $x(t)$. Use it to calculate the displacement during the interval $t = 0$ to $t = 5.0$ s. SSM

104 • Figure 2-41 shows the velocity of a particle versus time. (*a*) What is the magnitude, in meters, represented by the area of the shaded box? (*b*) Estimate the displacement of the particle for the two 1-s intervals, one beginning at $t = 1.0$ s and the other at $t = 2.0$ s. (*c*) Estimate the average velocity for the interval $1.0\,\text{s} \le t \le 3.0\,\text{s}$. (*d*) The equation of the curve is $v_x = (0.50\ \text{m/s}^3)t^2$. Find the displacement of the particle for the interval $1.0\,\text{s} \le t \le 3.0\,\text{s}$ by integration and compare this answer with your answer for Part (*b*). Is the average velocity equal to the mean of the initial and final velocities for this case?

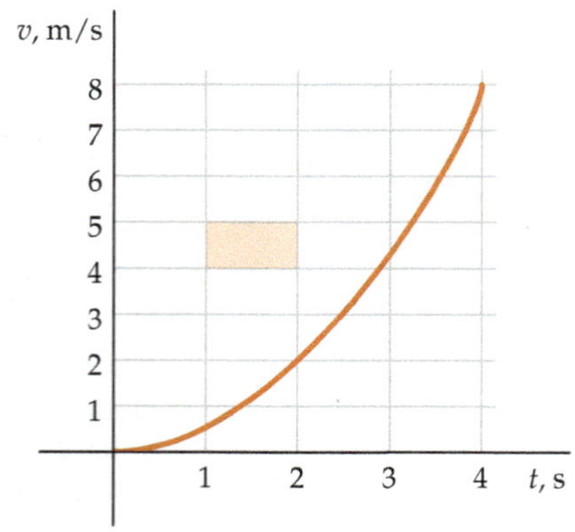

FIGURE 2-41 Problem 104

105 •• The velocity of a particle is given by $v_x = (7.0\ \text{m/s}^3)t^2 - 5.0\ \text{m/s}$. If the particle is at the origin at $t_0 = 0$, find the position function $x(t)$.

106 •• Consider the velocity graph in Figure 2-42. Assuming $x = 0$ at $t = 0$, write correct algebraic expressions for $x(t)$, $v_x(t)$, and $a_x(t)$ with appropriate numerical values inserted for all constants.

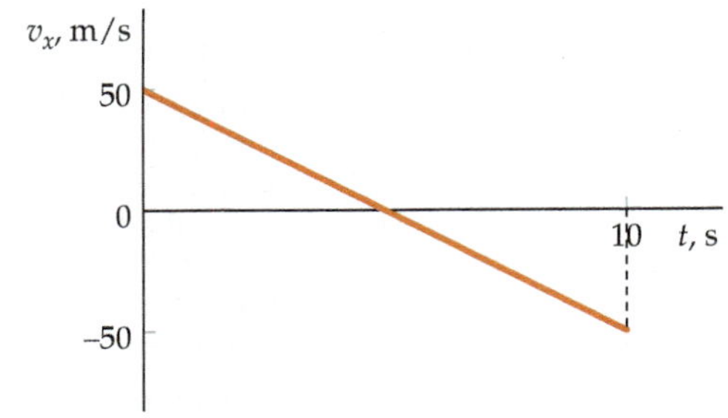

FIGURE 2-42 Problem 106

107 ••• Figure 2-43 shows the acceleration of a particle versus time. (*a*) What is the magnitude, in m/s, of the area of the shaded box? (*b*) The particle starts from rest at $t = 0$. Estimate the velocity at $t = 1.0$ s, 2.0 s, and 3.0 s by counting the boxes under the curve. (*c*) Sketch the curve v_x versus t from your results for Part (*b*); then estimate how far the particle travels in the interval $t = 0$ to $t = 3.0$ s.

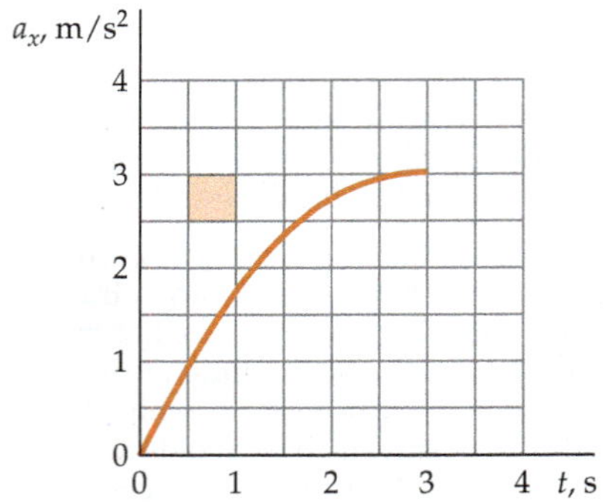

FIGURE 2-43 Problem 107

108 ••• Figure 2-44 is a graph of v_x versus t for a particle moving along a straight line. The position of the particle at time $t = 0$ is $x_0 = 5.0$ m. (*a*) Find x for various times t by counting boxes, and sketch x as a function of t. (*b*) Sketch a graph of the acceleration a_x as a function of the time t. (*c*) Determine the displacement of the particle between $t = 3.0$ s and 7.0 s.

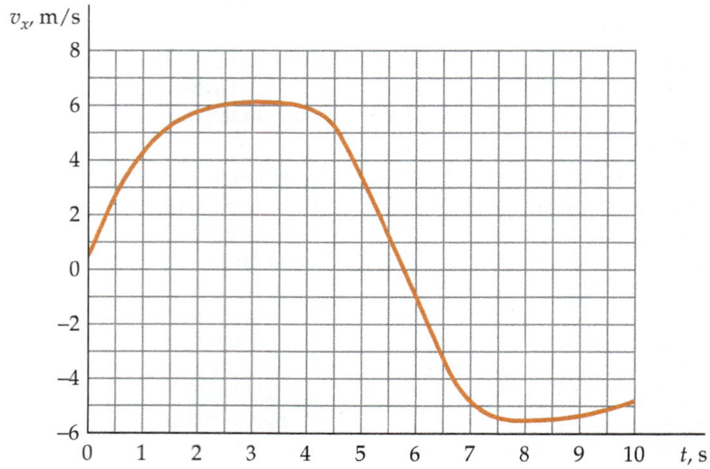

FIGURE 2-44 Problem 108

109 ••• **CONCEPTUAL** Figure 2-45 shows a plot of x versus t for an object moving along a straight line. For this motion, sketch graphs (using the same t axis) of (*a*) v_x as a function of t, and (*b*) a_x as a function of t. (*c*) Use your sketches to qualitatively compare the time(s) when the object is at its largest distance from the origin to the time(s) when its speed is greatest. Explain why the times are *not* the same. (*d*) Use your sketches to qualitatively compare the time(s) when the object is moving fastest to the time(s) when its acceleration is the largest. Explain why the times are *not* the same. SSM

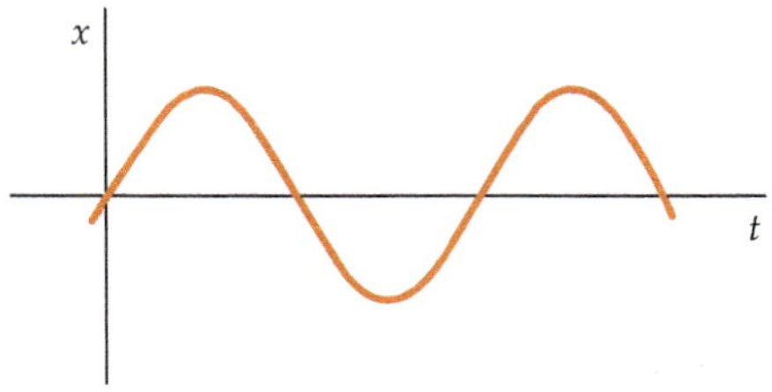

FIGURE 2-45 Problem 109

110 ••• **MULTISTEP** The acceleration of a certain rocket is given by $a_x = bt$, where b is a positive constant. (*a*) Find the position function $x(t)$ if $x = x_0$ and $v_x = v_{0x}$ at $t = 0$. (*b*) Find the position and velocity at $t = 5.0$ s if $x_0 = 0$, $v_{0x} = 0$ and $b = 3.0\ \text{m/s}^3$. (*c*) Compute the average velocity of the rocket between $t = 4.5$ s and 5.5 s at $t = 5.0$ s if $x_0 = 0$, $v_{0x} = 0$ and $b = 3.0\ \text{m/s}^3$. Compare this average velocity with the instantaneous velocity at $t = 5.0$ s.

111 ••• In the time interval from 0.0 s to 10.0 s, the acceleration of a particle traveling in a straight line is given by $a_x = (0.20\ \text{m/s}^3)t$. Let to the right be the $+x$ direction. The particle initially has a velocity to the right of 9.5 m/s and is located 5.0 m to the left of the origin. (*a*) Determine the velocity as a function of time during the interval; (*b*) determine the position as a function of time during the interval; (*c*) determine the average velocity between $t = 0.0$ s and 10.0 s, and compare it to the average of the instantaneous velocities at the start and ending times. Are these two averages equal? Explain. SSM

112 ••• Consider the motion of a particle that experiences a variable acceleration given by $a_x = a_{0x} + bt$, where a_{0x} and b are constants and $x = x_0$ and $v_x = v_{0x}$ at $t = 0$. (*a*) Find the instantaneous velocity as a function of time. (*b*) Find the position as a function of time. (*c*) Find the average velocity for the time interval with an initial time of zero and arbitrary final time t. (*d*) Compare the average of the initial and final velocities to your answer to Part (*c*). Are these two averages equal? Explain.

GENERAL PROBLEMS

113 ••• **CONTEXT-RICH** You are a student in a science class that is using the following apparatus to determine the value of g. Two photogates are used. (*Note:* You may be familiar with photogates in everyday living. You see them in the doorways of some stores. They are designed to ring a bell when someone interrupts the beam while walking through the door.) One photogate is located at the edge of a table that is 1.00 m above the floor, and the second photogate is located directly below the first, at a height 0.500 m above the floor. You are told to drop a marble through these gates, releasing it from rest a negligible distance above the upper gate. The upper gate starts a timer as the ball passes through its beam. The second photogate stops the timer when the ball passes through its beam. (*a*) Prove that the experimental magnitude of free-fall acceleration is given by $g_{exp} = (2\Delta y)/(\Delta t)^2$, where Δy is the vertical distance between the photogates and Δt is the fall time. (*b*) For your setup, what value of Δt would you expect to measure, assuming g_{exp} is the standard value ($9.81\ \text{m/s}^2$)? (*c*) During the experiment, a slight error is made. Instead of locating the first photogate even with the top of the table, your not-so-careful lab partner locates it 0.50 cm lower than the top of the table. However, she does manage to properly locate the second photogate at a height of 0.50 m above the floor. However, she releases the marble from the same height that it was released from when the photogate was 1.00 m above the floor. What value of g_{exp} will you and your partner determine? What percentage difference does this represent from the standard value of g?

114 ••• **MULTISTEP** The position of a body oscillating on a spring is given by $x = A \sin \omega t$, where A and ω (lower case Greek omega) are constants, $A = 5.0$ cm, and $\omega = 0.175\ \text{s}^{-1}$. (*a*) Plot x as a function of t for $0 \le t \le 36\,\text{s}$. (*b*) Measure the slope of your graph at $t = 0$ to find the velocity at this time. (*c*) Calculate the average velocity for a series of intervals, beginning at $t = 0$ and ending at $t = 6.0$, 3.0, 2.0, 1.0, 0.50, and 0.25 s. (*d*) Compute dx/dt to find the velocity at time $t = 0$. (*e*) Compare your results in Parts (*c*) and (*d*) and explain why your Part-(*c*) results approach your Part-(*d*) result.

115 ••• **CONCEPTUAL** Consider an object that is attached to a horizontally oscillating piston. The object moves with a velocity given by $v = B \sin(\omega t)$, where B and ω (lower case Greek omega) are constants and ω is in s^{-1}. (*a*) Explain why B is equal to the maximum speed v_{max}. (*b*) Determine the acceleration of the object as a function of time. Is the acceleration constant? (*c*) What is the maximum acceleration (magnitude) in terms of ω and v_{max}. (*d*) At $t = 0$, the object's position is known to be x_0. Determine the position as a function of time in terms of t, ω, x_0, and v_{max}. SSM

116 ••• Suppose the acceleration of a particle is a function of x, where $a_x(x) = (2.0\ \text{s}^{-2})x$. (*a*) If the velocity is zero when $x = 1.0$ m, what is the speed when $x = 3.0$ m? (*b*) How long does it take the particle to travel from $x = 1.0$ m to $x = 3.0$ m.

117 ••• A rock falls through water with a continuously decreasing acceleration. Assume that the rock's acceleration as a function of *velocity* has the form $a_y = g - bv_y$ where b is a positive constant. (The $+y$ direction is directly downward.) (*a*) What are the SI units of b? (*b*) Prove mathematically that if the rock is released from rest at time $t = 0$, the acceleration will depend exponentially on *time* according to $a_y(t) = ge^{-bt}$. (*c*) What is the terminal speed for the rock in terms of g and b? (See Problem 38 for an explanation of the phenomenon of *terminal speed*.) SSM

118 ••• A small rock sinking through water (see Problem 117) experiences an exponentially decreasing acceleration given by $a_y(t) = ge^{-bt}$, where b is a positive constant that depends on the shape and size of the rock and the physical properties of the water. Based upon this, find expressions for the velocity and position of the rock as functions of time. Assume that its initial position and velocity are both zero and that the $+y$ direction is directly downward.

119 ••• **SPREADSHEET** The acceleration of a skydiver jumping from an airplane is given by $a_y = g - bv_y^2$, where b is a positive constant that depends on the skydiver's cross-sectional area and the density of the surrounding atmosphere she is diving through. The $+y$ directions is directly downward. (*a*) If her initial speed is zero when stepping from a hovering helicopter, show that her speed as a function of time is given by $v_y(t) = v_t \tanh(t/T)$, where v_t is the terminal speed (see Problem 38) given by $v_t = \sqrt{g/b}$, and $T = v_t/g$ is a time-scale parameter. (*b*) What fraction of the terminal speed is the speed at $t = T$. (*c*) Use a **spreadsheet** program to graph $v_y(t)$ as a function of time, using a terminal speed of 56 m/s (use this value to calculate b and T). Does the resulting curve make sense?

120 ••• **APPROXIMATION** Imagine that you are standing at a wishing well, wishing that you knew how deep the surface of the water was. Cleverly, you make your wish. Then you take a penny from your pocket and drop it into the well. Exactly three seconds after you dropped the penny, you hear the sound it made when it struck the water. If the speed of sound is 343 m/s, how deep is the well? Neglect any effects due to air resistance.

121 ••• **CONTEXT-RICH** You are driving a car at the 25-mi/h speed limit when you observe the light at the intersection 65 m in front of you turn yellow. You know that at that particular intersection the light remains yellow for exactly 5.0 s before turning red. After you think for 1.0 s, you then accelerate the car at a constant rate. You somehow manage to pass your 4.5-m-long car completely through the 15.0-m-wide intersection just as the light turns red, thus narrowly avoiding a ticket for being in an intersection when the light is red. Immediately after passing through the intersection, you take your foot off the accelerator, relieved. However, down the road you are pulled over for speeding. You assume that you were ticketed for the speed of your car as it exited the intersection. Determine this speed and decide whether you should fight this ticket in court. Explain.

122 ••• For a spherical celestial object of radius R, the acceleration due to gravity g at a distance x from the *center* of the object is $g = g_0R^2/x^2$, where g_0 is the acceleration due to gravity at the object's surface and $x > R$. For the moon, take $g_0 = 1.63\ \text{m/s}^2$ and $R = 3200$ km. If a rock is released from rest at a height of $4R$ above the lunar surface, with what speed does the rock impact the moon? *Hint: Its acceleration is a function of position and increases as the object falls. So do not use constant acceleration free-fall equations, but go back to basics.*

CHAPTER 3

SAILBOATS DO NOT TRAVEL IN STRAIGHT LINES TO THEIR DESTINATIONS, BUT INSTEAD MUST "TACK" BACK AND FORTH ACROSS THE WIND. THIS BOAT MUST SAIL EAST, THEN SOUTH, AND THEN EAST AGAIN, IN ITS JOURNEY TO A SOUTHEASTERN PORT. *(PhotoDisc/Getty Images.)*

Motion in Two and Three Dimensions

? How can we calculate the boat's displacement and its average velocity? (See Example 3-1.)

The motion of a sailboat tacking into the wind or the path of a home-run ball as it flies out of a ballpark cannot be fully described using the equations we presented in Chapter 2. Instead, to describe these motions, we must extend the idea of motion in one dimension discussed in Chapter 2 to two and three dimensions. To do this, we must revisit the concept of vectors and look at how they can be used to analyze and describe motion in more than one dimension.

In this chapter, we will discuss the displacement, velocity, and acceleration vectors in further detail. In addition, we will discuss two specific types of motion: projectile motion and circular motion. The material in this chapter presumes you are familiar with the material that introduces vectors in Sections 6 and 7 of Chapter 1. You are encouraged to review these sections before proceeding in this chapter.

3-1 DISPLACEMENT, VELOCITY, AND ACCELERATION

In Chapter 2, the concepts of displacement, velocity, and acceleration were used to describe the motion of an object moving in a straight line. Now we use the concept of vectors to extend these characteristics of motion in two and three dimensions.

POSITION AND DISPLACEMENT VECTORS

The **position vector** of a particle is a vector drawn from the origin of a coordinate system to the location of the particle. For a particle in the xy plane at the point with coordinates (x, y), the position vector $\vec{r}$ is

$$\vec{r} = x\hat{i} + y\hat{j} \qquad \text{3-1}$$

DEFINITION—POSITION VECTOR

Note that x and y components of the position vector $\vec{r}$ are the Cartesian coordinates (Figure 3-1) of the particle.

Figure 3-2 shows the actual path or trajectory of the particle. At time t_1, the particle is at P_1, with position vector $\vec{r}_1$; by time t_2, the particle has moved to P_2, with position vector $\vec{r}_2$. The particle's change in position is the displacement vector $\Delta\vec{r}$:

$$\Delta\vec{r} = \vec{r}_2 - \vec{r}_1 \qquad \text{3-2}$$

DEFINITION—DISPLACEMENT VECTOR

Using unit vectors, we can rewrite this displacement as

$$\Delta\vec{r} = \vec{r}_2 - \vec{r}_1 = (x_2 - x_1)\hat{i} + (y_2 - y_1)\hat{j} = \Delta x\hat{i} + \Delta y\hat{j} \qquad \text{3-3}$$

VELOCITY VECTORS

Recall that average velocity is defined as displacement divided by the elapsed time. The result of the displacement vector divided by the elapsed time interval $\Delta t = t_2 - t_1$ is the **average-velocity vector:**

$$\vec{v}_{\text{av}} = \frac{\Delta\vec{r}}{\Delta t} \qquad \text{3-4}$$

DEFINITION—AVERAGE-VELOCITY VECTOR

The average velocity vector and the displacement vector are in the same direction.

The magnitude of the displacement vector is less than the distance traveled along the curve unless the particle moves along a straight line and never reverses its direction. However, if we consider smaller and smaller time intervals (Figure 3-3), the magnitude of the displacement approaches the distance along the curve, and the angle between $\Delta\vec{r}$ and the tangent to the curve at the beginning of the interval approaches zero. We define the **instantaneous-velocity vector** as the limit of the average-velocity vector as Δt approaches zero:

$$\vec{v} = \lim_{\Delta t \to 0} \frac{\Delta\vec{r}}{\Delta t} = \frac{d\vec{r}}{dt} \qquad \text{3-5}$$

DEFINITION—INSTANTANEOUS-VELOCITY VECTOR

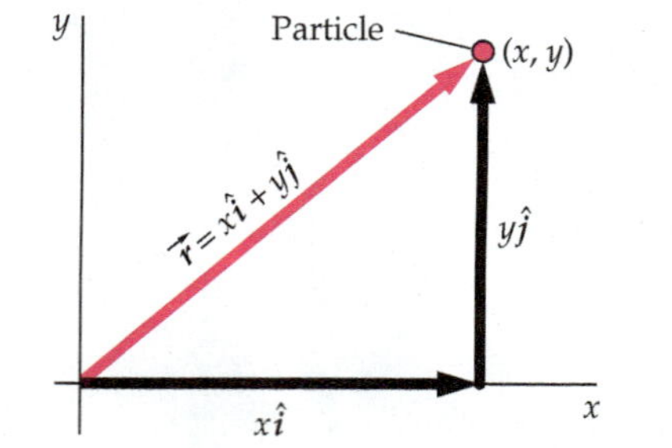

FIGURE 3-1 The x and y components of the position vector $\vec{r}$ for a particle are the x and y (Cartesian) coordinates of the particle.

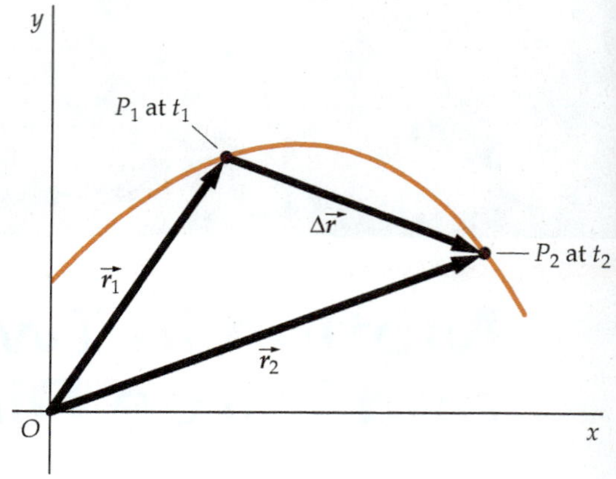

FIGURE 3-2 The displacement vector $\Delta\vec{r}$ is the difference in the position vectors, $\Delta\vec{r} = \vec{r}_2 - \vec{r}_1$. Equivalently, $\Delta\vec{r}$ is the vector that, when added to the initial position vector $\vec{r}_1$, yields the final position vector $\vec{r}_2$. That is, $\vec{r}_1 + \Delta\vec{r} = \vec{r}_2$.

! Do not confuse the trajectory in graphs of x versus y with the curve in the x-versus-t plots of Chapter 2.

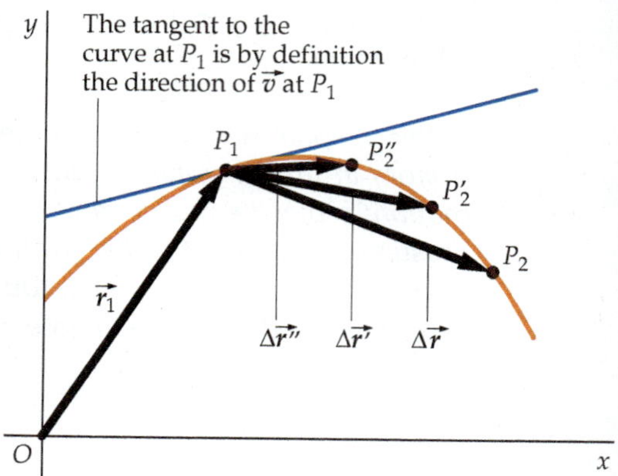

FIGURE 3-3 As the time interval decreases, the angle between direction of $\Delta\vec{r}$ and the tangent to the curve approaches zero.

The instantaneous-velocity vector is the derivative of the position vector with respect to time. Its magnitude is the speed and its direction is along the line tangent to the curve in the direction of motion of the particle.

To calculate the derivative in Equation 3-5, we write the position vectors in terms of their components (Equation 3-1):

$$\Delta\vec{r} = \vec{r}_2 - \vec{r}_1 = (x_2 - x_1)\hat{i} + (y_2 - y_1)\hat{j} = \Delta x\hat{i} + \Delta y\hat{j}$$

Then

$$\vec{v} = \lim_{\Delta t \to 0}\frac{\Delta\vec{r}}{\Delta t} = \lim_{\Delta t \to 0}\frac{\Delta x\hat{i} + \Delta y\hat{j}}{\Delta t} = \lim_{\Delta t \to 0}\left(\frac{\Delta x}{\Delta t}\right)\hat{i} + \lim_{\Delta t \to 0}\left(\frac{\Delta y}{\Delta t}\right)\hat{j}$$

or

$$\vec{v} = \frac{dx}{dt}\hat{i} + \frac{dy}{dt}\hat{j} = v_x\hat{i} + v_y\hat{j} \qquad 3\text{-}6$$

where $v_x = dx/dt$ and $v_y = dy/dt$ are the x and y components of the velocity.

The magnitude of the velocity vector is given by:

$$v = \sqrt{v_x^2 + v_y^2} \qquad 3\text{-}7$$

and the direction of the velocity is given by

$$\theta = \tan^{-1}\frac{v_y}{v_x} \qquad 3\text{-}8$$

! Do not trust your calculator to always give the correct value for θ when using Equation 3-8. Most calculators will return the correct value for θ if v_x is positive. If v_x is negative, however, you will need to add 180° (π rad) to the value returned by the calculator.

See **Math Tutorial** *for more information on* **Trigonometry**

Example 3-1 The Velocity of a Sailboat

A sailboat has coordinates $(x_1, y_1) = (130\text{ m}, 205\text{ m})$ at $t_1 = 60.0$ s. Two minutes later, at time t_2, it has coordinates $(x_2, y_2) = (110\text{ m}, 218\text{ m})$. (*a*) Find the average velocity $\vec{v}_{av}$ for this two-minute interval. Express $\vec{v}_{av}$ in terms of its rectangular components. (*b*) Find the magnitude and direction of this average velocity. (*c*) For $t \geq 20.0$ s, the position of a second sailboat as a function of time is $x(t) = b_1 + b_2 t$ and $y(t) = c_1 + c_2/t$, where $b_1 = 100$ m, $b_2 = 0.500$ m/s, $c_1 = 200$ m, and $c_2 = 360$ m · s. Find the instantaneous velocity as a function of time t, for $t \geq 20.0$ s.

PICTURE The initial and final positions of the first sailboat are given. Because the motion of the boat is in two dimensions, we need to express the displacement, average velocity, and instantaneous velocity as vectors. Then we can use Equations 3-5 through 3-8 to obtain the requested values.

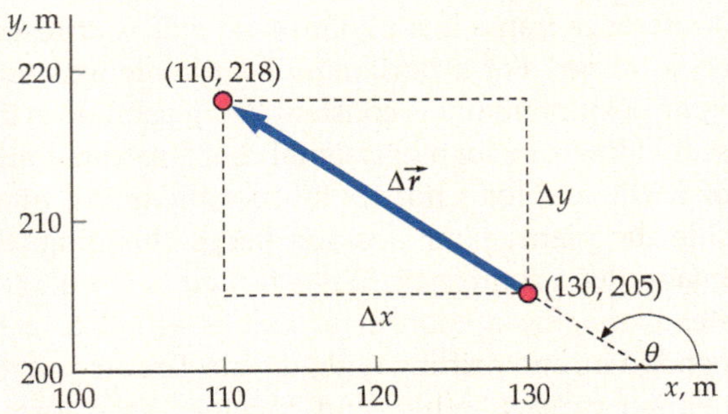

FIGURE 3-4

SOLVE

(*a*) 1. Draw a coordinate system (Figure 3-4) and draw the displacement of the sailboat. Draw the average-velocity vector (it and the displacement vector are in the same direction):

2. The x and y components of the average velocity $\vec{v}_{av}$ are calculated directly from their definitions:

$$\vec{v}_{av} = v_{x\,av}\hat{i} + v_{y\,av}\hat{j}$$

where

$$v_{x\,av} = \frac{\Delta x}{\Delta t} = \frac{110\text{ m} - 130\text{ m}}{120\text{ s}} = -0.167\text{ m/s}$$

$$v_{y\,av} = \frac{\Delta y}{\Delta t} = \frac{218\text{ m} - 205\text{ m}}{120\text{ s}} = 0.108\text{ m/s}$$

so

$$\vec{v}_{av} = \boxed{-(0.167\text{ m/s})\hat{i} + (0.108\text{ m/s})\hat{j}}$$

(*b*) 1. The magnitude of $\vec{v}_{av}$ is found from the Pythagorean theorem:

$$v_{av} = \sqrt{(v_{x\,av})^2 + (v_{y\,av})^2} = \boxed{0.199\text{ m/s}}$$

2. The ratio of $v_{y\,av}$ to $v_{x\,av}$ gives the tangent of the angle θ between $\vec{v}_{av}$ and the $+x$ direction (we add 180° to the value of −33.0° returned by the calculator because v_x is negative):

$$\tan\theta = \frac{v_{y\,av}}{v_{x\,av}}$$

so

$$\theta = \tan^{-1}\frac{v_{y\,av}}{v_{x\,av}} = \tan^{-1}\frac{0.108\text{ m/s}}{-0.167\text{ m/s}} = -33.0° + 180° = \boxed{147°}$$

(*c*) We find the instantaneous velocity $\vec{v}$ by calculating dx/dt and dy/dt:

$$\vec{v} = \frac{dx}{dt}\hat{i} + \frac{dy}{dt}\hat{j} = b_2\hat{i} - c_2t^{-2}\hat{j} = \boxed{(0.500\text{ m/s})\hat{i} - \frac{360\text{ m}\cdot\text{s}}{t^2}\hat{j}}$$

CHECK The magnitude of $\vec{v}_{av}$ is greater than the absolute value of either its x or its y component. With t in seconds, the units for the y component of $\vec{v}$ in Part (*c*) are $\text{m}\cdot\text{s}/\text{s}^2 = \text{m/s}$, which are appropriate units for velocity.

RELATIVE VELOCITY

Midair refueling. Each plane is nearly at rest relative to the other, though both are moving with very large velocities relative to Earth. *(Novastock/Dembinsky Photo Associates.)*

If you are sitting in an airplane that is moving with a velocity of 500 mi/h toward the east, your velocity is the same as that of the airplane. This velocity might be your velocity relative to the surface of Earth, or it might be your velocity relative to the air outside the airplane. (These two relative velocities would be very different if the plane were flying in a jet stream.) In addition, your velocity relative to the airplane itself is zero.

The surface of Earth, the air outside the plane, and the plane itself are frames of reference. A **frame of reference** (or reference frame) is an extended object or collection of objects whose parts are at rest relative to each other. To specify the velocity of an object requires that you specify the frame of reference that the velocity is relative to.

We use coordinate axes that are attached to reference frames to make position measurements. (A coordinate axis is said be attached to a reference frame if the coordinate axis is at rest relative to the reference frame.) For a horizontal coordinate axis attached to the plane, your position remains constant. (At least it does if you remain in your seat.) However, for a horizontal coordinate axis attached to the surface of Earth, and for a horizontal coordinate axis attached to the air outside the plane, your position keeps changing. (If you have trouble imagining a coordinate axis attached to the air outside the plane, instead imagine a coordinate axis attached to a balloon that is suspended in, and drifting with, the air. The air and the balloon are at rest relative to each other, and together they form a single reference frame.)

If a particle p moves with velocity $\vec{v}_{pA}$ relative to reference frame A, which is in turn moving with velocity $\vec{v}_{AB}$ relative to reference frame B, the velocity $\vec{v}_{pB}$ of the particle relative to reference frame B is related to $\vec{v}_{pA}$ and $\vec{v}_{AB}$ by

$$\vec{v}_{pB} = \vec{v}_{pA} + \vec{v}_{AB} \qquad 3\text{-}9$$

For example, if a person p is on a railroad car C that is moving with velocity $\vec{v}_{CG}$ relative to the ground G (Figure 3-5*a*), and the person is walking with velocity $\vec{v}_{pC}$ (Figure 3-5*b*) relative to the car, then the velocity of the person relative to the ground is the vector sum of these two velocities: $\vec{v}_{pG} = \vec{v}_{pC} + \vec{v}_{CG}$ (Figure 3-5*c*).

The velocity of object A relative to object B is equal in magnitude and opposite in direction to the velocity of object B relative to object A. For example, $\vec{v}_{pC}$ is equal to $-\vec{v}_{Cp}$, where $\vec{v}_{pC}$ is the velocity of the person relative to the car, and $\vec{v}_{Cp}$ is the velocity of the car relative to the person.

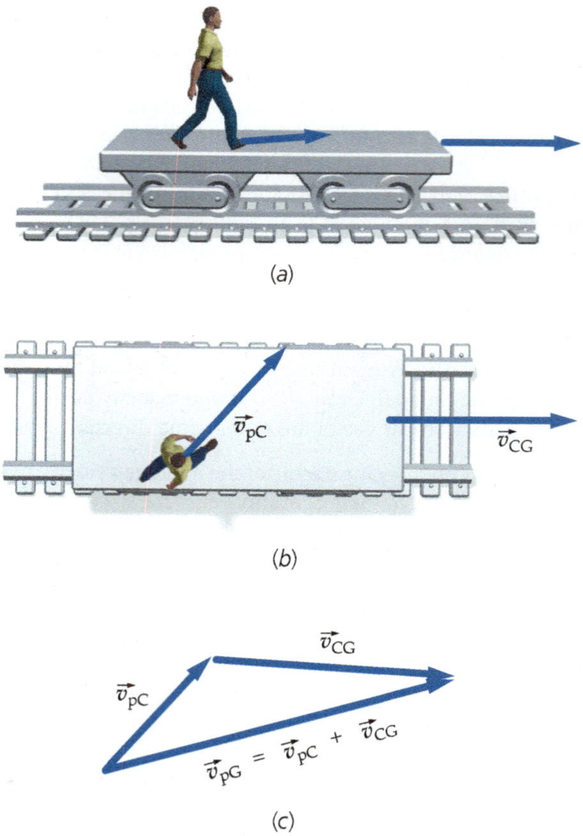

FIGURE 3-5 The velocity of the person relative to the ground is equal to the velocity of the person relative to the car plus the velocity of the car relative the ground.

PROBLEM-SOLVING STRATEGY

Relative Velocity

PICTURE The first step in solving a relative-velocity problem is to identify and label the relevant reference frames. Here, we will call them reference frame A and reference frame B.

SOLVE

1. Using $\vec{v}_{pB} = \vec{v}_{pA} + \vec{v}_{AB}$ (Equation 3-9), relate the velocity of the moving object (particle p) relative to frame A to the velocity of the particle relative to frame B.
2. Sketch a vector addition diagram for the equation $\vec{v}_{pB} = \vec{v}_{pA} + \vec{v}_{AB}$. Use the head-to-tail method of vector addition. Include coordinate axes on the sketch.
3. Solve for the desired quantity. Use trigonometry where appropriate.

CHECK Make sure that you solve for the velocity or position of the moving object relative to the proper reference frame.

! The order of the subscripts used when denoting relative velocity vectors is very important. When using relative velocity vectors, be very careful to write the subscripts in a consistent order.

Example 3-2 A Flying Plane

A pilot wishes to fly a plane due north relative to the ground. The airspeed of the plane is 200 km/h and the wind is blowing from west to east at 90 km/h. (*a*) In which direction should the plane head? (*b*) What is the ground speed of the plane?

PICTURE Because the wind is blowing toward the east, a plane headed due north will drift off course toward the east. To compensate for this crosswind, the plane must head west of due north. The velocity of the plane relative to the ground $\vec{v}_{pG}$ is equal to the velocity of the plane relative to the air $\vec{v}_{pA}$ plus the velocity of the air relative to the ground $\vec{v}_{AG}$.

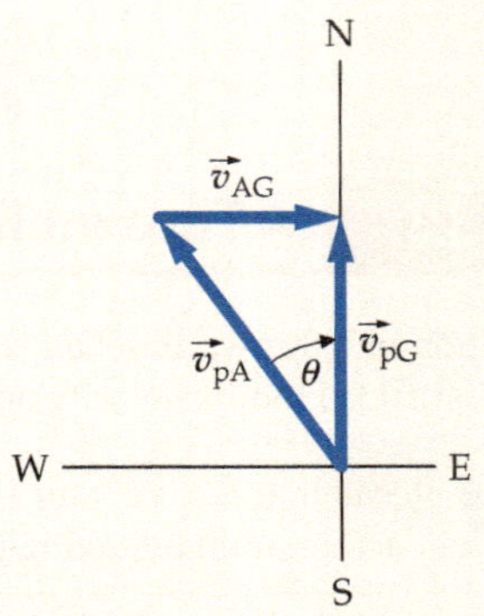

FIGURE 3-6

SOLVE

(a) 1. The velocity of the plane relative to the ground is given by Equation 3-9:

$$\vec{v}_{pG} = \vec{v}_{pA} + \vec{v}_{AG}$$

2. Make a velocity addition diagram (Figure 3-6) showing the addition of the vectors in step 1. Include direction axes:

3. The sine of the angle θ between the velocity of the plane relative to the air and due north equals the ratio of v_{AG} and v_{pA}:

$$\sin\theta = \frac{v_{AG}}{v_{pA}} = \frac{90\text{ km/h}}{200\text{ km/h}} = \frac{9}{20}$$

so

$$\theta = \sin^{-1}\frac{9}{20} = \boxed{27° \text{ west of north}}$$

(*b*) Because $\vec{v}_{AG}$ and $\vec{v}_{pG}$ are mutually perpendicular, we can use the Pythagorean theorem to find the magnitude of $\vec{v}_{pG}$:

$$v_{pA}^2 = v_{pG}^2 + v_{AG}^2$$

so

$$v_{pG} = \sqrt{v_{pA}^2 - v_{AG}^2}$$

$$= \sqrt{(200\text{ km/h})^2 - (90\text{ km/h})^2} = \boxed{180\text{ km/h}}$$

CHECK Heading directly into the 90 km/h wind would result in a ground speed of 200 km/h − 90 km/h = 110 km/h. The Part-(*b*) result of 180 km/h is greater than 110 km/h and less than 200 km/h, as expected.

ACCELERATION VECTORS

The **average-acceleration vector** is the ratio of the change in the instantaneous-velocity vector, $\Delta\vec{v}$, to the elapsed time interval Δt:

$$\vec{a}_{av} = \frac{\Delta\vec{v}}{\Delta t} \qquad 3\text{-}10$$

DEFINITION—AVERAGE-ACCELERATION VECTOR

The **instantaneous-acceleration vector** is the limit of this ratio as Δt approaches zero; in other words, it is the derivative of the velocity vector with respect to time:

$$\vec{a} = \lim_{\Delta t \to 0} \frac{\Delta\vec{v}}{\Delta t} = \frac{d\vec{v}}{dt} \qquad 3\text{-}11$$

DEFINITION—INSTANTANEOUS-ACCELERATION VECTOR

To calculate the instantaneous acceleration, we express $\vec{v}$ in rectangular coordinates:

$$\vec{v} = v_x\hat{i} + v_y\hat{j} + v_z\hat{k} = \frac{dx}{dt}\hat{i} + \frac{dy}{dt}\hat{j} + \frac{dz}{dt}\hat{k}$$

Then

$$\vec{a} = \frac{dv_x}{dt}\hat{i} + \frac{dv_y}{dt}\hat{j} + \frac{dv_z}{dt}\hat{k} = \frac{d^2x}{dt^2}\hat{i} + \frac{d^2y}{dt^2}\hat{j} + \frac{d^2z}{dt^2}\hat{k}$$

$$= a_x\hat{i} + a_y\hat{j} + a_z\hat{k} \qquad 3\text{-}12$$

where the components of $\vec{a}$ are

$$a_x = \frac{dv_x}{dt}, \quad a_y = \frac{dv_y}{dt}, \quad a_z = \frac{dv_z}{dt}.$$

Example 3-3 A Thrown Baseball

The position of a thrown baseball is given by $\vec{r} = [1.5\text{ m} + (12\text{ m/s})t]\hat{i} + [(16\text{ m/s})t - (4.9\text{ m/s}^2)t^2]\hat{j}$. Find its velocity and acceleration as functions of time.

PICTURE Recall that $\vec{r} = x\hat{i} + y\hat{j}$ (Equation 3-1). We can find the x and y components of the velocity and acceleration by taking the time derivatives of x and y.

SOLVE

1. Find the x and y components of $\vec{r}$:

$$x = 1.5\text{ m} + (12\text{ m/s})t$$
$$y = (16\text{ m/s})t - (4.9\text{ m/s}^2)t^2$$

2. The x and y components of the velocity are found by differentiating x and y:

$$v_x = \frac{dx}{dt} = 12\text{ m/s}$$
$$v_y = \frac{dy}{dt} = (16\text{ m/s}) - 2(4.9\text{ m/s}^2)t$$

3. We differentiate v_x and v_y to obtain the components of the acceleration:

$$a_x = \frac{dv_x}{dt} = 0$$
$$a_y = \frac{dv_y}{dt} = -9.8\text{ m/s}^2$$

4. In vector notation, the velocity and acceleration are

$$\vec{v} = \boxed{(12\text{ m/s})\hat{i} + [16\text{ m/s} - (9.8\text{ m/s}^2)t]\hat{j}}$$
$$\vec{a} = \boxed{(-9.8\text{ m/s}^2)\hat{j}}$$

CHECK The units that accompany the quantities for velocity and acceleration are m/s and m/s², respectively. Our step 4 results for velocity and acceleration have the correct units of m/s and m/s².

For a vector to be constant, both its magnitude and direction must remain constant. If either magnitude or direction changes, the vector changes. Thus, if a car rounds a curve in the road at constant speed, it is accelerating because the velocity is changing due to the change in direction of the velocity vector.

Example 3-4 Rounding a Curve

A car is traveling east at 60 km/h. It rounds a curve, and 5.0 s later it is traveling north at 60 km/h. Find the average acceleration of the car.

PICTURE We can calculate the average acceleration from its definition, $\vec{a}_{av} = \Delta\vec{v}/\Delta t$. To do this, we first calculate $\Delta\vec{v}$, which is the vector that when added to $\vec{v}_i$, results in $\vec{v}_f$.

SOLVE

1. The average acceleration is the change in velocity divided by the elapsed time. To find $\vec{a}_{av}$, we first find the change in velocity: $\vec{a}_{av} = \dfrac{\Delta\vec{v}}{\Delta t}$
2. To find $\Delta\vec{v}$, we first specify $\vec{v}_i$ and $\vec{v}_f$. Draw $\vec{v}_i$ and $\vec{v}_f$ (Figure 3-7*a*), and draw the vector addition diagram (Figure 3-7*b*) corresponding to $\vec{v}_f = \vec{v}_i + \Delta\vec{v}$:
3. The change in velocity is related to the initial and final velocities: $\vec{v}_f = \vec{v}_i + \Delta\vec{v}$
4. Substitute these results to find the average acceleration: $\vec{a}_{av} = \dfrac{\vec{v}_f - \vec{v}_i}{\Delta t} = \dfrac{60\text{ km/h}\,\hat{j} - 60\text{ km/h}\,\hat{i}}{5.0\text{ s}}$
5. Convert 60 km/h to meters per second: $60\text{ km/h} \times \dfrac{1\text{ h}}{3600\text{ s}} \times \dfrac{1000\text{ m}}{1\text{ km}} = 16.7\text{ m/s}$
6. Express the acceleration in meters per second squared: $\vec{a}_{av} = \dfrac{\vec{v}_f - \vec{v}_i}{\Delta t} = \dfrac{16.7\text{ m/s}\,\hat{j} - 16.7\text{ m/s}\,\hat{i}}{5.0\text{ s}}$ $= \boxed{-3.4\text{ m/s}^2\hat{i} + 3.4\text{ m/s}^2\hat{j}}$

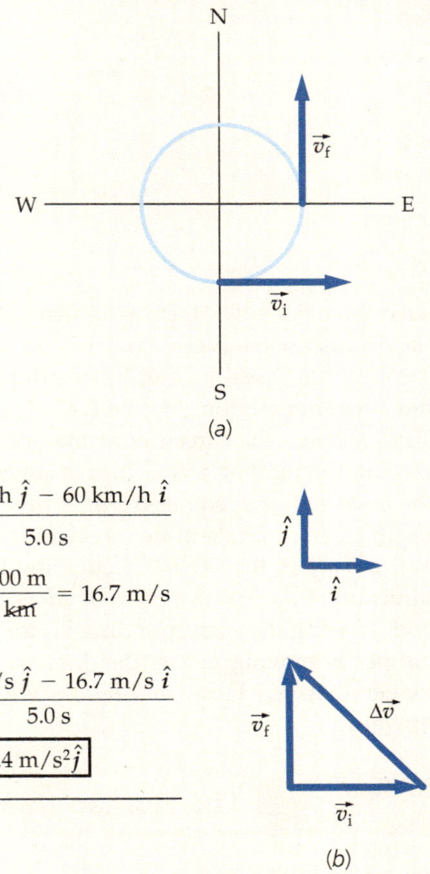

FIGURE 3-7

CHECK The eastward component of the velocity decreases from 60 km/h to zero, so we expect a negative acceleration component in the *x* direction. The northward component of the velocity increases from zero to 60 km/h, so we expect a positive acceleration component in the *y* direction. Our step 6 result meets both of these expectations.

TAKING IT FURTHER Note that the car is accelerating even though its speed remains constant.

PRACTICE PROBLEM 3-1 Find the magnitude and direction of the average acceleration vector.

The motion of an object traveling in a circle at constant speed is an example of motion in which the direction of the velocity changes even though its magnitude, the speed, remains constant.

> ! Do not assume that the acceleration of an object is zero just because the object is traveling at constant speed. For the acceleration to be zero, neither the magnitude nor the direction of the velocity vector can be changing.

THE DIRECTION OF THE ACCELERATION VECTOR

In the next few chapters, you will need to determine the direction of the acceleration vector from a description of the motion. To see how this is done, consider a bungee jumper as she slows down prior to reversing direction at the lowest point of her jump. To find the direction of her acceleration as she loses speed during the last stages of her descent, we draw a series of dots representing her position at

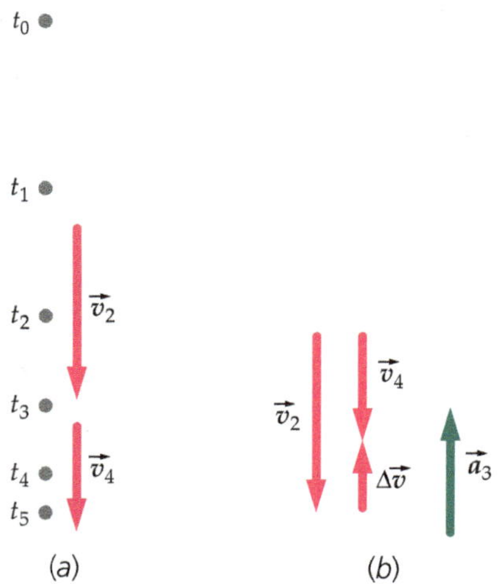

FIGURE 3-8 (*a*) A motion diagram of a bungee jumper losing speed as she descends. The dots are drawn at successive ticks of a clock. (*b*) We draw vectors $\vec{v}_2$ and $\vec{v}_4$ starting from the same point. Then, we draw $\Delta\vec{v}$ from the head of $\vec{v}_2$ to the head of $\vec{v}_4$ to obtain the graphical expression of the relation $\vec{v}_2 + \Delta\vec{v} = \vec{v}_4$. The acceleration $\vec{a}_3$ is in the same direction as $\Delta\vec{v}$.

CONCEPT CHECK 3-1

Figure 3-9 is a motion diagram of the bungee jumper before, during, and after time t_6, when she momentarily come to rest at the lowest point in her descent. During the part of her ascent shown, she is moving upward with increasing speed. Use this diagram to determine the direction of the jumper's acceleration (*a*) at time t_6 and (*b*) at time t_9.

successive ticks of a clock, as shown in Figure 3-8*a*. The faster she moves, the greater the distance she travels between ticks, and the greater the space between the dots in the diagram. Next we number the dots, starting with zero and increasing in the direction of her motion. At time t_0 she is at dot 0, at time t_1 she is at dot 1, and so forth. To determine the direction of the acceleration at time t_3, we draw vectors representing the jumper's velocities at times t_2 and t_4. The average acceleration during the interval t_2 to t_4 equals $\Delta\vec{v}/\Delta t$, where $\Delta\vec{v} = \vec{v}_4 - \vec{v}_2$ and $\Delta t = t_4 - t_2$. We use this result as an estimate of her acceleration at time t_3. That is, $\vec{a}_3 \approx \Delta\vec{v}/\Delta t$. Because $\vec{a}_3$ and $\Delta\vec{v}$ are in the same direction, by finding the direction of $\Delta\vec{v}$ we also find the direction of $\vec{a}_3$. The direction of $\Delta\vec{v}$ is obtained by using the relation $\vec{v}_2 + \Delta\vec{v} = \vec{v}_4$ and drawing the corresponding vector addition diagram (Figure 3-8*b*). Because the jumper is moving faster (the dots are farther apart) at t_2 than at t_4, we draw $\vec{v}_2$ longer than $\vec{v}_4$. From this figure, we can see that $\Delta\vec{v}$, and thus $\vec{a}_3$, is directed upward.

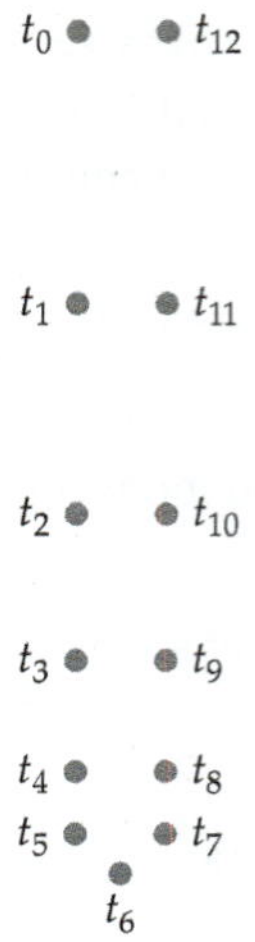

FIGURE 3-9 The dots for the bungee jumper's ascent are drawn to the right of those for her descent so that they do not overlap each other. Her motion, however, is straight down and then straight up.

Example 3-5 The Human Cannonball — *Context-Rich*

You are asked to substitute for an ill performer in a circus that is sponsored by your school. The job, should you choose to accept it, is to get shot out of a cannon. Never afraid to accept a challenge, you accept. The barrel of the cannon is inclined an angle of 60° above the horizontal. Your physics teacher offers you extra credit on the next exam if you successfully use a motion diagram to estimate the direction of your acceleration during the ascending portion of the flight.

PICTURE During the ascending portion of the flight, you travel in a curved path with decreasing speed. To estimate the direction of your acceleration you use $\vec{a}_{av} = \Delta\vec{v}/\Delta t$ and estimate the direction of $\Delta\vec{v}$. To estimate the direction of $\Delta\vec{v}$, we draw a motion diagram and then make a sketch of the relation $\vec{v}_i + \Delta\vec{v} = \vec{v}_f$.

SOLVE

1. Make a motion diagram (Figure 3-10*a*) of your motion during the ascending portion of the flight. Because your speed decreases as you ascend, the spacing between adjacent dots on your diagram decreases as you rise:
2. Pick a dot on the motion diagram and draw a velocity vector on the diagram for both the preceding and the following dot. These vectors should be drawn tangent to your trajectory.

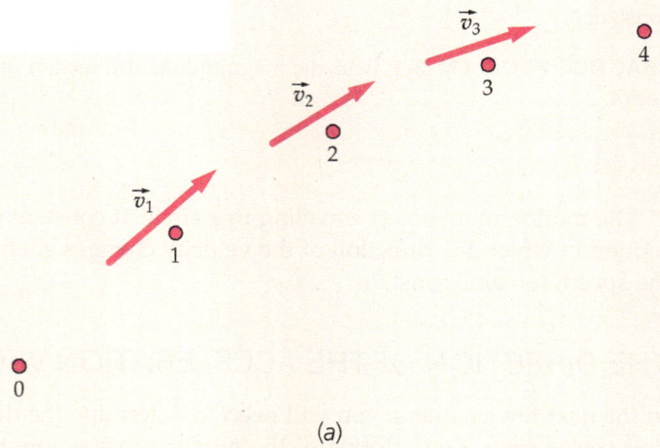

FIGURE 3-10

3. Draw the vector addition diagram (Figure 3-10*b*) of the relation $\vec{v}_i + \Delta\vec{v} = \vec{v}_f$. Begin by drawing the two velocity vectors from the same point. These vectors have the same magnitude and direction as the vectors drawn for step 2. Then, draw the $\Delta\vec{v}$ vector from the head of $\vec{v}_i$ to the head of $\vec{v}_f$.
4. Draw the estimated acceleration vector in the same direction as $\Delta\vec{v}$, but not the same length (because $\vec{a} \approx \Delta\vec{v}/\Delta t$).

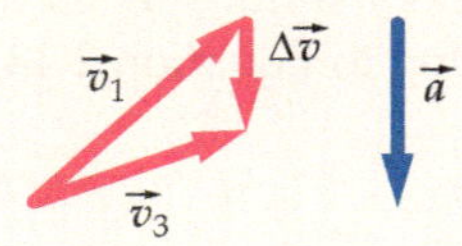

(*b*)

FIGURE 3.10 *(continued)*

CHECK During the ascent, the upward component of the velocity is decreasing, so we expect $\Delta\vec{v}$ to have a downward vertical component. Our step 3 result satisfies this expectation.

TAKING IT FURTHER The process of finding the direction of the acceleration using a motion diagram is not precise. Therefore, the result is an estimate of the direction of the acceleration, as opposed to a precise determination.

CONCEPT CHECK 3-2

Use a motion diagram to estimate the direction of the acceleration in Example 3-5 during the descending portion of your flight.

3-2 SPECIAL CASE 1: PROJECTILE MOTION

In a home run hit or a field goal kick, the ball follows a particular curved path through the air. This type of motion, known as **projectile motion,** occurs when an object (the projectile) is launched into the air and is allowed to move freely. The projectile might be a ball, a dart, water shooting out of a fountain, or even a human body during a long jump. If air resistance is negligible, then the projectile is said to be in free-fall. For objects in free-fall near the surface of Earth, the acceleration is the downward acceleration due to gravity.

Figure 3-11 shows a particle launched with initial speed v_0 at angle θ_0 above the horizontal. Let the launch point be at (x_0, y_0); y is positive upward and x is positive to the right. The initial velocity $\Delta\vec{v}_0$ then has components

$$v_{0x} = v_0 \cos\theta_0 \qquad \text{3-13}a$$

$$v_{0y} = v_0 \sin\theta_0 \qquad \text{3-13}b$$

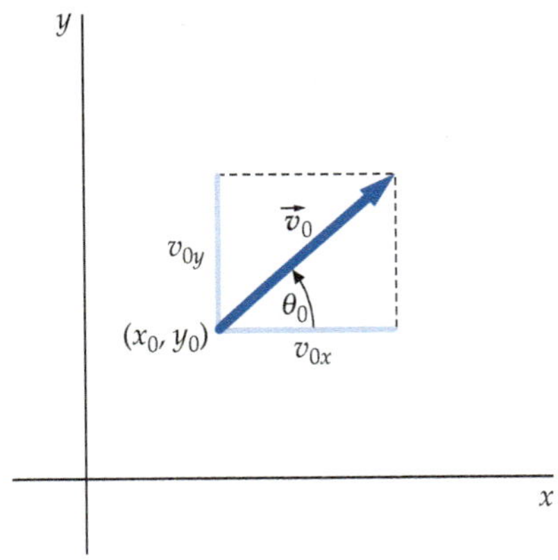

FIGURE 3-11 The components of $\vec{v}_0$ are $v_{0x} = v_0 \cos\theta_0$ and $v_{0y} = v_0 \sin\theta_0$, where θ_0 is the angle above the horizontal of $\vec{v}_0$.

In the absence of air resistance, the acceleration $\vec{a}$ is constant. The projectile has no horizontal acceleration, so the only acceleration is the free-fall acceleration $\vec{g}$, directed downward:

$$a_x = 0 \qquad \text{3-14}a$$

and

$$a_y = -g \qquad \text{3-14}b$$

Because the acceleration is constant, we can use the kinematic equations for constant acceleration presented in Chapter 2. The x component of the velocity $\vec{v}$ is constant because no horizontal acceleration exists:

$$v_x = v_{0x} \qquad \text{3-15}a$$

The y component of the velocity varies with time according to $v_y = v_{0y} + a_y t$ (Equation 2-12), with $a_y = -g$:

$$v_y = v_{0y} - gt \qquad \text{3-15}b$$

Notice that v_x does not depend on v_y and v_y does not depend on v_x: *The horizontal and vertical components of projectile motion are independent.* Dropping a ball from a desktop and projecting a second ball horizontally at the same time can demonstrate the independence of v_x and v_y, as shown in Figure 3-12. Notice that the two balls strike the floor simultaneously.

According to Equation 2-14, the displacements x and y are given by

$$x(t) = x_0 + v_{0x}t \qquad \text{3-16}a$$

$$y(t) = y_0 + v_{0y}t - \tfrac{1}{2}gt^2 \qquad \text{3-16}b$$

The notation $x(t)$ and $y(t)$ simply emphasizes that x and y are functions of time. If the y component of the initial velocity is known, the time t for which the particle is at height y can be found from Equation 3-16*b*. The horizontal position at that time can then be found using Equation 3-16*a*. (Equations 3-14 to 3-16 are expressed in vector form immediately preceding Example 3-10.)

The general equation for the path $y(x)$ of a projectile can be obtained from Equations 3-16 by eliminating the variable t. Choosing $x_0 = 0$ and $y_0 = 0$, we obtain $t = x/v_{0x}$ from Equation 3-16*a*. Substituting this into Equation 3-16*b* gives

$$y(x) = v_{0y}\left(\frac{x}{v_{0x}}\right) - \frac{1}{2}g\left(\frac{x}{v_{0x}}\right)^2 = \left(\frac{v_{0y}}{v_{0x}}\right)x - \left(\frac{g}{2v_{0x}^2}\right)x^2$$

Substituting for the velocity components using $v_{0x} = v_0 \cos\theta_0$ and $v_{0y} = v_0 \sin\theta_0$ yields

$$y(x) = (\tan\theta_0)x - \left(\frac{g}{2v_0^2\cos^2\theta_0}\right)x^2 \qquad \text{3-17}$$

PATH OF PROJECTILE

for the projectile's path. This equation is of the form $y = ax + bx^2$, which is the equation for a parabola passing through the origin. Figure 3-13 shows the path of a projectile with its velocity vector and components at several points. The path is for a projectile that impacts the ground at P. The horizontal distance $|\Delta x|$ between launch and impact at the same elevation is the **horizontal range** R.

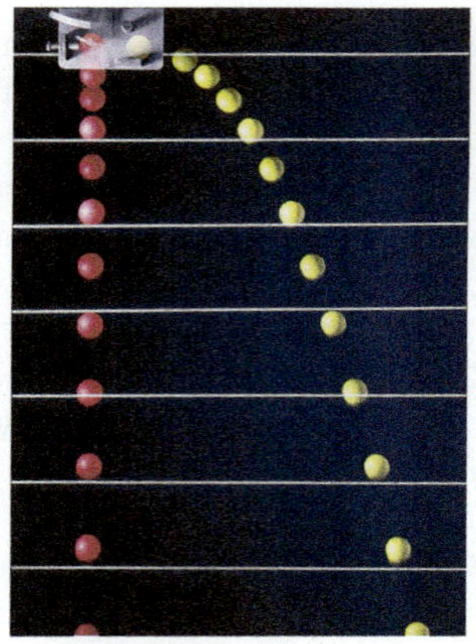

FIGURE 3-12 The red ball is released from rest at the instant the yellow ball rolls off the tabletop. The positions of the two balls are shown at successive equal time intervals. The vertical motion of the yellow ball is identical with the vertical motion of the red ball, thus demonstrating that the vertical motion of the yellow ball is independent of its horizontal motion. *(Richard Megna/Fundamental Photographs.)*

! Do not think that the velocity of a projectile is zero when the projectile is at the highest point in its trajectory. At the highest point in the trajectory v_y is zero, but the projectile may still be moving horizontally.

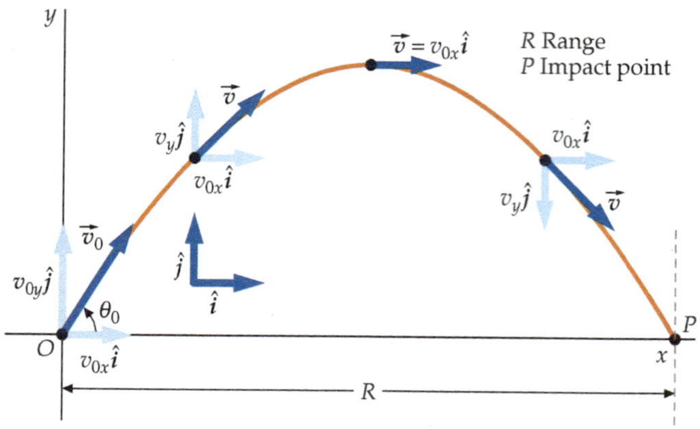

FIGURE 3-13 The path of a projectile, showing velocity components at different times.

Example 3-6 A Cap in the Air

A delighted physics graduate throws her cap into the air with an initial velocity of 24.5 m/s at 36.9° above the horizontal. The cap is later caught by another student. Find (*a*) the total time the cap is in the air, and (*b*) the total horizontal distance traveled. (Ignore effects of air resistance.)

PICTURE We choose the origin to be the initial position of the cap so that $x_0 = y_0 = 0$. We assume it is caught at the same height. The total time the cap is in the air is found by setting $y(t) = 0$ in $y(t) = y_0 + v_{0y}t - \frac{1}{2}gt^2$ (Equation 3-16*b*). We can then use this result in $x(t) = x_0 + v_{0x}t$ (Equation 3-16*a*) to find the total horizontal distance traveled.

SOLVE

(*a*) 1. Setting $y = 0$ in Equation 3-16*b*:

$$y = v_{0y}t - \tfrac{1}{2}gt^2$$
$$0 = t(v_{0y} - \tfrac{1}{2}gt)$$

2. There are two solutions for t:

$$t_1 = 0 \text{ (initial time)}$$
$$t_2 = \frac{2v_{0y}}{g}$$

3. Use trigonometry to relate v_{0y} to v_0 and θ_0 (see Figure 3-11):

$$v_{0y} = v_0 \sin\theta_0$$

4. Substitute for v_{0y} in the step 2 result to find the total time t_2:

$$t_2 = \frac{2v_{0y}}{g} = \frac{2v_0\sin\theta_0}{g} = \frac{2(24.5\text{ m/s})\sin 36.9°}{9.81\text{ m/s}^2} = \boxed{3.00\text{ s}}$$

(*b*) Use the value for the time in step 4 to calculate the total horizontal distance traveled:

$$x = v_{0x}t_2 = (v_0\cos\theta_0)t_2 = (24.5\text{ m/s})\cos 36.9°(3.00\text{ s}) = \boxed{58.8\text{ m}}$$

CHECK If the cap traveled at a constant speed of 24.5 m/s for 3.00 s it would have traveled a distance of 73.5 m. Because it was launched at an angle, its horizontal speed was less than 24.5 m/s, so we expect its distance traveled to be less than 73.5 m. Our Part-(*b*) result of 58.8 m meets this expectation.

TAKING IT FURTHER The vertical component of the initial velocity of the cap is 14.7 m/s, the same as that of the cap in Example 2-13 (Chapter 2), where the cap was thrown straight up with $v_0 = 14.7$ m/s. The time the cap is in the air is also the same as in Example 2-13. Figure 3-14 shows the height y versus t for the cap. This curve is identical to Figure 2-20*a* (Example 2-13) because the caps each have the same vertical acceleration and vertical velocity. Figure 3-14 can be reinterpreted as a graph of y versus x if its time scale is converted to a distance scale, as shown in the figure. This is accomplished by multiplying the time values by 19.6 m/s. This works because the cap moves at (24.5 m/s) cos 36.9° = 19.6 m/s horizontally. The curve y versus x is a parabola (as is the curve y versus t).

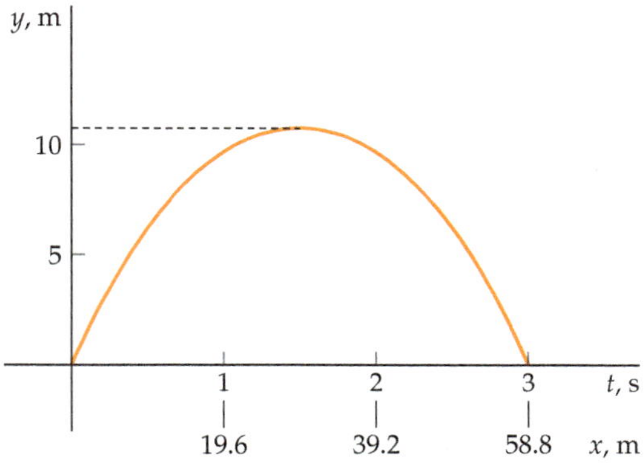

FIGURE 3-14 A plot of y versus t and of y versus x.

Example 3-7 A Supply Drop

A helicopter drops a supply package to flood victims on a raft on a swollen lake. When the package is released, the helicopter is 100 m directly above the raft and flying at a velocity of 25.0 m/s at an angle $\theta_0 = 36.9°$ above the horizontal. (*a*) How long is the package in the air? (*b*) How far from the raft does the package land? (*c*) If the helicopter continues at constant velocity, where is the helicopter when the package lands? (Ignore effects of air resistance.)

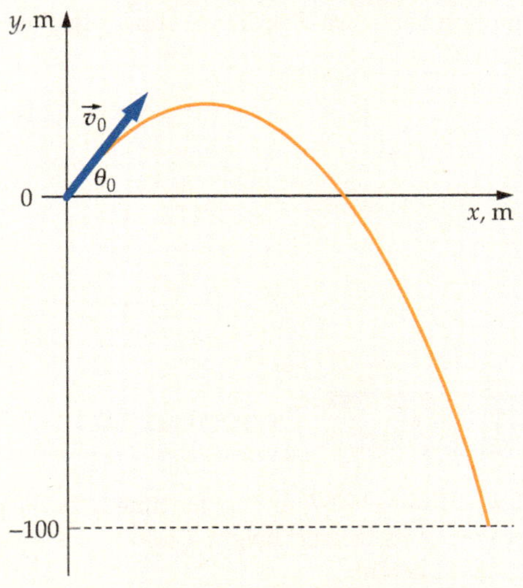

FIGURE 3-15 The parabola intersects the $y = -100$ m line twice, but only one of those times is greater than zero.

PICTURE The time in the air depends only on the vertical motion. Using $y(t) = y_0 + v_{0y}t - \frac{1}{2}gt^2$ (Equation 3-16*b*), you can solve for the time. Choose the origin to be at the location of the package when it is released. The initial velocity of the package is the velocity of the helicopter. The horizontal distance traveled by the package is given by $x(t) = v_{0x}t$ (Equation 3-16*a*), where t is the time the package is in the air.

SOLVE

(*a*) 1. Sketch the trajectory of the package during the time it is in the air. Include coordinate axes as shown in Figure 3-15:

2. To find the time of flight, write $y(t)$ for motion with constant acceleration, then set $y_0 = 0$ and $a_y = -g$ in the equation:

$$y = y_0 + v_{0y}t + \tfrac{1}{2}a_yt^2$$
$$y = 0 + v_{0y}t - \tfrac{1}{2}gt^2 = v_{0y}t - \tfrac{1}{2}gt^2$$

3. The solution to the quadratic equation $ax^2 + bx + c = 0$ is given by the quadratic formula:

$$x = \frac{-b \pm \sqrt{b^2 - 4ac}}{2a}$$

Using this, solve the quadratic equation from step 2 for t:

$$y = v_{0y}t - \tfrac{1}{2}gt^2$$

so

$$0 = \tfrac{1}{2}gt^2 - v_{0y}t + y$$

and

$$t = \frac{v_{0y} \pm \sqrt{v_{0y}^2 - 2gy}}{g}$$

4. Solve for the time when $y = -100$ m. First, solve for v_{0y}, then use the value for v_{0y} to find t.

$$v_{0y} = v_0 \sin\theta_0 = (25.0\ \text{m/s})\sin 36.9° = 15.0\ \text{m/s}$$

so

$$t = \frac{15.0\ \text{m/s} \pm \sqrt{(15.0\ \text{m/s})^2 - 2(9.81\ \text{m/s}^2)(-100\ \text{m})}}{9.81\ \text{m/s}^2}$$

so

$$t = -3.24\ \text{s} \quad \text{or} \quad t = 6.30\ \text{s}$$

Because the package is released at $t = 0$, the time of impact cannot be negative. Hence:

$$t = \boxed{6.30\ \text{s}}$$

(*b*) 1. At impact the package has traveled a horizontal distance x, where x is the horizontal velocity times the time of flight. First solve for the horizontal velocity:

$$v_{0x} = v_0 \cos\theta_0 = (25.0\ \text{m/s})\cos 36.9° = 20.0\ \text{m/s}$$

2. Next substitute for v_{0x} in $x = x_0 + v_{0x}t$ (Equation 3-16*a*) to find x.

$$x = v_{0x}t = (20.0\ \text{m/s})(6.30\ \text{s}) = \boxed{126\ \text{m}}$$

(*c*) The coordinates of the helicopter at the time of impact are

$$x_h = v_{0x}t = (20.0\ \text{m/s})(6.30\ \text{s}) = 126\ \text{m}$$

$$y_h = y_{h0} + v_{h0}t = 0 + (15.0\ \text{m/s})(6.30\ \text{s}) = 94.4\ \text{m}$$

At impact, the helicopter is

$\boxed{\text{194 m directly above the package}}$.

CHECK The helicopter is directly above the package when the package hits the water (and at all other times before then). This is because the horizontal velocities of the package and the helicopter were equal at release, and the horizontal velocities of both remain constant during flight.

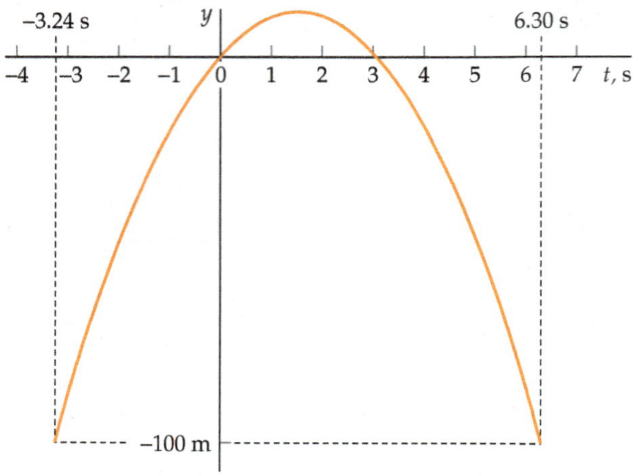

FIGURE 3-16

TAKING IT FURTHER The positive time is appropriate because it corresponds to a time after the package is dropped (which occurs at $t = 0$). The negative time is when the package would have been at $y = -100$ m if its motion had started earlier as shown in Figure 3-16.

Example 3-8 Dropping Supplies *Try It Yourself*

Using Example 3-7, find (*a*) the time t_1 for the package to reach its greatest height h above the water, (*b*) its greatest height h, and (*c*) the time t_2 for the package to fall to the water from its greatest height.

PICTURE The time t_1 is the time at which the vertical component of the velocity is zero. Using $v_y(t) = v_{0y} - gt$ (Equation 3-15*b*) solve for t_1.

SOLVE

Cover the column to the right and try these on your own before looking at the answers.

Steps	Answers
(*a*) 1. Write $v_y(t)$ for the package.	$v_y(t) = v_{0y} - gt$
2. Set $v_y(t_1) = 0$ and solve for t_1.	$t_1 = \boxed{1.53\text{ s}}$
(*b*) 1. Find $v_{y\,\text{av}}$ during the time the package is moving up.	$v_{y\,\text{av}} = 7.505\text{ m/s}$
2. Use $v_{y\,\text{av}}$ to find the distance traveled up. Then find h.	$\Delta y = 11.48\text{ m}$, so $h = \boxed{111\text{ m}}$
(*c*) Find the time for the package to fall a distance h.	$t_2 = \boxed{4.77\text{ s}}$

CHECK Note that $t_1 + t_2 = 6.30$ s, in agreement with Example 3-7. Also, note that t_1 is less than t_2. This is as expected because the package rises a distance of 12 m but falls a distance of 112 m.

PRACTICE PROBLEM 3-2 Solve Part (*b*) of Example 3-8 using $y(t)$ (Equation 3-16*b*) instead of finding $v_{y\,\text{av}}$.

HORIZONTAL RANGE OF A PROJECTILE

The horizontal range R of a projectile can be written in terms of its initial speed and initial angle above the horizontal. As in the preceding examples, we find the horizontal range by multiplying the x component of the velocity by the total time that the projectile is in the air. The total flight time T is obtained by setting $y = 0$ and $t = T$ in $y = v_0t - \frac{1}{2}gt^2$ (Equation 3-16*b*).

$$0 = v_{0y}T - \tfrac{1}{2}gT^2 \qquad T > 0$$

Dividing through by T gives

$$v_{0y} - \tfrac{1}{2}gT = 0$$

The flight time of the projectile is thus

$$T = \frac{2v_{0y}}{g} = \frac{2v_0}{g}\sin\theta_0$$

To find the horizontal range R, we substitute T for t in $x(t) = v_{0x}t$ (Equation 3-16*a*) to obtain

$$R = v_{0x}T = (v_0\cos\theta_0)\left(\frac{2v_0}{g}\sin\theta_0\right) = \frac{2v_0^2}{g}\sin\theta_0\cos\theta_0$$

This can be further simplified by using the trigonometric identity:

$$\sin 2\theta = 2\sin\theta\cos\theta$$

Thus,

$$R = \frac{v_0^2}{g}\sin 2\theta_0 \qquad \text{3-18}$$

HORIZONTAL RANGE OF A PROJECTILE

PRACTICE PROBLEM 3-3

Use Equation 3-18 to verify the answer Part (*b*) of Example 3-6.

Equation 3-18 is useful if you want to find the range for several projectiles that have equal initial speeds. For this case, this equation shows how the range depends on θ. Because the maximum value of $\sin 2\theta$ is 1, and because $\sin 2\theta = 1$ when $\theta = 45°$, the range is greatest when $\theta = 45°$. Figure 3-17 shows graphs of the vertical heights versus the horizontal distances for projectiles with an initial speed of 24.5 m/s and several different initial angles. The angles drawn are 45°, which has the maximum range, and pairs of angles at equal amounts above and below 45°. Notice that the paired angles have the same range. The green curve has an initial angle of 36.9°, as in Example 3-6.

In many practical applications, the initial and final elevations may not be equal, or other considerations are important. For example, in the shot put, the ball ends its flight when it hits the ground, but it is projected from an initial height of about 2 m above the ground. This condition causes the horizontal displacement to be at a maximum at an angle somewhat lower than 45°, as shown in Figure 3-18. Studies of the best shot-putters show that maximum horizontal displacement occurs at an initial angle of about 42°.

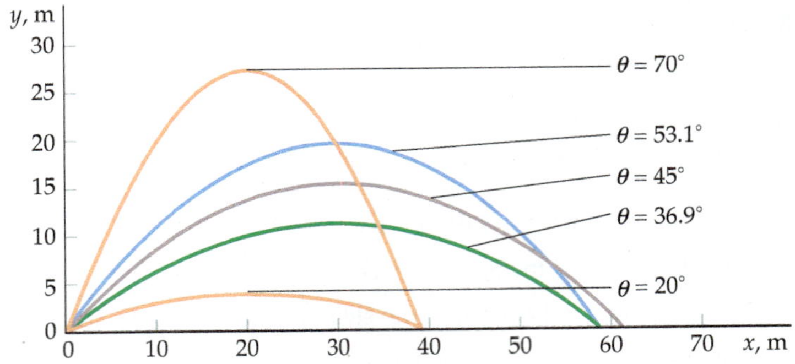

FIGURE 3-17 The initial speed is the same for each trajectory.

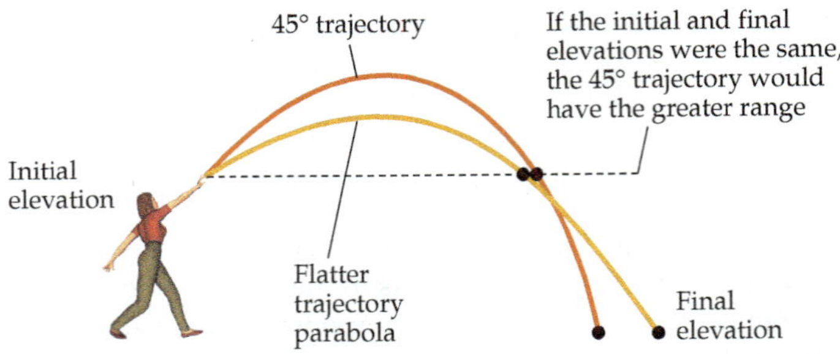

FIGURE 3-18 If a projectile lands at an elevation lower than the initial elevation, the maximum horizontal displacement is achieved when the projection angle is somewhat lower than 45°.

Example 3-9 To Catch a Thief

A police officer chases a master jewel thief across city rooftops. They are both running when they come to a gap between buildings that is 4.00 m wide and has a drop of 3.00 m (Figure 3-19). The thief, having studied a little physics, leaps at 5.00 m/s, at an angle of 45.0° above the horizontal, and clears the gap easily. The police officer did not study physics and thinks he should maximize his horizontal velocity, so he leaps horizontally at 5.00 m/s. (*a*) Does he clear the gap? (*b*) By how much does the thief clear the gap?

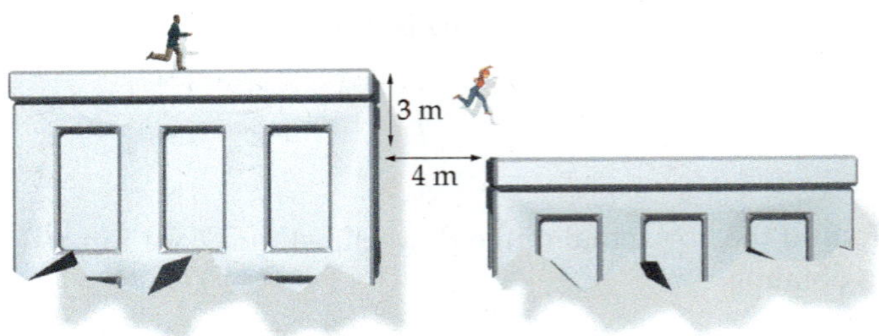

FIGURE 3-19

PICTURE Assuming they both clear the gap, the total time in the air depends only on the vertical aspects of the motion. Choose the origin at the launch point, with upward positive so that Equations 3-16 apply. Use Equation 3-16*b* for $y(t)$ and solve for the time when $y = -3.00$ m for $\theta_0 = 0$ and again for $\theta_0 = 45.0°$. The horizontal distances traveled are the values of x at these times.

SOLVE

(*a*) 1. Write $y(t)$ for the police officer and solve for t when $y = -3.00$ m.

$$y = y_0 + v_{y0}t - \tfrac{1}{2}gt^2 \quad -3.00\text{ m} = 0 + 0 - \tfrac{1}{2}(9.81\text{ m/s}^2)t^2$$

$$t = 0.782\text{ s}$$

2. Substitute this into the equation for $x(t)$ and find the horizontal distance traveled during this time.

$$x = x_0 + v_{x0}t$$

$$x = 0 + (5.00\text{ m/s})(0.782\text{ s})$$

$$x = \boxed{3.91\text{ m}}$$

Because 3.91 m is less than 4.00 m, it appears the police officer fails to make it across the gap between buildings.

(*b*) 1. Write $y(t)$ for the thief and solve for t when $y = -3.00$ m. $y(t)$ is a quadratic equation with two solutions, but only one of its solutions is acceptable.	$y = y_0 + v_{y0}t - \frac{1}{2}gt^2$ $-3.00\text{ m} = 0 + (5.00\text{ m/s})\sin 45.0° - \frac{1}{2}(9.81\text{ m/s}^2)t^2$ $t = -0.500$ s or $t = 1.22$ s She must land at a time after she leaps, so $t = 1.22$ s
2. Find the horizontal distance covered for the positive value of t.	$x = x_0 + v_{0x}t = 0 + (5.00\text{ m/s})\cos 45°(1.22\text{ s}) = 4.31\text{ m}$
3. Subtract 4.0 m from this distance.	$4.31\text{ m} - 4.00\text{ m} = \boxed{0.31\text{ m}}$

CHECK The policeman's horizontal speed remains 5.00 m/s during his flight. So, the policeman travels the 4.00 m to the next building in 4.00 m/(5.00 m/s) = 0.800 s. Because our Part-(*a*), step-1 result is less than 0.800 s, we know he falls below the second roof before reaching the second building—in agreement with our Part-(*a*), step-2 result.

TAKING IT FURTHER By modeling the policeman as a particle, we found that he was slightly below the second roof at impact. However, we cannot conclude that he did not complete the jump because he is not a particle. It is likely that he would raise his feet enough for them to clear the edge of the second roof.

PROJECTILE MOTION IN VECTOR FORM

For projectile motion, we have $a_x = 0$ and $a_y = -g$ (Equations 3-14*a* and 3-14*b*), where the $+y$ direction is directly upward. To express these equations in vector form, we multiply both sides of each equation by the appropriate unit vector and then add the two resulting equations. That is, $a_x\hat{i} = 0\hat{i}$ plus $a_y\hat{j} = -g\hat{j}$ gives

$$a_x\hat{i} + a_y\hat{j} = -g\hat{j} \quad \text{or} \quad \vec{a} = \vec{g} \qquad \text{3-14}c$$

where $\vec{g}$ is the free-fall acceleration vector. The magnitude of $\vec{g}$ is $g = 9.81\text{ m/s}^2$ (at sea level and at 45° latitude).

Combining equations $v_x = v_{0x}$ and $v_y = v_{0y} - gt$ in like manner gives

$$\vec{v} = \vec{v}_0 + \vec{g}t \quad (\text{or } \Delta\vec{v} = \vec{g}t) \qquad \text{3-15}c$$

where $\vec{v} = v_x\hat{i} + v_y\hat{j}$, $\vec{v}_0 = v_{0x}\hat{i} + v_{0y}\hat{j}$, and $\vec{g} = -g\hat{j}$. Repeating the process, this time for Equations $x(t) = x_0 + v_{0x}t$ and $y(t) = y_0 + v_{0y}t - \frac{1}{2}gt^2$, gives

$$\vec{r} = \vec{r}_0 + \vec{v}_0t + \tfrac{1}{2}\vec{g}t^2 \quad (\text{or } \Delta\vec{r} = \vec{v}_0t + \tfrac{1}{2}\vec{g}t^2) \qquad \text{3-16}c$$

where $\vec{r} = x\hat{i} + y\hat{j}$ and $\vec{r}_0 = x_0\hat{i} + y_0\hat{j}$. The vector forms of the kinematic equations (Equations 3-15*c* and 3-16*c*) are useful for solving a number of problems, including the following example.

Example 3-10 The Ranger and the Monkey

A park ranger with a tranquilizer dart gun intends to shoot a monkey hanging from a branch. The ranger points the barrel directly at the monkey, not realizing that the dart will follow a parabolic path that will pass below the present position of the creature. The monkey, seeing the gun discharge, immediately lets go of the branch and drops out of the tree, expecting to avoid the dart. (*a*) Show that the monkey will be hit regardless of the initial speed of the dart as long as this speed is great enough for the dart to travel the horizontal distance to the tree. Assume the reaction time of the monkey is negligible. (*b*) Let $\vec{v}_{d0}$ be the initial velocity of the dart relative to the monkey. Find the velocity of the dart *relative to the monkey* at an arbitrary time t during the dart's flight.

PICTURE In this example, both the monkey and the dart exhibit projectile motion. To show that the dart hits the monkey, we have to show that at some time t, the dart and the monkey have the same coordinates, regardless of the initial speed of the dart. To do this, we apply

Equation 3-16*c* to both the monkey and the dart. For Part (*b*), we can use Equation 3-15*c*, keeping in mind the relative reference frames.

SOLVE

(*a*) 1. Apply Equation 3-16*c* to the monkey at arbitrary time *t*:

$\Delta\vec{r}_{\rm m} = \frac{1}{2}\vec{g}t^2$

(The initial velocity of the monkey is zero.)

2. Apply Equation 3-16*c* to the dart at arbitrary time *t*:

$\Delta\vec{r}_{\rm d} = \vec{v}_{\rm d0}t + \frac{1}{2}\vec{g}t^2$

where $\vec{v}_{\rm d0}$ is the velocity of the dart as it leaves the gun.

3. Make a sketch of the monkey, the dart, and the gun, as shown in Figure 3-20. Show the dart and the monkey at their initial locations and at their locations a time *t* later. On the figure draw a vector representing each term in the step-1 and -2 results:

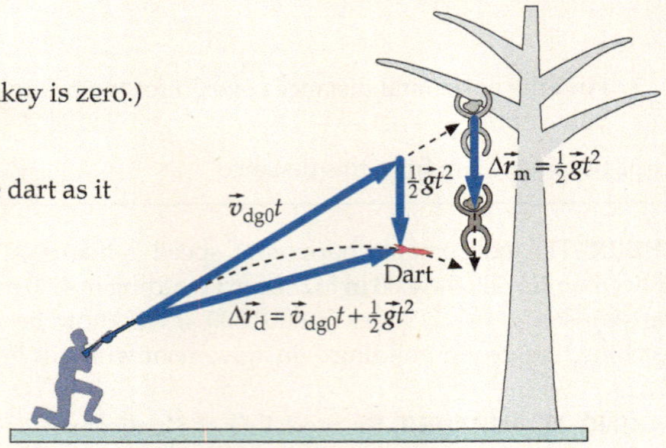

FIGURE 3-20

4. Note that at time *t* the dart and the monkey both are a distance $\frac{1}{2}gt^2$ below the line of sight of the gun:

The dart will strike the monkey when the dart reaches the monkey's line of fall.

(*b*) 1. The velocity of the dart relative to the monkey equals the velocity of the dart relative to the gun plus the velocity of the gun relative to the monkey:

$\vec{v}_{\rm dm} = \vec{v}_{\rm dg} + \vec{v}_{\rm gm}$

2. The velocity of the gun relative to the monkey is the negative of the velocity of the monkey relative to the gun:

$\vec{v}_{\rm dm} = \vec{v}_{\rm dg} - \vec{v}_{\rm mg}$

3. Using $\vec{v} = \vec{v}_0 + \vec{g}t$ (Equation 3-15*c*), express both the velocity of the dart relative to the gun and the velocity of the monkey relative to the gun:

$\vec{v}_{\rm dg} = \vec{v}_{\rm dg0} + \vec{g}t$

$\vec{v}_{\rm mg} = \vec{g}t$

4. Substitute these expressions into the Part-(*b*), step-2 result:

$\vec{v}_{\rm dm} = (\vec{v}_{\rm dg0} + \vec{g}t) - (\vec{g}t) = \boxed{\vec{v}_{\rm dg0}}$

CHECK The Part-(*a*), step-4 and the Part-(*b*), step-4 results are in agreement with each other. They agree that the dart will strike the monkey if the dart reaches the monkey's line of fall before the monkey lands on the ground.

TAKING IT FURTHER Relative to the falling monkey, the dart moves with constant speed $v_{\rm dg0}$ in a straight line. The dart strikes the monkey at time $t = L/v_{\rm d0}$, where L is the distance from the muzzle of the gun to the initial position of the monkey.

In a familiar lecture demonstration, a target is suspended by an electromagnet. When the dart leaves the gun, the circuit to the magnet is broken and the target falls. The demonstration is then repeated using a different initial velocity of the dart. For a large value of $v_{\rm dg0}$, the target is hit very near its original height, and for some lesser value of $v_{\rm dg0}$ it is hit just before it reaches the floor.

PRACTICE PROBLEM 3-4 A hockey puck at ice level is struck such that it misses the net and just clears the top of the Plexiglas wall of height $h = 2.80$ m. The flight time at the moment the puck clears the wall is $t_1 = 0.650$ s, and the horizontal position is $x_1 = 12.0$ m. (*a*) Find the initial speed and direction of the puck. (*b*) When does the puck reach its maximum height? (*c*) What is the maximum height of the puck?

3-3 SPECIAL CASE 2: CIRCULAR MOTION

Figure 3-21 shows a pendulum bob swinging back and forth in a vertical plane. The path of the bob is a segment of a circular path. Motion along a circular path, or a segment of a circular path, is called **circular motion.**

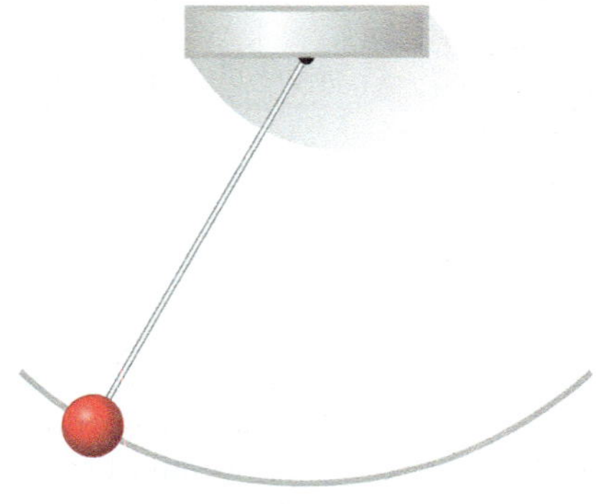

FIGURE 3-21 A pendulum bob swings along a circular arc centered at the point of support of the string.

Example 3-11 A Swinging Pendulum — Conceptual

Consider the motion of the pendulum bob shown in Figure 3-21. Using a motion diagram (see Example 3-5), find the direction of the acceleration vector when the bob is swinging from left to right and (*a*) on the descending portion of the path, (*b*) passing through the lowest point on the path, and (*c*) on the ascending portion of the path.

PICTURE As the bob descends, it both gains speed and changes direction. The acceleration is related to the change in velocity by $\vec{a} \approx \Delta\vec{v}/\Delta t$. The direction of the acceleration at a point can be estimated by constructing a vector addition diagram for the relation $\vec{v}_i + \Delta\vec{v} = \vec{v}_f$ to find the direction of $\Delta\vec{v}$, and thus the direction of the acceleration vector.

SOLVE

(*a*) 1. Make a motion diagram for a full left-to-right swing of the bob (Figure 3-22*a*). The spacing between dots is greatest at the lowest point where the speed is greatest.

2. Pick the dot at t_2 on the descending portion of the motion and draw a velocity vector on the diagram for both the preceding and the following dot (the dots at t_1 and t_3). The velocity vectors should be drawn tangent to the path and with lengths proportional to the speed (Figure 3-22*a*).

3. Draw the vector addition diagram (Figure 3-22*b*) for the relation $\vec{v}_i + \Delta\vec{v} = \vec{v}_f$. On this diagram draw the acceleration vector. Because $\vec{a} \approx \Delta\vec{v}/\Delta t$, $\vec{a}$ is in the same direction as $\Delta\vec{v}$.

(*b*) Repeat steps 2 and 3 (Figure 3-22*c*) for the point at t_4, the lowest point on the path.

(*c*) Repeat steps 2 and 3 (Figure 3-22*d*) for the point at t_6, a point on the ascending portion of the path.

CHECK At the lowest point (at t_4) the horizontal component of $\vec{v}$ is a maximum, so we expect the horizontal component of $\vec{a}$ to be zero. Near the lowest point, the upward component of $\vec{v}$ is negative just prior to $t = t_4$ and is positive just after $t = t_4$, so the upward component of $\vec{v}$ is increasing at $t = t_4$. This means we expect the upward component of $\vec{a}$ to be positive at $t = t_4$. The acceleration vector in Figure 3-22*c* is in agreement with both of these expectations.

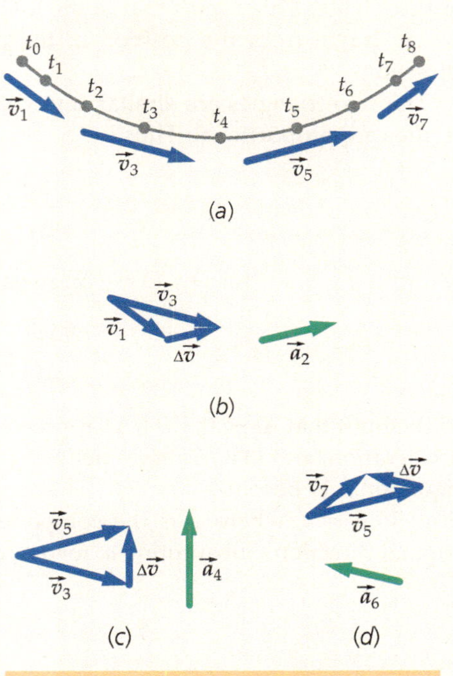

FIGURE 3-22

In Example 3-11, we saw that the acceleration vector is directed straight upward at the lowest point of the pendulum's swing (Figure 3-23)—toward point *P* at the center of the circle. Where the speed is increasing (on the descending portion), the acceleration vector has a component in the direction of the velocity vector as well as a component in the direction toward *P*. Where the speed is decreasing, the acceleration vector has a component in the direction opposite to the direction of the velocity vector, as well as a component in the direction toward *P*.

As a particle moves along a circular arc, the direction from the particle toward *P* (toward the center of the circle) is called the **centripetal direction** and the direction of the velocity vector is called the **tangential direction.** In Figure 3-23, the acceleration vector at the lowest point of the pendulum bob's path is in the centripetal direction. At virtually all other points along the path, the acceleration vector has both a tangential component and a centripetal component.

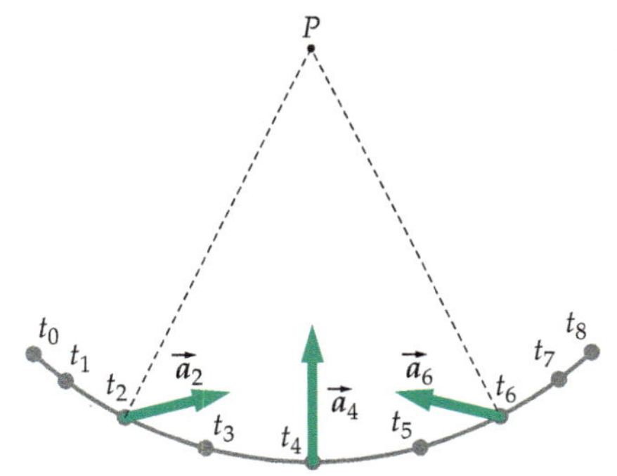

FIGURE 3-23

UNIFORM CIRCULAR MOTION

Motion in a circle at constant speed is called **uniform circular motion.** Even though the speed of a particle moving in uniform circular motion is constant, the moving particle is accelerating. To find an expression for the acceleration of a particle during uniform circular motion, we will extend the method used in Example 3-11 to relate the acceleration to the speed and the radius of the circle. The position and velocity vectors for a particle moving in a circle at constant speed are shown in Figure 3-24. The angle $\Delta\theta$ between $\vec{v}(t)$ and $\vec{v}(t + \Delta t)$ is equal to the angle between $\vec{r}(t)$ and $\vec{r}(t + \Delta t)$ because $\vec{r}$

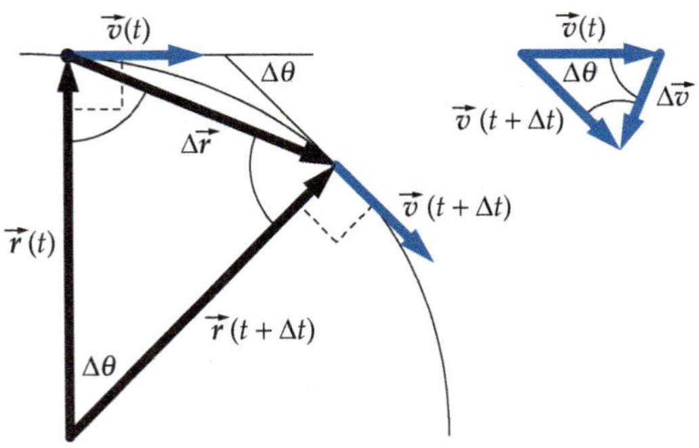

FIGURE 3-24

and $\vec{v}$ both rotate through the same angle $\Delta\theta$ during time Δt. An isosceles triangle is formed by the two velocity vectors and the $\Delta\vec{v}$ vector, and a second isosceles triangle is formed by the two position vectors and the $\Delta\vec{r}$ vector.

To find the direction of the acceleration vector we examine the triangle formed by the two velocity vectors and the $\Delta\vec{v}$ vector. The sum of the angles of any triangle is 180° and the base angles of any isosceles triangle are equal. In the limit that Δt approaches zero, $\Delta\theta$ also approaches zero, so in this limit the two base angles each approach 90°. This means that in the limit that $\Delta t \rightarrow 0$, $\Delta\vec{v}$ is perpendicular to $\vec{v}$. If $\Delta\vec{v}$ is drawn from the position of the particle, then it points in the centripetal direction.

The two triangles are similar, and corresponding lengths of similar geometric figures are proportional. Thus,

$$\frac{|\Delta\vec{v}|}{|\Delta\vec{r}|} = \frac{v}{r}$$

Multiplying both sides by $|\Delta\vec{r}|\,/\Delta t$ gives

$$\frac{|\Delta\vec{v}|}{\Delta t} = \frac{v}{r}\frac{|\Delta\vec{r}|}{\Delta t} \qquad \text{3-19}$$

In the limit that $\Delta t \rightarrow 0$, $|\Delta\vec{v}|/\Delta t$ approaches a, the magnitude of the instantaneous acceleration, and $|\Delta\vec{r}|/\Delta t$ approaches v (the speed). Thus, in the limit that $\Delta t \rightarrow 0$, Equation 3-19 becomes $a = v^2/r$. The acceleration vector is in the centripetal direction, so $a_c = a$, where a_c is the component of the acceleration vector in the centripetal direction. Substituting a_c for a, we have

$$a_c = \frac{v^2}{r} \qquad \text{3-20}$$

CENTRIPETAL ACCELERATION

Centripetal acceleration is the component of the acceleration vector in the centripetal direction. The motion of a particle moving in a circle with constant speed is often described in terms of the time T required for one complete revolution, called the **period.** During one period, the particle travels a distance of $2\pi r$ (where r is the radius of the circle), so its speed v is related to r and T by

$$v = \frac{2\pi r}{T} \qquad \text{3-21}$$

Example 3-12 A Satellite's Motion

A satellite moves at constant speed in a circular orbit about the center of Earth and near the surface of Earth. If the magnitude of its acceleration is $g = 9.81\ \text{m/s}^2$, find (*a*) its speed, and (*b*) the time for one complete revolution.

PICTURE Because the satellite orbits near the surface of Earth, we take the radius of the orbit to be 6370 km, the radius of Earth. Then, we can use Equations 3-20 and 3-21 to find the satellite's speed and the time for the satellite to make one complete revolution around Earth.

SOLVE

(*a*) Make a sketch of the satellite orbiting Earth in a low-Earth orbit (Figure 3-25). Include the velocity and acceleration vectors:

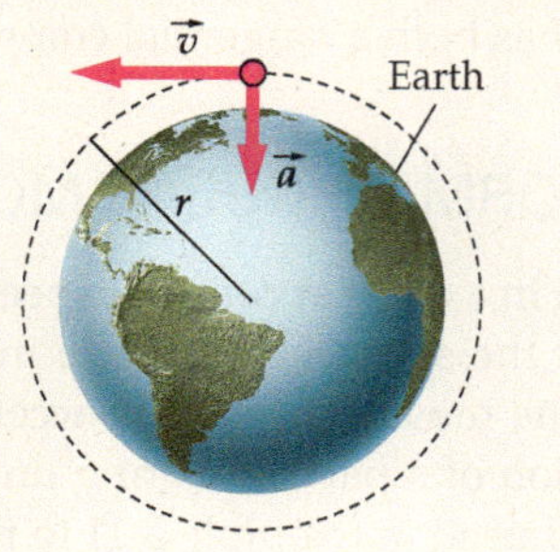

FIGURE 3-25 A satellite in a circular, low-Earth orbit.

Set the centripetal acceleration v^2/r equal to g and solve for the speed v:

$$a_c = \frac{v^2}{r} = g, \quad \text{so}$$

$$v = \sqrt{rg} = \sqrt{(6370\text{ km})(9.81\text{ m/s}^2)}$$

$$= \boxed{7.91\text{ km/s} = 17{,}700\text{ mi/h} = 4.91\text{ mi/s}}$$

(*b*) Use Equation 3-21 to get the period T:

$$T = \frac{2\pi r}{v} = \frac{2\pi(6370\text{ km})}{7.91\text{ km/s}} = \boxed{5060\text{ s} = 84.3\text{ min}}$$

CHECK It is well known that the orbital period for satellites orbiting just above the atmosphere of Earth is about 90 min, so the Part-(*b*) result of 84.3 min is close to what we would expect.

TAKING IT FURTHER For actual satellites in orbit a few hundred kilometers above Earth's surface, the orbital radius r is a few hundred kilometers greater than 6370 km. As a result, the acceleration is slightly less than 9.81 m/s² because of the decrease in the gravitational force with distance from the center of Earth. Orbits just above Earth's atmosphere are referred to as "low Earth orbits." Many satellites, including the Hubble telescope and the International Space Station, are in low Earth orbits. Information about these and other satellites can be found at www.nasa.gov.

PRACTICE PROBLEM 3-5 A car rounds a curve of radius 40 m at 48 km/h. What is its centripetal acceleration?

TANGENTIAL ACCELERATION

A particle moving in a circle with *varying speed* has a component of acceleration tangent to the circle, a_t, as well as the radially inward centripetal acceleration, v^2/r. For general motion along a curve, we can treat a portion of the curve as an arc of a circle (Figure 3-26). The particle then has centripetal acceleration $a_c = v^2/r$ toward the center of curvature, and if the speed v is changing, it has a tangential acceleration given by

$$a_t = \frac{dv}{dt} \qquad 3\text{-}22$$

TANGENTIAL ACCELERATION

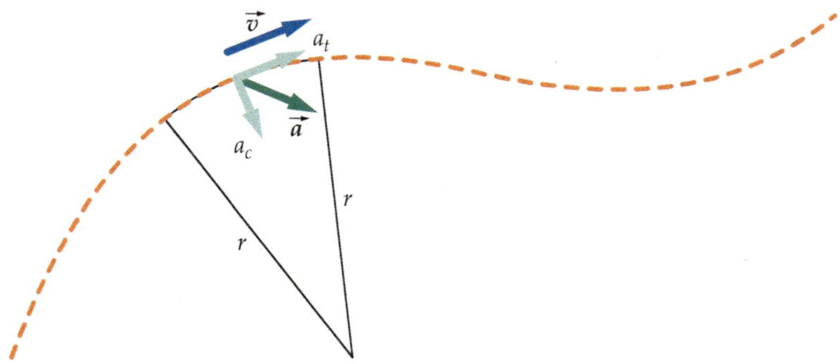

FIGURE 3-26 The tangential and centripetal acceleration components of a particle moving along a curved path.

PRACTICE PROBLEM 3-6

You are in a cart on a rollercoaster track that is entering a loop-the-loop. At the instant you are one fourth the way through the loop-the-loop your cart is going straight up at 20 m/s, and is losing speed at 5.0 m/s². The radius of curvature of the track is 25 m. What are your centripetal and tangential acceleration components at that instant?

Physics Spotlight

GPS: Vectors Calculated While You Move

The navigation systems used in automobiles obtain information from GPS satellites and use the information to calculate the position and velocity of the car. At times, they calculate the car's displacement vector using a procedure called *dead reckoning*. *(Andrew Fox/Corbis.)*

If you fly to another city and rent a car to reach your destination, you might rent a Global Positioning System (GPS) navigation computer with the car. Many people use GPS navigators, but not everyone knows that these computers constantly calculate vectors for you.

Twenty-four GPS satellites orbit Earth at an altitude of 11,000 mi.* At most times and places, at least three satellites are visible (above the horizon). In many cases, four or more satellites are visible. Each satellite broadcasts a continual stream that includes its identification, information about its orbit, and a time marker that is precise to one billionth of a second.† The satellite's internal clocks and orbits are checked by a ground station that can send correction information.

A GPS receiver listens for the signals from the satellites. When it can get a lock on three or more satellite signals, it calculates how far away each satellite is by the difference between the satellite's time marker and the time on the receiver's clock when the marker is detected. From the known orbits of each satellite and the distance to each satellite, the receiver can triangulate its position. A calculation from three satellites will give the longitude and latitude of the receiver. A calculation from four satellites will also give altitude.

But where do vectors come in? The receiver does not just triangulate its position once—that would give a point position. The receiver constantly listens for the satellites and calculates changes in the receiver position from changes in the triangulation results. It calculates any changes in distance and direction from the last known position. Within a very short time it has taken several readings, enough to calculate the velocity of your travel. The result? A speed in a particular direction—a velocity vector—is always part of the receiver's calculations.

But this vector is not there just to draw a pretty line on the screen for you. There are times when it is not possible to get a good reading on a receiver. Perhaps you have driven under a bridge or through a tunnel. If the GPS receiver is unable to lock onto a meaningful signal, it will start from your last known position. It will then use your last known velocity and direction to calculate a *dead reckoning*. It will assume that you are continuing in the same direction and at the same speed until it is able to get a reliable signal from enough satellites. Once it is able to receive good signals again, it will make corrections to your position and your course.

When GPS was pioneered, the time signal broadcast by the satellites was encoded with distortion, or *selective availability*, which could be removed only with decoding receivers enabled for defense purposes. The military could track location to within six meters, while civilians could track location to within only around 45 meters.** That coding was turned off in the year 2000. Theoretically, a GPS receiver would be able to tell your position down to the width of your finger,‡ given the right signals, and give equally precise and accurate measurements of your speed and direction—all from at least 11,000 miles away.

---

* The actual number of operational satellites varies. It is more than twenty-four, in case of satellite failure. "Block II Satellite Information." *ftp://tycho.usno.navy.mil/pub/gps/gpsb2.txt* United States Naval Observatory. March 2006.

† "GPS: The Role of Atomic Clocks—It Started with Basic Research." *http://www.beyonddiscovery.org/content/view.page.asp?I=464* Beyond Discovery. The National Academy of Sciences. March 2006.

** "Comparison of Positions With and Without Selective Availability: Full 24 Hour Data Sets." *http://www.ngs.noaa.gov/FGCS/info/sans_SA/compare/ERLA.htm* National Geodetic Survey. March, 2006.

‡ "Differential GPS: Advanced Concepts." *http://www.trimble.com/gps/advanced1.html* Trimble. March, 2006.

Summary

TOPIC	RELEVANT EQUATIONS AND REMARKS	
1. Kinematic Vectors		
Position vector	The position vector $\vec{r}$ points from the origin of the coordinate system to the particle.	
Instantaneous-velocity vector	The velocity vector $\vec{v}$ is the rate of change of the position vector. Its magnitude is the speed, and it points in the direction of motion. $$\vec{v} = \lim_{\Delta t \to 0} \frac{\Delta \vec{r}}{\Delta t} = \frac{d\vec{r}}{dt}$$	3-5
Instantaneous-acceleration vector	$$\vec{a} = \lim_{\Delta t \to 0} \frac{\Delta \vec{v}}{\Delta t} = \frac{d\vec{v}}{dt}$$	3-11
2. Relative Velocity	If a particle p moves with velocity $\vec{v}_{pA}$ relative to reference frame A, which is in turn moving with velocity $\vec{v}_{AB}$ relative to reference frame B, the velocity of p relative to B is $$\vec{v}_{pB} = \vec{v}_{pA} + \vec{v}_{AB}$$	3-9
3. Projectile Motion with No Air Resistance	The $+x$ direction is horizontal and the $+y$ direction is upward for the equations in this section.	
Independence of motion	In projectile motion, the horizontal and vertical motions are independent.	
Acceleration	$$a_x = 0 \quad \text{and} \quad a_y = -g$$	
Dependence on time	$$v_x(t) = v_{0x} + a_x t \quad \text{and} \quad v_y(t) = v_{0y} + a_y t$$ (2-12) $$\Delta x = v_{0x}t + \tfrac{1}{2}a_x t^2 \quad \text{and} \quad \Delta y = v_{0y}t + \tfrac{1}{2}a_y t^2$$ (2-14) where $v_{x0} = v_0 \cos\theta_0$ and $v_{y0} = v_0 \sin\theta_0$. Alternatively, $$\Delta\vec{v} = \vec{g}t \quad \text{and} \quad \Delta\vec{r} = \vec{v}_0 t + \tfrac{1}{2}\vec{g}t^2$$	2-12, 2-14, 3-15*c*, 3-16*c*
Horizontal displacement	The horizontal displacement is found by multiplying v_{0x} by the total time the projectile is in the air.	
4. Circular Motion		
Centripetal acceleration	$$a_c = \frac{v^2}{r}$$	3-20
Tangential acceleration	$$a_t = \frac{dv}{dt}$$ where v is the *speed*.	3-22
Period	$$v = \frac{2\pi r}{T}$$	3-21

Answers to Concept Checks

3-1 (*a*) upward, (*b*) upward

3-2 vertically downward

Answers to Practice Problems

3-1 $a_{av} = 4.7\ \text{m/s}^2$ at 45° west of north

3-2
$$\begin{aligned} y(t) &= y_0 + v_{0y}t - \tfrac{1}{2}gt^2 \\ &= 0 + (25.0\ \text{m/s})\sin 36.9°(1.43\ \text{s}) \\ &\quad + \tfrac{1}{2}(9.81\ \text{m/s}^2)(1.43\ \text{s})^2 \\ &= 11.48\ \text{m} \end{aligned}$$
$\therefore h = 111\ \text{m}$

3-3 $R = \dfrac{v_0^2}{g}\sin 2\theta_0 = \dfrac{(24.5\ \text{m/s})^2}{9.81\ \text{m/s}^2}\sin(2\times 36.9°) = 58.8°$

3-4 (*a*) $\vec{v}_0 = 20.0\ \text{m/s}$ at $\theta_0 = 22.0°$, (*b*) $t = 0.764$ s, (*c*) $v_{y\,av}t = 2.86$ m

3-5 $4.44\ \text{m/s}^2$

3-6 $a_c = 16\ \text{m/s}^2$ and $a_t = -5.0\ \text{m/s}^2$

Problems

In a few problems, you are given more data than you actually need; in a few other problems, you are required to supply data from your general knowledge, outside sources, or informed estimate.

Interpret as significant all digits in numerical values that have trailing zeros and no decimal points.

- • Single-concept, single-step, relatively easy
- •• Intermediate-level, may require synthesis of concepts
- ••• Challenging
- SSM Solution is in the *Student Solutions Manual*
- Consecutive problems that are shaded are paired problems.

CONCEPTUAL PROBLEMS

1 • Can the magnitude of the displacement of a particle be less than the distance traveled by the particle along its path? Can its magnitude be more than the distance traveled? Explain. SSM

2 • Give an example in which the distance traveled is a significant amount, yet the corresponding displacement is zero. Can the reverse be true? If so, give an example.

3 • What is the average velocity of a batter who hits a home run (from when he hits the ball to when he touches home plate after rounding the bases)?

4 • A baseball is hit so its initial velocity upon leaving the bat makes an angle of 30° above the horizontal. It leaves that bat at a height of 1.0 m above the ground and lands untouched for a single. During its flight, from just after it leaves the bat to just before it hits the ground, describe how the angle between its velocity and acceleration vectors changes. Neglect any effects due to air resistance.

5 • If an object is moving toward the west at some instant, in what direction is its acceleration? (*a*) north, (*b*) east, (*c*) west, (*d*) south, (*e*) may be any direction

6 • Two astronauts are working on the lunar surface to install a new telescope. The acceleration due to gravity on the moon is only 1.64 m/s². One astronaut tosses a wrench to the other astronaut, but the speed of throw is excessive and the wrench goes over her colleague's head. When the wrench is at the highest point of its trajectory, (*a*) its velocity and acceleration are both zero, (*b*) its velocity is zero but its acceleration is nonzero, (*c*) its velocity is nonzero but its acceleration is zero, (*d*) its velocity and acceleration are both nonzero, (*e*) insufficient information is given to choose between any of the previous choices.

7 • The velocity of a particle is directed toward the east while the acceleration is directed toward the northwest, as shown in Figure 3-27. The particle is (*a*) speeding up and turning toward the north, (*b*) speeding up and turning toward the south, (*c*) slowing down and turning toward the north, (*d*) slowing down and turning toward the south, (*e*) maintaining constant speed and turning toward the south.

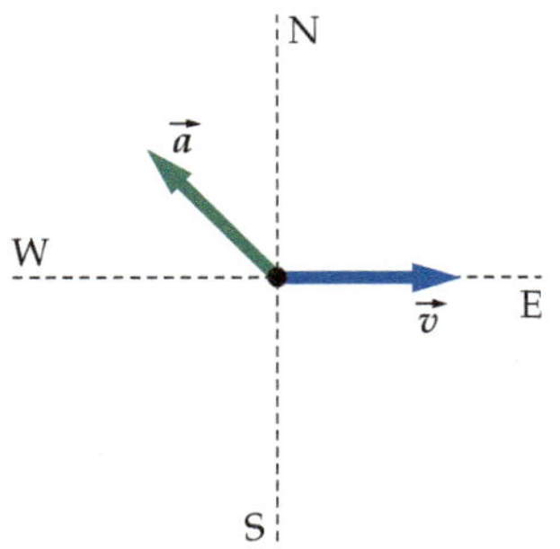

FIGURE 3-27
Problem 7

8 • Assume you know the position vectors of a particle at two points on its path, one earlier and one later. You also know the time it took the particle to move from one point to the other. Then, you can compute the particle's (*a*) average velocity, (*b*) average acceleration, (*c*) instantaneous velocity, and (*d*) instantaneous acceleration.

9 • Consider the path of a moving particle. (*a*) How is the velocity vector related geometrically to the path of the particle? (*b*) Sketch a curved path and draw the velocity vector for the particle for several positions along the path.

10 • The acceleration of a car is zero when it is (*a*) turning right at a constant speed, (*b*) driving up a long straight incline at constant speed, (*c*) traveling over the crest of a hill at constant speed, (*d*) bottoming out at the lowest point of a valley at constant speed, (*e*) speeding up as it descends a long straight decline.

11 • Give examples of motion in which the directions of the velocity and acceleration vectors are (*a*) opposite, (*b*) the same, and (*c*) mutually perpendicular. SSM

12 • How is it possible for a particle moving at constant speed to be accelerating? Can a particle with constant velocity be accelerating at the same time?

13 •• Imagine throwing a dart straight upward so that it sticks into the ceiling. After it leaves your hand, it steadily slows down as it rises before it sticks. (*a*) Draw the dart's velocity vector at times t_1 and t_2, where t_1 and t_2 occur after it leaves your hand but before it hits the ceiling and $t_2 - t_1$ is small. From your drawing, find the direction of the change in velocity $\Delta\vec{v} = \vec{v}_2 - \vec{v}_1$, and thus the direction of the acceleration vector. (*b*) After it has stuck in the ceiling for a few seconds, the dart falls down to the floor. As it falls it speeds up, of course, until it hits the floor. Repeat Part (*a*) to find the direction of its acceleration vector as it falls. (*c*) Now imagine tossing the dart horizontally. What is the direction of its acceleration vector after it leaves your hand, but before it strikes the floor? SSM

14 • As a bungee jumper approaches the lowest point in her descent, the rubber cord holding her stretches and she loses speed as she continues to move downward. Assuming that she is dropping straight down, make a motion diagram to find the direction of her acceleration vector as she slows down by drawing her velocity vectors at times t_1 and t_2, where t_1 and t_2 are two instants during the portion of her descent in which she is losing speed and $t_2 - t_1$ is small. From your drawing find the direction of the change in velocity $\Delta\vec{v} = \vec{v}_2 - \vec{v}_1$, and thus the direction of the acceleration vector.

15 • After reaching the lowest point in her jump at time t_{low}, a bungee jumper moves upward, gaining speed for a short time until gravity again dominates her motion. Draw her velocity vectors at times t_1 and t_2, where $t_2 - t_1$ is small and $t_1 < t_{low} < t_2$. From your drawing find the direction of the change in velocity $\Delta\vec{v} = \vec{v}_2 - \vec{v}_1$, and thus the direction of the acceleration vector at time t_{low}.

16 • A river is 0.76 km wide. The banks are straight and parallel (Figure 3-28). The current is 4.0 km/h and is parallel to the banks. A boat has a maximum speed of 4.0 km/h in still water. The pilot of the boat

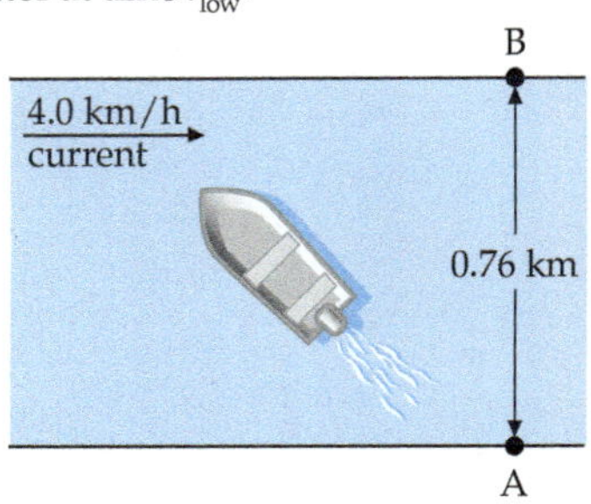

FIGURE 3-28
Problem 16

wishes to go on a straight line from A to B, where the line AB is perpendicular to the banks. The pilot should (*a*) head directly across the river, (*b*) head 53° upstream from the line AB, (*c*) head 37° upstream from the line AB, (*d*) give up—the trip from A to B is not possible with a boat of this limited speed, (*e*) do none of the above.

17 • During a heavy rain, the drops are falling at a constant velocity and at an angle of 10° west of the vertical. You are walking in the rain and notice that only the top surfaces of your clothes are getting wet. In what direction are you walking? Explain. SSM

18 • In Problem 17, what is your walking speed if the speed of the drops relative to the ground is 5.2 m/s?

19 • True or false (ignore any effects due to air resistance):

(*a*) When a projectile is fired horizontally, it takes the same amount of time to reach the ground as an identical projectile dropped from rest from the same height.
(*b*) When a projectile is fired from a certain height at an upward angle, it takes longer to reach the ground than does an identical projectile dropped from rest from the same height.
(*c*) When a projectile is fired horizontally from a certain height, it has a higher speed upon reaching the ground than does an identical projectile dropped from rest from the same height.

20 • A projectile is fired at 35° above the horizontal. Any effects due to air resistance are negligible. At the highest point in its trajectory, its speed is 20 m/s. The initial velocity had a horizontal component of (*a*) 0, (*b*) (20 m/s) cos 35°, (*c*) (20 m/s) sin 35°, (*d*) (20 m/s)/cos 35°, (*e*) 20 m/s.

21 • A projectile is fired at 35° above the horizontal. Any effects due to air resistance are negligible. The initial velocity of the projectile in Problem 20 has a vertical component that is (*a*) less than 20 m/s, (*b*) greater than 20 m/s, (*c*) equal to 20 m/s, (*d*) cannot be determined from the data given. SSM

22 • A projectile is fired at 35° above the horizontal. Any effects due to air resistance are negligible. The projectile lands at the same elevation of launch, so the vertical component of the impact velocity of the projectile is (*a*) the same as the vertical component of its initial velocity in both magnitude and sign, (*b*) the same as the vertical component of its initial velocity in magnitude but opposite in sign, (*c*) less than the vertical component of its initial velocity in magnitude but with the same sign, (*d*) less than the vertical component of its initial velocity in magnitude but with the opposite sign.

23 • Figure 3-29 represents the parabolic trajectory of a projectile going from A to E. Air resistance is negligible. What is the direction of the acceleration at point B? (*a*) up and to the right, (*b*) down and to the left, (*c*) straight up, (*d*) straight down, (*e*) the acceleration of the ball is zero

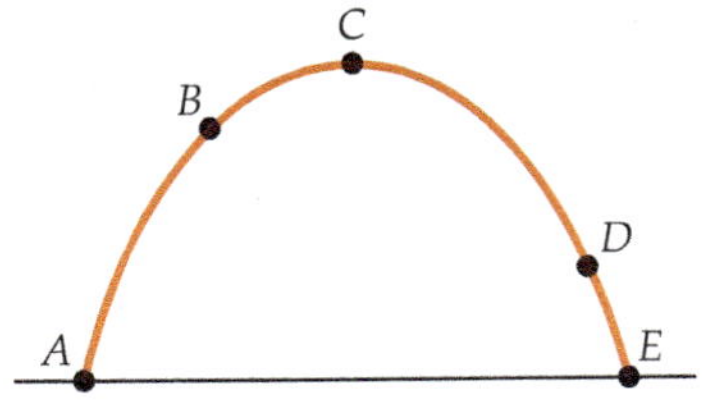

FIGURE 3-29 Problems 23 and 24

24 • Figure 3-29 represents the trajectory of a projectile going from *A* to *E*. Air resistance is negligible. (*a*) At which point(s) is the speed the greatest? (*b*) At which point(s) is the speed the least? (*c*) At which two points is the speed the same? Is the velocity also the same at these points?

25 • True or false:

(*a*) If an object's speed is constant, then its acceleration must be zero.
(*b*) If an object's acceleration is zero, then its speed must be constant.
(*c*) If an object's acceleration is zero, its velocity must be constant.
(*d*) If an object's speed is constant, then its velocity must be constant.
(*e*) If an object's velocity is constant, then its speed must be constant. SSM

26 • The initial and final velocities of a particle are as shown in Figure 3-30. Find the direction of the average acceleration.

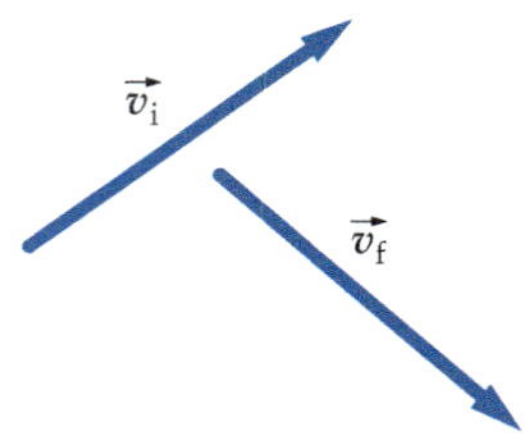

FIGURE 3-30 Problem 26

27 •• The automobile path shown in Figure 3-31 is made up of straight lines and arcs of circles. The automobile starts from rest at point A. After it reaches point B, it travels at constant speed until it reaches point E. It comes to rest at point F. (*a*) At the middle of each segment (AB, BC, CD, DE, and EF), what is the direction of the velocity vector? (*b*) At which of these points does the automobile have a nonzero acceleration? In those cases, what is the direction of the acceleration? (*c*) How do the magnitudes of the acceleration compare for segments BC and DE?

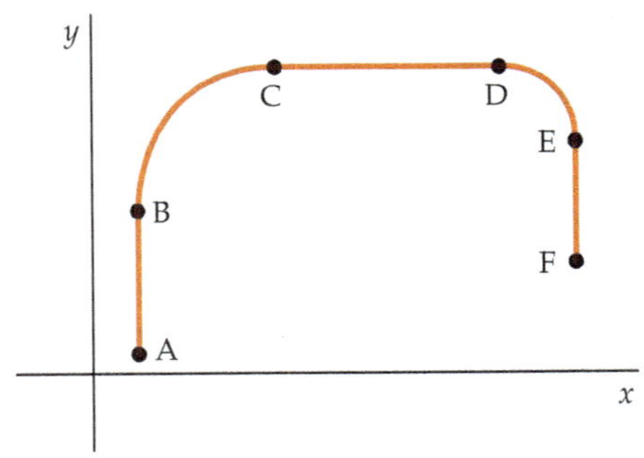

FIGURE 3-31 Problem 27

28 •• Two cannons are pointed directly toward each other, as shown in Figure 3-32. When fired, the cannonballs will follow the trajectories shown—P is the point where the trajectories cross each other. If we want the cannonballs to hit each other, should the gun crews fire cannon A first, cannon B first, or should they fire simultaneously? Ignore any effects due to air resistance.

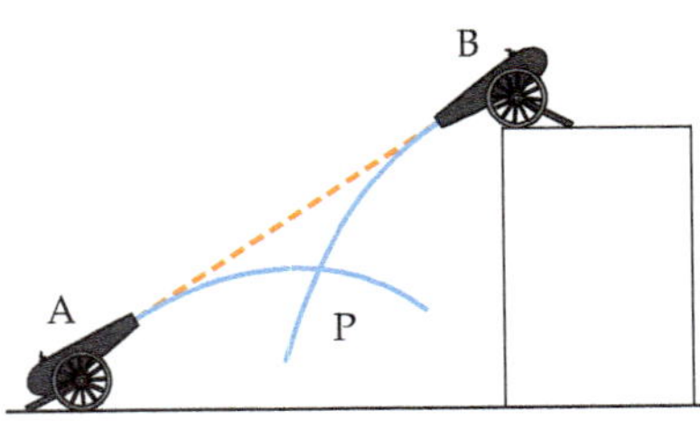

FIGURE 3-32 Problem 28

29 •• Galileo wrote the following in his *Dialogue concerning the two world systems:* "Shut yourself up . . . in the main cabin below decks on some large ship, and . . . hang up a bottle that empties drop by drop into a wide vessel beneath it. When you have observed [this] carefully . . . have the ship proceed with any speed you like, so long as the motion is uniform and not fluctuating this way and that. . . . The droplets will fall as before into the vessel beneath without dropping towards the stern, although while the drops are in the air the ship runs many spans." Explain this quotation.

30 •• A man swings a stone attached to a rope in a horizontal circle at constant speed. Figure 3-33 represents the path of the rock looking down from above. (*a*) Which of the vectors could represent the velocity of the stone? (*b*) Which could represent the acceleration?

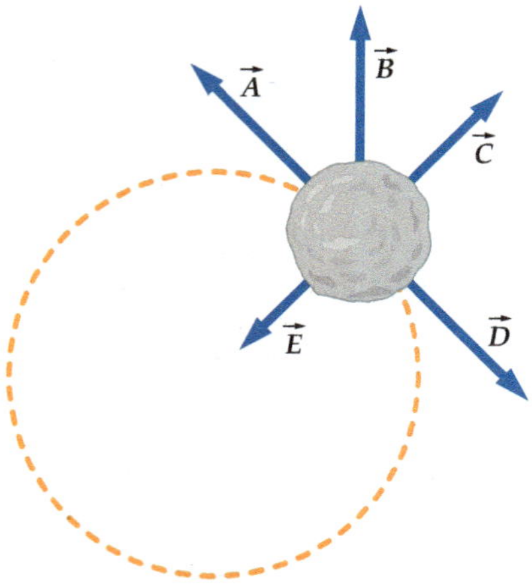

FIGURE 3-33 Problem 30

31 •• True or false:

(*a*) An object cannot move in a circle unless it has centripetal acceleration.
(*b*) An object cannot move in a circle unless it has tangential acceleration.
(*c*) An object moving in a circle cannot have a variable speed.
(*d*) An object moving in a circle cannot have a constant velocity.

32 •• Using a motion diagram, find the direction of the acceleration of the bob of a pendulum when the bob is at a point where it is just reversing its direction.

33 •• **CONTEXT-RICH** During your rookie bungee jump, your friend records your fall using a camcorder. By analyzing it frame by frame, he finds that the y component of your velocity is (recorded every 1/20 of a second) as follows:

t (s)	12.05	12.10	12.15	12.20	12.25	12.30	12.35	12.40	12.45
v_y (m/s)	−0.78	−0.69	−0.55	−0.35	−0.10	0.15	0.35	0.49	0.53

(*a*) Draw a motion diagram. Use it to find the direction and relative magnitude of your average acceleration for each of the eight successive 0.050-s time intervals in the table. (*b*) Comment on how the y component of your acceleration does or does not vary in sign and magnitude as you reverse your direction of motion. SSM

ESTIMATION AND APPROXIMATION

34 •• **CONTEXT-RICH** Estimate the speed in mph with which water comes out of a garden hose using your past observations of water coming out of garden hoses and your knowledge of projectile motion.

35 •• **CONTEXT-RICH** You won a contest to spend a day with a baseball team during spring training. You are allowed to try to hit some balls thrown by a pitcher. Estimate the acceleration during the hit of a fastball thrown by a major league pitcher when you hit the ball squarely—straight back at the pitcher. You will need to make reasonable choices for ball speeds, both just before and just after the ball is hit, and of the contact time of the ball with the bat.

36 •• Estimate how far you can throw a ball if you throw it (*a*) horizontally while standing on level ground, (*b*) at $\theta = 45°$ above horizontal while standing on level ground, (*c*) horizontally from the top of a building 12 m high, (*d*) at $\theta = 45°$ above horizontal from the top of a building 12 m high. Ignore any effects due to air resistance.

37 •• In 1978, Geoff Capes of Great Britain threw a heavy brick a horizontal distance of 44.5 m. Find the approximate speed of the brick at the highest point of its flight, neglecting any effects due to air resistance. Assume the brick landed at the same height it was launched.

POSITION, DISPLACEMENT, VELOCITY, AND ACCELERATION VECTORS

38 • A wall clock has a minute hand with a length of 0.50 m and an hour hand with a length of 0.25 m. Take the center of the clock as the origin, and use a Cartesian coordinate system with the positive x axis pointing to 3 o'clock and the positive y axis pointing to 12 o'clock. Using unit vectors $\hat{i}$ and $\hat{j}$, express the position vectors of the tip of the hour hand ($\vec{A}$) and the tip of the minute hand ($\vec{B}$) when the clock reads (*a*) 12:00, (*b*) 3:00, (*c*) 6:00, (*d*) 9:00.

39 • In Problem 38, find the displacements of the tip of each hand (that is, $\Delta\vec{A}$ and $\Delta\vec{B}$) when the time advances from 3:00 p.m. to 6:00 p.m. SSM

40 • In Problem 38, write the vector that describes the displacement of a fly if it quickly goes from the tip of the minute hand to the tip of the hour hand at 3:00 p.m.

41 • **CONCEPTUAL, APPROXIMATION** A bear, awakening from winter hibernation, staggers directly northeast for 12 m and then due east for 12 m. Show each displacement graphically and graphically determine the single displacement that will take the bear back to her cave to continue her hibernation.

42 • A scout walks 2.4 km due east from camp, then turns left and walks 2.4 km along the arc of a circle centered at the campsite, and finally walks 1.5 km directly toward the camp. (*a*) How far is the scout from camp at the end of his walk? (*b*) In what direction is the scout's position relative to the campsite? (*c*) What is the ratio of the final magnitude of the displacement to the total distance walked.

43 • The faces of a cubical storage cabinet in your garage have 3.0-m-long edges that are parallel to the xyz coordinate planes. The cube has one corner at the origin. A cockroach, on the hunt for crumbs of food, begins at that corner and walks along three edges until it is at the far corner. (*a*) Write the roach's displacement using the set of $\hat{i}$, $\hat{j}$, and $\hat{k}$ unit vectors, and (*b*) find the magnitude of its displacement. SSM

44 • **CONTEXT-RICH** You are the navigator of a ship at sea. You receive radio signals from two transmitters A and B, which are 100 km apart, one due south of the other. The direction finder shows you that transmitter A is at a heading of 30° south of east from the ship, while transmitter B is due east. Calculate the distance between your ship and transmitter B.

45 • A stationary radar operator determines that a ship is 10 km due south of him. An hour later the same ship is 20 km due southeast. If the ship moved at constant speed and always in the same direction, what was its velocity during this time?

46 • A particle's position coordinates (x, y) are (2.0 m, 3.0 m) at $t = 0$; (6.0 m, 7.0 m) at $t = 2.0$ s; and (13 m, 14 m) at $t = 5.0$ s. (*a*) Find the magnitude of the average velocity from $t = 0$ to $t = 2.0$ s. (*b*) Find the magnitude of the average velocity from $t = 0$ to $t = 5.0$ s.

47 • A particle moving at a velocity of 4.0 m/s in the $+x$ direction is given an acceleration of 3.0 m/s^2 in the $+y$ direction for 2.0 s. Find the final speed of the particle. SSM

48 • Initially, a swift-moving hawk is moving due west with a speed of 30 m/s; 5.0 s later it is moving due north with a speed of 20 m/s. (*a*) What are the magnitude and direction of $\Delta\vec{v}_{av}$ during this 5.0-s interval? (*b*) What are the magnitude and direction of $\vec{a}_{av}$ during this 5.0-s interval?

49 • At $t = 0$, a particle located at the origin has a velocity of 40 m/s at $\theta = 45°$. At $t = 3.0$ s, the particle is at $x = 100$ m and $y = 80$ m and has a velocity of 30 m/s at $\theta = 50°$. Calculate (*a*) the average velocity, and (*b*) the average acceleration of the particle during this 3.0-s interval.

50 •• At time zero, a particle is at $x = 4.0$ m and $y = 3.0$ m and has velocity $\vec{v} = (2.0\text{ m/s})\hat{i} + (-9.0\text{ m/s})\hat{j}$. The acceleration of the particle is constant and is given by $\vec{a} = (4.0\text{ m/s}^2)\hat{i} + (3.0\text{ m/s}^2)\hat{j}$. (*a*) Find the velocity at $t = 2.0$ s. (*b*) Express the position vector at $t = 4.0$ s in terms of $\hat{i}$ and $\hat{j}$. In addition, give the magnitude and direction of the position vector at this time.

51 •• A particle has a position vector given by $\vec{r} = (30t)\hat{i} + (40t - 5t^2)\hat{j}$, where r is in meters and t is in seconds. Find the instantaneous-velocity and instantaneous-acceleration vectors as functions of time t. SSM

52 •• A particle has a constant acceleration of $\vec{a} = (6.0\text{ m/s}^2)\hat{i} + (4.0\text{ m/s}^2)\hat{j}$. At time $t = 0$, the velocity is zero and the position vector is $\vec{r}_0 = (10\text{ m})\hat{i}$. (*a*) Find the velocity and position vectors as functions of time t. (*b*) Find the equation of the particle's path in the xy plane and sketch the path.

53 •• Starting from rest at a dock, a motor boat on a lake heads north while gaining speed at a constant 3.0 m/s^2 for 20 s. The boat then heads west and continues for 10 s at the speed that it had at 20 s. (*a*) What is the average velocity of the boat during the 30-s trip? (*b*) What is the average acceleration of the boat during the 30-s trip? (*c*) What is the displacement of the boat during the 30-s trip? SSM

54 •• Starting from rest at point A, you ride your motorcycle north to point B 75.0 m away, increasing speed at a steady rate of 2.00 m/s^2. You then gradually turn toward the east along a circular path of radius 50.0 m at constant speed from B to point C, until your direction of motion is due east at C. You then continue eastward, slowing at a steady rate of 1.00 m/s^2 until you come to rest at point D. (*a*) What is your average velocity and acceleration for the trip from A to D? (*b*) What is your displacement during your trip from A to C? (*c*) What distance did you travel for the entire trip from A to D?

RELATIVE VELOCITY

55 •• A plane flies at an airspeed of 250 km/h. A wind is blowing at 80 km/h toward the direction 60° east of north. (*a*) In what direction should the plane head in order to fly due north relative to the ground? (*b*) What is the speed of the plane relative to the ground?

56 •• A swimmer *heads* directly across a river, swimming at 1.6 m/s relative to the water. She arrives at a point 40 m downstream from the point directly across the river, which is 80 m wide. (*a*) What is the speed of the river current? (*b*) What is the swimmer's speed relative to the shore? (*c*) In what direction should the swimmer head in order to arrive at the point directly opposite her starting point?

57 •• A small plane departs from point A heading for an airport 520 km due north at point B. The airspeed of the plane is 240 km/h and there is a steady wind of 50 km/h blowing directly toward the southeast. Determine the proper heading for the plane and the time of flight. SSM

58 •• Two boat landings are 2.0 km apart on the same bank of a stream that flows at 1.4 km/h. A motorboat makes the round trip between the two landings in 50 min. What is the speed of the boat relative to the water?

59 •• **ENGINEERING APPLICATION, CONTEXT-RICH** During a radio-controlled model-airplane competition, each plane must fly from the center of a 1.0-km-radius circle to any point on the circle and back to the center. The winner is the plane that has the shortest round-trip time. The contestants are free to fly their planes along any route as long as the plane begins at the center, travels to the circle, and then returns to the center. On the day of the race, a steady wind blows out of the north at 5.0 m/s. Your plane can maintain an air speed of 15 m/s. Should you fly your plane upwind on the first leg and downwind on the trip back, or across the wind, flying east and then west? Optimize your chances by calculating the round-trip time for both routes using your knowledge of vectors and relative velocities. With this prerace calculation, you can determine the best route and have a major advantage over the competition!

60 •• **CONTEXT-RICH** You are piloting a small plane that can maintain an air speed of 150 kt (knots, or nautical miles per hour) and you want to fly due north (azimuth = 000°) relative to the ground. (*a*) If a wind of 30 kt is blowing from the east (azimuth = 090°), calculate the heading (azimuth) you must ask your copilot to maintain. (*b*) At that heading, what will be your ground speed?

61 •• Car A is traveling east at 20 m/s toward an intersection. As car A crosses the intersection, car B starts from rest 40 m north of the intersection and moves south steadily gaining speed at 2.0 m/s^2. Six seconds after A crosses the intersection find (*a*) the position of B relative to A, (*b*) the velocity of B relative to A, (*c*) the acceleration of B relative to A. *Hint: Let the unit vectors* $\hat{i}$ *and* $\hat{j}$ *be toward the east and north, respectively, and express your answers using* $\hat{i}$ *and* $\hat{j}$. SSM

62 •• While walking between gates at an airport, you notice a child running along a moving walkway. Estimating that the child runs at a constant speed of 2.5 m/s relative to the surface of the walkway, you decide to try to determine the speed of the walkway itself. You watch the child run on the entire 21-m walkway in one direction, immediately turn around, and run back to his starting point. The entire trip takes a total elapsed time of 22 s. Given this information, what is the speed of the moving walkway relative to the airport terminal?

63 •• Ben and Jack are shopping in a department store. Ben leaves Jack at the bottom of the escalator and walks east at a speed of 2.4 m/s. Jack gets on the escalator, which is inclined at an angle of 37° above the horizontal, and travels eastward and upward at a speed of 2.0 m/s. (*a*) What is the velocity of Ben relative to Jack? (*b*) At what speed should Jack walk up the escalator so that he is always directly above Ben (until he reaches the top)? SSM

64 ••• A juggler traveling in a train on level track throws a ball straight up, relative to the train, with a speed of 4.90 m/s. The train has a velocity of 20.0 m/s due east. As observed by the juggler, (*a*) what is the ball's total time of flight, and (*b*) what is the displacement of the ball during its rise? According to a friend standing on the ground next to the tracks, (*c*) what is the ball's initial speed, (*d*) what is the angle of the launch, and (*e*) what is the displacement of the ball during its rise?

CIRCULAR MOTION AND CENTRIPETAL ACCELERATION

65 • What is the magnitude of the acceleration of the tip of the minute hand of the clock in Problem 38? Express it as a fraction of the magnitude of free-fall acceleration *g*.

66 • **CONTEXT-RICH** You are designing a centrifuge to spin at a rate of 15,000 rev/min. (*a*) Calculate the maximum centripetal acceleration that a test-tube sample held in the centrifuge arm 15 cm from the rotation axis must withstand. (*b*) It takes 1 min, 15 s for the centrifuge to spin up to its maximum rate of revolution from rest. Calculate the *magnitude* of the tangential acceleration of the centrifuge while it is spinning up, assuming that the tangential acceleration is constant.

67 ••• Earth rotates on its axis once every 24 hours, so that objects on its surface execute uniform circular motion about the axis with a period of 24 hours. Consider only the effect of this rotation on the person on the surface. (Ignore Earth's orbital motion about the Sun.) (*a*) What is the speed and what is the magnitude of the acceleration of a person standing on the equator? (Express the magnitude of this acceleration as a percentage of *g*.) (*b*) What is the direction of the acceleration vector? (*c*) What is the speed and what is the magnitude of the acceleration of a person standing on the surface at 35°N latitude? (*d*) What is the angle between the direction of the acceleration of the person at 35°N and the direction of the acceleration of the person at the equator if both persons are at the same longitude? SSM

68 •• Determine the acceleration of the moon toward Earth, using values for its mean distance and orbital period from the Terrestrial and Astronomical Data table in this book. Assume a circular orbit. Express the acceleration as a fraction of *g*.

69 •• (*a*) What are the period and speed of a person on a carousel if the person has an acceleration with a magnitude of 0.80 m/s^2 when she is standing 4.0 m from the axis? (*b*) What are her acceleration magnitude and speed if she then moves to a distance of 2.0 m from the carousel center and the carousel keeps rotating with the same period?

70 ••• Pulsars are neutron stars that emit X rays and other radiation in such a way that we on Earth receive pulses of radiation from the pulsars at regular intervals equal to the period that they rotate. Some of these pulsars rotate with periods as short as 1 ms! The Crab Pulsar, located inside the Crab Nebula in the constellation Orion, has a period currently of length 33.085 ms. It is estimated to have an equatorial radius of 15 km, an average radius for a neutron star. (*a*) What is the value of the centripetal acceleration of an object on the surface at the equator of the pulsar? (*b*) Many pulsars are observed to have periods that lengthen slightly with time, a phenomenon called "spin down." The rate of slowing of the Crab Pulsar is 3.5×10^{-13} s per second, which implies that if this rate remains constant, the Crab Pulsar will stop spinning in 9.5×10^{10} s (about 3000 years from today). What is the tangential acceleration of an object on the equator of this neutron star?

71 ••• **BIOLOGICAL APPLICATION** Human blood contains plasma, platelets, and blood cells. To separate the plasma from other components, centrifugation is used. Effective centrifugation requires subjecting blood to an acceleration of 2000*g* or more. In this situation, assume that blood is contained in test tubes that are 15 cm long and are full of blood. These tubes ride in the centrifuge tilted at an angle of 45.0° above the horizontal (Figure 3-34). (*a*) What is the distance of a sample of blood from the rotation axis of a centrifuge rotating at 3500 rpm, if it has an acceleration of 2000*g*? (*b*) If the blood at the center of the tubes revolves around the rotation axis at the radius calculated in Part (*a*), calculate the accelerations experienced by the blood at each end of the test tube. Express all accelerations as multiples of *g*.

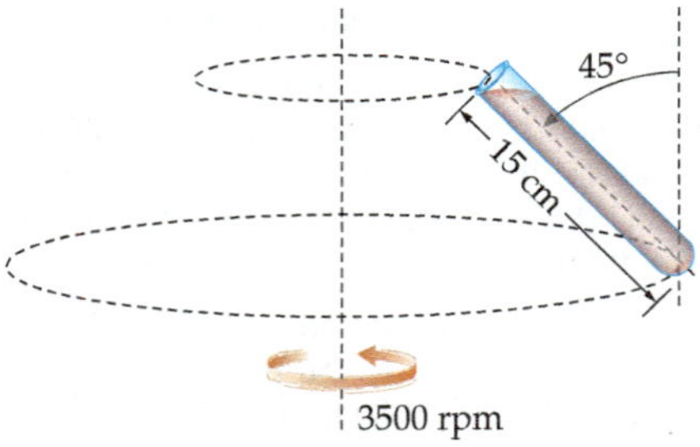

FIGURE 3-34 Problem 71

PROJECTILE MOTION AND PROJECTILE RANGE

72 • While trying out for the position of pitcher on your high school baseball team, you throw a fastball at 87 mi/h toward home plate, which is 18.4 m away. How far does the ball drop due to effects of gravity by the time it reaches home plate? (Ignore any effects due to air resistance.)

73 • A projectile is launched with speed v_0 at an angle of θ_0 above the horizontal. Find an expression for the maximum height it reaches above its starting point in terms of v_0, θ_0, and g. (Ignore any effects due to air resistance.)

74 •• A cannonball is fired with initial speed v_0 at an angle 30° above the horizontal from a height of 40 m above the ground. The projectile strikes the ground with a speed of $1.2v_0$. Find v_0. (Ignore any effects due to air resistance.)

75 •• In Figure 3-35, what is the minimum initial speed of the dart if it is to hit the monkey before the monkey hits the ground, which is 11.2 m below the initial position of the monkey, if $x = 50$ m and $h = 10$ m? (Ignore any effects due to air resistance.)

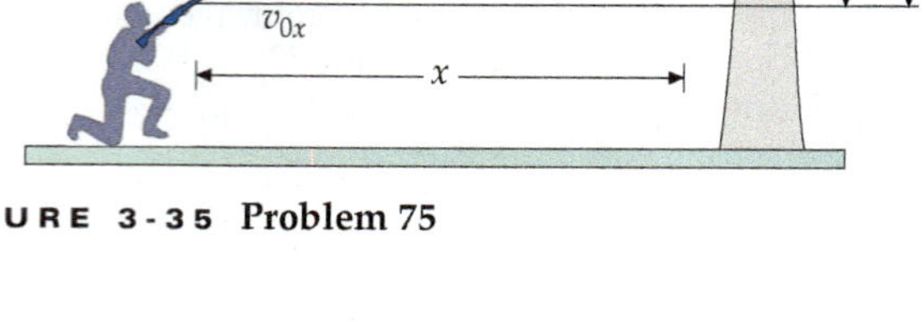

FIGURE 3-35 Problem 75

76 •• A projectile is launched from ground level with an initial speed of 53 m/s. Find the launch angle (the angle the initial velocity vector is above the horizontal) such that the maximum height of the projectile is equal to its horizontal range. (Ignore any effects due to air resistance.)

77 •• A ball launched from ground level lands 2.44 s later on a level field 40.0 m away from the launch point. Find the magnitude of the initial velocity vector and the angle it is above the horizontal. (Ignore any effects due to air resistance.) SSM

78 •• Consider a ball that is launched from ground level with initial speed v_0 at an angle θ_0 above the horizontal. If we consider its speed to be v at some height h above the ground, show that for a given value of h, v is independent of θ_0. (Ignore any effects due to air resistance.)

79 ••• At $\frac{1}{2}$ of its maximum height, the speed of a projectile is $\frac{3}{4}$ of its initial speed. What was its launch angle? (Ignore any effects due to air resistance.)

80 •• A cargo plane is flying horizontally at an altitude of 12 km with a speed of 900 km/h when a large crate falls out of the rear loading ramp. (Ignore any effects due to air resistance.) (*a*) How long does it take the crate to hit the ground? (*b*) How far horizontally is the crate from the point where it fell off when it hits the ground? (*c*) How far is the crate from the aircraft when the crate hits the ground, assuming that the plane continues to fly with the same velocity?

81 •• Wile E. Coyote (*Carnivorus hungribilus*) is chasing the Roadrunner (*Speedibus cantcatchmi*) yet again. While running down the road, they come to a deep gorge, 15.0 m straight across and 100 m deep. The Roadrunner launches himself across the gorge at a launch angle of 15° above the horizontal, and lands with 1.5 m to spare. (*a*) What was the Roadrunner's launch speed? (*b*) Wile E. Coyote launches himself across the gorge with the same initial speed, but at a different launch angle. To his horror, he is short of the other lip by 0.50 m. What was his launch angle? (Assume that it was less than 15°.) SSM

82 •• A cannon barrel is elevated 45° above the horizontal. It fires a ball with a speed of 300 m/s. (*a*) What height does the ball reach? (*b*) How long is the ball in the air? (*c*) What is the horizontal range of the cannon ball? (Ignore any effects due to air resistance.)

83 •• A stone thrown horizontally from the top of a 24-m tower hits the ground at a point 18 m from the base of the tower. (Ignore any effects due to air resistance.) (*a*) Find the speed with which the stone was thrown. (*b*) Find the speed of the stone just before it hits the ground. SSM

84 •• A projectile is fired into the air from the top of a 200-m cliff above a valley (Figure 3-36). Its initial velocity is 60 m/s at 60° above the horizontal. Where does the projectile land? (Ignore any effects due to air resistance.)

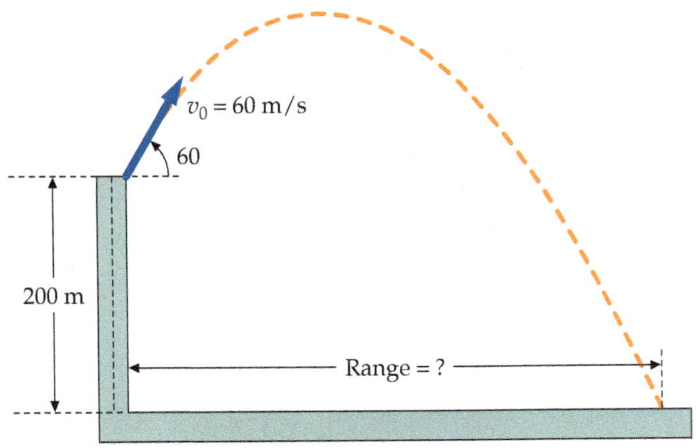

FIGURE 3-36 Problem 84

85 •• The range of a cannonball fired horizontally from a cliff is equal to the height of the cliff. What is the direction of the velocity vector of the projectile as it strikes the ground? (Ignore any effects due to air resistance.)

86 •• An archer fish launches a droplet of water from the surface of a small lake at an angle of 60° above the horizontal. He is aiming at a juicy spider sitting on a leaf 50 cm to the east and on a branch 25 cm above the water surface. The fish is trying to knock the spider into the water so that the fish can eat the spider. (*a*) What must the speed of the water droplet be for the fish to be successful? (*b*) When it hits the spider, is the droplet rising or falling?

87 •• **CONTEXT-RICH** You are trying out for the position of place-kicker on a professional football team. With the ball teed 50.0 m from the goalposts with a crossbar 3.05 m off the ground, you kick the ball at 25.0 m/s and 30° above the horizontal. (*a*) Is the field goal attempt good? (*b*) If so, by how much does it clear the bar? If not, by how much does it go under the bar? (*c*) How far behind the plane of the goalposts does the ball land? SSM

88 •• The speed of an arrow fired from a compound bow is about 45.0 m/s. (*a*) A Tartar archer sits astride his horse and launches an arrow into the air, elevating the bow at an angle of 10° above the horizontal. If the arrow is 2.25 m above the ground at launch, what is the arrow's horizontal range? Assume that the ground is level, and ignore any effects due to air resistance. (*b*) Now assume that his horse is at full gallop, moving in the same direction that the archer will fire the arrow. Also assume that the archer elevates the bow at the same elevation angle as in Part (*a*) and fires. If the horse's speed is 12.0 m/s, what is the arrow's horizontal range now?

89 The roof of a two-story house makes an angle of 30° with the horizontal. A ball rolling down the roof rolls off the edge at a speed of 5.0 m/s. The distance to the ground from that point is 7.0 m. (*a*) How long is the ball in the air? (*b*) How far from the base of the house does it land? (*c*) What is its speed and direction just before landing?

90 •• Compute $dR/d\theta_0$ from $R = (v_0^2/g)\sin(2\theta_0)$ and show that setting $dR/d\theta_0 = 0$ gives $\theta_0 = 45°$ for the maximum range.

91 •• In a science fiction short story written in the 1970s, Ben Bova described a conflict between two hypothetical colonies on the moon—one founded by the United States and the other by the USSR. In the story, colonists from each side started firing bullets at each other, only to find to their horror that their rifles had large enough muzzle velocities so that the bullets went into orbit. (*a*) If the magnitude of free-fall acceleration on the moon is 1.67 m/s^2, what is the maximum range of a rifle bullet with a muzzle velocity of 900 m/s? (Assume the curvature of the surface of the moon is negligible.) (*b*) What would the muzzle velocity have to be to send the bullet into a circular orbit just above the surface of the moon?

92 ••• On a level surface, a ball is launched from ground level at an angle of 55° above the horizontal, with an initial speed of 22 m/s. It lands on a hard surface, and bounces, reaching a peak height of 75% of the height it reached on its first arc. (Ignore any effects due to air resistance.) (*a*) What is the maximum height reached in its first parabolic arc? (*b*) How far horizontally from the launch point did it strike the ground the first time? (*c*) How far horizontally from the launch point did the ball strike the ground the second time? Assume the horizontal component of the velocity remains constant during the collision of the ball with the ground. *Hint: You cannot assume that the angle with which the ball leaves the ground after the first collision is the same as the initial launch angle.*

93 ••• In the text, we calculated the range for a projectile that lands at the same elevation from which it is fired as $R = (v_0^2/g)\sin 2\theta_0$. A golf ball hit from an elevated tee at 45.0 m/s and an angle of 35.0° lands on a green 20.0 m below the tee (Figure 3-37). (Ignore any effects due to air resistance.) *(a)* Calculate the range using the equation $R = (v_0^2/g)\sin 2\theta_0$ even though the ball is hit from an elevated tee. *(b)* Show that the range for the more general problem (Figure 3-37) is given by $R = \left(1 + \sqrt{1 - \frac{2gy}{v_0^2 \sin^2\theta_0}}\right)\frac{v_0^2}{2g}\sin 2\theta_0$, where y is the height of the green above the tee. That is, $y = -h$. *(c)* Compute the range using this formula. What is the percentage error in neglecting the elevation of the green?

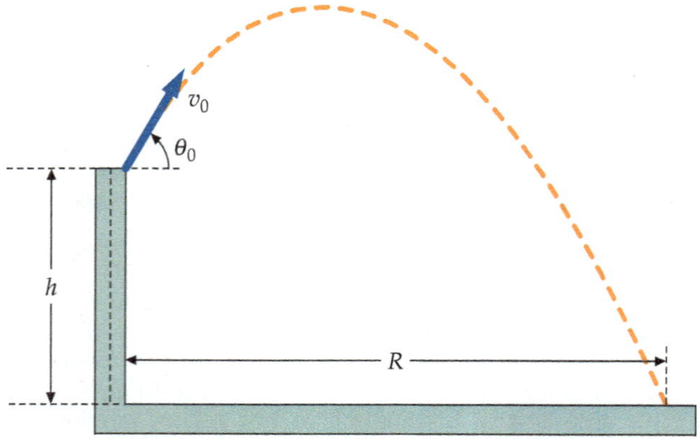

FIGURE 3-37 Problem 93

94 ••• **MULTISTEP** In the text, we calculated the range for a projectile that lands at the same elevation from which it is fired as $R = (v_0^2/g)\sin 2\theta_0$ if the effects of the air resistance are negligible. *(a)* Show that for the same conditions the change in the range for a small change in free-fall acceleration g, and the same initial speed and angle, is given by $\Delta R/R = -\Delta g/g$. *(b)* What would be the length of a homerun at a high altitude where g is 0.50% less than at sea level if the homerun at sea level traveled 400 ft?

95 ••• **MULTISTEP, APPROXIMATION** In the text, we calculated the range for a projectile that lands at the same elevation from which it is fired as $R = (v_0^2/g)\sin 2\theta_0$ if the effects of the air resistance are negligible. *(a)* Show that for the same conditions the change in the range for a small change in launch speed, and the same initial angle and free-fall acceleration, is given by $\Delta R/R = 2\Delta v_0/v_0$. *(b)* Suppose a projectile's range was 200 m. Use the formula in Part *(a)* to estimate its increase in range if the launch speed were increased by 20.0%. *(c)* Compare your answer in *(b)* to the increase in range by calculating the increase in range directly from $R = (v_0^2/g)\sin 2\theta_0$. If the results for Parts *(b)* and *(c)* are different, is the estimate too small or large, and why? SSM

96 ••• A projectile, fired with unknown initial velocity, lands 20.0 s later on the side of a hill, 3000 m away horizontally and 450 m vertically above its starting point. (Ignore any effects due to air resistance.) *(a)* What is the vertical component of its initial velocity? *(b)* What is the horizontal component of its initial velocity? *(c)* What was its maximum height above its launch point? *(d)* As it hit the hill, what speed did it have and what angle did its velocity make with the vertical?

97 ••• A projectile is launched over level ground at an initial elevation angle of θ. An observer standing at the launch site sees the projectile at the point of its highest elevation and measures the angle ϕ shown in Figure 3-38. Show that $\tan\phi = \frac{1}{2}\tan\theta$. (Ignore any effects due to air resistance.) SSM

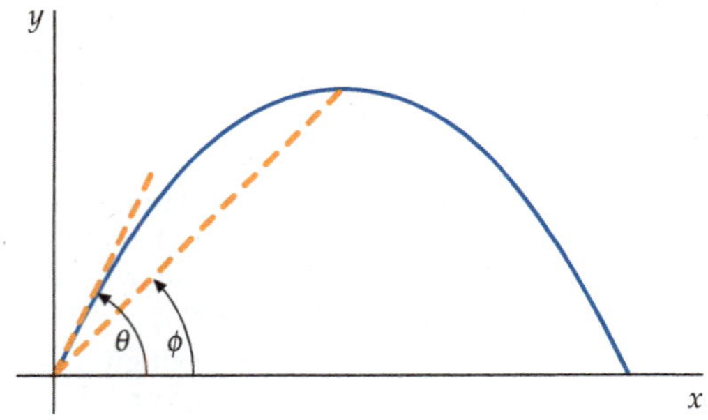

FIGURE 3-38 Problem 97

98 ••• A toy cannon is placed on a ramp that has a slope of angle ϕ. *(a)* If the cannonball is projected up the hill at an angle of θ_0 above the horizontal (Figure 3-39) and has a muzzle speed of v_0, show that the range R of the cannonball (as measured along the ramp) is given by

$$R = \frac{2v_0^2\cos^2\theta_0(\tan\theta_0 - \tan\phi)}{g\cos\phi}$$

Ignore any effects due to air resistance.

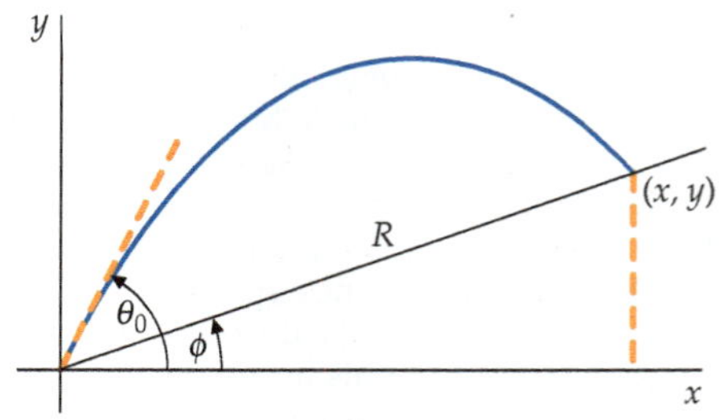

FIGURE 3-39 Problem 98

99 ••• A rock is thrown from the top of a 20-m-high building at an angle of 53° above the horizontal. *(a)* If the horizontal range of the throw is equal to the height of the building, with what speed was the rock thrown? *(b)* How long is it in the air? *(c)* What is the velocity of the rock just before it strikes the ground? (Ignore any effects due to air resistance.)

100 ••• A woman throws a ball at a vertical wall 4.0 m away (Figure 3-40). The ball is 2.0 m above ground when it leaves the woman's hand with an initial velocity of 14 m/s at 45°, as shown. When the ball hits the wall, the horizontal component of its velocity is reversed; the vertical component remains unchanged. *(a)* Where does the ball hit the ground? *(b)* How long was the ball in the air before it hit the wall? *(c)* Where did the ball hit the wall? *(d)* How long was the ball in the air after it left the wall? Ignore any effects due to air resistance.

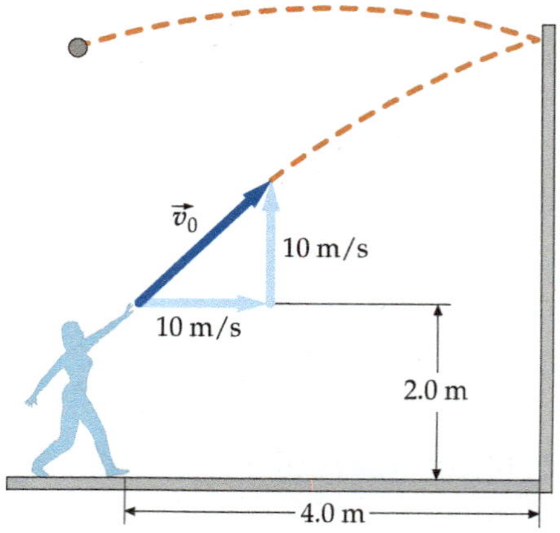

FIGURE 3-40 Problem 100

101 ••• **ENGINEERING APPLICATION** Catapults date from thousands of years ago, and were used historically to launch everything from stones to horses. During a battle in what is now Bavaria, inventive artillerymen from the united German clans launched giant *spaetzle* from their catapults toward a Roman fortification whose walls were 8.50 m high. The catapults launched the *spaetzle* projectiles from a height of 4.00 m above the ground and a distance of 38.0 m from the walls of the fortification at an angle of 60.0 degrees above the horizontal (Figure 3-41). If the projectiles were to hit the top of the wall, splattering the Roman soldiers atop the wall with pulverized pasta, (*a*) what launch speed was necessary? (*b*) How long were the *spaetzle* in the air? (*c*) At what speed did the projectiles hit the wall? Ignore any effects due to air resistance.

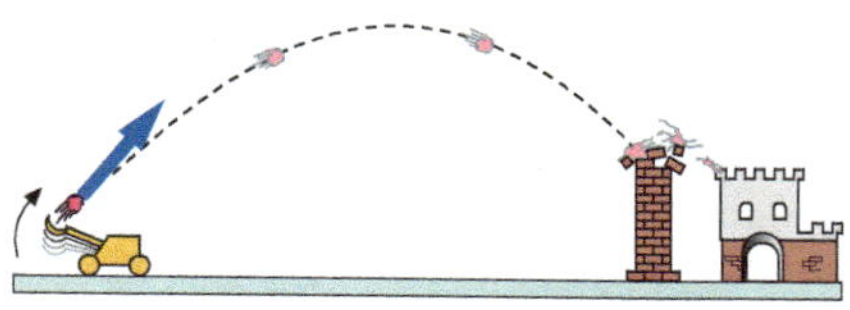

FIGURE 3-41 Problem 101

102 ••• The distance from the pitcher's mound to home plate is 18.4 m. The mound is 0.20 m above the level of the field. A pitcher throws a fastball with an initial speed of 37.5 m/s. At the moment the ball leaves the pitcher's hand, it is 2.30 m above the mound. (*a*) What should the angle between $\vec{v}_0$ and the horizontal be so that the ball crosses the plate 0.70 m above ground? (Ignore any effects due to air resistance.) (*b*) With what speed does the ball cross the plate? SSM

103 ••• You are watching your friend play hockey. During the course of the game, he strikes the puck in such a way that, when it is at its highest point, it just clears the surrounding 2.80-m-high Plexiglas wall that is 12.0 m away. Find (*a*) the vertical component of its initial velocity, (*b*) the time it takes to reach the wall, and (*c*) the horizontal component of its initial velocity, and its initial speed and angle. (Ignore any effects due to air resistance.)

104 ••• Carlos is on his trail bike, approaching a creek bed that is 7.0 m wide. A ramp with an incline of 10° has been built for daring people who try to jump the creek. Carlos is traveling at his bike's maximum speed, 40 km/h. (*a*) Should Carlos attempt the jump or brake hard? (*b*) What is the minimum speed a bike must have to make this jump? Assume equal elevations on either side of the creek. (Ignore any effects due to air resistance.)

105 ••• If a bullet that leaves the muzzle of a gun at 250 m/s is to hit a target 100 m away at the level of the muzzle (1.7 m above the level ground), the gun must be aimed at a point above the target. (*a*) How far above the target is that point? (*b*) How far behind the target will the bullet strike the ground? (Ignore any effects due to air resistance.) SSM

GENERAL PROBLEMS

106 •• During a do-it-yourself roof repair project, you are on the roof of your house and accidentally drop your hammer. The hammer then slides down the roof at constant speed of 4.0 m/s. The roof makes an angle of 30° with the horizontal, and its lowest point is 10 m from the ground. (*a*) How long after leaving the roof does the hammer hit the ground? (*b*) What is the horizontal distance traveled by the hammer between the instant it leaves the roof and the instant it hits the ground? (Ignore any effects due to air resistance.)

107 •• A squash ball typically rebounds from a surface with 25% of the speed with which it initially struck the surface. Suppose a squash ball is served in a shallow trajectory, from a height above the ground of 45 cm, at a launch angle of 6.0° degrees above the horizontal, and at a distance of 12 m from the front wall. (*a*) If it strikes the front wall exactly at the top of its parabolic trajectory, determine how high above the floor the ball strikes the wall. (*b*) How far horizontally from the wall does it strike the floor, after rebounding? (Ignore any effects due to air resistance.)

108 •• A football quarterback throws a pass at an angle of 36.5°. He releases the pass 3.50 m behind the line of scrimmage. His receiver left the line of scrimmage 2.50 s earlier, goes straight downfield at a constant speed of 7.50 m/s. In order that the pass land gently in the receiver's hands without the receiver breaking stride, with what speed must the quarterback throw the pass? Assume that the ball is released at the same height it is caught and that the receiver is straight downfield from the quarterback at the time of release. (Ignore any effects due to air resistance.)

109 •• Suppose a test pilot is able to safely withstand an acceleration of up to 5.0 times the acceleration due to gravity (that is, remain conscious and alert enough to fly). During the course of maneuvers, he is required to fly the plane in a horizontal circle at its top speed of 1900 mi/h. (*a*) What is the radius of the smallest circle he will be able to safely fly the plane in? (*b*) How long does it take him to go halfway around this minimum-radius circle?

110 •• A particle moves in the xy plane with constant acceleration. At $t = 0$ the particle is at $\vec{r}_1 = (4.0\text{ m})\hat{i} + (3.0\text{ m})\hat{j}$, with velocity $\vec{v}_1$. At $t = 2.0$ s, the particle has moved to $\vec{r}_2 = (10\text{ m})\hat{i} - (2.0\text{ m})\hat{j}$ and its velocity has changed to $\vec{v}_2 = (5.0\text{ m/s})\hat{i} - (6.0\text{ m/s})\hat{j}$. (*a*) Find $\vec{v}_1$. (*b*) What is the acceleration of the particle? (*c*) What is the velocity of the particle as a function of time? (*d*) What is the position vector of the particle as a function of time?

111 •• Plane A is flying due east at an air speed of 400 mph. Directly below, at a distance of 4000 ft, plane B is headed due north, flying at an air speed of 700 mph. Find the velocity vector of plane B relative to A. SSM

112 •• A diver steps off the cliffs at Acapulco, Mexico, 30.0 m above the surface of the water. At that moment, he activates his rocket-powered backpack horizontally, which gives him a constant horizontal acceleration of 5.00 m/s^2 but does not affect his vertical motion. (*a*) How long does he take to reach the surface of the water? (*b*) How far out from the base of the cliff does he enter the water, assuming the cliff is vertical? (*c*) Show that his flight path is a straight line. (Ignore any effects of air resistance.)

113 •• A small steel ball is projected horizontally off the top of a long flight of stairs. The initial speed of the ball is 3.0 m/s. Each step is 0.18 m high and 0.30 m wide. Which step does the ball strike first? SSM

114 •• Suppose you can throw a ball a maximum horizontal distance L when standing on level ground. How far can you throw it from the top of a building of height h if you throw it at (*a*) 0°, (*b*) 30°, (*c*) 45°? (Ignore any effects due to air resistance.)

115 ••• Darlene is a stunt motorcyclist in a traveling circus. For the climax of her show, she takes off from the ramp at angle θ_0, clears a ditch of width L, and lands on an elevated ramp (height h) on the other side (Figure 3-42). (*a*) For a given height h, find the minimum necessary takeoff speed v_{min} needed to make the jump successfully. (*b*) What is v_{min} for $L = 8.0$ m, $\theta = 30°$, and $h = 4.0$ m?

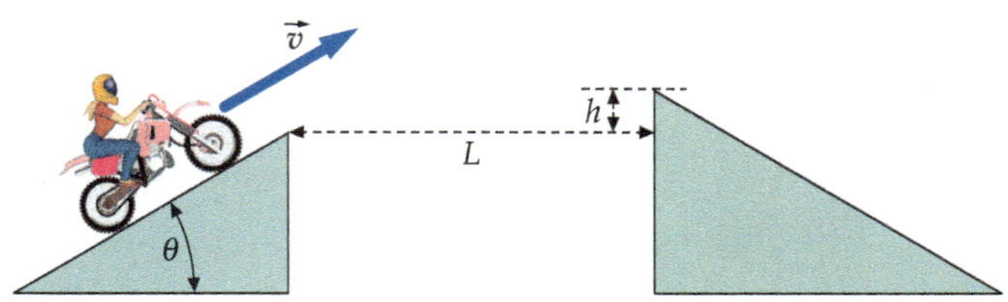

FIGURE 3-42 Problem 115

(*c*) Show that regardless of her takeoff speed, the maximum height of the platform is $h < L \tan \theta_0$. Interpret this result physically. (Neglect any effects due to air resistance and treat the rider and the bike as if they were a single particle.)

116 ••• A small boat is headed for a harbor 32 km directly northwest of its current position when it is suddenly engulfed in heavy fog. The captain maintains a compass bearing of northwest and a speed of 10 km/h relative to the water. The fog lifts 3.0 h later and the captain notes that he is now exactly 4.0 km south of the harbor. (*a*) What was the average velocity of the current during those 3.0 h? (*b*) In what direction should the boat have been heading to reach its destination along a straight course? (*c*) What would its travel time have been if it had followed a straight course?

117 •• Galileo showed that, if any effects due to air resistance are ignored, the ranges for projectiles (on a level field) whose angles of projection exceed or fall short of 45° by the same amount are equal. Prove Galileo's result. **SSM**

118 •• Two balls are thrown with equal speeds from the top of a cliff of height h. One ball is thrown at an angle of α above the horizontal. The other ball is thrown at an angle of β below the horizontal. Show that each ball strikes the ground with the same speed, and find that speed in terms of h and the initial speed v_0. (Ignore any effects due to air resistance.)

119 •• In his car, a driver tosses an egg vertically from chest height so that the peak of its path is just below the ceiling of the passenger compartment, which is 65 cm above his release point. He catches the egg at the same height that he releases it. If you are a roadside observer and measure the horizontal distance between catch and release points to be 19 m, (*a*) how fast is the car moving? (*b*) In your reference frame, at what angle above the horizontal was the egg thrown?

120 ••• A straight line is drawn on the surface of a 16-cm-radius turntable from the center to the perimeter. A bug crawls along this line from the center outward as the turntable spins counterclockwise at a constant 45 rpm. Its walking speed relative to the turntable is a steady 3.5 cm/s. Let its initial heading be in the positive x direction. As the bug reaches the edge of the turntable (still traveling at 3.5 cm/s radially, relative to the turntable), what are the x and y components of the velocity of the bug?

121 ••• On a windless day, a stunt pilot is flying his vintage World War I Sopwith Camel from Dubuque, Iowa, to Chicago, Illinois, for an air show. Unfortunately, he is unaware that his plane's ancient magnetic compass has a serious problem in that what it records as "north" is in fact 16.5° east of true north. At one moment during his flight, the airport in Chicago notifies him that he is, in reality, 150 km due west of the airport. He then turns due east, according to his plane's compass, and flies for 45 minutes at 150 km/h. At that point, he expects to see the airport and begin final descent. What is the plane's actual distance from Chicago and what should be the pilot's heading if he is to fly directly toward Chicago?

122 ••• **ENGINEERING APPLICATION, CONTEXT-RICH** A cargo plane in flight lost a package because somebody forgot to close the rear cargo doors. You are on the team of safety experts trying to analyze what happened. From the point of takeoff, while climbing to altitude, the airplane traveled in a straight line and at a constant speed of 275 mi/h at an angle of 37° above the horizontal. During the ascent, the package slid off the back ramp. You found the package in a field a distance of 7.5 km from the takeoff point. To complete the investigation you need to know exactly how long after takeoff the package slid off the back ramp of the plane. (Consider the sliding speed to be negligible.) Calculate the time at which the package fell off the back ramp. (Ignore any effects due to air resistance.)

CHAPTER 4

THIS AIRPLANE IS ACCELERATING AS IT HEADS DOWN THE RUNWAY BEFORE TAKEOFF. NEWTON'S LAWS RELATE AN OBJECT'S ACCELERATION TO ITS MASS AND THE FORCES ACTING ON IT. *(John Neubauer/FPG/Getty.)*

Newton's Laws

Now that we have studied how objects move in one, two, and three dimensions, we can ask the questions "Why do objects start to move?" and "What causes a moving object to change speed or change direction?"

These questions occupied the mind of Sir Isaac Newton, who was born in 1642, the year Galileo died. As a student at Cambridge, where he was later a mathematics professor, Newton studied the work of Galileo and Kepler. He wanted to figure out why planets move in ellipses at speeds dependent on their distance to the Sun, and even why the solar system stays together at all. During his lifetime, he developed both his law of gravitation, which we will examine in Chapter 12, and his three basic laws of motion that form the basis of classical mechanics.

Newton's laws relate the forces objects exert on each other, and relate any change in the motion of an object to the forces that act on it. Newton's laws of motion are the tools that enable us to analyze a wide range of mechanical phenomena. Although we may already have an intuitive idea of a force as a push or a pull, like that exerted by our muscles or by stretched rubber bands and springs, Newton's laws allow us to refine our understanding of forces.

In this chapter, we describe Newton's three laws of motion and begin using them to solve problems involving objects in motion and at rest.

? If you were a passenger on this plane, how might you use Newton's laws to determine the plane's acceleration? (See Example 4-9.)

4-1 NEWTON'S FIRST LAW: THE LAW OF INERTIA

Give a shove to a piece of ice on a counter top: It slides, and then slows to a stop. If the counter is wet, the ice will travel farther before stopping. A piece of dry ice (frozen carbon dioxide) riding on a cushion of carbon dioxide vapor slides quite far with little change in velocity. Before Galileo, it was thought that a steady force, such as a push or pull, was always needed to keep an object moving with constant velocity. But Galileo, and later Newton, recognized that the slowing of objects in everyday experience is due to the force of friction. If friction is reduced, the rate of slowing is reduced. A water slick or a cushion of gas is especially effective at reducing friction, allowing the object to slide a great distance with little change in velocity. Galileo reasoned that, if we could remove all external forces including friction from an object, then the velocity of the object would never change—a property of matter he described as **inertia.** This conclusion, restated by Newton as his first law, is also called the **law of inertia.**

Friction is greatly reduced by a cushion of air that supports the hovercraft. *(Jose Dupont/Explorer/Photo Researchers.)*

A modern wording of Newton's first law is

> **First law.** An object at rest stays at rest *unless* acted on by an external force. An object in motion continues to travel with constant speed in a straight line *unless* acted on by an external force.
>
> NEWTON'S FIRST LAW

INERTIAL REFERENCE FRAMES

Newton's first law makes no distinction between an object at rest and an object moving with constant (nonzero) velocity. Whether an object remains at rest or remains moving with constant velocity depends on the reference frame in which the object is observed. Suppose you are a passenger on an airplane that is flying along a straight path at constant altitude and you carefully place a tennis ball on your seat tray (which is horizontal). Relative to the plane, the ball will remain at rest as long as the plane continues to fly at constant velocity relative to the ground. Relative to the ground, the ball remains moving with the same velocity as the plane (Figure 4-1*a*).

Now, suppose that the pilot opens the throttle and the plane suddenly accelerates forward (relative to the ground). You will then observe that the ball on your tray suddenly starts to roll toward the rear of the plane, accelerating (relative to the plane) even though there is no horizontal force acting on it (Figure 4-1*b*). In this accelerating reference frame of the plane, Newton's first-law statement does not apply. Newton's first-law statement applies *only* in reference frames known as inertial reference frames. In fact, Newton's first law gives us the criterion for determining if a reference frame is an inertial frame:

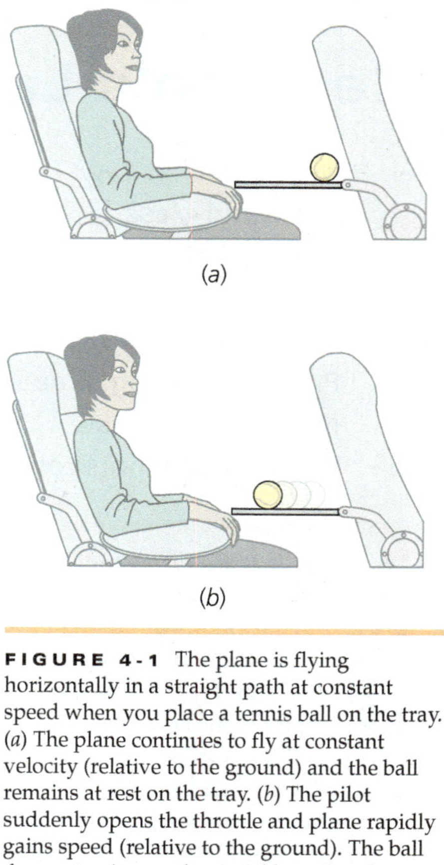

(*a*)

(*b*)

FIGURE 4-1 The plane is flying horizontally in a straight path at constant speed when you place a tennis ball on the tray. (*a*) The plane continues to fly at constant velocity (relative to the ground) and the ball remains at rest on the tray. (*b*) The pilot suddenly opens the throttle and plane rapidly gains speed (relative to the ground). The ball does not gain speed as quickly as the plane, so it accelerates (relative to the plane) toward the back of the plane.

> If no forces act on an object, any reference frame for which the acceleration of the object remains zero is an **inertial reference frame.**
>
> INERTIAL REFERENCE FRAME

Both the plane, when cruising at constant velocity, and the ground are, to a good approximation, inertial reference frames. Any reference frame moving with constant velocity relative to an inertial reference frame is also an inertial reference frame.

A reference frame attached to the ground is not quite an inertial reference frame because of the small acceleration of the ground due to the rotation of Earth and

> ! Newton's first-law statement applies *only* to inertial reference frames.

the small acceleration of Earth itself due to its revolution around the Sun. However, these accelerations are of the order of $0.01\,\text{m/s}^2$ or less, so to a good approximation, a reference frame attached to the surface of Earth is an inertial reference frame.

4-2 FORCE AND MASS

Using Newton's first law and the concept of inertial reference frames, we can define a **force** as an external influence or action on an object that causes the object to change velocity, that is, to accelerate relative to an inertial reference frame. (We assume that no other forces are acting on the object.) Force is a vector quantity. It has both magnitude (the size or strength of the force) and direction.

Forces are exerted on objects by other objects, and forces that are due to one object being physically touched by a second object are known as **contact forces.** Common examples of contact forces include hitting a ball with a bat, your hand pulling on a fishing line, your hands pushing on a shopping cart, and the force of friction between your sneakers and the floor. Note that in each case there is direct physical contact between the object applying the force and the object to which the force is applied. Other forces act on an object without direct physical contact with a second object. These forces, referred to as **action-at-a-distance forces,** include the gravitational force, the magnetic force, and the electric force.

THE FUNDAMENTAL INTERACTIONS OF NATURE

Forces are interactions between particles. Traditionally, physicists explain all interactions observed in nature in terms of four basic interactions that occur between elementary particles (see Figure 4-2):

1. The gravitational interaction—the long-range interaction between particles due to their mass. It is believed by some that the gravitational interaction involves the exchange of hypothetical particles called gravitons.
2. The electromagnetic interaction—the long-range interaction between electrically charged particles involving the exchange of photons.
3. The weak interaction—the extremely short-range interaction between subnuclear particles involving the exchange or production of W and Z bosons. The electromagnetic and weak interactions are now viewed as a single unified interaction called the electroweak interaction.
4. The strong interaction—the long-range interaction between hadrons, which themselves consist of quarks, that binds protons and neutrons together to form the atomic nuclei. It involves the exchange of mesons between hadrons, or gluons between quarks.

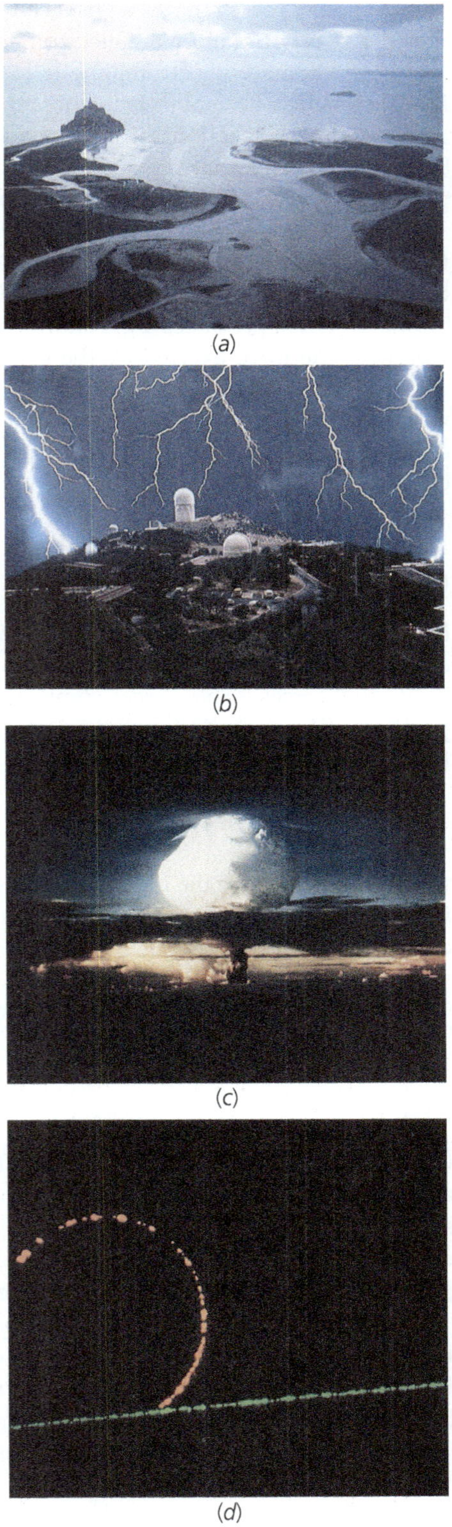

FIGURE 4-2 (*a*) The magnitude of the gravitational force between Earth and an object near Earth's surface is the weight of the object. The gravitational interaction between the Sun and the other planets is responsible for keeping the planets in their orbits around the Sun. Similarly, the gravitational interaction between Earth and the moon keeps the moon in its nearly circular orbit around Earth. The gravitational forces exerted by the moon and the Sun on the oceans of Earth are responsible for the diurnal and semidiurnal tides. Mont-Saint-Michel, France, shown in the photo, is an island when the tide is in. (*b*) The electromagnetic interaction includes both the electric and the magnetic forces. A familiar example of the electric interaction is the attraction between small bits of paper and a comb that is electrified after being run through hair. The lightning bolts above the Kitt Peak National Observatory, shown in the photo, are the result of the electromagnetic interaction. (*c*) The strong nuclear interaction between elementary particles called hadrons, which include protons and neutrons, the constituents of atomic nuclei. This interaction results from the interaction of quarks, which are the building blocks of hadrons, and is responsible for holding nuclei together. The hydrogen bomb explosion shown here illustrates the strong nuclear interaction. (*d*) The weak nuclear interaction between leptons (which include electrons and muons) and between hadrons (which include protons and neutrons). This false-color cloud chamber photograph illustrates the weak interaction between a cosmic ray muon (green) and an electron (red) knocked out of an atom. *((a) Cotton Coulson/Woodfin Camp and Assoc.; (b) Gary Ladd; (c) Los Alamos National Lab; (d) Science Photo Library/Photo Researchers.)*

The everyday forces that we observe between macroscopic objects are due to either the gravitational or the electromagnetic interactions. Contact forces, for example, are actually electromagnetic in origin. They are exerted between the surface molecules of the objects in contact. Action-at-a-distance forces are due to the fundamental interactions of gravity and electromagnetism. These two forces act between particles that are separated in space. Although Newton could not explain how forces act through empty space, later scientists introduced the concept of a field, which acts as an intermediary agent. For example, we consider the attraction of Earth by the Sun in two steps. The Sun creates a condition in space that we call the gravitational field. This field then exerts a force on Earth. Similarly, Earth produces a gravitational field that exerts a force on the Sun. Your weight is the force exerted by the gravitational field of Earth on you. When we study electricity and magnetism (Chapters 22–31) we will study electric fields, which are produced by electrical charges, and magnetic fields, which are produced by electrical charges in motion. The strong and weak interactions are discussed in Chapter 42.

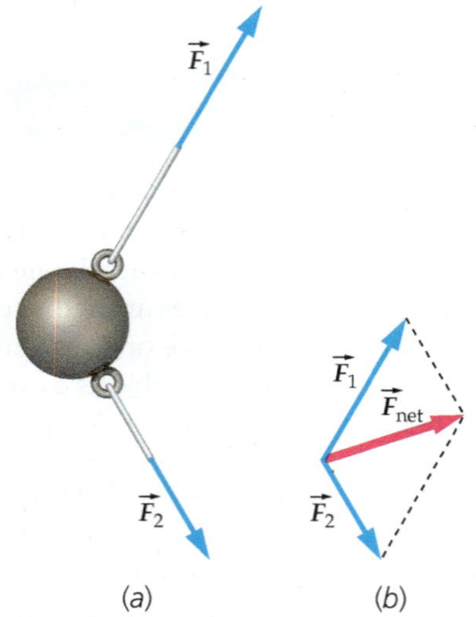

FIGURE 4-3 (*a*) The forces $\vec{F}_1$ and $\vec{F}_2$ pull on the sphere. (*b*) The effect of the two forces is as if a single force $\vec{F}_{net} = \vec{F}_1 + \vec{F}_2$ acts on the sphere instead of the two distinct forces $\vec{F}_1$ and $\vec{F}_2$.

COMBINING FORCES

If two or more individual forces simultaneously act on an object, the result is as if a single force, equal to the vector sum of the individual forces, acts in place of the individual forces. (That forces combine this way is called the **principle of superposition.**) The vector sum of the individual forces on an object is called the **net force** $\vec{F}_{net}$ on the object. That is,

$$\vec{F}_{net} = \vec{F}_1 + \vec{F}_2 + \cdots$$

where $\vec{F}_1, \vec{F}_2, \ldots$ are the individual forces. Figure 4-3 shows an object being pulled in two directions by ropes. The effect is as if a single force equal to the net force acts on the object.

The SI unit of force is the newton (N). The newton is defined in the next section. One newton is equal to the weight of a modest-sized apple.

CONCEPT CHECK 4-1

Is the net force an actual force?

MASS

Objects intrinsically resist being accelerated. Imagine kicking both a soccer ball and a bowling ball. The bowling ball resists being accelerated much more than does the soccer ball, as would be evidenced by your sore toes. This intrinsic property is called the object's **mass.** It is a measure of the object's inertia. The greater an object's mass, the more the object resists being accelerated.

As noted in Chapter 1, the object chosen as the international standard for mass is a platinum-iridium alloy cylinder carefully preserved at the International Bureau of Weights and Measures at Sèvres, France. The mass of this standard object is 1 **kilogram** (kg), the SI unit of mass.

A convenient standard unit for mass in atomic and nuclear physics is the **unified atomic mass unit** (u), which is defined as one-twelfth of the mass of the carbon-12 (^{12}C) atom. The unified atomic mass unit is related to the kilogram by

$$1\text{ u} = 1.660\,540 \times 10^{-27}\text{ kg}$$

The concept of mass is defined as a constant of proportionality in Newton's second law. To measure the mass of an object, we compare its mass with a standard mass, such as the 1-kg standard kept at Sèvres. The comparison is accomplished using Newton's second law, and a procedure for doing this is described in Section 4-3 immediately following Example 4-1.

4-3 NEWTON'S SECOND LAW

Newton's first law tells us what happens when there is *no* force acting on an object. But what happens when there are forces exerted on the object? Consider again a block of ice sliding with constant velocity on a smooth, *frictionless* surface. If you push on the ice, you exert a force $\vec{F}$ that causes the ice's velocity to change. The harder you push, the greater the resulting acceleration $\vec{a}$. The acceleration, $\vec{a}$, of any object is directly proportional to the net force $\vec{F}_{net}$ exerted on it, and the reciprocal of the mass of the object is the proportionality constant. In addition, the acceleration vector and the net force vector are in the same direction. Newton summarized these observations in his second law of motion:

> **Second law.** The acceleration of an object is directly proportional to the net force acting on it, and the reciprocal of the mass of the object is the constant of proportionality. Thus,
>
> $$\vec{a} = \frac{\vec{F}_{net}}{m}, \quad \text{where} \quad \vec{F}_{net} = \sum \vec{F} \qquad \text{4-1}$$
>
> NEWTON'S SECOND LAW

A net force on an object causes it to accelerate. It is a matter of cause and effect. The net force is the cause and the effect is the acceleration.*

A net force of 1 newton gives a 1-kg mass an acceleration of 1 m/s^2, so

$$1\,\text{N} = (1\,\text{kg})(1\,\text{m/s}^2) = 1\,\text{kg}\cdot\text{m/s}^2 \qquad \text{4-2}$$

Thus, a force of 2 N gives a 2-kg mass an acceleration of 1 m/s^2, and so on.

In the U.S. customary system, the unit of force is the *pound* (lb), where 1 lb ≈ 4.45 N,† and the unit of mass is the *slug*. The pound is defined to be the force required to produce an acceleration of 1 ft/s^2 on a mass of 1 slug:

$$1\,\text{lb} = 1\,\text{slug}\cdot\text{ft/s}^2$$

It follows that 1 slug ≈ 14.6 kg.

The Equation 4-1 is frequently expressed:

$$\vec{F}_{net} = m\vec{a}$$

and we will express it this way most of the time.

> **!** Newton's second-law statement, like Newton's first–law statement, can be applied *only* in inertial reference frames.

CONCEPT CHECK 4-2

Is $m\vec{a}$ a force?

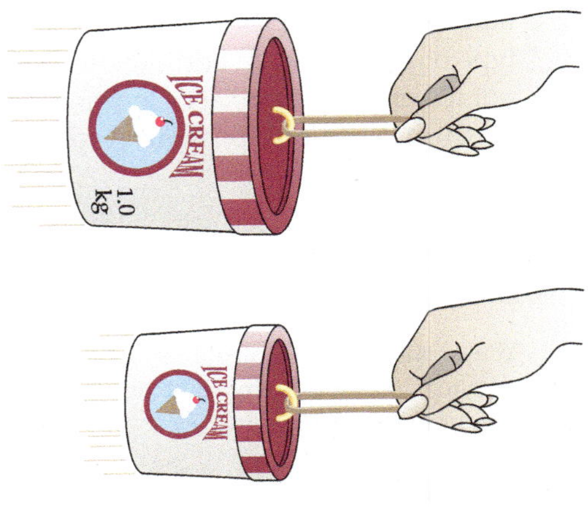

FIGURE 4-4

Example 4-1 A Sliding Ice-Cream Carton

A force exerted by a stretched rubber band (see Figure 4-4) produces an acceleration of 5.0 m/s^2 on an ice-cream carton of mass m_1 = 1.0 kg. When a force exerted by an identical rubber band stretched by the same amount is applied to a carton of ice cream of mass m_2, it produces an acceleration of 11 m/s^2. (*a*) What is the mass of the second carton of ice cream? (*b*) What is the magnitude of the force exerted by the rubber band on the carton?

PICTURE We can apply Newton's second law, $\Sigma\vec{F} = m\vec{a}$, to each object and solve for the mass of the ice-cream carton and the magnitude of the force. The magnitudes of the forces exerted by the rubber bands are equal.

SOLVE

(*a*) 1. Apply $\Sigma\vec{F} = m\vec{a}$ to each object. There is only one force on each object, and we only need to consider magnitudes of the vector quantities: $\quad F_1 = m_1a_1$ and $F_2 = m_2a_2$

---

* Newton's second law relates the net force and the acceleration. Not everyone agrees that the net force is the cause and the acceleration is the effect.

† The pound we are talking about is the pound force (i.e., 1 pound force is exactly equal to 4.448 221 615 260 5 N). There is also a pound mass, which is exactly equal to 0.453 592 37 kg.

2. Because the applied forces are equal in magnitude, the ratio of the masses is equal to the reciprocal of the ratio of the accelerations:

$$F_1 = F_2 \Rightarrow m_1 a_1 = m_2 a_2 \quad \text{and} \quad \frac{m_2}{m_1} = \frac{a_1}{a_2}$$

3. Solve for m_2 in terms of m_1, which is 1.0 kg:

$$m_2 = \frac{a_1}{a_2} m_1 = \frac{5.0\ \text{m/s}^2}{11\ \text{m/s}^2}(1.0\ \text{kg}) = \boxed{0.45\ \text{kg}}$$

(*b*) The magnitude F_1 is found by using the mass and acceleration of either object:

$$F_1 = m_1 a_1 = (1.0\ \text{kg})(5.0\ \text{m/s}^2) = \boxed{5.0\ \text{N}}$$

CHECK A mass of 0.45 kg is a plausible mass for a carton of ice cream. One kilogram weighs about 2.2 pounds. So, the carton weighs about one pound and is the size of a one-pint carton.

PRACTICE PROBLEM 4-1 A net force of 3.0 N produces an acceleration of 2.0 m/s² on an object of unknown mass. What is the mass of the object?

To describe mass quantitatively, we can apply identical forces to two masses and compare their accelerations. If a force of magnitude F produces acceleration of magnitude a_1 when applied to an object of mass m_1, and an identical force produces acceleration of magnitude a_2 when applied to an object of mass m_2, then $m_1 a_1 = m_2 a_2$ (or $m_2/m_1 = a_1/a_2$). That is

If $F_1 = F_2$ then

$$\frac{m_2}{m_1} = \frac{a_1}{a_2} \qquad \text{4-3}$$

COMPARING MASSES

This definition agrees with our intuitive idea of mass. If a force is applied to an object and a force of equal magnitude is applied to a second object, then the object with more mass will accelerate less. The ratio a_1/a_2 produced by forces of equal magnitude acting on two objects is independent of the magnitude, direction, or type of force used. In addition, mass is an intrinsic property of an object that does not depend on the object's location—it remains the same whether the object is on Earth, on the moon, or in deep space.

Example 4-2 A Walk in Space

Context-Rich

You're stranded in space away from your spaceship. Fortunately, you have a propulsion unit that provides a constant net force $\vec{F}$ for 3.0 s. After 3.0 s, you have moved 2.25 m. If your mass is 68 kg, find $\vec{F}$.

The propulsion unit (not shown) is pushing the astronaut to the right. *(NASA/Science Source/Photo Researchers.)*

PICTURE The force acting on you is constant, so your acceleration is also constant. We can use the kinematic equations of Chapters 2 and 3 to find $\vec{a}$, and then obtain the force from $\Sigma\vec{F} = m\vec{a}$. We choose the $+x$ direction to be in the direction of $\vec{F}$ (Figure 4-5), so $\vec{F} = F_x\hat{i}$ and $F_x = ma_x$.

SOLVE

1. To find the acceleration, we use Equation 2-14 with $v_0 = 0$:

$$\Delta x = v_0 t + \tfrac{1}{2}a_x t^2 = 0 + \tfrac{1}{2}a_x t^2$$

$$a_x = \frac{2\Delta x}{t^2} = \frac{2(2.25\ \text{m})}{(3.0\ \text{s})^2} = 0.50\ \text{m/s}^2$$

$$\vec{a} = a_x\hat{i} = 0.50\ \text{m/s}^2\hat{i}$$

2. Because $\vec{F}$ is the *net* force, $\Sigma\vec{F}_i = \vec{F}$. Therefore, we substitute $\vec{a} = 0.50\text{ m/s}^2\hat{i}$ and $m = 68\text{ kg}$ into this equation to find the force:

$$\vec{F} = ma_x\hat{i} = (68\text{ kg})(0.50\text{ m/s}^2)\hat{i} = \boxed{34\text{ N}\hat{i}}$$

CHECK The acceleration is 0.50 m/s^2, which is about 5% of $g = 9.81\text{ m/s}^2$. This value seems plausible. If the magnitude of the acceleration were equal to g you would move a lot farther than 2.25 m in 3 s.

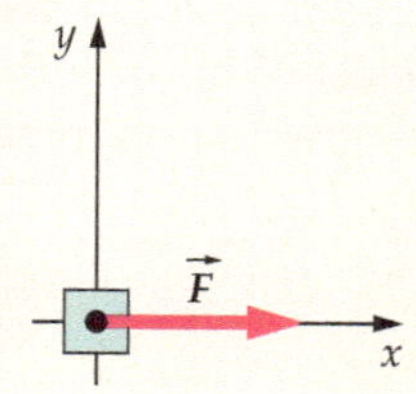

FIGURE 4-5

Example 4-3 A Particle Subjected to Two Forces

A particle of mass 0.400 kg is subjected simultaneously to two forces $\vec{F}_1 = -2.00\text{ N}\hat{i} - 4.00\text{ N}\hat{j}$ and $\vec{F}_2 = -2.60\text{ N}\hat{i} + 5.00\text{ N}\hat{j}$ (Figure 4-6). If the particle is at the origin and starts from rest at $t = 0$, find (*a*) its position $\vec{r}$ and (*b*) its velocity $\vec{v}$ at $t = 1.60\text{ s}$.

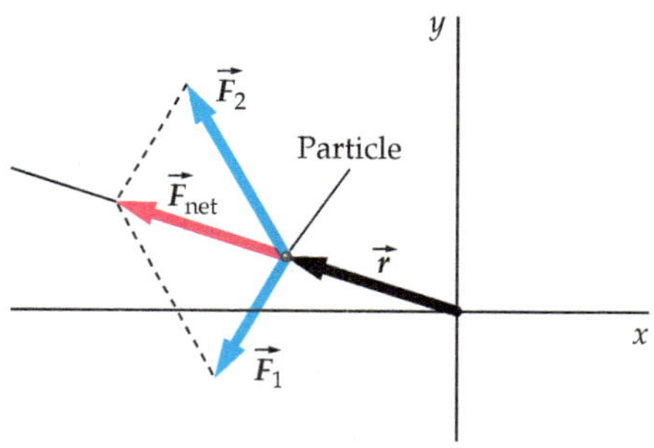

FIGURE 4-6 The acceleration is in the direction of the net force. The particle is released from rest at the origin. Following release, it moves in the direction of the net force, which is also the direction of the acceleration vector.

PICTURE Apply $\Sigma\vec{F} = m\vec{a}$ to find the acceleration. Once the acceleration is known, we can use the kinematic equations of Chapters 2 and 3 to determine the particle's position and velocity as functions of time.

SOLVE

(*a*) 1. Write the general equation for the position vector $\vec{r}$ as a function of time t for constant acceleration $\vec{a}$ in terms of $\vec{r}_0$, $\vec{v}_0$, and $\vec{a}$, and substitute $\vec{r}_0 = \vec{v}_0 = 0$.

$$\vec{r} = \vec{r}_0 + \vec{v}_0 t + \tfrac{1}{2}\vec{a}t^2 = 0 + 0 + \tfrac{1}{2}\vec{a}t^2 = \tfrac{1}{2}\vec{a}t^2$$

2. Use $\Sigma\vec{F} = m\vec{a}$ to write the acceleration $\vec{a}$ in terms of the resultant force $\Sigma\vec{F}$ and the mass m.

$$\vec{a} = \frac{\Sigma\vec{F}}{m}$$

3. Compute $\Sigma\vec{F}$ from the given forces.

$$\Sigma\vec{F} = \vec{F}_1 + \vec{F}_2 = (-2.00\text{ N}\,\hat{i} - 4.00\text{ N}\,\hat{j}) + (-2.60\text{ N}\,\hat{i} + 5.00\text{ N}\,\hat{j}) = -4.60\text{ N}\,\hat{i} + 1.00\text{ N}\,\hat{j}$$

4. Find the acceleration $\vec{a}$.

$$\vec{a} = \frac{\Sigma\vec{F}}{m} = -11.5\text{ m/s}^2\,\hat{i} + 2.50\text{ m/s}^2\,\hat{j}$$

5. Find the position $\vec{r}$ for a general time t.

$$\vec{r} = \tfrac{1}{2}\vec{a}t^2 = \tfrac{1}{2}a_x t^2\hat{i} + \tfrac{1}{2}a_y t^2\hat{j} = (-5.75\text{ m/s}^2\,\hat{i} + 1.25\text{ m/s}^2\,\hat{j})t^2$$

6. Find $\vec{r}$ at $t = 1.60\text{ s}$.

$$\vec{r} = \boxed{-14.7\text{ m}\,\hat{i} + 3.20\text{ m}\,\hat{j}}$$

(*b*) Write the velocity $\vec{v}$ by taking the time derivative of the step-5 result. Evaluate the velocity at $t = 1.6\text{ s}$.

$$\vec{v}(t) = \frac{d\vec{r}}{dt} = 2(-5.75\text{ m/s}^2\,\hat{i} + 1.25\text{ m/s}^2\,\hat{j})t$$

$$\vec{v}(1.6\text{ s}) = \boxed{-18.4\text{ m/s}\,\hat{i} + 4.00\text{ m/s}\,\hat{j}}$$

CHECK The position, velocity, acceleration, and net force vectors all have negative x components and positive y components. This is as expected for motion starting from rest at the origin and moving with constant acceleration.

4-4 THE FORCE DUE TO GRAVITY: WEIGHT

If you drop an object near Earth's surface, it accelerates toward Earth. If air resistance is negligible, all objects fall with the same acceleration, called the free-fall acceleration $\vec{g}$. The force causing this acceleration is the **gravitational force** ($\vec{F}_g$) exerted by Earth on the object. The *weight* of the object is the magnitude of the gravitational force on it. If the gravitational force is the *only* force acting on an object, the object is said to be in **freefall.** We can apply Newton's second law ($\Sigma\vec{F} = m\vec{a}$) to an object of mass m that is in freefall with acceleration $\vec{g}$ to obtain an expression for the gravitation force $\vec{F}_g$:

$$\vec{F}_g = m\vec{g} \tag{4-4}$$

WEIGHT

Because $\vec{g}$ is the same for all objects, it follows that the gravitational force on an object is proportional to its mass. Near Earth, the vector $\vec{g}$ is the gravitational force per unit mass exerted by the planet Earth on any object and is called the **gravitational field** of Earth. Near the surface of Earth, the magnitude of $\vec{g}$ has the value

$$g = 9.81\,\text{N/kg} = 9.81\,\text{m/s}^2 \tag{4-5}$$

When working problems in the U.S. customary system, we substitute F_g/g for mass m, where F_g is the magnitude of the gravitational force, in pounds, and g is the magnitude of the acceleration due to gravity in feet per second squared. Because $9.81\,\text{m} = 32.2\,\text{ft}$,

$$g = 32.2\,\text{ft/s}^2 \tag{4-6}$$

Careful measurements show that near Earth $\vec{g}$ varies with location. $\vec{g}$ points toward the center of Earth and, at points above the surface of Earth, the magnitude of $\vec{g}$ varies inversely with the square of the distance to the center of Earth. Thus, an object weighs slightly less at very high altitudes than it does at sea level. The gravitational field also varies slightly with latitude because Earth is not exactly spherical but is slightly flattened at the poles. Thus weight, unlike mass, is not an intrinsic property of an object. Although the weight of an object varies from place to place because of variations in g, these variations are too small to be noticed in most practical applications on or near the surface of Earth.

! Weight is not an intrinsic property of an object.

An example should help clarify the difference between mass and weight. Consider a bowling ball near the moon. Its weight is the magnitude of the gravitational force exerted on it by the moon, but that force is a mere sixth of the magnitude of the gravitational force exerted on the bowling ball when it is similarly positioned on Earth. The ball weighs about one-sixth as much on the moon, and lifting the ball on the moon requires one-sixth the force. However, because the mass of the ball is the same on the moon as on Earth, throwing the ball horizontally at a specified speed requires the same force on the moon as on Earth.

Although the weight of an object may vary from one place to another, at any particular location the weight of the object is proportional to its mass. Thus, we can conveniently compare the masses of two objects at a given location by comparing their weights.

Our sensation of the gravitational force on us comes from other forces that balance it. When you sit on a chair, you feel a force exerted by the chair that balances the gravitational force on you and prevents you from accelerating toward the floor. When you stand on a spring scale, your feet feel the force exerted by the scale. The scale is calibrated to read the magnitude of the force it exerts (by the compression of its springs) to balance the gravitational force on you. The magnitude of this force is called your **apparent weight.** If there is no force to balance your weight, as in free-fall, your apparent weight is zero. This condition, called **weightlessness,** is experienced by astronauts in orbiting satellites. A satellite in a circular orbit near the surface of Earth is accelerating toward Earth. The only force acting on the satellite is that of gravity, so it is in free-fall. Astronauts in the satellite are also in free-fall. The only force on them is the gravitational force on them, which produces the acceleration $\vec{g}$. Because there is no force balancing the force of gravity, the astronauts have zero apparent weight.

Example 4-4 An Accelerating Student

The net force acting on a 130-lb student has a magnitude of 25.0 lb. What is the magnitude of her acceleration?

PICTURE Apply Newton's second law and solve for the acceleration. The mass can be found from the student's weight.

SOLVE

According to Newton's second law, the student's acceleration is the force divided by her mass, and her mass is equal to her weight divided by g:

$$a = \frac{F_{net}}{m} = \frac{F_{net}}{F_g/g} = \frac{25.0 \text{ lb}}{(130 \text{ lb})/(32.2 \text{ ft/s}^2)} = \boxed{6.19 \text{ ft/s}^2}$$

CHECK The force is slightly less than one-fifth of her weight, so we expect the acceleration to be slightly less than one-fifth of g. $(32.2 \text{ ft/s}^2)/5 = 6.44 \text{ ft/s}^2$, and 6.19 ft/s^2 is slightly less than 6.44 ft/s^2, so the result is plausible.

TAKING IT FURTHER Rearranging the equation in the solution gives

$$m = \frac{F_{net}}{a} = \frac{F_g}{g}$$

This arrangement reveals that you can solve for her acceleration without first solving for the mass. For any object, the ratio of F_{net} to a equals the ratio of F_g to g.

PRACTICE PROBLEM 4-2 What force is needed to give an acceleration of 3.0 ft/s^2 to a 5.0-lb block?

4-5 CONTACT FORCES: SOLIDS, SPRINGS, AND STRINGS

Many forces are exerted by one body in contact with another. In this section, we will examine some of the more common contact forces.

SOLIDS

If a surface is pushed against, it pushes back. Consider the ladder leaning against a wall shown in Figure 4-7. At the region of contact, the ladder pushes against the wall with a horizontal force, compressing the distance between the molecules in the surface of the wall. Like mattress springs, the compressed molecules in the wall push back on the ladder with a horizontal force. Such a force, *perpendicular* to the contacting surfaces, is called a **normal force** (the word *normal* means perpendicular). The wall bends slightly in response to a load, though this is rarely noticeable to the unaided eye.

Normal forces can vary over a wide range of magnitudes. A horizontal tabletop, for instance, will exert an upward normal force on any object resting on it. As long as the table doesn't break, this normal force will balance the downward gravitational force on the object. Furthermore, if you press down on the object, the magnitude of the upward normal force exerted by the table will increase, countering the extra force, thus preventing the object from accelerating downward.

In addition, surfaces in contact can exert forces on each other that are *parallel* to the contacting surfaces. Consider the large block on the floor shown in Figure 4-8. If the block is pushed sideways with a gentle enough force, it will not slide. The surface of the floor exerts a force back on the block, opposing its tendency to slide in the direction of the push. However, if the block is pushed sideways with a sufficiently large force, it will start to slide. To keep the block sliding, it is necessary to continue to push it. If the sideways push is not sustained, the contact force will slow the motion of the box until it stops. A component of a contact force that opposes sliding, or

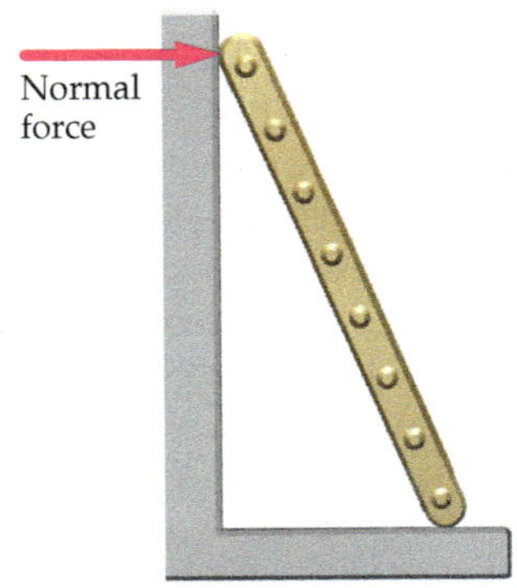

FIGURE 4-7 The wall supports the ladder by pushing on the ladder with a force normal to the wall.

FIGURE 4-8 The man is pushing on a block. The frictional force exerted by the floor on the block opposes its sliding motion or its tendency to slide.

the tendency to slide, is called a **frictional force;** it acts parallel to the contacting surfaces. (Frictional forces are treated more extensively in Chapter 5.)

A construction dumpster (Figure 4-9*a*) is situated on a road with a steep incline. Gravity pulls the dumpster downward, so to prevent the dumpster from moving, the road must exert an upward force $\vec{F}$ of equal magnitude on the dumpster (Figure 4-9*b*). The force $\vec{F}$ is a contact force by the road on the dumpster. A contact force such as this one is often thought of as two distinct forces, one, called the normal force $\vec{F}_n$, directed perpendicular to the road surface, and a second, called the frictional force $\vec{f}$, that is directed parallel to the road surface. The frictional force opposes any tendency of the dumpster to slide down the hill.

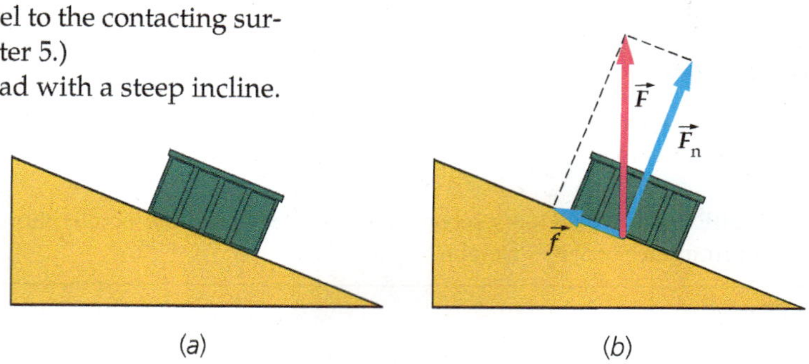

FIGURE 4-9 (*a*) A dumpster is parked on a steep incline. (*b*) The contact force by the road on the dumpster is represented either as the single force $\vec{F}$, or as a superposition of a normal force $\vec{F}_n$ and a frictional force $\vec{f}$.

SPRINGS

When a spring is stretched from its unstressed length by a distance x, the force it exerts is found experimentally to be

$$F_x = -kx \qquad \text{4-7}$$

HOOKE'S LAW

where the positive constant k, called the **force constant** (or spring constant), is a measure of the stiffness of the spring (Figure 4-10). A negative value of x means the spring has been compressed a distance $|x|$ from its unstressed length. The negative sign in Equation 4-7 signifies that when the spring is stretched (or compressed) in one direction, the force it exerts is in the opposite direction. This relation, known as **Hooke's law,** turns out to be quite important. An object at rest under the influence of forces that balance is said to be in *static equilibrium.* If a small displacement results in a net restoring force toward the equilibrium position, the equilibrium is called *stable equilibrium.* For small displacements, nearly all restoring forces obey Hooke's law.

The molecular force of attraction between atoms in a molecule or solid varies much like that of a spring. We can therefore use two masses on a spring to model a diatomic molecule, or a set of masses connected by springs to model a solid as shown in Figure 4-11.

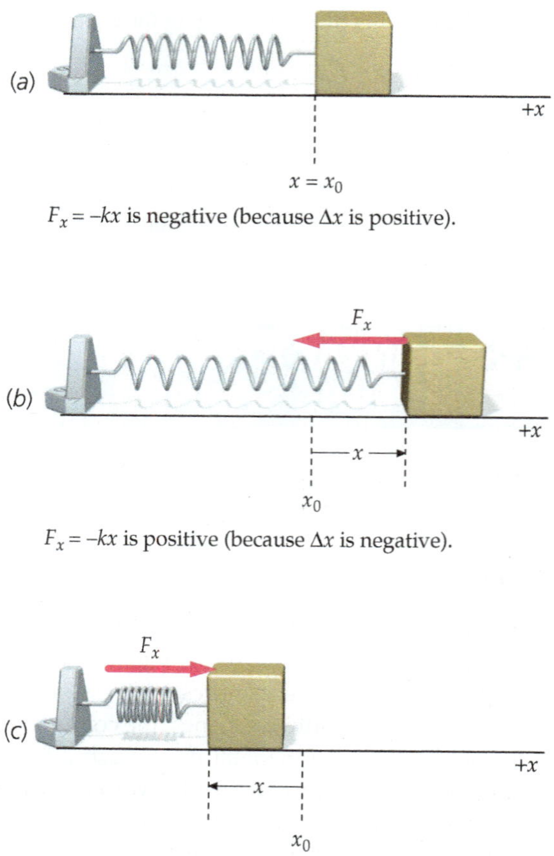

FIGURE 4-10 A horizontal spring. (*a*) When the spring is unstressed, it exerts no force on the block. (*b*) When the spring is stretched so that x is positive, it exerts a force of magnitude kx in the $-x$ direction. (*c*) When the spring is compressed so that x is negative, the spring exerts a force of magnitude $k|x|$ in the $+x$ direction.

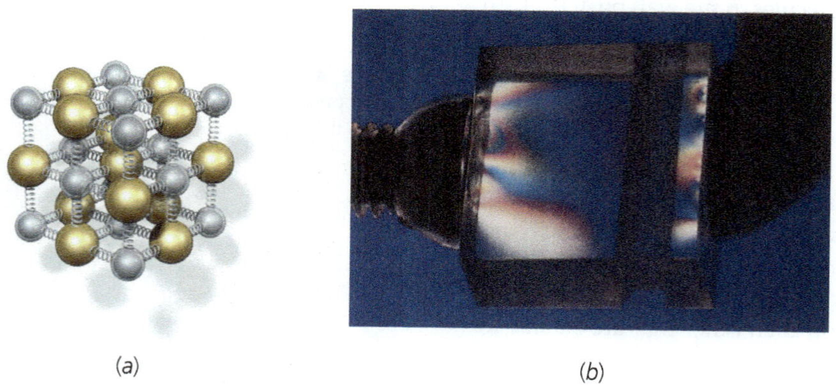

FIGURE 4-11 (*a*) Model of a solid consisting of atoms connected to each other by springs. The springs are very stiff (large force constant) so that when a weight is placed on the solid, its deformation is not visible. However, compression such as that produced by the clamp on the plastic block in (*b*) leads to stress patterns that are visible when viewed with polarized light. *((b) Fundamental Photographs.)*

Example 4-5 The Slam Dunk

A 110-kg basketball player hangs on the rim following a slam dunk (Figure 4-12). Prior to dropping to the floor, he hangs motionless with the front of the rim deflected down a distance of 15 cm. Assume the rim can be approximated by a spring and calculate the force constant k.

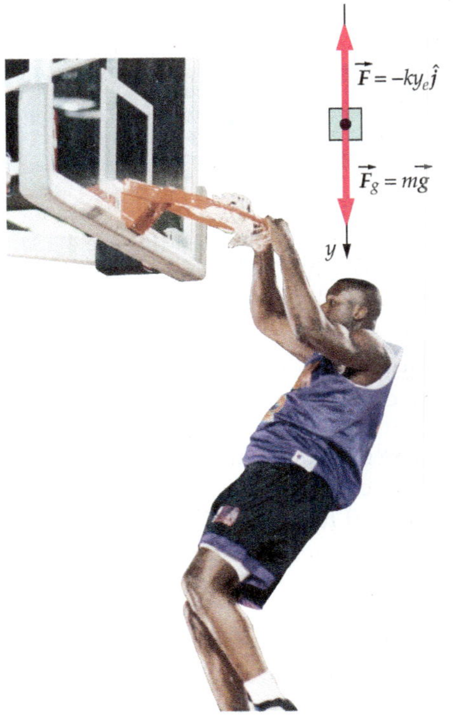

FIGURE 4-12 *(AFP-Getty Images.)*

PICTURE Because the acceleration of the player is zero, the net force exerted on him must also be zero. The upward force exerted by the rim balances his weight. Let $y = 0$ be the original position of the rim and choose down to be the $+y$ direction. Then the displacement of the rim y_e is positive, the weight $F_{gy} = mg$ is positive, and the force $F_y = -ky_e$ exerted by the rim is negative.

SOLVE

1. Apply $\Sigma F_y = m\vec{a}_y$ to the player. The acceleration of the player is zero:

$$\Sigma F_y = ma_y$$
$$F_{gy} + F_y = 0$$

2. Use Hooke's law (Equation 4-7) to find F_y:

$$F_y = -ky_e$$

3. Substitute expressions or values for the force components in step 1 and solve for k:

$$F_{gy} + F_y = 0$$
$$mg + (-ky_e) = 0$$
$$k = \frac{mg}{y_e} = \frac{(110\ \text{kg})(9.8\ \text{N/kg})}{0.15\ \text{m}}$$
$$= \boxed{7.2 \times 10^3\ \text{N/m}}$$

CHECK The weight of any object, in newtons, is almost ten times larger than the object's mass in kilograms. Thus, the weight is more than 1000 N. A deflection of only 0.10 m would mean k would be 10,000 N/m, so getting $k = 7200$ N/m for a deflection of 0.15 m seems about right.

TAKING IT FURTHER Although a basketball rim doesn't look much like a spring, the rim is sometimes suspended by a hinge with a spring that is distorted when the front of the rim is pulled down. As a result, the upward force the rim exerts on the player's hands is proportional to the rim front's displacement and oppositely directed. Note that we used N/kg for the units of g so that kg cancels, giving N/m for the units of k. We can use either 9.81 N/kg or 9.81 m/s^2 for g, whichever is more convenient, because 1 N/kg = 1 m/s^2.

PRACTICE PROBLEM 4-3 A 4.0-kg bunch of bananas is suspended motionless from a spring balance whose force constant is 300 N/m. By how much is the spring stretched?

PRACTICE PROBLEM 4-4 A spring of force constant 400 N/m is attached to a 3.0-kg block that rests on a horizontal air track that renders friction negligible. What extension of the spring is needed to give the block an acceleration of 4.0 m/s^2 upon release?

PRACTICE PROBLEM 4-5 An object of mass m oscillates at the end of an ideal spring of force constant k. The time for one complete oscillation is the period T. Assuming that T depends on m and k, use dimensional analysis to find the form of the relationship $T = f(m, k)$, ignoring numerical constants. This is most easily found by looking at the units. Note that the units of k are N/m = (kg · m/s^2)/m = kg/s^2, and the units of m are kg.

STRINGS

Strings (ropes) are used to pull things. We can think of a string as a spring with such a large force constant that the extension of the string is negligible. Strings are flexible, however, so unlike springs, they cannot push things. Instead, they flex or bend sharply. The magnitude of the force that one segment of a string exerts on an adjacent segment is called **tension,** T. It follows that if a string pulls on an object, the magnitude of the force on the object equals the tension. The concept of tension in a string or rope is further developed in Section 4-8.

4-6 PROBLEM SOLVING: FREE-BODY DIAGRAMS

Imagine a sled being pulled across icy ground by a sled dog. The dog pulls on a rope attached to the sled (Figure 4-13*a*) with a horizontal force causing the sled to gain speed. We can think of the sled and rope together as a single particle. What forces act on the sled-rope particle? Both the dog and the ice touch the sled-rope, so we know that the dog and the ice exert contact forces on it. We also know that Earth exerts a gravitational force on the sled-rope (the sled-rope's weight). Thus, a total of three forces act on the sled-rope (assuming that friction is negligible):

1. The gravitational force on the sled-rope $\vec{F}_g$.
2. The contact force $\vec{F}_n$ exerted by the ice on the runners. (Without friction, the contact force is directed normal to the ice.)
3. The contact force $\vec{F}$ exerted by the dog on the rope.

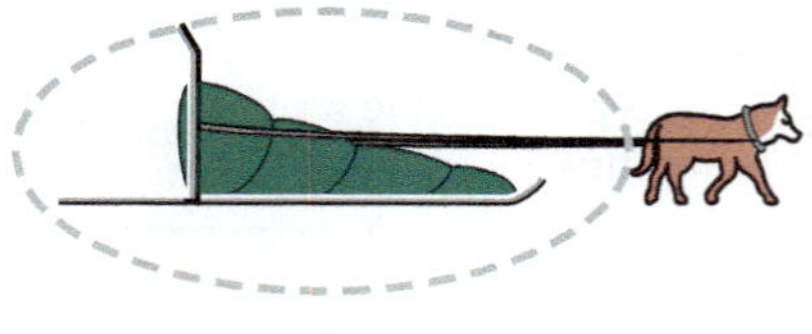

(*a*)

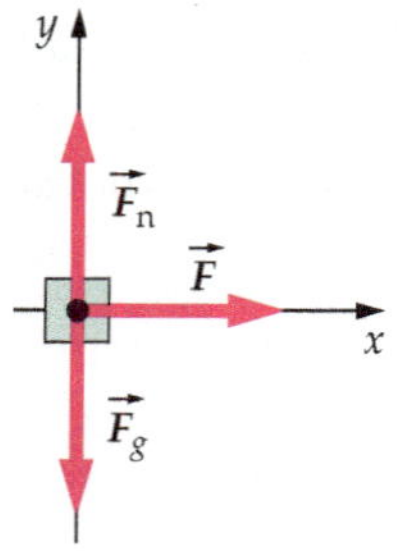

(*b*)

FIGURE 4-13 (*a*) A dog pulling a sled. The first step in problem solving is to isolate the system to be analyzed. In this case, the closed dashed curve represents the boundary between the sled-rope object and its surroundings. (*b*) The forces acting on the sled in Figure 4.13*a*.

A diagram that shows schematically all the forces acting on a system, such as Figure 4-13*b*, is called a **free-body diagram.** It is called a free-body diagram because the body (object) is drawn free from its surroundings.

Drawing the force vectors on a free-body diagram to scale requires that we first determine the direction of the acceleration vector using kinematic methods. We know the object is moving to the right with increasing speed. It follows from kinematics that its acceleration vector is in the direction that the velocity vector is changing—the forward direction. Note that $\vec{F}_n$ and $\vec{F}_g$ in the diagram have equal magnitudes. We know the magnitudes are equal because the vertical component of the acceleration is zero. As a qualitative check on the plausibility of our free-body diagram, we draw a vector-addition diagram (Figure 4-14) verifying that the vector sum of the forces is in the same direction as the acceleration vector.

We can now apply Newton's second law to determine the x and y components of the net force on the sled-rope particle. The x component of Newton's second law gives

$$\Sigma F_x = F_{nx} + F_{gx} + F_x = ma_x$$

$$0 + 0 + F = ma_x$$

or

$$a_x = \frac{F}{m}$$

The y component of Newton's second law gives

$$\Sigma F_y = F_{ny} + F_{gy} + F_y = ma_y$$

$$F_n - F_g + 0 = 0$$

or

$$F_n = F_g$$

Thus, the sled-rope particle has an acceleration in the $+x$ direction of F/m and the magnitude of the vertical force $\vec{F}_n$ exerted by the ice is $F_n = F_g = mg$.

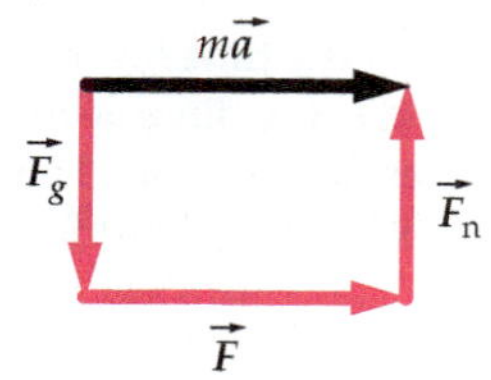

FIGURE 4-14 The vector sum of the forces in the free-body diagram is equal the mass times the acceleration vector.

PROBLEM-SOLVING STRATEGY

Applying Newton's Second Law

PICTURE Make sure you identify all of the forces acting on a particle. Then, determine the direction of the acceleration vector of the particle, if possible. Knowing the direction of the acceleration vector will help you choose the best coordinate axes for solving the problem.

SOLVE

1. Draw a neat diagram that includes the important features of the problem.
2. Isolate the object (particle) of interest, and identify each force that acts on it.
3. Draw a free-body diagram showing each of these forces.
4. Choose a suitable coordinate system. If the direction of the acceleration vector is known, choose a coordinate axis parallel to that direction. For objects sliding along a surface, choose one coordinate axis parallel to the surface and the other perpendicular to it.
5. Apply Newton's second law, $\Sigma\vec{F} = m\vec{a}$, usually in component form.
6. Solve the resulting equations for the unknowns.

CHECK Make sure your results have the correct units and seem plausible. Substituting extreme values into your symbolic solution is a good way to check your work for errors.

Example 4-6 A Dogsled Race

During your winter break, you enter a dogsled race in which students replace the dogs. Wearing cleats for traction, you begin the race by pulling on a rope attached to the sled with a force of 150 N at 25° above the horizontal. The mass of the sled–passenger–rope particle is 80 kg and there is negligible friction between the sled runners and the ice. Find (*a*) the acceleration of the sled and (*b*) the magnitude of the normal force exerted by the surface on the sled.

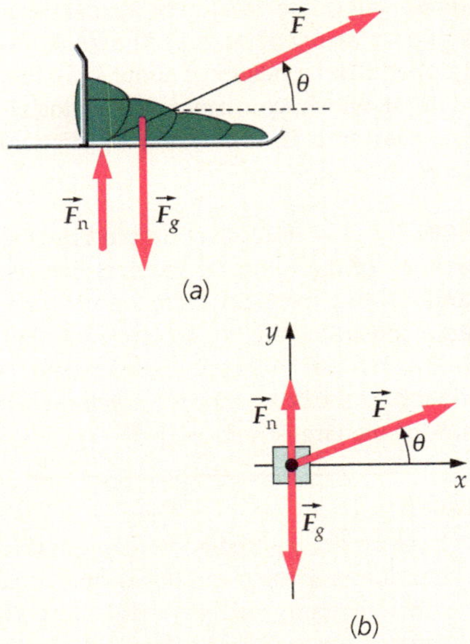

FIGURE 4-15

PICTURE Three forces act on the particle: its weight $\vec{F}_g$, which acts downward; the normal force $\vec{F}_n$, which acts upward; and the force with which you pull the rope $\vec{F}$, directed 25° above the horizontal. Because the forces are not all parallel to a single line, we study the system by applying Newton's second law to the x and y directions separately.

SOLVE

(*a*) 1. Sketch a free-body diagram (Figure 4-15*b*) of the sled-passenger-rope particle. Include a coordinate system with one of the coordinate axes in the direction of the acceleration. The particle moves to the right with increasing speed, so we know the acceleration is also to the right:

2. *Note:* Use the head-to-tail method of vector addition to verify that the sum of the forces on the free-body diagram can be in the direction of the acceleration (Figure 4-16):

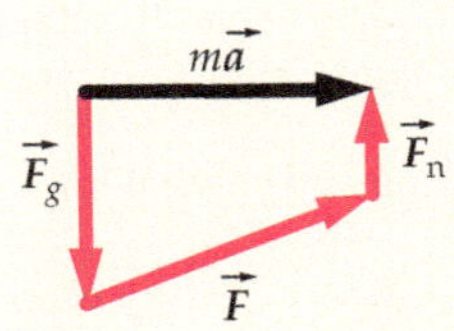

FIGURE 4-16 The vector sum of the forces in the free-body diagram is equal the mass times the acceleration vector.

3. Apply Newton's second law to the particle. Write out the equation in both vector and component form:

$$\vec{F}_n + \vec{F}_g + \vec{F} = m\vec{a} \quad \text{or}$$
$$F_{nx} + F_{gx} + F_x = ma_x$$
$$F_{ny} + F_{gy} + F_y = ma_y$$

4. Express the x components of $\vec{F}_n$, $\vec{F}_g$, and $\vec{F}$:

$$F_{nx} = 0, \quad F_{gx} = 0, \quad \text{and} \quad F_x = F\cos\theta$$

5. Substitute the step-4 results into the x component equation in step 3. Then solve for the acceleration a_x:

$$\Sigma F_x = 0 + 0 + F\cos\theta = ma_x \quad \text{so}$$
$$a_x = \frac{F\cos\theta}{m} = \frac{(150\text{ N})\cos 25°}{80\text{ kg}} = \boxed{1.7\text{ m/s}^2}$$

(*b*) 1. Express the y component of $\vec{a}$:

$$a_y = 0$$

2. Express the y components of $\vec{F}_n$, $\vec{F}_g$, and $\vec{F}$:

$$F_{ny} = F_n, \quad F_{gy} = -mg, \quad \text{and} \quad F_y = F\sin\theta$$

3. Substitute the Part-(*b*), steps-1 and -2 results into the y component equation in Part-(*a*), step-3. Then solve for F_n:

$$\Sigma F_y = F_n - mg + F\sin\theta = 0$$
$$F_n = mg - F\sin\theta$$
$$= (80\text{ kg})(9.81\text{ N/kg}) - (150\text{ N})\sin 25° = \boxed{7.2 \times 10^2\text{ N}}$$

CHECK Note that only the x component of $\vec{F}$, which is $F\cos\theta$, causes the object to accelerate. We expect the acceleration to be less if the rope is not horizontal. Also, we expect the normal force exerted by the ice to counter less than the full weight of the object because part of the weight is countered by the force exerted by the rope.

PRACTICE PROBLEM 4-6 If $\theta = 25°$, what is the maximum of the magnitude of the force $\vec{F}$ that can be applied to the rope without lifting the sled off the surface?

Example 4-7 Unloading a Truck

Context-Rich

You are working for a big delivery company, and must unload a large, fragile package from your truck, using a delivery ramp (Figure 4-17). If the downward component of the velocity of the package when it reaches the bottom of the ramp is greater than 2.50 m/s (2.50 m/s is the speed an object would have if it were dropped from a height of about 1 ft), the package will break. What is the largest angle at which you can safely unload? The ramp is 1.00 m high, has rollers (i.e., the ramp is approximately frictionless), and is inclined at an angle θ to the horizontal.

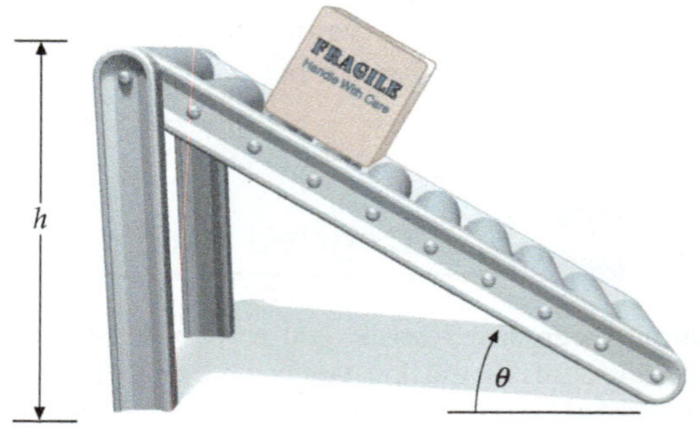

FIGURE 4-17

PICTURE Two forces act on the box, the gravitational force $\vec{F}_g$ and the normal force $\vec{F}_n$ of the ramp on the box. Because these forces are not antiparallel, they cannot sum to zero. So, there is a net force on the box causing it to accelerate. The ramp constrains the box to move parallel to its surface. We choose down the incline as the $+x$ direction. To determine the acceleration, we apply Newton's second law to the box. Once the acceleration is known, we can use kinematics to determine the largest safe angle.

SOLVE

1. First we draw a free-body diagram (Figure 4-18). Two forces act on the package, the gravitational force and the normal force. We choose the direction of the acceleration, down the ramp, as the $+x$ direction. (*Note:* The angle between $\vec{F}_g$ and the $-y$ direction is the same as the angle between the horizontal and the incline as we see from the free-body diagram.) We can also see that $F_{gx} = F_g \sin\theta$.

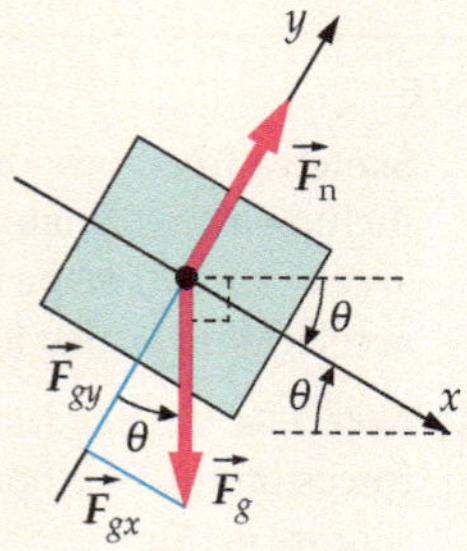

FIGURE 4-18

2. To find a_x we apply Newton's second law ($\Sigma F_x = ma_x$) to the package. (*Note:* $\vec{F}_n$ is perpendicular to the x axis and $F_g = mg$.)

$$F_{nx} + F_{gx} = ma_x \quad \text{where}$$
$$F_{nx} = 0 \quad \text{and} \quad F_{gx} = F_g \sin\theta = mg\sin\theta$$

3. Substituting and solving for the acceleration gives:

$$0 + mg\sin\theta = ma_x \quad \text{so} \quad a_x = g\sin\theta$$

4. Relate the downward component of the velocity of the box to its velocity component v_x in the x direction:

$$v_d = v_x \sin\theta$$

5. The velocity component v_x is related to the displacement Δx along the ramp by the kinematic equation:

$$v_x^2 = v_{0,x}^2 + 2a_x\,\Delta x$$

6. Substituting for a_x in the kinematic equation (step 5) and setting $v_{0,x}$ to zero gives:

$$v_x^2 = 2g\sin\theta\,\Delta x$$

7. From Figure 4-17, we can see that when Δx equals the length of the ramp, $\Delta x \sin\theta = h$, where h is the height of the ramp: $v_x^2 = 2gh$
8. Solve for v_d using the step-4 result and the expression for v_x from step 7: $v_d = \sqrt{2gh}\sin\theta$
9. Solve for the maximum angle: $2.50\text{ m/s} = \sqrt{2(9.81\text{ m/s}^2)(1.00\text{ m})}\sin\theta_{max}$

$\therefore\ \theta_{max} = \boxed{34.4°}$

CHECK At an angle of 34.4°, the downward component of the velocity will be slightly greater then half the speed that the box would have if it were dropped from a height of 1.00 m.

TAKING IT FURTHER The acceleration down the incline is constant and equal to $g\sin\theta$. In addition, the speed v at the bottom depends upon h but not upon the angle θ.

PRACTICE PROBLEM 4-7 Apply $\Sigma F_y = ma_y$ to the package and show that $F_n = mg\cos\theta$.

Example 4-8 Picture Hanging

Try It Yourself

A picture weighing 8.0 N is supported by two wires with tensions T_1 and T_2, as shown in Figure 4-19. Find each tension.

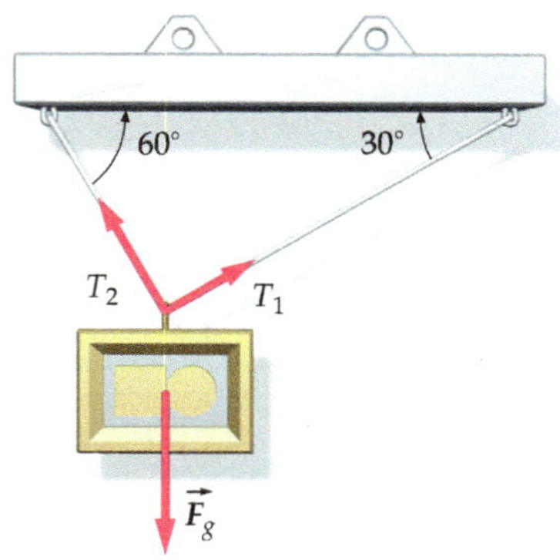

FIGURE 4-19

PICTURE Because the picture does not accelerate, the net force acting on it must be zero. The three forces acting on the picture (the gravitational force $\vec{F}_g$, and the tension forces $\vec{T}_1$ and $\vec{T}_2$) must therefore sum to zero.

SOLVE

Cover the column to the right and try these on your own before looking at the answers.

Steps:	Answers
1. Draw a free-body diagram for the picture (Figure 4-20). On your diagram show the x and y components of each tension force.	
2. Apply $\Sigma\vec{F} = m\vec{a}$ in vector form to the picture.	$\vec{T}_1 + \vec{T}_2 + \vec{F}_g = m\vec{a}$
3. Resolve each force into its x and y components. This gives you two equations for the two unknowns T_1 and T_2. The acceleration is zero.	$T_{1x} + T_{2x} + F_{gx} = 0$ $T_1\cos30° - T_2\cos60° + 0 = 0$ and $T_{1y} + T_{2y} + F_{gy} = 0$ $T_1\sin30° + T_2\sin60° - F_g = 0$
4. Solve the x component equation for T_2.	$T_2 = T_1\dfrac{\cos30°}{\cos60°}$
5. Substitute your result for T_2 (from step 4) into the y component equation and solve for T_1.	$T_1\sin30° + \left(T_1\dfrac{\cos30°}{\cos60°}\right)\sin60° - F_g = 0$ $T_1 = 0.50F_g = \boxed{4.0\text{ N}}$
6. Use your result for T_1 to find T_2.	$T_2 = T_1\dfrac{\cos30°}{\cos60°} = \boxed{6.9\text{ N}}$

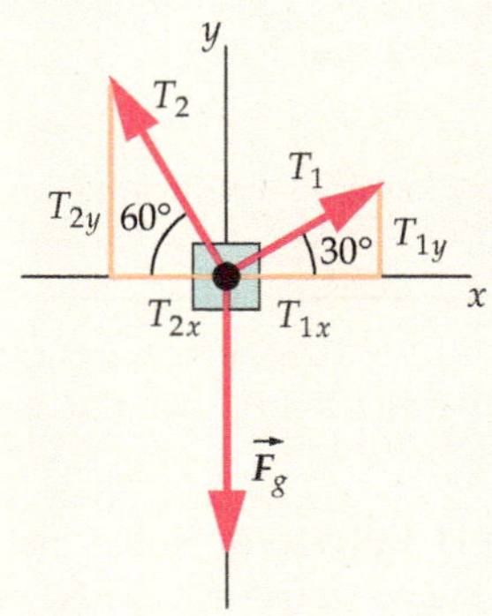

FIGURE 4-20

CHECK The more vertical of the two wires supports the greater share of the load, as you might expect. Also, we see that $T_1 + T_2 > 8$ N. The "extra" force is due to the wires pulling to the right and left.

Example 4-9 An Accelerating Jet Plane

As your jet plane speeds down the runway on takeoff, you decide to determine its acceleration, so you take out your yo-yo and note that when you suspend it, the string makes an angle of 22.0° with the vertical (Figure 4-21*a*). (*a*) What is the acceleration of the plane? (*b*) If the mass of the yo-yo is 40.0 g, what is the tension in the string?

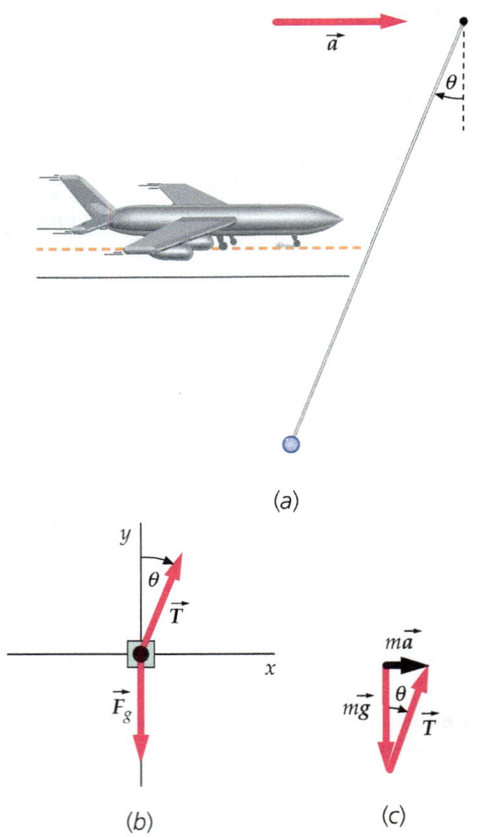

FIGURE 4-21

PICTURE Both the yo-yo and plane have the same acceleration. The net force on the yo-yo is in the direction of its acceleration—to the right. This force is supplied by the horizontal component of the tension force $\vec{T}$. The vertical component of $\vec{T}$ balances the gravitational force $\vec{F}_g$ on the yo-yo. We choose a coordinate system in which the $+x$ direction is in the direction of the acceleration vector $\vec{a}$ and the $+y$ direction is vertically upward. Writing Newton's second law for both the x and y directions gives two equations to determine the two unknowns, a and T.

SOLVE

(*a*) 1. Draw a free-body diagram for the yo-yo (Figure 4-21*b*). Choose the $+x$ direction to be the direction of the yo-yo's acceleration vector.

2. Apply $\Sigma F_x = ma_x$ to the yo-yo. Then simplify using trigonometry:

$$T_x + F_{gx} = ma_x$$

$$T\sin\theta + 0 = ma_x$$

or

$$T\sin\theta = ma_x$$

3. Apply $\Sigma F_y = ma_y$ to the yo-yo. Then, simplify using trigonometry (Figure 4-21) and $F_g = mg$. Since the acceleration is in the $+x$ direction, $a_y = 0$:

$$T_y + F_{gy} = ma_y$$

$$T\cos\theta - mg = 0$$

or

$$T\cos\theta = mg$$

4. Divide the step-2 result by the step 3 result and solve for the acceleration. Because the acceleration vector is in the $+x$ direction, $a = a_x$:

$$\frac{T\sin\theta}{T\cos\theta} = \frac{ma_x}{mg} \quad \text{so} \quad \tan\theta = \frac{a_x}{g} \quad \text{and}$$

$$a_x = g\tan\theta = (9.81\ \text{m/s}^2)\tan 22.0° = \boxed{3.96\ \text{m/s}^2}$$

(*b*) Using the step-3 result, solve for the tension:

$$T = \frac{mg}{\cos\theta} = \frac{(0.0400\ \text{kg})(9.81\ \text{m/s}^2)}{\cos 22.0°} = \boxed{0.423\ \text{N}}$$

CHECK At $\theta = 0$, $\cos\theta = 1$ and $\tan\theta = 0$. Substituting these values into the expressions in the last two steps of the solution gives and $a_x = 0$ and $T = mg$, as expected.

TAKING IT FURTHER Notice that for the Part-(*b*) result T is greater than the gravitational force on the yo-yo ($mg = 0.392$ N) because the cord not only keeps the yo-yo from falling but also accelerates it in the horizontal direction. Here we use the units m/s² for g (instead of N/kg) because we are calculating acceleration.

PRACTICE PROBLEM 4-8 For what acceleration magnitude a would the tension in the string be equal to 3.00 mg? What is θ in this case?

Our next example is the application of Newton's second law to objects that are at rest relative to a reference frame that is itself accelerating.

Example 4-10 "Weighing" Yourself in an Elevator

Suppose that your mass is 80 kg, and you are standing on a scale fastened to the floor of an elevator. The scale measures force and is calibrated in newtons. What does the scale read when (*a*) the elevator is rising with upward acceleration of magnitude a; (*b*) the elevator is descending with downward acceleration of magnitude a'; (*c*) the elevator is rising at 20 m/s and its speed is decreasing at a rate of $8.0\,\mathrm{m/s^2}$?

PICTURE The scale reading is the magnitude of the normal force exerted by the scale on you (Figure 4-22). Because you are at rest relative to the elevator, you and the elevator have the same acceleration. Two forces act on you: the downward force of gravity, $\vec{F}_g = m\vec{g}$, and the upward normal force from the scale, F_n. The sum of these forces gives you the observed acceleration. We choose upward to be the $+y$ direction.

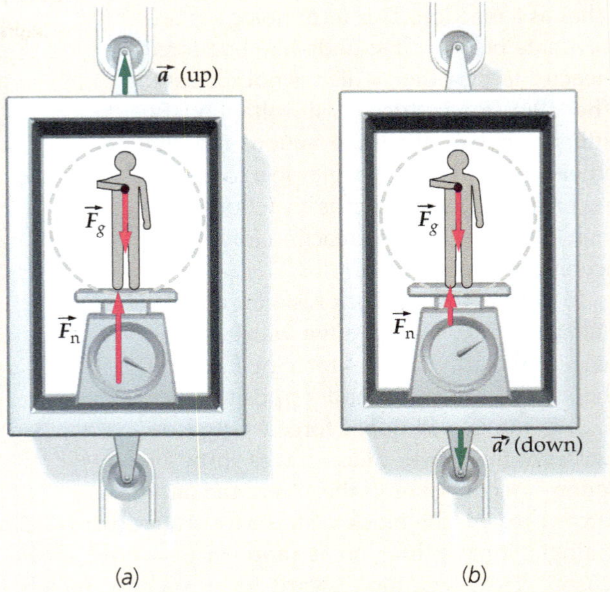

FIGURE 4-22

SOLVE

(*a*) 1. Draw a free-body diagram of yourself (Figure 4-23):

2. Apply $\Sigma F_y = ma_y$: $F_{ny} + F_{gy} = ma_y$

$F_n - mg = ma_y$

3. Solve for F_n. This is the reading on the scale (your apparent weight): $F_n = mg + ma_y = m(g + a_y)$

4. $a_y = +a$: $F_n = \boxed{m(g + a)}$

(*b*) $a_y = -a'$. Substitute for a_y in the Part-(*a*), step-3 result: $F_n = m(g + a_y) = \boxed{m(g - a')}$

(*c*) The velocity is positive but decreasing, so the acceleration is negative. Thus, $a_y = -8.0\,\mathrm{m/s^2}$. Substitute into the Part-(*a*), step-3 result: $F_n = m(g + a_y) = (80\text{ kg})(9.81\text{ m/s}^2 - 8.0\text{ m/s}^2)$

$= 144.8\text{ N} = \boxed{1.40 \times 10^2\text{ N}}$

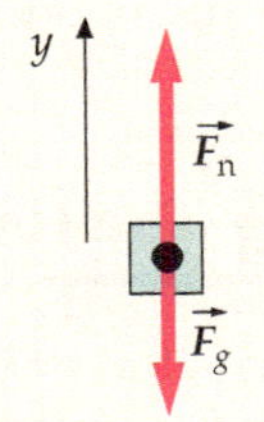

FIGURE 4-23

CHECK Independent of whether the elevator is ascending or descending, if its acceleration is upward you would expect to "feel heavier" and expect your apparent weight to be greater than mg. This is in keeping with the Part-(*a*) result. If its acceleration is downward you would expect to "feel lighter" and expect your apparent weight to be less than mg. The results for Parts (*b*) and (*c*) are in agreement with these expectations.

PRACTICE PROBLEM 4-9 A descending elevator comes to a stop with an acceleration of magnitude $4.00\,\mathrm{m/s^2}$. If your mass is 70.0 kg and you are standing on a force scale in the elevator, what does the scale read as the elevator is stopping?

4-7 NEWTON'S THIRD LAW

Newton's third law describes an important property of forces: forces always occur in pairs. For example, if a force is exerted on some object A, there must be another object B exerting the force. Newton's third law states that these forces are equal in magnitude and opposite in direction. That is, if object A exerts a force on object B, then B exerts an equally strong but oppositely directed force on A.

Third law. When two bodies interact, the force $\vec{F}_{BA}$ exerted by object B on object A is equal in magnitude and opposite in direction to the force $\vec{F}_{AB}$ exerted by object A on object B. Thus,

$$\vec{F}_{BA} = -\vec{F}_{AB} \qquad \text{4-8}$$

NEWTON'S THIRD LAW

Each pair of forces is called a Newton's third-law **(N3L) pair.** It is common to refer to one force in the pair as an action and the other as a reaction. This terminology is unfortunate because it sounds like one force "reacts" to the other, which is not the case. The two forces occur simultaneously. Either can be called the action and the other the reaction. If we refer to a force acting on a particular object as an action force, then the corresponding reaction force must act on a different object.

In Figure 4-24, a block rests on a table. The force $\vec{F}_{gEB}$ acting downward on the block is the gravitational force by Earth on the block. An equal and opposite force $\vec{F}_{gBE}$ is the gravitational force exerted on Earth by the block. These forces form an action–reaction pair. If they were the only forces present, the block would accelerate downward because it would have only a single force acting on it (and Earth would accelerate upward for the same reason). However, the upward force $\vec{F}_{nTB}$ by the table on the block balances the gravitational force on the block. In addition, there is a downward force $\vec{F}_{nBT}$ by the block on the table. The forces $\vec{F}_{nBT}$ and $\vec{F}_{nTB}$ form a Newton's third law pair and, thus, are equal and opposite.

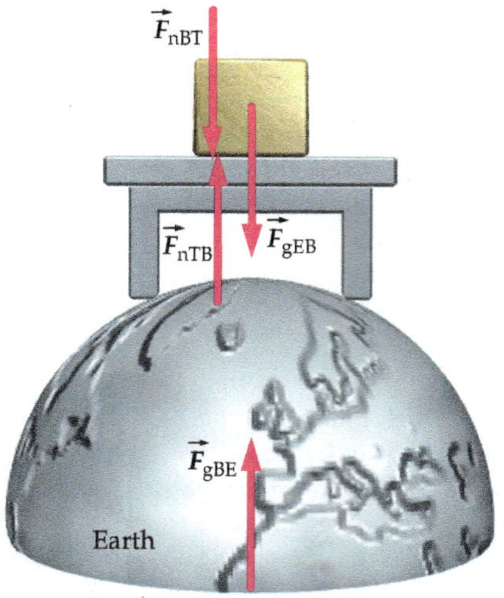

FIGURE 4-24

> Newton's third-law force pairs are *always* equal and opposite.

> No two external forces acting on *the same object* can ever constitute a Newton's third-law pair.

CONCEPT CHECK 4-3

Do the forces $\vec{F}_{gBE}$ and $\vec{F}_{nBT}$ in Figure 4-24 form a Newton's third-law pair?

Example 4-11 The Horse Before the Cart *Conceptual*

A horse refuses to pull a cart (Figure 4-25*a*). The horse reasons, "according to Newton's third law, whatever force I exert on the cart, the cart will exert an equal and opposite force on me, so the net force will be zero and I will have no chance of accelerating the cart." What is wrong with this reasoning?

PICTURE Because we are interested in the motion of the cart, we draw a simple diagram for it (Figure 4-25*b*). The force exerted by the horse on the cart is labeled $\vec{F}_{HC}$. (This force is actually exerted on the harness. Because the harness is attached to the cart, we consider it a part of the cart.) Other forces acting on the cart are the gravitational force of Earth on the cart $\vec{F}_{gEC}$, the normal force of the pavement on the cart $\vec{F}_{n\,PC}$ and the frictional force exerted by the pavement on the cart, labeled $\vec{f}_{PC}$.

SOLVE

1. Draw a free-body diagram for the cart (see Figure 4-25*c*). Because the cart does not accelerate vertically, the vertical forces must sum to zero. The horizontal forces are $\vec{F}_{HC}$ to the right and $\vec{f}_{PC}$ to the left. The cart will accelerate to the right if $|\vec{F}_{HC}|$ is greater than $|\vec{f}_{PC}|$.

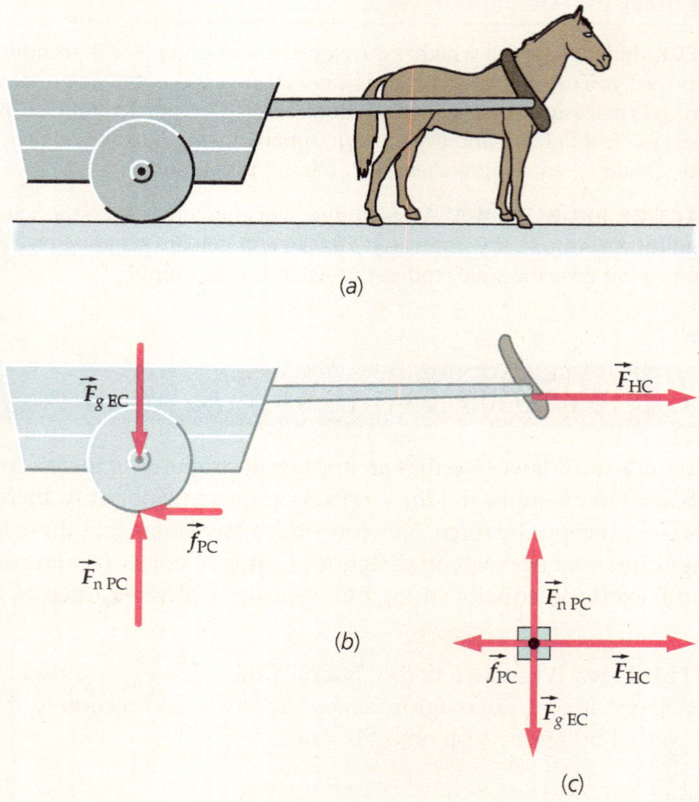

FIGURE 4-25

2. Note that the reaction force to $\vec{F}_{HC}$, which we call $\vec{F}_{CH}$, is exerted on the horse, not on the cart (Figure 4-25*d*). It has no effect on the motion of the cart, but it does affect the motion of the horse. If the horse is to accelerate to the right, there must be a force $\vec{F}_{PH}$ (to the right) exerted on the horse's hooves by the pavement that is greater in magnitude than $\vec{F}_{CH}$.

Because the reaction force to $\vec{F}_{HC}$ is exerted on the horse, it has no effect on the motion of the cart. This is the flaw in the horse's reasoning.

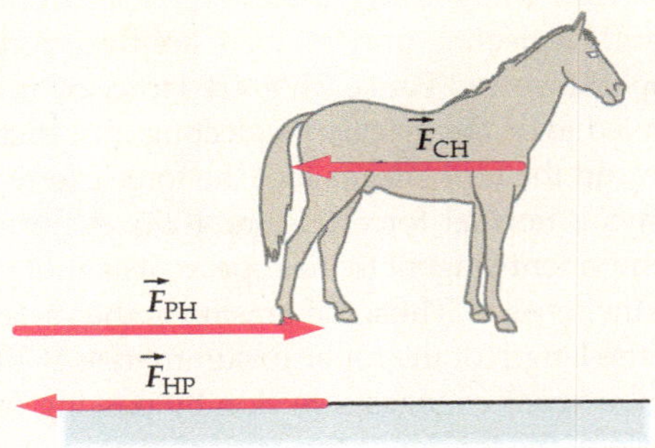

FIGURE 4-25 *(continued)*

CHECK All forces on the cart have C for a rightmost subscript, and all forces on the horse have H for a rightmost subscript. Thus, we have not drawn both forces of a Newton's third-law force pair on either the horse or the cart.

TAKING IT FURTHER This example illustrates the importance of drawing a simple diagram when solving mechanics problems. Had the horse done so, he would have seen that he need only push back hard against the pavement so that the pavement will push him forward.

CONCEPT CHECK 4-4

As you stand facing a friend, place your palms against your friend's palms and push. Can your friend exert a force on you if you do not exert a force back? Try it.

4-8 PROBLEM SOLVING: PROBLEMS WITH TWO OR MORE OBJECTS

In some problems, the motions of two (or more) objects are influenced by the interactions between the objects. For example, such objects might touch each other, or they might be connected to each other by a string or spring.

The tension in a string or rope is the magnitude of the force that one segment of the rope exerts on a neighboring segment. The tension can vary throughout the length of the rope. For a rope dangling from a girder at the ceiling of a gymnasium, the tension is greatest at points near the ceiling because a short segment of rope next to the ceiling has to support the weight of all the rope below it. For the problems in this book, however, you can almost always assume that the masses of strings and ropes are negligible, so variations in tension due to the weight of a string or rope can be neglected. Conveniently, this also means that you may assume that variations in the tension due to the acceleration of a string or rope can also be neglected.

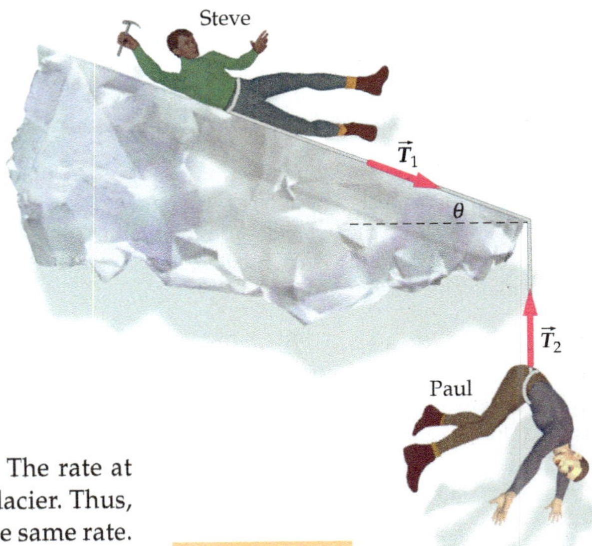

FIGURE 4-26

Consider, for example, the motion of Steve and Paul in Figure 4-26. The rate at which Paul descends equals the rate at which Steve slides along the glacier. Thus, their speeds remain equal. If Paul gains speed, Steve gains speed at the same rate. That is, their tangential accelerations remain equal. (The tangential acceleration of a particle is the component of the acceleration that is tangent to the path of the motion of the particle.)

The free-body diagram of a segment of the rope attached to Steve, where Δm_s is the segment's mass, is shown in Figure 4-27. Applying Newton's second law to the segment gives $T - T' = \Delta m_s a_x$. If the mass of the segment is negligible, then $T = T'$. To give a segment of negligible mass any finite acceleration, a net force of only a negligible magnitude is needed. (That is, only a negligible difference in tension is needed to give a rope segment of negligible mass any finite acceleration.)

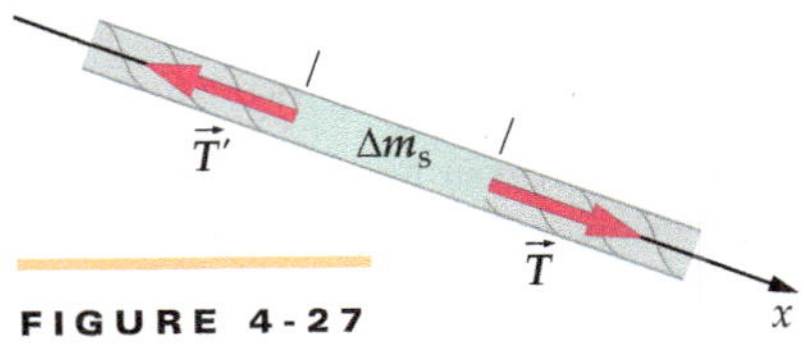

FIGURE 4-27

Next, we consider the entire rope connecting Steve and Paul. Neglecting gravity, there are three forces acting on the rope. Steve and Paul each exert a force on it, as does the ice at the edge of the glacier. Neglecting any friction between the ice and the rope means that the force exerted by the ice is always a normal force (Figure 4-28). A normal force has no component tangent to the rope, so it cannot produce a change in the tension. Thus, the tension is the same throughout the entire length of the rope. To summarize, if a taut rope of negligible mass changes direction by passing over a frictionless surface, the tension is the same throughout the rope. The following box summarizes the steps for solving such problems.

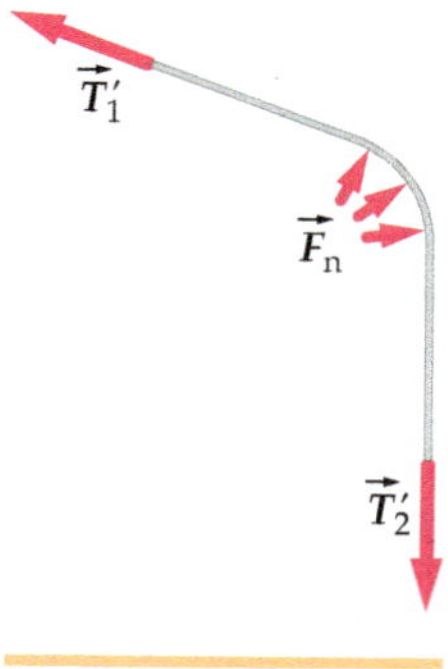

FIGURE 4-28

CONCEPT CHECK 4-5

Suppose that instead of passing over the edge of a glacier, the rope passed around a pulley with frictionless bearings as shown in Figure 4-29. Would the tension then be the same throughout the length of the rope?

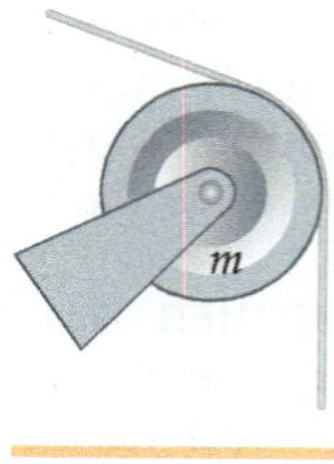

FIGURE 4-29

PROBLEM-SOLVING STRATEGY

Applying Newton's Laws to Problems with Two or More Objects

PICTURE Remember to draw a separate free-body diagram for each object. The unknowns can be obtained by solving simultaneous equations.

SOLVE

1. Draw a separate free-body diagram for each object. Use a separate coordinate system for each object. Remember, if two objects touch, the forces they exert on each other are equal and opposite (Newton's third law).
2. Apply Newton's second law to each object.
3. Solve the resultant equations, together with any equations describing interactions and constraints, for the unknown quantities.

CHECK Make sure your answer is consistent with the free-body diagrams that you have created.

Example 4-12 The Ice Climbers

Paul (mass m_P) accidentally falls off the edge of a glacier as shown in Figure 4-26. Fortunately, he is connected by a long rope to Steve (mass m_S), who has a climbing ax. Before Steve sets his ax to stop them, Steve slides without friction along the ice, attached by the rope to Paul. Assume there is no friction between the rope and the glacier. Find the acceleration of each person and the tension in the rope.

PICTURE The tension forces $\vec{T}_1$ and $\vec{T}_2$ have equal magnitudes because the rope is assumed massless and the glacial ice is assumed frictionless. The rope does not stretch or become slack, so Paul and Steve have equal speeds at all times. Their accelerations $\vec{a}_S$ and $\vec{a}_P$ must therefore be equal in magnitude (but not in direction). Steve accelerates down the face of the glacier whereas Paul accelerates straight downward. We can solve this problem by applying $\Sigma\vec{F} = m\vec{a}$ to each person, and then solving for the accelerations and the tension.

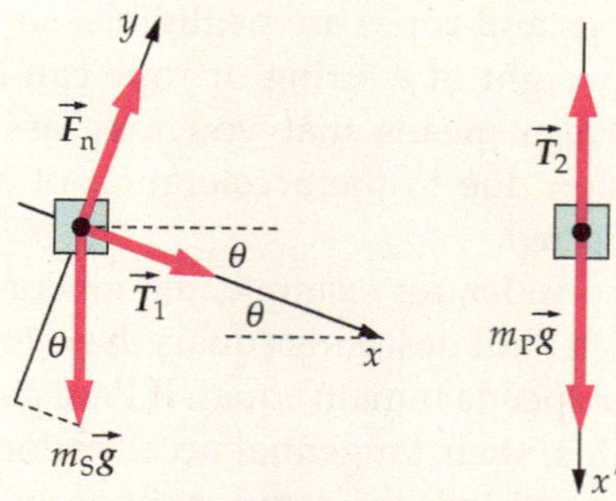

FIGURE 4-30

SOLVE

1. Draw separate free-body diagrams for Steve and Paul (Figure 4-30). Put axes x and y on Steve's diagram, choosing the direction of Steve's acceleration as the $+x$ direction. Choose the direction of Paul's acceleration as the $+x'$ direction.

2. Apply $\Sigma F_x = ma_x$ in the x direction to Steve:

$$F_{nx} + T_{1x} + m_S g_x = m_S a_{Sx}$$

3. Apply $\Sigma F_{x'} = ma_{x'}$ in the x' direction to Paul:

$$T_{2x'} + m_P g_{x'} = m_P a_{Px'}$$

4. Because they are connected by a taut rope that does not stretch, the accelerations of Paul and Steve are related. Express this relation:

$$a_{Px'} = a_{Sx} = a_t$$

a_t stands for the acceleration component in the *tangential* direction. (The direction of the motion.)

5. Because the rope is of negligible mass and slides over the ice with negligible friction, the forces $\vec{T}_1$ and $\vec{T}_2$ are simply related. Express this relation:
$$T_2 = T_1 = T$$

6. Substitute the steps-4 and -5 results into the step-2 and step-3 equations:
$$T + m_S g \sin\theta = m_S a_t$$
$$-T + m_P g = m_P a_t$$

7. Solve the step-6 equations for the acceleration by eliminating T and solving for a_t:
$$\boxed{a_t = \frac{m_S \sin\theta + m_P}{m_S + m_P} g}$$

8. Substitute the step-7 result into either step-6 equation and solve for T:
$$\boxed{T = \frac{m_S m_P}{m_S + m_P}(1 - \sin\theta)g}$$

CHECK If m_P is very much greater than m_S, we expect the acceleration to be approximately g and the tension to be approximately zero. Taking the limit as m_S approaches 0 does indeed give $a_t = g$ and $T = 0$ for this case. If m_P is much less than m_S, we expect the acceleration to be approximately $g \sin\theta$ (see step 3 of Example 4-7) and the tension to be zero. Taking the limit as m_P approaches 0 in steps 7 and 8, we indeed obtain at $a_t = g \sin\theta$ and $T = 0$. At an extreme value of the inclination ($\theta = 90°$) we again check our answers. Substituting $\theta = 90°$ in steps 7 and 8, we obtain $a_t = g$ and $T = 0$. This seems right since Steve and Paul would be in free-fall for $\theta = 90°$.

TAKING IT FURTHER In Step 1 we chose down the incline and straight down to be positive to keep the solution as simple as possible. With this choice, when Steve moves in the $+x$ direction (down the surface of the glacier), Paul moves in the $+x'$ direction (straight downward).

PRACTICE PROBLEM 4-10 (*a*) Find the acceleration if $\theta = 15°$ and if the masses are $m_S = 78\,\text{kg}$ and $m_P = 92\,\text{kg}$. (*b*) Find the acceleration if these two masses are interchanged.

Example 4-13 Building a Space Station

You are an astronaut constructing a space station, and you push on a box of mass m_1 with force $\vec{F}_{A1}$. The box is in direct contact with a second box of mass m_2 (Figure 4-31). (*a*) What is the acceleration of the boxes? (*b*) What is the magnitude of the force each box exerts on the other?

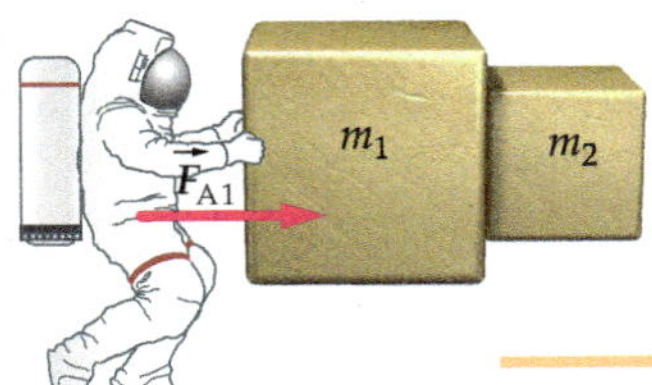

FIGURE 4-31

PICTURE Force $\vec{F}_{A1}$ is a contact force and only acts on box 1. Let $\vec{F}_{21}$ be the force exerted by box 2 on box 1, and $\vec{F}_{12}$ be the force exerted by box 1 on box 2. In accord with Newton's third law, these forces are equal and opposite ($\vec{F}_{21} = -\vec{F}_{12}$), so $F_{21} = F_{12}$. Apply Newton's second law to each box separately. The motions of the two boxes are identical, so the accelerations $\vec{a}_1$ and $\vec{a}_2$ are equal.

SOLVE

(*a*) 1. Draw free-body diagrams for the two boxes (Figure 4-32).

2. Apply $\Sigma\vec{F} = m\vec{a}$ to box 1.
$$F_{A1} - F_{21} = m_1 a_{1x}$$

3. Apply $\Sigma\vec{F} = m\vec{a}$ to box 2.
$$F_{12} = m_2 a_{2x}$$

4. Express both the relation between the two accelerations and the relation between the magnitudes of the forces the blocks exert on each other. The accelerations are equal because the speeds are equal at all times, so the rate of change of the speeds are equal. The forces are equal in magnitude because the forces constitute a N3L force pair:
$$a_{2x} = a_{1x} = a_x$$
$$F_{21} = F_{12} = F$$

5. Substitute these back into the step-2 and step-3 results and solve for a_x.
$$\boxed{a_x = \frac{F_{A1}}{m_1 + m_2}}$$

(*b*) Substitute your expression for a_x into either the step-2 or the step-3 result and solve for F.
$$\boxed{F = \frac{m_2}{m_1 + m_2} F_{A1}}$$

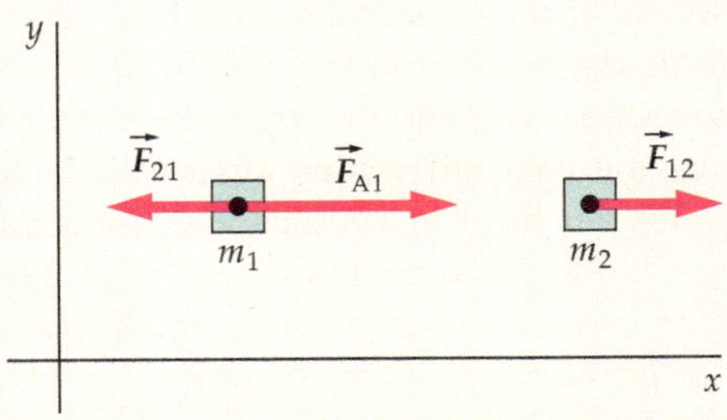

FIGURE 4-32

CHECK Note that the result in step 5 is the same as if the force $\vec{F}_{A1}$ had acted on a single mass equal to the sum of the masses of the two boxes. In fact, because the two boxes have the same acceleration, we can consider them to be a single particle with mass $m_1 + m_2$.

Physics Spotlight

Roller Coasters and the Need for Speed

Roller coasters have fascinated people since the spectacular Promenades Aeriennes (Aerial Walks) opened in Paris in 1817.* Until recently, though, ride designers were stuck with one major limitation—the ride needed to start at the top of a large hill.

In the 1970s, Anton Schwartzkopf, a German amusement park designer, was inspired by planes taking off from an aircraft carrier. In 1976, the Shuttle Loop roller coaster opened. A multiton weight was cranked to the top of a tower near the roller coaster. One end of a cable was attached to the weight, while the other was hooked to the coaster, in order to pull it. The weight dropped and quickly pulled the cable with the attached train of cars. In less than 3 seconds, the coaster train accelerated to 60 miles per hour.

At the same time, Schwartzkopf came up with a second catapult-style method of launching a coaster. A 5-ton flywheel was spun at high speed. A cable was connected to the coaster and the fly wheel. In less than 3 seconds, the coaster train, with up to 28 passengers, accelerated to speeds of 60 miles per hour. Both of these methods pioneered the use of catapult-style launches in roller coasters.†

Two new methods of launching roller coasters have allowed roller coasters to travel at even faster speeds. Intamin AG has created a *hydraulic*, or liquid-driven, system to pull the cable. The car alone for the Top Thrill Dragster weighs 12,000 pounds. Eighteen passengers are usually along for the ride, as well. The car is weighed as it passes over sensors, and a computer calculates just how fast the cable needs to go in order to catapult the car and passengers so they reach the top of the 420-foot first hill. Then, the liquid-filled motors quickly provide of up to 10,000 horsepower to wind the cable at up to 500 rpm, and to accelerate the coaster car to 120 miles per hour in 4 seconds.‡

Stan Checketts invented the first *pneumatic*, or compressed-air, roller coaster. The Thrust Air 2000™ is powered by a single, very large blast of air. The eight-passenger car is weighed as it passes over sensors. Then, four compressors swing into action. They pump air into a storage tank sitting at the base of a tower. The compressed air is measured into a shot tank, depending on the weight of the car. Finally, the air rapidly releases through a valve in the top of the tower, pushing against a piston that drives the catapult pulley system. The fully loaded car accelerates to 80 miles per hour in 1.8 seconds. A minimum of 40,000 pounds of thrust is needed to produce this acceleration. For comparison, a single F-15 jet engine is rated at a maximum 29,000 pounds of thrust.§ Roller coasters now are powered by a thrust more powerful than a jet engine. Something to think about the next time you pass by an amusement park.

The Hypersonic XLC at Paramount's King's Dominion Amusement Park, Virginia, the world's first compressed-airlaunched roller coaster, goes from 0 to 80 miles per hour in 1.8 seconds. *(Courtesy of King's Dominion Amusement Park.)*

* Cartmell, Robert, *The Incredible Scream Machine: A History of the Roller Coaster.* Bowling Green State University Popular Press, Bowling Green Ohio. 1987.

† "The Tidal Wave" http://www.greatamericaparks.com/tidalwave.html Marriott Great America Parks, 2006; Cartmell, op. cit.

‡ Hitchcox, Alan L. "Want Thrills? Go with Hydraulics." Hydraulics and Pneumatics, July 2005.

§ Goldman, Lea. "Newtonian Nightmare." *Forbes*, 7/23/2001. Vol. 168, Issue 2; "The F-100 Engine." http://www.pratt-whitney.com/prod_mil_f100.asp Pratt & Whitney, March 2006.

SUMMARY

1. Newton's laws of motion are fundamental laws of nature that serve as the basis for our understanding of mechanics.
2. Mass is an *intrinsic* property of an object.
3. Force is an important *derived* dynamic quantity.

TOPIC	RELEVANT EQUATIONS AND REMARKS
1. Newton's Laws	
First law	An object at rest stays at rest *unless* acted on by an external force. An object in motion continues to travel with constant velocity *unless* acted on by an external force. (Reference frames in which these statement hold are called inertial reference frames.)
Second law	The acceleration of an object is directly proportional to the net force acting on it. The reciprocal of the mass of the object is the proportionality constant. Thus $\vec{F}_{\text{net}} = m\vec{a}, \quad \text{where} \quad \vec{F}_{\text{net}} = \Sigma\vec{F}$ 4-1
Third law	When two bodies interact, the force $\vec{F}_{\text{BA}}$ exerted by object B on object A is equal in magnitude and opposite in direction to the force $\vec{F}_{\text{AB}}$ exerted by object A on object B: $\vec{F}_{\text{BA}} = -\vec{F}_{\text{AB}}$ 4-8
2. Inertial Reference Frames	Our statements of Newton's first and second laws are valid only in inertial reference frames. Any reference frame that is moving with constant velocity relative to an inertial reference frame is itself an inertial reference frame, and any reference frame that is accelerating relative to an inertial frame is not an inertial reference frame. Earth's surface is, to a good approximation, an inertial reference frame.
3. Force, Mass, and Weight	
Force	Force is defined in terms of the acceleration it produces on a given object. A force of 1 newton (N) is that force which produces an acceleration of 1 m/s^2 on a mass of 1 kilogram (kg).
Mass	Mass is an intrinsic property of an object. It is the measure of the inertial resistance of the object to acceleration. Mass does not depend on the location of the object. Applying identical forces to each of two objects and measuring their respective accelerations allows the masses of two objects to be compared. The ratio of the masses of the objects is equal to the inverse ratio of the accelerations produced: $\frac{m_2}{m_1} = \frac{a_1}{a_2}$
Gravitational Force	The gravitational force $\vec{F}_g$ on an object near the surface of Earth is the force of gravitational attraction exerted by Earth on the object. It is proportional to the gravitational field $\vec{g}$ (which is equal to the free-fall acceleration), and the mass m of the object is the proportionality constant: $\vec{F}_g = m\vec{g}$ 4-4 The weight of an object is the magnitude of the gravitational force on the object.
4. Fundamental Forces	All the forces observed in nature can be explained in terms of four basic interactions:
	1. The gravitational interaction
	2. The electromagnetic interaction
	3. The weak interaction*
	4. The strong nuclear interaction (also called the *hadronic force*)
5. Contact Forces	Contact forces of support and friction and those exerted by springs and strings are due to molecular forces that arise from the basic electromagnetic force.
Hooke's law	When an unstressed spring is compressed or extended by a small amount Δx, the restoring force it exerts is proportional to Δx: $F_x = -k\Delta x$ 4-7

* The electromagnetic and weak interactions are now viewed as the electroweak interaction.

Answers to Concept Checks

4-1 No, the net force is not an actual force. It is the vector sum of the actual forces.

4-2 No, it is the net force that causes the acceleration of the mass.

4-3 No, they do not.

4-4 No. Doing so would be contrary to Newton's third law.

4-5 No. Doing away with friction in the bearing is one thing, but the pulley still has mass. A difference in tension is needed in order to change the rate of rotation of the pulley wheel. Pulleys with non-negligible mass are studied in Chapter 8.

Answers to Practice Problems

4-1 1.5 kg

4-2 0.47 lb

4-3 13 cm

4-4 3.0 cm

4-5 $T = C\sqrt{m/k}$ where C is some dimensionless constant. The correct expression for the period, as we will see in Chapter 14, is $T = 2\pi\sqrt{m/k}$.

4-6 1.9 kN

4-7 Applying Newton's second law (for y components), we see from the free-body diagram (Figure 4-18) that $\Sigma F_y = ma_y \Rightarrow F_n - F_g\cos\theta = 0$, where we have used that a_y equals zero. Thus, $F_n = F_g\cos\theta$.

4-8 $a = 27.8\ \text{m/s}^2$, $\theta = 70.5°$

4-9 967 N

4-10 (*a*) $a_t = 0.66g$, (*b*) $a_t = 0.60g$

Problems

In a few problems, you are given more data than you actually need; in a few other problems, you are required to supply data from your general knowledge, outside sources, or informed estimate.

Interpret as significant all digits in numerical values that have trailing zeros and no decimal points.

For all problems, use $g = 9.81\ \text{m/s}^2$ for the free-fall acceleration due to gravity and neglect friction and air resistance unless instructed to do otherwise.

- • Single-concept, single-step, relatively easy
- •• Intermediate-level, may require synthesis of concepts
- ••• Challenging
- SSM Solution is in the *Student Solutions Manual*
- Consecutive problems that are shaded are paired problems.

CONCEPTUAL PROBLEMS

1 • While on a very smooth level transcontinental plane flight, your coffee cup sits motionless on your tray. Are there forces acting on the cup? If so, how do they differ from the forces that would be acting on the cup if it sat on your kitchen table at home?

2 • You are passing another car on a highway and determine that, relative to you, the car you pass has an acceleration $\vec{a}$ toward the west. However, the driver of the other car is maintaining a constant speed and direction relative to the road. Is the reference frame of your car an inertial one? If not, in which direction (east or west) is your car accelerating relative to the other car?

3 • **Context-Rich** You are riding in a limousine that has opaque windows that do not allow you to see outside. The car is on a flat horizontal plain, so the car can accelerate by speeding up, slowing down, or turning. Equipped with just a small heavy object on the end of a string, how can you use it to determine if the limousine is changing either speed or direction? Can you determine the limousine's velocity? SSM

4 •• If only a single nonzero force acts on an object, does the object accelerate relative to all inertial reference frames? Is it possible for such an object to have zero velocity in some inertial reference frame and not in another? If so, give a specific example.

5 •• A baseball is acted upon by a single known force. From this information alone, can you tell in which direction the baseball is moving relative to some reference frame? Explain.

6 •• A truck moves directly away from you at constant velocity (as observed by you while standing in the middle of the road). It follows that (*a*) no forces act on the truck, (*b*) a constant net force acts on the truck in the direction of its velocity, (*c*) the net force acting on the truck is zero, (*d*) the net force acting on the truck is its weight.

7 • **Engineering Application** Several space probes have been launched that are now far out in space. *Pioneer 10*, for example, was launched in the 1970s and is still moving away from the Sun and its planets. Is the mass of *Pioneer 10* changing? Which of the known fundamental forces continue to act on it? Does it have a net force on it?

8 •• **Engineering Application** Astronauts in apparent weightlessness during their stay on the International Space Station must carefully monitor their masses because significant loss of body mass is known to cause serious medical problems. Give an example of how you might design equipment to measure the mass of an astronaut on the orbiting space station.

9 •• **Context-Rich** You are riding in an elevator. Describe two situations in which your apparent weight is greater than your true weight. SSM

10 •• Suppose you are in a train moving at constant velocity relative to the ground. You toss a ball to your friend several seats in front of you. Use Newton's second law to explain why you cannot use your observations of the tossed ball to determine the train's velocity relative to the ground.

11 •• Explain why, of the fundamental interactions, gravitational interaction is the main concern in our everyday lives. One other on this list also plays an increasingly significant role in our rapidly advancing technology. Which one is that? Why are the others not obviously important?

12 •• Give an example of an object that has three forces acting on it, and (*a*) accelerates, (*b*) moves at constant (nonzero) velocity, and (*c*) remains at rest.

13 •• Suppose a block of mass m_1 rests on a block of mass m_2 and the combination rests on a table as shown in Figure 4-33. Tell the name of the force and its category (contact versus action-at-a-distance) for each of the following forces: (*a*) force exerted by m_1 on m_2, (*b*) force exerted by m_2 on m_1, (*c*) force exerted by m_2 on the table, (*d*) force exerted by the table on m_2, (*e*) force exerted by Earth on m_2. Which, if any, of these forces constitute a Newton's third-law pair of forces? SSM

FIGURE 4-33 Problem 13

14 •• CONTEXT-RICH You yank a fish you have just caught on your line upward from rest into your boat. Draw a free-body diagram of the fish after it has left the water and as it gains speed as it rises. In addition, tell the type (tension, spring, gravity, normal, friction, etc.) and category (contact versus action-at-a-distance) of each force on your diagram. Which, if any, pairs of the forces on your diagram constitute a Newton's third-law pair? Can you tell the relative magnitudes of the forces on your diagram from the information given? Explain.

15 •• If you gently set a fancy plate on the table, it will not break. However if you drop it from a height, it might very well break. Discuss the forces that act on the plate (as it contacts the table) in both these situations. Use kinematics and Newton's second law to describe what is different about the second situation that causes the plate to break?

16 •• For each of the following forces, give what produces it, what object it acts on, its direction, and the reaction force. (*a*) The force you exert on your briefcase as you hold it while standing at the bus stop. (*b*) The normal force on the soles of your feet as you stand barefooted on a horizontal wood floor. (*c*) The gravitational force on you as you stand on a horizontal floor. (*d*) The horizontal force exerted on a baseball by a bat as the ball is hit straight up the middle toward center field for a single.

17 •• For each case, identify the force (including its direction) that causes the acceleration. (*a*) A sprinter at the very start of the race. (*b*) A hockey puck skidding freely but slowly coming to rest on the ice. (*c*) A long fly ball at the top of its arc. (*d*) A bungee jumper at the very bottom of her descent.

18 • True or false:

(*a*) If two external forces that are both equal in magnitude and opposite in direction act on the same object, the two forces can never be a Newton's third-law pair.

(*b*) The two forces of a Newton's third-law pair are equal only if the objects involved are not accelerating.

19 •• An 80-kg man on ice skates is pushing his 40-kg son, also on skates, with a force of 100 N. Together, they move across the ice steadily gaining speed. (*a*) The force exerted by the boy on his father is (1) 200 N, (2) 100 N, (3) 50 N, or (4) 40 N. (*b*) How do the magnitudes of the two accelerations compare? (*c*) How do the directions of the two accelerations compare?

20 •• A girl holds a stone in her hand and can move it up or down or keep it still. True or false: (*a*) The force exerted by her hand on the rock is always the same magnitude as the force of gravity on the stone. (*b*) The force exerted by her hand on the rock is the reaction force to the force of gravity on the stone. (*c*) The force exerted by her hand on the stone is always the same magnitude as the force on her hand by the stone, but in the opposite direction. (*d*) If the girl moves her hand down at a constant speed, then her upward force on the stone is less than the force of gravity on the stone. (*e*) If the girl moves her hand downward but slows the stone to rest, then the force of the stone on the girl's hand is the same magnitude as the force of gravity on the stone.

21 •• A 2.5-kg object hangs at rest from a string attached to the ceiling. (*a*) Draw a free-body diagram of the object, indicate the reaction force to each force drawn and tell what object the reaction force acts on. (*b*) Draw a free-body diagram of the string, indicate the reaction force to each force drawn, and tell what object each reaction force acts on. Do not neglect the mass of the string. SSM

22 •• (*a*) Which of the free-body diagrams in Figure 4-34 represents a block sliding down a frictionless inclined surface? (*b*) For the correct diagram, label the forces and tell which are contact forces and which are action-at-a-distance forces. (*c*) For each force in the correct diagram, identify the reaction force, the object it acts on and its direction.

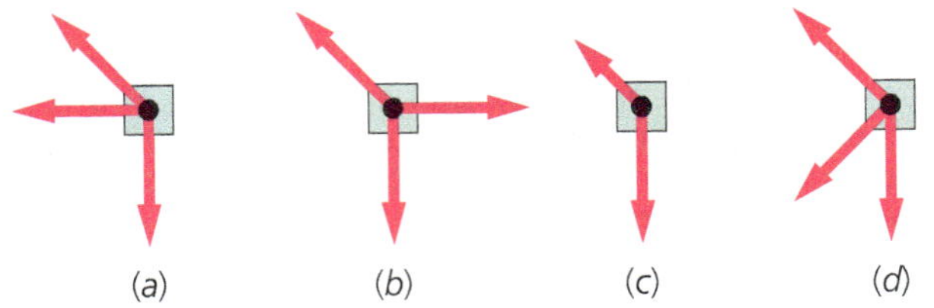

FIGURE 4-34 Problem 22

23 •• A wooden box on the floor is pressed against a compressed, horizontal spring that is attached to a wall. The horizontal floor beneath the box is frictionless. Draw the free-body diagram of the box in the following cases. (*a*) The box is held at rest against the compressed spring. (*b*) The force holding the box against the spring no longer exists, but the box is still in contact with the spring. (*c*) When the box no longer has contact with the spring.

24 •• Imagine yourself seated on a wheeled desk chair at your desk. Consider friction forces between the chair and the floor to be negligible. However, the friction forces between the desk and the floor are not negligible. When sitting at rest, you decide you need another cup of coffee. You push horizontally against the desk, and the chair rolls backward away from the desk. (*a*) Draw a free-body diagram of yourself during the push and clearly indicate which force was responsible for your acceleration. (*b*) What is the reaction force to the force that caused your acceleration? (*c*) Draw the free-body diagram of the desk and explain why it did not accelerate. Does this violate Newton's third law? Explain.

25 ••• The same (net) horizontal force F is applied for a fixed time interval Δt to each of two objects, having masses m_1 and m_2, that sit on a flat, frictionless surface. (Let $m_1 > m_2$.) (*a*) Assuming the two objects are initially at rest, what is the ratio of their accelerations during the time interval, in terms of F, m_1 and m_2? (*b*) What is the ratio of their speeds v_1 and v_2 at the end of the time interval? (*c*) How far apart are the two objects (and which is ahead) at the end of the time interval?

ESTIMATION AND APPROXIMATION

26 •• **CONCEPTUAL** Most cars have four springs attaching the body to the frame, one at each wheel position. Devise an experimental method of estimating the force constant of one of the springs using your known weight and the weights of several of your friends. Assume the 4 springs are identical. Use the method to estimate the force constant of your car's springs.

27 •• Estimate the force exerted on the goalie's glove by the puck when he catches a hard slap shot for a save. SSM

28 •• A baseball player slides into second base during a steal attempt. Assuming reasonable values for the length of the slide, the speed of the player at the beginning of the slide, and the speed of the player at the end of the slide, estimate the average force of friction acting on the player.

29 •• **ENGINEERING APPLICATION** A race car skidding out of control manages to slow down to 90 km/h before crashing head-on into a brick wall. Fortunately, the driver is wearing a safety harness. Using reasonable values for the mass of the driver and the stopping distance, estimate the average force exerted on the driver by the safety harness, including its direction. Neglect any effects of frictional forces on the driver by the seat.

NEWTON'S FIRST AND SECOND LAWS: MASS, INERTIA, AND FORCE

30 • A particle is traveling in a straight line at a constant speed of 25.0 m/s. Suddenly, a constant force of 15.0 N acts on it, bringing it to a stop in a distance of 62.5 m. (*a*) What is the direction of the force? (*b*) Determine the time it takes for the particle to come to a stop. (*c*) What is its mass?

31 • An object has an acceleration of 3.0 m/s^2 when a single force of magnitude F_0 acts on it. (*a*) What is the magnitude of its acceleration when the magnitude of this force is doubled? (*b*) A second object has an acceleration magnitude of 9.0 m/s^2 under the influence of a single force of magnitude F_0. What is the ratio of the mass of the second object to that of the first object? (*c*) If the two objects are glued together to form a composite object, what acceleration magnitude will a single force of magnitude F_0 acting on the composite object produce?

32 • A tugboat tows a ship with a constant force of magnitude F_1. The increase in the ship's speed during a 10-s interval is 4.0 km/h. When a second tugboat applies an additional constant force of magnitude F_2 in the same direction, the speed increases by 16 km/h during a 10-s interval. How do the magnitudes of F_1 and F_2 compare? (Neglect the effects of water resistance and air resistance.)

33 • A single constant force of magnitude 12 N acts on a particle of mass m. The particle starts from rest and travels in a straight line a distance of 18 m in 6.0 s. Find m.

34 • A net force of (6.0 N)$\hat{i}$ − (3.0 N)$\hat{j}$ acts on a 1.5 kg object. Find the acceleration $\vec{a}$.

35 •• A bullet of mass 1.80×10^{-3} kg moving at 500 m/s impacts a tree stump and penetrates 6.00 cm into the wood before coming to rest. (*a*) Assuming that the acceleration of the bullet is constant, find the force (including direction) exerted by the wood on the bullet. (*b*) If the same force acted on the bullet and it had the same impact speed but half the mass, how far would it penetrate into the wood? SSM

36 •• A cart on a horizontal, linear track has a fan attached to it. The cart is positioned at one end of the track, and the fan is turned on. Starting from rest, the cart takes 4.55 s to travel a distance of 1.50 m. The mass of the cart plus fan is 355 g. Assume that the cart travels with constant acceleration. (*a*) What is the net force exerted on the cart–fan combination? (*b*) Mass is added to the cart until the total mass of the cart–fan combination is 722 g, and the experiment is repeated. How long does it take for the cart, starting from rest, to travel 1.50 m now? Ignore the effects due to friction.

37 •• A horizontal force of magnitude F_0 causes an acceleration of magnitude 3.0 m/s^2 when it acts on an object of mass m sliding on a frictionless surface. Find the magnitude of the acceleration of the same object in the circumstances shown in Figure 4-35*a* and 4-35*b*.

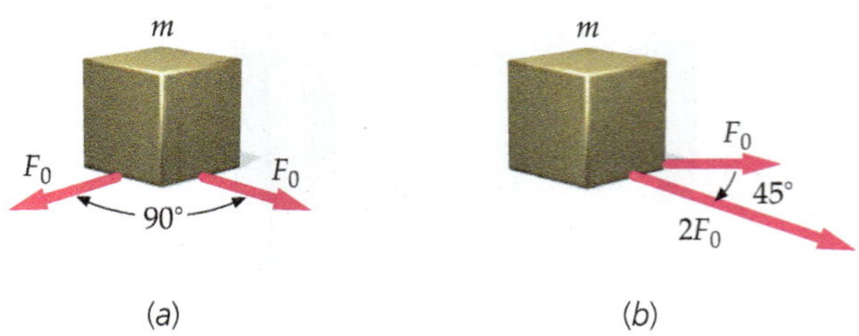

FIGURE 4-35 Problem 37

38 •• Al and Bert stand in the middle of a large frozen lake (frictionless surface). Al pushes on Bert with a force of 20 N for 1.5 s. Bert's mass is 100 kg. Assume that both are at rest before Al pushes Bert. (*a*) What is the speed that Bert reaches as he is pushed away from Al? (*b*) What speed does Al reach if his mass is 80 kg?

39 •• If you push a block whose mass is m_1 across a frictionless floor with a horizontal force of a magnitude F_0, the block has an acceleration of 12 m/s^2. If you push on a different block whose mass is m_2 with a horizontal force of magnitude F_0, its acceleration is 3.0 m/s^2. (*a*) What acceleration will a horizontal force of magnitude F_0 give to a single block with mass $m_2 - m_1$? (*b*) What acceleration will a horizontal force of magnitude F_0 give to a single block with mass $m_2 + m_1$?

40 •• **MULTISTEP** To drag a 75.0-kg log along the ground at constant velocity, your tractor has to pull it with a horizontal force of 250 N. (*a*) Draw the free-body diagram of the log. (*b*) Use Newton's laws to determine the force of friction on the log. (*c*) What is the normal force of the ground on the log? (*d*) What horizontal force must you exert if you want to give the log an acceleration of 2.00 m/s^2 assuming the force of friction does not change. Redraw the log's free-body diagram for this situation.

41 •• A 4.0-kg object is subjected to two constant forces, $\vec{F}_1 = (2.0\text{ N})\hat{i} + (-3.0\text{ N})\hat{j}$ and $\vec{F}_2 = (4.0\text{ N})\hat{i} - (11\text{ N})\hat{j}$. The object is at rest at the origin at time $t = 0$. (*a*) What is the object's acceleration? (*b*) What is its velocity at time $t = 3.0$ s? (*c*) Where is the object at time $t = 3.0$ s?

MASS AND WEIGHT

42 • On the moon, the acceleration due to the effect of gravity is only about 1/6 of that on Earth. An astronaut whose weight on Earth is 600 N travels to the lunar surface. His mass, as measured on the moon, will be (*a*) 600 kg, (*b*) 100 kg, (*c*) 61.2 kg, (*d*) 9.81 kg, (*e*) 360 kg.

43 • Find the weight of a 54-kg student in (*a*) newtons, and (*b*) pounds.

44 • Find the mass of a 165-lb engineer in kilograms.

45 •• **ENGINEERING APPLICATION** To train astronauts to work on the moon, where the free-fall acceleration is only about 1/6 of that on Earth, NASA submerges them in a tank of water. If an astronaut who is carrying a backpack, air conditioning unit, oxygen supply, and other equipment, has a total mass of 250 kg, determine the following quantities. (*a*) her weight including backpack, etc. on Earth, (*b*) her weight on the moon, (*c*) the required upward buoyancy force of the water during her training for the moon's environment on Earth. SSM

46 •• It is the year 2075 and space travel is common. A physics professor brings his favorite teaching demonstration with him to the moon. The apparatus consists of a very smooth (frictionless) horizontal table and an object to slide on it. On Earth, when the professor attaches a spring (force constant 50 N/m) to the object and pulls horizontally so the spring stretches 2.0 cm, the object accelerates at 1.5 m/s^2. (*a*) Draw the free-body diagram of the object and use it and Newton's laws to determine the object's mass. (*b*) What would the object's acceleration be under identical conditions on the moon?

FREE-BODY DIAGRAMS: STATIC EQUILIBRIUM

47 • **ENGINEERING APPLICATION, MULTISTEP** A 35.0-kg traffic light is supported by two wires as in Figure 4-36. (*a*) Draw the light's free-body diagram and use it to answer the following question qualitatively: Is the tension in wire 2 greater than or less than the tension in wire 1? (*b*) Verify your answer by applying Newton's laws and solving for the two tensions.

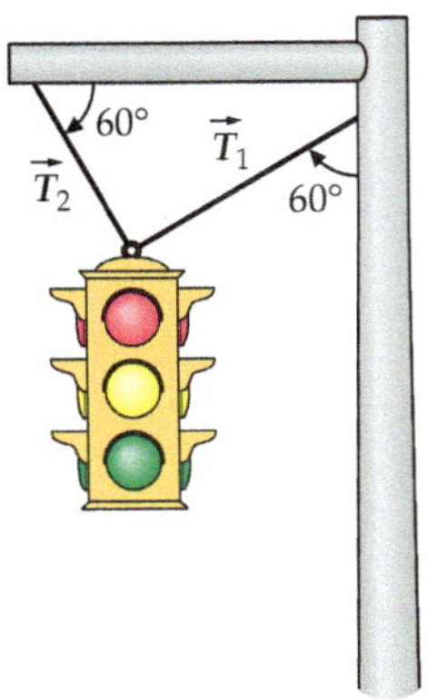

FIGURE 4-36 Problem 47

48 • A 42.6-kg lamp is hanging from wires as shown in Figure 4-37. The ring has negligible mass. The tension T_1 in the vertical wire is (*a*) 209 N, (*b*) 418 N, (*c*) 570 N, (*d*) 360 N, (*e*) 730 N.

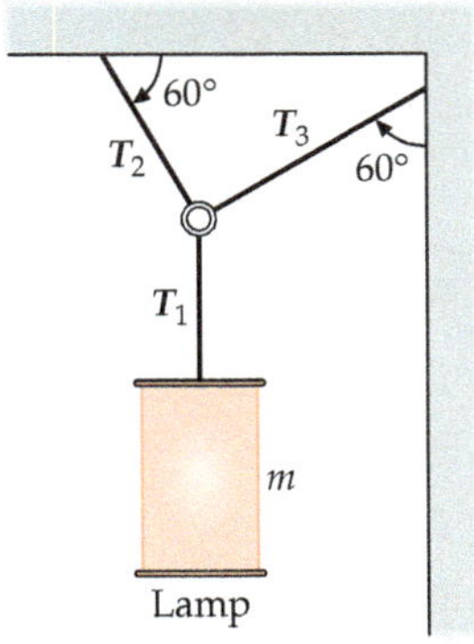

FIGURE 4-37 Problem 48

49 •• In Figure 4-38*a*, a 0.500-kg block is suspended at the midpoint of a 1.25-m-long string. The ends of the string are attached to the ceiling at points separated by 1.00 m. (*a*) What angle does the string make with the ceiling? (*b*) What is the tension in the string? (*c*) The 0.500-kg block is removed and two 0.250-kg blocks are attached to the string such that the lengths of the three string segments are equal (Figure 4-38*b*). What is the tension in each segment of the string? SSM

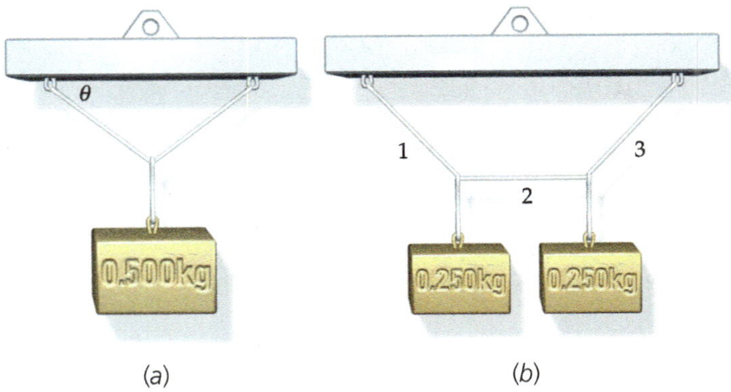

FIGURE 4-38 Problem 49

50 •• A ball weighing 100-N is shown suspended from a system of cords (Figure 4-39). What are the tensions in the horizontal and angled cords?

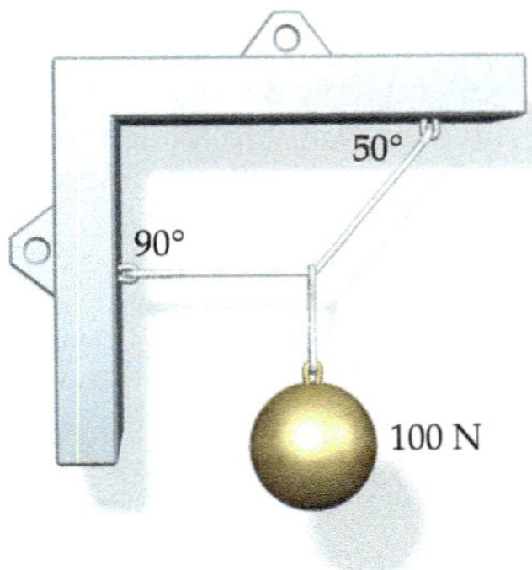

FIGURE 4-39 Problem 50

51 •• A 10-kg object on a frictionless table is subjected to two horizontal forces, $\vec{F}_1$ and $\vec{F}_2$, with magnitudes $F_1 = 20$ N and $F_2 = 30$ N, as shown in Figure 4-40. Find the third horizontal force $\vec{F}_3$ that must be applied so that the object is in static equilibrium. SSM

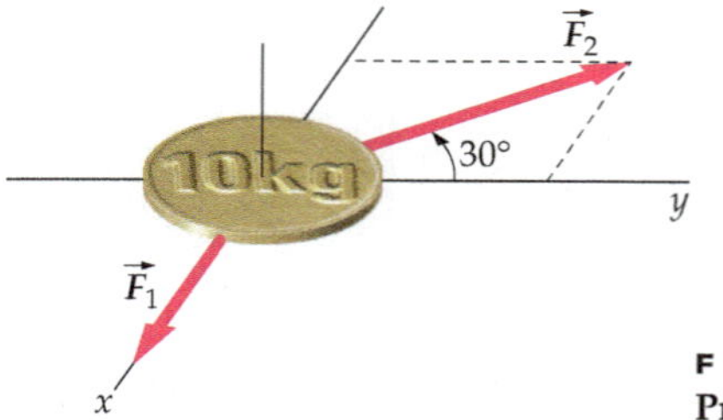

FIGURE 4-40 Problem 51

52 •• For the systems to be in equilibrium in Figure 4-41*a*, Figure 4-41*b*, and Figure 4-41*c*, find the unknown tensions and masses.

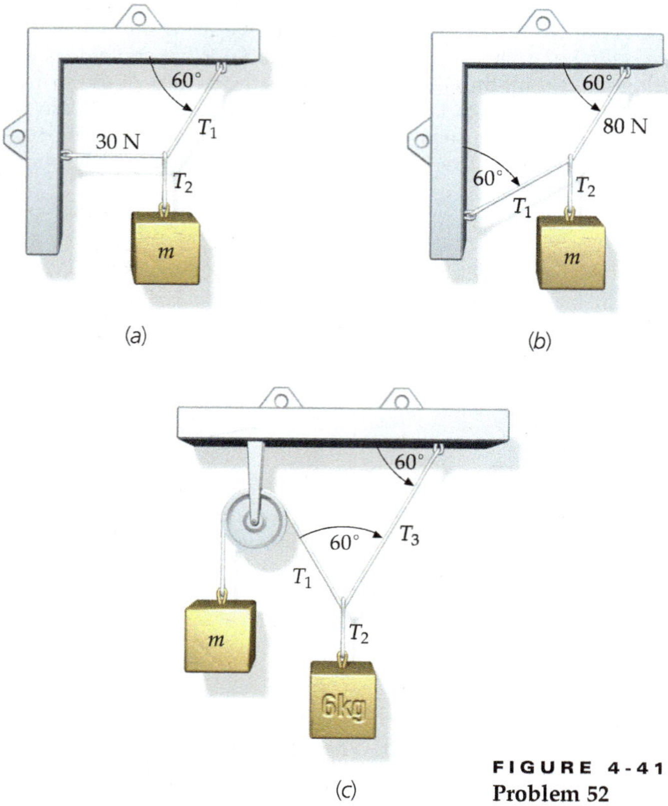

FIGURE 4-41 Problem 52

53 •• **ENGINEERING APPLICATION** Your car is stuck in a mud hole. You are alone, but you have a long, strong rope. Having studied physics, you tie the rope tautly to a telephone pole and pull on it sideways, as shown in Figure 4-42. (*a*) Find the force exerted by the rope on the car when the angle θ is 3.00° and you are pulling with a force of 400 N, but the car does not move. (*b*) How strong must the rope be if it takes a force of 600 N to move the car when θ is 4.00°?

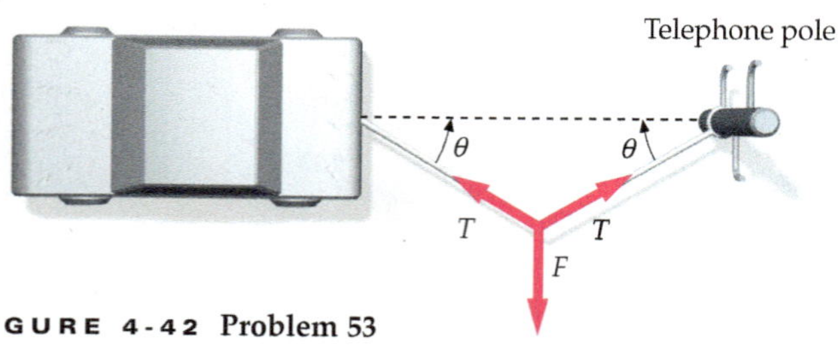

FIGURE 4-42 Problem 53

54 ••• **ENGINEERING APPLICATION, MULTISTEP** Balloon arches are often seen at festivals or celebrations; they are made by attaching helium-filled balloons to a rope that is fixed to the ground at each end. The lift from the balloons raises the structure into the arch shape. Figure 4-43*a* shows the geometry of such a structure: N balloons are attached at equally spaced intervals along a massless rope of length L, which is attached to two supports at its ends. Each balloon provides a lift force of magnitude F. The horizontal and vertical coordinates of the point on the rope where the ith balloon is attached are x_i and y_i, and T_i is the tension in the ith segment. (Note segment 0 is the segment between the point of attachment and the first balloon, and segment N is the segment between the last balloon and the other point of attachment). (*a*) Figure 4-43*b* shows a free-body diagram for the ith balloon. From this diagram, show that the horizontal component of the force T_i (call it T_H) is the same for all the string segments. (*b*) By considering the vertical component of the forces, use Newton's laws to derive the following relationship between the tension in the ith and $(i-1)$th segments: $T_{i-1}\sin\theta_{i-1} - T_i\sin\theta_i = F$. (*c*) Show that $\tan\theta_0 = -\tan\theta_{N+1} = NF/2T_H$. (*d*) From the diagram and the two expressions above, show that $\tan\theta_i = (N-2i)F/2T_H$ and that $y_i = \frac{L}{N+1}\sum_{j=0}^{i-1}\cos\theta_j$, $y_i = \frac{L}{N+1}\sum_{j=0}^{i-1}\sin\theta_j$.

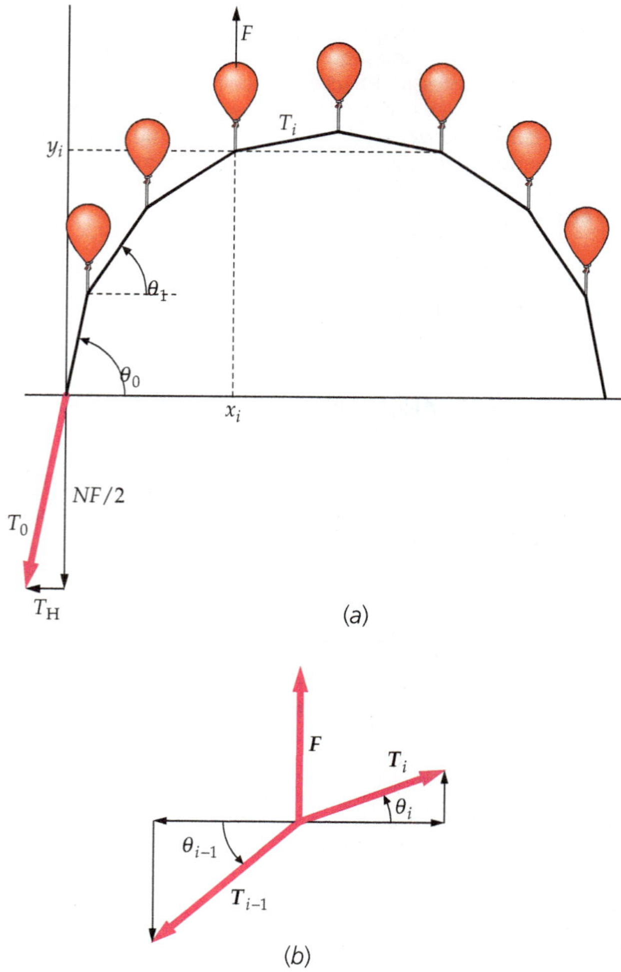

FIGURE 4-43 Problem 54

55 ••• **Engineering Application, Spreadsheet** (*a*) Consider a numerical solution to Problem 54. Write a **spreadsheet** program to make a graph of the shape of a balloon arch. Use the following parameters: $N = 10$ (balloons), each providing a lift force $F = 1.0$ N and each attached to a rope of length $L = 11$ m, with a horizontal component of tension $T_H = 10$ N. How far apart are the two points of attachment? How high is the arch at its highest point? (*b*) Note that we have not specified the spacing between the supports—it is determined by the other parameters. Vary T_H while keeping the other parameters the same until you create an arch that has a spacing of 8.0 m between the supports. What is T_H then? As you increase T_H, the arch should get flatter and more spread out. Does your spreadsheet model show this?

FREE-BODY DIAGRAMS: INCLINED PLANES AND THE NORMAL FORCE

56 • A large box whose mass is 20.0 kg rests on a frictionless floor. A mover pushes on the box with a force of 250 N at an angle 35.0° below the horizontal. Draw the box's free-body diagram and use it to determine the acceleration of the box.

57 • A 20.0-kg box rests on a frictionless ramp with a 15.0° slope. The mover pulls on a rope attached to the box to pull it up the incline (Figure 4-44). If the rope makes an angle of 40.0° with the horizontal, what is the smallest force F the mover will have to exert to move the box up the ramp?

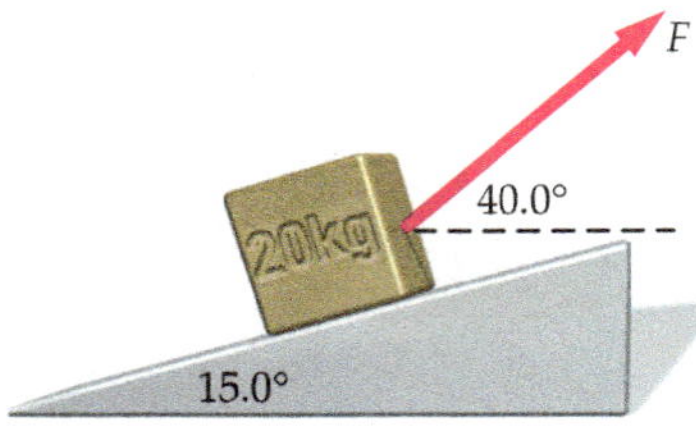

FIGURE 4-44 Problem 57

58 • In Figure 4-45, the objects are attached to spring scales calibrated in newtons. Give the reading(s) of the balance(s) in each case, assuming that both the scales and the strings are massless.

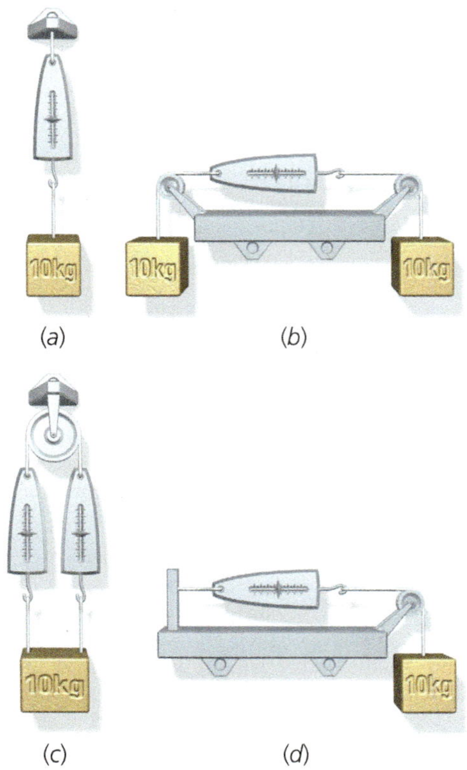

FIGURE 4-45 Problem 58

59 •• A box is held in position on a frictionless incline by a cable (Figure 4-46). (*a*) If $\theta = 60°$ and $m = 50$ kg, find the tension in the cable and the normal force exerted by the incline. (*b*) Find the tension as a function of θ and m, and check your result for plausibility in the special cases of $\theta = 0°$ and $\theta = 90°$.

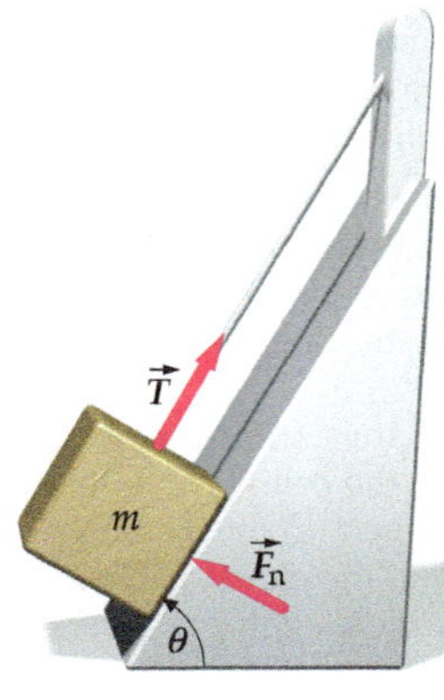

FIGURE 4-46 Problem 59

60 •• A horizontal force of 100 N pushes a 12-kg block up a frictionless incline that makes an angle of 25° with the horizontal. (*a*) What is the normal force that the incline exerts on the block? (*b*) What is the magnitude of acceleration of the block?

61 •• A 65-kg student weighs himself by standing on a force scale mounted on a skateboard that is rolling down an incline, as shown in Figure 4-47. Assume there is no friction so that the force exerted by the incline on the skateboard is normal to the incline. What is the reading on the scale if $\theta = 30°$? SSM

FIGURE 4-47 Problem 61

62 •• A block of mass m slides across a frictionless floor and then up a frictionless ramp (Figure 4-48). The angle of the ramp is θ and the speed of the block before it starts up the ramp is v_0. The block will slide up to some maximum height h above the floor before stopping. Show that h is independent of m and θ by deriving an expression for h in terms of v_0 and g.

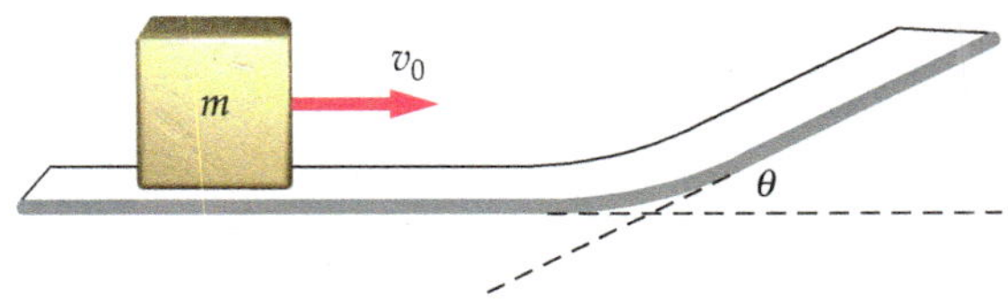

FIGURE 4-48 Problem 62

FREE-BODY DIAGRAMS: ELEVATORS

63 • **CONCEPTUAL** (*a*) Draw the free-body diagram (with accurate relative force magnitudes) for an object that is hung by a rope from the ceiling of an elevator that is ascending but slowing. (*b*) Repeat Part (*a*) but for the situation in which the elevator is descending and speeding up. (*c*) Can you tell the difference between the two diagrams? Explain why the diagrams do not tell anything about the object's velocity. SSM

64 • A 10.0-kg block is suspended from the ceiling of an elevator by a cord rated to withstand a tension of 150 N. Shortly after the elevator starts to ascend, the cord breaks. What was the minimum acceleration of the elevator when the cord broke?

65 •• A 2.0-kg block hangs from a spring scale calibrated in newtons that is attached to the ceiling of an elevator (Figure 4-49). What does the scale read when (*a*) the elevator is ascending with a constant speed of 30 m/s; (*b*) the elevator is descending with a constant speed of 30 m/s; (*c*) the elevator is ascending at 20 m/s and gaining speed at a rate of 3.0 m/s²? (*d*) Suppose that from $t = 0$ to $t = 5.0$ s, the elevator ascends at a constant speed of 10 m/s. Its speed is then steadily reduced to zero during the next 4.0 s, so that it is at rest at $t = 9.0$ s. Describe the reading of the scale during the interval $0 < t < 9.0$ s.

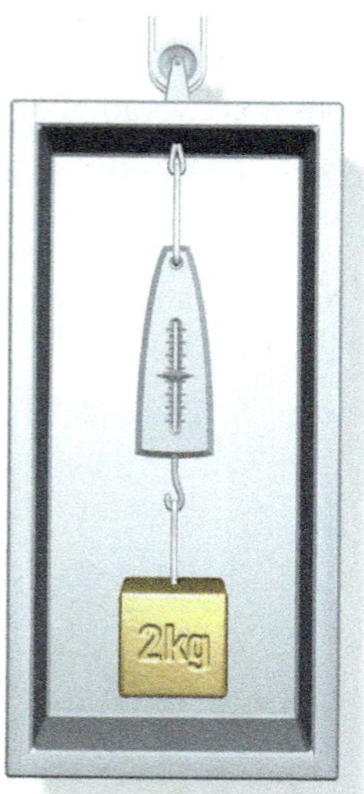

FIGURE 4-49 Problem 65

FREE-BODY DIAGRAMS: SEVERAL OBJECTS AND NEWTON'S THIRD LAW

66 •• **CONCEPTUAL** Two boxes of mass m_1 and m_2 connected by a massless string are being pulled along a horizontal frictionless surface by the tension force in a second string, as shown in Figure 4-50. (*a*) Draw the free-body diagram of both boxes separately and show that $T_1/T_2 = m_1/(m_1 + m_2)$. (*b*) Is this result plausible? Explain. Does your answer make sense both in the limit that $m_2/m_1 \gg 1$ and in the limit that $m_2/m_1 \ll 1$? Explain.

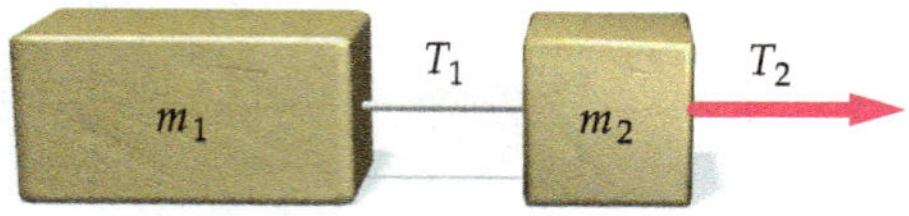

FIGURE 4-50 Problem 66

67 •• A box of mass $m_2 = 3.5$ kg rests on a frictionless horizontal shelf and is attached by strings to boxes of masses $m_1 = 1.5$ kg and $m_3 = 2.5$ kg as shown in Figure 4-51. Both pulleys are frictionless and massless. The system is released from rest. After it is released, find (*a*) the acceleration of each of the boxes, and (*b*) the tension in each string.

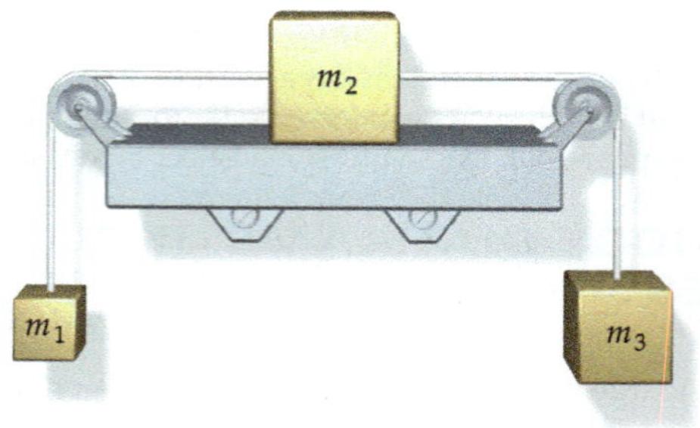

FIGURE 4-51 Problem 67

68 •• Two blocks are in contact on a frictionless horizontal surface. The blocks are accelerated by a single horizontal force $\vec{F}$ applied to one of them (Figure 4-52). Find the acceleration and the contact force of block 1 on block 2 (*a*) in terms of F, m_1 and m_2, and (*b*) for the specific values $F = 3.2$ N, $m_1 = 2.0$ kg and $m_2 = 6.0$ kg.

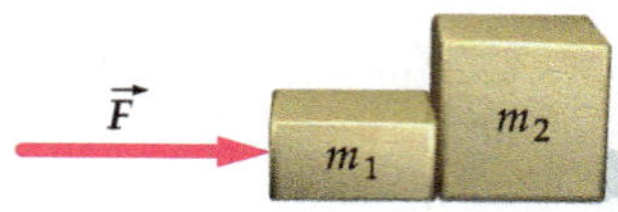

FIGURE 4-52 Problem 68

69 •• Repeat Problem 68, but with the two blocks interchanged. Are your answers for this problem the same as in Problem 68? Explain.

70 •• Two 100-kg boxes are dragged along a horizontal frictionless surface at a constant acceleration of 1.00 m/s², as shown in Figure 4-53. Each rope has a mass of 1.00 kg. Find the magnitude of the force $\vec{F}$ and the tension in the ropes at points A, B, and C.

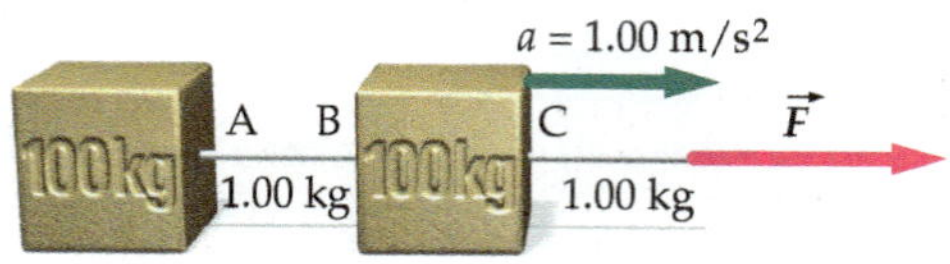

FIGURE 4-53 Problem 70

71 •• A block of mass m is being lifted vertically by a uniform rope of mass M and length L. The rope is being pulled upward by a force applied to its top end, and the rope and block are accelerating upward with an acceleration of magnitude a. Show that the tension in the rope at a distance x (where $x < L$) above the block is given by $(a + g)[m + (x/L)M]$. SSM

72 •• A chain consists of 5 links, each having a mass of 0.10 kg. The chain is being pulled upward by a force applied by your hand to its top link, giving the chain an upward acceleration of 2.5 m/s². Find (*a*) the force magnitude F exerted on the top link by your hand; (*b*) the net force on each link; and (*c*) the magnitude of the force that each link exerts on the link below it.

73 •• **MULTISTEP** A 40.0-kg object supported by a vertical rope. The rope, and thus the object, is then accelerated from rest upward so that it attains a speed of 3.50 m/s in 0.700 s. (*a*) Draw the object's free-body diagram with the relative lengths of the vectors showing the relative magnitudes of the forces. (*b*) Use the free-body diagram and Newton's laws to determine the tension in the rope. SSM

74 •• **ENGINEERING APPLICATION, MULTISTEP** A 15000-kg helicopter is lowering a 4000-kg truck to the ground by a cable of fixed length. The truck, helicopter, and cable are descending at 15.0 m/s and must be slowed to 5.00 m/s in the next 50.0 m of descent to prevent damaging the truck. Assume a constant rate of slowing. (*a*) Draw the free-body diagram of the truck. (*b*) Determine the tension in the cable. (*c*) Determine the lift force on the helicopter blades.

75 •• Two objects are connected by a massless string, as shown in Figure 4-54. The incline and the massless pulley are frictionless. Find the acceleration of the objects and the tension in the string (*a*) in terms of θ, m_1, and m_2, and for (*b*) $\theta = 30°$ and $m_1 = m_2 = 5.0\,\text{kg}$.

FIGURE 4-54 Problem 75

76 •• **ENGINEERING APPLICATION** During a stage production of *Peter Pan*, the 50-kg actress playing Peter has to fly in vertically (descend). To be in time with the music, she must, starting from rest, be lowered a distance of 3.2 m in 2.2 s at a constant acceleration. Backstage, a smooth surface sloped at 50° supports a counterweight of mass *m*, as shown in Figure 4-55. Show the calculations that the stage manager must perform to find (*a*) the mass of the counterweight that must be used and (*b*) the tension in the wire.

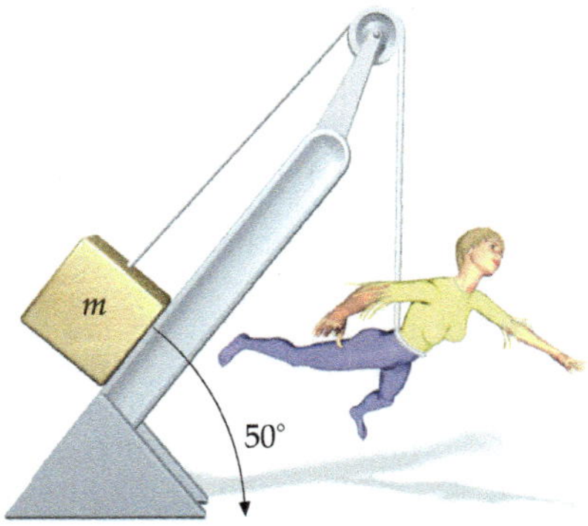

FIGURE 4-55 Problem 76

77 •• An 8.0-kg block and a 10-kg block, connected by a rope that passes over a frictionless peg, slide on frictionless incline, (Figure 4-56). (*a*) Find the acceleration of the blocks and the tension in the rope. (*b*) The two blocks are replaced by two others of masses m_1 and m_2 such that there is no acceleration. Find whatever information you can about the masses of these two new blocks.

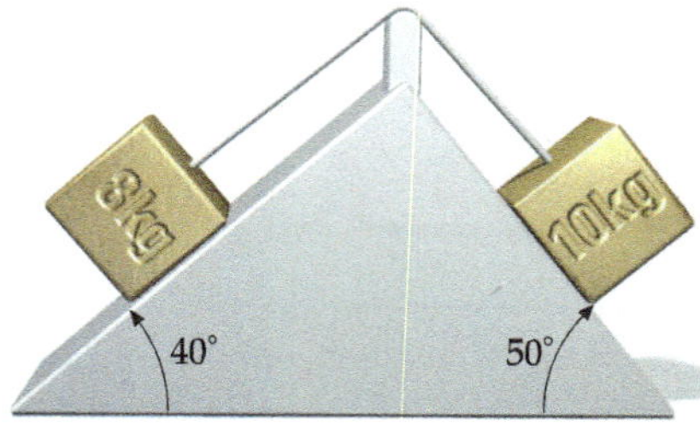

FIGURE 4-56 Problem 77

78 •• A heavy rope of length 5.0 m and mass 4.0 kg lies on a frictionless horizontal table. One end is attached to a 6.0-kg block. The other end of the rope is pulled by a constant horizontal 100-N force. (*a*) What is the acceleration of the system? (*b*) Give the tension in the rope as a function of position along the rope.

79 •• A 60-kg housepainter stands on a 15-kg aluminum platform. The platform is attached to a rope that passes through an overhead pulley, which allows the painter to raise herself and the platform (Figure 4-57). (*a*) With what force *F* must she pull down on the rope to accelerate herself and the platform upward at a rate of $0.80\,\text{m/s}^2$? (*b*) When her speed reaches 1.0 m/s, she pulls in such a way that she and the platform go up at a constant speed. What force is she exerting on the rope now? (Ignore the mass of the rope.) SSM

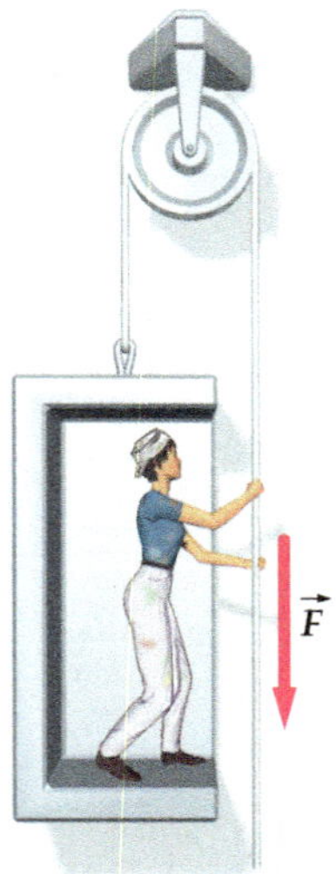

FIGURE 4-57 Problem 79

80 ••• Figure 4-58 shows a 20-kg block sliding on a 10-kg block. All surfaces are frictionless and the pulley is massless and frictionless. Find the acceleration of each block and the tension in the string that connects the blocks.

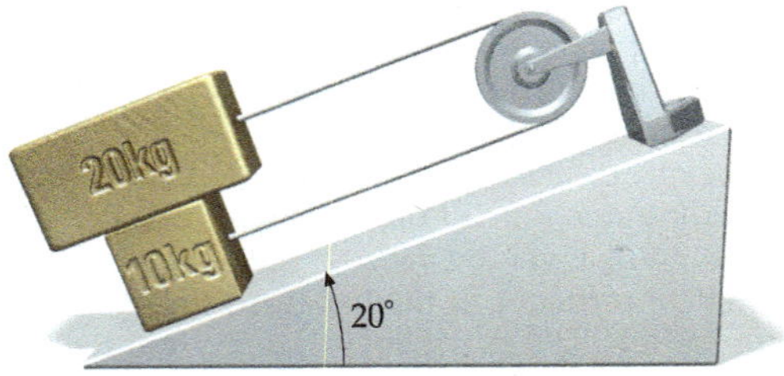

FIGURE 4-58 Problem 80

81 ••• A 20-kg block with a pulley attached slides along a frictionless ledge. It is connected by a massless string to a 5.0-kg block via the arrangement shown in Figure 4-59. Find (*a*) the acceleration of each block, and (*b*) the tension in the connecting string.

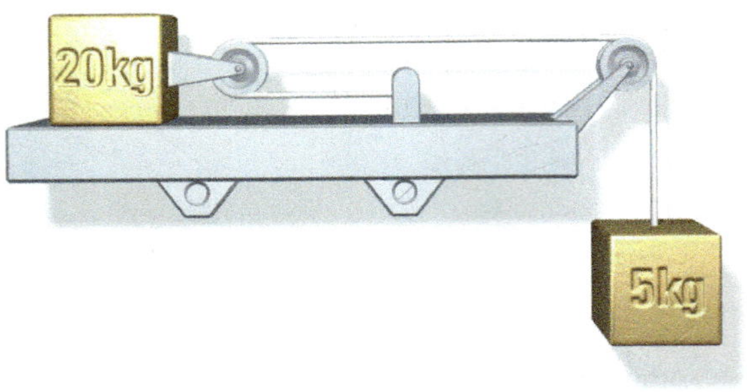

FIGURE 4-59 Problem 81

82 •• **MULTISTEP** The apparatus in Figure 4-60 is called an *Atwood's machine* and is used to measure the free-fall acceleration g by measuring the acceleration of the two blocks connected by a string over a pulley. Assume a massless, frictionless pulley and a massless string. (*a*) Draw the free-body diagram of each block. (*b*) Use the free-body diagrams and Newton's laws to show that the magnitude of the acceleration of either block and the tension in the string are $a = (m_1 - m_2)g/(m_1 + m_2)$ and $T = 2m_1m_2g/(m_1 + m_2)$. (*c*) Do these expressions give plausible results if $m_1 = m_2$, in the limit that $m_1 \gg m_2$ and in the limit that $m_1 \ll m_2$? Explain.

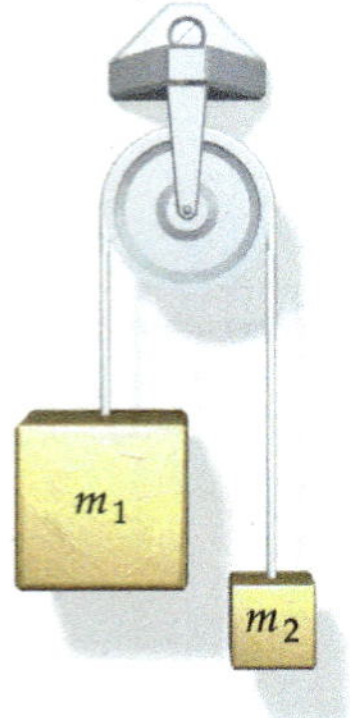

FIGURE 4-60 Problems 82 and 83

83 •• If one of the masses of the Atwood's machine in Figure 4-60 is 1.2 kg, what should be the other mass so that the displacement of either mass during the first second following release is 0.30 m? Assume a massless, frictionless pulley and a massless string.

84 ••• The acceleration of gravity g can be determined by measuring the time t it takes for a mass m_2 in an Atwood's machine described in Problem 82 to fall a distance L, starting from rest. (*a*) Using the results of Problem 82 (note the acceleration is constant), find an expression for g in terms of L, t, m_1, and m_2. (*b*) Show that a small error in the time measurement dt, will lead to an error in g by an amount dg given by $dg/g = -2dt/t$. (*c*) Assume that the only significant uncertainty in the experimental measurements is the time of fall. If $L = 3.00$ m and m_1 is 1.00 kg, find the value of m_2 such that g can be measured with an accuracy of ±5 percent with a time measurement that is accurate to ±0.1 s.

GENERAL PROBLEMS

85 •• A pebble of mass m rests on the block of mass m_2 of the ideal Atwood's machine in Figure 4-60. Find the force exerted by the pebble on the block of mass m_2.

86 •• A simple accelerometer can be made by suspending a small massive object from a string attached to a fixed point on an accelerating object. Suppose such an accelerometer is attached to point P on the ceiling of an automobile traveling in a straight line on a flat surface at constant acceleration. Due to the acceleration, the string will make an angle θ with the vertical. (*a*) Show that the magnitude of the acceleration a is related to the angle θ by $a = g\tan\theta$. (*b*) Suppose the automobile brakes steadily to rest from 50 km/h over a distance of 60 m. What angle will the string make with the vertical? Will the suspended object be positioned below and ahead or below and behind point P during the braking?

87 •• **ENGINEERING APPLICATION** The mast of a sailboat is supported at the bow and stern by stainless steel wires, the forestay and backstay, anchored 10 m apart (Figure 4-61). The 12.0-m-long mast weighs 800 N and stands vertically on the deck of the boat. The mast is positioned 3.60 m behind where the forestay is attached. The tension in the forestay is 500 N. Find the tension in the backstay and the force that the mast exerts on the deck. SSM

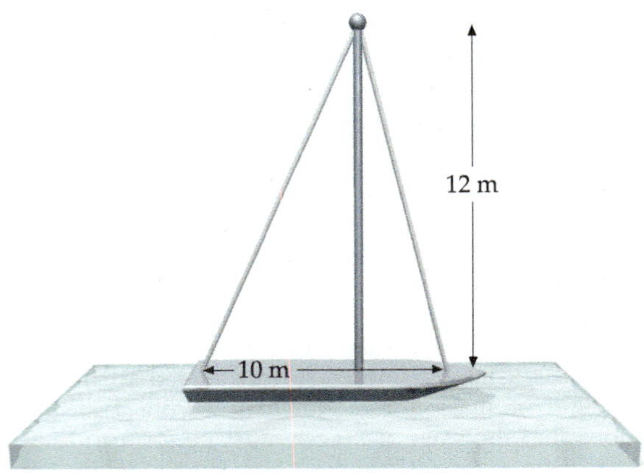

FIGURE 4-61 Problem 87

88 •• A 50-kg block is suspended from a uniform 1.5-m-long chain that is hanging from the ceiling. The mass of the chain itself is 20 kg. Determine the tension in the chain (*a*) at the point where the chain is attached to the block, (*b*) midway up the chain, and (*c*) at the point where the chain is attached to the ceiling.

89 •• The speed of the head of a red headed woodpecker reaches 5.5 m/s before impact with the tree. If the mass of the head is 0.060 kg and the average force on the head is 6.0 N, find (*a*) the acceleration of the head (assuming constant acceleration), (*b*) the depth of penetration into the tree, and (*c*) the time it takes for the head to come to a stop.

90 •• **MULTISTEP** A frictionless surface is inclined at an angle of 30.0° to the horizontal. A 270-g block on the ramp is attached to a 75.0-g block using a pulley, as shown in Figure 4-62. (*a*) Draw two free-body diagrams, one for the 270-g block and the other for the 75.0-g block. (*b*) Find the tension in the string and the acceleration of the 270-g block. (*c*) The 270-g block is released from rest.

FIGURE 4-62 Problem 90

How long does it take for it to slide a distance of 1.00 m along the surface? Will it slide up the incline, or down the incline?

91 •• A box of mass m_1 is pulled along a frictionless horizontal surface by a horizontal force $\vec{F}$ that is applied to the end of a rope of mass m_2 (see Figure 4-63). Neglect any sag of the rope. (*a*) Find the acceleration of the rope and block, assuming them to be one object. (*b*) What is the net force acting on the rope? (*c*) Find the tension in the rope at the point where it is attached to the block.

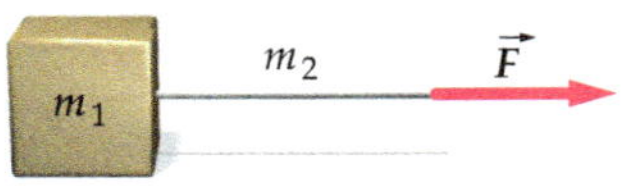

FIGURE 4-63 Problem 91

92 •• A 2.0-kg block rests on a frictionless wedge that has a 60° incline and an acceleration $\vec{a}$ to the right such that the mass remains stationary relative to the wedge (Figure 4-64). (*a*) Draw the free-body diagram of the block and use it to determine the magnitude of the acceleration. (*b*) What would happen if the wedge were given an acceleration larger than this value? Smaller than this value?

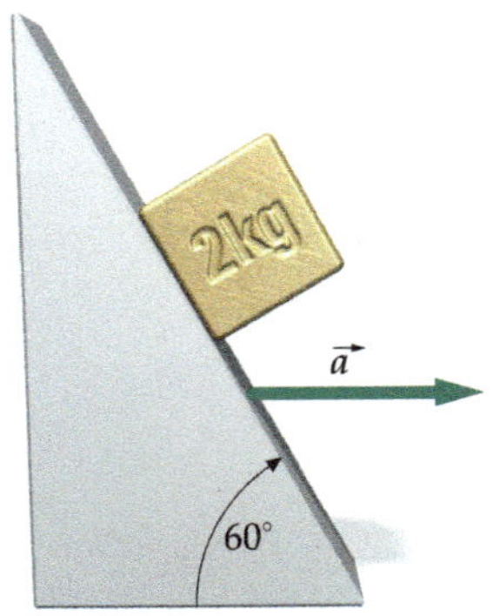

FIGURE 4-64 Problem 92

93 ••• The masses attached to each side of an ideal Atwood's machine consist of a stack of five washers, each of mass m, as shown in Figure 4-65. The tension in the string is T_0. When one of the washers is removed from the left side, the remaining washers accelerate and the tension decreases by 0.300 N. (*a*) Find m. (*b*) Find the new tension and the acceleration of each mass when a second washer is removed from the left side. SSM

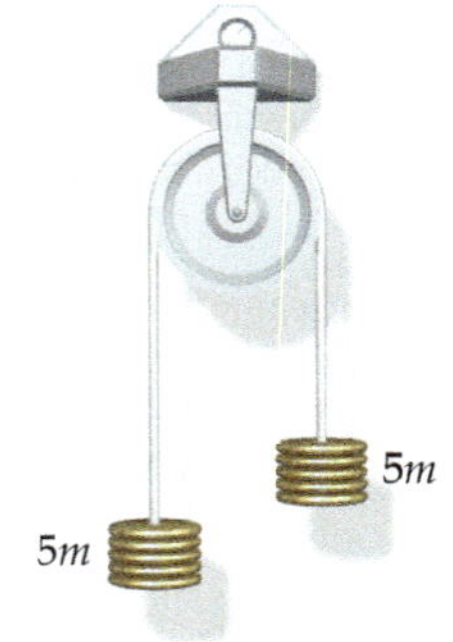

FIGURE 4-65 Problems 93 and 94

94 •• Consider the ideal Atwood's machine in Figure 4-65. When N washers are transferred from the left side to the right side, the right side descends 47.1 cm in 0.40 s. Find N.

95 •• Blocks of mass m and $2m$ are on a horizontal frictionless surface (Figure 4-66). The blocks are connected by a horizontal string. In addition, forces $\vec{F}_1$ and $\vec{F}_2$ are applied as shown. (*a*) If the forces shown are constant, find the tension in the connecting string. (*b*) If the magnitudes of the forces vary with time as $F_1 = Ct$ and $F_2 = 2Ct$, where C equals to 5.00 N/s and t is time, find the time t_0 at which the tension in the string equals to 10.0 N.

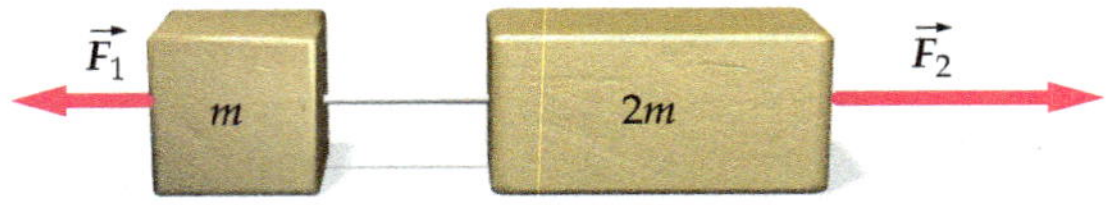

FIGURE 4-66 Problem 95

96 •• Elvis Presley has supposedly been sighted numerous times since his death on August 16, 1977. The following is a chart of what Elvis's weight would be if he were sighted on the surfaces of other objects in our solar system. Use the chart to determine: (*a*) Elvis's mass on Earth, (*b*) Elvis's mass on Pluto, and (*c*) the free-fall acceleration on Mars. (*d*) Compare the free-fall acceleration on Pluto to the free-fall acceleration on the moon.

Planet	Elvis's Weight (N)
Mercury	431
Venus	1031
Earth	1133
Mars	431
Jupiter	2880
Saturn	1222
Pluto	58
Moon	191

97 ••• **CONTEXT-RICH** As a prank, your friends have kidnapped you in your sleep, and transported you out onto the ice covering a local pond. When you wake up you are 30.0 m from the nearest shore. The ice is so slippery (i.e. frictionless) that you cannot seem to get yourself moving. You realize that you can use Newton's third law to your advantage, and choose to throw the heaviest thing you have, one boot, in order to get yourself moving. Take your weight to be 595 N. (*a*) What direction should you throw your boot so that you will most quickly reach the shore? (*b*) If you throw your 1.20-kg boot with an average force of 420 N, and the throw takes 0.600 s (the time interval over which you apply the force), what is the magnitude of the force that the boot exerts on you? (Assume constant acceleration.) (*c*) How long does it take you to reach shore, including the short time in which you were throwing the boot?

98 ••• The pulley of an ideal Atwood's machine is given an upward acceleration *a*, as shown in Figure 4-67. Find the acceleration of each mass and the tension in the string that connects them. The speeds of the two blocks are not equal in this situation.

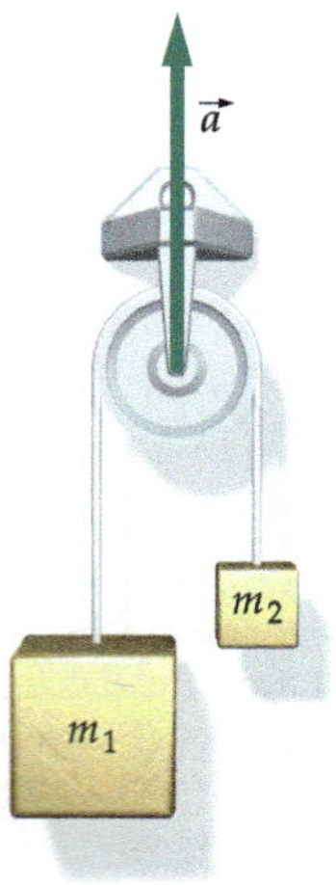

FIGURE 4-67 Problem 98

99 •• **ENGINEERING APPLICATION, CONTEXT-RICH, SPREADSHEET** You are working for an automotive magazine and putting a certain new automobile (mass 650 kg) through its paces. While accelerating from rest, its onboard computer records its velocity as a function of time as follows:

v_x (m/s):	0	10	20	30	40	50
t (s):	0	1.8	2.8	3.6	4.9	6.5

(*a*) Using a **spreadsheet,** find the average acceleration of the five time intervals and graph the velocity versus time and acceleration versus time for this car. (*b*) Where on the graph of velocity versus time is the net force on the car highest and lowest? Explain your reasoning. (*c*) What is the average net force on the car over the whole trip? (*d*) From the graph of velocity versus time, estimate the total distance covered by the car.

CHAPTER 5

DAYTONA INTERNATIONAL SPEEDWAY, THE "WORLD CENTER OF RACING," FEATURES A 2.5-MILE TRI-OVAL TRACK, WHICH HAS FOUR-STORY, 31-DEGREE-HIGH BANKED CURVES. IN THE DAYTONA 500 RACE THE STOCK CARS TRAVEL THROUGH THE CURVES AT SPEEDS CLOSE TO 200 MPH. SURPRISINGLY, THE FIERY CRASHES THAT THE DAYTONA 500 IS FAMOUS FOR, WITH THEIR ACCOMPANYING INJURIES AND FATALITIES, ARE USUALLY NOT CAUSED BY SKIDDING ON THE CURVES. *(PhotoDisc/Getty Images.)*

Additional Applications of Newton's Laws

5-1 Friction

5-2 Drag Forces

5-3 Motion Along a Curved Path

*5-4 Numerical Integration: Euler's Method

5-5 The Center of Mass

? What factors determine how fast a car can go through a curve without skidding? (See Example 5-12.)

In Chapter 4, we introduced Newton's laws and applied them to situations where action was restricted to straight-line motion and frictional forces were introduced. We now will consider some more general applications and how Newton's laws can be used to explain innumerable properties of the world in which we live.

In this chapter, we will extend the application of Newton's laws to motion along curved paths, and we will analyze the effects of resistive forces such as friction and air drag. We will also introduce the concept of the center of mass of a system of particles and show how modeling the system as a single particle located at the center of mass can result in being able to predict the bulk motion of such a system.

5-1 FRICTION

If you shove a book that is resting on a desktop, the book will probably skid across the desktop. If the desktop is long enough, the book will eventually skid to a stop. This happens because a frictional force is exerted by the desktop on the book in a direction opposite to the book's velocity. This force, which acts on the surface of the book in contact with the desktop, is known as a *frictional force*. Frictional forces are a necessary part of our lives. Without friction our ground-based transportation system, from walking to automobiles, could not function. Friction allows you to start walking, and once you are already moving, friction allows you to change either your speed or direction. Friction allows you to start, steer, and stop a car. Friction holds a nut on a screw, a nail in wood, and a knot in a piece of rope. However, as important as friction is, it is often not desirable. Friction causes wear whenever moving pieces of machinery are in contact, and large amounts of time and money are spent trying to reduce such effects.

Friction is a complex, incompletely understood phenomenon that arises from the attraction between the molecules of one surface and the molecules on a second surface in close contact. The nature of this attraction is electromagnetic—the same as the molecular bonding that holds an object together. This short-ranged attractive force becomes negligible at distances of only a few atomic diameters.

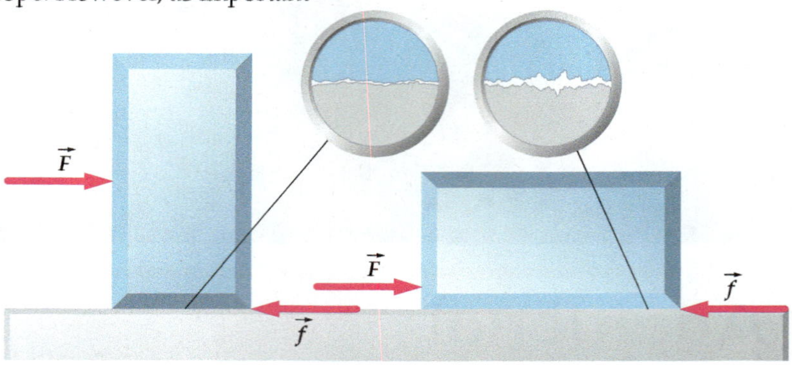

FIGURE 5-1 The microscopic area of contact between box and floor is only a small fraction of the macroscopic area of the box's bottom surface. The microscopic area of contact is proportional to the normal force exerted between the surfaces. If the box rests on its side, the macroscopic area is increased, but the force per unit area is decreased, so the microscopic area of contact is unchanged. Whether the box is upright or on its side, the same horizontal applied force F is required to keep it sliding at constant speed.

As shown in Figure 5-1, ordinary objects that look smooth and feel smooth are rough and bumpy at the microscopic (atomic) scale. This is the case even if the surfaces are highly polished. When surfaces come into contact, they touch only at prominences, called *asperities*, shown in Figure 5-1. The normal force exerted by a surface is exerted at the tips of these asperities where the normal force per unit area is very large, large enough to flatten the tips of the asperities. If the two surfaces are pressed together more strongly, the normal force increases and so does this flattening, resulting in a larger microscopic contact area. Under a wide range of conditions the microscopic area of contact is proportional to the normal force. The frictional force is proportional to the microscopic contact area; so, like the microscopic contact area, it is proportional to the normal force.

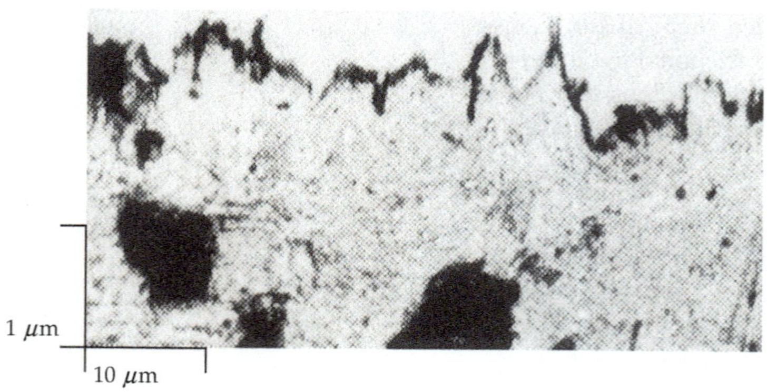

Magnified section of a polished steel surface showing surface irregularities. The irregularities are about $= 5 \times 10^{-7}$m high, a height that corresponds to several thousand atomic diameters. *(From F. P. Bowden and D. Tabor,* Lubrication of Solids, *Oxford University Press, 2000.)*

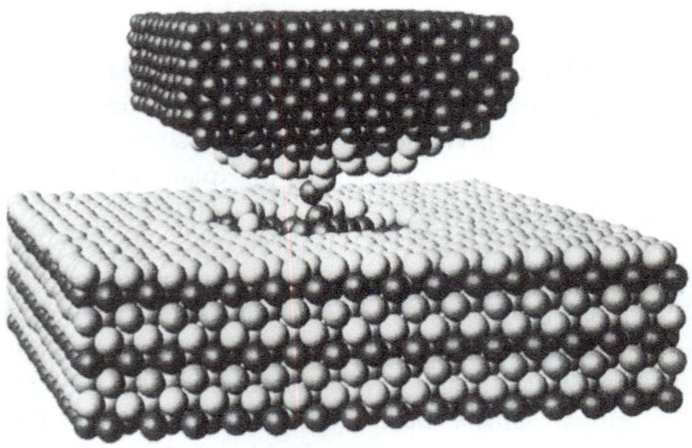

The computer graphic shows gold atoms (bottom) adhering to the fine point of a nickel probe (top) that has been in contact with the gold surface. *(Uzi Landman and David W. Leudtke/Georgia Institute of Technology.)*

STATIC FRICTION

Suppose you apply a small horizontal force $\vec{F}$ (Figure 5-2) to a large box resting on the floor. The box may not move noticeably because the force of static friction $\vec{f}_s$, exerted by the floor on the box, balances the force you apply. **Static friction** is the frictional force that acts when there is no sliding between the two surfaces in contact—it is the force that keeps the box from sliding. The force of static friction, which opposes the applied force on the box, can vary in magnitude from zero to some maximum value $f_{s\max}$, depending on how hard you push. That is, as you push on the box, the opposing force of static friction increases to remain equal in magnitude to the applied force until the magnitude of the applied force exceeds $f_{s\max}$. Data show that $f_{s\max}$ is proportional to the strength of the forces pressing the two surfaces together. That is, $f_{s\max}$ is proportional to the magnitude of the normal force exerted by one surface on the other:

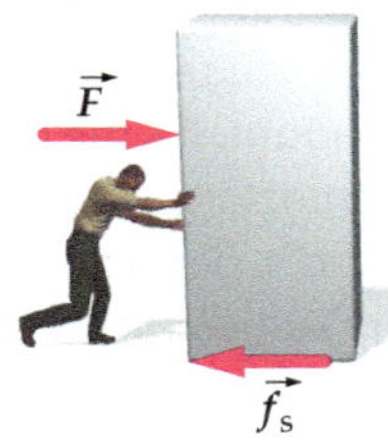

FIGURE 5-2

$$f_{s\max} = \mu_s F_n \qquad 5\text{-}1$$

STATIC FRICTION RELATION

where the proportionality constant μ_s, is the **coefficient of static friction.** This coefficient depends on what materials the surfaces in contact are made of as well as the temperatures of the surfaces. If you exert a horizontal force with a magnitude that is less than or equal to $f_{s\max}$ on the box, the static frictional force will just balance this horizontal force and the box will remain at rest. If you exert a horizontal force even slightly greater than $f_{s\max}$ on the box, then the box will begin to slide. Thus, we can write Equation 5-1 as:

$$f_s \leq \mu_s F_n \qquad 5\text{-}2$$

The direction of the static frictional force is such that it opposes the tendency of the box to slide.

! Equation 5-2 is an inequality because the magnitude of the force of static friction ranges from zero up to $f_{s\max}$.

! If the horizontal force you exert on the box is toward the left, then the static frictional force is toward the right. The static frictional force always opposes any tendency to slide.

KINETIC FRICTION

If you push the box in Figure 5-2 hard enough, it will slide across the floor. As it slides, the floor exerts a force of **kinetic friction** $\vec{f}_k$ (also called sliding friction) that opposes the motion. To keep the box sliding with constant velocity, you must exert a force on the box that is equal in magnitude and opposite in direction to the force of kinetic friction exerted by the floor.

Like the magnitude of a maximum static frictional force, the magnitude of a kinetic frictional force f_k is proportional to the microscopic contact area and the strength of the forces pressing the two surfaces together. That is, f_k is proportional to the normal force F_n one surface exerts on the other:

$$f_k = \mu_k F_n \qquad 5\text{-}3$$

KINETIC FRICTION RELATION

where the proportionality constant μ_k, is the **coefficient of kinetic friction.** The coefficient of kinetic friction depends on what materials the surfaces in contact are made of as well as the temperature of the contacting surfaces. Unlike static friction, the force of kinetic friction is independent of the magnitude of the applied horizontal force. Experiments show that μ_k is approximately constant for a wide range of speeds.

Figure 5-3 shows a plot of the frictional force exerted on the box by the floor as a function of the applied force. The force of friction balances the applied force until the box starts to slide, which occurs when the applied force exceeds $\mu_s F_n$ by an infinitesimal amount. As the box slides, the frictional force remains equal to $\mu_k F_n$. For any given contacting surfaces, μ_k is less than μ_s. This means you have to push harder to get the box to begin sliding than to keep it sliding at constant speed. Table 5-1 lists some approximate values of μ_s and μ_k for various pairs of surfaces.

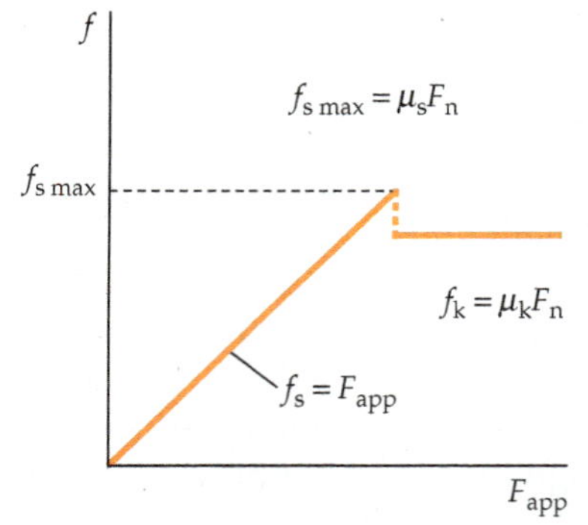

FIGURE 5-3

Table 5-1 Approximate Values of Frictional Coefficients

Materials	μ_s	μ_k
Steel on steel	0.7	0.6
Brass on steel	0.5	0.4
Copper on cast iron	1.1	0.3
Glass on glass	0.9	0.4
Teflon on Teflon	0.04	0.04
Teflon on steel	0.04	0.04
Rubber on concrete (dry)	1.0	0.80
Rubber on concrete (wet)	0.30	0.25
Waxed ski on snow (0°C)	0.10	0.05

ROLLING FRICTION

When a perfectly rigid wheel rolls *at constant speed* along a perfectly rigid horizontal road without slipping, no frictional force slows its motion. However, because real tires and roads continually deform (Figure 5-4) and because the tread and the road are continually peeled apart, in the real world the road exerts a force of **rolling friction** $\vec{f}_r$ that opposes the motion. To keep the wheel rolling with constant velocity, you must exert a force on the wheel that is equal in magnitude and opposite in direction to the force of rolling friction exerted on the wheel by the road.

The **coefficient of rolling friction** μ_r is the ratio of the magnitudes of the rolling frictional force f_r and the normal force F_n:

$$f_r = \mu_r F_n \quad \text{5-4}$$

ROLLING FRICTION RELATION

where μ_r depends on the nature of the surfaces in contact and the composition of the wheel and road. Typical values of μ_r are 0.01 to 0.02 for rubber tires on concrete and 0.001 to 0.002 for steel wheels on steel rails. Coefficients of rolling friction are typically less than coefficients of kinetic friction by one to two orders of magnitude. Rolling friction is considered to be negligible in this book, except where it is specifically stated that it is significant.

FIGURE 5-4 As the car moves down the highway, the rubber flexes radially inward where the tread initiates contact with the pavement, and flexes radially outward where the tread loses contact with the road. The tire is not perfectly elastic, so the forces exerted on the tread by the pavement that flex the tread inward are greater than those exerted on the tread by the pavement as the tread flexes back as it leaves the pavement. This imbalance of forces results in a force opposing the rolling of the tire. This force is called a rolling frictional force. The more the tire flexes, the greater the rolling frictional force.

SOLVING PROBLEMS INVOLVING STATIC, KINETIC, AND ROLLING FRICTION

The following examples illustrate how to solve problems involving static and kinetic friction. The guidelines for approaching these types of problems are as follows:

PROBLEM-SOLVING STRATEGY

Solving Problems Involving Friction

PICTURE Determine which types of friction are involved in solving a problem. Objects experience static friction when no sliding exists between the surfaces of the objects that are in contact. The force of static friction opposes the tendency of the surfaces to slide on each other. The maximum static frictional force f_{smax} is equal to the product of the normal force and the coefficient of static friction. If two surfaces are sliding against each other, they experience kinetic frictional forces (unless the problem states that one of the surfaces is frictionless). Rolling friction occurs because a rolling object and the surface that the object is rolling on continually deform and the object and the surface are continually peeling apart.

SOLVE

1. Construct a free-body diagram with the y axis normal to (and the x axis parallel to) the contacting surfaces. The direction of the frictional force is such that it opposes slipping, or the tendency to slip.
2. Apply $\Sigma F_y = ma_y$ and solve for the normal force F_n.
   If the friction is *kinetic* or *rolling*, relate the frictional and normal forces using $f_k = \mu_k F_n$ or $f_r = \mu_r F_n$, respectively.
   If the friction is *static*, relate the frictional and normal forces using $f_s \leq \mu_s F_n$ (or $f_{smax} = \mu_s F_n$).
3. Apply $\Sigma F_x = ma_x$ to the object and solve for the desired quantity.

CHECK In making sure that your answer makes sense, remember that coefficients of friction are dimensionless and that you must account for all forces (for example, tensions in ropes).

Example 5-1 A Game of Shuffleboard

A cruise-ship passenger uses a shuffleboard cue to push a shuffleboard disk of mass 0.40 kg horizontally along the deck so that the disk leaves the cue with a speed of 8.5 m/s. The disk then slides a distance of 8.0 m before coming to rest. Find the coefficient of kinetic friction between the disk and the deck.

PICTURE The force of kinetic friction is the only horizontal force acting on the disk after it separates from the cue. The acceleration is constant, because the frictional force is constant. We can find the acceleration using the constant-acceleration equations of Chapter 2 and relate the acceleration to μ_k using $\Sigma F_x = ma_x$.

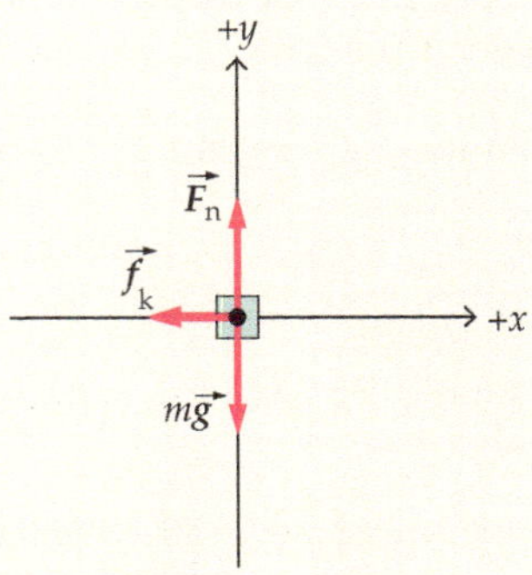

FIGURE 5-5

SOLVE

1. Draw a free-body diagram for the disk after it leaves the cue (Figure 5-5). Choose as the $+x$ direction the direction of the disk's velocity:
2. The coefficient of kinetic friction relates the magnitudes of the frictional and normal forces:

   $f_k = \mu_k F_n$

3. Apply $\Sigma F_y = ma_y$ to the disk. Solve for the normal force. Then, using the relationship from step 2, solve for the frictional force:

   $\Sigma F_y = ma_y$

   $F_n - mg = 0 \Rightarrow F_n = mg$

   so $f_k = \mu_k mg$

4. Apply $\Sigma F_x = ma_x$ to the disk. Using the step-3 result, solve for the acceleration:

   $\Sigma F_x = ma_x$

   $-f_k = ma_x$

   so $-\mu_k mg = ma_x$ so $a_x = -\mu_k g$

5. The acceleration is constant. Relate it to the total distance traveled and the initial velocity using $v_x^2 = v_{0x}^2 + 2a_x\Delta x$ (Equation 2-15). Using the step-4 result, solve for μ_k:

$$v_x^2 = v_{0x}^2 + 2a_x\,\Delta x \Rightarrow 0 = v_{0x}^2 - 2\mu_k g\,\Delta x$$

$$\text{so}\quad \mu_k = \frac{v_{0x}^2}{2g\,\Delta x} = \frac{(8.5\text{ m/s})^2}{2(9.81\text{ m/s}^2)(8.0\text{ m})} = \boxed{0.46}$$

CHECK The value obtained for μ_k is dimensionless and within the range of values for other materials listed in Table 5-1, so it is plausible.

TAKING IT FURTHER Note that the acceleration and the coefficient of friction are independent of the mass m. The greater the mass, the harder it is to stop the disk, but a greater mass is accompanied by a greater normal force, and thus a greater frictional force. The net result is that the mass has no effect on the acceleration (or the stopping distance).

Example 5-2 A Sliding Coin

A hardcover book (Figure 5-6) is resting on a tabletop with its front cover facing upward. You place a coin on this cover and very slowly open the book until the coin starts to slide. The angle θ_{max} (known as the *angle of repose*) is the angle the front cover makes with the horizontal just as the coin starts to slide. Find the coefficient of static friction μ_s between the book cover and the coin in terms of θ_{max}.

FIGURE 5-6 *(Ramón Rivera-Moret.)*

PICTURE The forces acting on the coin are the gravitational force $F_g = mg$, the normal force F_n, and the frictional force f. Because the coin is on the verge of sliding (but not yet sliding), the frictional force is a static frictional force directed up the incline. Because the coin remains stationary, its acceleration is zero. We use Newton's second law to relate this acceleration to the forces on the coin, and then solve for the fictional force.

SOLVE

1. Draw a free-body diagram for the coin when the book cover inclined at angle θ, where $\theta \leq \theta_{max}$ (Figure 5-7). Draw the y axis normal to the book cover:

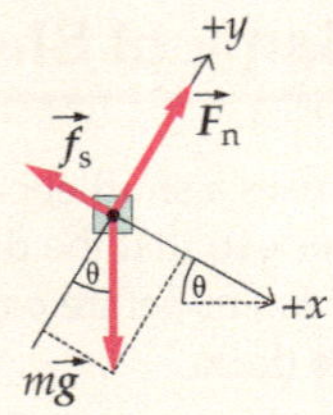

FIGURE 5-7

2. The coefficient of static friction relates the frictional and normal forces:

$$f_s \leq \mu_s F_n$$

3. We apply $\Sigma F_y = ma_y$ to the coin and solve for the normal force:

$$\Sigma F_y = ma_y$$
$$F_n - mg\cos\theta = 0 \Rightarrow F_n = mg\cos\theta$$

4. Substitute for F_n in $f_s \leq \mu_s F_n$ (Equation 5-1):

$$f_s \leq \mu_s F_n \Rightarrow f_s \leq \mu_s mg\cos\theta$$

5. Apply $\Sigma F_x = ma_x$ to the coin. Then solve for the friction force:

$$\Sigma F_x = ma_x$$
$$-f_s + mg\sin\theta = 0 \Rightarrow f_s = mg\sin\theta$$

6. Substituting $mg\sin\theta$ for f_s in the step-4 result gives:

$$mg\sin\theta \leq \mu_s mg\cos\theta \Rightarrow \tan\theta \leq \mu_s$$

7. θ_{max}, the largest angle satisfying the condition $\tan\theta \leq \mu_s$, is the largest angle such that the coin does not slide:

$$\boxed{\mu_s = \tan\theta_{max}}$$

CHECK The coefficient of friction is dimensionless, and so is the tangent function. Also, for $0 < \theta_{max} < 45°$, $\tan\theta_{max}$ is between zero and one. One would expect the coin to slide before the angle reached 45° and one would expect the coefficient of static friction to be between zero and one. Thus, the step-7 result is plausible.

PRACTICE PROBLEM 5-1 The coefficient of static friction between a car's tires and the road on a particular day is 0.70. What is the steepest angle of inclination of the road for which the car can be parked with all four wheels locked and not slide down the hill?

Erosion due to a stream cutting across a beach. Even though the edge weaves in and out, the angle of the slope remains constant. The angle of the slope is the angle of repose for the granular material. *(David R. Bailey.)*

CONCEPT CHECK 5-1

The car in Practice Problem 5-1 is parked at the steepest angle of inclination with all four wheels locked. Would the car slide down the incline if only two of the wheels are locked?

Example 5-3 Pulling a Sled

Two children sitting on a sled at rest in the snow ask you to pull them. You oblige by pulling on the sled's rope, which makes an angle of 40° with the horizontal (Figure 5-8). The children have a combined mass of 45 kg and the sled has a mass of 5.0 kg. The coefficients of static and kinetic friction are $\mu_s = 0.20$ and $\mu_k = 0.15$ and the sled is initially at rest. Find both the magnitude of the frictional force exerted by the snow on the sled and the acceleration of the children and sled if the tension in the rope is (*a*) 100 N and (*b*) 140 N.

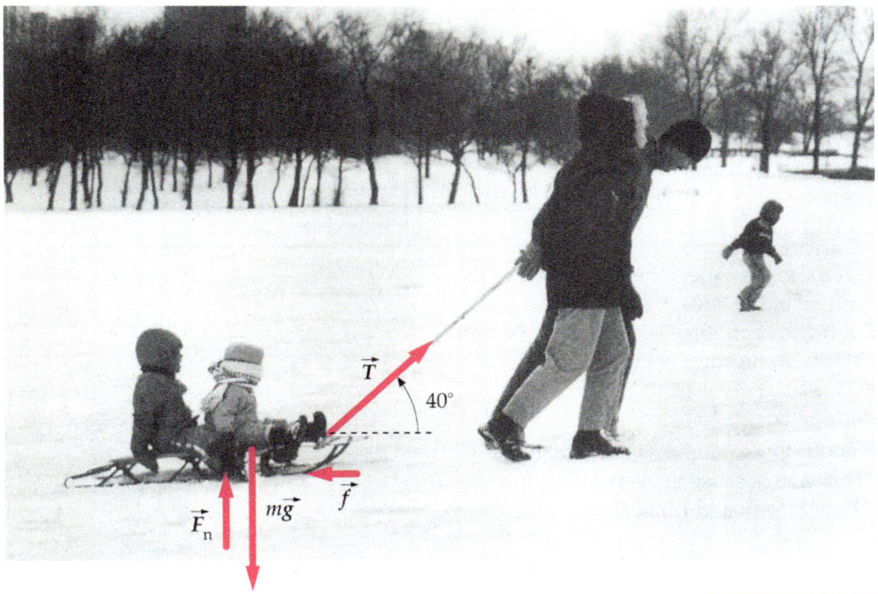

FIGURE 5-8 *(Jean-Claude LeJeune/Stock Boston.)*

PICTURE First, we need to find out whether the frictional force is static or kinetic. To do this, we see if the given tension forces satisfy the relation $f_s \le \mu_s F_n$. Once we have done that, we can select the correct expression for the frictional force, and solve the corresponding expression for f.

SOLVE

(*a*) 1. Draw a free-body diagram for the sled (Figure 5-9):

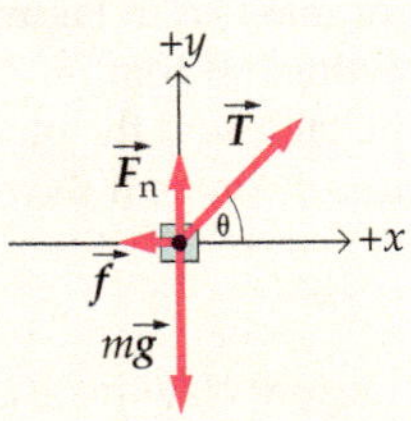

FIGURE 5-9

2. Write down the static friction relation. If this relation is satisfied, the sled does not begin sliding:

$$f_s \le \mu_s F_n$$

3. Apply $\Sigma F_y = ma_y$ to the sled and solve for the normal force:

$$\Sigma F_y = ma_y$$
$$F_n + T\sin\theta - mg = 0 \Rightarrow F_n = mg - T\sin\theta$$

4. Apply $\Sigma F_x = ma_x$ (with $a_x = 0$) to the sled and solve for the static frictional force:

$$\Sigma f_x = ma_x$$
$$-f_s + T\cos\theta = 0 \Rightarrow f_s = T\cos\theta$$

5. Substitute the step-4 and step-5 results into the step-2 result: $T\cos\theta \le \mu_s(mg - T\sin\theta)$

6. Check to see if the given tension of 100 N satisfies the nonslip condition (the step-3 inequality): $(100\text{ N})\cos 40° \le 0.20[(50\text{ kg})(9.81\text{ N/kg}) - (100\text{ N})\sin 40°]$

   $77\text{ N} \le 85\text{ N}$

   The inequality *is* satisfied, thus the sled *is not* sliding.

7. Because the sled is not sliding, the frictional force is that of static friction. To find the frictional force, use the step-3 expression for f_s: $a_x = 0$

   $f_s = T\cos\theta = (100\text{ N})\cos 40° = \boxed{77\text{ N}}$

(*b*) 1. Check the step-4 result from Part (*a*) with $T = 140$ N. If the relation is satisfied, the sled does not slide: $(140\text{ N})\cos 40° \le 0.20[(50\text{ kg})(9.81\text{ N/kg}) - (140\text{ N})\sin 40°]$

   $107\text{ N} \le 80\text{ N}$

   The inequality *is not* satisfied, thus the sled *is* sliding.

2. Because the sled is sliding, the friction is kinetic friction, where $f_k = \mu_k F_n$. In Part-(*a*), step 3 we applied $\Sigma F_y = ma_y$ to the sled and found $F_n = mg - T\sin\theta$. Using these results, solve for the kinetic frictional force:

$$f_k = \mu_k F_n$$
$$f_k = \mu_k(mg - T\sin\theta) = 0.15[(50\text{ kg})(9.81\text{ N/kg}) - (140\text{ N})\sin 40°] = \boxed{60\text{ N}}$$

3. Apply $\Sigma F_x = ma_x$ to the sled and solve for the frictional force. Then substitute the Part-(*b*), step-2 result for f_k and solve for the acceleration:

$$\Sigma F_x = ma_x$$
$$-f_k + T\cos\theta = ma_x \Rightarrow a_x = \frac{-f_k + T\cos\theta}{m}$$
$$\text{so} \quad a_x = \frac{(-60\text{ N}) + (140\text{ N})\cos 40°}{50\text{ kg}} = \boxed{0.94\text{ m/s}^2}$$

CHECK We expect a_x to be greater than or equal to zero, so we expect the magnitude of the frictional force to be less than or equal to the x component of the tension force. In Part (*a*) the magnitude of the frictional force and x component of the tension force both equal 77 N, and in Part (*b*) the magnitude of the frictional force equals 60 N and the x component of the tension force is $140\text{ N}\cos 40° = 107$ N.

TAKING IT FURTHER Note two important points about this example: (1) the normal force is less than the weight of the children and the sled. This is so because the vertical component of the tension helps the ground counter the gravitational force; and (2) in Part (*a*), the force of static friction is less than $\mu_s F_n$.

PRACTICE PROBLEM 5-2 What is the maximum force you can pull the rope at the specified angle without the sled beginning to slide?

Example 5-4 A Sliding Block

The block of mass m_2 in Figure 5-10 has been adjusted so that the block of mass m_1 is on the verge of sliding. (*a*) If $m_1 = 7.0$ kg and $m_2 = 5.0$ kg, what is the coefficient of static friction between the table and the block? (*b*) With a slight nudge, the blocks move with acceleration of magnitude a. Find a if the coefficient of kinetic friction between the table and the block is $\mu_k = 0.54$.

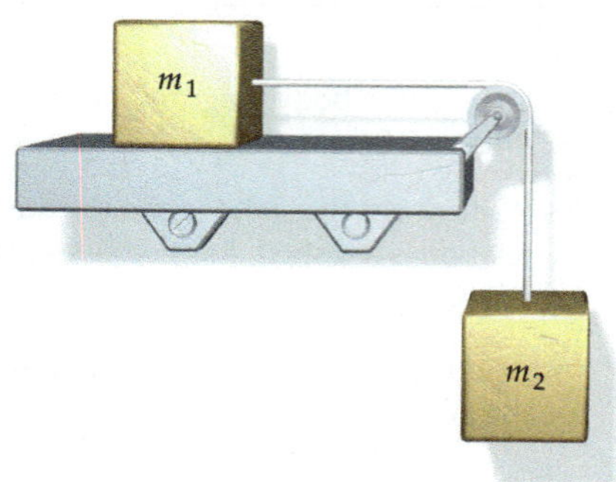

FIGURE 5-10

PICTURE Apply Newton's second law to each block. By neglecting the masses of both the rope and the pulley, and by neglecting friction in the pulley bearing, the tension has the same magnitude throughout the rope, so $T_1 = T_2 = T$. Because the rope remains taut but does not stretch, the accelerations have the same magnitude, so $a_1 = a_2 = a$.

To find the coefficient of static friction μ_s, as requested in Part (*a*), set the force of static friction on m_1 equal to its maximum value $f_{s\max} = \mu_s F_n$ and set the acceleration equal to zero.

SOLVE

(*a*) 1. Draw a free-body diagram for each block (Figure 5-11). Choose the $+x$ and $+x'$ directions to be the same as the directions of the accelerations of blocks 1 and 2, respectively. That is, the $+x$ direction is to the right and the $+x'$ direction is vertically downward:

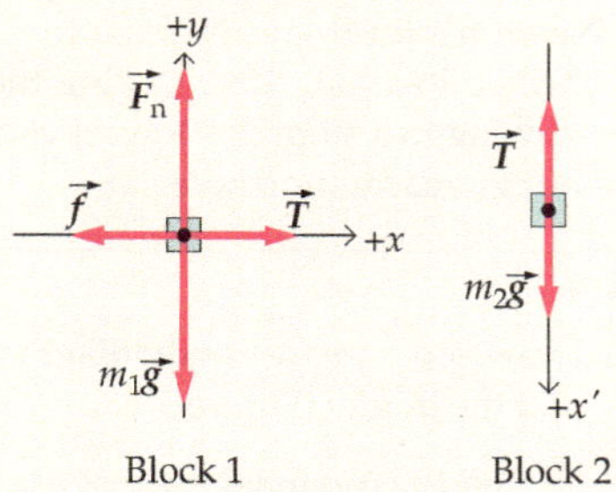

FIGURE 5-11

2. Apply $\Sigma F_y = ma_y$ to block 1 and solve for the normal force. Then solve for the static frictional force.

$\Sigma F_y = m_1 a_{1y}$

$F_n - m_1 g = 0 \Rightarrow F_n = m_1 g$

so

$f_{smax} = \mu_s F_n$ so $f_{smax} = \mu_s m_1 g$

3. Apply $\Sigma F_x = ma_x$ to block 1 and solve for the frictional force. Then substitute into the step-2 result.

$\Sigma F_x = m_1 a_{1x}$

$T - f_{smax} = 0 \Rightarrow T = f_{smax}$

so

$T = \mu_s m_1 g$

4. Apply $\Sigma F_x = ma_x$ to block 2 and solve for the tension. Then substitute into the step-3 result.

$\Sigma F_{x'} = m_2 a_{2x'} \Rightarrow m_2 g - T = 0$

so

$T = m_2 g$ and $m_2 g = \mu_s m_1 g$

5. Solve the step-4 result for μ_s.

$$\mu_s = \frac{m_2}{m_1} = \frac{5.0\text{ kg}}{7.0\text{ kg}} = \boxed{0.71}$$

(*b*) 1. During sliding, the frictional force is kinetic and the accelerations have the same magnitude a. Relate the kinetic frictional force f_k to the normal force. The normal force was found in step 2 of Part (*a*).

$f_k = \mu_k F_n$

so

$f_k = \mu_k m_1 g$

$|\vec{a}_1| = a_{1x} = a$ and $|\vec{a}_2| = a_{2x'} = a$

2. Apply $\Sigma F_x = ma_x$ to block 1. Then substitute for the frictional force using the result from step 1 of Part (*b*).

$\Sigma F_x = m_1 a_{1x} \Rightarrow T - f_k = m_1 a$

so

$T - \mu_k m_1 g = m_1 a$

3. Apply $\Sigma F_{x'} = ma_{x'}$ to block 2.

$\Sigma F_{x'} = m_2 a_{2x'} \Rightarrow m_2 g - T = m_2 a$

4. Add the equations in steps 2 and 3 of Part (*b*) and solve for a.

$$a = \frac{m_2 - \mu_k m_1}{m_1 + m_2} g = \boxed{1.0\text{ m/s}^2}$$

CHECK Note that if $m_1 = 0$ the expression for the acceleration reduces to $a = g$ as one would expect.

PRACTICE PROBLEM 5-3 What is the tension in the rope when the blocks are sliding?

Example 5-5 The Runaway Buggy

Context-Rich

A runaway baby buggy is sliding without friction across a frozen pond toward a hole in the ice (Figure 5-12). You race after the buggy on skates. As you grab it, you and the buggy are moving toward the hole at speed v_0. The kinetic coefficient of friction between your skates and the ice as you turn out the blades to brake is μ_k. D is the distance between the buggy and the hole at the instant you reach the buggy, m_B is the mass of the buggy (including its precious cargo), and m_Y is your mass. (*a*) What is the lowest value of D such that you stop the buggy before it reaches the hole in the ice? (*b*) What force do you exert on the buggy as you slow it?

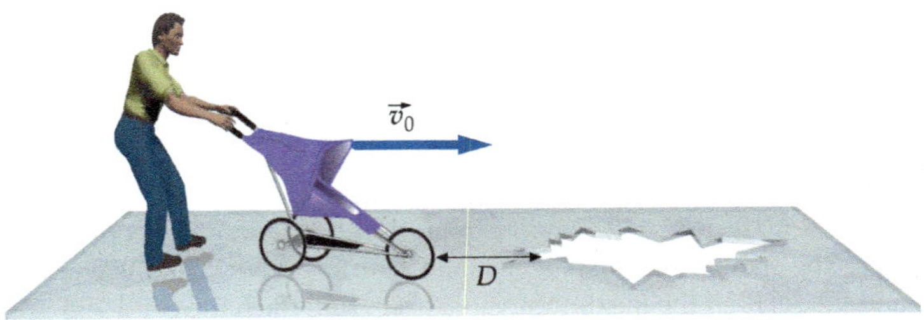

FIGURE 5-12

PICTURE Initially, you and the buggy are moving toward the hole with speed v_0, which we take to be in the $+x$ direction. If you exert a force $\vec{F}_{YB}$ on the buggy, the buggy, in accord with Newton's third law, exerts a force $\vec{F}_{BY}$ on you. Apply Newton's second law to determine the acceleration. After finding the acceleration, find the distance D the buggy travels while slowing to a stop. The lowest value of D is that for which your speed reaches zero just as the buggy reaches the hole.

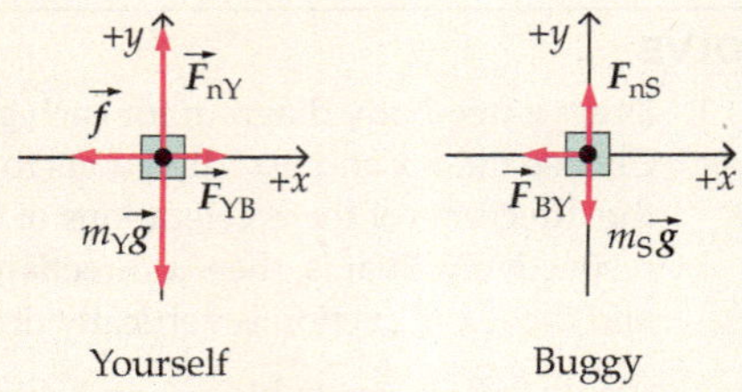

FIGURE 5-13

SOLVE

(*a*) 1. Draw separate free-body diagrams for yourself and the buggy (Figure 5-13).

2. To find the frictional force of the ice on you, you need to first find the normal force of the ice on you:
$$f_{IYk} = \mu_k F_{IYn}$$

3. Apply $\Sigma F_y = ma_y$ to yourself and solve first for the normal force and then for the frictional force:
$$\Sigma F_y = ma_y \Rightarrow F_{IYn} - m_Y g = 0 \quad (a_y = 0)$$
and $f_{IYk} = \mu_k F_{IYn}$ so $f_{IYk} = \mu_k m_Y g$

4. Apply $\Sigma F_x = ma_x$ to yourself. Then substitute in the step-3 result:
$$\Sigma F_x = ma_x \Rightarrow F_{BY} - f_{IYk} = m_Y a_x$$
so $F_{BY} - \mu_k m_Y g = m_Y a_x$

5. Apply $\Sigma F_x = ma_x$ to the buggy:
$$\Sigma F_x = m_B a_x \Rightarrow F_{YB} = m_B a_x$$

6. $\vec{F}_{BY}$ and $\vec{F}_{YB}$ form an N3L force pair, so they are equal in magnitude:
$$F_{BY} = F_{YB}$$

7. Add the step-4 and step-5 results and use $F_{BY} - F_{YB} = 0$ to simplify:
$$-F_{YB} + (F_{BY} - \mu_s m_Y g) = m_B a_x + m_Y a_x$$
$$0 - \mu_s m_Y g = m_B a_x + m_Y a_x$$

8. Solve the step-7 result for a_x:
$$a_x = -\frac{\mu_k m_Y}{m_Y + m_B} g \quad (a_x \text{ is negative, as expected.})$$

9. Substitute the step-8 result into a kinematic equation and solve for the magnitude of the displacement D:
$$v_x^2 = v_{0x}^2 + 2a_x \Delta x \Rightarrow 0 = v_0^2 + 2a_x D$$
so
$$D = \frac{-v_0^2}{2a_x} = \boxed{\left(1 + \frac{m_B}{m_Y}\right)\frac{v_0^2}{2\mu_k g}}$$

(*b*) F_{YB} can be found by combining the results for steps 5 and 8:
$$F_{YB} = m_B|a_x| = \boxed{\frac{\mu_k m_B g}{1 + (m_Y/m_B)}}$$

CHECK For large values of m_B/m_Y, D is large, as expected.

TAKING IT FURTHER The minimum value of D is proportional to v_0^2 and inversely proportional to μ_k. Figure 5-14 shows the stopping distance D versus initial velocity squared for values of m_B/m_Y equal to 0.1, 0.3, and 1.0, with $\mu_k = 0.5$. Note that the larger the mass ratio m_B/m_Y, the greater the distance D needed to stop for a given initial velocity. This is akin to braking to a stop in a car that is pulling a trailer that does not have its own brakes. The mass of the trailer increases the stopping distance for a given speed.

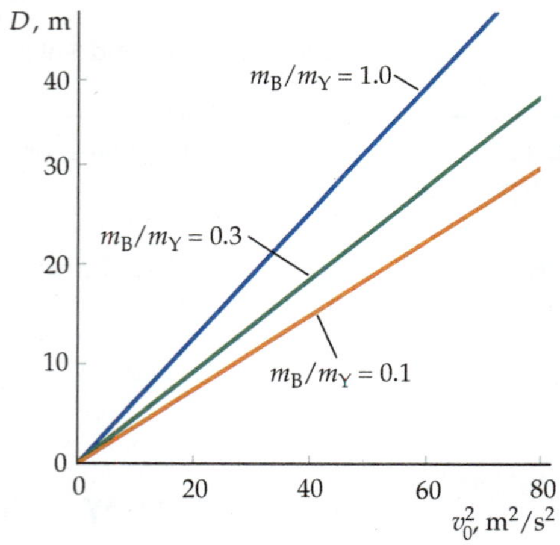

FIGURE 5-14

Example 5-6 Pulling a Child on a Toboggan

Try It Yourself

A child of mass m_C sits on a toboggan of mass m_T, which in turn sits on a frictionless frozen pond (Figure 5-15). The toboggan is pulled with a horizontal applied force $\vec{F}_{ap}$ as shown. The coefficients of static and sliding friction between the child and toboggan are μ_s and μ_k. (*a*) Find the maximum value of F_{ap} for which the child will not slide relative to the toboggan. (*b*) Find the acceleration of the toboggan and the acceleration of the child when F_{ap} is greater than this value.

FIGURE 5-15

PICTURE The only force accelerating the child forward is the frictional force exerted by the toboggan on the child. In Part (*a*) the challenge is to find F_{ap} when this frictional force is static and maximum. To do this, apply $\sum\vec{F} = m\vec{a}$ to the child and solve for the acceleration when the static frictional force is maximum. Then apply $\sum\vec{F} = m\vec{a}$ to the toboggan and solve for F_{ap}. In Part (*b*), we follow a parallel procedure. However, in Part (*b*) the minimum value of F_{ap} is given and we solve for the acceleration of the toboggan.

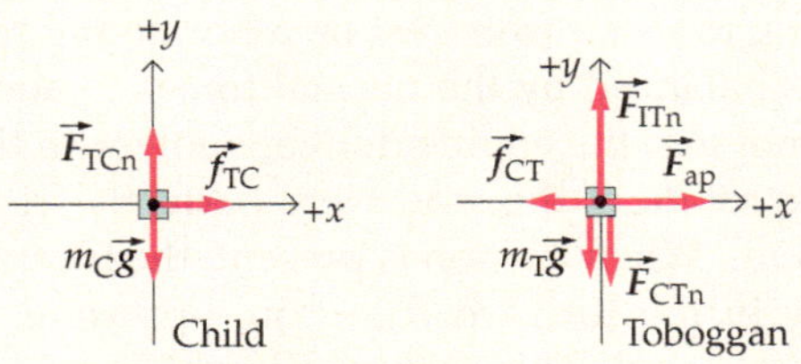

FIGURE 5-16 The force $\vec{F}_{ITn}$ is the normal force exerted by the ice on the toboggan.

SOLVE

Cover the column to the right and try these on your own before looking at the answers.

Steps	Answers
(*a*) 1. Draw one free-body diagram for the child and another for the toboggan (Figure 5-16).	
2. Apply $\sum F_x = ma_x$ to the toboggan:	$\sum F_{Tx} = m_T a_{Tx} \Rightarrow F_{ap} - f_{CTs\,max} = m_T a_{Tx}$
3. Apply $\sum F_y = ma_y$ to the child and solve for the normal force. Then apply $f_{x\,max} = \mu_s F_n$ and solve for the frictional force.	$\sum F_{Cy} = m_C a_y \Rightarrow F_{TCn} - m_C g = 0$ so $F_{TCn} = m_C g$ $f_{s\,max} = \mu F_n \Rightarrow f_{TCs\,max} = \mu_s F_{TCn}$ so $f_{TCs\,max} = \mu_s m_C g$
4. Apply $\sum F_x = ma_x$ to the child and solve for the acceleration.	$\sum F_{Cx} = m_C a_{Cx} \Rightarrow f_{TCs\,max} = m_C a_{Cx}$ so $\mu_s m_C g = m_C a_{Cx} \Rightarrow a_{Cx} = \mu_s g$
5. Equate the magnitudes of the forces in each N3L force pair appearing in the two free-body diagrams. In addition, express the relation between the accelerations due to the nonslipping constraint.	$F_{TCn} = F_{CTn}$ and $f_{TCs\,max} = f_{CTs\,max}$ and $a_{Cx} = a_{Tx} = a_x$
6. Substitute the step-4 and step-5 results into the step-2 result and solve for F_{ap}.	$F_{ap} - \mu_s m_C g = m_T \mu_s g$ so $F_{ap} = \boxed{(m_C + m_T)\mu_s g}$
(*b*) 1. Equate the magnitudes of each N3L force pair and express the relation between the accelerations if the child is slipping on the toboggan.	$F_{TCn} = F_{CTn} = F_n$ and $f_{TCk} = f_{CTk} = f_k$ and $a_{Cx} < a_{Tx}$
2. Solve for the magnitude of the kinetic frictional force using the result from step 3 of Part (*a*) for the normal force.	$f_k = \mu_k F_n$ so $f_k = \mu_k m_C g$
3. Apply $\sum F_x = ma_x$ to the child and solve for *her acceleration.*	$\sum F_{Cx} = m_C a_{Cx} \Rightarrow f_k = m_C a_{Cx}$ so $\mu_k m_C g = m_C a_{Cx} \Rightarrow a_{Cx} = \boxed{\mu_k g}$
4. Apply $\sum F_x = ma_x$ to the toboggan. Using the result from step 2 of Part (*b*), solve for *its acceleration.*	$\sum F_{Tx} = m_T a_{Tx} \Rightarrow F_{ap} - f_k = m_T a_{Tx}$ so $F_{ap} - \mu_k m_C g = m_T a_{Tx} \Rightarrow a_{Tx} = \boxed{\dfrac{F_{ap} - \mu_k m_C g}{m_T}}$

CHECK The Part-(*a*) result is what we expect if the child does not slip on the toboggan. If we model the child and the toboggan as a single particle and apply Newton's second law to it, we obtain $F_{ap} = (m_C + m_T)a_x$. If we substitute $\mu_s(m_C + m_T)g$ (our Part-(*a*), step-6 result) for F_{ap}, we get $\mu_s(m_C + m_T)g = (m_C + m_T)a_x$. Dividing both sides by the sum of the masses gives $a_x = \mu_s g$, our Part-(*a*), step-3 result. Thus, our Part (*a*) effort is consistent with modeling the sled and child as a single particle.

TAKING IT FURTHER In this example, the frictional force does not oppose the motion of the child, it causes it. Without friction, the girl would slip back relative to the toboggan. However, even though the child moves, or tends to move, backward (leftward) *relative to the toboggan,* she moves forward *relative to the ice.* Friction forces oppose relative motion, or tendency toward relative motion, between two surfaces in contact. In addition, relative to the child, the toboggan slips, or tends to slip, forward. The friction force on the toboggan is directed rearward, opposing this slipping forward, or tendency toward slipping forward.

! Note that frictional forces *do not always* oppose motion. However, frictional forces between contacting surfaces *do always* oppose relative motion, or the tendency toward relative motion, between the two contacting surfaces.

FRICTION, CARS, AND ANTILOCK BRAKES

Figure 5-17 shows the forces acting on a front-wheel-drive car that is just starting to move from rest on a horizontal road. The gravitational force F_g on the car is balanced by the normal forces F_n and F_n' exerted on the tires. To start the car moving, the engine delivers power to the axle that makes the front wheels start to rotate. If the road were perfectly frictionless, the front wheels would merely spin. When friction is present, the frictional force exerted by the road on the tires is in the forward direction, opposing the tendency of the tire surface to slip backward. This frictional force enables the acceleration needed for the car to start moving forward. If the power delivered by the engine is not so great that the surface of the tire slips on the surface of the road, the wheels will roll without slipping and that part of the tire tread touching the road remains at rest relative to the road. (The part in contact with the road continuously changes as the tire rolls.) The friction between the road and the tire tread is then that of static friction. The largest frictional force that the tire can exert on the road (and that the road can exert on the tire) is $\mu_s F_n$.

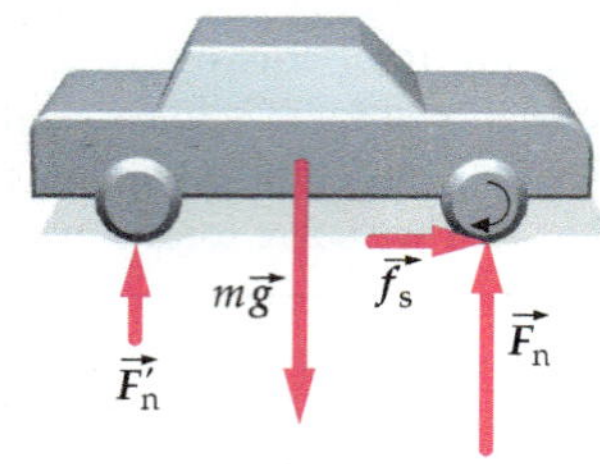

FIGURE 5-17 Forces acting on a car with front-wheel drive that is accelerating from rest. The normal force $\vec{F}_n$ on the front tires is usually larger than the normal force $\vec{F}_n'$ on the rear tires because typically the engine of the car is mounted near the front of the car. There is no drag force from the air shown because the car is just starting to move. There would be a rearward-directed rolling frictional force on all wheels, but that force has been ignored.

For a car moving in a straight line with speed v relative to the road, the center of each of its wheels also moves with speed v, as shown in Figure 5-18. If a wheel is rolling without slipping, its top is moving faster than v, whereas its bottom is moving slower than v. However, *relative to the car*, each point on the perimeter of the wheel moves in a circle with the same speed v. Moreover, the speed of the point on the tire momentarily in contact with the ground is zero *relative to the ground*. (Otherwise, the tire would be skidding along the road.)

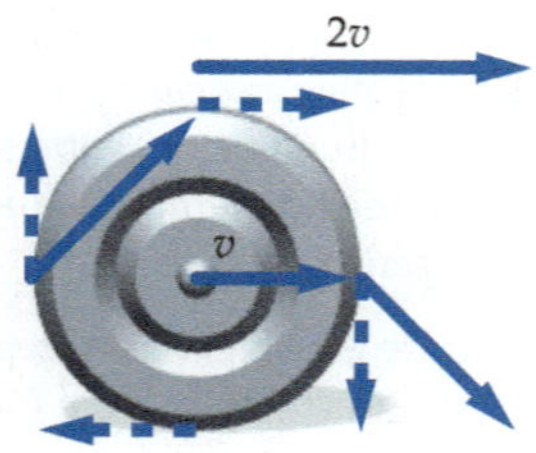

FIGURE 5-18 In this figure, dashed lines represent velocities relative to the body of the car; solid lines represent velocities relative to the ground.

If the engine supplies excessive power, the tire will slip and the wheels will spin. Then the force that accelerates the car is the force of kinetic friction, which is less than the maximum force of static friction. If we are stuck on ice or snow, our chances of getting free are better if we avoid spinning the wheels by using a light touch on the accelerator pedal. Similarly, when braking a car to a stop, the force exerted by the road on the tires may be either static friction or kinetic friction, depending on how hard the brakes are applied. If the brakes are applied so hard that the wheels lock, the tires will skid along the road and the stopping force will be that of kinetic friction. If the brakes are applied less forcefully, so that no slipping occurs between the tires and the road, the stopping force will be that of static friction.

When wheels do lock and tires skid, two undesirable things happen. The minimum stopping distance is increased and the ability to control the direction of the motion of the car is greatly diminished. Obviously, this loss of directional control can have dire consequences. Antilock braking systems (ABS) in cars are designed to prevent the wheels from locking even if the brakes are applied hard. These systems have wheel-speed sensors. If the control unit senses that a wheel is about to lock, the module signals the brake pressure modulator to drop, hold, and then restore the pressure to that wheel up to 15 times per second. This varying of pressure is much like "pumping" the brake, but with the ABS system, the wheel that is locking is the only one being pumped. This is called *threshold braking*. With threshold braking, maximum friction for stopping is maintained.

Example 5-7 The Effect of Antilock Brakes

A car is traveling at 30 m/s along a horizontal road. The coefficients of friction between the road and the tires are $\mu_s = 0.50$ and $\mu_k = 0.40$. How far does the car travel before stopping if (*a*) the car is braked with an antilock braking system so that threshold braking is sustained, and (*b*) the car is braked hard with no antilock braking system so that the wheels lock? (*Note:* Skidding heats up the tires and the coefficients of friction vary with changes in temperature. These temperature effects are neglected in this example.)

PICTURE The force that stops a car when it brakes without skidding is the force of static friction exerted by the road on the tires (Figure 5-19). We use Newton's second law to solve for the frictional force and the car's acceleration. Kinematics is then used to find the stopping distance.

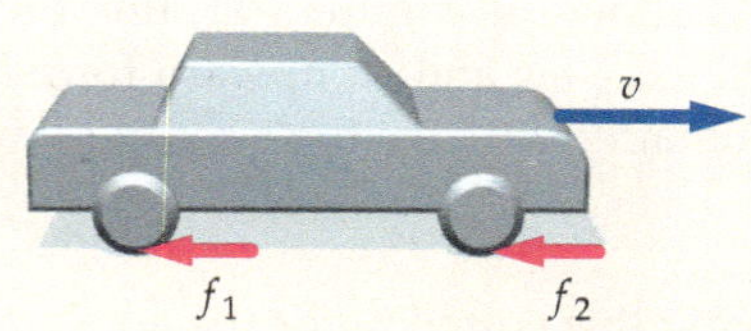

FIGURE 5-19

SOLVE

(*a*) 1. Draw a free-body diagram for the car (Figure 5-20). Treat all four wheels as if they were a single point of contact with the ground. Assume further that the brakes are applied to all four wheels. The $\vec{f}$ in the free-body diagram is the total of the frictional forces on individual wheels:

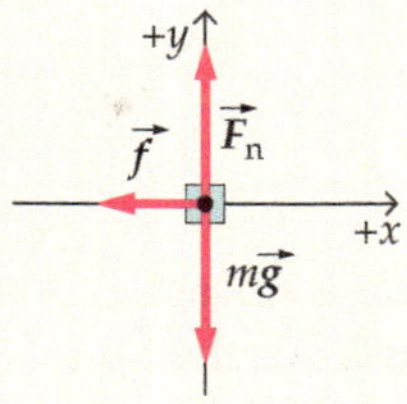

FIGURE 5-20

2. Assuming that the acceleration is constant, we use Equation 2-15 to relate the stopping distance Δx to the initial velocity v_{0x}:

$$v_x^2 = v_{0x}^2 + 2a_x\,\Delta x$$

$$\text{When } v_x = 0,\ \Delta x = -\frac{v_{0x}^2}{2a_x}$$

3. Apply $\Sigma F_y = ma_y$ to the car and solve for the normal force. Then apply $f_{x\max} = \mu_s F_n$ and solve for the frictional force:

$$\Sigma F_y = ma_y \ \Rightarrow\ F_n - mg = 0 \quad \text{so} \quad F_n = mg$$

$$f_{s\max} = \mu_s F_n \quad \text{so} \quad f_{s\max} = \mu_s mg$$

4. Apply $\Sigma F_x = ma_x$ to the car and solve for the acceleration:

$$\Sigma F_x = ma_x \ \Rightarrow\ -f_{s\max} = ma_x$$

Substituting $\mu_s mg$ for $f_{s\max}$ gives $-\mu_s mg = ma_x \ \Rightarrow\ a_x = -\mu_s g$

5. Substituting for a_x in the equation for Δx in step 2 gives the stopping distance:

$$\Delta x = -\frac{v_{0x}^2}{2a_x} = \frac{v_{0x}^2}{2\mu_s g} = \frac{(30\ \text{m/s})^2}{2(0.50)(9.81\ \text{m/s}^2)} = \boxed{0.92 \times 10^2\ \text{m}}$$

(*b*) 1. When the wheels lock, the force exerted by the road on the car is that of kinetic friction. Using reasoning similar to that in Part (*a*), we obtain for the acceleration:

$$a_x = -\mu_k g$$

2. The stopping distance is then:

$$\Delta x = -\frac{v_{0x}^2}{2a_x} = \frac{v_{0x}^2}{2\mu_k g} = \frac{(30\ \text{m/s})^2}{2(0.40)(9.81\ \text{m/s}^2)} = \boxed{1.1 \times 10^2\ \text{m}}$$

CHECK The calculated displacements are both positive as expected. In addition, the antilock brake system significantly shortens the stopping distance of the car as expected.

TAKING IT FURTHER Notice that the stopping distance is more than 20% greater when the wheels are locked. Also note that the stopping distance is independent of the car's mass—the stopping distance is the same for a subcompact car as for a large truck—provided the coefficients of friction are the same. The tires heat up dramatically when skidding occurs. This produces a change in μ_k that was not taken into account in this solution.

5-2 DRAG FORCES

When an object moves through a fluid such as air or water, the fluid exerts a **drag force** or retarding force that opposes the motion of the object. The drag force depends on the shape of the object, the properties of the fluid, and the speed of the object relative to the fluid. Unlike ordinary friction, the drag force increases as the speed of the object increases. At very low speeds, the drag force is approximately proportional to the speed of the object; at higher speeds, it is more nearly proportional to the square of the speed.

Consider an object dropped from rest and falling under the influence of the force of gravity, which we assume to be constant. The magnitude of the drag force is

$$F_d = bv^n \qquad \text{5-5}$$

DRAG FORCE RELATION

where b is a constant.

The drag force on the parachute slows this dragster. *(IHRA/Live Nation.)*

As shown in Figure 5-21, the forces acting on the object are a constant downward force mg and an upward force bv^n. If we take downward as the $+y$ direction, we obtain from Newton's second law

$$mg - bv^n = ma_y \tag{5-6}$$

Solving this equation for the acceleration gives:

$$a_y = g - \frac{b}{m}v^n \tag{5-7}$$

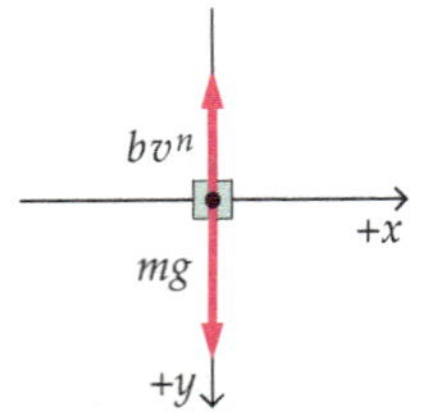

FIGURE 5-21 Free-body diagram showing forces on an object falling through the air.

The speed is zero at $t = 0$ (the instant the object is released), so at $t = 0$ the drag force is zero and the acceleration is g downward. As the speed of the object increases, the drag force increases so the acceleration decreases. Eventually, the speed is great enough for the magnitude of the drag force bv^n to approach the force of gravity mg. In this limit, the acceleration approaches zero and the speed approaches the **terminal speed** v_T. At terminal speed, the drag force balances the weight force and the acceleration is zero. Setting v equal to v_T and a_y equal to zero in Equation 5-6, we obtain

$$bv_T^n = mg$$

Solving for the terminal speed, we get

$$v_T = \left(\frac{mg}{b}\right)^{1/n} \tag{5-8}$$

The larger the constant b, the smaller the terminal speed. Cars are designed to minimize b to reduce the effect of wind resistance. A parachute, on the other hand, is designed to maximize b so that the terminal speed will be small. For example, the terminal speed of a skydiver before release of the parachute is about 200 km/h, which is about 60 m/s. When the parachute opens, the drag force rapidly increases, becoming greater than the force of gravity. As a result, the skydiver experiences an upward acceleration while falling; that is, the speed of the descending skydiver decreases. As the speed of the skydiver decreases, the drag force decreases and the speed approaches a new terminal speed of about 20 km/h.

(Stuart Williams/Dembinsky Photo Associates.)

(Joe McBride/Stone.)

Example 5-8 Terminal Speed

A 64.0-kg skydiver falls with a terminal speed of 180 km/h with her arms and legs outspread. (*a*) What is the magnitude of the upward drag force F_d on the skydiver? (*b*) If the drag force is equal to bv^2, what is the value of b?

PICTURE We use Newton's second law to solve for the drag force in Part (*a*), and then substitute in the appropriate values to find b in Part (*b*).

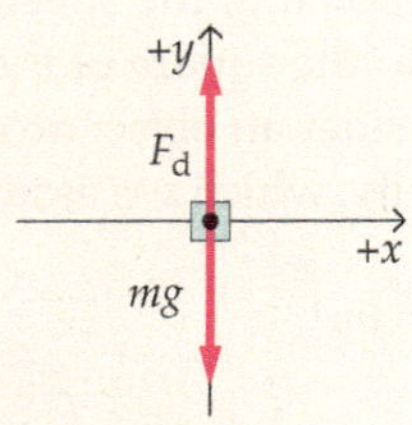

FIGURE 5-22

SOLVE

(*a*) 1. Draw a free-body diagram (Figure 5-22).

2. Apply $\Sigma F_y = ma_y$. Because the skydiver is moving with constant velocity, the acceleration is zero:

$$\Sigma F_y = ma_y \;\Rightarrow\; F_d - mg = 0$$

so $F_d = mg = (64.0\text{ kg})(9.81\text{ N/kg}) = \boxed{628\text{ N}}$

(*b*) 1. To find b we set $F_d = bv^2$:

$$F_d = mg = bv^2$$

$$\text{so} \quad b = \frac{mg}{v^2}$$

2. Convert the speed to m/s, then calculate b:

$$180\ \text{km/h} = \frac{180\ \text{km}}{1\ \text{h}} \times \frac{1\ \text{h}}{3.6\ \text{ks}} = 50.0\ \text{m/s}$$

$$b = \frac{(64.0\ \text{kg})(9.81\ \text{m/s}^2)}{(50.0\ \text{m/s})^2} = \boxed{0.251\ \text{kg/m}}$$

CHECK The units obtained for b are kg/m. To check that these units are correct, we multiply kg/m by $(\text{m/s})^2$ to get kg m/s^2. These units are units of force (because one newton is defined as a kg m/s^2). We expected these units to be units of force because the drag force is given by bv^2.

TAKING IT FURTHER The conversion factor 1 h/3.6 ks appears in the last step of the solution. This conversion factor is exact because 1 h is exactly equal to 3600 s. Consequently, in converting her speed from km/h to m/s, three-figure accuracy is maintained. This is so even though we did not write the conversion factor as 1.00 h/3.60 ks.

5-3 MOTION ALONG A CURVED PATH

Objects often do not move in a straight line: a car rounding a curve is an example, as is a satellite orbiting Earth.

Consider a satellite moving in a circular orbit around Earth as shown in Figure 5-23. At an altitude of 200 km, the gravitational force on the satellite is just slightly less than at Earth's surface. Why does the satellite not fall toward Earth? Actually, the satellite does "fall." But because the surface of Earth is curved, the satellite does not get closer to the surface of Earth. If the satellite were not accelerating, it would move from point P_1 to P_2 in some time t. Instead, it arrives at point P_2' on its circular orbit. In a sense, the satellite "falls" the distance h shown in Figure 5-23. If t is small, P_2 and P_2' are nearly on a radial line. In that case, we can calculate h from the right triangle of sides vt, r, and $r + h$: Because $r + h$ is the hypotenuse of the right triangle, the Pythagorean theorem gives:

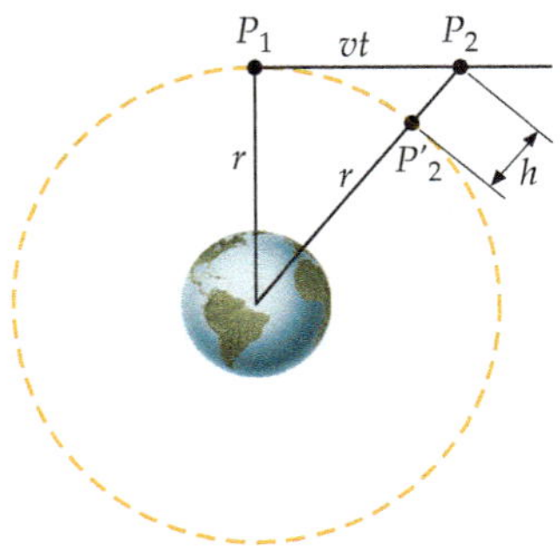

FIGURE 5-23 The satellite is moving with speed v in a circular orbit of radius r about Earth. If the satellite did not accelerate toward Earth, it would move from point P_1 to P_2 along a straight line. Because of its acceleration, it instead falls a distance h. For a sufficiently short time t, the acceleration is essentially constant, so $h = \frac{1}{2}at^2 = \frac{1}{2}(v^2/r)t^2$.

$$(r + h)^2 = (vt)^2 + r^2$$
$$r^2 + 2hr + h^2 = v^2t^2 + r^2$$

or

$$h(2r + h) = v^2t^2$$

For very short times, h will be much less than r, so we can neglect h compared with $2r$. Then

$$2rh \approx v^2t^2$$

or

$$h \approx \frac{1}{2}\left(\frac{v^2}{r}\right)t^2$$

Comparing this with the constant-acceleration expression $h = \frac{1}{2}at^2$, we see that the magnitude of the acceleration of the satellite is

$$a = \frac{v^2}{r}$$

which is the expression for centripetal acceleration established in Chapter 3. From Figure 3-24, we see this acceleration is directed toward the center of the circular orbit. By applying Newton's second law along the direction of the acceleration

See Math Tutorial for more information on Trigonometry

vector, we find that the magnitude of the net force causing the acceleration is related to the magnitude of the acceleration by:

$$F_{\text{net}} = m\frac{v^2}{r}$$

Figure 5-24*a* shows a ball on the end of a string, the other end of which is attached to a fixed support. The ball is traveling in a horizontal circle of radius r at constant speed v. Consequently, the acceleration of the ball has magnitude v^2/r.

As we saw in Chapter 3, a particle moving with constant speed v in a circle of radius r (Figure 5-24*a*) has an acceleration of magnitude $a = v^2/r$ directed toward the center of the circle (the centripetal direction). The net force acting on an object is always in the same direction as the acceleration vector, so the net force (Figure 5-24*b*) on an object moving in a circle at constant speed is also in the centripetal direction. A net force in the centripetal direction is sometimes referred to as the **centripetal force.** It may be due to a string, spring, or other contact force such as a normal or frictional force; it may be an action-at-a-distance type of force such as a gravitational force; or it may be any combination of these. It is always directed inward—toward the center of curvature of the path.

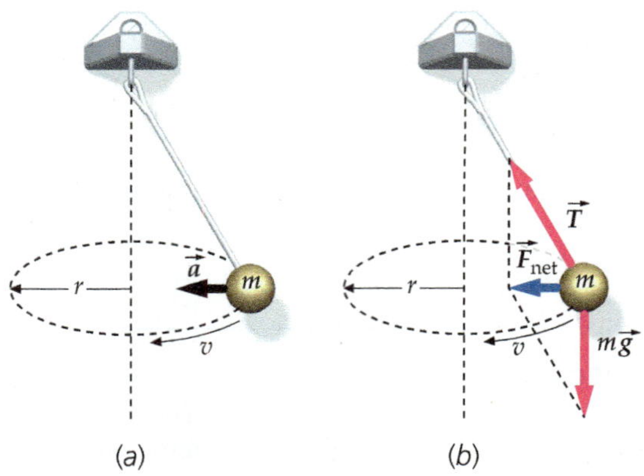

FIGURE 5-24 A ball suspended from a string moves in a horizontal circle at constant speed. (*a*) The acceleration vector is in the centripetal direction (toward the center of the circular path). Acceleration in the centripetal direction is called centripetal acceleration. (*b*) Two forces act on the ball, the tension force exerted by the string and the gravitational force. These two forces sum so that the net force is in the centripetal direction. A net force in the centripetal direction is sometimes referred to as the centripetal force.

! The centripetal force is not actually a force. It is merely a name for the component of net force toward the center of curvature of the path. Like the net force, the centripetal force does not belong on a free-body diagram. Only actual forces belong on free-body diagrams.

PROBLEM-SOLVING STRATEGY

Solving Motion Along A Curved Path Problems

PICTURE Remember that you should never label a force as a centripetal force on a free-body diagram. Instead you should label a force as a tension force, or a normal force, or a gravitational force, and so forth.

SOLVE

1. Draw a free-body diagram of the object. Include coordinate axes with the origin at a point of interest on the path. Draw one coordinate axis in the tangential direction (the direction of motion) and a second in the centripetal direction.
2. Apply $\Sigma F_c = ma_c$ and $\Sigma F_t = ma_t$ (Newton's second law in component form).
3. Substitute $a_c = v^2/r$ and $a_t = dv/dt$, where v is the speed.
4. If the object moves in a circle of radius r with constant speed v, use $v = 2\pi r/T$, where T is the time for one revolution.

CHECK Make sure that your answers are in accordance with the fact that the direction of the centripetal acceleration is always toward the center of curvature and perpendicular to the direction of the velocity vector.

Example 5-9 Swinging a Pail

You swing a pail containing mass m of water in a vertical circle of radius r (Figure 5-25). If the speed is v_{top} at the top of the circle, find (*a*) the force F_{PW} exerted by the pail on the water at the top of the circle, and (*b*) the minimum value of v_{top} for the water to remain in the pail. (*c*) What is the force exerted by the pail on the water at the bottom of the circle, where the speed of the pail is v_{bot}?

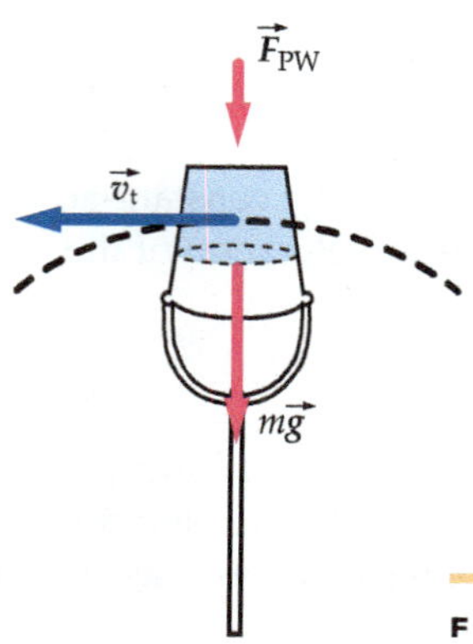

FIGURE 5-25

PICTURE When the bucket is at the top of its circle, both the force of gravity on the water and the contact force of the pail on the water are in the centripetal direction (downward). At the bottom of the circle, the contact force of the pail on the water must be greater than the force of gravity on the water to provide a net force in the centripetal direction (upward). We can apply Newton's second law to find the force exerted by the pail on the water at these points. Because the water moves in a circular path, it always has an acceleration component equal to v^2/r toward the center of the circle.

SOLVE

(*a*) 1. Draw free-body diagrams for the water at the top and bottom of the circle (Figure 5-26). Choose the $+r$ direction to be toward the center of the circle in each case.

2. Apply $\Sigma F_r = ma_r$ to the water as it passes through the top of the circle with speed v_{top}. Solve for the force F_{PW} exerted by the pail on the water:

$$\Sigma F_r = ma_r$$

$$F_{PW} + mg = m\frac{v_{top}^2}{r} \Rightarrow F_{PW} = \boxed{m\left(\frac{v_{top}^2}{r} - g\right)}$$

(*b*) The pail can push on the water, but not pull on it. The minimum force it can exert on the water is zero. Set $F_{PW} = 0$ and solve for the minimum speed:

$$0 = m\left(\frac{v_{top,min}^2}{r} - g\right) \Rightarrow v_{top,min} = \boxed{\sqrt{rg}}$$

(*c*) Apply $\Sigma F_r = ma_r$ to the water as it passes through the bottom of its path with speed v_{bot}. Solve for F_{PW}:

$$\Sigma F_r = ma_r$$

$$F_{PW} - mg = \frac{mv_{bot}^2}{r} \Rightarrow F_{PW} = \boxed{m\left(\frac{v_{bot}^2}{r} + g\right)}$$

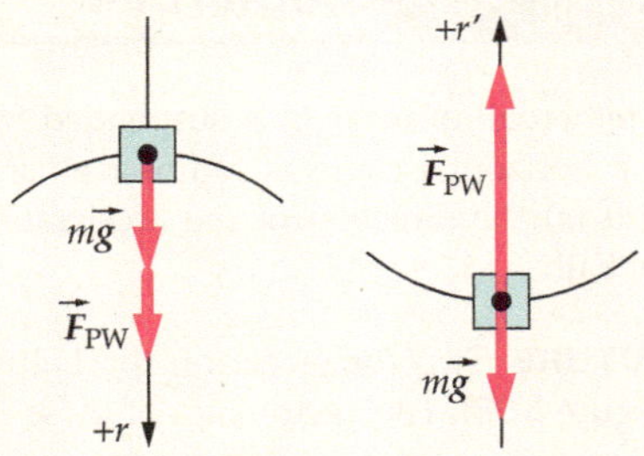

FIGURE 5-26

CHECK In the Part-(*c*) result, when $v_{bot} = 0$, $F_{PW} = mg$. This is as expected.

TAKING IT FURTHER Note that there is no arrow labeled "centripetal force" in the free-body diagram. Centripetal force is not a kind of force exerted by some agent; it is just the name for the component of the resultant force in the centripetal direction.

PRACTICE PROBLEM 5-4 Estimate (*a*) the minimum speed at the top of the circle and (*b*) the maximum period of revolution that will keep you from getting wet if you swing a pail of water in a vertical circle at constant speed.

Example 5-10 Playing Ape

You step off the limb of a tree clinging to a 30-m-long vine that is attached to another limb at the same height and 30 m distant. Assuming air resistance is negligible, how fast are you gaining speed at the instant the vine makes an angle of 25° with the vertical during your descent?

PICTURE Model this situation as a rope of negligible mass with one end attached to a limb and the other to a particle of mass m. Apply Newton's second law to the mass. The tangential acceleration is the rate of change of speed, so solve for the tangential acceleration.

SOLVE

1. Sketch a free-body diagram of the object (Figure 5-27). Let the $+r$ direction be toward the center of the path and the $+t$ direction be in the direction of the velocity:

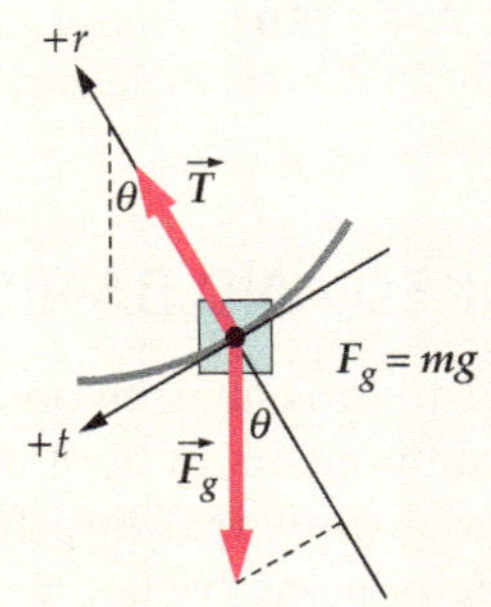

FIGURE 5-27 The $+t$ direction is in the direction of the velocity vector.

2. Apply $\Sigma F_t = ma_t$ and use the free-body diagram to find expressions for the force components:

$$\Sigma F_t = ma_t$$
$$T_t + F_{gt} = ma_t$$
$$0 + mg\sin\theta = ma_t$$

3. Solve for a_t:

$$a_t = g\sin\theta = (9.81\ \text{m/s}^2)\sin 25° = \boxed{4.1\ \text{m/s}}$$

CHECK You expect your rate of change of speed to be positive because your speed would be increasing as long as you are descending. In addition, you expect that you will not be gaining speed at a rate equal to or greater than $g = 9.81\ \text{m/s}^2$. The step-3 result meets these expectations.

PRACTICE PROBLEM 5-5 What will your rate of change of speed be at the instant the vine is vertical and you are passing through the lowest point in your arc?

Example 5-11 Tetherball — *Try It Yourself*

A tetherball of mass m is suspended from a length of rope and travels at constant speed v in a horizontal circle of radius r as shown. The rope makes an angle θ with the vertical. Find (*a*) the direction of the acceleration, (*b*) the tension in the rope, and (*c*) the speed of the ball.

PICTURE Two forces act on the ball; the gravitational force and the tension in the rope (Figure 5-28). The vector sum of these forces is in the direction of the acceleration vector.

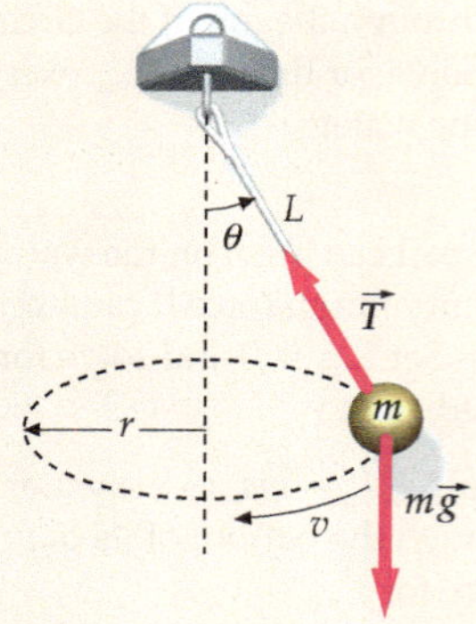

FIGURE 5-28

SOLVE

Cover the column to the right and try these on your own before looking at the answers.

Steps	Answers
(*a*) 1. The ball is moving in a horizontal circle at constant speed. The acceleration is in the centripetal direction.	The acceleration is horizontal and directed from the ball toward the center of the circle it is moving in.
(*b*) 1. Draw a free-body diagram for the ball (Figure 5-29). Choose as the $+x$ direction the direction of the ball's acceleration (toward the center of the circular path).	
2. Apply $\Sigma F_y = ma_y$ to the ball and solve for the tension T.	$\Sigma F_y = ma_y \Rightarrow T\cos\theta - mg = 0$ so $\boxed{T = \dfrac{mg}{\cos\theta}}$
(*c*) 1. Apply $\Sigma F_x = ma_x$ to the ball.	$\Sigma F_x = ma_x \Rightarrow T\sin\theta = m\dfrac{v^2}{r}$
2. Substitute $mg/\cos\theta$ for T and solve for v.	$\dfrac{mg}{\cos\theta}\sin\theta = m\dfrac{v^2}{r} \Rightarrow g\tan\theta = \dfrac{v^2}{r}$ so $v = \boxed{\sqrt{rg\tan\theta}}$

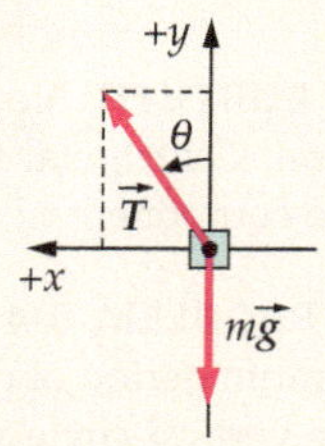

FIGURE 5-29

CHECK As $\theta \rightarrow 90°$, $\cos\theta \rightarrow 0$ and $\tan\theta \rightarrow \infty$. In the results for Parts (*b*) and (*c*), the expressions T and v both approach infinity as $\theta \rightarrow 90°$. This is what anyone who has played tetherball would expect. For θ to even approach 90°, the ball would have to move very fast.

TAKING IT FURTHER An object attached to a string and moving in a horizontal circle so that the string makes an angle θ with the vertical is called a *conical pendulum*.

UNBANKED AND BANKED CURVES

When a car rounds a curve on a horizontal road, the force components in both the centripetal and the tangential (forward) directions are provided by the force of static friction exerted by the road on the tires of the car. If the car travels at constant speed, then the forward component of the frictional force is balanced by the rearward-directed forces of air drag and rolling friction. The forward component of the static frictional force remains zero if air drag and rolling friction are both negligible and if the speed of the car remains constant.

If a curved road is not horizontal but banked, the normal force of the road will have a component in the centripetal direction. The *banking angle* is usually chosen so that no friction is needed for a car to complete the curve at the specified speed.

As a car rounds the curve, the tire is distorted by the frictional force exerted by the road. *(David Allio/Icon SMI/Corbis.)*

In 1993 a descent probe containing instruments went deep into the Jovian atmosphere—to the surface of Jupiter. The fully assembled probe was tested at accelerations up to 200*g*s in this large centrifuge at Sandia National Laboratories. *(Sandia National Laboratory.)*

Example 5-12 Rounding a Banked Curve

A curve of radius 30.0 m is banked at an angle θ. That is, the normal to the road surface makes an angle of 30.0° with the vertical. Find θ such that a car can round the curve at 40.0 km/h even if the road is coated with ice, making the road essentially frictionless.

PICTURE In this case, only two forces act on the car: the force of gravity and the normal force $\vec{F}_n$ (Figure 5-30). Because the road is banked, the normal force has a horizontal component that causes the car's centripetal acceleration. The vector sum of the two force vectors is in the direction of the acceleration. We can apply Newton's second law and then solve for θ.

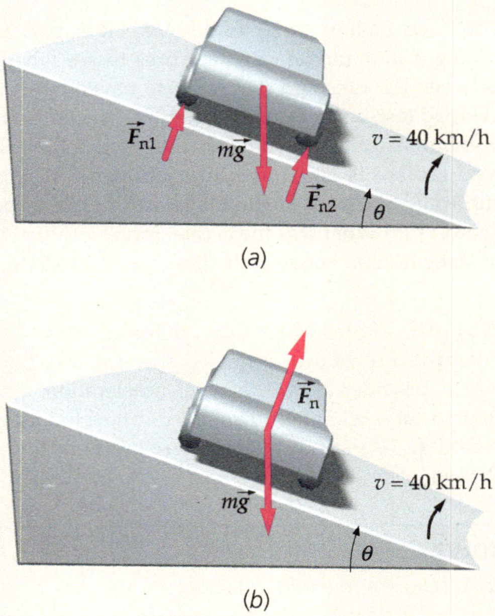

FIGURE 5-30

SOLVE

1. Draw a free-body diagram for the car (Figure 5-31)
2. Apply $\Sigma F_y = ma_y$ to the car:
   $$\Sigma F_y = ma_y$$
   $$F_n \cos\theta - mg = 0 \Rightarrow F_n \cos\theta = mg$$
3. Apply $\Sigma F_x = ma_x$ to the car.
   $$\Sigma F_x = ma_x \Rightarrow F_n \sin\theta = m\frac{v^2}{r}$$
4. Divide the step-3 result by the step-2 result, then solve for θ:
   $$\frac{\sin\theta}{\cos\theta} = \frac{mv^2}{rmg} \Rightarrow \tan\theta = \frac{v^2}{rg}$$
   $$\theta = \tan^{-1}\frac{[(40.0\text{ km/h})(1\text{ h}/3.6\text{ ks})]^2}{(30.0\text{ m})(9.81\text{ m/s}^2)}$$
   $$= \boxed{22.8°}$$

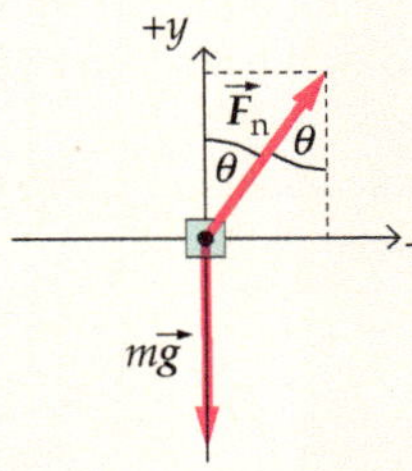

FIGURE 5-31

CHECK The banking angle is 22.8°. It is plausible because 30.0 m is a very small radius for a highway turn. For comparison, the turns at the Daytona International Speedway have a radius of 300 m and a banking angle of 31°.

TAKING IT FURTHER The banking angle θ depends on v and r, but not the mass m; θ increases with increasing v, and decreases with increasing r. When the banking angle, speed, and radius satisfy $\tan\theta = v^2/rg$, the car rounds the curve smoothly, with no tendency to slide either inward or outward. If the car speed is greater than $\sqrt{rg\tan\theta}$, the road must exert a static frictional force down the incline if the car is stay on the road. This force has a horizontal component, which provides the additional centripetal force needed to keep r from increasing. If the car speed is less than $\sqrt{rg\tan\theta}$, the road must exert a frictional force up the incline for the car is to stay on the road.

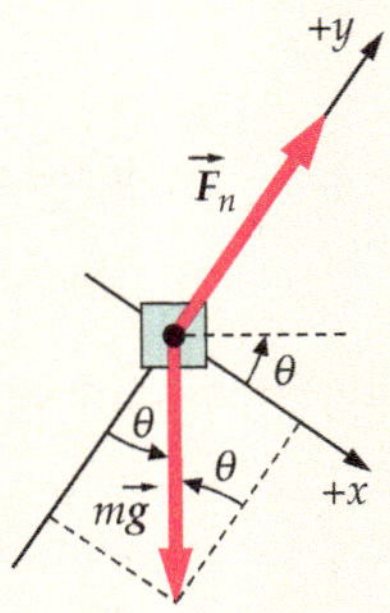

FIGURE 5-32

ALTERNATIVE SOLUTION In the preceding solution, we followed the guidelines to choose one of the coordinate axis directions to be the direction of the acceleration vector, the centripetal direction. However, the solution is no more difficult if we choose one of the axis directions to be down the incline. This choice is taken in the following solution.

1. Draw a free-body diagram for the car (Figure 5-32). The $+x$ direction is down the incline and the $+y$ direction is the normal direction.
2. Apply $\Sigma F_x = ma_x$ to the car:
   $$\Sigma F_x = ma_x \Rightarrow mg\sin\theta = ma_x$$
3. Draw a sketch and use trigonometry to obtain an expression for a_x in terms of a and θ (Figure 5-33):
   $$a_x = a\cos\theta$$
4. Substitute the step-3 result into the step-2 result. Then substitute v^2/r for a and solve for θ:
   $$mg\sin\theta = ma\cos\theta$$
   $$g\sin\theta = \frac{v^2}{r}\cos\theta$$
   $$\tan\theta = \frac{v^2}{rg} \Rightarrow \theta = \boxed{\tan^{-1}\frac{v^2}{rg}}$$

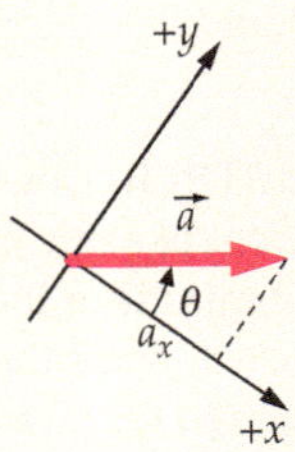

FIGURE 5-33

PRACTICE PROBLEM 5-6 Find the component of the acceleration normal to the road surface.

Example 5-13 A Road Test

Context-Rich

You have a summer job with NASCAR as part of an automobile tire testing team. You are testing a new model of racing tires to see whether or not the coefficient of static friction between the tires and dry concrete pavement is 0.90 as claimed by the manufacturer. In a skidpad test, a racecar is able to travel at constant speed in a circle of radius 45.7 m in 15.2 s without skidding. Assume air drag and rolling friction are negligible and assume that the road surface is horizontal. In a skidpad test a car travels in a circle on a flat, horizontal surface (a skidpad) at the maximum possible speed v without skidding. (*a*) What was its speed v? (*b*) What was the acceleration? (*c*) What was the minimum value for the coefficient of static friction between the tires and the road?

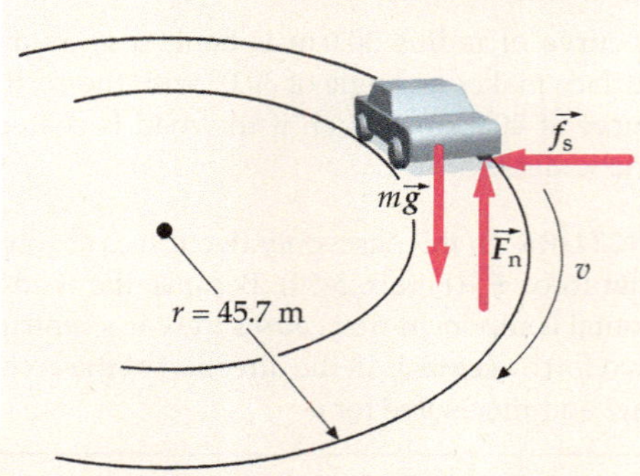

FIGURE 5-34

PICTURE Figure 5-34 shows the forces acting on the car. The normal force $\vec{F}_n$ balances the downward force due to gravity $\vec{F}_g = m\vec{g}$. The horizontal force is the force of static friction, which provides the centripetal acceleration. The faster the car travels, the greater the centripetal acceleration. The speed can be found from the circumference of the circle and the period T. This speed puts a lower limit on the maximum value of the coefficient of static friction.

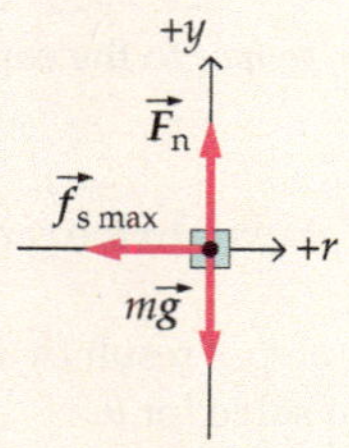

FIGURE 5-35

SOLVE

(*a*) 1. Draw a free-body diagram for the car (Figure 5-35). The $+r$ direction is away from the center of curvature.

2. The speed v is the circumference of the circle divided by the time required for one revolution:

$$v = \frac{2\pi r}{T} = \frac{2\pi(45.7\text{ m})}{15.2\text{ s}} = \boxed{18.9\text{ m/s}}$$

(*b*) Use v to calculate the centripetal and tangential accelerations:

$$a_c = \frac{v^2}{r} = \frac{(18.9\text{ m/s})^2}{(45.7\text{ m})} = \boxed{7.81\text{ m/s}^2}$$

$$a_t = \frac{dv}{dt} = \boxed{0}$$

The acceleration is 7.81 m/s² in the centripetal direction.

(*c*) 1. Apply $\Sigma F_y = ma_y$ to the car. Solve for the normal and maximum frictional force:

$$\Sigma F_y = ma_y$$

$$F_n - mg = 0 \quad \text{so} \quad F_n = mg$$

$$\text{and} \quad f_{s\max} = \mu_s F_n = \mu_s mg$$

2. Apply $\Sigma F_r = ma_r$ to the car:

$$\Sigma F_r = ma_r$$

$$-f_{s\max} = m\left(-\frac{v^2}{r}\right) \Rightarrow f_{s\max} = m\frac{v^2}{r}$$

3. Substitute from step 1 of Part (*c*) and solve for μ_s:

$$\mu_s mg = m\frac{v^2}{r} \Rightarrow \mu_s g = \frac{v^2}{r}$$

$$\mu_s = \frac{v^2}{rg} = \frac{(18.9\text{ m/s})^2}{(45.7\text{ m})(9.81\text{ m/s}^2)} = \boxed{0.796}$$

CHECK If μ_s were equal to 1.00, the inward force would be equal to mg and the centripetal acceleration would be g. Here μ_s is about 0.80 and the centripetal acceleration is about $0.80g$.

TAKING IT FURTHER Does the result of the skidpad test support the manufacturer's claim that the coefficient of static friction is 0.90? It does support the manufacturer's claim. In calculating the magnitude of the frictional force, we accounted for the frictional force needed to accelerate the car toward the center of curvature, but we neglected to account for the frictional force required to counter the effects of air drag and rolling friction. A speed of 18.9 m/s equals 42.3 mi/h, a speed at which air drag is definitely significant.

* 5-4 NUMERICAL INTEGRATION: EULER'S METHOD

If a particle moves under the influence of a *constant* force, its acceleration is constant and we can find its velocity and position from the constant-acceleration kinematic formulas in Chapter 2. But consider a particle moving through space where the force on it, and therefore its acceleration, depends on its position and its velocity. The position, velocity, and acceleration of the particle at one instant determine the position and velocity the next instant, which then determines its acceleration at that instant. The actual position, velocity, and acceleration of an object all change continuously with time. We can approximate this by replacing the continuous time variations with small time steps of duration Δt. The simplest approximation is to assume constant acceleration during each step. This approximation is called **Euler's method.** If the time interval is sufficiently short, the change in acceleration during the interval will be small and can be neglected.

Let x_0, v_{0x}, and a_{0x} be the known position, velocity, and acceleration of a particle at some initial time t_0. If we neglect any change in velocity during the time interval, the new position is given by

$$x_1 = x_0 + v_{0x}\,\Delta t$$

Similarly, if we assume constant acceleration during Δt, the velocity at time $t_1 = t_0 + \Delta t$ is given by

$$v_1 = v_{0x} + a_{0x}\,\Delta t$$

We can use the values x_1 and v_{1x} to compute the new acceleration a_{1x} using Newton's second law, and then use x_1, v_{1x}, and a_{1x} to compute x_2 and v_{2x}.

$$x_2 = x_1 + v_{1x}\,\Delta t$$
$$v_2 = v_1 + a_{1x}\,\Delta t$$

The connection between the position and velocity at time t_n and time $t_{n+1} = t_n + \Delta t$ is given by

$$x_{n+1} = x_n + v_{nx}\,\Delta t \qquad 5\text{-}9$$

and

$$v_{n+1} = v_{nx} + a_{nx}\,\Delta t \qquad 5\text{-}10$$

To find the velocity and position at some time t, we therefore divide the time interval $t - t_0$ into a large number of smaller intervals Δt and apply Equations 5-9 and 5-10, beginning at the initial time t_0. This involves a large number of simple, repetitive calculations that are most easily done on a computer. The technique of breaking the time interval into small steps and computing the acceleration, velocity, and position at each step using the values from the previous step is called *numerical integration*.

To illustrate the use of numerical integration, let us consider a problem in which a skydiver is dropped from rest at some height under the influences of both gravity and a drag force that is proportional to the square of the speed. We will find the velocity v_x and the distance traveled x as functions of time.

The equation describing the motion of an object of mass m dropped from rest is Equation 5-6 with $n = 2$:

$$mg - bv^2 = ma_x$$

where down is the positive direction. The acceleration is thus

$$a_x = g - \frac{b}{m}v^2 \qquad 5\text{-}11$$

	A	B	C	D
1	Δt =	0.5	s	
2	x0 =	0	m	
3	v0 =	0	m/s	
4	a0 =	9.81	m/s^2	
5	vt =	60	m/s	
6				
7	t	x	v	a
8	(s)	(m)	(m/s)	(m/s^2)
9	0.00	0.0	0.00	9.81
10	0.50	0.0	4.91	9.74
11	1.00	2.5	9.78	9.55
12	1.50	7.3	14.55	9.23
13	2.00	14.6	19.17	8.81
14	2.50	24.2	23.57	8.30
15	3.00	36.0	27.72	7.72
41	16.00	701.0	59.55	0.15
42	16.50	730.7	59.62	0.16
43	17.00	760.6	59.68	0.10
44	17.50	790.4	59.74	0.09
45	18.00	820.3	59.78	0.07
46	18.50	850.2	59.82	0.06
47	19.00	880.1	59.85	0.05
48	19.50	910.0	59.87	0.04
49	20.00	939.9	59.89	0.04
50				

(a)

	A	B	C	D
1	Δt =	0.5	s	
2	x0 =	0	m	
3	v0 =	0	m/s	
4	a0 =	9.81	m/s^2	
5	vt =	60	m/s	
6				
7	t	x	v	a
8	(s)	(m)	(m/s)	(m/s^2)
9	0	=B2	=B3	=$B$4*(1−C9^2/$B$5^2)
10	=A9+$B$1	=B9+C9*$B$1	=C9+D9*$B$1	=$B$4*(1−C10^2/$B$5^2)
11	=A10+$B$1	=B10+C10*$B$1	=C10+D10*$B$1	=$B$4*(1−C11^2/$B$5^2)
12	=A11+$B$1	=B11+C11*$B$1	=C11+D11*$B$1	=$B$4*(1−C12^2/$B$5^2)
13	=A12+$B$1	=B12+C12*$B$1	=C12+D12*$B$1	=$B$4*(1−C13^2/$B$5^2)
14	=A13+$B$1	=B13+C13*$B$1	=C13+D13*$B$1	=$B$4*(1−C14^2/$B$5^2)
15	=A14+$B$1	=B14+C14*$B$1	=C14+D14*$B$1	=$B$4*(1−C15^2/$B$5^2)
41	=A40+$B$1	=B40+C40*$B$1	=C40+D40*$B$1	=$B$4*(1−C41^2/$B$5^2)
42	=A41+$B$1	=B41+C41*$B$1	=C41+D41*$B$1	=$B$4*(1−C42^2/$B$5^2)
43	=A42+$B$1	=B42+C42*$B$1	=C42+D42*$B$1	=$B$4*(1−C43^2/$B$5^2)
44	=A43+$B$1	=B43+C43*$B$1	=C43+D43*$B$1	=$B$4*(1−C44^2/$B$5^2)
45	=A44+$B$1	=B44+C44*$B$1	=C44+D44*$B$1	=$B$4*(1−C45^2/$B$5^2)
46	=A45+$B$1	=B45+C45*$B$1	=C45+D45*$B$1	=$B$4*(1−C46^2/$B$5^2)
47	=A46+$B$1	=B46+C46*$B$1	=C46+D46*$B$1	=$B$4*(1−C47^2/$B$5^2)
48	=A47+$B$1	=B47+C47*$B$1	=C47+D47*$B$1	=$B$4*(1−C48^2/$B$5^2)
49	=A48+$B$1	=B48+C48*$B$1	=C48+D48*$B$1	=$B$4*(1+C49^2/$B$5^2)
50				

(b)

FIGURE 5-36 (*a*) Spreadsheet to compute the position and speed of a skydiver with air drag proportional to v^2. (*b*) The same Excel spreadsheet displaying the formulas rather than the values.

It is convenient to write the constant b/m in terms of the terminal speed v_T. Setting $a_x = 0$ in Equation 5-11, we obtain

$$0 = g - \frac{b}{m}v_T^2$$

$$\frac{b}{m} = \frac{g}{v_T^2}$$

Substituting g/v_T^2 for b/m in Equation 5-11 gives

$$a_x = g\left(1 - \frac{v^2}{v_T^2}\right) \qquad \text{5-12}$$

The acceleration at time t_n is calculated using the values x_n and v_{nx}.

To solve Equation 5-12 numerically, we need to use numerical values for g and v_t. A reasonable terminal speed for a skydiver is 60.0 m/s. If we choose $x_0 = 0$ for the initial position, the initial values are $x_0 = 0$, $v_0 = 0$, and $a_{0x} = g = 9.81\text{ m/s}^2$. To find the velocity v_x and position x after some time, say, $t = 20.0$ s, we divide the time interval $0 < t < 20.0$ s into many small intervals Δt and apply Equations 5-9, 5-10, and 5-12. We do this by using a computer spreadsheet (or by writing a computer program) as shown in Figure 5-36. This spreadsheet has $\Delta t = 0.5$ s, and the computed values at $t = 20$ s are $v = 59.89$ m/s and $x = 939.9$ m. Figure 5-37 shows graphs of v_x versus t and x versus t plotted from these data.

But how accurate are our computations? We can estimate the accuracy by running the program again using a smaller time interval. If we use $\Delta t = 0.25$ s, one-half of the value we originally used, we obtain $v = 59.86$ m/s and $x = 943.1$ m at $t = 20$ s. The difference in v is about 0.05 percent and that in x is about 0.3 percent. These are our estimates of the accuracy of the original computations.

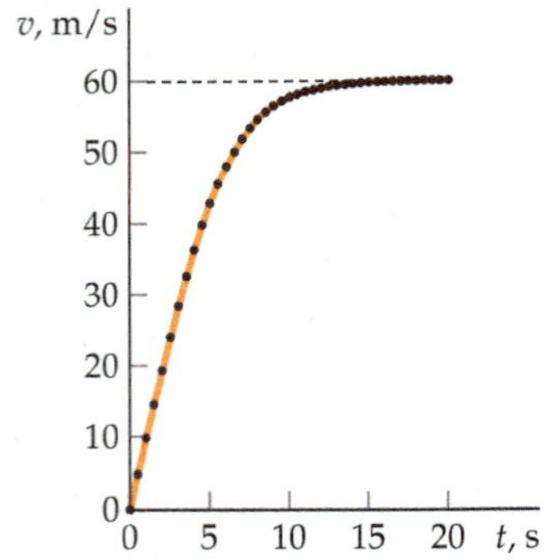

(a)

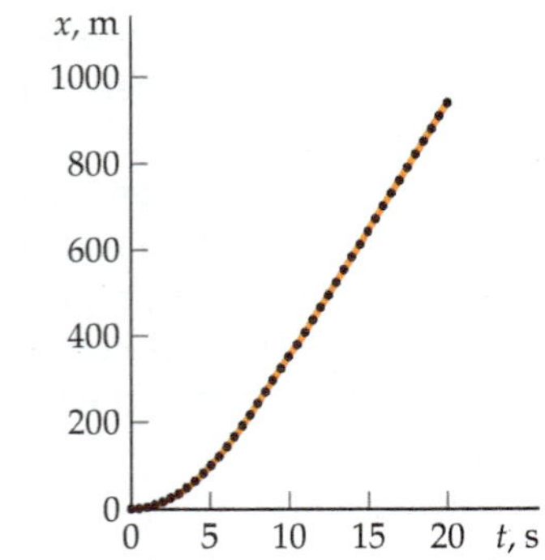

(b)

FIGURE 5-37 (*a*) Graph of v versus t for a skydiver, found by numerical integration using $\Delta t = 0.5$s. The horizontal dashed line is the terminal speed $v_t = 60$m/s. (*b*) Graph of x versus t using $\Delta t = 0.5$s.

Because the difference between the value of a_{avx} for some time interval Δt and the value of a_x at the beginning of the interval becomes smaller as the time interval becomes shorter, we might expect that it would be better to use very short time intervals, say, $\Delta t = 0.000\,000\,001$ s. But there are two reasons for not using extremely short intervals. First, the shorter the time interval, the larger the number of calculations that are required and the longer the program takes to run. Second, the computer keeps only a fixed number of digits at each step of the calculation, so that at each step there is a round-off error. These round-off errors add up. The larger the number of calculations, the more significant the total round-off errors become. When we first decrease the time interval, the accuracy improves because a_i more nearly approximates a_{av} for the interval. However, as the time interval is decreased further, the round-off errors build up and the accuracy of the computation decreases. A good rule of thumb to follow is to use no more than about 10^5 time intervals for the typical numerical integration.

5-5 THE CENTER OF MASS

If you throw a ball into the air, the ball follows a smooth parabolic path. But if you toss a baton in the air (Figure 5-38), the motion of the baton is more complicated. Each end of the baton moves in a different direction, and both ends move in a different way than the middle. However, if you look at the motion of the baton more closely, you will see that there is one point on the baton that moves in a parabolic path, even though the rest of the baton does not. This point, called the **center of mass,** moves as if all the baton's mass were concentrated at that point and all external forces were applied there.

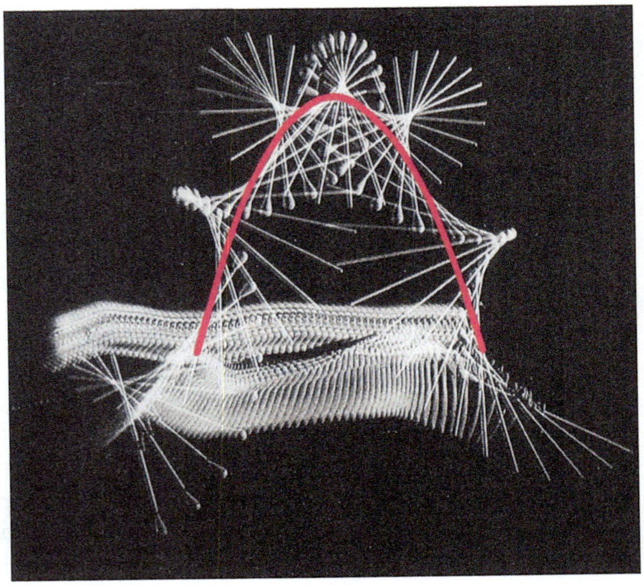

FIGURE 5-38 A multiflash photo of a baton thrown into the air. *(Estate of Harold E. Edgerton/Palm Press.)*

To determine the center of mass of an object, it is helpful to think of the object as a system of particles. Consider, for example, a simple system that consists of two point particles located on the x axis at positions x_1 and x_2 (Figure 5-39). If the particles have masses m_1 and m_2, then the center-of-mass is located on the x axis at position x_{cm}, defined by

$$Mx_{cm} = m_1x_1 + m_2x_2 \qquad \text{5-13}$$

where $M = m_1 + m_2$ is the total mass of the system. If we choose the position of the origin and the $+x$ direction such that the position of m_1 is at the origin and that of m_2 is on the positive x axis, then $x_1 = 0$ and $x_2 = d$, where d is the distance between the particles. The center of mass is then given by

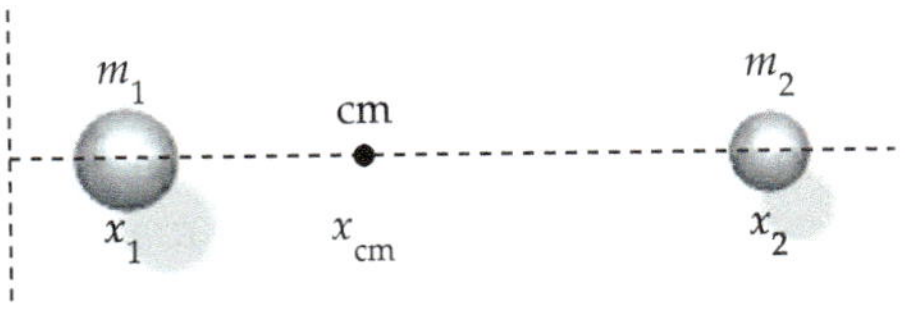

FIGURE 5-39

$$Mx_{cm} = m_1x_1 + m_2x_2 = m_1(0) + m_2d$$

$$x_{cm} = \frac{m_2}{M}d = \frac{m_2}{m_1 + m_2}d \qquad \text{5-14}$$

In the case of just two particles, the center of mass lies at some point on the line between the particles; if the particles have equal masses, then the center of mass is midway between them (Figure 5-40*a*). If the two particles are of unequal mass, then the center of mass is closer to the more massive particle (Figure 5-40*b*).

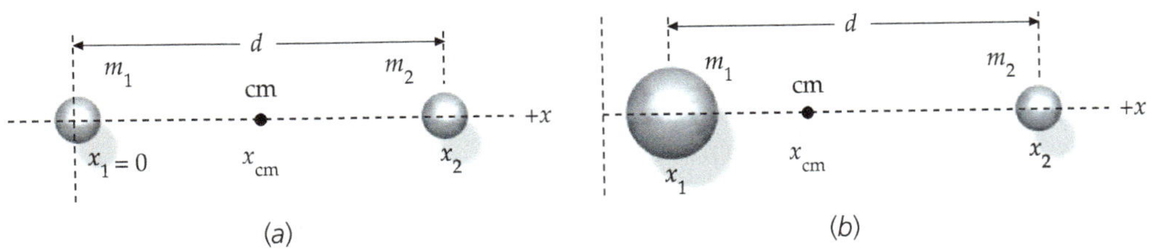

FIGURE 5-40

PRACTICE PROBLEM 5-7

A 4.0-kg mass is at the origin and a 2.0-kg mass is on the x axis at $x = 6.0$ cm. Find x_{cm}.

We can generalize from two particles in one dimension to a system of many particles in three dimensions. For N particles in three dimensions,

$$Mx_{cm} = m_1x_1 + m_2x_2 + m_3x_3 + \cdots + m_Nx_N$$

Using more concise notation, this is written

$$Mx_{cm} = \sum_i m_ix_i \qquad \text{5-15}$$

where again $M = \sum_i m_i$ is the total mass of the system. Similarly, in the y and z directions,

$$My_{cm} = \sum_i m_iy_i \qquad \text{5-16}$$

and

$$Mz_{cm} = \sum_i m_iz_i \qquad \text{5-17}$$

In vector notation, $\vec{r}_i = x_i\hat{i} + y_i\hat{j} + z_i\hat{k}$ is the position vector of the ith particle. The **position of the center of mass, $\vec{r}_{cm}$,** is defined by

$$M\vec{r}_{cm} = m_1\vec{r}_1 + m_2\vec{r}_2 + \cdots = \sum_i m_i\vec{r}_i \qquad \text{5-18}$$

DEFINITION: CENTER OF MASS

where $\vec{r}_{cm} = x_{cm}\hat{i} + y_{cm}\hat{j} + z_{cm}\hat{k}$.

Now let us consider extended objects, such as balls, baseball bats, and even cars. We can think of objects such as these as a system containing a very large number of particles, with a continuous distribution of mass. For highly symmetric objects, the center of mass is at the *center of symmetry*. For example, the center of mass of a uniform sphere or a uniform cylinder is located at its *geometric center*. For an object with a line or plane of symmetry, the center of mass lies somewhere along that line or plane. To find the position center of mass of an object, we replace the sum in Equation 5-18 with an integral:

$$M\vec{r}_{cm} = \int \vec{r}\,dm \qquad \text{5-19}$$

CENTER OF MASS, CONTINUOUS OBJECT

where dm is a small element of mass located at position $\vec{r}$, as shown in Figure 5-41. (We will examine in detail how this integral is set up after Example 5-15.)

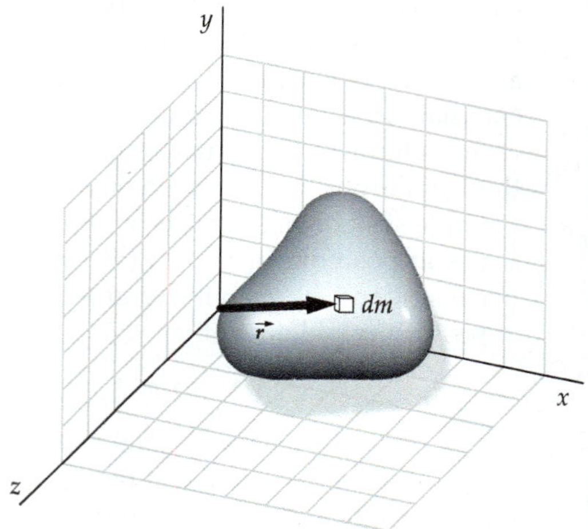

FIGURE 5-41 A mass element dm located at position $\vec{r}$ is used for finding the center of mass by integration.

PROBLEM-SOLVING STRATEGY

Solving Center-of-Mass Problems

PICTURE Determining centers of mass often simplifies determinations of the motions of an object or system of objects. Drawing a sketch of the object or system of objects is useful when trying to determine a center of mass.

SOLVE

1. Check the mass distribution for symmetry axes. If there are symmetry axes, the center of mass will be located on them. Use existing symmetry axes as coordinate axes where feasible.

2. Check to see if the mass distribution is composed of highly symmetric subsystems. If so, then calculate the centers of mass of the individual subsystems, and then calculate the center of mass of the system by treating each subsystem as a point particle at its center of mass.
3. If the system contains one or more point particles, place the origin at the location of a point particle. (If the ith particle is at the origin, then $\vec{r}_i = 0$.)

CHECK Make sure your center-of-mass determinations make sense. In many cases, the center of mass of an object is located near the more massive and larger part of the object. The center of mass of a multi-object system or an object such as a hoop may not be located within or on any object.

Example 5-14 The Center of Mass of a Water Molecule

A water molecule consists of an oxygen atom and two hydrogen atoms. An oxygen atom has a mass of 16.0 unified mass units (u), and each hydrogen atom has a mass of 1.00 u. The hydrogen atoms are each at an average distance of 96.0 pm (96.0×10^{-12} m) from the oxygen atom, and are separated from one another by an angle of 104.5°. Find the center of mass of a water molecule.

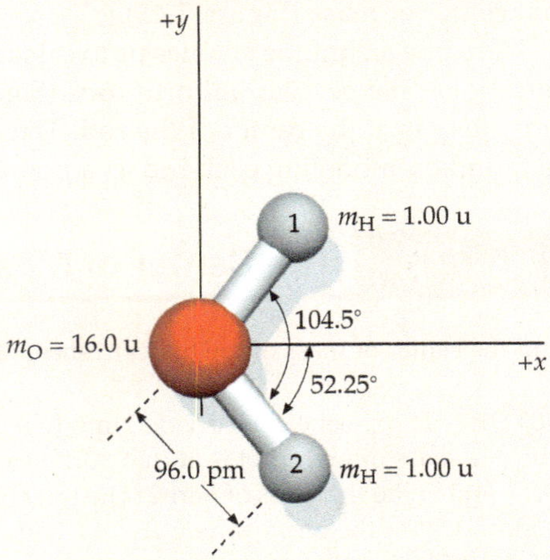

FIGURE 5-42

PICTURE We can simplify the calculation by selecting a coordinate system such that the origin is located at the oxygen atom, with the x axis bisecting the angle between the hydrogen atoms (Figure 5-42). Then, given the symmetries of the molecule, the center of mass will be on the x axis, and the line from the oxygen atom to each hydrogen atom will make an angle of 52.25°.

SOLVE

1. The location of the center of mass is given by its coordinates, x_{cm} and y_{cm} (Equations 5-15 and 5-16):

$$x_{cm} = \frac{\sum m_i x_i}{M}, \quad y_{cm} = \frac{\sum m_i y_i}{M}$$

2. Writing these out explicitly gives:

$$x_{cm} = \frac{m_{H1}x_{H1} + m_{H2}x_{H2} + m_O x_O}{m_{H1} + m_{H2} + m_O}$$

$$y_{cm} = \frac{m_{H1}y_{H1} + m_{H2}y_{H2} + m_O y_O}{m_{H1} + m_{H2} + m_O}$$

3. We have chosen the origin to be the location of the oxygen atom, so both the x and y coordinates of the oxygen atom are zero. The x and y coordinates of the hydrogen atoms are calculated from the 52.25° angle each hydrogen makes with the x axis:

$$x_O = y_O = 0$$
$$x_{H1} = 96.0 \text{ pm} \cos 52.25° = 58.8 \text{ pm}$$
$$x_{H2} = 96.0 \text{ pm} \cos(-52.25°) = 58.8 \text{ pm}$$
$$y_{H1} = 96.0 \text{ pm} \sin 52.25° = 75.9 \text{ pm}$$
$$y_{H2} = 96.0 \text{ pm} \sin(-52.25°) = -75.9 \text{ pm}$$

4. Substituting the coordinate and mass values into step 2 gives x_{cm}:

$$x_{cm} = \frac{(1.00 \text{ u})(58.8 \text{ pm}) + (1.00 \text{ u})(58.8 \text{ pm}) + (16.0 \text{ u})(0)}{1.00 \text{ u} + 1.00 \text{ u} + 16.0 \text{ u}} = 6.53 \text{ pm}$$

$$x_{cm} = \frac{(1.00 \text{ u})(75.9 \text{ pm}) + (1.00 \text{ u})(-75.9 \text{ pm}) + (16.0 \text{ u})(0)}{1.00 \text{ u} + 1.00 \text{ u} + 16.0 \text{ u}} = 0.00 \text{ pm}$$

5. The center of mass is on the x axis:

$$\vec{r}_{cm} = x_{cm}\hat{i} + y_{cm}\hat{j} = \boxed{6.53 \text{ pm } \hat{i} + 0.00\hat{j}}$$

CHECK That $y_{cm} = 0$ can be seen from the symmetry of the mass distribution. Also, the center of mass is very close to the relatively massive oxygen atom, as expected.

TAKING IT FURTHER The distance 96 pm is read "ninety-six picometers," where pico is the prefix for 10^{-12}.

Note that we could also have solved Example 5-14 by first finding the center of mass of just the two hydrogen atoms. For a system of three particles Equation 5-18 is

$$M\vec{r}_{\text{cm}} = m_1\vec{r}_1 + m_2\vec{r}_2 + m_3\vec{r}_3$$

The first two terms on the right side of this equation are related to the center of mass of the first two particles $\vec{r}\,'_{\text{cm}}$:

$$m_1\vec{r}_1 + m_2\vec{r}_2 = (m_1 + m_2)\vec{r}\,'_{\text{cm}}$$

The center of mass of the three-particle system can then be written

$$M\vec{r}_{\text{cm}} = (m_1 + m_2)\vec{r}\,'_{\text{cm}} + m_3\vec{r}_3$$

So we can first find the center of mass for two of the particles, the hydrogen atoms, for example, and then replace them with a single particle of total mass $m_1 + m_2$ at that center of mass (Figure 5-43).

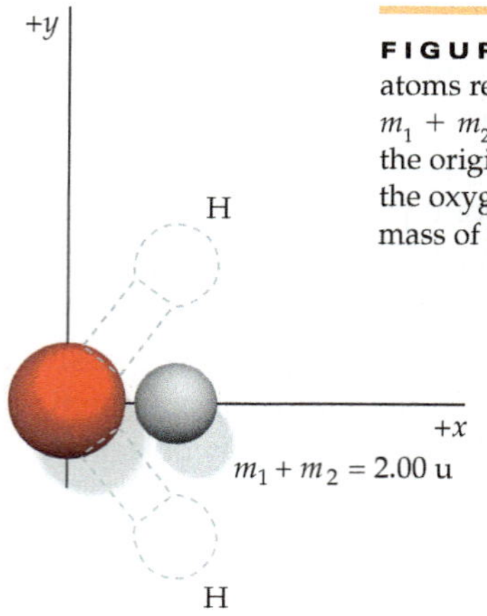

FIGURE 5-43 Example 5-14 with the two hydrogen atoms replaced by a single particle of mass $m_1 + m_2 = 2.00\,\text{u}$ on the x axis at the center of mass of the original atoms. The center of mass then falls between the oxygen atom at the origin and the calculated center of mass of the two hydrogen atoms.

The same technique enables us to calculate centers of mass for more complex systems, for instance, two uniform rods (Figure 5-44). The center of mass of each rod separately is at the center of the rod. The center of mass of the two-rod system can be found by modeling each rod as a point particle at its individual center of mass.

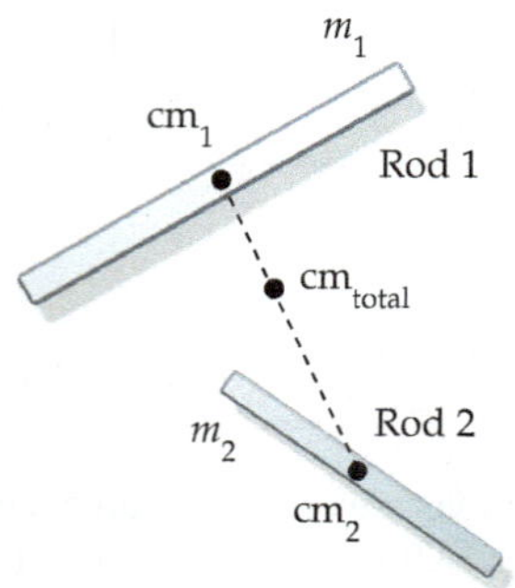

FIGURE 5-44

Example 5-15 The Center of Mass of a Plywood Sheet

Find the center of mass of the uniform sheet of plywood shown in Figure 5-45*a*.

PICTURE The sheet can be divided into two symmetrical parts (Figure 5-45*b*). The center of mass of each part is at that part's geometric center. Let m_1 be the mass of part 1 and m_2 be the mass of part 2. The total mass is $M = m_1 + m_2$. The masses are proportional to the areas, where A_1, A_2, and A are the areas of the pieces of mass m_1, m_2, and M, respectively.

SOLVE

Steps

1. Write the x and y coordinates of the center of mass in terms of m_1 and m_2.

$$x_{\text{cm}} = \frac{1}{M}(m_1 x_{\text{cm}1} + m_2 x_{\text{cm}2})$$

$$y_{\text{cm}} = \frac{1}{M}(m_1 y_{\text{cm}1} + m_2 y_{\text{cm}2})$$

2. Substitute area ratios for the mass ratios.

$$x_{\text{cm}} = \frac{A_1}{A}x_{\text{cm}1} + \frac{A_2}{A}x_{\text{cm}2}$$

$$y_{\text{cm}} = \frac{A_1}{A}y_{\text{cm}1} + \frac{A_2}{A}y_{\text{cm}2}$$

3. Calculate the areas and the ratios of the areas, using the values from Figure 5-45*b*.

$$A_1 = 0.32\text{ m}^2; \qquad A_2 = 0.040\text{ m}^2$$

$$\frac{A_1}{A} = \frac{8.0}{9.0} \qquad \frac{A_2}{A} = \frac{1.0}{9.0}$$

4. Write the x and y coordinates of the center-of-mass coordinates for each part by inspection of the figure.

$$x_{\text{cm}1} = 0.40\text{ m}, \quad y_{\text{cm}1} = 0.20\text{ m}$$

$$x_{\text{cm}2} = 0.70\text{ m}, \quad y_{\text{cm}2} = 0.50\text{ m}$$

5. Substitute these results into the step-2 result to calculate x_{cm} and y_{cm}.

$$x_{\text{cm}} = \boxed{0.43\text{ m}}, \quad y_{\text{cm}} = \boxed{0.23\text{ m}}$$

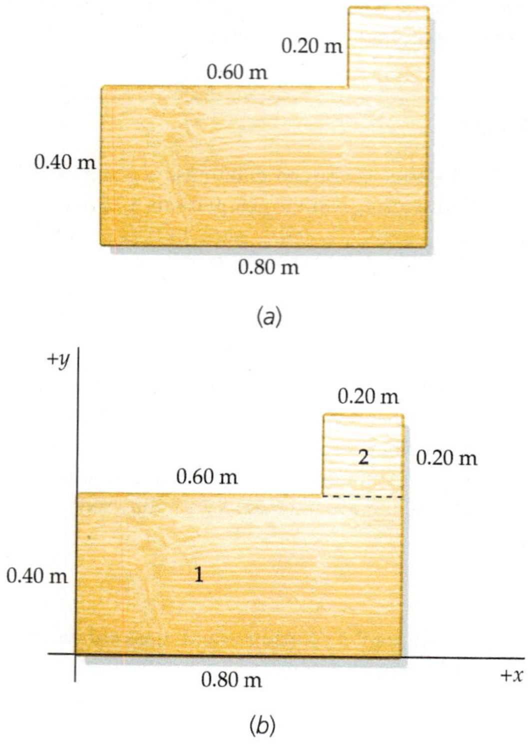

FIGURE 5-45

CHECK As expected, the center of mass of the system is very near the center of mass of part 1 (because $m_1 = 8m_2$.)

TAKING IT FURTHER Drawing an axis through the geometric centers of parts 1 and 2 and placing the origin at the geometric center of part 1, would have made locating the center of mass considerably easier.

* FINDING THE CENTER OF MASS BY INTEGRATION

In this section, we find the center of mass by integration (Equation 5-19):

$$\vec{r}_{\text{cm}} = \frac{1}{M}\int \vec{r}\, dm$$

We will start by finding the center of mass of a uniform thin rod to illustrate the technique for setting up the integration.

Uniform rod We first choose a coordinate system. A good choice for a coordinate system is one with an x axis through the length of the rod, with the origin at one end of the rod (Figure 5-46). Shown on the figure is a mass element dm of length dx a distance x from the origin. Equation 5-19 thus gives

$$\vec{r}_{\text{cm}} = \frac{1}{M}\int \vec{r}\, dm = \frac{1}{M}\int x\hat{i}\, dm$$

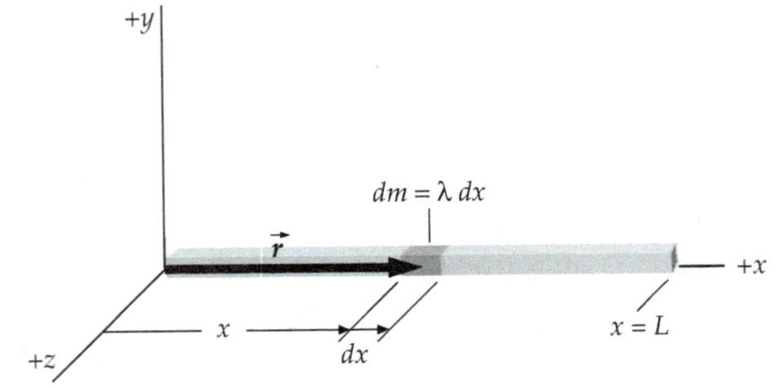

FIGURE 5-46

The mass is distributed on the x axis along the interval $0 \le x \le L$. To integrate dm along the mass distribution means the limits of the integral are 0 and L. (We integrate in the direction of increasing x.) The ratio dm/dx is the mass per unit length λ, so $dm = \lambda\, dx$:

$$\vec{r}_{\text{cm}} = \frac{1}{M}\hat{i}\int x\,dm = \frac{1}{M}\hat{i}\int_0^L x\lambda\,dx \qquad 5\text{-}20$$

where

$$M = \int dm = \int_0^L \lambda\, dx \qquad 5\text{-}21$$

Because the rod is uniform, λ is constant and can be factored from each of the integrals in Equations 5-20 and 5-21, giving

$$\vec{r}_{\text{cm}} = \frac{1}{M}\lambda\hat{i}\int_0^L x\,dx = \frac{1}{M}\lambda\hat{i}\frac{L^2}{2} \qquad 5\text{-}22$$

and

$$M = \lambda\int_0^L dx = \lambda L \qquad 5\text{-}23$$

Solving Equation 5-23 for λ gives $\lambda = M/L$. Thus, for a uniform rod the mass per unit length is equal to the total mass divided by the total length. Substituting λL for M in Equation 5-22, we complete the calculation and obtain the expected result

$$\vec{r}_{\text{cm}} = \frac{1}{\lambda L}\lambda\hat{i}\frac{L^2}{2} = \frac{1}{2}L\hat{i}$$

Semicircular hoop In calculating the center of mass of a uniform semicircular hoop of radius R, a good choice of coordinate axes is one with the origin at the center and with the y axis bisecting the semicircle (Figure 5-47). To find the center of mass, we use $M\vec{r}_{\text{cm}} = \int \vec{r}\,dm$ (Equation 5-19,) where $\vec{r} = x\hat{i} + y\hat{j}$. The semicircular mass distribution suggests using polar coordinates,* for which $x = r\cos\theta$ and $y = r\sin\theta$. The distance of the points on the semicircle from the origin is $r = R$. With these substitutions, we have

$$\vec{r}_{\text{cm}} = \frac{1}{M}\int (x\hat{i} + y\hat{j})\,dm = \frac{1}{M}\int R(\cos\theta\hat{i} + \sin\theta\hat{j})\,dm$$

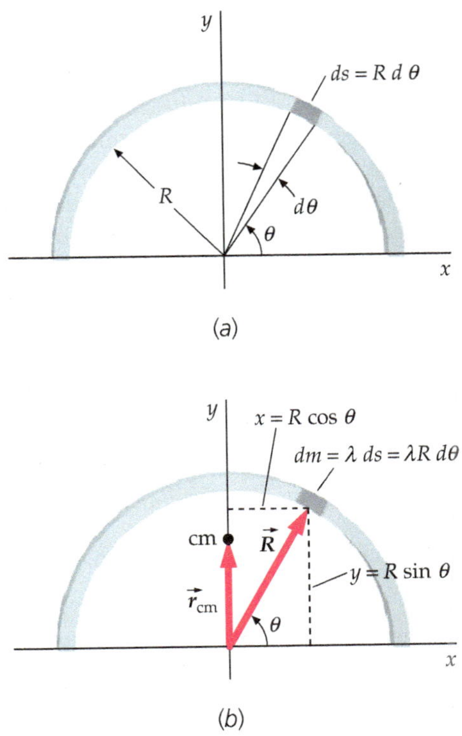

FIGURE 5-47 Geometry for calculating the center of mass of a semicircular hoop by integration.

* In polar coordinates the coordinates of a point are r and θ, were r is the magnitude of the position vector $\vec{r}$ and θ is the angle that the position vector makes with the $+x$ direction.

Next we express dm in terms of $d\theta$. First, the mass element dm has length $ds = R\,d\theta$, so

$$dm = \lambda\,ds = \lambda R\,d\theta$$

where $\lambda = dm/ds$ is the mass per unit length. Thus, we have

$$\vec{r}_{\text{cm}} = \frac{1}{M}\int R(\cos\theta\,\hat{i} + \sin\theta\,\hat{j})\lambda R\,d\theta$$

Evaluating this integral involves integrating dm along the semicircular mass distribution. This means that $0 \le \theta \le \pi$. Integrating in the direction of increasing θ, the integration limits go from 0 to π. That is,

$$\vec{r}_{\text{cm}} = \frac{1}{M}\int_0^{\pi} R(\cos\theta\hat{i} + \sin\theta\hat{j})\lambda R\,d\theta = \frac{\lambda R^2}{M}\left(\hat{i}\int_0^{\pi}\cos\theta\,d\theta + \hat{j}\int_0^{\pi}\sin\theta\,d\theta\right)$$

where we have used that an integral of a sum is the sum of the integrals. Because the hoop is uniform, we know that $M = \lambda\pi R$, where πR is the length of the semicircular arc. Substituting $M/(\pi R)$ for λ and integrating gives

$$\vec{r}_{\text{cm}} = \frac{R}{\pi}\left(\hat{i}\int_0^{\pi}\cos\theta\,d\theta + \hat{j}\int_0^{\pi}\sin\theta\,d\theta\right) = \frac{R}{\pi}\left(\hat{i}\sin\theta\Big|_0^{\pi} - \hat{j}\cos\theta\Big|_0^{\pi}\right) = \frac{2}{\pi}R\hat{j}$$

The center of mass is on the y axis a distance of $2R/\pi$ from the origin. Curiously, it is outside of the material of the semicircular hoop.

See **Math Tutorial** *for more information on* **Integrals**

MOTION OF THE CENTER OF MASS

The motion of any object or system of particles can be described in terms of the motion of the center of mass plus the motion of individual particles in the system relative to the center of mass. The multiple image photograph in Figure 5-48 shows a hammer thrown into the air. While the hammer is in the air, the center of mass follows a parabolic path, the same path that would be followed by a point particle. The other parts of the hammer rotate about this point as the hammer moves through the air.

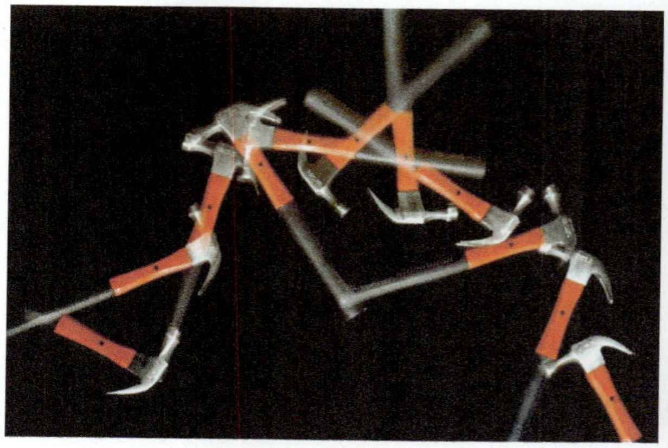

FIGURE 5-48 The center of mass (the black dot) of the hammer moves in a smooth parabolic path. *(Loren Winters/Visual Unlimited.)*

The motion of the center of mass for a system of particles is related to the net force on the system as a whole. We can show this by examining the motion of a system of n particles of total mass M. First, we find the velocity of the center-of-mass system by differentiating both sides of Equation 5-18 ($M\vec{r}_{\text{cm}} = \Sigma m_i\vec{r}_i$) with respect to time:

$$M\frac{d\vec{r}_{\text{cm}}}{dt} = m_1\frac{d\vec{r}_1}{dt} + m_2\frac{d\vec{r}_2}{dt} + \cdots = \sum_i m_i\frac{d\vec{r}_i}{dt}$$

Because the time derivative of position is velocity, this gives

$$M\vec{v}_{\text{cm}} = m_1\vec{v}_1 + m_2\vec{v}_2 + \cdots = \sum_i m_i\vec{v}_i \qquad \text{5-24}$$

Differentiating both sides again, we obtain the accelerations:

$$M\vec{a}_{\text{cm}} = m_1\vec{a}_1 + m_2\vec{a}_2 + \cdots = \sum_i m_i\vec{a}_i \qquad \text{5-25}$$

where $\vec{a}_i$ is the acceleration of the ith particle and $\vec{a}_{\text{cm}}$ is the acceleration of the center of mass. From Newton's second law, $m_i\vec{a}_i$ equals the sum of the forces acting on the ith particle, so

$$\sum_i m_i\vec{a}_i = \sum_i \vec{F}_i$$

where the sum on the right is the sum of all the forces acting on each and every particle in the system. Some of these forces are *internal* forces (exerted on a particle

Throwing the hammer. If the ball moves in a horizontal circle at constant speed, its acceleration vector points in the centripetal direction (toward the center of the circle). The net force on the ball is in the direction of the acceleration vector. The centripetal component of the net force is supplied by the person pulling the handle inward. *(Pete Saloutos/Corbis.)*

in the system by some other particle in the system) and others are *external* forces (exerted on a particle in the system by something external to the system). Thus,

$$M\vec{a}_{cm} = \sum_i \vec{F}_{i\,int} + \sum_i \vec{F}_{i\,ext} \qquad 5\text{-}26$$

According to Newton's third law, forces come in equal and opposite pairs. Therefore, for each internal force acting on a particle in the system there is an equal and opposite internal force acting on some other particle in the system. When we sum all the internal forces, each third-law force pair sums to zero, so $\sum \vec{F}_{i\,int} = 0$. Equation 5-22 then becomes

$$\vec{F}_{net\,ext} = \sum_i \vec{F}_{i\,ext} = M\vec{a}_{cm} \qquad 5\text{-}27$$

NEWTON'S SECOND LAW FOR A SYSTEM

That is, the net external force acting on the system equals the total mass M times the acceleration of the center of mass $\vec{a}_{cm}$. Thus,

The center of mass of a system moves like a particle of mass $M = \sum m_i$ under the influence of the net external force acting on the system.

This theorem is important because it describes the motion of the center of mass for *any* system of particles: The center of mass moves exactly like a single point particle of mass M acted on by only the external forces. The individual motion of a particle in the system is typically much more complex and is not described by Equation 5-27. The hammer thrown into the air in Figure 5-48 is an example. The only external force acting is gravity, so the center of mass of the hammer moves in a simple parabolic path, as would a point particle. However, Equation 5-27 does not describe the rotational motion of the head of the hammer about the center of mass.

If a system has a zero net external force acting on it, then $\vec{a}_{cm} = 0$. In this case the center of mass either remains at rest or moves with constant velocity. The internal forces and motion may be complex, but the motion of the center of mass is simple. Further, if the component of the net next force in a given direction, say the x direction, remains zero, then a_{cmx} remains zero and v_{cmx} remains constant. An example of this is a projectile in the absence of air drag. The net external force on the projectile is the gravitational force. This force acts straight downward, so its component in any horizontal direction remains zero. It follows that the horizontal component of the velocity of the center of mass remains constant.

CONCEPT CHECK 5-2

A cylinder rests on a sheet of paper on a table (Figure 5-49). You pull on the paper causing the paper to slide to the right. This results in the cylinder rolling leftward *relative to the paper*. How does the center of mass of the cylinder move *relative to the table*?

FIGURE 5-49

Example 5-16 An Exploding Projectile

A projectile is fired into the air over level ground on a trajectory that would result in it landing 55 m away. However, at its highest point it explodes into two fragments of equal mass. Immediately following the explosion one fragment has a momentary speed of zero and then falls straight down to the ground. Where does the other fragment land? Neglect air resistance.

PICTURE Let the projectile be the system. Then, the forces of the explosion are all internal forces. Because the only *external* force acting on the system is that due to gravity, the center of mass, which is midway between the two fragments, continues on its parabolic path as if there had been no explosion (Figure 5-50).

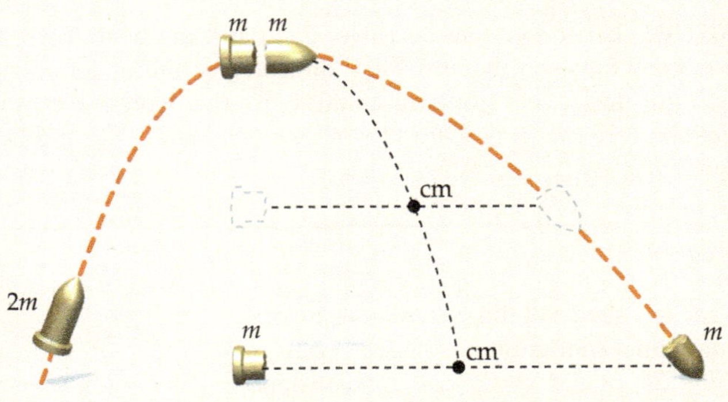

FIGURE 5-50

SOLVE

1. Let $x = 0$ be the initial position of the projectile. The landing positions x_1 and x_2 of the fragments are related to the final position of the center of mass by:

$$(2m)x_{cm} = mx_1 + mx_2$$
$$\text{or} \quad 2x_{cm} = x_1 + x_2$$

2. At impact, $x_{cm} = R$ and $x_1 = 0.5R$, where $R = 55$ m is the horizontal range for the unexploded projectile. Solve for x_2:

$$x_2 = 2x_{cm} - x_1 = 2R - 0.5R = 1.5R$$
$$= 1.5(55\text{ m}) = \boxed{83\text{ m}}$$

CHECK Fragment #1 was pushed backwards by the explosive forces, so fragment #2 was pushed forward by an equal but opposite force. As expected, fragment #2 impacts the ground at a distance farther from the launch point than the projectile would have impacted had it not exploded into two pieces.

TAKING IT FURTHER In Figure 5-51, height versus distance is plotted for exploding projectiles when fragment #1 has a horizontal velocity of half of the initial horizontal velocity. The center of mass follows a normal parabolic trajectory, as it did in the original example in which fragment #1 falls straight down. If both fragments have the same vertical component of velocity after the explosion, they land at the same time. If just after the explosion the vertical component of the velocity of one fragment is less than that of the other, the fragment with the smaller vertical-velocity component will hit the ground first. As soon as it does, the ground exerts a force on it and the net external force on the system is no longer just the gravitational force. From that instant on, our analysis is invalid.

PRACTICE PROBLEM 5-8 If the fragment that falls straight down has twice the mass of the other fragment, how far from the launch position does the lighter fragment land?

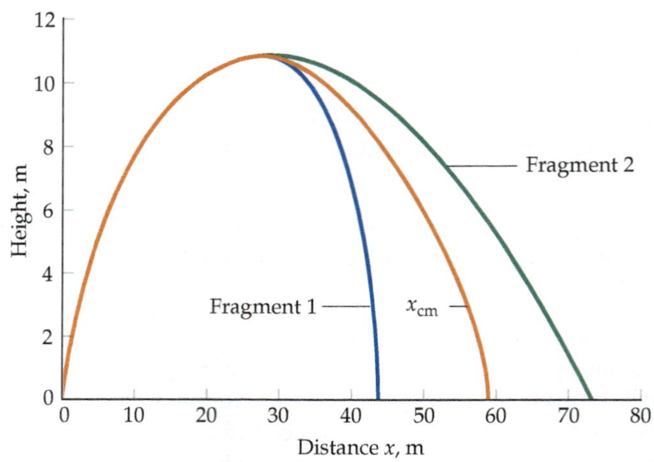

FIGURE 5-51

Example 5-17 Changing Places in a Rowboat

Pete (mass 80 kg) and Dave (mass 120 kg) are in a rowboat (mass 60 kg) on a calm lake. Dave is near the bow of the boat, rowing, and Pete is at the stern, 2.0 m from the center. Dave gets tired and stops rowing. Pete offers to row, so after the boat comes to rest they change places. How far does the boat move as Pete and Dave change places? (Neglect any horizontal force exerted by the water.)

PICTURE Let the system be Dave, Pete, and the boat. There are no *external* forces in the horizontal direction, so the center of mass does not move horizontally relative to the water. Flesh out Equation 5-15 ($Mx_{cm} = \Sigma m_i x_i$) both before and after Pete and Dave change places.

SOLVE

1. Make a sketch of the system in its initial and final configurations (Figure 5-52). Let $L = 2.0$ m and let $d = \Delta x_{boat}$, the distance the boat moves forward when Pete and Dave switch places:

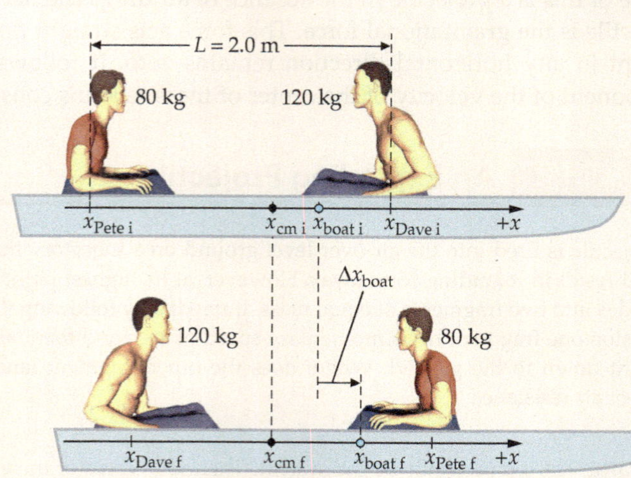

FIGURE 5-52 Pete and Dave changing places viewed from the reference frame of the water. The blue dot is the center of mass of the boat and the black dot is the center of mass of the Pete–Dave–boat system.

2. Flesh out $Mx_{cm} = \Sigma m_i x_i$ both before and after Pete and Dave change places. The coordinate axis measures positions in the reference frame of the water:

$$Mx_{cm\,i} = m_{Pete}x_{Pete\,i} + m_{Dave}x_{Dave\,i} + m_{boat}x_{boat\,i}$$

and

$$Mx_{cm\,f} = m_{Pete}x_{Pete\,f} + m_{Dave}x_{Dave\,f} + m_{boat}x_{boat\,f}$$

3. Subtract the third step-2 equation from the second step-2 equation. Then substitute 0 for Δx_{cm}, $d + L$ for Δx_{Pete}, $d - L$ for Δx_{Dave} and d for Δx_{boat}:

$$M\Delta x_{cm} = m_{Pete}\,\Delta x_{Pete} + m_{Dave}\,\Delta x_{Dave} + m_{boat}\,\Delta x_{boat}$$
$$0 = m_{Pete}(d + L) + m_{Dave}(d + L) + m_{boat}d$$

4. Solve for d:

$$d = \frac{(m_{Dave} - m_{Pete})}{m_{Dave} + m_{Pete} + m_{boat}}L = \frac{(120\text{ kg} - 80\text{ kg})}{120\text{ kg} + 80\text{ kg} + 60\text{ kg}}(2.0\text{ m}) = \boxed{0.31\text{ m}}$$

CHECK Dave's mass is greater than Pete's mass, so when they changed places their center of mass moved toward the stern of the boat. The boat's center of mass had to move in the opposite direction for the center of mass of the Dave–Pete–boat system to remain stationary. The step-4 result is the displacement of the boat. It is positive as expected.

Example 5-18 A Sliding Block

A wedge of mass m_2 sits at rest on a scale, as shown in Figure 5-53. A small block of mass m_1 slides down the frictionless incline of the wedge. Find the scale reading while the block slides. The wedge does not slide on the scale.

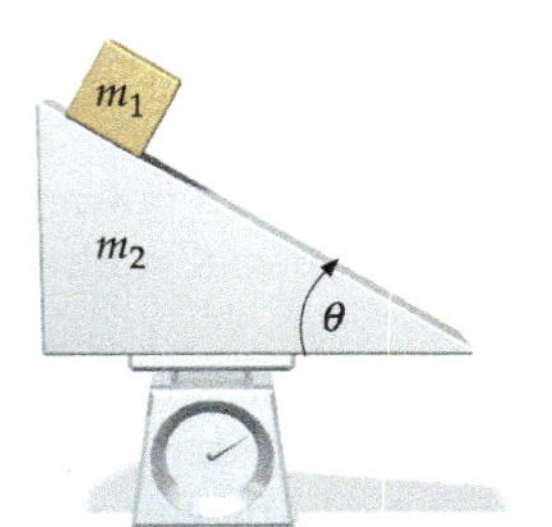

FIGURE 5-53

PICTURE We choose the wedge plus block to be the system. Because the block accelerates down the wedge, the center of mass of the system has acceleration components to the right and downward. The external forces on the system are the gravitational forces on the block and wedge, the static frictional force f_s of the scale on the wedge, and the normal force F_n exerted by the scale on the wedge. The scale reading is equal to the magnitude of F_n.

SOLVE

1. Draw a free-body diagram for the wedge–block system (Figure 5-54):

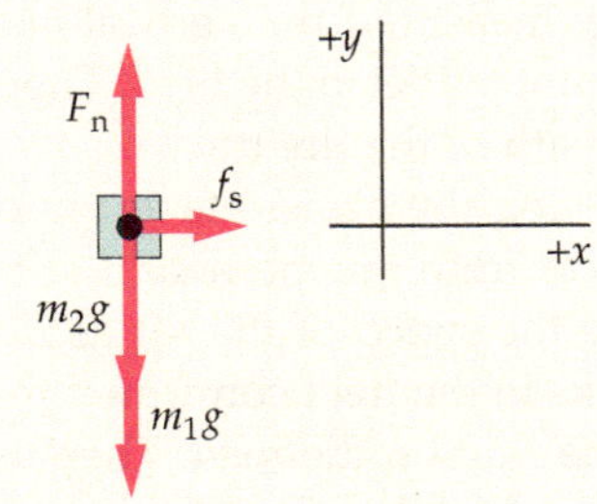

FIGURE 5-54

2. Write the vertical component of Newton's second law for the system and solve for F_n:

$$F_n - m_1g - m_2g = Ma_{cm\,y} = (m_1 + m_2)a_{cm\,y}$$
$$F_n = (m_1 + m_2)g + (m_1 + m_2)a_{cm\,y}$$

3. Using Equation 5-21, express $a_{cm\,y}$ in terms of the acceleration of the block a_{1y}:

$$Ma_{cm\,y} = m_1a_{1y} + m_2a_{2y}$$
$$(m_1 + m_2)a_{cm\,y} = m_1a_{1y} + 0$$
$$a_{cm\,y} = \frac{m_1}{m_1 + m_2}a_{1y}$$

4. From Example 4-7, a block sliding down a stationary frictionless incline has acceleration $g \sin\theta$ down the incline. Use trigonometry to find the y component of this acceleration and use it to find $a_{cm\,y}$:

so $a_{1y} = -a_1 \sin\theta$, where $a_1 = g\sin\theta$

$$a_{1y} = -(g\sin\theta)\sin\theta = -g\sin^2\theta$$

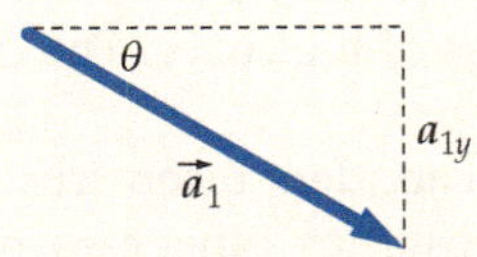

FIGURE 5-55

5. Substitute for a_{1y} in the step-3 result:

$$a_{cm\,y} = \frac{m_1}{m_1 + m_2}a_{1y} = -\frac{m_1}{m_1 + m_2}g\sin^2\theta$$

6. Substitute for $a_{cm\,y}$ in the step-2 result and solve for F_n:

$$F_n = (m_1 + m_2)g + (m_1 + m_2)a_{cm\,y}$$
$$= (m_1 + m_2)g - m_1g\sin^2\theta = [m_1(1 - \sin^2\theta) + m_2]g$$
$$= \boxed{(m_1\cos^2\theta + m_2)g}$$

CHECK For $\theta = 0$, $\cos\theta = 1$, the step-5 result is, as expected, equal to the sum of the two weights. For $\theta = 90°$, $\cos\theta = 0$, and the step-5 result is, as expected, equal to the weight of the wedge alone.

PRACTICE PROBLEM 5-9 Find the force component F_x exerted on the wedge by the scale.

Physics Spotlight

Accident Reconstruction—Measurements and Forces

A car being towed after an accident. *(Mikael Karlsson.)*

Four teenagers drove to a Halloween haunted house that was located out in the countryside. As their car started into a curve, the driver saw a deer in the middle of the road. Frantic swerving and braking put the car into a slide. It skidded off the edge of the gently banked curve, flew over the narrow ditch, and landed in a newly harvested field below the road, where it skidded to a stop in the loose dirt.

Thanks to airbags and safety belts, no one was killed. All were taken to the hospital. The car was towed. But the accident investigation was not finished until a question was answered—was the car speeding?

Accident reconstruction specialists investigated the scene, and used information about the physics of an accident to determine what happened before, during, and immediately after that accident.* After this accident, a police officer, a specialist hired by the driver's insurance company, and a specialist hired by the county road department all looked at the scene.

The first thing the specialists did was measure and photograph everything that might be pertinent to the accident. They measured the road, so that the curve's radius and bank angle could be calculated and compared to the information at the county roads office. They measured the tire marks on the road, and in the field. They used a *drag sled* to determine the coefficient of kinetic friction for the field.† They measured the vertical and horizontal distance from the edge of the road to the first marks in the field. They measured the angle of the road to horizontal along the path of the tire tracks.

Using the measurement information, they calculated a simplified trajectory of the car from the moment it left the road until it landed in the field. This trajectory gave the speed of the car as it left the roadway. Their calculations using the skid marks in the field confirmed that speed. Finally, they calculated the starting speed of the skid on the road. They used the coefficient of kinetic friction of the road, as it was clear that the wheels had been locked, and not spinning.

They concluded that the car had been under the speed limit for the road, but, like most cars, had been going faster than the marked advisory speed for the curve.‡ The county put up deer warning signs, and installed a guard rail along the outer edge of the curve. The driver was ticketed for failure to maintain control of the vehicle.

Not all accident reconstruction is so simple or straightforward. Many accidents involve obstacles, other cars, or tires of the wrong size for the vehicle. Others may involve looking at the physics inside the car to determine whether or not seat belts were worn, or who was driving. But all accident reconstructions start with measurements to determine the forces at work during the accident.

---

* The International Association of Accident Reconstruction Specialists. http://www.iaars.org/ March 2006.

† Marks, Christopher C. O'N., *Pavement Skid Resistance Measurement and Analysis in the Forensic Context.* International Conference on Surface Friction. 2005, Christchurch. http://www.surfacefriction.org.nz.

‡ Chowdhury, Mashrur A., Warren, Davey L., "Analysis of Advisory Speed Setting Criteria." *Public Roads*, 00333735, Dec. 91. Vol. 55, Issue 3.

SUMMARY

Friction and drag forces are complex phenomena empirically approximated by simple equations. For a particle to move in a curved path at constant speed the net force is directed toward the center of curvature. The center of mass of a system moves as if the system were a single point particle with the net force on the system acting on it.

TOPIC	RELEVANT EQUATIONS AND REMARKS	
1. **Friction**	Two objects in contact exert frictional forces on each other. These forces are parallel to the contacting surfaces and directed so as to oppose sliding or tendency to slide.	
Static friction	$f_s \leq \mu_s F_n$	5-2
	where F_n is the normal force of contact and μ_s is the coefficient of static friction.	
Kinetic friction	$f_k = \mu_k F_n$	5-3
	where μ_k is the coefficient of kinetic friction. The coefficient of kinetic friction is slightly less than the coefficient of static friction.	
Rolling friction	$f_r = \mu_r F_n$	5-4
	where μ_r is the coefficient of rolling friction.	
2. **Drag Forces**	When an object moves through a fluid, it experiences a drag force that opposes its motion. The drag force increases with increasing speed. If the body is dropped from rest, its speed increases. As it does, the magnitude of the drag force comes closer and closer to the magnitude of the force of gravity, so the net force, and thus the acceleration, approaches zero. As the acceleration approaches zero, the speed approaches a constant value called its terminal speed. The terminal speed depends on both the shape of the body and on the medium through which it falls.	
3. **Motion Along a Curved Path**	A particle moving along an arbitrary curve can be considered to be moving along a circular arc for a short time interval. Its instantaneous acceleration vector has a component $a_c = v^2/r$ toward the center of curvature of the arc and a component $a_t = dv/dt$ that is tangential to the arc. If the particle is moving along a circular path of radius r at constant speed v, $a_t = 0$ and the speed, radius, and period T are related by $2\pi r = vT$.	
4. ***Numerical Integration: Euler's Method**	To estimate the position x and velocity v at some time t, we first divide the interval from zero to t into a large number of small intervals, each of length Δt. The initial acceleration a_0 is then calculated from the initial position x_0 and velocity v_0. The position x_1 and velocity v_1 a time Δt later are estimated using the relations	
	$x_{n+1} = x_n + v_n \Delta t$	5-9
	and	
	$v_{n+1} = v_n + a_n \Delta t$	5-10
	with $n = 0$. The acceleration a_{n+1} is calculated using the values for x_{n+1} and v_{n+1} and the process is repeated. This continues until estimations for the position and velocity at time t are calculated.	
5. **Center of Mass**		
Center of mass for a system of particles	The center of mass of a system of particles is defined to be the point whose coordinates are given by:	
	$Mx_{cm} = \sum_i m_i x_i$	5-15
	$My_{cm} = \sum_i m_i y_i$	5-16
	$Mz_{cm} = \sum_i m_i z_i$	5-17

TOPIC	RELEVANT EQUATIONS AND REMARKS	
Center of mass for continuous objects	If the mass is continuously distributed, the center of mass is given by: $M\vec{r}_{cm} = \int \vec{r}\, dm$	5-19
Position, velocity, and acceleration for the center of mass of a system of particles	$M\vec{r}_{cm} = m_1\vec{r}_1 + m_2\vec{r}_2 + \cdots$	5-18
	$M\vec{v}_{cm} = m_1\vec{v}_1 + m_2\vec{v}_2 + \cdots$	5-20
	$M\vec{a}_{cm} = m_1\vec{a}_1 + m_2\vec{a}_2 + \cdots$	5-21
Newton's second law for a system	$\vec{F}_{net\,ext} = \sum_i \vec{F}_{i\,ext} = M\vec{a}_{cm}$	5-23

Answers to Concept Checks

5-1 Yes, the car would slide down the incline.

5-2 It accelerates to the right, because the net external force acting on the cylinder is the frictional force to the right exerted on it by the paper. Try it. The cylinder may *appear* to move to the left, because relative to the paper it rolls leftward. However, relative to the table, which serves as an inertial reference frame, it moves to the right. If you mark the table with the original position of the cylinder, you will observe the center of mass move to the right *during the time the cylinder remains in contact with the moving paper*.

Answers to Practice Problems

5-1 35°

5-2 1.1×10^2 N

5-3 $T = m_2(g - a) = 44$ N

5-4 (*a*) Assuming $r \approx 1$ m, we find $v_{t,\min} \approx 3$ m/s, (*b*) $T = 2\pi r/v \approx 2$ s

5-5 Zero. At that instant you are no longer gaining speed and not yet losing speed. Your rate of change of speed is momentarily zero.

5-6 1.60 m/s^2

5-7 $x_{cm} = 2.0$ cm

5-8 $2R = 1.1 \times 10^2$ m

5-9 $F_x = m_1 g \sin\theta \cos\theta$

PROBLEMS

In a few problems, you are given more data than you actually need; in a few other problems, you are required to supply data from your general knowledge, outside sources, or informed estimate.

Interpret as significant all digits in numerical values that have trailing zeros and no decimal points.

- • Single-concept, single-step, relatively easy
- •• Intermediate-level, may require synthesis of concepts
- ••• Challenging
- SSM Solution is in the *Student Solutions Manual*
- Consecutive problems that are shaded are paired problems.

CONCEPTUAL PROBLEMS

1 • Various objects lie on the bed of a truck that is moving along a straight horizontal road. If the truck gradually speeds up, what force acts on the objects to cause them to speed up too? Explain why some of the objects might stay stationary on the floor while others might slip backward on the floor. SSM

2 • Blocks made of the same material but differing in size lie on the bed of a truck that is moving along a straight horizontal road. All of the blocks will slide if the truck's acceleration is sufficiently great. How does the minimum acceleration at which a small block slips compare with the minimum acceleration at which a much heavier block slips?

3 • A block of mass m rests on a plane that is inclined at an angle θ with the horizontal. It follows that the coefficient of static friction between the block and plane is (*a*) $\mu_s \geq g$, (*b*) $\mu_s = \tan\theta$, (*c*) $\mu_s \leq \tan\theta$, (*d*) $\mu_s \geq \tan\theta$.

4 • A block of mass m is at rest on a plane that is inclined at an angle of 30° with the horizontal, as shown in Figure 5-56. Which of the following statements about the magnitude of the static frictional force f_s is necessarily true? (*a*) $f_s > mg$, (*b*) $f_s > mg\cos 30°$, (*c*) $f_s = mg\cos 30°$, (*d*) $f_s = mg\sin 30°$, (*e*) None of these statements is true.

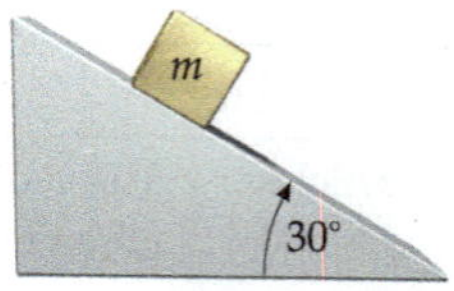

FIGURE 5-56
Problem 4

5 •• On an icy winter day, the coefficient of friction between the tires of a car and a roadway is reduced to one-quarter of its value on a dry day. As a result, the maximum speed $v_{\max\,dry}$ at which the car can safely negotiate a curve of radius R is reduced. The new value for this speed is (*a*) $v_{\max\,dry}$, (*b*) $0.71v_{\max\,dry}$, (*c*) $0.50v_{\max\,dry}$, (*d*) $0.25v_{\max\,dry}$, (*e*) reduced by an unknown amount depending on the car's mass.

6 •• If it is started properly on the frictionless inside surface of a cone (Figure 5-57), a block is capable of maintaining uniform circular motion. Draw the free-body diagram of the block and identify clearly which force (or forces, or force components) is responsible for the centripetal acceleration of the block.

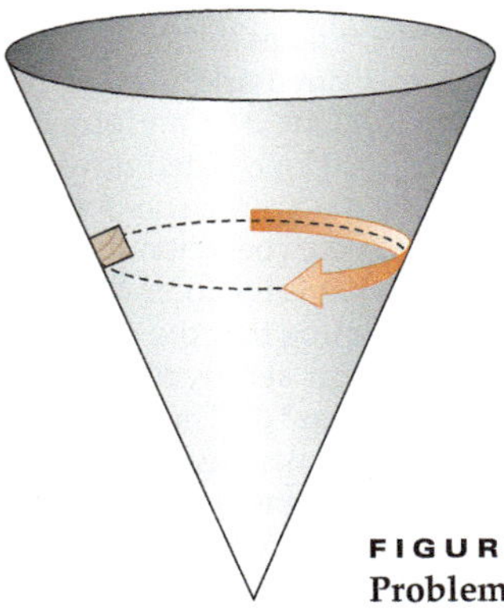

FIGURE 5-57
Problem 6

7 •• Here is an interesting experiment that you can perform at home: take a wooden block and rest it on the floor or some other flat surface. Attach a rubber band to the block and pull gently on the rubber band in the horizontal direction. Keep your hand moving at constant speed. At some point, the block will start moving, but it will not move smoothly. Instead, it will start moving, stop again, start moving again, stop again, and so on. Explain why the block moves this way. (The start-stop motion is sometimes called "stick-slip" motion.)

8 • Viewed from an inertial reference frame, an object is seen to be moving in a circle. Which, if any, of the following statements must be true. (*a*) A nonzero net force acts on the object. (*b*) The object cannot have a radially outward force acting on it. (*c*) At least one of the forces acting on the object must point directly toward the center of the circle.

9 •• A particle is traveling in a vertical circle at constant speed. One can conclude that the magnitude of its _____ is (are) constant. (*a*) velocity, (*b*) acceleration, (*c*) net force, (*d*) apparent weight

10 •• You place a lightweight piece of iron on a table and hold a small kitchen magnet above the iron at a distance of 1.00 cm. You find that the magnet cannot lift the iron, even though there is obviously a force between the iron and the magnet. Next, holding the magnet in one hand and the piece of iron in the other, with the magnet 1.00 cm above the iron, you simultaneously drop them from rest. As they fall, the magnet and the piece of iron bang into each other before hitting the floor. (*a*) Draw free-body diagrams illustrating all of the forces on the magnet and the iron for each demonstration. (*b*) Explain why the magnet and iron move closer together while they are falling, even though the magnet cannot lift the piece of iron when it is sitting on the table.

11 ••• The following question is an excellent "braintwister," invented by Boris Korsunsky:* Two identical blocks are attached by a massless string running over a pulley, as shown in Figure 5-58.

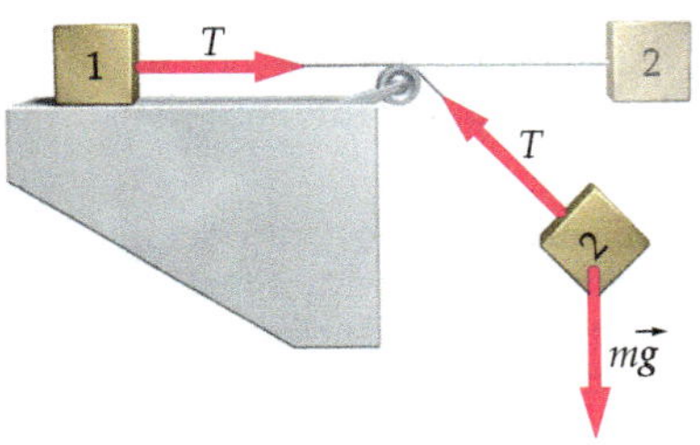

FIGURE 5-58
Problem 11

The rope initially runs over the pulley at the rope's midpoint, and the surface that block 1 rests on is frictionless. Blocks 1 and 2 are initially at rest when block 2 is released with the string taut and horizontal. Will block 1 hit the pulley before or after block 2 hits the wall? (Assume that the initial distance from block 1 to the pulley is the same as the initial distance from block 2 to the wall.) There is a very simple solution. SSM

12 •• In class, most professors do the following experiment while discussing the conditions under which air drag can be neglected while analyzing free-fall. First, a flat piece of paper and a small lead weight are dropped next to each other, and clearly the paper's acceleration is less than that of the lead weight. Then, the paper is crumpled into a small wad and the experiment repeated. Over the distance of a meter or two, it is clear the acceleration of the paper is now very close to that of the lead weight. To your dismay, the professor calls on you to explain why the paper's acceleration changed so dramatically. Repeat your explanation here!

13 •• CONTEXT-RICH Jim decides to attempt to set a record for terminal speed in skydiving. Using the knowledge he has gained from a physics course, he makes the following plans. He will be dropped from as high an altitude as possible (equipping himself with oxygen), on a warm day and go into a "knife" position, in which his body is pointed vertically down and his hands are pointed ahead. He will outfit himself with a special sleek helmet and rounded protective clothing. Explain how each of these factors helps Jim attain the record. SSM

14 •• CONTEXT-RICH You are sitting in the passenger seat in a car driving around a circular, horizontal, flat racetrack at a high speed. As you sit there, you "feel" a "force" pushing you toward the outside of the track. What is the true direction of the force acting on you, and where does it come from? (Assume that you do not slide across the seat.) Explain the *sensation* of an "outward force" on you in terms of the Newtonian perspective.

15 • The mass of the moon is only about 1% of that of Earth. Therefore, the force that keeps the moon in its orbit around Earth (*a*) is much smaller than the gravitational force exerted on the moon by Earth, (*b*) is much greater than the gravitational force exerted on the moon by Earth, (*c*) actually is the gravitational force exerted on the moon by Earth, (*d*) cannot be answered yet, because we have not yet studied Newton's law of gravity. SSM

16 •• A block is sliding on a frictionless surface along a loop-the-loop, as in Figure 5-59*a*. The block is moving fast enough so that it never loses contact with the track. Match the points along the track to the appropriate free-body diagrams in Figure 5-59*b*.

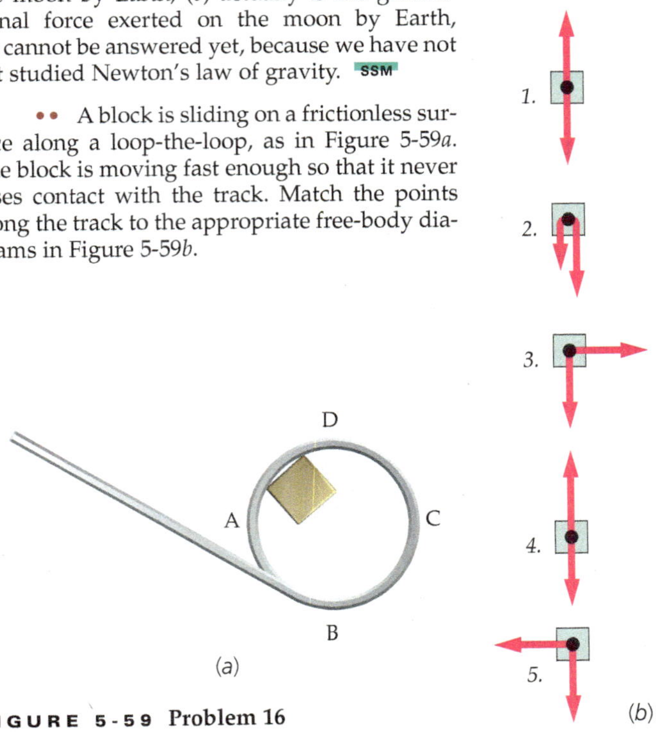

FIGURE 5-59 Problem 16

* Boris Korsunsky, "Braintwisters for Physics Students," *The Physics Teacher*, 33, 550 (1995).

17 •• (*a*) A rock and a feather held at the same height above the ground are simultaneously dropped. During the first few milliseconds following release, the drag force on the rock is smaller than the drag force on the feather, but later on during the fall the *opposite* is true. Explain. (*b*) In light of this result, explain how the rock's acceleration can be so obviously larger than that of the feather. *Hint: Draw a free-body diagram of each object.* SSM

18 •• Two pucks of masses m_1 and m_2 are lying on a frictionless table and are connected by a massless spring of force constant k. A horizontal force F_1 directed away from m_2 is then exerted on m_1. What is the magnitude of the resulting acceleration of the center of mass of the two-puck system? (*a*) F_1/m_1, (*b*) $F_1/(m_1 + m_2)$, (*c*) $(F_1 + kx)/(m_1 + m_2)$, where x is the amount the spring is stretched, (*d*) $(m_1 + m_2)F_1/m_1m_2$

19 •• The two pucks in Problem 18 lie unconnected on a frictionless table. A horizontal force F_1 directed away from m_2 is then exerted on m_1. How does the magnitude of the resulting acceleration of the center of mass of the two-puck system compare to the magnitude acceleration of m_1? Explain your reasoning.

20 •• If only external forces can cause the center of mass of a system of particles to accelerate, how can a car on level ground ever accelerate? We normally think of the car's engine as supplying the force needed to accelerate the car, but is this true? Where does the external force that accelerates the car come from?

21 •• When you push on the brake pedal to slow down a car, a brake pad is pressed against the rotor so that the friction of the pad slows the wheel's rotation. However, the friction of the pad against the rotor *cannot be* the force that slows the car down, because it is an internal force (both the rotor and the wheel are parts of the car, so any forces between them are purely internal to the system). What is the external force that slows down the car? Give a detailed explanation of how this force operates.

22 •• Give an example of each of the following: (*a*) a three-dimensional object that has no matter at its center of mass, (*b*) a solid object whose center of mass is outside of it, (*c*) a solid sphere whose center of mass does not lie at its geometrical center, (*d*) a long wooden stick whose center of mass does not lie at its middle.

23 •• **BIOLOGICAL APPLICATION** When you are standing upright, your center of mass is located within the volume of your body. However, as you bend over (say to pick up a package), its location changes. Approximately where is it when you are bent over at right angles and what change in your body caused the center of mass location to change? Explain. SSM

24 •• **ENGINEERING APPLICATION** Early on their three-day (one-way) trip to the moon, the Apollo team (late 1960s to early 1970s) would explosively separate the lunar ship from the third-stage booster (that provided the final "boost") while still fairly close to Earth. During the explosion, how did the velocity of each of the two pieces of the system change? How did the velocity of the center of mass of the system change?

25 •• You throw a boomerang and for a while it "flies" horizontally in a straight line at a constant speed, while spinning rapidly. Draw a series of pictures, as viewed vertically down from overhead, of the boomerang in different rotational positions as it moves parallel to the surface of Earth. On each picture, indicate the location of the boomerang's center of mass and connect the dots to trace the trajectory of its center of mass. What is the center of mass's acceleration during this part of the flight?

ESTIMATION AND APPROXIMATION

26 •• **ENGINEERING APPLICATION** To determine the aerodynamic drag on a car, automotive engineers often use the "coast-down" method. The car is driven on a long, flat road at some convenient speed (60 mi/h is typical), shifted into neutral, and allowed to coast to a stop. The time that it takes for the speed to drop by successive 5-mi/h intervals is measured and used to compute the net force slowing the car down. (*a*) One day, a group measured that a Toyota Tercel with a mass of 1020 kg coasted down from 60.0 mi/h to 55.0 mi/h in 3.92 s. Estimate the average net force slowing the car down in this speed range. (*b*) If the coefficient of rolling friction for this car is known to be 0.020, what is the force of rolling friction that is acting to slow it down? Assuming that the only two forces acting on the car are rolling friction and aerodynamic drag, what is the average drag force acting on the car? (*c*) The drag force has the form $\frac{1}{2}C\rho Av^2$, where A is the cross-sectional area of the car facing into the air, v is the car's speed, ρ is the density of air, and C is a dimensionless constant of order-of-magnitude 1. If the cross-sectional area of the car is 1.91 m^2, determine C from the data given. (The density of air is 1.21 kg/m^3; use 57.5 mi/h for the speed of the car in this computation.)

27 •• Using dimensional analysis, determine the units and dimensions of the constant b in the retarding force bv^n if (*a*) $n = 1$ and (*b*) $n = 2$. (*c*) Newton showed that the air resistance of a falling object with a circular cross section should be approximately $\frac{1}{2}\rho\pi r^2v^2$, where $\rho = 1.20$ kg/m^3, the density of air. Show that this is consistent with your dimensional analysis for part (*b*). (*d*) Find the terminal speed for a 56.0-kg skydiver; approximate his cross-sectional area as a disk of radius 0.30 m. The density of air near the surface of Earth is 1.20 kg/m^3. (*e*) The density of the atmosphere decreases with height above the surface of Earth; at a height of 8.0 km, the density is only 0.514 kg/m^3. What is the terminal velocity at this height? SSM

28 •• Estimate the terminal velocity of an average sized raindrop and a golf-ball-sized hailstone. *Hint: See Problems 26 and 27.*

29 •• Estimate the minimum coefficient of static friction needed between a car's tires and the pavement in order to complete a left turn at a city street intersection at the posted straight-ahead speed limit of 25 mph and on narrow inner-city streets. Comment on the wisdom of attempting such a turn at that speed.

30 •• Estimate the widest stance you can take when standing on a dry, icy surface. That is, how wide can you safely place your feet and not slip into an undesired "split"? Let the coefficient of static friction of rubber on ice be roughly 0.25.

FRICTION

31 • A block of mass m slides at constant speed down a plane inclined at an angle of θ with the horizontal. It follows that (*a*) $\mu_k = mg\sin\theta$, (*b*) $\mu_k = \tan\theta$, (*c*) $\mu_k = 1 - \cos\theta$, (*d*) $\mu_k = \cos\theta - \sin\theta$. SSM

32 • A block of wood is pulled at constant velocity by a horizontal string across a horizontal surface with a force of 20 N. The coefficient of kinetic friction between the surfaces is 0.3. The force of friction is (*a*) impossible to determine without knowing the mass of the block, (*b*) impossible to determine without knowing the speed of the block, (*c*) 0.30 N, (*d*) 6.0 N, (*e*) 20 N.

33 • A block weighing 20-N rests on a horizontal surface. The coefficients of static and kinetic friction between the surface and the

block are $\mu_s = 0.80$ and $\mu_k = 0.60$. A horizontal string is then attached to the block and a constant tension T is maintained in the string. What is the magnitude of the force of friction acting on the block if (*a*) $T = 15$ N, (*b*) $T = 20$ N? SSM

34 • A block of mass m is pulled at a constant velocity across a horizontal surface by a string as shown in Figure 5-60. The magnitude of the frictional force is (*a*) $\mu_k mg$, (*b*) $T\cos\theta$, (*c*) $\mu_k(T - mg)$, (*d*) $\mu_k T\sin\theta$, or (*e*) $\mu_k(mg - T\sin\theta)$.

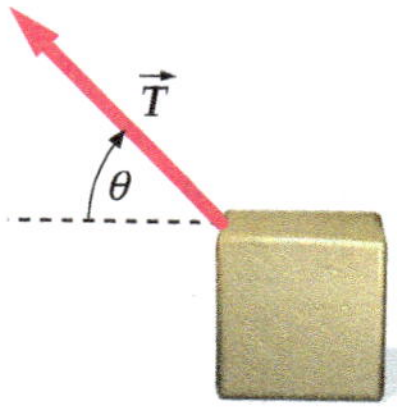

FIGURE 5-60 Problem 34

35 • A 100-kg crate rests on a thick-pile carpet. A weary worker then pushes on the crate with a horizontal force of 500 N. The coefficients of static and kinetic friction between the crate and the carpet are 0.600 and 0.400, respectively. Find the magnitude of the frictional force exerted by the carpet on the crate. SSM

36 • A box weighing 600 N is pushed along a horizontal floor at constant velocity with a force of 250 N parallel to the floor. What is the coefficient of kinetic friction between the box and the floor?

37 • The coefficient of static friction between the tires of a car and a horizontal road is 0.60. Neglecting air resistance and rolling friction, (*a*) what is the magnitude of the maximum acceleration of the car when it is braked? (*b*) What is the shortest distance in which the car can stop if it is initially traveling at 30 m/s? SSM

38 • The force that accelerates a car along a flat road is the frictional force exerted by the road on the car's tires. (*a*) Explain why the acceleration can be greater when the wheels do not slip. (*b*) If a car is to accelerate from 0 to 90 km/h in 12 s, what is the minimum coefficient of friction needed between the road and tires? Assume that the drive wheels support exactly half the weight of the car.

39 •• A 5.00-kg block is held at rest against a vertical wall by a horizontal force of 100 N. (*a*) What is the frictional force exerted by the wall on the block? (*b*) What is the minimum horizontal force needed to prevent the block from falling if the static coefficient of friction between the wall and the block is 0.400?

40 •• A tired and overloaded student is attempting to hold a large physics textbook wedged under his arm, as shown in Figure 5-61.

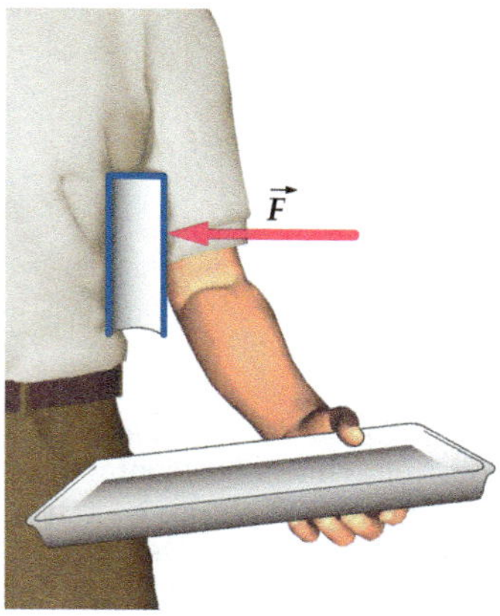

FIGURE 5-61 Problem 40

The textbook has a mass of 3.2 kg, while the coefficient of static friction of the textbook against the student's underarm is 0.320 and the coefficient of static friction of the book against the student's shirt is 0.160. (*a*) What is the minimum horizontal force that the student must apply to the textbook to prevent it from falling? (*b*) If the student can only exert a force of 61 N, what is the acceleration of the textbook as it slides from under his arm? The coefficient of kinetic friction of arm against textbook is 0.200, while that of shirt against textbook is 0.090.

41 •• **ENGINEERING APPLICATION** You are racing in a rally on a snowy day when the temperature is near the freezing point. The coefficient of static friction between a car's tires and an icy road is 0.080. Your crew boss is concerned about some of the hills on the course and wants you to think about switching to studded tires. To address the issue, he wants to compare the actual hill angles on the course to see which of them your car can negotiate. (*a*) What is the angle of the steepest incline that a vehicle with four-wheel drive can climb at constant speed? (*b*) Given that the hills are icy, what is the steepest possible hill angle for the same four-wheel drive car to descend at constant speed?

42 •• A 50-kg box that is resting on a level floor must be moved. The coefficient of static friction between the box and the floor is 0.60. One way to move the box is to push down on the box at an angle θ below the horizontal. Another method is to pull up on the box at an angle θ above the horizontal. (*a*) Explain why one method requires less force than the other. (*b*) Calculate the minimum force needed to move the box by each method if $\theta = 30°$ and compare the answer with the results when $\theta = 0°$.

43 •• A block of mass $m_1 = 250$ g is at rest on a plane that makes an angle of $\theta = 30°$ with the horizontal. The coefficient of kinetic friction between the block and the plane is 0.100. The block is attached to a second block of mass $m_2 = 200$ g that hangs freely by a string that passes over a frictionless, massless pulley (Figure 5-62). When the second block has fallen 30.0 cm, what will be its speed? SSM

FIGURE 5-62 Problems 43, 44, 45

44 •• In Figure 5-62, $m_1 = 4.0$ kg and the coefficient of static friction between the block and the incline is 0.40. (*a*) Find the range of possible values for m_2 for which the system will be in static equilibrium. (*b*) Find the frictional force on the 4.0-kg block if $m_2 = 1.0$ kg.

45 •• In Figure 5-62, $m_1 = 4.0$ kg, $m_2 = 5.0$ kg, and the coefficient of kinetic friction between the inclined plane and the 4.0-kg block is $\mu_k = 0.24$. Find the magnitude of the acceleration of the masses and the tension in the cord.

46 •• A 12-kg turtle rests on the bed of a zookeeper's truck, which is traveling down a country road at 55 mi/h. The zookeeper spots a deer in the road, and slows to a stop in 12 s. Assuming

constant acceleration, what is the minimum coefficient of static friction between the turtle and the truck bed surface that is needed to prevent the turtle from sliding?

47 •• A 150-g block is projected up a ramp with an initial speed of 7.0 m/s. The coefficient of kinetic friction between the ramp and the block is 0.23. (*a*) If the ramp is inclined 25° with the horizontal, how far along the surface of the ramp does the block slide before coming to a stop? (*b*) The block then slides back down the ramp. What is the minimum coefficient of static friction between the block and the ramp if the block is not to slide back down the ramp? SSM

48 •• An automobile is going up a 15° grade at a speed of 30 m/s. The coefficient of static friction between the tires and the road is 0.70. (*a*) What minimum distance does it take to stop the car? (*b*) What minimum distance would it take to stop if the car were going down the grade?

49 •• **ENGINEERING APPLICATION** A rear-wheel-drive car supports 40 percent of its weight on its two drive wheels and has a coefficient of static friction of 0.70 with a horizontal straight road. (*a*) Find the vehicle's maximum acceleration. (*b*) What is the shortest possible time in which this car can achieve a speed of 100 km/h? (Assume the engine can provide unlimited power.)

50 •• You and your best pal make a friendly bet that you can place a 2.0-kg box against the side of a cart, as in Figure 5-63, and that the box will not fall to the ground, even though you guarantee to use no hooks, ropes, fasteners, magnets, glue, or adhesives of any kind. When your friend accepts the bet, you begin pushing the cart in the direction shown in the figure. The coefficient of static friction between the box and the cart is 0.60. (*a*) Find the minimum acceleration for which you will win the bet. (*b*) What is the magnitude of the frictional force in this case? (*c*) Find the force of friction on the box if the acceleration is twice the minimum needed for the box not to fall. (*d*) Show that, for a box of any mass, the box will not fall if the magnitude of the forward acceleration is $a \geq g/\mu_s$, where μ_s is the coefficient of static friction.

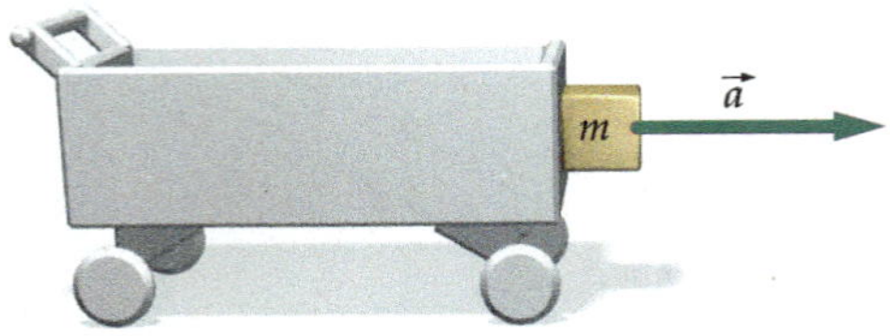

FIGURE 5-63 Problem 50

51 •• Two blocks attached by a string (Figure 5-64) slide down a 10° incline. Block 1 has mass $m_1 = 0.80$ kg and block 2 has mass $m_2 = 0.25$ kg. In addition, the kinetic coefficients of friction between the blocks and the incline are 0.30 for block 1 and 0.20 for block 2. Find (*a*) the magnitude of the acceleration of the blocks, and (*b*) the tension in the string.

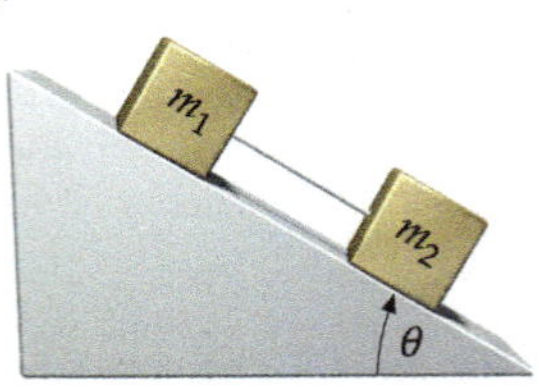

FIGURE 5-64 Problems 51 and 52

52 •• Two blocks of masses m_1 and m_2 are sliding down an incline as shown in Figure 5-64. They are connected by a massless *rod*. The coefficients of kinetic friction between the block and the surface are μ_1 for block 1 and μ_2 for block 2. (*a*) Determine the acceleration of the two blocks. (*b*) Determine the force that the rod exerts on each of the two blocks. Show that these forces are both 0 when $\mu_1 = \mu_2$ and give a simple, nonmathematical argument why this is true.

53 •• A block of mass m rests on a horizontal table (Figure 5-65). The block is pulled by a massless rope with a force $\vec{F}$ at an angle θ. The coefficient of static friction is 0.60. The minimum value of the force needed to move the block depends on the angle θ. (*a*) Discuss qualitatively how you would expect the magnitude of this force to depend on θ. (*b*) Compute the force for the angles $\theta = 0°, 10°, 20°, 30°, 40°, 50°$, and $60°$, and make a plot of F versus θ for $mg = 400$ N. From your plot, at what angle is it most efficient to apply the force to move the block? SSM

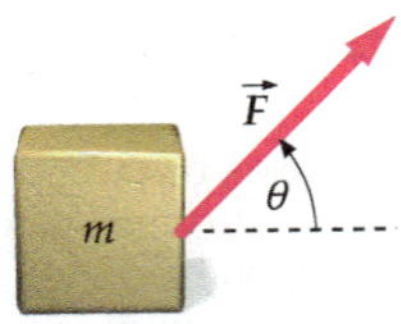

FIGURE 5-65 Problems 53 and 54

54 •• Consider the block in Figure 5-65. Show that, in general, the following results hold for a block of mass m resting on a horizontal surface whose coefficient of static friction is μ_s. (*a*) If you want to apply the *minimum* possible force to move the block, you should apply it with the force pulling upward at an angle $\theta = \tan^{-1}\mu_s$. (*b*) The minimum force necessary to start the block moving is $F_{min} = (\mu_s/\sqrt{1+\mu_s^2})mg$. (*c*) Once the block starts moving, if you want to apply the least possible force to keep it moving, should you keep the angle at which you are pulling the same, increase it, or decrease it?

55 •• Answer the questions in Problem 54, but for a force $\vec{F}$ that pushes down on the block at an angle θ *below* the horizontal.

56 •• A 100-kg mass is pulled along a frictionless surface by a horizontal force $\vec{F}$ such that its acceleration is $a_1 = 6.00\ \text{m/s}^2$ (Figure 5-66). A 20.0-kg mass slides along the top of the 100-kg mass and has an acceleration of $a_2 = 4.00\ \text{m/s}^2$. (It thus slides backward relative to the 100-kg mass.) (*a*) What is the frictional force exerted by the 100-kg mass on the 20.0-kg mass? (*b*) What is the net force acting on the 100-kg mass? What is the force F? (*c*) After the 20.0-kg mass falls off the 100-kg mass, what is the acceleration of the 100-kg mass? (Assume that the force F does not change.)

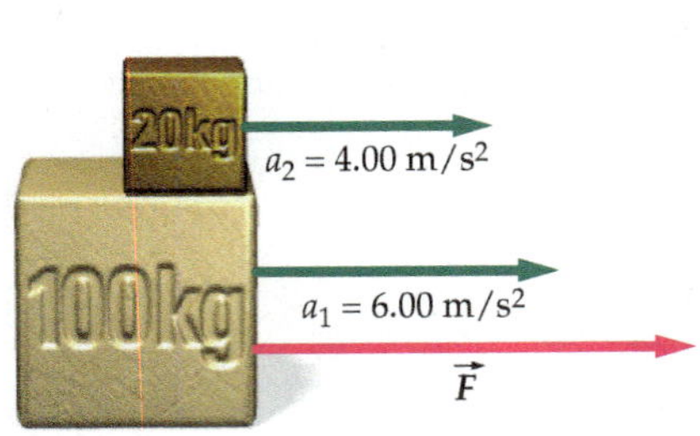

FIGURE 5-66 Problem 56

57 •• A 60-kg block slides along the top of a 100-kg block. The 60-kg block has an acceleration of 3.0 m/s² while a horizontal force of 320 N is applied to it, as shown in Figure 5-67. There is no friction between the 100-kg block and a horizontal frictionless surface, but there is friction between the two blocks. (*a*) Find the coefficient

FIGURE 5-67 Problem 57

of kinetic friction between the blocks. (*b*) Find the acceleration of the 100-kg block during the time that the 60-kg block remains in contact.

58 •• The coefficient of static friction between a rubber tire and the road surface is 0.85. What is the maximum acceleration of a 1000-kg four-wheel-drive truck if the road makes an angle of 12° with the horizontal and the truck is (*a*) climbing, and (*b*) descending?

59 •• A 2.0-kg block sits on a 4.0-kg block that is on a frictionless table (Figure 5-68). The coefficients of friction between the blocks are $\mu_s = 0.30$ and $\mu_k = 0.20$. (*a*) What is the maximum horizontal force F that can be applied to the 4.0-kg block if the 2.0-kg block is not to slip? (*b*) If F has half this value, find the acceleration of each block and the force of friction acting on each block. (*c*) If F is twice the value found in (*a*), find the acceleration of each block.

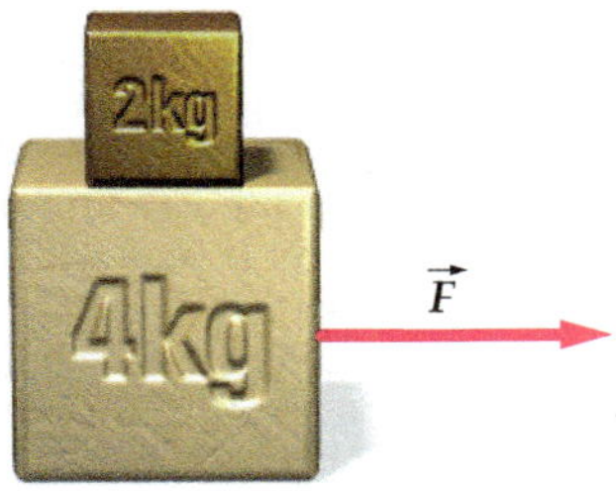

FIGURE 5-68 Problem 59

60 ••• A 10.0-kg block rests on a 5.0-kg bracket, as shown in Figure 5-69. The 5.0-kg bracket sits on a frictionless surface. The coefficients of friction between the 10.0-kg block and the bracket on which it rests are $\mu_s = 0.40$ and $\mu_k = 0.30$. (*a*) What is the maximum force F that can be applied if the 10.0-kg block is not to slide on the bracket? (*b*) What is the corresponding acceleration of the 5.0-kg bracket?

FIGURE 5-69 Problem 60

61 ••• You and your friends push a 75.0-kg greased pig up an aluminum slide at the county fair, starting from the low end of the slide. The coefficient of kinetic friction between the pig and the slide is 0.070. (*a*) All of you pushing together (parallel to the incline) manage to accelerate the pig from rest at the constant rate of 5.0 m/s^2 over a distance of 1.5 m, at which point you release the pig. The pig continues up the slide, reaching a maximum *vertical height* above its release point of 45 cm. What is the angle of inclination of the slide? (*b*) At the maximum height the pig turns around and begins to slip down the slide, how fast is it moving when it arrives at the low end of the slide?

62 ••• A 100-kg block on an inclined plane is attached to another block of mass m via a string, as in Figure 5-70. The coefficients of static and kinetic friction for the block and the incline are $\mu_s = 0.40$ and $\mu_k = 0.20$ and the plane is inclined 18° with horizontal. (*a*) Determine the range of values for m, the mass of the hanging block, for which the 100-kg block will not move unless disturbed, but if nudged, will slide down the incline. (*b*) Determine a range of values for m for which the 100-kg block will not move unless nudged, but if nudged will slide up the incline.

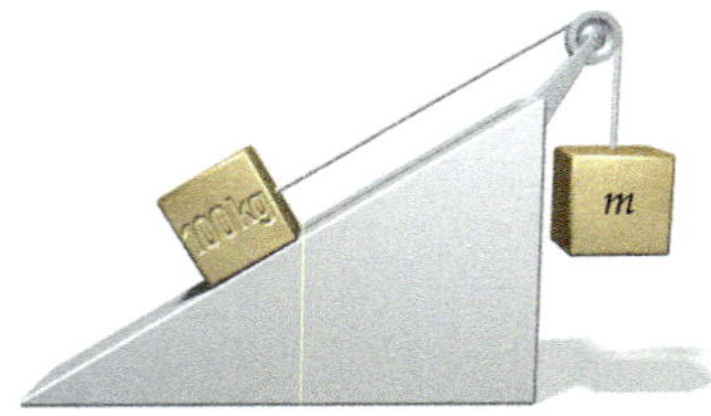

FIGURE 5-70 Problem 62

63 ••• A block of mass 0.50 kg rests on the inclined surface of a wedge of mass 2.0 kg, as in Figure 5-71. The wedge is acted on by a horizontal applied force $\vec{F}$ and slides on a frictionless surface. (*a*) If the coefficient of static friction between the wedge and the block is $\mu_s = 0.80$ and the wedge is inclined 35° with the horizontal, find the maximum and minimum values of the applied force for which the block does not slip. (*b*) Repeat part (*a*) with $\mu_s = 0.40$.

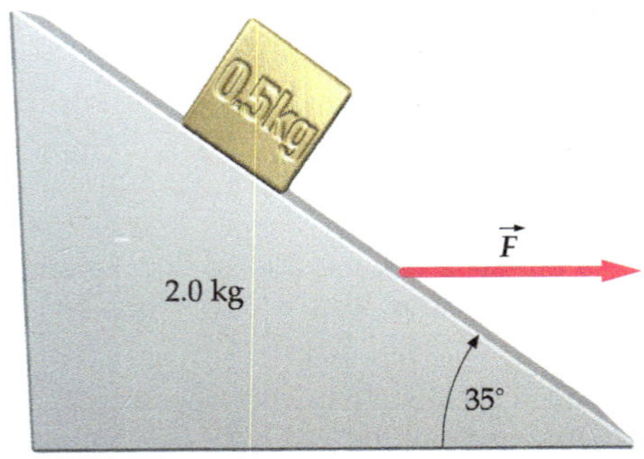

FIGURE 5-71 Problem 63

64 ••• SPREADSHEET In your physics lab, you and your lab partners push a block of wood with a mass of 10.0 kg (starting from rest), with a constant horizontal force of 70 N across a wooden floor. In the previous week's laboratory meeting, your group determined that the coefficient of kinetic friction was not exactly constant, but instead was found to vary with the object's speed according to $\mu_k = 0.11/(1 + 2.3 \times 10^{-4}v^2)^2$. Write a **spreadsheet** program using Euler's method to calculate and graph both the speed and the position of the block as a function of time from 0 to 10 s. Compare this result to the result you would get if you assumed the coefficient of kinetic friction had a constant value of 0.11.

65 ••• **MULTISTEP** In order to determine the coefficient of kinetic friction of a block of wood on a horizontal table surface, you are given the following assignment: Take the block of wood and give it an initial velocity across the surface of the table. Using a stopwatch, measure the time Δt it takes for the block to come to a stop and the total displacement Δx the block slides following the push. (*a*) Using Newton's laws and a free-body diagram of the block, show that the expression for the coefficient of kinetic friction is given by $\mu_k = 2\Delta x/[(\Delta t)^2 g]$. (*b*) If the block slides a distance of 1.37 m in 0.97 s, calculate μ_k. (*c*) What was the initial speed of the block?

66 •• **SPREADSHEET** (*a*) A block is sliding down an inclined plane. The coefficient of kinetic friction between the block and the plane is μ_k. Show that a graph of $a_x/\cos\theta$ versus $\tan\theta$ (where a_x is the acceleration down the incline and θ is the angle the plane is inclined with the horizontal) would be a straight line with slope g and intercept $-\mu_k g$. (*b*) The following data show the acceleration of a block sliding down an inclined plane as a function of the angle θ that the plane is inclined with the horizontal:*

θ (degrees)	Acceleration (m/s²)
25.0	1.69
27.0	2.10
29.0	2.41
31.0	2.89
33.0	3.18
35.0	3.49
37.0	3.78
39.0	4.15
41.0	4.33
43.0	4.72
45.0	5.11

Using a **spreadsheet** program, graph these data and fit a straight line to them to determine μ_k and g. What is the percentage difference between the obtained value of g and the commonly specified value of 9.81 m/s²?

DRAG FORCES

67 • A Ping-Pong ball has a mass of 2.3 g and a terminal speed of 9.0 m/s. The drag force is of the form bv^2. What is the value of b? SSM

68 • A small pollution particle settles toward Earth in still air. The terminal speed of the particle is 0.30 mm/s, the mass of the particle is 1.0×10^{-10} g and the drag force of the particle is of the form bv. What is the value of b?

69 •• A common classroom demonstration involves dropping basket-shaped coffee filters and measuring the time required for them to fall a given distance. A professor drops a single basket-shaped coffee filter from a height h above the floor, and records the time for the fall as Δt. How far will a stacked set of n identical filters fall during the same time interval Δt? Consider the filters to be so light that they instantaneously reach their terminal velocities. Assume a drag force that varies as the square of the speed and assume the filters are released oriented right-side up. SSM

70 •• A skydiver of mass 60.0 kg can slow herself to a constant speed of 90 km/h by orienting her body horizontally, looking straight down with arms and legs extended. In this position, she presents the maximum cross-sectional area and thus maximizes the air-drag force on her. (*a*) What is the magnitude of the drag force on the skydiver? (*b*) If the drag force is given by bv^2, what is the value of b? (*c*) At some instant she quickly flips into a "knife" position, orienting her body vertically with her arms straight down. Suppose this reduces the value of b to 55 percent of the value in Parts (*a*) and (*b*). What is her acceleration at the instant she achieves the "knife" position?

71 •• **ENGINEERING APPLICATION, CONTEXT-RICH** Your team of test engineers is to release the parking brake so an 800-kg car will roll down a very long 6.0 percent grade in preparation for a crash test at the bottom of the incline. (On a 6.0 percent grade the change in altitude is 6.0 percent of the horizontal distance traveled.) The total resistive force (air drag plus rolling friction) for this car has been previously established to be $F_d = 100\text{ N} + (1.2\text{ N}\cdot\text{s}^2/\text{m}^2)v^2$, where v is the speed of the car. What is the terminal speed for the car rolling down this grade?

72 ••• **APPROXIMATION** Small, slowly moving spherical particles experience a drag force given by Stokes' law: $F_d = 6\pi\eta r v$, where r is the radius of the particle, v is its speed, and η is the coefficient of viscosity of the fluid medium. (*a*) Estimate the terminal speed of a spherical pollution particle of radius 1.00×10^{-5} m and density of 2000 kg/m³. (*b*) Assuming that the air is still and that η is $1.80 \times 10^{-5}\text{ N}\cdot\text{s/m}^2$, estimate the time it takes for such a particle to fall from a height of 100 m.

73 ••• **ENGINEERING APPLICATION, CONTEXT-RICH** You have an environmental chemistry internship, and are in charge of a sample of air that contains pollution particles of the size and density given in Problem 72. You capture the sample in an 8.0-cm-long test tube. You then place the test tube in a centrifuge with the midpoint of the test tube 12 cm from the rotation axis of the centrifuge. You set the centrifuge to spin at 800 revolutions per minute. (*a*) Estimate the time you have to wait so that nearly all of the pollution particles settle to the end of the test tube. (*b*) Compare this to the time required for a pollution particle to fall 8.0 cm under the action of gravity and subject to the drag force given in Problem 72. SSM

MOTION ALONG A CURVED PATH

74 • A rigid rod with a 0.050-kg ball at one end rotates about the other end so the ball travels at constant speed in a vertical circle with a radius of 0.20 m. What is the maximum speed of the ball so that the force of the rod on the ball does not exceed 10 N?

75 • A 95-g stone is whirled in a horizontal circle on the end of an 85-cm-long string. The stone takes 1.2 s to make each complete revolution. Determine the angle that the string makes with the horizontal. SSM

76 •• A 0.20-kg stone is whirled in a horizontal circle on the end of an 0.80-m-long string. The string makes an angle of 20° with the horizontal. Determine the speed of the stone.

77 •• A 0.75-kg stone attached to a string is whirled in a horizontal circle of radius 35 cm as in the tetherball in Example 5-11. The string makes an angle of 30° with the vertical. (*a*) Find the speed of the stone. (*b*) Find the tension in the string.

78 •• **BIOLOGICAL APPLICATION** A pilot with a mass of 50 kg comes out of a vertical dive in a circular arc such that at the bottom of the arc her upward acceleration is 3.5g. (*a*) How does the magnitude

* Data taken from Dennis W. Phillips, "Science Friction Adventure–Part II," *The Physics Teacher*, 553 (1990).

of the force exerted by the airplane seat on the pilot at the bottom of the arc compare to her weight? (*b*) Use Newton's laws of motion to explain why the pilot might be subject to a blackout. This means that an above normal volume of blood "pools" in her lower limbs. How would an inertial reference frame observer describe the cause of the blood pooling?

79 •• An 80.0-kg airplane pilot pulls out of a dive by following, at a constant speed of 180 km/h, the arc of a circle whose radius is 300 m. (*a*) At the bottom of the circle, what are the direction and magnitude of his acceleration? (*b*) What is the net force acting on him at the bottom of the circle? (*c*) What is the force exerted on the pilot by the airplane seat?

80 •• An small object of mass m_1 moves in a circular path of radius r on a frictionless horizontal tabletop (Figure 5-72). It is attached to a string that passes through a small frictionless hole in the center of the table. A second object with a mass of m_2 is attached to the other end of the string. Derive an expression for r in terms of m_1, m_2, and the time T for one revolution.

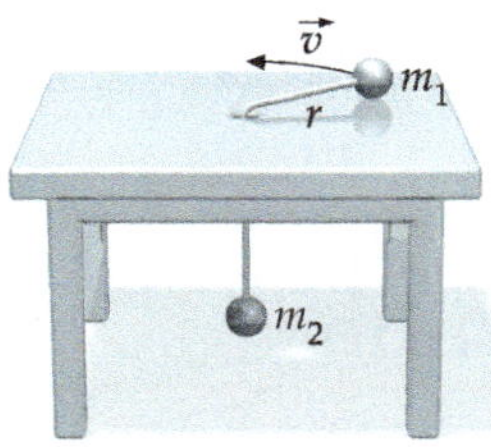

FIGURE 5-72 Problem 80

81 •• A block of mass m_1 is attached to a cord of length L_1, which is fixed at one end. The block moves in a horizontal circle on a frictionless tabletop. A second block of mass m_2 is attached to the first by a cord of length L_2 and also moves in a circle on the same frictionless tabletop, as shown in Figure 5-73. If the period of the motion is T, find the tension in each cord in terms of the given symbols. SSM

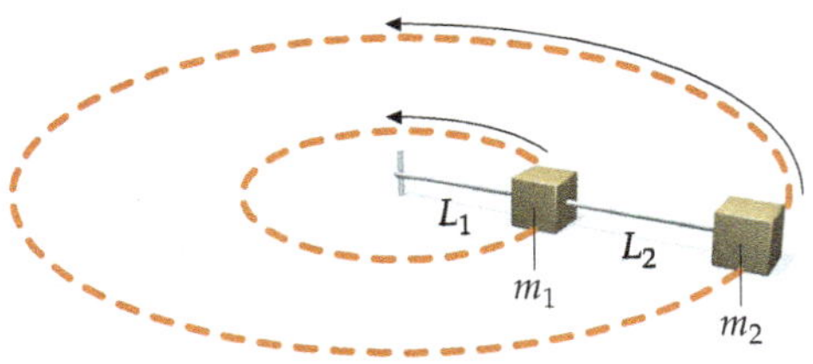

FIGURE 5-73 Problem 81

82 •• MULTISTEP A particle moves with constant speed in a circle of radius 4.0 cm. It takes 8.0 s to complete each revolution. (*a*) Draw the path of the particle to scale, and indicate the particle's position at 1.0-s intervals. (*b*) Sketch the displacement vectors for each interval. These vectors also indicate the directions for the average-velocity vectors for each interval. (*c*) Graphically find the magnitude of the change in the average velocity $|\Delta\vec{v}|$ for two consecutive 1-s intervals. Compare $|\Delta\vec{v}|/\Delta t$, measured in this way, with the magnitude of the instantaneous acceleration computed from $a_c = v^2/r$.

83 •• You are swinging your younger sister in a circle of radius 0.75 m, as shown in Figure 5-74. If her mass is 25 kg and you arrange it so she makes one revolution every 1.5 s, (*a*) what is the magnitude and direction of the force that must be exerted by you on her? (Model her as a point particle.) (*b*) What is the magnitude and direction of the force she exerts on you?

FIGURE 5-74 Problem 83 *(David de Lossy/The Image Bank.)*

84 •• The string of a conical pendulum is 50.0 cm long and the mass of the bob is 0.25 kg. (*a*) Find the angle between the string and the horizontal when the tension in the string is six times the weight of the bob. (*b*) Under those conditions, what is the period of the pendulum?

85 •• A 100-g coin sits on a horizontally rotating turntable. The turntable makes exactly 1.00 revolution each second. The coin is located 10 cm from the axis of rotation of the turntable. (*a*) What is the frictional force acting on the coin? (*b*) If the coin slides off the turntable when it is located more than 16.0 cm from the axis of rotation, what is the coefficient of static friction between the coin and the turntable?

86 •• A 0.25-kg tether ball is attached to a vertical pole by a 1.2-m cord. Assume the radius of the ball is negligible. If the ball moves in a horizontal circle with the cord making an angle of 20° with the vertical, (*a*) what is the tension in the cord? (*b*) What is the speed of the ball?

87 ••• A small bead with a mass of 100 g (Figure 5-75) slides without friction along a semicircular wire with a radius of 10 cm that rotates about a vertical axis at a rate of 2.0 revolutions per second. Find the value of θ for which the bead will remain stationary relative to the rotating wire.

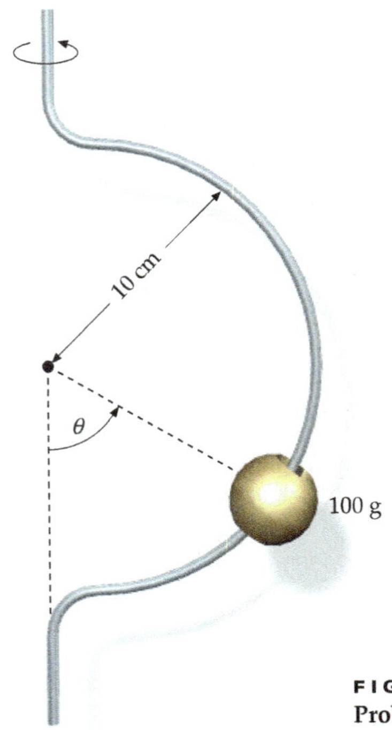

FIGURE 5-75 Problem 87

CENTRIPETAL FORCE

88 • A car speeds along the curved exit ramp of a freeway. The radius of the curve is 80.0 m. A 70.0-kg passenger holds the armrest of the car door with a 220-N force in order to keep from sliding across the front seat of the car. (Assume the exit ramp is not banked and ignore friction with the car seat.) What is the car's speed?

89 • The radius of curvature of the track at the top of a loop-the-loop on a roller-coaster ride is 12.0 m. At the top of the loop, the force that the seat exerts on a passenger of mass m is $0.40mg$. How fast is the roller-coaster car moving as it moves through the highest point of the loop. SSM

90 •• **ENGINEERING APPLICATION** On a runway of a decommissioned airport, a 2000-kg car travels at a constant speed of 100 km/h. At 100 km/h the air drag on the car is 500 N. Assume that rolling friction is negligible. (*a*) What is the force of static friction exerted on the car by the runway surface, and what is the minimum coefficient of static friction necessary for the car to sustain this speed? (*b*) The car continues to travel at 100 km/h, but now along a path with radius of curvature r. For what value of r will the angle between the static friction force vector and the velocity vector equal 45.0°, and for what value of r will it equal 88.0°? What is the minimum coefficient of static friction necessary for the car to hold this last radius of curvature without skidding?

91 •• Suppose you ride a bicycle in a 20-m-radius circle on a horizontal surface. The resultant force exerted by the surface on the bicycle (normal force plus frictional force) makes an angle of 15° with the vertical. (*a*) What is your speed? (*b*) If the frictional force on the bicycle is half its maximum possible value, what is the coefficient of static friction?

92 •• An airplane is flying in a horizontal circle at a speed of 480 km/h. The plane is banked for this turn, its wings tilted at an angle of 40° from the horizontal (Figure 5-76). Assume that a lift force acting perpendicular to the wings acts on the aircraft as it moves through the air. What is the radius of the circle in which the plane is flying?

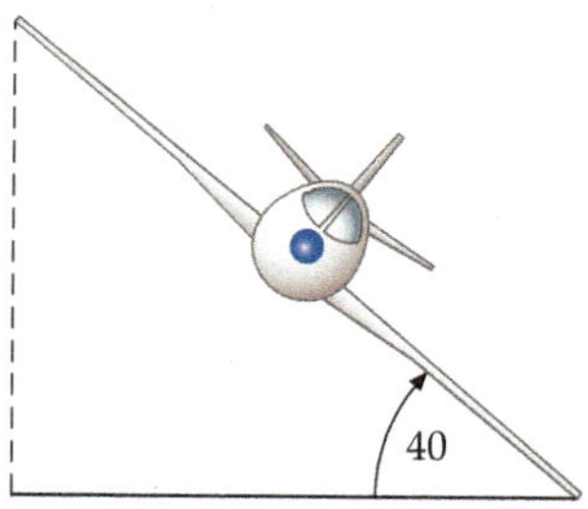

FIGURE 5-76 Problem 92

93 •• An automobile club plans to race a 750-kg car at the local racetrack. The car needs to be able to travel around several 160-m-radius curves at 90 km/h. What should the banking angle of the curves be so that the force of the pavement on the tires of the car is in the normal direction? *Hint: What does this requirement tell you about the frictional force?*

94 •• A curve of radius 150 m is banked at an angle of 10°. An 800-kg car negotiates the curve at 85 km/h without skidding. Neglect the effects of air drag and rolling friction. Find (*a*) the normal force exerted by the pavement on the tires, (*b*) the frictional force exerted by the pavement on the tires, (*c*) the minimum coefficient of static friction between the pavement and the tires.

95 •• On another occasion, the car in Problem 94 negotiates the curve at 38 km/h. Neglect the effects of air drag and rolling friction. Find (*a*) the normal force exerted on the tires by the pavement, and (*b*) the frictional force exerted on the tires by the pavement.

96 ••• **ENGINEERING APPLICATION** As a civil engineering intern during one of your summers in college, you are asked to design a curved section of roadway that meets the following conditions: When ice is on the road, and the coefficient of static friction between the road and rubber is 0.080, a car at rest must not slide into the ditch and a car traveling less than 60 km/h must not skid to the outside of the curve. Neglect the effects of air drag and rolling friction. What is the minimum radius of curvature of the curve and at what angle should the road be banked?

97 ••• **ENGINEERING APPLICATION** A curve of radius 30 m is banked so that a 950-kg car traveling at 40.0 km/h can round it even if the road is so icy that the coefficient of static friction is approximately zero. You are commissioned to tell the local police the range of speeds at which a car can travel around this curve without skidding. Neglect the effects of air drag and rolling friction. If the coefficient of static friction between the road and the tires is 0.300, what is the range of speeds you tell them?

*NUMERICAL INTEGRATION: EULER'S METHOD

*** 98** •• **SPREADSHEET, APPROXIMATION** You are riding in a hovering hot air balloon when you throw a baseball straight down with an initial speed of 35.0 km/h. The baseball falls with a terminal speed of 150 km/h. Assuming air drag is proportional to the speed squared, use Euler's method (**spreadsheet**) to estimate the speed of the ball after 10.0 s. What is the uncertainty in this estimate? You drop a second baseball, this one is released from rest. How long does it take for it to reach 99 percent of its terminal speed? How far does it fall during this time?

*** 99** •• **SPREADSHEET, APPROXIMATION** You throw a baseball straight up with an initial speed of 150 km/h. The ball's terminal speed when falling is also 150 km/h. (*a*) Use Euler's method (**spreadsheet**) to estimate its height 3.50 s after release. (*b*) What is the maximum height it reaches? (*c*) How long after release does it reach its maximum height? (*d*) How much later does it return to the ground? (*e*) Is the time the ball spends on the way up less than, the same as, or greater than the time it spends on the way down? SSM

*** 100** ••• **SPREADSHEET, APPROXIMATION** A 0.80-kg block on a horizontal frictionless surface is held against a massless spring, compressing it 30 cm. The force constant of the spring is 50 N/m. The block is released and the spring pushes it 30 cm. Use Euler's method (**spreadsheet**) with $\Delta t = 0.0050$ s to estimate the time it takes for the spring to push the block the 30 cm. How fast is the block moving at this time? What is the uncertainty in this speed?

FINDING THE CENTER OF MASS

101 • Three point masses of 2.0 kg each are located on the x axis. One is at the origin, another at $x = 0.20$ m, and another at $x = 0.50$ m. Find the center of mass of the system.

102 • On a weekend archeological dig, you discover an old club-ax that consists of a symmetrical 8.0-kg stone attached to the end of a uniform 2.5-kg stick. You measure the dimensions of the club-ax as shown in Figure 5-77. How far is the center of mass of the club-ax from the handle end of the club-ax?

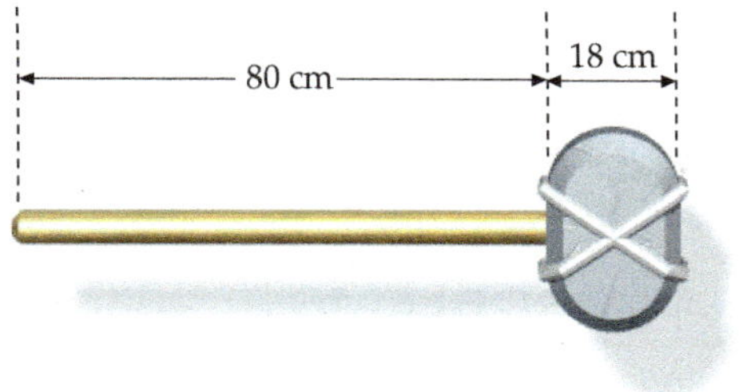

FIGURE 5-77 Problem 102

103 • Three balls A, B, and C, with masses of 3.0 kg, 1.0 kg, and 1.0 kg, respectively, are connected by massless rods, as shown in Figure 5-78. What are the coordinates of the center of mass of this system?

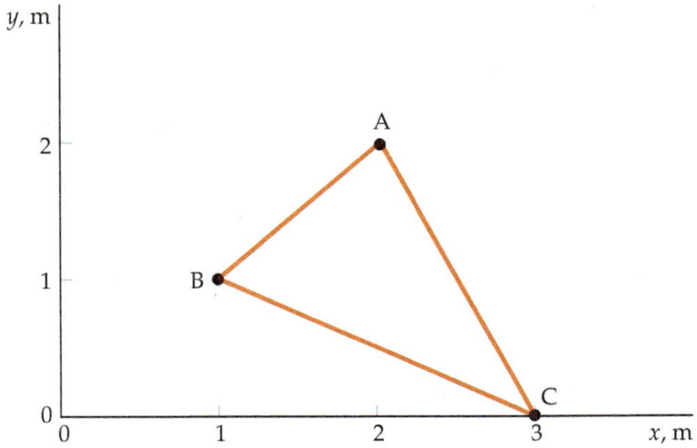

FIGURE 5-78 Problems 103 and 115

104 • By symmetry, locate the center of mass of a uniform sheet in the shape of an equilateral triangle with edges of length a. The triangle has one vertex on the y axis and the others at $(-a/2, 0)$ and $(+a/2, 0)$.

105 •• Find the center of mass of the uniform sheet of plywood in Figure 5-79. Consider this as a system of effectively two sheets, letting one have a "negative mass" to account for the cutout. Thus, one is a square sheet of 3.0-m edge length and mass m_1 and the second is a rectangular sheet measuring 1.0 m × 2.0 m with a mass of $-m_2$. Let the coordinate origin be at the lower left corner of the sheet. SSM

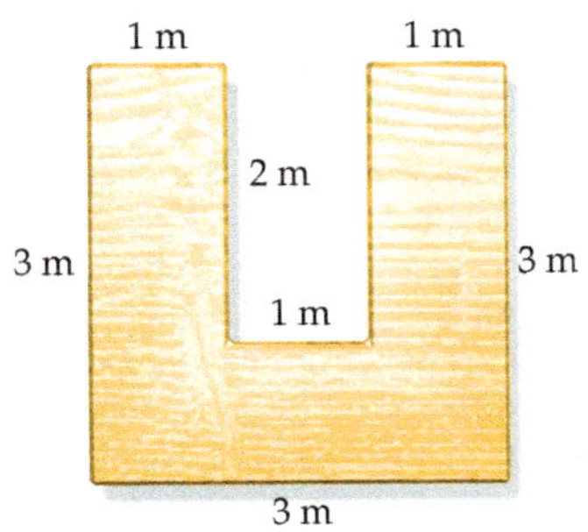

FIGURE 5-79 Problem 105

106 •• A can in the shape of a symmetrical cylinder with mass M and height H is filled with water. The initial mass of the water is M, the same mass as the can. A small hole is punched in the bottom of the can, and the water drains out. (*a*) If the height of the water in the can is x, what is the height of the center of mass of the can plus the water remaining in the can? (*b*) What is the minimum height of the center of mass as the water drains out?

107 •• Two identical thin uniform rods each of length L are glued together at the ends so that the angle at the joint is 90°. Determine the location of the center of mass (in terms of L) of this configuration relative to the origin taken to be at the joint. *Hint: You do not need the mass of the rods, but you should start by assuming a mass m and see that it cancels out.* SSM

108 ••• Repeat the analysis of Problem 107 with a general angle θ at the joint instead of 90°. Does your answer agree with the specific 90°-angle answer in Problem 107 if you set θ equal to 90°? Does your answer give plausible results for angles of zero and 180°?

* 109 •• **FINDING THE CENTER OF MASS BY INTEGRATION** Show that the center of mass of a uniform semicircular disk of radius R is at a point $4R/(3\pi)$ from the center of the circle.

* 110 •• Find the location of the center of mass of a nonuniform rod 0.40 m in length if its density varies linearly from 1.00 g/cm at one end to 5.00 g/cm at the other end. Specify the center-of-mass location relative to the less-massive end of the rod.

* 111 ••• You have a thin uniform wire bent into part of a circle that is described by a radius R and angle θ_m (see Figure 5-80). Show that the location of its center of mass is on the x axis and located a distance $x_{cm} = (R \sin\theta_m)/\theta_m$, where θ_m is expressed in radians. Check your answer by showing that this answer gives the physically expected limit for $\theta_m = 180°$. Verify that your answer gives you the result in the text (in the subsection Finding the Center of Mass by Integration) for the special case of $\theta_m = 90°$.

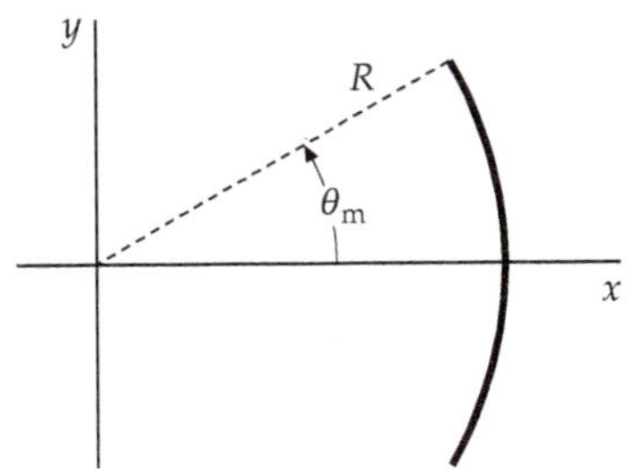

FIGURE 5-80 Problem 111

* 112 ••• A long, thin wire of length L has a linear mass density given by $A - Bx$, where A and B are positive constants and x is the distance from the more massive end. (*a*) A condition for this problem to be realistic is that $A > BL$. Explain why. (*b*) Determine x_{cm} in terms of L, A, and B. Does your answer makes sense if $B = 0$? Explain.

MOTION OF THE CENTER OF MASS

113 • Two 3.0-kg particles have velocities $\vec{v}_1 = (2.0\ \text{m/s})\hat{i} + (3.0\ \text{m/s})\hat{j}$ and $\vec{v}_2 = (4.0\ \text{m/s})\hat{i} - (6.0\ \text{m/s})\hat{j}$. Find the velocity of the center of mass of the system. SSM

114 • A 1500-kg car is moving westward with a speed of 20.0 m/s, and a 3000-kg truck is traveling east with a speed of 16.0 m/s. Find the velocity of the center of mass of the car–truck system.

115 • A force $\vec{F} = 12\,\text{N}\ \hat{i}$ is applied to the 3.0-kg ball in Figure 5-78 in Problem 103. (No forces act on the other two balls.) What is the acceleration of the center of mass of the three-ball system?

116 •• A block of mass m is attached to a string and suspended inside an otherwise empty box of mass M. The box rests on a scale that measures the system's weight. (*a*) If the string breaks, does the reading on the scale change? Explain your reasoning. (*b*) Assume that the string breaks and the mass m falls with constant acceleration g. Find the magnitude and direction of the acceleration of the center of mass of the box–block system. (*c*) Using the result from (*b*), determine the reading on the scale while m is in free-fall.

117 •• The bottom end of a massless, vertical spring of force constant k rests on a scale and the top end is attached to a massless cup, as in Figure 5-81. Place a ball of mass m_b gently into the cup and ease it down into an equilibrium position where it sits at rest in the cup. (*a*) Draw the separate free-body diagrams for the ball and the spring. (*b*) Show that in this situation, the spring compression d is given by $d = m_b g/k$. (*c*) What is the scale reading under these conditions? SSM

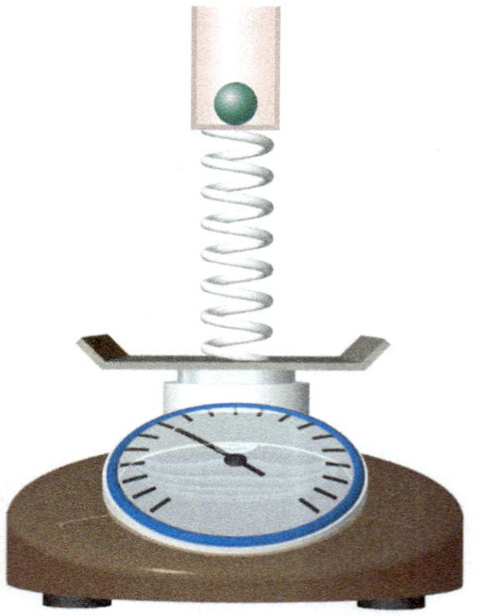

FIGURE 5-81 Problem 117

118 ••• In the Atwood's machine in Figure 5-82 the string passes over a fixed cylinder of mass m_c. The cylinder does not rotate. Instead, the string slides on its frictionless surface. (*a*) Find the acceleration of the center of mass of the two-block–cylinder-string system. (*b*) Use Newton's second law for systems to find the force F exerted by the support. (*c*) Find the tension T in the string connecting the blocks and show that $F = m_c g + 2T$.

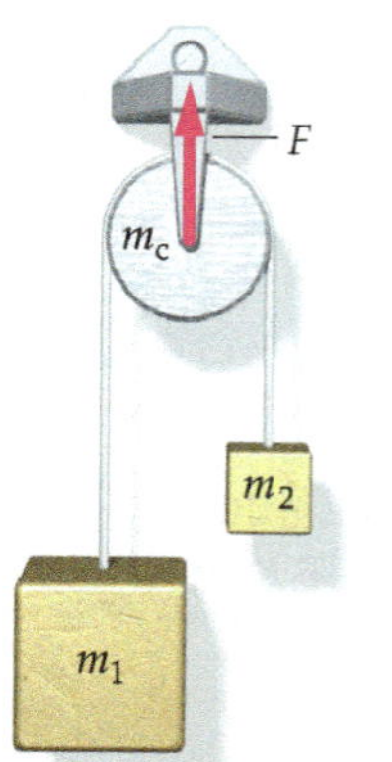

FIGURE 5-82 Problem 118

119 ••• Starting with the equilibrium situation in Problem 117, the whole system (scale, spring, cup, and ball) is now subjected to an upward acceleration of magnitude a (for example, in an elevator). Repeat the free-body diagrams and calculations in Problem 117?

GENERAL PROBLEMS

120 • In designing your new house in California, you are prepared for it to withstand a maximum horizontal acceleration of $0.50g$. What is the minimum coefficient of static friction between the floor and your prized Tuscan vase so that the vase does not slip on the floor under these conditions?

121 • A 4.5-kg block slides down an inclined plane that makes an angle of 28° with the horizontal. Starting from rest, the block slides a distance of 2.4 m in 5.2 s. Find the coefficient of kinetic friction between the block and plane.

122 •• You plan to fly a model airplane of mass 0.400 kg that is attached to a horizontal string. The plane will travel in a horizontal circle of radius 5.70 m. (Assume the weight of the plane is balanced by the upward "lift" force of the air on the wings of the plane.) The plane will make 1.20 revolutions every 4.00 s. (*a*) Find the speed at which you must fly the plane. (*b*) Find the force exerted on your hand as you hold the string (assume the string is massless).

123 •• **CONTEXT-RICH** Your moving company is to load a crate of books on a truck with the help of some planks that slope upward at 30°. The mass of the crate is 100 kg, and the coefficient of sliding friction between it and the planks is 0.500. You and your employees push *horizontally* with a combined net force $\vec{F}$. Once the crate has started to move, how large must F be in order to keep the crate moving at constant speed?

124 •• Three forces act on an object in static equilibrium (Figure 5-83). (*a*) If F_1, F_2, and F_3 represent the magnitudes of the forces acting on the object, show that $F_1/\sin\theta_{23} = F_2/\sin\theta_{31} = F_3/\sin\theta_{12}$. (*b*) Show that $F_1^2 = F_2^2 + F_3^2 + 2F_2F_3\cos\theta_{23}$.

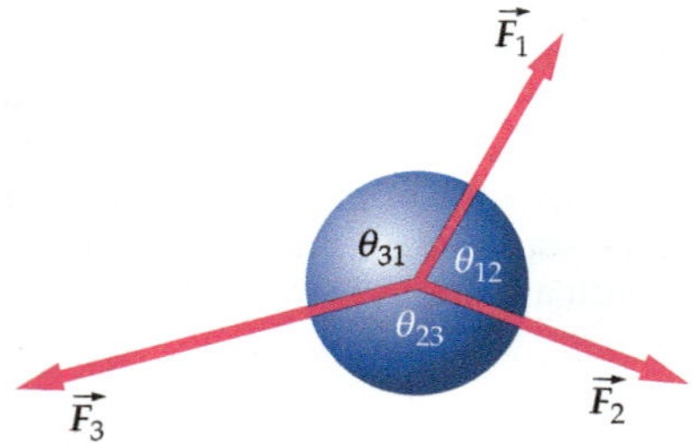

FIGURE 5-83 Problem 124

125 •• In a carnival ride, you sit on a seat in a compartment that rotates with constant speed in a vertical circle of radius 5.0 m. The ride is designed so your head always points toward the center of the circle. (*a*) If the ride completes one full circle in 2.0 s, find the direction and magnitude of your acceleration. (*b*) Find the slowest rate of rotation (in other words, the longest time T_m to complete one full circle) if the seat belt is to exert no force on you at the top of the ride.

126 •• A flat-topped toy cart moves on frictionless wheels, pulled by a rope under tension T. The mass of the cart is m_1. A load of mass m_2 rests on top of the cart with the coefficient of static friction μ_s between the cart and the load. The cart is pulled up a ramp that is inclined at angle θ above the horizontal. The rope is parallel to the ramp. What is the maximum tension T that can be applied without causing the load to slip?

127 ••• A sled weighing 200 N that is held in place by static friction, rests on a 15° incline. The coefficient of static friction between the sled and the incline is 0.50. (*a*) What is the magnitude of the normal force on the sled? (*b*) What is the magnitude of the static frictional force on the sled? (*c*) The sled is now pulled up the incline (Figure 5-84) at constant speed by a child walking up the incline ahead of the sled. The child weighs 500 N and pulls on the rope with a constant force of 100 N. The rope makes an angle of 30° with the incline and has negligible mass. What is the magnitude of the kinetic frictional force on the sled? (*d*) What is the coefficient of kinetic friction between the sled and the incline? (*e*) What is the magnitude of the force exerted on the child by the incline?

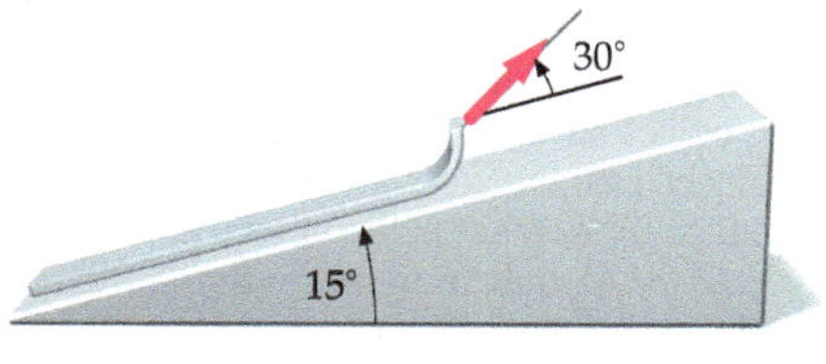

FIGURE 5-84 Problem 127

128 •• **ENGINEERING APPLICATION** In 1976, Gerard O'Neill proposed that large space stations be built for human habitation in orbit around Earth and the moon. Because prolonged free-fall has adverse medical effects, he proposed making the stations in the form of long cylinders and spinning them around the cylinder axis to provide the inhabitants with the sensation of gravity. One such O'Neill colony is to be built 5.0 miles long, with a diameter of 0.60 mi. A worker on the inside of the colony would experience a sense of "gravity," because he would be in an accelerated frame of reference due to the rotation. (*a*) Show that the "acceleration of gravity" experienced by the worker in the O'Neill colony is equal to his centripetal acceleration. *Hint: Consider someone "looking in" from outside the colony.* (*b*) If we assume that the space station is composed of several decks that are at varying distances (radii) from the axis of rotation, show that the "acceleration of gravity" becomes weaker the closer the worker gets to the axis. (*c*) How many revolutions per minute would this space station have to make to give an "acceleration of gravity" of 9.8 m/s^2 at the outermost edge of the station?

129 •• A child of mass m slides down a slide inclined at 30° in time t_1. The coefficient of kinetic friction between her and the slide is μ_k. She finds that if she sits on a small sled (also of mass m) with frictionless runners, she slides down the same slide in time $\frac{1}{2}t_1$. Find μ_k.

130 ••• The position of a particle of mass $m = 0.80$ kg as a function of time is given by $\vec{r} = x\hat{i} + y\hat{j} = (R\sin\omega t)\,\hat{i} + (R\cos\omega t)\,\hat{j}$, where $R = 4.0$ m and $\omega = 2\pi\ \mathrm{s}^{-1}$. (*a*) Show that the path of this particle is a circle of radius R, with its center at the origin of the xy plane. (*b*) Compute the velocity vector. Show that $v_x/v_y = -y/x$. (*c*) Compute the acceleration vector and show that it is directed toward the origin and has the magnitude v^2/R. (*d*) Find the magnitude and direction of the net force acting on the particle.

131 ••• **MULTISTEP** You are on an amusement park ride with your back against the wall of a spinning vertical cylinder. The floor falls away and you are held up by static friction. Assume your mass is 75 kg. (*a*) Draw a free-body diagram of yourself. (*b*) Use this diagram with Newton's laws to determine the force of friction on you. (*c*) If the radius of the cylinder is 4.0 m and the coefficient of static friction between you and the wall is 0.55. What is the minimum number of revolutions per minute necessary to prevent you from sliding down the wall? Does this answer hold only for you? Will other, more massive, patrons fall downward? Explain.

132 ••• An block of mass m_1 is on a horizontal table. The block is attached to a 2.5-kg block (m_2) by a light string that passes over a pulley at the edge of the table. The block of mass m_2 dangles 1.5 m above the ground (Figure 5-85). The string that connects them passes over a frictionless, massless pulley. This system is released from rest at $t = 0$ and the 2.5-kg block strikes the ground at $t = 0.82$ s. The system is now placed in its initial configuration and a 1.2-kg block is placed on top of the block of mass m_1. Released from rest, the 2.5-kg block now strikes the ground 1.3 s later. Determine the mass m_1 and the coefficient of kinetic friction between the block whose mass is m_1 and the table.

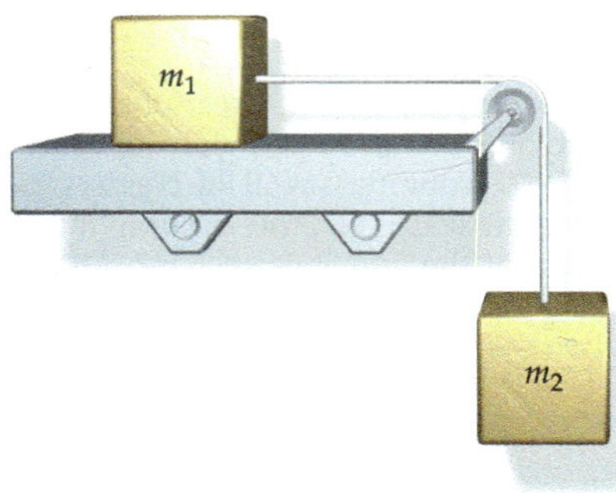

FIGURE 5-85 Problem 132

133 ••• Sally claims flying squirrels do not really fly; they jump and use the folds of skin that connect their forelegs and their back legs like a parachute to allow them to glide from tree to tree. Liz decides to test Sally's hypothesis by calculating the terminal speed of a falling outstretched flying squirrel. If the constant b in the drag force is proportional to the area of the object facing the air flow, use the results of Example 5-12 and some assumptions about the size of the squirrel to estimate its terminal (downward) speed. Is Sally's claim supported by Liz's calculation?

134 •• **BIOLOGICAL APPLICATION** After a parachutist jumps from an airplane (but before he pulls the rip cord to open his parachute), a downward speed of up to 180 km/h can be reached. When the parachute is finally opened, the drag force is increased by about a factor of 10, and this can create a large jolt on the jumper. Suppose this jumper falls at 180 km/h before opening his chute. (*a*) Determine the parachutist's acceleration when the chute is just opened, assuming his mass is 60 kg. (*b*) If rapid accelerations greater than 5.0g can harm the structure of the human body, is this a safe practice?

135 • Find the location of the center of mass of the Earth–moon system relative to the center of Earth. Is it inside or outside the surface of Earth?

136 •• A circular plate of radius R has a circular hole of radius $R/2$ cut out of it (Figure 5-86). Find the center of mass of the plate after the hole has been cut. *Hint: The plate can be modeled as two disks superimposed, with the hole modeled as a disk negative mass.*

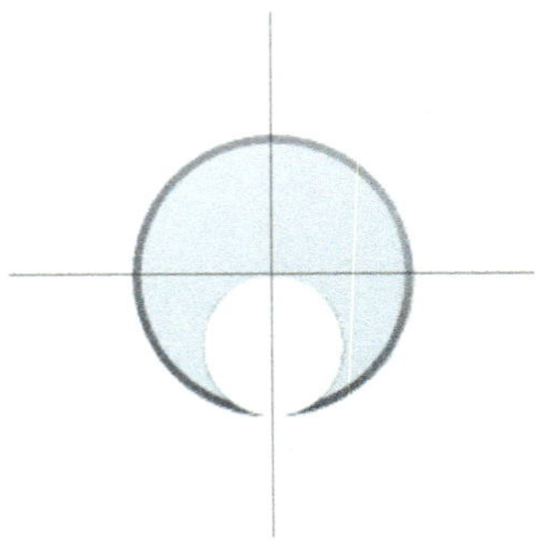

FIGURE 5-86 Problem 136

137 •• An unbalanced baton consists of a 50-cm-long uniform rod of mass 200 g. At one end there is a 10-cm-diameter uniform solid sphere of mass 500 g, and at the other end there is a 8.0-cm-diameter uniform solid sphere of mass 750 g. (The center-to-center distance between the spheres is 59 cm.) (*a*) Where, relative to the center of the light sphere, is the center of mass of this baton? (*b*) If this baton is tossed straight up (but spinning) so that its initial center of mass speed is 10.0 m/s, what is the velocity of the center of mass 1.5 s later? (*c*) What is the net external force on the baton while in the air? (*d*) What is the acceleration of the baton's center of mass 1.5 s following its release?

138 •• You are standing at the very rear of a 6.0-m-long, 120-kg raft that is at rest in a lake with its prow only 0.50 m from the end of the pier (Figure 5-87). Your mass is 60 kg. Neglect frictional forces between the raft and the water. (*a*) How far from the end of the pier is the center of mass of the you–raft system? (*b*) You walk to the front of the raft and then stop. How far from the end of the pier is the center of mass now? (*c*) When you are at the front of the raft, how far are you from the end of the pier?

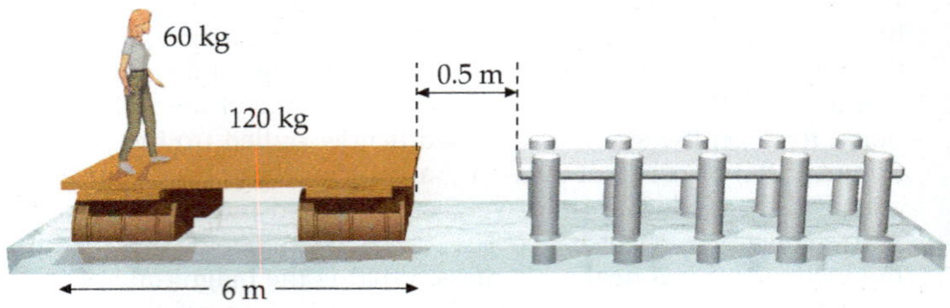

FIGURE 5-87 Problem 138

139 •• An Atwood's machine that has a frictionless massless pulley and massless strings has a 2.00-kg object hanging from one side and 4.00-kg object hanging from the other side. (*a*) What is the speed of each object 1.50 s after they are simultaneously released from rest? (*b*) At that time, what is the velocity of the center of mass of the two objects? (*c*) At that moment, what is the acceleration of the center of mass of the two objects?

CHAPTER 6

THE SNOW MELTS UNDER THE SKIES DUE TO THE KINETIC FRICTION BETWEEN THE SKIS AND THE SNOW. THE SKIER IS GLIDING DOWN THE MOUNTAIN ON A THIN LAYER OF LIQUID WATER. *(Courtesy of Rossignol Ski Company.)*

? How does the shape of the hill or the length of the path affect a skier's final speed at the finish line? (See Example 6-12.)

Work and Kinetic Energy

Thus far, we have analyzed motion by using concepts such as position, velocity, acceleration, and force. However, some types of motion are difficult to describe using Newton's laws directly. (A speed skier sliding down a curved slope is an example of this type of motion.) In this chapter and Chapter 7, we look at alternative methods for analyzing motion that involve two central concepts in science: energy and work. Unlike force, which is a vector physical quantity, energy and work are scalar physical quantities associated with particles and with systems of particles. As you will see, these new concepts provide powerful methods for solving a wide class of problems.

In this chapter, we explore the concept of work and how work is related to kinetic energy—the energy associated with the motion of objects. We also discuss the related concepts of power and center-of-mass work.

6-1 WORK DONE BY A CONSTANT FORCE

You may be used to thinking of work as anything that requires physical or mental exertion, such as studying for an exam, carrying a backpack, or riding a bike. But in physics, work is the transfer of energy by a force. If you stretch a spring by pulling on it with your hand (Figure 6-1), energy is transferred from you to the spring, and the energy transferred from you to the spring is equal to the work done by the force of your hand on the spring. The energy transferred to the spring can be evidenced if you let go of the spring and watch it rapidly contract and vibrate.

Work is a scalar quantity that can be positive, negative, or zero. The work done by object A on object B is positive if energy is transferred from A to B, and is negative if energy is transferred from B to A. If no energy is transferred, the work done is zero. In the case of you stretching a spring, the work done by you on the spring is positive because energy is transferred from you to the spring. However, suppose you move your hand so the spring slowly contracts to its unstressed state. During the contraction the spring loses energy—energy is transferred from the spring to you—and the work you do on the spring is negative.

It is commonly said that work is force times distance. Unfortunately, the statement "work is force times distance" is misleadingly simple. Work is done on an object by a force when the point of application of the force moves through a displacement. For a constant force, the work done equals the force component in the direction of the displacement times the magnitude of the displacement. For example, suppose you push a box along the ground with a constant horizontal force $\vec{F}$ in the direction of displacement $\Delta x \hat{i}$ (Figure 6-2*a*). Because the force acts on the box in the same direction as the displacement, the work W done by the force on the box is

$$W = F|\Delta x|$$

Now suppose you pull on a string attached to the box, such that force acts at an angle to the displacement, as shown in Figure 6-2*b*. In this case, the work done on the box by the force is given by the force component *in the direction* of the displacement times the magnitude of the displacement:

$$W = F_x \Delta x = F\cos\theta\,|\Delta x| \qquad 6\text{-}1$$

WORK BY A CONSTANT FORCE

where F is the magnitude of the constant force, $|\Delta x|$ is the magnitude of the displacement of the point of application of the force, and θ is the angle between the directions of the force and displacement vectors. The displacement of the point of application of the force is identical to the displacement of any other point on the box because the box is rigid and moves without rotating. If you raise or lower a box by exerting a force $\vec{F}$ on it, you are doing work on the box. Let up be the positive y direction and let $\Delta y \hat{j}$ be the displacement of the box. The work done by you on the box is positive if Δy and F_y have the same signs and negative if they have opposite signs. But if you are simply holding the box in a fixed position then, according to the definition of work, you are not doing work *on the box* because the Δy is zero (Figure 6-3). In this case, the work you do on the box is zero, even though you are applying a force.

The SI unit of work is the **joule** (J), which equals the product of a newton and a meter:

$$1\,\text{J} = 1\,\text{N}\cdot\text{m} \qquad 6\text{-}2$$

In the U.S. customary system, the unit of work is the foot-pound: 1 ft·lb = 1.356 J. Another convenient unit of work in atomic and nuclear physics is the **electron volt** (eV):

$$1\,\text{eV} = 1.602 \times 10^{-19}\,\text{J} \qquad 6\text{-}3$$

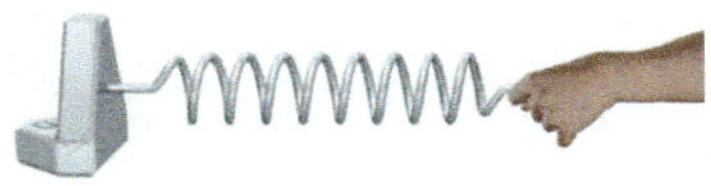

FIGURE 6-1 Energy is transferred from the person to the spring as the spring lengthens. The energy transferred is equal to the work done by the person on the spring.

CONCEPT CHECK 6-1

For the contraction of the spring described immediately before this concept check, is the work done by the spring on the person positive or negative?

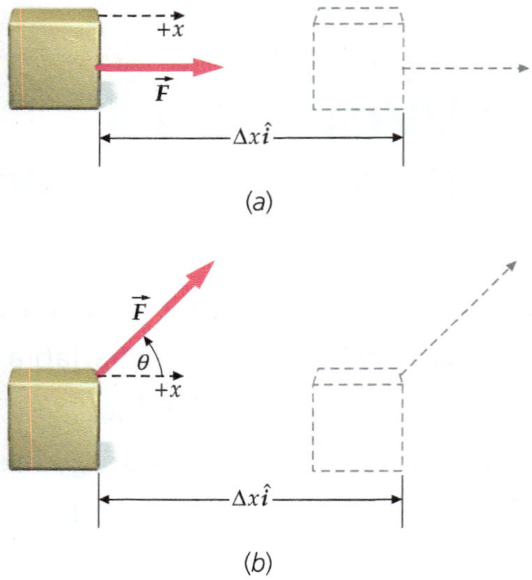

FIGURE 6-2

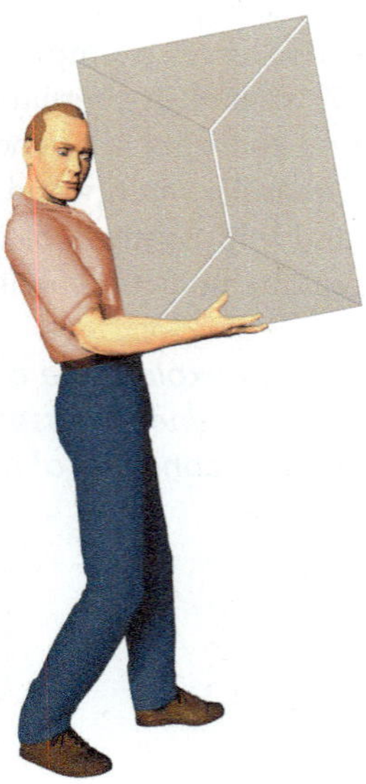

FIGURE 6-3

Commonly used multiples of eV are keV (10^3 eV) and MeV (10^6 eV). The work required to remove an electron from an atom is of the order of a few eV, whereas the work needed to remove a proton or a neutron from an atomic nucleus is of the order of several MeV.

> **PRACTICE PROBLEM 6-1**
> A force of 12 N is exerted on a box at an angle of $\theta = 20°$, as in Figure 6-2*b*. How much work is done by the force on the box as the box moves along the table a distance of 3.0 m?

If there are several forces that do work on *a system*, the total work is found by computing the work done by each force and adding each individual work together.

$$W_{\text{total}} = F_{1x}\,\Delta x_1 + F_{2x}\,\Delta x_2 + F_{3x}\,\Delta x_3 + \cdots \qquad 6\text{-}4$$

We model the system as a *particle* if the system moves so all the parts of the system undergo identical displacements. When several forces do work on such a *particle*, the displacements of the points of application of these forces are identical. Let the displacement of the point of application of any one of the forces be Δx. Then

$$W_{\text{total}} = F_{1x}\,\Delta x + F_{2x}\,\Delta x + \cdots = (F_{1x} + F_{2x} + \cdots)\Delta x = F_{\text{net}\,x}\,\Delta x \qquad 6\text{-}5$$

For a particle constrained to move along the x axis, the net force has only an x component. That is, $\vec{F}_{\text{net}} = F_{\text{net}\,x}\hat{i}$. Thus, for a particle, the x component of the net force times the displacement of any part of the object is equal to the total work done on the object.

Example 6-1 Loading with a Crane

A 3000-kg truck is to be loaded onto a ship by a crane that exerts an upward force of 31 kN on the truck. This force, which is strong enough to overcome the gravitational force and keep the truck moving upward, is applied over a distance of 2.0 m. Find (*a*) the work done on the truck by the crane, (*b*) the work done on the truck by gravity, and (*c*) the net work done on the truck.

PICTURE In Parts (*a*) and (*b*), the force acting on the truck is constant and the displacement is in a straight line, so we can use Equation 6-1, choosing the $+y$ direction as the direction of the displacement:

SOLVE

(*a*) 1. Sketch the truck at its initial and final positions, and choose the $+y$ direction to be the direction of the displacement (Figure 6-4):

2. Calculate the work done by the applied force:

$$W_{\text{app}} = F_{\text{app}\,y}\,\Delta y = (31\text{ kN})(2.0\text{ m}) = \boxed{62\text{ kJ}}$$

(*b*) Calculate the work done by the force of gravity: (*Note*: The vector $\vec{g}$ is directed downward and the $+y$ direction is upward. Consequently, $g_y = g\cos 180° = -g$.)

$$W_g = mg_y\,\Delta y = (3000\text{ kg})(-9.81\text{ N/kg})(2.0\text{ m}) = \boxed{-59\text{ kJ}}$$

(*c*) The net work done on the truck is the sum of the work done by each force:

$$W_{net} = W_{\text{app}\,y} + W_g = 62\text{ kJ} + (-59\text{ kJ}) = \boxed{3\text{ kJ}}$$

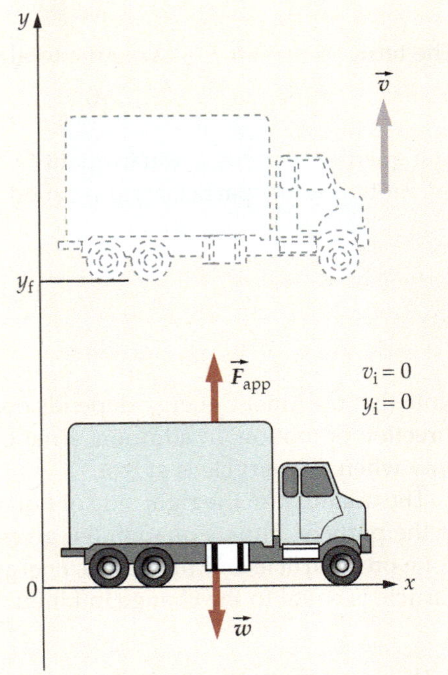

FIGURE 6-4

CHECK In Part (*a*), the force is applied in the same direction as the displacement, so we expect the work done on the truck to be positive. In Part (*b*), the force is applied in a direction opposite the displacement, so we expect the work done on the truck to be negative. Our results match these expectations.

TAKING IT FURTHER In Part (*c*), we also could have found the total work by first calculating the net force on the truck and then using Equation 6-5.

THE WORK–KINETIC-ENERGY THEOREM

Energy is one of the most important unifying concepts in science. All physical processes involve energy. The **energy** of a system is a measure of its ability to do work.

Different terms are used to describe the energy associated with different conditions or states. **Kinetic energy** is energy associated with motion. *Potential energy* is energy associated with the configuration of a system, such as the separation distance between two objects that attract each other. *Thermal energy* is associated with the random motion of the atoms, molecules, or ions within a system and is closely connected with the temperature of the system. In this chapter, we focus on kinetic energy. Potential energy and thermal energy are discussed in Chapter 7.

When forces do work on a particle, the result is a change in the energy associated with the motion of the particle—the kinetic energy. To evaluate the relationship between kinetic energy and work, let us look at what happens if a constant net force $\vec{F}_{net}$ acts on a particle of mass m that moves along the x axis. Applying Newton's second law, we see that

$$F_{net\,x} = ma_x$$

If the net force is constant, the acceleration is constant, and we can relate the displacement to the initial speed v_i and final speed v_f by using the constant-acceleration kinematic equation (Equation 2-16)

$$v_f^2 = v_i^2 + 2a_x\,\Delta x$$

Solving this for a_x gives

$$a_x = \frac{1}{2\Delta x}(v_f^2 - v_i^2)$$

Substituting for a_x in $F_{net\,x} = ma_x$ and then multiplying both sides by Δx, gives

$$F_{net\,x}\,\Delta x = \tfrac{1}{2}mv_f^2 - \tfrac{1}{2}mv_i^2$$

The term on the left $F_{net\,x}\,\Delta x$ is the total work done on the particle. Thus

$$W_{total} = \tfrac{1}{2}mv_f^2 - \tfrac{1}{2}mv_i^2 \qquad 6\text{-}6$$

The quantity $\frac{1}{2}mv^2$ is a scalar quantity that represents the energy associated with the motion of the particle, and is called the **kinetic energy** K of the particle:

$$K = \tfrac{1}{2}mv^2 \qquad 6\text{-}7$$

DEFINITION—KINETIC ENERGY

Note that the kinetic energy depends on only the particle's speed and mass, not its direction of motion. In addition, kinetic energy can never be negative, and is zero only when the particle is at rest.

The quantity on the right side of Equation 6-6 is the change in the kinetic energy of the particle. Thus, Equation 6-6 gives us a relationship between the total work done on a particle and the kinetic energy of the particle. The total work done on a particle is equal to the change in kinetic energy of the particle:

$$W_{total} = \Delta K \qquad 6\text{-}8$$

WORK–KINETIC-ENERGY THEOREM

This result is known as the **work–kinetic-energy theorem.** This theorem tells us that when W_{total} is positive, the kinetic energy increases, which means the particle is moving faster at the end of the displacement than at the beginning. When W_{total} is negative, the kinetic energy decreases. When W_{total} is zero, the kinetic energy does not change, which means the particle's speed is unchanged.

Because total work on a particle is equal to its change in kinetic energy, we can see that the units of energy are the same as those of work. Three commonly used units of energy are the joule (J), the foot-pound (ft-lb), and the electron volt (eV).

The derivation of the work–kinetic-energy theorem presented here is valid only if the net force remains constant. However, as you will see later in this chapter, this theorem is valid even when the net force varies and the motion is not along a straight line.

! Note that kinetic energy depends on the *speed* of the particle, not the velocity. If the velocity changes direction, but not magnitude, the kinetic energy remains the same.

PROBLEM-SOLVING STRATEGY

Solving Problems Involving Work and Kinetic Energy

PICTURE The way you choose the $+y$ direction or $+x$ direction can help you to easily solve a problem that involves work and kinetic energy.

SOLVE

1. Draw the particle first at its initial position and second at its final position. For convenience, the object can be represented as a dot or a box. Label the initial and final positions of the object.
2. Put one or more coordinate axes on the drawing.
3. Draw arrows for the initial and final velocities, and label them appropriately.
4. On the initial-position drawing of the particle, place a vector for each force acting on it. Accompany each vector with a suitable label.
5. Calculate the total work done on the particle by the forces and equate this total to the change in the particle's kinetic energy.

CHECK Make sure you pay attention to negative signs during your calculations. For example, values for work done can be positive or negative, depending on the direction of the displacement relative to the direction of the force.

Example 6-2 Force on an Electron

In a television picture tube*, electrons are accelerated by an electron gun. The force that accelerates the electron is an electric force due to the electric field in the gun. An electron is accelerated from rest by an electron gun to a kinetic energy of 2.5 keV over a distance of 2.5 cm. Find the force on the electron, assuming it to be both constant and in the direction of the electron's motion.

PICTURE The electron can be modeled as a particle. Its initial and final kinetic energies are both given, and the electric force is the only force acting on it. Apply the work–kinetic-energy theorem and solve for the force.

* A television picture tube is a type of cathode-ray tube.

SOLVE

1. Make a drawing of the electron in its initial and final positions. Include the displacement, the initial and final speeds, and the force (Figure 6-5):

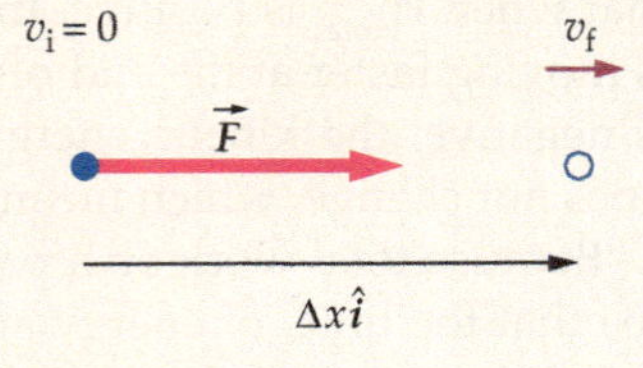

FIGURE 6-5

2. Set the work done equal to the change in kinetic energy:

$$W_{total} = \Delta K$$
$$F_x \Delta x = K_f - K_i$$

3. Solve for the force using the conversion factor $1.6 \times 10^{-19}\,\text{J} = 1.0\,\text{eV}$:

$$F_x = \frac{K_f - K_i}{\Delta x} = \frac{2500\,\text{eV} - 0}{0.025\,\text{m}} \times \frac{1.6 \times 10^{-19}\,\text{J}}{1.0\,\text{eV}}$$
$$= \boxed{1.6 \times 10^{-14}\,\text{N}}$$

CHECK The mass of an electron is only 9.1×10^{-31} kg. Thus, it is no surprise that such a small force would give it a large speed and thus a noticeable change in kinetic energy.

TAKING IT FURTHER (*a*) $1\,\text{J} = 1\,\text{N}\cdot\text{m}$, so $1\,\text{J/m} = 1\,\text{N}$. (*b*) 1 eV is the kinetic energy acquired by a particle of charge $-e$ (an electron, for example) when it is accelerated from the $-$ terminal to the $+$ terminal of a 1-V battery through a vacuum.

Example 6-3 A "Dogsled" Race

During your winter break you enter a "dogsled" race across a frozen lake. This is a race where each sled is pulled by a person, not by dogs. To get started you pull the sled (total mass 80 kg) with a force of 180 N at 40° above the horizontal. Find (*a*) the work you do, and (*b*) the final speed of the sled after it moves $\Delta x = 5.0$ m, assuming that it starts from rest and there is no friction.

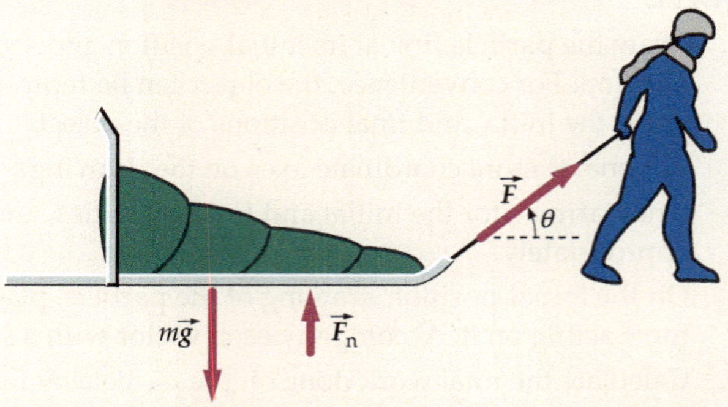

FIGURE 6-6

PICTURE The work done by you is $F_x \Delta x$, where we choose the direction of the displacement as the positive x direction. This is also the *total* work done on the sled because the other forces, mg and F_n, have no x components. The final speed of the sled can be found by applying the work–kinetic-energy theorem to the sled. Calculate the work done by each force on the sled (Figure 6-6) and equate the total work to the change in kinetic energy of the sled.

SOLVE

(*a*) 1. Sketch the sled both in its initial position and in its position after moving the 5.0 m. Draw the x axis in the direction of the motion (Figure 6-7).

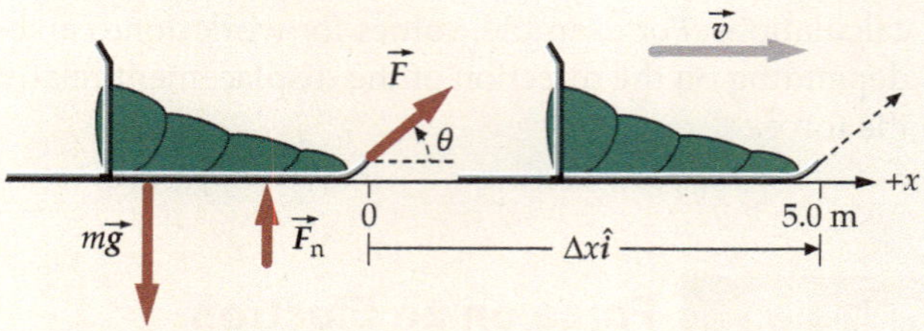

FIGURE 6-7

2. The work done by you on the sled is $F_x \Delta x$. This is the total work done on the sled. The other two forces each act perpendicular to the x direction (see Figure 6-7), so they do zero work:

$$W_{total} = W_{you} = F_x \Delta x = F \cos\theta\, \Delta x$$
$$= (180\,\text{N})(\cos 40°)(5.0\,\text{m}) = 689\,\text{J}$$
$$= \boxed{6.9 \times 10^2\,\text{J}}$$

(*b*) Apply the work–kinetic-energy theorem to the sled and solve for the final speed:

$$W_{total} = \tfrac{1}{2}mv_f^2 - \tfrac{1}{2}mv_i^2$$
$$v_f^2 = v_i^2 + \frac{2W_{total}}{m}$$
$$= 0 + \frac{2(689\,\text{J})}{80\,\text{kg}} = 17.2\,\text{m}^2/\text{s}^2$$
$$v_f = \sqrt{17.2\,\text{m}^2/\text{s}^2} = 4.151\,\text{m/s} = \boxed{4.2\,\text{m/s}}$$

CHECK In Part (*b*) we used that 1 J/kg = 1 m²/s². This is correct because

$$1\ \text{J/kg} = 1\ \text{N}\cdot\text{m/kg} = (1\ \text{kg}\cdot\text{m/s}^2)\cdot\text{m/kg} = 1\ \text{m}^2/\text{s}^2$$

TAKING IT FURTHER The square root of 17.2 is 4.147, which rounds off to 4.1. However, the correct answer to Part (*b*) is 4.2 m/s. It is correct because it is calculated by taking the square root of 17.235 999 970 178 (the value stored in my calculator after executing the calculation of v_f).

PRACTICE PROBLEM 6-2 What is the magnitude of the force you exert if the 80-kg sled starts with a speed of 2.0 m/s and its final speed is 4.5 m/s after you pull it through a distance of 5.0 m while keeping the angle at 40°?

6-2 WORK DONE BY A VARIABLE FORCE–STRAIGHT-LINE MOTION

See Math Tutorial *for more information on* Integrals

Many forces vary with position. For example, a stretched spring exerts a force proportional to the distance it is stretched. In addition, the gravitational force Earth exerts on a spaceship varies inversely with the square of the center-to-center distance between the two bodies. How can we calculate the work done by forces like these?

Figure 6-8 shows the plot of a *constant* force F_x as a function of position x. Notice that the work done by the force on a particle whose displacement is Δx is represented by the area under the force-versus-position curve—indicated by the shading in Figure 6-8. We can approximate a variable force by a series of essentially constant forces (Figure 6-9). For each small displacement interval Δx_i, the force is approximately constant. Therefore the work done is approximately equal to the *area* of the rectangle of height F_{xi} and width Δx_i. The work W done by a variable force is then equal to the sum of the areas of an increasingly large number of these rectangles in the limit that the width of each individual rectangle approaches zero:

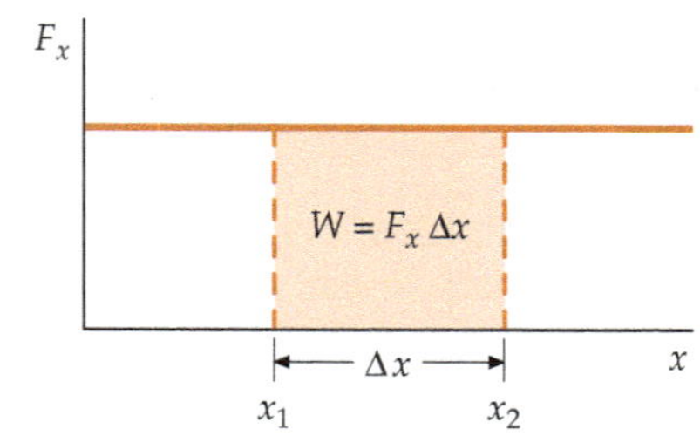

FIGURE 6-8 The work done by a constant force is represented graphically as the area under the F_x-versus-x curve.

$$W = \lim_{\Delta x_i \to 0} \sum_i F_{xi}\,\Delta x_i = \text{area under the } F_x\text{-versus-}x \text{ curve} \qquad 6\text{-}9$$

This limit is the integral of $F_x dx$ over the interval from x_1 to x_2. So the work done by a variable force F_x acting on a particle as it moves from x_1 to x_2 is

$$W = \int_{x_1}^{x_2} F_x dx = \text{area under the } F_x\text{-versus-}x \text{ curve} \qquad 6\text{-}10$$

WORK BY A VARIABLE FORCE—STRAIGHT-LINE MOTION

FIGURE 6-9 A variable force can be approximated by a series of constant forces over small intervals. The work done by the constant force in each interval is the area of the rectangle beneath the force curve. The sum of these rectangular areas is the sum of the work done by the set of constant forces that approximates the varying force. In the limit of infinitesimally small Δx_i, the sum of the areas of the rectangles equals the area under the complete force curve.

If the force plotted in Figure 6-9 is the net force on the particle, then each term $F_{xi}\,\Delta x_i$ in the sum in Equation 6-9 represents the total work done on the particle by a constant force as the particle undergoes the incremental displacement Δx_i. Thus, $F_{xi}\,\Delta x_i$ is equal to the change in kinetic energy ΔK_i of the particle during incremental displacement Δx_i (see Equation 6-8). In addition, the total change in the kinetic energy ΔK of the particle during the total displacement is equal to the sum of the incremental changes in kinetic energy. It follows that the total work W_{total} done on the particle for the total displacement equals the change in kinetic energy for the total displacement. Therefore, $W_{\text{total}} = \Delta K$ (Equation 6-8) holds for variable forces as well as for constant forces.

Example 6-4 Work Done by a Varying Force

A force $\vec{F} = F_x\hat{i}$ varies with x, as shown in Figure 6-10. Find the work done by the force on a particle as the particle moves from $x = 0.0$ m to $x = 6.0$ m.

PICTURE The work done is the area under the curve from $x = 0.0$ m to $x = 6.0$ m. Because the curve consists of straight-line segments, the easiest approach is to break the area into two segments, one consisting of a rectangle (area A_1) and the other consisting of a right triangle (A_2), and then use the geometric formulas for area to find the work. (The alternative approach is to set up and execute an integration, as is done in Example 6-5.)

SOLVE

1. We find the work done by calculating the area under the F_x-versus-x curve:
   $W = A_{\text{total}}$
2. This area is the sum of the two areas shown. The area of a triangle is one half the altitude times the base:
   $$W = A_{\text{total}} = A_1 + A_2$$
   $$= (5.0\text{ N})(4.0\text{ m}) + \tfrac{1}{2}(5.0\text{ N})(2.0\text{ m})$$
   $$= 20\text{ J} + 5.0\text{ J} = \boxed{25\text{ J}}$$

CHECK If the force were a constant 5.0 N over the entire 6.0 m, the work would be (5.0 N)(6.0 m) = 30 J. The step-2 result of 25 J is slightly less than the 30 J, as expected.

PRACTICE PROBLEM 6-3 The force shown is the only force that acts on a particle of mass 3.0 kg. If the particle starts from rest at $x = 0.0$ m, how fast is it moving when it reaches $x = 6.0$ m?

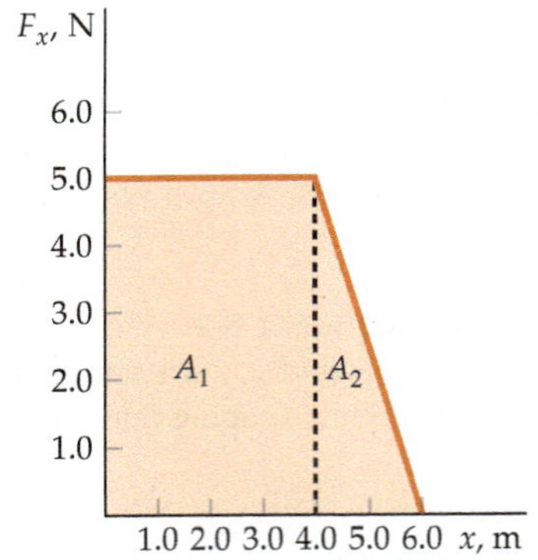

FIGURE 6-10

WORK DONE BY A SPRING THAT OBEYS HOOKE'S LAW

Figure 6-11 shows a block on a horizontal frictionless surface connected to a spring. If the spring is stretched or compressed, the spring exerts a force on the block. Recall from Equation 4-7 that the force exerted by the spring on the block is given by

$$F_x = -kx \qquad \text{(Hooke's law)} \qquad 6\text{-}11$$

where the k is a positive constant and x is the extension of the spring. If the spring is extended, then x is positive and the force component F_x is negative. If the spring is compressed, then x is negative and the force component F_x is positive.

Because the force varies with x, we can use Equation 6-10 to calculate the work done by the spring force on the block as the block undergoes a displacement from $x = x_i$ to $x = x_f$. (Besides the spring force, two other forces act on the block; the force of gravity, $m\vec{g}$, and the normal force of the table, $\vec{F}_n$. However, each of these forces does no work because neither has a component in the direction of the displacement. The only force that does work on the block is the spring force.) Substituting F_x from Equation 6-11 into Equation 6-10, we get

$$W_{\text{by spring}} = \int_{x_i}^{x_f} F_x dx = \int_{x_i}^{x_f} (-kx)\,dx = -k\int_{x_i}^{x_f} x\,dx = -k\left(\frac{x_f^2}{2} - \frac{x_i^2}{2}\right) \qquad 6\text{-}12$$

Rearranging this gives:

$$W_{\text{by spring}} = \tfrac{1}{2}kx_i^2 - \tfrac{1}{2}kx_f^2 \qquad 6\text{-}13$$

WORK BY A SPRING FORCE

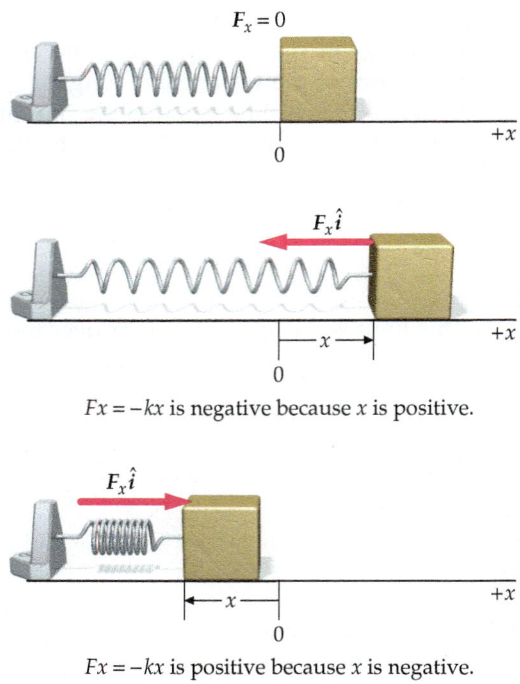

FIGURE 6-11 A horizontal spring. (*a*) When the spring is unstretched, it exerts no force on the block. (*b*) When the spring is stretched so that x is positive, it exerts a force of magnitude kx in the $-x$ direction. (*c*) When the spring is compressed so that x is negative, the spring exerts a force of magnitude $k|x|$ in the $+x$ direction.

The integral in Equation 6-12 can also be computed using geometry to calculate the area under the curve (Figure 6-12*a*). This gives

$$W_{\text{by spring}} = A_1 + A_2 = |A_1| - |A_2| = \tfrac{1}{2}kx_1^2 - \tfrac{1}{2}kx_2^2$$

which is identical to Equation 6-13.

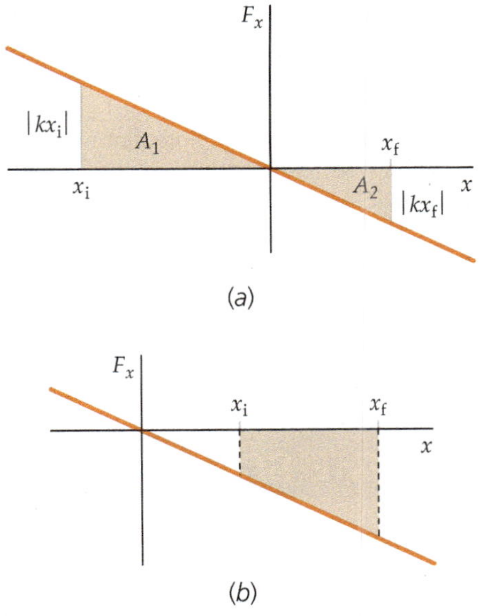

FIGURE 6-12

> **PRACTICE PROBLEM 6-4**
>
> Using geometry, calculate the area under the curve shown in Figure 6-12*b* and show you get an expression identical to that shown in Equation 6-13.

Suppose you pull on an initially relaxed spring (Figure 6-13), stretching it to a final extension x_f. How much work does the force exerted on the spring by your hand $\vec{F}_{SH}$ do? The force by your hand on the spring is equal to kx. (It is equal and opposite to the force by the spring on your hand.) As x increases from 0 to x_f, the force on the spring increases linearly from $F_{SHx} = 0$ to $F_{SHx} = kx_f$, and so has an average value* of $\frac{1}{2}kx_f$. The work done by this force is equal to the product of this average value and x_f. Thus, the work W done on the spring by your hand is given by

$$W = \tfrac{1}{2}kx_f^2$$

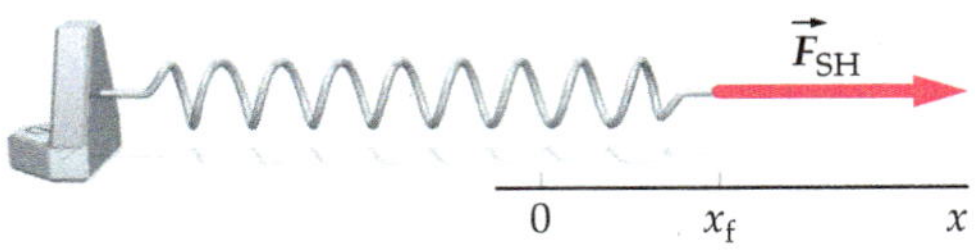

FIGURE 6-13

Example 6-5 Work Done on a Block by a Spring

A 4.0-kg block on a frictionless table is attached to a horizontal spring with $k = 400$ N/m. The spring is initially compressed 5.0 cm (Figure 6-14). Find (*a*) the work done on the block by the spring as the block moves from $x = x_1 = -5.0$ cm to its equilibrium position $x = x_2 = 0.0$ cm, and (*b*) the speed of the block at $x_2 = 0.0$ cm.

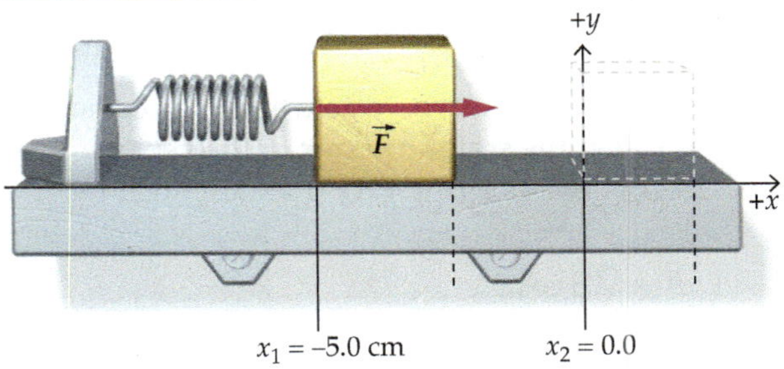

FIGURE 6-14

PICTURE Make a graph of F_x versus x. The work done on the block as it moves from x_1 to x_2 equals the area under the F_x-versus-x curve between these limits, shaded in Figure 6-15, which can be calculated by integrating the force over the distance. The work done equals the change in kinetic energy, which is simply its final kinetic energy because the initial kinetic energy is zero. The speed of the block at $x = 0.0$ cm is found from the kinetic energy of the block.

SOLVE

(*a*) The work W done on the block by the spring is the integral of $F_x dx$ from x_1 to x_2:

$$W = \int_{x_1}^{x_2} F_x\,dx = \int_{x_1}^{x_2} -kx\,dx = -k\int_{x_1}^{x_2} x\,dx$$

$$= -\tfrac{1}{2}kx^2\Big|_{x_1}^{x_2} = -\tfrac{1}{2}k(x_2^2 - x_1^2)$$

$$= -\tfrac{1}{2}(400\text{ N/m})[(0.000\text{ m})^2 - (0.050\text{ m})^2]$$

$$= \boxed{0.50\text{ J}}$$

(*b*) Apply the work–kinetic-energy theorem to the block and solve for v_2:

$$W_{\text{total}} = \tfrac{1}{2}mv_2^2 - \tfrac{1}{2}mv_1^2$$

so

$$v_2^2 = v_1^2 + \frac{2W_{\text{total}}}{m} = 0 + \frac{2(0.50\text{ J})}{4.0\text{ kg}} = 0.25\text{ m}^2/\text{s}^2$$

$$v_2 = \boxed{0.50\text{ m/s}}$$

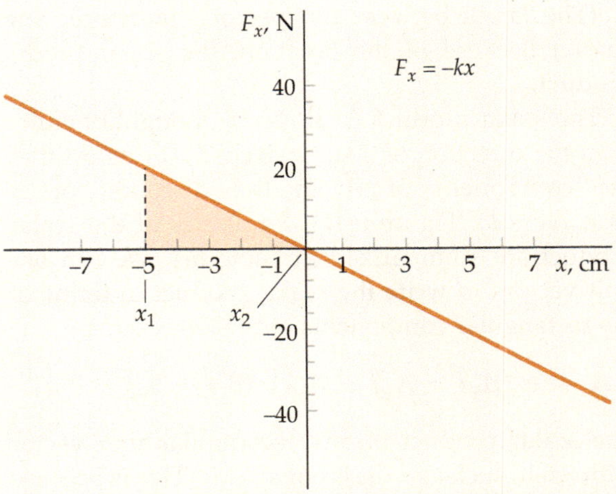

FIGURE 6-15

* Typically, an average value refers to an average over time. In this case, it refers to an average over position.

CHECK The work done is positive. The force and displacement are in the same direction, so this is as expected. The work is positive, so we expect the kinetic energy, and thus the speed, to increase. Our results fulfill this expectation.

TAKING IT FURTHER Note that we could *not* have solved this example by first applying Newton's second law to find the acceleration, and then using the constant-acceleration kinematic equations. This is so because the force exerted by the spring on the block, $F_x = -kx$, varies with position. Thus, the acceleration also varies with position. Therefore, the constant-acceleration kinematic equations are not in play.

PRACTICE PROBLEM 6-5 Find the speed of the 4.0-kg block when it reaches $x = 3.0$ cm if it starts from $x = 0.0$ cm with velocity $v_x = 0.50$ m/s.

6-3 THE SCALAR PRODUCT

Work is based on the component of force in the direction of an object's displacement. For straight-line motion, it is easy to calculate the component of the force in the direction of the displacement. However, in situations involving motion along a curved path, the force and the displacement can point in any direction. For these situations we can use a mathematical operation known as the *scalar* or *dot product* to determine the component of a given force in the direction of the displacement. The scalar product involves multiplying one vector by a second vector to produce a scalar.

Consider the particle moving along the arbitrary curve shown in Figure 6-16*a*. The component $F_{\parallel}$ in Figure 6-16*b* is related to the angle ϕ between the directions of $\vec{F}$ and $d\vec{\ell}$ by $F_{\parallel} = F\cos\phi$, so the work dW done by $\vec{F}$ for the displacement $d\vec{\ell}$ is

$$dW = F_{\parallel}d\ell = F\cos\phi\, d\ell$$

This combination of two vectors and the cosine of the angle between their directions is called the **scalar product** of the vectors. The scalar product of two general vectors $\vec{A}$ and $\vec{B}$ is written $\vec{A}\cdot\vec{B}$ and defined by

$$\vec{A}\cdot\vec{B} = AB\cos\phi \qquad \text{6-14}$$

DEFINITION—SCALAR PRODUCT

where A and B are the magnitudes of the vectors and ϕ is the angle between $\vec{A}$ and $\vec{B}$. (The "angle between two vectors" means the angle between their directions in space.) Because of the notation, the scalar product is also known as the **dot product.**

The scalar product $\vec{A}\cdot\vec{B}$ can be thought of either as A times the component of $\vec{B}$ in the direction of $\vec{A}$ ($A \times B\cos\phi$), or as B times the component of $\vec{A}$ in the direction of $\vec{B}$ ($B \times A\cos\phi$) (Figure 6-17). Properties of the scalar product are summarized in Table 6-1. We can use unit vectors to write the scalar product in terms of the rectangular components of the two vectors:

$$\vec{A}\cdot\vec{B} = (A_x\hat{i} + A_y\hat{j} + A_z\hat{k})\cdot(B_x\hat{i} + B_y\hat{j} + B_z\hat{k})$$

The scalar product of any rectangular unit vector with itself, such as $\hat{i}\cdot\hat{i}$, is equal to 1. (This is because $\hat{i}\cdot\hat{i} = |\hat{i}||\hat{i}|\cos(0) = 1 \times 1 \times \cos(0) = 1$.) Thus, a term like $A_x\hat{i}\cdot B_x\hat{i} = A_xB_x\hat{i}\cdot\hat{i}$ is equal to A_xB_x. Also, because the unit vectors $\hat{i}$, $\hat{j}$, and $\hat{k}$ are mutually perpendicular, the scalar product of one of them

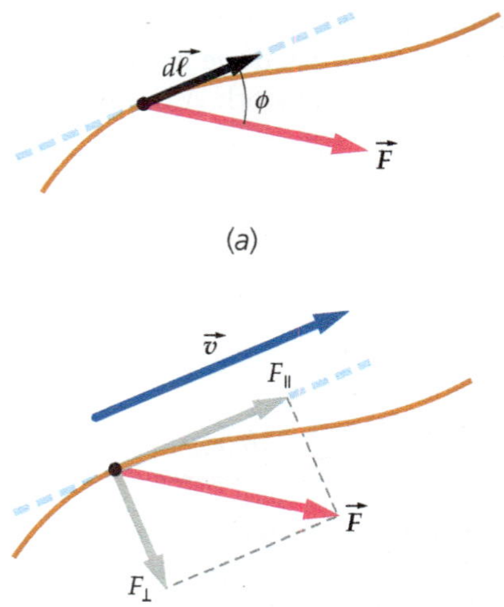

FIGURE 6-16 (*a*) A particle moving along an arbitrary curve in space. (*b*) The perpendicular component of the force $F_{\perp}$ changes the direction of the particle's motion, but not its speed. The tangential, or parallel, component $F_{\parallel}$ changes the particle's speed, but not its direction of motion. $F_{\parallel}$ is equal to the mass m times the tangential acceleration dv/dt. The parallel component of the force does work $F_{\parallel}d\ell$ and the perpendicular component does no work.

Table 6-1 Properties of Scalar Products

If	Then
$\vec{A}$ and $\vec{B}$ are perpendicular,	$\vec{A}\cdot\vec{B} = 0$ (because $\phi = 90°$, $\cos\phi = 0$)
$\vec{A}$ and $\vec{B}$ are parallel,	$\vec{A}\cdot\vec{B} = AB$ (because $\phi = 0°$, $\cos\phi = 1$)
$\vec{A}\cdot\vec{B} = 0$,	Either $\vec{A} = 0$ or $\vec{B} = 0$ or $\vec{A} \perp \vec{B}$
Furthermore,	
$\vec{A}\cdot\vec{A} = A^2$	Because $\vec{A}$ is parallel to itself
$\vec{A}\cdot\vec{B} = \vec{B}\cdot\vec{A}$	Commutative rule of multiplication
$(\vec{A} + \vec{B})\cdot\vec{C} = \vec{A}\cdot\vec{C} + \vec{B}\cdot\vec{C}$	Distributive rule of multiplication

with any other one of them, such as $\hat{i}\cdot\hat{j}$, is zero. (This is because $\hat{i}\cdot\hat{j} = |\hat{i}||\hat{j}|\cos(90°) = 1 \times 1 \times \cos(90°) = 0$.) Thus, any term like $A_x\hat{i}\cdot B_y\hat{j}$ (called a *cross term*) is equal to zero. The result is

$$\vec{A}\cdot\vec{B} = A_xB_x + A_yB_y + A_zB_z \qquad \text{6-15}$$

The component of a vector in a specific direction can be written as the scalar product of the vector and the unit vector in that direction. For example, the component A_x is found from

$$\vec{A}\cdot\hat{i} = (A_x\hat{i} + A_y\hat{j} + A_z\hat{k})\cdot\hat{i} = A_x \qquad \text{6-16}$$

This result provides an algebraic procedure for obtaining a component equation, given a vector equation. That is, multiplying both sides of the vector equation $\vec{A} + \vec{B} = \vec{C}$ by $\hat{i}$ gives $(\vec{A} + \vec{B})\cdot\hat{i} = \vec{C}\cdot\hat{i}$, which in turn gives $A_x + B_x = C_x$.

The rule for differentiating a dot product is

$$\frac{d}{dt}(\vec{A}\cdot\vec{B}) = \frac{d\vec{A}}{dt}\cdot\vec{B} + \vec{A}\cdot\frac{d\vec{B}}{dt} \qquad \text{6-17}$$

This rule is analogous with the rule for differentiating the product of two scalars. The rule for differentiating a dot product can be obtained by differentiating both sides of Equation 6-15.

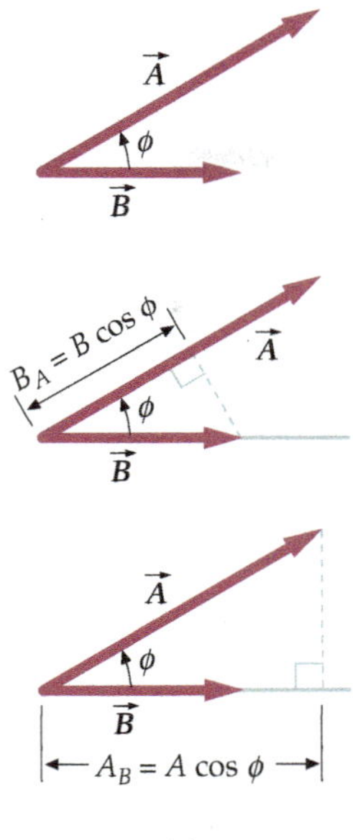

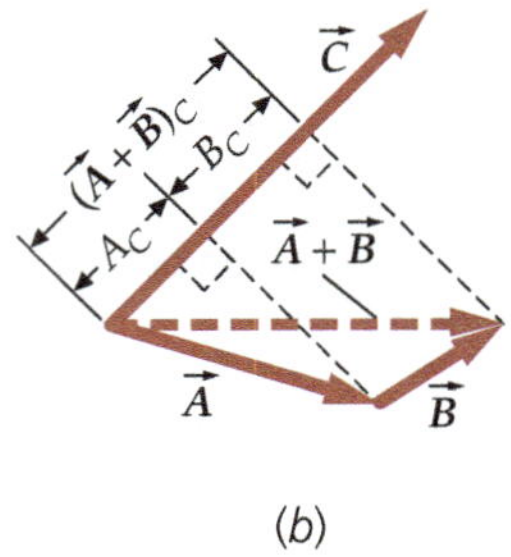

FIGURE 6-17 (*a*) The scalar product $\vec{A}\cdot\vec{B}$ is the product of A and the projection of $\vec{B}$ on $\vec{A}$, or the product of B and the projection of $\vec{A}$ on $\vec{B}$. That is, $\vec{A}\cdot\vec{B} = AB\cos\phi = AB_A = BA_B$. (*b*) $(\vec{A} + \vec{B})\cdot\vec{C}$ equals $(\vec{A} + \vec{B})_C C$ (the projection of $\vec{A} + \vec{B}$ in the direction of $\vec{C}$ times C). However, $(\vec{A} + \vec{B})_C = A_C + B_C$, so $(\vec{A} + \vec{B})\cdot\vec{C} = (A_C + B_C)C = A_CC + B_CC = \vec{A}\cdot\vec{C} + \vec{B}\cdot\vec{C}$. That is, for the scalar product, multiplication is distributive over addition.

Example 6-6 Using the Scalar Product

(*a*) Find the angle between the vectors $\vec{A} = (3.00\hat{i} + 2.00\hat{j})$ m and $\vec{B} = (4.00\hat{i} - 3.00\hat{j})$ m (Figure 6-18). (*b*) Find the component of $\vec{A}$ in the direction of $\vec{B}$.

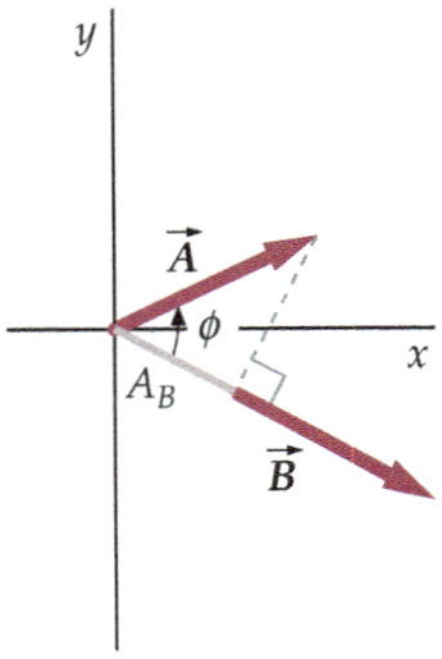

FIGURE 6-18

PICTURE For Part (a), we find the angle ϕ from the definition of the scalar product. Because we are given the components of the vectors, we first determine the scalar product and the values of A and B. Then, we use these values to determine the angle ϕ. For Part (*b*), the component of $\vec{A}$ in the direction of $\vec{B}$ is found from the scalar product $\vec{A}\cdot\hat{B}$, where $\hat{B} = \vec{B}/B$.

SOLVE

(*a*) 1. Write the scalar product of $\vec{A}$ and $\vec{B}$ in terms of A, B, and $\cos\phi$ and solve for $\cos\phi$:

$\vec{A}\cdot\vec{B} = AB\cos\phi$, so

$$\cos\phi = \frac{\vec{A}\cdot\vec{B}}{AB}$$

2. Find $\vec{A}\cdot\vec{B}$ from the components of $\vec{A}$ and $\vec{B}$:

$$\vec{A}\cdot\vec{B} = A_xB_x + A_yB_y = (3.00\text{ m})(4.00\text{ m}) + (2.00\text{ m})(-3.00\text{ m})$$
$$= 12.0\text{ m}^2 - 6.00\text{ m}^2 = 6.0\text{ m}^2$$

3. The magnitudes of the vectors are obtained from the scalar product of the vector with itself:

$$\vec{A}\cdot\vec{A} = A^2 = A_x^2 + A_y^2 = (3.00\text{ m})^2 + (2.00\text{ m})^2 = 13.0\text{ m}^2$$

so $A = \sqrt{13.0}$ m

and

$$\vec{B}\cdot\vec{B} = B^2 = B_x^2 + B_y^2 = (4.00\text{ m})^2 + (-3.00\text{ m})^2 = 25.0\text{ m}^2$$

so $B = 5.00$ m

4. Substitute these values into the equation in step 1 for $\cos\phi$ to find ϕ:

$$\cos\phi = \frac{\vec{A}\cdot\vec{B}}{AB} = \frac{6.0\text{ m}^2}{(\sqrt{13}\text{ m})(5.00\text{ m})} = 0.333$$

$$\phi = \boxed{71°}$$

(*b*) The component of $\vec{A}$ in the direction of $\vec{B}$ is the scalar product of $\vec{A}$ with the unit vector $\hat{B} = \vec{B}/B$:

$$A_B = \vec{A}\cdot\hat{B} = \vec{A}\cdot\frac{\vec{B}}{B} = \frac{\vec{A}\cdot\vec{B}}{B} = \frac{6.0\text{ m}^2}{5.00\text{ m}} = \boxed{1.2\text{ m}}$$

CHECK The component of $\vec{A}$ in the direction of $\vec{B}$ is $A\cos\phi = (\sqrt{13}\text{ m})\cos 71° = 1.2\text{ m}$. This answer verifies our Part-(*b*) result.

PRACTICE PROBLEM 6-6 (*a*) Find $\vec{A}\cdot\vec{B}$ for $\vec{A} = (3.0\hat{i} + 4.0\hat{j})\text{ m}$ and $\vec{B} = (2.0\hat{i} + 8.0\hat{j})\text{ m}$. (*b*) Find A, B, and the angle between $\vec{A}$ and $\vec{B}$ for these vectors.

WORK IN SCALAR-PRODUCT NOTATION

In scalar-product notation, the work dW done by a force $\vec{F}$ on a particle over an infinitesimal displacement $d\vec{\ell}$ is

$$dW = F_{\parallel}d\ell = F\cos\phi\, d\ell = \vec{F}\cdot d\vec{\ell} \qquad \text{6-18}$$

INCREMENTAL WORK

where $d\ell$ is the magnitude of $d\vec{\ell}$ and $F_{\parallel}$ is the component of $\vec{F}$ in the direction of $d\vec{\ell}$. The work done on the particle as it moves from point 1 to point 2 is

$$W = \int_1^2 \vec{F}\cdot d\vec{\ell} \qquad \text{6-19}$$

THE DEFINITION OF WORK

(If the force remains constant, the work can be expressed $W = \vec{F}\cdot\vec{\ell}$, where $\vec{\ell}$ is the displacement. In Chapter 3, the displacement is denoted $\Delta\vec{r} = \Delta x\hat{i} + \Delta y\hat{j}$; $\vec{\ell}$ and $\Delta\vec{r}$ are different symbols for the same quantity.)

When several forces $\vec{F}_i$ act on a particle whose displacement is $d\vec{\ell}$, the total work done on it is

$$dW_{\text{total}} = \vec{F}_1\cdot d\vec{\ell} + \vec{F}_2\cdot d\vec{\ell} + \cdots = (\vec{F}_1 + \vec{F}_2 + \cdots)\cdot d\vec{\ell} = (\Sigma\vec{F}_i)\cdot d\vec{\ell} \qquad \text{6-20}$$

Example 6-7 Pushing a Box

You push a box up a ramp using a constant horizontal 100-N force $\vec{F}$. For each distance of 5.00 m along the ramp, the box gains 3.00 m of height. Find the work done by $\vec{F}$ for each 5.00 m the box moves along the ramp (*a*) by directly computing the scalar product from the components of $\vec{F}$ and $\vec{\ell}$, where $\vec{\ell}$ is the displacement, (*b*) by multiplying the product of the magnitudes of $\vec{F}$ and $\vec{\ell}$ by $\cos\phi$, where ϕ is the angle between the direction of $\vec{F}$ and the direction of $\vec{\ell}$, (*c*) by finding $F_{\parallel}$ (the component of $\vec{F}$ in the direction of $\vec{\ell}$) and multiplying it by ℓ (the magnitude of $\vec{\ell}$), and (*d*) by finding $\ell_{\parallel}$ (the component of $\vec{\ell}$ in the direction of $\vec{F}$) and multiplying it by the magnitude of the force.

PICTURE Draw a sketch of the box in its initial and final positions. Place coordinate axes on the sketch with the x axis horizontal. Express the force and displacement vectors in component form and take the scalar product. Then find the component of the force in the direction of the displacement, and vice versa.

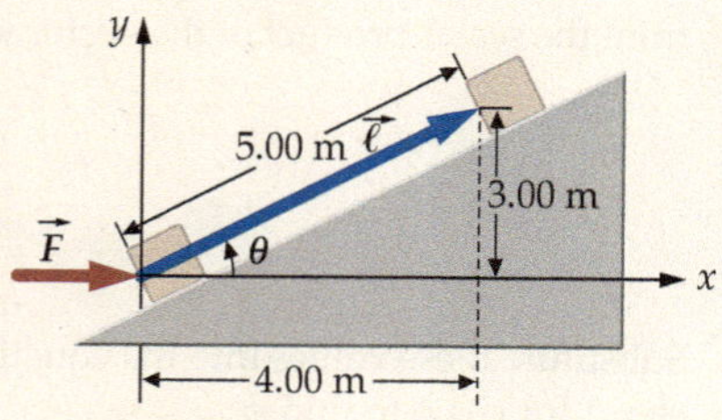

FIGURE 6-19

SOLVE

(*a*) 1. Draw a sketch of the situation (Figure 6-19).

2. Express $\vec{F}$ and $\vec{\ell}$ in component form and take the scalar product:

$$\vec{F} = (100\hat{i} + 0\hat{j})\text{N}$$

$$\vec{\ell} = (4.00\hat{i} + 3.00\hat{j})\text{m}$$

$$W = \vec{F}\cdot\vec{\ell} = F_x\Delta x + F_y\Delta y = (100\text{ N})(4.00\text{ m}) + 0(3.00\text{ m})$$

$$= \boxed{4.00\times 10^2\text{ J}}$$

(*b*) Calculate $F\ell\cos\phi$, where ϕ is the angle between the directions of the two vectors as shown. Equate this expression with the Part-(*a*) result and solve for $\cos\phi$. Then solve for the work:

$\vec{F}\cdot\vec{\ell} = F\ell\cos\phi$ and $\vec{F}\cdot\vec{\ell} = F_x\Delta x + F_y\Delta y$

so

$$\cos\phi = \frac{F_x\Delta x + F_y\Delta y}{F\ell} = \frac{(100\text{ N})(4.00\text{ m}) + 0}{(100\text{ N})(5.00\text{ m})} = 0.800$$

and

$$W = F\ell\cos\phi = (100\text{ N})(5.00\text{ m})0.800 = \boxed{4.00\times 10^2\text{ J}}$$

(*c*) Find $F_{\|}$ and multiply it by ℓ:

$$F_{\|} = F\cos\phi = (100\text{ N})0.800 = 80.0\text{ N}$$

$$W = F_{\|}\ell = (80.0\text{ N})(5.00\text{ m}) = \boxed{4.00\times 10^2\text{ J}}$$

(*d*) Multiply F and $\ell_{\|}$, where $\ell_{\|}$ is the component of $\vec{\ell}$ in the direction of $\vec{F}$:

$$\ell_{\|} = \ell\cos\phi = (5.00\text{ m})0.800 = 4.00\text{ m}$$

$$W = F\ell_{\|} = (100\text{ N})(4.00\text{ m}) = \boxed{4.00\times 10^2\text{ J}}$$

CHECK The four distinct calculations give the same result for the work.

TAKING IT FURTHER For this problem, computing the work is easiest using the procedure in Part (*a*). For other problems, the procedure in Part (*b*), Part (*c*), or Part (*d*) may be the easiest. You need to be competent in using all four procedures. (The more problem-solving tools you have at your disposal, the better.)

Example 6-8 A Displaced Particle

Try It Yourself

A particle undergoes a displacement $\vec{\ell} = (2.00\hat{i} - 5.00\hat{j})$ m. During this displacement a constant force $\vec{F} = (3.00\hat{i} + 4.00\hat{j})$ N acts on the particle. Find (*a*) the work done by the force, and (*b*) the component of the force in the direction of the displacement.

PICTURE The force is constant, so the work W can be found by computing $W = \vec{F}\cdot\vec{\ell} = F_x\Delta x + F_y\Delta y$. Combining this with the relation $\vec{F}\cdot\vec{\ell} = F_{\|}\ell$, we can find the component of $\vec{F}$ in the direction of the displacement.

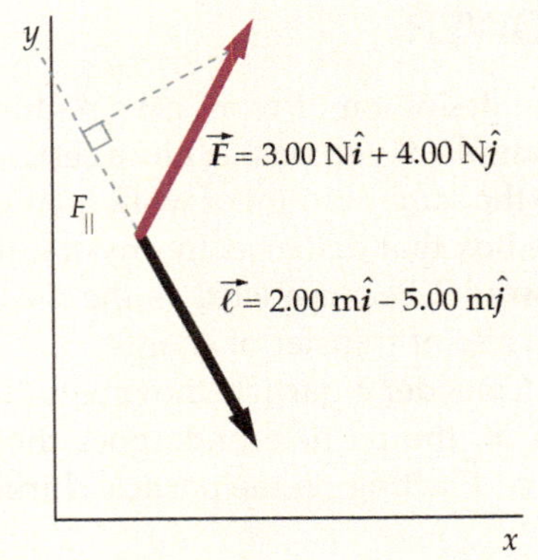

FIGURE 6-20

SOLVE

Cover the column to the right and try these on your own before looking at the answers.

Steps	Answers
(*a*) 1. Make a sketch showing $\vec{F}$, $\vec{\ell}$, and $F_{\|}$ (Figure 6-20).	
2. Compute the work done W.	$W = \vec{F}\cdot\vec{\ell} = \boxed{-14.0\text{ J}}$
(*b*) 1. Compute $\vec{\ell}\cdot\vec{\ell}$ and use your result to find ℓ.	$\vec{\ell}\cdot\vec{\ell} = 29.0\text{ m}^2$, so $\ell = \sqrt{29.0}$ m
2. Using $\vec{F}\cdot\vec{\ell} = F_{\|}\ell$, solve for $F_{\|}$.	$F_{\|} = \vec{F}\cdot\vec{\ell}/\ell = \boxed{-2.60\text{ N}}$

CHECK From Figure 6-20, we see that the angle between $\vec{F}$ and $\vec{\ell}$ is between 90° and 180°, so we expect both $F_{\|}$ and the work to be negative. Our results are in agreement with this expectation.

TAKING IT FURTHER Nowhere in the wording of either the problem statement or the solution of Example 6-8 does it say that the motion of the particle is along any particular path. Because the force is constant, the solution depends on the net displacement $\vec{\ell}$, but not on the path taken. The path could be a straight-line path or a curved path (Figure 6-21) and not a word in the solution would have to be changed.

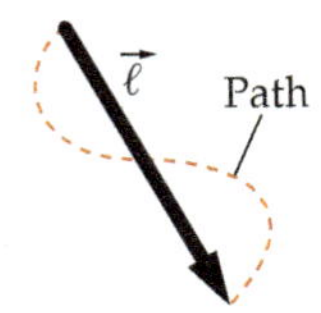

FIGURE 6-21

PRACTICE PROBLEM 6-7 Find the magnitude of $\vec{F}$ and the angle ϕ between $\vec{F}$ and $\vec{\ell}$.

Example 6-9 Differentiating a Scalar Product

Show that $\vec{a}\cdot\vec{v} = \frac{1}{2}d(v^2)/dt$, where $\vec{a}$ the acceleration, $\vec{v}$ the velocity, and v the speed.

PICTURE Note that $v^2 = \vec{v}\cdot\vec{v}$, so the rule for differentiating scalar products can be used here.

SOLVE

Apply the rule for differentiating scalar products (Equation 6-17) to the scalar product $\vec{v}\cdot\vec{v}$:

$$\frac{d}{dt}v^2 = \frac{d}{dt}(\vec{v}\cdot\vec{v}) = \frac{d\vec{v}}{dt}\cdot\vec{v} + \vec{v}\cdot\frac{d\vec{v}}{dt} = 2\frac{d\vec{v}}{dt}\cdot\vec{v} = 2\vec{a}\cdot\vec{v}$$

$$\text{so}\quad \vec{a}\cdot\vec{v} = \frac{1}{2}\frac{d}{dt}v^2$$

CHECK Speed v has dimensions of length over time, so dv^2/dt has dimensions of length squared over time cubed. Acceleration $\vec{a}$ has dimensions of length over time squared, so $\vec{a}\cdot\vec{v}$ has dimensions of length squared over time cubed. Thus, both sides of $\vec{a}\cdot\vec{v} = \frac{1}{2}d(v^2)/dt$ have the same dimensions (length squared over time cubed).

TAKING IT FURTHER This example involves only kinematic parameters, so the resulting relation is a strictly kinematic relation. The equation $\vec{a}\cdot\vec{v} = \frac{1}{2}d(v^2)/dt$ has unrestricted validity (unlike some kinematic equations we have studied that are valid only if the acceleration remains constant).

From Example 6-9 we have the kinematic relation

$$\vec{a}\cdot\vec{v} = \frac{1}{2}\frac{d}{dt}v^2 = \frac{d}{dt}\left(\frac{1}{2}v^2\right) \qquad \text{6-21}$$

In Section 6-4, this equation is used to derive the work–kinetic-energy theorem for particles moving along curved paths under the influence of forces that are not necessarily constant.

POWER

The definition of work says nothing about how long it takes to do the work. For example, if you push a box a certain distance up a hill with a constant velocity, you do the same amount of work on the box regardless of how long it takes you to push the box that distance. In physics, the rate at which a force does work is called the **power** P. Because work is the measure of energy transferred by a force, power is the rate of transfer of energy.

Consider a particle moving with instantaneous velocity $\vec{v}$. In a short time interval dt, the particle undergoes the displacement $d\vec{\ell} = \vec{v}\,dt$. The work done by a force $\vec{F}$ acting on the particle during this time interval is

$$dW = \vec{F}\cdot d\vec{\ell} = \vec{F}\cdot\vec{v}\,dt$$

The power is then

$$P = \frac{dW}{dt} = \vec{F}\cdot\vec{v} \qquad \text{6-22}$$

POWER BY A FORCE

Note the difference between power and work. Two motors that lift a given load a given distance expend the same amount of energy, but the power is greater for the force that does the work in the least time.

Like work and energy, power is a scalar quantity. The SI unit of power, one joule per second, is called a **watt** (W):

$$1\ \text{W} = 1\ \text{J/s}$$

In the U.S. customary system, the unit of energy is the foot-pound and the unit of power is the foot-pound per second. A commonly used multiple of this unit, called a horsepower (hp), is defined as

$$1 \text{ hp} = 550 \text{ ft} \cdot \text{lb/s} \approx 746 \text{ W}$$

The product of a unit of power and a unit of time is a unit of energy. Electric companies charge for energy, not power, usually by the kilowatt-hour (kW · h). A kilowatt-hour of energy is the energy transferred in 1 hour at the constant rate of 1 kilowatt, or

$$1 \text{ kW} \cdot \text{h} = (10^3 \text{ W})(3600 \text{ s}) = 3.6 \times 10^6 \text{ W} \cdot \text{s} = 3.6 \text{ MJ}$$

Example 6-10 The Power of a Motor

A small motor is used to operate a lift that raises a load of bricks weighing 500 N to a height of 10 m in 20 s (Figure 6-22) at constant speed. The lift weighs 300 N. What is the power output of the motor?

PICTURE Because the acceleration is zero, the magnitude of the upward force $\vec{F}$ exerted by the motor is equal to the weight of the lift plus the weight of the bricks. The rate the motor does work is the power.

SOLVE

The power is given by $\vec{F} \cdot \vec{v}$:

$$P = \vec{F} \cdot \vec{v} = Fv\cos\phi = Fv\cos(0) = Fv$$

$$= (800 \text{ N})\frac{10 \text{ m}}{20 \text{ s}} = \boxed{4.0 \times 10^2 \text{ W}}$$

CHECK The work done by the force is (800 N)(10 m) = 8000 J. This work took 20 s to do, so we expect the power to be the 8000 J/20 s = 4.0×10^2 W. Our result is in perfect agreement with this.

TAKING IT FURTHER (1) The lift could not actually operate at constant speed. The bricks and lift will have to initially be brought up to speed (because they are starting from rest.) The power output will exceed 400 W during this speedup interval. In addition, the power output will be less than 400 W as the lift slows to a stop at the top. The average power output of the motor during the lift is 400 W (and the power provided by the force of gravity is −400 W). (2) A power of 400 W is slightly more than $\frac{1}{2}$ hp.

PRACTICE PROBLEM 6-8 Find the average power output of the motor needed to raise the bricks and lift to a height of 10 m in 40 s. What is the work done by the force of the motor? What is the work done by the force of gravity?

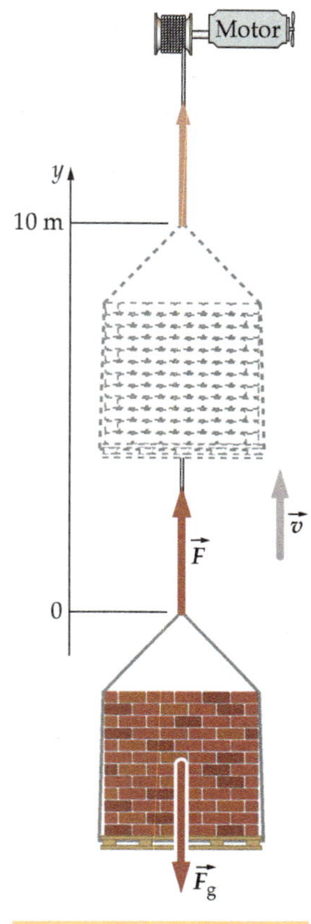

FIGURE 6-22

Example 6-11 Power and Kinetic Energy

Show that the power delivered by the net force acting on a particle equals the rate at which the kinetic energy of the particle is changing.

PICTURE The power delivered by the net force P_{net} equals $\vec{F}_{net} \cdot \vec{v}$. Show that $\vec{F}_{net} \cdot \vec{v} = dK/dt$, where $K = \frac{1}{2}mv^2$.

SOLVE

1. Substitute for $\vec{F}_{net}$ using Newton's second law:

$$\vec{F}_{net} \cdot \vec{v} = m\vec{a} \cdot \vec{v}$$

2. The product $\vec{a} \cdot \vec{v}$ is related to the time derivative of v^2 by $2\vec{a} \cdot \vec{v} = d(v^2)/dt$ (Equation 6–21):

$$\frac{d}{dt}v^2 = \frac{d}{dt}(\vec{v} \cdot \vec{v}) = 2\vec{a} \cdot \vec{v}$$

3. Substitute the step-2 result into the step-1 result: $\vec{F}_{net} \cdot \vec{v} = m\vec{a} \cdot \vec{v} = m\frac{1}{2}\frac{d}{dt}v^2$

4. The mass m is constant, so it and the fraction $\frac{1}{2}$ can be moved inside the argument of the derivative: $\vec{F}_{net} \cdot \vec{v} = \frac{d}{dt}\left(\frac{1}{2}mv^2\right)$

5. The argument of the derivative is the kinetic energy K: $\boxed{P_{net} = \vec{F}_{net} \cdot \vec{v} = \frac{dK}{dt}}$

CHECK The joule is the unit for energy, so dK/dt has units of joules per second, or watts. The watt is the unit for power, so $P_{net} = dK/dt$ is dimensionally consistent.

From Example 6-11 we have

$$P_{net} = \vec{F}_{net} \cdot \vec{v} = \frac{dK}{dt} \tag{6-23}$$

which relates the power delivered by the net force to the rate of change of kinetic energy of any object that can be modeled as a particle.

6-4 WORK–KINETIC-ENERGY THEOREM—CURVED PATHS

The work–energy theorem for motion along a curved path can be established by integrating both sides of $\vec{F}_{net} \cdot \vec{v} = dK/dt$ (Equation 6-23). Integrating both sides over time gives

$$\int_1^2 \vec{F}_{net} \cdot \vec{v}\,dt = \int_1^2 \frac{dK}{dt}dt \tag{6-24}$$

Because $d\vec{\ell} = \vec{v}\,dt$, where $d\vec{\ell}$ is the displacement during time dt, and because $(dK/dt)\,dt = dK$, Equation 6-24 can be expressed

$$\int_1^2 \vec{F}_{net} \cdot d\vec{\ell} = \int_1^2 dK$$

The integral on the left is the total work, W_{total}, done on the particle. The integral on the right can be evaluated, giving

$$\int_1^2 \vec{F}_{net} \cdot d\vec{\ell} = K_2 - K_1 \quad (\text{or } W_{total} = \Delta K) \tag{6-25}$$

WORK–KINETIC-ENERGY THEOREM

Equation 6-25 follows directly from Newton's second law of motion.

Example 6-12 Work Done on a Skier — *Context-Rich*

You and your friend are at a ski resort with two ski runs, a beginner's run and an expert's run. Both runs begin at the top of the ski lift and end at a finish line at the bottom of the same lift. Let h be the vertical descent for both runs. The beginner's run is longer and less steep than the expert's run. You and your friend, who is a much better skier than you, are testing some experimental frictionless skis. To make things interesting, you offer a wager that if she takes the expert's run and you take the beginner's run, her speed at the finish line will not be greater than your speed at the finish line. Forgetting that you are studying physics, she accepts the bet. The conditions are that you both start from rest at the top of the lift and both of you coast for the entire trip. Who wins the bet? (Assume air drag is negligible.)

(PhotoDisc/Getty.)

PICTURE Because you and your friend are coasting on the skis, you both can be modeled as particles. (The work–kinetic-energy theorem works only for particles.) Two forces act on each of you, a weight force and a normal force.

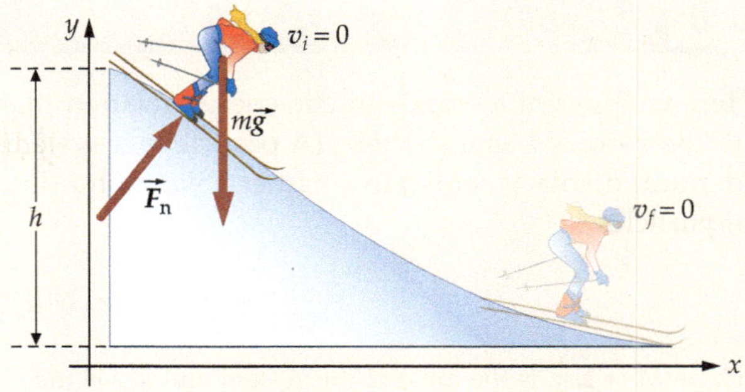

SOLVE

1. Make a sketch of yourself and draw the two force vectors on the sketch (Figure 6-23*a*). Also include coordinate axes. The work–kinetic-energy theorem, with $v_1 = 0$, relates the final speed v_f to the total work.

2. The final speed is related to the final kinetic energy, which in turn is related to the total work by the work–kinetic-energy theorem: $W_{total} = \frac{1}{2}mv_f^2 - \frac{1}{2}mv_i^2$

3. For each of you, the total work is the work done by the normal force plus the work done by the gravitational force: $W_{total} = W_n + W_g$

4. The force $m\vec{g}$ on you is constant, but the force $\vec{F}_n$ is not constant. First we calculate the work done by $\vec{F}_n$. Calculate the work dW_n done on you by $\vec{F}_n$ for an infinitesimal displacement $d\vec{\ell}$ (Figure 6-23*b*) at an arbitrary location along the run: $dW_n = \vec{F}_n \cdot d\vec{\ell} = F_n \cos\phi\, d\ell$

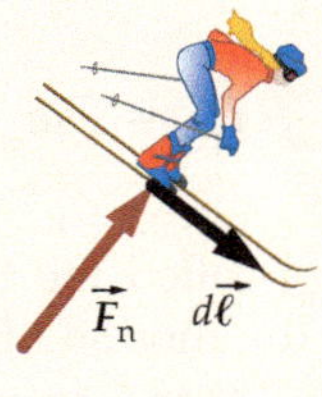

FIGURE 6-23

5. Find the angle ϕ between the directions of $\vec{F}_n$ and $d\vec{\ell}$. The displacement $d\vec{\ell}$ is tangent to the slope: $\phi = 90°$

6. Calculate the work done by $\vec{F}_n$ for the entire run: $W_n = \int F_n \cos 90°\, d\ell = \int (0)\, d\ell = 0$

7. The force of gravity $\vec{F}_g$ is constant, so the work done by gravity is $W_g = \vec{F}_g \cdot \vec{\ell}$, where $\vec{\ell}$ (Figure 6-24) is the net displacement from the top to the bottom of the lift:
$$W_g = m\vec{g} \cdot \vec{\ell} = -mg\hat{j} \cdot (\Delta x\hat{i} + \Delta y\hat{j}) = -mg\,\Delta y$$

8. The skier is descending the hill, so Δy is negative. From Figure 6-23*a*, we see that $\Delta y = -h$: $\Delta y = -h$

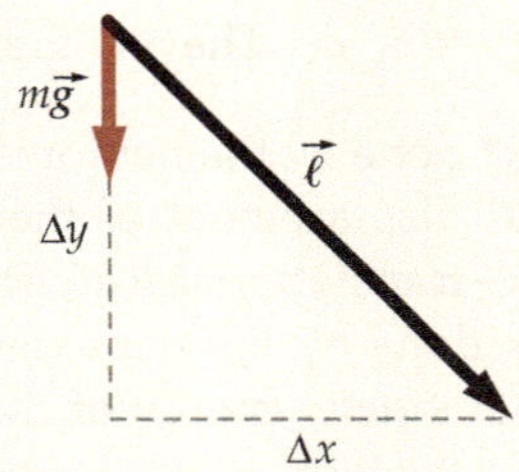

FIGURE 6-24

9. Substituting gives: $W_g = mgh$

10. Apply the work–kinetic-energy theorem to find v_f: $W_n + W_g = \Delta K$

11. The final speed depends only on h, which is the same for both runs. Both of you will have the same final speeds.
$$0 + mgh = \frac{1}{2}mv_f^2 - 0 \quad \text{so} \quad v_f = \sqrt{2gh}$$
YOU WIN! (The bet was that she would not be going faster than you.)

CHECK The force driving your motion is the gravitational force. This force is proportional to the mass, so the work done by it is proportional to the mass. Because the kinetic energy is also proportional to the mass, the mass cancels out of the work–kinetic-energy equation. Thus, we expect the final speed to be independent of mass. Our result is independent of mass as expected.

TAKING IT FURTHER Your friend on the steeper trail will cross the finish line in less time, but that was not the bet. What was shown here is that the work done by the gravitational force equals *mgh*. It does not depend upon the shape of the hill or upon the length of the path taken. It depends only upon the mass *m* and the vertical drop *h* between the starting point and the finishing point.

* 6-5 CENTER-OF-MASS WORK

Here we present a work–kinetic-energy relation that works for systems that cannot be modeled as a particle. (A particle is a system for which all parts undergo identical displacements.) In Chapter 5 we found (Equation 5-23) that for a system of particles

$$\vec{F}_{\text{net ext}} = \sum \vec{F}_{i\,\text{ext}} = M\vec{a}_{\text{cm}} \qquad \text{6-26}$$

where $M = \Sigma m_i$ is the mass of the system and $\vec{a}_{\text{cm}}$ is the acceleration of the center of mass. Equation 6-26 can be integrated to obtain a useful equation involving work and kinetic energy that can be applied to systems that cannot be modeled as a particle. First, we take the scalar product of both sides of Equation 6-26 with $\vec{v}_{\text{cm}}$ to obtain

$$\vec{F}_{\text{net ext}} \cdot \vec{v}_{\text{cm}} = M\vec{a}_{\text{cm}} \cdot \vec{v}_{\text{cm}} = \frac{d}{dt}\left(\tfrac{1}{2}Mv_{\text{cm}}^2\right) = \frac{dK_{\text{trans}}}{dt} \qquad \text{6-27}$$

where $K_{\text{trans}} = \frac{1}{2}Mv_{\text{cm}}^2$, called the **translational kinetic energy**, is the kinetic energy associated with the motion of the center of mass. Multiplying both sides of Equation 6-27 by dt and then integrating gives

$$\int_1^2 \vec{F}_{\text{net ext}} \cdot d\vec{\ell}_{\text{cm}} = \Delta K_{\text{trans}} \qquad \text{6-28}$$

CENTER-OF-MASS WORK–TRANSLATIONAL-KINETIC-ENERGY RELATION

where $d\vec{\ell}_{\text{cm}} = \vec{v}_{\text{cm}}dt$. The integral $\int_1^2 \vec{F}_{\text{net ext}} \cdot d\vec{\ell}_{\text{cm}}$ is referred to as the **center-of-mass work*** done by the net force on a system of particles, and $d\vec{\ell}_{\text{cm}} = \vec{v}_{\text{cm}}dt$ is the incremental displacement of the center of mass. Equation 6-28 is the **center-of-mass work–translational-kinetic-energy relation.** Stated in words "The center-of-mass work done by the net external force on a system equals the change in the translational kinetic energy of the system. Although Equation 6-28 looks like the equation for the work–kinetic-energy theorem (Equation 6-25), there are some important differences. The center-of-mass work–translational-kinetic-energy relation deals only with the displacement and speed of the center of mass of the system, so when using this relation, we ignore the motion of any part of the system relative to the center-of-mass reference frame. (A center-of-mass reference frame is a nonrotating reference frame† that moves with the center of mass.) This allows us to calculate the bulk motion of the system without knowing all the internal details of the system.

For a system that moves as a particle (with all parts having the same velocity) the center-of-mass work–translational-kinetic-energy relation reduces to the work–kinetic-energy theorem (Equation 6-25).

It is also sometimes useful to refer to the center-of-mass work done by a single force. The center-of-mass work W_{cm} done by any particular force $\vec{F}$ is given by

$$W_{\text{cm}} = \int_1^2 \vec{F} \cdot d\vec{\ell}_{\text{cm}} \qquad \text{6-29}$$

* Center-of-mass work is also called pseudowork.

† A nonrotating reference frame is a frame that is not rotating relative to an inertial reference frame.

Example 6-13 Two Pucks and a String

Two identical pucks on an air table are connected by a length of string (see Figure 6-25). The pucks, each of mass m, are initially at rest in the configuration shown. A constant force of magnitude F accelerates the system toward the right. After the point of application P of the force has moved a distance d, the pucks collide and stick together. What is the speed of the pucks immediately following the collision?

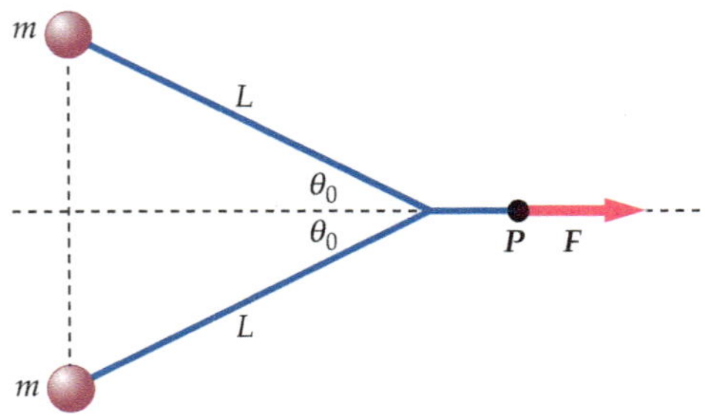

FIGURE 6-25

PICTURE Let the system be the two pucks and the string. Apply the center-of-mass work–kinetic-energy relation to the system. Following the collision the speed of each puck equals the speed of the center of mass. (The pucks can move without friction on the air table.)

SOLVE

1. Make a drawing showing the system initially, and after it has moved distance d (Figure 6-26):

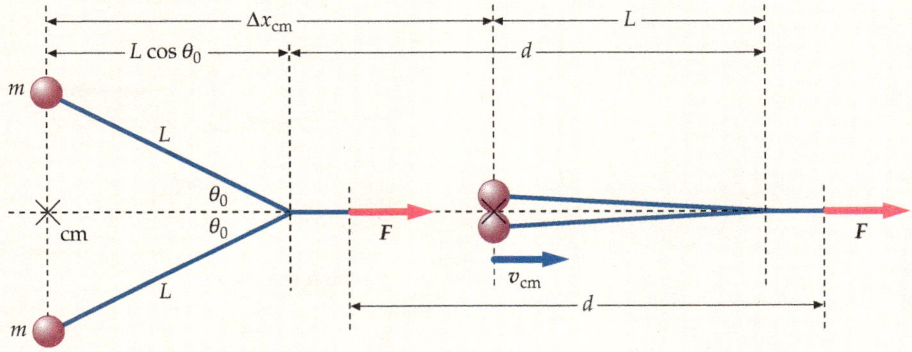

FIGURE 6-26 As the center of mass moves the distance Δx_{cm}, the point of application of the force $\vec{F}$ moves the distance d.

2. Apply the center-of-mass-work–translational-kinetic-energy relation to the system. The net force on the system is $\vec{F} = F\hat{i}$:

$$\int_i^f \vec{F}_{net\,ext} \cdot d\vec{\ell}_{cm} = \Delta K_{trans}$$

$$\int_i^f F\hat{i} \cdot dx_{cm}\hat{i} = K_{trans\,f} - K_{trans\,i}$$

$$F\int_i^f dx_{cm} = K_{trans\,f} - 0$$

$$F\Delta x_{cm} = \tfrac{1}{2}(2m)v_{cm}^2 = mv_{cm}^2$$

3. Find Δx_{cm} in terms of d and L. Figure 6-26 makes the calculation of Δx_{cm} fairly straightforward:

$$\Delta x_{cm} + L = L\cos\theta_0 + d$$

$$\text{so}\quad \Delta x_{cm} = d - L(1 - \cos\theta_0)$$

4. Substitute the step-3 result into the step-2 result and solve for v_{cm}:

$$F\Delta x_{cm} = mv_{cm}^2$$

$$F[d - L(1 - \cos\theta_0)] = mv_{cm}^2$$

$$\text{so}\quad v_{cm} = \boxed{\sqrt{\frac{F[d - L(1 - \cos\theta_0)]}{m}}}$$

CHECK If the initial angle θ_0 is zero, the system can be modeled as a particle and the work–kinetic-energy theorem can be used. This would give $Fd = \frac{1}{2}(2m)v^2 = mv^2$, or $v = \sqrt{Fd/m}$. Our step-4 result gives the very same expression for the speed if $\theta_0 = 0$.

TAKING IT FURTHER (1) In this example, the displacement of the center of mass Δx_{cm} is less than the displacement d of the point of application of the force $\vec{F}$. As a result, the center-of-mass work done by the force is less than the work Fd done by the force. (2) The pucks lose kinetic energy when they collide and stick together. This energy appears as some other form of energy, such as thermal energy. The conservation of energy is discussed further in Chapter 7.

Example 6-14 Stopping Distance

To avoid an accident, the driver of a 1000-kg car moving at 90 km/h on a straight horizontal road steps on the brakes with maximum force. The antilock braking system (ABS) is not working, so the wheels lock and the tires skid as the car comes to a stop. The kinetic coefficient of friction between the road and the tires is 0.80. How far does the car travel during the skid?

PICTURE The car cannot be modeled as a particle. The points of application of the kinetic frictional forces are the parts of the tires in contact with the road surface. The high points of the contacting surfaces alternatively stick and slip. Therefore, the car cannot be modeled as a particle during the skid. The center-of-mass work–translational-kinetic-energy relation applied to the car enables us to calculate the stopping distance.

SOLVE

1. Write the center-of-mass work–translational-kinetic-energy relation. We need to solve for the displacement of the center of mass of the car:

$$\int_1^2 \vec{F}_{\text{net ext}} \cdot d\vec{\ell}_{\text{cm}} = \Delta K_{\text{trans}}$$

2. Draw a free-body diagram of the car during the skid (Figure 6-27):

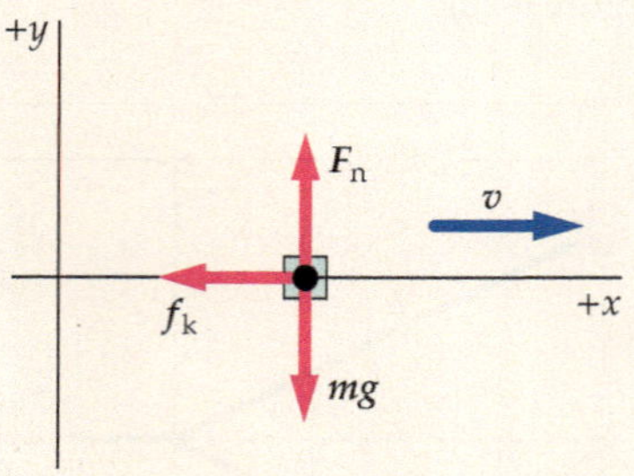

FIGURE 6-27

3. The vertical acceleration is zero, so the normal force and the gravitational force sum to zero. The net external force on the car is equal to the frictional force. Solve for the net force on the car:

$$\vec{F}_{\text{net}} = \vec{F}_{\text{n}} + m\vec{g} + \vec{f}_{\text{k}} = \vec{f}_{\text{k}}$$

so

$$F_{\text{net}} = f_{\text{k}} = \mu_{\text{k}} F_{\text{n}} = \mu_{\text{k}} mg$$

$$\vec{F}_{\text{net}} = -\mu_{\text{k}} mg\hat{i}$$

4. Apply the center-of-mass work–translational-kinetic-energy relation to the car:

$$\int_1^2 \vec{F}_{\text{net}} \cdot d\vec{\ell}_{\text{cm}} = \Delta K_{\text{trans}}$$

$$\int_1^2 -\mu_{\text{s}} mg\hat{i} \cdot dx_{\text{cm}}\hat{i} = K_{\text{trans}\,2} - K_{\text{trans}\,1}$$

$$-\mu_{\text{s}} mg \int_1^2 dx_{\text{cm}} = 0 - K_{\text{trans}\,1}$$

$$-\mu_{\text{s}} mg\,(x_{\text{cm}\,2} - x_{\text{cm}\,1}) = -\tfrac{1}{2} m v_{\text{cm}\,1}^2$$

5. Solve for the displacement, but first convert the initial speed from km/h to m/s:

$$v_{\text{cm}\,1} = 90\ \text{km/h} \cdot \frac{1\ \text{h}}{(3.6\ \text{ks})} = 25\ \text{m/s}$$

so

$$\Delta x_{\text{cm}} = x_{\text{cm}\,2} - x_{\text{cm}\,1} = \frac{v_{\text{cm}\,1}^2}{2\mu_{\text{k}} g}$$

$$\Delta x_{\text{cm}} = \frac{(25\ \text{m/s})^2}{2 \cdot (0.80)(9.81\ \text{m/s}^2)} = \boxed{40\ \text{m}}$$

CHECK We would expect the stopping distance to increase with initial speed, and to decrease with increasing coefficient of friction. The step-5 expression for Δx_{cm} meets these expectations.

TAKING IT FURTHER The translational kinetic energy of the car is dissipated as thermal energy of the tires and of the pavement. The dissipation of kinetic energy into thermal energy by kinetic friction is discussed further in Chapter 7.

Physics Spotlight

Coasters and Baggage and Work (Oh My!)

Baggage transfer methods at some major airports have a lot in common with roller coasters. High rates of change of acceleration for long periods of time are bad for both coaster passengers and baggage items. Both must move swiftly without unwanted jerking and halting.

Some roller coaster cars (and some baggage carriers) gain kinetic energy because of the work done on them by constant forces exerted on them by banks of *linear induction motors* (LIMs). A LIM is an electromagnetic method of providing force without moving parts.* The main reason for the use of linear induction motors is the flexibility of applying force at calculated locations during the travel of the coaster train or the baggage buggy. The roller coaster and the baggage buggies run on tracks that use sensors to determine the speed of the vehicles, and communicate that speed to the controllers for the motors. The LIMs can be turned off when the vehicle has reached the right speed. In both cases, some LIMs are also wired to act as brakes on the vehicles, exerting forces on them in opposition to their direction of travel.

The speeds of the buggies transporting luggage between terminals at Heathrow International Airport are controlled by LIMs. *(Vanderlande Industries.)*

Speed—The Ride is a roller coaster launched from the NASCAR Café in the Sahara Hotel and Casino in Las Vegas. The design firm, Ingenieurbuero Stengel GmbH, specified 88 motors in three locations along the track. The first bank of motors launches the coaster train. The 6-car, 24-passenger coaster train is smoothly accelerated to 45 mi/h in 2.0 s. It swoops around a corner and plunges 25 ft underground before rising and going through a clothoid loop-the-loop.† After it goes through the loop-the-loop, forces exerted on it by the second bank of LIMs quadruple its kinetic energy in 2.0 s.‡ The roller coaster glides along Las Vegas Boulevard, and races two hundred feet up a near-vertical incline. For safety's sake, a series of LIMs near the top of the incline can slow the train, if necessary. The coaster train then runs backward through the entire coaster route. As it returns to the station, the LIMs in the station act as brakes, and bring the train to a stop.

Other than the forces from the LIMs, the forces acting on the coaster train are gravity, friction, and the normal force. Each of the cars in the coaster train travels over the same path, although the starting and ending points for each car are not the same. The maximum acceleration of any passenger is $3.5g$. This is not excessive—the momentary acceleration caused by being hit on the head with a pillow can go above $20g$.§

Heathrow International Airport often transfers luggage between Terminals One and Four. The terminals are more than 1.0 km apart and are separated by a runway. Each piece of luggage is loaded onto a small buggy that rides on rails. (The speeds of the buggies are controlled by LIMs mounted on the tracks.) The buggy goes down a steep incline to reach the level of a tunnel, 20 m underground. It travels through the tunnel at 30 km/h, and is kept at that speed by regularly spaced LIMs. At the end of the tunnel, the buggy climbs up into the appropriate floor of the other terminal. When you transfer between flights at a large airport, remember that your luggage may well be going on its own special ride.

* "Whoa! Linear motors blast Vegas coaster straight up." *Machine Design*, May 4, 2000. Vol. 28; "Sectors" EI-WHS http://www.eiwhs.co.uk/sectors.asp April 2006; "Baggage Handling Case Study." Force Engineering http://www.force.co.uk/bagcase.htm, April 2006; "Leisure Rides." Force Engineering, http://www.force.co.uk/leishome.htm April 2006.

† "Roller coaster constructor Werner Stengel receives honorary doctorate at Göteborg University." Göteborg University Faculty of Science. http://www2.science.gu.se/english/werner_stengel.shtml April 2006.

‡ "Speed Facts." Sahara Hotel and Casino, http://www.saharavegas.com/thrills/facts.html April 2006.

§ Exponent Failure Analysis Associates. *Investigation of Amusement Park and Roller Coaster Injury Likelihood and Severity:* 48. http://www.emerson-associates.com/safety/articles/ExponentReport.pdf April 2006.

SUMMARY

1. Work, kinetic energy, and power are important derived dynamic quantities.
2. The work–kinetic-energy theorem is an important relation derived from Newton's laws applied to a particle. (In this context, a particle is a perfectly rigid object that moves without rotating.)
3. The scalar product of vectors is a mathematical definition that is useful throughout physics.

TOPIC	RELEVANT EQUATIONS AND REMARKS	
1. Work	$W = \int_1^2 \vec{F} \cdot d\vec{\ell} = \int_1^2 F_{\parallel} d\ell$ (definition)	
Constant force	$W = \vec{F} \cdot \vec{\ell} = F_{\parallel}\ell = F\ell_{\parallel} = F\ell \cos\theta$	
Constant force—straight-line motion	$W = F_x \Delta x = F\lvert\Delta x\rvert \cos\theta$	
Variable force—straight-line motion	$W = \int_{x_1}^{x_2} F_x dx =$ area under the F_x-versus-x curve	
2. Kinetic Energy	$K = \frac{1}{2}mv^2$ (definition)	
3. Work–Kinetic-Energy Theorem	$W_{\text{total}} = \Delta K = \frac{1}{2}mv_{\text{f}}^2 - \frac{1}{2}mv_{\text{i}}^2$	
4. Scalar or Dot Product	$\vec{A} \cdot \vec{B} = AB\cos\phi$ (definition)	
In terms of components	$\vec{A} \cdot \vec{B} = A_x B_x + A_y B_y + A_z B_z$	
Unit vector times vector	$\vec{A} \cdot \hat{i} = A_x$	
Derivative product rule	$\frac{d}{dt}(\vec{A} \cdot \vec{B}) = \frac{d\vec{A}}{dt} \cdot \vec{B} + \vec{A} \cdot \frac{d\vec{B}}{dt}$	
5. Power	$P = \frac{dW}{dt} = \vec{F} \cdot \vec{v}$	
6. Center-of-Mass Work–Translational-Kinetic-Energy Relation	$\int_1^2 \vec{F}_{\text{net ext}} \cdot d\vec{\ell}_{\text{cm}} = \Delta K_{\text{trans}}$ This relation is a useful problem-solving tool if for systems that cannot be modeled as a particle.	6-2
Center-of-Mass Work	$W_{\text{cm}} = \int_1^2 \vec{F} \cdot d\vec{\ell}_{\text{cm}}$	6-3
Translational Kinetic Energy	$K_{\text{trans}} = \frac{1}{2}Mv_{\text{cm}}^2$, where $M = \Sigma m_i$	

Answers to Concept Checks

6-1 The work being done by the force is negative.

Answers to Practice Problems

6-1 34 J

6-2 1.7×10^2 N

6-3 4.1 m/s

6-4 The region of interest is below the x axis, so the "area under the curve" is negative. The "area under the curve" is $-(\lvert A_1\rvert - \lvert A_2\rvert)$, where A_1 and A_2 are shown in Figure 6-28. The work done by the spring is equal to

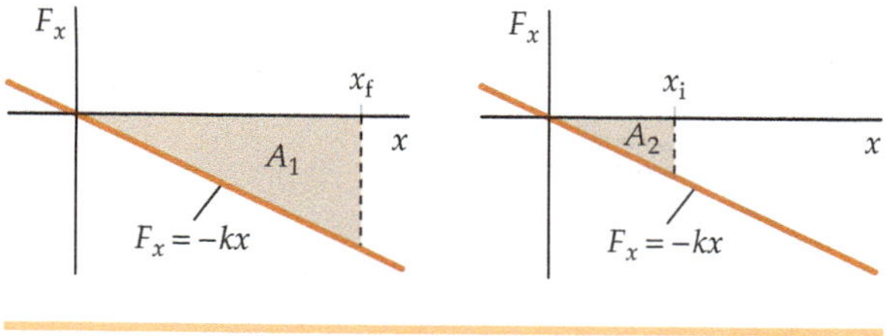

FIGURE 6-28

the "area under the curve," and the area of a triangle is one-half the altitude times the base. Thus,

$$W_{\text{by spring}} = -(|A_1| - |A_2|) = -(\tfrac{1}{2}\cdot kx_f \cdot x_f - \tfrac{1}{2}\cdot kx_i \cdot x_i) = \boxed{\tfrac{1}{2}kx_i^2 - \tfrac{1}{2}kx_f^2}$$

which is identical to Equation 6-13.

6-5 0.40 m/s

6-6 (*a*) 38 m^2, (*b*) $A = 5.0$ m, $B = 8.2$ m, $\phi = 23°$

6-7 $F = 5.00$ N, $\phi = 121°$

6-8 $P = 2.0 \times 10^2$ W, $W = 8.0 \times 10^3$ J, $W = -8.0 \times 10^3$ J

PROBLEMS

In a few problems, you are given more data than you actually need; in a few other problems, you are required to supply data from your general knowledge, outside sources, or informed estimates.

Interpret as significant all digits in numerical values that have trailing zeros and no decimal points.

For all problems, use g = 9.81 m/s² for the free-fall acceleration due to gravity and neglect friction and air resistance unless instructed to do otherwise.

- • Single-concept, single-step, relatively easy
- •• Intermediate-level, may require synthesis of concepts
- ••• Challenging
- SSM Solution is in the *Student Solutions Manual*
- Consecutive problems that are shaded are paired problems.

CONCEPTUAL PROBLEMS

1 • True or false: (*a*) If the net or total work done on a particle was not zero, then its speed must have changed. (*b*) If the net or total work done on a particle was not zero, then its velocity must have changed. (*c*) If the net or total work done on a particle was not zero, then its direction of motion could not have changed. (*d*) No work is done by the forces acting on a particle if it remains at rest. (*e*) A force that is always perpendicular to the velocity of a particle never does work on the particle.

2 • You push a heavy box in a straight line along the top of a rough horizontal table. The box starts at rest and ends at rest. Describe the work done on it (including sign) by each force acting on it and the net work done on it.

3 • You are riding on a Ferris wheel that is rotating at constant speed. True or false: During any fraction of a revolution: (*a*) None of the forces acting on you does work on you. (*b*) The total work done by all forces acting on you is zero. (*c*) There is zero net force on you. (*d*) You are accelerating.

4 • By what factor does the kinetic energy of a particle change if its speed is doubled but its mass is cut in half?

5 • Give an example of a particle that has constant kinetic energy but is accelerating. Can a non-accelerating particle have a changing kinetic energy? If so, give an example.

6 • An particle initially has kinetic energy K. Later it is found to be moving in the opposite direction with three times its initial speed. What is the kinetic energy now? (*a*) K, (*b*) $3K$, (*c*) $23K$, (*d*) $9K$, (*e*) $-9K$

7 • How does the work required to stretch a spring 2.0 cm from its unstressed length compare with the work required to stretch it 1.0 cm from its unstressed length? SSM

8 • A spring is first stretched 2.0 cm from its unstressed length. It is then stretched an additional 2.0 cm. How does the work required for the second stretch compare to the work required for the first stretch (give a ratio of second to first)?

9 • The dimension of power is (*a*) $M\cdot L^2\cdot T^2$, (*b*) $M\cdot L^2/T$, (*c*) $M\cdot L^2/T^2$, (*d*) $M\cdot L^2/T^3$.

10 • Show that the SI units of the force constant of a spring can be written as kg/s^2.

11 • True or false: (*a*) The gravitational force cannot do work on an object, because it is not a contact force. (*b*) Static friction can never do work on an object. (*c*) As a negatively charged electron is removed from a positively charged nucleus, the force on the electron does work that as a positive value. (*d*) If a particle is moving along a circular path, the total work being done on it is necessarily zero.

12 •• A hockey puck has an initial velocity in the $+x$ direction on a horizontal sheet of ice. Qualitatively sketch the force-versus-position graph for the (constant) horizontal force that would need to act on the puck to bring it to rest. Assume that the puck is located at $x = 0$ when the force begins to act. Show that the sign of the area under the curve agrees with the sign of the change in the puck's kinetic energy and interpret this in terms of the work–kinetic-energy theorem.

13 •• True or false: (*a*) The scalar product cannot have units. (*b*) If the scalar product of two nonzero vectors is zero, then they are parallel. (*c*) If the scalar product of two nonzero vectors is equal to the product of their magnitudes, then the two vectors are parallel. (*d*) As an object slides up an incline, the sign of the scalar product of the force of gravity on it and its displacement is negative. SSM

14 •• (*a*) Must the scalar product of two perpendicular unit vectors always be zero? If not, give an example. (*b*) An object has a velocity $\vec{v}$ at some instant. Interpret $\sqrt{\vec{v} \cdot \vec{v}}$ physically. (*c*) A ball rolls off a horizontal table. What is the scalar product between its velocity and its acceleration the instant after it leaves the table? Explain. (*d*) In Part (*c*), what is the sign of the scalar product of its velocity and acceleration the instant before it impacts the floor?

15 •• You lift a package vertically upward a distance L in time Δt. You then lift a second package that has twice the mass of the first package vertically upward the same distance while providing the same power as required for the first package. How much time does lifting the second package take (answer in terms of Δt)?

16 •• There are lasers that output more than 1.0 GW of power. A typical large modern electric generation plant typically outputs 1.0 GW of electrical power. Does this mean the laser outputs a huge amount of energy? Explain. *Hint: These high-power lasers are pulsed on and off, so they are not outputting power for very long time intervals.*

17 •• You are driving a car that accelerates from rest on a level road without spinning its wheels. Use the center-of-mass work–translational-kinetic-energy relation and free-body diagrams to clearly explain which force (or forces) is (are) directly responsible for the gain in translational kinetic energy of both you and the car. *Hint: The relation refers to external forces only, so the car's engine is not the answer. Pick your "system" correctly for each case.* SSM

ESTIMATION AND APPROXIMATION

18 •• (*a*) Estimate the work done on you by gravity as you take an elevator from the ground floor to the top of the Empire State Building, a building 102 stories high. (*b*) Estimate the amount of work the normal force of the floor did on you. *Hint: The answer is not zero.* (*c*) Estimate the average power of the force of gravity.

19 •• **ENGINEERING APPLICATION, CONTEXT-RICH** The nearest stars, apart from the Sun, are light-years away from Earth. If we are to investigate these stars, our space ships will have to travel at an appreciable fraction of the speed of light. (*a*) You are in charge of estimating the energy required to accelerate a 10,000-kg capsule from rest to 10 percent of the speed of light in one year. What is the minimum amount of energy that is required? Note that at velocities approaching the speed of light, the kinetic energy formula $\frac{1}{2}mv^2$ is not correct. However, it gives a value that is within 1% of the correct value for speeds up to 10% of the speed of light. (*b*) Compare your estimate to the amount of energy that the United States uses in a year (about 5×10^{20} J). (*c*) Estimate the minimum average power required of the propulsion system.

20 •• The mass of the Space Shuttle orbiter is about 8×10^4 kg and the period of its orbit is 90 min. Estimate the kinetic energy of the orbiter and the work done on it by gravity between launch and orbit. (Although the force of gravity decreases with altitude, this effect is small in low-Earth orbit. Use this fact to make the necessary approximation; you do not need to do an integral.) The orbits are about 250 miles above the surface of Earth.

21 • **CONTEXT-RICH** Ten inches of snow have fallen during the night, and you must shovel out your 50-ft-long driveway (Figure 6-29). Estimate how much work you do on the snow by completing this task. Make a plausible guess of any value(s) needed (the width of the driveway, for example), and state the basis for each guess.

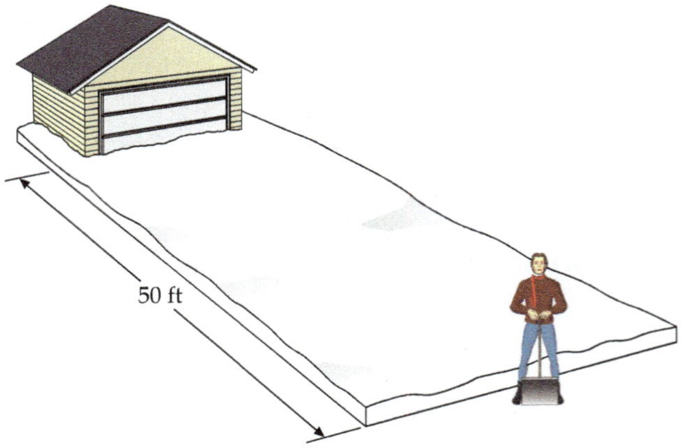

FIGURE 6-29 Problem 21

WORK, KINETIC ENERGY, AND APPLICATIONS

22 • A 15-g piece of space junk has a speed of 1.2 km/s. (*a*) What is its kinetic energy? (*b*) What is its kinetic energy if its speed is halved? (*c*) What is its kinetic energy if its speed is doubled?

23 • Find the kinetic energy of (*a*) a 0.145-kg baseball moving with a speed of 45.0 m/s, and (*b*) a 60.0-kg jogger running at a steady pace of 9.00 min/mi.

24 • A 6.0-kg box is raised a distance of 3.0 m from rest by a vertical applied force of 80 N. Find (*a*) the work done on the box by the applied force, (*b*) the work done on the box by gravity, and (*c*) the final kinetic energy of the box.

25 • A constant 80-N force acts on a 5.0 kg box. The box initially is moving at 20 m/s in the direction of the force, and 3.0 s later the box is moving at 68 m/s. Determine both the work done by this force and the average power delivered by the force during the 3.0-s interval.

26 •• You run a race with a friend. At first you each have the same kinetic energy, but she is running faster than you are. When you increase your speed by 25 percent, you are running at the same speed she is. If your mass is 85 kg, what is her mass?

27 •• A 3.0-kg particle moving along the x axis has a velocity of +2.0 m/s as it passes through the origin. It is subjected to a single force, F_x, that varies with position, as shown in Figure 6-30. (*a*) What is the kinetic energy of the particle as it passes through the origin? (*b*) How much work is done by the force as the particle moves from $x = 0.0$ m to $x = 4.0$ m? (*c*) What is the speed of the particle when it is at $x = 4.0$ m? SSM

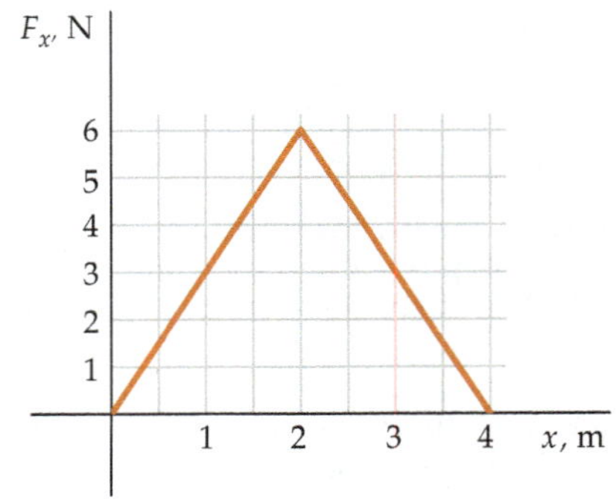

FIGURE 6-30 Problem 27

28 •• A 3.0-kg object moving along the x axis has a velocity of 2.4 m/s as it passes through the origin. It is acted on by a single force, F_x, that varies with x, as shown in Figure 6-31. (*a*) Find the work done by the force from $x = 0.0$ m to $x = 2.0$ m. (*b*) What is the kinetic energy of the object at $x = 2.0$ m? (*c*) What is the speed of the object at $x = 2.0$ m? (*d*) What is the work done on the object from $x = 0.0$ to $x = 4.0$ m? (*e*) What is the speed of the object at $x = 4.0$ m?

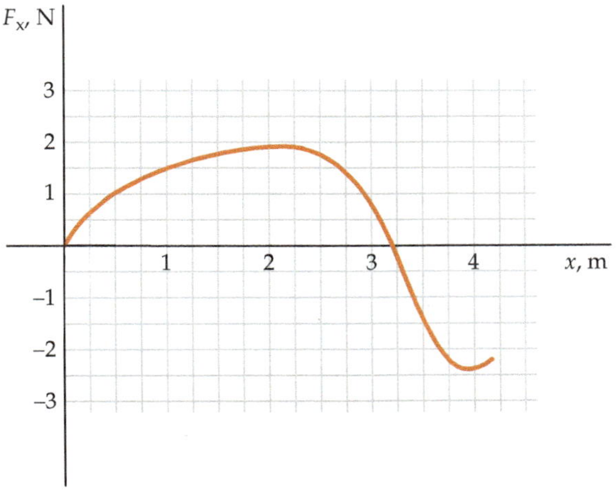

FIGURE 6-31 Problem 28

29 •• One end of a light spring (force constant k) is attached to the ceiling, the other end is attached to an object of mass m. The spring initially is vertical and unstressed. You then "ease the object down" to an equilibrium position a distance h below its initial position. Next, you repeat this experiment, but instead of easing the object down, you release it, with the result that it falls a distance H below the initial position before momentarily stopping. (*a*) Show that $h = mg/k$. (*b*) Use the work–kinetic-energy theorem to show that $H = 2h$. Try this experiment on your own.

30 •• A force F_x acts on a particle that has a mass of 1.5 kg. The force is related to the position x of the particle by the formula $F_x = Cx^3$, where $C = 0.50$ if x is in meters and F_x is in newtons. (*a*) What are the SI units of C? (*b*) Find the work done by this force as the particle moves from $x = 3.0$ m to $x = 1.5$ m. (*c*) At $x = 3.0$ m, the force points opposite the direction of the particle's velocity (speed is 12.0 m/s). What is its speed at $x = 1.5$ m? Can you tell its direction of motion at $x = 1.5$ m using only the work–kinetic-energy theorem? Explain.

31 •• You have a vacation cabin that has a nearby solar (black) water container used to provide a warm outdoor shower. For a few days last summer, your pump went out and you had to personally haul the water up the 4.0 m from the pond to the tank. Suppose your bucket has a mass of 5.0 kg and holds 15.0 kg of water when it is full. However, the bucket has a hole in it, and as you moved it vertically at a constant speed v, water leaked out at a constant rate. By the time you reached the top, only 5.0 kg of water remained. (*a*) Write an expression for the mass of the bucket plus water as a function of the height above the pond surface. (*b*) Find the work done by you on the bucket for each 5.0 kg of water delivered to the tank.

32 •• A 6.0-kg block slides 1.5 m down a frictionless incline that makes an angle of 60° with the horizontal. (*a*) Draw the free-body diagram of the block, and find the work done by each force when the block slides 1.5 m (measured along the incline). (*b*) What is the total work done on the block? (*c*) What is the speed of the block after it has slid 1.5 m, if it starts from rest? (*d*) What is its speed after 1.5 m, if it starts with an initial speed of 2.0 m/s?

33 •• **ENGINEERING APPLICATION** You are designing a jungle-vine–swinging sequence for the latest Tarzan movie. To determine his speed at the low point of the swing and to make sure it does not exceed mandatory safety limits, you decide to model the system of Tarzan + vine as a pendulum. Assume your model consists of a particle (Tarzan, mass 100 kg) hanging from a light string (the vine) of length ℓ attached to a support. The angle between the vertical and the string is written as ϕ. (*a*) Draw a free-body diagram for the object on the end of the string (Tarzan on the vine). (*b*) An infinitesimal distance along the arc (along which the object travels) is $\ell d\phi$. Write an expression for the total work dW_{total} done on the particle as it traverses that distance for an arbitrary angle ϕ. (*c*) If the $\ell = 7.0$ m, and if the particle starts from rest at an angle 50°, determine the particle's kinetic energy and speed at the low point of the swing using the work–kinetic-energy theorem. SSM

34 •• *Simple machines* are frequently used for reducing the amount of force that must be supplied to perform a task such as lifting a heavy weight. Such machines include the screw, block-and-tackle systems, and levers, but the simplest of the simple machines is the inclined plane. In Figure 6-32, you are raising a heavy box to the height of the truck bed by pushing it up an inclined plane (a ramp). (*a*) The *mechanical advantage MA* of the inclined plane is defined as the ratio of the magnitude of the force it would take to lift the block straight up (at constant speed) to the magnitude of the force it would take to push it up the ramp (at constant speed). If the plane is frictionless, show that $MA = 1/\sin\theta = L/H$, where H is the height of the truck bed and L is the length of the ramp. (*b*) Show that the work you do by moving the block into the truck is the same whether you lift it straight up or push it up the frictionless ramp.

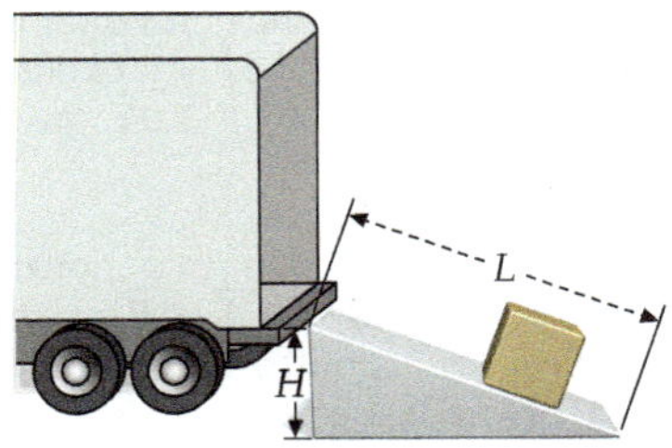

FIGURE 6-32 Problem 34

35 •• Particle a has mass m, is initially located on the positive x axis at $x = x_0$ and is subject to a repulsive force F_x from particle b. The location of particle b is fixed at the origin. The force F_x is inversely proportional to the square of the distance x between the particles. That is, $F_x = A/x^2$, where A *is a positive constant*. Particle a is released from rest and allowed to move under the influence of the force. Find an expression for the work done by the force on a as a function of x. Find both the kinetic energy and speed of a as x approaches infinity.

36 • You exert a force of magnitude F on the free end of the rope. (*a*) If the load moves up a distance h, through what distance does the point of application of the force move? (*b*) How much work is done by the rope on the load? (*c*) How much work do you do on the rope? (*d*) The mechanical advantage (defined in Problem 34) of this system is the ratio F/F_g, where F_g is the weight of the load. What is this mechanical advantage?

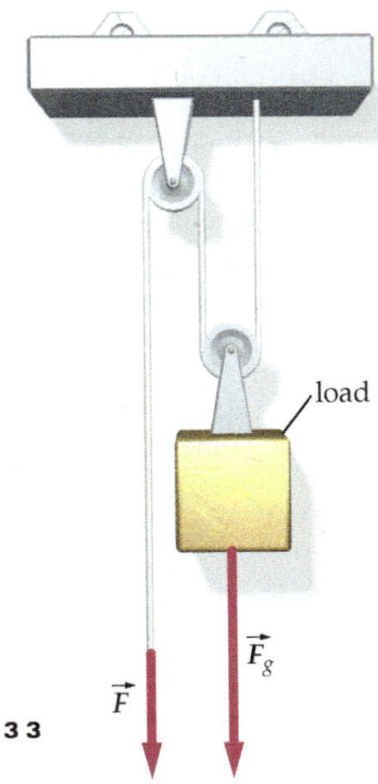

FIGURE 6-33 Problem 36

SCALAR (DOT) PRODUCTS

37 • What is the angle between the vectors $\vec{A}$ and $\vec{B}$ if $\vec{A}\cdot\vec{B} = -AB$?

38 • Two vectors $\vec{A}$ and $\vec{B}$ each have magnitudes of 6.0 m and the angle between their directions is 60°. Find $\vec{A}\cdot\vec{B}$.

39 • Find $\vec{A}\cdot\vec{B}$ for the following vectors: (*a*) $\vec{A} = 3\hat{i} - 6\hat{j}$, $\vec{B} = -4\hat{i} + 2\hat{j}$; (*b*) $\vec{A} = 5\hat{i} + 5\hat{j}$, $\vec{B} = 2\hat{i} - 4\hat{j}$; and (*c*) $\vec{A} = 6\hat{i} + 4\hat{j}$, $\vec{B} = 4\hat{i} - 6\hat{j}$.

40 • Find the angles between the vectors $\vec{A}$ and $\vec{B}$ given: (*a*) $\vec{A} = 3\hat{i} - 6\hat{j}$, $\vec{B} = -4\hat{i} + 2\hat{j}$; (*b*) $\vec{A} = 5\hat{i} + 5\hat{j}$, $\vec{B} = 2\hat{i} - 4\hat{j}$; and (*c*) $\vec{A} = 6\hat{i} + 4\hat{j}$, $\vec{B} = 4\hat{i} - 6\hat{j}$.

41 • A 2.0-kg particle is given a displacement of $\Delta\vec{r} = (3.0\text{ m})\hat{i} + (3.0\text{ m})\hat{j} + (-2.0\text{ m})\hat{k}$. During the displacement, a constant force $\vec{F} = (2.0\text{ N})\hat{i} - (1.0\text{ N})\hat{j} + (1.0\text{ N})\hat{k}$ acts on the particle. (*a*) Find the work done by $\vec{F}$ for this displacement. (*b*) Find the component of $\vec{F}$ in the direction of this displacement.

42 •• (*a*) Find the unit vector that is in the same direction as the vector $\vec{A} = 2.0\hat{i} - 1.0\hat{j} - 1.0\hat{k}$. (*b*) Find the component of the vector $\vec{A} = 2.0\hat{i} - 1.0\hat{j} - 1.0\hat{k}$ in the direction of the vector $\vec{B} = 3.0\hat{i} + 4.0\hat{j}$.

43 •• (*a*) Given two nonzero vectors $\vec{A}$ and $\vec{B}$, show that if $|\vec{A} + \vec{B}| = |\vec{A} - \vec{B}|$, then $\vec{A} \perp \vec{B}$. (*b*) Given a vector $\vec{A} = 4\hat{i} - 3\hat{j}$, find a vector in the *xy* plane that is perpendicular to $\vec{A}$ and has a magnitude of 10. Is this the only vector that satisfies the specified requirements? Explain. SSM

44 •• Unit vectors $\hat{A}$ and $\hat{B}$ are in the *xy* plane. They make angles of θ_1 and θ_2, respectively, with the $+x$ axis. (*a*) Use trigonometry to find the *x* and *y* components of the two vectors directly. (Your answer should be in terms of the angles.) (*b*) By considering the scalar product of $\hat{A}$ and $\hat{B}$, show that $\cos(\theta_1 - \theta_2) = \cos\theta_1\cos\theta_2 + \sin\theta_1\sin\theta_2$.

45 •• In Chapter 8, we shall introduce a new vector for a particle, called its *linear momentum*, symbolized by $\vec{p}$. Mathematically, it is related to the mass *m* and velocity $\vec{v}$ of the particle by $\vec{p} = m\vec{v}$. (*a*) Show that the particle's kinetic energy *K* can be expressed as $K = \dfrac{\vec{p}\cdot\vec{p}}{2m}$. (*b*) Compute the linear momentum of a particle of mass 2.5 kg that is moving at a speed of 15 m/s at an angle of 25° clockwise from the $+x$ axis in the *xy* plane. (*c*) Compute its kinetic energy using both $K = \frac{1}{2}mv^2$ and $K = \dfrac{\vec{p}\cdot\vec{p}}{2m}$ and verify that they give the same result.

46 ••• (*a*) Let $\vec{A}$ be a constant vector in the *xy* plane with its tail at the origin. Let $\vec{r} = x\hat{i} + y\hat{j}$ be a vector in the *xy* plane that satisfies the relation $\vec{A}\cdot\vec{r} = 1$. Show that the points with coordinates (*x*, *y*) lie on a straight line. (*b*) If $\vec{A} = 2\hat{i} - 3\hat{j}$, find the slope and *y* intercept of the line. (*c*) If we now let $\vec{A}$ and $\vec{r}$ be vectors in three-dimensional space, show that the relation $\vec{A}\cdot\vec{r} = 1$ specifies a plane.

47 ••• A particle moves in a circle that is centered at the origin and the magnitude of its position vector $\vec{r}$ is constant. (*a*) Differentiate $\vec{r}\cdot\vec{r} = r^2 =$ constant with respect to time to show that $\vec{v}\cdot\vec{r} = 0$, and therefore $\vec{v} \perp \vec{r}$. (*b*) Differentiate $\vec{v}\cdot\vec{r} = 0$ with respect to time and show that $\vec{a}\cdot\vec{r} + v^2 = 0$, and therefore $a_r = -v^2/r$. (*c*) Differentiate $\vec{v}\cdot\vec{v} = v^2$ with respect to time to show that $\vec{a}\cdot\hat{v} = dv/dt$, and therefore $a_t = dv/dt$. SSM

WORK AND POWER

48 • Force A does 5.0 J of work in 10 s. Force B does 3.0 J of work in 5.0 s. Which force delivers greater power, A or B? Explain.

49 • MULTISTEP A single force of 5.0 N in the $+x$ direction acts on an 8.0-kg object. (*a*) If the object starts from rest at $x = 0$ at time $t = 0$, write an expression for the power delivered by this force as a function of time. (*b*) What is the power delivered by this force at time $t = 3.0$ s?

50 • Find the power delivered by a force $\vec{F}$ acting on a particle that moves with a velocity $\vec{v}$, where (*a*) $\vec{F} = (4.0\text{ N})\hat{i} + (3.0\text{ N})\hat{k}$ and $\vec{v} = (6.0\text{ m/s})\hat{i}$; (*b*) $\vec{F} = (6.0\text{ N})\hat{i} - (5.0\text{ N})\hat{j}$ and $\vec{v} = -(5.0\text{ m/s})\hat{i} + (4.0\text{ m/s})\hat{j}$; and (*c*) $\vec{F} = (3.0\text{ N})\hat{i} + (6.0\text{ N})\hat{j}$ and $\vec{v} = (2.0\text{ m/s})\hat{i} + (3.0\text{ m/s})\hat{j}$.

51 •• ENGINEERING APPLICATION You are in charge of installing a small food-service elevator (called a *dumbwaiter* in the food industry) in a campus cafeteria. The elevator is connected by a pulley system to a motor, as shown in Figure 6-34. The motor raises and lowers the dumbwaiter. The mass of the dumbwaiter is 35 kg. In operation, it moves at a speed of 0.35 m/s upward, without accelerating (except for a brief initial period, which we can neglect, just after the motor is turned on). Electric motors typically have an efficiency of 78%. If you purchase a motor with an efficiency of 78%, what minimum power rating should the motor have? Assume that the pulleys are frictionless. SSM

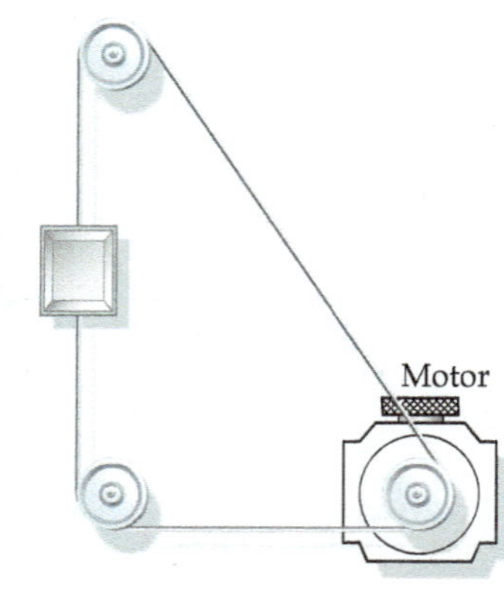

FIGURE 6-34 Problem 51

52 •• A cannon placed at the edge of a cliff of height *H* fires a cannonball directly upward with an initial speed v_0. The cannonball rises, falls back down (missing the cannon by a small margin), and lands at the foot of the cliff. Neglecting air resistance, calculate the velocity $\vec{v}$ as a function of time, and show explicitly that the integral of $\vec{F}_{net}\cdot\vec{v}$ over the time that the cannonball spends in flight is equal to the change in the kinetic energy of the cannonball over the same time.

53 •• A particle of mass *m* moves from rest at $t = 0$ under the influence of a single constant force $\vec{F}$. Show that the power delivered by the force at any time *t* is $P = F^2t/m$.

54 •• A 7.5-kg box is being lifted by means of a light rope that is threaded through a single, light, frictionless pulley that is attached to the ceiling. (*a*) If the box is being lifted at a *constant speed* of 2.0 m/s, what is the power delivered by the person pulling on the rope? (*b*) If the box is lifted, at *constant acceleration,* from rest on the floor to a height of 1.5 m above the floor in 0.42 s, what average power is delivered by the person pulling on the rope?

* CENTER OF MASS WORK AND CENTER OF MASS TRANSLATIONAL KINETIC ENERGY

55 ••• ENGINEERING APPLICATION, CONTEXT-RICH, SPREADSHEET You have been asked to test drive a car and study its actual performance relative to its specifications. This particular car's engine is rated at 164 hp. This value is the *peak* rating, which means that it is capable, at most, of providing energy at the rate of 164 hp to the drive wheels. You determine that the car's mass (including test equipment and driver on board) is 1220 kg. (*a*) When cruising at a constant 55.0 mi/h, your onboard engine-monitoring computer determines that the engine is producing 13.5 hp. From previous coasting experiments, it has been determined that the coefficient of rolling friction on the car is 0.0150. Assume that the drag force on the car varies as the square of the car's speed. That is, $F_d = Cv^2$. (*a*) What is the value of the constant, *C*? (*b*) Considering the peak power, what is the maximum speed (to the nearest 1 mi/h) that you would expect the car could attain? (This problem can be done by hand analytically, but it can be done more easily and quickly using a **graphing calculator** or **spreadsheet**.)

56 •• **CONTEXT-RICH, CONCEPTUAL** As you drive your car along a country road at night, a deer jumps out of the woods and stands in the middle of the road ahead of you. This occurs just as you are passing from a 55-mi/h zone to a 50-mi/h zone. At the 50-mi/h speed-limit sign, you slam on the car's brakes, causing them to lock up, and skid to a stop inches from the startled deer. As you breathe a sigh of relief, you hear the sound of a police siren. The policeman proceeds to write you a ticket for driving 56 mi/h in 50-mi/h zone. Because of your preparation in physics, you are able to use the 25-m-long skid marks that your car left behind as evidence that you were not speeding. What evidence do you present? In formulating your answer, you will need to know the coefficient of kinetic friction between automobile tires and dry concrete (see Table 5-1).

GENERAL PROBLEMS

57 • **APPROXIMATION** In February 2002, a total of 60.7 billion kW·h of electrical energy was generated by nuclear power plants in the United States. At that time, the population of the United States was about 287 million people. If the average American has a mass of 60 kg, and if 25% of the entire energy output of all nuclear power plants was diverted to supplying energy for a single giant elevator, estimate the height *h* to which the entire population of the country could be lifted by the elevator. In your calculations, assume also that *g* is constant over the entire height *h*.

58 • **ENGINEERING APPLICATION** One of the most powerful cranes in the world operates in Switzerland. It can slowly raise a 6000-t load to a height of 12.0 m. (Note that 1 t = one tonne is sometimes called a metric ton. It is a unit of mass, not force, and is equal to 1000 kg.) (*a*) How much work is done by the crane during this lift? (*b*) If it takes 1.00 min to lift this load to this height at constant velocity, and the crane is 20 percent efficient, find the total (gross) power rating of the crane.

59 • In Austria, there once was a 5.6-km-long ski lift. It took about 60 min for a gondola to travel up its length. If there were 12 gondolas going up, each with a cargo of mass 550 kg, and if there were 12 empty gondolas going down, and the angle of ascent was 30°, estimate the power *P* the engine needed to deliver in order to operate the ski lift.

60 •• **ENGINEERING APPLICATION** To complete your master's degree in physics, your advisor has you design a small, linear accelerator capable of emitting protons, each with a kinetic energy of 10.0 keV. (The mass of a single proton is 1.67×10^{-27} kg.) In addition, 1.00×10^9 protons per second must reach the target at the end of the 1.50-m-long accelerator. (*a*) What the average power must be delivered to the stream of protons? (*b*) What force (assumed constant) must be applied to each proton? (*c*) What speed does each proton attain just before it strikes the target, assuming the protons start from rest?

61 ••• The four strings pass over the bridge of a violin, as shown in Figure 6-35. The strings make and angle of 72.0° with the normal to the plane of the instrument on either side of the bridge. The resulting total normal force pressing the bridge into the violin is 1.00×10^3 N. The length of the strings from the bridge to the peg to which each is attached is 32.6 cm. (*a*) Determine the tension in the strings, assuming the tension is the same for each string. (*b*) One of the strings is plucked out a distance of 4.00 mm, as shown. Make a free-body diagram showing all of the forces acting on the segment of the string in contact with the finger (not shown), and determine the force pulling the segment back to its equilibrium position. Assume the tension in the string remains constant during the pluck. (*c*) Determine the work done on the string in plucking it out that distance. Remember that the net force pulling the string back to its equilibrium position is changing as the string is being pulled out, but assume that the magnitudes of the tension forces remain constant.

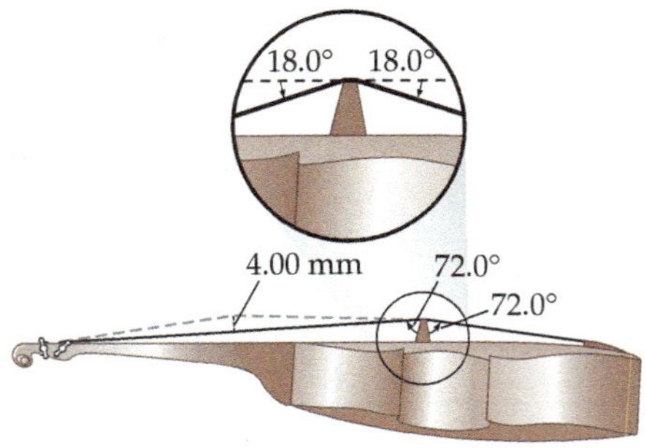

FIGURE 6-35 Problem 61

62 •• The magnitude of the single force acting on a particle of mass *m* is given by $F = bx^2$, where *b* is a constant. The particle starts from rest. After it travels a distance *L*, determine its (*a*) kinetic energy and (*b*) speed.

63 •• A single horizontal force in the $+x$ direction acts on a cart of mass *m*. The cart starts from rest at $x = 0$, and the speed of the cart increases with *x* as $v = Cx$, where *C* is a constant. (*a*) Find the force acting on the cart as a function of *x*. (*b*) Find the work done by the force in moving the cart from $x = 0$ to $x = x_1$. **SSM**

64 ••• A force $\vec{F} = (2.0\ \text{N/m}^2)x^2\hat{i}$ is applied to a particle initially at rest in the *xy* plane. Find the work done by this force on the particle and the final speed of the particle as it moves along a path that is (*a*) in a straight line from point (2.0 m, 2.0 m) to point (2.0 m, 7.0 m) and (*b*) in a straight line from point (2.0 m, 2.0 m) to point (5.0 m, 6.0 m). The given force is the only force doing work on the particle.

65 •• A particle of mass *m* moves along the *x* axis. Its position varies with time according to $x = 2t^3 - 4t^2$, where *x* is in meters and *t* is in seconds. Find (*a*) the velocity and acceleration of the particle as functions of *t*, (*b*) the power delivered to the particle as a function of *t*, and (*c*) the work done by the net force from $t = 0$ to $t = t_1$.

66 •• A 3.0-kg particle starts from rest at $x = 0.050$ m and moves along the *x* axis under the influence of a single force $F_x = 6.0 + 4.0x - 3.0x^2$, where F_x is in newtons and *x* is in meters. (*a*) Find the work done by the force as the particle moves from $x = 0.050$ m to $x = 3.0$ m. (*b*) Find the power delivered to the particle as it passes through the point $x = 3.0$ m.

67 •• The initial kinetic energy imparted to a 0.0200-kg bullet is 1200 J. (*a*) Assuming it accelerated down a 1.00-m-long rifle barrel, estimate the average power delivered to it during the firing. (*b*) Neglecting air resistance, find the range of this projectile when it is fired at an angle such that the range equals the maximum height attained.

68 •• The force F_x acting on a 0.500-kg particle is shown as a function of *x* in Figure 6-36. (*a*) From the graph, calculate the work done by the force when the particle moves from $x = 0.00$ to the following values of *x*: -4.00, -3.00, -2.00, -1.00, $+1.00$, $+2.00$,

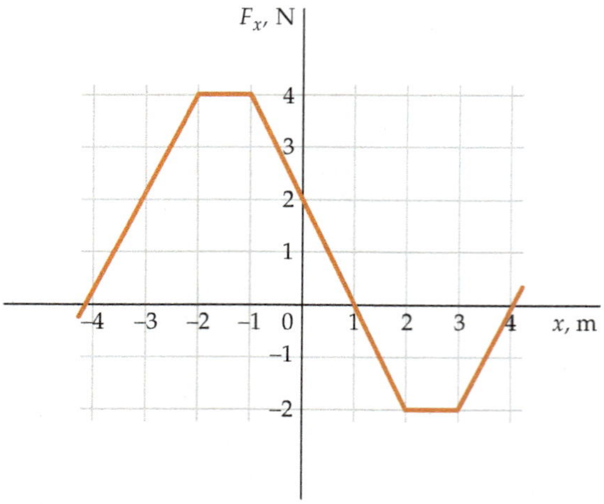

FIGURE 6-36 Problem 68

+3.00, and +4.00 m. (*b*) If it starts with a velocity of 2.00 m/s in the +x direction, how far will the particle go in that direction before stopping?

69 •• (*a*) Repeat Problem 68(*a*) for the force F_x shown in Figure 6-37. (*b*) If the object starts at the origin moving to the right with a kinetic energy of 25.0 J, how much kinetic energy does it have at x = 4.00 m.

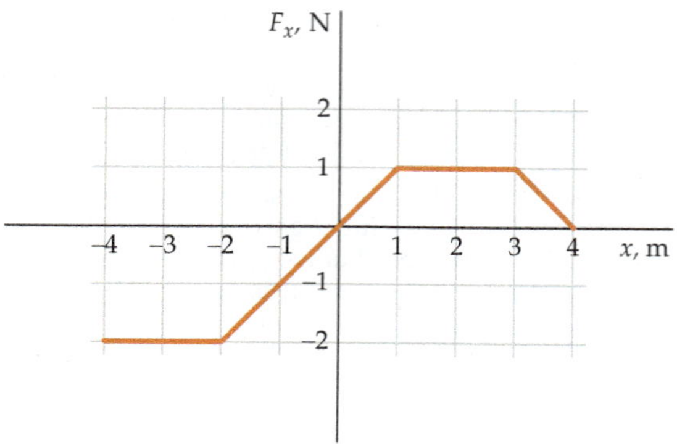

FIGURE 6-37 Problem 69

70 •• A box of mass M is at rest at the bottom of a frictionless inclined plane (Figure 6-38). The box is attached to a string that pulls with a constant tension T. (*a*) Find the work done by the tension T as the box moves through a distance x along the plane. (*b*) Find the speed of the box as a function of x. (*c*) Determine the power delivered by the tension in the string as a function of x.

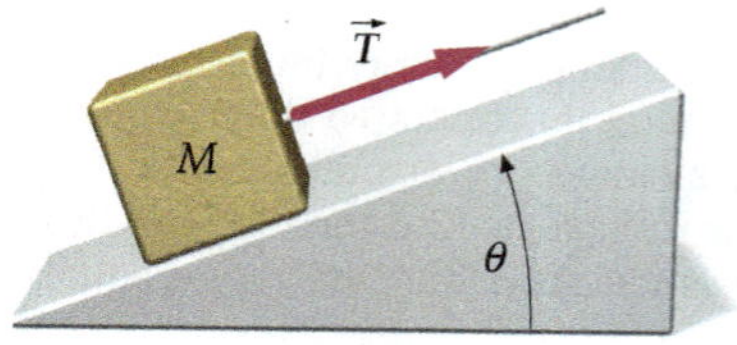

FIGURE 6-38 Problem 70

71 ••• A force acting on a particle in the xy plane at coordinates (x, y) is given by $\vec{F} = (F_0/r)(y\hat{i} - x\hat{j})$, where F_0 is a positive constant and r is the distance of the particle from the origin. (*a*) Show that the magnitude of this force is F_0 and that its direction is perpendicular to $\vec{r} = x\hat{i} + y\hat{j}$. (*b*) Find the work done by this force on a particle that moves once around a circle of radius 5.0 m that is centered at the origin. SSM

72 ••• A force acting on a 2.0-kg particle in the xy plane at coordinates (x, y) is given by $\vec{F} = -(b/r^3)(x\hat{i} + y\hat{j})$, where b is a positive constant and r is the distance from the origin. (*a*) Show that the magnitude of the force is inversely proportional to r^2, and that its direction is antiparallel (opposite) to the radius vector $\vec{r} = x\hat{i} + y\hat{j}$. (*b*) If $b = 3.0\ \text{N}\cdot\text{m}^2$, find the work done by this force as the particle moves from (2.0 m, 0.0 m), to (5.0 m, 0.0 m) along a straight-line path. (*c*) Find the work done by this force on a particle moving once around a circle of radius r = 7.0 m that is centered at the origin.

73 ••• A block of mass m on a horizontal frictionless tabletop is attached by a swivel to a spring that is attached to the ceiling (Figure 6-39). The vertical distance between the top of the block and the ceiling is y_0, and the horizontal position is x. When the block is at $x = 0$, the spring, which has force constant k, is completely unstressed. (*a*) What is F_x, the x component of the force on the block due to the spring, as a function of x? (*b*) Show that F_x is proportional to x^3 for sufficiently small values of $|x|$. (*c*) If the block is released from rest at $x = x_0$, where $|x_0| \ll y_0$, what is its speed when it reaches $x = 0$?

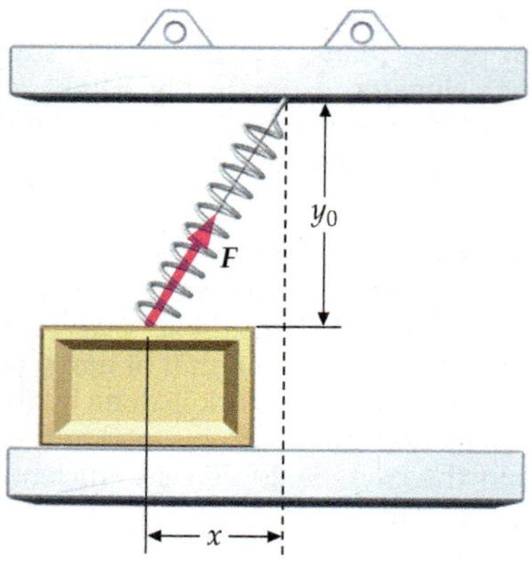

FIGURE 6-39 Problem 73

74 •• Two horses pull a large crate at constant speed across a barn floor by means of two light steel cables. A large box of mass 250 kg sits on the crate (Figure 6-40). As the horses pull, the cables are parallel to the horizontal floor. The coefficient of friction between the crate and the barn floor is 0.25. (*a*) What is the work done by each horse if the box is moved a distance of 25 m? (*b*) What is the tension in each cable if the angle between each cable and the direction the crate moves is 15°?

FIGURE 6-40 Problem 74

CHAPTER 7

AS A ROLLER COASTER RUSHES THROUGH ITS TWISTS AND TURNS, ENERGY IS TRANSFERRED IN DIFFERENT WAYS. ELECTRICAL POTENTIAL ENERGY PURCHASED FROM THE POWER COMPANY IS TRANSFORMED INTO GRAVITATIONAL POTENTIAL ENERGY AS THE CARS AND PASSENGERS ARE RAISED TO THE HIGHEST POINT OF THE RIDE. AS THE ROLLER-COASTER CARS PLUMMET DOWN THE STEEP INCLINE, THIS GRAVITATIONAL POTENTIAL ENERGY IS TRANSFORMED INTO KINETIC ENERGY AND THERMAL ENERGY—INCREASING THE TEMPERATURE OF BOTH THE CARS AND THEIR SURROUNDINGS BY A SMALL AMOUNT. *(Michael S. Lewis/Corbis.)*

Conservation of Energy

When work is done by one system on another, energy is transferred between the two systems. For example, when you pull a sled, energy from you goes partly into the kinetic energy of the sled and partly into the thermal energy that arises from the friction between the sled and the snow. At the same time, the internal chemical energy of your body decreases. The net result is the transfer of the internal chemical energy of your body to the external kinetic energy of the sled plus the thermal energy of the sled and the snow. This transfer of energy highlights one of the most important principles in science, *the law of conservation of energy,* which states that the total energy of a system and its surroundings does not change. Whenever the energy of a system changes, we can account for the change by the appearance or disappearance of energy somewhere else.

? How can we use the concept of energy transformation to determine how high the cars must be when they start their descent for them to complete the loop-the-loop? (See Example 7-8.)

In this chapter, we continue the study of energy begun in Chapter 6 by describing and applying the law of conservation of energy and examining the energy associated with several different states, including potential energy and thermal energy. We also discuss that energy changes for a system are often not continuous, but occur in discrete "bundles" or "lumps" called quanta. *Although a quantum of energy in a macroscopic system typically is so small that it goes unnoticed, its presence has profound consequences for microscopic systems such as atoms and molecules.*

7-1 POTENTIAL ENERGY

In Chapter 6, we showed that the total work done on a *particle* equals the change in its kinetic energy. However, sometimes a particle is part of a *system* consisting of two or more particles, and we need to examine the external work done on the system.* Often, the energy transferred to such a system by the work done by external forces on the system does not go into increasing the total kinetic energy *of the system*. Instead, the energy transferred is stored as **potential energy**—energy associated with the relative positions of different parts of the system. The configuration of a system is the way the different parts of the system are positioned relative to each other. Potential energy is an energy associated with system configuration, whereas kinetic energy is an energy associated with motion.

For example, consider a pile driver whose driver is suspended some distance h above a pile (a long, slender column). When the driver is released, it falls—gaining kinetic energy until it smashes into the pile, driving the pile into the ground. The driver is then raised back up to its initial height and released again. Each time the driver is raised from its lowest to its highest position, a gravitational force does work on it given by $-mgh$, where m is the mass of the driver. A second force, the force provided by the lifting agent, acts on the driver. As the driver is raised, the force exerted by the lifting agent does work on the driver that has a positive value. During the raising of the driver, these two work values sum to zero. We know they sum to zero because during the lift, the driver can be modeled as a particle, so the work–kinetic-energy theorem (Equation 6-8) tells us that the total work done on the driver is equal to the change in its kinetic energy—which is zero.

Consider lifting a barbell of mass m to a height h. The barbell starts at rest and ends at rest, so the net change in the kinetic energy of the barbell is zero. The barbell can be modeled as a particle during the lift, so the work–kinetic-energy theorem tells us that the total work done on the barbell is zero. There are two forces on the barbell, the force of gravity and the force of your hands. The gravitational force on the barbell is $m\vec{g}$, and the work done on the barbell by this force during the lift is $-mgh$. Because we know that the total work done on the barbell is zero, it follows that the work done on the barbell by the force of your hands is $+mgh$.

Consider the barbell and planet Earth to be a *system* of two particles (Figure 7-1). (You are not part of this system.) The external forces acting *on the Earth–barbell system* are the three forces exerted on it by you. These forces are the contact force by your hands *on the barbell*, the contact force by your feet *on the floor*, and the gravitational force by you *on Earth*. The gravitational force on Earth by you is equal and opposite to the gravitational force on you by Earth. (The gravitational forces you and the barbell attract each other with are negligible.) The barbell moves through a displacement of one or two meters, but displacements of the floor and planet Earth are negligibly small, so the force exerted on the barbell by your hands is the only one of the three external forces that does work on the Earth–barbell system. Thus, the total work done on this system by all three external forces is $+mgh$ (the work done on the barbell by your hands). The energy transferred to the system by this work is stored as *gravitational potential energy*, energy associated with the position of the barbell relative to Earth (energy associated with the height of the barbell above the floor).

Another system that stores energy associated with its configuration is a spring. If you stretch or compress a spring, energy associated with the length of the spring is stored as *elastic potential energy*. Consider the spring shown in Figure 7-2 as the system. You compress the spring, pushing it with equal and opposite forces $\vec{F}_1$ and $\vec{F}_2$. These forces sum to zero, so the net force on the spring remains zero. Thus, there is no change in the kinetic energy of the spring. The energy transfer associated with the work you do on the spring is stored not as kinetic energy, but as

* Systems of particles are discussed more thoroughly in Chapter 8.

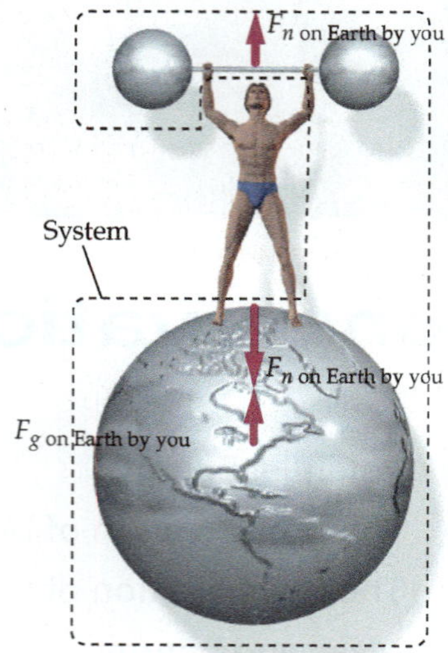

FIGURE 7-1

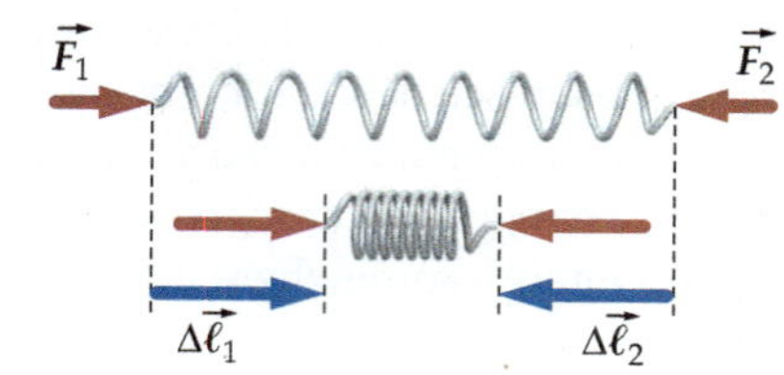

FIGURE 7-2 The spring is compressed by external forces $\vec{F}_1$ and $\vec{F}_2$. Both forces do work on the spring as they compress it. These work values are positive, so the elastic potential energy of the spring increases as it is compressed.

elastic potential energy. The configuration of this system has been changed, as evidenced by the change in the length of the spring. The total work done on the spring is positive because both $\vec{F}_1$ and $\vec{F}_2$ do positive work. (The work done by $\vec{F}_1$ is positive because both $\vec{F}_1$ and the displacement $\Delta\vec{\ell}_1$ are in the same direction. The same can be said for $\vec{F}_2$ and $\Delta\vec{\ell}_2$.)

CONSERVATIVE AND NONCONSERVATIVE FORCES

When you ride a ski lift to the top of a hill of height h, the work done on you by gravity is $-mgh$, where m is your mass. When you ski down the hill to the bottom, the work done by gravity is $+mgh$, independent of the shape of the hill (as you saw in Example 6-12). The total work done on you by gravity during the round-trip up and down the hill is zero and is independent of the path you take. In a situation such as this, where the total work done on an object by a force depends on only the initial and final positions of the object, and not the path taken, the force doing the work is called a **conservative force.**

> The work done by a conservative force on a particle is independent of the path taken as the particle moves from one point to another.
>
> DEFINITION—CONSERVATIVE FORCE

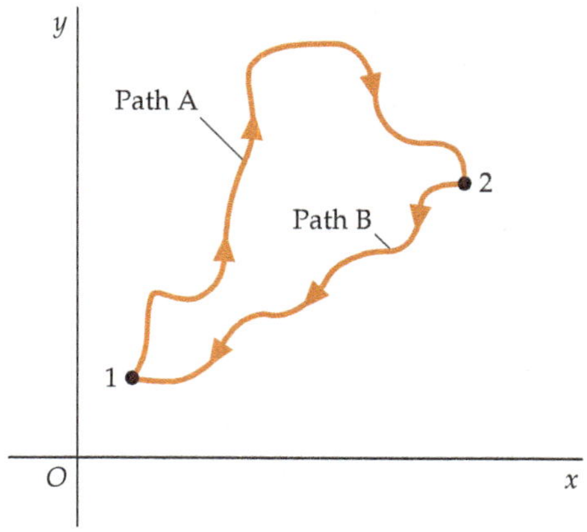

FIGURE 7-3 Two paths in space connecting the points 1 and 2. If the work done by a conservative force along path A from 1 to 2 is $+W$, the work done on the return trip along path B must be $-W$, because the roundtrip work is zero. When traversing path B from 1 to 2, the force is the same at each point, but the displacement is opposite to that when going on path B from 2 to 1. Thus, the work done along path B from 1 to 2 must also be W. It follows that the work done as a particle goes from point 1 to 2 is the same along any path connecting the two points.

From Figure 7-3 we see that this definition implies that:

> A force is conservative if the work it does on a particle is zero when the particle moves around *any* closed path, returning to its initial position.
>
> ALTERNATIVE DEFINITION—CONSERVATIVE FORCE

In the ski-lift example, the force of gravity, exerted by Earth on you, is a conservative force, because the total work done by gravity on you during the round-trip is zero, independent of the path you take. Both the gravitational force on an object and the force exerted by a spring of negligible mass on an object are conservative forces. (If a spring's mass is negligible then its kinetic energy is also negligible.) Any spring in this book has negligible mass unless otherwise stated.

Not all forces are conservative. A force is said to be **nonconservative** if it does not meet the definition for conservative forces. Suppose, for example, that you push a block across a table along a straight line from point A to point B and back again, so that the block ends up at A, its original position. Friction opposes the block's motion, so the force you push on the block with is in the direction of motion and the value of the work done by the push is positive on both legs of the round-trip. The total work done by the push does *not* equal zero. Thus, the push is an example of a nonconservative force.

As another example, consider the force $\vec{F}$ a donkey exerts on a pole as the donkey pulls the pole around the circle at constant speed. As the donkey walks, $\vec{F}$ is continuously doing work whose value is positive. The point of application (point P) of $\vec{F}$ returns to the same position each time the donkey completes one pass around the circle, so the work done by $\vec{F}$ is not equal to zero each time P completes one trip around a closed path (the circle). We can thus conclude that $\vec{F}$ is a nonconservative force.

A machine for pumping water. The donkey exerts force $\vec{F}$ on the pole at P, the point of application of the force. *(O. Alamany and E. Vicens/Corbis.)*

If the work done around *any* particular closed path is not zero, we can conclude that the force is nonconservative. However, we can conclude the force is conservative only if the work is zero around *all* possible closed paths. Because there are infinitely many possible closed paths, it is impossible to calculate the work done for each one. Therefore, finding a single closed path along which the work done by a particular force *is not zero* is sufficient for showing that the force is nonconservative, but is of limited use when investigating whether a given force is conservative. In more advanced physics courses, more sophisticated mathematical methods for evaluating whether a force is conservative are studied.

Example 7-1 Integral Around a Closed Path

To calculate the work done by a force $\vec{F}$ around a closed curve (path) C, we evaluate $\oint_C \vec{F}\cdot d\vec{\ell}$, where the circle on the integral sign means that the integration is evaluated for one complete trip around C. For $\vec{F} = Ax\hat{i}$, calculate $\oint_C \vec{F}\cdot d\vec{\ell}$ for the path C shown in Figure 7-4.

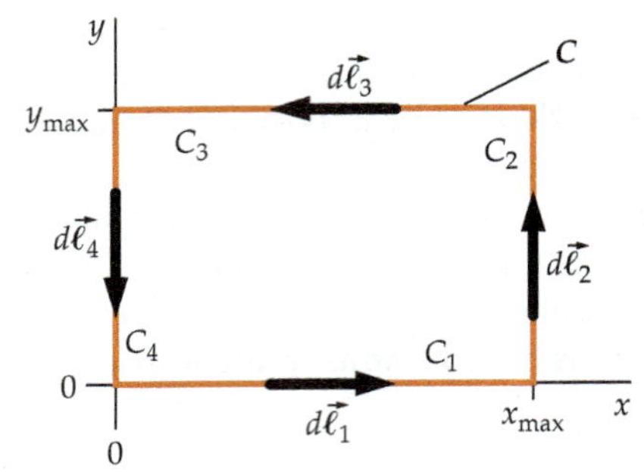

FIGURE 7-4

PICTURE The path C consists of four straight segments. Evaluate $d\vec{\ell} = dx\hat{i} + dy\hat{j}$ on each segment and calculate $\int \vec{F}\cdot d\vec{\ell}$ separately for each of the four segments.

SOLVE

1. The integral around C is equal to the sum of the integrals along the segments that make up C:

$$\oint_C \vec{F}\cdot d\vec{\ell} = \int_{C_1} \vec{F}\cdot d\vec{\ell}_1 + \int_{C_2} \vec{F}\cdot d\vec{\ell}_2 + \int_{C_3} \vec{F}\cdot d\vec{\ell}_3 + \int_{C_4} \vec{F}\cdot d\vec{\ell}_4$$

2. On C_1, $dy = 0$, so $d\vec{\ell}_1 = dx\hat{i}$:

$$\int_{C_1} \vec{F}\cdot d\vec{\ell}_1 = \int_0^{x_{max}} Ax\hat{i}\cdot dx\hat{i} = A\int_0^{x_{max}} x\,dx = \tfrac{1}{2}Ax_{max}^2$$

3. On C_2, $dx = 0$ and $x = x_{max}$, so $d\vec{\ell}_2 = dy\hat{j}$ and $\vec{F} = Ax_{max}\hat{i}$:

$$\int_{C_2} \vec{F}\cdot d\vec{\ell}_2 = \int_0^{y_{max}} Ax_{max}\hat{i}\cdot dy\hat{j} = Ax_{max}\int_0^{y_{max}} \hat{i}\cdot\hat{j}\,dy = 0$$

($\hat{i}\cdot\hat{j} = 0$ because $\hat{i}$ and $\hat{j}$ are perpendicular.)

4. On C_3, $dy = 0$, so $d\vec{\ell}_3 = dx\hat{i}$:

$$\int_{C_3} \vec{F}\cdot d\vec{\ell}_3 = \int_{x_{max}}^{0} Ax\hat{i}\cdot dx\hat{i} = -A\int_0^{x_{max}} x\,dx = -\tfrac{1}{2}Ax_{max}^2$$

5. On C_4, $dx = 0$ and $x = 0$, so $d\vec{\ell}_4 = dy\hat{j}$ and $\vec{F} = 0$:

$$\int_{C_4} \vec{F}\cdot d\vec{\ell}_4 = \int_{y_{max}}^{0} 0\hat{i}\cdot dy\hat{j} = 0$$

6. Add the step-2, -3, -4, and -5 results:

$$\oint_C \vec{F}\cdot d\vec{\ell} = \tfrac{1}{2}Ax_{max}^2 + 0 - \tfrac{1}{2}Ax_{max}^2 + 0 = \boxed{0}$$

CHECK The force is described by Hooke's law (the force for a spring). Thus, it is conservative, so the integral of this force around any closed path is zero.

TAKING IT FURTHER The negative sign in step 4 appeared because the integration limits were reversed.

PRACTICE PROBLEM 7-1 For $\vec{F} = Bxy\hat{i}$, calculate $\oint_C \vec{F}\cdot d\vec{\ell}$ for the path C shown in Figure 7-4.

POTENTIAL-ENERGY FUNCTIONS

The work done by a conservative force on a particle does not depend on the path, but it does depend on the endpoints of the path. We can use this property to define the **potential-energy function** U that is associated with a conservative force. Let us return to the ski-lift example once again. Now consider yourself and Earth to be a *two-particle system.* (The ski lift is not part of this system.) When a ski lift raises you to the top of the hill, it does work mgh on the you–Earth system. This work is stored as the gravitational potential energy of the you–Earth system. When you ski down the hill, this potential energy is converted to the kinetic energy of your motion. Note that when you ski down the hill, the work done by gravity *decreases* the potential energy of the system. We define the potential-energy function U such that the work done by a conservative force equals the decrease in the potential-energy function:

$$W = \int_1^2 \vec{F} \cdot d\vec{\ell} = -\Delta U$$

or

$$\Delta U = U_2 - U_1 = -\int_1^2 \vec{F} \cdot d\vec{\ell} \qquad \text{7-1}a$$

DEFINITION—POTENTIAL-ENERGY FUNCTION

This equation gives the change in potential energy due to a change in the configuration of the system as an object moves from point 1 to point 2.

For an infinitesimal displacement, $d\vec{\ell}$, the change in potential energy is given by

$$dU = -\vec{F} \cdot d\vec{\ell} \qquad \text{7-1}b$$

See Math Tutorial *for more information on* Integrals

Gravitational potential energy Using Equation 7-1*b* we can calculate the potential-energy function associated with the gravitational force near the surface of Earth. For the force $\vec{F} = -mg\hat{j}$, we have

$$dU = -\vec{F} \cdot d\vec{\ell} = -(-mg\hat{j}) \cdot (dx\hat{i} + dy\hat{j} + dz\hat{k}) = +mg\,dy$$

where we have exploited the fact that $\hat{j} \cdot \hat{i} = \hat{j} \cdot \hat{k} = 0$ and $\hat{j} \cdot \hat{j} = 1$. Integrating, we obtain

$$U = \int mg\,dy = mgy + U_0$$

$$U = U_0 + mgy \qquad \text{7-2}$$

GRAVITATIONAL POTENTIAL ENERGY NEAR EARTH'S SURFACE

where U_0, the arbitrary constant of integration, is the value of the potential energy at $y = 0$. Because only a change in the potential energy is defined, the actual value of U is not important. For example, if the gravitational potential energy of the Earth–skier system is chosen to be zero when the skier is at the bottom of the hill, its value when the skier is at a height h above that level is mgh. Or we could choose the potential energy to be zero when the skier is at point P halfway down the ski slope, in which case its value at any other point would be mgy, where y is the height of the skier above point P. On the lower half of the slope, the potential energy would then be negative.

! We are free to choose U to be zero at any convenient reference point.

PRACTICE PROBLEM 7-2

A 55-kg window washer stands on a platform 8.0 m above the ground. What is the potential energy U of the window-washer–Earth system if (*a*) U is chosen to be zero on the ground, (*b*) U is chosen to be zero 4.0 m above the ground, and (*c*) U is chosen to be zero 10 m above the ground?

Example 7-2 A Falling Bottle

A 0.350-kg bottle falls from rest from a shelf that is 1.75 m above the floor. Find the potential energy of the bottle–Earth system when the bottle is on the shelf and just before impact with the floor. Find the kinetic energy of the bottle just before impact.

PICTURE The work done by Earth on the bottle as it falls equals the negative of the change in the potential energy of the bottle–Earth system. Knowing the work, we can use the work–kinetic-energy theorem to find the kinetic energy.

SOLVE

1. Make a sketch showing the bottle on the shelf and again when it is about to impact the floor (Figure 7-5). Choose the potential energy of the bottle–Earth system to be zero when the bottle is on the floor, and place a y axis on the sketch with the origin at floor level:

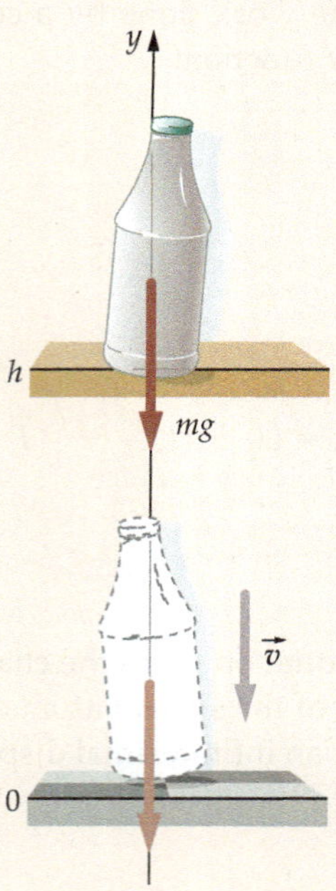

FIGURE 7-5

2. The only force doing work on the falling bottle is the force of gravity, so $W_{\text{total}} = W_g$. Apply the work–kinetic-energy theorem to the falling bottle:

$$W_{\text{total}} = W_g = \Delta K$$

3. The gravitational force exerted by Earth on the falling bottle is internal to the bottle–Earth system. It is also a conservative force, so the work done by it equals the negative of the change in the potential energy of the system:

$$W_g = -\Delta U = -(U_f - U_i) = -(mgy_f - mgy_i)$$
$$= mg(y_i - y_f) = mg(h - 0) = mgh$$

4. Substitute the step-3 result into the step-2 result and solve for the final kinetic energy. The original kinetic energy is zero:

$$mgh = \Delta K$$
$$mgh = K_f - K_i$$
$$K_f = K_i + mgh$$
$$= 0 + (0.350\text{ kg})(9.81\text{ N/kg})(1.75\text{ m})$$
$$= 6.01\text{ N}\cdot\text{m} = \boxed{6.01\text{ J}}$$

CHECK The units of the answer in step 4 are units of energy, because $1\text{ N}\cdot\text{m} = 1\text{ J}$.

TAKING IT FURTHER Potential energy is associated with the configuration of a *system of particles*, but we sometimes have systems, such as the bottle–Earth system in this example, in which only one particle moves (Earth's motion is negligible). For brevity, then, we sometimes refer to the potential energy of the bottle–Earth system as simply the potential energy of the bottle.

The gravitational potential energy of a system of particles in a uniform gravitational field is the same as if the entire mass of the system were concentrated at the system's center of mass. For such a system, let h_i be the height of the ith particle above some reference level. The gravitational potential energy of the system is then

$$U_g = \sum_i m_i g h_i = g \sum_i m_i h_i$$

where the sum is over all the particles in the system. By definition of the center of mass, the height of the center of mass of the system is given by

$$Mh_{\text{cm}} = \sum_i m_i h_i, \qquad \text{where } M = \sum_i m_i$$

Substituting Mh_{cm} for $\Sigma m_i h_i$ gives

$$U_g = Mgh_{cm} \qquad 7\text{-}3$$

GRAVITATIONAL POTENTIAL ENERGY OF A SYSTEM

Elastic potential energy Another example of a conservative force is that of a stretched (or compressed) spring of negligible mass. Suppose you pull a block attached to a spring from its equilibrium position at $x = 0$ to a new position at $x = x_1$ (Figure 7-6). The work done by the spring on the block is negative because the force exerted by the spring on the block and the displacement of the block are oppositely directed. If we then release the block, the force of the spring does positive work on the block as the block accelerates back toward its initial position. The total work done on the block by the spring as the block moves from $x = 0$ to $x = x_1$, and then back to $x = 0$, is zero. This result is independent of the size of x_1 (as long as the stretching is not so great as to exceed the elastic limit of the spring). The force exerted by the spring is therefore a conservative force. We can calculate the potential-energy function associated with this force from Equation 7-1*b*:

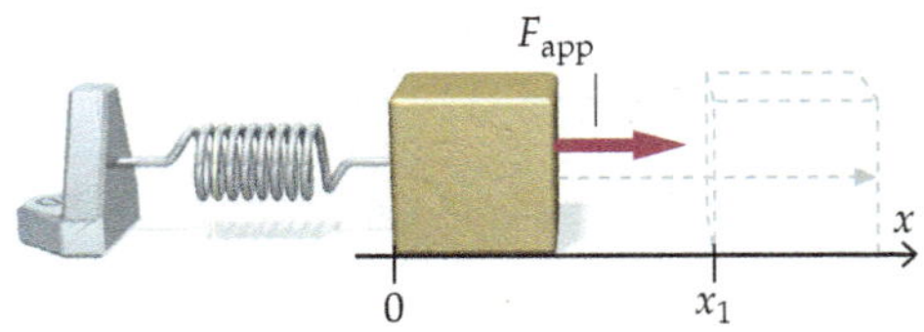

FIGURE 7-6 The applied force F_{app} pulls the block to the right, stretching the spring by x_1.

$$dU = -\vec{F} \cdot d\vec{\ell} = -F_x\,dx = -(-kx)dx = +kx\,dx$$

Then

$$U = \int kx\,dx = \frac{1}{2}kx^2 + U_0$$

where U_0 is the potential energy when $x = 0$, that is, when the spring is unstressed. Choosing U_0 to be zero gives

$$U = \tfrac{1}{2}kx^2 \qquad 7\text{-}4$$

POTENTIAL ENERGY OF A SPRING

The formula $U = \frac{1}{2}kx^2$ for the potential energy of a spring requires that the spring is relaxed if $x = 0$. Thus, the location of the point where $x = 0$ is not arbitrary when using the potential energy function $U = \frac{1}{2}kx^2$.

When we pull the block from $x = 0$ to $x = x_1$, the agent doing the pulling must exert an applied force on the block. If the block starts from rest at $x = 0$ and ends at rest at $x = x_1$, the change in its kinetic energy is zero. The work–energy theorem then tells us that the total work done on the block is zero. That is, $W_{app} + W_{spring} = 0$, or

$$W_{app} = -W_{spring} = \Delta U_{spring} = \tfrac{1}{2}kx_1^2 - 0 = \tfrac{1}{2}kx_1^2$$

The energy transferred from the agent doing the pulling to the block–spring system is equal to W_{app} and is stored as potential energy in the spring.

PRACTICE PROBLEM 7-3

A suspension spring on a Toyota Prius has a force constant of 11,000 N/m. How much energy is transferred to one of these springs when, starting from its relaxed length, it is compressed 30.0 cm?

Example 7-3 Potential Energy of a Basketball Player

A system consists of a 110-kg basketball player, the rim of a basketball hoop, and Earth. Assume that the potential energy of this system is zero when the player is standing on the floor and the rim is horizontal. Find the total potential energy of this system when the player is hanging on the front of the rim (much like that shown in Figure 7-7). Also, assume that the center of mass of the player is 0.80 m above the floor when he is standing and 1.30 m above the floor when he is hanging. The force constant of the rim is 7.2 kN/m and the front of the rim is displaced downward a distance of 15 cm. Assume the mass of the rim is negligible.

FIGURE 7-7 *(Elio Castoria/APF/Getty Images.)*

PICTURE When the player changes position from standing on the floor to hanging on the rim, the total change in potential energy is the change in gravitational potential energy plus the change in elastic potential energy stored in the stressed rim, whose potential energy can be measured just as if it were a spring: $U_s = \frac{1}{2}kx^2$. Choose 0.80 m above the floor as the reference point where $U_g = 0$.

SOLVE

1. Sketch the system, first in the initial configuration and again in the final configuration (Figure 7-8):

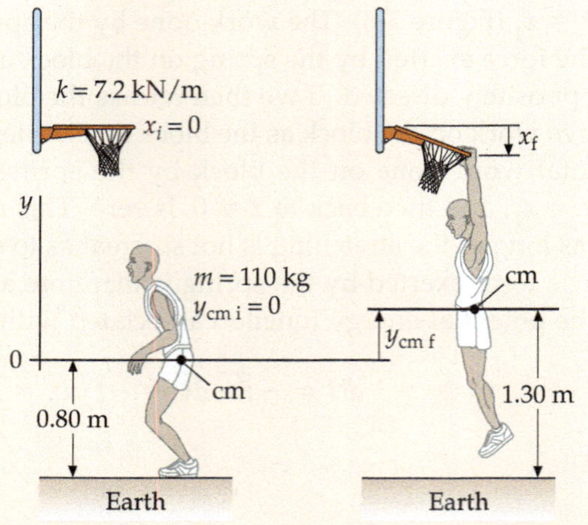

FIGURE 7-8 A basketball player jumps, grabs hold of the rim, and dangles from it.

2. The gravitational potential energy reference point where $U_g = 0$ is 0.80 m above the floor. Thus, $U_{gi} = 0$. The initial total potential energy equals zero:

$$U_{gi} = mgy_{cm\,i} = mg(0) = 0$$
$$U_{si} = \tfrac{1}{2}kx_i^2 = \tfrac{1}{2}k(0)^2 = 0$$
$$U_i = U_{gi} + U_{si} = 0$$

3. The total final potential energy is the sum of final gravitational potential energy and the final elastic potential energy of the rim:

$$U_f = U_{gf} + U_{sf} = mgy_{cm\,f} + \tfrac{1}{2}kx_f^2$$
$$= (110\text{ kg})(9.81\text{ N/kg})(0.50\text{ m}) + \tfrac{1}{2}(7.2\text{ kN/m})(0.15\text{ m})^2$$
$$= 540\text{ N}\cdot\text{m} + 81\text{ N}\cdot\text{m} = \boxed{6.2 \times 10^2\text{ J}}$$

CHECK The units check out if we use the definition of the joule. That definition is $1\text{ J} = 1\text{ N}\cdot\text{m}$.

TAKING IT FURTHER The front of the rim and player oscillate vertically immediately after the player grabs. However, they eventually come to rest with the front of the rim 15 cm below its initial position. The total potential energy is a minimum when the system is in equilibrium (Figure 7-9). Why this is so is explained near the end of Section 7-2.

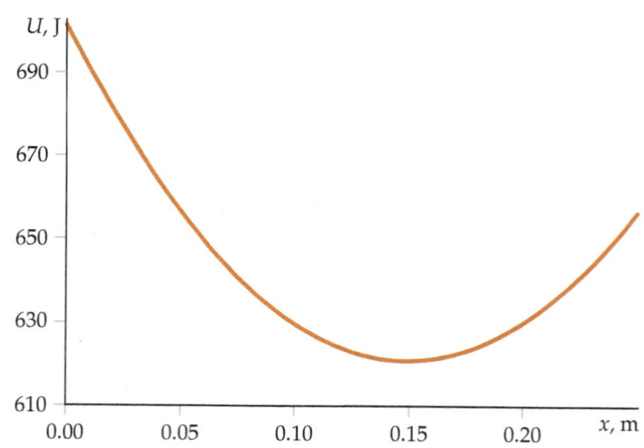

FIGURE 7-9 The graph shows the total potential energy $U = U_g + U_s$ as a function of the downward displacement of the rim.

PRACTICE PROBLEM 7-4 A 3.0-kg block is hung vertically from a spring with a force constant of 600 N/m. (*a*) By how much is the spring stretched? (*b*) How much potential energy is stored in the spring?

7-2 THE CONSERVATION OF MECHANICAL ENERGY

We are now ready to look at the relationship between kinetic energy and potential energy. Recall that the total work done on each particle in a system equals the change in the kinetic energy ΔK_i of that particle, so the total work done by all the forces W_{total} equals the change in the total kinetic energy of the system ΔK_{sys}:

$$W_{total} = \sum \Delta K_i = \Delta K_{sys} \qquad \text{7-5}$$

Two sets of forces do work on a particle in a system: the external forces and the internal forces. Each internal force is either conservative or nonconservative. The total work done by all forces equals the work done by all external forces W_{ext}, plus the work done by all internal nonconservative forces W_{nc}, plus that done by all internal conservative forces W_c:

$$W_{total} = W_{ext} + W_{nc} + W_c$$

Rearranging gives

$$W_{ext} + W_{nc} = W_{total} - W_c$$

The negative of the total work done by all the conservative internal forces $-W_c$ equals the change in the potential energy of the system ΔU_{sys}

$$-W_c = \Delta U_{sys} \qquad \text{7-6}$$

Substituting from Equations 7-5 and 7-6, we have

$$W_{ext} + W_{nc} = \Delta K_{sys} + \Delta U_{sys} \qquad \text{7-7}$$

The right side of this equation can be simplified as

$$\Delta K_{sys} + \Delta U_{sys} = \Delta(K_{sys} + U_{sys}) \qquad \text{7-8}$$

The sum of the kinetic energy K_{sys} and the potential energy U_{sys} is called the **total mechanical energy** E_{mech}:

$$E_{mech} = K_{sys} + U_{sys} \qquad \text{7-9}$$

DEFINITION—TOTAL MECHANICAL ENERGY

Combining Equations 7-8 and 7-9, and then substituting into Equation 7-7 gives

$$W_{ext} = \Delta E_{mech} - W_{nc} \qquad \text{7-10}$$

WORK–ENERGY THEOREM FOR SYSTEMS

The mechanical energy of a system of particles is conserved (E_{mech} = constant) if the total work done by all external forces and by all internal nonconservative forces is zero.

$$E_{mech} = K_{sys} + U_{sys} = \text{constant} \qquad \text{7-11}$$

CONSERVATION OF MECHANICAL ENERGY

This is **conservation of mechanical energy,** and is the origin of the expression "conservative force."

If $E_{\text{mech i}} = K_i + U_i$ is the initial mechanical energy of the system and $E_{\text{mech f}} = K_f + U_f$ is the final mechanical energy of the system, conservation of mechanical energy implies that

$$E_{\text{mech f}} = E_{\text{mech i}} \quad (\text{or } K_f + U_f = K_i + U_i) \tag{7-12}$$

In other words, when the mechanical energy of a system is conserved, we can relate the final mechanical energy of the system to the initial mechanical energy of the system without considering the intermediate motion and the work done by the forces involved. Therefore, conservation of mechanical energy allows us to solve problems that might be difficult to solve using Newton's laws directly.

APPLICATIONS

Suppose that you are on skis and, starting at rest from a height h_i above the bottom of a hill, you coast down the hill. Assuming that friction and air drag are negligible, what is your speed as you pass by a marker on the hill a height h above the bottom of the hill?

The mechanical energy of the Earth–skier system is conserved because the only force doing work is the internal, conservative force of gravity. If we choose $U = 0$ at the bottom of the hill, the initial potential energy is mgh_i. This energy is also the total mechanical energy because the initial kinetic energy is zero. Thus,

$$E_{\text{mech i}} = K_i + U_i = 0 + mgh_i$$

As you pass the marker, the potential energy is mgh and the speed is v. Hence,

$$E_{\text{mech f}} = K_f + U_f = \tfrac{1}{2}mv^2 + mgh$$

Setting $E_{\text{mech f}} = E_{\text{mech i}}$, we find

$$\tfrac{1}{2}mv^2 + mgh = mgh_i$$

Solving for v gives

$$v = \sqrt{2g(h_i - h)}$$

Your speed is the same as if you had undergone free-fall, falling straight down through a distance $h_i - h$. However, by skiing down the hill, you travel a greater distance and take more time than you would if you were in free-fall and falling straight down.

PROBLEM-SOLVING STRATEGY

Solving Problems Involving Mechanical Energy

PICTURE Identify a system that includes the object (or objects) of interest and any other objects that interact with the object of interest by either a conservative or a kinetic-frictional force.

SOLVE

1. Make a sketch of the system and include labels. Include a coordinate axis (or axes) and show the system in its initial and final configurations. (Showing an intermediate configuration is often helpful also.) Objects may be represented as dots, just as is done in free-body diagrams.
2. Identify any external forces acting on the system that do work, and any internal nonconservative forces that do work. Also identify any internal conservative forces that do work.
3. Apply Equation 7-10 (the work–energy theorem for systems). For each internal conservative force doing work use a potential-energy function to represent the work done.

CHECK Make sure that you have accounted for the work done by all conservative and nonconservative forces in determining your answer.

Example 7-4 Kicking a Ball

Standing near the edge of the roof of a 12-m-high building, you kick a ball with an initial speed of $v_i = 16$ m/s at an angle of 60° above the horizontal. Neglecting effects due to air resistance, find (*a*) how high above the height of the building the ball rises, and (*b*) its speed just before it hits the ground.

PICTURE We choose the ball and Earth as the system. We consider this system during the interval from just after the kick to just before impact with the ground. No external forces do work on the system, and no internal nonconservative forces do work, so the mechanical energy of the system is conserved. At the top of its flight, the ball is moving horizontally with a speed v_{top}, equal to the horizontal component of its initial velocity v_{ix}. We choose $y = 0$ at the roof of the building.

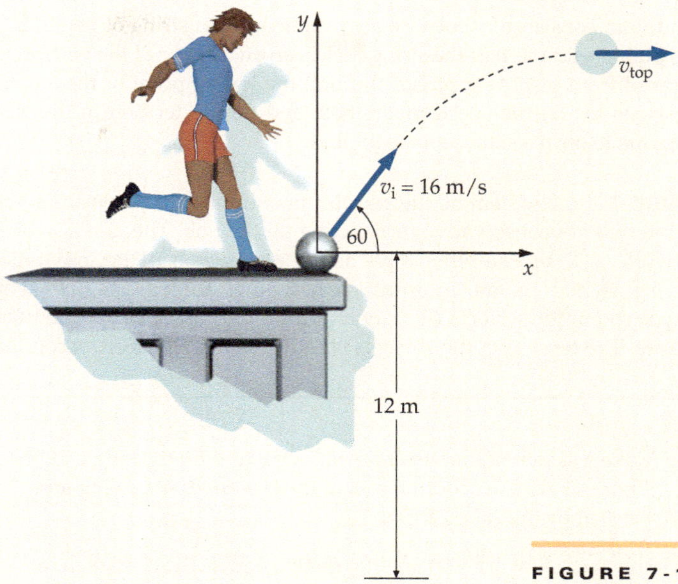

FIGURE 7-10

SOLVE

(*a*) 1. Make a sketch (Figure 7-10) of the trajectory. Include coordinate axes and show the initial position of the ball and its position at the top of its flight. Choose $y = 0$ at the roof of the building:

2. Apply the work–energy equation for systems. Choose the ball and Earth as the system. Following the kick and before impact with the ground no external forces do work and no nonconservative forces do work (we are neglecting the effects of air resistance):

$$W_{ext} = \Delta E_{mech} - W_{nc}$$
$$0 = \Delta E_{mech} - 0$$
$$\therefore E_{mech\,f} = E_{mech\,i}$$

3. The gravitational force does work on the system. We account for this work using the gravitational potential energy function mgy:

$$E_{mech\,top} = E_{mech\,i}$$
$$\tfrac{1}{2}mv_{top}^2 + mgy_{top} = \tfrac{1}{2}mv_i^2 + mgy_i$$
$$\tfrac{1}{2}mv_{top}^2 + mgh_{top} = \tfrac{1}{2}mv_i^2 + 0$$

4. Conservation of mechanical energy relates the height y_{top} above the roof of the building to the initial speed v_i and the speed at the top of its flight v_{top}:

$$E_{mech\,top} = E_{mech\,i}$$
$$\tfrac{1}{2}mv_{top}^2 + mgy_{top} = \tfrac{1}{2}mv_i^2 + mgy_i$$
$$\tfrac{1}{2}mv_{top}^2 + mgy_{top} = \tfrac{1}{2}mv_i^2 + 0$$

5. Solve for y_{top}:

$$y_{top} = \frac{v_i^2 - v_{top}^2}{2g}$$

6. The velocity at the top of its flight equals the x component of its initial velocity:

$$v_{top} = v_{ix} = v_i \cos\theta$$

7. Substitute the step-3 result into the step-2 result and solve for y_{top}:

$$y_{top} = \frac{v_i^2 - v_{top}^2}{2g} = \frac{v_i^2 - v_i^2\cos^2\theta}{2g} = \frac{v_i^2(1 - \cos^2\theta)}{2g}$$
$$= \frac{(16\text{ m/s})^2(1 - \cos^2 60°)}{2(9.81\text{ m s}^2)} = \boxed{9.8\text{ m}}$$

(*b*) 1. If v_f is the speed of the ball just before it hits the ground (where $y = y_f = -12$ m), its energy is expressed:

$$E_{mech\,f} = \tfrac{1}{2}mv_f^2 + mgy_f$$

2. Set the final mechanical energy equal to the initial mechanical energy:

$$\tfrac{1}{2}mv_f^2 + mgy_f = \tfrac{1}{2}mv_i^2 + 0$$

3. Solve for v_f, and set $y_f = -12$ m to find the final velocity:

$$v_f = \sqrt{v_i^2 - 2gy_f}$$
$$= \sqrt{(16\text{ m/s})^2 - 2(9.81\text{ m/s}^2)(-12\text{ m})}$$
$$= \boxed{22\text{ m/s}}$$

CHECK We would expect that the higher the building, the greater the speed at impact. The expression for v_f in step 3 of Part (*b*) reflects this expectation.

Example 7-5 A Pendulum

A pendulum consists of a bob of mass m attached to a string of length L. The bob is pulled aside so that the string makes an angle θ_0 with the vertical, and is released from rest. As it passes through the lowest point of the arc, find expressions for (*a*) the speed of the bob, and (*b*) the tension in the string. Effects due to air resistance are negligible.

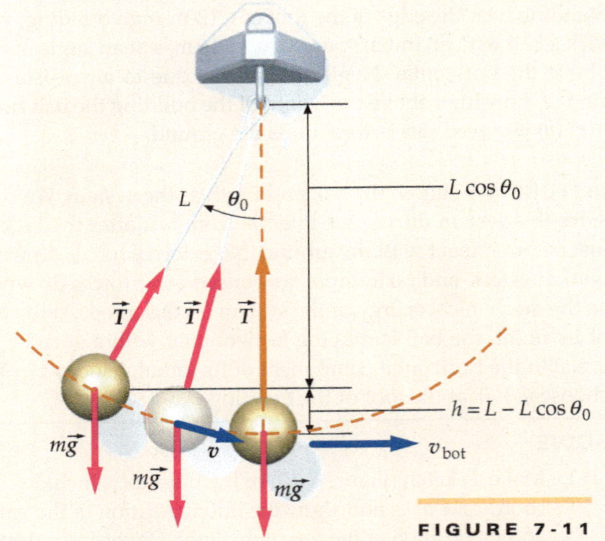

FIGURE 7-11

PICTURE Let the system be the pendulum and Earth. The tension force $\vec{T}$ is an internal nonconservative force acting on the bob. The rate at which $\vec{T}$ does work is $\vec{T}\cdot\vec{v}$. The other force acting on the bob is the gravitational force $m\vec{g}$, which is an internal conservative force. Use the work–energy theorem for systems (Equation 7-10) to find the speed at the bottom of the arc. The tension in the string is obtained using Newton's second law.

SOLVE

(*a*) 1. Make a sketch of the system in its initial and final configurations (Figure 7-11). We choose $y = 0$ at the bottom of the swing and $y = h$ at the initial position:

2. The external work done on the system equals the change in its mechanical energy minus the work done by internal nonconservative forces (Equation 7-10):

$$W_{ext} = \Delta E_{mech} - W_{nc}$$

3. There are no external forces acting on the system. The tension force is an internal nonconservative force:

$$W_{ext} = 0$$

$$W_{nc} = \int_1^2 \vec{T}\cdot d\vec{\ell}$$

4. The displacement increment $d\vec{\ell}$ equals the velocity times the time increment dt. Substitute into the step-3 result. The tension is perpendicular to the velocity, so $\vec{T}\cdot\vec{v} = 0$:

$$d\vec{\ell} = \vec{v}\,dt$$

$$\text{so}\quad W_{nc} = \int_1^2 \vec{T}\cdot d\vec{\ell} = \int_1^2 \vec{T}\cdot\vec{v}\,dt = 0$$

5. Substitute for W_{ext} and W_{nc} in the step-2 result. The bob initially is at rest:

$$W_{ext} = \Delta E_{mech} - W_{nc}$$

$$0 = \Delta E_{mech} - 0$$

$$\Delta E_{mech} = 0$$

6. Apply conservation of mechanical energy. The bob initially is at rest:

$$E_{mech\,f} = E_{mech\,i}$$

$$\tfrac{1}{2}mv_f^2 + mgy_f = \tfrac{1}{2}mv_i^2 + mgy_i$$

$$\tfrac{1}{2}mv_{bot}^2 + 0 = 0 + mgh$$

7. Conservation of mechanical energy thus relates the speed v_{bot} to the initial height $y_i = h$:

$$\tfrac{1}{2}mv_{bot}^2 = mgh$$

8. Solve for the speed v_{bot}:

$$v_{bot} = \sqrt{2gh}$$

9. To express speed in terms of the initial angle θ_0, we need to relate h to θ_0. This relation is illustrated in Figure 7-11:

$$L = L\cos\theta_0 + h$$

$$\text{so}\quad h = L - L\cos\theta_0 = L(1 - \cos\theta_0)$$

10. Substitute this value for h to express the speed at the bottom in terms of θ_0:

$$v_{bot} = \boxed{\sqrt{2gL(1 - \cos\theta_0)}}$$

(*b*) 1. When the bob is at the bottom of the circle, the forces on it are $m\vec{g}$ and $\vec{T}$. Apply $\Sigma F_y = ma_y$:

$$T - mg = ma_y$$

2. At the bottom, the bob has an acceleration v_{bot}^2/L in the centripetal direction (toward the center of the circle), which is upward:

$$a_y = \frac{v_{bot}^2}{L} = \frac{2gL(1 - \cos\theta_0)}{L} = 2g(1 - \cos\theta_0)$$

3. Substitute for a_y in the Part-(*b*), step-1 result and solve for T:

$$T = mg + ma_y = m(g + a_y) = m[g + 2g(1 - \cos\theta_0)]$$

$$= \boxed{(3 - 2\cos\theta_0)mg}$$

CHECK (1) The tension at the bottom is greater than the weight of the bob because the bob is accelerating upward. (2) Step 3 in Part (*b*) shows that for $\theta_0 = 0$, $T = mg$, the expected result for a stationary bob hanging from a string.

TAKING IT FURTHER (1) The rate at which a force does work is given by $\vec{F}\cdot\vec{v}$ (Equation 6-22). Step 4 of Part (*a*) reveals that the rate at which the tension force is doing work is zero. *Any* force that remains perpendicular to the velocity does zero work. (2) Step 8 of Part (*a*) shows that the speed at the bottom is the same as if the bob had dropped in free-fall from a height h. (3) The speed of the bob at the bottom of the arc can also be found using Newton's laws directly, but such a solution is more challenging because the tangential acceleration a_t varies with position, and therefore with time, so the constant-acceleration formulas do not apply. (4) If the string had not been included in the system then the W_{ext} would equal the work done by the tension force and W_{nc} would equal zero because there would be no internal nonconservative forces. The results would be identical.

Example 7-6 A Block Pushing a Spring

Try It Yourself

A 2.0-kg block on a frictionless horizontal surface is pushed against a spring that has a force constant of 500 N/m, compressing the spring by 20 cm. The block is then released, and the force of the spring accelerates the block as the spring decompresses. The block then glides along the surface and then up a frictionless incline of angle 45°. How far up the incline does the block travel before momentarily coming to rest?

PICTURE Let the system include the block, the spring, Earth, the horizontal surface, the ramp, and the wall to which the spring is attached. After the block is released there are no external forces on this system. The only forces that do work are the force exerted by the spring on the block and the force of gravity, both of which are conservative. Thus, the total mechanical energy of the system is conserved. Find the maximum height h from the conservation of mechanical energy, and then the maximum distance up the incline s from $\sin 45° = h/s$.

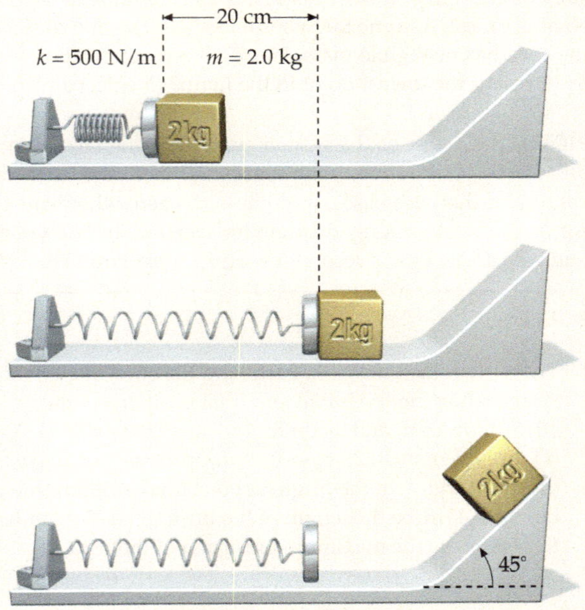

FIGURE 7-12

SOLVE

Cover the column to the right and try these on your own before looking at the answers.

Steps	Answers
1. Choose the block, the spring, Earth, the horizontal surface, the ramp, and the wall to which the spring is attached. Sketch the system with both the initial and final configurations (Figure 7-12).	
2. Apply the work–energy theorem for systems. Following release no external forces act on the system and no internal nonconservative forces do work on the system:	$W_{ext} = \Delta E_{mech} - W_{nc}$ $0 = \Delta E_{mech} - 0$ $\therefore E_{mech\,f} = E_{mech\,i}$
3. Write the initial mechanical energy in terms of the compression distance x_i.	$E_{mech\,i} = U_{s\,i} + U_{g\,i} + K_i = \frac{1}{2}kx_i^2 + 0 + 0$
4. Write the final mechanical energy in terms of the height h.	$E_{mech\,f} = U_{s\,f} + U_{g\,f} + K_f = 0 + mgh + 0$
5. Substitute into the step-3 result and solve for h.	$mgh = \frac{1}{2}kx_i^2$ $h = \dfrac{kx_i^2}{2mg} = 0.51\text{ m}$
6. Find the distance s from h and the angle of inclination (Figure 7-13).	$h = s \times \sin\theta$ $s = \boxed{0.72\text{ m}}$

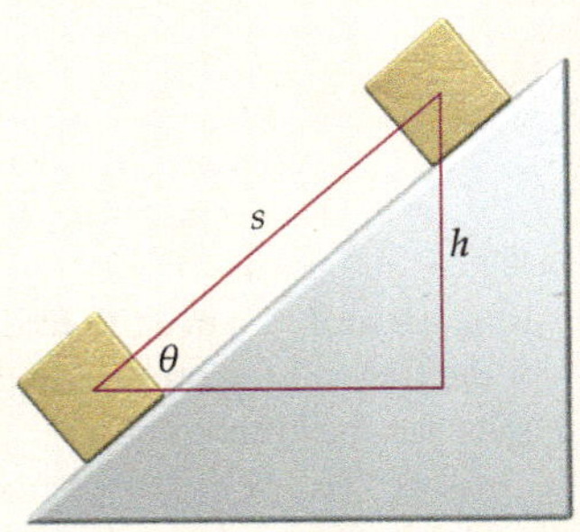

FIGURE 7-13

CHECK The expression for h in step 5 is plausible. It tells us by inspection that increasing x_i results in a larger maximum height, and using a larger mass results in a smaller maximum height.

TAKING IT FURTHER (1) In this problem, the initial mechanical energy of the system is the potential energy of the spring. This energy is transformed first into kinetic energy and then into gravitational potential energy. (2) The normal force $\vec{F}_n$ on the block always acts at right angles to the velocity, so $\vec{F}_n \cdot \vec{v} = 0$ at all times.

PRACTICE PROBLEM 7-5 Find the speed of the block just as it leaves the spring.

PRACTICE PROBLEM 7-6 How much work did the normal force on the block do?

Example 7-7 Bungee Jumping *Context-Rich*

You jump off a platform 134 m above the Nevis River. After you have free-fallen for the first 40 m, the bungee cord attached to your ankles starts to stretch. (The unstretched length of the cord is 40 m.) You continue to descend another 80 m before coming to rest. Assume that your mass is 100 kg, the cord follows Hooke's law, and the cord has negligible mass. What is your acceleration when you are momentarily at rest at the lowest point in the jump? (Neglect air drag.)

PICTURE Choose as the system everything mentioned in the problem statement, plus Earth. As you fall, your speed first increases, then reaches some maximum value, and then decreases until it is again zero when you are at your lowest point. Apply the work–energy theorem for systems. To find your acceleration at the bottom we use Newton's second law ($\Sigma F_x = ma_x$) and Hooke's law ($F_x = -kx$).

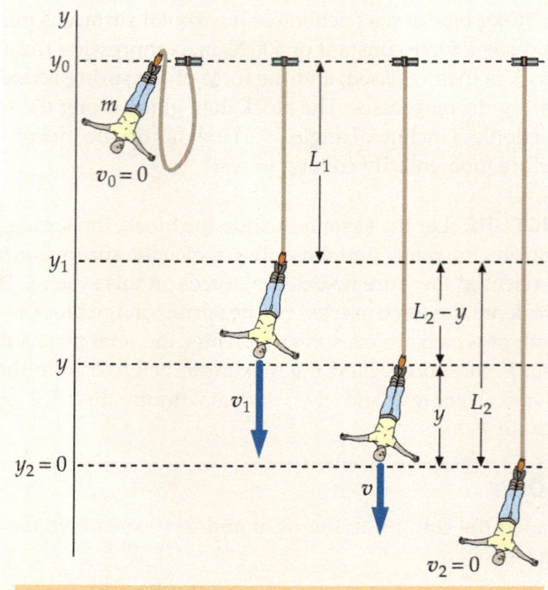

FIGURE 7-14

SOLVE

1. The system includes you, Earth and the cord. Sketch the system showing the initial and final positions of the first 40 m of descent, and again for the next 80 m of descent (Figure 7-14). Include a y axis with up as the positive y direction and with the origin at your final position (lowest). Let $L_1 = 40$ m be the length of the unstressed cord and let $L_2 = 80$ m be the maximum extension of the cord.

2. Apply the work–energy theorem for systems. There are no external forces and no internal nonconservative forces doing work:

$$W_{ext} = \Delta E_{mech} - W_{nc}$$
$$0 = \Delta E_{mech} - 0$$
$$\therefore E_{mech\,f} = E_{mech\,i}$$

3. Apply the step-2 result to the part of the descent that the cord is stretching. The extension of the cord is $L_2 - y$:

$$E_{mech\,3} = E_{mech\,2}$$
$$U_{g3} + U_{s3} + K_3 = U_{g2} + U_{s2} + K_2$$
$$mgy_3 + \tfrac{1}{2}k(L_2 - y_3)^2 + \tfrac{1}{2}mv_3^2 = mgy_2 + \tfrac{1}{2}ky_2^2 + \tfrac{1}{2}mv_2^2$$
$$0 + \tfrac{1}{2}kL_2^2 + 0 = mgL_2 + 0 + \tfrac{1}{2}mv_2^2$$
$$\tfrac{1}{2}kL_2^2 = mgL_2 + \tfrac{1}{2}mv_2^2$$

4. To solve for k, we need to find the kinetic energy at the end of the free-fall region. Apply the step-2 result again and solve for the kinetic energy:

$$E_{mech\,2} = E_{mech\,1}$$
$$U_{g2} + K_2 = U_{g1} + K_1$$
$$mgy_2 + \tfrac{1}{2}mv_2^2 = mgy_1 + \tfrac{1}{2}mv_1^2$$
$$mgL_2 + \tfrac{1}{2}mv_2^2 = mg(L_1 + L_2) + 0$$
$$\tfrac{1}{2}mv_2^2 = mgL_1$$

5. Substitute the step-4 result into the step-3 result and solve for k:

$$\tfrac{1}{2}kL_2^2 = mgL_2 + mgL_1$$

$$k = \frac{2mg(L_2 + L_1)}{L_2^2}$$

6. Apply Newton's second law when you are at the lowest point. First we construct a free-body diagram (Figure 7-15):

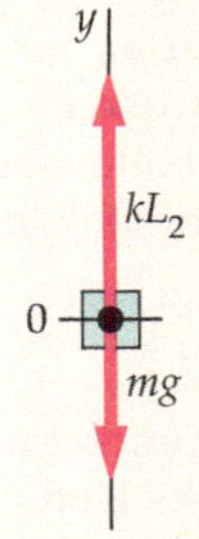

FIGURE 7-15

7. Apply Newton's second law and solve for the acceleration. Use the expression for k from step 5:

$$\Sigma F_y = ma_y$$

$$-mg + kL_2 = ma_y$$

$$a_y = -g + k\frac{L_2}{m} = -g + \frac{2mg(L_2 + L_1)}{L_2^2}\frac{L_2}{m}$$

$$= g\left(1 + 2\frac{L_1}{L_2}\right) = g\left(1 + 2\frac{40}{80}\right) = \boxed{2.0g}$$

CHECK We expect the acceleration at the bottom to be upward (the $+y$ direction) and our result is in agreement with that. Any time the velocity reverses directions, immediately following the reversal the velocity vector and the acceleration vector are in the same direction.

PRACTICE PROBLEM 7-7 As you fall, you gain speed until the upward pull of the cord equals the downward pull of gravity. What is your height above the lowest point when you achieve maximum speed?

Example 7-8 Back to the Future

Context-Rich

Imagine that you have time-traveled back to the late 1800s and are watching your great-great-great-grandparents on their honeymoon taking a ride on the Flip Flap Railway, a Coney Island roller coaster with a circular loop-the-loop. The car they are in is about to enter the loop-the-loop when a 100-lb sack of sand falls from a construction-site platform and lands in the back seat of the car. No one is hurt, but the impact causes the car to lose 25 percent of its speed. The car started from rest at a point 2 times as high as the top of the circular loop. Neglect losses due to friction or air drag. Will their car make it over the top of the loop-the-loop without falling off?

PICTURE Let the system be the car, its contents, the track (including the loop-the-loop), and Earth. The car has to have enough speed at the top of the loop to maintain contact with the track. We can use the work–energy theorem for systems to determine the speed just before the sandbag hits the car, and we can use it again to determine the speed the car has at the top of the loop. Then we can use Newton's second law to determine the magnitude of the normal force, if any, exerted on the car by the track.

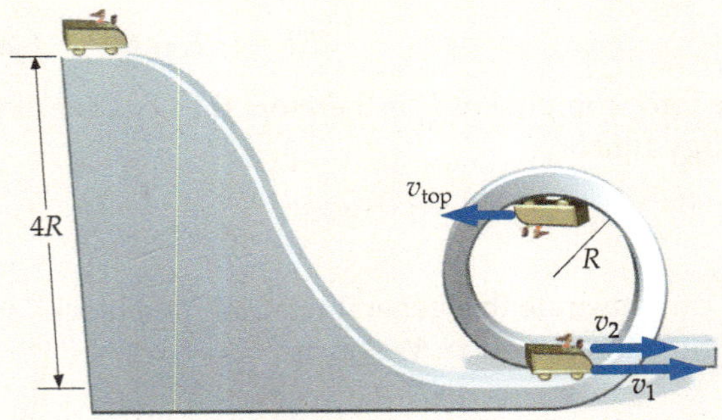

FIGURE 7-16

SOLVE

1. Choose the system to be the car, its contents, the track, and Earth. Draw a picture of the car and track, with the car at the starting point, at the bottom of the track, and again at the top of the loop (Figure 7-16):

2. Apply Newton's second law to relate the speed at the top of the loop to the normal force:

$$F_n + mg = m\frac{v_{top}^2}{R}$$

3. Apply the work–energy theorem to the interval prior to impact. There are no external forces and no internal nonconservative forces do work. Find the speed just prior to impact. Measuring heights from the bottom of the loop, the initial height of 4R, where R is the radius of the loop, is two times the height of the top of the loop:

$$W_{ext} = \Delta E_{mech} - W_{nc}$$
$$0 = \Delta E_{mech} - 0$$
$$\therefore E_{mech\,f} = E_{mech\,i}$$
$$U_0 + K_0 = U_1 + K_1$$
$$mg\,4R + 0 = 0 + \tfrac{1}{2}mv_1^2$$
$$\text{so}\quad v_1 = \sqrt{8Rg}$$

4. The impact with the sandbag results in a 25 percent decrease in speed. Find the speed after impact:

$$v_2 = 0.75v_1 = 0.75\sqrt{8Rg}$$

5. Apply the work–energy theorem to the interval following impact. Find the speed at the top of the loop-the-loop:

$$U_{top} + K_{top} = U_2 + K_2$$
$$mg2R + \tfrac{1}{2}mv_{top}^2 = 0 + \tfrac{1}{2}m(0.75^2 \cdot 8Rg)$$
$$\text{so}\quad v_{top}^2 = (0.75^2 \cdot 8 - 4)Rg = 0.5Rg$$

6. Substituting for v_{top}^2 in the step-2 result gives:

$$F_n + mg = m\frac{0.5Rg}{R}$$
$$F_n + mg = 0.5mg$$

7. Solve for F_n:

$$F_n = -0.5mg$$

8. F_n is the magnitude of the normal force. It cannot be negative:

Oops! The car has left the track.

CHECK A loss of 25 percent of your speed means losing almost 44 percent of your kinetic energy. The speed is the same as would be attained if the car started from rest at a height of $0.56 \times 4R = 1.12 \times 2R$ (12 percent higher than the height of the top of the loop). We should not be too surprised to find that the car left the track.

TAKING IT FURTHER Fortunately, there were safety devices to prevent the cars from falling, so your ancestors likely would have survived. The biggest concern for riders on the Flip Flap Railway was a broken neck. The Flip Flap Railway subjected riders to accelerations of up to 12g's during the loop-the-loop and was the last of the circular loop-the-loop roller coasters. Loop-the-loops on modern rides are higher than they are wide.

POTENTIAL ENERGY AND EQUILIBRIUM

We can gain insight into the motion of a system by looking at a graph of its potential energy versus the position of a particle in that system. For simplicity, we restrict our consideration to a particle constrained to move along a straight line—the x axis. To create such a graph, we first must find the relationship between the potential energy function and the force acting on the particle. Consider a conservative force $\vec{F} = F_x\hat{i}$ acting on the particle. Substituting this into Equation 7-1b gives

$$dU = -\vec{F}\cdot d\vec{\ell} = -F_x dx$$

The force component F_x is therefore the negative of the derivative* of the potential-energy function:

$$F_x = -\frac{dU}{dx} \qquad 7\text{-}13$$

We can illustrate this general relation for a block–spring system by differentiating the function $U = \tfrac{1}{2}kx^2$. We obtain

$$F_x = -\frac{dU}{dx} = -\frac{d}{dx}\left(\frac{1}{2}kx^2\right) = -kx$$

Figure 7-17 shows a plot of $U = \tfrac{1}{2}kx^2$ versus x for a block and spring. The derivative of this function is represented graphically as the slope of the tangent line

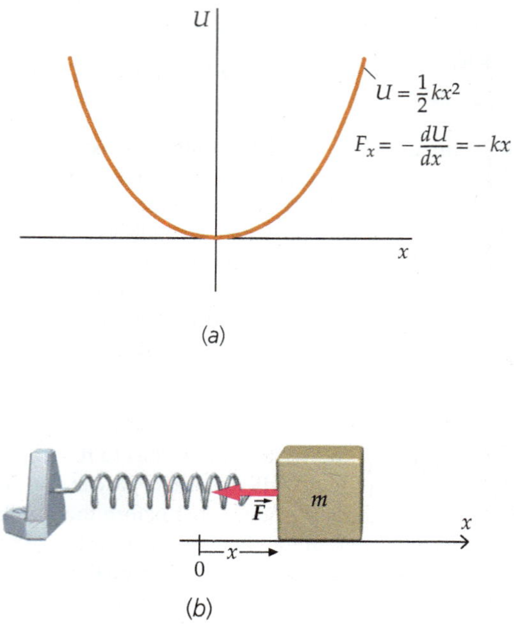

FIGURE 7-17 (a) Plot of the potential-energy function U versus x for an object on a spring. A minimum in the potential-energy curve is a point of stable equilibrium. Displacement in either direction results in a force directed toward the equilibrium position. (b) The applied force $\vec{F}_{app}$ pulls the block to the right, stretching the spring by x_1.

* The derivative in Equation 7-13 is replaced by the partial derivative with respect to x if the motion is not restricted to the x axis.

to the curve. The force is thus equal to the negative of the slope of the tangent line to the curve. At $x = 0$, the force $F_x = -dU/dx$ is zero and the block is in *equilibrium*, assuming no other forces are acting on it.

When x is positive in Figure 7-17*a*, the slope is positive and the force F_x is negative. When x is negative, the slope is negative and the force F_x is positive. In either case, the force is in the direction that will accelerate the block in the direction of decreasing potential energy. If the block is displaced slightly from $x = 0$, the force is directed back toward $x = 0$. The equilibrium at $x = 0$ is thus **stable equilibrium,** because a small displacement results in a restoring force that accelerates the particle back toward its equilibrium position.

! The potential energy function is a minimum at a point of stable equilibrium.

In stable equilibrium, a small displacement in any direction results in a restoring force that accelerates the particle back toward its equilibrium position.

CONDITION FOR STABLE EQUILIBRIUM

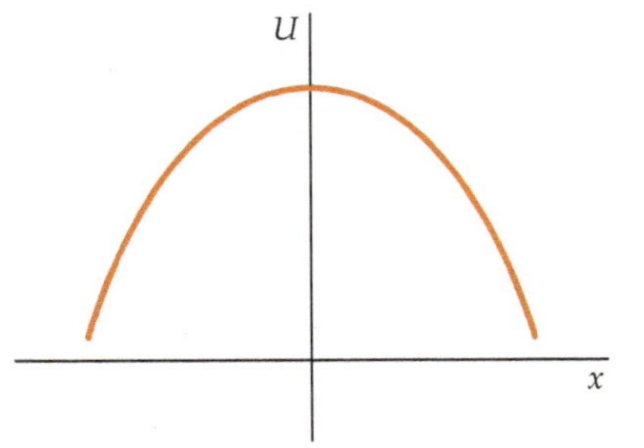

FIGURE 7-18 A particle with a potential energy shown by this potential-energy curve will be in unstable equilibrium at $x = 0$ because a displacement from $x = 0$ results in a force directed away from the equilibrium position.

Figure 7-18 shows a potential-energy curve with a maximum rather than a minimum at the equilibrium point $x = 0$. Such a curve could represent the potential energy of a space ship at the point between Earth and the moon where the gravitational pull on the ship by Earth is equal to the gravitational pull on the ship by the moon. (We are neglecting any gravitational pull from the Sun.) For this curve, when x is positive, the slope is negative and the force F_x is positive, and when x is negative, the slope is positive and the force F_x is negative. Again, the force is in the direction that will accelerate the particle toward lower potential energy, but this time the force is away from the equilibrium position. The maximum at $x = 0$ in Figure 7-18 is a point of **unstable equilibrium** because a small displacement results in a force that accelerates the particle away from its equilibrium position.

In unstable equilibrium, a small displacement results in a force that accelerates the particle away from its equilibrium position.

CONDITION FOR UNSTABLE EQUILIBRIUM

Figure 7-19 shows a potential-energy curve that is flat in the region near $x = 0$. No force acts on a particle at $x = 0$, and hence the particle is at equilibrium; furthermore, there will be no resulting force if the particle is displaced slightly in either direction. This is an example of **neutral equilibrium.**

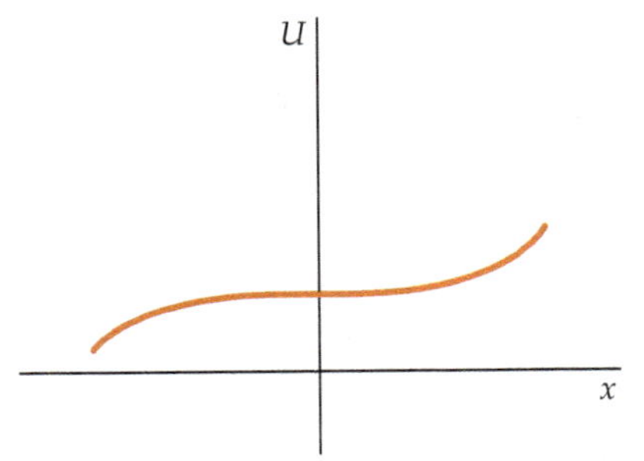

FIGURE 7-19 Neutral equilibrium. The force $F_x = -dU/dx$ is zero at $x = 0$ and at neighboring points, so displacement away from $x = 0$ results in no force, and the system remains in equilibrium.

In neutral equilibrium, a small displacement in any direction results in zero force and the particle remains in equilibrium.

CONDITION FOR NEUTRAL EQUILIBRIUM

Example 7-9 Force and the Potential-Energy Function

In the region $-a < x < a$ the force on a particle is represented by the potential-energy function

$$U = -b\left(\frac{1}{a+x} + \frac{1}{a-x}\right)$$

where a and b are positive constants. (*a*) Find the force F_x in the region $-a < x < a$. (*b*) At what value of x is the force zero? (*c*) At the location where the force equals zero, is the equilibrium stable or unstable?

PICTURE The force is the negative of the derivative of the potential-energy function. The equilibrium is stable where the potential-energy function is a minimum and it is unstable where the potential-energy function is a maximum.

SOLVE

(*a*) Compute $F_x = -dU/dx$.

$$F_x = -\frac{d}{dx}\left[-b\left(\frac{1}{(a+x)} + \frac{1}{(a-x)}\right)\right] = \boxed{-b\left(\frac{1}{(a+x)^2} - \frac{1}{(a-x)^2}\right)}$$

(*b*) Set F_x equal to zero and solve for x.

$$F_x = 0 \text{ at } \boxed{x = 0}$$

(*c*) Compute d^2U/dx^2. If it is positive at the equilibrium position, then U is a minimum and the equilibrium is stable. If it is negative, then U is a maximum and the equilibrium is unstable.

$$\frac{d^2U}{dx^2} = -2b\left(\frac{1}{(a+x)^3} + \frac{1}{(a-x)^3}\right)$$

$$\text{At } x = 0,\quad \frac{d^2U}{dx^2} = \frac{-4b}{a^3} < 0$$

Thus, $\boxed{\text{unstable}}$ equilibrium.

CHECK If U is expressed in joules and x and a are expressed in meters, then b must be expressed in joule·meters and F_x must be expressed in newtons. Our Part-(*a*) result shows that F_x has the same units as those of Part (*b*) divided by m^2. That is, our expression for F_x has units of $J \cdot m/m^2 = J/m$. Because $1\,J = 1\,N \cdot m$, our expression for F_x has units of newtons. Consequently, our Part-(*a*) result is dimensionally correct, and therefore is plausible.

TAKING IT FURTHER The potential-energy function in this example is for a particle under the influence of the gravitational forces exerted by two identical fixed-point masses, one at $x = -a$, the other at $x = +a$. The particle is located on the line joining the masses. Midway between the two masses the net force on the particle is zero. Otherwise, it is toward the closest mass.

We can use the result that a position of stable equilibrium is a potential-energy minimum to locate the center of mass experimentally. For example, two objects connected by a light rod will balance if the pivot is at the center of mass (Figure 7-20). If we pivot the system at any other point, the system will rotate until the gravitational potential energy is at a minimum, which occurs when the center of mass is at its lowest possible point directly below the pivot (Figure 7-21). (The gravitational potential energy of a system is given by $U_g = mgh_{cm}$ [Equation 7-3].)

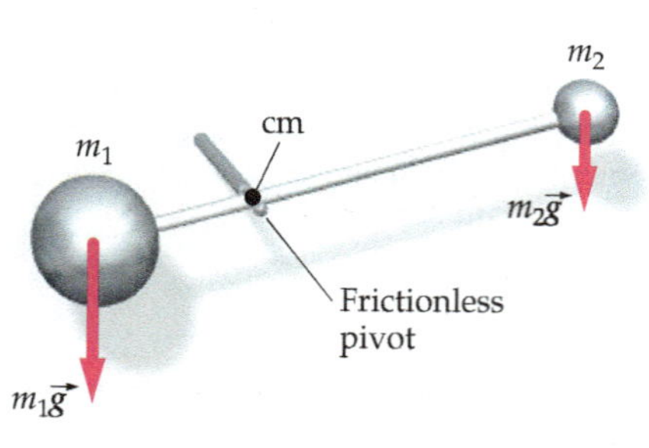

FIGURE 7-20

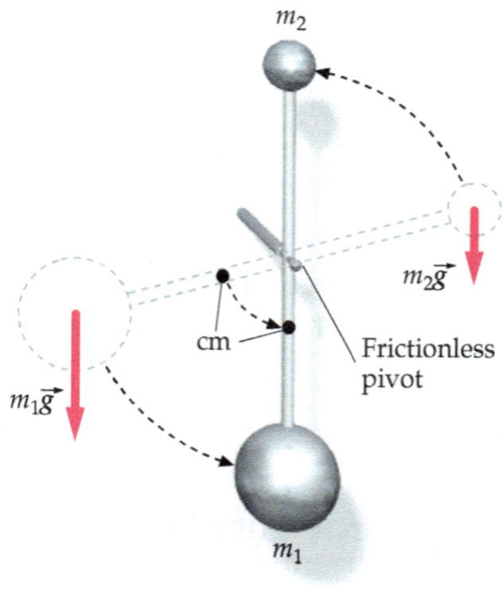

FIGURE 7-21

If we suspend any irregular object from a pivot, the object will hang so that its center of mass lies somewhere on the vertical line drawn directly downward from the pivot. Now suspend the object from another point and note where the vertical line through the pivot point now passes. The center of mass will lie at the intersection of the two lines (Figure 7-22).

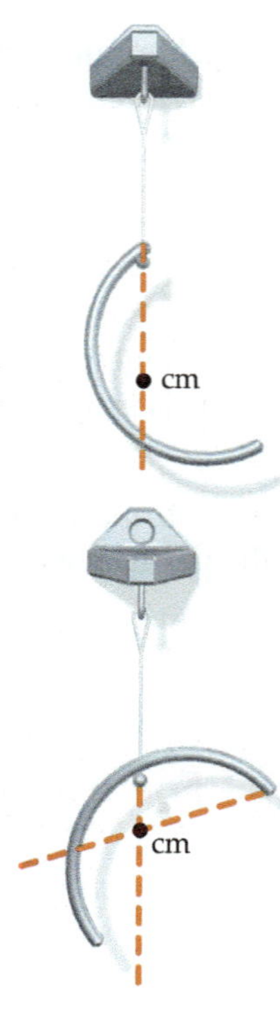

FIGURE 7-22 The center of mass of an irregular object can be found by suspending it first from one point and then from a second point.

For a system for which mechanical energy remains constant, graphs that plot both potential energy U and mechanical energy E are often useful. For example, Figure 7-23 is a plot of the potential-energy function

$$U = b\left(\frac{1}{a+x} + \frac{1}{a-x}\right)$$

which is the negative of the potential-energy function used in Example 7-9. Figure 7-23 contains plots of both this potential-energy function and the total mechanical energy E. The kinetic energy K for a specified value of x is represented by the distance that the total mechanical-energy line is above the potential-energy curve because $K = E - U$.

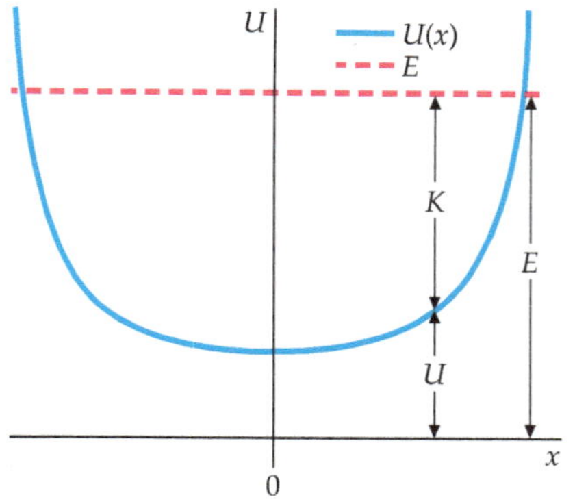

FIGURE 7-23 The potential energy U and the total mechanical energy E are plotted versus x. The sum of the kinetic energy K and the potential energy equals the total mechanical energy. That is, $K = E - U$.

7-3 THE CONSERVATION OF ENERGY

In the macroscopic world, dissipative nonconservative forces, such as kinetic friction, are always present to some extent. Such forces tend to decrease the mechanical energy of a system. However, any such decrease in mechanical energy is accompanied by a corresponding increase in thermal energy. (Consider that excessive automotive braking sometimes causes the temperature of the rotors or brake drums to increase to the point that the metal warps.) Another type of nonconservative force is that involved in the deformations of objects. When you bend a metal coat hanger back and forth, you do work on the coat hanger, but the work you do does not appear as mechanical energy. Instead, the coat hanger becomes warm. The work done in deforming the hanger is dissipated as thermal energy. Similarly, when a falling ball of modeling clay lands on the floor (thud), it becomes warmer as it deforms. The dissipated kinetic energy appears as thermal energy. For the clay–floor–Earth system, the total energy is the sum of the thermal energy and the mechanical energy. The total energy of the system is conserved even though neither the total mechanical energy nor the total thermal energy is individually conserved.

A third type of nonconservative force is associated with chemical reactions. When we include systems in which chemical reactions take place, the sum of mechanical energy plus thermal energy is not conserved. For example, suppose that you begin running from rest. You initially have no kinetic energy. When you begin to run, chemical energy stored in certain molecules in your muscles is transformed into kinetic energy and thermal energy. It is possible to identify and measure the chemical energy that is transformed into kinetic energy and thermal energy. In this case, the sum of mechanical energy, thermal energy, and chemical energy is conserved.

Even when thermal energy and chemical energy are included, the total energy of the system does not always remain constant, because energy can be converted to radiation energy, such as sound waves or electromagnetic waves. But *the increase or decrease in the total energy of a system can always be accounted for by the disappearance or appearance of energy outside the system.* This experimental result, known as the **law of conservation of energy,** is one of the most important laws in all of science. Let E_{sys} be the total energy of a given system, E_{in} be the energy that enters the system, and E_{out} be the energy that leaves the system. The law of conservation of energy then states

$$E_{in} - E_{out} = \Delta E_{sys} \qquad 7\text{-}14$$

LAW OF CONSERVATION OF ENERGY

Alternatively,

> The total energy of the universe is constant. Energy can be converted from one form to another, or transmitted from one region to another, but energy can never be created or destroyed.
>
> LAW OF CONSERVATION OF ENERGY

The total energy E of many systems from everyday life can be accounted for completely by mechanical energy E_{mech}, thermal energy E_{therm}, and chemical energy E_{chem}. To be comprehensive and include other possible forms of energy, such as electromagnetic or nuclear energy, we include E_{other}, and write

$$E_{\text{sys}} = E_{\text{mech}} + E_{\text{therm}} + E_{\text{chem}} + E_{\text{other}} \tag{7-15}$$

THE WORK–ENERGY THEOREM

One way energy is transferred into or out of a system is for work to be done on the system by agents outside the system. In situations where this is the only method of energy transfer into or out of a system, the law of conservation of energy is expressed as:

$$W_{\text{ext}} = \Delta E_{\text{sys}} = \Delta E_{\text{mech}} + \Delta E_{\text{therm}} + \Delta E_{\text{chem}} + \Delta E_{\text{other}} \tag{7-16}$$

WORK-ENERGY THEOREM

where W_{ext} is the work done on the system by external forces and ΔE_{sys} is the change in the system's total energy. This **work–energy theorem** for systems, which we call simply the work–energy theorem, is a powerful tool for studying a wide variety of systems. Note that if the system is just a single particle, its energy can only be kinetic. In that case, the work–energy theorem (Equation 7-16) reduces to the work–kinetic-energy theorem (Equation 6-8) studied in Chapter 6.

There are two methods for transferring energy into or out of a system. The second method is called heat. Heat is the transfer of energy due to a temperature difference. Exchanges of energy due to a temperature difference between a system and its surroundings are discussed in Chapter 18. In this chapter, the transfer of energy by heat is assumed to be negligible.

Example 7-10 Falling Clay — *Conceptual*

A ball of modeling clay with mass m is released from rest from a height h and falls to the perfectly rigid floor (thud). Discuss the application of the law of conservation of energy to (*a*) the system consisting of the clay ball alone, and (*b*) the system consisting of Earth, the floor, and the clay ball.

PICTURE Two forces act on the clay ball following its release: the force of gravity and the contact force of the floor. Because the floor does not move (it is rigid), the contact force it exerts on the clay ball does no work. There are no chemical or other energy changes, so we can neglect ΔE_{chem} and ΔE_{other}. (We neglect the sound energy radiated when the clay ball hits the floor.) Thus, the only energy transferred to or from the clay ball is the work done by the force of gravity.

SOLVE

(*a*) 1. Write the work–energy theorem for the clay ball:

$$W_{\text{ext}} = \Delta E_{\text{sys}} = \Delta E_{\text{mech}} + \Delta E_{\text{therm}} + \Delta E_{\text{chem}} + \Delta E_{\text{other}}$$

$$W_{\text{ext}} = \Delta E_{\text{sys}} = \Delta E_{\text{mech}} + E_{\text{therm}}$$

2. The two external forces on the system (the clay ball) are the force of gravity and the normal force exerted by the floor on the clay ball. However, the part of the ball in contact with the floor does not move, so the normal force on the ball by the floor does no work. Thus, the only work done on the clay ball is done by the force of gravity on the ball:
   $W_{ext} = mgh$
3. Because the clay ball alone is our system, its mechanical energy is entirely kinetic, which is zero both initially and finally. Thus, the change in mechanical energy is zero:
   $\Delta E_{mech} = 0$
4. Substitute mgh for W_{ext} and 0 for ΔE_{mech} in step 1:
   $W_{ext} = \Delta E_{mech} + E_{therm}$
   $mgh = 0 + \Delta E_{therm}$
   so $\boxed{\Delta E_{therm} = mgh}$
   (*Note:* If the floor were not perfectly rigid, the increase in thermal energy would be shared by the ball and the floor.)

(*b*) 1. No external forces act on the clay ball–Earth–floor system (the force of gravity and the force of the floor are now internal to the system), so there is no external work done:
   $W_{ext} = 0$
2. Write the work–energy theorem with $W_{ext} = 0$:
   $W_{ext} = \Delta E_{sys} = \Delta E_{mech} + \Delta E_{therm}$
   $0 = \Delta E_{mech} + \Delta E_{therm}$
3. The initial mechanical energy of the clay ball–Earth system is the initial gravitational potential energy. The final mechanical energy is zero:
   $E_{mech\,i} = mgh$
   $E_{mech\,f} = 0$
4. The change in mechanical energy of the clay ball–Earth system is thus:
   $\Delta E_{mech} = 0 - mgh = -mgh$
5. The work–energy theorem thus gives the same result as in Part (*a*):
   $\Delta E_{therm} = \boxed{-\Delta E_{mech} = mgh}$

CHECK The Part-(*a*) and Part-(*b*) results are the same—that the thermal energy of the system increases by mgh. This is as one would expect.

TAKING IT FURTHER In Part (*a*), energy is transferred to the ball by the work done on it by the force of gravity. This energy appears as the kinetic energy of the ball before it impacts the floor and as thermal energy after impact. The ball warms slightly and the energy is eventually transferred to the surroundings. In Part (*b*), no energy is transferred to the ball–Earth–floor system. The original potential energy of the system is converted to kinetic energy of the ball just before it hits, and then into thermal energy.

PROBLEMS INVOLVING KINETIC FRICTION

When surfaces slide across each other, kinetic friction decreases the mechanical energy of the system and increases the thermal energy. Consider a block that begins with initial velocity v_i and slides along a board that is on a frictionless surface (Figure 7-24). The board is initially at rest. We choose the block and board to be our system, and $\Delta E_{chem} = \Delta E_{other} = 0$. No external work is done on this system. The work–energy theorem gives

$$0 = \Delta E_{mech} + \Delta E_{therm} \qquad 7\text{-}17$$

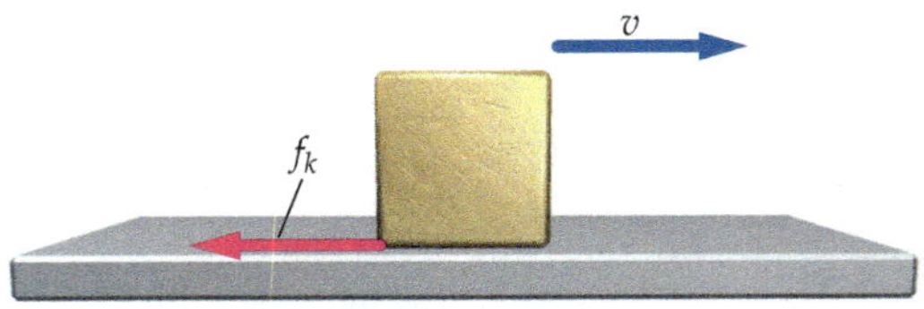

FIGURE 7-24

The change in mechanical energy is given by

$$\Delta E_{mech} = \Delta K_{block} + \Delta K_{board} = (\tfrac{1}{2}mv_f^2 - \tfrac{1}{2}mv_i^2) + (\tfrac{1}{2}MV_f^2 - 0) \qquad 7\text{-}18$$

where m is the mass of the block, M is the mass of the board, v is the speed of the block, and V is the speed of the board. We can relate this change in mechanical energy to the kinetic frictional force. If f_k is the magnitude of the frictional force on either the block or the board, Newton's second law applied to the block gives

$$-f_k = ma_x$$

where a_x is the acceleration of the block. Multiplying both sides by the displacement of the block Δx, we obtain

$$-f_k \Delta x = ma\Delta x \qquad 7\text{-}19$$

Solving the constant-acceleration formula $2a_x \Delta x = v_f^2 - v_i^2$ for $a_x \Delta x$ and substituting into Equation 7-19 gives

$$-f_k \Delta x = ma_x \Delta x = m(\tfrac{1}{2}v_f^2 - \tfrac{1}{2}v_i^2) = \tfrac{1}{2}mv_f^2 - \tfrac{1}{2}mv_i^2 \qquad 7\text{-}20$$

Equation 7-20 is just the center-of-mass work–translational-kinetic-energy relation (Equation 6-27) applied to the block. Applying this same relation to the board gives

$$f_k \Delta X = MA_X \Delta X = M(\tfrac{1}{2}V_f^2 - \tfrac{1}{2}V_i^2) = \tfrac{1}{2}MV_f^2 - 0 \qquad 7\text{-}21$$

where ΔX and A_x are the displacement and acceleration of the board. Adding Equations 7-20 and 7-21 gives

$$-f_k(\Delta x - \Delta X) = (\tfrac{1}{2}mv_f^2 - \tfrac{1}{2}mv_i^2) + \tfrac{1}{2}MV_f^2 \qquad 7\text{-}22$$

We note that $\Delta x - \Delta X$ is the distance s_{rel} that the block slides relative to the board, and that the right side of Equation 7-22 is the change in mechanical energy ΔE_{mech} of the block–board system. Substituting into Equation 7-22 gives

$$-f_k s_{rel} = \Delta E_{mech} \qquad 7\text{-}23$$

The decrease in mechanical energy of the block–board system is accompanied by a corresponding increase in the thermal energy of the system. This thermal energy appears both on the bottom surface of the block and on the upper surface of the board. Substituting ΔE_{therm} for $-\Delta E_{mech}$ (Equation 7-17), we obtain

$$f_k s_{rel} = \Delta E_{therm} \qquad 7\text{-}24$$

ENERGY DISSIPATED BY KINETIC FRICTION

where s_{rel} is the distance one contacting surface slides relative to the other contacting surface. Because the distance s_{rel} is the same in all frames of reference, Equation 7-24 is valid in all frames of reference, independent of whether they are inertial frames of reference or not.

Substituting this result into the work–energy theorem (with $\Delta E_{chem} = \Delta E_{other} = 0$), we obtain

$$W_{ext} = \Delta E_{mech} + \Delta E_{therm} = \Delta E_{mech} + f_k s_{rel} \qquad 7\text{-}25$$

WORK–ENERGY THEOREM WITH FRICTION

Example 7-11 Pushing a Box

A 4.0-kg box is initially at rest on a horizontal tabletop. You push the box a distance of 3.0 m along the tabletop with a horizontal force of 25 N. The coefficient of kinetic friction between the box and tabletop is 0.35. Find *(a)* the external work done on the block–table system, *(b)* the energy dissipated by friction, *(c)* the final kinetic energy of the box, and *(d)* the speed of the box.

FIGURE 7-25

PICTURE The box plus table is the system (Figure 7-25). You are external to this system, so the force you push with is an external force. The final speed of the box is found from its final kinetic energy, which we find using the work–energy theorem with $\Delta E_{chem} = 0$ and $\Delta E_{therm} = f_k s_{rel}$. The energy of the system is increased by the external work. Some of the energy increase is kinetic energy and some is thermal energy.

SOLVE

(*a*) Four external forces are acting on the system. However, only one of them does work. The total external work done is the product of the push force and the distance traveled:

$$\Sigma W_{\text{ext}} = W_{\text{by you on block}} + W_{\text{by gravity on block}} + W_{\text{by gravity on table}} + W_{\text{by floor on table}}$$
$$= F_{\text{push}}\,\Delta x + 0 + 0 + 0 = (25\text{ N})(3.0\text{ m})$$
$$= \boxed{75\text{ J}}$$

(*b*) The energy dissipated by friction is $f_k\,\Delta x$ (the magnitude of the normal force equals mg):

$$\Delta E_{\text{therm}} = f_k\,\Delta x = \mu_k F_n\,\Delta x = \mu_k mg\,\Delta x$$
$$= (0.35)(4.0\text{ kg})(9.81\text{ N/kg})(3.0\text{ m})$$
$$= \boxed{41\text{ J}}$$

(*c*) 1. Apply the work–energy theorem to find the final kinetic energy:

$$W_{\text{ext}} = \Delta E_{\text{mech}} + \Delta E_{\text{therm}}$$

2. No internal conservative forces do work, so the change in potential energy ΔU is zero. Thus, the change in mechanical energy equals the change in kinetic energy:

$$\Delta E_{\text{mech}} = \Delta U + \Delta K = 0 + (K_f - 0) = K_f$$

3. Substitute this into the step-1 result, then use the values from Parts (*a*) and (*b*) to find K_f:

$$W_{\text{ext}} = K_f + \Delta E_{\text{therm}}$$
so $$K_f = W_{\text{ext}} - \Delta E_{\text{therm}}$$
$$= 75\text{ J} - 41\text{ J} = \boxed{34\text{ J}}$$

(*d*) The final speed of the box is related to its kinetic energy. Solve for the final speed of the box:

$$K_f = \tfrac{1}{2}mv_f^2$$
so $$v_f = \sqrt{\frac{2K_f}{m}} = \sqrt{\frac{2(34\text{ J})}{4.0\text{ kg}}} = \boxed{4.1\text{ m/s}}$$

CHECK Part of the energy transferred to the system by the pusher (you) ends up as kinetic energy and some of the energy ends up as thermal energy. As expected, the change in thermal energy (Part (*b*)) is positive and is less than the work done by the external force (Part (*a*)).

Example 7-12 A Moving Sled

Try It Yourself

A sled is coasting on a horizontal snow-covered surface with an initial speed of 4.0 m/s. If the coefficient of friction between the sled and the snow is 0.14, how far will the sled travel before coming to rest?

PICTURE We choose the sled and snow as our system and then apply the work–energy theorem.

SOLVE

Cover the column to the right and try these on your own before looking at the answers.

Steps / **Answers**

1. Sketch the system in its initial and final configurations (Figure 7-26).

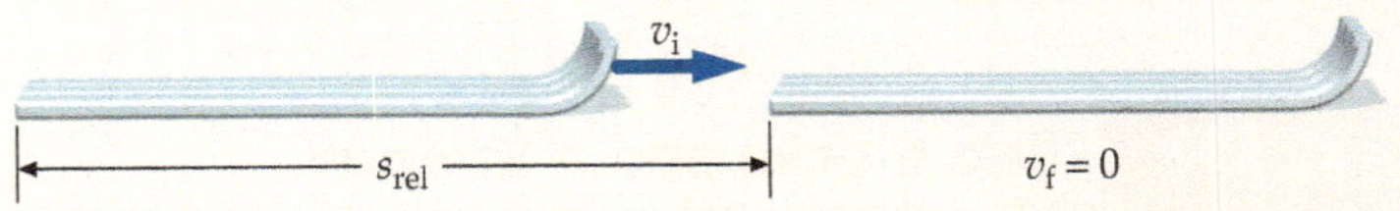

FIGURE 7-26

2. Apply the work–energy theorem. Relate the change in thermal energy to the frictional force.

$$W_{\text{ext}} = \Delta E_{\text{mech}} + \Delta E_{\text{therm}}$$
$$= (\Delta U + \Delta K) + f_k s_{\text{rel}}$$

3. Solve for f_k. The normal force is equal to mg.

$$f_k = \mu_k F_n = \mu_k mg$$

4. There are no external forces doing work on the system and there are no internal conservative forces doing work. Use these observations to eliminate two terms from the step-2 result.

$$W_{\text{ext}} = 0 \text{ and } \Delta U = 0$$
so $$W_{\text{ext}} = \Delta U + \Delta K + f_k s_{\text{rel}}$$
$$0 = 0 + \Delta K + \mu_k mg\, s_{\text{rel}}$$

5. Express the change in kinetic energy in terms of the mass and the initial speed, and solve for s_{rel}.

$$s_{\text{rel}} = \frac{v^2}{2\mu_k g} = \boxed{5.8\text{ m}}$$

CHECK The expression for the displacement in step 5 is dimensionally correct. The coefficient of friction μ_k is dimensionless, and v^2/g has the dimension of length.

Example 7-13 A Playground Slide

A child of mass 40 kg goes down an 8.0-m-long slide inclined at 30° with the horizontal. The coefficient of kinetic friction between the child and the slide is 0.35. If the child starts from rest at the top of the slide, how fast is he traveling when he reaches the bottom?

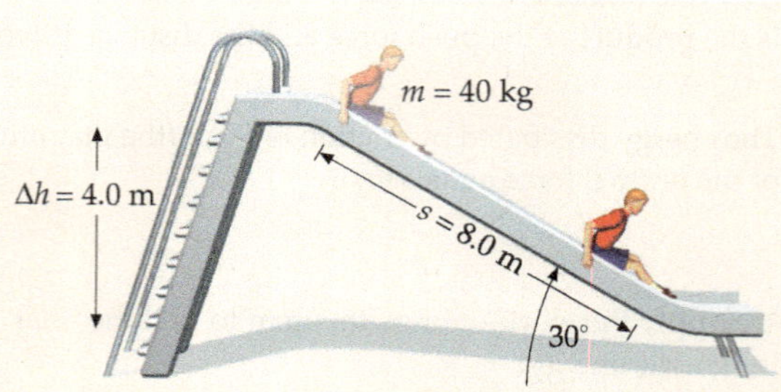

FIGURE 7-27

PICTURE As the child travels down the slide, some of his initial potential energy is converted into kinetic energy and, due to friction, some into thermal energy. We choose the child–slide–Earth as our system and apply the conservation of energy theorem.

SOLVE

1. Make a sketch of the child–slide–Earth system, showing both its initial and final configurations (Figure 7-27).
2. Write out the conservation-of-energy equation: $W_{\text{ext}} = \Delta E_{\text{mech}} + \Delta E_{\text{therm}} = (\Delta U + \Delta K) + f_k s_{\text{rel}}$
3. The initial kinetic energy is zero. The speed at the bottom is related to the final kinetic energy: $\Delta K = K_f - 0 = \frac{1}{2}mv_f^2$
4. There are no external forces acting on the system: $W_{\text{ext}} = 0$
5. The change in potential energy is related to the change in height Δh (which is negative): $\Delta U = mg\,\Delta h$
6. To find f_k we apply Newton's second law to the child. First we draw a free-body diagram (Figure 7-28):

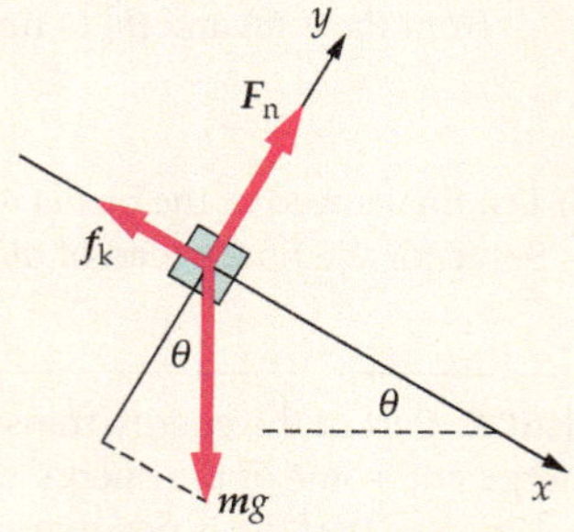

FIGURE 7-28

7. Next, we apply Newton's second law. The normal component of the acceleration is zero. To find F_n we take components in the normal direction. Then we solve for f_k using $f_k = \mu_k F_n$:
   $F_n - mg\cos\theta = 0$
   so $f_k = \mu_k F_n = \mu_k mg\cos\theta$
8. We use trigonometry to relate $s = s_{\text{rel}}$ to Δh: $|\Delta h| = s\sin\theta$
9. Substituting into the step-2 result gives:
   $0 = mg\,\Delta h + \frac{1}{2}mv_f^2 + f_k s = -mgs\sin\theta + \frac{1}{2}mv_f^2 + \mu_k mg\cos\theta\, s$
10. Solving for v_f gives:
    $$v_f^2 = 2gs(\sin\theta - \mu_k\cos\theta) = 2(9.81\text{ m/s}^2)(8.0\text{ m})(\sin 30° - 0.35\cos 30°)$$
    $$= 30.9\text{ m}^2/\text{s}^2$$
    so $v_f = \boxed{5.6\text{ m/s}}$

CHECK Note that, as expected, the expression for v_f^2 in step 10 is independent of the mass of the child. This is expected because all forces acting on the child are proportional to the mass m.

PRACTICE PROBLEM 7-8 Use the bottom end of the slide as the reference level where the potential energy is zero. For the Earth–child–slide system, calculate (*a*) the initial mechanical energy, (*b*) the final mechanical energy, and (*c*) the energy dissipated by friction.

Example 7-14 Two Blocks and a Spring

A 4.0-kg block hangs by a light string that passes over a massless, frictionless pulley and is connected to a 6.0-kg block that rests on a shelf. The coefficient of kinetic friction is 0.20. The 6.0-kg block is pushed against a spring, compressing it 30 cm. The spring has a force constant of 180 N/m. Find the speed of the blocks after the 6.0-kg block is released and the 4.0-kg block has fallen a distance of 40 cm. (Assume the 6.0-kg block is initially 40 cm or more from the pulley.)

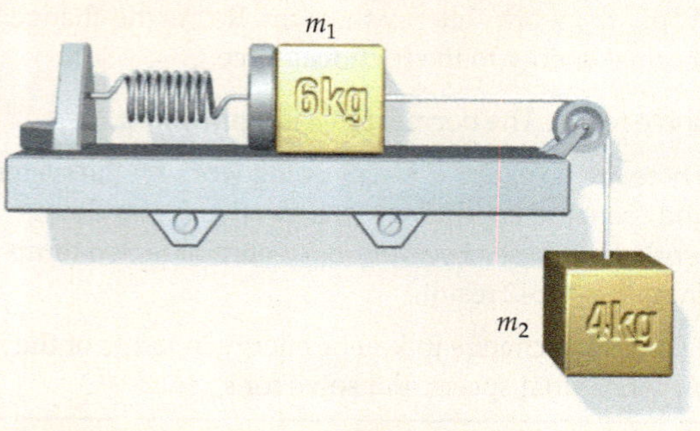

FIGURE 7-29 The system is everything shown plus Earth.

PICTURE The speed of the blocks is obtained from their final kinetic energy. Consider the system to be everything shown in Figure 7-29 plus Earth. This system has both gravitational and elastic potential energy. Apply the work–energy theorem to find the kinetic energy of the blocks. Then, use the kinetic energy of the blocks to solve for their speed.

3. The angular frequency is found from the frequency, which is found from the speed and wavelength:

$$\omega = 2\pi f \quad \text{and} \quad v = f\lambda,$$

$$\text{so} \quad \omega = 2\pi\frac{v}{\lambda} = 2\pi\frac{47.4\ \text{m/s}}{0.25\ \text{m}} = 1190\ \text{rad/s}$$

$$= \boxed{1200\ \text{rad/s}}$$

(*b*) The average total energy of a harmonic wave on the string is given by $(\Delta E)_{\text{av}} = \frac{1}{2}\mu\omega^2 A^2 \Delta x$ (Equation 15-23):

$$(\Delta E)_{\text{av}} = \frac{1}{2}\mu\omega^2 A^2\,\Delta x = \frac{1}{2}\frac{m}{L}\omega^2 A^2\,\Delta x$$

$$= \frac{1}{2}\frac{0.32\ \text{kg}}{60\ \text{m}}(1190\ \text{s}^{-1})^2(0.012\ \text{m})^2(15\ \text{m})$$

$$= 8.19\ \text{J} = \boxed{8.2\ \text{J}}$$

CHECK The units for the average energy in the Part-(*b*) result are given by

$$1\frac{\text{kg}\cdot\text{s}^{-2}\,\text{m}^3}{\text{m}} = 1\frac{\text{kg}\cdot\text{m}^2}{\text{s}^2} = 1\ \text{N}\cdot\text{m} = 1\ \text{J}$$

where we have used that $1\ \text{N} = 1\ \text{kg}\cdot\text{m/s}^2$. The units work out, so the Part-(*b*) result is plausible.

PRACTICE PROBLEM 15-4 Calculate the average rate at which energy is transmitted along the string.

HARMONIC SOUND WAVES

Harmonic sound waves can be generated by a tuning fork or loudspeaker that is vibrating with simple harmonic motion. The vibrating source causes the air molecules next to it to oscillate with simple harmonic motion about their equilibrium positions. These molecules collide with neighboring molecules, causing them to oscillate, which in turn collide with their neighboring molecules, causing them to oscillate, and so forth, thereby propagating the sound wave. Equation 15-15 describes a harmonic sound wave if the wave function $y(x,t)$ is replaced by $s(x,t)$, which represents the displacements of the molecules from their equilibrium positions. Thus,

$$s(x,t) = s_0 \sin(kx - \omega t) \qquad 15\text{-}24$$

These displacements are along the direction of propagation of the wave, and lead to variations in the density and pressure of the air. Figure 15-11 shows the displacement of air molecules and the density variations caused by a sound wave at some fixed time. The pressure is maximum where the density is maximum. We see from this figure

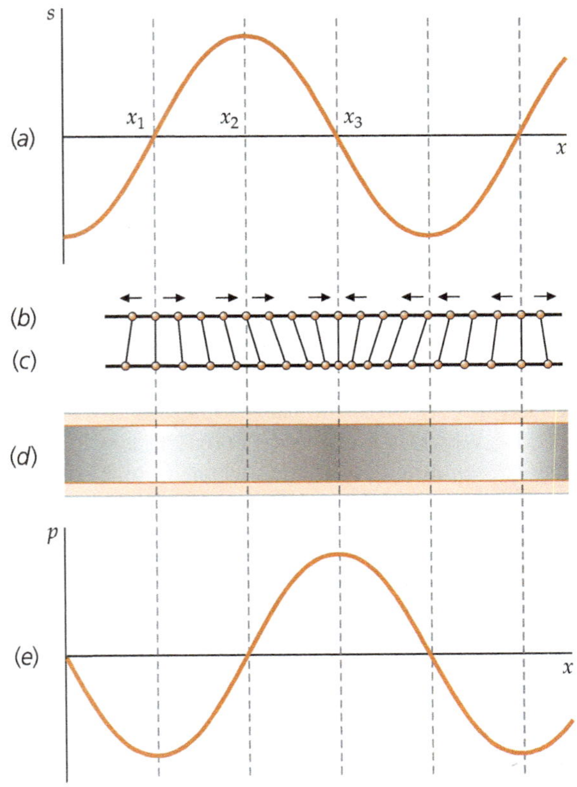

FIGURE 15-11 (*a*) Displacement from equilibrium of air molecules in a harmonic sound wave versus position at some instant. Points x_1 and x_3 are points of zero displacement. (*b*) Some representative molecules equally spaced at their equilibrium positions 1/4 cycle earlier. The arrows indicate the directions of their velocities at that instant. (*c*) Molecules near points x_1, x_2, and x_3 after the sound wave arrives. Just to the left of x_1, the displacement is negative, indicating that the gas molecules are displaced to the left, away from point x_1, at this time. Just to the right of x_1, the displacement is positive, indicating that the molecules are displaced to the right, which is again away from point x_1. So at point x_1, the density is a minimum because the gas molecules on both sides are displaced away from that point. At point x_3, the density is a maximum because the molecules on both sides of that point are displaced toward point x_3. At point x_2, the density does not change because the gas molecules on both sides of that point have equal displacements in the same direction. (*d*) Density of the air at this instant. The density is maximum at x_3 and minimum at x_1, which are both points of zero displacement. It is equal to the equilibrium value at point x_2, which is a maximum in displacement. (*e*) Pressure change, which is proportional to the density change, versus position. The pressure change and displacement (position change) are 90° out of phase.

that the density wave, and thus the pressure wave, is 90° out of phase with the displacement wave. (In the arguments of sine or cosine functions, we will always express phase angles in radians. However, in verbal descriptions, we usually say that "two waves are 90° out of phase" rather than "two waves are out of phase by $\pi/2$ rad.") Where the displacement s is zero, the density, and thus the pressure, is either maximum or minimum, and where the displacement is a maximum or a minimum, the density, and thus the pressure, is equal to its equilibrium value. A displacement wave given by Equation 15-24 thus implies a pressure wave given by

$$p = p_0 \sin\left(kx - \omega t - \frac{\pi}{2}\right) = -p_0 \cos(kx - \omega t) \qquad 15\text{-}25$$

where p stands for the pressure minus the local equilibrium pressure, and p_0, the maximum value of p, is called the pressure amplitude. It can be shown that the pressure amplitude p_0 is related to the displacement amplitude s_0 by

$$p_0 = \rho\omega v s_0 \qquad 15\text{-}26$$

where v is the speed of propagation and ρ is the equilibrium density of the gas. Thus, as a harmonic sound wave travels through air, the displacement of air molecules, the pressure, and the density all vary sinusoidally with the frequency of the vibrating source.

PRACTICE PROBLEM 15-5

Sound of frequencies from about 20 Hz to about 20,000 Hz are audible to humans (although many people have rather limited hearing above 15,000 Hz). If the speed of sound in air is 343 m/s, what are the wavelengths that correspond to the highest and lowest audible frequencies?

Energy of sound waves The average energy of a harmonic sound wave in a volume element ΔV is given by Equation 15-23 with A replaced by s_0 and $\mu\,\Delta x$, replaced by $\rho\,\Delta V$, where ρ is the equilibrium density of the medium.

$$(\Delta E)_{\text{av}} = \tfrac{1}{2}\rho\omega^2 s_0^2\,\Delta V \qquad 15\text{-}27$$

The energy per unit volume is the average energy density η_{av}:

$$\eta_{\text{av}} = \frac{\Delta E_{\text{av}}}{\Delta V} = \frac{1}{2}\rho\omega^2 s_0^2 \qquad 15\text{-}28$$

where η is the lowercase Greek letter eta.

ELECTROMAGNETIC WAVES

Electromagnetic waves include light, radio waves, X rays, gamma rays, and microwaves, among others. The various types of electromagnetic waves differ only in wavelength and frequency. Unlike mechanical waves, electromagnetic waves do not require a medium for propagation. They travel through a vacuum with speed c, which is a universal constant, $c \approx 3.00 \times 10^8$ m/s. The wave function for electromagnetic waves is an electric field $\vec{E}(x,t)$ associated with the wave. (Electric fields are introduced in Chapter 21. A wave equation, similar to those for string waves and sound waves, is derived from the laws of electricity and magnetism in Chapter 30.) The electric field is perpendicular to the direction of propagation, so electromagnetic waves are transverse waves.

Electromagnetic waves are produced when free electric charges accelerate or when electrons bound to atoms and molecules make transitions to lower energy states. Radio waves, which have frequencies of about 1 MHz for AM and 100 MHz for FM, are produced by macroscopic electric currents oscillating in radio antennas. The frequency of the emitted waves equals the frequency of oscillation of the charges. Light waves, which have frequencies of the order of 10^{14} Hz, are generally produced by atomic or molecular transitions involving bound electrons. The spectrum of electromagnetic waves is discussed in Chapter 31.

(David Sacks/ The Image Bank/ Getty.)

15-3 WAVES IN THREE DIMENSIONS

Figure 15-12 shows two-dimensional circular waves on the surface of water in a ripple tank. These waves are generated by drops of water striking the surface. The wave crests form concentric circles called **wavefronts.** For a point source of sound, the waves move out in three dimensions, and the wavefronts are concentric spherical surfaces.

FIGURE 15-12 Circular wavefronts diverging from a point source in a ripple tank. *(PhotoDisc/Getty Images.)*

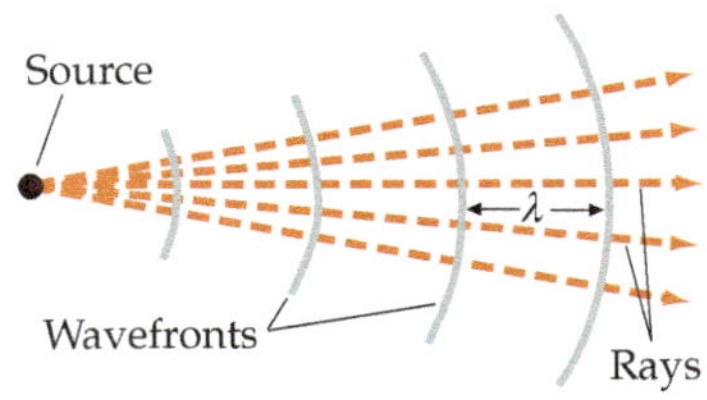

FIGURE 15-13 The motion of wavefronts can be represented by rays drawn perpendicular to the wavefronts. For a point source, the rays are radial lines diverging from the source.

The motion of any set of wavefronts can be indicated by rays, which are directed lines perpendicular to the wavefronts (Figure 15-13). For circular or spherical waves, the rays are radial lines.

In a homogeneous medium, such as air at constant density, the wavefronts travels in straight lines in the direction of the rays, much like a beam of particles. At a great distance from a point source, a sufficiently small section of the wavefront can be approximated by a flat surface (a plane), and the rays are approximately parallel lines; such a wave is called a **plane wave** (Figure 15-14). The two-dimensional analog of a plane wave is a *line wave,* which is a small part of a circular wavefront at a great distance from the source. Line waves can also be produced in a ripple tank by a line source, as in Figure 15-15.

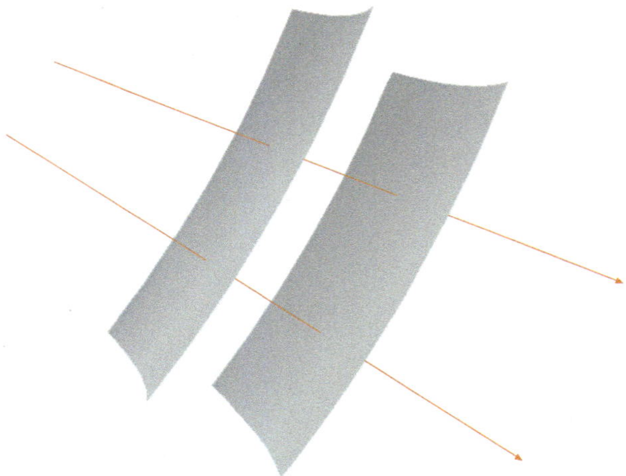

FIGURE 15-14 Plane waves. At great distances from a point source, the wavefronts are approximately parallel planes, and the rays are approximately parallel lines perpendicular to the wavefronts.

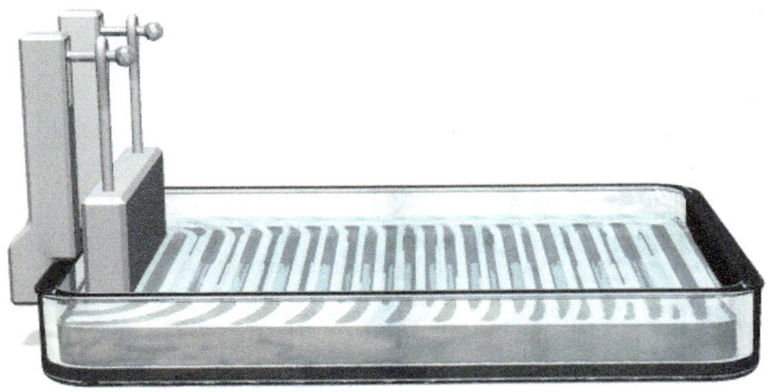

FIGURE 15-15 A two-dimensional analog of a plane wave can be generated in a ripple tank by a flat board that oscillates up and down in the water to produce the wavefronts, which are straight lines.

WAVE INTENSITY

If a point source emits waves uniformly in all directions, then the energy at a distance r from the source is distributed uniformly on a spherical surface of radius r and area $A = 4\pi r^2$. If P_{av} is the average power emitted by the source, then the average power per unit area at a distance r from the source is $P_{av}/(4\pi r^2)$. The average power per unit area that is incident perpendicular to the direction of propagation is called the **intensity:**

$$I = \frac{P_{av}}{A} \tag{15-29}$$

INTENSITY DEFINED

The SI units of intensity are watts per square meter (W/m^2). At a distance r from a point source, the intensity is

$$I = \frac{P_{av}}{4\pi r^2} \tag{15-30}$$

INTENSITY DUE TO A POINT SOURCE

The intensity of a three-dimensional wave varies inversely with the square of the distance from a point source.

There is a simple relation between the intensity of a wave and the energy density in the medium through which it propagates. Figure 15-16 shows a spherical wave that has just reached the radius r_1. The volume inside the radius r_1 contains energy because the particles in that region are oscillating. The region outside r_1 contains no energy because the wave has not yet reached it. After a short time Δt, the wave moves out a short distance $\Delta r = v\,\Delta t$ past r_1. The average energy in the spherical shell of surface area A, thickness $v\,\Delta t$, and volume $\Delta V = A\,\Delta r = Av\,\Delta t$ is

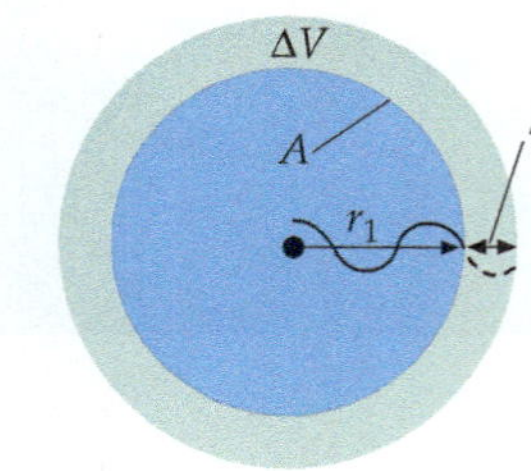

Volume of shell = $\Delta V = A\,\Delta r = Av\,\Delta t$

FIGURE 15-16

$$(\Delta E)_{av} = \eta_{av}\,\Delta V = \eta_{av} Av\,\Delta t$$

The rate of transfer of energy is the power passing into the shell. The average incident power is

$$P_{av} = \frac{(\Delta E)_{av}}{\Delta t} = \eta_{av} Av$$

and the intensity of the wave is

$$I = \frac{P_{av}}{A} = \eta_{av} v \tag{15-31}$$

Thus, the intensity equals the product of the wave speed v and the average energy density η_{av}. Substituting $\eta_{av} = \frac{1}{2}\rho\omega^2 s_0^2$ from Equation 15-28 for the energy density in a harmonic sound wave, we obtain

$$I = \eta_{av} v = \frac{1}{2}\rho\omega^2 s_0^2 v = \frac{1}{2}\frac{p_0^2}{\rho v} \tag{15-32}$$

where we have used $s_0 = p_0/(\rho\omega v)$ from Equation 15-26. This result—that the intensity of a sound wave is proportional to the square of the amplitude—is a general property of harmonic waves.

The human ear can accommodate a large range of sound-wave intensities, from about 10^{-12} W/m^2 (which is usually taken to be the threshold of hearing) to about 1 W/m^2 (an intensity great enough to stimulate pain in most people). The pressure

Sound waves from a telephone handset spreading out in the air. The waves have been made visible by sweeping out the space in front of the handset with a light source whose brightness is controlled by a microphone. *(From Winston E. Kock, Lasers and Holography, 1978, Dover Publications, New York.)*

2. Divide μ_2 by μ_1 and use the given information that $v_1 = 2v_2$:

$$\frac{\mu_2}{\mu_1} = \frac{v_1^2}{v_2^2} = \frac{(2v_2)^2}{v_2^2} = \boxed{4}$$

(*c*) 1. Write expressions for the incident, reflected, and transmitted power using $P_{av} = \frac{1}{2}\mu v\omega^2 A^2$ (Equation 15-22):

$$P_{\text{in av}} = \tfrac{1}{2}\mu_1\omega^2 A^2 v_1$$

$$P_{\text{r av}} = \tfrac{1}{2}\mu_1\omega^2 A_r^2 v_1$$

$$P_{\text{t av}} = \tfrac{1}{2}\mu_2\omega^2 A_t^2 v_2$$

2. Substitute the Part-(*a*) results into the expressions for the reflected and transmitted power:

$$P_{\text{r av}} = \tfrac{1}{2}\mu_1\omega^2\left(-\frac{1}{3}A\right)^2 v_1 = \frac{1}{18}\mu_1\omega^2 A^2 v_1$$

$$P_{\text{t av}} = \tfrac{1}{2}\mu_2\omega^2\left(\frac{2}{3}A\right)^2 v_2 = \frac{2}{9}\mu_2\omega^2 A^2 v_2$$

3. Obtain expressions for P_r/P_{in} and for P_t/P_{in}:

$$\frac{P_{\text{r av}}}{P_{\text{in av}}} = \frac{\frac{1}{18}\mu_1\omega^2 A^2 v_1}{\frac{1}{2}\mu_1\omega^2 A^2 v_1} = \boxed{\frac{1}{9}}$$

$$\frac{P_{\text{t av}}}{P_{\text{in av}}} = \frac{\frac{2}{9}\mu_2\omega^2 A^2 v_2}{\frac{1}{2}\mu_1\omega^2 A^2 v_1} = \frac{4}{9}\frac{\mu_2}{\mu_1}\frac{v_2}{v_1}$$

4. Simplify using the Part-(*b*) result and the given relation $v_1 = 2v_2$:

$$\frac{P_{\text{t av}}}{P_{\text{in av}}} = \frac{4}{9}4\frac{v_2}{2v_2} = \boxed{\frac{8}{9}}$$

CHECK The fractional power reflected plus the fractional power transmitted equals one, as one would expect.

TAKING IT FURTHER The reflected wave is inverted relative to the incident wave, so it is 180° out of phase with it. A negative amplitude corresponds to a phase shift of 180°.

PRACTICE PROBLEM 15-6 Repeat Example 15-10 except with $v_2 = 2v_1$.

Energy conservation gives another relation between the reflection and transmission coefficients. This relation, established in Problem 15-70 is given by

$$1 = r^2 + \frac{v_1}{v_2}\tau^2 \qquad \text{15-36}$$

where r^2 is the fraction of the incident power that is reflected and $(v_1/v_2)\tau^2$ is the fraction that is transmitted.

PRACTICE PROBLEM 15-7

Show that the values of r and τ for the wires in Example 15-10 satisfy Equation 15-36.

In three dimensions, a boundary between two regions of differing wave speed is a surface. Figure 15-20 shows a ray incident on such a boundary surface. This example could be an ultrasonic pressure wave in air striking a solid or liquid surface. The reflected ray makes an angle with the normal to the surface equal to that of the incident ray, as shown.

The transmitted ray is bent toward or away from the normal—depending on whether the wave speed in the second medium is less or greater than that in the incident medium. The bending of the transmitted ray is called **refraction.** When the wave speed in the second medium is greater than that in the incident medium (as occurs when a light wave in glass or water is refracted into the air), the ray

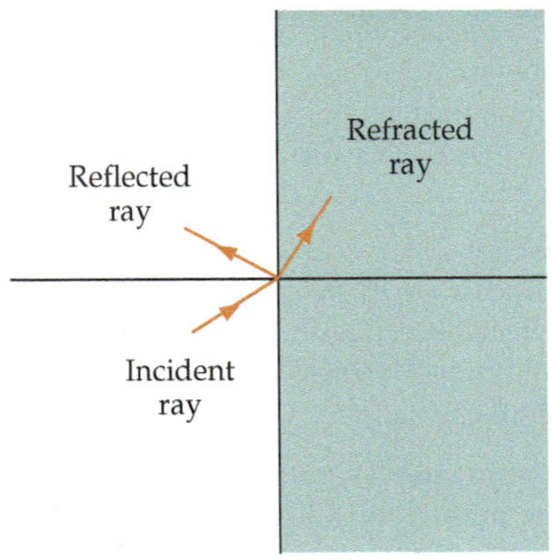

FIGURE 15-20 A wave striking a boundary surface between two media in which the wave speed differs. Part of the wave is reflected and part is transmitted. The change in direction of the transmitted (refracted) ray is called refraction.

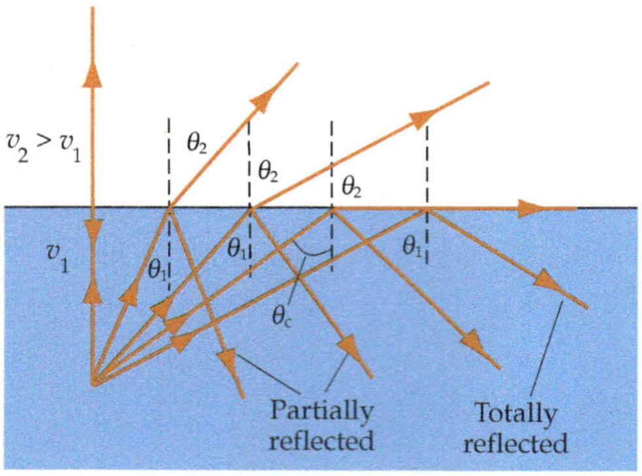

FIGURE 15-21 Light from a source in the water is bent away from the normal when it enters the air. For angles of incidence above a critical angle θ_C, there is no transmitted ray, a condition known as total internal reflection.

describing the direction of propagation is bent away from the normal, as shown in Figure 15-21. As the angle of incidence is increased, the angle of refraction increases, until a critical angle of incidence is reached for which the angle of refraction is 90°. For incident angles greater than the critical angle, there is no refracted ray, a phenomenon known as **total internal reflection.**

The amount of energy reflected from a surface depends on the surface. Rigid flat walls, floors, and ceilings make good reflectors for sound waves, whereas porous and less rigid materials, such as cloth in draperies and furniture coverings, absorb much of the incident sound. The reflection of sound waves plays an important role in the design of a lecture hall, a library, or a music auditorium. If a lecture hall has many flat reflecting surfaces, speech is difficult to understand because of the many echoes that simultaneously arrive at the listener's ear. Absorbent material is often placed on the walls and ceiling to reduce such reflections. In a concert hall, a reflecting shell is placed behind the orchestra, and reflecting panels are hung from the ceiling to reflect and direct the sound back toward the listeners.

(Courtesy of Davies Symphony Hall.)

Example 15-11 Balloon Hearing Aid — *Conceptual*

A popular physics demonstration uses a weather balloon filled with carbon dioxide. If you place the balloon between yourself and a sound source, the sound gets louder. Why is that?

PICTURE The molar mass of carbon dioxide is greater than the effective molar mass of air. Thus, sound travels faster in air than it does in carbon dioxide at atmospheric pressure. To "see" why the sound gets louder when the balloon is between you and the source of the sound, draw a diagram of the rays of sound as they pass through the balloon. The rays will refract (bend) when they are transmitted through a surface where the speed of sound changes.

SOLVE

1. Trace a ray from the source of the sound through the top half of the balloon (Figure 15-22*a*). The ray will refract toward the normal as it enters the balloon, and away from the normal as it exits the balloon:
2. Repeat step 1 for four or five additional rays, including some that go through the lower half of the balloon (Figure 15-22*b*).
3. Use the diagram to explain why the sound is louder when the balloon is between you and the source of sound: The sound is loudest in the region where the rays intersect.

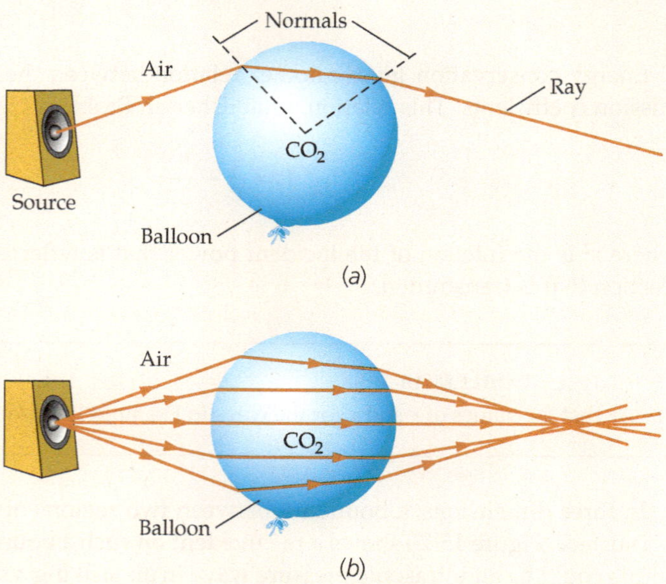

FIGURE 15-22

CHECK The balloon is to sound as a magnifying glass is to light. In glass, light travels more slowly than it does in air, just as in CO_2 sound travels more slowly than it does in air.

DIFFRACTION

If a wavefront is partially blocked by an obstacle, the unblocked part of the wavefront bends behind the obstacle. This bending of the wavefronts is called **diffraction.** Almost all of the diffraction occurs for that part of the wavefront that passes within a few wavelengths of the edge of the obstacle. For the parts of the wavefront that pass farther than a few wavelengths from the edge, diffraction is negligible and the wave propagates in straight lines in the direction of the incident rays. When wavefronts encounter a barrier with an aperture (hole) only a few wavelengths across, the part of the wavefronts passing through the aperture all pass within a few wavelengths of an edge. Thus, flat wavefronts bend and spread out and become spherical or circular (Figure 15-23). In contrast, for a beam of *particles* falling upon a barrier with an aperture, the part of the beam passing through the aperture does so with no change in the direction of the particles (Figure 15-24). Diffraction is one of the key characteristics that distinguish waves from particles. We will discuss how diffraction arises when we study the interference and diffraction of light in Chapter 35.

FIGURE 15-23 Plane waves in a ripple tank meeting a barrier with an opening that is approximately one wavelength wide. Beyond the barrier are circular waves that are concentric about the opening, much as if there were a point source at the opening. *(Fundamental Photographers.)*

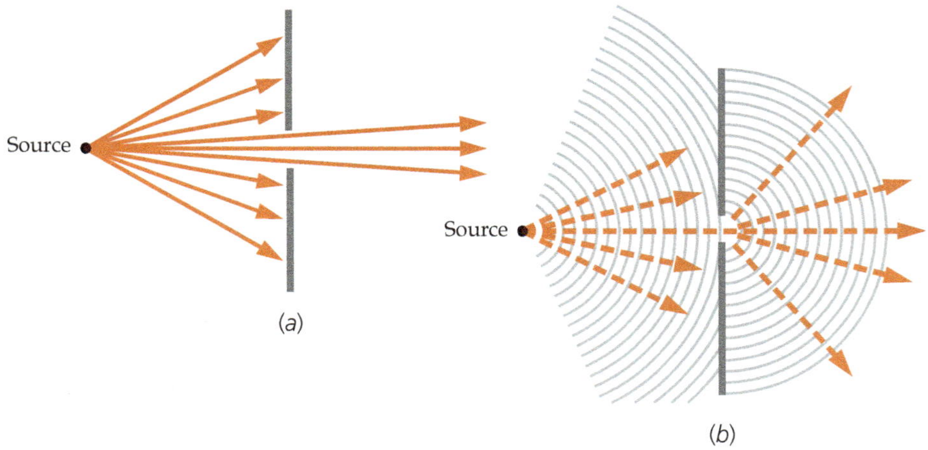

FIGURE 15-24 Comparison of particles and waves passing through a narrow opening in a barrier. (*a*) Transmitted particles are confined to a narrow-angle beam. (*b*) Transmitted waves spread out (radiate widely) from the aperture, which acts like a point source of circular waves.

Although waves passing through an aperture always bend, or diffract, to some extent, the amount of diffraction depends on whether the wavelength is small or large relative to the width of the aperture. If the wavelength is greater than or equal to the width aperture, as in Figure 15-25, the diffraction effects are large, and the waves spread out as they pass through the aperture—as if the waves were originating from a point source. On the other hand, if the wavelength is small relative to the aperture, the effect of diffraction is small, as shown in Figure 15-25. Near the edges of the aperture the wavefronts are distorted and the waves appear to bend slightly. For the most part, however, the wavefronts are not affected and the waves propagate in straight lines, much like a beam of particles. The approximation that waves propagate in straight lines in the direction of the rays with no diffraction is known as the **ray approximation.** Wavefronts are distorted *near* the edges of any obstacle blocking part of the wavefronts. By *near* we mean within a few wavelengths of the edges.

Because the wavelengths of audible sound (which range from a few centimeters to several meters) are generally large compared with apertures and obstacles (doors or windows, and people, for example), diffraction of sound waves is a phenomenon that is regularly observed. On the other hand, the wavelengths of visible light 4×10^{-7} to 7×10^{-7} m) are so small compared with the size of ordinary objects and apertures that the diffraction of light is not easily noticed; light appears to travel in straight lines. Nevertheless, the diffraction of light is an important phenomenon that we study in detail in Chapter 35.

Diffraction places a limitation on how accurately small objects can be located by reflecting waves off them and on how well details of the objects can be resolved. Waves are not reflected appreciably from objects smaller than the wavelength, so detail cannot be observed on a scale smaller than the wavelength used. If waves of wavelength λ are used to locate an object, then its position can be known only to within an uncertainty of one wavelength.

FIGURE 15-25 Plane waves in a ripple tank meeting a barrier with an opening width that is large compared to λ. The wave continues in the forward direction, with only a small amount of spreading into the regions to either side of the opening. *(Fundamental Photographers.)*

Sound waves with frequencies above 20,000 Hz are called **ultrasonic waves.** Because of their very small wavelengths, narrow beams of ultrasonic waves can be sent out and reflected from small objects. Bats can emit and detect frequencies up to about 120 kHz, corresponding to a wavelength of 2.8 mm, which they use to locate small prey such as moths. Echolocation systems, called sonar (from *so*und and *na*vigation *r*anging), are used to detect the outlines of submerged objects with sound waves. The frequency used by commercially available fish finders ranges from about 25 to 200 kHz, and porpoises produce echolocation clicks in the same frequency range. In medicine, ultrasonic waves are used for diagnostic purposes. Ultrasonic waves are passed through the human body and information about the frequency and intensity of the transmitted and reflected waves is processed to construct a three-dimensional picture of the body's interior, called a sonogram.

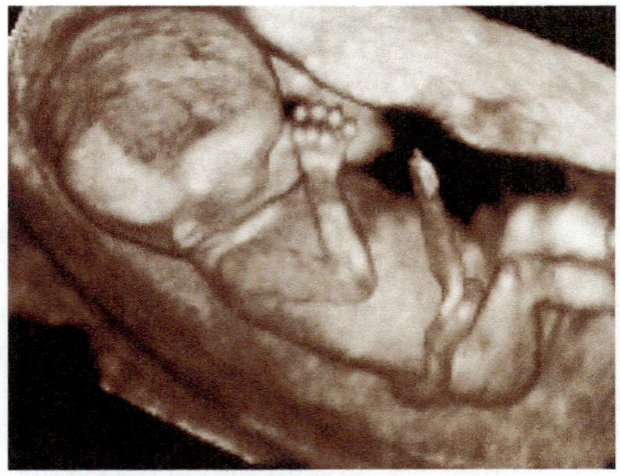

(GE Medical Systems/Photo Researchers, Inc.)

15-5 THE DOPPLER EFFECT

If a wave source and a receiver are moving relative to each other, the received frequency is not the same as the frequency of the source. If they are moving closer together, the received frequency is greater than the source frequency; and if they are moving farther apart, the received frequency is less than the source frequency. This is called the **Doppler effect.** A familiar example is the drop in pitch of the sound of the horn of an approaching car as the car passes by—and then recedes.

In the following discussion, all motions are relative to the medium. Consider the source moving with speed u_s, shown in Figure 15-26*a* and *b*, and a stationary receiver. The source has frequency f_s (and period $T_s = 1/f_s$). The received frequency f_r, the number of wave crests passing the receiver per unit time, is related to the wavelength λ (the distance between successive crests) and wave speed v by

$$f_r\lambda = v \quad \text{(stationary receiver)} \qquad 15\text{-}37$$

A wave crest leaves the source at time t_1 (Figure15-26*c*) and the next wave crest leaves the source at time t_2. The time between these two events is $T_s = t_2 - t_1$, and during this time the source and the crest leaving the source at time t_1 travel distances u_sT_s and vT_s, respectively. Consequently, at time t_2, the distance between the source and the crest leaving at time t_1 equals the wavelength λ. Behind the source $\lambda = \lambda_b = (v + u_s)T_s$, and in front of the source $\lambda = \lambda_f = (v - u_s)T_s$, provided $u_s < v$. (If $u_s \geq v$, no wavefronts reach the region ahead of the source.) We can express both λ_b and λ_f as

$$\lambda = (v \pm u_s)T_s = \frac{v \pm u_s}{f_s} \qquad 15\text{-}38$$

where the minus sign is used if $\lambda = \lambda_f$ and the plus sign is used if $\lambda = \lambda_b$. We have substituted $1/f_s$ for T_s. Substituting for λ in Equation 15-37 and rearranging gives

$$f_r = \frac{v}{\lambda} = \frac{v}{v \pm u_s} f_s \quad \text{(stationary receiver)} \qquad 15\text{-}39$$

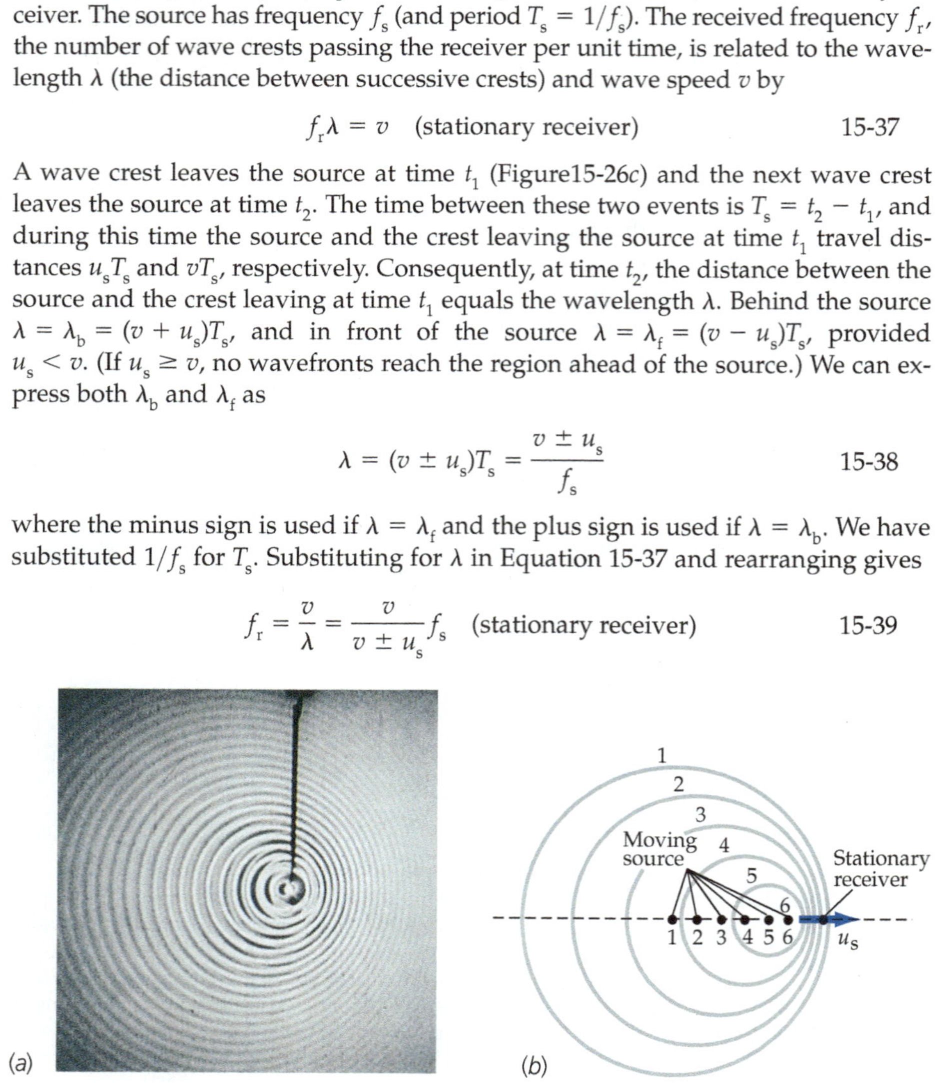

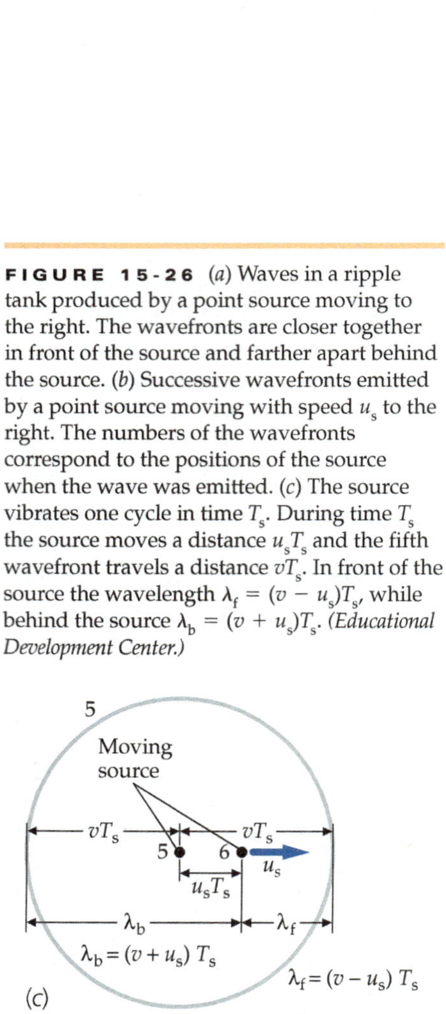

FIGURE 15-26 (*a*) Waves in a ripple tank produced by a point source moving to the right. The wavefronts are closer together in front of the source and farther apart behind the source. (*b*) Successive wavefronts emitted by a point source moving with speed u_s to the right. The numbers of the wavefronts correspond to the positions of the source when the wave was emitted. (*c*) The source vibrates one cycle in time T_s. During time T_s the source moves a distance u_sT_s and the fifth wavefront travels a distance vT_s. In front of the source the wavelength $\lambda_f = (v - u_s)T_s$, while behind the source $\lambda_b = (v + u_s)T_s$. *(Educational Development Center.)*

When the receiver moves relative to the medium, the received frequency is different simply because the receiver moves past more or fewer wave crests in a given time. Let T_r denote the time between arrivals of successive crests for a receiver moving with speed u_r. Then, during the time between the arrivals of two successive crests, each crest will have traveled a distance vT_r, and during the same time the receiver will have traveled a distance u_rT_r. If the receiver moves in the direction opposite to that of the wave (Figure 15-27), then during time T_r, the distance a crest moves plus the distance the receiver moves equals the wavelength. That is, $vT_r + u_rT_r = \lambda$, or $T_r = \lambda/(v + u_r)$. [If the receiver moves in the same direction as the wave, then $vT_r - \lambda = u_rT_r$, so $T_r = \lambda/(v - u_r)$]. Because $f_r = \lambda/T_r$, we have

$$f_r = \frac{1}{T_r} = \frac{v \pm u_r}{\lambda} \qquad \text{15-40}$$

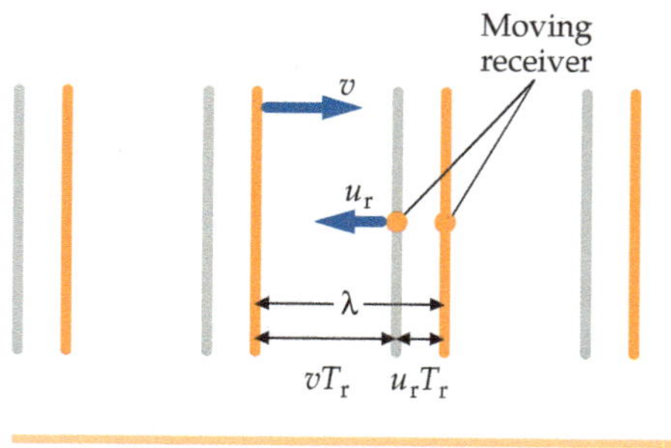

FIGURE 15-27 The time between arrivals of wave crests at the receiver is T_r. The wave crests are represented by orange lines when a wave crest reaches the receiver, and they are represented by grey lines when the next crest reaches the receiver. During time T_r the receiver travels the distance u_rT_r, while the wave crest travels the distance vT_r.

where, if the receiver moves in the same direction as the wave, the received frequency is lower, so we choose the negative sign. If the receiver moves in the direction opposite to that of the wave, the frequency is higher, so we choose the positive sign. Substituting for λ from Equation 15-38, we obtain

$$f_r = \frac{v \pm u_r}{v \pm u_s} f_s \qquad \text{15-41}a$$

The correct choices for the plus or minus signs are most easily determined by remembering that the frequency tends to increase both when the source moves toward the receiver and when the receiver moves toward the source. For example, if the receiver is moving toward the source, the plus sign is selected in the numerator, which tends to increase the received frequency; if the source is moving away from the receiver, the plus sign is selected in the denominator so that the equation predicts a decrease in the received frequency. Equation 15-41*a* appears more symmetric, and thus is easier to remember, if expressed in the form

$$\frac{f_r}{v \pm u_r} = \frac{f_s}{v \pm u_s} \qquad \text{15-41}b$$

It can be shown (see Problem 83) that if both u_s and u_r are much smaller than the wave speed v, then the shift in frequency $\Delta f = f_r - f_s$ is given approximately by

$$\frac{\Delta f}{f_s} \approx \pm \frac{u}{v} \quad (u \ll v) \qquad \text{15-42}$$

where $u = u_s \pm u_r$ is the speed of the source relative to the receiver.

In a reference frame in which the medium is moving (for example, the reference frame of the ground if air is the medium and if there is a wind blowing), the wave speed v is replaced by $v' = v \pm u_w$, where u_w is the speed of the wind relative to the ground.

! Equations 15-37 through 15-42 are valid only in the reference frame of the propagating medium.

PROBLEM-SOLVING STRATEGY

Solving Problems Involving Doppler Shift

PICTURE Solving problems involving the Doppler shift means using the equation

$$f_r = \frac{v \pm u_r}{v \pm u_s} f_s$$

(Equation 15-41*a*).

SOLVE

1. Find the speed of the source u_s and of the receiver u_r in the reference frame of the propagating medium.
2. Find the directions of the motions of the source and receiver in the same reference frame.
3. Substitute values into Equation 15-41*a*. Both the source moving toward the receiver and the receiver moving toward the source tend to increase the received frequency. Thus, if the source is moving toward the receiver,

choose the minus sign in the denominator, and if the receiver is moving toward the source, choose the plus sign in the numerator.

4. If the wave bounces off a reflector before reaching the receiver, treat the reflector first as a receiver and apply Equation 15-41*a*, then treat the reflector as a source and apply Equation 15-41*a* once again.

CHECK If the distance between a source and receiver is decreasing, then the received frequency f_r is higher than the source frequency f_s. If this distance is increasing, then f_r is lower than f_s.

Example 15-12 Sounding the Horn

The frequency of a car horn is 400 Hz. If the horn is honked as the car moves with a speed $u_s = 34$ m/s (about 122 km/h) through still air toward a stationary receiver, find (*a*) the wavelength of the sound passing the receiver, and (*b*) the frequency received. Take the speed of sound in air to be 343 m/s. (*c*) Find the wavelength of the sound passing the receiver and find the frequency received if the car is stationary as the horn is honked and a receiver moves with a speed $u_r = 34$ m/s toward the car.

PICTURE (*a*) The waves in front of the source are compressed, so we use the minus sign in $\lambda = (v \pm u_s)/f_s$ (Equation 15-38). (*b*) We calculate the received frequency using $f_r = [(v \pm u_r)/(v \pm u_s)]f_s$ (Equation 15-41*a*). (*c*) For a moving receiver, we use the same equations as in Parts (*a*) and (*b*).

SOLVE

(*a*) Using Equation 15-38, calculate the wavelength in front of the car. In front of the source the wavelength is shorter, so choose the sign accordingly:

$$\lambda = \frac{v - u_s}{f_s} = \frac{343\text{ m/s} - 34\text{ m/s}}{400\text{ Hz}} = 0.773\text{ m} = \boxed{0.77\text{ m}}$$

(*b*) Using Equation 15-41*a* with $u_r = 0$, solve for the received frequency:

$$f_r = \frac{v \pm u_r}{v \pm u_s}f_s = \frac{v}{v - u_s}f_s = \left(\frac{343}{343 - 34}\right)(400\text{ Hz}) = 444\text{ Hz} = \boxed{440\text{ Hz}}$$

(*c*) 1. Using Equation 15-38 with $u_s = 0$, calculate the wavelength in front of the source:

$$\lambda = \frac{v \pm u_s}{f_s} = \frac{343\text{ m/s}}{400\text{ Hz}} = 0.858\text{ m} = \boxed{0.86\text{ m}}$$

2. The received frequency is given by Equation 15-41*a* with $u_s = 0$. The source is approaching the receiver, so the frequency is shifted upward. Choose the sign accordingly:

$$f_r = \frac{v \pm u_r}{v \pm u_s}f_s = \frac{v + u_r}{v}f_s = \left(1 + \frac{u_r}{v}\right)f_s = \left(1 + \frac{34}{343}\right)(400\text{ Hz}) = \boxed{440\text{ Hz}}$$

CHECK The receiver is moving at about 10% of the speed of sound and the frequency received is about 10% higher than the frequency of the source, which is plausible. (Caution, though, this works only if the source is at rest.)

TAKING IT FURTHER The frequency f_r can also be obtained using Equation 15-40.

PRACTICE PROBLEM 15-8 As a train moving at 90 km/h is approaching a stationary listener on a windless day, it sounds its horn, which has a frequency of 630 Hz. (*a*) What is the wavelength of the sound waves in front of the train? (*b*) What frequency is heard by the listener? (Use 343 m/s for the speed of sound.)

Example 15-13 The Speed of the Wave *Context-Rich*

You work for an insurance company. An asteroid crashing into the ocean generates a tsunami. When the waves strikes land a 10-m-high wave does a lot of damage. Your boss wants to know how fast the big waves were moving. Knowing that you took a physics course, he asks you to find out. All you have to go on is the audio from a tape recorder that was found in a tree after the waves receded. The audio on the tape has a siren in the background, and in between blasts from the local warning siren is a faint echo of the siren. You measure the frequencies of the sound produced by the siren and its echo and find the siren had a frequency of 4000 Hz, but the echo had a frequency of 4080 Hz. How fast was the big wave approaching?

PICTURE You check with the weather service and find there was no wind at the time the tsunami hit. In addition, the reported temperature is 20°C, so the speed of sound was 343 m/s. First, apply the Doppler-effect equation (Equation 15-41*a*) to calculate the frequency of the sound received by the tsunami in terms of the speed u of the big wave. Apply the equation again, this time considering the big wave as the source of the sound and the tape recorder as the receiver. Assume that the tape recorder was not moving.

SOLVE

1. Apply the Doppler-effect equation with $u_s = 0$ to relate the frequency f_r received by the big wave to the speed u of the big wave:

$$f_r = \frac{v \pm u_r}{v \pm u_s} f_s = \frac{v + u}{v} f_s \qquad f_r = \frac{v + u}{v} f_s$$

2. Apply the Doppler-effect equation, this time with $u_r = 0$, to relate the frequency f'_r received by the tape recorder to the speed of the big wave. Use the step-1 result f_r as the frequency of the big wave as a source of sound:

$$f'_r = \frac{v \pm u_r}{v \pm u_s} f'_s = \frac{v}{v - u} f_r \qquad f'_r = \frac{v}{v - u} f_r$$

3. We now have two equations and two unknowns. Substitute the step-1 result into the step-2 result and simplify:

$$f'_r = \frac{v}{v - u} f_r = \frac{v}{v - u}\frac{v + u}{v} f_s \qquad f'_r = \frac{v + u}{v - u} f_s$$

4. Solve for the speed u:

$$u = \frac{f'_r - f_s}{f'_r - f_s} v = \frac{4400\text{ Hz} - 4000\text{ Hz}}{4400\text{ Hz} + 4000\text{ Hz}} 343\text{ m/s} = \boxed{16.3\text{ m/s}}$$

CHECK Sixteen meters per second is about twice as fast as a person can sprint under ideal conditions. Having seen video of tsunamis striking the shore, this seems like a plausible speed.

Another familiar example of the Doppler effect is the radar used by police to measure the speed of a car. Electromagnetic waves emitted by the radar transmitter strike the moving car. The car acts as both a moving receiver and a moving source as the waves reflect off it back to the radar receiver. Equation 15-41*a* is not valid for electromagnetic waves. Electromagnetic waves require the use of the relativistic Doppler-effect formula. (The relativistic Doppler effect is discussed following Example 15-14.) It turns out that if $u \ll c$, where c is the speed of light, Equation 15-42 is valid for electromagnetic waves.

Example 15-14 Police Radar

Try It Yourself

The radar unit in a police car sends out electromagnetic waves that travel at the speed of light c. The electric current in the antenna of the radar unit oscillates at frequency f_s. The waves reflect from a speeding car moving away from the police car at speed u relative to the police car. There is a frequency difference of Δf between f_s and f'_x, the frequency received at the police car. Find u in terms of f_s and Δf.

PICTURE The radar wave strikes the speeding car at frequency f_r. This frequency is less than f_s because the car is moving away from the source. The frequency shift is given by $\Delta f/f = \pm u/v$ (Equation 15-42) with $v = c$. The car then acts as a moving source emitting waves of frequency f_r. The police unit detects waves of frequency $f'_x < f_r$ because the source (the speeding car) is moving away from the police car. The frequency difference is $f'_x - f_s$.

SOLVE

Cover the column to the right and try these on your own before looking at the answers.

Steps	Answers
1. The radar unit must be able to determine the speed based only on what it transmits and what it detects.	The radar unit must determine u in terms of f_s and f'_r. We solve $\Delta f/f = \pm u/v$ (Equation 15-42) for u in terms of f_s and $\Delta f = f'_r - f_s$.
2. The frequency difference Δf is the frequency difference $\Delta f_1 = f_r - f_s$ plus the frequency difference $\Delta f_2 = f - f_r$.	$\Delta f = \Delta f_1 + \Delta f_2$
3. Using Equation 15-42 with $v = c$, substitute for the frequency differences in step 2.	$\Delta f = -\frac{u}{c} f_s - \frac{u}{c} f_r = -\frac{u}{c}(f_s + f_r)$
4. Again using Equation 15-42, solve for f_r in terms of f_s.	$\frac{\Delta f_1}{f_s} = -\frac{u}{c}$ so $f_r = \left(1 - \frac{u}{c}\right) f_s$

5. Substitute your step-4 result into your step-3 result and simplify.

$$\Delta f = -\frac{u}{c}\left(2 - \frac{u}{c}\right)f_s$$

6. Compared to 2, u/c is negligible. Use this to simplify the step-5 result and solve for u in terms of Δf and f_s.

$$\Delta f \approx -2\frac{u}{c}f_s \quad \text{so} \quad u = -\frac{\Delta f}{2f_s}c = \boxed{\frac{|\Delta f|}{2f_s}c}$$

CHECK The step-6 result is a dimensionless ratio times the speed of light, so it has the correct dimensions for a speed. Thus, the dimensions of the step-6 result are plausible.

TAKING IT FURTHER The difference in frequency between two waves of nearly equal frequency is easy to detect because the two waves interfere to produce a wave whose amplitude oscillates with frequency $|\Delta f|$, which is called the beat frequency. Interference and beats are discussed in Chapter 16.

PRACTICE PROBLEM 15-9 Calculate Δf if $f_s = 1.50 \times 10^9$ Hz, $c = 3.00 \times 10^8$ m/s, and $u = 50.0$ m/s.

The Doppler shift and relativity We see from Example 15-12 (and Equations 15-39, 15-40, and 15-41) that the magnitude of the Doppler shift in frequency depends on whether it is the source or the receiver that is moving relative to the medium. For sound, these two situations are physically different. For example, if you move relative to still air, you feel air rushing past you. In your reference frame, there is a wind. For sound waves in air, therefore, we can tell whether the source or receiver is moving by noting if there is a wind in the reference frame of the source or the receiver. However, light and other electromagnetic waves propagate through empty space in which there is no propagating medium. There is no "wind" to tell us whether the source or receiver is moving. According to Einstein's theory of relativity, absolute motion cannot be detected, and all observers measure the same speed c for light, independent of their motion relative to the source. Thus Equation 15-41 cannot be correct for the Doppler shift for light. Two modifications must be made in calculating the relativistic Doppler effect for light. First, the speed of waves passing a receiver is c, which is independent of the motion of the receiver. Second, the time interval between the emission of successive wave crests, which is $T_s = 1/f_s$ in the reference frame of the source, is different in the reference frame of the receiver when the two reference frames are in relative motion, because of relativistic time dilation and length contraction (Equations R-9 and R-3). (We discuss the relativistic Doppler effect in Chapter 39.) The result is that the frequency received depends only on the relative speed of approach (or recession) u, and is related to the frequency emitted by

$$f_r = \sqrt{\frac{c \pm u}{c \pm u}}f_s \qquad \text{15-43}$$

Choose the signs that give an up-shift in frequency when the source and receiver are approaching, and vice versa. Again, when $u \ll c$, $\Delta f/f_s \approx \pm u/c$, as given by Equation 15-42.

(a)

(b)

(a) Shock waves from a supersonic airplane. *(Sandia National Laboratory.)* (b) Shock waves produced by a bullet traversing a helium balloon. *(Estate of Harold E. Edgerton/ Palm Press Inc.)*

SHOCK WAVES

During our derivations of the Doppler-shift expressions, we assumed that the speed u of the source was less than the wave speed v. If a source moves with speed greater than the wave speed, then there will be no waves in front of the source. Instead, the waves pile up behind the source to form a shock wave. In the case of sound waves, this shock wave is heard as a sonic boom when it arrives at the receiver.

Figure 15-28 shows a source originally at point P_1 moving to the right with speed u. After some time t, the wave emitted from point P_1 has traveled a distance vt. The source has traveled a distance ut and will be at point P_2. The line from

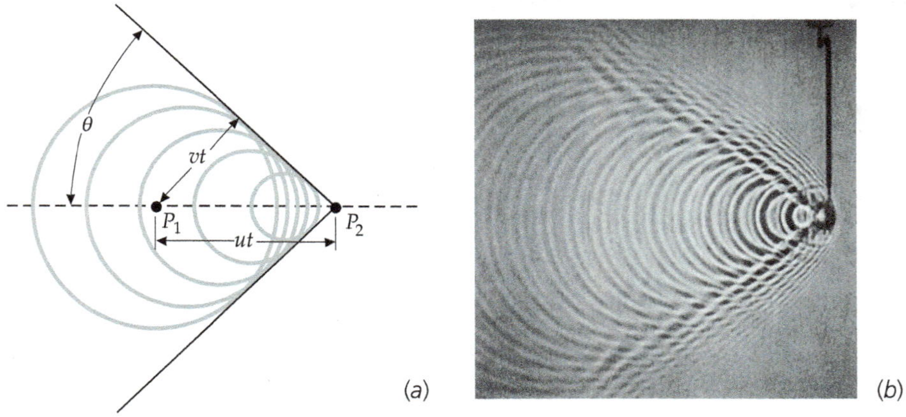

FIGURE 15-28 (*a*) Source moving with a speed u that is greater than the wave speed v. The envelope of the wavefronts forms a cone with the source at the apex. (*b*) Waves in a ripple tank produced by a source moving with a speed $u > v$. *(Educational Development Center.)*

this new position of the source to the wavefront emitted when the source was at P_1 makes an angle θ, called the **Mach angle,** with the path of the source, given by

$$\sin\theta = \frac{vt}{ut} = \frac{v}{u} \qquad 15\text{-}44$$

Thus, the shock wave is confined to a cone that narrows as u increases. The ratio of the source speed u to the wave speed v is called the **Mach number:**

$$\text{Mach number} = \frac{u}{v} \qquad 15\text{-}45$$

Equation 15-44 also applies to the electromagnetic radiation called *Cerenkov radiation,* which is given off when a charged particle moves in a medium with speed u that is greater than the speed of light v in that medium. (According to the special theory of relativity, it is impossible for a particle to move faster than c, the speed of light in vacuum. In a medium such as glass however, electrons and other particles can move faster than the speed of light in that medium.) The blue glow surrounding the fuel elements of a nuclear reactor is an example of Cerenkov radiation.

Example 15-15 A Sonic Boom — Try It Yourself

A supersonic plane flying due east at an altitude of 15 km passes directly over point P. The sonic boom is heard at point P when the plane is 22 km east of point P. What is the speed of the supersonic plane?

PICTURE The speed of the plane is related to the sine of the Mach angle (Equation 15-2). Draw a picture so the sine of the Mach angle can be calculated.

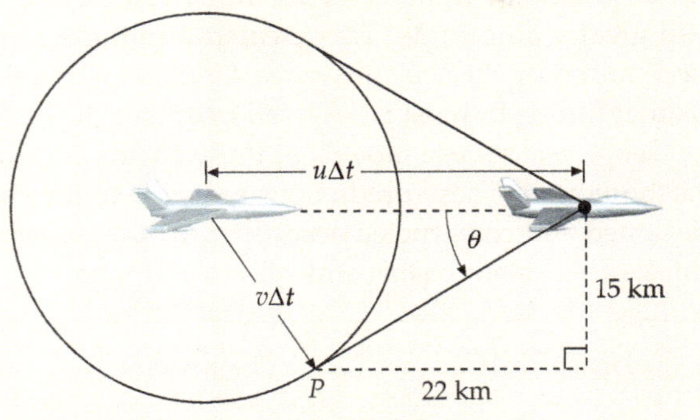

FIGURE 15-29 In the time that the plane moves distance $u\,\Delta t$, the sound moves distance $v\,\Delta t$.

SOLVE

Cover the column to the right and try these on your own before looking at the answers.

Steps	Answers
1. Sketch the position of the plane (Figure 15-29) both at the instant the sonic boom is heard at point P and at the instant that sound was produced. Label the distance the sound travels $v\,\Delta t$, and the distance the plane travels $u\,\Delta t$.	
2. From your sketch and Equation 15-44, calculate u:	$\tan\theta = \dfrac{15\text{ km}}{22\text{ km}}$ so $\theta = 34.3°$ $\sin\theta = \dfrac{v\,\Delta t}{u\,\Delta t} = \dfrac{v}{u}$ so $u = \dfrac{v}{\sin\theta} = 609\text{ m/s} = \boxed{610\text{ m/s}}$

CHECK The speed of sound is 343 m/s, so 610 m/s is plausible for a supersonic speed.

Physics Spotlight

All Shook Up: Sediment Basins and Earthquake Resonance

On April 18, 1906, the city of San Francisco was devastated by a powerful earthquake. All the buildings in the lowest area collapsed. These buildings were built on water-logged *unconsolidated sediments*—loose gravel, sand, dirt, and clay. Some buildings even sank one or more stories into the ground as the shaking liquefied the loose sediment. Buildings up on rocky Nob Hill and Russian Hill fared better.

The damage to buildings constructed on water-logged loose gravel, sand, dirt and clay is greater than the damage to those constructed on hard rock. *(Roger Ressmeyer/CORBIS.)*

Cities located on unconsolidated sediments and near major faults are more vulnerable to earthquake damage than those that are not. If they are partly surrounded by rocky hills or mountains, the danger increases. Vulnerable cities include Seattle,[*,†] Istanbul,[‡] Rome,[#] Los Angeles,[°] San Francisco,[§] and Taipei.[¶]

Unconsolidated sediments are at much greater risk for seismic shaking than rock. When an earthquake occurs, some of the earthquake's energy is sent forth in seismic waves. These waves cause the ground to vibrate over a wide range of frequencies. In solid rock, the waves vibrate and have relatively small amplitudes. The looser the rock or sediment, the more the propagation speed decreases and the amplitude increases.[**] In loose gravel, the waves vibrate more slowly and have greater amplitude. In water-logged sediments, the waves vibrate still more slowly and have much greater amplitude. If you tap a bowl of gelatin sharply on the side, you can hear the sound of tapping the bowl. If it is a metal or glass bowl, the sound will have a frequency of hundreds of hertz. But the gelatin attenuates and scatters the higher frequencies, and resonates at lower frequencies. The same principle underlies the vulnerability of many cities to earthquakes.[††]

Unfortunately, the resonance frequencies of most buildings are closer to the resonance frequencies of the seismic waves in loose sediments.[‡‡] Thus, not only do the sediments vibrate with greater amplitude, but they vibrate most strongly at frequencies that are closer to the resonance frequencies of buildings. This problem was clearly recognized in the State report on the San Francisco earthquake of 1906.[##] Buildings located in areas of unconsolidated sediment were far more damaged than those that were located on higher, firmer ground.

The situation is worst for those cities that are built on sediments that are partially surrounded by areas of hard rock. There the waves resonate within the basin of sediments with large amplitudes. This was true in 1906, when the town of Santa Rosa sustained great damage, even though it was farther from the epicenter of the quake than other, less damaged towns. Santa Rosa is located within its own sediment basin, surrounded by rock.[°°] The basin resonance causes the sediments to vibrate with still greater amplitude. This greater amplitude creates greater damage. Usually, the damage is from the horizontal acceleration caused by the seismic waves. Until the stringent earthquake codes of the 1970s, buildings were not built to withstand horizontal forces. In most cities, well over half the buildings date from before the stringent codes were adopted.

Geophysicists use models of these basins and their sediments to predict areas that are likely to sustain high damage from earthquakes.[§§] These predictions are used to improve codes or to require that bridges,[¶¶] breakwaters,[***] and buildings[†††] are designed and constructed according to current best practices for hazard reduction. The next time you shake a bowl of gelatin, think of sediment basins and seismic damage.

---

* Chang, S., et al., "Expected Ground Failure," Paper presented at the Seattle Fault Earthquake Scenario Conference, 2005. Seattle: Earthquake Engineering Research Institute. http://seattlescenario.eeri.org/presentations/Ch%202%20Ground%20Failure%20-%20Chang.pdf

† Pierepiekarz, M. et al., "Buildings," Paper presented at the Seattle Fault Earthquake Scenario Conference, 2005. Seattle: Earthquake Engineering Research Institute. http://seattlescenario.eeri.org/presentations/Ch%205%20Buildings%20-%20Pierepiekarz.pdf

‡ Erdik, M. *Earthquake Vulnerability of Buildings and a Mitigation Strategy: Case of Istanbul.* World Bank. http://info.worldbank.org/etools/docs/library/114715/istanbul03/docs/istanbul03/06erdik3-n%5B1%5D.pdf as of June 2006.

Perkins, S. "Rome at Risk: Seismic Shaking Could Be Long and Destructive," *Science News*, Feb. 25, 2006, 115.

° Perkins, S., "Portrait of Destruction," *Science News*, May 21, 2005, 325.

§ Zoback, M. L., "The 1906 Earthquake—Lessons Learned, Lessons Forgotten, and Future Directions," Paper presented at the American Geophysical Union Meeting, San Francisco, Dec. 5–9, 2005. http://www.ucmp.berkeley.edu/museum/events/shortcourse2006/zoback/

¶ Altenburger, E., "Earthquake Hazards in Taiwan—The September 1999 Chichi Earthquake," *FOCUS on Geography*, Winter 2004, 1–8.

** O'Connell, D. R.H., "Replications of Apparent Nonlinear Seismic Response with Linear Wave Propagation Models," *Science*, Mar. 26, 1999, Vol. 283, No. 5410, p. 2045–2050.

†† Page, R. A., Blume, J. A., and Joyner, W. B., "Earthquake Shaking and Damage to Buildings." *Science*, Aug. 22, 1975, Vol. 189., No. 4203, p. 601–608.

‡‡ Seed, H. B., et al., "Soil Conditions and Building Damage in 1967 Caracas Earthquake," *Journal of Soil Mechanics Division of American Society of Civil Engineers*, 1972, Vol. 98, No. 8, 787–806.

Lawson, A., et al., *Report of the State Earthquake Investigation Commission*. 1908. Washington, DC: Carnegie Institution.

°° Sloan, D., "Portrait of a Tectonic Landscape," *Bay Nature*, Spring 2006, Vol. 6, No. 2, 24–27.

§§ United States Geological Survey, "1906 Ground Motion Simulations," Earthquake Hazards Program. http://earthquake.usgs.gov/regional/nca/1906/simulations/, as of June 2006.

¶¶ Treyger, S., Jones, M., and Orsolini, G., "Suspending the Big One," *Roads and Bridges*, May 2004, 22–25.

*** Banijamali, B., "Rubble Mounds Feel the Rumble," *Dredging and Port Construction*, June 2005, 35–41.

††† Gonchar, J., "One Project, but Many Seismic Solutions," *Architectural Record*, May 2006, 167–174.

Summary

1. In wave motion, energy and momentum are transported from one point in space to another without the transport of matter.
2. The relation $v = f\lambda$ holds for all harmonic waves.

TOPIC	RELEVANT EQUATIONS AND REMARKS	
1. Transverse and Longitudinal Waves	In transverse waves, such as waves on a string, the disturbance is perpendicular to the direction of propagation. In longitudinal waves, such as sound waves, the disturbance is along the direction of propagation.	
2. Speed of Waves	The wave speed v is independent of the motion of the wave source. The speed of a wave relative to the medium depends on the density and elastic properties of the medium.	
Waves on a string	$v = \sqrt{F_T/\mu}$	15-3
Sound waves	$v = \sqrt{B/\rho}$	15-4
Sound waves in a gas	$v = \sqrt{\gamma RT/M}$	15-5
	where T is the absolute temperature,	
	$T = t_C + 273.15$	15-6
	R is the universal gas constant,	
	$R = 8.314\ \text{J}/(\text{mol}\cdot\text{K})$	15-7
	M is the molar mass of the gas, which for air is 29.0×10^{-3} kg/mol, and γ is a constant that depends on the kind of gas. For a diatomic gas such as air, $\gamma = 7/5$. For a monatomic gas such as helium, $\gamma = 5/3$.	
Electromagnetic waves	The speed of electromagnetic waves in vacuum is a universal constant $c = 3.00 \times 10^8$ m/s	
*** 3. Wave Equation**	$\frac{\partial^2 y}{\partial x^2} = \frac{1}{v^2}\frac{\partial^2 y}{\partial t^2}$	15-10*b*
4. Harmonic Waves		
Wave function	$y(x,t) = A\sin(kx \pm \omega t)$	15-15
	where A is the amplitude, k is the wave number, and ω is the angular frequency. Use the minus sign for a wave traveling in the $+x$ direction, and the plus sign for a wave traveling in the $-x$ direction.	
Wave number	$k = \frac{2\pi}{\lambda}$	15-14
Angular frequency	$\omega = 2\pi f = \frac{2\pi}{T}$	15-17
Speed	$v = f\lambda = \omega/k$	15-12, 15-16
Energy	The energy in a harmonic wave is proportional to the square of the amplitude.	
Power for harmonic waves on a string	$P_{av} = \frac{1}{2}\mu v\omega^2 A^2$	15-22
5. Harmonic Sound Waves	Sound waves can be considered to be either displacement waves or pressure waves. The human ear is sensitive to sound waves of frequencies from about 20 Hz to 20 kHz. In a harmonic sound wave, the pressure and displacement are 90° out of phase.	
Amplitudes	The pressure and displacement amplitudes are related by	
	$p_0 = \rho\omega v s_0$	15-26
	where ρ is the density of the medium.	
Energy density	$\eta_{av} = \frac{(\Delta E)_{av}}{\Delta V} = \frac{1}{2}\rho\omega^2 s_0^2$	15-28

TOPIC	RELEVANT EQUATIONS AND REMARKS	
6. Intensity	The intensity of a wave is the average power per unit area. $$I = \frac{P_{av}}{A}$$	15-29
Average intensity I of a sound wave	$$I = \eta_{av}v = \frac{1}{2}\rho\omega^2 s_0^2 v = \frac{1}{2}\frac{p_0^2}{\rho v}$$	15-32
*Intensity level β in dB	Sound intensity levels are measured on a logarithmic scale. $$\beta = (10\text{ dB})\log\frac{I}{I_0}$$ where $I_0 = 10^{-12}\text{ W/m}^2$ is taken as the threshold of hearing.	15-33
7. Reflection and Refraction	When a wave is incident on a boundary surface that separates two regions of differing wave speed, part of the wave is reflected and part is transmitted.	
	The reflection and transmission coefficients are $$r = \frac{v_2 - v_1}{v_2 + v_1} \quad \text{and} \quad \tau = \frac{2v_2}{v_2 + v_1}$$	15-35
8. Diffraction	If a wavefront is partially blocked by an obstacle, the unblocked part of the wavefront diffracts (bends) into the region behind the obstacle.	
Ray approximation	If a wavefront is partially blocked by an obstacle, almost all of the diffraction occurs for that part of the wavefront that passes within a few wavelengths of the edge. For those parts of the wavefront that pass farther from the edge than a few wavelengths, diffraction is negligible and the wave propagates in straight lines in the direction of the incident rays.	
9. Doppler Effect	When a sound source and receiver are in relative motion, the received frequency f_r is higher than the frequency of the source f_s if their separation is decreasing, and lower if their separation is increasing.	
Moving source	$$\lambda = \frac{v \pm u_s}{f_s}$$	15-38[4]
Moving receiver	$$f_r = \frac{v \pm u_r}{\lambda}$$	15-40[3]
Either source or receiver moving	$$f_r = \frac{v \pm u_r}{v \pm u_s}f_s \quad \text{or} \quad \frac{f_r}{v \pm u_r} = \frac{f_s}{v \pm u_s}$$ Choose the signs that give an up-shift in frequency for an approaching source or receiver, and vice versa.	15-41[3]
Small speeds of source or receiver	$$\frac{\Delta f}{f_s} \approx \pm\frac{u}{v} \quad (u \ll v), \quad \text{where} \quad u_s = u_s \pm u_r,$$	15-42[3]
Relativistic Doppler shift	$$f_r = \sqrt{\frac{c \pm u}{c \pm u}}f_s$$ Choose the signs that give an up-shift in frequency for an approaching source or receiver, and vice versa.	15-43
10. Shock Waves	When the source speed is greater than the wave speed, the waves behind the source are confined to a cone of angle θ given by	
Mach angle	$$\sin\theta = \frac{v}{u}$$	15-44
Mach number	$$\text{Mach number} = \frac{u}{v}$$	15-45

SOLVE

Cover the column to the right and try these on your own before looking at the answers.

Steps	Answers
1. Find the equivalent resistance R_{eq} of the 4.0-Ω and 12-Ω resistors in parallel.	$R_{eq} = 3.0\ \Omega$
2. Find the equivalent resistance R'_{eq} of R_{eq} in series with the 5.0-Ω resistor.	$R'_{eq} = 8.0\ \Omega$
3. Find the equivalent resistance of R'_{eq} in parallel with the 24-Ω resistor.	$R''_{eq} = \boxed{6.0\ \Omega}$

CHECK As expected for parallel combinations, the step-1 and step-3 results are less than the resistance of either of the two resistors in parallel. In addition, the step-2 result is greater than either of the two resistors in series, as expected for series combinations.

Example 25-13 Blowing the Fuse *Context-Rich*

You are making a snack for you and some friends to help you get ready for a full night of studying. You decide that coffee, toast, and popcorn would be a good start. You turn on the toaster, place some popcorn in the microwave, and turn on the microwave. Because your apartment is in an older building, you know that the fuse may blow when you turn on too many appliances. Should you start the coffeemaker? You look on the appliances and find that the toaster has a rating of 900 W, the microwave has a rating of 1200 W, and the coffeemaker has a rating of 600 W. You know that your apartment uses 20-A fuses.

PICTURE We can assume that household circuits are wired in parallel, because plugging in one device does not affect others that are in the circuit. Household voltage in the United States is 120 V. (We can neglect the fact that it is not dc.) If we can determine the current through each device, we can add up the total current in the circuit and see how it compares to the fuse current.

SOLVE

1. The power delivered to a device is the current multiplied by the voltage. That is, $P = IV$. Solve for the current for each device:

$$I_{toaster} = \frac{P_{toaster}}{V} = \frac{900\ \text{W}}{120\ \text{V}} = 7.5\ \text{A}$$

$$I_{m\text{-}wave} = \frac{P_{m\text{-}wave}}{V} = \frac{1200\ \text{W}}{120\ \text{V}} = 10.0\ \text{A}$$

$$I_{c\text{-}maker} = \frac{P_{c\text{-}maker}}{V} = \frac{600\ \text{W}}{120\ \text{V}} = 5.0\ \text{A}$$

2. The current through the fuse is the sum of these currents: $I_{fuse} = 22.5\ \text{A}$

3. A current this large is above the 20-A rating of the fuse: **Your guests will have to wait on the coffee.**

CHECK The maximum possible power that can be delivered by a 120-V circuit that has a 20-A fuse is $P_{max} = I_{max}V = (20\ \text{A})(120\ \text{V}) = 2400\ \text{W}$. The total power needed to run the three appliances simultaneously is 900 W + 1200 W + 600 W = 2700 W, which is 300 W more than the maximum that the circuit can deliver.

TAKING IT FURTHER We have assumed that the apartment has only one circuit, and thus only one fuse. Typically, there are several circuits, each fused separately. The coffeemaker can be plugged into an outlet that is on a different circuit than the outlet for the toaster and microwave without a fuse blowing.

25-5 KIRCHHOFF'S RULES

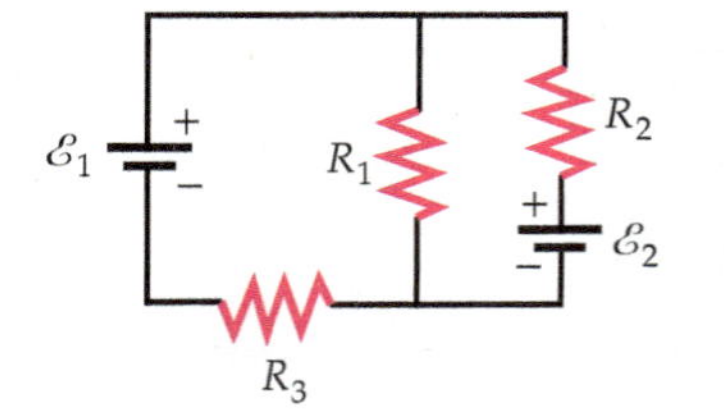

FIGURE 25-25 An example of a circuit that cannot be analyzed by replacing combinations of resistors in series or parallel with their equivalent resistances. The potential drops across R_1 and R_2 are not equal because of the emf source $\mathcal{E}_2$, so these resistors are not in parallel. (Parallel resistors would be connected together at both ends.) The resistors do not have the same current, so they are not in series.

There are many circuits, such as the circuit shown in Figure 25-25, that cannot be analyzed by merely replacing combinations of resistors by an equivalent resistance. The two resistors R_1 and R_2 in this circuit appear as though they might be in parallel, but they are not. The potential drop is not the same across both resistors because of the presence of the emf source $\mathcal{E}_2$ in series with R_2. Nor are R_1 and R_2 in series, because the wire connecting them has a branch point—they do not have the same current due to the way they are connected.

Two rules, called **Kirchhoff's rules,** apply to this circuit and to any other circuit:

1. When any closed loop is traversed, the algebraic sum of the changes in potential around the loop must equal zero.
2. At any junction (branch point) in a circuit where the current can divide, the sum of the currents into the junction must equal the sum of the currents out of the junction.

KIRCHHOFF'S RULES

Kirchhoff's first rule, called the **loop rule,** was introduced in Chapter 24. This rule follows directly from the presence of a conservative field $\vec{E}$.* To say $\vec{E}$ is conservative means that

$$\oint_C \vec{E} \cdot d\vec{r} = 0 \qquad 25\text{-}27$$

where the integral is taken around any closed curve C. Changes in potential ΔV and $\vec{E}$ are related by $\Delta V = V_b - V_a = -\int_a^b \vec{E} \cdot d\vec{r}$. Thus, Equation 25-27 implies that the sum of the changes in potential (the sum of the ΔVs) around any closed path equals zero.

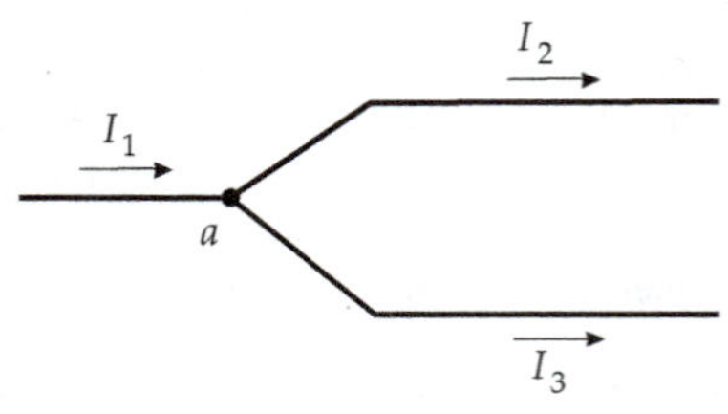

FIGURE 25-26 Illustration of Kirchhoff's junction rule. The current I_1 into point a equals the sum $I_2 + I_3$ of the currents out of point a.

Kirchhoff's second rule, called the **junction rule,** follows from the conservation of charge. Figure 25-26 shows the junction of three wires carrying currents I_1, I_2, and I_3. Because charge does not originate or accumulate at this point, the conservation of charge implies the junction rule, which for this case gives

$$I_1 = I_2 + I_3 \qquad 25\text{-}28$$

Charges do accumulate on the surfaces of conductors. However, it would require a very large surface area, such as the surface area of some capacitor plates, to accumulate a significant amount of charge. The surface areas of the conductors that are used in common circuits are much too small to accumulate a significant amount of charge.

SINGLE-LOOP CIRCUITS

As an example of using Kirchhoff's loop rule, consider the circuit shown in Figure 25-27, which contains two batteries that have internal resistances r_1 and r_2 and three external resistors. We wish to find the current in terms of the emfs and resistances.

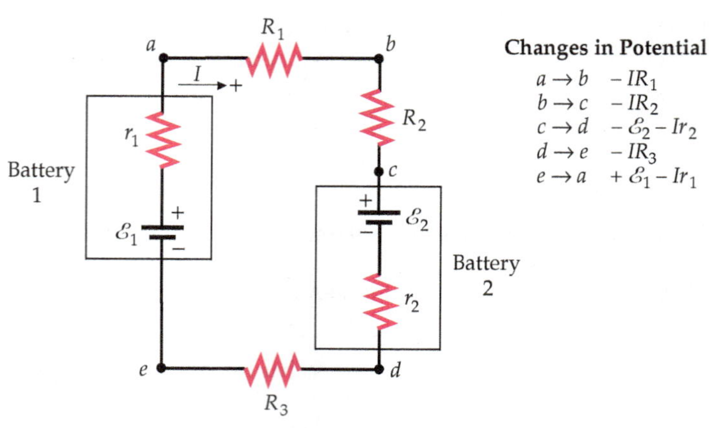

FIGURE 25-27 Circuit containing two batteries and three external resistors.

* There is also such a thing as a nonconservative electric field that is discussed in Chapter 28. The resultant electric field is the superposition of the conservative electric field and the nonconservative electric field. Kirchhoff's loop rule applies only to the conservative part of the electric field.

We choose clockwise as positive, as indicated by the arrow with the plus sign next to it in Figure 25-27. We then apply Kirchhoff's loop rule as we traverse the circuit in the positive direction, beginning at point a. Note that we encounter a potential drop as we traverse the source of emf between points c and d and we encounter a potential rise as we traverse the source of emf between e and a. Assuming that I is positive, we encounter a potential drop as we traverse each resistor. Beginning at point a, we obtain from Kirchhoff's loop rule

$$(V_b - V_a) + (V_c - V_b) + (V_d - V_c) + (V_e - V_d) + (V_a - V_e) = 0$$

Expressing the changes in potential in terms of the current, the emfs, and the resistances gives

$$(-IR_1) + (-IR_2) + (-\mathcal{E}_2 - Ir_2) + (-IR_3) + (\mathcal{E}_1 - Ir_1) = 0$$

Solving for the current I, we obtain

$$I = \frac{\mathcal{E}_1 - \mathcal{E}_2}{R_1 + R_2 + R_3 + r_1 + r_2} \qquad \text{25-29}$$

If $\mathcal{E}_2$ is greater than $\mathcal{E}_1$, we get a negative value for the current I, indicating that the current is in the negative (counterclockwise) direction.

For this example, we assume that $\mathcal{E}_1$ is greater than $\mathcal{E}_2$, so the current is positive. In addition, we model the current as positive charge carriers that move clockwise around the circuit. (The actual charge carriers are negatively charged electrons traveling counterclockwise.) Then, charge flows through battery 2 from the high-potential end to the low-potential end. Therefore, a positive charge ΔQ moving through battery 2 from point c to point d loses potential energy $\Delta Q\mathcal{E}_2$ (in addition to any energy dissipated within the battery due to the internal resistance r_2). If battery 2 is a rechargeable battery, much of this lost potential energy is stored in the battery as chemical energy, which means that battery 2 is *charging*.

The analysis of a circuit is usually simplified if we define the potential to equal zero at a convenient point in the circuit. Then we calculate the potential at the other points relative to it. Because only potential differences are important, any point in a circuit can be chosen to have zero potential. In many circuits, however, one point is connected to a rod that is driven into the ground. Such a point is said to be grounded or put to earth, and the potential is defined to be zero at that point. However, in an automobile the negative terminal of the battery is connected to the engine block by a heavy cable (called a grounding cable), and the point where the cable is connected to the engine block is referred to as ground. In the following example, we choose point e in the figure to be at zero potential. This is indicated by the ground symbol ⏚ at point e.

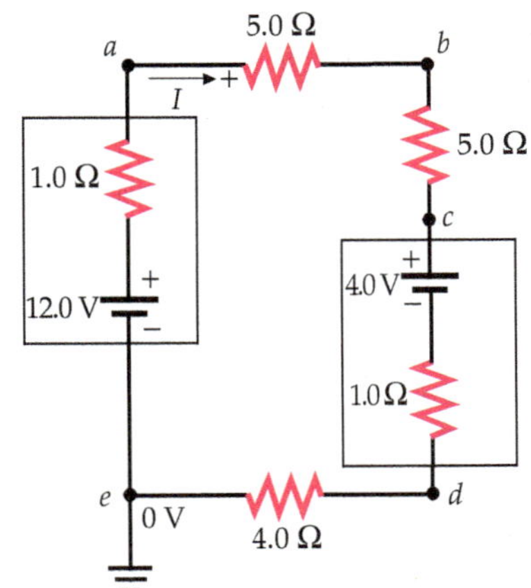

FIGURE 25-28

Example 25-14 Finding the Potential

Suppose the elements in the circuit in Figure 25-28 have the values $\mathcal{E}_1 = 12.0$ V, $\mathcal{E}_2 = 4.0$ V, $r_1 = r_2 = 1.0\ \Omega$, $R_1 = R_2 = 5.0\ \Omega$, and $R_3 = 4.0\ \Omega$. (*a*) Find the potentials at points a through e in the figure, assuming that the potential at point e is zero. (*b*) Discuss the energy transfers in the circuit.

PICTURE To find the potential differences, we first need to find the current I in the circuit. The potential drop across each resistor is equal to the product IR. To discuss the energy transfers, we calculate the power delivered to or supplied by each element using Equations 25-14 and 25-15.

SOLVE

(*a*) 1. The current I in the circuit is found using Equation 25-29:

$$I = \frac{12.0\text{ V} - 4.0\text{ V}}{5.0\ \Omega + 5.0\ \Omega + 4.0\ \Omega + 1.0\ \Omega + 1.0\ \Omega} = \frac{8.0\text{ V}}{16\ \Omega} = 0.50\text{ A}$$

2. We now find the potential at each labeled point in the circuit:

$$V_a = V_e + \mathcal{E}_1 - Ir_1 = 0 + 12.0\text{ V} - (0.50\text{ A})(1.0\ \Omega) = \boxed{11.5\text{ V}}$$

$$V_b = V_a - IR_1 = 11.5\text{ V} - (0.50\text{ A})(5.0\ \Omega) = \boxed{9.0\text{ V}}$$

$$V_c = V_b - IR_2 = 9.0\text{ V} - (0.50\text{ A})(5.0\ \Omega) = \boxed{6.5\text{ V}}$$

$$V_d = V_c - \mathcal{E}_2 - Ir_2 = 6.5\text{ V} - 4.0\text{ V} - (0.50\text{ A})(1.0\ \Omega) = \boxed{2.0\text{ V}}$$

$$V_e = V_d + IR_3 = 2.0\text{ V} - (0.50\text{ A})(4.0\ \Omega) = \boxed{0.0\text{ V}}$$

(*b*) 1. First, calculate the power supplied by the chemical reactions in the emf source that has emf $\mathcal{E}_1$:

$$P_{\mathcal{E}_1} = \mathcal{E}_1 I = (12.0\text{ V})(0.50\text{ A}) = 6.0\text{ W}$$

2. Part of this power is delivered to the resistors, both inside and outside the batteries:

$$P_R = I^2R_1 + I^2R_2 + I^2R_3 + I^2r_1 + I^2r_2$$
$$= (0.50\text{ A})^2(5.0\ \Omega + 5.0\ \Omega + 4.0\ \Omega + 1.0\ \Omega + 1.0\ \Omega) = 4.0\text{ W}$$

3. The remaining 2.0 W of power goes into charging battery 2:

$$P_{\mathcal{E}_2} = \mathcal{E}_2 I = (4.0\text{ V})(0.50\text{ A}) = 2.0\text{ W}$$

4. The rate at which potential energy is being delivered in the circuit is

$$P = P_R + P_{\mathcal{E}_2} = 6.0\text{ W}$$

CHECK The rate at which the 12-V battery converts chemical energy to electrical potential energy (6.0 W) is equal to the rate at which the 4.0-V battery converts electrical potential energy to chemical energy (2.0 W) plus the rate at which potential energy is dissipated (4.0 W).

Note that the terminal voltage of the battery that is being charged in Example 25-14 is $V_c - V_d = 4.5$ V, which is greater than the emf of the battery. If the same 4.0-V battery were used to deliver 0.50 A to an external circuit, its terminal voltage would be 3.5 V (again assuming that its internal resistance is 1.0 Ω). If the internal resistance is very small, the terminal voltage of a battery is nearly equal to its emf, whether the battery is delivering energy to an external circuit or is being charged. Some real batteries, such as those used in automobiles, are nearly reversible and can easily be recharged. Other types of batteries are not reversible. If you attempt to recharge one of those batteries by driving current through it from its positive to its negative terminal, virtually all of the energy will be dissipated into thermal energy rather than being transformed into the chemical energy of the battery.

Example 25-15 Jump-Starting a Car

A fully charged* car battery is to be connected by jumper cables to a discharged car battery in order to charge it. (*a*) To which terminal of the discharged battery should the positive terminal of the charged battery be connected? (*b*) Assume that the charged battery has an emf of $\mathcal{E}_1 = 12.0$ V, the discharged battery has an emf of $\mathcal{E}_2 = 11.0$ V, the internal resistances of the batteries are $r_1 = r_2 = 0.020\ \Omega$, and the resistance of the jumper cables is $R = 0.010\ \Omega$. What will be the charging current? (*c*) What will be the current if the batteries are connected incorrectly?

PICTURE For Part (*a*) the batteries should be connected so the initially discharged battery is charged. To calculate the current apply Kirchhoff's loop rule.

* Batteries do not store charge. A *fully charged* battery is one with a maximum amount of stored chemical energy.

SOLVE

(*a*) To charge the discharged battery, we connect positive terminal to positive terminal and negative terminal to negative terminal, to drive current through the discharged battery from the positive terminal to the negative terminal (Figure 25-29):

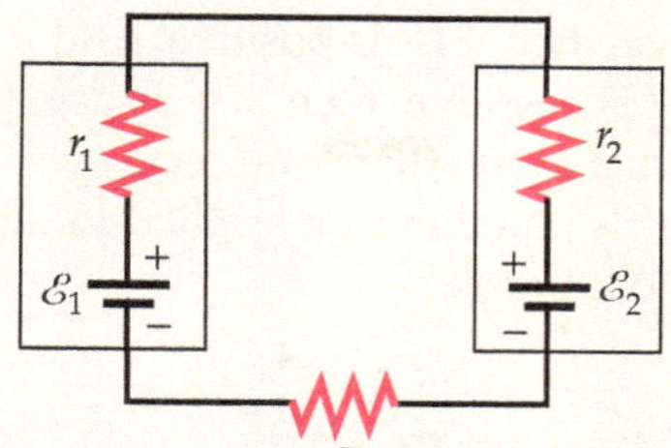

FIGURE 25-29

(*b*) Use Kirchhoff's loop rule to find the charging current:

$$\mathcal{E}_1 - Ir_1 - Ir_2 - \mathcal{E}_2 - IR = 0$$

so

$$I = \frac{\mathcal{E}_1 - \mathcal{E}_2}{R + r_1 + r_2} = \frac{12.0\text{ V} - 11.0\text{ V}}{0.050\ \Omega} = \boxed{20\text{ A}}$$

(*c*) When the batteries are connected incorrectly, positive terminals to negative terminals, the emfs add:

$$\mathcal{E}_1 - Ir_1 + \mathcal{E}_2 - Ir_2 - IR = 0$$

so

$$I = \frac{\mathcal{E}_1 + \mathcal{E}_2}{R + r_1 + r_2} = \frac{12.0\text{ V} + 11.0\text{ V}}{0.050\ \Omega} = \boxed{460\text{ A}}$$

CHECK If the batteries are connected incorrectly, as shown in Figure 25-30, the current is very large and the batteries could explode—producing a shower of boiling battery acid.

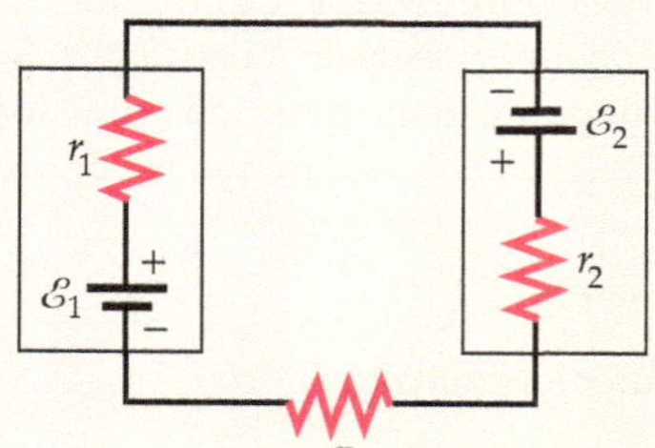

FIGURE 25-30 Two batteries connected incorrectly—dangerous!

MULTILOOP *CIRCUITS*

In multiloop circuits, often the direction of the current in one or more branches of the circuit are not obvious. Fortunately, Kirchhoff's rules do not require that we know these directions initially. In fact, the opposite is true. Kirchhoff's rules enable us to determine the directions of the currents. To accomplish this, for each branch of the circuit we arbitrarily assign a positive direction along the branch, and we indicate this assignment by placing a corresponding arrow on the circuit diagram (Figure 25-31). If the current density in the branch is in this positive direction, then when we solve for this current we will get a positive value. However, if the current density is opposite to the assigned positive direction, when we solve for the current we will get a negative value. In a resistor, an electric field within the resistor causes the current, and the current is in the same direction as the electric field. Because the electric field always points in the direction of decreasing potential, we know that in a resistor the direction of the current is also the direction of decreasing potential. Therefore, any time we traverse a resistor in the direction of the current, the change in potential is negative, and vice versa. Here is the rule:

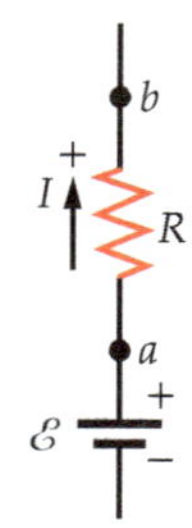

FIGURE 25-31 It is not known whether the current I has a positive or a negative value. Whether it is positive or negative, $V_b - V_a = -IR$. If the current is in the positive direction, then I is positive and $-IR$ is negative. If the current is in the negative direction, however, then I is negative and $-IR$ is positive.

> For each branch of a circuit, we draw an arrow to indicate the positive direction for that branch. Then, if we traverse a resistor in the direction of the arrow, the change in potential ΔV is equal to $-IR$ (and if we traverse a resistor in the direction opposite the direction of the arrow, ΔV is equal to $+IR$).
>
> SIGN RULE FOR THE CHANGE IN POTENTIAL ACROSS A RESISTOR

If we traverse a resistor in the positive direction, and if I is positive, then $-IR$ is negative. This is as expected, because in a resistor, the current is always in the direction of decreasing potential. However, if we traverse a resistor in the positive direction, and if I is negative, then $-IR$ is positive. Similarly, if we traverse a resistor in the negative direction, and if I is positive, then $+IR$ is positive. And if we traverse a resistor in the negative direction and if I is negative, then $+IR$ is negative.

To analyze circuits containing more than one loop, we need to use both of Kirchhoff's rules, with Kirchhoff's junction rule applied to junctions (points where the current splits into two or more parts).

Example 25-16 Applying Kirchhoff's Rules

(*a*) Find the current in each branch of the circuit shown in Figure 25-32. (*b*) Find the energy dissipated in the 4.0-Ω resistor in 3.0 s.

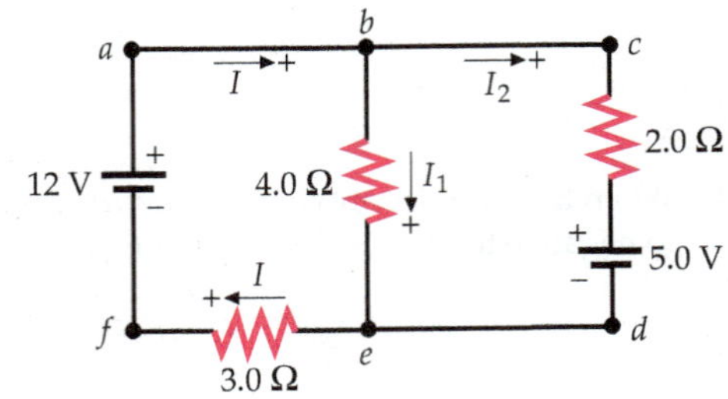

FIGURE 25-32

PICTURE There are three branch currents, I, I_1, and I_2, to be determined, so we need three equations. One equation comes from applying the junction rule to point *b*. (We can also apply the junction rule to point *e*, the only other junction in the circuit, but it gives exactly the same information.) The other two relations are obtained by applying the loop rule. There are three loops in the circuit: the two interior loops, *abefa* and *bcdeb*, and the exterior loop, *abcdefa*. We can use any two of these loops—the third will give redundant information. There is at least one direction arrow on each branch in Figure 25-32. Each direction arrow indicates the positive direction for that branch. If our analysis results in a negative value for a branch current, then that current is in the direction opposite to the direction arrow for that branch.

SOLVE

(*a*) 1. Apply the junction rule to point *b*:

$$I = I_1 + I_2$$

2. Apply the loop rule to the outer loop, *abcdefa*:

$$-(2.0\ \Omega)I_2 - 5.0\ \text{V} - (3.0\ \Omega)(I_1 + I_2) + 12\ \text{V} = 0$$

3. Divide the above equation by 1 Ω, recalling that (1 V)/(1 Ω) = 1 A, then simplify:

$$7.0\ \text{A} - 3.0I_1 - 5.0I_2 = 0$$

4. For the third condition, apply the loop rule to the loop on the right, *bcdeb*:

$$-(2.0\ \Omega)I_2 - 5.0\ \text{V} + (4.0\ \Omega)I_1 = 0 \quad \text{so} \quad -5.0\ \text{A} + 4.0I_1 - 2.0I_2 = 0$$

5. The results for steps 3 and 4 can be combined to solve for I_1 and I_2. To do so, first multiply the result for step 3 by 2, and then multiply the result for step 4 by −5:

$$14\ \text{A} - 6.0I_1 - 10I_2 = 0$$
$$25\ \text{A} - 20I_1 + 10I_2 = 0$$

6. Add the equations in step 5 to eliminate I_2, then solve for I_1:

$$39\ \text{A} - 26I_1 = 0$$
$$I_1 = \frac{39\ \text{A}}{26} = \boxed{1.5\ \text{A}}$$

7. Substitute I_1 in the results for step 3 or 4 to solve for I_2:

$$7.0\ \text{A} - 3.0(1.5\ \text{A}) - 5.0I_2 = 0$$
$$I_2 = \frac{2.5\ \text{A}}{5.0} = \boxed{0.50\ \text{A}}$$

8. Finally, I_1 and I_2 determine I using the equation in step 1:

$$I = I_1 + I_2 = 1.5\ \text{A} + 0.50\ \text{A} = \boxed{2.0\ \text{A}}$$

(*b*) 1. The power delivered to the 4.0-Ω resistor is found using $P = I_1^2R$:

$$P = I_1^2R = (1.5\ \text{A})^2(4.0\ \Omega) = 9.0\ \text{W}$$

2. The total energy dissipated in the 4.0-Ω resistor during time Δt is $W = P\,\Delta t$. In this case, $\Delta t = 3.0$ s:

$$W = P\,\Delta t = (9.0\ \text{W})(3.0\ \text{s}) = \boxed{27\ \text{J}}$$

CHECK In Figure 25-33, we have chosen the potential to be zero at point f, and we have labeled the currents and the potentials at the other points. Note that $V_b - V_e = 6.0$ V and $V_e - V_f = 6.0$ V. Applying the loop rule to the loop on the left gives +12 V − 6.0 V − 6.0 V = 0.

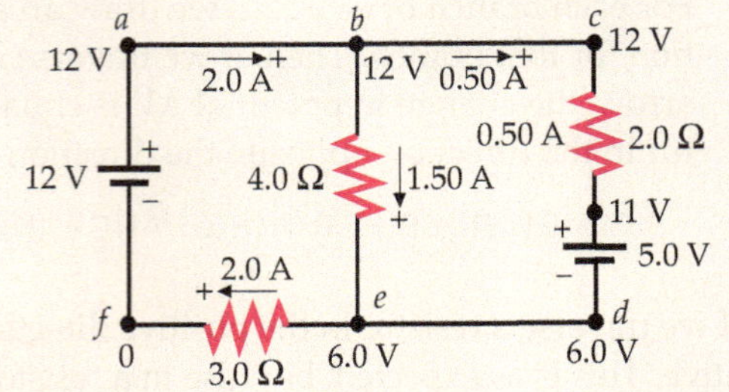

FIGURE 25-33

TAKING IT FURTHER Applying the loop rule to the loop on the left, *abefa*, gives $12\text{ V} - (4.0\ \Omega)I_1 - (3.0\ \Omega)(I_1 + I_2) = 0$, or $12\text{ A} - 7.0I_1 - 3.0I_2 = 0$. Note that this is just the result for step 3 minus the result for step 4 and hence contains no new information, as expected.

PRACTICE PROBLEM 25-9 Find I_1 for the case in which the 3.0-Ω resistor approaches (*a*) zero resistance and (*b*) infinite resistance.

Example 25-16 illustrates the general methods for the analysis of multiloop circuits. These methods are listed in the following problem-solving strategy.

PROBLEM-SOLVING STRATEGY

Method for Analyzing Multiloop Circuits

PICTURE Draw a sketch of the circuit.

SOLVE

1. Replace any series or parallel resistor combinations or capacitor combinations with their equivalent values.
2. Repeat step 1 as many times as possible.
3. Next, assign a positive direction for each branch of the circuit and indicate this direction with an arrow. Label the current in each branch. Add a plus sign and a minus sign to indicate the high-potential terminal and low-potential terminal of each source of emf.
4. Apply the junction rule to all but one of the junctions.
5. Apply the loop rule to the different loops until the total number of independent equations equals the total number of unknowns. When traversing a resistor in the positive direction, the change in potential equals $-IR$. When traversing a battery from the negative terminal to the positive terminal, the change in potential equals $\mathcal{E} \pm Ir$.
6. Solve the equations to obtain the values of the unknowns.

CHECK Check your results by assigning a potential of zero to one point in the circuit and use the values of the currents found to determine the potentials at other points in the circuit.

Example 25-17 A Three-Branch Circuit

(*a*) Find the current in each branch of the circuit shown in Figure 25-34. (*b*) Assign $V = 0$ to point *c* and then find the potential at each other point *a* through *f*.

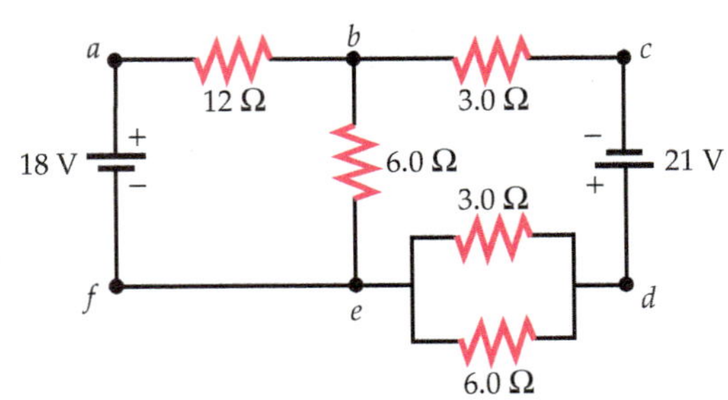

FIGURE 25-34

PICTURE First, replace the two parallel resistors by an equivalent resistance. Second, assign a positive direction to each branch and indicate each choice with an arrow. Third, place a plus sign and a minus sign at the high-potential and low-potential terminals of each battery. Assign a symbol for the current in each branch. These branch currents can then be found by applying the junction rule at either junction *b* or junction *e* and by applying the loop rule twice.

SOLVE

(*a*) 1. Find the equivalent resistance of the 3.0-Ω and 6.0-Ω parallel resistors:

$R_{eq} = 2.0\ \Omega$

2. Redraw the circuit (Figure 25-35) with the 2.0-Ω resistor in place of the parallel combination. Place an arrow on each branch indicating your direction-sign assignments. Let I be the current in branch with the 18-V battery, let I_1 be the current in the 6.0-Ω resistor, and let I_2 be the current in branch with the 21-V battery:

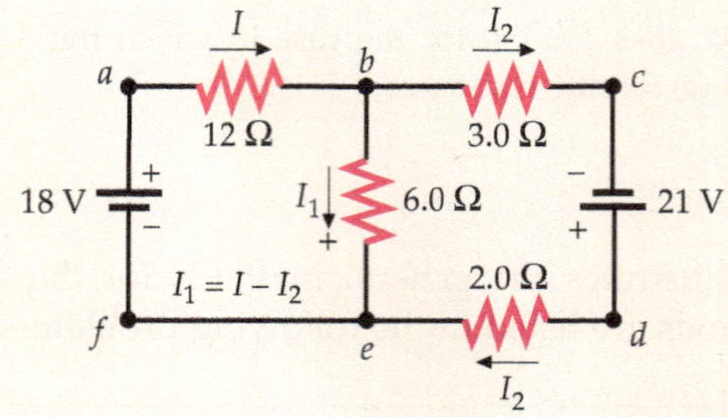

FIGURE 25-35

3. Apply the junction rule at point *b*:

$I = I_1 + I_2$

4. Apply Kirchhoff's loop rule to loop *abefa* to obtain an equation involving I and I_1:

$18\ \text{V} - (12\ \Omega)I - (6.0\ \Omega)I_1 = 0$

5. Simplify the equation from step 4 (dividing both sides by 6.0 Ω):

$3.0\ \text{A} - 2.0I - 1.0I_1 = 0$

6. Apply Kirchhoff's loop rule to loop *bcdeb*:

$-(3.0\ \Omega)I_2 + 21\ \text{V} - (2.0\ \Omega)I_2 + (6.0\ \Omega)I_1 = 0$

7. Simplify the equation in step 6 (dividing both sides by 1.0 Ω):

$21\ \text{A} + 6.0I_1 - 5.0I_2 = 0$

8. Solve the simultaneous equations (from steps 3, 5, and 7) for I, I_1, and I_2. One way to do this calculation is first substitute $I_1 + I_2$ for I in the step 5 equation to obtain $3.0\ \text{A} - 3.0I_1 - 2.0I_2 = 0$. This equation and the step 7 equation constitute two equations and two unknowns. Solve for the currents:

$I_1 = \boxed{-1.0\ \text{A}}$ $I_2 = \boxed{3.0\ \text{A}}$ $I = \boxed{2.0\ \text{A}}$

9. Use $V = I_2R_{eq}$ to find the potential drop across the parallel combination consisting of the 3.0-Ω and 6.0-Ω resistors:

$V = 6.0\ \text{V}$

10. Use the result of step 9 and Ohm's law to find the current in each of the parallel resistors;

$I_{3\,\Omega} = \boxed{2.0\ \text{A}}$ $I_{6\,\Omega} = \boxed{1.0\ \text{A}}$

(*b*) Redraw Figure 25-35 showing the direction and value of the current in each branch of the circuit (Figure 25-36). Begin with $V = 0$ at point *c* and calculate the potential at points *d*, *e*, *f*, *a*, and *b*:

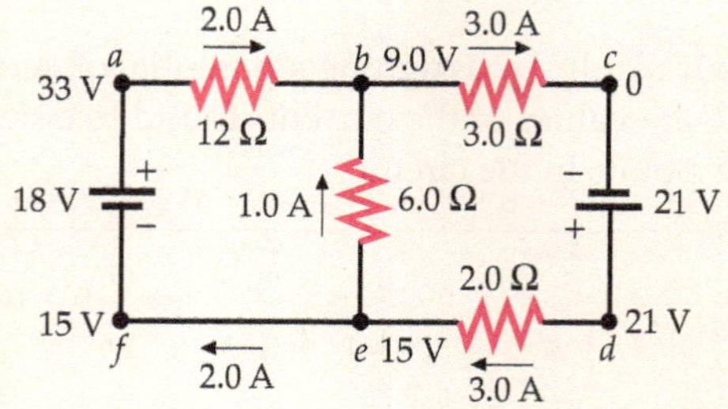

FIGURE 25-36

$V_d = V_c + 21\ \text{V} = 0 + 21\ \text{V} = \boxed{21\ \text{V}}$

$V_e = V_d - (3.0\ \text{A})(2.0\ \Omega) = 21\ \text{V} - 6.0\ \text{V} = \boxed{15\ \text{V}}$

$V_f = V_e = \boxed{15\ \text{V}}$

$V_a = V_f + 18\ \text{V} = 15\ \text{V} + 18\ \text{V} = \boxed{33\ \text{V}}$

$V_b = V_a - (2.0\ \text{A})(12.0\ \Omega) = 33\ \text{V} - 24\ \text{V} = \boxed{9\ \text{V}}$

CHECK From point *b* to point *c* the potential drops by $(3.0\ \text{A})(3.0\ \Omega) = 9.0\ \text{V}$, which gives $V_c = 0$, as assumed. From point *e* to point *b* the potential drops by $(1.0\ \text{A})(6.0\ \Omega) = 6.0\ \text{V}$, so $V_b = V_e - 6.0\ \text{V} = 15\ \text{V} - 6.0\ \text{V} = 9\ \text{V}$.

AMMETERS, VOLTMETERS, AND OHMMETERS

The devices that measure current, potential difference, and resistance are called **ammeters, voltmeters,** and **ohmmeters,** respectively. Often, all three of these meters are included in a single *multimeter* that can be switched from one use to another. You might use a voltmeter to measure the terminal voltage of your car battery and an ohmmeter to measure the resistance of some electrical device at home (for example, a toaster or lightbulb) when you suspect a short circuit or a broken wire.

To measure the current in a resistor in a simple circuit, we place an ammeter in series with the resistor, as shown in Figure 25-37, so that the current is the same in the ammeter and the resistor. Because the ammeter has a very low (but finite) resistance, the current in the circuit decreases very slightly when the ammeter is inserted. Ideally, the ammeter should have a negligibly small resistance so that the current to be measured is only negligibly affected.

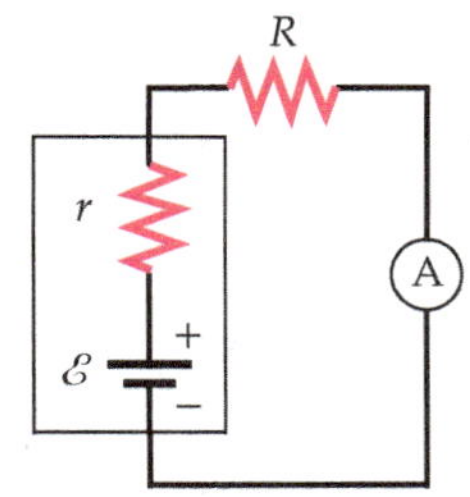

FIGURE 25-37 To measure the current in a resistor R, an ammeter A (circled) is placed in series with the resistor so that it carries the same current as the resistor.

The potential difference across a resistor is measured by placing a voltmeter across the resistor (in parallel with the resistor), as shown in Figure 25-38, so that the potential drop is the same across both the voltmeter and the resistor. The voltmeter reduces the resistance between points a and b, thus increasing the total current in the circuit and changing the potential drop across the resistor. A voltmeter should have an extremely high resistance so that its effect on the current in the circuit is negligible.

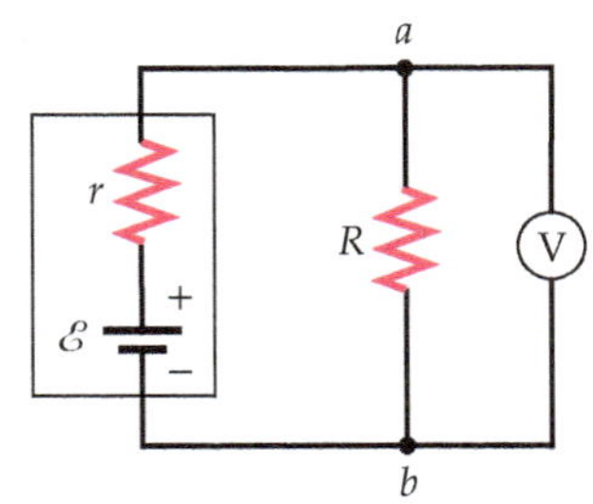

FIGURE 25-38 To measure the potential drop across a resistor, a voltmeter V (circled) is placed in parallel with the resistor so that the potential difference across the voltmeter is equal to the potential difference across the resistor.

The principal component of many commonly used ammeters and voltmeters is a **galvanometer,** a device that detects small currents passing through it. The galvanometer is designed so that the scale reading is proportional to the current passing through. The type of galvanometer used in many student laboratories consists of a coil of wire in the magnetic field of a permanent magnet. When there is current in the coil, the magnetic field exerts a torque on the coil, which causes the coil to rotate. A pointer attached to the coil indicates the reading on a scale. The coil itself contributes some resistance when the galvanometer is placed within a circuit.

To construct an ammeter from a galvanometer, we place a small resistor called a **shunt resistor** in *parallel* with the galvanometer. The resistance of the shunt resistor is usually much smaller than the resistance of the galvanometer so that the majority of the current is carried by the shunt resistor. The equivalent resistance of the ammeter is then approximately equal to the resistance of the shunt resistor, which is much smaller than the internal resistance of the galvanometer alone. To construct a voltmeter, we place a resistor that has a large resistance in *series* with the galvanometer so that the equivalent resistance of the voltmeter is much larger than the resistance of the galvanometer coil alone. Figure 25-39 illustrates the construction of an ammeter and voltmeter from a galvanometer. The resistance of the galvanometer R_g is shown separately in these schematic drawings, but it is actually part of the galvanometer.

A simple ohmmeter consists of a battery connected in series with a galvanometer and a resistor, as shown in Figure 25-40*a*.

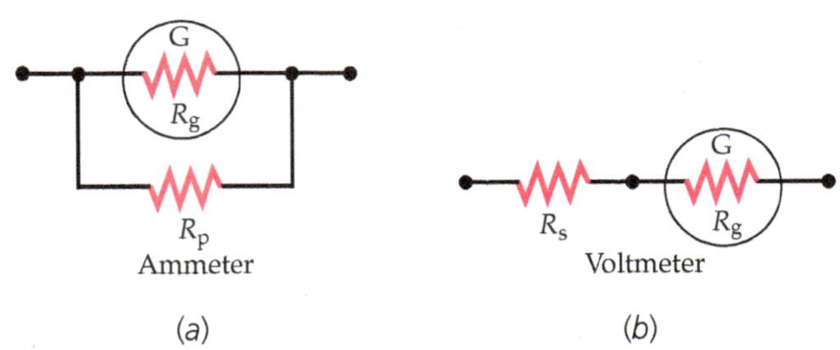

FIGURE 25-39 (*a*) An ammeter consists of a galvanometer G (circled) whose resistance is R_g and a small parallel resistance R_p. (*b*) A voltmeter consists of a galvanometer G (circled) and a large series resistance R_s.

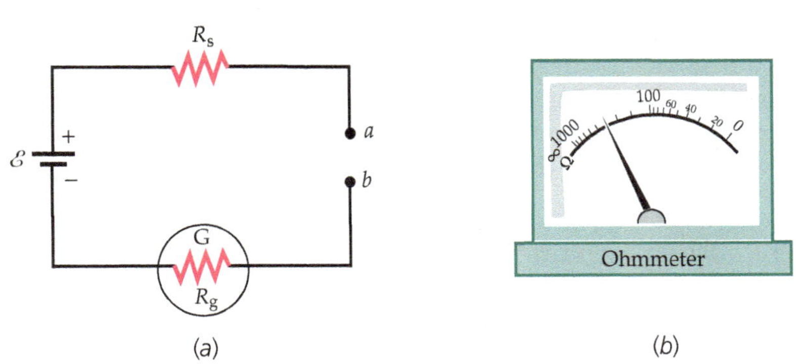

FIGURE 25-40 (*a*) An ohmmeter consists of a battery connected in series with a galvanometer and a resistor R_s, which is chosen so that the galvanometer gives full-scale deflection when points a and b are shorted. (*b*) When a resistor R is connected between terminals a and b, the galvanometer needle deflects by an amount that depends on the value of R. The galvanometer scale is calibrated to give a readout in ohms.

The resistance R_s is chosen so that when the terminals a and b are shorted (put in electrical contact with negligible resistance between them), the current through the galvanometer gives a full-scale deflection. Thus, a full-scale deflection indicates no resistance between terminals a and b. A zero deflection indicates an infinite resistance between the terminals. When the terminals are connected across an unknown resistance R, the current through the galvanometer depends on R, so the scale can be calibrated to give a direct reading of R, as shown in Figure 25-40*b*. Because an ohmmeter sends a current through the resistance to be measured, some caution must be exercised when using this instrument. For example, you would not want to try to measure the resistance of a sensitive galvanometer with an ohmmeter, because the current provided by the battery in the ohmmeter could possibly damage the galvanometer.

25-6 *RC* CIRCUITS

A circuit containing a resistor and a capacitor is called an ***RC* circuit.** The current in an *RC* circuit is in a single direction, as in all dc circuits, but the magnitude of the current varies with time. A practical example of an *RC* circuit is the circuit in the flash attachment of a camera. Before a flash photograph is taken, a battery in the flash attachment charges the capacitor through a resistor. When the capacitor is fully charged, the light is ready to flash. When the picture is taken, the capacitor discharges through the flashbulb. The battery then recharges the capacitor, and a short time later the flash is ready for another picture. Using Kirchhoff's rules, we can obtain equations for the charge Q and the current I as functions of time for both the charging and discharging of a capacitor through a resistor.

DISCHARGING A CAPACITOR

Figure 25-41 shows a capacitor that has an initial charge $+Q_0$ on the upper plate and an initial charge $-Q_0$ on the lower plate. The capacitor is connected to a resistor R and a switch S, which is initially open. The potential difference across the capacitor is initially $V_0 = Q_0/C$, where C is the capacitance.

We close the switch at time $t = 0$. Because there is now a potential difference across the resistor, there is also a current in the resistor. This initial current is

$$I_0 = \frac{V_0}{R} = \frac{Q_0}{RC} \quad \text{25-30}$$

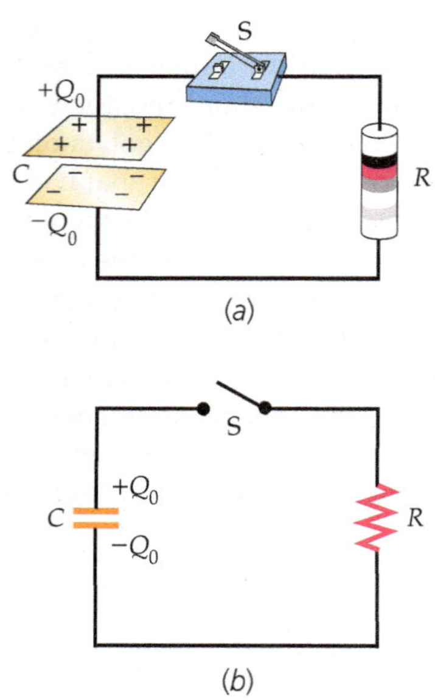

FIGURE 25-41 (*a*) A parallel-plate capacitor in series with a switch S and a resistor R. (*b*) A circuit diagram for Figure 25-41*a*.

The current is the rate of flow of positive charge from the positive plate of the capacitor to the negative plate through the resistor. (We model the charge carriers as being positively charged, when actually they are negatively charged electrons.) As time passes, the charge on the capacitor decreases. If we choose the positive direction to be clockwise, then the current equals the rate of decrease of the charge. If Q is the charge on the upper plate of the capacitor at time t, the current is related to Q by

$$I = -\frac{dQ}{dt} \quad \text{25-31}$$

(The minus sign is needed because while Q decreases, dQ/dt is negative.*) Traversing the circuit in the clockwise direction, we encounter a potential drop IR across the resistor and a potential increase Q/C across the capacitor. Thus, Kirchhoff's loop rule gives

$$\frac{Q}{C} - IR = 0 \quad \text{25-32}$$

* If the positive direction were chosen to be counterclockwise, then the sign in Equation 25-31 would be a plus sign.

where Q and I, both functions of time, are related by Equation 25-31. Substituting $-dQ/dt$ for I in Equation 25-32, we have

$$\frac{Q}{C} + R\frac{dQ}{dt} = 0$$

or

$$\frac{dQ}{dt} = -\frac{1}{RC}Q \qquad 25\text{-}33$$

To solve this equation, we first separate the variables Q and t by multiplying both sides of the equation by dt/Q, and then we integrate both sides. Multiplying both sides by dt/Q, we obtain

$$\frac{dQ}{Q} = -\frac{1}{RC}dt \qquad 25\text{-}34$$

The variables Q and t are now in separate terms. Integrating from Q_0 at $t = 0$ to Q' at time t' gives

$$\int_{Q_0}^{Q'} \frac{dQ}{Q} = -\frac{1}{RC}\int_0^{t'} dt$$

so

$$\ln\frac{Q'}{Q_0} = -\frac{t'}{RC}$$

Because t' is arbitrary, we can replace t' with t, and then $Q' = Q(t)$. Solving for $Q(t)$ gives

$$Q(t) = Q_0e^{-t/(RC)} = Q_0e^{-t/\tau} \qquad 25\text{-}35$$

where τ, called the **time constant,** is the time it takes for the charge to decrease by a factor of e^{-1}:

$$\tau = RC \qquad 25\text{-}36$$

DEFINITION—TIME CONSTANT

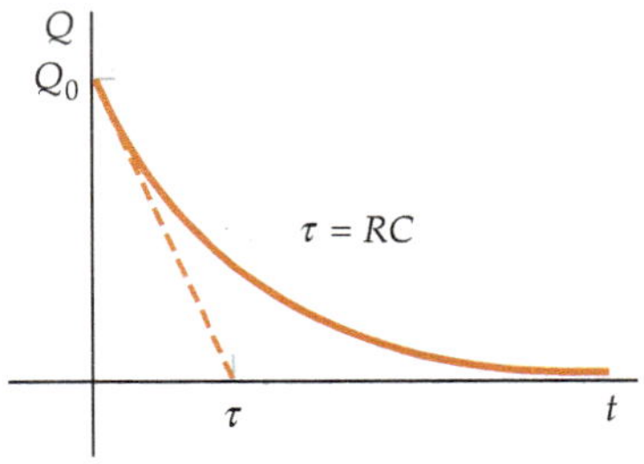

FIGURE 25-42 Plot of the charge on the capacitor versus time for the circuit shown in Figure 25-41. The switch is closed at time $t = 0$. The time constant $\tau = RC$ is the time it takes for the charge to decrease by a factor of e^{-1}. (The time constant is also the time it would take the capacitor to discharge fully if its discharge rate remains constant, as indicated by the dashed line.)

See **Math Tutorial** *for more information on*
Exponential Functions

Figure 25-42 shows the charge on the capacitor in the circuit of Figure 25-41 as a function of time. After a time $t = \tau$, the charge is $Q = e^{-1}Q_0 = 0.37Q_0$. After a time $t = 2\tau$, the charge is $Q = e^{-2}Q_0 = 0.135Q_0$, and so forth. After a time equal to several time constants, the charge Q is negligible. This type of decrease, which is called an **exponential decrease,** is very common in nature. It occurs whenever the rate at which a quantity decreases is proportional to the quantity itself.*

The decrease in the charge on a capacitor can be likened to the decrease in the amount of water in a bucket that has vertical sides and a small hole in the bottom. (The rate at which the water flows out of the hole is proportional to the difference in pressure of the water on either side of the hole, which is in turn proportional to the amount of water still in the bucket.)

The current is obtained by differentiating Equation 25-35

$$I = -\frac{dQ}{dt} = \frac{Q_0}{RC}e^{-t/(RC)}$$

Substituting, using Equation 25-30, we obtain

$$I = I_0e^{-t/\tau} \qquad 25\text{-}37$$

where $I_0 = V_0/R = Q_0/(RC)$ is the initial current. The current as a function of time is shown in Figure 25-43. As with the charge, the current decreases exponentially with time constant $\tau = RC$.

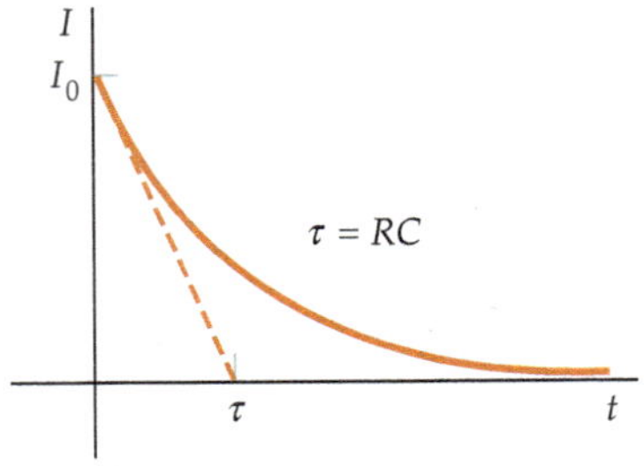

FIGURE 25-43 Plot of the current versus time for the circuit in Figure 25-41. The curve has the same shape as that in Figure 25-42. If the rate of decrease of the current remains constant, the current would reach zero after one time constant, as indicated by the dashed line.

* We encountered exponential decreases in Chapter 14 when we studied the damped oscillator.

Example 25-18 Discharging a Capacitor

A 4.0-μF capacitor is charged to 24 V and then connected across a 200-Ω resistor. Find (*a*) the initial charge on the capacitor, (*b*) the initial current through the 200-Ω resistor, (*c*) the time constant, and (*d*) the charge on the capacitor after 4.0 ms.

PICTURE The circuit diagram is the same as the circuit diagram shown in Figure 25-41.

SOLVE

(*a*) The initial charge is related to the capacitance and voltage:

$$Q_0 = CV_0 = (4.0\ \mu\text{F})(24\ \text{V}) = \boxed{96\ \mu\text{C}}$$

(*b*) The initial current is the initial voltage divided by the resistance:

$$I_0 = \frac{V_0}{R} = \frac{24\ \text{V}}{200\ \Omega} = \boxed{0.12\ \text{A}}$$

(*c*) The time constant is RC:

$$\tau = RC = (200\ \Omega)(4.0\ \mu\text{F}) = 800\ \mu\text{s} = \boxed{0.80\ \text{ms}}$$

(*d*) Substitute $t = 4.0$ ms into Equation 25-35 to find the charge on the capacitor at that time:

$$Q = Q_0 e^{-t/\tau} = (96\ \mu\text{C})e^{-(4.0\ \text{ms})/(0.80\ \text{ms})}$$

$$= (96\ \mu\text{C})e^{-5} = \boxed{0.65\ \mu\text{C}}$$

CHECK At the initial current of $I_0 = 0.12$ A it would take $Q_0/I_0 = 96\ \mu\text{C}/0.12\ \text{A} = 0.80$ ms for the capacitor to fully discharge. Because the current decreases exponentially during discharge, it is not surprising that it takes 4.0 ms for the charge to lose 99.3% of its initial charge.

TAKING IT FURTHER After five time constants, Q is less than 1 percent of its initial value.

PRACTICE PROBLEM 25-10 Find the current through the 200-Ω resistor at $t = 4.0$ ms.

CHARGING A CAPACITOR

Figure 25-44*a* shows a circuit for charging a capacitor. The capacitor is initially uncharged. The switch S, originally open, is closed at time $t = 0$. Charge immediately begins to flow through the battery (Figure 25-44*b*). If the charge on the rightmost plate of the capacitor at time t is Q, the current in the circuit is I, and clockwise is positive, then Kirchhoff's loop rule gives

$$\mathcal{E} - IR - \frac{Q}{C} = 0 \tag{25-38}$$

By inspecting this equation, we can see that when Q is zero (at $t = 0$) the current is $I = I_0 = \mathcal{E}/R$. The charge then increases and the current decreases. The charge approaches a maximum value of $Q_f = C\mathcal{E}$ as the current I approaches zero, as can also be seen from Equation 25-38.

In this circuit, we have chosen the positive direction so if I is positive Q is increasing. Thus,

$$I = +\frac{dQ}{dt}$$

Substituting dQ/dt for I in Equation 25-38 gives

$$\mathcal{E} - R\frac{dQ}{dt} - \frac{Q}{C} = 0 \tag{25-39}$$

Equation 25-39 can be solved in the same way that we solved Equation 25-33.

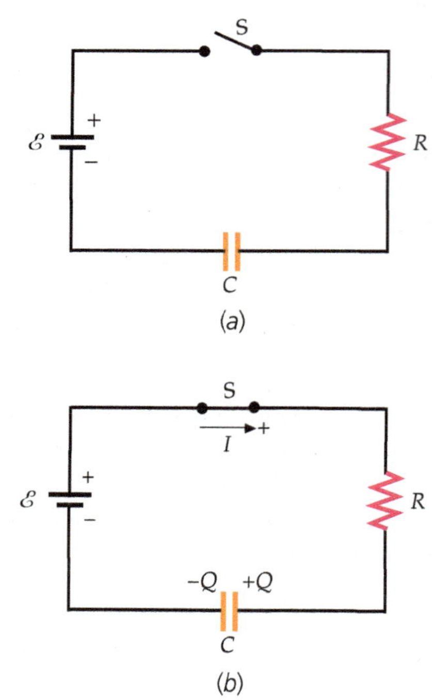

FIGURE 25-44 (*a*) A circuit for charging a capacitor to a potential difference $\mathcal{E}$. (*b*) After the switch is closed, there is both a current through and a potential drop across the resistor and both a charge on and a potential drop across the capacitor.

The details are left as a problem (see Problem 101). The result is

$$Q = C\mathcal{E}[1 - e^{-t/(RC)}] = Q_f(1 - e^{-t/\tau}) \tag{25-40}$$

where $Q_f = C\mathcal{E}$ is the final charge. The current is obtained from $I = dQ/dt$:

$$I = \frac{dQ}{dt} = C\mathcal{E}\left[-\frac{-1}{RC}e^{-t/(RC)}\right] = \frac{\mathcal{E}}{R}e^{-t/(RC)}$$

or

$$I = \frac{\mathcal{E}}{R}e^{-t/(RC)} = I_0 e^{-t/\tau} \tag{25-41}$$

where the initial current is $I_0 = \mathcal{E}/R$.

Figure 25-45 and Figure 25-46 show the charge and the current as functions of time.

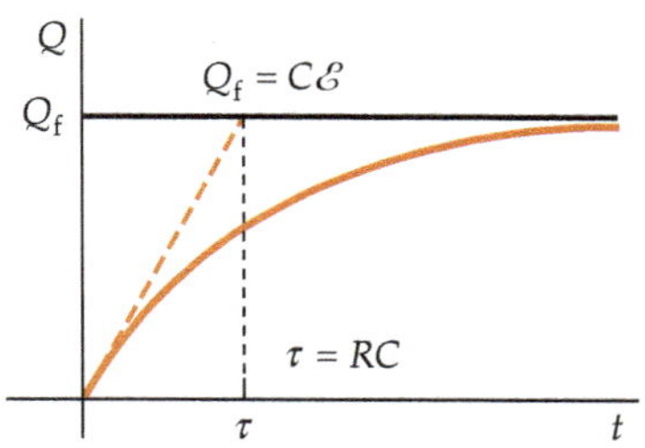

FIGURE 25-45 Plot of the charge on the capacitor versus time for the charging circuit of Figure 25-44 after the switch is closed (at $t = 0$). After a time $t = \tau = RC$, the charge on the capacitor is $0.63C\mathcal{E}$, where $C\mathcal{E}$ is its final charge. If the charging rate were constant, the capacitor would be fully charged after a time $t = \tau$.

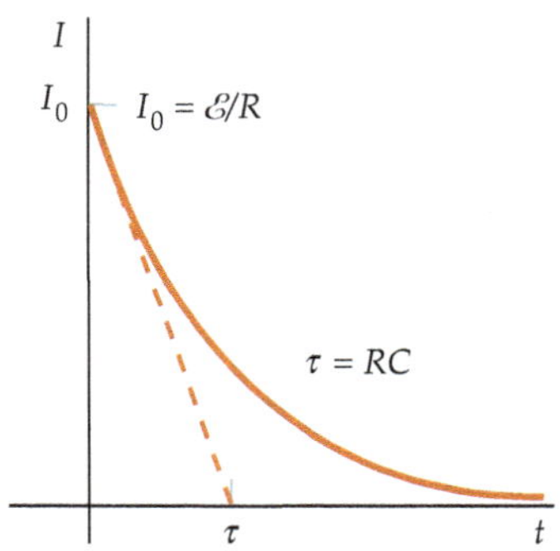

FIGURE 25-46 Plot of the current versus time for the charging circuit of Figure 25-44. The current is initially $\mathcal{E}/R$, and the current decreases exponentially with time.

PRACTICE PROBLEM 25-11

Show that Equation 25-40 does indeed satisfy Equation 25-39 by substituting the expressions for Q and dQ/dt into Equation 25-39.

PRACTICE PROBLEM 25-12

What fraction of the maximum charge is on the charging capacitor after a time $t = 2\tau$?

Example 25-19 Charging a Capacitor

Try It Yourself

A 6.0-V battery that has a negligible internal resistance is used to charge a 2.0-μF capacitor through a 100-Ω resistor. Find (*a*) the initial current, (*b*) the final charge on the capacitor, (*c*) the time required for the charge to reach 90 percent of its final value, and (*d*) the charge when the current is half its initial value.

PICTURE The charge initially is zero so the voltage across the resistor is equal to the emf of the battery. Apply Ohm's law to the resistor and solve for the current. After a long time, the current is zero so the voltage across the capacitor is equal to the emf of the battery. Apply the definition of capacitance and solve for the charge. Use Equation 25-40 to relate the charge to the time, and use Kirchhoff's loop rule to relate the charge to the current.

SOLVE

Cover the column to the right and try these on your own before looking at the answers.

Steps	Answers
(*a*) Find the initial current from $I_0 = \mathcal{E}/R$.	$I_0 = 0.060\text{ A} = \boxed{60\text{ mA}}$
(*b*) Find the final charge from $Q = C\mathcal{E}$.	$Q_f = \boxed{12\ \mu\text{C}}$
(*c*) Set $Q = 0.90Q_f$ in Equation 25-40 and solve for t. (First solve for $e^{t/\tau}$, then take the natural log of both sides and solve for t.)	$t = 2.3\tau = \boxed{0.46\text{ ms}}$
(*d*) 1. Apply Kirchhoff's loop rule to the circuit using Figure 25-44*b*.	$\mathcal{E} - IR - \frac{Q}{C} = 0$
2. Set $I = I_0/2$ and solve for Q.	$Q = \frac{Q_f}{2} = \boxed{6.0\ \mu\text{C}}$

CHECK The answer to Part (*d*) can be obtained by first solving for t using Equation 25-41, then substituting that time into Equation 25-40 and solving for Q. However, using the loop rule is certainly the more direct approach.

Example 25-20 Finding Values at Short and Long Times

The 6.0-μF capacitor in the circuit shown in Figure 25-47 is initially uncharged. Find the current through the 4.0-Ω resistor and the current through the 8.0-Ω resistor (*a*) immediately after the switch is closed and (*b*) a long time after the switch is closed. (*c*) Find the charge on the capacitor a long time after the switch is closed.

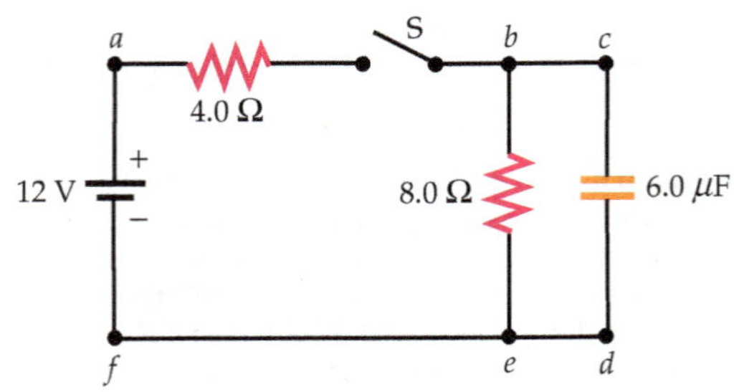

FIGURE 25-47

PICTURE Because the capacitor is initially uncharged (and the 4.0-Ω resistor limits the current through the battery), the initial potential difference across the capacitor is zero. The capacitor and the 8.0-Ω resistor are connected in parallel, and the difference in potential across each is the same. Thus, the initial potential difference across the 8.0-Ω resistor is also zero. The positive direction for the branch that has the battery is up the page, and the positive direction for the other two branches is down the page. Let Q be the charge on the upper plate of the capacitor.

SOLVE

(*a*) The charge on the capacitor initially is zero. Apply the loop rule to the outer loop and solve for the current through the 4.0-Ω resistor. Apply the loop rule to the loop containing the 8.0-Ω resistor and the capacitor and solve for the current through the 8.0-Ω resistor.

$$12\text{ V} - (4.0\ \Omega)I_{4\Omega} - \frac{0}{C} = 0$$

$$I_{4\Omega} = \boxed{3.0\text{ A}}$$

$$I_{8\Omega}(8\ \Omega) - \frac{0}{C} = 0$$

$$I_{8\Omega} = \boxed{0}$$

(*b*) After a long time, the capacitor is fully charged (no more charge flows onto its plates) and the current through both resistors is the same. Apply the loop rule to the loop on the left and solve for the current:

$$12\text{ V} - (4.0\ \Omega)I_f - (8.0\ \Omega)I_f = 0$$

$$I_f = \boxed{1.0\text{ A}}$$

(*c*) The potential difference across the 8.0-Ω resistor and the capacitor are equal. Use this to solve for Q_f:

$$I_f(8.0\ \Omega) = \frac{Q_f}{C}$$

$$Q_f = (1.0\text{ A})(8.0\ \Omega)(6.0\ \mu\text{F}) = \boxed{48\ \mu\text{C}}$$

CHECK The analysis of this circuit at the extreme times when the capacitor is either uncharged or fully charged is simple. When the capacitor is uncharged, it acts like a good conductor (a short circuit) between points c and d; that is, the circuit is the same as the one shown in Figure 25-48*a* where we have replaced the capacitor by a wire of zero resistance. When the capacitor is fully charged, it acts like an open switch, as shown in Figure 25-48*b*.

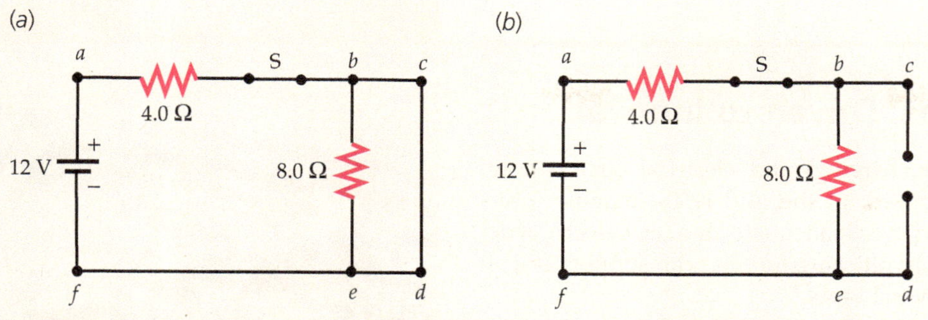

FIGURE 25-48

ENERGY CONSERVATION IN CHARGING A CAPACITOR

During the charging process, a total charge $Q_f = \mathcal{E}C$ flows through the battery. The battery therefore does work

$$W = Q_f\mathcal{E} = C\mathcal{E}^2$$

Half of this work is accounted for by the energy stored in the capacitor (see Equation 24-8):

$$U = \tfrac{1}{2}Q_f\mathcal{E}$$

We now show that the other half of work done by the battery is dissipated as thermal energy by the resistance of the circuit. The rate at which energy is dissipated by the resistance R is

$$\frac{dW_R}{dt} = I^2R$$

where $I = (\mathcal{E}/R)e^{-t/(RC)}$ (Equation 25-41). Substituting for I we have

$$\frac{dW_R}{dt} = \left(\frac{\mathcal{E}}{R}e^{-t/(RC)}\right)^2 R = \frac{\mathcal{E}^2}{R}e^{-2t/(RC)}$$

We find the total energy dissipated by integrating from $t = 0$ to $t = \infty$:

$$W_R = \int_0^\infty \frac{\mathcal{E}^2}{R}e^{-2t/(RC)}\,dt = \frac{\mathcal{E}^2}{R}\int_0^\infty e^{-at}\,dt$$

where $a = 2/(RC)$. Thus,

$$W_R = \frac{\mathcal{E}^2}{R}\frac{e^{-at}}{-a}\bigg|_0^\infty = -\frac{\mathcal{E}^2}{Ra}(0 - 1) = \frac{\mathcal{E}^2}{R}\frac{1}{a} = \frac{\mathcal{E}^2}{R}\frac{RC}{2}$$

The total amount of Joule heating is thus

$$W_R = \frac{1}{2}\mathcal{E}^2C = \frac{1}{2}Q_f\mathcal{E}$$

where $Q_f = \mathcal{E}C$. This result is independent of the resistance R. Thus, when a capacitor is charged through a resistor by a constant source of emf, half the energy provided by the source of emf is stored in the capacitor and half goes into thermal energy. This thermal energy includes the energy that is dissipated by the internal resistance of the source of emf.

Physics Spotlight

Vehicle Electrical Systems: Driven to Innovation

(Graham Harrison/Alamy.)

During the 1930s, 6-volt batteries (7-volt charging) and electrical circuits were standard for automobiles in the United States. In the mid-1950s, manufacturers around the world recognized that this battery was inadequate for a car's electrical demands, and changed to 12-volt batteries (14-volt charging) that can support 14-volt electrical systems.[*] The changeover took several years.[†]

In the mid-1960s, demands on a car's electrical system included the starter, ignition, lights, radio, and air-conditioning in luxury cars.[‡] Today, the electrical system and electronics of a car[#] can include crash sensors, automatic braking systems, seat motors, power steering, power braking, windshield washers with intermittent timing, video-based entertainment systems, engine controllers, cruise control, and window motors. Some high-end luxury cars put further demands on their electrical systems with electronic throttle control, radar to detect the distance to objects,[°] electronic stability and suspension control, and heated seats.[§] The power requirement for an automobile is 1.5 to 2.0 kW today, and is expected to increase to 3.0 to 3.5 kW or more in the near future.[¶] The electrical system and electronics of a car now account for more than 20 percent of the manufacturing cost of the average car.[**]

Because demands on the electrical system of cars are expected to increase still further,[††] many people have suggested that it would be a good idea to upgrade the electrical system of cars to a 36-volt battery and a 42-volt system. (Because power is the product of voltage and current, this means that when the voltage increased, the current would decrease in order to supply the same power.) Many people were excited about the prospect of using smaller wires and lighter wiring harnesses to carry power to all the electrical devices.[‡‡] In addition, the higher voltage would mean smaller and lighter starter motors and alternators.

But switching over to a 42-volt system is proving to be more difficult than expected. Although a concept car was built with a 42-volt system, it was made from nonstandard customized parts.[##] In a 14-V system, a connection that has vibrated loose will not arc persistently over a gap that is about a millimeter. In a 42-V system, the same loose connection will arc, which creates the danger of an electrical fire.[°°] At 42 V, more expensive electrical connectors are required. By mid-2005, several manufacturers had confirmed that they were not interested in using 42-volt systems within the next few years.[§§,¶¶] A consortium of researchers continues to look at issues involving 42-volt automotive systems.[***] When the economics of a switch to the 42-volt systems make sense, then 42-volt electrical systems will be put into mass-production automobiles.

---

* Ribbens, W. B., *Understanding Automotive Electronics, 6th ed.* New York: Newnes (Elsevier), 2003.

† Corbett, B., "No Flick of the Switch." *Ward's Auto World,* April 2001, Vol. 37, No. 4, p. 50.

‡ Ribbens, W. B., op. cit.

\# *Automotive Electronics Handbook,* R. Jurgen, ed., New York: McGraw-Hill, 1995.

° Allen, R., "New Technologies Make Roads Safer… One Smart Car at a Time." *Electronic Design,* Jun. 29, 2006, pp. 41–44.

§ "The 2007 S600 Sedan," Mercedes-Benz, http://www.mbusa.com/models/features/specs/overview.do?modelCode=S600V&class=07_S As of Sept. 2006.

¶ Masrur, M. A., Monroe, J., Patel, R., and Garg, V. K., "42-volt Electrical Power System for Military Vehicles—Comparison with Commercial Automotive Systems," *Vehicular Technology Conference,* 2002. Proceedings, VTC 2002-Fall, 2002 IEEE 56th, Vol. 3, pp. 1846–1850.

** Marsh, D., "LIN Simplifies and Standardizes In-Vehicle Networks." *Electronic Design News,* Apr. 8, 2005, pp. 29+.

†† Huber, P. W., and Mills, M. P., "The End of the M. E.?" *Mechanical Engineering,* May 2005, pp. 26–29.

‡‡ Truett, R., "42-Volt Systems Boost Fuel Economy Efforts." *Automotive News,* Oct. 21, 2001, Vol. 77, No. 6008, p. 6i.

\#\# "No-Compromise Mild Hybrid Car Engine Has a Promising Future." *Asia-Pacific Engineer,* Jun. 1, 2003. http://www.engineerlive.com/asiapacific-engineer/automotive-design/1603/nocompromise-mild-hybrid-car-engine-has-a-promising-future.thtml As of Sept. 2006.

°° Moran, T., "42-Volt Challenges: Arcs and Sparks." *Automotive News,* Mar. 12, 2001, Vol. 75, No. 5920, p. 8.

§§ Kelly, K., "DC Dumps 42-Volts." *Ward's AutoWorld,* Jun. 2004, p. 9.

¶¶ Crain, K., "Let's Step Back, Rethink Technology." *Automotive News,* Jan. 3, 2005, Vol. 79, No. 6128, p. 12.

*** MIT/Industry Consortium on Advanced Automotive Electrical/Electronic Components and Systems. "Consortium Research Units." http://lees-web.mit.edu/public/Public%20Documents/Research_Units_and_Deliverables.pdf As of Sept. 2006.

Summary

1. Ohm's law is an empirical law that holds only for certain materials.
2. Current, resistance, and emf are important *defined* quantities.
3. Kirchhoff's rules follow from the conservation of charge and the conservative nature of the electric field.

TOPIC	RELEVANT EQUATIONS AND REMARKS	
1. Electric Current	Electric current is the rate of flow of electric charge through a cross-sectional area. $$I = \frac{\Delta Q}{\Delta t}$$ in the limit that Δt approaches zero.	25-1
Drift velocity	In a conducting wire, electric current is the result of the slow drift of negatively charged electrons that are accelerated by an electric field in the wire and then collide with the lattice ions. Typical drift speeds of electrons in wires are of the order of a few millimeters per second. For mobile charges moving in the positive direction, $$I = qnAv_d$$ where $q = -e$, n is the number density of free electrons, A is the cross-sectional area, and v_d is the drift speed.	25-3
Current density	The current density $\vec{J}$ is related to the drift velocity by $$\vec{J} = qn\vec{v}_d$$ The current I through a cross-sectional surface is the flux of the current density through the surface.	25-4
2. Resistance		
Definition of resistance	$$R = \frac{V}{I}$$	25-7
Resistivity, ρ	$$R = \rho\frac{L}{A}$$	25-10
Temperature coefficient of resistivity, α	$$\alpha = \frac{(\rho - \rho_0)/\rho_0}{T - T_0}$$	25-12
3. Ohm's Law	For ohmic materials, the resistance does not depend on either the current or the potential drop: $$V = IR,\ R \text{ constant}$$	25-9
4. Power		
Supplied to a device or segment	$$P = IV$$	25-13
Delivered to a resistor	$$P = IV = I^2R = \frac{V^2}{R}$$	25-14
5. Emf		
Source of emf	A device that supplies electrical energy to a circuit.	
Power supplied by an ideal emf source	$$P = I\mathcal{E}$$	25-15
6. Battery		
Ideal	An ideal battery is a source of emf that maintains a constant potential difference between its two terminals, independent of the current through the battery.	
Real	A real battery can be considered as an ideal battery in series with a small resistance, called its internal resistance.	
Terminal voltage	$$V_a - V_b = \mathcal{E} - Ir$$ where in the battery the positive direction is the direction of increasing potential.	25-16

TOPIC	RELEVANT EQUATIONS AND REMARKS	
Total energy stored	$E_{\text{stored}} = Q\mathcal{E}$	25-18
7. Equivalent Resistance		
Resistors in series	$R_{\text{eq}} = R_1 + R_2 + R_3 + \ldots$	25-20
Resistors in parallel	$\frac{1}{R_{\text{eq}}} = \frac{1}{R_1} + \frac{1}{R_1} + \frac{1}{R_3} + \ldots$	25-25
8. Kirchhoff's Rules	1. When any closed loop is traversed, the algebraic sum of the changes in potential around the loop must equal zero. 2. At any junction (branch point) in a circuit where the current can divide, the sum of the currents into the junction must equal the sum of the currents out of the junction.	
9. Measuring Devices		
Ammeter	An ammeter is a very low resistance device that is placed in series with a circuit element to measure the current in the element.	
Voltmeter	A voltmeter is a very high resistance device that is placed in parallel with a circuit element to measure the potential difference across the element.	
Ohmmeter	An ohmmeter is a device containing a battery connected in series with a galvanometer and a resistor that is used to measure the resistance of a circuit element placed across its terminals.	
10. Discharging a Capacitor		
Charge on the capacitor	$Q(t) = Q_0 e^{-t/(RC)} = Q_0 e^{-t/\tau}$	25-35
Current in the circuit	$I = -\frac{dQ}{dt} = \frac{V_0}{R} e^{-t/(RC)} = I_0 e^{-t/\tau}$	25-37
Time constant	$\tau = RC$	25-36
11. Charging a Capacitor		
Charge on the capacitor	$Q = C\mathcal{E}[1 - e^{-t/(RC)}] = Q_f(1 - e^{-t/\tau})$	25-40
Current in the circuit	$I = \frac{\mathcal{E}}{R} e^{-t/(RC)} = I_0 e^{-t/\tau}$	25-41

Answers to Concept Checks

25-1 (*a*) The current is larger just after the switch is thrown because the bulb filament is a metal and is relatively cool, so its resistance is less than it will be when the bulb has been running for a while. Lower resistance means higher current. (*b*) Energy from the battery initially is delivered to the filament at a greater rate than the relatively cool filament releases heat. After a while, energy from the battery is delivered to the filament at the same rate that the now hot filament releases heat. Under these conditions the filament's temperature, and thus its resistance, remains constant.

Answers to Practice Problems

25-1 7.9 h

25-2 14 000

25-3 4.5 V

25-4 2.4 m

25-5 The color bands are colored, from top to bottom, brown, orange, blue, red, and brown. The value of the resistance is 13.6 kΩ and the tolerance is 1%.

25-6 (*a*) 45 W, (*b*) 270 J

25-7	(*a*) 6.0 Ω, (*b*) 1.3 Ω
25-8	(*a*) $R'_{eq} = 2.0\ \Omega$; (*b*) $I = 9.0$ A; (*c*) $V_2 = 18$ V, $V_0 = 0$, $V_{12} = 0$; (*d*) $I_2 = 9.0$ A, $I_0 = 9.0$ A, $I_{12} = 0$
25-9	(*a*) 3.0 A (*b*) 0.83 A
25-10	0.81 mA
25-12	0.86

Problems

In a few problems, you are given more data than you actually need; in a few other problems, you are required to supply data from your general knowledge, outside sources, or informed estimate.

Interpret as significant all digits in numerical values that have trailing zeros and no decimal points.

- • Single-concept, single-step, relatively easy
- •• Intermediate-level, may require synthesis of concepts
- ••• Challenging
- SSM Solution is in the *Student Solutions Manual*
- Consecutive problems that are shaded are paired problems.

CONCEPTUAL PROBLEMS

1 • In our study of electrostatics, we concluded that no electric field exists within the material of a conductor in electrostatic equilibrium. Why can we discuss electric fields within the material of conductors in this chapter?

2 • Figure 25-12 shows a mechanical analog of a simple electric circuit. Devise another mechanical analog in which the current is represented by a flow of water instead of marbles. In the water circuit, what would be analogous to the battery? What would be analogous to the wire? What would be analogous to the resistor?

3 • Wires A and B are both made of copper. The wires are connected in series, so we know they carry the same current. However, the diameter of wire A is twice the diameter of wire B. Which wire has the higher number density of charge carriers (number per unit charge)? (*a*) A, (*b*) B, (*c*) They have the same number density of charge carriers.

4 • The diameters of copper wires A and B are equal. The current carried by wire A is twice the current carried by wire B. In which wire do the charge carriers have the higher drift speed? (*a*) A, (*b*) B, (*c*) They have the same drift speed.

5 • Wire A and wire B are identical copper wires. The current carried by wire A is twice the current carried by wire B. Which wire has the higher current density? (*a*) A, (*b*) B, (*c*) They have the same current density. (*d*) none of the above

6 •• Consider a metal wire that has each end connected to a different terminal of the same battery. Your friend argues that no matter how long the wire is, the drift speed of the charge carriers in the wire is the same. Evaluate your friend's claim.

7 • In a resistor, the direction of the current must always be in the "downhill" direction, that is, in the direction of decreasing electric potential. Is it also the case that in a battery, the direction of the current must always be the "downhill"? Explain your answer.

8 • Discuss the distinction between an emf and a potential difference.

9 • Wire A and wire B are made of the same material and have the same length. The diameter of wire A is twice the diameter of wire B. If the resistance of wire B is R, then what is the resistance of wire A? (Neglet any effects that temperature may have on resistance.) (*a*) R, (*b*) $2R$, (*c*) $R/2$, (*d*) $4R$, (*e*) $R/4$

10 • Two cylindrical copper wires have the same mass. Wire A is twice as long as wire B. (Neglet any effects that temperature may have on resistance.) Their resistances are related by (*a*) $R_A = 8R_B$, (*b*) $R_A = 4R_B$, (*c*) $R_A = 2R_B$, (*d*) $R_A = R_B$

11 • If the current in a resistor is I, the power delivered to the resistor is P. If the current in the resistor is increased to $3I$, what is the power then delivered to the resistor? (Assume the resistance of the resistor does not change.) (*a*) P, (*b*) $3P$, (*c*) $P/3$, (*d*) $9P$, (*e*) $P/9$

12 • If the potential drop across a resistor is V, the power delivered to the resistor is P. If the potential drop is increased to $2V$, what is the power delivered to the resistor then equal to? (*a*) P, (*b*) $2P$, (*c*) $4P$, (*d*) $P/2$, (*e*) $P/4$

13 • A heater consists of a variable resistor (a resistor whose resistance can be varied) connected across an ideal voltage supply. (An ideal voltage supply is one that has a constant emf and a negligible internal resistance.) To increase the heat output, should you decrease the resistance or increase the resistance? Explain your answer. SSM

14 • One resistor has a resistance R_1 and another resistor has a resistance R_2. The resistors are connected in parallel. If $R_1 \gg R_2$, the equivalent resistance of the combination is approximately (*a*) R_1, (*b*) R_2, (*c*) 0, (*d*) infinity.

15 • One resistor has a resistance R_1 and another resistor has a resistance R_2. The resistors are connected in series. If $R_1 \gg R_2$, the equivalent resistance of the combination is approximately (*a*) R_1, (*b*) R_2, (*c*) 0, (*d*) infinity.

16 • A parallel combination consisting of resistors A and B is connected across the terminals of a battery. The resistor A has twice the resistance of resistor B. If the current carried by resistor A is I, then what is the current carried by resistor B? (*a*) I, (*b*) $2I$, (*c*) $I/2$, (*d*) $4I$, (*e*) $I/4$

17 • A series combination consisting of resistors A and B is connected across the terminals of a battery. The resistor A has twice the resistance of resistor B. If the current carried by resistor A is I, then what is the current carried by resistor B? (*a*) I, (*b*) $2I$, (*c*) $I/2$, (*d*) $4I$, (*e*) $I/4$

18 • Kirchhoff's junction rule is considered to be a consequence of (*a*) conservation of charge, (*b*) conservation of energy, (*c*) Newton's laws, (*d*) Coulomb's law, (*e*) quantization of charge.

19 • True or false:

(*a*) An ideal voltmeter has a zero internal resistance.

(*b*) An ideal ammeter has a zero internal resistance.

(*c*) An ideal voltage source has a zero internal resistance.

20 • Before you and your classmates run an experiment, your professor lectures about safety. She reminds you that to measure the voltage across a resistor you connect a voltmeter in parallel with the resistor, and to measure the current in a resistor you connect an ammeter in series with the resistor. She also states that connecting a voltmeter in series with a resistor will not measure the voltage across the resistor, but also cannot do any damage to the circuit or the instrument. In addition, connecting an ammeter in parallel with a resistor will not measure the current in the resistor, but could cause significant damage to the circuit and the instrument. Explain why connecting a voltmeter in series with a resistor causes no damage while connecting an ammeter in parallel with a resistor can cause significant damage.

21 • The capacitor C in Figure 25-49 is initially uncharged. Just after the switch S is closed, (*a*) the voltage across C equals $\mathcal{E}$, (*b*) the voltage across R equals $\mathcal{E}$, (*c*) the current in the circuit is zero, (*d*) both (*a*) and (*c*) are correct.

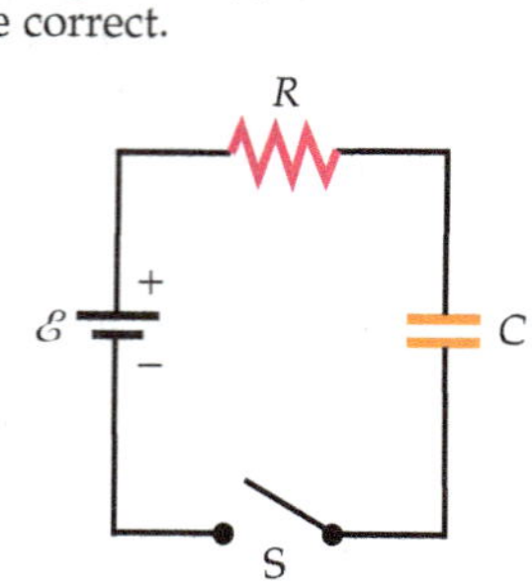

FIGURE 25-49 Problems 21 and 24

22 •• A capacitor is discharging through a resistor. If it takes a time T for the charge on a capacitor to drop to half its initial value, how long (in terms of T) does it take for the stored energy to drop to half its initial value?

23 •• In Figure 25-50, the values of the resistances are related as follows: $R_2 = R_3 = 2R_1$. If power P is delivered to R_1, what is the power delivered to R_2 and R_3? SSM

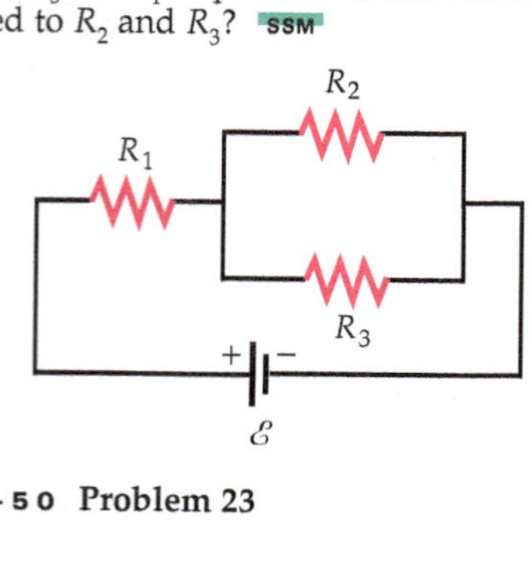

FIGURE 25-50 Problem 23

24 •• The capacitor in Figure 25-49 is initially uncharged. The switch S is closed and remains closed for a very long time. During this time, (*a*) the energy supplied by the battery is $\frac{1}{2}C\mathcal{E}^2$, (*b*) the energy dissipated in the resistor is $\frac{1}{2}C\mathcal{E}^2$, (*c*) the energy in the resistor is dissipated at a constant rate, (*d*) the total charge passing through the resistor is $\frac{1}{2}C\mathcal{E}$.

ESTIMATION AND APPROXIMATION

25 •• It is not a good idea to stick the ends of a metal paper clip into the two rectangular slots of a household electrical wall outlet in the United States. Explain why by estimating the current that a paper clip would carry until either the fuse blows or the breaker trips.

26 •• (*a*) Estimate the resistance of an automobile jumper cable. (*b*) Look up the current required to start a typical car. At that current, what is the potential drop that occurs across the jumper cable? (*c*) How much power is dissipated in the jumper cable when it carries that current?

27 •• **ENGINEERING APPLICATION, CONTEXT-RICH** Your manager wants you to design a new superinsulated water heater for the residential market. A coil of Nichrome wire is to be used as the heating element. Estimate the length of wire required. *Hint: You will need to determine the size of a typical water heater and a reasonable time period for creating hot water.*

28 •• A compact fluorescent lightbulb costs about \$6.00 and has a typical lifetime of 10 000 h. The bulb uses 20 W of power but produces illumination equivalent to that of a 75-W incandescent bulb. An incandescent bulb costs about \$1.50 and has a typical lifetime of 1000 h. Your family wonders whether it should buy fluorescent lightbulbs. Estimate the amount of money your household would save each year by using compact fluorescent lightbulbs instead of the incandescent bulbs.

29 •• **CONTEXT-RICH** The wires in a house must be large enough in diameter so that they do not get hot enough to start a fire. While working for a building contractor during the summer, you are involved in remodeling a house. The local building code states that the Joule heating of the wire used in houses should not exceed 2.0 W/m. Estimate the maximum gauge of the copper wire that you can use during the rewiring of the house with 20-A circuits.

30 •• A laser diode used in making a laser pointer is a highly nonlinear circuit element. Its behavior is as follows. For any voltage drop across it that is less than about 2.30 V, it behaves as if it has an infinite internal resistance, but for voltages higher than 2.30 V it has a very low internal resistance—effectively zero. (*a*) A laser pointer is made by putting two 1.55-V watch batteries in series across the laser diode. If the batteries each have an internal resistance between 1.00 Ω and 1.50 Ω, estimate the current in the laser diode. (*b*) About half of the power delivered to the laser diode goes into radiant energy. Using this fact, estimate the power of the laser beam, and compare this value to typical quoted values of about 3.00 mW. (*c*) If the batteries each have a capacity of 20.0 mA · h (i.e., they can deliver a constant current of 20.0 mA for approximately 1 hour before discharging), estimate how long one can continuously operate the laser pointer before replacing the batteries.

CURRENT, CURRENT DENSITY, DRIFT SPEED, AND THE MOTION OF CHARGES

31 • A 10-gauge copper wire carries a current equal to 20 A. Assuming copper has one free electron per atom, calculate the drift speed of the free electrons in the wire. SSM

32 •• A thin nonconducting ring that has a radius a and a linear charge density λ rotates with angular speed ω about an axis through its center and perpendicular to the plane of the ring. Find the current of the ring.

33 •• A length of 10-gauge copper wire and a length of 14-gauge copper wire are welded together end to end. The wires carry a current of 15 A. (*a*) If there is one free electron for each copper atom in each wire, find the drift speed of the electrons in each wire. (*b*) What is the ratio of the magnitude of the current density in the length of 10-gauge wire to the magnitude of the current density in the length of 14-gauge wire? SSM

34 •• An accelerator produces a beam of protons with a circular cross section that is 2.0 mm in diameter and has a current of 1.0 mA. The current density is uniformly distributed throughout the beam. The kinetic energy of each proton is 20 MeV. The beam strikes a metal target and is absorbed by the target. (*a*) What is the number density of the protons in the beam? (*b*) How many protons strike the target each minute? (*c*) What is the magnitude of the current density in this beam?

35 •• In one of the colliding beams of a planned proton *supercollider,* the protons are moving at nearly the speed of light and the beam current is 5.00 mA. The current density is uniformly distributed throughout the beam. (*a*) How many protons are there per meter of length of the beam? (*b*) If the cross-sectional area of the beam is $1.00 \times 10^{-6}\ \text{m}^2$, what is the number density of protons? (*c*) What is the magnitude of the current density in this beam? SSM

36 •• **CONTEXT-RICH** The *solar wind* consists of protons from the Sun moving toward Earth (the wind actually consists of about 95% protons). The number density of protons at a distance from the Sun equal to the orbital radius of Earth is about 7.0 protons per cubic centimeter. Your research team monitors a satellite that is in orbit around the Sun at a distance from the Sun equal to Earth's orbital radius. You are in charge of the satellite's *mass spectrometer,* an instrument used to measure the composition and intensity of the solar wind. The aperture of your spectrometer is a circle of radius 25 cm. The rate of collection of protons by the spectrometer is such that they constitute a measured current of 85 nA. What is the speed of the protons in the solar wind? (Assume the protons enter the aperture at normal incidence.)

37 •• A gold wire has a 0.10-mm-diameter cross section. Opposite ends of this wire are connected to the terminals of a 1.5-V battery. If the length of the wire is 7.5 cm, how much time, on average, is required for electrons leaving the negative terminal of the battery to reach the positive terminal? Assume the resistivity of gold is $2.44 \times 10^{-8}\ \Omega \cdot \text{m}$.

RESISTANCE, RESISTIVITY, AND OHM'S LAW

Note: **In this section, assume the resistors are ohmic (constant resistance) unless stated otherwise.**

38 • A 10-m-long wire has a resistance equal to 0.20 Ω and carries a current equal to 5.0 A. (*a*) What is the potential difference across the entire length of the wire? (*b*) What is the electric-field strength in the wire?

39 • A potential difference of 100 V across the terminals of a resistor produces a current of 3.00 A in the resistor. (*a*) What is the resistance of the resistor? (*b*) What is the current in the resistor when the potential difference is only 25.0 V? (Assume the resistance of the resistor remains constant.) SSM

40 • A block of carbon is 3.0 cm long and has a square cross-section whose sides are 0.50 cm long. A potential difference of 8.4 V is maintained across its length. (*a*) What is the resistance of the block? (*b*) What is the current in this resistor?

41 • An extension cord consists of a pair of 30-m-long 16-gauge copper wires. What is the potential difference that must be applied across one of the wires if it is to carry a current of 5.0 A? SSM

42 • (*a*) How long is a 14-gauge copper wire that has a resistance of 12.0 Ω? (*b*) How much current will it carry if a 120-V potential difference is applied across its length?

43 • A cylinder of glass is 1.00 cm long and has a resistivity of $1.01 \times 10^{12}\ \Omega \cdot \text{m}$. What length of copper wire that has the same cross-sectional area will have the same resistance as the glass cylinder?

44 •• **ENGINEERING APPLICATION** While remodeling your garage, you need to temporarily splice, end to end, an 80-m-long copper wire that is 1.00 mm in diameter with a 49-m-long aluminum wire that has the same diameter. The maximum current in the wires is 2.00 A. (*a*) Find the potential drop across each wire of this system when the current is 2.00 A. (*b*) Find the electric field in each wire when the current is 2.00 A.

45 •• A 1.00-m-long wire has a resistance equal to 0.300 Ω. A second wire made of identical material has a length of 2.00 m and a mass equal to the mass of the first wire. What is the resistance of the second wire? SSM

46 •• A 10-gauge copper wire can safely carry currents up to 30.0 A. (*a*) What is the resistance of a 100-m length of the wire? (*b*) What is the electric field in the wire when the current is 30.0 A? (*c*) How long does it take for an electron to travel 100 m in the wire when the current is 30.0 A?

47 •• A cube of copper has edges that are 2.00 cm long. If copper in the cube is drawn to form a length of 14-gauge wire, what will the resistance of the length of wire be? Assume the density of the copper does not change.

48 ••• Find an expression for the resistance between the ends of the half-ring shown in Figure 25-51. The resistivity of the material constituting the half-ring is ρ. *Hint: Model the half-ring as a parallel combination of a large number of thin half-rings. Assume the current is uniformly distributed on a cross section of the half-ring.*

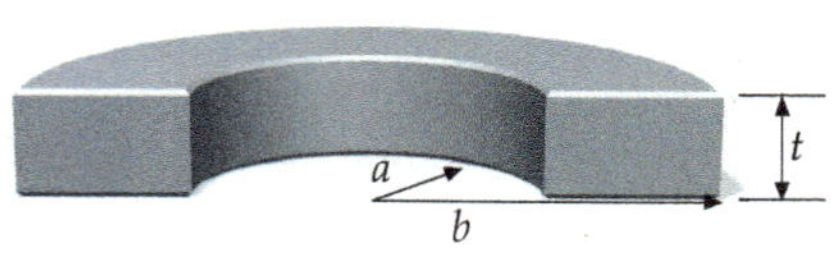

FIGURE 25-51 Problem 48

49 ••• Consider a wire of length L in the shape of a truncated cone. The radius of the wire varies with distance x from the narrow end according to $r = a + [(b - a)/L]x$, where $0 < x < L$. Derive an expression for the resistance of this wire in terms of its length L, radius a, radius b, and resistivity ρ. *Hint: Model the wire as a series combination of a large number of thin disks. Assume the current is uniformly distributed on a cross section of the cone.* SSM

50 ••• The space between two metallic concentric spherical shells is filled with a material that has a resistivity of $3.50 \times 10^{-5}\ \Omega \cdot \text{m}$. If the inner metal shell has an outer radius of 1.50 cm and the outer metal shell has an inner radius of 5.00 cm, what is the resistance between the conductors? *Hint: Model the material as a series combination of a large number of thin spherical shells.*

51 ••• The space between two metallic coaxial cylinders that have the same length L is completely filled with a nonmetallic material having a resistivity ρ. The inner metal shell has an outer radius a and the outer metal shell has an inner radius b. (*a*) What is the resistance between the two cylinders? *Hint: Model the material as a series combination of a large number of thin cylindrical shells.* (*b*) Find the current between the two metallic cylinders if $\rho = 30.0\ \Omega \cdot \text{m}$, $a = 1.50$ cm, $b = 2.50$ cm, $L = 50.0$ cm, and a potential difference of 10.0 V is maintained between the two cylinders.

TEMPERATURE DEPENDENCE OF RESISTANCE

52 • A tungsten rod is 50 cm long and has a square cross section that has 1.0-mm-long edges. (*a*) What is its resistance at 20°C? (*b*) What is its resistance at 40°C?

53 • At what temperature will the resistance of a copper wire be 10 percent greater than its resistance at 20°C? SSM

54 •• **ENGINEERING APPLICATION** You have a toaster that uses a Nichrome wire as a heating element. You need to determine the temperature of the Nichrome wire under operating conditions. First, you measure the resistance of the heating element at 20°C and find it to be 80.0 Ω. Then you measure the current immediately after you plug the toaster into a wall outlet—before the temperature of the Nichrome wire increases significantly. You find this startup current to be 8.70 A. When the heating element reaches its operating temperature, you measure the current to be 7.00 A. Use your data to determine the maximum operating temperature of the heating element.

55 •• **ENGINEERING APPLICATION** Your electric space heater has a Nichrome heating element that has a resistance of 8.00 Ω at 20.0°C. When 120 V is applied, the electric current heats the Nichrome wire to 1000°C. (*a*) What is the initial current in the heating element at 20.0°C? (*b*) What is the resistance of the heating element at 1000°C? (*c*) What is the operating power of this heater?

56 •• A 10.0-Ω Nichrome resistor is wired into an electronic circuit using copper leads (wires) that have diameters equal to 0.600 mm. The copper leads have a total length of 50.0 cm. (*a*) What additional resistance is due to the copper leads? (*b*) What percentage error in the total resistance is produced by neglecting the resistance of the copper leads? (*c*) What change in temperature would produce a change in resistance of the Nichrome wire equal to the resistance of the copper leads? Assume that the Nichrome section is the only one whose temperature is changed.

57 ••• A wire that has a cross-sectional area A, a length L_1, a resistivity ρ_1, and a temperature coefficient α_1 is connected end to end to a second wire that has the same cross-sectional area, a length L_2, a resistivity ρ_2, and a temperature coefficient α_2, so that the wires carry the same current. (*a*) Show that if $\rho_1 L_1 \alpha_1 + \rho_2 L_2 \alpha_2 = 0$, then the total resistance is independent of temperature for small temperature changes. (*b*) If one wire is made of carbon and the other wire is made of copper, find the ratio of their lengths for which the total resistance is approximately independent of temperature. SSM

58 ••• The resistivity of tungsten increases approximately linearly with temperature from 56.0 nΩ·m at 293 K to 1.10 μΩ·m at 3500 K. A lightbulb is powered by a 100-V dc power supply. Under those operating conditions the temperature of the tungsten filament is 2500 K, the length of the filament is equal to 5.00 cm, and the power delivered to the filament is 40 W. Estimate (*a*) the resistance of the filament and (*b*) the diameter of the filament.

59 ••• A 5.00-V lightbulb used in an electronics class has a carbon filament that has a length of 3.00 cm and a diameter of 40.0 μm. At temperatures between 500 K and 700 K, the resistivity of the carbon used in making small lightbulb filaments is about $3.00 \times 10^{-5}\ \Omega \cdot \text{m}$. (*a*) Assuming that the bulb is a perfect blackbody radiator, calculate the temperature of the filament under operating conditions. (*b*) One concern about carbon filament bulbs, unlike tungsten filament bulbs, is that the resistivity of carbon decreases with increasing temperature. Explain why this decrease in resistivity is a concern.

ENERGY IN ELECTRIC CIRCUITS

60 • A 1.00-kW heater is designed to operate at 240 V. (*a*) What is the heater's resistance and what is the current in the wires that supply power to the heater? (*b*) What is the power delivered to the heater if it operates at 120 V? Assume that its resistance remains the same.

61 • A battery has an emf of 12 V. How much work does it do in 5.0 s if it delivers a current of 3.0 A?

62 • An automotive battery has an emf of 12.0 V. When supplying power to the starter motor, the current in the battery is 20.0 A and the terminal voltage of the battery is 11.4 V. What is the internal resistance of the battery?

63 • (*a*) How much power is delivered by the battery in Problem 62 due to the chemical reactions within the battery when the current in the battery is 20 A? (*b*) How much of this power is delivered to the starter when the current in the battery is 20 A? (*c*) By how much does the chemical energy of the battery decrease if the current in the starter is 20 A for 7.0 s? (*d*) How much energy is dissipated in the battery during those 7.0 seconds? SSM

64 • A battery that has an emf of 6.0 V and an internal resistance of 0.30 Ω is connected to a variable resistor with resistance R. Find the current and power delivered by the battery when R is (*a*) 0, (*b*) 5.0 Ω, (*c*) 10 Ω, and (*d*) infinite.

65 •• **ENGINEERING APPLICATION, CONTEXT-RICH** A 12.0-V automobile battery that has a negligible internal resistance can deliver a total charge of 160 A·h. (*a*) What is the amount of energy stored in the battery? (*b*) After studying all night for a calculus test, you try to drive to class to take the test. However, you find that the car's battery is "dead" because you had left the headlights on! Assuming the battery was able to produce current at a constant rate until it died, how long were your lights on? Assume the pair of headlights together operates at a power of 150 W.

66 •• **ENGINEERING APPLICATION** The measured current in a circuit in your uncle's house is 12.5 A. In this circuit, the only appliance that is on is a space heater that is being used to heat the bathroom. A pair of 12-gauge copper wires carries the current from the supply panel in your basement to the wall outlet in the bathroom, a distance of 30.0 m. You measure the voltage at the supply panel to be exactly 120 V. What is the voltage at the wall outlet in the bathroom that the space heater is connected to?

67 •• **ENGINEERING APPLICATION** A lightweight electric car is powered by a series combination of ten 12.0-V batteries, each having negligible internal resistance. Each battery can deliver a charge of 160 A·h before needing to be recharged. At a speed of 80.0 km/h, the average force due to air drag and rolling friction is 1.20 kN. (*a*) What must be the minimum power delivered by the electric motor if the car is to travel at a speed of 80.0 km/h? (*b*) What is the

total charge, in coulombs, that can be delivered by the series combination of ten batteries before recharging is required? (*c*) What is the total electrical energy delivered by the ten batteries before recharging? (*d*) How far can the car travel (at 80.0 km/h) before the batteries must be recharged? (*e*) What is the cost per kilometer if the cost of recharging the batteries is 9.00 cents per kilowatt-hour? SSM

68 ••• A 100-W heater is designed to operate with an applied voltage of 120 V. (*a*) What is the heater's resistance, and what current does the heater carry? (*b*) Show that if the potential difference V across the heater changes by a small amount ΔV, the power P changes by a small amount ΔP, where $\Delta P/P \approx 2\,\Delta V/V$. *Hint: Approximate the changes by modeling them as differentials, and assume the resistance is constant.* (*c*) Using the Part-(*b*) result, find the approximate power delivered to the heater if the potential difference is decreased to 115 V. Compare your result to the exact answer.

COMBINATIONS OF RESISTORS

69 • If the potential drop from point a to point b (Figure 25-52) is 12.0 V, find the current in each resistor. SSM

70 • If the potential drop between point a and point b (Figure 25-53) is 12.0 V, find the current in each resistor.

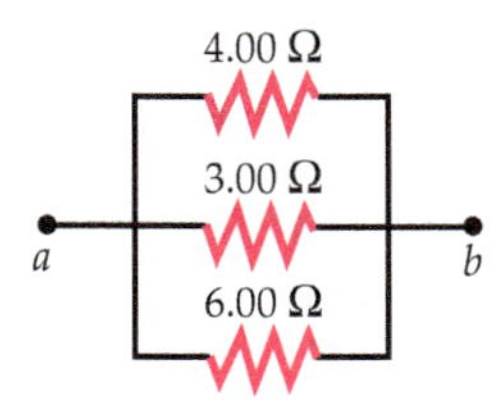

FIGURE 25-52 Problem 69

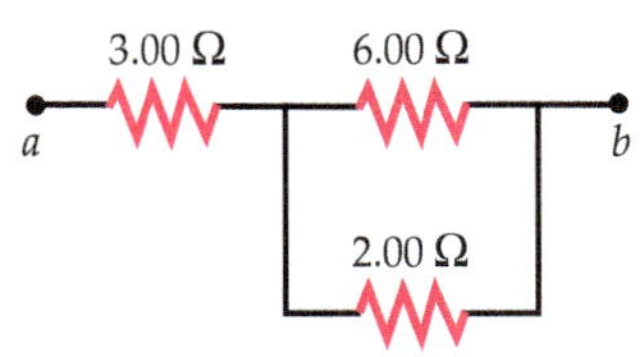

FIGURE 25-53 Problem 70

71 • (*a*) Show that the equivalent resistance between point a and point b in Figure 25-54 is R. (*b*) How would adding a fifth resistor that has resistance R between point c and point d affect the equivalent resistance between point a and point b?

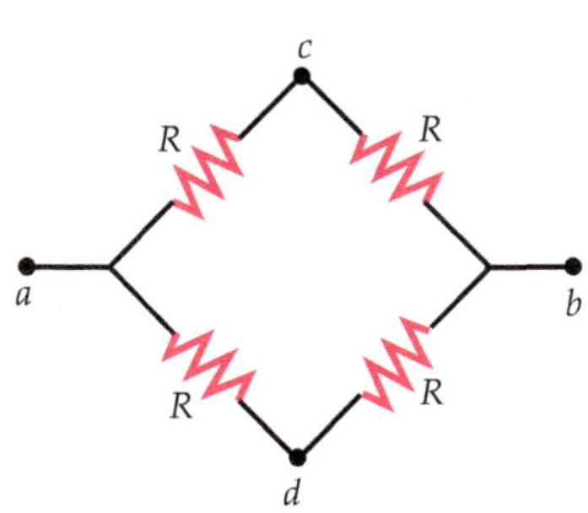

FIGURE 25-54 Problem 71

72 •• The battery in Figure 25-55 has negligible internal resistance. Find (*a*) the current in each resistor and (*b*) the power delivered by the battery.

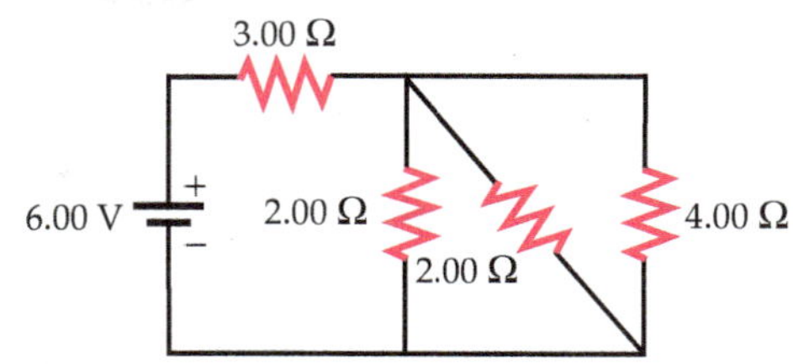

FIGURE 25-55 Problem 72

73 •• A 5.00-V power supply has an internal resistance of 50.0 Ω. What is the smallest resistor that can be put in series with the power supply so that the voltage drop across the resistor is larger than 4.50 V? SSM

74 •• **ENGINEERING APPLICATION** You have been handed an unknown battery. Using your multimeter, you determine that when a 5.00-Ω resistor is connected across the battery's terminals, the current in the battery is 0.500 A. When this resistor is replaced by an 11.0-Ω resistor, the current drops to 0.250 A. From those data, find (*a*) the emf and (*b*) internal resistance of the battery.

75 •• (*a*) Find the equivalent resistance between point a and point b in Figure 25-56. (*b*) If the potential drop between point a and point b is 12.0 V, find the current in each resistor.

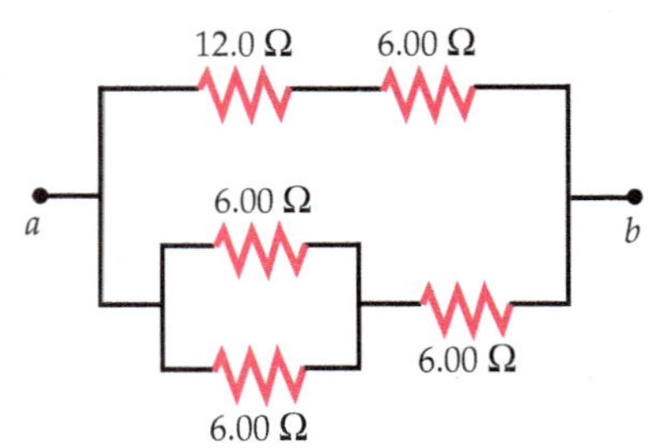

FIGURE 25-56 Problem 75

76 •• (*a*) Find the equivalent resistance between point a and point b in Figure 25-57. (*b*) If the potential drop between point a and point b is 12.0 V, find the current in each resistor.

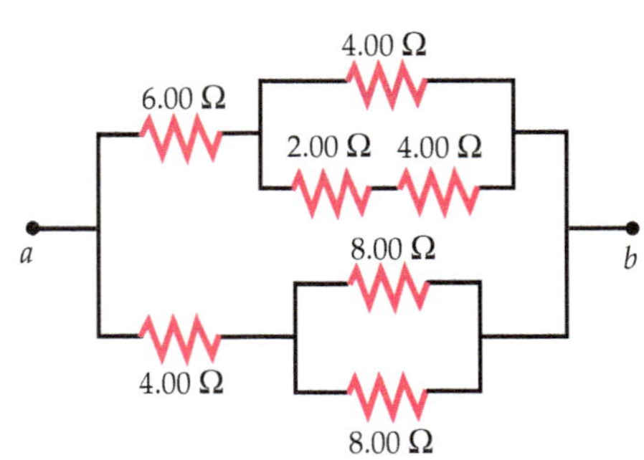

FIGURE 25-57 Problem 76

77 •• A length of wire has a resistance of 120 Ω. The wire is cut into pieces that have the same length, and then the wires are connected in parallel. The resistance of the parallel arrangement is 1.88 Ω. Find the number of pieces into which the wire was cut. SSM

78 •• A parallel combination of an 8.00-Ω resistor and a resistor of unknown resistance is connected in series with a 16.0-Ω resistor and an ideal battery. The circuit is disassembled and the three resistors are then connected in series with each other and the same battery. In both arrangements, the current through the 8.00-Ω resistor is the same. What is the resistance of the unknown resistor?

79 •• For the network shown in Figure 25-58, let R_{ab} denote the equivalent resistance between terminals a and b. Find (*a*) R_3, so that $R_{ab} = R_1$; (*b*) R_2, so that $R_{ab} = R_3$; and (*c*) R_1, so that $R_{ab} = R_1$.

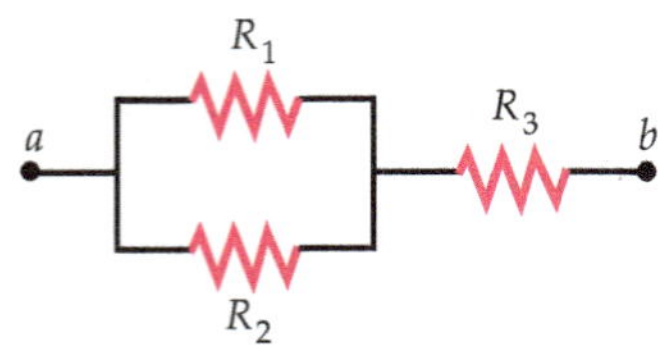

FIGURE 25-58 Problems 79 and 80

80 •• Check your results for Problem 79 using the following specific values: (*a*) $R_1 = 4.00\ \Omega$, $R_2 = 6.00\ \Omega$; (*b*) $R_1 = 4.00\ \Omega$, $R_3 = 3.00\ \Omega$; and (*c*) $R_2 = 6.00\ \Omega$, $R_3 = 3.00\ \Omega$.

KIRCHHOFF'S RULES

Note: **While simpler circuits in this section can be solved using the concepts of equivalent parallel and series resistor combinations, the intent is to gain practice using Kirchhoff's rules. Use them to solve all the problems in this section.**

81 • In Figure 25-59, the battery's emf is 6.00 V and R is 0.500 Ω. The rate of Joule heating in R is 8.00 W. (*a*) What is the current in the circuit? (*b*) What is the potential difference across R? (*c*) What is the resistance r? SSM

82 • The batteries in the circuit in Figure 25-60 have negligible internal resistance. (*a*) Find the current using Kirchhoff's loop rule. (*b*) Find the power delivered to or supplied by each battery. (*c*) Find the rate of Joule heating in each resistor.

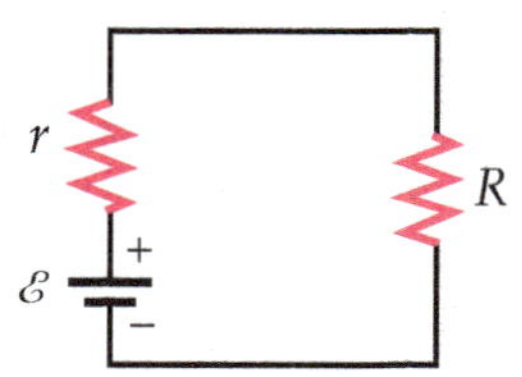

FIGURE 25-59 Problem 81

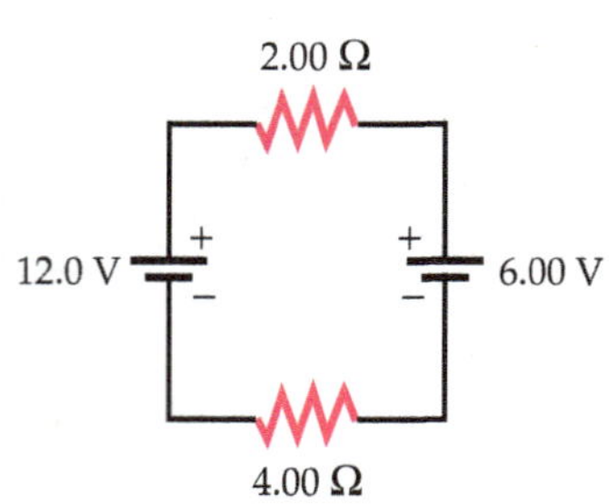

FIGURE 25-60 Problem 82

83 •• **ENGINEERING APPLICATION** An old car battery that has an emf of $\mathcal{E}_1 = 11.4$ V and an internal resistance of 50.0 mΩ is connected to a 2.00-Ω resistor. In an attempt to recharge the battery, you connect a second battery that has an emf of $\mathcal{E}_2 = 12.6$ V and an internal resistance of 10.0 mΩ in parallel with the first battery and the resistor with a pair of jumper cables. (*a*) Draw a diagram of the circuit. (*b*) Find the current in each branch of the circuit. (*c*) Find the power supplied by the second battery and discuss where that power is delivered. Assume that the emfs and internal resistances of both batteries remain constant.

84 •• In the circuit in Figure 25-61, the reading of the ammeter is the same when both switches are open and when both switches are closed. What is the unknown resistance R?

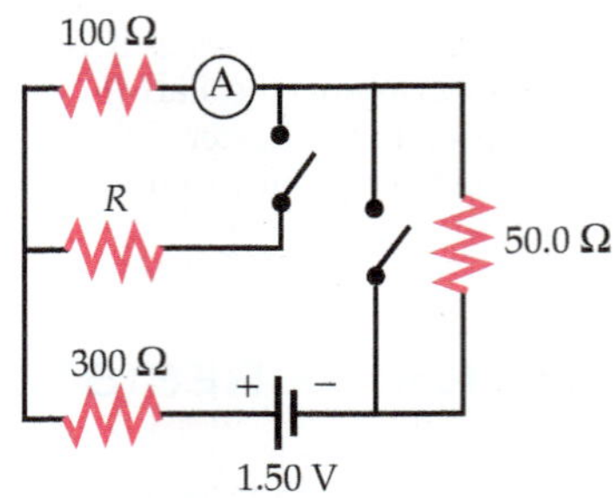

FIGURE 25-61 Problem 84

85 •• In the circuit shown in Figure 25-62, the batteries have negligible internal resistance. Find (*a*) the current in each branch of the circuit, (*b*) the potential difference between point a and point b, and (*c*) the power supplied by each battery. SSM

86 •• In the circuit shown in Figure 25-63, the batteries have negligible internal resistance. Find (*a*) the current in each branch of the circuit, (*b*) the potential difference between point a and point b, and (*c*) the power supplied by each battery.

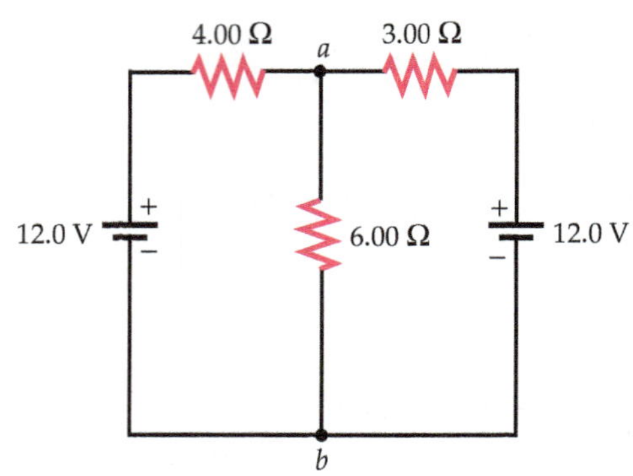

FIGURE 25-62 Problem 85

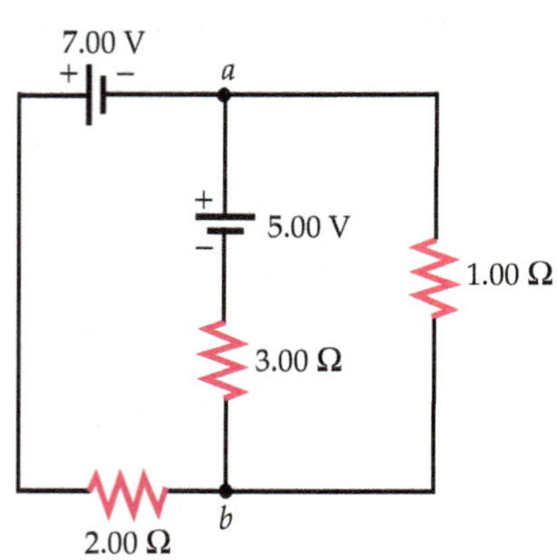

FIGURE 25-63 Problem 86

87 ••• Two identical batteries, each having an emf $\mathcal{E}$ and an internal resistance r, can be connected across a resistance R with the batteries connected either in series or in parallel. In each situation, determine explicitly whether the power supplied to R is greater when R is less than r or when R is greater than r.

88 •• **ENGINEERING APPLICATION** The circuit fragment shown in Figure 25-64 is called a *voltage divider.* (*a*) If R_{load} is not attached, show that $V_{\text{out}} = VR_2/(R_1 + R_2)$. (*b*) If $R_1 = R_2 = 10\ \text{k}\Omega$, what is the smallest value of R_{load} that can be used so that V_{out} drops by less than 10 percent from its unloaded value? (V_{out} is measured with respect to ground.)

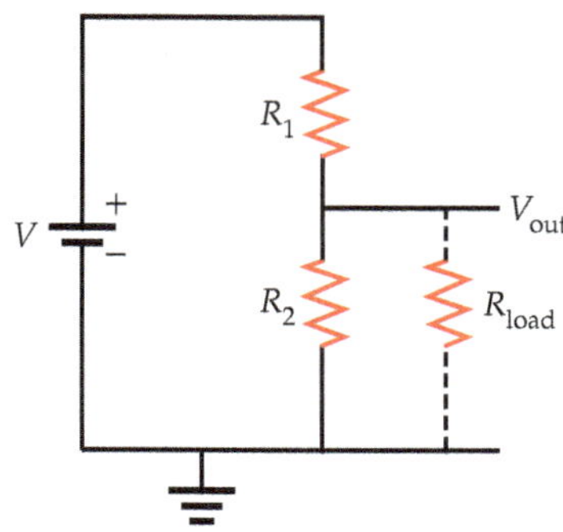

FIGURE 25-64 Problem 88

89 ••• For the circuit shown in Figure 25-65, find the potential difference between point *a* and point *b*. SSM

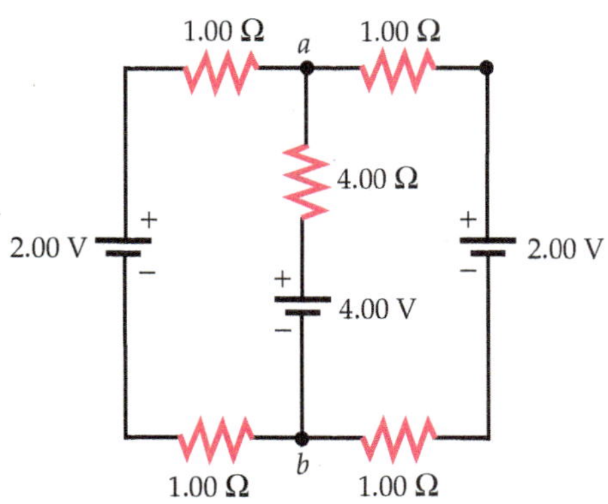

FIGURE 25-65 Problem 89

90 ••• For the circuit shown in Figure 25-66, find (*a*) the current in each resistor, (*b*) the power supplied by each source of emf, and (*c*) the power delivered to each resistor.

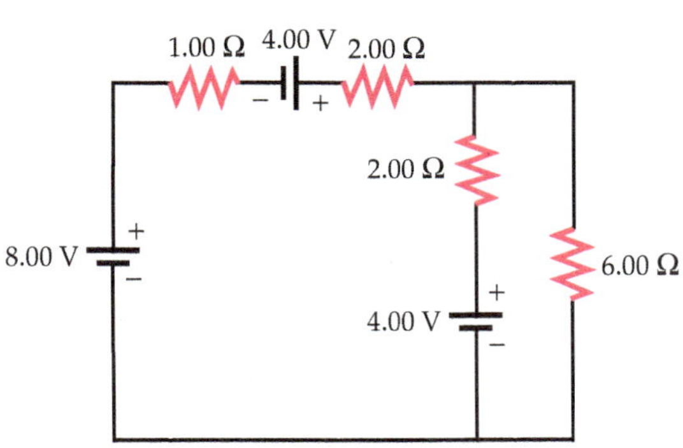

FIGURE 25-66 Problem 90

AMMETERS AND VOLTMETERS

91 •• The voltmeter shown in Figure 25-67 can be modeled as an ideal voltmeter (a voltmeter that has an infinite internal resistance) in parallel with a 10.0-MΩ resistor. Calculate the reading on the voltmeter when (*a*) $R = 1.00\ \text{k}\Omega$, (*b*) $R = 10.0\ \text{k}\Omega$, (*c*) $R = 1.00\ \text{M}\Omega$, (*d*) $R = 10.0\ \text{M}\Omega$, and (*e*) $R = 100\ \text{M}\Omega$. (*f*) What is the largest value of R possible if the measured voltage is to be within 10 percent of the *true* voltage (i.e., the voltage drop across R without the voltmeter in place)? SSM

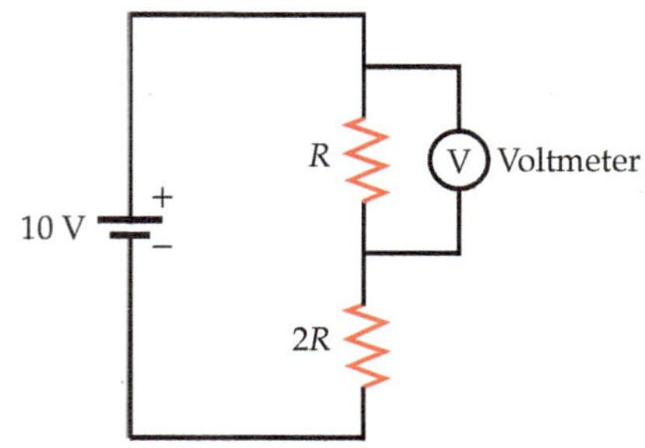

FIGURE 25-67 Problem 91

92 •• You are given a D'Arsonval galvanometer that will deflect full scale if a current of 50.0 μA runs through the galvanometer. At that current, there is a voltage drop of 0.250 V across the meter. What is the internal resistance of the galvanometer?

93 •• You are given a D'Arsonval galvanometer that will deflect full scale if a current of 50.0 μA runs through the galvanometer. At that current, there is a voltage drop of 0.250 V across the meter. You wish to use the galvanometer to construct an ammeter that can measure currents up to 100 mA. Show that this can be done by placing a resistor in parallel with the meter, and find the value of its resistance.

94 •• You are given a D'Arsonval galvanometer that will deflect full scale if a current of 50.0 μA runs through the galvanometer. At that current, there is a voltage drop of 0.250 V across the meter. You wish to use the galvanometer to construct a voltmeter that can measure potential differences up to 10.0 V. Show that this can be done by placing a large resistance in series with the meter movement, and find the resistance needed.

RC CIRCUITS

95 • For the circuit shown in Figure 25-68, $C = 6.00\ \mu\text{F}$, $\mathcal{E} = 100\ \text{V}$, and $R = 500\ \Omega$. After having been at contact *a* for a long time, the switch throw is rotated to contact *b*. (*a*) What is the charge on the upper plate of the capacitor just as the switch throw is moved to contact *b*? (*b*) What is the current just after the switch throw is rotated to contact *b*? (*c*) What is the time constant of this circuit? (*d*) How much charge is on the upper plate of the capacitor 6.00 ms after the switch throw is rotated to contact *b*?

96 • At $t = 0$ the switch throw in Figure 25-68 is rotated to contact *b* after having been at contact *a* for a long time. (*a*) How much energy is stored in the capacitor at $t = 0$? (*b*) For $t > 0$, find the energy stored in the capacitor as a function of time. (*c*) Sketch a plot of the energy stored in the capacitor versus time t.

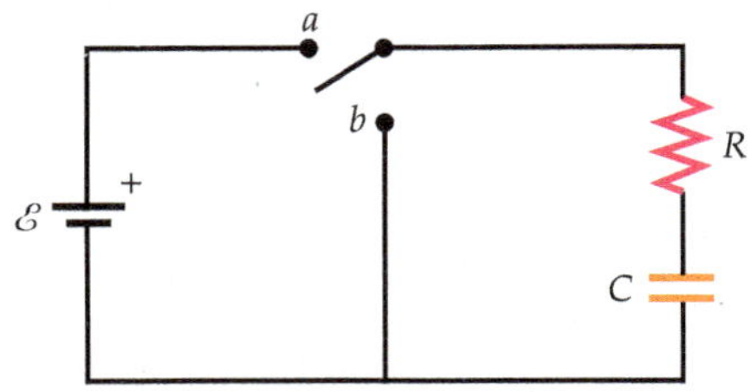

FIGURE 25-68 Problems 95, 96, and 98

97 •• In the circuit in Figure 25-69, the emf equals 50.0 V and the capacitance equals 2.00 μF. Switch S is opened after having been closed for a long time, and 4.00 s later the voltage drop across the resistor is 20.0 V. Find the resistance of the resistor. SSM

98 •• For the circuit shown in Figure 25-68, $C = 0.120\ \mu$F and $\mathcal{E} = 100$ V. The switch throw is rotated to contact b after having been at contact a for a long time, and 4.00 s later the potential difference across the capacitor is equal to $\frac{1}{2}\mathcal{E}$. What is the value of R?

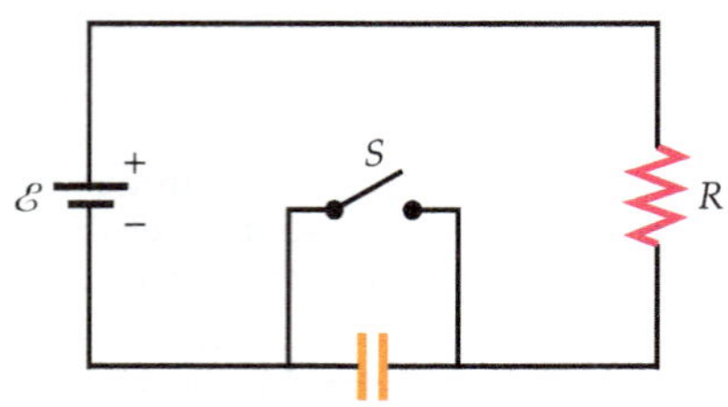

FIGURE 25-69 Problems 97 and 99

99 •• In the circuit in Figure 25-69, the emf equals 6.00 V and has negligible internal resistance. The capacitance equals 1.50 μF and the resistance equals 2.00 MΩ. Switch S has been closed for a long time. Switch S is opened. After a time interval equal to one time constant of the circuit has elapsed, find (*a*) the charge on the capacitor plate on the right, (*b*) the rate at which the charge is increasing, (*c*) the current, (*d*) the power supplied by the battery, (*e*) the power delivered to the resistor, and (*f*) the rate at which the energy stored in the capacitor is increasing.

100 •• A constant charge of 1.00 mC is on the positively charged plate of the 5.00-μF capacitor shown in Figure 25-70. Find (*a*) the battery current and (*b*) the resistances R_1, R_2, and R_3.

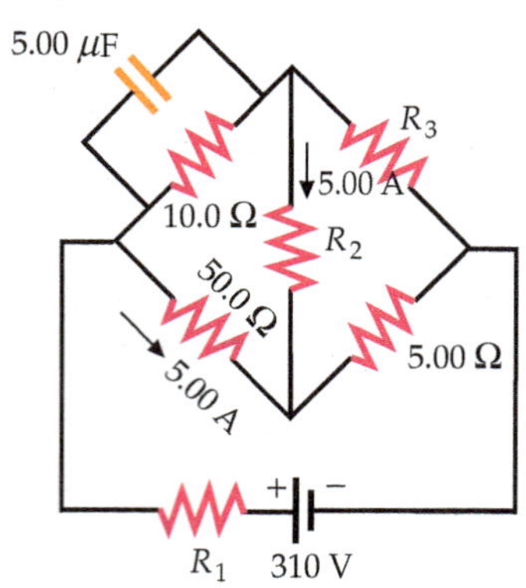

FIGURE 25-70 Problem 100

101 •• Show that Equation 25-39 can be rearranged and written as $\dfrac{dQ}{\mathcal{E}C - Q} = \dfrac{dt}{RC}$. Integrate this equation to derive the solution given by Equation 25-40.

102 •• Switch S, shown in Figure 25-71, is closed after having been open for a long time. (*a*) What is the initial value of the battery current just after switch S is closed? (*b*) What is the battery current a long time after switch S is closed? (*c*) What are the charges on the plates of the capacitors a long time after switch S is closed? (*d*) Switch S is reopened. What are the charges on the plates of the capacitors a long time after switch S is reopened?

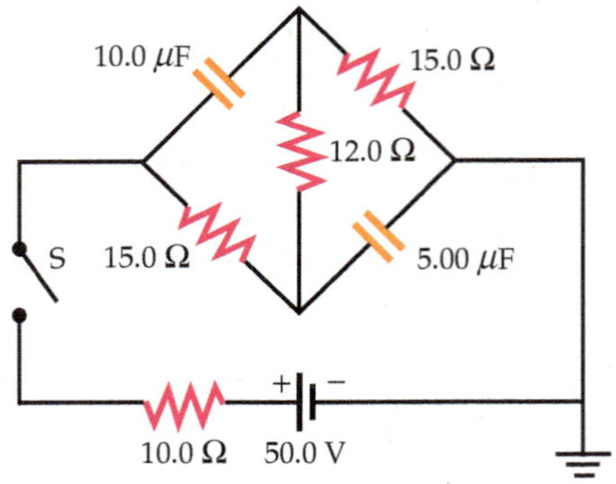

FIGURE 25-71 Problem 102

103 ••• For the circuit shown in Figure 25-72, switch S has been open for a long time. At time $t = 0$ the switch is then closed. (*a*) What is the battery current just after switch S is closed? (*b*) What is the battery current a long time after switch S is closed? (*c*) What is the current in the 600-Ω resistor as a function of time?

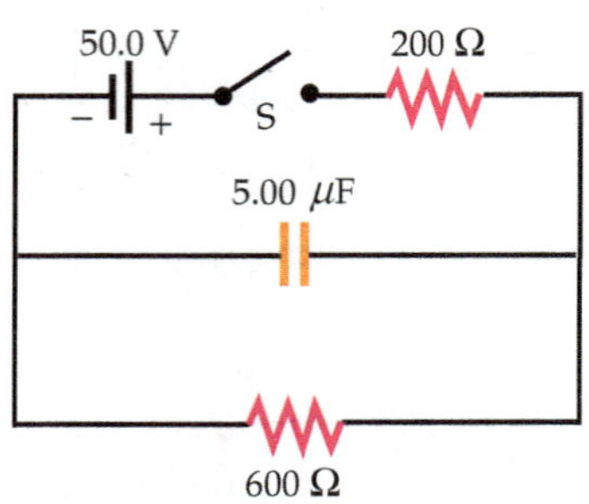

FIGURE 25-72 Problem 103

104 ••• For the circuit shown in Figure 25-73, switch S has been open for a long time. At time $t = 0$ the switch is then closed. (*a*) What is the battery current just after switch S is closed? (*b*) What is the battery current a long time after switch S is closed? (*c*) The switch has been closed for a long time. At time $t = 0$ the switch is then opened. Find the current through the 600-kΩ resistor as a function of time.

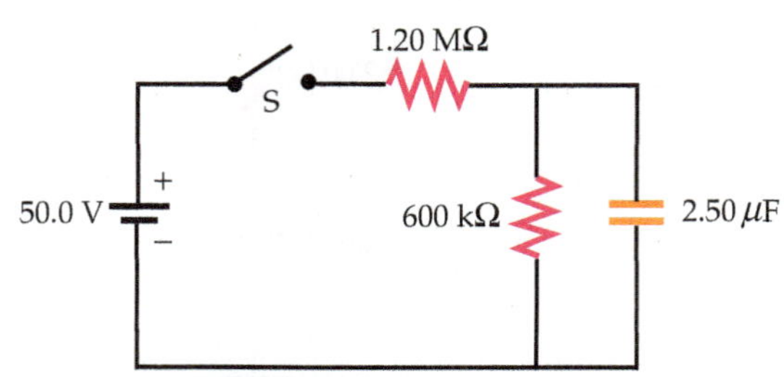

FIGURE 25-73 Problem 104

105 ••• In the circuit shown in Figure 25-74, the capacitor has a capacitance of 2.50 μF and the resistor has a resistance of 0.500 MΩ. Before the switch is closed, the potential drop across the capacitor is 12.0 V, as shown. Switch S is closed at $t = 0$. (*a*) What is the current immediately after switch S is closed? (*b*) At what time t is the voltage across the capacitor 24.0 V? SSM

106 ••• Repeat Problem 105 if the initial polarity of the capacitor is opposite to that shown in Figure 25-74.

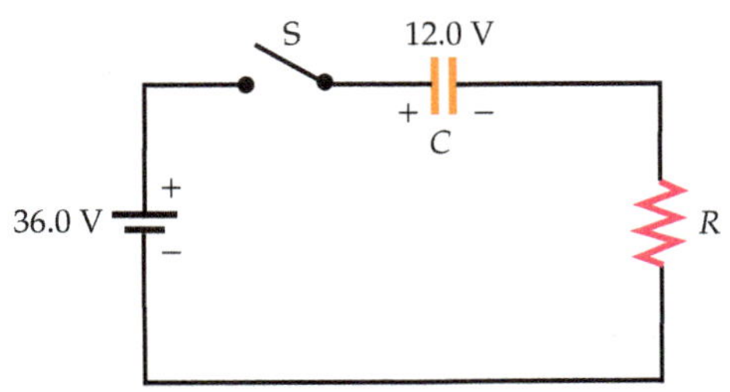

FIGURE 25-74 Problems 105 and 106

GENERAL PROBLEMS

107 •• In Figure 25-75, $R_1 = 4.00\ \Omega$, $R_2 = 6.00\ \Omega$, $R_3 = 12.0\ \Omega$, and the battery emf is 12.0 V. Denote the currents through the resistors as I_1, I_2, and I_3, respectively. (*a*) Decide which of the following inequalities holds for the circuit. Explain your answer conceptually. (1) $I_1 > I_2 > I_3$, (2) $I_2 = I_3$, (3) $I_3 > I_2$, (4) None of the above (*b*) To verify that your answer to Part (*a*) is correct, calculate all three currents. SSM

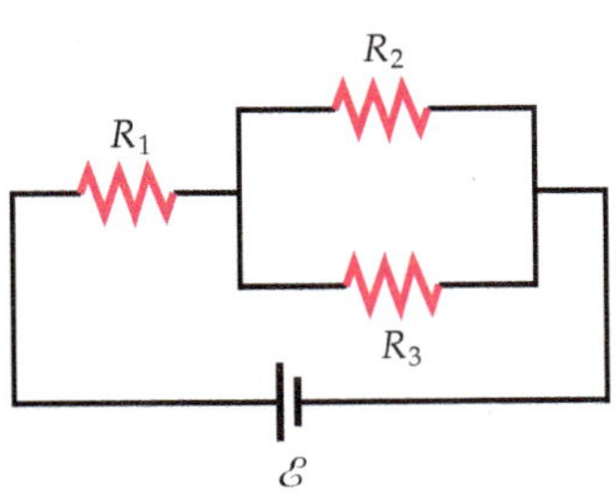

FIGURE 25-75 Problem 107

108 •• A 120 V, 25.0-W lightbulb is connected in series with a 120 V, 100-W lightbulb and a potential difference of 120 V is placed across the combination. Assume the bulbs have constant resistance. (*a*) Which bulb should be brighter under those conditions? Explain your answer conceptually. *Hint: What does the phrase "25.0-W lightbulb" mean? That is, under what conditions is 25 W of power delivered to the bulb?* (*b*) Determine the power delivered to each bulb under those conditions. Do your results support your answer to Part (*a*)?

109 •• The circuit shown in Figure 25-76 is a *Wheatstone bridge*, and the variable resistor is being used as a slide-wire potentiometer. The resistance R_0 is known. This "bridge" is used to determine an unknown resistance R_x. The resistances R_1 and R_2 comprise a wire 1.00 m long. Point *a* is a sliding contact that is moved along the wire to vary the resistances. Resistance R_1 is proportional to the distance from the left end of the wire (labeled 0.00 cm) to point *a*, and R_2 is proportional to the distance from point *a* to the right end of the wire (labeled 100 cm). The sum of R_1 and R_2 remains constant. When points *a* and *b* are at the same potential, there is no current in the galvanometer and the bridge is said to be balanced. (Because the galvanometer is used to detect the absence of a current, it is called a *null detector.*) If the fixed resistance $R_0 = 200\ \Omega$, find the unknown resistance R_x if (*a*) the bridge balances at the 18.0-cm mark, (*b*) the bridge balances at the 60.0-cm mark, and (*c*) the bridge balances at the 95.0-cm mark.

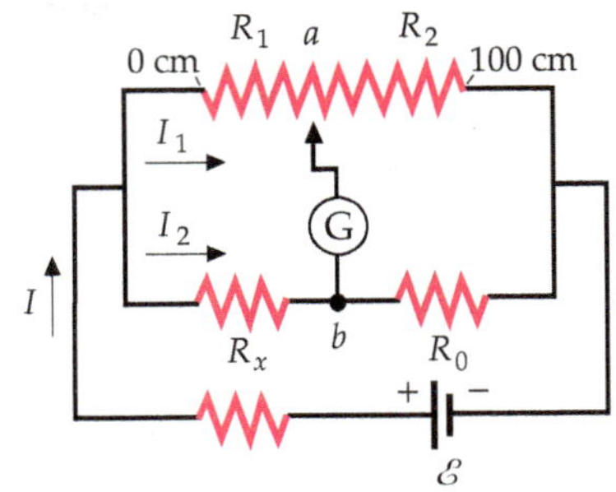

FIGURE 25-76 Problems 109 and 110

110 •• For the Wheatstone bridge in Problem 109, suppose the bridge balances at the 98.0-cm mark. (*a*) What is the unknown resistance? (*b*) What is the percentage error in the measured value of R_x if there is an error of 2.00 mm in the location of the balance point? (*c*) To what value should R_0 be changed to so that the balance point for the unknown resistor will be nearer the 50.0-cm mark? (*d*) If the balance point is at the 50.0-cm mark, what is the percentage error in the measured value of R_x if there is an error of 2.00 mm in the location of the balance point?

111 •• You are running an experiment that uses an accelerator that produces a 3.50-μA proton beam. Each proton in the beam has 60.0 MeV of kinetic energy. The protons impinge on, and come to rest inside, a 50.0-g copper target within a vacuum chamber. You are concerned that the target will get too hot and melt the solder on some connecting wires that are crucial to the experiment. (*a*) Determine the number of protons that strike the target per second. (*b*) Find the amount of energy delivered to the target each second. (*c*) Determine how much time elapses before the target temperature increases to 300°C. (Neglect any heat released by the target.) SSM

112 •• The belt of a Van de Graaff generator carries a surface charge density of 5.00 mC/m^2. The belt is 0.500 m wide and moves at 20.0 m/s. (*a*) What current does the belt carry? (*b*) If the potential of the dome of the generator is 100 kV above ground, what is the minimum power of the motor needed to drive the belt?

113 •• **ENGINEERING APPLICATION** Large conventional electromagnets use water cooling to prevent excessive heating of the magnet coils. A large laboratory electromagnet has a current equal to 100 A when a voltage of 240 V is applied to the terminals of the energizing coils. To cool the coils, water at an initial temperature of 15°C is circulated around the coils. How many liters of water must circulate by the coils each second if the temperature of the coils is not to exceed 50°C?

114 ••• (*a*) Give support to the assertion that a leaky capacitor (one for which the resistance of the dielectric is finite) can be modeled as a capacitor that has an infinite resistance in parallel with a resistor. (*b*) Show that the time constant for discharging the capacitor is given by $\tau = \kappa \mathcal{E}_0 \rho$. (For simplicity, assume the capacitor is a parallel-plate variety filled completely with a leaky dielectric.) (*c*) Mica has a dielectric constant equal to about 5.0 and a resistivity equal to about $9.0 \times 10^{13}\ \Omega \cdot \text{m}$. Calculate the time it takes for the charge of a mica-filled capacitor to decrease to 10 percent of its initial value.

115 ••• **ENGINEERING APPLICATION** Figure 25-77 shows the basis of the sweep circuit used in an oscilloscope. Switch S is an electronic switch that closes whenever the potential across the switch increases to a value V_c and opens when the potential across the switch decreases to 0.200 V. The emf $\mathcal{E}$, which is much greater than V_c, charges the capacitor C through a resistor R_1. The resistor R_2 represents the small but finite resistance of the electronic switch. In a typical circuit, $\mathcal{E} = 800$ V, $V_c = 4.20$ V, $R_2 = 1.00$ mΩ, $R_1 = 0.500$ MΩ, and $C = 20.0$ nF. (*a*) What is the time constant for charging of the capacitor C? (*b*) Show that as the potential across switch S increases from 0.200 V to 4.20 V, the potential across the capacitor increases almost linearly with time. *Hint: Use the approximation* $e^x \approx 1 + x$, *for* $|x| \ll 1$. (*This approximation of* e^x *can be derived using the differential approximation.*) (*c*) What should the value of R_1 be changed to so that the capacitor charges from 0.200 V to 4.20 V in 0.100 s? (*d*) How much time elapses during the discharge of the capacitor when switch S closes? (*e*) At what average rate is energy delivered to the resistor R_1 during charging and to the switch resistance R_2 during discharge?

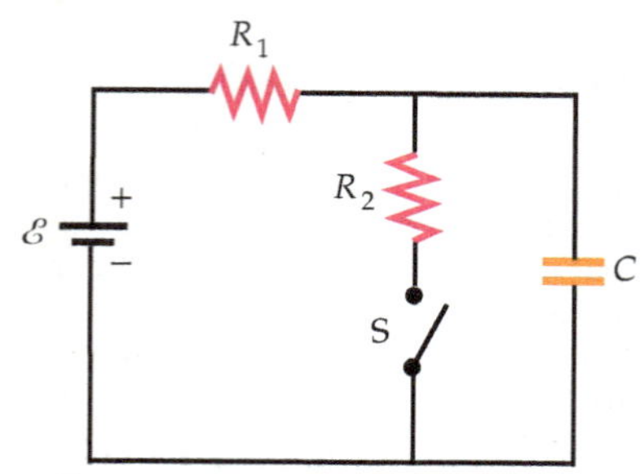

FIGURE 25-77 Problem 115

116 ••• In the circuit shown in Figure 25-78, $R_1 = 2.00$ MΩ, $R_2 = 5.00$ MΩ, and $C = 1.00$ μF. The capacitor is initially without charge on either plate. At $t = 0$, switch S is closed, and at $t = 2.00$ s switch S is opened. (*a*) Sketch a graph of the voltage across C and the current in R_2 between $t = 0$ and $t = 10.0$ s. (*b*) Find the voltage across the capacitor at $t = 2.00$ s and at $t = 8.00$ s.

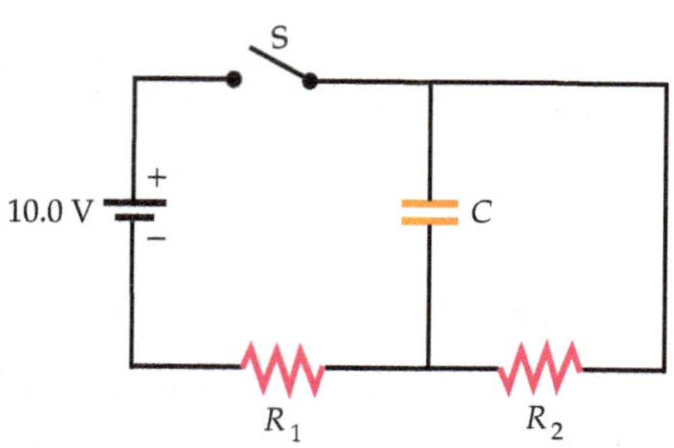

FIGURE 25-78 Problem 116

117 ••• Two batteries that have emfs $\mathcal{E}_1$ and $\mathcal{E}_2$ and internal resistances r_1 and r_2 are connected in parallel. Prove that if a resistor of resistance R is connected in parallel with combination, the optimal load resistance (the value of R at which maximum power is delivered) is given by $R = r_1r_2/(r_1 + r_2)$.

118 ••• Capacitors C_1 and C_2 are connected to a resistor of resistance R and an ideal battery that has $\mathcal{E}$ as shown in Figure 25-79. The throw of switch S is initially at contact *a* and both capacitors are without charge. The throw is then rotated to contact *b* and left there for a long time. Finally, at time $t = 0$, the throw is returned to contact *a*. (*a*) Quantitatively compare the total energy stored in the two capacitors at $t = 0$ and a long time later. (*b*) Find the current

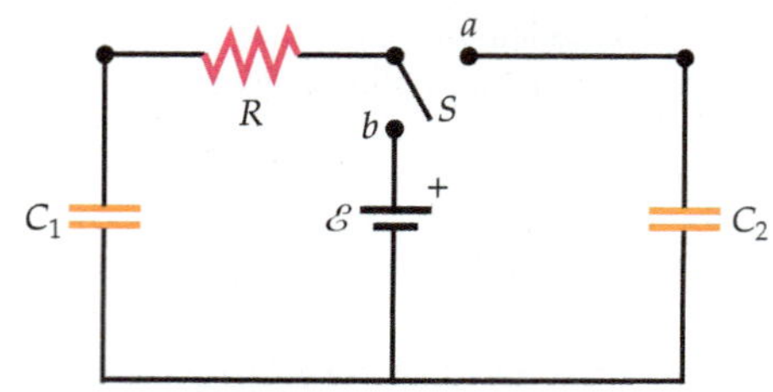

FIGURE 25-79 Problem 118

through R as a function of t for $t > 0$. (*c*) Find the energy delivered to the resistor as a function of t for $t > 0$. (*d*) Find the total energy dissipated in the resistor after $t = 0$ and compare it with the loss of stored energy found in Part (*a*).

119 ••• (*a*) Calculate the equivalent resistance (in terms of R, the resistance of each individual resistor) between points *a* and *b* for the infinite ladder of resistors shown in Figure 25-80 assuming the resistors are identical. That is, assuming $R = R_1 = R_2$. (*b*) Repeat Part (*a*) but do not assume that $R_1 = R_2$ and express your answer in terms of R_1 and R_2. (*c*) Check your results by showing that your result from Part (*b*) agrees with your result from Part (*a*) if you substitute R for both R_1 and R_2.

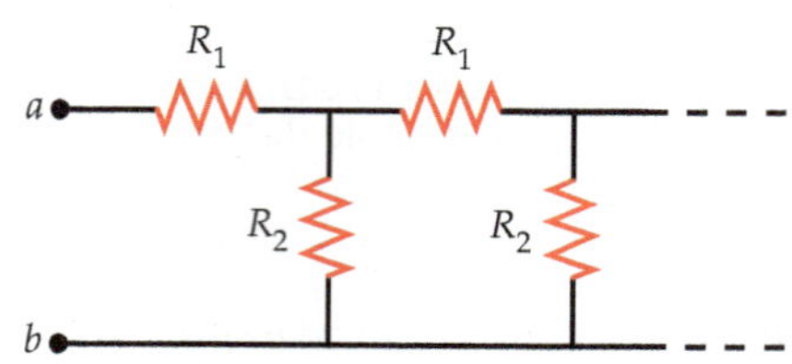

FIGURE 25-80 Problem 119

120 ••• A graph of current as a function of voltage for an Esaki diode is shown in Figure 25-81. (*a*) Make a graph of the differential resistance of the diode as a function of voltage. The differential resistance R_d of a circuit element is defined as $R_d = dV/dI$, where V is the voltage drop across the element and I is the current in the element. (*b*) At what value of the voltage drop does the differential resistance become negative? (*c*) What is the maximum differential resistance for this diode in the range shown and at what voltage does it occur? (*d*) Are there any places in the voltage range shown where the diode exhibits a differential resistance equal to zero? If so, under what value(s) of the voltage does this (do these) occur?

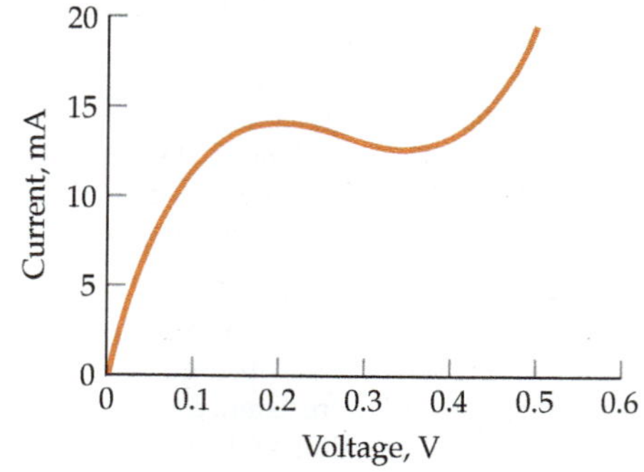

FIGURE 25-81 Problem 120

CHAPTER 26

THE AURORA BOREALIS APPEARS WHEN THE SOLAR WIND, CHARGED PARTICLES PRODUCED BY NUCLEAR FUSION REACTIONS IN THE SUN, BECOMES TRAPPED BY EARTH'S MAGNETIC FIELD. *(Atlas Photo Bank/Photo Researchers, Inc.)*

The Magnetic Field

? How does Earth's magnetic field act on subatomic particles? (See Example 26-1.)

More than 2000 years ago, the Greeks were aware that a certain type of stone (now called magnetite) attracts pieces of iron, and written references exist which describe the use of magnets for navigation dating from the twelfth century.

In 1269, Pierre de Maricourt discovered that a needle laid at various positions on a spherical natural magnet orients itself along lines that pass through points at opposite ends of the sphere. He called these points the *poles of the magnet*. Subsequently, many experimenters noted that every magnet of any shape has two poles, called the north and the south pole, where the force exerted by the magnet is strongest. It was also noted that the *like poles* of two magnets repel each other and the *unlike poles* of two magnets attract each other.

In 1600, William Gilbert discovered that Earth is a natural magnet and has magnetic poles near the north and south geographic poles. Because the north pole of a compass needle points toward the south pole of a given magnet, what we call the north pole of Earth is actually a south magnetic pole, as illustrated in Figure 26-1. Thus, the north and south poles of a magnet are sometimes referred to as the north-seeking and south-seeking poles, respectively.

Although electric charges and magnetic poles are similar in many respects, there is an important difference: Magnetic poles always occur in pairs. When a magnet is broken in half, equal and opposite poles appear at either side of the break point. The result is two magnets, each with a north and a south pole.

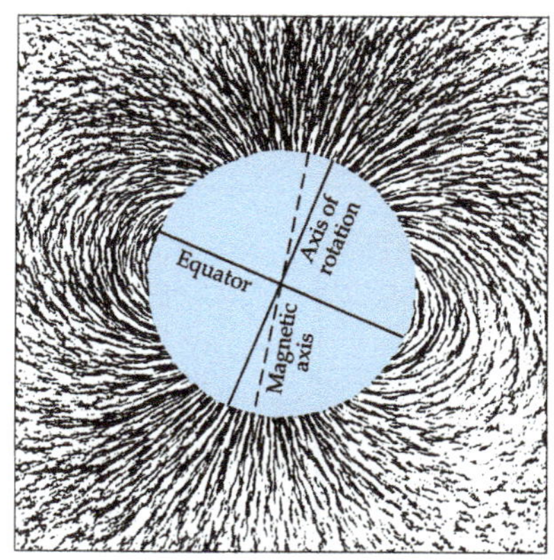

FIGURE 26-1 Magnetic field lines of Earth depicted by iron filings around a uniformly magnetized sphere. The field lines exit from the north magnetic pole, which is near the south geographic pole, and enter the south magnetic pole, which is near the north geographic pole.

There has long been speculation about the existence of an isolated magnetic pole, and in recent years considerable experimental effort has been made to find such an object. Thus far, there is no conclusive evidence that an isolated magnetic pole exists.

In this chapter, we consider the effects of a given magnetic field on moving charges and on wires carrying currents. The sources of magnetic fields are discussed in the next chapter.

26-1 THE FORCE EXERTED BY A MAGNETIC FIELD

The existence of a magnetic field $\vec{B}$ at some point in space can be demonstrated using a compass needle. If there is a magnetic field, the needle will align itself in the direction of the field.*

It has been experimentally observed that, when a particle that has charge q and velocity $\vec{v}$ is in a region with a magnetic field $\vec{B}$, a force acts on the particle that is proportional to q, to v, to B, and to the sine of the angle between the directions of $\vec{v}$ and $\vec{B}$. Surprisingly, the force is perpendicular to both the velocity and the magnetic field. These experimental results can be summarized as follows: When a particle that has a charge q and a velocity $\vec{v}$ is in a region with a magnetic field $\vec{B}$, the magnetic force $\vec{F}$ on the particle is

$$\vec{F} = q\vec{v} \times \vec{B} \tag{26-1}$$

MAGNETIC FORCE ON A MOVING CHARGED PARTICLE

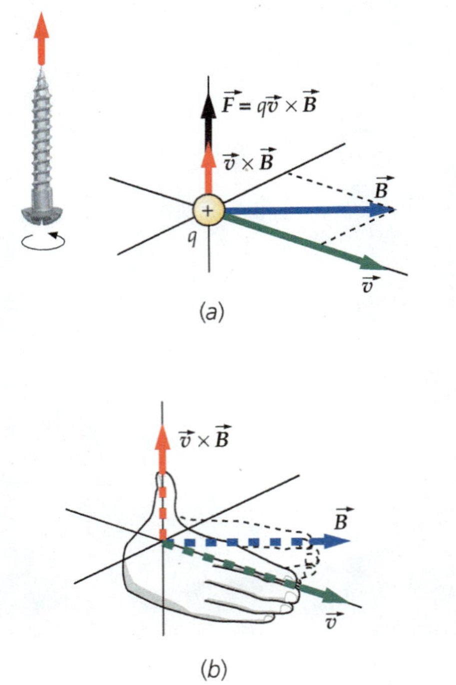

FIGURE 26-2 Right-hand rule for determining the direction of a force exerted on a charged particle moving in a magnetic field. If q is positive, then $\vec{F}$, is in the same direction as $\vec{v} \times \vec{B}$. (*a*) The vector product $\vec{v} \times \vec{B}$ is perpendicular to both $\vec{v}$ and $\vec{B}$ and is in the direction of the advance of a right-hand-threaded screw if turned in the same direction as to rotate $\vec{v}$ into $\vec{B}$. (*b*) If the fingers of the right hand are in the direction of $\vec{v}$ so that they can be curled toward $\vec{B}$, the thumb points in the direction of $\vec{v} \times \vec{B}$.

Because $\vec{F}$ is perpendicular to both $\vec{v}$ and $\vec{B}$, $\vec{F}$ is perpendicular to the plane defined by these two vectors. The direction of $\vec{v} \times \vec{B}$ is given by the right-hand rule as $\vec{v}$ is rotated into $\vec{B}$, as illustrated in Figure 26-2. If q is positive, then $\vec{F}$ is in the same direction as $\vec{v} \times \vec{B}$.

Examples of the direction of the forces exerted on moving charged particles when the magnetic field vector $\vec{B}$ is in the vertical direction are shown in Figure 26-3.

The direction of any particular magnetic field $\vec{B}$ can be found experimentally by measuring $\vec{F}$ and $\vec{v}$ for several velocities in different directions and then applying Equation 26-1.

Equation 26-1 defines the **magnetic field** $\vec{B}$ in terms of the force exerted on a moving charged particle. The SI unit of magnetic field is the **tesla** (T). A particle that has a charge of one coulomb and is moving with a velocity of one meter per second perpendicular to a magnetic field of one tesla experiences a force of one newton:

$$1\ \text{T} = 1\frac{\text{N}}{\text{C}\cdot\text{m/s}} = 1\ \text{N/(A}\cdot\text{m)} \tag{26-2}$$

FIGURE 26-3 Part (*a*) and (*b*) show the direction of the magnetic force on a positively charged particle moving with velocity $\vec{v}$ in a magnetic field $\vec{B}$. In Concept Check 26-2 (see page 889), you are asked to find the sign of the charge on the particle shown in part (*c*) of this figure.

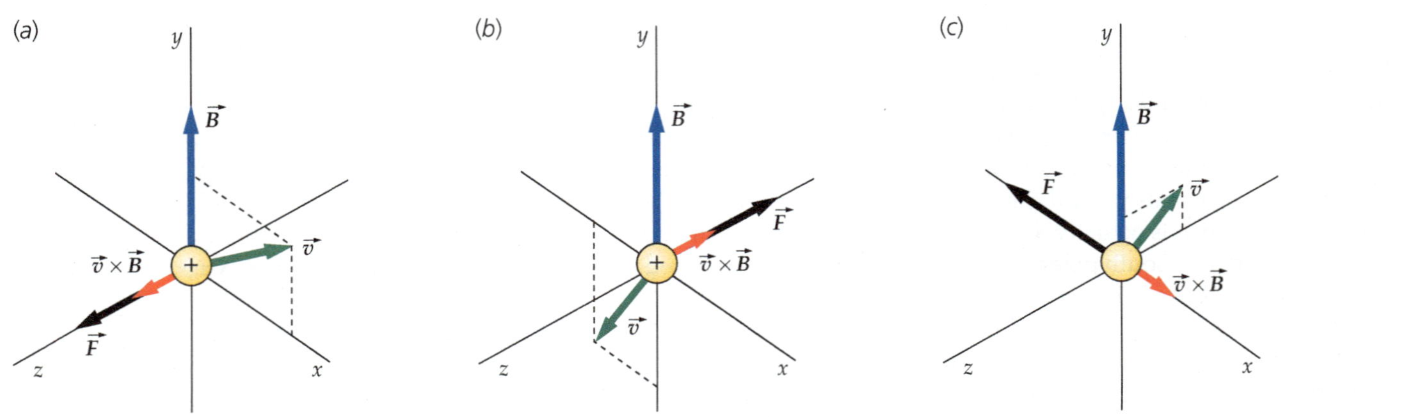

* Compass needles are suspended so they remain horizontal. This results in the compass needle aligning itself in the horizontal component of the magnetic field. A compass needle with an unrestricted suspension would align itself in the magnetic field.

CONCEPT CHECK 26-1

The direction of any magnetic field $\vec{B}$ is specified as the direction that the north pole of a compass needle points toward when the needle is aligned in the field. Suppose that the direction of the magnetic field $\vec{B}$ were instead specified as the direction pointed toward by the south pole of a compass needle aligned in the field. Would the right-hand rule shown in Figure 26-2 then give the direction of the magnetic force on the moving positive charge, or would a left-hand rule be required? Explain your answer.

CONCEPT CHECK 26-2

The particle in Figure 26-3(*c*) (*a*) is positively charged, (*b*) is negatively charged, (*c*) could be either positively or negatively charged. Explain your answer.

Like the farad, the tesla is a large unit. The magnetic field strength of Earth has a magnitude somewhat less than 10^{-4} T on Earth's surface. The magnetic field strengths near powerful permanent magnets are about 0.1 T to 0.5 T, and powerful laboratory and industrial electromagnets produce fields of 1 T to 2 T. Fields greater than 10 T are extremely difficult to produce because the resulting magnetic forces will either tear the magnets apart or crush the magnets. A commonly used unit, derived from the cgs system, is the **gauss** (G), which is related to the tesla as follows:

$$1\ \text{G} = 10^{-4}\ \text{T} \qquad 26\text{-}3$$

DEFINITION—GAUSS

Because magnetic fields are often given in gauss, which is not an SI unit, you need to remember to convert from gauss to teslas when making calculations.

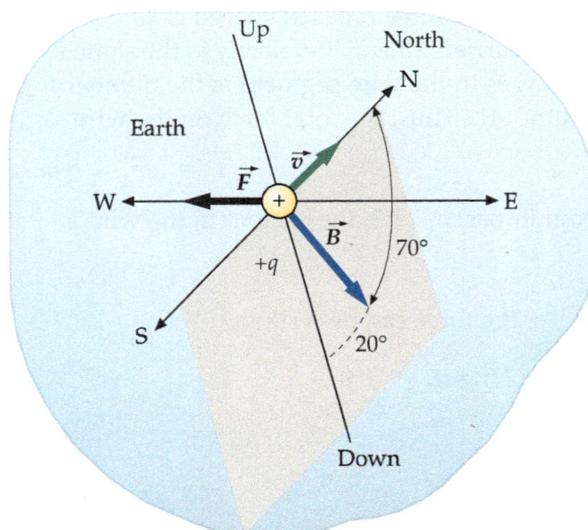

FIGURE 26-4

Example 26-1 Force on a Proton Going North

The magnetic field strength of Earth is measured at a point on the surface to have a magnitude of about 0.6 G and is tilted downward in the northern hemisphere, making an angle of about 70° with the horizontal, as shown in Figure 26-4. (Earth's magnetic field varies from place to place. These data are approximately correct for the central United States.) A proton ($q = +e$) is moving horizontally in the northward direction with speed $v = 1.0 \times 10^7$ m/s. Calculate the magnetic force on the proton (*a*) using $F = qvB \sin\theta$ and (*b*) by first expressing $\vec{v}$ and $\vec{B}$ in terms of the unit vectors $\hat{i}$, $\hat{j}$, and $\hat{k}$, and then computing $\vec{F} = q\vec{v} \times \vec{B}$.

PICTURE Let the x and y directions be to the east and to the north, respectively, and let the z direction be vertically upward (Figure 26-5). The velocity vector is then in the $+y$ direction.

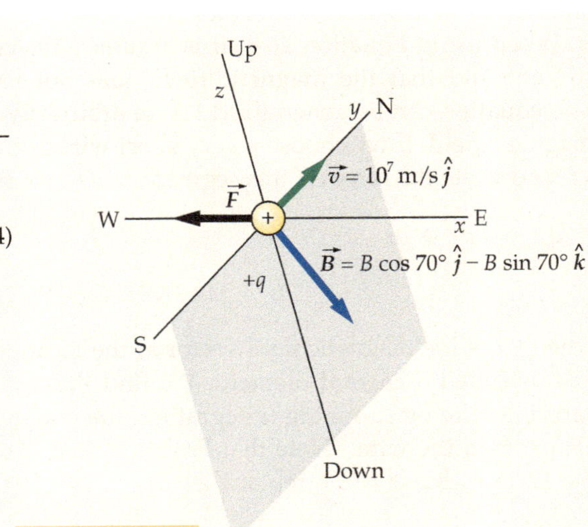

FIGURE 26-5

SOLVE

(*a*) Calculate $F = qvB \sin\theta$ using $\theta = 70°$. From Figure 26-4, we see that the direction of the force is westward.

$$F = qvB \sin 70°$$
$$= (1.6 \times 10^{-19}\ \text{C})(10 \times 10^{6}\ \text{m/s})(0.6 \times 10^{-4}\ \text{T})(0.94)$$
$$= \boxed{9.0 \times 10^{-17}\ \text{N}}$$

(*b*) 1. The magnetic force is the vector product of $q\vec{v}$ and $\vec{B}$:

$$\vec{F} = q\vec{v} \times \vec{B}$$

2. Express $\vec{v}$ and $\vec{B}$ in terms of their components:

$$\vec{v} = v_y\hat{j}$$
$$\vec{B} = B_y\hat{j} + B_z\hat{k}$$

3. Write $\vec{F} = q\vec{v} \times \vec{B}$ in terms of these components:

$$\vec{F} = q\vec{v} \times \vec{B} = q(v_y\hat{j}) \times (B_y\hat{j} + B_z\hat{k})$$
$$= qv_yB_y(\hat{j} \times \hat{j}) + qv_yB_z(\hat{j} \times \hat{k}) = qv_yB_z\hat{i}$$

4. Evaluate $\vec{F}$:

$$\vec{F} = qv(-B\sin\theta)\hat{i}$$
$$= -(1.6 \times 10^{-19}\text{C})(10^7\,\text{m/s})(0.6 \times 10^{-4}\,\text{T})\sin 70^\circ\,\hat{i}$$
$$= \boxed{-9.0 \times 10^{-17}\,\text{N}\hat{i}}$$

CHECK The Part-(*a*) result is equal to the magnitude of the Part-(*b*) result.

TAKING IT FURTHER Note that the direction of $\hat{i}$ is to the east, so the force is directed to the west as shown in Figure 26-5.

PRACTICE PROBLEM 26-1 Find the force on a proton moving with velocity $\vec{v} = 4 \times 10^6\,\text{m/s}\,\hat{i}$ in a magnetic field $\vec{B} = 2.0\,\text{T}\hat{k}$.

When a current-carrying wire is in a region that has a magnetic field, there is a force on the wire that is equal to the sum of the magnetic forces on the individual charge carriers in the wire. Figure 26-6 shows a short segment of wire that has cross-sectional area A, length L, and current I. If the wire is in a magnetic field $\vec{B}$, the magnetic force on each charge is $q\vec{v}_d \times \vec{B}$, where $\vec{v}_d$ is the drift velocity of the charge carriers (the drift velocity is the same as the average velocity). The number of charges in the wire segment is the number n per unit volume multiplied by the volume AL. Thus, the total force on the wire segment is

$$\vec{F} = (q\vec{v}_d \times \vec{B})nAL$$

From Equation 25-3, the current in the wire is

$$I = nqv_dA$$

Hence, the force can be written

$$\vec{F} = I\vec{L} \times \vec{B} \qquad 26\text{-}4$$

MAGNETIC FORCE ON A STRAIGHT SEGMENT OF CURRENT-CARRYING WIRE

where $\vec{L}$ is a vector whose magnitude is the length of the segment and whose direction is the same as that of the current.* For the current in the $+x$ direction (Figure 26-7) and the magnetic field vector at the segment in the xy plane, the force on the wire is in the $+z$ direction.

When using Equation 26-4, it is assumed that the wire segment is straight and that the magnetic field does not vary over its length. The equation can be generalized for an arbitrarily shaped wire in any magnetic field. If we choose a very short wire segment that has length $d\vec{\ell}$ and write the force on this segment as $d\vec{F}$, we have

$$d\vec{F} = I\,d\vec{\ell} \times \vec{B} \qquad 26\text{-}5$$

MAGNETIC FORCE ON A CURRENT ELEMENT

where $\vec{B}$ is the magnetic field vector at the location of the segment. The quantity $I\,d\vec{\ell}$ is called a **current element.** We find the total magnetic force on a current-carrying wire by summing (integrating) the magnetic forces due to all the current elements in the wire. (Note that Equation 26-5 is the same as Equation 26-1 with the current element $I\,d\vec{\ell}$ replacing $q\vec{v}$.)

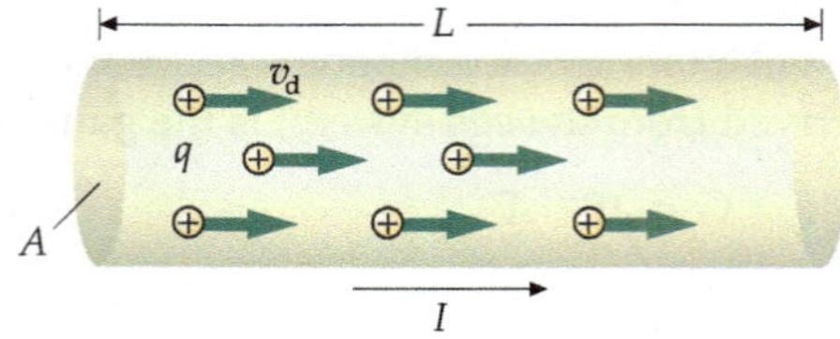

FIGURE 26-6 Wire segment that has a length L and carries a current I. If the wire is in a magnetic field $\vec{B}$, there will be a force on each charge carrier resulting in a force on the wire.

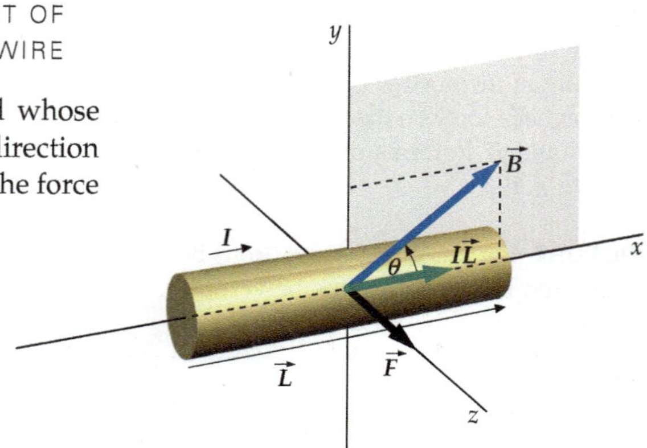

FIGURE 26-7 Magnetic force on a current-carrying segment of wire in a magnetic field. The current is in the x direction, and the magnetic field is in the xy plane and makes an angle θ with the $+x$ direction. The force $\vec{F}$ is in the $+z$ direction, perpendicular to both $\vec{B}$ and $\vec{L}$, and has magnitude $ILB\sin\theta$.

* By the direction of the current we mean the direction of the current-density vector $\vec{J}$.

Just as the electric field $\vec{E}$ can be represented by electric field lines, the magnetic field $\vec{B}$ can be represented by **magnetic field lines.** In both cases, the direction of the field is indicated by the direction of the field lines and the magnitude of the field is indicated by the density (number per unit area) of the lines on surface perpendicular to the lines. There are, however, two important differences between electric field lines and magnetic field lines:

1. Electric field lines are in the direction of the electric force on a positive charge, but the magnetic field lines are perpendicular to the magnetic force on a moving charge.
2. Electric field lines begin on positive charges and end on negative charges; magnetic field lines neither begin nor end.

Figure 26-8 shows the magnetic field lines both inside and outside a bar magnet.

! Do not think the field lines for the magnetic field of a magnet begin on magnetic south poles and end on magnetic north poles. In reality, they neither begin nor end. Instead they enter the magnet at one end and exit the magnet at the other end.

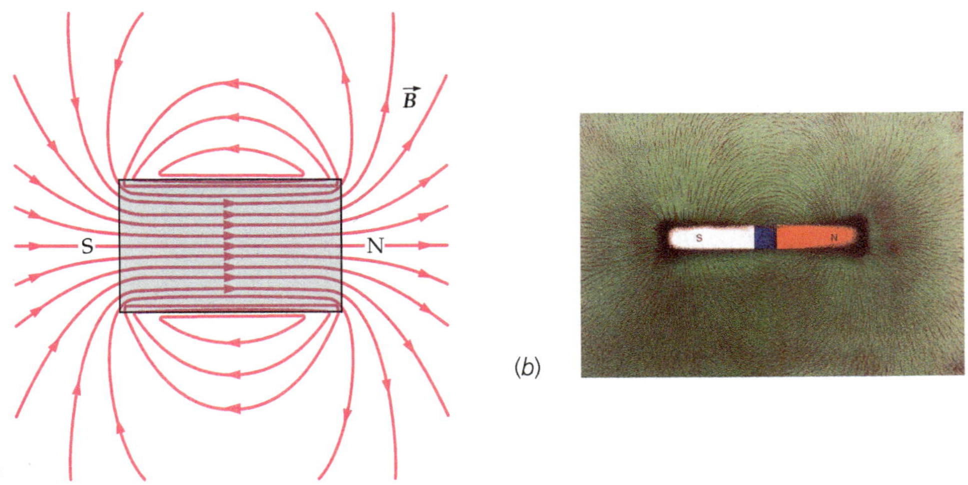

FIGURE 26-8 (*a*) Magnetic field lines inside and outside a bar magnet. The lines emerge from the north pole and enter the south pole, but they have no beginning or end. Instead, they form closed loops. (*b*) Magnetic field lines outside a bar magnet as indicated by iron filings. *(© 1995 Tom Pantages.)*

Example 26-2 Force on a Straight Wire

A 3.0-mm-long segment of wire carries a current of 3.0 A in the $+x$ direction. It lies in a magnetic field of magnitude 0.020 T that is in the xy plane and makes an angle of 30° with the $+x$ direction, as shown in Figure 26-9. What is the magnetic force exerted on the wire segment?

FIGURE 26-9

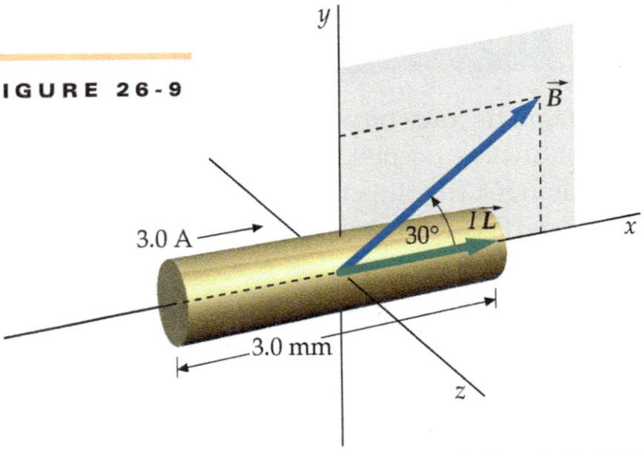

PICTURE The magnetic force is in the direction of $\vec{L} \times \vec{B}$, which we see from Figure 26-9 is in the $+z$ direction.

SOLVE

1. The magnetic force is given by Equation 26-4:

$$\vec{F} = I\vec{L} \times \vec{B} = ILB \sin 30^\circ\, \hat{k}$$
$$= (3.0\text{ A})(0.0030\text{ m})(0.020\text{ T})(\sin 30^\circ)\hat{k}$$
$$= \boxed{9.0 \times 10^{-5}\text{ N}\hat{k}}$$

CHECK The force is perpendicular to the wire, as expected.

Example 26-3 Force on a Bent Wire

A wire bent into a semicircular loop of radius R lies in the xy plane. It carries a current I from point a to point b, as shown in Figure 26-10. Throughout the region there is a uniform magnetic field $\vec{B} = B\hat{k}$ that is perpendicular to the plane of the loop. Find the magnetic force acting on the semicircular loop section of the wire.

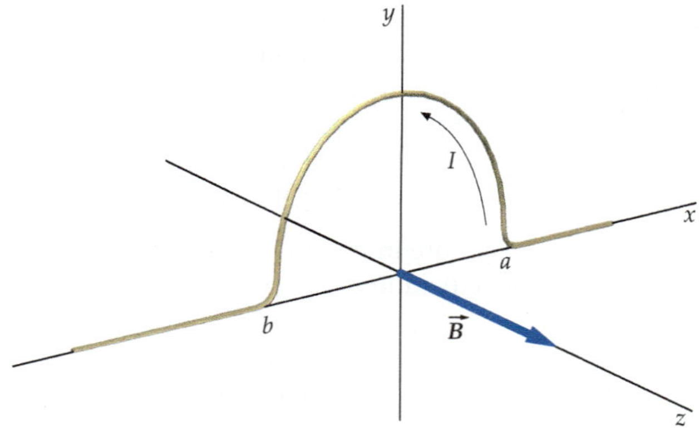

FIGURE 26-10

PICTURE The magnetic force $d\vec{F}$ is exerted on a segment of the semicircular wire that lies in the xy plane, as shown in Figure 26-11. We find the total magnetic force by expressing the x and y components of $d\vec{F}$ in terms of θ and integrating them separately from $\theta = 0$ to $\theta = \pi$.

SOLVE

1. Write the force $d\vec{F}$ on a current element $I\,d\vec{\ell}$.
$$d\vec{F} = I\,d\vec{\ell} \times \vec{B}$$
2. Express $d\vec{\ell}$ in terms of the unit vectors $\hat{i}$ and $\hat{j}$:
$$d\vec{\ell} = -d\ell\,\sin\theta\hat{i} + d\ell\,\cos\theta\hat{j}$$
3. Compute $I\,d\vec{\ell}$ using $d\ell = R\,d\theta$ and $\vec{B} = B\hat{k}$:
$$\begin{aligned} d\vec{F} &= I\,d\vec{\ell} \times \vec{B} \\ &= I(-R\sin\theta\,d\theta\hat{i} + R\cos\theta\,d\theta\hat{j}) \times B\hat{k} \\ &= IRB\sin\theta\,d\theta\hat{j} + IRB\cos\theta\,d\theta\hat{i} \end{aligned}$$
4. Integrate each component of $d\vec{F}$ from $\theta = 0$ to $\theta = \pi$.
$$\begin{aligned} \vec{F} &= \int d\vec{F} = IRB\hat{i}\int_0^{\pi}\cos\theta\,d\theta + IRB\hat{j}\int_0^{\pi}\sin\theta\,d\theta \\ &= IRB\hat{i}(0) + IRB\hat{j}(2) = \boxed{2IRB\hat{j}} \end{aligned}$$

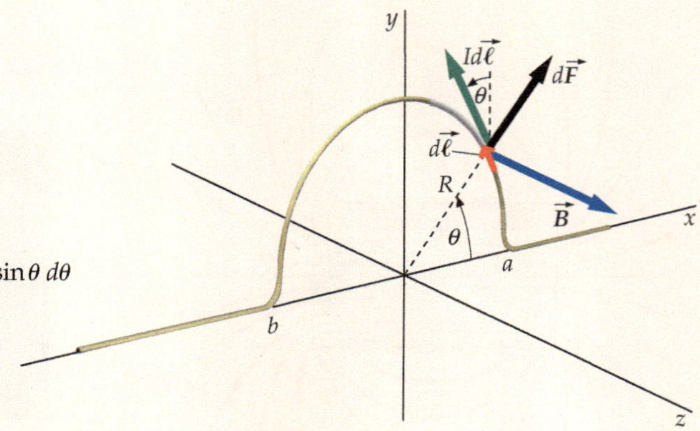

FIGURE 26-11

CHECK The result that the x component of $\vec{F}$ is zero can be seen from symmetry. For the right half of the loop, $d\vec{F}$ tilts to the right; for the left half of the loop, $d\vec{F}$ tilts to the left.

TAKING IT FURTHER The net force on the semicircular wire is the same as if the semicircle were replaced by a straight-line segment of length $2R$ connecting points a and b. (This is a general result that is derived in Problem 26.)

26-2 MOTION OF A POINT CHARGE IN A MAGNETIC FIELD

The magnetic force on a charged particle moving through a region with a magnetic field is always perpendicular to the velocity of the particle. The magnetic force thus changes the direction of the velocity but not the magnitude of the velocity (the speed). Therefore, *magnetic forces do no work on particles and do not change their kinetic energy.*

In the special case where the velocity of a charged particle is perpendicular to a uniform magnetic field, as shown in Figure 26-12, the particle moves in a circular orbit. The magnetic force provides the force in the centripetal direction that is necessary for circular motion. We can use Newton's second law to relate the radius of the circle to the magnetic field and the speed of the particle. If the velocity is $\vec{v}$, the magnetic force on a particle that has charge q is given by $\vec{F} = q\vec{v} \times \vec{B}$. The magnitude of the net force is equal to qvB, because $\vec{v}$ and $\vec{B}$ are perpendicular.

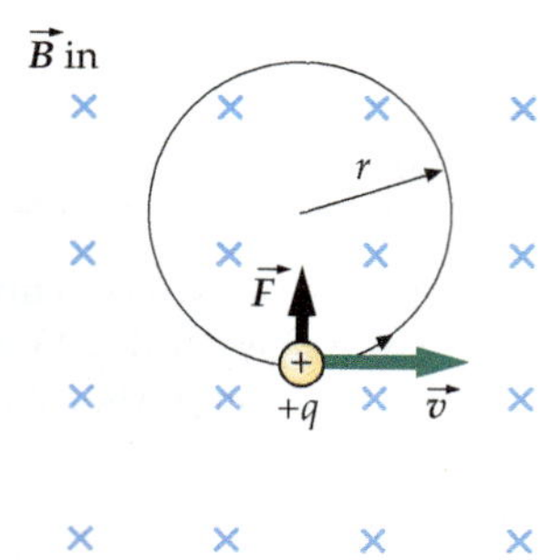

FIGURE 26-12 Charged particle moving in a plane perpendicular to a uniform magnetic field. The magnetic field is into the page as indicated by the crosses. (Each cross represents the tail feathers of an arrow. A field out of the plane of the page would be indicated by dots, each dot representing the point of an arrow.) The magnetic force is perpendicular to the velocity of the particle, causing it to move in a circular orbit.

Newton's second law gives

$$F = ma$$

$$qvB = m\frac{v^2}{r}$$

or

$$r = \frac{mv}{qB} \qquad 26\text{-}6$$

where m is the mass of the particle.

The period of the circular motion is the time it takes the particle to travel once around the circumference of the circle. The period is related to the speed by

$$T = \frac{2\pi r}{v}$$

Substituting $mv/(qB)$ for r (Equation 26-6), we obtain the period of the particle's circular orbit, which is called the **cyclotron period:**

$$T = \frac{2\pi(mv/qB)}{v} = \frac{2\pi m}{qB} \qquad 26\text{-}7$$

CYCLOTRON PERIOD

The frequency of the circular motion, called the **cyclotron frequency**, is the reciprocal of the period:

$$f = \frac{1}{T} = \frac{qB}{2\pi m} \quad \text{so} \quad \omega = 2\pi f = \frac{q}{m}B \qquad 26\text{-}8$$

CYCLOTRON FREQUENCY

Note that the period and the frequency given by Equations 26-7 and 26-8 depend on the charge-to-mass ratio q/m, but the period and the frequency are independent of the velocity v or the radius r. Two important applications of the circular motion of charged particles in a uniform magnetic field, the mass spectrometer and the cyclotron, are discussed later in this section.

(a) (b)

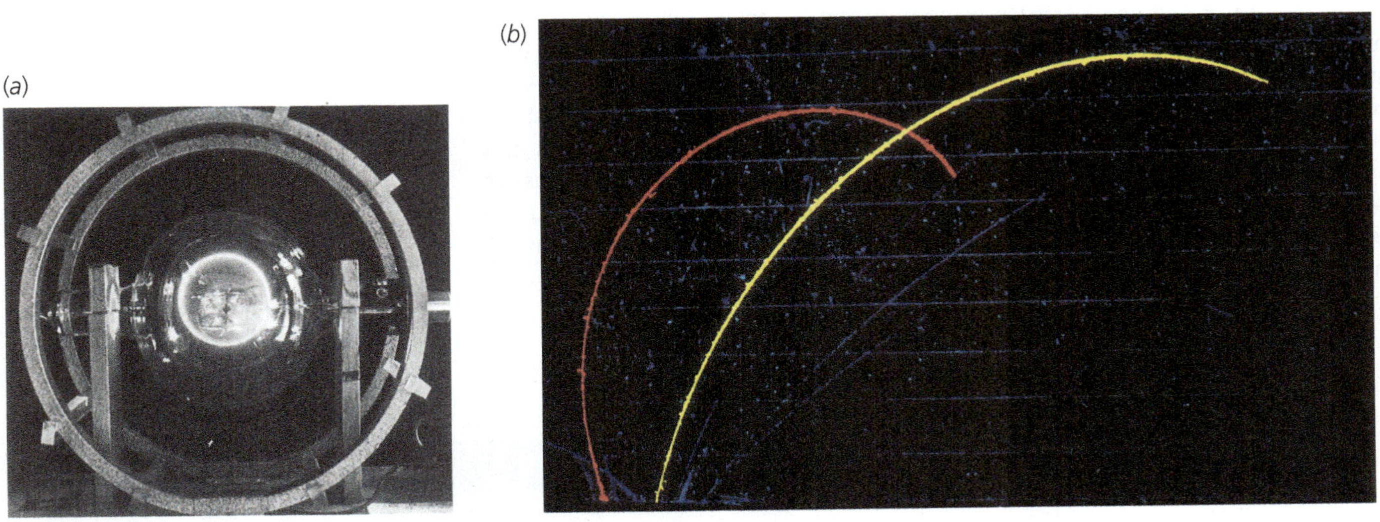

(*a*) Circular path of electrons moving in the magnetic field produced by the current in two large coils. The electrons ionize the dilute gas in the tube, causing it to give off a glow that indicates the path of the beam. (*b*) False-color photograph showing tracks of a 1.6-MeV proton (red) and a 7-MeVα particle (yellow) in a cloud chamber. The radius of curvature is proportional to the momentum and inversely proportional to the charge of the particle. For these energies, the momentum of the α particle, which has twice the charge of the proton, is about four times that of the proton and so its radius of curvature is greater. ((*a*) *Larry Langrill.* (*b*) *© Lawrence Berkeley Laboratory/Science Photo Library.*)

Example 26-4 Cyclotron Period

A proton has a mass equal to 1.67×10^{-27} kg, has a charge equal to 1.60×10^{-19} C, and moves in a circle of radius $r = 21.0$ cm perpendicular to a magnetic field equal to 4000 G. Find (*a*) the speed of the proton and (*b*) the period of the motion.

PICTURE Apply Newton's second law to find the speed, and use distance equals speed multiplied by time to find the period.

SOLVE

(*a*) 1. Apply Newton's second law ($F = ma$):

$$F = ma \quad \Rightarrow \quad qvB = m\frac{v^2}{r}$$

2. Solve for the speed:

$$v = \frac{rqB}{m} = \frac{(0.210\text{ m})(1.60 \times 10^{-19}\text{ C})(0.400\text{ T})}{1.67 \times 10^{-27}\text{ kg}}$$

$$= \boxed{8.05 \times 10^6\text{ m/s} = 0.0268c}$$

(*b*) Use distance equals speed multiplied by time and solve for the period:

$$2\pi r = vT$$

so

$$T = \frac{2\pi r}{v} = \frac{2\pi(0.210\text{ m})}{(8.05 \times 10^6\text{ m/s})} = 1.64 \times 10^{-7}\text{ s} = \boxed{164\text{ ns}}$$

TAKING IT FURTHER The radius of the circular orbit is proportional to the speed, but the period of the orbit is independent of both the speed and radius.

Suppose that a charged particle is in a region that has a uniform magnetic field and is moving with a velocity that is not perpendicular to $\vec{B}$. There is no magnetic force component, and thus no acceleration component, parallel to $\vec{B}$, so the component of the velocity that is parallel to $\vec{B}$ remains constant. The magnetic force on the particle is perpendicular to $\vec{B}$, so the change in motion of the particle due to this force is the same as that just discussed. The path of the particle is thus a helix, as shown in Figure 26-13.

The motion of charged particles in nonuniform magnetic fields can be quite complex. Figure 26-14 shows a *magnetic bottle,* an interesting magnetic field configuration in which the field is weak at the center and strong at both ends. A detailed analysis of the motion of a charged particle in such a field shows that the particle spirals around the field lines and becomes trapped, oscillating back and forth between points P_1 and P_2 in the figure. Such magnetic field configurations are used to confine dense beams of charged particles, called plasmas, in nuclear fusion research. A similar phenomenon is the oscillation of ions back and forth between Earth's magnetic poles in the Van Allen belts (Figure 26-15).

(a)

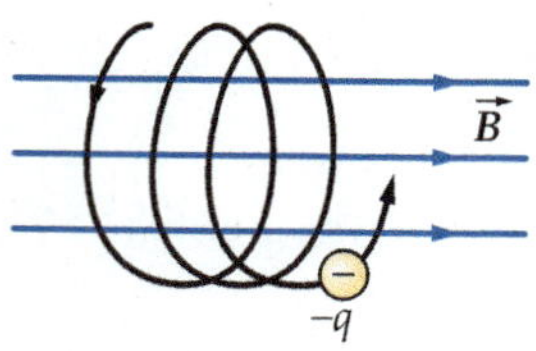

(b)

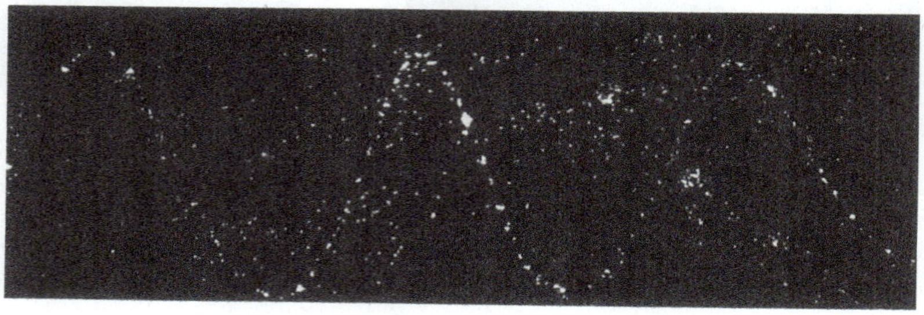

FIGURE 26-13 (*a*) When a particle has a velocity component parallel to a magnetic field as well as a velocity component perpendicular to the magnetic field the particle moves in a helical path around the field lines. (*b*) Cloud-chamber photograph of the helical path of an electron moving in a magnetic field. The path of the electron is made visible by the condensation of water droplets in the cloud chamber. *(Carl E. Nielson.)*

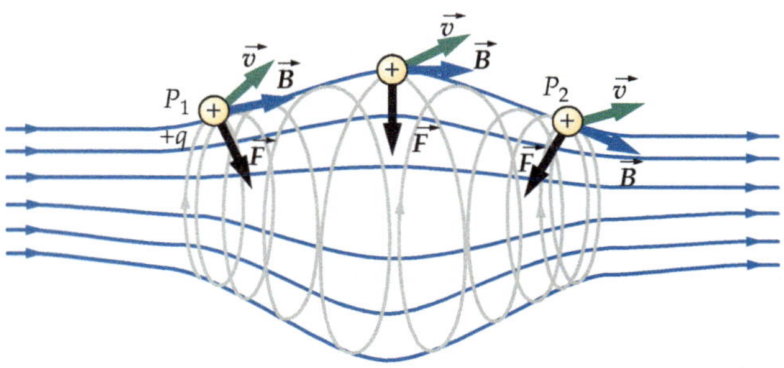

FIGURE 26-14 Magnetic bottle. When a charged particle moves in such a field, which is strong at both ends and weak in the middle, the particle becomes trapped and moves back and forth, spiraling around the field lines.

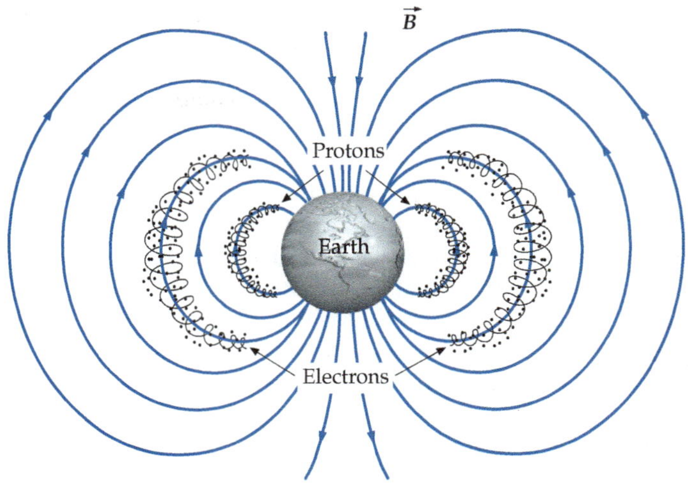

FIGURE 26-15 Van Allen belts. Protons (inner belts) and electrons (outer belts) are trapped in Earth's magnetic field and spiral around the field lines between the north and south poles.

*THE VELOCITY SELECTOR

The magnetic force on a charged particle moving in a uniform magnetic field can be balanced by an electric force if the magnitudes and directions of the magnetic field and the electric field are properly chosen. Because the electric force is in the direction of the electric field (for particles with positive charge) and the magnetic force is perpendicular to the magnetic field, the electric and magnetic fields in the region through which the particle is moving must be perpendicular to each other if the forces are to balance. Such a region is said to have **crossed fields.**

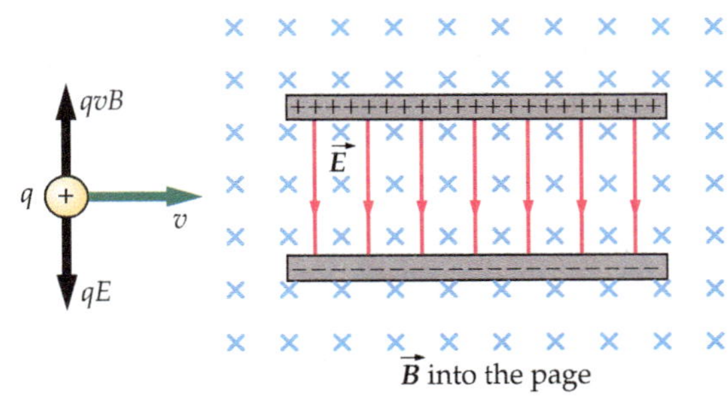

FIGURE 26-16 Crossed electric and magnetic fields. When a particle that has a positive charge moves to the right, the particle experiences a downward electric force and an upward magnetic force. These forces balance if the speed of the particle is related to the field strengths by $vB = E$.

Figure 26-16 shows a region of space between the plates of a capacitor where there is an electric field and a perpendicular magnetic field (produced by a magnet that has one pole on each side of this sheet of paper). Consider a particle that has charge q entering this space from the left. The net force on the particle is

$$\vec{F} = q\vec{E} + q\vec{v} \times \vec{B}$$

If q is positive, the electric force of magnitude qE is down the page and the magnetic force of magnitude qvB is up the page. If the charge is negative, the direction of each of these forces is reversed. The two forces balance if $qE = qvB$ that is, if

$$v = \frac{E}{B} \qquad 26\text{-}9$$

For given magnitudes of the electric and magnetic fields, the forces balance only for particles that have the exact speed given by Equation 26-9. Any particle that has this speed, regardless of its mass or charge, will traverse the space undeflected. A particle that has a greater speed will be deflected toward the direction of the magnetic force, and a particle that has a lesser speed will be deflected in the direction of the electric force. This arrangement of fields is often used as a **velocity selector,** which is a device that allows only particles with the speed specified by Equation 26-9 to pass.

PRACTICE PROBLEM 26-2

A proton is moving in the $+x$ direction in a region of crossed fields where $\vec{E} = 2.00 \times 10^5\ \text{N/C}\ \hat{k}$ and $\vec{B} = 0.300\ \text{T}\ \hat{j}$. (*a*) What is the speed of the proton if it is not deflected? (*b*) If the proton moves with twice this speed, in which direction will it be deflected?

*THOMSON'S MEASUREMENT OF q/m FOR ELECTRONS

An example of the use of crossed electric and magnetic fields is the famous experiment performed by J. J. Thomson in 1897 where he showed that the rays of a cathode-ray tube can be deflected by electric and magnetic fields, indicating that they must consist of charged particles. By measuring the deflections of these particles, Thomson showed that all the particles have the same charge-to-mass ratio q/m. He also showed that particles that have this charge-to-mass ratio can be obtained using any material for a source, which means that these particles, now called electrons, are a fundamental constituent of all matter.

Figure 26-17 shows a schematic diagram of the cathode-ray tube Thomson used. Electrons are emitted from the cathode C, which is at a negative potential relative to the potential at slits A and B. An electric field in the direction from A toward C accelerates the electrons, and some of the electrons pass through slits A and B into a field-free region. The electrons then enter the electric field between the capacitor plates D and F that is perpendicular to the velocity of the electrons. This field accelerates the electrons vertically for the short time that they are between the plates. The electrons are deflected and strike the phosphorescent screen S at the far right side of the tube at some deflection Δy from the point at which they strike when there is no electric field between the plates. The screen glows where the electrons strike the screen, indicating the location of the beam. The speed of the electrons v_0 is determined by introducing a magnetic field $\vec{B}$ between the plates in a direction that is perpendicular to both the electric field and the initial velocity of the electrons. The magnitude of $\vec{B}$ is adjusted until the beam is not deflected. The speed is then found from Equation 26-9.

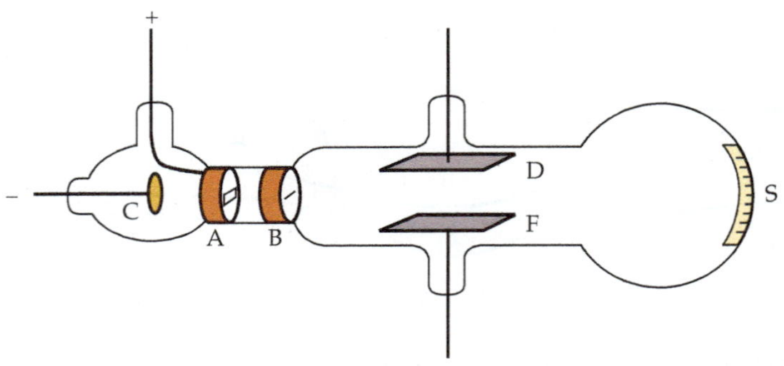

FIGURE 26-17 Thomson's tube for measuring q/m for the particles of cathode rays (electrons). Electrons from the cathode C pass through the slits at A and B and strike a phosphorescent screen S. The beam can be deflected by an electric field between plates D and F or by a magnetic field (not shown).

With the magnetic field turned off, the beam is deflected by an amount Δy, which consists of two parts: the deflection Δy_1, which occurs while the electrons are between the plates, and the deflection Δy_2, which occurs after the electrons leave the region between the plates (Figure 26-18).

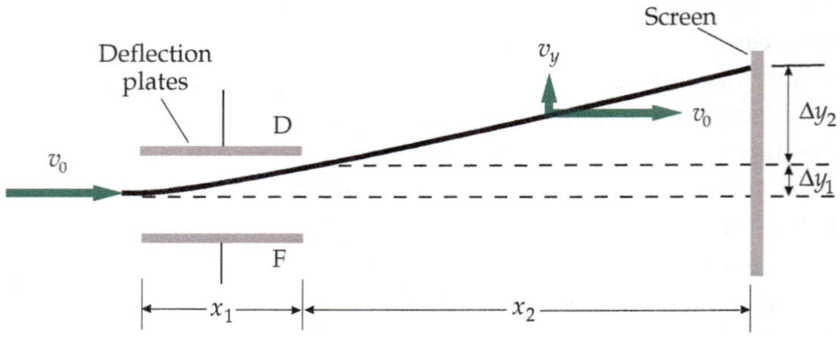

FIGURE 26-18 The total deflection of the beam in the J. J. Thomson experiments consists of the deflection Δy_1 while the electrons are between the plates plus the deflection Δy_2 that occurs in the field-free region between the plates and the screen.

Let x_1 be the horizontal distance across the deflection plates D and F. If the electron is moving horizontally with speed v_0 when it enters the region between the plates, the time spent between the plates is $t_1 = x_1/v_0$, and the vertical velocity when it leaves the plates is

$$v_y = a_y t_1 = \frac{qE_y}{m} t_1 = \frac{qE_y}{m}\frac{x_1}{v_0}$$

where E_y is the upward component of the electric field between the plates. The deflection in this region is

$$\Delta y_1 = \frac{1}{2} a_y t_1^2 = \frac{1}{2}\frac{qE_y}{m}\left(\frac{x_1}{v_0}\right)^2$$

The electron then travels an additional horizontal distance x_2 in the field-free region from the deflection plates to the screen. Because the velocity of the electron is constant in this region, the time to reach the screen is $t_2 = x_2/v_0$, and the additional vertical deflection is

$$\Delta y_2 = v_y t_2 = \frac{qE_y}{m}\frac{x_1}{v_0}\frac{x_2}{v_0}$$

The total deflection at the screen is therefore

$$\Delta y = \Delta y_1 + \Delta y_2 = \frac{1}{2}\frac{qE_y}{mv_0^2}x_1^2 + \frac{qE_y}{mv_0^2}x_1 x_2 \qquad \text{26-10}$$

The measured deflection Δy can be used to determine the charge-to-mass ratio, q/m, from Equation 26-10.

Example 26-5 Electron Beam Deflection

Electrons pass undeflected through the plates of Thomson's apparatus when the electric field is 3000 V/m and there is a crossed magnetic field of 0.140 mT. If the plates are 4.00 cm long and the ends of the plates are 30.0 cm from the screen, find the deflection on the screen when the magnetic field is turned off.

PICTURE The mass and charge of the electron are known: $m = 9.11 \times 10^{-31}$ kg and $q = -e = -1.60 \times 10^{-19}$ C. The speed of the electron can be found from the ratio of the magnetic and electric fields.

SOLVE

1. The total deflection of the electron is given by Equation 26-10:

$$\Delta y = \Delta y_1 + \Delta y_2 = \frac{1}{2}\frac{qE_y}{mv_0^2}x_1^2 + \frac{qE_y}{mv_0^2}x_1 x_2$$

2. The speed v_0 equals E/B:

$$v_0 = \frac{E}{B} = \frac{3000\ \text{V/m}}{1.40 \times 10^{-4}\ \text{T}} = 2.14 \times 10^7\ \text{m/s}$$

3. Substitute the value for v_0 determined in step 2, the given value of E, and the known values for m and q into Equation 26-10 to find Δy:

$$\Delta y_1 = \frac{1}{2}\frac{(-1.60 \times 10^{-19}\ \text{C})(-3000\ \text{V/m})}{(9.11 \times 10^{-31}\ \text{kg})(2.14 \times 10^7\ \text{m/s})^2}(0.0400\ \text{m})^2 = 9.20 \times 10^{-4}\ \text{m}$$

$$\Delta y_2 = \frac{(-1.60 \times 10^{-19}\ \text{C})(-3000\ \text{V/m})}{(9.11 \times 10^{-31}\ \text{kg})(2.14 \times 10^7\ \text{m/s})^2}(0.0400\ \text{m})(0.300\ \text{m}) = 1.38 \times 10^{-2}\ \text{m}$$

$$\Delta y = \Delta y_1 + \Delta y_2 = 9.20 \times 10^{-4}\ \text{m} + 1.38 \times 10^{-2}\ \text{m} = 0.92\ \text{mm} + 13.8\ \text{mm} = \boxed{14.7\ \text{mm}}$$

CHECK As expected, Δy_2 is an order of magnitude greater than Δy_1. This was expected because the distance from the plates to the screen is an order of magnitude greater than the length of the plates.

*THE MASS SPECTROMETER

The **mass spectrometer,** first designed by Francis William Aston in 1919, was developed as a means of measuring the masses of isotopes. Such measurements are important in determining both the presence of isotopes and their abundance in nature. On Earth, for example, naturally occurring magnesium has been found to consist of 78.7 percent ^{24}Mg, 10.1 percent ^{25}Mg, and 11.2 percent ^{26}Mg. These isotopes have masses in the approximate ratio 24:25:26.

Figure 26-19 shows a simple schematic drawing of a mass spectrometer. Positive ions are formed by bombarding atoms with X rays or a beam of electrons. (Electrons are knocked out of the atoms by the X rays or bombarding electrons to form positive ions.) The ions are accelerated by an electric field and enter a uniform magnetic field. If the positive ions start from rest and move through a potential difference ΔV, the ions' kinetic energy when they enter the magnetic field equals their loss in potential energy, $q|\Delta V|$:

$$\tfrac{1}{2}mv^2 = q|\Delta V| \qquad \text{26-11}$$

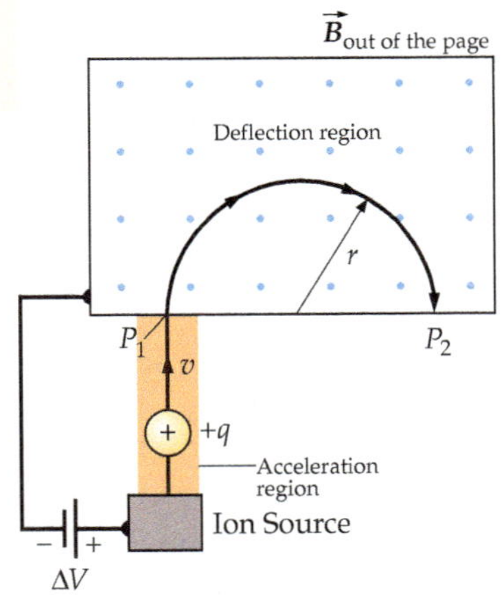

FIGURE 26-19 Schematic drawing of a mass spectrometer. Positive ions from an ion source are accelerated through a potential difference ΔV and enter a uniform magnetic field at P_1. The magnetic field is out of the plane of the page as indicated by the dots. The ions are bent into a circular arc and emerge at P_2. The radius r of the circle varies with the mass of the ion.

The ions move in a semicircle of radius r given by Equation 26-6, $r = mv/qB$, and strike a photographic plate at point P_2, a distance $2r$ from the point P_1 where the ions entered the magnetic field.

The speed v can be eliminated from Equations 26-6 and 26-11 to find m/q in terms of the known quantities V, B, and r. We first solve Equation 26-6 for v and square each term, which gives

$$v^2 = \frac{r^2q^2B^2}{m^2}$$

Substituting this expression for v^2 into Equation 26-11, we obtain

$$\frac{1}{2}m\left(\frac{r^2q^2B^2}{m^2}\right) = q|\Delta V|$$

Simplifying this equation and solving for m/q, we obtain

$$\frac{m}{q} = \frac{B^2r^2}{2|\Delta V|} \tag{26-12}$$

In Aston's original mass spectrometer, mass differences could be measured to a precision of about 1 part in 10,000. The precision has been improved by introducing a velocity selector between the ion source and the magnet, which increases the degree of accuracy with which the speeds of the incoming ions can be determined.

Example 26-6 Separating Isotopes of Nickel

A ^{58}Ni ion that has a charge equal to $+e$ and a mass equal to 9.62×10^{-26} kg is accelerated through a potential drop of 3.00 kV and deflected in a magnetic field of 0.120 T. (*a*) Find the radius of curvature of the orbit of the ion. (*b*) Find the difference in the radii of curvature of ^{58}Ni ions and ^{60}Ni ions. (Assume that the mass ratio is 58:60.)

PICTURE The radius of curvature r can be found using Equation 26-12. Using the mass dependence of r, we can find the radius of curvature for the orbit of the ^{60}Ni ions from the radius of curvature for the orbit of the ^{58}Ni ions, and then take the difference between the two radii.

SOLVE

(*a*) Solve Equation 26-12 for r:

$$r = \sqrt{\frac{2m|\Delta V|}{qB^2}} = \left[\frac{2(9.62 \times 10^{-26}\text{ kg})(3000\text{ V})}{(1.60 \times 10^{-19}\text{ C})(0.120\text{ T})^2}\right]^{1/2} = \boxed{0.501\text{ m}}$$

(*b*) 1. Let r_1 and r_2 be the radius of the orbit of the ^{58}Ni ion and the ^{60}Ni ion, respectively. Use the result in Part (*a*) to find the ratio of r_2 to r_1:

$$\frac{r_2}{r_1} = \sqrt{\frac{m_2}{m_1}} = \sqrt{\frac{60}{58}} = 1.017$$

2. Use the result of the previous step to calculate r_2 for ^{60}Ni:

$$r_2 = 1.017r_1 = (1.017)(0.501\text{ m}) = 0.510\text{ m}$$

3. The difference in orbital radii is $r_2 - r_1$:

$$r_2 - r_1 = 0.510\text{ m} - 0.501\text{ m} = \boxed{9\text{ mm}}$$

CHECK The difference in the orbital radii is less than 2 percent of the radius of curvature of either orbit. This result is expected for two ions whose masses differ by less than 4 percent.

THE CYCLOTRON

The cyclotron was invented by E. O. Lawrence and M. S. Livingston in 1934 to accelerate particles, such as protons or deuterons, to large kinetic energies.* The high-energy particles are used to bombard atomic nuclei, causing nuclear reactions that

* A deuteron is the nucleus of heavy hydrogen, ^{2}H, which consists of a proton and neutron tightly bound together.

are then studied to obtain information about nuclei. High-energy protons and deuterons are also used to produce radioactive materials and for medical purposes.

Figure 26-20 is a schematic drawing of a cyclotron. The particles move in two semicircular metal containers called *dees* (because they are the shape of the letter "D"). The dees are housed in a vacuum chamber that is in a region with a uniform magnetic field provided by an electromagnet. The region in which the particles move must be evacuated so that the particles will not be scattered in collisions with air molecules. A potential difference ΔV, which alternates in time with a period T, is maintained between the dees. The period is chosen to be the cyclotron period $T = 2\pi m/(qB)$ (Equation 26-7). The potential difference creates an electric field across the gap between the dees. At the same time, there is no electric field within each dee because the metal dees act as shields.

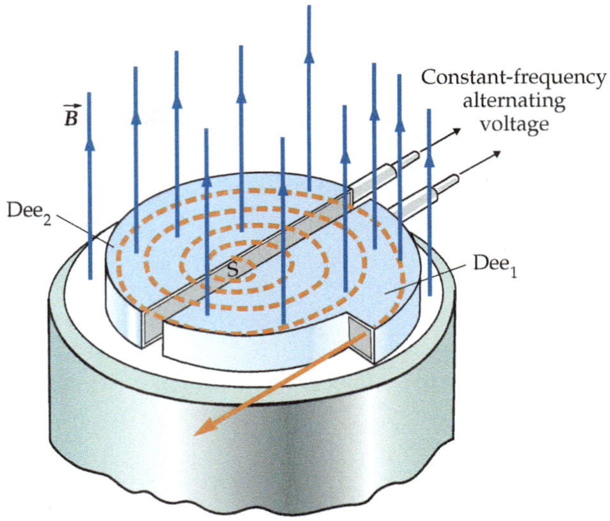

FIGURE 26-20 Schematic drawing of a cyclotron. The upper-pole face of the magnet has been omitted. Charged particles, such as protons, are accelerated from a source S at the center by the potential difference across the gap between the dees. When the charged particles arrive at the gap again the potential difference has changed sign so they are again accelerated across the gap and move in a larger circle. The potential difference across the gap alternates with the cyclotron frequency of the particle, which is independent of the radius of the circle.

Positively charged particles are initially injected into dee$_1$ with a small velocity from an ion source S near the center of the dees. They move in a semicircle in dee$_1$ and arrive at the gap between dee$_1$ and dee$_2$ after a time $\frac{1}{2}T$. The potential is adjusted so that dee$_1$ is at a higher potential than dee$_2$ when the particles arrive at the gap between them. Each particle is therefore accelerated across the gap by the electric field and gains kinetic energy equal to $q\Delta V$.

Because the particle now has more kinetic energy, the particle moves in a semicircle of larger radius in dee$_2$. It arrives at the gap again after a time $\frac{1}{2}T$, because the period is independent of the particle's speed. By this time, the potential difference between the dees has been reversed so that dee$_2$ is now at the higher potential. Once more the particle is accelerated across the gap and gains additional kinetic energy equal to $q\Delta V$. Each time the particle arrives at the gap, it is accelerated and gains kinetic energy equal to $q\Delta V$. Thus, the particle moves in larger and larger semicircular orbits until it eventually leaves the magnetic field. In the typical cyclotron, each particle may make 50 to 100 revolutions and exit with energies of up to several hundred megaelectron volts.

The kinetic energy of a particle leaving a cyclotron can be calculated by setting r in Equation 26-6 equal to the maximum radius of the dees and solving the equation for v:

$$r = \frac{mv}{qB} \quad \Rightarrow \quad v = \frac{qBr}{m}$$

Then

$$K = \frac{1}{2}mv^2 = \frac{1}{2}\left(\frac{q^2B^2}{m}\right)r^2 \qquad 26\text{-}13$$

Example 26-7 Energy of Accelerated Proton

A cyclotron for accelerating protons has a magnetic field of 0.150 T and a maximum radius of 0.500 m. (*a*) What is the cyclotron frequency? (*b*) What is the kinetic energy of the protons when they emerge?

PICTURE Apply Newton's second law ($F = ma$) with $F = |q\vec{v} \times \vec{B}|$. Use $v = r\omega$ and solve for the frequency and the speed.

SOLVE

(*a*) 1. Apply $F = ma$, where F is the magnetic force and a is the centripetal acceleration. Substitute ωr for v and solve for ω:

$$F = ma$$

$$qvB = m\frac{v^2}{r}$$

$$q\omega rB = m\frac{\omega^2 r^2}{r}$$

$$\omega = \frac{qB}{m} = \frac{(1.60 \times 10^{-19}\ \text{C})(0.150\ \text{T})}{1.67 \times 10^{-27}\ \text{kg}}$$

$$= 1.44 \times 10^7\ \text{rad/s}$$

2. Use $2\pi f = \omega$ to calculate the frequency in cycles per second (hertz):

$$f = \frac{\omega}{2\pi} = \frac{1.44 \times 10^7\ \text{rad/s}}{2\pi\ \text{rad}}$$

$$= 2.29 \times 10^6\ \text{Hz} = \boxed{2.29\ \text{MHz}}$$

(*b*) 1. Calculate the kinetic energy:

$$K = \frac{1}{2}mv^2 = \frac{1}{2}m\omega^2 r^2$$

$$= \frac{1}{2}(1.67 \times 10^{-27}\ \text{kg})(1.44 \times 10^7\ \text{rad/s})^2(0.500\ \text{m})^2$$

$$= 4.33 \times 10^{-14}\ \text{J}$$

2. The energies of protons and other elementary particles are usually expressed in electron volts. Use $1\ \text{eV} = 1.60 \times 10^{-19}\ \text{J}$ to convert to eV:

$$K = 4.33 \times 10^{-14}\ \text{J} \times \frac{1\ \text{eV}}{1.60 \times 10^{-19}\ \text{J}} = \boxed{271\ \text{keV}}$$

CHECK The exit speed of the proton is $v = r\omega = (0.500\ \text{m})(1.44 \times 10^7\ \text{rad/s}) = 7.20 \times 10^6\ \text{m/s}$. The speed of light is $3.00 \times 10^8\ \text{m/s}$. Our calculated value of $1.44 \times 10^7\ \text{rad/s}$ for the angular frequency is plausible because it is a high speed that is less than ten percent of the speed of light.

26-3 TORQUES ON CURRENT LOOPS AND MAGNETS

A current-carrying loop experiences no net force in a uniform magnetic field, but it does experience a net torque. The orientation of the loop can be described conveniently by a unit vector $\hat{n}$ that is normal to the plane of the loop, as illustrated in Figure 26-21. If the fingers of the right hand curl around the loop in the direction of the current, the thumb points in the direction of $\hat{n}$.

Figure 26-22 shows the forces exerted by a uniform magnetic field on a rectangular current-carrying loop whose vector $\hat{n}$ makes an angle θ with the direction of the magnetic field $\vec{B}$. The net force on the loop is zero. The forces $\vec{F}_1$ and $\vec{F}_2$ have the magnitude

$$F_1 = F_2 = IaB$$

The forces form a couple, so the torque they exert is the same about any point. Point P in Figure 26-22 is a convenient point about which to compute the torque. The magnitude of the torque is

$$\tau = F_2 b \sin\theta = IaBb \sin\theta = IAB \sin\theta$$

where $A = ab$ is the area of the loop. For a loop that has N turns, the torque has the magnitude

$$\tau = NIAB \sin\theta$$

This torque tends to twist the loop so that $\hat{n}$ is in the same direction as $\vec{B}$.

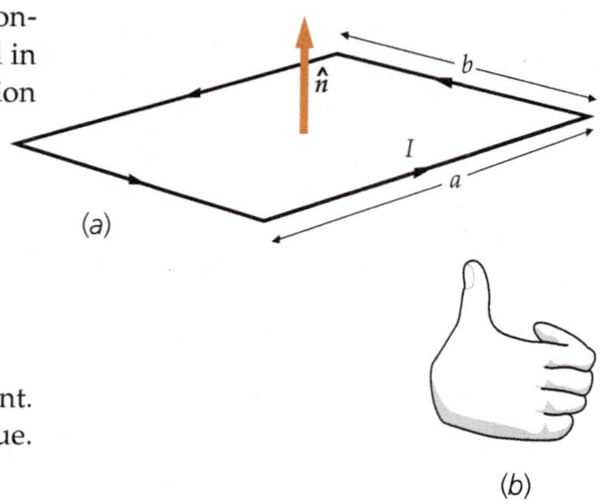

FIGURE 26-21 (*a*) The orientation of a current loop is described by the unit vector $\hat{n}$ perpendicular to the plane of the loop. (*b*) Right-hand rule for determining the direction of $\hat{n}$. If the fingers of the right hand curl around the loop in the direction of the current, the thumb points in the direction of $\hat{n}$.

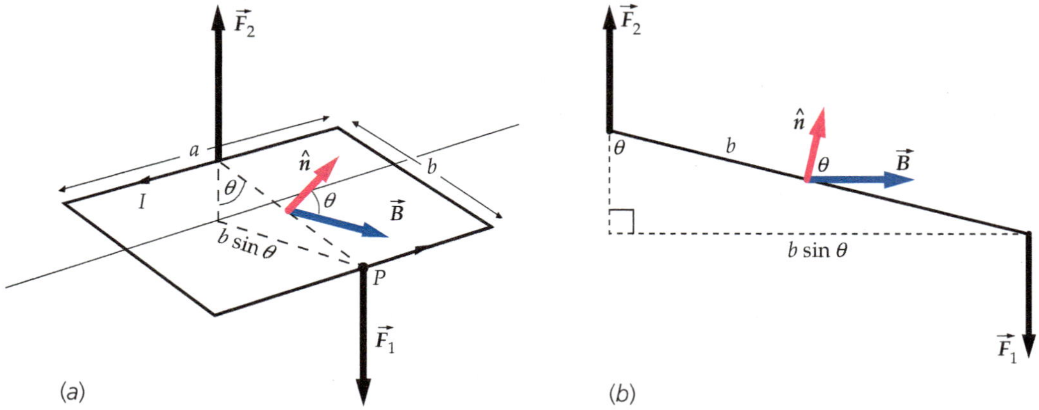

FIGURE 26-22 (*a*) Rectangular current loop whose unit normal $\hat{n}$ makes an angle θ with a uniform magnetic field $\vec{B}$. (*b*) An edge-on view of the current loop. The torque on the loop has magnitude $IAB \sin\theta$ and is in the direction such that $\hat{n}$ tends to rotate so as to align itself with $\vec{B}$.

The torque can be written conveniently in terms of the **magnetic dipole moment** $\vec{\mu}$ (also referred to simply as the **magnetic moment**) of the current loop, which is defined as

$$\vec{\mu} = NIA\hat{n} \qquad \text{26-14}$$

MAGNETIC DIPOLE MOMENT OF A CURRENT LOOP

The SI unit of magnetic moment is the ampere-square meter ($A \cdot m^2$). In terms of the magnetic dipole moment, the torque on the current loop is given by

$$\vec{\tau} = \vec{\mu} \times \vec{B} \qquad \text{26-15}$$

TORQUE ON A CURRENT LOOP

Equation 26-15, which we have derived for a rectangular loop, holds in general for a loop of any shape that lies in a single plane. The torque on any such loop is the vector product of the magnetic moment $\vec{\mu}$ of the loop and the magnetic field $\vec{B}$, where the magnetic moment (Figure 26-23) is defined as a vector that has a magnitude equal to NIA and has the same direction as $\hat{n}$. Comparing Equation 26-15 with Equation 21-11 ($\vec{\tau} = \vec{p} \times \vec{E}$) for the torque on an electric dipole, we see that the expression for the torque on a magnetic dipole in a magnetic field has the same form as that for the torque on an electric dipole in an electric field.

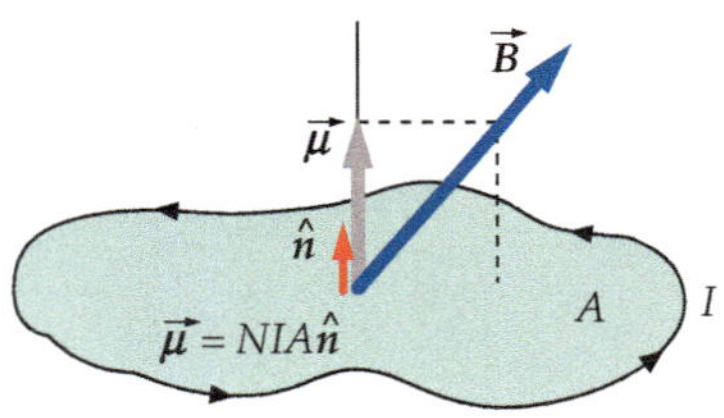

FIGURE 26-23 A flat current loop of arbitrary shape is described by its magnetic moment $\vec{\mu} = NIA\hat{n}$. In a magnetic field $\vec{B}$, the loop experiences a torque $\vec{\mu} \times \vec{B}$.

Example 26-8 Torque on a Current Loop

A circular loop has a radius equal to 2.00 cm, has 10 turns of wire, and carries a current equal to 3.00 A. The axis of the loop makes an angle of 30.0° with a magnetic field of 8000 G. Find the magnitude of the torque on the loop.

PICTURE The torque on a current loop is given by $\vec{\tau} = \vec{\mu} \times \vec{B}$ (Equation 26-15) where $\vec{\mu} = NIA\hat{n}$ (Equation 26-14).

SOLVE

The magnitude of the torque is given by Equation 26-15:

$$\tau = |\vec{\mu} \times \vec{B}| = \mu B \sin\theta = NIAB \sin\theta$$
$$= (10.0)(3.00\ \text{A})\pi(0.0200\ \text{m})^2(0.800\ \text{T}) \sin 30.0°$$
$$= \boxed{1.51 \times 10^{-2}\ \text{N} \cdot \text{m}}$$

CHECK From $\vec{F} = I\vec{L} \times \vec{B}$ (Equation 26-4) we can see that the SI unit for magnetic field (the tesla) must have units of $N/(A \cdot m)$. With this in mind, one can see by inspection that the units for the right-hand side of the equation in the solution work out to $N \cdot m$, which are SI units for torque.

Example 26-9 Tilting a Loop — Try It Yourself

A circular wire loop that has a radius R, a mass m, and a current I lies on a horizontal surface (Figure 26-24). There is a horizontal magnetic field $\vec{B}$. How large can the current I be before one edge of the loop will lift off the surface?

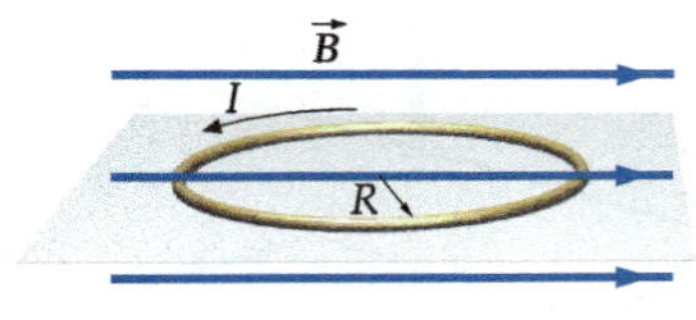

FIGURE 26-24

PICTURE The loop (Figure 26-25) will start to rotate when the magnitude of the net torque on the loop is greater than zero. To eliminate the torque due to the normal force, we calculate torques about the point of contact between the surface and the loop. The magnetic torque is given by $\vec{\tau} = \vec{\mu} \times \vec{B}$. The magnetic torque is the same about any point because the magnetic torque consists of couples. The lever arm for the gravitational torque is the radius of the loop.

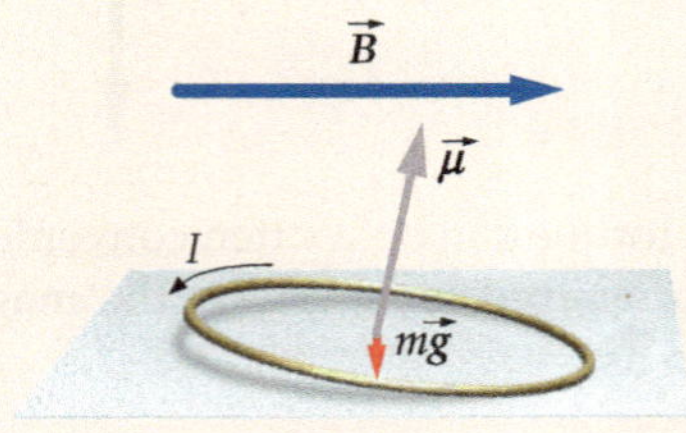

FIGURE 26-25

SOLVE

Cover the column to the right and try these on your own before looking at the answers.

Steps	Answers
1. Find the magnitude of the magnetic torque acting on the loop.	$\tau_m = \mu B \sin(90°) = I\pi R^2 B$
2. Find the magnitude of the gravitational torque exerted on the loop.	$\tau_g = mgR$
3. Equate the magnitudes of the torques and solve for the current I.	$I = \boxed{\dfrac{mg}{\pi RB}}$

CHECK The current is directly proportional to the mass for constant B, which makes sense. The larger the mass, the more current is needed to start to rotate the ring.

POTENTIAL ENERGY OF A MAGNETIC DIPOLE IN A MAGNETIC FIELD

When a torque is exerted on a rotating object, work is done. When a magnetic dipole is rotated through an angle $d\theta$, the work done is

$$dW = -\tau\, d\theta = -\mu B \sin\theta\, d\theta$$

where θ is the angle between $\vec{\mu}$ and $\vec{B}$. The minus sign arises because the magnetic torque tends to decrease θ. Setting this work equal to the decrease in potential energy U, we have

$$dU = -dW = +\mu B \sin\theta\, d\theta$$

Integrating, we obtain

$$U = -\mu B \cos\theta + U_0$$

We choose the potential energy to be zero when $\theta = 90°$. Then $U_0 = 0$ and the potential energy of the dipole is given by

$$U = -\mu B \cos\theta = -\vec{\mu} \cdot \vec{B} \qquad 26\text{-}16$$

POTENTIAL ENERGY OF A MAGNETIC DIPOLE

Equation 26-16 gives the potential energy of a magnetic dipole at an angle θ to the direction of a magnetic field.

Example 26-10 Torque on a Coil

A square 12-turn coil has an edge-length equal to 40.0 cm and carries a current of 3.00 A. It lies in the $z = 0$ plane, as shown in a uniform magnetic field $\vec{B} = 0.300\text{ T}\,\hat{i} + 0.400\text{ T}\,\hat{k}$. The current is counterclockwise when viewed from a point on the positive z axis. Find (*a*) the magnetic moment of the coil and (*b*) the torque exerted on the coil. (*c*) Find the potential energy of the coil.

PICTURE From Figure 26-26, we see that the magnetic moment of the loop is in the $+z$ direction.

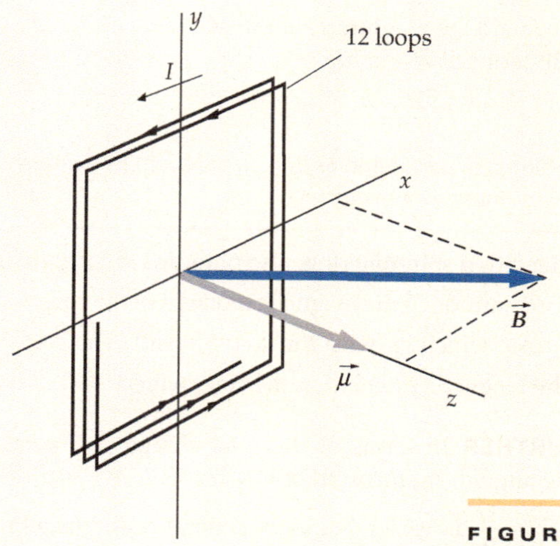

FIGURE 26-26

SOLVE

(a) Calculate the magnetic moment of the loop:

$$\vec{\mu} = NIA\hat{k} = (12)(3.00\text{ A})(0.400\text{ m})^2\hat{k}$$
$$= \boxed{5.76\text{ A}\cdot\text{m}^2\,\hat{k}}$$

(b) The torque on the current loop is given by Equation 26-15:

$$\vec{\tau} = \vec{\mu} \times \vec{B}$$
$$= (5.76\text{ A}\cdot\text{m}^2\,\hat{k}) \times (0.300\text{ T}\,\hat{i} + 0.400\text{ T}\,\hat{k})$$
$$= \boxed{1.73\text{ N}\cdot\text{m}\,\hat{j}}$$

(c) The potential energy is the negative dot product of $\vec{\mu}$ and $\vec{B}$:

$$U = -\vec{\mu}\cdot\vec{B}$$
$$= -(5.76\text{ A}\cdot\text{m}^2\hat{k})\cdot(0.300\text{ T}\,\hat{i} + 0.400\text{ T}\,\hat{k})$$
$$= \boxed{-2.30\text{ J}}$$

CHECK The torque in the Part-(*b*) result is perpendicular to both the magnetic moment vector and the magnetic field vector, as is expected for a vector product.

PRACTICE PROBLEM 26-3 The potential energy of a current-carrying coil in a uniform magnetic field $\vec{B}$ is equal to zero when the angle between the magnetic dipole moment of the coil $\vec{\mu}$ and the magnetic field is 90°. Calculate the potential energy of the system if the coil is oriented so $\vec{B}$ and $\vec{\mu}$ are (*a*) in the same direction and (*b*) in opposite directions.

When a permanent magnet, such as a compass needle or a bar magnet, is placed in a region where there is a magnetic field $\vec{B}$, the field exerts a torque on the magnet that tends to rotate the magnet so that it lines up with the field. (This effect also occurs with previously unmagnetized iron filings, which become magnetized in the presence of a field $\vec{B}$.) The bar magnet is characterized by a magnetic moment $\vec{\mu}$, a vector that points in the same direction as an arrow drawn from the south pole of the magnet to the north pole of the magnet. A short bar magnet thus behaves like a current loop.

Example 26-11 $\vec{\mu}$ of a Spinning Disk

A thin nonconducting disk that has a mass m, a radius a, and a uniform surface charge per unit area σ spins with angular velocity $\vec{\omega}$ about an axis through the center of the disk and perpendicular to the plane of the disk. Find the magnetic moment of the spinning disk.

PICTURE We find the magnetic moment of a circular element that has a radius R and a width dR and integrate (Figure 26-27). The charge on the element is $dq = \sigma\, dA = \sigma 2\pi R\, dR$. If the charge is positive, the magnetic moment is in the direction of $\vec{\omega}$, so we need only calculate its magnitude.

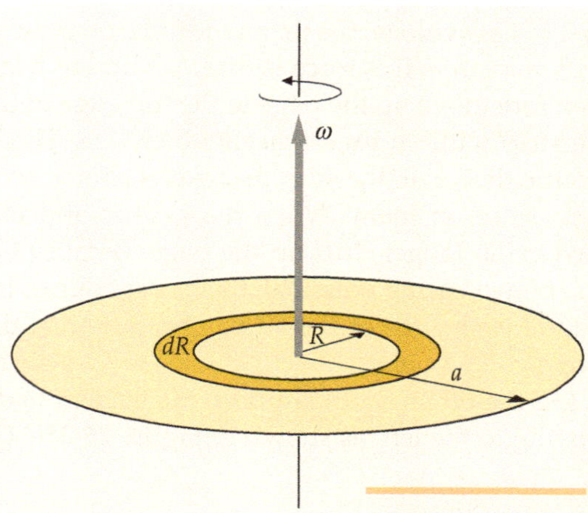

FIGURE 26-27

SOLVE

1. The magnitude of the magnetic moment of the strip shown is the current multiplied by the area of the loop:

$$d\mu = A\,dI = \pi R^2\,dI$$

2. The current in the strip is the total charge dq on the strip divided by the period T. During one period the charge dq passes by a point not rotating with the strip. The period is equal to the reciprocal of the frequency f of rotation $1/T = f = \omega/(2\pi)$:

$$dI = \frac{dq}{T} = \frac{\omega}{2\pi}dq = \frac{\omega}{2\pi}\sigma\,dA$$
$$= \frac{\omega}{2\pi}\sigma 2\pi R\,dR = \sigma\omega R\,dR$$

3. Substitute to obtain the magnitude of the magnetic moment of the strip $d\mu$ in terms of R and dR:

$$d\mu = \pi R^2\,dI = \pi R^2\sigma\omega R\,dR = \pi\sigma\omega R^3\,dR$$

4. Integrate from $R = 0$ to $R = a$:

$$\mu = \int_0^a \pi\sigma\omega R^3\,dR = \frac{1}{4}\pi\sigma\omega a^4$$

5. Use the fact that $\vec{\mu}$ is parallel to $\vec{\omega}$ (if σ is positive) to express the magnetic moment as a vector:

$$\vec{\mu} = \boxed{\tfrac{1}{4}\pi\sigma a^4\vec{\omega}}$$

CHECK Consider a thin spinning ring, also of radius a, carrying the same charge, $Q = \sigma\pi a^2$, as the disk. The magnitude of the magnetic moment of the ring is given by $\mu = IA = \frac{Q}{T}\pi a^2 = \frac{\sigma\pi a^2}{2\pi/\omega}\pi a^2 = \frac{1}{2}\pi\sigma a^4\omega$, which is twice the step-5 result. The step-5 result is smaller than the magnitude of the magnetic moment of the ring, which is what one would expect.

TAKING IT FURTHER In terms of the total charge $Q = \sigma\pi a^2$, the magnetic moment is $\vec{\mu} = \frac{1}{4}Qa^2\vec{\omega}$. The angular momentum of the disk is $\vec{L} = (\frac{1}{2}ma^2)\vec{\omega}$, so the magnetic moment can be written $\vec{\mu} = \frac{Q}{2m}\vec{L}$, which is a more general result. (See Problem 57.)

26-4 THE HALL EFFECT

As we have seen, charges moving in a region where there is a magnetic field each experience a force perpendicular to their motion. When these charges are traveling in a conducting wire, they will be pushed to one side of the wire. This results in a separation of charge in the wire—a phenomenon called the **Hall effect.** This phenomenon allows us to determine the sign of the charge on the charge carriers and the number of charge carriers per unit volume n in a conductor. The Hall effect also provides a convenient method for measuring magnetic fields.

Figure 26-28 shows two conducting strips; each conducting strip carries a current I to the right because the left sides of the strips are connected to the positive terminal of a battery and the right sides are connected to the negative terminal. A magnetic field $\vec{B}$ is directed into the paper. Let us suppose the current in the strip is due to positively charged particles moving to the right, as shown in Figure 26-28*a*. On average, the magnetic force on these particles is $q\vec{v}_d \times \vec{B}$ (where $\vec{v}_d$ is the drift velocity). This force is directed up the page. The positively charged particles therefore move up the page to the top edge of the strip, leaving the bottom edge of the strip with an excess negative charge. This separation of charge produces an electric field $\vec{E}$ in the strip that exerts a force on the particles that opposes the magnetic force on them. When the electric and magnetic forces balance, the charge carriers no longer drift up the page. Because the electric field points in the direction of decreasing potential, the upper edge of the strip is at a higher potential than is the lower edge of the strip. This potential difference can be measured using a sensitive voltmeter.

On the other hand, suppose the current is due to negatively charged particles moving to the left, as shown in Figure 26-28*b*. (The negatively charged particles in the strip must move to the left because the current, as before, is to the right.) The magnetic force $q\vec{v}_d \times \vec{B}$ is again up the page, because the signs of both q and the direction of $\vec{v}_d$ have been reversed. Again the carriers are forced to the upper edge of the strip, but the upper edge of the strip now carries a negative charge (because the charge carriers are negative) and the lower edge of the strip now carries a positive charge.

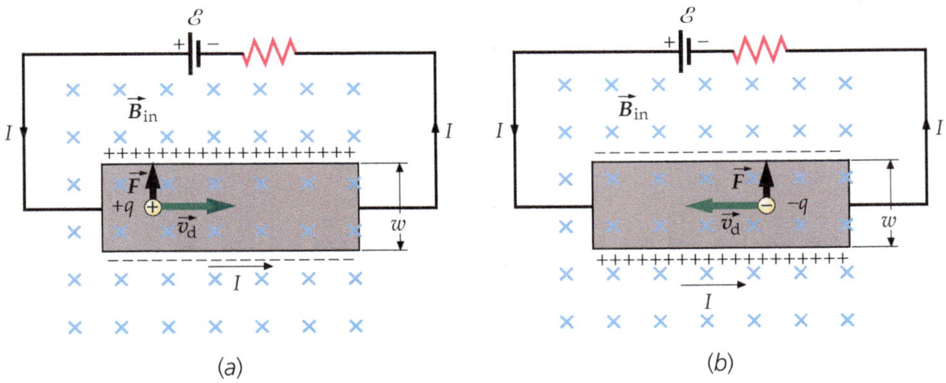

FIGURE 26-28 The Hall effect. The magnetic field is directed into the plane of the page as indicated by the crosses. The magnetic force on a charged particle is upward for a current to the right whether the current is due to (*a*) positive particles moving to the right or (*b*) negative particles moving to the left.

A measurement of the sign of the potential difference between the upper and lower parts of the strip tells us the sign of the charge carriers. In semiconductors, the charge carriers may be negative electrons or positive "holes." A measurement of the sign of the potential difference tells us which are dominant for a particular semiconductor. For a metal strip, we find that the upper edge of the strip in Figure 26-28*b* is at a lower potential than is the lower edge of the strip—which means that the upper part must carry a negative charge. Thus, Figure 26-28*b* is the correct illustration of the current in a metal strip. It was a measurement like this which led to the discovery that the charge carriers in metals are negatively charged.

The potential difference between the top of the strip and the bottom of the strip is called the **Hall voltage**. We can calculate the magnitude of the Hall voltage in terms of the drift velocity. The magnitude of the magnetic force on the charge carriers in the strip is qv_dB. This magnetic force is balanced by the electrostatic force of magnitude qE_H, where E_H is the electric field due to the charge separation. Thus, we have $E_H = v_dB$. If the width of the strip is w, the potential difference is E_Hw. The Hall voltage is therefore

$$V_H = E_Hw = v_dBw \tag{26-17}$$

PRACTICE PROBLEM 26-4

A conducting strip of width $w = 2.0$ cm is placed in a magnetic field of 0.80 T. The Hall voltage is measured to be 0.64 μV. Calculate the drift velocity of the electrons.

Because the drift velocity for ordinary currents is very small, we can see from Equation 26-17 that the Hall voltage is very small for ordinary-sized strips and magnetic fields. From measurements of the Hall voltage for a strip of a given size, we can determine the number of charge carriers per unit volume in the strip.

The magnitude of the current is given by Equation 26-3:

$$|I| = |q|nv_{\mathrm{d}}A$$

where A is the cross-sectional area of the strip. For a strip of width w and thickness t, the cross-sectional area is $A = wt$. Because the charge carriers are electrons, the quantity $|q|$ is the charge on one electron e. The number density of charge carriers n is thus given by

$$n = \frac{|I|}{A|q|v_{\mathrm{d}}} = \frac{|I|}{wtev_{\mathrm{d}}} \qquad \text{26-18}$$

Substituting V_{H}/B for $v_{\mathrm{d}}w$ (Equation 26-17), we have

$$n = \frac{|I|B}{teV_{\mathrm{H}}} \qquad \text{26-19}$$

Example 26-12 Charge Carrier Number Density in Silver

A silver slab has a thickness equal to 1.00 mm, a width equal to 1.50 cm, and a current equal to 2.50 A in a region where there is a magnetic field of magnitude 1.25 T perpendicular to the slab. The Hall voltage is measured to be 0.334 μV. (*a*) Calculate the number density of the charge carriers. (*b*) Calculate the number density of atoms in silver, which has a mass density of $\rho = 10.5\ \mathrm{g/cm^3}$ and a molar mass of $M = 107.9\ \mathrm{g/mol}$, and compare the number density of atoms in silver with the Part-(*a*) result.

PICTURE We can use Equation 26-19 to find the number density of charge carriers. The number density of atoms can be obtained from knowledge of the density and the molar mass.

SOLVE

(*a*) Substitute numerical values into Equation 26-19 to find n:

$$n = \frac{|I|B}{teV_{\mathrm{H}}} = \frac{(2.50\ \mathrm{A})(1.25\ \mathrm{T})}{(1.00 \times 10^{-3}\ \mathrm{m})(1.60 \times 10^{-19}\ \mathrm{C})(3.34 \times 10^{-7}\ \mathrm{V})}$$

$$= \boxed{5.85 \times 10^{28}\ \mathrm{electrons/m^3}}$$

(*b*) 1. The number of atoms per unit volume is $\rho N_{\mathrm{A}}/M$:

$$n_{\mathrm{a}} = \rho\frac{N_{\mathrm{A}}}{M} = (10.5\ \mathrm{g/cm^3})\frac{6.02 \times 10^{23}\ \mathrm{atoms/mol}}{107.9\ \mathrm{g/mol}}$$

$$= \boxed{5.86 \times 10^{22}\ \mathrm{atoms/cm^3} = 5.86 \times 10^{28}\ \mathrm{atoms/m^3}}$$

2. Compare the Part-(*b*), step-1 result with the Part-(*a*) result:

These results indicate that the number of charge carriers in silver is very nearly one per atom.

CHECK We should expect the number density of charge carriers and the number density of atoms in a metal to be the same order of magnitude. Our results validate that expectation.

The Hall voltage provides a convenient method for measuring magnetic fields. If we rearrange Equation 26-19, we can write for the Hall voltage

$$V_{\mathrm{H}} = \frac{|I|}{nte}B \qquad \text{26-20}$$

A given strip can be calibrated by measuring the Hall voltage for a given current in a known magnetic field. The strip can then be used to measure an unknown magnetic field B by measuring the Hall voltage for a given current.

*THE QUANTUM HALL EFFECTS

According to Equation 26-20, the Hall voltage should increase linearly with magnetic field strength B for a given current in a given slab. In 1980, while studying the Hall effect in semiconductors at very low temperatures and very large magnetic fields, Klaus von Klitzing discovered that a plot of V_H versus B resulted in a series of plateaus, as shown in Figure 26-29, rather than a straight line. That is, the Hall voltage is quantized. For the discovery of the integer quantum Hall effect, von Klitzing won the Nobel Prize in Physics in 1985.

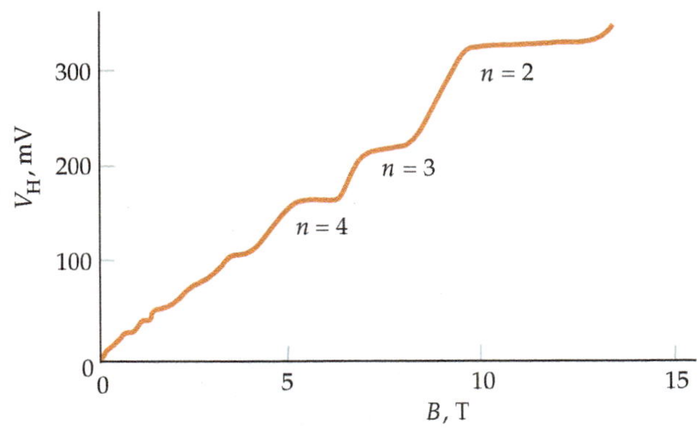

FIGURE 26-29 A plot of the Hall voltage versus applied magnetic field shows plateaus, indicating that the Hall voltage is quantized. The data were taken at a temperature of 1.39 K with the current I held fixed at 25.52 μA.

In the theory of the integer quantum Hall effect, the Hall resistance, defined as $R_H = V_H/I$, can take on only the values

$$R_H = \frac{V_H}{I} = \frac{R_K}{n} \qquad n = 1, 2, 3, \ldots \qquad \text{26-21}$$

where n is an integer, and R_K, called the **von Klitzing constant,** is related to the fundamental electric charge e and Planck's constant h by

$$R_K = \frac{h}{e^2} \qquad \text{26-22}$$

Because the von Klitzing constant can be measured to an accuracy of a few parts per billion, the quantum Hall effect is now used to define a standard of resistance. As of January 1990, the **ohm** is defined in terms of the conventional value* of the von Klitzing constant R_{K-90}, which has the value

$$R_{K-90} = 25\,812.8076\ \Omega\ \text{(exact)} \qquad \text{26-23}$$

In 1982, it was observed that under certain special conditions the Hall resistance is given by Equation 26-22 but with the integer n replaced by a series of rational fractions. This is called the *fractional quantum Hall effect*. For the discovery and explanation of the fractional quantum Hall effect, American professors Laughlin, Störmer, and Tsui won the Nobel Prize in Physics in 1998.

* The value of R_{K-90} differs only slightly from that of R_K. The currently used value of the von Klitzing constant is $R_K = (25\,812.807\,572 \pm 0.000\,095)\ \Omega$.

Physics Spotlight

Earth and the Sun—Magnetic Changes

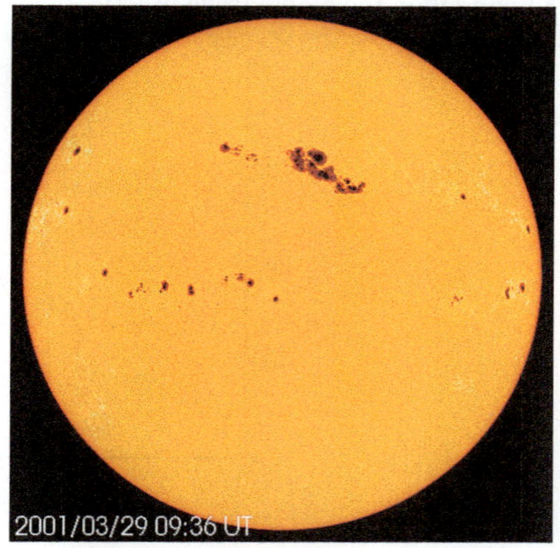

Sunspots are regions where the magnetic field strength is very high. They are darker than the surrounding surface because the temperature in the sunspot is cooler than the temperature of the surrounding area. *(SOHO/NASA.)*

The magnetic fields of the Sun and Earth have been measured almost constantly in recent years by satellite and ground-based magnetic observatories.[*] Geologists and physicists have collaborated to study the paleomagnetic fields of both Earth[†] and the Sun.[‡] The paleomagnetic studies and the ongoing observations show that the magnetic fields of the Earth and Sun are continuously changing.

Earth's magnetic field has been used as a navigational aid for over 900 years.[#] Navigators were soon aware that magnetic north does not coincide with celestial north, and that the magnetic declination (the difference in direction between magnetic north and celestial north) varied from place to place. Measurements of magnetic declination taken in the same places dating from the sixteenth century[°] showed that the apparent location of magnetic north varied with time at the same place.[§] These measurements are the first evidence that Earth's magnetic field is dynamic.

In the 1960s, drill cores showed many layers of magnetic reversals in volcanic rocks.[¶] It became clear that Earth's magnetic field reverses around every 200000 years, but there have been durations of over six million years during which there were no geomagnetic reversals. Immediately surrounding the reversal, the record shows that the field strength decreases, reverses, and then increases over a period of a few thousand years.[**] The last geomagnetic reversal was 700000 years ago. Lately, Earth's magnetic field strength has been decreasing. From 1840 to the present, Earth's magnetic field has decreased by 15 nT/y,[††] which is a decrease of 3% per century, and reconstruction of data from ships' logs shows a decrease of about 2 nT/y from 1590 to 1840.

In the early twentieth century, G. E. Hale noted that sunspots, which had been observed for hundreds of years, had magnetic fields. He demonstrated that during a 22-year sunspot cycle, the Sun's magnetic field gradually decreased, reversed, increased, and returned back to the original configuration.[‡‡] Sunspots themselves have been measured with a magnetic field strength in excess of 200 mT.[##] Recent observation has shown that sunspots are magnetically powered vortices in the Sun. Although the surface of the Sun has an apparent average field of 0.10 mT in regions without sunspots, small areas of such regions have magnetic strengths varying from below 20 mT up to 100 mT.[°°]

The solar wind, which consists of charged sub-atomic particles ejected from the Sun at around 400 km/s,[§§] carries a magnetic field. Satellite data show that the interplanetary magnetic field is complex and dynamic.[¶¶,***] Near Earth, the strength of the interplanetary magnetic field varies between 1 and 37 nT. Sometimes, the Sun ejects a large burst of charged particles. When a large burst arrives at Earth's magnetic field, it causes a magnetic storm that can block radio communications and cause widespread power blackouts. The *Voyager 1* spacecraft was more than 94 AU from the Sun when it measured the strength of the interplanetary magnetic field as 0.03 nT.[†††,‡‡‡] The solar wind still carries a measurable magnetic field well beyond the orbit of Pluto.

---

* "Geomagnetic Frequently Asked Questions." United States National Geophysical Data Center, National Oceanic and Atmospheric Administration. http://www.ngdc.noaa.gov/seg/geomag/faqgeom.shtml As of Sept., 2006.

† Yamazaki, T., and Oda, H., "Orbital Influence on Earth's Magnetic Field: 100,000-Year Periodicity in Inclination." *Science*, Mar. 29, 2002, Vol. 294 pp. 2435–2437.

‡ Solanki, S. K., et al., "11,000 Year Sunspot Number Reconstruction." *IBGP Pages/World Data Center for Paleoclimatology Data Contribution Series #2005-015*. 2005. ftp://ftp.ncdc.noaa.gov/pub/data/paleo/climate-forcing/solar_variability/solanki2004-ssn.txt As of Sept., 2006.

\# Hellemans, A., and Bunch, B., *The Timetables of Science*. New York: Simon and Schuster, 1988. p. 75.

° Kono, M., "Ships' Logs and Archeomagnetism." *Science*, May 12, 2006, Vol. 312, pp. 865–66.

§ Hermanus Magnetic Observatory, "Detailed History." http://www.hmo.ac.za/detailed-history.html As of Sept., 2006.

¶ Dunn, J. R., et al., "Paleomagnetic Study of a Reversal of the Earth's Magnetic Field." *Science*, May 21, 1971, Vol. 172, pp. 840–844.

** Merrill, R. T., and McFadden, P. L., "Geomagnetic Polarity Transitions." *Reviews of Geophysics*, May 1999, Vol. 37, No. 2, pp. 201–226.

†† Gubbins, D., Jones, A. L., and Finlay, C., "Fall in Earth's Magnetic Field is Erratic." *Science*, May 12, 2006, Vol 312, pp. 900–902.

‡‡ Abbot, C. G., "Sun-Spots and Weather." *Science*, Dec. 8, 1933, Vol. 78, pp. 518–519.

\#\# Liang, H.-F., Zhao, H.-J., and Xiang, F.-Y., "Vector Magnetic Field Measurement of NOAA AR 10197." *Chinese Journal of Astronomy and Astrophysics*, Aug. 2006, Vol. 6, No. 4, pp. 470–476.

°° Lin, H., and Rimmele, T., "The Granular Magnetic Fields of the Quiet Sun." *The Astrophysical Journal*, Mar. 20, 1999, Vol. 514, Pt. 1, pp. 448–455.

§§ Hathaway, D., "The Solar Wind." *Solar Physics*, Marshall Space Flight Center, NASA http://solarscience.msfc. nasa.gov/SolarWind.shtml. Jun. 1, 2006, As of Oct., 2006.

¶¶ Smith, E. J., et al., "The Sun and Heliosphere at Solar Maximum." *Science*, Nov. 14, 2003, Vol. 302, pp. 1165–1168.

*** Arnold, N., and Lyons, A., "Granta MIST: Meeting Report." *Astronomy and Geophysics*, Aug. 2006, Vol. 46, pp. 4.18–4.21.

††† Gurnett, D. A., and Kurth, W. S., "Electron Plasma Oscillations Upstream of the Solar Wind Termination Shock." *Science*, Sept. 23, 2005, Vol. 309, pp. 2025–2027.

‡‡‡ Burlaga, L. F., et al., "Crossing the Termination Shock into the Helosheath: Magnetic Fields." *Science*, Sept. 23, 2005, Vol. 309, pp. 2027–2029.

Summary

1. The magnetic field describes the condition in space in which moving charges experience a force perpendicular to their velocity.
2. The magnetic force is part of the electromagnetic interaction, one of the three known fundamental interactions in nature.
3. The magnitude and direction of a magnetic field $\vec{B}$ are defined by the formula $\vec{F} = q\vec{v} \times \vec{B}$, where $\vec{F}$ is the force exerted on a particle with charge q moving with velocity $\vec{v}$.

TOPIC	RELEVANT EQUATIONS AND REMARKS	
1. Magnetic Force		
On a moving charge	$\vec{F} = q\vec{v} \times \vec{B}$	26-1
On a current element	$d\vec{F} = I\,d\vec{\ell} \times \vec{B}$	26-5
Unit of the magnetic field	The SI unit of magnetic fields is the tesla (T). A commonly used unit is the gauss (G), which is related to the tesla by $1\text{ G} = 10^{-4}\text{ T}$	26-3
2. Motion of Point Charges	A particle of mass m and charge q moving with speed v in a plane perpendicular to a uniform magnetic field moves in a circular orbit. The period and frequency of the circular motion are independent of the radius of the orbit and of the speed of the particle.	
Newton's second law	$qvB = m\dfrac{v^2}{r}$	26-6
Cyclotron period	$T = \dfrac{2\pi m}{qB}$	26-7
Cyclotron frequency	$f = \dfrac{1}{T} = \dfrac{qB}{2\pi m}$	26-8
*Velocity selector	A velocity selector consists of crossed electric and magnetic fields so that the electric and magnetic forces balance for a particle moving with speed v. $v = \dfrac{E}{B}$	26-9
*Thomson's measurement of q/m	The deflection of a charged particle in an electric field depends on the speed of the particle and is proportional to the charge-to-mass ratio q/m of the particle. J. J. Thomson used crossed electric and magnetic fields to measure the speed of cathode rays and then measured q/m for these particles by deflecting them in an electric field. He showed that all cathode rays consist of particles which all have the same charge-to-mass ratio. These particles are now called electrons.	
*Mass spectrometer	The mass-to-charge ratio of an ion of known speed can be determined by measuring the radius of the circular path taken by the ion in a known magnetic field.	
3. Current Loops		
Magnetic dipole moment	$\vec{\mu} = NIA\hat{n}$	26-14
Torque	$\vec{\tau} = \vec{\mu} \times \vec{B}$	26-15
Potential energy of a magnetic dipole	$U = -\vec{\mu} \cdot \vec{B}$	26-16
Net force	The net force on a current loop in a *uniform* magnetic field is zero.	

TOPIC	RELEVANT EQUATIONS AND REMARKS	
4. The Hall Effect	When a conducting strip carrying a current is placed in a magnetic field, the magnetic force on the charge carriers causes a separation of charge called the Hall effect. This results in a voltage V_H, called the Hall voltage. The sign of the charge carriers can be determined from a measurement of the sign of the Hall voltage, and the number of carriers per unit volume can be determined from the magnitude of V_H.	
Hall voltage	$V_H = E_H w = v_d B w = \frac{\lvert I \rvert}{nte} B$	26-17, 26-20
*Quantum Hall effects	Measurements at very low temperatures in very large magnetic fields indicate that the Hall resistance $R_H = V_H/I$ is quantized and can take on only the values given by $R_H = \frac{V_H}{I} = \frac{R_K}{n} \quad n = 1, 2, 3, \ldots$	26-21
*Conventional von Klitzing constant (definition of ohm)	$R_{K-90} = 25\,812.8076\ \Omega$ (exact)	26-23

Answers to Concept Checks

26-1 A left-hand rule is one way to answer the question. The definition for the direction of $\vec{B}$ is a convention. If the definition for the direction of $\vec{B}$ were changed as described in the question statement, a correct force law could be written $\vec{F} = q\vec{v} \otimes \vec{B}$, where the symbol $\otimes$ denotes the same operation as the symbol $\times$, except the product denoted by $\otimes$ requires using the left-hand rule instead of the right-hand rule. Alternatively, the force law could be revised to $\vec{F} = \vec{B} \times q\vec{v}$ and then you could stay with the right-hand rule.

26-2 (*b*) Negatively charged. The force $\vec{F}$ and the vector $\vec{v} \times \vec{B}$ are in opposite directions only if the particle is negatively charged. This is consistent with the relation $\vec{F} = q\vec{v} \times \vec{B}$.

Answers to Practice Problems

26-1 $-1.3 \times 10^{-12}\ \mathrm{N}\ \hat{j}$

26-2 (*a*) 667 km/s, (*b*) in the $-z$ direction

26-3 (*a*) -2.88 J. Note that this potential energy is lower than the potential energy calculated in the example. (The potential energy is lowest when $\vec{\mu}$ and $\vec{B}$ are in the same direction.) (*b*) $+2.88$ J

26-4 4.0×10^{-5} m/s

Problems

In a few problems, you are given more data than you actually need; in a few other problems, you are required to supply data from your general knowledge, outside sources, or informed estimate.

Interpret as significant all digits in numerical values that have trailing zeros and no decimal points.

- • Single-concept, single-step, relatively easy
- •• Intermediate-level, may require synthesis of concepts
- ••• Challenging
- SSM Solution is in the *Student Solutions Manual*
- Consecutive problems that are shaded are paired problems.

CONCEPTUAL PROBLEMS

1 • When the axis of a cathode-ray tube is horizontal in a region in which there is a magnetic field that is directed vertically upward, the electrons emitted from the cathode follow one of the dashed paths to the face of the tube in Figure 26-30. The correct path is (*a*) 1, (*b*) 2, (*c*) 3, (*d*) 4, (*e*) 5. SSM

2 •• We define the direction of the electric field to be the same as the direction of the force on a positive test charge. Why then do we *not* define the direction of the magnetic field to be the same as the direction of the magnetic force on a moving positive test charge?

3 • A *flicker bulb* is a lightbulb that has a long, thin flexible filament. It is meant to be plugged into an ac outlet that delivers current at a frequency of 60 Hz. There is a small permanent magnet

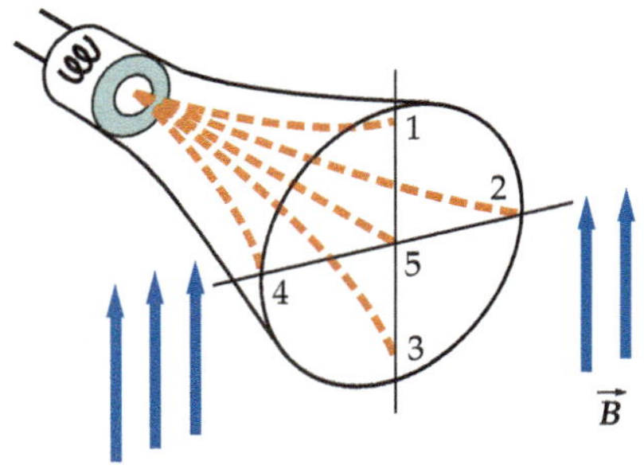

FIGURE 26-30 Problem 1

inside the bulb. When the bulb is plugged in the filament oscillates back and forth. At what frequency does it oscillate? Explain your answer. SSM

4 • In a cyclotron, the potential difference between the dees oscillates with a period given by $T = 2\pi m/(qB)$. Show that the expression to the right of the equal sign has units of seconds if q, B, and m have units of coulombs, teslas, and kilograms, respectively.

5 • A ^{7}Li nucleus has a charge equal to $+3e$ and a mass that is equal to the mass of seven protons. A ^{7}Li nucleus and a proton are both moving perpendicular to a uniform magnetic field $\vec{B}$. The magnitude of the momentum of the proton is equal to the magnitude of the momentum of the nucleus. The path of the proton has a radius of curvature equal to R_p and the path of the ^{7}Li nucleus has a radius of curvature equal to R_{Li}. The ratio R_p/R_{Li} is closest to (*a*) 3/1, (*b*) 1/3, (*c*) 1/7, (*d*) 7/1, (*e*) 3/7, (*f*) 7/3.

6 • An electron moving in the $+x$ direction enters a region that has a uniform magnetic field in the $+y$ direction. When the electron enters this region, it will (*a*) be deflected toward the $+y$ direction, (*b*) be deflected toward the $-y$ direction, (*c*) be deflected toward the $+z$ direction, (*d*) be deflected toward the $-z$ direction, (*e*) continue undeflected in the $+x$ direction.

7 • In a velocity selector, the speed of the undeflected charged particle is given by the ratio of the magnitude of the electric field to the magnitude of the magnetic field. Show that E/B in fact does have the units of m/s if E and B are in units of volts per meter and teslas, respectively. SSM

8 • How are the properties of magnetic field lines similar to the properties of electric field lines? How are they different?

9 • True or false:

(*a*) The magnetic moment of a bar magnet points from its north pole to its south pole.
(*b*) Inside the material of a bar magnet, the magnetic field due to the bar magnet points from the magnet's south pole toward its north pole.
(*c*) If a current loop simultaneously has its current doubled and its area cut in half, then the magnitude of its magnetic moment remains the same.
(*d*) The maximum torque on a current loop placed in a magnetic field occurs when the plane of the loop is perpendicular to the direction of the magnetic field.

10 •• Show that the von Klitzing constant, h/e^2, gives the SI unit for resistance (the ohm) if h and e are in units of joule-seconds and coulombs, respectively.

11 ••• The theory of relativity states that no law of physics can be described using the absolute velocity of an object, which is in fact impossible to define due to a lack of an absolute reference frame. Instead, the behavior of interacting objects can only be described by the relative velocities between the objects. New physical insights result from this idea. For example, in Figure 26-31 a magnet moving at high speed relative to some observer passes by an electron that is at rest relative to the same observer. Explain why you are sure that a force must be acting on the electron. In what direction will the force point at the instant the north pole of the magnet passes directly underneath the electron? Explain your answer.

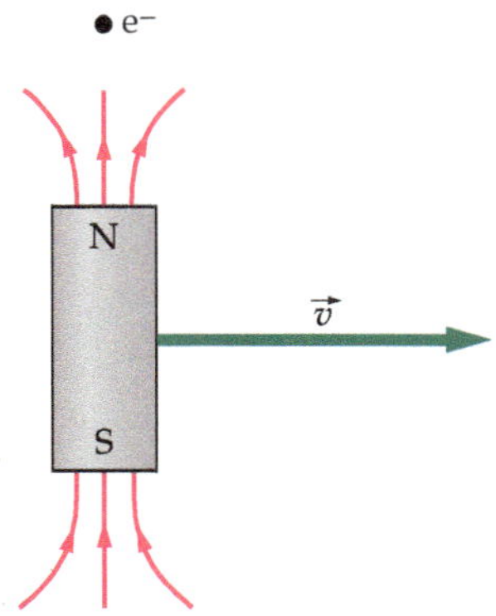

FIGURE 26-31 Problem 11

ESTIMATION AND APPROXIMATION

12 • Estimate the maximum magnetic force per meter that Earth's magnetic field could exert on a current-carrying wire in a 20-A circuit in your house.

13 •• CONTEXT-RICH Your friend wants to be a magician and intends to use Earth's magnetic field to suspend a current-carrying wire above the stage. He asks you to estimate the minimum current needed to suspend the wire just above Earth's surface at the equator (where Earth's magnetic field is horizontal). Assume the wire has a linear mass density of 10 g/m. Would you advise him to proceed with his plans for this act?

THE FORCE EXERTED BY A MAGNETIC FIELD

14 • Find the magnetic force on a proton moving in the $+x$ direction at a speed of 0.446 Mm/s in a uniform magnetic field of 1.75 T in the $+z$ direction.

15 • A point particle has a charge equal to -3.64 nC and a velocity equal to 2.75×10^3 m/s $\hat{i}$. Find the force on the charge if the magnetic field is (*a*) 0.38 T $\hat{j}$, (*b*) 0.75 T $\hat{i}$ + 0.75 T $\hat{j}$, (*c*) 0.65 T $\hat{i}$, and (*d*) 0.75 T $\hat{i}$ + 0.75 T $\hat{k}$.

16 • A uniform magnetic field equal to 1.48 T $\hat{k}$ is in the $+z$ direction. Find the force exerted by the field on a proton if the velocity of the proton is (*a*) 2.7 km/s $\hat{i}$, (*b*) 3.7 km/s $\hat{j}$, (*c*) 6.8 km/s $\hat{k}$, and (*d*) 4.0 km/s $\hat{i}$ + 3.0 km/s $\hat{j}$.

17 • A straight wire segment that is 2.0 m long makes an angle of 30° with a uniform 0.37-T magnetic field. Find the magnitude of the force on the wire if the wire has a current of 2.6 A.

18 • A straight segment of a current-carrying wire has a current element $I\vec{L}$, where $I = 2.7$ A and $\vec{L} = 3.0$ cm $\hat{i}$ + 4.0 cm $\hat{j}$. The segment is in a region with a uniform magnetic field given by 1.3 T $\hat{i}$. Find the force on the segment of wire.

19 • What is the force on an electron that has a velocity equal to 2.0×10^6 m/s $\hat{i} - 3.0 \times 10^6$ m/s $\hat{j}$ when it is in a region with a magnetic field given by 0.80 T $\hat{i}$ + 0.60 T $\hat{j}$ − 0.40 T $\hat{k}$?

20 •• The section of wire shown in Figure 26-32 carries a current equal to 1.8 A from a to b. The segment is in a region that has a magnetic field whose value is $1.2\text{ T}\,\hat{k}$. Find the total force on the wire and show that the total force is the same as if the wire were in the form of a straight wire directly from a to b and carrying the same current.

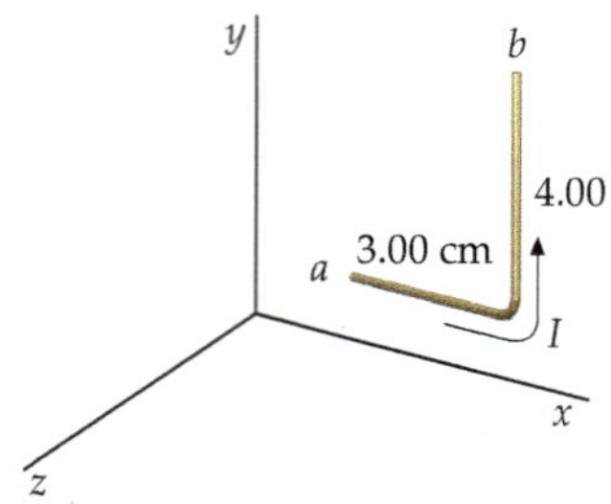

FIGURE 26-32 Problem 20

21 •• A straight, stiff, horizontal 25-cm-long wire that has a mass equal to 50 g is connected to a source of emf by light, flexible leads. A magnetic field of 1.33 T is horizontal and perpendicular to the wire. Find the current necessary to "float" the wire, that is, when the wire is released from rest it remains at rest.

22 •• **ENGINEERING APPLICATION** In your physics laboratory class, you have constructed a simple *gaussmeter* for measuring the horizontal component of magnetic fields. The setup consists of a stiff 50-cm wire that hangs vertically from a conducting pivot so that its free end makes contact with a pool of mercury in a dish below (Figure 26-33). The mercury provides an electrical contact without constraining the movement of the wire. The wire has a mass of 5.0 g and conducts a current downward. (*a*) What is the equilibrium angular displacement of the wire from vertical if the horizontal component of the magnetic field is 0.040 T and the current is 0.20 A? (*b*) What is the sensitivity of this gaussmeter? That is, what is the ratio of the output to the input (in radians per tesla)?

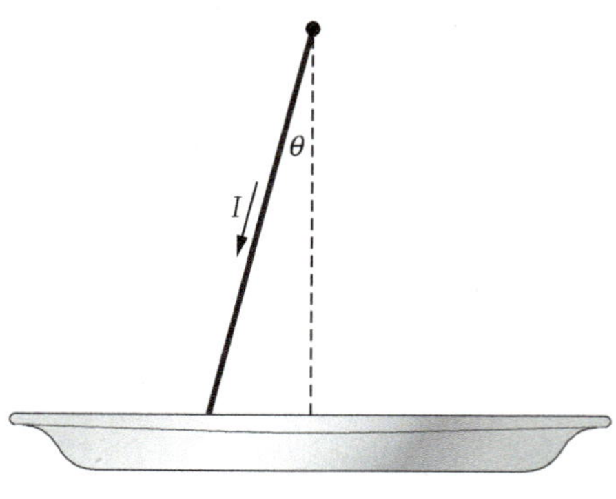

FIGURE 26-33 Problem 22

23 •• A 10-cm-long straight wire is parallel with the x axis and carries a current of 2.0 A in the $+x$ direction. The force on this wire due to the presence of a magnetic field $\vec{B}$ is $3.0\text{ N}\,\hat{j} + 2.0\text{ N}\,\hat{k}$. If this wire is rotated so that it is parallel with the y axis with the current in the $+y$ direction, the force on the wire becomes $-3.0\text{ N}\,\hat{j} - 2.0\text{ N}\,\hat{k}$. Determine the magnetic field $\vec{B}$. SSM

24 •• A 10-cm-long straight wire is parallel with the z axis and carries a current of 4.0 A in the $+z$ direction. The force on this wire due to a uniform magnetic field $\vec{B}$ is $-0.20\text{ N}\,\hat{i} + 0.20\text{ N}\,\hat{j}$. If this wire is rotated so that it is parallel with the x axis with the current in the $+x$ direction, the force on the wire becomes $0.20\,\hat{k}\text{ N}$. Find $\vec{B}$.

25 •• A current-carrying wire is bent into a closed semicircular loop of radius R that lies in the xy plane (Figure 26-34). The wire is in a uniform magnetic field that is in the $+z$ direction, as shown. Verify that the force acting on the loop is zero. SSM

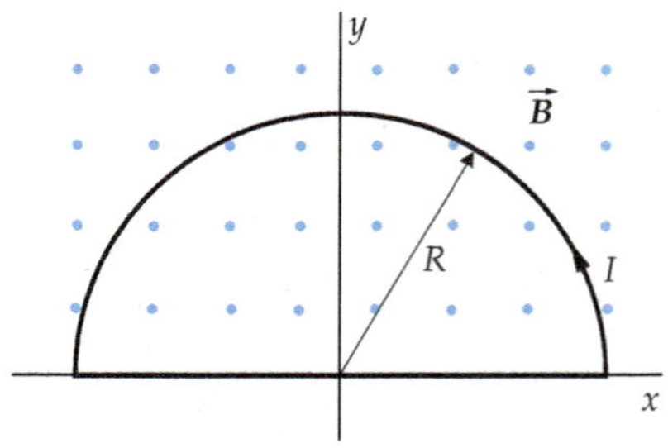

FIGURE 26-34 Problem 25

26 ••• A wire bent in some arbitrary shape carries a current I. The wire is in a region with a uniform magnetic field $\vec{B}$. Show that the total force on the part of the wire from some arbitrary point on the wire (designated as a) to some other arbitrary point on the wire (designated as b) is $\vec{F} = I\vec{L} \times \vec{B}$, where $\vec{L}$ is the vector from point a to point b. In other words, show that the force on an arbitrary section of the bent wire is the same as the force on a straight section wire carrying the same current and connecting the two endpoints of the arbitrary section.

MOTION OF A POINT CHARGE IN A MAGNETIC FIELD

27 • A proton moves in a 65-cm-radius circular orbit that is perpendicular to a uniform magnetic field of magnitude 0.75 T. (*a*) What is the orbital period for the motion? (*b*) What is the speed of the proton? (*c*) What is the kinetic energy of the proton? SSM

28 • A 4.5-keV electron (an electron that has a kinetic energy equal to 4.5 keV) moves in a circular orbit that is perpendicular to a magnetic field of 0.325 T. (*a*) Find the radius of the orbit. (*b*) Find the frequency and period of the orbital motion.

29 •• A proton, deuteron, and an alpha particle in a region with a uniform magnetic field each follow circular paths that have the same radius. The deuteron has a charge that is equal to the charge a proton has, and an alpha particle has a charge that is equal to twice the charge a proton has. Assume that $m_\alpha = 2m_d = 4m_p$. Compare (*a*) their speeds, (*b*) their kinetic energies, and (*c*) the magnitudes of their angular momenta about the centers of the orbits.

30 •• A particle has a charge q, a mass m, a linear momentum of magnitude p, and a kinetic energy K. The particle moves in a circular orbit of radius R perpendicular to a uniform magnetic field $\vec{B}$. Show that (*a*) $p = BqR$ and (*b*) $K = \frac{1}{2}B^2q^2R^2/m$.

31 •• A beam of particles with velocity $\vec{v}$ enters a region that has a uniform magnetic field $\vec{B}$ in the $+x$ direction. Show that when the x component of the displacement of one of the particles is $2\pi(m/qB)v\cos\theta$, where θ is the angle between $\vec{v}$ and $\vec{B}$, the velocity of the particle is in the same direction as it was when the particle entered the field. SSM

32 •• A proton that has a speed equal to 1.00×10^6 m/s enters a region with a uniform magnetic field that has a magnitude of 0.800 T and points into the page, as shown in Figure 26-35. The proton enters the region at an angle $\theta = 60°$. Find the exit angle ϕ and the distance d.

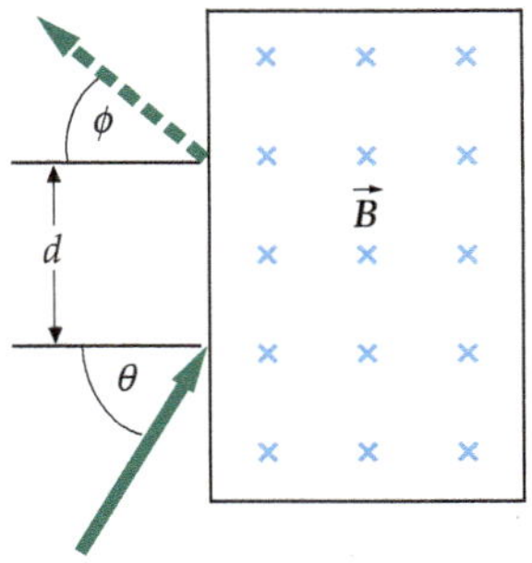

FIGURE 26-35
Problems 32 and 33

33 •• Suppose that in Figure 26-35, the magnetic field has a magnitude of 60 mT, the distance d is 40 cm, and θ is 24°. Find the speed v at which a particle enters the region and the exit angle ϕ if the particle is (*a*) a proton and (*b*) a deuteron. Assume that $m_d = 2m_p$. SSM

34 •• The galactic magnetic field in some region of interstellar space has a magnitude of 1.00×10^{-9} T. A particle of interstellar dust has a mass of 10.0 μg and a total charge of 0.300 nC. How many years does it take for the particle to complete a revolution of the circular orbit caused by its interaction with the magnetic field?

APPLICATIONS OF THE MAGNETIC FORCE ACTING ON CHARGED PARTICLES

35 • A velocity selector has a magnetic field that has a magnitude equal to 0.28 T and is perpendicular to an electric field that has a magnitude equal to 0.46 MV/m. (*a*) What must the speed of a particle be for that particle to pass through the velocity selector undeflected? What kinetic energy must (*b*) protons and (*c*) electrons have in order to pass through the velocity selector undeflected? SSM

36 •• A beam of protons is moving in the $+x$ direction with a speed of 12.4 km/s through a region in which the electric field is perpendicular to the magnetic field. The beam is not deflected in this region. (*a*) If the magnetic field has a magnitude of 0.85 T and points in the $+y$ direction, find the magnitude and direction of the electric field. (*b*) Would electrons that have the same velocity as the protons be deflected by these fields? If so, in what direction would they be deflected? If not, why not?

37 •• The plates of a Thomson q/m apparatus are 6.00 cm long and are separated by 1.20 cm. The end of the plates is 30.0 cm from the tube screen. The kinetic energy of the electrons is 2.80 keV. If a potential difference of 25.0 V is applied across the deflection plates, by how much will the point where the beam of electrons strikes the screen be displaced?

38 •• Chlorine has two stable isotopes, ^{35}Cl and ^{37}Cl. Chlorine gas which consists of singly ionized ions is to be separated into its isotopic components using a mass spectrometer. The magnetic field strength in the spectrometer is 1.2 T. What is the minimum value of the potential difference through which these ions must be accelerated so that the separation between them, after they complete their semicircular path, is 1.4 cm?

39 •• In a mass spectrometer, a singly ionized ^{24}Mg ion has a mass equal to 3.983×10^{-26} kg and is accelerated through a 2.50-kV potential difference. It then enters a region where it is deflected by a magnetic field of 557 G. (*a*) Find the radius of curvature of the ion's orbit. (*b*) What is the difference in the orbital radii of the ^{26}Mg and ^{24}Mg ions? Assume that their mass ratio is 26:24. SSM

40 •• A beam of singly ionized ^{6}Li and ^{7}Li ions passes through a velocity selector and enters a region of uniform magnetic field with a velocity that is perpendicular to the direction of the field. If the diameter of the orbit of the ^{6}Li ions is 15 cm, what is the diameter of the orbit for ^{7}Li ions? Assume their mass ratio is 7:6.

41 •• Using Example 26-6, determine the time required for a ^{58}Ni ion and a ^{60}Ni ion to complete the semicircular path.

42 •• Before entering a mass spectrometer, ions pass through a velocity selector consisting of parallel plates that are separated by 2.0 mm and have a potential difference of 160 V. The magnetic field strength is 0.42 T in the region between the plates. The magnetic field strength in the mass spectrometer is 1.2 T. Find (*a*) the speed of the ions entering the mass spectrometer and (*b*) the difference in the diameters of the orbits of singly ionized ^{238}U and ^{235}U. The mass of a ^{235}U ion is 3.903×10^{-25} kg.

43 •• A cyclotron for accelerating protons has a magnetic field strength of 1.4 T and a radius of 0.70 m. (*a*) What is the cyclotron's frequency? (*b*) Find the kinetic energy of the protons when they emerge. (*c*) How will your answers change if deuterons are used instead of protons? SSM

44 •• A certain cyclotron that has a magnetic field whose magnitude is 1.8 T is designed to accelerate protons to a kinetic energy of 25 MeV. (*a*) What is the cyclotron frequency for this cyclotron? (*b*) What must the minimum radius of the magnet be to achieve this energy? (*c*) If the alternating potential difference applied to the dees has a maximum value of 50 kV, how many revolutions must the protons make before emerging with kinetic energies of 25 MeV?

45 •• Show that for a given cyclotron the cyclotron frequency for accelerating deuterons is the same as the frequency for accelerating alpha particles and is half the frequency for accelerating protons in the same magnetic field. The deuteron has a charge that is equal to the charge a proton has, and an alpha particle has a charge that is equal to twice the charge a proton has. Assume that $m_\alpha = 2m_d = 4m_p$.

46 ••• Show that the radius of the orbit of a charged particle in a cyclotron is proportional to the square root of the number of orbits completed.

TORQUES ON CURRENT LOOPS, MAGNETS, AND MAGNETIC MOMENTS

47 • A small circular coil consisting of 20 turns of wire lies in a region with a uniform magnetic field whose magnitude is 0.50 T. The arrangement is such that the normal to the plane of the coil makes an angle of 60° with the direction of the magnetic field. The radius of the coil is 4.0 cm, and the wire carries a current of 3.0 A. (*a*) What is the magnitude of the magnetic moment of the coil? (*b*) What is the magnitude of the torque exerted on the coil? SSM

48 • What is the maximum torque on a 400-turn circular coil of radius 0.75 cm that carries a current of 1.6 mA and is in a region with a uniform magnetic field of 0.25 T?

49 • A current-carrying wire is in the shape of a square of edge length 6.0 cm. The square lies in the $z = 0$ plane. The wire carries a current of 2.5 A. What is the magnitude of the torque on the wire if it is in a region with a uniform magnetic field of magnitude 0.30 T that points in the (*a*) $+z$ direction and (*b*) $+x$ direction? SSM

50 • A current-carrying wire is in the shape of an equilateral triangle of edge length 8.0 cm. The triangle lies in the $z = 0$ plane. The wire carries a current of 2.5 A. What is the magnitude of the torque on the wire if it is in a region with a uniform magnetic field of magnitude 0.30 T that points in the (*a*) $+z$ direction and (*b*) $+x$ direction?

51 •• A rigid wire is in the shape of a square of edge length L. The square has mass m and the wire carries current I. The square lies on a flat horizontal surface in a region where there is a magnetic field of magnitude B that is parallel to two edges of the square. What is the minimum value of B so that one edge of the square will lift off the surface?

52 •• A rectangular current-carrying 50-turn coil, as shown in Figure 26-36, is pivoted about the z axis. (*a*) If the wires in the $z = 0$ plane make an angle $\theta = 37°$ with the y axis, what angle does the magnetic moment of the coil make with the unit vector $\hat{i}$? (*b*) Write an expression for $\hat{n}$ in terms of the unit vectors $\hat{i}$ and $\hat{j}$, where $\hat{n}$ is a unit vector in the direction of the magnetic moment. (*c*) What is the magnetic moment of the coil? (*d*) Find the torque on the coil when there is a uniform magnetic field $\vec{B} = 1.5\text{ T}\,\hat{j}$ in the region occupied by the coil. (*e*) Find the potential energy of the coil in this field. (The potential energy is zero when $\theta = 0$.)

FIGURE 26-36
Problems 52 and 53

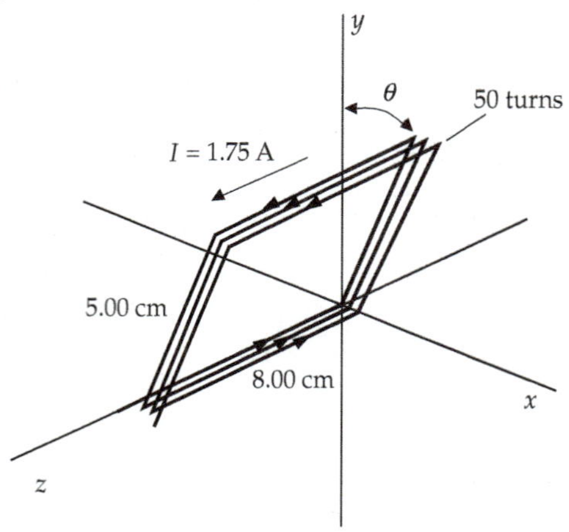

53 •• For the coil in Problem 52 the magnetic field is now $\vec{B} = 2.0\text{ T}\,\hat{j}$. Find the torque exerted on the coil when $\hat{n}$ is equal to (*a*) $\hat{i}$, (*b*) $\hat{j}$, (*c*) $-\hat{j}$, and (*d*) $(\hat{i} + \hat{j})/\sqrt{2}$. SSM

54 •• A small bar magnet has a length equal to 6.8 cm and its magnetic moment is aligned with a uniform magnetic field of magnitude 0.040 T. The bar magnet is then rotated through an angle of 60° about an axis perpendicular to its length. The observed torque on the bar magnet has a magnitude of 0.10 N · m. (*a*) Find the magnetic moment of the magnet. (*b*) Find the potential energy of the magnet.

55 •• A wire loop consists of two semicircles connected by straight segments (Figure 26-37). The inner and outer radii are 0.30 m and 0.50 m, respectively. A current of 1.5 A is in this wire and the current in the outer semicircle is in the clockwise direction. What is the magnetic moment of this current loop?

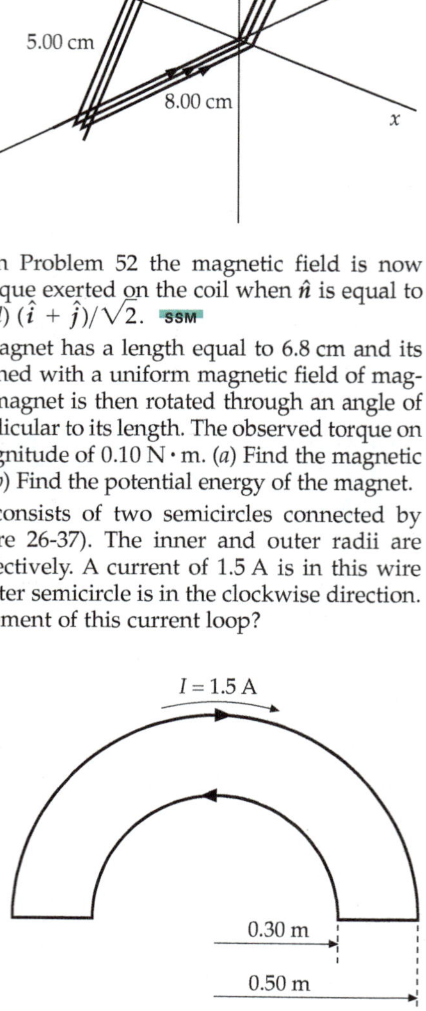

FIGURE 26-37
Problem 55

56 •• A wire of length L is wound into a circular coil that has N turns. Show that when the wire carries a current I, the magnetic moment of the coil has a magnitude given by $IL^2/(4\pi N)$.

57 •• A particle that has a charge q and a mass m moves with angular velocity ω in a circular path of radius r. (*a*) Show that the average current created by this moving particle is $\omega q/(2\pi)$ and that the magnetic moment of its orbit has a magnitude of $\frac{1}{2}q\omega r^2$. (*b*) Show that the angular momentum of this particle has the magnitude of $mr^2\omega$ and that the magnetic moment and angular momentum vectors are related by $\vec{\mu} = \frac{1}{2}(q/m)\,\vec{L}$, where $\vec{L}$ is the angular momentum about the center of the circle. SSM

58 ••• A uniformly charged nonconducting cylindrical shell (Figure 26-38) has length L, inner and outer radii R_i and R_o, respectively, a charge density ρ, and an angular velocity ω about its axis. Derive an expression for the magnetic moment of the cylinder.

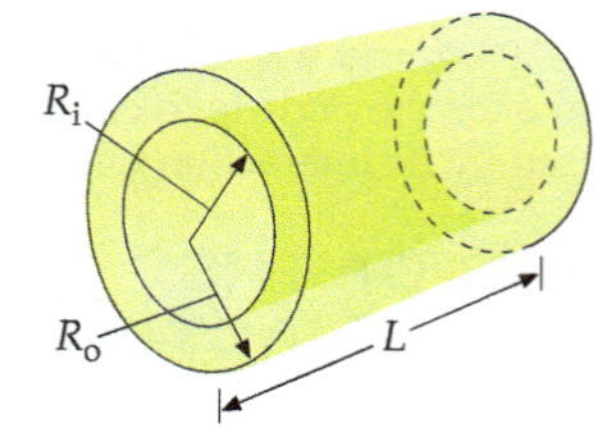

FIGURE 26-38 **Problem 58**

59 ••• A uniform nonconducting thin rod of mass m and length L has a uniform charge per unit length λ and rotates with angular speed ω about an axis through one end and perpendicular to the rod. (*a*) Consider a small segment of the rod of length dx and charge $dq = \lambda\,dr$ at a distance r from the pivot (Figure 26-39). Show that the average current created by this moving segment is $\omega\,dq/(2\pi)$ and show that the magnetic moment of this segment is $\frac{1}{2}\lambda\omega r^2\,dx$. (*b*) Use this to show that the magnitude of the magnetic moment of the rod is $\frac{1}{6}\lambda\omega L^3$. (*c*) Show that the magnetic moment $\vec{\mu}$ and angular momentum $\vec{L}$ are related by $\vec{\mu} = \frac{1}{2}(Q/m)\vec{L}$, where Q is the total charge on the rod. SSM

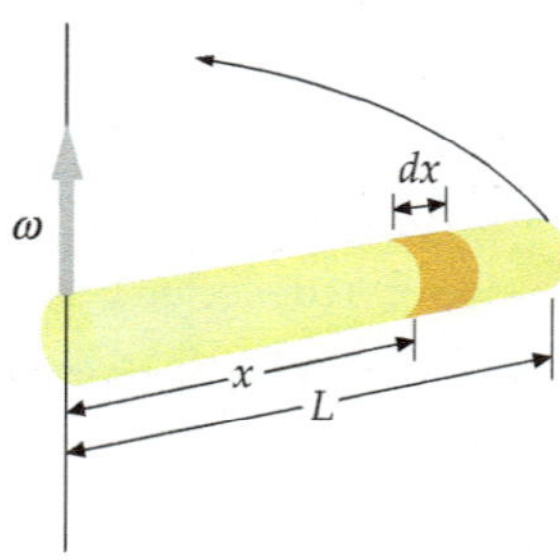

FIGURE 26-39
Problem 59

60 ••• A nonuniform, nonconducting thin disk of mass m, radius R, and total charge Q has a charge per unit area σ that varies as $\sigma_0 r/R$ and a mass per unit area σ_m that is given by $(m/Q)\sigma$. The disk rotates with angular speed ω about its central axis. (*a*) Show that the magnetic moment of the disk has a magnitude $\frac{1}{5}\pi\omega\sigma_0 R^4$, which can be alternatively rewritten as $\frac{3}{10}\omega QR^2$. (*b*) Show that the magnetic moment $\vec{\mu}$ and angular momentum $\vec{L}$ are related by $\vec{\mu} = \frac{1}{2}(Q/m)\vec{L}$.

61 ••• A spherical shell of radius R carries a constant surface charge density σ. The shell rotates about its diameter with angular speed ω. Find the magnitude of the magnetic moment of the rotating spherical shell. SSM

62 ••• A uniform, solid, uniformly charged sphere of radius R has a volume charge density ρ. The sphere rotates about an axis through its center with angular speed ω. Find the magnitude of the magnetic moment of the rotating sphere.

63 ••• A uniform, thin, uniformly charged disk of mass m, radius R, and uniform surface charge density σ rotates with angular speed ω about an axis through its center and perpendicular to the disk (Figure 26-40). The disk is in a region with a uniform magnetic field $\vec{B}$ that makes an angle θ with the rotation axis. Calculate (*a*) the magnitude of the torque exerted on the disk by the magnetic field and (*b*) the precession frequency of the disk in the magnetic field.

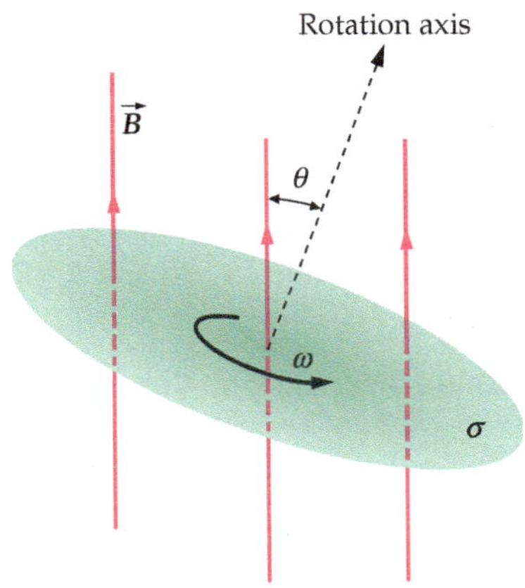

FIGURE 26-40 Problem 63

THE HALL EFFECT

64 • A metal strip that is 2.00 cm wide and 0.100 cm thick carries a current of 20.0 A in a region with a uniform magnetic field of 2.00 T, as shown in Figure 26-41. The Hall voltage is measured to be 4.27 μV. (*a*) Calculate the drift speed of the free electrons in the strip. (*b*) Find the number density of the free electrons in the strip. (*c*) Is point *a* or point *b* at the higher potential? Explain your answer.

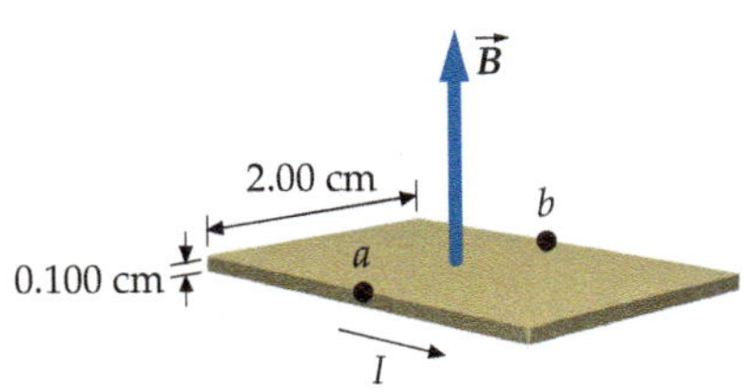

FIGURE 26-41 Problems 64 and 65

65 •• The number density of free electrons in copper is 8.47×10^{22} electrons per cubic centimeter. If the metal strip in Figure 26-41 is copper and the current is 10.0 A, find (*a*) the drift speed v_d and (*b*) the potential difference $V_a - V_b$. Assume that the magnetic field strength is 2.00 T. SSM

66 •• **ENGINEERING APPLICATION** A copper strip has 8.47×10^{22} free electrons per cubic centimeter, is 2.00 cm wide, is 0.100 cm thick, and is used to measure the magnitudes of unknown magnetic fields that are perpendicular to it. Find the magnitude B when the current is 20.0 A and the Hall voltage is (*a*) 2.00 μV, (*b*) 5.25 μV, and (*c*) 8.00 μV.

67 •• **BIOLOGICAL APPLICATION** Because blood contains ions, moving blood in the presence of a magnetic field develops a Hall voltage across the diameter of an artery. A large artery that has a diameter of 0.85 cm can have blood flowing through it with a maximum speed of 0.60 m/s. If a section of the artery is in a magnetic field of 0.20 T, what is the maximum potential difference across the diameter of the artery?

68 •• The Hall coefficient R_H is a property of conducting material (just as resistivity is). It is defined as $R_H = E_y/(J_x B_z)$, where J_x is the x component of the current density in the material, B_z is the z component of the magnetic field, and E_y is the y component of the resulting Hall electric field. Show that the Hall coefficient is equal to $1/(nq)$, where q is the charge of the charge carriers ($-e$ if they are electrons). (The Hall coefficients of monovalent metals, such as copper, silver, and sodium, are therefore negative.)

69 •• Aluminum has a density of 2.7×10^3 kg/m^3 and a molar mass of 27 g/mol. The Hall coefficient of aluminum is $R = -0.30 \times 10^{-10}$ m^3/C. (See Problem 68 for the definition of R.) What is the number of conduction electrons per aluminum atom? SSM

GENERAL PROBLEMS

70 • A long wire parallel to the x axis carries a current of 6.50 A in the $+x$ direction. The wire occupies a region that has a uniform magnetic field $\vec{B} = 1.35\text{ T}\,\hat{j}$. Find the magnetic force per unit length on the wire.

71 • An alpha particle (charge $+2e$) travels in a circular path of radius 0.50 m in a region with a magnetic field whose magnitude is 0.10 T. Find (*a*) the period, (*b*) the speed, and (*c*) the kinetic energy (in electron volts) of the alpha particle. (The mass of an alpha particle is 6.65×10^{-27} kg.)

72 •• The pole strength q_m of a bar magnet is defined by $\vec{\mu} = q_m \vec{\ell}$, where $\vec{\mu}$ is the magnetic moment of the magnet and $\vec{\ell}$ is the position of the north-pole end of the magnet relative to the south-pole end. Show that the torque exerted on a bar magnet in a uniform magnetic field $\vec{B}$ is the same as if a force $+q_m\vec{B}$ is exerted on the north pole of the magnet and a force $-q_m\vec{B}$ is exerted on the south pole.

73 •• A particle of mass m and charge q enters a region where there is a uniform magnetic field $\vec{B}$ parallel with the x axis. The initial velocity of the particle is $\vec{v} = v_{0x}\hat{i} + v_{0y}\hat{j}$, so the particle moves in a helix. (*a*) Show that the radius of the helix is $r = mv_{0y}/qB$. (*b*) Show that the particle takes a time $\Delta t = 2\pi m/qB$ to complete each turn of the helix. (*c*) What is the x component of the displacement of the particle during the time given in Part (*b*)? SSM

74 •• A metal cross-bar of mass m rides on a parallel pair of long horizontal conducting rails separated by a distance L and connected to a device that supplies constant current I to the circuit, as shown in Figure 26-42. The circuit is in a region with a uniform magnetic field $\vec{B}$ whose direction is vertically downward. There is no friction and the bar starts from rest at $t = 0$. (*a*) In which direction will the bar start to move? (*b*) Show that at time t the bar has a speed of $(BIL/m)t$.

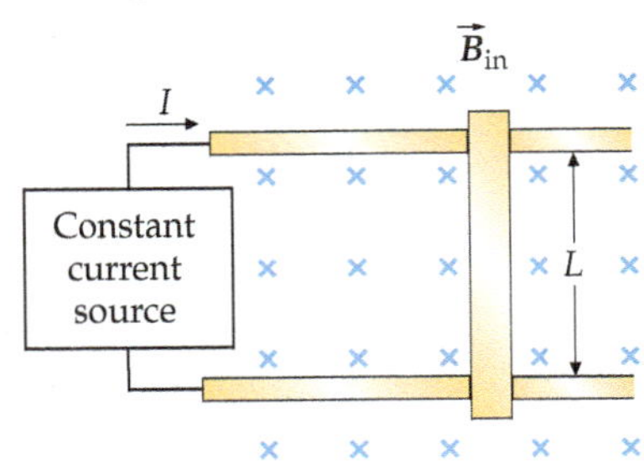

FIGURE 26-42 Problems 74 and 75

75 •• Assume that the rails Problem 74 are frictionless but tilted upward so that they make an angle θ with the horizontal, and with the current source attached to the low end of the rails. The magnetic field is still directed vertically downward. (*a*) What minimum value of B is needed to keep the bar from sliding down the rails? (*b*) What is the acceleration of the bar if B is twice the value found in Part (*a*)? SSM

76 •• A long, narrow bar magnet that has magnetic moment $\vec{\mu}$ parallel to its long axis is suspended at its center as a frictionless compass needle. When placed in region with a horizontal magnetic field $\vec{B}$, the needle lines up with the field. If it is displaced by a small angle θ, show that the needle will oscillate about its equilibrium position with frequency $f = \frac{1}{2\pi}\sqrt{\frac{\mu B}{I}}$, where I is the moment of inertia of the needle about the point of suspension.

77 •• A straight 20-m-long conducting wire is parallel to the y axis and is moving in the $+x$ direction with a speed of 20 m/s in a region that has a magnetic field given by 0.50 T $\hat{k}$. (*a*) A magnetic force acting on the conduction electrons leaves one end negatively charged due to an electron surplus and the other end positively charged due to an electron deficit. This charge separation process continues until the electric field due to the accumulated positive and negative charges exerts forces on the remaining conduction electrons that exactly balance the magnetic forces acting on them. Find the magnitude and direction of this electric field in the steady-state situation. (*b*) Which end of the wire is positively charged and which end is negatively charged? (*c*) Suppose the moving wire is 2.0 m long. What is the potential difference between its two ends due to this electric field?

78 ••• A circular loop of wire that has a mass m and a constant current I is in a region with a uniform magnetic field. It is initially in equilibrium and its magnetic moment is aligned with the magnetic field. The loop is given a small angular displacement about an axis through its center and perpendicular to the magnetic field and then released. What is the period of the subsequent motion? (Assume that the only torque exerted on the loop is due to the magnetic field and that there are no other forces acting on the loop.)

79 ••• A small bar magnet has a magnetic moment $\vec{\mu}$ that makes an angle θ with the x axis. The magnet is in a region that has a *nonuniform* magnetic field given by $\vec{B} = B_x(x)\hat{i} + B_y(y)\hat{j}$. Using $F_x = -\partial U/\partial x$, $F_y = -\partial U/\partial y$, and $F_z = -\partial U/\partial z$, show that there is a net magnetic force on the magnet that is given by

$$\vec{F} = \mu_x \frac{\partial B_x}{\partial x}\hat{i} + \mu_y \frac{\partial B_y}{\partial y}\hat{j}$$

80 •• A proton, a deuteron, and an alpha particle all have the same kinetic energy. They are moving in a region with a uniform magnetic field that is perpendicular to each of their velocities. Let R_p, R_d, and R_α be the radii of their circular orbits, respectively. The deuteron has a charge that is equal to the charge a proton has, and an alpha particle has a charge that is equal to twice the charge a proton has. Find the ratios R_d/R_p and R_α/R_p. Assume that $m_\alpha = 2m_d = 4m_p$.

81 ••• **ENGINEERING APPLICATION, CONTEXT-RICH** Your forensic chemistry group, working closely with local law enforcement agencies, has acquired a mass spectrometer similar to that discussed in the text. It employs a uniform magnetic field that has a magnitude of 0.75 T. To calibrate the mass spectrometer, you decide to measure the masses of various carbon isotopes by measuring the position of impact of the various singly ionized carbon ions that have entered the spectrometer with a kinetic energy of 25 keV. A wire chamber with position sensitivity of 0.50 mm is part of the apparatus. What will be the limit on its mass resolution (in kg) for ions in this mass range, that is, those whose mass is on the order of that of a carbon atom?

CHAPTER 27

THESE COILS AT THE KETTERING MAGNETICS LABORATORY AT OAKLAND UNIVERSITY ARE CALLED HELMHOLTZ COILS. THEY ARE USED TO CANCEL EARTH'S MAGNETIC FIELD AND TO PROVIDE A UNIFORM MAGNETIC FIELD IN A SMALL REGION OF SPACE FOR STUDYING THE MAGNETIC PROPERTIES OF MATTER. *(Bob Williamson, Oakland University, Rochester, Michigan.)*

Sources of the Magnetic Field

? Do you know how to calculate the magnitude of the magnetic field of a current-carrying coil? (See Example 27-2.)

As we discussed in Chapter 26, an awareness of the power of permanent magnets has been around since the year 1000. However, the study of magnets as they relate to electricity did not occur until 1819 when a Danish physicist, Hans Christian Oersted, discovered that a compass needle is deflected by an electric current. Just a month after Oersted's discovery, Jean-Baptiste Biot and Félix Savart announced the results of their measurements of the torque on a magnet near a long, current-carrying wire and they analyzed these results in terms of the magnetic field produced by each element of the current. André-Marie Ampère did additional experiments and showed that current elements also experience a force in the presence of a magnetic field and that two current elements exert forces on each other.

In this chapter, we begin by considering the magnetic field produced by a single moving charge and by the moving charges in a current element. We then calculate the magnetic fields produced by some common current configurations, such as a straight-wire segment; a long, straight wire; a current loop; and a solenoid. Next we discuss Ampère's law. Finally, we consider the magnetic properties of matter.

27-1 THE MAGNETIC FIELD OF MOVING POINT CHARGES

When a point charge q moves with velocity $\vec{v}$, the moving point charge produces a magnetic field $\vec{B}$ in space, given by*

$$\vec{B} = \frac{\mu_0}{4\pi}\frac{q\vec{v}\times\hat{r}}{r^2} \qquad 27\text{-}1$$

MAGNETIC FIELD OF A MOVING POINT CHARGE

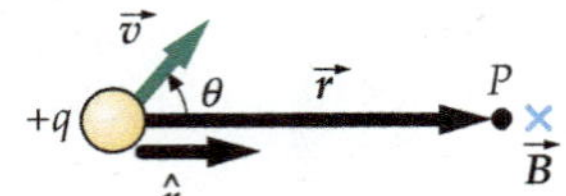

FIGURE 27-1 A positive point charge q moving with velocity $\vec{v}$ produces a magnetic field $\vec{B}$ at a field point P. The magnetic field at P is in the direction of $\vec{v}\times\hat{r}$, where $\vec{v}$ is the velocity of the point charge and $\hat{r}$ is the unit vector pointing from the charge to the field point. The field varies inversely as the square of the distance from the charge to the field point and is proportional to the sine of the angle between $\vec{v}$ and $\hat{r}$. (The blue × at the field point indicates that the direction of the field is into the page.)

where $\hat{r}$ is a unit vector (see Figure 27-1) that points to the field point P from the charge q moving with velocity $\vec{v}$, and μ_0 is a constant of proportionality called the **magnetic constant (permeability of free space),**† which, by definition, has the exact value

$$\mu_0 = 4\pi \times 10^{-7}\,\text{T}\cdot\text{m/A} = 4\pi \times 10^{-7}\,\text{N/A}^2 \qquad 27\text{-}2$$

The units of μ_0 are such that B is in teslas when q is in coulombs, v is in meters per second, and r is in meters. The unit N/A^2 comes from the fact that $1\text{ T} = 1\text{ N}\cdot\text{s/(C}\cdot\text{m)}$. The constant $1/(4\pi)$ is arbitrarily included in Equation 27-1 so that the factor 4π does not appear in Ampère's law (Equation 27-16), which we will study in Section 27-4.

Example 27-1 Magnetic Field of a Moving Point Charge

A point particle that has charge $q = 4.5\,\mu\text{C}$ is moving with velocity $\vec{v} = 3.0\text{ m/s}\,\hat{i}$ along the line $y = 3.0$ m in the $z = 0$ plane. Find the magnetic field at the origin produced by this charge when the charge is at the point $x = -4.0$ m, $y = 3.0$ m, as shown in Figure 27-2.

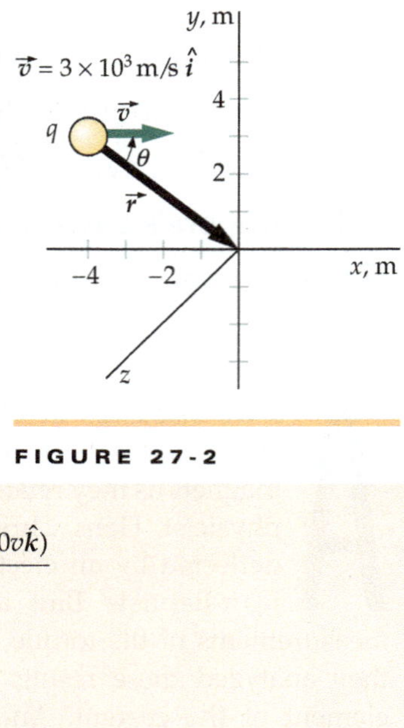

FIGURE 27-2

PICTURE The magnetic field associated with a moving charged particle is given by Equation 27-1.

SOLVE

1. The magnetic field is given by Equation 27-1:

$$\vec{B} = \frac{\mu_0}{4\pi}\frac{q\vec{v}\times\hat{r}}{r^2} \quad \text{where} \quad \vec{v} = v\hat{i}$$

2. Find $\hat{r}$ and r from Figure 27-2 and write $\hat{r}$ in terms of $\hat{i}$ and $\hat{j}$:

$$\vec{r} = 4.0\text{ m}\,\hat{i} - 3.0\text{ m}\,\hat{j}$$

$$r = \sqrt{4.0^2 + 3.0^2}\text{ m} = 5.0\text{ m}$$

$$\hat{r} = \frac{\vec{r}}{r} = \frac{4.0\text{ m}\,\hat{i} - 3.0\text{ m}\,\hat{j}}{5.0\text{ m}} = 0.80\,\hat{i} - 0.60\,\hat{j}$$

3. Substitute the above results in Equation 27-1:

$$\vec{B} = \frac{\mu_0}{4\pi}\frac{q\vec{v}\times\hat{r}}{r^2} = \frac{\mu_0}{4\pi}\frac{q(v\hat{i})\times(0.80\hat{i} - 0.60\hat{j})}{r^2} = \frac{\mu_0}{4\pi}\frac{q(-0.60v\hat{k})}{r^2}$$

$$= -(10^{-7}\,\text{T}\cdot\text{m/A})\frac{(4.5\times10^{-6}\text{ C})(0.60)(3.0\text{ m/s})}{(5.0\text{ m})^2}\hat{k}$$

$$= \boxed{-3.2\times10^{-14}\text{ T}\,\hat{k}}$$

CHECK It is also possible to obtain $\vec{B}$ without finding an explicit expression for the unit vector $\hat{r}$. From Figure 27-2 we note that $\vec{v}\times\hat{r}$ is in the $-z$ direction. In addition, the magnitude of $\vec{v}\times\hat{r}$ is $v\sin\theta$, where $\sin\theta = (3.0\text{ m})/(5.0\text{ m}) = 0.60$. Combining these results, we have $\vec{v}\times\hat{r} = v\sin\theta(-\hat{k}) = -v(0.60)\hat{k}$, in agreement with our result in line 1 of step 3.

PRACTICE PROBLEM 27-1 At the same instant, find the magnetic field on the y axis both at $y = 3.0$ m and at $y = 6.0$ m.

* This expression is used only for speeds much much less than the speed of light.

† Some care must be taken not to confuse the constant μ_0 with the magnitude of the magnetic moment vector $\vec{\mu}$.

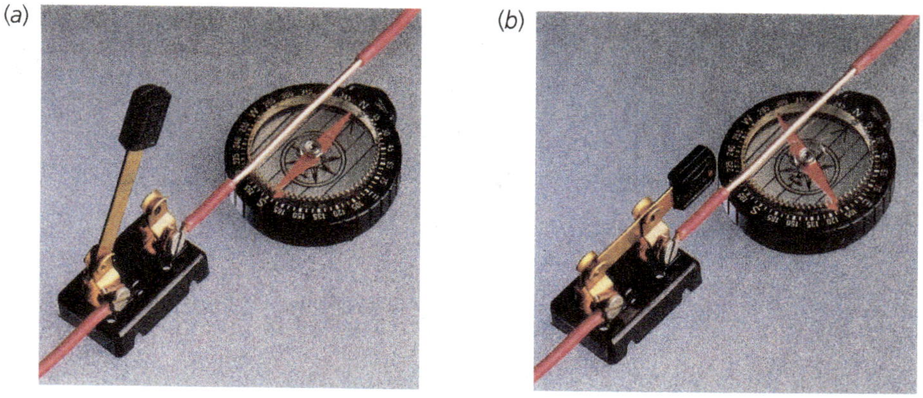

Oersted's experiment. (*a*) When no current exists in the wire, the compass needle points north. (*b*) When the wire carries a current, the compass needle is deflected in the direction of the resultant magnetic field. The current in the wire is directed upward, from left to right. The insulation has been stripped from the wire to improve the contrast of the photograph. (*© 1990 Richard Menga/Fundamental Photographs.*)

27-2 THE MAGNETIC FIELD OF CURRENTS: THE BIOT–SAVART LAW

In the previous chapter, we extended our discussion of forces on point charges to forces on current elements by substituting $q\vec{v}$ with the current element $I\,d\vec{\ell}$. We do the same for the magnetic field produced by a current element. The magnetic field $d\vec{B}$ produced by a current element $I\,d\vec{\ell}$ is given by Equation 27-1, where $q\vec{v}$ is replaced by $I\,d\vec{\ell}$:

$$d\vec{B} = \frac{\mu_0}{4\pi}\frac{I\,d\vec{\ell} \times \hat{r}}{r^2} \qquad \text{27-3}$$

BIOT–SAVART LAW

Equation 27-3, known as the **Biot–Savart law,** was also deduced by Ampère. The Biot–Savart law and Equation 27-1 are analogous to Coulomb's law for the electric field of a point charge. The source of the magnetic field is a moving charge $q\vec{v}$ or a current element $I\,d\vec{\ell}$, just as the charge q is the source of the electrostatic field. The magnetic field decreases with the square of the distance from the moving charge or current element, just as the electric field decreases with the square of the distance from a point charge. However, the directional aspects of the electric and magnetic fields are quite different. Whereas the electric field points in the radial direction $\hat{r}$ from the point charge to the field point (for a positive charge), the magnetic field is perpendicular to both $\hat{r}$ and $\vec{v}$, in the case of a point charge, or $\hat{r}$ and $d\vec{\ell}$ in the case of a current element. At a point along the line of a current element, such as point P_2 in Figure 27-3, the magnetic field due to that current element is zero. (Equation 27-3 gives $d\vec{B} = 0$ if $d\vec{\ell}$ and $\hat{r}$ are either parallel or antiparallel.)

The magnetic field due to the total current in a circuit can be calculated by using the Biot–Savart law to find the field due to each current element, and then summing (integrating) over all the current elements in the circuit. This calculation is challenging for all but the simplest circuit geometries.

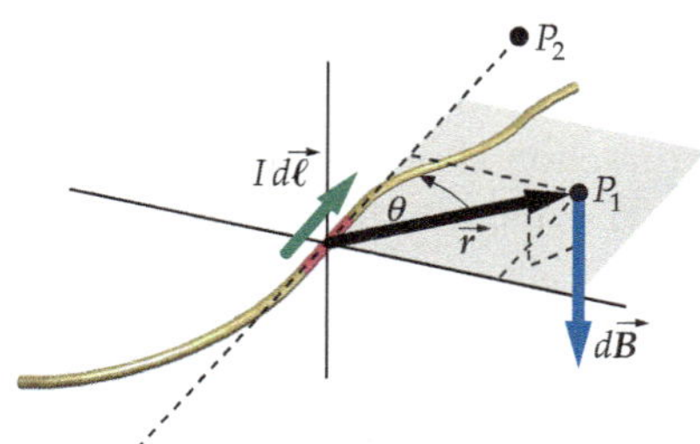

FIGURE 27-3 The current element $I\,d\vec{\ell}$ produces a magnetic field $d\vec{B}$ at point P_1 that is in the direction of $I\,d\vec{\ell} \times \hat{r}$, and thus perpendicular to both $d\vec{\ell}$ and $\hat{r}$. The current element produces no magnetic field at point P_2, which is along the line of $d\vec{\ell}$.

$\vec{B}$ DUE TO A CURRENT LOOP

Figure 27-4 shows a current element $I\,d\vec{\ell}$ of a current loop of radius R and the unit vector $\hat{r}$ that is directed from the element to the center of the loop. The magnetic field at the center of the loop due to this element is directed along the axis of the loop, and its magnitude, obtained by taking the magnitude of both sides of Equation 27-3, is given by

$$dB = \frac{\mu_0}{4\pi}\frac{I\,d\ell\sin\theta}{R^2} \qquad \text{27-4}$$

where θ is the angle between $d\vec{\ell}$ and $\hat{r}$, which is 90° for each current element, so $\sin\theta = 1$. The resultant magnetic field due to all the current elements in the loop is found by integrating Equation 27-4 over all the current elements in the loop.

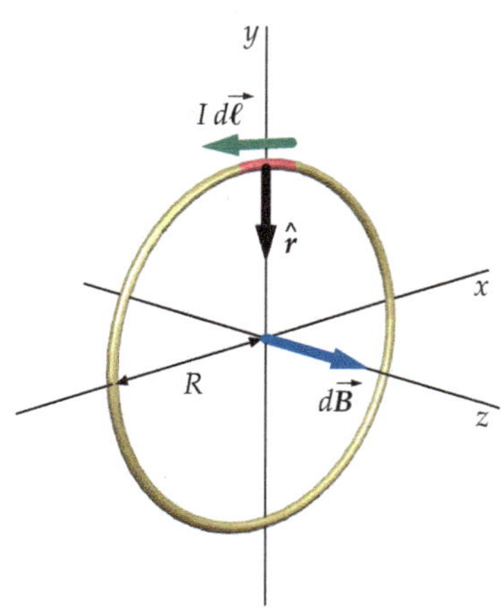

FIGURE 27-4 Current element for calculating the magnetic field at the center of a circular current loop. Each element produces a magnetic field that is directed along the axis of the loop.

Because I and R are the same for all elements, we obtain

$$B = \frac{\mu_0}{4\pi}\frac{I}{R^2}\oint d\ell$$

The integral of $d\ell$ around the complete loop gives the total length $2\pi R$, the circumference of the loop. The magnetic field due to the entire loop is thus

$$B = \frac{\mu_0}{4\pi}\frac{I}{R^2}2\pi R = \frac{\mu_0 I}{2R} \qquad \text{27-5}$$

B AT THE CENTER OF A CURRENT LOOP

PRACTICE PROBLEM 27-2

Find the current in a circular loop of radius 8.0 cm that will produce a magnetic field of 0.20 mT at the center of the loop.

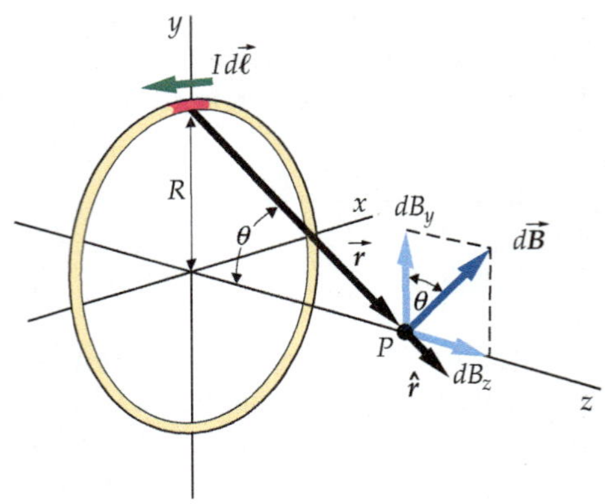

FIGURE 27-5 Geometry for calculating the magnetic field at a point on the axis of a circular current loop.

Figure 27-5 shows the geometry for calculating the magnetic field at a point on the axis of a circular current loop a distance z from the circular loop's center. We first consider the current element at the top of the loop. Here, as everywhere on the loop, $I\,d\vec{\ell}$ is tangent to the loop and perpendicular to the vector $\vec{r}$ from the current element to the field point P. The magnetic field $d\vec{B}$ due to this element is in the direction shown in the figure, perpendicular to $\hat{r}$ and also perpendicular to $I\,d\vec{\ell}$. The magnitude of $d\vec{B}$ is

$$|d\vec{B}| = \frac{\mu_0}{4\pi}\frac{I|d\hat{\ell}\times\hat{r}|}{r^2} = \frac{\mu_0}{4\pi}\frac{I\,d\ell}{(z^2+R^2)}$$

where we have used the facts that $r^2 = z^2 + R^2$ and that $d\vec{\ell}$ and $\hat{r}$ are perpendicular, so $|d\vec{\ell}\times\hat{r}| = d\ell$.

When we sum around all the current elements in the loop, the components of $d\vec{B}$ perpendicular to the axis of the loop, such as dB_y in Figure 27-5, sum to zero, which leave only the components dB_z that are parallel to the axis. We thus compute only the z component of the field. From Figure 27-5, we have

$$dB_z = dB\sin\theta = \left(\frac{\mu_0}{4\pi}\frac{I\,d\ell}{(z^2+R^2)}\right)\left(\frac{R}{\sqrt{z^2+R^2}}\right) = \frac{\mu_0}{4\pi}\frac{IR\,d\ell}{(z^2+R^2)^{3/2}}$$

To find the field due to the entire loop of current, we integrate dB_z around the loop:

$$B_z = \oint dB_z = \oint \frac{\mu_0}{4\pi}\frac{IR}{(z^2+R^2)^{3/2}}d\ell$$

Because neither z nor R varies as we sum over the elements in the loop, we can remove those quantities from the integral. Then,

$$B_z = \frac{\mu_0}{4\pi}\frac{IR}{(z^2+R^2)^{3/2}}\oint d\ell$$

The integral of $d\ell$ around the loop gives $2\pi R$. Thus,

$$B_z = \frac{\mu_0}{4\pi}\frac{IR}{(z^2+R^2)^{3/2}}2\pi R = \frac{\mu_0}{4\pi}\frac{2\pi R^2 I}{(z^2+R^2)^{3/2}} \qquad \text{27-6}$$

B ON THE AXIS OF A CURRENT LOOP

PRACTICE PROBLEM 27-3

Show that Equation 27-6 reduces to $B_z = \frac{1}{2}\mu_0 I/R$ (Equation 27-5) at the center of the loop.

At great distances from the loop, $|z|$ is much greater than R, so $(z^2 + R^2)^{3/2} \approx (z^2)^{3/2} = |z|^3$. Then,

$$B_z = \frac{\mu_0}{4\pi}\frac{2I\pi R^2}{|z|^3}$$

or

$$B_z = \frac{\mu_0}{4\pi}\frac{2\mu}{|z|^3} \qquad 27\text{-}7$$

MAGNETIC-DIPOLE FIELD ON THE AXIS OF THE DIPOLE

where $\mu = I\pi R^2$ is the magnitude of the magnetic moment of the loop. Note the similarity of this expression and the electric field on the axis of an electric dipole whose electric dipole moment has magnitude p (Equation 21-10):

$$E_z = \frac{1}{4\pi\epsilon_0}\frac{2p}{|z|^3}$$

Although it has not been demonstrated, our result that a current loop produces a magnetic dipole field far away holds in general for any point whether it is on the axis of the loop or off of the axis of the loop. Thus, a current loop behaves as a magnetic dipole because it experiences a torque $\vec{\mu} \times \vec{B}$ when placed in an external magnetic field (as was shown in Chapter 26) and it also produces a magnetic dipole field at field points far from the current loop. Figure 27-6 shows the magnetic field lines for a current loop.

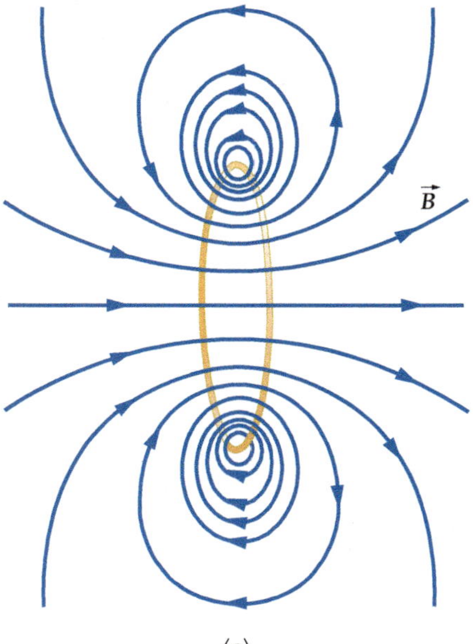

(a)

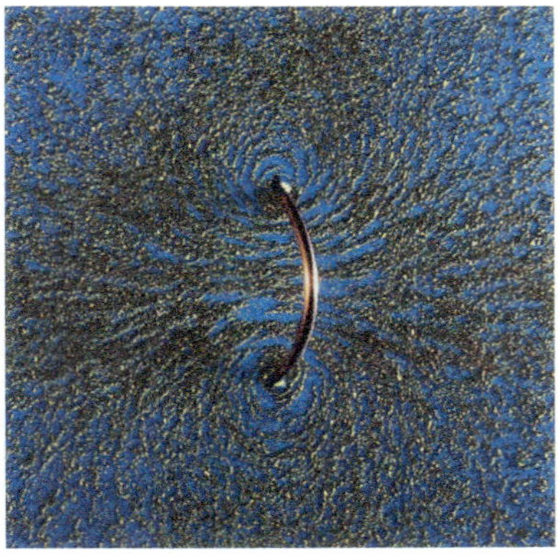

(b)

FIGURE 27-6 (*a*) The magnetic field lines of a circular current loop. (*b*) The magnetic field lines of a circular current loop indicated by iron filings. *(© 1990 Richard Menga/Fundamental Photographs.)*

Example 27-2 Find $\vec{B}$ on Axis of Coil

A circular coil has a radius equal to 5.00 cm, has 12 turns, lies in the $z = 0$ plane, and is centered at the origin. It carries a current of 4.00 A, and the magnetic moment of the coil is in the $+z$ direction. Using Equation 27-6, find the magnetic field on the z axis at (*a*) $z = 0$, (*b*) $z = 15.0$ cm, and (*c*) $z = 3.00$ m. (*d*) Using Equation 27-7, find the magnetic field on the z axis at $z = 3.00$ m.

PICTURE The magnetic field due to a loop that has N turns is N times that due to a single turn. (*a*) At $z = 0$, $B = \frac{1}{2}\mu_0 N/R$ (from Equation 27-5). Equation 27-6 gives the magnetic field on the axis due to the current in a single turn. Far from the loop, as in Part (*c*), the field can be found using Equation 27-7. In that case, because we have N loops, the magnetic moment is $\mu = NIA$, where $A = \pi R^2$.

SOLVE

(*a*) B_z at the center is N times that given by Equation 27-5 for a single turn of the coil:

$$B_z = \frac{\mu_0 NI}{2R} = (4\pi \times 10^{-7}\ \text{T}\cdot\text{m/A})\frac{(12)(4.00\ \text{A})}{2(0.0500\ \text{m})} = \boxed{6.03 \times 10^{-4}\ \text{T}}$$

(*b*) B_z on the axis is N times that given by Equation 27-6:

$$B_z = \frac{\mu_0}{4\pi}\frac{2\pi R^2 NI}{(z^2 + R^2)^{3/2}} = (10^{-7}\ \text{T}\cdot\text{m/A})\frac{2\pi(0.0500\ \text{m})^2(12)(4.00\ \text{A})}{\left[(0.1500\ \text{m})^2 + (0.0500\ \text{m})^2\right]^{3/2}} = \boxed{1.91 \times 10^{-5}\ \text{T}}$$

(*c*) Use Equation 27-6 again:

$$B_z = \frac{\mu_0}{4\pi}\frac{2\pi R^2 NI}{(z^2 + R^2)^{3/2}} = (10^{-7}\ \text{T}\cdot\text{m/A})\frac{2\pi(0.0500\ \text{m})^2(12)(4.00\ \text{A})}{\left[(3.00\ \text{m})^2 + (0.0500\ \text{m})^2\right]^{3/2}} = \boxed{2.79 \times 10^{-9}\ \text{T}}$$

(*d*) 1. Because 3.00 m is much greater than the radius $R = 0.0500$ m, we can use Equation 27-7 for the magnetic field far from the loop:

$$B_z = \frac{\mu_0}{4\pi}\frac{2\mu}{|z|^3}$$

2. The magnitude of the magnetic moment of the loop is N/A:

$$\mu = NI\pi R^2 = (12)(4.00\ \text{A})\,\pi\,(0.0500\ \text{m})^2 = 0.377\ \text{A}\cdot\text{m}^2$$

3. Substitute $\mu = 0.377\ \text{A}\cdot\text{m}^2$ and $z = 3.00$ m into the expression for B_z in step 1:

$$B_z = \frac{\mu_0}{4\pi}\frac{2\mu}{|z|^3} = (10^{-7}\ \text{T}\cdot\text{m/A})\frac{2(0.377\ \text{A}\cdot\text{m}^2)}{(3.00\ \text{m})^3} = \boxed{2.79 \times 10^{-9}\ \text{T}}$$

CHECK In Part (*d*) $z = 60R$, so we were able to use an approximation that is valid for $z \gg R$. The result differs from the exact value, calculated in Part (*c*), by less than one part in 279.

Example 27-3 Calculating the Amount of Mobile Charge

In the coil described in Example 27-2, the current is 4.00 A. Assuming the drift speed is 1.40×10^{-4} m/s, find the number of coulombs of mobile charge (free electrons) in the wire. (The drift speed for a wire carrying a current of 1 A was found to be 3.5×10^{-5} m/s in Example 25-1.)

PICTURE The amount of moving charge Q in the wire is the product of the rate at which charge enters one end of the wire and the time it takes the charge to travel the length of the wire. The rate at which charge enters one end of the wire is the current I, and the time for the charge to travel the length L of the wire is L/v_d, where v_d is the drift speed.

SOLVE

1. The amount of moving charge is the product of the current and the time for a charge carrier to travel the length of the wire:

$$Q = I\,\Delta t$$

2. The drift speed is the length of the wire divided by the time:

$$v_d = \frac{L}{\Delta t}$$

3. The length L is the number of turns multiplied by the length per turn. Also, we solve the step-2 result for the time:

$$L = N2\pi R = (12)2\pi(0.0500\text{ m}) = 3.77\text{ m}$$

and

$$\Delta t = \frac{L}{v_d} = \frac{3.77\text{ m}}{1.40 \times 10^{-4}\text{ m/s}} = 2.69 \times 10^4\text{ s}$$

4. Solve the step-1 result for the amount of moving charge in the wire:

$$Q = I\,\Delta t = (4.00\text{ A})(2.69 \times 10^4\text{ s})$$
$$= \boxed{1.08 \times 10^5\text{ C}}$$

CHECK There is approximately one conduction electron for each atom in a metal. If the wire is made of copper (with an molar mass equal to 63.5 g/mol), it is plausible that 3.77 m of the wire would have a mass approximately equal to 63.5 g. Thus, we estimate about one mole of copper exists in the wire. That means the number of conduction electrons in the wire is approximately equal to Avogadro's number. The total charge Q carried by these electrons is the number of electrons multiplied by the charge per electron. That is, $Q = -N_A e = -(6.02 \times 10^{23})(1.60 \times 10^{-19}\text{ C}) = -0.965 \times 10^5\text{ C}$. The magnitude of this result is very close to our step-4 result.

TAKING IT FURTHER The current consists of about 10^5 C of moving charges. In comparison to the amount of charge stored in an ordinary capacitor, this amount is an enormous amount of charge.

Example 27-4 Torque on a Bar Magnet — *Try It Yourself*

A small bar magnet that has a magnetic moment whose magnitude is $0.0300\text{ A}\cdot\text{m}^2$ is placed at the center of the coil of Example 27-2 so that its magnetic moment vector lies in the $x = 0$ plane and makes an angle of 30° with the $+z$ direction as shown. Neglecting any variation in $\vec{B}$ over the region occupied by the magnet, find the torque on the magnet.

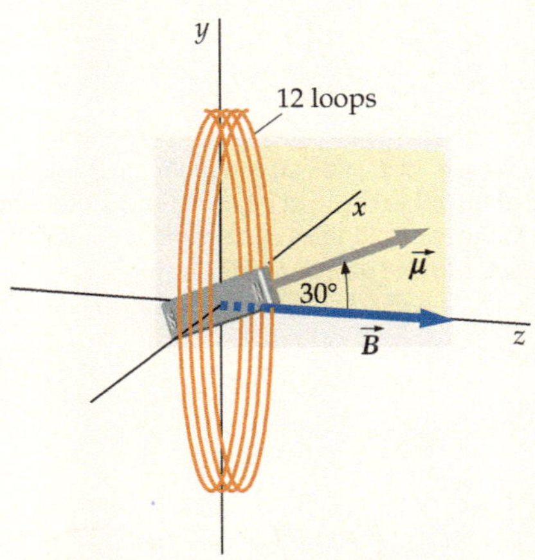

FIGURE 27-7

PICTURE The torque on a magnetic moment is given by $\vec{\tau} = \vec{\mu} \times \vec{B}$. $\vec{B}$ is in the $+z$ direction, so you can use the right-hand rule to show that $\vec{\mu} \times \vec{B}$ is in the $+x$ direction (Figure 27-7).

SOLVE

Cover the column to the right and try these on your own before looking at the answers.

Steps	Answers
1. Compute the magnitude of the torque from $\vec{\tau} = \vec{\mu} \times \vec{B}$.	$\tau = 9.04 \times 10^{-6}\text{ N}\cdot\text{m}$
2. Indicate the direction using a unit vector.	$\vec{\tau} = \boxed{(9.04 \times 10^{-6}\text{ N}\cdot\text{m})\hat{i}}$

CHECK We expect the torque to tend to align the magnetic moment with the magnetic field. Thus, a torque vector in the $+x$ direction is as expected.

$\vec{B}$ DUE TO A CURRENT IN A SOLENOID

A **solenoid** is a conducting wire wound into a helix of closely spaced turns, as illustrated in Figure 27-8. A solenoid is used to produce a strong, uniform magnetic field in the region surrounded by its loops. The solenoid's role in magnetism is analogous to that of the parallel-plate capacitor, which produces a strong, uniform electric field between its plates.

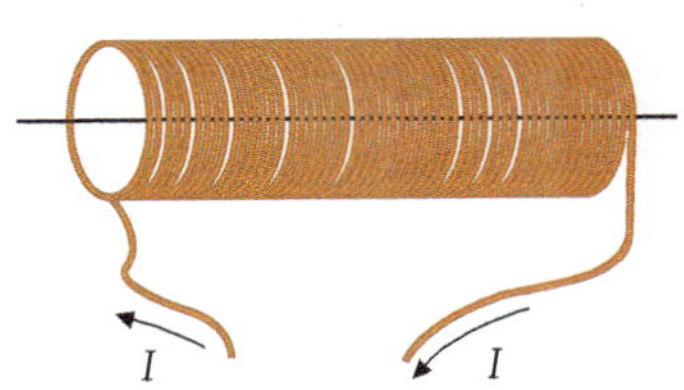

FIGURE 27-8 A tightly wound solenoid can be considered as a set of circular current loops placed side by side that carry the same current. The solenoid produces a uniform magnetic field inside the loops and distant from the ends of the solenoid.

The magnetic field of a solenoid is essentially that of a set of N identical current loops placed side by side. Figure 27-9 shows the magnetic field lines for two such loops.

Figure 27-10*a* shows the magnetic field lines for a tightly wound solenoid. Inside the solenoid and away from the ends, the field lines are approximately parallel to the axis and are closely and uniformly spaced, indicating a strong, uniform magnetic field. Outside the solenoid (above and below it) the lines are much less dense. In addition, the field lines become farther apart as you move away from either end of the solenoid. Comparing this figure with Figure 27-10*b*, we see that the magnetic field of a solenoid, both inside and outside the solenoid, are virtually identical to the magnetic field of a bar magnet of the same size and shape as the solenoid. In Figure 27-10*c*, iron filings align with the field of a current-carrying solenoid.

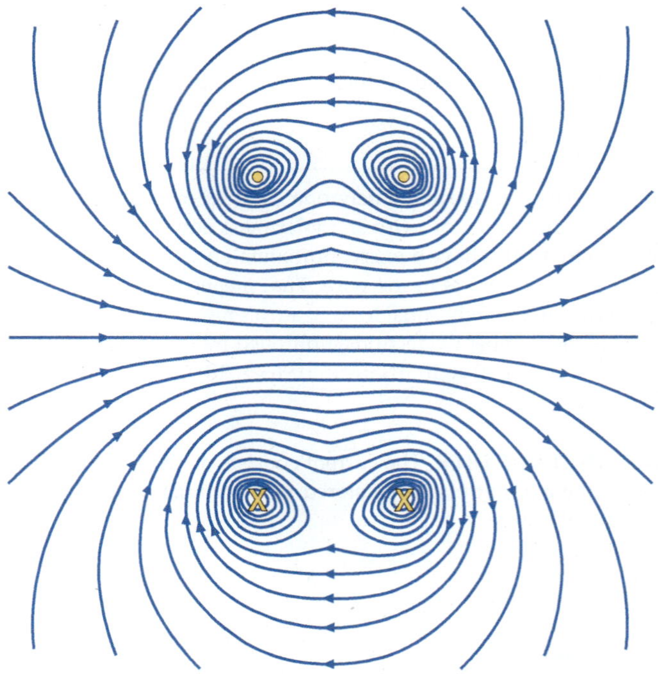

FIGURE 27-9 Magnetic field lines due to two identical coaxial loops carrying the same current. The points where the loops intersect the plane of the page are each marked by an × where the current enters the page and by a dot • where the current emerges from the page. In the region between the loops near the axis the magnetic fields of the individual loops superpose, so the resultant field is strong and surprisingly uniform. The region where the field is uniform is greatest if the planes of the two loops are separated by a distance equal to the radius of the loops.

FIGURE 27-10 (*a*) Magnetic field lines of a solenoid. The lines are identical to those of a bar magnet of the same shape, as in (*b*). (*c*) Magnetic field lines of a solenoid shown by iron filings. *(© 1990 Richard Menga/Fundamental Photographs.)*

(*a*)

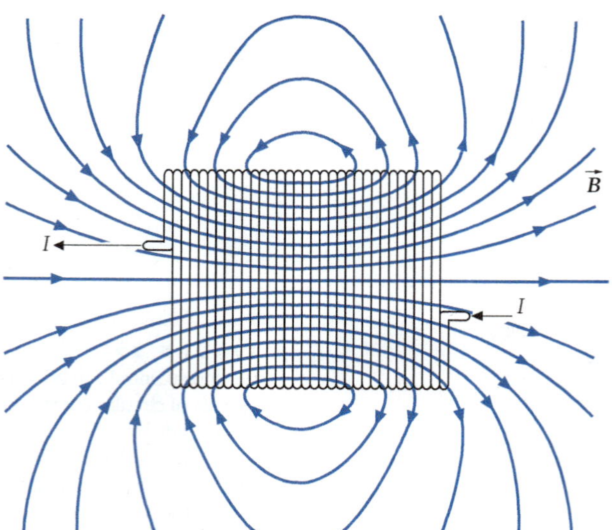

(*b*)

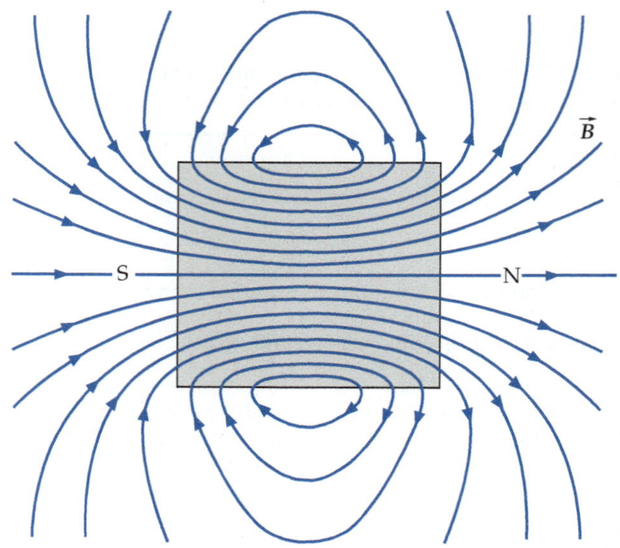

(*c*)

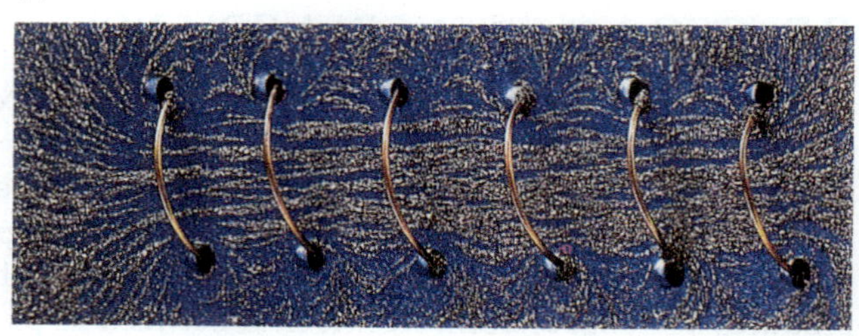

Example 27-5 The Field of a Long, Tightly Wound Solenoid — *Conceptual*

The previous paragraph asserts that magnetic field inside and far from the ends of a long, tightly wound, current-carrying solenoid is uniform and parallel with the axis of the solenoid, and that the magnetic field is zero outside the solenoid. Validate this assertion by modeling the solenoid as a tightly packed stack of current-carrying rings, and by using the field line drawing of a single current-carrying ring (Figure 27-11).

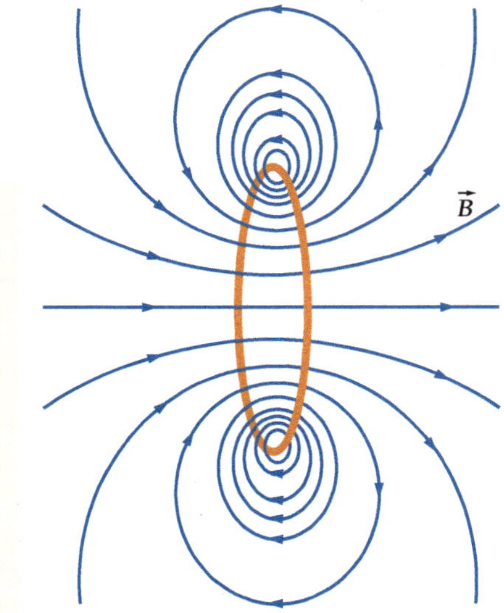

FIGURE 27-11

PICTURE Figure 27-12 shows three equally spaced current-carrying rings representing three loops of a long, tightly wound solenoid. At each of points A, B, and C, where A is just inside loop 2, B is just outside loop 2, and C is at the center of loop 2, sketch the three magnetic field vectors due to the three loops shown. Use the field line drawing of a single current-carrying ring (Figure 27-11) to obtain the directions and relative magnitudes of the three fields. Using your sketch, present an argument that the resultant magnetic fields at points A and B are equal in magnitude and are parallel with the axis of the solenoid. Using your sketch, present an argument that the resultant magnetic field at point C is zero.

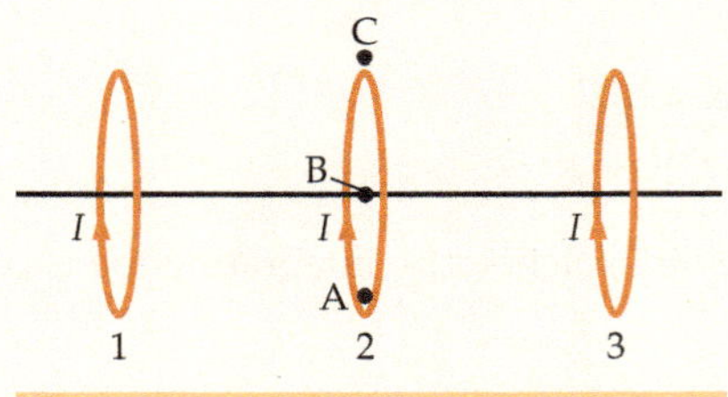

FIGURE 27-12

SOLVE

1. At point A, sketch the magnetic-field vectors $\vec{B}_1$, $\vec{B}_2$, and $\vec{B}_3$ due to the currents in loops 1, 2, and 3, respectively (Figure 27-13). Use Figure 27-11 for guidance:

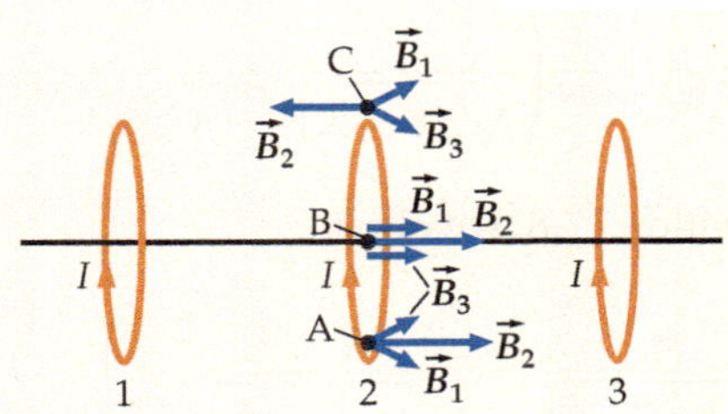

FIGURE 27-13

2. The magnitude of the magnetic field is greater where the field lines are closer together. An examination of the field lines in Figure 27-11 reveals that magnitude of the field $\vec{B}_2$ (due to loop 2) at point A is greater than it is at point B:

The magnitude of the field $\vec{B}_2$ (due to loop 2) is greater at point A than the magnitude of $\vec{B}_2$ is at point B. However, because $\vec{B}_1$, $\vec{B}_2$ and $\vec{B}_3$ are in the same direction, it is feasible that the resultant field ($\vec{B}_1 + \vec{B}_2 + \vec{B}_3$) at point B has the same magnitude as the resultant magnetic field at A.

3. An examination of the field lines in Figure 27-11 reveals that in the plane of the current-carrying ring, at points outside the ring the magnetic field is in the opposite direction than it is at points inside the ring:

At point C, the direction of $\vec{B}_2$ is to the left and the direction of $\vec{B}_1 + \vec{B}_3$ is to the right. In addition, additional loops in the solenoid that are near loops 1 and 3 will produce additional magnetic fields at C whose vector sum is to the right. It is feasible, therefore, that the resultant magnetic field at C is zero.

TAKING IT FURTHER The arguments presented in this example only hold for those sections of the solenoid located far from either end of the solenoid. Suppose loop 2 in Figure 27-13 is not near the center of a long solenoid, but is the last loop at the right end of the solenoid. Then, loop 3 would be absent from the picture, and the three vectors labeled $\vec{B}_3$ would be absent as well.

Consider a solenoid that has a length L, consists of N turns, and carries a current I. We choose the axis of the solenoid to be the z axis, with the left end at $z = z_1$ and the right end at $z = z_2$, as shown in Figure 27-14. We will calculate the magnetic field at field point P on the z axis a distance z from the origin. The figure shows an element of the solenoid of length dz' at a distance z' from the origin. If $n = N/L$ is the number of turns per unit length, there are $n\,dz'$ turns of wire in this element,

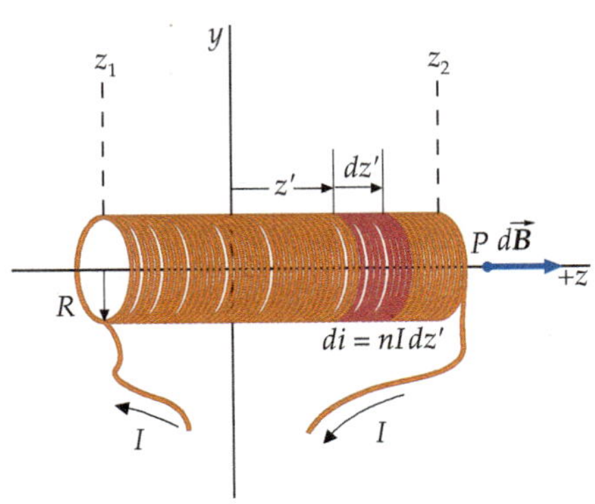

FIGURE 27-14 Geometry for calculating the magnetic field inside a solenoid on its axis. The number of turns in the element dz' is $n\,dz'$, where $n = N/L$ is the number of turns per unit length. The element dz' is treated as a current loop carrying a current $di = nI\,dz'$.

with each turn carrying a current I. The element is thus equivalent to a single loop carrying a current $di = nI\,dz'$. The magnetic field at a point on the z axis due to a loop at the origin carrying current di is given by Equation 27-6:

$$dB_z = \tfrac{1}{2}\mu_0 \frac{R^2\,di}{(z^2 + R^2)^{3/2}}$$

where z is the distance between the loop and the field point. For a loop at $z = z'$ carrying current $di = nI\,dz'$, the distance between the loop and the field point P is $z - z'$, so

$$dB_z = \tfrac{1}{2}\mu_0 \frac{R^2 nI\,dz'}{\left[(z - z')^2 + R^2\right]^{3/2}}$$

We find the magnetic field at P due to the entire solenoid by integrating the expression from $z' = z_1$ to $z' = z_2$:

$$B_z = \tfrac{1}{2}\mu_0 nIR^2 \int_{z_1}^{z_2} \frac{dz'}{\left[(z - z')^2 + R^2\right]^{3/2}} \qquad \text{27-8}$$

The integral in Equation 27-8 can be evaluated using trigonometric substitution with $z - z' = R\tan\theta$. Also, the integral can be looked up in standard tables of integrals. The integral's value is

$$\int_{z_1}^{z_2} \frac{dz'}{\left[(z - z')^2 + R^2\right]^{3/2}} = \frac{1}{R^2}\left(\frac{z - z_1}{\sqrt{(z - z_1)^2 + R^2}} - \frac{z - z_2}{\sqrt{(z - z_2)^2 + R^2}}\right)$$

See Math Tutorial *for more information on* Integrals

Substituting this into Equation 27-8, we obtain

$$B_z(z) = \tfrac{1}{2}\mu_0 nI\left(\frac{z - z_1}{\sqrt{(z - z_1)^2 + R^2}} - \frac{z - z_2}{\sqrt{(z - z_2)^2 + R^2}}\right) \qquad \text{27-9}$$

B_z ON THE AXIS OF A SOLENOID

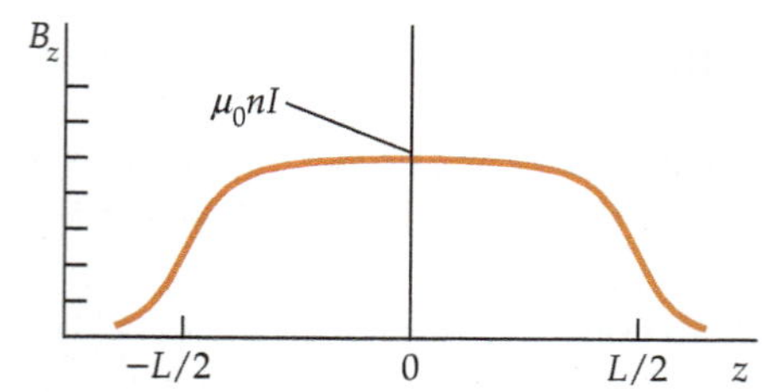

FIGURE 27-15 Graph of the magnetic field on the axis inside a solenoid versus the position z on the axis. The field inside the solenoid is nearly constant except near the ends. The length L of the solenoid is ten times longer than the radius.

A solenoid is called a long solenoid if its length L is much greater than its radius R. Inside and far from the ends of a long solenoid, the fraction on the left in the parentheses approaches +1 and the fraction on the right approaches −1. This means the expression in the parentheses tends toward +2. Thus, in the region inside and far from either end of the solenoid the magnetic field is given by

$$B_z = \mu_0 nI \qquad \text{27-10}$$

B_z INSIDE A LONG SOLENOID

To evaluate B_z at the right end of the solenoid we use Equation 27-9 with $z = z_2$. This gives $B_z(z_2) = \tfrac{1}{2}\mu_0 nI\, L/\sqrt{L^2 + R^2}$, where $L = z_2 - z_1$. Then, if $L \gg R$, the ratio $L/\sqrt{L^2 + R^2}$ becomes very close to one, so $B_z(z_2) \approx \tfrac{1}{2}\mu_0 nI$. Thus, B_z at either end of a long solenoid is half the value of B at points deep inside the solenoid (distant from either end). Figure 27-15 gives a plot of the magnetic field on the axis of a solenoid versus position z on the axis (with the origin at the center of the solenoid). The approximation that the field is uniform (independent of the position) along the axis is good, except for near the ends.

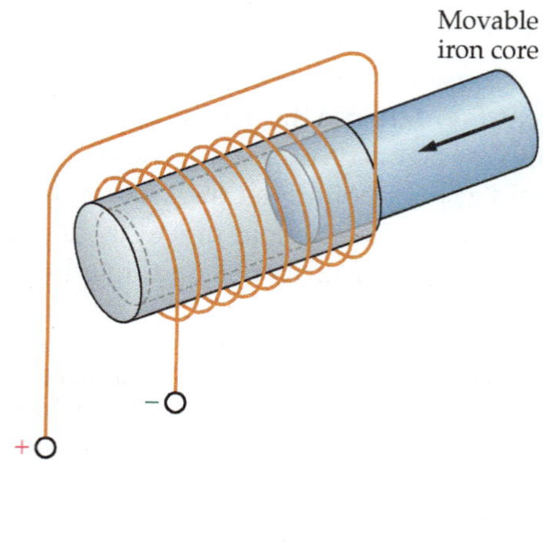

FIGURE 27-16 An automotive starter solenoid. When the solenoid is energized, its magnetic field pulls in the iron core. This engages gears that connect the starter motor to the flywheel of the engine. Once the current to the solenoid is interrupted, a spring disengages the gears and pushes the iron core to the right.

Example 27-6 $\vec{B}$ at Center of a Solenoid

Find the magnetic field at the center of a solenoid that has a length equal to 20.0 cm, a radius equal to 1.40 cm, 600 turns, and a current equal to 4.00 A.

PICTURE To find B on the axis of the solenoid we apply Equation 27-9 with the origin at the center of the solenoid.

SOLVE

1. We will calculate the field exactly, using Equation 27-9:

$$B_z(z) = \tfrac{1}{2}\mu_0 nI\left(\frac{z - z_1}{\sqrt{(z - z_1)^2 + R^2}} - \frac{z - z_2}{\sqrt{(z - z_2)^2 + R^2}}\right)$$

2. To find the magnetic field at the middle of the solenoid, we choose the middle of the solenoid as the origin. Then we set $z = 0$, $z_1 = -\frac{1}{2}L$, and $z_2 = \frac{1}{2}L$:

$$B_z(0) = \tfrac{1}{2}\mu_0 nI\left(\frac{0 - \left(-\frac{1}{2}L\right)}{\sqrt{\left[0 - \left(-\frac{1}{2}L\right)\right]^2 + R^2}} - \frac{0 - \left(\frac{1}{2}L\right)}{\sqrt{\left[\left(0 - \left(\frac{1}{2}L\right)\right)^2\right] + R^2}}\right)$$

$$= \tfrac{1}{2}\mu_0 nI\frac{L}{\sqrt{\frac{1}{4}L^2 + R^2}} = \mu_0 nI\frac{L}{\sqrt{L^2 + 4R^2}}$$

3. Substituting in the given information, we have:

$$\frac{L}{\sqrt{L^2 + 4R^2}} = \frac{20.0\text{ cm}}{\sqrt{(20.0\text{ cm})^2 + 4(1.40\text{ cm})^2}} = 0.990$$

$$B_z(0) = 0.990\,\mu_0 nI$$

$$= 0.990(4\pi \times 10^{-7}\,\text{T}\cdot\text{m/A})\frac{600}{0.200\text{ m}}(4.00\text{ A})$$

$$= \boxed{1.50 \times 10^{-2}\text{ T}}$$

CHECK Note that the approximation obtained using Equation 27-10 amounts to replacing 0.990 by 1 in step 3. Doing so gives a result that differs from the step-3 result by only one percent. This result is expected for a solenoid whose length to radius ratio is 20 cm/1.4 cm ≈ 14.

PRACTICE PROBLEM 27-4 Calculate B on the axis a distance halfway between the center and one end of the solenoid. Compare this result with the step-3 result.

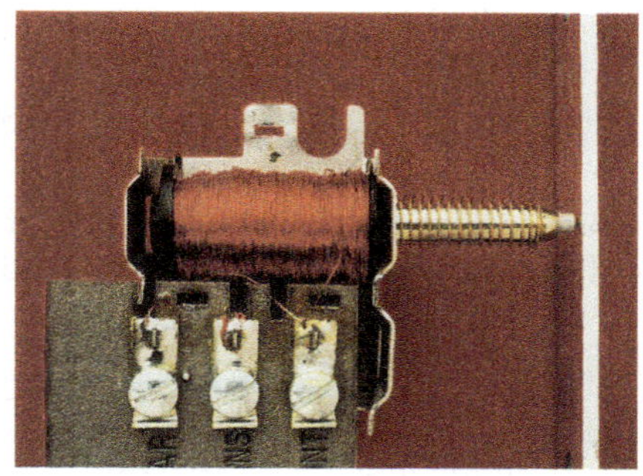

A cross section of a doorbell. When the solenoid is energized, its magnetic field pulls on the plunger, causing it to strike the bell (not shown). The spring returns the plunger to its normal position. *(© Bruce Iverson.)*

$\vec{B}$ DUE TO A CURRENT IN A STRAIGHT WIRE

Figure 27-17 shows the geometry for calculating the magnetic field $\vec{B}$ at a point P due to the current in the straight-wire segment shown. We choose R to be the perpendicular distance from the wire to point P, and we choose the x axis to be along the wire with $x = 0$ at the projection of P onto the x axis.

A typical current element $I\,d\vec{\ell}$ at a distance x from the origin is shown. The vector $\vec{r}$ points from the element to the field point P. The direction of the magnetic field at P due to this element is the direction of $I\,d\vec{\ell} \times \hat{r}$, which is out of the paper. Note that at P, the magnetic fields due to all the current elements of the wire are in

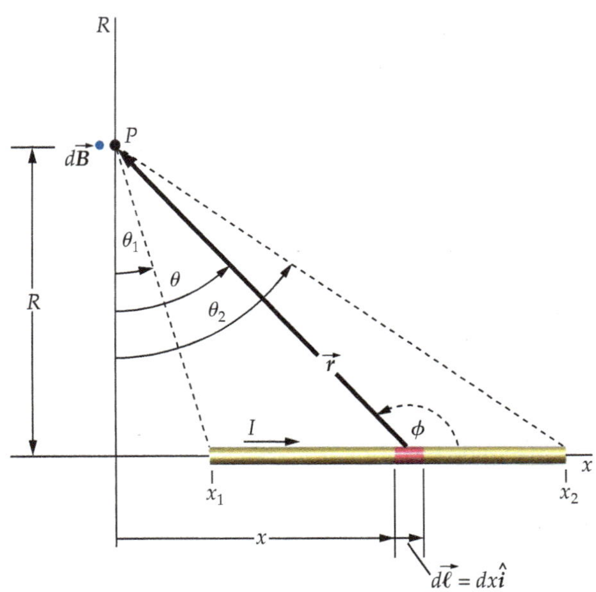

FIGURE 27-17 Geometry for calculating the magnetic field at point P due to a straight current segment. Each element of the segment contributes to the total magnetic field at point P, which is directed out of the paper. The result is expressed in terms of the angles θ_1 and θ_2.

this same direction. Thus, we need to compute only the magnitude of the field. The field due to the current element shown has the magnitude (Equation 27-3)

$$dB = \frac{\mu_0}{4\pi}\frac{Idx}{r^2}\sin\phi$$

It is more convenient to write this in terms of θ rather than ϕ:

$$dB = \frac{\mu_0}{4\pi}\frac{I\,dx}{r^2}\cos\theta \qquad 27\text{-}11$$

To sum over all the current elements, we need to relate the variables θ, r, and x. It turns out to be easiest to express x and r in terms of θ. We have

$$x = R\tan\theta$$

Then, taking the differential of each side with R as a constant gives

$$dx = R\sec^2\theta\,d\theta = R\frac{r^2}{R^2}d\theta = \frac{r^2}{R}d\theta$$

where we have used $\sec\theta = r/R$. Substituting this expression for dx into Equation 27-11, we obtain

$$dB = \frac{\mu_0}{4\pi}\frac{I}{r^2}\frac{r^2\,d\theta}{R}\cos\theta = \frac{\mu_0}{4\pi}\frac{I}{R}\cos\theta\,d\theta$$

We sum over these elements by integrating from $\theta = \theta_1$ to $\theta = \theta_2$, where θ_1 and θ_2 are shown in Figure 27-17. This calculation gives

$$B = \int_{\theta_1}^{\theta_2}\frac{\mu_0}{4\pi}\frac{I}{R}\cos\theta\,d\theta = \frac{\mu_0}{4\pi}\frac{I}{R}\int_{\theta_1}^{\theta_2}\cos\theta\,d\theta$$

Evaluating the integral, we obtain

$$B = \frac{\mu_0}{4\pi}\frac{I}{R}(\sin\theta_2 - \sin\theta_1) \qquad 27\text{-}12$$

B DUE TO A STRAIGHT-WIRE SEGMENT

This result gives the magnetic field due to any straight, current-carrying wire segment in terms of the perpendicular distance R and θ_1 and θ_2, which are the angles subtended at the field point by the ends of the wire. If the length of the wire approaches infinity in both directions, θ_2 approaches $+90°$ and θ_1 approaches $-90°$. The result for such a very long wire is obtained from Equation 27-12, by setting $\theta_1 = -90°$ and $\theta_2 = +90°$:

$$B = \frac{\mu_0}{4\pi}\frac{2I}{R} \qquad 27\text{-}13$$

B DUE TO AN INFINITELY LONG, STRAIGHT WIRE

At any point in space, the magnetic field lines of a long, straight, current-carrying wire are tangent to a circle of radius R about the wire, where R is the perpendicular distance from the wire to the field point. The direction of $\vec{B}$ can be determined by applying the right-hand rule, as shown in Figure 27-18*a*. The magnetic field lines thus encircle the wire, as shown in Figure 27-18*b*.

The result expressed by Equation 27-13 was found experimentally by Biot and Savart in 1820. From their analysis, Biot and Savart were able to discover the expression given in Equation 27-3 for the magnetic field due to a current element.

(a)

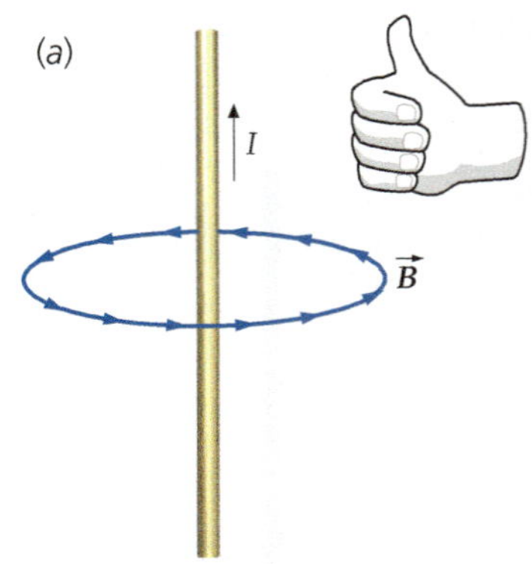

(b)

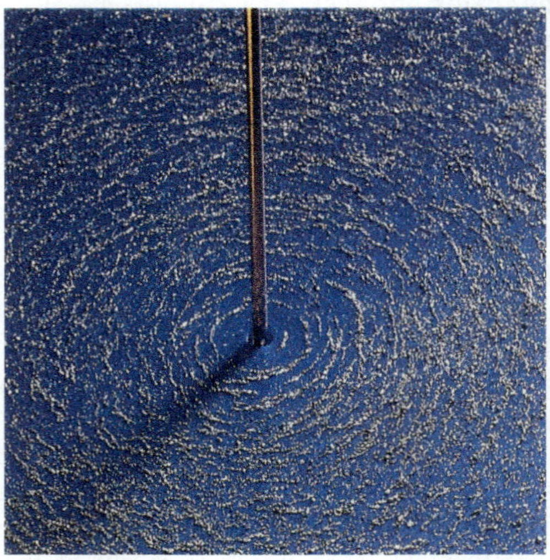

FIGURE 27-18 (*a*) Right-hand rule for determining the direction of the magnetic field due to a long, straight, current-carrying wire. The magnetic field lines encircle the wire in the direction of the fingers of the right hand when the thumb points in the direction of the current. (*b*) Magnetic field lines due to a long wire, which is indicated by iron filings. (*© 1990 Richard Menga/Fundamental Photographs.*)

Example 27-7 $\vec{B}$ at Center of Square Current Loop

Find the magnetic field at the center of a square current loop that has an edge length L, which is equal to 50 cm, and carries a current equal to 1.5 A.

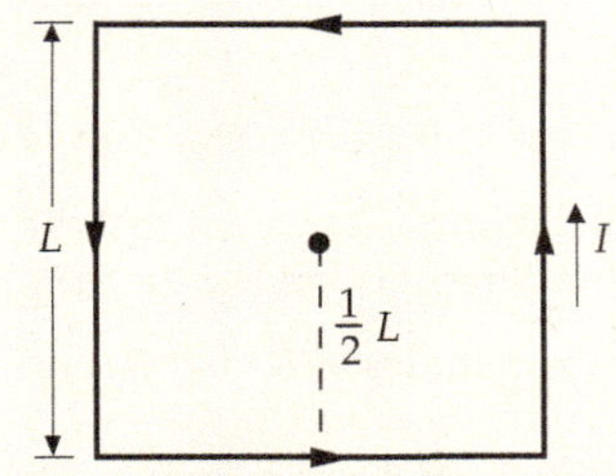

FIGURE 27-19

PICTURE The magnetic field at the center of the loop is the sum of the contributions from each of the four sides of the loop. From Figure 27-19, we can see that each side of the loop produces a field of equal magnitude pointing out of the page. Thus, we use Equation 27-12 for a given side, then multiply by 4 for the total field.

SOLVE

1. The magnitude of the resultant field is 4 times the magnitude of the field B_s due to a single side:

$$B = 4B_s$$

2. Calculate the magnitude of the magnetic field B_s due to a single side of the loop. Note from the figure that $R = \frac{1}{2}L$ and $\theta_1 = -45°$ and $\theta_2 = +45°$:

$$B_s = \frac{\mu_0}{4\pi}\frac{I}{R}(\sin\theta_2 - \sin\theta_1) = \frac{\mu_0}{4\pi}\frac{I}{\frac{1}{2}L}[\sin(+45°) - \sin(-45°)]$$

$$= (10^{-7}\,\mathrm{T\cdot m/A})\frac{1.5\ \mathrm{A}}{0.25\ \mathrm{m}}2\sin 45° = 8.5 \times 10^{-7}\,\mathrm{T}$$

3. Multiply this value by 4 to find the magnitude of the resultant field:

$$B = 4B_s = 4(8.5 \times 10^{-7}\,\mathrm{T}) = \boxed{3.4 \times 10^{-6}\,\mathrm{T}}$$

CHECK Practice Problem 27-5 serves as a check.

PRACTICE PROBLEM 27-5 Compare the magnetic field at the center of a circular current loop of radius R with the magnetic field at the center of a square current loop of side $L = 2R$ carrying the same current. Which is larger?

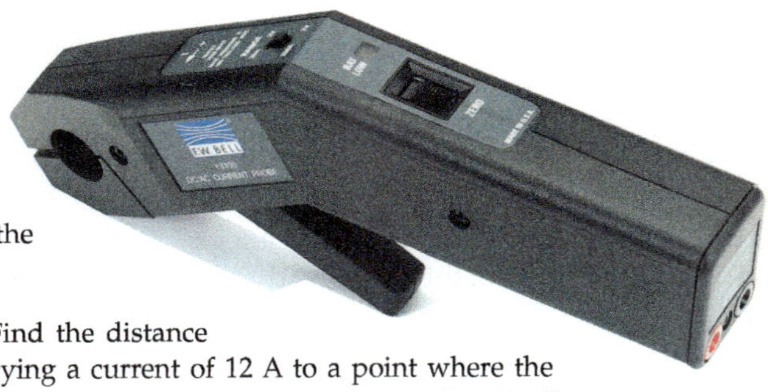

A current gun used to measure electric current. The jaws of the current gun clamp around a current-carrying wire without touching the wire. The magnetic field produced by the wire is measured with a Hall-effect device mounted in the current gun. The Hall-effect device puts out a voltage proportional to the magnetic field, which in turn is proportional to the current in the wire. *(Courtesy of F. W. Bell.)*

PRACTICE PROBLEM 27-6 Find the distance from a long, straight wire carrying a current of 12 A to a point where the magnetic field due to the current in the wire has a magnitude equal to 60 μT.

Example 27-8 $\vec{B}$ Due to Two Parallel Wires

A long, straight wire carries a current of 1.7 A in the $+z$ direction and lies along the line $x = -3.0$ cm, $y = 0$. A second such wire carries a current of 1.7 A in the $+z$ direction and lies along the line $x = +3.0$ cm, $y = 0$, as shown in Figure 27-20. Find the magnetic field at a point P on the y axis at $y = 6.0$ cm.

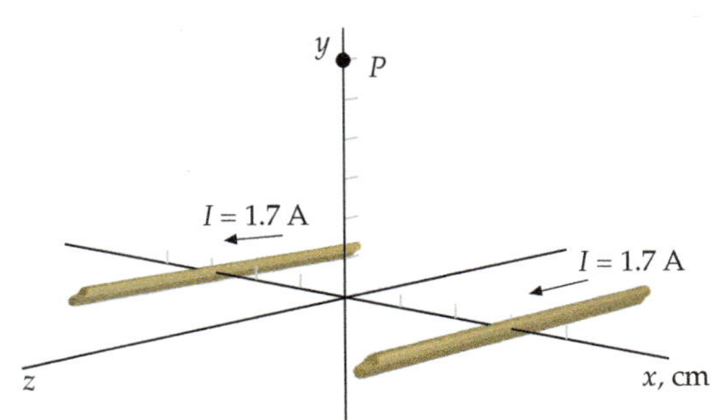

FIGURE 27-20

PICTURE The magnetic field at point P is the vector sum of the field $\vec{B}_L$ due to the wire on the left in Figure 27-21, and the field $\vec{B}_R$ due to the wire on the right. Because each wire carries the same current and each wire is the same distance from point P, the magnitudes B_L and B_R are equal. $\vec{B}_L$ is perpendicular to the radius from the left wire to point P, and $\vec{B}_R$ is perpendicular to the radius from the right wire to the point P.

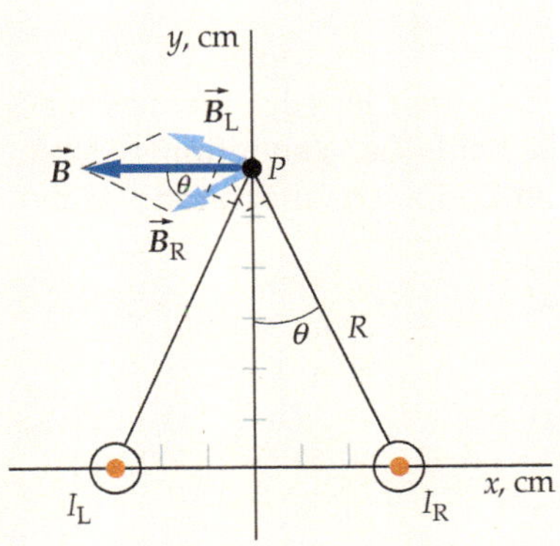

FIGURE 27-21

SOLVE

1. The field at P is the vector sum of the fields $\vec{B}_L$ and $\vec{B}_R$: $\vec{B} = \vec{B}_L + \vec{B}_R$
2. From Figure 27-21 we see that the resultant magnetic field is in the $-x$ direction and has the magnitude $2B_L \cos\theta$. $\vec{B} = -2B_L \cos\theta\, \hat{i}$
3. The magnitudes of $\vec{B}_L$ and $\vec{B}_R$ are given by Equation 27-13: $B_L = B_R = \frac{\mu_0}{4\pi}\frac{2I}{R}$
4. R is the radial distance from each wire to the point P. We find R from the figure and substitute R into the expression for B_L and B_R: $R = \sqrt{(3.0\text{ cm})^2 + (6.0\text{ cm})^2} = 6.7\text{ cm}$ so $B_L = B_R = (10^{-7}\text{ T}\cdot\text{m/A})\frac{2(1.7\text{ A})}{0.067\text{ m}} = 5.07 \times 10^{-6}\text{ T}$
5. We obtain $\cos\theta$ from the figure: $\cos\theta = \frac{6.0\text{ cm}}{R} = \frac{6.0\text{ cm}}{6.7\text{ cm}} = 0.894$
6. Substitute the values of $\cos\theta$ and B_L into the equation in step 2 for $\vec{B}$: $\vec{B} = -2(5.07 \times 10^{-6}\text{ T})(0.894)\hat{i} = \boxed{-9.1 \times 10^{-6}\text{ T}\,\hat{i}}$

CHECK The magnitude of the step-6 result is less than twice the step-4 result, which is expected because the vectors being added are not parallel.

PRACTICE PROBLEM 27-7 Find $\vec{B}$ at the origin.

PRACTICE PROBLEM 27-8 Find $\vec{B}$ at the origin assuming that the direction of the current is reversed in the wire along the line $x = 3.0$ cm, $y = 0$.

MAGNETIC FORCE BETWEEN PARALLEL WIRES

We can use Equation 27-13 for the magnetic field due to a long, straight, current-carrying wire and $d\vec{F} = I\, d\vec{\ell} \times \vec{B}$ (Equation 26-5) for the force exerted by a magnetic field on a segment of a current-carrying wire to find the magnetic force exerted by one long, straight, current-carrying wire on another. Figure 27-22 shows two long parallel wires carrying currents in the same direction. We consider the force on a segment $d\vec{\ell}_2$ carrying current I_2, as shown. The magnetic field $\vec{B}_1$ at this segment due to current I_1 is perpendicular to the segment $d\vec{\ell}_2$, as shown. This is true for all current elements along wire 2. That the force, given by $d\vec{F}_{12} = I_2\, d\vec{\ell}_2 \times \vec{B}_1$, on current element $I_2\, d\vec{\ell}_2$ is directed toward wire 1 can be revealed by applying the right hand rule. Similarly, a current segment $I_1\, d\vec{\ell}_1$ will experience a magnetic force directed toward current I_2, due to a magnetic field $\vec{B}_2$ arising from current I_2, given by $d\vec{F}_{21} = I_1\, d\vec{\ell}_1 \times \vec{B}_2$. Thus, two parallel currents attract each other. If one of the currents is reversed, the forces will be reversed, so two antiparallel currents will repel each other. The attraction or repulsion of parallel or antiparallel currents was discovered experimentally by Ampère one week after he heard of Oersted's discovery of the effect of a current on a compass needle.

The magnitude of the magnetic force on the current element $I_2\, d\vec{\ell}_2$ is

$$dF_{12} = |I_2\, d\vec{\ell}_2 \times \vec{B}_1|$$

Because the magnetic field at current element $I_2\, d\vec{\ell}_2$ is perpendicular to the current element, we have

$$dF_{12} = I_2\, d\ell_2 B_1$$

If the distance R between the wires is much less than their lengths, the magnetic field at $I_2\, d\vec{\ell}_2$ due to current I_1 will approximate the field due to an infinitely long, current-carrying wire, which is given by Equation 27-13. The magnitude of the force on the segment $I_2\, d\vec{\ell}_2$ is therefore

$$dF_{12} = I_2\, d\ell_2 \frac{\mu_0 I_1}{2\pi R}$$

and the force per unit length is

$$\frac{dF_{12}}{d\ell_2} = \frac{\mu_0}{2\pi}\frac{I_1 I_2}{R} \qquad 27\text{-}14$$

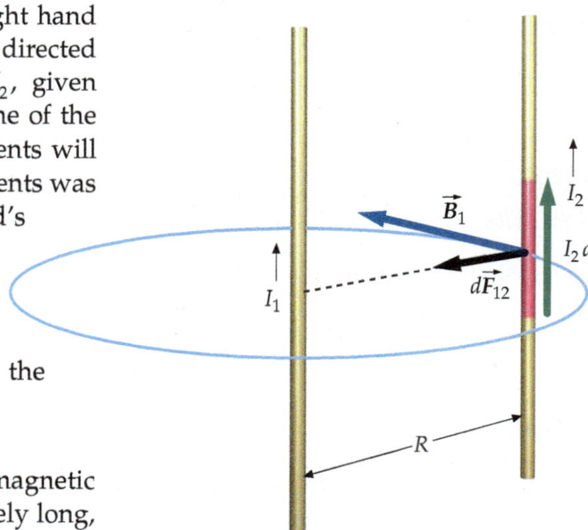

FIGURE 27-22 Two long, straight, wires carrying parallel currents. The magnetic field $\vec{B}_1$ due to current I_1 is perpendicular to current I_2. The force on current I_2 is toward current I_1. There is an equal and opposite force exerted by current I_2 on I_1. The current-carrying wires thus attract each other.

In Chapter 21, the coulomb was defined in terms of the ampere, but the definition of the ampere was deferred. The ampere is defined as follows:

> The ampere is that constant current which, if maintained in two straight, parallel conductors of infinite length and of negligible circular cross sections placed one meter apart in a vacuum, would produce a force between the conductors equal to 2×10^{-7} newton per meter of length.
>
> DEFINITION—AMPERE

That definition of the ampere allows the unit of current (and therefore the unit of electric charge) to be determined by a mechanical measurement. In practice, currents much closer together than 1 m are used so that the force can be measured more accurately.

Figure 27-23 shows a **current balance,** which is a device that can be used to calibrate an ammeter from the definition of the ampere. The upper conductor, directly above the lower conductor, is pivoted on knife-edge contacts and is balanced so that the wires (or conducting rods) are a small distance apart. The conductors are connected in series to carry the same current but in opposite directions so that the currents will repel each other. Weights are placed on the upper conductor until it balances again at the original separation. The force of repulsion is thus determined by measuring the total weight required to balance the upper conductor.

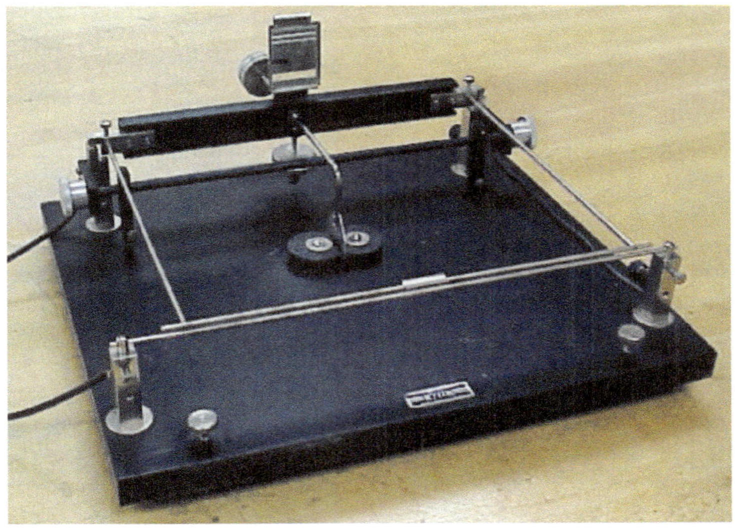

(a)

FIGURE 27-23 (*a*) A picture of a current balance used in a general physics lab. (*b*) A schematic diagram of a current balance. The two parallel rods in front carry equal but oppositely directed currents and therefore repel each other. The force of repulsion is balanced by weights placed on the upper rod, which is part of a rectangle that is balanced on knife edges at the back. The mirror on top is used to reflect a beam of laser light to accurately determine the position of the upper rod. *(Photo by Gene Mosca.)*

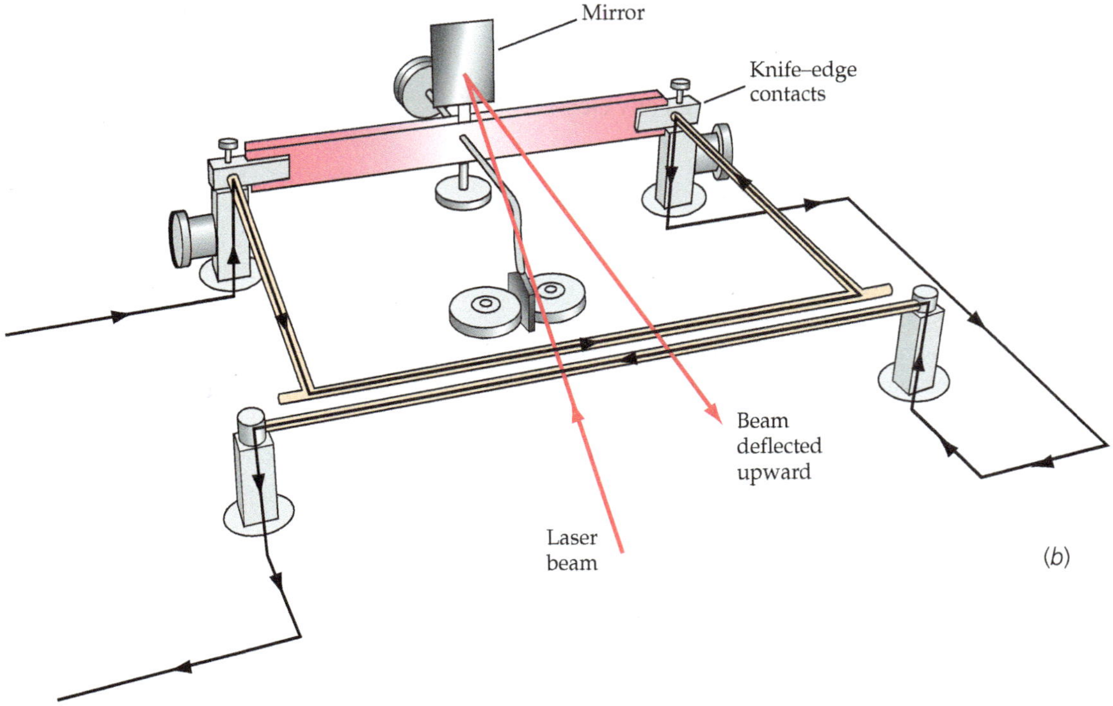

(b)

Example 27-9 Balancing the Magnetic Force

Try It Yourself

Two straight 50.0-cm-long rods have central axes that are 1.50 mm apart in a current balance, and carry currents of 15.0 A each in opposite directions. What mass must be placed on the upper rod to balance the magnetic force of repulsion?

PICTURE Equation 27-14 gives the magnitude of the magnetic force per unit length exerted by the lower rod on the upper rod. Find this force for a rod of length L and set it equal to the weight mg.

SOLVE

Cover the column to the right and try these on your own before looking at the answers.

Steps	Answers
1. Set the weight mg equal to the magnetic force of repulsion of the rods.	$mg = \frac{\mu_0}{2\pi}\frac{I_1 I_2}{R}L$
2. Solve for the mass m.	$m = 1.53 \times 10^{-3}\text{ kg} = \boxed{1.53\text{ g}}$

TAKING IT FURTHER Because only 1.53 g is required to balance the system, we see that the magnetic force between two straight current-carrying wires is relatively small, even for currents as large as 15.0 A separated by only 1.50 mm.

27-3 GAUSS'S LAW FOR MAGNETISM

The magnetic field lines shown in Figure 27-6, Figure 27-9, and Figure 27-10 differ from electric field lines because the lines of $\vec{B}$ form closed curves, whereas lines of $\vec{E}$ begin and end on electric charges. The magnetic equivalent of an electric charge is a magnetic pole, such as appears to be at the ends of a bar magnet. Magnetic field lines appear to diverge from the north-pole end of a bar magnet (Figure 27-10*b*) and appear to converge on the south-pole end. Inside the magnet, however, the magnetic field lines neither diverge from a point near the north-pole end, nor do they converge on a point near the south-pole end. Instead, the magnetic field lines pass through the bar magnet from the south-pole end to the north-pole end, as shown in Figure 27-10*b*. If a Gaussian surface encloses one end of a bar magnet, the number of magnetic field lines that penetrate the surface from the inside is exactly equal to the number of magnetic field lines that penetrate the surface from the outside. That is, the net flux $\phi_{\text{m net}}$ of the magnetic field $\vec{B}$ through any closed surface S is always zero.*

$$\phi_{\text{m net}} = \oint_S \vec{B}\cdot\hat{n}\,dA = \oint_S B_{\text{n}}\,dA = 0 \qquad 27\text{-}15$$

GAUSS'S LAW FOR MAGNETISM

where B_{n} is the component of $\vec{B}$ normal to surface S at area element dA. The definition of the magnetic flux ϕ_{m} is exactly analogous to the definition of electric flux, with $\vec{B}$ replacing $\vec{E}$. This result is called *Gauss's law for magnetism*. It is the mathematical statement that no point in space exists from which magnetic field lines diverge, or to which magnetic field lines converge. That is, isolated magnetic poles do not exist.† The fundamental unit of magnetism is the magnetic dipole.

* Recall that the net flux of the electric field is a measure of the net number of field lines that leave a closed surface and is equal to $Q_{\text{inside}}/\epsilon_0$.

† The existence of magnetic monopoles is a subject of great debate, and the search for magnetic monopoles remains active. To date, however, none have been discovered.

Figure 27-24 compares the field lines of $\vec{B}$ for a magnetic dipole with the field lines of $\vec{E}$ for an electric dipole. Note that far from the dipoles the field lines are identical. But inside the dipole, the field lines of $\vec{E}$ are directed opposite to the field lines of $\vec{B}$. The field lines of $\vec{E}$ diverge from the positive charge and converge to the negative charge, whereas the field lines of $\vec{B}$ are continuous loops.

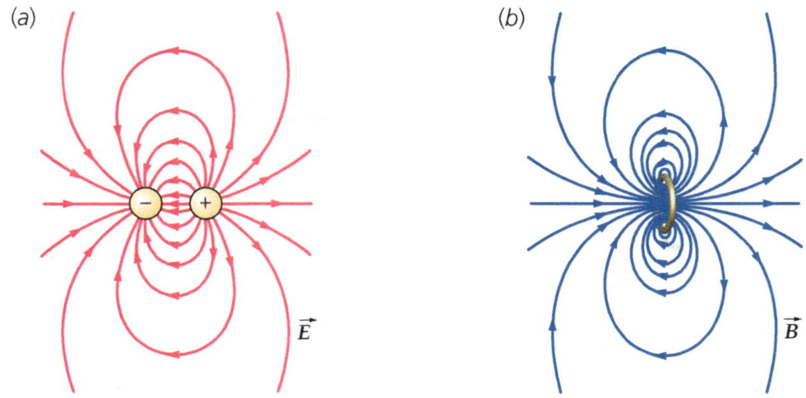

FIGURE 27-24 (*a*) Electric field lines of an electric dipole. (*b*) Magnetic field lines of a magnetic dipole. Far from the dipoles, the field lines are identical. In the region between the charges in (*a*), the electric field lines are approximately opposite the direction of the dipole moment, whereas inside the loop in (*b*), the magnetic field lines are approximately parallel to the direction of the dipole moment.

27-4 AMPÈRE'S LAW

In Chapter 22, we found that for highly symmetric charge distributions we could calculate the electric field more easily using Gauss's law than Coulomb's law. A similar situation exists in magnetism. Ampère's law relates the tangential component B_t of the magnetic field summed (integrated) around a closed curve C to the current I_C that passes through any surface bounded by C. This law can be used to obtain an expression for the magnetic field in situations that have a high degree of symmetry. In mathematical form, **Ampère's law** is

$$\oint_C B_t\, d\ell = \oint_C \vec{B}\cdot d\vec{\ell} = \mu_0 I_C \quad C \text{ is any closed curve} \qquad 27\text{-}16$$

AMPÈRE'S LAW

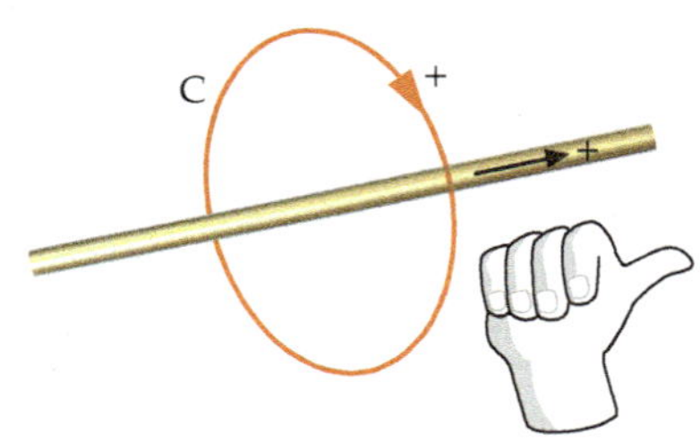

FIGURE 27-25 The positive direction for the path integral for Ampère's law is related to the positive direction for the current passing through the surface by a right-hand rule.

where I_C is the net current that penetrates any surface S bounded by the curve C. The positive tangential direction for the path integral along C is related to the choice for the positive direction for the current I_C through S by the right-hand rule shown in Figure 27-25. Ampère's law holds *as long as the currents are steady and continuous.* This means the current does not change in time and that charge is not accumulating anywhere. Ampère's law is useful in calculating the magnetic field $\vec{B}$ in situations that have a high degree of symmetry so that the line integral $\oint_C \vec{B}\cdot d\vec{\ell}$ can be written as $B\oint_C d\ell$ (the product of B and some distance). The integral $\oint_C \vec{B}\cdot d\vec{\ell}$ is called a **circulation integral.** More specifically, $\oint_C \vec{B}\cdot d\vec{\ell}$ is called the circulation of $\vec{B}$ around curve C. Ampère's law and Gauss's law are both of considerable theoretical importance, and both laws hold whether there is symmetry or there is no symmetry. If there is no symmetry, neither law is very useful in calculating electric or magnetic fields.

! Ampère's law holds as long as the currents are steady and continuous.

The simplest application of Ampère's law is to find the magnetic field due to the current in an infinitely long, straight wire. Figure 27-26 shows a circular curve C around a long wire with its center at the wire. We know the direction of the magnetic field due to each current element is tangent to this circle from the Biot–Savart law. Assuming that the magnetic field $\vec{B}$ is tangent to this circle, is in the same direction as $d\vec{\ell}$, and has the same magnitude B at any point on the circle, Ampère's law ($\oint_C B_t\, d\ell = \mu_0 I_C$) then gives

$$B\oint_C d\ell = \mu_0 I_C$$

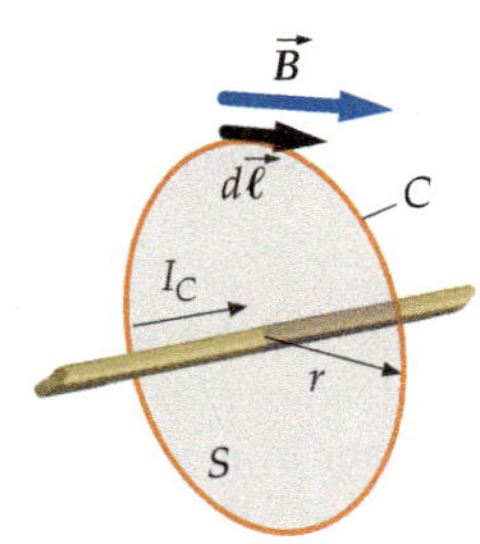

FIGURE 27-26 Geometry for calculating the magnetic field of a long, straight, current-carrying wire using Ampère's law. The magnetic field is tangent to the circle and the magnitude of the magnetic field is the same everywhere in the circle.

where $B = B_t$. We can factor B out of the integral because B has the same value everywhere on the circle. The integral of $d\ell$ around the circle equals $2\pi R$ (the circumference of the circle). The current I_C is the current I in the wire. We thus obtain $B2\pi R = \mu_0 I$, or

$$B = \frac{\mu_0 I}{2\pi R}$$

which is Equation 27-13.

Example 27-10 The Direction of the Magnetic Field — Conceptual

A long, straight cylindrical shell carries a current. Show that the direction of the magnetic field due to the current in the shell is tangent to a circle coaxial with the shell (Figure 27-27).

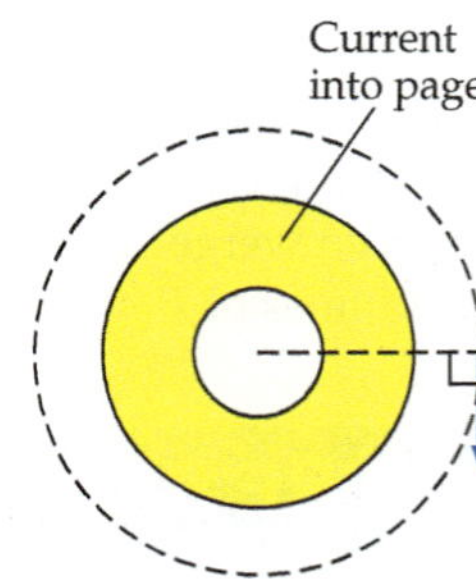

FIGURE 27-27

PICTURE Model the cylindrical shell as a bundle of thin, long, straight wires, each parallel to the central axis and each carrying a small fraction of the total current. Pick a field point P at an arbitrary location. Divide the shell in half with an imaginary plane that contains both P and the central axis of the shell. Using the right-hand rule (Figure 27-25), find the direction of the magnetic field at P due to the current in one of the thin wires in the model. Identify the symmetric counterpart to the thin wire in the other half of the shell. The counterpart is the wire equidistant from the plane and opposite the initially identified wire. Find the direction of the magnetic field at P due to the current in the counterpart thin wire. The direction of the magnetic field at P is midway between the directions of the magnetic fields due to currents in the thin wire and its counterpart.

SOLVE

1. Choose a field point P. Use the right-hand rule (Figure 27-25) to find the directions of the magnetic fields at P due to the current in a thin wire and the current in its symmetric counterpart. Draw a sketch of these two wires and their magnetic fields at the field point (Figure 27-28). Also show the sum of the two magnetic fields:

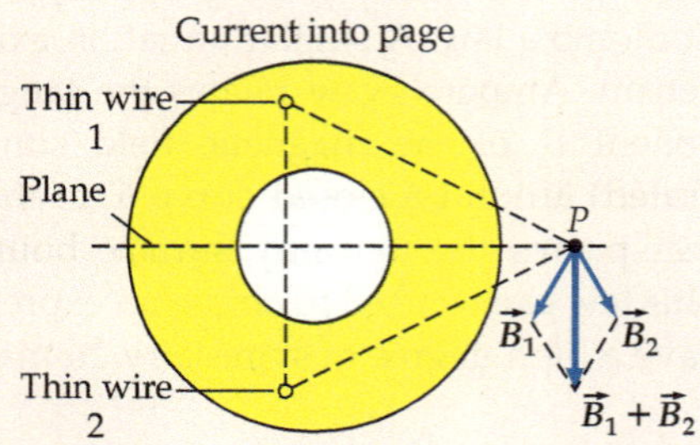

FIGURE 27-28

2. The resultant magnetic field at P is the sum of the magnetic fields due to all the thin wires that make up the cylindrical shell:

   The resultant magnetic field at P is in the same direction as the sum $\vec{B}_1 + \vec{B}_2$. This is because the sum of the magnetic fields due to the currents in any thin wire and its symmetric counterpart will point in the same direction as $\vec{B}_1 + \vec{B}_2$.

3. If the field point P is inside the shell, the magnetic field at P due to the currents in the thin wires to the right of P (Figure 27-29) will point in the opposite direction of those to the left of P (Figure 27-28):

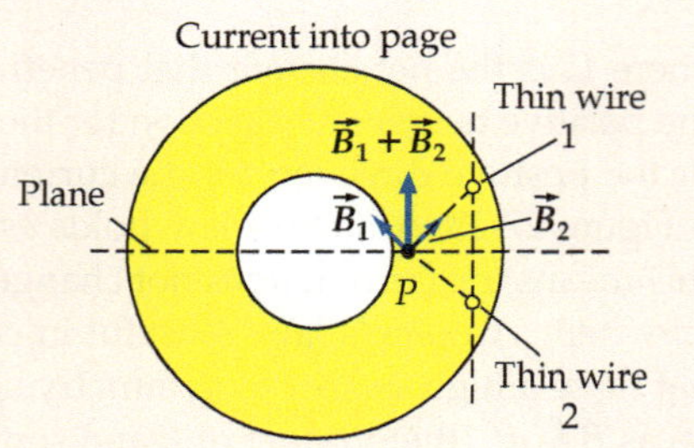

FIGURE 27-29

Example 27-11 $\vec{B}$ Inside and Outside a Wire

A long, straight wire has a radius R and carries a current I that is uniformly distributed over the circular cross section of the wire. Find the magnetic field both outside the wire and inside the wire.

PICTURE We can use Ampère's law to calculate $\vec{B}$ because of the high degree of symmetry. At a distance r (Figure 27-30), we know that $\vec{B}$ is tangent to the circle of radius r about the wire and $\vec{B}$ is constant in magnitude everywhere on the circle. The expression for the current through the surface S that is bounded by C depends on whether r is less than or greater than the radius of the wire R.

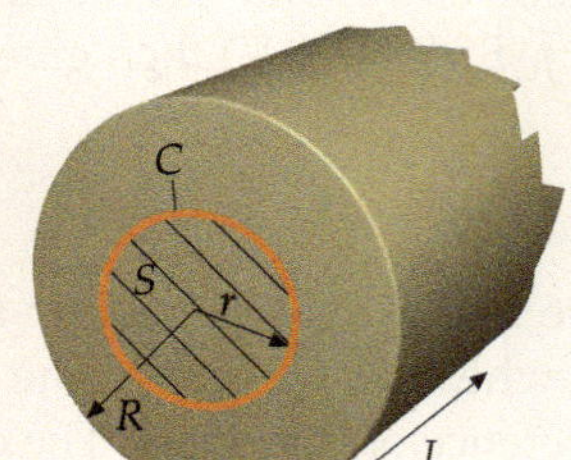

FIGURE 27-30

SOLVE

1. Ampère's law is used to relate the circulation of $\vec{B}$ around curve C to the current passing through the surface S bounded by C:

$$\oint_C \vec{B}\cdot d\vec{\ell} = \mu_0 I_C$$

2. Evaluate the circulation of $\vec{B}$ around a circle of radius r that is coaxial with the wire:

$$\oint_C \vec{B}\cdot d\vec{\ell} = B\oint_C d\ell = B2\pi r$$

3. Substitute into Ampère's law and solve for B:

$$B2\pi r = \mu_0 I_C$$

so

$$B = \frac{\mu_0 I_C}{2\pi r}$$

4. Outside the wire, $r > R$, and the total current passes through the surface bounded by C:

$$I_C = I$$

or

$$\boxed{B = \frac{\mu_0}{2\pi}\frac{I}{r} \qquad r \geq R}$$

5. Inside the wire, $r < R$. Assume that the current is distributed uniformly to solve for I_C. Solve for B:

$$\frac{I_C}{\pi r^2} = \frac{I}{\pi R^2}$$

or

$$\left(I_C = \frac{r^2}{R^2} I\right)$$

so

$$B = \frac{\mu_0}{2\pi}\frac{I_C}{r} = \frac{\mu_0}{2\pi}\frac{(r^2/R^2)I}{r} = \boxed{\frac{\mu_0 I}{2\pi R^2} r \leq R}$$

CHECK The step 4 and step 5 results give the same expression for B for $r = R$, as expected.

TAKING IT FURTHER Inside the wire, the field increases with distance from the center of the wire. Figure 27-31 shows the graph of B versus r for this example.

We see from Example 27-11 that the magnetic field due to a current uniformly distributed over a wire of radius R is given by

$$B = \begin{cases} \dfrac{\mu_0 I}{2\pi R^2} r & r \leq R \\[2ex] \dfrac{\mu_0}{2\pi}\dfrac{I}{r} & r \geq R \end{cases} \qquad 27\text{-}17$$

B FOR A LONG STRAIGHT WIRE

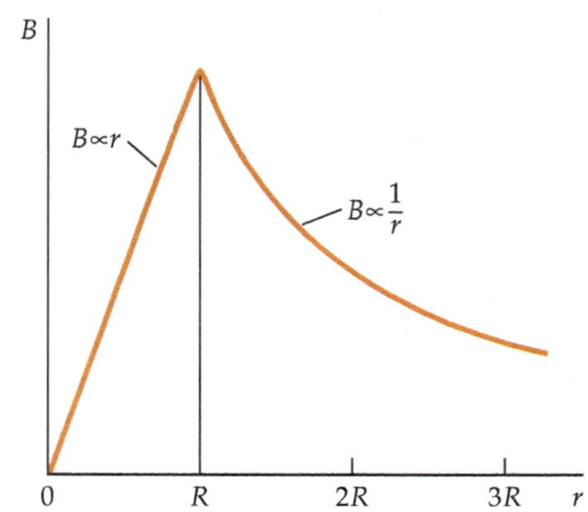

FIGURE 27-31

For the next application of Ampère's law, we calculate the magnetic field of a tightly wound **toroid,** which consists of loops of wire wound around a doughnut-shaped form, as shown in Figure 27-32. There are N turns of wire, each carrying a current I. To calculate B, we evaluate the line integral $\oint_C \vec{B} \cdot d\vec{\ell}$ around a circle of radius r that is coaxial with the toroid and is inside its loops. By symmetry, $\vec{B}$ is tangent to this circle and constant in magnitude at every point on the circle. Then,

$$\oint_C \vec{B} \cdot d\vec{\ell} = B2\pi r = \mu_0 I_C$$

Let a and b be the inner and outer radii of the toroid, respectively. The total current through the surface S bounded by a circle of radius r for $a < r < b$ is $I_C = NI$. Ampère's law then gives

$$\oint_C \vec{B} \cdot d\vec{\ell} = \mu_0 I_C \quad \text{or} \quad (B2\pi r = \mu_0 NI)$$

or

$$B = \frac{\mu_0 NI}{2\pi r} \qquad a < r < b \qquad 27\text{-}18$$

B INSIDE A TIGHTLY WOUND TOROID

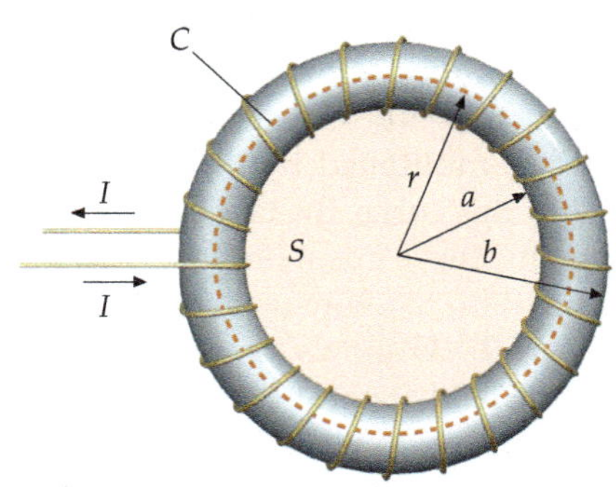

FIGURE 27-32 A toroid consists of loops of wire wound around an imaginary doughnut-shaped form. The magnetic field at any distance r can be found by applying Ampère's law to the circle of radius r. The surface S is bounded by curve C. The wire penetrates S once for each turn.

If r is less than a, there is no current through the surface S. If r is greater than b, the total current through S is zero because for each turn of the wire the current penetrates the surface S twice (Figure 27-33), once going into the page and once coming out of the page. Thus, the magnetic field is zero for both the region $r < a$ and the region $r > b$:

$$B = 0, \qquad r < a \text{ or } r > b$$

The magnetic field intensity inside the toroid is not uniform but decreases with increasing r. However, if $b - a$ (the diameter of the loops of the coil) is much less than $2b$, the variation in r from $r = a$ to $r = b$ is small, so the variation of B inside the loops is also small.

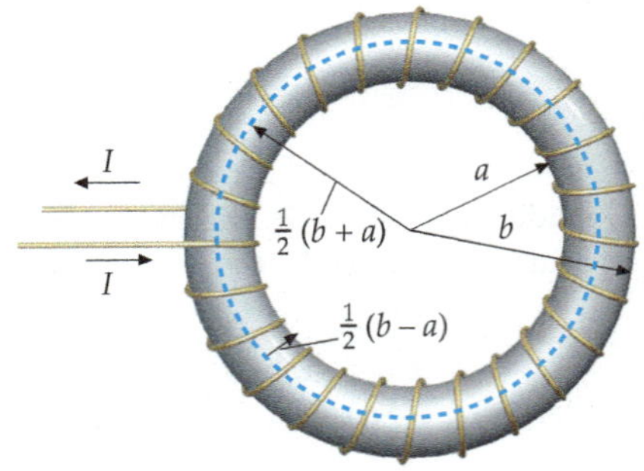

FIGURE 27-33 The toroid has mean radius $r = \frac{1}{2}(b + a)$, where a and b are the inner and outer radii of the toroid. Each turn of the wire is a circle of radius $\frac{1}{2}(b - a)$.

(a)

(b)

(*a*) The Tokamak fusion-test reactor is a large toroid that produces a magnetic field for confining charged particles. Coils containing over 10 km of water-cooled copper wire carry a pulsed current, which has a peak value of 73,000 A and produces a magnetic field of 5.2 T for about 3 s. (*b*) Inspection of the assembly of the Tokamak reactor from inside the toroid. *(Courtesy of Princeton University Plasma Physics Laboratory.)*

LIMITATIONS OF AMPÈRE'S LAW

Ampère's law is useful for calculating the magnetic field only when there is both a steady and continuous current and a high degree of symmetry. Consider the current loop shown in Figure 27-34. According to Ampère's law, the line integral $\oint_C \vec{B} \cdot d\vec{\ell} = \oint_C B_t \, d\ell$ around a curve, such as curve C in the figure, equals μ_0 multiplied by the current I in the loop. Although Ampère's law is valid for this curve, the tangential component of magnetic field B_t is not constant along any curve encircling the current. Thus, there is not enough symmetry in this situation to allow us to evaluate the integral $\oint_C B_t \, d\ell$ and solve for B_t.

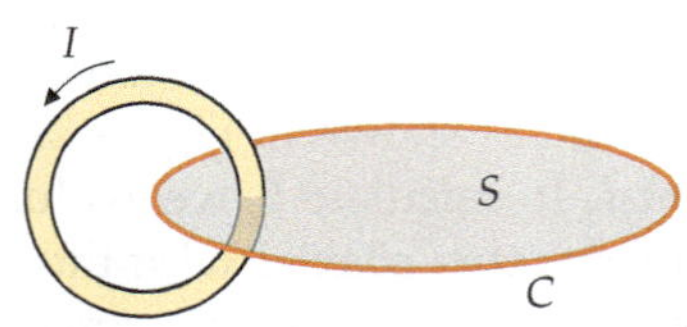

FIGURE 27-34 Ampère's law holds for the curve C encircling the current in the circular loop, but it is not useful for finding B_t because B_t cannot be factored out of the circulation integral.

Figure 27-35 shows a finite current segment of length ℓ. We wish to find the magnetic field at point P, which is equidistant from the ends of the segment and at a distance r from the center of the segment. A direct application of Ampère's law gives

$$B = \frac{\mu_0}{2\pi}\frac{I}{r}$$

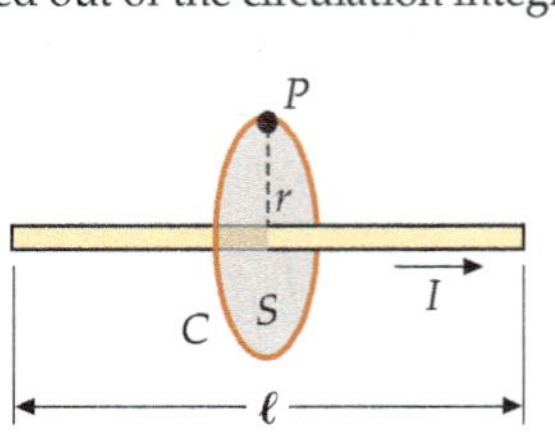

FIGURE 27-35 The application of Ampère's law to find the magnetic field on the bisector of a finite current segment gives an incorrect result.

This result is the same as for an infinitely long wire, because the same symmetry arguments apply. It does not agree with the result obtained from the Biot–Savart law, which depends on the length of the current segment and which agrees with experiment. If the current segment is just one part of a continuous circuit carrying a current, as shown in Figure 27-36, Ampère's law for curve C is valid, but it cannot be used to find the magnetic field at point P because there is insufficient symmetry.

In Figure 27-37, the current in the segment arises from a small spherical conductor that has an initial charge $+Q$ at the left of the segment and another small spherical conductor at the right that has a charge $-Q$. When they are connected, a current $I = -dQ/dt$ exists in the segment for a short time, until the spheres are uncharged. For this case, we *do* have the symmetry needed to assume that $\vec{B}$ is tangential to the curve and $\vec{B}$ is constant in magnitude along the curve. For a situation like this, in which the current is discontinuous in space, Ampère's law is not valid. In Chapter 30, we will see how Maxwell was able to modify Ampère's law so that it holds for all currents. When Maxwell's generalized form of Ampère's law is used to calculate the magnetic field for a current segment, such as the current segment shown in Figure 27-37, the result agrees with the result found from the Biot–Savart law.

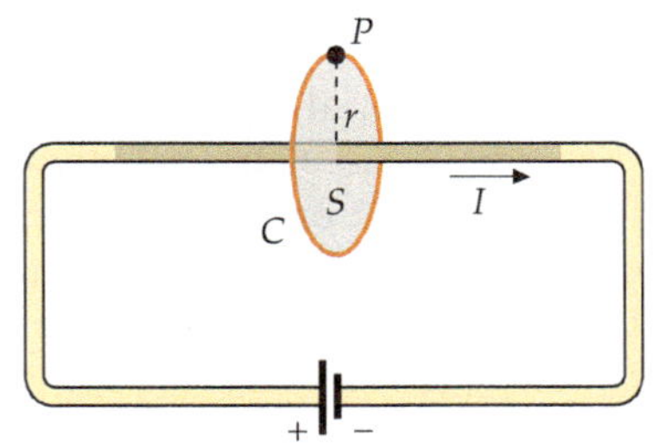

FIGURE 27-36 If the current segment in Figure 27-34 is part of a complete circuit, Ampère's law for the curve C is valid, but there is not enough symmetry to use Ampère's law to find the magnetic field at point P.

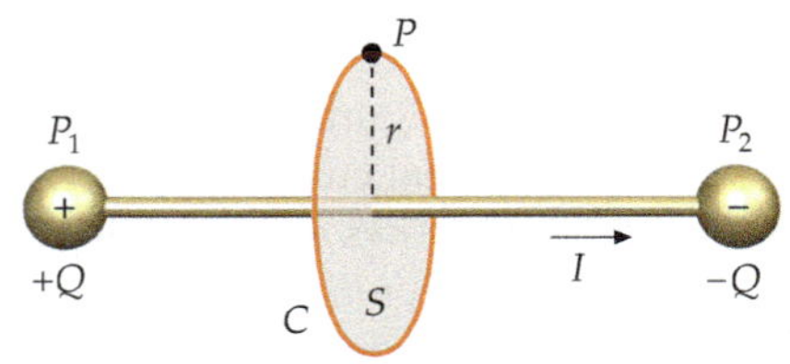

FIGURE 27-37 If the current segment in Figure 27-35 is due to a momentary flow of charge from a small conductor on the left to a small conductor on the right, there is enough symmetry to use Ampère's law to compute the magnetic field at P, but Ampère's law is not valid because the current is not steady.

27-5 MAGNETISM IN MATTER

Atoms have magnetic dipole moments due to the motion of their electrons and due to the intrinsic magnetic dipole moment associated with the spin of the electrons. Unlike the situation with electric dipoles, the alignment of magnetic dipoles parallel to an external magnetic field tends to *increase* the field. We can see this difference by comparing the electric field lines of an electric dipole with the magnetic field lines of a magnetic dipole, such as a small current loop, as was shown in Figure 27-24. Far from the dipoles, the field lines are identical. However, between the charges of the electric dipole, the electric field lines are opposite the direction of the dipole moment, whereas inside the current loop, the magnetic field lines are parallel to the magnetic dipole moment. Thus, inside a magnetically polarized material, the magnetic dipoles create a magnetic field that is parallel to the magnetic dipole moment vectors.

Materials fall into three categories—**paramagnetic, ferromagnetic,** and **diamagnetic**—according to the behavior of their magnetic moments in an external magnetic field. Paramagnetism arises from the partial alignment of the electron spins (in metals) or from the atomic or molecular magnetic moments by an applied magnetic field in the direction of the field. In paramagnetic materials, the magnetic dipoles do not interact strongly with each other and are normally randomly oriented. In the presence of an applied magnetic field, the dipoles are partially aligned in the direction of the field, thereby increasing the field. However, in external magnetic fields of ordinary strength at ordinary temperatures, only a very small fraction of the atoms are aligned because thermal motion tends to randomize their orientation. The increase in the total magnetic field is therefore very small. Ferromagnetism is much more complicated. Because of a strong interaction between neighboring magnetic dipoles, a high degree of alignment occurs even in weak external magnetic fields, which causes a very large increase in the total field. Even when there is no external magnetic field, a ferromagnetic material may have its magnetic dipoles aligned, as in permanent magnets. Diamagnetism arises from the orbital magnetic dipole moments induced by an applied magnetic field. These magnetic moments are opposite the direction of the applied magnetic field, thereby decreasing the field. This effect actually occurs in all materials; however, because the induced magnetic moments are very small compared to the permanent magnetic moments, diamagnetism is often masked by paramagnetic or ferromagnetic effects. Diamagnetism is thus observed only in materials whose atoms have no permanent magnetic moments.

MAGNETIZATION AND MAGNETIC SUSCEPTIBILITY

When some material is placed in a strong magnetic field, such as that of a solenoid, the magnetic field of the solenoid tends to align the magnetic dipole moments (either permanent or induced) inside the material and the material is said to be magnetized. We describe a magnetized material by its **magnetization** $\vec{M}$**,** which is defined as the net magnetic dipole moment per unit volume of the material:

$$\vec{M} = \frac{d\vec{\mu}}{dV} \tag{27-19}$$

Long before we had any understanding of atomic or molecular structure, Ampère proposed a model of magnetism in which the magnetization of materials is due to microscopic current loops inside the magnetized material. We now know that these current loops are a classical model for the orbital motion and spin of the electrons in atoms. Consider a cylinder of magnetized material. Figure 27-38 shows atomic current loops in the cylinder aligned with their magnetic moments along the axis of the cylinder. Because of cancellation of neighboring current loops, the net current at any point inside the material is zero, leaving a net current on the surface of the material (Figure 27-39). This surface current, called an **amperian current,** is similar to the real current in the windings of the solenoid.

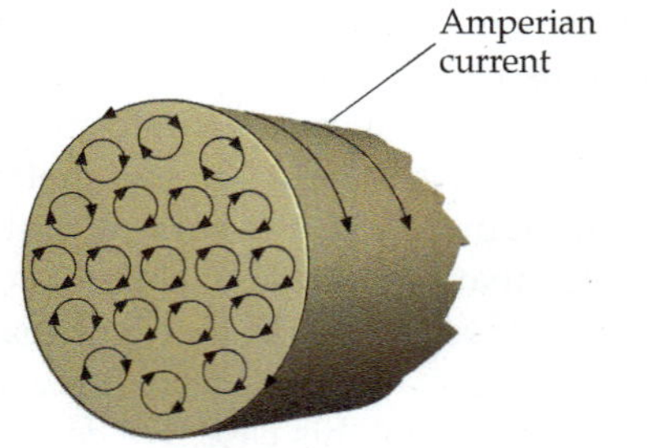

FIGURE 27-38 A model of atomic current loops in which all the atomic dipoles are parallel to the axis of the cylinder. The net current at any point inside the material is zero due to cancellation of neighboring atoms. The result is a surface current similar to that of a solenoid.

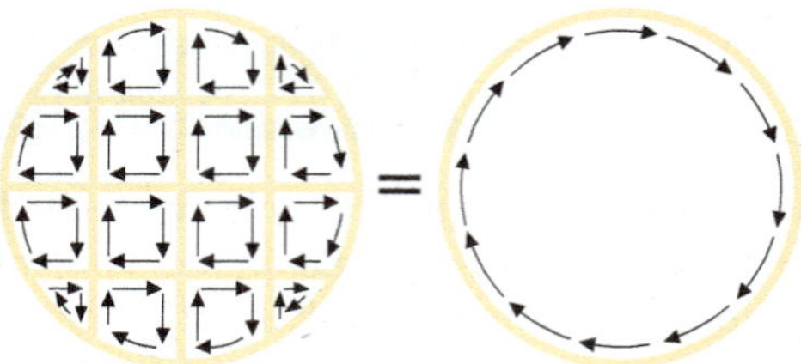

FIGURE 27-39 The currents in the adjacent current loops in the interior of a uniformly magnetized material cancel, leaving only a surface current. Cancellation occurs at every interior point independent of the shape of the very small loops.

Figure 27-40 shows a short cylinder of cross-sectional area A, length $d\ell$, and volume $dV = A\,d\ell$. Let di be the amperian current on the curved surface of the disk. The magnitude of the magnetic dipole moment of the disk is the same as that of a current loop that has an area A and carries a current di:

$$d\mu = A\,di$$

The magnitude of the magnetization of the disk is the magnetic moment per unit volume:

$$M = \frac{d\mu}{dV} = \frac{A\,di}{A\,d\ell} = \frac{di}{d\ell} \tag{27-20}$$

Thus, the magnitude of the magnetization vector is the amperian current per unit length along the surface of the magnetized material. We see from this result that the SI units for magnetization M are amperes per meter.

Consider a cylinder that has a uniform magnetization $\vec{M}$ parallel to its axis. The effect of the magnetization is the same as if the cylinder carried a surface current per unit length of magnitude M. This current is similar to the current carried by a tightly wound solenoid. For a solenoid, the current per unit length is nI, where n is the number of turns per unit length and I is the current in each turn. The magnitude of the magnetic field B_m inside the cylinder and far from its ends is thus given by $B = \mu_0 nI$ (Equation 27-10) for a solenoid with nI replaced by M:

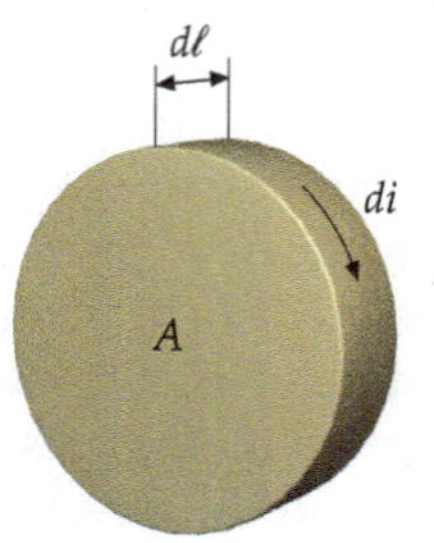

FIGURE 27-40 A disk element for relating the magnetization M to the surface current per unit length $di/d\ell$.

$$B_m = \mu_0 M \tag{27-21}$$

Suppose we place a cylinder of magnetic material inside a long solenoid that has n turns per unit length and carries a current I. The applied field of the solenoid $\vec{B}_{app}$ ($B_{app} = \mu_0 nI$) magnetizes the material so that it has a magnetization $\vec{M}$. The resultant magnetic field at a point inside the solenoid and far from its ends due to the current in the solenoid plus the magnetized material is

$$\vec{B} = \vec{B}_{app} + \mu_0\vec{M} \tag{27-22}$$

For paramagnetic and ferromagnetic materials, $\vec{M}$ is in the same direction as $\vec{B}_{app}$; for diamagnetic materials, $\vec{M}$ is opposite to $\vec{B}_{app}$. For paramagnetic and diamagnetic materials, the magnetization is found to be proportional to the applied magnetic field that produces the alignment of the magnetic dipoles in the material. We can thus write

$$\vec{M} = \chi_m \frac{\vec{B}_{app}}{\mu_0} \tag{27-23}$$

where the proportionality constant χ_m is a dimensionless number called the **magnetic susceptibility.** Equation 27-22 is then

$$\vec{B} = \vec{B}_{app} + \mu_0\vec{M} = \vec{B}_{app}(1 + \chi_m) = K_m\vec{B}_{app} \qquad 27\text{-}24$$

where

$$K_m = 1 + \chi_m \qquad 27\text{-}25$$

is called the **relative permeability** of the material. For paramagnetic materials, χ_m is a small positive number that depends on temperature. For diamagnetic materials (other than superconductors), it is a small negative constant independent of temperature. Table 27-1 lists the magnetic susceptibility of various paramagnetic and diamagnetic materials. We see that the magnetic susceptibility for the solids listed is of the order of 10^{-5}, and $K_m \approx 1$.

The magnetization of ferromagnetic materials, which we discuss shortly, is much more complicated. The relative permeability K_m defined as the ratio B/B_{app} is not constant and has maximum values ranging from 5000 to 100,000. In the case of permanent magnets, K_m is not even defined because such materials exhibit magnetization even in the absence of an applied field.

Table 27-1 Magnetic Susceptibility of Various Materials at

Material	χ_m
Aluminum	2.3×10^{-5}
Bismuth	-1.66×10^{-5}
Copper	-0.98×10^{-5}
Diamond	-2.2×10^{-5}
Gold	-3.6×10^{-5}
Magnesium	1.2×10^{-5}
Mercury	-3.2×10^{-5}
Silver	-2.6×10^{-5}
Sodium	-0.24×10^{-5}
Titanium	7.06×10^{-5}
Tungsten	6.8×10^{-5}
Hydrogen (1 atm)	-9.9×10^{-9}
Carbon dioxide (1 atm)	-2.3×10^{-9}
Nitrogen (1 atm)	-5.0×10^{-9}
Oxygen (1 atm)	2090×10^{-9}

ATOMIC MAGNETIC MOMENTS

The magnetization of a paramagnetic or ferromagnetic material can be related to the permanent magnetic moments of the individual atoms or electrons of the material. The orbital magnetic moment of an atomic electron can be derived semiclassically, even though it is quantum mechanical in origin. Consider a particle of mass m and charge q moving with speed v in a circle of radius r, as shown in Figure 27-41. The magnitude of the angular momentum of the particle about the center of the circle is

$$L = mvr \qquad 27\text{-}26$$

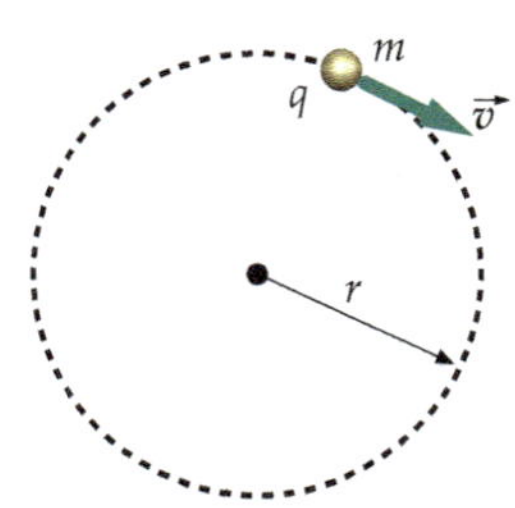

FIGURE 27-41 A particle of charge q and mass m moving with speed v in a circle of radius r. The angular momentum is directed into the paper and has a magnitude mvr. The magnetic moment is directed into the paper (if q is positive) and has a magnitude $\frac{1}{2}qvr$.

The magnitude of the magnetic moment is the product of the current and the area of the circle:

$$\mu = IA = I\pi r^2$$

If T is the time for the charge to complete one revolution, the current (the charge passing a point per unit time) is q/T. Because the period T is the distance $2\pi r$ divided by the velocity v, the current is

$$I = \frac{q}{T} = \frac{qv}{2\pi r}$$

The magnetic moment is then

$$\mu = IA = \frac{qv}{2\pi r}\pi r^2 = \tfrac{1}{2}qvr \qquad 27\text{-}27$$

Using $vr = L/m$ from Equation 27-26, we have for the magnetic moment

$$\mu = \frac{q}{2m}L$$

If the charge q is positive, the angular momentum and magnetic moment are in the same direction. We can therefore write

$$\vec{\mu} = \frac{q}{2m}\vec{L} \qquad 27\text{-}28$$

CLASSICAL RELATION BETWEEN MAGNETIC MOMENT AND ANGULAR MOMENTUM

Equation 27-28 is the general classical relation between magnetic moment and angular momentum. It also holds in the quantum theory of the atom for orbital angular momentum, but the equation does not hold for the intrinsic spin angular momentum of the electron. For electron spin, the magnetic moment is twice that predicted by the equation.* The extra factor of 2, which is accounted for by quantum theory, has no analog in classical mechanics.

Because angular momentum is quantized, the magnetic moment of an atom is also quantized. The quantum of angular momentum is $\hbar = h/(2\pi)$, where h is Planck's constant, so we express the magnetic moment in terms of $\vec{L}/\hbar$:

$$\vec{\boldsymbol{\mu}} = \frac{q\hbar}{2m}\frac{\vec{L}}{\hbar}$$

For an electron, $m = m_e$ and $q = -e$, so the magnetic moment of the electron due to its orbital motion is

$$\vec{\boldsymbol{\mu}}_\ell = -\frac{e\hbar}{2m_e}\frac{\vec{L}}{\hbar} = -\mu_B\frac{\vec{L}}{\hbar} \qquad \text{27-29}$$

MAGNETIC MOMENT DUE TO THE ORBITAL MOTION OF AN ELECTRON

where

$$\mu_B = \frac{e\hbar}{2m_e} = 9.27 \times 10^{-24}\,\mathrm{A\cdot m^2} = 9.27 \times 10^{-24}\,\mathrm{J/T}$$

$$= 5.79 \times 10^{-5}\,\mathrm{eV/T} \qquad \text{27-30}$$

BOHR MAGNETON

is the quantum unit of magnetic moment called a **Bohr magneton.** The magnetic moment of an electron due to its intrinsic spin angular momentum $\vec{S}$ is

$$\vec{\boldsymbol{\mu}}_s = -2 \times \frac{e\hbar}{2m_e}\frac{\vec{S}}{\hbar} = -2\mu_B\frac{\vec{S}}{\hbar} \qquad \text{27-31}$$

MAGNETIC MOMENT DUE TO ELECTRON SPIN

Although the calculation of the magnetic moment of any atom is a complicated problem in quantum theory, the result for all electrons, according to both theory and experiment, is that the magnetic moment is of the order of a few Bohr magnetons. In addition, any atom that has an angular momentum equal to zero has a net magnetic moment equal to zero. (This topic is further discussed in Chapter 36.)

If all the atoms in a sample of material have their magnetic moments aligned, the magnetic moment per unit volume of the sample is the product of the number of atoms per unit volume n and the magnetic moment μ of each atom. For this extreme case, the **saturation magnetization** M_s is

$$M_s = n\mu \qquad \text{27-32}$$

The number of atoms per unit volume can be found from the molar mass M, the density ρ of the material, and Avogadro's number N_A:

$$n = \frac{N_A\,(\text{atoms/mol})}{M\,(\text{kg/mol})}\rho(\text{kg/m}^3) \qquad \text{27-33}$$

* This result and the phenomenon of electron spin was predicted in 1927 by Paul Dirac, who combined special relativity and quantum mechanics into a relativistic wave equation called the Dirac equation. Precise measurements indicate that the magnetic moment of the electron due to its spin is 2.00232 times that predicted by Equation 27-28. The fact that the intrinsic magnetic moment of the electron is approximately twice what we would expect makes it clear that the simple model of the electron as a spinning ball is not to be taken literally.

Example 27-12 Saturation Magnetization for Iron

Find the saturation magnetization of a sample of iron and find the magnetic field it produces deep within the sample. Assume that each iron atom has a magnetic moment of 1 Bohr magneton.

PICTURE We find the number of atoms per unit volume from the density of iron, $\rho = 7.87 \times 10^3$ kg/m^3, and its molar mass $M = 55.8 \times 10^{-3}$ kg/mol.

SOLVE

1. The saturation magnetization is the product of the number of atoms per unit volume and the magnetic moment of each molecule:

$$M_s = n\mu$$

2. Calculate the number of atoms per unit volume from Avogadro's number, the molar mass, and the density:

$$n = \frac{N_A}{M}\rho = \frac{6.02 \times 10^{23}\ \text{atoms/mol}}{55.8 \times 10^{-3}\ \text{kg/mol}}(7.87 \times 10^3\ \text{kg/m}^3)$$
$$= 8.49 \times 10^{28}\ \text{atoms/m}^3$$

3. Substitute this result and $\mu = 1$ Bohr magneton to calculate the saturation magnetization:

$$M_s = n\mu$$
$$= (8.49 \times 10^{28}\ \text{atoms/m}^3)(9.27 \times 10^{-24}\ \text{A}\cdot\text{m}^2)$$
$$= \boxed{7.88 \times 10^5\ \text{A/m}}$$

4. The magnetic field on the axis inside and far from the ends of a long iron cylinder resulting from this maximum magnetization is given by $B = \mu_0 M_s$:

$$B = \mu_0 M_s$$
$$= (4\pi \times 10^{-7}\ \text{T}\cdot\text{A})(7.86 \times 10^5\ \text{A/m})$$
$$= \boxed{0.990\ \text{T} \approx 1\ \text{T}}$$

CHECK The step-4 result of $B \approx 1$ T is a large magnetic field. This result is as expected for the saturation magnetic field inside a ferromagnetic material.

TAKING IT FURTHER The measured saturation magnetic field of annealed iron is about 2.16 T, indicating that the magnetic moment of an iron atom is slightly greater than 2 Bohr magnetons. This magnetic moment is due mainly to the spins of two unpaired electrons in the iron atom.

*PARAMAGNETISM

Paramagnetism occurs in materials whose atoms have permanent magnetic moments that interact with each other only very weakly, resulting in a very small, positive magnetic susceptibility χ_m. When there is no external magnetic field, these magnetic moments are randomly oriented. In the presence of an external magnetic field, the magnetic moments tend to line up parallel to the field, but this is counteracted by the tendency for the magnetic moments to be randomly oriented due to thermal motion. The degree to which the moments line up with the field depends on the strength of the field and on the temperature. This degree of alignment usually is small because the energy of a magnetic moment in an external magnetic field is typically much smaller than the thermal energy of an atom of the material, which is of the order of kT, where k is Boltzmann's constant and T is the absolute temperature.

The potential energy of a magnetic dipole of moment $\vec{\mu}$ in an external magnetic field $\vec{B}$ is given by Equation 27-16:

$$U = -\mu B\cos\theta = -\vec{\mu}\cdot\vec{B}$$

The potential energy U_{min} when the moment and the field are parallel ($\theta = 0$) is thus lower than the potential energy U_{max} when the moment and the field are antiparallel ($\theta = 180°$) by $2\mu B$. For a typical atomic magnetic moment of 1 Bohr magneton and a typical strong magnetic field of 1 T, the difference in these potential energies is

$$\Delta U = 2\mu_B B = 2(5.79 \times 10^{-5}\ \text{eV/T})(1\ \text{T}) = 1.16 \times 10^{-4}\ \text{eV}$$

Liquid oxygen, which is paramagnetic, is attracted by the magnetic field of a permanent magnet. A net force is exerted on the magnetic dipoles because the magnetic field is not uniform. *(J. F. Allen, St. Andrews University, Scotland.)*

At a normal temperature of $T = 300$ K, the typical thermal energy kT is

$$kT = (8.62 \times 10^{-5}\ \text{eV/K})(300\ \text{K}) = 2.59 \times 10^{-2}\ \text{eV}$$

which is more than 200 times greater than $2\mu_B B$. Thus, even in a very strong magnetic field of 1 T, most of the magnetic moments will be randomly oriented because of thermal motions (unless the temperature is very low).

Figure 27-42 shows a plot of the magnetization M versus an applied external magnetic field B_{app} at a given temperature. In very strong applied fields, nearly all the magnetic moments are aligned with the field and $M \approx M_s$. (For magnetic fields attainable in the laboratory, this can occur only for very low temperatures.) When $B_{app} = 0$, $M = 0$, indicating that the orientation of the moments is completely random. In weak fields, the magnetization is approximately proportional to the applied field, as indicated by the orange dashed line in the figure. In this region, the magnetization is given by

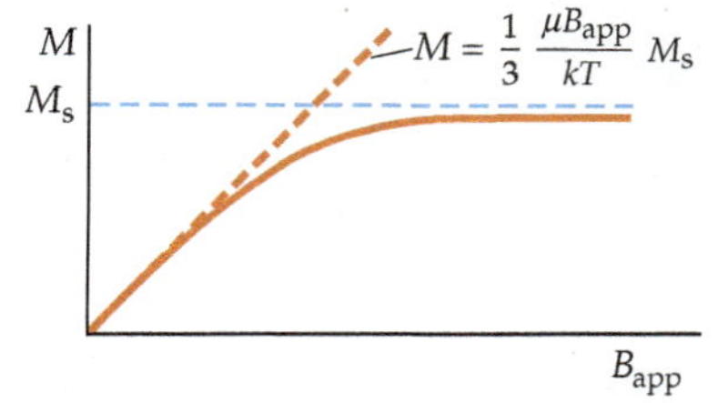

FIGURE 27-42 Plot of magnetization M versus an applied magnetic field B_{app}. In very strong fields, the magnetization approaches the saturation value M_s. This can be achieved only at very low temperatures. In weak fields, the magnetization is approximately proportional to B_{app}, a result known as Curie's law.

$$M = \frac{1}{3}\frac{\mu B_{app}}{kT} M_s \qquad 27\text{-}34$$

CURIE'S LAW

Note that $\mu B_{app}/(kT)$ is the ratio of the maximum energy of a dipole in the magnetic field to the characteristic thermal energy. The result that the magnetization varies inversely with the absolute temperature was discovered experimentally by Pierre Curie and is known as **Curie's law.**

Example 27-13 Applying Curie's Law

If $\mu = \mu_B$, at what temperature will the magnetization be 1.00 percent of the saturation magnetization in an applied magnetic field of 1.00 T?

PICTURE Using Equation 27-34, solve for the temperature when M/M_s equals 0.0100.

SOLVE

1. Curie's law relates M, T, M_s, and B_{app}: $\quad M = \frac{1}{3}\frac{\mu B_{app}}{kT} M_s$

2. Solve for T using $\mu = \mu_B$ and $M/M_s = 0.0100$: $\quad T = \frac{\mu_B B_{app}}{3k}\frac{M_s}{M} = \frac{(5.79 \times 10^{-5}\ \text{eV/T})(1.00\ \text{T})}{3(8.62 \times 10^{-5}\ \text{eV/K})} 100$

$$= \boxed{22.4\ \text{K}}$$

CHECK The step-2 result is greater than absolute zero, as expected.

TAKING IT FURTHER From this example, we see that even in a strong applied magnetic field of 1.00 T, the magnetization is less than 1.00 percent of saturation at temperatures above 22.4 K.

PRACTICE PROBLEM 27-9 If $\mu = \mu_B$, what fraction of the saturation magnetization is M at 300 K for an external magnetic field of 1.5 T?

*FERROMAGNETISM

Ferromagnetism occurs in pure iron, cobalt, and nickel as well as in alloys of those metals with each other. It also occurs in gadolinium, dysprosium, and a few compounds. Ferromagnetism arises from a strong interaction between the electrons in a partially full band in a metal or between the localized electrons that form magnetic moments on neighboring atoms. This interaction, called the **exchange interaction,** lowers the energy of a pair of electrons with parallel spins.

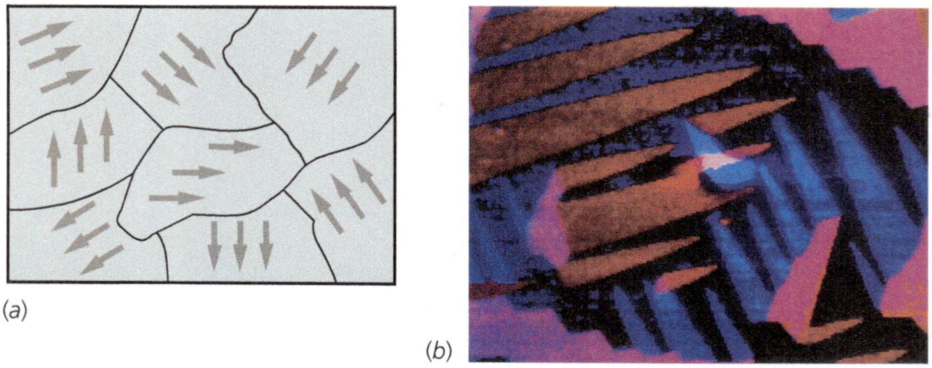
(a) (b)

FIGURE 27-43 (*a*) Schematic illustration of ferromagnetic domains. Within a domain, the magnetic dipoles are aligned, but the direction of alignment varies from domain to domain so that the net magnetic moment is zero. A small external magnetic field may cause the enlargement of those domains that are aligned parallel to the field (at the expense of the neighboring domains), or it may cause the alignment within a domain to rotate. In either case, the result is a net magnetic moment parallel to the field. (*b*) Magnetic domains on the surface of a 97 percent Fe–3 percent Si crystal observed using a scanning electron microscope with polarization analysis. The four colors indicate four possible domain orientations. *(Robert J. Celotta, National Institute of Standards and Technology.)*

A Canadian quarter that is attracted by a magnet. Canadian coins often contain significant amounts of nickel, which is ferromagnetic. *(Photo by Gene Mosca.)*

A chunk of magnetite (lodestone) attracts the needle of a compass. *(© Paul Silverman/Fundamental Photographs.)*

Ferromagnetic materials have very large positive values of magnetic susceptibility χ_m (as measured under conditions described, which follow). In samples of these substances, a small external magnetic field can produce a very large degree of alignment of the atomic magnetic dipole moments. In some cases, the alignment can persist even when the external magnetizing field is removed. This alignment persists because the magnetic dipole moments exert strong forces on their neighbors so that over a small region of space the moments are aligned with each other even when there is no external field. The region of space over which the magnetic dipole moments are aligned is called a **magnetic domain.** The size of a domain is usually microscopic. Within the domain, all the permanent atomic magnetic moments are aligned, but the direction of alignment varies from domain to domain so that the net magnetic moment of a macroscopic piece of ferromagnetic material is zero in the normal state. Figure 27-43 illustrates this situation. The dipole forces that produce this alignment are predicted by quantum theory and cannot be explained with classical physics. At temperatures above a critical temperature, called the **Curie temperature,** thermal agitation is great enough to break up this alignment and ferromagnetic materials become paramagnetic.

When an external magnetic field is applied, the boundaries of the domains may shift or the direction of alignment within a domain may change so that there is a net macroscopic magnetic moment in the direction of the applied field. Because the degree of alignment is large for even a small external field, the magnetic field produced in the material by the dipoles is often much greater than the external field.

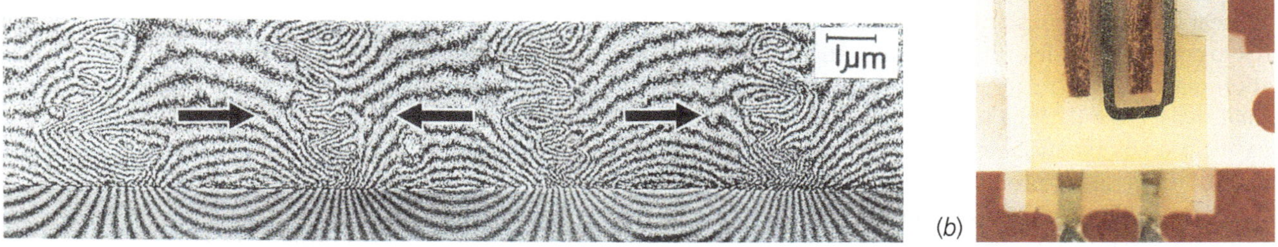

(a) (b)

(*a*) Magnetic field lines on a cobalt magnetic recording tape. The solid arrows indicate the encoded magnetic bits. (*b*) Cross section of a magnetic tape recording head. Current from an audio amplifier is sent to wires around a magnetic core in the recording head where it produces a magnetic field. When the tape passes over a gap in the core of the recording head, the fringing magnetic field encodes information on the tape. *((a) Akira Tonomura, Hitachi Advanced Research Library, Hatomaya, Japan; (b) © Bruce Iverson.)*

Let us consider what happens when we magnetize a long iron rod by placing it inside a solenoid and gradually increase the current in the solenoid windings. We assume that the rod and the solenoid are long enough to permit us to neglect end effects. Because the induced magnetic moments are in the same direction as the applied field, $\vec{B}_{app}$ and $\vec{M}$ are in the same direction. Then,

$$B = B_{app} + \mu_0 M = \mu_0 nI + \mu_0 M \qquad 27\text{-}35$$

In ferromagnetic materials, the magnetic field $\mu_0 M$ due to the magnetic moments is often greater than the magnetizing field B_{app} by a factor of several thousand.

Figure 27-44 shows a plot of B versus the magnetizing field B_{app}. As the current is gradually increased from zero, B increases from zero along the part of the curve from the origin O to point P_1. The flattening of this curve near point P_1 indicates that the magnetization M is approaching its saturation value M_s, at which all the atomic magnetic moments are aligned. Above saturation, B increases only because the magnetizing field $B_{app} = \mu_0 nI$ increases. When B_{app} is gradually decreased from point P_1, there is not a corresponding decrease in the magnetization. The shift of the domains in a ferromagnetic material is not completely reversible, and some magnetization remains even when B_{app} is reduced to zero, as indicated in the figure. This effect is called **hysteresis,** from the Greek word *hysteros* meaning later or behind, and the curve in Figure 27-44 is called a **hysteresis curve.** The value of the magnetic field at point P_4 when B_{app} is zero is called the **remnant field** B_{rem}. At that point, the iron rod is a permanent magnet. If the current in the solenoid is now reversed so that B_{app} is in the opposite direction, the magnetic field B is gradually brought to zero at point c. The remaining part of the hysteresis curve is obtained by further increasing the current in the opposite direction until point P_2 is reached, which corresponds to saturation in the opposite direction, and then decreasing the current to zero at point P_3 and increasing it again in its original direction.

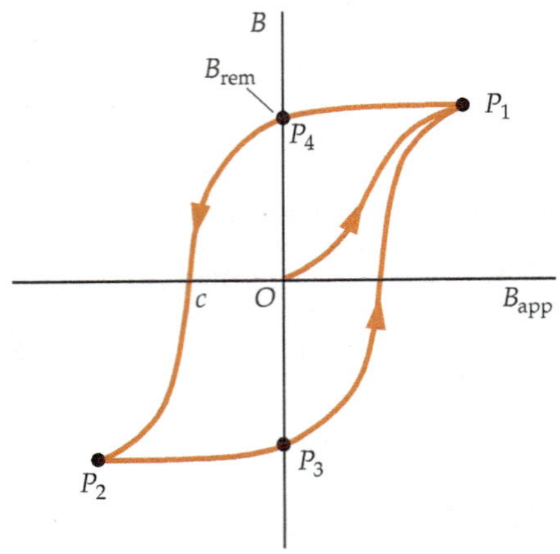

FIGURE 27-44 Plot of B versus the applied magnetizing field B_{app}. The outer curve is called a hysteresis curve. The field B_{rem} is called the remnant field. It remains when the applied field returns to zero.

Because the magnetization M depends on the previous history of the material, and because it can have a large value even when the applied field is zero, it is not simply related to the applied field B_{app}. However, if we confined ourselves to that part of the magnetization curve from the origin to point P_1 in Figure 27-44, $\vec{B}_{app}$ and $\vec{M}$ are parallel and M is zero when B_{app} is zero. We can then define the magnetic susceptibility as in Equation 27-23,

$$M = \chi_m \frac{B_{app}}{\mu_0}$$

and

$$B = B_{app} + \mu_0 M = B_{app}(1 + \chi_m) = K_m \mu_0 nI = \mu nI \qquad 27\text{-}36$$

where

$$\mu = (1 + \chi_m)\mu_0 = K_m \mu_0 \qquad 27\text{-}37$$

is called the **permeability** of the material. [For paramagnetic and diamagnetic materials, χ_m is much less than 1 so the permeability μ and the magnetic constant (permeability of empty space) μ_0 are very nearly equal.]

Because B does not vary linearly with B_{app}, as can be seen from Figure 27-44, the relative permeability is not constant. The maximum value of K_m occurs at a magnetization that is considerably less than the saturation magnetization. Table 27-2 lists the saturation magnetic field $\mu_0 M_s$ and the maximum values of K_m for some ferromagnetic materials. Note that the maximum values of K_m are much greater than 1.

The area enclosed by the hysteresis curve is proportional to the energy dissipated as heat in the irreversible process of magnetizing and demagnetizing. If the hysteresis effect is small, so that the area inside the curve is small, indicating a small energy loss, the material is called **magnetically soft.** Soft iron (chemically pure iron) is an example. The hysteresis curve for a magnetically soft material is shown in Figure 27-45. Here the remnant field B_{rem} is nearly zero, and the energy loss per cycle is small. Magnetically soft materials are used for transformer cores to allow the magnetic field B to change without incurring large energy losses as the field changes.

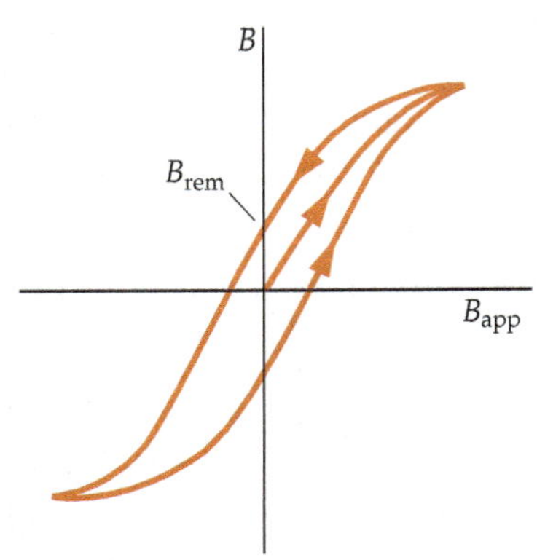

FIGURE 27-45 Hysteresis curve for a magnetically soft material. The remnant field is very small compared with the remnant field for a magnetically hard material such as that shown in Figure 27-44.

Table 27-2 Maximum Values of and for Some Ferromagnetic Materials

Material	$\mu_0 M_s$, T	K_m
Iron (annealed)	2.16	5500
Iron–silicon (96 percent Fe, 4 percent Si)	1.95	7000
Permalloy (55 percent Fe, 45 percent Ni)	1.60	25,000
Mu-metal (77 percent Ni, 16 percent Fe, 5 percent Cu, 2 percent Cr)	0.65	100,000

On the other hand, a large remnant field is desirable in a permanent magnet. **Magnetically hard** materials, such as carbon steel, the alloy Alnico 5, and the rare earths samarium and neodymium (samarium–cobalt and neodymium–iron–boron) are used for permanent magnets.

(*a*) An extremely high-capacity, hard-disk drive for magnetic storage of information, capable of storing over 250 gigabytes of information. (*b*) A magnetic test pattern on a hard disk, magnified 2400 times. The light and dark regions correspond to oppositely directed magnetic fields. The smooth region just outside the pattern is a region of the disk that has been erased just prior to writing. (*(a) © 2003 Western Digital Corporation. All rights reserved. (b) Tom Chang/ IBM Storage Systems Division, San Jose, CA.)*

Example 27-14 Solenoid with Iron Core

A long solenoid has 12 turns per centimeter and a core of soft iron. When the current is 0.500 A, the magnetic field inside the iron core is 1.36 T. Find (*a*) the applied field B_{app}, (*b*) the relative permeability K_m, and (*c*) the magnetization M.

PICTURE The applied field is just that of a long solenoid given by $B_{app} = \mu_0 nI$. Because the total magnetic field is given, we can find the relative permeability from its definition ($K_m = B/B_{app}$) and we can find M from $B = B_{app} + \mu_0 M$.

SOLVE

(*a*) The applied field is given by Equation 27-10:

$$B_{app} = \mu_0 nI = (4\pi \times 10^{-7}\ \text{T}\cdot\text{m/A})(1200\ \text{m}^{-1})(0.500\ \text{A}) = \boxed{7.54 \times 10^{-4}\ \text{T}}$$

(*b*) The relative permeability is the ratio of B to B_{app}:

$$K_m = \frac{B}{B_{app}} = \frac{1.36\ \text{T}}{7.54 \times 10^{-4}\ \text{T}} = \boxed{1.80 \times 10^3}$$

(*c*) The magnetization M is found from Equation 27-35:

$$\mu_0 M = B - B_{\text{app}}$$

$$= 1.36\text{ T} - 7.54 \times 10^{-4}\text{ T} \approx B = 1.36\text{ T}$$

$$M = \frac{B}{\mu_0} = \frac{1.36\text{ T}}{4\pi \times 10^{-7}\text{ T}\cdot\text{m/A}} = \boxed{1.08 \times 10^6\text{ A/m}}$$

CHECK Table 27-2 gives 5500 for the maximum value for K_{m}. Our Part-(*b*) result is less than this maximum value, as expected.

TAKING IT FURTHER The applied magnetic field of 7.54×10^{-4} T is a negligible fraction of the total field of 1.36 T.

*DIAMAGNETISM

Diamagnetic materials are those materials that have very small negative values of magnetic susceptibility χ_{m}. Diamagnetism was discovered by Michael Faraday in 1845 when Faraday found that a piece of bismuth is repelled by either pole of a magnet, indicating that the external field of the magnet induces a magnetic moment in bismuth in the direction opposite the field.

We can understand this effect qualitatively from Figure 27-46, which shows two positive charges moving in circular orbits with the same speed but in opposite directions. Their magnetic moments are in opposite directions and therefore cancel.* In the presence of an external magnetic field $\vec{B}$ directed into the paper, the charges experience an extra force $q\vec{v} \times \vec{B}$, which is along the radial direction. For the charge on the left, this extra force is inward, increasing the centripetal force. If the charge is to remain in the same circular orbit, it must speed up so that mv^2/r equals the total centripetal force.† Its magnetic moment, which is out of the page, is thus increased. For the charge on the right, the additional force is outward, so the particle must slow down to maintain its circular orbit. Its magnetic moment, which is into the page, is decreased. In each case, the *change* in the magnetic moment of the charges is in the direction out of the page, opposite that of the external applied field. Because the permanent magnetic moments of the two charges are equal and oppositely directed they add to zero, leaving only the induced magnetic moments, which are both in the direction opposite to the direction of the applied magnetic field.

A material is diamagnetic if its atoms have zero net angular momentum and therefore no permanent magnetic moment. (The net angular momentum of an atom depends on the electronic structure of the atom, which is presented in Chapter 36.) The induced magnetic moments that cause diamagnetism have magnitudes of the order of 10^{-5} Bohr magnetons. Because this is much smaller than the permanent magnetic moments of the atoms of paramagnetic or ferromagnetic materials, the diamagnetic effect in these atoms is masked by the alignment of their permanent magnetic moments. However, because this alignment decreases with temperature, all materials are theoretically diamagnetic at sufficiently high temperatures.

When a superconductor is placed in an external magnetic field, electric currents are induced on the superconductor's surface so that the net magnetic field in the superconductor is zero. Consider a superconducting rod inside a solenoid of n turns per unit length. When the solenoid is connected to a source of emf so that it carries a current I, the magnetic field due to the solenoid is $\mu_0 nI$. A surface current of $-nI$ per unit length is induced on the superconducting rod that cancels out the field due to the solenoid so that the net field inside the superconductor is zero. From Equation 27-24,

$$\vec{B} = \vec{B}_{\text{app}}(1 + \chi_{\text{m}}) = 0$$

so

$$\chi_{\text{m}} = -1$$

A superconductor is thus a perfect diamagnet with a magnetic susceptibility of −1.

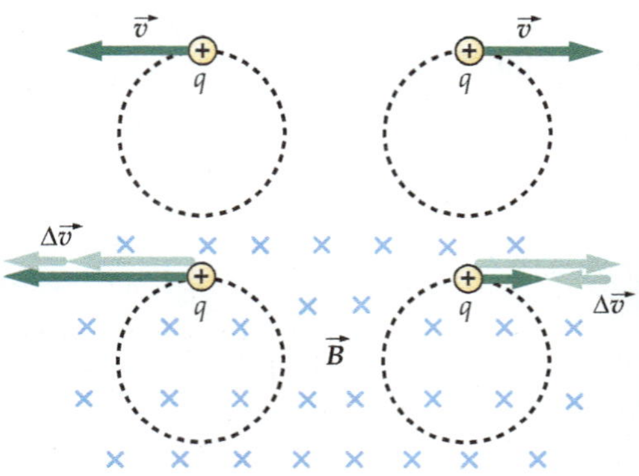

FIGURE 27-46 (*a*) A positive charge moving counterclockwise in a circle has its magnetic moment directed out of the paper. When an external magnetic field directed into the page is turned on, the magnetic force increases the centripetal force so the speed of the particle must increase. The change in the magnetic moment is out of the page. (*b*) A positive charge moving clockwise in a circle has its magnetic moment directed into the page. When an external magnetic field directed into the page is turned on, the magnetic force decreases the centripetal force so the speed of the particle must decrease. As in (*a*), the change in the magnetic moment is directed out of the page.

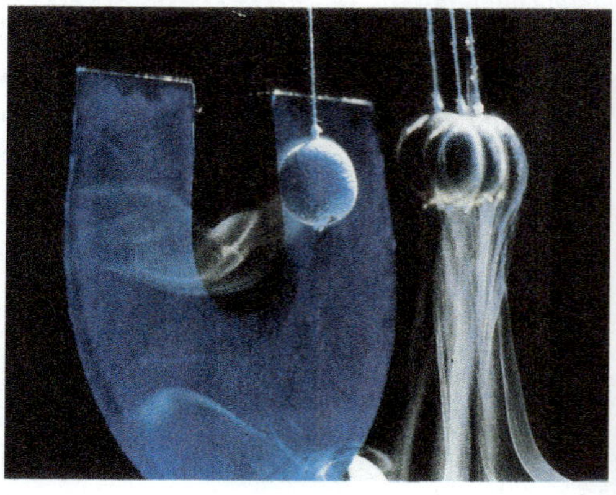

A superconductor is a perfect diamagnet. Here the superconducting pendulum bob is repelled by the permanent magnet. (*© Bill Pierce/ Time Magazine, Inc.*)

* It is simpler to consider positive charges even though negatively charged electrons provide the magnetic moments in matter.

† The electron speeds up because of an electric field induced by the changing magnetic field, an effect called induction, which we discuss in Chapter 28.

Physics Spotlight

Solenoids at Work

Why solenoids? Unlike gears, solenoids do not depend on friction for transferring motion, which means that solenoid-based movement is less likely to wear out machine parts. Solenoid valves, switches, and actuators are all based on the same principle—a central core in the solenoid is moved when current exists in the solenoid coils. Solenoid valves to control the flow of liquids and gases are the most popular mechanical use of solenoids. Some solenoid valves are opened directly by the motion of solenoid cores. When the solenoids turn off, springs return the valves to the off position.* Other solenoid valves, known as pilot-operated valves, use solenoid cores as switches for pistons that have very large ports, or use the motion of solenoid cores to open small pilot ports, which cause enough of a pressure differential in the main fluid line to open the main valve port.†

The world's largest superconducting solenoid magnet first reached its full field strength of 4 T in December 2006. Weighing in at over 10,000 tonnes, the magnet is built around a 6-m diameter, 13-m long superconducting solenoid. The solenoid, which is at CERN, will be used as part of a muon detector. *(CERN.)*

Because the cost of replacing a small valve is large due to lost time on a manufacturing line, manufacturers often choose solenoid valves in these lines.‡ Some solenoid valves have been rated at several million cycles.# Solenoid valves also have been designed for use in very challenging environments. Solenoid valves can operate in corrosive°,§ areas and in areas that have explosive atmospheres.¶ Landscaping and irrigation applications** require solenoids that can operate outdoors. Solenoids are increasingly used for complex automated manufacturing processes.††

Because of their high reliability, long duty life, and low power draw compared to strictly mechanical systems, solenoids are used in robotics, aviation, and automotive applications. In robotics, solenoids often control air valves. In automotive applications, some solenoids control fluid pressure in the transmission, while others control automatic door locks.

One shortcoming of solenoids is that they are liable to overheating if they are overpowered‡‡ or if the solenoid continually receives the power that is required to originally turn the solenoid on.## Overheating can fuse solenoid coils, shut down manufacturing lines, keep engines from starting, and even start fires. Because of these problems, designers are very careful to fit the solenoid to the use and the circuit.

Not all solenoids are used in mechanical applications. Some of the most powerful solenoids on Earth are used to provide large uniform magnetic fields for particle physics experiments. Many of these solenoids use cryogenically cooled superconductors to reach their full strength without overheating. A 5-T superconducting solenoid is set up at the Deutsches Elektronen-Synchrotron, or DESY. This magnet produces a magnetic field of 5.25 T when it carries currents of up to 1000 A. For the superconductors in the coil to work most efficiently, as well as to prevent overheating, the DESY solenoid must be cooled to 4.4 K.°° In Cessy, France, the largest superconducting solenoid magnet on Earth, the Compact Muon Solenoid, is scheduled to begin operating in November 2007.§§ The coils of the solenoid contain 1947 km of superconducting niobium/titanium strands, and the solenoid has an internal diameter of 6.0 m. When it is cooled to 4.5 K, a current of more than 56 kA produces a magnetic field of 4 T.¶¶ Whether very large solenoids for particle physics or miniature solenoids for chemical plants, the reliability and predictability of solenoids are prized assets.

* Hargraves, D., "Solenoid Valves: Operation, Selection, and Application." *Air Conditioning, Heating, & Refrigeration News*, Apr. 5, 1999, pp. 26–28.

† Zdobinski, D., Mudd, W., and Byrne, G., "Understanding Applications, Uses, Key to Solenoid Valve Selection." *Plant Engineering*, Jun. 2006, pp. 65–68.

‡ Heney, P. J., "Wide Variety of Solenoid Valves Available to Designers." *Hydraulics and Pneumatics*, Sept. 1998, Vol. 51, No. 9, pp. 51–56.

\# "Updated Solenoid Survives 20 Million Cycles." *Machine Design*, Aug. 23, 2001, p. 54.

° "Direct-Acting Solenoid Valves." *Design News*, Jun. 5, 2006, pp. 83–84.

§ "Solenoid Valve Handles Acids." *Manufacturing Chemist*, Jul. 1996, Vol. 67, No. 7, p. 51.

¶ "Solenoid Valve for Hazardous Areas." *Offshore*, Nov. 1998, Vol. 58, No. 11, p. 216.

** Mentzer, T., "Control Gets 'Smart'." *Landscape Management*, Jan. 2000, Vol. 39, No. 1, pp. 38+.

†† Mervartova, K., Martinez Calatayud, J., and Catala Icardo, M., "A Fully Automated Assembly Using Solenoid Valves for the Photodegradation and Chemiluminometric Determination of the Herbicide Chlorsulfuron." *Analytical Letters*, Jan. 2005, Vol. 38, No. 1, pp. 179–194.

‡‡ Zdobinski, D., Mudd, W., and Byrne, G., op. cit.

\## Nakhe, S. V., "Smart Solenoid Driver Reduces Power Loss." *Electronic Design*, Oct. 13, 2005, Vol. 53, No. 22, pp. 62–64.

°° Gadwinkel, E., et al., "Cryogenics for a 5 Tesla Superconducting Solenoid with Large Aperture at DESY." *CP170, Advances in Cryogenic Engineering: Transactions of the Cryogenic Engineering Conference—CEC, Vol. 49, AIP Conference Proceedings*, 2004, Vol. 710, Issue 1, pp. 719–725.

§§ Science Daily, "World's Largest Superconducting Solenoid Magnet Reaches Full Field." *Science Daily*, Sept. 26, 2006. http://www.sciencedaily.com/releases/2006/09/060925075001.htm As of Oct. 2006.

¶¶ Blau, B., and Pauss, F., "Superconducting Magnet: ETH Zürich and Superconductor Manufacture for CMS." *CMS Info, CERN*, Apr. 2003, http://cmsinfo.cern.ch/outreach/CMSdocuments/MagnetBrochure/MagnetBrochure.pdf As of Oct. 2006.

Summary

1. Magnetic fields arise from moving charges, and therefore from currents.
2. The Biot–Savart law describes the magnetic field produced by a current element.
3. Ampère's law relates the line integral of the magnetic field along some closed curve to the current that passes through any surface bounded by the curve.
4. The magnetization vector $\vec{M}$ describes the magnetic dipole moment per unit volume of matter.
5. The classical relation $\vec{\mu} = [q/(2m)]\vec{L}$ is derived from the definitions of angular momentum and magnetic moment.
6. The Bohr magneton is a convenient unit for atomic and nuclear magnetic moments.

TOPIC	RELEVANT EQUATIONS AND REMARKS	
1. Magnetic Field $\vec{B}$		
Due to a moving point charge	$\vec{B} = \frac{\mu_0}{4\pi}\frac{q\vec{v} \times \hat{r}}{r^2}$	27-1
	where $\hat{r}$ is a unit vector that points to the field point P from the charge q moving with velocity $\vec{v}$, and μ_0 is a constant of proportionality called the magnetic constant (the permeability of empty space):	
	$\mu_0 = 4\pi \times 10^{-7}\,\mathrm{T\cdot m/A} = 4\pi \times 10^{-7}\,\mathrm{N/A^2}$	27-2
Due to a current element (Biot–Savart law)	$d\vec{B} = \frac{\mu_0}{4\pi}\frac{I\,d\vec{\ell} \times \hat{r}}{r^2}$	27-3
On the axis of a current loop	$B_z = \frac{\mu_0}{4\pi}\frac{2\pi R^2 I}{(z^2 + R^2)^{3/2}}$	27-6
Inside a long solenoid, far from the ends	$B_z = \mu_0 nI$	27-10
	where n is the number of turns per unit length.	
Due to a straight-wire segment	$B = \frac{\mu_0}{4\pi}\frac{I}{R}(\sin\theta_2 - \sin\theta_1)$	27-12
	where R is the perpendicular distance to the wire and θ_1 and θ_2 are the angles subtended at the field point by the ends of the wire.	
Due to a long, straight wire	Use Equation 27-12 with $\theta_2 = 90°$ and $\theta_2 = -90°$, or derive using Ampere's law.	
	The direction of $\vec{B}$ is such that the magnetic field lines of $\vec{B}$ encircle the wire in the direction of the fingers of the right hand if the thumb points in the direction of the current.	
Inside the loops of a tightly wound toroid	$B = \frac{\mu_0}{2\pi}\frac{NI}{r}$	27-18
2. Magnetic Field Lines	Magnetic lines neither begin nor end. Either they form closed loops or they continue indefinitely.	
3. Gauss's Law for Magnetism	$\phi_{\mathrm{m\,net}} = \oint_S \vec{B}\cdot\hat{n}\,dA = \oint_S B_n\,dA = 0$	27-15
4. Magnetic Poles	Magnetic poles always occur in north–south pairs. Isolated magnetic poles have not been found.	
5. Ampère's Law	$\oint_C \vec{B}\cdot d\vec{\ell} = \oint_C B_t\,d\ell = \mu_0 I_C$	27-16
	where C is any closed curve.	

TOPIC	RELEVANT EQUATIONS AND REMARKS	
Validity of Ampère's law	Ampère's law is valid only if the currents are steady and continuous. It can be used to derive expressions for the magnetic field for situations with a high degree of symmetry, such as a long, straight, current-carrying wire or a long, tightly wound solenoid.	
6. Magnetism in Matter	Matter can be classified as paramagnetic, ferromagnetic, or diamagnetic.	
Magnetization	A magnetized material is described by its magnetization vector $\vec{M}$, which is defined as the magnetic dipole moment per unit volume of the material: $$\vec{M} = \frac{d\vec{\mu}}{dV}$$	27-19
	The magnetic field due to a uniformly magnetized cylinder is the same as if the cylinder carried a current per unit length of magnitude M on its surface. This current, which is due to the intrinsic motion of the atomic charges in the cylinder, is called an amperian current.	
7. $\vec{B}$ in Magnetic Materials	$$\vec{B} = \vec{B}_{\text{app}} + \mu_0 \vec{M}$$	27-22
Magnetic susceptibility χ_m	$$\vec{M} = \chi_m \frac{\vec{B}_{\text{app}}}{\mu_0}$$	27-23
	For paramagnetic materials, χ_m is a small positive number that depends on temperature. For diamagnetic materials (other than superconductors), it is a small negative constant independent of temperature. For superconductors, $\chi_m = -1$. For ferromagnetic materials, the magnetization depends not only on the magnetizing current but also on the past history of the material.	
Relative permeability	$$\vec{B} = K_m \vec{B}_{\text{app}}$$	27-24
	where $$K_m = 1 + \chi_m$$	27-25
8. Atomic Magnetic Moments	$$\vec{\mu} = \frac{q}{2m}\vec{L}$$	27-28
	where $\vec{L}$ is the orbital angular momentum of the particle.	
Bohr magneton	$$\mu_B = \frac{e\hbar}{2m_e} = 9.27 \times 10^{-24}\ \text{A}\cdot\text{m}^2$$ $$= 9.27 \times 10^{-24}\ \text{J/T} = 5.79 \times 10^{-5}\ \text{eV/T}$$	27-30
Due to the orbital motion of an electron	$$\vec{\mu}_\ell = -\mu_B \frac{\vec{L}}{\hbar}$$	27-29
Due to electron spin	$$\vec{\mu}_s = -2\mu_B \frac{\vec{S}}{\hbar}$$	27-31
***9. Paramagnetism**	Paramagnetic materials have permanent atomic magnetic moments that have random directions in the absence of an applied magnetic field. In an applied field these dipoles are aligned with the field to some degree, producing a small contribution to the total field that adds to the applied field. The degree of alignment is small except in very strong fields and at very low temperatures. At ordinary temperatures, thermal motion tends to maintain the random directions of the magnetic moments.	
Curie's law	In weak fields, the magnetization is approximately proportional to the applied field and inversely proportional to the absolute temperature. $$M = \frac{1}{3}\frac{\mu B_{\text{app}}}{kT} M_s$$	27-34

TOPIC	RELEVANT EQUATIONS AND REMARKS
*10. **Ferromagnetism**	Ferromagnetic materials have small regions of space called magnetic domains in which all the permanent atomic magnetic moments are aligned. When the material is unmagnetized, the direction of alignment in one domain is independent of that in another domain so that no net magnetic field is produced. When the material is magnetized, the domains of a ferromagnetic material are aligned, producing a very strong contribution to the magnetic field. This alignment can persist in magnetically hard materials, even when the external field is removed, thus leading to permanent magnets.
*11. **Diamagnetism**	Diamagnetic materials are those materials in which the magnetic moments of all electrons in each atom cancel, leaving each atom with a zero magnetic moment in the absence of an external field. In an applied magnetic field, a very small magnetic moment is induced that tends to weaken the field. This effect is independent of temperature. Superconductors are diamagnetic with a magnetic susceptibility equal to -1.

Answers to Practice Problems

27-1	$\vec{B} = 0, \vec{B} = 3.2 \times 10^{-14}\ \text{T}\ \hat{k}$
27-2	25 A
27-4	1.48×10^{-2} T. This is about 2 percent less than the step-3 result.
27-5	B at the center is larger for the circle.
27-6	$R = 4.0$ cm
27-7	0
27-8	$\vec{B} = 2.3 \times 10^{-5}\ \text{T}\ \hat{j}$
27-9	$M/M_s = 1.12 \times 10^{-3}$

Problems

In a few problems, you are given more data than you actually need; in a few other problems, you are required to supply data from your general knowledge, outside sources, or informed estimate.

Interpret as significant all digits in numerical values that have trailing zeros and no decimal points.

- • Single-concept, single-step, relatively easy
- •• Intermediate-level, may require synthesis of concepts
- ••• Challenging
- SSM Solution is in the *Student Solutions Manual*
- Consecutive problems that are shaded are paired problems.

CONCEPTUAL PROBLEMS

1 • Sketch the field lines for the electric dipole and the magnetic dipole shown in Figure 27-47. How do the field lines differ in appearance close to the center of each dipole?

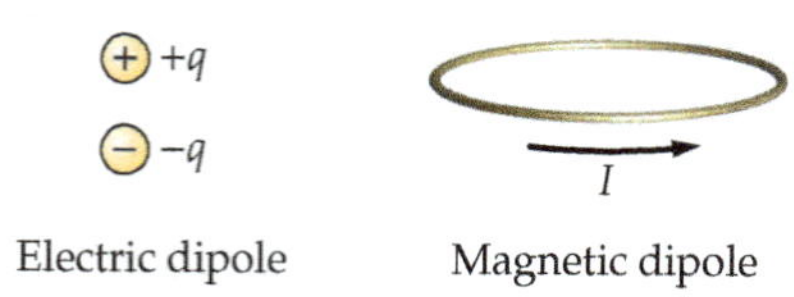

FIGURE 27-47 Problem 1

2 • Two wires lie in the plane of the page and carry equal currents in opposite directions, as shown in Figure 27-48. At a point midway between the wires, the magnetic field is (*a*) zero, (*b*) into the page, (*c*) out of the page, (*d*) toward the top or bottom of the page, (*e*) toward one of the two wires.

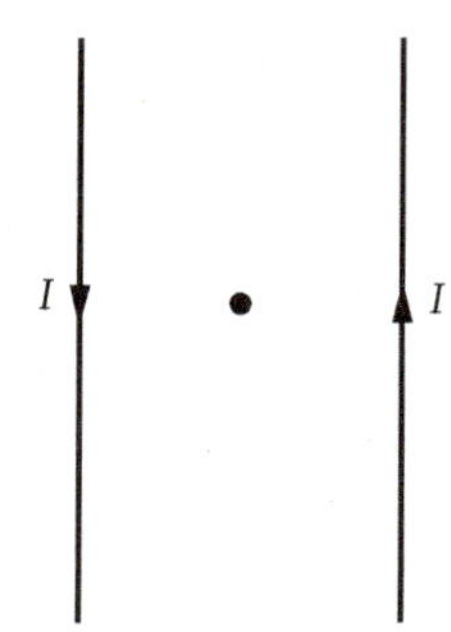

FIGURE 27-48 Problem 2

3 • Parallel wires 1 and 2 carry currents I_1 and I_2, respectively, where $I_2 = 2I_1$. The two currents are in the same direction. The magnitudes of the magnetic force by current 1 on wire 2

and by current 2 on wire 1 are F_{12} and F_{21}, respectively. These magnitudes are related by (*a*) $F_{12} = F_{21}$, (*b*) $F_{21} = 2F_{12}$, (*c*) $2F_{21} = F_{12}$, (*d*) $F_{21} = 4F_{12}$, (*e*) $4F_{21} = F_{12}$.

4 • Make a field-line sketch of the magnetic field due to the currents in the pair of identical coaxial coils shown in Figure 27-49. Consider two cases: (*a*) the currents in the coils have the same magnitude and have the same direction and (*b*) the currents in the coils have the same magnitude and have the opposite directions.

FIGURE 27-49
Problem 4

5 • Discuss the differences and similarities between Gauss's law for magnetism and Gauss's law for electricity. SSM

6 • Explain how you would modify Gauss's law if scientists discovered that single, isolated magnetic poles actually existed.

7 • You are facing directly into one end of a long solenoid and the magnetic field inside of the solenoid points away from you. From your perspective, is the direction of the current in the solenoid coils clockwise or counterclockwise? Explain your answer. SSM

8 • Opposite ends of a helical metal spring are connected to the terminals of a battery. Do the spacings between the coils of the spring tend to increase, decrease, or remain the same when the battery is connected? Explain your answer.

9 • The current density is constant and uniform in a long, straight wire that has a circular cross section. True or false:

(*a*) The magnitude of the magnetic field produced by the wire is greatest at the surface of the wire.
(*b*) The magnetic field strength in the region surrounding the wire varies inversely with the square of the distance from the wire's central axis.
(*c*) The magnetic field is zero at all points on the wire's central axis.
(*d*) The magnitude of the magnetic field inside the wire increases linearly with the distance from the wire's central axis.

10 • If the magnetic susceptibility of a material is positive, (*a*) paramagnetic effects or ferromagnetic effects must be greater than diamagnetic effects, (*b*) diamagnetic effects must be greater than paramagnetic effects, (*c*) diamagnetic effects must be greater than ferromagnetic effects, (*d*) ferromagnetic effects must be greater than paramagnetic effects, (*e*) paramagnetic effects must be greater than ferromagnetic effects.

11 • Of the four gases listed in Table 27-1, which are diamagnetic and which are paramagnetic? SSM

12 • When a current is passed through the wire in Figure 27-50, will the wire tend to bunch up or will it tend to form a circle? Explain your answer.

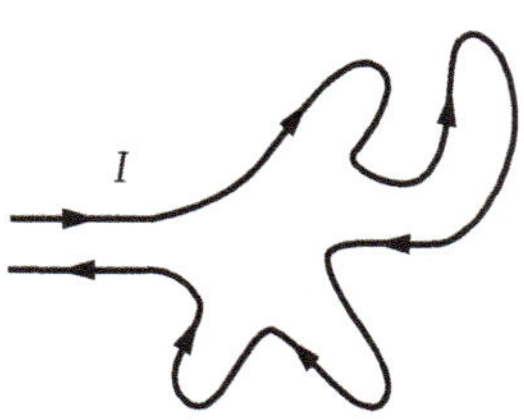

FIGURE 27-50
Problem 12

THE MAGNETIC FIELD OF MOVING POINT CHARGES

13 • At time $t = 0$, a particle has a charge of 12 μC, is located in the $z = 0$ plane at $x = 0, y = 2.0$ m, and has a velocity equal to 30 m/s $\hat{i}$. Find the magnetic field in the $z = 0$ plane at (*a*) the origin, (*b*) $x = 0, y = 1.0$ m, (*c*) $x = 0, y = 3.0$ m, and (*d*) $x = 0, y = 4.0$ m. SSM

14 • At time $t = 0$, a particle has a charge of 12 μC, is located in the $z = 0$ plane at $x = 0, y = 2.0$ m, and has a velocity equal to 30 m/s $\hat{i}$. Find the magnetic field in the $z = 0$ plane at (*a*) $x = 1.0$ m, $y = 3.0$ m, (*b*) $x = 2.0$ m, $y = 2.0$ m, and (*c*) $x = 2.0$ m, $y = 3.0$ m.

15 • A proton has a velocity of 1.0×10^2 m/s $\hat{i} + 2.0 \times 10^2$ m/s $\hat{j}$ and is located in the $z = 0$ plane at $x = 3.0$ m, $y = 4.0$ m at some time $t = T$. Find the magnetic field in the $z = 0$ plane at (*a*) $x = 2.0$ m, $y = 2.0$ m, (*b*) $x = 6.0$ m, $y = 4.0$ m, and (*c*) $x = 3.0$ m, $y = 6.0$ m.

16 •• In a pre–quantum-mechanical model of the hydrogen atom, an electron orbits a proton at a radius of 5.29×10^{-11} m. According to this model, what is the magnitude of the magnetic field at the proton due to the orbital motion of the electron? Neglect any motion of the proton.

17 •• Two equal point charges q are, at some instant, located at (0, 0, 0) and at (0, b, 0). They are both moving with speed v in the $+x$ direction (assume $v \ll c$). Find the ratio of the magnitude of the magnetic force to the magnitude of the electric force on each charge.

THE MAGNETIC FIELD USING THE BIOT–SAVART LAW

18 • A small current element at the origin has a length of 2.0 mm and carries a current of 2.0 A in the $+z$ direction. Find the magnetic field due to the current element (*a*) on the x axis at $x = 3.0$ m, (*b*) on the x axis at $x = -6.0$ m, (*c*) on the z axis at $z = 3.0$ m, and (*d*) on the y axis at $y = 3.0$ m.

19 • A small current element at the origin has a length of 2.0 mm and carries a current of 2.0 A in the $+z$ direction. Find the magnitude and direction of the magnetic field due to this current element at the point (0, 3.0 m, 4.0 m). SSM

20 • A small current element at the origin has a length of 2.0 mm and carries a current of 2.0 A in the $+z$ direction. Find the magnitude of the magnetic field due to this current element and indicate its direction on a diagram at (*a*) $x = 2.0$ m, $y = 4.0$ m, $z = 0$ and (*b*) $x = 2.0$ m, $y = 0$, $z = 4.0$ m.

THE MAGNETIC FIELD DUE TO CURRENT LOOPS AND COILS

21 • A single conducting loop has a radius equal to 3.0 cm and carries a current equal to 2.6 A. What is the magnitude of the magnetic field on the line through the center of the loop and perpendicular to the plane of the loop (*a*) at the center of the loop, (*b*) 1.0 cm from the center, (*c*) 2.0 cm from the center, and (*d*) 35 cm from the center?

22 ••• **SPREADSHEET** A pair of identical coils, each having a radius of 30 cm, are separated by a distance equal to their radii, that is, 30 cm. Called *Helmholtz coils*, they are coaxial and carry equal currents in directions such that their axial fields are in the same direction. A feature of Helmholtz coils is that the resultant magnetic field in the region between the coils is very uniform. Assume the current in each is 15 A and there are 250 turns for each coil. Using a spreadsheet program, calculate and graph the magnetic field as a function of z, the distance from the center of the coils along the common axis, for -30 cm $< z < +30$ cm. Over what range of z does the field vary by less than 20%?

23 ••• A pair of Helmholtz coils that have radii R have their axes along the z axis (see Problem 22). One coil is in the $z = -\frac{1}{2}R$ plane and the second coil is in the $z = +\frac{1}{2}R$ plane. Show that on the z axis at $z = 0$ $dB_z/dz = 0$, $d^2B_z/dz^2 = 0$, and $d^3B_z/dz^3 = 0$. (*Note:* These results show that the magnitude and direction of the magnetic field in the region to either side of the midpoint are approximately equal to the magnitude and direction of the magnetic field at the midpoint.)

24 ••• **ENGINEERING APPLICATION** *Anti-Helmholtz* coils are used in many physics applications, such as laser cooling and trapping, where a field with a uniform gradient is desired. These coils have the same construction as a Helmholtz coil, except that the currents have opposite directions, so that the axial fields are in opposite directions, and the coil separation is $\sqrt{3}R$ rather than R. Graph the magnetic field as a function of z, the axial distance from the center of the coils, for an anti-Helmholtz coil using the same parameters as in Problem 22. Over what interval of the z axis is dB_z/dz within one percent of its value at the midpoint between the coils?

THE MAGNETIC FIELD DUE TO STRAIGHT-LINE CURRENTS

Problems 25 to 30 refer to Figure 27-51 which shows two long, straight wires in the xy plane and parallel to the x axis. One wire is at $y = -6.0$ cm and the other wire is at $y = +6.0$ cm. The current in each wire is 20 A.

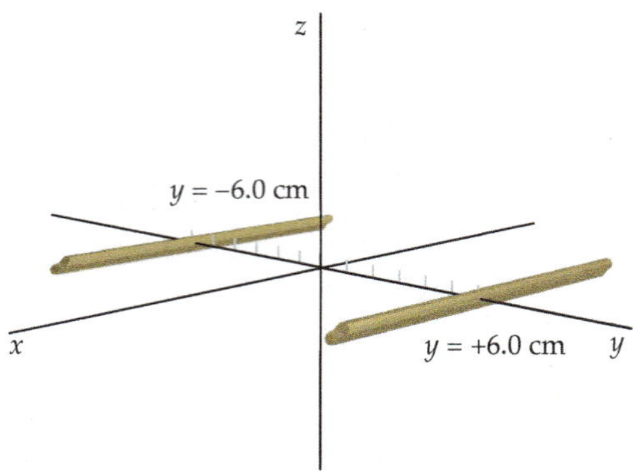

FIGURE 27-51 Problems 25–30

25 •• If the currents are both in the $-x$ direction, find the magnetic field at the following points on the y axis: (*a*) $y = -3.0$ cm, (*b*) $y = 0$, (*c*) $y = +3.0$ cm, and (*d*) $y = +9.0$ cm. SSM

26 •• **SPREADSHEET** Using a spreadsheet program or graphing calculator, graph B_z versus y when both currents are in the $-x$ direction.

27 •• The current in the wire at $y = -6.0$ cm is in the $-x$ direction and the current in the wire at $y = +6.0$ cm is in the $+x$ direction. Find the magnetic field at the following points on the y axis: (*a*) $y = -3.0$ cm, (*b*) $y = 0$, (*c*) $y = +3.0$ cm, and (*d*) $y = +9.0$ cm.

28 •• **SPREADSHEET** The current in the wire at $y = -6.0$ cm is in the $+x$ direction and the current in the wire at $y = +6.0$ cm is in the $-x$ direction. Using a spreadsheet program or graphing calculator, graph B_z versus y.

29 • Find the magnetic field on the z axis at $z = +8.0$ cm if (*a*) the currents are both in the $-x$ direction and (*b*) the current in the wire at $y = -6.0$ cm is in the $-x$ direction and the current in the wire at $y = +6.0$ cm is in the $+x$ direction.

30 • Find the magnitude of the force per unit length exerted by one wire on the other.

31 • Two long, straight parallel wires 8.6 cm apart carry equal currents. The wires repel each other with a force per unit length of 3.6 nN/m. (*a*) Are the currents parallel or antiparallel? Explain your answer. (*b*) Determine the current in each wire.

32 •• The current in the wire shown in Figure 27-52 is 8.0 A. Find the magnetic field at point P.

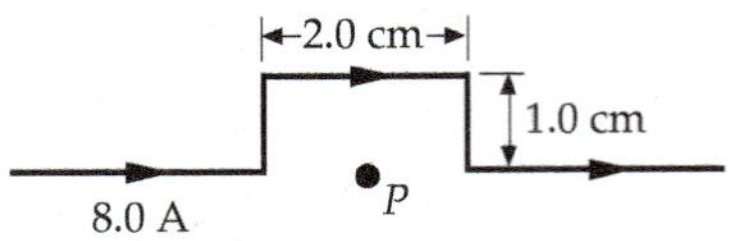

FIGURE 27-52 Problem 32

33 •• As a student technician, you are preparing a lecture demonstration on "magnetic suspension." You have a 16-cm-long straight, rigid wire that will be suspended by flexible conductive lightweight leads above a long, straight wire. Currents that are equal but are in opposite directions will be established in the two wires so the 16-cm wire "floats" a distance h above the long wire with no tension in its suspension leads. If the mass of the 16-cm wire is 14 g and if h (the distance between the central axes of the two wires) is 1.5 mm, what should their common current be? SSM

34 •• Three long, parallel straight wires pass through the vertices of an equilateral triangle that has sides equal to 10 cm, as shown in Figure 27-53. A dot indicates that the direction of the current is out of the page and a cross indicates that the direction of the current is into the page. If each current is 15 A, find (*a*) the magnetic field at the location of the upper wire due to the currents in the two lower wires and (*b*) the force per unit length on the upper wire.

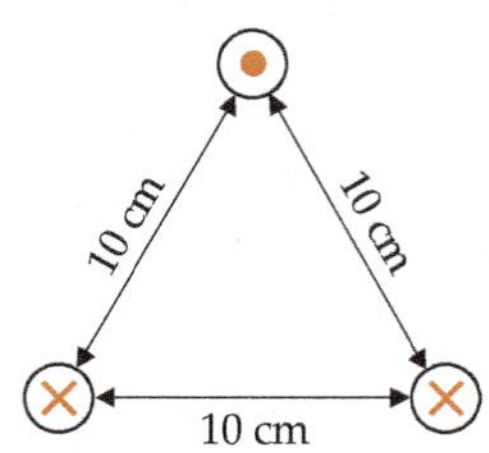

FIGURE 27-53 Problems 34 and 35

35 •• Rework Problem 34 with the current in the lower right corner of Figure 27-53 reversed.

36 •• An infinitely long wire lies along the x axis and carries current I in the $+x$ direction. A second infinitely long wire lies along the y axis and carries current I in the $+y$ direction. At what points in the $z = 0$ plane is the resultant magnetic field zero?

37 •• An infinitely long wire lies along the z axis and carries a current of 20 A in the $+z$ direction. A second infinitely long wire is parallel to the z axis and intersects the x axis at $x = 10.0$ cm.

(*a*) Find the current in the second wire if the magnetic field is zero at (2.0 cm, 0, 0). (*b*) What is the magnetic field at (5.0 cm, 0, 0)? SSM

38 •• Three long parallel wires are at the corners of a square, as shown in Figure 27-54. The wires each carry a current I. Find the magnetic field at the unoccupied corner of the square when (*a*) all the currents are into the page, (*b*) I_1 and I_3 are into the page and I_2 is out, and (*c*) I_1 and I_2 are into the page and I_3 is out. Your answers should be in terms of I and L.

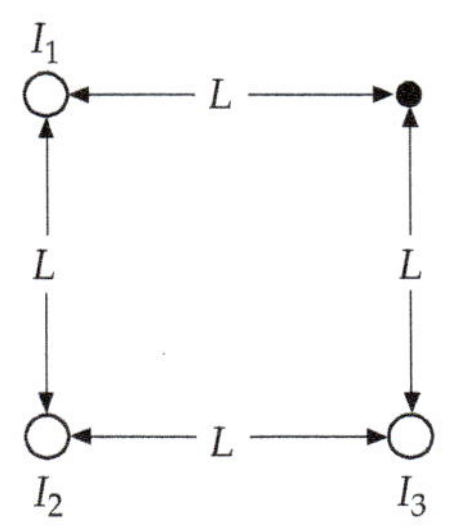

FIGURE 27-54 Problem 38

39 •• Four long, straight parallel wires each carry current I. In a plane perpendicular to the wires, the wires are at the corners of a square of side length a. Find the magnitude of the force per unit length on one of the wires if (*a*) all the currents are in the same direction and (*b*) the currents in the wires at adjacent corners are oppositely directed. SSM

40 •• Five long, straight, current-carrying wires are parallel to the z axis, and each carries a current I in the $+z$ direction. The wires each are a distance R from the z axis. Two of the wires intersect the x axis, one at $x = R$ and the other at $x = -R$. Another wire intersects the y axis at $y = R$. One of the remaining wires intersects the $z = 0$ plane at the point $(R/\sqrt{2}, R/\sqrt{2})$ and the last remaining wire intersects the $z = 0$ plane at the point $(-R/\sqrt{2}, R/\sqrt{2})$. Find the magnetic field on the z axis.

MAGNETIC FIELD DUE TO A CURRENT-CARRYING SOLENOID

41 •• A solenoid that has length 30 cm, radius 1.2 cm, and 300 turns carries a current of 2.6 A. Find the magnitude of the magnetic field on the axis of the solenoid (*a*) at the center of the solenoid and (*b*) at one end of the solenoid. SSM

42 • A solenoid is 2.7 m long, has a radius of 0.85 cm, and has 600 turns. It carries a current I of 2.5 A. What is the magnitude of the magnetic field B inside the solenoid and far from either end?

43 •• A solenoid has n turns per unit length, has a radius R, and carries a current I. Its axis coincides with the z axis—one end at $z = -\frac{1}{2}\ell$ and the other end at $z = +\frac{1}{2}\ell$. Show that the magnitude of the magnetic field at a point on the z axis is given by $B = \frac{1}{2}\mu_0 nI(\cos\theta_1 - \cos\theta_2)$, where the angles are related to the geometry by $\cos\theta_1 = (z + \frac{1}{2}\ell)/\sqrt{(z + \frac{1}{2}\ell)^2 + R^2}$ and $\cos\theta_2 = (z - \frac{1}{2}\ell)/\sqrt{(z - \frac{1}{2}\ell)^2 + R^2}$.

44 ••• In Problem 43, an expression for the magnitude of the magnetic field along the axis of a solenoid is given. For $z \gg \ell$ and $z \gg R$, the angles θ_1 and θ_2 are very small, so the small-angle approximations $\cos\theta \approx 1 - \frac{1}{2}\theta^2$ and $\sin\theta \approx \tan\theta \approx \theta$ are highly accurate. (*a*) Draw a diagram and use it to show that, for these conditions, the angles can be approximated as $\theta_1 \approx R/(z + \frac{1}{2}\ell)$ and $\theta_2 \approx R/(z - \frac{1}{2}\ell)$. (*b*) Using these approximations, show that the magnetic field at points on the z axis where $z \gg \ell$ can be written as $B = \frac{\mu_0}{4\pi}\left(\frac{q_m}{r_2^2} - \frac{q_m}{r_1^2}\right)$ where $r_2 = z - \frac{1}{2}\ell$ is the distance to the near end of the solenoid, $r_1 = z + \frac{1}{2}\ell$ is the distance to the far end, and the quantity q_m is defined by $q_m = nI\pi R^2 = \mu/\ell$, where $\mu = NI\pi R^2$ is the magnitude of the magnetic moment of the solenoid.

USING AMPÈRE'S LAW

45 • A long, straight, thin-walled cylindrical shell of radius R carries a current I parallel to the central axis of the shell. Find the magnetic field (including direction) both inside and outside the shell. SSM

46 • In Figure 27-55, one current is 8.0 A into the page, the other current is 8.0 A out of the page, and each curve is a circular path. (*a*) Find $\oint_C \vec{B} \cdot d\vec{\ell}$ for each path, assuming that each integral is to be evaluated in the counterclockwise direction. (*b*) Which path, if any, can be used to find the combined magnetic field of the currents?

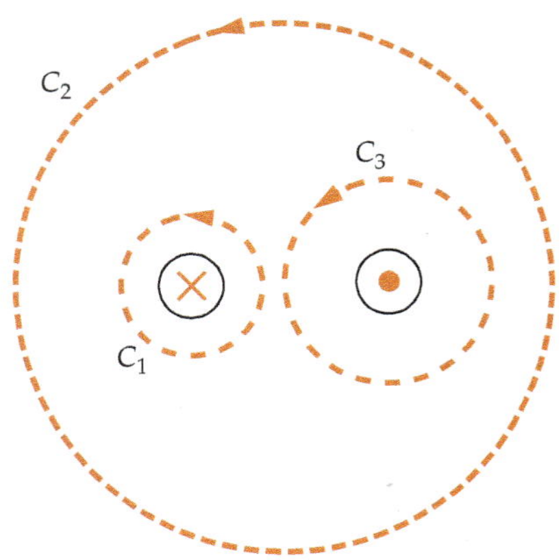

FIGURE 27-55 Problem 46

47 •• Show that a uniform magnetic field that has no fringing field, such as that shown in Figure 27-56, is impossible because it violates Ampère's law. Do this calculation by applying Ampère's law to the rectangular curve shown by the dashed lines. SSM

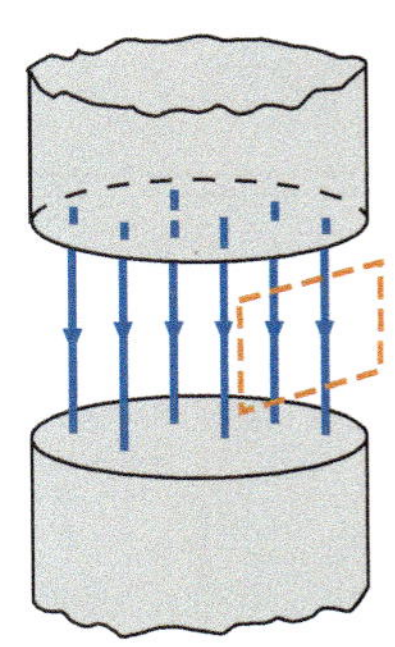

FIGURE 27-56 Problem 47

48 •• **SPREADSHEET** A coaxial cable consists of a solid conducting cylinder that has a radius equal to 1.00 mm and a conducting cylindrical shell that has an inner radius equal to 2.00 mm and an outer radius equal to 3.00 mm. The solid cylinder carries a current of 15.0 A parallel to the central axis. The cylindrical shell carries a current of 15.0 A in the opposite direction. Assume that the current densities are uniformly distributed in both conductors. (*a*) Using a spreadsheet program or graphing calculator, graph the magnitude of the magnetic field as a function of the radial distance r from the central axis for $0 < R < 3.00$ mm. (*b*) What is the magnitude of the field for $R > 3.00$ mm?

49 •• A long cylindrical shell has an inner radius a and an outer radius b and carries a current I parallel to the central axis. Assume that within the material of the shell the current density is uniformly distributed. Find an expression for the magnitude of the magnetic field for (*a*) $0 < R < a$, (*b*) $a < R < b$, and (*c*) $R > b$. SSM

50 •• Figure 27-57 shows a solenoid that has n turns per unit length and carries a current I. Apply Ampère's law to the rectangular curve shown in the figure to derive an expression for the magnitude of the magnetic field. Assume that inside the solenoid the magnetic field is uniform and parallel with the central axis, and that outside the solenoid there is no magnetic field.

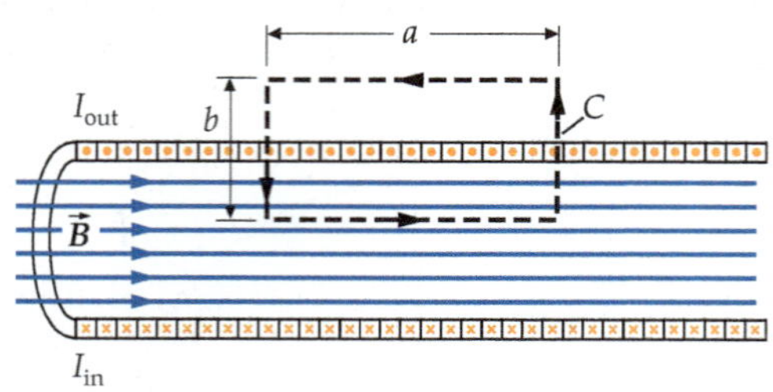

FIGURE 27-57 Problem 50

51 •• A tightly wound 1000-turn toroid has an inner radius of 1.00 cm and an outer radius of 2.00 cm, and carries a current of 1.50 A. The toroid is centered at the origin with the centers of the individual turns in the $z = 0$ plane. In the $z = 0$ plane: (*a*) What is the magnetic field strength at a distance of 1.10 cm from the origin? (*b*) What is the magnetic field strength at a distance of 1.50 cm from the origin? SSM

52 ••• A thin conducting sheet in the $z = 0$ plane carries current in the $-x$ direction (Figure 27-58*a*). The sheet extends indefinitely in all directions and the current is uniformly distributed throughout the sheet. To find the direction of the magnetic field at point P consider the field due only to the currents I_1 and I_2 in the two narrow strips shown. The strips are identical, so $I_1 = I_2$. (*a*) What is the direction of the magnetic field at point P due to just I_1 and I_2? Explain your answer using a sketch. (*b*) What is the direction of the magnetic field at point P due to the *entire* sheet? Explain your answer. (*c*) What is the direction of the field at a point to the right of point P (where $y \neq 0$)? Explain your answer. (*d*) What is the direction of the field at a point below the sheet (where $z < 0$)? Explain your answer using a sketch. (*e*) Apply Ampère's law to the rectangular curve (Figure 27-58*b*) to show that the magnetic field strength at point P is given by $B = \frac{1}{2}\mu_0\lambda$, where $\lambda = dI/dy$ is the current per unit length along the y axis.

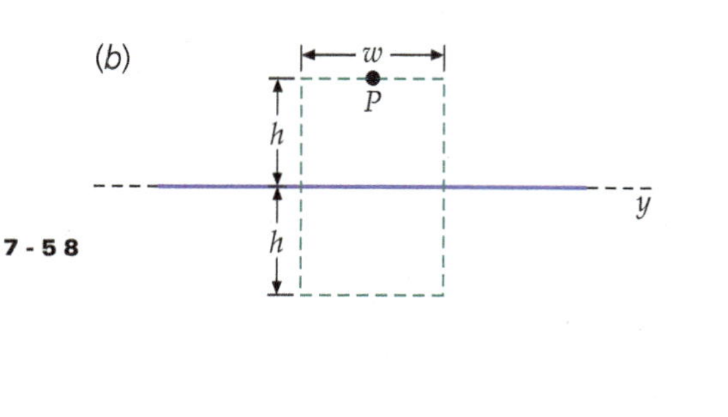

FIGURE 27-58 Problem 52

MAGNETIZATION AND MAGNETIC SUSCEPTIBILITY

53 • A tightly wound solenoid is 20.0 cm long, has 400 turns, and carries a current of 4.00 A so that its axial field is in the $+z$ direction. Find B and B_{app} at the center when (*a*) there is no core in the solenoid and (*b*) there is a soft iron core that has a magnetization of 1.2×10^6 A/m. SSM

54 • A long tungsten-core solenoid carries a current. (*a*) If the core is removed while the current is held constant, does the magnetic field strength in the region inside the solenoid decrease or increase? (*b*) By what percentage does the magnetic field strength in the region inside the solenoid decrease or increase?

55 • As a liquid fills the interior volume of a solenoid that carries a constant current, the magnetic field inside the solenoid *decreases* by 0.0040 percent. Determine the magnetic susceptibility of the liquid.

56 • A long thin solenoid carrying a current of 10 A has 50 turns per centimeter of length. What is the magnetic field strength in the region occupied by the interior of the solenoid when the interior is (*a*) a vacuum, (*b*) filled with aluminum, and (*c*) filled with silver?

57 •• A cylinder of iron, initially unmagnetized, is cooled to 4.00 K. What is the magnetization of the cylinder at that temperature due to the influence of Earth's magnetic field of 0.300 G? Assume a magnetic moment of 2.00 Bohr magnetons per atom. SSM

58 •• A cylinder of silver at a temperature of 77 K has a magnetization equal to 0.075% of its saturation magnetization. Assume a magnetic moment of one Bohr magneton per atom. The density of silver is 1.05×10^4 kg/m^3. (*a*) What value of applied magnetic field parallel to the central axis of the cylinder is required to reach this magnetization? (*b*) What is the magnetic field strength at the center of the cylinder?

59 •• During a solid-state physics lab, you are handed a cylindrically shaped sample of unknown magnetic material. You and your lab partners place the sample in a long solenoid that has n turns per unit length and a current I. The values for magnetic field B

within the material versus nI are given below. Use these values to plot B versus B_{app} and K_m versus nI, where B_{app} is the field due to the current I and K_m is the relative permeability of the sample.

nI, A/m	0	50	100	150	200	500	1000	10 000
B, T	0	0.04	0.67	1.00	1.2	1.4	1.6	1.7

ATOMIC MAGNETIC MOMENTS

60 •• Nickel has a density of 8.70 g/cm^3 and a molar mass of 58.7 g/mol. Nickel's saturation magnetization is 0.610 T. Calculate the magnetic moment of a nickel atom in Bohr magnetons.

61 •• Repeat Problem 60 for cobalt, which has a density of 8.90 g/cm^3, a molar mass of 58.9 g/mol, and a saturation magnetization of 1.79 T.

*PARAMAGNETISM

62 • Show that Curie's law predicts that the magnetic susceptibility of a paramagnetic substance is given by $\chi_m = \mu_0 M_s/(3kT)$.

63 •• In a simple model of paramagnetism, we can consider that some fraction f of the atoms have their magnetic moments aligned with the external magnetic field and that the rest of the atoms are randomly oriented and therefore do not contribute to the magnetic field. (*a*) Use this model and Curie's law to show that at temperature T and external magnetic field B, the fraction of aligned atoms f is given by $\mu B/(3kT)$. (*b*) Calculate this fraction for a sample temperature of 300 K and an external field of 1.00 T. Assume that μ has a value of 1.00 Bohr magneton.

64 •• Assume that the magnetic moment of an aluminum atom is 1.00 Bohr magneton. The density of aluminum is 2.70 g/cm^3 and its molar mass is 27.0 g/mol. (*a*) Calculate the value of the saturation magnetization and the saturation magnetic field for aluminum. (*b*) Use the result of Problem 62 to calculate the magnetic susceptibility at 300 K. (*c*) Explain why the result for Part (*b*) is larger than the value listed in Table 27-1.

65 •• A toroid has N turns, carries a current I, has a mean radius R, and has a cross-sectional radius r, where $r \ll R$ (Figure 27-59). When the toroid is filled with material, it is called a *Rowland ring*. Find B_{app} and B in such a ring, assuming a magnetization that is everywhere parallel to $\vec{B}_{app}$. SSM

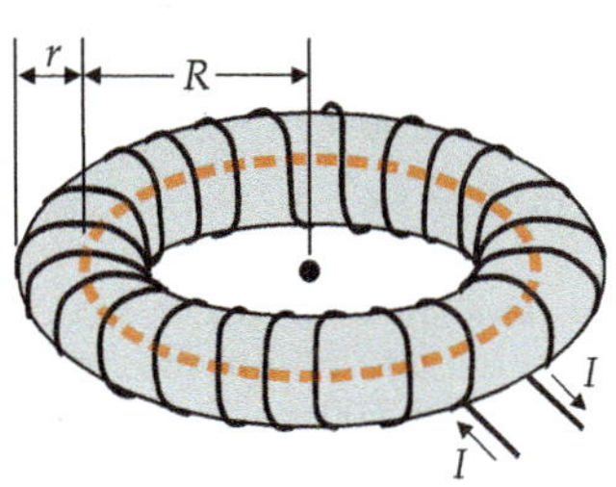

FIGURE 27-59 Problems 65 and 73

66 •• A toroid is filled with liquid oxygen that has a magnetic susceptibility of 4.00×10^{-3}. The toroid has 2000 turns and carries a current of 15.0 A. Its mean radius is 20.0 cm, and the radius of its cross section is 8.00 mm. (*a*) What is the magnetization? (*b*) What is the magnetic field? (*c*) What is the percentage change in the magnetic field produced by the liquid oxygen?

67 •• The centers of the turns of a toroid form a circle with a radius of 14.0 cm. The cross-sectional area of each turn is 3.00 cm^2. It is wound with 5278 turns of fine wire, and the wire carries a current of 4.00 A. The core is filled with a paramagnetic material of magnetic susceptibility 2.90×10^{-4}. (*a*) What is the magnitude of the magnetic field within the substance? (*b*) What is the magnitude of the magnetization? (*c*) What would the magnitude of the magnetic field be if there were no paramagnetic core present?

*FERROMAGNETISM

68 • For annealed iron, the relative permeability K_m has its maximum value of approximately 5500 at $B_{app} = 1.57 \times 10^{-4}$ T. Find the magnitude of the magnetization and magnetic field in annealed iron when K_m is maximum.

69 •• The saturation magnetization for annealed iron occurs when $B_{app} = 0.201$ T. Find the permeability and the relative permeability of annealed iron at saturation. (See Table 27-2.) SSM

70 •• The *coercive force* (which is a misnomer because it is really a magnetic field value) is defined as the applied magnetic field needed to bring the magnetic field back to zero along the hysteresis curve (which is point c in Figure 27-44). For a certain permanent bar magnet, the coercive force is known to be 5.53×10^{-2} T. The bar magnet is to be demagnetized by placing it inside a 15.0-cm-long solenoid that has 600 turns. What minimum current is needed in the solenoid to demagnetize the magnet?

71 •• A long thin solenoid has 50 turns/cm and carries a current of 2.00 A. The solenoid is filled with iron and the magnetic field is measured to be 1.72 T. (*a*) Neglecting end effects, what is the magnitude of the applied magnetic field? (*b*) What is the magnetization? (*c*) What is the relative permeability?

72 •• When the current in Problem 71 is 0.200 A, the magnetic field is measured to be 1.58 T. (*a*) Neglecting end effects, what is the applied magnetic field? (*b*) What is the magnetization? (*c*) What is the relative permeability?

73 •• A toroid has N turns, carries a current I, has a mean radius R, and has a cross-sectional radius r, where $r \ll R$ (Figure 27-59). The core of the toroid is filled with iron. When the current is 10.0 A, the magnetic field in the region where the iron is has a magnitude of 1.80 T. (*a*) What is the magnetization? (*b*) Find the values for the relative permeability, the permeability, and the magnetic susceptibility for this iron sample. SSM

74 • The centers of the turns of a toroid form a circle with a radius of 14.0 cm. The cross-sectional area of each turn is 3.00 cm^2. It is wound with 5278 turns of fine wire, and the wire carries a current of 0.200 A. The core is filled with soft iron, which has a relative permeability of 500. What is the magnetic field strength in the core?

75 ••• A long, straight wire that has a radius of 1.00 mm is coated with an insulating ferromagnetic material that has a thickness of 3.00 mm and a relative magnetic permeability of 400. The coated wire is in air and the wire itself is nonmagnetic. The wire carries a current of 40.0 A. (*a*) Find the magnetic field in the region occupied by the inside of the wire as a function of the perpendicular distance, r, from the central axis of the wire. (*b*) Find the magnetic field in the region occupied by the inside of the ferromagnetic material as a function of the perpendicular distance, r, from the central axis of the wire. (c) Find the magnetic field in the region surrounding the wire and coating as a function of the perpendicular distance, r, from the central axis of the wire. (d) What must the magnitudes and directions of the amperian currents be on the surfaces of the ferromagnetic material to account for the magnetic fields observed?

GENERAL PROBLEMS

76 • Find the magnetic field at point P in Figure 27-60.

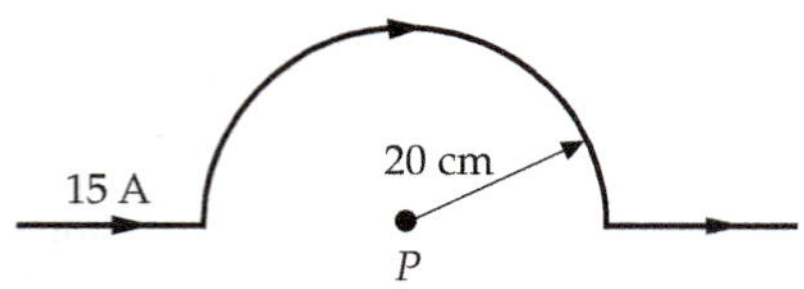

FIGURE 27-60 Problem 76

77 • Using Figure 27-61, find the magnetic field (in terms of the parameters given in the figure) at point P, the common center of the two arcs. SSM

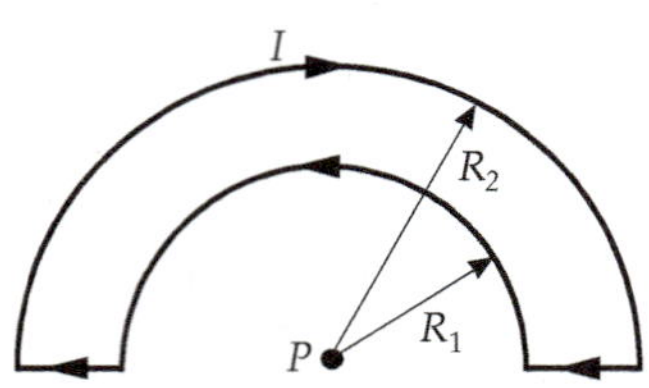

FIGURE 27-61 Problem 77

78 •• A wire of length ℓ is wound into a circular coil of N turns and carries a current I. Show that the magnetic field strength in the region occupied by the center of the coil is given by $\mu_0 \pi N^2 I/\ell$.

79 •• A very long wire carrying a current I is bent into the shape shown in Figure 27-62. Find the magnetic field at point P.

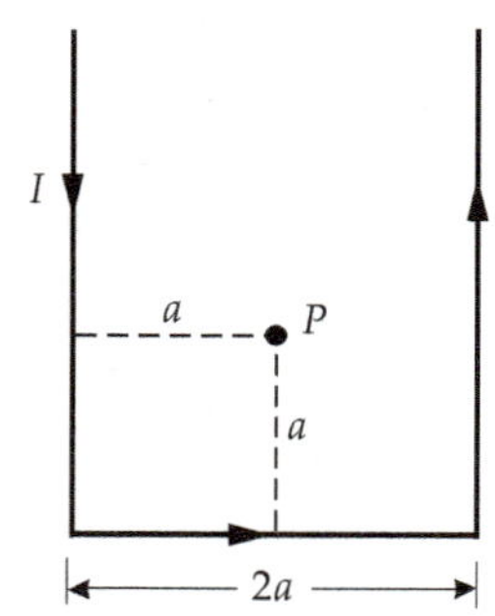

FIGURE 27-62 Problem 79

80 •• A power cable carrying 50 A is 2.0 m below Earth's surface, but the cable's direction and precise position are unknown. Explain how you could locate the cable using a compass. Assume that you are at the equator, where Earth's magnetic field is horizontal and 0.700 G due north.

81 •• A long, straight wire carries a current of 20.0 A, as shown in Figure 27-63. A rectangular coil that has two sides parallel to the straight wire has sides that are 5.00 cm long and 10.0 cm long. The side nearest to the wire is 2.00 cm from the wire. The coil carries a current of 5.00 A. (a) Find the force on each segment of the rectangular coil due to the current in the long, straight wire. (b) What is the net force on the coil? SSM

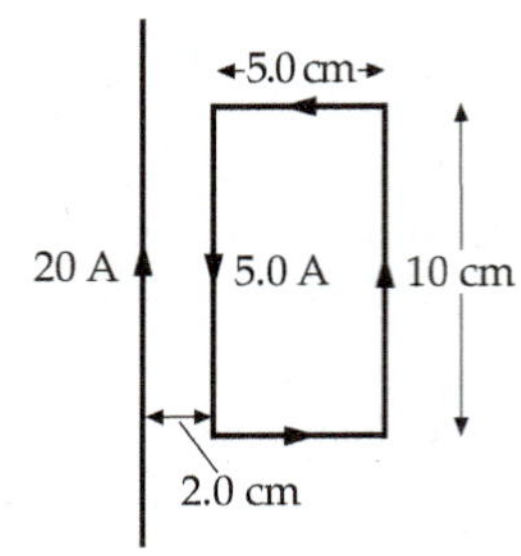

FIGURE 27-63 Problem 81

82 •• The closed loop shown in Figure 27-64 carries a current of 8.0 A in the counterclockwise direction. The radius of the outer arc is 0.60 m and that of the inner arc is 0.40 m. Find the magnetic field at point P.

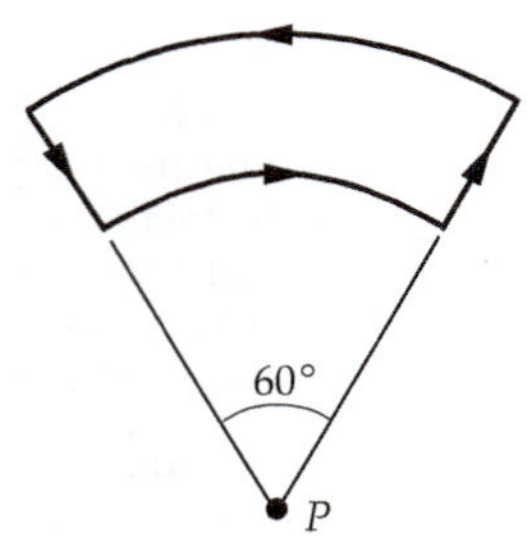

FIGURE 27-64 Problem 82

83 •• A closed circuit consists of two semicircles of radii 40 cm and 20 cm that are connected by straight segments, as shown in Figure 27-65. A current of 3.0 A exists in this circuit and has a clockwise direction. Find the magnetic field at point P.

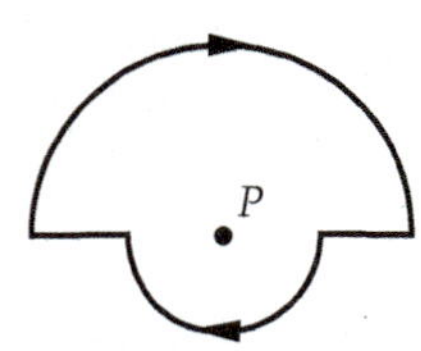

FIGURE 27-65 Problem 83

84 •• A very long, straight wire carries a current of 20.0 A. An electron outside the wire is 1.00 cm from the central axis of the wire and is moving with a speed of 5.00×10^6 m/s. Find the force on the electron when it moves (a) directly away from the wire, (b) parallel to the wire in the direction of the current, and (c) perpendicular to the central axis of wire and tangent to a circle that is coaxial with the wire.

85 •• SPREADSHEET A current of 5.00 A is uniformly distributed over the cross section of a long, straight wire of radius $R_0 = 2.55$ mm. Using a spreadsheet program, graph the magnetic field strength as a function of R (the distance from the central axis of the wire) for $0 \leq R \leq 10R_0$.

86 •• A 50-turn coil of radius 10.0 cm carries a current of 4.00 A and a concentric 20-turn coil of radius 0.500 cm carries a current of 1.00 A. The planes of the two coils are perpendicular. Find the magnitude of the torque exerted by the large coil on the small coil. (Neglect any variation in magnetic field due to the current in the large coil over the region occupied by the small coil.)

87 •• The magnetic needle of a compass is a uniform rod with a length of 3.00 cm, a radius of 0.850 mm, and a density of 7.96×10^3 kg/m^3. The needle is free to rotate in a horizontal plane, where the horizontal component of Earth's magnetic field is 0.600 G. When disturbed slightly, the compass executes simple harmonic motion about its midpoint with a frequency of 1.40 Hz. (*a*) What is the magnetic dipole moment of the needle? (*b*) What is the magnetization of the needle? (*c*) What is the amperian current on the surface of the needle?

88 •• A relatively inexpensive ammeter, called a *tangent galvanometer*, can be made using Earth's magnetic field. A plane circular coil that has N turns and a radius R is oriented so the magnetic field B_c it produces in the center of the coil is either east or west. A compass is placed at the center of the coil. When there is no current in the coil, assume the compass needle points due north. When there is a current in the coil (I), the compass needle points in the direction of the resultant magnetic field at an angle θ to the north. Show that the current I is related to θ and to the horizontal component of Earth's magnetic field B_e by $I = \dfrac{2RB_e}{\mu_0 N}\tan\theta$.

89 •• Earth's magnetic field is about 0.600 G at the magnetic poles, and is pointed vertically downward at the magnetic pole in the northern hemisphere. If the magnetic field were due to an electric current circulating in a loop at the radius of the inner iron core of Earth (approximately 1300 km), (*a*) what would be the magnitude of the current required? (*b*) What direction would this current have–the same as Earth's spin, or opposite? Explain your answer.

90 •• A long, narrow bar magnet has its magnetic moment $\vec{\mu}$ parallel to its long axis and is suspended at its center—in essence becoming a frictionless compass needle. When the magnet is placed in a magnetic field $\vec{B}$, it lines up with the field. If it is displaced by a small angle and released, show that the magnet will oscillate about its equilibrium position with frequency given by $\dfrac{1}{2\pi}\sqrt{\dfrac{\mu B}{I}}$, where I is the moment of inertia about the point of suspension.

91 •• An infinitely long, straight wire is bent, as shown in Figure 27-66. The circular portion has a radius of 10.0 cm and its center is a distance r from the straight part. Find the value of r such that the magnetic field at the center of the circular portion is zero.

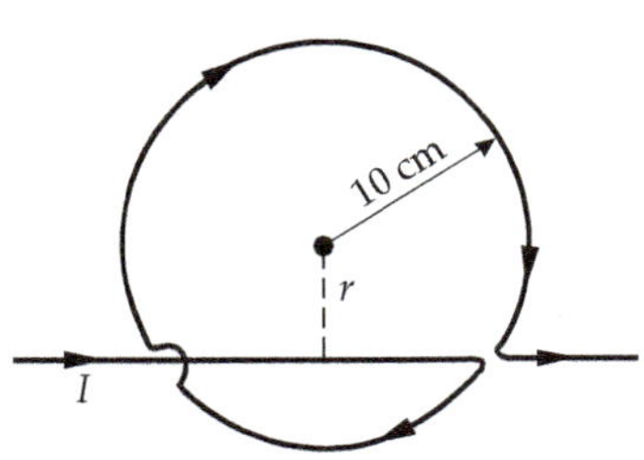

FIGURE 27-66 Problem 91

92 •• (*a*) Find the magnetic field strength at point P on the perpendicular bisector of a wire segment carrying current I, as shown in Figure 27-67. (*b*) Use your result from Part (*a*) to find the magnetic field strength at the center of a regular polygon of N sides. (*c*) Show that when N is very large, your result approaches that for the magnetic field strength at the center of a circle.

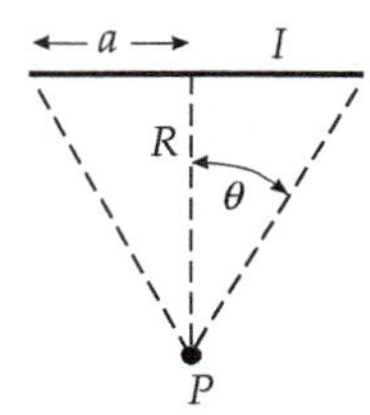

FIGURE 27-67 Problem 92

93 •• The current in a long cylindrical conductor of radius 10 cm varies with distance from the axis of the cylinder according to the relation $I(r) = (50\text{ A/m})r$. Find the magnetic field at the following perpendicular distances from the wire's central axis: (*a*) 5.0 cm, (*b*) 10 cm, and (*c*) 20 cm.

94 •• Figure 27-68 shows a square loop that has 20-cm-long sides and is in the $z = 0$ plane with its center at the origin. The loop carries a current of 5.0 A. An infinitely long wire that is parallel to the x axis and carries a current of 10 A intersects the z axis at $z = 10$ cm. The directions of the currents are shown in the figure. (*a*) Find the net torque on the loop. (*b*) Find the net force on the loop.

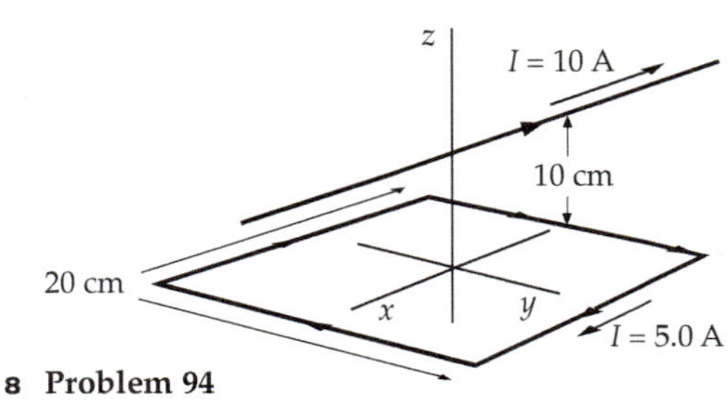

FIGURE 27-68 Problem 94

95 •• A current balance is constructed in the following way: A straight 10.0-cm-long section of wire is placed on top of the pan of an electronic balance (Figure 27-69). This section of wire is connected in series with a power supply and a long straight horizontal section of wire that is parallel to it and positioned directly above it. The distance between the central axes of the two wires is 2.00 cm. The power supply provides a current in the wires. When the power supply is switched on, the reading on the balance increases by 5.00 mg. What is the current in the wire? SSM

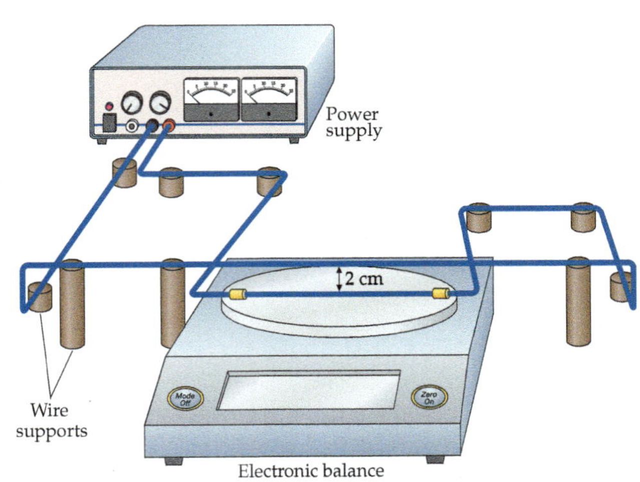

FIGURE 27-69 Problem 95

96 •• Consider the current balance of Problem 95. If the sensitivity of the balance is 0.100 mg, what is the minimum current detectable using this current balance?

97 ••• A nonconducting disk has radius R, carries a uniform surface charge density σ, and rotates with angular speed ω. (*a*) Consider an annular strip that has a radius r, a width dr, and a charge dq. Show that the current (dI) produced by this rotating strip is given by $\omega\sigma r\,dr$. (*b*) Use your result from Part (*a*) to show that the magnetic field strength at the center of the disk is given by the expression $\frac{1}{2}\mu_0\sigma\omega R$. (*c*) Use your result from Part (*a*) to find an expression for the magnetic field strength at a point on the central axis of the disk a distance z from its center. SSM

98 ••• A square loop that has sides of length ℓ lies in the $z = 0$ plane with its center at the origin. The loop carries a current I. (*a*) Derive an expression for the magnetic field strength at any point on the z axis. (*b*) Show that for z much larger than ℓ, your result from Part (*a*) becomes $B \approx \mu\mu_0/(2\pi z^3)$, where μ is the magnitude of the magnetic moment of the loop.

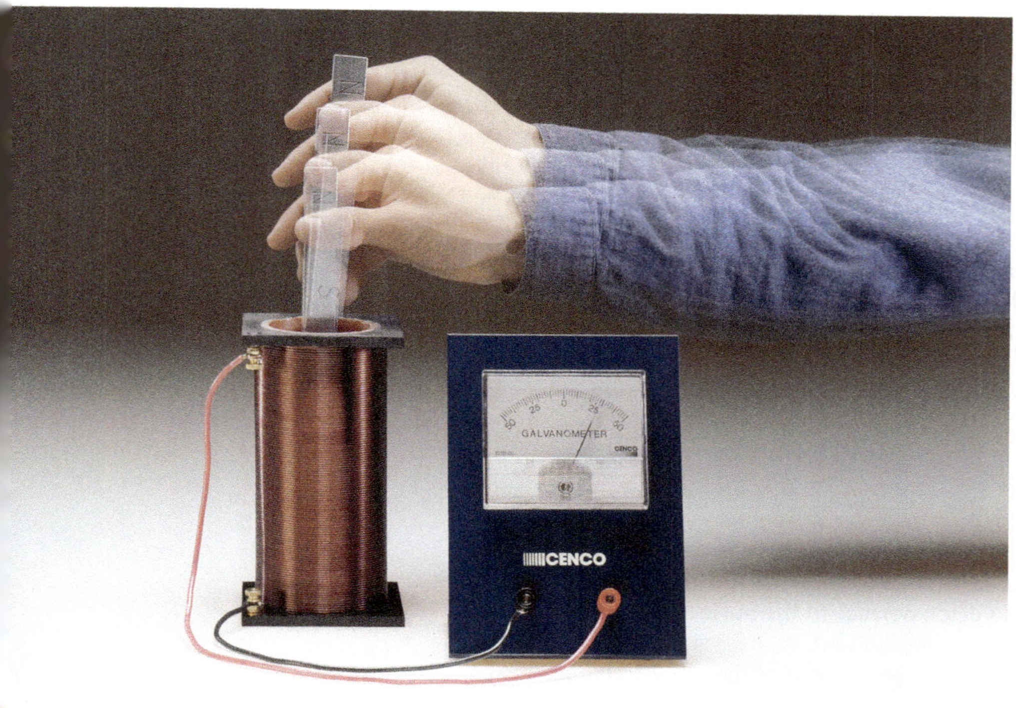

CHAPTER 28

DEMONSTRATION OF INDUCED EMF. WHEN THE MAGNET IS MOVING TOWARD OR AWAY FROM THE COIL, AN EMF IS INDUCED IN THE COIL, AS SHOWN BY THE GALVANOMETER'S DEFLECTION. NO DEFLECTION IS OBSERVED WHEN THE MAGNET IS STATIONARY. *(Richard Megna/ Fundamental Photographs.)*

? How do you calculate the magnitude of an induced emf in a coil? (See Example 28-2.)

Magnetic Induction

In the early 1830s, Michael Faraday in England and Joseph Henry in the United States independently discovered that in a *changing* magnetic field a changing magnetic flux through a surface bounded by a closed stationary loop of wire induces a current in the wire. The emfs and currents caused by such changing magnetic fluxes are called **induced emfs** and **induced currents.** The process itself is referred to as **induction.** Faraday and Henry also discovered that in a *static* magnetic field a changing magnetic flux through a surface bounded by a moving loop of wire induces an emf in the wire. An emf caused by the motion of a conductor in a region with a magnetic field is called a **motional emf.**

When you pull the plug of an electric cord from its socket, you sometimes observe a small spark. Before the cord is disconnected, the cord carries a current that produces a magnetic field encircling the current. When the cord is disconnected, the current abruptly ceases and the magnetic field encircling the cord collapses.

This changing magnetic field induces an emf that tends to maintain the original current, resulting in a spark at the points of the disconnect. Once the magnetic field collapses to zero it is no longer changing, and the induced emf is zero.

This chapter will explore the various methods of magnetic induction, all of which can be summarized by a single relation known as Faraday's law. Faraday's law relates the induced emf in a circuit to the rate of change in magnetic flux through the circuit. (The magnetic flux *through the circuit refers to the flux of the magnetic field through a surface bounded by the circuit.)*

28-1 MAGNETIC FLUX

The flux of any vector field through a surface is calculated in the same way as the flux of an electric field through a surface (Section 22-2). Let dA be an element of area on the surface S, and let $\hat{n}$ be a unit vector normal to the surface element of area dA (Figure 28-1). If $\hat{n}$ is a normal to a surface element, then so is $-\hat{n}$, so there are two directions normal to any surface element, and which of the two directions is selected for the direction of $\hat{n}$ is optional. However, the sign of the flux does depend on the choice for the direction of $\hat{n}$. The magnetic flux ϕ_m through S is

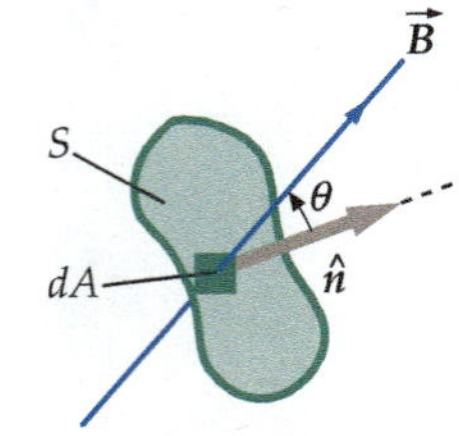

FIGURE 28-1 When $\vec{B}$ makes an angle θ with the normal to the surface S, the flux through the area dA is $\vec{B}\cdot\hat{n}\,dA = B\,dA\cos\theta$, where dA is the area of the surface element.

$$\phi_m = \int_S \vec{B}\cdot\hat{n}\,dA = \int_S B_n\,dA \qquad 28\text{-}1$$

MAGNETIC FLUX

The unit of magnetic flux is that of magnetic field strength multiplied by area, namely, the tesla-meter-squared, which is called a **weber** (Wb):

$$1\ \text{Wb} = 1\ \text{T}\cdot\text{m}^2 \qquad 28\text{-}2$$

Because B is proportional to the number of field lines per unit area, the magnetic flux is proportional to the number of field lines through an element of area.

PRACTICE PROBLEM 28-1

Show that one weber per second is one volt.

If the surface is flat and has an area A, and if $\vec{B}$ is uniform (has the same magnitude and direction) everywhere on the surface, the magnetic flux through the surface is

$$\phi_m = \vec{B}\cdot\hat{n}A = BA\cos\theta = B_nA \qquad 28\text{-}3$$

where θ is the angle between the direction of $\vec{B}$ and the direction of $\hat{n}$. We shall refer to the direction of $\hat{n}$ as the positive normal direction. We are often interested in the flux through a surface bounded by a coil that has several turns of wire. If the coil has N turns, the flux through the surface is N multiplied by the flux through each turn (Figure 28-2). That is,

$$\phi_m = NBA\cos\theta \qquad 28\text{-}4$$

where A is the area of the flat surface bounded by a single turn. (*Note:* Only a closed curve can actually bound a surface. A single turn of a multiturn coil is not closed, so a single turn cannot actually bound a surface. However, if a coil is tightly wound a single turn is almost closed, and A is the area of the flat surface that it almost bounds.)

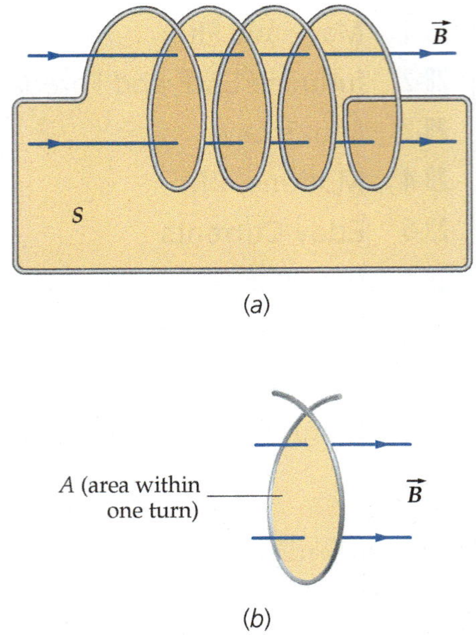

FIGURE 28-2 (*a*) The flux through the surface S bounded by a coil that has N turns is proportional to the number of field lines penetrating the surface. The coil shown has 4 turns. For the two field lines shown, each line penetrates the surface S four times, once for each turn, so the flux through S is four times greater than the flux through the surface "bounded" by a single turn of the coil. The coil shown is not tightly wound so the surface S can be seen better. (*b*) The area A of the flat surface is (almost) bounded by a single turn.

Example 28-1 Flux through a Solenoid

Find the magnetic flux through a solenoid that is 40 cm long, has a radius of 2.5 cm, has 600 turns, and carries a current of 7.5 A.

PICTURE The magnetic field $\vec{B}$ inside this long solenoid is uniform and parallel with the axis of the solenoid. (We are neglecting the end effects.) It is therefore perpendicular to the plane of each loop of the solenoid. Therefore, to find the flux we need to find B inside the solenoid and then multiply B by NA.

SOLVE

1. The magnetic flux is the product of the number of turns, the magnetic field strength, and the area bounded by one turn (Equation 28-4):
   $$\phi_m = NBA$$
2. The magnetic field inside the solenoid is given by $B = \mu_0 nI$ (Equation 27-10), where $n = N/\ell$ is the number of turns per unit length:
   $$\phi_m = N\mu_0 nIA = N\mu_0 \frac{N}{\ell}IA = \frac{\mu_0 N^2 IA}{\ell}$$
3. Express the area A in terms of its radius:
   $$A = \pi r^2$$
4. Substitute the given values to calculate the flux:
   $$\phi_m = \frac{\mu_0 N^2 I\pi r^2}{\ell}$$
   $$= \frac{(4\pi \times 10^{-7}\,\mathrm{T\cdot m/A})(600)^2(7.5\,\mathrm{A})\pi(0.025\,\mathrm{m})^2}{0.40\,\mathrm{m}}$$
   $$= \boxed{1.67 \times 10^{-2}\,\mathrm{Wb}}$$

CHECK The units in line 2 of step 4 work out to $\mathrm{T\cdot m^2}$, and a weber is defined as $1\ \mathrm{T\cdot m^2}$. These are the correct units for magnetic flux.

TAKING IT FURTHER Note that because $\phi_m = NBA$ and B is proportional to the number of turns N, ϕ_m is proportional to N^2.

28-2 INDUCED EMF AND FARADAY'S LAW

Experiments by Faraday, Henry, and others showed that if the magnetic flux through a surface bounded by a wire (a conducting path) changes, an emf equal in magnitude to the rate of change of the flux is induced in the wire. We usually detect the emf by observing a current in the conductor, but the emf around the boundary of the surface exists even if the conducting path does not exist or is incomplete (not closed) and no current exists. In previous chapters, we considered emfs that were localized in a specific part of a circuit, such as between the terminals of a battery. However, induced emfs can be distributed throughout a circuit.

The magnetic flux ϕ_m through a flat surface of area A in a uniform magnetic field $\vec{B}$ is given by $\phi_m = BA\cos\theta$ (Equation 28-3), where θ is the angle between $\vec{B}$ and the normal to the surface. The flux can be changed by increasing or decreasing B, by increasing or decreasing A, or by changing the angle θ. If the magnetic field is due to a permanent magnet, the magnitude of the magnetic field can be increased or decreased by moving a permanent magnet toward or away from the surface. If the magnetic field is due to a current in a circuit, the magnitude of the magnetic field can be increased or decreased by increasing or decreasing the current. The flux through the surface can also be changed by varying the angle θ. To vary θ, we can change either the orientation of the surface or the direction of the magnetic field. In each case, if along the perimeter of the surface there is a conducting path, such as a metal wire, an emf $\mathcal{E}$ is induced along the path that is equal in magnitude to the rate of change of the magnetic flux through the surface. That is,

$$\mathcal{E} = -\frac{d\phi_m}{dt} \qquad \text{28-5}$$

FARADAY'S LAW

This result is known as **Faraday's law.** The minus sign in Faraday's law has to do with the direction of the induced emf (clockwise or counterclockwise), which is addressed later in this section.

Figure 28-3 shows a single stationary loop of wire in a magnetic field. The flux through the loop is changing because the magnetic field strength on surface S is increasing, so an emf is induced in the loop. Because emf is the work done per unit charge, we know there must be forces exerted on the mobile charges that are doing work on these charges. Magnetic forces can do no work; therefore, we cannot attribute the emf to the work done by magnetic forces. It is electric forces associated with a nonconservative electric field $\vec{E}_{nc}$ doing the work on the mobile charges. The line integral of this electric field around a complete circuit equals the work done per unit charge, which is equal to the induced emf in the circuit.

The electric fields that we studied in earlier chapters resulted from static electric charges. Such electric fields are conservative, meaning that their circulation about any closed path C is zero. (The circulation of a vector field $\vec{E}$ about a closed path C is defined as $\oint_C \vec{E} \cdot d\vec{\ell}$.) However, the electric field associated with a changing magnetic field is nonconservative. Its circulation about C is equal to the induced emf in the loop of wire. The circulation of the electric field is equal to the negative of the rate of change of the magnetic flux through any surface S bounded by C:

$$\mathcal{E} = \oint_C \vec{E}_{nc} \cdot d\vec{\ell} = -\frac{d}{dt}\int_S \vec{B} \cdot \hat{n}\, dA = -\frac{d\phi_m}{dt} \qquad \text{28-6}$$

INDUCED EMF FOR A STATIONARY CIRCUIT IN A CHANGING MAGNETIC FIELD

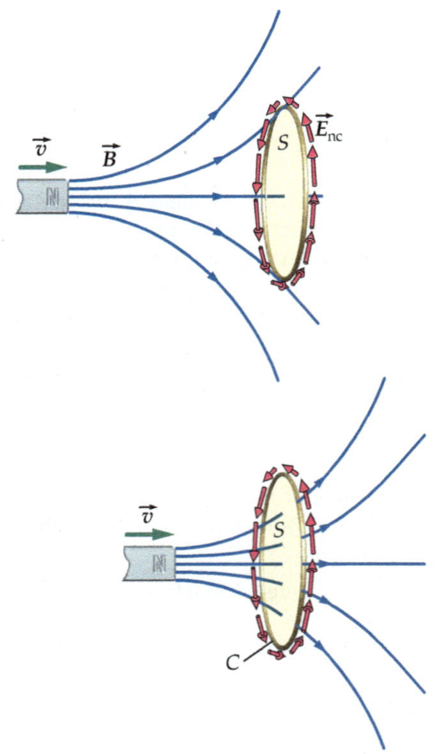

FIGURE 28-3 If the magnetic flux through the stationary wire loop is changing, an emf is induced in the loop. The emf is distributed throughout the loop, which is due to a nonconservative electric field $\vec{E}_{nc}$ tangent to the wire. The closed path C is within the material of the conducting loop.

Example 28-2 Induced EMF in a Circular Coil I

A uniform magnetic field makes an angle of 30.0° with the axis of a circular coil that has 300 turns and a radius equal to 4.00 cm. The magnitude of the magnetic field increases at a rate of 85.0 T/s while its direction remains fixed. Find the magnitude of the induced emf in the coil.

PICTURE The induced emf equals the number of turns N multiplied by the rate of change of the flux through a single turn. Because $\vec{B}$ is uniform, the flux through each turn is simply $\phi_m = BA\cos\theta$, where $A = \pi r^2$ is the area of the circle bounded by one turn of the coil.

SOLVE

1. The magnitude of the induced emf is given by Faraday's law: $\mathcal{E} = -\frac{d\phi_m}{dt}$
2. For a uniform field, the flux is: $\phi_m = N\vec{B}\cdot\hat{n}A = NBA\cos\theta$
3. Substitute this expression for ϕ_m and calculate $\mathcal{E}$: $\mathcal{E} = -\frac{d\phi_m}{dt} = -\frac{d}{dt}(NBA\cos\theta) = -N\pi r^2\cos\theta\frac{dB}{dt}$
$= (300)\pi(0.0400\text{ m})^2\cos 30.0°(85.0\text{ T/s}) = -111\text{ V}$
$\mathcal{E} = \boxed{111\text{ V}}$

CHECK Line 2 of step 3 has units of $\text{T}\cdot\text{m}^2/\text{s}$, where $1\text{ T}\cdot\text{m}^2/\text{s} = 1\text{ Wb/s} = 1$ volt. [Use the formula $\vec{F} = q\vec{v}\times\vec{B}$ as a reminder that $1\text{ N} = 1\text{ C}\cdot\text{m}\cdot\text{T/s}$, so $1\text{ T} = 1\text{ N}\cdot\text{s}/(\text{C}\cdot\text{m})$.]

PRACTICE PROBLEM 28-2 If the resistance of the coil is 200 Ω, what is the induced current?

Example 28-3 Induced EMF in a Circular Coil II

Try It Yourself

An 80.0-turn coil that has a radius equal to 5.00 cm and a resistance equal to 30.0 Ω sits in a region that has a uniform magnetic field normal to the plane of the coil. At what rate must the magnitude of the magnetic field change to produce a current of 4.00 A in the coil?

PICTURE The number of turns, multiplied by the rate of change of the magnetic flux through a surface bounded by a single turn, is equal to the negative of the induced emf using Faraday's law. The emf in the coil equals IR.

SOLVE

Cover the column to the right and try these on your own before looking at the answers.

Steps	Answers
1. Write the magnetic flux in terms of B, N, and the radius r, and solve for B.	$\phi_m = NBA = NB\pi r^2$ $B = \dfrac{\phi_m}{N\pi r^2}$
2. Take the time derivative of B.	$\dfrac{dB}{dt} = \dfrac{1}{N\pi r^2}\dfrac{d\phi_m}{dt}$
3. Use Faraday's law to relate the rate of change of the flux to the emf.	$\mathcal{E} = -\dfrac{d\phi_m}{dt}$
4. Calculate the magnitude of the emf in the coil from the current and resistance of the coil.	$\lvert\mathcal{E}\rvert = IR = 120\text{ V}$
5. Substitute numerical values of $\mathcal{E}$, N, and r to calculate $\lvert dB/dt\rvert$.	$\left\lvert\dfrac{dB}{dt}\right\rvert = \dfrac{1}{N\pi r^2}\lvert\mathcal{E}\rvert = \boxed{191\text{ T/s}}$

A sign convention allows us to use the minus sign in Faraday's law to find the direction of the induced emf. According to this convention, the positive tangential direction along the integration path C is related to the direction of the unit normal $\hat{n}$ on the surface S bounded by C by a right-hand rule (Figure 28-4). By placing your right thumb in the direction of $\hat{n}$, the fingers of your hand curl in the positive tangential direction on C. If $d\phi_m/dt$ is positive, then in accord with Faraday's law (Equation 28-6), $\mathcal{E}$ is in the negative tangential direction. (The direction of $\mathcal{E}$ can also be determined via Lenz's law, which is discussed in Section 28-3.)

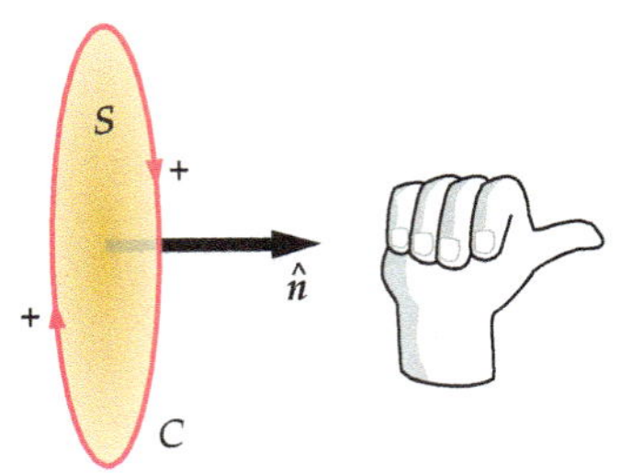

FIGURE 28-4 By placing your right thumb in the direction of $\hat{n}$ on the surface S, the fingers of your hand curl in the positive tangential direction on C.

Example 28-4 Induced Nonconservative Electric Field

A magnetic field $\vec{B}$ is perpendicular to the plane of the page. $\vec{B}$ is uniform throughout a circular region that has a radius R, as shown in Figure 28-5. Outside this region, B equals zero. The direction of $\vec{B}$ remains fixed and the rate of change of B is dB/dt. What are the magnitude and direction of the induced electric field in the plane of the page (*a*) a distance $r < R$ from the center of the circular region and (*b*) a distance $r > R$ from the center, where $B = 0$?

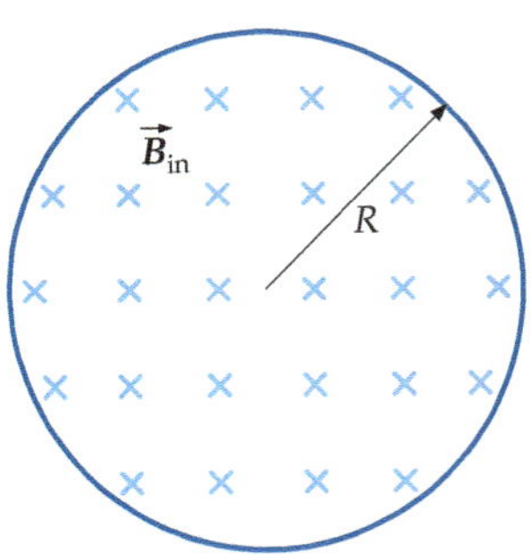

FIGURE 28-5

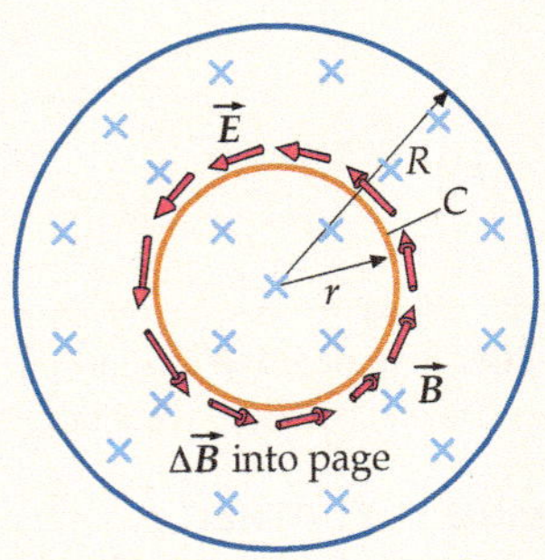

FIGURE 28-6

PICTURE The magnetic field $\vec{B}$ is into the page and uniform over a circular region of radius R, as shown in Figure 28-6. As B increases or decreases, the magnetic flux through a surface bounded by closed curve C also changes, and an emf $\mathcal{E} = \oint_C \vec{E}\cdot d\vec{\ell}$ is induced around C. The induced electric field is found by applying $\oint_C \vec{E}\cdot d\vec{\ell} = -d\phi_m/dt$ (Equation 28-6). To take advantage of the system's symmetry, we choose C to be a circular curve of radius r and then evaluate the line integral. By symmetry, $\vec{E}$ is tangent to circle C and has the same magnitude at any point on the circle. We will assign into the page as the direction of $\hat{n}$. The sign convention then tells us that the positive tangential direction is clockwise. We then calculate the magnetic flux ϕ_m, take its time derivative, and solve for E_t (the tangential component of $\vec{E}$).

SOLVE

(*a*) 1. The $\vec{E}$ and $\vec{B}$ fields are related by Equation 28-6:

$$\oint_C \vec{E}\cdot d\vec{\ell} = -\frac{d\phi_m}{dt}$$

where

$$\phi_m = \int_S \vec{B}\cdot\hat{n}\,dA$$

2. E_t (the tangential component of $\vec{E}$) is found from the line integral for a circle of radius $r < R$. $\vec{E}$ is tangent to the circle and has the same magnitude at all points on the circle:

$$\oint_C \vec{E}\cdot d\vec{\ell} = \oint_C E_t\,d\ell = E_t\oint_C d\ell = E_t 2\pi r$$

3. For $r < R$, $\vec{B}$ is uniform on the flat surface S bounded by the circle C. We choose into the page as the direction of $\hat{n}$. Because $\vec{B}$ is also into the page, the flux through S is simply BA:

$$\phi_m = \int_S \vec{B}\cdot\hat{n}\,dA = \int_S B_n\,dA = B_n\int_S dA = BA = B\pi r^2$$

4. Calculate the time derivative of ϕ_m:

$$\frac{d\phi_m}{dt} = \frac{d}{dt}(B\pi r^2) = \frac{dB}{dt}\pi r^2$$

5. Substitute the step 2 and step 4 results into the step 1 result and solve for E_t:

$$E_t 2\pi r = -\frac{dB}{dt}\pi r^2$$

so

$$\boxed{E_t = -\frac{r}{2}\frac{dB}{dt} \quad r < R}$$

6. For the choice of the direction of $\hat{n}$ in step 3, the positive tangential direction is clockwise:

E_t is negative, so the direction of $\vec{E}$ is **counterclockwise**.

(*b*) 1. For a circle of radius $r > R$ (the region where the magnetic field is zero), the line integral is the same as before:

$$\oint_C \vec{E}\cdot d\vec{\ell} = E_t 2\pi r$$

2. Because $B = 0$ for $r > R$, the magnetic flux through S is $B\pi R^2$:

$$\phi_m = B\pi R^2$$

3. Apply Faraday's law to find E_t:

$$E_t 2\pi r = -\frac{dB}{dt}\pi R^2$$

so

$$\boxed{E_t = -\frac{R^2}{2r}\frac{dB}{dt} \quad r > R}$$

E_t is negative, so the direction of $\vec{E}$ is **counterclockwise**.

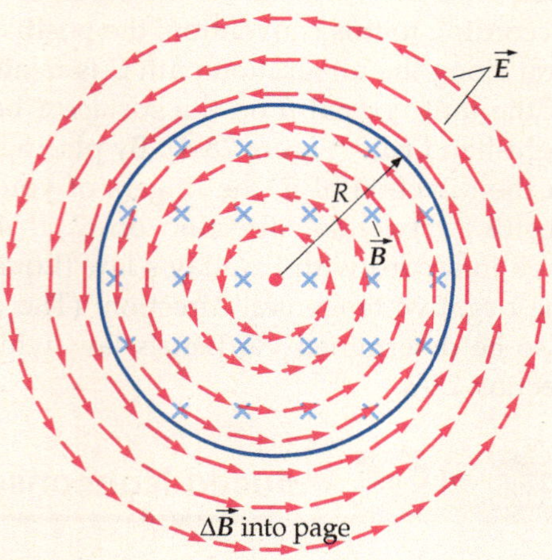

FIGURE 28-7 The magnetic field is into the page and increasing in magnitude. The induced electric field is counterclockwise.

CHECK The positive tangential direction is clockwise. When $d\phi_m/dt$ is positive, E_t is negative. The electric field direction is counterclockwise, as shown in Figure 28-7.

TAKING IT FURTHER Note that the electric field in this example is produced by a changing magnetic field rather than by electric charges. Note also that $\vec{E}$, and thus an emf, exists along any closed curve bounding the area through which the magnetic flux is changing, whether there is a wire or circuit along the curve or there is not.

! Note also that $\vec{E}$, and thus the emf, exists along any closed curve bounding the area through which the magnetic flux is changing, whether there is a wire or circuit along the curve or there is not.

28-3 LENZ'S LAW

The minus sign in Faraday's law has to do with the direction of the induced emf. This can be obtained by applying the sign convention described in the previous section, or by applying a general physical principle known as **Lenz's law:**

> The induced emf is in such a direction as to oppose, or tend to oppose, the change that produces it.
>
> LENZ'S LAW

Note that Lenz's law does not specify just what kind of change causes the induced emf and current. The statement of Lenz's law is purposely left vague to cover a variety of conditions, which we will now illustrate.

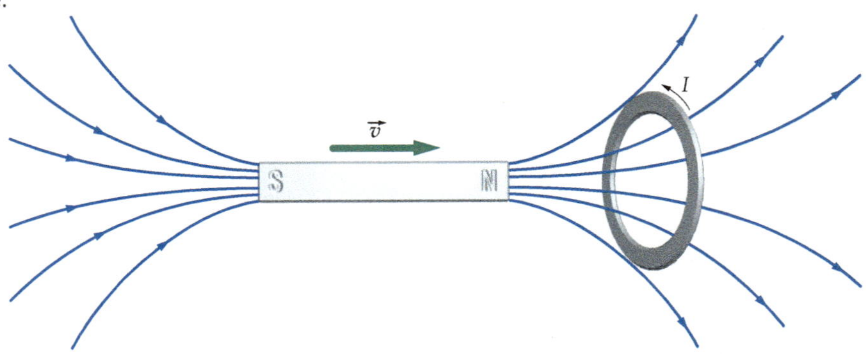

FIGURE 28-8 When the bar magnet is moving to the right, toward the loop, the emf induced in the loop produces an induced current in the direction shown. The magnetic field due to this induced current in the loop produces a magnetic field that exerts a force on the bar magnet opposing its motion to the right.

Figure 28-8 shows a bar magnet moving toward a conducting loop. It is the motion of the bar magnet to the right that induces an emf and current in the loop. Lenz's law tells us that this induced emf and current must be in a direction to oppose the motion of the bar magnet. That is, the current induced in the loop produces a magnetic field of its own, and this magnetic field must exert a force to the left on the approaching bar magnet. Figure 28-9 shows the induced magnetic moment of the current loop when the magnet is moving toward it. The loop acts like a small magnet with its north pole to the left and its south pole to the right. Because like poles repel, the induced magnetic moment of the loop repels the bar magnet; that is, it opposes its motion toward the loop. This result means the direction of the induced current in the loop must be as shown in Figure 28-9.

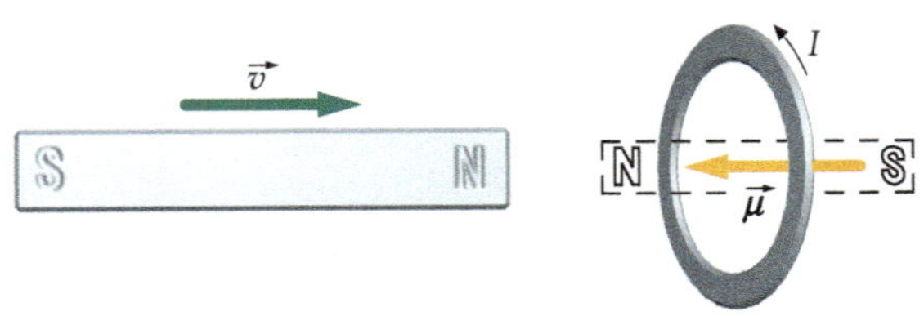

FIGURE 28-9 The magnetic moment of the loop $\vec{\mu}$ (shown in outline as if it were a bar magnet) due to the induced current is such as to oppose the motion of the bar magnet. The bar magnet is moving toward the loop, so the induced magnetic moment repels the bar magnet.

Suppose the induced current in the loop shown in Figure 28-9 was opposite to the direction shown. Then there would be a magnetic force toward the right on the approaching bar magnet, causing the bar magnet to gain speed. This gain in speed would cause an increase in the induced current, which in turn would cause the force on the bar magnet to increase, and so on. This result is too good to be true. Any time we nudge a bar magnet toward a conducting loop it would move toward the loop with ever increasing speed and with no significant effort on our part. Were this situation to occur, it would be a violation of energy conservation. The reality, however, is that energy is conserved, and Lenz's law is consistent with this reality.

An alternative statement of Lenz's law in terms of magnetic flux is frequently of use. This statement is

> When a magnetic flux through a surface changes, the magnetic field due to any induced current produces a flux of its own—through the same surface and opposite in sign to the initial change in flux.
>
> ALTERNATIVE STATEMENT OF LENZ'S LAW

For an example which shows how this alternative statement is applied, see Example 28-5.

Example 28-5 Lenz's Law and Induced Current

Using the alternative statement of Lenz's law, find the direction of the induced current in the loop shown in Figure 28-8.

PICTURE Use the alternative statement of Lenz's law to determine the direction of the magnetic field due to the current induced in the loop. When a magnetic flux through a surface changes, the magnetic field due to any induced current produces a flux of its own—through the same surface and opposite in sign to the initial change in flux. Then use a right-hand rule to determine the direction of the induced current.

SOLVE

1. Draw a sketch of the loop bounding the flat surface S (Figure 28-10). On surface S draw the vector $\Delta\vec{B}_1$, which is the change in the magnetic field $\vec{B}_1$ of the approaching bar magnet on S:

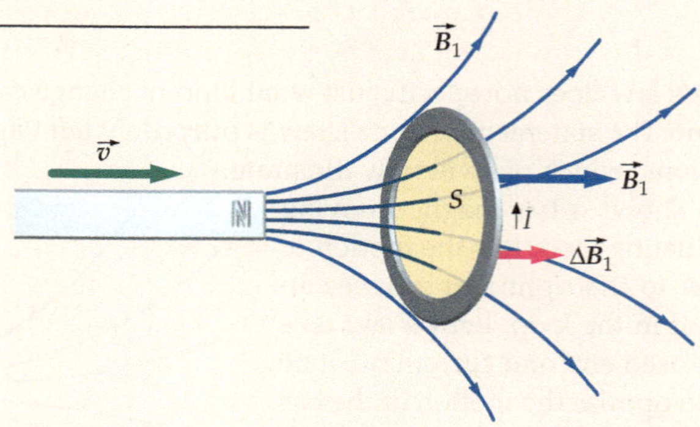

FIGURE 28-10

2. On the sketch draw the vector $\vec{B}_2$, which is the magnetic field of the current induced in the loop (Figure 28-11). Because $\vec{B}_2$ was initially zero, $\vec{B}_2$ is in the same direction as $\Delta\vec{B}_2$. Use the alternative statement of Lenz's law to determine the direction of $\vec{B}_2$. $\vec{B}_2$ and $\Delta\vec{B}_1$ must penetrate S in opposite directions for the change in the flux of $\vec{B}_2$ to be opposite in sign to the change in flux of $\vec{B}_1$:

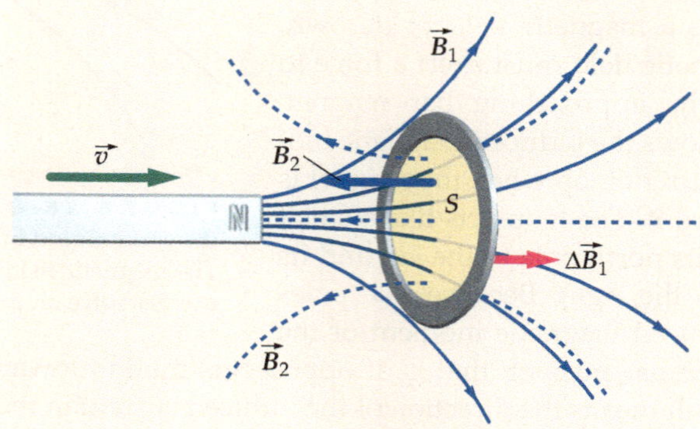

FIGURE 28-11

3. Using the right-hand rule and the direction of $\vec{B}_2$, determine the direction of the current induced in the loop (Figure 28-12):

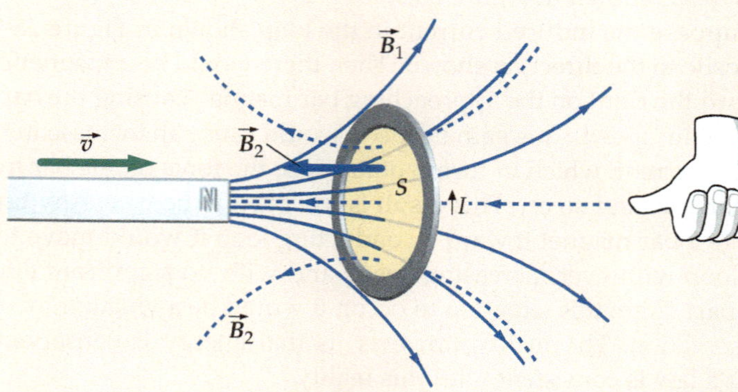

FIGURE 28-12

CHECK The step-3 result gives the same direction as was obtained on page 965 using the initial statement of Lenz's law.

CONCEPT CHECK 28-1

Using the alternative statement of Lenz's law, find the direction of the induced current in the loop shown in Figure 28-8 if the magnet is moving to the left (away from the loop).

In Figure 28-13, the bar magnet is at rest and the loop is moving away from the magnet. The induced current and magnetic moment are shown in the figure. In this case, the bar magnet attracts the loop, thus opposing the motion of the loop as required by Lenz's law.

In Figure 28-14, when the current in circuit 1 is changing, there is a changing flux through circuit 2. Suppose that the switch S in circuit 1 is initially open so that there is no current in the circuit (Figure 28-14*a*). When we close the switch (Figure 28-14*b*), the current in circuit 1 does not instantaneously reach its steady value $\mathcal{E}_1/R_1$ but takes some time to change from zero to that value. During the time the current is increasing, the flux through circuit 2 is changing and a current is induced in circuit 2 in the direction shown. When the current in circuit 1 reaches its steady value, the flux through circuit 2 is no longer changing, so there is no longer an induced current in circuit 2. An induced current in circuit 2 in the opposite direction appears briefly when the switch in circuit 1 is opened (Figure 28-14*c*) and the current in circuit 1 is decreasing to zero. It is important to understand that there is an induced emf *only while the flux is changing*. The emf does not depend on the magnitude of the flux itself, but only on its rate of change. If there is a large steady flux through a circuit, there is no induced emf.

FIGURE 28-13 When the loop is moving away from the stationary bar magnet, the bar magnet attracts the magnetic moment of the loop, again opposing the relative motion.

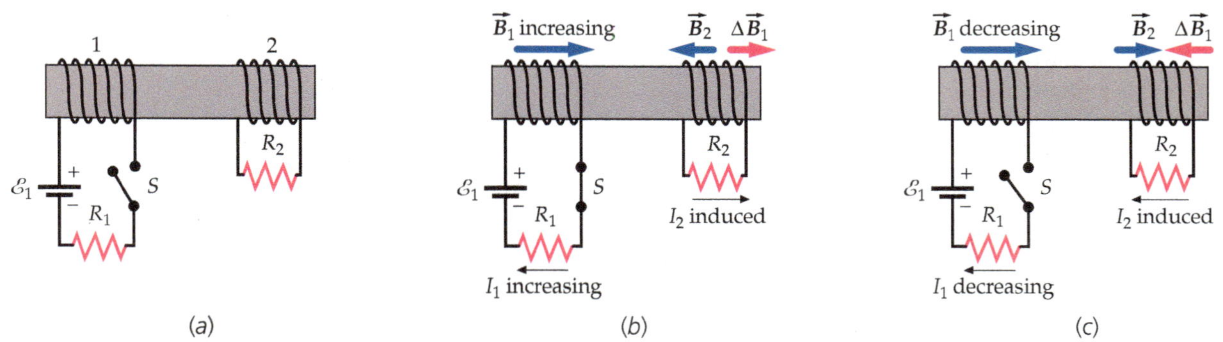

FIGURE 28-14 (*a*) Two adjacent circuits. (*b*) Just after the switch is closed, I_1 is increasing in the direction shown. The changing flux through circuit 2 induces the current I_2. The flux through circuit 2 due to I_2 opposes the change in flux due to I_1. (*c*) As the switch is opened, I_1 decreases and the flux through circuit 2 changes. The induced current I_2 then tends to maintain the flux through circuit 2.

For our next example, we consider the single isolated circuit shown in Figure 28-15. If there is a current in the circuit, there is a magnetic flux through the coil due to its own current. If the current is changing, the flux through the coil is changing and there is an induced emf in the circuit while the flux is changing. This *self-induced emf* opposes the change in the current. It is therefore called a **back emf.** Because of this self-induced emf, the current in a circuit cannot jump instantaneously from zero to some finite value or from some finite value to zero. Henry first noticed this effect when he experimented with a circuit consisting of many turns of a wire like that in Figure 28-15. This arrangement gives a large flux through the circuit for even a small current. Joseph Henry noticed a spark across the switch when he tried to break the circuit. Such a spark is due to the large induced emf that occurs when the current varies rapidly, as during the opening of the switch. In this case, the induced emf is directed so as to maintain the original current. The large induced emf produces a large potential difference across the switch as it is opened. The electric field between the contacts of the switch is large enough to produce dielectric breakdown in the surrounding air. When dielectric breakdown occurs, the air conducts electric current in the form of a spark.

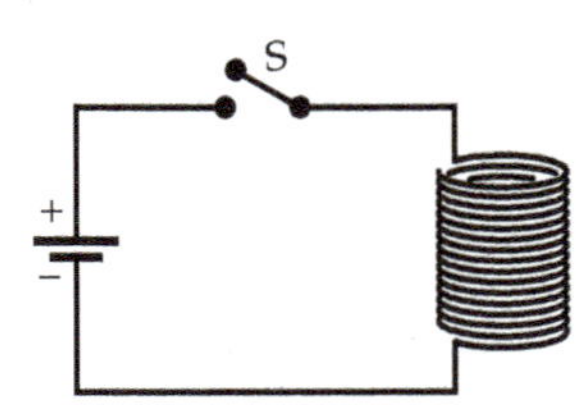

FIGURE 28-15 The coil that has many turns of wire yields a large flux for a given current in the circuit. Thus, when the current changes, there is a large emf induced in the coil opposing the change.

Example 28-6 Lenz's Law and a Moving Coil

A rectangular coil has $N = 80$ turns and each turn has a width $a = 20.0$ cm and a length $b = 30.0$ cm. Half the coil is located in a region that has a magnetic field of magnitude $B = 0.800$ T directed into the page (Figure 28-16). The resistance R of the coil is 30.0 Ω. Find the magnitude and direction of the induced current if the coil is moved with a speed of 2.00 m/s (*a*) to the right, (*b*) up the page, and (*c*) down the page.

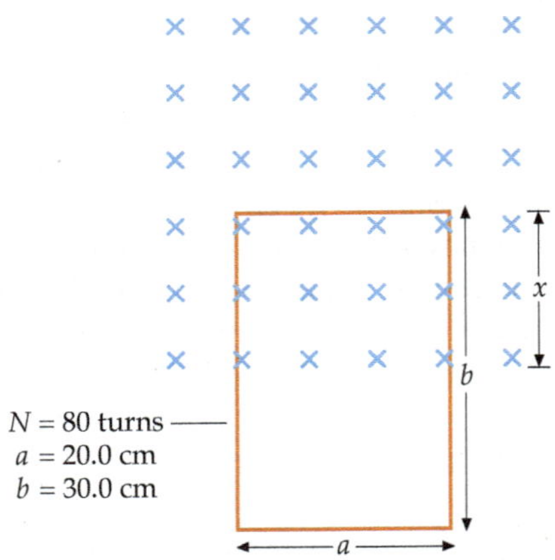

FIGURE 28-16

PICTURE The induced current equals the induced emf divided by the resistance. We can calculate the emf induced in the circuit as the coil moves by calculating the rate of change of the flux through the coil. The direction of the induced current is found from Lenz's law.

SOLVE

(*a*) 1. The induced current equals the emf divided by the resistance:

$$I = \frac{\mathcal{E}}{R}$$

2. The induced emf and the magnetic flux are related by Faraday's law:

$$\mathcal{E} = -\frac{d\phi_m}{dt}$$

3. The flux through the surface bounded by the coil is N multiplied by the flux through each turn of the coil. We choose into the page as the direction of $\hat{n}$. The flux through the surface S bounded by a single turn is Bax:

$$\phi_m = N\vec{B}\cdot\hat{n}A$$
$$= N[Bax - (0)a(b - x)] = NBax$$

4. When the coil is moving to the right (or to the left), x does not change and the flux does not change (until the coil leaves the region of the magnetic field). The current is therefore zero:

$$\mathcal{E} = -\frac{d\phi_m}{dt} = 0$$

so

$$\boxed{I = 0}$$

(*b*) 1. Compute the rate of change of the flux when the coil is moving up the page. In this case x is increasing, so dx/dt is positive:

$$\frac{d\phi_m}{dt} = \frac{d}{dt}(NBax) = NBa\frac{dx}{dt}$$

2. The derivative dx/dt is equal to the speed of the coil.

$$|I| = \frac{|\mathcal{E}|}{R} = \frac{NBa|dx/dt|}{R} = \frac{(80)(0.800\text{ T})(0.200\text{ m})(2.00\text{ m/s})}{30.0\ \Omega}$$
$$= 0.853\text{ A}$$

3. As the coil moves up the page, the flux of $\vec{B}$ through S is increasing. The induced current must produce a magnetic field whose flux through S decreases as x increases. That would be a magnetic field whose dot product with $\hat{n}$ is negative. Such a magnetic field is directed out of the page on S. To produce a magnetic field in this direction the induced current must be counterclockwise:

$$\boxed{I = 0.853\text{ A, counterclockwise}}$$

(*c*) As the coil moves down the page, the flux of $\vec{B}$ through S is decreasing. The induced current must produce a magnetic field whose flux through S increases as x decreases. That would be a magnetic field whose dot product with $\hat{n}$ is positive. Such a magnetic field is directed into the page on S. To produce a magnetic field in this direction the induced current must be clockwise:

$$\boxed{I = 0.853\text{ A, clockwise}}$$

CHECK In Part (*b*) the motion up the page causes the current to be induced, so the direction of the induced current must result in a force opposing the upward motion. Applying $\vec{F} = I\vec{L} \times \vec{B}$ (Equation 26-4) to the upper section of the loop gives a force down the page if the current in the loop is counterclockwise. This is in agreement with our Part-(*b*) result.

TAKING IT FURTHER In this example the magnetic field is static, so no nonconservative electric field exists. Thus, the emf is not the work done by a nonconservative electric field. The cause of this emf is examined in the next section.

28-4 MOTIONAL EMF

The emf induced in a conductor due to its motion in a region in which there exists a magnetic field is called motional emf. More generally,

> Motional emf is any emf induced by the motion of a conductor in a region in which there exists a magnetic field.
>
> DEFINITION—MOTIONAL EMF

Example 28-7 Total Charge through a Flipped Coil

A small coil of N turns has its plane perpendicular to a uniform static magnetic field $\vec{B}$, as shown in Figure 28-17. The coil is connected to a current integrator (C.I.), which is a device used to measure the total charge passing through the coil. Find the charge passing through the coil if the coil is rotated through 180° about the axis shown.

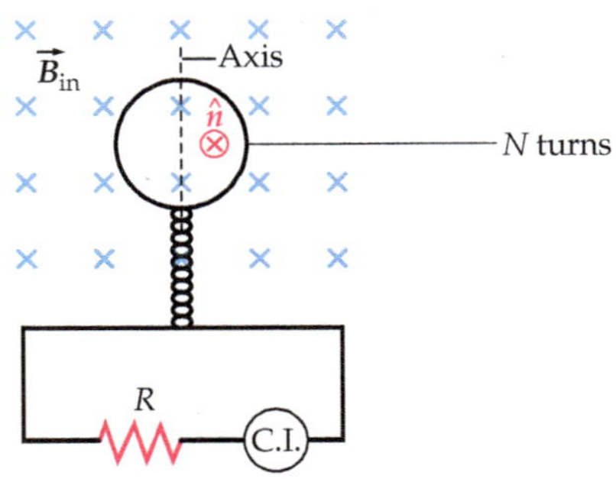

FIGURE 28-17

PICTURE When the coil in Figure 28-17 is rotated, the magnetic flux through the coil changes, causing an induced emf $\mathcal{E}$. The emf in turn causes a current $I = \mathcal{E}/R$, where R is the total resistance of the circuit. Because $I = dq/dt$, we can find the charge Q passing through the integrator by integrating I; that is, $Q = \int I\,dt$.

SOLVE

1. The increment of charge dq equals the current I multiplied by the increment of time dt:

$$dq = I\,dt$$

2. The emf $\mathcal{E}$ is related to I by Ohm's law:

$$\mathcal{E} = RI$$

so

$$\mathcal{E}\,dt = RI\,dt$$

3. The emf is related to the flux ϕ_m by Faraday's law:

$$\mathcal{E} = -\frac{d\phi_m}{dt}$$

or

$$\mathcal{E}\,dt = -d\phi_m$$

4. Substitute $-d\phi_m$ for $\mathcal{E}\,dt$ and dq for $I\,dt$ in the step 2 result and solve for dq:

$$-d\phi_m = R\,dq$$

so

$$dq = -\frac{1}{R}d\phi_m$$

5. Integrate to find the total charge Q:

$$Q = \int_0^Q dq = -\frac{1}{R}\int_{\phi_{mi}}^{\phi_{mf}} d\phi_m = -\frac{1}{R}(\phi_{mf} - \phi_{mi}) = -\frac{\Delta\phi_m}{R}$$

6. The flux through the coil is $\phi_m = n\vec{B}\cdot\hat{n}A$, where $\hat{n}$ is the normal to the flat surface bounded by the coil (Figure 28-18). The normal initially is directed into the page. When the coil rotates, so does the surface and its normal. Find the change in ϕ_m when the coil rotates 180°:

$$\Delta\phi_m = \phi_{mf} - \phi_{mi} = N\vec{B}\cdot\hat{n}_f A - N\vec{B}\cdot\hat{n}_i A$$
$$= NA(\vec{B}\cdot\hat{n}_f - \vec{B}\cdot\hat{n}_i) = NA[(-B) - (+B)] = -2NBA$$

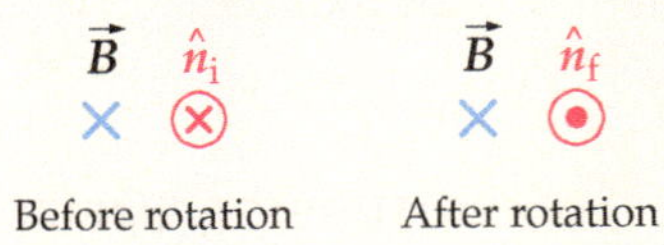

FIGURE 28-18

7. Combining the results from the previous two steps yields Q:

$$Q = \boxed{\frac{2NBA}{R}}$$

TAKING IT FURTHER Note that the charge Q does not depend on whether or not the coil is rotated slowly or quickly—all that matters is the change in the magnetic flux through the coil. A coil used in this way is called a *flip coil*. It is used to measure magnetic fields. For example, if the current integrator (C.I.) measures a total charge Q passing through the coil when it is flipped, the magnetic field strength is given by $B = RQ/(2NA)$, which can be obtained directly from the step-7 result.

PRACTICE PROBLEM 28-3 A flip coil has 40 turns, a radius of 3.00 cm, and a resistance of 16.0 Ω, and the plane of the coil is initially perpendicular to a static, uniform 0.500-T magnetic field. If the coil is rotated 90° about an axis perpendicular to the magnetic field, how much charge passes through the coil?

Figure 28-19 shows a thin conducting rod sliding to the right along conducting rails that are connected by a resistor. A uniform magnetic field $\vec{B}$ is directed into the page.

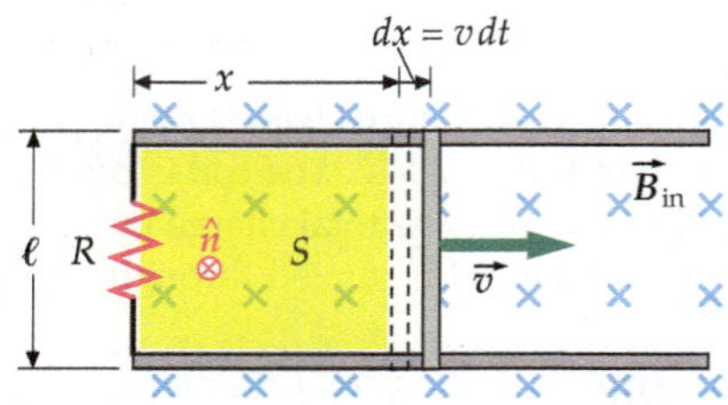

FIGURE 28-19 A conducting rod sliding on conducting rails in a magnetic field. As the rod moves to the right, the area of the surface S increases, so the magnetic flux through S into the paper increases. An emf of magnitude $B\ell v$ is induced in the circuit, inducing a counterclockwise current that produces flux through the surface S directed out of the paper opposing the change in flux due to the motion of the rod.

Consider the magnetic flux through the flat surface S bounded by the circuit. Let the normal $\hat{n}$ to the surface be into the page. As the rod moves to the right, the surface S increases, as does the magnetic flux through S. Thus, an emf is induced in the circuit. Let ℓ be the separation of the rails and x be the distance from the left end of the rails to the rod. The area of surface S is then ℓx, and the magnetic flux through S is

$$\phi_{\rm m} = \vec{B}\cdot\hat{n}A = B_nA = B\ell x$$

Taking the time derivative of both sides gives

$$\frac{d\phi_{\rm m}}{dt} = B\ell\frac{dx}{dt} = B\ell v$$

where $v = dx/dt$ is the speed of the rod. The emf induced in this circuit is therefore

$$\mathcal{E} = -\frac{d\phi_{\rm m}}{dt} = -B\ell v$$

where the negative sign tells us that the emf is in the negative tangential direction. Put your right thumb in the direction of $\hat{n}$ (into the page) and your fingers will curl in the positive tangential direction (clockwise). Thus, the induced emf is counterclockwise.

We can check this result (the direction of the induced emf) using Lenz's law. It is the motion of the rod to the right that produces the induced current, so the magnetic force on this rod due to the induced current must be to the left. The magnetic force on a current-carrying conductor is given by $I\vec{L}\times\vec{B}$ (Equation 26-4), where $\vec{L}$ is in the direction of the current. If $\vec{L}$ is up the page and $\vec{B}$ is into the page, the force is to the left, which affirms our previous result (that the induced emf is counterclockwise). If the rod is given some initial velocity $\vec{v}$ to the right and is then released, the force due to the induced current slows the rod until it stops. To maintain the motion of the rod, an external force pushing the rod to the right must be maintained.

A second check on the direction of the induced emf and current is implemented by considering the direction of the magnetic force on the charge carriers moving to the right with the rod. The charge carriers move rightward with the same velocity $\vec{v}$ as the rod, so the charge carriers experience a magnetic force $\vec{F} = q\vec{v}\times\vec{B}$. If q is positive that force is upward, which means the induced emf is counterclockwise.

$$|\mathcal{E}| = B\ell v \qquad 28\text{-}7$$

MAGNITUDE OF EMF FOR A ROD MOVING PERPENDICULAR TO BOTH THE LENGTH OF THE ROD AND $\vec{B}$

(If the magnetic field is not normal to the plane of the circuit, the B in Equation 28-7 should be replaced with the component of B normal to the plane of the circuit.)

Figure 28-20 shows a positive charge carrier in a conducting rod that is moving at constant speed through a uniform magnetic field directed into the paper. Because the charge carrier is moving horizontally with the rod, there is an upward magnetic force on the charge carrier of magnitude qvB. Responding to that force, the charge carriers in the rod move upward, producing a net positive charge at the top of the rod and leaving a net negative charge at the bottom of the rod. The charge carriers continue to move upward until the electric field $\vec{E}_{\parallel}$ produced by the separated charges exerts a downward force of magnitude $qE_{\parallel}$ on the separated charges, which balances the upward magnetic force qvB. In equilibrium, the magnitude of this electric field in the rod is

$$E_{\parallel} = vB$$

The direction of this electric field is parallel to the rod, directed downward. The associated potential difference across the length ℓ of the rod is

$$\Delta V = E_{\parallel}\ell = vB\ell$$

with the potential being higher at the top. That is, when there is no current through the rod, the potential difference across the length of the rod equals $vB\ell$ (the motional emf). When there is a current I through the rod, the potential difference is

$$\Delta V = vB\ell - Ir \qquad 28\text{-}8$$

where r is the resistance of the rod.

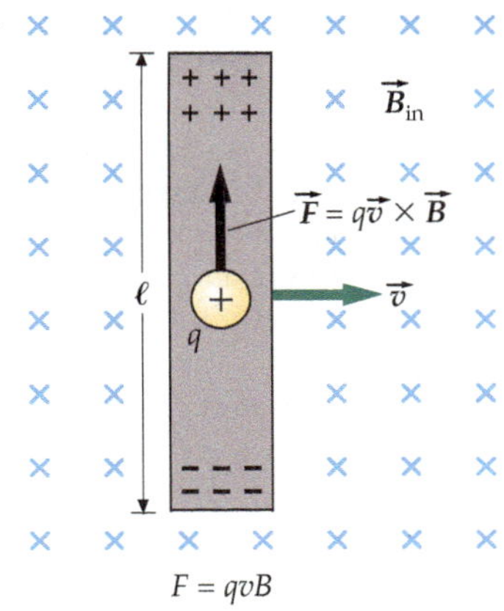

FIGURE 28-20 A positive charge carrier in a conducting rod that is moving through a magnetic field experiences a magnetic force that has an upward component. Some of these charge carriers move to the top of the rod, leaving the bottom of the rod negative. The charge separation produces a downward electric field of magnitude $E_{\parallel} = vB$ in the rod. Thus, the potential at the top of the rod is greater than the potential at the bottom of the rod by $E_{\parallel}\ell = vB\ell$.

PRACTICE PROBLEM 28-4

A rod 40 cm long moves at 12 m/s in a plane perpendicular to a magnetic field of 0.30 T. The rod's velocity is perpendicular to its length. Find the emf induced in the rod.

Example 28-8 A U-Shaped Conductor and a Sliding Rod — *Try It Yourself*

Using Figure 28-19, let $B = 0.600$ T, $v = 8.00$ m/s, $\ell = 15.0$ cm, and $R = 25.0\ \Omega$; assume that the resistances of the rod and the rails are negligible. Find (*a*) the induced emf in the circuit, (*b*) the current in the circuit, (*c*) the force needed to move the rod with constant velocity, and (*d*) the power dissipated in the resistor.

SOLVE

Cover the column to the right and try these on your own before looking at the answers.

Steps	Answers
1. Calculate the induced emf from Equation 28-7.	$\mathcal{E} = Bv\ell = \boxed{0.720\ \text{V}}$
2. Find the current from Ohm's law.	$I = \dfrac{\mathcal{E}}{R} = \boxed{28.8\ \text{mA}}$
3. The force needed to move the rod with constant velocity is equal and opposite to the force exerted by the magnetic field on the rod, which has the magnitude $IB\ell$ (Equation 26-4). Calculate the magnitude of this force.	$F = IB\ell = \boxed{2.59\ \text{mN}}$
4. Find the power dissipated in the resistor.	$P = I^2R = \boxed{20.7\ \text{mW}}$

CHECK Using $P = Fv$, we confirm that the power is 20.7 mW.

TAKING IT FURTHER The potential at the top of the rod is greater than the potential at the bottom of the rod by the emf.

Example 28-9 Magnetic Drag

A rod that has a mass m slides on frictionless conducting rails in a region that has a static uniform magnetic field $\vec{B}$ directed into the page (Figure 28-21). An external agent is pushing the rod, maintaining its motion to the right at constant speed v_0. At time $t = 0$, the agent abruptly stops pushing and the rod continues forward while being slowed by the magnetic force. Find the speed v of the rod as a function of time.

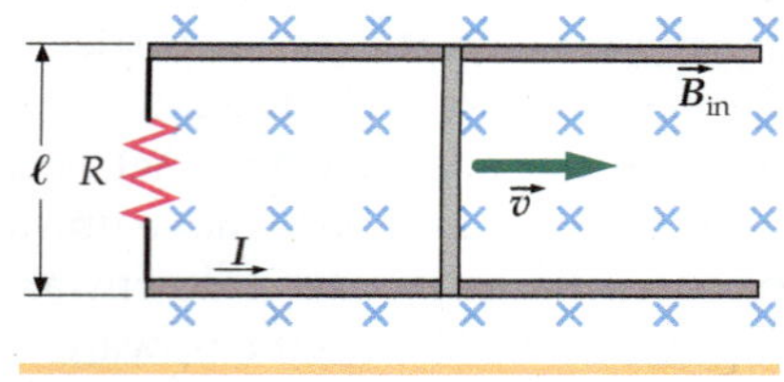

FIGURE 28-21

PICTURE The speed of the rod changes because a magnetic force acts on the induced current. The motion of the rod through a magnetic field induces an emf $\mathcal{E} = B\ell v$ and, therefore, a current in the rod, $I = \mathcal{E}/R$. This result causes a magnetic force to act on the rod, $F = I\ell B$ (Equation 26-4). With the force known, we apply Newton's second law to find the speed as a function of time. Take the $+x$ direction as being to the right.

SOLVE

1. Apply Newton's second law to the rod:
$$F_x = ma_x = m\frac{dv}{dt}$$

2. The force exerted on the rod is the magnetic force (Equation 26-4), which is proportional to the current and in the $-x$ direction, as shown in Figure 28-21:
$$F_x = -I\ell B$$

3. The current equals the motional emf divided by the resistance of the rod:
$$I = \frac{\mathcal{E}}{R} = \frac{B\ell v}{R}$$

4. Combining these results, we find the x component of the magnetic force exerted on the rod:
$$F_x = -IB\ell = -\frac{B\ell v}{R}B\ell = -\frac{B^2\ell^2 v}{R}$$

5. Newton's second law then gives:
$$-\frac{B^2\ell^2 v}{R} = m\frac{dv}{dt}$$

6. Separate the variables, then integrate the velocity from v_0 to v_f and the time from 0 to t_f:
$$\frac{dv}{v} = -\frac{B^2\ell^2}{mR}dt$$
$$\int_{v_0}^{v_f}\frac{dv}{v} = -\frac{B^2\ell^2}{mR}\int_0^{t_f}dt$$
$$\ln\frac{v_f}{v_0} = -\frac{B^2\ell^2}{mR}t_f$$

7. Let $v = v_f$ and $t = t_f$, then solve for v:
$$v = \boxed{v_0 e^{-t/\tau} \quad \text{where } \tau = \frac{mR}{B^2\ell^2}}$$

CHECK The kinetic energy of the rod is transformed into thermal energy in the resistor. To conserve energy the kinetic energy of the rod must decrease, which means its speed must decrease. The step 7 result is in agreement with the conservation of energy.

TAKING IT FURTHER If the force were constant, the rod's speed would decrease linearly with time. However, because the force is proportional to the rod's speed, as found in step 4, the force is large initially but the force decreases as the speed decreases. In principle, the rod never stops moving. Even so, the rod travels only a finite distance.

GENERATORS AND MOTORS

Most electrical energy used today is produced by electric generators in the form of alternating current (ac). A simple **generator** of alternating current is a coil rotating in a uniform magnetic field as shown in Figure 28-22. The ends of the coil are connected to rings called slip rings that rotate with the coil. Electrical contact is made with the coil by stationary graphite brushes in contact with the rings.

When the normal to the plane of the coil $\hat{n}$ makes an angle θ with a uniform magnetic field $\vec{B}$, as shown in the figure, the magnetic flux through the coil is

$$\phi_m = NBA\cos\theta \qquad 28\text{-}9$$

where N is the number of turns in the coil and A is the area of the flat surface bounded by the coil. When the coil is mechanically rotated, the flux through it will change, and an emf will be induced in the coil according to Faraday's law. If the initial angle between $\hat{n}$ and $\vec{B}$ is zero, then the angle at some later time t is given by

$$\theta = \omega t$$

where ω is the angular frequency of rotation. Substituting this expression for θ into Equation 28-9, we obtain

$$\phi_m = NBA\cos\omega t = NBA\cos 2\pi f t$$

The emf in the coil will then be

$$\mathcal{E} = -\frac{d\phi_m}{dt} = -NBA\frac{d}{dt}\cos\omega t = \omega NBA\sin\omega t \qquad 28\text{-}10$$

This can be written

$$\mathcal{E} = \mathcal{E}_{max}\sin\omega t$$

where

$$\mathcal{E}_{max} = \omega NBA$$

is the maximum value of the emf. We can thus produce a sinusoidal emf in a coil by rotating it with constant frequency in a magnetic field. In this source of emf, the mechanical energy of the rotating coil is converted into electric energy. The mechanical energy usually comes from a waterfall or a steam turbine. Although practical generators are considerably more complicated, they work on the same principle that an alternating emf is produced in a coil rotating in a magnetic field, and they are designed so that the emf produced is sinusoidal.

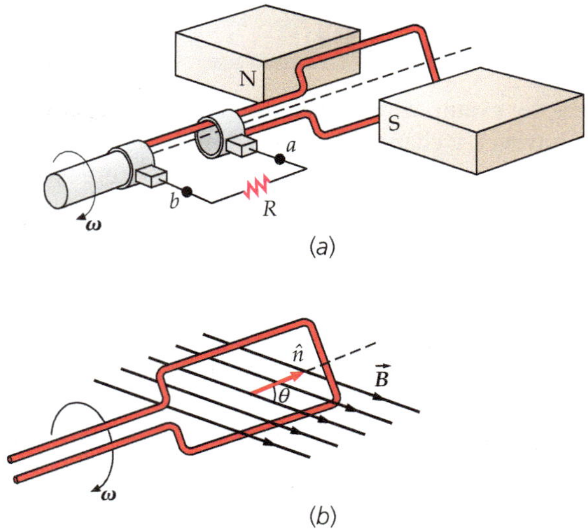

FIGURE 28-22 (*a*) An ac generator. A coil rotating with constant angular speed ω in a magnetic field $\vec{B}$ generates a sinusoidal emf. Energy from a waterfall or a steam turbine is used to rotate the coil to produce electrical energy. The emf is supplied to an external circuit by the brushes in contact with the rings. (*b*) At this instant, the normal to the plane of the coil makes an angle θ with the magnetic field and the flux is equal to $BA\sin\theta$.

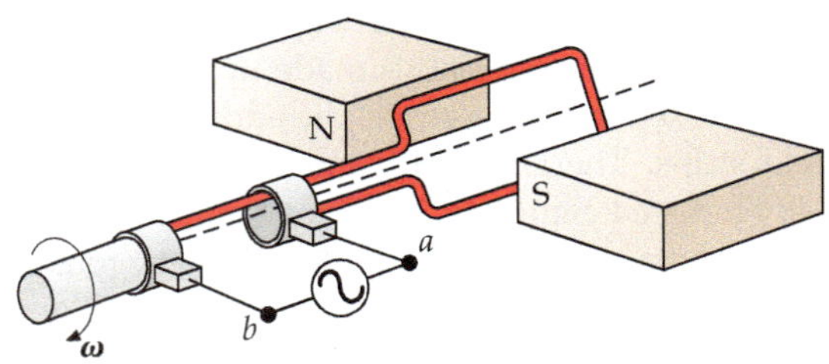

FIGURE 28-23 When alternating current is supplied to the coil of Figure 28-22, the coil becomes a motor. As the coil rotates, a back emf is generated, limiting the current.

The same coil in a magnetic field that can be used to generate an alternating emf can also be used as an ac **motor.** Instead of mechanically rotating the coil to generate an emf, we apply an alternating current to the coil from another ac generator as shown in Figure 28-23. (In circuit diagrams, an ac generator is represented by the symbol ⊙.) A current loop in a magnetic field experiences a torque that tends to rotate the loop such that its magnetic moment $\vec{\mu}$ points in the direction of $\vec{B}$ and the plane of the loop is perpendicular to $\vec{B}$. If direct current were supplied to the coil in Figure 28-23, the torque on the coil would change directions when the coil rotates past its equilibrium position, which is when the plane of the coil is vertical in the figure. The coil would then oscillate about its equilibrium position, eventually coming to rest there with its plane vertical. However, if the direction of the current is reversed just as the coil passes the vertical position, the torque does not change direction but continues to rotate the coil in the same direction. As the coil rotates in the magnetic field, a back emf is generated that tends to oppose the current. When the motor is first turned on, there is no back emf and the current is very large, being limited only by the resistance in the circuit. As the motor begins to rotate, the back emf increases and the current decreases.

PRACTICE PROBLEM 28-5

A 250-turn coil has an area per turn of 3.0 cm^2. If it rotates at 60 rev/s in a 0.40-T magnetic field at 60 Hz, what is the maximum emf in the coil?

CONCEPT CHECK 28-2

When a generator delivers electric energy to a circuit, where does the energy come from?

28-5 EDDY CURRENTS

In the examples we have discussed, currents were induced in thin wires or rods. However, a changing flux often induces circulating currents, which are called *eddy currents*, in a piece of bulk metal like the core of a transformer. The heat produced by such current constitutes a power loss in the transformer. Consider a conducting slab between the pole faces of an electromagnet (Figure 28-24). If the magnetic field $\vec{B}$ between the pole faces is changing with time (as it will if the current in the magnet windings is alternating current), the flux through any closed loop in the slab, such as through the curve C indicated in the figure, will change. Consequently, there will be an induced emf around C. Because path C is in a conductor, the emf will drive currents in the conductor.

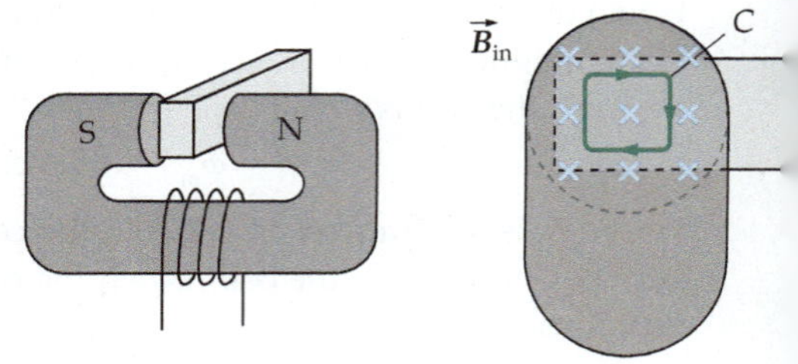

FIGURE 28-24 Eddy currents. When the magnetic field through a metal slab is changing, an emf is induced in any closed loop in the metal, such as loop C. The induced emfs drive currents, which are called eddy currents.

The existence of eddy currents can be demonstrated by pulling a copper or aluminum sheet through the region between the poles of a strong permanent magnet (Figure 28-25). Part of the area enclosed by curve C in the figure is in the magnetic field, and part of the area enclosed by curve C is outside the magnetic field. As the sheet is pulled to the right, the flux through this curve decreases (assuming that into the paper is the positive normal direction). A clockwise emf is induced around this curve. This emf drives a current that is directed upward in the region between the pole faces, and the magnetic field exerts a force on this current to the left opposing motion of the sheet. You can feel this drag force on the sheet if you pull a conducting sheet rapidly through a region that has a strong magnetic field.

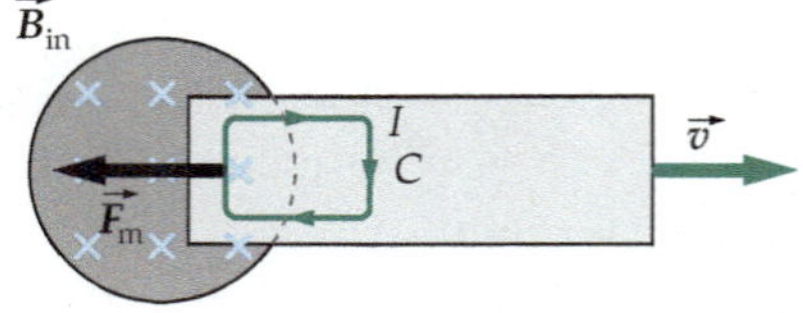

FIGURE 28-25 Demonstration of eddy currents. When the metal sheet is pulled to the right, there is a magnetic force to the left on the induced current opposing the motion.

Eddy currents are frequently undesirable because power is lost due to Joule heating by the current, and this dissipated energy must be transferred to the environment. The power loss can be reduced by increasing the resistance of the possible paths for the eddy currents, as shown in Figure 28-26*a*. Here the conducting slab is laminated; that is, the conducting slab is made up of small strips glued together. Because insulating glue separates the strips, the eddy currents are essentially confined to the individual strips. The large eddy-current loops are broken up, and the power loss is greatly reduced. Similarly, if there are cuts in the sheet, as shown in Figure 28-26*b*, the eddy currents are lessened and the magnetic force is greatly reduced.

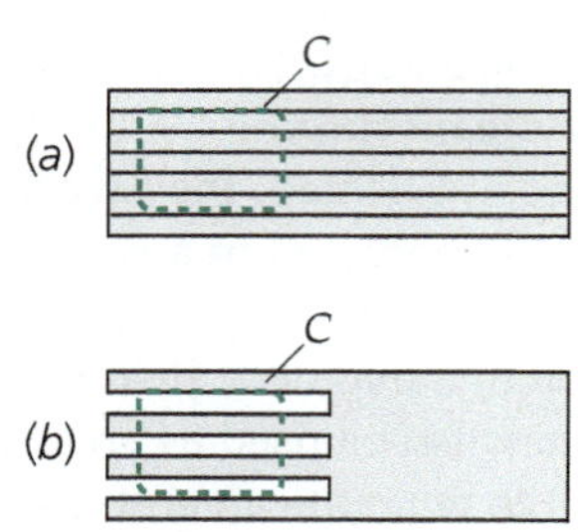

FIGURE 28-26 Disrupting the conduction paths in the metal slab can reduce the eddy current. (*a*) If the slab is constructed from strips of metal glued together, the insulating glue between the slabs increases the resistance of the closed loop C. (*b*) Slots cut into the metal slab also reduce the eddy current.

Eddy currents are not always undesirable. For example, eddy currents are often used to damp unwanted oscillations. With no damping present, sensitive mechanical balance scales that are used to measure small masses might oscillate back and forth around their equilibrium reading many times. Such scales are usually designed so that a small sheet of aluminum (or some other metal) moves between the poles of a permanent magnet as the scales oscillate. The resulting eddy currents dampen the oscillations so that equilibrium is quickly reached. Eddy currents also play a role in the magnetic braking systems of some rapid transit cars. A large electromagnet is positioned in the vehicle over the rails. If the magnet is energized by a current in its windings, eddy currents are induced in the rails by the motion of the magnet and the magnetic forces provide a drag force on the magnet that slows the car.

28-6 INDUCTANCE

SELF-INDUCTANCE

Consider a coil carrying a current I. The current in the coil produces a magnetic field $\vec{B}$ that varies from point to point, but at each point in space the value of B is proportional to I. The magnetic flux of $\vec{B}$ through the coil is therefore also proportional to I:

$$\phi_m = LI \qquad 28\text{-}11$$

DEFINITION—SELF-INDUCTANCE

where L, the proportionality constant, is called the self-inductance of the coil. The self-inductance depends on the geometric shape of the coil. The SI unit of inductance is the **henry** (H). From Equation 28-11, we can see that the unit of inductance equals the unit of flux divided by the unit of current:

$$1\text{ H} = 1\text{ Wb/A} = 1\text{ T}\cdot\text{m}^2/\text{A}$$

In principle, the self-inductance of any coil or circuit can be calculated by assuming a current I, calculating $\vec{B}$ at every point on a surface bounded by the coil, calculating the flux ϕ_m, and using $L = \phi_m/I$. In actual practice, the calculation is often very challenging. However, the self-inductance of a long, tightly wound solenoid can be calculated directly. The magnetic flux through a long, thin solenoid is given by NBA, where $B = \mu_0 nI$, N is the number of turns, n is the number of turns per unit length, I is the current and A is the area per turn. Thus, the magnetic flux through the coil is

$$\phi_m = NBA = \mu_0 N(\mu_0 nI)A = \frac{\mu_0 N^2 IA}{\ell} = \mu_0 n^2 IA\ell \qquad \text{28-12}$$

where ℓ is the length of the solenoid. As expected, the flux is proportional to the current. The proportionality constant is the self-inductance L:

$$L = \frac{\phi_m}{I} = \mu_0 n^2 A\ell \qquad \text{28-13}$$

SELF-INDUCTANCE OF A LONG SOLENOID

The self-inductance of a long solenoid is proportional to the square of the number of turns per unit length n and to the volume $A\ell$. Thus, like capacitance, self-inductance depends only on geometric factors.* From the dimensions of Equation 28-13, we can see that μ_0 can be expressed in henrys per meter:

$$\mu_0 = 4\pi \times 10^{-7}\text{ H/m}$$

Example 28-10 Self-Inductance of a Solenoid

Find the self-inductance of a solenoid of length 10.0 cm, area 5.00 cm^2, and 100 turns.

PICTURE We can calculate the self-inductance in henrys from Equation 28-13.

SOLVE

1. L is given by Equation 28-13: $L = \mu_0 n^2 A\ell$

2. Convert the given quantities to SI units:
$\ell = 10.0\text{ cm} = 0.100\text{ m}$
$A = 5.00\text{ cm}^2 = 5.00 \times 10^{-4}\text{ m}^2$
$n = N/\ell = (100\text{ turns})/(0.100\text{ m}) = 1000\text{ turns/m}$
$\mu_0 = 4\pi \times 10^{-7}\text{ H/m}$

3. Substitute the given quantities:
$$L = \mu_0 n^2 A\ell$$
$$= (4\pi \times 10^{-7}\text{ H/m})(1000\text{ turns/m})^2(5.00 \times 10^{-4}\text{ m}^2)(0.100\text{ m})$$
$$= \boxed{6.28 \times 10^{-5}\text{ H}}$$

CHECK The inductance of a solenoid that does not have a soft iron core is expected to be a small fraction of a henry. That is the case for the solenoid in this example.

---

* If the inductor has a material in the core, the self-inductance also depends on the properties of the material.

When the current in a circuit is changing, the magnetic flux due to the current is also changing, so an emf is induced in the circuit. Because the self-inductance L of a circuit is constant, the rate of change of the flux is related to the rate of change of the current by

$$\frac{d\phi_{\text{m}}}{dt} = \frac{d(LI)}{dt} = L\frac{dI}{dt}$$

According to Faraday's law, we have

$$\mathcal{E} = -\frac{d\phi_{\text{m}}}{dt} = -L\frac{dI}{dt} \qquad 28\text{-}14$$

SELF-INDUCED EMF

Thus, the self-induced emf is proportional to the rate of change of the current. Because of the negative sign in Equation 28-14, the self-induced emf is often called a back emf. A coil or solenoid that has enough turns to have a significant self-inductance is called an **inductor.** In circuits, it is denoted by the symbol . Typically, the self-inductance of the rest of the circuit is negligible in comparison to the self-inductance of a coil or solenoid. The potential difference across an inductor is given by

$$\Delta V = \mathcal{E} - Ir = -L\frac{dI}{dt} - Ir \qquad 28\text{-}15$$

POTENTIAL DIFFERENCE ACROSS AN INDUCTOR

where r is the internal resistance of the inductor.* For an ideal inductor, $r = 0$.

PRACTICE PROBLEM 28-6

At what rate must the current in the solenoid of Example 28-10 change in order to induce a back emf of 20.0 V?

MUTUAL INDUCTANCE

When two or more circuits are close to each other, as in Figure 28-27, the magnetic flux through one circuit depends not only on the current in that circuit but also on the current in the nearby circuits. Let I_1 be the current in circuit 1, on the left in Figure 28-27, and let I_2 be the current in circuit 2, on the right. The magnetic field $\vec{B}$ at surface S_2 is the superposition of $\vec{B}_1$ due to I_1 and $\vec{B}_2$ due to I_2, where B_1 is proportional to I_1 and B_2 is proportional to I_2. We can therefore write the flux of $\vec{B}_1$ through circuit 2, $\phi_{\text{m}12}$, as

$$\phi_{\text{m}12} = M_{12}I_1 \qquad 28\text{-}16a$$

DEFINITION—MUTUAL INDUCTANCE

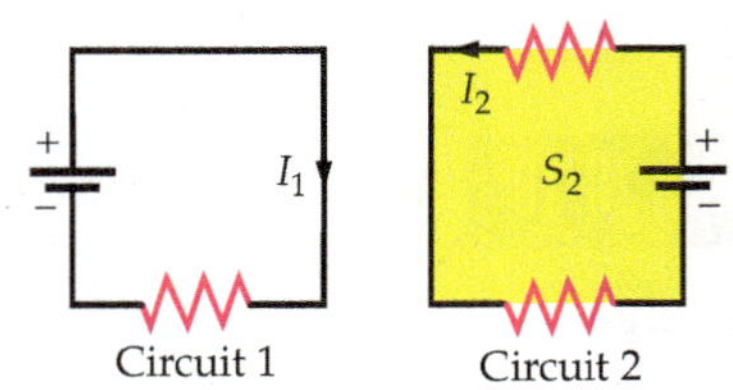

FIGURE 28-27 Two adjacent circuits. The magnetic field on S_2 is partly due to current I_1 and partly due to current I_2. The flux through S_2 is the sum of two terms, one proportional to I_1 and the other to I_2.

where M_{12} is called the **mutual inductance** of the two circuits. The mutual inductance depends on the geometrical arrangement of the two circuits. For instance, if the circuits are far apart, the flux of $\vec{B}_1$ through circuit 2 will be small and the mutual inductance will be small. (The net flux $\phi_{\text{m}2}$ of $\vec{B} = \vec{B}_1 + \vec{B}_2$ through circuit 2 is given by $\phi_{\text{m}2} = \phi_{\text{m}12} + \phi_{\text{m}22}$.) An equation similar to Equation 28-16a can be written for the flux of $\vec{B}_2$ through circuit 1:

$$\phi_{\text{m}21} = M_{21}I_2 \qquad 28\text{-}16b$$

We can calculate the mutual inductance for two tightly wound coaxial solenoids like the solenoids shown in Figure 28-28. Let ℓ be the length of both solenoids, and let the inner solenoid have N_1 turns and radius r_1 and the outer solenoid have N_2 turns and radius r_2. We will first calculate the mutual inductance M_{12} by assuming that the inner solenoid carries a current I_1 and finding the magnetic flux $\phi_{\text{m}2}$ due to this current through the outer solenoid.

* If the inductor has an iron core, the internal resistance includes properties of the core.

The magnetic field $\vec{B}_1$ due to the current in the inner solenoid is uniform in the space inside the inner solenoid and has magnitude

$$B_1 = \mu_0(N_1/\ell)I_1 = \mu_0 n_1 I_1 \quad r < r_1 \qquad \text{28-17}$$

and outside the inner solenoid this magnetic field B_1 is essentially zero. The flux of $\vec{B}_1$ through the outer solenoid is therefore

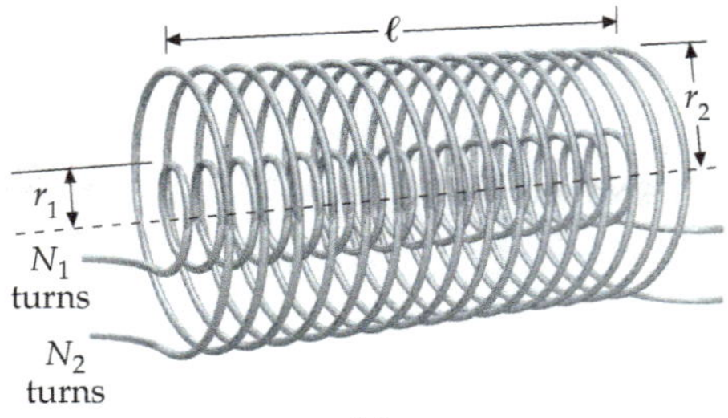

(a)

(b)

FIGURE 28-28 *(a)* A long, narrow solenoid inside a second solenoid of the same length. A current in either solenoid produces a magnetic flux in the other. *(b)* A tesla coil illustrating the geometry of the wires in Figure 28-28*a*. Such a device functions as a transformer.* Here, low-voltage alternating current in the outer winding is transformed into a higher-voltage alternating current in the inner winding. The emf induced in the inner coil by the field of the charging current in the outer coil is high enough to light the bulb above the coils. *((b) © Michael Holford, Collection of the Science Museum, London.)*

$$\phi_{m2} = N_2 B_1(\pi r_1^2) = n_2 \ell B_1(\pi r_1^2) = \mu_0 n_2 n_1 \ell(\pi r_1^2) I_1$$

Note that the area used to compute the flux through the outer solenoid is not the area of the surface bounded by a loop of that solenoid, πr_2^2, but rather is the area of the surface bounded by a loop of the inner solenoid, πr_1^2. This is because the magnetic field due to the inner solenoid is zero outside the inner solenoid. The mutual inductance M_{12} is thus

$$M_{12} = \frac{\phi_{m12}}{I_1} = \mu_0 n_2 n_1 \ell \pi r_1^2 \qquad \text{28-18}$$

PRACTICE PROBLEM 28-7

Calculate the mutual inductance M_{21} of the coaxial solenoids of Figure 28-28 by finding the flux through the inner solenoid due to a current I_2 in the outer solenoid.

Note the result of Practice Problem 28-7 reveals that $M_{12} = M_{21}$. It can be shown that this is a general result. Therefore in the future we will drop the subscripts for mutual inductance and simply write M.

28-7 MAGNETIC ENERGY

An inductor stores magnetic energy, just as a capacitor stores electrical energy. Consider the circuit shown in Figure 28-29, which consists of an ideal inductor that has an inductance L and a resistor that has a resistance R in series with an ideal battery that has emf $\mathcal{E}_0$ and a switch S. We assume that R and L are the resistance and inductance of the entire circuit. The switch is initially open, so no current exists in the circuit. A short time after the switch is closed, there are a current I in the circuit, a potential difference $-IR$ across the resistor, and a potential difference $-L\,dI/dt$ across the inductor. (For an ideal inductor, the difference in potential across the inductor equals the back emf, which was given in Equation 28-14.) Applying Kirchhoff's loop rule to this circuit gives

$$\mathcal{E}_0 - IR - L\frac{dI}{dt} = 0 \qquad \text{28-19}$$

If we multiply each term by the current I and rearrange, we obtain

$$\mathcal{E}_0 I = I^2 R + LI\frac{dI}{dt} \qquad \text{28-20}$$

The term $\mathcal{E}_0 I$ is the rate at which electrical potential energy is supplied by the battery. The term I^2R is the rate at which potential energy is delivered to the resistor. (It is also the rate at which potential energy is dissipated by the resistance in the circuit.) The term $LI\,dI/dt$ is the rate at which potential energy is delivered to the inductor, so if U_m is the energy stored in the inductor, then

$$\frac{dU_m}{dt} = LI\frac{dI}{dt}$$

which implies

$$dU_m = LI\,dI$$

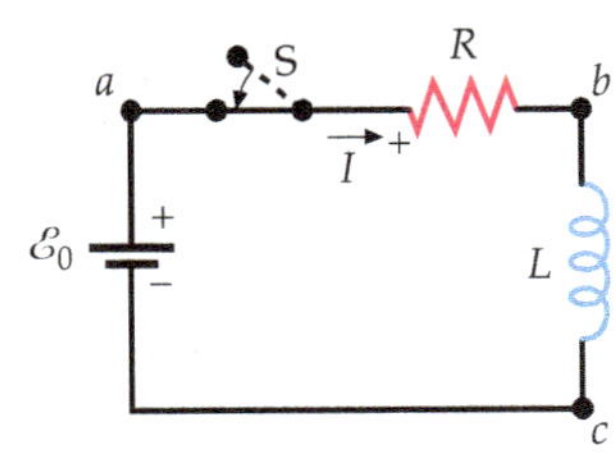

FIGURE 28-29 Just after the switch S is closed in this circuit, the current begins to increase and a back emf of magnitude $L\,dI/dt$ is induced in the inductor. The potential drop across the resistor IR plus the potential drop across the inductor $(L\,dI/dt)$ equals the emf of the battery $(\mathcal{E}_0)$.

* The transformer is discussed in Chapter 29.

Integrating this equation, we obtain

$$U_m = \frac{1}{2}LI^2 + C$$

where C is a constant of integration. To evaluate C, we set U_m equal to zero when I is equal to zero. The energy stored in an inductor carrying a current I is thus given by

$$U_m = \frac{1}{2}LI^2 \qquad \text{28-21}$$

ENERGY STORED IN AN INDUCTOR

When a current is produced in an inductor, a magnetic field is created in the region in and around the inductor coil. We can think of the energy stored in an inductor as energy stored in this magnetic field. For the special case of a long, thin solenoid, the magnetic field strength is zero except for the region inside the inductor, where it is given by

$$B = \mu_0 nI$$

The self-inductance of a long, thin solenoid is given by Equation 28-13:

$$L = \mu_0 n^2 A\ell$$

where A is the cross-sectional area and ℓ is the length. Substituting $B/(\mu_0 n)$ for I and $\mu_0 n^2 A\ell$ for L in Equation 28-21, we obtain

$$U_m = \frac{1}{2}LI^2 = \frac{1}{2}\mu_0 n^2 A\ell\left(\frac{B}{\mu_0 n}\right)^2 = \frac{B^2}{2\mu_0}A\ell$$

The quantity $A\ell$ is the volume of the space within the solenoid containing the magnetic field. The energy per unit volume is the **magnetic energy density** u_m:

$$u_m = \frac{B^2}{2\mu_0} \qquad \text{28-22}$$

MAGNETIC ENERGY DENSITY

Although we derived this expression by considering the special case of the magnetic field in a long solenoid, it is a general result. Whenever a magnetic field exists in space, the magnetic energy per unit volume is given by Equation 28-22. Note the similarity to the energy density in a region where there is an electric field (Equation 24-9):

$$u_e = \frac{1}{2}\epsilon_0 E^2$$

Example 28-11 Electromagnetic Energy Density

A certain region of space has a uniform magnetic field of 0.0200 T and a uniform electric field of 2.50×10^6 N/C. Find (*a*) the total electromagnetic energy density in the region, and (*b*) the energy in a cubical box of edge length $\ell = 12.0$ cm.

PICTURE The total energy density u is the sum of the electrical and magnetic energy densities, $u = u_e + u_m$. The energy in a volume $\mathcal{V}$ is given by $U = u\mathcal{V}$.

SOLVE

(*a*) 1. Calculate the electrical energy density:

$$u_e = \frac{1}{2}\epsilon_0 E^2$$

$$= \frac{1}{2}(8.85 \times 10^{-12}\ \text{C}^2/\text{N}\cdot\text{m}^2)(2.50 \times 10^6\ \text{N/C})^2$$

$$= 27.7\ \text{J/m}^3$$

2. Calculate the magnetic energy density:

$$u_m = \frac{B^2}{2\mu_0} = \frac{(0.0200\ \text{T})^2}{2(4\pi \times 10^{-7}\ \text{N/A}^2)} = 159\ \text{J/m}^3$$

3. The total energy density is the sum of the above two contributions:

$$u = u_e + u_m = 27.7\ \text{J/m}^3 + 159\ \text{J/m}^3 = \boxed{187\ \text{J/m}^3}$$

(*b*) The total energy in the box is $U = u\mathcal{V}$, where $\mathcal{V} = \ell^3$ is the volume of the box:

$$U = u\mathcal{V} = u\ell^3 = (187\ \text{J/m}^3)(0.120\ \text{m})^3 = \boxed{0.323\ \text{J}}$$

*28-8 *RL* CIRCUITS

A circuit containing a resistor and an inductor, such as that shown in Figure 28-29, is called an ***RL* circuit.** Because all circuits have resistance and self-inductance at room temperature, the analysis of an *RL* circuit can be applied to some extent to all circuits.*

For the circuit shown in Figure 28-29, application of Kirchhoff's loop rule gave us

$$\mathcal{E}_0 - IR - L\frac{dI}{dt} = 0 \qquad \text{28-19}$$

Let us look at some general features of this equation. First, the sum $IR + L\,dI/dt$ equals the emf of the battery, which is constant. Immediately after we close the switch in the circuit, the current is still zero, so IR is zero, and $L\,dI/dt$ equals the emf of the battery, $\mathcal{E}_0$. Setting $I = 0$ in Equation 28-19, we get

$$\left.\frac{dI}{dt}\right|_{I=0} = \frac{\mathcal{E}_0}{L} \qquad \text{28-23}$$

As the current increases, IR increases and dI/dt decreases. Note that the current cannot abruptly jump from zero to some final value as it would if the inductance L were zero. When the inductance L is greater than zero dI/dt is finite, and therefore the current must be continuous in time. After a short time, the current has reached a positive value I, and the rate of change of the current is

$$\frac{dI}{dt} = \frac{\mathcal{E}_0 - IR}{L}$$

At this time the current is still increasing, but its rate of increase is less than it was at $t = 0$. The final value I_f of the current can be obtained by setting dI/dt equal to zero in Equation 28-19:

$$I_f = \frac{\mathcal{E}_0}{R} \qquad \text{28-24}$$

Figure 28-30 shows the current in this circuit as a function of time. This figure is the same as that for the charge on a capacitor as a function of time when the capacitor is being charged in an *RC* circuit (Figure 25-45).

Equation 28-19 is of the same form as Equation 25-38 for the charging of a capacitor—and can be solved in the same way (by separating variables and integrating). The result is

$$I = \frac{\mathcal{E}_0}{R}(1 - e^{-(R/L)t}) = I_f(1 - e^{-t/\tau}) \qquad \text{28-25}$$

where $I_f = \mathcal{E}_0/R$ is the current as $t \rightarrow \infty$, and

$$\tau = \frac{L}{R} \qquad \text{28-26}$$

is the **time constant** of the circuit. The larger the self-inductance L or the smaller the resistance R, the longer it takes for the current to reach any specified fraction of its final current I_f.

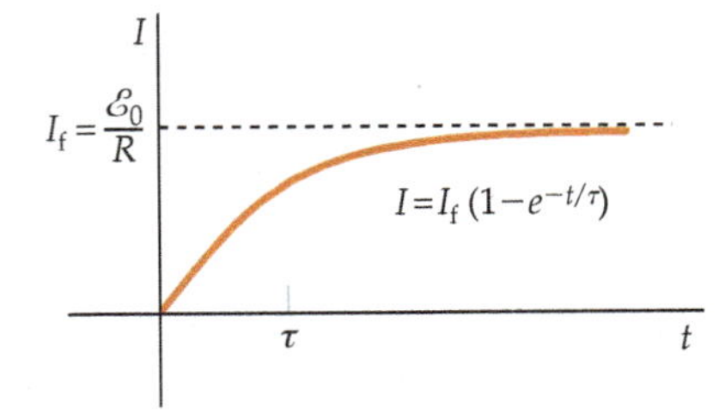

FIGURE 28-30 Current versus time in an *RL* circuit. At a time $t = \tau = L/R$, the current is at 63 percent of its maximum value $\mathcal{E}_0/R$.

* All circuits also have some capacitance between parts of the circuits at different potentials. We will consider the effects of capacitance in Chapter 29 when we study ac circuits. Here we will neglect capacitance to simplify the analysis and to focus on the effects of inductance.

Example 28-12 Energizing a Coil

A coil that has a self-inductance equal to 5.00 mH and a resistance equal to 15.0 Ω is placed across the terminals of a 12.0-V battery that has a negligible internal resistance. (*a*) What is the final current? (*b*) What is the time constant? (*c*) How many time constants does it take for the current to reach 99.0 percent of its final value?

PICTURE The final current is the current when $dI/dt = 0$. The current as a function of time is given by Equation 28-25, $I = I_f(1 - e^{-t/\tau})$, where $\tau = L/R$.

SOLVE

(*a*) Using Equation 28-19, set dI/dt equal to zero to find the final current, I_f:

$$\mathcal{E}_0 - IR - L\frac{dI}{dt} = 0$$

$$\mathcal{E}_0 - I_f R - 0 = 0$$

$$I_f = \frac{\mathcal{E}_0}{R} = \frac{12.0\text{ V}}{15.0\ \Omega} = \boxed{0.800\text{ A}}$$

(*b*) Calculate the time constant τ.

$$\tau = \frac{L}{R} = \frac{5.00 \times 10^{-3}\text{H}}{15.0\ \Omega} = \boxed{333\ \mu\text{s}}$$

(*c*) Use Equation 28-25 and calculate the time t for $I = 0.990I_f$:

$$I = I_f(1 - e^{-t/\tau}),$$

so

$$e^{-t/\tau} = \left(1 - \frac{I}{I_f}\right)$$

Taking the logarithm of both sides gives

$$-\frac{t}{\tau} = \ln\left(1 - \frac{I}{I_f}\right)$$

Thus,

$$t = -\tau \ln\left(1 - \frac{I}{I_f}\right) = -\tau \ln(1 - 0.990)$$

$$= -\tau \ln(0.010) = \tau \ln 100 = \boxed{4.61\tau}$$

CHECK In five time constants, the current is within one percent of its final value. This is consistent with the results of Example 25-18 where we found that after five time constants the charge on a discharging capacitor was less that one percent of its initial charge.

PRACTICE PROBLEM 28-8

How much energy is stored in this inductor when the final current has been attained?

In Figure 28-31, the circuit has a make-before-break switch (shown in Figure 28-32) that allows us to remove the battery from the circuit without interrupting the current through the inductor. The resistor R_1 protects the battery so that the battery is not shorted when the switch is thrown. If the switch throw is in position *e*, the battery, the inductor, and the two resistors are connected in series and the current builds up in the circuit as just discussed, except that the total resistance is now $R_1 + R$ and the final current is $\mathcal{E}_0/(R + R_1)$. Suppose that the throw has been in position *e* for a long time, so that the current remains at its final value, which we will call I_0. At time $t = 0$, we rapidly move the throw from position *e* to position *f*. With the throw at *f*, the current is zero in the branch with the battery and R_1. We now have a closed single-loop circuit (loop *abcdfa*) that has a resistor and an inductor carrying an initial current I_0. Applying Kirchoff's loop rule to this circuit gives

$$-IR - L\frac{dI}{dt} = 0$$

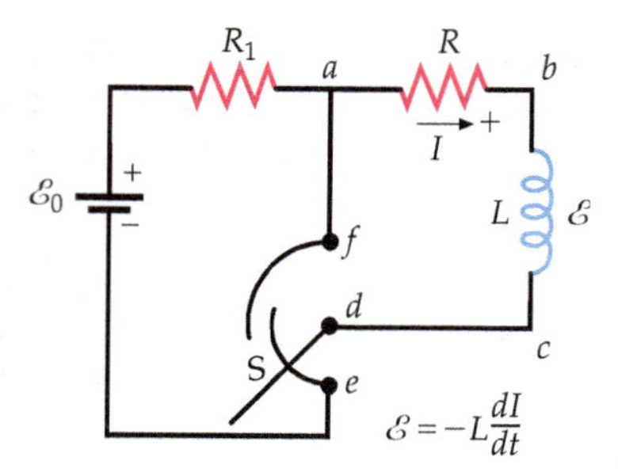

FIGURE 28-31 An *RL* circuit that has a make-before-break switch so that the battery can be removed from the circuit without interrupting the current through the inductor. The current in the inductor reaches its steady-state value with the switch pole in position *e*. The pole is then rapidly moved to position *f*.

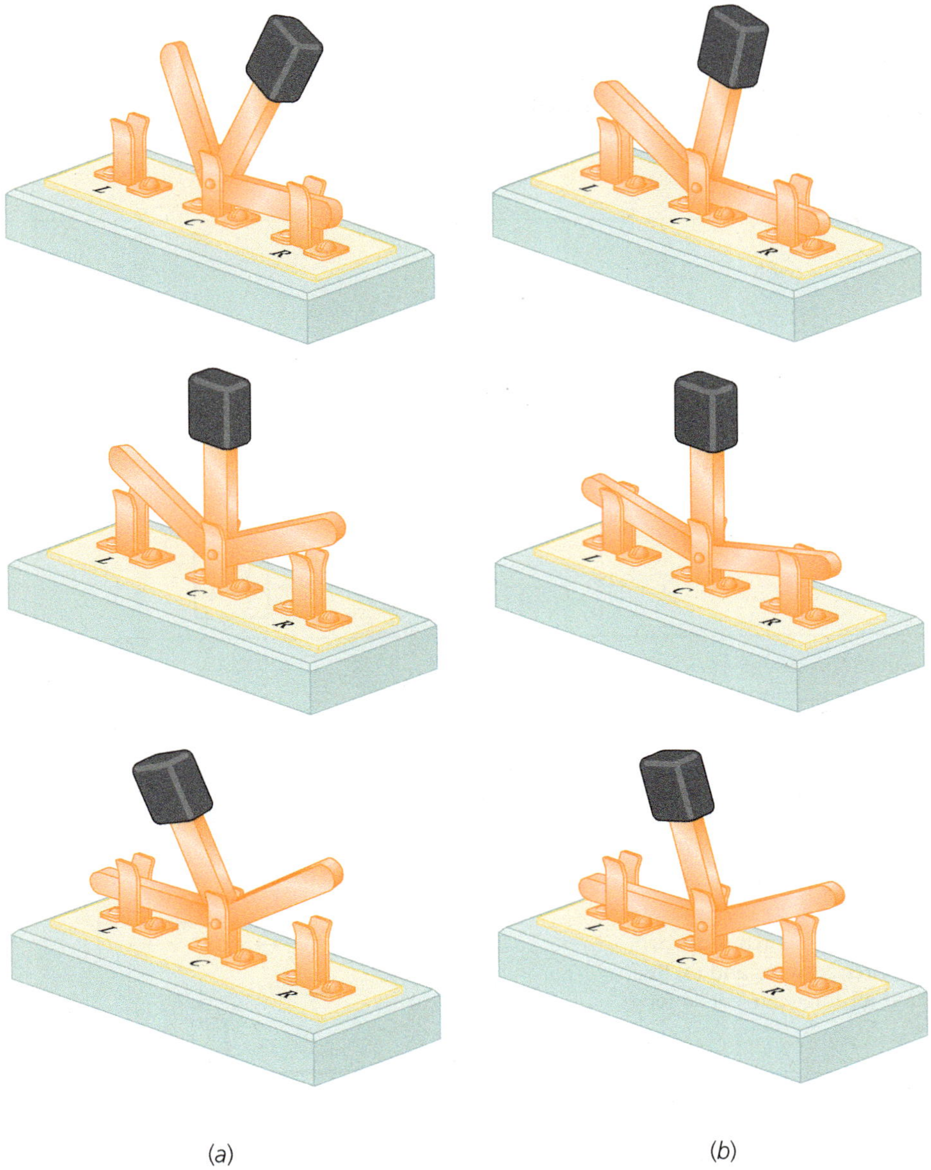

FIGURE 28-32 (*a*) The standard single-pole, double-throw switch is a break-before-make switch. That is, it breaks the first contact before making the second contact. (*b*) In a make-before-break, single-pole, double-throw switch, the throw makes the second contact before breaking the first contact. With the throw in the middle position, the throw is simultaneously in electrical contact with contacts *L* and *R*.

Rearranging this equation to separate the variables I and t gives

$$\frac{dI}{I} = -\frac{R}{L}\,dt \qquad 28\text{-}27$$

(Equation 28-27 is of the same form as Equation 25-34 for the discharge of a capacitor.) Integrating and then solving for I gives

$$I = I_0 e^{-t/\tau} \qquad 28\text{-}28$$

where $\tau = L/R$ is the time constant. Figure 28-33 shows the current as a function of time.

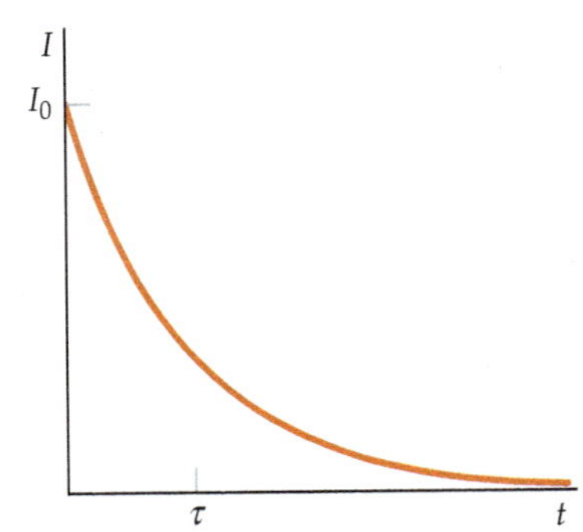

FIGURE 28-33 Current versus time for the circuit in Figure 28-31. The current decreases exponentially with time.

> **PRACTICE PROBLEM 28-9**
>
> What is the time constant of a closed single-loop circuit that has a resistance equal to 85 Ω and an inductance equal to 6.0 mH?

Example 28-13 Energy Dissipated

Find the total energy dissipated in the resistor R, as shown in Figure 28-31, when the current in the circuit decreases from its initial value of I_0 to 0.

PICTURE The rate of energy dissipation is equal to I^2R.

SOLVE

1. The rate of energy dissipation is I^2R: $\frac{dU}{dt} = I^2R$

2. The total energy U dissipated in the resistor is the integral of $P\,dt$ from $t = 0$ to $t = \infty$: $U = \int_0^\infty I^2R\,dt$

3. The current I is given by Equation 28-28: $I = I_0 e^{-(R/L)t}$

4. Substitute this current into the integral: $U = \int_0^\infty I^2R\,dt = \int_0^\infty I_0^2 e^{-2(R/L)t}R\,dt = I_0^2 R \int_0^\infty e^{-2(R/L)t}\,dt$

5. The integration can be done by substituting $x = 2Rt/L$: $U = I_0^2 R \frac{e^{-2(R/L)t}}{-2(R/L)}\bigg|_0^\infty = I_0^2 R \frac{-L}{2R}(0 - 1) = \boxed{\frac{1}{2}LI_0^2}$

CHECK The total amount of energy dissipated equals the energy $\frac{1}{2}LI_0^2$ originally stored in the inductor. (The energy stored in an inductor is $\frac{1}{2}LI^2$ (Equation 28-21.)

Example 28-14 Initial Currents and Final Currents

For the circuit shown in Figure 28-34, find the currents I_1, I_2, and I_3 (*a*) immediately after switch S is closed and (*b*) a long time after switch S has been closed. After the switch has been closed for a long time the switch is opened. Immediately after the switch is opened (*c*) find the three currents and (*d*) find the potential drop across the 20-Ω resistor. (*e*) Find all three currents a long time after switch S was opened.

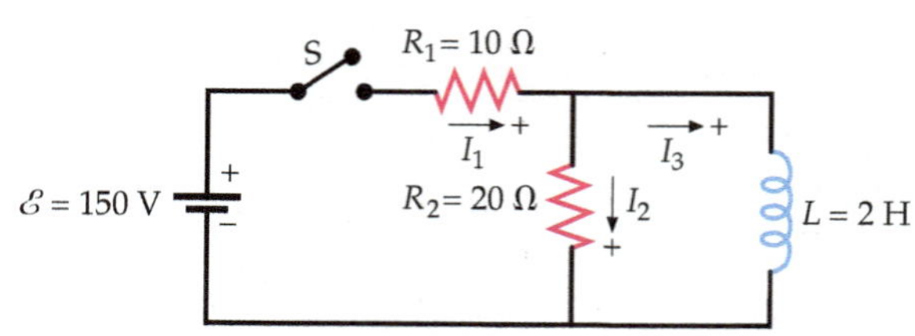

FIGURE 28-34

PICTURE (*a*) We simplify our calculations by using the fact that the current in an inductor cannot change abruptly. Thus, because the current in the inductor is zero before the switch is closed, the current in the inductor must be zero just after the switch is closed. (*b*) When the current reaches its final value dI/dt equals zero, so there is no potential drop across the inductor. The inductor thus acts like a short circuit; that is, the inductor acts like a wire with zero resistance. (*c*) Immediately after the switch is opened, the current in the inductor is the same as it was just before the switch was opened. (*d*) A long time after the switch is opened, all the currents must be zero.

SOLVE

(*a*) 1. The switch is just closed. The current through the inductor is zero, just as it was before the switch was closed. Apply the junction rule to relate I_1 and I_2:

$I_3 = \boxed{0}$

$I_1 = I_2 + I_3$

so

$I_1 = I_2$

2. The current in the left loop is obtained by applying the loop rule to the loop on the left:

$\mathcal{E} - I_1R_1 - I_1R_2 = 0$

so

$I_1 = I_2 = \frac{\mathcal{E}}{R_1 + R_2} = \frac{150\text{ V}}{10\ \Omega + 20\ \Omega} = \boxed{5.0\text{ A}}$

(*b*) 1. After a long time, the currents are steady and the inductor acts like a short circuit, so the potential drop across R_2 is zero. Apply the loop rule to the loop on the right and solve for I_2:

$-L\frac{dI_3}{dt} + I_2R_2 = 0$

$0 + I_2R_2 = 0 \quad \Rightarrow \quad I_2 = \boxed{0}$

2. Apply the loop rule to the loop on the left and solve for I_1:

$$\mathcal{E} - I_1R_1 - I_2R_2 = 0$$
$$\mathcal{E} - I_1R_1 - 0 = 0$$

so

$$I_1 = \frac{\mathcal{E}}{R_1} = \frac{150\ \text{V}}{10\ \Omega} = \boxed{15\ \text{A}}$$

3. Apply the junction rule and solve for I_3:

$$I_1 = I_2 + I_3$$
$$15\ \text{A} = 0 + I_3$$

so

$$I_3 = \boxed{15\ \text{A}}$$

(*c*) When the switch is reopened, I_1 *"instantaneously"* becomes zero. The current I_3 in the inductor changes continuously, so at that "instant" $I_3 = 15$ A. Apply the junction rule and solve for I_2:

$$I_3 = \boxed{15\ \text{A}}$$
$$I_1 = I_2 + I_3$$

so

$$I_2 = I_1 - I_3 = 0 - 15\ \text{A} = \boxed{-15\ \text{A}}$$

(*d*) Apply Ohm's law to find the potential drop across R_2:

$$V = I_2R_2 = (15\ \text{A})(20\ \Omega) = \boxed{300\ \text{V}}$$

(*e*) A long time after the switch is opened, all the currents must equal zero.

$$I_1 = I_2 = I_3 = \boxed{0}$$

TAKING IT FURTHER Were you surprised to find the potential drop across R_2 in Part (*d*) to be larger than the emf of the battery? This potential drop is equal to the emf of the inductor.

PRACTICE PROBLEM 28-10 Suppose $R_2 = 200\ \Omega$ and the switch has been closed for a long time. What is the potential drop across R_2 immediately after the switch is then opened?

28-9 MAGNETIC PROPERTIES OF SUPERCONDUCTORS

A superconductor has a resistivity equal to zero below a critical temperature T_c, which varies from material to material. In the presence of a magnetic field $\vec{B}$, the critical temperature is lower than the critical temperature is when there is no magnetic field. As the magnetic field increases, the critical temperature decreases. If the magnetic field strength is greater than some critical field strength B_c, superconductivity does not exist at any temperature.

*MEISSNER EFFECT

As a superconductor in a region that has a magnetic field is cooled below its critical temperature, the magnetic field in the region within the superconducting material becomes zero (Figure 28-35). This effect was discovered by Walter Meissner and Robert Ochsenfeld in 1933 and is now known as the **Meissner effect.** The magnetic field becomes zero because superconducting currents induced on the surface of the superconductor produce a

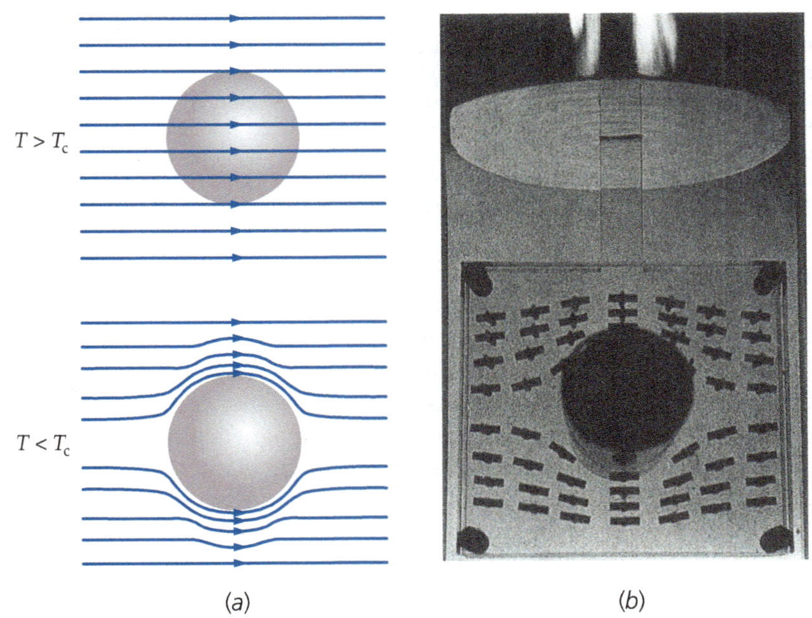

FIGURE 28-35 (*a*) The Meissner effect in a superconducting solid sphere cooled in a constant applied magnetic field. As the temperature drops below the critical temperature T_c, the magnetic field inside the sphere becomes zero. (*b*) Demonstration of the Meissner effect. A superconducting tin cylinder is situated with its axis perpendicular to a horizontal magnetic field. The directions of the field lines are indicated by weakly magnetized compass needles mounted in a Lucite sandwich so that they are free to turn. (*A. Leitner/Rensselaer Polytechnic Institute.*)

second magnetic field that cancels out the applied one. The magnetic levitation (see the photo) results from the repulsion between the permanent magnet producing the applied field and the magnetic field produced by the currents induced in the superconductor.

Only certain superconductors, called **type I superconductors,** exhibit the complete Meissner effect. Figure 28-36*a* shows a plot of the magnetization M multiplied by μ_0 versus the applied magnetic field B_{app} for a type I superconductor. For a magnetic field less than the critical field strength B_c, the magnetic field $\mu_0 M$ induced in the superconductor is equal and opposite to the applied magnetic field. The values of B_c for type I superconductors are always too small for such materials to be useful in the coils of a superconducting magnet.

Other materials, known as **type II superconductors,** have a magnetization curve similar to that in Figure 28-36*b*. Such materials are usually alloys or metals that have large resistivities in the normal state. Type II superconductors exhibit the electrical properties of superconductors, except for the Meissner effect, up to the critical field B_{c2}, which may be several hundred times the typical values of critical fields for type I superconductors. For example, the alloy Nb_3Ge has a critical field $B_{c2} = 34$ T. Such materials can be used for high-field superconducting magnets. Below the critical field B_{c1}, the behavior of a type II superconductor is the same as that of a type I superconductor.

The disk is a superconductor. The magnetic levitation results from the repulsion between the permanent magnet producing the applied field and the magnetic field produced by the currents induced in the superconductor. *(© Palmer/Kane, Inc./CORBIS.)*

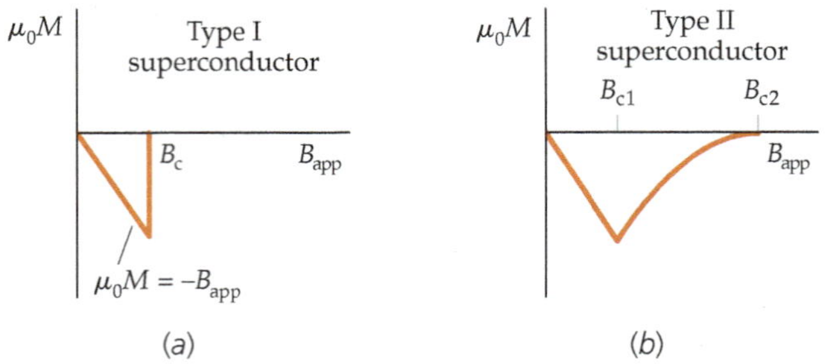

FIGURE 28-36 Plots of μ_0 multiplied by the magnetization M versus applied magnetic field for type I and type II superconductors. (*a*) In a type I superconductor, the resultant magnetic field is zero below a critical applied field B_c because the field due to induced currents on the surface of the superconductor exactly cancels the applied field. Above the critical field, the material is a normal conductor and the magnetization is too small to be seen on this scale. (*b*) In a type II superconductor, the magnetic field starts to penetrate the superconductor at a field B_{c1}, but the material remains superconducting up to a field B_{c2}, after which the material becomes a normal conductor.

*FLUX QUANTIZATION

Consider a superconducting ring that has an area A and carries a current. A magnetic flux $\phi_m = B_n A$ can exist through the flat surface S bounded by the ring due to the current in the ring and due also perhaps to other currents external to the ring. According to Equation 28-6, if the flux of the magnetic field through S changes, an electric field will be induced in the ring whose circulation is proportional to the rate of change of the flux. But no electric field can exist in a superconducting ring because the ring has no resistance and a finite electric field would drive an infinite current. The flux through S is thus fixed and cannot change.

Another effect, which results from the quantum-mechanical treatment of superconductivity, is that the total flux through surface S is quantized and is given by

$$\phi_m = n\frac{h}{2e} \quad n = 1, 2, 3, \ldots \qquad 28\text{-}29$$

The smallest unit of flux, called a **fluxon,** is

$$\phi_0 = \frac{h}{2e} = 2.0678 \times 10^{-15}\ \text{T} \cdot \text{m}^2 \qquad 28\text{-}30$$

Physics Spotlight

The Promise of Superconductors

The researcher is filling tubes with high-temperature superconducting powder in order to make wire. *(Courtesy of Department of Energy.)*

In 1986, a pair of IBM researchers tested a metal oxide—a ceramic—and found it was superconductive at 23 K.[*] Researchers worldwide then began testing different ceramics for superconductivity. By early 1987, a high-temperature superconductor (HTS) ceramic had been found. It is superconducting at 90 K—warm enough to be cooled by liquid nitrogen rather than liquid helium.[†] Superconducting ceramics were also found that could carry very large currents. The popular press assumed that room-temperature superconductors would be discovered. Books written in the late 1980s discussed the possibilities of superconductor-assisted levitating trains, superconductor-enabled computers, transfer of power over long distances without large losses due to resistance, and even superconductor-assisted satellite lasers.[‡, #]

Unfortunately, room-temperature superconductors have not been reliably observed. Furthermore, high-temperature ceramic superconductors are difficult to work with.[°] They are brittle and cannot be connected easily to wires, so several ways of depositing the superconducting ceramics on other surfaces had to be invented. In addition, if the boundaries between the tiny ceramic grains are not properly oriented or if the layers are too thick, the ceramic is not superconductive.[§]

These difficulties have, however, slowly been overcome. HTS are now used in a growing number of applications. Superconducting quantum interference detectors, or SQUIDs, use interruptions[¶] in superconductivity to detect extremely small amounts of energy. They are used in extraordinarily sensitive metal detectors,[**] light detectors,[††] and even to detect magnetic fields in the nervous systems of newborns.[‡‡] HTS have been tested in short lengths of electrical cable that have been nitrogen-cooled and carry large currents[##] and in fine superconducting wires.[°°]

Superconductors become resistive conductors when they carry large currents, which can be advantageous for long-distance power distribution systems. When short circuits occur in electrical circuits, the current rapidly increases unless the circuit is protected by a fuse or circuit breaker. Without the protections, the large currents can damage equipment and cause fires. Superconducting current limiters are being developed[§§] to protect electrical distribution networks from those excessively large currents.[¶¶]

In 2001, Japanese researchers discovered that magnesium diboride, MgB_2, is superconductive at 39 K, much warmer than any other metallic superconductor. Unlike other metallic superconductors, it can be cooled by liquid neon instead of the more costly liquid helium. Because MgB_2 is a metallic alloy, it is easily made into wire.[***] Pure MgB_2 does develop resistance at a lower critical current than other metallic superconductors, so it is not currently used for high-current applications.[†††] Researchers are studying the additions of small amounts of other elements to improve the characteristics of MgB_2.[‡‡‡]

* Yamazaki, S., "Superconducting Ceramics," *United States Patent* 7,1112,556 *B1*. September 26, 2006.
† Chu, C. W., "Superconductivity Above 90 K." *Proceedings of the National Academy of Sciences*, Jul. 1987, Vol. 84, pp. 4681–4682.
‡ Asimov, I., *How Did We Find Out About Superconductivity?* New York: Walker and Company, 1988, pp. 57–62.
\# Lampton, C. E., *Superconductors*. Hillside, New Jersey: Enslow, 1989, pp. 7–8, 53–69.
° Pool, R., "Superconductors' Material Problems." *Science*, Apr. 1, 1988, Vol. 240, No. 4848, pp. 25–27.
§ Service, R. F., "YBCO Confronts Life in the Slow Lane." *Science*, Feb. 1, 2002, Vol. 295, p. 787.
¶ Irwin, K. D., "Seeing with Superconductors." *Scientific American*, Nov. 2006, pp. 86–94.
** Bick, M., et al., "A SQUID-Based Metal Detector-Comparison to Coil and X-Ray Systems." *Superconducting Science and Technology*, Jan. 18, 2005, Vol. 18, pp. 346–351.
†† "Color Video Streaming from Space" *Machine Design*, May 25, 2006, p. 40.
‡‡ Draganova, R., et al., "Sound Frequency Change Detection in Fetuses and Newborns, a Magnetoencephalographic Study." *Neuroimage*, Nov. 1, 2005, Vol. 28, No. 2, pp. 354–361.
\#\# Malozemoff, A. P., Mannhart, J., and Scalapino, D., "High-Temperature Cuprate Superconductors Get to Work." *Physics Today*, April 2005, pp. 41–47.
°° Kang, S., "High-Performance, High-T_C Superconducting Wires." *Science*, Mar. 31, 2006, Vol. 311, pp. 1911–1914.
§§ Malozemoff, A. P., Mannhart, J., and Scalapino, D., op. cit.
¶¶ Meerovich, V., and Sokolovsky, V., "Experimental Study of a Transformer with Superconducting Elements for Fault Current Limitation and Energy Redistribution." *Cryogenics*, Aug. 2005, Vol. 45, No. 8, pp. 572–577.
*** Service, R., "MgB_2 Trades Performance for a Shot at the Real World." *Science*, Feb. 1, 2002, Vol. 295, pp. 786–788.
††† Canfield, P., and Bud'ko, S., "Low-Temperature Superconductivity Is Warming Up." *Scientific American*, Apr. 2005, pp. 80–87.
‡‡‡ Senkowicz, B. J., et al., "Atmospheric Conditions and Their Effect on Ball-Milled Magnesium Diboride." *Superconductor Science and Technology*, Oct. 2006, Vol. 19, pp. 1173–1177.

Summary

1. Faraday's law and Lenz's law are fundamental laws of physics.
2. Self-inductance is a property of a circuit element that relates the flux through the element to the current.

TOPIC	RELEVANT EQUATIONS AND REMARKS	
1. Magnetic Flux ϕ_m		
General definition	$\phi_m = \int_S \vec{B}\cdot\hat{n}\,dA$	28-1
Uniform field, flat surface bounded by coil of N turns	$\phi_m = NBA\cos\theta$ where A is the area of the flat surface bounded by a single turn.	28-4
Units	$1\text{ Wb} = 1\text{ T}\cdot\text{m}^2$	28-2
Due to current in a circuit	$\phi_m = LI$	28-11
Due to current in two circuits	$\phi_{m1} = L_1I_1 + MI_2$ $\phi_{m2} = L_2I_2 + MI_1$	28-16
2. EMF		
Faraday's law (includes both induction and motional emf)	$\mathcal{E} = -\dfrac{d\phi_m}{dt}$	28-5
Induction (time-varying magnetic field, C stationary)	$\mathcal{E} = \oint_C \vec{E}\cdot d\vec{\ell}$	28-6
Rod moving perpendicular to both its length and $\vec{B}$	$\lvert\mathcal{E}\rvert = vB\ell$	28-7
Self-induced (back emf)	$\mathcal{E} = -L\dfrac{dI}{dt}$	28-14
3. Faraday's Law	$\mathcal{E} = -\dfrac{d\phi_m}{dt}$	28-5
4. Lenz's Law	The induced emf and induced current are in such a direction as to oppose, or tend to oppose, the change that produces them.	
Alternative statement	When a magnetic flux through a surface changes, the magnetic field due to any induced current produces a flux of its own—through the same surface and opposite in sign to the change in flux.	
5. Inductance		
Self-inductance	$L = \dfrac{\phi_m}{I}$	28-11
Self-inductance of a solenoid	$L = \mu_0 n^2 A\ell$	28-13

TOPIC	RELEVANT EQUATIONS AND REMARKS	
Mutual inductance	$M = \frac{\phi_{m21}}{I_1} = \frac{\phi_{m12}}{I_2}$	28-18
Units and constants	$1\text{ H} = 1\text{ Wb/A} = 1\text{ T}\cdot\text{m}^2/\text{A}$ $\mu_0 = 4\pi \times 10^{-7}\text{ H/m}$	
6. Magnetic Energy		
Energy stored in an inductor	$U_m = \frac{1}{2}LI^2$	28-21
Energy density in a magnetic field	$u_m = \frac{B^2}{2\mu_0}$	28-22
***7. *RL* Circuits**		
Potential difference across an inductor	$\Delta V = \mathcal{E} - Ir = -L\frac{dI}{dt} - Ir$ where r is the internal resistance of the inductor. For an ideal inductor $r = 0$.	28-15
Energizing an inductor with a battery	In a single-loop circuit consisting of a resistor that has a resistance R, an inductor that has a self-inductance L, and a battery that has an emf $\mathcal{E}_0$, the current does not reach its maximum value I_f instantaneously, but rather takes some time to build up. If the current is initially zero, its value at some later time t is given by $I = \frac{\mathcal{E}_0}{R}(1 - e^{-t/\tau}) = I_f(1 - e^{-t/\tau})$	28-25
Time constant τ	$\tau = \frac{L}{R}$	28-26
De-energizing an inductor	In a single-loop circuit consisting of a resistor that has a resistance R and an inductor that has a self-inductance L, the current does not drop to zero through a resistor instantaneously, but rather takes some time to decrease. If the current is initially I_0, its value at some later time t is given by $I = I_0 e^{-t/\tau}$	28-28

Answers to Concept Checks

28-1 Opposite to the direction shown in Figure 28-12.

28-2 The external agent turning the coil does work on the coil. The energy comes from the external agent.

Answers to Practice Problems

28-2 0.555 A

28-3 3.53 mC

28-4 1.4 V

28-5 11 V

28-6 3.18×10^5 A/s

28-7 $M_{12} = \mu_0 n_2 n_1 \ell \pi r_1^2$

28-8 $U_m = \frac{1}{2}LI_f^2 = 1.60 \times 10^{-3}$ J

28-9 71 μs

28-10 3.0 kV

Problems

In a few problems, you are given more data than you actually need; in a few other problems, you are required to supply data from your general knowledge, outside sources, or informed estimate.

Interpret as significant all digits in numerical values that have trailing zeros and no decimal points.

- • Single-concept, single-step, relatively easy
- •• Intermediate-level, may require synthesis of concepts
- ••• Challenging
- SSM Solution is in the *Student Solutions Manual*
- Consecutive problems that are shaded are paired problems.

CONCEPTUAL PROBLEMS

1 • (*a*) The magnetic equator is a line on the surface of Earth on which Earth's magnetic field is horizontal. At the magnetic equator, how would you orient a flat sheet of paper so as to create the maximum magnitude of magnetic flux through it? (*b*) How about the minimum magnitude of magnetic flux? SSM

2 • At one of Earth's magnetic poles, how would you orient a flat sheet of paper so as to create the maximum magnitude of magnetic flux through it?

3 • Show that the following combination of SI units is equivalent to the volt: $\mathrm{T \cdot m^2/s}$ SSM

4 • Show that the following combination of SI units is equivalent to the ohm: $\mathrm{Wb/(A \cdot s)}$

5 • A current is induced in a conducting loop that lies in a horizontal plane, and the induced current is clockwise when viewed from above. Which of the following statements could be true? (*a*) A constant magnetic field is directed vertically downward. (*b*) A constant magnetic field is directed vertically upward. (*c*) A magnetic field whose magnitude is increasing is directed vertically downward. (*d*) A magnetic field whose magnitude is decreasing is directed vertically downward. (*e*) A magnetic field whose magnitude is decreasing is directed vertically upward. SSM

6 • Give the direction of the induced current in the circuit, shown on the right in Figure 28-37, when the resistance in the circuit on the left is suddenly (*a*) increased and (*b*) decreased. Explain your answer.

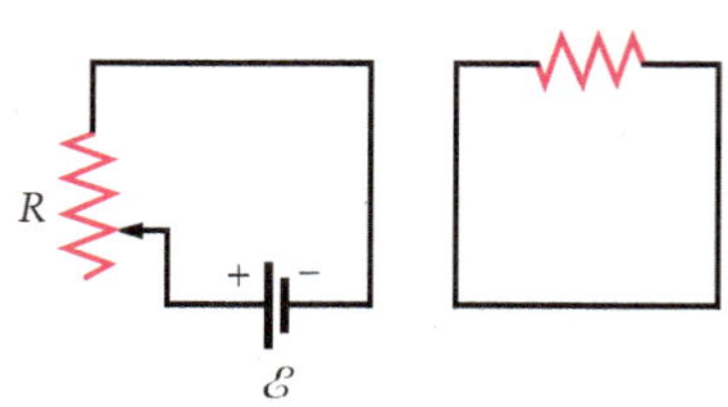

FIGURE 28-37 Problem 6

7 • The planes of the two circular loops in Figure 28-38 are parallel. As viewed from the left, a counterclockwise current exists in loop A. If the magnitude of the current in loop A is increasing, what is the direction of the current induced in loop B? Do the loops attract or repel each other? Explain your answer. SSM

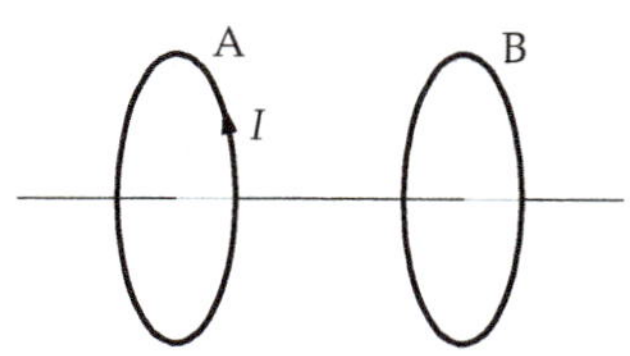

FIGURE 28-38 Problem 7

8 • A bar magnet moves with constant velocity along the axis of a loop, as shown in Figure 28-39. (*a*) Make a graph of the magnetic flux through the loop as a function of time. Indicate on the graph when the magnet is halfway through the loop by designating this time t_1. Choose the direction of the normal to the flat surface bounded by the loop to be to the right. (*b*) Make a graph of the induced current in the loop as a function of time. Choose the positive direction for the current to be clockwise as viewed from the left.

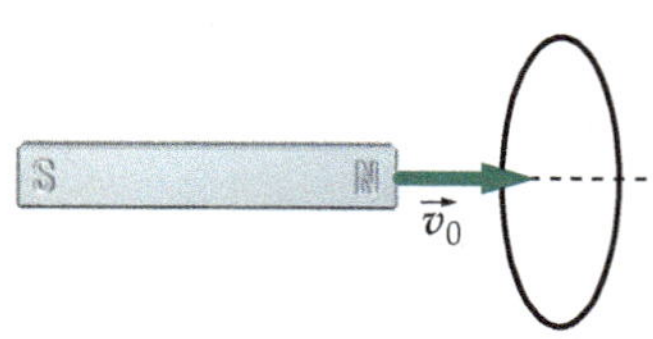

FIGURE 28-39 Problem 8

9 • A bar magnet is mounted on the end of a coiled spring and is oscillating in simple harmonic motion along the axis of a loop, as shown in Figure 28-40. The magnet is in its equilibrium position when its midpoint is in the plane of the loop. (*a*) Make a graph of the magnetic flux through the loop as a function of time. Indicate when the magnet is halfway through the loop by designating these times t_1 and t_2. (*b*) Make a graph of the induced current in the loop as a function of time, choosing the current to be positive when it is clockwise as viewed from above.

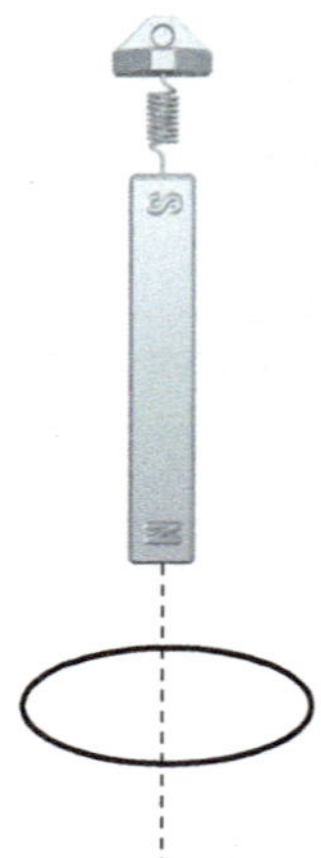

FIGURE 28-40 Problem 9

10 • A pendulum is fabricated from a thin, flat piece of aluminum. At the bottom of its arc, it passes between the poles of a strong permanent magnet. In Figure 28-41*a*, the metal sheet is continuous, whereas in Figure 28-41*b*, there are slots in the sheet. When released from the same angle, the pendulum that has slots swings back and forth many times, but the pendulum that does not have slots stops swinging after no more than one complete oscillation. Explain why.

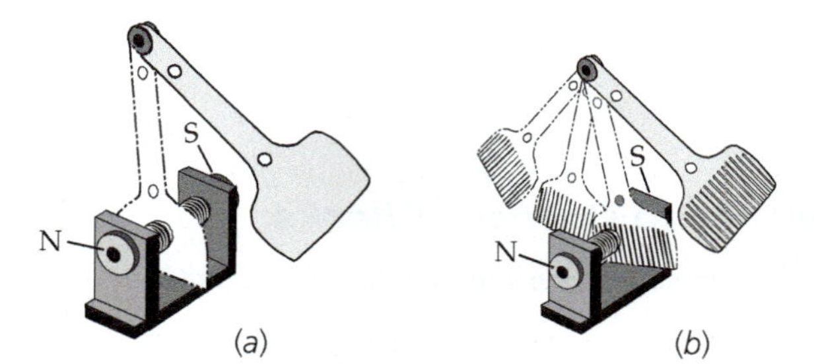

FIGURE 28-41 Problem 10 *(Courtesy of PASCO Scientific Co.)*

11 • A bar magnet is dropped inside a long, vertical tube. If the tube is made of metal, the magnet quickly approaches a terminal speed, but if the tube is made of cardboard, the magnet falls with constant acceleration. Explain why the magnet falls differently in the metal tube than it does in the cardboard tube.

12 • A small square wire loop lies in the plane of this page, and a constant magnetic field is directed into the page. The loop is moving to the right, which is the $+x$ direction. Find the direction of the induced current, if any, in the loop if (*a*) the magnetic field is uniform, (*b*) the magnetic field strength increases as x increases, and (*c*) the magnetic field strength decreases as x increases.

13 • If the current in an inductor doubles, the energy stored in the inductor will (*a*) remain the same, (*b*) double, (*c*) quadruple, (*d*) halve.

14 • Two solenoids are equal in length and radius, and the cores of both are identical cylinders of iron. However, solenoid A has three times the number of turns per unit length as solenoid B. (*a*) Which solenoid has the larger self-inductance? (*b*) What is the ratio of the self-inductance of solenoid A to the self-inductance of solenoid B?

15 • True or false:

(*a*) The induced emf in a circuit is equal to the negative of the magnetic flux through the circuit.
(*b*) There can be a nonzero induced emf at an instant when the flux through the circuit is equal to zero.
(*c*) The self-inductance of a solenoid is proportional to the rate of change of the current in the solenoid.
(*d*) The magnetic energy density at some point in space is proportional to the square of the magnitude of the magnetic field at that point.
(*e*) The inductance of a solenoid is proportional to the current in it. SSM

ESTIMATION AND APPROXIMATION

16 • CONTEXT-RICH Your baseball teammates, having just studied this chapter, are concerned about generating enough voltage to shock them while swinging aluminum bats at fastballs. Estimate the maximum possible motional emf measured between the ends of an aluminum baseball bat during a swing. Do you think your team should switch to wooden bats to avoid electrocution?

17 • Compare the energy density stored in Earth's electric field near its surface to that stored in Earth's magnetic field near its surface.

18 •• A physics teacher does the following emf demonstration. She has two students hold a long wire connected to a voltmeter. The wire is held slack, so that it sags with a large arc in it. When she says, "Start," the students begin rotating the wire as if they were playing jump rope. The students stand 3.0 m apart, and the sag in the wire is about 1.5 m. The motional emf from the "jump rope" is then measured on the voltmeter. (*a*) Estimate a reasonable value for the maximum angular speed at which the students can rotate the wire. (*b*) From this, estimate the maximum motional emf in the wire. *Hint: What field is involved in creating the induced emf?*

19 •• (*a*) Estimate the maximum possible motional emf between the wingtips of a typical commercial airliner in flight. (*b*) Estimate the magnitude of the electric field between the wingtips.

MAGNETIC FLUX

20 • A uniform magnetic field of magnitude 0.200 T is in the $+x$ direction. A square coil that has 5.00-cm-long sides has a single turn and makes an angle θ with the z axis, as shown in Figure 28-42. Find the magnetic flux through the coil when θ is (*a*) 0°, (*b*) 30°, (*c*) 60°, and (*d*) 90°.

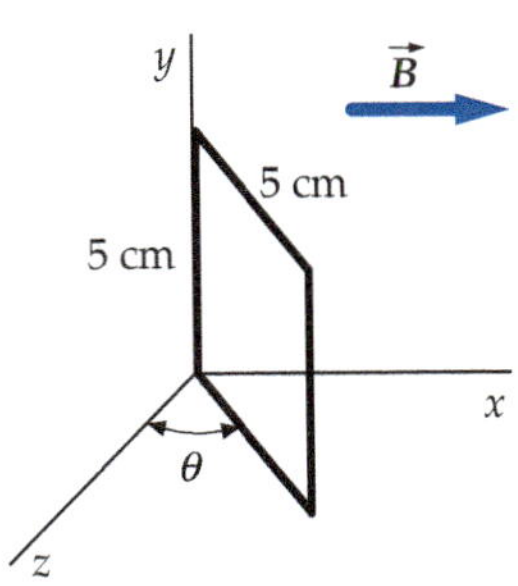

FIGURE 28-42 Problem 20

21 • A circular coil has 25 turns and a radius of 5.0 cm. It is at the equator, where Earth's magnetic field is 0.70 G, north. The axis of the coil is the line that passes through the center of the coil and is perpendicular to the plane of the coil. Find the magnetic flux through the coil when the axis of the coil is (*a*) vertical, (*b*) horizontal with the axis pointing north, (*c*) horizontal with the axis pointing east, and (*d*) horizontal with the axis making an angle of 30° with north. SSM

22 • A magnetic field of 1.2 T is perpendicular to the plane of a 14-turn square coil with sides 5.0 cm long. (*a*) Find the magnetic flux through the coil. (*b*) Find the magnetic flux through the coil if the magnetic field makes an angle of 60° with the normal to the plane of the coil.

23 • A uniform magnetic field $\vec{B}$ is perpendicular to the base of a hemisphere of radius R. Calculate the magnetic flux (in terms of B and R) through the spherical surface of the hemisphere.

24 • Find the magnetic flux through a 400-turn solenoid that has a length equal to 25.0 cm, has a radius equal to 1.00 cm, and carries a current of 3.00 A.

25 • Find the magnetic flux through a 800-turn solenoid that has a length equal to 30.0 cm, has a radius equal to 1.00 cm, and carries a current of 2.00 A.

26 •• A circular coil has 15.0 turns, has a radius of 4.00 cm, and is in a uniform magnetic field of 4.00 kG in the $+x$ direction. Find the flux through the coil when the unit normal to the plane of the coil is (*a*) $\hat{i}$, (*b*) $\hat{j}$, (*c*) $(\hat{i} + \hat{j})/\sqrt{2}$, (*d*) $\hat{k}$, and (*e*) $0.60\hat{i} + 0.80\hat{j}$.

27 •• A long solenoid has n turns per unit length, has a radius R_1, and carries a current I. A circular coil with radius R_2 and with N total turns is coaxial with the solenoid and equidistant from its ends. (*a*) Find the magnetic flux through the coil if $R_2 > R_1$. (*b*) Find the magnetic flux through the coil if $R_2 < R_1$. SSM

28 ••• (*a*) Compute the magnetic flux through the rectangular loop shown in Figure 28-43. (*b*) Evaluate your answer for $a = 5.0$ cm, $b = 10$ cm, $d = 2.0$ cm, and $I = 20$ A.

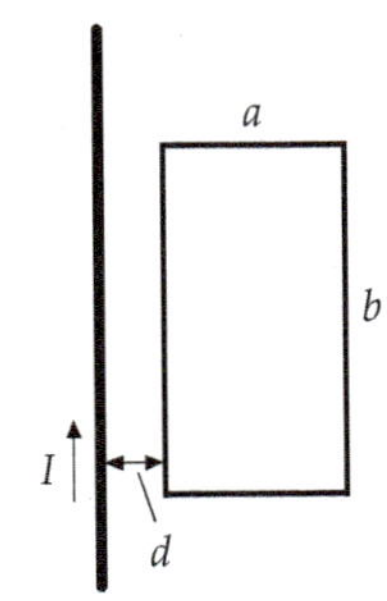

FIGURE 28-43 Problem 28

29 ••• A long, cylindrical conductor with a radius R and a length L carries a current I. Find the magnetic flux per unit length through the area indicated in Figure 28-44.

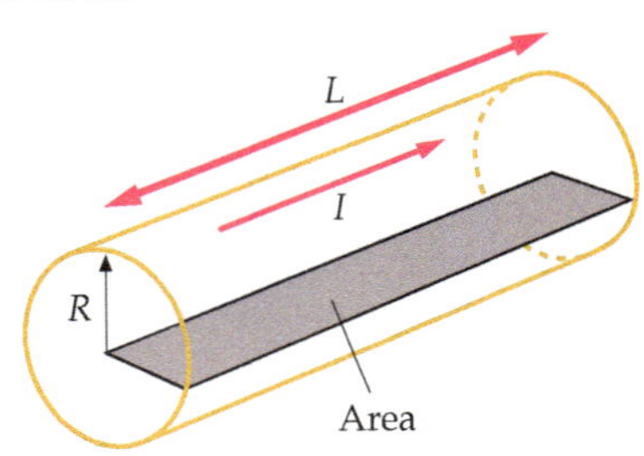

FIGURE 28-44 Problem 29

INDUCED EMF AND FARADAY'S LAW

30 • The flux through a loop is given by $\phi_m = (0.10t^2 - 0.40t)$, where ϕ_m is in webers and t is in seconds. (*a*) Find the induced emf as a function of time. (*b*) Find both ϕ_m and $\mathcal{E}$ at $t = 0$, $t = 2.0$ s, $t = 4.0$ s, and $t = 6.0$ s.

31 • The flux through a loop is given by $\phi_m = (0.10t^2 - 0.40t)$, where ϕ_m is in webers and t is in seconds. (*a*) Sketch graphs of magnetic flux and induced emf as a function of time. (*b*) At what time(s) is the flux minimum? What is the induced emf at that (those) time(s)? (*c*) At what time(s) is the flux zero? What is (are) the induced emf(s) at those time(s)?

32 • A solenoid that has a length equal to 25.0 cm, a radius equal to 0.800 cm, and 400 turns is in a region where a magnetic field of 600 G exists and makes an angle of 50° with the axis of the solenoid. (*a*) Find the magnetic flux through the solenoid. (*b*) Find the magnitude of the average emf induced in the solenoid if the magnetic field is reduced to zero in 1.40 s.

33 •• A 100-turn circular coil has a diameter of 2.00 cm and a resistance of 50.0 Ω, and the two ends of the coil are connected together. The plane of the coil is perpendicular to a uniform magnetic field of magnitude 1.00 T. The direction of the field is reversed. (*a*) Find the total charge that passes through a cross section of the wire. If the reversal takes 0.100 s, find (*b*) the average current and (*c*) the average emf during the reversal. SSM

34 •• At the equator, a 1000-turn coil that has a cross-sectional area of 300 cm^2 and a resistance of 15.0 Ω is aligned so that its plane is perpendicular to Earth's magnetic field of 0.700 G. (*a*) If the coil is flipped over in 0.350 s, what is the average induced current in it during the 0.350 s? (*b*) How much charge flows through a cross section of the coil wire during the 0.350 s?

35 •• **ENGINEERING APPLICATION** A *current integrator* measures the current as a function of time and integrates (adds) the current to find the total charge passing through it. (Because $I = dq/dt$, the integrator calculates the integral of the current or $Q = \int I\,dt$.) A circular coil that has 300 turns and a radius equal to 5.00 cm is connected to such an instrument. The total resistance of the circuit is 20.0 Ω. The plane of the coil is originally aligned perpendicular to Earth's magnetic field at some point. When the coil is rotated through 90° about an axis that is in the plane of the coil a charge of 9.40 μC passes through the current integrator. Calculate the magnitude of Earth's magnetic field at that point.

MOTIONAL EMF

36 • A 30.0-cm-long rod moves steady at 8.00 m/s in a plane that is perpendicular to a magnetic field of 500 G. The velocity of the rod is perpendicular to its length. Find (*a*) the magnetic force on an electron in the rod, (*b*) the electrostatic field in the rod, and (*c*) the potential difference between the ends of the rod.

37 • A 30.0-cm-long rod moves in a plane that is perpendicular to a magnetic field of 500 G. The velocity of the rod is perpendicular to its length. Find the speed of the rod if the potential difference between the ends is 6.00 V.

38 •• In Figure 28-45, let the magnetic field strength be 0.80 T, the rod speed be 10 m/s, the rod length be 20 cm, and the resistance of the resistor be 2.0 Ω. (The resistance of the rod and rails are negligible.) Find (*a*) the induced emf in the circuit, (*b*) the induced current in the circuit (including direction), and (*c*) the force needed to move the rod with constant speed (assuming negligible friction). Find (*d*) the power delivered by the force found in Part (*c*) and (*e*) the rate of Joule heating in the resistor.

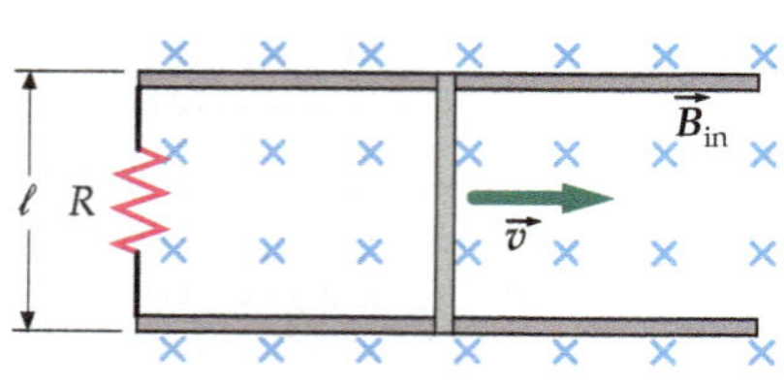

FIGURE 28-45 Problem 38

39 •• A 10-cm by 5.0-cm rectangular loop (Figure 28-46) that has a resistance equal to 2.5 Ω moves at a constant speed of 2.4 cm/s through a region that has a uniform 1.7-T magnetic field directed out of the page as shown. The front of the loop enters the field region at time $t = 0$. (*a*) Graph the flux through the loop as a function of time. (*b*) Graph the induced emf and the current in the loop as functions of time. Neglect any self-inductance of the loop and construct your graphs to include the interval $0 \le t \le 16$ s.

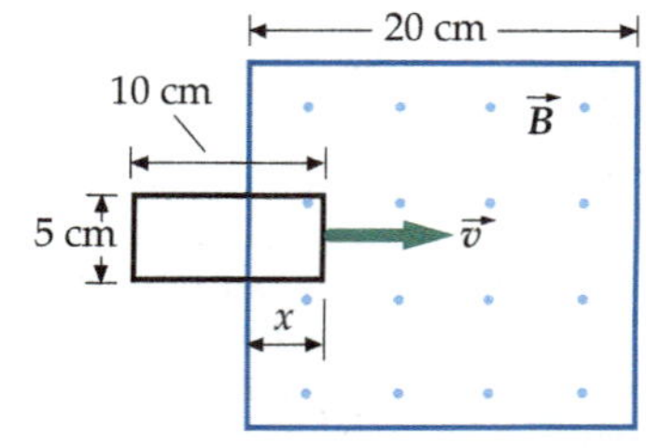

FIGURE 28-46 Problem 39

40 •• A uniform 1.2-T magnetic field is in the $+z$ direction. A conducting rod of length 15 cm lies parallel to the y axis and oscillates in the x direction with displacement given by $x = (2.0\text{ cm})\cos(120\pi t)$, where 120π has units of rad/s. (*a*) Find an expression for potential difference between the ends of the rod as a function of time? (*b*) What is the maximum potential difference between the ends the rod?

41 •• In Figure 28-47, the rod has a mass m, and a resistance R. The rails are horizontal, frictionless, and have negligible resistances. The distance between the rails is ℓ. An ideal battery that has an emf $\mathcal{E}$ is connected between points a and b so that the current in the rod is downward. The rod is released from rest at $t = 0$. (*a*) Derive an expression for the force on the rod as a function of the speed. (*b*) Show that the speed of the rod approaches a terminal speed and find an expression for the terminal speed. (*c*) What is the current when the rod is moving at its terminal speed? SSM

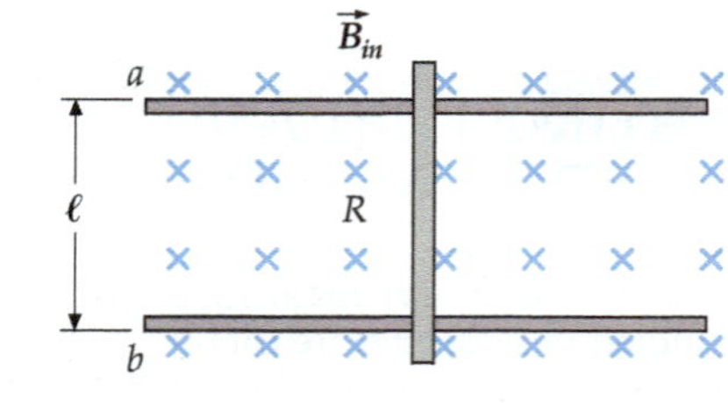

FIGURE 28-47 Problem 41

42 • A uniform magnetic field is established perpendicular to the plane of a loop that has a radius equal to 5.00 cm and a resistance equal to 0.400 Ω. The magnitude of the field is increasing at a rate of 40.0 mT/s. Find (*a*) the magnitude of the induced emf in the loop, (*b*) the induced current in the loop, and (*c*) the rate of Joule heating in the loop.

43 •• In Figure 28-48, a conducting rod that has a mass m and a negligible resistance is free to slide without friction along two parallel frictionless rails that have negligible resistances separated by a distance ℓ and connected by a resistance R. The rails are attached to a long inclined plane that makes an angle θ with the horizontal. There is a magnetic field directed upward as shown. (*a*) Show that there is a retarding force directed up the incline given by $F = (B^2\ell^2 v\cos^2\theta)/R$. (*b*) Show that the terminal speed of the rod is $v_t = mgR\sin\theta/(B^2\ell^2\cos^2\theta)$.

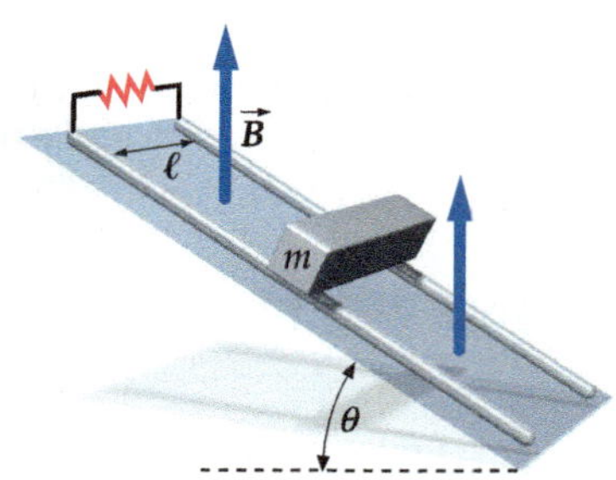

FIGURE 28-48 Problem 43

44 ••• A conducting rod of length ℓ rotates at constant angular speed ω about one end, in a plane perpendicular to a uniform magnetic field B (Figure 28-49). (*a*) Show that the potential difference between the ends of the rod is $\frac{1}{2}B\omega\ell^2$. (*b*) Let the angle θ between the rotating rod and the dashed line be given by $\theta = \omega t$. Show that the area of the pie-shaped region swept out by the rod during time t is $\frac{1}{2}\ell^2\theta$. (*d*) Compute the flux ϕ_m through that area, and apply $\mathcal{E} = -d\phi_m/dt$ (Faraday's law) to show that the motional emf is given by $\frac{1}{2}B\omega\ell^2$.

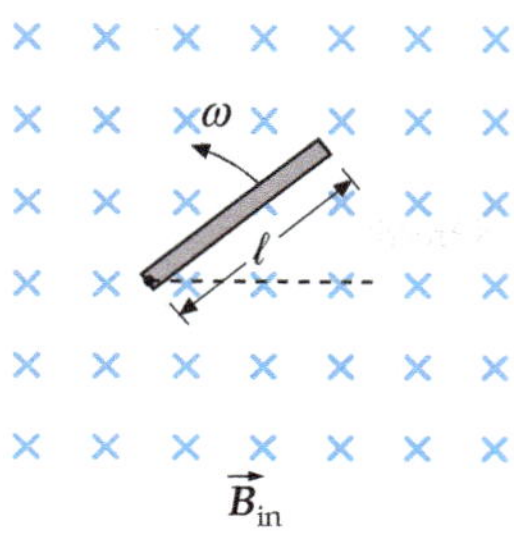

FIGURE 28-49 Problem 44

GENERATORS AND MOTORS

45 • A 2.00-cm by 1.50-cm rectangular coil has 300 turns and rotates in a region that has magnetic field of 0.400 T. (*a*) What is the maximum emf generated when the coil rotates at 60 rev/s? (*b*) What must its angular speed be to generate a maximum emf of 110 V? SSM

46 • The coil of Problem 45 rotates at 60 rev/s in a magnetic field. If the maximum emf generated by the coil is 24 V, what is the magnitude of the magnetic field?

INDUCTANCE

47 • When the current in an 8.00-H coil is equal to 3.00 A and is increasing at 200 A/s, find (*a*) the magnetic flux through the coil and (*b*) the induced emf in the coil.

48 •• A 300-turn solenoid has a radius equal to 2.00 cm and a length equal to 25.0 cm; a 1000-turn solenoid has a radius equal to 5.00 cm and is also 25.0 cm long. The two solenoids are coaxial, with one nested completely inside the other. What is their mutual inductance?

49 •• An insulated wire that has a resistance of 18.0 Ω/m and a length of 9.00 m will be used to construct a resistor. First, the wire is bent in half and then the doubled wire is wound on a cylindrical form (Figure 28-50) to create a 25.0-cm-long helix that has a diameter equal to 2.00 cm. Find both the resistance and the inductance of this wire-wound resistor. SSM

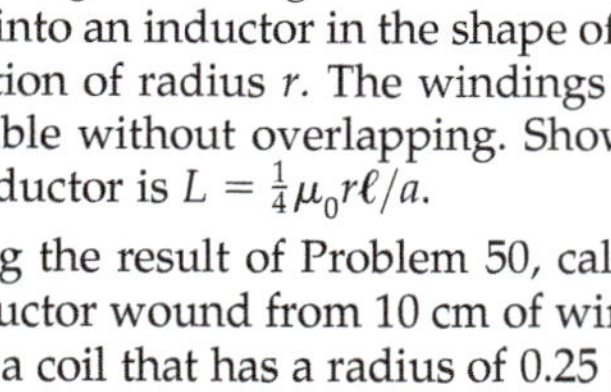
FIGURE 28-50 Problem 49

50 •• You are given a length ℓ of wire that has radius a and are told to wind it into an inductor in the shape of a helix that has a circular cross section of radius r. The windings are to be as close together as possible without overlapping. Show that the self-inductance of this inductor is $L = \frac{1}{4}\mu_0 r\ell/a$.

51 • Using the result of Problem 50, calculate the self-inductance of an inductor wound from 10 cm of wire that has a diameter of 1.0 mm into a coil that has a radius of 0.25 cm.

52 ••• In Figure 28-51, circuit 2 has a total resistance of 300 Ω. After switch S is closed, the current in Circuit 1 increases—reaching a value of 5.00 A after a long time. A charge of 200 μC passes through the galvanometer in Circuit 2 during the time that the current in Circuit 1 is increasing. What is the mutual inductance between the two coils?

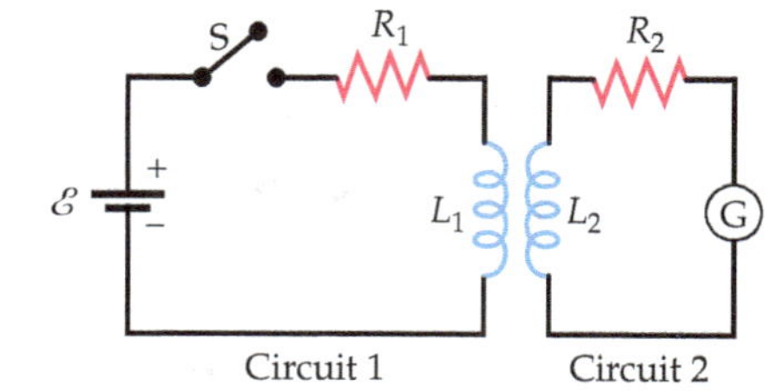

FIGURE 28-51 Problem 52

53 ••• Show that the inductance of a toroid of rectangular cross section, as shown in Figure 28-52, is given by $L = \dfrac{\mu_0 N^2 H \ln(b/a)}{2\pi}$, where N is the total number of turns, a is the inside radius, b is the outside radius, and H is the height of the toroid. SSM

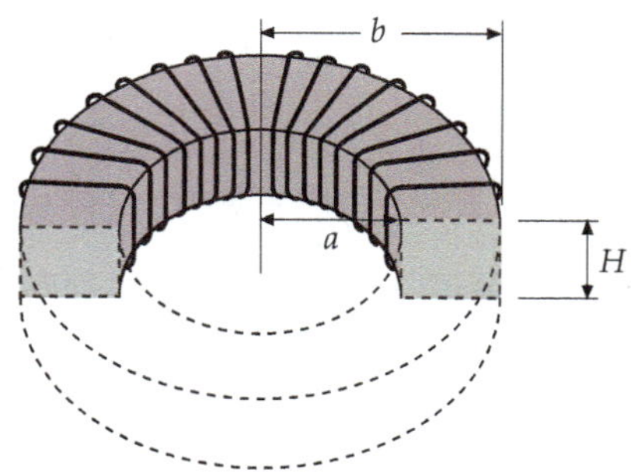

FIGURE 28-52 Problem 53

MAGNETIC ENERGY

54 • A coil that has a self-inductance of 2.00 H and a resistance of 12.0 Ω is connected to an ideal 24.0-V battery. (*a*) What is the steady-state current? (*b*) How much energy is stored in the inductor when the steady-state current is established?

55 • In a plane electromagnetic wave, the magnitudes of the electric fields and magnetic fields are related by $E = cB$, where $c = 1/\sqrt{\epsilon_0\mu_0}$ is the speed of light. Show that when $E = cB$ the electric and the magnetic energy densities are equal. SSM

56 •• A 2000-turn solenoid has a cross-sectional area equal to 4.0 cm² and a length equal to 30 cm. The solenoid carries a current of 4.0 A. (*a*) Calculate the magnetic energy stored in the solenoid using $U = \frac{1}{2}LI^2$, where $L = \mu_0 n^2 A\ell$. (*b*) Divide your answer in Part (*a*) by the volume of the region inside the solenoid to find the magnetic energy per unit volume in the solenoid. (*c*) Check your Part-(*b*) result by computing the magnetic energy density from $u_m = B^2/(2\mu_0)$ where $B = \mu_0 nI$.

57 •• A long cylindrical wire has a radius equal to 2.0 cm and carries a current of 80 A uniformly distributed over its cross-sectional area. Find the magnetic energy per unit length within the wire.

58 •• A toroid that has a mean radius equal to 25.0 cm and circular loops with radii equal to 2.00 cm is wound with a superconducting wire. The wire has a length equal to 1000 m and carries a current of 400 A. (*a*) What is the number of turns of the wire? (*b*) What is the magnetic field strength and magnetic energy density at the mean radius? (*c*) Estimate the total energy stored in this toroid by assuming that the energy density is uniformly distributed in the region inside the toroid.

**RL* CIRCUITS

59 • A circuit consists of a coil that has a resistance equal to 8.00 Ω and a self-inductance equal to 4.00 mH, an open switch, and an ideal 100-V battery—all connected in series. At $t = 0$ the switch is closed. Find the current and its rate of change at times (*a*) $t = 0$, (*b*) $t = 0.100$ ms, (*c*) $t = 0.500$ ms, and (*d*) $t = 1.00$ ms. SSM

60 • In the circuit shown in Figure 28-53, the throw of the make-before-break switch has been at contact a for a long time and the current in the 1.00 mH coil is equal to 2.00 A. At $t = 0$ the throw is quickly moved to contact b. The total resistance $R + r$ of the coil and the resistor is 10.0 Ω. Find the current when (*a*) $t = 0.500$ ms and (*b*) $t = 10.0$ ms.

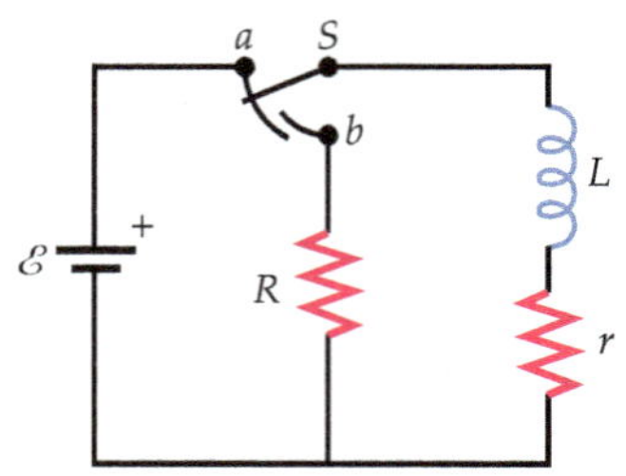

FIGURE 28-53 Problem 60

61 •• In the circuit shown in Figure 28-54, let $\mathcal{E}_0 = 12.0$ V, $R = 3.00$ Ω, and $L = 0.600$ H. The switch, which was initially open, is closed at time $t = 0$. At time $t = 0.500$ s, find (*a*) the rate at which the battery supplies energy, (*b*) the rate of Joule heating in the resistor, and (*c*) the rate at which energy is being stored in the inductor. SSM

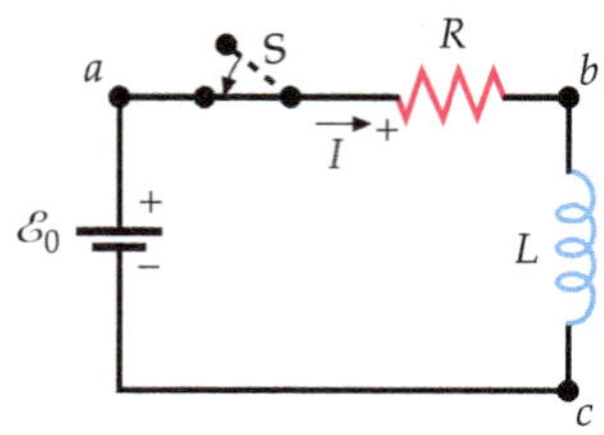

FIGURE 28-54 Problems 61, 62 and 69

62 •• How many time constants must elapse before the current in an RL circuit (Figure 28-54) that is initially zero reaches (*a*) 90 percent, (*b*) 99 percent, and (*c*) 99.9 percent of its steady-state value?

63 •• A circuit consists of a 4.00-mH coil, a 150-Ω resistor, a 12.0-V ideal battery, and an open switch—all connected in series. After the switch is closed: (*a*) What is the initial rate of increase of the current? (*b*) What is the rate of increase of the current when the current is equal to half its steady-state value? (*c*) What is the steady-state value of the current? (*d*) How long does it take for the current to reach 99 percent of its steady state value? SSM

64 •• A circuit consists of a large electromagnet that has an inductance of 50.0 H and a resistance of 8.00 Ω, a dc 250-V power source, and an open switch—all connected in series. How long after the switch is closed is the current equal to (*a*) 10 A and (*b*) 30 A.

65 •• **SPREADSHEET** Given the circuit shown in Figure 28-55, assume that the inductor has negligible internal resistance and that the switch S has been closed for a long time so that a steady current exists in the inductor. (*a*) Find the battery current, the current in the 100 Ω resistor, and the current in the inductor. (*b*) Find the potential drop across the inductor immediately after the switch S is opened. (*c*) Using a spreadsheet program, make graphs of the current in the inductor and the potential drop across the inductor as functions of time, for the period during which the switch is open. SSM

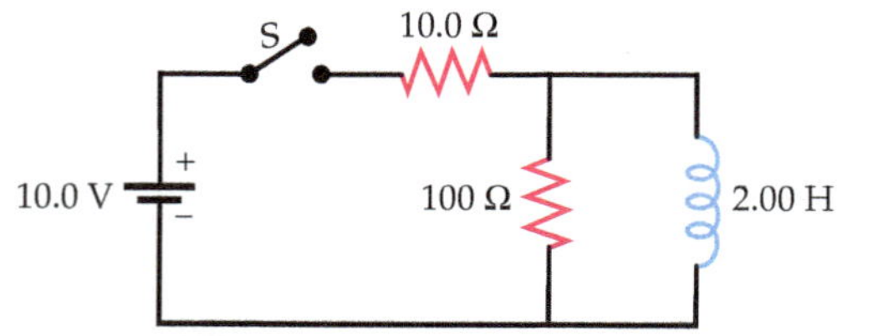

FIGURE 28-55 Problem 65

66 •• Given the circuit shown in Figure 28-56, the inductor has negligible internal resistance and the switch S has been open for a long time. The switch is then closed. (*a*) Find the current in the battery, the current in the 100-Ω resistor, and the current in the inductor immediately after the switch is closed. (*b*) Find the current in the battery, the current in the 100-Ω resistor, and the current in the inductor a long time after the switch is closed. After being closed for a long time the switch is now opened. (*c*) Find the current in the battery, the current in the 100-Ω resistor, and the current in the inductor immediately after the switch is opened. (*d*) Find the current in the battery, the current in the 100-Ω resistor, and the current in the inductor after the switch is opened for a long time.

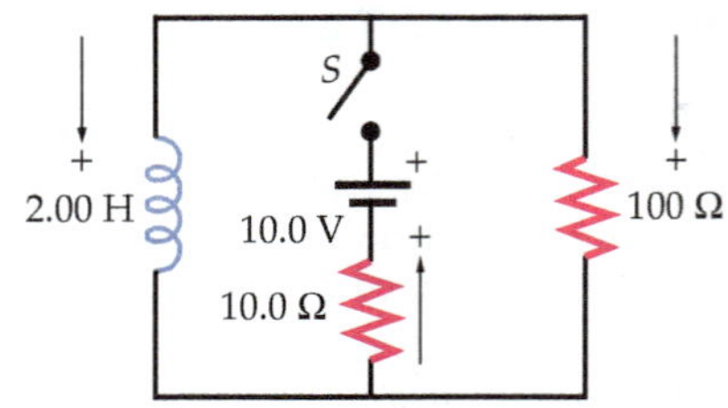

FIGURE 28-56 Problem 66

67 •• An inductor, two resistors, a make-before-break switch, and a battery are connected as shown in Figure 28-57. The switch throw has been at contact e for a long time and the current in the inductor is 2.5 A. Then, at $t = 0$, the throw is quickly moved to contact f. During the next 45 ms the current in the inductor drops to 1.5 A. (*a*) What is the time constant for this circuit? (*b*) If the resistance R is equal to 0.40 Ω, what is the value of the inductance L?

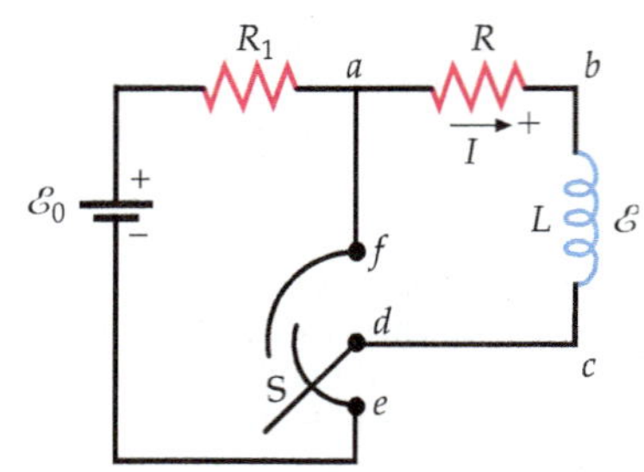

FIGURE 28-57 Problem 67

68 •• A circuit consists of a coil that has a self-inductance equal to 5.00 mH and an internal resistance equal to 15.0 Ω, an ideal 12.0-V battery, and an open switch—all connected in series (Figure 28-58). At $t = 0$ the switch is closed. Find the time when the rate at which energy is dissipated in the coil equals the rate at which magnetic energy is stored in the coil.

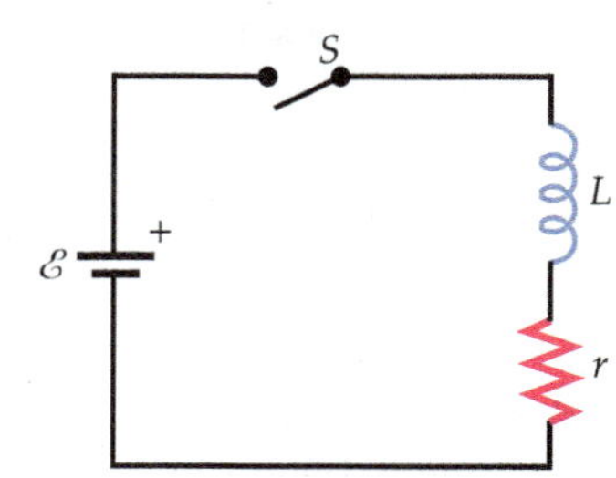

FIGURE 28-58 Problem 68

69 ••• In the circuit shown in Figure 28-54, let $\mathcal{E}_0 = 12.0$ V, $R = 3.00$ Ω, and $L = 0.600$ H. The switch is closed at time $t = 0$. During the time from $t = 0$ to $t = L/R$, find (*a*) the amount of energy supplied by the battery, (*b*) the amount of energy dissipated in the resistor, and (*c*) the amount of energy delivered to the inductor. *Hint: Find the energy transfer rates as functions of time and integrate.* SSM

GENERAL PROBLEMS

70 • A 100-turn coil has a radius of 4.00 cm and a resistance of 25.0 Ω. (*a*) The coil is in a uniform magnetic field that is perpendicular to the plane of the coil. What rate of change of the magnetic field strength will induce a current of 4.00 A in the coil? (*b*) What rate of change of the magnetic field strength is required if the magnetic field makes an angle of 20° with the normal to the plane of the coil?

71 •• **ENGINEERING APPLICATION** Figure 28-59 shows a schematic drawing of an *ac generator*. The basic generator consists of a rectangular loop of dimensions *a* and *b* and has *N* turns connected to *slip rings*. The loop rotates (driven by a gasoline engine) at an angular speed of ω in a uniform magnetic field $\vec{B}$. (*a*) Show that the induced potential difference between the two slip rings is given by $\mathcal{E} = NBab\omega \sin \omega t$. (*b*) If $a = 2.00$ cm, $b = 4.00$ cm, $N = 250$, and $B = 0.200$ T, at what angular frequency ω must the coil rotate to generate an emf whose maximum value is 100 V? SSM

72 •• **ENGINEERING APPLICATION** Prior to 1960, magnetic field strengths were usually measured by a *rotating coil gaussmeter*. The device uses a small multiturn coil rotating at a high speed on an axis perpendicular to the magnetic field. The coil is connected to an ac voltmeter by means of slip rings, like those shown in Figure 28-59. In one specific design, the rotating coil has 400 turns and an area of 1.40 cm^2. The coil rotates at 180 rev/min. If the magnetic field strength is 0.450 T, find the maximum induced emf in the coil and the orientation of the normal to the plane of the coil relative to the field for which the maximum induced emf occurs.

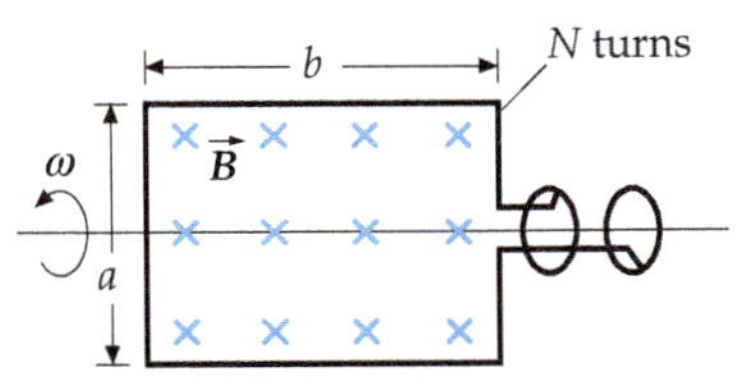

FIGURE 28-59 Problems 71 and 72

73 •• Show that the equivalent self-inductance for two inductors that have self-inductances L_1 and L_2 and are connected in series is given by $L_{eq} = L_1 + L_2$ if there is no flux linkage between the two inductors. (Saying there is no flux linkage between them is equivalent to saying that the mutual inductance between them is zero.)

74 •• Show that the equivalent self-inductance for two in ductors that have self-inductances L_1 and L_2 and are connected in parallel is given by $\frac{1}{L_{eq}} = \frac{1}{L_1} + \frac{1}{L_2}$ if there is no flux linkage between the two inductors. (Saying there is no flux linkage between them is equivalent to saying that the mutual inductance between them is equal to zero.)

75 •• A circuit consists of a 12-V battery, a switch, and a lightbulb—all connected in series. It is known that the lightbulb requires a minimum current of 0.10 A in order to produce a visible glow. In the circuit, that particular bulb draws 2.0 W when the switch has been closed for a long time. Now, an inductor is put in series with the bulb and the rest of the circuit. If the lightbulb begins to glow 3.5 ms after the switch is closed, how large is the self-inductance of the inductor? Ignore any heating time of the filament and assume the glow is observed as soon as the current in the filament reaches the 0.10-A threshold.

76 •• Your friend decides to generate electrical power by rotating a 100 000-turn coil of wire around an axis in the plane of the coil and through its center. The coil is perpendicular to Earth's magnetic field in a region where the field strength is equal to 0.300 G. The loops of the coil have a radius of 25.0 cm, and the coil has negligible resistance. (*a*) If your friend turns the coil at a rate of 150 rev/s, what peak current will exist in a 1500-Ω resistor that is connected across the terminals of the coil? (*b*) The average of the square of the current will equal half of the square of the peak current. What will be the average power delivered to the resistor? Is this an economical way to generate power? *Hint: Energy has to be expended to keep the coil rotating.*

77 •• Figure 28-60*a* shows an experiment designed to measure the acceleration due to gravity. A large plastic tube is encircled by a wire, which is arranged in single loops separated by a distance of 10 cm. A strong magnet is dropped through the top of the loop. As the magnet falls through each loop the voltage rises; then the voltage rapidly falls through zero to a large negative value and then returns to zero. The shape of the voltage signal is shown in Figure 28-60*b*. (*a*) Explain the basic physics behind the generation of this voltage pulse. (*b*) Explain why the tube cannot be made of a conductive material. (*c*) Qualitatively explain the *shape* of the voltage signal in Figure 28-60*b*. (*d*) The times at which the voltage crosses zero as the magnet falls through each loop in succession are given in the table on the next page. Use these data to calculate a value for *g*. SSM

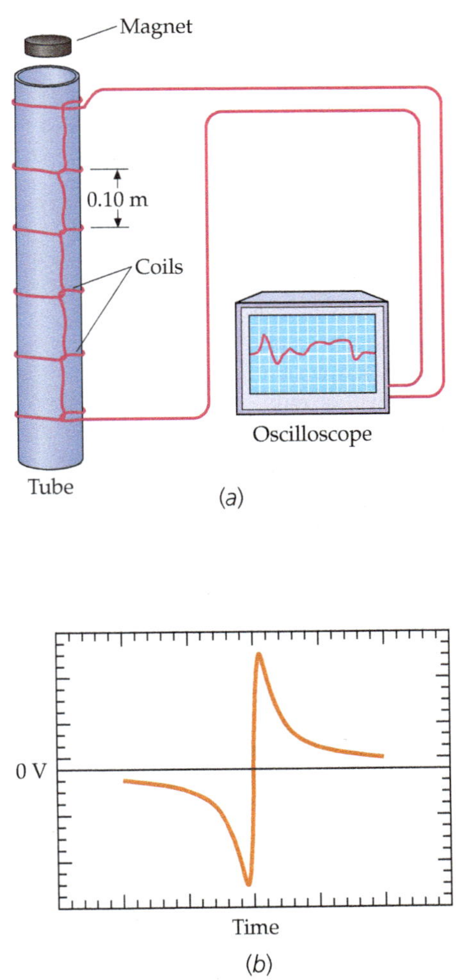

FIGURE 28-60 Problem 77

Loop Number	Zero Crossing Time (s)
1	0.011189
2	0.063133
3	0.10874
4	0.14703
5	0.18052
6	0.21025
7	0.23851
8	0.26363
9	0.28853
10	0.31144
11	0.33494
12	0.35476
13	0.37592
14	0.39107

78 •• The rectangular coil shown in Figure 28-61 has 80 turns, is 25 cm wide, is 30 cm long, and is located in a magnetic field of 0.14 T directed out of the page, as shown. Only half of the coil is in the region of the magnetic field. The resistance of the coil is 24 Ω. Find the magnitude and the direction of the induced current if the coil is moving with a velocity of 2.0 m/s (*a*) to the right, (*b*) up the page, (*c*) to the left, and (*d*) down the page.

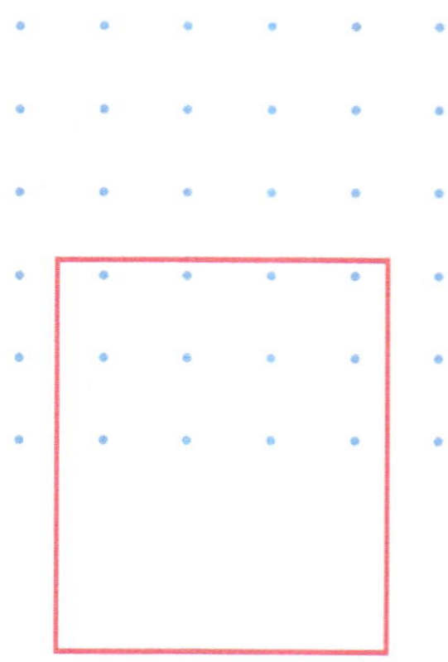

FIGURE 28-61 Problem 78

79 •• A long solenoid has n turns per unit length and carries a current that varies with time according to $I = I_0 \sin \omega t$. The solenoid has a circular cross section of radius R. Find the induced electric field, at points near the plane equidistant from the ends of the solenoid, as a function of both the time t and the perpendicular distance r from the axis of the solenoid for (*a*) $r < R$ and (*b*) $r > R$. SSM

80 ••• A coaxial cable consists of two very thin-walled conducting cylinders of radii r_1 and r_2 (Figure 28-62). The currents in the inner and outer cylinders are equal in magnitude but opposite in direction. (*a*) Use Ampère's law to find the magnetic field as a function of the perpendicular distance r from the central axis of the cable for (1) $0 < r < r_1$, (2) $r_1 < r < r_2$, and (3) $r > r_2$. (*b*) Show that the magnetic energy density in the region between the cylinders is given by $u_m = \frac{1}{2}(\mu_0/4\pi)I^2/(\pi r^2)$. (*c*) Show that the total magnetic energy in a cable volume of length ℓ is given by $U_m = (\mu_0/4\pi)I^2\ell \ln(r_2/r_1)$. (*d*) Use the result in Part (*c*) and the relationship between magnetic energy, current, and inductance to show that the self-inductance per unit length of the cable arrangement is given by $L/\ell = (\mu_0/2\pi)\ln(r_2/r_1)$.

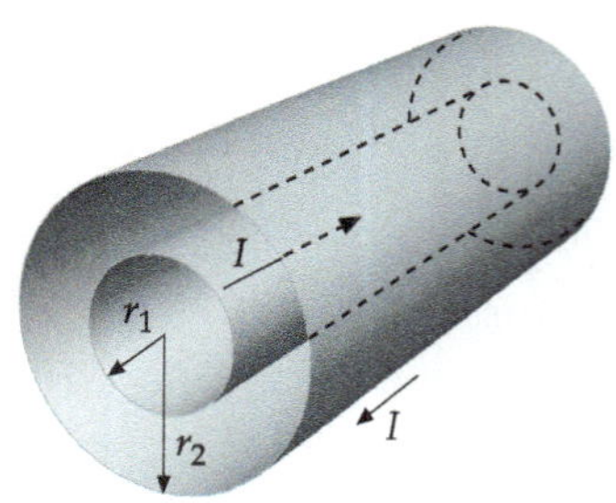

FIGURE 28-62 Problem 80

81 ••• A coaxial cable consists of two very thin-walled conducting cylinders of radii r_1 and r_2 (Figure 28-63). The currents in the inner and outer cylinders are equal in magnitude but opposite in direction. Compute the flux through a rectangular area of sides ℓ and $r_2 - r_1$ between the conductors shown in Figure 28-63. Use the relationship between flux and current ($\phi_m = LI$) to show that the self-inductance per unit length of the cable is given by $L/\ell = (\mu_0/2\pi)\ln(r_2/r_1)$.

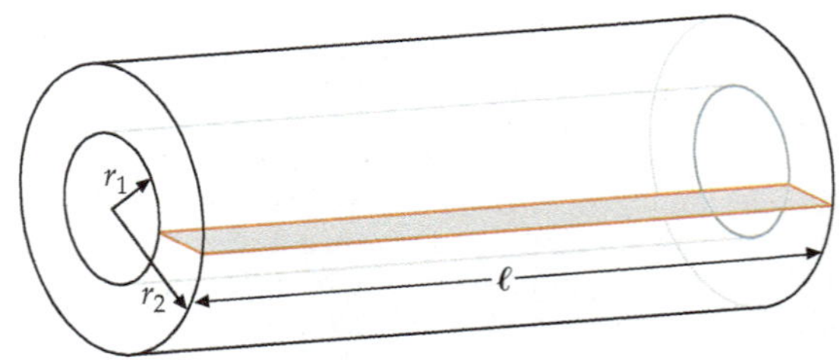

FIGURE 28-63 Problem 81

82 ••• **SPREADSHEET** Figure 28-64 shows a rectangular loop of wire that is 0.300 m wide, is 1.50 m long, and lies in the vertical plane which is perpendicular to a region that has a uniform magnetic field. The magnitude of the uniform magnetic field is 0.400 T and the direction of the magnetic field is into the page. The portion of the loop not in the magnetic field is 0.100 m long. The resistance of the loop is 0.200 Ω and its mass is 50.0 g. The loop is released from rest at $t = 0$. (*a*) What are the magnitude and direction of the induced current when the loop has a downward speed v? (*b*) What is the force that acts on the loop as a result of the current? (*c*) What is the net force acting on the loop? (*d*) Write out Newton's second law for the loop. (*e*) Obtain an expression for the speed of the loop as a function of time. (*f*) Integrate the expression obtained in Part (*e*) to find the distance the loop falls as a function of time. (*g*) Using a spreadsheet program, make a graph of the position of the loop as a function of time (letting $t = 0$ at the start) for values of y between 0 m and 1.40 m (i.e., when the loop leaves the magnetic field). (*h*) At what time does the loop completely leave the field region? Compare this to the time it would have taken if there were no field.

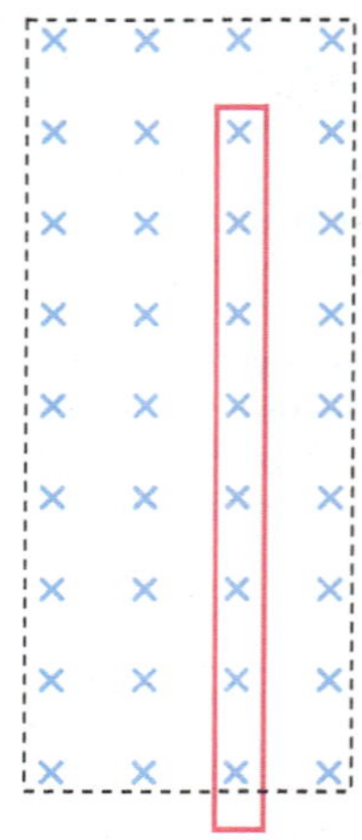

FIGURE 28-64 Problem 82

83 ••• A coil that has N turns and an area A is suspended from the ceiling by a wire that provides a linear restoring torque that has a torsion constant κ. The two ends of the coil are connected to each other, the coil has resistance R, and the moment of inertia of the coil is I. The plane of the coil is vertical and parallel to a uniform horizontal magnetic field $\vec{B}$ when the wire is not twisted (i.e., $\theta = 0$). The coil is twisted about a vertical axis through its center by a small angle θ_0 and released. The coil then undergoes damped harmonic oscillation. Show that its angle with its equilibrium position will vary with time according to $\theta(t) = \theta_0 e^{-t/2\tau} \cos \omega' t$, where $\tau = RI/(NBA)^2$, $\omega = \sqrt{\kappa/I}$ and $\omega' = \omega_0\sqrt{1 - (2\omega_0\tau)^{-2}}$.

CHAPTER 29

THIS LISTENER DIALS IN HER FAVORITE RADIO STATION. THIS CHANGES THE RESONANT FREQUENCY OF AN OSCILLATING ELECTRIC CIRCUIT WITHIN THE TUNER, SO ONLY THE STATION SHE SELECTS IS AMPLIFIED. *(© Roger Ressmeyer/Corbis.)*

Alternating-Current Circuits

? What component of the circuit is modified as she turns the dial? (See Example 29-11.)

More than 99 percent of the electrical energy used today is produced by electrical generators in the form of alternating current, which has a great advantage over direct current. Electrical energy can be distributed over large regions at very high voltages and low currents to reduce energy losses due to Joule heating. With alternating current, electrical energy can then be transformed, with almost no energy loss, to lower and safer voltages and correspondingly higher currents for local distribution and use.* The transformer that accomplishes these changes in potential difference and current works on the basis of magnetic induction. In North America, power is delivered by a sinusoidal current that has a frequency equal to 60 Hz. Devices such as radios, television sets, and microwave ovens detect or generate alternating currents of much higher frequencies.

Alternating current is produced by motional emf or by magnetic induction in an ac generator, which is designed to provide a sinusoidal emf.

* High voltage direct current is sometimes used to transmit electrical power between one point and another distant point. However, alternating current is always used to transmit power from one point to two or more distant points.

In this chapter, we will see that when the generator output is sinusoidal, the current in an inductor, a capacitor, or a resistor is also sinusoidal, although it is generally not in phase with the generator's emf. When the emf and current are both sinusoidal, their maximum values are proportional. The study of sinusoidal currents is particularly important because even currents that are not sinusoidal can be analyzed in terms of sinusoidal components using Fourier analysis.

29-1 ALTERNATING CURRENT IN A RESISTOR

Figure 29-1 shows a simple **ac generator.** An analysis of such a generator is presented in Chapter 28. The emf of such a generator is given by the equation immediately following Equation 28-10:

$$\mathcal{E} = \mathcal{E}_{\text{peak}} \cos \omega t \tag{29-1}$$

where ω is the angular speed of the coil. (Equation 28-10 has the emf proportional to $\sin \omega t$ rather than $\cos \omega t$. The distinction between the two is the choice of when $t = 0$.) If the N-turn coil has area A, and if the magnetic field is uniform and has magnitude B, the peak emf is given by ωNBA. Although practical generators are considerably more complicated, they all produce a sinusoidal emf either by induction or by motional emf. In circuit diagrams, an ac generator is represented by the symbol ⊙.

Figure 29-2 shows a simple ac circuit that consists of an ideal generator and a resistor. (A generator is ideal if its internal resistance, its self-inductance, and its capacitance are negligible.) The voltage drop across the resistor V_R is equal to the emf $\mathcal{E}$ of the generator. If the generator produces an emf given by Equation 29-1, we have

$$V_R = V_{R\,\text{peak}} \cos \omega t$$

Applying Ohm's law, we have

$$V_R = IR \tag{29-2}$$

Thus,

$$V_{R\,\text{peak}} \cos \omega t = IR \tag{29-3}$$

so the current in the resistor is

$$I = \frac{V_{R\,\text{peak}}}{R} \cos \omega t = I_{\text{peak}} \cos \omega t \tag{29-4}$$

where

$$I_{\text{peak}} = \frac{V_{R\,\text{peak}}}{R} \tag{29-5}$$

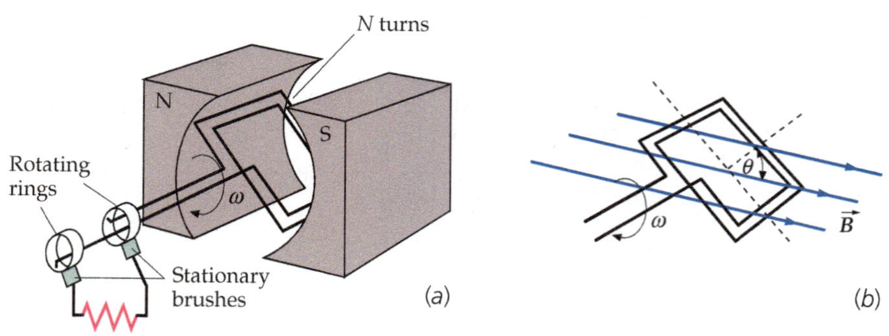

FIGURE 29-1 (*a*) An ac generator. A coil rotating with constant angular frequency ω in a static magnetic field $\vec{B}$ generates a sinusoidal emf. Energy from falling water or from a steam turbine is used to rotate the coil to produce electrical energy. The emf is supplied to an external circuit by the brushes that are in contact with the rings. (*b*) At this instant, the normal to the plane of the coil makes an angle θ with the magnetic field, and the flux through the each turn of the coil is $BA \cos \theta$.

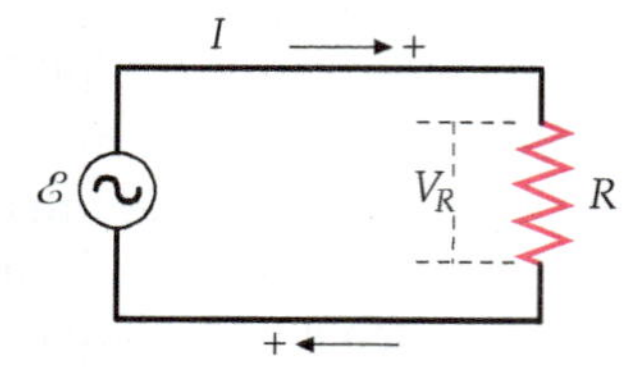

FIGURE 29-2 An ac generator in series with a resistor R.

(a) (b)

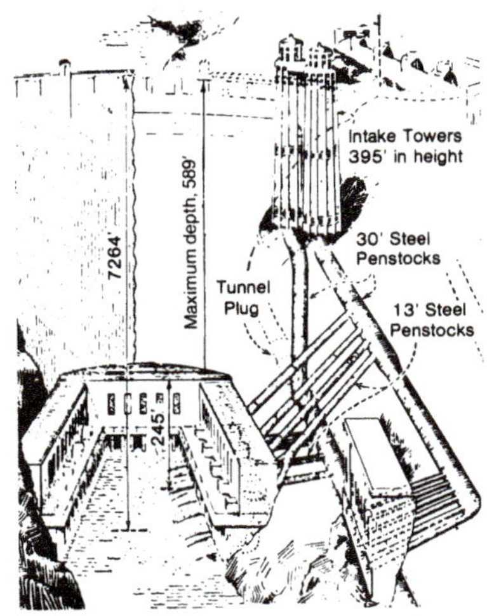

(c)

(*a*) The mechanical energy of falling water drives turbines (*b*) for the generation of electricity. (*c*) Schematic drawing of the Hoover Dam showing the intake towers and pipes (penstocks) that carry the water to the generators below. *((a) Courtesy of U.S. Department of the Interior, Department of Reclamation. (b) © Lee Langum/ Photo Researchers, Inc.)*

Note that the current through the resistor is in phase with the potential drop across the resistor, as shown in Figure 29-3.

The power delivered to the resistor varies with time. Its instantaneous value is

$$P = I^2R = (I_{\text{peak}} \cos \omega t)^2 R = I^2_{\text{peak}} R \cos^2 \omega t \qquad \text{29-6}$$

Figure 29-4 shows the power as a function of time. The power varies from zero to its peak value I^2_{peak}, as shown. We are usually interested in the average power over one or more complete cycles:

$$P_{\text{av}} = (I^2R)_{\text{av}} = I^2_{\text{peak}} R (\cos^2 \omega t)_{\text{av}}$$

The average value of $\cos^2 \omega t$ over one or more complete periods is $\frac{1}{2}$. This result can be seen from the identity $\cos^2 \omega t + \sin^2 \omega t = 1$. A plot of $\sin^2 \omega t$ looks the same as a plot of $\cos^2 \omega t$ except that the plot is shifted by 90°. Both have the same average value over one or more complete periods, and because their sum is 1, the average value of each must be $\frac{1}{2}$. The average power dissipated in the resistor is thus

$$P_{\text{av}} = (I^2R)_{\text{av}} = \frac{1}{2} I^2_{\text{peak}} R \qquad \text{29-7}$$

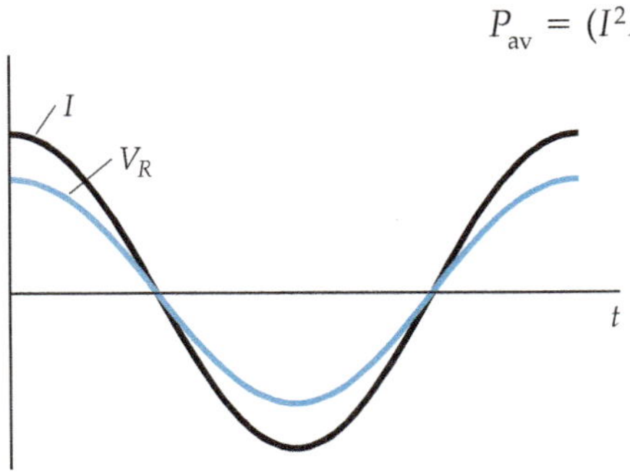

FIGURE 29-3 The voltage drop across a resistor is in phase with the current.

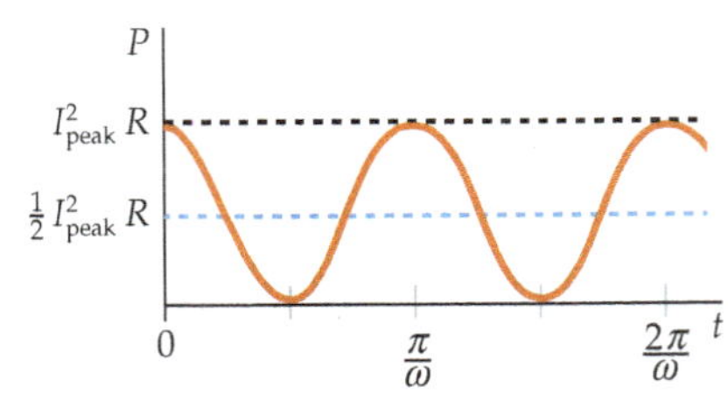

FIGURE 29-4 Plot of the power delivered to the resistor shown in Figure 29-2 versus time. The power varies from zero to a peak value $I^2_{\text{peak}} R$. The average power is half the peak power.

ROOT-MEAN-SQUARE VALUES

Most ac ammeters and voltmeters are designed to measure the **root-mean-square (rms) values** of current and potential difference. The rms value I_{rms} of a current is defined by

$$I_{\text{rms}} = \sqrt{(I^2)_{\text{av}}} \qquad \text{29-8}$$

DEFINITION—RMS CURRENT

For a sinusoidal current, the average value of I^2 is

$$(I^2)_{\text{av}} = \left[(I_{\text{peak}}\cos\omega t)^2\right]_{\text{av}} = \frac{1}{2}I^2_{\text{peak}}$$

Substituting $\frac{1}{2}I^2_{\text{peak}}$ for $(I^2)_{\text{av}}$ in Equation 29-8, we obtain

$$I_{\text{rms}} = \frac{1}{\sqrt{2}}I_{\text{peak}} \approx 0.707I_{\text{peak}} \tag{29-9}$$

RMS VALUE RELATED TO PEAK VALUE

The rms value of *any quantity* that varies sinusoidally with time is equal to the peak value of that quantity divided by $\sqrt{2}$.

Substituting $(I_{\text{rms}})^2$ for $\frac{1}{2}I^2_{\text{peak}}$ in Equation 29-7, we obtain for the average power delivered to the resistor

$$P_{\text{av}} = (I_{\text{rms}})^2R \tag{29-10}$$

The rms current equals the steady dc current that would produce the same Joule heating as the actual ac current.

> ! The rms current equals the steady dc current that would produce the same Joule heating as the actual ac current.

For the simple circuit in Figure 29-2, the average power delivered by the generator is

$$P_{\text{av}} = (\mathcal{E}I)_{\text{av}} = \left[(\mathcal{E}_{\text{peak}}\cos\omega t)(I_{\text{peak}}\cos\omega t)\right]_{\text{av}} = \mathcal{E}_{\text{peak}}I_{\text{peak}}(\cos^2\omega t)_{\text{av}}$$

or

$$P_{\text{av}} = \tfrac{1}{2}\mathcal{E}_{\text{peak}}I_{\text{peak}}$$

Using $I_{\text{rms}} = I_{\text{peak}}/\sqrt{2}$ and $\mathcal{E}_{\text{rms}} = \mathcal{E}_{\text{peak}}/\sqrt{2}$, this can be written

$$P_{\text{av}} = \mathcal{E}_{\text{rms}}I_{\text{rms}} \tag{29-11}$$

AVERAGE POWER DELIVERED BY A GENERATOR

The rms current is related to the rms potential drop in the same way that the peak current is related to the peak potential drop. We can see this by dividing each side of Equation 29-5 by $\sqrt{2}$ and substituting for I_{peak} and $V_{R\,\text{rms}}$ using $I_{\text{rms}} = I_{\text{peak}}/\sqrt{2}$ and $V_{R\,\text{rms}} = V_{R\,\text{peak}}/\sqrt{2}$.

$$I_{\text{rms}} = \frac{V_{R\,\text{rms}}}{R} \tag{29-12}$$

Equations 29-10, 29-11, and 29-12 are of the same form as the corresponding equations for direct-current circuits; however, I is replaced by I_{rms} and V_R is replaced by $V_{R\,\text{rms}}$. We can therefore calculate the power input and the heat generated using the same equations that we used for direct current, if we use rms values for the current and potential drop.

> **PRACTICE PROBLEM 29-1**
>
> The sinusoidal potential drop across a 12-Ω resistor has a peak value of 48 V. Find (*a*) the rms current, (*b*) the average power, and (*c*) the maximum power delivered to the resistor.

The ac power supplied to domestic wall outlets and light fixtures in the United States has an rms potential difference of 120 V at a frequency of 60 Hz. This potential difference is maintained, independent of the current. If you plug a 1600-W space heater into a wall outlet it will draw a current of

$$I_{\text{rms}} = \frac{P_{\text{av}}}{V_{\text{rms}}} = \frac{1600\text{ W}}{120\text{ V}} = 13.3\text{ A}$$

All appliances plugged into the outlets of a single 120-V circuit are connected in parallel. If you plug a 500-W toaster into another outlet of the same circuit, it will draw a current of 500 W/120 V = 4.17 A, and the total current through the parallel combination will be 17.5 A. Typical household wall outlets are rated at 15 A and are part of a circuit using wires rated at either 15 A or 20 A, with each circuit having several outlets. A total current greater than the rated current for the wiring is likely to overheat the wiring and is a fire hazard. Each circuit is therefore equipped with a circuit breaker (or a fuse in older houses) that trips (or blows) when the total current exceeds the 15-A or 20-A rating.

High-power domestic appliances, such as electric clothes dryers, kitchen ranges, and hot water heaters, typically require power delivered at 240 V rms. For a given power requirement, only half as much current is required at 240 V as at 120 V, but 240 V is more likely to deliver a fatal shock or to start a fire than 120 V.

Example 29-1 Sawtooth Waveform

Find (*a*) the average current and (*b*) the rms current for the sawtooth waveform shown in Figure 29-5. In the region $0 < t < T$, the current is given by $I = (I_0/T)t$.

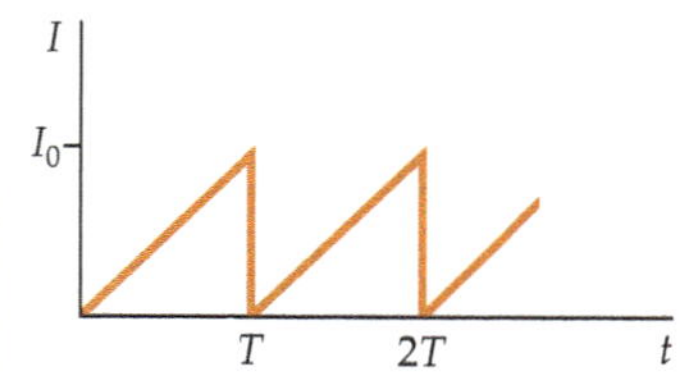

FIGURE 29-5

PICTURE The average of any quantity over a time interval T is the integral of the quantity over the interval divided by T. We use this to find both the average current, I_{av}, and the average of the square of the current, $(I^2)_{av}$.

SOLVE

(*a*) Calculate I_{av} by integrating I from $t = 0$ to $t = T$ and dividing by T:

$$I_{av} = \frac{1}{T}\int_0^T I\,dt = \frac{1}{T}\int_0^T \frac{I_0}{T}t\,dt = \frac{I_0}{T^2}\frac{T^2}{2} = \boxed{\frac{1}{2}I_0}$$

(*b*) 1. Find $(I^2)_{av}$ by integrating I^2:

$$(I^2)_{av} = \frac{1}{T}\int_0^T I^2\,dt = \frac{1}{T}\left(\frac{I_0}{T}\right)^2\int_0^T t^2\,dt = \frac{I_0^2}{T^3}\frac{T^3}{3} = \frac{1}{3}I_0^2$$

2. The rms current is the square root of $(I^2)_{av}$:

$$I_{rms} = \sqrt{(I^2)_{av}} = \boxed{\frac{I_0}{\sqrt{3}}}$$

CHECK Both the average current and the rms current are less than I_0, as expected.

29-2 ALTERNATING-CURRENT CIRCUITS

Alternating current behaves differently than direct current in inductors and capacitors. When a capacitor becomes fully charged in a dc circuit, the capacitor blocks the current; that is, the capacitor acts like an open circuit. However, if the current alternates, charge continually flows onto the plates and off of the plates of the capacitor. We will see that at high frequencies, a capacitor hardly impedes the current at all. That is, the capacitor acts like a short circuit. Conversely, an induction coil that has a low internal resistance is essentially a short circuit for direct current; however, when the current is changing, a back emf is generated in an inductor that is proportional to dI/dt. At high frequencies, the back emf is large and the inductor acts like an open circuit.

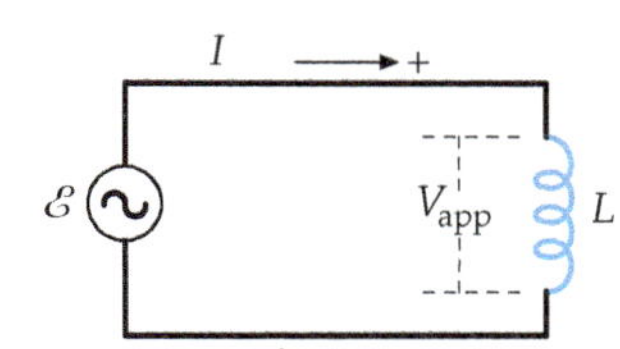

FIGURE 29-6 An ac generator in series with an inductor L. The arrow indicates the positive direction along the wire. Note that for a positive value of dI/dt, the potential drop V_L across the inductor is positive.

INDUCTORS IN ALTERNATING-CURRENT CIRCUITS

Figure 29-6 shows an induction coil in series with an ac generator. When the current changes in the inductor, a back emf equal to $L\,dI/dt$ is generated due to the changing flux. Usually this back emf is much greater than the Ir drop due to the resistance r of the coil, so we normally neglect the resistance of the coil.

The potential drop across the inductor V_L is then given by

$$V_L = L\frac{dI}{dt} \qquad 29\text{-}13$$

POTENTIAL DROP ACROSS AN IDEAL INDUCTOR

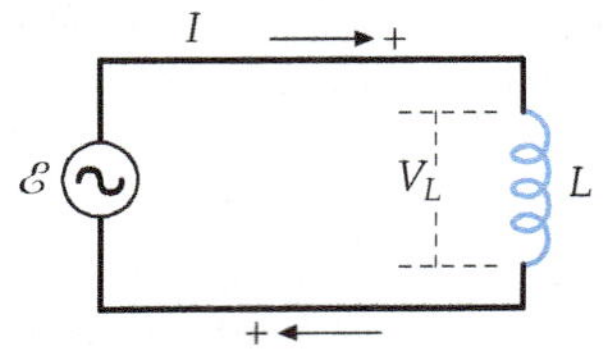

FIGURE 29-6 *(repeated)* An ac generator in series with an inductor L. The arrow indicates the positive direction along the wire. Note that for a positive value of dI/dt, the potential drop V_L across the inductor is positive.

In this circuit, the potential drop V_L across the inductor equals the emf $\mathcal{E}$ of the generator. That is,

$$V_L = \mathcal{E} = \mathcal{E}_{max}\cos\omega t = V_{L\,peak}\cos\omega t$$

where $V_{L\,peak} = \mathcal{E}_{peak}$. Substituting for V_L in Equation 29-13 gives

$$V_{L\,peak}\cos\omega t = L\frac{dI}{dt} \qquad 29\text{-}14$$

Rearranging, we obtain

$$dI = \frac{V_{L\,peak}}{L}\cos\omega t\,dt \qquad 29\text{-}15$$

We solve for the current I by integrating both sides of the equation:

$$I = \frac{V_{L\,peak}}{L}\int\cos\omega t\,dt = \frac{V_{L\,peak}}{\omega L}\sin\omega t + C \qquad 29\text{-}16$$

where the constant of integration C is the dc component of the current. Setting the dc component of the current to be zero, we have

$$I = \frac{V_{L\,peak}}{\omega L}\sin\omega t = I_{peak}\sin\omega t \qquad 29\text{-}17$$

where

$$I_{peak} = \frac{V_{L\,peak}}{\omega L} \qquad 29\text{-}18$$

The potential drop $V_L = V_{L\,peak}\cos\omega t$ across the inductor is 90° out of phase with the current $I = I_{peak}\sin\omega t$. From Figure 29-7, which shows I and V_L as functions of time, we can see that the peak value of the potential drop occurs $\frac{1}{4}T$ earlier in time than the corresponding peak value of the current, where T is the period. The potential drop across an inductor is said to *lead the current by* 90°. We can also conceptually understand this result. The potential drop across the inductor its equal to the emf induced in it. When I is zero but decreasing, dI/dt is at its minimum, which is negative, so the emf induced in the inductor is at its maximum. One-quarter cycle later, I is maximum. At this time, dI/dt is zero, so V_L is zero. Using the trigonometric identity $\sin\theta = \cos(\theta - \frac{\pi}{2})$, Equation 29-17 for the current can be written

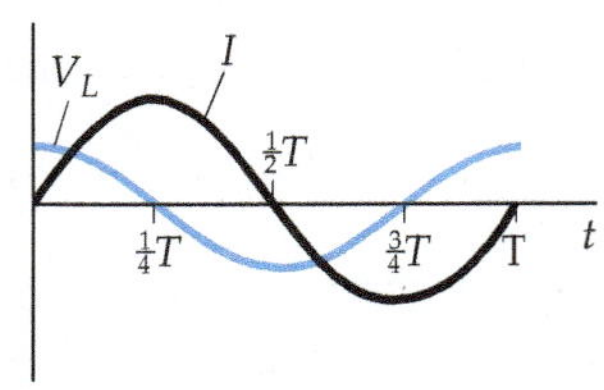

FIGURE 29-7 Current and potential drop across the inductor shown in Figure 29-6 as functions of time. The maximum potential drop occurs one-fourth period before the maximum current. Thus, the potential drop is said to lead the current by one-fourth period or 90°.

$$I = I_{peak}\cos(\omega t - \tfrac{\pi}{2}) \qquad 29\text{-}19$$

The relation between the peak current and the peak potential drop (or between the rms current and rms potential drop) for an inductor can be written in a form similar to $I_{rms} = V_{R\,rms}/R$ (Equation 29-12). From Equation 29-18, we have

$$I_{peak} = \frac{V_{L\,peak}}{\omega L} = \frac{V_{L\,peak}}{X_L} \qquad 29\text{-}20$$

where

$$X_L = \omega L \qquad 29\text{-}21$$

DEFINITION—INDUCTIVE REACTANCE

is called the **inductive reactance.** Because $I_{\text{rms}} = I_{\text{peak}}/\sqrt{2}$ and $V_{L\,\text{rms}} = V_{L\,\text{peak}}/\sqrt{2}$, the rms current is given by

$$I_{\text{rms}} = \frac{V_{L\,\text{rms}}}{X_L} \qquad \text{29-22}$$

Like resistance, inductive reactance has units of ohms. As we can see from Equation 29-22, the larger the reactance for a given rms potential drop, the smaller the rms current. Unlike resistance, the inductive reactance depends on the frequency—the greater the frequency, the greater the reactance.

The *instantaneous* power delivered to the inductor is

$$P = V_L I = (V_{L\,\text{peak}} \cos \omega t)(I_{\text{peak}} \sin \omega t) = V_{L\,\text{peak}} I_{\text{peak}} \cos \omega t \sin \omega t$$

The *average* power delivered to the inductor is zero. We can see this by using the trigonometric identity

$$2 \cos \omega t \sin \omega t = \sin 2\omega t$$

The value of $\sin 2\omega t$ oscillates twice during each cycle of the current, and so is negative as often as it is positive. Thus, when averaged over an integral number of cycles, no energy is delivered to an inductor. (If the internal resistance r of the inductor is not negligible, then the average power delivered is equal to $(I_{\text{rms}})^2 r$.)

Example 29-2 Inductive Reactance

The potential drop across a 40.0-mH inductor is sinusoidal and has an rms potential drop of 120 V. Find the inductive reactance and the rms current when the frequency is (*a*) 60.0 Hz and (*b*) 2000 Hz.

PICTURE We calculate the inductive reactance at each frequency and use Equation 29-20 to find the peak current.

SOLVE

(*a*) 1. The peak current equals the rms potential drop divided by the inductive reactance:

$$I_{\text{rms}} = \frac{V_{L\,\text{rms}}}{X_L}$$

2. Compute the inductive reactance at 60.0 Hz:

$$X_{L1} = \omega_1 L = 2\pi f_1 L = (2\pi)(60.0\text{ Hz})(40.0 \times 10^{-3}\text{ H}) = \boxed{15.1\ \Omega}$$

3. Use this value of X_L to compute the rms current at 60.0 Hz:

$$I_{1\,\text{rms}} = \frac{120\text{ V}}{15.1\ \Omega} = \boxed{7.95\text{ A}}$$

(*b*) 1. Compute the inductive reactance at 2000 Hz:

$$X_{L2} = \omega_2 L = 2\pi f_2 L = (2\pi)(2000\text{ Hz})(40.0 \times 10^{-3}\text{ H}) = \boxed{503\ \Omega}$$

2. Use this value of X_L to compute the rms current at 2000 Hz:

$$I_{2\,\text{rms}} = \frac{120\text{ V}}{503\ \Omega} = \boxed{0.239\text{ A}}$$

CHECK The rms current at 2000 Hz is about 3 percent of the rms current at 60.0 Hz. This result is expected because we expect the inductor to behave more and more like an open circuit as the frequency increases.

CAPACITORS IN ALTERNATING-CURRENT CIRCUITS

When a capacitor is connected across the terminals of an ac generator (Figure 29-8), the voltage drop across the capacitor is

$$V_C = \frac{Q}{C} \qquad 29\text{-}23$$

where Q is the charge on the upper plate of the capacitor.

In this circuit, the potential drop V_C across the capacitor equals the emf $\mathcal{E}$ of the generator. That is,

$$V_C = \mathcal{E}_{\text{peak}} \cos \omega t = V_{C\,\text{peak}} \cos \omega t$$

where $V_{C\,\text{peak}} = \mathcal{E}_{\text{peak}}$. Substituting for V_C in Equation 29-23 and solving for Q gives

$$Q = V_C C = V_{C\,\text{peak}} C \cos \omega t = Q_{\text{peak}} \cos \omega t$$

The current is

$$I = \frac{dQ}{dt} = -\omega Q_{\text{peak}} \sin \omega t = -I_{\text{peak}} \sin \omega t$$

where

$$I_{\text{peak}} = \omega Q_{\text{peak}} \qquad 29\text{-}24$$

Using the trigonometric identity $\sin\theta = -\cos(\theta + \frac{\pi}{2})$, where $\theta = \omega t$, we obtain

$$I = -\omega Q_{\text{peak}} \sin \omega t = I_{\text{peak}} \cos(\omega t + \tfrac{\pi}{2}) \qquad 29\text{-}25$$

The voltage drop V_C across the capacitor is in phase with the charge Q (Equation 29-23), so as with the inductor, the voltage drop across the capacitor is 90° out of phase with the current in the circuit. From Figure 29-9, we see that the maximum value of the potential drop occurs 90° or one-fourth period later in time than the maximum value of the current. Thus, *the potential drop across a capacitor lags the current by* 90°. We can also understand this result in another way. The charge Q is proportional to the potential drop V_C, so the maximum value of $dQ/dt = I$ occurs when the charge Q, and therefore V_C, is zero. As the charge on the capacitor plate increases the current decreases, until one-fourth period later the charge Q, and therefore V_C, is a maximum and the current is zero. The current then becomes negative as the charge Q decreases.

We can relate the current to the potential drop in a form similar to $I_{\text{rms}} = V_{R\,\text{rms}}/R$ (Equation 29-5) for a resistor. From Equation 29-24, we have

$$I_{\text{peak}} = \omega Q_{\text{peak}} = \omega C V_{C\,\text{peak}} = \frac{V_{C\,\text{peak}}}{1/(\omega C)} = \frac{V_{C\,\text{peak}}}{X_C}$$

and, similarly,

$$I_{\text{rms}} = \frac{V_{C\,\text{rms}}}{X_C} \qquad 29\text{-}26$$

where

$$X_C = \frac{1}{\omega C} \qquad 29\text{-}27$$

DEFINITION—CAPACITIVE REACTANCE

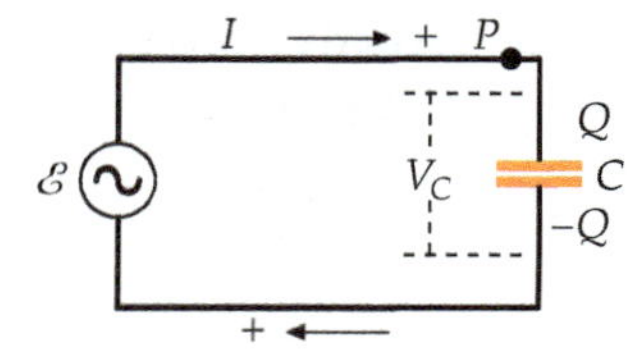

FIGURE 29-8 An ac generator in series with a capacitor C. The positive direction along the circuit is such that when the current is positive the charge Q on the upper capacitor plate is increasing, so the current is related to the charge by $I = +dQ/dt$.

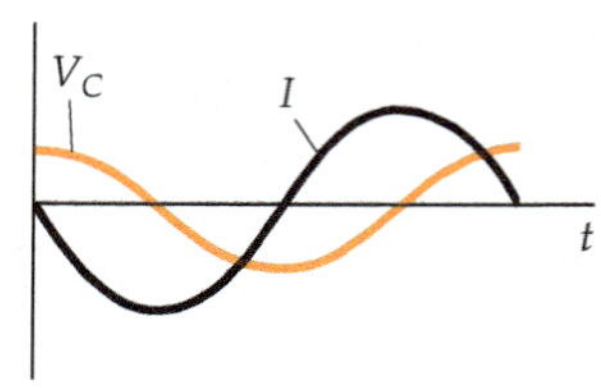

FIGURE 29-9 Current and potential drop across the capacitor shown in Figure 29-8 versus time. The maximum potential drop occurs one-fourth period after the maximum current. Thus, the potential drop is said to lag the current by 90°.

is called the **capacitive reactance** of the circuit. Like resistance and inductive reactance, capacitive reactance has units of ohms and, like inductive reactance, capacitive reactance depends on the frequency of the current. In this case, the greater the frequency, the smaller the reactance. The average power delivered to a capacitor in an ac circuit is zero, as it is for an inductor. This is so because the potential drop is proportional to $\cos \omega t$ and the current is proportional to $\sin \omega t$, and $(\cos \omega t \sin \omega t)_{av} = 0$. Thus, like inductors with no resistance, capacitors dissipate no energy.

Because charge cannot pass across the space between the plates of a capacitor, it may seem strange that there is a continuing alternating current in the circuit shown in Figure 29-8. Suppose we choose the time to be zero at the instant that the voltage drop V_C across the capacitor is both zero and increasing. (At this same instant, the charge Q on the upper plate of the capacitor is also both zero and increasing.) As V_C then increases, positive charge flows off the lower plate and onto the upper plate, and Q reaches its maximum value Q_{peak} a quarter period later. After Q reaches its maximum value Q continues to change, reaching zero at the half-period point, $-Q_{peak}$ at the three-quarter-period point, and zero (again) at the completion of the cycle at the full-period point. The charge Q_{peak} flows past point P (see Figure 29-8) on the wire each quarter period. If we double the frequency, we halve the period. Thus, if we double the frequency we halve the time for the charge Q_{peak} to flow past point P on the wire, so we have doubled the current amplitude I_{peak}. Hence, the greater the frequency, the less the capacitor impedes the flow of charge.

Example 29-3 Capacitive Reactance

A 20.0-μF capacitor is placed across an ac generator that applies a potential drop which has an amplitude (peak value) of 100 V. Find the capacitive reactance and the current amplitude when the frequency is (*a*) 60.0 Hz and (*b*) 6000 Hz.

PICTURE The capacitive reactance is $X_C = 1/(\omega C)$ and the peak current is $I_{peak} = V_{C\,peak}/X_C$.

SOLVE

(*a*) Calculate the capacitive reactance at 60.0 Hz and use this value to find the peak current at 60.0 Hz:

$$X_{C1} = \frac{1}{\omega_1 C} = \frac{1}{2\pi f_1 C} = \frac{1}{2\pi(60.0\text{ Hz})(20.0 \times 10^{-6}\text{ F})} = \boxed{133\ \Omega}$$

$$I_{1\,peak} = \frac{V_{C\,peak}}{X_{C1}} = \frac{100\text{ V}}{133\ \Omega} = \boxed{0.752\text{ A}}$$

(*b*) Calculate the capacitive reactance at 6000 Hz and use this value to find the peak current at 6000 Hz:

$$X_{C2} = \frac{1}{\omega_2 C} = \frac{1}{2\pi f_2 C} = \frac{1}{2\pi(6000\text{ Hz})(20.0 \times 10^{-6}\text{ F})} = \boxed{1.33\ \Omega}$$

$$I_{2\,peak} = \frac{V_{C\,peak}}{X_{C2}} = \frac{100\text{ V}}{1.33\ \Omega} = \boxed{75.2\text{ A}}$$

CHECK The current at 60.0 Hz is about 1 percent of the current at 6000 Hz. This result is expected because we expect the capacitor to be more like an open circuit at lower frequencies.

TAKING IT FURTHER Note that the capacitive reactance is inversely proportional to the frequency, so increasing the frequency by two orders of magnitude decreases the reactance by two orders of magnitude. The current is directly proportional to the frequency, as expected.

* 29-3 THE TRANSFORMER

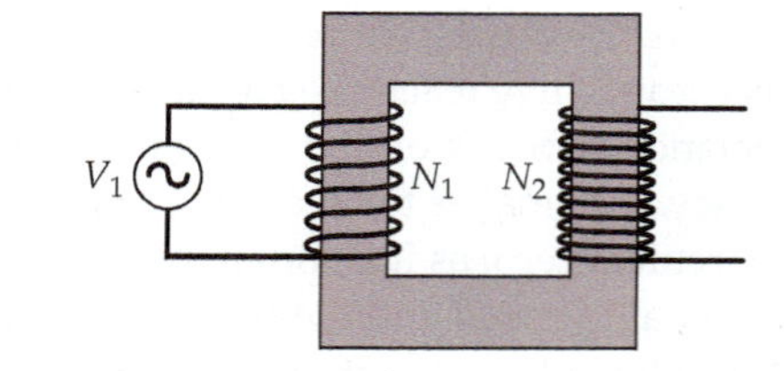

FIGURE 29-10 A transformer that has N_1 turns in the primary and N_2 turns in the secondary.

A transformer is a device used to increase or decrease the voltage in a circuit without an appreciable loss of power. Figure 29-10 shows a simple transformer consisting of two wire coils around a common iron core. The coil carrying the input power is called the primary, and the other coil is called the secondary. Either coil of a transformer can be used for the **primary** or **secondary.** The transformer operates on the principle that an alternating current in one circuit induces an alternating emf in a nearby circuit due to the mutual inductance of the two circuits. The iron core increases the magnetic field for a given current and guides its direction so that the flux linkage between the coils approaches 100%. (At 100% flux linkage, all the magnetic field lines through one coil also go through the other coil.) If no power were lost, the product of the potential drop across and the current in the secondary windings would equal the product of the potential drop across and the current in the primary windings. Thus, if the potential difference across the secondary coil is larger than the potential difference across the primary coil, the current in the secondary coil is smaller than the current in the primary coil, and vice versa. Power losses arise because of Joule heating, both in the two coils and in the iron core currents,* and from hysteresis in the iron core. We will neglect those losses and consider an ideal transformer of 100 percent efficiency, for which all of the power supplied to the primary coil appears in the secondary coil. Actual power distribution transformers often have efficiencies of 98 percent or more.

Consider a transformer with a potential drop V_1 across the primary coil of N_1 turns; the secondary coil of N_2 turns is an open circuit. Because of the iron core, there is a large flux through each coil even when the magnetizing current I_m in the primary circuit is very small. (The magnetizing current is the current in the primary when the secondary circuit is open.) We can ignore the resistances of the coils, which are negligible in comparison with their inductive reactances. The primary circuit is then a simple circuit consisting of an ac generator and a pure inductance, like that discussed in Section 29-2. The magnetizing current in the primary coil and the voltage drop across the primary coil are out of phase by 90°, and the average power dissipated in the primary coil is zero. If ϕ_{turn} is the magnetic flux per turn of the primary coil, the potential drop across the primary coil is equal to the back emf, so

$$V_1 = N_1\frac{d\phi_{turn}}{dt} \qquad \text{29-28}$$

If there is no flux leakage out of the iron core, the flux through each turn is the same for both coils. Thus, the total flux through the secondary coil is $N_2\phi_{turn}$, and the potential difference across the secondary coil is

$$V_2 = N_2\frac{d\phi_{turn}}{dt} \qquad \text{29-29}$$

Comparing Equations 29-28 and 29-29, we can see that

$$V_2 = \frac{N_2}{N_1}V_1 \qquad \text{29-30}$$

If N_2 is greater than N_1, the potential difference across the secondary coil is greater than the potential drop across the primary coil, and the transformer is called a *step-up transformer*. If N_2 is less than N_1, the potential difference across the secondary coil is less than the potential drop across the primary coil, and the transformer is called a *step-down transformer*.

(a)

(b)

(c)

(*a*) A power box that has a transformer for stepping down voltage for distribution to homes. (*b*) A suburban power substation where transformers step down voltage from high-voltage transmission lines. (*c*) A 9-volt ac plug-in transformer. ((*a*) *Yaov/Phototake.* (*b*) *Daniel S. Brody/Stock Boston.* (*c*) *Ramòn Rivera Moret.*)

* The induced currents, called eddy currents, can be greatly reduced by using a core of laminated metal to break up current paths.

When we put a resistance R, called a *load resistance*, across the secondary coil, there will then be a current I_2 in the secondary circuit that is in phase with the potential drop V_2 across the resistance. This current sets up an additional flux ϕ'_{turn} through each turn that is proportional to N_2I_2. This flux opposes the original flux set up by the original magnetizing current I_m in the primary. However, the potential drop across the primary coil is determined by the generator emf, which is unaffected by the secondary circuit. According to Equation 29-29, the flux in the iron core must change at the original rate; that is, the total flux in the iron core must be the same as when no load exists across the secondary. The primary coil thus draws an additional current I_1 to maintain the original flux ϕ_{turn}. The flux through each turn produced by this additional current is proportional to N_1I_1. Because this flux equals $-\phi'_{\text{turn}}$, the additional current I_1 in the primary is related to the current I_2 in the secondary by

$$N_1I_1 = -N_2I_2 \qquad \text{29-31}$$

The currents are 180° out of phase and produce counteracting fluxes. Because I_2 is in phase with V_2, the additional current I_1 is in phase with the potential drop across the primary circuit. The power input from the generator is $V_{1\,\text{rms}}I_{1\,\text{rms}}$, and the power output is $V_{2\,\text{rms}}I_{2\,\text{rms}}$. (The magnetizing current does not contribute to the power input because it is 90° out of phase with the generator voltage.) If there are no losses,

$$V_{1\,\text{rms}}I_{1\,\text{rms}} = V_{2\,\text{rms}}I_{2\,\text{rms}} \qquad \text{29-32}$$

In most cases, the additional current in the primary I_1 is much greater than the original magnetizing current I_m that is drawn from the generator when no load exists. This can be demonstrated by putting a lightbulb in series with the primary coil. The lightbulb is much brighter when there is a load across the secondary circuit than when the secondary circuit is open. If I_m can be neglected, Equation 29-32 relates the total currents in the primary and secondary circuits.

Example 29-4 Doorbell Transformer

A doorbell requires 0.40 A rms of alternating current at 6.0 V rms. It is connected to a transformer whose primary has 2000 turns and is connected to a 120-V-rms ac line. (*a*) How many turns should there be in the secondary? (*b*) What is the current in the primary?

PICTURE We can find the number of turns from the turns ratio, which equals the voltage ratio. The primary current can be found by equating the power out to the power in.

SOLVE

(*a*) The turns ratio can be obtained from Equation 29-30. Solve for the number of turns in the secondary, N_2:

$$\frac{N_2}{N_1} = \frac{V_2}{V_1}$$

so

$$N_2 = \frac{V_{2\,\text{rms}}}{V_{1\,\text{rms}}}N_1 = \frac{6.0\text{ V}}{120\text{ V}}2000\text{ turns} = \boxed{100\text{ turns}}$$

(*b*) Because we are assuming 100 percent efficiency in power transmission, the input and output currents are related by Equation 29-32. Solve for the current in the primary, I_1:

$$V_{2\,\text{rms}}I_{2\,\text{rms}} = V_{1\,\text{rms}}I_{1\,\text{rms}}$$

so

$$I_{1\,\text{rms}} = \frac{V_{2\,\text{rms}}}{V_{1\,\text{rms}}}I_{2\,\text{rms}} = \frac{6.0\text{ V}}{120\text{ V}}(0.40\text{ A}) = \boxed{0.020\text{ A}}$$

CHECK To step-down the voltage requires fewer turns in the secondary than in the primary. In addition, a transformer that steps down the voltage steps up the current. Our results reflect both of these attributes.

An important use of transformers is in the transmission and distribution of electrical power. To minimize the I^2R heat loss (Joule heating) in transmission lines, it is economical to use a high voltage and a low current. On the other hand, safety and other considerations require that power be delivered to consumers at lower voltages and therefore with higher currents. Suppose, for example, that each person in a city with a population of 50000 uses consumes electrical energy at a rate of 1.2 kW. (The per capita rate of consumption of electrical energy in the United States is actually somewhat higher than this value.) At 120 V, the current required for each person would be

$$I = \frac{1200\text{ W}}{120\text{ V}} = 10\text{ A}$$

The total current for 50000 people would then be 500000 A. The transport of such a current from a power-plant generator to a city many kilometers away would require conductors of enormous thickness, and the I^2R power loss would be substantial. Rather than transmit the power at 120 V, step-up transformers are used at power plants to step up the voltage to values as great as 600000 V. For this voltage, the current needed is only

$$I = \frac{120\text{ V}}{600\,000\text{ V}}(500\,000\text{ A}) = 100\text{ A}$$

To reduce the voltage to a safer level for transport within a city, power substations are located just outside the city to step down the voltage to a safer value, such as 10000 V. Transformers in boxes attached to the power poles outside each house again step down the voltage to 120 V (or 240 V) for distribution to the house. Because of the ease of stepping the voltage up or down with transformers, alternating current rather than direct current is in common use.

Example 29-5 Transmission Losses

A transmission line has a resistance of 0.020 Ω/km. Calculate the power loss due to Joule heating if 200 kW of power is transmitted from a power generator to a city 10 km away at (*a*) 240 V rms and (*b*) 4.4 kV rms.

PICTURE First, note that the total resistance of 10 km of wire is $R = (0.020\ \Omega/\text{km})(10\text{ km}) = 0.20\ \Omega$. In each case, begin by finding the current needed to transmit 200 kW using $P = IV$, then find the power loss using $(I_{\text{rms}})^2R$. In the solution, the voltages and currents are rms values and the power is the average power.

SOLVE

(*a*) 1. Find the current needed to transmit 200 kW of power at 240 V:

$$I = \frac{P}{V} = \frac{200\text{ kW}}{240\text{ V}} = 833\text{ A}$$

2. Calculate the power loss:

$$I^2R = (833\text{ A})^2(0.20\ \Omega) = \boxed{1.4 \times 10^2\text{ kW}}$$

(*b*) 1. Now, find the current needed to transmit 200 kW of power at 4.4 kV:

$$I = \frac{P}{V} = \frac{200\text{ kW}}{4.4\text{ kV}} = 45.4\text{ A}$$

2. Calculate the power loss:

$$I^2R = (45.4\text{ A})^2(0.20\ \Omega) = \boxed{0.41\text{ kW}}$$

CHECK The power loss at 4.4 kV is less than one percent of the power loss at 240 V. This result is consistent with the reason for stepping up the voltage for transmission.

TAKING IT FURTHER Note that with a transmission voltage of 240 V, almost 70 percent of the power is wasted through heat loss. In addition, there is an *IR* (voltage) drop across the transmission line of 167 V, so the power is delivered at only 73 V. However, with transmission at 4.4 kV, only about 0.2 percent of the power is lost during transmission and there is an *IR* drop across the transmission line of only 9 V, so the power is delivered with only a 0.2 percent voltage drop.

* 29-4 *LC* AND *RLC* CIRCUITS WITHOUT A GENERATOR

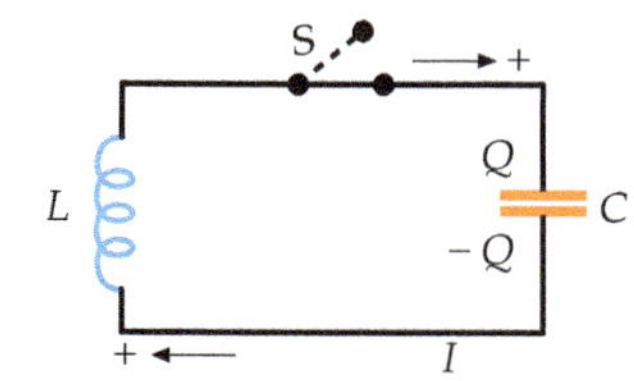

FIGURE 29-11 An LC circuit. When the switch is closed, the initially charged capacitor discharges through the inductor, producing a back emf.

Figure 29-11 shows a simple circuit that has inductance and capacitance but has no resistance. Such a circuit is called an ***LC* circuit.** We assume that the upper capacitor plate carries an initial positive charge Q_0 and that the switch is initially open. After the switch is closed at $t = 0$, the charge begins to flow through the inductor. Let Q be the charge on the upper plate of the capacitor and let the positive direction around the circuit be clockwise, as shown. Then,

$$I = +\frac{dQ}{dt}$$

Applying Kirchhoff's loop rule to the circuit, we have

$$L\frac{dI}{dt} + \frac{Q}{C} = 0 \qquad 29\text{-}33$$

Substituting dQ/dt for I gives

$$L\frac{d^2Q}{dt^2} + \frac{1}{C}Q = 0 \qquad 29\text{-}34$$

This equation is of the same form as Equation 14-2 for the acceleration of a mass on a spring:

$$m\frac{d^2x}{dt^2} + kx = 0$$

The behavior of an LC circuit is thus analogous to that of a mass on a spring—L analogous to the mass m, Q analogous to the position x, and $1/C$ analogous to the spring constant k. Also, the current I is analogous to the velocity v, because $\omega = dx/dt$ and $I = dQ/dt$. In mechanics, the mass of an object describes the inertia of the object. The greater the mass, the more opposition there is to change the velocity of the object. Similarly, the inductance L can be thought of as the inertia of an ac circuit. The greater the inductance, the more opposition there is to changes in the current I.

If we divide each term in Equation 29-34 by L and rearrange, we obtain

$$\frac{d^2Q}{dt^2} = -\frac{1}{LC}Q \qquad 29\text{-}35$$

which is analogous to

$$\frac{d^2x}{dt^2} = -\frac{k}{m}x \qquad 29\text{-}36$$

In Chapter 14, we found that we could write the solution of Equation 29-36 for simple harmonic motion in the form

$$x = A\cos(\omega t - \delta)$$

where $\omega = \sqrt{k/m}$ is the angular frequency, A is the displacement amplitude, and δ is the phase constant, which depends on the initial conditions. The solution to Equation 29-35 is thus

$$Q = A\cos(\omega t - \delta)$$

with

$$\omega = \frac{1}{\sqrt{LC}} \qquad 29\text{-}37$$

The current I is found by differentiating:

$$I = \frac{dQ}{dt} = -\omega A \sin(\omega t - \delta)$$

If we choose our initial conditions to be $Q = Q_{\text{peak}}$ and $I = 0$ at $t = 0$, the phase constant δ is zero and $A = Q_{\text{peak}}$. Our solutions are then

$$Q = Q_{\text{peak}} \cos \omega t \qquad \text{29-38}$$

and

$$I = -\omega Q_{\text{peak}} \sin \omega t = -I_{\text{peak}} \sin \omega t \qquad \text{29-39}$$

where $I_{\text{peak}} = \omega Q_{\text{peak}}$.

Figure 29-12 shows graphs of Q and I versus time. The charge oscillates between the values $+Q_{\text{peak}}$ and $-Q_{\text{peak}}$ with angular frequency $\omega = 1/\sqrt{LC}$. The current oscillates between $+\omega Q_{\text{peak}}$ and $-\omega Q_{\text{peak}}$ with the same frequency. Also, the charge lags behind the current by 90° (see Problem 29-33). The current is maximum when the charge is zero and the current is zero when the charge is maximum.

In our study of the oscillations of a mass on a spring, we found that the total energy is constant, and that the total energy oscillates between potential energy and kinetic energy. We also have two kinds of energy in the LC circuit—electric energy and magnetic energy. The electric energy stored in the capacitor is

$$U_{\text{e}} = \frac{1}{2}QV_C = \frac{1}{2}\frac{Q^2}{C}$$

Substituting $Q_{\text{peak}} \cos \omega t$ for Q, we have for the electric energy

$$U_{\text{e}} = \frac{1}{2}\frac{Q_{\text{peak}}^2}{C}\cos^2\omega t \qquad \text{29-40}$$

The electric energy oscillates between its maximum value $Q_{\text{peak}}^2/(2C)$ and zero at an angular frequency of 2ω (see Problem 29-33). The magnetic energy stored in the inductor is

$$U_{\text{m}} = \frac{1}{2}LI^2 \qquad \text{29-41}$$

Substituting $I = -\omega Q_{\text{peak}} \sin \omega t$ (Equation 29-39), we get

$$U_{\text{m}} = \frac{1}{2}L\omega^2 Q_{\text{peak}}^2 \sin^2\omega t = \frac{1}{2}\frac{Q_{\text{peak}}^2}{C}\sin^2\omega t \qquad \text{29-42}$$

where we have used $\omega^2 = 1/LC$ (Equation 29-37). The magnetic energy also oscillates between its maximum value of $Q_{\text{peak}}^2/2C$ and zero at an angular frequency of 2ω. The sum of the electrostatic energy and the magnetic energy is the total energy, which is constant in time:

$$U_{\text{total}} = U_{\text{e}} + U_{\text{m}} = \frac{1}{2}\frac{Q_{\text{peak}}^2}{C}\cos^2\omega t + \frac{1}{2}\frac{Q_{\text{peak}}^2}{C}\sin^2\omega t = \frac{1}{2}\frac{Q_{\text{peak}}^2}{C}$$

This sum equals the energy initially stored on the capacitor.

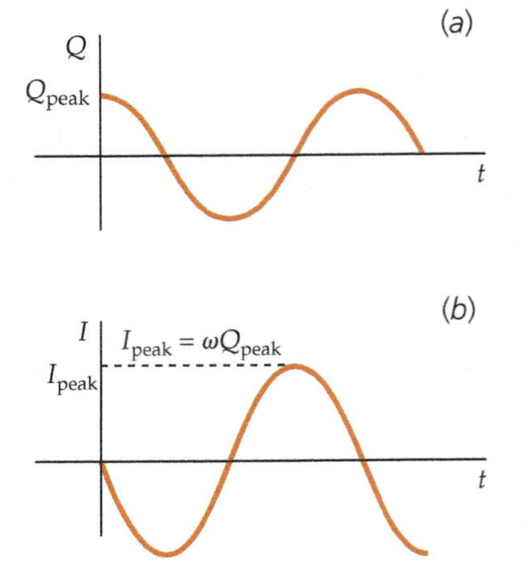

FIGURE 29-12 Graphs of (*a*) Q versus t and (*b*) I versus t for the LC circuit shown in Figure 29-11.

Example 29-6 *LC* Oscillator

A 2.0-μF capacitor is charged to 20 V and the capacitor is then connected across a 6.0-μH inductor. (*a*) What is the frequency of oscillation? (*b*) What is the peak value of the current?

PICTURE In (*b*), the current is maximum when dQ/dt is maximum, so the current amplitude is ωQ_{peak}. Also, $Q = Q_{\text{peak}}$ when $V = V_{\text{peak}}$, where V is the voltage across the capacitor.

SOLVE

(*a*) The frequency of oscillation depends only on the values of the capacitance and the inductance:

$$f = \frac{\omega}{2\pi} = \frac{1}{2\pi\sqrt{LC}} = \frac{1}{2\pi\sqrt{(6.0 \times 10^{-6}\,\text{H})(2.0 \times 10^{-6}\,\text{F})}}$$

$$= \boxed{4.6 \times 10^4\ \text{Hz}}$$

(*b*) 1. The peak value of the current is related to the peak value of the charge:

$$I_{\text{peak}} = \omega Q_{\text{peak}} = \frac{Q_{\text{peak}}}{\sqrt{LC}}$$

2. The peak charge on the capacitor is related to the peak potential drop across the capacitor:

$$Q_{\text{peak}} = CV_{\text{peak}}$$

3. Substitute CV_{peak} for Q_{peak} and calculate I_{peak}:

$$I_{\text{peak}} = \frac{CV_{\text{peak}}}{\sqrt{LC}} = \frac{V_{\text{peak}}}{\sqrt{L/C}}$$

$$= \frac{(20\ \text{V})}{\sqrt{(6.0\ \mu\text{H})/(2.0\ \mu\text{F})}} = \boxed{12\ \text{A}}$$

PRACTICE PROBLEM 29-2 A 5.0-μF capacitor is charged and is then discharged through an inductor. What should the value of the inductance be so that the current oscillates with frequency 8.0 kHz?

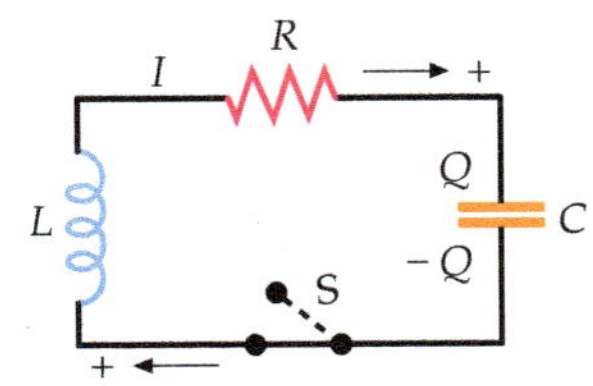

FIGURE 29-13 An *RLC* circuit.

If we include a resistor in series with the capacitor and the inductor, as in Figure 29-13, we have an ***RLC* circuit.** Kirchhoff's loop rule gives

$$L\frac{dI}{dt} + IR + \frac{Q}{C} = 0 \qquad \text{29-43}a$$

or

$$L\frac{d^2Q}{dt^2} + R\frac{dQ}{dt} + \frac{1}{C}Q = 0 \qquad \text{29-43}b$$

where we have used $I = dQ/dt$ as before. Equations 29-43*a* and 29-43*b* are analogous to the equation for a damped harmonic oscillator (see Equation 14-38):

$$m\frac{d^2x}{dt^2} + b\frac{dx}{dt} + kx = 0$$

The first term, $L\,dI/dt = L\,d^2Q/dt^2$, is analogous to the mass multiplied by the acceleration, $m\,dv/dt = m\,d^2x/dt^2$; the second term, $IR = R\,dQ/dt$, is analogous to the damping term, $bv = b\,dx/dt$; and the third term, Q/C, is analogous to the restoring force kx. In the oscillation of a mass on a spring, the damping constant b leads to a dissipation of energy. In an *RLC* circuit, the resistance R is analogous to the damping constant b and leads to a dissipation of energy.

If the resistance is small, the charge and the current oscillate with (angular) frequency* that is very nearly equal to $\omega_0 = 1/\sqrt{LC}$, which is called the **natural frequency** of the circuit, but the oscillations are damped. We can understand this qualitatively from energy considerations. If we multiply each term in Equation 29-43*a* by the current I, we obtain

$$LI\frac{dI}{dt} + I^2R + I\frac{Q}{C} = 0 \qquad \text{29-44}$$

* As in Chapter 14 when we discussed mechanical oscillations, we usually omit the word *angular* when the omission will not cause confusion.

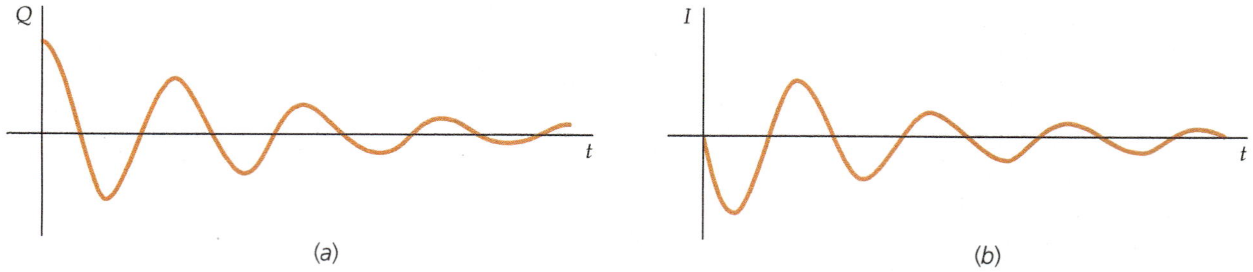

FIGURE 29-14 Graphs of (*a*) Q versus t and (*b*) I versus t for the RLC circuit shown in Figure 29-13 when the value of R is small enough so that the oscillations are underdamped.

The magnetic energy in the inductor is given by $\frac{1}{2}LI^2$ (see Equation 28-21). Note that

$$\frac{d\left(\frac{1}{2}LI^2\right)}{dt} = LI\frac{dI}{dt}$$

where $LI\,dI/dt$ is the first term in Equation 29-44. If $LI\,dI/dt$ is positive, it equals the rate at which electrical potential energy is transformed into magnetic energy. If $LI\,dI/dt$ is negative, it equals the rate at which magnetic energy is transformed back into electrical potential energy. Note that $LI\,dI/dt$ is positive or negative depending on whether I and dI/dt have the same sign or different signs. The second term in Equation 29-44 is I^2R, the rate at which electrical potential energy is dissipated in the resistor. I^2R is never negative. Note that

$$\frac{d\left(\frac{1}{2}Q^2/C\right)}{dt} = \frac{Q}{C}\frac{dQ}{dt} = I\frac{Q}{C}$$

where IQ/C is the third term in Equation 29-44. This result is the rate of change of the electric potential energy of the capacitor, which may be positive or negative. The sum of the electric and magnetic energies is not constant for this circuit because energy is continually dissipated in the resistor. Figure 29-14 shows graphs of Q versus t and I versus t for a small resistance R in an RLC circuit. If we increase R, the oscillations become more heavily damped until a critical value of R is reached for which not even one oscillation exists. Figure 29-15 shows a graph of Q versus t in an RLC circuit when the value of R is greater than the critical damping value.

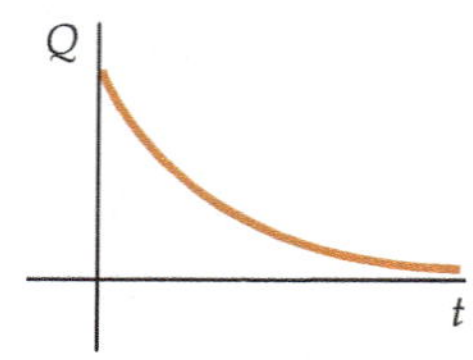

FIGURE 29-15 A graph of Q versus t for the RLC circuit shown in Figure 29-13 when the value of R is so large that the oscillations are overdamped.

* 29-5 PHASORS

Until this point, the circuits considered contained an ideal ac generator and only a single passive element (for example, resistor, inductor, or capacitor). In those circuits, the potential drop across the passive element equaled the emf of the generator. In circuits that contain an ideal ac generator and two or more additional elements connected in series, the sum of the potential drops across the elements at a given instant is equal to the generator emf at that instant; this is the same as with dc circuits. However, in series ac circuits the potential drops typically are not in phase, so the sum of their rms values does not equal the rms value of the generator emf.

Two-dimensional vectors, which are called **phasors,** can represent the phase relations between the current and the potential drops across resistors, capacitors, or inductors. In Figure 29-16, the potential drop across a resistor V_R is represented by a vector $\vec{V}_R$ that has magnitude $I_{\text{peak}}R$ and makes an angle θ with the x axis. This potential drop is in phase with the current. The current in a steady-state ac circuit varies with time, as

$$I = I_{\text{peak}}\cos\theta = I_{\text{peak}}\cos(\omega t - \delta) \qquad 29\text{-}45$$

where ω is the angular frequency and δ is some phase constant. The potential drop across a resistor is then given by

$$V_R = IR = I_{\text{peak}}R\cos(\omega t - \delta) \qquad 29\text{-}46$$

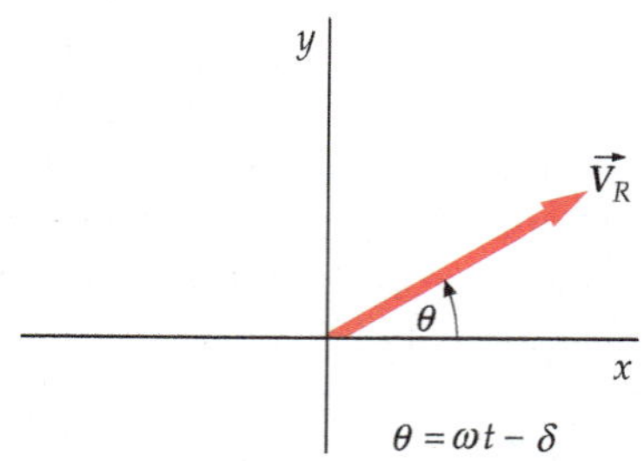

FIGURE 29-16 The potential drop across a resistor can be represented by a vector $\vec{V}_R$, which is called a phasor, that has magnitude $I_{\text{peak}}R$ and makes an angle $\theta = \omega t - \delta$ with the x axis. The phasor rotates with an angular frequency ω. The potential drop $V_R = IR$ is the x component of $\vec{V}_R$.

The potential drop across a resistor is thus equal to the x component of the phasor vector $\vec{V}_R$, which rotates counterclockwise with an angular frequency ω. The current I may be written as the x component of a phasor $\vec{I}$ having the same direction as $\vec{V}_R$.

When several components are connected together in a series combination, their potential drops add. When several components are connected in parallel, their currents add. Unfortunately, adding sines or cosines of different amplitudes and phases algebraically is awkward. It is much easier to do this by vector addition.*

Let us look at how phasors are used. Any ac current or any potential drop is written in the form $A\cos(\omega t - \delta)$, which in turn is treated as A_x, the x component of a phasor that makes an angle $(\omega t - \delta)$ with the $+x$ direction. Instead of adding two potential drops or currents algebraically, as $A\cos(\omega t - \delta_1) + B\cos(\omega t - \delta_2)$, we represent the quantities as phasors $\vec{A}$ and $\vec{B}$ and find the phasor sum $\vec{C} = \vec{A} + \vec{B}$ geometrically. The resultant potential drop or current is then the x component of the resultant phasor, $C_x = A_x + B_x$. The geometric representation conveniently shows the relative amplitudes and phases of the phasors.

Consider an ac circuit that contains an inductor L, a capacitor C, and a resistor R connected in series. They all carry the same current, which is represented as the x component of the current phasor $\vec{I}$. The potential drop across the resistor V_R is represented by a phasor $\vec{V}_R$ that has magnitude $I_{\text{peak}}R$ and is in phase with the current phasor $\vec{I}$. The potential drop across the inductor V_L is represented by a phasor $\vec{V}_L$ that has magnitude $I_{\text{peak}}X_L$ and leads the current phasor $\vec{I}$ by 90°. Similarly, the potential drop across the capacitor V_C is represented by a phasor $\vec{V}_C$ that has magnitude $I_{\text{peak}}X_C$ and lags $\vec{I}$ by 90°. Figure 29-17 shows the phasors $\vec{V}_R$, $\vec{V}_L$, $\vec{V}_C$, and $\vec{V}_{\text{app}}$, where the x component of V_{app} is the potential drop across the series combination. The phasors all rotate counterclockwise with an angular frequency ω. At any instant in time, the instantaneous value of the potential drop across any of these elements equals the x component of the corresponding phasor.

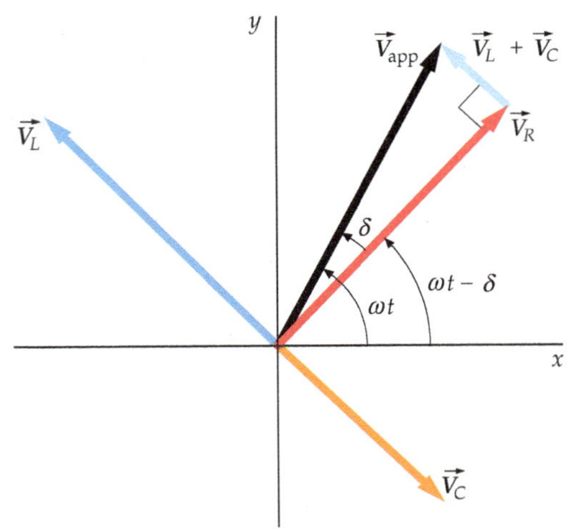

FIGURE 29-17 Phasor representations of the potential drops V_R, V_L, and V_C. Each vector rotates in the counterclockwise direction with an angular frequency ω. At any instant, the potential drop across an element equals the x component of the corresponding phasor, and the potential drop V_{app} across the *RLC*-series combination, which equals the sum of the potential drops, equals the x component of the vector sum $\vec{V}_R + \vec{V}_L + \vec{V}_C$.

* 29-6 DRIVEN *RLC* CIRCUITS

SERIES *RLC* CIRCUIT

Figure 29-18 shows a series *RLC* circuit being sinusoidally driven by an ac generator. If the potential drop applied by the generator to the series *RLC* combination is $V_{\text{app}} = V_{\text{app peak}}\cos\omega t$, applying Kirchhoff's loop rule gives

$$V_{\text{app peak}}\cos\omega t - L\frac{dI}{dt} - IR - \frac{Q}{C} = 0$$

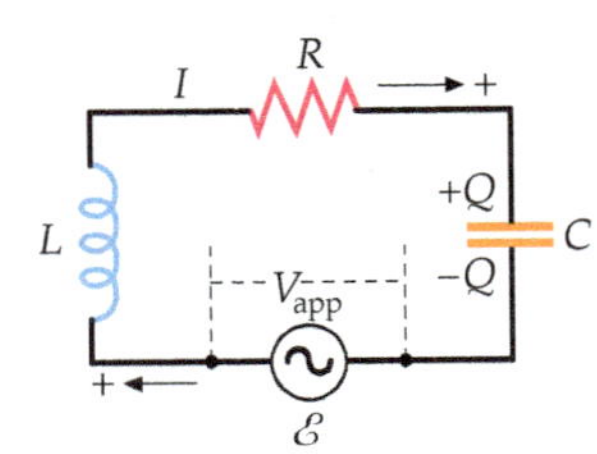

FIGURE 29-18 A series *RLC* circuit with an ac generator.

Using $I = dQ/dt$ and rearranging, we obtain

$$L\frac{d^2Q}{dt^2} + R\frac{dQ}{dt} + \frac{1}{C}Q = V_{\text{app peak}}\cos\omega t \qquad \text{29-47}$$

This equation is analogous to Equation 14-53 for the forced oscillation of a mass on a spring:

$$m\frac{d^2x}{dt^2} + b\frac{dx}{dt} + m\omega_0^2 x = F_0\cos\omega t$$

(In Equation 14-53, the force constant k was written in terms of the mass m and the natural angular frequency ω_0 using $k = m\omega_0^2$. The capacitance in Equation 29-47 could be similarly written in terms of L and the natural angular frequency using $1/C = L\omega_0^2$.)

---

* It is also easier to do using complex numbers.

We will discuss the solution of Equation 29-47 qualitatively as we did with Equation 14-53 for the forced oscillator. The current in the circuit consists of a transient current that depends on the initial conditions (for example, the initial phase of the generator and the initial charge on the capacitor) and a steady-state current that does not depend on the initial conditions. We will ignore the transient current, which decreases exponentially with time and is eventually negligible, and concentrate on the steady-state current. The steady-state current obtained by solving Equation 29-47 is

$$I = I_{\text{peak}} \cos(\omega t - \delta) \qquad 29\text{-}48$$

where the phase angle δ is given by

$$\tan \delta = \frac{X_L - X_C}{R} \qquad 29\text{-}49$$

PHASE CONSTANT FOR A SERIES *RLC* CIRCUIT

The peak current is

$$I_{\text{peak}} = \frac{V_{\text{app peak}}}{\sqrt{R^2 + (X_L - X_C)^2}} = \frac{V_{\text{app peak}}}{Z} \qquad 29\text{-}50$$

PEAK CURRENT IN A SERIES *RLC* CIRCUIT

where

$$Z = \sqrt{R^2 + (X_L - X_C)^2} \qquad 29\text{-}51$$

IMPEDANCE OF A SERIES *RLC* CIRCUIT

The quantity $X_L - X_C$ is called the **total reactance,** and Z is called the **impedance.** Combining these results, we have

$$I = \frac{V_{\text{app peak}}}{Z} \cos(\omega t - \delta) \qquad 29\text{-}52$$

Equation 29-52 can also be obtained from a simple diagram using vectors called phasors. Figure 29-19 shows the phasors representing the potential drops across the resistance, the inductance, and the capacitance. The x component of each of these vectors equals the instantaneous potential drop across the corresponding element. Because the sum of the x components equals the x component of the sum, the sum of the x components equals the sum of the potential drops across these elements, which by Kirchhoff's loop rule equals the instantaneous applied potential drop.

If we represent the potential drop applied across the series combination $V_{\text{app}} = V_{\text{app peak}} \cos \omega t$ as a phasor $\vec{V}_{\text{app}}$ that has the magnitude $V_{\text{app peak}}$, we have

$$\vec{V}_{\text{app}} = \vec{V}_R + \vec{V}_L + \vec{V}_C \qquad 29\text{-}53$$

In terms of the magnitudes,

$$V_{\text{app peak}} = |\vec{V}_R + \vec{V}_L + \vec{V}_C| = \sqrt{V_{R\,\text{peak}}^2 + (V_{L\,\text{peak}} - V_{C\,\text{peak}})^2}$$

But $V_R = I_{\text{peak}} R$, $V_L = I_{\text{peak}} X_L$, and $V_C = I_{\text{peak}} X_C$. Thus,

$$V_{\text{app peak}} = I_{\text{peak}} \sqrt{R^2 + (X_L - X_C)^2} = I_{\text{peak}} Z$$

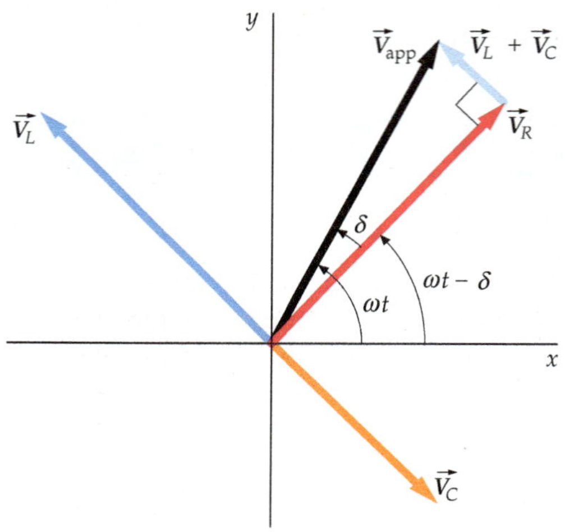

FIGURE 29-19 Phase relations among potential drops in a series *RLC* circuit. The potential drop across the resistor is in phase with the current. The potential drop across the inductor V_L leads the current by 90°. The potential drop across the capacitor lags the current by 90°. The sum of the vectors representing the potential drops gives a vector at an angle δ with the current representing the applied emf. For the case shown here, V_L is greater than V_C, and the current lags the applied potential drop by δ.

The phasor $\vec{V}_{app}$ makes an angle δ with $\vec{V}_R$, as shown in Figure 29-19. From the figure, we can see that

$$\tan\delta = \frac{|\vec{V}_L + \vec{V}_C|}{|\vec{V}_R|} = \frac{I_{peak}X_L - I_{peak}X_C}{I_{peak}R} = \frac{X_L - X_C}{R}$$

in agreement with Equation 29-49. Because $\vec{V}_{app}$ makes an angle ωt with the x axis, $\vec{V}_R$ makes an angle $\omega t - \delta$ with the x axis. This applied potential drop is in phase with the current, which is therefore given by

$$I = I_{peak}\cos(\omega t - \delta) = \frac{V_{app\ peak}}{Z}\cos(\omega t - \delta)$$

This is Equation 29-52. The relation between the impedance Z, the resistance R, and the total reactance $X_L - X_C$ is best remembered by using the right triangle shown in Figure 29-20.

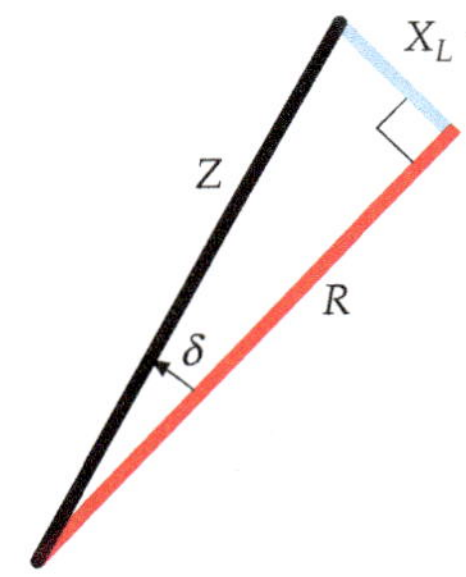

FIGURE 29-20 A right triangle relating capacitive and inductive reactance, resistance, impedance, and the phase angle in an *RLC* circuit.

RESONANCE

When X_L and X_C are equal, the total reactance is zero, and the impedance Z has its smallest value R. Then I_{peak} has its greatest value and the phase angle δ is zero, which means that the current is in phase with the applied potential drop. Let ω_{res} be the value of ω for which X_L and X_C are equal. It is obtained from

$$X_L = X_C$$

$$\omega_{res}L = \frac{1}{\omega_{res}C}$$

or

$$\omega_{res} = \frac{1}{\sqrt{LC}}$$

which equals the natural frequency ω_0. When the frequency of the applied potential drop equals the natural frequency ω_0, the impedance is smallest, I_{peak} is greatest, and the circuit is said to be at **resonance.** The natural frequency ω_0 is therefore also called the **resonance frequency.** This resonance condition in a driven *RLC* circuit is similar to that in a driven simple harmonic oscillator.

Because neither an inductor nor a capacitor dissipates energy, the average power delivered to a series *RLC* circuit is the average power supplied to the resistor. The instantaneous power supplied to the resistor is

$$P = I^2R = [I_{peak}\cos(\omega t - \delta)]^2R$$

Averaging over one or more cycles and using $(\cos^2\theta)_{av} = \frac{1}{2}$, we obtain for the average power

$$P_{av} = \frac{1}{2}I_{peak}^2R = (I_{rms})^2R \qquad 29\text{-}54$$

Using $R/Z = \cos\delta$ from Figure 29-20 and $I_{peak} = V_{app\ peak}/Z$, this can be written

$$P_{av} = \frac{1}{2}V_{app\ peak}I_{peak}\cos\delta = V_{app\ rms}I_{rms}\cos\delta \qquad 29\text{-}55$$

The quantity $\cos\delta$ is called the **power factor** of the *RLC* circuit. At resonance, δ is zero, and the power factor is 1.

The power can also be expressed as a function of the angular frequency ω. Using $I_{rms} = V_{app\ rms}/Z$, Equation 29-54 becomes

$$P_{av} = (I_{rms})^2 = (V_{app\ rms})^2\frac{R}{Z^2}$$

From the definition of impedance Z, we have

$$Z^2 = (X_L - X_C)^2 + R^2 = \left(\omega L - \frac{1}{\omega C}\right)^2 + R^2$$
$$= \frac{L^2}{\omega^2}\left(\omega^2 - \frac{1}{LC}\right)^2 + R^2$$
$$= \frac{L^2}{\omega^2}(\omega^2 - \omega_0^2)^2 + R^2$$

where we have used $\omega_0 = 1/\sqrt{LC}$. Using this expression for Z^2, we obtain the average power as a function of ω:

$$P_{av} = \frac{(V_{app\ rms})^2 R\omega^2}{L^2(\omega^2 - \omega_0^2)^2 + \omega^2 R^2} \qquad 29\text{-}56$$

Figure 29-21 shows the average power supplied by the generator to the series combination as a function of generator frequency for two different values of the resistance R. These curves, called **resonance curves,** are the same as the power-versus-frequency curves for a driven damped oscillator (see Section 14-5). The average power is greatest when the generator frequency equals the resonance frequency. When the resistance is small, the resonance curve is narrow; when the resistance is large, the resonance curve is broad. A resonance curve can be characterized by the **resonance width** $\Delta\omega$. As shown in Figure 29-21, the resonance width is the frequency difference between the two points on the curve where the power is half its maximum value. When the width is small compared with the resonance frequency, the resonance is sharp; that is, the resonance curve is narrow.

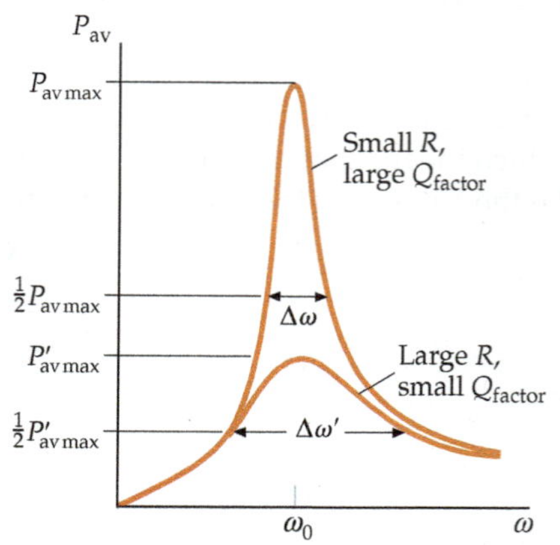

FIGURE 29-21 Plot of average power versus frequency for a series RLC circuit. The power is maximum when the frequency of the generator ω equals the natural frequency of the circuit $\omega_0 = 1/\sqrt{LC}$. If the resistance is small, the Q factor is large and the resonance is sharp. The resonance width $\Delta\omega$ of the curves is measured between points where the power is half its maximum value.

The Q factor for a mechanical oscillator is defined as $Q_{factor} = \omega_0 m/b$ (Equations 14-42 and 14-45), where m is the mass and b is the damping constant. We then saw that for weakly-damped oscillator $Q_{factor} = 2\pi E/|\Delta E|$, where E is the total energy of the system at the beginning of a cycle and ΔE is the energy dissipated during the cycle. The **Q factor** for an RLC circuit can be defined in a similar way. Because L is analogous to the mass m and R is analogous to the damping constant b, the Q factor for an RLC circuit that has a small resistance is given by

$$Q_{factor} = 2\pi\left(\frac{E}{|\Delta E|}\right)_{cycle} = \frac{\omega_0 L}{R} \qquad 29\text{-}57$$

When the resonance curve is reasonably narrow (that is, when Q is greater than about 2 or 3), the Q factor can be approximated by

$$Q_{factor} = \frac{\omega_0}{\Delta\omega} = \frac{f_0}{\Delta f} \qquad 29\text{-}58$$

Q FACTOR FOR AN RLC CIRCUIT

Resonance circuits are used in radio receivers, where the resonance frequency of the circuit is varied either by varying the capacitance or the inductance. Resonance occurs when the natural frequency of the circuit equals one of the frequencies of the radio waves picked up at the antenna. At resonance, a relatively large current exists in the antenna circuit. If the Q factor of the circuit is sufficiently high, currents due to other station frequencies off resonance will be negligible compared with those currents due to the station frequency to which the circuit is tuned.

Example 29-7 Driven Series *RLC* Circuit

A series *RLC* combination that has $L = 2.0\text{ H}$, $C = 2.0\ \mu\text{F}$, and $R = 20\ \Omega$ is driven by an ideal generator that has a peak emf of 100 V and a frequency which can be varied. Find (*a*) the resonance frequency f_0, (*b*) the *Q* factor, (*c*) the width of the resonance Δf, and (*d*) the current amplitude at resonance.

PICTURE The resonance frequency is found from $\omega_0 = 1/\sqrt{LC}$ and the *Q* factor is found from $Q_{\text{factor}} = \omega_0 L/R$ (Equation 29-57).

SOLVE

(*a*) The resonance frequency is $f_0 = \omega_0/2\pi$:

$$f_0 = \frac{\omega_0}{2\pi} = \frac{1}{2\pi\sqrt{LC}}$$

$$= \frac{1}{2\pi\sqrt{(2.0\text{ H})(2.0\times 10^{-6}\text{ F})}} = \boxed{80\text{ Hz}}$$

(*b*) Use this result to calculate Q_{factor}:

$$Q_{\text{factor}} = \frac{\omega_0 L}{R} = \frac{2\pi(80\text{ Hz})(2.0\text{ H})}{20\ \Omega} = \boxed{50}$$

(*c*) Use the value of Q_{factor} to find the width of the resonance Δf, where $Q_{\text{factor}} = f_0/\Delta f$:

$$\Delta f = \frac{f_0}{Q_{\text{factor}}} = \frac{80\text{ Hz}}{50} = \boxed{1.6\text{ Hz}}$$

(*d*) At resonance, the impedance is equal to *R* and I_{peak} is $V_{\text{app peak}}/R$:

$$I_{\text{peak}} = \frac{V_{\text{app peak}}}{R} = \frac{\mathcal{E}_{\text{peak}}}{R} = \frac{100\text{ V}}{20\ \Omega} = \boxed{5.0\text{ A}}$$

CHECK At resonance the inductive and capacitances each equal $X_L = \omega_0 L = 2\pi(80\text{ Hz})(2.0\text{ H}) = 1.0\text{ k}\Omega$. The resistance is given as $R = 20\ \Omega$. Because the resistance is much smaller than the inductive reactance at resonance, we expect the *Q* factor to be high and the resonance to be sharp. The Part-(*b*) and -(*c*) results meet this expectation.

TAKING IT FURTHER The width of 1.6 Hz is about 2.0 percent of the resonance frequency of 80 Hz, so the resonance peak is quite sharp.

Example 29-8 Driven Series *RLC* Circuit Current, Phase, and Power

Try It Yourself

If the generator in Example 29-7 has a frequency of 60 Hz, find (*a*) the peak current, (*b*) the phase constant δ, (*c*) the power factor, and (*d*) the average power delivered.

PICTURE The peak current is the peak applied potential drop divided by the total impedance of the series combination. The phase angle δ is found from $\tan\delta = (X_L - X_C)/R$. You can use either Equation 29-54 or Equation 29-55 to find the average power delivered.

SOLVE

Cover the column to the right and try these on your own before looking at the answers.

Steps	Answers
(*a*) 1. Write the peak current in terms of $V_{\text{app peak}}$ and the impedance.	$I_{\text{peak}} = \frac{V_{\text{app peak}}}{Z} = \frac{\mathcal{E}_{\text{peak}}}{Z}$
2. Calculate the capacitive and inductive reactances and the total reactance.	$X_C = 1326\ \Omega$, $X_L = 754\ \Omega$ so $X_L - X_C = -572\ \Omega$
3. Calculate the total impedance *Z*.	$Z = 573\ \Omega$
4. Use the results of steps 2 and 3 to calculate I_{peak}.	$I_{\text{peak}} = \boxed{0.17\text{ A}}$
(*b*) Use the results of Part (*a*), steps 2 and 3 to calculate δ.	$\delta = \tan^{-1}\frac{X_L - X_C}{R} = \boxed{-88.0^\circ}$

(*c*) Use your value of δ to compute the power factor. $\cos\delta = \boxed{0.035}$

(*d*) Calculate the average power delivered from Equation 29-54. $P_{av} = \frac{1}{2}I^2_{peak}R = \boxed{0.29\text{ W}}$

CHECK To check our result for the average power using the power factor found in Part (*c*), we have $P_{av} = \frac{1}{2}V_{app\,peak}I_{peak}\cos\delta = \frac{1}{2}\mathcal{E}_{peak}I_{peak}\cos\delta = 0.29\text{ W}$. This result is in agreement with our result for Part (*d*).

TAKING IT FURTHER The frequency of 60 Hz is well below the resonance frequency of 80 Hz. (Recall that the width of the resonance peak, calculated in Example 29-7, is only 1.6 Hz.) As a result, the total reactance is much greater in magnitude than the resistance. This result is always the case far from resonance. Similarly, a peak current of 0.17 A is much less than the peak current at resonance, which was found to be 5.0 A in Example 29-7. Finally, we see from Figure 29-19 that a negative phase angle δ means that the current leads the applied potential drop.

CONCEPT CHECK 29-1

A circuit consists of an ideal constant-frequency generator, a resistor, a capacitor and an inductor with a moveable soft-iron core—all connected in series. You notice that if you nudge the soft-iron core a bit deeper into the induction coil, the rms current increases slightly. Prior to the nudge, the resonance frequency of the circuit was (*a*) below the generator frequency, (*b*) equal to the generator frequency, (*c*) above the generator frequency.

Example 29-9 Driven Series *RLC* Circuit at Resonance — *Try It Yourself*

Find the peak potential drop across the resistor, the inductor, and the capacitor at resonance for the circuit in Example 29-7.

PICTURE The peak potential drop across the resistor is I_{peak} multiplied by R. Similarly, the peak potential drop across the inductor or capacitor is I_{peak} multiplied by the appropriate reactance. We found that at resonance $I_{peak} = 5.0$ A and $f_0 = 80$ Hz in Example 29-7.

SOLVE

Cover the column to the right and try these on your own before looking at the answers.

Steps	Answers
1. Calculate $V_{R\,peak} = I_{peak}R$.	$V_{R\,peak} = I_{peak}R = \boxed{100\text{ V}}$
2. Express $V_{L\,peak}$ in terms of I_{peak} and X_L.	$V_{L\,peak} = I_{peak}X_L = I_{peak}\omega_0 L = \boxed{5.0\text{ kV}}$
3. Express $V_{C\,peak}$ in terms of I_{peak} and X_C.	$V_{C\,peak} = I_{peak}X_C = \dfrac{I_{peak}}{\omega_0 C} = \boxed{5.0\text{ kV}}$

CHECK The inductive and capacitive reactances are equal, as we would expect at resonance. (We solved for the resonance frequency by setting them equal.)

TAKING IT FURTHER The phasor diagram for the potential drops across the resistor, capacitor, and inductor is shown in Figure 29-22. The peak potential drop across the resistor is a relatively safe 100 V, equal to the peak emf of the generator. However, the peak potential drops across the inductor and the capacitor are a dangerously high 5.0 kV. These potential drops are 180° out of phase. At resonance, the potential drop across the inductor at any instant is the negative of the potential drop across the capacitor, so they always sum to zero, leaving the potential drop across the resistor equal to the emf in the circuit.

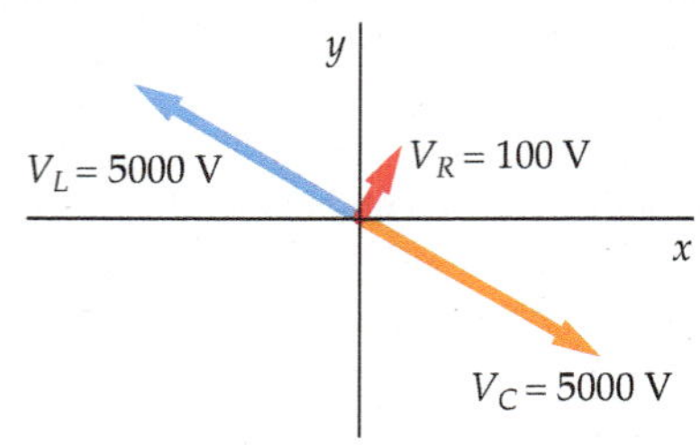

FIGURE 29-22

Example 29-10 *RC* Low-Pass Filter

A resistor R and capacitor C are in series with an ideal generator, as shown in Figure 29-23. The generator applies a potential difference across the *RC* combination given by $V_{app} = \sqrt{2}\,V_{app\,rms}\cos\omega t$. Find the rms potential difference across the capacitor $V_{out\,rms}$ as a function of frequency ω.

PICTURE The rms potential difference across the capacitor is the product of the rms current and the capacitive reactance. The rms current is found from the potential difference applied by the generator and the impedance of the series *RC* combination.

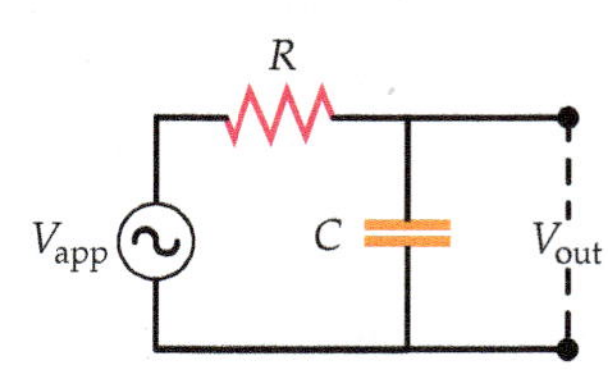

FIGURE 29-23 The peak output voltage decreases as frequency increases.

SOLVE

1. The potential drop across the capacitor is I_{rms} multiplied by X_C: $V_{out\,rms} = I_{rms}X_C$

2. The rms current depends on the applied rms potential difference and the impedance: $I_{rms} = \dfrac{V_{app\,rms}}{Z}$

3. In this circuit, only R and X_C contribute to the total impedance: $Z = \sqrt{R^2 + X_C^2}$

4. Substitute those expressions and $X_C = 1/(\omega C)$ to find the output rms potential difference:

$$V_{out\,rms} = I_{rms}X_C = \frac{V_{app\,rms}X_C}{\sqrt{R^2 + X_C^2}} = \frac{V_{app\,rms}}{\sqrt{1 + \dfrac{R^2}{X_C^2}}} = \boxed{\frac{V_{app\,rms}}{\sqrt{1 + (\omega RC)^2}}}$$

CHECK The dimensions of our step 4 result are correct. The dimension of ω is 1/T and the dimension of RC is T, so the product ωRC is dimensionless.

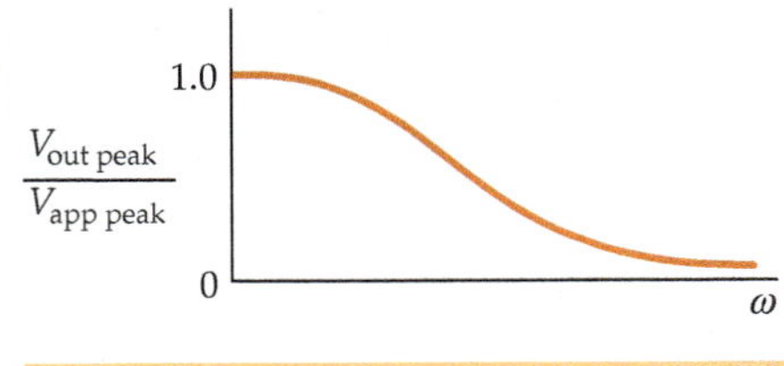

FIGURE 29-24

TAKING IT FURTHER This circuit is called an *RC low-pass filter*, because it transmits low frequencies with greater amplitude than high frequencies. In fact, the output potential difference equals the potential difference applied by the generator in the limit that $\omega \rightarrow 0$, but also approaches zero for $\omega \rightarrow \infty$, as shown in the graph of the ratio of output potential difference to applied potential difference in Figure 29-24.

PRACTICE PROBLEM 29-3 Find the output potential difference for this circuit if the capacitor is replaced by an inductor L.

Example 29-11 An FM Tuner

Context-Rich

You have been tinkering with building a radio tuner using your new knowledge of physics. You know that the FM dial gives its frequencies in megahertz, and you would like to determine what percentage of change in an inductor would allow you to tune for the whole FM range. You decide to start at midrange and determine a percent increase and decrease needed for inductance. A variable inductor is usually an iron-core solenoid, and the inductance is increased by further inserting the core. The FM dial goes from 88 MHz to 108 MHz.

PICTURE We can relate inductance to the resonant frequency with $\omega = 2\pi f$ and $\omega = 1\sqrt{LC}$. Then, if we find the percent change in frequency, we can determine the percent change in inductance. The capacitance C does not vary.

SOLVE

1. The resonant angular frequency ω is related to the inductance L:

$\omega = 1/\sqrt{LC}$

and

$\omega = 2\pi f$

so

$$f = \frac{1}{2\pi\sqrt{LC}}$$

2. Solving for L gives:

$L = a/f^2$

where

$a = (4\pi^2 C)^{-1}$

3. Express the fractional change in L in terms of the frequencies. When L is maximum, f is minimum and vice versa. The middle frequency f_{mid} is halfway between the maximum and minimum frequency and L_{mid} is the inductance when $f = f_{mid}$:

$$\frac{\Delta L}{L} = \frac{L_{max} - L_{min}}{L_{mid}} = \frac{a/f_{max}^2 - a/f_{min}^2}{a/f_{mid}^2}$$

$$= f_{mid}^2\left(\frac{1}{f_{max}^2} - \frac{1}{f_{min}^2}\right) = 98^2\left(\frac{1}{108^2} - \frac{1}{88^2}\right)$$

$$= -0.417$$

4. The minus sign is not relevant, except as an indication that when the inductance increases the resonant frequency decreases. Express the step 3 result as a percentage:

The inductance varies by about $\boxed{\text{42 percent}}$.

A shipboard radio, circa 1920. Exposed at the operator's left are the inductance coils and capacitor plates of the tuning circuit. (*© George H. Clark Radioana Collection-Archive Center, National Museum of American History.*)

PARALLEL RLC CIRCUIT

Figure 29-25 shows a resistor R, a capacitor C, and an inductor L connected in parallel across an ac generator. The total current I from the generator divides into three currents: the current I_R in the resistor, the current I_C in the capacitor, and the current I_L in the inductor. The instantaneous potential drop V_{app} is the same across each element. The current in the resistor is in phase with the potential drop and the phasor $\vec{I}_R$ has magnitude V_{peak}/R. Because the potential drop across an inductor *leads* the current in the inductor by 90°, I_L *lags* the potential drop by 90°, and the phasor $\vec{I}_L$ has magnitude V_{peak}/X_L. Similarly, I_C leads the potential drop by 90° and the phasor $\vec{I}_C$ has magnitude V_{peak}/X_C. These currents are represented by phasors in Figure 29-26. The total current I is the x component of the vector sum of the individual current phasors as shown in the figure. The magnitude of the total current phasor is

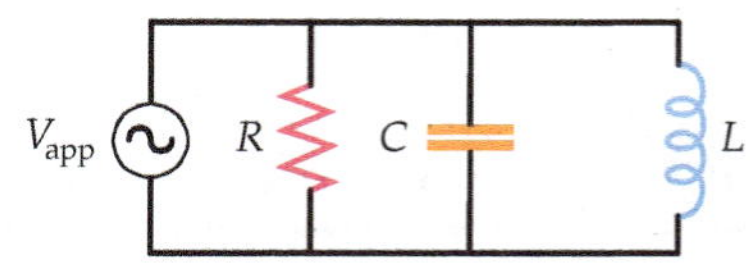

FIGURE 29-25 A parallel *RLC* circuit.

$$I = \sqrt{I_R^2 + (I_L - I_C)^2} = \sqrt{\left(\frac{V_{peak}}{R}\right)^2 + \left(\frac{V_{peak}}{X_L} - \frac{V_{peak}}{X_C}\right)^2} = \frac{V_{peak}}{Z} \qquad \text{29-59}$$

where the total impedance Z is related to the resistance and the capacitive and inductive reactances by

$$\frac{1}{Z^2} = \frac{1}{R^2} + \left(\frac{1}{X_L} - \frac{1}{X_C}\right)^2 \qquad \text{29-60}$$

At resonance, the currents in the inductor and capacitor are 180° out of phase, so the total current is a minimum and is just the current in the resistor. We see from Equation 29-59 that this occurs if Z is maximum, so $1/Z$ is minimum. Then, we see from Equation 29-60 that if $X_L = X_C$, $1/Z$ has its minimum value $1/R$. Equating X_L with X_C and solving for ω obtains the resonant frequency, which equals the natural frequency $\omega_0 = 1/\sqrt{LC}$.

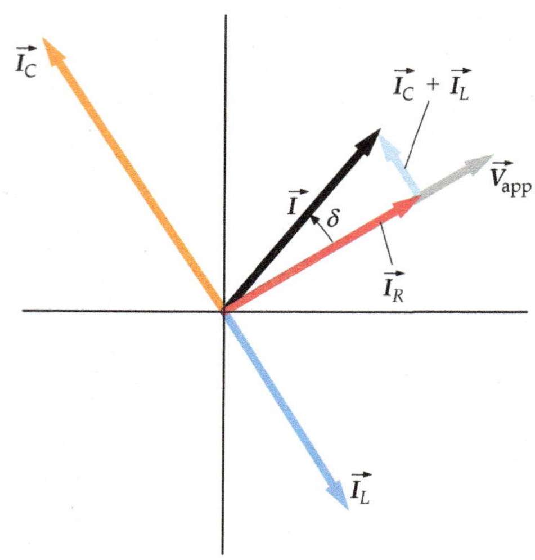

FIGURE 29-26 A phasor diagram for the currents in the parallel *RLC* circuit shown in Figure 29-25. The potential drop is the same across each element. The current in the resistor is in phase with the potential drop. The current in the capacitor leads the potential drop by 90° and the current in the inductor lags the potential drop by 90°. The phase difference δ between the total current and the potential drop depends on the relative magnitudes of the currents, which depend on the values of the resistance and of the capacitive and inductive reactances.

Physics Spotlight

The Electric Grid: Power to the People

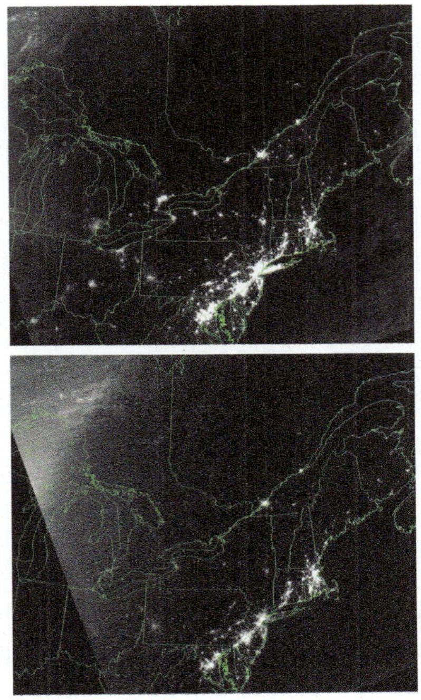

The pair of satellite images shows how a power failure affected many American and Canadian cities during the evening of Thursday, August 14, 2003. The top image was taken 20 hours before the blackout while the bottom image was taken 7 hours after the blackout. *(Courtesy of Chris Elridge/U.S. Airforce.)*

All around the world, people depend on electrical distribution systems, or grids, for reliable electric power. Generators, substations that have transformers, and high-voltage transmission lines are all necessary to efficiently move electrical energy from one place to another.[*] As of 2002, over 150 000 miles of high-voltage ac transmission lines and more than 10 000 transmission substations were part of the grid in the United States alone.[†] Grids are growing in complexity[‡] all over the world.[#, °] Unfortunately, as grids grow, so do the number of possible points of failure.

Most electrical grid failures are small scale, caused by local weather, by equipment failure,[§] or even by animals.[¶] But these failures give clues to the causes of large-scale power outages. Power surges within transformers and lines are the primary cause of local outages. Damage is prevented from propagating, or cascading, by relay switches that close off the line and act as surge suppressors for the system as a whole. Rarely, local outages are caused when there is much more demand than local generators can supply.

Sometimes the same mechanisms meant to prevent damage from a local outage can cause the outage to cascade. On November 9, 1965, a relay switch tripped at a hydroelectric plant in southern Ontario. The current from that line was shunted to five other transmission lines, which caused the relay switches for those lines to trip. Because of this dramatic reduction in load, the generators sped up, which meant that the power they provided was out of phase with power from other providers.[**] Over the course of a few minutes, relay after relay tripped, and many generators were shut down as they were isolated from their loads. Within four seconds, relays had tripped throughout the northeastern United States. Within a few minutes, generators were taken off-line, and over 30 million people were left without electricity for several hours.

That blackout prompted the formation of the National Electric Reliability Council.[††] Measures put in place for coordination of electric loads[‡‡, ##] have prevented many large blackouts, but blackouts have still happened. In July 1977, a lightning strike to transmission lines in New York tripped relays, and because of the slow response of the system operator,[°°] New York City was left without power for three days.[§§] On August 14, 2003, an unfortunate combination of high demand, a transmission line shorting against an untrimmed tree, and inadequate communications led to a blackout in the northeast United States and Canada that left 50 million people without power, some for days.[¶¶]

To prevent future outages, technical improvements to the grid are being actively sought. One improvement is software able to monitor and control portions of the grid with speed and flexibility.[***] Other improvements may include higher capacity transmission lines, improved transformers, and more responsive maintenance programs.[†††, ‡‡‡]

---

* *The Electricity Delivery System.* United States Department of Energy, Office of Electricity Delivery and Energy Reliability, Feb. 2006. http://www.energetics.com/gridworks/pdfs/factsheet.pdf As of Nov. 2006.

† Ibid.

‡ Harris, J. L., et al., "Peak Demand and Energy Projection Bandwidths 2005–2014 Regional and National." National Energy Reliability Council, Sept. 14, 2005. ftp://www.nerc.com/pub/sys/all_updl/docs/pubs/Final_NERC_2005-2014_REGIONAL_BANDWIDTH_REPORT.pdf As of Nov. 2006.

\# "Towards National Power Grid." [sic] Power Grid Corporation of India Limited. http://www.powergridindia.com/pgnew/01-0001-003.asp As of Nov. 2006.

° Chow, J., Kopp, R., and Portney, P., "Energy Resources and Global Development." *Science*, Nov. 28, 2003, Vol. 302, pp. 1528–1531.

§ Chowdhury, A., et al., "MAPP Bulk Transmission System Outage Report." Mid-Continent Area Power Pool, Jun. 2001. http://www.mapp.org/assets/pdf/BTOR19_1.PDF As of Nov. 2006.

¶ Orso, J., "Bangor Hit with Power Outage." *La Crosse Tribune*, Jul. 16, 2006.

** U.S. Federal Power Commission, "Northeast Power Failure: November 9 and 10, 1965." Washington, DC: U.S. Government Printing Office. At http://blackout.gmu.edu/archive/pdf/fpc_65.pdf As of Nov. 2006.

†† Central Maine Power Company, "The Great Northeast Blackout of 1965." http://www.cmpco.com/about/system/blackout.html As of Nov. 2006.

‡‡ California Independent System Operator, "Load Reduction Programs." California Independent System Operator Procedure E-502, Mar. 15, 2005. http://www.caiso.com/docs/2000/06/15/200006151111359621.pdf As of Nov. 2006.

\## "Emergency Manual Load Shedding." California Independent System Operator Procedure E-502, Feb. 17, 2006. http://www.caiso.com/docs/1998/12/02/1998120218100812000.pdf As of Nov. 2006.

°° Boffey, P. M., "Investigators Agree N. Y. Blackout of 1977 Could Have Been Avoided." *Science*, Sept. 15, 1978, Vol. 201, No. 4360, pp. 994–998.

§§ Metz, W. D., " New York Blackout: Weak Links Tie Con Ed to Neighboring Utilities." *Science*, Jul. 29, 1977, Vol. 197, No. 4302, pp. 441–442.

¶¶ U.S.–Canada Power System Outage Task Force, "Final Report on the August 14, 2003 Blackout in the United States and Canada: Causes and Recommendations." ftp://www.nerc.com/pub/sys/all_updl/docs/blackout/ch1-3.pdf As of Nov. 2006.

*** Brown, E., "Creating Stability in a World of Unstable Electricity Distribution." *Logos*, Argonne National Laboratories, Spring 2004, Vol. 22, No. 1. At http://www.anl.gov/Media_Center/logos22-1/electricity.htm As of Nov. 2006.

††† Office of Electric Transmission and Distribution, "GridWorks Multi-Year Plan." United States Department of Energy. http://www.oe.energy.gov/DocumentsandMedia/multiyear-plan_final.pdf As of Nov. 2006.

‡‡‡ U.S.–Canada Power System Outage Task Force, "The August 14, 2003 Blackout One Year Later: Actions Taken in the United States and Canada to Reduce Blackout Risk." Natural Resources Canada and the U.S. Department of Energy, Aug. 13, 2004. ftp://www.nerc.com/pub/sys/all_updl/docs/blackout/Blackout-OneYearLater(PRINT).pdf As of Nov. 2006.

Summary

1. Reactance is a frequency-dependent property of capacitors and inductors that is analogous to the resistance of a resistor.
2. Impedance is a frequency-dependent property of an ac circuit or circuit loop that is analogous to the resistance in a dc circuit.
3. Phasors are two-dimensional vectors that allow us to picture the phase relations in a circuit.
4. Resonance occurs when the frequency of the generator equals the natural frequency of the oscillating circuit.

TOPIC	RELEVANT EQUATIONS AND REMARKS	
1. Alternating-Current Generators	An ac generator is a device that transforms mechanical energy into electrical energy. This transformation can be accomplished by using the mechanical energy to either rotate a conducting coil in a magnetic field or to rotate a magnet in a conducting coil.	
Emf generated	$\mathcal{E} = \mathcal{E}_{peak}\cos(\omega t + \delta)$	29-1
2. Current		
Rms current	$I_{rms} = \sqrt{(I^2)_{av}}$	29-8
Rms current and peak current	$I_{rms} = \frac{1}{\sqrt{2}} I_{peak}$	29-9
For a resistor	$I_{rms} = \frac{V_{R\,rms}}{R}$ potential drop and current in phase	29-12
For an inductor	$I_{rms} = \frac{V_{L\,rms}}{\omega L} = \frac{V_{L\,rms}}{X_L}$ potential drop leads current by 90°	29-22
For a capacitor	$I_{rms} = \frac{V_{C\,rms}}{1/\omega C} = \frac{V_{C\,rms}}{X_C}$ potential drop lags current by 90°	29-26
3. Reactance		
Inductive reactance	$X_L = \omega L$	29-21
Capacitive reactance	$X_C = \frac{1}{\omega C}$	29-27
4. Average Power		
To a resistor	$P_{av} = V_{R\,rms} I_{rms} = (I_{rms})^2 R$	29-10, 29-12
To an inductor or to a capacitor	$P_{av} = 0$	
5. *Transformers	A transformer is a device used to increase or decrease the voltage in a circuit without an appreciable loss in power. For a transformer with N_1 turns in the primary and N_2 turns in the secondary, the potential difference across the secondary coil is related to the potential drop across the primary coil by $V_2 = \frac{N_2}{N_1} V_1$	29-30

TOPIC	RELEVANT EQUATIONS AND REMARKS	
	If there are no power losses, $$V_{1\,\text{rms}}I_{1\,\text{rms}} = V_{2\,\text{rms}}I_{2\,\text{rms}}$$	29-32
6. *LC and RLC Series Circuits	If a capacitor is discharged through an inductor, the charge and the voltage on the capacitor oscillate with angular frequency $$\omega = \frac{1}{\sqrt{LC}}$$ The current in the inductor oscillates with the same frequency, but it is out of phase with the charge by 90°. The energy oscillates between electric energy in the capacitor and magnetic energy in the inductor. If the circuit also has resistance, the oscillations are damped because energy is dissipated in the resistor.	29-37
7. *Phasors	Phasors are two-dimensional vectors that represent the current $\vec{I}$, the potential drop across a resistor $\vec{V}_R$, the potential drop across a capacitor $\vec{V}_C$, and the potential drop across an inductor $\vec{V}_L$ in an ac circuit. These phasors rotate in the counterclockwise direction with an angular velocity that is equal to the angular frequency ω of the current. $\vec{V}_R$ is in phase with the current, $\vec{V}_L$ leads the current by 90°, and $\vec{V}_C$ lags the current by 90°. The x component of each phasor equals the magnitude of the current or the corresponding potential drop at any instant.	
8. *Driven Series RLC Circuit		
Applied potential drop	$$V_{\text{app}} = V_{\text{app peak}} \cos \omega t$$	
Current	$$I = \frac{V_{\text{app peak}}}{Z} \cos(\omega t - \delta)$$	29-52
Impedance Z	$$Z = \sqrt{R^2 + (X_L - X_C)^2}$$	29-51
Phase angle δ	$$\tan \delta = \frac{X_L - X_C}{R}$$	29-49
Average power	$$P_{\text{av}} = (I_{\text{rms}})^2 R = V_{\text{app rms}} I_{\text{rms}} \cos \delta = \frac{(V_{\text{app rms}})^2 R \omega^2}{L^2(\omega^2 - \omega_0^2)^2 + \omega^2 R^2}$$	29-54, 29-55, 29-56
Power factor	The quantity $\cos \delta$ in Equation 29-55 is called the power factor of the RLC circuit. At resonance, δ is zero, the power factor is 1, and $$P_{\text{av}} = V_{\text{app rms}} I_{\text{rms}}$$	
Resonance	When the rms current is maximum, the circuit is said to be at resonance. The conditions for resonance are $$Z = \sqrt{R^2 + (X_L - X_C)^2} = R$$ $X_L = X_C$, so $$\omega = \omega_0 = \frac{1}{\sqrt{LC}} \quad \text{and} \quad \delta = 0$$	
9. *Q Factor	The sharpness of the resonance curve is described by the Q factor: $$Q_{\text{factor}} = \frac{\omega_0 L}{R}$$	29-57
	When the resonance curve is reasonably narrow, the Q factor can be approximated by $$Q_{\text{factor}} = \frac{\omega_0}{\Delta \omega} = \frac{f_0}{\Delta f}$$	29-58

Answers to Concept Check

29-1 (*c*)

Answers to Practice Problems

29-1 (*a*) 2.8 A, (*b*) 96 W, (*c*) 1.9×10^2 W

29-2 79 μH

29-3 $V_{\text{out rms}} = V_{\text{in rms}}/\sqrt{1 + (R/L)^2/\omega^2}$. This circuit is a *high-pass filter*.

Problems

In a few problems, you are given more data than you actually need; in a few other problems, you are required to supply data from your general knowledge, outside sources, or informed estimate.

Interpret as significant all digits in numerical values that have trailing zeros and no decimal points.

- • Single-concept, single-step, relatively easy
- •• Intermediate-level, may require synthesis of concepts
- ••• Challenging
- SSM Solution is in the *Student Solutions Manual*
- Consecutive problems that are shaded are paired problems.

CONCEPTUAL PROBLEMS

1 • A coil in an ac generator rotates at 60 Hz. How much time elapses between successive peak emf values of the coil?

2 • If the rms voltage in an ac circuit is doubled, the peak voltage is (*a*) doubled, (*b*) halved, (*c*) increased by a factor of $\sqrt{2}$, (*d*) not changed.

3 • If the frequency in the circuit shown in Figure 29-27 is doubled, the inductance of the inductor will (*a*) double, (*b*) not change, (*c*) halve, (*d*) quadruple. SSM

4 • If the frequency in the circuit shown in Figure 29-27 is doubled, the inductive reactance of the inductor will (*a*) double, (*b*) not change, (*c*) halve, (*d*) quadruple.

FIGURE 29-27 Problems 3 and 4

5 • If the frequency in the circuit in Figure 29-28 is doubled, the capacitive reactance of the circuit will (*a*) double, (*b*) not change, (*c*) halve, (*d*) quadruple.

FIGURE 29-28 Problem 5

6 • (*a*) In a circuit consisting solely of an ac generator and an ideal inductor, are there any time intervals when the inductor receives energy from the generator? If so, when? Explain your answer. (*b*) Are there any time intervals when the inductor supplies energy back to the generator? If so when? Explain your answer.

7 • (*a*) In a circuit consisting of a generator and a capacitor, are there any time intervals when the capacitor receives energy from the generator? If so, when? Explain your answer. (*b*) Are there any time intervals when the capacitor supplies power to the generator? If so, when? Explain your answer. SSM

8 • (*a*) Show that the SI unit of inductance multiplied by the SI unit of capacitance is equivalent to seconds squared. (*b*) Show that the SI unit of inductance divided by the SI unit of resistance is equivalent to seconds.

9 • Suppose you increase the rotation rate of the coil in the generator shown in the simple ac circuit in Figure 29-29. Then the rms current (*a*) increases, (*b*) does not change, (*c*) may increase or decrease depending on the magnitude of the original frequency, (*d*) may increase or decrease depending on the magnitude of the resistance, (*e*) decreases. SSM

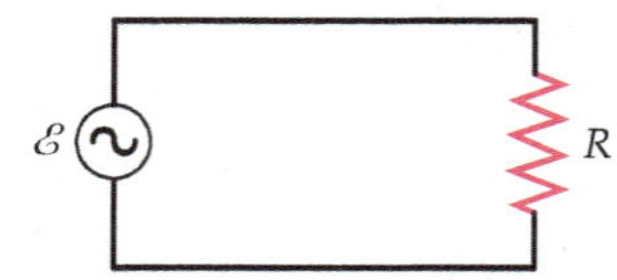

FIGURE 29-29 Problem 9

10 • If the inductance value is tripled in a circuit consisting solely of a variable inductor and a variable capacitor, how would you have to change the capacitance so that the natural frequency of the circuit is unchanged? (*a*) Triple the capacitance. (*b*) Decrease the capacitance to one-third of its original value. (*c*) You should not change the capacitance. (*d*) You cannot determine how to change the capacitance from the data given.

11 • Consider a circuit consisting solely of an ideal inductor and an ideal capacitor. How does the maximum energy stored in the capacitor compare to the maximum value stored in the inductor? (*a*) They are the same and each is equal to the total energy stored in the circuit. (*b*) They are the same and each is equal to half of the total energy stored in the circuit. (*c*) The maximum energy stored in the capacitor is larger than the maximum energy stored in the inductor. (*d*) The maximum energy stored in the inductor is larger than the maximum energy stored in the capacitor. (*e*) You cannot compare the maximum energies based on the data given because the ratio of the maximum energies depends on the actual capacitance and inductance values. SSM

12 • True or false:

(*a*) A driven series *RLC* circuit that has a high *Q* factor has a narrow resonance curve.

(*b*) A circuit consists solely of a resistor, an inductor, and a capacitor, all connected in series. If the resistance of the resistor is doubled, the natural frequency of the circuit remains the same.

(*c*) At resonance, the impedance of a driven series *RLC* combination equals the resistance *R*.

(*d*) At resonance, the current in a driven series *RLC* circuit is in phase with the voltage applied to the combination.

13 • True or false:

(*a*) Near resonance, the power factor of a driven series *RLC* circuit is close to zero.

(*b*) The power factor of a driven series *RLC* circuit does not depend on the value of the resistance.

(*c*) The resonance frequency of a driven series *RLC* circuit does not depend on the value of the resistance.

(*d*) At resonance, the peak current of a driven series *RLC* circuit does not depend on the capacitance or the inductance.

(*e*) For frequencies below the resonant frequency, the capacitive reactance of a driven series *RLC* circuit is larger than the inductive reactance.

(*f*) For frequencies below the resonant frequency of a driven series *RLC* circuit, the phase of the current leads (is ahead of) the phase of the applied voltage.

14 • You may have noticed that sometimes two radio stations can be heard when your receiver is tuned to a specific frequency. This situation often occurs when you are driving and are between two cities. Explain how this situation can occur.

15 • True or false:

(*a*) At frequencies much higher than or much lower than the resonant frequency of a driven series *RLC* circuit, the power factor is close to zero.

(*b*) The larger the resonance width of a driven series *RLC* circuit is, the larger the *Q* factor for the circuit is.

(*c*) The larger the resistance of a driven series *RLC* circuit is, the larger the resonance width for the circuit is.

16 • An ideal transformer has N_1 turns on its primary and N_2 turns on its secondary. The average power delivered to a load resistance *R* connected across the secondary is P_2 when the primary rms voltage is V_1. The rms current in the primary windings can then be expressed as (*a*) P_2/V_1, (*b*) $(N_1/N_2)(P_2/V_1)$, (*c*) $(N_2/N_1)(P_2/V_1)$, (*d*) $(N_2/N_1)2(P_2/V_1)$.

17 • True or false:

(*a*) A transformer is used to change frequency.

(*b*) A transformer is used to change voltage.

(*c*) If a transformer steps up the current, it must step down the voltage.

(*d*) A step-up transformer steps down the current.

(*e*) The standard household wall-outlet voltage in Europe is 220 V, about twice that used in the United States. If a European traveler wants her hair dryer to work properly in the United States, she should use a transformer that has more windings in its secondary coil than in its primary coil.

(*f*) The standard household wall-outlet voltage in Europe is 220 V, about twice that used in the United States. If an American traveler wants his electric razor to work properly in Europe, he should use a transformer that steps up the current. SSM

ESTIMATION AND APPROXIMATION

18 •• **ENGINEERING APPLICATION** The impedances of motors, transformers, and electromagnets include both resistance and inductive reactance. Suppose that the phase of the current to a large industrial plant lags the phase of the applied voltage by 25° when the plant is under full operation and using 2.3 MW of power. The power is supplied to the plant from a substation 4.5 km from the plant; the 60 Hz rms line voltage at the plant is 40 kV. The resistance of the transmission line from the substation to the plant is 5.2 Ω. The cost per kilowatt-hour to the company that owns the plant is \$0.14, and the plant pays only for the actual energy used. (*a*) Estimate the resistance and inductive reactance of the plant's total load. (*b*) Estimate the rms current in the power lines and the rms voltage at the substation. (*c*) How much power is lost in transmission? (*d*) Suppose that the phase at which the current lags the phase of the applied voltage is reduced to 18° by adding a bank of capacitors in series with the load. How much money would be saved by the electric utility during one month of operation, assuming the plant operates at full capacity for 16 h each day? (*e*) What must be the capacitance of this bank of capacitors to achieve this change in phase angle?

ALTERNATING CURRENT IN RESISTORS, INDUCTORS, AND CAPACITORS

19 • A 100-W lightbulb is screwed into a standard 120-V-rms. socket. Find (*a*) the rms current, (*b*) the peak current, and (*c*) the peak power. SSM

20 • A circuit breaker is rated for a current of 15 A rms at a voltage of 120 V rms. (*a*) What is the largest value of peak current that the breaker can carry? (*b*) What is the maximum average power that can be supplied by this circuit?

21 • What is the reactance of a 1.00-μH inductor at (*a*) 60 Hz, (*b*) 600 Hz, and (*c*) 6.00 kHz? SSM

22 • An inductor has a reactance of 100 Ω at 80 Hz. (*a*) What is its inductance? (*b*) What is its reactance at 160 Hz?

23 • At what frequency would the reactance of a 10-μF capacitor equal the reactance of a 1.0-mH inductor?

24 • What is the reactance of a 1.00-nF capacitor at (*a*) 60.0 Hz, (*b*) 6.00 kHz, and (*c*) 6.00 MHz?

25 • A 20-Hz ac generator that produces a peak emf of 10 V is connected to a 20-μF capacitor. Find (*a*) the peak current and (*b*) the rms current. SSM

26 • At what frequency is the reactance of a 10-μF capacitor (*a*) 1.00 Ω, (*b*) 100 Ω, and (*c*) 10.0 mΩ?

27 •• A circuit consists of two ideal ac generators and a 25-Ω resistor, all connected in series. The potential difference across the terminals of one of the generators is given by $V_1 = (5.0\ \text{V})\cos(\omega t - \alpha)$, and the potential difference across the terminals of the other generator is given by $V_2 = (5.0\ \text{V})\cos(\omega t + \alpha)$, where $\alpha = \pi/6$. (*a*) Use Kirchhoff's loop rule and a trigonometric identity to find the peak current in the circuit. (*b*) Use a phasor diagram to find the peak current in the circuit. (*c*) Find the current in the resistor if $\alpha = \pi/4$ and the amplitude of V_2 is increased from 5.0 V to 7.0 V.

*UNDRIVEN CIRCUITS CONTAINING CAPACITORS, RESISTORS, AND INDUCTORS

28 • (*a*) Show that $1/\sqrt{LC}$ has units of inverse seconds by substituting SI units for inductance and capacitance into the expression. (*b*) Show that $\omega_0 L/R$ (the expression for the *Q* factor) is dimensionless by substituting SI units for angular frequency, inductance, and resistance into the expression.

29 • (*a*) What is the period of oscillation of an *LC* circuit consisting of a 2.0-mH coil and a 20-μF capacitor? (*b*) A circuit that oscillates consists solely of an 80-μF capacitor and a variable ideal inductor. What inductance is needed in order to tune this circuit to oscillate at 60 Hz? SSM

30 • An *LC* circuit has capacitance C_0 and inductance *L*. A second *LC* circuit has capacitance $\frac{1}{2}C_0$ and inductance $2L$, and a third *LC* circuit has capacitance $2C_0$ and inductance $\frac{1}{2}L$. (*a*) Show that each circuit oscillates with the same frequency. (*b*) In which circuit would the peak current be greatest if the peak voltage across the capacitor in each circuit were the same?

31 •• A 5.0-μF capacitor is charged to 30 V and is then connected across an ideal 10-mH inductor. (*a*) How much energy is stored in the system? (*b*) What is the frequency of oscillation of the circuit? (*c*) What is the peak current in the circuit?

32 •• A coil with internal resistance can be modeled as a resistor and an ideal inductor in series. Assume that the coil has an internal resistance of 1.00 Ω and an inductance of 400 mH. A 2.00-μF capacitor is charged to 24.0 V and is then connected across coil. (*a*) What is the initial voltage across the coil? (*b*) How much energy is dissipated in the circuit before the oscillations die out? (*c*) What is the frequency of oscillation of the circuit? (Assume the internal resistance is sufficiently small that it has no impact on the frequency of the circuit.) (*d*) What is the quality factor of the circuit?

33 •• An inductor and a capacitor are connected, as shown in Figure 29-30. Initially, the switch is open, and the left plate of the capacitor has charge Q_0. The switch is then closed. (*a*) Plot both Q versus t and I versus t on the same graph, and explain how it can be seen from these two plots that the current leads the charge by 90°. (*b*) The expressions for the charge and for the current are given by Equations 29-38 and 29-39, respectively. Use trigonometry and algebra to show that the current leads the charge by 90°. SSM

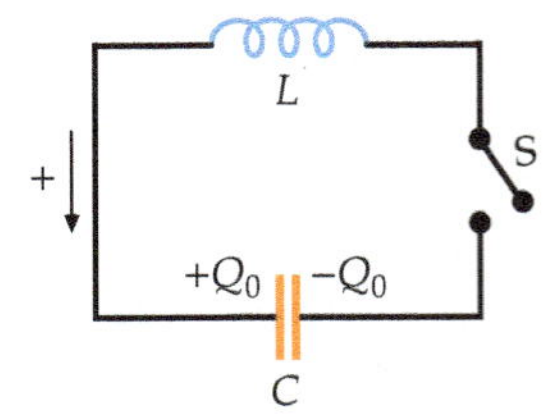

FIGURE 29-30 Problem 33

DRIVEN *RL* CIRCUITS

34 •• A circuit consists of a resistor, an ideal 1.4-H inductor, and an ideal 60-Hz generator, all connected in series. The rms voltage across the resistor is 30 V and the rms voltage across the inductor is 40 V. (*a*) What is the resistance of the resistor? (*b*) What is the peak emf of the generator?

35 •• A coil that has a resistance of 80.0 Ω has an impedance of 200 Ω when driven at a frequency of 1.00 kHz. What is the inductance of the coil? SSM

36 •• **Engineering Application** A two-conductor transmission line simultaneously carries a superposition of two voltage signals, so the potential difference between the two conductors is given by $V = V_1 + V_2$, where $V_1 = (10.0 \text{ V}) \cos(\omega_1 t)$ and $V_2 = (10.0 \text{ V}) \cos(\omega_2 t)$, where $\omega_1 = 100$ rad/s and $\omega_2 = 10\,000$ rad/s. A 1.00-H inductor and a 1.00-kΩ shunt resistor are inserted into the transmission line as shown in Figure 29-31. (Assume that the output is connected to a load that draws only an insignificant amount of current.) (*a*) What is the voltage (V_{out}) at the output of the transmission line? (*b*) What is the ratio of the low-frequency amplitude to the high-frequency amplitude at the output?

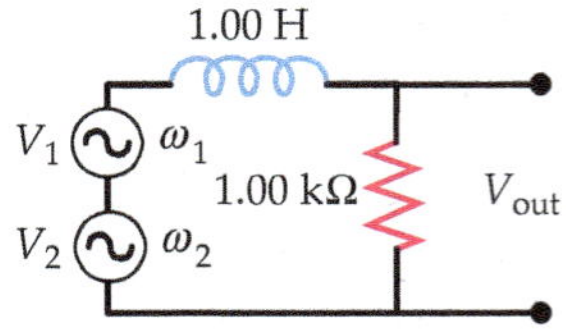

FIGURE 29-31 Problem 36

37 •• A coil is connected to a 120-V-rms, 60-Hz line. The average power supplied to the coil is 60 W, and the rms current is 1.5 A. Find (*a*) the power factor, (*b*) the resistance of the coil, and (*c*) the inductance of the coil. (*d*) Does the current lag or lead the voltage? Explain your answer. (*e*) Support your answer to Part (*d*) by determining the phase angle.

38 •• A 36-mH inductor that has a resistance of 40 Ω is connected to an ideal ac voltage source whose output is given by $\mathcal{E} = (345 \text{ V}) \cos(150\pi t)$, where t is in seconds. Determine (*a*) the peak current in the circuit, (*b*) the peak and rms voltages across the inductor, (*c*) the average power dissipation, and (*d*) the peak and average magnetic energy stored in the inductor.

39 •• A coil that has a resistance R and an inductance L has a power factor equal to 0.866 when driven at a frequency of 60 Hz. What is the coil's power factor if it is driven at 240 Hz? SSM

40 •• A resistor and an inductor are connected in parallel across an ideal ac voltage source whose output is given by $\mathcal{E} = \mathcal{E}_{max} \cos \omega t$ as shown in Figure 29-32. Show that (*a*) the current in the resistor is given by $I_R = (\mathcal{E}_{peak}/R) \cos \omega t$, (*b*) the current in the inductor is given by $I_L = (\mathcal{E}_{peak}/X_L) \cos(\omega t - 90°)$, and (*c*) the current in the voltage source is given by $I = I_R + I_L = I_{peak} \cos(\omega t - \delta)$, where $I_{peak} = \mathcal{E}_{peak}/Z$.

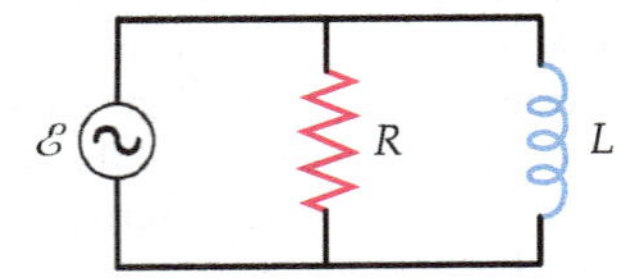

FIGURE 29-32 Problem 40

41 •• Figure 29-33 shows a load resistor that has a resistance of $R_L = 20.0$ Ω connected to a high-pass filter consisting of an inductor that has an inductance $L = 3.20$-mH inductor and a resistor that has resistance $R = 4.00$ Ω. The output of the ideal ac generator is given by $\mathcal{E} = (100 \text{ V}) \cos(2\pi ft)$. Find the rms currents in all three branches of the circuit if the driving frequency is (*a*) 500 Hz and (*b*) 2000 Hz. Find the fraction of the total average power supplied by the ac generator that is delivered to the load resistor if the frequency is (*c*) 500 Hz and (*d*) 2000 Hz. SSM

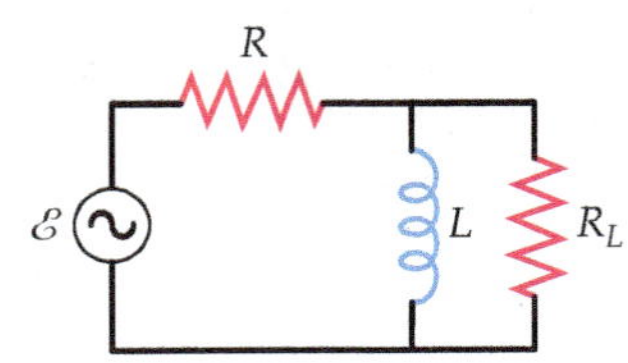

FIGURE 29-33 Problem 41

42 •• An ideal ac voltage source whose emf $\mathcal{E}_1$ is given by $(20 \text{ V}) \cos(2\pi ft)$ and an ideal battery whose emf $\mathcal{E}_2$ is 16 V are connected to a combination of two resistors and an inductor (Figure 29-34), where $R_1 = 10$ Ω, $R_2 = 8.0$ Ω, and $L = 6.0$ mH. Find the average power delivered to each resistor if the driving frequency is (*a*) 100 Hz, (*b*) 200 Hz, and (*c*) 800 Hz.

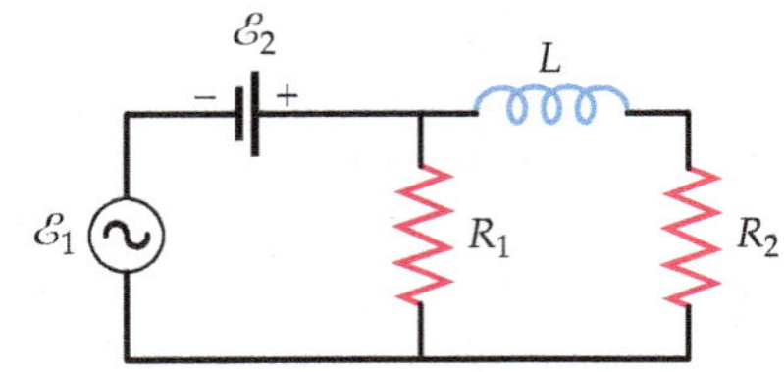

FIGURE 29-34 Problem 42

43 •• An ac circuit contains a resistor and an ideal inductor connected in series. The rms voltage drop across the series combination is 100 V and the rms voltage drop across the inductor alone is 80 V. What is the rms voltage drop across the resistor?

FILTERS AND RECTIFIERS

44 •• **Engineering Application** The circuit shown in Figure 29-35 is called an *RC high-pass filter* because it transmits input voltage signals that have high frequencies with greater amplitude than it transmits input voltage signals that have low frequencies. If the input voltage is given by $V_{in} = V_{in\,peak} \cos \omega t$, show that the output voltage is $V_{out} = V_H \cos(\omega t - \delta)$ where $V_H = V_{in\,peak}/\sqrt{1 + (\omega RC)^{-2}}$.

(Assume that the output is connected to a load that draws only an insignificant amount of current.) Show that this result justifies calling this circuit a high-pass filter.

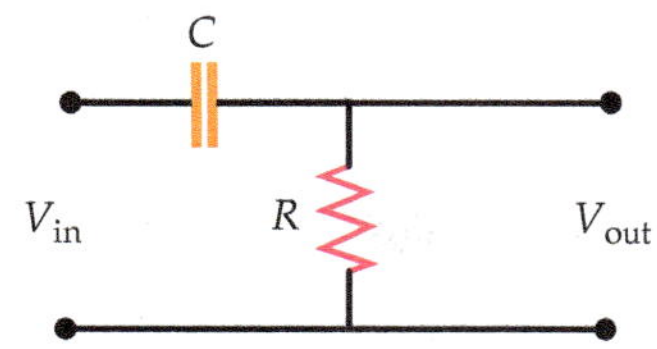

FIGURE 29-35 Problem 44

45 •• (*a*) Find an expression for the phase constant δ in Problem 44 in terms of ω, *R*, and *C*. (*b*) What is the value of δ in the limit that $\omega \to 0$? (*c*) What is the value of δ in the limit that $\omega \to \infty$? (*d*) Explain your answers to Parts (*b*) and (*c*).

46 •• **SPREADSHEET** Assume that in Problem 44, $R = 20\ \text{k}\Omega$ and $C = 15\ \text{nF}$. (*a*) At what frequency is $V_H = \frac{1}{\sqrt{2}} V_{\text{in peak}}$? That particular frequency is known as the 3-dB frequency, or f_{3dB}, for the circuit. (*b*) Using a spreadsheet program, make a graph of $\log_{10}(V_H)$ versus $\log_{10}(f)$, where *f* is the frequency. Make sure that the scale extends from at least 10% of the 3-dB frequency to ten times the 3-dB frequency. (*c*) Make a graph of δ versus $\log_{10}(f)$ for the same range of frequencies as in Part (*b*). What is the value of the phase constant when the frequency is equal to the 3-dB frequency?

47 •• A slowly varying voltage signal *V*(*t*) is applied to the input of the high-pass filter of Problem 44. Slowly varying means that during one time constant (equal to *RC*) there is no significant change in the voltage signal. Show that under those conditions the output voltage is proportional to the time derivative of *V*(*t*). This situation is known as a *differentiation circuit*. SSM

48 •• We can describe the output from the high-pass filter from Problem 44 using a decibel scale: $\beta = (20\ \text{dB})\log_{10}(V_H/V_{\text{in peak}})$, where β is the output in decibels. Show that for $V_H = \frac{1}{\sqrt{2}} V_{\text{in peak}}$, $\beta = 3.0\ \text{dB}$. The frequency at which $V_H = \frac{1}{\sqrt{2}} V_{\text{in peak}}$ is known as f_{3dB} (the 3-dB frequency). Show that for $f \ll f_{3dB}$, the output β drops by 6 dB if the frequency *f* is halved.

49 •• Show that the average power dissipated in the resistor of the high-pass filter of Problem 44 is given by

$$P_{av} = \frac{V_{\text{in peak}}^2}{2R[1 + (\omega RC)^{-2}]}.$$ SSM

50 •• One application of the high-pass filter of Problem 44 is as a noise filter for electronic circuits (a filter that blocks out low-frequency noise). Using a resistance value of 20 kΩ, find a value for the capacitance for the high-pass filter that attenuates a 60-Hz input voltage signal by a factor of 10, that is, so $V_H = \frac{1}{10} V_{\text{in peak}}$.

51 •• **ENGINEERING APPLICATION** The circuit shown in Figure 29-36 is an example of a low-pass filter. (Assume that the output is connected to a load that draws only an insignificant amount of current.) (*a*) If the input voltage is given by $V_{in} = V_{\text{in peak}} \cos \omega t$, show that the output voltage is $V_{out} = V_L \cos(\omega t - \delta)$ where $V_L = V_{\text{in peak}}/\sqrt{1 + (\omega RC)^2}$. (*b*) Discuss the trend of the output voltage in the limiting cases $\omega \to 0$ and $\omega \to \infty$. SSM

52 •• (*a*) Find an expression for the phase angle δ for the low-pass filter of Problem 51 in terms of ω, *R*, and *C*. (*b*) Find the value of δ in the limit that $\omega \to 0$ and in the limit that $\omega \to \infty$. Explain your answer.

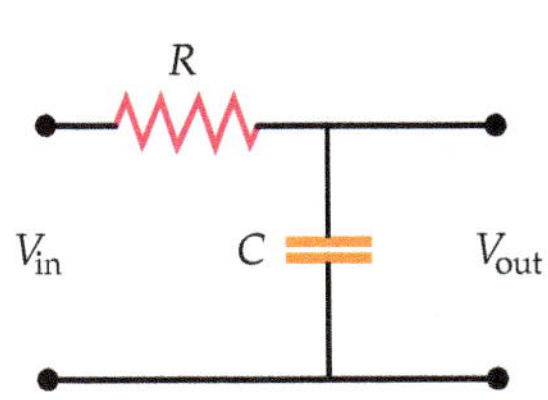

FIGURE 29-36 Problems 51 and 52

53 •• **SPREADSHEET** Using a spreadsheet program, make a graph of V_L versus input frequency *f* and a graph of phase angle δ versus input frequency for the low-pass filter of Problems 51 and 52. Use a resistance value of 10 kΩ and a capacitance value of 5.0 nF.

54 ••• A rapidly varying voltage signal *V*(*t*) is applied to the input of the low-pass filter of Problem 51. Rapidly varying means that during one time constant (equal to *RC*) there are significant changes in the voltage signal. Show that under those conditions the output voltage is proportional to the integral of *V*(*t*) with respect to time. This situation is known as an *integration circuit*.

55 ••• **ENGINEERING APPLICATION** The circuit shown in Figure 29-37 is a *trap filter*. (Assume that the output is connected to a load that draws only an insignificant amount of current.) (*a*) Show that the *trap filter* acts to reject signals in a band of frequencies centered at $\omega = 1/\sqrt{LC}$. (*b*) How does the width of the frequency band rejected depend on the resistance *R*? SSM

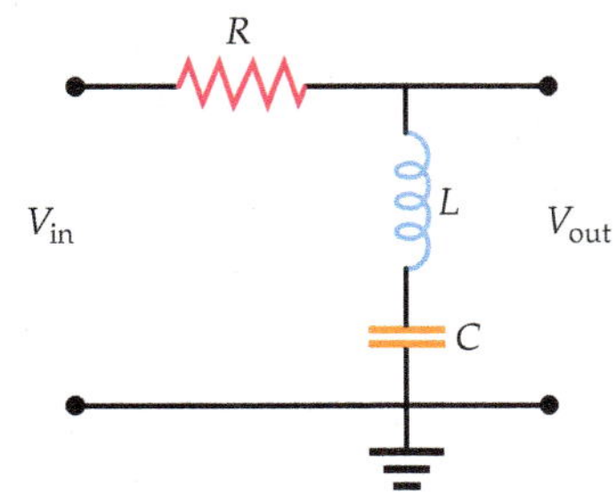

FIGURE 29-37 Problem 55

56 ••• **ENGINEERING APPLICATION** A half-wave rectifier for transforming an ac voltage into a dc voltage is shown in Figure 29-38. The diode in the figure can be thought of as a one-way valve for current. It allows current to pass in the forward direction (the direction of the arrowhead) only when V_{in} is at a higher electric potential than V_{out} by 0.60 V (i.e., whenever $V_{in} - V_{out} \geq +0.60\ \text{V}$). The resistance of the diode is effectively infinite when $V_{in} - V_{out}$ is less than +0.60 V. Plot two cycles of both input and output voltages as a function of time, on the same graph, assuming the input voltage is given by $V_{in} = V_{\text{in peak}} \cos \omega t$.

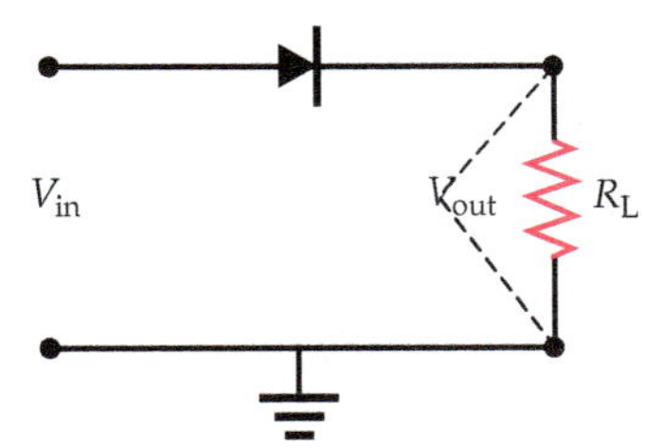

FIGURE 29-38 Problem 56

57 ••• **ENGINEERING APPLICATION** The output of the rectifier of Problem 56 can be further filtered by putting its output through a low-pass filter as shown in Figure 29-39*a*. The resulting output is a dc voltage with a small ac component (ripple) shown in Figure 29-39*b*. If the input frequency is 60 Hz and the load resistance is 1.00 kΩ, find the value for the capacitance so that the output voltage varies by less than 50 percent of the mean value over one cycle.

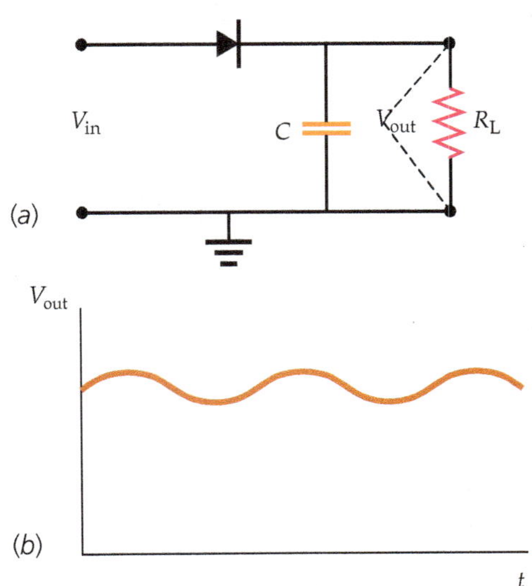

FIGURE 29-39 Problem 57

DRIVEN *LC* CIRCUITS

58 •• The generator voltage in Figure 29-40 is given by $\mathcal{E} = (100\text{ V})\cos(2\pi ft)$. (*a*) For each branch, what is the peak current and what is the phase of the current relative to the phase of the generator voltage? (*b*) At the resonance frequency there is no current in the generator. What is the angular frequency at resonance? (*c*) At the resonance frequency, find the current in the inductor and the current in the capacitor. Express your results as functions of time. (*d*) Draw a phasor diagram showing the phasors for the applied voltage, the generator current, the capacitor current, and the inductor current for the case where the frequency is higher than the resonance frequency.

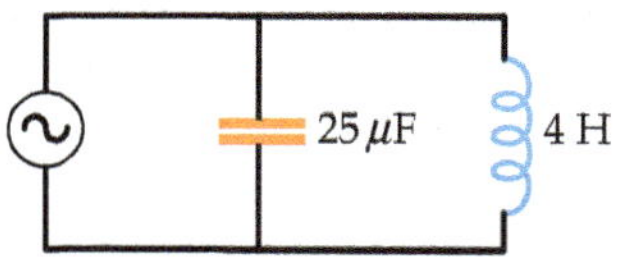

FIGURE 29-40 Problem 58

59 •• A circuit consists of an ideal ac generator, a capacitor, and an ideal inductor, all connected in series. The charge on the capacitor is given by $Q = (15\ \mu\text{C})\cos(\omega t + \frac{\pi}{4})$, where $\omega = 1250\text{ rad/s}$. (*a*) Find the current in the circuit as a function of time. (*b*) Find the capacitance if the inductance is 28 mH. (*c*) Write expressions for the electrical energy U_e, the magnetic energy U_m, and the total energy U as functions of time.

60 ••• **ENGINEERING APPLICATION** One method for determining the compressibility of a dielectric material uses a driven *LC* circuit that has a parallel-plate capacitor. The dielectric is inserted between the plates and the change in resonance frequency is determined as the capacitor plates are subjected to a compressive stress. In one such arrangement, the resonance frequency is 120 MHz when a dielectric of thickness 0.100 cm and dielectric constant $\kappa = 6.80$ is placed between the plates. Under a compressive stress of 800 atm, the resonance frequency decreases to 116 MHz. Find the Young's modulus of the dielectric material. (Assume that the dielectric constant does not change with pressure.)

61 ••• Figure 29-41 shows an inductor in series with a parallel-plate capacitor. The capacitor has a width w of 20 cm and a gap of 2.0 mm. A dielectric that has a dielectric constant of 4.8 can be slid in and out of the gap. The inductor has an inductance of 2.0 mH. When half the dielectric is between the capacitor plates (when $x = \frac{1}{2}w$), the resonant frequency of this combination is 90 MHz. (*a*) What is the capacitance of the capacitor without the dielectric? (*b*) Find the resonance frequency as a function of x for $0 \leq x \leq w$.

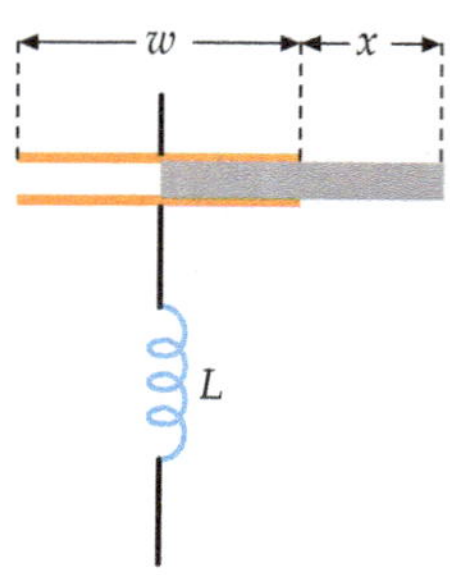

FIGURE 29-41 Problem 61

DRIVEN *RLC* CIRCUITS

62 • A circuit consists of an ideal ac generator, a 20-μF capacitor, and an 80-Ω resistor, all connected in series. The output of the generator has a peak emf of 20 V, and the armature of the generator rotates at 400 rad/s. Find (*a*) the power factor, (*b*) the rms current, and (*c*) the average power supplied by the generator.

63 •• Show that the expression $P_{av} = R\mathcal{E}_{rms}^2/Z^2$ gives the correct result for a circuit containing only an ideal ac generator and (*a*) a resistor, (*b*) a capacitor, and (*c*) an inductor. In the expression $P_{av} = R\mathcal{E}_{rms}^2/Z^2$, P_{av} is the average power supplied by the generator, $\mathcal{E}_{rms}$ is the root-mean-square value of the emf of the generator, R is the resistance, C is the capacitance, and L is the inductance. [In Part (*a*), $C = L = 0$; in Part (*b*), $R = L = 0$; and in Part (*c*), $R = C = 0$.] SSM

64 •• A series *RLC* circuit that has an inductance of 10 mH, a capacitance of 2.0 μF, and a resistance of 5.0 Ω is driven by an ideal ac voltage source that has a peak emf of 100 V. Find (*a*) the resonant frequency and (*b*) the root-mean-square current at resonance. When the frequency is 8000 rad/s, find (*c*) the capacitive and inductive reactances, (*d*) the impedance, (*e*) the root-mean-square current, and (*f*) the phase angle.

65 •• Find (*a*) the Q factor and (*b*) the resonance width (in hertz) for the circuit in Problem 64. (*c*) What is the power factor when $\omega = 8000\text{ rad/s}$? SSM

66 •• **ENGINEERING APPLICATION** FM radio stations typically operate at frequencies separated by 0.20 MHz. Thus, when your radio is tuned to a station operating at a frequency of 100.1 MHz, the resonance width of the receiver circuit should be much smaller than 0.20 MHz, so that you do not receive signal from stations operating at adjacent frequencies. Assume your receiving circuit has a resonance width of 0.050 MHz. When tuned in to that particular station, what is the Q factor of your circuit?

67 •• A coil is connected to a 60-Hz ac generator with a peak emf equal to 100 V. At this frequency, the coil has an impedance of 10 Ω and a reactance of 8.0 Ω. (*a*) What is the peak current in the coil? (*b*) What is the phase angle between the current and the applied voltage? (*c*) A capacitor is put in series with the coil and the generator. What capacitance is required so that the current is in phase with the generator emf? (*d*) What is the peak voltage measured across the capacitor?

68 •• An ideal 0.25-H inductor and a capacitor are connected in series with an ideal 60-Hz generator. A digital voltmeter is used to measure the rms voltages across the inductor and capacitor independently. The voltmeter reading across the capacitor is 75 V and that across the inductor is 50 V. (*a*) Find the capacitance and the rms current in the circuit. (*b*) What is the rms voltage across the series combination of the capacitor and the inductor?

69 •• In the circuit shown in Figure 29-42 the ideal generator produces an rms voltage of 115 V when operated at 60 Hz. What is the rms voltage between points (*a*) *A* and *B*, (*b*) *B* and *C*, (*c*) *C* and *D*, (*d*) *A* and *C*, and (*e*) *B* and *D*? SSM

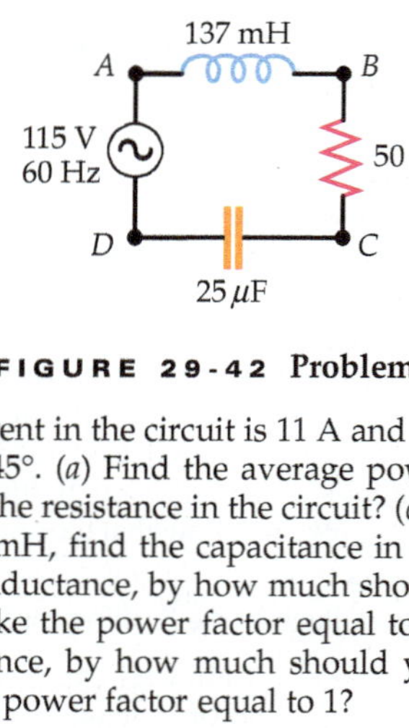

FIGURE 29-42 Problem 69

70 •• When an *RLC* series circuit is connected to a 120-V-rms, 60-Hz line, the rms current in the circuit is 11 A and the current leads the line voltage by 45°. (*a*) Find the average power supplied to the circuit. (*b*) What is the resistance in the circuit? (*c*) If the inductance in the circuit is 50 mH, find the capacitance in the circuit. (*d*) Without changing the inductance, by how much should you change the capacitance to make the power factor equal to 1? (*e*) Without changing the capacitance, by how much should you change the inductance to make the power factor equal to 1?

71 •• **SPREADSHEET** Plot the circuit impedance versus the angular frequency for each of the following circuits: (*a*) a driven series *LR* circuit, (*b*) a driven series *RC* circuit, and (*c*) a driven series *RLC* circuit.

72 •• In a driven series *RLC* circuit, the ideal generator has a peak emf equal to 200 V, the resistance is 60.0 Ω, and the capacitance is 8.00 μF. The inductance can be varied from 8.00 mH to 40.0 mH by the insertion of an iron core in the solenoid. The angular frequency of the generator is 2500 rad/s. If the capacitor voltage is not to exceed 150 V, find (*a*) the peak current and (*b*) the range of inductances that is safe to use.

73 •• A certain electrical device draws an rms current of 10 A at an average power of 720 W when connected to a 120-V-rms, 60-Hz power line. (*a*) What is the impedance of the device? (*b*) What series combination of resistance and reactance would have the same impedance as this device? (*c*) If the current leads the emf, is the reactance inductive or capacitive?

74 •• A method for measuring inductance is to connect the inductor in series with a known capacitance, a known resistance, an ac ammeter, and a variable-frequency signal generator. The frequency of the signal generator is varied and the emf is kept constant until the current is maximum. (*a*) If the capacitance is 10 μF, the peak emf is 10 V, the resistance is 100 Ω, and the rms current in the circuit is maximum when the driving frequency is 5000 rad/s, what is the value of the inductance? (*b*) What is the maximum rms current?

75 •• A resistor and a capacitor are connected in parallel across an ac generator (Figure 29-43) that has an emf given by $\mathcal{E} = \mathcal{E}_{peak} \cos \omega t$. (*a*) Show that the current in the resistor is given by $I_R = (\mathcal{E}_{peak}/R) \cos \omega t$. (*b*) Show that the current in the capacitor branch is given by $I_C = (\mathcal{E}_{peak}/X_C) \cos(\omega t + 90°)$.(*c*) Show that the current in the generator is given by $I = I_{peak} \cos(\omega t + \delta)$, where $\tan \delta = R/X_C$ and $I_{peak} = \mathcal{E}_{peak}/Z$.

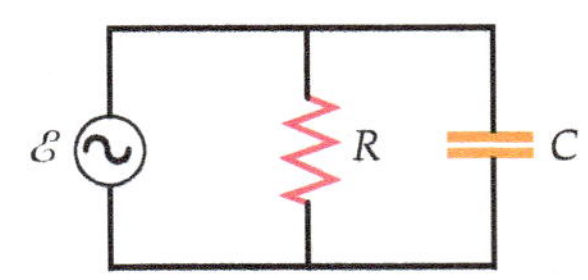

FIGURE 29-43 Problem 75

76 ••• Figure 29-44 shows a plot of average power P_{av} versus generator frequency ω for a series *RLC* circuit driven by an ac generator. The average power P_{av} is given by Equation 29-56. The full width at half-maximum, $\Delta\omega$, is the width of the resonance curve between the two points, where P_{av} is one-half its maximum value. Show that for a sharply peaked resonance, $\Delta\omega \approx R/L$ and that $Q_{factor} \approx \omega_0/\Delta\omega$ (Equation 29-58). *Hint: The half-power points occur when the denominator of Equation 29-56 is equal to twice the value it has at resonance; that is, when* $L^2(\omega^2 - \omega_0^2)^2 + \omega^2R^2 \approx +\omega_0^2R^2$. *Let* ω_1 *and* ω_2 *be the solutions of this equation. Then, show that* $\Delta\omega = \omega_2 - \omega_1 \approx R/L$.

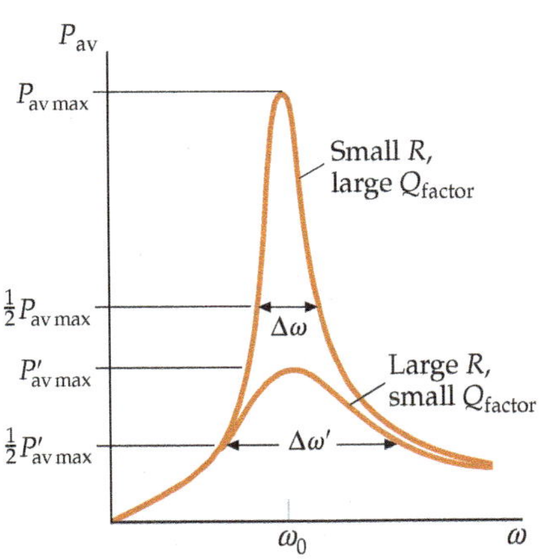

FIGURE 29-44 Problem 76

77 ••• Show by direct substitution that $L\frac{d^2Q}{dt^2} + R\frac{dQ}{dt} + \frac{1}{C}Q = 0$ (Equation 29-43*b*) is satisfied by $Q = Q_0e^{-t/\tau} \cos \omega' t$, where $\tau = 2L/R$, $\omega' = \sqrt{1/(LC) - 1/\tau^2}$, and Q_0 is the charge on the capacitor at $t = 0$.

78 ••• **ENGINEERING APPLICATION** One method for measuring the magnetic susceptibility of a sample uses an *LC* circuit consisting of an air-core solenoid and a capacitor. The resonant frequency of the circuit without the sample is determined and then measured again with the sample inserted in the solenoid. Suppose you have a solenoid that is 4.00 cm long, is 3.00 mm in diameter, and has 400 turns of fine wire. You have a sample that is inserted in the solenoid and completely fills the air space. Neglect end effects. (*a*) Calculate the inductance of the empty solenoid. (*b*) What value for the capacitance of the capacitor should you choose so that the resonance frequency of the circuit without a sample is exactly 6.0000 MHz? (*c*) When a sample is inserted in the solenoid, you determine that the resonance frequency drops to 5.9989 MHz. Use your data to determine the sample's susceptibility.

*THE TRANSFORMER

79 • A rms voltage of 24 V is required for a device whose impedance is 12 Ω. (*a*) What should the turns ratio of a transformer be, so that the device can be operated from a 120-V-rms line? (*b*) Suppose the transformer is accidentally connected in reverse with the secondary winding across the 120-V-rms line and the 12-Ω load across the primary. How much rms current will then be in the primary winding? SSM

80 • A transformer has 400 turns in the primary and 8 turns in the secondary. (*a*) Is this a step-up or a step-down transformer? (*b*) If the primary is connected to a 120-V-rms voltage source, what is the open-circuit rms voltage across the secondary? (*c*) If the primary rms current is 0.100 A, what is the secondary rms current, assuming negligible magnetization current and no power loss?

81 • The primary of a step-down transformer has 250 turns and is connected to a 120-V-rms line. The secondary is to supply 20 A rms at 9.0 V rms. Find (*a*) the rms current in the primary and (*b*) the number of turns in the secondary, assuming 100 percent efficiency.

82 •• An audio oscillator (ac source) that has an internal resistance of 2000 Ω and an open-circuit rms output voltage of 12.0 V is to be used to drive a loudspeaker coil that has a resistance of 8.00 Ω. (*a*) What should be the ratio of primary to secondary turns of a transformer so that maximum average power is transferred to the speaker? (*b*) Suppose a second identical speaker is connected in parallel with the first speaker. How much average power is then supplied to the two speakers combined?

83 • The distribution circuit of a residential power line is operated at 2000 V rms. This voltage must be reduced to 240 V rms for use within residences. If the secondary side of the transformer has 400 turns, how many turns are in the primary?

GENERAL PROBLEMS

84 •• A resistor that has a resistance R carries a current given by $(5.0\text{ A}) \sin 2\pi ft + (7.0\text{ A}) \sin 4\pi ft$, where $f = 60$ Hz. (*a*) What is the rms current in the resistor? (*b*) If $R = 12$ Ω, what is the average power delivered to the resistor? (*c*) What is the rms voltage across the resistor?

85 •• Figure 29-45 shows the voltage versus time for a *square-wave* voltage source. If $V_0 = 12$ V, (*a*) what is the rms voltage of this source? (*b*) If this alternating waveform is rectified by eliminating the negative voltages, so that only the positive voltages remain, what is the new rms voltage? SSM

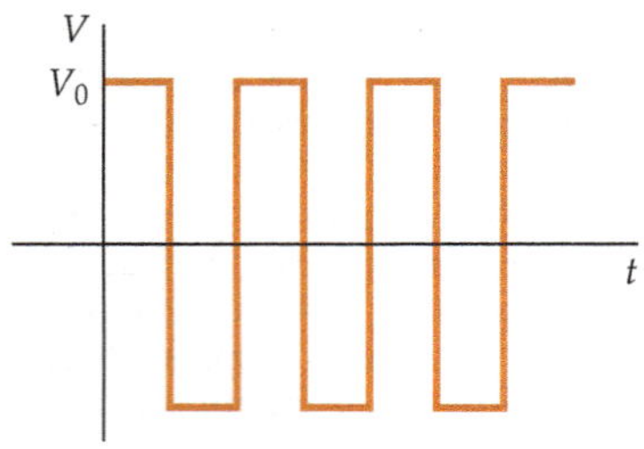

FIGURE 29-45 Problem 85

86 •• What are the average values and rms values of current for the two current waveforms shown in Figure 29-46?

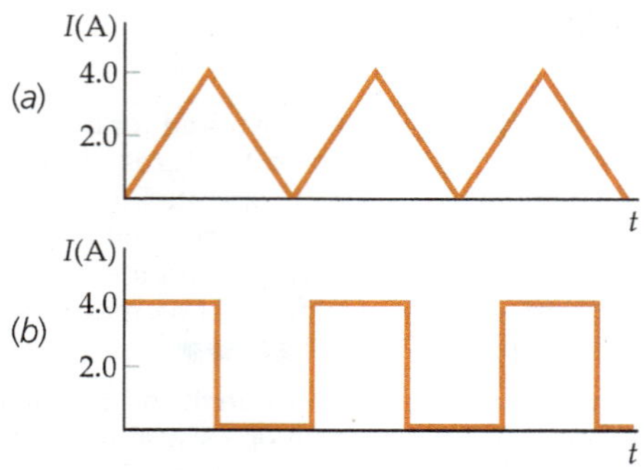

FIGURE 29-46 Problem 86

87 •• In the circuit shown in Figure 29-47, $\mathcal{E}_1 = (20\text{ V})\cos 2\pi ft$, where $f = 180$ Hz; $\mathcal{E}_2 = 18$ V; and $R = 36\ \Omega$. Find the maximum, minimum, average, and rms values of the current in the resistor.

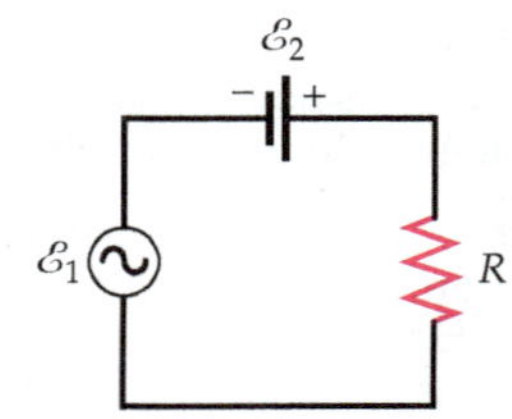

FIGURE 29-47 Problems 87, 88, and 89

88 •• Repeat Problem 87 if the resistor is replaced by a 2.0-μF capacitor.

89 ••• A circuit consists of an ac generator, a capacitor and an ideal inductor—all connected in series. The emf of the generator is given by $\mathcal{E}_{\text{peak}}\cos\omega t$. (*a*) Show that the charge on the capacitor obeys the equation $L\dfrac{d^2Q}{dt^2} + \dfrac{Q}{C} = \mathcal{E}_{\text{peak}}\cos\omega t$. (*b*) Show by direct substitution that this equation is satisfied by $Q = Q_{\text{peak}}\cos\omega t$ where $Q_{\text{peak}} = -\dfrac{\mathcal{E}_{\text{peak}}}{L(\omega^2 - \omega_0^2)}$. (*c*) Show that the current can be written as $I = I_{\text{peak}}\cos(\omega t - \delta)$, where $I_{\text{peak}} = \dfrac{\omega\mathcal{E}_{\text{peak}}}{L|\omega^2 - \omega_0^2|} = \dfrac{\mathcal{E}_{\text{peak}}}{|X_L - X_C|}$, $\delta = -90°$ for $\omega < \omega_0$, and $\delta = 90°$ for $\omega > \omega_0$, where ω_0 is the resonance frequency. SSM

CHAPTER 30

SET ACROSS THE DESERT NEAR SOCORRO, NEW MEXICO, THE NATIONAL RADIO ASTRONOMY OBSERVATORY'S VERY LARGE ARRAY IS A SYSTEM OF 27 RADIO ANTENNAS SET IN A Y-SHAPED CONFIGURATION. BECAUSE THE INFORMATION GATHERED FROM THE ARRAY IS COMBINED ELECTRONICALLY, THE INSTRUMENT HAS A RESOLUTION THAT IS 22 MILES WIDE. *(NRAO/AUI.)*

Maxwell's Equations and Electromagnetic Waves

Maxwell's equations, first proposed by the great Scottish physicist James Clerk Maxwell, relate the electric and magnetic field vectors $\vec{E}$ and $\vec{B}$ and their sources, which are electric charges and currents. These equations summarize the experimental laws of electricity and magnetism—the laws of Coulomb, Gauss, Biot–Savart, Ampère, and Faraday. These experimental laws hold in general except for Ampère's law, which applies only to steady continuous currents.

In this chapter, we will see how Maxwell was able to generalize Ampère's law with the invention of the displacement current (Section 30-1). Maxwell was then able to show that the generalized laws of electricity and magnetism imply the existence of electromagnetic waves.

? Did you ever wonder whether a radio antenna generates a wave equally in all directions?

(See Example 30-5.)

30-1 MAXWELL'S DISPLACEMENT CURRENT

Maxwell's equations play a role in classical electromagnetism analogous to the role of Newton's laws in classical mechanics. In principle, all problems in classical electricity and magnetism can be solved using Maxwell's equations, just as all problems in classical mechanics can be solved by using Newton's laws. Maxwell's equations are considerably more complicated than Newton's laws, however, and their application to most problems involves mathematics beyond the scope of this book. Nevertheless, Maxwell's equations are of great theoretical importance. For example, Maxwell showed that these equations can be combined to yield a wave equation for the electric and magnetic field vectors $\vec{E}$ and $\vec{B}$. Such **electromagnetic waves** are generated by accelerating charges (for example, the charges in an alternating current in an antenna). Electromagnetic waves were first produced in the laboratory by Heinrich Hertz in 1887. Maxwell showed that his equations predicted the speed of electromagnetic waves in free space to be

$$c = \frac{1}{\sqrt{\mu_0 \epsilon_0}} \qquad 30\text{-}1$$

THE SPEED OF ELECTROMAGNETIC WAVES

where ϵ_0, the electric constant, is the constant appearing in Coulomb's and Gauss's laws and μ_0, the magnetic constant, is the constant appearing in the Biot–Savart law and Ampère's law. Maxwell noticed with great excitement the coincidence that the measure for the speed of light equaled $1/\sqrt{\mu_0 \epsilon_0}$, and Maxwell correctly surmised that light itself is an electromagnetic wave. Today, the value of c is defined as $2.997\,924\,58 \times 10^8$ m/s, the value of μ_0 is defined as $4\pi \times 10^{-7}$ N/A^2, and the value of ϵ_0 is defined by Equation 30-1.

Ampère's law (Equation 27-16) relates the line integral of the magnetic field around some closed curve C to the current that passes through any surface bounded by that curve:

$$\oint_C \vec{B} \cdot d\vec{\ell} = \mu_0 I_S \qquad \text{for any closed curve } C \qquad 30\text{-}2$$

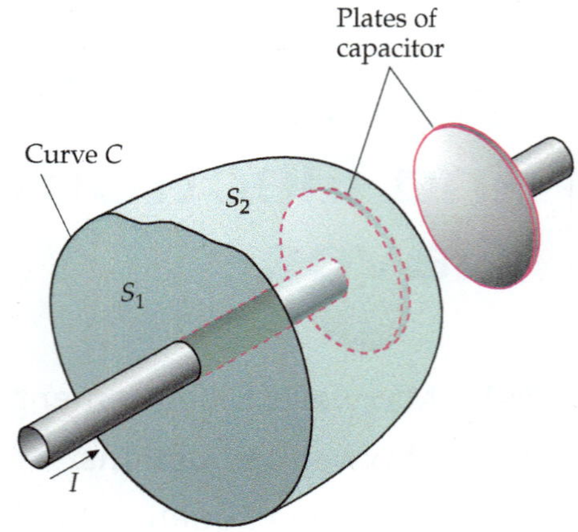

FIGURE 30-1 Two surfaces S_1 and S_2 bounded by the same curve C. The current I passes through surface S_1 but not through surface S_2. Ampère's law, which relates the line integral of the magnetic field around the curve C to the total current passing through any surface bounded by C, is not valid when the current is not continuous, as when it stops at the capacitor plate here.

Maxwell recognized a flaw in Ampère's law. Figure 30-1 shows two different surfaces, S_1 and S_2, bounded by the same curve C, which encircles a current carrying wire that is connected to a capacitor plate. The current through surface S_1 is I, but no current exists through surface S_2 because the charge stops on the capacitor plate. Thus, ambiguity exists in the phrase "the current through any surface bounded by the curve." Such a problem arises when the current is not continuous.

Maxwell showed that the law can be generalized to include all situations if the current I in the equation is replaced by the sum of the current I and another term I_d, called **Maxwell's displacement current,** defined as

$$I_d = \epsilon_0 \frac{d\phi_e}{dt} \qquad 30\text{-}3$$

DEFINITION—DISPLACEMENT CURRENT

where ϕ_e is the flux of the electric field through the same surface bounded by the curve C. The generalized form of Ampère's law is then

$$\oint_C \vec{B} \cdot d\vec{\ell} = \mu_0(I + I_d) = \mu_0 I + \mu_0 \epsilon_0 \frac{d\phi_e}{dt} \qquad 30\text{-}4$$

GENERALIZED FORM OF AMPÈRE'S LAW

We can understand this generalization by considering Figure 30-1 again. Let us call the sum $I + I_d$* the generalized current. According to the argument just stated, the same generalized current must cross any surface bounded by the curve C. The surfaces S_1 and S_2 together form a single closed surface. Thus, the sum of the generalized currents into the region enclosed by the two surfaces S_1 and S_2, is equal to the sum of the generalized currents out of the region. If a net current I into the enclosed region exists, an equal net displacement current I_d out of the enclosed region must exist. In the enclosed region in the figure, a net current I into the region exists that increases the charge Q_{inside} within the region:

$$I = \frac{dQ_{\text{inside}}}{dt}$$

The net flux of the electric field out of the enclosed region is related to the charge enclosed by Gauss's law:

$$\phi_{\text{e net}} = \oint_S E_n\, dA = \frac{1}{\epsilon_0} Q_{\text{inside}}$$

Solving for the charge gives

$$Q_{\text{inside}} = \epsilon_0 \phi_{\text{e net}}$$

and taking the derivative of each side gives

$$\frac{dQ_{\text{inside}}}{dt} = \epsilon_0 \frac{d\phi_{\text{e net}}}{dt}$$

The rate of increase of the enclosed charge is thus proportional to the rate of increase of the net flux of the electric field out of the region:

$$\frac{dQ_{\text{inside}}}{dt} = \epsilon_0 \frac{d\phi_{\text{e net}}}{dt} = I_d$$

Thus, the net current into the volume equals the net displacement current out of the volume. The generalized current is thus continuous, and this is *always* the case.

It is interesting to compare Equation 30-4 to Equation 28-6:

$$\mathcal{E} = \oint_C \vec{E} \cdot d\vec{\ell} = -\frac{d\phi_m}{dt} = -\int_S \frac{\partial B_n}{\partial t}\, dA \qquad 30\text{-}5$$

FARADAY'S LAW

which in this chapter will be referred to as Faraday's law. (Equation 30-5 is a restricted form of Faraday's law, a form that does include emfs associated with time-varying magnetic fields, but does not include emfs associated with moving conductors.) According to Faraday's law, a changing magnetic flux produces an electric field whose line integral around a closed curve is proportional to the rate of change of magnetic flux through any surface bounded by the curve. Maxwell's modification of Ampère's law shows that a changing electric flux produces a magnetic field whose line integral around a curve is proportional to the rate of change of the electric flux. We thus have the interesting reciprocal result that a changing magnetic field produces an electric field (Faraday's law) and a changing electric field produces a magnetic field (generalized form of Ampère's law). Note, no magnetic analog of a current I exists. This is consistent with the observation that the magnetic monopole, the magnetic analog of an electric charge, does not exist.†

---

* In more advanced treatments, the generalized current is taken as the sum of a conduction current and a displacement current, where the conduction current is attributed to the motion of free (delocalized) charge carriers, and the displacement current is what is referred to in this book as the displacement current and a term associated with the motion of bound (localized) charge carriers.

† The question of the existence of magnetic monopoles has theoretical importance. Numerous attempts to observe magnetic monopoles have been made but to date no one has been unambiguously successful.

Example 30-1 Calculating Displacement Current

A parallel-plate capacitor has closely spaced circular plates of radius R. The current I in the wires connected to the plates is 2.5 A, as shown in Figure 30-2. Compute the displacement current I_d through surface S passing between the plates by directly computing the rate of change of the flux of $\vec{E}$ through surface S.

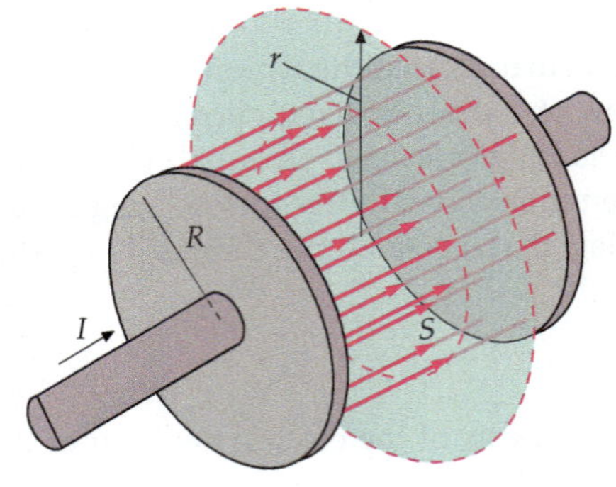

FIGURE 30-2 The surface S between the capacitor plates is penetrated by electric field lines. The charge Q on the positively charged plate is increasing at 2.5 C/s = 2.5 A. The distance between the plates is not drawn to scale. The plates are much closer together than the plates shown in the figure.

PICTURE The displacement current is $I_d = \epsilon_0\, d\phi_e/dt$, where ϕ_e is the electric flux through the surface between the plates. Because the parallel plates are closely spaced, the electric field in the region between the plates is uniform and perpendicular to the plates. Outside the region between the plates the electric field is negligible. Thus, the electric flux is simply $\phi_e = EA$, where E is the electric field between the plates and A is the plate area.

SOLVE

1. The displacement current is found by taking the time derivative of the electric flux:
$$I_d = \epsilon_0 \frac{d\phi_e}{dt}$$
2. The flux equals the electric field magnitude multiplied by the plate area:
$$\phi_e = EA$$
3. The electric field is proportional to the charge density on the plates, which we treat as uniformly distributed:
$$E = \frac{\sigma}{\epsilon_0} = \frac{Q/A}{\epsilon_0}$$
4. Substitute these results to calculate I_d:
$$I_d = \epsilon_0 \frac{d(EA)}{dt} = \epsilon_0 A\frac{dE}{dt} = \epsilon_0 A \frac{d}{dt}\left(\frac{Q}{A\epsilon_0}\right)$$
$$= \frac{dQ}{dt} = \boxed{2.5\text{ A}}$$

CHECK The step-4 result is equal to the current in the wires, as expected.

Example 30-2 Calculating $\vec{B}$ from Displacement Current

The circular plates in Example 30-1 have a radius of $R = 3.0$ cm. Find the magnetic field strength B at a point between the plates a distance $r = 2.0$ cm from the axis through the centers of the plates when the current into the positive plate is 2.5 A.

PICTURE We find B from the generalized form of Ampère's law (Equation 30-4). We chose a circular path C of radius $r = 2.0$ cm about the centerline joining the plates, as shown in Figure 30-3. We then calculate the displacement current through the surface S bounded by C. By symmetry, $\vec{B}$ is tangent to C and has the same magnitude everywhere on C.

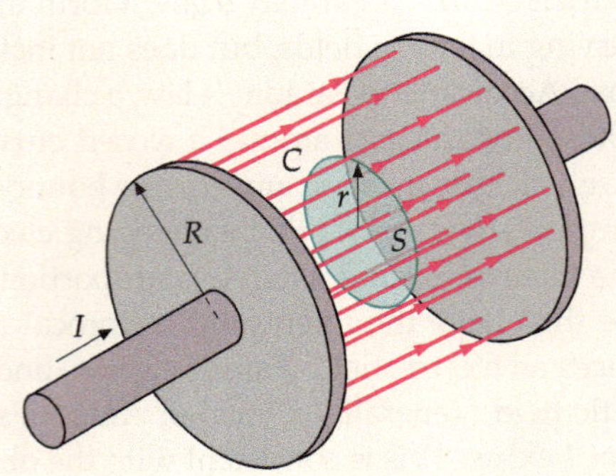

FIGURE 30-3 The distance between the plates is not drawn to scale. The plates are much closer together than they appear.

SOLVE

1. We find B from the generalized form of Ampère's law:
$$\oint_C \vec{B}\cdot d\vec{\ell} = \mu_0(I + I_d)$$
where
$$I_d = \epsilon_0 \frac{d\phi_e}{dt}$$
2. The line integral is B multiplied by the circumference of the circle:
$$\oint_C \vec{B}\cdot d\vec{\ell} = B\cdot 2\pi r$$

3. Because no charges are moving through the surface S, $I = 0$. Thus, the generalized current through S is just the displacement current:

$$\oint_C \vec{B}\cdot d\vec{\ell} = \mu_0 I + \mu_0\epsilon_0\frac{d\phi_e}{dt}$$

$$B\cdot 2\pi r = 0 + \mu_0\epsilon_0\frac{d\phi_e}{dt}$$

4. The electric flux through S equals the product of the uniform field strength E and the area A of the flat surface S bounded by the curve C, and E is equal to σ/ϵ_0:

$$\phi_e = AE = \pi r^2 E = \pi r^2\frac{\sigma}{\epsilon_0}$$

$$= \pi r^2\frac{Q}{\epsilon_0\pi R^2} = \frac{Qr^2}{\epsilon_0 R^2}$$

5. Substitute these results into step 3 and solve for B:

$$B\cdot 2\pi r = \mu_0\epsilon_0\frac{d}{dt}\left(\frac{Qr^2}{\epsilon_0 R^2}\right) = \mu_0\frac{r^2}{R^2}\frac{dQ}{dt}$$

$$B = \frac{\mu_0}{2\pi}\frac{r}{R^2}\frac{dQ}{dt} = \frac{\mu_0}{2\pi}\frac{r}{R^2}I$$

$$= (2\times 10^{-7}\,\text{T}\cdot\text{m/A})\frac{0.02\text{ m}}{(0.03\text{ m})^2}(2.5\text{ A})$$

$$= \boxed{1.11\times 10^{-5}\text{ T}}$$

30-2 MAXWELL'S EQUATIONS

Maxwell's equations are

$$\oint_S E_n\,dA = \frac{1}{\epsilon_0}Q_{\text{inside}} \qquad \text{30-6}a$$

GAUSS'S LAW

$$\oint_S B_n\,dA = 0 \qquad \text{30-6}b$$

GAUSS'S LAW FOR MAGNETISM

$$\oint_C \vec{E}\cdot d\vec{\ell} = -\frac{d}{dt}\int_S B_n\,dA = -\int_S\frac{\partial B_n}{\partial t}dA \qquad \text{30-6}c$$

FARADAY'S LAW

$$\oint_C \vec{B}\cdot d\vec{\ell} = \mu_0(I + I_d),\text{ where } I_d = \epsilon_0\int_S\frac{\partial E_n}{\partial t}dA \qquad \text{30-6}d$$

AMPÈRE'S LAW

MAXWELL'S EQUATIONS*

Gauss's law (Equation 30-6a) states that the flux of the electric field through any closed surface equals $1/\epsilon_0$ multiplied by the net charge inside the surface. As discussed in Chapter 22, Gauss's law implies that the electric field $\vec{E}$ due to a point charge varies inversely as the square of the distance from the charge. This law describes how electric field lines diverge from a positive charge and converge on a negative charge. Its experimental basis is Coulomb's law.

* In all four of Maxwell's equations, the integration paths C and the integration surfaces S are at rest and the integrations take place at an instant in time.

Gauss's law for magnetism (Equation 30-6*b*) states that the flux of the magnetic field $\vec{B}$ through *any* closed surface is zero. This equation describes the experimental observation that magnetic field lines do not diverge from any point in space or converge to any point in space; that is, it implies that isolated magnetic poles do not exist.

Faraday's law (Equation 30-6*c*) states that the line integral of the electric field $\vec{E}$ around any closed curve C equals the negative of the rate of change of the flux of the magnetic field $\vec{B}$ through any surface S bounded by curve C. (S is not a closed surface, so the magnetic flux through S is not necessarily zero.) Faraday's law describes how electric field lines encircle any area through which the magnetic flux is changing, and it relates the electric field vector $\vec{E}$ to the rate of change of the magnetic field vector $\vec{B}$.

Ampère's law modified to include Maxwell's displacement current (Equation 30-6*d*) states that the line integral of the magnetic field $\vec{B}$ around any closed curve C equals μ_0 multiplied by the sum of the current I through any surface S bounded by the curve and the displacement current I_d through the same surface. This law describes how the magnetic field lines encircle an area through which a current or a displacement current is passing.

In Section 30-3, we show how wave equations for both the electric field $\vec{E}$ and the magnetic field $\vec{B}$ can be derived from Maxwell's equations.

30-3 THE WAVE EQUATION FOR ELECTROMAGNETIC WAVES

In Section 15-1, we saw that waves on a string obey a partial differential equation called the **wave equation:**

$$\frac{\partial^2 y(x,t)}{\partial x^2} = \frac{1}{v^2}\frac{\partial^2 y(x,t)}{\partial t^2} \qquad \text{30-7}$$

where $y(x, t)$ is the wave function, which for string waves is the displacement of the string. The velocity of the wave is given by $v = \sqrt{F_T/\mu}$, where F_T is the tension and μ is the linear mass density. The general solution to this equation is

$$y(x,t) = y_1(x - vt) + y_2(x + vt)$$

where y_1 and y_2 are functions of $x - vt$ and $x + vt$, respectively. The general solution functions can be expressed as a superposition of harmonic wave functions of the form

$$y(x,t) = y_0 \sin(kx - \omega t) \qquad \text{and} \qquad y(x,t) = y_0 \sin(kx + \omega t)$$

where $k = 2\pi/\lambda$ is the wave number and $\omega = 2\pi f$ is the angular frequency.

Maxwell's equations imply that $\vec{E}$ and $\vec{B}$ obey wave equations similar to Equation 30-7. We consider only empty space (space in which no charges or currents exist) and we assume that the electric and magnetic fields $\vec{E}$ and $\vec{B}$ are functions of time and one space coordinate only, which we will take to be the x coordinate. Such a wave is called a **plane wave,** because $\vec{E}$ and $\vec{B}$ are uniform throughout any plane perpendicular to the x axis. For a plane electromagnetic wave traveling parallel to the x axis, the x components of the fields are zero, so the vectors $\vec{E}$ and $\vec{B}$ are perpendicular to the x axis and each obeys the wave equation:

$$\frac{\partial^2 \vec{E}}{\partial x^2} = \frac{1}{c^2}\frac{\partial^2 \vec{E}}{\partial t^2} \qquad \text{30-8a}$$

WAVE EQUATION FOR $\vec{E}$

$$\frac{\partial^2 \vec{B}}{\partial x^2} = \frac{1}{c^2}\frac{\partial^2 \vec{B}}{\partial t^2} \qquad 30\text{-}8b$$

WAVE EQUATION FOR $\vec{B}$

where $c = 1/\sqrt{\mu_0 \epsilon_0}$ is the speed of the waves. (*Note:* Dimensional analysis helps in remembering these equations. For each equation, the numerators on both sides are the same and the denominators on both sides have the dimension of length squared.)

DERIVATION OF THE WAVE EQUATION

We can relate the space derivative of one of the field vectors to the time derivative of the other field vector by applying Faraday's law (Equation 30-6*c*) and the modified version of Ampère's law (Equation 30-6*d*) to appropriately chosen curves in space. We first relate the space derivative of E_y to the time derivative of B_z by applying Equation 30-6*c* (Faraday's law) to the rectangular curve of sides Δx and Δy lying in the xy plane (Figure 30-4). The circulation of $\vec{E}$ around C, for small Δx and Δy, is given by

$$\oint_C \vec{E}\cdot d\vec{\ell} = E_y(x_2)\,\Delta y - E_y(x_1)\,\Delta y = [E_y(x_2) - E_y(x_1)]\,\Delta y$$

where $E_y(x_1)$ is the value of E_y at $x = x_1$ and $E_y(x_2)$ is the value of E_y at $x = x_2$. The contributions of the type $E_x \Delta x$ from the top and bottom of this curve are zero because $E_x = 0$. Because Δx is very small (compared to the wavelength), we can approximate the difference in E_y on the left and right sides of this curve (at x_1 and at x_2) by

$$E_y(x_2) - E_y(x_1) = \Delta E_y \approx \frac{\partial E_y}{\partial x}\Delta x$$

Then

$$\oint_C \vec{E}\cdot d\vec{\ell} \approx \frac{\partial E_y}{\partial x}\Delta x\,\Delta y$$

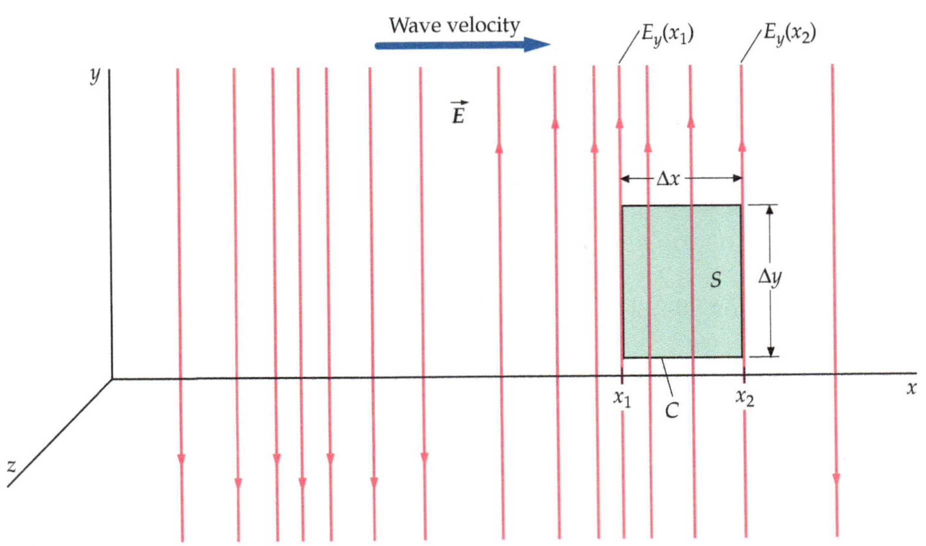

FIGURE 30-4 A rectangular curve in the xy plane for the derivation of Equation 30-9.

Faraday's law is

$$\oint_C \vec{E}\cdot d\vec{\ell} = -\int_S \frac{\partial B_n}{\partial t}dA$$

The time derivative of the flux of $\partial B_n/\partial t$ through the rectangular surface bounded by this curve is given by

$$\int_S B_n\, dA \approx \frac{\partial B_z}{\partial t}\Delta x\,\Delta y$$

Faraday's law then gives

$$\frac{\partial E_y}{\partial x}\Delta x\,\Delta y = -\frac{\partial B_z}{\partial t}\Delta x\,\Delta y$$

or

$$\frac{\partial E_y}{\partial x} = -\frac{\partial B_z}{\partial t} \qquad \text{30-9}$$

Equation 30-9 implies that if there is a component of the electric field E_y that depends on x, there must be a component of the magnetic field B_z that depends on time or, conversely, if there is a component of the magnetic field B_z that depends on time, there must be a component of the electric field E_y that depends on x. We can get a similar equation relating the space derivative of the magnetic field B_z to the time derivative of the electric field E_y by applying Ampère's law (Equation 30-6*d*) to the curve of sides Δx and Δz in the xz plane shown in Figure 30-5.

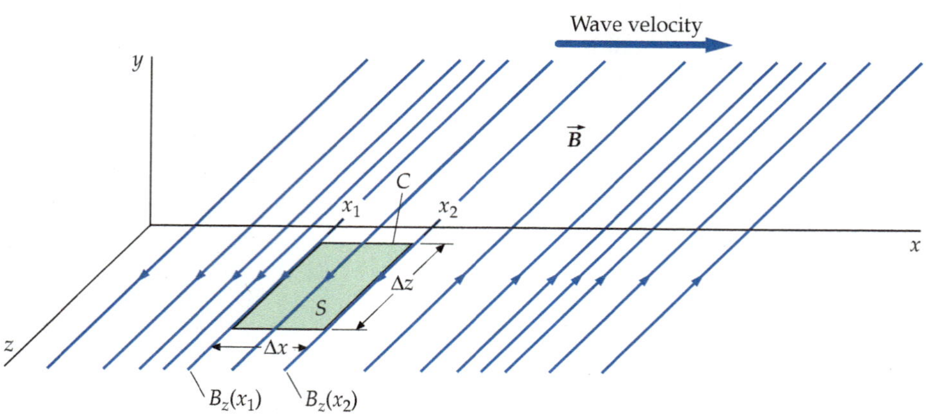

FIGURE 30-5 A rectangular curve in the $y = 0$ plane for the derivation of Equation 30-10.

For the case of no currents ($I = 0$), Equation 30-6*d* is

$$\oint_C \vec{B}\cdot d\vec{\ell} = \mu_0\epsilon_0\int_S \frac{\partial E_n}{\partial t}dA$$

The details of this calculation are similar to those for Equation 30-9. The result is

$$\frac{\partial B_z}{\partial x} = -\mu_0\epsilon_0\frac{\partial E_y}{\partial t} \qquad \text{30-10}$$

We can eliminate either B_z or E_y from Equations 30-9 and 30-10 by differentiating both sides of either equation with respect to either x or t. If we differentiate both sides of Equation 30-9 with respect to x, we obtain

$$\frac{\partial}{\partial x}\left(\frac{\partial E_y}{\partial x}\right) = -\frac{\partial}{\partial x}\left(\frac{\partial B_z}{\partial t}\right)$$

Interchanging the order of the time and space derivatives on the term to the right of the equal sign gives

$$\frac{\partial^2 E_y}{\partial x^2} = -\frac{\partial}{\partial t}\left(\frac{\partial B_z}{\partial x}\right)$$

Using Equation 30-10, we substitute for $\partial B_z/\partial x$ to obtain

$$\frac{\partial^2 E_y}{\partial x^2} = -\frac{\partial}{\partial t}\left(-\mu_0\epsilon_0\frac{\partial E_y}{\partial t}\right)$$

which yields the wave equation

$$\frac{\partial^2 E_y}{\partial x^2} = \mu_0\epsilon_0\frac{\partial^2 E_y}{\partial t^2} \qquad \text{30-11}$$

Comparing Equation 30-11 with Equation 30-7, we see that E_y obeys a wave equation for waves with speed $v = 1/\sqrt{\mu_0\epsilon_0}$, which is Equation 30-1.

If we had instead chosen to eliminate E_y from Equations 30-9 and 30-10 (by differentiating Equation 30-9 with respect to t, for example), we would have obtained an equation identical to Equation 30-11 except with B_z replacing E_y. We can thus see that both the electric field E_y and the magnetic field B_z obey a wave equation for waves traveling with the velocity $1/\sqrt{\mu_0\epsilon_0}$. By substituting the measured values for μ_0 and ϵ_0, Maxwell showed that the value of $1/\sqrt{\mu_0\epsilon_0}$ is equal to the measured value for speed of light.

By following the same line of reasoning as used above, and applying Equation 30-6*c* (Faraday's law) to the curve in the *xz* plane (Figure 30-5), we would obtain

$$\frac{\partial E_z}{\partial x} = \frac{\partial B_y}{\partial t} \qquad \text{30-12}$$

Similarly, the application of Equation 30-6*d* to the curve in the *xy* plane (Figure 30-4) gives

$$\frac{\partial B_y}{\partial x} = \mu_0\epsilon_0\frac{\partial E_z}{\partial t} \qquad \text{30-13}$$

We can use these results to show that, for a wave propagating in the x direction, the components E_z and B_y also obey the wave equation.

To show that the magnetic field B_z is in phase with the electric field E_y, consider the harmonic wave function of the form

$$E_y = E_0 \sin(kx - \omega t) \qquad \text{30-14}$$

If we substitute this solution into Equation 30-9, we have

$$\frac{\partial B_z}{\partial t} = -\frac{\partial E_y}{\partial x} = -kE_0\cos(kx - \omega t)$$

To solve for B_z, we take the integral of $\partial B_z/\partial t$ with respect to time. Doing so yields

$$B_z = \int \frac{\partial B_z}{\partial t}\,dt = \frac{k}{\omega}E_0\sin(kx - \omega t) + f(x) \qquad \text{30-15}$$

where $f(x)$ is an arbitrary function of x.

PRACTICE PROBLEM 30-1

Verify Equation 30-15 by showing that $\frac{\partial}{\partial t}\left[\frac{k}{\omega}E_0\sin(kx - \omega t) + f(x)\right]$ is equal to $-kE_0\cos(kx - \omega t)$.

We next substitute the solution (Equation 30-14) into Equation 30-10 and obtain

$$\frac{\partial B_z}{\partial x} = -\mu_0\epsilon_0\frac{\partial E_y}{\partial t} = \omega\mu_0\epsilon_0 E_0 \cos(kx - \omega t)$$

Solving for B_z gives

$$B_z = \int \frac{\partial B_z}{\partial x}\,dx = \frac{\omega\mu_0\epsilon_0}{k}E_0 \sin(kx - \omega t) + g(t) \qquad \text{30-16}$$

where $g(t)$ is an arbitrary function of time. Equating the right sides of Equations 30-15 and 30-16 gives

$$\frac{k}{\omega}E_0 \sin(kx - \omega t) + f(x) = \frac{\omega\mu_0\epsilon_0}{k}E_0 \sin(kx - \omega t) + g(t)$$

Substituting c for ω/k and $1/c^2$ for $\mu_0\epsilon_0$ gives

$$\frac{1}{c}E_0 \sin(kx - \omega t) + f(x) = \frac{1}{c}E_0 \sin(kx - \omega t) + g(t)$$

which implies $f(x) = g(t)$ for all values of x and t. These remain equal only if $f(x) = g(t) =$ constant (independent of both x and t). Thus, Equation 30-15 becomes

$$B_z = \frac{k}{\omega}E_0 \sin(kx - \omega t) + \text{constant} = B_0 \sin(kx - \omega t) \qquad \text{30-17}$$

where $B_0 = (k/\omega)E_0 = (1/c)E_0$. The integration constant was dropped because it plays no part in the wave. It merely allows for the presence of a static uniform magnetic field. Because the electric and magnetic fields oscillate in phase and have the same frequency, we have the general result that the magnitude of the electric field is c multiplied by the magnitude of the magnetic field for an electromagnetic wave:

$$E = cB \qquad \text{30-18}$$

The direction of propagation of an electromagnetic wave is always the direction of the vector product $\vec{E} \times \vec{B}$. For the wave described in the preceding discussion, the electric and magnetic fields are given by $\vec{E} = E_0 \sin(kx - \omega t)\hat{j}$ and $\vec{B} = B_0 \sin(kx - \omega t)\hat{k}$. Thus,

$$\vec{E} \times \vec{B} = [E_0 \sin(kx - \omega t)\hat{j}] \times [B_0 \sin(kx - \omega t)\hat{k}] = E_0B_0 \sin^2(kx - \omega t)\hat{i}$$

The term on the right is a vector in the $+x$ direction, so we have verified that $\vec{E} \times \vec{B}$ is in the direction of propagation for this electromagnetic wave.

We see that Maxwell's equations imply wave equations 30-8*a* and 30-8*b* for the electric and magnetic fields; and that if E_y varies harmonically, as in Equation 30-14, the magnetic field B_z is in phase with E_y and has an amplitude related to the amplitude of E_y by $B_z = E_y/c$. The electric and magnetic fields are perpendicular to each other and to the direction of the wave propagation.

> **!** The direction of propagation of an electromagnetic wave is always the direction of the vector product $\vec{E} \times \vec{B}$.

Example 30-3 $\vec{B}(y, t)$ for a Linearly Polarized Plane Wave

The expression for the electric field of a certain electromagnetic wave is given by $\vec{E}(y, t) = E_0 \sin(ky + \omega t)\hat{k}$. (*a*) What is the direction of propagation of the wave? (*b*) What is the corresponding expression for the magnetic field on the wave?

PICTURE The argument of the sine function gives the direction of propagation. $\vec{B}$ is perpendicular to both $\vec{E}$ and to the direction of propagation. $\vec{B}$ and $\vec{E}$ are in phase and $\vec{E} \times \vec{B}$ is in the direction of propagation.

SOLVE

(*a*) The argument of the sine function $(ky + \omega t)$ tells us the direction of propagation:

The direction of propagation is the $-y$ direction, which is the direction of $-\hat{j}$.

(*b*) 1. $\vec{B}$ is in phase with $\vec{E}$ and is perpendicular to both $\vec{E}$ and the direction of propagation $-\hat{j}$. (That is, $\vec{B}$ is perpendicular to both $\hat{j}$ and $\hat{k}$.) This result means:

Either $\vec{B}(y,t) = +B_0 \sin(ky + \omega t)\hat{i}$ or

$\vec{B}(y,t) = -B_0 \sin(ky + \omega t)\hat{i}$

2. $\vec{E} \times \vec{B}$ is in the direction of propagation, $-\hat{j}$. Assume $\vec{B}(x,t) = +B_0 \sin(ky + \omega t)\hat{i}$ and calculate the product $\vec{E} \times \vec{B}$:

$$\begin{aligned}\vec{E} \times \vec{B} &= E_0 \sin(ky + \omega t)\hat{k} \times B_0 \sin(ky + \omega t)\hat{i} \\ &= E_0B_0 \sin^2(ky + \omega t)(\hat{k} \times \hat{i}) \\ &= E_0B_0 \sin^2(ky + \omega t)\hat{j}\end{aligned}$$

3. The step-2 result contradicts the reality that the direction of propagation is the $-y$ direction. Take the cross product $\vec{E} \times \vec{B}$ with the other choice for the expression for the magnetic field:

$$\begin{aligned}\vec{E} \times \vec{B} &= E_0 \sin(ky + \omega t)\hat{k} \times (-B_0) \sin(ky + \omega t)\hat{i} \\ &= E_0(-B_0) \sin^2(ky + \omega t)(\hat{k} \times \hat{i}) \\ &= -E_0B_0 \sin^2(ky + \omega t)\,\hat{j}\end{aligned}$$

4. The step-3 result is in the direction of propagation. The correct expression for the magnetic field is:

$$\boxed{\vec{B}(y,t) = -B_0 \sin(ky + \omega t)\hat{i}}$$

where $B_0 = E_0/c$ (Equation 30-18).

CHECK The step-4 result is perpendicular to both $\vec{E}$ and to the direction of propagation, as expected.

Example 30-4 $\vec{B}(x, t)$ for a Circular Polarized Plane Wave

The expression for the electric field of a certain electromagnetic wave is given by $\vec{E}(x,t) = E_0 \sin(kx - \omega t)\hat{j} + E_0 \cos(kx - \omega t)\hat{k}$. (*a*) Find the corresponding magnetic field of the same wave. (*b*) Compute $\vec{E} \cdot \vec{B}$ and $\vec{E} \times \vec{B}$.

PICTURE We can solve this example by using the principle of superposition. The given electric field is the superposition of two fields, the one given in Equation 30-14 and the other given by $E_0 \cos(kx - \omega t)\hat{k}$.

SOLVE

(*a*) 1. From the arguments of the trigonometric functions we can see that the direction of propagation is the $+x$ direction:

The wave is traveling in the $+x$ direction.

2. The given electric field can be considered as the superposition of $\vec{E}_1 = E_0 \sin(kx - \omega t)\hat{j}$ and $\vec{E}_2 = E_0 \cos(kx - \omega t)\hat{k}$. Find the magnetic fields $\vec{B}_1$ and $\vec{B}_2$ associated with these electric fields, respectively. Use the procedure followed in Example 30-3:

For $\vec{E}_1 = E_0 \sin(kx - \omega t)\hat{j}$, $\vec{B}_1 = B_0 \sin(kx - \omega t)\hat{k}$ where $B_0 = E_0/c$ (Equation 30-18),

and

For $\vec{E}_2 = E_0 \cos(kx - \omega t)\hat{k}$, $\vec{B}_2 = -B_0 \cos(kx - \omega t)\hat{j}$ where $B_0 = E_0/c$.

3. The superposition of magnetic fields gives the resultant magnetic field:

$$\boxed{\begin{aligned}\vec{B}(x,t) &= \vec{B}_1 + \vec{B}_2 \\ &= B_0 \sin(kx - \omega t)\hat{k} - B_0 \cos(kx - \omega t)\hat{j} \\ &\text{where} \\ B_0 &= E_0/c\end{aligned}}$$

(*b*) 1. Let $\theta = kx - \omega t$ to simplify the notation and calculate $\vec{E} \cdot \vec{B}$:

$$\begin{aligned}\vec{E} \cdot \vec{B} &= (E_0 \sin\theta\,\hat{j} + E_0 \cos\theta\,\hat{k}) \cdot (B_0 \sin\theta\,\hat{k} - B_0 \cos\theta\,\hat{j}) \\ &= E_0B_0 \sin^2\theta\,\hat{j} \cdot \hat{k} - E_0B_0 \sin\theta \cos\theta\,\hat{j} \cdot \hat{j} \\ &\quad + E_0B_0 \cos\theta \sin\theta\,\hat{k} \cdot \hat{k} - E_0B_0 \cos^2\theta\,\hat{k} \cdot \hat{j} \\ &= 0 - E_0B_0 \sin\theta \cos\theta + E_0B_0 \cos\theta \sin\theta - 0 = \boxed{0}\end{aligned}$$

2. Calculate $\vec{E} \times \vec{B}$:

$$\begin{aligned}\vec{E} \times \vec{B} &= (E_0 \sin\theta\, \hat{j} + E_0 \cos\theta\, \hat{k}) \times (-B_0 \cos\theta\, \hat{j} + B_0 \sin\theta\, \hat{k}) \\ &= -E_0B_0 \sin\theta \cos\theta\, (\hat{j} \times \hat{j}) + E_0B_0 \sin^2\theta\, (\hat{j} \times \hat{k}) \\ &\quad - E_0B_0 \cos^2\theta\, (\hat{k} \times \hat{j}) + E_0B_0 \cos\theta \sin\theta\, (\hat{k} \times \hat{k}) \\ &= 0 + E_0B_0 \sin^2\theta\, \hat{i} + E_0B_0 \cos^2\theta\, \hat{i} + 0 = \boxed{E_0B_0\hat{i}}\end{aligned}$$

CHECK The Part (*b*), step 1 result verifies that $\vec{E}$ and $\vec{B}$ are perpendicular to one another and the the Part (*b*), step 2 result verifies that the $+x$ direction is the direction of propagation.

TAKING IT FURTHER This type of electromagnetic wave is said to be *circularly polarized*. At a fixed value of x, both $\vec{E}$ and $\vec{B}$ rotate in a circle with angular frequency ω.

PRACTICE PROBLEM 30-2 Calculate $\vec{E} \cdot \vec{E}$ and $\vec{B} \cdot \vec{B}$. Note that the fields $\vec{E}$ and $\vec{B}$ are constant in magnitude.

30-4 ELECTROMAGNETIC RADIATION

Figure 30-6 shows the electric and magnetic field vectors of an electromagnetic wave. The electric and magnetic fields are perpendicular to each other and perpendicular to the direction of propagation of the wave. Electromagnetic waves are thus transverse waves. The electric and magnetic fields are in phase and, at each point in space and at each instant in time, their magnitudes are related by

$$E = cB \qquad \text{30-18}$$

where $c = 1/\sqrt{\mu_0\epsilon_0}$ is the speed of the wave. The direction of propagation of an electromagnetic wave is the direction of the cross product $\vec{E} \times \vec{B}$.

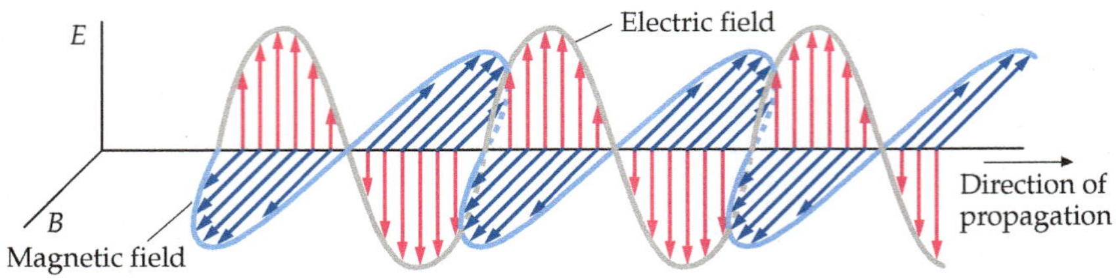

FIGURE 30-6 The electric and magnetic field vectors in an electromagnetic wave. The fields are in phase, perpendicular to each other, and perpendicular to the direction of propagation of the wave.

THE ELECTROMAGNETIC SPECTRUM

The various types of electromagnetic waves (for example, radio waves and gamma rays) differ only in wavelength and frequency, which are related according to the equation $f\lambda = c$. Table 30-1 lists the **electromagnetic spectrum** and the names usually associated with the various frequency and wavelength ranges. These ranges are often not well defined and sometimes overlap. For example, electromagnetic waves that have wavelengths of approximately 0.1 nm are usually called X rays, but if the electromagnetic waves originate from nuclear radioactivity, they are called gamma rays.

The human eye is sensitive to electromagnetic radiation that has wavelengths between 400 and 780 nm,* which is the range called **visible light.** The shortest wavelengths of visible light are those of violet light and the longest wavelengths are those of red light. Electromagnetic waves that have wavelengths shorter than 400 nm but longer than 10 nm (the longest wavelength in the X-rays region) are called **ultraviolet rays. Infrared waves** have wavelengths longer than 780 nm but

* Light whose wavelength is between 700 and 780 nm can only be seen under special circumstances that include the light intensity being very high.

Table 30-1 The Electromagnetic Spectrum

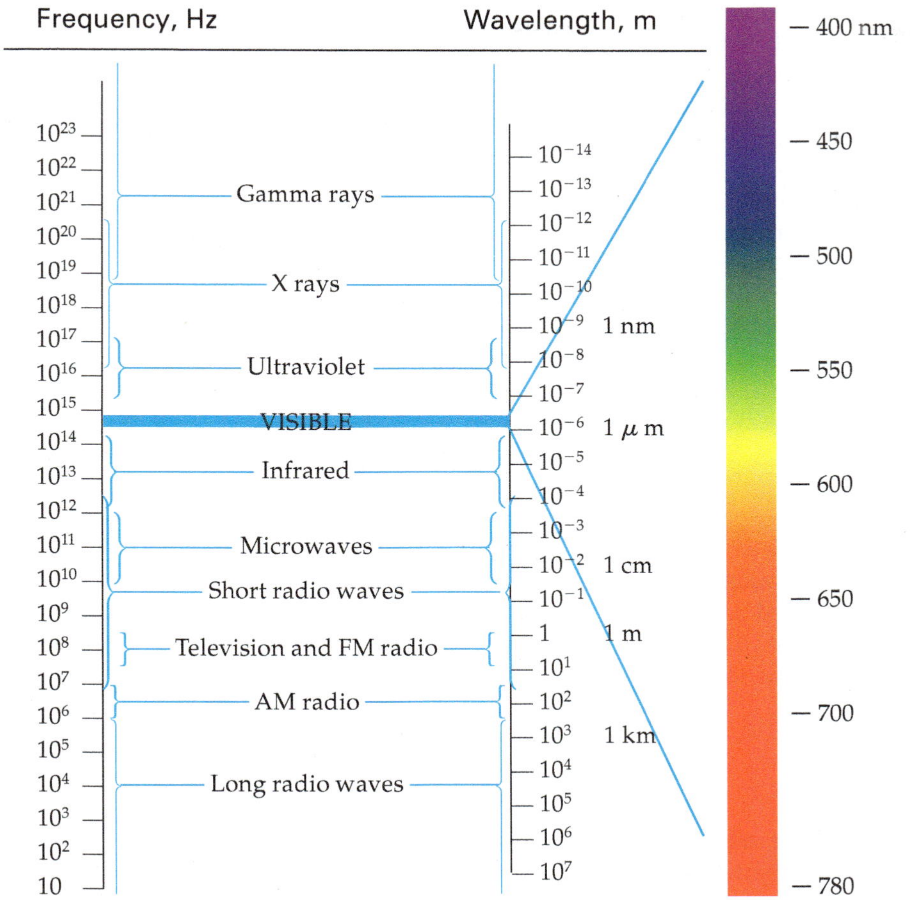

shorter than 100 μm. Heat radiation emitted by objects at temperatures in the range of room temperature is in the infrared region of the electromagnetic spectrum. There are no limits on the wavelengths of electromagnetic radiation; that is, all wavelengths (or frequencies) are theoretically possible.

The differences in wavelengths of the various kinds of electromagnetic waves have important physical consequences. As you know, the behavior of waves depends strongly on the relative sizes of the wavelengths and the objects or apertures (openings) the waves encounter. Because the wavelengths of visible light are in the rather narrow range between 400 and 780 nm, they are much smaller than most obstacles. Thus, the ray approximation (introduced in Section 15-4) is often valid. The wavelength and frequency are also important in determining the kinds of interactions between electromagnetic waves and matter. X rays, for example, have very short wavelengths and high frequencies. They easily penetrate many materials that are opaque to lower-frequency light waves, which are absorbed by the materials. Microwaves have wavelengths between 1 mm and 30 cm. Wavelengths in that range are used to heat food in microwave ovens. The main mechanism of this heating is that molecules which have large dipole moments align themselves in the electric field of the radiation. This electric field flips its direction at twice the frequency of the radiation, so the polar molecules must rotate rapidly to keep up with the alternating electric field. These rapidly rotating molecules bump into surrounding molecules—causing them to heat up. Bluetooth and other wireless local-area-network protocols use wavelengths in the microwave region.

PRODUCTION OF ELECTROMAGNETIC WAVES

Electromagnetic waves are produced when free charges accelerate or when electrons bound to atoms and molecules make transitions to lower energy states. Radio waves, which have frequencies between 550 and 1600 kHz for AM and between 88 and 108 MHz for FM, are produced by electric currents oscillating in radio transmission antennas. The frequency of the emitted waves equals the frequency of oscillation of the charges.

A continuous spectrum of X rays is produced by the deceleration of electrons when they crash into a metal target. The radiation produced is called **bremsstrahlung** (which means "braking radiation" in the German language). Accompanying the broad, continuous bremsstrahlung spectrum is a discrete spectrum of X-ray lines produced by transitions of core electrons in the atoms of the target material.

Synchrotron radiation arises from the circular orbital motion of charged particles (usually electrons or positrons) in nuclear accelerators called synchrotrons. Originally considered a nuisance by accelerator scientists, X rays produced by synchrotrons are now used as a medical diagnostic tool because of the ease of manipulating the beams with reflection and diffraction optics. Synchrotron radiation is also emitted by charged particles trapped in magnetic fields associated with stars and galaxies. It is believed that most low-frequency radio waves reaching Earth from outer space originate as synchrotron radiation.

Heat is radiated by thermally excited molecular motion. The spectrum of heat radiation is the blackbody radiation spectrum discussed in Section 20-4.

Light waves, which have frequencies of the order of 10^{14} Hz, are generally produced by transitions of bound atomic charges. We discuss sources of light waves in Chapter 31.

ELECTRIC DIPOLE RADIATION

Figure 30-7 is a schematic drawing of an electric-dipole radio antenna that consists of two conducting rods connected to an alternating-current generator. At time $t = 0$ (Figure 30-7*a*), the ends of the rods are charged, and an electric field parallel to the rod exists near the rod. A magnetic field also exists, which is not shown, encircling the rods due to the current in the rods. The fluctuations in these fields move out away from the rods with the speed of light. After one-fourth period, at $t = T/4$ (Figure 30-7*b*), the rods are uncharged, and the electric field near the rod is zero. At $t = T/2$ (Figure 30-7*c*), the rods are again charged, but the charges are opposite those at $t = 0$. The electric and magnetic fields at a great distance from the antenna are quite different from the fields near the antenna. Far from the antenna, the electric and magnetic fields oscillate in phase with simple harmonic motion, perpendicular to each other and to the direction of propagation of the wave. Figure 30-8 shows the electric and magnetic fields far from an electric dipole antenna.

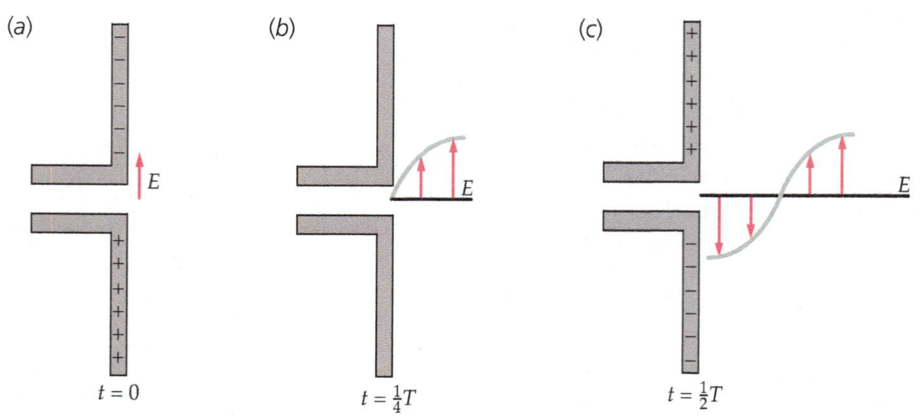

FIGURE 30-7 An electric dipole radio antenna for radiating electromagnetic waves. Alternating current is supplied to the antenna by a generator (not shown). The fluctuations in the electric field due to the fluctuations in the charges in the antenna propagate outward at the speed of light. There is also a fluctuating magnetic field (not shown) perpendicular to the paper due to the current in the antenna.

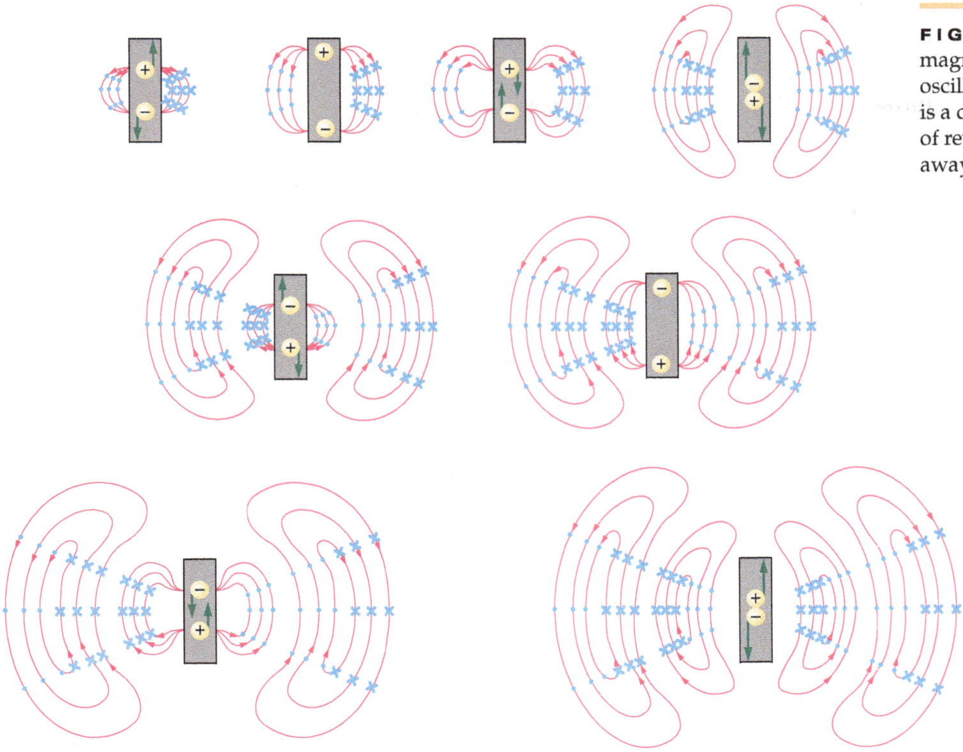

FIGURE 30-8 Electric field lines (in red) and magnetic field lines (in blue) produced by an oscillating electric dipole. Each magnetic field line is a circle with the long axis of the dipole as its axis of revolution. The cross product $\vec{E} \times \vec{B}$ is directed away from the dipole at all points.

Electromagnetic waves of radio or television frequencies can be detected by an electric dipole antenna placed parallel to the electric field of the incoming wave, so that it induces an alternating current in the antenna (Figure 30-9). These electromagnetic waves can also be detected by a loop antenna placed perpendicular to the magnetic field, so that the changing magnetic flux through the loop induces a current in the loop (Figure 30-10). Electromagnetic waves of frequencies in the visible light range are detected by the eye or by photographic film, both of which are mainly sensitive to the electric field.

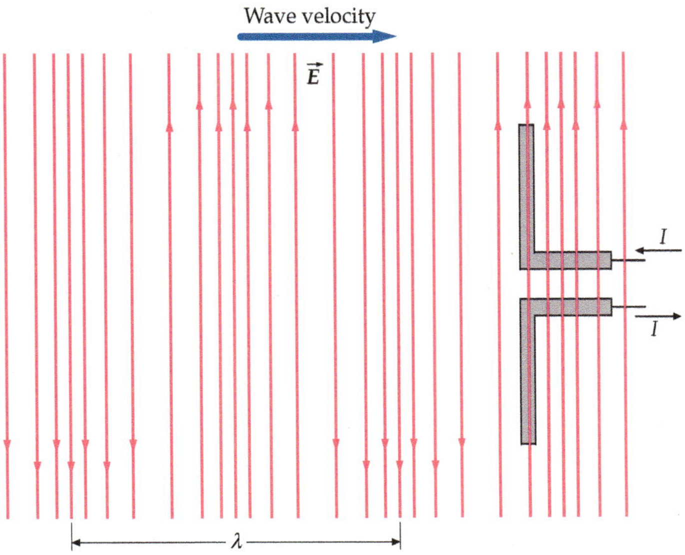

FIGURE 30-9 An electric dipole antenna for detecting electromagnetic waves. The alternating electric field of the incoming wave produces an alternating current in the antenna. The magnetic field lines (not shown) are perpendicular to the plane of the page.

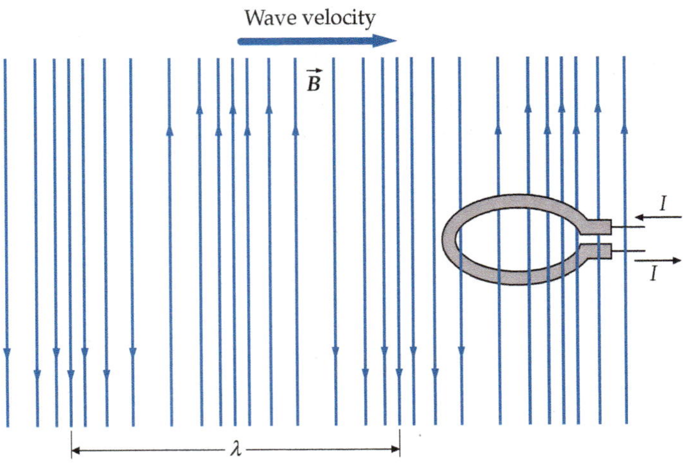

FIGURE 30-10 Loop antenna for detecting electromagnetic radiation. The alternating magnetic flux through the loop due to the magnetic field of the radiation induces an alternating current in the loop. The electric field lines (not shown) are perpendicular to the plane of the page.

The radiation from a dipole antenna, such as that shown in Figure 30-7, is called electric dipole radiation. Many electromagnetic waves exhibit the characteristics of electric dipole radiation. An important feature of this type of radiation is that the intensity of the electromagnetic waves radiated by a dipole antenna is zero along the axis of the antenna and maximum in the radial direction (away from the axis). If the dipole is in the y direction with its center at the origin, as in Figure 30-11, the intensity is zero along the y axis and maximum in the xz plane. In the direction of a line making an angle θ with the y axis, the intensity is proportional to $\sin^2\theta$.

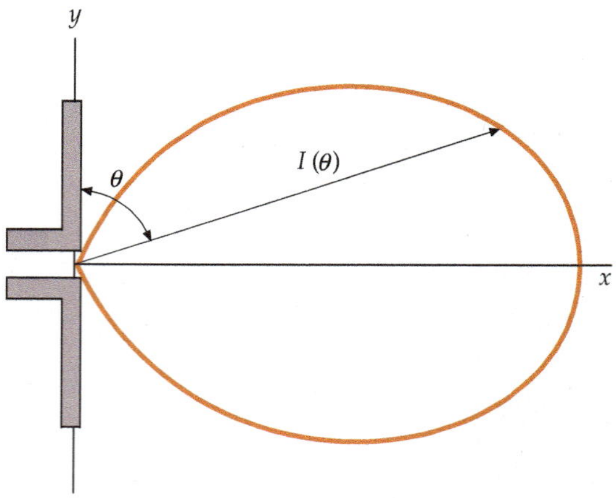

FIGURE 30-11 Polar plot of the intensity of electromagnetic radiation from an electric dipole antenna versus angle. The intensity $I(\theta)$ is proportional to the length of the arrow. The intensity is maximum perpendicular to the antenna (at $\theta = 90°$) and minimum along the antenna at $\theta = 0$ or $\theta = 180°$.

Example 30-5 EMF Induced in a Loop Antenna

A loop antenna consisting of a single 10.0-cm radius loop of wire is used to detect electromagnetic waves for which $E_{rms} = 0.150$ V/m. If the plane of the loop is perpendicular to the magnetic field, find the rms emf induced in the loop when the frequency of the plane wave is (*a*) 600 kHz and (*b*) 60.0 MHz.

PICTURE The induced emf in the wire is related to the rate of change of the magnetic flux through the loop by Faraday's law (Equation 30-5). Using Equation 30-18, we can obtain the rms value of the magnetic field from the given rms value of the electric field.

SOLVE

(*a*) 1. Faraday's law relates the emf to the rate of change of the magnetic flux through the flat stationary surface bounded by the loop:

$$\mathcal{E} = -\frac{d\phi_m}{dt}$$

2. The wavelength of a 600-kHz wave traveling at speed c is $\lambda = c/f = 500$ m. Over the flat surface bounded by the 10-cm radius loop $\vec{B}$ is quite uniform.

$$\phi_m = BA = \pi r^2 B \quad \text{so} \quad \mathcal{E} = -\frac{d\phi_m}{dt} = -\pi r^2 \frac{\partial B}{\partial t}$$

and

$$\mathcal{E}_{rms} = \pi r^2 \left(\frac{\partial B}{\partial t}\right)_{rms}$$

3. Compute $\partial B_{rms}/\partial t$ from a sinusoidal B:

$$B = B_0 \sin(kx - \omega t)$$

$$\frac{\partial B}{\partial t} = -\omega B_0 \cos(kx - \omega t)$$

4. Calculate the rms value of $\partial B/\partial t$. The rms value of any sinusoidal function of time equals $1/\sqrt{2}$, and the peak value divided by $\sqrt{2}$ equals the rms value:

$$\left(\frac{\partial B}{\partial t}\right)_{rms} = \omega B_0[-\cos(kx - \omega t)]_{rms} = \omega B_0 \frac{1}{\sqrt{2}} = \omega B_{rms}$$

5. Using $E = cB$ (Equation 30-18), relate the rms value of $\partial B/\partial t$ to E_{rms}:

$$E = cB$$

so

$$B_{rms} = \frac{E_{rms}}{c}$$

6. Substituting into the step-3 result gives:

$$\left(\frac{\partial B}{\partial t}\right)_{rms} = \omega B_{rms} = \omega \frac{E_{rms}}{c} = \frac{2\pi f}{c} E_{rms}$$

7. Substituting the step-6 result into the step-2 result, calculate $\mathcal{E}_{rms}$ at $f = 600$ kHz:

$$\mathcal{E}_{rms} = \pi r^2 \left(\frac{\partial B}{\partial t}\right)_{rms} = \pi r^2 \frac{2\pi f}{c} E_{rms}$$

$$= \pi(0.100\text{ m})^2 \frac{2\pi(6.00 \times 10^5\text{ Hz})}{3.00 \times 10^8\text{ m/s}}(0.150\text{ V/m})$$

$$= \boxed{5.92 \times 10^{-5}\text{ V} = 59.2\ \mu\text{V}}$$

(*b*) The induced emf is proportional to the frequency [Part (*a*), step 4], so at 60 MHz it will be 100 times greater than at 600 kHz:

$$\mathcal{E}_{rms} = (100)(5.92 \times 10^{-5}\ \text{V}) = 0.00592\ \text{V}$$
$$= \boxed{5.92\ \text{mV}}$$

CHECK Step 7 of Part (*a*) shows that $\mathcal{E}_{rms}$ increases with increases in frequency, E_{rms}, and area. These results are all expected.

TAKING IT FURTHER For Part (*b*) the frequency is 60.0 MHz, so $\lambda = c/f = 5.00$ m. $\vec{B}$ is not as uniform over the surface bounded by the 10-cm radius loop when $\lambda = 5.00$ m as it is when $\lambda = 500$ m, as in Part (*a*). However, $\vec{B}$ on the surface when $\lambda = 5.00$ m is sufficiently uniform that the Part-(*b*) result is sufficiently accurate for most purposes.

ENERGY AND MOMENTUM IN AN ELECTROMAGNETIC WAVE

Like other waves, electromagnetic waves carry energy and momentum. The energy carried is described by the intensity, which is the average power per unit area incident on a surface perpendicular to the direction of propagation. The momentum per unit time per unit area carried by an electromagnetic wave is called the **radiation pressure.**

Intensity Consider an electromagnetic wave traveling toward the right and a cylindrical region that has a length L, a cross-sectional area A, and its central axis from left to right. The average amount of electromagnetic energy U_{av} within this region equals $u_{av}\mathcal{V}$ where u_{av} is the average energy density and $\mathcal{V} = LA$ is the volume of the cylindrical region. In the time it takes the electromagnetic wave to travel the distance L, all of the electromagnetic energy equal to $u_{av}LA$ passes through the right end of the region. The time Δt for the wave to travel the distance L is L/c, so the power P_{av} (the energy per unit time) passing out the right end of the region is

$$P_{av} = U_{av}/\Delta t = u_{av}LA/(L/c) = u_{av}Ac$$

and the intensity I (the average power per unit area) is

$$I = P_{av}/A = u_{av}c$$

The total energy density in the wave u is the sum of the electric and magnetic energy densities. The electric energy density u_e (Equation 24-9) and magnetic energy density u_m (Equation 28-22) are given by

$$u_e = \frac{1}{2}\epsilon_0 E^2 \quad \text{and} \quad u_m = \frac{B^2}{2\mu_0}$$

At each point in a region where there is an electromagnetic wave in free space, E equals cB, so we can express the magnetic energy density in terms of the electric field:

$$u_m = \frac{B^2}{2\mu_0} = \frac{(E/c)^2}{2\mu_0} = \frac{E^2}{2\mu_0 c^2} = \frac{1}{2}\epsilon_0 E^2$$

where we have used $\mu_0\epsilon_0 = 1/c^2$. Thus, the electric and magnetic energy densities are equal. Using $E = cB$, we may express the total energy density in several useful ways:

$$u = u_e + u_m = \epsilon_0 E^2 = \frac{B^2}{\mu_0} = \frac{EB}{\mu_0 c} \qquad 30\text{-}19$$

ENERGY DENSITY IN AN ELECTROMAGNETIC WAVE

To compute the average energy density, we replace the instantaneous fields E and B by their rms values $E_{rms} = \frac{1}{\sqrt{2}}E_0$ and $B_{rms} = \frac{1}{\sqrt{2}}B_0$, where E_0 and B_0 are the maximum values of the fields. The intensity is then

$$I = u_{av}c = \frac{E_{rms}B_{rms}}{\mu_0} = \frac{1}{2}\frac{E_0B_0}{\mu_0} = |\vec{S}|_{av} \tag{30-20}$$

INTENSITY OF AN ELECTROMAGNETIC WAVE

where the vector

$$\vec{S} = \frac{\vec{E} \times \vec{B}}{\mu_0} \tag{30-21}$$

DEFINITION—POYNTING VECTOR

is called the **Poynting vector** after its discoverer, John Poynting. The average magnitude of $\vec{S}$ is the intensity of the wave, and the direction of $\vec{S}$ is the direction of propagation of the wave.

Radiation pressure We now show by a simple example that an electromagnetic wave carries momentum. Consider a plane wave traveling in the $+x$ direction that is incident on a stationary charge, as shown in Figure 30-12. Let $\vec{E}$ be in the $+y$ direction and $\vec{B}$ in the $+z$ direction, and neglect the time dependence of the fields. The particle experiences a force $q\vec{E}$ in the $+y$ direction and is thus accelerated by the electric field. At any time t, the velocity in the $+y$ direction is

$$v_y = a_y t = \frac{qE}{m}t$$

After a short time t_1, the charge has acquired kinetic energy equal to

$$K = \frac{1}{2}mv_y^2 = \frac{1}{2}\frac{mq^2E^2t_1^2}{m^2} = \frac{1}{2}\frac{q^2E^2}{m}t_1^2 \tag{30-22}$$

When the charge is moving in the y direction, it experiences a magnetic force

$$\vec{F}_m = q\vec{v} \times \vec{B} = qv_y\,\hat{j} \times B\,\hat{k} = qv_yB\,\hat{i} = \frac{q^2EB}{m}t\,\hat{i}$$

Note that this force is in the direction of propagation of the wave. Using $dp_x = F_x\,dt$, we find for the momentum p_x transferred by the wave to the particle in time t_1:

$$p_x = \int_0^{t_1} F_x\,dt = \int_0^{t_1}\frac{q^2EB}{m}t\,dt = \frac{1}{2}\frac{q^2EB}{m}t_1^2$$

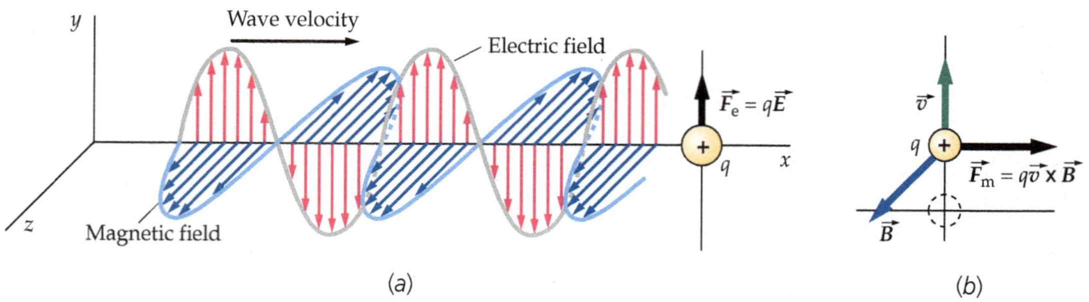

FIGURE 30-12 An electromagnetic wave incident on a point charge that is initially at rest on the x axis. (*a*) The electric force $q\vec{E}$ accelerates the charge in the $+y$ direction. (*b*) When the velocity $\vec{v}$ of the charge is in the $+y$ direction, the magnetic force $q\vec{v} \times \vec{B}$ accelerates the charge in the direction of propagation (the $+x$ direction) of the wave.

If we use $B = E/c$, this becomes

$$p_x = \frac{1}{c}\left(\frac{1}{2}\frac{q^2E^2}{m}t_1^2\right) \qquad \text{30-23}$$

Comparing Equations 30-22 and 30-23, we see that the momentum acquired by the charge in the direction of the wave is $1/c$ multiplied by the energy. Although our simple calculation was not rigorous, the results are correct. The magnitude of the momentum carried by an electromagnetic wave is $1/c$ multiplied by the energy carried by the wave:

$$p = \frac{U}{c} \qquad \text{30-24}$$

MOMENTUM AND ENERGY IN AN ELECTROMAGNETIC WAVE

Because the intensity is the energy per unit area per unit time, the intensity divided by c is the momentum carried by the wave per unit area per unit time. The momentum carried per unit time is a force. The intensity divided by c is thus a force per unit area, which is a pressure. This pressure is the radiation pressure P_r:

$$P_r = \frac{I}{c} \qquad \text{30-25}$$

RADIATION PRESSURE AND INTENSITY

We can relate the radiation pressure to the electric or magnetic fields by using Equation 30-20 to relate I to E and B, and Equation 30-18 to eliminate either E or B:

$$P_r = \frac{I}{c} = \frac{E_0B_0}{2\mu_0 c} = \frac{E_{rms}B_{rms}}{\mu_0 c} = \frac{E_0^2}{2\mu_0 c^2} = \frac{B_0^2}{2\mu_0} \qquad \text{30-26}$$

RADIATION PRESSURE IN TERMS OF E AND B

Consider an electromagnetic wave incident normally on some surface. If the surface absorbs energy U from the electromagnetic wave, it also absorbs momentum p given by Equation 30-24, and the pressure exerted on the surface equals the radiation pressure. If the wave is reflected, the momentum transferred is $2p$ because the wave now carries momentum in the opposite direction. The pressure exerted on the surface by the wave is then twice that given by Equation 30-26.

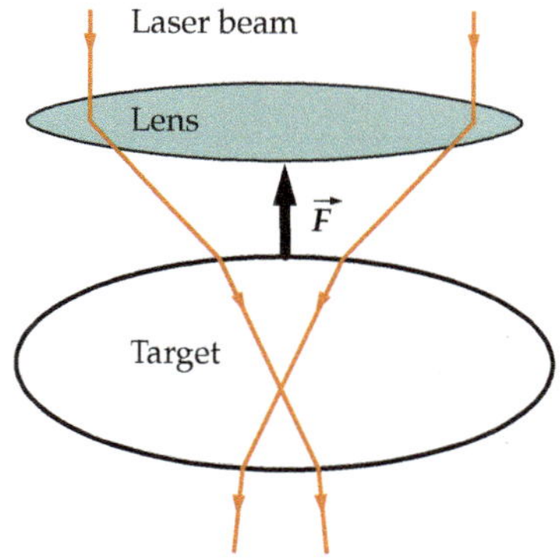

"Laser tweezers" make use of the momentum carried by electromagnetic waves to manipulate targets on a molecular scale. The two rays shown are refracted as they pass through a transparent target, such as a biological cell or, on an even smaller scale, a tiny transparent bead attached to a large molecule within a cell. At each refraction, the rays are bent downward, which increases the downward component of momentum of the rays. The target thus exerts a downward force on the laser beams, and the laser beams exert an upward force on the target, which pulls the target toward the laser source. The force is typically of the order of piconewtons. Laser tweezers have been used to accomplish such astonishing feats as stretching out coiled molecules of DNA.

Example 30-6 Radiation Pressure 3.0 m from a Lightbulb

A lightbulb emits spherical electromagnetic waves uniformly in all directions. Find (*a*) the intensity, (*b*) the radiation pressure, and (*c*) the electric and magnetic field magnitudes at a distance of 3.0 m from the lightbulb, assuming that 50 W of electromagnetic radiation is emitted.

PICTURE At a distance r from the lightbulb, the energy is spread uniformly over the surface of a sphere of radius r—an area equal to $4\pi r^2$. The intensity is the power divided by the area. The radiation pressure can then be found from $P_r = I/c$.

SOLVE

(*a*) 1. Divide the power output by the area to find the intensity: $I = \dfrac{50\text{ W}}{4\pi r^2}$

2. Substitute $r = 3.0$ m: $I = \dfrac{50\text{ W}}{4\pi(3.0\text{ m})^2} = \boxed{0.44\text{ W/m}^2}$

(*b*) The radiation pressure is the intensity divided by the speed of light:

$$P_r = \frac{I}{c} = \frac{0.44\ \text{W/m}^2}{3.00 \times 10^8\ \text{m/s}} = \boxed{1.5 \times 10^{-9}\ \text{Pa}}$$

(*c*) 1. B_0 is related to P_r by Equation 30-26:

$$B_0 = \sqrt{2\mu_0 P_r}$$
$$= [2(4\pi \times 10^{-7}\ \text{T}\cdot\text{m/A})(1.5 \times 10^{-9}\ \text{Pa})]^{1/2}$$
$$= 6.1 \times 10^{-8}\ \text{T}$$

2. The maximum value of the electric field E_0 is c multiplied by B_0:

$$E_0 = cB_0 = (3.00 \times 10^8\ \text{m/s})(6.1 \times 10^{-8}\ \text{T})$$
$$= 18\ \text{V/m}$$

3. The electric and magnetic field magnitudes at that point are of the form:

$$\boxed{\begin{aligned} &E = E_0 \sin\omega t \quad \text{and} \quad B = B_0 \sin\omega t \\ &\text{with } E_0 = 18\ \text{V/m} \\ &\text{and } B_0 = 6.1 \times 10^{-8}\ \text{T} \end{aligned}}$$

CHECK Our Part-(*b*) result is a very small pressure. (It is fourteen orders of magnitude less than atmospheric pressure.) We do not perceive any pressure on us by the light from a lightbulb, so a very small pressure is as expected.

TAKING IT FURTHER Only about 2 percent of the power consumed by incandescent bulbs is transformed into visible light.

Example 30-7 A Laser Rocket

Context-Rich

You are stranded in space a distance of 20 m from your spaceship. You carry a 1.0-kW laser. If your total mass, including your space suit and laser, is 95 kg, how long will it take you to reach the spaceship if you point the laser beam directly away from the ship?

PICTURE The laser emits light, which carries with it momentum. By momentum conservation, you are given an equal and opposite momentum toward the spaceship. The momentum carried by light is $p = U/c$, where U is the energy of the light. If the power of the laser is $P = dU/dt$, then the rate of change of momentum produced by the laser is $dp/dt = (dU/dt)/c = P/c$. This force is the force exerted on you, which is constant.

SOLVE

1. The time taken is related to the distance and the acceleration. We assume that you are initially at rest relative to the spaceship:

$$x = \frac{1}{2}at^2 \quad t = \sqrt{\frac{2x}{a}}$$

2. Your acceleration is the force divided by your mass, and the force is the power divided by c:

$$a = \frac{F}{m} = \frac{P/c}{m} = \frac{P}{mc}$$

3. Use $x = \frac{1}{2}at^2$ to calculate the time t:

$$t = \sqrt{\frac{2x}{a}} = \sqrt{\frac{2xmc}{P}}$$
$$= \sqrt{\frac{2(20\ \text{m})(95\ \text{kg})(3.00 \times 10^8\ \text{m/s})}{1000\ \text{W}}}$$
$$= 3.38 \times 10^4\ \text{s} = \boxed{9.4\ \text{h}}$$

CHECK You expect the time to be long because you know from experience that the pressure from a lightbulb is very small. The step-3 result is as expected.

TAKING IT FURTHER Note that the acceleration is extremely small—only about $10^{-9}\,g$. Your speed when you reach the spaceship would be $v = at = 1.2$ mm/s, which is practically imperceptible.

PRACTICE PROBLEM 30-3 How long would it take you to reach the spaceship if you took off one of your shoelaces and threw it as fast as you could in the direction opposite the ship? (To answer this, you must first estimate the mass of the shoelace and the maximum speed that you can throw the shoelace.) Compare this time with the step-3 result.

Physics Spotlight

Wireless: Sharing the Spectrum

Marconi's transmitter. This transmitter was used in 1901 during the first transatlantic radio broadcast. *(Cornwall, UK, to Newfoundland, USA.)*

One day in March 1998,[*] the remote heart monitors at Baylor University Medical Center and Dallas Methodist Hospital stopped working. WFFA, a Dallas television station, was testing its new digital broadcast system on its licensed frequency. The heart monitors, longtime unlicensed low-power users of the same frequency, were overwhelmed by the test. No patients were harmed, and the station stopped further testing until the hospitals could replace their monitors with ones that used different frequencies.[†] In 2000, the Wireless Medical Telemetry Service established a set of licensed frequencies for medical monitoring devices.[‡]

When Guglielmo Marconi transmitted signals on his wireless telegraph in 1896, he used a spark-gap transmitter.[#] The sparks produced electromagnetic radiation at frequencies over a range of 5 or more orders of magnitude (from a few kilohertz up to 2 GHz). When more than one wireless telegraph was transmitting in the same area, they had to take turns and follow rules. One reckless operator could wreck communications for an entire area.[°]

In 1903, the International Telegraph Union began to study the problems of radiotelegraphy. In 1906, the first Radio Telegraph Convention, signed in Berlin, assigned the frequency of 500 kHz to maritime distress signals.[§] Ships were directed to use less than 1 kW of power, unless they were over 300 km from the nearest shore station.[¶] (These communications were still broadband, but the transmitted power spectrum peaked at 500 kHz.) The first practical alternative to spark-gap transmitters was a continuous wave circuit invented by Edwin Armstrong in 1912.[**] Also in 1912, the International Radiotelegraph Convention issued the first table of frequency allocations,[††] but spark-gap transmitters were still plentiful and could overwhelm local and regional communications.[‡‡]

Radio transmissions became concentrated around narrowing frequency bands. In 1927, national bodies were established to coordinate use of the electromagnetic spectrum.[##] By 1934, the international body was renamed the International Telecommunication Union.[°°] The Federal Communications Commission regulates the radio frequency portions of the spectrum in the United States.[§§] Since then, the ITU and the FCC have cooperated in international frequency allocation with other regulatory bodies around the world.

As new services have been added, changes and accommodations have been made to allocations of the frequency spectrum. These changes are not always made globally. For instance, in the United States, cell phone service frequencies are at 850 and 1900 MHz. In many other countries, cell phone bands are at 900 and 1800 MHz.[¶¶]

No matter what the power, devices that have the potential of emitting electromagnetic interference have to be certified that they do not disturb the spectrum beyond their frequency allocation outside of a small radius.[***] Many applications share frequency bands that do not have exclusive licenses granted to any one application. For instance, microwave ovens, wireless computer devices, and some cordless telephones all operate at frequencies close to 2.4 GHz.[†††] Low-power applications that are not the licensed users of a frequency sometimes need frequencies of their own, as demonstrated by the new medical telemetry bands. Occasionally, wireless receivers in an area pick up broadband interference caused by the intermittent sparking of a short circuit. In essence, the malfunctioning electrical equipment has become an unlicensed spark-gap transmitter.

---

* "Wireless Medical Telemetry—Electromagnetic Interference." *United States Food and Drug Administration Center for Devices and Radiological Health*, Sept. 1, 2002. http://www.fda.gov/cdrh/emc/wmt-emi.html#1 As of Nov. 2006.

† McClain, J. P., "Time to Upgrade." *American Society for Healthcare Engineering*. www.ashe.org/ashe/wmts/pdfs/timetoupgrade.pdf As of Nov. 2006.

‡ Federal Communications Commission. *FCC-00211*. Washington, DC: United States Federal Communications Commission, Jun. 12, 2000. http://www.fcc.gov/Bureaus/Engineering_Technology/Orders/2000/fcc00211.doc As of Nov. 2006.

\# Thomson, E., "The Field of Experimental Research." *Science*, Aug. 25, 1899, Vol. X, No. 243, pp. 236–245.

° Pitts, A., "Backgrounder: What Is Amateur Radio?" *American Radio Relay League*, Oct. 4, 2004. http://www.arrl.org/pio/bwhatis.html As of Nov. 2006.

§ "ARRL Granted Experimental License for 500 kHz Research by Radio Amateurs." *American Radio Relay League*, Sep. 15, 2006. http://www.arrl.org/news/stories/2006/09/15/104/ As of Nov. 2006.

¶ "Service Regulations Affixed to the International Wireless Convention." *United States Early Radio History*. http://earlyradiohistory.us/1906conv.htm#SR As of Nov. 2006.

** Lewis, T., *Empire of the Air.* New York: HarperCollins, 1991, pp. 70–74.

†† "History." *International Telecommunication Union*, Nov. 15, 2004. http://www.itu.int/aboutitu/overview/history.html As of Nov. 2006.

‡‡ Lapin, G. D., "Lessons Learned about Frequency Sharing in the Amateur Radio Service." *American Radio Relay League*. http://www.arrl.org/tis/info/HTML/plc/files/ Lessons%20Learned%20About%20Frequency%20Sharing%20in%20the%20Amateur%20Radio%20Service%20Rev%202.ppt As of Nov. 2006.

\#\# *Radio Act of 1927.* United States Public Law 632. Feb. 23, 1927. Available at http://showcase.netins.net/web/akline/ pdf/1927act.pdf As of Nov. 2006.

°° "History." *International Telecommunication Union*, Nov. 15, 2004. http://www.itu.int/aboutitu/overview/history.html As of Nov. 2006.

§§ "About the FCC." *United States Federal Communications Commission*, Sept. 26, 2006. http://www.fcc.gov/aboutus.html As of Nov. 2006.

¶¶ Luna, N. "Globetrotting with Cell Phones Tricky but Not Impossible." *The Orange County Register*, May 4, 2005.

*** "Rule 47 CFR Part 15." *United States Federal Register*, Washington, DC: Aug. 14, 2006. http://www.fcc.gov/oet/info/rules/part15/part15-8-14-06.pdf As of Nov. 2006.

††† Lowe, M., "Muting Microwaves." *Appliance Design*, Jan. 2006, Vol. 54, No. 1, pp. 74–75.

Summary

1. Maxwell's equations summarize the fundamental laws of physics that govern electricity and magnetism.
2. Electromagnetic waves include light, radio waves, television waves, X rays, gamma rays, microwaves, and others.

TOPIC	RELEVANT EQUATIONS AND REMARKS	
1. Maxwell's Displacement Current	Ampère's law can be generalized to apply to currents that are not steady (and not continuous) if the current I is replaced by $I + I_d$, where I_d is Maxwell's displacement current: $$I_d = \epsilon_0 \frac{d\phi_e}{dt}$$	30-3
Generalized form of Ampère's law	$$\oint_C \vec{B} \cdot d\vec{\ell} = \mu_0(I + I_d) = \mu_0 I + \mu_0\epsilon_0 \frac{d\phi_e}{dt}$$	30-4
2. Maxwell's Equations	The laws of electricity and magnetism are summarized by Maxwell's equations.	
Gauss's law	$$\oint_S E_n\, dA = \frac{1}{\epsilon_0} Q_{\text{inside}}$$	30-6*a*
Gauss's law for magnetism (isolated magnetic poles do not exist)	$$\oint_S B_n\, dA = 0$$	30-6*b*
Faraday's law (form that does not include motional emf)	$$\oint_C \vec{E} \cdot d\vec{\ell} = -\frac{d}{dt}\int_S B_n\, dA = -\int_S \frac{\partial B_n}{\partial t} dA$$	30-6*c*
Ampère's law modified	$$\oint_C \vec{B} \cdot d\vec{\ell} = \mu_0(I + I_d) = \mu_0 I + \mu_0\epsilon_0 \int_S \frac{\partial E_n}{\partial t} dA$$	30-6*d*
3. The Wave Equation for Electromagnetic Waves	Maxwell's equations imply that the electric and magnetic field vectors in free space obey a wave equation. $$\frac{\partial^2 \vec{E}}{\partial x^2} = \frac{1}{c^2}\frac{\partial^2 \vec{E}}{\partial t^2}$$ $$\frac{\partial^2 \vec{B}}{\partial x^2} = \frac{1}{c^2}\frac{\partial^2 \vec{B}}{\partial t^2}$$	30-8*a* 30-8*b*
4. Electromagnetic Waves	In an electromagnetic wave, the electric and magnetic field vectors are perpendicular to each other and to the direction of propagation. Their magnitudes are related by $$E = cB$$ The vector product $\vec{E} \times \vec{B}$ is in the direction of propagation.	30-18
Wave speed	$$c = \frac{1}{\sqrt{\mu_0\epsilon_0}} = 3.00 \times 10^8 \text{ m/s}$$	30-1
Electromagnetic spectrum	The various types of electromagnetic waves—light, radio waves, X rays, gamma rays, microwaves, and others—differ only in wavelength and frequency. The human eye is sensitive to the range from about 400 nm to 780 nm.	
Electric dipole radiation	Electromagnetic waves are produced when free electric charges accelerate. Oscillating charges in an electric dipole antenna radiate electromagnetic waves with an intensity that is greatest in directions perpendicular to the antenna. There is no radiated intensity along the long axis of the antenna. Perpendicular to the antenna and far away from it, the electric field of the electromagnetic wave is parallel to the antenna.	
Energy density in an electromagnetic wave	$$u = u_e + u_m = \epsilon_0 E^2 = \frac{B^2}{\mu_0} = \frac{EB}{\mu_0 c}$$	30-19

TOPIC	RELEVANT EQUATIONS AND REMARKS	
Intensity of an electromagnetic wave	$I = u_{av}c = \frac{E_{rms}B_{rms}}{\mu_0} = \frac{1}{2}\frac{E_0B_0}{\mu_0} = \lvert\vec{S}\rvert_{av}$	30-20
Poynting vector	$\vec{S} = \frac{\vec{E}\times\vec{B}}{\mu_0}$	30-21
Momentum and energy in an electromagnetic wave	$p = \frac{U}{c}$	30-24
Radiation pressure and intensity	$P_r = \frac{I}{c}$	30-25

Answers to Practice Problems

30-2 $\vec{E}\cdot\vec{E} = E_0^2$ and $\vec{B}\cdot\vec{B} = B_0^2$

30-3 About 5 h for a 10-g shoelace thrown at 10 m/s. Light beam propulsion takes almost twice as long as shoelace propulsion.

Problems

In a few problems, you are given more data than you actually need; in a few other problems, you are required to supply data from your general knowledge, outside sources, or informed estimate.

Interpret as significant all digits in numerical values that have trailing zeros and no decimal points.

- • Single-concept, single-step, relatively easy
- •• Intermediate-level, may require synthesis of concepts
- ••• Challenging
- SSM Solution is in the *Student Solutions Manual*
- Consecutive problems that are shaded are paired problems.

CONCEPTUAL PROBLEMS

1 • True or false:

(*a*) The displacement current has different units than the conduction current.
(*b*) Displacement current only exists if the electric field in the region is changing with time.
(*c*) In an oscillating *LC* circuit, no displacement current exists between the capacitor plates when the capacitor is momentarily fully charged.
(*d*) In an oscillating *LC* circuit, no displacement current exists between the capacitor plates when the capacitor is momentarily uncharged. SSM

2 • Using SI units, show that $\epsilon_0\, d\phi_e/dt$ has units of current.

3 • True or false:

(*a*) Maxwell's equations apply only to electric and magnetic fields that are constant over time.
(*b*) The electromagnetic wave equation can be derived from Maxwell's equations.
(*c*) Electromagnetic waves are transverse waves.
(*d*) The electric and magnetic fields of an electromagnetic wave in free space are in phase. SSM

4 • Theorists have speculated about the existence of *magnetic monopoles*, and several experimental searches for such monopoles have occurred. Suppose magnetic monopoles were found and that the magnetic field at a distance r from a monopole of strength q_m is given by $B = (\mu_0/4\pi)q_m/r^2$. Modify the Gauss's law for magnetism equation to be consistent with such a discovery.

5 • (*a*) For each of the following pairs of electromagnetic waves, which has the higher frequency: (1) visible light or X rays, (2) green light or red light, (3) infrared waves or red light. (*b*) For each of the following pairs of electromagnetic waves, which has the longer wavelength: (1) visible light or microwaves, (2) green light or ultraviolet light, (3) gamma rays or ultraviolet light.

6 • The detection of radio waves can be accomplished with either an electric dipole antenna or a loop antenna. True or false:

(*a*) The electric dipole antenna works according to the Faraday's law.
(*b*) If a linearly polarized radio wave is approaching you head on such that its electric field oscillates vertically, to best detect this wave the normal to a loop antenna's plane should be oriented so that it points either right or left.
(*c*) If a linearly polarized radio wave is approaching you such that its electric field oscillates in a horizontal plane, to best detect this wave using a dipole antenna the antenna should be oriented vertically.

7 • A transmitter emits electromagnetic waves using an electric dipole antenna oriented vertically. (*a*) A receiver to detect the waves also uses an electric dipole antenna that is one mile from the transmitting antenna and at the same altitude. How should the receiver's electric dipole antenna be oriented for optimum signal reception? (*b*) A receiver to detect these waves uses a loop antenna that is one mile from the transmitting antenna and at the same altitude. How should the loop antenna be oriented for optimum signal reception?

8 • Show that the expression $(\vec{E}\times\vec{B})/\mu_0$ for the Poynting vector $\vec{S}$ (Equation 30-21) has units of watts per square meter (the SI units for electromagnetic wave intensity).

9 • If a red light beam, a green light beam, and a violet light beam, all traveling in empty space, have the same intensity, which light beam carries more momentum? (*a*) the red light beam, (*b*) the green light beam, (*c*) the violet light beam, (*d*) They all have the same momentum. (*e*) You cannot determine which beam carries the most momentum from the data given. SSM

10 • If a red light plane wave, a green light plane wave, and a violet light plane wave, all traveling in empty space, have the same intensity, which wave has the largest peak electric field? (*a*) the red light wave, (*b*) the green light wave, (*c*) the violet light wave, (*d*) They all have the same peak electric field. (*e*) You cannot determine the largest peak electric field from the data given.

11 • Two sinusoidal plane electromagnetic waves are identical except that wave A has a peak electric field that is three times the peak electric field of wave B. How do their intensities compare? (*a*) $I_A = \frac{1}{3}I_B$, (*b*) $I_A = \frac{1}{9}I_B$, (*c*) $I_A = 3I_B$, (*d*) $I_A = 9I_B$, (*e*) You cannot determine how their intensities compare from the data given.

ESTIMATION AND APPROXIMATION

12 •• **ENGINEERING APPLICATION** In *laser cooling and trapping*, the forces associated with radiation pressure are used to slow down atoms from thermal speeds of hundreds of meters per second at room temperature to speeds of a few meters per second or slower. An isolated atom will absorb only radiation of specific frequencies. If the frequency of the laser-beam radiation is tuned so that the target atoms will absorb the radiation, then the radiation is absorbed during a process called *resonant absorption*. The cross-sectional area of the atom for resonant absorption is approximately equal to λ^2, where λ is the wavelength of the laser light. (*a*) Estimate the acceleration of a rubidium atom (molar mass 85 g/mol) in a laser beam whose wavelength is 780 nm and intensity is 10 W/m^2. (*b*) About how long would it take such a light beam to slow a rubidium atom in a gas at room temperature (300 K) to near-zero speed?

13 •• **ENGINEERING APPLICATION** One of the first successful satellites launched by the United States in the 1950s was essentially a large spherical (aluminized) Mylar balloon from which radio signals were reflected. After several orbits around Earth, scientists noticed that the orbit itself was changing with time. They eventually determined that radiation pressure from sunlight was causing the orbit of this object to change—a phenomenon not taken into account in planning the mission. Estimate the ratio of the radiation-pressure force by the sunlight on the satellite to the gravitational force by Earth's gravity on the satellite. SSM

14 •• Some science fiction writers have described solar sails that could propel interstellar spaceships. Imagine a giant sail on a spacecraft subjected to radiation pressure from our Sun. (*a*) Explain why this arrangement works better if the sail is highly reflective rather than highly absorptive. (*b*) If the sail is assumed highly reflective, show that the force exerted by the sunlight on the spacecraft's sail is given by $P_SA/(4\pi r^2c)$ where P_S is the power output of the Sun (3.8×10^{26} W), A is the surface area of the sail, r is the distance from the Sun, and c is the speed of light. (Assume the area of the sail is much larger than the area of the spacecraft so that all the force is due to radiation pressure on the sail only.) (*c*) Using a reasonable value for A, compare the force on the spacecraft due to the radiation pressure and the force on the spacecraft due to the gravitational force of the Sun on the spacecraft. Does the result imply that such a system will work? Explain your answer.

MAXWELL'S DISPLACEMENT CURRENT

15 • A parallel-plate capacitor has circular plates and no dielectric between the plates. Each plate has a radius equal to 2.3 cm and the plates are separated by 1.1 mm. Charge is flowing onto the upper plate (and off of the lower plate) at a rate of 5.0 A. (*a*) Find the rate of change of the electric field strength in the region between the plates. (*b*) Compute the displacement current in the region between the plates and show that it equals 5.0 A. SSM

16 • In a region of space, the electric field varies with time as (0.050 N/C) $\sin\omega t$, where $\omega = 2000$ rad/s. Find the peak displacement current through a surface that is perpendicular to the electric field and has an area equal to 1.00 m^2.

17 •• For Problem 15, show that the magnetic field strength between the plates a distance r from the axis through the centers of both plates is given by $B = (1.9 \times 10^{-3}\ \text{T/m})r$.

18 •• The capacitors referred to in this problem have only empty space between the plates. (*a*) Show that a parallel-plate capacitor has a displacement current in the region between its plates that is given by $I_d = C\,dV/dt$, where C is the capacitance and V is the potential difference between the plates. (*b*) A 5.00-nF parallel-plate capacitor is connected to an ideal ac generator so the potential difference between the plates is given by $V = V_0 \cos\omega t$, where $V_0 = 3.00$ V and $\omega = 500\pi$ rad/s. Find the displacement current in the region between the plates as a function of time.

19 •• There is a current of 10 A in a resistor that is connected in series with a parallel-plate capacitor. The plates of the capacitor have an area of 0.50 m^2, and no dielectric exists between the plates. (*a*) What is the displacement current between the plates? (*b*) What is the rate of change of the electric field strength between the plates? (*c*) Find the value of the line integral $\oint_C \vec{B} \cdot d\vec{\ell}$, where the integration path C is a 10-cm-radius circle that lies in a plane that is parallel with the plates and is completely within the region between them. SSM

20 ••• Demonstrate the validity of the generalized form of Ampère's law (Equation 30-4) by showing that it gives the same result as the Biot–Savart law (Equation 27-3) in a specified situation. Figure 30-13 shows two momentarily equal but opposite point charges ($+Q$ and $-Q$) on the x axis at $x = -a$ and $x = +a$, respectively. At the same instant there is a current I in the wire connecting them, as shown. Point P is on the y axis at $y = R$. (*a*) Use the Biot–Savart law to show that the magnitude of the magnetic field at point P is given by $B = \dfrac{\mu_0 Ia}{2\pi R}\dfrac{1}{\sqrt{R^2 + a^2}}$. (*b*) Now consider a circular strip of radius r and width dr in the $x = 0$ plane that has its center at the origin. Show that the flux of the electric field through this strip is given by $E_x\,dA = \dfrac{Q}{\epsilon_0(r^2 + a^2)^{3/2}}\pi r\,dr$. (*c*) Use the result from Part (*b*) to show that the total electric flux ϕ_e through a circular surface S of radius R is given by $\phi_e = \dfrac{Q}{\epsilon_0}\left(1 - \dfrac{a}{\sqrt{a^2 + R^2}}\right)$. (*d*) Find the displacement current I_d through S, and show that $I + I_d = I\dfrac{a}{\sqrt{a^2 + R^2}}$. (*e*) Finally, show that the generalized form of Ampère's law (Equation 30-4) gives the same result for the magnetic field as found in Part (*a*).

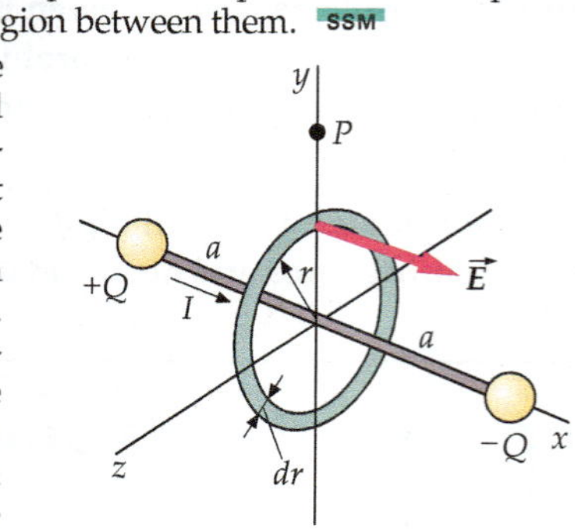

FIGURE 30-13 Problem 20

MAXWELL'S EQUATIONS AND THE ELECTROMAGNETIC SPECTRUM

21 • The color of the dominant light from the Sun is in the yellow-green region of the visible spectrum. Estimate the wavelength and frequency of the dominant light emitted by our Sun. *Hint: See Table 30-1.*

22 • (*a*) What is the frequency of microwave radiation that has a 3.00-cm-long wavelength? (*b*) Using Table 30-1, estimate the ratio of the shortest wavelength of green light to the shortest wavelength of red light.

23 • (*a*) What is the frequency of an X ray that has a 0.100-nm-long wavelength? (*b*) The human eye is sensitive to light that has a wavelength equal to 550 nm. What are the color and frequency of this light? Comment on how this answer compares to your answer for Problem 21.

ELECTRIC DIPOLE RADIATION

Note: **All of the problems in this section are based on the following information. Refer to Figure 30-11. It can be shown that the intensity of radiation from a radiating electric dipole at a field point far from the antenna is proportional to $\sin^2\theta/r^2$, where θ is the angle between the electric dipole moment vector $\vec{p}$ and the position vector $\vec{r}$ of the field point relative to the center of the antenna. The pattern of radiation from this type of antenna is independent of the azimuthal angle, that is, you can rotate the pattern about the antenna axis and it does not change shape.**

24 •• Suppose a radiating electric dipole lies along the z axis. Let I_1 be the intensity of the radiation at a distance of 10 m and at angle of 90°. Find the intensity (in terms of I_1) at (*a*) a distance of 30 m and an angle of 90°, (*b*) a distance of 10 m and an angle of 45°, and (*c*) a distance of 20 m and an angle of 30°.

25 •• (*a*) For the situation described in Problem 24, at what angle is the intensity at a distance of 5.0 m equal to I_1? (*b*) At what distance is the intensity equal to I_1 when $\theta = 45°$?

26 •• **ENGINEERING APPLICATION, CONTEXT-RICH** You and your engineering crew are in charge of setting up a wireless telephone network for a village in a mountainous region. The transmitting antenna of one station is an electric dipole antenna located atop a mountain 2.00 km above sea level. There is a nearby mountain that is 4.00 km from the antenna and is also 2.00 km above sea level. At that location, one member of the crew measures the intensity of the signal to be 4.00×10^{-12} W/m². What should be the intensity of the signal at the village that is located at sea level and 1.50 km from the transmitter?

27 ••• **ENGINEERING APPLICATION** A radio station that uses a vertical electric dipole antenna broadcasts at a frequency of 1.20 MHz and has a total power output of 500 kW. Calculate the intensity of the signal at a horizontal distance of 120 km from the station. SSM

28 ••• **ENGINEERING APPLICATION** Regulations require that licensed radio stations have limits on their broadcast power so as to avoid interference with signals from distant stations. You are in charge of checking compliance with the law. At a distance of 30.0 km from a radio station that broadcasts from a single vertical electric dipole antenna at a frequency of 800 kHz, the intensity of the electromagnetic wave is 2.00×10^{-13} W/m². What is the total power radiated by the station?

29 ••• **ENGINEERING APPLICATION** A small private plane approaching an airport is flying at an altitude of 2.50 km above sea level. As a flight controller at the airport, you know your system uses a vertical electric dipole antenna to transmit 100 W at 24.0 MHz. What is the intensity of the signal at the plane's receiving antenna when the plane is 4.00 km from the airport? Assume the airport is at sea level.

ENERGY AND MOMENTUM IN AN ELECTROMAGNETIC WAVE

30 • An electromagnetic wave has an intensity of 100 W/m². Find its (*a*) rms electric field strength, and (*b*) rms magnetic field strength.

31 • The amplitude of an electromagnetic wave's electric field is 400 V/m. Find the wave's (*a*) rms electric field strength, (*b*) rms magnetic field strength, (*c*) intensity, and (*d*) radiation pressure (P_r). SSM

32 • The rms value of an electromagnetic wave's electric field strength is 400 V/m. Find the wave's (*a*) rms magnetic field strength, (*b*) average energy density, and (*c*) intensity.

33 •• (*a*) An electromagnetic wave that has an intensity equal to 200 W/m² is normal to a black 20 cm by 30 cm rectangular card that absorbs 100 percent of the wave. Find the force exerted on the card by the radiation. (*b*) Find the force exerted by the same wave if the card reflects 100 percent of the wave.

34 •• Find the force exerted by the electromagnetic wave on the card in Part (*b*) of Problem 33 if both the incident and the reflected waves are at angles of 30° to the normal.

35 • (*a*) For a given distance from a radiating electric dipole, at what angle (expressed as θ and measured from the dipole axis) is the intensity equal to 50 percent of the maximum intensity? (*b*) At what angle θ is the intensity equal to 1 percent of the maximum intensity? SSM

36 •• A laser pulse has an energy of 20.0 J and a beam radius of 2.00 mm. The pulse duration is 10.0 ns and the energy density is uniformly distributed within the pulse. (*a*) What is the spatial length of the pulse? (*b*) What is the energy density within the pulse? (*c*) Find the rms values of the electric and magnetic fields in the pulse.

37 •• An electromagnetic plane wave has an electric field that is parallel to the y axis, and has a Poynting vector given by $\vec{S}(x, t) = (100\ \text{W/m}^2)\cos^2(kx - \omega t)\hat{i}$, where x is in meters, $k = 10.0$ rad/m, $\omega = 3.00 \times 10^9$ rad/s, and t is in seconds. (*a*) What is the direction of propagation of the wave? (*b*) Find the wavelength and frequency of the wave. (*c*) Find the electric and magnetic fields of the wave as functions of x and t. SSM

38 •• A parallel-plate capacitor is being charged. The capacitor consists of a pair of identical circular parallel plates that each have a radius b and a separation distance d. (*a*) Show that the displacement current in the capacitor gap has the same value as the conduction current in the capacitor leads. (*b*) What is the direction of the Poynting vector in the region between the capacitor plates? (*c*) Find an expression for the Poynting vector in this region and show that its flux into the region between the plates is equal to the rate of change of the energy stored in the capacitor.

39 •• A pulsed laser fires a 1000-MW pulse that has a 200-ns duration at a small object that has a mass equal to 10.0 mg and is suspended by a fine fiber that is 4.00 cm long. If the radiation is completely absorbed by the object, what is the maximum angle of deflection of this pendulum? (Think of the system as a ballistic pendulum and assume the small object was hanging vertically before the radiation hit it.) SSM

40 •• The mirrors used in a particular type of laser are 99.99% reflecting. (*a*) If the laser has an average output power of 15 W, what is the average power of the radiation incident on one of the mirrors? (*b*) What is the force due to radiation pressure on one of the mirrors?

41 •• (*a*) Estimate the force on Earth due to the pressure of the radiation on Earth by the Sun, and compare this force to the gravitational force of the Sun on Earth. (At Earth's orbit, the intensity of sunlight is 1.37 kW/m².) (*b*) Repeat Part (*a*) for Mars which is at an average distance of 2.28×10^8 km from the Sun and has a radius of 3.40×10^3 km. (*c*) Which planet has the larger ratio of radiation pressure to gravitational attraction? SSM

THE WAVE EQUATION FOR ELECTROMAGNETIC WAVES

42 • Show by direct substitution that Equation 30-8*a* is satisfied by the wave function $E_y = E_0 \sin(kx - \omega t) = E_0 \sin k(x - ct)$ where $c = \omega/k$.

43 • Use the values of μ_0 and ϵ_0 in SI units to compute $1/\sqrt{\epsilon_0\mu_0}$ and show that it is equal to 3.00×10^8 m/s.

44 •• (*a*) Use Maxwell's equations to show for a plane wave, in which $\vec{E}$ and $\vec{B}$ are independent of y and z, that $\frac{\partial E_z}{\partial x} = \frac{\partial B_y}{\partial t}$ and $\frac{\partial B_y}{\partial x} = \mu_0\epsilon_0\frac{\partial E_z}{\partial t}$. (*b*) Show that E_z and B_y also satisfy the wave equation.

45 •• Show that any function of the form $y(x, t) = f(x - vt)$ or $y(x, t) = g(x + vt)$ satisfies the wave equation (Equation 30-7). SSM

GENERAL PROBLEMS

46 • An electromagnetic wave has a frequency of 100 MHz and is traveling in a vacuum. The magnetic field is given by $\vec{B}(z, t) = (1.00 \times 10^{-8}\text{ T})\cos(kz - \omega t)\hat{i}$. (*a*) Find the wavelength and the direction of propagation of this wave. (*b*) Find the electric field vector $\vec{E}(z, t)$. (*c*) Determine the Poynting vector, and use it to find the intensity of the wave.

47 •• **ENGINEERING APPLICATION** A circular loop of wire can be used to detect electromagnetic waves. Suppose the signal strength from a 100-MHz FM radio station 100 km distant is 4.0 μW/m^2, and suppose the signal is vertically polarized. What is the maximum rms voltage induced in your antenna, assuming your antenna is a 10.0-cm-radius loop? SSM

48 •• **ENGINEERING APPLICATION** The electric field strength from a radio station some distance from the electric dipole transmitting antenna is given by $(1.00 \times 10^{-4}\text{ N/C})\cos[(1.00 \times 10^6\text{ rad/s})t]$. (*a*) What peak voltage is picked up on a 50.0-cm-long wire oriented parallel with the electric field direction? (*b*) What is the maximum voltage that can be induced by this electromagnetic wave in a conducting loop of radius 20.0 cm, and what orientation of the loop does this require?

49 ••• A parallel-plate capacitor has circular plates of radius a that are separated by a distance d. In the gap between the two plates is a thin straight wire of resistance R that connects the centers of the two plates. A time-varying voltage given by $V_0 \sin\omega t$ is applied across the plates. (*a*) What is the current drawn by the capacitor? (*b*) What is the magnetic field as a function of the radial distance r from the centerline within the capacitor plates? (*c*) What is the phase angle between the current drawn by the capacitor and the applied voltage?

50 •• A 20-kW beam of electromagnetic radiation is normal to a surface that reflects 50 percent of the radiation. What is the force exerted by the radiation on the surface?

51 •• The electric fields of two harmonic electromagnetic waves of angular frequency ω_1 and ω_2 are given by $\vec{E}_1 = E_{10}\cos(k_1x - \omega_1t)\hat{j}$ and by $\vec{E}_2 = E_{20}\cos(k_2x - \omega_2t + \delta)\hat{j}$. For the resultant of these two waves, find (*a*) the instantaneous Poynting vector and (*b*) the time-averaged Poynting vector. (*c*) Repeat Parts (*a*) and (*b*) if the direction of propagation of the second wave is reversed so that $\vec{E}_2 = E_{20}\cos(k_2x + \omega_2t + \delta)\hat{j}$. SSM

52 •• Show that $\frac{\partial B_z}{\partial x} = -\mu_0\epsilon_0\frac{\partial E_y}{\partial t}$ (Equation 30-10) follows from $\oint_C \vec{B}\cdot d\vec{\ell} = \mu_0\epsilon_0\int_S \frac{\partial E_n}{\partial t}dA$ (Equation 30-6*d* and where $I = 0$) by integrating along a suitable curve C and over a suitable surface S in a manner that parallels the derivation of Equation 30-9.

53 •• For your backpacking excursions, you have purchased a radio capable of detecting a signal as weak as 1.00×10^{-14} W/m^2. This radio has a 2000-turn coil antenna that has a radius of 1.00 cm wound on an iron core that increases the magnetic field by a factor of 200. The broadcast frequency of the radio station is 1400 kHz. (*a*) What is the peak magnetic field strength of an electromagnetic wave of this minimum intensity? (*b*) What is the peak emf that it is capable of inducing in the antenna? (*c*) What would be the peak emf induced in a straight 2.00-m-long metal wire oriented parallel to the direction of the electric field?

54 •• The intensity of the sunlight striking Earth's upper atmosphere is 1.37 kW/m^2. (*a*) Find the rms values of the magnetic and electric fields of this light. (*b*) Find the average power output of the Sun. (*c*) Find the intensity and the radiation pressure at the surface of the Sun.

55 ••• A conductor in the shape of a long solid cylinder that has a length L, a radius a, and a resistivity ρ carries a steady current I that is uniformly distributed over its cross section. (*a*) Use Ohm's law to relate the electric field $\vec{E}$ in the conductor to I, ρ, and a. (*b*) Find the magnetic field $\vec{B}$ just outside the conductor. (*c*) Use the results from Part (*a*) and Part (*b*) to compute the Poynting vector $\vec{S} = (\vec{E} \times \vec{B})/\mu_0$ at $r = a$ (the edge of the conductor). In what direction is $\vec{S}$? (*d*) Find the flux $\oint S_n\,dA$ through the surface of the cylinder, and use the flux to show that the rate of energy flow into the conductor equals I^2R, where R is the resistance of the cylinder. SSM

56 ••• A long solenoid that has n turns per unit length carries a current that increases linearly with time. The solenoid has radius R, length L, and the current I in the windings is given by $I = at$. (*a*) Find the induced electric field at a distance $r < R$ from the central axis of the solenoid. (*b*) Find the magnitude and direction of the Poynting vector $\vec{S}$ at $r = R$ (just inside the solenoid windings). (*c*) Calculate the flux $\oint S_n\,dA$ into the region inside the solenoid, and show that the flux equals the rate of increase of the magnetic energy inside the solenoid.

57 ••• Small particles are blown out of the solar system by the radiation pressure of sunlight. Assume that each particle is spherical, has a radius r, has a density of 1.00 g/cm^3, and absorbs all the radiation in a cross-sectional area of πr^2. Assume the particles are located at some distance d from the Sun, which has a total power output of 3.83×10^{26} W. (*a*) What is the critical value for the radius r of the particle for which the radiation force of repulsion just balances the gravitational force of attraction to the Sun? (*b*) Do particles that have radii larger than the critical value get ejected from the solar system, or is it only particles that have radii smaller than the critical value that get ejected? Explain your answer.

58 ••• When an electromagnetic wave at normal incidence on a perfectly conducting surface is reflected, the electric field of the reflected wave at the reflecting surface is equal and opposite to the electric field of the incident wave at the reflecting surface. (*a*) Explain why this assertion is valid. (*b*) Show that the superposition of incident and reflected waves results in a standing wave. (*c*) Are the magnetic fields of the incident waves and reflected waves at the reflecting surface equal and opposite as well? Explain your answer.

59 ••• An intense point source of light radiates 1.00 MW isotropically (uniformly in all directions). The source is located 1.00 m above an infinite, perfectly reflecting plane. Determine the force that the radiation pressure exerts on the plane. SSM

PART V LIGHT

CHAPTER 31

LIGHT IS TRANSMITTED BY TOTAL INTERNAL REFLECTION THROUGH TINY GLASS FIBERS. *(© James L. Amos/Corbis.)*

? How large must the angle of incidence of the light on the wall of the tube be so that no light escapes? (See Example 31-4.)

Properties of Light

The human eye is sensitive to electromagnetic radiation with wavelengths from approximately 400 nm to 700 nm.* The shortest wavelengths in the visible spectrum correspond to violet light and the longest to red light. The perceived colors of light are the result of the physiological and psychological responses of the eyes and brain to the different frequencies of visible light. Although the correspondence between perceived color and frequency is quite good, there are many interesting deviations. For example, a mixture of red light and green light is perceived by the eyes and brain as yellow—even in the absence of light in the yellow region of the spectrum.

* Wavelengths as short as 380 nm and as long as 780 nm can be seen by some individuals.

In this chapter, we study how light is produced; how its speed is measured; and how light is scattered, reflected, refracted, and polarized.

31-1 THE SPEED OF LIGHT

Prior to the seventeenth century the speed of visible light was thought by many people to be infinite, and an effort to measure the speed of visible light was made by Galileo. He and a partner stood on hilltops about three kilometers apart, each with a lantern and a shutter to cover it. Galileo proposed to measure the time it took for visible light to travel back and forth between the experimenters. First, one would uncover his lantern, and when the other saw the light he would uncover his lantern. The time between the first partner's uncovering his lantern and his seeing the light from the other lantern would be the time it took for light to travel back and forth between the experimenters. Though this method is sound in principle, the speed of light is so great that the time interval to be measured is much smaller than variations in human response time, so Galileo was unable to obtain a value for the speed of light.

The first indication of the true magnitude of the speed of light came from astronomical observations of the period of Io, one of the moons of Jupiter. This period is determined by measuring the time between eclipses of Io. An eclipse occurs when Io enters the region behind Jupiter where no direct sunlight reaches. The eclipse period is about 42.5 h, but measurements made when Earth is moving away from Jupiter along path *ABC* in Figure 31-1 give a greater time for this period than do measurements made when Earth is moving toward Jupiter along path *CDA* in the figure. Because these measurements differ from the average value by only about 15 s, the discrepancies were difficult to measure accurately. In 1675, the astronomer Ole Römer attributed these discrepancies to the fact that the speed of light is finite, and that during the 42.5 h between eclipses of Jupiter's moon, the distance between Earth and Jupiter changes, making the path for the light longer or shorter. Römer devised the following method for measuring the cumulative effect of these discrepancies. Jupiter is moving much more slowly than Earth, so we can neglect its motion. When Earth is at point *A*, nearest to Jupiter, the distance between Earth and Jupiter is changing negligibly. The period of Io's eclipse is measured, providing the time between the beginnings of successive eclipses. Based on this measurement, the number of eclipses during 6 months is computed, and the time when an eclipse should begin a half-year later when Earth is at point *C* is predicted. When Earth is actually at point *C*, the observed beginning of the eclipse is about 16.6 min later than predicted. This is the time it takes light to travel a distance equal to the diameter of Earth's orbit. This calculation neglects the distance traveled by Jupiter toward Earth. However, because the orbital speed of Jupiter is so much slower than that of Earth, the distance Jupiter moves toward (or away from) Earth during the 6 months is much less than the diameter of Earth's orbit.

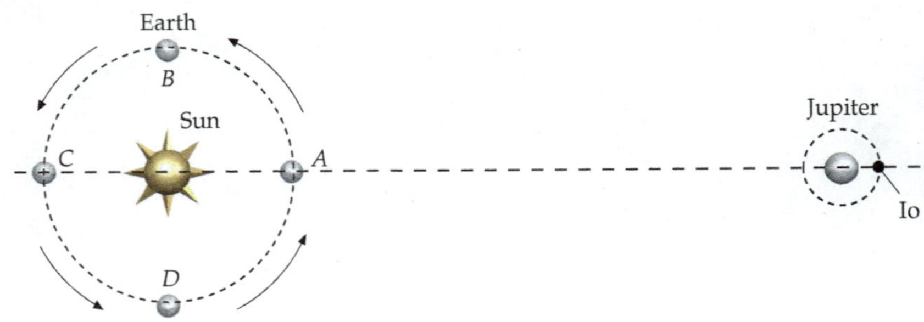

FIGURE 31-1 Römer's method of measuring the speed of light. The time between eclipses of Jupiter's moon Io appears to be greater when Earth is moving along path *ABC* than when Earth is moving along path *CDA*. The difference is due to the time it takes light to travel the distance traveled by Earth along the line of sight during one period of Io.

> **PRACTICE PROBLEM 31-1**
>
> Calculate (*a*) the distance traveled by Earth between successive eclipses of Io and (*b*) the speed of light, given that the time between successive eclipses is 15 s longer than average when Earth is moving directly away from Jupiter.

The French physicist Armand Fizeau made the first nonastronomical measurement of the speed of visible light in 1849. On a hill in Paris, Fizeau placed a light source and a system of lenses arranged so that the light reflected from a semitransparent mirror was focused on a gap in a toothed wheel, as shown in Figure 31-2.

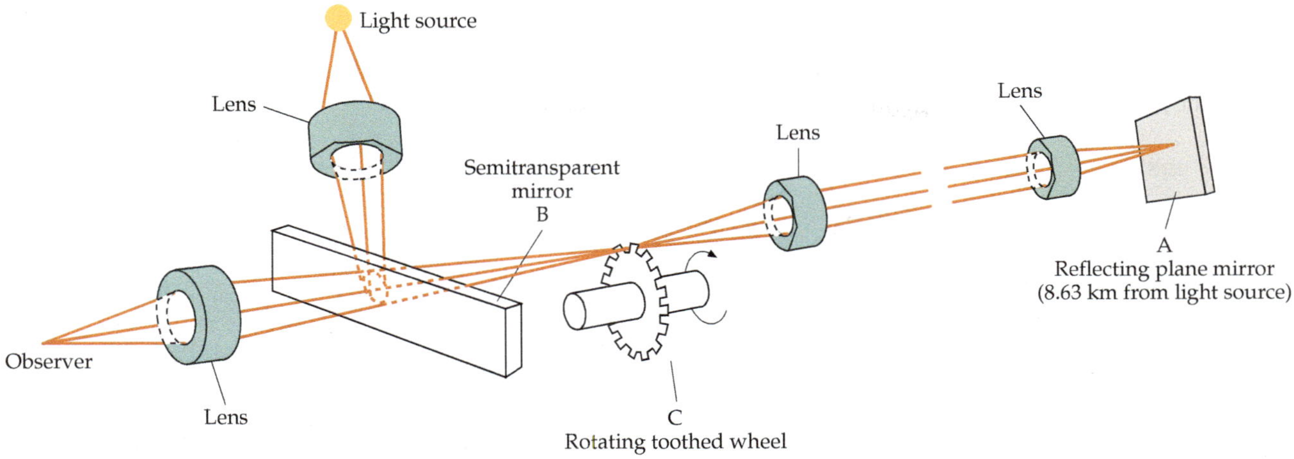

FIGURE 31-2 Fizeau's method of measuring the speed of light. Light from the source is reflected by mirror B and is transmitted through a gap in the toothed wheel to mirror A. The speed of light is determined by measuring the angular speed of the wheel that will permit the reflected light to pass through the next gap in the toothed wheel so that an image of the source is observed.

On a distant hill (about 8.63 km away) Fizeau placed a mirror to reflect the light back, to be viewed by an observer as shown. The toothed wheel was rotated, and the speed of rotation was varied. At low speeds of rotation, no light was visible because the light that passed through a gap in the rotating wheel and was reflected back by the mirror was obstructed by the next tooth of the wheel. The speed of rotation was then increased. The light suddenly became visible when the rotation speed was such that the reflected light passed through the next gap in the wheel. The time for the wheel to rotate through the angle between successive gaps equals the time for the light to make the round trip to the distant mirror and back.

Fizeau's method was improved upon by Jean Foucault, who replaced the toothed wheel with a rotating mirror, as shown in Figure 31-3. Light that strikes the rotating mirror is reflected toward a distant fixed mirror, where it is reflected back toward the rotating mirror. The light is then reflected toward the telescope by the rotating mirror. During the time taken for the light to travel from the rotating mirror to the distant fixed mirror and back, the mirror rotates through a small angle θ. By measuring the angle θ, the time for the light to travel to the distant mirror and back is determined. In approximately 1850, Foucault measured the speed of light in air and in water, and he showed that the speed of light in water is less than the speed of light in air. Using essentially the same method, the American physicist Albert Michelson made more precise measurements of the speed of light in approximately 1880. A half-century later, Michelson made even more precise measurements of the speed of light, using an octagonal rotating mirror (Figure 31-4). In these measurements, the mirror rotates through one-eighth of a turn during the time it takes for the light to travel to the fixed mirror and back. The rotation rate is varied until another face of the mirror is in the right position for the reflected light to enter the telescope.

FIGURE 31-3 Simplified drawing of Foucault's method of measuring the speed of light.

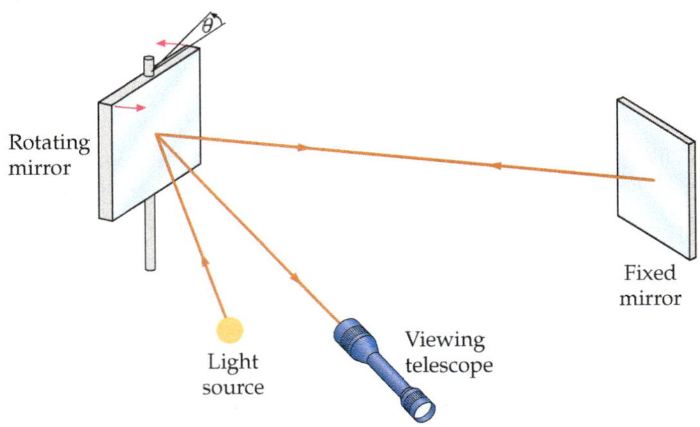

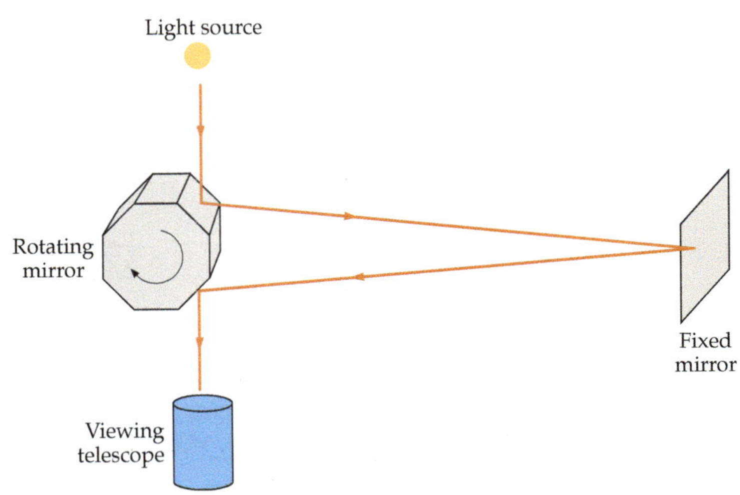

FIGURE 31-4 Simplified drawing of Michelson's method of measuring the speed of light at Mt. Wilson in the late 1920s.

Another method of determining the speed of light involves the measurement of the electrical constants ϵ_0 and μ_0 to determine c from $c = 1/\sqrt{\epsilon_0\mu_0}$.

The various methods we have discussed for measuring the speed of light are all in agreement. Today, the speed of light is defined to be exactly

$$c = 299\,792\,458 \text{ m/s} \tag{31-1}$$

DEFINITION—SPEED OF LIGHT

and the standard unit of length, the meter, is defined in terms of this speed and the standard unit of time. The meter is the distance light travels (in a vacuum) in 1/299 792 458 s. The value 3.00×10^8 m/s for the speed of light is accurate enough for nearly all calculations in this book. The speed of radio waves and all other electromagnetic waves (in a vacuum) is the same as the speed of visible light.

Example 31-1 The Speed of Light

What is the speed of light in feet per nanosecond?

PICTURE This is an exercise in unit conversions. There are ~30 cm = 0.30 m in 1.0 ft.

SOLVE

1. Convert m/s to ft/ns: $c = 3.0 \times 10^8 \text{ m/s} \times \left(\frac{1.0 \text{ ft}}{0.30 \text{ m}}\right) \times \left(\frac{1.0 \text{ s}}{10^9 \text{ ns}}\right) = \boxed{1.0 \text{ ft/ns}}$

Example 31-2 Fizeau's Determination of *c*

You are attempting to reproduce Fizeau's determination of the speed of light. Using a wheel that has 720 teeth, light is observed when the wheel rotates at 22.3 rev/s. If the distance from the wheel to the distant mirror is 8.63 km, what value do these numbers give for the speed of light?

PICTURE The time taken for the light to travel from the wheel to the mirror and back is the time for the wheel to rotate one Nth of a revolution, where $N = 720$ is the total number of teeth.

SOLVE

1. The speed is the distance divided by the time. The distance from the wheel to the mirror is L: $c = \frac{2L}{\Delta t}$
2. The angular displacement equals the angular speed multiplied by the time: $\Delta\theta = \omega\,\Delta t$
3. Solve for the time: $\Delta t = \frac{\Delta\theta}{\omega}$
4. Substitute for Δt and solve for c: $c = \frac{2L\omega}{\Delta\theta} = \frac{2(8.63 \times 10^3 \text{ m})(22.3 \text{ rev/s})}{\frac{1}{720}\text{ rev}} = \boxed{2.77 \times 10^8 \text{ m/s}}$

CHECK This result is slightly more than 7 percent too low for the speed of light. However, because it is only 7 percent off, it is a plausible answer.

PRACTICE PROBLEM 31-2 Space travelers on the moon use electromagnetic waves to communicate with the space control center on Earth. Use $c = 3.00 \times 10^8$ m/s to calculate the time delay for their signal to reach Earth, which is 3.84×10^8 m away.

CONCEPT CHECK 31-1

The step-4 result to Example 31-2 is 7 percent too low. What might account for this discrepancy between this measured value and the known value for the speed of light? An error in counting the number of teeth, in measuring the angular speed, or in measuring the distance to the mirror are possible sources of error, but they are unlikely sources of error. There is a more likely source of error. Can you find it?

Large distances are often given in terms of the distance traveled by light in a given time. For example, the distance to the Sun is 8.33 light-minutes, written 8.33 $c \cdot$ min. A light-year is the distance light travels in one year. We can easily find a conversion factor between light-years and meters. The number of seconds in one year is

$$1\text{ y} = 1\text{ y} \times \frac{365.24\text{ d}}{1\text{ y}} \times \frac{24\text{ h}}{1\text{ d}} \times \frac{3600\text{ s}}{1\text{ h}} = 3.156 \times 10^7\text{ s}$$

(*Note:* There are approximately π multiplied by 10^7 seconds per year, which is the mechanism by which some individuals remember the approximate value of the conversion.) The number of meters in one light-year is

$$1\, c \cdot y = (2.998 \times 10^8\text{ m/s})(3.156 \times 10^7\text{ s}) = 9.46 \times 10^{15}\text{ m} \qquad 31\text{-}2$$

31-2 THE PROPAGATION OF LIGHT

The propagation of light is governed by the wave equation discussed in Chapter 30. But long before Maxwell's theory of electromagnetic waves, the propagation of light and other waves was described empirically by two interesting and very different principles attributed to the Dutch physicist Christian Huygens (1629–1695) and the French mathematician Pierre de Fermat (1601–1665).

HUYGENS'S CONSTRUCTION

Figure 31-5 shows a portion of a spherical wavefront emanating from a point source. The wavefront is the locus of points of constant phase. If the radius of the wavefront is r at time t, its radius at time $t + \Delta t$ is $r + c\,\Delta t$, where c is the speed of the wave. However, if a part of the wave is blocked by some obstacle or if the wave passes through a different medium, as in Figure 31-6, the determination of the new wavefront position at time $t + \Delta t$ is much more difficult. The propagation of any wavefront through space can be described using a geometric construction invented by Huygens in approximately 1678, which is now known as **Huygens's construction** or **Huygens's principle:**

> Each point on a primary wavefront serves as the source of spherical secondary wavelets that advance at the wavespeed for the propagating medium. The primary wavefront at some later time is the envelope of these wavelets.
>
> HUYGENS'S CONSTRUCTION

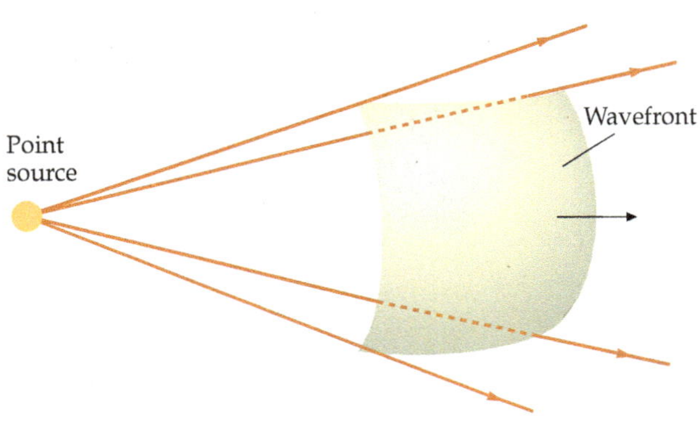

FIGURE 31-5 Spherical wavefront from a point source.

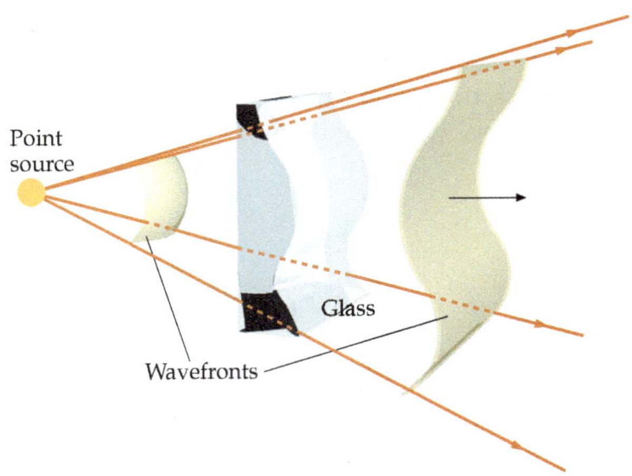

FIGURE 31-6 Wavefront from a point source before and after passing through a piece of glass of varied thickness.

Figure 31-7 shows the application of Huygens's construction to the propagation of a plane wave and the propagation of a spherical wave. Of course, if each point on a wavefront were really a point source, there would be waves in the backward direction as well. Huygens ignored those back waves.

Huygens's construction was later modified by Augustin Fresnel, so that the new wavefront was calculated from the old wavefront by superposition of the wavelets considering their relative amplitudes and phases. Kirchhoff later showed that the Huygens–Fresnel construction was a consequence of the wave equation (Equation 30-8*a*), thus putting it on a firm mathematical basis. Kirchhoff showed that the intensity of each wavelet depends on the angle and is zero at 180° (the backward direction).

We will use Huygens's construction to derive the laws of reflection and refraction in Section 31-5. In Chapter 33, we apply Huygens's construction with Fresnel's modification to calculate the diffraction pattern of a single slit. Because the wavelength of light is so small, we can often use the ray approximation to describe its propagation.

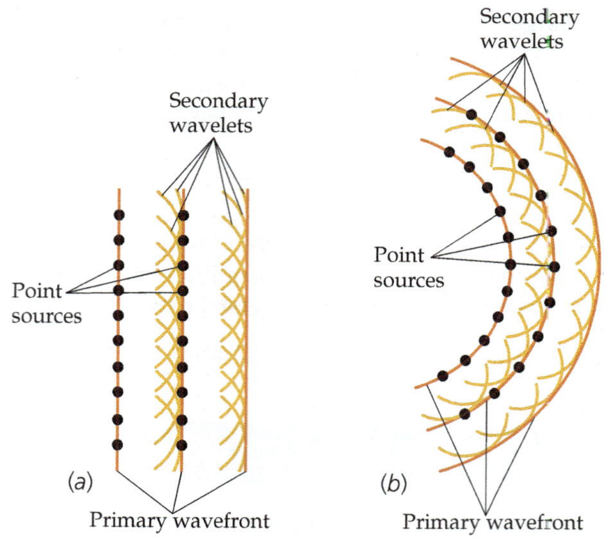

FIGURE 31-7 Huygens's construction for the propagation from left to right of a primary wavefront of (*a*) a plane wave and (*b*) an outgoing spherical, or circular, wave.

FERMAT'S PRINCIPLE

The propagation of light can also be described by Fermat's principle:

> The path taken by light traveling from one point to another is such that the time of travel is a minimum. That is, light travels along the path of least time.*
>
> FERMAT'S PRINCIPLE

The path of least time is not necessarily the path of shortest distance. For example, suppose you are a lifeguard at one end of a pool and a person needs immediate assistance at the far end of the pool. You could get to the person by swimming the entire length of the pool, but you would get to the person more quickly if you ran along the perimeter of the pool and entered the water only when you are near the person.

In Section 31-5, we will use Fermat's principle to derive the laws of reflection and refraction.

31-3 REFLECTION AND REFRACTION

The speed of light in a transparent medium such as air, water, or glass is less than the speed $c = 3 \times 10^8$ m/s in vacuum.† A transparent medium is characterized by an **index of refraction**, n, which is defined as the ratio of the speed of light in a vacuum, c, to the speed in the medium, v:

$$n = \frac{c}{v} \qquad 31\text{-}3$$

DEFINITION—INDEX OF REFRACTION

* A more complete and general statement is that the time of travel is stationary with respect to variations in path; that is, if t is expressed in terms of some parameter x, the path taken will be such that $dt/dx = 0$. The important characteristic of a stationary path is that the time taken along nearby paths will be approximately the same as that along the stationary path.

† It is not the case that the wave speed is never greater than c. In certain materials it is greater than c. However, this does not mean that information can travel at speeds greater than c.

For water, $n = 1.33$, whereas for glass n ranges from approximately 1.50 to 1.66, depending on the type of glass. Diamond has a very high index of refraction—approximately 2.4. The index of refraction of air is approximately 1.0003, so for most purposes we can assume that the speed of light in air is the same as the speed of light in vacuum.

When a beam of light strikes a boundary surface separating two different media, such as an air–glass interface, part of the light energy is reflected and part of the light energy enters the second medium. If the incident light is not perpendicular to the surface, then the transmitted beam is not parallel to the incident beam. The change in direction of the transmitted ray is called **refraction.** Figure 31-8 shows a light ray striking a smooth air–glass interface. The angle θ_1 between the incident ray and the normal (the line perpendicular to the surface) is called the **angle of incidence,** and the plane containing the incident ray and the normal is called the **plane of incidence.** The reflected ray lies in the plane of incidence and makes an angle θ_1' with the normal that is equal to the angle of incidence as shown in the figure:

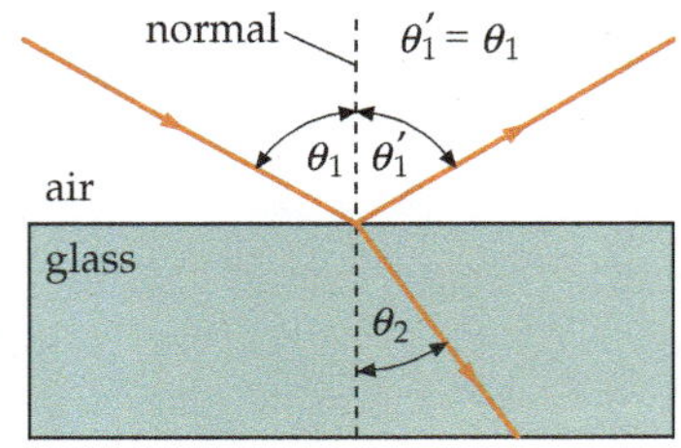

FIGURE 31-8 The angle of reflection θ_1' equals the angle of incidence θ_1. The angle of refraction θ_2 is less than the angle of incidence if the light speed in the second medium is less than that in the incident medium.

See **Math Tutorial** *for more information on* **Trigonometry**

$$\theta_1' = \theta_1 \qquad \text{31-4}$$

LAW OF REFLECTION

This result is known as the **law of reflection.** The law of reflection holds for any type of wave. Figure 31-9 illustrates the law of reflection for rays of light and for wavefronts of ultrasonic waves.

The ray that enters the glass in Figure 31-8 is called the *refracted ray*, and the angle θ_2 is called the angle of refraction. When a wave crosses a boundary where the wave speed is reduced, as in the case of light entering glass from air, the angle of refraction is less than the angle of incidence θ_1, as shown in Figure 31-8; that is, the refracted ray is bent toward the normal. If, on the other hand, the light beam originates in the glass and is refracted into the air, then the refracted ray is bent away from the normal.

The angle of refraction θ_2 depends on the angle of incidence and on the relative speed of light waves in the two mediums. If v_1 is the wave speed in the incident medium and v_2 is the wave speed in the transmission medium, the angles of incidence and refraction are related by

$$\frac{1}{v_1}\sin\theta_1 = \frac{1}{v_2}\sin\theta_2 \qquad \text{31-5}a$$

Equation 31-5*a* holds for the refraction of any kind of wave incident on a boundary separating two media.

The indices of refraction of the two media are n_1 and n_2. Combining Equations 31-3 and 31-5*a* gives

$$n_1 \sin\theta_1 = n_2 \sin\theta_2 \qquad \text{31-5}b$$

SNELL'S LAW OF REFRACTION

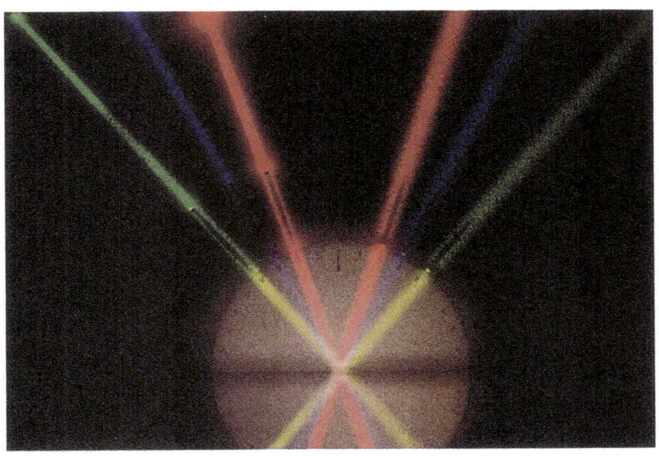

(*a*)

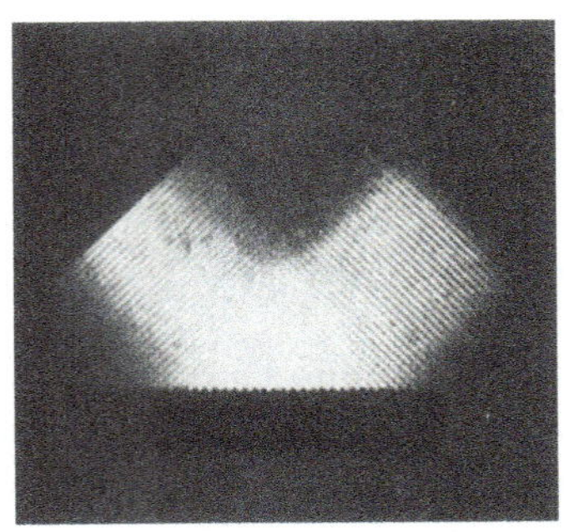

(*b*)

FIGURE 31-9 (*a*) Light rays reflecting from an air–glass interface showing equal angles of incidence and reflection. (*b*) Ultrasonic plane waves in water reflecting from a steel plate. *((a) Ken Kay/Fundamental Photographs. (b) Courtesy Battelle-Northwest Laboratories.)*

Reflection and refraction of a beam of light incident on a glass slab. *(Richard Megna/Fundamental Photographs.)*

This result was discovered experimentally in 1621 by the Dutch scientist Willebrord Snell and is known as **Snell's law** or the **law of refraction.** It was independently discovered a few years later by the French mathematician and philosopher René Descartes.

PHYSICAL MECHANISMS FOR REFLECTION AND REFRACTION

The physical mechanism of the reflection and refraction of light can be understood in terms of the absorption and reradiation of the light by the atoms in the reflecting or refracting medium. When light traveling in air strikes a glass surface, the atoms in the glass absorb the light and reradiate it at the same frequency in all directions. The waves radiated backward by the glass atoms interfere constructively at an angle equal to the angle of incidence to produce the reflected wave.

The transmitted wave is the result of the interference of the incident wave and the wave produced by the absorption and reradiation of light energy by the atoms in the medium. For light entering glass from air, a phase lag exists between the reradiated wave and the incident wave. Therefore, a phase lag also exists between the resultant wave and the incident wave. This phase lag means that the position of a wave crest of the transmitted wave is retarded relative to the position of a wave crest of the incident wave in the medium. As a result, a transmitted wave crest does not travel as far in a given time as the original incident wave crest; that is, the wave speed of the transmitted wave is less than that of the incident wave. The index of refraction is therefore greater than 1. The frequency of the light in the second medium is the same as the frequency of the incident light—the atoms absorb and reradiate the light at the same frequency—but the wave speed is different, so the wavelength of the transmitted light is different from that of the incident light. If λ is the wavelength of light in a vacuum, then $\lambda f = c$, and if λ_n is the wavelength in a medium that has an index of refraction n in which the wave has speed v, then $\lambda_n f = v$. Combining these two relations gives $\lambda/\lambda_n = c/v$, or

$$\lambda_n = \frac{\lambda}{c/v} = \frac{\lambda}{n} \qquad 31\text{-}6$$

SPECULAR REFLECTION AND DIFFUSE REFLECTION

Figure 31-10*a* shows a bundle of light rays from a point source P that are reflected from a flat surface. After reflection, the rays diverge exactly as if they came from a point P' behind the surface. (The point P' is called the *image point*. We will study the formation of images by reflecting and refracting surfaces in the next chapter.) When the rays enter the eye, they cannot be distinguished from rays actually diverging from a source at P'.

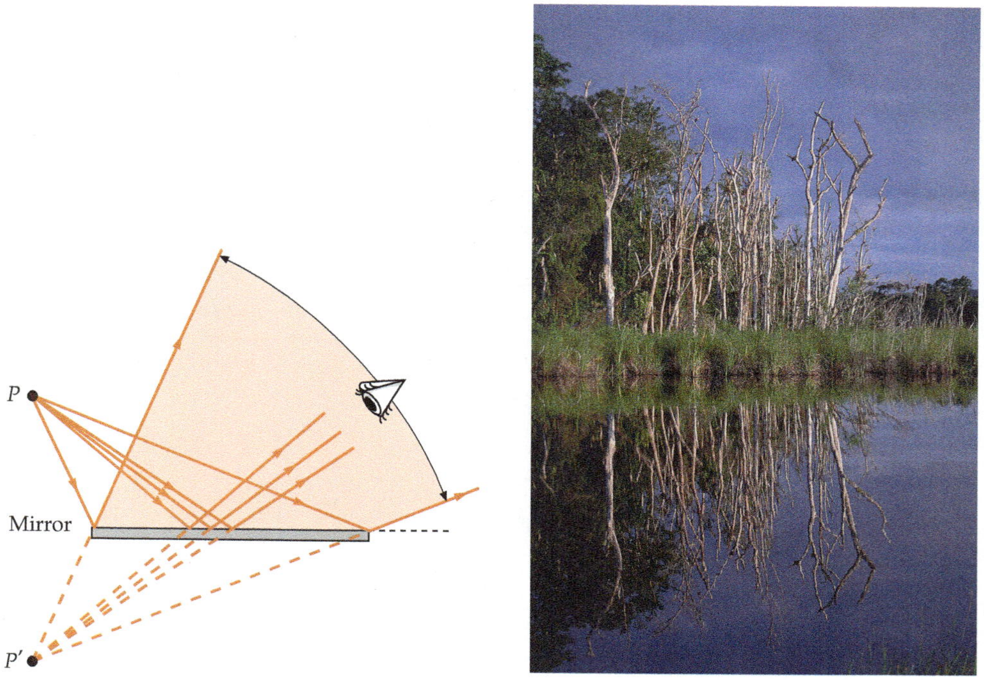

(a) (b)

FIGURE 31-10 (*a*) Specular reflection from a smooth surface. (*b*) Specular reflection of trees from water. *(Macduff Everton/Corbis.)*

Reflection from a smooth surface is called **specular reflection.** It differs from **diffuse reflection,** which is illustrated in Figure 31-11. Here, because the surface is rough, the rays from a point reflect in random directions and do not diverge from any point, so no image exists. The reflection of light from the page of this book is diffuse reflection. The glass used in picture frames is sometimes ground slightly to give diffuse reflection and thereby cut down on glare from the light used to illuminate the picture. Diffuse reflection from the surface of a road allows you to see the road when you are driving at night because some of the light from your headlights reflects back toward you. In wet weather the reflection is mostly specular; therefore, little light is reflected back toward you, which makes the road difficult to see.

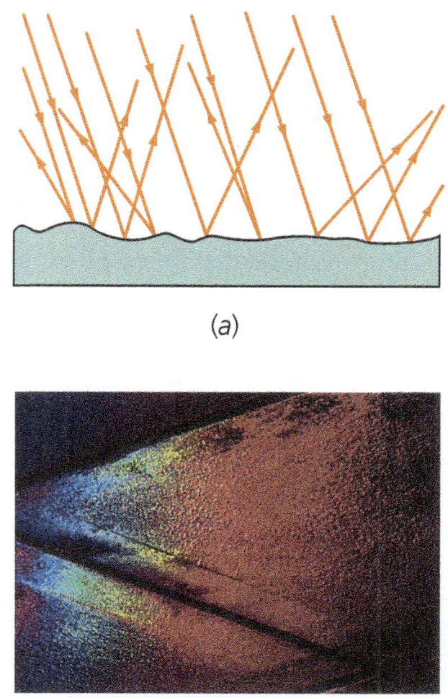

(a) (b)

FIGURE 31-11 (*a*) Diffuse reflection from a rough surface. (*b*) Diffuse reflection of colored lights from a sidewalk. *((b) Pete Saloutos/The Stock Market.)*

RELATIVE INTENSITY OF REFLECTED AND TRANSMITTED LIGHT

The fraction of light energy reflected at a boundary, such as an air–glass interface, depends in a complicated way on the angle of incidence, the orientation of the electric field vector associated with the wave, and the indices of refraction of the two media. For the special case of normal incidence ($\theta_1 = \theta_1' = 0$), the reflected intensity can be shown to be

$$I = \left(\frac{n_1 - n_2}{n_1 + n_2}\right)^2 I_0 \qquad \text{31-7}$$

where I_0 is the incident intensity and n_1 and n_2 are the indices of refraction of the two media.* For a typical case of reflection from a clean air–glass interface for which $n_1 = 1$ and $n_2 = 1.5$, Equation 31-7 gives $I = I_0/25$. Only about 4 percent of the energy is reflected; the remainder of the energy is transmitted.

* An equation for waves on a string that is similar to Equation 31-7 is presented in Section 4 of Chapter 15.

Example 31-3 Refraction from Air to Water

Light traveling in air enters water with an angle of incidence of 45.0°. If the index of refraction of water is 1.33, what is the angle of refraction?

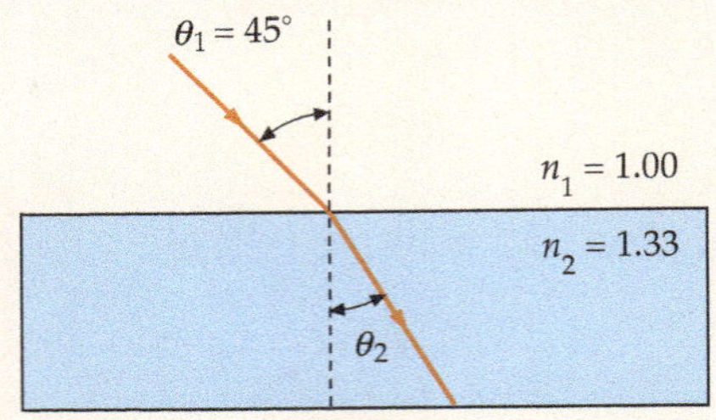

FIGURE 31-12

PICTURE The angle of refraction is found using Snell's law of refraction. Let subscripts 1 and 2 refer to the air and water, respectively. Then $n_1 = 1.00$, $\theta_1 = 45.0°$, $n_2 = 1.33$, and θ_2 is the angle of refraction (Figure 31-12).

SOLVE

1. Use Snell's law of refraction to solve for $\sin\theta_2$, the sine of the angle of refraction:

$$n_1 \sin\theta_1 = n_2 \sin\theta_2$$

so

$$\sin\theta_2 = \frac{n_1}{n_2}\sin\theta_1$$

2. Find the angle whose sine is 0.532:

$$\theta_2 = \sin^{-1}\left(\frac{n_1}{n_2}\sin\theta_1\right) = \sin^{-1}\left(\frac{1.00}{1.33}\sin 45.0°\right)$$

$$= \sin^{-1}(0.532) = \boxed{32.1°}$$

CHECK When entering a medium in which light travels more slowly, light bends toward the normal; so we expect θ_2 to be less than θ_1. The step-2 result meets this expectation.

TAKING IT FURTHER Note that the light is bent toward the normal as the light travels into the medium with the larger index of refraction.

TOTAL INTERNAL REFLECTION

Figure 31-13 shows a point source in glass and rays striking the glass–air interface at various angles. All the rays not perpendicular to the interface are bent away from the normal. As the angle of incidence is increased, the angle of refraction increases until a critical angle of incidence θ_c is reached for which the angle of refraction is 90°. For incident angles greater than the critical angle, no refracted ray exists. All the energy is reflected. This phenomenon is called **total internal reflection.** The critical angle can be found in terms of the indices of refraction of the two media by solving Equation 31-5*b* ($n_1 \sin\theta_1 = n_2 \sin\theta_2$) for $\sin\theta_1$ and setting θ_2 equal to 90°.

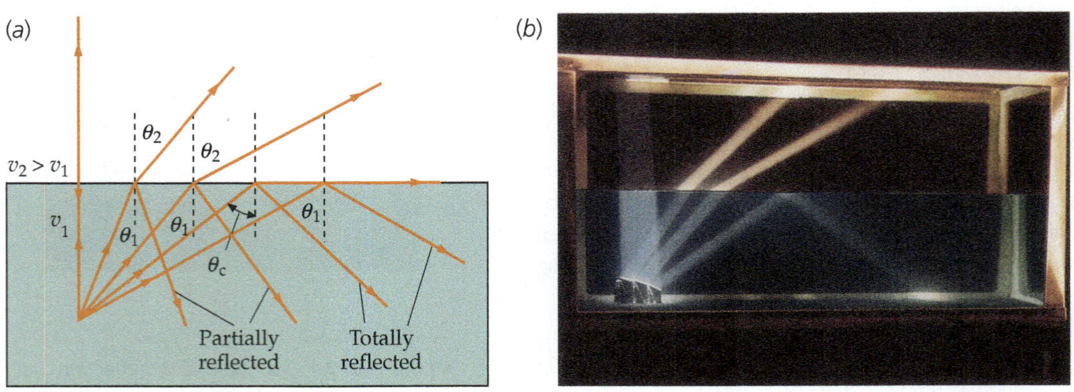

FIGURE 31-13 (*a*) Total internal reflection. As the angle of incidence is increased, the angle of refraction is increased until, at a critical angle of incidence θ_c, the angle of refraction is 90°. For angles of incidence greater than the critical angle, there is no refracted ray. (*b*) A photograph of refraction and total internal reflection from a water–air interface. *(Ken Kay/Fundamental Photographs.)*

That is,

$$\sin\theta_c = \frac{n_2}{n_1}\sin 90° = \frac{n_2}{n_1} \qquad 31\text{-}8$$

CRITICAL ANGLE FOR TOTAL INTERNAL REFLECTION

Note that total internal reflection occurs only when the incident light is in the medium that has the higher index of refraction. Mathematically, if n_2 is greater than n_1, Snell's law of refraction cannot be satisfied because there is no angle whose sine is greater than 1.

Example 31-4 Total Internal Reflection

Try It Yourself

A particular glass has an index of refraction of $n = 1.50$. What is the critical angle for total internal reflection for light leaving the glass and entering air, for which $n = 1.00$?

PICTURE Apply the law of refraction (Equation 31-5*b*) with the angle of refraction equal to 90°.

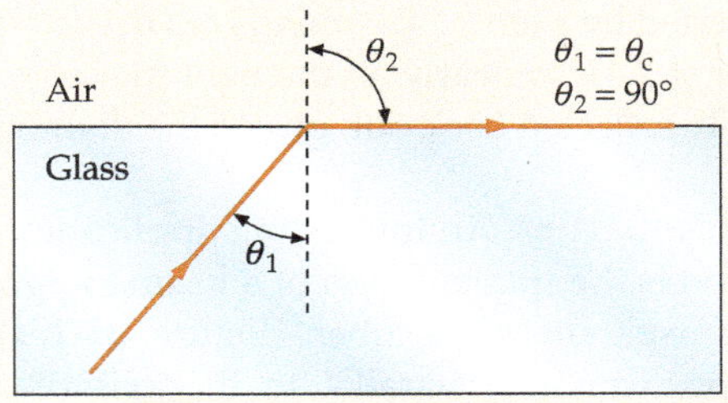

FIGURE 31-14

SOLVE

Cover the column to the right and try these on your own before looking at the answers.

Steps	Answers
1. Make a diagram (Figure 31-14) showing the incident and refracted rays. For the critical angle, the angle of refraction is 90°.	
2. Apply the law of refraction (Equation 31-5*b*). The critical angle is the angle of incidence.	$\theta_c = \boxed{41.8°}$

CHECK Figure 31-13*b* shows that the critical angle for a water-air boundary is a bit larger than 45°. Because the index of refraction for glass is somewhat larger than the index for water, we expect a critical angle for a glass-water boundary to be a bit less than 45°. An angle of 41.8° meets this expectation.

Example 31-5 How Deep Are You?

Context-Rich

You are spending time at the pool. While under the water, you look up and notice that you see objects above water level in a circle of light of radius approximately 2.0 m, and the rest of your vision is the color of the sides of the pool. How deep are you below the surface of the water?

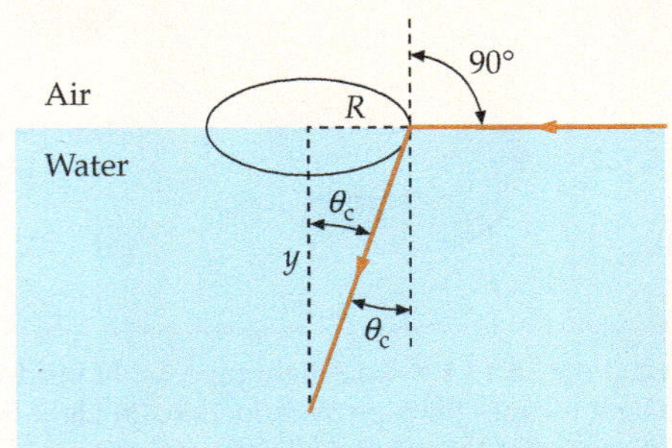

FIGURE 31-15

PICTURE We can determine the depth of the pool from the radius of the light and the angle at which the light is entering your eye from the edge of the circle. At the edge of the circle the light is entering the water at 90.0°, so the angle of refraction at the air–water surface is the critical angle for total internal reflection at the water–air surface. From Figure 31-15, we see that the depth y is related to this angle and the radius of the circle R by $\tan\theta_c = R/y$. The critical angle is found from Equation 31-8 with $n_2 = 1.00$ and $n_1 = 1.33$.

SOLVE

1. The depth y is related to the radius of the circle R and the critical angle θ_c:
$$\tan\theta_c = R/y$$

2. Solve for the depth y:
$$y = \frac{R}{\tan\theta_c}$$

3. Find the critical angle for total internal reflection at a water–air surface:
$$\sin\theta_c = \frac{n_2}{n_1} = \frac{1.00}{1.33} = 0.752$$
$$\theta_c = 48.8°$$

4. Solve for the depth y:
$$y = \frac{R}{\tan\theta_c} = \frac{2.0\text{ m}}{\tan 48.8°} = \boxed{1.7\text{ m}}$$

CHECK The step-4 result seems like a plausible value. Most swimming pools are at least that deep.

Figure 31-16*a* shows light reaching one of the short sides of a 45–45–90° glass prism at normal incidence. If the index of refraction of the glass is 1.5, the critical angle for total internal reflection is 41.8°, as we calculated in Example 31-4. Because the angle of incidence of the ray on the glass–air interface is 45°, the light will be totally reflected and will exit perpendicular to the other face of the prism, as shown. In Figure 31-16*b*, the light is incident perpendicular to the hypotenuse of the prism and is totally reflected twice so that it emerges at 180° to its original direction. Prisms are used to change the directions of light rays. In binoculars, two prisms are used on each side. These prisms reflect the light, thus shortening the required length, and reinvert the image (first inverted by a lens).* Diamonds have a very large index of refraction ($n \approx 2.4$), so nearly all the light that enters a diamond is eventually reflected back out, giving the diamond its sparkle.

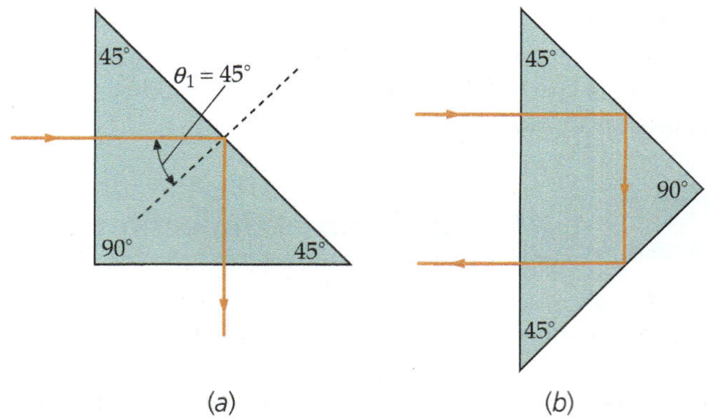

FIGURE 31-16 (*a*) Light entering through one of the short sides of a 45–45–90° glass prism is totally reflected. (*b*) Light entering through the long side of the prism is totally reflected twice.

Fiber optics An interesting application of total internal reflection is the transmission of a beam of light down a long, narrow, transparent glass fiber (Figure 31-17*a*). If the beam begins approximately parallel to the axis of the fiber, it will strike the walls of the fiber at angles greater than the critical angle (if the bends in the fiber are not too sharp) and no light energy will be lost through the walls of the fiber. A bundle of such fibers can be used for imaging, as illustrated in Figure 31-17*b*.

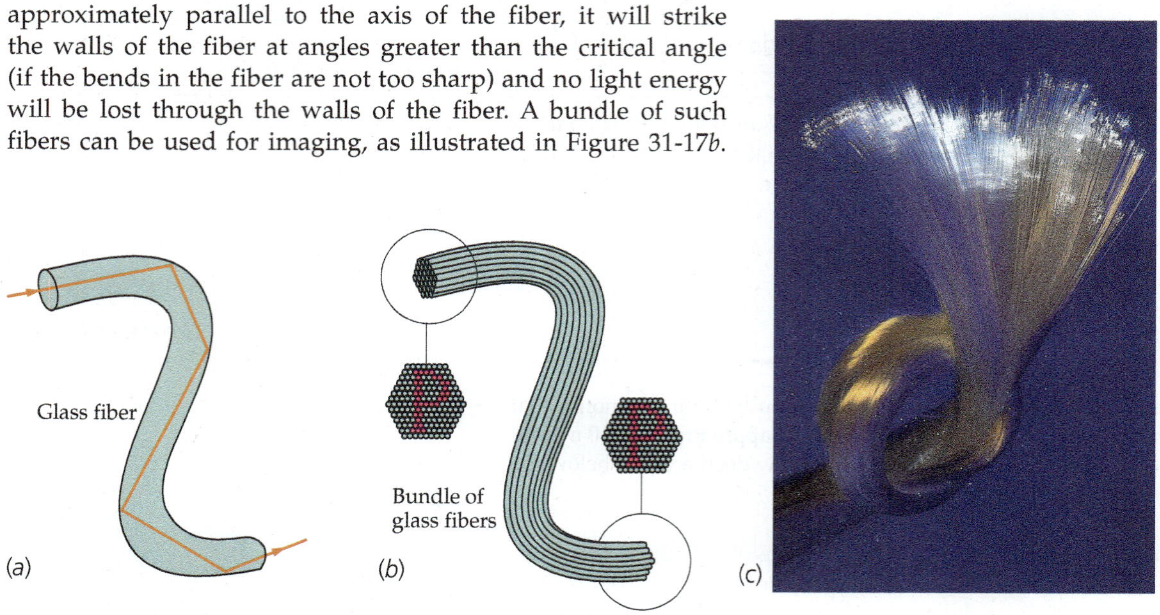

FIGURE 31-17 (*a*) A light pipe. Light inside the pipe is always incident at an angle greater than the critical angle, so no light escapes the pipe by refraction. (*b*) Light from the object is transported by a bundle of glass fibers to form an image of the object at the other end of the pipe. (*c*) Light emerging from a bundle of glass fibers. *((c) Ted Horowitz/The Stock Market.)*

Fiber optics has many applications in medicine and in communications. In medicine, light is transmitted along tiny fibers to visually probe various internal organs without surgery. In communications, the rate at which information can be transmitted is related to the signal frequency. A transmission system using light of frequencies of the order of 10^{14} Hz can transmit information at a much greater rate than one using radio waves, which have frequencies of the order of 10^6 Hz. In telecommunication systems, a single glass fiber that is the thickness of a human hair can transmit audio or video information equivalent to 32 000 voices speaking simultaneously.

* The image produced by the objective lens of a telescope is discussed in Section 32-4.

MIRAGES

When the index of refraction of a medium changes gradually, the refraction is continuous, leading to a gradual bending of the light. An interesting example of this is the formation of a mirage. On a hot and sunny day, the surface of exposed rocks, pavement, and sand often gets very hot. In this case there is often a layer of air near the ground that is warmer, and therefore less dense, than the air just above it. The speed of any light wave is slightly greater in this less dense layer, so a light beam passing from the cooler layer into the warmer layer is bent. Figure 31-18*a* shows the light from a tree when all the surrounding air is at the same temperature. The wavefronts are spherical, and the rays are straight lines. In Figure 31-18*b*, the air near the ground is warmer, resulting in the wavefronts traveling faster there. The portions of the wavefronts near the hot ground get ahead of the higher portions, creating a nonspherical wavefront and causing a curving of the rays. Thus, the ray shown initially heading for the ground is bent upward. As a result, the viewer sees an image of the tree looking as if it were reflected off a water surface on the ground. When driving on a hot sunny day, you may have noticed apparent wet spots on the highway ahead that disappear as you approach them. These mirages are due to the refraction of light from the sky by a layer of air that has been heated due to its proximity to the hot pavement.

(*a*)

(*b*)

(*a*) In this demonstration at the Naval Research Laboratory, a combination of laser sources generates different colors that excite adjacent fiber sensor elements, leading to a separation of the information as indicated by the separation of the colors. (*b*) The tip of a light guide preform is softened by heat and drawn into a long, tiny fiber. The colors in the preform indicate a layered structure of differing compositions, which is retained in the fiber. *((a) Dan Boyd/Courtesy Naval Research Laboratory. (b) Courtesy AT&T Archives.)*

(*a*)

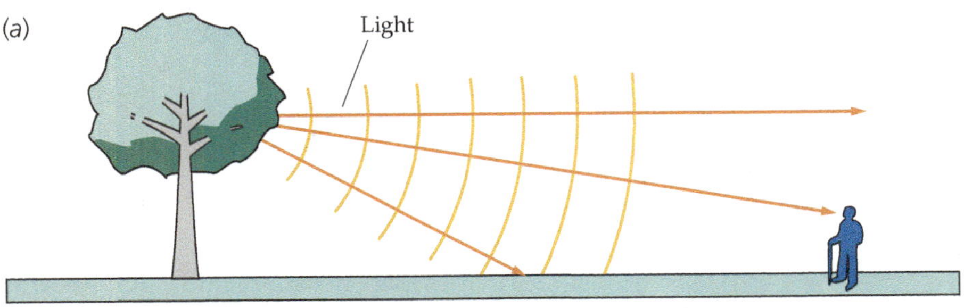

(*b*)

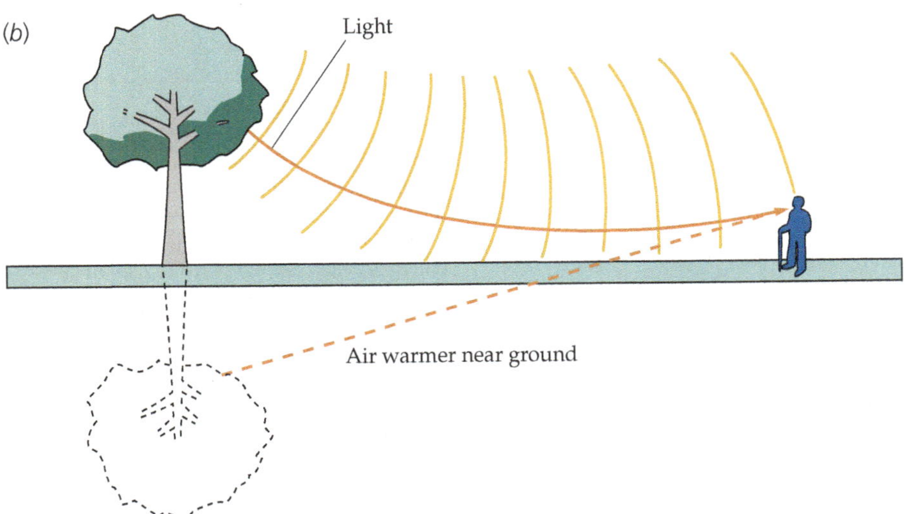

(*c*)

FIGURE 31-18 A mirage. (*a*) When the air is at a uniform temperature, the wavefronts of the light from the tree are spherical. (*b*) When the air near the ground is warmer, the wavefronts are not spherical and the light from the tree is continuously refracted into a curved path. (*c*) Apparent reflections of motorcycles on a hot road. *(Robert Greenler.)*

DISPERSION

The index of refraction of a material has a slight dependence on wavelength. For many materials, n decreases slightly as the wavelength increases, as shown in Figure 31-19. The dependence of the index of refraction on wavelength (and therefore on frequency) is called **dispersion.** When a beam of white light is incident at some angle on the surface of a glass prism, the angle of refraction (which is measured relative to the normal) for the shorter wavelengths is slightly smaller than the angle of refraction for the longer wavelengths. The light of shorter wavelength (toward the violet end of the spectrum) is therefore bent more toward the normal than that of longer wavelength. The beam of white light is thus spread out or dispersed into its component colors or wavelengths (Figure 31-20).

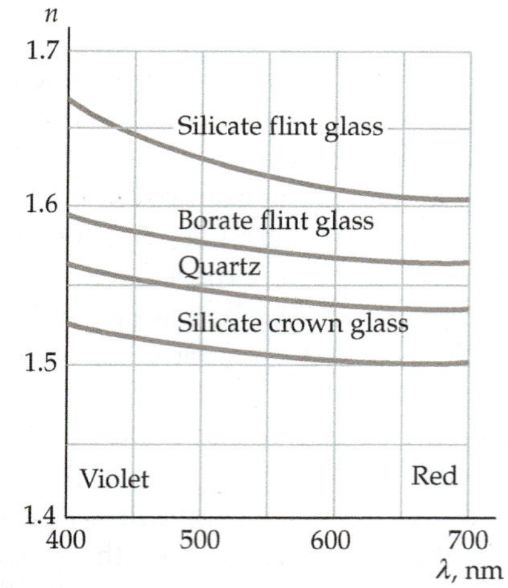

FIGURE 31-19 Index of refraction versus wavelength for various materials.

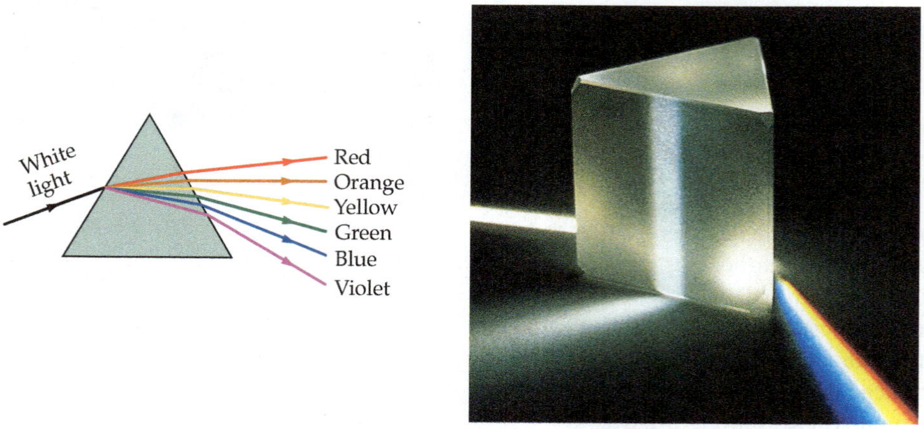

FIGURE 31-20 A beam of white light incident on a glass prism is dispersed into its component colors. The index of refraction decreases as the wavelength increases so that the longer wavelengths (red) are bent less than the shorter wavelengths (blue). *(David Parker/Science Photo Library/Photo Researchers.)*

Rainbows The rainbow is a familiar example of dispersion, in this case the dispersion of sunlight. Figure 31-21 is a diagram originally drawn by Descartes, showing parallel rays of light from the Sun entering a spherical water drop. First, the rays are refracted as they enter the drop. The rays are then reflected from the water–air interface on the other side of the drop and finally are refracted again as they leave the drop.

From Figure 31-21, we can see that the angle made by the emerging rays and the diameter (along ray 1) reaches a maximum around ray 7 and then decreases. The concentration of rays emerging at approximately the maximum angle gives rise to the rainbow. By construction, using the law of refraction, Descartes showed that the maximum angle is about 42°. To observe a rainbow, we must therefore look at the water drops at an angle of 42° relative to the line back to the Sun, as shown in Figure 31-22. The angular radius of the rainbow is therefore 42°.

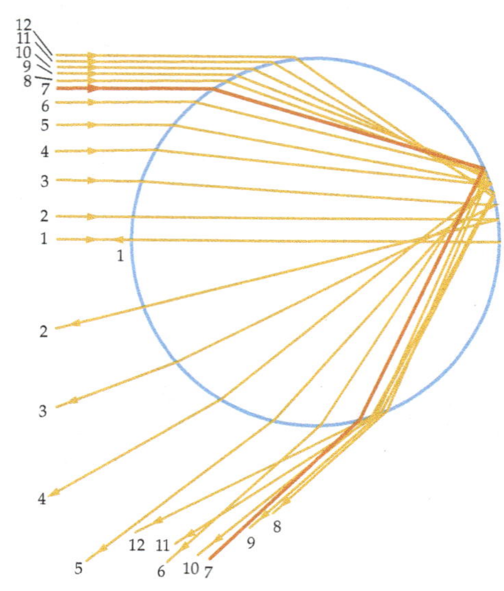

FIGURE 31-21 Descartes's construction of parallel rays of light entering a spherical water drop. Ray 1 enters the drop along a diameter and is reflected back along its incident path. Ray 2 enters slightly above the diameter and emerges below the diameter at a small angle with the diameter. The rays entering farther and farther away from the diameter emerge at greater and greater angles up to ray 7, shown as the heavy line. Rays entering above ray 7 emerge at smaller and smaller angles with the diameter.

The separation of the colors in the rainbow results from the fact that the index of refraction of water depends slightly on the wavelength of light. The angular radius of the rainbow will therefore depend slightly on the wavelength of the light. The observed rainbow is made up of light rays from many different droplets of water (Figure 31-23). The color seen at a particular angular radius corresponds to the wavelength of light that allows the light to reach the eye from the droplets at that angular radius. Because n_{water} is smaller for red light than for blue light, the red part of the rainbow is at a slightly greater angular radius than the blue part of the rainbow, so red is at the outer side of the rainbow.

When a light ray strikes a surface separating water and air, part of the light is reflected and part of the light is refracted. A secondary rainbow results from the light rays that are reflected twice within a droplet (Figure 31-24). The secondary rainbow has an angular radius of 51°, and its color sequence is the reverse of that of the primary rainbow; that is, the violet is on the outside in the secondary rainbow. Because of the small fraction of light reflected from a water–air interface, the secondary rainbow is considerably fainter than the primary rainbow.

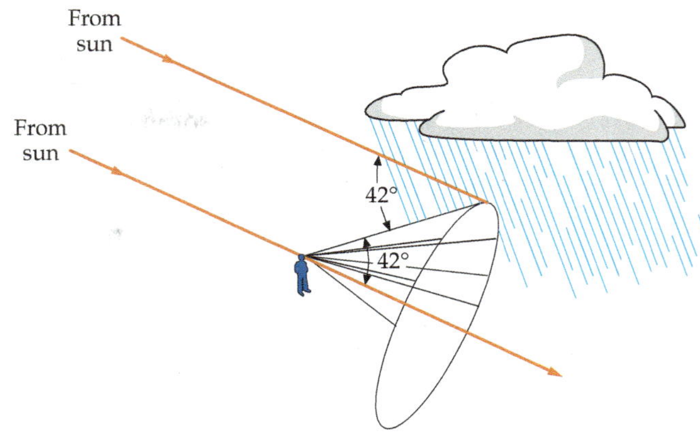

FIGURE 31-22 A rainbow is viewed at an angle of 42° from the line to the Sun, as predicted by Descartes's construction, as shown in Figure 31-21.

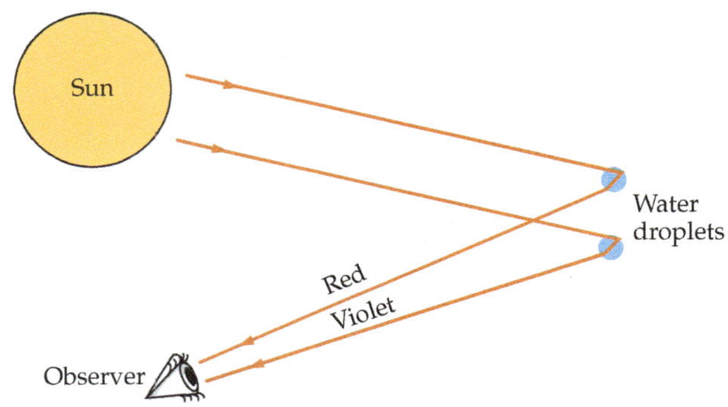

FIGURE 31-23 The rainbow results from light rays from many different water droplets.

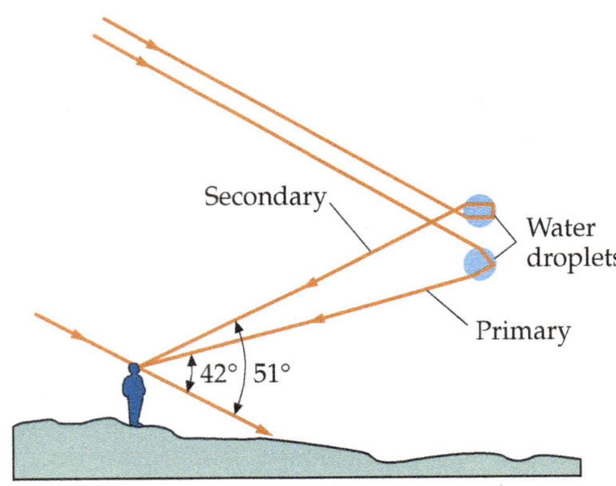

FIGURE 31-24 The secondary rainbow results from light rays that are reflected twice within a water droplet.

(*a*) This 22° halo around the Sun results from refraction by hexagonal ice crystals that are randomly oriented in the upper atmosphere. (*b*) When the ice crystals are not randomly oriented but are falling with their flat bases horizontal, only parts of the halo to the left and to the right of the Sun, called *sun dogs*, are seen. ((*a*) *Robert Greenler.* (*b*) *Giovanni DeAmici, NSF, Lawrence Berkeley Laboratory.*)

*Calculating the angular radius of the rainbow

We can calculate the angular radius of the rainbow from the laws of reflection and refraction. Figure 31-25 shows a ray of light incident on a spherical water droplet at point A. The angle of refraction θ_2 is related to the angle of incidence θ_1 by Snell's law of refraction:

$$n_{\text{air}} \sin\theta_1 = n_{\text{water}} \sin\theta_2 \qquad 31\text{-}9$$

Point P in Figure 31-25 is the intersection of the line of the incident ray and the line of the emerging ray. The angle ϕ_d is called the angle of deviation of the ray, and ϕ_d and 2β form a straight angle. Thus,

$$\phi_d + 2\beta = \pi \qquad 31\text{-}10$$

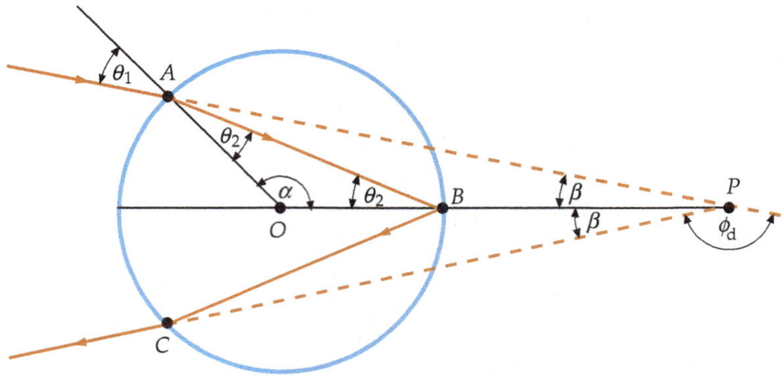

FIGURE 31-25 Light ray incident on a spherical water drop. The refracted ray strikes the back of the water droplet at point B. It makes an angle θ_2 with the radial line OB and is reflected at an equal angle. The ray is refracted again at point C, where it leaves the droplet.

We wish to relate the angle of deviation ϕ_d to the angle of incidence θ_1. From the triangle AOB, we have

$$2\theta_2 + \alpha = \pi \qquad 31\text{-}11$$

Similarly, from the triangle AOP, we have

$$\theta_1 + \beta + \alpha = \pi \qquad 31\text{-}12$$

Eliminating α from Equations 31-11 and 31-12 and solving for β gives

$$\beta = \pi - \theta_1 - \alpha = \pi - \theta_1 - (\pi - 2\theta_2) = 2\theta_2 - \theta_1$$

Substituting this value for β into Equation 31-10 gives the angle of deviation:

$$\phi_d = \pi - 2\beta = \pi - 4\theta_2 + 2\theta_1 \qquad 31\text{-}13$$

Equation 31-13 can be combined with Equation 31-9 to eliminate θ_2 and give the angle of deviation ϕ_d in terms of the angle of incidence θ_1:

$$\phi_d = \pi + 2\theta_1 - 4\sin^{-1}\left(\frac{n_{\text{air}}}{n_{\text{water}}}\sin\theta_1\right) \qquad 31\text{-}14$$

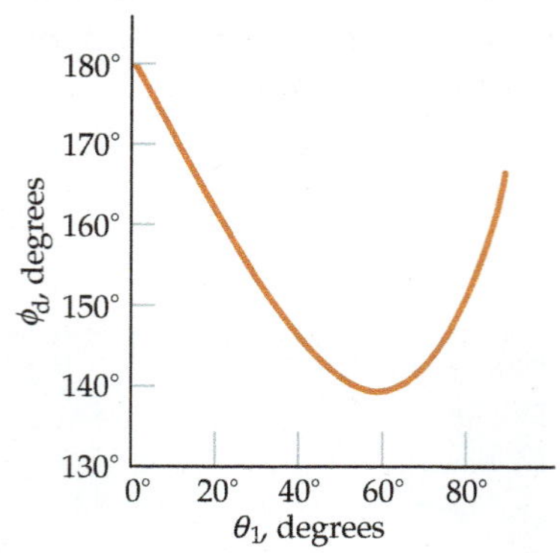

FIGURE 31-26 Plot of the angle of deviation ϕ_d as a function of incident angle θ_1. The angle of deviation has its minimum value of 138° when the angle of incidence is 60°. Because $d\phi_d/d\theta_1 = 0$ at minimum deviation, the deviation of rays with incident angles slightly less or slightly greater than 60° will be approximately the same.

Figure 31-26 shows a plot of ϕ_d versus θ_1. The angle of deviation ϕ_d has its minimum value when $\theta_1 \approx 60°$. At an angle of incidence of 60°, the angle of deviation is $\phi_{d,\text{min}} = 138°$. This angle is called the **angle of minimum deviation.** At incident angles that are slightly greater or slightly smaller than 60°, the angle of deviation is approximately the same. Therefore, the intensity of the light reflected by the water droplet will be a maximum at the angle of minimum deviation. We can see from Figure 31-25 that the maximum value of β corresponds to the minimum value of ϕ_d. Thus, the angular radius of the intensity maximum, given by $2\beta_{\text{max}}$, is

$$2\beta_{\text{max}} = \pi - \phi_{d,\text{min}} = 180° - 138° = 42° \qquad 31\text{-}15$$

The index of refraction of water varies slightly with wavelength. Thus, for each wavelength (color), the intensity maximum occurs at an angular radius slightly different than that of neighboring wavelengths.

31-4 POLARIZATION

In an electromagnetic wave, the direction of the electric field is perpendicular to the direction of propagation of the wave. If the electric field remains parallel to a line perpendicular to the direction of propagation, the wave is said to be **linearly polarized.** A wave produced by an electric dipole antenna is polarized with the electric field vector at any field point remaining in the plane containing the field point and the antenna axis. Waves produced by numerous sources are usually not polarized. An incandescent light source, for example, contains millions of atoms acting independently. The electric field for such a wave can be resolved into x and y components that vary randomly, because there is no correlation between the individual atoms producing the light.

The polarization of electromagnetic waves can be demonstrated with microwaves, which have wavelengths on the order of centimeters. In a typical microwave generator, polarized waves are radiated by an electric dipole antenna. In Figure 31-27, the electric dipole antenna is vertical, so the electric field vector $\vec{E}$ of the horizontally radiated waves is also vertical. An absorber can be made of a screen of parallel straight wires. When the wires are vertical, as in Figure 31-27*a*, the electric field parallel to the wires sets up currents in the wires and energy is absorbed. When the wires are horizontal and therefore perpendicular to $\vec{E}$, as in Figure 31-27*b*, no currents are set up and the waves are transmitted.

There are four phenomena that produce polarized electromagnetic waves from unpolarized waves: (1) absorption, (2) reflection, (3) scattering, and (4) birefringence (also called double refraction), each of which is examined in the upcoming sections.

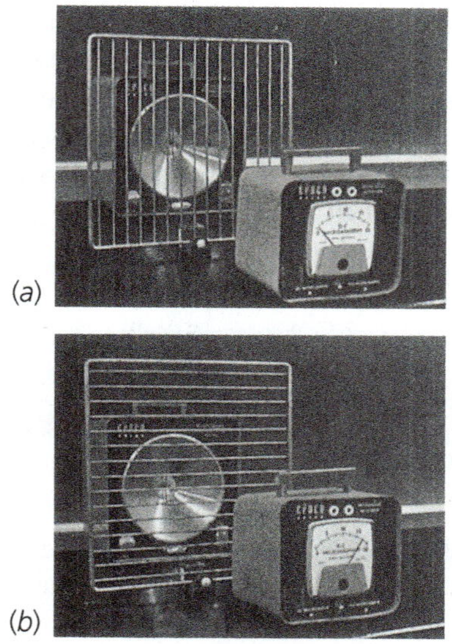

FIGURE 31-27 Demonstration showing the polarization of microwaves. The electric field of the microwaves is vertical, parallel to the vertical dipole antenna. (*a*) When the metal wires of the absorber are vertical, electric currents are set up in the wires and energy is absorbed, as indicated by the low reading on the microwave detector. (*b*) When the wires are horizontal, no currents are set up and the microwaves are transmitted, as indicated by the high reading on the detector. *(Larry Langrill.)*

POLARIZATION BY ABSORPTION

Several naturally occurring crystals, when cut into appropriate shapes, absorb and transmit light differently depending on the polarization of the light. These crystals can be used to produce linearly polarized light. In 1938, E. H. Land invented a simple commercial polarizing sheet called Polaroid. This material contains long-chain hydrocarbon molecules that are aligned when the sheet is stretched in one direction during the manufacturing process. These chains become conducting at optical frequencies when the sheet is dipped in a solution containing iodine. When light is incident with its electric field vector parallel to the chains, electric currents are set up along the chains, and the light energy is absorbed, just as the microwaves are absorbed by the wires in Figure 31-27. If the electric field is perpendicular to the chains, the light is transmitted. The direction perpendicular to the chains is called the **transmission axis.** We will make the simplifying assumption that all the light is transmitted when the electric field is parallel to the transmission axis and all the light is absorbed when it is perpendicular to the transmission axis. (In reality, polarizing sheets absorb some of the light, even when the electric field is parallel to the transmission axis.)

Consider an unpolarized light beam incident on a polarizing sheet with its transmission axis along the x direction, as shown in Figure 31-28. The beam is incident on a second polarizing sheet, the analyzer, whose transmission axis makes an angle θ with the x axis. If E is the electric field amplitude of the incident beam, the component parallel with the transmission axis is $E_{\|} = E\cos\theta$, and the component perpendicular to the transmission axis is $E_{\perp} = E\sin\theta$. The sheet absorbs $E_{\perp}$ and transmits $E_{\|}$, so the transmitted beam has an electric field amplitude of $E_{\|} = E\cos\theta$ and is linearly polarized in the direction of the transmission axis. Because the intensity of light is proportional to the square of the magnitude of the electric field amplitude, the intensity I of light transmitted by the sheet is given by

$$I = I_0\cos^2\theta \qquad 31\text{-}16$$

LAW OF MALUS

where I_0 is the intensity of the incident beam. If we have an incident beam of unpolarized light of intensity I_0 incident on a polarizing sheet, the direction of the incident electric field varies from location to location on the sheet, and at each location it fluctuates in time. At each location, the average value of $\cos^2\theta$ is one-half, so applying Equation 31-16 gives $I = I_0(\cos^2\theta)_{\text{av}} = \frac{1}{2}I_0$, where I is the intensity of the transmitted beam.

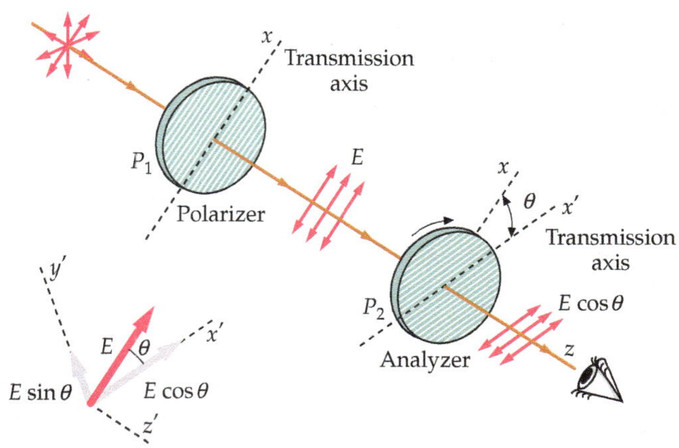

FIGURE 31-28 A vertically polarized beam is incident on a polarizing sheet with its transmission axis x' making an angle θ with the vertical. Only the component $E\cos\theta$ is transmitted through the second sheet, and the transmitted beam is linearly polarized in the direction of the transmission axis x'. If the intensity between the sheets is I_0, the intensity transmitted by the second sheet is $I_0\cos^2\theta$.

When two polarizing elements are placed in succession in a beam of unpolarized light, the first polarizing element is called the **polarizer** and the second polarizing element is called the **analyzer.** If the polarizer and the analyzer are crossed, that is, if their transmission axes are perpendicular to each other, no light gets through. Equation 31-16 is known as the **law of Malus** after its discoverer, E. L. Malus (1775–1812). It applies to any two polarizing elements whose transmission axes make an angle θ with each other.

(a)

(b)

(*a*) Crossed polarizers block out all of the light. (*b*) In a liquid crystal display, the crystal is between crossed polarizers. Light incident on the crystal is transmitted because the crystal rotates the direction of polarization of the light 90°. The light is reflected back out through the crystal by a mirror behind the crystal, and a uniform background is seen. When a voltage is applied across a small segment of the crystal, the polarization is not rotated, so no light is transmitted and the segment appears black. *((a) Fundamental Photographs. (b) 1990 PAR/NYC, Inc./Photo by Elizabeth Algieri.)*

Example 31-6 Intensity Transmitted

Unpolarized light of intensity 3.0 W/m² is incident on two polarizing sheets whose transmission axes make an angle of 60° (Figure 31-29). What is the intensity of light transmitted by the second sheet?

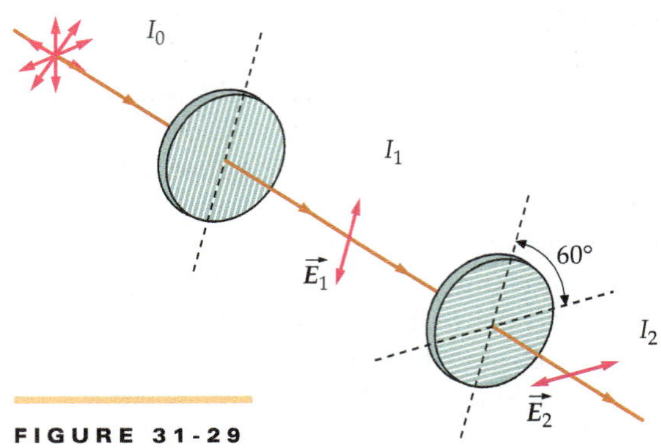

FIGURE 31-29

PICTURE The incident light is unpolarized, so the intensity transmitted by the first polarizing sheet is half the incident intensity. The second sheet further reduces the intensity by a factor of $\cos^2\theta$, with $\theta = 60°$.

SOLVE

1. The intensity I_1 transmitted by the first sheet is half the intensity I_0 of unpolarized light incident on the first sheet: $I_1 = \frac{1}{2}I_0$
2. The intensity I_2 transmitted by the second sheet is related to the intensity I_1 of the light incident on the second sheet by Equation 31-16: $I_2 = I_1\cos^2\theta$
3. Combine these results and substitute the given data: $I_2 = \frac{1}{2}I_0\cos^2 60° = \frac{1}{2}(3.0\ \mathrm{W/m^2})(0.500)^2 = \boxed{0.38\ \mathrm{W/m^2}}$

CHECK The first polarizer cuts the intensity in half, so we should expect the intensity transmitted through both sheets to be less than half the incident intensity of 3.0 W/m². The step-3 result meets this expectation.

TAKING IT FURTHER Note that the second sheet rotates the plane of polarization by 60°.

POLARIZATION BY REFLECTION

When unpolarized light is reflected from a plane surface boundary between two transparent media, such as air and glass or air and water, the reflected light is partially polarized. The degree of polarization depends on the angle of incidence and on the ratio of the wave speeds in the two media. For a certain angle of incidence called the polarizing angle θ_p, the reflected light is completely polarized. At the polarizing angle, the reflected and refracted rays are perpendicular to each other. David Brewster (1781–1868), a Scottish scientist and an inventor of numerous instruments (including the kaleidoscope), discovered this experimentally in 1812. The polarizing angle is also referred to as the Brewster angle.

Figure 31-30 shows light incident at the polarizing angle θ_p for which the reflected light is completely polarized. The electric field of the incident light can be resolved into components parallel and perpendicular to the plane of incidence. The reflected light is linearly polarized with its electric field perpendicular to the plane of incidence. We can relate the polarizing angle to the indices of refraction of the media using Snell's law (the law of refraction). If n_1 is the index of refraction of the first medium and n_2 is the index of refraction of the second medium, the law of refraction gives

$$n_1 \sin\theta_p = n_2 \sin\theta_2$$

where θ_2 is the angle of refraction. From Figure 31-30, we can see that the sum of the angle of reflection and the angle of refraction is 90°. Because the angle of reflection equals the angle of incidence, we have

$$\theta_2 = 90° - \theta_p$$

Then

$$n_1 \sin\theta_p = n_2 \sin(90° - \theta_p) = n_2 \cos\theta_p$$

or

$$\tan\theta_p = \frac{n_2}{n_1} \qquad \text{31-17}$$

POLARIZING ANGLE

Although the reflected light is completely polarized for this angle of incidence, the transmitted light is only partially polarized (because only a small fraction of the incident light is reflected). If the incident light itself is polarized with the electric field in the plane of incidence, no reflected light exists when the angle of incidence is θ_p. We can qualitatively understand this result by using Figure 31-31. If we consider the charges in the atoms next to the surface of the second medium to be driven, by the electric field of the refracted light, to oscillate parallel to the direction of electric field, no reflected ray can exist because for an electric dipole antenna, no energy is radiated along the line of oscillation. (Each of the oscillating atoms is a small electric dipole antenna.)

Because of the polarization of reflected light, sunglasses that contain a polarizing sheet can be very effective in cutting out glare. If light is reflected from a horizontal surface, such as a lake surface or snow on the ground, the electric field of the reflected light will be predominantly horizontal. Polarizing sunglasses with a vertical transmission axis will then reduce glare by absorbing much of the reflected light. If you have polarizing sunglasses, you can observe this effect by looking through the glasses at reflected light and then rotating the glasses 90°; much more of the light will be transmitted.

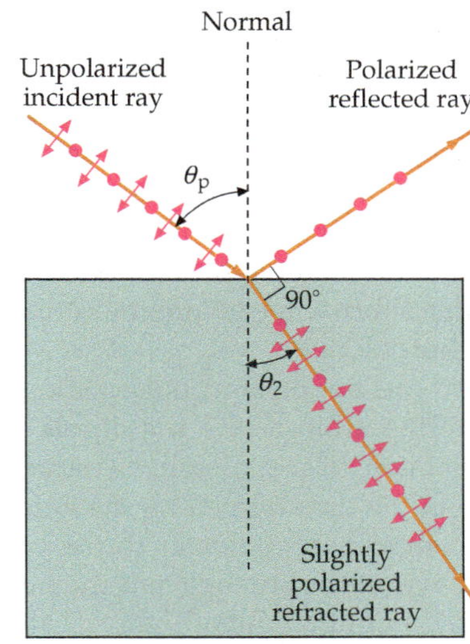

FIGURE 31-30 Polarization by reflection. The incident wave is unpolarized and has components of the electric field parallel to the plane of incidence (arrows) and components perpendicular to that plane (dots). For incidence at the polarizing angle, the reflected wave is completely polarized, with its electric field perpendicular to the plane of incidence.

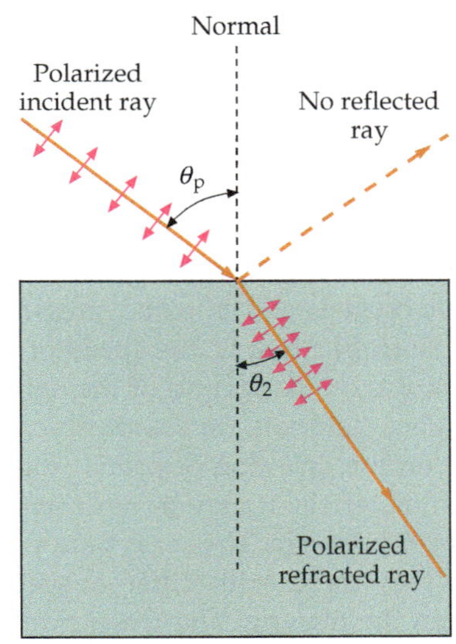

FIGURE 31-31 Polarized light incident at the polarizing angle. When the incident light is polarized with $\vec{E}$ in the plane of incidence, there is no reflected ray.

POLARIZATION BY SCATTERING

The phenomenon of absorption and reradiation is called **scattering.** Scattering can be demonstrated by passing a light beam through a container of water to which a small amount of powdered milk has been added. The milk particles scatter the light, making the light beam visible. Similarly, laser beams can be made visible by introducing chalk or smoke particles into the air to scatter the light. A familiar example of light scattering is that from air molecules, which tend to scatter short wavelengths more than long wavelengths, thereby giving the sky its blue color.

We can understand polarization by scattering if we think of the charges in a scattering atom as electric dipole antennas that radiate waves that have maximum intensities in directions perpendicular to the antenna axes and intensities of zero in the direction along the antenna axis. The electric field vector of the scattered light perpendicular to the direction of propagation is in the plane of the long axis of the antenna and the field point. Figure 31-32 shows a beam of unpolarized light that initially travels along the z axis, striking a particle at the origin. The electric field in the light beam has components in both the x and y directions perpendicular to the direction of motion of the light beam. These fields set up oscillations of the charges within the molecule in the $z = 0$ plane, and no oscillation exists along the z direction. These oscillations can be thought of as a superposition of an oscillation along the x axis and another along the y axis, and each of these oscillations produce dipole radiation. Thus, the oscillation along the x axis produces no radiation along the x axis, which means the light radiated along the x axis is produced only by the oscillation along the y axis. It follows that the light radiated along the x axis is polarized with its electric field parallel with the y axis. There is nothing special about the choice of axes for this discussion, so the result can be generalized. That is, the light scattered in a direction perpendicular to the incident light beam is polarized with its electric field perpendicular to both the incident beam and the direction of propagation of the scattered light. This can be seen easily by examining the scattered light with a piece of polarizing sheet.

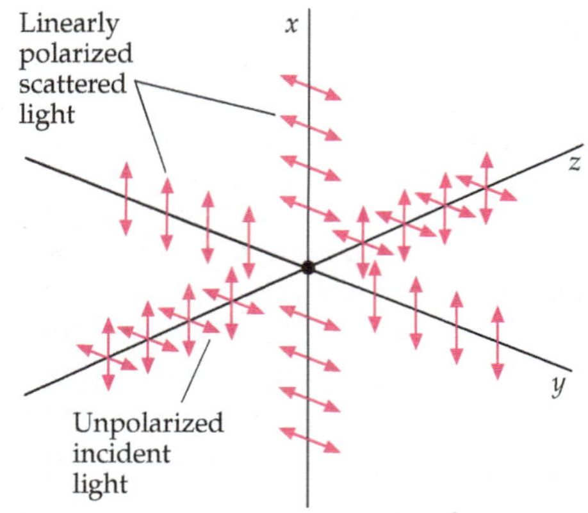

FIGURE 31-32 Polarization by scattering. Unpolarized light propagating in the $+z$ direction is incident on a scattering center at the origin. The light scattered in the $z = 0$ plane along the $\pm x$ direction is polarized parallel with the y axis (and the light scattered in the $\pm y$ direction is polarized parallel with the x axis).

POLARIZATION BY BIREFRINGENCE

Birefringence is a complicated phenomenon that occurs in calcite and other noncubic crystals and in some stressed plastics, such as cellophane. Most materials are **isotropic,** that is, the speed of light passing through the material is independent of the polarization of the light. Because of their microscopic structure, birefringent materials are **anisotropic.** The speed of light depends on the polarization and on the direction of propagation of the light. When a light ray is incident on such materials, it may be separated into two rays called the *ordinary ray* and the *extraordinary ray*. These rays are polarized in mutually perpendicular directions, and they travel with different speeds. Depending on the relative orientation between the material and the incident light beam, the two rays may also travel in different directions.

There is one particular direction in a birefringent material in which both rays propagate with the same speed. This direction is called the **optic axis** of the material. (The optic axis is actually a *direction* rather than a line in the material.) Nothing unusual happens when light travels in the direction of the optic axis. However, when light is incident at an angle to the optic axis, as shown in Figure 31-33, the

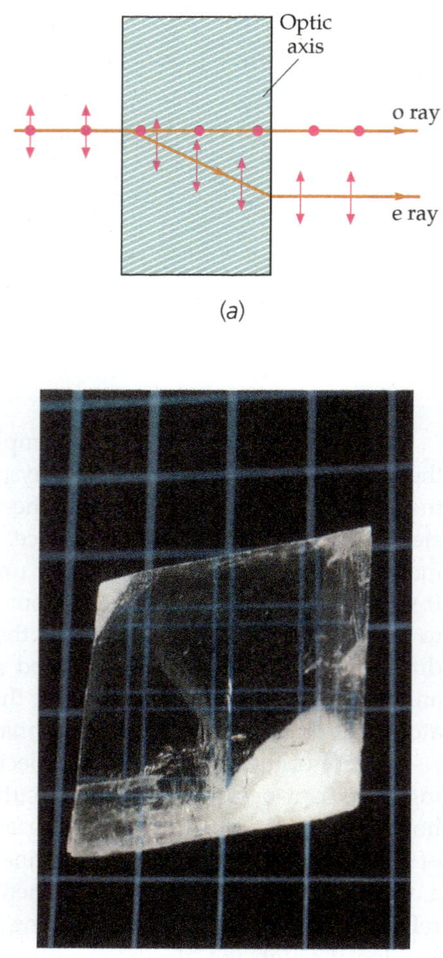

FIGURE 31-33 (*a*) A narrow beam of light incident on a birefringent crystal such as calcite is split into two beams, called the ordinary ray (o ray) and the extraordinary ray (e ray), that have mutually perpendicular polarizations. If the crystal is rotated, the extraordinary ray rotates in space. (*b*) A double image of the cross-hatching is produced by this birefringent crystal of calcium carbonate. *(Paul Silverman Photographs.)*

rays travel in different directions and emerge separated in space. If the material is rotated, the extraordinary ray (the e ray in the figure) revolves in space around the ordinary ray (o ray).

If light is incident on a birefringent plate perpendicular to its crystal face and perpendicular to the optic axis, the two rays travel in the same direction but at different speeds. The number of wavelengths in the two rays in the plate is different because the wavelengths ($\lambda = v/f$) of the rays differ. The rays emerge with a phase difference that depends on the thickness of the plate and on the wavelength of the incident light. In a **quarter-wave plate,** the thickness is such that a 90° phase difference exists between the waves of a particular wavelength when they emerge. In a **half-wave plate,** the rays emerge with a phase difference of 180°.

Suppose that the incident light is linearly polarized so that the electric field vector is at 45° to the optic axis, as illustrated in Figure 31-34. The ordinary and extraordinary rays start out in phase and have equal amplitudes. With a quarter-wave plate, the waves emerge with a phase difference of 90°, so the resultant electric field has components $E_x = E_0 \sin \omega t$ and $E_y = E_0 \sin(\omega t + 90°) = E_0 \cos \omega t$. The electric field vector thus rotates in a circle and has constant magnitude. Such a wave is said to be **circularly polarized.**

With a half-wave plate, the waves that emerge have a phase difference of 180°, so the resultant electric field is linearly polarized with components $E_x = E_0 \sin \omega t$ and $E_y = E_0 \sin(\omega t + 180°) = -E_0 \sin \omega t$. The net effect is that the direction of polarization of the wave is rotated by 90° relative to that of the incident light, as shown in Figure 31-35.

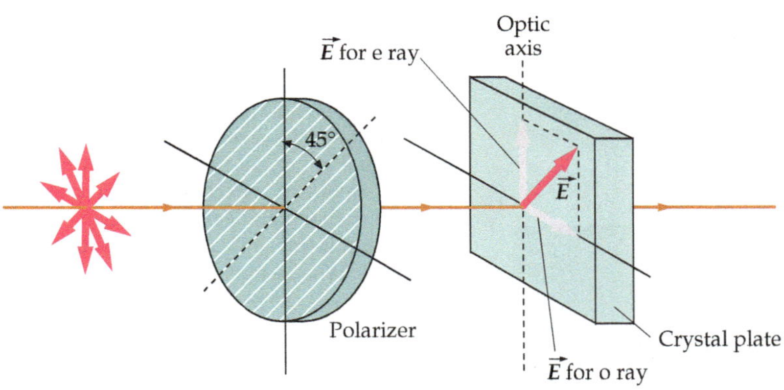

FIGURE 31-34 Polarized light emerging from the polarizer is incident on a birefringent crystal so that the electric field vector makes a 45° angle with the optic axis, which is perpendicular to the light beam. The ordinary and extraordinary rays travel in the same direction but at different speeds. The polarization of the emerging light depends on the thickness of the crystal and the wavelength of the light.

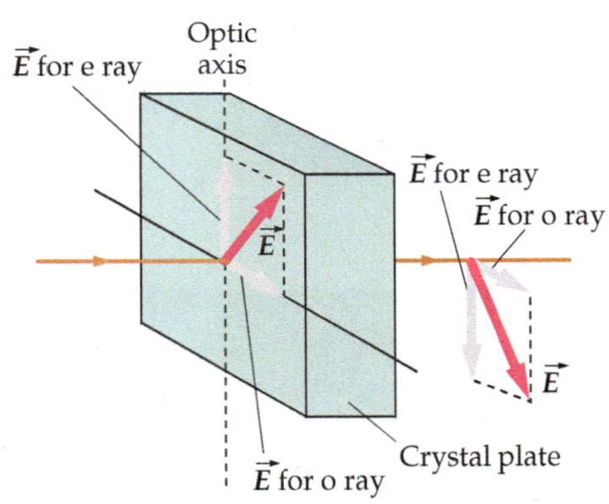

FIGURE 31-35 If the birefringent crystal in Figure 31-34 is a half-wave plate, and if the electric field vector of the incident light makes an angle of 45° with the optic axis, then the direction of polarization of the emerging light is rotated by 90°.

Interesting and beautiful patterns can be observed by placing birefringent materials, such as cellophane or stressed plastic, between two polarizing sheets with their transmission axes perpendicular to each other. Ordinarily, no light is transmitted through crossed polarizing sheets. However, if we place a birefringent material between the crossed polarizing sheets, the material acts as a half-wave plate for light of a certain color depending on the material's thickness. The direction of polarization is rotated and some light gets through both sheets. Various glasses and plastics become birefringent when under stress. The stress patterns can be observed when the material is placed between crossed polarizing sheets.

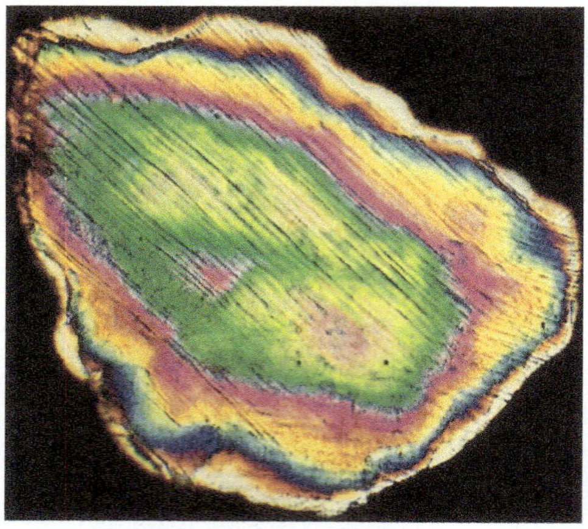

(a)

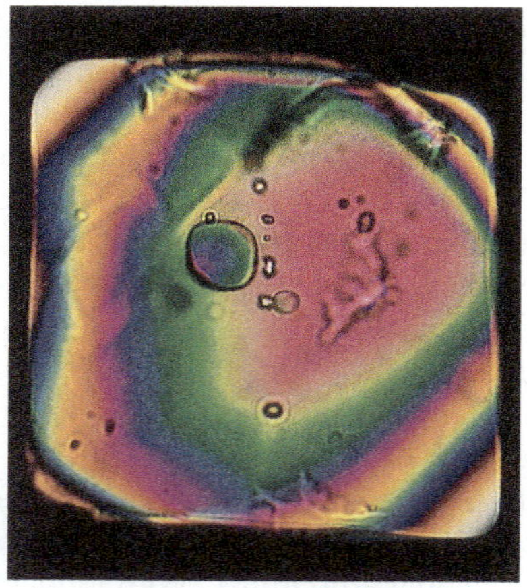

(b)

(c)

(d)

(e)

When the transmission axes of two polarizing sheets are perpendicular, the polarizers are said to be crossed and no light is transmitted. However, many materials are birefringent or become so under stress. Such materials rotate the direction of polarization of the light so that light of a particular wavelength is transmitted through both polarizers. When a birefringent material is viewed between crossed polarizers, information about its internal structure is revealed. (*a*) A shocked quartz grain from the site of a meteorite crater. The layered structure, evidenced by the parallel lines, arises from the shock of the impact of the meteor. (*b*) A grain of quartz typically found in silicic volcanic rocks. No shock lines are seen. (*c*) Thin sections of an ice core from the Antarctic ice sheet reveal bubbles of trapped CO_2, which appear amber-colored. The sample was taken from a depth of 194 m, corresponding to air trapped 1600 years ago, whereas the sample in (*d*) is from a depth of 56 m, corresponding to air trapped 450 years ago. Ice core measurements have replaced the less reliable technique of analyzing carbon in tree rings to compare current atmospheric CO_2 levels with those of the recent past. (*e*) Robert Mark of the Princeton School of Architecture examines the stress patterns in a plastic model of the nave structure of Chartres Cathedral. *((a, b) Glen A. Izett, US Geological Survey. (c, d) Dr. Anthony J Gow/Cold Regions Research and Engineering Laboratory, Hanover New Hampshire. (e) Sepp Seitz/Woodfin Camp and Associates.)*

31-5 DERIVATION OF THE LAWS OF REFLECTION AND REFRACTION

The laws of reflection and refraction can be derived from either Huygens's principle or Fermat's principle.

HUYGENS'S CONSTRUCTION

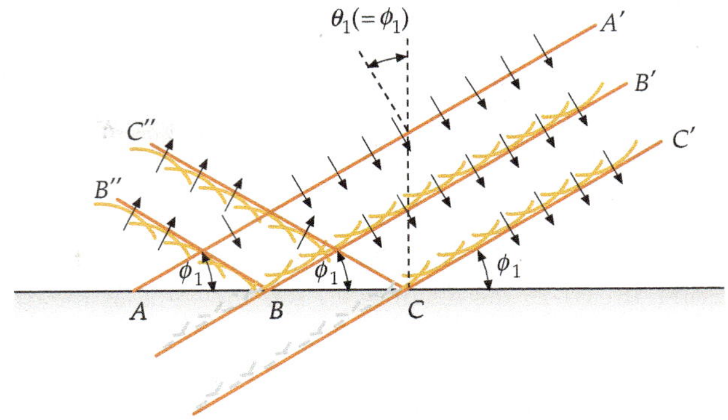

FIGURE 31-36 Plane wave reflected at a plane mirror. The angle θ_1 between the incident ray and the normal to the mirror is the angle of incidence. It is equal to the angle ϕ_1 between the incident wavefront and the mirror.

Reflection Figure 31-36 shows a plane wavefront AA' striking a mirror at point A. As can be seen from the figure, the angle ϕ_1 between the wavefront and the mirror is the same as the angle of incidence θ_1, which is the angle between the normal to the mirror and the rays (which are perpendicular to the wavefronts). According to Huygens's construction, each point on a given wavefront can be considered a point source of secondary wavelets. The position of the wavefront after a time t is found by constructing wavelets of radius ct with their centers on the wavefront AA'. Wavelets that have not yet reached the mirror form the portion of the new wavefront BB'. Wavelets that have already reached the mirror are reflected and form the portion of the new wavefront $B''B$. By a similar construction, the wavefront $C''C$ is obtained from the Huygens's wavelets originating on the wavefront $B''B$. Figure 31-37 is an enlargement of a portion of Figure 31-36 showing AP, which is part of the initial position of the wavefront. During the time t, the wavelet from point P reaches the mirror at point B, and the wavelet from point A reaches point B''. The reflected wavefront $B''B$ makes an angle ϕ_1' with the mirror that is equal to the angle of reflection θ_1' between the reflected ray and the normal to the mirror. The triangles $AB''B$ and APB are both right triangles that have a common side AB and equal sides $AB'' = BP = ct$. Hence, these triangles are congruent, and the angles ϕ_1 and ϕ_1' are equal, implying that the angle of reflection θ_1' equals the angle of incidence θ_1.

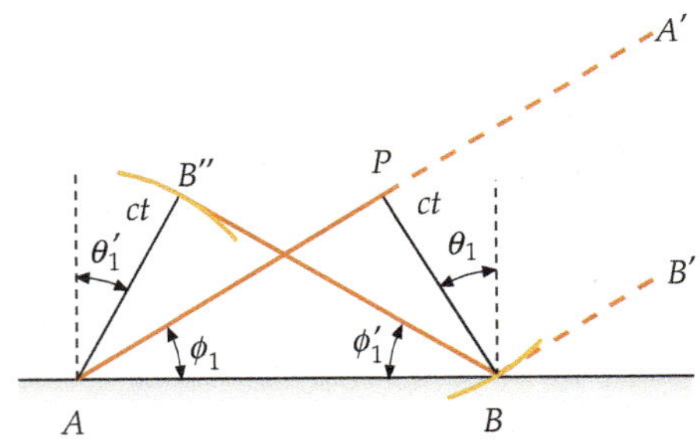

FIGURE 31-37 Geometry of Huygens's construction for the calculation of the law of reflection. The wavefront AP initially strikes the mirror at point A. After a time t, the Huygens wavelet from P strikes the mirror at point B, and the Huygens wavelet from point A reaches point B''.

Refraction Figure 31-38 shows a plane wave incident on an air–glass interface. We apply Huygens's construction to find the wavefront in the transmitted wave. Line AP indicates a portion of the wavefront in medium 1 that strikes the glass surface at an angle ϕ_1. In time t, the wavelet from P travels the distance v_1t and reaches the point B on the line AB separating the two media, while the wavelet from point A travels a shorter distance v_2t into the second medium. The new wavefront BB' is not parallel to the original wavefront AP because the speeds v_1 and v_2 are different. From the triangle APB,

$$\sin\phi_1 = \frac{v_1t}{AB}$$

or

$$AB = \frac{v_1t}{\sin\phi_1} = \frac{v_1t}{\sin\theta_1}$$

using the fact that the angle ϕ_1 equals the angle of incidence θ_1. Similarly, from triangle $AB'B$,

$$\sin\phi_2 = \frac{v_2t}{AB}$$

or

$$AB = \frac{v_2t}{\sin\phi_2} = \frac{v_2t}{\sin\theta_2}$$

where $\theta_2 = \phi_2$ is the angle of refraction. Equating the reciprocals of the two values for AB, we obtain

$$\frac{1}{v_1}\sin\theta_1 = \frac{1}{v_2}\sin\theta_2 \qquad \text{31-18}$$

Multiplying both sides by c, and then substituting n_1 for c/v_1 and n_2 for c/v_2, we obtain $n_1\sin\theta_1 = n_2\sin\theta_2$, which is Snell's law.

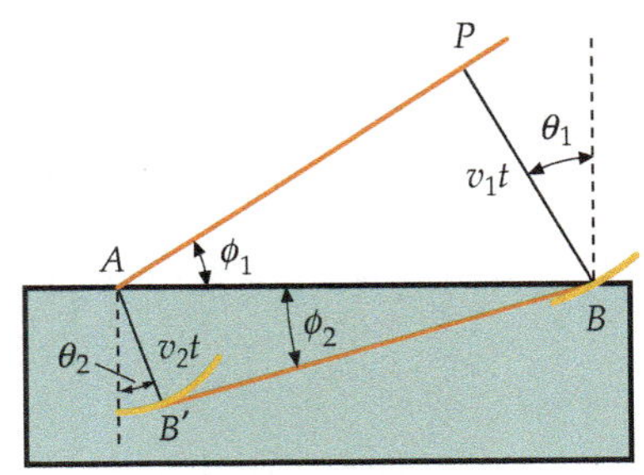

FIGURE 31-38 Application of Huygens's principle to the refraction of plane waves at the surface separating a medium in which the wave speed is v_1 from a medium in which the wave speed v_2 is less than v_1. The angle of refraction θ_2 is less than the angle of incidence θ_1.

FERMAT'S PRINCIPLE

Reflection Figure 31-39 shows two paths in which light leaves point A, strikes the plane surface, which we can consider to be a mirror, and travels to point B. The problem for the application of Fermat's principle to reflection can be stated as follows: At what point P in the figure must the light strike the mirror so that it will travel from point A to point B in the least time? Because the light is traveling in the same medium for this problem, the time will be minimum when the distance is minimum. In Figure 31-39 the distance APB is the same as the distance $A'PB$, where point A' lies along the perpendicular from A to the mirror and is equidistant behind the mirror. As we vary point P, the distance $A'PB$ is least when the points A', P, and B lie on a straight line. We can see from the figure that this occurs when the angle of incidence equals the angle of reflection.

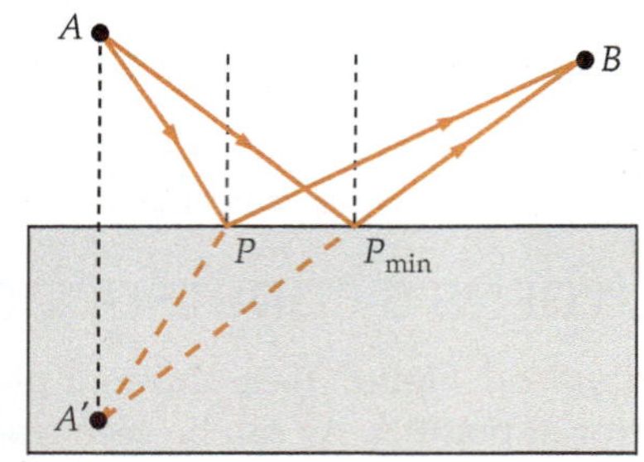

FIGURE 31-39 Geometry for deriving the law of reflection from Fermat's principle. The time it takes for the light to travel from point A to the surface and then on to point B is a minimum for the light striking the surface at point P_{min}.

Refraction The derivation of Snell's law of refraction from Fermat's principle is slightly more complicated. Figure 31-40 shows several possible paths for light traveling from point A in air to point B in glass. Point P_1 is on the straight line between A and B, but this path is not the one for the shortest travel time because light travels at a slower speed in the glass. If we move slightly to the right of P_1, the total path length is longer, but the distance traveled in the slower medium is less than for the path through P_1. It is not apparent from the figure which path is the path of least time. However, it is not surprising that a path slightly to the right of the straight-line path takes less time because the time gained by traveling a shorter distance in the glass more than compensates for the time lost traveling a longer distance in the air. As we move the point of intersection of the possible path to the right of point P_1, the total time of travel from point A to point B decreases until we reach a minimum at point P_{min}. Beyond this point, the time saved by traveling a shorter distance in the glass is not enough to compensate for the additional time spent by traveling the greater distance in the air.

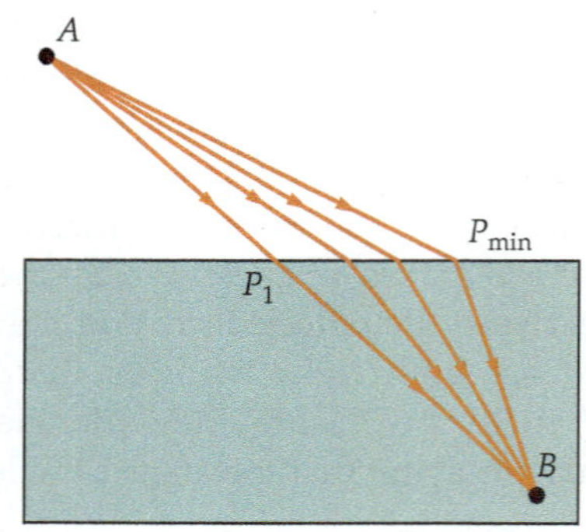

FIGURE 31-40 Geometry for deriving Snell's law from Fermat's principle. The point P_{min} is the point at which light must strike the glass in order that the travel time from point A to point B is a minimum.

Figure 31-41 shows the geometry for finding the path of least time. If L_1 is the distance traveled in medium 1 that has an index of refraction n_1, and L_2 is the distance traveled in medium 2 that has an index of refraction n_2, the time for light to travel the total path AB is

$$t = \frac{L_1}{v_1} + \frac{L_2}{v_2} = \frac{L_1}{c/n_1} + \frac{L_2}{c/n_2} = \frac{n_1 L_1}{c} + \frac{n_2 L_2}{c} \qquad \text{31-19}$$

We wish to find the point P_{min} for which this time is a minimum. We do this by expressing the time in terms of a single parameter x, as shown in the figure, indicating the position of point P_{min}. In terms of the distance x,

$$L_1^2 = a^2 + x^2 \quad \text{and} \quad L_2^2 = b^2 + (d - x)^2 \qquad \text{31-20}$$

Figure 31-42 shows the time t as a function of x. At the value of x for which the time is a minimum, the slope of the graph of t versus x is zero:

$$\frac{dt}{dx} = 0$$

Differentiating each term in Equation 31-19 with respect to x and setting the result equal to zero, we obtain

$$\frac{dt}{dx} = \frac{1}{c}\left(n_1 \frac{dL_1}{dx} + n_2 \frac{dL_2}{dx}\right) = 0 \qquad \text{31-21}$$

We can compute these derivatives from Equations 31-20. We have

$$2L_1 \frac{dL_1}{dx} = 2x \quad \text{or} \quad \frac{dL_1}{dx} = \frac{x}{L_1}$$

where x/L_1 is just $\sin\theta_1$, and where θ_1 is the angle of incidence. Thus,

$$\frac{dL_1}{dx} = \sin\theta_1 \qquad \text{31-22}$$

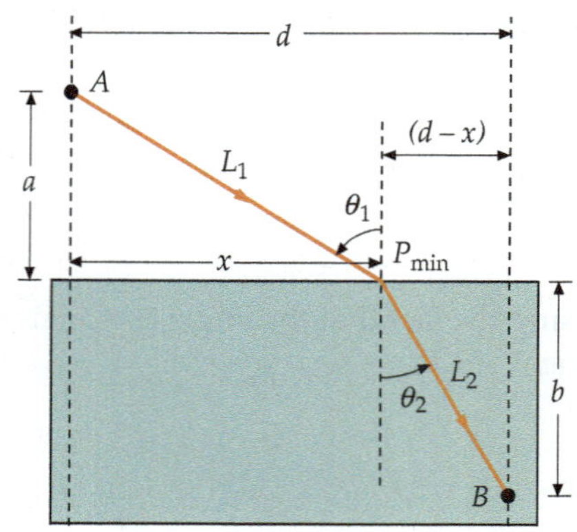

FIGURE 31-41 Geometry for calculating the minimum time in the derivation of Snell's law from Fermat's principle.

Similarly,

$$2L_2\frac{dL_2}{dx} = 2(d - x)(-1)$$

or

$$\frac{dL_2}{dx} = -\frac{d - x}{L_2} = -\sin\theta_2 \qquad 31\text{-}23$$

where θ_2 is the angle of refraction. From Equation 31-21,

$$n_1\frac{dL_1}{dx} + n_2\frac{dL_2}{dx} = 0 \qquad 31\text{-}24$$

Substituting the results of Equations 31-22 and 31-23 for dL_1/dx and dL_2/dx gives

$$n_1 \sin\theta_1 + n_2(-\sin\theta_2) = 0$$

or

$$n_1 \sin\theta_1 = n_2 \sin\theta_2$$

which is Snell's law.

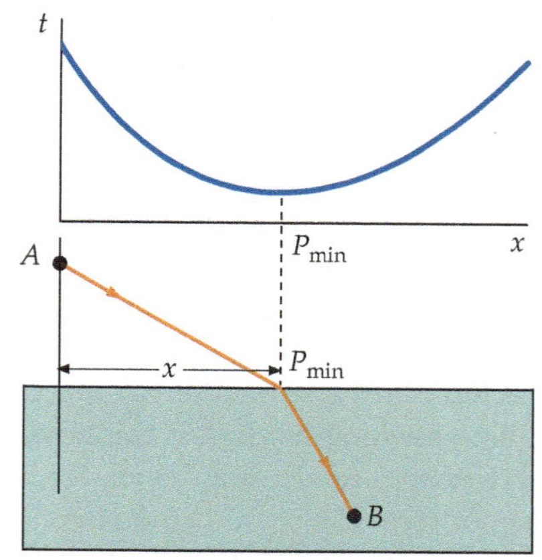

FIGURE 31-42 Graph of the time it takes for light to travel from point A to point B versus x, measured along the refracting surface. The time is a minimum at the point at which the angles of incidence and refraction obey Snell's law.

31-6 WAVE–PARTICLE DUALITY

The wave nature of light was first demonstrated by Thomas Young, who in 1801 observed the interference pattern of two coherent light sources produced by illuminating a pair of narrow, parallel slits with a single source. (Young's experiment is presented in Section 3 of Chapter 33.) The wave theory of light culminated in 1860 with Maxwell's prediction of electromagnetic waves. The particle nature of light was first proposed by Albert Einstein in 1905 in his explanation of the photoelectric effect.* A particle of light called a **photon** has energy E that is related to the frequency f and wavelength λ of the light wave by the Einstein equation

$$E = hf = \frac{hc}{\lambda} \qquad 31\text{-}25$$

EINSTEIN'S EQUATION FOR PHOTON ENERGY

where c is the speed of light and h is Planck's constant:

$$h = 6.626 \times 10^{-34}\ \text{J}\cdot\text{s} = 4.136 \times 10^{-15}\ \text{eV}\cdot\text{s}$$

Because energies are often given in electron volts and wavelengths are given in nanometers, it is convenient to express the combination hc in eV·nm. We have

$$hc = (4.1357 \times 10^{-15}\ \text{eV}\cdot\text{s})(2.9979 \times 10^{8}\ \text{m/s}) = 1.2398 \times 10^{-6}\ \text{eV}\cdot\text{m}$$

or

$$hc = 1240\ \text{eV}\cdot\text{nm} \qquad 31\text{-}26$$

The propagation of light is governed by its wave properties, whereas the exchange of energy between light and matter is governed by its particle properties. This wave–particle duality is a general property of nature. For example, the propagation of electrons (and other so-called particles) is also governed by wave properties, whereas the exchange of energy between the electrons and other particles is governed by particle properties.

* The photoelectric effect is discussed in Chapter 34.

31-7 LIGHT SPECTRA

Newton was the first to recognize that white light is a mixture of light of all colors of approximately equal intensity. He demonstrated this by letting sunlight fall on a glass prism and observing the spectrum of the refracted light (Figure 31-43). Because the angle of refraction produced by a glass prism depends on the wavelength of the light, the refracted beam is spread out in space into its component colors or wavelengths, like a rainbow. Figure 31-44 shows a spectroscope, which is a device for analyzing the spectra of light sources. Light from the source passes through a narrow slit, traverses a lens that makes the beam parallel, and falls on a glass prism, where it is refracted twice (once as it enters the glass and again as it leaves the glass). The refracted beam is viewed with a telescope, which is mounted on a rotating platform so that the angle of the refracted beam, which depends on the wavelength of the light, can be measured. The spectrum of the light source can thus be analyzed in terms of its component wavelengths. The spectrum of sunlight has a continuous range of wavelengths and is therefore called a **continuous spectrum.** The light emitted by the atoms in low-pressure gases, such as mercury atoms in a fluorescent lamp, contains only a discrete set of wavelengths. Each wavelength emitted by the source produces a separate image of the collimating slit in the spectroscope. Such a spectrum is called a **line spectrum.** The continuous visible spectrum and the line spectra from several elements are shown in the photograph.

FIGURE 31-43 Newton demonstrating the spectrum of sunlight with a glass prism. *(Corbis/Bettmann.)*

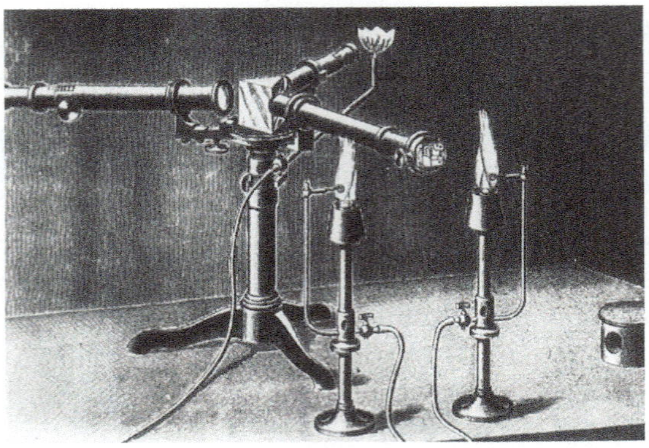

FIGURE 31-44 A late nineteenth-century spectroscope belonging to Gustav Kirchhoff. Modern student spectroscopes usually have the same general design. *(Corbis/Bettmann.)*

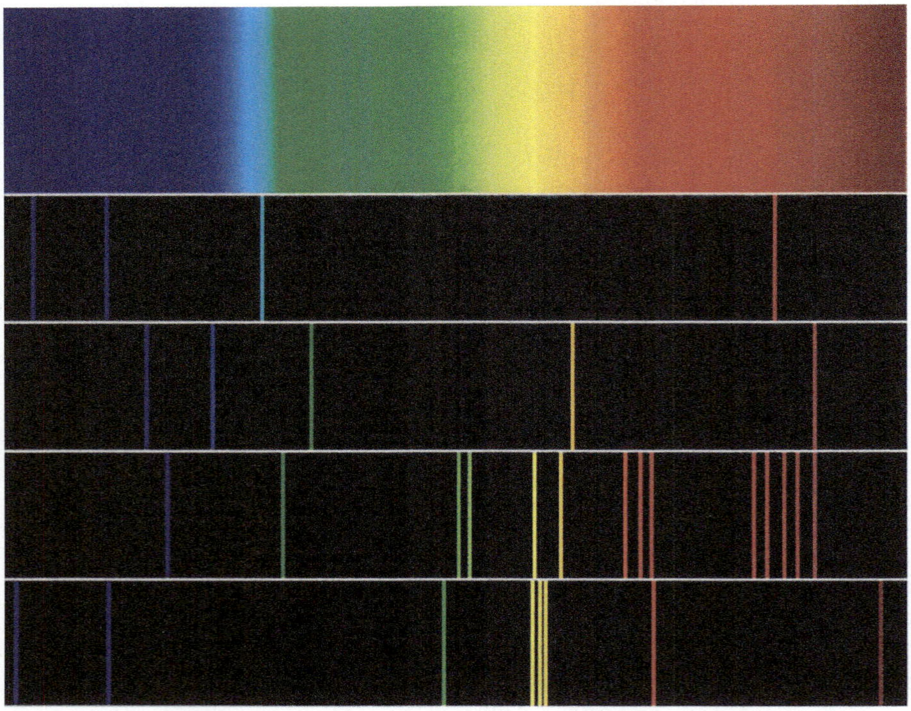

The continuous visible spectrum (top) and the line spectra of (from top to bottom) hydrogen, helium, barium, and mercury. *(Adapted from Eastman Kodak and Wabash Instrument Corporation.)*

* 31-8 SOURCES OF LIGHT

LINE SPECTRA

The most common sources of visible light are transitions of the valence electrons in atoms. Normally, an atom is in its ground state where its electrons are at their lowest allowed energy levels consistent with the exclusion principle. (The exclusion principle, which was first proposed by Wolfgang Pauli in 1925 to explain the electronic structure of atoms, states that no two electrons in an atom can have the same quantum state.) The lowest energy electrons are tightly bound to the nucleus, forming a stable core of electrons. The one or two electrons in the highest energy states are much less tightly bound to the nucleus and are relatively easily excited to vacant higher energy states. These outer electrons are responsible for the energy changes in the atom that result in the emission or absorption of visible light.

When an atom collides with another atom or with a free electron, or when the atom absorbs electromagnetic energy, the valence electrons can be excited to higher energy states. After a time of approximately 10 ns (1 ns = 10^{-9} s), the valence electrons spontaneously make transitions to lower energy states with the emission of a photon. This process, called **spontaneous emission,** is random; the photons emitted from two different atoms are not correlated. The emitted light is thus incoherent. By conservation of energy, the energy of an emitted photon is the energy difference $|\Delta E|$ between the initial state and the final state of the atom. The frequency of the light wave is related to the energy by the Einstein equation, $|\Delta E| = hf$ (Equation 31-25). The wavelength of the emitted light is then

$$\lambda = \frac{c}{f} = \frac{hc}{hf} = \frac{hc}{|\Delta E|} \qquad \text{31-27}$$

The photon energies corresponding to shortest wavelengths (400 nm) and longest (700 nm) wavelengths in the visible spectrum are

$$E_{400\text{ nm}} = \frac{hc}{\lambda} = \frac{1240\text{ eV}\cdot\text{nm}}{400\text{ nm}} = 3.10\text{ eV} \qquad \text{31-28}a$$

and

$$E_{700\text{ nm}} = \frac{hc}{\lambda} = \frac{1240\text{ eV}\cdot\text{nm}}{700\text{ nm}} = 1.77\text{ eV} \qquad \text{31-28}b$$

Because the energy levels in atoms form a discrete set, the emission spectrum of light from single atoms or from atoms in low-pressure gases consists of a set of sharp discrete lines that are characteristic of the element. These narrow lines are broadened somewhat by Doppler shifts, due to the motion of the atom relative to the observer and by collisions with other atoms; generally, however, if the gas density is low enough, the lines are narrow and well separated from one another. The study of the line spectra of hydrogen and other atoms led to the first understanding of the energy levels of atoms.

Continuous spectra When atoms are close together and interact strongly, as in liquids and solids, the energy levels of the individual atoms are spread out into energy bands, resulting in essentially continuous bands of energy levels. When the bands overlap, as they often do, the result is a continuous spectrum of possible energies and a continuous emission spectrum. In an incandescent material such as a hot metal filament, electrons are randomly accelerated by frequent collisions, resulting in a broad spectrum of thermal radiation. The rate at which an object radiates thermal energy is proportional to the fourth power of its absolute temperature.*

* This is known as the Stefan–Boltzmann law. This property and other properties of thermal radiation, such as Wien's displacement law, are discussed more fully in Section 20-4.

The radiation emitted by an object at temperatures below approximately 600°C is concentrated in the infrared and is not visible. As an object is heated, the energy radiated extends to shorter and shorter wavelengths. Between approximately 600°C and 700°C, enough of the radiated energy is in the visible spectrum for the object to glow a dull red. At higher and higher temperatures, the object becomes bright red and then white. For a given temperature, the wavelength λ_{peak} at which the emitted power is a maximum varies inversely with the temperature, a result known as Wien's displacement law. The surface of the Sun at $T = 6000$ K emits a continuous spectrum of approximately constant intensity over the visible range of wavelengths.

ABSORPTION, SCATTERING, SPONTANEOUS EMISSION, AND STIMULATED EMISSION

When radiation is emitted, an atom (the words atom and molecule are used interchangeably in this section) makes a transition from an excited state to a state of lower energy; when radiation is absorbed, an atom makes a transition from a lower state to a higher state. When atoms are irradiated with a continuous spectrum of radiation, the transmitted spectrum shows dark lines corresponding to the absorption of light at discrete wavelengths. The absorption spectra of atoms were the first line spectra observed. Because atoms at normal temperatures are in either their ground states or low-energy excited states, only transitions from a ground state (or a near ground state) to a more highly excited state are observed. Thus, absorption spectra usually have far fewer lines than do emission spectra.

Figure 31-45 illustrates several interesting phenomena that can occur when a photon is incident on an atom. In Figure 31-45*a*, the energy of the incoming photon is too small to excite the atom to an excited state, so the atom remains in its ground state and the photon is said to be scattered. Because the incoming and outgoing or scattered photons have the same energy, the scattering is said to be elastic. If the wavelength of the incident light is large compared with the size of the atom, the scattering can be described in terms of classical electromagnetic theory and is called **Rayleigh scattering** after Lord Rayleigh, who worked out the theory in 1871. The probability of Rayleigh scattering varies as $1/\lambda^4$. This means that blue light is scattered much more readily than red light, which accounts for the bluish color of the sky. The removal of blue light by Rayleigh scattering also accounts for some of the reddish color of the light transmitted through the atmosphere at sunrise and sunset.

Inelastic scattering, also called **Raman scattering,** occurs when an incident photon that has just the right amount of energy is absorbed and the atom undergoes a transition to a more energetic state. Then, the atom emits a photon as it undergoes a transition to a less energetic state, whose energy differs from that of the initial state. If the energy of the scattered photon hf' is less than that of the incident photon hf (Figure 31-45*b*), it is called **Stokes Raman scattering.** If the energy of the scattered photon is greater than that of the incident photon (Figure 31-45*c*), it is called **anti-Stokes Raman scattering.**

In Figure 31-45*d*, the energy of the incident photon is just equal to the difference in energy between the initial state and a more energetic state. The atom absorbs the photon and makes a transition to the more excited state in a process called **resonance absorption.**

In Figure 31-45*e*, an atom in an excited state spontaneously undergoes a transition to a less energetic state, in a process called **spontaneous emission.** Often an atom in an excited state undergoes transitions to one or more intermediate states as it returns to the ground state. A common example occurs when an atom is excited by ultraviolet light and emits visible light as it returns to its ground state by multiple transitions. This process, often called **fluorescence,** occurs in a thin film lining on the inside of the glass tubes of fluorescent lightbulbs. Because the lifetime of a typical excited atomic energy state is of the order of 10 ns, this process appears to occur instantaneously. However, some excited states have much longer lifetimes—of the

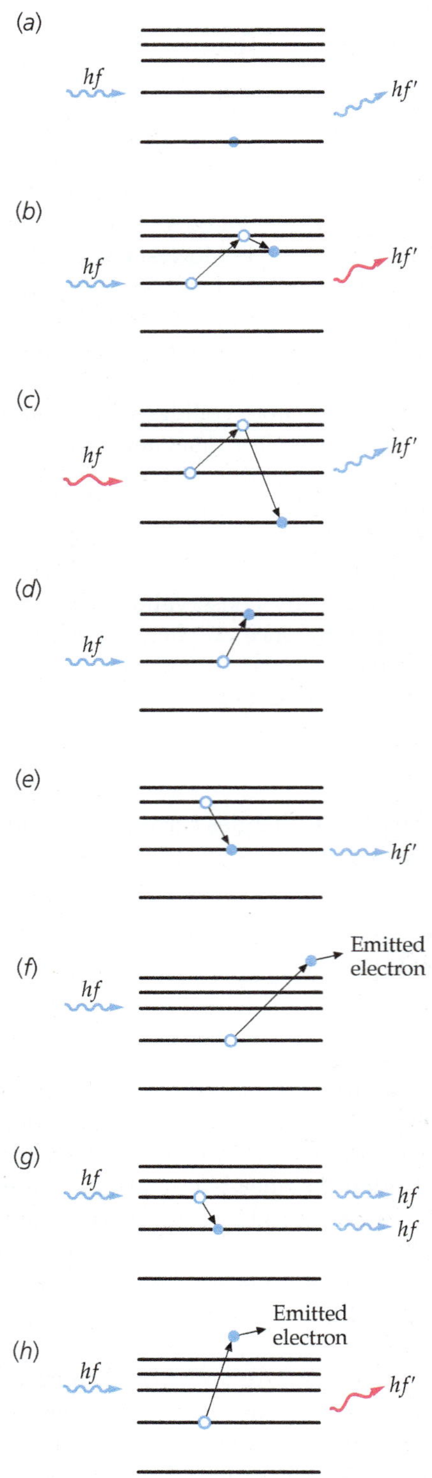

FIGURE 31-45 Photon–atom and photon–molecule interactions. (*a*) Elastic scattering. (*b*) Stokes Raman scattering. (*c*) Anti-Stokes Raman scattering. (*d*) Resonance absorption. (*e*) Spontaneous emission. (*f*) Photoelectric effect. (*g*) Stimulated emission. (*h*) Compton scattering.

order of milliseconds or occasionally seconds or even minutes. Such a state is called a **metastable state.** Materials that have very long-lived metastable states and emit light long after the original excitation are called **phosphorescent materials.**

Figure 31-45*f* illustrates the photoelectric effect, in which the absorption of the photon ionizes the atom by causing the emission of an electron. Figure 31-45*g* illustrates **stimulated emission.** This process occurs if the atom is initially in an excited state of energy E_H, and the energy of the incident photon is equal to $E_H - E_L$, where E_H and E_L are the energies of higher and lower energy states, respectively. In this case, the oscillating electromagnetic field associated with the incident photon can stimulate the excited atom, which then emits a photon in the same direction as the incident photon and in phase with it. The photons from the stimulated atoms can stimulate the emission of additional photons propagating in the same direction with the same phase. This process amplifies the initially emitted photon, yielding a beam of light originating from different atoms that is coherent. As a result, interference of the light from a large number of atoms can easily be observed.

Figure 31-45*h* illustrates **Compton scattering,** which occurs if the energy of the incident photon is much greater than the ionization energy. Note that in Compton scattering, a photon is absorbed and a photon is emitted, whereas in the photoelectric effect, a photon is absorbed with none emitted.

(*a*)

(*b*)

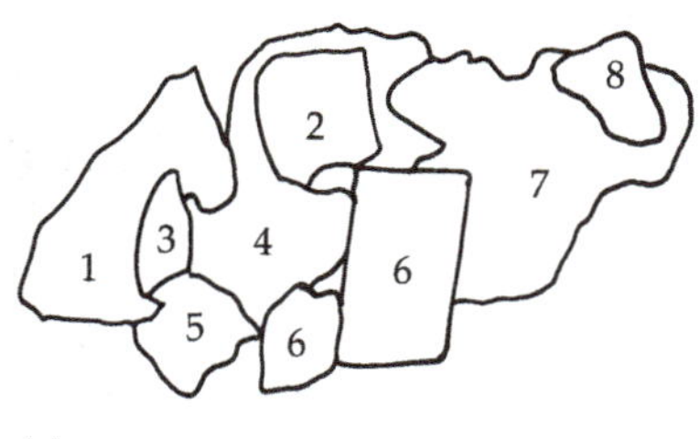

(*c*)

A collection of minerals in (*a*) daylight and in (*b*) ultraviolet light (sometimes called *black light*). Identified by number in the schematic (*c*), they are 1, powerllite; 2, willemite; 3, scheelite; 4, calcite; 5, calcite and willemite composite; 6, optical calcite; 7, willemite; and 8, opal. The change in color is due to the minerals fluorescing under the ultraviolet light. In optical calcite, both fluorescence and phosphorescence occur. *(Paul Silverman/Fundamental Photographs.)*

Example 31-7 Resonant Absorption and Emission

The energy level E_1 of the first excited state of a potassium atom is 1.62 eV above the energy level E_0 of the ground state. The energy levels E_2 and E_3 of the second and third excited states of a potassium atom are 2.61 eV and 3.07 eV, respectively, above the ground state energy E_0. (*a*) What is the longest wavelength of radiation that can be absorbed by a potassium atom in its ground state? Calculate the wavelength of the emitted photon when the atom makes a transition from (*b*) the third excited state (E_3) to the ground state and from (*c*) the third excited state (E_3) to the second excited state (E_2).

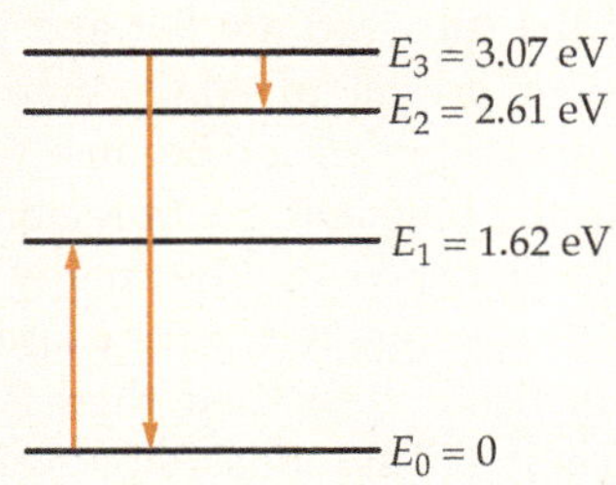

FIGURE 31-46

PICTURE The ground state and the first three excited energy levels are shown in Figure 31-46. (*a*) Because the wavelength is related to the energy of a photon by $\lambda = hc/\Delta E$, longer wavelengths correspond to smaller energy differences, and the smallest energy difference for a transition originating at the ground state is from the ground state to the first excited state. (*b*) The wavelengths of the photons emitted when the atom transitions to lower energy states are related to the energy differences by $\lambda = hc/|\Delta E|$.

SOLVE

(*a*) Calculate the wavelength of radiation absorbed in a transition from the ground state to the first excited state:

$$\lambda = \frac{hc}{\Delta E} = \frac{hc}{E_1 - E_0} = \frac{1240\ \text{eV}\cdot\text{nm}}{1.62\ \text{eV} - 0} = \boxed{765\ \text{nm}}$$

(*b*) For the transition from E_3 to the ground state, the photon energy is $E_3 - E_0 = E_3$. Calculate the wavelength of radiation emitted in this transition:

$$\lambda = \frac{hc}{|\Delta E|} = \frac{hc}{E_3 - E_0} = \frac{1240\ \text{eV}\cdot\text{nm}}{3.07\ \text{eV} - 0} = \boxed{404\ \text{nm}}$$

(*c*) For the transition from E_3 to E_2, the photon energy is $E_3 - E_2$. Calculate the wavelength of radiation emitted in this transition:

$$\lambda = \frac{hc}{|\Delta E|} = \frac{hc}{E_3 - E_2} = \frac{1240 \text{ eV}\cdot\text{nm}}{3.07 \text{ eV} - 2.61 \text{ eV}} = \boxed{2.70\ \mu\text{m}}$$

CHECK The Part-(*b*) result is smaller than the Part-(*a*) result. This result is expected because the more energy a photon has, the shorter the wavelength.

TAKING IT FURTHER The wavelength of radiation emitted in the transition from E_1 to the ground state E_0 is 765 nm, the same as the wavelength for radiation absorbed in the transition from the ground state to E_1. This transition and the transition from E_3 to the ground state both result in photons in the visible spectrum. The photon emitted during the transition from E_3 to E_2 is in the infrared region of the electromagnetic spectrum.

LASERS

The *laser* (*l*ight *a*mplification by *s*timulated *e*mission of *r*adiation) is a device that produces a strong beam of coherent photons by stimulated emission. Consider a system consisting of atoms that have a ground state of energy E_0 and an excited metastable state of energy E_1. If the atoms are irradiated by photons of energy $E_1 - E_0$, those atoms in the ground state can absorb a photon and make the transition to state E_1, whereas those atoms already in the excited state may be stimulated to decay back to the ground state. The relative probabilities of absorption and stimulated emission, first worked out by Einstein, are equal. Ordinarily, nearly all the atoms of the system at normal temperature will initially be in the ground state, so absorption will be the main effect. To produce more stimulated-emission transitions than absorption transitions, we must arrange to have more atoms in the excited state than in the ground state. This condition, called *population inversion*, can be achieved by a method called optical pumping in which atoms are *pumped* up to levels of energy greater than E_1 by the absorption of an intense auxiliary radiation. The atoms then decay down to state E_1 either by spontaneous emission or by nonradiative transitions, such as those transitions due to collisions.

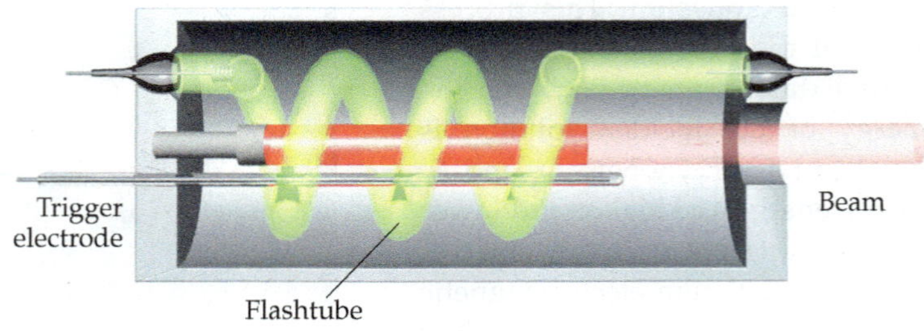

FIGURE 31-47 Schematic diagram of the first ruby laser.

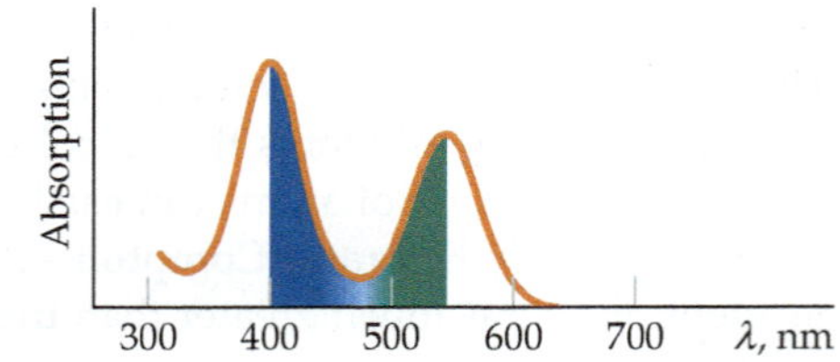

FIGURE 31-48 Absorption versus wavelength for Cr^{3+} in ruby. Ruby appears red because of the strong absorption of green and blue light by the chromium ions.

Figure 31-47 shows a schematic diagram of the first laser, a ruby laser built by Theodore Maiman in 1960. The laser consists of a ruby rod a few centimeters long surrounded by a helical gaseous flashtube that emits a broad spectrum of light. The ends of the ruby rod are flat and perpendicular to the axis of the rod. Ruby is a transparent crystal of Al_2O_3 that has a small amount (about 0.05 percent) of chromium. It appears red because the chromium ions (Cr^{3+}) have strong absorption bands in the blue and green regions of the visible spectrum, as shown in Figure 31-48. The energy levels of chromium—important for the operation of a ruby laser–are shown in Figure 31-49. When the flashtube is fired, there is an intense burst of light that lasts several milliseconds. Photon absorption excites many of the chromium ions to the bands of energy levels indicated by the shading in Figure 31-49. The excited chromium ions then rapidly drop down to a closely spaced pair of metastable states labeled E_1 in the figure. The metastable states are approximately 1.79 eV

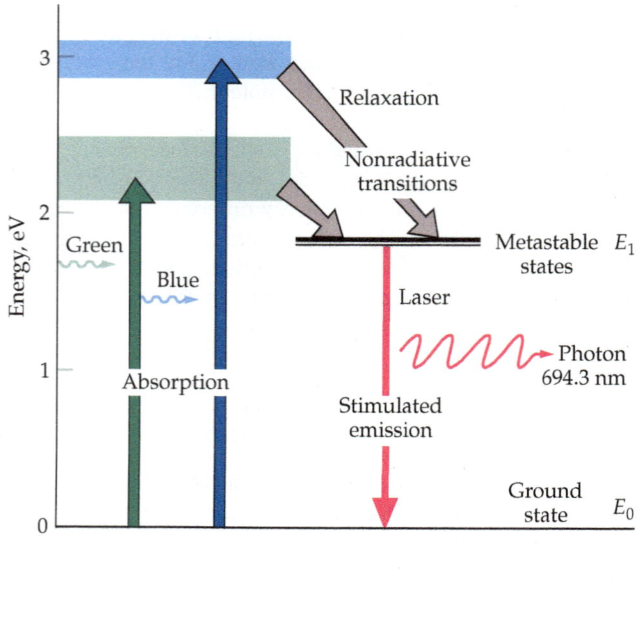

FIGURE 31-49 Energy levels in a ruby laser. To make the population of the metastable states greater than that of the ground state, the ruby crystal is subjected to intense radiation that contains energy in the green and blue wavelengths. This excites atoms from the ground state to the bands of energy levels indicated by the shading, from which the atoms decay to the metastable states by nonradiative transitions. Then, by stimulated emission, the atoms undergo the transition from the metastable states to the ground state.

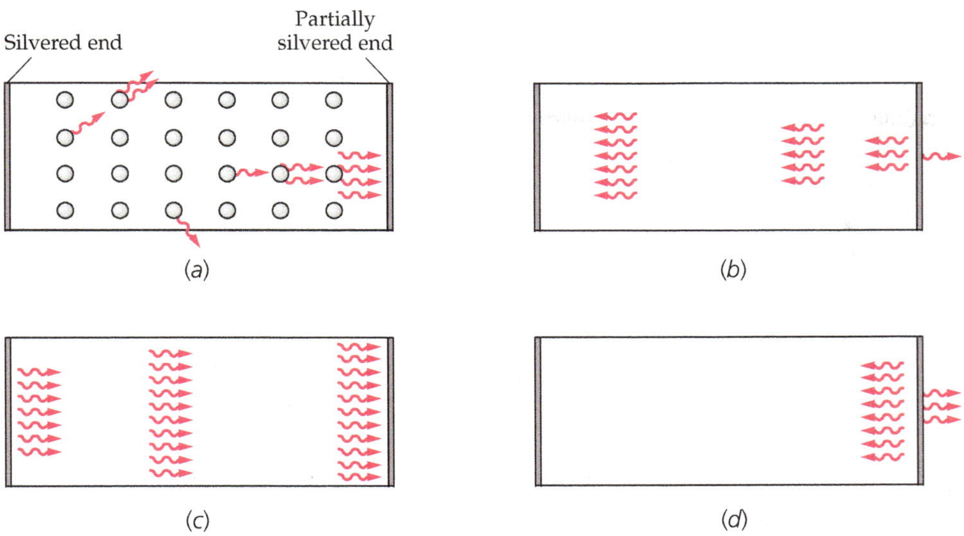

FIGURE 31-50 Buildup of a photon beam in a laser. (*a*) When irradiated, some atoms spontaneously emit photons, some of which travel to the right and stimulate other atoms to emit photons parallel to the axis of the crystal. (*b*) Of the four photons that strike the right face, one is transmitted and three are reflected. As the reflected photons traverse the lasing material, they stimulate other atoms to emit photons, and the beam builds up. By the time the beam reaches the right face again (*c*), it comprises many photons. (*d*) Some of the photons are transmitted, and the rest of the photons are reflected.

above the ground state. The expected lifetime for a chromium ion to remain in one of the metastable states is about 5 ms, after which the chromium ion spontaneously emits a photon and decays to the ground state. A millisecond is a long time for an atomic process. Consequently, if the flash is intense enough, the number of chromium ions populating the two metastable states will exceed the population of chromium ions in the ground state. It follows that during the time the flashtube is firing, the populations of ions in the ground state and in the metastable states are inverted. When the chromium ions in the state E_1 decay to the ground state by spontaneous emission, they emit photons of energy 1.79 eV and wavelength 694.3 nm. The photons have just the right energy to stimulate chromium ions in the metastable states to emit photons of the same energy (and wavelength) as they undergo the transition to the ground state. The photons also have just the right energy to stimulate chromium ions in the ground state to absorb a photon as they undergo the transition to one of the metastable states. These processes are competing processes, but the stimulated emission process dominates as long as the population of chromium ions in the metastable states exceeds the population in the ground state.

In the ruby laser, one end of the crystal is fully silvered, so it is 100 percent* reflecting; the other end of the crystal, called the output coupler, is partially silvered, leaving it about 85 percent reflecting. When photons traveling parallel to the axis of the crystal strike the silvered ends, all are reflected from the back face and 85 percent are reflected from the front face, with 15 percent of the photons escaping through the partially silvered front face. During each pass through the crystal, the photons stimulate more and more atoms so that an intense beam is emitted from the partially silvered end (Figure 31-50). Because the duration of each flash of the flashtube is between two and three seconds, the laser beam is produced in pulses lasting a few milliseconds. Modern ruby lasers generate intense light beams with energies ranging from 50 J to 100 J per pulse. The beam can have a diameter as small as 1 mm and an angular divergence as small as 0.25 mrad to about 7 mrad.

Population inversion is achieved somewhat differently in the continuous helium–neon laser. The energy levels of helium and neon that are important for the operation of the laser are shown in Figure 31-51. Helium has an excited energy state $E_{1\,\mathrm{He}}$ that is 20.61 eV above its ground state. Helium atoms are excited to state $E_{1\,\mathrm{He}}$ by an electric discharge. Neon has an excited state $E_{2\,\mathrm{Ne}}$ that is 20.66 eV above its ground state. This is just 0.05 eV above the first excited state of helium. The neon atoms are excited to state $E_{2\,\mathrm{Ne}}$ by collisions with excited helium atoms. The kinetic energy of the helium atoms provides the extra 0.05 eV of energy needed to excite the neon atoms. Another excited state of neon $E_{1\,\mathrm{Ne}}$ exists that is 18.70 eV above its ground state and 1.96 eV below state $E_{2\,\mathrm{Ne}}$. Because state $E_{1\,\mathrm{Ne}}$ is normally unoccupied, population

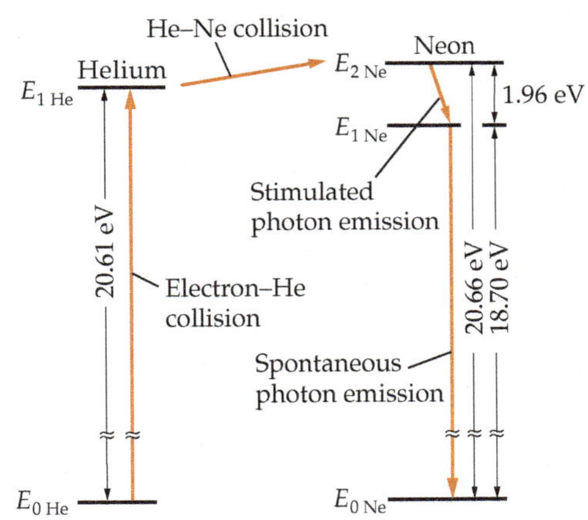

FIGURE 31-51 Energy levels of helium and neon that are important for the helium–neon laser. The helium atoms are excited by electrical discharge to an energy state 20.61 eV above the ground state. They collide with neon atoms, exciting some neon atoms to an energy state 20.66 eV above the ground state. Population inversion is thus achieved between this level and one 1.96 eV below it. The spontaneous emission of photons of energy 1.96 eV stimulates other atoms in the upper state to emit photons of energy 1.96 eV.

* In actuality, the light is only 99.7 percent reflected by a "fully silvered" end. In addition, the reflective coating consists of multiple dielectric layers, not silver.

inversion between states $E_{2\,\text{Ne}}$ and $E_{1\,\text{Ne}}$ is obtained immediately. The stimulated emission that occurs between these states results in photons of energy 1.96 eV and wavelength 632.8 nm, which produces a bright red light. After stimulated emission, the neon atoms in state $E_{1\,\text{Ne}}$ decay to the ground state $E_{0\,\text{Ne}}$ by spontaneous emission.

Note that stimulated emission involves transitions between two excited states of the neon atom in the helium–neon laser, whereas stimulated emission involves transitions between an excited state and the ground state of the chromium ion in the ruby laser. For stimulated emission between an excited state and a ground state, population inversion is difficult to achieve because more than half the atoms in the ground state must be excited. However, for stimulated emission between two excited states, population inversion is easily achieved because the state after stimulated emission is not the ground state but an excited state that is normally unpopulated.

Figure 31-52 shows a schematic diagram of a helium–neon laser commonly used for physics demonstrations. The helium–neon laser consists of a gas tube that contains 15 percent helium gas and 85 percent neon gas. A totally reflecting flat mirror is mounted at one end of the gas tube and a 99 percent reflecting concave mirror is placed at the other end of the gas tube. The concave mirror focuses parallel light at the flat mirror and also acts as a lens that transmits part of the light, so that the light emerges as a parallel beam.

A laser beam is coherent, very narrow, and intense. Its coherence makes the laser beam useful in the production of holograms, which we discuss in Chapter 33. The precise direction and small angular spread of the laser beam make it useful as a surgical tool for destroying cancer cells or reattaching a detached retina. Lasers are also used by surveyors for precise alignment over large distances. Distances can be accurately measured by reflecting a laser pulse from a mirror and measuring the time the pulse takes to travel to the mirror and back. The distance to the moon has been measured to within a few centimeters using an array of mirrors placed on the moon for that purpose. Laser beams are also used in fusion research. An intense laser pulse is focused on tiny pellets of deuterium–tritium in a combustion chamber. The beam heats the pellets to temperatures of the order of 10^8 K in a very short time, causing the deuterium and tritium to fuse and release energy.

Laser technology is advancing so quickly that it is possible to mention only a few of the recent developments. In addition to the ruby laser, many other solid-state lasers exist that have output wavelengths which range from approximately 170 nm to 3900 nm. Lasers that generate more than 1 kW of continuous power have been constructed. Pulsed lasers can now deliver nanosecond pulses of power exceeding 10^{14} W. Various gas lasers can now produce beams of wavelengths that range from the far infrared to the ultraviolet. Semiconductor lasers (also known as diode lasers or junction lasers) have shrunk in just 10 years from the size of a pinhead to mere billionths of a meter. Liquid lasers that use chemical dyes can be tuned over a range of wavelengths (approximately 70 nm for continuous lasers and more than 170 nm for pulsed lasers). A relatively new laser, the free-electron laser, extracts light energy from a beam of free electrons moving through a spatially varying magnetic field.

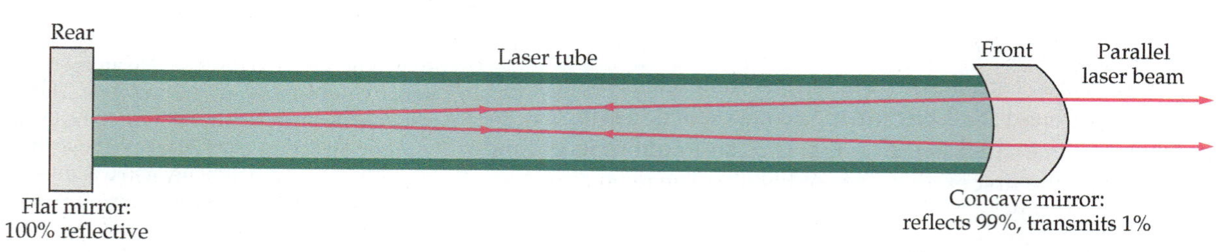

FIGURE 31-52 Schematic drawing of a helium–neon laser. The use of a concave mirror rather than a second plane mirror makes the alignment of the mirrors less critical than it is for the ruby laser. The concave mirror on the right also serves as a lens that focuses the emitted light into a parallel beam.

The free-electron laser has the potential to have very high power and large efficiency and can be tuned over a large range of wavelengths. There appears to be no limit to the variety and uses of modern lasers.

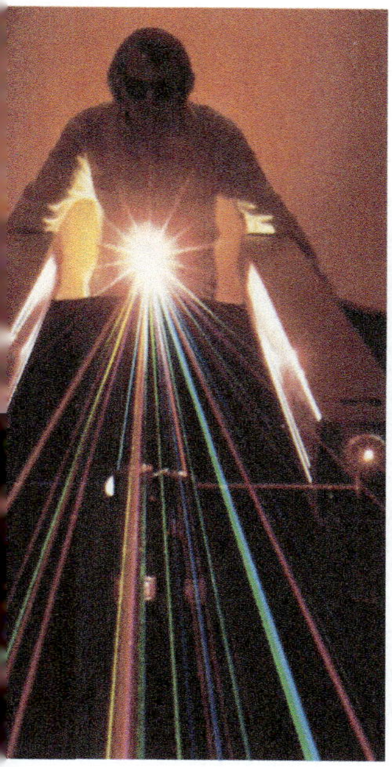

(*b*)

(*c*)

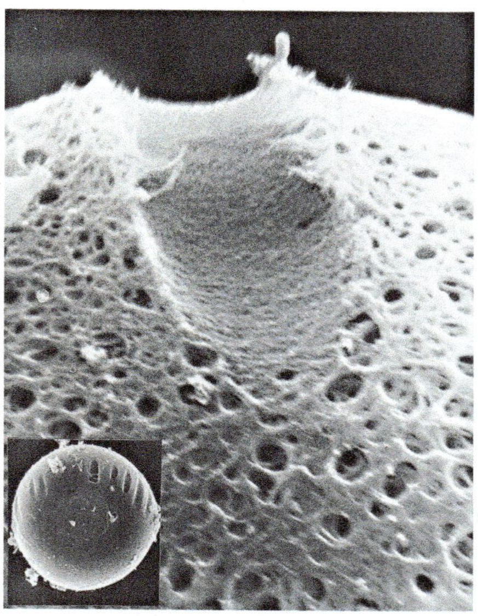

(*d*)

(*e*)

(*a*) Beams from a krypton laser and an argon laser, split into their component wavelengths. In these gas lasers, krypton and argon atoms have been stripped of multiple electrons, forming positive ions. The light-emitting energy transitions occur when excited electrons in the ions decay from one upper energy level to another. Here, several energy transitions are occurring at once, each corresponding to emitted light of a different wavelength. (*b*) A femtosecond pulsed laser. By a technique known as *modelocking*, different excited modes within a laser's cavity can be made to interfere with one another and create a series of ultrashort pulses, which are picoseconds long, that correspond to the time it takes light to bounce back and forth once within the cavity. Ultrashort pulses have been used as probes to study the behavior of atoms and molecules during chemical reactions. (*c*) A carbon dioxide laser takes just 2 minutes to cut out a steel saw blade. (*d*) A groove etched in the zona pellucida (protective outer covering) of a mouse egg by a *laser scissor* facilitates implantation. This technique has already been applied in human fertility therapies. Several effects contribute to the ability of the finely focused laser to cut on such a delicate scale—photon absorption may heat the target, break molecular bonds, or drive chemical reactions. (*e*) The so-called nanolasers shown are semiconductor disks mere microns in diameter and fractions of a micron in width. These tiny lasers work like their larger counterparts. Exploiting quantum effects that prevail on the microscopic scale, nanolasers promise great efficiency and they are being explored as ultrafast, low-energy switching devices. *((a, c) Chuck O'rear/West Light. (b) Courtesy of Ahmed H. Zewail. (d) Michael W. Berns/Scientific American. (e) David Scharf.)*

Physics Spotlight

Optical Tweezers and Vortices: Light at Work

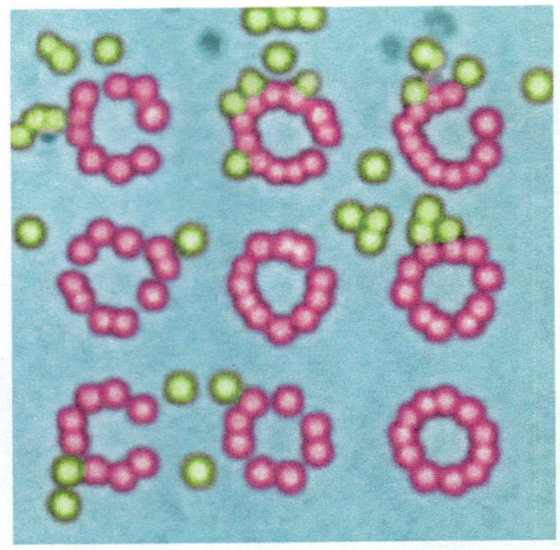

Silicon spheres in water are caught in a three by three array of optical vortices. The optical vortices trap the spheres and exert torques on them. *(Courtesy David G. Grier, from E. Curtis, B. A. Koss and D. G. Grier, "Dynamic holographic optical tweezers," Optics Communications 207, 169-175 (2002).)*

Light pressure has been used to measure the force exerted by biological molecules,[*] to unfold and refold proteins,[†] and even to assist in the trapping and study of atoms.[‡] Using the radiation pressure of light to hold microscopic particles in place is called *optical trapping.* Some optical traps, often called *optical tweezers,* can move and manipulate particles.

In the 1970s, a research group led by Arthur Ashkin at Bell Laboratories used radiation pressure of light to levitate droplets of water from 1 to 40 micrometers in diameter.[#] After many years of experimentation, this group demonstrated that a single laser could control the position of a virus in a solution on a microscope slide.[°] Molecular biologists and microbiologists quickly began to use optical tweezers in their studies.

Often, optical trapping is done using lasers that transmit light which has a wavelength near 1000 nanometers,[§] because many biological materials are relatively transparent to such near-infrared wavelengths. The liquid used for holding biological specimens absorbs scattered light near this wavelength.[¶,**] (This means that the trapped object is less likely to be cooked by the light.) Other wavelengths of light may be used, depending on the items to be trapped. The force used in optical traps for examining biological molecules is a few piconewtons.[††]

Optical trapping works both by light pressure and by taking advantage of the gradient of light intensity of a tightly focused laser beam. If a beam of light shines through a small translucent spherical object within the beam, the light will refract. The average of the pressure of refraction of an intense beam will act to keep the object centered in the beam. The more tightly the beam of light is focused, the closer a particle is trapped along the beam near the focus by the strength of the gradient of light intensity.[‡‡,##] This allows for positional control of an object in three dimensions. During the study of biological molecules, a molecule is usually attached to a polystyrene sphere that can be anywhere from 100 nm to 2 μm in diameter. Moving the sphere allows the molecule to be stretched, folded, and brought into focus with the help of optical tweezers. Much larger objects, such as entire cells, can also be moved by optical tweezers.[°°]

Specialized digital lenses can give laser light carefully calculated twists. The twisted light is called an *optical vortex.* Although they have many other potential uses, optical vortices can be used as specialized optical tweezers that have angular momentum.[§§] Different twists have different amounts of angular momentum and can be used to spin and rotate particles. Optical vortices have even been used to spin particles into one another to combine particles.

Physicists at the University of Chicago have come up with a method of generating hundreds of different optical tweezers from the same beam by passing laser beams through a digitally controlled lens.[¶¶] These tweezers can include optical vortices that exert different amounts of torque on particles. The holographic optical tweezers (HOT) method of creating optical vortices has been patented for manipulating particles, and for pumping, mixing, and sorting fluids and objects on a microscopic scale.[***,†††] Makers of miniature machines are enthusiastic about this technology because, unlike delicate tiny machines, light does not wear out.[‡‡‡]

* Mehta, A. D., et al., "Single-Molecule Biomechanics with Optical Methods." *Science*, Mar. 12, 1999, Vol. 283, No. 5408, pp. 1689–1695.

† Cecconi, C., et al., "Direct Observation of the Three-State Folding of a Single Protein Molecule." *Science*, Sept. 23, 2005, Vol. 309, No. 5743, pp. 2057–2060.

‡ Nagel, B., "Presentation Speech for 1997 Nobel Prize in Physics." *Les Prix Nobel. The Nobel Prizes 1997*, Ed. Tore Frängsmyr [Nobel Foundation], Stockholm, 1998, at http://nobelprize.org/nobel_prizes/physics/laureates/1997/presentation-speech.html As of Dec. 2006.

\# Ashkin, A., and Dziedzic, J., M., "Optical Levitation of Liquid Drops by Radiation Pressure." *Science*, Mar. 21, 1975, Vol. 187, No. 4181, pp. 1073–1075.

° Ashkin, A., and Dziedzic, J., M., "Optical Trapping and Manipulation of Viruses and Bacteria." *Science*, Mar. 20, 1987, Vol. 235, No. 4795, pp. 1517–1520.

§ Mohanty, S. K., Dasgupta, R., and Gupta, P. K., "Three-Dimensional Orientation of Microscopic Objects Using Combined Elliptical and Point Optical Tweezers." *Applied Physics B*, Dec. 2005, Vol. 81, No. 8, pp. 1063–1066.

¶ Molloy, J. E., and Padgett, M. E., "Lights, Action: Optical Tweezers." *Contemporary Physics*, Jul./Aug. 2002, Vol. 43, No. 4, pp. 241–258.

** Block, S. M., "Construction of Optical Tweezers." *Cells: A Laboratory Manual. Vol. 2, Sect. 7*, Eds. D. L. Spector, R. D. Goldman, and L. A. Leinwand. Cold Spring Harbor: Cold Spring Harbor Laboratory Press, 1998. At http://www.cshlpress.com/chap_cells.tpl#intro As of Dec. 2006.

†† Mehta, A. D., et al., op. cit.

‡‡ Block, S. M., op. cit.

\#\# Molloy, J. E., and Padgett, M. J., op. cit.

°° Pool, R., "Trapping with Optical Tweezers." *Science*, Aug. 26, 1988, Vol. 241, No. 4869, p. 1042.

§§ Dholakia, K., Spalding, G., and MacDonald, M., "Optical Tweezers: The Next Generation." *Physics World*, Oct. 2002, Vol. 15, No. 9, pp. 31–35.

¶¶ Curtis, J. E., Koss, B. A., and Grier, D. G., "Dynamic Holographic Optical Tweezers." *Optics Communications*, 2002, Vol. 207, pp. 169–175

*** Curtis, J. E., Koss, B. A., and Grier, D. G., *Use of Multiple Optical Vortices for Pumping, Mixing and Sorting.* U.S. Patent 6,858,833 B2. Feb. 22, 2005.

††† Curtis, J. E., Koss, B. A., and Grier, D. G., *Multiple Optical Vortices for Manipulating Particles* U.S. Patent 6,995,351 B2. Feb. 7, 2006.

‡‡‡ "Motors of Light May Resolve Dilemma of How to Power MEMS." *Small Times*, April 4, 2004. At http://www.smalltimes.com/document_display.cfm?document_id=5796 As of May 2006.

Summary

TOPIC	RELEVANT EQUATIONS AND REMARKS	
1. Speed of Light	The SI unit of length, the meter, is defined so that the speed of light in vacuum is exactly $c = 299\,792\,458\ \text{m/s}$	31-1
v in a transparent medium	$v = \frac{c}{n}$ where n is the index of refraction.	31-3
2. Reflection and Refraction	When light is incident on a surface separating two media in which the speed of light differs, part of the light energy is transmitted and part of the light energy is reflected.	
Law of reflection	The reflected ray lies in the plane of incidence and makes an angle θ_1' with the normal that is equal to the angle of incidence. $\theta_1' = \theta_1$	31-4
Reflected intensity, normal incidence	$I = \left(\frac{n_1 - n_2}{n_1 + n_2}\right)^2 I_0$	31-7
Index of refraction	$n = \frac{c}{v}$	31-3
Law of refraction (Snell's law)	$n_1 \sin\theta_1 = n_2 \sin\theta_2$	31-5*b*
Total internal reflection	When light is traveling in a medium that has an index of refraction n_1 and is incident on the boundary of a second medium that has a lower index of refraction $n_2 < n_1$, the light is totally reflected if the angle of incidence is greater than the critical angle θ_c given by	
Critical angle	$n_1 \sin\theta_c = n_2 \sin 90^\circ \qquad n_1 > n_2$	31-8
Dispersion	The speed of light in a medium, and therefore the index of refraction of that medium, depends on the wavelength of light. Because of dispersion, a beam of white light incident on a refracting prism is dispersed into its component colors. Similarly, the reflection and refraction of sunlight by raindrops produce a rainbow.	
3. Polarization	Transverse waves can be polarized. The four phenomena that produce polarized electromagnetic waves from unpolarized waves are (1) absorption, (2) scattering, (3) reflection, and (4) birefringence.	
Malus's law	When two polarizers have their transmission axes at an angle θ, the intensity transmitted by the second polarizer is reduced by the factor $\cos^2\theta$: $I = I_0 \cos^2\theta$	31-16
4. Huygens's Construction	Each point on a primary wavefront serves as the source of spherical secondary wavelets that advance with a speed and frequency equal to that of the primary wave. The primary wavefront at some later time is the envelope of these wavelets.	
5. Wave-Particle Duality	Light propagates like a wave, but interacts with matter like a particle.	
Photon energy	$E = hf = \frac{hc}{\lambda}$	31-25
Planck's constant	$h = 6.626 \times 10^{-34}\ \text{J}\cdot\text{s} = 4.136 \times 10^{-15}\ \text{eV}\cdot\text{s}$	
hc	$hc = 1240\ \text{eV}\cdot\text{nm}$	31-26

TOPIC	RELEVANT EQUATIONS AND REMARKS
6. Emission of Light	Light is emitted when a valence electron makes a transition from an excited state to a state of lower energy.
Line spectra	Atoms in dilute gases emit a discrete set of wavelengths called a line spectrum. The photon energy $E = hf = hc/\lambda$ equals the difference in energy of the initial and final states of the atom.
Continuous spectra	Atoms in high-density gases, liquids, or solids have continuous bands of energy levels, so they emit a continuous spectrum of light. Thermal radiation is visible if the temperature of the emitting object is above approximately 600°C.
Spontaneous emission	An atom in an excited state will spontaneously make a transition to a lower state with the emission of a photon. This process is random, with a characteristic lifetime of about 10^{-8} s. The photons from two or more atoms are not correlated, so the light is incoherent.
Stimulated emission	Stimulated emission occurs if an atom is initially in an excited state and a photon of energy equal to the energy difference between that state and a lower state is incident on the atom. The oscillating electromagnetic field of the incident photon stimulates the excited atom to emit another photon in the same direction and in phase with the incident photon. The emitted light is coherent with the incident light.
7. Visible Light	The human eye is sensitive to electromagnetic radiation that has wavelengths from approximately 400 nm (violet) to 700 nm (red). The photon energies range from approximately 1.8 eV to 3.1 eV. A uniform distribution of wavelengths, such as the wavelengths emitted by the Sun, appears white to our eyes.
8. Lasers	A laser produces an intense, coherent, and narrow beam of photons as the result of stimulated emission. The operation of a laser depends on population inversion, in which there are more atoms in an excited state than in the ground state or a lower state.

Answers to Concept Check

31-1 There are 720 teeth, but there are also 720 gaps, so the width of a tooth is less than $\frac{1}{720}$ of the circumference of the wheel. Consequently, the wheel actually has to rotate through less than $\frac{1}{720}$ rev for the light from the distant mirror to be again observed.

Answers to Practice Problems

31-1 (*a*) 4.57×10^6 km (*b*) 3.05×10^8 m/s

31-2 1.28 s each way

Problems

In a few problems, you are given more data than you actually need; in a few other problems, you are required to supply data from your general knowledge, outside sources, or informed estimate.

Interpret as significant all digits in numerical values that have trailing zeros and no decimal points.

- • Single-concept, single-step, relatively easy
- •• Intermediate-level, may require synthesis of concepts
- ••• Challenging
- SSM Solution is in the *Student Solutions Manual*
- Consecutive problems that are shaded are paired problems.

CONCEPTUAL PROBLEMS

1 • A ray of light reflects from a plane mirror. The angle between the incoming ray and the reflected ray is 70°. What is the angle of reflection? (*a*) 70°, (*b*) 140°, (*c*) 35°, (*d*) Not enough information is given to determine the reflection angle. SSM

2 • A ray of light in air is incident on the surface of a piece of glass. The angle between the normal to the surface and the incident ray is 40°, and the angle between the normal and the refracted ray is 28°. What is the angle between the incident ray and the refracted ray? (*a*) 12°, (*b*) 28°, (*c*) 40°, (*d*) 68°

3 • **Engineering Application** During a physics experiment, you are measuring refractive indices of different transparent materials using a red helium–neon laser beam. For a given angle of incidence, the beam has an angle of refraction equal to 28° in material A, and an angle of refraction equal to 26° in material B. Which material has the larger index of refraction? *(a)* A, *(b)* B, *(c)* The indices of refraction are the same. *(d)* You cannot determine the relative magnitudes of the indices of refraction from the data given.

4 • A ray of light passes from air into water, striking the surface of the water at an angle of incidence of 45°. Which, if any, of the following four quantities changes as the light enters the water: *(a)* wavelength, *(b)* frequency, *(c)* speed of propagation, *(d)* direction of propagation, *(e)* none of the above?

5 • Earth's atmosphere decreases in density as the altitude increases. As a consequence, the index of refraction of the atmosphere also decreases as altitude increases. Explain how one can see the Sun when it is below the horizon. (The horizon is the extension of a plane that is tangent to the Earth's surface.) Why does the setting Sun appear flattened?

6 • A physics student playing pocket billiards wants to strike her cue ball so that it hits a cushion and then hits the eight ball squarely. She chooses several points on the cushion and then measures the distances from each point to the cue ball and to the eight ball. She aims at the point for which the sum of these distances is least. *(a)* Will her cue ball hit the eight ball? *(b)* How is her method related to Fermat's principle? Neglect any effects due to ball rotation.

7 • A swimmer at point *S* in Figure 31-53 develops a leg cramp while swimming near the shore of a calm lake and calls for help. A lifeguard at point *L* hears the call. The lifeguard can run 9.0 m/s and swim 3.0 m/s. She knows physics and chooses a path that will take the least time to reach the swimmer. Which of the paths shown in the figure does the lifeguard take? SSM

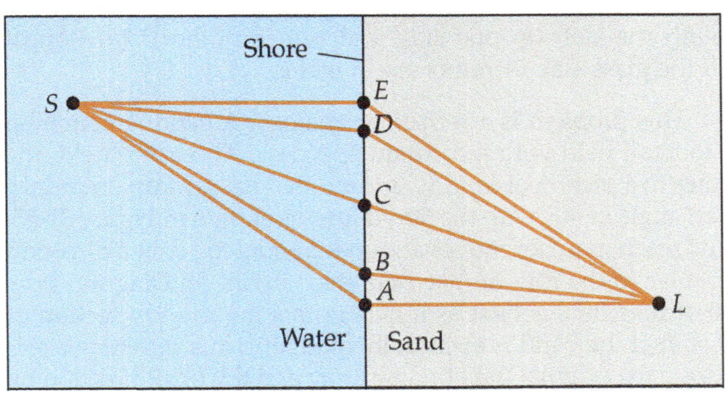

FIGURE 31-53 Problem 7

8 • Material A has a higher index of refraction than material B. Which material has the larger critical angle for total internal reflection when the material is in air? *(a)* A, *(b)* B, *(c)* The angles are the same. *(d)* You cannot compare the angles based on the data given.

9 • **Biological Application** A human eye perceives color using a structure which is called a *cone* that is located on the retina. Three types of molecules compose these cones and each type of molecule absorbs either red, green, or blue light by resonance absorption. Use this fact to explain why the color of an object that appears blue in air appears blue underwater, in spite of the fact that the wavelength of the light is shortened in accordance with Equation 31-6. SSM

10 • Let θ be the angle between the transmission axes of two polarizing sheets. Unpolarized light of intensity I is incident on the first sheet. What is the intensity of the light transmitted through both sheets? *(a)* $I\cos^2\theta$, *(b)* $(I\cos^2\theta)/2$, *(c)* $(I\cos^2\theta)/4$, *(d)* $I\cos\theta$, *(e)* $(I\cos\theta)/4$, *(f)* none of the above

11 •• Draw a diagram to explain how Polaroid sunglasses reduce glare from sunlight reflected from a smooth horizontal surface, such as the surface found on a pool of water. Your diagram should clearly indicate the direction of polarization of the light as it propagates from the Sun to the reflecting surface and then through the sunglasses into the eye. SSM

12 • **Biological Application** Why is it far less dangerous to stand in front of an intense beam of red light than in front of a very low-intensity beam of gamma rays?

13 • Three energy states of an atom are A, B, and C. State B is 2.0 eV above state A and state C is 3.00 eV above state B. Which atomic transition results in the emission of the shortest wavelength of light? *(a)* B → A, *(b)* C → B, *(c)* C → A, *(d)* A → C

14 • In Problem 13, if the atom is initially in state A, which transition results in the emission of the longest wavelength of light? *(a)* A → B, *(b)* B → C, *(c)* A → C, *(d)* B → A

15 • What role does the helium play in a helium–neon laser? SSM

16 • When a beam of visible white light that passes through a gas of atomic hydrogen at room temperature is viewed with a spectroscope, dark lines are observed at the wavelengths of the hydrogen atom emission series. The atoms that participate in the resonance absorption then emit light of same wavelength as they return to the ground state. Explain why the observed spectrum nevertheless exhibits pronounced dark lines.

17 • Which of the following types of light would have the highest energy photons? *(a)* red, *(b)* infrared, *(c)* blue, *(d)* ultraviolet SSM

ESTIMATION AND APPROXIMATION

18 • Estimate the time required for light to make the round trip during Galileo's experiment to measure the speed of light. Compare the time of the round trip to typical human response times. How accurate do you think this experiment is?

19 • Estimate the time delay in receiving a light on your retina when you are wearing eyeglasses compared to when you are not wearing your eyeglasses.

20 •• **Biological Application** Estimate the number of photons that enter your eye if you look for a tenth of a second at the Sun. What energy is absorbed by your eye during that time, assuming that all the photons are absorbed? The total power output of the Sun is 4.2×10^2 W.

21 •• Römer was observing the eclipses of Jupiter's moon Io with the hope that they would serve as a highly accurate clock that would be independent of longitude. (Prior to GPS, such a clock was needed for accurate navigation.) Io eclipses (enters the umbra of Jupiter's shadow) every 42.5 h. Assuming an eclipse of Io is observed on Earth on June 1 at midnight when Earth is at location *A*

(as shown in Figure 31-54), predict the expected time of observation of an eclipse one-quarter of a year later when Earth is at location *B*, assuming (*a*) the speed of light is infinite and (*b*) the speed of light is 2.998×10^8 m/s.

22 •• If the angle of incidence is small enough, the small angle approximation $\sin\theta \approx \theta$ may be used to simplify Snell's law of refraction. Determine the maximum value of the angle that would make the value for the angle differ by no more than one percent from the value for the sine of the angle. (This approximation will be used in connection with image formation by spherical surfaces in Chapter 32.)

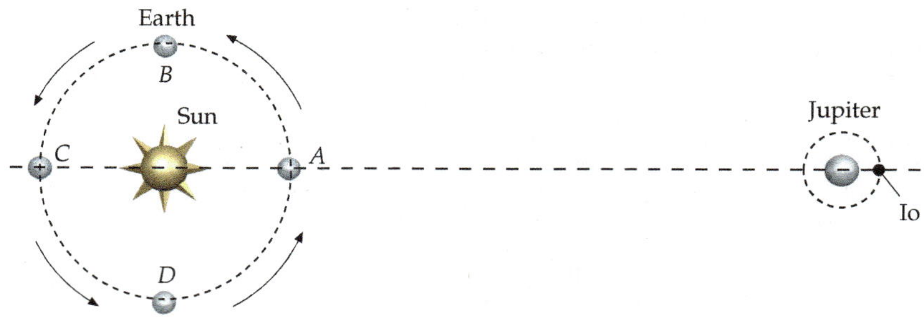

FIGURE 31-54 Problem 21

THE SPEED OF LIGHT

23 • Mission Control sends a brief wake-up call to astronauts in a spaceship that is far from Earth. At a time 5.0 s after the call is sent, Mission Control can hear the groans of the astronauts. How far from Earth is the spaceship? (*a*) 7.5×10^8 m, (*b*) 15×10^8 m, (*c*) 30×10^8 m, (*d*) 45×10^8 m, (*e*) The spaceship is on the moon.

24 •• **ENGINEERING APPLICATION** The distance from a point on the surface of Earth to a point on the surface of the moon is measured by aiming a laser light beam at a reflector on the surface of the moon and measuring the time required for the light to make a round trip. The uncertainty in the measured distance Δx is related to the uncertainty in the measured time Δt by $\Delta x = \frac{1}{2}c\,\Delta t$. If the time intervals can be measured to ± 1.00 ns, (*a*) find the uncertainty of the distance. (*b*) Estimate the percentage uncertainty in the distance.

25 •• Ole Römer discovered the finiteness of the speed of light by observing Jupiter's moons. Approximately how sensitive would the timing apparatus need to be in order to detect a shift in the predicted time of the moon's eclipses that occur when the moon happens to be at perigee (3.63×10^5 km) and those that occur when the moon is at apogee (4.06×10^5 km)? Assume that an instrument should be able to measure to at least one-tenth the magnitude of the effect it is to measure. SSM

REFLECTION AND REFRACTION

26 • Calculate the fraction of light energy reflected from an air–water interface at normal incidence.

27 • A ray of light is incident on one of two mirrors that are set at right angles to each other. The plane of incidence is perpendicular to both mirrors. Show that after reflecting from each mirror, the ray will emerge traveling in the direction opposite to the incident direction, regardless of the angle of incidence.

28 •• **SPREADSHEET** (*a*) A ray of light in air is incident on an air–water interface. Using a spreadsheet or graphing program, plot the angle of refraction as a function of the angle of incidence from 0° to 90°. (*b*) Repeat Part (*a*), but for a ray of light in water that is incident on a water–air interface. [For Part (*b*), there is no reflected ray for angles of incidence that are greater than the critical angle.]

29 •• The red light from a helium–neon laser has a wavelength of 632.8 nm in air. Find the (*a*) speed, (*b*) wavelength, and (*c*) frequency of helium–neon laser light in air, water, and glass. (The glass has an index of refraction equal to 1.50.)

30 •• The index of refraction for silicate flint glass is 1.66 for violet light that has a wavelength in air equal to 400 nm and 1.61 for red light that has a wavelength in air equal to 700 nm. A ray of 700-nm-wavelength red light and a ray of 400-nm-wavelength violet light both have angles of refraction equal to 30° upon entering the glass from air. (*a*) Which is greater, the angle of incidence of the ray of red light or the angle of incidence of the ray of violet light? Explain your answer. (*b*) What is the difference between the angles of incidence of the two rays?

31 •• A slab of glass that has an index of refraction of 1.50 is submerged in water that has an index of refraction of 1.33. Light in the water is incident on the glass. Find the angle of refraction if the angle of incidence is (*a*) 60°, (*b*) 45°, and (*c*) 30°. SSM

32 •• Repeat Problem 31 for a beam of light initially in the glass that is incident on the glass–water interface at the same angles.

33 •• A beam of light in air strikes a glass slab at normal incidence. The glass slab has an index of refraction of 1.50. (*a*) Approximately what percentage of the incident light intensity is transmitted through the slab (in one side and out the other)? (*b*) Repeat Part (*a*) if the glass slab is immersed in water.

34 •• This problem is a refraction analogy. A band is marching down a football field with a constant speed v_1. About midfield, the band comes to a section of muddy ground that has a sharp boundary making an angle of 30° with the 50-yd line, as shown in Figure 31-55. In the mud, each marcher moves at a speed equal to $\frac{1}{2}v_1$ in a direction perpendicular to the row of marchers they are in. (*a*) Diagram how each line of marchers is bent as it encounters the muddy section of the field so that the band is eventually marching in a different direction. Indicate the original direction by a ray and the final direction by a second ray. (*b*) Find the angles between these rays and the line normal to the boundary. Is their direction of motion "bent" toward the normal or away from it? Explain your answer in terms of refraction.

35 •• In Figure 31-56, light is initially in a medium that has an

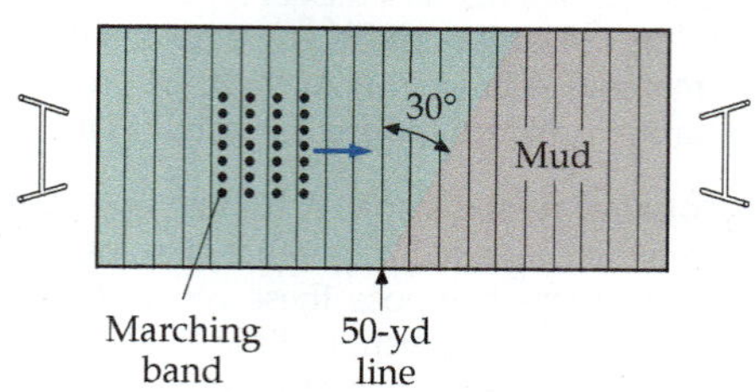

FIGURE 31-55 Problem 34

index of refraction n_1. It is incident at angle θ_1 on the surface of a liquid that has an index of refraction n_2. The light passes through the layer of liquid and enters glass that has an index of refraction n_3. If θ_3 is the angle of refraction in the glass, show that $n_1 \sin\theta_1 = n_3 \sin\theta_3$. That is, show that the second medium can be neglected when finding the angle of refraction in the third medium. SSM

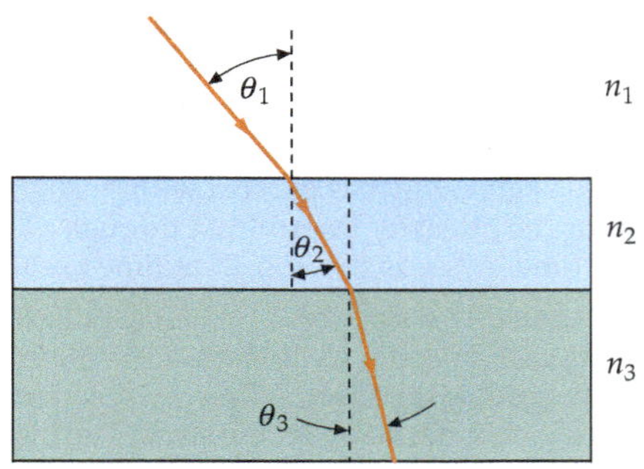

FIGURE 31-56 Problem 35

36 •• On a safari, you are spearfishing while wading in a river. You observe a fish gliding by you. If your line of sight to the fish is 64.0° below the horizontal in air, and assuming the spear follows a straight-line path through the air and water after it is released, determine the angle below the horizontal that you should aim your spear gun in order to catch dinner. Assume the spear gun barrel is 1.50 m above the water surface, the fish is 1.20 m below the surface, and the spear travels in a straight line all the way to the fish.

37 ••• You are standing on the edge of a swimming pool and looking directly across at the opposite side. You notice that the bottom edge of the opposite side of the pool appears to be at an angle of 28° below the horizontal. However, when you sit on the pool edge, the bottom edge of the opposite side of the pool appears to be at an angle of only 14° below the horizontal. Use these observations to determine the width and depth of the pool. *Hint: You will need to estimate the height of your eyes above the surface of the water when standing and sitting.*

38 ••• Figure 31-57 shows a beam of light incident on a glass plate of thickness d and index of refraction n. (*a*) Find the angle of incidence so that the separation b between the ray reflected from the top surface and the ray reflected from the bottom surface and exiting the top surface is a maximum. (*b*) What is this angle of incidence if the index of refraction of the glass is 1.60? (*c*) What is the separation of the two beams if the thickness of the glass plate is 4.0 cm?

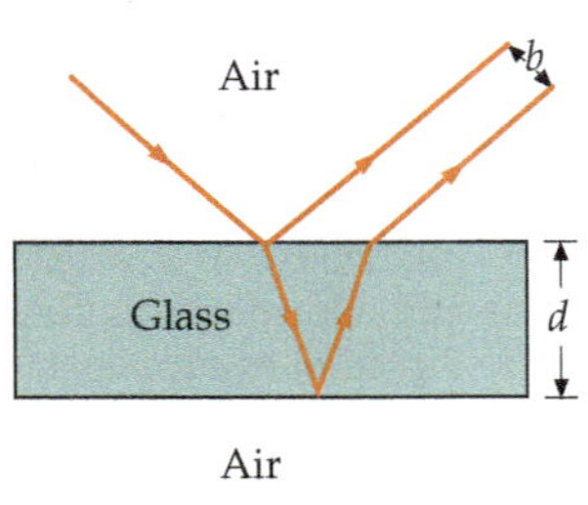

FIGURE 31-57 Problems 38 and 48

TOTAL INTERNAL REFLECTION

39 • What is the critical angle for light traveling in water that is incident on a water–air interface? SSM

40 • A glass surface ($n = 1.50$) has a layer of water ($n = 1.33$) on it. Light in the glass is incident on the glass–water interface. Find the critical angle for total internal reflection.

41 • A point source of light is located 5.0 m below the surface of a large pool of water. Find the area of the largest circle on the pool's surface through which light coming directly from the source can emerge.

42 •• Light traveling in air strikes the largest face of an isosceles-right-triangle prism at normal incidence. What is the speed of light in this prism if the prism is just barely able to produce total internal reflection?

43 •• A point source of light is located at the bottom of a steel tank, and an opaque circular card of radius 6.00 cm is placed horizontally over it. A transparent fluid is gently added to the tank so that the card floats on the fluid surface with its center directly above the light source. No light is seen by an observer above the surface until the fluid is 5.00 cm deep. What is the index of refraction of the fluid?

44 •• **ENGINEERING APPLICATION** An optical fiber allows rays of light to propagate long distances by using total internal reflection. Optical fibers are used extensively in medicine and in digital communications. As shown in Figure 31-58 the fiber consists of a core material that has an index of refraction n_2 and radius b surrounded by a cladding material that has an index of refraction $n_3 < n_2$. The *numerical aperture* of the fiber is defined as $\sin\theta_1$, where θ_1 is the angle of incidence of a ray of light that impinges on the center of the end of the fiber and then reflects off the core-cladding interface just at the critical angle. Using the figure as a guide, show that the numerical aperture is given by $\sin\theta_1 = \sqrt{n_2^2 - n_3^2}$ assuming the ray is initially in air. *Hint: Use of the Pythagorean theorem may be required.*

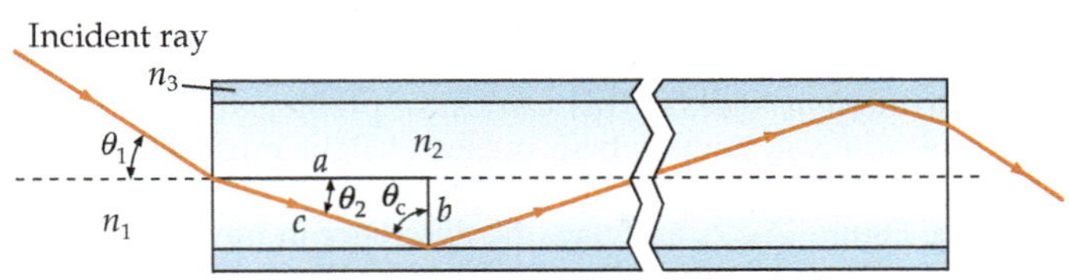

FIGURE 31-58 Problems 44, 45, and 46

45 •• **ENGINEERING APPLICATION** Find the maximum angle of incidence θ_1 of a ray that would propagate through an optical fiber that has a core index of refraction of 1.492, a core radius of 50.00 μm, and a cladding index of 1.489. See Problem 44. SSM

46 •• **ENGINEERING APPLICATION** Calculate the difference in time needed for two pulses of light to travel down 15.0 km of the fiber that is described in Problem 44. Assume that one pulse enters the fiber at normal incidence and the second pulse enters the fiber at the maximum angle of incidence calculated in Problem 45. In fiber optics, this effect is known as *modal dispersion*.

47 ••• Investigate how a thin film of water on a glass surface affects the critical angle for total reflection. Use $n = 1.50$ for glass and $n = 1.33$ for water. (*a*) What is the critical angle for total internal reflection at the glass–water interface? (*b*) Does a range of incident angles exist such that the angles are greater than θ_c for glass-to-air refraction and for which the light rays will leave the glass, travel through the water, and then pass into the air?

48 •• A laser beam is incident on a plate of glass that is 3.0 cm thick (Figure 31-57). The glass has an index of refraction of 1.5 and the angle of incidence is 40°. The top and bottom surfaces of the glass are parallel. What is the distance b between the beam formed by reflection off the top surface of the glass and the beam reflected off the bottom surface of the glass.

DISPERSION

49 • A beam of light strikes the plane surface of silicate flint glass at an angle of incidence of 45°. The index of refraction of the glass varies with wavelength (see Figure 31-59). How much smaller is the angle of refraction for violet light of wavelength 400 nm than the angle of refraction for red light of wavelength 700 nm?

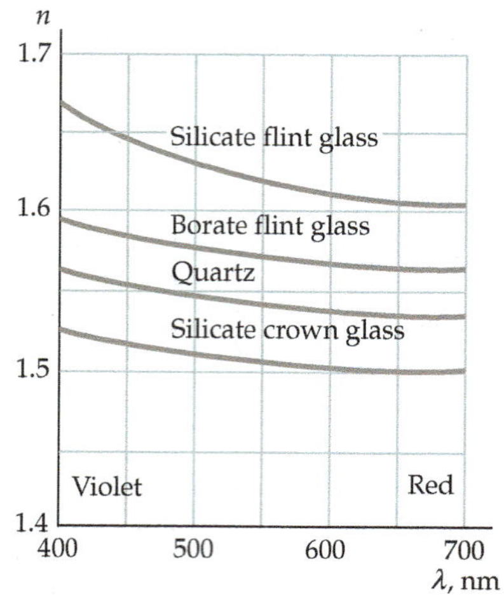

FIGURE 31-59 Problems 49, 50, 73, and 77

50 •• **ENGINEERING APPLICATION** In many transparent materials, dispersion causes different colors (wavelengths) of light to travel at different speeds. This can cause problems in fiber-optic communications systems where pulses of light must travel very long distances in glass. Assuming a fiber is made of silicate crown glass (see Figure 31-59), calculate the difference in travel times that two short pulses of light take to travel 15.0 km in the fiber if the first pulse has a wavelength of 700 nm and the second pulse has a wavelength of 500 nm.

POLARIZATION

51 • What is the polarizing angle for light in air that is incident on (*a*) water ($n = 1.33$) and (*b*) glass ($n = 1.50$)? SSM

52 • Light that is horizontally polarized is incident on a polarizing sheet. It is observed that only 15 percent of the intensity of the incident light is transmitted through the sheet. What angle does the transmission axis of the sheet make with the horizontal?

53 • Two polarizing sheets have their transmission axes crossed so that no light gets through. A third sheet is inserted between the first two so that its transmission axis makes an angle θ with the transmission axis of the first sheet. Unpolarized light of intensity I_0 is incident on the first sheet. Find the intensity of the light transmitted through all three sheets if (*a*) $\theta = 45°$ and (*b*) $\theta = 30°$.

54 • A horizontal 5.0 mW laser beam that is vertically polarized is incident on a polarizing sheet that is oriented with its transmission axis vertical. Behind the first sheet is a second sheet that is oriented so that its transmission axis makes an angle of 27° with respect to vertical. What is the power of the beam transmitted through the second sheet?

55 •• The polarizing angle for light in air that is incident on a certain substance is 60°. (*a*) What is the angle of refraction of light incident at this angle? (*b*) What is the index of refraction of this substance? SSM

56 •• Two polarizing sheets have their transmission axes crossed so that no light is transmitted. A third sheet is inserted so that its transmission axis makes an angle θ with the transmission axis of the first sheet. (*a*) Derive an expression for the intensity of the transmitted light as a function of θ. (*b*) Show that the intensity transmitted through all three sheets is maximum when $\theta = 45°$.

57 •• If the middle polarizing sheet in Problem 56 is rotating at an angular speed ω about an axis parallel with the light beam, find an expression for the intensity transmitted through all three sheets as a function of time. Assume that $\theta = 0°$ at time $t = 0$.

58 •• **SPREADSHEET** A stack of $N + 1$ ideal polarizing sheets is arranged so that each sheet is rotated by an angle of $\pi/(2N)$ rad with respect to the preceding sheet. A linearly polarized light wave of intensity I_0 is incident normally on the stack. The incident light is polarized along the transmission axis of the first sheet and is therefore perpendicular to the transmission axis of the last sheet in the stack. (*a*) Show that the intensity of the light transmitted through the entire stack is given by $I_0 \cos^{2N}[\pi/(2N)]$. (*b*) Using a spreadsheet or graphing program, plot the transmitted intensity as a function of N for values of N from 2 to 100. (*c*) What is the direction of polarization of the transmitted beam in each case?

59 •• **SPREADSHEET, ENGINEERING APPLICATION** The device described in Problem 58 could serve as a *polarization rotator*, which changes the linear plane of polarization from one direction to another. The efficiency of such a device is measured by taking the ratio of the output intensity at the desired polarization to the input intensity. The result of Problem 58 suggests that the highest efficiency is achieved by using a large value for the number N. A small amount of intensity is lost regardless of the input polarization when using a real polarizer. For each polarizer, assume the transmitted intensity is 98 percent of the amount predicted by the law of Malus and use a **spreadsheet** or graphing program to determine the optimum number of sheets you should use to rotate the polarization 90°. SSM

60 •• Show mathematically that a linearly polarized wave can be thought of as a superposition of a right and a left circularly polarized wave.

61 •• Suppose that the middle sheet in Problem 53 is replaced by two polarizing sheets. If the angles between the transmission axes of the second, third and fourth sheets in the stack make angles of 30°, 60°, and 90°, respectively, with the transmission axis of the first sheet, (*a*) what is the intensity of the transmitted light? (*b*) How does this intensity compare with the intensity obtained in Part (*a*) of Problem 53?

62 •• Show that the electric field of a circularly polarized wave propagating parallel to the x axis can be expressed by $\vec{E} = E_0 \sin(kx + \omega t)\hat{j} + E_0 \cos(kx + \omega t)\hat{k}$.

63 •• A circularly polarized wave is said to be *right circularly polarized* if the electric and magnetic fields rotate clockwise when viewed along the direction of propagation and *left circularly polarized* if the fields rotate counterclockwise. (*a*) What is the sense of the circular polarization for the wave described by the expression in Problem 62? (*b*) What would be the expression for the electric field of a circularly polarized wave traveling in the same direction as the wave in Problem 60, but with the fields rotating in the opposite sense? SSM

SOURCES OF LIGHT

64 • A helium–neon laser emits light that has a wavelength equal to 632.8 nm and has a power output of 4.00 mW. How many photons are emitted per second by this laser?

65 •• The first excited state of an atom of a gas is 2.85 eV above the ground state. (*a*) What is the maximum wavelength of radiation for resonance absorption by atoms of the gas that are in the ground state? (*b*) If the gas is irradiated with monochromatic light that has a wavelength of 320 nm, what is the wavelength of the Raman scattered light?

66 •• A gas is irradiated with monochromatic ultraviolet light that has a wavelength of 368 nm. Scattered light that has a wavelength equal to 368 nm is observed, and scattered light that has a wavelength of 658 nm is also observed. Assuming that the gas atoms were in their ground state prior to irradiation, find the energy difference between the ground state and the excited state obtained by the irradiation.

67 •• Sodium has excited states 2.11 eV, 3.20 eV, and 4.35 eV above the ground state. Assume that the atoms of the gas are all in the ground state prior to irradiation. (*a*) What is the maximum wavelength of radiation that will result in resonance fluorescence? What is the wavelength of the fluorescent radiation? (*b*) What wavelength will result in excitation of the state 4.35 eV above the ground state? If that state is excited, what are the possible wavelengths of resonance fluorescence that might be observed? SSM

68 •• Singly ionized helium is a hydrogen-like atom that has a nuclear charge of $+2e$. Its energy levels are given by $E_n = -4E_0/n^2$, where $n = 1, 2, \ldots$ and $E_0 = 13.6$ eV. If a beam of visible white light is sent through a gas of singly ionized helium, at what wavelengths will dark lines be found in the spectrum of the transmitted radiation? (Assume that the ions of the gas are all in the state with energy E_1 prior to irradiation.)

69 • A pulse from a ruby laser has an average power of 10 MW and lasts 1.5 ns. (*a*) What is the total energy of the pulse? (*b*) How many photons are emitted in the pulse? SSM

GENERAL PROBLEMS

70 • A beam of red light that has a wavelength of 700 nm in air travels in water. (*a*) What is the wavelength in water? (*b*) Does a swimmer underwater observe the same color or a different color for this light?

71 • The critical angle for total internal reflection for a substance is 48°. What is the polarizing angle for the substance? SSM

72 •• Show that when a flat mirror is rotated through an angle θ about an axis in the plane of the mirror, a reflected beam of light (from a fixed incident beam) that is perpendicular to the rotation axis is rotated through 2θ.

73 •• Use Figure 31-59 to calculate the critical angles for light initially in silicate flint glass that is incident on a glass–air interface if the light is (*a*) violet light of wavelength 400 nm and (*b*) red light of wavelength 700 nm. SSM

74 •• Light is incident on a slab of transparent material at an angle θ_1, as shown in Figure 31-60. The slab has a thickness t and an index of refraction n. Show that $n = \sin(\theta_1)/\sin[\tan^{-1}(d/t)]$, where d is the distance shown in the figure.

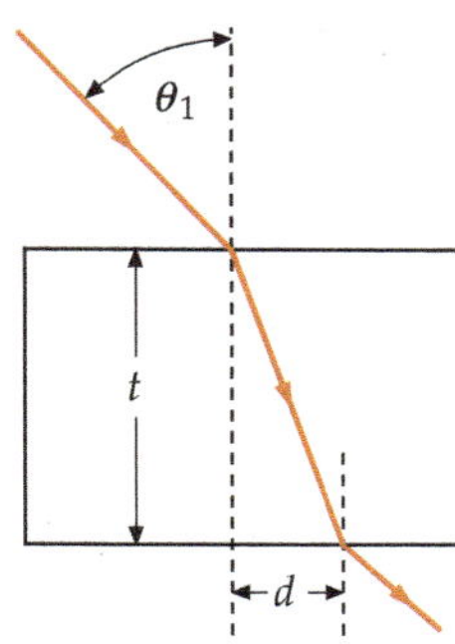

FIGURE 31-60 Problem 74

75 •• A ray of light begins at the point (−2.00 m, 2.00 m, 0.00 m), strikes a mirror in the $y = 0$ plane at some point $(x, 0, 0)$, and reflects through the point (2.00 m, 6.00 m, 0.00 m). (*a*) Find the value of x that makes the total distance traveled by the ray a minimum. (*b*) What is the angle of incidence on the reflecting plane? (*c*) What is the angle of reflection?

76 •• **ENGINEERING APPLICATION** To produce a polarized laser beam, a plate of transparent material (Figure 31-61) is placed in the laser cavity and oriented so the light strikes it at the polarizing angle. Such a plate is called a Brewster window. Show that if θ_{P1} is the polarizing angle for the n_1 to n_2 interface, then θ_{P2} is the polarizing angle for the n_2 to n_1 interface.

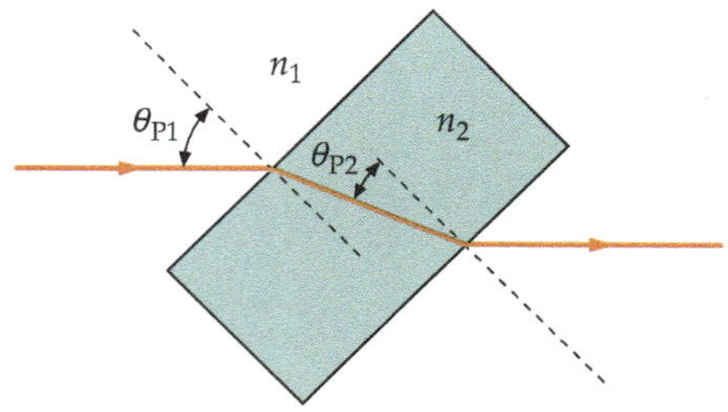

FIGURE 31-61 Problem 76

77 •• From the data provided in Figure 31-59, calculate the polarization angle for an air–glass interface, using light of wavelength 550 nm in each of the four types of glass shown. SSM

78 •• A light ray passes through a prism with an apex angle of α, as shown in Figure 31-62. The ray and the bisector of the apex angle intersect at right angles. Show that the *angle of deviation* δ is related to the apex angle and the index of refraction of the prism material by $\sin\left[\frac{1}{2}(\alpha + \delta)\right] = n \sin\left(\frac{1}{2}\alpha\right)$.

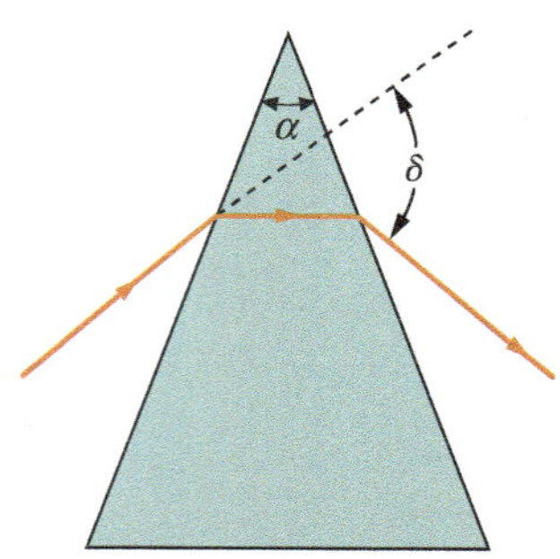

FIGURE 31-62 Problems 78 and 84

79 •• (*a*) For light rays inside a transparent medium that is surrounded by a vacuum, show that the polarizing angle and the critical angle for total internal reflection satisfy $\tan\theta_p = \sin\theta_c$. (*b*) Which angle is larger, the polarizing angle or the critical angle for total internal reflection? SSM

80 •• Light in air is incident on the surface of a transparent substance at an angle of 58° with the normal. The reflected and refracted rays are observed to be mutually perpendicular. (*a*) What is the index of refraction of the transparent substance? (*b*) What is the critical angle for total internal reflection in this substance?

81 •• A light ray in dense flint glass that has an index of refraction of 1.655 is incident on the glass surface. An unknown liquid condenses on the surface of the glass. Total internal reflection on the glass–liquid interface occurs for a minimum angle of incidence on the glass–liquid interface of 53.7°. (*a*) What is the refractive index of the unknown liquid? (*b*) If the liquid is removed, what is the minimum angle of incidence for total internal reflection? (*c*) For the angle of incidence found in Part (*b*), what is the angle of refraction of the ray into the liquid film? Does a ray emerge from the liquid film into the air above? Assume the glass and liquid have parallel planar surfaces.

82 ••• (*a*) Show that for normally incident light, the intensity transmitted through a glass slab that has an index of refraction of n and is surrounded by air is approximately given by $I_T = I_0[4n/(n+1)^2]^2$. (*b*) Use the Part-(*a*) result to find the ratio of the transmitted intensity to the incident intensity through N parallel slabs of glass for light of normal incidence. (*c*) How many slabs of a glass that has an index of refraction of 1.5 are required to reduce the intensity to 10 percent of the incident intensity?

83 ••• Equation 31-14 gives the relation between the angle of deviation ϕ_d of a light ray incident on a spherical drop of water in terms of the incident angle θ_1 and the index of refraction of water. (*a*) Assume that $n_{air} = 1$ and derive an expression for $d\phi_d/d\theta_1$. *Hint: If* $y = \sin^{-1}x$, *then* $dy/dx = (1 - x^2)^{-1/2}$. (*b*) Use the result to show that the angle of incidence for minimum deviation θ_{1m} is given by $\cos\theta_{1m} = \sqrt{\frac{1}{3}(n^2 - 1)}$. (*c*) The index of refraction for a certain red light in water is 1.3318 and the index of refraction for a certain blue light in water is 1.3435. Use the result of Part (*a*) to find the angular separation of these colors in the primary rainbow.

84 ••• Show that the angle of deviation δ is a minimum if the angle of incidence is such that the ray and the bisector of the apex angle α (Figure 31-62) intersect at right angles.

CHAPTER 32

THIS IMAGE IS OF A FEMALE MOSQUITO AND WAS PRODUCED BY A MICROSCOPE. *(Nuridsany & Perennou/ Photo Researchers.)*

Optical Images

? How do you determine the magnifying power of a compound microscope? (See Example 32-15.)

Because the wavelength of light is very small compared with most obstacles and openings, diffraction—the bending of waves around corners—is often negligible, and the ray approximation, in which waves are considered to propagate in straight lines, accurately describes observations.

In this chapter, we apply the laws of reflection and refraction to the formation of images by mirrors and lenses.

32-1 MIRRORS

PLANE MIRRORS

Figure 32-1 shows a bundle of light rays emanating from a point source P and reflected from a plane mirror. After reflection, the rays diverge exactly as if they came from a point P' behind the plane of the mirror. The point P' is called the **image** of the **object** P. When the reflected rays enter the eye, they cannot be distinguished from rays diverging from a source at P' when no mirror is present. This image at P' is called a **virtual image,** virtual because the light does not really emanate from it.

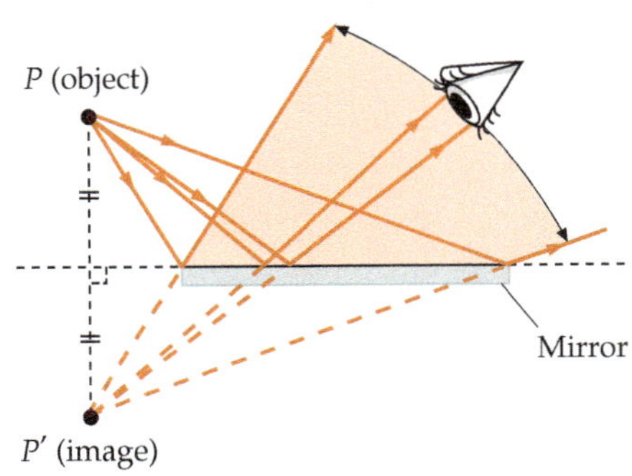

FIGURE 32-1 Image formed by a plane mirror. The rays from point P that strike the mirror and enter the eye appear to come from the image point P' behind the mirror. The image can be seen by an eye located anywhere in the shaded region.

The plane of the mirror is the perpendicular bisector of the line from the object point P to the image point P' as shown. The image can be seen by an eye located anywhere in the shaded region indicated, in which a straight line from the image to the eye passes through the mirror. The object need not be directly in front of the mirror. As long as the object is not located behind the plane of the mirror, some location exists at which the eye can be positioned so as to view the image.

If you stand in front of a mirror and hold up your right hand, the image you see is neither magnified nor reduced, but it looks like a left hand (Figure 32-2). This right-to-left reversal is a result of **depth inversion**—the hand is transformed from a right hand to a left hand because the front and the back of the hand are reversed by the mirror. Depth inversion is also illustrated in Figure 32-3. Figure 32-4 shows the image of a simple rectangular coordinate system. The mirror transforms a right-handed coordinate system, for which $\hat{i} \times \hat{j} = \hat{k}$, into a left-handed coordinate system, for which $\hat{i} \times \hat{j} = -\hat{k}$.

Figure 32-5 shows an arrow of height y standing parallel to a plane mirror a distance s from the mirror. We can locate the image of the tip of the arrowhead (and of any other point on the arrow) by drawing two rays. One ray, drawn perpendicular to the mirror, hits the mirror at point A and is reflected back onto itself. The other ray, making an angle θ with the normal to the mirror, is reflected, making an equal angle θ with the x axis. The extension of these two rays back behind the mirror locates the image of the arrowhead, as shown by the dashed lines in the figure. We can see from this figure that the image is the same distance behind the mirror as the object is in front of the mirror, and that the image is upright (the image of the arrow points in the same direction as the object) and is the same size as the object.

The formation of multiple images by two flat mirrors whose planes make an angle with each other is illustrated in Figure 32-6. We frequently see multiple images like this in clothing stores that provide mirrors for this purpose. The light from source point P that is reflected from mirror 1 strikes mirror 2 just as if it came from the image point P'_1. The image P'_1 is the object for mirror 2. Its image is behind the plane of mirror 2 at point P''_{12}. This image will be formed as long as the image point P'_1 is in front of the plane of mirror 2. The image at point P'_2 is due to rays from P that reflect directly from mirror 2. Because P'_2 is behind the plane of mirror 1, it cannot serve as an object point for an additonal image in mirror 1. The image at point P'_2 cannot serve as an object for mirror 1 because the geometry dictates that none of the rays from P that reflect directly from mirror 2 can then strike mirror 1.

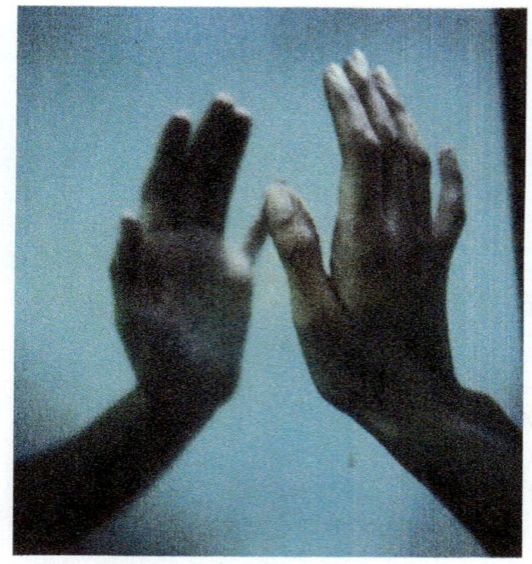

FIGURE 32-2 The image of a right hand in a plane mirror is transformed to a left hand. This right-to-left reversal is a result of depth inversion. *(Demitrios Zangos.)*

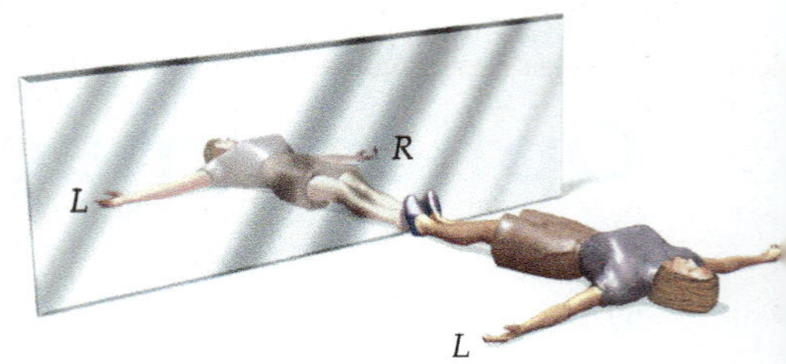

FIGURE 32-3 A person lying down with her feet against the mirror. The image is depth inverted.

Do not make the mistake of thinking that a mirror is an inversion in the left to right direction. Instead, it is an inversion in the front to back direction.

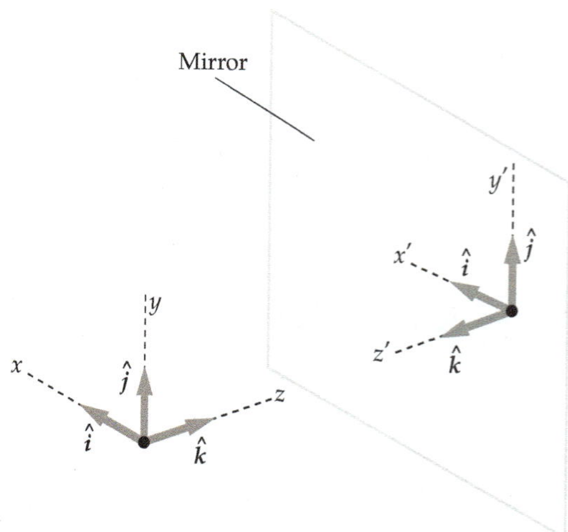

FIGURE 32-4 Image of a rectangular coordinate system in a plane mirror. The arrow along the z axis is reversed in the image. The image of the original right-handed coordinate system, for which $\hat{i} \times \hat{j} = \hat{k}$, is a left-handed coordinate system, for which $\hat{i} \times \hat{j} = -\hat{k}$.

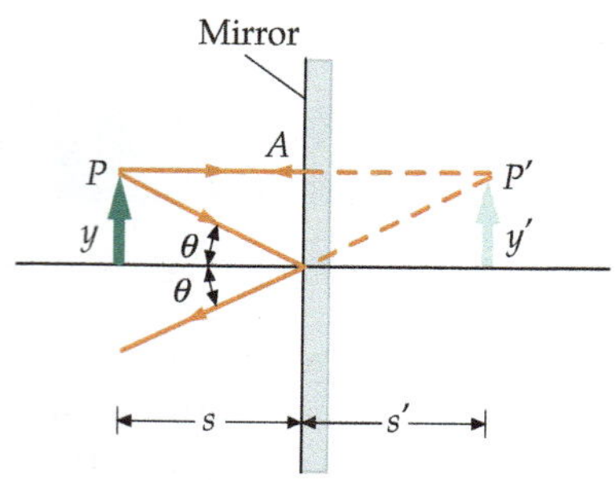

FIGURE 32-5 Ray diagram for locating the image of an arrow in a plane mirror.

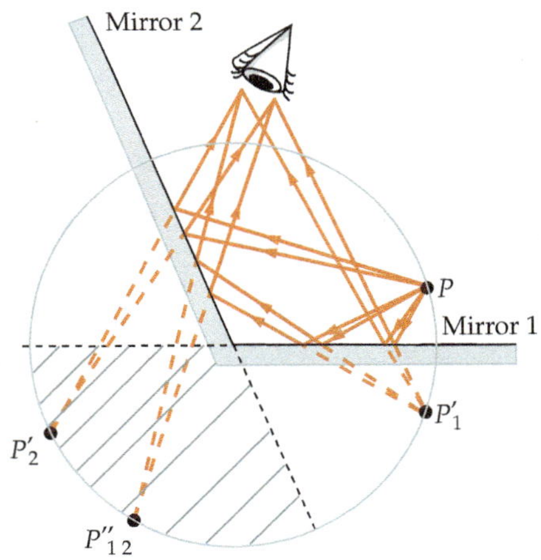

FIGURE 32-6 Images formed by two plane mirrors. P'_1 is the image of the object P in mirror 1, and P'_2 is the image of the object in mirror 2. Point P''_{12} is the image of P'_1 in mirror 2, which is seen when light rays from the object reflect first from mirror 1 and then from mirror 2. The image P'_2 does not have an image in mirror 1 because it is behind that mirror.

> **CONCEPT CHECK 32-1**
>
> Show that a source point and all consequent image points formed by two flat mirrors are equidistant from the intersection of the planes of the two mirrors. (The circle shown in Figure 32-6 is equidistant from such an intersection.)

> **CONCEPT CHECK 32-2**
>
> Which of the images of himself can Ben see?

An alternative way of stating this is that because P'_2 is behind the plane of mirror 1, the image at P'_2 cannot serve as an object for mirror 1. The number of images formed by two mirrors depends on the angle between the mirrors and the position of the object.

Suppose your friend Ben is standing at point P (Figure 32-6) and is wearing a sweatshirt with BEN printed on it. In addition, suppose Ben is facing the intersection of the two mirrors and waving his right hand. Also, suppose that you are standing at the location of the eye. You can see an image of Ben at all three image locations. For the images at P'_1 and P'_2, Ben is waving his left hand and the printing on his sweatshirt appears as ИƎB. However, for the image at P''_{12} Ben is waving his right hand and the printing appears as BEN. For the image at P''_{12} depth inversion occurs twice, once for each reflection, so the result is as if no depth inversion occurs.

Figure 32-7 illustrates the fact that a horizontal ray reflected from two mutually perpendicular vertical mirrors is reflected back along a parallel path no matter what angle the ray makes with the mirrors. If three mirrors are mutually perpendicular to each other, like the sides of an inside corner of a box, any ray incident on any one of the mirrors from any direction is reflected back on a path parallel to the incident ray. A set of three mirrors arranged in this manner is called a corner-cube reflector. An array of corner-cube reflectors was placed on the moon in the Sea of Tranquility by the *Apollo 11* astronauts in 1969. A laser beam from Earth that is directed at the mirrors is reflected back to the same place on Earth. Such a beam has been used to measure the distance from the laser to the mirrors to within a few centimeters by measuring the time it takes for the light to travel to the mirrors and return.

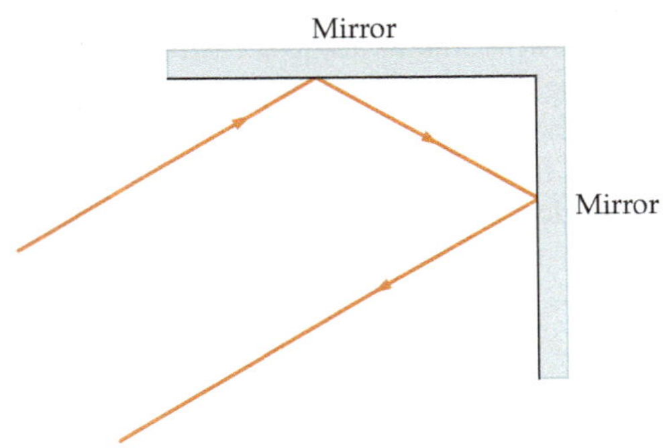

FIGURE 32-7 A ray striking one of two perpendicular plane mirrors is reflected from the second mirror in the direction opposite the original direction for any angle of incidence. The plane of the rays is perpendicular to both mirrors.

SPHERICAL MIRRORS

Figure 32-8 shows a bundle of rays from a point source P on the axis of a concave spherical mirror reflecting from the mirror and converging at point P'. (A concave mirror is shaped like a cave when you look into it.) The rays then diverge from point P', just as if there were an object at the point P'. This image is called a **real image,** because light really does emanate from the image point. The image can be seen by an eye at the left of the image looking into the mirror. It can also be observed on a small viewing screen* or on a small piece of photographic film placed at the image point.

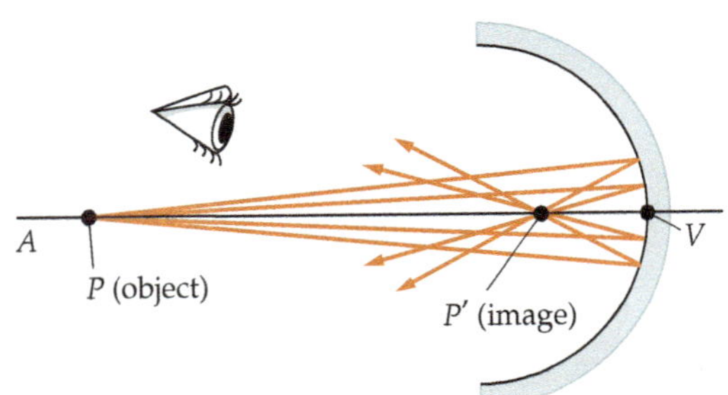

FIGURE 32-8 Rays from a point object P on the axis AV of a concave spherical mirror form an image at P'. The image is sharp if the rays strike the mirror near the axis and if the rays are almost parallel with the axis.

* A viewing screen must produce either diffuse reflection or diffuse transmission of the light. Ground glass is commonly used for this purpose. The screen must be relatively small, so that some of the light from the source reaches the mirror without being blocked by the screen.

A virtual image, such as that formed by a plane mirror as discussed in the previous section, cannot be observed on a screen at the image point because no light from the object point exists at the image point. Despite this distinction between real and virtual images, the eye makes no distinction between them. The light rays diverging from a real image and those appearing to diverge from a virtual image are the same to the eye.

From Figure 32-9, we can see that only rays that strike the spherical mirror at points near the axis (line AV) are reflected through the image point. Rays that are almost parallel with the axis and are near to the axis are called **paraxial rays.** Rays that strike the mirror at points far from the axis upon reflection pass near the image point, but not through it. Such rays cause the image to appear blurred, an effect called **spherical aberration.** The image can be sharpened by blocking off all but the central part of the mirror, so that rays far from the axis do not strike it. The image is then sharper, but its brightness is reduced because less light is reflected to the image point.

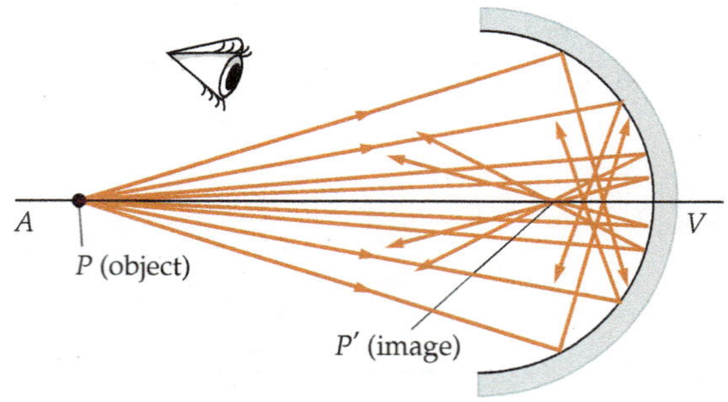

FIGURE 32-9 Spherical aberration of a mirror. Nonparaxial rays that strike the mirror at points far from the axis AV are not reflected through the image point P' formed by the paraxial rays. The nonparaxial rays blur the image.

In order to describe image formation, we wish to obtain an equation relating the position of the image point to the position of the object point. To do this, we draw two rays (Figure 32-10*a*) from an arbitrarily positioned object point P. One ray passes through point C, the center of curvature of the mirror, and the other ray strikes point A, an arbitrarily positioned point on the mirror. The image point P' is where these two rays intersect after reflecting off the mirror. Using the law of reflection, we obtain the location of P'. The ray passing through point C strikes the mirror at normal incidence, so the ray reflects back upon itself. The ray striking the mirror at A makes angle θ with the normal, so, as shown, the reflected ray also makes angle θ with the normal. (Any line normal to a spherical surface passes through the center of curvature.) The image distance s' and object distance s are measured from the plane tangent to the mirror at its vertex V. The angle β is an exterior angle to the triangle PAC, therefore, $\beta = \alpha + \theta$. Similarly, from the triangle PAP', $\gamma = \alpha + 2\theta$. Eliminating θ from these equations gives

$$\alpha + \gamma = 2\beta$$

See **Math Tutorial** *for more information on* **Geometry and Trigonometry**

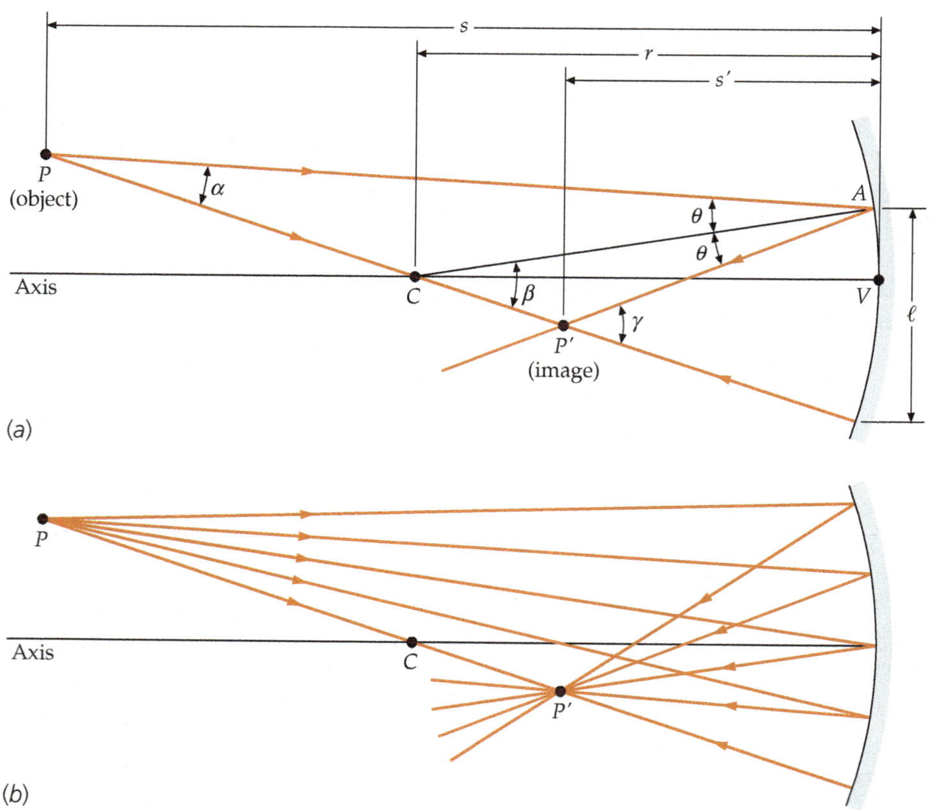

FIGURE 32-10 (a) Geometry for calculating the image distance s' from the object distance s and the radius of curvature r. The angle β is an exterior angle to the triangle PAC; therefore, $\beta = \alpha + \theta$. Similarly, from the triangle PAP', $\gamma = \alpha + 2\theta$. Eliminating θ from these equations gives $2\beta = \alpha + \gamma$. Equation 32-1 follows directly, if we assume the following small-angle approximations: $\alpha \approx \ell/s$, $\beta \approx \ell/r$, and $\gamma \approx \ell/s'$. (b) All paraxial rays from object point P pass through image point P' after reflecting off the mirror.

By assuming all rays are paraxial, we can substitute using the small-angle approximations: $\alpha \approx \ell/s$, $\beta \approx \ell/r$, and $\gamma \approx \ell/s'$. Equation 32-1 follows directly:

$$\frac{1}{s} + \frac{1}{s'} = \frac{2}{r} \qquad 32\text{-}1$$

This equation relates the object and image distances with the radius of curvature. The striking thing about this equation is that it has absolutely no information about the location of point A. Therefore, the equation is valid for *any* choice for the location of point A, as long as point A is on the surface of the mirror and all rays are paraxial. That is, as shown in Figure 32-10*b*, *all* paraxial rays emanating from an object point will, upon reflection, pass through a *single* image point.

Equation 32-1 specifies the image position in terms of its distance from the mirror. We now specify the image position in terms of its distance from the axis. We first draw a single ray (Figure 32-11) that reflects off the mirror at its vertex. The two right triangles formed are similar. Corresponding sides of similar triangles are equal, so

$$\frac{y'}{y} = -\frac{s'}{s} \qquad 32\text{-}2$$

The minus sign takes into account that y'/y is negative as P and P' are on opposite sides of the axis. Thus, if y is positive, y' is negative and if y is negative, y' is positive.

PRACTICE PROBLEM 32-1

For the image point and object point shown in Figure 32-11, show that

$$\frac{y'}{y} = -\frac{r/2}{s - (r/2)}$$

Hint: Solve Equation 32-1 for s′ and substitute your result into Equation 32-2.

When the object distance is large compared with the radius of curvature of the mirror, the term $1/s$ in Equation 32-1 is much smaller than $2/r$ and can be neglected. That is, as $s \to \infty$, $s' \to \frac{1}{2}r$, where s' is the image distance. This distance is called the **focal length** f of the mirror, and the plane on which parallel rays incident on the mirror are focused is called the **focal plane.** The intersection of the axis with the focal plane is called the **focal point** F, as illustrated in Figure 32-12*a*. (Again, only paraxial rays are focused at a single point.)

$$f = \tfrac{1}{2}r \qquad 32\text{-}3$$

FOCAL LENGTH FOR A MIRROR

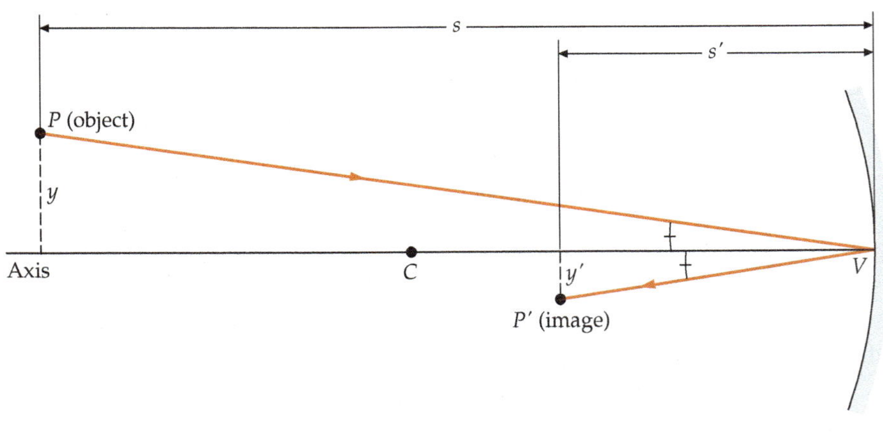

FIGURE 32-11 Geometry for calculating the position y' of the image point with respect to its distance from the axis.

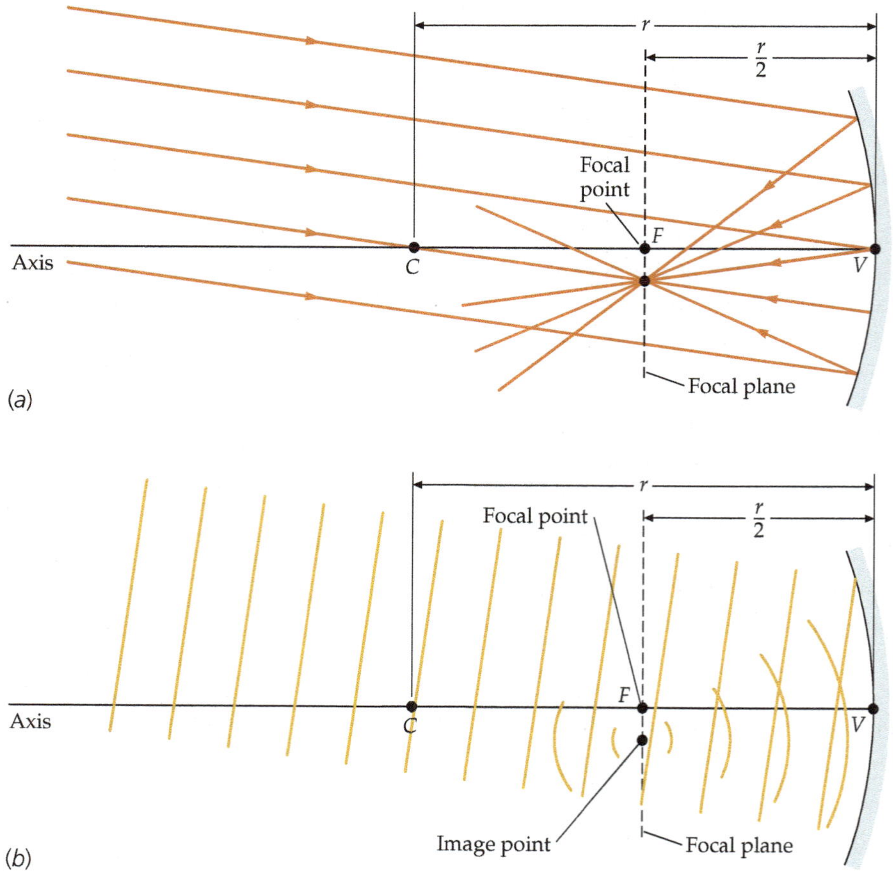

FIGURE 32-12 (*a*) Parallel rays strike a concave mirror and are reflected to a point on the focal plane a distance $\frac{1}{2}r$ to the left of the mirror. (*b*) The incoming wavefronts are planes; upon reflection, they become spherical wavefronts that converge to, and then diverge from, the image point.

PRACTICE PROBLEM 32-2

Show that solving Equation 32-1 for s' gives

$$s' = \frac{r}{2 - \frac{r}{s}}$$

Then show that as $s \rightarrow \infty$, $s' \rightarrow \frac{1}{2}r$.

The focal length of a spherical mirror is half the radius of curvature. In terms of the focal length f, Equation 32-1 is

$$\frac{1}{s} + \frac{1}{s'} = \frac{1}{f} \qquad \text{32-4}$$

MIRROR EQUATION

Equation 32-4 is called the **mirror equation.**

When an object point is very far from the mirror, the rays reaching the mirror are approximately parallel, and the wavefronts are approximately planes (Figure 32-12*b*). In Figure 32-12*b*, note that the last part of each wavefront to reflect from the concave mirror surface is the part just below the vertex V. This results in a spherical wavefront upon reflection. Figure 32-13 shows both the wavefronts and the rays for plane waves striking a convex mirror. In this case, the central part of the wavefront strikes the mirror first, and the reflected waves appear to come from the focal point behind the mirror.

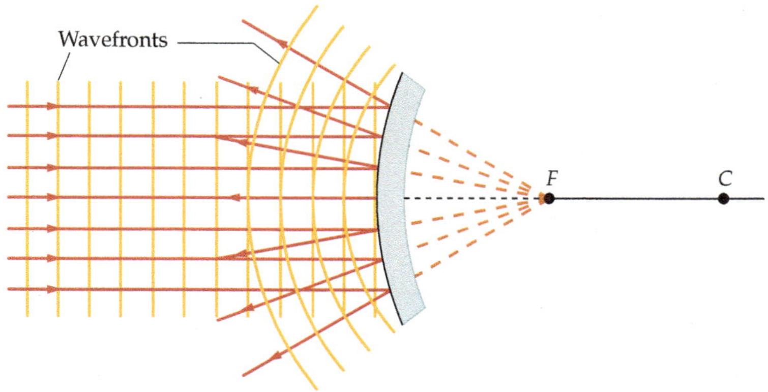

FIGURE 32-13 Reflection of plane waves from a convex mirror. The outgoing wavefronts are spherical, as if emanating from the focal point F behind the mirror. The rays are normal to the wavefronts and appear to diverge from F.

Figure 32-14 illustrates a property of waves called **reversibility.** If we reverse the direction of a reflected ray, the law of reflection assures us that the reflected ray will be along the original incoming ray, but in the opposite direction. (Reversibility also holds for refracted rays, which are discussed in later sections.) Thus, if we have a real image of an object formed by a reflecting (or refracting) surface, we can place an object at the image point and a new, real image will be formed at the position of the original object.

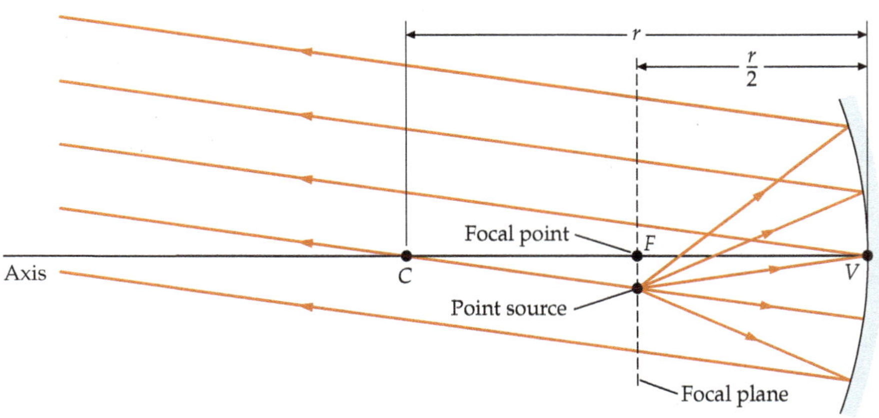

FIGURE 32-14 Reversibility. Rays diverging from a point source on the focal plane of a concave mirror are reflected from the mirror as parallel rays. The rays follow the same paths as in Figure 32-12*a* but in the reverse direction.

Example 32-1 Image in a Concave Mirror

A point source is 12 cm from a concave mirror and 3.0 cm above the axis of the mirror. The radius of curvature of the mirror is 6.0 cm. Find (*a*) the focal length of the mirror and (*b*) the image distance. (*c*) Find the position of the image relative to the axis.

PICTURE The focal length of a spherical mirror is half the radius of curvature. Once the focal length is known, the image distance can be found using the mirror equation (Equation 32-4), and the distance of the image from the axis can be found using Equation 32-2. The image distance from the mirror is the distance from the plane tangent to the mirror at its vertex.

SOLVE

(*a*) The focal length is half the radius of curvature:

$$f = \tfrac{1}{2}r = \tfrac{1}{2}(6.0\text{ cm}) = \boxed{3.0\text{ cm}}$$

(*b*) 1. Use the mirror equation to find a relation for the image distance s':

$$\frac{1}{s} + \frac{1}{s'} = \frac{1}{f}$$

or

$$\frac{1}{12\text{ cm}} + \frac{1}{s'} = \frac{1}{3.0\text{ cm}}$$

2. Solve for s':

$$\frac{1}{s'} = \frac{4}{12\text{ cm}} - \frac{1}{12\text{ cm}} = \frac{3}{12\text{ cm}}$$

$$s' = \boxed{4.0\text{ cm}}$$

(*c*) 1. Use Equation 32-2 to find the distance y' of image from the axis:

$$\frac{y'}{y} = -\frac{s'}{s}$$

2. Solve for y':

$$y' = -\frac{s'}{s}y = -\frac{4.0\text{ cm}}{12\text{ cm}}(3.0\text{ cm})$$

$$= \boxed{-1.0\text{ cm}}$$

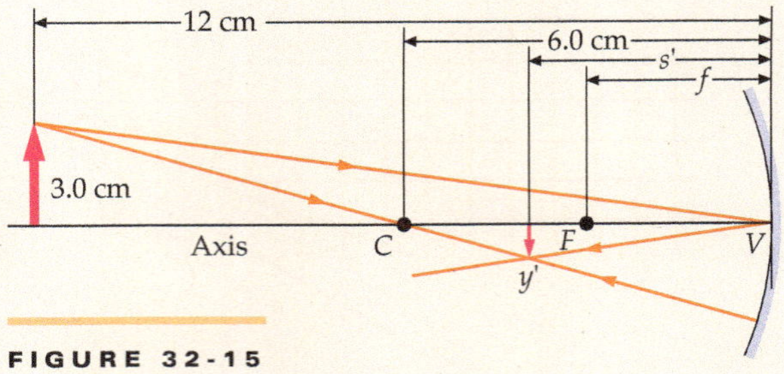

FIGURE 32-15

CHECK In Figure 32-15, two rays from the tip of the arrow have been drawn to locate the corresponding point on the image. By choosing to draw the ray that passes through *C* and the ray that reflects off the mirror at *V*, we have chosen to draw two rays that are easy to trace. From this figure, we can see that the results of the solution are very plausible.

PRACTICE PROBLEM 32-3 A concave mirror has a focal length of 4.0 cm. (*a*) What is the mirror's radius of curvature? (*b*) Find the image distance for an object 2.0 cm from the mirror.

CONCEPT CHECK 32-3

What is the radius of curvature of a plane mirror?

RAY DIAGRAMS FOR MIRRORS

A useful method to locate images is by geometric construction of a **ray diagram,** as illustrated in Figure 32-16, where the object is a human figure perpendicular to the axis a distance *s* from the mirror. By the judicious choice of rays from the head of the figure, we can quickly locate the image. Of the infinitely many rays, there are three rays, **principal rays,** that are particularly convenient to use:

1. The **parallel ray,** drawn parallel to the axis. This ray is reflected through the focal point.
2. The **focal ray,** drawn through the focal point. This ray is reflected parallel to the axis.
3. The **radial ray,** drawn through the center of curvature. This ray strikes the mirror perpendicular to its surface and is thus reflected back on itself.

PRINCIPAL RAYS FOR A MIRROR

These rays are shown in Figure 32-16. The intersection of any two paraxial rays locates the image point of the head. The three principal rays are easier to draw than any of the other rays. Typically, you draw two of the principal rays to locate the image, and then draw the third principal ray as a check to verify the result. Ray diagrams are best drawn with the mirror replaced by a straight line that extends as far as necessary to intercept the rays, as shown in Figure 32-17. (Note that the image in this case is real, inverted, and smaller than the object.)

When the object is between the mirror and its focal point, the rays from an object point reflected from the mirror do not converge but appear to diverge from a point behind the mirror,

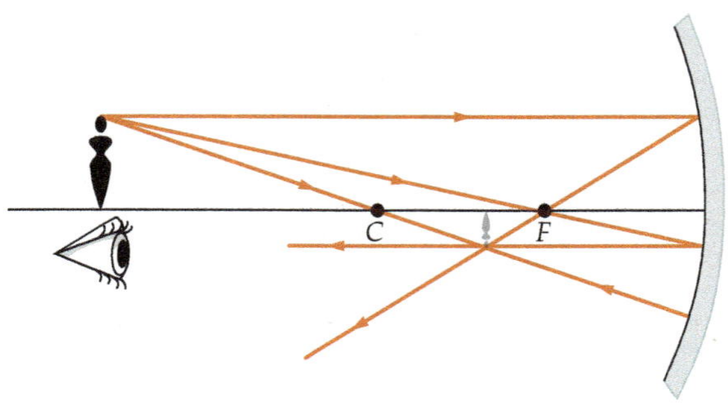

FIGURE 32-16 Ray diagram for the location of the image by geometric construction.

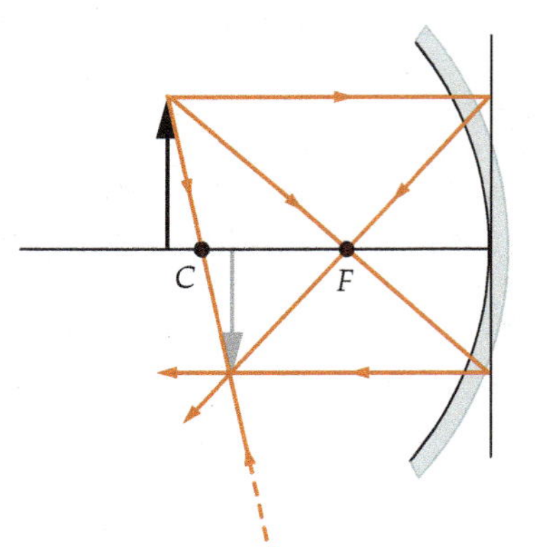

FIGURE 32-17 Ray diagrams are easier to construct if the curved surface is replaced by a plane tangent to the surface at the vertex.

as illustrated in Figure 32-18. In this case, the image is virtual and upright (*upright* meaning not inverted relative to the object). For an object between the mirror and the focal point, s is less than $r/2$, so the image distance s' calculated from Equation 32-1 turns out to be negative. We can apply Equations 32-1, 32-2, 32-3, and 32-4 to this case and to convex mirrors if we adopt a convenient sign convention. Whether the mirror is convex or concave, real images can be formed only in front of the mirror, that is, on the same side of the mirror as the reflected light (and the object). Virtual images are formed behind the mirror where there is no actual light from the object. Our sign convention is as follows:

1. s is positive if the object is on the incident-light side of the mirror.
2. s' is positive if the image is on the reflected-light side of the mirror.
3. r (and thus f) is positive if the mirror is concave so the center of curvature is on the reflected-light side of the mirror.

SIGN CONVENTIONS FOR REFLECTION

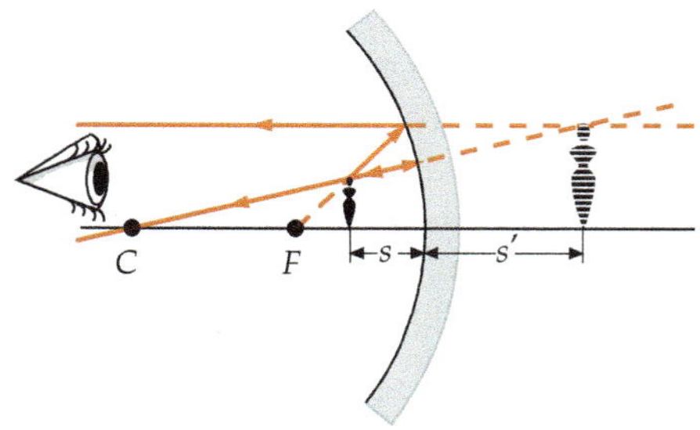

FIGURE 32-18 A virtual image is formed by a concave mirror when the object is inside the focal point. Here the image is located by the radial ray, which is reflected back on itself, and the focal ray, which is reflected parallel to the axis. The two reflected rays appear to diverge from an image point behind the mirror. This image point is found by constructing extensions to the reflected rays.

The incident-light side and the reflected-light side are, of course, the same side of the mirror. The parameters s, s', r, and f are all positive if a real object* is in front of a concave mirror that forms a real image. A parameter is negative if it does not meet the stated condition for being positive.

The ratio of the image height to the object height is defined as the **lateral magnification** m of the image. From Figure 32-19 and Equation 32-2, we see that the lateral magnification is

$$m = \frac{y'}{y} = -\frac{s'}{s} \qquad \text{32-5}$$

LATERAL MAGNIFICATION

A negative magnification, which occurs when both s and s' are positive, indicates that the image is inverted.

For plane mirrors, the radius of curvature is infinite. The focal length given by Equation 32-3 is then also infinite. Equation 32-4 then gives $s' = -s$, indicating that the image is behind the mirror at a distance equal to the object distance. The magnification given by Equation 32-5 is then $+1$, indicating that the image is upright and the same size as the object.

Although the preceding equations, coupled with our sign conventions, are relatively easy to use, we often need to know only the approximate location and magnification of the image and whether it is real or virtual and upright or inverted. This knowledge is usually easiest to obtain by constructing a ray diagram. It is always a good idea to use both this graphical method and the algebraic method to locate an image, so that one method serves as a check on the results of the other.

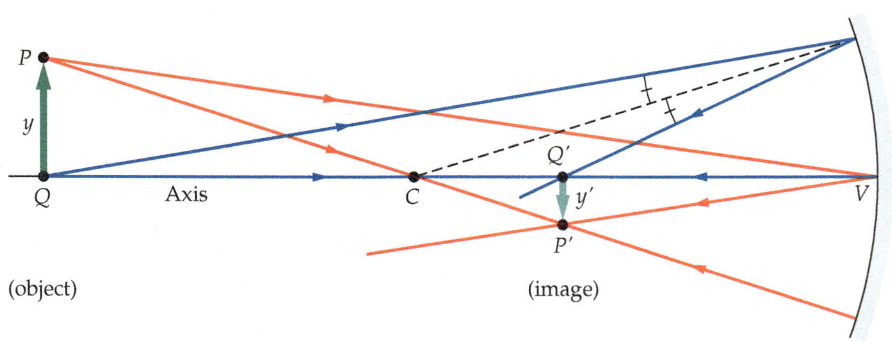

FIGURE 32-19 Geometry for showing the lateral magnification. Rays from the top of the object at P, upon reflection, intersect at P'; and rays from the bottom of the object at Q intersect at Q', where points P and P' have vertical positions y and y', respectively. The lateral magnification m is given by the ratio y'/y. In accord with Equation 32-2, $y'/y = -s'/s$. The minus sign results from the fact that y'/y is negative when s and s' are both positive. A negative m means the image is inverted.

* An object is real if it is on the same side of the mirror as the incident light.

Convex mirrors Figure 32-20 shows a ray diagram for an object in front of a convex mirror. The central ray heading toward the center of curvature C is perpendicular to the mirror and is reflected back on itself. The parallel ray is reflected as if it came from the focal point F behind the mirror. The focal ray (not shown) would be drawn toward the focal point and would be reflected parallel to the axis. We can see from the figure that the image is behind the mirror and is therefore virtual. The image is also upright and smaller than the object.

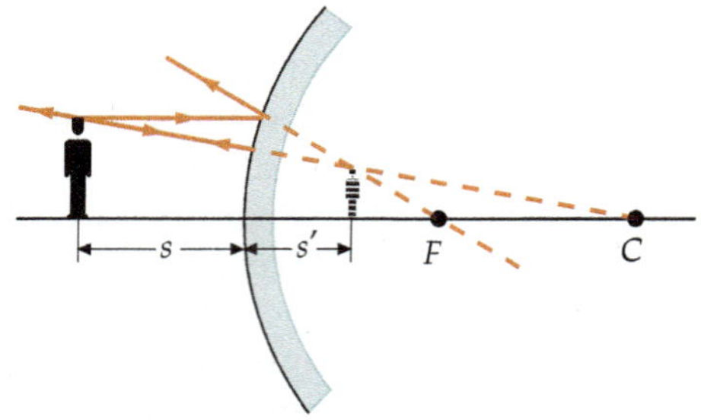

FIGURE 32-20 Ray diagram for an object in front of a convex mirror.

Example 32-2 Image in a Convex Mirror

An object that is 2.0 cm high is 10 cm from a convex mirror that has a radius of curvature equal to 10 cm. (*a*) Locate the image and (*b*) find the height of the image.

PICTURE The ray diagram for this problem is the same as shown in Figure 32-20. From the figure, we see that the image is upright, virtual, and smaller than the object. To find the exact location and height of the image, we use the mirror equation, with $s = 10$ cm and $r = -10$ cm.

SOLVE

(*a*) 1. The image distance s' is related to the object distance s and the focal length f by the mirror equation:

$$\frac{1}{s} + \frac{1}{s'} = \frac{1}{f}$$

2. Calculate the focal length of the mirror:

$$f = \tfrac{1}{2}r = \tfrac{1}{2}(-10\text{ cm}) = -5.0\text{ cm}$$

3. Substitute $s = 10$ cm and $f = -5.0$ cm into the mirror equation to find the image distance:

$$\frac{1}{10\text{ cm}} + \frac{1}{s'} = \frac{1}{-5.0\text{ cm}}$$

4. Solve for s':

$$s' = \boxed{-3.3\text{ cm}}$$

(*b*) 1. The height of the image is m multiplied by the height of the object:

$$y' = my$$

2. Calculate the magnification m:

$$m = -\frac{s'}{s} = -\frac{-3.3\text{ cm}}{10\text{ cm}} = +0.33$$

3. Use m to find the height of the image:

$$y' = my = (0.33)(2.0\text{ cm}) = \boxed{0.67\text{ cm}}$$

CHECK The image distance is negative, indicating a virtual image behind the mirror. The magnification is positive and less than one, indicating that the image is upright and smaller than the object. The results are in agreement with the information obtained from the ray diagram (Figure 32-21).

PRACTICE PROBLEM 32-4 Find the image distance and lateral magnification for an object 5.0 cm away from the mirror in Example 32-2.

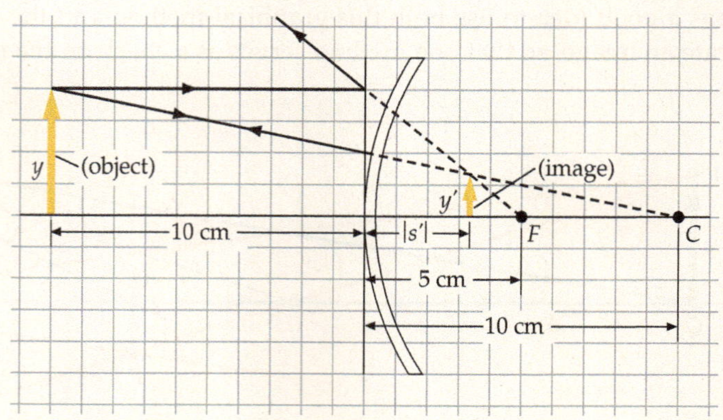

FIGURE 32-21

Example 32-3 Determining the Range

Context-Rich

You have a part-time job at Pleasant Hills Golf Course. The fairway of the 16th hole is horizontal for the first 50 yd and then goes down a not-too-steep hill (Figure 32-22), so the people on the tee cannot see the party in front. To prevent people from driving off the tee into the party in front, a convex mirror is mounted on a pole, enabling golfers on the tee to see whether or not the party in front is out of range.* Your boss says that a range finder that works by triangulation could be placed facing the mirror, so the golfers could measure how far the image of the party in front is behind the mirror. Then the golfers could be given a chart telling them how far the next party is from the tee. Your boss knows you are taking a physics course, so he asks you to calculate the distance of the image behind the mirror if the next party is 250 yd from the tee. The radius of curvature of the mirror has a magnitude of 20.0 yd.

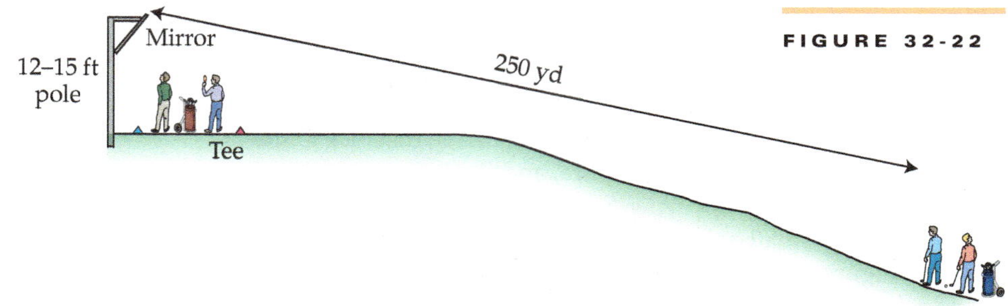

FIGURE 32-22

PICTURE The image distance is related to the object distance by the mirror formula, and the focal length of the lens is half the radius of curvature.

SOLVE

1. Use the mirror equation. For a convex mirror, the radius of curvature is negative:

$$\frac{1}{s} + \frac{1}{s'} = \frac{1}{f}$$

and

$$f = \frac{2}{r}$$

so

$$\frac{1}{250\text{ yd}} + \frac{1}{s'} = \frac{2}{-20.0\text{ yd}}$$

2. The image is 9.62 yd behind the mirror:

$$s' = \boxed{-9.62\text{ yd}}$$

CHECK That the step-2 result is negative is as expected. That is, it was expected that the image would be behind the mirror.

PRACTICE PROBLEM 32-5 What is the distance to the party in front if the image is 9.75 yd behind the mirror?

(a)

(b)

(*a*) A convex mirror resting on paper that has equally spaced parallel stripes. Note the large number of lines imaged in a small space and the reduction in size and distortion in shape of the image. (*b*) A convex mirror is used for security in a store. *(Richard Megna/Fundamental Photographs.)*

* The mirror was not replaced after it was knocked down during a storm a couple of years ago.

32-2 LENSES

IMAGES FORMED BY REFRACTION

One end of a long transparent cylinder is machined and polished to form a convex spherical surface. Figure 32-23 illustrates the formation of an image by refraction at such a surface. Suppose the cylinder is submerged in a transparent liquid that has an index of refraction n_1, and suppose the cylinder is made of a plastic material that has an index of refraction n_2, where n_2 is greater than n_1. Again, only in the paraxial limit do rays from an object point converge to one point. An equation relating the image distance to the object distance, the radius of curvature, and the indices of refraction can be derived by applying Snell's law of refraction to the rays and using small-angle approximations. The geometry is shown in Figure 32-24. The angles θ_1 and θ_2 are related by Snell's law of refraction: $n_1 \sin\theta_1 = n_2 \sin\theta_2$. Using the small-angle approximation ($\sin\theta = \theta$), Snell's law becomes $n_1\theta_1 = n_2\theta_2$. From triangle ACP', we have $\beta = \theta_2 + \gamma = (n_1/n_2)\theta_1 + \gamma$; and from triangle PAC, we have $\theta_1 = \alpha + \beta$. Eliminating θ_1 from these two equations gives $n_1\alpha + n_2\gamma = (n_2 - n_1)\beta$. Substituting for α, β and γ and using the small-angle approximations $\alpha \approx \ell/s$, $\beta \approx \ell/r$, and $\gamma = \ell/s'$ gives

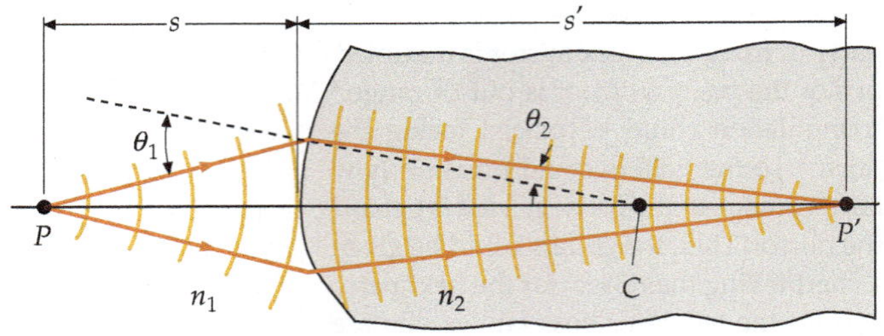

FIGURE 32-23 Image formed by refraction at a spherical surface between two media where the waves move slower in the second medium.

$$\frac{n_1}{s} + \frac{n_2}{s'} = \frac{n_2 - n_1}{r} \tag{32-6}$$

REFRACTION AT A SINGLE SURFACE

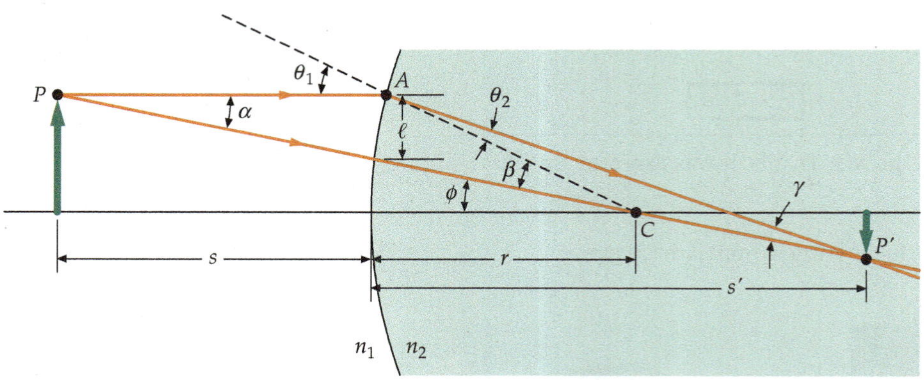

FIGURE 32-24 Geometry for relating the image position to the object position for refraction at a single spherical surface.

In refraction, real images are formed in back of the surface, which we will call the refracted-light side, whereas virtual images occur on the incident-light side, in front of the surface. The sign conventions we use for refraction are similar to those for reflection:

1. s is positive for objects on the incident-light side of the surface.
2. s' is positive for images on the refracted-light side of the surface.
3. r is positive if the center of curvature is on the refracted-light side.

SIGN CONVENTIONS FOR REFRACTION*

* The sign convention of choice for advanced work on optical design is the Cartesian sign convention. It can readily be found on the Internet.

We see that parameters s, s', and r are all positive if a real object is in front of a convex refracting surface that forms a real image. A parameter is negative if it does not meet the stated condition for being positive.

Example 32-4 Magnification by a Refracting Surface *Try It Yourself*

Derive an expression for the magnification $m = y'/y$ of an image formed by a spherical refracting surface.

PICTURE The magnification is the ratio of y' to y. Using Figure 32-19 and Figure 32-24 as guides, draw a ray diagram suitable for this derivation. The heights are related to the tangents of the angles θ_1 and θ_2, as shown in Figure 32-25. The angles are related by Snell's law. For paraxial rays, you can use the approximations $\tan\theta \approx \sin\theta \approx \theta$, and $\cos\theta \approx 1$.

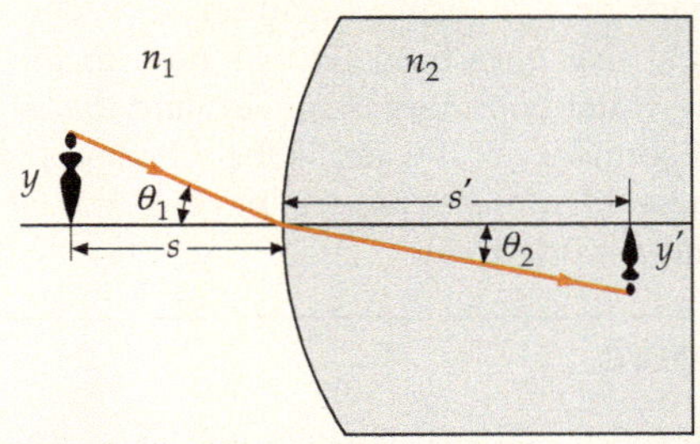

FIGURE 32-25

SOLVE

Cover the column to the right and try these on your own before looking at the answers.

Steps:	Answers
1. Using Figure 32-19 and Figure 32-24 as guides, draw a ray diagram suitable for this derivation. The drawing should include an object, a real image, a refracting surface, and an axis. Then draw an incident ray from the top of the object to the intersection of the axis with the refracting surface, and draw the refracted ray to the corresponding image point (Figure 32-25).	
2. Write expressions for $\tan\theta_1$ and $\tan\theta_2$ in terms of the heights y and $-y'$ and the object and image distances s and s'. (Because y' is negative, use $-y'$, so that $\tan\theta_2$ is positive.)	$\tan\theta_1 = \frac{y}{s}$; $\tan\theta_2 = \frac{-y'}{s'}$
3. Apply the small-angle approximation $\tan\theta \approx \theta$ to your expressions.	$\theta_1 = \frac{y}{s}$; $\theta_2 = \frac{-y'}{s'}$
4. Write Snell's law of refraction relating the angles θ_1 and θ_2 using the small-angle approximation $\sin\theta \approx \theta$.	$n_1 \sin\theta_1 = n_2 \sin\theta_2$ $n_1\theta_1 = n_2\theta_2$
5. Substitute the expressions for θ_1 and θ_2 found in step 3.	$n_1\left(\frac{y}{s}\right) = n_2\left(\frac{-y'}{s'}\right)$
6. Solve for the magnification $m = y'/y$.	$m = \frac{y'}{y} = -\frac{n_1 s'}{n_2 s}$

CHECK The step-6 result for the lateral magnification is dimensionless, as expected.

We see from Example 32-4 that the magnification due to refraction at a spherical surface is

$$m = \frac{y'}{y} = -\frac{n_1 s'}{n_2 s} \qquad 32\text{-}7$$

MAGNIFICATION FOR A REFRACTING BOUNDARY

Example 32-5 Image Seen from a Goldfish Bowl

Goldie the goldfish is in a 15.0-cm-radius spherical bowl of water that has an index of refraction of 1.33. Fluffy the cat is sitting on the table with her nose 10.0 cm from the surface of the bowl (Figure 32-26). The light from Fluffy's nose is refracted by the air–water boundary to form an image. Find (*a*) the image distance and (*b*) the magnification of the image of Fluffy's nose. Neglect any effect of the bowl's thin glass wall.

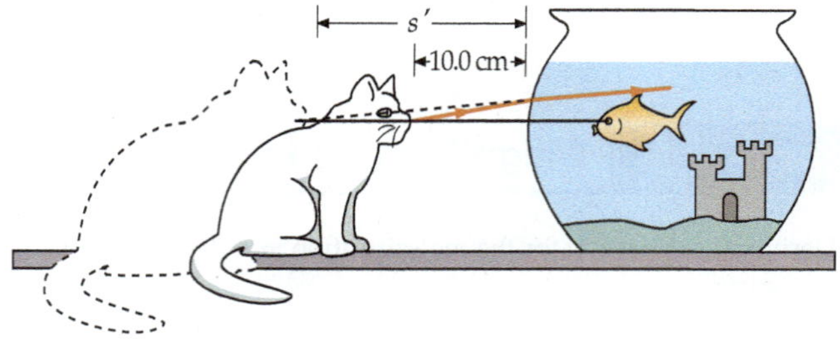

FIGURE 32-26 Goldie sees Fluffy's image farther from the bowl than Fluffy actually is.

PICTURE We find the image distance s' using Equation 32-6 and the magnification using Equation 32-7. Because we are interested in light that goes from Fluffy's nose to the bowl, it follows that the air–water boundary is convex, and that air is the incident-light side of boundary and water is the refracted-light side of boundary. With these identifications, we have $n_1 = 1.00$, $n_2 = 1.33$, $s = +10.0$ cm, and $r = +15.0$ cm.

SOLVE

(*a*) 1. The equation relating the object distance to the image distance is Equation 32-6:

$$\frac{n_1}{s} + \frac{n_2}{s'} = \frac{n_2 - n_1}{r}$$

2. Identify and assign signs to the parameters in the previous step:

$n_1 = 1.00$, $n_2 = 1.33$, $s = +10.0$ cm, and $r = +15.0$ cm

3. Substitute numerical values and solve for s':

$$\frac{1.00}{10.0\text{ cm}} + \frac{1.33}{s'} = \frac{1.33 - 1.00}{15.0\text{ cm}}$$

so

$$s' = \boxed{-17.1\text{ cm}}$$

(*b*) Substitute numerical values into Equation 32-7 to find the magnification m:

$$m = -\frac{n_1 s'}{n_2 s} = -\frac{(1.00)(-17.1\text{ cm})}{(1.33)(10.0\text{ cm})} = \boxed{1.29}$$

CHECK Because s' is negative, the image is virtual; that is, the image is on the side of the refracting surface opposite the refracted light, as shown in Figure 32-26. The fish, Goldie, would see Fluffy to be slightly farther away ($|s'| > s$) than she actually is, and larger ($|m| > 1$) than she actually is. That m is positive indicates the image is upright.

PRACTICE PROBLEM 32-6 If Goldie is 7.5 cm from the side of the bowl nearest Fluffy, find (*a*) the location and (*b*) the magnification of Goldie's image, as seen by Fluffy.

PRACTICE PROBLEM 32-7 The bowl is replaced by an aquarium that has flat sides and Goldie is 7.5 cm from the side where Fluffy is. Use Equation 32-6 to find the location of the image of Goldie that Fluffy sees.

Example 32-6 Image Seen from an Overhead Branch

During the summer months, Goldie the fish spends much of her time in a small pond in her owner's backyard. While enjoying a rest at the bottom of the 1.00-m-deep pond, Goldie is being watched by Fluffy the cat, who is perched on a tree limb 3.00 m above the surface of the pond. How far below the surface is the image of the fish that Fluffy sees? (The index of refraction of water is 1.33.)

PICTURE The surface of the pond is a spherical refracting surface that has a radius of curvature equal to the radius of Earth. (Neglecting the curvature of the surface of Earth, we use $r \approx \infty$.) Thus, Equation 32-6 applies. Because the light reaching Fluffy originates in the water, use $n_1 = 1.33$ and $n_2 = 1.00$.

SOLVE

1. Draw a picture of the situation. Label the object distance and the indices of refraction of the media. Goldie is the object (Figure 32-27):

2. Using Equation 32-6, relate the image position s' to the other relevant parameters:

$$\frac{n_1}{s} + \frac{n_2}{s'} = \frac{n_2 - n_1}{r}$$

3. The refracting surface is virtually flat. Using $r = \infty$, solve for s':

$$s' = -\frac{n_2}{n_1}s$$

4. Using the given values $n_1 = 1.33$, $n_2 = 1.00$ and $s = 1.00$ m, substitute to obtain s':

$$s' = -\frac{1}{1.33}(1.00 \text{ m}) = -0.752 \text{ m}$$

That the image is negative means that the image is on the side of the surface opposite the refracted light. That is, it is 0.752 m below the surface.

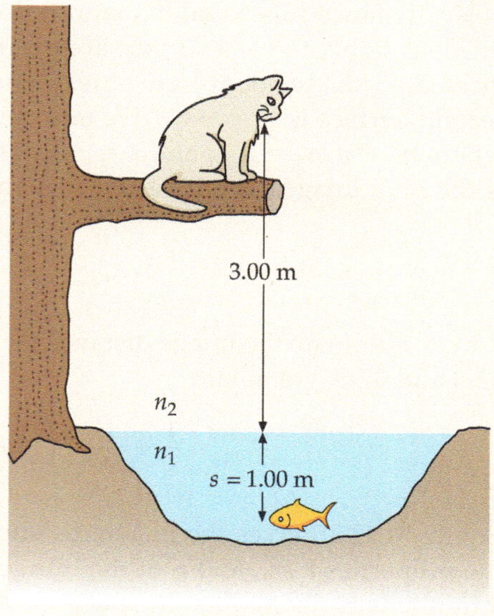

FIGURE 32-27

CHECK The image is between the location of the fish and the surface of the water. This result is expected. Recall that if you dip an oar into the water at an angle, the part of the oar that is underwater appears above where you know it has to be.

TAKING IT FURTHER (1) This image can be seen at the calculated position only when the object is viewed from directly overhead, or nearly so. From that observation point the rays are paraxial, a condition necessary for Equation 32-6 to be valid. If Fluffy is standing on the edge of the pond, the rays will not satisfy the paraxial approximation and Equation 32-6 will not correctly predict the location of the image. (2) The distance (n_2/n_1) multiplied by s is called the apparent depth of the submerged object. If $n_2 = 1$, the apparent depth equals s/n_1.

CONCEPT CHECK 32-4

Draw a ray diagram for the image of Goldie, as described in Example 32-6. That is, draw several rays diverging from an object point P on Goldie, and show that after refracting the rays appear to diverge from an image point P' somewhat above the object point.

THIN LENSES

An important application of Equation 32-6 for refraction at a single surface is finding the position of the image formed by a lens. This determination is done by considering the refraction at each surface of the lens separately to derive an equation relating the image distance to the object distance, the radius of curvature of each surface of the lens, and the index of refraction of the lens material.

We will consider a thin lens that has an index of refraction n and air on both sides. Let the radii of curvature of the surfaces of the lens be r_1 and r_2. If an object is at a distance s from the first surface (and therefore from the lens), the distance s_1' of the image due to refraction at the first surface can be found using Equation 32-6:

$$\frac{n_{\text{air}}}{s} + \frac{n}{s_1'} = \frac{n - n_{\text{air}}}{r_1} \qquad 32\text{-}8$$

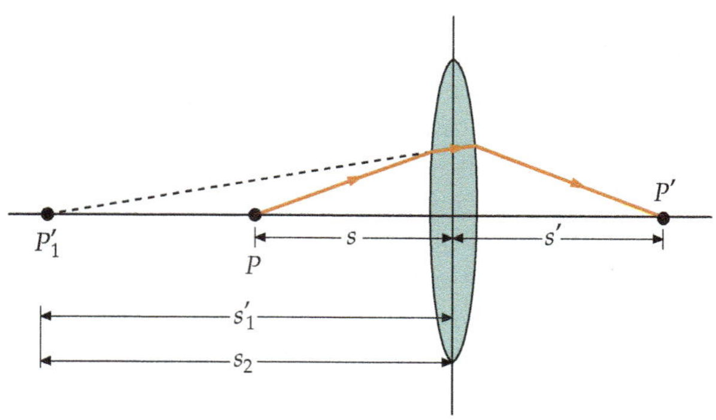

FIGURE 32-28 Refraction occurs at both surfaces of a lens. Here, the refraction at the first surface leads to a virtual image at P_1'. The rays strike the second surface as if they came from P_1'. Image distances are negative when the image is on the incident-light side of the surface, whereas object distances are positive for objects located on that side. Thus, $s_2(\approx -s_1')$ is the object distance for the second surface of the lens.

The light refracted at the first surface is again refracted at the second surface. Figure 32-28 shows the case when the image distance s_1' for the first surface is negative, indicating a virtual image to the left of the surface. Rays in the glass refracted from the first surface diverge as if they came from the image point P_1'. The rays strike the second surface at the same angles as if there were an object at image point P_1'. The image for the first surface therefore becomes the object for the second surface. Because the lens is of negligible thickness, the object distance s_2 is equal in magnitude to s_1'. Object distances for objects on the incident-light side of a surface are positive, whereas image distances for images located on the incident-light side are negative. Thus, the object distance for the second surface is $s_2' = -s_1'$.* We now write Equation 32-6 for the second surface, where $n_1 = n$, $n_2 = n_{air}$, and $s = -s_1'$. The image distance for the second surface is the final image distance s' for the lens:

$$\frac{n}{-s_1'} + \frac{n_{air}}{s'} = \frac{n_{air} - n}{r_2} \tag{32-9}$$

We can eliminate the image distance for the first surface $-s_1'$ by adding Equations 32-8 and 32-9. We obtain

$$\frac{1}{s} + \frac{1}{s'} = \left(\frac{n}{n_{air}} - 1\right)\left(\frac{1}{r_1} - \frac{1}{r_2}\right) \tag{32-10}$$

Equation 32-10 gives the image distance s' in terms of the object distance s and the properties r_1, r_2, and n of the thin lens. As with mirrors, the focal length f of a thin lens is defined as the image distance when the object distance is infinite. Setting s equal to infinity and writing f for the image distance s', we obtain

$$\frac{1}{f} = \left(\frac{n}{n_{air}} - 1\right)\left(\frac{1}{r_1} - \frac{1}{r_2}\right) \tag{32-11}$$

LENS-MAKER'S EQUATION

Equation 32-12 is called the **lens-maker's** equation. Substituting $1/f$ for the right side of Equation 32-10, we obtain

$$\frac{1}{s} + \frac{1}{s'} = \frac{1}{f} \tag{32-12}$$

THIN-LENS EQUATION

This **thin-lens equation** is the same as the mirror equation (Equation 32-4). Recall, however, that the sign conventions for refraction are somewhat different from the sign conventions for reflection. For refraction, the image distance s' is positive when the image is on the refracted-light side of the refracting surface(s), that is, when it is on the side opposite the incident-light side. The sign of the focal length of a lens (see Equation 32-11) is determined by the sign convention for a single refracting boundary. That is, r is positive if the center of curvature is on the same side of the surface as the refracted light. For a lens like that shown in Figure 32-28, r_1 is positive and r_2 is negative, so f is positive.

Figure 32-29*a* shows the wavefronts of plane waves incident on a double convex lens. The central part of the wavefront strikes the lens first. Because the wave speed in the lens is less than that in air (assuming $n > 1$), the central part of the wavefront lags behind the outer parts, resulting in a spherical wavefront that converges at the focal point F'. The rays for this situation are shown in Figure 32-29*b*. Such a lens is called a **converging lens.** Because its focal length as calculated from Equation 32-2 is positive, it is also called a **positive lens.**

---

* If s_1' were positive, the rays would be converging as they strike the second surface. The object for the second surface would then be a virtual object located to the right of the second surface. This object would be a virtual object. Again, $s_2 = -s_1'$.

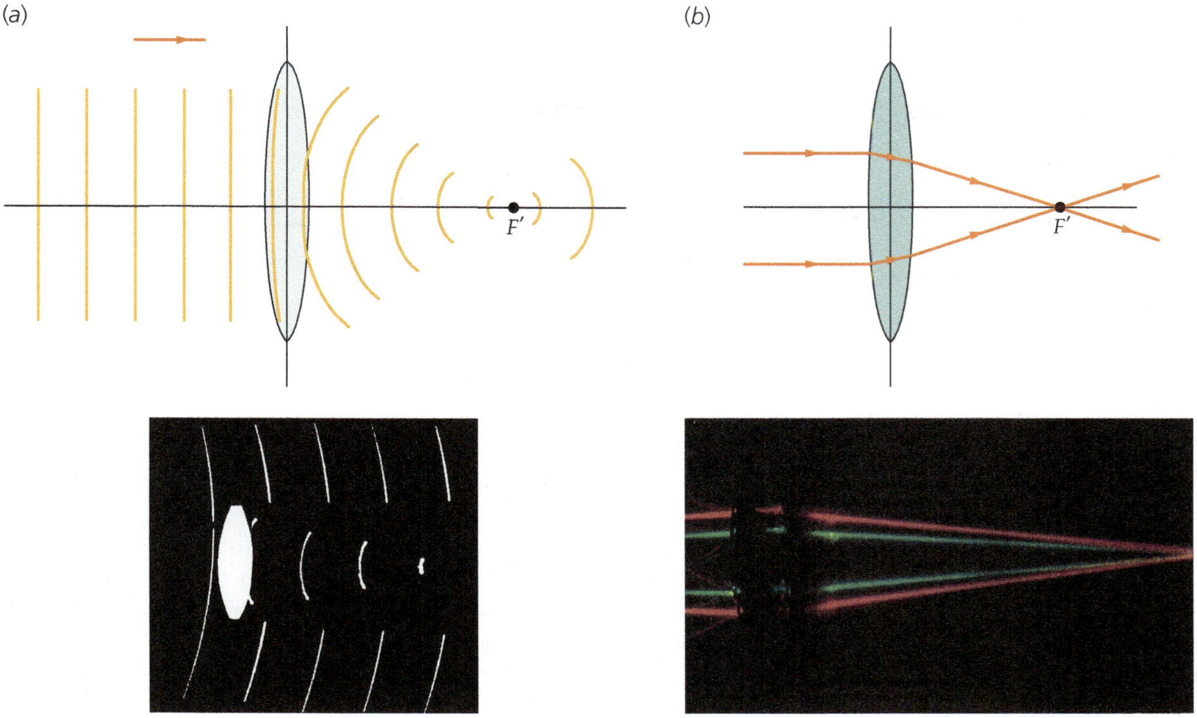

FIGURE 32-29 (*a*) *Top:* Wavefronts for plane waves striking a converging lens. The central part of the wavefront is retarded more by the lens than the outer part, resulting in a spherical wave that converges at the focal point F'. *Bottom:* Wavefronts passing through a lens, shown by a photographic technique called *light-in-flight recording* that uses a pulsed laser to make a hologram of the wavefronts of light. (*b*) *Top:* Rays for plane waves striking a converging lens. The rays are bent at each surface and converge at the focal point. *Bottom:* A photograph of rays focused by a converging lens. *((a) Nils Abramson, (b) Fundamental Photographs.)*

Any lens that is thicker in the middle than at the edges is a converging lens (providing that the index of refraction of the lens is greater than that of the surrounding medium). Figure 32-30 shows the wavefronts and rays for plane waves incident on a double concave lens. In this case, the outer part of the wavefronts lag behind the central parts, resulting in outgoing spherical waves that diverge from a focal point on the incident-light side of the lens. The focal length of this lens is negative. Any lens that is thinner in the middle than at the edges is a **diverging,** or **negative,** lens.

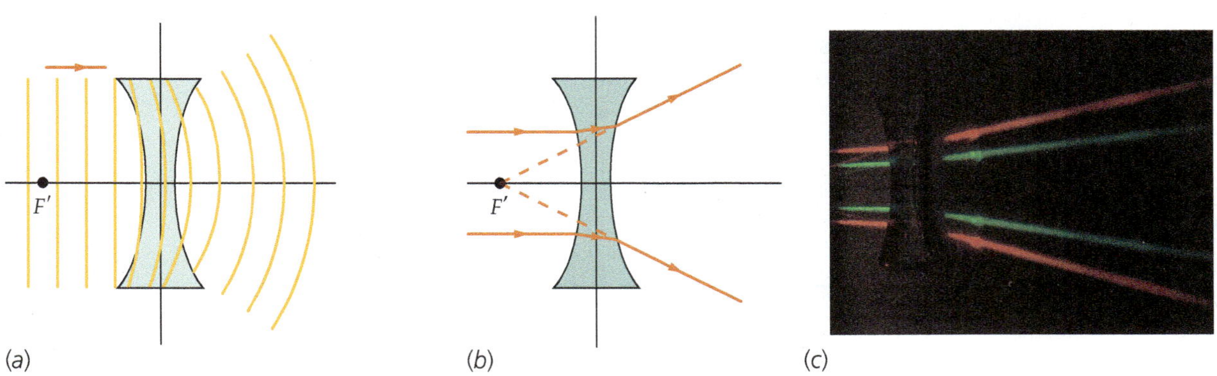

FIGURE 32-30 (*a*) Wavefronts for plane waves striking a diverging lens. Here, the outer part of the wavefront is retarded more than the central part, resulting in a spherical wave that diverges as it moves out, as if it came from the focal point F' to the left of the lens. (*b*) Rays for plane waves striking the same diverging lens. The rays are bent outward and diverge, as if they came from the focal point F'. (*c*) A photograph of rays passing through a diverging lens. *(Fundamental Photographs.)*

Example 32-7 The Lens-Maker's Formula

A double convex, thin glass lens that has an index of refraction $n = 1.50$ has radii of curvature whose magnitudes are 10 cm and 15 cm, as shown in Figure 32-31. Find its focal length in air.

PICTURE We can find the focal length using the lens-maker's equation (Equation 32-11). Here, light is incident on the surface that has the smaller radius of curvature. The center of curvature C_1 of this surface is on the refracted-light side of the lens; thus, $r_1 = +10$ cm. For the second surface, the center of curvature C_2 is on the incident-light side; therefore, $r_2 = -15$ cm.

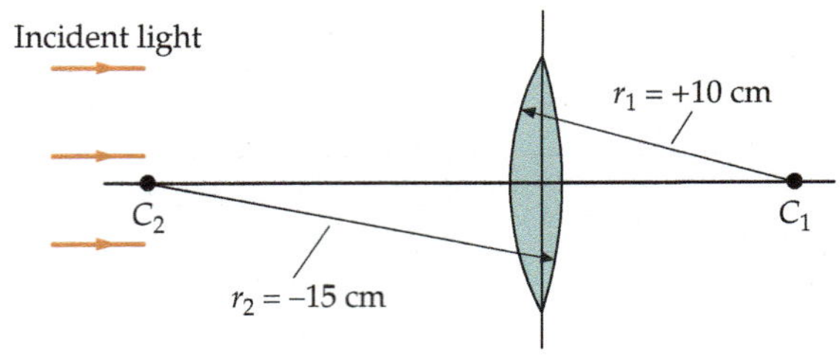

FIGURE 32-31

SOLVE

Numerical substitution in Equation 32-11 yields the focal length f:

$$\frac{1}{f} = \left(\frac{n}{n_{air}} - 1\right)\left(\frac{1}{r_1} - \frac{1}{r_2}\right)$$

$$= \left(\frac{1.50}{1.00} - 1\right)\left(\frac{1}{10\text{ cm}} - \frac{1}{-15\text{ cm}}\right) = 0.50\left(\frac{5.0}{30\text{ cm}}\right)$$

$$\therefore f = \boxed{12\text{ cm}}$$

CHECK The calculated focal length is positive, as expected. The lens is thicker in the middle than at the edges, so the focal length is expected to be positive.

PRACTICE PROBLEM 32-8 A double convex thin lens has an index of refraction $n = 1.6$ and its surfaces have radii of curvature of equal magnitude. If the focal length of the lens is 15 cm, what is the magnitude of the radii of curvature of the surfaces?

PRACTICE PROBLEM 32-9 Show that if you reverse the direction of the incoming light for the lens shown in Example 32-7, so that the incoming light is incident on the surface that has the greater radius of curvature, you get the same result for the focal length.

If parallel light strikes the lens of Example 32-7 from the left, it is focused at a point 12 cm to the right of the lens; whereas if parallel light strikes the lens from the right, it is focused at 12 cm to the left of the lens. Both of these points are focal points of the lens. Using the reversibility property of light rays, we can see that light diverging from a focal point and striking a lens will leave the lens as a parallel beam, as shown in Figure 32-32. Incident rays parallel to the axis emerge directed either toward or away from the **first focal point** F. Incident rays directed either toward or away from the **second focal point** F' emerge parallel with the axis. For a converging lens, the first focal point is on the incident-light side and the second focal point is on the refracted-light side. (For a diverging lens the opposite is true.) If parallel light is incident on the lens at a small angle with the axis, as in Figure 32-33, it is focused at a point in the **focal plane** a distance f from the lens.

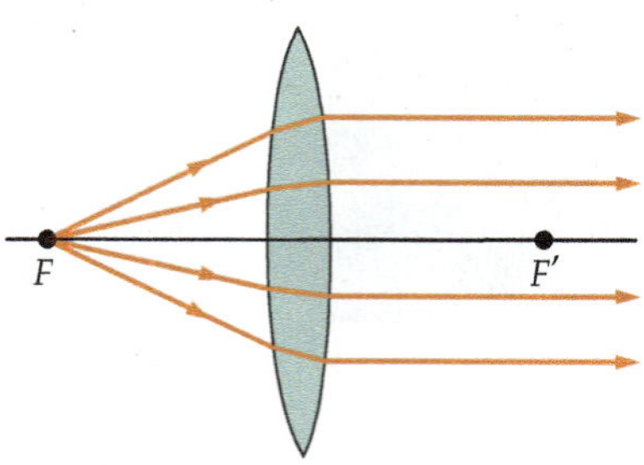

FIGURE 32-32 Light rays diverging from the focal point of a positive lens emerge parallel to the axis.

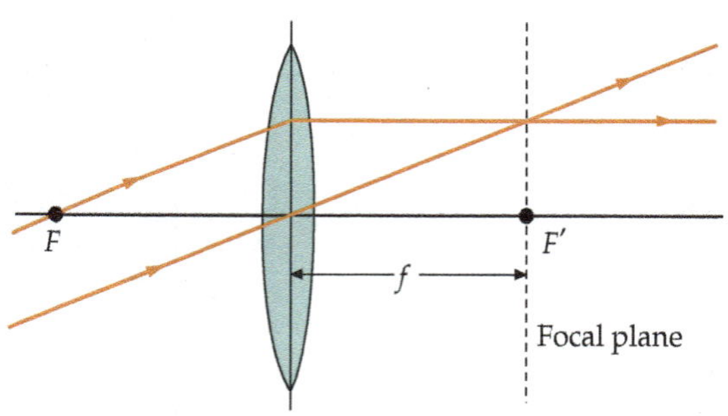

FIGURE 32-33 Parallel rays incident on the lens at an angle to its axis are focused at a point in the focal plane of the lens.

The reciprocal of the focal length is called the **power of a lens.** When the focal length is expressed in meters, the power is given in reciprocal meters, called **diopters** (D):

$$P = \frac{1}{f} \qquad 32\text{-}13$$

The power of a lens measures its ability to focus parallel light at a short distance from the lens. The shorter the focal length, the greater the power. For example, a lens that has a focal length of 25 cm = 0.25 m has a power of 4.0 D. A lens that has a focal length of 10 cm = 0.10 m has a power of 10 D. Because the focal length of a diverging lens is negative, its power is negative.

Example 32-8 Power of a Lens

The lens shown in Figure 32-34 has an index of refraction of 1.50 and radii of curvature whose magnitudes are 10.0 cm and 13.0 cm. Find (*a*) its focal length in air and (*b*) its power.

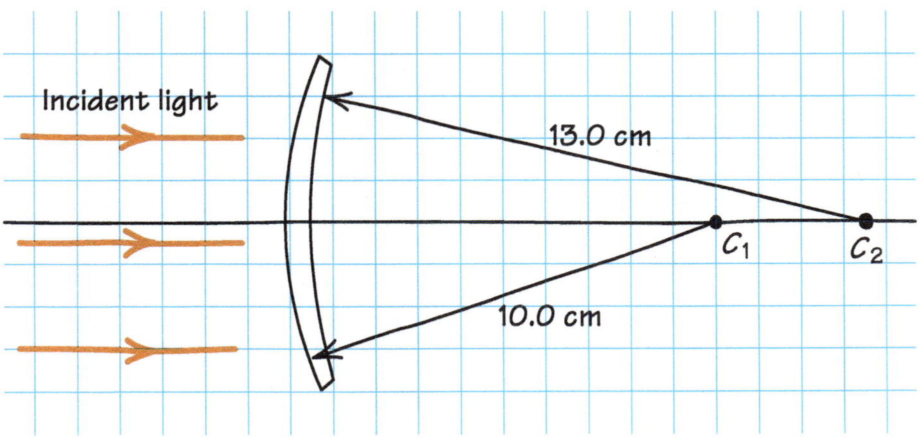

FIGURE 32-34

PICTURE For the orientation of the lens relative to the incident light shown in Figure 32-34, the radius of curvature of the first surface is $r_1 = +10.0$ cm and that of the second surface is $r_2 = +13.0$ cm.

SOLVE

(*a*) Calculate f from the lens-maker's equation using the given value of n and the values of r_1 and r_2 for the orientation shown:

$$\frac{1}{f} = \left(\frac{n}{n_{\text{air}}} - 1\right)\left(\frac{1}{r_1} - \frac{1}{r_2}\right)$$

$$= \left(\frac{1.50}{1.00} - 1\right)\left(\frac{1}{10.0\text{ cm}} - \frac{1}{13.0\text{ cm}}\right)$$

$$\therefore f = \boxed{87\text{ cm}}$$

(*b*) The power is the reciprocal of the focal length expressed in meters:

$$P = \frac{1}{f} = \frac{1}{0.867\text{ m}} = \boxed{1.2\text{ D}}$$

CHECK The values for the focal length and the power are positive. That is what is expected for a lens that is thicker in the middle than at the edges.

TAKING IT FURTHER We obtain the same result no matter which surface the light strikes first.

During laboratory experiments involving lenses, it is usually much easier to measure the focal length than to measure the radii of curvature of the surfaces.

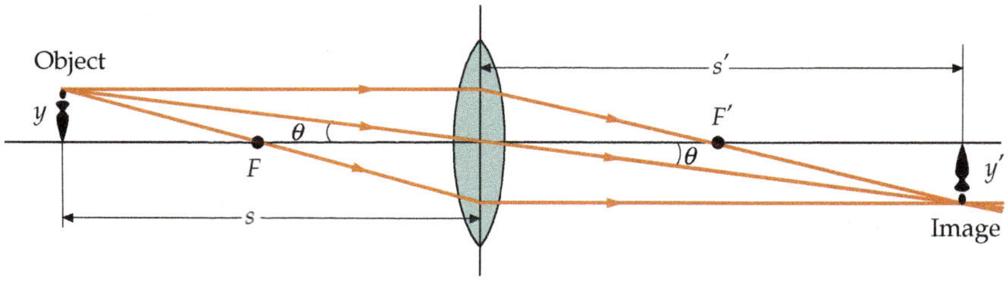

FIGURE 32-35 Ray diagram for a thin converging lens. We draw the rays as if all the bending of light takes place at the central plane. The ray through the center is undeflected because the lens surfaces there are parallel and close together.

RAY DIAGRAMS FOR LENSES

As with images formed by mirrors, it is convenient to locate the images of lenses by graphical methods. Figure 32-35 illustrates the graphical method for a thin converging lens. In the thin lens approximation, we consider the rays to bend at the plane through the center of the lens and perpendicular to the optic axis. The three principal rays are as follows:

1. The **parallel ray,** drawn parallel to the axis. The emerging ray is directed toward the second focal point of the lens.
2. The **central ray,** drawn through the center (the vertex) of the lens. This ray is undeflected. (The faces of the lens are parallel at the center, so the ray emerges in the same direction but displaced slightly. Because the lens is thin, the displacement is negligible.)
3. The **focal ray,** drawn through the first focal point.* This ray emerges parallel to the axis.

PRINCIPAL RAYS FOR A THIN LENS

The weight and bulk of a large-diameter lens can be reduced by constructing the lens from annular segments at different angles so that light from a point is refracted by the segments into a parallel beam. Such an arrangement is called a Fresnel lens. Several Fresnel lenses are used in this lighthouse to produce intense parallel beams of light from a source at the focal point of the lenses. The illuminated surface of an overhead projector is a Fresnel lens. *(Bohdan Hrynewych/Stock Boston.)*

These three rays converge to the image point, as shown in Figure 32-35. In this case, the image is real and inverted. From the figure, we have $\tan\theta = y/s = -y'/s'$. The lateral magnification is then

$$m = \frac{y'}{y} = -\frac{s'}{s} \qquad \text{32-14}$$

This expression is the same as the expression for mirrors. Again, a negative magnification indicates that the image is inverted. The ray diagram for a diverging lens is shown in Figure 32-36.

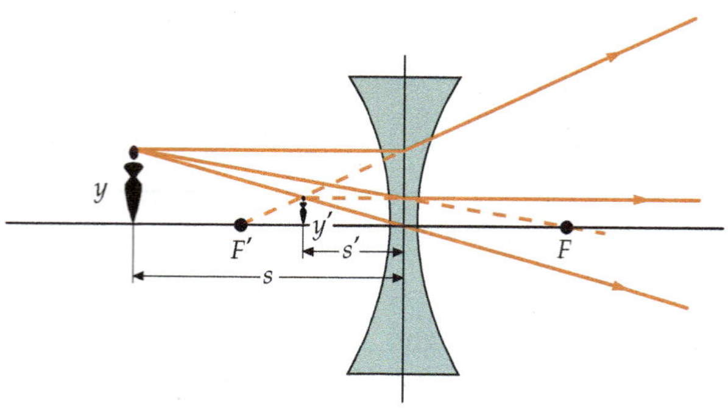

FIGURE 32-36 Ray diagram for a diverging lens. The parallel ray is bent away from the axis, as if it came from the second focal point F'. The ray toward the first focal point F emerges parallel to the axis. For a diverging lens the first focal point F is on the refracted-light side of the lens.

* The focal ray is drawn toward the first focal point F for a diverging lens.

Example 32-9 Image Formed by a Lens

An object that is 1.2 cm high is placed 4.0 cm from a double convex lens that has a focal length of 12 cm. Locate the image both graphically and algebraically, state whether the image is real or virtual, and find its height. Place an eye on the figure positioned and oriented so as to view the image.

PICTURE Locate the image by graphical methods. That means by drawing the three principal rays. The eye is positioned and oriented so the light from the image enters the eye.

SOLVE

1. Draw the parallel ray. This ray leaves the object parallel to the axis, then is bent by the lens to pass through the second focal point, F' (Figure 32-37):

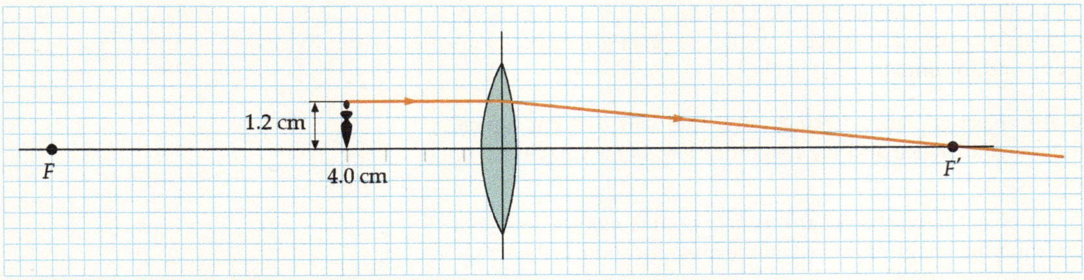

FIGURE 32-37

2. Draw the central ray, which passes undeflected through the center of the lens. Because the two rays are diverging on the refracted-light side, we extend them back to the incident-light side to find the image (Figure 32-38 here):

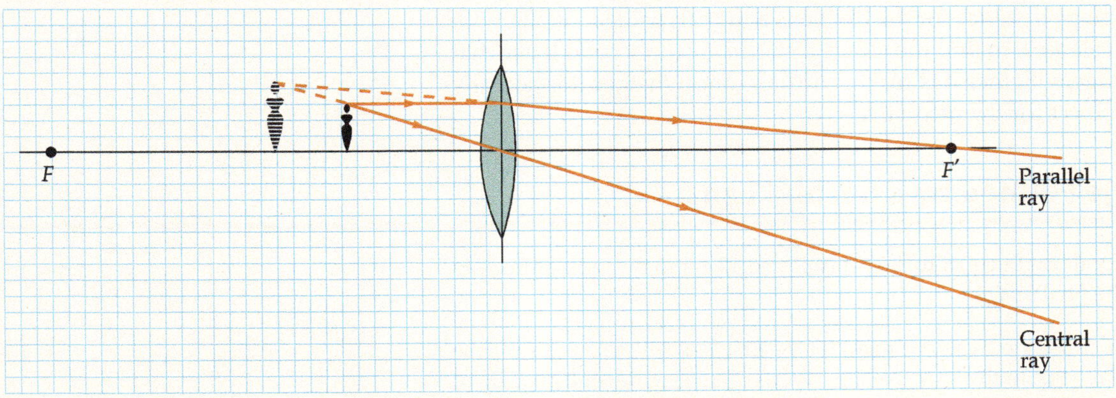

FIGURE 32-38

3. As a check, we also draw the focal ray. This ray leaves the object on a line passing through the first focal point, then emerges parallel to the axis. Note that the image is virtual, upright, and enlarged (Figure 32-39):

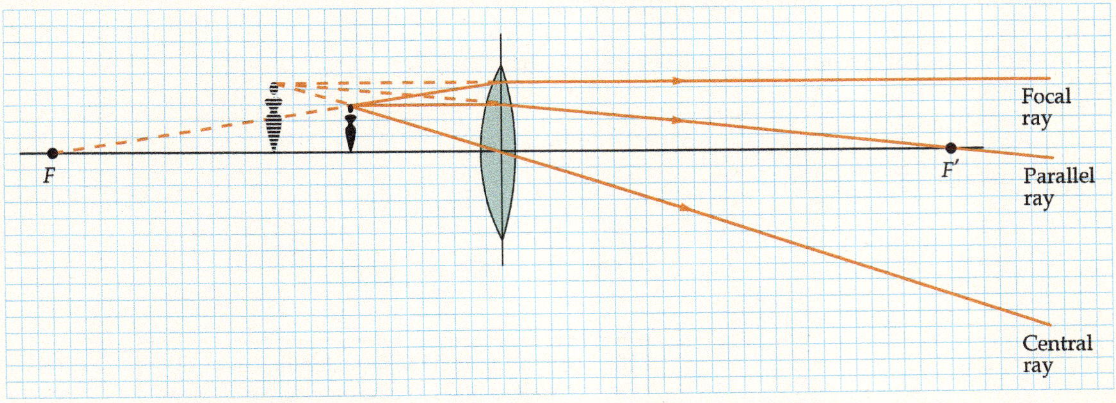

FIGURE 32-39

4. The eye must be positioned so the light from the image enters the eye.

5. We now verify the results of the ray diagram algebraically. First, find the image distance using Equation 32-12:

$$\frac{1}{4.0\text{ cm}} + \frac{1}{s'} = \frac{1}{12\text{ cm}}$$

$$\frac{1}{s'} = \frac{1}{12\text{ cm}} - \frac{1}{4.0\text{ cm}} = -\frac{1}{6.0\text{ cm}}$$

$$s' = -6.0\text{ cm}$$

6. The height of the image is found from the height of the object and the magnification:

$$h' = mh$$

7. The magnification m is given by Equation 32-14:

$$m = -\frac{s'}{s} = -\frac{-6.0\text{ cm}}{4.0\text{ cm}} = \boxed{+1.5}$$

8. Using this result, we find the height of the image, h':

$$h' = mh = (1.5)(1.2\text{ cm}) = \boxed{1.8\text{ cm}}$$

CHECK Note the agreement between the algebraic and ray diagram results. Algebraically, we find that the image is 6.0 cm from the lens on the incident-light side (because $s' < 0$); that is, the image is 2.0 cm to the left of the object. Because $m > 0$, it follows that the image is upright, and because $m > 1$, the image is enlarged. It is good practice to process lens problems both graphically and algebraically and to compare the results.

PRACTICE PROBLEM 32-10 An object is placed 15 cm from a thin lens that has a focal length equal to 10 cm. Find the image distance and the magnification. Is the image real or virtual? Is the image upright or inverted?

PRACTICE PROBLEM 32-11 An object is placed 5.0 cm from a double convex lens that has a focal length equal to 10 cm. Find the image distance and the magnification. Is the image real or virtual? Is the image upright or inverted?

COMBINATIONS OF LENSES

If we have two or more thin lenses, we can find the final image produced by the system by finding the image distance for the first lens and then using it, along with the distance between the lenses, to find the object distance for the second lens. That is, we consider each image, whether it is real or virtual—and whether it is actually formed or not—as the object for the next lens.

Example 32-10 Image Formed by a Second Lens

A second lens that has a focal length equal to +6 cm is placed 12 cm to the right of the lens in Example 32-9. Locate the final image.

PICTURE The principal rays used to locate the image of the first lens will not necessarily be principal rays for the second lens. In this example, however, we have chosen the position of the second lens (Figure 32-40*a*) so that the parallel ray for the first lens turns out to be the central ray for the second lens. Also, the focal ray for the first lens emerges parallel to the axis and is therefore the parallel ray for the second lens. If additional principal rays for the second lens are needed, we simply draw them from the image formed by the first lens. For example, in Figure 32-40*b* we added such a ray, drawn from the first image through the first focal point F_2 of the second lens.

Algebraically we use $s_2 = 18$ cm, because the first image is 6 cm to the left of the first lens and therefore 18 cm to the left of the second lens.

SOLVE

1. Use $s_2 = 18$ cm and $f = 6$ cm to calculate s_2':

$$\frac{1}{s_2} + \frac{1}{s_2'} = \frac{1}{f_2}$$

$$\frac{1}{18\text{ cm}} + \frac{1}{s_2'} = \frac{1}{6\text{ cm}}$$

$$s_2' = 9\text{ cm}$$

> The final image is on the refracted-light side of the second lens and is 9 cm from the second lens.

2.

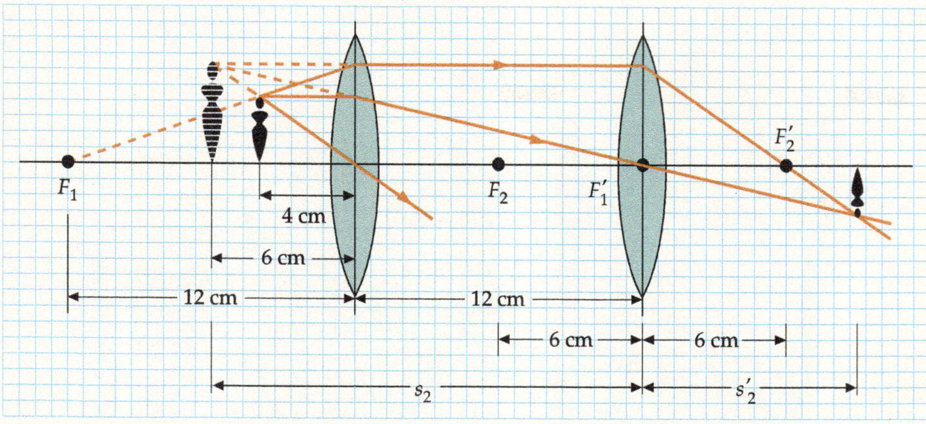

FIGURE 32-40*a*

CHECK The ray diagram in step 2 checks the calculation in step 1. In addition, to check the ray diagram in step 2, we draw the focal ray for the second lens.

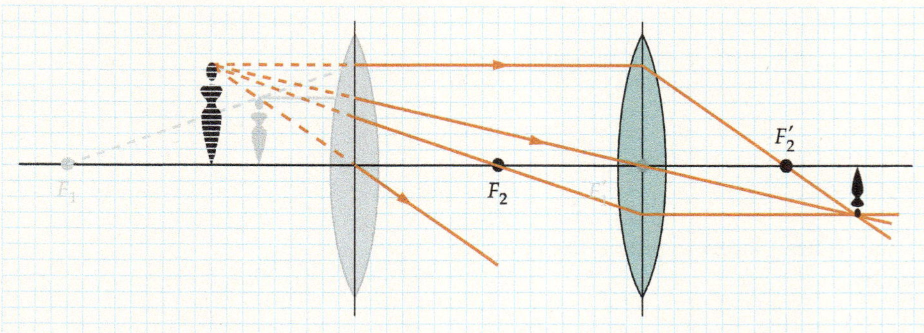

FIGURE 32-40*b*

Example 32-11 A Combination of Two Lenses

Try It Yourself

Two lenses, each having a focal length equal to 10 cm, are 15 cm apart. Find the final image location for an object 15 cm from one of the lenses and 30 cm from the other.

PICTURE Use a ray diagram to find the location of the image formed by lens 1. When these rays strike lens 2 they are further refracted, leading to the final image. More accurate results are obtained algebraically using the thin-lens equation for both lens 1 and lens 2.

SOLVE

Cover the column to the right and try these on your own before looking at the answers.

Steps

1. Draw the (*a*) parallel, (*b*) central, and (*c*) focal rays for lens 1 (Figure 32-41). If lens 2 did not alter these rays, they would form the image I_1.

Answers

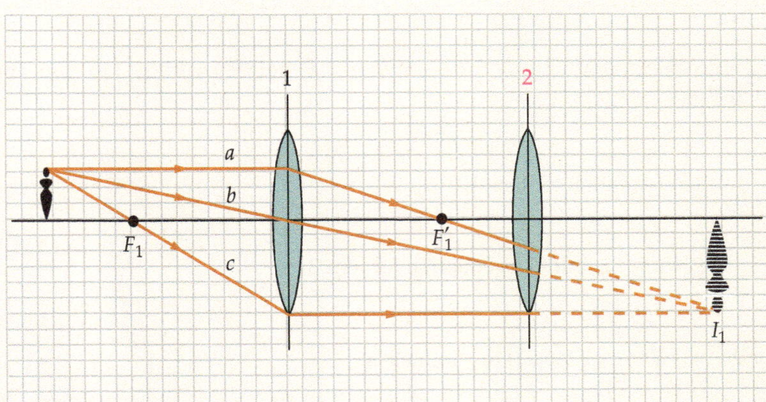

FIGURE 32-41

2. To locate the final image, add three principal rays (d, e, and f) for lens 2. The intersection of these rays gives the location of image I_2 (Figure 32-42).

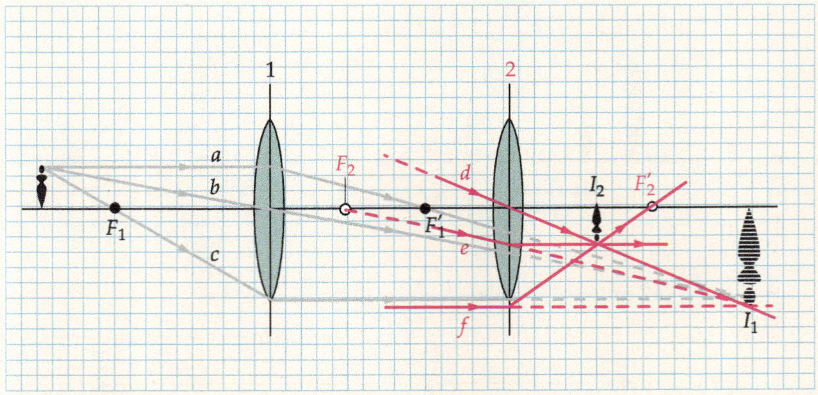

FIGURE 32-42

3. To proceed algebraically, use the thin-lens equation to find the image distance s_1' produced by lens 1. — $s_1' = 30$ cm

4. For lens 2, the image, I_1 is 15 cm from the lens on the refracted-light side; hence, $s_2 = -15$ cm. Use this to find the final image distance s_2'. — $\boxed{s_2' = 6 \text{ cm}}$

CHECK From the step-2 ray diagram we see that the final image is about six tenths of the focal length of lens 2. The focal length of lens 2 is 10 cm, so step-2 result and the step-4 result are in agreement.

COMPOUND LENSES

When two thin lenses of focal lengths f_1 and f_2 are placed together, the effective focal length of the combination f_{eff} is given by

$$\frac{1}{f_{\text{eff}}} = \frac{1}{f_1} + \frac{1}{f_2} \qquad 32\text{-}15$$

as is shown in the following example (Example 32-12). The power of two lenses in contact is given by

$$P_{\text{eff}} = P_1 + P_2 \qquad 32\text{-}16$$

Example 32-12 Two Lenses in Contact

Try It Yourself

For two lenses very close together, derive the relation $\dfrac{1}{f_{\text{eff}}} = \dfrac{1}{f_1} + \dfrac{1}{f_2}$.

PICTURE Apply the thin-lens equation to each lens using the fact that the distance between the lenses is negligible, so the object distance for the second lens is the negative of the image distance for the first lens.

SOLVE

Cover the column to the right and try these on your own before looking at the answers.

Steps	Answers
1. Write the thin-lens equation for lens 1.	$\dfrac{1}{s_1} + \dfrac{1}{s_1'} = \dfrac{1}{f_1}$
2. Using $s_2 = -s_1'$, write the thin-lens equation for lens 2.	$\dfrac{1}{-s_1'} + \dfrac{1}{s_2'} = \dfrac{1}{f_2}$
3. Add your two resulting equations to eliminate s_1'.	$\boxed{\dfrac{1}{s_1} + \dfrac{1}{s_2'} = \dfrac{1}{f_1} + \dfrac{1}{f_2} = \dfrac{1}{f_{\text{eff}}}}$

* 32-3 ABERRATIONS

When all the rays from a point object are not focused at a single image point, the resultant blurring of the image is called **aberration.** Figure 32-43 shows rays from a point source on the axis traversing a thin lens with spherical surfaces. Rays that strike the lens far from the axis are bent much more than are the rays near the axis, with the result that not all the rays are focused at a single point. Instead, the image appears as a circular disk at point C, where the diameter is minimum. This type of aberration in a lens is called **spherical aberration;** it is the same as the spherical aberration of mirrors discussed in Section 32-1. Similar but more complicated aberrations called *coma* and *astigmatism* occur when objects are off axis. The aberration in the shape of the image of an extended object that occurs, because the magnification depends on the distance of the object point from the axis, is called **distortion.** We will not discuss these aberrations further, except to point out that they do not arise from any defect in the lens or mirror but instead result from the application of the laws of refraction and reflection to spherical surfaces. These aberrations are not evident in our simple equations, because we used small-angle approximations in the derivation of the equations.

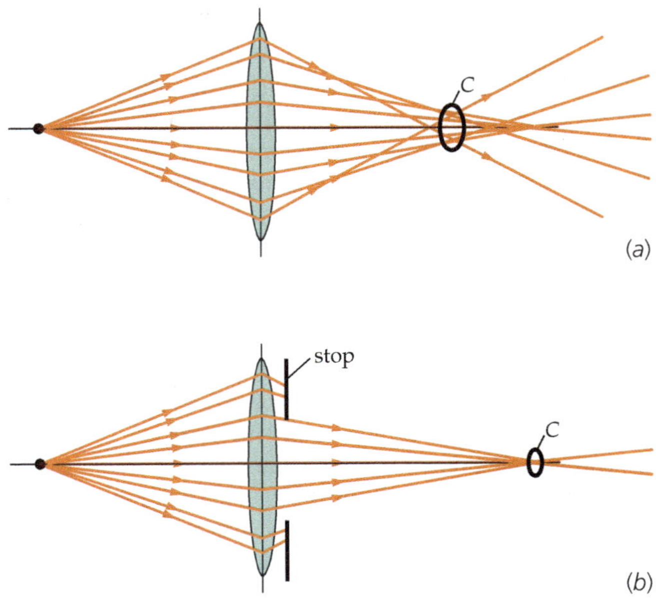

FIGURE 32-43 Spherical aberration in a lens. (*a*) Rays from a point object on the axis are not focused at a point. (*b*) Spherical aberration can be reduced by using a stop to block off the outer parts of the lens, but this also reduces the amount of light reaching the image.

Some aberrations can be eliminated or partially corrected by using nonspherical surfaces for mirrors or lenses, but nonspherical surfaces are usually much more difficult and costly to produce than spherical surfaces. One example of a nonspherical reflecting surface is the parabolic mirror illustrated in Figure 32-44. Rays that are parallel to the axis of a parabolic surface are reflected and focused at a common point, no matter how far the rays are from the axis. Parabolic reflecting surfaces are sometimes used in large astronomical telescopes, which need a large reflecting surface to gather as much light as possible to make the image as intense as possible (reflecting telescopes are described in the upcoming optional Section 32-4). Satellite dishes use parabolic surfaces to focus microwaves from communications satellites. A parabolic surface can also be used in a searchlight to produce a parallel beam of light from a small source placed at the focal point of the surface.

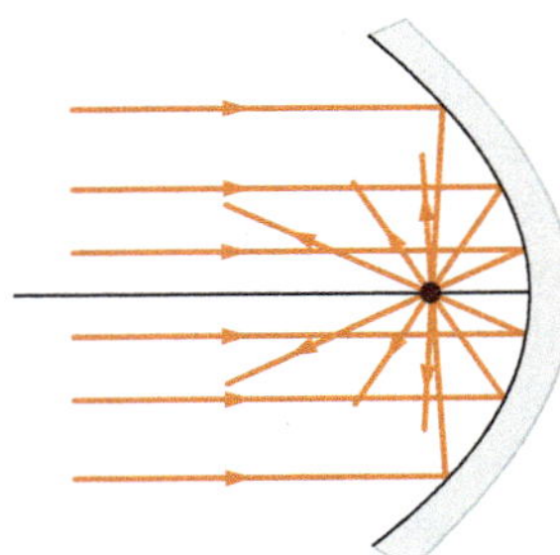

FIGURE 32-44 A parabolic mirror focuses all rays parallel to the axis to a single point with no spherical aberration.

An important aberration found with lenses but not found with mirrors is **chromatic aberration,** which is due to variations in the index of refraction with wavelength. From Equation 32-11, we can see that the focal length of a lens depends on its index of refraction and is therefore different for different wavelengths. Because n is slightly greater for blue light than for red light, the focal length for blue light will be shorter than the focal length for red light.

Chromatic aberration and other aberrations can be partially corrected by using combinations of lenses instead of a single lens. For example, a positive lens and a negative lens of greater focal length can be used together to produce a converging lens system that has much less chromatic aberration than a single lens of the same focal length. A high quality camera lens typically contains six elements to correct the various aberrations that are present.

* 32-4 OPTICAL INSTRUMENTS

*THE EYE

The optical system of prime importance is the eye, which is shown in Figure 32-45. Light enters the eye through a variable aperture, the pupil. The light is converged by the cornea, with assistance from the lens, to form an image on the retina, which has a film of nerve fibers covering the back surface of the eye. The retina contains tiny sensing structures called *rods* and *cones*, which detect the light and transmit the information along the optic nerve to the brain. The shape of the crystalline lens can be altered slightly by the action of the ciliary muscle. When the eye is focused on an object far away, the muscle is relaxed and the cornea-lens system has its maximum focal length, about 2.5 cm, which is the distance from the cornea to the retina. When the object is brought closer to the eye, the ciliary muscle increases the curvature of the lens slightly, thereby decreasing its focal length, so that the image is again focused on the retina. This process is called *accommodation*. If the object is too close to the eye, the lens cannot focus the light on the retina and the image is blurred. The closest point for which the lens can focus the image on the retina is called the **near point.** The distance from the eye to the near point varies greatly from one person to another and changes with age. At 10 years, the near point may be as close as 7 cm, whereas at 60 years it may recede to 200 cm because of the loss of flexibility of the lens. The standard value taken for the near point is 25 cm.

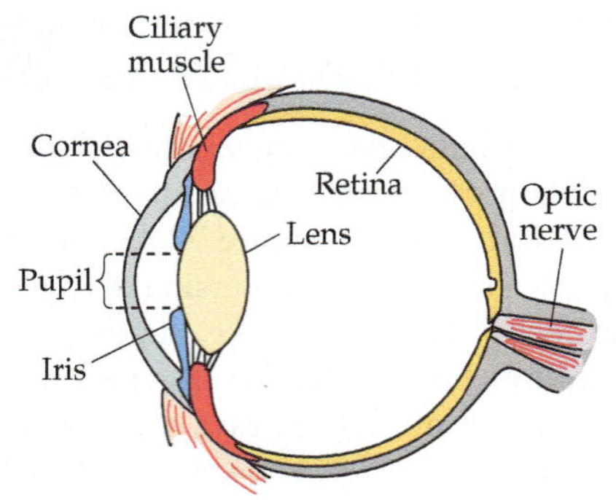

FIGURE 32-45 The human eye. The amount of light entering the eye is controlled by the iris, which regulates the size of the pupil. The lens thickness is controlled by the ciliary muscle. The cornea and lens together converge the light to focus the image on the retina, which contains approximately 125 million receptors, called rods and cones, and approximately 1 million optic-nerve fibers.

If the eye underconverges, resulting in the images being focused behind the retina, the person is said to be farsighted. A farsighted person can see distant objects clearly where little convergence is required, but has trouble seeing close objects clearly. Farsightedness is corrected with a converging (positive) lens (Figure 32-46).

On the other hand, the eye of a nearsighted person overconverges and focuses light from distant objects in front of the retina. A nearsighted person can see nearby objects clearly (objects for which the widely diverging incident rays can be focused on the retina), but has trouble seeing distant objects clearly. Nearsightedness is corrected with a diverging (negative) lens (Figure 32-47).

Another common defect of vision is astigmatism, which is caused by the cornea being not quite spherical but having a different curvature in one plane than in another. This results in a blurring of the image of a point object into a short line. Astigmatism is corrected by glasses using lenses of cylindrical rather than spherical shape.

The *apparent size* of an object is determined by the actual size of the image on the retina. The larger the image on the retina, the larger the apparent size and the greater the number of rods and cones that are illuminated by the light from the object. From Figure 32-48, we see that the size of the image on the retina is greater

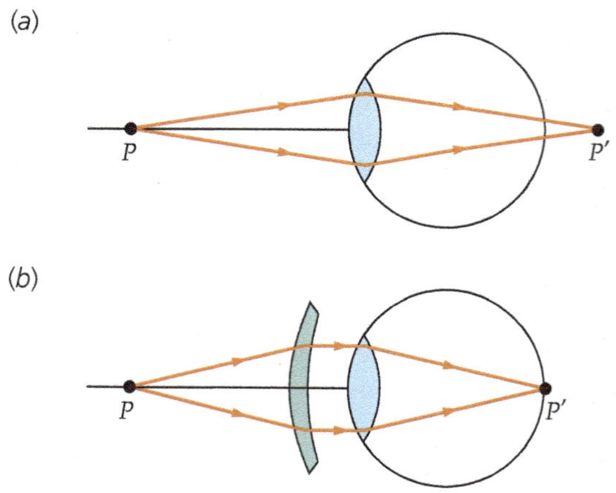

FIGURE 32-46 (*a*) A farsighted eye focuses rays from a nearby object to a point behind the retina. (*b*) A converging lens corrects this defect by bringing the image onto the retina. These diagrams, and those following, are drawn as if all the focusing of the eye is done at the cornea; in fact, the lens and cornea system act more like a spherical refracting surface than a thin lens.

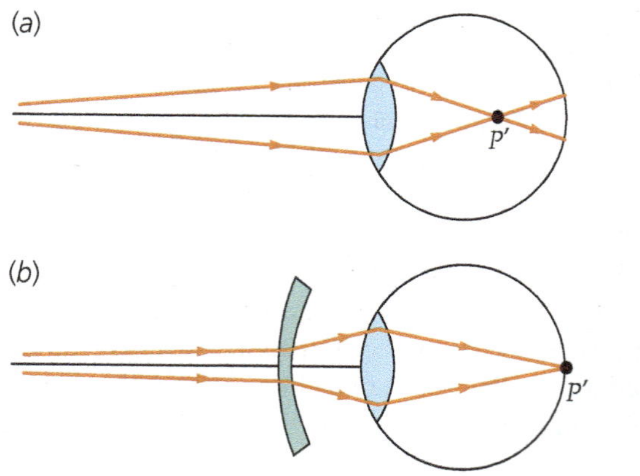

FIGURE 32-47 (*a*) A nearsighted eye focuses rays from a distant object to a point in front of the retina. (*b*) A diverging lens corrects this defect.

when the object is close than it is when the object is far away. The apparent size of an object is thus greater when the object is closer to the eye. The image size is proportional to the angle θ subtended by the object at the eye. For Figure 32-48,

$$\phi = \frac{y'}{2.5\text{ cm}} \quad \text{and} \quad \theta \approx \frac{y}{s} \qquad 32\text{-}17$$

for small angles. Applying the law of refraction gives $n_{air} \sin\theta = n \sin\phi$, where $n_{air} = 1.00$ and n is the refractive index inside the eye. For small angles this becomes

$$\theta \approx n\phi \qquad 32\text{-}18$$

Combining Equations 32-17 and 32-18 gives

$$\frac{y}{s} \approx n\frac{y'}{2.5\text{ cm}} \quad \text{or} \quad y' \approx \frac{2.5\text{ cm}}{n}\frac{y}{s} \qquad 32\text{-}19$$

The size of the image on the retina is proportional to the size of the object and inversely proportional to the distance from the cornea to the object. Because the near point is the closest point to the eye for which a sharp image can be formed on the retina, the distance from the cornea to the near point is called the *distance of most distinct vision.*

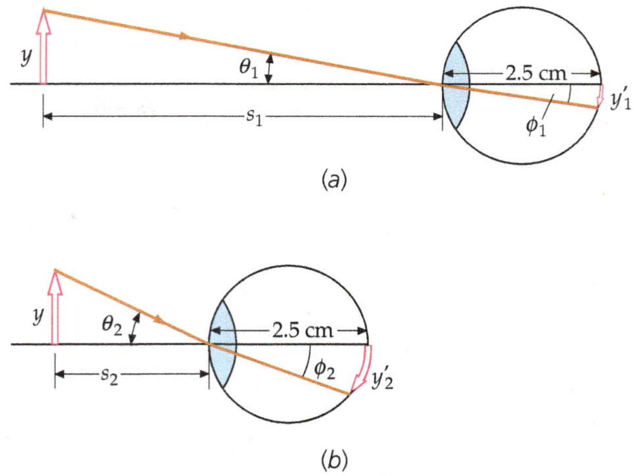

FIGURE 32-48 (*a*) A distant object of height y looks small because the image on the retina is small. (*b*) When the same object is closer, it looks larger because the image on the retina is larger.

Example 32-13 Reading Glasses

The near-point distance of a person's eye is 75 cm. Using a reading glasses lens placed a negligible distance from the eye, the near-point distance of the lens-eye system is 25 cm. That is, the image on the retina is blurry if an object is placed less than 25 cm in front of the lens. (*a*) What power is the reading glasses lens and (*b*) what is the lateral magnification of the image formed by the lens? (*c*) Which produces the bigger image on the retina, (1) the object at the near point of, and viewed by, the unaided eye or (2) the object at the near point of the lens-eye system and viewed through the lens that is immediately in front of the eye?

PICTURE A near-point distance of the lens-eye system of 25 cm means the lens forms a virtual image 75 cm in front of the lens if an object is placed 25 cm in front of the lens. Figure 32-49*a* shows a diagram of an object 25 cm from a converging lens that produces a virtual, upright image at $s' = -75$ cm. Figure 32-49*b* shows the image on the retina formed by the focusing power of the eye.

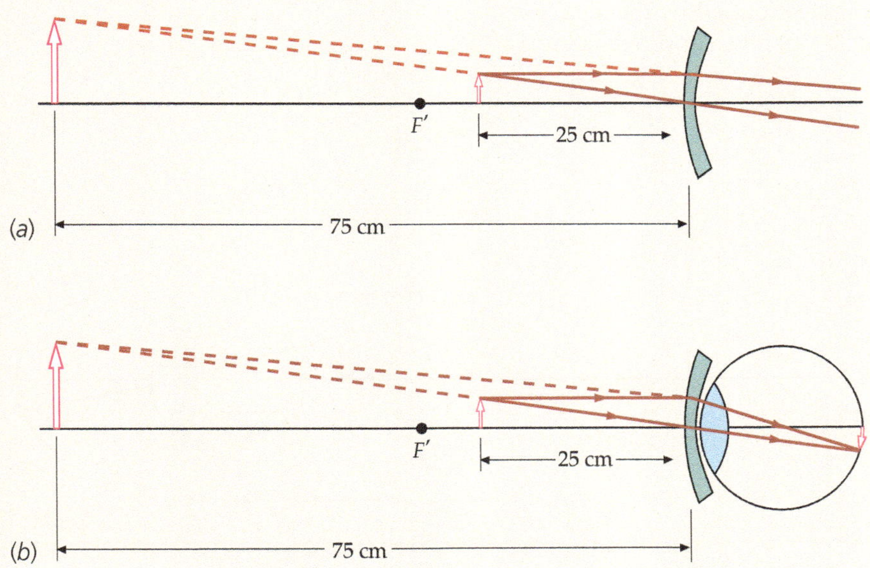

FIGURE 32-49

SOLVE

(*a*) Use the thin-lens equation, $s = 25$ cm, and $s' = -75$ cm to calculate the power, $1/f$:

$$\frac{1}{f} = \frac{1}{s} + \frac{1}{s'} = \frac{1}{25\text{ cm}} + \frac{1}{-75\text{ cm}}$$

$$= \frac{2}{75\text{ cm}} = \frac{2}{0.75\text{ m}} = \boxed{2.7\text{ D}}$$

(*b*) Using $m = -s'/s$, find m:

$$m = -\frac{s'}{s} = -\frac{-75\text{ cm}}{25\text{ cm}} = \boxed{3.0}$$

(*c*) In both cases, the rays entering the eye appear to diverge from points on an image 75 cm in front of the eye. However, with the lens in place, the image 75 cm in front of the eye is larger than the object by a factor of 3:

$\boxed{\text{Option 2}}$

TAKING IT FURTHER (1) If your near point is 75 cm, you are farsighted. To read a book you must hold it at least 75 cm from your eye to be able to focus on the print. The image of the print on your retina is then very small. The reading glasses lens produces an image also 75 cm from your eye, and this image is three times larger than the actual print. Thus, looking through the lens, the image of the print on the retina is larger by a factor of 3. (2) In this example, the distance from the lens to the eye was negligible. The results are slightly different if this distance is not negligible, which is factored into the calculations.

PRACTICE PROBLEM 32-12 Calculate the power of the eye for which the near point is 75 cm and the cornea-retina distance is 2.5 cm, and calculate the combined power of the lens and eye when they are in contact. Compare this result with the power of a lens for which $s' = 2.5$ cm, when $s = 25$ cm.

*THE SIMPLE MAGNIFIER

We saw in Example 32-13 that the *apparent* size of an object can be increased by using a converging lens placed next to the eye. A converging lens is called a **simple magnifier** if it is placed next to the eye and if the object is placed closer to the lens than its focal length, as was the case for the lens in Example 32-13. In that example, the lens formed a virtual image at the near point of the eye, the same location that the object must be placed for best viewing by the unaided eye. So, with the lens in place, the magnitude of the image distance $|s'|$ was greater than the object distance s, so the image seen by the eye is magnified by $m = |s'|/s$. If the actual height of the object is y, then the height y' of the image formed by the lens is my. To the eye, this image subtends an angle θ (Figure 32-50) given approximately by

$$\theta = \frac{my}{|s'|} = m\frac{y}{|s'|} = \frac{|s'|}{s}\frac{y}{|s'|} = \frac{y}{s}$$

which is the *very same angle* the object would subtend if the lens were removed while the object and the eye are left in place. That is, the apparent size of the image seen by the eye through the lens is the same as the apparent size of the object that would be seen by the eye were the lens removed (assuming the eye could focus at that distance). Thus, the apparent size of the object seen through the lens is inversely proportional to the distance from the object to the eye with the lens in place. The smaller s is, the larger the subtended angle θ and the larger the apparent size of the object.

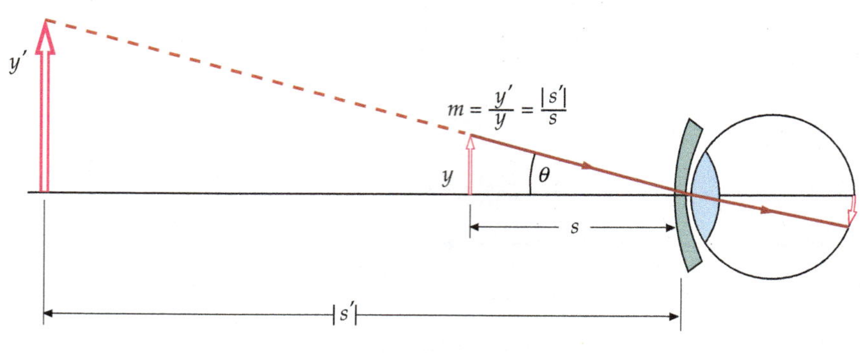

FIGURE 32-50

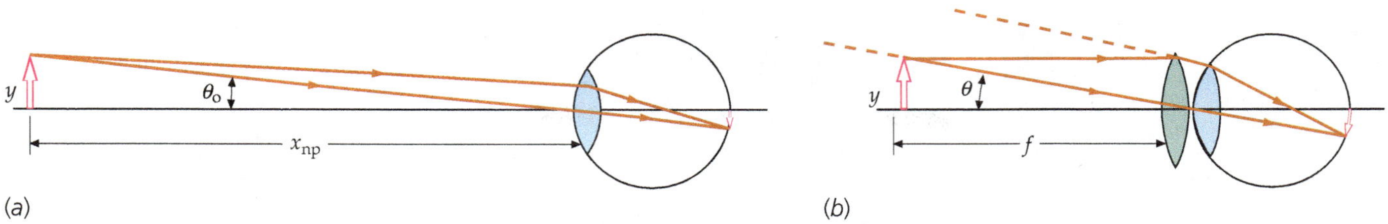

FIGURE 32-51 (*a*) An object at the near point subtends an angle θ_o at the naked eye. (*b*) When the object is at the focal point of the converging lens, the rays emerge from the lens parallel and enter the eye as if they came from an object a very large distance away. The image can thus be viewed at infinity by the relaxed eye. When f is less than the near-point distance, the converging lens allows the object to be brought closer to the eye. This increases the angle subtended by the object to θ, thereby increasing the size of the image on the retina.

In Figure 32-51*a*, a small object of height y is at the near point of the eye at a distance x_{np}. The angle subtended, θ_o, is given approximately by

$$\theta_o = \frac{y}{x_{np}}$$

In Figure 32-51*b*, a converging lens of focal length f that is smaller than x_{np} is placed a negligible distance in front of the eye, and the object is placed in the focal plane of the lens. The rays emerge from the lens parallel, indicating that the image is located an infinite distance in front of the lens. The parallel rays are focused by the relaxed eye on the retina. The angle subtended by this image is equal to the angle subtended by the object (assuming that the lens is a negligible distance from the eye). The angle subtended by the object is approximately

$$\theta = \frac{y}{f}$$

The ratio θ/θ_o is called the *angular magnification* or *magnifying power* M of the lens:

$$M = \frac{\theta}{\theta_o} = \frac{x_{np}}{f} \qquad 32\text{-}20$$

Simple magnifiers are used as eyepieces (called oculars) in microscopes and telescopes to view a real image formed by another lens or lens system. To correct aberrations, combinations of lenses that result in a short positive focal length may be used in place of a single lens, but the principle of the simple magnifier is the same.

Example 32-14 Angular Magnification of a Simple Magnifier

Try It Yourself

A person who has a near point of 25 cm uses a 40-D lens as a simple magnifier. What angular magnification is obtained?

PICTURE The angular magnification is found from the focal length f (Equation 32-20), which is the reciprocal of the power.

SOLVE

Cover the column to the right and try these on your own before looking at the answers.

Steps	Answers
1. Calculate the focal length of the lens using $P = 1/f$ (Equation 32-13).	$f = 2.5$ cm
2. Use your result from step 1 and incorporate the result into Equation 32-20 to calculate the angular magnification.	$M = \boxed{10}$

TAKING IT FURTHER Looking through the lens, the object appears 10 times larger because it can be placed at 2.5 cm rather than at 25 cm from the eye, thus increasing the size of the image on the retina tenfold.

PRACTICE PROBLEM 32-13 What is the angular magnification in this example if the near point of the person is 30 cm rather than 25 cm?

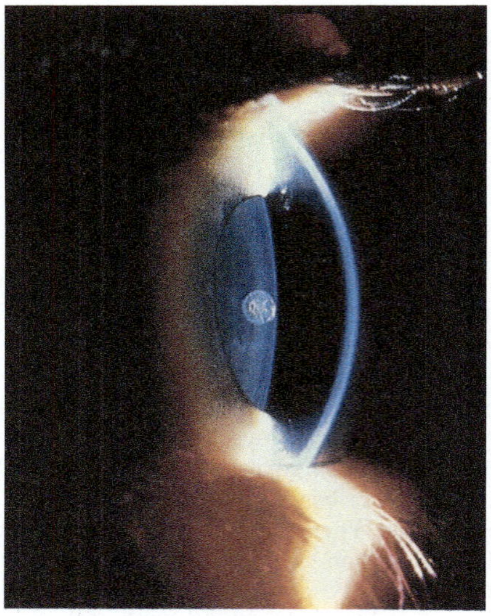

(a)

(b)

(*a*) The human eye in profile. (*b*) The lens of the eye is kept in place by the ciliary muscle (shown here in the upper left), which rings the lens. When the ciliary muscle contracts, the lens tends to bulge. The greater lens curvature enables the eye to focus on nearby objects. (*c*) Some of the 120 million rods and 7 million cones in the eye, magnified approximately 5000 times. The rods (the more slender of the two) are more sensitive in dim light, whereas the cones are more sensitive to color. The rods and cones form the bottom layer of the retina and are covered by nerve cells, blood vessels, and supporting cells. Most of the light entering the eye is reflected or absorbed before reaching the rods and cones. The light that does reach them triggers electrical impulses along nerve fibers that ultimately reach the brain. (*d*) A neural net used in the vision system of certain robots. Loosely modeled on the human eye, it contains 1920 sensors. *((a), (b) and (c) Lennart Nilsson, (d) Courtesy IMEC and University of Pennsylvania Department of Electrical Engineering.)*

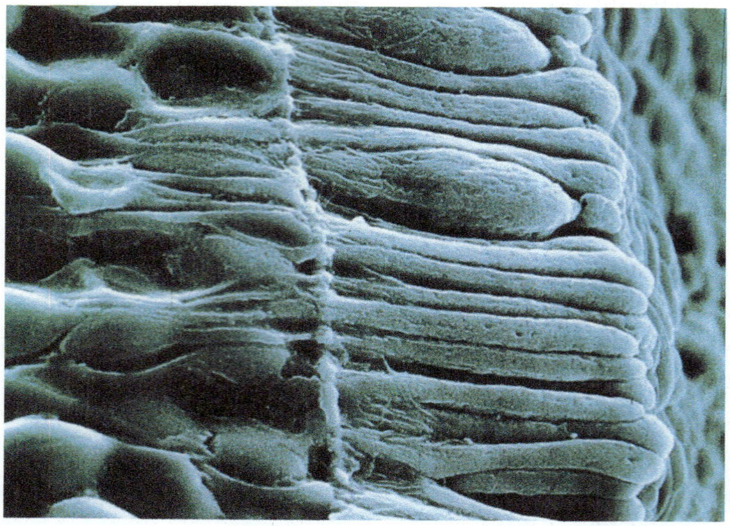

(c)

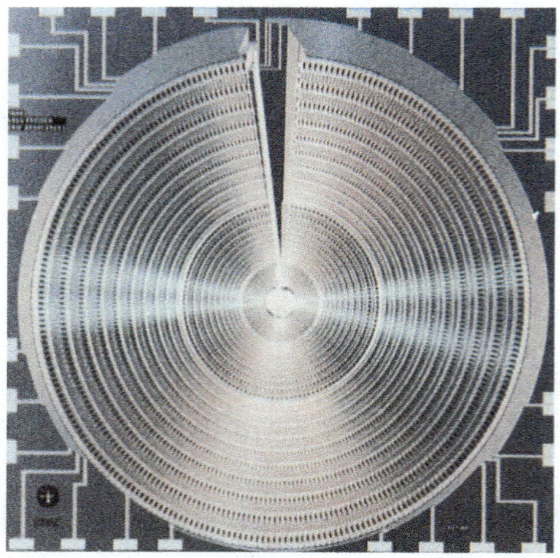

(d)

*THE COMPOUND MICROSCOPE

The compound microscope (Figure 32-52) is used to look at very small objects at short distances. In its simplest form, it consists of two converging lenses. The lens nearest the object, called the **objective,** forms a real image of the object. This image is enlarged and inverted. The lens nearest the eye, called the **eyepiece** or **ocular,** is used as a simple magnifier to view the image formed by the objective. The eyepiece is placed so that the image formed by the objective falls at the first focal point of the eyepiece. The light from each point on the object thus emerges from the eyepiece as a parallel beam, as if it were coming from a point a great distance in front of the eye. (This is commonly called *viewing the image at infinity*.)

The distance between the second focal point of the objective and the first focal point of the eyepiece is called the **tube length** L. The tube length is fixed at 16 cm. The object is placed just outside the first focal point of the objective so that an enlarged image is formed at the first focal point of the eyepiece a distance $L + f_o$ from the objective, where f_o is the focal length of the objective. From Figure 32-52, $\tan\beta = y/f_o = -y'/L$. The lateral magnification of the objective is therefore

$$m_o = \frac{y'}{y} = -\frac{L}{f_o} \qquad 32\text{-}21$$

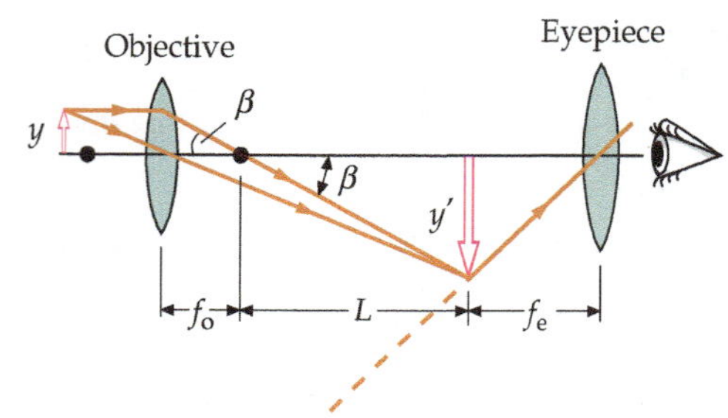

FIGURE 32-52 Schematic diagram of a compound microscope consisting of two positive lenses, the objective of focal length f_o and the eyepiece of focal length f_e. The real image of the object formed by the objective is viewed by the eyepiece, which acts as a simple magnifier. The final image is at infinity.

The angular magnification of the eyepiece (from Equation 32-20) is

$$M_e = \frac{x_{np}}{f_e}$$

where x_{np} is the near-point distance of the viewer and f_e is the focal length of the eyepiece. The magnifying power of the compound microscope is the product of the lateral magnification of the objective and the angular magnification of the eyepiece:

$$M = m_o M_e = -\frac{L}{f_o}\frac{x_{np}}{f_e} \qquad \text{32-22}$$

MAGNIFYING POWER OF A MICROSCOPE

Example 32-15 The Compound Microscope

A microscope has an objective lens that has a focal length equal to 1.2 cm and an eyepiece that has a focal length equal to 2.0 cm. These lenses are separated by 20.0 cm. (*a*) Find the magnifying power if the near point of the viewer is 25.0 cm. (*b*) Where should the object be placed if the final image is to be viewed at infinity?

PICTURE To calculate the magnifying power, we use Equation 32-22. To calculate the object distance for the objective, we use the lens equation.

SOLVE

(*a*) 1. The magnifying power is given by Equation 32-22:
$$M = -\frac{L}{f_o}\frac{x_{np}}{f_e}$$

2. The tube length L is the distance between the lenses minus the focal distances:
$$L = 20.0\text{ cm} - 2.0\text{ cm} - 1.2\text{ cm} = 16.8\text{ cm}$$

3. Substitute this value for L and the given values of x_{np}, f_o, and f_e to calculate M:
$$M = -\frac{L}{f_o}\frac{x_{np}}{f_e} = -\frac{16.8\text{ cm}}{1.2\text{ cm}}\frac{25.0\text{ cm}}{2.0\text{ cm}} = \boxed{-180}$$

(*b*) 1. Calculate the object distance s in terms of the image distance for the objective s' and the focal length f_o:
$$\frac{1}{s} + \frac{1}{s'} = \frac{1}{f_o}$$

2. From Figure 32-52, the image distance for the image of the objective is $f_o + L$:
$$s' = f_o + L = 1.2\text{ cm} + 16.8\text{ cm} = 18.0\text{ cm}$$

3. Substitute to calculate s:
$$\frac{1}{s} + \frac{1}{18.0\text{ cm}} = \frac{1}{1.2\text{ cm}}$$
$$s = \boxed{1.3\text{ cm}}$$

CHECK The magnifying power is very large, which is the purpose of a compound microscope.

TAKING IT FURTHER The object should thus be placed at 1.3 cm from the objective or 0.1 cm outside its first focal point.

*THE TELESCOPE

A telescope is used to view objects that are far away and are often large. The telescope works by creating a real image of the object that is much closer than the object. The astronomical telescope, illustrated schematically in Figure 32-53, consists of two converging lenses—an objective lens that forms a real, inverted image and an eyepiece that is used as a simple magnifier to view that image. Because the object is very far away, the image of the objective lies in the focal plane of the objective, and the image distance equals the focal length f_o. The image formed by the objective is much smaller than the object (because the object distance is

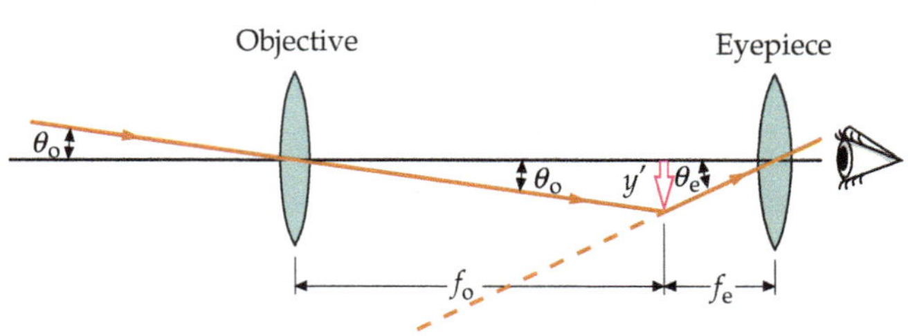

FIGURE 32-53 Schematic diagram of an astronomical telescope. The objective lens forms a real, inverted image of a distant object near its second focal point, which coincides with the first focal point of the eyepiece. The eyepiece serves as a simple magnifier to view the image.

much larger than the focal length of the objective). For example, if we are looking at the moon, the image of the moon formed by the objective is much smaller than the moon itself. The function of the objective is not to magnify the object, but to produce an image that is close enough to us so it can be magnified by the eyepiece acting as a simple magnifier. The eyepiece is placed a distance f_e from the image, where f_e is the focal length of the eyepiece, so the final image can be viewed at infinity. Because this image is at the second focal plane of the objective and at the first focal plane of the eyepiece, the objective lens and the eyepiece must be separated by the sum of the focal lengths of the objective lens and eyepiece, $f_o + f_e$.

The magnifying power of the telescope is the angular magnification θ_e/θ_o, where θ_e is the angle subtended by the virtual image produced by the eyepiece as viewed through the eyepiece, and θ_o is the angle subtended by the object when it is viewed directly by the unaided eye. The angle θ_o is the same as that subtended by the object at the objective shown in Figure 32-53. (The distance from a distant object, such as the moon, to the objective is essentially the same as the distance to the eye.) From this figure, we can see that

$$\tan\theta_o = \frac{y}{s} = -\frac{y'}{f_o} \approx \theta_o$$

where we have used the small-angle approximation $\tan\theta \approx \theta$. The angle θ_e in the figure is that subtended by the image at infinity formed by the eyepiece:

$$\tan\theta_e = \frac{y'}{f_e} \approx \theta_e$$

Because y' is negative, θ_e is negative, indicating that the image is inverted. The magnifying power of the telescope is then

$$M = \frac{\theta_e}{\theta_o} = -\frac{f_o}{f_e} \qquad \text{32-23}$$

MAGNIFYING POWER OF A TELESCOPE

From Equation 32-23, we can see that a large magnifying power is obtained with an objective of large focal length and an eyepiece of short focal length.

PRACTICE PROBLEM 32-14

The world's largest refracting telescope is at the Yerkes Observatory of the University of Chicago at Williams Bay, Wisconsin. The telescope's objective has a diameter of 1.02 m and a focal length of 19.5 m. The focal length of the eyepiece is 10.0 cm. What is its magnifying power?

The main consideration with an astronomical telescope is not its magnifying power but its light-gathering power, which depends on the size of the objective. The larger the objective, the brighter the image. Very large lenses without aberrations are difficult to produce. In addition, there are mechanical problems in supporting very large, heavy lenses by their edges. A reflecting telescope (Figure 32-54 and Figure 32-55) uses a concave mirror instead of a lens for its objective. The mirror offers several advantages. For example, a mirror does not produce chromatic aberration.

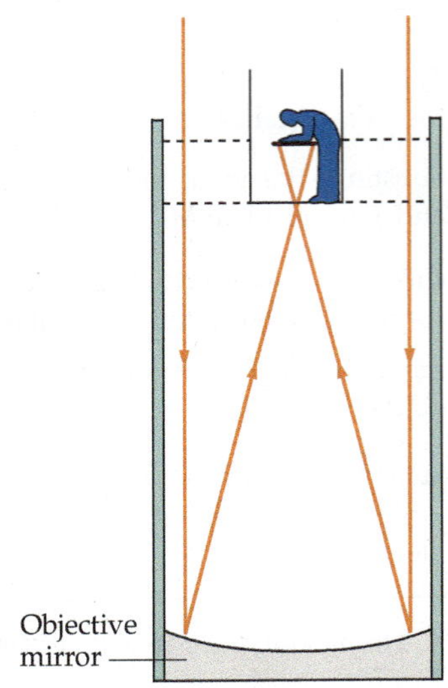

FIGURE 32-54 A reflecting telescope uses a concave mirror instead of a lens for its objective. Because the viewer compartment blocks off some of the incoming light, the arrangement shown here is used only in telescopes with very large objective mirrors.

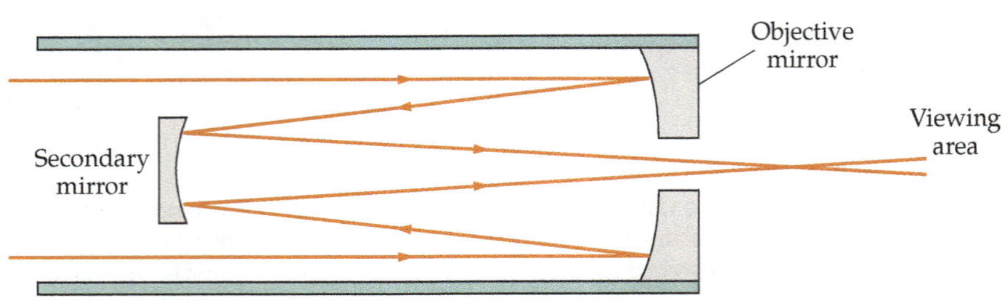

FIGURE 32-55 This reflecting telescope has a secondary mirror to redirect the light through a small hole in the objective mirror, thus providing more room for auxiliary instruments in the viewing area.

In addition, mechanical support is much simpler, because the mirror weighs far less than a lens of equivalent optical quality, and the mirror can be supported over its entire back surface. In modern Earth-based telescopes, the objective mirror consists of several dozen adaptive mirror segments that can be adjusted individually to correct for minute variations in gravitational stress when the telescope is tilted, and to compensate for thermal expansions and contractions and other changes caused by climatic conditions. In addition, they can adjust to nullify the distortions produced by atmospheric fluctuations.

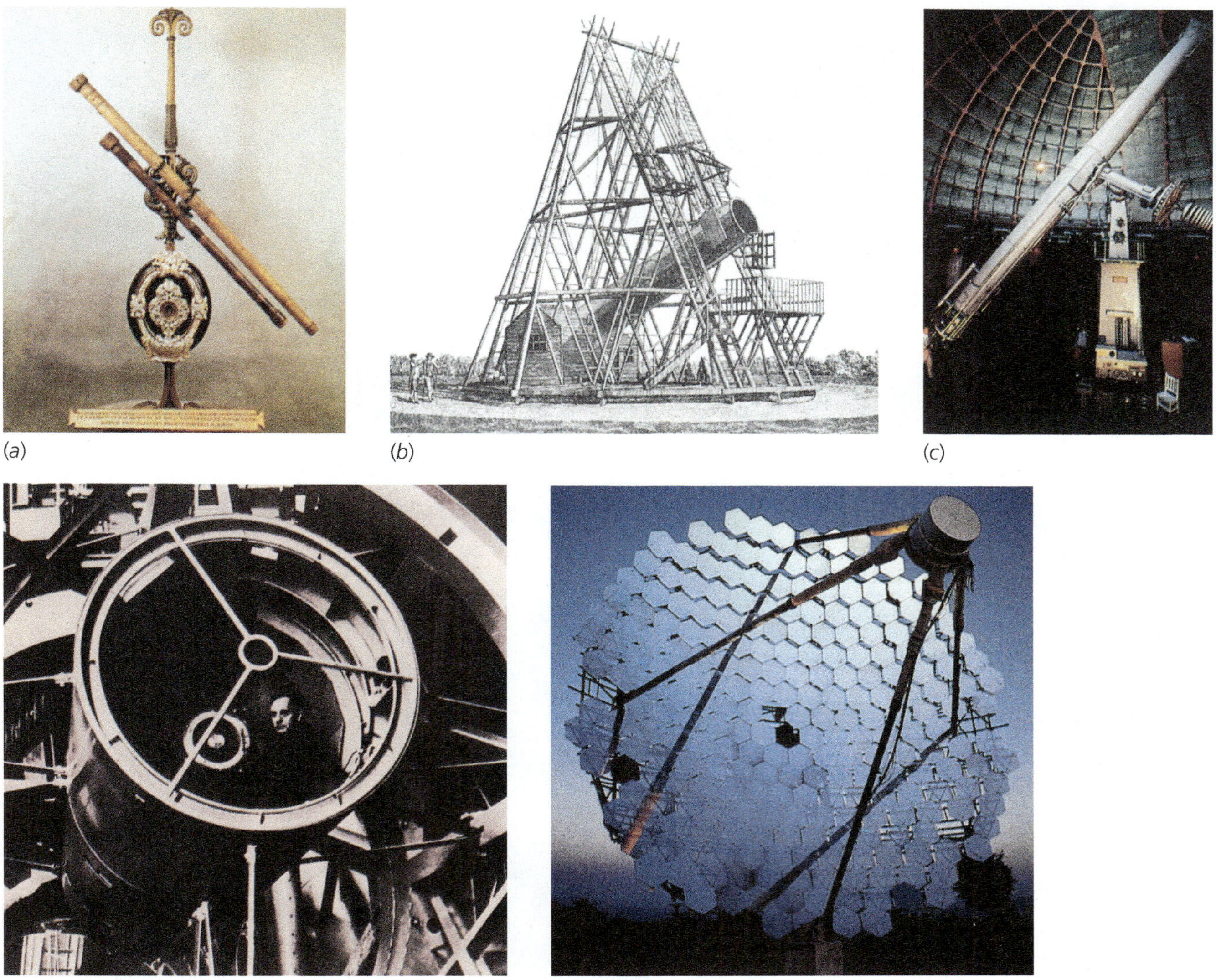

(a) (b) (c) (d) (e)

Astronomy at optical wavelengths began with Galileo approximately 400 years ago. In the twentieth century, astronomers began to explore the electromagnetic spectrum at other wavelengths; beginning with radio astronomy in the 1940s, satellite-based X-ray astronomy in the early 1960s, and more recently, ultraviolet, infrared, and gamma-ray astronomy. (*a*) Galileo's seventeenth-century telescope, with which he discovered mountains on the moon, sunspots, Saturn's rings, and the bands and moons of Jupiter. (*b*) An engraving of the reflector telescope built in the 1780s and used by the great astronomer Friedrich Wilhelm Herschel, who was the first to observe galaxies outside our own. (*c*) Because it is difficult to make large, flaw-free lenses, refractor telescopes like this 91.4-cm telescope at Lick Observatory have been superseded in light-gathering power by reflector telescopes. (*d*) The great astronomer Edwin Powell Hubble, who discovered the apparent expansion of the universe, is shown seated in the observer's cage of the 5.08-m Hale reflecting telescope, which is large enough for the observer to sit at the prime focus itself. (*e*) This 10-m optical reflector at the Whipple Observatory in southern Arizona is the largest instrument designed exclusively for use in gamma-ray astronomy. High-energy gamma rays of unknown origin strike the upper atmosphere and create cascades of particles. Among these particles are high-energy electrons that emit Cerenkov radiation observable from the ground. According to one hypothesis, high-energy gamma rays are emitted as matter is accelerated toward ultradense rotating stars called pulsars. *((a) Scala/Art Resource, (b) Royal Astronomical Society Library, (c) Lick Observatory, courtesy of the University of California Regents, (d) California Institute of Technology, (e) Gary Ladd.)*

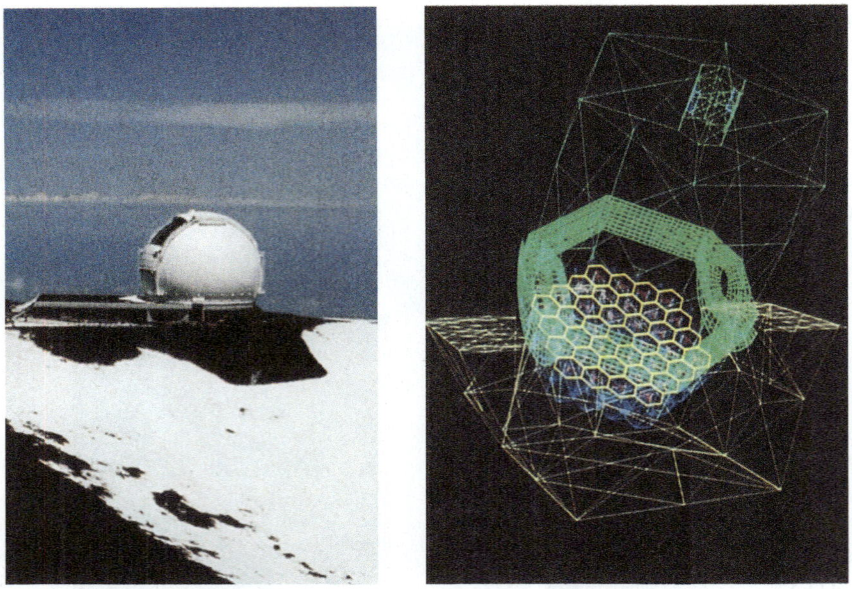

(a) (b)

(c)

(*a*) The Keck Observatory, atop the inactive volcano of Mauna Kea, Hawaii, houses the world's largest optical telescope. The clear, dry air and lack of light pollution make the remote heights of Mauna Kea an ideal site for astronomical observations. (*b*) The Keck telescope is composed of 36 hexagonal mirror segments performing together as if they were a single mirror 10 m wide—roughly twice as large as the largest single-mirror telescope presently in operation. (*c*) Beneath each Keck mirror is a system of computer-controlled sensors and motor-driven actuators that can continuously vary the mirror's shape. These variations, which are sensitive to within 100 nm, enable the system to correct for variations in the alignments of the segments due to minute variations in gravitational stress when the telescope is tilted and to compensate for thermal expansions and contractions and fluctuations caused by gusts of wind on the mountaintop. *(California Association for Research in Astronomy.)*

The Hubble Space Telescope is high above the atmospheric turbulence that limits the ability of ground-based telescopes to resolve images at optical wavelengths. *(NASA.)*

Physics Spotlight

Eye Surgery: New Lenses for Old

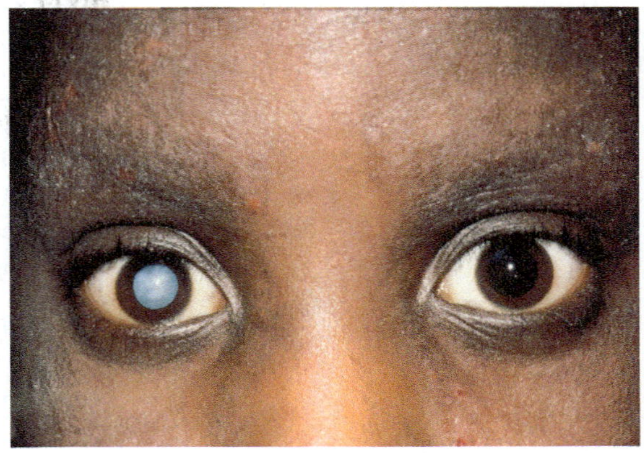

Unfortunately, this person has a cataract affecting the entire lens of her right eye. The presence of the cataract is associated with the severe eczema that can be seen on her forehead. *(Western Ophthalmic Hospital/Photo Researchers.)*

The first documented eye surgery took place over 2000 years ago and was performed by a surgeon in India known today as Susruta.[*, †] The lens of the human eye is susceptible to clouding with age and exposure to ultraviolet light or with exposure to some diseases or chemicals. This opacity is called a cataract. In the *Susruta Samhita*, the surgeon wrote of diagnosing and removing cataract-clouded lenses.

In 2005, more than two million cataract surgeries were performed in the United States.[‡] Most involved removing the lens and implanting an artificial lens inside the eye. Although permanent *i*ntra*o*cular *l*enses (IOLs) were pioneered in 1951,[#] they did not gain widespread acceptance until the 1980s. Before then, patients had to wear strong external corrective lenses (eyeglasses or contact lenses after the 1960s) after cataract surgery.

Cataract patients who have had IOLs implanted have had to wear external corrective lenses, because the IOLs are fixed focus lenses. However, several different IOLs that have variable focus have been recently introduced.[°, §] These accommodative, multifocal lenses are flexed and focused by the eye muscles. Some patients have had such good results with accommodative IOLs that they do not need supplemental glasses.

Many people without cataracts wear external corrective lenses. In the United States alone, over 150 million people spend at least $15 billion on external corrective lenses each year.[¶] In addition, nearly a million get refractive surgery each year that promises to reduce their need for external corrective lenses.[**] Refractive surgery reshapes the cornea to minimize refractive errors. It was pioneered in the 1930s, but did not become popular until techniques that use lasers were developed in the late 1980s.[††] Refractive surgery can be done by using chemicals or mechanical abrasion, but it is most commonly done by using lasers that vaporize minute pieces of the cornea. Some refractive surgeries involve cutting a thin flap in the outer layers of the cornea, shaping the stroma, the thickest layer of the cornea underneath, and then replacing the flap. Other surgeries just involve shaping the outer surface of the cornea.[‡‡]

Refractive surgeries work best on patients who have medium to low refractive error. In those patients, however, the results are promising. Up to 72% of the patients who have low to moderate nearsightedness ended up with postsurgical uncorrected vision that is considered 20/20, or normal.[##] Recently, customized wavefront-guided laser surgery has been used to minimize refractive errors that are not fixed by earlier techniques of eye surgery.[°°] A light is bounced off of the retina through many locations on the cornea. The returning wavefront is mapped, and very tiny refractive errors are located for elimination. With custom wavefront-guided laser surgery, up to 89% of patients ended up with uncorrected 20/20 vision.

But even with the best surgeons, there can be complications with refractive surgeries.[§§, ¶¶] The most common is dry eyes, but in up to 2% of patients vision can actually be worse after surgery.[***] Complications are more likely in people who have large refractive errors, and refractive surgery is not recommended for them. But those people may also eventually be able to not wear external corrective lenses. Recently, accommodative IOLs have been tested on people who have high refractive errors. Many were able to have near-normal vision after surgery without additional corrective lenses.[†††]

---

* Raju, V. K., "Susruta of Ancient India." *Indian Journal of Ophthalmology*, Feb. 2003, Vol. 51, No. 2, pp. 119–122. http://www.ijo.in/article.asp?issn=0301-4738;year=2003;volume=51;issue=2;spage=119;epage=122;aulast=Raju As of Nov. 2006.

† Hellemans, A., and Bunch, B., *The Timetables of Science*. New York: Simon & Schuster, 1988, p. 28.

‡ American Academy of Ophthalmology, "Industry News." *Academy Express*, Vol. 5, No. 14, http://www.aao.org/news/academy_express/20060405.cfm#asc As of Nov. 2006.

\# Apple, D. J., "Sir Harold Ridley." *Journal of Cataract and Refractive Surgery*, Mar. 2004, Vol. 30, No. 3, pp. 47–52.

° Cummings, J. S., et al., "Clinical Evaluation of the Crystalens AT-45 Accommodating Intraocular Lens: Results of the U.S. Food and Drug Administration Clinical Trial." *Journal of Cataract and Refractive Surgery*, May 2006, Vol. 32, No. 5, pp. 812–825.

§ Charters, L., "Dual-Optic IOL Effective Answer to Presbyopia." *Ophthalmology Times*, Feb. 1, 2006, pp. 20–21.

¶ Rados, C., "A Focus on Vision." *FDA Consumer*, Jul.–Aug. 2006, pp. 10–17.

** American Academy of Ophthalmology, "Industry News." *Academy Express*, Vol. 5, No. 14, http://www.aao.org/news/academy_express/20060405.cfm#asc As of Nov. 2006.

†† Kornmehl, E., "The Start of Something Big." *Ophthalmology Times*, Nov. 1, 2006, Vol. 31, No. 21, p. 24.

‡‡ Sakimoto, T., Rosenblatt, M., and Azar, D., "Laser Eye Surgery for Refractive Errors." *The Lancet*, Apr. 29, 2006, Vol. 367, No. 9520, pp. 1432–1447.

\#\# Sakimoto, T., Rosenblatt, M., and Azar, D., op. cit.

°° Mackenzie, D., "Coming Soon: 'Wavefront Eye Surgery'?" *Science*, Mar. 14, 2003, Vol. 299, No. 5613, p. 1655.

§§ Potter, J., "Do What's Right When Refractive Surgery Goes Wrong." *Review of Optometry*, Oct. 15, 2006, pp. 52–62.

¶¶ Guttman, C., "DLK a Lifelong Risk in Post-LASIK eyes." *Ophthalmology Times*, Nov. 1, 2006, Vol. 31, No. 21, pp. 1+.

*** Sakimoto, T., Rosenblatt, M., and Azar, D., op. cit.

††† Charters, L., "Accommodating IOL Improves Vision for High Refractive Errors in Analysis." *Ophthalmology Times*, Jun. 1, 2006, Vol. 31, No. 11, p. 43.

Summary

TOPIC	RELEVANT EQUATIONS AND REMARKS	
1. Virtual and Real Images and Objects		
Images	An image is *real* if actual light rays converge to each image point. This can occur on the reflected-light side of a mirror or on the refracted-light side of a thin lens or refracting surface. An image is *virtual* if only the extensions of the actual light rays converge to each image point. This can occur behind a mirror or on the incident-light side of a lens or refracting surface.	
Objects	A real object is either a physical object or a real image. An object is *real* if actual light rays diverge from each object point. This can occur only on the incident-light side of a mirror, lens, or refracting surface. An object is *virtual* if only extensions of actual light rays diverge from each object point. This can occur only behind a mirror or on the refracted-light side of a lens or refracting surface.	
2. Spherical Mirrors		
Focal length	The focal length is the image distance when the object is at infinity, so the incident light is parallel to the axis.	
Mirror equation (for locating an image)	$$\frac{1}{s}+\frac{1}{s'}=\frac{1}{f}$$	32-4
	where	
	$$f=\frac{r}{2}$$	32-3
Lateral magnification	$$m=\frac{y'}{y}=-\frac{s'}{s}$$	32-5
Ray diagrams	Images can be located by a ray diagram using any two paraxial rays. The parallel, focal, and radial rays are the easiest to draw: 1. The parallel ray, drawn parallel to the axis, is reflected through the focal point. 2. The focal ray, drawn through the focal point, is reflected parallel to the axis. 3. The radial ray, drawn through the center of curvature, is reflected back on itself.	
Sign conventions for reflection	1. s is positive if the object is on the incident-light side of the mirror. 2. s' is positive if the image is on the reflected-light side of the mirror. 3. r and f are positive if the mirror is concave so the center of curvature is on the reflected-light side of the mirror.	
3. Images Formed by Refraction		
Refraction at a single surface	$$\frac{n_1}{s}+\frac{n_2}{s'}=\frac{n_2-n_1}{r}$$	32-6
	where n_1 is the index of refraction of the medium on the incident-light side of the surface.	
Magnification	$$m=\frac{y'}{y}=-\frac{n_1 s'}{n_2 s}$$	32-7
Sign conventions for refraction	1. s is positive for objects on the incident-light side of the surface.	
	2. s' is positive for images on the refracted-light side of the surface.	
	3. r is positive if the center of curvature is on the refracted-light side of the surface.	

TOPIC	RELEVANT EQUATIONS AND REMARKS
4. Thin Lenses	
Focal length (lens-maker's equation)	$\frac{1}{f} = \left(\frac{n}{n_{\text{air}}} - 1\right)\left(\frac{1}{r_1} - \frac{1}{r_2}\right)$ 32-11 A positive lens ($f > 0$) is a converging lens. A negative lens ($f < 0$) is a diverging lens.
First and second focal points	Incident rays parallel to the axis emerge directed either toward or away from the *first focal point F′*. Incident rays directed either toward or away from the *second focal point F* emerge parallel with the axis.
Power	$P = \frac{1}{f}$ 32-13
Thin-lens equation (for locating image)	$\frac{1}{s} + \frac{1}{s'} = \frac{1}{f}$ 32-12
Magnification	$m = \frac{y'}{y} = -\frac{s'}{s}$ 32-14
Ray diagrams	Images can be located by a ray diagram using any two paraxial rays. The parallel, central, and focal rays are the easiest to draw: 1. The parallel ray, drawn parallel to the axis, emerges directed toward (or away from) the second focal point. 2. The central ray, drawn through the center of the lens, is not deflected. 3. The focal ray, drawn through (or toward) the first focal point, emerges parallel to the axis.
Sign conventions for lenses	The sign conventions are the same as for refraction at a spherical surface.
5. *Aberrations	Blurring of the image of a single object point is called aberration. Spherical aberration results from the fact that a spherical surface focuses only paraxial rays (those that travel close to the axis) to a single point. Nonparaxial rays are focused at nearby points depending on the angle made with and distance from the axis. Spherical aberration can be reduced by blocking the rays farthest from the axis. This, of course, reduces the amount of light reaching the image. Chromatic aberration, which occurs with lenses but not mirrors, results from the variation in the index of refraction with wavelength. Lens aberrations are most commonly reduced by using a series of lens elements.
6. *The Eye	The cornea–lens system of the eye focuses light on the retina, where it is sensed by the rods and cones that send information along the optic nerve to the brain. When the eye is relaxed, the focal length of the cornea-lens system is about 2.5 cm, which is the distance to the retina. When objects are brought near the eye, the lens changes shape to decrease the overall focal length so that the image remains focused on the retina. The closest distance for which the image can be focused on the retina is called the near point, typically about 25 cm. The apparent size of an object depends on the size of the image on the retina. The closer the object, the larger the image on the retina and therefore the larger the apparent size of the object.
7. *The Simple Magnifier	A simple magnifier consists of a lens with a positive focal length that is smaller than the near point.
Magnifying power (angular magnification)	$M = \frac{\theta}{\theta_o} = \frac{x_{\text{np}}}{f}$ 32-20
8. *The Compound Microscope	The compound microscope is used to look at very small objects that are nearby. It consists of two converging lenses (or lens systems), an objective and an eyepiece. The object to be viewed is placed just outside the focal point of the objective, which forms an enlarged image of the object at the focal plane of the eyepiece. The eyepiece acts as a simple magnifier to view the final image.

TOPIC	RELEVANT EQUATIONS AND REMARKS	
Magnifying power (angular magnification)	$M = m_o M_e = -\frac{L}{f_o}\frac{x_{np}}{f_e}$ where L is the tube length, the distance between the second focal point of the objective and the first focal point of the eyepiece.	32-22
9. *The Telescope	The telescope is used to view objects that are far away. The objective of the telescope forms a real image of the object that is much smaller than the object but much closer. The eyepiece is then used as a simple magnifier to view the image. A reflecting telescope uses a mirror for its objective.	
Magnifying power (angular magnification)	$M = \frac{\theta_e}{\theta_o} = -\frac{f_o}{f_e}$	32-23

Answers to Concept Checks

32-2 Ben can see only the image at P'_1.

32-3 Infinity

32-4

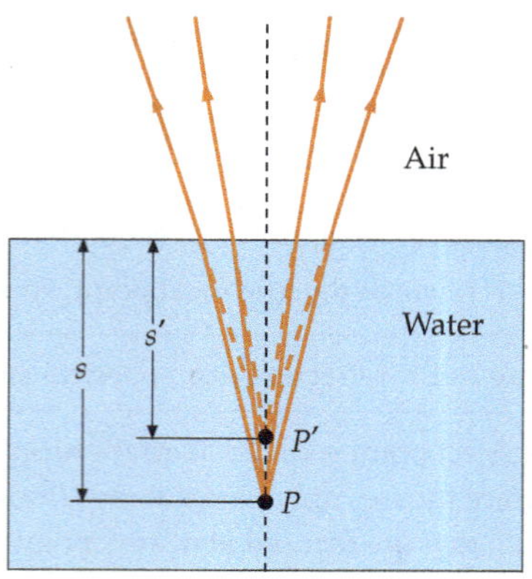

FIGURE 32-56 Ray diagram for the image of an object in water as viewed from directly overhead. The depth of the image is less than the depth of the object.

Answers to Practice Problems

32-3	(*a*) 8.0 cm (*b*) $s' = -4.0$ cm
32-4	$s' = -2.5$ cm, $m = +0.50$; the image is upright, virtual, and reduced in size.
32-5	390 yd
32-6	(*a*) $s' = -6.44$ cm and (*b*) $m = 1.14$. Fluffy sees Goldie to be 1.1 cm closer and 14 percent larger than she actually is.
32-7	The image is 5.6 cm from the near side of the aquarium.
32-8	18 cm
32-10	$s' = 30$ cm, $m = -2.0$; real, inverted
32-11	$s' = -10$ cm, $m = 2.0$; virtual, upright
32-12	$P_{eye} = 41.33$ D; $P_c = 41.33$ D $+ 2.67$ D $= 44$ D; the two powers are equal
32-13	$M = 12$
32-14	$M = -195$

Problems

In a few problems, you are given more data than you actually need; in a few other problems, you are required to supply data from your general knowledge, outside sources, or informed estimate.

Interpret as significant all digits in numerical values that have trailing zeros and no decimal points.

- • Single-concept, single-step, relatively easy
- •• Intermediate-level, may require synthesis of concepts
- ••• Challenging
- SSM Solution is in the *Student Solutions Manual*
- Consecutive problems that are shaded are paired problems.

CONCEPTUAL PROBLEMS

1 • Can a virtual image be photographed? If so, give an example. If not, explain why.

2 • Suppose the x, y, and z axes of a 3-D coordinate system are painted different colors. One photograph is taken of the coordinate system and another is taken of its image in a plane mirror. Is it possible to tell that one of the photographs is of a mirror image? Or could both photographs be of the real coordinate system from different angles?

3 • True or false
(*a*) The virtual image formed by a concave mirror is always smaller than the object.
(*b*) A concave mirror always forms a virtual image.
(*c*) A convex mirror never forms a real image of a real object.
(*d*) A concave mirror never forms an enlarged real image of an object. SSM

4 • An ant is crawling along the axis of a concave mirror that has radius of curvature R. At what object distances, if any, will the mirror produce (*a*) an upright image, (*b*) a virtual image, (*c*) an image smaller than the object, and (*d*) an image larger than the object?

5 • An ant is crawling along the axis of a convex mirror that has radius of curvature R. At what object distances, if any, will the mirror produce (*a*) an upright image, (*b*) a virtual image, (*c*) an image smaller than the object, and (*d*) an image larger than the object? SSM

6 •• Convex mirrors are often used for rearview mirrors on cars and trucks to give a wide-angle view. "Warning, objects are closer than they appear" is written below the mirrors. Yet, according to a ray diagram, the image distance for distant objects is much shorter than the object distance. Why then do they appear more distant?

7 • As an ant on the axis of a concave mirror crawls from a great distance to the focal point of a concave mirror, the image of the ant moves (*a*) from a great distance toward the focal point and is always real, (*b*) from the focal point to a great distance from the mirror and is always real, (*c*) from the focal point to the center of curvature of the mirror and is always real, (*d*) from the focal point to a great distance from the mirror and changes from a real image to a virtual image.

8 • A kingfisher bird that is perched on a branch a few feet above the water is viewed by a scuba diver submerged beneath the surface of the water directly below the bird. Does the bird appear to the diver to be closer to or farther from the surface than the actual bird? Explain your answer using a ray diagram.

9 • An object is placed on the axis of a diverging lens whose focal length has a magnitude of 10 cm. The distance from the object to the lens is 40 cm. The image is (*a*) real, inverted, and diminished; (*b*) real, inverted, and enlarged; (*c*) virtual, inverted, and diminished; (*d*) virtual, upright, and diminished; (*e*) virtual, upright, and enlarged.

10 • If an object is placed between the focal point of a converging lens and the optical center of the lens, the image is (*a*) real, inverted, and enlarged; (*b*) virtual, upright, and diminished; (*c*) virtual, upright, and enlarged; (*d*) real, inverted, and diminished.

11 • A converging lens is made of glass that has an index of refraction of 1.6. When the lens is in air, its focal length is 30 cm. When the lens is immersed in water, its focal length (*a*) is greater than 30 cm, (*b*) is between zero and 30 cm, (*c*) is equal to 30 cm, (*d*) has a negative value.

12 • True or false:

(*a*) A virtual image cannot be displayed on a screen.
(*b*) A negative image distance implies that the image is virtual.
(*c*) All rays parallel to the axis of a spherical mirror are reflected through a single point.
(*d*) A diverging lens cannot form a real image from a real object.
(*e*) The image distance for a converging lens is always positive.

13 • **BIOLOGICAL APPLICATION** Both the human eye and the digital camera work by forming real images on light-sensitive surfaces. The eye forms a real image on the retina and the camera forms a real image on a CCD array. Explain the difference between the ways in which these two systems *accommodate*. That is, the difference between how an eye adjusts and how a camera adjusts (or can be adjusted) to form a focused image for objects at both large and short distances from the camera.

14 • **BIOLOGICAL APPLICATION** If an object is 25 cm in front of the naked eye of a farsighted person, an image (*a*) would be formed behind the retina if it were not for the fact that that the light is blocked (by the back of the eyeball) and the corrective contact lens should be convex, (*b*) would be formed behind the retina if it were not for the fact that that the light is blocked (by the back of the eyeball) and the corrective contact lens should be concave, (*c*) is formed in front of the retina and the corrective contact lens should be convex, (*d*) is formed in front of the retina and the corrective contact lens should be concave.

15 •• Explain the following statement: A microscope is an object magnifier, but a telescope is an angle magnifier. *Hint: Take a look at the ray diagram for each magnifier and use it to explain the difference in adjectives.*

ESTIMATION AND APPROXIMATION

16 • Estimate the location and size of the image of your face when you hold a shiny new tablespoon a foot in front of your face and with the convex side toward you.

17 • Estimate the focal length of the "mirror" produced by the surface of the water in the reflection pool in front of the Lincoln Memorial on a still night.

18 •• Estimate the maximum value that could be obtained for the magnifying power of a simple magnifier, using Equation 32-20. *Hint: Think about the smallest focal length lens that could be made from glass and still be used as a magnifier.*

PLANE MIRRORS

19 • The image of the object point P in Figure 32-57 is viewed by an eye, as shown. Draw two rays from the object point that reflect from the mirror and enter the eye. If the object point and the mirror are fixed in their locations, indicate the range of locations where the eye can be positioned and still see the image of the object point. SSM

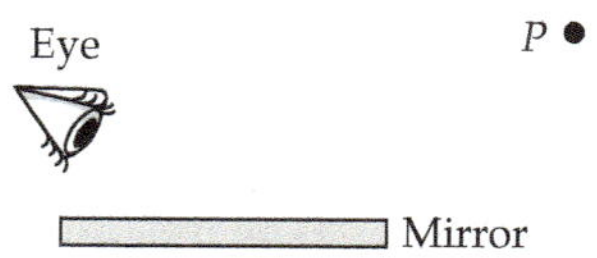

FIGURE 32-57 Problem 19

20 • You are 1.62 m tall and want to be able to see your full image in a vertical plane mirror. (*a*) What is the minimum height of the mirror that will meet your needs? (*b*) How far above the floor should the bottom of mirror in (*a*) be placed, assuming that the top of your head is 14 cm above your eye level? Use a ray diagram to explain your answer.

21 •• (*a*) Two plane mirrors make an angle of 90°. The light from a point object that is arbitrarily positioned in front of the mirrors produces images at three locations. For each image location, draw two rays from the object that, after one or two reflections, appear to come from the image location. (*b*) Two plane mirrors make an angle of 60° with each other. Draw a sketch to show the location of all the images formed of an object on the bisector of the angle between the mirrors. (*c*) Repeat Part (*b*) for an angle of 120°. SSM

22 •• Show that the mirror equation (Equation 32-4 where $f = r/2$) yields the correct image distance and magnification for a plane mirror.

23 •• When two plane mirrors are parallel, such as on opposite walls in a barber shop, multiple images arise because each image in one mirror serves as an object for the other mirror. An object is placed between parallel mirrors separated by 30 cm. The object is 10 cm in front of the left mirror and 20 cm in front of the right mirror. (*a*) Find the distance from the left mirror to the first four images in that mirror. (*b*) Find the distance from the right mirror to the first four images in that mirror. (*c*) Explain why each more distant image becomes fainter and fainter.

SPHERICAL MIRRORS

24 •• A concave mirror has a radius of curvature equal to 24 cm. Use ray diagrams to locate the image, if it exists, for an object near the axis at distances of (*a*) 55 cm, (*b*) 24 cm, (*c*) 12 cm, and (*d*) 8.0 cm from the mirror. For each case, state whether the image is real or virtual; upright or inverted; and enlarged, reduced, or the same size as the object.

25 • (*a*) Use the mirror equation (Equation 32-4 where $f = r/2$) to calculate the image distances for the object distances and mirror of Problem 24. (*b*) Calculate the magnification for each given object distance. SSM

26 • A convex mirror has a radius of curvature that has a magnitude equal to 24 cm. Use ray diagrams to locate the image, if it exists, for an object near the axis at distances of (*a*) 55 cm, (*b*) 24 cm, (*c*) 12 cm, (*d*) 8.0 cm, and (*e*) 1.0 cm from the mirror. For each case, state whether the image is real or virtual; upright or inverted; and enlarged, reduced, or the same size as the object.

27 • (*a*) Use the mirror equation (Equation 32-4 where $f = r/2$) to calculate the image distances for the object distances and mirror of Problem 26. (*b*) Calculate the magnification for each given object distance.

28 •• Use the mirror equation (Equation 32-4 where $f = r/2$) to prove that a convex mirror cannot form a real image of a real object, no matter where the object is placed.

29 • A dentist wants a small mirror that will produce an upright image that has a magnification of 5.5 when the mirror is located 2.1 cm from a tooth. (*a*) Should the mirror be concave or convex? (*b*) What should the radius of curvature of the mirror be? SSM

30 •• CONTEXT-RICH Convex mirrors are used in many stores to provide a wide angle of surveillance for a reasonable mirror size. Your summer job is at a local convenience store that uses the mirror shown in Figure 32-58. This setup allows you (or the clerk) to survey the entire store when you are 5.0 m from the mirror. The mirror has a radius of curvature equal to 1.2 m. Assume all rays are paraxial. (*a*) If a customer is 10 m from the mirror, how far from the mirror is his image? (*b*) Is the image in front of or behind the mirror? (*c*) If the customer is 2.0 m tall, how tall is his image?

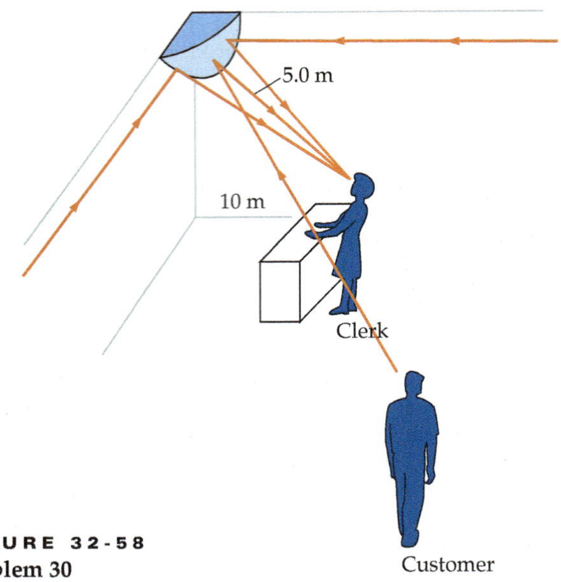

FIGURE 32-58
Problem 30

31 •• A certain telescope uses a concave spherical mirror that has a radius equal to 8.0 m. Find the location and diameter of the image of the moon formed by this mirror. The moon has a diameter of 3.5×10^6 m and is 3.8×10^8 m from Earth.

32 •• A piece of a thin spherical shell that has a radius of curvature of 100 cm is silvered on both sides. The concave side of the piece forms a real image 75 cm from the piece. The piece is then turned around so that its convex side faces the object. The piece is moved so that the image is now 35 cm from the piece on the concave side. (*a*) How far was the piece moved? (*b*) Was it moved toward the object or away from the object?

33 •• Two light rays parallel to the optic axis of a concave mirror strike that mirror as shown in Figure 32-59. The mirror has a radius of curvature equal to 5.0 m. They then strike a small spherical mirror that is 2.0 m from the large mirror. The light rays finally meet at the vertex of the large mirror. (*Note:* The small mirror is shown as planar, so as not to give away the answer, but it is *not* actually planar.) (*a*) What is the radius of curvature of the small mirror? (*b*) Is that mirror convex or concave? Explain your answer.

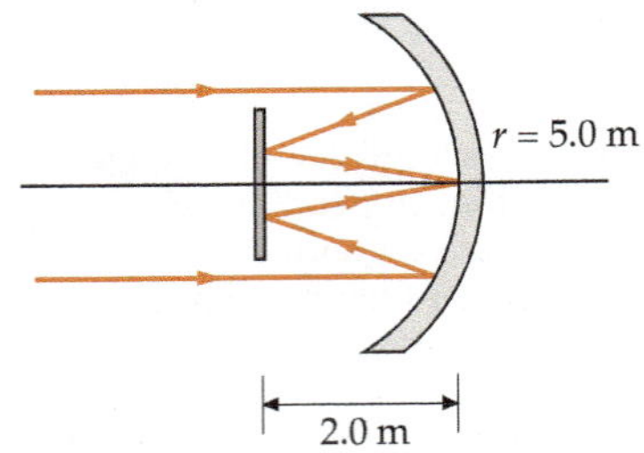

FIGURE 32-59
Problem 33

IMAGES FORMED BY REFRACTION

34 •• A very long 1.75-cm-diameter glass rod has one end ground and polished to a convex spherical surface that has a 7.20-cm radius. The glass material has an index of refraction of 1.68. (*a*) A point object in air is on the axis of the rod and 30.0 cm from the spherical surface. Find the location of the image and state whether the image is real or virtual. (*b*) Repeat Part (*a*) for a point object in air, on the axis, and 5.00 cm from the spherical surface. Draw a ray diagram for each case.

35 • A fish is 10 cm from the front surface of a spherical fish bowl of radius 20 cm. (*a*) How far behind the surface of the bowl does the fish appear to someone viewing the fish from in front of the bowl? (*b*) By what distance does the fish's apparent location change (relative to the front surface of the bowl) when it swims away to 30 cm from the front surface? SSM

36 •• A very long 1.75-cm-diameter glass rod has one end ground and polished to a concave spherical surface that has a 7.20-cm radius. The glass material has an index of refraction of 1.68. A point object in air is on the axis of the rod and 15.0 cm from the spherical surface. Find the location of the image and state whether the image is real or virtual. Draw a ray diagram.

37 •• Repeat Problem 34 for when the glass rod and the object are immersed in water, and (*a*) the object is 6.00 cm from the spherical surface, and (*b*) the object is 12.0 cm from the spherical surface. SSM

38 •• Repeat Problem 36 for when the glass rod and the object are immersed in water and the object is 20 cm from the spherical surface.

39 •• A rod that is 96.0 cm long is made of glass that has an index of refraction equal to 1.60. The rod has its ends ground to convex spherical surfaces that have radii equal to 8.00 cm and 16.0 cm. An object is in air on the long axis of the rod 20.0 cm from the end that has the 8.00-cm radius. (*a*) Find the image distance due to refraction at the 8.00-cm-radius surface. (*b*) Find the position of the final image due to refraction at both surfaces. (*c*) Is the final image real or virtual?

40 •• Repeat Problem 39 for an object in air on the axis of the glass rod 20.0 cm from the end that has the 16.0-cm radius.

THIN LENSES AND THE LENS-MAKER'S EQUATION

41 • A double concave lens that has an index of refraction equal to 1.45 has radii whose magnitudes are equal to 30.0 cm and 25.0 cm. An object is located 80.0 cm to the left of the lens. Find (*a*) the focal length of the lens, (*b*) the location of the image, and (*c*) the magnification of the image. (*d*) Is the image real or virtual? Is the image upright or inverted? SSM

42 • The following thin lenses are made of glass that has an index of refraction equal to 1.60. Make a sketch of each lens and find each focal length in air: (*a*) $r_1 = 20.0$ cm and $r_2 = 10.0$ cm, (*b*) $r_1 = 10.0$ cm and $r_2 = 20.0$ cm, and (*c*) $r_1 = -10.0$ cm and $r_2 = -20.0$ cm.

43 • The following four thin lenses are made of glass that has an index of refraction of 1.5. The radii given are *magnitudes*. Make a sketch of each lens and find each focal length in air: (*a*) double-convex that has radii of curvature equal to 15 cm and 26 cm, (*b*) plano-convex that has a radius of curvature equal to 15 cm, (*c*) double concave that has radii of curvature equal to 15 cm, and (*d*) plano-concave that has a radius of curvature equal to 26 cm.

44 • Find the focal length of a glass lens that has an index of refraction equal to 1.62, a concave surface that has a radius of curvature of magnitude 100 cm, and a convex surface that has a radius of curvature of magnitude 40.0 cm.

45 •• (*a*) An object that is 3.00 cm high is placed 25.0 cm in front of a thin lens that has a power equal to 10.0 D. Draw a ray diagram to find the position and the size of the image and check your results using the thin-lens equation. (*b*) Repeat Part (*a*) if the object is placed 20.0 cm in front of the lens. (*c*) Repeat Part (*a*) for an object placed 20.0 cm in front of a thin lens that has a power equal to −10.0 D. SSM

46 •• The lens-maker's equation has three design parameters. They consist of the index of refraction of the lens and the radii of curvature for its two surfaces. Thus, there are many ways to design a lens that has a particular focal length in air. Use the lens-maker's equation to design three different thin converging lenses, each having a focal length of 27.0 cm and each made from glass that has an index of refraction of 1.60. Sketch each of your designs.

47 •• Repeat Problem 46, but for a diverging lens that has a focal length in air of the same magnitude.

48 •• (*a*) What is meant by a negative object distance? Describe a specific situation in which a negative object distance can occur. (*b*) Find the image distance and the magnification for a thin lens in air when the object distance is −20 cm and the lens is a converging lens that has a focal length of 20 cm. Describe the image—is it virtual or real, upright or inverted? (*c*) Repeat Part (*b*) if the object distance is, instead, −10 cm, and the lens is diverging and has a focal length (magnitude) of 30 cm.

49 •• Two converging lenses, each having a focal length equal to 10 cm, are separated by 35 cm. An object is 20 cm to the left of the first lens. (*a*) Find the position of the final image using both a ray diagram and the thin-lens equation. (*b*) Is the final image real or virtual? Is the final image upright or inverted? (*c*) What is the overall lateral magnification? SSM

50 •• Repeat Problem 49, but with the second lens replaced by a diverging lens that has a focal length equal to −15 cm.

51 •• (*a*) Show that to obtain a magnification of magnitude $|m|$ using a converging thin lens of focal length f, the object distance must be equal to $(1 + |m|^{-1})f$. (*b*) You want to use a digital camera which has a lens whose focal length is 50.0 mm to take a picture of a person 1.75 m tall. How far from the camera lens should you have that person stand so that the image size on the light-receiving sensors of your camera is 24.0 mm?

52 •• **SPREADSHEET** A converging lens has a focal length of 12.0 cm. (*a*) Using a spreadsheet program or graphing calculator, plot the image distance as a function of the object distance, for object distances ranging from $1.10f$ to $10.0f$, where f is the focal length. (*b*) On the same graph used in Part (*a*), but using a different y axis, plot the magnification of the lens as a function of the object distance. (*c*) What type of image is produced for this range of object distances—real or virtual, upright or inverted? (*d*) Discuss the significance of any asymptotic limits your graphs have.

53 •• **SPREADSHEET** A converging lens has a focal length of 12.0 cm. (*a*) Using a spreadsheet program or graphing calculator, plot the image distance as a function of the object distance, for object distances ranging from $0.010f$ to $0.90f$, where f is the focal length. (*b*) On the same graph used in Part (*a*), but using a different y axis, plot the magnification of the lens as a function of the object distance. (*c*) What type of image is produced for this range of object distances—real or virtual, upright or inverted? (*d*) Discuss the significance of any asymptotic limits your graphs have.

54 •• An object is 15.0 cm in front of a converging lens that has a focal length equal to 15.0 cm. A second converging lens that also has a focal length equal to 15.0 cm is located 20.0 cm in back of the first. (*a*) Find the location of the final image and describe its properties (for example, real and inverted) and (*b*) draw a ray diagram to corroborate your answers to Part (*a*).

55 •• An object is 15.0 cm in front of a converging lens that has a focal length equal to 15.0 cm. A diverging lens that has a focal length whose magnitude is equal to 15.0 cm is located 20.0 cm in back of the first. (*a*) Find the location of the final image and describe its properties (for example, real and inverted) and (*b*) draw a ray diagram to corroborate your answers to Part (*a*). SSM

56 ••• In a convenient form of the thin-lens equation used by Newton, the object and image distances x and x' are measured from the focal points F and F', and not from the center of the lens. (*a*) Indicate x and x' on a sketch of a lens and show that if $x = s - f$ and $x' = s' - f$, the thin-lens equation (Equation 32-12) can be rewritten as $xx' = f^2$. (*b*) Show that the lateral magnification is given by $m = -x'/f = -f/x$.

57 ••• In *Bessel's method* for finding the focal length f of a lens, an object and a screen are separated by distance L, where $L > 4f$. It is then possible to place the lens at either of two locations, both between the object and the screen, so that there is an image of the object on the screen, in one case magnified and in the other case reduced. Show that if the distance between those two lens locations is D, then the focal length is given by $f = \frac{1}{4}(L^2 - D^2)/L$. *Hint: Refer to Figure 32-60.* SSM

58 ••• **ENGINEERING APPLICATION, CONTEXT-RICH** You are working for an optician during the summer. The optician needs to measure an unknown focal length and you suggest using *Bessel's method* (see Problem 57), which you used during a physics lab. You set the object-to-image distance at 1.70 m. The lens position is adjusted to get a sharp image on the screen. A second sharp image is found when the lens is moved a distance of 72 cm from its first location. (*a*) Sketch the ray diagram for the two locations. (*b*) Find the focal length of the lens using Bessel's method. (*c*) What are the two locations of the lens with respect to the object? (*d*) What are the magnifications of the images when the lens is in the two locations?

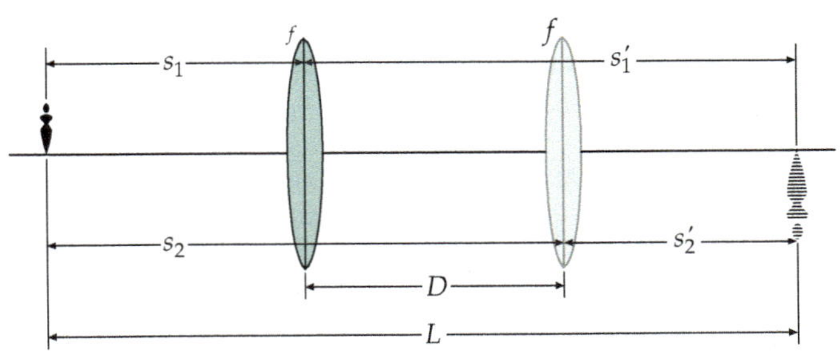

FIGURE 32-60 Problems 57 and 58

59 ••• An object is 17.5 cm to the left of a lens that has a focal length of +8.50 cm. A second lens, which has a focal length of −30.0 cm, is 5.00 cm to the right of the first lens. (*a*) Find the distance between the object and the final image formed by the second lens. (*b*) What is the overall magnification? (*c*) Is the final image real or virtual? Is the final image upright or inverted?

*ABERRATIONS

60 • Chromatic aberration is a common defect of (*a*) concave and convex lenses, (*b*) concave lenses only, (*c*) concave and convex mirrors, (*d*) all lenses and mirrors.

61 • **ENGINEERING APPLICATION** Discuss some of the reasons why most telescopes that are used by astronomers are reflecting rather than refracting telescopes.

62 • A symmetric double-convex lens has radii of curvature equal to 10.0 cm. It is made from glass that has an index of refraction equal to 1.530 for blue light and equal to 1.470 for red light. Find the focal length of this lens for (*a*) red light and (*b*) blue light.

*THE EYE

63 • **BIOLOGICAL APPLICATION** Find the change in the focal length of the eye when an object originally at 3.0 m is brought to 30 cm from the eye.

64 • **BIOLOGICAL APPLICATION** A farsighted person requires lenses that have powers equal to 1.75 D to read comfortably from a book that is 25.0 cm from his eyes. What is that person's near point without the lenses?

65 • **BIOLOGICAL APPLICATION** If two point objects close together are to be seen as two distinct objects, the images must fall on the retina on two different cones that are not adjacent. That is, there must be an unactivated cone between them. The separation of the cones is about 1.00 μm. Model the eye as a uniform 2.50-cm-diameter sphere that has a refractive index of 1.34. (*a*) What is the smallest angle the two points can subtend? (See Figure 32-61.) (*b*) How close together can two points be if they are 20.0 m from the eye? SSM

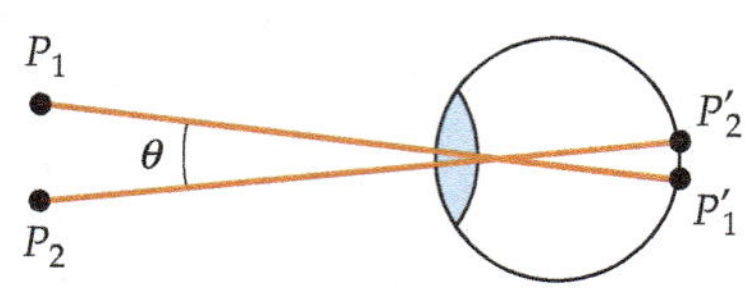

FIGURE 32-61 Problem 65

66 • **BIOLOGICAL APPLICATION** Suppose the eye were designed like a camera that has a lens of fixed focal length equal to 2.50 cm that could move toward or away from the retina, has air on both sides of the lens, and has no cornea. Approximately how far would the lens have to move to focus the image of an object 25.0 cm from the eye onto the retina? *Hint: Find the distance from the retina to the image behind it for an object at* 25.0 cm.

***Note:* Problems 67 through 69 refer to the model of the eye shown in Figure 32-62.**

67 •• **BIOLOGICAL APPLICATION** A simple model for the eye is a lens that has a variable power P located a fixed distance d in front of a screen, with the space between the lens and the screen filled by air. This "eye" can focus for all values of object distance s such that $x_{np} \leq s \leq x_{fp}$ where the subscriptson the variables refer to "near point" and "far point," respectively. This "eye" is said to be normal if it can focus on very distant objects. (*a*) Show that for a normal "eye" of this type, the required minimum value of P is given by $P_{min} = 1/d$. (*b*) Show that the maximum value of P is given by $P_{max} = 1/x_{np} + 1/d$. (*c*) The difference between the maximum and minimum powers, symbolized by A, is defined as $A = P_{max} - P_{min}$ and is called the *accommodation*. Find the minimum power and accommodation for this model eye that has a screen distance of 2.50 cm, a far point distance of infinity, and a near point distance of 25.0 cm. SSM

68 •• **BIOLOGICAL APPLICATION** (This problem refers to the model eye described in Problem 67.) In an eye that exhibits nearsightedness, the eye cannot focus on distant objects. (*a*) Show that for a nearsighted model eye capable of focusing out to a maximum distance x_{fp}, the minimum value of the power P is greater than that of a normal eye (that has a far point at infinity) and is given by $P_{min} = 1/x_{fp} + 1/d$. (*b*) To correct for nearsightedness, a contact lens may be placed directly in front of the lens of the model eye. What power contact lens would be needed to correct the vision of a nearsighted model eye that has a far-point distance of 50.0 cm?

69 •• **BIOLOGICAL APPLICATION** (This problem refers to the model eye described in Problem 67.) In an eye that exhibits farsightedness, the eye may be able to focus on distant objects but cannot focus on close objects. (*a*) Show that for a farsighted model eye capable of focusing only as close as a distance x'_{np}, the maximum value of the power P is given by $P_{max} = 1/x'_{np} + 1/d$. (*b*) Show that, compared to a model eye capable of focusing as close as a distance x_{np} (where $x_{np} < x'_{np}$), the maximum power of the farsighted lens is too small by $1/x_{np} - 1/x'_{np}$. (*c*) What power contact lens would be needed to correct the vision of a farsighted model eye, with $x'_{np} = 150$ cm, so that the eye may focus on objects as close as 15 cm?

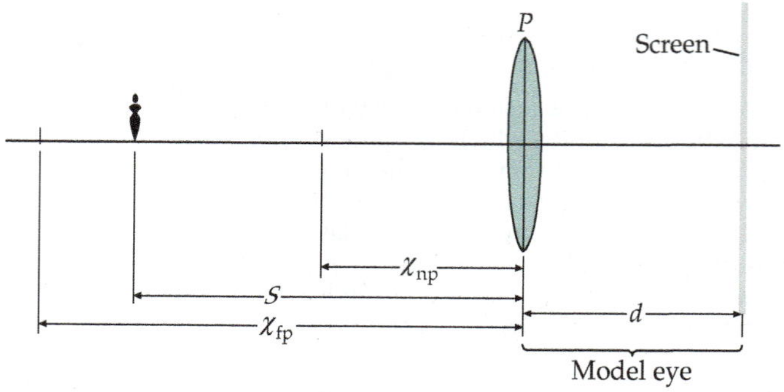

FIGURE 32-62 Problems 67, 68, and 69

70 •• **BIOLOGICAL APPLICATION** A person who has a near point of 80 cm needs to read from a computer screen that is only 45 cm from her eye. (*a*) Find the focal length of the lenses in reading glasses that will produce an image of the screen at a distance of 80 cm from her eye. (*b*) What is the power of the lenses?

71 •• **BIOLOGICAL APPLICATION** A nearsighted person cannot focus clearly on objects that are more distant than 2.25 m from her eye. What power lenses are required for her to see distant objects clearly?

72 •• **BIOLOGICAL APPLICATION** Because the index of refraction of the lens of the eye is not very different from that of the surrounding material, most of the refraction takes place at the cornea, where the index changes abruptly from 1.00 (air) to approximately 1.38. (*a*) Modeling the cornea, aqueous humor, lens and vitreous humor as a transparent homogeneous solid sphere that has an index of refraction of 1.38, calculate the sphere's radius if it focuses parallel light on the retina a distance 2.50 cm away. (*b*) Do you expect your result to be larger or smaller than the actual radius of the cornea? Explain your answer.

73 •• **Biological Application** The near point of a certain person's eyes is 80 cm. Reading glasses are prescribed so that he can read a book at 25 cm from his eye. The glasses are 2.0 cm from the eye. What diopter lenses should be used in the glasses?

74 ••• **Biological Application** At age 45, a person is fitted for reading glasses that have a power equal to 2.10 D in order to read at 25 cm. By the time she reaches the age of 55, she discovers herself holding her newspaper at a distance of 40 cm in order to see it clearly with her glasses on. (*a*) Where was her near point at age 45? (*b*) Where is her near point at age 55? (*c*) What power is now required for the lenses of her reading glasses so that she can again read at 25 cm? Assume the glasses are placed 2.2 cm from her eyes.

*THE SIMPLE MAGNIFIER

75 • What is the magnifying power of a lens that has a focal length equal to 7.0 cm when the image is viewed at infinity by a person whose near point is at 35 cm? SSM

76 •• A lens that has a focal length equal to 6.0 cm is used as a simple magnifier by one person whose near point is 25 cm and by another person whose near point is 40 cm. (*a*) What is the effective magnifying power of the lens for each person? (*b*) Compare the sizes of the images on the retinas when each person looks at the same object with the magnifier.

77 •• In your botany class, you examine a leaf using a convex 12-D lens as a simple magnifier. What is the angular magnification of the leaf if the image formed by the lens is (*a*) at infinity and (*b*) at 25 cm? SSM

*THE MICROSCOPE

78 •• Your laboratory microscope objective has a focal length of 17.0 mm. It forms an image of a tiny specimen at 16.0 cm from its second focal point. (*a*) How far from the objective is the specimen located? (*b*) What is the magnifying power for you if your near point distance is 25.0 cm and the focal length of the eyepiece is 51.0 mm?

79 •• A microscope has an objective that has a focal length equal to 8.5 mm. The eyepiece provides an angular magnification of 10 for a person whose near point distance is 25 cm. The tube length is 16 cm. (*a*) What is the lateral magnification of the objective? (*b*) What is the magnifying power of the microscope? SSM

80 •• A crude, symmetric handheld microscope consists of two 20-D lenses fastened at the ends of a tube 30 cm long. (*a*) What is the tube length of this microscope? (*b*) What is the lateral magnification of the objective? (*c*) What is the magnifying power of the microscope? (*d*) How far from the objective should the object be placed?

81 •• A compound microscope has an objective lens that has a power of 45 D and an eyepiece that has a power of 80 D. The lenses are separated by 28 cm. Assuming that the final image formed by the microscope is 25 cm from the eye, what is the magnifying power?

82 ••• A microscope has a magnifying power of 600. The eyepiece has an angular magnification of 15.0. The objective lens is 22.0 cm from the eyepiece. Calculate (*a*) the focal length of the eyepiece, (*b*) the location of the object so that it is in focus for a normal relaxed eye, and (*c*) the focal length of the objective lens.

*THE TELESCOPE

83 • You have a simple telescope that has an objective which has a focal length of 100 cm and an eyepiece which has a focal length of 5.00 cm. You are using it to look at the moon, which subtends an angle of about 9.00 mrad. (*a*) What is the diameter of the image formed by the objective? (*b*) What angle is subtended by the image formed at infinity by the eyepiece? (*c*) What is the magnifying power of your telescope? SSM

84 • The objective lens of the refracting telescope at the Yerkes Observatory has a focal length of 19.5 m. The moon subtends an angle of about 9.00 mrad. When the telescope is used to look at the moon, what is the diameter of the image of the moon formed by the objective?

85 •• The 200-in (5.10-m) diameter mirror of the reflecting telescope at Mt. Palomar has a focal length of 16.8 m. (*a*) By what factor is the light-gathering power increased over the 40.0-in (1.02-m) diameter refracting lens of the Yerkes Observatory telescope? (*b*) If the focal length of the eyepiece is 1.25 cm, what is the magnifying power of the 200-in telescope?

86 •• An astronomical telescope has a magnifying power of 7.0. The two lenses are 32 cm apart. Find the focal length of each lens.

87 •• A disadvantage of the astronomical telescope for terrestrial use (for example, at a football game) is that the image is inverted. A Galilean telescope uses a converging lens as its objective, but a diverging lens as its eyepiece. The image formed by the objective is at the second focal point of the eyepiece (the focal point on the refracted-light side of the eyepiece), so that the final image is virtual, upright, and at infinity. (*a*) Show that the magnifying power is given by $M = -f_o/f_e$, where f_o is the focal length of the objective and f_e is that of the eyepiece (which is negative). (*b*) Draw a ray diagram to show that the final image is indeed virtual, upright, and at infinity. SSM

88 •• A Galilean telescope (see Problem 87) is designed so that the final image is at the near point, which is 25 cm (rather than at infinity). The focal length of the objective is 100 cm and the focal length of the eyepiece is −5.0 cm. (*a*) If the object distance is 30.0 m, where is the image of the objective? (*b*) What is the object distance for the eyepiece so that the final image is at the near point? (*c*) How far apart are the lenses? (*d*) If the object height is 1.5 m, what is the height of the final image? What is the angular magnification?

89 ••• If you look into the wrong end of a telescope, that is, into the objective, you will see distant objects reduced in angular size. For a refracting telescope that has an objective with a focal length equal to 2.25 m and an eyepiece with a focal length equal to 1.50 cm, by what factor is the angular size of the object changed?

GENERAL PROBLEMS

90 • To focus a camera, the distance between the lens and the image-sensing surface is varied. A wide-angle lens of a camera has a focal length of 28 mm. By how much must the lens move to change from focusing on an object at infinity to an object at a distance of 5.00 m in front of the camera?

91 • A converging lens that has a focal length equal to 10 cm is used to obtain an image that is twice the height of the object. Find the object and image distances if (*a*) the image is to be upright and (*b*) the image is to be inverted. Draw a ray diagram for each case.

92 •• You are given two converging lenses that have focal lengths of 75 mm and 25 mm. (*a*) Show how the lenses should be arranged to form a telescope. State which lens to use as the objective, which lens to use as the eyepiece, how far apart to place the lenses, and what angular magnification you expect. (*b*) Draw a ray diagram to show how rays from a distant object are refracted by the two lenses.

93 •• (*a*) Show how the same two lenses in Problem 92 should be arranged to form a compound microscope that has a tube length of 160 mm. State which lens to use as the objective, which lens to use as the eyepiece, how far apart to place the lenses, and what

overall magnification you expect to get, assuming the user has a near point of 25 cm. (*b*) Draw a ray diagram to show how rays from a close object are refracted by the lenses. SSM

94 •• **CONTEXT-RICH** On a vacation, you are scuba diving and using a diving mask that has a face plate that bulges outward with a radius of curvature of 0.50 m. As a result, a convex spherical surface exists between the water and the air in the mask. A fish swims by you 2.5 m in front of your mask. (*a*) How far in front of the mask does the fish appear to be? (*b*) What is the lateral magnification of the image of the fish?

95 •• A 35-mm digital camera has a rectangular array of CCDs (light sensors) that is 24 mm by 36 mm. It is used to take a picture of a person 175 cm tall so that the image just fills the height (24 mm) of the CCD array. How far should the person stand from the camera if the focal length of the lens is 50 mm? SSM

96 •• A 35-mm camera that has interchangeable lenses is used to take a picture of a hawk that has a wingspan of 2.0 m. The hawk is 30 m away. What would be the ideal focal length of the lens used so that the image of the wings just fills the width of the light-sensitive area of the camera, which is 36 mm?

97 •• An object is placed 12.0 cm in front of a lens that has a focal length equal to 10.0 cm. A second lens that has a focal length equal to 12.5 cm is placed 20.0 cm in back of the first lens. (*a*) Find the position of the final image. (*b*) What is the magnification of the image? (*c*) Sketch a ray diagram showing the final image.

98 •• (*a*) Show that if f_a is the focal length of a thin lens in air, its focal length in water is given by $f_w = -(n_w/n_a)(n - n_a)/(n - n_w)f_a$, where n_w is the index of refraction of water, n is that of the lens material and n_a is that of air. (*b*) Calculate the focal length in air and in water of a double concave lens that has an index of refraction of 1.50 and radii of magnitudes 30 cm and 35 cm.

99 •• While parked in your car, you see a jogger in your rear view mirror, which is convex and has a radius of curvature whose magnitude is equal to 2.00 m. The jogger is 5.00 m from the mirror and is approaching at 3.50 m/s. How fast is the image of the jogger moving relative to the mirror?

100 •• A 2.00-cm-thick layer of water ($n = 1.33$) floats on top of a 4.00-cm-thick layer of carbon tetrachloride ($n = 1.46$) in a tank. How far below the top surface of the water does the bottom of the tank appear, according to an observer looking down from above at normal incidence?

101 ••• An object is 15.0 cm in front of a thin converging lens that has a focal length equal to 10.0 cm. A concave mirror that has a radius equal to 10.0 cm is 25.0 cm in back of the lens. (*a*) Find the position of the final image formed by the mirror–lens combination. (*b*) Is the image real or virtual? Is the image upright or inverted? (*c*) On a diagram, show where your eye must be to see this image. SSM

102 ••• When a bright light source is placed 30 cm in front of a lens, there is an upright image 7.5 cm from the lens. There is also a faint inverted image 6.0 cm from the lens on the incident-light side due to reflection from the front surface of the lens. When the lens is turned around, this weaker, inverted image is 10 cm in front of the lens. Find the index of refraction of the lens.

103 ••• A concave mirror that has a radius of curvature equal to 50.0 cm is oriented with its axis vertical. The mirror is filled with water that has an index of refraction equal to 1.33 and a maximum depth of 1.00 cm. At what height above the vertex of the mirror must an object be placed so that its image is at the same position as the object?

104 ••• The concave side of a lens has a radius of curvature that has a magnitude equal to 17.0 cm, and the convex side of the lens has a radius of curvature that has a magnitude equal to 8.00 cm. The focal length of the lens in air is 27.5 cm. When the lens is placed in a liquid that has an unknown index of refraction, the focal length increases to 109 cm. What is the index of refraction of the liquid?

105 ••• A solid glass ball of radius 10.0 cm has an index of refraction equal to 1.500. The right half of the ball is silvered so that it acts as a concave mirror (Figure 32-63). Find the position of the final image formed for an object located at (*a*) 40.0 cm and (*b*) 30.0 cm to the left of the center of the ball.

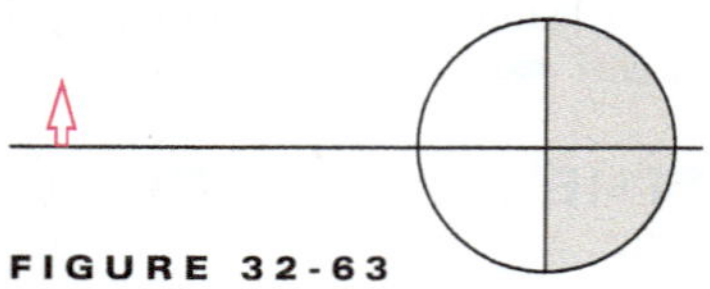

FIGURE 32-63 Problem 105

106 ••• (*a*) Show that a small change dn in the index of refraction of a lens material produces a small change in the focal length df given approximately by $df/f = -dn/(n - n_{air})$. (*b*) Use this result to estimate the focal length of a thin lens for blue light, for which $n = 1.530$, if the focal length for red light, for which $n = 1.470$, is 20.0 cm.

107 ••• The lateral magnification of a spherical mirror or a thin lens is given by $m = -s'/s$. Show that for objects of small horizontal extent, the longitudinal magnification is approximately $-m^2$. *Hint: Show that* $ds'/ds = -s'^2/s^2$. SSM

CHAPTER 33

WHITE LIGHT IS REFLECTED OFF A SOAP BUBBLE. WHEN LIGHT OF ONE WAVELENGTH IS INCIDENT ON A THIN SOAP-AND-WATER FILM, LIGHT IS REFLECTED FROM BOTH THE FRONT AND THE BACK SURFACES OF THE FILM. IF THE ORDER OF MAGNITUDE OF THE THICKNESS OF THE FILM IS ONE WAVELENGTH OF THE LIGHT, THE TWO REFLECTED LIGHT WAVES INTERFERE. IF THE TWO REFLECTED WAVES ARE 180° OUT OF PHASE, THE REFLECTED WAVES INTERFERE DESTRUCTIVELY, SO THE NET RESULT IS THAT NO LIGHT IS REFLECTED. IF WHITE LIGHT, WHICH CONTAINS A CONTINUUM OF WAVELENGTHS, IS INCIDENT ON THE THIN FILM, THEN THE REFLECTED WAVES WILL INTERFERE DESTRUCTIVELY ONLY FOR CERTAIN WAVELENGTHS, AND FOR OTHER WAVELENGTHS THEY WILL INTERFERE CONSTRUCTIVELY. THIS PROCESS PRODUCES THE COLORED FRINGES THAT YOU SEE IN THE SOAP BUBBLE. *(Aaron Haupt/ Photo Researchers.)*

? Have you ever wondered if the phenomenon that produces the bands that you see in the light reflected off a soap bubble has any practical applications? (See Example 33-2.)

Interference and Diffraction

Interference and diffraction are the important phenomena that distinguish waves from particles.* Interference is the formation of a lasting intensity pattern by two or more waves that superpose in space. Diffraction is the bending of waves around corners that occurs when a portion of a wavefront is cut off by a barrier or obstacle.

In this chapter, we will see how the pattern of the resulting wave can be calculated by treating each point on the original wavefront as a point source, according to Huygens's principle, and calculating the interference pattern resulting from these sources.

* Before you study this chapter, you may wish to review Chapter 15 and Chapter 16, where the general topics of interference and diffraction of waves are first discussed.

33-1 PHASE DIFFERENCE AND COHERENCE

When two harmonic sinusoidal waves of the same frequency and wavelength but of different phase combine, the resultant wave is a harmonic wave whose amplitude depends on the phase difference. If the phase difference is zero, or an integer multiplied by 360°, the waves are in phase and interfere constructively. The resultant amplitude equals the sum of the two individual amplitudes, and the intensity (which is proportional to the square of the amplitude) is maximum. (If the amplitudes are equal and the waves are in phase, the intensity is four times that of either individual wave.) If the phase difference is 180° or any odd integer multiplied by 180°, the waves are out of phase and interfere destructively. The resultant amplitude is then the difference between the two individual amplitudes, and the intensity is a minimum. (If the amplitudes are equal and the waves are 180° out of phase, the intensity is zero.)

A phase difference between two waves is often the result of a difference in path lengths. When a light wave reflects from a thin transparent film, such as a soap bubble, the reflected light is a superposition of the light reflected from the front surface of the film and the light reflected from the back surface of the film. The additional distance traveled by the light reflected from the back surface is called the path-length difference between the two reflected waves. A path-length difference of one wavelength produces a phase difference of 360°, which is equivalent to no phase difference at all. A path-length difference of one-half wavelength produces a 180° phase difference. In general, a path-length difference of Δr contributes a phase difference δ given by

$$\delta = \frac{\Delta r}{\lambda} 2\pi = \frac{\Delta r}{\lambda} 360^\circ \qquad 33\text{-}1$$

PHASE DIFFERENCE DUE TO A PATH-LENGTH DIFFERENCE

Example 33-1 Phase Difference

(*a*) What is the minimum path-length difference that will produce a phase difference of 180° for light of wavelength 800 nm? (*b*) What phase difference will that path-length difference produce in light of wavelength 700 nm?

PICTURE The phase difference is to 360° as the path-length difference is to the wavelength.

SOLVE

(*a*) The phase difference δ is to 360° as the path-length difference Δr is to the wavelength λ. We know that $\lambda = 800$ nm and $\delta = 180°$:

$$\frac{\delta}{360^\circ} = \frac{\Delta r}{\lambda}$$

$$\Delta r = \frac{\delta}{360^\circ}\lambda = \frac{180^\circ}{360^\circ}(800\text{ nm}) = \boxed{400\text{ nm}}$$

(*b*) Set $\lambda = 700$ nm, $\Delta r = 400$ nm, and solve for δ:

$$\delta = \frac{\Delta r}{\lambda}360^\circ = \frac{400\text{ nm}}{700\text{ nm}}360^\circ = \boxed{206^\circ = 3.59\text{ rad}}$$

CHECK The Part-(*b*) result is somewhat larger than 180°. This result is expected because 400 nm is longer than half of the 700-nm wavelength.

Another cause of phase difference is the 180° phase change a wave sometimes undergoes upon reflection from a surface. This phase change is analogous to the inversion of a pulse on a string when it reflects from a point where the density suddenly increases, such as when a light string is attached to a heavier string or rope. The inversion of the reflected pulse is equivalent to a phase change of 180° for a sinusoidal wave (which can be thought of as a series of pulses). When light traveling in air strikes the surface of a medium in which light travels more slowly, such as glass or water, there is a 180° phase change in the reflected light.

When light is traveling in the liquid wall of a soap bubble, there is no phase change in the light reflected from the surface between the liquid and the air. This situation is analogous to the reflection without inversion of a pulse on a heavy string at a point where the heavy string is attached to a lighter string.

> If light traveling in one medium strikes the surface of a medium in which light travels more slowly, there is a 180° phase change in the reflected light.
>
> PHASE DIFFERENCE DUE TO REFLECTION

As we saw in Chapter 16, interference of waves is observed when two or more coherent waves overlap. Interference of overlapping waves from two sources is not observed unless the sources are coherent. Because the light from each source is usually the result of millions of atoms radiating independently, the phase difference between the waves from such sources fluctuates randomly many times per second, so two light sources are usually not coherent. Coherence in optics is often achieved by splitting the light beam from a single source into two or more beams that can then be combined to produce an interference pattern. The light beam can be split by reflecting the light from the two surfaces of a thin film (Section 33-2), by diffracting the beam through two small openings or slits in an opaque barrier (Section 33-3), or by using a single point source and its image in a plane mirror for the two sources (Section 33-3). Today, lasers are the most important sources of coherent light in the laboratory.

Light from an ideal monochromatic source is a sinusoidal wave of infinite duration, and light from certain lasers approaches this ideal. However, light from conventional *monochromatic* sources, such as gas discharge tubes designed for this purpose, consists of packets of sinusoidal light that are only a few million wavelengths long. The light from such a source consists of many such packets, each approximately the same length. The packets have essentially the same wavelength, but the packets differ in phase in a random manner. The length of the individual packets is called the **coherence length** of the light, and the time it takes one of the packets to pass a point in space is the **coherence time.** The light emitted by a gas discharge tube designed to produce monochromatic light has a coherence length of only a few millimeters. By comparison, some highly stable lasers produce light that has a coherence length many kilometers long.

33-2 INTERFERENCE IN THIN FILMS

You have probably noticed the colored bands in a soap bubble or in the film on the surface of oily water. These bands are due to the interference of light reflected from the top and bottom surfaces of the film. The different colors arise because of variations in the thickness of the film, causing interference for different wavelengths at different points.

When waves traveling in a medium cross a surface where the wave speed changes, part of the wave is reflected and part is transmitted. In addition, the reflected wave undergoes a 180° phase shift upon reflection if the transmitted wave travels at a slower speed than do the incident and reflected waves. (This 180° phase shift is established for waves on a string in Section 15-4 of Chapter 15.) The reflected wave does not undergo a phase shift upon reflection if the transmitted wave travels at a faster speed than do the incident and reflected waves.

Consider a thin film of water (such as a small section of a soap bubble) of uniform thickness viewed at small angles with the normal, as shown in Figure 33-1. Part of the light is reflected from the upper air–water interface where it undergoes a 180° phase change. Most of the light enters the film and part of it is reflected by the bottom water–air interface. There is no phase change in this reflected light. If the light is nearly perpendicular to the surfaces, both the light reflected from the top surface and the light reflected from the bottom surface can enter the eye. The path-length

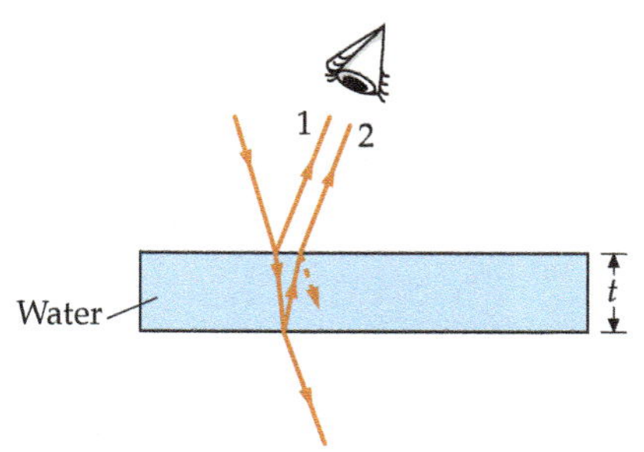

FIGURE 33-1 Light rays reflected from the top and bottom surfaces of a thin film are coherent because both rays come from the same source. If the light is incident almost normally, the two reflected rays will be very close to each other and will produce interference.

difference between these two rays is $2t$, where t is the thickness of the film. This path-length difference produces a phase difference of $(2t/\lambda')360°$, where $\lambda' = \lambda/n$ is the wavelength of the light in the film, n is the index of refraction of the film, and λ is the wavelength of the light in vacuum. The total phase difference between the two rays is thus 180° plus the phase difference due to the path-length difference. Destructive interference occurs when the path-length difference $2t$ is zero or a whole number of wavelengths λ' (in the film). Constructive interference occurs when the path-length difference is an odd number of half-wavelengths.

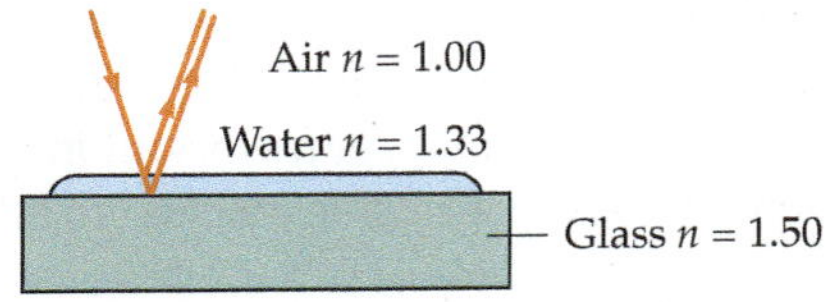

FIGURE 33-2 The interference of light reflected from a thin film of water resting on a glass surface. In this case, both rays undergo a change in phase of 180° upon reflection.

When a thin film of water lies on a glass surface, as in Figure 33-2, the ray that reflects from the lower water–glass interface also undergoes a 180° phase change, because the index of refraction of glass (approximately 1.50) is greater than that of water (approximately 1.33). Thus, both the rays shown in the figure have undergone a 180° phase change upon reflection. The phase difference δ between these rays is due solely to the path-length difference and is given by $\delta = (2t/\lambda')360°$.

When a thin film of varying thickness is viewed with monochromatic light, such as the yellow light from a sodium lamp, alternating bright and dark bands or lines, called **interference fringes,** are observed. The distance between a bright fringe and a dark fringe is that distance over which the film's thickness t changes so that the path-length difference $2t$ changes by $\lambda'/2$. Figure 33-3*a* shows the interference pattern observed when light is reflected from an air film between a spherical glass surface and a plane glass surface in contact. These circular interference fringes are known as **Newton's rings.** Typical rays reflected at the top and bottom of the air film are shown in Figure 33-3*b*. Near the point of contact of the surfaces, where the path-length difference between the ray reflected from the upper glass–air interface and the ray reflected from the lower air–glass interface is approximately zero (it is small compared with the wavelength of light) the interference is destructive because of the 180° phase shift of the ray reflected from the lower air–glass interface. This central region in Figure 33-3*a* is therefore dark. The first bright fringe occurs at the radius at which the path-length difference is $\lambda/2$, which contributes a phase difference of 180°. This adds to the phase shift due to reflection to produce a total phase difference of 360°, which is equivalent to a zero phase difference. The second dark region occurs at the radius at which the path-length difference is λ, and so on.

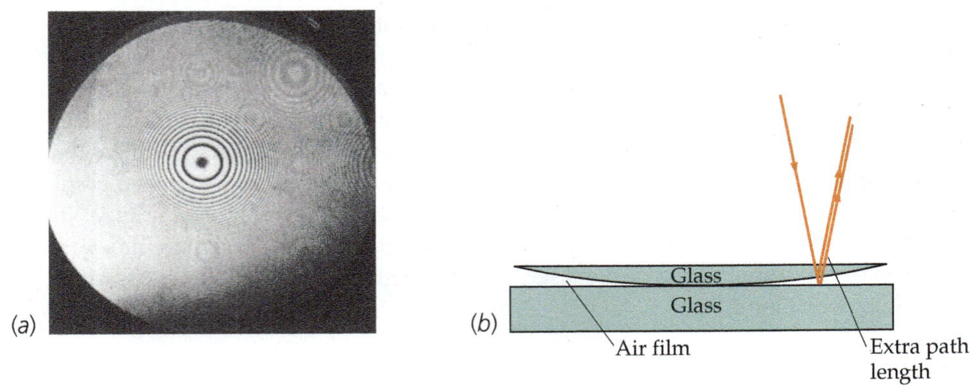

FIGURE 33-3 (*a*) Newton's rings observed when light is reflected from a thin film of air between a plane glass surface and a spherical glass surface. At the center, the thickness of the air film is negligible and the interference is destructive because of the 180° phase change of one of the rays upon reflection. (*b*) Glass surfaces for the observation of Newton's rings shown in Figure 33-3*a*. The thin film in this case is the film of air between the glass surfaces. *(Courtesy of Bausch & Lomb.)*

Example 33-2 A Wedge of Air

A wedge-shaped film of air is made by placing a small slip of paper between the edges of two flat pieces of glass, as shown in Figure 33-4. Light of wavelength 500 nm is incident normally on the glass, and interference fringes are observed by reflection. If the angle θ made by the plates is 3.0×10^{-4} rad (0.017°), how many dark interference fringes per centimeter are observed?

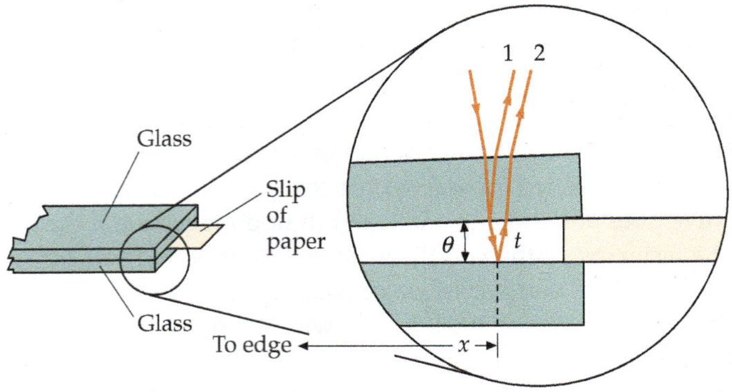

FIGURE 33-4 The angle θ, which is less than 0.02°, is exaggerated. The incoming and outgoing rays are virtually perpendicular to all air–glass interfaces.

PICTURE We find the number of fringes per centimeter by finding the horizontal distance x to the mth fringe and solving for m/x. Because the ray reflected from the bottom plate undergoes a 180° phase shift, the point of contact (where the path-length difference is zero) will be dark. The mth dark fringe after the contact point occurs when $2t = m\lambda'$, where $\lambda' = \lambda$ is the wavelength in the air film, and t is the plate separation at x, as shown in Figure 33-4. Because the angle θ is small, we can use the small-angle approximation $\theta \approx \tan\theta = t/x$.

SOLVE

1. The mth dark fringe from the contact point occurs when the path-length difference $2t$ equals m wavelengths:
$$2t = m\lambda' = m\lambda$$
$$m = \frac{2t}{\lambda}$$

2. The thickness t is related to the angle θ:
$$\theta = \frac{t}{x}$$

3. Substitute $t = x\theta$ into the equation for m:
$$m = \frac{2x\theta}{\lambda}$$

4. Calculate m/x:
$$\frac{m}{x} = \frac{2\theta}{\lambda} = \frac{2(3.0 \times 10^{-4})}{5.0 \times 10^{-7}\,\text{m}} = 1200\,\text{m}^{-1} = \boxed{12\,\text{cm}^{-1}}$$

CHECK The expression for the number of dark fringes per unit length in step 4 shows that the number per centimeter would decrease if light of a longer wavelength is used. This result is as expected.

TAKING IT FURTHER We observe 12 dark fringes per centimeter. In practice, the number of fringes per centimeter, which is easy to count, can be used to determine the angle. Note that if the angle of the wedge is increased, the fringes become more closely spaced.

PRACTICE PROBLEM 33-1 How many dark fringes per centimeter are observed if light of wavelength 650 nm is used?

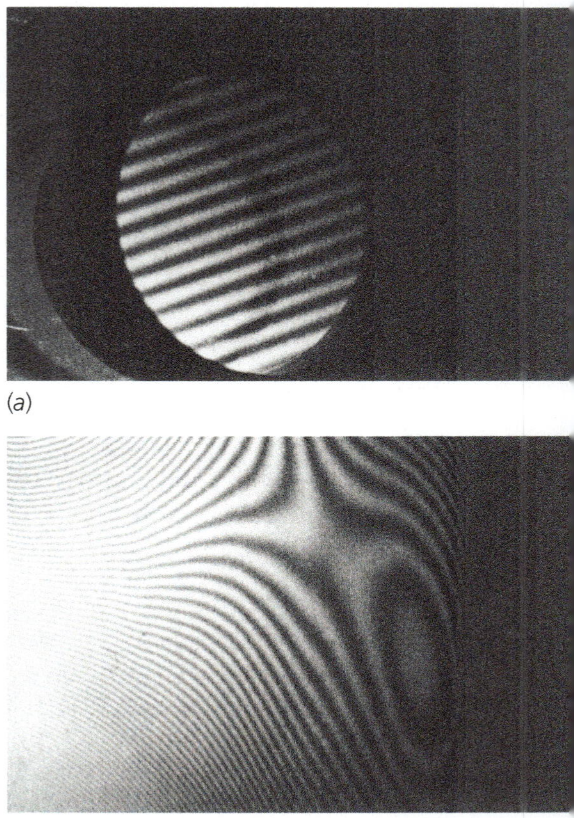
(a)

(b)

FIGURE 33-5 (*a*) Straight-line fringes from a wedge-shaped film of air, like that shown in Figure 33-4. The straightness of the fringes indicates that the glass plates are optically flat. (*b*) Fringes from a wedge-shaped film of air between glass plates that are not optically flat. *(Courtesy T. A. Wiggins.)*

Figure 33-5*a* shows interference fringes produced by a wedge-shaped air film between two flat glass plates, as in Example 33-2. Plates that produce straight fringes, such as those in Figure 33-5*a*, are said to be **optically flat.** To be optically flat, a surface must be flat to within a small fraction of a wavelength. A similar wedge-shaped air film formed by two ordinary glass plates yields the irregular fringe pattern in Figure 33-5*b*, which indicates that these plates are not optically flat.

One application of interference effects in thin films is in nonreflecting lenses, which are made by coating the surface of a lens with a thin film of a material that has an index of refraction equal to approximately 1.38, which is between the index of refraction of glass and that of air. The intensities of the light reflected from the top and bottom surfaces of the film are approximately equal, and because the reflected rays undergo a 180° phase change at both surfaces there is no phase difference due to reflection between the two rays. The thickness of the film is chosen to be $\frac{1}{4}\lambda' = \frac{1}{4}\lambda n$, where λ is the wavelength, in vacuum, that is in the middle of the visible spectrum, so that there is a phase change of 180° due to the path-length difference of $\lambda'/2$ for light of normal incidence. Reflection from the coated surface is thus minimized, which means that transmission through the surface is maximized.

33-3 TWO-SLIT INTERFERENCE PATTERN

Interference patterns of light from two or more sources can be observed only if the sources are coherent. The interference in thin films discussed previously can be observed because the two beams come from the same light source but are separated by reflection. In Thomas Young's famous 1801 experiment, in which he demonstrated the wave nature of light, two coherent light sources are produced by illuminating two very narrow parallel slits using a single light source. We saw in Chapter 15 that when a wave encounters a barrier that has a very small opening, the opening acts as a point source of waves (Figure 33-6).

During Young's experiment, diffraction causes each slit to act as a line source (which is equivalent to a point source in two dimensions). The interference pattern is observed on a screen far from the slits (Figure 33-7*a*). At very large distances from the slits, the lines from the two slits to some point P on the screen are approximately parallel, and the path-length difference is approximately $d \sin\theta$, where d is the separation of the slits, as shown in Figure 33-7*b*. When the path-length difference is equal to an integral number of wavelengths, the interference is constructive. We thus have interference maxima at an angle θ_m given by

$$d \sin\theta_m = m\lambda \qquad m = 0, 1, 2, \ldots \tag{33-2}$$

TWO-SLIT INTERFERENCE MAXIMA

where m is called the **order number.** The interference minima occur at

$$d \sin\theta_m = \left(m - \tfrac{1}{2}\right)\lambda \qquad m = 1, 2, 3, \ldots \tag{33-3}$$

TWO-SLIT INTERFERENCE MINIMA

The phase difference δ at a point P is related to the path-length difference $d \sin\theta$ by

$$\delta = \frac{\Delta r}{\lambda}2\pi = \frac{d\sin\theta}{\lambda}2\pi \tag{33-4}$$

We can relate the distance y_m measured along the screen from the central point to the mth bright fringe (see Figure 33-7*b*) to the distance L from the slits to the screen:

$$\tan\theta_m = \frac{y_m}{L}$$

For small angles, $\tan\theta \approx \sin\theta$. Substituting y_m/L for $\sin\theta_m$ in Equation 33-2 and solving for y_m gives

$$y_m = m\frac{\lambda L}{d} \tag{33-5}$$

From this result, we see that for small angles the fringes are equally spaced on the screen.

FIGURE 33-6 Plane water waves in a ripple tank encountering a barrier that has a small opening. The waves to the right of the barrier are circular waves that are concentric about the opening, just as if there were a point source at the opening. *(From PSSC Physics, 2nd Edition, 1965. D. C. Heath & Co. and Education Development Center, Newton MA.)*

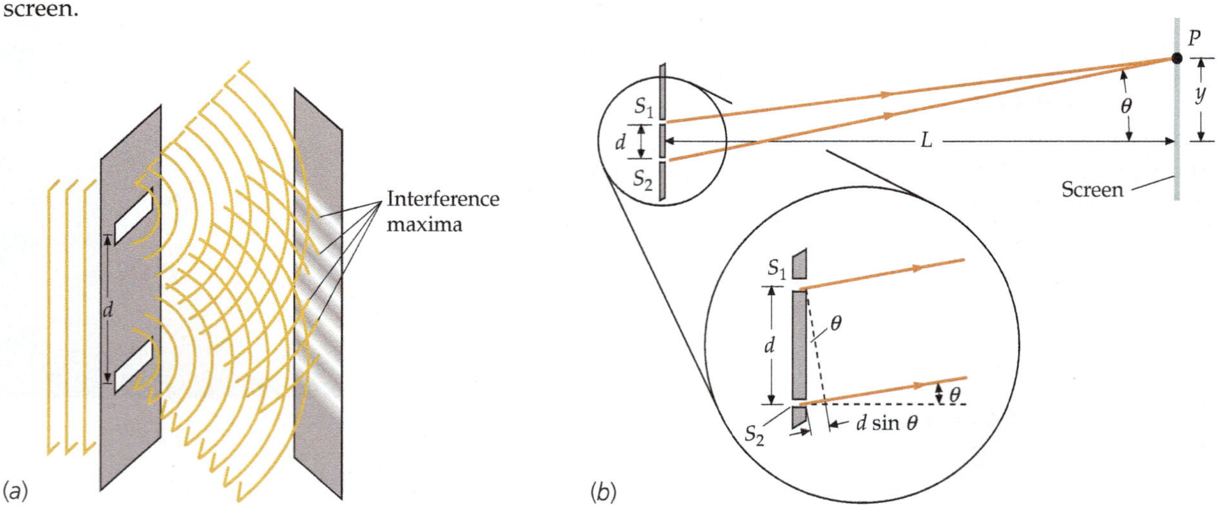

FIGURE 33-7 (*a*) Two slits act as coherent sources of light for the observation of interference in Young's experiment. Cylindrical waves from the slits overlap and produce an interference pattern on a screen. (*b*) Geometry for relating the distance y measured along the screen to L and θ. When the screen is very far away compared with the slit separation, the rays from the slits to a point on the screen are approximately parallel, and the path-length difference between the two rays is $d \sin\theta$.

Example 33-3 Fringe Spacing from Slit Spacing — *Try It Yourself*

Two narrow slits separated by 1.50 mm are illuminated by yellow light from a sodium lamp that has a wavelength equal to 589 nm. Find the spacing of the bright fringes observed on a screen 3.00 m away.

PICTURE The distance y_m measured along the screen to the mth bright fringe is given by Equation 33-5, where $L = 3.00$ m, $d = 1.50$ mm, and $\lambda = 589$ nm.

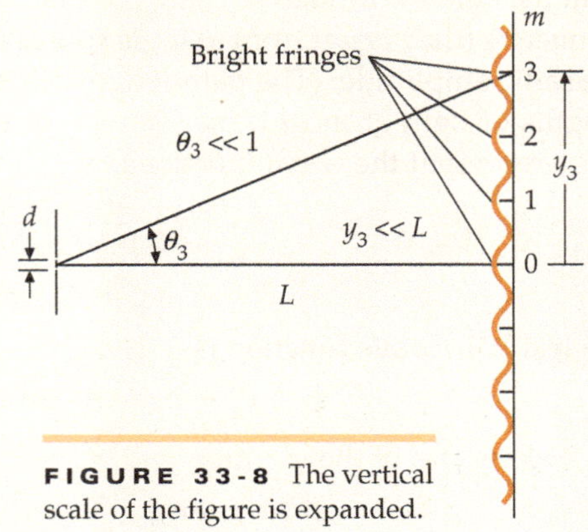

FIGURE 33-8 The vertical scale of the figure is expanded.

SOLVE

Cover the column to the right and try these on your own before looking at the answers.

Steps	Answers
1. Make a sketch of the situation (Figure 33-8).	
2. The fringe spacing is the distance between the mth bright fringe and the $(m + 1)$th bright fringe. Using the sketch, obtain an expression for the spacing between fringes.	fringe spacing $= y_{m+1} - y_m$
3. Apply Equation 33-5 to the mth and $(m + 1)$th fringe.	$y_m = m\frac{\lambda L}{d}$ and $y_{m+1} = (m+1)\frac{\lambda L}{d}$
4. Substitute into the step-2 result and simplify.	$y_{m+1} - y_m = \frac{\lambda L}{d}$
5. Substitute into the step-4 result and solve for the fringe spacing.	fringe spacing $= \boxed{1.18 \text{ mm}}$

TAKING IT FURTHER The fringes are uniformly spaced only to the degree that the small-angle approximation is valid, that is, to the degree that $\lambda/d \ll 1$. In this example, $\lambda/d = (589 \text{ nm})/(1.50 \text{ mm}) \approx 0.0004$.

See **Math Tutorial** *for more information on* **Trigonometry**

Example 33-4 How Many Fringes? — *Conceptual*

Two narrow slits are illuminated by monochromatic light. If the distance between the slits is equal to 2.75 wavelengths, what is the maximum number of bright fringes that can be seen on a screen? (*a*) 1, (*b*) 2, (*c*) 3, (*d*) 4, (*e*) 5, (*f*) 6 or more

PICTURE A bright fringe (constructive interference) exists at points on the screen for which the distance to the two slits differs by an integer multiplied by the wavelength. However, the maximum difference in distance possible is equal to the distance between the two slits.

SOLVE

1. Find the maximum difference in distance from points on the screen to the two slits: — At all points on the screen, the difference in distance from the two slits is 2.75 wavelengths or less.
2. A bright fringe (constructive interference) exists at points on the screen for which the distance to the two slits differs by an integer multiplied by the wavelength: — Bright fringes exist on the screen at places where the difference in distance to the slits is 2 wavelengths, 1 wavelength, or zero wavelengths.
3. Count up the bright fringes. There is the central maximum and two on either side of the central maximum: — $\boxed{(e)\ 5}$

CONCEPT CHECK 33-1

What is the maximum number of dark fringes that can be seen on a screen?

CALCULATION OF INTENSITY

To calculate the intensity of the light on the screen at a general point P, we need to add two harmonic wave functions that differ in phase.* The wave functions for electromagnetic waves are the electric field vectors. Let E_1 be the electric field at some point P on the screen due to the waves from slit 1, and let E_2 be the electric field at that point due to waves from slit 2. Because the angles of interest are small, we can treat the fields as though they are parallel. Both electric fields oscillate with the same frequency (they result from a single source that illuminates both slits) and they have the same amplitude. (The path-length difference is only of the order of a few wavelengths of light at most.) They have a phase difference δ given by Equation 33-4. If we represent the wave functions by

$$E_1 = A_0 \sin \omega t$$

and

$$E_2 = A_0 \sin(\omega t + \delta)$$

the resultant wave function is

$$E = E_1 + E_2 = A_0 \sin \omega t + A_0 \sin(\omega t + \delta) \qquad \text{33-6}$$

By making use of the identity

$$\sin \alpha + \sin \beta = 2 \cos \tfrac{1}{2}(\alpha - \beta) \sin \tfrac{1}{2}(\alpha + \beta)$$

the resultant wave function is given by

$$E = \left[2A_0 \cos \tfrac{1}{2}\delta\right] \sin\left(\omega t + \tfrac{1}{2}\delta\right) \qquad \text{33-7}$$

The amplitude of the resultant wave is thus $2A_0 \cos \frac{1}{2}\delta$. It has its maximum value of $2A_0$ when the waves are in phase and is zero when they are 180° out of phase. Because the intensity is proportional to the square of the amplitude, the intensity at any point P is

$$I = 4I_0 \cos^2 \tfrac{1}{2}\delta \qquad \text{33-8}$$

INTENSITY IN TERMS OF PHASE DIFFERENCE

where I_0 is the intensity of the light reaching the screen from either slit separately. The phase angle δ is related to the position on the screen by $\delta = (d \sin \theta/\lambda)2\pi$ (Equation 33-4).

Figure 33-9*a* shows the intensity pattern as seen on a screen. A graph of the intensity as a function of $\sin \theta$ is shown in Figure 33-9*b*. For small θ, this graph is equivalent to a plot of intensity versus y (because $y = L \tan \theta \approx L \sin \theta$). The intensity I_0 is the intensity from each slit separately. The dashed line in Figure 33-9*b* shows the average intensity $2I_0$, which is the result of averaging over a distance containing many interference maxima and minima. This is the intensity that would arise from the two sources if they acted independently without interference, that is, if they were not coherent. Then, the phase difference between the two sources would fluctuate randomly, so that only the average intensity would be observed.

Figure 33-10 shows another method of producing the two-slit interference pattern, an arrangement known as **Lloyd's mirror.** A monochromatic horizontal line source is placed at a distance $\frac{1}{2}d$ above the plane of a mirror. Light striking the screen directly from the source interferes with the light that is reflected from the mirror. The reflected light can be considered to come from the virtual image of the line source formed by the mirror. Because of the 180° change in phase upon reflection at the mirror, the interference pattern is that of two coherent line sources that differ in phase by 180°. The pattern is the same as that shown in Figure 33-9 for two slits, except that the maxima and minima are interchanged. Constructive interference occurs at points for which the path-length difference is a half-wavelength or any odd number of half-wavelengths. At those points, the 180° phase difference due to the path-length difference combines with the 180° phase difference of the sources to produce constructive interference.

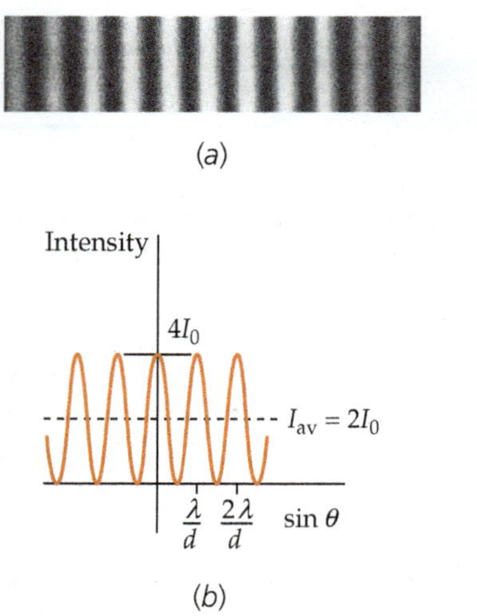

FIGURE 33-9 (*a*) The interference pattern observed on a screen far away from the two slits shown in Figure 33-7. (*b*) Plot of intensity versus $\sin\theta$. The maximum intensity is $4I_0$, where I_0 is the intensity due to each slit separately. The average intensity (dashed line) is $2I_0$. *(Courtesy of Michael Cagnet.)*

* We did this in Chapter 16 where we first discussed the superposition of two waves.

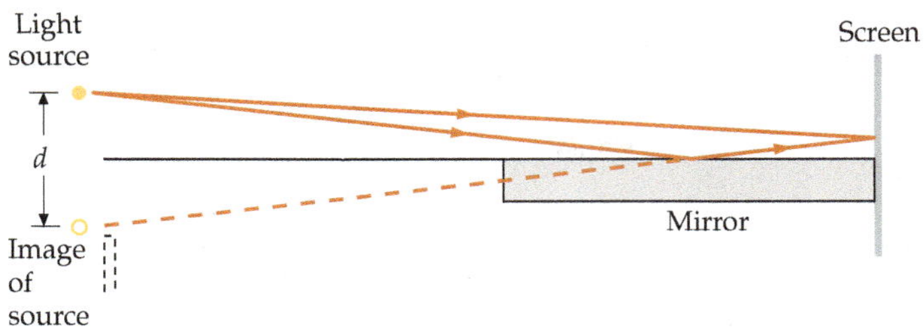

FIGURE 33-10 Lloyd's mirror for producing a two-slit interference pattern. The two sources (the source and its virtual image) are coherent and are 180° out of phase.

PRACTICE PROBLEM 33-2

A point source of light ($\lambda = 589$ nm) is placed 0.40 mm above the surface of a glass mirror. Interference fringes are observed on a screen 6.0 m away, and the interference is between the light reflected from the front surface of the glass and the light traveling from the source directly to the screen. Find the spacing of the fringes on the screen.

The physics of Lloyd's mirror was used in the early days of radio astronomy to determine the location of distant radio sources on the celestial sphere. A radio-wave receiver was placed on a cliff overlooking the sea, and the surface of the sea served as the mirror.

33-4 DIFFRACTION PATTERN OF A SINGLE SLIT

In our discussion of the interference patterns produced by two or more slits, we assumed that the slits were very narrow so that we could consider the slits to be line sources of cylindrical waves, which in our two-dimensional diagrams are point sources of circular waves. We could therefore assume that the value of the intensity due to one slit acting alone was the same (I_0) at any point P on the screen, independent of the angle θ made between the ray to point P and the normal line between the slit and the screen. When the slit is not narrow, the intensity on a screen far away is not independent of angle but decreases as the angle increases. Consider a slit of width a. Figure 33-11 shows the intensity pattern on a screen far away from the slit of width a as a function of $\sin\theta$. We can see that the intensity is maximum in the forward direction ($\sin\theta = 0$) and decreases to zero at an angle that depends on the slit width a and the wavelength λ.

Most of the light intensity is concentrated in the broad **central diffraction maximum,** although there are minor secondary maxima bands on either side of the central maximum. The first zeroes in the intensity occur at angles specified by

$$\sin\theta_1 = \lambda/a \qquad 33\text{-}9$$

Note that for a given wavelength λ, Equation 33-9 describes how variations in the slit width result in variations in the angular width of the central maximum. If we *increase* the slit width a, the angle θ_1 at which the intensity first becomes zero *decreases*, giving a more narrow central diffraction maximum. Conversely, if we *decrease* the slit width, the angle of the first zero *increases*, giving a wider central diffraction maximum. When a is smaller than λ, then $\sin\theta_1$ would have to exceed 1

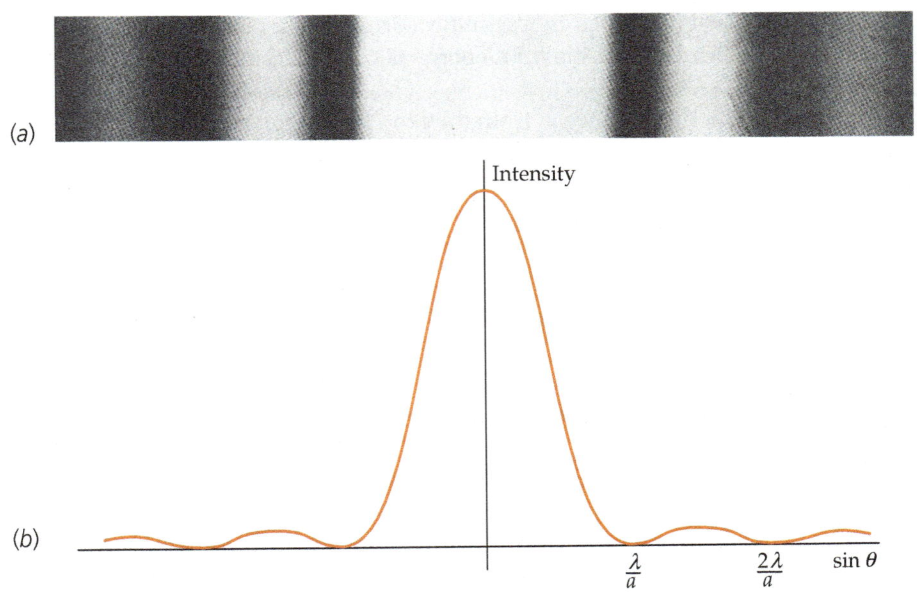

FIGURE 33-11 (*a*) Diffraction pattern of a single slit as observed on a screen far away. (*b*) Plot of intensity versus $\sin\theta$ for the pattern in Figure 33-11*a*. *(Courtesy of Michael Cagnet.)*

to satisfy Equation 33-9. Thus, for a less than λ, there are no points of zero intensity in the pattern, and the slit acts as a line source (a point source in two dimensions) radiating light energy essentially equal in all directions.

Multiplying both sides of Equation 33-9 by $a/2$ gives

$$\tfrac{1}{2}a\sin\theta_1 = \tfrac{1}{2}\lambda \qquad 33\text{-}10$$

The quantity $\frac{1}{2}a\sin\theta_1$ is the path-length difference between a light ray leaving the middle of the upper half of the slit and one leaving the middle of the lower half of the slit. We see that the first diffraction *minimum* occurs when these two rays are 180° out of phase, that is, when their path-length difference equals a half-wavelength. We can understand this result by considering each point on a wavefront to be a point source of light in accordance with Huygens's principle. In Figure 33-12, we have placed a line of dots on the wavefront at the slit to represent these point sources schematically. Suppose, for example, that we have 100 such dots and that we look at an angle θ_1 for which $a\sin\theta_1 = \lambda$. Let us consider the slit to be divided into two halves, with sources 1 through 50 in the upper half and sources 51 through 100 in the lower half. When the path-length difference between the middle of the upper half and the middle of the lower half of the slit equals a half-wavelength, the path-length difference between source 1 (the first source in the upper half) and source 51 (the first source in the lower half) is also $\frac{1}{2}\lambda$. The waves from those two sources will be out of phase by 180° and will thus cancel. Similarly, waves from the second source in each region (source 2 and source 52) will cancel. Continuing this argument, we can see that the waves from each pair of sources separated by $a/2$ will cancel. Thus, there will be no light energy at that angle. We can extend this argument to the second and third minima in the diffraction pattern of Figure 33-11. At an angle θ_2 where $a\sin\theta_2 = 2\lambda$, we can divide the slit into four regions, two regions for the top half and two regions for the bottom half. Using this same argument, the light intensity from the top half is zero because of the cancellation of pairs of sources; similarly, the light intensity from the bottom half is zero. The general expression for the points of zero intensity in the diffraction pattern of a single slit is thus

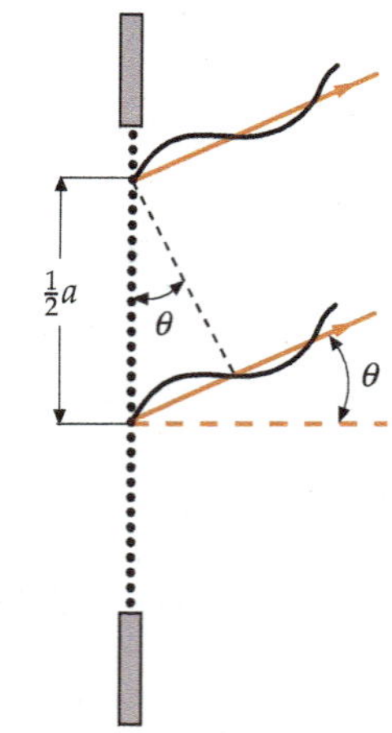

FIGURE 33-12 A single slit is represented by a large number of point sources of equal amplitude. At the first diffraction minimum of a single slit, the waves from each point source in the upper half of the slit are 180° out of phase with the wave from the point source a distance $a/2$ lower in the slit. Thus, the interference from each such pair of point sources is destructive.

$$a\sin\theta_m = m\lambda \qquad m = 1, 2, 3, \ldots \qquad 33\text{-}11$$

POINTS OF ZERO INTENSITY FOR A SINGLE-SLIT DIFFRACTION PATTERN

Usually, we are just interested in the first occurrence of a minimum in the light intensity because nearly all of the light energy is contained in the central diffraction maximum.

In Figure 33-13, the distance y_1 from the central maximum to the first diffraction minimum is related to the angle θ_1 and the distance L from the slit to the screen by

$$\tan\theta_1 = \frac{y_1}{L}$$

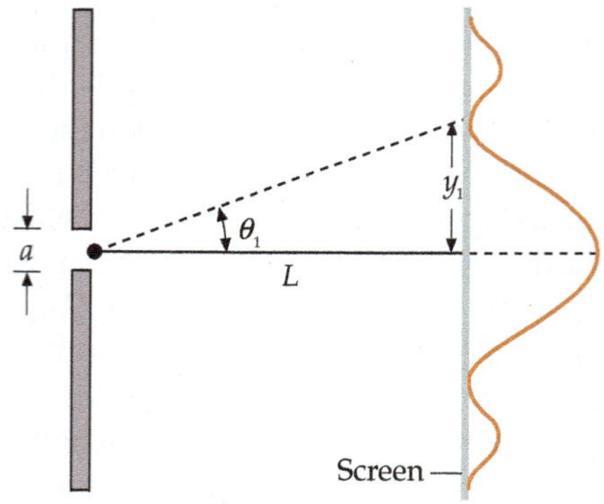

FIGURE 33-13 The distance y_1 measured along the screen from the central maximum to the first diffraction minimum is related to the angle θ_1 by $\tan\theta_1 = y_1/L$, where L is the distance to the screen.

Example 33-5 Width of the Central Diffraction Maximum

During a lecture demonstration of single-slit diffraction, a laser beam that has a wavelength equal to 700 nm passes through a vertical slit 0.20 mm wide and hits a screen 6.0 m away. Find the width of the central diffraction maximum on the screen; that is, find the distance between the first minimum on the left and the first minimum on the right of the central maximum.

PICTURE Referring to Figure 33-13, the width of the central diffraction maximum is $2y_1$.

SOLVE

1. The half-width of the central maximum y_1 is related to the angle θ_1 by: $\qquad \tan\theta_1 = \dfrac{y_1}{L}$

2. The angle θ_1 is related to the slit width a by Equation 33-11: $\sin\theta_1 = \lambda/a$

3. Solve the step-2 result for θ_1, substitute into the step-1 result, and solve for $2y_1$:

$$2y_1 = 2L\tan\theta_1 = 2L\tan\left(\sin^{-1}\frac{\lambda}{a}\right)$$
$$= 2(6.0\text{ m})\tan\left(\sin^{-1}\frac{700\times10^{-9}\text{ m}}{0.00020\text{ m}}\right)$$
$$= 4.2\times10^{-2}\text{ m} = \boxed{4.2\text{ cm}}$$

CHECK Because $\sin\theta_1 = \lambda/a = (700\text{ nm})/(0.20\text{ mm}) = 0.0035$, we can use the small-angle approximation to evaluate $2y_1$. In this approximation, $\sin\theta_1 = \tan\theta_1$, so $\lambda/a = y_1/L$ and $2y_1 = 2L\lambda/a = 2(6.0\text{ m})(700\text{ nm})/(0.20\text{ mm}) = 4.2$ cm. (This approximate value is in agreement with the exact value to within 0.0006 percent.)

INTERFERENCE–DIFFRACTION PATTERN OF TWO SLITS

When there are two or more slits, the intensity pattern on a screen far away is a combination of the single-slit diffraction pattern of the individual slits and the multiple-slit interference pattern we have studied. Figure 33-14 shows the intensity pattern on a screen far from two slits whose separation d is $10a$, where a is the width of each slit. The pattern is the same as the two-slit pattern that has very narrow slits (Figure 33-11) except that it is modulated by the single-slit diffraction pattern; that is, the intensity due to each slit separately is now not constant but decreases with angle, as shown in Figure 33-14*b*.

Note that the central diffraction maximum in Figure 33-14 has 19 interference maxima—the central interference maximum and 9 maxima on either side. The tenth interference maximum on either side of the central one is at the angle θ_{10}, given by $\sin\theta_{10} = 10\lambda/d = \lambda/a$, because $d = 10a$. This coincides with the position of the first diffraction minimum, so this interference maximum is not seen. At these points, the light from the two slits would be in phase and would interfere constructively, but there is no light coming from either slit because the points are at diffraction minima of each slit. In general, we can see that if $m = d/a$, the mth interference maximum will fall at the first diffraction minimum. Because the mth fringe is not seen, there will be $m - 1$ fringes on each side of the central fringe for a total of N fringes in the central maximum, where N is given by

$$N = 2(m-1) + 1 = 2m - 1 \qquad 33\text{-}12$$

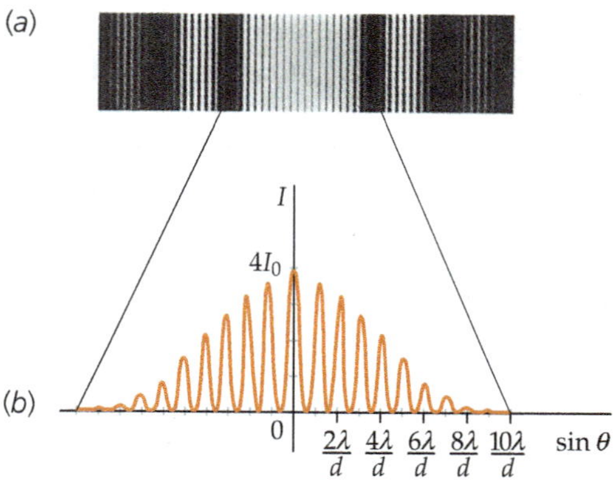

FIGURE 33-14 (*a*) Interference–diffraction pattern for two slits whose separation d is equal to 10 times their width a. The tenth interference maximum on either side of the central interference maximum is missing because it falls at the first diffraction minimum. (*b*) Plot of intensity versus $\sin\theta$ for the central band of the pattern in Figure 33-14*a*. (*Courtesy of Michael Cagnet.*)

Example 33-6 Interference and Diffraction

Two slits that each have a width $a = 0.015$ mm are separated by a distance $d = 0.060$ mm and are illuminated by light of wavelength $\lambda = 650$ nm. How many bright fringes are seen in the central diffraction maximum?

PICTURE We need to find the value of m for which the mth interference maximum coincides with the first diffraction minimum. Then there will be $N = 2m - 1$ fringes in the central maximum.

SOLVE

1. Relate the angle θ_1 of the first diffraction minimum to the slit width a: $\sin\theta_1 = \dfrac{\lambda}{a}$ (first diffraction minimum)

2. Relate the angle θ_m of the mth interference maxima to the slit separation d: $\sin\theta_m = \dfrac{m\lambda}{d}$ (mth interference maxima)

3. Set the angles equal and solve for m:

$$\frac{m\lambda}{d} = \frac{\lambda}{a}$$

$$m = \frac{d}{a} = \frac{0.060\text{ mm}}{0.015\text{ mm}} = 4.0$$

4. The first diffraction minimum coincides with the fourth bright fringe. Therefore, there are 3 bright fringes visible on either side of the central diffraction maximum. These 6 maxima, plus the central interference maximum, combine for a total of 7 bright fringes in the central diffraction maximum:

$$N = \boxed{7\text{ bright fringes}}$$

* 33-5 USING PHASORS TO ADD HARMONIC WAVES

To calculate the interference pattern produced by three, four, or more coherent light sources and to calculate the diffraction pattern of a single slit, we need to combine several harmonic waves of the same frequency that differ in phase. A simple geometric interpretation of harmonic wave functions leads to a method of adding harmonic waves of the same frequency by geometric construction.

Let the wave functions for two waves at some point be $E_1 = A_1 \sin\alpha$ and $E_2 = A_2 \sin(\alpha + \delta)$, where $\alpha = \omega t$. Our problem is then to find the sum:

$$E_1 + E_2 = A_1 \sin\alpha + A_2 \sin(\alpha + \delta)$$

We can represent each wave function by the y component of a two-dimensional vector, as shown in Figure 33-15. The geometric method of addition is based on the fact that the y component of the sum of two or more vectors equals the sum of the y components of the vectors, as illustrated in the figure. The wave function E_1 is represented by the y component of the vector $\vec{A}_1$. As the time continues on, this vector rotates in the xy plane with angular frequency ω. The vector $\vec{A}_1$ is called a **phasor.** (We encountered phasors in our study of ac circuits in Section 29-5.) The wave function E_2 is the y component of a phasor of magnitude A_2 that makes an angle $\alpha + \delta$ with the x axis. By the laws of vector addition, the sum of the y components of the individual phasors equals the y component of the resultant phasor $\vec{A}$, as shown in Figure 33-15. The y component of the resultant phasor, $A \sin(\alpha + \delta')$, is a harmonic wave function that is the sum of the two original wave functions. That is,

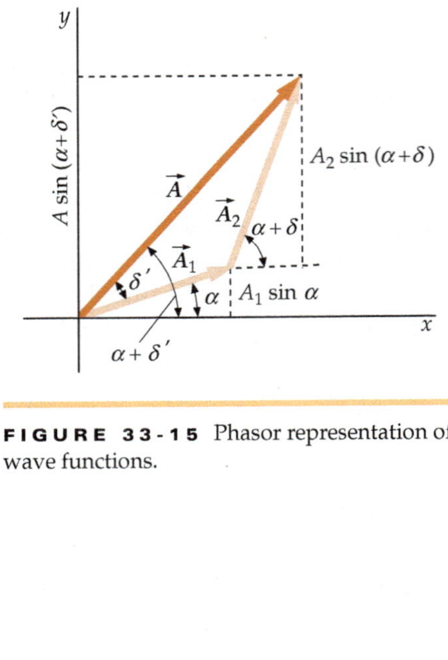

FIGURE 33-15 Phasor representation of wave functions.

$$A_1 \sin\alpha + A_2 \sin(\alpha + \delta) = A \sin(\alpha + \delta') \qquad \text{33-13}$$

where A (the amplitude of the resultant wave) and δ' (the phase of the resultant wave relative to the phase of the first wave) are found by adding the phasors representing the waves. As time varies, α varies. The phasors representing the two wave functions and the resultant phasor representing the resultant wave function rotate in space, but their relative positions do not change because they all rotate with the same angular velocity ω.

Example 33-7 Wave Superposition Using Phasors — *Try It Yourself*

Use the phasor method of addition to derive $E = \left[2A_0 \cos\frac{1}{2}\delta\right] \sin\left(\omega t + \frac{1}{2}\delta\right)$ (Equation 33-7) for the superposition of two waves of the same amplitude.

PICTURE Represent the waves $y_1 = A_0 \sin\alpha$ and $y_2 = A_0 \sin(\alpha + \delta)$ by vectors (phasors) of length A_0 making an angle δ with one another. The resultant wave $y_r = A \sin(\alpha + \delta')$ is represented by the sum of these vectors, which form an isosceles triangle, as shown in Figure 33-16.

SOLVE

Cover the column to the right and try these on your own before looking at the answers.

Steps	Answers
1. Relate δ and δ' using the theorem: "An external angle to a triangle is equal to the sum of the two non-adjacent internal angles."	$\delta' + \delta' = \delta$
2. Solve for δ'.	$\delta' = \frac{1}{2}\delta$
3. Write $\cos\delta'$ in terms of A and A_0.	$\cos\delta' = \dfrac{\frac{1}{2}A}{A_0}$
4. Solve for A in terms of δ.	$A = 2A_0\cos\delta' = 2A_0\cos\frac{1}{2}\delta$
5. Use your results for A and δ' to write the resultant wave function.	$y_r = A\sin(\alpha + \delta')$ $= \boxed{\left[2A_0\cos\frac{1}{2}\delta\right]\sin\left(\alpha + \frac{1}{2}\delta\right)}$

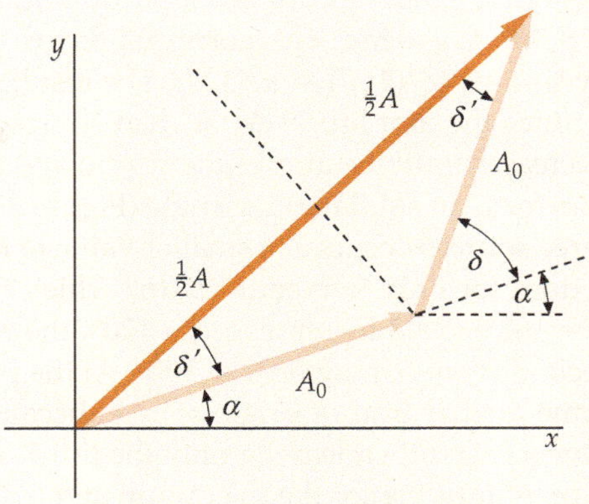

FIGURE 33-16

CHECK The step-5 result is identical to Equation 33-7 (see Problem statement).

PRACTICE PROBLEM 33-3 Find the amplitude and phase constant of the resultant wave function produced by the superposition of the two waves $E_1 = (4.0\text{ V/m})\sin(\omega t)$ and $E_2 = (3.0\text{ V/m})\sin(\omega t + 90°)$.

*THE INTERFERENCE PATTERN OF THREE OR MORE EQUALLY SPACED SOURCES

We can apply the phasor method of addition to calculate the interference pattern of three or more coherent sources that are equally spaced and in phase. We are most interested in the location of the interference maxima and minima. Figure 33-17 illustrates the case of three such sources. The geometry is the same as for two sources. At a great distance from the sources, the rays from the sources to a point P on the screen are approximately parallel. The path-length difference between the first and second source is then $d\sin\theta$, as before, and the path-length difference between the first and third source is $2d\sin\theta$. The wave at point P is the sum of the three waves. Let $\alpha = \omega t$ be the phase of the first wave at point P. We thus have the problem of adding three waves of the form

$$E_1 = A_0\sin\alpha$$
$$E_2 = A_0\sin(\alpha + \delta) \qquad 33\text{-}14$$
$$E_3 = A_0\sin(\alpha + 2\delta)$$

where

$$\delta = \frac{2\pi}{\lambda}d\sin\theta \approx \frac{2\pi}{\lambda}\frac{yd}{L} \qquad 33\text{-}15$$

as in the two-slit problem.

At $\theta = 0$, $\delta = 0$, so all the waves are in phase. The amplitude of the resultant wave is 3 times that of each individual wave and the intensity is 9 times that due to each source acting separately. As the angle θ increases from $\theta = 0$, the phase angle δ increases and the intensity decreases. The position $\theta = 0$ is thus a position of maximum intensity.

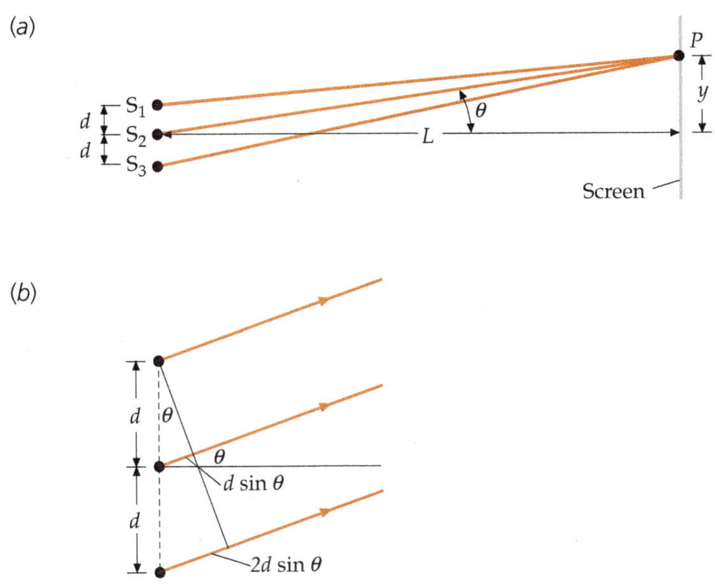

FIGURE 33-17 Geometry for calculating the intensity pattern far away from three equally spaced, coherent sources that are in phase.

Figure 33-18 shows the phasor addition of three waves for a phase angle $\delta = 30° = \pi/6$ rad. This corresponds to a point P on the screen for which θ is given by $\sin\theta = \lambda\delta/(2\pi d) = \lambda/(12d)$. The resultant amplitude A is considerably less than 3 times the amplitude A_0 of each source. As δ increases, the resultant amplitude decreases until the amplitude is zero at $\delta = 120°$. For this value of δ, the three phasors form an equilateral triangle (Figure 33-19). This first interference minimum for three sources occurs at a smaller value of δ (and therefore at a smaller angle θ) than it does for only two sources (for which the first interference minimum occurs at $\delta = 180°$). As δ increases from 120°, the resultant amplitude increases, reaching a secondary maximum at $\delta = 180°$. At the phase angle $\delta = 180°$, the amplitude is the same as that from a single source, because the waves from the first two sources cancel each other, leaving only the third. The intensity of the secondary maximum is one-ninth that of the maximum at $\theta = 0$. As δ increases beyond 180°, the amplitude again decreases and is zero at $\delta = 180° + 60° = 240°$. For δ greater than 240°, the amplitude increases and is again 3 times that of each source when $\delta = 360°$. This phase angle corresponds to a path-length difference of 1 wavelength for the waves from the first two sources and 2 wavelengths for the waves from the first and third sources. Hence, the three waves are in phase at this point. The largest maxima, called the principal maxima, are at the same positions as for just two sources, which are those points corresponding to the angles θ given by

$$d \sin\theta_{\rm m} = m\lambda \quad m = 0, 1, 2, \ldots \tag{33-16}$$

These maxima are stronger and narrower than those for two sources. They occur at points for which the path-length difference between adjacent sources is zero or an integral number of wavelengths.

These results can be generalized to more than three sources. For four coherent sources that are equally spaced and in phase, the principal interference maxima are again given by Equation 33-16, but the maxima are even more intense, they are narrower, and there are two small secondary maxima between each pair of principal maxima. At $\theta = 0$, the intensity is 16 times that due to a single source. The first interference minimum occurs when δ is 90°, as can be seen from the phasor diagram of Figure 33-20. The first secondary maximum is near $\delta = 132°$. The intensity of the secondary maximum is about one-fourteenth that of the central maximum. There is another minimum at $\delta = 180°$, another secondary maximum near $\delta = 228°$, and another minimum at $\delta = 270°$ before the next principal maximum at $\delta = 360°$.

Figure 33-21 shows the intensity patterns for two, three, and four equally spaced coherent sources. Figure 33-22 shows a graph of I/I_0, where I_0 is the intensity due to each source acting separately. For three sources, there is a very small secondary maximum between each pair of principal maxima, and the principal maxima are sharper and more intense than those due to just two sources. For four sources, there are two small secondary maxima between each pair of principal maxima, and the principal maxima are even more narrow and intense.

From this discussion, we can see that as we increase the number of sources, the intensity becomes more and more concentrated in the principal maxima given by Equation 33-16, and these maxima become narrower. For N sources, the intensity of the principal maxima is N^2 times that due to a single source. The first minimum occurs at a phase angle of $\delta = 360°/N$, for which the N phasors form a closed polygon of N sides. There are $N - 2$ secondary maxima between each pair of principal maxima. These secondary maxima are very weak compared with the principal maxima.

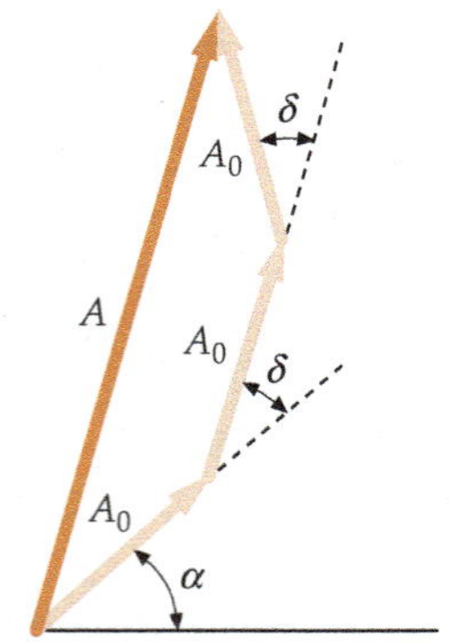

FIGURE 33-18 Phasor diagram for determining the resultant amplitude A due to three waves, each of amplitude A_0, that have phase differences of δ and 2δ due to path-length differences of $d \sin\theta$ and $2d \sin\theta$. The angle $\alpha = \omega t$ varies with time, but this does not affect the calculation of A.

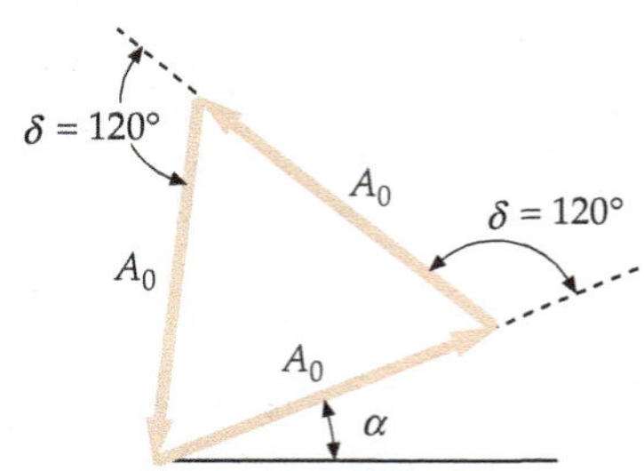

FIGURE 33-19 The resultant amplitude for the waves from three sources is zero when δ is 120°. This interference minimum occurs at a smaller angle θ than does the first minimum for two sources, which occurs when δ is 180°.

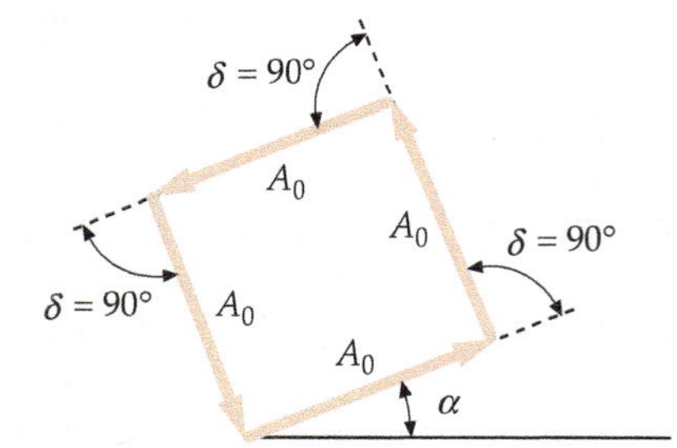

FIGURE 33-20 Phasor diagram for the first minimum for four coherent sources that are equally spaced and in phase. The amplitude is zero when the phase difference of the waves from adjacent sources is 90°.

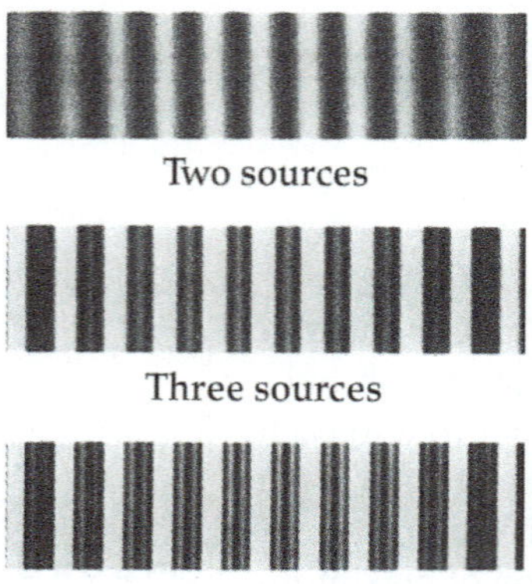

FIGURE 33-21 Intensity patterns for two, three, and four coherent sources that are equally spaced and in phase. There is a secondary maximum between each pair of principal maxima for three sources, and two secondary maxima between each pair of principal maxima for four sources. *(Courtesy of Michael Cagnet.)*

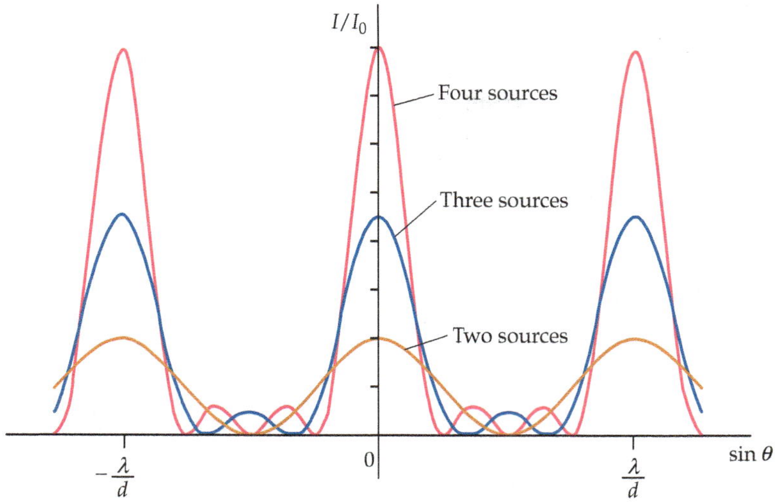

FIGURE 33-22 Plot of relative intensity versus $\sin\theta$ for two, three, and four coherent sources that are equally spaced and in phase.

As the number of sources is increased, the principal maxima become sharper and more intense, and the intensities of the secondary maxima become negligible compared to those of the principal maxima.

*CALCULATING THE SINGLE-SLIT DIFFRACTION PATTERN

We now use the phasor method for the addition of harmonic waves to calculate the intensity pattern shown in Figure 33-11. We assume that the slit of width a is divided into N equal intervals and that there is a point source of waves at the midpoint of each interval (Figure 33-23). If d is the distance between two adjacent sources and a is the width of the opening, we have $d = a/N$. Because the screen on which we are calculating the intensity is far from the sources, the rays from the sources to a point P on the screen are approximately parallel. The path-length difference between any two adjacent sources is $\delta \sin\theta$, and the phase difference δ is related to the path-length difference by

$$\delta = \frac{d\sin\theta}{\lambda}2\pi$$

If A_0 is the amplitude due to a single source, the amplitude at the central maximum, where $\theta = 0$ and all the waves are in phase, is $A_{max} = NA_0$ (Figure 33-24).

We can find the amplitude at some other point at an angle θ by using the phasor method for the addition of harmonic waves. As in the addition of two, three, or four waves, the intensity is zero at any point where the phasors representing the waves form a closed polygon. In this case, the polygon has N sides (Figure 33-25). At the first minimum, the wave from the first source just below the top of the opening and the wave from the source just below the middle of the opening are 180° out of phase. In this case, the waves from the source near the top of the opening differ

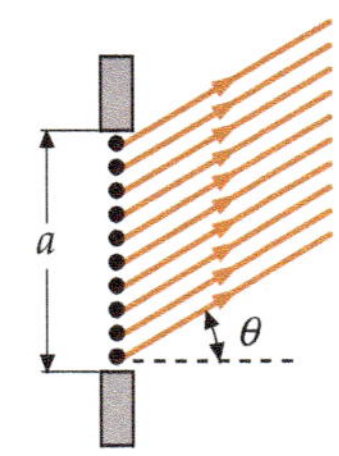

FIGURE 33-23 Diagram for calculating the diffraction pattern far away from a narrow slit. The slit width a is assumed to contain a large number of in-phase, equally spaced point sources separated by a distance d. The rays from the sources to a point far away are approximately parallel. The path-length difference for the waves from adjacent sources is $d\sin\theta$.

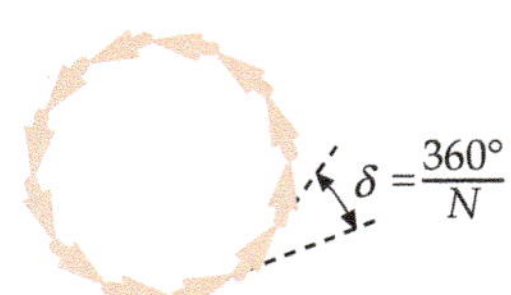

FIGURE 33-25 Phasor diagram for calculating the first minimum in the single-slit diffraction pattern. When the waves from the N sources completely cancel, the N phasors form a closed polygon. The phase difference between the waves from adjacent sources is then $\delta = 360°/N$. When N is very large, the waves from the first and last sources are approximately in phase.

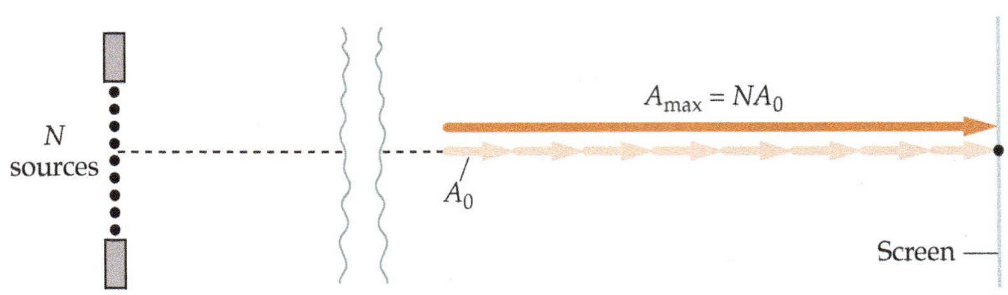

FIGURE 33-24 A single slit is represented by N sources, each of amplitude A_0. At the central maximum point, where $\theta = 0$, the waves from the sources add in phase, giving a resultant amplitude $A_{max} = NA_0$.

from those from the bottom of the opening by nearly 360°. [The phase difference is, in fact, 360° − (360°/N).] Thus, if the number of sources is very large, 360°/N is negligible and we get complete cancellation if the waves from the first and last sources are out of phase by 360°, corresponding to a path-length difference of one wavelength, in agreement with Equation 33-11.

We will now calculate the amplitude at a general point at which the waves from two adjacent sources differ in phase by δ. Figure 33-26 shows the phasor diagram for the addition of N waves, where the subsequent waves differ in phase from the first wave by $\delta, 2\delta, \ldots, (N-1)\delta$. When N is very large and δ is very small, the phasor diagram approximates the arc of a circle. The resultant amplitude A is the length of the chord of this arc. We will calculate this resultant amplitude in terms of the phase difference ϕ between the first wave and the last wave. From Figure 33-26, we have

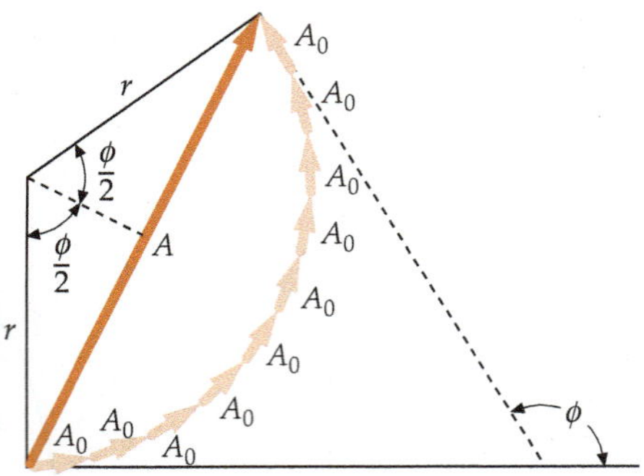

FIGURE 33-26 Phasor diagram for calculating the resultant amplitude due to the waves from N sources in terms of the phase difference ϕ between the wave from the first source just below the top of the slit and the wave from the last source just above the bottom of the slit. When N is very large, the resultant amplitude A is the chord of a circular arc of length $NA_0 = A_{max}$.

$$\sin \tfrac{1}{2}\phi = \frac{A/2}{r}$$

or

$$A = 2r \sin \tfrac{1}{2}\phi \qquad 33\text{-}17$$

where r is the radius of the arc. Because the length of the arc is $A_{max} = NA_0$ and the angle subtended is ϕ, we have

$$\phi = \frac{A_{max}}{r} \qquad 33\text{-}18$$

or

$$r = \frac{A_{max}}{\phi}$$

Substituting this into Equation 33-17 gives

$$A = \frac{2A_{max}}{\phi}\sin \tfrac{1}{2}\phi = A_{max}\frac{\sin \frac{1}{2}\phi}{\frac{1}{2}\phi}$$

Because the amplitude at the center of the central maximum ($\theta = 0$) is A_{max}, the ratio of the intensity at any other point to that at the center of the central maximum is

$$\frac{I}{I_0} = \frac{A^2}{A_{max}^2} = \left(\frac{\sin \frac{1}{2}\phi}{\frac{1}{2}\phi}\right)^2$$

or

$$I = I_0\left(\frac{\sin \frac{1}{2}\phi}{\frac{1}{2}\phi}\right)^2 \qquad 33\text{-}19$$

INTENSITY FOR A SINGLE-SLIT DIFFRACTION PATTERN

The phase difference ϕ between the first and last waves is related to the path-length difference $a \sin\theta$ between the top and bottom of the opening by

$$\phi = \frac{a \sin\theta}{\lambda}2\pi \qquad 33\text{-}20$$

Equation 33-19 and Equation 33-20 describe the intensity pattern shown in Figure 33-11. The first minimum occurs at $a \sin\theta = \lambda$, which is the point where the waves from the middle of the upper half and the middle of the lower half of the slit have a path-length difference of $\lambda/2$ and are 180° out of phase. The second minimum occurs at $a \sin\theta = 2\lambda$, where the waves from the upper half of the upper half of the slit and those from the lower half of the upper half of the slit have a path-length difference of $\lambda/2$ and are 180° out of phase.

There is a secondary maximum approximately midway between the first and second minima at $a \sin\theta \approx \frac{3}{2}\lambda$. Figure 33-27 shows the phasor diagram for determining the approximate intensity of this secondary maximum. The phase difference ϕ between the first and last waves is approximately $2\pi + \pi$. The phasors thus complete $1\frac{1}{2}$ circles. The resultant amplitude is the diameter of a circle that has a circumference which is two-thirds the total length A_{max}. If $C = \frac{2}{3}A_{max}$ is the circumference, the diameter A is

$$A = \frac{C}{\pi} = \frac{\frac{2}{3}A_{max}}{\pi} = \frac{2}{3\pi}A_{max}$$

and

$$A^2 = \frac{4}{9\pi^2}A_{max}^2$$

The intensity at this point is

$$I = \frac{4}{9\pi^2}I_0 = \frac{1}{22.2}I_0 \qquad 33\text{-}21$$

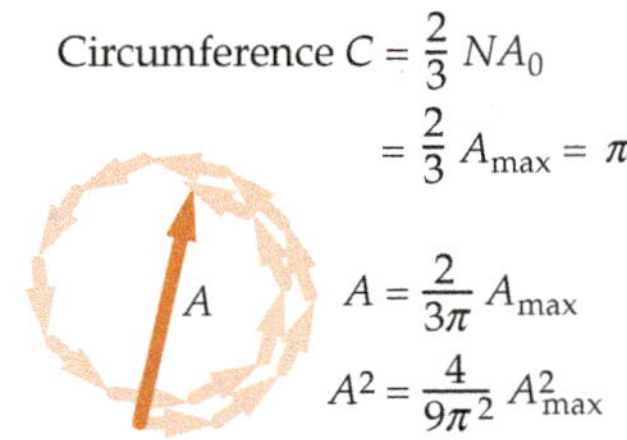

FIGURE 33-27 Phasor diagram for calculating the approximate amplitude of the first secondary maximum of the single-slit diffraction pattern. The secondary maximum occurs near the midpoint between the first and second minima when the N phasors complete $1\frac{1}{2}$ circles.

*CALCULATING THE INTERFERENCE–DIFFRACTION PATTERN OF MULTIPLE SLITS

The intensity of the two-slit interference–diffraction pattern can be calculated from the two-slit pattern (Equation 33-8) where the intensity of each slit (I_0 in that equation) is replaced by the diffraction pattern intensity due to each slit, I, given by Equation 33-19. The intensity for the two-slit interference–diffraction pattern is thus

$$I = 4I_0\left(\frac{\sin\frac{1}{2}\phi}{\frac{1}{2}\phi}\right)^2 \cos^2\tfrac{1}{2}\delta \qquad 33\text{-}22$$

INTERFERENCE–DIFFRACTION INTENSITY FOR TWO SLITS

where ϕ is the difference in phase between rays from the top and bottom of each slit, which is related to the width of each slit by

$$\phi = \frac{a\sin\theta}{\lambda}2\pi$$

and δ is the difference in phase between rays from the centers of two adjacent slits, which is related to the slit separation by

$$\delta = \frac{d\sin\theta}{\lambda}2\pi$$

In Equation 33-22, the intensity I_0 is the intensity at $\theta = 0$ due to one slit alone.

Example 33-8 Five-Slit Interference–Diffraction Pattern

Find the interference–diffraction intensity pattern for five equally spaced slits, where a is the width of each slit and d is the distance between adjacent slits.

PICTURE First, find the interference intensity pattern for the five slits, assuming no angular variations in the intensity due to diffraction. To do this, first construct a phasor diagram to find the amplitude of the resultant wave in an arbitrary direction θ. Intensity is proportional to the square of the amplitude. Next, correct for the variation of intensity with θ by using the single-slit diffraction pattern intensity relation (Equation 33-20 and Equation 33-20).

SOLVE

1. The diffraction pattern intensity I' due to a slit of width a is given by Equation 33-19 and Equation 33-20:

$$I' = I_0\left(\frac{\sin\frac{1}{2}\phi}{\frac{1}{2}\phi}\right)^2$$

where

$$\phi = \frac{2\pi}{\lambda}a\sin\theta$$

2. The interference pattern intensity I is proportional to the square of the amplitude A of the superposition of the wave functions for the light from the five slits:

$$I \propto A^2$$

where

$$A\sin(\alpha + \delta') = A_0\sin\alpha + A_0\sin(\alpha + \delta) + A_0\sin(\alpha + 2\delta) + A_0\sin(\alpha + 3\delta) + A_0\sin(\alpha + 4\delta)$$

and where $\alpha = \omega t$ and $\delta = \dfrac{d\sin\theta}{\lambda}2\pi$

3. To solve for A, we construct a phasor diagram (Figure 33-28). The amplitude A equals the sum of the projections of the individual phasors onto the resultant phasor:

$$\delta' = \beta + \delta$$

so

$$\beta = \delta' - \delta = 2\delta - \delta = \delta$$

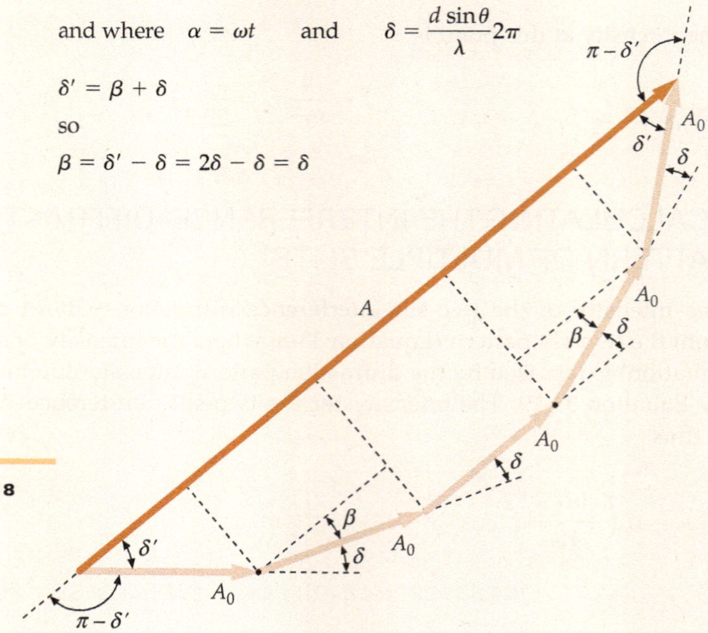

FIGURE 33-28

4. To find δ', we add the exterior angles. The sum of the exterior angles equals 2π (if you walk the perimeter of any polygon you rotate through the sum of the exterior angles, and you rotate through 2π radians):

$$2(\pi - \delta') + 4\delta = 2\pi \quad \Rightarrow \quad \delta' = 2\delta$$

5. Solve for A from the figure:

$$A = 2A_0\cos\delta' + 2A_0\cos\beta + A_0$$

6. Substitute for δ' using the step-4 result, and substitute for β using the relation $\beta = \delta$. (That β and δ are equal follows from the theorem "If two parallel lines are cut by a transversal, the interior and exterior angles on the same side of the transversal are equal.")

$$A = A_0(2\cos 2\delta + 2\cos\delta + 1)$$

7. Square both sides to relate the intensities. Recall, I' is the intensity from a single slit, and A_0 is the amplitude from a single slit:

$$A^2 = A_0^2(2\cos 2\delta + 2\cos\delta + 1)^2$$

so

$$I = I'(2\cos 2\delta + 2\cos\delta + 1)^2$$

8. Substitute for I' using the step-1 result:

$$\boxed{I = I_0\left(\frac{\sin\frac{1}{2}\phi}{\frac{1}{2}\phi}\right)^2(2\cos 2\delta + 2\cos\delta + 1)^2}$$

where $\phi = \dfrac{a\sin\theta}{\lambda}2\pi$ and $\delta = \dfrac{d\sin\theta}{\lambda}2\pi$

CHECK If $\theta = 0$, both $\phi = 0$ and $\delta = 0$. So, for $\theta = 0$, step 5 becomes $A = 5A_0$ and step 8 becomes $I = 5^2I_0 = 25I_0$ as expected.

33-6 FRAUNHOFER AND FRESNEL DIFFRACTION

Diffraction patterns, like the single-slit pattern shown in Figure 33-11, that are observed at points for which the rays from an aperture or an obstacle are nearly parallel are called **Fraunhofer diffraction patterns.** Fraunhofer patterns can be observed at great distances from the obstacle or the aperture so that the rays reaching any point are approximately parallel, or they can be observed using a lens to focus parallel rays on a viewing screen placed in the focal plane of the lens.

The diffraction pattern observed near an aperture or an obstacle is called a **Fresnel diffraction pattern.** Because the rays from an aperture or an obstacle close to a screen cannot be considered parallel, Fresnel diffraction is much more difficult to analyze. Figure 33-29 illustrates the difference between the Fresnel and the Fraunhofer patterns for a single slit.*

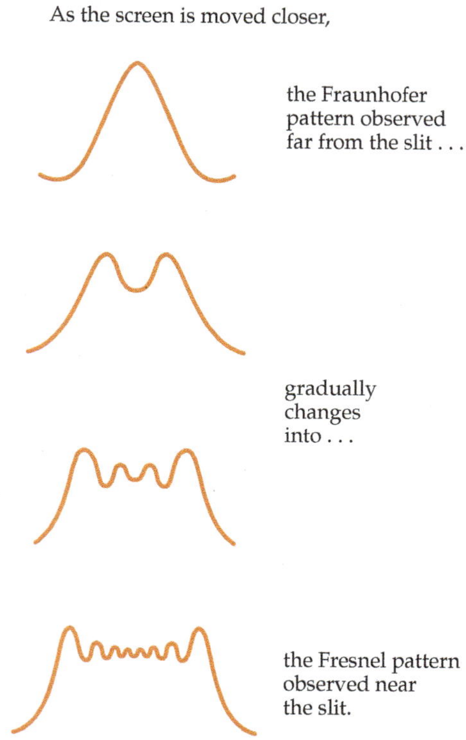

FIGURE 33-29 Diffraction patterns for a single slit at various screen distances.

Figure 33-30*a* shows the Fresnel diffraction pattern of an opaque disk. Note the bright spot at the center of the pattern caused by the constructive interference of the light waves diffracted from the edge of the disk. This pattern is of some historical interest. In an attempt to discredit Augustin Fresnel's wave theory of light, Siméon Poisson pointed out that it predicted a bright spot at the center of the shadow, which he assumed was a ridiculous contradiction of fact. However, Fresnel immediately demonstrated experimentally that such a spot does, in fact, exist. This demonstration convinced many doubters of the validity of the wave theory of light. The Fresnel diffraction pattern of a circular aperture is shown in Figure 33-30*b*. Comparing this with the pattern of the opaque disk in Figure 33-30*a*, we can see that the two patterns are complements of each other.

Figure 33-31*a* shows the Fresnel diffraction pattern of a straightedge illuminated by light from a point source. A graph of the intensity versus distance (measured along a line perpendicular to the edge) is shown in Figure 33-31*b*. The light intensity does not fall abruptly to zero in the geometric shadow, but it decreases rapidly and is negligible within a few wavelengths of the edge. The Fresnel diffraction pattern

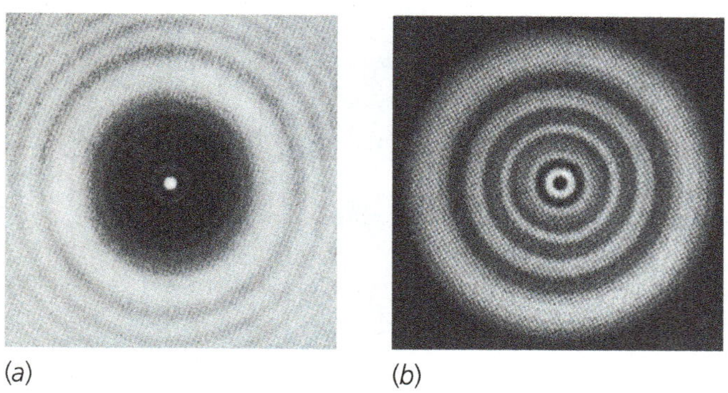

FIGURE 33-30 (*a*) The Fresnel diffraction pattern of an opaque disk. At the center of the shadow, the light waves diffracted from the edge of the disk are in phase and produce a bright spot called the *Poisson spot.* (*b*) The Fresnel diffraction pattern of a circular aperture. Compare this with Figure 33-30*a*. *((a) and (b) M. Cagnet, M. Françon, J. C. Thrierr, Atlas of Optical Phenomena.)*

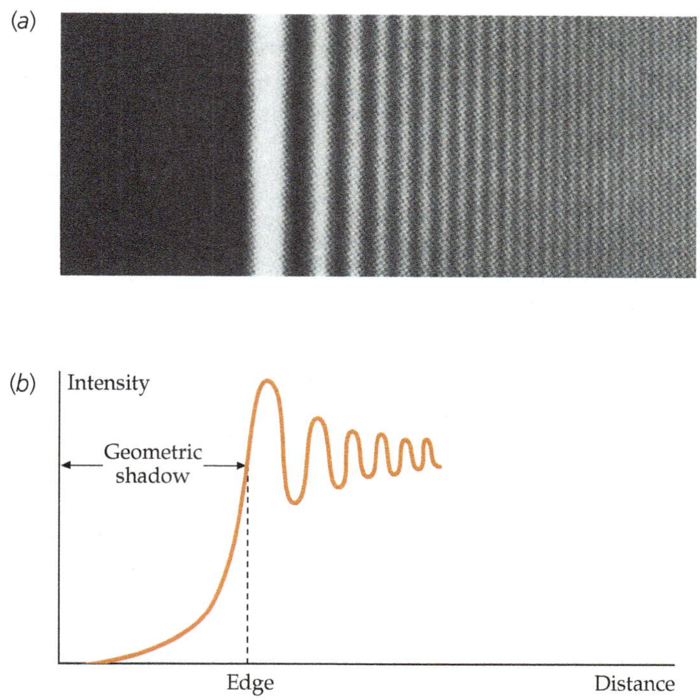

FIGURE 33-31 (*a*) The Fresnel diffraction of a straightedge. (*b*) A graph of intensity versus distance along a line perpendicular to the edge. *(Courtesy Battelle-Northwest Laboratories.)*

* See Richard E. Haskel, "A Simple Experiment on Fresnel Diffraction," *American Journal of Physics* 38 (1970): 1039.

of a rectangular aperture is shown in Figure 33-32. These patterns cannot be seen using extended light sources like an ordinary lightbulb, because the dark fringes of the pattern produced by light from one point on the source overlap the bright fringes of the pattern produced by light from another point.

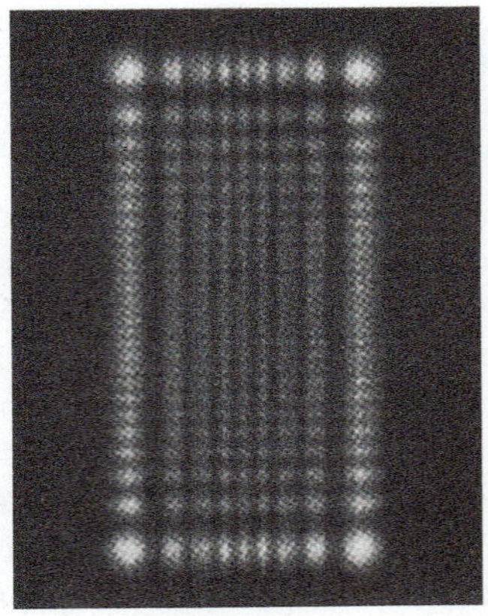

FIGURE 33-32 The Fresnel diffraction pattern of a rectangular aperture. *(Courtesy of Michael Cagnet.)*

33-7 DIFFRACTION AND RESOLUTION

Diffraction due to a circular aperture has important implications for the resolution of many optical instruments. Figure 33-33 shows the Fraunhofer diffraction pattern of a circular aperture. The angle θ subtended by the first diffraction minimum is related to the wavelength and the diameter of the opening D by

$$\sin\theta = 1.22\frac{\lambda}{D} \qquad 33\text{-}23$$

Equation 33-23 is similar to Equation 33-9 except for the factor 1.22, which arises from the mathematical analysis, and is similar to the equation for a single slit but more complicated because of the circular geometry. In many applications, the angle θ is small, so $\sin\theta$ can be replaced by θ. The first diffraction minimum is then at an angle θ given by

$$\theta \approx 1.22\frac{\lambda}{D} \qquad 33\text{-}24$$

FIGURE 33-33 The Fraunhofer diffraction pattern of a circular aperture. *(Courtesy of Michael Cagnet.)*

Figure 33-34 shows two point sources that subtend an angle α at a circular aperture far from the sources. The intensities of the Fraunhofer diffraction pattern are also indicated in this figure. If α is much greater than $1.22\lambda/D$, the sources will be seen as two sources. However, as α is decreased, the overlap of the diffraction patterns increases, and it becomes difficult to distinguish the two sources from one source. At the critical angular separation, α_c, given by

$$\alpha_c = 1.22\frac{\lambda}{D} \qquad 33\text{-}25$$

the first minimum of the diffraction pattern of one source falls on the central maximum of the other source. These objects are said to be just resolved by **Rayleigh's criterion for resolution.** Figure 33-35 shows the diffraction patterns for two sources when α is greater than the critical angle for resolution and when α is just equal to the critical angle for resolution.

Equation 33-25 has many applications. The *resolving power* of an optical instrument, such as a microscope or telescope, is the ability of the instrument to resolve two objects that are close together. The images of the objects tend to overlap because of diffraction at the entrance aperture of the instrument. We can see from Equation 33-25 that the resolving power can be increased either by increasing the diameter D of the lens (or mirror) or by decreasing the wavelength λ. Astronomical telescopes use large objective lenses or mirrors to increase their resolution as well as to increase their light-gathering power. An array of 27 radio antennas (Figure 33-36) mounted on rails can be configured to form a single telescope that has a resolution distance D of 36 km (22 mi). In a microscope, a film of transparent oil that has an index of refraction

(a)

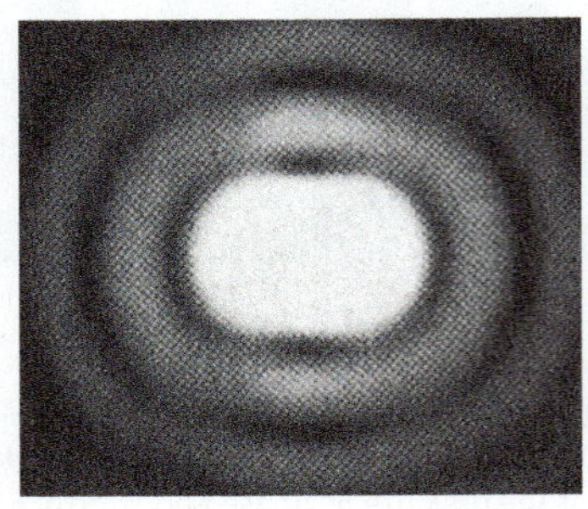

(b)

FIGURE 33-35 The diffraction patterns for a circular aperture and two incoherent point sources when (*a*) α is a factor of 2 or so greater than $\alpha_c = 1.22\lambda/D$ and (*b*) when α is equal to the limit of resolution, $\alpha_c = 1.22\lambda/D$. *((a) and (b) Courtesy of Michael Cagnet.)*

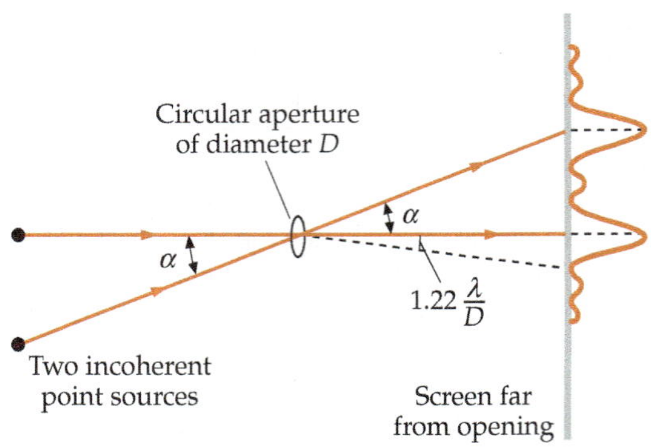

FIGURE 33-34 Two distant sources that subtend an angle α. If α is much greater than $1.22\lambda/D$, where λ is the wavelength of light and D is the diameter of the aperture, the diffraction patterns have little overlap and the sources are easily seen as two distinct sources. If α is not much greater than $1.22\lambda/D$, the overlap of the diffraction patterns makes it difficult to distinguish two sources from one.

of approximately 1.55 is sometimes used under the objective to decrease the wavelength of the light ($\lambda' = \lambda/n$). The wavelength can be reduced further by using ultraviolet light and photographic film; however, ordinary glass is opaque to ultraviolet light, so the lenses in an ultraviolet microscope must be made from quartz or fluorite. To obtain very high resolutions, electron microscopes are used—microscopes that use electrons rather than light. The wavelengths of electrons vary inversely with the square root of their kinetic energy and can be made as small as desired.*

FIGURE 33-36 The very large array (VLA) of radio antennas is located near Socorro, New Mexico. The 25-m-diameter antennas are mounted on rails, which can be arranged in several configurations, and can be extended over a diameter of 36 km. The data from the antennas are combined electronically, so the array is really a single high-resolution telescope. *(Courtesy of National Radio Astronomy Observatory/Associated Universities, Inc./National Science Foundation. Photographer: Kelly Gatlin. Digital composite: Patricia Smiley.)*

Example 33-9 Physics in the Library — *Context-Rich*

While studying in the library, you lean back in your chair and ponder the small holes you notice in the ceiling tiles. You notice that the holes are approximately 5.0 mm apart. You can clearly see the holes directly above you, about 2 m up, but the tiles far away do not appear to have the holes. You wonder if the reason you cannot see the distant holes is because they are not within the criteria for resolution established by Rayleigh. Is this a feasible explanation for the disappearance of the holes? You notice the holes disappear about 20 m from you.

PICTURE We will need to make assumptions about the situation. If we use Equation 33-25, we will need to know the wavelength of light and the aperture diameter. Assuming our pupil is the aperture, we can assume approximately 5.0 mm for the diameter. (This is the value used in this physics textbook.) The wavelength of the light is probably about 500 nm or so.

SOLVE

1. The angular limit for resolution by the eye depends on the ratio of the wavelength and the diameter of the pupil:
$$\theta_c \approx 1.22\frac{\lambda}{D}$$
2. The angle subtended by two holes depends on their separation distance d and their distance L from your eye:
$$\theta \approx \frac{d}{L}$$
3. Equating the two angles and putting in the numbers give:
$$\frac{d}{L} \approx 1.22\frac{\lambda}{D}$$
$$\frac{5.0\text{ mm}}{L} \approx 1.22\frac{500\text{ nm}}{5.0\text{ mm}}$$
4. Solving for L gives:
$$L = 41\text{ m}$$
5. By a factor of 2, 41 m is too large. However, you are suspect of the value given for the pupil diameter in your physics textbook. You know the pupil is smaller when the light is bright, and the library ceiling is very bright and colored white. An online search for eye pupil diameter soon yields the information you need. The pupil diameter ranges from 2 to 3 mm up to 7 mm:

Sucess. If the pupil diameter is 2.5 mm, the value of L is 20 m.

It is instructive to compare the limitation on resolution of the eye due to diffraction, as seen in Example 33-9, with the limitation on resolution due to the separation of the receptors (cones) on the retina. To be seen as two distinct objects, the images of the objects must fall on the retina on two nonadjacent cones. (See Problem 65 in Chapter 32.) Because the retina is about 2.5 cm from the cornea, the distance y on the retina corresponding to an angular separation of 1.5×10^{-4} rad is found from

$$\alpha_c = 1.5 \times 10^{-4}\text{ rad} = \frac{y}{2.5\text{ cm}}$$

or

$$y = 3.8 \times 10^{-4}\text{ cm} = 3.8 \times 10^{-6}\text{ m} = 3.8\ \mu\text{m}$$

The actual separation of the cones in the fovea centralis, where the cones are the most tightly packed, is about 1 μm. Outside this region, they are about 3 μm to 5 μm apart.

CONCEPT CHECK 33-2

True or False:
Fraunhofer diffraction is a limiting case of Fresnel diffraction.

* The wave properties of electrons are discussed in Chapter 34.

* 33-8 DIFFRACTION GRATINGS

A widely used tool for measuring the wavelength of light is the **diffraction grating,** which consists of a large number of equally spaced lines or slits on a flat surface. Such a grating can be made by cutting parallel, equally spaced grooves on a glass or metal plate with a precision ruling machine. With a reflection grating, light is reflected from the ridges between the lines or grooves. Phonograph records and compact disks exhibit some of the properties of reflection gratings. In a transmission grating, the light passes through the clear gaps between the rulings. Inexpensive, optically produced plastic gratings that have 10 000 or more slits per centimeter are common items in teaching laboratories. The spacing of the slits in a grating that has 10 000 slits per centimeter is $d = (1\text{ cm})/10\,000\text{ slits} = 10^{-4}\text{ cm slit}$.

Compact disks act as reflection gratings. *(Kevin R. Morris/Corbis.)*

Consider a plane wave of monochromatic light that is incident normally on a transmission grating (Figure 33-37). Assume that the width of each slit is very small so that it produces a widely diffracted beam. The interference pattern produced on a screen a large distance from the grating is due to a large number of coherent, equally spaced light sources. Suppose we have N slits that have separation d between adjacent slits. At $\theta = 0$, the light from each slit is in phase with that from all the other slits, so the amplitude of the wave is NA_0, where A_0 is the amplitude from each slit, and the intensity is $N^2 I_0$, where I_0 is the intensity due to a single slit alone. At an angle θ_1, where $d \sin\theta_1 = \lambda_1$, the path-length difference between any two successive slits is λ_1, so again the light from each slit is in phase with that from all the other slits and the intensity is $N^2 I_0$. The interference maxima are thus at angles θ given by

$$d \sin\theta_m = m\lambda \qquad m = 0, 1, 2, \ldots \qquad 33\text{-}26$$

The positions of the interference maxima do not depend on the number of sources, but the more sources there are, the sharper (narrower) and more intense the maxima will be.

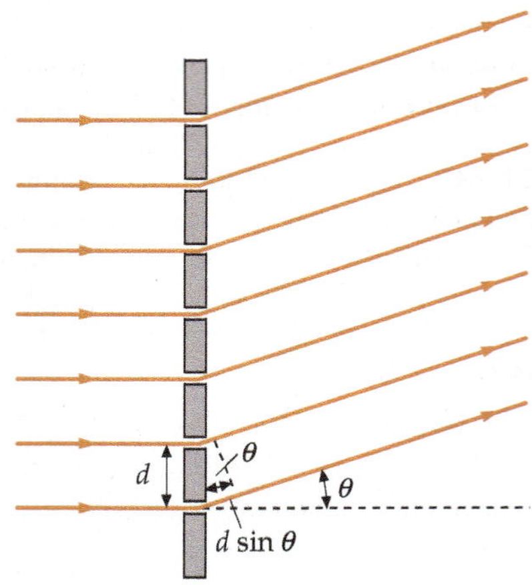

FIGURE 33-37 Light incident normally on a diffraction grating. At an angle θ, the path-length difference between rays from adjacent slits is $d \sin\theta$.

To see that the interference maxima will be sharper when there are many slits, consider the case of N illuminated slits, where N is large ($N \gg 1$). The distance from the first slit to the Nth slit is $(N-1)d \approx Nd$. When the path-length difference for the light from the first slit and that from the Nth slit is λ, the resulting intensity will be zero because the light from any two slits separated by $\frac{1}{2}Nd$ interferes destructively. (We saw this in our discussion of single-slit diffraction in Section 33-4.) Because the first and Nth slits are separated by approximately Nd, the intensity will be zero at angle θ_{min} given by

$$Nd \sin\theta_{\text{min}} = \lambda$$

so

$$\theta_{\text{min}} \approx \sin\theta_{\text{min}} = \frac{\lambda}{Nd}$$

The angular width of the interference maximum, which is equal to $2\theta_{\text{min}}$, is thus inversely proportional to N. Therefore, the greater the number of illuminated slits N, the sharper the maximum. Because the intensity in the maximum is proportional to $N^2 I_0$, the intensity in the maximum multiplied by the width of the maximum is proportional to NI_0. The intensity multiplied by the width is a measure of power per unit length in the maximum.

Figure 33-38*a* shows a student spectroscope that uses a diffraction grating to analyze light. In student laboratories, the light source is typically a glass tube containing atoms of a gas (for example, helium or sodium vapor) that are excited by a bombardment of electrons accelerated by high voltage across the tube. The light emitted by such a source contains only certain wavelengths that are characteristic of the atoms in the source. Light from the source passes through a narrow collimating slit and is made parallel by a lens. Parallel light from the lens is incident on the grating. Instead of falling on a screen a large distance away, the parallel light from the grating is focused by a telescope and viewed by the eye. The telescope is

(a) 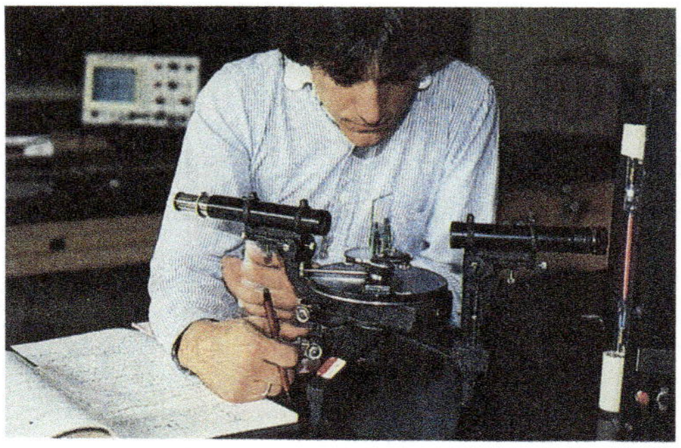 (b) 

mounted on a rotating platform that has been calibrated so that the angle θ can be measured. In the forward direction ($\theta = 0$), the central maximum for all wavelengths is seen. If light of a particular wavelength λ is emitted by the source, the first interference maximum is seen at the angle θ given by $d \sin\theta_m = m\lambda$ (Equation 33-26) with $m = 1$. Each wavelength emitted by the source produces a separate image of the collimating slit in the spectroscope called a **spectral line.** The set of lines corresponding to $m = 1$ is called the **first-order spectrum.** The **second-order spectrum** corresponds to $m = 2$ for each wavelength. Higher orders may be seen, providing the angle θ given by $d \sin\theta_m = m\lambda$ is less than 90°. Depending on the wavelengths, the orders may be mixed; that is, the third-order line for one wavelength may occur at a smaller value of θ than does the second-order line for another wavelength. If the spacing of the slits in the grating is known, the wavelengths emitted by the source can be determined by measuring the angles.

FIGURE 33-38 (*a*) A typical student spectroscope. Light from a collimating slit near the source is made parallel by a lens and falls on a grating. The diffracted light is viewed with a telescope at an angle that can be accurately measured. (*b*) Aerial view of the very large array (VLA) radio telescope in New Mexico. Radio signals from distant galaxies add constructively when Equation 33-26 is satisfied, where *d* is the distance between two adjacent telescopes. *((a) Clarence Bennett/ Oakland University, Rochester, Michigan. (b) NRAO/AUI/Science Photo Library/ Photo Researchers.)*

Example 33-10 Sodium D Lines

Sodium light is incident on a diffraction grating with 12 000 lines per centimeter. At what angles will the two yellow lines (called the sodium D lines) of wavelengths 589.00 nm and 589.59 nm be seen in the first order?

PICTURE Apply $d \sin\theta_m = m\lambda$ to each wavelength, with $m = 1$ and $d = (1/12\,000)$ cm.

SOLVE

1. The angle θ_m is given by $d \sin\theta_m = m\lambda$ with $m = 1$:
$$\sin\theta_1 = \frac{\lambda}{d}$$

2. Calculate θ_1 for $\lambda = 589.00$ nm:
$$\theta_1 = \sin^{-1}\left[\frac{589.00 \times 10^{-9}\text{ m}}{(1/12\,000)\text{ cm}} \times \left(\frac{100\text{ cm}}{1\text{ m}}\right)\right] = \boxed{44.98°}$$

3. Repeat the calculation for $\lambda = 589.59$ nm:
$$\theta_1 = \sin^{-1}\left[\frac{589.59 \times 10^{-9}\text{ m}}{(1/12\,000)\text{ cm}} \times \left(\frac{100\text{ cm}}{1\text{ m}}\right)\right] = \boxed{45.03°}$$

CHECK The first-order intensity maximum for the longer wavelength appears at the larger angle, as expected.

PRACTICE PROBLEM 33-4 Find the angles for the first-order intensity maxima of the two yellow lines if the grating has 15 000 lines per centimeter.

An important feature of a spectroscope is its ability to resolve spectral lines of two nearly equal wavelengths λ_1 and λ_2. For example, the two prominent yellow lines in the spectrum of sodium have wavelengths 589.00 and 589.59 nm. These can be seen as two separate wavelengths if their interference maxima do not overlap. According to Rayleigh's criterion for resolution, these wavelengths are resolved if the angular separation of their interference maxima is greater than the angular separation between an interference maximum and the first interference minimum on either side of it. The **resolving power** of a diffraction grating is defined to be $\lambda/|\Delta\lambda|$, where $|\Delta\lambda|$ is the smallest observable difference between two nearby wavelengths, each approximately equal to λ, that may be resolved. The resolving power is proportional to the number of slits illuminated because the more slits illuminated, the sharper the interference maxima. The resolving power R can be shown to be

$$R = \frac{\lambda}{|\Delta\lambda|} = mN \qquad 33\text{-}27$$

where N is the number of illuminated slits and m is the order number (see Problem 78). We can see from Equation 33-27 that to resolve the two yellow lines in the first order ($m = 1$) of the sodium spectrum the resolving power must be

$$R = 1 \times \frac{589.00\text{ nm}}{589.59\text{ nm} - 589.00\text{ nm}} = 998$$

Thus, to resolve the two yellow sodium lines in the first order, we need a grating containing 998 or more slits in the area illuminated by the light.

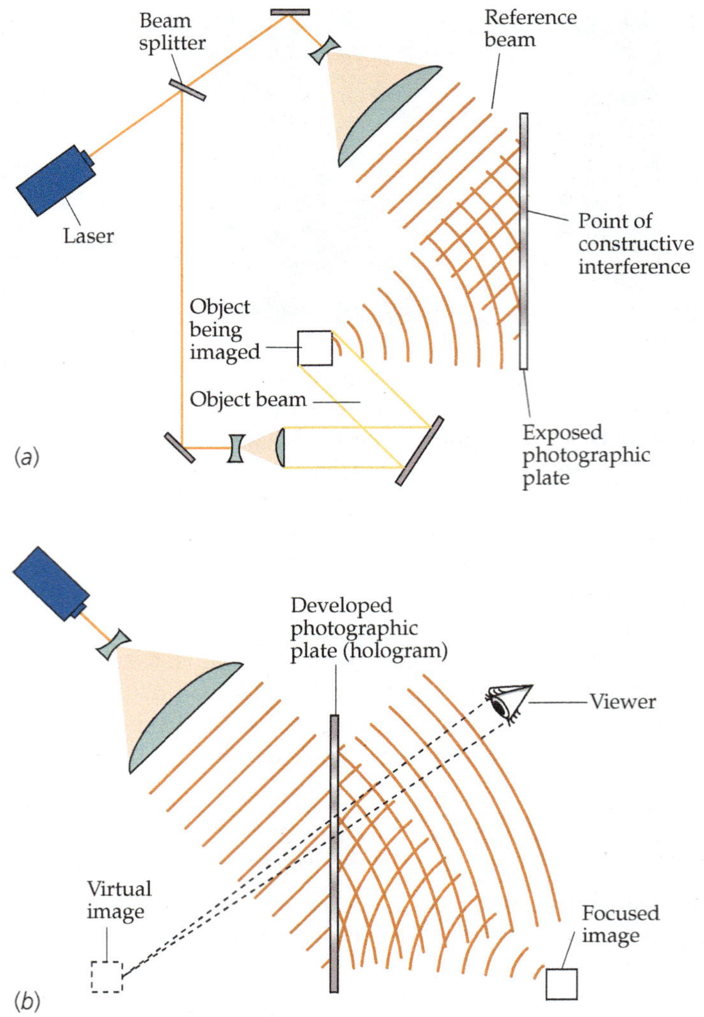

FIGURE 33-39 (*a*) The production of a hologram. The interference pattern produced by the reference beam and object beam is recorded on a photographic film. (*b*) When the film is developed and illuminated by coherent laser light, a three-dimensional image is seen.

*HOLOGRAMS

An interesting application of diffraction gratings is the production of a three-dimensional photograph called a **hologram** (Figure 33-39). In an ordinary photograph, the intensity of reflected light from an object is focused on a light-sensitive surface. As a result, a two-dimensional image is recorded. In a hologram, a beam from a laser is split into two beams, a reference beam and an object beam. The object beam reflects from the object to be photographed, and the interference pattern between it and the reference beam is recorded on a transparent film coated with a photosensitive emulsion. This can be done because the laser beam is coherent so that the relative phase difference between the reference beam and the object beam can be kept constant during the exposure. The film can be used to produce a holographic image after the emulsion is developed (chemically processed). The interference fringes on the film act as a diffraction grating. When the developed film is illuminated with a laser, a three-dimensional holographic image of the object is produced.

Holograms that you see on credit cards or postage stamps, called rainbow holograms, are more complicated. A horizontal strip of the original hologram is used to make a second hologram. The three-dimensional image can be seen as the viewer moves from side to side, but if viewed using monochromatic light, the image disappears when the viewer's eyes move above or below the slit image. When viewed using white light, the image is seen in different colors as the viewer moves in the vertical direction.

(a)

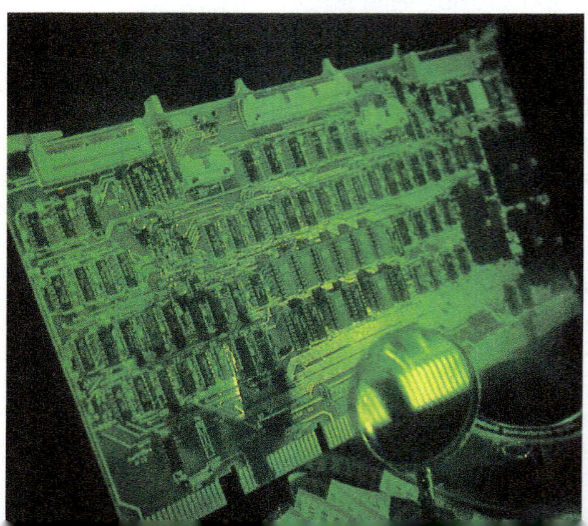

(b)

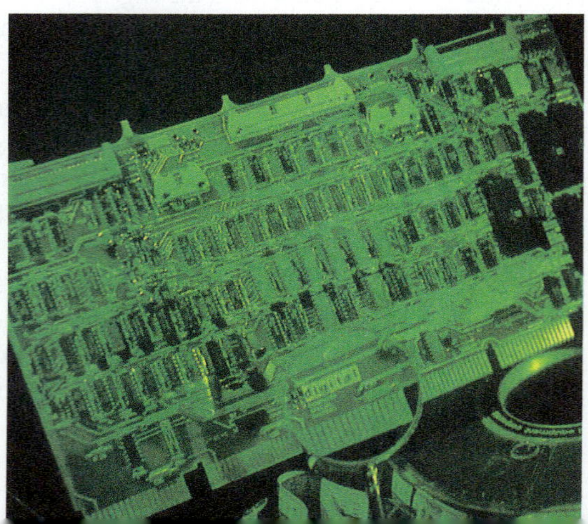

A hologram viewed from two different angles. Note that different parts of the circuit board appear behind the front magnifying lens. (*© 1981 by Ronald R. Erickson, Hologram by Nicklaus Phillips, 1978, for Digital Equipment Corporation.*)

Physics Spotlight

Holograms: Guided Interference

The machine is using heat to emboss holograms on plastic cards (perhaps credit cards). Such holograms serve as both a security feature and aesthetic enrichment. *(Pascal Goetgheluck/Photo Researchers.)*

Holography was invented by Dennis Gabor in 1948 when he tried to improve the resolution of electron microscopy.[*] He reconstructed wavefronts using interference on the photographic plate to make a picture that contained phase information as well as intensity information. He named this type of imagery holography, after the Greek words for "whole" and "writing," for he felt that including the phase information gave a complete picture.[†]

It was extremely difficult to create those first few holograms, and they did not achieve the desired resolution. He used mercury vapor lamps as a light source. The light was highly monochromatic, but incoherent. (The phase of the light fluctuated randomly.) A decade or so later, after the laser was invented, the use of coherent laser light made holography practical for many purposes.

Embossed holograms are frequently used because they are inexpensive. Embossed holograms are made by hot-stamping a metallized plastic film[‡] with a die that is a negative copy of the extremely shallow (around 0.3–0.5 micron deep) interference lines present in a hologram.[#] The plastic film is then a duplicate of the very tiny interference lines in the original hologram. When light shines through the film and reflects from the metallic backing, the holographic image is reconstructed. Almost all embossed holograms are rainbow holograms—able to be viewed without a laser. Creating the master of a rainbow hologram is a complex process involving multiple exposures at precise angles.[°]

Embossed holograms are highly visible, easy to recognize, and difficult to forge.[§] Because they can take the place of paper labels or be added to paper or plastic, they are used on credit cards, pharmaceutical packaging, currency, and traveler's checks as a quick method for authentication.[¶,**]

In January 1999, the Ford Motor Company used a series of digital holograms to create a 10 foot by 4 foot hologram of a concept car. The holograms were printed directly from computer design data.[††] Digital holography is now used to help physicians visualize the results of either computed tomography scans or magnetic resonance scans.[‡‡] The output from a series of MR or CT slices is collected, digitally processed, and then printed onto a single hologram, which can be viewed on a portable viewer. The resulting hologram allows surgeons to prepare for difficult surgeries[##] and may also have biomedical and industrial engineering applications.[°°] Digital holography is beginning to be used in holographic video applications.[§§]

Holograms have also been used as substitutes for traditional lenses. Holographic optical elements allow smaller and more compact displays to be built. Heads-up displays for airplane pilots are created using holographic optical elements.[¶¶] An extremely compact system that uses digitally calculated holograms as the optical element has been tested for use as a cell-phone based projector.[***] The use of holograms as optical elements and in optical data storage depends on advances in materials that are lightweight, tough, and have the desired optical properties.[†††]

Recently, holograms have been used to measure the electrostatic potential[‡‡‡] and magnetic fields[###] of very small objects. They have also been used to create higher resolution optics for X-ray lenses.[°°°] More than fifty years after holograms were invented, they are used to improve the resolution of microscopic images.

---

* Gabor, D., "Nobel Lecture." *Nobel Prize Lectures, 1971*, Dec. 11, 1971, at http://nobelprize.org/nobel_prizes/physics/laureates/1971/gabor-lecture.pdf As of Nov. 2006.

† Scanlon, L., "The Whole Picture." *Technology Review*, Dec. 2002/Jan. 2003, Vol. 105, No. 10, p. 88.

‡ Ruschmann, H. W., "Apparatus for Embossing Holograms on Metallized Thermoplastic Films." *United States Patent 4,547,141*, Oct. 15, 1985.

\# Abraham, N. C., "Optical Data Storage Disc." *United States Patent 5,452,282*, Sept. 19, 1995.

° Benton, S., Houde-Walter, W., and Mingace, Jr., H., "Methods of Making Holographic Images." *United States Patent 4,415,225.*, Nov. 15, 1983.

§ Cross, L., "Brand Security." *Graphic Arts Monthly*, Jan. 2006, Vol. 78, No. 1, pp. 32–33.

¶ "MasterCard Renews Hologram Contract." *American Banker*, Mar. 3, 2003, Vol. 168, No. 44, p. 18.

** Miller, H. I., "Fear and Pharmaceutical Failure." *The Washington Times*, Oct. 5, 2006, p. A16.

†† Mahoney, D. P., "Ford Drives Holography Development." *Computer Graphics World*, Feb. 1999, Vol. 22, No. 2, pp. 12–13.

‡‡ Samudhram, A., "Digital Holography Opens New Frontiers." *New Straits Times (Malaysia)*, Nov. 23, 2000, p. 2W.

\## Penrod, S., "3D Imaging Assisting Surgeons in Separation Surgery." *Local News*, KSL, Salt Lake City, Aug. 7, 2006. At http://www.ksl.com/?nid=148&sid=408002 As of Dec. 2006.

°° Liu, C., Yan, C., and Gao, S., "Digital Holographic Method for Tomography Reconstruction." *Applied Physics Letters*, Feb. 9, 2004, Vol. 84, No. 6, pp. 1010–1012.

§§ Freedman, D. H., "Holograms in Motion." *Technology Review*, Nov. 2002, Vol. 105, No. 9, pp. 48–55.

¶¶ Stevens, T., "Holograms: More than Pretty Pictures." *Industry Week*, Oct. 4, 1993, Vol. 242, No. 19, pp. 34–46.

*** Buckley, E., "Miniature Projectors Based on LBO Technology." *SID Mobile Displays Conference*, San Diego: Oct. 3–5. 2006. At http://www.lightblueoptics.com/images/news/SID_Mobile_Displays_2006.pdf As of Nov. 2006.

††† Huang, G. T., "Holographic Memory." *Technology Review*, Sept. 2005, Vol. 108, No. 9, pp. 64–67.

‡‡‡ Chou, L.-J., Chang, M.-T., and Chueh, Y.-L., "Electron Holography for Improved Measurement of Microfields in Nanoelectrode Assemblies." *Applied Physics Letters*, Jul. 10, 2006, Vol. 89, No. 2, Letter 023112, 3 pp.

\### Nepijko, S., and Wiesendanger, R., "Studies of Magnetic Properties of Small Particles by Electron Holography." *Applied Physics A, Materials Science and Processing*, 1997, Vol. 65, No. 4/5, pp. 361–366.

°°° "Solak, H. H., David, C., and Gobrecht, J., "Fabrication of High-Resolution Zone Plates with Wideband Extreme-Ultraviolet Holography." *Applied Physics Letters*, Oct. 4, 2004, Vol. 85, No. 14, pp. 2700–2702.

Summary

TOPIC	RELEVANT EQUATIONS AND REMARKS	
1. Interference	Two superposing light waves interfere if their phase difference remains constant for a time long enough to observe. They interfere constructively if their phase difference is zero or an integer multiplied by 360°. They interfere destructively if their phase difference is 180° or an odd integer multiplied by 180°.	
Phase difference due to a path-length difference	$\delta = \frac{\Delta r}{\lambda} 2\pi$	33-1
Phase difference due to reflection	A phase difference of 180° is introduced when a light wave is reflected from a boundary between two media for which the wave speed is greater on the incident-wave side of the boundary.	
Thin films	The interference of light waves reflected from the front and back surfaces of a thin film produces interference fringes, commonly observed in soap films or oil films. The difference in phase between the two reflected waves results from the path-length difference of twice the thickness of the film plus any phase change due to reflection of one or both of the rays.	
Two slits	The path-length difference at an angle θ on a screen far away from two narrow slits separated by a distance d is $d \sin\theta$. If the intensity due to each slit separately is I_0, the intensity at points of constructive interference is $4I_0$, and the intensity at points of destructive interference is zero.	
Two slit interference maxima (sources in phase)	$d \sin\theta_m = m\lambda \qquad m = 0, 1, 2, \ldots$	33-2
Two slit interference maxima (sources 180° out of phase)	$d \sin\theta_m = (m - \frac{1}{2})\lambda \qquad m = 1, 2, 3, \ldots$	33-3
2. Diffraction	Diffraction occurs whenever a portion of a wavefront is limited by an obstacle or an aperture. The intensity of light at any point in space can be computed using Huygens's construction by taking each point on the wavefront to be a point source and computing the resulting interference pattern.	
Fraunhofer patterns	Fraunhofer patterns are observed at great distances from the obstacle or aperture so that the rays reaching any point are approximately parallel, or they can be observed using a lens to focus parallel rays on a viewing screen placed in the focal plane of the lens.	
Fresnel patterns	Fresnel patterns are observed at points not necessarily far from the source.	
Single slit	When light is incident on a single slit of width a, the intensity pattern on a screen far away shows a broad central diffraction maximum that decreases to zero at an angle θ_1 given by	
	$\sin\theta_1 = \frac{\lambda}{a}$	33-9
	The width of the central maximum is inversely proportional to the width of the slit. The zeros in the single-slit diffraction pattern occur at angles given by	
	$a \sin\theta_m = m\lambda \qquad m = 1, 2, 3, \ldots$	33-11
	The maxima on either side of the central maximum have intensities that are much smaller than the intensity of the central maximum.	
Two slits	The interference-diffraction pattern of two slits is the two-slit interference pattern modulated by the single-slit diffraction pattern.	
Resolution of two sources	When light from two point sources that are close together passes through an aperture, the diffraction patterns of the sources may overlap. If the overlap is too great, the two sources cannot be resolved as two separate sources. When the central diffraction maximum of one source falls at the diffraction minimum of the other source, the two sources are said to be just resolved by Rayleigh's criterion for resolution. For a circular aperture of diameter D, the critical angular separation of two sources for resolution by Rayleigh's criterion is given by	
Rayleigh's criterion	$\alpha_c = 1.22\frac{\lambda}{D}$	33-25

TOPIC	RELEVANT EQUATIONS AND REMARKS
*Gratings	A diffraction grating consisting of a large number of equally spaced lines or slits is used to measure the wavelength of light emitted by a source. The positions of the mth-order interference maxima from a grating are at angles given by $d \sin\theta_m = m\lambda \qquad m = 0, 1, 2, \ldots$ (33-26). The resolving power of a grating is $R = \frac{\lambda}{\lvert\Delta\lambda\rvert} = mN$ (33-27), where N is the number of slits of the grating that are illuminated and m is the order number.
3. *Phasors	Two or more harmonic waves can be added by representing each wave as the y component of a two-dimensional vector called a phasor. The phase difference between two harmonic waves is represented as the angle between the phasors.

Answers to Concept Checks

33-1	6
33-2	True. Fresnel diffraction is the name that describes the observations when the observing screen is any distance from the source of the diffraction. Fraunhofer diffraction is the name that describes the observations in the limit that the observing screen is far from the source of the diffraction.

Answers to Practice Problems

33-1	9.2 cm^{-1}
33-2	4.4 mm
33-3	$A = 5.0$ V/m, $\delta = 37°$
33-4	62.07° and 62.18°

Problems

In a few problems, you are given more data than you actually need; in a few other problems, you are required to supply data from your general knowledge, outside sources, or informed estimate.

Interpret as significant all digits in numerical values that have trailing zeros and no decimal points.

- • Single-concept, single-step, relatively easy
- •• Intermediate-level, may require synthesis of concepts
- ••• Challenging
- SSM Solution is in the *Student Solutions Manual*
- Consecutive problems that are shaded are paired problems.

CONCEPTUAL PROBLEMS

1 • A phase difference due to a path-length difference is observed for monochromatic visible light. Which phase difference requires the least (minimum) path-length difference? (*a*) 90°,(*b*) 180°, (*c*) 270°,(*d*) The answer depends on the wavelength of the light.

2 • Which of the following pairs of light sources are coherent: (*a*) two candles, (*b*) one point source and its image in a plane mirror, (*c*) two pinholes uniformly illuminated by the same point source, (*d*) two headlights of a car, (*e*) two images of a point source due to reflection from the front and back surfaces of a soap film?

3 • The spacing between Newton's rings decreases rapidly as the diameter of the rings increases. Explain why this result occurs. SSM

4 • If the angle of a wedge-shaped air film such as the angle in Example 33-2 is too large, fringes are not observed. Why?

5 • Why must a film that is used to observe interference colors be thin?

6 • A loop of wire is dipped in soapy water and held up so that the soap film is vertical. (*a*) Viewed by reflection using white light, the top of the film appears black. Explain why. (*b*) Below the black region are colored bands. Is the first band red or violet?

7 • A two-slit interference pattern is formed using monochromatic laser light that has a wavelength of 640 nm. At the second maximum from the central maximum, what is the path-length difference between the light coming from each of the slits? (*a*) 640 nm, (*b*) 320 nm, (*c*) 960 nm, (*d*) 1280 nm SSM

8 • A two-slit interference pattern is formed using monochromatic laser light that has a wavelength of 640 nm. At the first minimum from the central maximum, what is the path-length difference between the light coming from each of the slits? (*a*) 640 nm, (*b*) 320 nm, (*c*) 960 nm, (*d*) 1280 nm

9 • A two-slit interference pattern is formed using monochromatic laser light that has a wavelength of 450 nm. What happens to the distance between the first maximum and the central maximum as the two slits are moved closer together? (*a*) The distance increases. (*b*) The distance decreases. (*c*) The distance remains the same.

10 • A two-slit interference pattern is formed using two different monochromatic lasers, one green and one red. Which color light has its first maximum closer to the central maximum? (*a*) green, (*b*) red, (*c*) Both maxima are at the same location.

11 • A single-slit diffraction pattern is formed using monochromatic laser light that has a wavelength of 450 nm. What happens to the distance between the first maximum and the central maximum as the slit is made narrower? (*a*) The distance increases. (*b*) The distance decreases. (*c*) The distance remains the same.

12 • Equation 33-2, which is $d \sin\theta_m = m\lambda$, and Equation 33-11, which is $a \sin\theta_m = m\lambda$, are sometimes confused. For each equation, define the symbols and explain the equation's application.

13 • When a diffraction grating is illuminated by white light, the first-order maximum of green light (*a*) is closer to the central maximum than the first-order maximum of red light, (*b*) is closer to the central maximum than the first-order maximum of blue light, (*c*) overlaps the second-order maximum of red light, (*d*) overlaps the second-order maximum of blue light.

14 • A double-slit interference experiment is set up in a chamber that can be evacuated. Using light from a helium–neon laser, an interference pattern is observed when the chamber is open to air. As the chamber is evacuated, one will note that (*a*) the interference fringes remain fixed, (*b*) the interference fringes move closer together, (*c*) the interference fringes move farther apart, (*d*) the interference fringes disappear completely.

15 • True or false:

(*a*) When waves interfere destructively, the energy is converted into heat.
(*b*) Interference patterns are observed only if the relative phases of the waves that superpose remain constant.
(*c*) In the Fraunhofer diffraction pattern for a single slit, the narrower the slit, the wider the central maximum of the diffraction pattern.
(*d*) A circular aperture can produce both a Fraunhofer diffraction pattern and a Fresnel diffraction pattern.
(*e*) The ability to resolve two point sources depends on the wavelength of the light. SSM

16 • You observe two very closely spaced sources of white light through a circular opening using various filters. Which color filter is most likely to prevent your resolving the images on your retinas as coming from two distinct sources? (*a*) red, (*b*) yellow, (*c*) green, (*d*) blue, (*e*) The filter choice is irrelevant.

17 •• Explain why the ability to distinguish the two headlights of an oncoming car at a given distance is easier for a human eye at night than during the daytime. Assume the headlights of the oncoming car are on both during both daytime and nighttime hours.

ESTIMATION AND APPROXIMATION

18 • It is claimed that the Great Wall of China is the only human-made object that can be seen from space using no equipment. Check to see if this claim is true, based on the resolving power of the human eye. Assume the observers are in a low-Earth orbit that has an altitude of about 250 km.

19 •• (*a*) Estimate how close an approaching car at night on a flat, straight part of a highway must be before its headlights can be distinguished from the single headlight of a motorcycle. (*b*) Estimate how far ahead of you a car is if its two red taillights merge to look as if they were one. SSM

20 •• A small loudspeaker is located at a large distance to the east from you. The loudspeaker is driven by a sinusoidal current whose frequency can be varied. Estimate the lowest frequency for which one of your ears would receive the sound waves exactly out of phase with that received by your other ear when you are facing north.

21 •• Estimate the maximum distance at which a binary star system could be resolvable by the human eye. Assume the two stars are about fifty times farther apart than Earth and the Sun are. Neglect any atmospheric effects. (A test similar to this "eye test" was used in ancient Rome to test for eyesight acuity before entering the army. A person who had normal eyesight could just barely resolve two well-known stars that appear close in the sky. Anyone who could not tell there were two stars failed the test.) SSM

PHASE DIFFERENCE AND COHERENCE

22 • Light that has a wavelength of 500 nm is incident normally on a film of water 1.00 μm thick. (*a*) What is the wavelength of the light in the water? (*b*) How many wavelengths are contained in the distance $2t$, where t is the thickness of the film? (*c*) The film has air on both sides. What is the phase difference between the wave reflected from the front surface and the wave reflected from the back surface in the region where the two reflected waves superpose?

23 •• Two coherent microwave sources both produce waves that each have a wavelength equal to 1.50 cm. The sources are located in the $z = 0$ plane, one at $x = 0$, $y = 15.0$ cm and the other at $x = 3.00$ cm, $y = 14.0$ cm. If the sources are in phase, find the difference in phase between the two waves for a receiver located at the origin. SSM

INTERFERENCE IN THIN FILMS

24 • A wedge-shaped film of air is made by placing a small slip of paper between the edges of two flat plates of glass. Light that has a wavelength of 700 nm is incident normally on the glass plates, and interference fringes are observed by reflection. (*a*) Is the first fringe near the point of contact of the plates dark or bright? Why? (*b*) If there are five dark fringes per centimeter, what is the angle of the wedge?

25 •• The diameters of fine fibers can be accurately measured using interference patterns. Two optically flat pieces of glass that each have a length L are arranged with the fiber between them, as shown in Figure 33-40. The setup is illuminated by monochromatic light, and the resulting interference fringes are observed. Suppose that L is 20.0 cm and that yellow sodium light (590 nm) is used for illumination. If 19 bright fringes are seen along this 20.0-cm distance, what are the limits on the diameter of the fiber? *Hint: The nineteenth fringe might not be right at the end, but you do not see a twentieth fringe at all.* SSM

26 •• Light that has a wavelength equal to 600 nm is used to illuminate two glass plates at normal incidence. The plates are 22 cm in length, touch at one end, and are separated at the other end by a wire that has a radius equal to 0.025 mm. How many bright fringes appear along the total length of the plates?

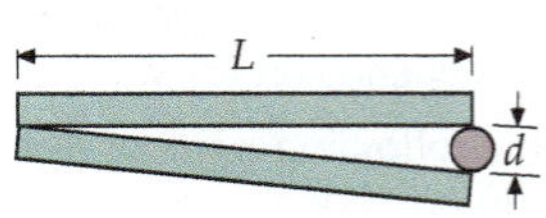

FIGURE 33-40 Problem 25

27 •• A thin film having an index of refraction of 1.50 is surrounded by air. It is illuminated normally by white light. Analysis of the reflected light shows that the wavelengths 360, 450, and 602 nm are the only missing wavelengths in or near the visible portion of the spectrum. That is, for those wavelengths, there is destructive interference. (*a*) What is the thickness of the film? (*b*) What visible wavelengths are brightest in the reflected interference pattern? (*c*) If this film were resting on glass that has an index of refraction of 1.60, what wavelengths in the visible spectrum would be missing from the reflected light?

28 •• A drop of oil (refractive index of 1.22) floats on water (refractive index of 1.33). When reflected light is observed from above, as shown in Figure 33-41, what is the thickness of the drop at

the point where the second red fringe, counting from the edge of the drop, is observed? Assume red light has a wavelength of 650 nm.

FIGURE 33-41 Problem 28

29 •• A film of oil that has an index of refraction of 1.45 rests on an optically flat piece of glass that has an index of refraction of 1.60. When illuminated by white light at normal incidence, light of wavelengths 690 nm and 460 nm is predominant in the reflected light. Determine the thickness of the oil film. SSM

30 •• A film of oil that has an index of refraction equal to 1.45 floats on water. When illuminated by white light at normal incidence, light of wavelengths 700 nm and 500 nm is predominant in the reflected light. Determine the thickness of the oil film.

NEWTON'S RINGS

31 •• A Newton's ring apparatus consists of a plano-convex glass lens that has a radius of curvature R and rests on a flat glass plate, as shown in Figure 33-42. The thin film is air of variable thickness. The apparatus is illuminated from above by light from a sodium lamp that has a wavelength equal to 590 nm. The pattern is viewed by reflected light. (*a*) Show that for a thickness t the condition for a bright (constructive) interference ring is $2t = \left(m + \frac{1}{2}\right)\lambda$, where $m = 0, 1, 2, \ldots$. (*b*) Show that for $t \ll R$, the radius r of a fringe is related to t by $r = \sqrt{2tR}$. (*c*) For a radius of curvature of 10.0 m and a lens diameter of 4.00 cm, how many bright fringes would you see in the reflected light? (*d*) What would be the diameter of the sixth bright fringe? (*e*) If the glass used in the apparatus has an index of refraction $n = 1.50$ and water replaces the air between the two pieces of glass, explain qualitatively the changes that will take place in the bright-fringe pattern. SSM

32 •• A plano-convex glass lens of radius of curvature 2.00 m rests on an optically flat glass plate. The arrangement is illuminated from above using monochromatic light that has a 520-nm wavelength. The indices of refraction of the lens and plate are 1.60. Determine the radii of the first and second bright fringe from the center in the reflected light.

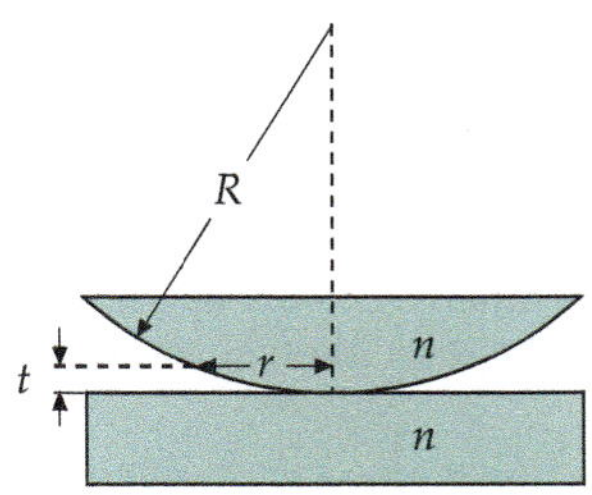

FIGURE 33-42 Problem 31

33 ••• Suppose that before the lens of Problem 32 is placed on the plate, a film of oil that has a refractive index equal to 1.82 is deposited on the plate. What will then be the radii of the first and second bright fringes?

TWO-SLIT INTERFERENCE PATTERNS

34 • Two narrow slits separated by 1.00 mm are illuminated by light that has a 600-nm wavelength, and the interference pattern is viewed on a screen 2.00 m away. Calculate the number of bright fringes per centimeter on the screen in the region near the center fringe.

35 • Using a conventional two-slit apparatus and light that has a 589-nm wavelength, 28 bright fringes per centimeter are observed near the center of a screen 3.00 m away. What is the slit separation? SSM

36 • Light that has a 633-nm wavelength and is from a helium–neon laser is shone normal to a plane having two slits. The first interference maximum is 82 cm from the central maximum on a screen 12 m away. (*a*) Find the separation of the slits. (*b*) How many interference maxima is it, in principle, possible to observe?

37 •• Two narrow slits are separated by a distance d. Their interference pattern is to be observed on a screen a large distance L away. (*a*) Calculate the spacing between successive maxima near the center fringe for light that has a 500-nm wavelength, when L is 1.00 m and d is 1.00 cm. (*b*) Would you expect to be able to observe the interference of light on the screen for this situation? (*c*) How close together should the slits be placed for the maxima to be separated by 1.00 mm for this wavelength and screen distance?

38 •• Light is incident at an angle ϕ with the normal to a vertical plane that has two slits of separation d (Figure 33-43). Show that the interference maxima are located at angles θ_m given by $\sin\theta_m + \sin\phi = m\lambda/d$.

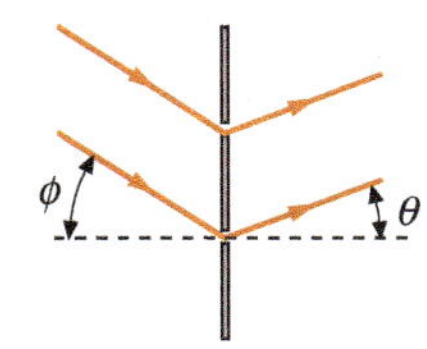

FIGURE 33-43 Problems 38 and 39

39 •• White light falls at an angle of 30° to the normal of a plane that has a pair of slits separated by 2.50 μm. What visible wavelengths give a bright interference maximum in the transmitted light in the direction normal to the plane? (See Problem 38.) SSM

40 •• Two small loudspeakers are separated by a distance of 5.0 cm, as shown in Figure 33-44. The speakers are driven in phase with a sine wave signal of frequency 10 kHz. A small microphone is placed a distance 1.00 m away from the speakers on the axis running through the middle of the two speakers, and the microphone is then moved perpendicular to the axis. Where does the microphone record the first minimum and the first maximum of the interference pattern from the speakers? The speed of sound in air is 343 m/s.

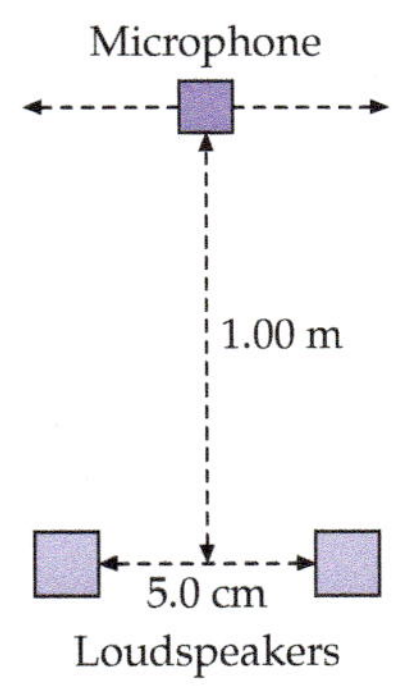

FIGURE 33-44 Problem 40

DIFFRACTION PATTERN OF A SINGLE SLIT

41 • Light that has a 600-nm wavelength is incident on a long narrow slit. Find the angle of the first diffraction minimum if the width of the slit is (*a*) 1.0 mm, (*b*) 0.10 mm, and (*c*) 0.010 mm.

42 • Plane microwaves are incident on a thin metal sheet that has a long, narrow slit of width 5.0 cm in it. The microwave radiation strikes the sheet at normal incidence. The first diffraction minimum is observed at $\theta = 37°$. What is the wavelength of the microwaves?

43 ••• Measuring the distance to the moon (lunar ranging) is routinely done by firing short-pulse lasers and measuring the time it takes for the pulses to reflect back from the moon. A pulse is fired from Earth. To send the pulse out, the pulse is expanded so that it fills the aperture of a 6.00-in-diameter telescope. Assuming the only thing spreading the beam out is diffraction and that the light wavelength is 500 nm, how large will the beam be when it reaches the moon, 3.82×10^5 km away? SSM

INTERFERENCE–DIFFRACTION PATTERN OF TWO SLITS

44 • How many interference maxima will be contained in the central diffraction maximum in the interference–diffraction pattern of two slits if the separation of the slits is exactly 5 times their width? How many will there be if the slit separation is an integral multiple of the slit width (that is, $d = na$ for any value of n)?

45 •• A two-slit Fraunhofer interference–diffraction pattern is observed using light that has a wavelength equal to 500 nm. The slits have a separation of 0.100 mm and an unknown width. (*a*) Find the width if the fifth interference maximum is at the same angle as the first diffraction minimum. (*b*) For that case, how many bright interference fringes will be seen in the central diffraction maximum? SSM

46 •• A two-slit Fraunhofer interference–diffraction pattern is observed using light that has a wavelength equal to 700 nm. The slits have widths of 0.010 mm and are separated by 0.20 mm. How many bright fringes will be seen in the central diffraction maximum?

47 •• Suppose that the *central* diffraction maximum for two slits has 17 interference fringes for some wavelength of light. How many interference fringes would you expect in the diffraction maximum adjacent to one side of the central diffraction maximum?

48 •• Light that has a wavelength equal to 550 nm illuminates two slits that both have widths equal to 0.030 mm and separations equal to 0.15 mm. (*a*) How many interference maxima fall within the full width of the central diffraction maximum? (*b*) What is the ratio of the intensity of the third interference maximum to one side of the center interference maximum to the intensity of the center interference maximum?

*USING PHASORS TO ADD HARMONIC WAVES

49 • Find the resultant of the two waves whose electric fields at a given location vary with time as follows: $\vec{E}_1 = 2.0A_0 \sin\omega t\hat{i}$ and $\vec{E}_2 = 3.0A_0 \sin\left(\omega t + \frac{3}{2}\pi\right)\hat{i}$. SSM

50 • Find the resultant of the two waves whose electric fields at a given location vary with time as follows: $\vec{E}_1 = 4.0A_0 \sin\omega t\hat{i}$ and $\vec{E}_2 = 3.0A_0 \sin\left(\omega t + \frac{1}{6}\pi\right)\hat{i}$.

51 •• Monochromatic light is incident on a sheet with a long narrow slit (Figure 33-45). Let I_0 be the intensity at the central maximum of the diffraction pattern on a distant screen, and let I be the intensity at the second intensity maximum from the central intensity maximum. The distance from this second intensity maximum to the far edge of the slit is longer than the distance from the second intensity maximum to the near edge of the slit by approximately 2.5 wavelengths. What is the ratio of I to I_0?

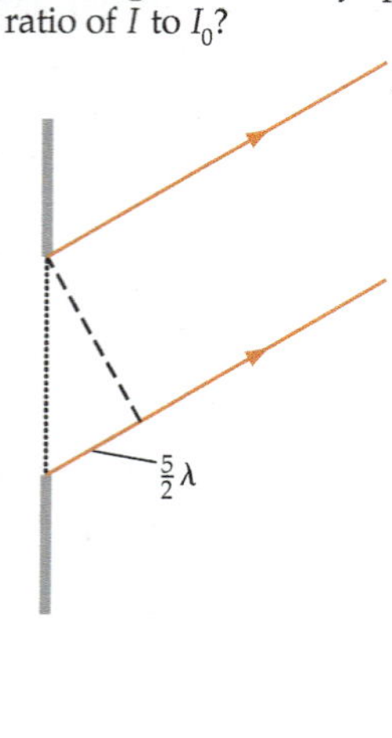

FIGURE 33-45
Problem 51

52 •• Monochromatic light is incident on a sheet that has three long narrow parallel equally spaced slits a distance d apart. (*a*) Show that the positions of the interference minima on a screen a large distance L away from the sheet that has the three equally spaced slits (spacing d, with $d \gg \lambda$) are given approximately by $y_m = m\lambda L/3d$, where $m = 1, 2, 4, 5, 7, 8, 10, \ldots$, that is, m is *not* a multiple of 3. (*b*) For a screen distance of 1.00 m, a light wavelength of 500 nm, and a source spacing of 0.100 mm, calculate the width of the principal interference maxima (the distance between successive minima) for three sources.

53 •• Monochromatic light is incident on a sheet that has four long narrow parallel equally spaced slits a distance d apart. (*a*) Show that the positions of the interference minima on a screen a large distance L away from the sheet that has the four equally spaced slits (spacing d, with $d \gg \lambda$) are given approximately by $y_m = m\lambda L/4d$, where $m = 1, 2, 3, 5, 6, 7, 9, 10, \ldots$, that is, m is *not* a multiple of 4. (*b*) For a screen distance of 2.00 m, a light wavelength of 600 nm, and a source spacing of 0.100 mm, calculate the width of the principal interference maxima (the distance between successive minima) for four sources. Compare the width with that for two sources with the same spacing. SSM

54 •• Light that has a wavelength equal to 480 nm falls normally on four slits. Each slit is 2.00 μm wide and the center-to-center separation between it and the next slit is 6.00 μm. (*a*) Find the angular width of the central intensity maximum of the single-slit diffraction pattern on a distant screen. (*b*) Find the angular position of all interference intensity maxima that lie inside the central diffraction maximum. (*c*) Find the angular width of the central interference maximum. That is, find the angle between the first interference intensity minima on either side of the central interference maximum. (*d*) Sketch the relative intensity as a function of the sine of the angle.

55 ••• Three slits, each separated from its neighbor by 60.0 μm, are illuminated at the central intensity maximum by a coherent light source that has a wavelength equal to 550 nm. The slits are extremely narrow. A screen is located 2.50 m from the slits. The intensity is 50.0 mW/m². Consider a location 1.72 cm from the central maximum. (*a*) Draw a phasor diagram suitable for the addition of the three harmonic waves at that location. (*b*) From the phasor diagram, calculate the intensity of light at that location. SSM

56 ••• In single-slit Fraunhofer diffraction, the intensity pattern (Figure 33-11) consists of a broad central maximum with a sequence of secondary maxima to either side of the central maximum. The intensity is given by $I = I_0\left(\frac{\sin\frac{1}{2}\phi}{\frac{1}{2}\phi}\right)^2$, where ϕ is the phase difference between the wavelets arriving from the opposite edges of the slits. Calculate the values of ϕ for the first three seconday maxima to one side of the central maximum by finding the values of ϕ for which $dI/d\phi$ is equal to zero. Check your results by comparing your answers with approximate values for ϕ of 3π, 5π and 7π. (That these values for ϕ are approximatly correct at the secondary intensity maxima is discussed in the discussion surrounding Figure 33-27.)

DIFFRACTION AND RESOLUTION

57 • Light that has a wavelength equal to 700 nm is incident on a pinhole of diameter 0.100 mm. (*a*) What is the angle between the central maximum and the first diffraction minimum for a Fraunhofer diffraction pattern? (*b*) What is the distance between the central maximum and the first diffraction minimum on a screen 8.00 m away? SSM

58 • Two sources of light that both have wavelengths equal to 700 nm are 10.0 m away from the pinhole of Problem 57. How far apart must the sources be for their diffraction patterns to be resolved by Rayleigh's criterion?

59 • Two sources of light that both have wavelengths equal to 700 nm are separated by a horizontal distance x. They are 5.00 m from a vertical slit of width 0.500 mm. What is the smallest value of x for which the diffraction pattern of the sources can be resolved by Rayleigh's criterion?

60 •• The ceiling of your lecture hall is probably covered with acoustic tile, which has small holes separated by about 6.0 mm. (*a*) Using light that has a wavelength of 500 nm, how far could you be from this tile and still resolve the holes? Assume the diameter of the pupil of each of your eyes is about 5.0 mm. (*b*) Could you resolve the holes better using red light or violet light? Explain your answer.

61 •• The telescope on Mount Palomar has a diameter of 200 in. Suppose a double star were 4.00 light-years away. Under ideal conditions, what must be the minimum separation of the two stars for their images to be resolved using light that has a wavelength equal to 550 nm? SSM

62 •• The star Mizar in Ursa Major is a binary system of stars that have nearly equal magnitudes. The angular separation between the two stars is 14 seconds of arc. What is the minimum diameter of the pupil that allows resolution of the two stars using light that has a wavelength equal to 550 nm?

*DIFFRACTION GRATINGS

63 • A diffraction grating that has 2000 slits per centimeter is used to measure the wavelengths emitted by hydrogen gas. (*a*) At what angles in the first-order spectrum would you expect to find the two violet lines that have wavelengths of 434 nm and 410 nm? (*b*) What are the angles if the grating has 15 000 slits per centimeter? SSM

64 • Using a diffraction grating that has 2000 slits per centimeter, two lines in the first-order hydrogen spectrum are found at angles of 9.72×10^{-2} rad and 1.32×10^{-1} rad. What are the wavelengths of the lines?

65 • The colors of many butterfly wings and beetle carapaces are due to the effects of diffraction. The *Morpho* butterfly has structural elements on its wings that effectively act as a diffraction grating that has an 880-nm spacing. At what angle will the first diffraction maximum occur for normally incident light diffracted by the butterfly's wings? Assume the light is blue and has a wavelength of 440 nm.

66 •• A diffraction grating that has 2000 slits per centimeter is used to analyze the spectrum of mercury. (*a*) Find the angular separation in the first-order spectrum of the two lines that have wavelengths equal to 579 nm and 577 nm. (*b*) How wide must the beam on the grating be for the lines to be resolved?

67 •• A diffraction grating that has 4800 lines per centimeter is illuminated at normal incidence using white light (wavelength range of 400 nm to 700 nm). How many orders of complete spectra can one observe in the transmitted light? Do any of these orders overlap? If so, describe the overlapping regions. SSM

68 •• A square diffraction grating that has an area of 25.0 cm^2 has a resolution of 22 000 in the fourth order. At what angle should you look to see a wavelength of 510 nm in the fourth order?

69 •• Sodium light that has a wavelength equal to 589 nm falls normally on a 2.00-cm-square diffraction grating ruled with 4000 lines per centimeter. The Fraunhofer diffraction pattern is projected onto a screen a distance of 1.50 m from the grating by a 1.50-m-focal-length lens that is placed immediately in front of the grating. Find (*a*) the distances of the first and second order intensity maxima from the central intensity maximum, (*b*) the width of the central maximum, and (*c*) the resolution in the first order. (Assume the entire grating is illuminated.)

70 •• The spectrum of neon is exceptionally rich in the visible region. Among the many lines are two lines at wavelengths of 519.313 nm and 519.322 nm. If light from a neon discharge tube is normally incident on a transmission grating that has 8400 lines per centimeter and the spectrum is observed in second order, what must be the width of the grating that is illuminated, so that the two lines can be resolved?

71 •• Mercury has several stable isotopes, among them ^{198}Hg and ^{202}Hg. The strong spectral line of mercury, at about 546.07 nm, is a composite of spectral lines from the various mercury isotopes. The wavelengths of the line for ^{198}Hg and ^{202}Hg are 546.07532 nm and 546.07355 nm, respectively. What must be the resolving power of a grating capable of resolving the two isotopic lines in the third-order spectrum? If the grating is illuminated over a 2.00-cm-wide region, what must be the number of lines per centimeter of the grating? SSM

72 ••• A diffraction grating has n lines per unit length. Show that the angular separation ($\Delta\theta$) of two lines of wavelengths λ and $\lambda + \Delta\lambda$ is approximately $\Delta\theta = \Delta\lambda/\sqrt{\frac{1}{n^2m^2} - \lambda^2}$, where m is the order number.

73 ••• For a diffraction grating in which all the surfaces are normal to the incident radiation, most of the energy goes into the zeroth order, which is useless from a spectroscopic point of view, because in zeroth order all the wavelengths are at 0°. Therefore, modern reflection gratings have shaped, or *blazed*, grooves, as shown in Figure 33-46. This shifts the specular reflection, which contains most of the energy, from the zeroth order to some higher order. (*a*) Calculate the *blaze angle* ϕ_m in terms of the groove separation d, the wavelength λ, and the order number m in which specular reflection is to occur for $m = 1, 2, \ldots$. (*b*) Calculate the proper blaze angle for the specular reflection to occur in the second order for light of wavelength 450 nm incident on a grating with 10 000 lines per centimeter. SSM

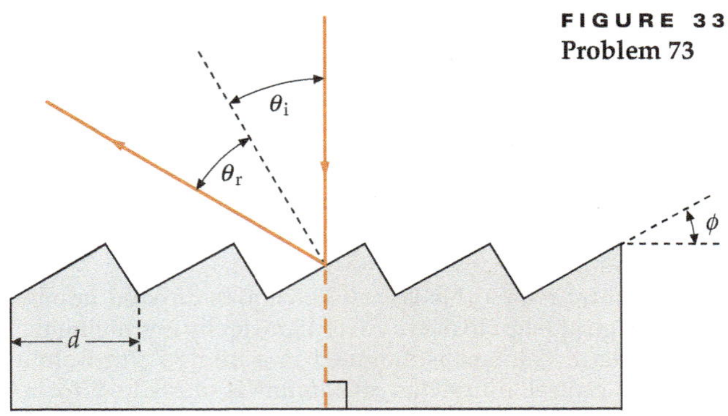

FIGURE 33-46 Problem 73

74 ••• In this problem, you will derive the relation $R = \lambda/|\Delta\lambda| = mN$ (Equation 33-27) for the resolving power of a diffraction grating having N slits separated by a distance d. To do this, you will calculate the angular separation between the intensity maximum and intensity minimum for some wavelength λ and set it equal to the angular separation of the mth-order maximum for two nearby wavelengths. (*a*) First show that the phase difference ϕ waves between the waves from two adjacent slits is given by $\phi = \frac{2\pi d}{\lambda}\sin\theta$. (*b*) Next, differentiate that expression to show that a small change in angle $d\theta$ results in a change in phase of $d\phi$ given by $d\phi = \frac{2\pi d}{\lambda}\cos\theta\, d\theta$. (*c*) Then for N slits, the angular separation between an interference maximum and an interference minimum corresponds to a phase change of $d\phi = 2\pi/N$. Use that to show that the angular separation $d\theta$ between the intensity maximum and intensity minimum for some wavelength λ is given by $d\theta = \frac{\lambda}{Nd\cos\theta}$. (*d*) Next, use the fact that the angle of the mth-order interference maximum for wavelength λ is specified by $d\sin\theta = m\lambda$ (Equation 33-26). Compute the differential of each side of the equation to show that angular separation of the mth-order

maximum for two nearly equal wavelengths differing by $d\lambda$ is given by $d\theta = \frac{m\,d\lambda}{d\cos\theta}$. (*e*) According to Rayleigh's criterion, two wavelengths will be resolved in the mth order if the angular separation of the wavelengths, given by the Part-(*d*) result, equals the angular separation of the interference maximum and the interference minimum given by the Part-(*c*) result. Use this to arrive at $R = \lambda/|\Delta\lambda| = mN$ (Equation 33-27) for the resolving power of a grating.

GENERAL PROBLEMS

75 • Naturally occuring coronas (brightly colored rings) are sometimes seen around the moon or the Sun when viewed through a thin cloud. (Warning: When viewing a Sun corona, be sure that the entire Sun is blocked by the edge of a building, a tree, or a traffic pole to safeguard your eyes.) These coronas are due to diffraction of light by small water droplets in the cloud. A typical angular diameter for a coronal ring is about 10°. From this, estimate the size of the water droplets in the cloud. Assume that the water droplets can be modeled as opaque disks that have the same radius as the droplet, and that the Fraunhofer diffraction pattern from an opaque disk is the same as the pattern from an aperture of the same diameter. (This last statement is known as *Babinet's principle*.) SSM

76 • An artificial corona (see Problem 75) can be made by placing a suspension of polystyrene microspheres in water. Polystyrene microspheres are small, uniform spheres that are made of plastic and have indices of refraction equal to 1.59. Assuming that the water has an index of refraction equal to 1.33, what is the angular diameter of such an artificial corona if 5.00-μm-diameter particles are illuminated by light from a helium–neon laser that has a wavelength in air of 632.8 nm?

77 • Coronas (see Problem 75) can be caused by pollen grains, typically of birch or pine. Such grains are irregular in shape, but they can be treated as if they had an average diameter of about 25 μm. What is the angular diameter (in degrees) of the corona for blue light? What is the angular diameter (in degrees) of the corona for red light?

78 • Light from a He–Ne laser (632.8 nm) is directed upon a human hair, in an attempt to measure its diameter by examining the diffraction pattern. The hair is mounted in a frame 7.5 m from a wall, and the central diffraction maximum is measured to be 14.6 cm wide. What is the diameter of the hair? (The diffraction pattern of a hair that has a diameter d is the same as the diffraction pattern of a single slit that has a width $a = d$. See Babinet's principle discussed in Problem 75.)

79 • A long, narrow horizontal slit lies 1.00 μm above a plane mirror, which is in the horizontal plane. The interference pattern produced by the slit and its image is viewed on a screen 1.00 m from the slit. The wavelength of the light is 600 nm. (*a*) Find the distance from the mirror to the first maximum. (*b*) How many dark bands per centimeter are seen on the screen? SSM

80 • A radio telescope is situated at the edge of a lake. The telescope is looking at light from a radio galaxy that is just rising over the horizon. If the height of the antenna is 20 m above the surface of the lake, at what angle above the horizon will the radio galaxy be when the telescope is centered in the first intensity interference maximum of the radio waves? Assume the wavelength of the radio waves is 20 cm. *Hint: The interference is caused by the light reflecting off the lake and remember that this reflection will result in a 180° phase shift.*

81 • The diameter of the radio telescope at Arecibo, Puerto Rico, is 300 m. What is the smallest angular separation of two objects that this telescope can detect when it is tuned to detect microwaves of 3.2-cm wavelength?

82 •• A thin layer of a transparent material that has an index of refraction of 1.30 is used as a nonreflective coating on the surface of glass that has an index of refraction of 1.50. What should the minimum thickness of the material be for the material to be nonreflecting for light that has a wavelength of 600 nm?

83 •• A *Fabry–Perot interferometer* (Figure 33-47) consists of two parallel, half-silvered mirrors that face each other and are separated by a small distance a. A half-silvered mirror is one that transmits 50% of the incident intensity and reflects 50% of the incident intensity. Show that when light is incident on the interferometer at an angle of incidence θ, the transmitted light will have maximum intensity when $2a = m\lambda\cos\theta$. SSM

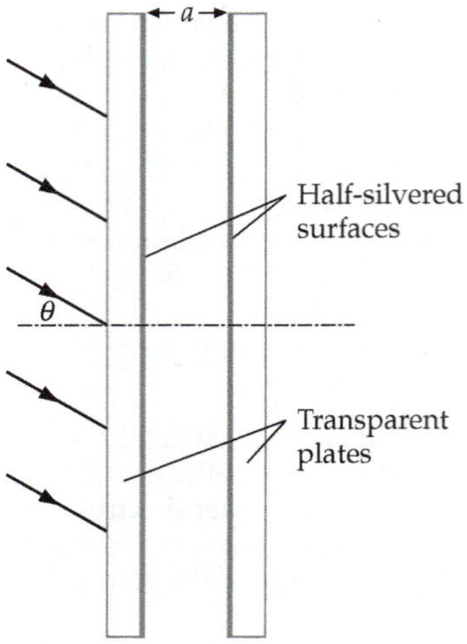

FIGURE 33-47 Problem 83

84 •• A mica sheet that is 1.20 μm thick is suspended in air. In reflected light, there are gaps in the visible spectrum at 421, 474, 542, and 633 nm. Find the index of refraction of the mica sheet.

85 •• A camera lens is made of glass that has an index of refraction of 1.60. This lens is coated with a magnesium fluoride film (index of refraction equal to 1.38) to enhance its light transmission. The purpose of this film is to produce zero reflection for light that has a wavelength of 540 nm. Treat the lens surface as a flat plane and the film as a uniformly thick flat film. (*a*) What minimum thickness of this film will accomplish its objective? (*b*) Would there be destructive interference for any other visible wavelengths? (*c*) By what factor would the reflection for light of 400 nm wavelength be reduced by the presence of this film? Neglect the variation in the reflected light amplitudes from the two surfaces. SSM

86 •• In a pinhole camera, the image is fuzzy because of geometry (rays arrive at the film after passing through different parts of the pinhole) and because of diffraction. As the pinhole is made smaller, the fuzziness due to geometry is reduced, but the fuzziness due to diffraction is increased. The optimum size of the pinhole for the sharpest possible image occurs when the spread due to diffraction equals the spread due to the geometric effects of the pinhole. Estimate the optimum size of the pinhole if the distance from the pinhole to the film is 10.0 cm and the wavelength of the light is 550 nm.

87 •• The Impressionist painter Georges Seurat used a technique called *pointillism*, in which his paintings are composed of small, closely spaced dots of pure color, each about 2.0 mm in diameter. The illusion of the colors blending together smoothly is produced in the eyes of the viewer by diffraction effects. Calculate the minimum viewing distance for this effect to work properly. Use the wavelength of visible light that requires the *greatest* distance between dots, so that you are sure the effect will work for *all* visible wavelengths. Assume the pupil of the eye has a diameter of 3.0 mm. SSM

PART VI MODERN PHYSICS: Quantum Mechanics, Relativity, and the Structure of Matter

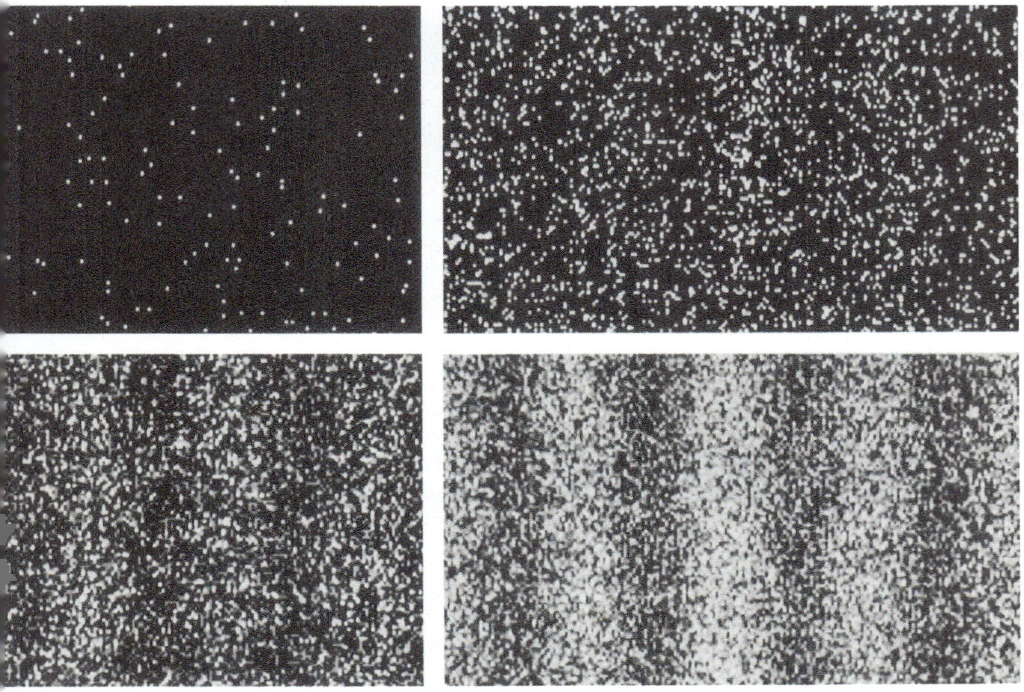

CHAPTER 34

ELECTRON INTERFERENCE PATTERN PRODUCED BY ELECTRONS INCIDENT ON A BARRIER CONTAINING TWO SLITS: (*A*) 10 ELECTRONS, (*B*) 100 ELECTRONS, (*C*) 3000 ELECTRONS, AND (*D*) 70,000 ELECTRONS. THE MAXIMA AND MINIMA DEMONSTRATE THE WAVE NATURE OF THE ELECTRON AS IT TRAVERSES THE SLITS. INDIVIDUAL DOTS ON THE SCREEN INDICATE THE PARTICLE NATURE OF THE ELECTRON AS IT EXCHANGES ENERGY WITH THE DETECTOR. THE PATTERN IS THE SAME WHETHER ELECTRONS OR PHOTONS (PARTICLES OF LIGHT) ARE USED. *(Courtesy of Akira Tononmura, Advanced Research Laboratory, Hitachi, Ltd.)*

Wave–Particle Duality and Quantum Physics

? How do you calculate the wavelength of an electron? (See Example 34-4.)

At the beginning of the twentieth century, it was thought that sound, light, and other electromagnetic radiation (such as radio waves) were waves, and electrons, protons, atoms, and similar units were particles. The first 30 years of that century revealed startling developments in theoretical and experimental physics, such as the finding that light actually exchanges energy in discrete lumps or quanta, just like particles, and the finding that an electron exhibits diffraction and interference as it propagates through space, just like a wave. The fact that light exchanges energy like a particle implies that light energy is not continuous but is *quantized*. Similarly, the wave nature of the electron, along with the fact that the standing wave condition requires a discrete set of frequencies, implies that the energy of an electron in a confined region of space is not continuous, but is quantized to a discrete set of values.

In this chapter, we begin by discussing some basic properties of light and electrons, examining their wave and particle characteristics. We then consider some of the detailed properties of matter waves, showing, in particular, how standing waves imply the quantization of energy. Finally, we discuss some of the important features of the theory of quantum physics, which was developed in the 1920s and which has been extremely successful in describing nature. Quantum physics is now the basis of our understanding of both atomic and subatomic systems and systems that have very low temperatures.

34-1 WAVES AND PARTICLES

We have seen that the propagation of waves through space is quite different from the propagation of particles. Waves bend around corners (diffraction) and interfere with one another, producing interference patterns. If a wave encounters a small aperture, the wave spreads out on the other side as if the aperture were a point source. The propagation of particles is quite unlike the propagation of waves. Particles travel in straight lines until they collide with something, after which the particles again travel in straight lines. If two particle beams meet in space, they never produce an interference pattern.

Particles and waves also exchange energy differently. Particles exchange energy in collisions that occur at specific points in space and in time. The energy of waves, on the other hand, is spread out in space and deposited continuously as the wave fronts interact with matter.

Sometimes the propagation of a wave cannot be distinguished from the propagation of a beam of particles. If the wavelength λ is very small compared to distances from the edges of objects, diffraction effects are negligible and the wave travels in straight lines. Also, interference maxima and minima are so close together in space as to be unobservable. The result is that the wave interacts with a detector, like a beam of numerous small particles each exchanging a small amount of energy; the exchange cannot distinguish particles from waves. For example, you do not observe the individual air molecules bouncing off your face if the wind blows on it. Instead, the interaction of billions of particles appears to be continuous, like that of a wave.

34-2 LIGHT: FROM NEWTON TO MAXWELL

The question of whether light consists of a beam of particles or waves in motion is one of the most interesting in the history of science (see Chapter 31). Isaac Newton used a particle theory of light to explain the laws of reflection and refraction; however, for refraction, Newton needed to assume that light travels faster in water or glass than in air, an assumption later shown to be incorrect. The chief early proponents of the wave theory were Robert Hooke and Christian Huygens, who explained refraction by assuming that light travels more slowly in glass or water than it does in air.* Newton favored the theory that light consists of particles and does not consist of waves because, in his time, light was believed to travel through a medium only in straight lines—diffraction had not yet been observed.

Because of Newton's great reputation and authority, his particle theory of light was accepted for more than a century. Then, in 1801, Thomas Young demonstrated the wave nature of light in a famous experiment in which two coherent light sources are produced by illuminating a pair of narrow, parallel slits with a single source (Figure 34-1*a*).† In Chapter 33, we saw that when light encounters a small opening, the opening acts as a point source of waves (Figure 33-7). In Young's experiment, each slit acts as a line source, which is equivalent to a point source in two dimensions.‡ The interference pattern is observed on a screen placed behind the slits. Interference maxima occur at angles so that the path difference is an integral number of wavelengths. Similarly, interference minima occur if the path difference is one-half the wavelength or any odd number of half wavelengths. Figure 34-1*b* shows the intensity pattern as seen on the screen. Remember that if two coherent waves of equal intensity I_0 meet in space, the result can be a wave of intensity $4I_0$ (constructive interference), an intensity of zero (destructive interference), or a wave of intensity between zero and $4I_0$, depending on the phase difference between the waves at the observation point. Young's experiment and many other experiments demonstrate that light propagates like a wave.

(*a*)

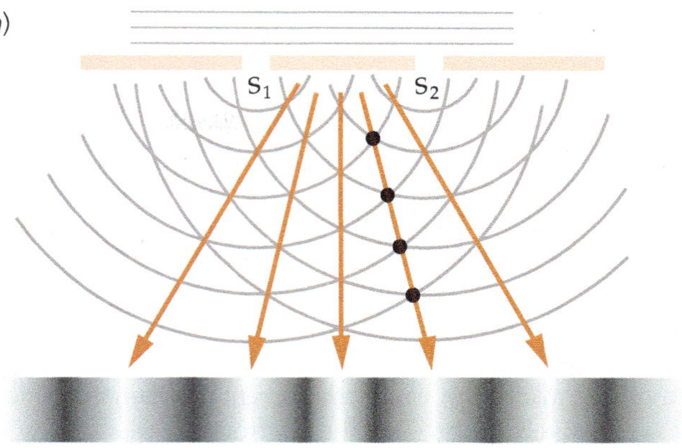

(*b*)

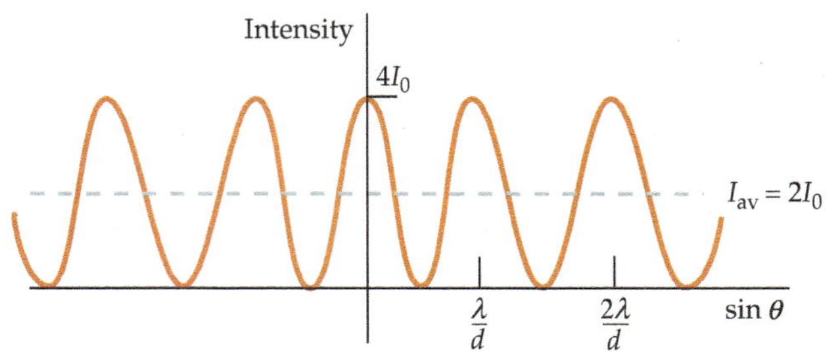

FIGURE 34-1 (*a*) Two slits act as coherent sources of light for the observation of interference in Young's experiment. Cylindrical waves from the slits overlap and produce an interference pattern on a screen far away. (*b*) The intensity pattern produced in Figure 34-1*a*. The intensity is maximum at points where the path difference is an even number of half wavelengths, and the intensity is zero where the path difference is an odd number of half wavelengths.

In the early nineteenth century, the French physicist Augustin Fresnel (1788–1827) performed extensive experiments on interference and diffraction and put the wave theory on a rigorous mathematical basis. Fresnel showed that the observed straight-line propagation of light is a result of the very short wavelengths of visible light.

The classical wave theory of light culminated in 1860 when James Clerk Maxwell published his mathematical theory of electromagnetism. This theory yielded a wave equation that predicted the existence of electromagnetic waves that propagate with a speed that can be calculated from the laws of electricity and magnetism.# The fact that the result of this calculation was $c \approx 3 \times 10^8$ m/s, the same as the speed of light, suggested to Maxwell that light is an electromagnetic wave. The eye is sensitive to electromagnetic waves that have wavelengths in the range from approximately 400 nm (1 nm = 10^{-9} m) to approximately 700 nm. This range is called *visible light*. Other electromagnetic waves (for example, microwave waves, radio waves, and X rays) differ from visible light waves only in wavelength and in frequency.

34-3 THE PARTICLE NATURE OF LIGHT: PHOTONS

The diffraction of light and the existence of an interference pattern in the two-slit experiment give clear evidence that light has wave properties. However, early in the twentieth century, it was found that light energy comes in discrete amounts.

* See Section 5 of Chapter 31.

† See Section 3 of Chapter 33.

‡ See Section 4 of Chapter 33.

\# See Section 3 of Chapter 30

THE PHOTOELECTRIC EFFECT

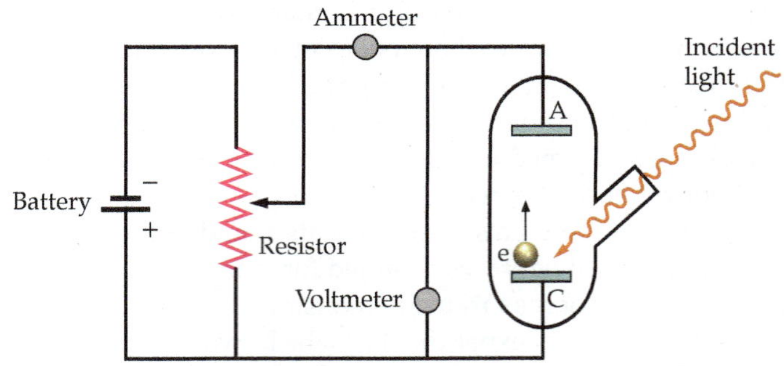

FIGURE 34-2 A schematic drawing of the apparatus for studying the photoelectric effect. Light of a single frequency enters an evacuated chamber and strikes the cathode C, which then ejects electrons (electron in figure is not drawn to scale). The current in the ammeter measures the number of these electrons that reach the anode A per unit time. The anode is made electrically negative with respect to the cathode to repel the electrons. Only those electrons that have enough initial kinetic energy to overcome the repulsion can reach the anode. The voltage between the two plates is slowly increased until the current becomes zero, which happens when even the most energetic electrons do not make it to plate A.

The quantum nature of light and the quantization of energy were suggested by Albert Einstein in 1905 in his explanation of the photoelectric effect. Einstein's work marked the beginning of quantum theory, and for his work, Einstein received the Nobel Prize in Physics. Figure 34-2 shows a schematic diagram of the basic apparatus for studying the photoelectric effect. Light of a single frequency enters an evacuated chamber and falls on a clean metal surface C (C for cathode), causing electrons to be emitted. Some of these electrons strike the second metal plate A (A for anode), constituting an electric current between the plates. Plate A is negatively charged, so the electrons are repelled by it, with only the most energetic electrons reaching plate A. The maximum kinetic energy of the emitted electrons is measured by slowly increasing the voltage until the current becomes zero. Experiments give the surprising result that the maximum kinetic energy of the emitted electrons is *independent of the intensity* of the incident light. Classically, we would expect that increasing the rate at which light energy falls on the metal surface would increase the energy absorbed by individual electrons and, therefore, would increase the maximum kinetic energy of the electrons emitted. Experiments show that this classical result does not happen. The maximum kinetic energy of the emitted electrons is the same for a given wavelength of incident light, no matter how intense the light is. Einstein demonstrated that this experimental result can be explained if light energy is quantized in small bundles called **photons.** The energy E of each photon is given by

$$E = hf = \frac{hc}{\lambda} \qquad \text{34-1}$$

EINSTEIN EQUATION FOR PHOTON ENERGY

where f is the frequency, and h is a constant now known as **Planck's constant.*** The measured value of this constant is

$$h = 6.626 \times 10^{-34}\,\text{J}\cdot\text{s} = 4.136 \times 10^{-15}\,\text{eV}\cdot\text{s} \qquad \text{34-2}$$

PLANCK'S CONSTANT

Equation 34-1 is sometimes called the **Einstein equation.**

A light beam consists of a beam of particles—photons—each having energy hf. The intensity (power per unit area) of a monochromatic light beam is the number of photons per unit area per unit of time, multiplied by the energy per photon. The interaction of the light beam with the metal surface consists of collisions between photons and electrons. During each of these collisions, the photon gives all its energy to an electron and the photon no longer exists. The electron is emitted from the surface after it receives the energy from a single photon. If the intensity of light is increased, more photons fall on the surface per unit time, and more electrons are emitted per unit time. However, each photon still has the same energy hf, so the energy absorbed by each electron is unchanged.

* In 1900, the German physicist Max Planck introduced this constant to explain discrepancies between the theoretical curves and experimental data on the spectrum of blackbody radiation. Planck also assumed that the radiation was emitted and absorbed by a blackbody in quanta of energy hf, but he considered his assumption to be just a computational device rather than a fundamental property of electromagnetic radiation. (Blackbody radiation is discussed in Chapter 20.)

If ϕ is the minimum energy necessary to remove an electron from a metal surface, the maximum kinetic energy of the electrons emitted is given by

$$K_{max} = \left(\tfrac{1}{2}mv^2\right)_{max} = hf - \phi \qquad 34\text{-}3$$

EINSTEIN'S PHOTOELECTRIC EQUATION

where f is the frequency of the photons. The quantity ϕ, called the **work function,** is a characteristic of the particular metal. (Some electrons will have kinetic energies less than $hf - \phi$, because of the loss of energy from traveling through the metal.)

According to Einstein's photoelectric equation, a plot of K_{max} versus frequency f should be a straight line that has the slope h. This was a bold prediction, because, at the time, no evidence existed that Planck's constant had any application outside of blackbody radiation. In addition, there was no experimental data on K_{max} versus frequency f, because no one before had even suspected that the frequency of the light was related to K_{max}. This prediction was difficult to verify experimentally, but careful experiments by R. A. Millikan approximately 10 years later showed that Einstein's equation was correct. Figure 34-3 shows a plot of Millikan's data.

Photons that have frequencies less than a **threshold frequency** f_t, and therefore have wavelengths greater than a **threshold wavelength** $\lambda_t = c/f_t$, do not have enough energy to eject an electron from a particular metal. The threshold frequency and the corresponding threshold wavelength can be related to the work function ϕ by setting the maximum kinetic energy of the electrons equal to zero in Equation 34-3. Then

$$\phi = hf_t = \frac{hc}{\lambda_t} \qquad 34\text{-}4$$

Work functions for metals are typically a few electron volts. Because wavelengths are usually given in nanometers and energies in electron volts, it is useful to have the value of hc in electron volt–nanometers:

$$hc = (4.1357 \times 10^{-15}\ \text{eV}\cdot\text{s})(2.9979 \times 10^{8}\ \text{m/s}) = 1.240 \times 10^{-6}\ \text{eV}\cdot\text{m}$$

or

$$hc = 1240\ \text{eV}\cdot\text{nm} \qquad 34\text{-}5$$

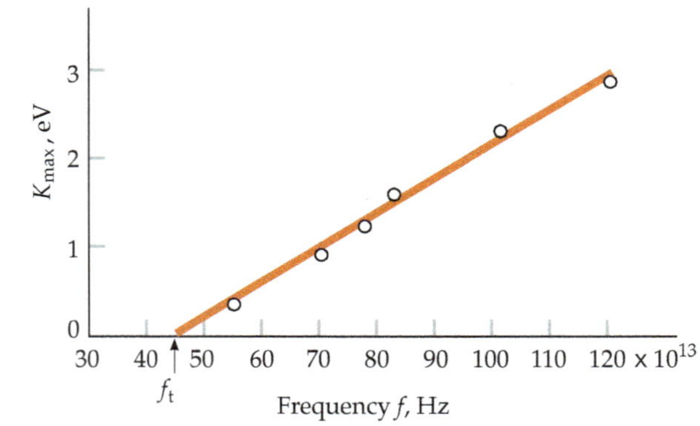

FIGURE 34-3 Millikan's data for the maximum kinetic energy K_{max} versus frequency f for the photoelectric effect. The data fall on a straight line that has a slope h, as predicted by Einstein approximately a decade before the experiment was performed.

Example 34-1 Photon Energies for Visible Light

Calculate the photon energies for light that has a wavelength equal to 400 nm (violet) and light that has a wavelength equal to 700 nm (red). (The wavelengths of 400 nm and 700 nm are the approximate wavelengths for the two extremes of the visible light spectrum.)

PICTURE Photon energies are related to photon frequencies and wavelengths by $E = hf = hc/\lambda$ (Equation 34-1).

SOLVE

1. The energy is related to the wavelength by Equation 34-1: $E = hf = \dfrac{hc}{\lambda}$

2. For $\lambda = 400$ nm, the energy is $E = \dfrac{hc}{\lambda} = \dfrac{1240\ \text{eV}\cdot\text{nm}}{400\ \text{nm}} = \boxed{3.10\ \text{eV}}$

3. For $\lambda = 700$ nm, the energy is $E = \dfrac{hc}{\lambda} = \dfrac{1240\ \text{eV}\cdot\text{nm}}{700\ \text{nm}} = \boxed{1.77\ \text{eV}}$

CHECK The shorter the wavelength of light, the greater the energy and 3.10 eV for 400 nm is greater than 1.77 eV for 700 nm.

TAKING IT FURTHER We can see from these calculations that visible light has photons that have energies which range from approximately 1.8 eV to 3.1 eV. X rays, which have much shorter wavelengths, have photons that have energies of the order of keV. Gamma rays emitted by nuclei have even shorter wavelengths and photons that have energies of the order of MeV.

PRACTICE PROBLEM 34-1 Find the energy of a photon corresponding to electromagnetic radiation in the FM radio band of wavelength 3.00 m.

PRACTICE PROBLEM 34-2 Find the wavelength of a photon whose energy is (*a*) 0.100 eV, (*b*) 1.00 keV, and (*c*) 1.00 MeV.

Example 34-2 The Number of Photons per Second in Sunlight

Try It Yourself

The intensity of sunlight at Earth's surface is approximately 1400 W/m². Assuming the average photon energy is 2.00 eV (corresponding to a wavelength of approximately 600 nm), calculate the number of photons that strike an area of 1.00 cm² each second.

PICTURE The intensity (power per unit area) is given, as is the area. From these given quantities, we can calculate the power, which is the energy per unit time.

SOLVE

Cover the column to the right and try these on your own before looking at the answers.

Steps	Answers
1. The energy ΔE is related to the number N of photons and the energy per photon $hf = 2.00$ eV:	$\Delta E = Nhf$
2. The intensity I (power per unit area) and the area A are given, so we can find the power:	$I = \frac{P}{A}$
3. Knowing the power (energy per unit time) and the time, we can find the energy:	$\Delta E = P\Delta t$
4. Combine the results from steps 1–3 and solve for N (take care to get the units to cancel):	$N = \boxed{4.38 \times 10^{17}}$

CHECK This is an enormous number of photons. However, in everyday situations we do not notice that the energy of sunlight arrives in discrete amounts. Thus, an enormous number is expected.

PRACTICE PROBLEM 34-3 Calculate the photon density (in photons per cubic centimeter) of the sunlight in Example 34-2. The number arriving on an area of 1.00 cm² in one second is the number in a column whose cross section is 1.00 cm² and whose height is the distance light travels in one second.

COMPTON SCATTERING

The first use of the photon concept was to explain the results of photoelectric-effect experiments. In the photoelectric effect, all the energy of the photon is transferred to an electron. However, in Compton scattering only some of the energy of the photon is transferred to an electron. The photon concept was also used by Arthur H. Compton to explain the results of his measurements of the scattering of X rays by free electrons in 1923. According to classical theory, if an electromagnetic wave of frequency f_i is incident on material containing free charges, the charges will oscillate with this frequency and reradiate electromagnetic waves of the same frequency.

Compton considered these reradiated waves as scattered photons, and he pointed out that if the scattering process were a collision between a photon and an electron (Figure 34-4), the electron would recoil and thus absorb energy. The scattered photon would then have less energy, and therefore a lower frequency and longer wavelength, than the incident photon.

According to classical electromagnetic theory (see Section 30-3), the energy and momentum of an electromagnetic wave are related by

$$E = pc \qquad \text{34-6}$$

The momentum of a photon is thus related to its wavelength λ by $p = E/c = hf/c = h/\lambda$.

$$p = \frac{h}{\lambda} \qquad \text{34-7}$$

MOMENTUM OF A PHOTON

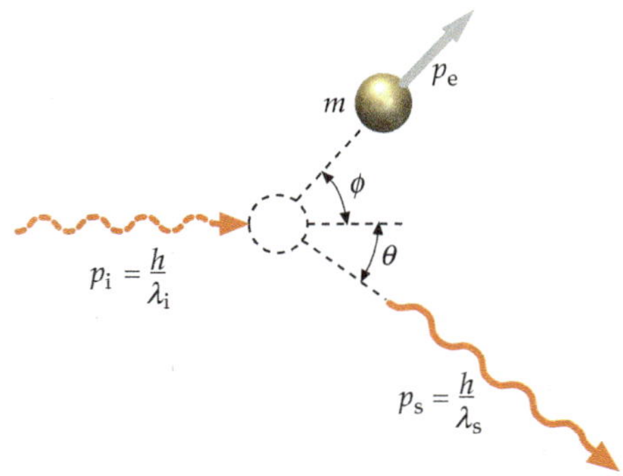

FIGURE 34-4 The scattering of light by an electron is considered as a collision of a photon of momentum h/λ_i and a stationary electron. The scattered photon has less energy and therefore has a longer wavelength than does the incident photon.

Compton applied the laws of conservation of momentum and energy to the collision of a photon and an electron to calculate the momentum p_s and thus the wavelength $\lambda_s = h/p_s$ of the scattered photon (see Figure 34-4). Applying conservation of momentum to the collision gives

$$\vec{p}_i = \vec{p}_s + \vec{p}_e \qquad \text{34-8}$$

where $\vec{p}_i$ is the momentum of the incident photon and $\vec{p}_e$ is the momentum of the electron after the collision. The initial momentum of the electron is zero. Rearranging Equation 34-8, we have $\vec{p}_e = \vec{p}_i - \vec{p}_s$. Taking the dot product of each side with itself gives

$$p_e^2 = p_i^2 + p_s^2 - 2p_ip_s\cos\theta \qquad \text{34-9}$$

where θ is the angle the direction of motion of the scattered photon makes with the direction of motion of the incident photon. Because the kinetic energy of the electron after the collision can be a significant fraction of the rest energy of an electron, the relativistic expression relating the total energy E of the electron to its momentum is used (see Chapter R). This expression (Equation R-17) is

$$E = \sqrt{p_e^2c^2 + (m_ec^2)^2}$$

where m_e is the mass of the electron. Applying conservation of energy to the collision gives

$$p_ic + m_ec^2 = p_sc + \sqrt{p_e^2c^2 + (m_ec^2)^2} \qquad \text{34-10}$$

where pc (Equation 34-6) has been used to express the energies of the photons. Eliminating p_e^2 from Equations 34-9 and 34-10 gives

$$\frac{1}{p_s} - \frac{1}{p_i} = \frac{1}{m_ec}(1 - \cos\theta)$$

and substituting for p_i and p_s, using Equation 34-7, gives

$$\lambda_s - \lambda_i = \frac{h}{m_ec}(1 - \cos\theta) \qquad \text{34-11}$$

COMPTON EQUATION

The increase in wavelength is independent of the wavelength λ_i of the incident photon. The quantity $h/(m_ec)$ has dimensions of length and is called the *Compton wavelength* λ_C. Its value is

$$\lambda_C = \frac{h}{m_ec} = \frac{hc}{m_ec^2} = \frac{1240\ \text{eV}\cdot\text{nm}}{5.110 \times 10^5\ \text{eV}} = 2.426 \times 10^{-12}\ \text{m} = 2.426\ \text{pm} \qquad \text{34-12}$$

Because $\lambda_s - \lambda_i$ is small, it is difficult to observe unless λ_i is so small that the fractional change $(\lambda_s - \lambda_i)/\lambda_i$ is appreciable.

Compton used X rays that have wavelengths equal to 71.1 pm (1 pm $= 10^{-12}$ m $= 10^{-3}$ nm). The energy of a photon of this wavelength is $E = hc/\lambda = (1240 \text{ eV}\cdot\text{nm})/(0.0711 \text{ nm}) = 17.4$ keV. The electrons in the experiment can be considered essentially free because the energy of the X rays is much greater than the binding energies of the valence electrons in atoms (which are of the order of a few eV). Compton's measurements of $\lambda_s - \lambda_i$ as a function of scattering angle θ agreed with Equation 34-11, thereby confirming the correctness of the photon concept (the particle nature of light).

Example 34-3 Finding the Increase in Wavelength

An X-ray photon of wavelength 6.00 pm makes a head-on collision with an electron, so that the scattered photon goes in a direction opposite to that of the incident photon. The electron is initially at rest. (*a*) How much longer is the wavelength of the scattered photon than the wavelength of the incident photon? (*b*) What is the kinetic energy of the recoiling electron?

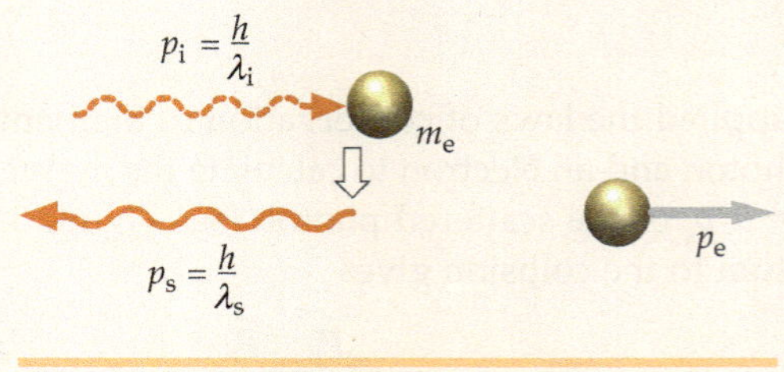

FIGURE 34-5

PICTURE We can calculate the increase in wavelength, and thus the new wavelength, from the Compton equation (Equation 34-11). We then use the new wavelength to find the energy of the scattered photon and then to find the kinetic energy of the recoiling electron from conservation of energy (Figure 34-5).

SOLVE

(*a*) Use Equation 34-11 to calculate the increase in wavelength:

$$\Delta\lambda = \lambda_s - \lambda_i = \frac{h}{m_e c}(1 - \cos\theta)$$

$$= (2.43 \text{ pm})(1 - \cos 180°) = \boxed{4.86 \text{ pm}}$$

(*b*) 1. The kinetic energy of the recoiling electron equals the energy of the incident photon E_i minus the energy of the scattered photon E_s:

$$K_e = E_i - E_s = hf_i - hf_s = \frac{hc}{\lambda_i} - \frac{hc}{\lambda_s}$$

2. Calculate λ_s from the given wavelength of the incident photon and the change found in Part (*a*):

$$\lambda_s = \lambda_i + \Delta\lambda = 6.00 \text{ pm} + 4.86 \text{ pm}$$
$$= 10.86 \text{ pm}$$

3. Substitute the values of λ_i and λ_s into the Part (*b*), step-1 result to find the energy of the recoiling electron:

$$K_e = \frac{hc}{\lambda_i} - \frac{hc}{\lambda_s}$$

$$= \frac{1240 \text{ eV}\cdot\text{nm}}{6.00 \text{ pm}} - \frac{1240 \text{ eV}\cdot\text{nm}}{10.86 \text{ pm}}$$

$$= \frac{1.240 \text{ keV}\cdot\text{nm}}{6.00 \times 10^{-3} \text{ nm}} - \frac{1.240 \text{ keV}\cdot\text{nm}}{10.86 \times 10^{-3} \text{ nm}}$$

$$= 207 \text{ keV} - 114 \text{ keV} = \boxed{93 \text{ keV}}$$

TAKING IT FURTHER The kinetic energy of the scattered electron is 93 keV and the rest energy of an electron is 511 keV, so the kinetic energy is 18 percent of the rest energy. Thus, the nonrelativistic formula for the kinetic energy ($\frac{1}{2}m_e v^2$) is not valid.

PRACTICE PROBLEM 34-4 What is the speed of the scattered electron given by the nonrelativistic formula for the kinetic energy ($\frac{1}{2}m_e v^2$)?

34-4 ENERGY QUANTIZATION IN ATOMS

Ordinary white light has a continuous spectrum; that is, it contains *all* the wavelengths in the visible spectrum. But if atoms in a gas at low pressure are excited by an electric discharge, they emit light of specific wavelengths that are characteristic of the element or the compound. Because the energy of a photon is related to its wavelength by $E = hf = hc/\lambda$, a discrete set of wavelengths implies a discrete set of energies. Conservation of energy then implies that if an atom absorbs a photon, its internal energy increases by a discrete amount, an amount equal to the energy of the photon. (It also implies that if an atom emits a photon, its internal energy decreases by a discrete amount that is equal to the energy of the photon.) In 1913, this led Niels Bohr to postulate that the internal energy of an atom can have only a discrete set of values. That is, the internal energy of an atom is quantized. If an excited atom radiates light of frequency f, the atom makes a transition from one allowed level to another level that has less energy by $|\Delta E| = hf$. Bohr was able to construct a semiclassical model of the hydrogen atom that had a discrete set of energy levels consistent with the observed spectrum of emitted light.* However, the *reason* for the quantization of energy levels in atoms and other systems remained a mystery until the wave nature of electrons was discovered a decade later.

34-5 ELECTRONS AND MATTER WAVES

In 1897, J. J. Thomson showed that the rays of a cathode-ray tube (Figure 34-6) can be deflected by electric and magnetic fields and therefore must consist of electrically charged particles. By measuring the deflections of these particles, Thomson showed that all the particles have the same charge-to-mass ratio q/m. He also showed that particles with this charge-to-mass ratio can be obtained using any material for the cathode, which means that these particles, now called electrons, are a fundamental constituent of all matter.

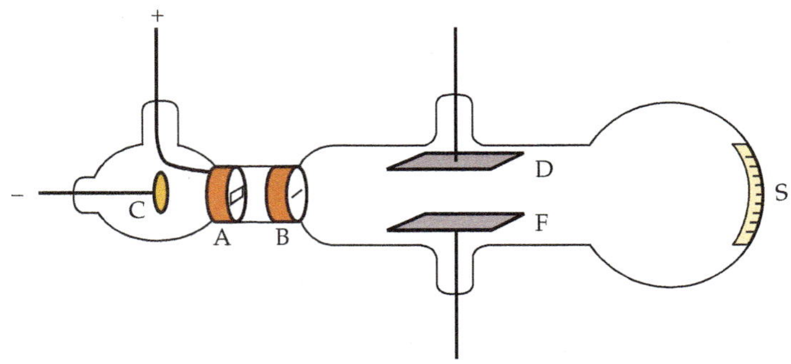

FIGURE 34-6 Schematic diagram of the cathode-ray tube Thomson used to measure q/m for the particles that comprise cathode rays (electrons). Electrons from the cathode C pass through the slits at A and B and strike a phosphorescent screen S. The beam can be deflected by an electric field between plates D and F or by a magnetic field (not shown).

THE DE BROGLIE HYPOTHESIS

Because light seems to have both wave and particle properties, it is natural to ask whether matter (for example, electrons and protons) might also have both wave and particle characteristics. In 1924, a French physics student, Louis de Broglie, suggested this idea in his doctoral dissertation. de Broglie's work was highly speculative, because no evidence existed at that time of any wave aspects of matter.

For the wavelength of electron waves, de Broglie chose

$$\lambda = \frac{h}{p} \tag{34-13}$$

DE BROGLIE RELATION FOR THE WAVELENGTH OF ELECTRON WAVES

where p is the momentum of the electron. Note that this is the same as Equation 34-7 for a photon. For the frequency of electron waves, de Broglie chose the Einstein equation relating the frequency and energy of a photon.

$$f = \frac{E}{h} \tag{34-14}$$

DE BROGLIE RELATION FOR THE FREQUENCY OF ELECTRON WAVES

* The Bohr model is reviewed in Chapter 36.

These equations are thought to apply to all matter. However, for macroscopic objects, the wavelengths calculated from Equation 34-13 are so small that it is impossible to observe the usual wave properties of interference or diffraction. Even a dust particle that has a mass as small as 1 μg is much too massive for any wave characteristics to be noticed, as we see in the following example.

Example 34-4 The de Broglie Wavelength — *Try It Yourself*

Find the de Broglie wavelength of a 1.00×10^{-6} g particle moving with a speed of 1.00×10^{-6} m/s.

PICTURE The wavelength λ and the momentum p of a particle are related by $\lambda = h/p$.

SOLVE

Cover the column to the right and try this on your own before looking at the answers.

Steps	Answers
Write the definition of the de Broglie wavelength and substitute the given data.	$\lambda = \frac{h}{p} = \frac{h}{mv} = \frac{6.63 \times 10^{-34}\,\text{J}\cdot\text{s}}{(1.00 \times 10^{-9}\,\text{kg})(1.00 \times 10^{-6}\,\text{m/s})} = \boxed{6.63 \times 10^{-19}\,\text{m}}$

CHECK As expected, this wavelength, which is four or five orders of magnitude smaller than the diameter of an atomic nucleus, is too small to be observed.

Because the wavelength found in Example 34-4 is so small, much smaller than any possible apertures or obstacles, diffraction or interference of such waves cannot be observed. In fact, the propagation of waves of very small wavelengths is indistinguishable from the propagation of particles. The momentum of the particle in Example 34-4 is only 10^{-15} kg·m/s. A macroscopic particle that has a greater momentum would have an even smaller de Broglie wavelength. We therefore do not observe the wave properties of such macroscopic objects as baseballs and billiard balls.

PRACTICE PROBLEM 34-5

Find the de Broglie wavelength of a baseball of mass 0.17 kg moving at 100 km/h.

The situation is different for low-energy electrons and other subatomic particles. Consider a particle with kinetic energy K. Its momentum is found from

$$K = \frac{p^2}{2m}$$

or

$$p = \sqrt{2mK}$$

Its wavelength is then

$$\lambda = \frac{h}{p} = \frac{h}{\sqrt{2mK}}$$

If we multiply the numerator and the denominator by c, we obtain

$$\lambda = \frac{hc}{\sqrt{2mc^2K}} = \frac{1240\,\text{eV}\cdot\text{nm}}{\sqrt{2mc^2K}} \qquad \text{34-15}$$

WAVELENGTH ASSOCIATED WITH A PARTICLE OF MASS m

where we have used $hc = 1240$ eV·nm. For electrons, $mc^2 = 0.5110$ MeV. Then,

$$\lambda = \frac{1240 \text{ eV} \cdot \text{nm}}{\sqrt{2mc^2K}} = \frac{1240 \text{ eV} \cdot \text{nm}}{\sqrt{2(0.5110 \times 10^6 \text{ eV})K}}$$

or

$$\lambda = \frac{1.226}{\sqrt{K}} \text{nm} \qquad (K \text{ in electron volts}) \qquad 34\text{-}16$$

ELECTRON WAVELENGTH

Equation 34-15 and Equation 34-16 do not hold for relativistic particles whose kinetic energies are a significant fraction of their rest energies mc^2. (Rest energies are discussed in Chapter 7 and in Chapter R.)

PRACTICE PROBLEM 34-6

Find the wavelength of an electron whose kinetic energy is 10.0 eV.

ELECTRON INTERFERENCE AND DIFFRACTION

The observation of diffraction and interference of electron waves would provide the crucial test of the existence of wave properties of electrons. This observation was first seen serendipitously in 1927 by C. J. Davisson and L. H. Germer as they were studying electron scattering from a nickel target at the Bell Telephone Laboratories. After heating the target to remove an oxide coating that had accumulated during an accidental break in the vacuum system, they found that the scattered electron intensity as a function of the scattering angle showed maxima and minima. Their target had crystallized, and they had observed electron diffraction by accident. Davisson and Germer then prepared a target consisting of a single crystal of nickel and investigated this phenomenon extensively. Figure 34-7*a* illustrates their experiment. Electrons from an electron gun are directed at a crystal and detected at some angle ϕ that can be varied. Figure 34-7*b* shows a typical pattern observed. There is a strong scattering maximum at an angle of 50°. The angle for maximum scattering of waves from a crystal depends on the wavelength of the waves and the spacing of the atoms in the crystal. Using the known spacing of atoms in their crystal, Davisson and Germer calculated the wavelength that could

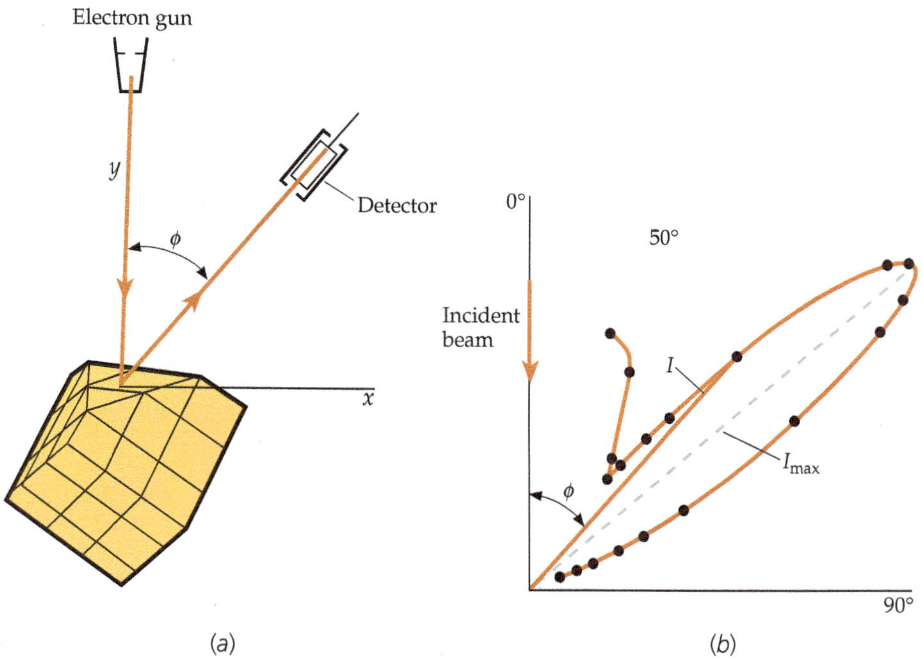

FIGURE 34-7 The Davisson–Germer experiment. (*a*) Electrons are scattered from a nickel crystal into a detector. (*b*) A polar plot of the intensity I of scattered electrons versus scattering angle. The maximum intensity I_{max} is at the angle predicted by the diffraction of waves of wavelength λ given by the de Broglie formula.

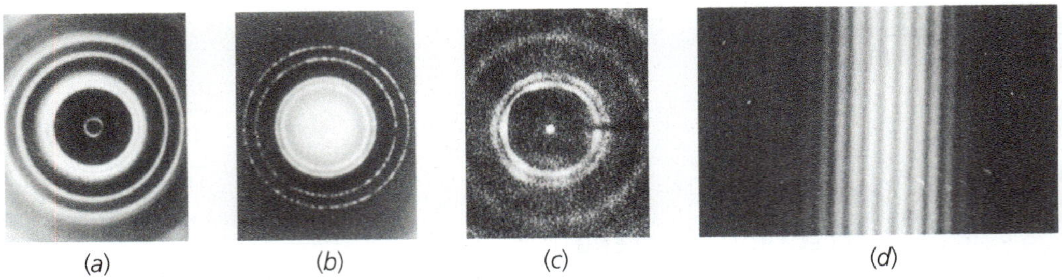

FIGURE 34-8 (*a*) The diffraction pattern produced by X rays of wavelength 0.071 nm on an aluminum foil target. (*b*) The diffraction pattern produced by 600-eV electrons (λ = 0.050 nm) on an aluminum foil target. (*c*) The diffraction of 0.0568 eV neutrons (λ = 0.12 nm) incident on a copper foil. (*d*) A two-slit electron diffraction–interference pattern. (*(a) and (b) PSSC Physics, 2nd ed., 1965. D.C. Heath & Co., and Education Development Center, Inc., Newton, MA, (c) C.G. Shull, (d) Claus Jönsson.)*

produce such a maximum and found that it agreed with the de Broglie equation (Equation 34-16) for the electron energy they were using. By varying the energy of the incident electrons, they could vary the electron wavelengths and produce maxima and minima at different locations in the diffraction patterns. In all cases, the measured wavelengths agreed with de Broglie's hypothesis.

Another demonstration of the wave nature of electrons was provided in the same year by G. P. Thomson (son of J. J. Thomson) who observed electron diffraction in the transmission of electrons through thin metal foils. A metal foil consists of tiny, randomly oriented crystals. The diffraction pattern resulting from such a foil is a set of concentric circles. Figure 34-8*a* and Figure 34-8*b* show the diffraction pattern observed using X rays and electrons on an aluminum foil target. Figure 34-8*c* shows the diffraction patterns of neutrons on a copper foil target. Note the similarity of the patterns. The diffraction of hydrogen and helium atoms was observed in 1930. In all cases, the measured wavelengths agree with the de Broglie predictions. Figure 34-8*d* shows a diffraction pattern produced by electrons incident on two narrow slits. This experiment is equivalent to Young's famous double-slit experiment with light. The pattern is identical to the pattern observed with photons of the same wavelength. (Compare with Figure 34-1.)

Shortly after the wave properties of the electron were demonstrated, it was suggested that electrons rather than light might be used to *see* small objects. As discussed in Chapter 33, reflected waves or transmitted waves can resolve details of objects only if the details are larger than the wavelength of the reflected wave. Beams of electrons, which can be focused by electric and magnetic fields, can have very small wavelengths—much shorter than visible light. Today, the electron microscope (Figure 34-9) is an important research tool used to visualize specimens at scales far smaller than those possible with a light microscope.

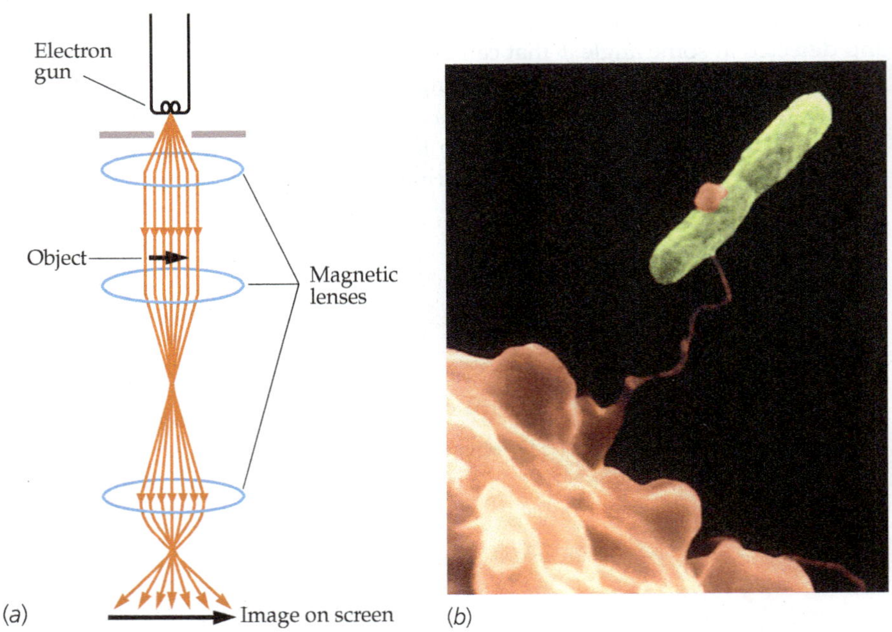

FIGURE 34-9 (*a*) An electron microscope. Electrons from a heated filament (the electron gun) are accelerated by a large potential difference. The electron beam is made parallel by a magnetic focusing lens. The electrons strike a thin target and are then focused by a second magnetic lens. The third magnetic lens projects the electron beam onto a fluorescent screen to produce the image. (*b*) An electron micrograph of an amoeba (*Hartmannella vermiformis*) that uses an extended pseudopod to entrap a bacterium (*Legionella pneumophila*). (*(b) CDC/Dr. Barry S. Fields.)*

STANDING WAVES AND ENERGY QUANTIZATION

Given that electrons have wavelike properties, it should be possible to produce standing electron waves. If energy is associated with the frequency of a standing wave, as in $E = hf$ (Equation 34-14), then standing waves imply quantized energies.

The idea that the discrete energy states in atoms could be explained by standing waves led to the development of a detailed mathematical theory known as quantum theory, quantum mechanics, or wave mechanics by Erwin Schrödinger and others in 1926. In this theory, the electron is described by a wave function ψ that obeys a wave equation called the **Schrödinger equation.** The form of the Schrödinger equation of a particular system depends on the forces acting on the particle, which are described by the potential energy functions associated with those forces. In Chapter 35, we discuss this equation, which is somewhat similar to the classical wave equations for sound or for light. Schrödinger solved the standing wave problem for the hydrogen atom, the simple harmonic oscillator, and other systems of interest. He found that the allowed frequencies, combined with $E = hf$, resulted in the set of energy levels found experimentally for the hydrogen atom, thereby demonstrating that quantum theory provides a general method of finding the quantized energy levels for a given system. Quantum theory is the basis for our modern understanding of the world—from the inner workings of the atomic nucleus to the radiation spectra of distant galaxies.

! Do not think energy is always quantized. It is not unless the system is bound. The energy of a system consisting of a proton and an electron is quantized only if the electron is bound to the proton—as it is in the hydrogen atom. If the electron is not bound to the proton, then the energy of the system is not quantized.

34-6 THE INTERPRETATION OF THE WAVE FUNCTION

The wave function for waves on a string is the string displacement y. The wave function for sound waves can be either the displacement of the air molecules s, or the pressure P. The wave function for electromagnetic waves is the electric field $\vec{E}$ and the magnetic field $\vec{B}$. What is the wave function for electron waves? The symbol we use for this wave function is ψ (the Greek letter psi). When Schrödinger published his wave equation, neither he nor anyone else knew just how to interpret the wave function ψ. We can get a hint about how to interpret ψ by considering the quantization of light waves. For classical waves, such as sound or light, the energy per unit volume in the wave is proportional to the square of the wave function. Because the energy of a light wave is quantized, the energy per unit volume is proportional to the number of photons per unit volume. We might therefore expect the square of the photon's wave function to be proportional to the number of photons per unit volume in a light wave. But suppose we have a very low-energy source of light that emits just one photon at a time. In any unit volume, there is either one photon or none. The square of the wave function must then describe the *probability* of finding a photon in some unit volume.

The Schrödinger equation describes a single particle. The square of the wave function for a particle must then describe the *probability density*, which is the probability per unit volume, of finding the particle at a location. The probability of finding the particle in some volume element must also be proportional to the size of the volume element dV. Thus, in one dimension, the probability of finding a particle in a region of length dx at the position x is $\psi^2(x)\,dx$. If we call this probability $P(x)\,dx$, where $P(x)$ is the **probability density,** we have

$$P(x) = \psi^2(x) \qquad \text{34-17}$$

PROBABILITY DENSITY

Generally the wave function depends on time as well as position, and is written $\psi(x,t)$. However, for standing waves, the probability density is independent of time. Because we will be concerned mostly with standing waves in this chapter, we omit the time dependence of the wave function and write it $\psi(x)$ or just ψ.

The probability of finding the particle either in the region between x_1 and $x_1 + dx$ or in the region between x_2 and $x_2 + dx$ is the sum of the separate probabilities $P(x_1)\,dx + P(x_2)\,dx$. If we have a particle at all, the probability of finding the particle somewhere must be 1. Then the sum of the probabilities over all the possible values of x must equal 1. That is,

$$\int_{-\infty}^{\infty} \psi^2\,dx = 1 \qquad \text{34-18}$$

NORMALIZATION CONDITION

Equation 34-18 is called the **normalization condition.** If ψ is to satisfy the normalization condition, it must approach zero as $|x|$ approaches infinity. This condition places restrictions on the possible solutions of the Schrödinger equation. There are mathematical solutions to the Schrödinger equation that do not approach zero as $|x|$ approaches infinity. However, these solutions are not acceptable as wave functions.

Example 34-5 Probability Calculation for a Classical Particle

It is known that a classical point particle moves back and forth with constant speed between two walls at $x = 0$ and $x = 8.0$ cm (Figure 34-10). No additional information about the location of the particle is known. (*a*) What is the probability density $P(x)$? (*b*) What is the probability of finding the particle at the point where x equals exactly 2 cm? (*c*) What is the probability of finding the particle between $x = 3.0$ cm and $x = 3.4$ cm?

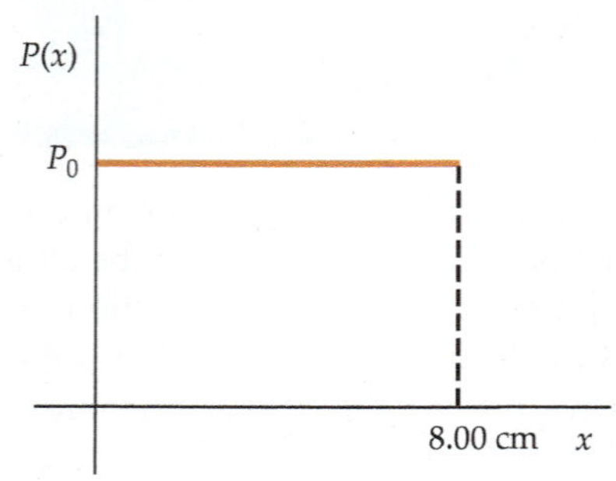

FIGURE 34-10 The probability function $P(x)$.

PICTURE We do not know the initial position of the particle. Because the particle moves with constant speed, it is equally likely to be anywhere in the region $0 < x < 8.0$ cm. The probability density $P(x)$ is therefore independent of x, for $0 < x < 8.0$ cm, and zero outside of this range. We can find $P(x)$, for $0 < x < 8.0$ cm, by normalization, that is, by requiring that the probability that the particle is somewhere between $x = 0$ and $x = 8.0$ cm is 1.

SOLVE

(*a*) 1. The probability density $P(x)$ is uniform between the walls and zero elsewhere:

$$P(x) = \begin{cases} 0 & x < 0 \\ P_0 & 0 < x < 8.0\text{ cm} \\ 0 & x > 8.0\text{ cm} \end{cases}$$

2. Apply the normalization condition:

$$\int_{-\infty}^{+\infty} P(x)\,dx = \int_{-\infty}^{0} P(x)\,dx + \int_{0}^{8.0\text{ cm}} P(x)\,dx + \int_{8.0\text{ cm}}^{\infty} P(x)\,dx$$

$$= 0 + \int_{0}^{8.0\text{ cm}} P_0\,dx + 0 = P_0\,(8.0\text{ cm}) = 1$$

3. Solve for P_0:

$$\boxed{P_0 = \frac{1}{8.0\text{ cm}}}$$

(*b*) On the interval $0 < x < 8.0$ cm, the probability of finding the particle in some range Δx is proportional to $P_0\Delta x = \Delta x/(8\text{ cm})$. The probability of finding the particle at the point $x = 2$ cm is zero because Δx is zero (no range exists). Alternatively, because an infinite number of points exists between $x = 0$ and $x = 8$ cm, and the particle is equally likely to be at any point, the chance that the particle will be at any one particular point must be zero.

The probability of finding the particle at the point where x equals exactly 2 cm is 0.

(c) Because the probability density is uniform, the probability of a particle being in some range Δx in the region $0 < x < 8.0$ cm is $P_0\Delta x$. The probability of the particle being in the region $3.0 \text{ cm} < x < 3.4 \text{ cm}$ is thus:

$$P_0\Delta x = \left(\frac{1}{8.0 \text{ cm}}\right)0.4 \text{ cm} = \boxed{0.05}$$

CHECK The length of the interval $3.0 \text{ cm} < x < 3.4 \text{ cm}$ is 0.4 cm, which is 5 percent of $L = 8.0$ cm. Because the particle is moving at constant speed v, we expect it to be in the interval $3.0 \text{ cm} < x < 3.4 \text{ cm}$ during 5 percent of the time, provided the total time is much longer than the time L/v (the time required for the particle to travel 8.0 cm). Our Part-(c) result meets this expectation.

34-7 WAVE–PARTICLE DUALITY

We have seen that light, which we ordinarily think of as a wave, exhibits particle properties when it interacts with matter, as in the photoelectric effect or in Compton scattering. Electrons, which we usually think of as particles, exhibit the wave properties of interference and diffraction when they pass near the edges of obstacles. All carriers of momentum and energy (for example, electrons, atoms, or photons) exhibit both wave and particle characteristics. It might be tempting to say that an electron, for example, is both a wave and a particle, but what does this mean? In classical physics, the concepts of waves and particles are mutually exclusive. A **classical particle** behaves like a piece of shot; it can be localized and scattered, it exchanges energy suddenly at a point in space, and it obeys the laws of conservation of energy and momentum in collisions. It does *not* exhibit interference or diffraction. A **classical wave,** on the other hand, behaves like a sound or light wave; it exhibits diffraction and interference, and its energy is spread out continuously in space and time. A classical wave and a classical particle are mutually exclusive. Nothing can be both a classical particle and a classical wave at the same time.

After Thomas Young observed the two-slit interference pattern by using light in 1801, light was thought to be a classical wave. On the other hand, the electrons discovered by J. J. Thomson were thought to be classical particles. We now know that these classical concepts of waves and particles do not adequately describe the complete behavior of any phenomenon.

Everything propagates like a wave and exchanges energy like a particle.

Often the concepts of the classical particle and the classical wave give the same results. If the wavelength is very small, diffraction effects are negligible, so the waves travel in straight lines like classical particles. Also, interference is not seen for waves of very short wavelength, because the interference fringes are too closely spaced to be observed. It then makes no difference which concept we use. If diffraction is negligible, we can think of light as a wave propagating along rays, as in geometrical optics, or as a beam of photon particles. Similarly, we can think of an electron as a wave propagating in straight lines along rays or, more commonly, as a particle.

We can also use either the wave or particle concept to describe exchanges of energy if we have a large number of particles and we are interested only in the average values of energy and momentum exchanges.

THE TWO-SLIT EXPERIMENT REVISITED

The wave–particle duality of nature is illustrated by the analysis of the experiment in which a single electron is incident on a barrier that has two slits. The analysis is virtually the same whether we use an electron or a photon (light). To describe the propagation of an electron, we must use wave theory. Let us assume the source is a point source, such as a needle point, so we have spherical waves spreading out from the source. After passing through the two slits, the wavefronts spread out—as if each slit were a source of wavefronts. The wave function ψ at a point on a screen or film far from the slits depends on the difference in path lengths from the source to the point, one path through one slit, and the other path through the other slit. At points on the screen for which the difference in path lengths is either zero or an integral number of wavelengths, the amplitude of the wave function is a maximum. Because the probability of detecting the electron is proportional to ψ^2, the electron is very likely to arrive near these points. At points for which the path difference is an odd number of half wavelengths, the wave function ψ is zero, so the electron is very unlikely to arrive near these points. The chapter opening photos show the interference pattern produced by 10 electrons, 100 electrons, 3000 electrons, and 70,000 electrons. Note that, although the electron propagates through the slits like a wave, the electron interacts with the screen at a single point—like a particle.

THE UNCERTAINTY PRINCIPLE

An important principle consistent with the wave–particle duality of nature is the **uncertainty principle.** It states that, in principle, it is impossible to simultaneously measure both the position and the momentum of a particle with unlimited precision. A common way to measure the position of an object is to look at the object by using light. If we do this, we scatter light from the object and determine the position by the direction of the scattered light. If we use light of wavelength λ, we can measure the position x only to an uncertainty Δx of the order of λ because of diffraction effects.

$$\Delta x \sim \lambda$$

To reduce the uncertainty in position, we therefore use light of very short wavelength, perhaps even X rays. In principle, there is no limit to the accuracy of such a position measurement, because there is no limit on how small the wavelength λ can be.

We can determine the momentum p_x of the object if we know the mass and can determine its velocity. The momentum of the object can be found by measuring the object's position at two nearby times and computing its velocity. If we use light of wavelength λ, the photons carry momentum h/λ. If these photons are scattered by the object we are looking at, the scattering changes the momentum of the object in an uncontrollable way. Each photon carries momentum h/λ, so the uncertainty in the momentum Δp_x of the object is of the order of h/λ:

$$\Delta p_x \sim \frac{h}{\lambda}$$

If the wavelength of the radiation is small, the momentum of each photon will be large and the momentum measurement will have a large uncertainty. Reducing the intensity of light cannot eliminate this uncertainty; such a reduction merely reduces the number of photons in the beam. To see the object, we must scatter at least one photon. Therefore, the uncertainty in the momentum measurement of the object will be large if λ is small, and the uncertainty in the position measurement of the object will be large if λ is large.

Of course, we could always look at the objects by scattering electrons instead of photons, but the same difficulty remains. If we use low-momentum electrons to reduce the uncertainty in the momentum measurement, we have a large uncertainty in the position measurement because of diffraction of the electrons. The relation between the wavelength and momentum $\lambda = h/p_x$ is the same for electrons as it is for photons.

The product of the intrinsic uncertainties in position and momentum is

$$\Delta x \Delta p_x \sim \lambda \times \frac{h}{\lambda} = h$$

This relation between the uncertainties in position and momentum is called the uncertainty principle. If we define precisely what we mean by uncertainties in measurement, we can give a precise statement of the uncertainty principle. If Δx and Δp are defined to be the standard deviations in the measurements of position and momentum, it can be shown that their product must be greater than or equal to $\hbar/2$.

$$\Delta x \Delta p_x \geq \tfrac{1}{2}\hbar \qquad \text{34-19}$$

where $\hbar = h/2\pi$.*

Equation 34-19 provides a statement of the uncertainty principle first given by Werner Heisenberg in 1927. In practice, the experimental uncertainties are usually much greater than the intrinsic lower limit that results from wave–particle duality.

34-8 A PARTICLE IN A BOX

We can illustrate many of the important features of quantum physics without solving the Schrödinger equation by considering a simple problem of a particle of mass m confined to a one-dimensional box of length L, like the particle in Example 34-5. This can be considered a crude description of an electron that is constrained to be within an atom or a proton that is constrained to be within a nucleus. If a classical particle bounces back and forth between the walls of the box, the particle's energy and momentum can have any values. However, according to quantum theory, the particle is described by a wave function ψ, whose square describes the probability of finding the particle in some region. Because we are assuming that the particle is indeed inside the box, the wave function must be zero everywhere outside the box. If the box is between $x = 0$ and $x = L$, we have

$$\psi = 0 \qquad \text{for } x \leq 0 \text{ and for } x \geq L$$

In particular, if we assume the wave function to be continuous everywhere, it must be zero at the end points of the box $x = 0$ and $x = L$. This is the same condition as the condition for standing waves on a string fixed at $x = 0$ and $x = L$, and the results are the same. The allowed wavelengths for a particle in the box are those where the length L equals an integral number of half wavelengths (Figure 34-11).

$$L = n\frac{\lambda_n}{2} \qquad n = 1, 2, 3, \ldots \qquad \text{34-20}$$

STANDING-WAVE CONDITION FOR A PARTICLE IN A BOX OF LENGTH L

The total energy E of the particle is its kinetic energy

$$E = \frac{1}{2}mv^2 = \frac{p^2}{2m}$$

Substituting the de Broglie relation $p_n = h/\lambda_n$,

$$E_n = \frac{p_n^2}{2m} = \frac{(h/\lambda_n)^2}{2m} = \frac{h^2}{2m\lambda_n^2}$$

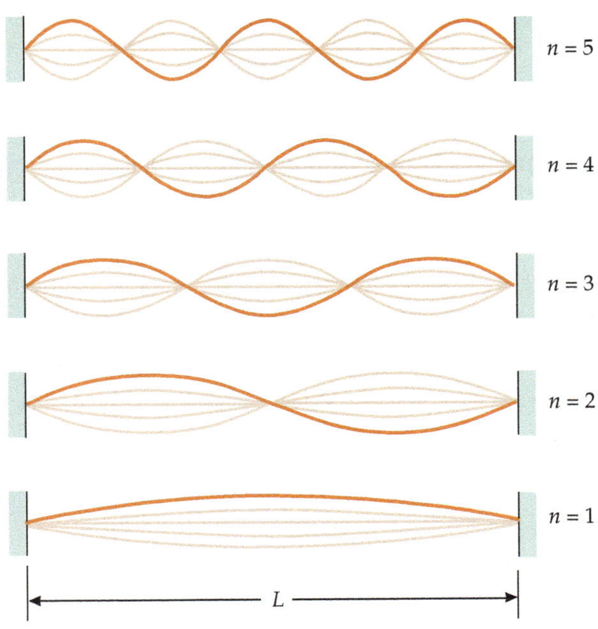

FIGURE 34-11 Standing waves on a string fixed at both ends. The standing-wave condition is the same as for standing electron waves in a box.

* The combination $h/2\pi$ occurs so often it is given a special symbol, somewhat analogous to giving the special symbol ω for $2\pi f$, which occurs often in oscillations.

Then the standing-wave condition $\lambda_n = 2L/n$ gives the allowed energies.

$$E_n = n^2\frac{h^2}{8mL^2} = n^2E_1 \qquad 34\text{-}21$$

ALLOWED ENERGIES FOR A PARTICLE IN A BOX

where

$$E_1 = \frac{h^2}{8mL^2} \qquad 34\text{-}22$$

GROUND-STATE ENERGY FOR A PARTICLE IN A BOX

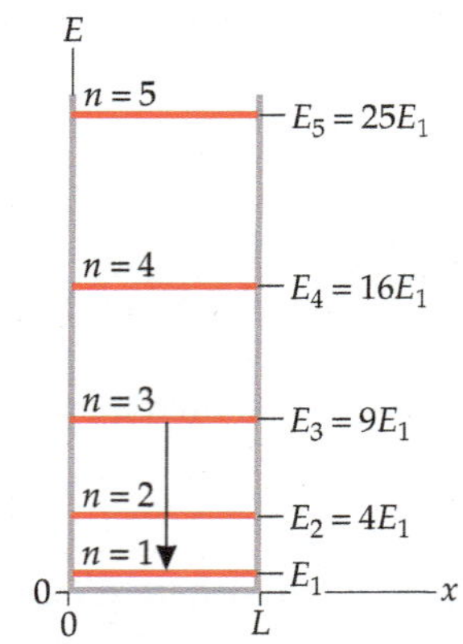

FIGURE 34-12 Energy-level diagram for a particle in a box. Classically, a particle can have any energy value. Quantum mechanically, only those energy values given by Equation 34-21 are allowed. A transition between the state $n = 3$ and the ground state $n = 1$ is indicated by the vertical arrow.

is the energy of the lowest state, which is the ground state.

The condition $\psi = 0$ at $x = 0$ and $x = L$ is called a **boundary condition.** Boundary conditions in quantum theory lead to energy quantization. Figure 34-12 shows the energy-level diagram for a particle in a box. Note that the lowest energy is not zero. This result is a general feature of quantum theory. If a particle is confined to some region of space, the particle has a minimum kinetic energy, called the **zero-point energy** that is greater than zero. The smaller the region of space the particle is confined to, the greater its zero-point energy. In Equation 34-22, this is indicated by the fact that E_1 varies as $1/L^2$.

If an electron is confined (bound to an atom) in some energy state E_i, the electron can make a transition to another energy state E_f by the emission of a photon if E_f is less than E_i. (If E_f is greater than E_i, the system absorbs a photon.) The transition from state 3 to the ground state is indicated in Figure 34-12 by the vertical arrow. The frequency of the emitted photon is found from the conservation of energy*

$$hf = E_i - E_f \qquad 34\text{-}23$$

The wavelength of the photon is then

$$\lambda = \frac{c}{f} = \frac{hc}{E_i - E_f} \qquad 34\text{-}24$$

STANDING-WAVE FUNCTIONS

The amplitude of a vibrating string that is fixed at $x = 0$ and $x = L$ is given by Equation 16-15:

$$A_n(x) = A_n \sin k_n x \qquad n = 1, 2, 3, \ldots$$

where A_n is a constant, $k_n = 2\pi/\lambda_n$ is the wave number, and $\lambda_n = 2L/n$. The wave functions for a particle in a box (which can be obtained by solving the Schrödinger equation, as we will see in Chapter 35) are the same:

$$\psi_n(x) = A_n \sin k_n x \qquad n = 1, 2, 3, \ldots$$

where $k_n = 2\pi/\lambda_n$. Using $\lambda_n = 2L/n$, we have

$$k_n = \frac{2\pi}{\lambda_n} = \frac{2\pi}{2L/n} = \frac{n\pi}{L}$$

The wave functions can thus be written

$$\psi_n(x) = A_n \sin\left(n\pi\frac{x}{L}\right)$$

* This equation was first proposed by Niels Bohr in his semiclassical model of the hydrogen atom in 1913, about 10 years before de Broglie's suggestion that electrons have wave properties. The Bohr model is presented in Chapter 36.

The constant A_n is determined by the normalization condition (Equation 34-18):

$$\int_{-\infty}^{\infty} \psi^2\,dx = \int_0^L A_n^2 \sin^2\!\left(n\pi\frac{x}{L}\right) dx = 1$$

Note that we need integrate only from $x = 0$ to $x = L$ because $\psi(x)$ is zero everywhere else. The result of evaluating the integral and solving for A_n is

$$A_n = \sqrt{\frac{2}{L}}$$

which is independent of n. The normalized standing-wave functions for a particle in a box are thus

$$\psi_n(x) = \sqrt{\frac{2}{L}}\sin\!\left(n\pi\frac{x}{L}\right) \qquad n = 1, 2, 3, \ldots \qquad 34\text{-}25$$

STANDING-WAVE FUNCTIONS FOR A PARTICLE IN A BOX

The standing-wave functions for $n = 1$, $n = 2$, and $n = 3$ are shown in Figure 34-13.

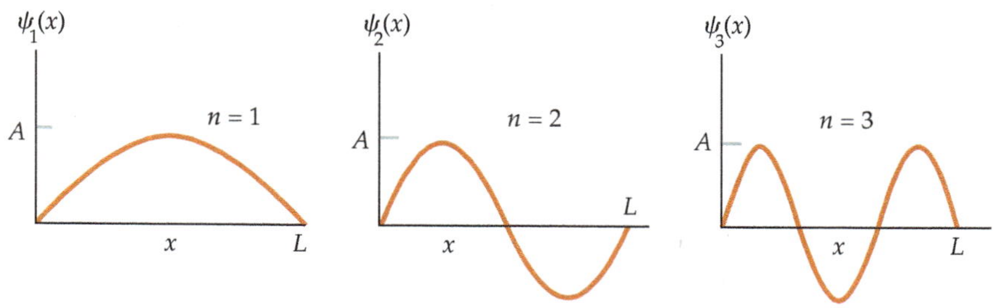

FIGURE 34-13 Standing-wave functions for $n = 1$, $n = 2$, and $n = 3$.

The number n is called a **quantum number.** It characterizes the wave function for a particular state and for the energy of that state. In our one-dimensional problem, a quantum number arises from the boundary condition on the wave function that it must be zero at $x = 0$ and $x = L$. In three-dimensional problems, three quantum numbers arise, one associated with a boundary condition in each dimension.

Figure 34-14 shows plots of ψ^2 for the ground state $n = 1$, the first excited state $n = 2$, the second excited state $n = 3$, and the state $n = 10$. In the ground state, the particle is most likely to be found near the center of the box, as indicated by

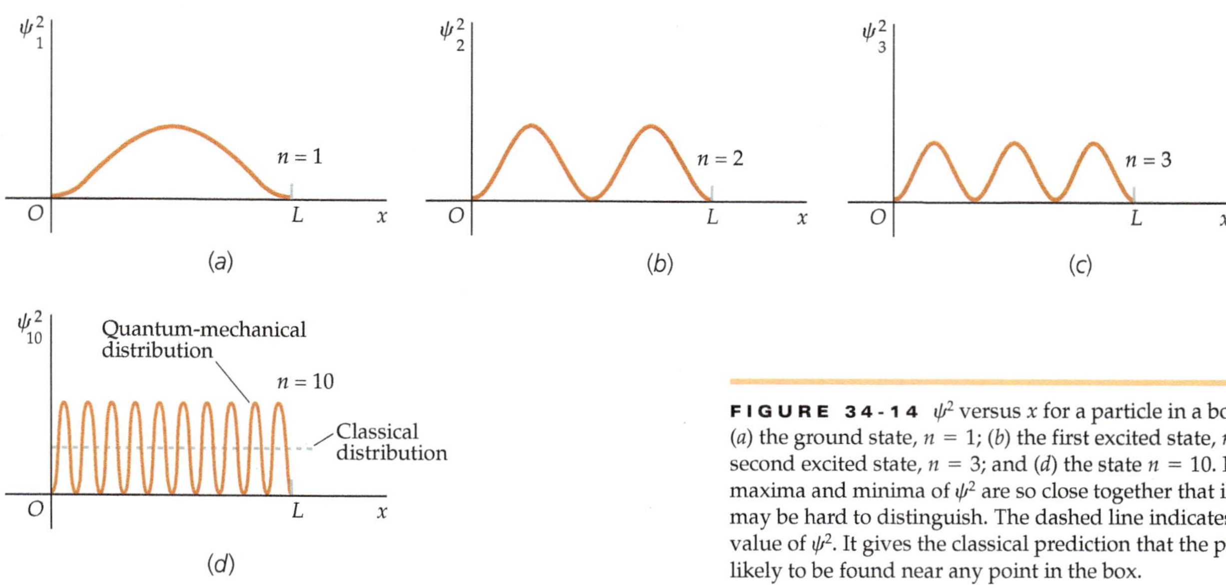

FIGURE 34-14 ψ^2 versus x for a particle in a box of length L for (*a*) the ground state, $n = 1$; (*b*) the first excited state, $n = 2$; (*c*) the second excited state, $n = 3$; and (*d*) the state $n = 10$. For $n = 10$, the maxima and minima of ψ^2 are so close together that individual maxima may be hard to distinguish. The dashed line indicates the average value of ψ^2. It gives the classical prediction that the particle is equally likely to be found near any point in the box.

the maximum value of ψ^2 at $x = L/2$. In the first excited state, the particle is least likely to be found near the center of the box because ψ^2 is small near $x = L/2$. For very large values of n, the maxima and minima of ψ^2 are very close together, as illustrated for $n = 10$. The average value of ψ^2 is indicated in this figure by the dashed line. For very large values of n, the maxima and minima are so closely spaced that ψ^2 cannot be distinguished from its average value. The fact that $(\psi^2)_{av}$ is constant across the whole box means that the particle is equally likely to be found anywhere in the box—the same as in the classical result. This is an example of **Bohr's correspondence principle:**

> In the limit of very large quantum numbers, the classical calculation and the quantum calculation must yield the same results.
>
> BOHR'S CORRESPONDENCE PRINCIPLE

The region of very large quantum numbers is also the region of very large energies. For large energies, the percentage change in energy between adjacent quantum states is very small, so energy quantization is not important (see Problem 71).

We are so accustomed to thinking of the electron as a classical particle that we tend to think of an electron in a box as a particle bouncing back and forth between the walls. But the probability distributions shown in Figure 34-14 are stationary; that is, they do not depend on time. A better picture for an electron in a bound state is a cloud of charge that has the charge density proportional to ψ^2. Figure 34-14 can then be thought of as plots of the charge density versus x for the various states. In the ground state, $n = 1$, the electron cloud is centered in the middle of the box and is spread out over most of the box, as indicated in Figure 34-14*a*. In the first excited state, $n = 2$, the charge density of the electron cloud has two maxima, as indicated in Figure 34-14*b*. For very large values of n, there are many closely spaced maxima and minima in the charge density resulting in an average charge density that is approximately uniform throughout the box. This electron-cloud picture of an electron is very useful in understanding the structure of atoms and molecules. However, it should be noted that whenever an electron is observed to interact with matter or radiation, it is always observed as a whole unit charge.

Example 34-6 Photon Emission by a Particle in a Box

An electron is in a one-dimensional box of length 0.100 nm. (*a*) Find the ground-state energy. (*b*) Find the energies of the four lowest-energy states that have energies above the ground-state energy, and then sketch an energy-level diagram. (*c*) Find the wavelength of the photon emitted for each transition from the state $n = 3$ to a lower-energy state.

PICTURE For Part (*a*) the ground state is the $n = 1$ state, and $E_1 = h^2/8mL^2$ (Equation 34-22). For Part (*b*), the energies are given by $E_n = n^2E_1$ (Equation 34-21), where, $n = 2, 3, 4,$ and 5. For Part (*c*), the photon wavelengths are given by $\lambda = hc/(E_i - E_f)$ (Equation 34-24).

SOLVE

(*a*) Use $hc = 1240\ \text{eV}\cdot\text{nm}$ and $mc^2 = 0.5110$ MeV to calculate E_1:

$$E_1 = \frac{h^2}{8mL^2} = \frac{(hc)^2}{8(mc^2)L^2} = \frac{(1240\ \text{eV}\cdot\text{nm})^2}{8(5.110\times10^5\ \text{eV})(0.100\ \text{nm})^2} = \boxed{37.6\ \text{eV}}$$

(*b*) 1. Calculate $E_n = n^2E_1$ for $n = 2, 3, 4,$ and 5:

$$E_2 = (2)^2(37.6\ \text{eV}) = \boxed{150\ \text{eV}}$$

$$E_3 = (3)^2(37.6\ \text{eV}) = \boxed{338\ \text{eV}}$$

$$E_4 = (4)^2(37.6\ \text{eV}) = \boxed{602\ \text{eV}}$$

$$E_5 = (5)^2(37.6\ \text{eV}) = \boxed{940\ \text{eV}}$$

2. Sketch an energy-level diagram using the values for the five energy states (Figure 34-15).

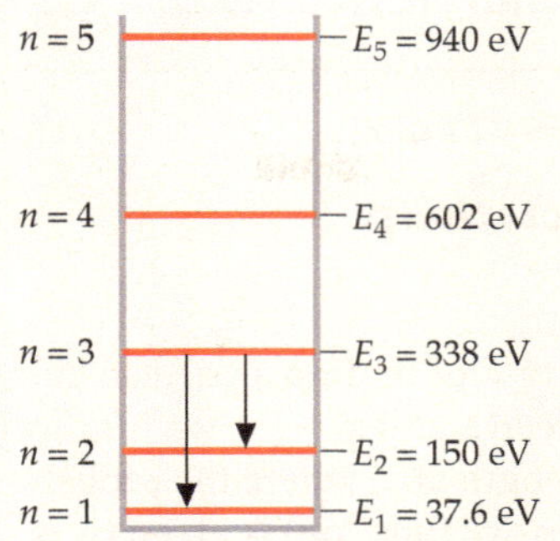

FIGURE 34-15

(*c*) 1. Use the energies found in Part (*b*) to calculate the wavelength for a transition from state 3 to state 2:

$$\lambda = \frac{hc}{E_3 - E_2} = \frac{1240 \text{ eV} \cdot \text{nm}}{338 \text{ eV} - 150 \text{ eV}} = \boxed{6.60 \text{ nm}}$$

2. Then use the energies in Part (*a*) and Part (*b*) to calculate the wavelength for a transition from state 3 to state 1:

$$\lambda = \frac{hc}{E_3 - E_1} = \frac{1240 \text{ eV} \cdot \text{nm}}{338 \text{ eV} - 37.6 \text{ eV}} = \boxed{4.13 \text{ nm}}$$

CHECK The wavelength of the photon emitted during the transition from the $n = 3$ to the $n = 1$ state is shorter than the wavelength of the photon emitted during the transition from the $n = 3$ to the $n = 2$ state. This result is expected—the greater the energy of the photon the shorter its wavelength.

TAKING IT FURTHER The energy-level diagram is shown in Figure 34-15. The transitions from $n = 3$ to $n = 2$ and from $n = 3$ to $n = 1$ are indicated by the vertical arrows. The ground-state energy of 37.6 eV is on the same order of magnitude as the kinetic energy of the electron in the ground state of the hydrogen atom, which is 13.6 eV. In the hydrogen atom, the electron has potential energy of −27.2 eV in the ground state, giving it a total ground-state energy (potential energy plus kinetic energy) of −13.6 eV.

PRACTICE PROBLEM 34-7 Calculate the wavelength of the photon emitted if the electron in the box makes a transition from $n = 4$ to $n = 3$.

34-9 EXPECTATION VALUES

The solution of a classical mechanics problem is typically specified by giving the position of a particle as a function of time. But the wave nature of matter prevents us from doing this for microscopic systems. The most that we can know is the relative probability of measuring a certain value of the position x. If we measure the position for a large number of identical systems, we get a range of values corresponding to the probability distribution. The average value of x obtained from such measurements is called the **expectation value** and is written $\langle x \rangle$. The expectation value of x is the same as the average value of x that we would expect to obtain from a measurement of the positions of a large number of particles that have the same wave function $\psi(x)$.

Because $\psi^2(x)\,dx$ is the probability of finding a particle in the region dx, the expectation value of x is

$$\langle x \rangle = \int_{-\infty}^{+\infty} x\psi^2(x)\,dx \qquad 34\text{-}26$$

EXPECTATION VALUE OF *x* DEFINED

See **Math Tutorial** *for more information on* **Integrals**

The expectation value of any function $F(x)$ is given by

$$\langle F(x) \rangle = \int_{-\infty}^{+\infty} F(x)\psi^2(x)\,dx \qquad 34\text{-}27$$

EXPECTATION VALUE OF *F*(*x*) DEFINED

CALCULATING PROBABILITIES AND EXPECTATION VALUES

PROBLEM-SOLVING STRATEGY

Probabilities and Expectations

SOLVE

1. To calculate the probability P of finding a particle in the region of infinitesimal length between x and $x + dx$, we multiply the length dx by the probability per unit length at x, where the probability per unit length (called the probability density function) is given by ψ^2.
2. To calculate the probability P of finding a particle in the region $x_1 < x < x_2$, we, in principle, divide the region into an infinite number of regions of infinitesimal length dx, calculate the probability P of finding the particle in each infinitesimal length, and then sum the probabilities. That is, we evaluate the integral $\int_{x_1}^{x_2} \psi^2\, dx$.
3. To calculate the expected value of a function $F(x)$, we evaluate the integral $\int_{-\infty}^{+\infty} F(x)\psi^2(x)\, dx$. The result of this calculation is called the expected value of $F(x)$.

The problem of a particle in a box allows us to illustrate the calculation of the probability of finding the particle in various regions of the box and the expectation values for various energy states. We give two examples, using the wave functions given by Equation 34-25.

Example 34-7 The Probability of the Particle Being Found in a Specified Region of a Box

A particle in a one-dimensional box of length L is in the ground state. Find the probability of finding the particle (a) in the region that has a length $\Delta x = 0.01L$ and is centered at $x = \frac{1}{2}L$ and (b) in the region $0 < x < \frac{1}{4}L$.

PICTURE The probability P of finding the particle in some infinitesimal range dx is $\psi^2\, dx$. For a particle in the nth state, the wave function is given by $\psi_n = \sqrt{2/L}\,\sin(n\pi x/L)$ (Equation34-25). For a particle in the ground state, $n = 1$; and ψ_1^2 is illustrated in Figure 34-14. The probability of finding x in some region is just the area under this curve for the region. For Part (a) , the region is $\Delta x = 0.01L$, centered at $x = L/2$, and the area under the ψ_1^2 versus x curve is shown in Figure 34-16a. This area is $\sim\psi_1^2\Delta x$. For Part (b) , the region is $0 < x < L/4$, and the area under the curve is shown in Figure 34-16b. To calculate this area, we must integrate ψ_1^2 from $x = 0$ to $x = L/4$.

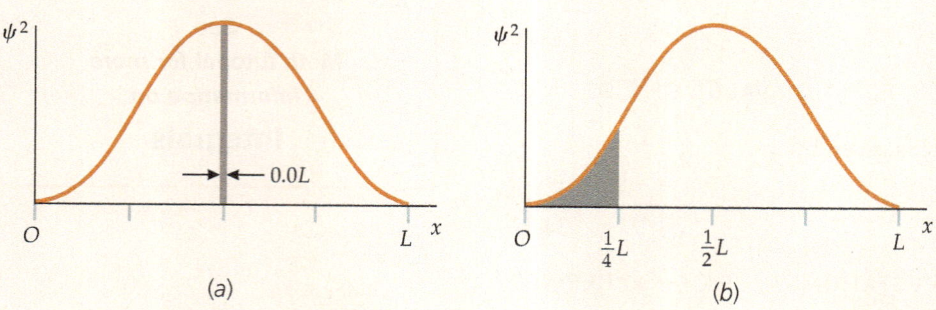

FIGURE 34-16

SOLVE

(*a*) 1. The probability of finding the particle is the area under the curve shown in Figure 34-16*a*. To calculate this area, we need to calculate the height of curve at $x = \frac{1}{2}L$:

$$\psi(x) = \psi_1(x) = \sqrt{\frac{2}{L}}\sin\left(\pi\frac{x}{L}\right)$$

so

$$\psi^2(\tfrac{1}{2}L) = \frac{2}{L}\sin^2\frac{\pi}{2} = \frac{2}{L}$$

2. The area is the height multiplied by the width, and the width is $\Delta x = 0.01L$:

$$P = \psi^2(\tfrac{1}{2}L)\Delta x = \frac{2}{L} \times 0.01L = \boxed{0.02}$$

(*b*) 1. The probability of finding the particle is the area under the curve shown in Figure 34-16*b*. To calculate this area, we need to integrate from $x = 0$ to $x = L/4$:

$$P = \int_0^{L/4} \psi^2(x)\,dx = \int_0^{L/4} \frac{2}{L}\sin^2\frac{\pi x}{L}\,dx$$

2. The integral can be evaluated a number of ways. If a table of integrals is used, a change in the integration variable in required. Changing the integration variable to $\theta = \pi x/L$ gives:

$$P = \frac{2}{\pi}\int_0^{\pi/4} \sin^2\theta\,d\theta$$

3. The integral can be found in tables:

$$\int_0^{\pi/4} \sin^2\theta\,d\theta = \left(\frac{\theta}{2} - \frac{\sin 2\theta}{4}\right)\Big|_0^{\pi/4} = \left(\frac{\pi}{8} - \frac{1}{4}\right)$$

4. Use the result from Part (*b*), step 3 to calculate the probability:

$$P = \frac{2}{\pi}\left(\frac{\pi}{8} - \frac{1}{4}\right) = \boxed{0.091}$$

CHECK If ψ_1^2 were uniformly distributed on the interval $0 < x < L$, the step-4 result would be 0.25. However, instead of ψ_1^2 being uniformly distributed, it is relatively small on the interval $0 < x < \frac{1}{4}L$, so a step-4 result that is less than 0.25 is expected.

TAKING IT FURTHER An integral was not necessary for Part (*a*) because the area of interest could be well approximated by a rectangle of height ψ^2 and width Δx. The chance of finding the particle in the region $\Delta x = 0.01L$ at $x = \frac{1}{2}L$ is approximately 2 percent. The chance of finding the particle in the region $0 < x < \frac{1}{4}L$ is about 9.1 percent.

CONCEPT CHECK 34-1

A fair six-sided die has the number 1 printed on four faces and the number 6 printed on the other two faces. What is the probability that a 1 comes up when the die is thrown? *Hint: The probability that a specific value comes up for one throw is the fraction of the throws that that value comes up after a large number of throws.*

Example 34-8 Calculating Expectation Values

Find (*a*) $\langle x\rangle$ and (*b*) $\langle x^2\rangle$ for a particle in its ground state in a box of length L.

PICTURE We use $\langle F(x)\rangle = \int F(x)\psi^2(x)\,dx$, with $\psi_n(x) = \sqrt{\frac{2}{L}}\sin\frac{n\pi x}{L}$.

SOLVE

(*a*) 1. Write $\langle x\rangle$ using the ground-state wave function given by Equation 34-25, with $n = 1$:

$$\langle x\rangle = \int_{-\infty}^{+\infty} x\psi^2(x)\,dx = \frac{2}{L}\int_0^L x\sin^2\left(\frac{\pi x}{L}\right)dx$$

2. To evaluate this integral by using a table of integrals, first change the integration variable to $\theta = \pi x/L$:

$$\langle x\rangle = \frac{2}{L}\left(\frac{L}{\pi}\right)^2\int_0^{\pi} \theta\sin^2\theta\,d\theta = \frac{2L}{\pi^2}\int_0^{\pi} \theta\sin^2\theta\,d\theta$$

3. The table of integrals gives:

$$\int_0^{\pi} \theta\sin^2\theta\,d\theta = \left[\frac{\theta^2}{4} - \frac{\theta\sin 2\theta}{4} - \frac{\cos 2\theta}{8}\right]_0^{\pi} = \frac{\pi^2}{4}$$

4. Substitute this value into the expression in step 2:

$$\langle x\rangle = \frac{2L}{\pi^2}\int_0^{\pi} \theta\sin^2\theta\,d\theta = \frac{2L}{\pi^2}\frac{\pi^2}{4} = \boxed{\frac{L}{2}}$$

(*b*) 1. Repeat step 1 and step 2 of Part (*a*) for $\langle x^2 \rangle$:

$$\langle x^2 \rangle = \int_{-\infty}^{+\infty} x^2 \psi^2(x)\, dx = \int_0^L x^2 \frac{2}{L} \sin^2(\pi x/L)\, dx$$
$$= \frac{2}{L}\left(\frac{L}{\pi}\right)^3 \int_0^\pi \theta^2 \sin^2\theta\, d\theta = \frac{2L^2}{\pi^3} \int_0^\pi \theta^2 \sin^2\theta\, d\theta$$

2. Evaluating the integral using a table of integrals gives:

$$\int_0^\pi \theta^2 \sin^2\theta\, d\theta = \left[\frac{\theta^3}{6} - \left(\frac{\theta^2}{4} - \frac{1}{8}\right)\sin 2\theta - \frac{\theta \cos 2\theta}{4}\right]\Bigg|_0^\pi$$
$$= \frac{\pi^3}{6} - \frac{\pi}{4}$$

3. Substitute this value into the expression in step 1 of Part (*b*):

$$\langle x^2 \rangle = \frac{2L^2}{\pi^3}\left(\frac{\pi^3}{6} - \frac{\pi}{4}\right) = \left(\frac{1}{3} - \frac{1}{2\pi^2}\right)L^2 = \boxed{0.283L^2}$$

CHECK The expectation value of x is $L/2$, as we would expect, because the probability distribution is symmetric about the midpoint of the box.

TAKING IT FURTHER Note that $\langle x^2 \rangle$ is greater than $\langle x \rangle^2$.

CONCEPT CHECK 34-2

A fair six-sided die has the number 1 printed on four faces and the number 6 printed on the other two faces. Let N be the number that comes up when the die is thrown. What is the expectation value of N? What is the expectation value of N^2? *Hint: The expectation value of a quantity is the average value of that quantity after a large number of throws.*

34-10 ENERGY QUANTIZATION IN OTHER SYSTEMS

The quantized energies of a system are generally determined by solving the Schrödinger equation for that system. The form of the Schrödinger equation depends on the potential energy of the particle. The potential energy for a one-dimensional box from $x = 0$ to $x = L$ is shown in Figure 34-17. This potential energy function is called an **infinite square-well potential,** and it is described mathematically by

$$U(x) = \begin{cases} \infty & x < 0 \\ 0 & 0 < x < L \\ \infty & x > L \end{cases} \qquad 34\text{-}28$$

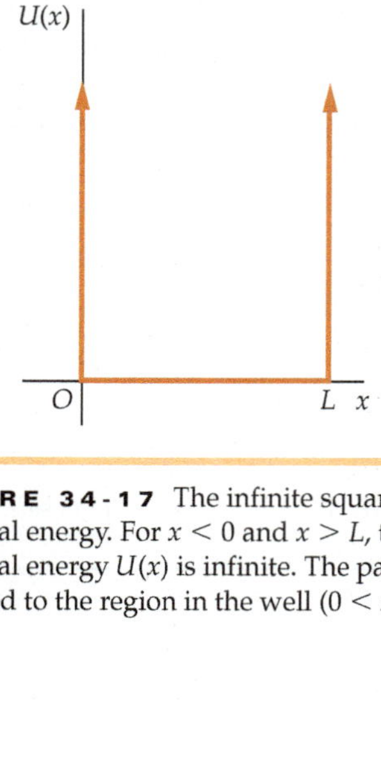

FIGURE 34-17 The infinite square-well potential energy. For $x < 0$ and $x > L$, the potential energy $U(x)$ is infinite. The particle is confined to the region in the well ($0 < x < L$).

The particle moves freely inside the box, so the potential energy is uniform. For convenience, we choose the value of this potential energy to be zero. Outside the box the potential energy is infinite, so the particle cannot exist outside the box no matter what its energy. We did not need to solve the Schrödinger equation for this potential because the wave functions and quantized frequencies are the same as for a string fixed at both ends, which we studied in Chapter 16. Although this problem seems artificial, actually it is useful for some physical problems, such as a neutron that is constrained to a nucleus that has a large number of protons and neutrons.

THE HARMONIC OSCILLATOR

More realistic than the particle in a box is the harmonic oscillator, which applies to an object of mass m on a spring that has a force constant k or to any system undergoing small oscillations about a stable equilibrium. Figure 34-18 shows the potential energy function

$$U(x) = \tfrac{1}{2}kx^2 = \tfrac{1}{2}m\omega_0^2 x^2$$

where $\omega_0 = \sqrt{k/m}$ is the natural frequency of the oscillator. Classically, the object oscillates between $x = +A$ and $x = -A$. Its total energy is $E = \frac{1}{2}m\omega_0^2 A^2$, which can have any nonnegative value, including zero.

In quantum theory, the particle is represented by the wave function $\psi(x)$, which is determined by solving the Schrödinger equation for this potential. Normalizable wave functions $\psi_n(x)$ occur only for discrete values of the energy E_n given by

$$E_n = (n + \tfrac{1}{2})hf_0 \qquad n = 0, 1, 2, 3, \ldots \qquad 34\text{-}29$$

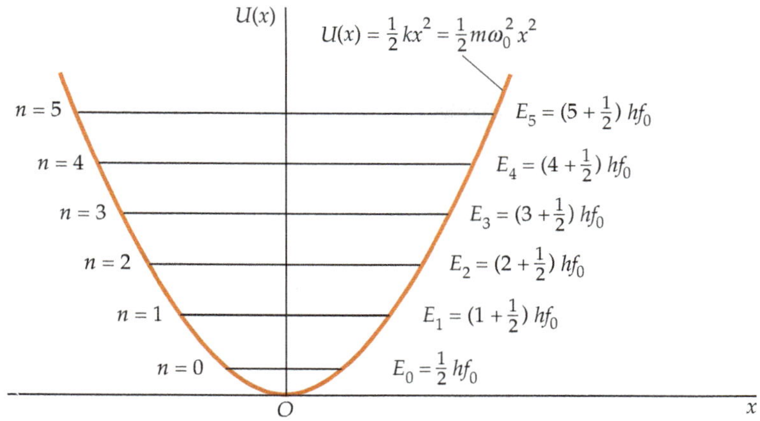

FIGURE 34-18 Harmonic oscillator potential energy function. The allowed energy levels are indicated by the equally spaced horizontal lines. Also, $\omega_0 = 2\pi f_0$.

where $f_0 = \omega_0/2\pi$ is the classical frequency of the oscillator. Note that the energy levels of a harmonic oscillator are evenly spaced with separation hf_0, as indicated in Figure 34-18. Compare this with the uneven spacing of the energy levels for the particle in a box, as shown in Figure 34-12. If a harmonic oscillator makes a transition from energy level n to the next lowest energy level $n - 1$, the frequency f of the photon emitted is given by $hf = E_i - E_f$ (Equation 34-23). Applying this equation gives

$$hf = E_n - E_{n-1} = (n + \tfrac{1}{2})hf_0 - (n - 1 + \tfrac{1}{2})hf_0 = hf_0$$

The frequency f of the emitted photon is therefore equal to the classical frequency f_0 of the oscillator.

THE HYDROGEN ATOM

In the hydrogen atom, an electron is bound to a proton by the electrostatic force of attraction (discussed in Chapter 21). This force varies inversely as the square of the separation distance (exactly like the gravitational attraction of Earth and the Sun). The potential energy of the electron–proton system therefore varies inversely with separation distance (Equation 23-9). As in the case of gravitational potential energy, the potential energy of the electron–proton system is chosen to be zero if the electron is an infinite distance from the proton. Then for all finite distances, the potential energy is negative. Like the case of an object orbiting Earth, the electron–proton system is a bound system if its total energy is negative. Like the energies of a particle in a box and of a harmonic oscillator, the energies are described by a quantum number n. As we will see in Chapter 36, the allowed energies of the hydrogen atom are given by

$$E_n = -\frac{13.6\text{ eV}}{n^2} \qquad n = 1, 2, 3, \ldots \tag{34-30}$$

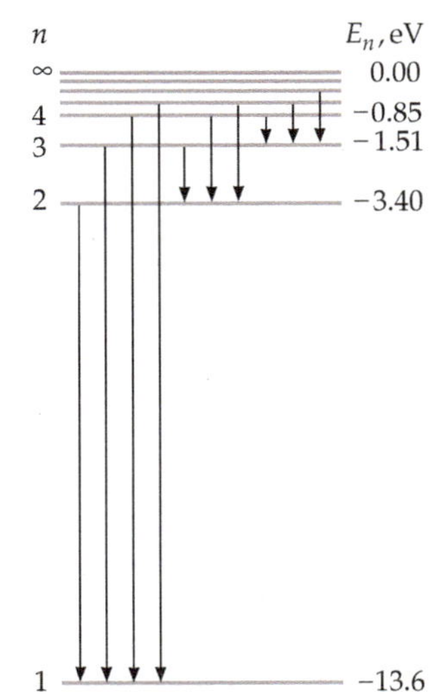

FIGURE 34-19 Energy-level diagram for the hydrogen atom. The energy of the ground state is −13.6 eV. As n approaches infinity, the energy approaches 0, which is the highest energy state for which an electron is bound to the nucleus.

The lowest energy corresponds to $n = 1$. The ground-state energy is thus −13.6 eV. The energy of the first excited state is $-(13.6\text{ eV})/2^2 = -3.40$ eV. Figure 34-19 shows the energy-level diagram for the hydrogen atom. The vertical arrows indicate transitions from a higher state to a lower state that accompany the emission of electromagnetic radiation. Only those transitions ending at the first excited state ($n = 2$) involve energy differences in the range of visible light of 1.77 eV to 3.10 eV, as calculated in Example 34-1.

Other atoms are more complicated than the hydrogen atom, but their energy levels are similar in many ways to those of hydrogen. Their ground-state energies are of the order of −1 eV to −10 eV, and many transitions involve energies corresponding to photons in the visible range.

! Do not think an electron orbits a proton in a classical orbit like Earth's orbit around the Sun. It doesn't.

TOPIC	RELEVANT EQUATIONS AND REMARKS
5. Wave–Particle Duality	Photons, electrons, neutrons, and all other carriers of momentum and energy exhibit both wave and particle properties. Everything propagates like a classical wave, exhibiting diffraction and interference, but exchanges energy in discrete lumps like a classical particle. Because the wavelength of macroscopic objects is so small, diffraction and interference are not observed. Also, if a macroscopic amount of energy is exchanged, so many quanta are involved that the particle nature of the energy is not evident.
6. Uncertainty Principle	The wave–particle duality of nature leads to the uncertainty principle, which states that the product of the uncertainty in a measurement of position and the uncertainty in a measurement of momentum must be greater than or equal to $\frac{1}{2}\hbar$, where $\hbar$ is Planck's constant divided by 2π. $$\Delta x\,\Delta p_x \geq \tfrac{1}{2}\hbar \qquad 34\text{-}19$$

Answers to Concept Checks

34-1 2/3

34-2 $\langle N\rangle = 8/3$ $\langle N^2\rangle = 38/3$

Answers to Practice Problems

34-1 4.13×10^{-7} eV

34-2 (*a*) 12.4 μm, (*b*) 1.24 nm, (*c*) 1.24 pm

34-3 $1.46 \times 10^{7}\ \text{cm}^{-3}$

34-4 0.6*c*

34-5 1.4×10^{-34} m

34-6 0.388 nm. From this result, we see that a 10-eV electron has a de Broglie wavelength of about 0.4 nm. This quantity is of the same order of magnitude as the size of the atom and the spacing of atoms in a crystal.

34-7 4.70 nm

Problems

In a few problems, you are given more data than you actually need; in a few other problems, you are required to supply data from your general knowledge, outside sources, or informed estimate.

Interpret as significant all digits in numerical values that have trailing zeros and no decimal points.

- • Single-concept, single-step, relatively easy
- •• Intermediate-level, may require synthesis of concepts
- ••• Challenging
- SSM Solution is in the *Student Solutions Manual*
- Consecutive problems that are shaded are paired problems.

CONCEPTUAL PROBLEMS

1 • The quantized character of electromagnetic radiation is observed by (*a*) the Young double-slit experiment, (*b*) diffraction of light by a small aperture, (*c*) the photoelectric effect, (*d*) the J. J. Thomson cathode-ray experiment. SSM

2 •• Two monochromatic light sources, A and B, emit the same number of photons per second. The wavelength of A is $\lambda_A = 400$ nm and the wavelength of B is $\lambda_B = 600$ nm. The power radiated by source B (*a*) is equal to the power of source A, (*b*) is less than the power of source A, (*c*) is greater than the power of source A, (*d*) cannot be compared to the power from source A using the available data.

3 • The work function of a surface is ϕ. The threshold wavelength for emission of photoelectrons from the surface is equal to (*a*) hc/ϕ, (*b*) ϕ/hf, (*c*) hf/ϕ, (*d*) none of the above. SSM

4 •• When light of wavelength λ_1 is incident on a certain photoelectric cathode, no electrons are emitted, no matter how intense the incident light is. Yet, when light of wavelength $\lambda_2 < \lambda_1$ is incident, electrons are emitted, even when the incident light has low intensity. Explain this observation.

5 • True or false: (*a*) The wavelength of an electron's matter wave varies inversely with the momentum of the electron. (*b*) Electrons can undergo diffraction. (*c*) Neutrons can undergo diffraction.

6 • If the wavelength of an electron is equal to the wavelength of a proton, then (*a*) the speed of the proton is greater than the speed of the electron, (*b*) the speeds of the proton and the electron are equal, (*c*) the speed of the proton is less than the speed of the electron, (*d*) the energy of the proton is greater than the energy of the electron, (*e*) both (*a*) and (*d*) are correct.

7 • A proton and an electron have equal kinetic energies. It follows that the wavelength of the proton is (*a*) greater than the wavelength of the electron, (*b*) equal to the wavelength of the electron, (*c*) less than the wavelength of the electron.

8 • The parameter x represents the position of a particle. Can the expectation value of x ever have a value such that the probability density function $P(x)$ is zero? Give a specific example.

9 •• It was once believed that if two identical experiments are done on identical systems under the same conditions, the results must be identical. Explain how this statement can be modified so that it is consistent with quantum physics.

10 •• A six-sided die has the numeral 1 painted on three sides and the numeral 2 painted on the other three sides. (*a*) What is the probability of a 1 coming up when the die is thrown? (*b*) What is the expectation value of the numeral that comes up when the die is thrown? (*c*) What is the expectation value of the cube of the numeral that comes up when the die is thrown?

ESTIMATION AND APPROXIMATION

11 •• During an advanced physics lab, students measure the Compton wavelength, λ_C. The students obtain the following wavelength shifts $\lambda_s - \lambda_i$ as a function of scattering angle θ.

θ	45°	75°	90°	135°	180°
$\lambda_s - \lambda_i$	0.647 pm	1.67 pm	2.45 pm	3.98 pm	4.95 pm

Use their data to estimate the value for the Compton wavelength. Compare this number with the accepted value. SSM

12 •• **SPREADSHEET** Students in a physics lab are trying to determine the value of Planck's constant h, using a photoelectric apparatus similar to the one shown in Figure 34-2. The students are using a helium–neon laser that has a tunable wavelength as the light source. The data that the students obtain for the maximum electron kinetic energies are

λ	544 nm	594 nm	604 mn	612 nm	633 mn
K_{max}	0.360 eV	0.199 eV	0.156 eV	0.117 eV	0.062 eV

(*a*) Using a spreadsheet program or graphing calculator, plot K_{max} versus light frequency. (*b*) Use the graph to estimate the value of Planck's constant. (*Note:* You may wish to use a feature of your spreadsheet program or graphing calculator to obtain the best straight-line fit to the data.) (*c*) Compare your result with the accepted value for Planck's constant.

13 •• **SPREADSHEET** The cathode that was used by the students in the experiment described in Problem 12 is constructed from one of the following metals:

Metal	Tungsten	Silver	Potassium	Cesium
Work function	4.58 eV	2.4 eV	2.1 eV	1.9 eV

Determine which metal composes the cathode by using the same data given in Problem 12. (*a*) Using a spreadsheet program or graphing calculator, plot K_{max} versus frequency. (*b*) Use the graph to estimate the value of the work function based on the students' data. (*Note:* You may wish to use a feature of your spreadsheet program or graphing calculator to obtain the best straight-line fit to the data.) (*c*) Which metal was most likely used for the cathode in their experiment?

THE PARTICLE NATURE OF LIGHT: PHOTONS

14 • Find the photon energy in electron volts for light of wavelength (*a*) 450 nm, (*b*) 550 nm, and (*c*) 650 nm.

15 • Find the photon energy in electron volts for an electromagnetic wave of frequency (*a*) 100 MHz in the FM radio band and (*b*) 900 kHz in the AM radio band.

16 • What are the frequencies of photons that have the following energies: (*a*) 1.00 eV, (*b*) 1.00 keV, and (*c*) 1.00 MeV?

17 • Find the photon energy in electron volts if the wavelength is (*a*) 0.100 nm (about 1 atomic diameter) and (*b*) 1.00 fm (1 fm = 10^{-15} m, about 1 nuclear diameter).

18 •• The wavelength of red light emitted by a 3.00-mW helium–neon laser is 633 nm. If the diameter of the laser beam is 1.00 mm, what is the density of photons in the beam? Assume that the intensity is uniformly distributed across the beam.

19 • **ENGINEERING APPLICATION** Lasers used in a telecommunications network typically produce light that has a wavelength near 1.55 μm. How many photons per second are being transmitted if such a laser has an output power of 2.50 mW? SSM

THE PHOTOELECTRIC EFFECT

20 • The work function for tungsten is 4.58 eV. (*a*) Find the threshold frequency and wavelength for the photoelectric effect to occur when monochromatic electromagnetic radiation is incident on the surface of a sample of tungsten. Find the maximum kinetic energy of the electrons if the wavelength of the incident light is (*b*) 200 nm and (*c*) 250 nm.

21 • When monochromatic ultraviolet light that has a wavelength equal to 300 nm is incident on a sample of potassium, the emitted electrons have maximum kinetic energy of 2.03 eV. (*a*) What is the energy of an incident photon? (*b*) What is the work function for potassium? (*c*) What would be the maximum kinetic energy of the electrons if the incident electromagnetic radiation had a wavelength of 430 nm? (*d*) What is the maximum wavelength of incident electromagnetic radiation that will result in the photoelectric emission of electrons by a sample of potassium?

22 • The maximum wavelength of electromagnetic radiation that will result in the photoelectric emission of electrons from a sample of silver is 262 nm. (*a*) Find the work function for silver. (*b*) Find the maximum kinetic energy of the electrons if the incident radiation has a wavelength of 175 nm.

23 • The work function for cesium is 1.90 eV. (*a*) Find the minimum frequency and maximum wavelength of electromagnetic radiation that will result in the photoelectric emission of electrons from a sample of cesium. Find the maximum kinetic energy of the electrons if the wavelength of the incident radiation is (*b*) 250 nm and (*c*) 350 nm.

24 •• When a surface is illuminated with electromagnetic radiation of wavelength 780 nm, the maximum kinetic energy of the emitted electrons is 0.37 eV. What is the maximum kinetic energy if the surface is illuminated using radiation of wavelength 410 nm?

COMPTON SCATTERING

25 • Find the shift in wavelength of photons scattered by free stationary electrons at $\theta = 60°$. (Assume that the electrons are initially moving with negligible speed and are virtually free of (unattached to) any atoms or molecules.)

26 • When photons are scattered by electrons in a carbon sample, the shift in wavelength is 0.33 pm. Find the scattering angle. (Assume that the electrons are initially moving with negligible speed and are virtually free of (unattached to) any atoms or molecules.)

27 • The photons in a monochromatic beam are scattered by electrons. The wavelength of the photons that are scattered at an angle of 135° with the direction of the incident photon beam is 2.3 percent less than the wavelength of the incident photons. What is the wavelength of the incident photons?

28 • Compton used photons of wavelength 0.0711 nm. (*a*) What is the energy of one of those photons? (*b*) What is the wavelength of the photons scattered in the direction opposite to the direction of the incident photons? (*c*) What is the energy of the photon scattered in that direction?

29 • For the photons used by Compton (see Problem 28), find the momentum of the incident photon and the momentum of the photon scattered in the direction opposite to the direction of the incident photons. Use the conservation of momentum to find the momentum of the recoil electron in this case.

30 •• A beam of photons that have a wavelength equal to 6.00 pm is scattered by electrons initially at rest. A photon in the beam is scattered in a direction perpendicular to the direction of the incident beam. (*a*) What is the change in wavelength of the photon? (*b*) What is the kinetic energy of the electron?

ELECTRONS AND MATTER WAVES

31 • An electron is moving at 2.5×10^5 m/s. Find the electron's wavelength.

32 • An electron has a wavelength of 200 nm. Find (*a*) the magnitude of its momentum and (*b*) its kinetic energy.

33 •• An electron, a proton, and an alpha particle each have a kinetic energy of 150 keV. Find (*a*) the magnitudes of their momenta and (*b*) their de Broglie wavelengths.

34 • A neutron in a reactor has a kinetic energy of approximately 0.020 eV. Calculate the wavelength of the neutron.

35 • Find the wavelength of a proton that has a kinetic energy of 2.00 MeV.

36 • What is the kinetic energy of a proton whose wavelength is (*a*) 1.00 nm and (*b*) 1.00 fm?

37 • The kinetic energy of the electrons in the electron beam in a run of Davisson and Germer's experiment was 54 eV. Calculate the wavelength of the electrons in the beam.

38 • The distance between Li^+ and Cl^- ions in a LiCl crystal is 0.257 nm. Find the energy of electrons that have a wavelength equal to that spacing.

39 • An electron microscope uses electrons that have energies equal to 70 keV. Find the wavelength of the electrons. SSM

40 • What is the wavelength of a neutron that has a speed of 1.00×10^6 m/s?

A PARTICLE IN A BOX

41 •• (*a*) Find the energy of the ground state ($n = 1$) and the first two excited states of a neutron in a one-dimensional box of length $L = 1.00 \times 10^{-15}$ m = 1.00 fm (about the diameter of an atomic nucleus). Make an energy-level diagram for the system. Calculate the wavelength of electromagnetic radiation emitted when the neutron makes a transition from (*b*) $n = 2$ to $n = 1$, (*c*) $n = 3$ to $n = 2$, and (*d*) $n = 3$ to $n = 1$.

42 •• (*a*) Find the energy of the ground state ($n = 1$) and the first two excited states of a neutron in a one-dimensional box of length 0.200 nm (about the diameter of a H_2 molecule). Calculate the wavelength of electromagnetic radiation emitted when the neutron makes a transition from (*b*) $n = 2$ to $n = 1$, (*c*) $n = 3$ to $n = 2$, and (*d*) $n = 3$ to $n = 1$.

CALCULATING PROBABILITIES AND EXPECTATION VALUES

43 •• A particle is in the ground state of a one-dimensional box that has length L. (The box has one end at the origin and the other end on the positive x axis.) Determine the probability of finding the particle in the interval of length $\Delta x = 0.002L$ and centered at (*a*) $\frac{1}{4}x = L$, (*b*) $x = \frac{1}{2}L$, and (*c*) $x = \frac{3}{4}L$. (Because Δx is very small you need not do any integration.)

44 •• A particle is in the second excited state ($n = 3$) of a one-dimensional box that has length L. (The box has one end at the origin and the other end on the positive x axis.) Determine the probability of finding the particle in the interval of length $\Delta x = 0.002L$ and centered at (*a*) $x = \frac{1}{3}L$, (*b*) $x = \frac{1}{2}L$, and (*c*) $x = \frac{2}{3}L$. (Because Δx is very small you need not do any integration.)

45 •• A particle is in the first excited ($n = 2$) state of a one-dimensional box that has length L. (The box has one end at the origin and the other end on the positive x axis.) Find (*a*) $\langle x \rangle$ and (*b*) $\langle x^2 \rangle$.

46 •• A particle in a one-dimensional box that has length L is in the first excited state ($n = 2$). (The box has one end at the origin and the other end on the positive x axis.) (*a*) Sketch $\psi^2(x)$ versus x for this state. (*b*) What is the expectation value $\langle x \rangle$ for this state? (*c*) What is the probability of finding the particle in some small region dx centered at $x = L/2$? (*d*) Are your answers for Part (*b*) and Part (*c*) contradictory? If not, explain why your answers are not contradictory.

47 •• A particle of mass m has a wave function given by $\psi(x) = Ae^{-|x|/a}$, where A and a are positive constants. (*a*) Find the normalization constant A. (*b*) Calculate the probability of finding the particle in the region $-a \leq x \leq a$.

48 •• A one-dimensional box is on the x axis in the region of $0 \leq x \leq L$. A particle in this box is in its ground state. Calculate the probability that the particle will be found in the region (*a*) $0 < x < \frac{1}{2}L$, (*b*) $0 < x < \frac{1}{3}L$, and (*c*) $0 < x < \frac{3}{4}L$.

49 •• A one-dimensional box is on the x axis in the region of $0 \leq x \leq L$. A particle in this box is in its first excited state. Calculate the probability that the particle will be found in the region (*a*) $0 < x < \frac{1}{2}L$, (*b*) $0 < x < \frac{1}{3}L$, and (*c*) $0 < x < \frac{3}{4}L$.

50 •• The classical probability distribution function for a particle in a one-dimensional box on the x axis in the region of $0 < x < L$ is given by $P(x) = 1/L$. Use this expression to show that $\langle x \rangle = \frac{1}{2}L$ and $\langle x^2 \rangle = \frac{1}{3}L^2$ for a classical particle in the box.

51 •• A one-dimensional box is on the x axis in the region of $0 \leq x \leq L$. (*a*) The wave functions for a particle in the box are given by

$$\psi_n(x) = \sqrt{\frac{2}{L}} \sin \frac{n\pi x}{L} \qquad n = 1, 2, 3, \ldots$$

For a particle in the nth state, show that $\langle x \rangle = \frac{1}{2}L$ and $\langle x^2 \rangle = L^2/3 - L^2/(2n^2\pi^2)$. (*b*) Compare these expressions for $\langle x \rangle$ and $\langle x^2 \rangle$, for $n \gg 1$, with the expressions for $\langle x \rangle$ and $\langle x^2 \rangle$ for the classical distribution of Problem 50.

52 •• **SPREADSHEET** (*a*) Use a spreadsheet program or graphing calculator to plot $\langle x^2 \rangle$ as a function of the quantum number n for the particle in the box described in Problem 48 and for values of n from 1 to 100. Assume $L = 1.00$ m for your graph. Refer to Problem 51. (*b*) Comment on the significance of any asymptotic limits that your graph shows.

53 •• The wave functions for a particle of mass m in a one-dimensional box of length L *centered at the origin* (so that the ends are at $x = \pm\frac{1}{2}L$) are given by

$$\psi(x) = \sqrt{\frac{2}{L}} \cos \frac{n\pi x}{L} \qquad n = 1, 3, 5, 7, \ldots$$

and

$$\psi(x) = \sqrt{\frac{2}{L}} \sin \frac{n\pi x}{L} \qquad n = 2, 4, 6, 8, \ldots$$

Calculate $\langle x \rangle$ and $\langle x^2 \rangle$ for the ground state ($n = 1$).

54 •• Calculate $\langle x \rangle$ and $\langle x^2 \rangle$ for the first excited state ($n = 2$) of the box described in Problem 53.

GENERAL PROBLEMS

55 • Photons in a uniform 4.00-cm-diameter light beam have wavelengths equal to 400 nm and the beam has an intensity of 100 W/m^2. (*a*) What is the energy of each photon in the beam? (*b*) How much energy strikes an area of 1.00 cm^2 perpendicular to the beam in 1.00 s? (*c*) How many photons strike this area in 1.00 s? SSM

56 • A 1-μg particle is moving with a speed of approximately 1 mm/s in a one-dimensional box that has a length equal to 1 cm. Calculate the approximate value of the quantum number n of the state occupied by the particle.

57 • (*a*) For the particle and box of Problem 56, find Δx and Δp_x, assuming that these uncertainties are given by $\Delta x/L = 0.01$ percent and $\Delta p_x/p_x = 0.01$ percent. (*b*) What is $(\Delta x\,\Delta p_x)/\hbar$?

58 • In 1987, a laser at Los Alamos National Laboratory produced a flash that lasted 1×10^{-12} s and had a power of 5×10^{15} W. Estimate the number of emitted photons, assuming they all had wavelengths equal to 400 nm.

59 • **ENGINEERING APPLICATION** You cannot "see" anything smaller than the wavelength of the wave used to make the observation. What is the minimum energy of an electron needed in an electron microscope to "see" an atom that has a diameter of about 0.1 nm?

60 • A common flea that has a mass of 0.008 g can jump vertically as high as 20 cm. Estimate the wavelength for the flea immediately after takeoff.

61 •• **BIOLOGICAL APPLICATION** A 100-W source radiates light of wavelength 600 nm uniformly in all directions. An eye that has been adapted to the dark has a 7-mm-diameter pupil and can detect the light if at least 20 photons per second enter the pupil. How far from the source can the light be detected under these rather extreme conditions? SSM

62 •• **BIOLOGICAL APPLICATION** The diameter of the pupil of an eye under room-light conditions is approximately 5 mm. Find the intensity of light that has a wavelength equal to 600 nm so that 1 photon per second passes through the pupil.

63 •• A 100-W incandescent lightbulb radiates 2.6 W of visible light uniformly in all directions. (*a*) Find the intensity of the light from the bulb at a distance of 1.5 m. (*b*) If the average wavelength of the visible light is 650 nm, and counting only those photons in the visible spectrum, find the number of photons per second that strike a surface that has an area equal to 1.0 cm^2, is oriented so that the line to the bulb is perpendicular to the surface, and is a distance of 1.5 m from the bulb.

64 •• When light of wavelength λ_1 is incident on the cathode of a photoelectric tube, the maximum kinetic energy of the emitted electrons is 1.8 eV. If the wavelength is reduced to $\frac{1}{2}\lambda_1$, the maximum kinetic energy of the emitted electrons is 5.5 eV. Find the work function ϕ of the cathode material.

65 •• An incident photon of energy E_i undergoes Compton scattering at an angle of θ. Show that the energy E_s of the scattered photon is given by

$$E_s = \frac{E_i}{1 + (E_i/m_e c^2)(1 - \cos\theta)}$$

66 •• A particle is confined to a one-dimensional box. While the particle makes a transition from the state n to the state $n - 1$, radiation of 114.8 nm is emitted. While the particle makes the transition from the state $n - 1$ to the state $n - 2$, radiation of wavelength 147 nm is emitted. The ground-state energy of the particle is 1.2 eV. Determine n.

67 •• The Pauli exclusion principle states that no more than one electron may occupy a particular quantum state at a time. Electrons intrinsically occupy two spin states. Therefore, if we wish to model an atom as a collection of electrons trapped in a one-dimensional box, no more than two electrons in the box can have the same value of the quantum number n. Calculate the energy that the most energetic electron(s) would have for the uranium atom that has an atomic number 92. Assume the box has a length of 0.050 nm and the electrons are in the lowest possible energy states. How does this energy compare to the rest energy of the electron? SSM

68 •• A beam of electrons that each have the same kinetic energy illuminates a pair of slits separated by a distance 54 nm. The beam forms bright and dark fringes on a screen located a distance 1.5 m beyond the two slits. The arrangement is otherwise identical to that used in the optical two-slit interference experiment described in Chapter 33 and in Figure 33-7 and the fringes have the appearance shown in Figure 34-8*d*. The bright fringes are found to be separated by a distance of 0.68 mm. What is the kinetic energy of the electrons in the beam?

69 •• When a surface is illuminated by light of wavelength λ, the maximum kinetic energy of the emitted electrons is 1.20 eV. If the wavelength $\lambda' = 0.800\lambda$ is used, the maximum kinetic energy increases to 1.76 eV. For wavelength $\lambda' = 0.600\lambda$, the maximum kinetic energy of the emitted electrons is 2.676 eV. Determine the work function of the surface and the wavelength λ.

70 •• A simple pendulum has a length equal to 1.0 m and has a bob that has a mass equal to 0.30 kg. The energy of this oscillator is quantized, and the allowed values of the energy are given by $E_n = \left(n + \frac{1}{2}\right)hf_0$, where n is an integer and f_0 is the frequency of the pendulum. (*a*) Find n if the angular amplitude is 1.0°. (*b*) Find n such that E_{n+1} exceeds E_n by 0.010 percent.

71 •• (*a*) Show that for large n, the fractional difference in energy between state n and state $n + 1$ for a particle in a one-dimensional box is given approximately by

$$(E_{n+1} - E_n)/E_n \approx 2/n$$

(*b*) What is the approximate percentage energy difference between the states $n_1 = 1000$ and $n_2 = 1001$? (*c*) Comment on how this result is related to Bohr's correspondence principle. SSM

72 •• A mode-locked, titanium–sapphire laser has a wavelength of 850 nm and produces 100 million pulses of light each second. Each pulse has a duration of 125 femtoseconds (1 fs = 10^{-15} s) and consists of 5×10^9 photons. What is the average power produced by the laser?

73 •• This problem estimates the time lag in the photoelectric effect that is expected classically but not observed. Let the intensity of the incident radiation falling on an atom be 0.010 W/m^2. (*a*) If the area presented by an atom is 0.010 nm^2, find the energy per second falling on an atom. (*b*) If the work function is 2.0 eV, how long would it take for this much energy to fall on the atom if the radiation energy was distributed uniformly rather than in compact packets (photons)?

CHAPTER 35

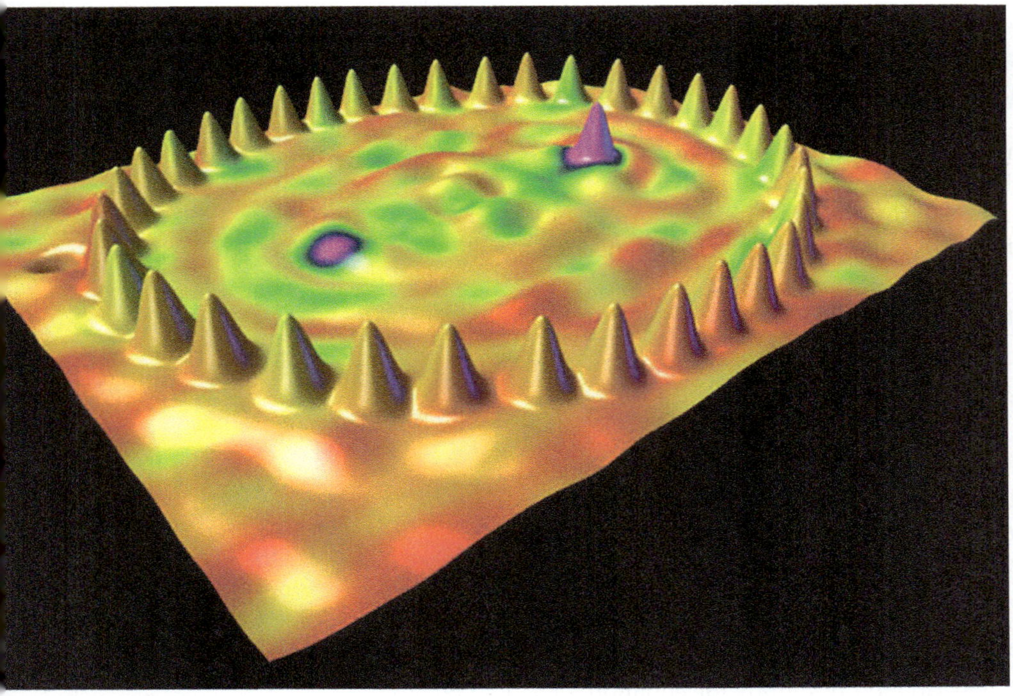

A QUANTUM MIRAGE. THE SCANNING TUNNELING MICROSCOPE (STM) ALLOWS ONE TO PUSH INDIVIDUAL ATOMS AROUND ON A SURFACE AND TO IMAGE THEM. ESPECIALLY INTRIGUING ARE IMAGES OF QUANTUM CORRALS, WHICH ARE CIRCULAR OR ELLIPTICAL ARRANGEMENTS ON A SURFACE INSIDE OF WHICH THE WAVES CORRESPONDING TO ELECTRONS NEAR THE SUBSTRATE SURFACE CAN BE REVEALED. THIS IMAGE COMES FROM IBM, WHERE PHYSICISTS PLACED THIRTY-SIX COBALT ATOMS IN AN ELLIPTICAL "STONEHENGE" PATTERN ON A COPPER SURFACE. AN EXTRA MAGNETIC COBALT ATOM WAS PLACED AT ONE OF THE TWO FOCI OF THE ELLIPSE, CAUSING VISIBLE INTERACTIONS WITH THE SURFACE ELECTRON WAVES. BUT THE WAVES ALSO SEEM TO BE INTERACTING WITH A PHANTOM COBALT ATOM AT THE OTHER FOCUS, AN ATOM THAT IS NOT REALLY THERE. *(Courtesy of IBM and the IBM Almadin Laboratories.)*

Applications of the Schrödinger Equation

In Chapter 34, we found that electrons and other particles have wave properties and are described by wave functions in the form $\Psi(x, t)$. We also mentioned that the wave function is a solution of the Schrödinger equation, and we discussed some solutions qualitatively without reference to the equation itself. In particular, we showed how the standing-wave conditions lead to quantization of energy for a particle confined to a one-dimensional box.

In this chapter we continue our discussion of the material introduced in Chapter 34. We discuss the Schrödinger equation and apply the equation to the particle-in-the-box problem and to several other situations in which a particle is confined to a region of space to illustrate how boundary conditions lead to energy quantization. We then show how the Schrödinger equation leads to barrier penetration and discuss the extension of the Schrödinger equation to more than one dimension and to more than one particle.

? Could the phantom cobalt atom described above be caused by reflections of waves from the corral of cobalt atoms? (See Section 35-4).

35-1 THE SCHRÖDINGER EQUATION

Like the classical wave equation (Equation 15-10*b*), the Schrödinger equation is a partial differential equation in space and time. Like Newton's laws of motion, the Schrödinger equation cannot be derived. Its validity, like that of Newton's laws, lies in its agreement with experiment. In one dimension, the Schrödinger equation is*

$$-\frac{\hbar^2}{2m}\frac{\partial^2\Psi(x,t)}{\partial x^2} + U\Psi(x,t) = i\hbar\frac{\partial\Psi(x,t)}{\partial t} \qquad 35\text{-}1$$

TIME-DEPENDENT SCHRÖDINGER EQUATION

where U is the potential energy function and $\Psi(x, t)$ is a wave function. Equation 35-1 is called the **time-dependent Schrödinger equation.** Unlike the classical wave equation, it relates the *second* space derivative of the wave function to the *first* time derivative of the wave function, and it contains the imaginary number $i = \sqrt{-1}$. The wave functions that are solutions of this equation are not necessarily real. $\Psi(x, t)$ is not a measurable function like the classical wave functions for sound or electromagnetic waves. The probability of finding a particle in some region of space dx certainly has a real value, though. We can modify slightly the equation for probability density given in Chapter 34 (Equation 34-17) to determine the probability of finding a particle in some region dx

See **Math Tutorial** *for more information on* **Complex Numbers**

$$P(x,t)\,dx = |\Psi(x,t)|^2\,dx = \Psi^*\Psi\,dx \qquad 35\text{-}2$$

where Ψ^*, the complex conjugate of Ψ, is identical to Ψ, except that $-i$ is substituted for i wherever i appears in the expression for Ψ.†

In classical mechanics, the standing-wave solutions to the wave equation (Equation 16-16) are of great interest and value. Not surprisingly, standing-wave solutions to the Schrödinger wave equation are also of great interest and value. The wave function for the standing-wave motion of a uniform taut string is $A\sin(kx)\cos(\omega t + \delta)$, which is representative of all standing waves. A standing wave function can always be expressed as a function of position multiplied by a function of time, where the function of time is one that varies sinusoidally with time. Standing-wave solutions to the one-dimensional Schrödinger wave equation are thus expressed

$$\Psi(x,t) = \psi(x)e^{-i\omega t} \qquad 35\text{-}3$$

where $e^{-i\omega t} = \cos(\omega t) - i\sin(\omega t)$. The right side of Equation 35-1 is then

$$i\hbar\frac{\partial\Psi(x,t)}{\partial t} = i\hbar(-i\omega)\psi(x)e^{-i\omega t} = \hbar\omega\psi(x)e^{-i\omega t} = E\psi(x)e^{-i\omega t}$$

where $E = \hbar\omega$ is the energy of the particle.

The Schrödinger wave equation has standing-wave solutions only if the potential energy function U depends on position x alone. Substituting $\psi(x)e^{-i\omega t}$ into Equation 35-1 and canceling the common factor $e^{-i\omega t}$, we obtain an equation for $\psi(x)$, called the **time-independent Schrödinger equation:**

$$-\frac{\hbar^2}{2m}\frac{d^2\psi(x)}{dx^2} + U(x)\psi(x) = E\psi(x) \qquad 35\text{-}4$$

TIME-INDEPENDENT SCHRÖDINGER EQUATION

* Although we simply state the Schrödinger equation, Schrödinger himself had a vast knowledge of classical wave theory that led him to this equation.

† Every complex number can be written in the form $z = a + bi$, where a and b are real numbers and $i = \sqrt{-1}$. The complex conjugate of z is $z^* = a - bi$, so $z^*z = (a + bi)(a - bi) = a^2 + b^2 = |z|^2$.

where we have written U as $U(x)$ to emphasize that while U may depend on position, U does not depend on time. The function $U(x)$ represents the interaction between the environment and the particle being observed. Different environments require different expressions for the potential energy function U in the Schrödinger equation.

The calculation of the allowed energy levels in a system involves only the time-independent Schrödinger equation, whereas finding the probabilities of transition between these levels requires the solution of the time-dependent equation. In this book, we will be concerned only with the time-independent Schrödinger equation (Equation 35-4).

The solution of Equation 35-4 depends on the form of the potential energy function $U(x)$. When $U(x)$ is such that the particle is confined to some region of space, only certain discrete energies E_n give solutions ψ_n that can satisfy the normalization condition (Equation 34-18):

$$\int_{-\infty}^{\infty} |\psi_n|^2 dx = 1$$

The complete time-dependent wave functions are then given, from Equation 35-3, by

$$\Psi_n(x, t) = \psi_n(x) e^{-i\omega_n t} = \psi_n(x) e^{-i(E_n/\hbar)t} \qquad 35\text{-}5$$

A PARTICLE IN AN INFINITE SQUARE-WELL POTENTIAL

We will illustrate the use of the time-independent Schrödinger equation by solving it for the problem of a particle in a box. The potential energy for a one-dimensional box from $x = 0$ to $x = L$ is shown in Figure 35-1. It is called an **infinite square-well potential** and is described mathematically by

$$U(x) = \begin{cases} \infty & x < 0 \\ 0 & 0 < x < L \\ \infty & x > L \end{cases} \qquad 35\text{-}6$$

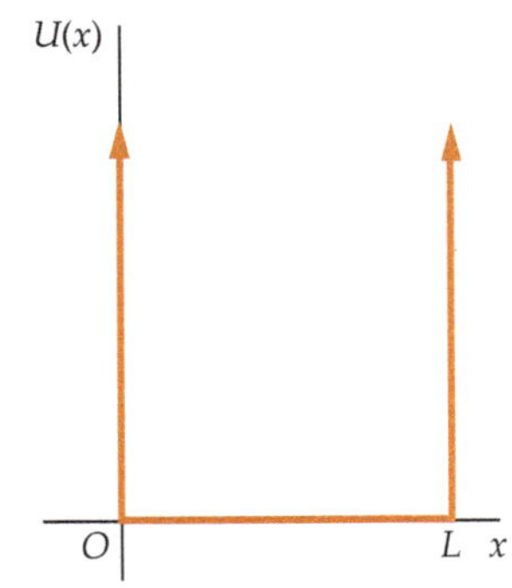

FIGURE 35-1 The infinite square-well potential energy function. For both $x < 0$ and $x > L$, the potential energy $U(x)$ is infinite. The particle is confined to the region in the well $0 < x < L$.

Inside the box, the potential energy is zero, whereas outside the box it is infinite. Because we require the particle to be in the box, we have $\psi(x) = 0$ everywhere outside the box. We then need to solve the Schrödinger equation inside the box for wave functions $\psi(x)$ that must be zero at $x = 0$ and at $x = L$.

Inside the box $U(x) = 0$, so the Schrödinger equation is written

$$-\frac{\hbar^2}{2m}\frac{d^2\psi(x)}{dx^2} = E\psi(x)$$

or

$$\frac{d^2\psi(x)}{dx^2} + k^2\psi(x) = 0 \qquad 35\text{-}7$$

where

$$k^2 = \frac{2mE}{\hbar^2} \qquad 35\text{-}8$$

The general solution of Equation 35-7 can be written as

$$\psi(x) = A \sin kx + B \cos kx \qquad 35\text{-}9$$

where A and B are constants. At $x = 0$, we have

$$\psi(0) = A \sin(0) + B\cos(0) = 0 + B$$

Summary

1. All phenomena propagate like waves and interact like particles.
2. The quantum of light is called a photon and has energy $E = hf$, where h is Planck's constant.
3. The relation between wavelength and momentum of electrons, photons, and other particles is given by the de Broglie relation $\lambda = h/p$.
4. Energy quantization in bound systems arises from standing-wave conditions, which are equivalent to boundary conditions on the wave function.
5. The uncertainty principle is a fundamental law of nature that places theoretical restrictions on the precision of a simultaneous measurement of the position and momentum of a particle. It follows from the general properties of waves.

TOPIC	RELEVANT EQUATIONS AND REMARKS	
1. Constants and Values		
Planck's constant	$h = 6.626 \times 10^{-34}\ \text{J}\cdot\text{s} = 4.136 \times 10^{-15}\ \text{eV}\cdot\text{s}$	34-2
hc	$hc = 1240\ \text{eV}\cdot\text{nm}$	34-5
2. The Particle Nature of Light: Photons	Energy is quantized.	
Photon energy and momentum	$E = hf$ and $E = pc$	34-1 and 34-6
3. Frequency–Wavelength (Energy–Momentum) Relations		
Photons and material particles (de Broglie relations)	$E = hf$ and $p = \dfrac{h}{\lambda}$	34-14 and 34-13
Nonrelativistic particles	$K = \dfrac{p^2}{2m}$ so $\lambda = \dfrac{hc}{\sqrt{2mc^2K}}$	34-15
Photoelectric effect	$K_{max} = \left(\frac{1}{2}mv^2\right)_{max} = hf - \phi$ where ϕ is the work function of the cathode.	34-3
Compton scattering	$\lambda_s - \lambda_i = \dfrac{h}{m_e c}(1 - \cos\theta) = \lambda_C(1 - \cos\theta) = 2.426\ \text{pm}\,(1 - \cos\theta)$	34-11
4. Quantum Mechanics	The state of a particle, such as an electron, is described by its wave function ψ, which is the solution of the Schrödinger wave equation.	
Probability density	The probability of finding the particle in some region of space dx is given by $P(x) = \psi^2(x)\,dx$	34-17
Normalization condition	$\int_{-\infty}^{\infty} \psi^2\,dx = 1$	34-18
Quantum number	The wave function for a particular energy state is characterized by a quantum number n. In three dimensions there are three quantum numbers—one associated with a boundary condition in each dimension.	
Expectation value	The expectation value of x is the same as the average value of x that we would expect to obtain from a measurement of the positions of a large number of particles with the same wave function $\psi(x)$. $\langle x \rangle = \int_{-\infty}^{+\infty} x\psi^2(x)\,dx$ $\langle F(x) \rangle = \int_{-\infty}^{+\infty} F(x)\psi^2(x)\,dx$	34-26 34-27

The boundary condition $\psi(x) = 0$ at $x = 0$ thus gives $B = 0$, and Equation 35-9 becomes

$$\psi(x) = A \sin kx \qquad 35\text{-}10$$

The wave function is thus a sine wave where the wavelength λ is related to the wave number k in the usual way, $\lambda = 2\pi/k$. The boundary condition $\psi(x) = 0$ at $x = L$ restricts the possible values of k and therefore the values of the wavelength λ, and (from Equation 35-8) the energy $E = \frac{1}{2}\hbar^2k^2/m$. We have

$$\psi(L) = A \sin kL = 0 \qquad 35\text{-}11$$

This condition is satisfied if kL is π or any integer multiplied by π, that is, if k is restricted to the values k_n given by

$$k_n = n\frac{\pi}{L} \qquad n = 1, 2, 3, \ldots \qquad 35\text{-}12$$

The condition (Equation 35-11) is also satisfied for $n = 0$. The function $\psi(x) = A \sin 0 = 0$ for all values of x, in the interval $0 < x < L$, is also a solution to the wave equation. However, if the wave function has a value of zero everywhere inside the box, then the box is empty. Furthermore, the wave function cannot be normalized and cannot be a wave function for a particle. Substituting $n\pi/L$ for k_n into Equation 35-8 and solving for E gives us the allowed energy values:

$$E_n = \frac{\hbar^2 k_n^2}{2m} = \frac{\hbar^2}{2m}\left(n\frac{\pi}{L}\right)^2 = n^2\left(\frac{h^2}{8mL^2}\right) = n^2E_1 \qquad 35\text{-}13$$

where

$$E_1 = \frac{h^2}{8mL^2} \qquad 35\text{-}14$$

Equation 35-14 is the same as Equation 34-22, which we obtained by fitting an integral number of half-wavelengths into the box.

For each value of n, there is wave function $\psi_n(x)$ given by

$$\psi_n(x) = \begin{cases} 0 & x < 0 \\ A_n \sin\dfrac{n\pi x}{L} & 0 < x < L \\ 0 & x > L \end{cases} \qquad 35\text{-}15$$

which is the same as Equation 34-25, where the constant $A_n = \sqrt{2/L}$ is determined by normalization.*

CONCEPT CHECK 35-1

The function below is not an acceptable wave function. What makes the function unacceptable?

$$\psi_n(x) = A_n \sin\frac{n\pi x}{L} \qquad -\infty < x < \infty$$

35-2 A PARTICLE IN A FINITE SQUARE WELL

The quantization of energy that we found for a particle in an infinite square well is a result that follows from the general solution of the Schrödinger equation for any particle confined to some region of space. We will illustrate this by considering the qualitative behavior of the wave function for a slightly more general potential energy function, the finite square well, which is shown in Figure 35-2. This potential energy function is described mathematically by

$$U(x) = \begin{cases} U_0 & x < 0 \\ 0 & 0 < x < L \\ U_0 & x > L \end{cases} \qquad 35\text{-}16$$

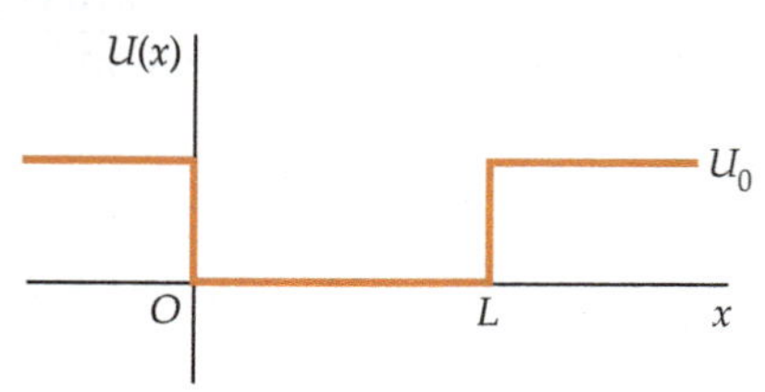

FIGURE 35-2 The finite square-well potential energy function.

* See Equation 34-18.

This potential energy function is discontinuous at $x = 0$ and $x = L$, but it is finite everywhere. The solutions of the Schrödinger equation for this type of potential energy function depend on whether the total energy E is greater or less than U_0. We will not discuss the case of $E > U_0$, except to remark that in that case the particle is not confined and any value of the energy is allowed. That is, there is no energy quantization when $E > U_0$. Here, we assume that $0 \le E < U_0$.

Inside the well, $U(x) = 0$, and the time-independent Schrödinger equation is the same as for the infinite well (Equation 35-7):

$$-\frac{\hbar^2}{2m}\frac{d^2\psi(x)}{dx^2} = E\psi(x) \qquad 0 \le x \le L$$

or

$$\frac{d^2\psi(x)}{dx^2} + k^2\psi(x) = 0$$

where $k^2 = 2mE/\hbar^2$. The general solution is of the form

$$\psi(x) = A\sin kx + B\cos kx$$

In this case, $\psi(x)$ is not required to be zero at $x = 0$ (the particle is not required to be inside the box), so B is not zero. Outside the well, the time-independent Schrödinger equation is

$$-\frac{\hbar^2}{2m}\frac{d^2\psi(x)}{dx^2} + U_0\psi(x) = E\psi(x) \qquad x < 0 \text{ and } x > L$$

or

$$\frac{d^2\psi(x)}{dx^2} - \alpha^2\psi(x) = 0 \qquad \text{35-17}$$

where

$$\alpha^2 = \frac{2m}{\hbar^2}(U_0 - E) \qquad U_0 > E \qquad \text{35-18}$$

The wave functions and allowed energies for the particle can be found by solving Equation 35-17 for $\psi(x)$ outside the well and then requiring that both $\psi(x)$ and $d\psi(x)/dx$ be continuous at the boundaries $x = 0$ and $x = L$. The solution of Equation 35-17 is not difficult [in the region $x > L$, it is of the form $\psi(x) = Ce^{-\alpha x}$ and in the region $x < 0$ it is of the form $\psi(x) = Ce^{+\alpha x}$], but applying the boundary conditions involves much tedious algebra and is not important for our purpose. The important feature of Equation 35-17 is that $d^2\psi/dx^2$ has the same sign as ψ. Thus, if ψ is positive, $d^2\psi/dx^2$ is also positive and the wave function curves away from the axis as x approaches either $+\infty$ or $-\infty$, as shown in Figure 35-3*a*. Similarly, if ψ is negative, $d^2\psi/dx^2$ is negative and ψ again curves away from the axis as x approaches either $+\infty$ or $-\infty$, as shown in Figure 35-3*b*. This behavior is very different from the behavior inside the well, where ψ and $d^2\psi/dx^2$ have opposite signs so that ψ always curves toward the axis like a sine or cosine function. Because of this behavior outside the well, for most values of the energy E in Equation 35-17, $\psi(x)$ becomes infinite as x approaches $\pm\infty$; that is, most wave functions $\psi(x)$ are not well behaved outside the well. Though they satisfy the Schrödinger equation, such functions are not proper wave functions because they cannot be normalized. The solutions of the Schrödinger equation are well behaved (that is, they approach 0 as $|x|$ becomes very large) only for certain values of the energy. These energy values are the allowed energies for the finite square well.

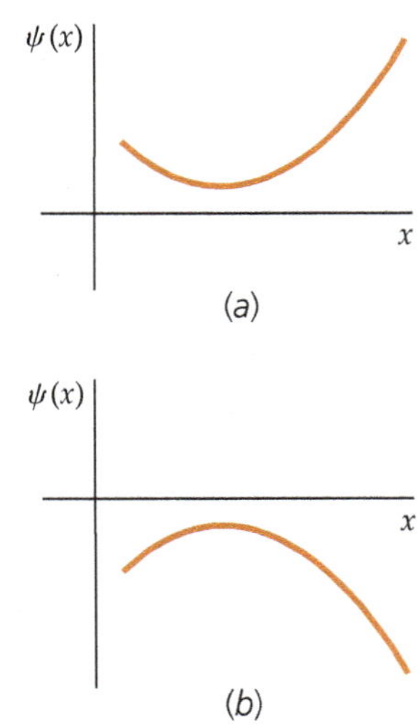

FIGURE 35-3 (*a*) A function ψ that has both a positive value and a positive concavity throughout the region shown. (Concavity is the sign of $d\psi/dx$.) (*b*) A function ψ that has both a negative value and a negative concavity throughout the region shown.

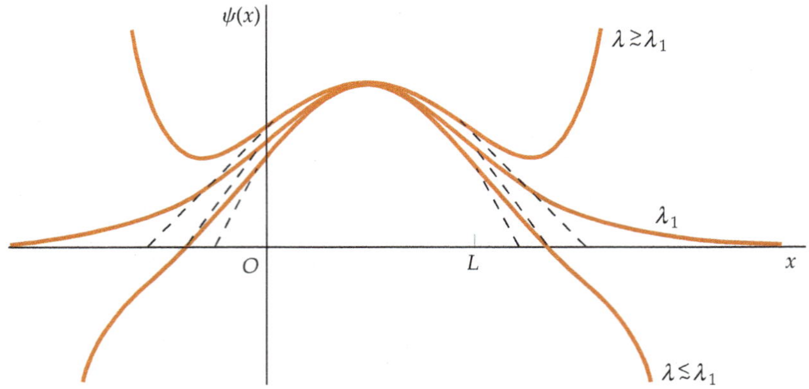

FIGURE 35-4 Functions satisfying the Schrödinger equation that have a wavelength λ that is almost equal to the wavelength λ_1, which is the wavelength that corresponds to the ground-state energy $E_1 = \hbar^2/2m\lambda_1^2$ in the finite well. If λ is slightly greater than λ_1, the function approaches plus infinity as $|x|$ approaches infinity, like the function in Figure 35-3*a*. At the critical wavelength λ_1, the function and its slope approach zero together as $|x|$ approaches infinity. If λ is slightly less than λ_1, the function crosses the x axis while the slope is still negative. The slope then becomes more negative because its rate of change $d^2\psi/dx^2$ is now negative. This function approaches negative infinity as $|x|$ approaches infinity.

Figure 35-4 shows a well-behaved wave function, a wave function that has a wavelength λ_1 inside the well that corresponds to the ground-state energy. The behavior of the wave functions corresponding to nearby wavelengths and energies is also shown. Figure 35-5 shows the wave functions and probability distributions for the ground state and first two excited states. From this figure, we can see that the wavelengths inside the well are slightly longer than the corresponding wavelengths for the infinite well (Figure 34-14), so the corresponding energies are slightly less than those for the infinite well. Another feature of the finite-well problem is that there are only a finite number of allowed energies. For very small values of U_0, there is only one allowed energy.

Note that the wave function penetrates beyond the edges of the well at $x = L$ and $x = 0$, indicating that there is some small probability of finding the particle in the region in which its total energy E is less than its potential energy U_0. This region is called the *classically forbidden region* because the kinetic energy, $E - U_0$, would be negative when $U_0 > E$. Because negative kinetic energy has no meaning in classical physics, it is interesting to speculate on the result of an attempt to observe the particle in the classically forbidden region. It can be shown from the uncertainty principle that if an attempt is made to localize the particle in the classically forbidden region, such a measurement introduces an uncertainty in the momentum of the particle corresponding to a minimum kinetic energy that is greater than $U_0 - E$. This is just great enough to prevent us from measuring a negative kinetic energy. The penetration of the wave function into a classically forbidden region does have important consequences in barrier penetration, which will be discussed in Section 35-4.

Much of our discussion of the finite-well problem applies to any problem in which $E > U(x)$ in some region and $E < U(x)$ outside that region, as we see in the next section.

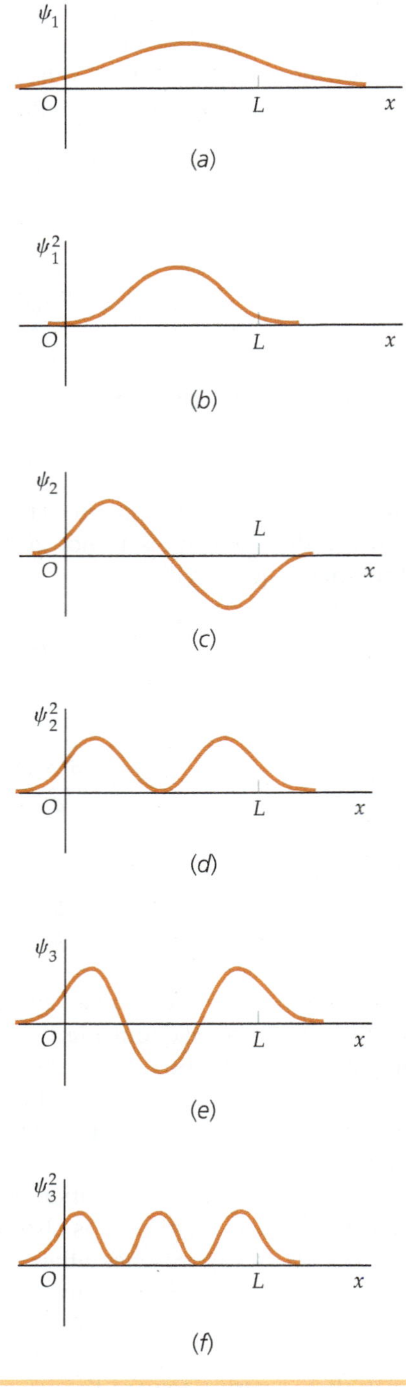

FIGURE 35-5 Graphs of the wave functions $\psi_n(x)$ and probability distributions $\psi^2(x)$ for $n = 1$, $n = 2$, and $n = 3$ for the finite square well. Compare these graphs with those of Figure 34-14 for the infinite square well, where the wave functions are zero at $x = 0$ and $x = L$. The wavelengths here are slightly longer than the corresponding wavelengths for the infinite well, so the allowed energies are somewhat smaller.

35-3 THE HARMONIC OSCILLATOR

The potential energy for a particle that has mass m and is attached to a spring that has force constant k is

$$U(x) = \tfrac{1}{2}kx^2 = \tfrac{1}{2}m\omega_0^2 x^2 \qquad 35\text{-}19$$

where $\omega_0 = \sqrt{k/m}$ is the natural frequency of the oscillator. Classically, the object oscillates between $x = +A$ and $x = -A$. The object's total energy is $E = \frac{1}{2}m\omega_0^2A^2$, which can have any positive value or zero.

This potential energy function, shown in Figure 35-6, applies to virtually any system undergoing small oscillations about a position of stable equilibrium. For example, it could apply to the oscillations of the atoms of a diatomic molecule (such as H_2 or HCl) in which the atoms are oscillating about their equilibrium positions. Between the classical turning points $(-A < x < A)$, the total energy is greater than the potential energy, and the Schrödinger equation can be written

$$\frac{d^2\psi(x)}{dx^2} = -k^2\psi(x) \qquad -A < x < A \qquad \text{35-20}$$

where $k^2 = (2m/\hbar^2)[E - U(x)]$ now depends on x. The solutions of this equation are no longer simple sine or cosine functions because the wave number $k = 2\pi/\lambda$ now varies with x; but because $d^2\psi/dx^2$ and ψ have opposite signs throughout the region $-A < x < A$, ψ will always curve toward the axis and the solutions will oscillate.

Outside the classical turning points $(|x| > A)$, the potential energy is greater than the total energy and the Schrödinger equation is similar to Equation 35-17:

$$\frac{d^2\psi(x)}{dx^2} - \alpha^2\psi(x) = 0 \qquad |x| > A \qquad \text{35-21}$$

except that here $\alpha^2 = (2m/\hbar^2)[U(x) - E] > 0$, where $U(x) > E$, depends on x. For $|x| > A$, $d^2\psi/dx^2$ and ψ have the same sign, so ψ will curve away from the axis and there will be only certain values of E for which solutions exist that approach zero as x approaches infinity.

For the harmonic oscillator potential energy function, the Schrödinger equation is

$$-\frac{\hbar^2}{2m}\frac{d^2\psi(x)}{dx^2} + \tfrac{1}{2}m\omega_0^2x^2\psi(x) = E\psi(x) \qquad \text{35-22}$$

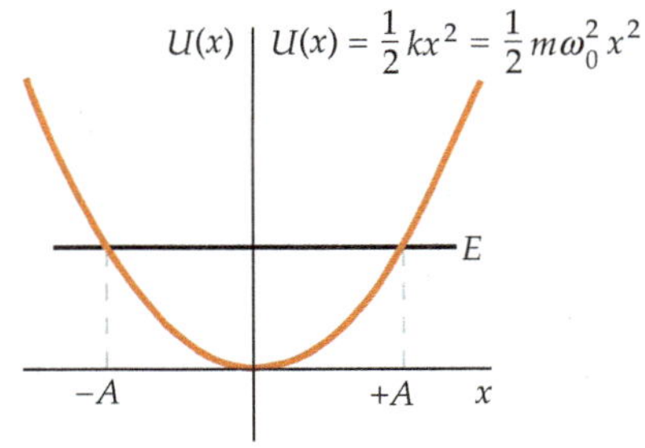

FIGURE 35-6 Harmonic oscillator potential.

WAVE FUNCTIONS AND ENERGY LEVELS

Rather than pursue a general solution to the Schrödinger equation for this system, we simply present the solution for the ground state and the first excited state.

The ground-state wave function $\psi_0(x)$ is found to be a Gaussian function centered at the origin:

$$\psi_0(x) = A_0e^{-ax^2} \qquad \text{35-23}$$

where A_0 and a are positive constants. This wave function and the wave function for the first excited state are shown in Figure 35-7.

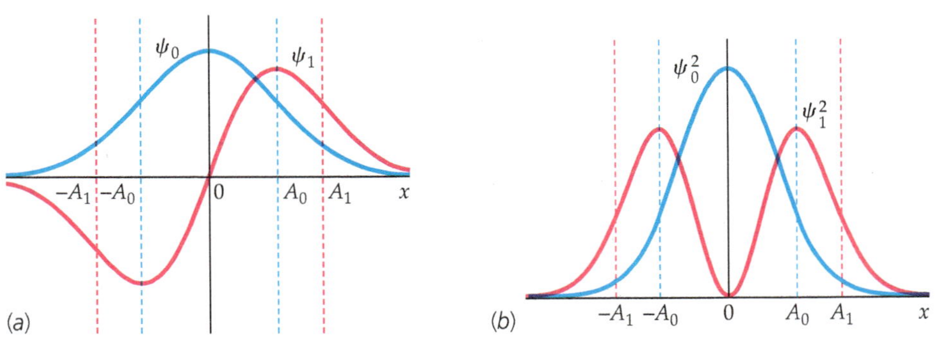

FIGURE 35-7 (*a*) The ground-state ψ_0 and first excited state ψ_1 wave functions for the harmonic oscillator potential. Classically, the motion of a harmonic oscillator with the ground-state energy E_0 would be restricted to the region $-A_0 \le x \le +A_0$ and the motion of a harmonic oscillator with the first excited state energy E_1 would be restricted to the region $-A_1 \le x \le +A_1$. (*b*) The ground-state ψ_0^2 and first excited state ψ_1^2 probability density functions for the harmonic oscillator potential.

Example 35-1 Verifying the Ground-State Wave Function

Verify that $\psi_0(x) = A_0e^{-ax^2}$, where A_0 and a are positive constants, is a solution of the Schrödinger equation for the harmonic oscillator.

PICTURE We calculate the second derivative of ψ_0 with respect to x and substitute into Equation 35-22. Because this expression is the ground-state wave function, we write E_0 for the energy E.

SOLVE

1. Compute $d\psi_0/dx$:

$$\frac{d\psi_0(x)}{dx} = \frac{d}{dx}(A_0e^{-ax^2}) = -2axA_0e^{-ax^2}$$

2. Compute $d^2\psi_0/dx^2$:

$$\frac{d^2\psi_0(x)}{dx^2} = -2aA_0e^{-ax^2} + 4a^2x^2A_0e^{-ax^2}$$

$$= (4a^2x^2 - 2a)A_0e^{-ax^2}$$

3. Substitute into the Schrödinger equation (Equation 35-22):

$$-\frac{\hbar^2}{2m}\frac{d^2\psi(x)}{dx^2} + \frac{1}{2}m\omega_0^2x^2\psi(x) = E\psi(x)$$

$$-\frac{\hbar^2}{2m}(4a^2x^2 - 2a)A_0e^{-ax^2} + \frac{1}{2}m\omega_0^2x^2A_0e^{-ax^2} = E_0A_0e^{-ax^2}$$

4. Cancel the common factor $A_0e^{-ax^2}$ and show the result in standard polynomial form:

$$-\frac{\hbar^2}{2m}(4a^2x^2 - 2a) + \frac{1}{2}m\omega_0^2x^2 = E_0$$

so

$$\left(\frac{1}{2}m\omega_0^2 - \frac{2\hbar^2a^2}{m}\right)x^2 + \left(\frac{\hbar^2a}{m} - E_0\right) = 0$$

5. The equation in step 4 must hold for all x. Set $x = 0$ and solve for E_0:

$$0 + \left(\frac{\hbar^2a}{m} - E_0\right) = 0$$

so

$$E_0 = \frac{\hbar^2a}{m}$$

6. Substitute this result for E_0 into the equation in step 4 and simplify:

$$\left(-\frac{2\hbar^2a^2}{m} + \frac{1}{2}m\omega_0^2\right)x^2 + 0 = 0$$

7. It follows that the coefficient of x^2 must equal zero:

$$-\frac{2\hbar^2a^2}{m} + \frac{1}{2}m\omega_0^2 = 0$$

8. Solve for a:

$$a = \frac{m\omega_0}{2\hbar}$$

9. Substitute this result into the equation for E_0 in step 5:

$$E_0 = \frac{\hbar^2a}{m} = \frac{1}{2}\hbar\omega_0$$

> We have shown that the given function, $\psi_0(x) = A_0e^{-ax^2}$, satisfies the Schrödinger equation for any value of A_0, as long as the energy is given by $E_0 = \frac{1}{2}\hbar\omega_0$.

CHECK Planck's constant has units of joules multiplied by seconds, and angular frequency has units of reciprocal seconds, so the step-9 expression $\frac{1}{2}\hbar\omega_0$ has the dimensions of energy, as expected.

TAKING IT FURTHER The step-4 equation is a polynomial that is equal to zero. A theorem that would have simplified the solution is "If a polynomial is equal to zero over a continuous range of values of x, then each of the polynomial coefficients is equal to zero. For example, if $Ax^3 + Bx^2 + Cx + D = 0$ on the interval $1 < x < 2$, then $A = B = C = D = 0$."

We see from this example that the ground-state energy is given by

$$E_0 = \frac{\hbar^2 a}{m} = \frac{1}{2}\hbar\omega_0 \qquad 35\text{-}24$$

The first excited state has a node in the center of the potential well, just as with the particle in a box.* The wave function $\psi_1(x)$ is

$$\psi_1(x) = A_1 x e^{-ax^2} \qquad 35\text{-}25$$

where $a = \frac{1}{2}m\omega_0/\hbar$, as in Example 35-1. This function is also shown in Figure 35-7. Substituting $\psi_1(x)$ into the Schrödinger equation, as was done for $\psi_0(x)$ in Example 35-1, yields the energy of the first excited state,

$$E_1 = \tfrac{3}{2}\hbar\omega_0$$

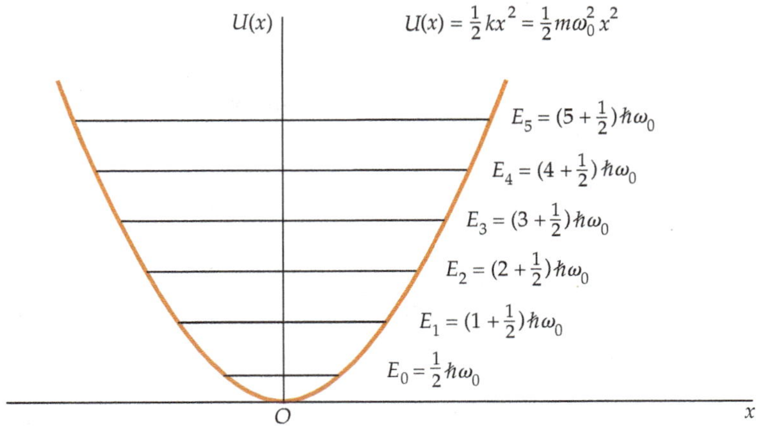

FIGURE 35-8 Energy levels in the harmonic oscillator potential.

In general, the energy of the nth excited state of the harmonic oscillator is

$$E_n = \left(n + \tfrac{1}{2}\right)\hbar\omega_0 \qquad n = 0, 1, 2, \ldots \qquad 35\text{-}26$$

as indicated in Figure 35-8. The fact that the energy levels are evenly spaced by the amount $\hbar\omega_0$ is a peculiarity of the harmonic oscillator potential. As we saw in Chapter 34, the energy levels for a particle in a box, or for the hydrogen atom, are not evenly spaced. The precise spacing of energy levels is closely tied to the particular form of the potential energy function.

35-4 REFLECTION AND TRANSMISSION OF ELECTRON WAVES: BARRIER PENETRATION

In Sections 35-2 and 35-3, we were concerned with bound-state problems in which the potential energy is larger than the total energy for large values of $|x|$. In this section, we consider some simple examples of unbound states for which E is greater than $U(x)$. For these problems, $d^2\psi/dx^2$ and ψ have opposite signs, so $\psi(x)$ curves toward the axis and does not become infinite as x approaches either $+\infty$ or $-\infty$.

STEP POTENTIAL

Consider a particle of energy E moving in a region in which the potential energy is the step function

$$U(x) = \begin{cases} 0 & x < 0 \\ U_0 & x > 0 \end{cases}$$

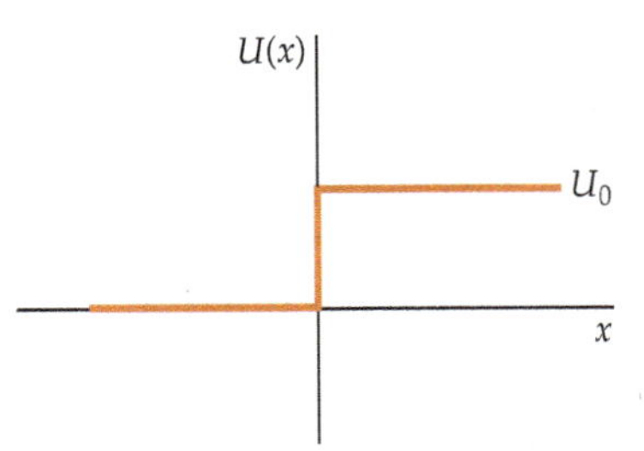

FIGURE 35-9 Step potential. A classical particle that is incident from the left and has total energy $E > U_0$ is always transmitted. The change in potential energy at $x = 0$ merely provides an impulsive force that reduces the speed of the particle. A wave incident from the left is partially transmitted and partially reflected because the wavelength changes abruptly at $x = 0$.

as shown in Figure 35-9. We are interested in what happens when a particle moving from left to right encounters the step.

The classical answer is simple. To the left of the step, the particle moves with a speed $v = \sqrt{2E/m}$. At $x = 0$, an impulsive force acts on the particle. If the initial energy E is less than U_0, the particle will be turned around and will then move to the left at its original speed; that is, the particle will be reflected by the step.

* Each higher-energy state has one additional node in the wave function.

If E is greater than U_0, the particle will continue to move to the right but with reduced speed given by $v = \sqrt{2(E - U_0)/m}$. We can picture this classical problem as a ball rolling along a level surface and coming to a steep hill of height h given by $mgh = U_0$. If the initial kinetic energy of the ball is less than mgh, the ball will roll part way up the hill and then back down and to the left along the lower surface at its original speed. If E is greater than mgh, the ball will roll up the hill and proceed to the right at a lesser speed.

The quantum-mechanical result is similar when E is less than U_0. Figure 35-10 shows the wave function for the case $E < U_0$. The wave function does not go to zero at $x = 0$ but rather decays exponentially, like the wave function for the bound state in a finite square-well problem. The wave penetrates slightly into the classically forbidden region $x > 0$, but it is eventually completely reflected. This problem is somewhat similar to that of total internal reflection in optics.

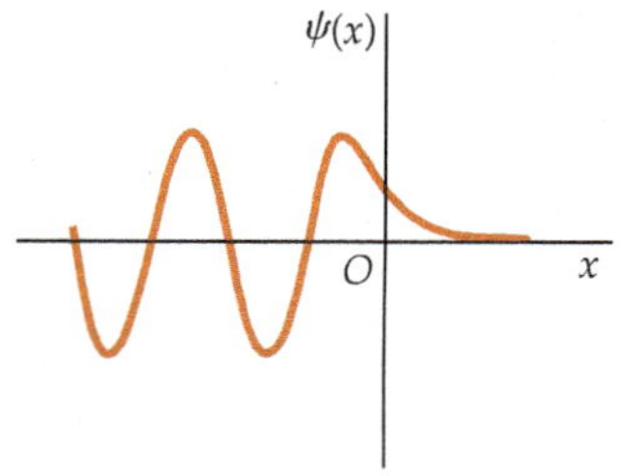

FIGURE 35-10 When the total energy E is less than U_0, the wave function penetrates slightly into the region $x > 0$. However, the probability of reflection for this case is 1, so no energy is transmitted.

For $E > U_0$, the quantum-mechanical result differs markedly from the classical result. At $x = 0$, the wavelength changes abruptly from $\lambda_1 = h/p_1 = h/\sqrt{2mE}$ to $\lambda_2 = h/p_2 = h/\sqrt{2m(E - U_0)}$. We know from our study of waves that when the wavelength changes suddenly, part of the wave is reflected and part of the wave is transmitted. Because the motion of an electron (or other particle) is governed by a wave equation, the electron sometimes will be transmitted and sometimes will be reflected. The probabilities of reflection and transmission can be calculated by solving the Schrödinger equation in each region of space and comparing the amplitudes of the transmitted waves and reflected waves with the amplitudes of the incident wave. These calculations and their results are similar to finding the fraction of light reflected from an air–glass interface. If R is the probability of reflection, called the reflection coefficient, this calculation gives

$$R = \frac{(k_1 - k_2)^2}{(k_1 + k_2)^2} \qquad 35\text{-}27$$

where k_1 is the wave number for the incident wave and k_2 is the wave number for the transmitted wave. This result is the same as the result in optics for the reflection of light at normal incidence from the boundary between two media having different indexes of refraction n (Equation 31-17). The probability of transmission T, called the **transmission coefficient,** can be calculated from the reflection coefficient, because the probability of transmission plus the probability of reflection must equal 1:

$$T + R = 1 \qquad 35\text{-}28$$

Example 35-2 Reflection and Transmission at a Step Barrier

A particle that has kinetic energy E_0 and is traveling in a region in which the potential energy is zero is incident on a potential-energy barrier of height $U_0 = 0.20E_0$. Find the probability that the particle will be reflected.

PICTURE We need to calculate the wave numbers k_1 and k_2 and use them to calculate the reflection coefficient R from Equation 35-27. The wave numbers are related to the momentum by the de Broglie relation $p = h/\lambda$ (Equation 34-13), where $k = 2\pi/\lambda$. Combining these two equations gives $p = \hbar k$. Thus, the kinetic energy K is related to the wave number by $K = \frac{1}{2}p^2/m = \frac{1}{2}\hbar^2k^2/m$.

SOLVE

1. The probability of reflection is the reflection coefficient:

$$R = \frac{(k_2 - k_1)^2}{(k_1 + k_2)^2}$$

2. Calculate k_1 from the initial kinetic energy E_0:

$$E_0 = \frac{\hbar^2 k_1^2}{2m}$$

$$k_1 = \sqrt{2mE_0/\hbar^2}$$

3. Relate the final kinetic energy K_2 to the initial kinetic energy E_0 and the potential energy U_0 in the region $x > 0$:

$$K_2 = E_0 - U_0 = E_0 - 0.2E_0 = 0.8E_0$$

4. Relate k_2 to the final kinetic energy K_2 and solve for k_2:

$$K_2 = \frac{\hbar k_2^2}{2m}$$

so

$$k_2 = \sqrt{2mK_2/\hbar^2} = \sqrt{2m(0.8E_0)/\hbar^2}$$
$$= \sqrt{0.80}\sqrt{2mE_0/\hbar^2}$$

5. Substitute these values into Equation 35-27 to calculate R:

$$R = \frac{(k_1 - k_2)^2}{(k_1 + k_2)^2} = \left(\frac{1 - \sqrt{0.80}}{1 + \sqrt{0.80}}\right)^2 = \boxed{0.0031}$$

CHECK Classically, the particle would not be reflected by such a low barrier. The step-5 result gives a probability of 0.31 percent that the particle will be reflected. Such a low probability approaches being in agreement with our classical expectations.

TAKING IT FURTHER The probability of reflection is only 0.31 percent. This probability is small because the barrier height reduces the kinetic energy by only 20 percent. Because k is proportional to the square root of the kinetic energy, the wave number and therefore the wavelength is changed by only 10 percent.

PRACTICE PROBLEM 35-1 Express the index of refraction n of light in terms of the wave number k and the frequency ω, and show that the expression $(n_1 - n_2)^2/(n_1 + n_2)^2$ (Equation 31-7) for the reflection coefficient of light at normal incidence is the same as Equation 35-27. *Hint: Express the index of refraction n of light in terms of the wave number k and the angular frequency* ω.

In quantum mechanics, a localized particle is represented by a wave packet, which has a maximum at the most probable position of the particle. Figure 35-11 shows a wave packet representing a particle of energy E incident on a step potential of height U_0, which is less than E. After the encounter, there are two wave packets. The relative heights of the transmitted packet and reflected packet indicate the relative probabilities of transmission and reflection. For the situation shown here, E is much greater than U_0, and the probability of transmission is much greater than that of reflection.

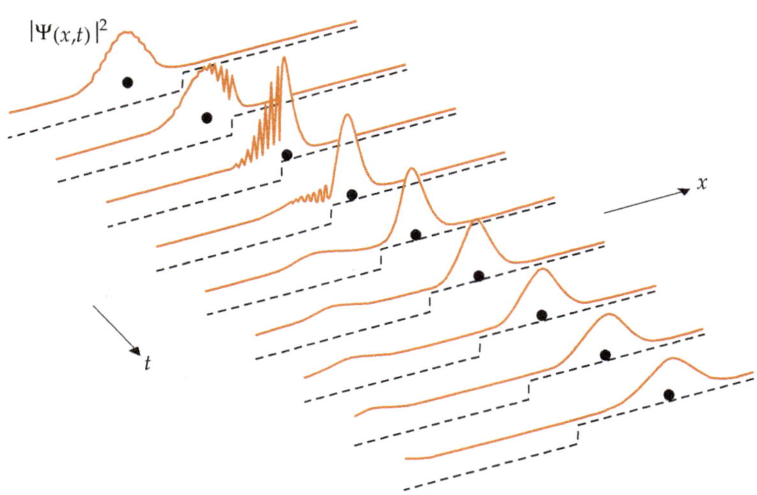

FIGURE 35-11 Time development of a one-dimensional wave packet representing a particle incident on a step potential for $E > U_0$. The position of a classical particle is indicated by the dot. Note that part of the packet is transmitted and part is reflected.

BARRIER PENETRATION

Figure 35-12*a* shows a rectangular potential-energy barrier of height U_0 and width a given by

$$U(x) = \begin{cases} 0 & x < 0 \\ U_0 & 0 < x < a \\ 0 & x > a \end{cases}$$

We consider a particle of energy E, which is slightly less than U_0, that is incident on the barrier from the left. Classically, the particle would always be reflected. However, a wave incident from the left does not decrease immediately to zero at the barrier, but it will instead decay exponentially in the classically forbidden region $0 < x < a$. On reaching the far wall of the barrier ($x = a$), the wave function must join smoothly to a sinusoidal wave function to the right of the barrier, as shown in Figure 35-12*b* This implies that there is some probability of the particle (which is represented by the wave function) being found on the far side of the barrier even though, classically, it should never pass through the barrier. For the case in which the quantity αa [where $\alpha^2 = 2m(U_0 - E)/\hbar^2$] is much greater than 1 the transmission coefficient T is equal to $e^{-2\alpha a}$:

> **!** Do not think it might be possible to detect a particle in the classically forbidden region. It cannot. A proof shows this claim to be a consequence of the uncertainty principle.

$$T = e^{-2\alpha a} \qquad \alpha a \gg 1 \tag{35-29}$$

TRANSMISSION THROUGH A BARRIER

The probability of penetration of the barrier thus decreases exponentially with both the barrier thickness a and the square root of the relative barrier height $(U_0 - E)$. This phenomenon, the penetrating of a classically forbidden region, is called **quantum tunneling.**

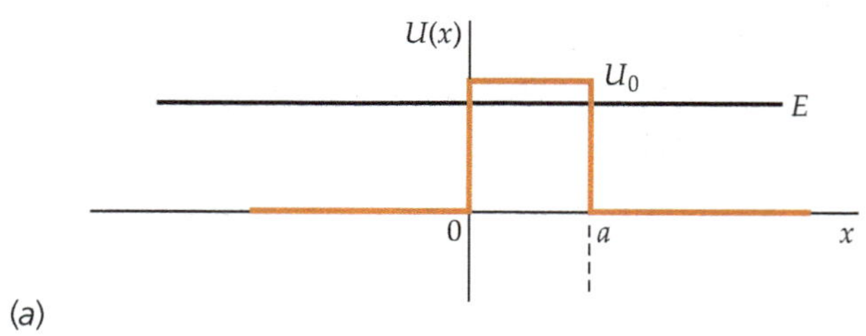

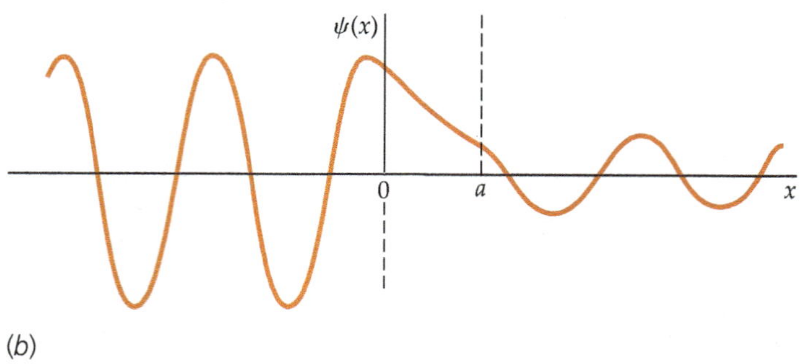

FIGURE 35-12 (*a*) A rectangular potential-energy barrier. (*b*) The penetration of the barrier by a wave that has a total energy less than the barrier energy. Part of the wave is transmitted by the barrier even though, classically, the particle cannot enter the region $0 < x < a$ in which the potential energy is greater than the total energy. To the left of the barrier, there is both an incident wave and a reflected wave. These waves form a resultant wave so that ψ is a superposition of a standing wave and a traveling wave (traveling toward the barrier). Only the transmitted wave exists in the region $x > a$, and it is traveling away from the barrier.

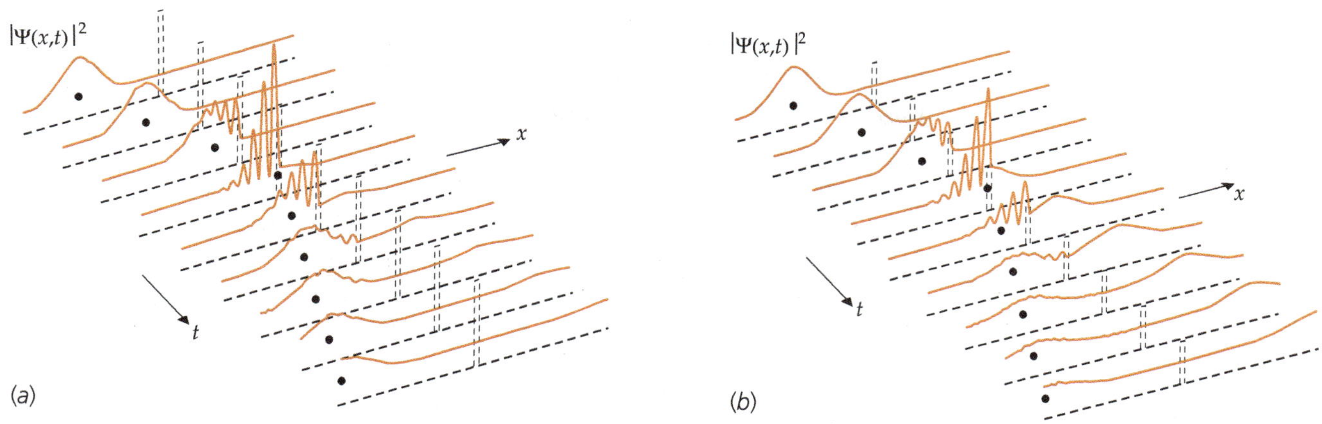

FIGURE 35-13 Barrier penetration. (*a*) The same particle incident on a barrier of height much greater than the energy of the particle. A very small part of the packet tunnels through the barrier. In both drawings, the position of a classical particle is indicated by a dot. (*b*) A wave packet representing a particle incident on a barrier of height just slightly greater than the energy of the particle. For this particular choice of energies, the probability of transmission is approximately equal to the probability of reflection, as indicated by the relative sizes of the transmitted and reflected packets.

Figure 35-13*a* shows a wave packet incident on a potential-energy barrier of height U_0 that is considerably greater than the energy of the particle. The probability of penetration is very small, as indicated by the relative sizes of the reflected and transmitted packets. In Figure 35-13*b*, the barrier is just slightly greater than the energy of the particle. In this case, the probability of penetration is about the same as the probability of reflection. Figure 35-14 shows a particle incident on two potential-energy barriers of height just slightly greater than the energy of the particle.

As we have mentioned, the penetration of a barrier is not unique to quantum mechanics. When light is totally reflected from a glass–air interface, the light wave can penetrate the air barrier if a second piece of glass is brought within a few

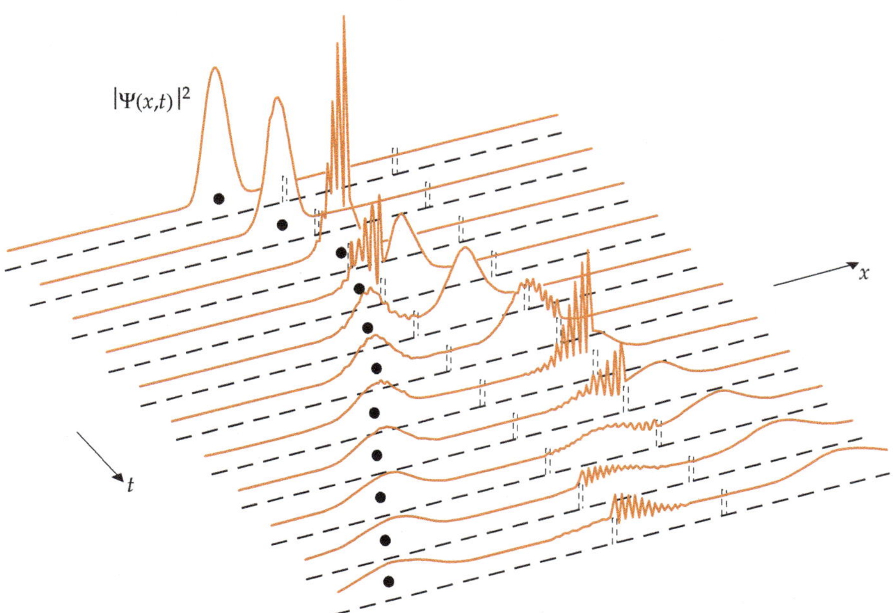

FIGURE 35-14 A wave packet representing a particle incident on two barriers. At each encounter, part of the packet is transmitted and part reflected, resulting in part of the packet being trapped between the barriers for some time.

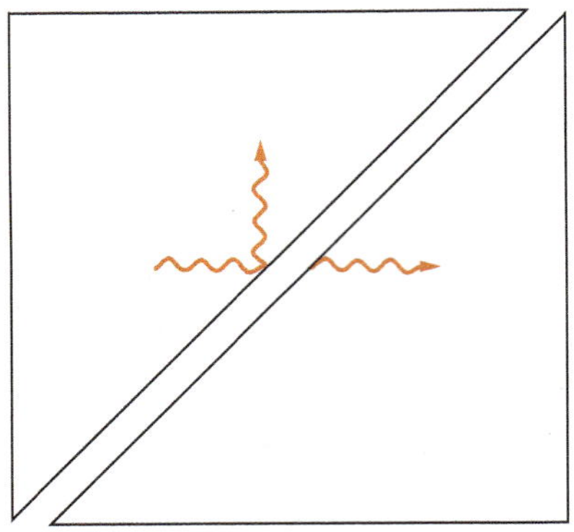

FIGURE 35-15 The penetration of an optical barrier. If the second prism is close enough to the first, part of the wave penetrates the air barrier even when the angle of incidence in the first prism is greater than the critical angle.

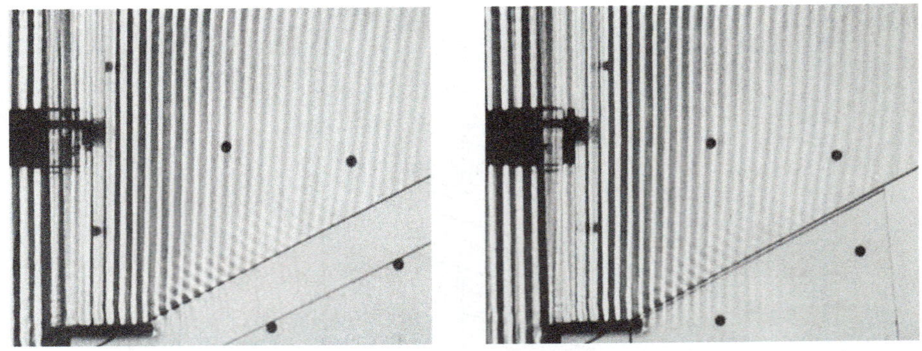

(a) (b)

FIGURE 35-16 The penetration of a barrier by water waves in a ripple tank. In Figure 35-16*a*, the waves are totally reflected from a gap of deeper water. When the gap is very narrow, as in Figure 35-16*b*, a transmitted wave appears. The dark circles are spacers that are used to support the prisms from below. *(Education Development Center.)*

wavelengths of the first. This effect can be demonstrated with a laser beam and two 45° prisms (Figure 35-15). Similarly, water waves in a ripple tank can penetrate a gap of deep water (Figure 35-16).

The theory of barrier penetration was used by George Gamow in 1928 to explain the enormous variation in the half-lives for α decay of radioactive nuclei. (Alpha particles are emitted from atoms during radioactive decay and consist of two protons and two neutrons tightly bound together.) In general, the smaller the energy of the emitted α particle, the longer the half-life of the particle is. The energies of α particles from natural radioactive sources range from approximately 4 MeV to 7 MeV, whereas the half-lives range from approximately 10^{-5} seconds to 10^{10} years. Gamow represented a radioactive nucleus by a potential well of finite depth containing an α particle, as shown in Figure 35-17. Without knowing very much about the nuclear force that is exerted by the nucleus on the particle, Gamow represented it by a square well. Just outside the well, the α particle that has a charge of $+2e$ is repelled by the nucleus that has a charge $+Ze$, where Ze is the remaining nuclear charge. This force is represented by the Coulomb potential energy $+k(2e)(Ze)/r$. The energy E is the measured kinetic energy of the emitted α particle, because when it is far from the nucleus its potential energy is zero. After the α particle is formed from the radioactive nucleus, it bounces back and forth inside the nucleus, hitting the barrier at the nuclear radius R. Each time the α particle strikes the barrier, some small probability exists of the particle penetrating the barrier and appearing outside the nucleus. We can see from Figure 35-17 that a small increase in E reduces the relative height of the barrier $U - E$ and also the barrier's thickness. Because the probability of penetration is so sensitive to the barrier thickness and relative height, a small increase in E leads to a large increase in the probability of transmission and therefore to a shorter lifetime. Gamow was able to derive an expression for the half-life as a function of E that is in excellent agreement with experimental results.

In the **scanning tunneling microscope (STM),** developed in the 1980s, a thin space between the surface of a sample and the tip of a tiny needle-like probe (Figure 35-18) acts as a potential-energy barrier to electrons bound in the sample. (The height of the barrier is the work function of the surface.) A small voltage applied between the probe and the sample causes the electrons to *tunnel* through the vacuum separating the tip of the probe and the surface of the sample if the surfaces are close enough together. The tunneling current is extremely sensitive to the size of the gap between the probe and sample. A constant tunneling current is maintained as the probe scans (travels along) the surface by a feedback

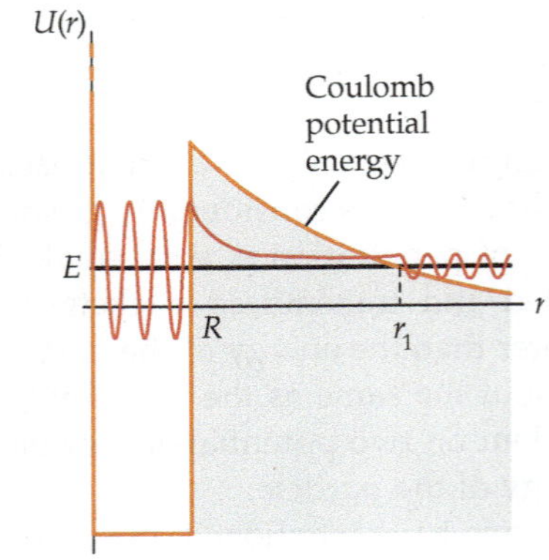

FIGURE 35-17 Model of a potential energy function for an α particle of a radioactive nucleus. The strong attractive nuclear force when r is approximately equal to the nuclear radius R can be approximately described by the potential well shown. The nuclear force is negligible outside the nucleus, and the potential there is given by Coulomb's law, $U(r) = +k(2e)(Ze)/r$, where Ze is the nuclear charge and $2e$ is the charge of the α particle. The wave function of the alpha particle, shown in red, is placed on the graph.

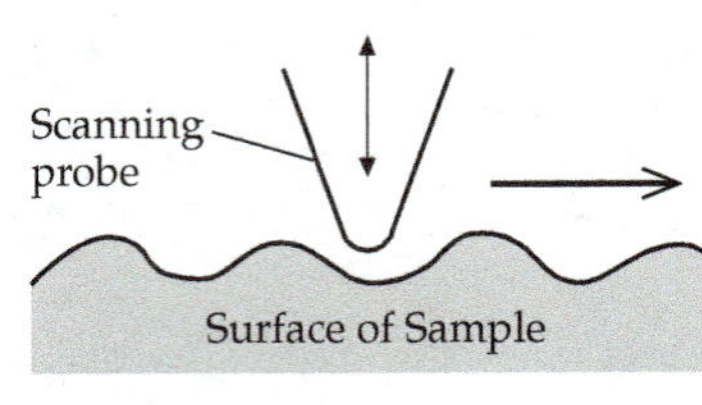

FIGURE 35-18 The tiny probe of a scanning tunneling microscope travels along the surface of a sample. A constant potential difference is maintained between the probe and the sample and electrons tunnel through the potential-energy barrier at the surface. A feedback mechanism maintains a constant tunneling current by moving the probe up and down as it travels along the surface.

mechanism that moves the probe up and down (farther from or closer to the surface). The surface of the sample is mapped out by the tracking of the motions of the probe. In this way, the surface features of the sample can be measured with a resolution of the order of the size of an atom.

35-5 THE SCHRÖDINGER EQUATION IN THREE DIMENSIONS

The one-dimensional time-independent Schrödinger equation is easily extended to three dimensions. In rectangular coordinates, it is

$$-\frac{\hbar^2}{2m}\left(\frac{\partial^2\psi}{\partial x^2}+\frac{\partial^2\psi}{\partial y^2}+\frac{\partial^2\psi}{\partial z^2}\right)+U\psi=E\psi \qquad 35\text{-}30$$

where the wave function ψ and the potential energy U are generally functions of all three coordinates, x, y, and z. To illustrate some of the features of problems in three dimensions, we consider a particle in a three-dimensional infinite square well given by $U(x, y, z) = 0$ for $0 < x < L$, $0 < y < L$, and $0 < z < L$. Outside this cubical region, $U(x, y, z) = \infty$. For this problem, the wave function must be zero at the edges of the well.

There are standard methods in partial differential equations for solving Equation 35-30. We can guess the form of the solution from our knowledge of probability. For a one-dimensional box along the x axis, we have found the probability that a particle is in the region between x and $x + dx$ to be $A_1^2 \sin^2(k_1x)\,dx$ (from Equation 35-10), where A_1 is a normalization constant and $k_1 = n\pi/L$ is the wave number. Similarly, for a box along the y axis, the probability of a particle being in a region between y and $y + dy$ is $A_2^2\sin^2(k_2y)dy$. The probability of two independent events occurring is the product of the probabilities of each event occurring.* So the probability of a particle being in region between x and $x + dx$ *and* in region between y and $y + dy$ is $A_1^2\sin^2(k_1x)\,dx\, A_2^2\sin^2(k_2y)\,dy = A_1^2\sin^2(k_1x)\, A_2^2\sin^2(k_2y)\,dx\,dy$. The probability of a particle being in the region between x and $x + dx$, y and $y + dy$, and z and $z + dz$ is $\psi^2(x, y, z)\,dx\,dy\,dz$, where $\psi(x, y, z)$ is the solution of Equation 35-30. This solution is of the form

$$\psi(x, y, z) = A\sin(k_1x)\sin(k_2y)\sin(k_3z) \qquad 35\text{-}31$$

where the constant A is determined by normalization. Inserting this solution into Equation 35-30, we obtain for the energy

$$E=\frac{\hbar^2}{2m}(k_1^2+k_2^2+k_3^2)$$

which is equivalent to $E = \frac{1}{2}(p_x^2 + p_y^2 + p_z^2)/m$, where $p_x = \hbar k_1$, $p_y = \hbar k_2$, and $p_z = \hbar k_3$. The wave function (Equation 35-31) will be zero at $x = L$ if $k_1 = n_1\pi/L$, where n_1 is an integer. Similarly, the wave function will be zero at $y = L$ if $k_2 = n_2\pi/L$, and the wave function will be zero at $z = L$ if $k_3 = n_3\pi/L$. (It is also zero at $x = 0$, $y = 0$, and $z = 0$.) The energy is thus quantized to the values

$$E_{n_1n_2n_3}=\frac{\hbar^2\pi^2}{2mL^2}(n_1^2+n_2^2+n_3^2)=E_1(n_1^2+n_2^2+n_3^2) \qquad 35\text{-}32$$

where n_1, n_2, and n_3 are positive integers and $E_1 = \hbar^2\pi^2/(2mL^2)$ is the ground-state energy of the one-dimensional well. Note that the energy and wave function are characterized by three quantum numbers, each arising from the boundary conditions for one of the coordinates x, y, and z.

* For example, if you throw two dice, the probability of the first die coming up 6 is 1/6 and the probability of the second die coming up an odd number is 1/2. The probability of the first die coming up 6 *and* the second die coming up an odd number is (1/6)(1/2) = 1/12.

The lowest energy state (the ground state) for the cubical well occurs when $n_1 = n_2 = n_3 = 1$ and has the value

$$E_{111} = \frac{3\hbar^2\pi^2}{2mL^2} = 3E_1$$

The first excited energy level can be obtained in three different ways: $n_1 = 2$, $n_2 = n_3 = 1$; $n_2 = 2$, $n_1 = n_3 = 1$; or $n_3 = 2$, $n_1 = n_2 = 1$. Each has a different wave function. For example, the wave function for $n_1 = 2$ and $n_2 = n_3 = 1$ is

$$\psi_{211} = A \sin\frac{2\pi x}{L} \sin\frac{\pi y}{L} \sin\frac{\pi z}{L} \qquad 35\text{-}33$$

There are thus three different quantum states as described by the three different wave functions corresponding to the same energy level. An energy level with which more than one wave function is associated is said to be **degenerate.** In this case, there is threefold degeneracy. Degeneracy is related to the spatial symmetry of the system. If, for example, we consider a noncubic well, where $U = 0$ for $0 < x < L_1$, $0 < y < L_2$, and $0 < z < L_3$, the boundary conditions at the edges would lead to the quantum conditions $k_1L_1 = n_1\pi$, $k_2L_2 = n_2\pi$, and $k_3L_3 = n_3\pi$, and the total energy would be

$$E_{n_1n_2n_3} = \frac{\hbar^2\pi^2}{2m}\left(\frac{n_1^2}{L_1^2} + \frac{n_2^2}{L_2^2} + \frac{n_3^2}{L_3^2}\right) \qquad 35\text{-}34$$

These energy levels are not degenerate if L_1, L_2, and L_3 are all different. Figure 35-19 shows the energy levels for the ground state and first two excited states for an infinite cubic well in which the excited states are degenerate and for a noncubic infinite well in which L_1, L_2, and L_3 are all slightly different so that the excited levels are slightly split apart and the degeneracy is removed. The ground state is the state where the quantum numbers n_1, n_2, and n_3 all equal 1. None of the three quantum numbers can be zero. If any one of n_1, n_2, and n_3 were zero, the corresponding wave number k would also equal zero and the corresponding wave function (Equation 35-31) would equal zero for all values of x, y, and z.

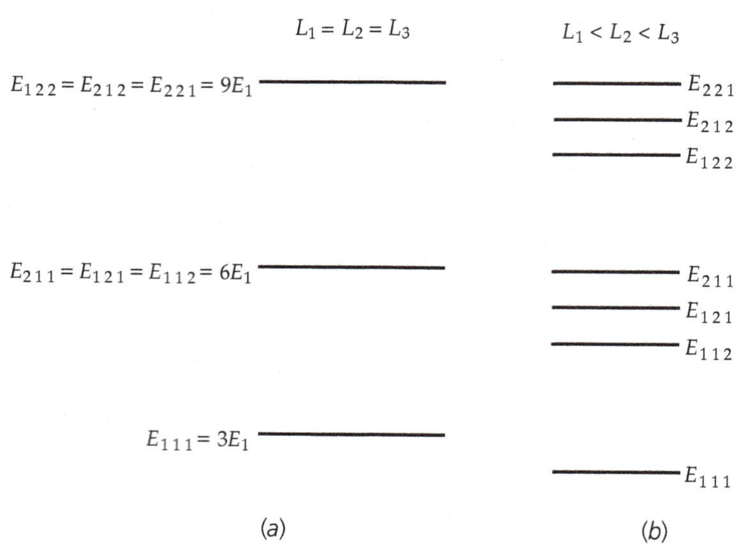

FIGURE 35-19 Energy-level diagrams for (*a*) a cubic infinite well and (*b*) a noncubic infinite well. In Figure 35-19*a* the energy levels are degenerate; that is, there are two or more wave functions having the same energy. The degeneracy is removed when the symmetry of the potential is removed, as in Figure 35-19*b*.

Example 35-3 Energy Levels for a Particle in a Three-Dimensional Box

A particle is in a three-dimensional box where $L_3 = L_2 = 2L_1$. Give the quantum numbers n_1, n_2, and n_3 that correspond to the state(s) in each of the seven lowest energy levels of this box.

PICTURE We can use Equation 35-34 to write the energy in terms of L_1 and the quantum numbers n_1, n_2, and n_3. Then we can find by inspection the values of the quantum numbers that give the lowest energies.

SOLVE

1. The energy of a level is given by Equation 35-34:

$$E_{n_1n_2n_3} = \frac{\hbar^2\pi^2}{2m}\left(\frac{n_1^2}{L_1^2} + \frac{n_2^2}{L_2^2} + \frac{n_3^2}{L_3^2}\right)$$

2. Factor out $1/L_1^2$:

$$E_{n_1n_2n_3} = \frac{\hbar^2\pi^2}{2m}\left(\frac{n_1^2}{L_1^2} + \frac{n_2^2}{4L_1^2} + \frac{n_3^2}{4L_1^2}\right) = \frac{\hbar^2\pi^2}{8mL_1^2}(4n_1^2 + n_2^2 + n_3^2)$$

3. The lowest energy is E_{111}:

$$E_{111} = E_1(4 \cdot 1^2 + 1^2 + 1^2) = \boxed{6E_1} \quad \text{(1st)}$$

where $E_1 = \hbar^2\pi^2/8mL_1^2$.

4. The energy increases the least when we increase n_2 or n_3. Trying various values of the quantum numbers:

$$E_{121} = E_{112} = E_1(4 \cdot 1^2 + 2^2 + 1^2) = \boxed{9E_1} \quad \text{(2nd)}$$

$$E_{122} = E_1(4 \cdot 1^2 + 2^2 + 2^2) = \boxed{12E_1} \quad \text{(3rd)}$$

$$E_{131} = E_{113} = E_1(4 \cdot 1^2 + 3^2 + 1^2) = \boxed{14E_1} \quad \text{(4th)}$$

$$E_{132} = E_{123} = E_1(4 \cdot 1^2 + 3^2 + 2^2) = \boxed{17E_1} \quad \text{(5th)}$$

$$E_{211} = E_1(4 \cdot 2^2 + 1^2 + 1^2) = \boxed{18E_1} \quad \text{(6th)}$$

$$\left.\begin{aligned} E_{221} = E_{212} &= E_1(4 \cdot 2^2 + 2^2 + 1^2) = \boxed{21E_1} \\ E_{141} = E_{114} &= E_1(4 \cdot 1^2 + 4^2 + 1^2) = \boxed{21E_1} \end{aligned}\right\} \text{(7th)}$$

CHECK Because two of the lengths are equal, degenerate energy levels are expected. Our results meet this expectation.

TAKING IT FURTHER Energies E_{221} and E_{212} are exactly equal because L_2 and L_3 are exactly equal. However, energies E_{221} and E_{141} are exactly equal because L_1 is exactly half of L_2.

PRACTICE PROBLEM 35-2 Find the quantum numbers and energies of the next two energy levels in step 4.

Example 35-4 Wave Functions for a Particle in a Three-Dimensional Box

Try It Yourself

Write the degenerate wave functions for the fourth and fifth excited states (the 5th and 6th levels) of the results in step 4 of Example 35-3.

PICTURE Use $\psi(x, y, z) = A\sin(k_1x)\sin(k_2y)\sin(k_3z)$ (a generalized version of Equation 35-31) with $k_i = n_i\pi/L_i$.

SOLVE

Cover the column to the right and try these on your own before looking at the answers.

Steps	Answers
Write the wave functions corresponding to the energies E_{131} and E_{113}	$\psi_{131} = A \sin\frac{\pi x}{L_1} \sin\frac{3\pi y}{2L_1} \sin\frac{\pi z}{2L_1}$ $\psi_{113} = A \sin\frac{\pi x}{L_1} \sin\frac{\pi y}{2L_1} \sin\frac{3\pi z}{2L_1}$

35-6 THE SCHRÖDINGER EQUATION FOR TWO IDENTICAL PARTICLES

Our discussion of quantum mechanics has thus far been limited to situations in which a single particle moves in some force field characterized by a potential energy function U. The most important physical problem of this type is the hydrogen atom, in which a single electron moves in the Coulomb potential of a proton. This problem is actually a two-body problem, because the proton also moves in the field of the electron. However, the motion of the much more massive proton requires only a very small correction to the energy of the atom that is easily made in both classical and quantum mechanics. When we consider more complicated problems, such as the helium atom, we must apply quantum mechanics to two or more electrons moving in an external field. Such problems are complicated, not only by the interaction of the electrons with each other, but also by the fact that the electrons are identical.

The interaction of two electrons with each other is electromagnetic and is essentially the same as the classical interaction of two charged particles. The Schrödinger equation for an atom that has two or more electrons cannot be solved exactly, so approximation methods must be used. This situation is not very different from the situation in which a classical expression describes three or more particles. However, the complications arising from the identity of electrons are purely quantum mechanical and have no classical counterpart. They are due to the fact that it is impossible to keep track of which electron is which. Classically, identical particles can be identified by their positions, which in principle can be determined with unlimited accuracy. This is impossible quantum mechanically because of the uncertainty principle. Figure 35-20 offers a schematic illustration of the problem.

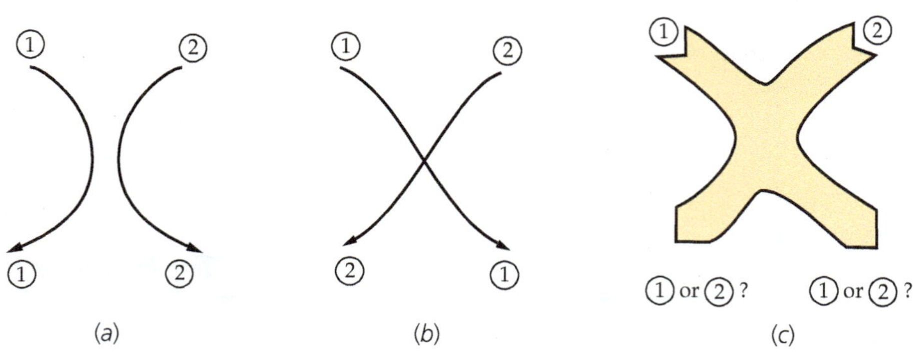

FIGURE 35-20 (*a*, *b*) Two possible classical electron paths. If electrons were classical particles, they could be distinguished by the paths followed. (*c*) However, because of the quantum-mechanical wave properties of electrons, the paths are spread out, as indicated by the shaded region. It is impossible to distinguish which electron is which after they separate.

The indistinguishability of identical particles has important consequences. For instance, consider the very simple case of two identical, noninteracting particles in a one-dimensional infinite square well. The time-independent Schrödinger equation for two particles, each of mass m, is

$$-\frac{\hbar^2}{2m}\frac{\partial^2\psi(x_1, x_2)}{\partial x_1^2} - \frac{\hbar^2}{2m}\frac{\partial^2\psi(x_1, x_2)}{\partial x_2^2} + U\psi(x_1, x_2) = E\psi(x_1, x_2) \qquad \text{35-35}$$

where x_1 and x_2 are the coordinates of the two particles. If the particles interact, the potential energy U is described by terms that have both x_1 and x_2 and cannot be separated into separate terms having only x_1 or x_2. For example, the electrostatic repulsion of two electrons in one dimension is represented by the potential energy function $ke^2/|x_2 - x_1|$. However, if the particles do not interact (as we are assuming here), we can write $U = U_1(x_1) + U_2(x_2)$. For the infinite square well, we need only solve the Schrödinger equation inside the well where $U = 0$ and require that the wave function be zero at the walls of the well. With $U = 0$, Equation 35-35 looks just like the expression for a particle in a two-dimensional well (Equation 35-30, but no z terms exist and y is replaced by x_2).

Solutions of this equation can be written in the form*

$$\psi_{nm} = \psi_n(x_1)\psi_m(x_2) \qquad \text{35-36}$$

where ψ_n and ψ_m are the single-particle wave functions for a particle in an infinite well and n and m are the quantum numbers of particles 1 and 2, respectively. For example, for $n = 1$ and $m = 2$, the wave function is

$$\psi_{12} = A \sin\frac{\pi x_1}{L} \sin\frac{2\pi x_2}{L} \qquad \text{35-37}$$

The probability of finding particle 1 in the region between $x = x_1$ and $x = x_1 + dx_1$ and particle 2 in the region between $x = x_2$ and $x = x_2 + dx_2$ is $\psi_{nm}^2(x_1, x_2)\, dx_1\, dx_2$, which is just the product of the separate probabilities $\psi_n^2(x_1)\, dx_1$ and $\psi_m^2(x_2)\, dx_2$. However, even though we have labeled the particles 1 and 2, we cannot distinguish which is between x_1 and $x_1 + dx_1$ and which is between x_2 and $x_2 + dx_2$ if they are identical. The mathematical descriptions of identical particles must be the same if we interchange the labels. The probability density $\psi^2(x_1, x_2)$ must therefore be the same as $\psi^2(x_2, x_1)$:

$$\psi^2(x_2, x_1) = \psi^2(x_1, x_2) \qquad \text{35-38}$$

Equation 35-38 is satisfied if $\psi(x_2, x_1)$ is either **symmetric** or **antisymmetric** on the exchange of particles—that is, if either

$$\psi(x_2, x_1) = \psi(x_1, x_2) \quad \text{symmetric} \qquad \text{35-39}$$

or

$$\psi(x_2, x_1) = -\psi(x_1, x_2) \quad \text{antisymmetric} \qquad \text{35-40}$$

Note that the wave functions given by Equations 35-36 and 35-37 are neither symmetric nor antisymmetric. If we interchange x_1 and x_2 in these wave functions, we get a different wave function, which implies that the particles can be distinguished.

---

* Again, this result can be obtained by solving Equation 35-35, but it also can be understood in terms of our knowledge of probability. The probability of electron 1 being between $x = x_1$ and $x = x_1 + dx_1$ and electron 2 being between $x = x_2$ and $x = x_2 + dx_2$ is the product of the individual probabilities.

We can find symmetric and antisymmetric wave functions that are solutions of the Schrödinger equation by adding or subtracting ψ_{nm} and ψ_{mn}. Adding them, we obtain

$$\psi_S = A'[\psi_n(x_1)\psi_m(x_2) + \psi_n(x_2)\psi_m(x_1)] \quad \text{symmetric} \qquad 35\text{-}41$$

and subtracting them, we obtain

$$\psi_A = A'[\psi_n(x_1)\psi_m(x_2) - \psi_n(x_2)\psi_m(x_1)] \quad \text{antisymmetric} \qquad 35\text{-}42$$

For example, the symmetric and antisymmetric wave functions for the first excited state of two identical particles in an infinite square well would be

$$\psi_S = A'\left(\sin\frac{\pi x_1}{L}\sin\frac{2\pi x_2}{L} + \sin\frac{\pi x_2}{L}\sin\frac{2\pi x_1}{L}\right) \qquad 35\text{-}43$$

and

$$\psi_A = A'\left(\sin\frac{\pi x_1}{L}\sin\frac{2\pi x_2}{L} - \sin\frac{\pi x_2}{L}\sin\frac{2\pi x_1}{L}\right) \qquad 35\text{-}44$$

There is an important difference between antisymmetric and symmetric wave functions. If $n = m$, the antisymmetric wave function is identically zero for all values of x_1 and x_2, whereas the symmetric wave function is not. Thus, if the wave function describing two identical particles is antisymmetric, the quantum numbers n and m of two particles cannot be the same. The idea that no two electrons in an atom can occupy the same quantum state, and thus have the same quantum numbers, was first stated by Wolfgang Pauli in 1925. This idea was soon generalized to include systems other than atoms and particles other than electrons. For example, no two protons of a nucleus can occupy the same quantum state, and no two neutrons of a nucleus can occupy the same quantum state. Electrons, protons, neutrons, neutrinos, and quarks all have a spin quantum number s equal to one-half, and all particles that have half-integer spin are called fermions. The two allowed values for the secondary spin quantum number m_s are plus and minus one half. Pauli's idea is called the **Pauli exclusion principle:**

> No two identical fermions can simultaneously occupy the same quantum state.
>
> PAULI EXCLUSION PRINCIPLE

The wave function for two or more identical fermions must be an antisymmetric wave function. Other particles (for example, α particles, deuterons, photons, and mesons) have integer spin and symmetric wave functions. These particles are called **bosons.**

A wave function that is a solution to the multiparticle time-independent wave equation (Equation 35-35) is called a spatial state. A bound system that contains fermions has either one or two identical fermions in each occupied spatial state. However, for a bound system that contain bosons there is no limit to the number of identical bosons in each spatial state.

Summary

1. The Schrödinger equation is a partial differential equation that relates the second space derivative of a wave function to its first time derivative. Wave functions that describe physical situations are solutions of this differential equation.
2. Because a wave function must satisfy the normalization condition, it must be well behaved; this means, among other things, that it must approach zero as $|x|$ approaches infinity. For bound systems such as a particle in a box, a simple harmonic oscillator, or an electron in an atom, this requirement leads to energy quantization.
3. The well-behaved wave functions for bound systems describe standing waves.

TOPIC — **RELEVANT EQUATIONS AND REMARKS**

1. Time-Independent Schrödinger Equation

$$-\frac{\hbar^2}{2m}\frac{d^2\psi(x)}{dx^2} + U(x)\psi(x) = E\psi(x) \qquad 35\text{-}4$$

Allowable solutions

In addition to satisfying the Schrödinger equation, a wave function $\psi(x)$ must be continuous and must have a continuous first derivative $d\psi/dx$.* Because the probability of finding an electron somewhere must be 1, the wave function must obey the normalization condition

$$\int_{-\infty}^{\infty} |\psi|^2 dx = 1$$

This condition implies the boundary condition that ψ must approach 0 as $|x|$ approaches ∞. Such boundary conditions lead to the quantization of energy.

2. Confined Particles

When the total energy E is greater than the potential energy $U(x)$ in some region (the classically allowed region) and less than $U(x)$ outside that region, the wave function ψ oscillates within the classically allowed region and $|\psi|$ decreases exponentially outside that region. The wave function approaches zero as $|x|$ approaches ∞ only for certain values of the total energy E. The energy is thus quantized.

In a finite square well

In a finite well of height U_0, there are only a finite number of allowed energies, and these are slightly less than the corresponding energies in an infinite well.

In the simple harmonic oscillator

For the oscillator with potential energy function $U(x) = \frac{1}{2}m\omega_0^2 x^2$, the allowed energies are equally spaced and given by

$$E_n = \left(n + \tfrac{1}{2}\right)\hbar\omega_0 \qquad n = 0, 1, 2, \ldots \qquad 35\text{-}26$$

The ground-state wave function is given by

$$\psi_0(x) = A_0 e^{-ax^2} \qquad 35\text{-}23$$

where A_0 is the normalization constant and

$$a = \tfrac{1}{2}m\omega_0/\hbar.$$

3. Reflection and Barrier Penetration

When the potential changes abruptly over a small distance, a particle may be reflected even though $E > U(x)$. A particle may penetrate a region in which $E < U(x)$. Reflection and penetration of matter waves are similar to those for other kinds of waves.

4. The Schrödinger Equation in Three Dimensions

The wave function for a particle in a three-dimensional box can be written

$$\psi(x, y, z) = \psi_1(x)\psi_2(y)\psi_3(z)$$

where ψ_1, ψ_2, and ψ_3 are wave functions for a one-dimensional box.

Degeneracy

When more than one wave function is associated with the same energy level, the energy level is said to be degenerate. Energy-level degeneracy occurs because of spatial symmetry.

* An exception to this claim is for the infinite well potential (where U is equal to zero inside the well and to infinity outside the well). For this potential function $d\psi/dx$ is not continuous at the boundary of the well (see Figure 34-17).

TOPIC	RELEVANT EQUATIONS AND REMARKS
5. **The Schrödinger Equation for Two Identical Particles**	A wave function that describes two identical particles must be either symmetric or antisymmetric when the coordinates of the particles are exchanged. Fermions (which include electrons, protons, and neutrons) are described by antisymmetric wave functions and obey the Pauli exclusion principle, which states that no two identical particles can simultaneously have the same values for their quantum number. Bosons (which include α particles, deuterons, photons, and mesons) have symmetric wave functions and do not obey the Pauli exclusion principle.

Answers to Concept Checks

35-1 The wave function cannot be normalized.

Answers to Practice Problems

35-2 $E_{133} = 22E_1, E_{142} = E_{124} = E_{222} = 24E_1$

Problems

In a few problems, you are given more data than you actually need; in a few other problems, you are required to supply data from your general knowledge, outside sources, or informed estimate.

Interpret as significant all digits in numerical values that have trailing zeros and no decimal points.

- • Single-concept, single-step, relatively easy
- •• Intermediate-level, may require synthesis of concepts
- ••• Challenging
- SSM Solution is in the *Student Solutions Manual*
- Consecutive problems that are shaded are paired problems.

CONCEPTUAL PROBLEMS

1 • Sketch (*a*) the wave function and (*b*) the probability density function for the $n = 5$ state of the finite square-well potential.

2 • Sketch (*a*) the wave function and (*b*) the probability density function for the $n = 4$ state of the finite square-well potential.

THE SCHRÖDINGER EQUATION

3 •• Show that if $\psi_1(x)$ and $\psi_2(x)$ are each solutions to the time-independent Schrödinger equation (Equation 35-4), then $\psi_3(x) = \psi_1(x) + \psi_2(x)$ is also a solution. This result, known as the superposition principle, applies to the solutions of all linear equations.

THE HARMONIC OSCILLATOR

4 •• The harmonic oscillator problem may be used to describe the vibrations of molecules. For example, the hydrogen molecule H_2 is found to have equally spaced vibrational energy levels separated by 8.7×10^{-20} J. What value of the force constant of the spring would be needed to get this energy spacing, assuming that half the molecule can be modeled as a hydrogen atom attached to one end of a spring that has its other end fixed? *Hint: The spacing for the energy levels of this half-molecule would be half of the spacing for the energy levels of the complete molecule. In addition, the force constant of a spring is inversely proportional to its relaxed length, so if half of the spring has force constant k, the entire spring has a force constant that is equal to* $\frac{1}{2}k$.

5 •• Use the procedure of Example 35-1 to verify that the energy of the first excited state of the harmonic oscillator is $E_1 = \frac{3}{2}\hbar\omega_0$. (*Note:* Rather than solve for a again, use the step-8 result $a = \frac{1}{2}m\omega_0/\hbar$ obtained in Example 35-1). SSM

6 •• Show that the expectation value $\langle x\rangle = \int_{-\infty}^{\infty} x|\psi|^2\,dx$ is zero for both the ground state and the first excited state of the harmonic oscillator.

7 •• Verify that the normalization constant A_0 in the ground-state harmonic-oscillator wave function $\psi_0(x) = A_0e^{-ax^2}$ (Equation 35-23) is given by $A_0 = (2m\omega_0/h)^{1/4}$.

8 •• Using the result of Problem 7, show that for the ground state of the harmonic oscillator $\langle x^2\rangle = \int x^2|\psi|^2\,dx = \hbar/(2m\omega_0) = 1/(4a)$. Use this result to show that the average potential energy equals half the total energy.

9 •• The quantity $\sqrt{\langle x^2\rangle - \langle x\rangle^2}$ is a measure of the average spread in the location of a particle. (*a*) Consider an electron trapped in a harmonic oscillator potential. Its lowest energy level is found to be 2.1×10^{-4} eV. Calculate $\sqrt{\langle x^2\rangle - \langle x\rangle^2}$ for this electron. (See Problems 6 and 8.) (*b*) Now consider an electron trapped in an infinite square-well potential. If the width of the well is equal to $\sqrt{\langle x^2\rangle - \langle x\rangle^2}$, what would be the lowest energy level for this electron?

10 ••• Classically, the average kinetic energy of the harmonic oscillator equals the average potential energy. Assume that this result is also true for the quantum-mechanical harmonic oscillator, and use this result, along with the result of Problem 8, to determine the expectation value of p_x^2 (where $p_x = mv_x$) for the ground state of the one-dimensional harmonic oscillator.

11 ••• We know that for the classical harmonic oscillator, $p_{x\,\text{av}} = 0$. It can be shown that for the quantum-mechanical harmonic oscillator, $\langle p_x\rangle = 0$. Use the results of Problems 6, 8, and 10 to determine the uncertainty product $\Delta x\,\Delta p_x$ for the ground state of the harmonic oscillator. The uncertainties are defined by $(\Delta x)^2 = \langle(x - \langle x\rangle)^2\rangle$ and $(\Delta p_x)^2 = \langle(p_x - \langle p_x\rangle)^2\rangle$.

REFLECTION AND TRANSMISSION OF ELECTRON WAVES: BARRIER PENETRATION

12 •• A particle of energy E approaches a step barrier of height U_0. What should be the ratio E/U_0 so that the reflection coefficient is $\frac{1}{2}$?

13 •• **SPREADSHEET** A particle that has mass m is traveling in the direction of increasing x. The potential energy of the particle is equal to zero everywhere in the region $x < 0$ and is equal to U_0 everywhere in the region $x > 0$, where $U_0 > 0$. (*a*) Show that if the total energy is $E = \alpha U_0$, where $\alpha \geq 1$, then the wave number k_2 in the region $x > 0$ is given by $k_2 = k_1\sqrt{(\alpha - 1)/\alpha}$, where k_1 is the wave number in the region $x < 0$. (b) Using a spreadsheet program or graphing calculator, graph the reflection coefficient R and the transmission coefficient T as functions of α, for $1 \leq \alpha \leq 5$. SSM

14 •• Suppose that the potential energy in Problem 13 is equal to zero everywhere in the region $x < 0$ and is equal to $-U_0$ everywhere in the region $x > 0$, where $U_0 > 0$. The wave number for the incident particle is again k_1, and the total energy is $2U_0$. (*a*) What is the wave number for the particle in the region where $x > 0$? (*b*) What is the reflection coefficient R? (*c*) What is the transmission coefficient T? (*d*) If each of one million particles that are in the region $x < 0$ are traveling with wave number k_1 in the direction of increasing x and are incident upon the potential energy drop at $x = 0$, how many of these particles are expected to continue along in the direction of increasing x? How does this compare with the classical prediction?

15 •• A 10-eV electron (an electron with a kinetic energy of 10 eV) is incident on a potential-energy barrier that has a height equal to 25 eV and a width equal to 1.0 nm. (*a*) Use Equation 35-29 to calculate the order of magnitude of the probability that the electron will tunnel through the barrier. (*b*) Repeat your calculation for a width of 0.10 nm. SSM

16 •• Use Equation 35-29 to calculate the order of magnitude of the probability that a proton will tunnel out of a nucleus in one collision with the nuclear barrier if the proton has an energy 6.0 MeV below the top of the potential-energy barrier and the barrier thickness is 1.2×10^{-15} m.

17 ••• To understand how a small change in α-particle energy can dramatically change the probability of the α particle tunneling from a nucleus, consider an α particle emitted by a uranium nucleus ($Z = 92$). (*a*) Referring to Figure 35-17, calculate the center-to-center distance of closest approach r_1 that α particles that have kinetic energies of 4.0 MeV and 7.0 MeV could make to the uranium nucleus. (*b*) Use the result from Part (*a*) to calculate the relative transmission coefficient $e^{-2\alpha a}$ for the same α particles. (*Note:* The actual half-lives of uranium nuclei vary over nine orders of magnitude. Your calculation will show a smaller range than this; however, to find half-life, you must also include the frequency with which the α particle strikes the barrier.)

THE SCHRÖDINGER EQUATION IN THREE DIMENSIONS

18 •• (*a*) A particle is confined to a three-dimensional box that has sides L_1, $L_2 = 2L_1$, and $L_3 = 3L_1$. Give the quantum numbers n_1, n_2 and n_3 that correspond to the ten lowest-energy quantum states of this box. *Hint: A spreadsheet can be helpful.* (*b*) What quantum numbers, if any, correspond to degenerate energy levels? (*c*) Give a wave function for the fifth excited state. (There are only five states that have energy levels below the energy level of the fifth excited state.)

19 •• (*a*) A particle is confined to a three-dimensional box that has sides L_1, $L_2 = 2L_1$, and $L_3 = 4L_1$. Give the quantum numbers n_1, n_2 and n_3 that correspond to the ten lowest-energy quantum states of this box. *Hint: A spreadsheet can be helpful.* (*b*) What combinations of these quantum numbers, if any, correspond to degenerate energy levels? (*c*) Give the wave function for the fourth excited energy state. (There are only four states that have energy levels below the energy level of the fourth excited state.)

20 • A particle moves in a potential well given by $U(x, y, z) = 0$ for $-L/2 < x < L/2, 0 < y < L$, and $0 < z < L$; $U = \infty$ outside these ranges. (*a*) Write an expression for the ground-state wave function for the particle. (*b*) How do the allowed energies compare with those for a well having $U = 0$ for $0 < x < L$, rather than for $-L/2 < x < L/2$? Explain your answer.

21 •• A particle is constrained to the two-dimensional region defined by $0 \leq x \leq L$ and $0 \leq y \leq L$ and moves freely throughout that region. (*a*) Find the wave functions that meet these conditions and are solutions of the Schrödinger equation. (*b*) Find the energies that correspond to the wave functions in Part (*a*). (*c*) Find the quantum numbers of the two lowest states that have the same energy (that are degenerate). (*d*) Find the quantum numbers of the three lowest states that have the same energy.

THE SCHRÖDINGER EQUATION FOR TWO IDENTICAL PARTICLES

22 • Show that the two-particle wave function $\psi_{12} = A\sin(\pi x_1/L)\sin(2\pi x_2/L)$, $0 < x_1 < L$ and $0 < x_2 < L$, (Equation 35-37), is a solution of

$$-\frac{\hbar^2}{2m}\frac{\partial^2\psi(x_1, x_2)}{\partial x_1^2} - \frac{\hbar^2}{2m}\frac{\partial^2\psi(x_1, x_2)}{\partial x_2^2} + U\psi(x_1, x_2) = E\psi(x_1, x_2)$$

(Equation 35-35), if $U(x_1, x_2) = 0$, and find the energy of the state represented by this wave function.

23 • What is the ground-state energy of ten noninteracting bosons in a one-dimensional box of length L?

24 •• What is the ground-state energy of seven identical noninteracting fermions in a one-dimensional box of length L? (Because the quantum number associated with spin can have two values, each spatial state can be occupied by two fermions.)

ORTHOGONALITY OF WAVE FUNCTIONS

The integral of two functions over some space interval is somewhat analogous to the dot product of two vectors. If this integral is zero, the functions are said to be orthogonal, which is analogous to two vectors being perpendicular. The following problems illustrate the general principle that any two wave functions corresponding to different energy levels in the same potential are orthogonal. A general hint for all these problems is that the integral $\int_{x_1}^{x_2} f(x)\,dx$ *is equal to zero if* x_1 *is equal to* $-x_2$ *and if* $f(x)$ *is equal to* $-f(-x)$.

25 •• Show that the ground-state and the first excited state wave functions of the harmonic oscillator are orthogonal; that is, show that $\int_{-\infty}^{\infty} \psi_0(x)\psi_1(x)\,dx = 0$.

26 •• The wave function for the state $n = 2$ of the harmonic oscillator is $\psi_2(x) = A_2\left(2ax^2 - \frac{1}{2}\right)e^{-ax^2}$, where A_2 is the normalization constant and a is a positive constant. Show that the wave functions for the states $n = 1$ and $n = 2$ of the harmonic oscillator are orthogonal.

27 •• For the wave functions

$$\psi_n(x) = A\sin(n\pi x/L) \qquad n = 1, 2, 3, \ldots$$

corresponding to a particle in an infinite square-well potential from 0 to L, show that $\int_0^L \psi_m(x)\psi_n(x)\,dx = 0$ for all positive integers m and n, where $m \neq n$; that is, show that the wave functions are orthogonal.

GENERAL PROBLEMS

28 •• Consider a particle in an infinite one-dimensional box that has a length L and is centered at the origin. (*a*) What are the values of $\psi_1(0)$ and $\psi_2(0)$? (*b*) What are the values of $\langle x\rangle$ for the $n = 1$ and $n = 2$ states? (*c*) Evaluate $\langle x^2\rangle$ for the $n = 1$ and $n = 2$ states.

29 •• Eight identical noninteracting fermions are confined to an infinite two-dimensional square box of side length L. Determine the energies of the three lowest energy states. (See Problem 22.) SSM

30 •• A particle is confined to a two-dimensional box defined by the following boundary conditions: $U(x, y) = 0$ for $-L/2 \leq x \leq L/2$ and $-3L/2 \leq y \leq 3L/2$, and $U(x, y) = \infty$ outside these ranges. (*a*) Determine the energies of the three lowest energy states. Are any of these states degenerate? (*b*) Identify the quantum numbers of the two lowest energy degenerate states and determine the energy of these states.

31 ••• The classical probability distribution function for a particle in an infinite one-dimensional well of length L is $P = 1/L$. (See Example 34-5.) (*a*) Show that the classical expectation value of x^2 for a particle in an infinite one-dimensional well of length L that is centered at the origin is $L^2/12$. (*b*) Find the quantum expectation value of x^2 for the nth state of a particle in the one-dimensional box and show that it approaches the classical limit $L^2/12$ as n approaches infinity.

32 •• Show that Equations 35-27 and 35-28 imply that the transmission coefficient for particles of energy E incident on a step barrier $U_0 < E$ is given by

$$T = \frac{4k_1k_2}{(k_1 + k_2)^2} = \frac{4r}{(1 + r)^2}$$

where $r = k_2/k_1$.

33 •• (*a*) Show that for the case of a particle of energy E incident on a step barrier $U_0 < E$, the wave numbers k_1 and k_2 are related by

$$\frac{k_2}{k_1} = r = \sqrt{1 - \frac{U_0}{E}}$$

(*b*) Use this result to show that $R = (1 - r)^2/(1 + r)^2$.

34 •• **SPREADSHEET** (*a*) Using a spreadsheet program or graphing calculator and the results of Problem 32 and Problem 33, graph the transmission coefficient T and reflection coefficient R as a function of incident energy E for values of E ranging from $E = U_0$ to $E = 10.0U_0$. (*b*) What limiting values do your graphs indicate?

35 ••• The wave function for the state $n = 2$ of the harmonic oscillator is $\psi_2(x) = A_2(2ax^2 - \frac{1}{2})e^{-ax^2}$, where $a = \frac{1}{2}m\omega_0/\hbar$. Determine the normalization constant A_2.

36 ••• Consider the time-independent, one-dimensional Schrödinger equation when the potential function is symmetric about the origin, that is, when $U(x)$ is even.* (*a*) Show that if $\psi(x)$ is a solution of the Schrödinger equation and has energy E, then $\psi(-x)$ is also a solution that has the same energy E. Furthermore, $\psi(x)$ and $\psi(-x)$ can differ by only a multiplicative constant. (*b*) Write $\psi(x) = C\psi(-x)$, and show that $C = \pm 1$. Note that $C = +1$ means that $\psi(x)$ is an even function of x, and $C = -1$ means that $\psi(x)$ is an odd function of x.

37 ••• In this problem, you will derive the ground-state energy of the harmonic oscillator using the precise form of the uncertainty principle, $\Delta x\,\Delta p_x \geq \hbar/2$, where Δx and Δp_x are defined to be the standard deviations $(\Delta x)^2 = \langle(x - \langle x\rangle)^2\rangle$ and $(\Delta p_x)^2 = \langle(p_x - \langle p_x\rangle)^2\rangle$.

Proceed as follows:

1. Write the total classical energy in terms of the position x and momentum p_x using $U(x) = \frac{1}{2}m\omega_0^2x^2$ and $K = \frac{1}{2}p_x^2/m$.
2. Show that $(\Delta x)^2 = \langle(x - \langle x\rangle)^2\rangle = \langle x^2\rangle - \langle x\rangle^2$ and $(\Delta p_x)^2 = \langle(p_x - \langle p_x\rangle)^2\rangle = \langle p_x^2\rangle - \langle p_x\rangle^2$. *Hint: See Equations 17-34a and 17-34b.*
3. Use the symmetry of the potential energy function to argue that $\langle x\rangle$ and $\langle p_x\rangle$ must be zero, so that $(\Delta x)^2 = \langle x^2\rangle$ and $(\Delta p_x)^2 = \langle p_x^2\rangle$.
4. Assume that $\Delta p_x \Delta x = \hbar/2$ to eliminate $\langle p_x^2\rangle$ from the average energy $\langle E\rangle = \langle \frac{1}{2}p_x^2/m + \frac{1}{2}m\omega_0^2x^2\rangle = \frac{1}{2}\langle p_x^2\rangle/m + \frac{1}{2}m\omega_0^2\langle x^2\rangle$ and write $\langle E\rangle$ as $\langle E\rangle = \hbar^2/(8mZ) + \frac{1}{2}m\omega^2Z$, where $Z = \langle x^2\rangle$.
5. Set $d\langle E\rangle/dZ = 0$ to find the value of Z for which $\langle E\rangle$ is a minimum.
6. Show that the minimum energy is given by $\langle E\rangle_{min} = +\frac{1}{2}\hbar\omega_0$. SSM

38 ••• A particle that has mass m and is near Earth's surface, at which $z = 0$, can be described by the potential energy function $U = mgz$ in the region $z > 0$, and by $U = \infty$ in the region $z < 0$. Sketch a graph of $U(z)$ versus z. For some positive value of total energy E, indicate the classically allowed region on the graph and plot the classical kinetic energy versus z on the graph. The Schrödinger equation for this problem is quite difficult to solve. Using arguments similar to those in Section 35-2 about the concavity of the wave function as given by the Schrödinger equation, sketch the shape of the wave function for the ground state and for the first two excited states.

* A function $f(x)$ is even if $f(x) = f(-x)$ for all x, and a function $f(x)$ is odd if $f(x) = -f(-x)$ for all x.

CHAPTER 36

AT A DISTANCE OF 6000 LIGHTYEARS FROM EARTH, THE STAR CLUSTER RCW 38 IS A RELATIVELY CLOSE STAR-FORMING REGION. THIS IMAGE COVERS AN AREA ABOUT 5 LIGHTYEARS ACROSS AND CONTAINS THOUSANDS OF HOT, VERY YOUNG STARS FORMED LESS THAN A MILLION YEARS AGO. X RAYS FROM THE HOT UPPER ATMOSPHERES OF 190 OF THESE STARS WERE DETECTED BY CHANDRA, AN X-RAY OBSERVATORY ORBITING EARTH. THE MECHANISMS GENERATING THESE X RAYS ARE NOT KNOWN. ON EARTH, X-RAY MACHINES PRODUCE X RAYS BY BOMBARDING A TARGET WITH HIGH-ENERGY ELECTRONS. THE ATOMIC NUMBER OF THE ATOMS THAT MAKE UP THE TARGET CAN BE DETERMINED BY ANALYZING THE RESULTING X-RAY SPECTRA. *(NASA/CXC/CfA/S. Wolk et al.)*

Atoms

? How is the atomic number obtained from the spectral analysis? (See Example 36-8.)

One hundred eleven chemical elements have been discovered, and several additional chemical elements recently have been reported but not authenticated. Each element is characterized by an atom that has a number of protons Z and an equal number of electrons. The number of protons Z is called the **atomic number.** The atom that has the fewest protons is called hydrogen (H) and has $Z = 1$. A helium (He) atom has two protons ($Z = 2$), a lithium (Li) atom has three protons ($Z = 3$), and so forth. Nearly all the mass of an atom is concentrated in its tiny nucleus, which is made up of protons and neutrons. An atom's nuclear radius is approximately 1 fm to 10 fm (1 fm = 10^{-15} m) and the radius of an atom is approximately 0.1 nm = 100 000 fm.

The chemical properties and physical properties of an element are determined by the number and arrangement of the electrons in an atom of the element. Because each proton has a positive charge $+e$, the nucleus has a total positive charge $+Ze$. The electrons are negatively charged ($-e$), so they are attracted to the nucleus and repelled by each other. Because electrons and protons have equal but opposite charges and an atom has equal numbers of electrons and protons, atoms are electrically neutral. Atoms that lose or gain one or more electrons are then electrically charged and are called *ions*.

We will begin our study of atoms by discussing the Bohr model, a semiclassical model developed by Niels Bohr in 1913 to explain the electromagnetic spectra produced by hydrogen atoms. Although this pre-quantum mechanics model has many shortcomings, it provides a useful framework for the discussion of atomic phenomena. After discussing the Bohr model, we will then apply our knowledge of quantum mechanics from Chapter 35 to give a much more productive model of the hydrogen atom. We will then discuss the structure of other atoms and the periodic table of the elements. Finally, we will discuss both optical and X-ray spectra.

36-1 THE ATOM

ATOMIC SPECTRA

By the beginning of the twentieth century, a large body of data had been collected on the emission of light by atoms in a gas when the atoms are excited by an electric discharge. Viewed through a spectroscope that has a narrow-slit aperture, light that has been emitted by atoms of a particular element appears as a discrete set of lines of different colors or wavelengths. The spacing and relative intensities of the lines are characteristic of the element. The wavelengths of these spectral lines could be accurately determined, and much effort went into finding regularities in the spectra. Figure 36-1 shows line spectra for hydrogen and for mercury.

In 1885, Johann Balmer determined that the wavelengths of the lines in the visible spectrum of hydrogen can be represented by the formula

$$\lambda = (364.6\text{ nm})\frac{m^2}{m^2 - 4} \qquad m = 3, 4, 5, \ldots \tag{36-1}$$

Balmer suggested that this expression might be a special case of a more general expression that would be applicable to the spectra of other elements. Such an expression, found by Johannes R. Rydberg and Walter Ritz and known as the **Rydberg–Ritz formula,** gives the reciprocal wavelength as

$$\frac{1}{\lambda} = R\left(\frac{1}{n_2^2} - \frac{1}{n_1^2}\right) \tag{36-2}$$

(a)

(b)

FIGURE 36-1 (*a*) Line spectrum of hydrogen and (*b*) line spectrum of mercury. *((a) and (b) adapted from Eastern Kodak and Wabash Instrument Corporation.)*

where n_1 and n_2 are integers, $n_1 > n_2$, and R is the Rydberg constant. The Rydberg constant does vary slightly, and in a regular way, from element to element. For hydrogen, R has the value

$$R_H = 1.097776 \times 10^7\ m^{-1}$$

RYDBERG CONSTANT FOR HYDROGEN

The Rydberg–Ritz formula gives the wavelengths for all the lines in the spectra of hydrogen, as well as alkali elements such as lithium and sodium. The hydrogen Balmer series given by Equation 36-1 is also given by Equation 36-2, with $R = R_H$, $n_2 = 2$, and $n_1 = m$.

Many attempts were made to construct a model of the atom that would yield these formulas for an atom's radiation spectrum. The most popular model, created by J. J. Thomson, considered various arrangements of electrons embedded in some kind of fluid that had most of the mass of the atom and had enough positive charge to make the atom electrically neutral. Thomson's model, called the "plum pudding" model, is illustrated in Figure 36-2. Because classical electromagnetic theory predicted that a charge oscillating with frequency f would radiate electromagnetic energy of that frequency, Thomson searched for configurations that were stable and that had normal modes of vibration of frequencies equal to those of the spectrum of the atom. A difficulty of this model and all other models was that, according to classical physics, electric forces alone cannot produce stable equilibrium. Thomson was unsuccessful in finding a model that predicted the observed frequencies for any atom.

The Thomson model was essentially ruled out by a set of experiments by H. W. Geiger and E. Marsden, under the supervision of E. Rutherford in approximately 1911, in which alpha particles from radioactive radium were scattered by atoms in a gold foil. Rutherford showed that the number of alpha particles scattered at large angles could not be accounted for by an atom in which the positive charge was distributed throughout the atom (known to be about 0.1 nm in diameter). Instead, the results suggested that the positive charge and most of the mass of an atom is concentrated in a very small region, now called the nucleus, which has a diameter of the order of 10^{-6} nm = 1 fm.

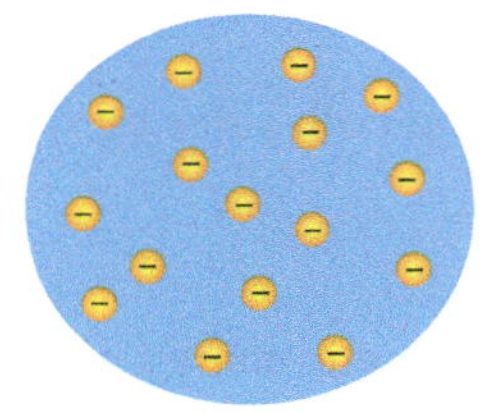

FIGURE 36-2 J. J. Thomson's plum pudding model of the atom. In this model the electrons, which have a negative charge, are embedded in a fluid of positive charge. For a given configuration in such a system, the resonance frequencies of oscillations of the electrons can be calculated. According to classical theory, the atom should radiate light of frequency equal to the frequency of oscillation of the electrons. Thomson could not find any configuration that would give frequencies in agreement with the measured frequencies of the spectrum of any atom.

36-2 THE BOHR MODEL OF THE HYDROGEN ATOM

Niels Bohr, working in the Rutherford laboratory in 1912, proposed a model of the hydrogen atom that extended the work of Planck, Einstein, and Rutherford and successfully predicted the observed spectra. According to Bohr's model, the electron of the hydrogen atom moves in either a circular or an elliptical orbit around the positive nucleus according to Coulomb's law and classical mechanics like the planets orbit the Sun. For simplicity, Bohr chose a circular orbit, as shown in Figure 36-3.

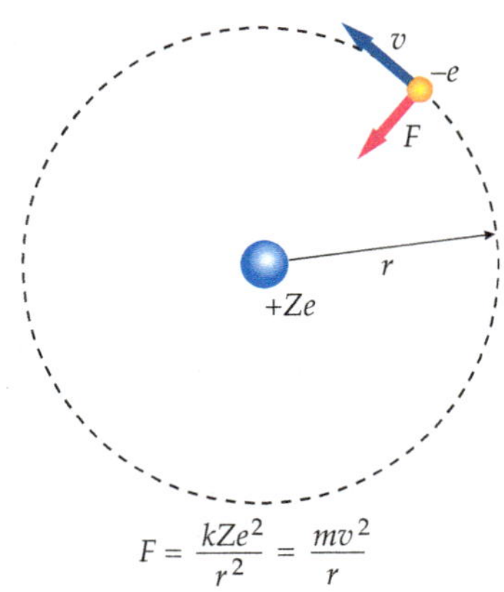

FIGURE 36-3 Electron of charge $-e$ traveling in a circular orbit of radius r around the nuclear charge $+Ze$. The attractive electrical force kZe^2/r^2 keeps the electron in its orbit.

ENERGY FOR A CIRCULAR ORBIT

Consider an electron of charge $-e$ moving in a circular orbit of radius r about a positive charge $+Ze$ such as the nucleus of a hydrogen atom ($Z = 1$) or of a singly ionized helium atom ($Z = 2$). The total energy of the electron can be related to the radius of the orbit. The potential energy of the electron of charge $-e$ at a distance r from a positive charge Ze is

$$U = \frac{kq_1q_2}{r} = \frac{k(Ze)(-e)}{r} = -\frac{kZe^2}{r} \qquad \text{36-3}$$

where k is the Coulomb constant. The kinetic energy K can be obtained as a function of r by using Newton's second law, $F_{\text{net}} = ma$. Setting the Coulomb attractive force equal to the mass multiplied by the centripetal acceleration gives

$$\frac{kZe^2}{r^2} = m\frac{v^2}{r} \qquad \text{36-4a}$$

Multiplying both sides by $r/2$ gives

$$K = \frac{1}{2}mv^2 = \frac{1}{2}\frac{kZe^2}{r} \qquad \text{36-4b}$$

Thus, the kinetic energy and the potential energy vary inversely with r. Note that the magnitude of the potential energy is twice that of the kinetic energy:

$$U = -2K \qquad \text{36-5}$$

Equation 36-5 a general result for particles orbiting under the influence of forces that vary inversely with the square of the distance from a fixed point. [It also holds for circular orbits in a gravitational field (see Example 11-6 in Section 11-3)]. The total energy is the sum of the kinetic energy and the potential energy:

$$E = K + U = \frac{1}{2}\frac{kZe^2}{r} - \frac{kZe^2}{r}$$

or

$$E = -\frac{1}{2}\frac{kZe^2}{r} \qquad \text{36-6}$$

ENERGY IN A CIRCULAR ORBIT FOR A $1/r^2$ FORCE

Although mechanical stability is achieved because the Coulomb attractive force provides the centripetal force necessary for the electron to remain in orbit, classical *electromagnetic* theory says that such an atom would be electrically unstable. The atom would be unstable because the electron must accelerate when moving in a circle and therefore radiate electromagnetic energy of frequency equal to that of its motion. According to the classical theory, such an atom would quickly collapse as the electron spiraled into the nucleus and radiated away the energy.

BOHR'S POSTULATES

Bohr circumvented the difficulty of the collapsing atom by *postulating* that only certain orbits, called stationary states, are allowed and that an atom with an electron in one of these orbits does not radiate. An atom radiates only when the electron makes a transition from one allowed orbit (stationary state) to another.

The electron in the hydrogen atom can move only in certain nonradiating, circular orbits called stationary states.

BOHR'S FIRST POSTULATE—NONRADIATING ORBITS

Bohr's second postulate relates the frequency of radiation to the energies of the stationary states. If E_i and E_f are the initial and final energies of the atom, the frequency of the emitted radiation during a transition is given by

$$f = \frac{E_i - E_f}{h} \qquad \text{36-7}$$

BOHR'S SECOND POSTULATE—PHOTON FREQUENCY FROM ENERGY CONSERVATION where h is Planck's constant. This postulate is equivalent to the assumption of conservation of energy when a photon of energy hf is emitted. Combining Equation 36-6 and Equation 36-7, we obtain for the frequency

$$f = \frac{E_i - E_f}{h} = \frac{1}{2}\frac{kZe^2}{h}\left(\frac{1}{r_f} - \frac{1}{r_i}\right) \qquad \text{36-8}$$

where r_i and r_f are the radii of the initial and final orbits.

To obtain the frequencies implied by the Rydberg–Ritz formula, $f = c/\lambda = cR(1/n_2^2 - 1/n_1^2)$, it is evident that the radii of stable orbits must be proportional to the squares of integers. Bohr searched for a quantum condition for the radii of the stable orbits that would yield this result. After much trial and error, Bohr found that he could obtain the desired result if he postulated that the magnitude of the angular momentum of the electron in a stable orbit equals an integer multiplied by $\hbar$. Because the magnitude of the angular momentum of a circular orbit is just mvr, this postulate is

$$mv_n r_n = n\hbar \qquad n = 1, 2, 3, \ldots \qquad \text{36-9}$$

BOHR'S THIRD POSTULATE—QUANTIZED ANGULAR MOMENTUM

where $\hbar = h/2\pi = 1.055 \times 10^{-34}\ \text{J}\cdot\text{s} = 6.582 \times 10^{-16}\ \text{eV}\cdot\text{s}$.

Equation 36-9 relates the speed v_n to the radius r_n of the orbit that has angular momentum $n\hbar$. Equation 36-4*a* gives us another equation relating the speed to the radius:

$$\frac{kZe^2}{r_n^2} = m\frac{v_n^2}{r_n}$$

or

$$v_n^2 = \frac{kZe^2}{mr_n} \qquad \text{36-10}$$

We can determine r_n by first solving for v_n in Equation 36-9. Squaring the result then gives

$$v_n^2 = n^2\frac{\hbar^2}{m^2r_n^2}$$

Equating this expression for v_n^2 with the expression given by Equation 36-10, we get

$$n^2\frac{\hbar^2}{m^2r_n^2} = \frac{kZe^2}{mr_n}$$

Solving for r_n, we obtain

$$r_n = n^2\frac{\hbar^2}{mkZe^2} = n^2\frac{a_0}{Z} \qquad \text{36-11}$$

RADII OF THE BOHR ORBITS

where a_0 is called the **first Bohr radius.** According to the Bohr model, a_0 is the orbital radius of the electron in a hydrogen atom that has $n = 1$.

$$a_0 = \frac{\hbar^2}{mke^2} = \frac{\epsilon_0 h^2}{\pi me^2} = 0.0529\ \text{nm} \qquad \text{36-12}$$

FIRST BOHR RADIUS

Substituting the expressions for r_n in Equation 36-11 into Equation 36-8 for the frequency gives

$$f = \frac{1}{2}\frac{kZe^2}{h}\left(\frac{1}{r_f} - \frac{1}{r_i}\right) = Z^2\frac{mk^2e^4}{4\pi\hbar^3}\left(\frac{1}{n_f^2} - \frac{1}{n_i^2}\right) \qquad 36\text{-}13$$

If we compare this expression where $Z = 1$ and $f = c/\lambda$ with the empirical Rydberg–Ritz formula (Equation 36-2), we obtain for the Rydberg constant

$$R = \frac{mk^2e^4}{4\pi c\hbar^3} = \frac{me^4}{8\epsilon_0^2 ch^3} \qquad 36\text{-}14$$

Using the values of m, e, c, k, and $\hbar$ known in 1913, Bohr calculated R and found his result to agree (within the limits of the uncertainties of the constants) with the value obtained from spectroscopy.

Example 36-1 Standing-Wave Condition Implies Quantization of Angular Momentum

For waves in a circle, the standing-wave condition is that there is an integral number of wavelengths in the circumference. That is, $n\lambda_n = 2\pi r_n$, where $n = 1, 2, 3$, and so on. Show that this condition for electron waves implies quantization of angular momentum.

PICTURE The wavelength and the momentum are related by the de Broglie relation $p = h/\lambda$ (Equation 34-13). Using this relation and the standing wave condition $n\lambda_n = 2\pi r_n$ to relate the momentum to the radius.

SOLVE

1. Write the standing-wave condition: $n\lambda_n = 2\pi r_n$
2. Use the de Broglie relation (Equation 34-13) to relate the momentum p to λ_n: $p = \frac{h}{\lambda_n} = \frac{nh}{2\pi r_n} = n\frac{\hbar}{r_n}$
3. Solve for pr_n. The angular momentum of an electron in a circular orbit is $mvr_n = pr_n$, where $p = mv$: $pr_n = \boxed{mvr_n = n\hbar}$

CHECK The step-3 result is the check. It is what the problem statement asks us to show.

ENERGY LEVELS

The total mechanical energy of the electron in the hydrogen atom is related to the radius of the circular orbit by Equation 36-6. If we substitute the quantized values of r as given by Equation 36-11, we obtain

$$E_n = -\frac{1}{2}\frac{kZe^2}{r_n} = -\frac{1}{2}\frac{kZ^2e^2}{n^2a_0} = -\frac{1}{2}\frac{mk^2Z^2e^4}{n^2\hbar^2}$$

or

$$E_n = -Z^2\frac{E_0}{n^2} \qquad 36\text{-}15$$

ENERGY LEVELS IN THE HYDROGEN ATOM

where

$$E_0 = \frac{mk^2e^4}{2\hbar^2} = \frac{1}{2}\frac{ke^2}{a_0} = 13.6\text{ eV} \qquad 36\text{-}16$$

The energies E_n corresponding to $Z = 1$ are the quantized allowed energies for the hydrogen atom.

Transitions between these allowed energies result in the emission or absorption of a photon whose frequency is given by $f = (E_i - E_f)/h$, and whose wavelength is

$$\lambda = \frac{c}{f} = \frac{hc}{E_i - E_f} \qquad 36\text{-}17$$

As we found in Chapter 34, it is convenient to have the value of hc in electron-volt nanometers:

$$hc = 1240 \text{ eV}\cdot\text{nm} \qquad 36\text{-}18$$

Because the energies are quantized, the frequencies and wavelengths of the radiation emitted by the hydrogen atom are quantized in agreement with the observed line spectrum.

Figure 36-4 shows the energy-level diagram for hydrogen. The energy of the hydrogen atom in the ground state is $E_1 = -13.6$ eV. As n approaches infinity the energy approaches zero. The process of removing an electron from an atom is called **ionization,** and the minimum amount of energy required to remove the electron is the **ionization energy.** The ionization energy of the ground-state hydrogen atom, which is also its **binding energy,** is 13.6 eV. A few transitions from a higher state to a lower state are indicated in Figure 36-4. When Bohr published his model of the hydrogen atom, the Balmer series (corresponding to $n_f = 2$ and $n_i = 3, 4, 5, \ldots$) and the Paschen series (corresponding to $n_f = 3$ and $n_i = 4, 5, 6, \ldots$) were known. In 1916, T. Lyman found the series corresponding to $n_f = 1$. F. Brackett in 1922 and H. A. Pfund in 1924 found the series corresponding to $n_f = 4$ and $n_f = 5$, respectively. Only the Balmer series lies in the visible portion of the electromagnetic spectrum.

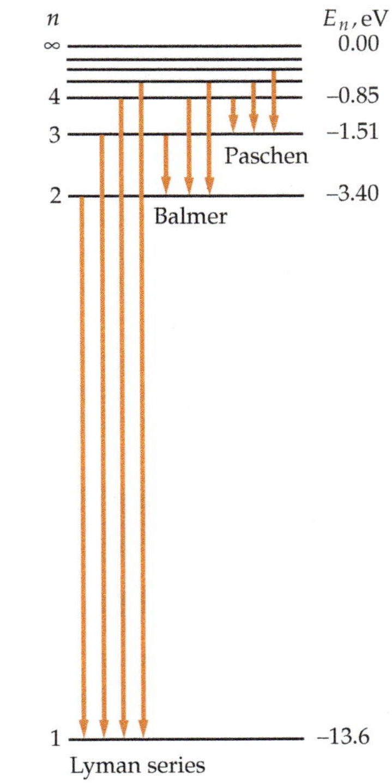

FIGURE 36-4 Energy-level diagram for hydrogen showing the first few transitions in each of the Lyman, Balmer, and Paschen series. The energies of the levels are given by Equation 36-15.

! Do not think the λ in Equation 36-17 is the wavelength of the electron. It is not. It is the wavelength of the emitted or absorbed photon.

Example 36-2 Longest Wavelength in the Lyman Series

Find (*a*) the energy and (*b*) the wavelength of the spectral line that has the longest wavelength in the Lyman series.

PICTURE For the Lyman series $n_f = 1$. From Figure 36-4, we can see that the Lyman series corresponds to transitions ending at the ground-state energy, $E_f = E_1 = -13.6$ eV. Because the photon wavelength λ varies inversely with energy, the transition that has the longest wavelength is the transition that has the lowest energy, which is from the first excited state $n = 2$ to the ground state $n = 1$.

SOLVE

1. The energy of the photon is the difference in the energies of the initial and final atomic states:

$$E_{\text{photon}} = \Delta E_{\text{atom}} = E_i - E_f$$
$$= E_2 - E_1 = \frac{-13.6 \text{ eV}}{2^2} - \frac{-13.6 \text{ eV}}{1^2}$$
$$= -3.40 \text{ eV} + 13.6 \text{ eV} = \boxed{10.2 \text{ eV}}$$

2. The wavelength of the photon is

$$\lambda = \frac{hc}{E_2 - E_1} = \frac{1240 \text{ eV}\cdot\text{nm}}{10.2 \text{ eV}} = \boxed{122 \text{ nm}}$$

CHECK The step-1 result of 10.2 eV is less than 13.6 eV (the binding energy of ground-state hydrogen). This result is expected.

TAKING IT FURTHER This photon has a wavelength that corresponds to the ultraviolet region of the electromagnetic spectrum. Because all the other lines in the Lyman series have even greater energies and shorter wavelengths, the Lyman series is completely in the ultraviolet region.

PRACTICE PROBLEM 36-1 Find the shortest wavelength for a line in the Lyman series.

Despite its spectacular successes, the Bohr model of the hydrogen atom had many shortcomings. There was no justification for the postulates of stationary states or for the quantization of angular momentum other than the fact that these postulates led to energy levels that agreed with spectroscopic data. Furthermore, attempts to apply the model to atoms that have more electrons and protons had little success. The quantum-mechanical model resolves these difficulties. The stationary states of the Bohr model are replaced by the standing-wave solutions of the Schrödinger equation analogous to the standing electron waves for a particle in a box discussed in Chapter 34 and Chapter 35. Energy quantization is a direct consequence of the standing-wave solutions of the Schrödinger equation. For hydrogen, these quantized energies agree with those obtained from the Bohr model and with experiment. The quantization of angular momentum that had to be postulated in the Bohr model is predicted by the quantum theory.

36-3 QUANTUM THEORY OF ATOMS

THE SCHRÖDINGER EQUATION IN SPHERICAL COORDINATES

In quantum theory, an electron in an atom is described by its wave function ψ. The probability of finding the electron in some volume dV of space equals the square of the absolute value of the wave function $|\psi|^2$ multiplied by dV. Boundary conditions on the wave function lead to the quantization of the wavelengths and frequencies and thereby to the quantization of the electron energy.

Consider a single electron of mass m moving in three dimensions in a region in which the potential energy is U. The time-independent Schrödinger equation for such a particle is given by Equation 35-30:

$$-\frac{\hbar^2}{2m}\left(\frac{\partial^2\psi}{\partial x^2}+\frac{\partial^2\psi}{\partial y^2}+\frac{\partial^2\psi}{\partial z^2}\right)+U(x,y,z)\psi=E\psi \qquad \text{36-19}$$

For a single isolated atom, the potential energy U depends only on the radial distance $r=\sqrt{x^2+y^2+z^2}$ of the electron from the center of the nucleus. The problem is then most conveniently treated using the spherical coordinates r, θ, and ϕ, which are related to the rectangular coordinates x, y, and z by

$$\begin{aligned} z &= r\cos\theta \\ x &= r\sin\theta\cos\phi \\ y &= r\sin\theta\sin\phi \end{aligned} \qquad \text{36-20}$$

These relations are shown in Figure 36-5. The transformation of Equation 36-19 from rectangular to spherical coordinates is a straightforward, but tedious, calculation, which we will omit. The result of this transformation can be found in Problem 42. We will discuss qualitatively some of the interesting features of the wave functions that satisfy this equation.

The transformed version of Equation 36-19 can be solved using the technique called separation of variables. This is accomplished by expressing the wave function $\psi(r,\theta,\phi)$ as a product of three functions, each of which is a function of only one of the three spherical coordinates:

$$\psi(r,\theta,\phi)=R(r)f(\theta)g(\phi) \qquad \text{36-21}$$

where R is a function that depends only on the radial coordinate r, f is a function that depends only on the polar coordinate θ, and g is a function that depends only on the azimuth coordinate ϕ. When this form of $\psi(r,\theta,\phi)$ is substituted into the Schrödinger equation, the Schrödinger equation can be transformed into three ordinary differential equations, one for $R(r)$, one for $f(\theta)$, and one for $g(\phi)$. The potential energy $U(r)$ appears only in the equation for $R(r)$, which is called the **radial equation.**

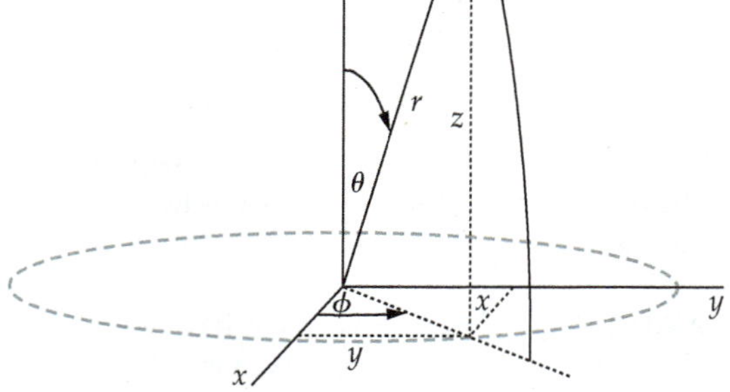

FIGURE 36-5 Geometric relations between spherical coordinates and rectangular coordinates.

Because the potential energy depends only on the coordinate r, the potential energy has no effect on the solutions of the equations for $f(\theta)$ and $g(\phi)$, and therefore has no effect on the angular dependence of the wave function $\psi(r, \theta, \phi)$. These solutions are applicable to *any* problem in which the potential energy depends only on r.

QUANTUM NUMBERS IN SPHERICAL COORDINATES

In three dimensions, the requirement that the wave function be continuous and normalizable introduces three quantum numbers, one associated with each spatial dimension. In spherical coordinates, the quantum number associated with r is labeled n, that associated with θ is labeled ℓ, and that associated with ϕ is labeled m_ℓ.* The quantum numbers n_1, n_2, and n_3 that we found in Chapter 35 for a particle in a three-dimensional square well in rectangular coordinates x, y, and z were independent of one another, but the quantum numbers associated with wave functions in spherical coordinates are dependent on each other. The possible values of these quantum numbers are

$$n = 1, 2, 3, \ldots$$
$$\ell = 0, 1, 2, 3, \ldots, n - 1$$
$$m_\ell = -\ell, -\ell + 1, -\ell + 2, \ldots, 0, \ldots, \ell - 2, \ell - 1, \ell \qquad 36\text{-}22$$

QUANTUM NUMBERS IN SPHERICAL COORDINATES

That is, n can be any positive integer; ℓ can be 0 or any positive integer up to $n - 1$; and m_ℓ can have $2\ell + 1$ possible values, ranging from $-\ell$ to $+\ell$ in integral steps.

The number n is called the **principal quantum number.** It is associated with the dependence of the wave function on the distance r and therefore with the probability of finding the electron at various distances from the nucleus. The quantum numbers ℓ and m_ℓ are associated with the orbital angular momentum of the electron and with the angular dependence of the electron wave function. The quantum number ℓ is called the **orbital quantum number.** The magnitude L of the orbital angular momentum $\vec{L}$ is related to the orbital quantum number ℓ by

$$L = \sqrt{\ell(\ell + 1)}\hbar \qquad 36\text{-}23$$

The quantum number m_ℓ is called the **magnetic quantum number.** It is related to the component of the orbital angular momentum along some direction in space. All spatial directions are equivalent for an isolated atom, but placing the atom in a magnetic field results in the direction of the magnetic field being separated out from the other directions. The convention is that the $+z$ direction is chosen for the magnetic-field direction. Then the z component of the orbital angular momentum of the electron is given by the quantum condition

$$L_z = m_\ell \hbar \qquad 36\text{-}24$$

This quantum condition arises from the boundary condition on the azimuth coordinate ϕ that the probability of finding the electron at some arbitrary angle ϕ_1 must be the same as that of finding the electron at angle $\phi_2 = \phi_1 + 2\pi$, because these two values of ϕ represent the same point in space.

If we measure the angular momentum of the electron in units of $\hbar$, we see that the magnitude of the orbital angular momentum is quantized to the value $\sqrt{\ell(\ell + 1)}$ units, and that its component along any direction can have only the $2\ell + 1$ values ranging from $-\ell$ to $+\ell$ units. Figure 36-6 shows a vector-model diagram illustrating the possible orientations of the angular-momentum vector for $\ell = 2$. Note that only specific values of θ are allowed; that is, the directions in space are quantized.

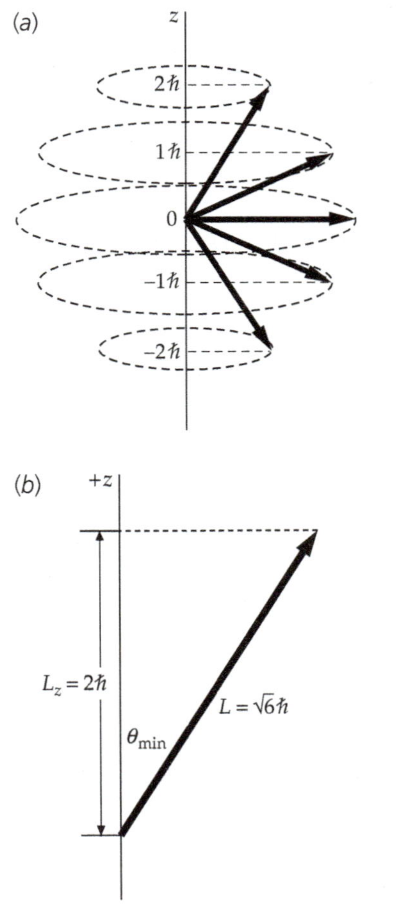

FIGURE 36-6 (*a*) Vector-model diagram illustrating the possible values of the z component of the orbital angular-momentum vector for the case $\ell = 2$. The magnitude of the orbital angular momentum is $L = \hbar\sqrt{\ell(\ell + 1)} = \hbar\sqrt{2(2 + 1)} = \hbar\sqrt{6}$. (*b*) The values of the z component of the orbital angular-momentum vector for the case $\ell = 2$ and $m_\ell = 2$. The value of the z component of the orbital angular momentum is $L_z = 2\hbar$.

* For simplicity, m_ℓ is sometimes written as m.

Example 36-3 The Directions of the Angular Momentum

If the orbital angular momentum is characterized by the quantum number $\ell = 2$, what are the possible values of L_z, and what is the smallest possible angle between $\vec{L}$ and the direction of increasing z?

PICTURE The possible orientations of $\vec{L}$ and the z axis are shown in Figure 36-6. The direction of increasing z is the direction of the external magnetic field in the vicinity of the atom.

SOLVE

1. Write the possible values of L_z:

$$\boxed{L_z = m_\ell \hbar \quad \text{where} \quad m_\ell = -2, -1, 0, 1, 2}$$

2. Express the angle θ between $\vec{L}$ and the $+z$ direction in terms of L and L_z:

$$\cos\theta = \frac{L_z}{L} = \frac{m_\ell \hbar}{\sqrt{\ell(\ell+1)}\hbar} = \frac{m_\ell}{\sqrt{\ell(\ell+1)}}$$

3. The smallest angle occurs when $\ell = 2$ and $m_\ell = 2$:

$$\cos\theta_{\text{min}} = \frac{2}{\sqrt{2(2+1)}} = \frac{2}{\sqrt{6}} = 0.816$$

$$\theta_{\text{min}} = \boxed{35.3°}$$

CHECK The angle in Figure 36-6*b* looks to be between about 30° and 40°, so the step-3 result of 35.3° is plausible.

TAKING IT FURTHER We note the somewhat strange result that the orbital angular-momentum vector cannot be parallel to the z axis.

PRACTICE PROBLEM 36-2 An atom is in a region that has a magnetic field. An electron in the atom has an angular momentum characterized by the quantum number $\ell = 4$. What are the possible values of m_ℓ for this electron?

36-4 QUANTUM THEORY OF THE HYDROGEN ATOM

We can treat the simplest atom, the hydrogen atom, as a stationary nucleus (a proton) and a single moving particle, an electron, which has linear momentum p and kinetic energy $p^2/2m$. The potential energy $U(r)$ due to the electrostatic attraction between the electron and the proton* is

$$U(r) = \frac{kZe^2}{r} \qquad \text{36-25}$$

For this potential-energy function, the Schrödinger equation can be solved exactly. In the lowest energy state, which is the ground state, the principal quantum number n has the value 1, ℓ is 0, and m_ℓ is 0.

ENERGY LEVELS

The allowed energies of the hydrogen atom that result from the solution of the Schrödinger equation are

$$E_n = -Z^2\frac{E_0}{n^2} \qquad n = 1, 2, 3, \ldots \qquad \text{36-26}$$

ENERGY LEVELS FOR HYDROGEN

* We include the factor Z, which is 1 for hydrogen, so that we can apply our results to other one-electron "atoms," such as singly ionized helium He^+, for which $Z = 2$.

where

$$E_0 = \frac{mk^2e^4}{2\hbar^2} = 13.6\text{ eV} \qquad 36\text{-}27$$

These energies are the same as those obtained using the Bohr model. Note that the energies E_n are negative, indicating that the electron is bound to the nucleus (thus the term *bound state*), and that the energies depend only on the principal quantum number n. The fact that the energy does not depend on the orbital quantum number ℓ is a peculiarity of the inverse-square force and holds only for an inverse r potential such as Equation 36-25. For atoms having multiple electrons, the interaction of the electrons leads to a dependence of the energy on ℓ. In general, the lower the value of ℓ, the lower the energy for such atoms. Because there is usually no preferred direction in space, the energy for any atom does not ordinarily depend on the magnetic quantum number m_ℓ, which is related to the z component of the angular momentum. However, the energy does depend on m_ℓ if the atom is in a magnetic field.

Figure 36-7 shows an energy-level diagram for hydrogen. This diagram is similar to Figure 36-4, except that the states which have the same value of n but different values of ℓ are shown separately. These states (called *terms*) are referred to by giving the value of n along with a code letter: s for $\ell = 0$, p for $\ell = 1$, d for $\ell = 2$, and f for $\ell = 3$.* (Lowercase letters s, p, d, f, and so on, are used to specify the orbital angular momentum of an individual electron, whereas uppercase letters S, P, D, F, and so on, are used to identify the orbital angular momentum for the entire multielectron atom. For a single-electron atom, like hydrogen, either uppercase or lowercase letters will suffice.) When an atom makes a transition from one allowed energy state to another, electromagnetic radiation in the form of a photon is emitted or absorbed. Such transitions result in spectral lines that are characteristic of the atom. The transitions obey the **selection rules:**

$$\Delta m_\ell = -1, 0, \text{ or } +1$$
$$\Delta\ell = -1 \text{ or } +1 \qquad 36\text{-}28$$

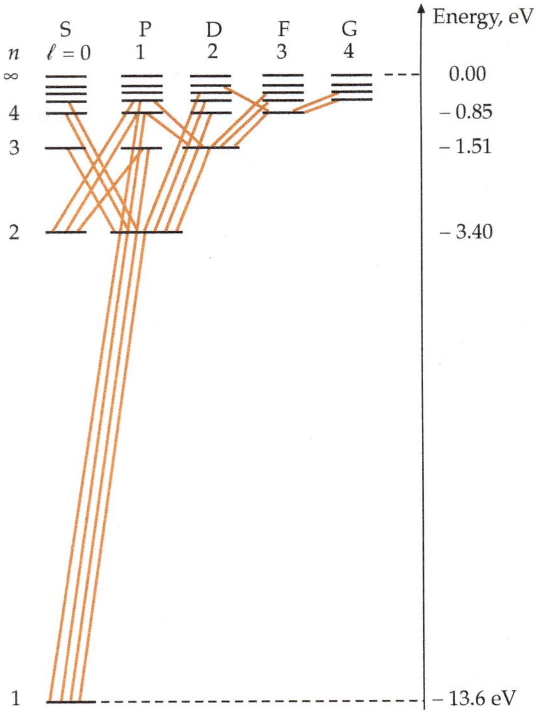

FIGURE 36-7 Energy-level diagram for hydrogen. The diagonal lines show transitions that involve emission or absorption of radiation and obey the selection rule $\Delta\ell = \pm 1$. States that have the same value of n but with different values of ℓ have the same energy $-E_0/n^2$, where $E_0 = 13.6$ eV as in the Bohr model.

These selection rules are related to the conservation of angular momentum and to the fact that the photon itself has an intrinsic angular momentum that has a maximum angular-momentum component of $1\hbar$ in any direction. The wavelengths of the light emitted by hydrogen (and by other atoms) are related to the energy levels by

$$hf = \frac{hc}{\lambda} = E_\text{i} - E_\text{f} \qquad 36\text{-}29$$

where E_i and E_f are the energies of the initial and final states.

WAVE FUNCTIONS AND PROBABILITY DENSITIES

The solutions of the Schrödinger equation in spherical coordinates are characterized by the quantum numbers n, ℓ, and m_ℓ, and are written $\psi_{n\ell m_\ell}$. The principal quantum number n can take on any of the values 1, 2, 3, ... In addition, for each value of n, ℓ can take on any of the values 0, 1, ..., $n - 1$, and for each value of ℓ, m_ℓ can take on any of the values $-\ell, -\ell + 1, -\ell + 2, \ldots, +\ell$. Thus, for any given value of n, there are n possible values of ℓ, and for any given value of ℓ, there are $2\ell + 1$ possible values of m_ℓ. For hydrogen, the energy depends only on n, so there are generally many different wave functions that correspond to the same

* These code letters for the values of ℓ are remnants of spectroscopists' descriptions of various spectral lines as *sharp, principal, diffuse,* and *fundamental*. For values greater than 3, the letters follow alphabetically, starting with g for $\ell = 4$.

energy (except at the lowest energy level, for which $n = 1$ and therefore ℓ and m_ℓ must both equal 0). These energy levels are therefore degenerate (see Section 35-5). The origins of this degeneracy are the $1/r$ dependence of the potential energy and the fact that, in the absence of any external fields, there is no preferred direction in space.*

The ground state In the lowest energy state, the ground state of hydrogen, the principal quantum number n has the value 1, ℓ is 0, and m_ℓ is 0. The energy is −13.6 eV, and the angular momentum is zero. (In the Bohr model of the atom the angular momentum in the ground state is equal to $\hbar$, not zero.) The wave function for the ground state is

$$\psi_{100} = C_{100}e^{-Zr/a_0} \qquad 36\text{-}30$$

where

$$a_0 = \frac{\hbar^2}{mke^2} = 0.0529 \text{ nm}$$

is the first Bohr radius and C_{100} is a constant that is determined by normalization. In three dimensions, the normalization condition is

$$\int |\psi|^2 dV = 1$$

where dV is a volume element and the integration is performed over all space. In spherical coordinates, the volume element (Figure 36-8) is

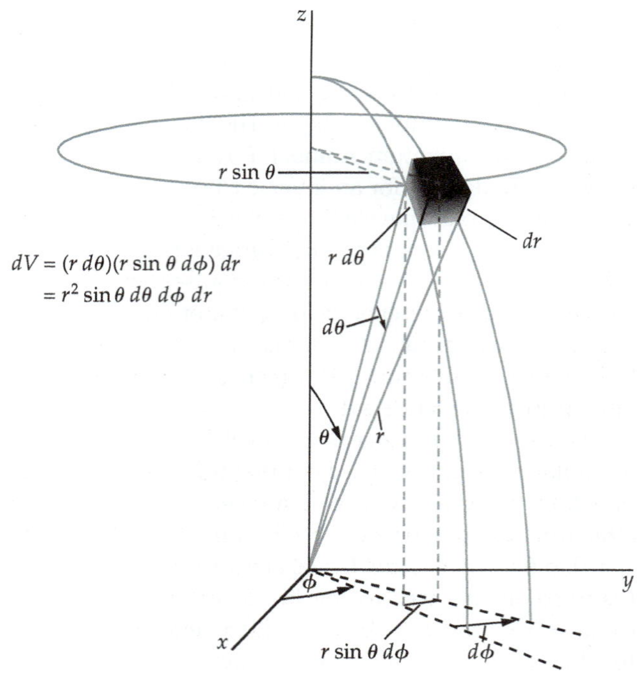

FIGURE 36-8 Volume element in spherical coordinates.

$$dV = (r\sin\theta\, d\phi)(r\, d\theta)dr = r^2 \sin\theta\, d\theta\, d\phi\, dr$$

We integrate over all space by integrating over ϕ from $\phi = 0$ to $\phi = 2\pi$, over θ from $\theta = 0$ to $\theta = \pi$, and over r from $r = 0$ to $r = \infty$. The normalization condition is thus

$$\int |\psi|^2 dV = \int_0^\infty \left[\int_0^\pi \left(\int_0^{2\pi} |\psi|^2 r^2 \sin\theta\, d\phi\right) d\theta\right] dr$$

$$= \int_0^\infty \left[\int_0^\pi \left(\int_0^{2\pi} C_{100}^2 e^{-2Zr/a_0} r^2 \sin\theta\, d\phi\right) d\theta\right] dr = 1$$

Because there is no θ or ϕ dependence in ψ_{100}, the triple integral can be factored into the product of three integrals. This gives

$$\int |\psi|^2 dV = \left(\int_0^{2\pi} d\phi\right)\left(\int_0^\pi \sin\theta\, d\theta\right)\left(\int_0^\infty C_{100}^2 e^{-2Zr/a_0} r^2\, dr\right)$$

$$= 2\pi \cdot 2 \cdot C_{100}^2 \left(\int_0^\infty e^{-2Zr/a_0} r^2\, dr\right) = 1$$

The remaining integral is of the form $\int_0^\infty x^n e^{-ax}\, dx$, where n a positive integer and $a > 0$. Using successive integration-by-parts operations† yields the result

$$\int_0^\infty x^n e^{-ax}\, dx = \frac{n!}{a^{n+1}}$$

so

$$\int_0^\infty r^2 e^{-2Zr/a_0}\, dr = \frac{a_0^3}{4Z^3}$$

See Math Tutorial *for more information on* **Integrals**

* If spin, relativistic effects, the spin of the nucleus, and quantum electrodynamics are considered, the degeneracy is broken.

† This integral can also be looked up in a table of integrals.

then

$$4\pi C_{100}^2\left(\frac{a_0^3}{4Z^3}\right) = 1$$

so

$$C_{100} = \frac{1}{\sqrt{\pi}}\left(\frac{Z}{a_0}\right)^{3/2} \qquad 36\text{-}31$$

The normalized ground-state wave function is thus

$$\psi_{100} = \frac{1}{\sqrt{\pi}}\left(\frac{Z}{a_0}\right)^{3/2} e^{-Zr/a_0} \qquad 36\text{-}32$$

The probability of finding the electron in a volume dV is $|\psi|^2\,dV$. The probability density $|\psi|^2$ is shown in Figure 36-9. Note that this probability density is spherically symmetric; that is, the probability density depends only on r, and is independent of θ or ϕ. The probability density is maximum at the origin.

We are more often interested in the probability of finding the electron at some radial distance r between r and $r + dr$. This radial probability $P(r)\,dr$ is the probability density $|\psi|^2$ multiplied by the volume of the spherical shell of thickness dr, which is $dV = 4\pi r^2\,dr$. The probability of finding the electron in the range from r to $r + dr$ is thus $P(r)\,dr = |\psi|^2\,4\pi r^2\,dr$, and the **radial probability density** is

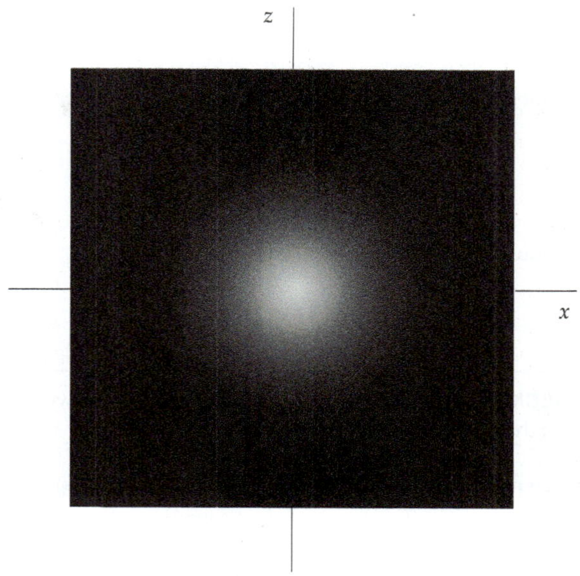

FIGURE 36-9 Computer-generated picture of the probability density $|\psi|^2$ for the ground state of hydrogen. The quantity $-e|\psi|^2$ can be thought of as the electron charge density in the atom. The density is spherically symmetric, is greatest at the origin, and decreases exponentially with r.

$$P(r) = 4\pi r^2|\psi|^2 \qquad 36\text{-}33$$

RADIAL PROBABILITY DENSITY

For the hydrogen atom in the ground state, the radial probability density is

$$P(r) = 4\pi r^2|\psi|^2 = 4\pi C_{100}^2 r^2 e^{-2Zr/a_0} = 4\left(\frac{Z}{a_0}\right)^3 r^2 e^{-2Zr/a_0} \qquad 36\text{-}34$$

Figure 36-10 shows the radial probability density $P(r)$ as a function of r. The maximum value of $P(r)$ occurs at $r = a_0/Z$, which for $Z = 1$ is the first Bohr radius. In contrast to the Bohr model, in which the electron stays in a well-defined orbit at $r = a_0$, we see that it is possible for the electron to be found at any distance from the nucleus. However, the most probable distance is a_0 (assuming $Z = 1$), and the chance of finding the electron at a much different distance is small. It is often useful to think of the electron in an atom as a charged cloud of charge density $\rho = -e|\psi|^2$, but we should remember that when it interacts with matter, an electron is always observed as a single charge.

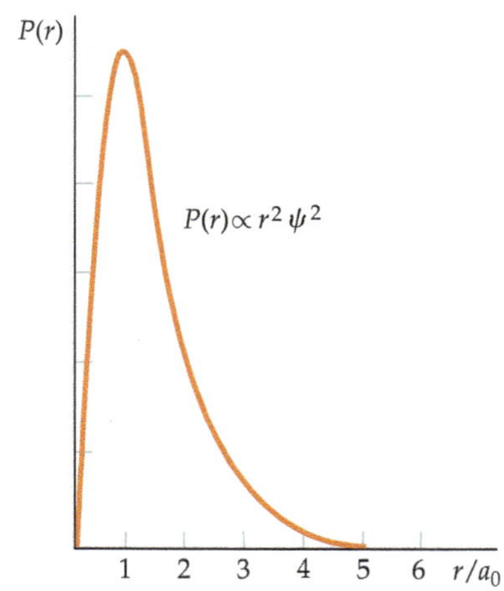

FIGURE 36-10 Radial probability density $P(r)$ versus r/a_0 for the ground state of the hydrogen atom. $P(r)$ is proportional to $r^2\psi^2$. The value of r for which $P(r)$ is maximum is the most probable distance $r = a_0$.

Example 36-4 Probability That the Electron Is in a Thin Spherical Shell

Consider a hydrogen atom that is in the ground state. Estimate the probability of finding the electron in a thin spherical shell of inner radius r and outer radius $r + \Delta r$, where $\Delta r = 0.06a_0$ at (*a*) $r = a_0$ and (*b*) $r = 2a_0$.

PICTURE Because Δr is so small compared to r, the variation in the radial probability density $P(r)$ in the shell can be neglected. The probability of finding the electron in some small range Δr is then $P(r)\,\Delta r$.

SOLVE

1. Substitute $Z = 1$ and $r = a_0$ into Equation 36-34:

$$P(r)\Delta r = \left[4\left(\frac{1}{a_0}\right)^3 r^2 e^{-2r/a_0}\right]\Delta r$$

$$P(a_0)(0.06a_0) = \left[4\left(\frac{1}{a_0}\right)^3 a_0^2 e^{-2}\right](0.06a_0) = \boxed{0.0325}$$

2. Substitute $Z = 1$ and $r = 2a_0$ into Equation 36-34:

$$P(r)\Delta r = \left[4\left(\frac{1}{a_0}\right)^3 r^2 e^{-2r/a_0}\right]\Delta r$$

$$P(2a_0)(0.06a_0) = \left[4\left(\frac{1}{a_0}\right)^3 4a_0^2 e^{-4}\right](0.06a_0) = \boxed{0.0176}$$

CHECK The probability of finding the electron between $r = a_0$ and $r = a_0 + 0.06a_0$ is larger than the probability of finding the particle between $r = 2a_0$ and $r = 2a_0 + 0.06a_0$, as expected.

TAKING IT FURTHER The volume of the spherical shell that has an inner radius $2a_0$ and outer radius $2a_0 + 0.06a_0$ is almost four times larger than the volume of the spherical shell that has an inner radius a_0 and outer radius $a_0 + 0.06a_0$. In spite of this there is approximately a 3 percent chance of finding the electron in this range at $r = a_0$, but at $r = 2a_0$ the chance is slightly less than 2 percent.

The first excited state In the first excited state of a hydrogen atom, n is equal to 2 and ℓ can equal either 0 or 1. For $\ell = 0$, $m_\ell = 0$, and we again have a spherically symmetric wave function, this time given by

$$\psi_{200} = C_{200}\left(2 - \frac{Zr}{a_0}\right)e^{-Zr/(2a_0)} \tag{36-35}$$

For $\ell = 1$, m_ℓ can be $+1$, 0, or -1. The corresponding wave functions are

$$\psi_{210} = C_{210}\frac{Zr}{a_0}e^{-Zr/(2a_0)}\cos\theta \tag{36-36}$$

$$\psi_{21\pm1} = C_{211}\frac{Zr}{a_0}e^{-Zr/(2a_0)}\sin\theta\, e^{\pm i\phi} \tag{36-37}$$

where C_{200}, C_{210}, and C_{211} are normalization constants. The probability densities are given by

$$\psi_{200}^2 = C_{200}^2\left(2 - \frac{Zr}{a_0}\right)^2 e^{-Zr/a_0} \tag{36-38}$$

$$\psi_{210}^2 = C_{210}^2\left(\frac{Zr}{a_0}\right)^2 e^{-Zr/a_0}\cos^2\theta \tag{36-39}$$

$$|\psi_{21\pm1}|^2 = C_{211}^2\left(\frac{Zr}{a_0}\right)^2 e^{-Zr/a_0}\sin^2\theta \tag{36-40}$$

The wave functions and probability densities for $\ell \neq 0$ are not spherically symmetric, but instead depend on the angle θ. The probability densities do not depend on ϕ. Figure 36-11 shows the probability density $|\psi|^2$ for $n = 2$, $\ell = 0$, and $m_\ell = 0$ (Figure 36-11*a*); for $n = 2$, $\ell = 1$, and $m_\ell = 0$ (Figure 36-11*b*); and for $n = 2$, $\ell = 1$, and $m_\ell = \pm1$ (Figure 36-11*c*). An important feature of these plots is that the electron cloud is spherically symmetric for $\ell = 0$ and is not spherically symmetric for $\ell \neq 0$. These angular distributions of the electron charge density depend only on the values of ℓ and m_ℓ and not on the radial part of the wave function. Similar charge distributions for the valence electrons of more complicated atoms play an important role in the chemistry of molecular bonding. (Electrons in the outermost shell are called **valence electrons.**)

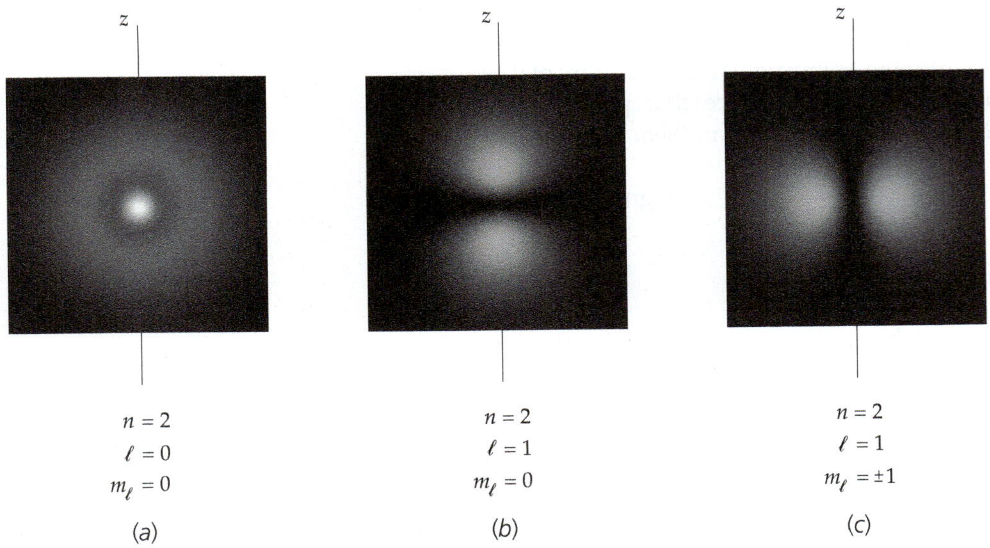

FIGURE 36-11 Computer-generated picture of the probability densities $|\psi|^2$ for the electron in the $n = 2$ states of hydrogen. All three images represent figures of revolution about the z axis. (*a*) For $\ell = 0$, $|\psi|^2$ is spherically symmetric. (*b*) For $\ell = 1$ and $m_\ell = 0$, $|\psi|^2$ is proportional to $\cos^2\theta$. (*c*) For $\ell = 1$ and $m_\ell = +1$ or -1, $|\psi|^2$ is proportional to $\sin^2\theta$.

Figure 36-12 shows the probability density for finding the electron at a distance r from the nucleus for $n = 2$, when $\ell = 1$ and when $\ell = 0$. We can see from the figure that this radial probability density depends on ℓ as well as on n.

For $n = 1$, we found that the most likely distance between the electron and the nucleus is a_0, which is the first Bohr radius, whereas for $n = 2$ and $\ell = 1$, the most probable distance between the electron and the nucleus is $4a_0$. These are the orbital radii for the first and second Bohr orbits (Equation 36-11). For $n = 3$ (and $\ell = 2$),* the most likely distance between the electron and nucleus is $9a_0$, which is the radius of the third Bohr orbit.

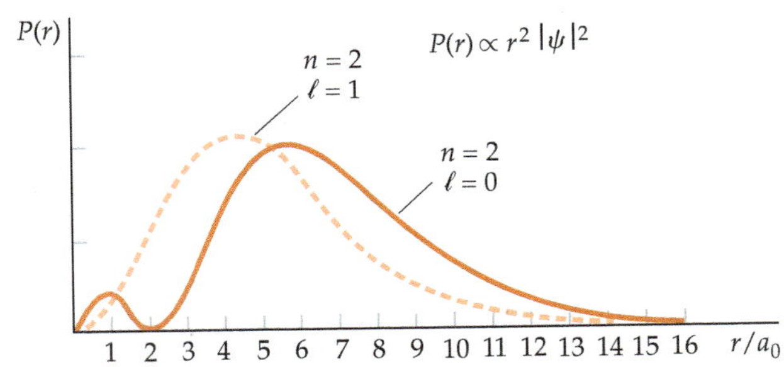

FIGURE 36-12 Radial probability density $P(r)$ versus r/a_0 for the $n = 2$ states of hydrogen. For $\ell = 1$, $P(r)$ is maximum at the Bohr value $r_2 = 2^2 a_0$. For $\ell = 0$, there is a maximum near this value and a much smaller maximum near the origin.

36-5 THE SPIN–ORBIT EFFECT AND FINE STRUCTURE

The orbital magnetic moment of an electron in an atom can be derived semiclassically, even though it is quantum mechanical in origin.† Consider a particle of mass m and charge q moving with speed v in a circle of radius r. The magnitude of the orbital angular momentum of the particle is $L = mvr$, and the magnitude of the magnetic moment is the product of the current and the area of the circle $\mu = IA = I\pi r^2$. If T is the time for the charge to complete one revolution, the current (charge passing a point per unit time) is q/T. Because the period T is the distance $2\pi r$ divided by the velocity v, the current is $I = q/T = qv/(2\pi r)$. The magnetic moment is then

$$\mu = IA = \frac{qv}{2\pi r}\pi r^2 = \frac{1}{2}qvr = \frac{q}{2m}L$$

where we have substituted L/m for vr. If the charge q is positive, the orbital angular momentum and orbital magnetic moment vectors are in the same direction. We can therefore write

$$\vec{\mu} = \frac{q}{2m}\vec{L} \qquad 36\text{-}41$$

* The correspondence with the Bohr model is closest for the maximum value of ℓ, which is $n - 1$.

† This topic was first presented in Section 27-5.

Equation 36-41 is the general classical relation between magnetic moment and angular momentum. It also holds in the quantum theory of the atom for orbital angular momentum, but not for the intrinsic spin angular momentum of the electron. For electron spin, the magnetic moment is twice that predicted by Equation 36-41.* The extra factor of 2 is a result from quantum theory that has no analog in classical mechanics.

The quantum of angular momentum is $\hbar$, so we express the magnetic moment in terms of $\vec{L}/\hbar$:

$$\vec{\mu} = \frac{q\hbar}{2m}\frac{\vec{L}}{\hbar}$$

For an electron, $m = m_e$ and $q = -e$, so the magnetic moment of the electron due to its orbital motion is

$$\vec{\mu}_\ell = -\frac{e\hbar}{2m_e}\frac{\vec{L}}{\hbar} = -\mu_B\frac{\vec{L}}{\hbar}$$

where $\mu_B = e\hbar/(2m_e) = 5.79 \times 10^{-5}$ eV/T is the quantum unit of magnetic moment called a Bohr magneton. The magnetic moment of an electron due to its intrinsic spin angular momentum $\vec{S}$ is

$$\vec{\mu}_S = -2\frac{e\hbar}{2m_e}\frac{\vec{S}}{\hbar} = -2\mu_B\frac{\vec{S}}{\hbar}$$

In general, an electron in an atom has both orbital angular momentum characterized by the quantum number ℓ and spin angular momentum characterized by the quantum number s. Analogous classical systems that have two kinds of angular momentum are Earth, which is spinning about its axis of rotation in addition to revolving about the Sun, and a precessing gyroscope that has angular momentum of precession in addition to its spin. The total angular momentum $\vec{J}$ is the sum of the orbital angular momentum $\vec{L}$ and the spin angular momentum $\vec{S}$, where

$$\vec{J} = \vec{L} + \vec{S} \qquad \text{36-42}$$

Classically $\vec{J}$ is an important quantity because the resultant torque on a system equals the rate of change of the total angular momentum, and in the case of only central forces, the total angular momentum is conserved. For a classical system, the direction of the total angular momentum $\vec{J}$ is without restrictions and the magnitude of $\vec{J}$ can take on any value between $J_{max} = L + S$ and $J_{min} = |L - S|$. In quantum mechanics, however, the directions of both $\vec{L}$ and $\vec{S}$ are more restricted and the magnitudes L and S are both quantized. Furthermore, like $\vec{L}$ and $\vec{S}$, the direction of the total angular momentum $\vec{J}$ is restricted and the magnitude of $\vec{J}$ is quantized. For an electron that has an orbital angular momentum characterized by the quantum number ℓ and spin $s = \frac{1}{2}$, the total angular-momentum magnitude J is equal to $\sqrt{j(j+1)}\hbar$, where the quantum number j is given by

$$j = +\tfrac{1}{2} \quad \text{if} \quad \ell = 0$$

and either

$$j = \ell + \tfrac{1}{2} \quad \text{or} \quad j = \ell - \tfrac{1}{2} \quad \text{if} \quad \ell > 0 \qquad \text{36-43}$$

Figure 36-13 is a vector model illustrating the two possible combinations $j = \frac{3}{2}$ and $j = \frac{1}{2}$ for the case of $\ell = 1$. The lengths of the vectors are proportional to $\sqrt{\ell(\ell+1)}\hbar$, $\sqrt{s(s+1)}\hbar$, and $\sqrt{j(j+1)}\hbar$. The spin angular momentum and the orbital angular momentum are said to be *parallel* when $j = \ell + s$ and *antiparallel* when $j = \ell - s$.

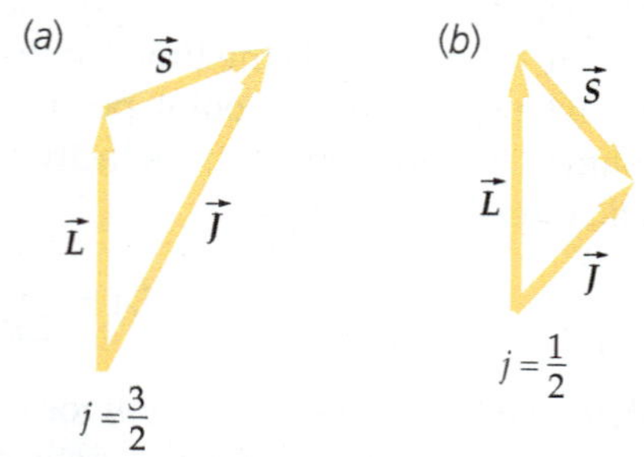

FIGURE 36-13 Vector diagrams illustrating the addition of orbital angular momentum and spin angular momentum for the case $\ell = 1$ and $s = \frac{1}{2}$. There are two possible values of the quantum number for the total angular momentum: $j = \ell + s = \frac{3}{2}$ and $j = \ell - s = \frac{1}{2}$.

* This result and the phenomenon of electron spin itself, was predicted in 1927 by Paul Dirac, who combined special relativity and quantum mechanics into a relativistic wave equation called the *Dirac equation*. Precise measurements indicate that the magnetic moment of the electron due to its spin is 2.00232 times that predicted by Equation 36-42. The fact that the intrinsic magnetic moment of the electron is approximately twice what we would expect makes it clear that the simple model of the electron as a spinning ball is not to be taken literally.

Atomic states that have the same n and ℓ values but with different j values have slightly different energies because of the interaction of the spin of the electron with its orbital motion. This effect is called the **spin–orbit effect.** The resulting splitting of spectral lines is called **fine-structure splitting.**

In the notation $n\ell_j$, the ground state of the hydrogen atom is written $1s_{1/2}$, where the 1 indicates that $n = 1$, the s indicates that $\ell = 0$, and the 1/2 indicates that $j = \frac{1}{2}$. The $n = 2$ states can have either $\ell = 0$ or $\ell = 1$, and the $\ell = 1$ state can have either $j = \frac{3}{2}$ or $j = \frac{1}{2}$. These states are thus denoted by $2s_{1/2}$, $2p_{3/2}$, and $2p_{1/2}$. Because of the spin–orbit effect, the $2p_{3/2}$ and $2p_{1/2}$ states have slightly different energies resulting in the fine-structure splitting of the transitions $2p_{3/2} \rightarrow 2s_{1/2}$ and $2p_{1/2} \rightarrow 2s_{1/2}$.

We can understand the spin–orbit effect qualitatively from a simple Bohr-model picture, as shown in Figure 36-14. In this figure, the electron moves in a circular orbit around a fixed proton. In Figure 36-14*a*, the orbital angular momentum $\vec{L}$ is up. In an inertial reference frame in which the electron is momentarily at rest (see Figure 36-14*b*), the proton is moving at right angles to the line connecting the proton and the electron. The moving proton produces a magnetic field $\vec{B}$ at the position of the electron. The direction of $\vec{B}$ is up, parallel to $\vec{L}$. The energy of the electron depends on its spin because of the magnetic moment $\vec{\mu}_s$ associated with the electron's spin. The energy is lowest when $\vec{\mu}_s$ is parallel to $\vec{B}$ and the energy is highest when it is antiparallel. This energy is given by (Equation 36-16).

$$U = -\vec{\mu}_s \cdot \vec{B} = -\mu_{s_z} B \approx -\mu_B B \qquad 36\text{-}44^*$$

Because $\vec{\mu}_s$ is directed opposite to its spin (because the electron has a negative charge), the energy is lowest when the spin $\vec{S}$ is antiparallel to $\vec{B}$ and thus to $\vec{L}$. The energy of the $2p_{1/2}$ state in hydrogen, in which $\vec{L}$ and $\vec{S}$ are antiparallel (Figure 36-15), is therefore slightly lower than that of the $2p_{3/2}$ state, in which $\vec{L}$ and $\vec{S}$ are parallel.

* Transferring the energy of the dipole to the frame of the proton gives a factor of 2, which is included in this result.

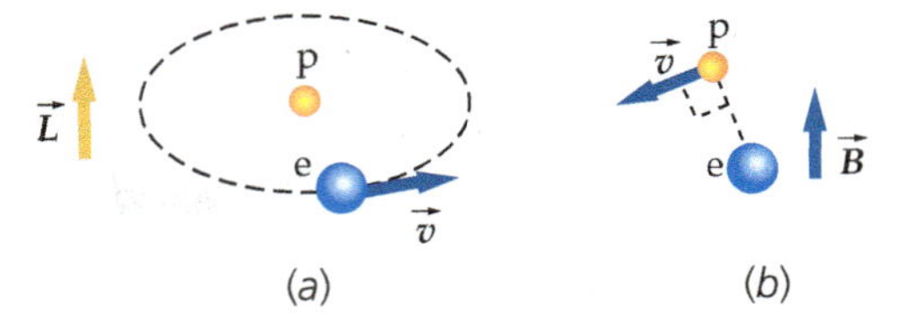

FIGURE 36-14 (*a*) An electron moving about a proton in a circular orbit in the horizontal plane with angular momentum $\vec{L}$ up. (*b*) In an inertial reference frame in which the electron is momentarily at rest there is, at the location of the electron, a magnetic field $\vec{B}$ due to the motion of the proton that is also directed up. When the electron spin $\vec{S}$ is parallel to $\vec{L}$, its magnetic moment $\vec{\mu}_s$ is antiparallel to $\vec{L}$ and $\vec{B}$, so the spin–orbit energy is at its greatest.

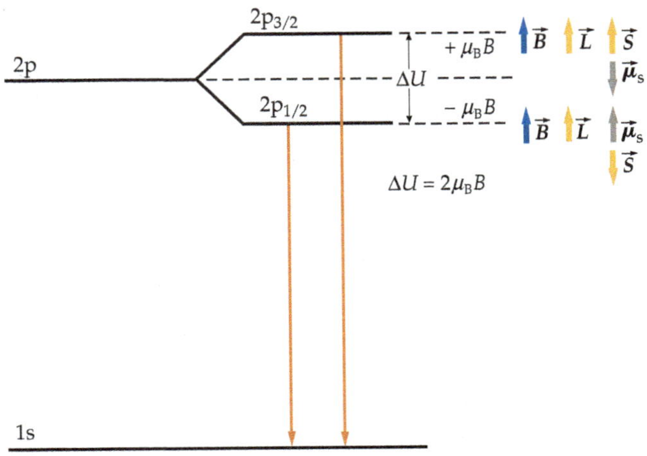

FIGURE 36-15 Fine-structure energy-level diagram. On the left, the levels in the absence of a magnetic field are shown. The effect of the field is shown on the right. Because of the spin-orbit interaction, the magnetic field splits the 2p level into two energy levels, with the $j = \frac{3}{2}$ level having slightly greater energy than the $j = \frac{1}{2}$ level. The spectral line due to the transition $2p \rightarrow 1s$ is therefore split into two lines of slightly different wavelengths.

Example 36-5 Determining *B* by Fine-Structure Splitting

As a consequence of fine-structure splitting, the energies of the $2p_{3/2}$ and $2p_{1/2}$ levels in hydrogen differ by 4.5×10^{-5} eV. If the 2p electron sees an internal magnetic field of magnitude B, the spin–orbit energy splitting will be of the order of $\Delta E = 2\mu_B B$, where μ_B is the Bohr magneton. From this, estimate the magnetic field that the 2p electron in hydrogen experiences.

PICTURE Use the equation $\Delta E = 2\mu_B B$ along with the given value of the energy difference and known value of μ_B.

SOLVE

1. Write the spin–orbit energy splitting in terms of the magnetic moment:

$$\Delta E = 2\mu_B B$$

where

$$\Delta E = 4.5 \times 10^{-5}\ \text{eV}$$

2. Solve for the magnetic field B:

$$B = \frac{\Delta E}{2\mu_B} = \frac{4.5 \times 10^{-5}\ \text{eV}}{2(5.79 \times 10^{-5}\ \text{eV/T})} = \boxed{0.389\ \text{T}}$$

36-6 THE PERIODIC TABLE

For atoms that have more than one electron, the Schrödinger equation cannot be solved exactly. However, powerful approximation methods allow us to calculate the energy levels of the atoms and wave functions of the electrons to a high degree of accuracy. As a first approximation, the Z electrons in an atom are assumed to be noninteracting. The Schrödinger equation can then be solved, and the resulting wave functions used to calculate the interaction of the electrons, which in turn can be used to better approximate the wave functions. Because the spin of an electron can have two possible components along an axis, there is an additional quantum number m_s, which can have the possible values $+\frac{1}{2}$ or $-\frac{1}{2}$. The state of each electron is thus described by the four quantum numbers n, ℓ, m_ℓ, and m_s, and such states are called **stationary states.** The energy of the electron is determined mainly by the principal quantum number n (which is related to the radial dependence of the wave function) and by the orbital angular-momentum quantum number ℓ. Generally, the lower the values of n, the lower the energy; and for a given value of n, the lower the value of ℓ, the lower the energy. The dependence of the energy on ℓ is due to the interaction of the electrons in the atom with each other. In hydrogen, of course, there is only one electron, and the energy is independent of ℓ. The specification of n, ℓ, m_ℓ, and m_s for each electron in an atom is called the **electron configuration.** Customarily, ℓ is specified according to the same code used to label the states of the hydrogen atom rather than by its numerical value. The code is

	s	p	d	f	g	h
ℓ value	0	1	2	3	4	5

The n values are sometimes referred to as shells, which are identified by another letter code: $n = 1$ denotes the K shell*; $n = 2$, the L shell; and so on.

The electron configuration of atoms is constrained by the Pauli exclusion principle—no two electrons in an atom can be in the same quantum state. That is, no two electrons can have the same set of values for the quantum numbers n, ℓ, m_ℓ, and m_s. Using the exclusion principle and the restrictions on the quantum numbers discussed in the previous sections (n is a positive integer, ℓ is an integer that ranges from 0 to $n - 1$, m_ℓ can have $2\ell + 1$ values from $-\ell$ to ℓ in integral steps, and m_s can be either $+\frac{1}{2}$ or $-\frac{1}{2}$), we can understand much of the structure of the periodic table.

We have already discussed the lightest element, hydrogen, which has just one electron. In the ground (lowest energy) state, the electron has $n = 1$ and $\ell = 0$, with $m_\ell = 0$ and $m_s = +\frac{1}{2}$ or $-\frac{1}{2}$. We call this a 1s electron. The 1 signifies that $n = 1$, and the s signifies that $\ell = 0$.

Electrons of atoms whose atomic numbers are greater than 1 will have states that will give the lowest total energy consistent with the Pauli exclusion principle.

CONCEPT CHECK 36-1

The following table lists candidates for the quantum numbers of an electron in an atom. Which of these candidates are not found in nature?

	n	ℓ	m_ℓ	m_s
(*a*)	2	2	−1	$+\frac{1}{2}$
(*b*)	3	2	−1	$+\frac{1}{2}$
(*c*)	2	−1	−1	$-\frac{1}{2}$
(*d*)	3	0	1	$+\frac{1}{2}$
(*e*)	3	1	1	$+\frac{1}{2}$

HELIUM ($Z = 2$)

The next element after hydrogen in the periodic table is helium ($Z = 2$); a helium atom has two electrons. In its ground state, both electrons are in the K shell, where $n = 1$, $\ell = 0$, and $m_\ell = 0$; one electron has $m_s = +\frac{1}{2}$ and the other has $m_s = -\frac{1}{2}$. This configuration is lower in energy than any other two-electron configuration. The resultant spin of the two electrons is zero. Because the orbital angular momentum is also zero, the total angular momentum is zero. The electron configuration for helium is written $1s^2$. The 1 signifies that $n = 1$, the s signifies that $\ell = 0$, and the superscript 2 signifies that there are two electrons in this state. Because ℓ

* The designation of the $n = 1$ shell as K is usually found when dealing with X-ray levels where the final shell in an inner electron transition is labeled as K, L, M, and so on.

can be only 0 for $n = 1$, these two electrons fill the K ($n = 1$) shell. The energy required to remove the most loosely bound electron from an atom in the ground state is called the **first ionization energy.** For a helium atom, the first ionization energy is 24.6 eV, which is relatively large. Helium is therefore basically inert.

Example 36-6 Electron Interaction Energy in Helium

(*a*) Use the measured first ionization energy to calculate the energy of interaction of the two electrons in the ground state of the helium atom. (*b*) Use your result to estimate the average separation of the two electrons.

PICTURE The energy of one electron in the ground state of helium is given by $E_n = -Z^2E_0/n^2$ (Equation 36-26), where $n = 1$ and $Z = 2$. If the electrons did not interact, the energy of the second electron would also be E_1, the same as that of the first electron. Thus, for an atom that has noninteracting electrons, the ionization energy of the first electron removed would be $|E_1|$ and the ground-state energy would be $E_{non} = 2E_1$. This is represented by the lowest level in Figure 36-16. Because of the interaction energy, the ground-state energy is greater than $2E_1$, which is represented by the higher level labeled E_g in the figure. The measured first ionization energy of helium is 24.6 eV. When we add $E_{ion} = 24.6$ eV to ionize He, we obtain ionized helium, written He^+, which has just one electron and therefore energy E_1.

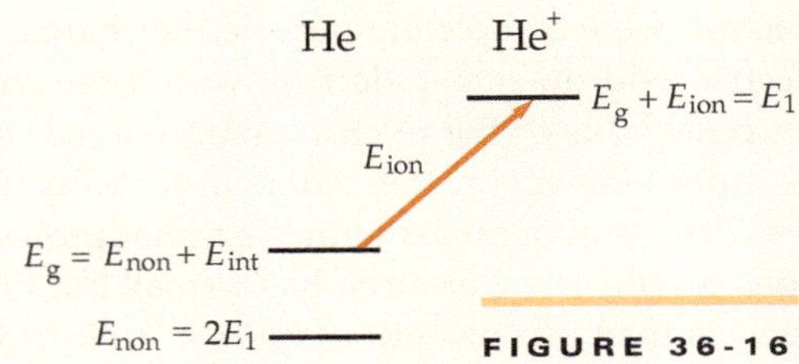

FIGURE 36-16

SOLVE

(*a*) 1. The sum of the energy of interaction E_{int} and the energy of two noninteracting electrons E_{non} equals the ground-state energy of helium:

$$E_{int} + E_{non} = E_g$$

2. Solve for E_{int} and substitute $E_{non} = 2E_1$:

$$E_{int} = E_g - E_{non} = E_g - 2E_1$$

3. Use Equation 36-26 to calculate the energy E_1 of one electron in the ground state:

$$E_n = -Z^2\frac{E_0}{n^2}$$

so

$$E_1 = -(2)^2\frac{13.6\text{ eV}}{1^2} = -54.4\text{ eV}$$

4. Substitute this value for E_1:

$$E_{int} = E_g - 2E_1 = E_g - 2(-54.4\text{ eV}) = E_g + 108.8\text{ eV}$$

5. The sum of the ground-state energy of He E_g and the ionization energy E_{ion} equals the ground-state energy of He^+, which is E_1:

$$E_g + E_{ion} = E_1 = -54.4\text{ eV}$$

6. Substitute $E_{ion} = 24.6$ eV to calculate E_g:

$$E_g = E_1 - E_{ion} = -54.4\text{ eV} - 24.6\text{ eV} = -79.0\text{ eV}$$

7. Substitute this result for E_g to obtain E_{int}:

$$E_{int} = E_g + 108.8\text{ eV} = -79.0\text{ eV} + 108.8\text{ eV} = \boxed{29.8\text{ eV}}$$

(*b*) 1. The energy of interaction of two electrons separated by distance r_s is the potential energy:

$$U = +\frac{ke^2}{r_s}$$

2. Set U equal to 29.8 eV, and solve for r_s. It is convenient to express r in terms of a_0, the radius of the first Bohr orbit in hydrogen, and to use $E_0 = ke^2/(2a_0) = 13.6$ eV (Equation 36-16):

$$r_s = \frac{ke^2}{U} = \frac{ke^2}{a_0}\frac{a_0}{U} = 2\frac{ke^2}{2a_0}\frac{a_0}{U} = 2\frac{E_0}{U}a_0 = 2\frac{13.6\text{ eV}}{29.8\text{ eV}}a_0 = \boxed{0.913a_0}$$

CHECK This separation is approximately the size of the diameter d_1 of the first Bohr orbit for an electron in helium, which is $d_1 = 2r_1 = 2a_0/Z = a_0$.

LITHIUM ($Z = 3$)

The next element, lithium, has an atom that has three electrons. Because the K shell ($n = 1$) of a ground-state lithium atom is completely filled with two electrons, the third electron occupies a higher energy shell. The next lowest energy shell after $n = 1$ is the $n = 2$ or L shell. This $n = 2$ electron has a greater probability of being much farther from the nucleus than the two $n = 1$ electrons. It is most likely to be found at a radius near that of the second Bohr orbit, which is four times the radius of the first Bohr orbit.

The nuclear charge is partially screened from the $n = 2$ electron by the two $n = 1$ electrons. Recall that the electric field outside a spherically symmetric charge density is the same as if all the charge were at the center of the sphere. If the $n = 2$ electron were completely outside the charge cloud of the two $n = 1$ electrons, the electric field the $n = 2$ electron would see would be that of a single charge $+e$ at the center due to the nuclear charge of $+3e$ and the charge $-2e$ of the two $n = 1$ electrons. However, the radial probability distribution (Equation 36-33) of the $n = 2$ electron overlaps with the radial probability distributions of the $n = 1$ electrons, so the $n = 2$ electron has a small but finite probability of being closer to the nucleus than one or both of the $n = 1$ electrons. Because of this, the effective nuclear charge $Z'e$ seen by the $n = 2$ electron is somewhat greater than $+1e$. Consequently, the energy of the $n = 2$ electron at a distance r from a point charge $+Z'e$ is given by Equation 36-6, where the nuclear charge $+Z$ replaced by $+Z'$.

$$E = -\frac{1}{2}\frac{kZ'e^2}{r} \tag{36-45}$$

The greater the overlap of the radial probability distributions of a higher energy electron with lower energy electrons, the greater is the effective nuclear charge $Z'e$ seen by the higher energy electron and the lower is the energy of the higher energy electron. Because the overlap is greater for ℓ values closer to zero (see Figure 36-12), the energy of the $n = 2$ electron in lithium is lower for the s state ($\ell = 0$) than for the p state ($\ell = 1$). The electron configuration of a lithium atom in the ground state is therefore $1s^2 2s$. The first ionization energy of a lithium atom is only 5.39 eV. Because its $n = 2$ electron is so loosely bound to the atom, lithium is very active chemically. It behaves like a one-electron atom, similar to hydrogen.

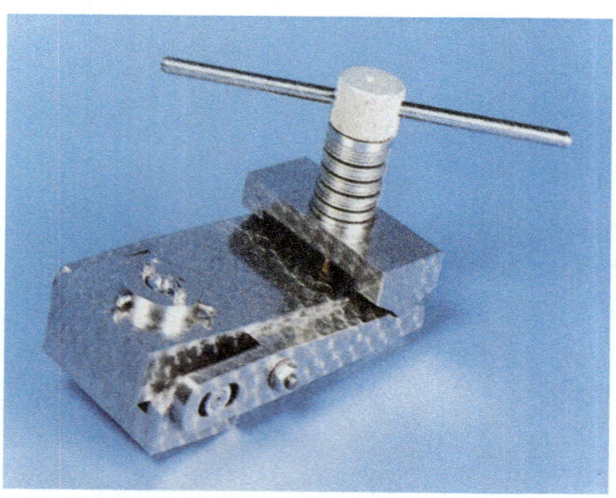

(a)

(b)

(*a*) A diamond anvil cell, in which the facets of two diamonds (approximately 1 mm^2 each) are used to compress a sample substance, subjecting it to very high pressure. (*b*) Samarium monosulfide (SmS) is normally a black, dull-looking semiconductor. When it is subjected to pressure above 7000 atm, an electron from the 4f state moves into the 5d state. The resulting compound glitters like gold and behaves like a metal. *((a) and (b) A. Jayaraman/AT&T Bell Labs.)*

Example 36-7 Effective Nuclear Charge for an Electron of a Lithium Atom

Suppose the radial probability distribution of the $n = 2$ electron in the lithium atom in the ground state did not overlap the probability distribution of the two $n = 1$ electrons; the nuclear charge would be shielded by the two $n = 1$ electrons and the effective nuclear charge would be $Z'e$, where $Z' = 1$. Then the energy of the $n = 2$ electron would be $-(13.6 \text{ eV})/2^2 = -3.4$ eV. However, the first ionization energy of lithium is 5.39 eV, not 3.4 eV. Use this fact to calculate the effective nuclear charge Z' seen by the $n = 2$ electron in lithium.

PICTURE Because the $n = 2$ electron is in the $n = 2$ shell, we will take $r = 4a_0$ for its average distance from the nucleus. We can then calculate Z' from Equation 36-45. Because r is given in terms of a_0, it will be convenient to use the fact that $E_0 = ke^2/(2a_0) = 13.6$ eV (Equation 36-16).

SOLVE

1. Equation 36-45 relates the energy of the $n = 2$ electron to its average distance r from the nucleus and the effective nuclear charge Z':

$$E = -\frac{1}{2}\frac{kZ'e^2}{r}$$

2. Substitute the given values $r = 4a_0$ and $E = -5.39$ eV:

$$-5.39 \text{ eV} = -\frac{1}{2}\frac{kZ'e^2}{4a_0}$$

3. Use $ke^2/(2a_0) = E_0 = 13.6$ eV and solve for Z':

$$-5.39 \text{ eV} = -\frac{Z'}{4}\frac{ke^2}{2a_0} = -\frac{Z'}{4}(13.6 \text{ eV})$$

so

$$Z' = 4\frac{5.39 \text{ eV}}{13.\ 6 \text{ eV}} = \boxed{1.59}$$

CHECK We expected Z' to be larger than one and certainly less than 3. Our step-3 result meets these expectations.

TAKING IT FURTHER This calculation is interesting but not very rigorous. We essentially used the radius ($r = 4a_0$) for the circular orbit from the semiclassical Bohr model and the measured first ionization energy to calculate the effective nuclear charge seen by the $n = 2$ electron. We know, of course, that this $n = 2$ electron does not move in a circular orbit of constant radius, but is better represented by a probability density $|\psi|^2$ that overlaps the probability distributions of the $n = 1$ electrons.

Hydrogen

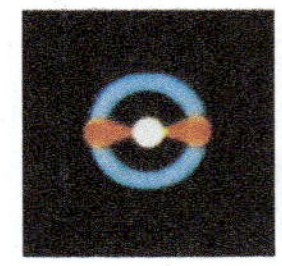

Carbon

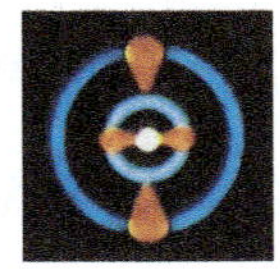

Silicon

Iron

Silver

Europium

A schematic depiction of the electron configurations in atoms. The spherically symmetric s states can have 2 electrons and are shown as white and blue. The dumbbell-shaped p states can have up to 6 electrons and are shown as orange. The d states can have up to 10 electrons and are shown as yellow-green. The f states can have up to 14 electrons and are shown as purple. *(David Parker/Photo Researchers.)*

BERYLLIUM ($Z = 4$)

The energy of the beryllium atom is a minimum if both $n = 2$ electrons are in the 2s state. There can be two electrons that have $n = 2$, $\ell = 0$, and $m_\ell = 0$ because of the two possible values for the spin quantum number m_s. The configuration of a beryllium atom is thus $1s^22s^2$.

BORON TO NEON ($Z = 5$ TO $Z = 10$)

If the 2s subshell of a ground-state boron atom is filled, the fifth electron must be in the next available (lowest energy) subshell, which is the 2p subshell, where $n = 2$ and $\ell = 1$. Because there are three possible values of m_ℓ (+1, 0, and −1) and two values of m_s for each value of m_ℓ, there can be six electrons in this subshell.

The electron configuration for a boron atom is $1s^22s^22p$. The electron configurations for carbon atoms (Z = 6) to neon atoms (Z = 10) differ from that for boron atoms only in the number of electrons in the 2p subshell. The first ionization energy increases with Z for these elements, reaching the value of 21.6 eV for the last element in the group, neon. A neon atom has the maximum number of electrons allowed in the $n = 2$ shell and its electron configuration is $1s^22s^22p^6$. Because of its very large first ionization energy, neon, like helium, basically is chemically inert. The atom whose atomic number is one less than neon's atomic number is fluorine, which has an unoccupied electron state in the 2p subshell; that is, a fluorine atom can have one more electron in its 2p subshell. Fluorine readily combines with elements such as lithium that have one electron in its highest energy shell (that is, an electron in an unfilled highest energy shell of an atom in the ground state). Lithium, for example, will donate its single valence electron to the fluorine atom to make an F^- ion and a Li^+ ion. These ions then bond together to form lithium fluoride.

SODIUM TO ARGON ($Z = 11$ TO $Z = 18$)

The eleventh electron of a ground-state sodium atom must be the $n = 3$ shell. Because this electron is shielded from the nucleus by $n = 2$ and $n = 1$ electrons, it is weakly bound in the sodium (Z = 11) atom. The first ionization energy of sodium is only 5.14 eV. Sodium atoms therefore combine readily with atoms such as fluorine. With $n = 3$, the value of ℓ can be 0, 1, or 2. A 3s electron has a lower energy than a 3p or 3d electron because its probability density overlap with the probability densities of $n = 2$ and $n = 1$ electrons is greatest. This energy difference between subshells of the same n value becomes greater as the number of electrons increases. The electron configuration of a sodium atom is $1s^22s^22p^63s^1$. For elements whose atoms have larger values of Z, the 3s subshell and then the 3p subshell are occupied. These two subshells can accommodate 2 + 6 = 8 electrons. The configuration of an argon (Z = 18) atom is $1s^22s^22p^63s^23p^6$. One might expect the nineteenth electron of potassium would occupy the third subshell (the d subshell where $\ell = 2$), but the overlap effect is now so strong that the energy of the nineteenth electron is lower in the 4s subshell than in the 3d subshell. There is thus another large energy difference between the eighteenth and nineteenth electrons of a potassium atom, and so an argon atom, with its full 3p subshell, is basically stable and inert.

ELEMENTS WITH Z > 18

The nineteenth electron in a potassium (Z = 19) atom and the twentieth electron in a calcium (Z = 20) atom occupy the 4s subshell rather than the 3d subshell. The electron configurations of the atoms of the next ten elements, scandium (Z = 21) through zinc (Z = 30), differ only in the number of electrons in the 3d shell, except for a chromium (Z = 24) atom and a copper (Z = 29) atom, each of which has only one 4s electron. These ten elements are called **transition elements.**

Figure 36-17 shows a plot of the first ionization energies versus Z for Z = 1 to Z = 60. The peaks in first ionization energy at Z = 2, 10, 18, 36, and 54 mark a filled shell or subshell. Table 36-1 gives the ground-state electron configurations of atoms up to atomic number 111.

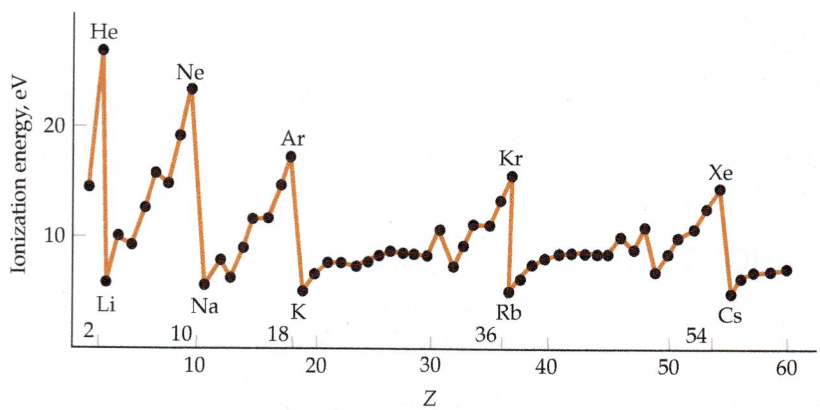

FIGURE 36-17 First ionization energies versus Z for Z = 1 to Z = 60. The first ionization energy increases with Z until a shell is filled at Z = 2, 10, 18, 36, and 54. An atom that has a filled shell and a single valence electron, such as sodium (Z = 11), has a very low ionization energy because the valence electron is shielded by the core electrons.

Table 36-1 Electron Configurations of the Atoms in Their Ground States
For some of the rare-earth elements and the heavy elements
the configurations are not firmly established.

		Shell (n):	K (1)	L (2)		M (3)			N (4)				O (5)				P (6)			Q (7)
			s	s	p	s	p	d	s	p	d	f	s	p	d	f	s	p	d	s
Z	Element	Subshell (ℓ):	(0)	(0)	(1)	(0)	(1)	(2)	(0)	(1)	(2)	(3)	(0)	(1)	(2)	(3)	(0)	(1)	(2)	(1)
1	H	hydrogen	1																	
2	He	helium	2																	
3	Li	lithium	2	1																
4	Be	beryllium	2	2																
5	B	boron	2	2	1															
6	C	carbon	2	2	2															
7	N	nitrogen	2	2	3															
8	O	oxygen	2	2	4															
9	F	fluorine	2	2	5															
10	Ne	neon	2	2	6															
11	Na	sodium	2	2	6	1														
12	Mg	magnesium	2	2	6	2														
13	Al	aluminum	2	2	6	2	1													
14	Si	silicon	2	2	6	2	2													
15	P	phosphorus	2	2	6	2	3													
16	S	sulfur	2	2	6	2	4													
17	Cl	chlorine	2	2	6	2	5													
18	Ar	argon	2	2	6	2	6													
19	K	potassium	2	2	6	2	6	.	1											
20	Ca	calcium	2	2	6	2	6	.	2											
21	Sc	scandium	2	2	6	2	6	1	2											
22	Ti	titanium	2	2	6	2	6	2	2											
23	V	vanadium	2	2	6	2	6	3	2											
24	Cr	chromium	2	2	6	2	6	5	1											
25	Mn	manganese	2	2	6	2	6	5	2											
26	Fe	iron	2	2	6	2	6	6	2											
27	Co	cobalt	2	2	6	2	6	7	2											
28	Ni	nickel	2	2	6	2	6	8	2											
29	Cu	copper	2	2	6	2	6	10	1											
30	Zn	zinc	2	2	6	2	6	10	2											
31	Ga	gallium	2	2	6	2	6	10	2	1										
32	Ge	germanium	2	2	6	2	6	10	2	2										
33	As	arsenic	2	2	6	2	6	10	2	3										
34	Se	selenium	2	2	6	2	6	10	2	4										
35	Br	bromine	2	2	6	2	6	10	2	5										
36	Kr	krypton	2	2	6	2	6	10	2	6										
37	Rb	rubidium	2	2	6	2	6	10	2	6	.	.	1							
38	Sr	strontium	2	2	6	2	6	10	2	6	.	.	2							

Continued on next page

Table 36-1 Continued

		Shell (n):	K (1)	L (2)		M (3)			N (4)				O (5)				P (6)			Q (7)
Z	Element	Subshell (ℓ):	s (0)	s (0)	p (1)	s (0)	p (1)	d (2)	s (0)	p (1)	d (2)	f (3)	s (0)	p (1)	d (2)	f (3)	s (0)	p (1)	d (2)	s (1)
39	Y	yttrium	2	2	6	2	6	10	2	6	1	.	2							
40	Zr	zirconium	2	2	6	2	6	10	2	6	2	.	2							
41	Nb	niobium	2	2	6	2	6	10	2	6	4	.	1							
42	Mo	molybdenum	2	2	6	2	6	10	2	6	5	.	1							
43	Tc	technetium	2	2	6	2	6	10	2	6	6	.	1							
44	Ru	ruthenium	2	2	6	2	6	10	2	6	7	.	1							
45	Rh	rhodium	2	2	6	2	6	10	2	6	8	.	1							
46	Pd	palladium	2	2	6	2	6	10	2	6	10	.	.							
47	Ag	silver	2	2	6	2	6	10	2	6	10	.	1							
48	Cd	cadmium	2	2	6	2	6	10	2	6	10	.	2							
49	In	indium	2	2	6	2	6	10	2	6	10	.	2	1						
50	Sn	tin	2	2	6	2	6	10	2	6	10	.	2	2						
51	Sb	antimony	2	2	6	2	6	10	2	6	10	.	2	3						
52	Te	tellurium	2	2	6	2	6	10	2	6	10	.	2	4						
53	I	iodine	2	2	6	2	6	10	2	6	10	.	2	5						
54	Xe	xenon	2	2	6	2	6	10	2	6	10	.	2	6						
55	Cs	cesium	2	2	6	2	6	10	2	6	10	.	2	6	.	.	1			
56	Ba	barium	2	2	6	2	6	10	2	6	10	.	2	6	.	.	2			
57	La	lanthanum	2	2	6	2	6	10	2	6	10	.	2	6	1	.	2			
58	Ce	cerium	2	2	6	2	6	10	2	6	10	1	2	6	1	.	2			
59	Pr	praseodymium	2	2	6	2	6	10	2	6	10	3	2	6	.	.	2			
60	Nd	neodymium	2	2	6	2	6	10	2	6	10	4	2	6	.	.	2			
61	Pm	promethium	2	2	6	2	6	10	2	6	10	5	2	6	.	.	2			
62	Sm	samarium	2	2	6	2	6	10	2	6	10	6	2	6	.	.	2			
63	Eu	europium	2	2	6	2	6	10	2	6	10	7	2	6	.	.	2			
64	Gd	gadolinium	2	2	6	2	6	10	2	6	10	7	2	6	1	.	2			
65	Tb	terbium	2	2	6	2	6	10	2	6	10	9	2	6	.	.	2			
66	Dy	dysprosium	2	2	6	2	6	10	2	6	10	10	2	6	.	.	2			
67	Ho	holmium	2	2	6	2	6	10	2	6	10	11	2	6	.	.	2			
68	Er	erbium	2	2	6	2	6	10	2	6	10	12	2	6	.	.	2			
69	Tm	thulium	2	2	6	2	6	10	2	6	10	13	2	6	.	.	2			
70	Yb	ytterbium	2	2	6	2	6	10	2	6	10	14	2	6	.	.	2			
71	Lu	lutetium	2	2	6	2	6	10	2	6	10	14	2	6	1	.	2			
72	Hf	hafnium	2	2	6	2	6	10	2	6	10	14	2	6	2	.	2			
73	Ta	tantalum	2	2	6	2	6	10	2	6	10	14	2	6	3	.	2			
74	W	tungsten (wolfram)	2	2	6	2	6	10	2	6	10	14	2	6	4	.	2			
75	Re	rhenium	2	2	6	2	6	10	2	6	10	14	2	6	5	.	2			
76	Os	osmium	2	2	6	2	6	10	2	6	10	14	2	6	6	.	2			
77	Ir	iridium	2	2	6	2	6	10	2	6	10	14	2	6	7	.	2			
78	Pt	platinum	2	2	6	2	6	10	2	6	10	14	2	6	9	.	1			
79	Au	gold	2	2	6	2	6	10	2	6	10	14	2	6	10	.	1			

Continued on next page

Table 36-1 Continued

		Shell (*n*):	*K* (1)	*L* (2)		*M* (3)			*N* (4)				*O* (5)				*P* (6)			*Q* (7)
			s	s	p	s	p	d	s	p	d	f	s	p	d	f	s	p	d	s
Z	Element	Subshell (ℓ):	(0)	(0)	(1)	(0)	(1)	(2)	(0)	(1)	(2)	(3)	(0)	(1)	(2)	(3)	(0)	(1)	(2)	(1)
80	Hg	mercury	2	2	6	2	6	10	2	6	10	14	2	6	10	.	2			
81	Tl	thallium	2	2	6	2	6	10	2	6	10	14	2	6	10	.	2	1		
82	Pb	lead	2	2	6	2	6	10	2	6	10	14	2	6	10	.	2	2		
83	Bi	bismuth	2	2	6	2	6	10	2	6	10	14	2	6	10	.	2	3		
84	Po	polonium	2	2	6	2	6	10	2	6	10	14	2	6	10	.	2	4		
85	At	astatine	2	2	6	2	6	10	2	6	10	14	2	6	10	.	2	5		
86	Rn	radon	2	2	6	2	6	10	2	6	10	14	2	6	10	.	2	6		
87	Fr	francium	2	2	6	2	6	10	2	6	10	14	2	6	10	.	2	6	.	1
88	Ra	radium	2	2	6	2	6	10	2	6	10	14	2	6	10	.	2	6	.	2
89	Ac	actinium	2	2	6	2	6	10	2	6	10	14	2	6	10	.	2	6	1	2
90	Th	thorium	2	2	6	2	6	10	2	6	10	14	2	6	10	.	2	6	2	2
91	Pa	protactinium	2	2	6	2	6	10	2	6	10	14	2	6	10	2	2	6	1	2
92	U	uranium	2	2	6	2	6	10	2	6	10	14	2	6	10	3	2	6	1	2
93	Np	neptunium	2	2	6	2	6	10	2	6	10	14	2	6	10	4	2	6	1	2
94	Pu	plutonium	2	2	6	2	6	10	2	6	10	14	2	6	10	6	2	6	.	2
95	Am	americium	2	2	6	2	6	10	2	6	10	14	2	6	10	7	2	6	.	2
96	Cm	curium	2	2	6	2	6	10	2	6	10	14	2	6	10	7	2	6	1	2
97	Bk	berkelium	2	2	6	2	6	10	2	6	10	14	2	6	10	9	2	6	.	2
98	Cf	californium	2	2	6	2	6	10	2	6	10	14	2	6	10	10	2	6	.	2
99	Es	einsteinium	2	2	6	2	6	10	2	6	10	14	2	6	10	11	2	6	.	2
100	Fm	fermium	2	2	6	2	6	10	2	6	10	14	2	6	10	12	2	6	.	2
101	Md	mendelevium	2	2	6	2	6	10	2	6	10	14	2	6	10	13	2	6	.	2
102	No	nobelium	2	2	6	2	6	10	2	6	10	14	2	6	10	14	2	6	.	2
103	Lr	lawrencium	2	2	6	2	6	10	2	6	10	14	2	6	10	14	2	6	1	2
104	Rf	rutherfordium	2	2	6	2	6	10	2	6	10	14	2	6	10	14	2	6	2	2
105	Db	dubnium	2	2	6	2	6	10	2	6	10	14	2	6	10	14	2	6	3	2
106	Sg	seaborgium	2	2	6	2	6	10	2	6	10	14	2	6	10	14	2	6	4	2
107	Bh	bohrium	2	2	6	2	6	10	2	6	10	14	2	6	10	14	2	6	5	2
108	Hs	hassium	2	2	6	2	6	10	2	6	10	14	2	6	10	14	2	6	6	2
109	Mt	meitnerium	2	2	6	2	6	10	2	6	10	14	2	6	10	14	2	6	7	2
110	Ds	darmstadtium	2	2	6	2	6	10	2	6	10	14	2	6	10	14	2	6	9	1
111	Rg	roentgenium	2	2	6	2	6	10	2	6	10	14	2	6	10	14	2	6	10	1

36-7 OPTICAL SPECTRA AND X-RAY SPECTRA

When an atom is in an excited state (when it is in an energy state above the ground state), it makes transitions to lower energy states, and in doing so emits electromagnetic radiation. The wavelength of the electromagnetic radiation emitted is related to the initial and final states by the Bohr formula (Equation 36-17), $\lambda = hc/(E_i - E_f)$, where E_i and E_f are the initial and final energies and h is Planck's constant. The atom can be excited to a higher energy state by bombarding the atom with a beam of electrons, as in a spectral tube that has a high voltage across it. Because the excited energy states of an atom form a discrete (rather than continuous) set, only certain wavelengths are emitted. These wavelengths of the emitted radiation constitute the emission spectrum of the atom.

A neon sign outside a Chinatown restaurant in Paris. Neon atoms in the tube are excited by an electron current passing through the tube. The excited neon atoms emit light in the visible range as they decay toward their ground states. The colors of neon signs result from the characteristic red-orange spectrum of neon plus the color of the glass tube itself. *(Robert Landau/Westlight.)*

OPTICAL SPECTRA

To understand atomic spectra we need to understand the excited states of the atom. The situation for an atom that has many electrons is, in general, much more complicated than that of a hydrogen atom that has just one electron. An excited state of the atom may involve a change in the state occupied by any one of the electrons, or even two or more electrons. Fortunately, in most cases, an excited state of an atom involves the excitation of just one of the electrons in the atom. The energies of excitation of the valence electrons of an atom are of the order of a few electron volts. Transitions involving these electrons result in photons in or near the visible or **optical spectrum.** (Recall that the energies of visible photons range from approximately 1.5 eV to 3 eV.) The excitation energies can often be calculated from a simple model in which the atom is pictured as a single electron plus a stable core consisting of the nucleus plus the other electrons. This model works particularly well for the alkali metals: Li, Na, K, Rb, and Cs. These elements are in the first column of the periodic table. The optical spectra of these elements are similar to the optical spectra of hydrogen.

Figure 36-18 shows an energy-level diagram for the optical transitions of a sodium atom, whose electrons form a neon core plus one electron. Because the total spin angular momentum of the core adds up to zero, the spin of each state of the sodium atom is $\frac{1}{2}$ (the spin of the valence electron). Because of the spin–orbit effect, the atomic states for which $J = L - \frac{1}{2}$ have a slightly different energy than those for which $J = L + \frac{1}{2}$ (except for states with $L = 0$). Each state (except for the $L = 0$ states) is therefore split into two states, called a doublet. The doublet splitting is very small and not evident on the energy scale of this diagram. The usual spectroscopic notation is that the states of these atoms are labeled with a superscript given by $2S + 1$, followed by a letter denoting the orbital angular momentum, followed by a subscript denoting the total angular momentum J. For states that have a total spin angular momentum $S = \frac{1}{2}$ the superscript is 2, indicating the state is a doublet. Thus, $^2P_{3/2}$, read as "doublet P three halves," denotes a state in which $L = 1$ and $J = \frac{3}{2}$. (The $L = 0$, or S, states are customarily labeled as if they were doublets even though they are not.) For a sodium atom in the first excited state, the electron is excited from the 3s level to the 3p level, which is approximately 2.1 eV above the ground state. The energy difference between the $P_{3/2}$ and $P_{1/2}$ states due to the spin–orbit effect is about 0.002 eV. Transitions from these states to the ground state give the familiar sodium yellow doublet:

$$3p(^2P_{1/2}) \rightarrow 3s(^2S_{1/2}) \qquad \lambda = 589.6 \text{ nm}$$
$$3p(^2P_{3/2}) \rightarrow 3s(^2S_{1/2}) \qquad \lambda = 589.0 \text{ nm}$$

The energy levels and spectra of other alkali metal atoms are similar to those for sodium. The optical spectrum for atoms such as helium, beryllium, and magnesium that have two valence electrons is considerably more complex because of the interaction of the two electrons.

X-RAY SPECTRA

X rays are usually produced in the laboratory by bombarding a target element with a high-energy beam of electrons in an X-ray tube. The result (Figure 36-19) consists of a continuous spectrum that depends only on the energy of the bombarding electrons and a line spectrum that is characteristic of the target element. The characteristic spectrum results from excitation of the core electrons in the target element.

The energy needed to excite a core electron—for example, an electron in the $n = 1$ state (K shell)—is much greater than the energy required to excite a valence

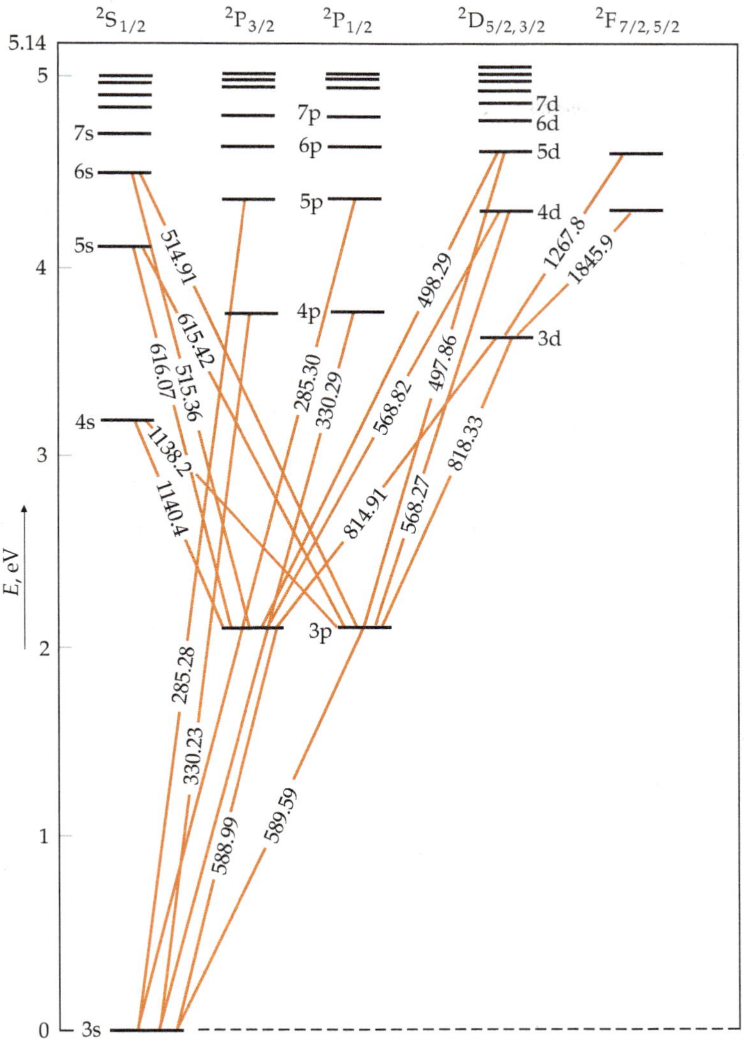

FIGURE 36-18 Energy-level diagram for sodium. The diagonal lines show observed optical transitions, where wavelengths are given in nanometers. The energy of the ground state has been chosen as the zero point for the scale on the left.

electron. A core electron cannot be excited to any of the filled states (for example, the $n = 2$ states in an atom with $Z \geq 10$) because of the exclusion principle. The energy required to excite a core electron to an unoccupied state is typically of the order of several thousand electron volts. If an electron is knocked out of the $n = 1$ shell (the K shell), there is a vacancy left in that shell. The vacancy can be filled if an electron in a higher energy shell makes a transition into the K shell. The photons emitted by electrons making such transitions also have energies of the order of several thousand electron volts and produce the sharp peaks in the X-ray spectrum, as shown in Figure 36-19. The K_α spectral line arises from transitions from the $n = 2$ shell (the L shell) to the $n = 1$ shell (the K shell). The K_β spectral line arises from transitions from the $n = 3$ shell to the $n = 1$ shell. These and other lines arising from transitions ending at the $n = 1$ shell make up the K series of the characteristic X-ray spectrum of the target element. Similarly, a second series, the L series, is produced by transitions from higher energy states to a vacated place in the $n = 2$ (L) shell. The letters K, L, M, and so on, designate the final shell of the electron making the transition and the series α, β, and so on, designates the number of shells above the final shell for the initial state of the electron.

In 1913, the English physicist Henry Moseley measured the wavelengths of the characteristic K_α X-ray spectra for approximately 40 elements. Using this data, Moseley showed that a plot of $\lambda^{-1/2}$ versus the order in which the elements appeared in the periodic table resulted in a straight line (with a few gaps and a few outliers). From his data, Moseley was able to accurately determine the atomic

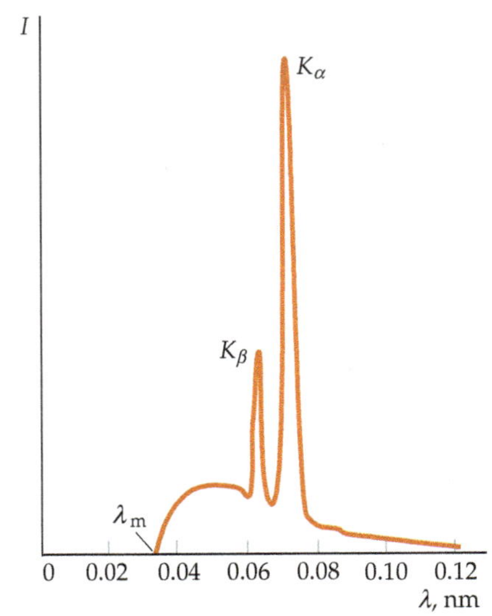

FIGURE 36-19 X-ray spectrum of molybdenum. The sharp peaks labeled K_α and K_β are characteristic of the element. The cutoff wavelength λ_m is independent of the target element and is related to the voltage V of the X-ray tube by $\lambda_m = hc/eV$.

number Z for each known element, and to predict the existence of some elements that were later discovered. The equation of the straight line of his plot is given by

$$\frac{1}{\sqrt{\lambda_{K_\alpha}}} = a(Z - 1)$$

The work of Bohr and Moseley can be combined to obtain an equation relating the wavelength of the emitted photon and the atomic number. According to the Bohr model of a single-electron atom (see Equation 36-13), the wavelength of the emitted photon when the electron makes the transition from $n = 2$ to $n = 1$ is given by

$$\frac{1}{\lambda} = Z^2 \frac{E_0}{hc}\left(1 - \frac{1}{2^2}\right)$$

where $E_0 = 13.6$ eV is the binding energy of the ground-state hydrogen atom. Taking the square root of both sides gives

$$\frac{1}{\sqrt{\lambda_{K_\alpha}}} = \left[\frac{E_0}{hc}\left(1 - \frac{1}{2^2}\right)\right]^{1/2} Z$$

Moseley's equation and this equation are in agreement if $Z - 1$ is substituted for Z in Bohr's equation and if $a = 3E_0/(4hc)$. This result raises the question, why a factor of $Z - 1$ instead of a factor of Z? Part of the explanation is that the formula from the Bohr theory ignores the shielding of the nuclear charge. In a multielectron atom, electrons in the $n = 2$ states are electrically shielded from the nuclear charge by the two electrons in the $n = 1$ state, so the $n = 2$ state electrons are attracted by an effective nuclear charge of about $(Z - 2)e$. However, when there is only one electron in the K shell, the $n = 2$ electrons are attracted by an effective nuclear charge of about $(Z - 1)e$. When an electron from state n drops into the vacated state in the $n = 1$ shell, a photon of energy $E_n - E_1$ is emitted. For $n = 2$, the wavelength of this photon is

$$\lambda_{K_\alpha} = \frac{hc}{(Z - 1)^2 E_0\left(1 - \frac{1}{2^2}\right)} \qquad \text{36-46}$$

which is obtained from the previous equation with $Z - 1$ substituted for Z.

Example 36-8 Identifying the Element from the K_α X-Ray Line

The wavelength of the K_α X-ray line for a certain element is $\lambda = 0.0721$ nm. What is the element?

PICTURE The K_α line corresponds to a transition from $n = 2$ to $n = 1$. The wavelength is related to the atomic number Z by Equation 36-46.

SOLVE

1. Solve Equation 36-46 for $(Z - 1)^2$:

$$\lambda_{K_\alpha} = \frac{hc}{(Z - 1)^2 E_0\left(1 - \frac{1}{2^2}\right)}$$

so

$$(Z - 1)^2 = \frac{4hc}{3\lambda_{K_\alpha} E_0}$$

2. Substitute the given data and solve for Z:

$$(Z - 1)^2 = \frac{4(1240 \text{ eV}\cdot\text{nm})}{3(0.0721 \text{ nm})(13.6 \text{ eV})} = 1686$$

so

$$Z = 1 + \sqrt{1686} = 42.06$$

3. Because Z is an integer, we round to the nearest integer:

$$Z = 42$$

The element is molybdenum.

CHECK The naturally occurring atom that has the largest atomic number is uranium, which has an atomic number $Z = 92$. That our step-3 result is greater than 0 and less than 93 is as expected.

Summary

1. The Bohr model is important because it was the first model to succeed at explaining the discrete optical spectrum of atoms in terms of the quantization of energy. It has been superceded by the quantum-mechanical model.
2. The quantum theory of atoms results from the application of the Schrödinger equation to a bound system consisting of nucleus of charge $+Ze$ and Z electrons of charge $-e$.
3. For the hydrogen atom, an atom that consists of one proton and one electron, the time-independent Schrödinger equation can be solved exactly to obtain the wave functions ψ, which depend on the quantum numbers n, ℓ, m_ℓ, and m_s.
4. The electron configuration of atoms is governed by the Pauli exclusion principle—no two electrons in an atom can have the same set of values for the quantum numbers n, ℓ, m_ℓ, and m_s. Using the exclusion principle and the restrictions on the quantum numbers, we can understand much of the structure of the periodic table.

TOPIC	RELEVANT EQUATIONS AND REMARKS	
1. The Bohr Model of the Hydrogen Atom		
Postulates for the hydrogen atom		
Nonradiating orbits	The idea that an electron moves in a circular nonradiating orbit around the proton.	
Photon frequency from energy conservation	$f = \frac{E_i - E_f}{h}$	36-7
Quantized angular momentum	$L_n = mv_n r_n = n\hbar \qquad n = 1, 2, 3, \ldots$	36-9
First Bohr radius	$a_0 = \frac{\hbar^2}{mke^2} = 0.0529\text{ nm}$	36-12
Radii of the Bohr orbits	$r_n = n^2\frac{a_0}{Z}$	36-11
Energy levels in hydrogen-like atoms	$E_n = -Z^2\frac{E_0}{n^2}$	36-15
	where $E_0 = -\frac{mk^2e^4}{2\hbar^2} = \frac{1}{2}\frac{ke^2}{a_0} = 13.6\text{ eV}$	36-16
Wavelengths emitted by the hydrogen atom	$\lambda = \frac{c}{f} = \frac{hc}{E_i - E_f} = \frac{1240\text{ eV}\cdot\text{nm}}{E_i - E_f}$	36-17, 36-18
2. Quantum Theory of Atoms	The electron is described by a wave function ψ that is a solution of the Schrödinger equation. Energy quantization arises from standing-wave conditions. ψ is described by the principal, orbital, and magnetic quantum numbers n, ℓ, and m_ℓ, and the spin quantum number $m_s = \pm\frac{1}{2}$.	
Time-independent Schrödinger equation	$-\frac{\hbar^2}{2m}\left(\frac{\partial^2\psi}{\partial x^2} + \frac{\partial^2\psi}{\partial y^2} + \frac{\partial^2\psi}{\partial z^2}\right) + U\psi = E\psi$	36-19
For an isolated atom, the solutions can be written as products of functions of r, θ, and ϕ separately	$\psi(r, \theta, \phi) = R(r)f(\theta)g(\phi)$	36-21
Quantum numbers in spherical coordinates		
Principal quantum number	$n = 1, 2, 3, \ldots$	
Orbital quantum number	$\ell = 0, 1, 2, 3, \ldots, n - 1$	
Magnetic quantum number	$m_\ell = -\ell, (-\ell + 1), \ldots, 0, \ldots, (\ell + 1), \ell$	36-22
Orbital angular momentum	$L = \sqrt{\ell(\ell + 1)}\hbar$	36-23

TOPIC	RELEVANT EQUATIONS AND REMARKS	
z component of orbital angular momentum	$L_z = m_\ell \hbar$	36-24
3. Quantum Theory of the Hydrogen Atom		
Energy levels for hydrogen-like atoms (same as for the Bohr model)	$E_n = -Z^2 \frac{E_0}{n^2} \quad n = 1, 2, 3, \ldots$	36-26
	where $E_0 = -\frac{mk^2e^4}{2\hbar^2} = 13.6\text{ eV}$	36-27
Wavelengths emitted by the hydrogen atom (same as for Bohr model)	$\lambda = \frac{c}{f} = \frac{hc}{E_i - E_f} = \frac{1240\text{ eV}\cdot\text{nm}}{E_i - E_f}$	36-17, 36-18
Wave functions		
The ground state	$\psi_{100} = C_{100}e^{-Zr/a_0} = \frac{1}{\sqrt{\pi}}\left(\frac{Z}{a_0}\right)^{3/2} e^{-Zr/a_0}$	36-30, 36-32
The first excited state	$\psi_{200} = C_{200}\left(2 - \frac{Zr}{a_0}\right)e^{-Zr/2a_0}$	36-35
	$\psi_{210} = C_{210}\frac{Zr}{a_0}e^{-Zr/2a_0}\cos\theta$	36-36
	$\psi_{21\pm1} = C_{21\pm1}\frac{Zr}{a_0}e^{-Zr/2a_0}\sin\theta\, e^{\pm i\phi}$	36-37
Probability densities	For $\ell = 0$, $\lvert\psi\rvert^2$ is spherically symmetric. For $\ell > 0$, $\lvert\psi\rvert^2$ depends on the angle θ.	
Radial probability density	$P(r) = 4\pi r^2 \lvert\psi\rvert^2$ The radial probability density is maximum at the distances corresponding roughly to the Bohr orbits.	36-33
4. The Spin–Orbit Effect and Fine Structure	The total angular momentum of an electron in an atom is a combination of the orbital angular momentum and spin angular momentum. It is characterized by the quantum number j, which can be either $\lvert\ell - \frac{1}{2}\rvert$ or $\ell + \frac{1}{2}$. Because of the interaction of the orbital and spin magnetic moments, the state $j = \lvert\ell - \frac{1}{2}\rvert$ has lower energy than the state $j = \ell + \frac{1}{2}$, for $\ell > 0$. This small splitting of the energy states gives rise to a small splitting of the spectral lines called fine structure.	
5. The Periodic Table	An atom of an element has Z electrons, where Z is the atomic number of the element. For an atom in the ground state, the electrons are in those states that will give the lowest energy consistent with the Pauli exclusion principle. The state of an atom is described by its electron configuration, which gives the values of n and ℓ for each electron. The ℓ values are specified by a code: s p d f g h ℓ value 0 1 2 3 4 5	
Pauli exclusion principle	No two electrons in an atom can have the same set of values for the quantum numbers n, ℓ, m_ℓ, and m_s.	
6. Atomic Spectra	Atomic spectra include optical spectra and X-ray spectra. Optical spectra result from transitions between energy levels of a single valence electron moving in the field of the nucleus and core electrons of the atom. Characteristic X-ray spectra result from the excitation of a core electron and the subsequent filling of the vacancy by other electrons in the atom.	
Selection rules	Transitions between energy states with the emission of a photon are governed by the following selection rules $\Delta m_\ell = 0 \quad \text{or} \quad \Delta m_\ell = \pm 1$ $\Delta\ell = \pm 1$	36-28

Answers to Concept Checks

36-1 (*a*), (*c*), and (*d*)

Answers to Practice Problems

36-1 91.2 nm

36-2 $-4, -3, -2, -1, 0, 1, 2, 3, 4$

Problems

In a few problems, you are given more data than you actually need; in a few other problems, you are required to supply data from your general knowledge, outside sources, or informed estimate.

Interpret as significant all digits in numerical values that have trailing zeros and no decimal points.

• Single-concept, single-step, relatively easy

•• Intermediate-level, may require synthesis of concepts

••• Challenging

SSM Solution is in the *Student Solutions Manual*

Consecutive problems that are shaded are paired problems.

CONCEPTUAL PROBLEMS

1 • For the hydrogen atom, as n increases, does the spacing of adjacent energy levels on an energy-level diagram increase or decrease? SSM

2 • The energy of the ground state of doubly ionized lithium ($Z = 3$) is ______, where $E_0 = 13.6$ eV. (*a*) $-9E_0$, (*b*) $-3E_0$, (*c*) $-E_0/3$, (*d*) $-E_0/9$

3 • Bohr's quantum condition on electron orbits requires (*a*) that the orbital angular momentum of the electron about the hydrogen nucleus equals $n\hbar$, where n is an integer, (*b*) that no more than one electron occupy a given state, (*c*) that the electrons spiral into the nucleus while radiating electromagnetic waves, (*d*) that the energies of an electron in a hydrogen atom be equal to nE_0, where E_0 is a constant and n is an integer, (*e*) none of the above.

4 • According to the Bohr model, if an electron moves to a larger orbit, does the electron's total energy increase or decrease? Does the electron's kinetic energy increase or decrease?

5 • According to the Bohr model, the kinetic energy of the electron in the ground state of hydrogen is E_0, where $E_0 = 13.6$ eV. The kinetic energy of the electron in the state $n = 2$ is (*a*) $4E_0$, (*b*) $2E_0$, (*c*) $E_0/2$, (*d*) $E_0/4$.

6 • According to the Bohr model, the radius of the $n = 1$ orbit in the hydrogen atom is $a_0 = 0.053$ nm. What is the radius of the $n = 5$ orbit? (*a*) $25a_0$, (*b*) $5a_0$, (*c*) a_0, (*d*) $a_0/5$, (*e*) $a_0/25$

7 • For the principal quantum number $n = 4$, how many different values can the orbital quantum number ℓ have? (*a*) 4, (*b*) 3, (*c*) 7, (*d*) 16, (*e*) 25 SSM

8 • For the principal quantum number $n = 4$, how many different combinations of ℓ and m_ℓ can occur? (*a*) 4, (*b*) 3, (*c*) 7, (*d*) 16, (*e*) 25

9 •• Why is the energy of the 3s state considerably lower than the energy of the 3p state for sodium, whereas in hydrogen 3s and 3p states have essentially the same energy? SSM

10 • The d state of an electron configuration corresponds to (*a*) $n = 2$, (*b*) $\ell = 3$, (*c*) $\ell = 2$, (*d*) $n = 3$, (*e*) $\ell = 0$.

11 •• Why are three quantum numbers inadequate to describe the states of the electrons in atoms that have more than one electron?

12 •• Group the following six atoms—potassium, calcium, titanium, chromium, manganese, and copper—according to their ground-state electron configurations for the $n = 4$ shell.

13 • What element has the electron configuration (*a*) $1s^22s^22p^63s^23p^3$ and (*b*) $1s^22s^22p^63s^23p^63d^54s^1$?

14 • For the principal quantum number $n = 3$, what are the possible combinations of the quantum numbers ℓ and m_ℓ?

15 • An electron in the L shell means that the electron is represented by (*a*) $\ell = 0$, (*b*) $\ell = 1$, (*c*) $n = 1$, (*d*) $n = 2$, or (*e*) $m_\ell = 2$.

16 •• The Bohr model and the quantum-mechanical model of the hydrogen atom give the same results for the energy levels. Discuss the advantages and disadvantages of each model.

17 •• The Sommerfeld–Hosser displacement theorem states that the optical spectrum of any atom is very similar to the spectrum of the singly charged positive ion of the element immediately following it in the periodic table. Discuss why this theorem is accurate.

18 • Using the triplet of numbers (n, ℓ, m_ℓ) to represent an electron that has the principal quantum number n, orbital quantum number ℓ, and magnetic quantum number m_ℓ, which of the following transitions is allowed? (*a*) $(5, 2, 2) \rightarrow (3, 1, -2)$, (*b*) $(2, 1, 0) \rightarrow (3, 0, 0)$, (*c*) $(4, 3, -2) \rightarrow (3, 2, -1)$, (*d*) $(1, 0, 0) \rightarrow (2, 1, -1)$, (*e*) $(2, 1, 0) \rightarrow (3, 1, 0)$

19 •• The Ritz combination principle states that for any atom, one can find different spectral lines λ_1, λ_2, λ_3, and λ_4, so that $1/\lambda_1 + 1/\lambda_2 = 1/\lambda_3 + 1/\lambda_4$. Show why this is true using an energy-level diagram. SSM

ESTIMATION AND APPROXIMATION

20 •• (*a*) We can define a thermal de Broglie wavelength λ_T for an atom in a gas at temperature T as being the de Broglie wavelength for an atom moving at the rms speed appropriate to that temperature. (The average kinetic energy of an atom is equal to $\frac{3}{2}kT$, where k is the Boltzmann constant. Use this value to calculate the rms speed of the atoms.) Show that $\lambda_T = h/\sqrt{3mkT}$ where m is the mass of the atom. (*b*) Cooled atoms can form a Bose *condensate* (a new state of matter) when their thermal de Broglie wavelength becomes larger than the average interatomic spacing. From this criterion, estimate the temperature needed to create a Bose condensate in a gas of ^{85}Rb atoms whose number density is 10^{12} atoms/cm^3.

21 •• In laser cooling and trapping, a beam of atoms traveling in one direction is slowed by interaction with an intense laser beam in the opposite direction. The photons scatter off the atoms by resonance absorption, a process by which the incident photon is absorbed by the atom, and a short time later a photon of equal energy is emitted in a random direction. The net result of a single such scattering event is a transfer of momentum to the atom in a direction opposite to the motion of the atom, followed by a second transfer of momentum to the atom in a random direction. Thus, during photon absorption the atom loses speed, but during photon emission the change in speed of the atom is, on average, zero (because the directions of the emitted photons are random). An analogy often made to this process is that of slowing down a bowling ball by bouncing ping-pong balls off of it. (*a*) Given that the typical photon energy used in these experiments is about 1 eV, and that the typical kinetic energy of an atom in the beam is the typical kinetic energy of the atoms in a gas that has a temperature of about 500 K (a typical temperature for an oven that produces an atomic beam), estimate the number of photon–atom collisions that are required to bring an atom to rest. (The average kinetic energy of an atom is equal to $\frac{3}{2}kT$, where k is the Boltzmann constant and T is the temperature. Use this to estimate the speed of the atoms.) (*b*) Compare the Part-(*a*) result with the number of ping-pong ball–bowling ball collisions that are required to bring the bowling ball to rest. (Assume the typical speed of the incident ping-pong balls are all equal to the initial speed of the bowling ball.) (*c*) ^{85}Rb is a type of atom often used during cooling experiments. The wavelength of the light resonant with the cooling transition of the atoms is $\lambda = 780.24$ nm. Estimate the number of photons needed to slow down an ^{85}Rb atom from a typical thermal velocity of 300 m/s to a stop. SSM

THE BOHR MODEL OF THE HYDROGEN ATOM

22 • The first Bohr radius is given by $a_0 = \hbar^2/(mke^2) = 0.0529$ nm (Equation 36-12). Use the known values of the constants in the equation to show that a_0 is equal to 0.0529 nm.

23 • The longest wavelength in the Lyman series for the hydrogen atom was calculated in Example 36-2. Find the wavelengths for the transitions (*a*) $n_i = 3$ to $n_f = 1$ and (*b*) $n_i = 4$ to $n_f = 1$.

24 • Find the photon energies for the three longest wavelengths in the Balmer series for the hydrogen atom, and calculate the three wavelengths.

25 •• Find the photon energy and wavelength for the series limit (shortest wavelength) in the Paschen series ($n_f = 3$) for the hydrogen atom. (*b*) Calculate the wavelength for the three longest wavelengths in Paschen series.

26 •• (*a*) Find the photon energy and wavelength for the series limit (shortest wavelength) in the Brackett series ($n_f = 4$) for the hydrogen atom. (*b*) Calculate the wavelength for the three longest wavelengths in Brackett series.

27 ••• In the center-of-mass reference frame of a hydrogen atom, the electron and nucleus have momenta that have equal magnitudes p and opposite directions. (*a*) Using the Bohr model, show that the total kinetic energy of the electron and nucleus can be written $K = p^2/(2\mu)$, where $\mu = m_eM/(M + m_e)$ is called the reduced mass, m_e is the mass of the electron, and M is the mass of the nucleus. (*b*) For the equations for the Bohr model of the atom, the motion of the nucleus can be taken into account by replacing the mass of the electron with the reduced mass. Use Equation 36-14 to calculate the Rydberg constant for a hydrogen atom that has a nucleus of mass $M = m_p$. Find the approximate value of the Rydberg constant by letting M go to infinity in the reduced mass formula. To how many figures does this approximate value agree with the actual value? (*c*) Find the percentage correction for the ground-state energy of the hydrogen atom by using the reduced mass in Equation 36-16.

Note: In general, the reduced mass for a two-body problem with masses m_1 and m_2 is given by

$$\mu = \frac{m_1 m_2}{m_1 + m_2}$$ SSM

28 •• The Pickering series of the spectrum of He^+ (singly ionized helium) consists of spectral lines due to transitions to the $n = 4$ state of He^+. Every other line of the Pickering series is very close to a spectral line in the Balmer series for hydrogen transitions to $n = 2$. (*a*) Show that this statement is accurate. (*b*) Calculate the wavelength of the photon during a transition from the $n = 6$ level to the $n = 4$ level of He^+, and show that it corresponds to one of the Balmer series lines.

QUANTUM NUMBERS IN SPHERICAL COORDINATES

29 • For an electron in an atom that has an orbital quantum number $\ell = 1$, find (*a*) the magnitude of the angular momentum L and (*b*) the possible values of the magnetic quantum number m_ℓ. (*c*) Draw a vector diagram to scale showing the possible orientations of $\vec{L}$ relative to the $+z$ direction.

30 • For an electron in an atom that has an orbital quantum number $\ell = 3$, find (*a*) the magnitude of the angular momentum L and (*b*) the possible values of m_ℓ. (*c*) Draw a vector diagram to scale showing the possible orientations of $\vec{L}$ relative to the $+z$ direction.

31 • An electron in an atom has principal quantum number $n = 3$. (*a*) What are the possible values of ℓ? (*b*) What are the possible combinations of ℓ and m_ℓ? (*c*) Using the fact that there are two quantum states for each combination of ℓ and m_ℓ because of electron spin, find the total number of electron states for $n = 3$.

32 • In an atom, find the total number of electron states that have (*a*) $n = 4$ and (*b*) $n = 2$. (See Problem 31.)

33 •• Find the minimum value of the angle θ between $\vec{L}$ and the $+z$ direction for an electron in an atom that has (*a*) $\ell = 1$, (*b*) $\ell = 4$, and (*c*) $\ell = 50$. SSM

34 •• What are the possible values of n and m_ℓ for an electron in an atom that has (*a*) $\ell = 3$, (*b*) $\ell = 4$, and (*c*) $\ell = 0$?

35 •• For an electron in an atom that is in an $\ell = 2$ state, find (*a*) the magnitude of the angular momentum squared L^2, (*b*) the maximum value of L_z^2, and (*c*) the smallest value of $L_x^2 + L_y^2$.

QUANTUM THEORY OF THE HYDROGEN ATOM

36 • For the ground state of the hydrogen atom, find the values of (*a*) $\psi(r)$ at $r = a_0$, (*b*) $\psi^2(r)$ at $r = a_0$, and (*c*) the radial probability density $P(r)$ at $r = a_0$. Give your answers in terms of a_0.

37 • (*a*) If electron spin is not included, how many different wave functions are there corresponding to the first excited energy level $n = 2$ for a hydrogen atom? (*b*) Specify the quantum numbers for each of these wave functions. SSM

38 •• For the ground state of the hydrogen atom, calculate the probability of finding the electron in the region between r and $r + \Delta r$, where $\Delta r = 0.03a_0$ and (*a*) $r = a_0$ and (*b*) $r = 2a_0$.

39 •• The value of the constant C_{200} in the equation

$$\psi_{200} = C_{200}\left(2 - \frac{Zr}{a_0}\right)e^{-Zr/(2a_0)}$$

(Equation 36-35) is given by

$$C_{200} = \frac{1}{4\sqrt{2\pi}}\left(\frac{Z}{a_0}\right)^{3/2}$$

Find the values of (*a*) $\psi(r)$ at $r = a_0$, (*b*) $\psi^2(r)$ at $r = a_0$, and (*c*) the radial probability density $P(r)$ at $r = a_0$ for the state $n = 2$, $\ell = 0$, and $m_\ell = 0$ of a hydrogen atom. Give your answers in terms of a_0.

40 ••• Show that the radial probability density for the $n = 2$, $\ell = 1$, and $m_\ell = 0$ state of a one-electron atom can be written as $P(r) = A\cos^2\theta\, r^4 e^{-Zr/a_0}$, where A is a constant.

41 ••• Calculate the probability of finding the electron in the region between r and $r + \Delta r$, where $\Delta r = 0.02a_0$ and (*a*) $r = a_0$ and (*b*) $r = 2a_0$ for the state $n = 2$, $\ell = 0$, and $m_\ell = 0$ in hydrogen. (See Problem 39 for the value of C_{200}.)

42 •• Show that the ground-state hydrogen atom wave function $\psi_{100} = \pi^{-1/2}(Z/a_0)^{3/2}e^{-Zr/a_0}$ (Equation 36-32) is a solution to Schrödinger's equation in spherical coordinates:

$$\frac{-\hbar}{2mr^2}\left\{\frac{\partial}{\partial r}\left(r^2\frac{\partial\psi}{\partial r}\right) + \left[\frac{1}{\sin\theta}\frac{\partial}{\partial\theta}\left(\sin\theta\frac{\partial\psi}{\partial\theta}\right) + \frac{1}{\sin^2\theta}\frac{\partial^2\psi}{\partial\phi^2}\right]\right\} + U(r)\psi = E\psi$$

where $U(r) = kZe^2/r$ (Equation 36-25).

43 •• Show by dimensional analysis that the expression for the hydrogen atom ground-state energy given by $E_0 = \frac{1}{2}mk^2e^4/\hbar^2$ (Equation 36-27) has the dimensions of energy.

44 •• By dimensional analysis, show that the expression for the first Bohr radius given by $a_0 = \hbar^2/(mke^2)$ (Equation 36-12) has the dimensions of length.

45 •• The radial probability distribution function for a one-electron atom in its ground state can be written $P(r) = Cr^2e^{-2Zr/a_0}$, where C is a constant. Show that $P(r)$ has its maximum value at $r = a_0/Z$.

46 ••• Show that the number of states in the hydrogen atom for a given n is $2n^2$.

47 ••• Calculate the probability that the electron in the ground state of a hydrogen atom is in the region $0 < r < a_0$.

THE SPIN-ORBIT EFFECT AND FINE STRUCTURE

48 • The potential energy of a magnetic moment in an external magnetic field is given by $U = -\vec{\mu}\cdot\vec{B}$. (*a*) Calculate the difference in energy between the two possible orientations of an electron in a magnetic field $\vec{B} = 1.50\ \text{T}\hat{k}$. (*b*) If the electrons are bombarded with photons of energy equal to that energy difference, "spin flip" transitions can be induced. Find the wavelength of the photons needed for such transitions. This phenomenon is called *electron spin resonance*.

49 • The total angular momentum of a hydrogen atom in a certain excited state has the quantum number $j = \frac{1}{2}$. What can you say about the value of the orbital angular-momentum quantum number ℓ?

50 • A hydrogen atom is in the state $n = 3$, $\ell = 2$. What are the possible values of j?

51 • Using a scaled vector diagram, show how the orbital angular momentum $\vec{L}$ combines with the spin angular momentum $\vec{S}$ to produce the two possible values of total angular momentum $\vec{J}$ for the $\ell = 3$ state of the hydrogen atom.

THE PERIODIC TABLE

52 • The total number of states of a hydrogen atom that has principal quantum number $n = 4$ is (*a*) 4, (*b*) 16, (*c*) 32, (*d*) 36, (*e*) 48.

53 • How many of the eight electrons in an oxygen atom in the ground state are in a p state? (*a*) 0, (*b*) 2, (*c*) 4, (*d*) 6, (*e*) 8

54 • Write the ground-state electron configuration of (*a*) an atom of carbon and (*b*) an atom of oxygen.

55 • Give the possible values of the *z* component of the orbital angular momentum of (*a*) a d electron and (*b*) an f electron.

OPTICAL SPECTRA AND X-RAY SPECTRA

56 • The optical spectra of atoms that have two electrons in the same highest energy shell are similar, but they are quite different from the spectra of atoms that have just one electron in the highest energy shell because of the interaction of the two electrons. Group the elements according to similar spectra: lithium, beryllium, sodium, magnesium, potassium, calcium, chromium, nickel, cesium, and barium.

57 • Write down the possible electron configurations for the first excited state of (*a*) a hydrogen atom, (*b*) a sodium atom, and (*c*) a helium atom.

58 • Indicate which of the following atoms should have optical spectra similar to a hydrogen atom and which of the following atoms should have optical spectra similar to a helium atom: Li, Ca, Ti, Rb, Hg, Ag, Cd, Ba, Fr, and Ra.

59 • (*a*) Calculate the next two longest wavelengths in the K series (after the K_α line) of molybdenum. (*b*) What is the wavelength of the shortest wavelength in this series?

60 • The wavelength of the K_α line for a certain element is 0.3368 nm. What is the element?

61 • Calculate the wavelength of the K_α line in (*a*) a magnesium ($Z = 12$) atom and (*b*) a copper ($Z = 29$) atom.

GENERAL PROBLEMS

62 • What is the energy of the shortest wavelength photon emitted by the hydrogen atom?

63 • The wavelength of a spectral line of hydrogen is 97.254 nm. Identify the transition that results in this line, assuming that the transition is to the ground state.

64 •• The wavelength of a spectral line of hydrogen is 1093.8 nm. Identify the transition that results in this line.

65 •• Spectral lines of the following wavelengths are emitted by a singly ionized helium atom: 164 nm, 230.6 nm, and 541 nm. Identify the transitions that result in those spectral lines.

66 •• The combination of physical constants $\alpha = e^2k/\hbar c$, where k is the Coulomb constant, is known as the *fine-structure constant*. It appears in numerous relations in atomic physics. (*a*) Show that α is dimensionless. (*b*) Show that in the Bohr model of the hydrogen atom $v_n = c\alpha/n$, where v_n is the speed of the electron in the state of quantum number n.

67 •• The wavelengths of the photons emitted by a potassium atom corresponding to transitions from the $4P_{3/2}$ and $4P_{1/2}$ states to the ground state are 766.41 nm and 769.90 nm. (*a*) Calculate the energies of the photons in electron volts. (*b*) The difference in the energies of the photons equals the difference in energy ΔE between the $4P_{3/2}$ and $4P_{1/2}$ states in potassium. Calculate ΔE. (*c*) Estimate the magnetic field that the 4p electron in potassium experiences.

68 •• To observe the characteristic K lines of the X-ray spectrum, one of the $n = 1$ electrons must be ejected from the atom. This is generally accomplished by bombarding the target material with electrons of sufficient energy to eject this tightly bound electron. What is the minimum energy required to observe the K lines of (*a*) a tungsten atom, (*b*) a molybdenum atom, and (*c*) a copper atom?

69 •• We are often interested in finding the quantity ke^2/r in electron volts when r is given in nanometers. Show that $ke^2 = 1.44\ \text{eV}\cdot\text{nm}$. SSM

70 •• The *positron* is a particle that has the same mass as the electron and carries a charge equal to $+e$. *Positronium* is a bound state of an electron–positron combination. (*a*) Calculate the energies of the five lowest energy states of positronium using the reduced mass, as given in Problem 27. (*b*) Do transitions between any of the levels found in Part (*a*) fall in the visible range of wavelengths? If so, which transitions are they?

71 • In 1947, Lamb and Retherford showed that there is a very small energy difference between the $2S_{1/2}$ and the $2P_{1/2}$ states of the hydrogen atom. They measured this difference essentially by causing transitions between the two states using very long wavelength electromagnetic radiation. The energy difference (the Lamb shift) is 4.372×10^{-6} eV and is explained by quantum electrodynamics as being due to fluctuations in the energy level of the vacuum. (*a*) What is the frequency of a photon whose energy is equal to the Lamb shift energy? (*b*) What is the wavelength of that photon? In what spectral region does it belong?

72 • A Rydberg atom is one in which an electron is in a *very* high excited state ($n \approx 40$ or higher). Such atoms are useful for experiments that probe the transition from quantum-mechanical behavior to classical. Furthermore, these excited states have extremely long lifetimes (i.e., the electron will stay in this high excited state for a very long time). A hydrogen atom is in the $n = 45$ state. (*a*) What is the ionization energy of the atom when it is in that state? (*b*) What is the energy level separation (in electron volts) between that state and the $n = 44$ state? (*c*) What is the wavelength of a photon resonant with the transition between these two states? (*d*) What is the radius of the atom when it is in the $n = 45$ state?

73 •• The deuteron, the nucleus of deuterium (heavy hydrogen), was first recognized from the spectrum of hydrogen. The deuteron has a mass that is approximately twice the mass of the proton. (*a*) Calculate the Rydberg constant for hydrogen and for deuterium using the reduced mass as given in Problem 27. (*b*) Using the result obtained in Part (*a*), determine the difference between the longest wavelength Balmer line of hydrogen (protium) and the longest wavelength Balmer line of deuterium.

74 •• The muonium atom is a hydrogen atom that has the electron replaced by a μ^- particle. The μ^- has a mass 207 times as great as the electron. (*a*) Calculate the energies of the five lowest energy levels of muonium using the reduced mass as given in Problem 27. (*b*) Do transitions between any of the levels found in Part (*a*) fall in the visible range of wavelengths (for example, between $\lambda = 700$ nm and $\lambda = 400$ nm)? If so, which transitions are they?

75 •• The triton, a nucleus consisting of a proton and two neutrons, is unstable and has a half-life of approximately 12 years. Tritium is an atom consisting of an electron and a triton. (*a*) Calculate the Rydberg constant of tritium using the reduced mass as given in Problem 27. (*b*) Determine the difference between the longest wavelength of the Balmer lines of tritium and the longest wavelength of the Balmer lines of deuterium (see Problem 73). In addition, (*c*) determine the difference between the longest wavelength of the Balmer lines of tritium and the longest wavelength of the Balmer lines of hydrogen (protium).

CHAPTER 37

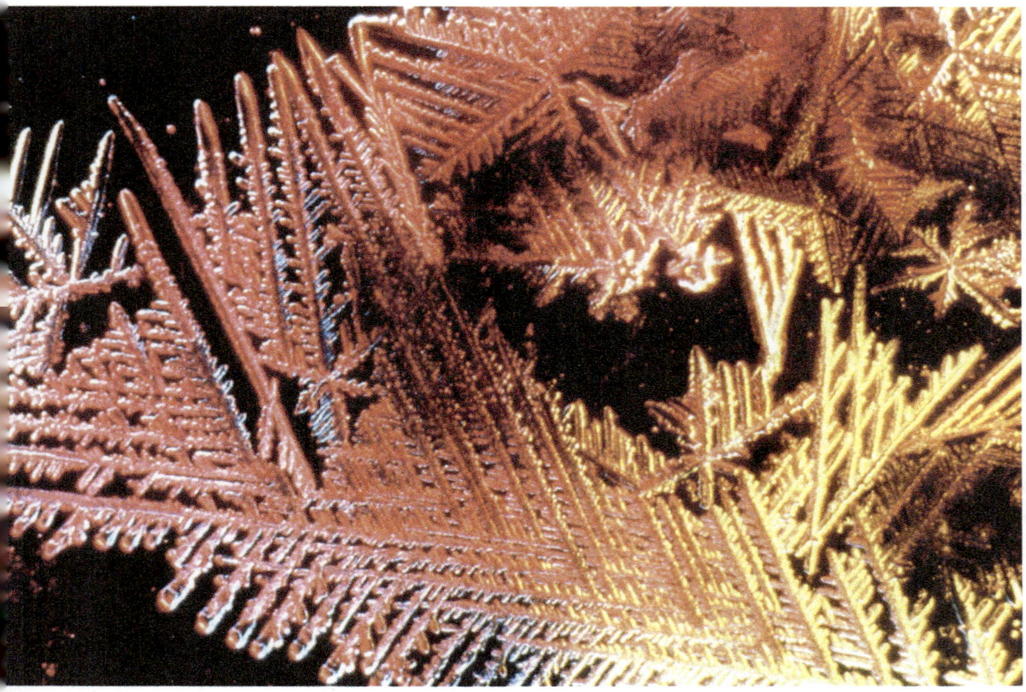

A MICROGRAPH OF SODIUM FLUORIDE CRYSTALS. SODIUM FLUORIDE IS OFTEN ADDED TO PUBLIC WATER SUPPLIES AS A TOOTH-DECAY PREVENTATIVE. *(National Institutes of Health/Photo Researchers.)*

Molecules

? How much energy is needed to form sodium fluoride? (See Example 37-1).

Most atoms bond together to form molecules or solids. Molecules may exist as separate entities, as in gaseous O_2 or N_2, or they may bond together to form liquids or solids. A molecule is the smallest constituent of a substance that retains its chemical properties.

In this chapter, we use our understanding of quantum mechanics to discuss bonding and the energy levels and spectra of diatomic molecules. Much of our discussion will be qualitative because, as in atomic physics, the quantum-mechanical calculations are very difficult.

37-1 BONDING

Consider a hydrogen molecule (H_2). We can think of H_2 either as two H atoms joined together or as a quantum-mechanical system of two protons and two electrons. The latter picture is more useful in this case because neither of the electrons in the H_2 molecule is confined to the region surrounding either one of the two protons. Instead, each electron is equally shared by both protons. For more complicated molecules, however, an intermediate picture is useful. For example, the fluorine molecule F_2 consists of 18 protons and 18 electrons, but only two of the electrons take part in the bonding. We therefore can consider this molecule as two F^+ ions and two electrons that belong to the molecule as a whole. The molecular wave functions for the bonding electrons are called **molecular orbitals.** In many cases, these molecular wave functions can be constructed from combinations of the atomic wave functions with which we are familiar.

The two principal types of bonds responsible for the formation of solids and molecules are the ionic bond and the covalent bond. Other types of bonds that are important in the bonding of liquids and solids are van der Waals bonds, metallic bonds, and hydrogen bonds. In many cases, bonding is a mixture of these mechanisms.

THE IONIC BOND

The simplest type of bond is the **ionic bond,** which is found in salts such as sodium chloride (NaCl). The sodium atom has one 3s electron outside a stable ten-electron core. The first ionization energy of sodium is the energy needed to remove the 3s electron from an isolated sodium atom. This energy is just 5.14 eV (see Figure 36-18). The removal of this electron results in an isolated positive ion that has its $n = 1$ and $n = 2$ electron shells filled. A chlorine atom has 17 electrons, and so is one electron short of having its first three shells filled. A measure of the energy released when an isolated atom gains one electron is called its **electron affinity;** a chlorine atom releases 3.62 eV of energy when it acquires an electron to form a Cl^- ion. Thus, the chlorine atom is said to have an electron affinity of -3.62 eV. The acquisition of one electron by chlorine results in a negative ion that has a filled outer electron shell. Thus, the formation of a Na^+ ion and a Cl^- ion by the donation of one electron of sodium to chlorine requires only $5.14\text{ eV} - 3.62\text{ eV} = 1.52\text{ eV}$ at infinite separation. The electrostatic potential energy U_e of the two ions when they are a distance r apart is $-ke^2/r$. When the separation of the ions is less than approximately 0.95 nm, the negative potential energy of attraction is of greater magnitude than the 1.52 eV of energy needed to create the ions. Thus, at separation distances less than 0.95 nm, it is energetically favorable (the total energy of the system is reduced) for the sodium atom to donate an electron to the chlorine atom to form NaCl.

Because the electrostatic attraction increases as the ions get closer together, it might seem that equilibrium could not exist. However, when the separation of the ions is very small, there is a strong repulsion that can be described by quantum mechanics and the exclusion principle. This repulsion is also responsible for the repulsion of the atoms in all molecules and ions (except H_2).* We can understand it qualitatively as follows. When the ions are very far apart, the probability distribution for a core electron in one of the ions does not overlap the probability distribution of any electron in the other ion. We can distinguish the electrons by the ion to which they belong. This means that electrons in the two ions can have the same quantum numbers because they occupy different regions of space. However, as the distance between the ions decreases, the probability distributions of the core electrons begin to overlap; that is, the electrons in the two ions begin to occupy the same region of space. Some of these electrons must go into higher energy quantum states as described by the exclusion principle.† But energy is required to shift the electrons into higher energy quantum states. This increase in energy when the ions are pushed closer together is equivalent to the repulsion energy of the ions. It is not a sudden process. The energy states of the electrons change gradually as the ions are brought together. A sketch of the potential energy $U(r)$ of the Na^+ and Cl^- ions versus separation distance r is shown in Figure 37-1. The energy is lowest at an

---

* In H_2, the repulsion is simply that of the two positively charged protons.

† Recall from our discussion in Chapter 35 that the exclusion principle is related to the fact that the wave function for two identical electrons is antisymmetric on the exchange of the electrons and that an antisymmetric wave function for two electrons with the same quantum numbers is zero if the space coordinates of the electrons are the same.

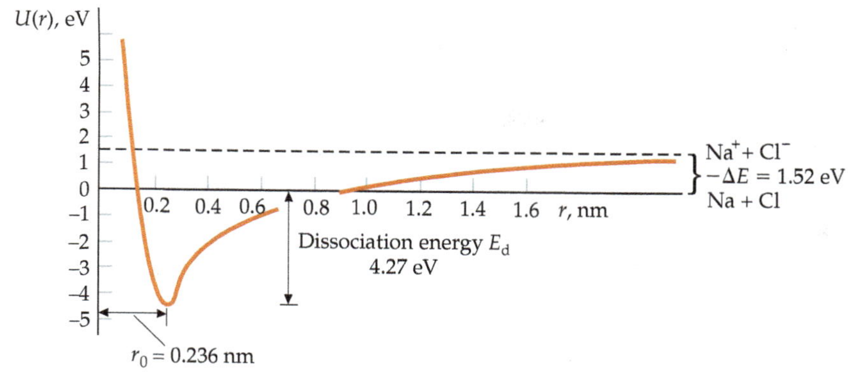

FIGURE 37-1 Potential energy for Na^+ and Cl^- ions as a function of separation distance r. The energy at infinite separation is chosen to be 1.52 eV, corresponding to the energy $-\Delta E$ needed to form the ions from atoms. The minimum energy is at the equilibrium separation $r_0 = 0.236$ nm for the ions.

equilibrium separation r_0 of approximately 0.236 nm. At smaller separations, the energy increases steeply. The energy required to separate the ions and form sodium and chlorine atoms is called the **dissociation energy** E_d, which is approximately 4.27 eV for NaCl.

The equilibrium separation distance of the gaseous NaCl, which can be obtained by evaporating solid NaCl, is 0.236 nm. Normally, NaCl exists as a solid in a cubic crystal structure, where the Na^+ and Cl^- ions are at the alternate corners of a cube. The separation of the Na^+ and Cl^- ions in a crystal is approximately 0.28 nm, which is somewhat larger than the 0.236 nm separation for the gaseous NaCl. Because of the presence of neighboring ions of opposite charge, the electrostatic energy per ion pair is lower when the ions are in a crystal.

Example 37-1 The Energy of Sodium Fluoride

The electron affinity of fluorine is −3.40 eV and the equilibrium separation of sodium fluoride (NaF) is 0.193 nm. (*a*) How much energy is needed to form Na^+ and F^- ions from sodium and fluorine atoms? (*b*) What is the electrostatic potential energy of the Na^+ and F^- ions at their equilibrium separation? (*c*) The dissociation energy of NaF is 5.38 eV. What is the energy due to repulsion of the ions at the equilibrium separation?

PICTURE (*a*) The energy ΔE needed to form Na^+ and F^- ions from the sodium and fluorine atoms is the sum of the first ionization energy of sodium (5.14 eV) and the electron affinity of fluorine. (*b*) The electrostatic potential energy, where $U = 0$ at infinity, is $U_e = -ke^2/r$. (*c*) If we choose the potential energy at infinity to be ΔE, the total potential energy is $U_{tot} = U_e + \Delta E + U_{rep}$, where U_{rep} is the energy of repulsion, which is found by setting the dissociation energy equal to $-U_{tot}$.

SOLVE

(*a*) Calculate the energy needed to form Na^+ and F^- ions from the sodium and fluorine atoms (see the Picture section):

$$\Delta E = 5.14\text{ eV} - 3.40\text{ eV} = \boxed{1.74\text{ eV}}$$

(*b*) Calculate the electrostatic potential energy at the equilibrium separation of $r = 0.193$ nm:

$$U_e = -\frac{ke^2}{r}$$

$$= -\frac{(8.99 \times 10^9\text{ N}\cdot\text{m}^2/\text{C}^2)(1.60 \times 10^{-19}\text{ C})^2}{1.93 \times 10^{-10}\text{ m}}$$

$$= -1.19 \times 10^{-18}\text{ J} = \boxed{-7.45\text{ eV}}$$

(*c*) The dissociation energy equals the negative of the total potential energy:

$$E_d = -U_{tot} = -(U_e + \Delta E + U_{rep})$$

so

$$U_{rep} = -(E_d + \Delta E + U_e)$$

$$= -(5.38\text{ eV} + 1.74\text{ eV} - 7.45\text{ eV}) = \boxed{0.33\text{ eV}}$$

CHECK The Part-(*c*) result is greater than zero as expected.

THE COVALENT BOND

A completely different mechanism, the **covalent bond,** is responsible for the bonding of identical or similar atoms to form molecules such as gaseous hydrogen (H_2), nitrogen (N_2), and carbon monoxide (CO). If we calculate the energy needed to form H^+ and H^- ions by the transfer of an electron from one atom to the other and then add this energy to the electrostatic potential energy, we find that there is no separation distance for which the total energy is negative. The bond thus cannot be ionic. Instead, the attraction of two hydrogen atoms can only be explained quantum-mechanically. The decrease in energy when two hydrogen atoms approach each other is due to the sharing of the two electrons by both atoms, which can be explained using the symmetry properties of the wave functions of electrons.

We can gain some insight into covalent bonding by considering a simple, one-dimensional quantum-mechanics problem of two identical finite square wells. We first consider a single electron that is equally likely to be in either well. Because the wells are identical, the probability distribution, which is proportional to $|\psi^2|$, must be symmetric about the midpoint between the wells. Then ψ must be either symmetric or antisymmetric with respect to the two wells. The two possibilities for the ground state are shown in Figure 37-2*a* for the case in which the wells are far apart and in Figure 37-2*b* for the case in which the wells are close together. An important feature of Figure 37-2*b* is that in the region between the wells the symmetric wave function is large and the antisymmetric wave function is small.

Now consider adding a second electron to the two wells. We saw in Section 6 of Chapter 35 that the wave functions for particles that obey the exclusion principle are antisymmetric on exchange of the particles. Thus, the total wave function for the two electrons must be antisymmetric on exchange of the electrons. Note that exchanging the electrons while keeping the wells in place is equivalent to keeping the electrons in place and exchanging the wells. The total wave function for two electrons can be written as a spatial expression and an expression for spin. So, an antisymmetric wave function can be the product of a symmetric spatial expression and an antisymmetric expression for spin or of a symmetric expression for spin and an antisymmetric spatial expression.

To understand the symmetry of the total wave function, we must therefore understand the symmetry of the expression for spin of the wave function. The spin of a single electron can have two possible values for its quantum number m_S: $m_S = +\frac{1}{2}$, which we call spin up, or $m_S = -\frac{1}{2}$, which we call spin down. We will use arrows to designate the spin wave function for a single electron: $\uparrow_1$ or $\uparrow_2$ for electron 1 or electron 2 that both are spin up and $\downarrow_1$ or $\downarrow_2$ for electron 1 or electron 2 that are both spin down. The total spin quantum number for two electrons can be $S = 1$, where $m_S = +1, 0,$ or -1; or $S = 0$, where $m_S = 0$. We use $\phi_{S\, m_S}$ to denote the spin wave function for two electrons. The spin state $\phi_{1\,+1}$, corresponding to $S = 1$ and $m_S = +1$, can be written

$$\phi_{1\,+1} = \uparrow_1\uparrow_2 \qquad S = 1, m_S = +1 \tag{37-1}$$

Similarly, the spin state for $S = 1, m_S = -1$ is

$$\phi_{1\,-1} = \downarrow_1\downarrow_2 \qquad S = 1, m_S = -1 \tag{37-2}$$

Note that both of these states are symmetric upon exchange of the electrons. The spin state corresponding to $S = 1$ and $m_S = 0$ is not quite so obvious. It turns out to be proportional to

$$\phi_{10} = \uparrow_1\downarrow_2 + \uparrow_2\downarrow_1 \qquad S = 1, m_S = 0 \tag{37-3}$$

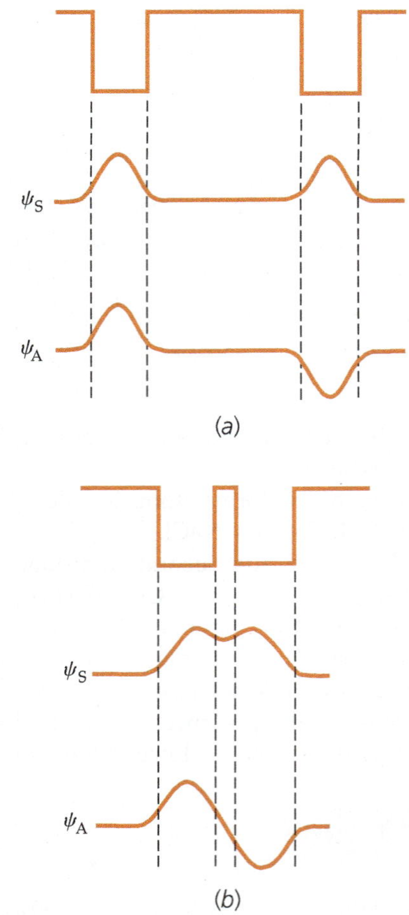

FIGURE 37-2 (*a*) Two square wells far apart. The electron wave function can be either symmetric (ψ_S) or antisymmetric (ψ_A) in space. The probability distributions and energies are the same for the two wave functions when the wells are far apart. (*b*) Two square wells that are close together. Between the wells, the antisymmetric space wave function is approximately zero, whereas the symmetric space wave function is quite large.

This spin state is also symmetric upon exchange of the electrons. The spin state for two electrons with antiparallel spins ($S = 0$) is

$$\phi_{00} = \uparrow_1\downarrow_2 - \uparrow_2\downarrow_1 \qquad S = 0, m_S = 0 \qquad 37\text{-}4$$

This spin state is antisymmetric upon exchange of electrons.

We thus have the important result that the spin part of the wave function is symmetric for parallel spins ($S = 1$) and antisymmetric for antiparallel spins ($S = 0$). Because the total wave function is the product of the spatial expression and the expression for spin, we have the following important result:

> For the total wave function of two electrons to be antisymmetric, the spatial part of the wave function must be antisymmetric for parallel spins ($S = 1$) and symmetric for antiparallel spins ($S = 0$).
>
> SPIN ALIGNMENT AND WAVE-FUNCTION SYMMETRY

We can now consider the problem of two hydrogen atoms. Figure 37-3*a* shows a spatially symmetric wave function ψ_S and a spatially antisymmetric wave function ψ_A for two hydrogen atoms that are far apart, and Figure 37-3*b* shows the same two wave functions for two hydrogen atoms that are close together. The squares of these two wave functions are shown in Figure 37-3*c*. Note that the probability distribution $|\psi|^2$ in the region between the protons is large for the symmetric wave function and small for the antisymmetric wave function. Thus, when the spatial part of the wave function is symmetric ($S = 0$), the electrons are often found in the region between the protons. The electron cloud, as shown in the upper part of Figure 37-3*c*, is concentrated in the space between the protons and the protons are bound together by this negatively charged cloud. Conversely, when the spatial part of the wave function is antisymmetric ($S = 1$), the electrons spend little time between the protons and the atoms do not bond together to form a molecule. In this case, the electron cloud is not concentrated in the space between the protons, as shown in the lower part of Figure 37-3*c*.

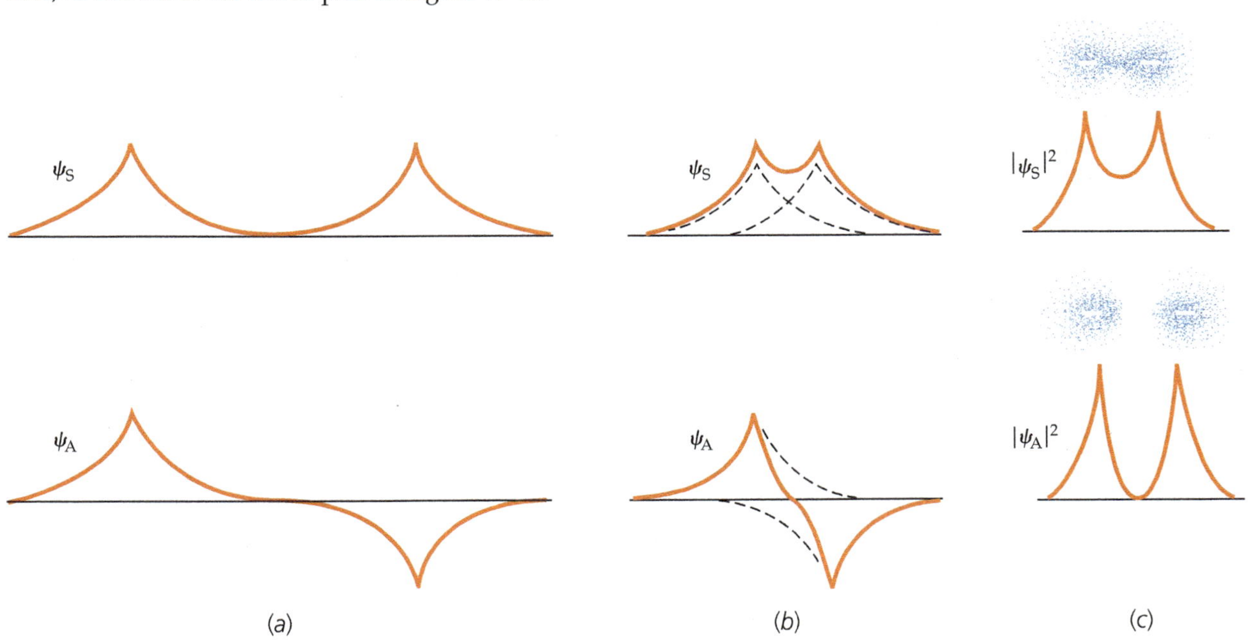

FIGURE 37-3 One-dimensional symmetric and antisymmetric wave functions for two hydrogen atoms (*a*) far apart and (*b*) close together. (*c*) Electron probability distributions ($|\psi|^2$) for the wave functions in Figure 37-3*b*. For the symmetric wave function, the electron charge density is large between the protons. This negative charge density holds the protons together in the hydrogen molecule H_2. For the antisymmetric wave function, the electron charge density is not large between the protons.

The total electrostatic potential energy for the H_2 molecule consists of the positive energy of repulsion of the two electrons and the negative potential energy of attraction of each electron for each proton. Figure 37-4 shows the electrostatic potential energy function U_S for two hydrogen atoms versus separation for the case in which the spatial part of the electron wave function is symmetric, and the electrostatic potential energy function U_A for the case in which the spatial part of the wave function is antisymmetric. We can see that the potential energy for the symmetric state is lower than the potential energy for the antisymmetric state and that the shape of the potential energy curve for the symmetric state is similar to the shape of the potential energy curve for ionic bonding (Figure 37-1). The equilibrium separation for H_2 is $r_0 = 0.074$ nm, and the binding energy is 4.52 eV. For the antisymmetric state, the potential energy is never negative and there is no bonding.

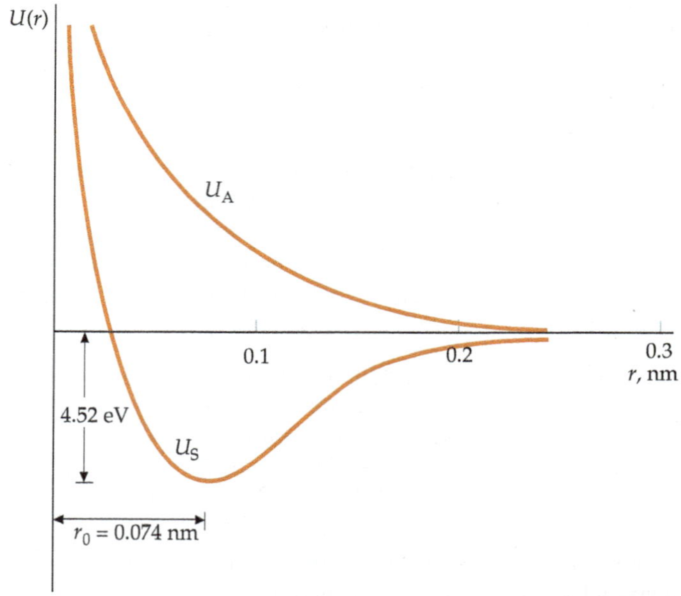

FIGURE 37-4 Potential energy versus separation for two hydrogen atoms. The curve labeled U_S is for a wave function that has a symmetric expression for the spatial part and the curve labeled U_A is for a wave function that has an antisymmetric expression for the spatial part.

We can now see why three hydrogen atoms do not bond to form H_3. If a third hydrogen atom is brought near an H_2 molecule, the third electron cannot be in a 1s state and have its spin antiparallel to the spin of both of the other electrons. If that electron is in an antisymmetric spatial state with respect to exchange with one of the electrons, the repulsion of this atom is greater than the attraction of the other. As the three atoms are pushed together, the third electron is, in effect, forced into a higher quantum-energy state according to the exclusion principle. The bond between two hydrogen atoms is called a **saturated bond** because there is no room for another electron. The two shared electrons essentially fill the 1s states of both atoms.

We can also see why two helium atoms do not normally bond together to form the He_2 molecule. There are no valence electrons that can be shared. The electrons in the filled shells are forced into higher energy states when the two atoms are brought together. At low temperatures or high pressures, helium atoms do bond together due to van der Waals forces, which we will discuss next. This bonding is so weak that at atmospheric pressure helium boils at 4 K, and it does not form a solid at any temperature unless the pressure is greater than about 20 atm.

> ! Don't think all bonds between atoms are partly ionic. They are not. Bonds between two identical atoms are always 100 percent covalent.

When two identical atoms bond, as in O_2 or N_2, the bonding is purely covalent. However, the bonding of two dissimilar atoms is often a mixture of covalent and ionic bonding. Even in NaCl, the electron donated by sodium to chlorine has some probability of being at the sodium atom because its wave function in the vicinity of the sodium atom, while small, is not zero. Thus, this electron is partially shared in a covalent bond, although this bonding is only a small part of the total bond, which is mainly ionic.

A measure of the degree to which a bond is ionic or covalent can be obtained from the electric dipole moment of the molecule or ionic unit. For example, if the bonding in NaCl were purely ionic, the center of positive charge would be at the Na^+ ion and the center of negative charge would be at the Cl^- ion. The electric dipole moment would have the magnitude

$$p_{\text{ionic}} = er_0 \qquad 37\text{-}5$$

where $r_0 = 2.36 \times 10^{-10}$ m is the equilibrium separation of the ions. Thus, the dipole moment of NaCl would be (from Figure 37-1)

$$p_{\text{ionic}} = er_0$$
$$= (1.60 \times 10^{-19}\ \text{C})(2.36 \times 10^{-10}\ \text{m}) = 3.78 \times 10^{-29}\ \text{C}\cdot\text{m}$$

The actual measured electric dipole moment of NaCl is

$$p_{\text{measured}} = 3.00 \times 10^{-29}\ \text{C} \cdot \text{m}$$

We can define the ratio of p_{measured} to p_{ionic} as the fractional amount of ionic bonding. For NaCl, this ratio is $3.00/3.78 = 0.79$. Thus, the bonding in NaCl is about 79 percent ionic.

PRACTICE PROBLEM 37-1

The equilibrium separation of HCl is 0.128 nm and its measured electric dipole moment is 3.60×10^{-30} C · m. What is the percentage of ionic bonding in HCl?

OTHER BONDING TYPES

The van der Waals Bond Any two separated molecules will be attracted to one another by electrostatic forces called *van der Waals forces.* So will any two atoms that do not form ionic or covalent bonds. The **van der Waals bonds** due to these forces are much weaker than the bonds already discussed. At high enough temperatures, these forces are not strong enough to overcome the motion of the atoms or molecules due to thermal energy. At sufficiently low temperatures, these motions become negligible and the van der Waals forces will cause virtually all substances to condense into a liquid and then a solid form.* The van der Waals forces arise from the interaction of the instantaneous electric dipole moments of the molecules.

Figure 37-5 shows how two polar molecules—molecules that have *permanent* electric dipole moments, such as H_2O—can bond. The electric field due to the dipole moment of one molecule orients the other molecule so that the two dipole moments attract. Nonpolar molecules also attract other nonpolar molecules by the van der Waals forces. Although nonpolar molecules have zero electric dipole moments on the average, they have instantaneous dipole moments that are generally not zero because of fluctuations in the positions of the charges. When two nonpolar molecules are near each other, the fluctuations in the instantaneous dipole moments tend to become correlated so as to produce attraction. This is illustrated in Figure 37-6.

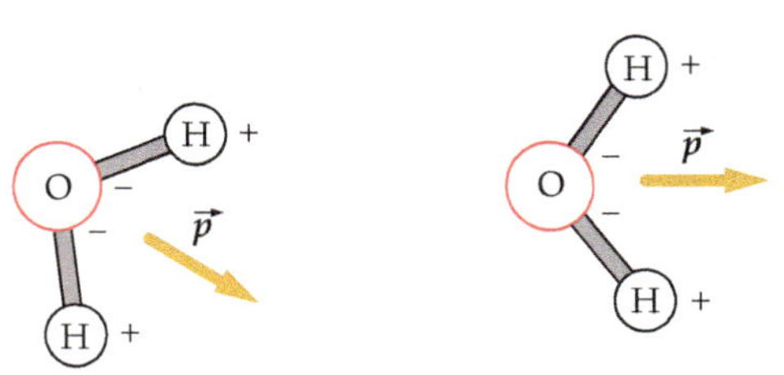

FIGURE 37-5 Bonding of H_2O molecules due to the attraction of the electric dipoles. The dipole moment of each molecule is indicated by $\vec{p}$. The electric field of one dipole orients the other dipole so the two dipole moments tend to be parallel. When the dipole moments are approximately parallel, the center of negative charge of one molecule is closer to the center of positive charge of the other molecule than it is to the center of the negative charge, so the molecules attract.

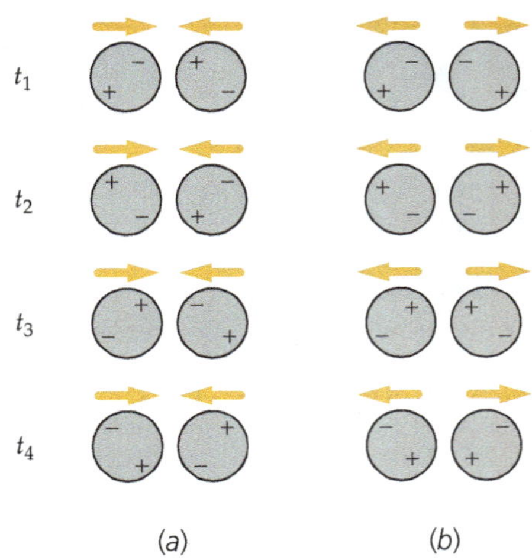

FIGURE 37-6 van der Waals attraction of molecules that have zero permanent dipole moments. (*a*) Possible orientations of instantaneous dipole moments at different times leading to attraction. (*b*) Possible orientations leading to repulsion. The electric field of the instantaneous dipole moment of one molecule tends to polarize the other molecule; thus the orientations leading to attraction (Figure 37-6*a*) are much more likely than those leading to repulsion (Figure 37-6*b*).

* Helium is the only element that does not solidify at any temperature at atmospheric pressure.

The hydrogen bond Another bonding mechanism of great importance is the hydrogen bond, which is formed by the sharing of a proton (the nucleus of the hydrogen atom) between two atoms, frequently two oxygen atoms. This sharing of a proton is similar to the sharing of electrons responsible for the covalent bond already discussed. It is facilitated by the small mass of the proton and by the absence of core electrons in hydrogen. The hydrogen bond often holds groups of molecules together and is responsible for the cross-linking that allows giant biological molecules and polymers to hold their fixed shapes. The well-known helical structure of DNA is due to hydrogen-bond linkages across turns of the helix (Figure 37-7).

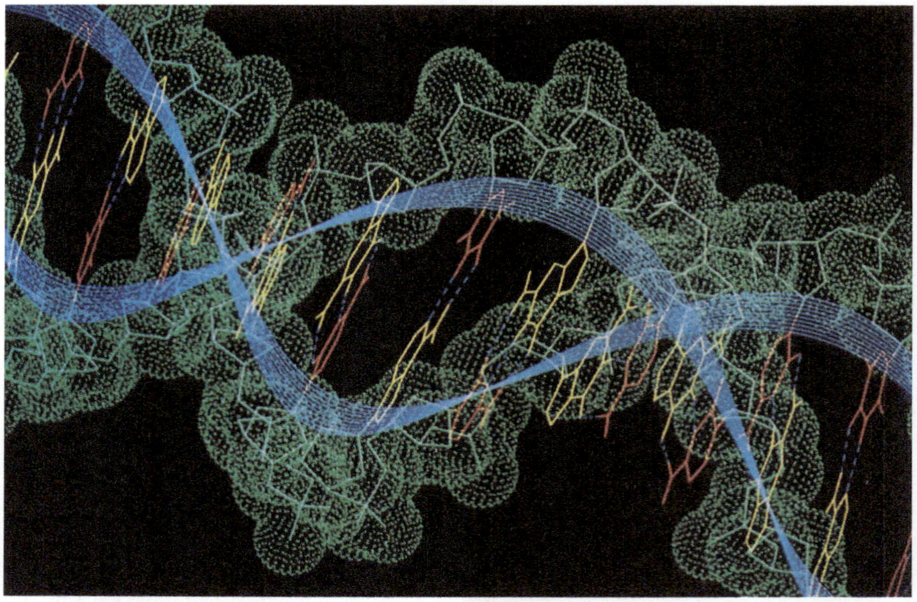

FIGURE 37-7 The DNA molecule. *(© Will and Demi McIntire/Photo Researchers.)*

(*a*) The discoverers of the structure of DNA. James Watson at left and Francis Crick are shown with their model of part of a DNA molecule in 1953. Crick and Watson met at the Cavendish Laboratory, Cambridge, in 1951. Their work on the structure of DNA was performed with a knowledge of Chargaff's ratios of the bases in DNA and some access to the X-ray crystallography of Maurice Wilkins and Rosalind Franklin at King's College London. Combining all of this work led to the deduction that DNA exists as a double helix, thus to its structure. Crick, Watson, and Wilkins shared the 1962 Nobel Prize for Physiology or Medicine; Franklin died from cancer in 1958. *((a) Norman Collection for the History of Molecular Biology.)*

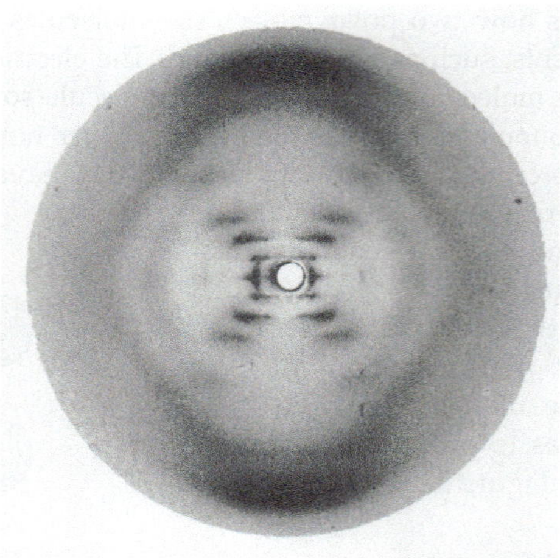

(*b*) X-ray diffraction pattern of the B form of DNA. Rosalind Franklin's colleague Maurice Wilkins, without obtaining her permission, made available to Watson and Crick her then unpublished X-ray diffraction pattern of the B form of DNA, which was crucial evidence for the helical structure. In his account of this discovery, Watson wrote: "The instant I saw the picture, my mouth fell open and pulse began to race. . . . The black cross of reflections which dominated the picture could arise only from a helical structure. . . . Mere inspection of the X-ray picture gave several of the vital helical parameters" (from Stent, Gunther, *The Double Helix*, New York: Norton, 1980). *((b) © A. Barrington Brown/Photo Researchers, NY.)*

The metallic bond In a metal, two atoms do not bond together by exchanging or sharing an electron to form a molecule. Instead, each valence electron is shared by many atoms. The bonding is thus distributed throughout the entire metal. A metal can be thought of as a lattice of positive ions held together by essentially free electrons that roam throughout the solid. In the quantum-mechanical picture, these free electrons form a cloud of negative charge density between the positively charged lattice ions that holds the ions together. In this respect, the metallic bond is somewhat similar to the covalent bond. However, with the metallic bond, there are far more than just two atoms involved, and the negative charge is distributed uniformly throughout the volume of the metal. The number of free electrons per lattice ion varies from metal to metal but is of the order of one free electron per ion.

* 37-2 POLYATOMIC MOLECULES

Molecules that have more than two atoms range from relatively simple molecules such as water, which has a molecular mass number of 18, to such giants as proteins and DNA, which can have molecular mass numbers of hundreds of thousands up to many millions. As with diatomic molecules, the structure of polyatomic molecules can be understood by applying basic quantum mechanics to the bonding of individual atoms. The bonding mechanisms for most polyatomic molecules are the covalent bond and the hydrogen bond. We will discuss only some of the simplest polyatomic molecules—H_2O, NH_3, and CH_4—to illustrate both the simplicity and complexity of the application of quantum mechanics to molecular bonding.

The basic requirement for the sharing of electrons in a covalent bond is that the wave functions of the valence electrons in the individual atoms must overlap as much as possible. As our first example, we will consider the water molecule. The ground-state configuration of the oxygen atom is $1s^2 2s^2 2p^4$. The 1s and 2s electrons are in filled shells and do not contribute to the bonding. The 2p shell has room for six electrons, two in each of the three spatial states (orbitals) corresponding to $\ell = 1$. In an isolated atom, we describe these spatial states by the hydrogen-like wave functions corresponding to $\ell = 1$ and $m_\ell = +1, 0,$ and -1. Because the energy is the same for these three spatial states, we could equally well use any linear combination of these wave functions. When an atom participates in molecular bonding, certain combinations of these atomic wave functions are important. These combinations are called **p_x, p_y, and p_z atomic orbitals.** The angular dependence of these orbitals is

$$p_x \propto \sin\theta \cos\phi \qquad 37\text{-}6$$

$$p_y \propto \cos\theta \cos\phi \qquad 37\text{-}7$$

$$p_z \propto \cos\phi \qquad 37\text{-}8$$

The electron charge distribution for these orbitals is maximum along the x, y, or z axis, respectively, as shown in Figure 37-8.

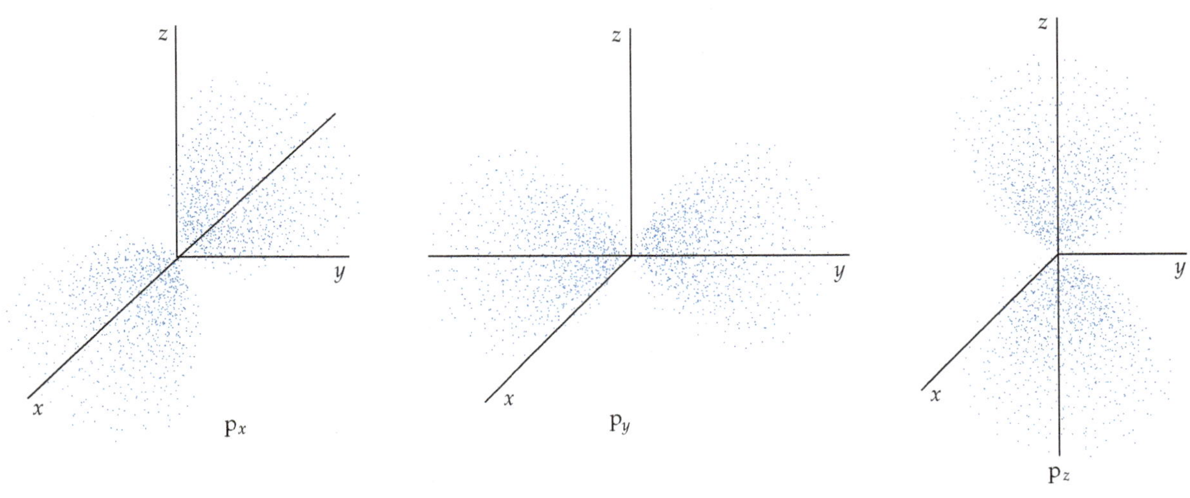

FIGURE 37-8 Computer-generated dot plot illustrating the spatial dependence of the electron charge distribution in the p_x, p_y, and p_z atomic orbitals.

For the oxygen in an H_2O molecule, maximum overlap of the electron wave functions occurs when two of the four 2p electrons are in one of the atomic orbitals (for this example, assume the p_z orbital) with their spins antiparallel, the third 2p electron is in a second orbital (the p_x orbital), and the fourth 2p electron is in the third orbital (the p_y orbital). Each of the unpaired electrons (in the p_x and p_y orbitals for this example) forms a bond with the electron of a hydrogen atom, as shown in Figure 37-9. Because of the repulsion of the two hydrogen atoms, the angle between the O—H bonds is actually greater than 90°. The effect of this repulsion can be calculated, and the result is in agreement with the measured angle of 104.5°.

Similar reasoning leads to an understanding of the bonding in NH_3 (not shown). In the ground state, nitrogen has three electrons in the 2p state. When these three electrons are in the p_x, p_y, and p_z atomic orbitals, they bond to the electrons of hydrogen atoms. Again, because of the repulsion of the hydrogen atoms, the angles between the bonds are somewhat larger than 90°.

The bonding of carbon atoms is somewhat more complicated. Carbon forms single, double and triple bonds, leading to a great diversity in the kinds of organic molecules. The ground-state configuration of carbon is $1s^22s^22p^2$. From our previous discussion, we might expect carbon to be divalent—that is, bonding only through its two 2p electrons—with the two bonds forming at approximately 90°. However, one of the most important features of the chemistry of carbon is that tetravalent carbon compounds, such as CH_4, are overwhelmingly favored.

The observed valence of 4 for carbon comes about in an interesting way. One of the first excited states of carbon occurs when a 2s electron is excited to a 2p state, giving a configuration of $1s^22s^12p^3$. In this excited state, we can have four unpaired electrons, one each in the 2s, $2p_x$, $2p_y$, and $2p_z$ atomic orbitals. We might expect there to be three similar bonds corresponding to the three p orbitals and one different bond corresponding to the s orbital. However, when carbon forms tetravalent bonds, these four atomic orbitals become mixed and form four new *equivalent* molecular orbitals called **hybrid orbitals.** This mixing of atomic orbitals, called hybridization, is among the most important features involved in the physics of complex molecular bonds. Figure 37-10 shows the tetrahedral structure of the methane molecule (CH_4), and Figure 37-11 shows the structure of the ethane molecule (CH_3—CH_3), which is similar to two joined methane molecules in which one of the C—H bonds is replaced with a C—C bond.

Carbon orbitals can also hybridize such that the s, p_x, and p_y orbitals combine to form three hybrid orbitals that are in the xy plane and form bonds that are 120° apart (the p_z orbital does not participate in bonding). An example of this configuration is graphite, in which the bonds in the xy plane provide the strongly layered structure characteristic of the material.

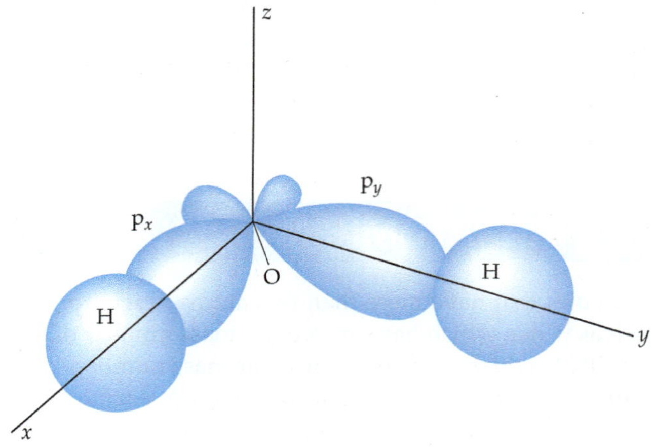

FIGURE 37-9 Electron charge distribution in the H_2O molecule.

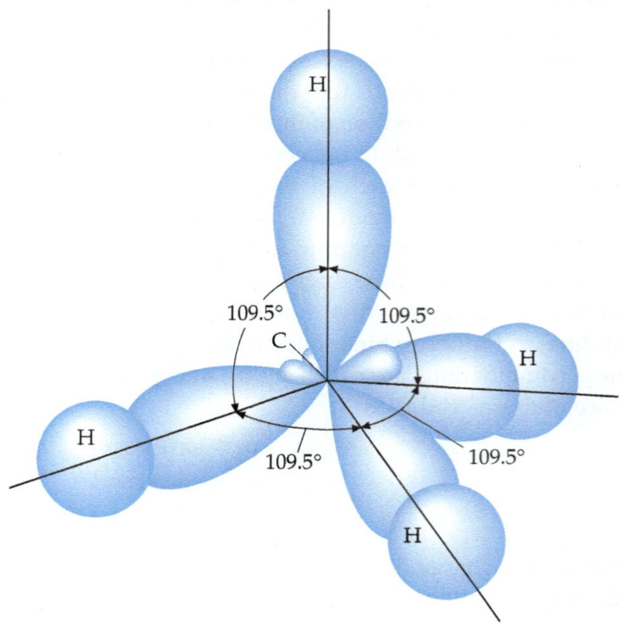

FIGURE 37-10 Electron charge distribution in the CH_4 (methane) molecule.

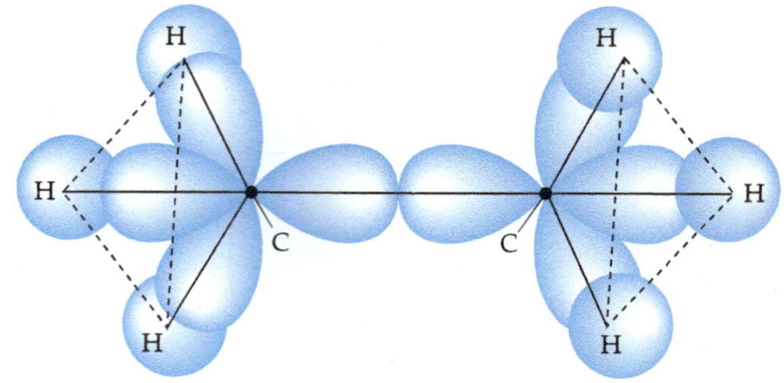

FIGURE 37-11 Electron charge distribution in the CH_3—CH_3 (ethane) molecule.

37-3 ENERGY LEVELS AND SPECTRA OF DIATOMIC MOLECULES

As is the case with an atom, a molecule often emits electromagnetic radiation when it makes a transition from an excited energy state to a state of lower energy. Conversely, a molecule can absorb radiation and make a transition from a lower energy state to a higher energy state. The study of molecular emission and absorption spectra thus provides us with information about the energy states of molecules. For simplicity, we will consider only diatomic molecules here.

The internal energy of a molecule can be conveniently separated into three parts: electronic, due to the excitation of the electrons of the molecule; vibrational, due to the oscillations of the atoms of the molecule; and rotational, due to the rotation of the molecule about its center of mass. The magnitudes of these energies are sufficiently different that they can be treated separately. The energies due to the electronic excitations of a molecule are typically of the order of magnitude of 1 eV, the same as for the electronic excitations of an atom. The energies of the vibrations of the atoms and of the rotation of the molecule are much smaller than the electronic excitation energy.

ROTATIONAL ENERGY LEVELS

Figure 37-12 shows a simple schematic model of a diatomic molecule consisting of particles that have masses of m_1 and m_2, are separated by a distance r and are rotating about the center of mass. Classically, the kinetic energy of rotation (Equation 9-11) is

$$E = \tfrac{1}{2}I\omega^2 \qquad 37\text{-}9$$

where I is the moment of inertia and ω is the angular speed of the rotation motion. If we write this in terms of the angular momentum $L = I\omega$, we have

$$E = \frac{(I\omega)^2}{2I} = \frac{L^2}{2I} \qquad 37\text{-}10$$

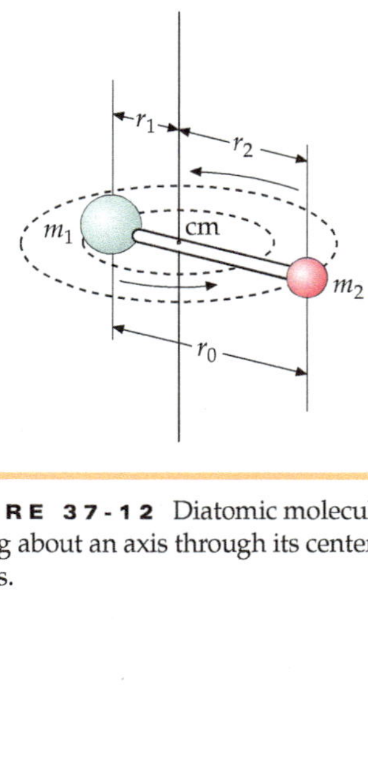

FIGURE 37-12 Diatomic molecule rotating about an axis through its center of mass.

The solution of the Schrödinger equation for rotation leads to quantization of the angular momentum with values given by

$$L^2 = \ell(\ell + 1)\hbar^2 \qquad \ell = 0, 1, 2, \ldots \qquad 37\text{-}11$$

where ℓ is the **rotational quantum number.** This is the same quantum condition on angular momentum that holds for the orbital angular momentum of an electron in an atom. Note, however, that L in Equation 37-10 refers to the angular momentum of the entire molecule rotating about its center of mass. The energy levels of a rotating molecule are therefore given by

$$E_\ell = \frac{\ell(\ell + 1)\hbar^2}{2I} = \ell(\ell + 1)E_{0r} \qquad \ell = 0, 1, 2, \ldots \qquad 37\text{-}12$$

ROTATIONAL ENERGY LEVELS

where E_{0r} is the characteristic rotational energy of a particular molecule, which is inversely proportional to its moment of inertia:

$$E_{0r} = \frac{\hbar^2}{2I} \qquad 37\text{-}13$$

CHARACTERISTIC ROTATIONAL ENERGY

A measurement of the rotational energy of a molecule from its rotational spectrum can be used to determine the moment of inertia of the molecule, which can then be used to find the separation of the atoms in the molecule. The moment of inertia about an axis through the center of mass of a diatomic molecule (see Figure 37-12) is

$$I = m_1 r_1^2 + m_2 r_2^2$$

Using $m_1 r_1 = m_2 r_2$, where r_1 is the distance of atom 1 from the center of mass, r_2 is the distance of atom 2 from the center of mass, and $r_0 = r_1 + r_2$, we can write the moment of inertia (see Problem 26) as

$$I = \mu r_0^2 \tag{37-14}$$

where μ, called the **reduced mass,** is

$$\mu = \frac{m_1 m_2}{m_1 + m_2} \tag{37-15}$$

DEFINITION—REDUCED MASS

If the masses are equal ($m_1 = m_2 = m$), as in H_2 and O_2, the reduced mass is $\mu = \frac{1}{2}m$ and

$$I = \tfrac{1}{2} m r_0^2 \tag{37-16}$$

A unit of mass convenient for discussing atomic and molecular masses is the **unified atomic mass unit,** u, which is defined as one-twelfth the mass of the carbon-12 (^{12}C) atom. The mass of one ^{12}C atom is thus 12 u. The mass of an atom in unified mass units is therefore numerically equal to the molar mass of the atom in grams. The unified mass unit is related to the gram and kilogram by

$$1\ \text{u} = \frac{1\ \text{g}}{N_A} = \frac{10^{-3}\ \text{kg}}{6.0221 \times 10^{23}} = 1.6606 \times 10^{-27}\ \text{kg} \tag{37-17}$$

where N_A is Avogadro's number.

Example 37-2 The Reduced Mass of a Diatomic Molecule

Find the reduced mass of the HCl molecule.

PICTURE We find the masses of the hydrogen and chlorine atoms in Appendix C* and use the definition of reduced mass (Equation 37-15).

SOLVE

1. The reduced mass μ is related to the individual masses m_H and m_{Cl}: $\mu = \dfrac{m_H m_{Cl}}{m_H + m_{Cl}}$
2. Find the masses in the periodic table: $m_H = 1.01\ \text{u}, \quad m_{Cl} = 35.5\ \text{u}$
3. Substitute to calculate the reduced mass: $\mu = \dfrac{m_H m_{Cl}}{m_H + m_{Cl}} = \dfrac{(1.01\ \text{u})(35.5\ \text{u})}{1.01\ \text{u} + 35.5\ \text{u}} = \boxed{0.982\ \text{u}}$

CHECK The formula for reduced mass is identical to the formula for the equivalent resistance for two resistors in parallel. As expected, the reduced mass is less than either mass.

TAKING IT FURTHER When one atom of a diatomic molecule is much more massive than the other, the center of mass of the molecule is approximately at the center of the more massive atom and the reduced mass is approximately equal to the mass of the lighter atom.

* The masses in the tables are weighted according to the natural isotopic distribution. Thus, the mass of carbon is given as 12.011 rather than 12.000 because natural carbon consists of about 98.9 percent ^{12}C and 1.1 percent ^{13}C. Similarly, natural chlorine consists of about 76 percent ^{35}Cl and 24 percent ^{37}Cl.

Example 37-3 Rotational Kinetic Energy of a Molecule

Estimate the characteristic rotational kinetic energy of an O_2 molecule, assuming that the separation of the atoms is 0.100 nm.

PICTURE The characteristic rotational kinetic energy is given by $E_{0r} = \hbar/(2I)$ (Equation 37-13), where I is the moment of inertia. The moment of inertia is given by $I = \mu r_0^2$ (Equation 37-14), where μ is the reduced mass and r_0 is the average center-to-center separation of the atomic nuclei.

SOLVE

1. The characteristic rotational energy is inversely proportional to the moment of inertia: $$E_{0r} = \frac{\hbar^2}{2I}$$
2. Calculate the moment of inertia: $$I = \mu r_0^2 = \tfrac{1}{2} m r_0^2$$
3. Substitute this expression for I into the expression for E_{0r}: $$E_{0r} = \frac{\hbar^2}{m r_0^2}$$
4. Use $m = 16$ u for the mass of oxygen to calculate E_{0r}: $$E_{0r} = \frac{\hbar^2}{m r_0^2} = \frac{(1.055 \times 10^{-34}\,\text{J}\cdot\text{s})^2}{(16\,\text{u})(10^{-10}\,\text{m})^2} \times \left(\frac{1\,\text{u}}{1.66 \times 10^{-27}\,\text{kg}}\right)$$ $$= 4.19 \times 10^{-23}\,\text{J} = \boxed{2.62 \times 10^{-4}\,\text{eV}}$$

CHECK As expected, the characteristic rotational kinetic energy is small compared with 1 eV (a typical electronic excitation energy).

We can see from Example 37-3 that the rotational energy levels are several orders of magnitude smaller than energy levels due to electron excitation. Transitions within a given set of rotational energy levels yield photons in the microwave region of the electromagnetic spectrum. The rotational energies are also small compared with the typical thermal energy kT at normal temperatures. For $T = 300$ K, for example, kT is about 2.6×10^{-2} eV, which is approximately 100 times the characteristic rotational energy as calculated in Example 37-3 and approximately 1 percent of the typical electronic energy. Thus, at ordinary temperatures, a molecule can be easily excited to the lower rotational energy levels by collisions with other molecules. But such collisions cannot excite the molecule to its electronic energy levels above the ground state.

CONCEPT CHECK 37-1

At room temperature, the molecules of a diatomic gas undergo transitions between rotational states, but the atoms of a monatomic gas do not. Why?

VIBRATIONAL ENERGY LEVELS

The quantization of energy in a simple harmonic oscillator was one of the first problems solved by Schrödinger in his paper proposing his wave equation. Solving the Schrödinger equation for a simple harmonic oscillator gives

$$E_\nu = (\nu + \tfrac{1}{2})hf \qquad \nu = 0, 1, 2, \ldots \qquad \text{37-18}$$

VIBRATIONAL ENERGY LEVELS

where f is the frequency of the oscillator and ν (lowercase Greek nu) is the **vibrational quantum number.*** An interesting feature of this result is that the energy levels are equally spaced with intervals equal to hf. The frequency of vibration of a diatomic molecule can be related to the force exerted by one atom on the other. Consider two objects of mass m_1 and m_2 connected by a spring of force constant k_F.

* We use ν here rather than n so as not to confuse the vibrational quantum number with the principal quantum number n for electronic energy levels.

The frequency of oscillation of this system (see Problem 32) can be shown to be

$$f = \frac{1}{2\pi}\sqrt{\frac{k_F}{\mu}} \qquad 37\text{-}19$$

where μ is the reduced mass given by Equation 37-15. The effective force constant k_F of a diatomic molecule can thus be determined from a measurement of the frequency of oscillation of the molecule.

A selection rule on transitions between vibrational states (of the same electronic state) requires that the vibrational quantum number ν can change only by ± 1, so the energy of a photon emitted by such a transition is hf and the frequency of the photon is f, the same as the frequency of vibration. There is a similar selection rule that ℓ must change by ± 1 for transitions between rotational states.

A typical measured frequency of a transition between vibrational states is 5×10^{13} Hz, which gives

$$E \approx hf = (4.14 \times 10^{-15}\ \text{eV}\cdot\text{s})(5.0 \times 10^{13}\ \text{s}^{-1}) = 0.2\ \text{eV}$$

and is an estimate for the order of magnitude of vibrational energies. This typical vibrational energy is approximately 1000 times greater than the typical rotational energy E_{0r} of the O_2 molecule we found in Example 37-3 and about 8 times greater than the typical thermal energy $kT = 0.026$ eV at $T = 300$ K. Thus, the vibrational levels are almost never excited by molecular collisions at ordinary temperatures.

Example 37-4 Determining the Force Constant

The frequency of vibration of the CO molecule is 6.42×10^{13} Hz. What is the effective force constant for this molecule?

PICTURE We use $2\pi f = \sqrt{k_F/\mu}$ (Equation 37-19) to relate k_F to the frequency and the reduced mass, and calculate μ from its definition.

SOLVE

1. The effective force constant is related to the frequency and reduced mass by Equation 37-19:

$$f = \frac{1}{2\pi}\sqrt{\frac{k_F}{\mu}}$$

$$k_F = (2\pi f)^2\mu$$

2. Calculate the reduced mass using 12 u for the mass of the carbon atom and 16 u for the mass of the oxygen atom:

$$\mu = \frac{m_1 m_2}{m_1 + m_2} = \frac{(12\ \text{u})(16\ \text{u})}{12\ \text{u} + 16\ \text{u}} = 6.86\ \text{u}$$

3. Substitute this value of μ into the equation for k_F in step 1 and convert to SI units:

$$\begin{aligned} k_F &= (2\pi f)^2\mu \\ &= 4\pi^2(6.42 \times 10^{13}\ \text{Hz})^2(6.86\ \text{u}) \\ &= 1.12 \times 10^{30}\ \text{u/s}^2 \times \left(\frac{1.66 \times 10^{-27}\ \text{kg}}{1\ \text{u}}\right) \\ &= \boxed{1.85 \times 10^{3}\ \text{N/m}} \end{aligned}$$

CHECK From Newton's second law we know that 1 kg m/s² = 1 N, so the units of kg/s² that remain after canceling out the u's in step 3 are equal to N/m, which is what is expected for the force constant of a "spring."

EMISSION SPECTRA

Figure 37-13 shows schematically some electronic, vibrational, and rotational energy levels of a diatomic molecule. The vibrational levels are labeled with the quantum number ν and the rotational levels are labeled with ℓ. The lower vibrational levels are evenly spaced, with $\Delta E = hf$. For higher vibrational levels, the approximation that the vibration is simple harmonic is not valid and the levels are not quite evenly spaced.

Note that the potential energy curves representing the force between the two atoms in the molecule do not have exactly the same shape for the electronic ground and excited states. This implies that the fundamental frequency of vibration f is different for different electronic states. For transitions between vibrational states of different electronic states, the selection rule $\Delta\nu = \pm 1$ does not hold. Such transitions result in the emission of photons of wavelengths in or near the visible spectrum, so the emission spectrum of a molecule for electronic transitions is also sometimes called the optical spectrum.

The spacing of the rotational levels increases with increasing values of ℓ. Because the energies of rotation are so much smaller than those of vibrational excitation or electronic excitation of a molecule, molecular rotation shows up in optical spectra as a fine splitting of the spectral lines. When this fine structure is not resolved, the spectrum appears as bands, as shown in Figure 37-14*a*. Close inspection of these bands reveals that they have a fine structure due to the rotational energy levels, as shown in the enlargement in Figure 37-14*c*.

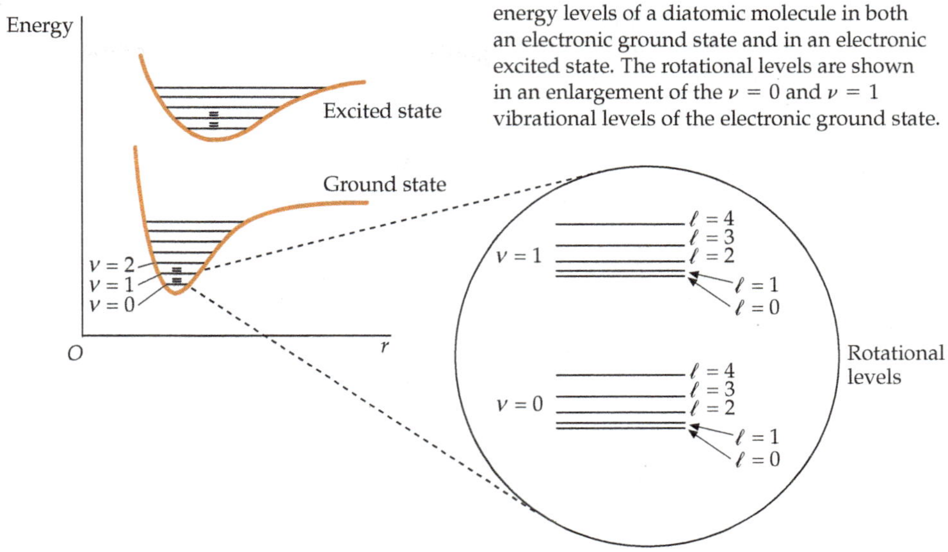

FIGURE 37-13 Vibrational and rotational energy levels of a diatomic molecule in both an electronic ground state and in an electronic excited state. The rotational levels are shown in an enlargement of the $\nu = 0$ and $\nu = 1$ vibrational levels of the electronic ground state.

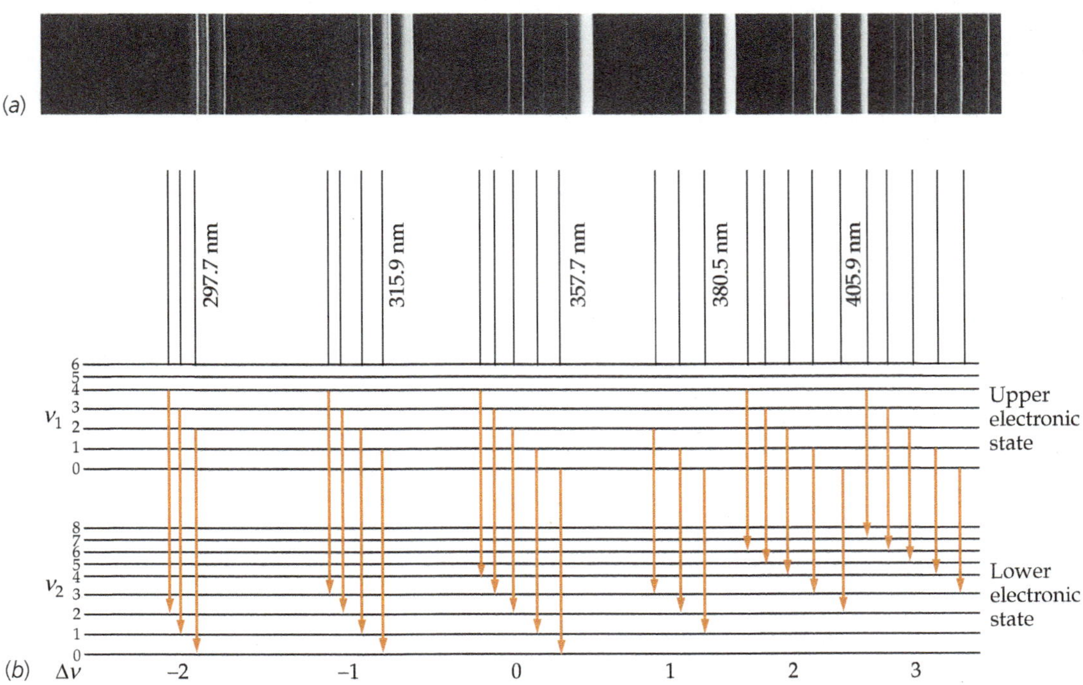

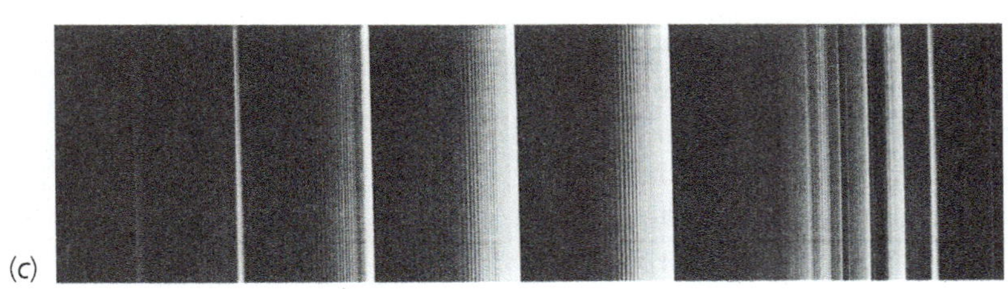

FIGURE 37-14 (*a*) Part of the emission spectrum of N_2. The spectral lines are due to transitions between the vibrational levels of two electronic states, as indicated in the energy level diagram (*b*). (*c*) An enlargement of part of Figure 37-14*a* shows that the apparent lines are in fact bands with structure caused by rotational levels. *(Courtesy of Dr. J. A. Marquissee.)*

ABSORPTION SPECTRA

Much molecular spectroscopy is done using infrared absorption techniques in which only the vibrational and rotational energy levels of the ground-state electronic level are excited. For ordinary temperatures, the vibrational energies are sufficiently large in comparison with the thermal energy kT that most of the molecules are in the lowest vibrational state $\nu = 0$, for which the energy is $E_0 = \frac{1}{2}hf$. The transition from $\nu = 0$ to $\nu = 1$ is the predominant transition in absorption. However, at room temperature the rotational energies are much less than the thermal energy kT. Thus, a number of the rotational energy states are occupied. If the molecule is originally in a vibrational state characterized by $\nu = 0$ and a rotational state characterized by the quantum number ℓ, the molecule's initial energy is

$$E_\ell = \tfrac{1}{2}hf + \ell(\ell + 1)E_{0r} \qquad \text{37-20}$$

where E_{0r} is given by Equation 37-13. From this state, two transitions are permitted by the selection rules. For a transition to the next higher vibrational state $\nu = 1$ and a rotational state characterized by $\ell + 1$, the final energy is

$$E_{\ell+1} = \tfrac{3}{2}hf + (\ell + 1)(\ell + 2)E_{0r} \qquad \text{37-21}$$

For a transition to the next higher vibrational state and to a rotational state characterized by $\ell - 1$, the final energy is

$$E_{\ell-1} = \tfrac{3}{2}hf + (\ell - 1)\ell E_{0r} \qquad \text{37-22}$$

The energy differences therefore are

$$\Delta E_{\ell\to\ell+1} = E_{\ell+1} - E_\ell = hf + 2(\ell + 1)E_{0r} \qquad \ell = 0, 1, 2, \ldots \qquad \text{37-23}$$

and

$$\Delta E_{\ell\to\ell-1} = E_{\ell-1} - E_\ell = hf - 2\ell E_{0r} \qquad \ell = 1, 2, 3, \ldots \qquad \text{37-24}$$

(In Equation 37-24, ℓ begins at $\ell = 1$ rather than at $\ell = 0$ because from $\ell = 0$ only the transition $\ell \to \ell + 1$ can occur.) Figure 37-15 illustrates these transitions. The frequencies of these transitions are given by

$$f_{\ell\to\ell+1} = \frac{\Delta E_{\ell\to\ell+1}}{h} = f + \frac{2(\ell + 1)E_{0r}}{h} \qquad \ell = 0, 1, 2, \ldots \qquad \text{37-25}$$

and

$$f_{\ell\to\ell-1} = \frac{\Delta E_{\ell\to\ell-1}}{h} = f - \frac{2\ell E_{0r}}{h} \qquad \ell = 1, 2, 3, \ldots \qquad \text{37-26}$$

The frequencies for the transitions $\ell \to \ell + 1$ are thus $f + 2(E_{0r}/h)$, $f + 4(E_{0r}/h)$, $f + 6(E_{0r}/h)$, and so forth; those corresponding to the transition $\ell \to \ell - 1$ are $f - 2(E_{0r}/h)$, $f - 4(E_{0r}/h)$, $f - 6(E_{0r}/h)$, and so forth. We thus expect the absorption spectrum to contain frequencies equally spaced by $2E_{0r}/h$ except for a gap of $4E_{0r}/h$ at the vibrational frequency f, as shown in Figure 37-16. A measurement of the position of the gap gives f and a measurement of the spacing of the absorption peaks gives E_{0r}, which is inversely proportional to the moment of inertia of the molecule.

Figure 37-17 shows the absorption spectrum of HCl. The double-peak structure results from the fact that chlorine occurs naturally in two isotopes, ^{35}Cl and ^{37}Cl, which gives HCl with two different moments of inertia. If all the rotational levels were equally populated initially, we would expect the intensities of each absorption line to be equal. However, the population of a rotational level is proportional to the degeneracy of the level, that is, to the number of states with the same value of ℓ, which is $2\ell + 1$, and to the Boltzmann factor $e^{-E/kT}$, where E is the energy of

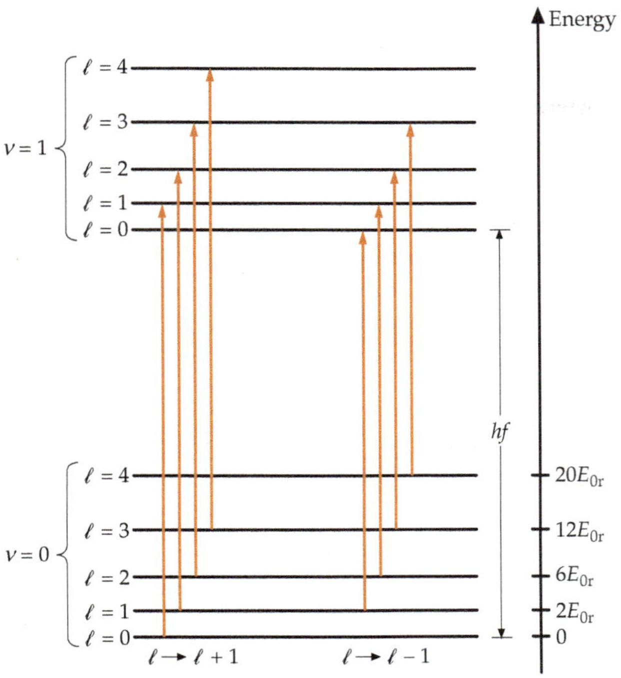

FIGURE 37-15 Absorptive transitions between the lowest vibrational states $\nu = 0$ and $\nu = 1$ in a diatomic molecule. These transitions obey the selection rule $\Delta\ell \pm 1$ and fall into two bands. The energies of the $\ell \to \ell + 1$ band are $hf + 2E_{0r}$, $hf + 4E_{0r}$, $hf + 6E_{0r}$, and so forth; whereas the energies of the $\ell \to \ell - 1$ band are $hf - 2E_{0r}$, $hf - 4E_{0r}$, $hf - 6E_{0r}$, and so forth.

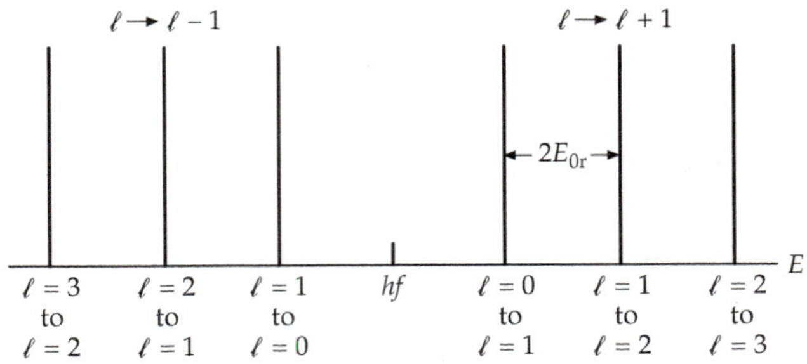

FIGURE 37-16 Expected absorption spectrum of a diatomic molecule. The right branch corresponds to transitions $\ell \to \ell + 1$ and the left branch corresponds to the transitions $\ell \to \ell - 1$. The lines are equally spaced by $2E_{0r}$. The energy midway between the branches is hf, where f is the frequency of vibration of the molecule.

the state. (The Boltzmann factor is presented in Chapter 17.) For low values of ℓ, the population increases slightly because of the degeneracy factor, whereas for higher values of ℓ, the population decreases because of the Boltzmann factor. The intensities of the absorption lines therefore increase with ℓ for low values of ℓ and then decrease with ℓ for high values of ℓ, as can be seen from the figure.

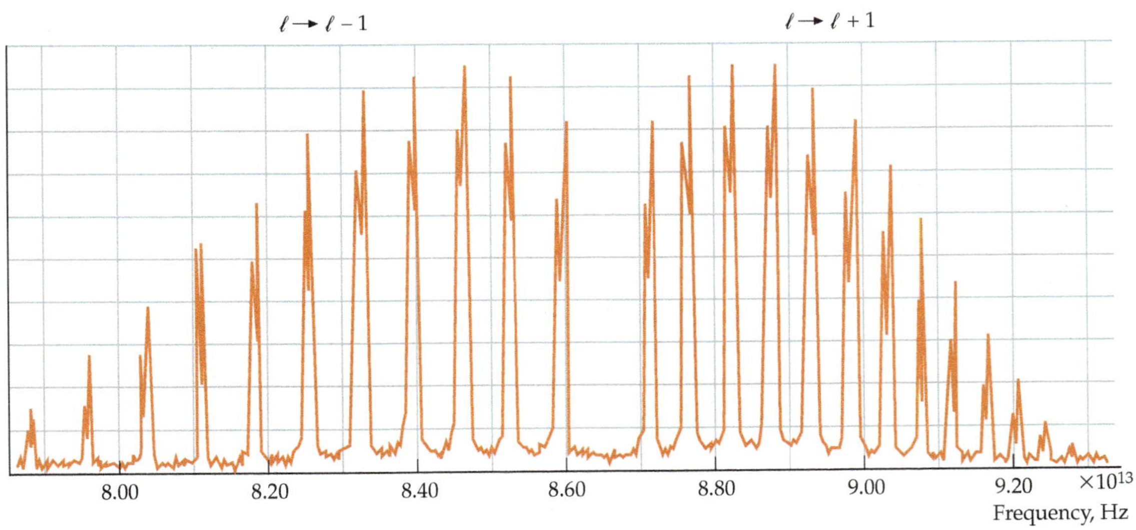

FIGURE 37-17 Absorption spectrum of the diatomic molecule HCl. The double-peak structure results from the two isotopes of chlorine, ^{35}Cl (abundance 75.5 percent) and ^{37}Cl (abundance 24.5 percent). The intensities of the peaks vary because the population of the initial state depends on ℓ.

Summary

1. Atoms are usually found in nature bonded to form molecules or in the lattices of crystalline solids.
2. Ionic bonnds and covalent bonds are the principal mechanisms responsible for forming molecules. Metallic bonds and van der Waals bonds are important in the formation of solids and liquids. Hydrogen bonds enable large biological molecules to maintain their shape.
3. Like atoms, molecules emit electromagnetic radiation when making a transition from a higher energy state to a lower energy state. The internal energy of a molecule can be separated into three parts: electronic, vibrational, and rotational energy.

TOPIC	RELEVANT EQUATIONS AND REMARKS	
1. Molecular Bonding		
Ionic	Ionic bonds result when an electron is transferred from one atom to another, resulting in a positive ion and a negative ion that bond together.	
Covalent	The covalent bond is the sharing of one or more electrons by atoms.	
van der Waals	The van der Waals bonds are weak bonds that result from the interaction of the instantaneous electric dipole moments of molecules.	
Hydrogen	The hydrogen bond results from the sharing of a proton of the hydrogen atom by other atoms.	
Metallic	In the metallic bond, the positive lattice ions of the metal are held together by a cloud of negative charge composed of free electrons.	
Mixed	A diatomic molecule formed from two identical atoms, such as O_2, must bond by covalent bonding. The bonding of two nonidentical atoms is often a mixture of covalent and ionic bonding. The percentage of ionic bonding can be found from the ratio of the magnitude of the measured electric dipole moment to the magnitude of the ionic electric dipole moment defined by	
	$p_{\text{ionic}} = er_0$	37-5
	where r_0 is the equilibrium separation of the ions.	
2. *Polyatomic Molecules	The shapes of such polyatomic molecules as H_2O and NH_3 can be understood from the spatial distribution of the atomic-orbital or molecular-orbital wave functions. The tetravalent nature of the carbon atom is a result of the hybridization of the 2s and 2p atomic orbitals.	
3. Diatomic Molecules		
Moment of inertia	$I = \mu r_0^2$	37-14
	where $\mu = \dfrac{m_1 m_2}{m_1 + m_2}$	37-15
	r_0 is the equilibrium separation, and μ is the reduced mass.	
Rotational energy levels	$E_\ell = \ell(\ell+1)E_{0r}$ where $E_{0r} = \hbar/2\pi$ and $\ell = 0, 1, 2, \ldots$	37-12
Vibrational energy levels	$E_\nu = \left(\nu + \frac{1}{2}\right)hf \quad \nu = 0, 1, 2, \ldots$	37-18
Effective force constant k_F	$f = \dfrac{1}{2\pi}\sqrt{\dfrac{k_F}{\mu}}$	37-19
4. Molecular Spectra	The optical spectra of molecules have a band structure due to transitions between rotational levels. Information about the structure and bonding of a molecule can be found from its rotational and vibrational absorption spectrum involving transitions from one vibrational-rotational level to another. These transitions obey the selection rules	
	$\Delta\nu = \pm 1 \quad \Delta\ell = \pm 1$	

Answers to Practice Problems

37-1 17.6 percent

Answers to Concept Check

37-1 The moment of inertia of an atom is much much less than the moment of inertia of a diatomic molecule, so the amount of energy needed to change the rotational state of a single atom is much much larger than the amount needed for a diatomic molecule. At 300 K, the required energy is not available by way of collisions between atoms.

Problems

In a few problems, you are given more data than you actually need; in a few other problems, you are required to supply data from your general knowledge, outside sources, or informed estimate.

Interpret as significant all digits in numerical values that have trailing zeros and no decimal points.

- • Single-concept, single-step, relatively easy
- •• Intermediate-level, may require synthesis of concepts
- ••• Challenging
- SSM Solution is in the *Student Solutions Manual*
- Consecutive problems that are shaded are paired problems.

CONCEPTUAL PROBLEMS

1 • Would you expect NaCl to be polar or nonpolar? SSM

2 • Would you expect N_2 to be polar or nonpolar?

3 • Does neon naturally occur as Ne or Ne_2? Explain your answer.

4 • What type of bonding mechanism would you expect for atoms of (*a*) HF, (*b*) KBr, (*c*) N_2, (*d*) Ag in solid silver?

5 •• The elements in the far right column of the periodic table are sometimes called noble gases, both because they are gases under a wide range of conditions and because atoms of these elements almost never react with other atoms to form molecules or ionic compounds. However, atoms of noble gases can react if the resulting molecule is formed in an electronic excited state. An example is ArF. When it is formed in the excited state, it is written ArF* and is called an excimer (for *ex*cited d*imer*). Refer to Figure 37-13 and discuss how a diagram for the electronic, vibrational, and rotation energy levels of ArF and ArF* would look in which the ArF ground state is unstable and the ArF* excited state is stable. (*Note:* Excimers are used in certain kinds of lasers.) SSM

6 • Find other atoms that have the same subshell electron configurations in their two highest energy orbitals as carbon atoms do. Would you expect the same type of hybridization for these orbitals as for carbon?

7 • How does the value of the effective force constant calculated for a CO molecule in Example 37-4 compare with the value of the force constant of the suspension springs on a typical automobile, which is about 1.5 kN/m?

8 • Explain why the moment of inertia of a diatomic molecule increases slightly with increasing angular momentum.

9 • Why would you expect the separation distance between the two protons to be larger in a H_2^+ ion than in a H_2 molecule?

10 • At room temperature an atom typically absorb radiation only from the ground state, whereas a diatomic molecule typically absorbs radiation from many different rotational states. Why?

11 •• The vibrational energy levels of diatomic molecules are described by a single vibrational frequency f that is the frequency of vibration of the two atoms of the molecule along the line through their centers. Would you expect to see one or more than one vibrational frequency in molecules that have three or more atoms? Consider in particular a water molecule H_2O (Figure 37-9).

ESTIMATION AND APPROXIMATION

12 •• The potential energy for a diatomic molecule has a minimum as shown in Figure 37-13. Near this minimum, the graph for the energy as a function of distance between the atoms may be approximated as a parabola, leading to the harmonic oscillator model for the vibrating molecule. An improved approximation is called the anharmonic oscillator and leads to a modification of the expression for the energy $E_\nu = (\nu + \frac{1}{2})hf$, where $\nu = 0, 1, 2, \ldots$ (Equation 37-18). The modified expression for energy is $E_\nu = (\nu + \frac{1}{2})hf - (\nu + \frac{1}{2})^2 hf\alpha$, where $\nu = 0, 1, 2, \ldots$. For an O_2 molecule, the constants have the values $f = 4.74 \times 10^{13}\,\text{s}^{-1}$ and $\alpha = 7.6 \times 10^{-3}$. Use this formula to estimate the smallest value of the quantum number ν for which the modified expression differs from the original expression by 10 percent.

13 •• To understand why quantum mechanics is not needed to describe many macroscopic systems, estimate the rotational energy quantum number ℓ and spacing between adjacent energy levels for a baseball ($m \sim 300$ g, $r \sim 3$ cm) spinning about its own axis at 20 rev/min. *Hint: Pick ℓ so the quantum energy formula $E_\ell = \ell(\ell + 1)\hbar^2/(2I)$, where $\ell = 0, 1, 2, \ldots$ (Equation 37-12) gives the correct energy for the given system. Then find the energy increase for the next highest energy level.*

14 •• Estimate the quantum number ν and spacing between adjacent energy levels for a 1.0-kg mass attached to spring. The spring has a force constant equal to 1200 N/m and the mass-spring system is vibrating with an amplitude of 3.0 cm. *Hint: Pick ν so that the quantum energy formula $E_\nu = (\nu + \frac{1}{2})hf$, where $\nu = 0, 1, 2, \ldots$ (Equation 37-18) gives the correct energy for the given system. Then find the energy increase for the next highest energy level.*

MOLECULAR BONDING

15 • Calculate the separation of Na^+ and Cl^- ions for which the potential energy of a single ionic unit (one Na^+ ion and one Cl^- ion) is -1.52 eV.

16 • The equilibrium separation of the atoms in a HF molecule is 0.0917 nm and the measured electric dipole moment of the molecule is 6.40×10^{-30} C·m. What percentage of the HF bond is ionic?

17 •• The dissociation energy of RbF is 5.12 eV, and the equilibrium separation of RbF is 0.227 nm. The electron affinity of a fluorine atom is -3.40 eV and the ionization energy of rubidium is 4.18 eV. Determine the core-repulsion energy of RbF.

18 •• The equilibrium separation of the K^+ and Cl^- ions in KCl is about 0.267 nm. (*a*) Calculate the potential energy of attraction of the ions. Assume that the ions are point charges at this separation. (*b*) The ionization energy of potassium is 4.34 eV and the electron affinity of chlorine is -3.62 eV. Calculate a value for the dissociation energy using the assumption that the energy of repulsion is negligible. (See Figure 37-1.) (*c*) The measured dissociation energy is 4.49 eV. What is the energy due to repulsion of the ions at the equilibrium separation?

19 •• Indicate an approximate value for the average value of the separation distance r for two vibrational levels on the potential energy curve for a diatomic molecule (one of the curves in Figure 37-13). Your teacher claims that the increase in r_{av} with increases in vibration energy explains why solids expand when heated. Do you agree? If so, give an argument supporting this claim. If not, give an argument opposing this claim.

20 •• Calculate the potential energy of attraction between the Na^+ and Cl^- ions at the equilibrium separation $r_0 = 0.236$ nm. Compare this result with the dissociation energy given in Figure 37-1. What is the energy due to repulsion of the ions at the equilibrium separation?

21 •• The equilibrium separation of the K^+ and F^- ions in KF is about 0.217 nm. (*a*) Calculate the potential energy of attraction of these ions. Assume that the ions are point charges at this separation. (*b*) The ionization energy of potassium is 4.34 eV and the electron affinity of fluorine is −3.40 eV. Find the dissociation energy by neglecting any energy of repulsion. (*c*) The measured dissociation energy is 5.07 eV. Calculate the energy due to repulsion of the ions at the equilibrium separation.

ENERGY LEVELS OF SPECTRA OF DIATOMIC MOLECULES

22 • The characteristic rotational energy E_{0r} for the rotation of a N_2 molecule is 2.48×10^{-4} eV. Using this value, find the separation distance of the 2 nitrogen atoms.

23 • The separation of the two oxygen atoms in a molecule of O_2 is actually slightly greater than the 0.100 nm used in Example 37-3. Furthermore, the characteristic energy of rotation E_{0r} for O_2 is 1.78×10^{-4} eV rather than the result obtained in that example. Use this value to calculate the separation distance of the two oxygen atoms. SSM

24 •• Show that the reduced mass of a diatomic molecule is always smaller than the mass of the molecule. Calculate the reduced mass for (*a*) H_2, (*b*) N_2, (*c*) CO, and (*d*) HCl. Express your answers in unified atomic mass units.

25 •• A CO molecule has a binding energy of approximately 11 eV. Find the vibrational quantum number ν that corresponds to 11 eV. (If a CO molecule actually had this much vibrational energy, it would "shake" apart.)

26 •• Derive the equation $I = \mu r_0^2$ (Equation 37-14) for the moment of inertia in terms of the reduced mass of a diatomic molecule.

27 •• The equilibrium separation between the atoms of a LiH molecule is 0.16 nm. Determine the energy separation between the $\ell = 3$ and $\ell = 2$ rotational levels of the diatomic molecule. SSM

28 •• The equilibrium separation of the K^+ and Cl^- ions in KCl is about 0.267 nm. Use this value together with the reduced mass of KCl to calculate the characteristic rotational energy E_{0r} (Equation 37-13) of KCl.

29 •• The central frequency for the absorption band of HCl shown in Figure 37-17 is at 8.66×10^{13} Hz, and the absorption peaks to either side of the central frequency are separated by about 6×10^{11} Hz. Use this information to find (*a*) the lowest (zero-point) vibrational energy for HCl, (*b*) the moment of inertia of HCl, and (*c*) the equilibrium separation of the two atoms.

30 •• Calculate the effective force constant for HCl from its reduced mass and from the fundamental vibrational frequency obtained from Figure 37-17.

31 •• The equilibrium separation between the atoms of a CO molecule is 0.113 nm. For a molecule, such as CO, that has a permanent electric dipole moment, radiative transitions obeying the selection rule $\Delta\ell = \pm 1$ between two rotational energy levels of the same vibrational level are allowed. (That is, the selection rule $\Delta\nu = \pm 1$ does not hold.) (*a*) Find the moment of inertia of CO and calculate the characteristic rotational energy E_{0r} (in eV). (*b*) Make an energy-level diagram for the rotational levels from $\ell = 0$ to $\ell = 5$ for some vibrational level. Label the energies in electron volts, starting with $E = 0$ for $\ell = 0$. Indicate on your diagram the transitions that obey $\Delta\ell = -1$, and calculate the energies of the photons emitted. (*c*) Find the wavelength of the photons emitted during each transition in (*b*). In what region of the electromagnetic spectrum are those photons? SSM

32 •• Two objects, one of mass m_1 and the other of mass m_2, are connected to opposite ends of a spring of force constant k_F. The objects are released from rest with the spring compressed. (*a*) Show that when the spring is extended and the object of mass m_1 is a distance Δr_1 from its equilibrium position in the center-of-mass reference frame, the force exerted by the spring is given by $F = -k_F(m_1/\mu)\Delta r_1$, where μ is the reduced mass. (*b*) Show that the frequency of oscillation f is related to k_F and μ by $2\pi f = \sqrt{k_F/\mu}$.

33 ••• Calculate the reduced masses μ for $H^{35}Cl$ and $H^{37}Cl$ molecules and the fractional difference $\Delta\mu/\mu$. Show that the mixture of isotopes in HCl leads to a fractional difference in the frequency of a transition from one rotational state to another given by $\Delta f/f = -\Delta\mu/\mu$. Compute $\Delta f/f$ and compare your result with Figure 37-17.

GENERAL PROBLEMS

34 • Show that when one atom of a diatomic molecule is much more massive than the other the reduced mass is approximately equal to the mass of the lighter atom.

35 •• The equilibrium separation between the nuclei of a CO molecule is 0.113 nm. Determine the energy difference between the $\ell = 2$ and $\ell = 1$ rotational energy levels of the molecule.

36 •• The effective force constant for a HF molecule is 970 N/m. Find the frequency of vibration for the molecule.

37 •• The frequency of vibration of a NO molecule is 5.63×10^{13} Hz. Find the effective force constant for NO.

38 •• The effective force constant of the hydrogen bond in a H_2 molecule is 580 N/m. Obtain the energies of the four lowest vibrational levels of the H_2, HD, and D_2 molecules, where H is protium and D is deuterium, and find the wavelengths of photons resulting from transitions between adjacent vibrational levels for each of the molecules.

39 •• The potential energy between two atoms in a molecule separated by a distance r can often be described rather well by the Lenard-Jones (or 6-12) potential function, which can be written as

$$U = U_0\left[\left(\frac{a}{r}\right)^{12} - 2\left(\frac{a}{r}\right)^{6}\right]$$

where U_0 and a are constants. Find the equilibrium separation r_0 in terms of a. *Hint: At the equilibrium separation the potential energy is a minimum.* Find U_{min}, the value of U when $r = r_0$. Use Figure 37-4 to obtain numerical values of r_0 and U_0 for a H_2 molecule, and express your answers in nanometers and electron volts.

40 •• In this problem, you are to determine how the van der Waals force between a polar molecule and a nonpolar molecule depends on the distance between the molecules. Let the polar molecule be at the origin and let its dipole moment be in the $+x$ direction. In addition, let the nonpolar molecule be on the x axis a distance x away. (*a*) How does the electric field strength due to an electric dipole vary with the distance from the dipole in a given direction? (*b*) Use (1) that the potential energy U of an electric dipole of dipole moment $\vec{p}$ in an electric field $\vec{E}$ can be expressed as $U = -\vec{p} \cdot \vec{E}$, and (2) that the induced dipole moment $\vec{p}'$ of the nonpolar molecule is in the direction of $\vec{E}$, and that p' is proportional to E, to determine how the potential energy of interaction of the two molecules depends on the separation distance x. (*c*) Using $F_x = -dU/dx$, determine how the force between the two molecules depends on distance.

41 •• Find the dependence of the force on separation distance between the two polar molecules described in Problem 40.

42 •• Use the infrared absorption spectrum of HCl in Figure 37-17 to obtain (*a*) the characteristic rotational energy E_{0r} (in eV) and (*b*) the vibrational frequency f and the vibrational energy hf (in eV).

43 • The dissociation energy is sometimes expressed in kilocalories per mole (kcal/mol). (*a*) Find the relation between the units eV/molecule and kcal/mol. (*b*) Find the dissociation energy of NaCl in kcal/mol.

CHAPTER 38

IT IS WELL KNOWN THAT ARSENIC IS A POISON. IT IS LESS WELL KNOWN THAT SILICON CRYSTALS THAT HAVE SMALL CONCENTRATIONS OF ARSENIC ATOMS HAVE A MUCH LOWER RESISTIVITY THAN DO CRYSTALS THAT ARE 100 PERCENT SILICON. *(The Natural History Museum/Alamy.)*

Solids

? Do you know how many atoms of arsenic it takes to increase the charge-carrier density by a factor of 5 million? (See Example 38-7.)

The first microscopic model of electric conduction in metals was proposed by Paul K. Drude in 1900 and developed by Hendrik A. Lorentz about 1909. This model successfully predicts that the current is proportional to the potential drop (Ohm's law) and relates the resistivity of conductors to the mean speed and the mean free path* of the free electrons within the conductor. However, when mean speed and mean free path are interpreted classically, there is a disagreement between the calculated values and the measured values of the resistivity, and a similar disagreement between the predicted temperature dependence and the observed temperature dependence that resistivity values have. Thus, the classical theory fails to adequately describe the resistivity of metals. Furthermore, the classical theory says nothing about the most striking

* The mean free path is the average distance traveled between collisions.

property of solids, namely, that some substances are conductors, others are insulators, and still others are semiconductors, which are substances whose resistivity falls between that of conductors and insulators.

When mean speed and mean free path are interpreted using quantum theory, both the magnitude and the temperature dependence of the resistivity are correctly predicted. In addition, quantum theory allows us to determine if a substance will be a conductor, an insulator, or a semiconductor.

In this chapter, we use our understanding of quantum mechanics to discuss the structure of solids and solid-state semiconducting devices. Much of our discussion will be qualitative because, as in atomic physics, the quantum-mechanical calculations are mathematically sophisticated.

38-1 THE STRUCTURE OF SOLIDS

The three phases of matter we observe everyday—gas, liquid, and solid—result from the relative strengths of the attractive forces between atoms and molecules and the thermal energies of the particles. Molecules and atoms in the gas phase have relatively large thermal kinetic energies, and such particles have little influence on one another except during their frequent but brief collisions. (By using the term thermal kinetic energies, we mean the kinetic energies of the molecules and atoms in the center-of-mass reference frame of the gas.) At sufficiently low temperatures, van der Waals forces will cause practically every substance to condense into a liquid and then into a solid. In liquids, the molecules or atoms are close enough—and their thermal kinetic energies are low enough—that they can develop a temporary **short-range order.** As their thermal kinetic energies are further reduced, the molecules or atoms form solids, which are characterized by a lasting order.

If a liquid is cooled slowly so that the kinetic energy of its molecules is reduced slowly, the molecules (or atoms or ions) may arrange themselves in a regular crystalline array, producing the maximum number of bonds and leading to a minimum potential energy. However, if the liquid is cooled rapidly so that its internal energy is removed before the molecules have a chance to arrange themselves, the solid formed is often not crystalline or the arrangement is not regular. Such a solid is called an **amorphous solid.** It displays short-range order but not the long-range order (the order over many molecular, atomic, or ionic diameters) that is characteristic of a crystal. Glass is a typical amorphous solid. A characteristic result of the long-range ordering of a crystal is that it has a well-defined melting point, whereas an amorphous solid merely softens as its temperature is increased. Many substances may solidify into either an amorphous state or a crystalline state depending on how the substances are prepared; others exist only in one such state or the other.

Most common solids are polycrystalline; that is, they consist of many single crystals that meet at *grain boundaries.* The size of a single crystal is typically a fraction of a millimeter. However, large single crystals do occur naturally and can be produced artificially. The most important property of a single crystal is the symmetry and regularity of its structure. It can be thought of as having a single unit structure that is repeated throughout the crystal. This smallest unit of a crystal is called the **unit cell;** its structure depends on the type of bonding—ionic, covalent, metallic, hydrogen, van der Waals—between the atoms, ions, or molecules. If more than one kind of atom is present, the structure will also depend on the relative sizes of the atoms.

Figure 38-1 shows the structure unit cell of crystalline sodium chloride (NaCl). The Na^+ and Cl^- ions are spherically symmetric, and the Cl^- ion is approximately twice as large as the Na^+ ion. The minimum potential energy for this crystal occurs when an ion of either kind has six nearest neighbors of the other kind. This structure is called *face-centered-cubic* (fcc). Note that the Na^+ and Cl^- ions in solid NaCl are *not* paired into NaCl molecules.

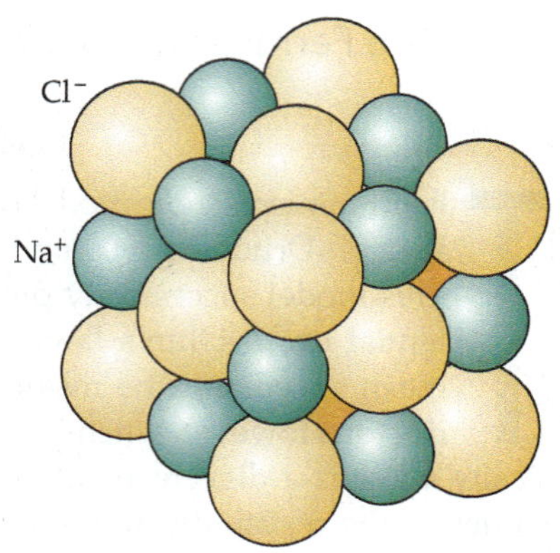

FIGURE 38-1 Face-centered-cubic structure of the NaCl crystal.

The net attractive part of the potential energy of an ion in a crystal can be written

$$U_{\text{att}} = -\alpha\frac{ke^2}{r} \qquad 38\text{-}1$$

where r is the (center-to-center) separation distance between neighboring ions (0.281 nm for the Na^+ and Cl^- ions in crystalline NaCl) and α, called the **Madelung constant,** depends on the geometry of the crystal. If only the six nearest neighbors of each ion in a face-centered-cubic crystalline structure were important, α would be six. However, in addition to the six neighbors of the opposite charge at a distance r, there are twelve ions of the same charge at a distance $\sqrt{2}r$, eight ions of opposite charge at a distance $\sqrt{3}r$, and so on. The Madelung constant is thus an infinite sum:

$$\alpha = 6 - \frac{12}{\sqrt{2}} + \frac{8}{\sqrt{3}} - \dots \qquad 38\text{-}2$$

The value of the Madelung constant for face-centered-cubic structures is $\alpha = 1.7476$.*

* A large number of terms are needed to calculate the Madelung constant accurately because the sum converges very slowly.

(a)

Crystal structure. (*a*) The hexagonal symmetry of a snowflake arises from a hexagonal symmetry in its lattice of hydrogen atoms and oxygen atoms. (*b*) NaCl (salt) crystals, magnified approximately thirty times. The crystals are built up from a cubic lattice of sodium and chloride ions. In the absence of impurities, an exact cubic crystal is formed. This (false-color) scanning electron micrograph shows that in practice the basic cube is often disrupted by dislocations, giving rise to crystals that have a wide variety of shapes. The underlying cubic symmetry, though, remains evident. (*c*) A crystal of quartz (SiO_2, silicon dioxide), the most abundant and widespread mineral on Earth. If molten quartz solidifies without crystallizing, glass is formed. (*d*) A soldering iron tip, ground down to reveal the copper core within its iron sheath. Visible in the iron is its underlying microcrystalline structure. *((a) Richard Waters 2/89 p. 52 Discover. (b) © Dr. Jeremy Burgess/Science Photo Library/Photo Researchers. (c) © Thomas R. Taylor/Photo Researchers. (d) Courtesy the AT&T Archives.)*

(b)

(c)

(d)

When Na^+ and Cl^- ions are very close together, they repel each other because of the overlap of their electron orbitals and the exclusion-principle repulsion discussed in Section 37-1. A simple empirical expression for the potential energy associated with this repulsion that works fairly well is

$$U_{\text{rep}} = \frac{A}{r^n}$$

where A and n are constants. The total potential energy of an ion is then

$$U = -\alpha\frac{ke^2}{r} + \frac{A}{r^n} \qquad \text{38-3}$$

The equilibrium separation $r = r_0$ is that at which the force $F = -dU/dr$ is zero. Differentiating and setting $dU/dr = 0$ at $r = r_0$, we obtain

$$A = \frac{\alpha ke^2 r_0^{n-1}}{n} \qquad \text{38-4}$$

Substituting for A in Equation 38-3 gives

$$U = -\alpha\frac{ke^2}{r_0}\left[\frac{r_0}{r} - \frac{1}{n}\left(\frac{r_0}{r}\right)^n\right] \qquad \text{38-5}$$

At $r = r_0$, we have

$$U(r_0) = -\alpha\frac{ke^2}{r_0}\left(1 - \frac{1}{n}\right) \qquad \text{38-6}$$

If we know the equilibrium separation r_0, the value of n can be found approximately from the *dissociation energy* of the crystal, which is the energy needed to break up the crystal into atoms.

Example 38-1 Separation Distance between Na^+ and Cl^- in NaCl

Calculate the equilibrium separation r_0 for NaCl from the measured density of NaCl, which is $\rho = 2.16\ \text{g/cm}^3$.

PICTURE We consider each ion to occupy a cubic volume of side r_0. The mass of 1 mol of NaCl is 58.4 g, which is the sum of the molar masses of sodium and chlorine. There are $2N_A$ ions in 1 mol of NaCl, where $N_A = 6.02 \times 10^{23}$ is Avogadro's number.

SOLVE

1. We consider each ion to occupy a cubic volume of side r_0. The volume v of one mole of NaCl equals the number of ions multiplied by the volume per ion:

$$v = 2N_A r_0^3$$

2. Relate r_0 to the density ρ and the molar mass M of NaCl:

$$\rho = \frac{M}{v} = \frac{M}{2N_A r_0^3}$$

3. Solve for r_0^3 and substitute the known values:

$$r_0^3 = \frac{M}{2N_A\rho} = \frac{58.4\ \text{g}}{2(6.02 \times 10^{23})(2.16\ \text{g/cm}^3)} = 2.25 \times 10^{-23}\ \text{cm}^3$$

so

$$r_0 = 2.82 \times 10^{-8}\ \text{cm} = \boxed{0.282\ \text{nm}}$$

CHECK In Chapter 36, we found the diameter of the hydrogen atom in the ground state to be about 0.11 nm. Our step 3 result is less than three times larger. Thus, $r_0 = 0.282$ nm is plausible.

The measured dissociation energy of NaCl is 770 kJ/mol. Using $1\ \text{eV} = 1.602 \times 10^{-19}$ J and the fact that 1 mol of NaCl has N_A pairs of ions, we can express the dissociation energy in electron volts per ion pair. The conversion between electron volts per ion pair and kilojoules per mole is

$$1\frac{\text{eV}}{\text{ion pair}} \times \frac{6.022 \times 10^{23}\ \text{ion pairs}}{1\ \text{mol}} \times \frac{1.602 \times 10^{-19}\ \text{J}}{1\ \text{eV}}$$

The result is

$$1\frac{\text{eV}}{\text{ion pair}} = 96.47\frac{\text{kJ}}{\text{mol}} \qquad 38\text{-}7$$

Thus, 770 kJ/mol = 7.98 eV per ion pair. Substituting −7.98 eV for $U(r_0)$, 0.282 nm for r_0, and 1.75 for α in Equation 38-6, we can solve for n. The result is $n = 9.35 \approx 9$.

Most ionic crystals, such as LiF, KF, KCl, KI, and AgCl, have a face-centered-cubic structure. Some elemental solids that have fcc structure are silver, aluminum, gold, calcium, copper, nickel, and lead.

Figure 38-2 shows the structure of CsCl, which is called the *body-centered-cubic* (bcc) structure. In this structure, each ion has eight nearest neighbor ions of the opposite charge. The Madelung constant for these crystals is 1.7627. Elemental solids that have bcc structure include barium, cesium, iron, potassium, lithium, molybdenum, and sodium.

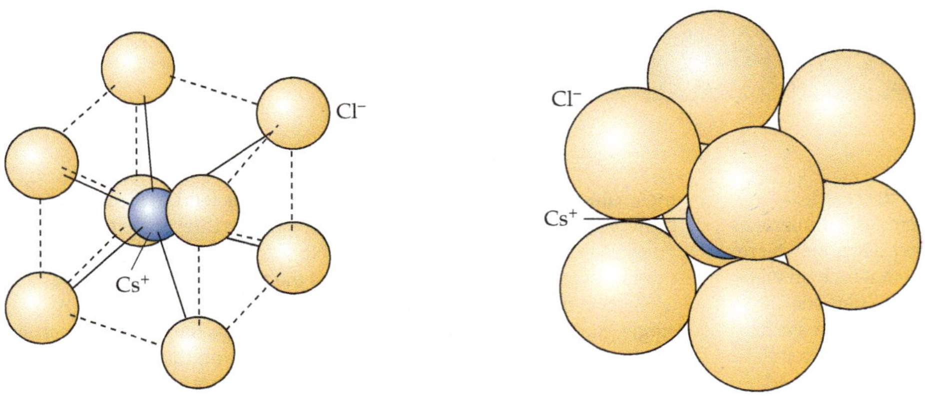

FIGURE 38-2 Body-centered-cubic structure of the CsCl crystal.

Figure 38-3 shows another important crystal structure: the *hexagonal close-packed* (hcp) structure. This structure is obtained by stacking identical spheres, such as bowling balls. In the first layer, each ball touches six others; thus, the name *hexagonal*. In the next layer, each ball fits into a triangular depression of the first layer. In the third layer, each ball fits into a triangular depression of the second layer, so it lies directly over a ball in the first layer. Elemental solids that have hcp structure include beryllium, cadmium, cerium, magnesium, osmium, and zinc.

For solids that have covalent bonding, the crystal structure is determined by the configuration of the bonds. Figure 38-4 illustrates the diamond structure of carbon, in which each atom is bonded to four other atoms as a result of hybridization, which is discussed in Section 37-2. This configuration is also the structure of germanium and silicon.

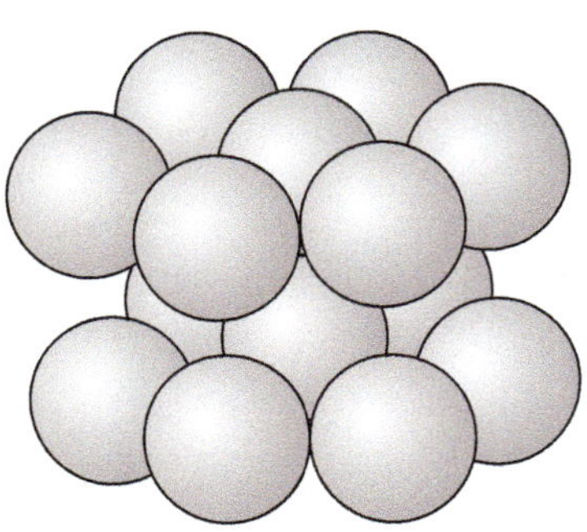

FIGURE 38-3 Hexagonal close-packed crystal structure.

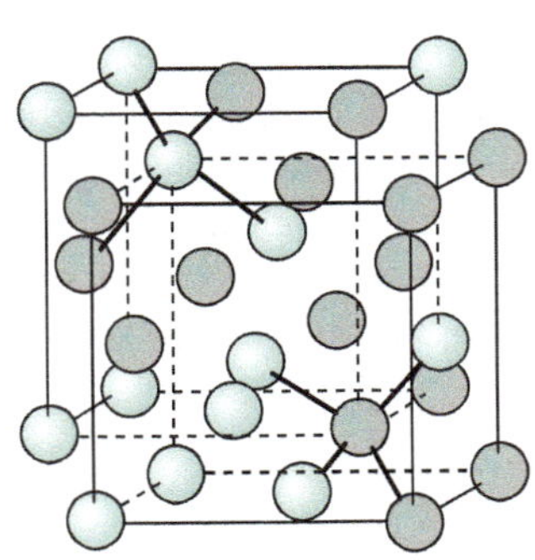

FIGURE 38-4 Diamond crystal structure. This structure can be considered to be a combination of two interpenetrating face-centered-cubic structures.

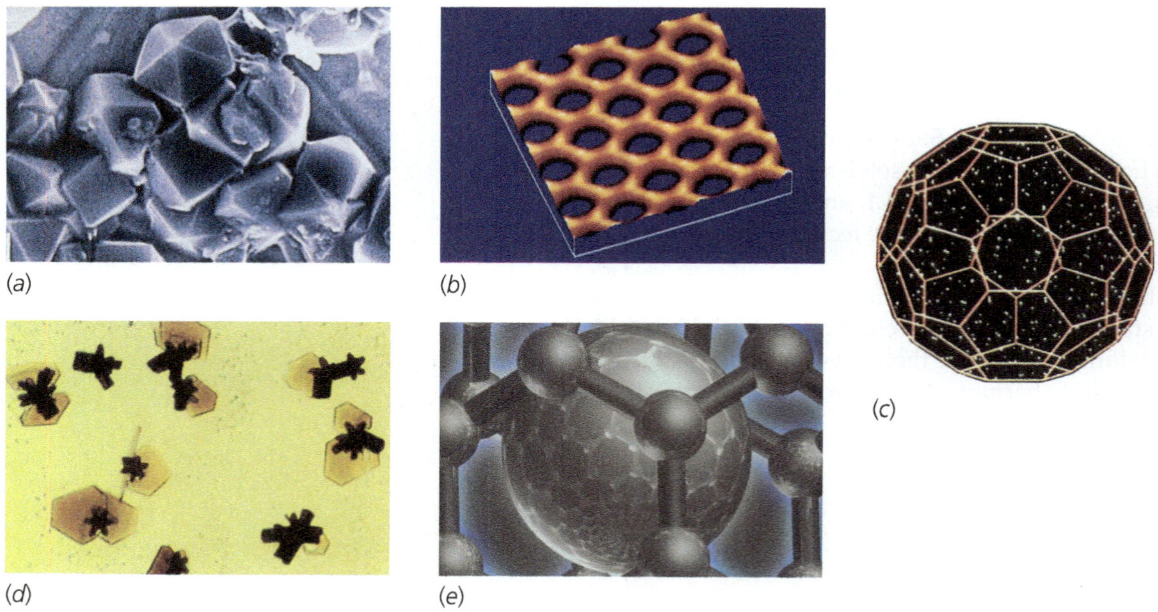

Carbon exists in three well-defined crystalline forms: diamond, graphite, and fullerenes (short for "buckminsterfullerenes"). Fullerenes were discovered in 1985. The forms differ in how the carbon atoms are packed together in a lattice. A fourth form of carbon, in which no well-defined crystalline form exists, is common charcoal. (*a*) Synthetic diamonds, magnified approximately 75,000 times. In diamond, each carbon atom is centered in a tetrahedron of four other carbon atoms. The strength of these bonds accounts for the hardness of a diamond. (*b*) An atomic-force micrograph of graphite. In graphite, carbon atoms are arranged in sheets, where each sheet is made up of atoms in hexagonal rings. The sheets slide easily across one another, a property that allows graphite to function as a lubricant. (*c*) A single sheet of carbon rings can be closed on itself if certain rings are allowed to be pentagonal, instead of hexagonal. A computer-generated image of the smallest such structure, C_{60}, is shown here. Each of the sixty vertices corresponds to a carbon atom; twenty of the faces are hexagons and twelve of the faces are pentagons. The same geometric pattern is encountered in a soccer ball. (*d*) Fullerene crystals, in which C_{60} molecules are close-packed. The smaller crystals tend to form thin brownish platelets; larger crystals are usually rodlike in shape. Fullerenes exist in which more than sixty carbon atoms appear. In the crystals shown here, about one-sixth of the molecules are C_{70}. (*e*) Carbon nanotubes have very interesting electrical properties. A single graphite sheet is a semimetal, which means that it has properties intermediate between those of semiconductors and those of metals. When a graphite sheet is rolled into a nanotube, not only do the carbon atoms have to line up around the circumference of the tube, but the wave functions of the electrons must also match up. This boundary-matching requirement places restrictions on these wave functions, which affects the motion of the electrons. Depending on exactly how the tube is rolled up, the nanotube can be either a semiconductor or a metal. *((a) Chris Kovach 3/91 p. 69 Discover. (b) Srinivas Manne, University of California, Santa Barbara. (c) Dr. F. A. Quiocho and J. S. Spurlino/Howard Hughes Medical Institute, Baylor College of Medicine. (d) W. Krätschmer/Max-Planck-Institute for Nuclear Physics. (e) © Kenneth Weard/BioGrafx/Science Source/Photo Researchers.)*

38-2 A MICROSCOPIC PICTURE OF CONDUCTION

We consider a metal as a regular three-dimensional lattice of ions filling some volume V and having a large number N of electrons that are free to move throughout the whole metal. The number of free electrons in a metal is approximately one to four electrons per atom. In the absence of an electric field, the free electrons move about the metal randomly, much the way gas molecules move about in a container.

The current in a conducting wire segment is proportional to the voltage drop across the segment:

$$I = \frac{V}{R} \qquad (\text{or } V = IR)$$

The resistance R is proportional to the length L of the wire segment and inversely proportional to the cross-sectional area A:

$$R = \rho\frac{L}{A}$$

where ρ is the resistivity. Substituting $\rho L/A$ for R, and EL for V, we can write the current in terms of the electric field strength E and the resistivity. We have

$$I = \frac{V}{R} = \frac{EL}{\rho L/A} = \frac{1}{\rho}EA$$

Dividing both sides by the area A gives $I/A = (1/\rho)E$, or $J = (1/\rho)E$, where $J = I/A$ is the magnitude of the **current density** vector $\vec{J}$. The current density vector is defined as

$$\vec{J} = qn\vec{v}_{\rm d} \qquad \text{38-8}$$

DEFINITION—CURRENT DENSITY

where q, n, and $\vec{v}_{\rm d}$ are the charge, the number density, and the drift velocity of the charge carrier. (This follows from Equation 25-3.) In vector form, the relation between the current density and the electric field is

$$\vec{J} = \frac{1}{\rho}\vec{E} \qquad \text{38-9}$$

This relation is the point form of Ohm's law. The reciprocal of the resistivity is called the **conductivity.**

According to Ohm's law, the resistivity is independent of both the current density and the electric field $\vec{E}$. Combining Equations 38-8 and 38-9 gives

$$-en_{\rm e}\vec{v}_{\rm d} = \frac{1}{\rho}\vec{E} \qquad \text{38-10}$$

where $-e$ and $n_{\rm e}$ have been substituted for q and n, respectively. According to Equation 38-10, the drift velocity $\vec{v}_{\rm d}$ is proportional to $\vec{E}$.

In the presence of an electric field, a free electron experiences a force $-e\vec{E}$. If this were the only force acting, the electron would have a constant acceleration $-e\vec{E}/m_{\rm e}$. However, Equation 38-10 implies a steady-state situation with a constant drift velocity that is proportional to the field $\vec{E}$. In the microscopic model, it is assumed that a free electron is accelerated for a short time and then makes a collision with a lattice ion. The velocity of the electron immediately after the collision is completely unrelated to the drift velocity. The justification for this assumption is that the magnitude of the drift velocity is extremely small compared with the speeds associated with the thermal kinetic energies of the free electrons.

For a typical free electron, its velocity a time t after its last collision is $\vec{v}_0 - (-e\vec{E}/m_{\rm e})t$, where $\vec{v}_0$ is its velocity immediately after that collision. Because the direction of $\vec{v}_0$ is random, it does not contribute to the average velocity of the electrons. Thus, the average velocity or drift velocity of the electrons is

$$\vec{v}_{\rm d} = -\frac{e\vec{E}}{m_{\rm e}}\tau \qquad \text{38-11}$$

where τ is the average time since the last collision. Substituting for $\vec{v}_{\rm d}$ in Equation 38-10, we obtain

$$-n_{\rm e}e\left(\frac{e\vec{E}}{m_{\rm e}}\tau\right) = \frac{1}{\rho}\vec{E}$$

so

$$\rho = \frac{m_{\rm e}}{n_{\rm e}e^2\tau} \qquad \text{38-12}$$

The time τ, called the **collision time,** is also the average time between collisions.*

* It is tempting but incorrect to think that if τ is the average time between collisions, the average time since its last collision is $\frac{1}{2}\tau$ rather than τ. If you find this confusing, you may take comfort in the fact that Drude used the incorrect result $\frac{1}{2}\tau$ in his original work.

The average distance an electron travels between collisions is $v_{av}\tau$, which is called the mean free path λ:

$$\lambda = v_{av}\tau \qquad 38\text{-}13$$

where v_{av} is the mean speed of the electrons. (The mean speed is many orders of magnitude greater than the drift speed.) In terms of the mean free path and the mean speed, the resistivity is

$$\rho = \frac{m_e v_{av}}{n_e e^2 \lambda} \qquad 38\text{-}14$$

RESISTIVITY IN TERMS OF v_{AV} AND λ

According to Ohm's law, the resistivity ρ is independent of the electric field $\vec{E}$. Because m_e, n_e, and e are constants, the only quantities that could possibly depend on $\vec{E}$ are the mean speed v_{av} and the mean free path λ. Let us examine these quantities to see if they can possibly depend on the applied field $\vec{E}$.

CLASSICAL INTERPRETATION OF v_{av} AND λ

Classically, at $T = 0$ all the free electrons in a conductor should have zero kinetic energy. As the conductor is heated, the lattice ions acquire an average kinetic energy of $\frac{3}{2}kT$, which is imparted to the free electrons by the collisions between the electrons and the ions. (This is a result of the equipartition theorem studied in Chapters 17 and 18.) The free electrons would then have a Maxwell–Boltzmann distribution just like a gas of molecules. In equilibrium, the electrons would be expected to have a mean kinetic energy of $\frac{3}{2}kT$, which at ordinary temperatures (~300 K) is approximately 0.04 eV. At $T = 300$ K, their root-mean-square (rms) speed,* which is slightly greater than the mean speed, is

$$v_{av} \approx v_{rms} = \sqrt{\frac{3kT}{m_e}} = \sqrt{\frac{3(1.38 \times 10^{-23}\,\text{J/K})(300\,\text{K})}{9.11 \times 10^{-31}\,\text{kg}}} \qquad 38\text{-}15$$

$$= 1.17 \times 10^5\ \text{m/s}$$

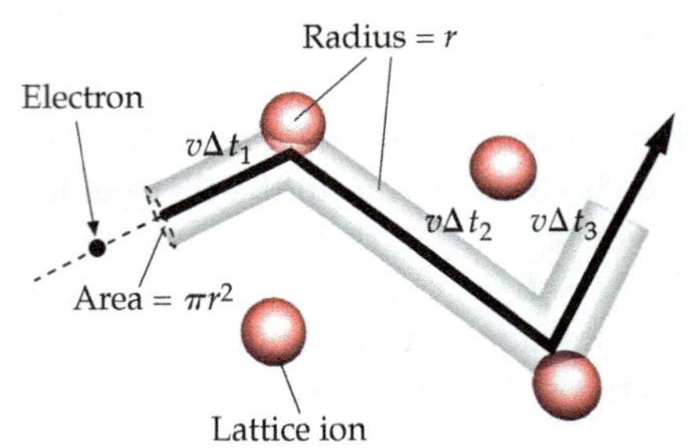

FIGURE 38-5 Model of an electron moving through the lattice ions of a conductor. The electron, which is considered to be a point particle, collides with an ion if it comes within a distance r of the center of the ion, where r is the radius of the ion. If the electron speed is v, it collides in time Δt with all the ions whose centers are in the volume $\pi r^2 v\Delta t$. While this picture is in accord with the classical Drude model for conduction in metals, it is in conflict with the current quantum-mechanical model presented later in this chapter.

Note that this is about nine orders of magnitude greater than the typical drift speed of 3.5×10^{-5} m/s, which was calculated in Example 25-1. The very small drift speed caused by the electric field therefore has essentially no effect on the very large mean speed of the electrons, so v_{av} in Equation 38-14 cannot depend on the electric field $\vec{E}$.

The mean free path is related classically to the size of the lattice ions in the conductor and to the number of ions per unit volume. Consider one electron moving with speed v through a region of stationary ions that are assumed to be hard spheres (Figure 38-5). Assume the size of the electron is negligible. The electron will collide with an ion if it comes within a distance r from the center of the ion, where r is the radius of the ion. During some time interval Δt_1, the electron moves a distance vt_1. If there is an ion whose center is in the cylindrical volume $\pi r^2 v\Delta t_1$, the electron will collide with the ion. The electron will then change directions and collide with another ion in time Δt_2 if the center of the ion is in the volume $\pi r^2 vt_2$. Thus, in the total time $\Delta t = \Delta t_1 + \Delta t_2 + \ldots$, the electron will collide with the number of ions whose centers are in the volume $\pi r^2 v\Delta t$. The number of ions in this volume is $n_{ion}\pi r^2 v\Delta t$, where n_{ion} is the number of ions per unit volume.

* See Equation 17-21.

The total path length divided by the number of collisions is the mean free path:

$$\lambda = \frac{v\Delta t}{n_{\text{ion}}\pi r^2 v\Delta t} = \frac{1}{n_{\text{ion}}\pi r^2} = \frac{1}{n_{\text{ion}}A} \qquad \text{38-16}$$

where $A = \pi r^2$ is the cross-sectional area of a lattice ion.

SUCCESSES AND FAILURES OF THE CLASSICAL MODEL

Neither n_{ion} nor r depends on the electric field $\vec{E}$, so λ also does not depend on $\vec{E}$. v_{av} and λ do not depend on $\vec{E}$ according to their classical interpretations, so the resistivity ρ does not depend on $\vec{E}$ in accordance with Ohm's law. However, the classical theory gives an incorrect temperature dependence for the resistivity. Because λ depends only on the radius and the number density of the lattice ions, the only quantity in Equation 38-14 that depends on temperature in the classical theory is v_{av}, which is proportional to $\sqrt{T}$. But experiments show that ρ varies linearly with temperature. Furthermore, when ρ is calculated at $T = 300$ K using the Maxwell–Boltzmann distribution for v_{av} and Equation 38-16 for λ, the calculated result is about six times greater than the measured value.

The classical theory of conduction fails because electrons are not classical particles. The wave nature of the electrons must be considered. Because of the wave properties of electrons and the constraints described by the exclusion principle (to be discussed in the following section), the energy distribution of the free electrons in a metal is not even approximately given by the Maxwell–Boltzmann distribution. Furthermore, the collision of an electron with a lattice ion is not similar to the collision of a baseball with a tree. Instead, it involves the scattering of electron waves by the lattice. To understand the quantum theory of conduction, we need a qualitative understanding of the energy distribution of free electrons in a metal. This will also help us understand the origin of contact potentials between two dissimilar metals in contact and the contribution of free electrons to the heat capacity of metals.

38-3 FREE ELECTRONS IN A SOLID

One may want to consider free electrons in a metal to be an *electron gas* in a metal. However, molecules in an ordinary gas, such as air, obey the classical Maxwell–Boltzmann energy distribution, but the free electrons in a metal do not. Instead, they obey a quantum energy distribution called the *Fermi–Dirac distribution.* The main features of a free electron can be understood by considering the electron in a metal to be a particle in a box, a problem whose one-dimensional version we studied extensively in Chapter 34. We discuss the main features of a free electron semiquantitatively in this section and leave the details of the Fermi–Dirac distribution to Section 38-9.

ENERGY QUANTIZATION IN A BOX

In Chapter 34, we found that the wavelength associated with an electron of momentum p is given by the de Broglie relation:

$$\lambda = \frac{h}{p} \qquad \text{38-17}$$

where h is Planck's constant. When a particle is confined to a finite region of space, such as a box, only certain wavelengths λ_n, where $n = 1, 2, \ldots$, that are specified by standing-wave conditions are allowed. For a one-dimensional box of length L, the standing-wave condition is

$$n\frac{\lambda_n}{2} = L \qquad n = 1, 2, \ldots \qquad \text{38-18}$$

This results in the quantization of energy:

$$E_n = \frac{p_n^2}{2m} = \frac{(h/\lambda_n)^2}{2m} = \frac{h^2}{2m}\frac{1}{\lambda_n^2} = \frac{h^2}{2m}\frac{1}{(2L/n)^2}$$

or

$$E_n = n^2 E_1 \qquad \text{38-19}$$

where $E_1 = h^2/(8mL^2)$. The spatial wave function for the nth state is given by

$$\psi_n(x) = \sqrt{\frac{2}{L}}\sin(n\pi x/L) \qquad \text{38-20}$$

The quantum number n characterizes the wave function for a particular state and the energy of that state. In three-dimensional problems, three quantum numbers arise, one associated with each dimension.

THE EXCLUSION PRINCIPLE

The distribution of electrons among the possible energy states is described by the exclusion principle, which states that no two electrons in an atom can be in the same quantum state; that is, they cannot have the same set of values for their quantum numbers. The exclusion principle applies to all "spin one-half" particles (fermions), which include electrons, protons, and neutrons. These particles have a *spin* quantum number m_s which has two possible values, $+\frac{1}{2}$ and $-\frac{1}{2}$. The quantum state of a particle is characterized by the spin quantum number m_s and the quantum numbers associated with the spatial part of the wave function. Because the spin quantum numbers have just two possible values, the exclusion principle can be stated in terms of the spatial states:

> There can be at most two electrons with the same set of values for their *spatial* quantum numbers.
>
> EXCLUSION PRINCIPLE IN TERMS OF SPATIAL STATES

When there are more than two electrons in a system, such as an atom, only two can be in the lowest energy state. The third and fourth electrons must go into the second-lowest state, and so on.

Example 38-2 Boson-System Energy versus Fermion-System Energy

Compare the total energy of the ground state of five identical bosons of mass m in a one-dimensional box with the total energy of the ground state of five identical fermions of mass m in the same box.

PICTURE The ground state is the lowest possible energy state. The energy levels in a one-dimensional box are given by $E_n = n^2E_1$, where $E_1 = h^2/(8mL^2)$. (This is in accord with Equation 38-19.) The lowest energy for five bosons occurs when all the bosons are in the state $n = 1$, as shown in Figure 38-6*a*. For fermions, the lowest state occurs when two fermions are in the state $n = 1$, two fermions are in the state $n = 2$, and one fermion is in the state $n = 3$, as shown in Figure 38-6*b*.

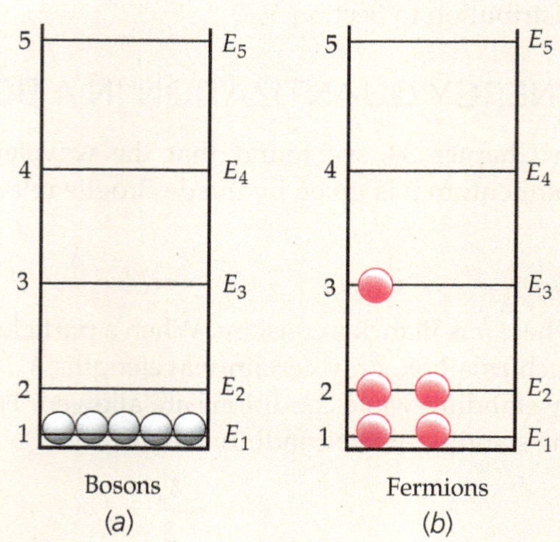

FIGURE 38-6

SOLVE

1. The energy of five bosons in the state $n = 1$ is: $E = 5E_1$

2. The energy of two fermions in the state $n = 1$, two fermions in the state $n = 2$, and one fermion in the state $n = 3$ is: $E = 2E_1 + 2E_2 + 1E_3 = 2E_1 + 2(2)^2E_1 + 1(3)^2E_1$
$= 2E_1 + 8E_1 + 9E_1 = 19E_1$

3. Compare the total energies: The five identical fermions have 3.8 times the total energy of the five identical bosons.

CHECK The fact that fermions must have different quantum states has a large effect on the total energy of a multiple-particle system, as expected.

THE FERMI ENERGY

When there are many electrons in a box, at $T = 0$ the electrons will occupy the lowest energy states consistent with the exclusion principle. If we have N electrons, we can put two electrons in the lowest energy level, two electrons in the next lowest energy level, and so on. The N electrons thus fill the lowest $N/2$ energy levels (Figure 38-7). The energy of the last filled (or half-filled) level at $T = 0$ is called the Fermi energy E_F. If the electrons moved in a one-dimensional box, the Fermi energy would be given by Equation 38-19, with $n = N/2$:

$$E_F = \left(\frac{N}{2}\right)^2 \frac{h^2}{8m_eL^2} = \frac{h^2}{32m_e}\left(\frac{N}{L}\right)^2 \qquad 38\text{-}21$$

FERMI ENERGY AT $T = 0$ IN ONE DIMENSION

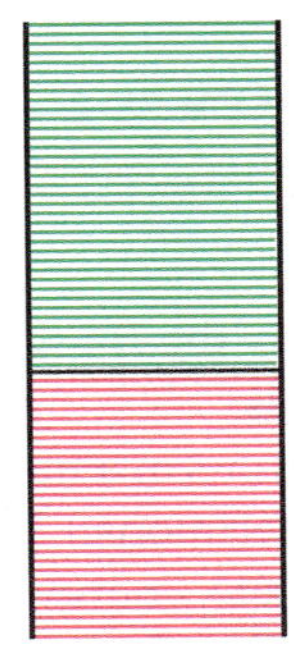

FIGURE 38-7 At $T = 0$ the electrons fill up the allowed energy states to the Fermi energy E_F. The levels are so closely spaced that they can be assumed to be continuous.

In a one-dimensional box, the Fermi energy depends on the number of free electrons per unit length of the box.

PRACTICE PROBLEM 38-1

Suppose there is an ion, and therefore a free electron, every 0.100 nm in a one-dimensional box. Calculate the Fermi energy. *Hint: Write Equation* 38-21 *as*

$$E_F = \frac{(hc)^2}{32m_ec^2}\left(\frac{N}{L}\right)^2 = \frac{(1240\text{ eV}\cdot\text{nm})^2}{32(0.511\text{ MeV})}\left(\frac{N}{L}\right)^2$$

In our model of conduction, the free electrons move in a *three-dimensional* box of volume V. The derivation of the Fermi energy in three dimensions is somewhat difficult, so we will just give the result. In three dimensions, the Fermi energy at $T = 0$ is

$$E_F = \frac{h^2}{8m_e}\left(\frac{3N}{\pi V}\right)^{2/3} \qquad 38\text{-}22a$$

FERMI ENERGY AT $T = 0$ IN THREE DIMENSIONS

The Fermi energy depends on the number density of free electrons N/V. Substituting numerical values for the constants gives

$$E_F = (0.3646\text{ eV}\cdot\text{nm}^2)\left(\frac{N}{V}\right)^{2/3} \qquad 38\text{-}22b$$

FERMI ENERGY AT $T = 0$ IN THREE DIMENSIONS

Example 38-3 The Fermi Energy for Copper

The number density for electrons in copper was calculated in Example 25-1 and found to be $84.7/\text{nm}^3$. Calculate the Fermi energy at $T = 0$ for copper.

PICTURE The Fermi energy is given by Equations 38-22.

SOLVE

1. The Fermi energy is given by Equation 38-22*b*: $E_F = (0.3646\ \text{eV}\cdot\text{nm}^2)\left(\frac{N}{V}\right)^{2/3}$

2. Substitute the given number density for copper: $E_F = (0.3646\ \text{eV}\cdot\text{nm}^2)(84.7/\text{nm}^3)^{2/3} = \boxed{7.03\ \text{eV}}$

CHECK The Fermi energy (the step-2 result) is much greater than kT at room temperatures as expected. For example, at $T = 300$ K, kT is only about 0.026 eV.

PRACTICE PROBLEM 38-2 Use Equation 38-22*b* to calculate the Fermi energy at $T = 0$ for gold, which has a free-electron number density of $59.0/\text{nm}^3$.

Table 38-1 lists the free-electron number densities and Fermi energies at $T = 0$ for several metals.

The free electrons in a metal are sometimes referred to as a Fermi gas. (They constitute a gas of fermions.) The average energy of a free electron can be calculated from the complete energy distribution of the electrons, which is discussed in Section 38-9. At $T = 0$, the average energy turns out to be

$$E_{av} = \tfrac{3}{5}E_F \tag{38-23}$$

AVERAGE ENERGY OF ELECTRONS IN A FERMI GAS AT $T = 0$

Table 38-1 Free-Electron Number Densities* and Fermi Energies at $T = 0$ for Selected Elements

	Element	N/V, electrons/nm³	E_F, eV
Al	Aluminum	181	11.7
Ag	Silver	58.6	5.50
Au	Gold	59.0	5.53
Cu	Copper	84.7	7.03
Fe	Iron	170	11.2
K	Potassium	14.0	2.11
Li	Lithium	47.0	4.75
Mg	Magnesium	86.0	7.11
Mn	Manganese	165	11.0
Na	Sodium	26.5	3.24
Sn	Tin	148	10.2
Zn	Zinc	132	9.46

* Number densities are measured using the Hall effect, discussed in Section 26-4.

For copper, E_{av} is approximately 4 eV. This average energy is huge compared with thermal energies of about $kT \approx 0.026$ eV at a temperature of $T = 300$ K. This result is very different from the classical Maxwell–Boltzmann distribution result that at $T = 0, E = 0$, and that at some temperature T, E is of the same order as kT.

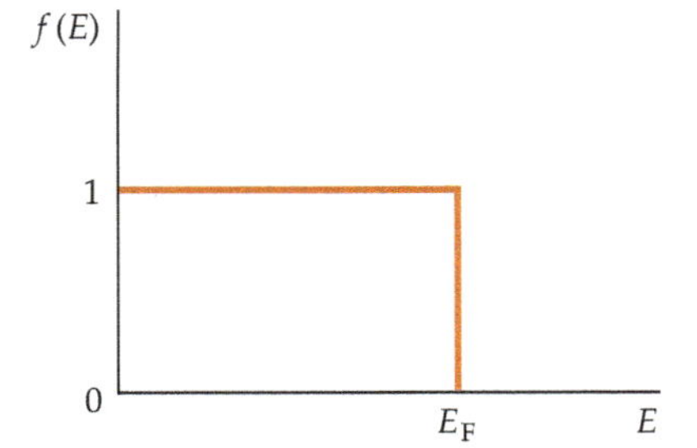

FIGURE 38-8 Fermi factor versus energy at $T = 0$.

THE FERMI FACTOR AT $T = 0$

The probability of an energy state being occupied is called the **Fermi factor,** $f(E)$. At $T = 0$ all the states below E_F are filled, whereas all those above that energy are empty, as shown in Figure 38-8. Thus, at $T = 0$ the Fermi factor is simply

$$f(E) = \begin{cases} 1 & E < E_F \\ 0 & E > E_F \end{cases} \qquad \text{38-24}$$

THE FERMI FACTOR FOR $T > 0$

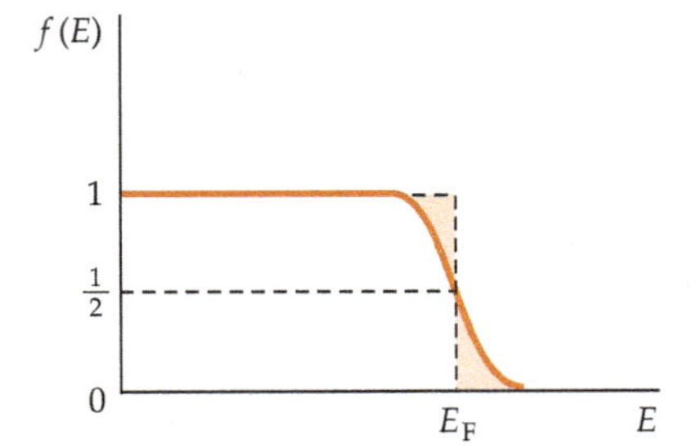

FIGURE 38-9 The Fermi factor for some temperature T. Some electrons that have energies near the Fermi energy are excited, as indicated by the shaded regions. The Fermi energy E_F is that value of E for which $f(E) = \frac{1}{2}$.

At temperatures greater than $T = 0$, some electrons will occupy higher energy states because of thermal energy gained during collisions with the lattice. However, an electron cannot move to a higher or lower state unless it is unoccupied. Because the kinetic energy of the lattice ions is of the order of kT, electrons cannot gain much more energy than kT in collisions with the lattice ions. Therefore, only those electrons that have energies within about kT of the Fermi energy can gain energy as the temperature is increased. At 300 K, kT is only 0.026 eV, so the exclusion principle prevents all but a very few electrons near the top of the energy distribution from gaining energy through random collisions with the lattice ions. Figure 38-9 shows a plot of the Fermi factor for some temperature T. Because for $T > 0$ there is no distinct energy that separates filled levels from unfilled levels, the definition of the Fermi energy must be slightly modified. At temperature T, the Fermi energy is defined to be the energy of the energy state for which the probability of being occupied is $\frac{1}{2}$. For all but extremely high temperatures, the difference between the Fermi energy at temperature T and the Fermi energy at temperature $T = 0$ is very small.

The **Fermi temperature** T_F is defined by

$$kT_F = E_F \qquad \text{38-25}$$

For temperatures much lower than the Fermi temperature, the average energy of the lattice ions will be much less than the Fermi energy, and the electron energy distribution will not differ greatly from that at $T = 0$.

Example 38-4 The Fermi Temperature for Copper

Find the Fermi temperature for copper.

PICTURE We use Equation 38-25 to find the Fermi temperature. The Fermi energy for copper at $T = 0$, calculated in Example 38-3, is 7.03 eV.

SOLVE

Use $E_F = 7.03$ eV and $k = 8.617 \times 10^{-5}$ eV/K in Equation 38-25:

$$T_F = \frac{E_F}{k} = \frac{7.03 \text{ eV}}{8.617 \times 10^{-5} \text{ eV/K}} = \boxed{81\,600 \text{ K}}$$

CHECK The Fermi temperature is very high, as expected.

TAKING IT FURTHER We can see from this example that the Fermi temperature of copper is much greater than any temperature T for which copper remains a solid.

Because an electric field in a conductor accelerates all of the conduction electrons together, the exclusion principle does not prevent the free electrons in filled states from participating in conduction. Figure 38-10 shows the Fermi factor in one dimension versus *velocity* for an ordinary temperature. The factor is approximately 1 for velocities v_x in the range $-u_F < v_x < u_F$, where the Fermi speed u_F is related to the Fermi energy by $E_F = \frac{1}{2}mu_F^2$. Then

$$u_F = \sqrt{\frac{2E_F}{m_e}} \qquad 38\text{-}26$$

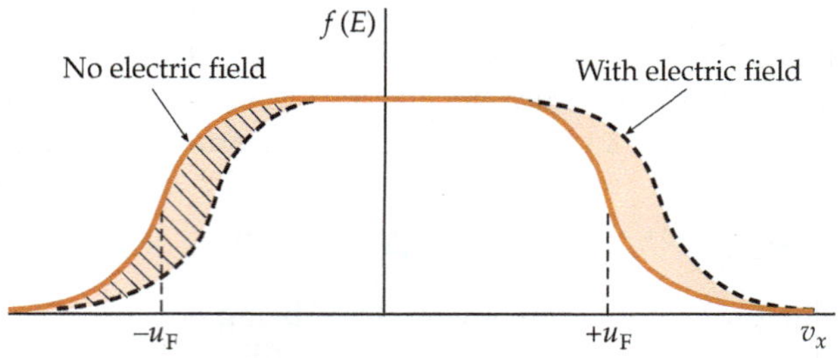

FIGURE 38-10 Fermi factor versus velocity in one dimension with no electric field (solid) and with an electric field in the $-x$ direction (dashed). The difference is greatly exaggerated.

Example 38-5 The Fermi Speed for Copper

Calculate the Fermi speed for copper.

PICTURE We use Equation 38-26 to find the Fermi speed. The Fermi energy for copper at $T = 0$, calculated in Example 38-3, is 7.03 eV.

SOLVE

Use Equation 38-26 with $E_F = 7.03$ eV:

$$u_F = \sqrt{\frac{2(7.03\text{ eV})}{9.11 \times 10^{-31}\text{ kg}}\left(\frac{1.60 \times 10^{-19}\text{ J}}{1\text{ eV}}\right)} = \boxed{1.57 \times 10^6\text{ m/s.}}$$

CHECK As expected, the result (the Fermi speed for copper) is high, but less than the speed of light.

The dashed curve in Figure 38-10 shows the Fermi factor after the electric field has been acting for some time t. Although all of the free electrons have their velocities shifted in the direction opposite to the electric field, the net effect is equivalent to shifting only the electrons near the Fermi energy.

CONTACT POTENTIAL

When two different metals are placed in contact, a potential difference $V_{contact}$ called the **contact potential** develops between them. The contact potential depends on both the work functions of the two metals, ϕ_1 and ϕ_2 (we encountered work functions when the photoelectric effect was introduced in Chapter 34), and the Fermi energies of the two metals. When the metals are in contact, the total energy of the system is lowered if electrons near the boundary move from the metal that has the higher Fermi energy into the metal that has the lower Fermi energy until the Fermi energies of the two metals are the same, as shown in Figure 38-11. When equilibrium is established, the metal that has the lower initial Fermi energy is negatively charged and the other metal is positively charged, so that between them there is a potential difference $V_{contact}$ given by

$$V_{contact} = \frac{\phi_1 - \phi_2}{e} \qquad 38\text{-}27$$

Table 38-2 lists the work functions for several metals.

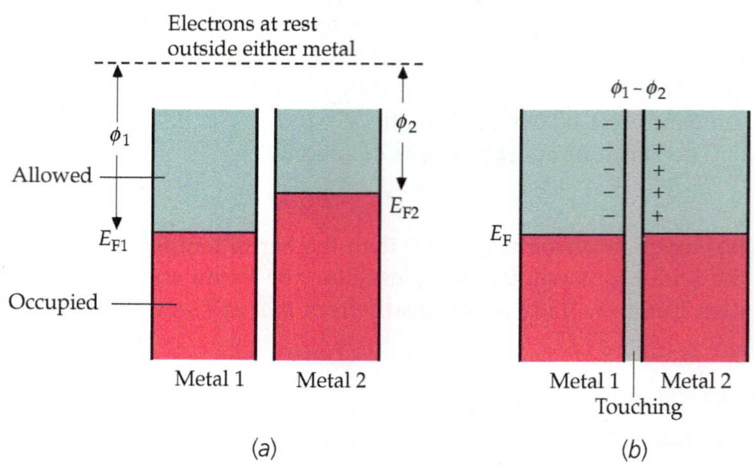

FIGURE 38-11 (*a*) Energy levels for two different metals that have different Fermi energies E_F and work functions ϕ. The work function is the difference between the energy of an electron at rest outside the metal and the Fermi energy within the metal. (*b*) When the metals are in contact, electrons flow from the metal that initially has the higher Fermi energy to the metal that initially has the lower Fermi energy until the Fermi energies are equal.

Table 38-2 Work Functions for Some Metals

	Metal	ϕ, eV		Metal	ϕ, eV
Ag	Silver	4.7	K	Potassium	2.1
Au	Gold	4.8	Mn	Manganese	3.8
Ca	Calcium	3.2	Na	Sodium	2.3
Cu	Copper	4.1	Ni	Nickel	5.2

Example 38-6 Contact Potential between Silver and Tungsten

The threshold wavelength for the photoelectric effect is 271 nm for tungsten and 262 nm for silver. What is the contact potential developed when silver and tungsten are placed in contact?

PICTURE The contact potential is proportional to the difference in the work functions for the two metals (Equation 38-27). The work function ϕ can be found from the given threshold wavelengths using $\phi = hc/\lambda_t$ (Equation 34-4).

SOLVE

1. The contact potential is given by Equation 38-27:

$$V_{\text{contact}} = \frac{\phi_1 - \phi_2}{e}$$

2. The work function is related to the threshold wavelength (Equation 34-4):

$$\phi = \frac{hc}{\lambda_t}$$

3. Substitute $\lambda_t = 271$ nm for tungsten (the symbol for tungsten is W):

$$\phi_W = \frac{hc}{\lambda_t} = \frac{1240\text{ eV}\cdot\text{nm}}{271\text{ nm}} = 4.58\text{ eV}$$

4. Substitute $\lambda_t = 262$ nm for silver:

$$\phi_{Ag} = \frac{1240\text{ eV}\cdot\text{nm}}{262\text{ nm}} = 4.73\text{ eV}$$

5. The contact potential is thus:

$$V_{\text{contact}} = \frac{\phi_{Ag} - \phi_W}{e} = 4.73\text{ V} - 4.58\text{ V} = \boxed{0.15\text{ V}}$$

CHECK As expected, the contact potential is small (less than one volt). You do not get large potential differences just by putting two metals in contact.

HEAT CAPACITY DUE TO ELECTRONS IN A METAL

The quantum-mechanical description of the electron distribution in metals allows us to understand why the contribution of the free electrons to the heat capacity of a metal is much less that of the ions. According to the classical equipartition theorem, the energy of the lattice ions in n moles of a solid is $3nRT$, and thus the molar specific heat is $c' = 3R$, where R is the universal gas constant (see Section 18-7). In a metal, the number of free electrons is approximately equal to the number of lattice ions. If these electrons obey the classical equipartition theorem, they should have an energy of $\frac{3}{2}nRT$ and contribute an additional $\frac{3}{2}R$ to the molar specific heat. But measured heat capacities of metals are just slightly greater than those of insulators. We can understand this result because at some temperature T, only those electrons that have energies near the Fermi energy can be excited by random collisions with the lattice ions. The number of those electrons is of the order of $(kT/E_F)N$, where N is the total number of free electrons. The energy of those electrons is increased from that at $T = 0$ by an amount that is of the order of kT. So the total increase in thermal energy is of the order of $(kT/E_F)N \times kT$.

We can thus express the energy of N electrons at temperature T as

$$E = NE_{av}(0) + \alpha N \frac{kT}{E_F} kT \qquad \text{38-28}$$

where $E_{av}(0)$ is the average energy at $T = 0$ and α is a constant that we expect to be of the order of 1 if our reasoning is correct. The calculation of α is quite challenging. The result is $\alpha = \pi^2/4$. Using this result and writing E_F in terms of the Fermi temperature, $E_F = kT_F$, we obtain the following for the contribution of the free electrons to the heat capacity at constant volume:

$$C_V = \frac{dE}{dT} = 2\alpha Nk\frac{kT}{E_F} = \frac{1}{2}\pi^2 nR\frac{T}{T_F}$$

where we have written Nk in terms of the gas constant R ($R = Nk/n$). The molar specific heat at constant volume is then

$$c'_V = \frac{1}{2}\pi^2 R\frac{T}{T_F} \qquad \text{38-29}$$

We can see that because of the large value of T_F, the contribution of the free electrons is a small fraction of R at ordinary temperatures. Because $T_F = 81\,600$ K for copper, the molar specific heat of the free electrons at $T = 300$ K is

$$c'_V = \frac{1}{2}\pi^2\frac{300\text{ K}}{81\,600\text{ K}}R \approx 0.02R$$

which is in good agreement with the experiment.

38-4 QUANTUM THEORY OF ELECTRICAL CONDUCTION

We can use Equation 38-14 for the resistivity if we use the Fermi speed u_F (Equation 38-26) in place of v_{av}:

$$\rho = \frac{m_e u_F}{n_e e^2 \lambda} \qquad \text{38-30}$$

We now have two problems. First, because the Fermi speed u_F is approximately independent of temperature, the resistivity given by Equation 38-30 is independent of temperature unless the mean free path should depend on the temperature. The second problem concerns magnitudes. As mentioned earlier, the classical expression for resistivity using v_{av} calculated from the Maxwell–Boltzmann distribution gives values that are about 6 times too large at $T = 300$ K. Because the Fermi speed u_F is about 16 times the Maxwell-Boltzmann value of v_{av}, the magnitude of ρ predicted by Equation 38-30 will be approximately 100 times greater than the experimentally determined value. The resolution of both of these problems lies in the calculation of the mean free path λ.

THE SCATTERING OF ELECTRON WAVES

In Equation 38-16 for the classical mean free path $\lambda = 1/(n_{ion}A)$, the quantity $A = \pi r^2$ is the cross-sectional area of the lattice ion as seen by an electron. In the quantum calculation, the mean free path is related to the scattering of electron waves by the crystal lattice. Detailed calculations show that, for a *perfectly* ordered crystal, $\lambda = \infty$; that is, there is no scattering of the electron waves. The scattering of electron waves arises because of *imperfections* in the crystal lattice, which have nothing to do with the actual cross-sectional area A of the lattice ions.

According to the quantum theory of electron scattering, A depends merely on *deviations* of the lattice ions from a perfectly ordered array and not on the size of the ions. The most common causes of such deviations are thermal vibrations of the lattice ions or impurities.

We can use $\lambda = 1/(n_{ion}A)$ for the mean free path if we reinterpret the area A. Figure 38-12 compares the classical picture and the quantum picture of this area. In the quantum picture, the lattice ions are points that have no size but present an area $A = \pi r_0^2$, where r_0 is the amplitude of thermal vibrations. In Chapter 14, we saw that the energy of vibration in simple harmonic motion is proportional to the square of the amplitude, which is πr_0^2. Thus, the effective area A is proportional to the energy of vibration of the lattice ions. From the equipartition theorem,* we know that the average energy of vibration is proportional to kT. Thus, A is proportional to T, and λ is proportional to $1/T$. Then the resistivity given by Equation 38-14 is proportional to T, in agreement with experiment.

The effective area A due to thermal vibrations can be calculated, and the results give values for the resistivity that are in agreement with experiments. At $T = 300$ K, for example, the effective area turns out to be about 100 times smaller than the actual cross-sectional area of a lattice ion. We see, therefore, that the free-electron model of metals gives a good account of electrical conduction if the classical mean speed v_{av} is replaced by the Fermi speed u_F and if the collisions between electrons and the lattice ions are interpreted in terms of the scattering of electron waves, for which only deviations from a perfectly ordered lattice are important.

The presence of impurities in a metal also causes deviations from perfect regularity in the crystal lattice. The effects of impurities on resistivity are approximately independent of temperature. The resistivity of a metal containing impurities can be written $\rho = \rho_t + \rho_i$, where ρ_t is the resistivity due to the thermal motion of the lattice ions and ρ_i is the resistivity due to impurities. Figure 38-13 shows typical resistance versus temperature curves for metals with impurities. As the absolute temperature approaches zero, the resistivity due to thermal motion approaches zero, and the total resistivity approaches the resistivity due to impurities, which is constant.

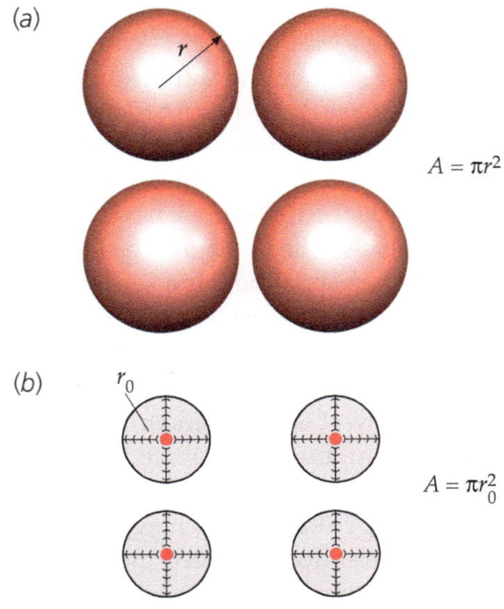

FIGURE 38-12 (*a*) Classical picture of the lattice ions as spherical balls of radius r that each present an area πr^2 to the electrons. (*b*) Quantum-mechanical picture of the lattice ions as points that are vibrating in three dimensions. The area presented to the electrons is πr_0^2, where r_0 is the amplitude of oscillation of the ions.

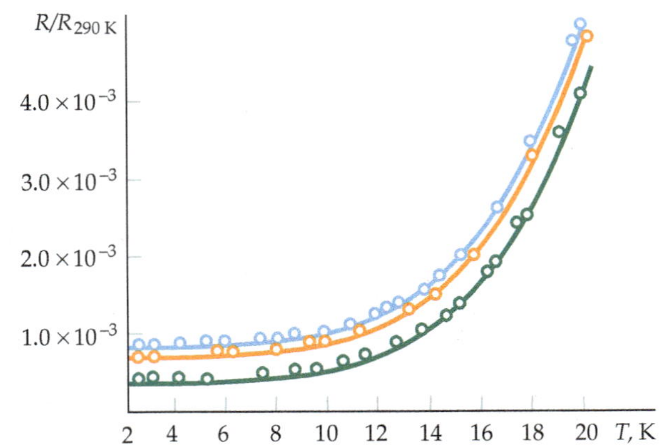

FIGURE 38-13 Relative resistance versus temperature for three samples of sodium. The three curves have the same temperature dependence but different magnitudes because of differing amounts of impurities in the samples.

38-5 BAND THEORY OF SOLIDS

Resistivities vary enormously between insulators and conductors. For a typical insulator, such as quartz, $\rho \sim 10^{16}\ \Omega\cdot\text{m}$, whereas for a typical conductor, $\rho \sim 10^{-8}\ \Omega\cdot\text{m}$. The reason for this enormous variation is the variation in the number density of free electrons n_e. To understand this variation, we consider the effect of the lattice on the electron energy levels.

We begin by considering the energy levels of the individual atoms as they are brought together. The allowed energy levels in an isolated atom are often far apart. For example, in hydrogen, the lowest allowed energy $E_1 = -13.6$ eV is 10.2 eV below the next lowest allowed energy $E_2 = (-13.6\text{ eV})/4 = -3.4$ eV.† Let us consider two identical atoms and focus our attention on one particular energy level. When the atoms are far apart, the energy of a particular level is the same for each atom. As the atoms are brought closer together, the energy level for each atom changes because of the influence of the other atom. As a result, the level splits into two levels of slightly different energies for the two-atom system. If we bring three atoms close together,

* The equipartition theorem does hold for the lattice ions, which obey the Maxwell–Boltzmann energy distribution.

† The energy levels in hydrogen are discussed in Chapter 36.

a particular energy level splits into three separate levels of slightly different energies. Figure 38-14 shows the energy splitting of two energy levels for six atoms as a function of the separation of the atoms.

If we have N identical atoms, a particular energy level in the isolated atom splits into N different, closely spaced energy levels when the atoms are close together. In a macroscopic solid, N is very large—of the order of 10^{23}—so each energy level splits into a very large number of levels called a **band.** The levels are spaced almost continuously within the band. There is a separate band of levels for each particular energy level of the isolated atom. The bands may be widely separated in energy, they may be close together, or they may even overlap, depending on the kind of atom and the type of bonding in the solid.

The lowest energy bands, corresponding to the lowest energy levels of the atoms in the lattice, are filled with electrons that are bound to the individual atoms. The electrons that can take part in conduction occupy the higher energy bands. The highest energy band that contains electrons is called the **valence band.** The valence band may be completely filled with electrons or only partially filled, depending on the kind of atom and the type of bonding in the solid.

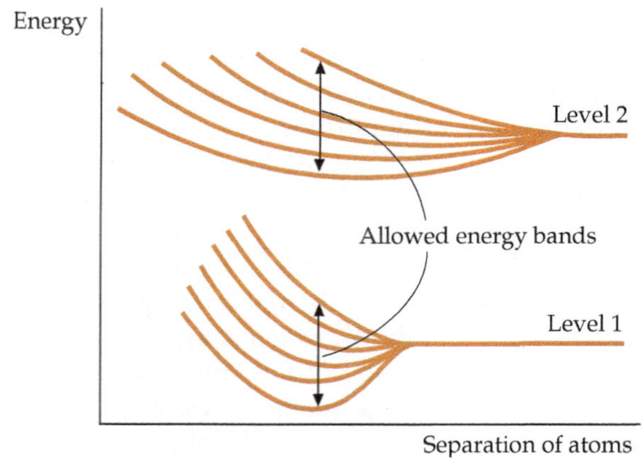

FIGURE 38-14 Energy splitting of two energy levels for six atoms as a function of the separation of the atoms. When there are many atoms, each level splits into a near-continuum of levels called a band.

We can now understand why some solids are conductors and why others are insulators. If the valence band is only partially filled, there are many available empty energy states in the band, and the electrons in the band can easily be raised to a higher energy state by an electric field. Accordingly, this substance is a good conductor. If the valence band is filled and there is a large energy gap between it and the next available band, an applied electric field may be too weak to excite an electron from the upper energy levels of the filled band across the large gap into the energy levels of the empty band, so the substance is an insulator. The lowest band in which there are unoccupied states is called the **conduction band.** In a conductor, the valence band is only partially filled, so the valence band is also the conduction band. An energy gap between allowed bands is called a **forbidden energy band.**

The band structure for a conductor, such as copper, is shown in Figure 38-15*a*. The lower bands (not shown) are filled with the lower energy electrons of the atoms. The valence band is only about half-filled. When an electric field is established in the conductor, the electrons in the conduction band are accelerated, which means that their energies are increased. This is consistent with the exclusion principle because there are many empty energy states just above those occupied by electrons in this band. These electrons are thus the conduction electrons.

Figure 38-15*b* shows the band structure for magnesium, which is also a conductor. In this case, the highest occupied band is completely filled, but there is an empty band above it that overlaps it. The two bands thus form a combined valence–conduction band that is only partially filled.

Figure 38-15*c* shows the band structure for a typical insulator. At $T = 0$ K, the valence band is completely filled. The next energy band having empty energy states, the conduction band, is separated from the valence band by a large energy gap.

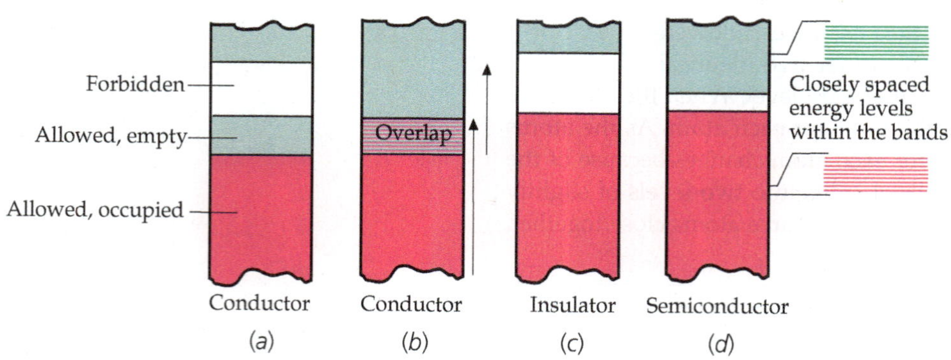

FIGURE 38-15 Four possible band structures for a solid. (*a*) A typical conductor. The valence band is also the conduction band. It is only partially filled, so electrons can be easily excited to nearby energy states. (*b*) A conductor in which the valence band overlaps a conduction band above it. (*c*) A typical insulator. There is a forbidden band that has a large energy gap between the filled valence band and the conduction band. (*d*) A semiconductor. The energy gap between the filled valence band and the conduction band is very small, so some electrons are excited to the conduction band at normal temperatures, leaving holes in the valence band.

At $T = 0$, the conduction band is empty. At ordinary temperatures, a few electrons can be excited to states in that band, but most cannot be excited to states because the energy gap is large compared with the energy an electron might obtain by thermal excitation. Very few electrons can be thermally excited to the nearly empty conduction band, even at fairly high temperatures. When an electric field of ordinary magnitude is established in the solid, electrons cannot be accelerated because there are no empty energy states at nearby energies. We describe this by saying that there are no free electrons. The small conductivity that is observed is due to the very few electrons that are thermally excited into the nearly empty conduction band. When an electric field applied to an insulator is sufficiently strong to cause an electron to be excited across the energy gap to the empty band, dielectric breakdown occurs.

In some substances, the energy gap between the filled valence band and the empty conduction band is very small, as shown in Figure 38-15*d*. At $T = 0$, there are no electrons in the conduction band and the material is an insulator. At ordinary temperatures, however, there are an appreciable number of electrons in the conduction band due to thermal excitation. Such a material is called an **intrinsic semiconductor.** For typical intrinsic semiconductors, such as silicon and germanium, the energy gap is only about 1 eV. In the presence of an electric field, the electrons in the conduction band can be accelerated because there are empty states nearby. Also, for each electron in the conduction band there is a vacancy, or **hole,** in the nearly filled valence band. In the presence of an electric field, electrons in this band can also be excited to a vacant energy level. This contributes to the electric current and is most easily described as the motion of a hole in the direction of the field and opposite to the motion of the electrons. The hole thus acts like a positive charge. To visualize the conduction of holes, think of a two-lane, one-way road that has one lane completely filled with parked cars and the other lane empty. If a car moves out of the completely filled lane into the empty lane, it can move ahead freely. As the other cars move up to occupy the vacated space, the vacated space propagates backward in the direction opposite the motion of the cars. Both the forward motion of the car in the nearly empty lane and the backward propagation of the empty space contribute to a net forward propagation of the cars.

An interesting characteristic of semiconductors is that the resistivity of the substance decreases as the temperature increases, which is contrary to the case for normal conductors. The reason is that as the temperature increases, the number of free electrons increases because there are more electrons in the conduction band. The number of holes in the valence band also increases, of course. In semiconductors, the effect of the increase in the number of charge carriers, both electrons and holes, exceeds the effect of the increase in resistivity due to the increased scattering of the electrons by the lattice ions due to thermal vibrations. Semiconductors therefore have a negative temperature coefficient of resistivity.

38-6 SEMICONDUCTORS

The semiconducting property of intrinsic semiconductors makes them useful as a basis for electronic circuit components whose resistivity can be controlled by application of an external voltage or current. Most such *solid-state devices,* however, such as the semiconductor diode and the transistor, make use of **impurity semiconductors,** which are created through the controlled addition of certain impurities to intrinsic semiconductors. This process is called **doping.** Figure 38-16*a* is a schematic illustration of silicon doped with a small amount of arsenic so that the arsenic atoms replace a few of the silicon atoms in the crystal lattice. The conduction band of pure silicon is virtually empty at ordinary temperatures, so pure silicon is a poor conductor of electricity. However, arsenic has five valence electrons rather than the four valence electrons of silicon. Four of these electrons take part in bonds with the four neighboring silicon atoms, and the fifth electron is very

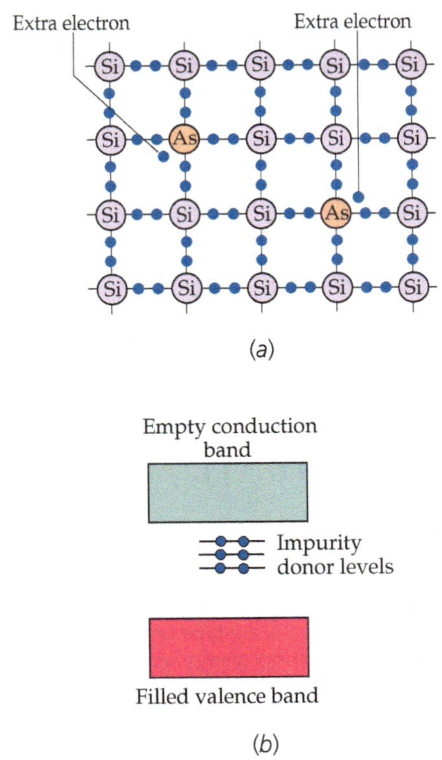

FIGURE 38-16 (*a*) A two-dimensional schematic illustration of silicon doped with arsenic. Because arsenic has five valence electrons, there is an extra, weakly bound electron that is easily excited to the conduction band, where it can contribute to electrical conduction. (*b*) Band structure of an *n*-type semiconductor, such as silicon doped with arsenic. The impurity atoms provide filled energy levels that are just below the conduction band. These levels donate electrons to the conduction band.

loosely bound to the atom. This extra electron occupies an energy level that is just slightly below the conduction band in the solid, and it is easily excited into the conduction band, where it can contribute to electrical conduction.

The effect on the band structure of a silicon crystal achieved by doping it with arsenic is shown in Figure 38-16*b*. The levels shown just below the conduction band are due to the extra electrons of the arsenic atoms. These levels are called **donor levels** because they donate electrons to the conduction band without leaving holes in the valence band. Such a semiconductor is called an ***n*-type semiconductor** because the major charge carriers are negatively charged electrons. The conductivity of a doped semiconductor can be controlled by controlling the amount of impurity added. The addition of just one part per million can increase the conductivity by several orders of magnitude.

Another type of impurity semiconductor can be made by replacing a silicon atom with a gallium atom, which has three valence electrons (Figure 38-17*a*). The gallium atom accepts electrons from the valence band to complete its four covalent bonds, thus creating a hole in the valence band. The effect on the band structure of silicon achieved by doping it with gallium is shown in Figure 38-17*b*. The empty levels shown just above the valence band are due to the holes from the ionized gallium atoms. These levels are called **acceptor levels** because they accept electrons from the filled valence band when those electrons are thermally excited to a higher energy state. This creates holes in the valence band that are free to propagate in the direction of an electric field. Such a semiconductor is called a ***p*-type semiconductor** because the charge carriers are positively charged holes. The fact that conduction is due to the motion of positively charged holes can be verified by the Hall effect. (The Hall effect is discussed in Chapter 26.)

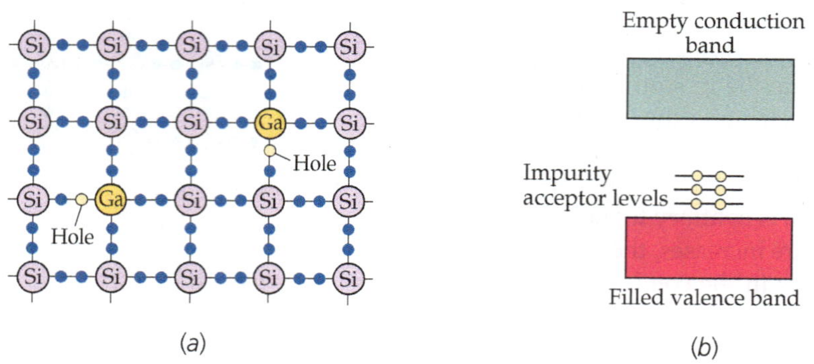

FIGURE 38-17 (*a*) A two-dimensional schematic illustration of silicon doped with gallium. Because gallium has only three valence electrons, there is a hole in one of its bonds. As electrons move into the hole the hole moves about, contributing to the conduction of electrical current. (*b*) Band structure of a *p*-type semiconductor, such as silicon doped with gallium. The impurity atoms provide empty energy levels just above the filled valence band that accept electrons from the valence band.

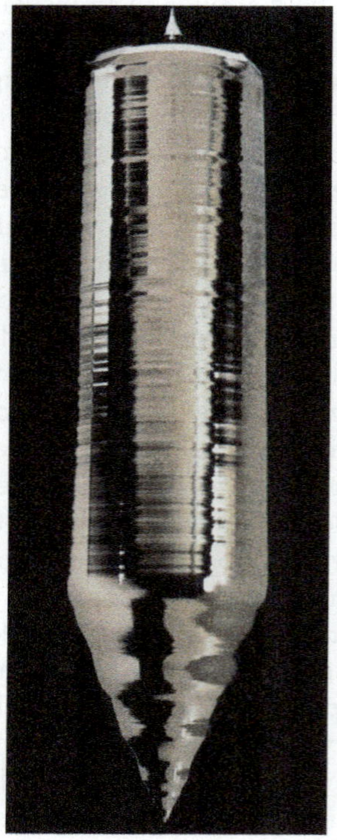

Synthetic crystal silicon is produced beginning with a raw material containing silicon (for instance, common beach sand), separating out the silicon, and melting it. From a seed crystal, the molten silicon grows into a cylindrical crystal, such as the one shown here. The crystals (typically about 1.3 m long) are formed under highly controlled conditions to ensure that they are flawless and the crystals are then sliced into thousands of thin wafers onto which the layers of an integrated circuit are etched. *(Museum of Modern Art.)*

Example 38-7 Number Density of Free Electrons in Arsenic-Doped Silicon

Try It Yourself

The number of free electrons in pure silicon is approximately 10^{10} electrons/cm^3 at ordinary temperatures. If one silicon atom out of every 10^6 atoms is replaced by an arsenic atom, how many free electrons per cubic centimeter are there? (The density of silicon is 2.33 g/cm^3 and its molar mass is 28.1 g/mol.)

PICTURE The number of silicon atoms per cubic centimeter, n_{Si}, can be found from $n_{\text{Si}} = \rho N_A/M$. Then, because each arsenic atom contributes one free electron, the number of electrons contributed by the arsenic atoms is $10^{-6}\, n_{\text{Si}}$.

SOLVE

Cover the column to the right and try these on your own before looking at the answers.

Steps	Answers
1. Calculate the number of silicon atoms per cubic centimeter.	$n_{Si} = \frac{\rho N_A}{M} = \frac{(2.33\ g/cm^3)(6.02 \times 10^{23}\ atoms/mol)}{28.1\ g/mol} = 4.99 \times 10^{22}\ atoms/cm^3$
2. Multiply by 10^{-6} to obtain the number of arsenic atoms per cubic centimeter, which equals the added number of free electrons per cubic centimeter.	$n_{As} = 10^{-6} n_{Si} = 4.99 \times 10^{16}\ atoms/cm^3$
3. The number of free electrons per cubic centimeter is equal to the number of arsenic atoms per cubic centimeter plus 1×10^{-10} (the number of silicon atoms per cubic centimeter).	$n_e = n_{As} + 1 \times 10^{-10} n_{Si} = 4.99 \times 10^{16}\ cm^{-3} + 1 \times 10^{10}\ cm^{-3} \approx \boxed{5 \times 10^{16}\ electrons/cm^3}$

CHECK As expected, the step-3 result is less than the number density of silicon atoms and more than the number density of conduction electrons in pure silicon.

TAKING IT FURTHER Because silicon has so few free electrons per atom, the number density of conduction electrons is increased by a factor of approximately 5 million per cubic centimeter by doping silicon with just one arsenic atom per million silicon atoms.

PRACTICE PROBLEM 38-3 How many free electrons are there per silicon atom in pure silicon?

* 38-7 SEMICONDUCTOR JUNCTIONS AND DEVICES

Semiconductor devices such as diodes and transistors make use of n-type semiconductors and p-type semiconductors joined together, as shown in Figure 38-18. In practice, the two types of semiconductors are often incorporated into a single silicon crystal doped with donor impurities on one side and acceptor impurities on the other side. The region in which the semiconductor changes from a p-type semiconductor to an n-type semiconductor is called a ***pn* junction.**

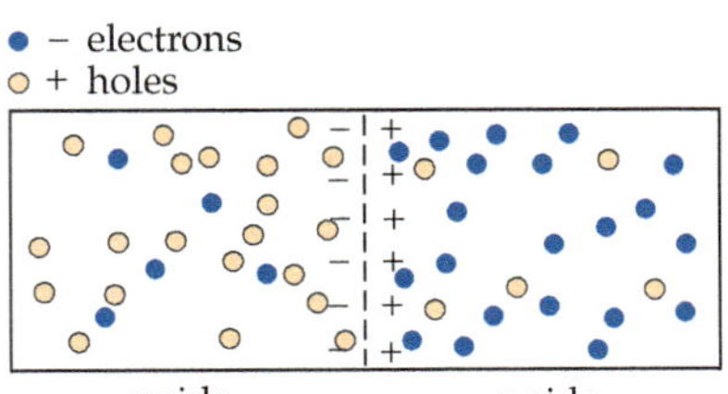

FIGURE 38-18 A *pn* junction. Because of the difference in their concentrations on either side of the *pn* junction, holes diffuse from the *p* side to the *n* side, and electrons diffuse from the *n* side to the *p* side. As a result, there is a double layer of charge at the junction, with the *p* side being negative and the *n* side being positive.

When an n-type semiconductor and a p-type semiconductor are placed in contact, the initially unequal concentrations of electrons and holes result in the diffusion of electrons across the junction from the n side to the p side and holes from the p side to the n side until equilibrium is established. The result of this diffusion is a net transport of positive charge from the p side to the n side. Unlike the case when two different metals are in contact, the electrons cannot travel very far from the junction region because the semiconductor is not a particularly good conductor. The diffusion of electrons and holes therefore creates a double layer of charge at the junction similar to that on a parallel-plate capacitor. There is, thus, a potential difference V across the junction, which tends to inhibit further diffusion. In equilibrium, the n side which has a net positive charge will be at a higher potential than the p side which has a net negative charge. In the junction region, between the charge layers, there will be very few charge carriers of either type, so the junction region has a high resistance. Figure 38-19 shows the energy-level diagram for a pn junction. The junction region is also called the **depletion region** because it has been depleted of charge carriers.

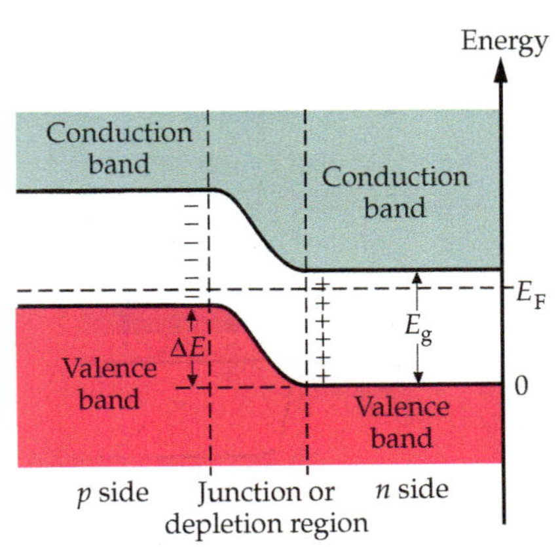

FIGURE 38-19 Electron energy levels for a *pn* junction.

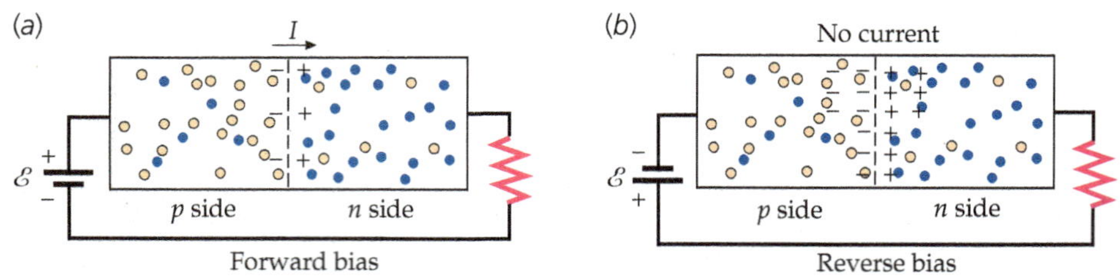

FIGURE 38-20 A *pn*-junction diode. (*a*) Forward-biased *pn* junction. The applied potential difference enhances the diffusion of holes from the *p* side to the *n* side and of electrons from the *n* side to the *p* side, resulting in a current *I*. (*b*) Reverse-biased *pn* junction. The applied potential difference inhibits the further diffusion of holes and electrons across the junction, so there is no current.

*DIODES

In Figure 38-20, an external potential difference has been applied across a *pn* junction by connecting a battery and a resistor to the semiconductor. When the positive terminal of the battery is connected to the *p* side of the junction, as shown in Figure 38-20*a*, the junction is said to be **forward biased.** Forward biasing lowers the potential across the junction. The diffusion of electrons and holes is thereby increased as they attempt to reestablish equilibrium, resulting in a current in the circuit.

If the positive terminal of the battery is connected to the *n* side of the junction, as shown in Figure 38-20*b*, the junction is said to be **reverse biased.** Reverse biasing tends to increase the potential difference across the junction, thereby further inhibiting diffusion. Figure 38-21 shows a plot of current versus voltage for a typical semiconductor junction. Essentially, the junction conducts only in one direction for applied voltages greater than the breakdown voltage. A single-junction semiconductor device is called a **diode.*** Diodes have many uses. One use is to convert alternating current into direct current, a process called *rectification*.

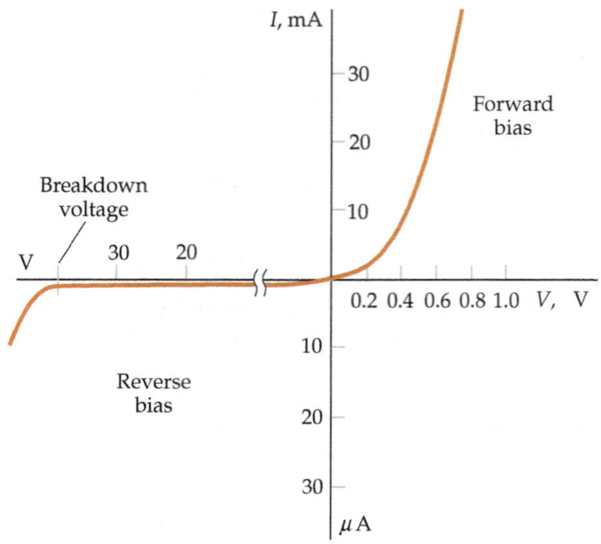

FIGURE 38-21 Plot of current versus applied voltage across a *pn* junction. Note the different scales on both axes for the forward and reverse bias conditions.

Note that the current in Figure 38-21 suddenly increases in magnitude at extreme values of reverse bias. In such large electric fields, electrons are stripped from their atomic bonds and accelerated across the junction. These electrons, in turn, cause others to break loose. This effect is called **avalanche breakdown.** Although such a breakdown can be disastrous in a circuit where it is not intended, the fact that it occurs at a sharply defined voltage makes it of use in a special voltage reference standard known as a **Zener diode.** Zener diodes are also used to protect devices from excessively high voltages.

An interesting effect, one that we discuss only qualitatively, occurs if both the *n* side and the *p* side of a *pn*-junction diode are so heavily doped that the donors on the *n* side provide so many electrons that the lower part of the conduction band is practically filled, and the acceptors on the *p* side accept so many electrons that the upper part of the valence band is nearly empty. Figure 38-22*a* shows the energy-level

FIGURE 38-22 Electron energy levels for a heavily doped *pn*-junction tunnel diode. (*a*) With no bias voltage, some electrons tunnel in each direction. (*b*) With a small bias voltage, the tunneling current is enhanced in one direction, making a sizable contribution to the net current. (*c*) With further increases in the bias voltage, the tunneling current decreases dramatically.

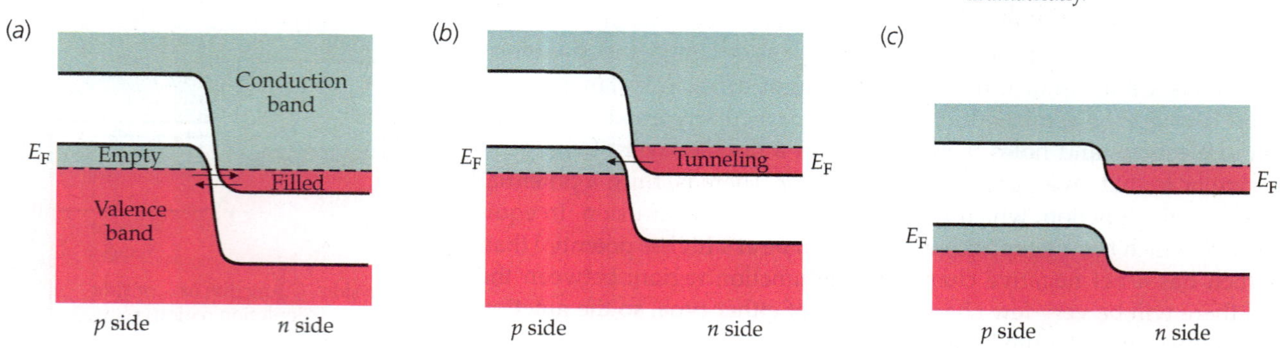

* The name *diode* originates from a vacuum tube device consisting of just two electrodes that also conducts electric current in one direction only.

diagram for this situation. Because the depletion region is now so narrow, electrons can easily penetrate the potential barrier across the junction and tunnel to the other side. The flow of electrons through the barrier is called a **tunneling current,** and such a heavily doped diode is called a **tunnel diode.**

At equilibrium where there is no bias, there is an equal tunneling current in each direction. When a small bias voltage is applied across the junction, the energy-level diagram is as shown in Figure 38-22*b*, and the tunneling of electrons from the *n* side to the *p* side is increased, whereas the tunneling of electrons in the opposite direction is decreased. This tunneling current, in addition to the usual current due to diffusion, results in a considerable net current. When the bias voltage is increased slightly, the energy-level diagram is as shown in Figure 38-22*c*, and the tunneling current is decreased. Although the diffusion current is increased, the net current is decreased. At large bias voltages, the tunneling current is completely negligible, and the total current increases with increasing bias voltage due to diffusion, as in an ordinary *pn*-junction diode. Figure 38-23 shows the current-versus-voltage curve for a tunnel diode. Such diodes are used in electric circuits because of their very fast response time. When operated near the peak in the current-versus-voltage curve, a small change in bias voltage results in a large change in the current.

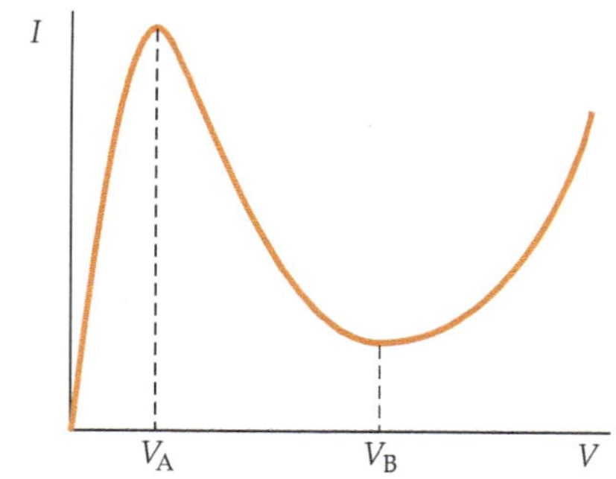

FIGURE 38-23 Current versus applied (bias) voltage V for a tunnel diode. For $V < V_A$, an increase in the bias voltage V enhances tunneling. For $V_A < V < V_B$, an increase in the bias voltage inhibits tunneling. For $V > V_B$, the tunneling is negligible, and the diode behaves like an ordinary *pn*-junction diode.

Another use for the *pn*-junction semiconductor is the **solar cell,** which is illustrated schematically in Figure 38-24. When a photon of energy greater than the gap energy (1.1 eV in silicon) strikes the *p*-type region, it can excite an electron from the valence band into the conduction band, leaving a hole in the valence band. This region is already rich in holes. Some of the electrons created by the photons will recombine with holes, but some will migrate to the junction. From there, they are accelerated into the *n*-type region by the electric field between the double layer of charge. This creates an excess negative charge in the *n*-type region and an excess positive charge in the *p*-type region. The result is a potential difference between the two regions, which in practice is approximately 0.6 V. If a load resistance is connected across the two regions, a charge flows through the resistance. Some of the incident light energy is thus converted into electrical energy. The current in the resistor is proportional to the rate of arrival of incident photons, which is in turn proportional to the intensity of the incident light.

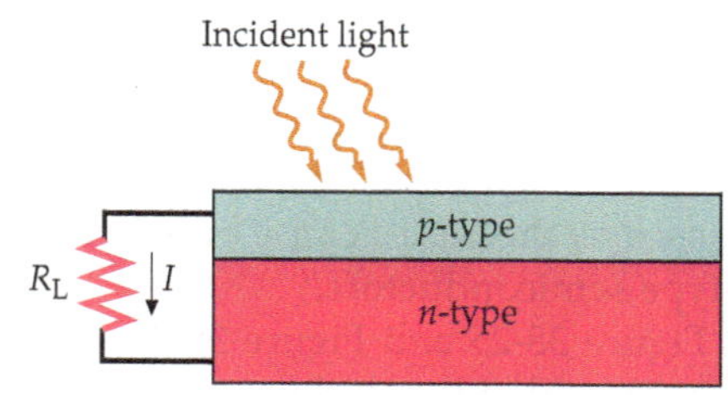

FIGURE 38-24 A *pn*-junction semiconductor as a solar cell. When light strikes the *p*-type region, electron-hole pairs are created, resulting in a current through the load resistance R_L.

There are many other applications of semiconductors with *pn* junctions. Particle detectors, called **surface-barrier detectors,** consist of a *pn*-junction semiconductor that has a large reverse bias so that there is ordinarily no current. When a high-energy particle, such as an electron, passes through the semiconductor, it creates many electron–hole pairs as it loses energy. The resulting current pulse signals the passage of the particle. **Light-emitting diodes** (LEDs) are *pn*-junction semiconductors that have large forward biases that produce large excess concentrations of electrons on the *p* sides and holes on the *n* sides of the junctions. Under these conditions, an LED emits light as the electrons and holes recombine. This is essentially the reverse of the process that occurs in a solar cell, in which electron–hole pairs are created by the absorption of light. LEDs are commonly used as warning indicators and as sources of infrared light beams.

A light-emitting diode (LED). *(© C. Falco/Photo Researchers.)*

*TRANSISTORS

The transistor, a semiconducting device that is used to produce a desired output signal in response to an input signal, was invented in 1948 by William Shockley, John Bardeen, and Walter Brattain and has revolutionized the electronics industry and our everyday world. A *simple bipolar junction transistor** consists of three distinct semiconductor regions called the **emitter,** the **base,** and the **collector.** The base is a very thin region of one type of semiconductor sandwiched between two regions of the opposite type. The emitter semiconductor is much more heavily

* Besides the bipolar junction transistor, there are other categories of transistors, notably, the field–effect transistor.

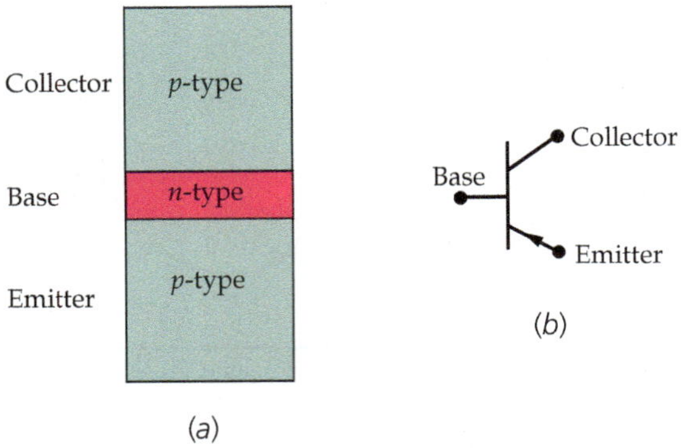

FIGURE 38-25 A *pnp* transistor. (*a*) The heavily doped emitter emits holes that pass through the thin base to the collector. (*b*) Symbol for a *pnp* transistor in a circuit. The arrow points in the direction of the conventional current, which is the same as that of the emitted holes.

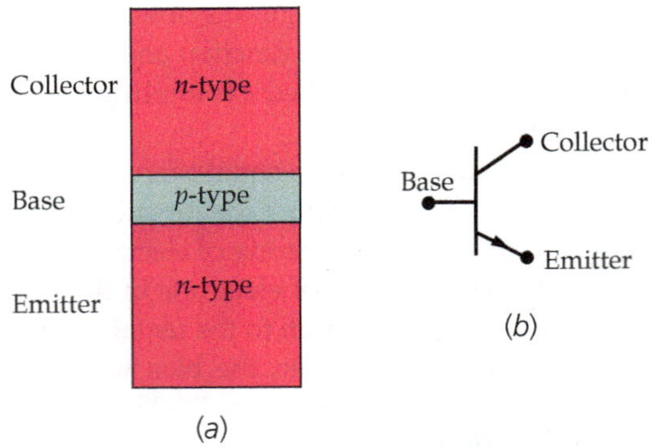

FIGURE 38-26 An *npn* transistor. (*a*) The heavily doped emitter emits electrons that pass through the thin base to the collector. (*b*) Symbol for an *npn* transistor. The arrow points in the direction of the conventional current, which is opposite the direction of the emitted electrons.

doped than either the base or the collector. In an *npn* transistor, the emitter and collector are *n*-type semiconductors and the base is a *p*-type semiconductor; in a *pnp* transistor, the base is an *n*-type semiconductor and the emitter and collector are *p*-type semiconductors.

Figure 38-25 and Figure 38-26 show, respectively, a *npn* transistor and an *npn* transistor, along with the symbols used to represent each transistor in circuit diagrams. We see that either transistor consists of two *pn* junctions. We will discuss the operation of a *npn* transistor. The operation of an *npn* transistor is similar.

In the normal operation of a *pnp* transistor, the emitter-base junction is forward biased, and the base-collector junction is reverse biased, as shown in Figure 38-27. The heavily doped *p*-type emitter emits holes that flow toward the emitter-base junction. This flow constitutes the emitter current I_e. Because the base is very thin, most of the holes flow across the base into the collector. This flow in the collector constitutes a current I_c. However, some of the holes recombine in the base producing a positive charge that inhibits the further flow of charge. To prevent this, some of the holes that do not reach the collector are drawn off the base as a base current I_b in a wire connected to the base. In Figure 38-27, therefore, I_c is almost but not quite equal to I_e, and I_b is much smaller than either I_c or I_e. It is customary to express I_c as

$$I_c = \beta I_b \qquad 38\text{-}31$$

where β is called the current gain of the transistor. Transistors can be designed to have values of β as low as ten or as high as several hundred.

Figure 38-28 shows a simple *pnp* transistor used as an amplifier. A small, time-varying input voltage v_s is connected in series with a constant bias voltage V_{eb}. The base current is then the sum of a steady current I_b produced by the bias voltage V_{eb} and a time-varying current i_b due to the signal voltage v_s. Because v_s may at any instant be either positive or negative, the bias voltage V_{eb} must be large enough to ensure that there is always a forward bias on the emitter-base junction. The collector current will consist of two parts: a constant direct current $I_c = \beta I_b$ and a time-varyng current $i_c = \beta i_b$. We thus have a current amplifier in which the time-varying output current i_c is β multiplied by the input current i_b. In such an amplifier, the steady currents I_c and I_b, although essential to the operation of the transistor, are usually not of interest. The input signal voltage v_s is related to the base current by Ohm's law:

$$i_b = \frac{v_s}{R_b + r_b} \qquad 38\text{-}32$$

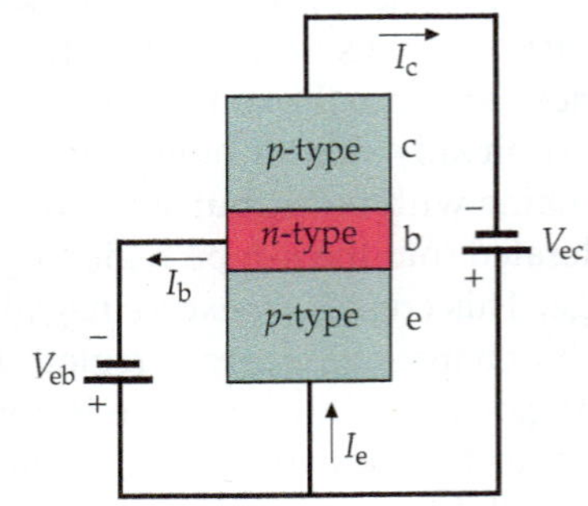

FIGURE 38-27 A *pnp* transistor biased for normal operation. Holes from the emitter can easily diffuse across the base, which is only tens of nanometers thick. Most of the holes flow to the collector, producing the current I_c.

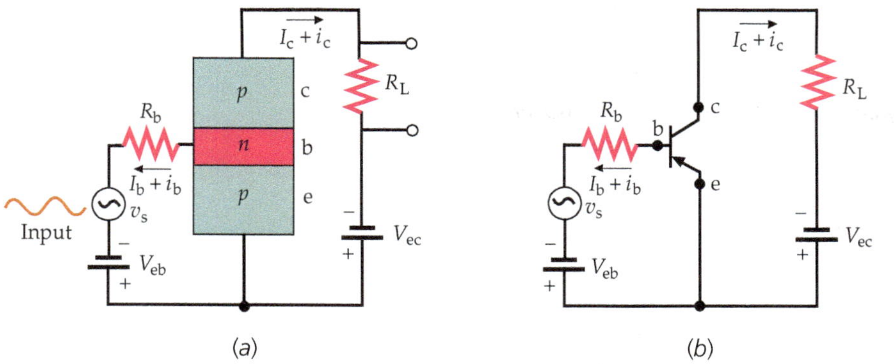

FIGURE 38-28 (*a*) A *pnp* transistor used as an amplifier. A small change i_b in the base current results in a large change i_c in the collector current. Thus, a small signal in the base circuit results in a large signal across the load resistor R_L in the collector circuit. (*b*) The same circuit as in Figure 38-28*a* with the conventional symbol for the transistor.

where r_b is the internal resistance of that part of the transistor between the base and emitter. Similarly, the collector current i_c produces a time-varying voltage v_L across the output or load resistance R_L given by

$$v_L = i_c R_L \qquad 38\text{-}33$$

Using Equation 38-31 and Equation 38-32, we have

$$i_c = \beta i_b = \beta \frac{v_s}{R_b + r_b} \qquad 38\text{-}34$$

The output voltage is thus related to the input voltage by

$$v_L = \beta \frac{v_s}{R_b + r_b} R_L = \beta \frac{R_L}{R_b + r_b} v_s \qquad 38\text{-}35$$

The ratio of the output voltage to the input voltage is the **voltage gain** of the amplifier:

$$\text{Voltage gain} = \frac{v_L}{v_s} = \beta \frac{R_L}{R_b + r_b} \qquad 38\text{-}36$$

A typical amplifier (for example, in a tape player) has several transistors, similar to the one shown in Figure 38-28, connected in series so that the output of one transistor serves as the input for the next. Thus, the very small voltage fluctuations produced by the motion of the magnetic tape past the pickup heads controls the large amounts of power required to drive the loudspeakers. The power delivered to the speakers is supplied by the dc sources connected to each transistor.

The technology of semiconductors extends well beyond individual transistors and diodes. Many of the electronic devices we use every day, such as laptop computers and the processors that govern the operation of vehicles and appliances, rely on large-scale integration of many transistors and other circuit components on a single chip. Large-scale integration combined with advanced concepts in semiconductor theory has created remarkable new instruments for scientific research.

38-8 SUPERCONDUCTIVITY

There are some substances for which the resistivity suddenly drops to zero below a certain temperature T_c, which is called the **critical temperature.** This amazing phenomenon, called **superconductivity,** was discovered in 1911 by the Dutch physicist H. Kamerlingh Onnes, who developed a technique for liquefying helium (boiling point equal to 4.2 K) and used his technique to explore the properties of substances at temperatures in that range. Figure 38-29 shows Onnes's plot of the resistance of mercury versus temperature. The critical temperature for mercury is approximately the same as the boiling point of helium, which is 4.2 K. Critical temperatures for other superconducting elements range from less than 0.1 K for hafnium and iridium to 9.2 K for niobium. The temperature range for superconductors is much higher for a number of metallic compounds. For example, the superconducting alloy Nb_3Ge, discovered in 1973, has a critical temperature of 25 K, which was the highest known until 1986, when the discoveries of J. Georg Bednorz and K. Alexander Müller launched the era of high-temperature superconductors, now defined as materials that exhibit superconductivity at temperatures above 77 K (the temperature at which nitrogen boils). The highest temperature at which superconductivity has been demonstrated, using thallium-doped $HgBa_2Ca_2Cu_3O_8$ + delta, is 138 K at atmospheric pressure. At extremely high pressures, some materials exhibit superconductivity at temperatures as high as 164 K.

The resistivity of a superconductor is zero. There can be a current in a superconductor even when there is no emf in the superconducting circuit. Indeed, in superconducting rings in which there was no electric field, steady currents have been observed to persist for years without apparent loss. Despite the cost and inconvenience of refrigeration using expensive liquid helium, many superconducting magnets have been built using superconducting materials, because such magnets require no power expenditure to maintain the large current needed to produce a large magnetic field.

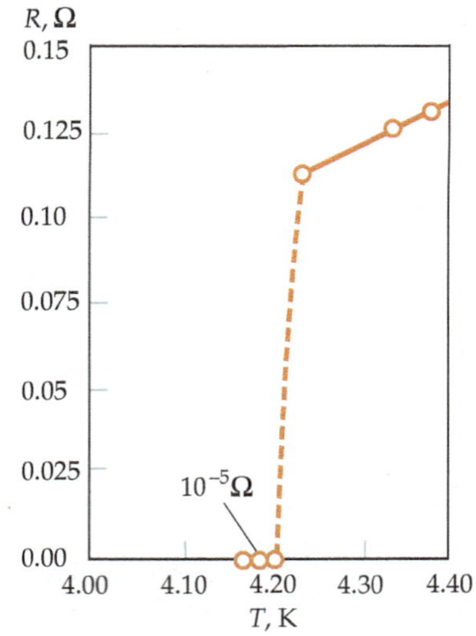

FIGURE 38-29 Plot by H. Kamerlingh Onnes of the resistance of mercury versus temperature, showing the sudden decrease at the critical temperature of $T = 4.2$ K.

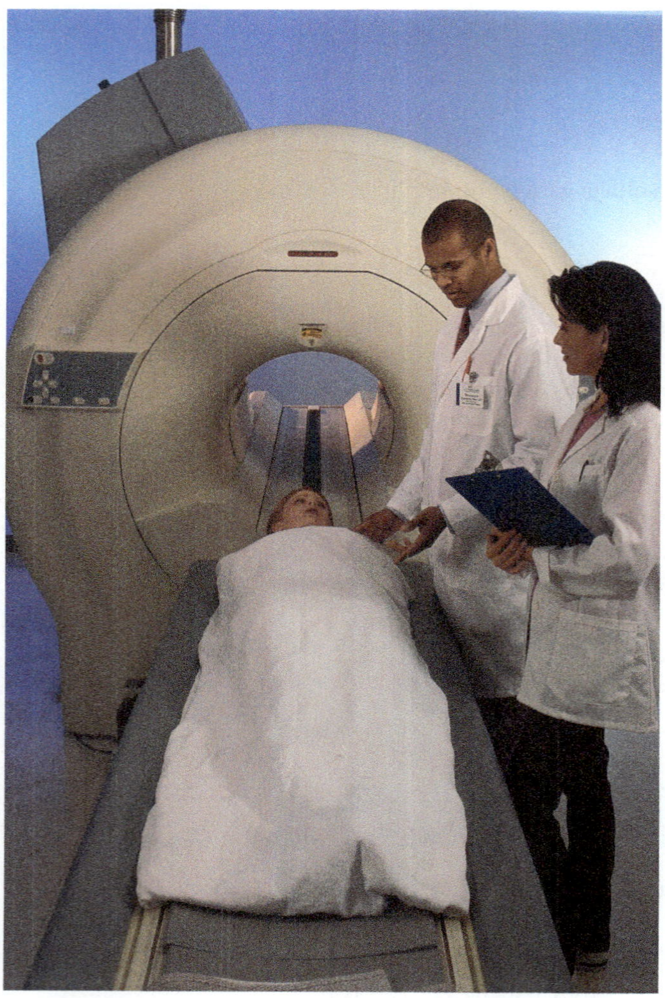

The wires for the magnetic field of a magnetic resonance imaging (MRI) machine carry large currents. To keep the wires from overheating, they are maintained at superconducting temperatures. To accomplish this, they are immersed in liquid helium. *(Corbis.)*

The discovery of high-temperature superconductors has revolutionized the study of superconductivity because relatively inexpensive liquid nitrogen, which boils at 77 K, can be used for a coolant. However, many problems, such as brittleness and the toxicity of the materials, make these new superconductors difficult to use. The search continues for new materials that will be superconductors at even higher temperatures.

THE BCS THEORY

It had been recognized for some time that low temperature superconductivity is due to a collective action of the conducting electrons. In 1957, John Bardeen, Leon Cooper, and Robert Schrieffer published a successful theory of low temperature superconductivity now known by the initials of the inventors as the **BCS theory.** According to this theory, the electrons in a superconductor are coupled in pairs at low temperatures. The coupling comes about because of the interaction between electrons and the crystal lattice. One electron interacts with the lattice and perturbs it. The perturbed lattice interacts with another electron in such a way that there is an attraction between the two electrons that at low temperatures can exceed the Coulomb repulsion between them. The electrons form a bound state called a **Cooper pair.** The electrons in a Cooper pair have equal and opposite spins, so they form a system with zero spin. Each Cooper pair acts as a *single particle* with zero spin, in other words, as a boson. Bosons do not obey the exclusion principle. Any number of Cooper pairs may be in the same quantum state with the same energy. In the ground state of a superconductor (at $T = 0$), all the conduction electrons are in Cooper pairs and all the Cooper pairs are in the same energy state. In the superconducting state, the Cooper pairs are correlated so that they act collectively. An electric current can be produced in a superconductor because all of the electrons in this collective state move together. But energy cannot be dissipated by individual collisions of electron and lattice ions unless the temperature is high enough to break the binding of the Cooper pairs. The required energy is called the *superconducting energy gap* E_g. In the BCS theory, this energy at zero temperature is related to the critical temperature by

$$E_g = \tfrac{7}{2}kT_c \qquad \text{38-37}$$

The energy gap can be determined by measuring the current across a junction between a normal metal and a superconductor as a function of voltage. Consider two metals separated by a layer of insulating material, such as aluminum oxide, that is only a few nanometers thick. The insulating material between the metals forms a barrier that prevents most electrons from traversing the junction. However, waves can tunnel through a barrier if the barrier is not too thick, even if the energy of the wave is less than that of the barrier.

When the materials on either side of the gap are normal nonsuperconducting metals, the current resulting from the tunneling of electrons through the insulating layer obeys Ohm's law for low applied voltages (Figure 38-30*a*). When one of the metals is a normal metal and the other is a superconductor, there is no current (at absolute zero) unless the applied voltage V is greater than a critical voltage $V_c = E_g/(2e)$, where E_g is the superconductor energy gap. Figure 38-30*b* shows the plot of current versus voltage for this situation. The current escalates rapidly when the energy $2eV$ absorbed by a Cooper pair traversing the barrier approaches $E_g = 2eV_c$, the minimum energy needed to break up the pair. (The small current visible in Figure 38-30*b* before the critical voltage is reached is present because at any temperature above absolute zero some of the electrons in the superconductor are thermally excited above the energy gap and are therefore not paired.) At voltages slightly above V_c, the current-versus-voltage curve becomes that for a normal metal. The superconducting energy gap can thus be measured by measuring the average voltage for the transition region.

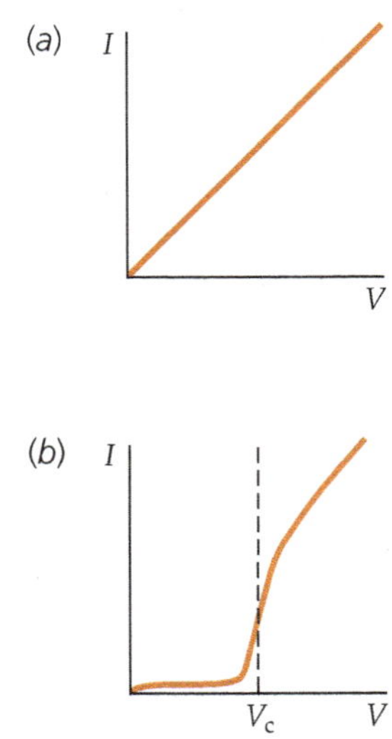

FIGURE 38-30 Tunneling current versus voltage for a junction of two metals separated by a thin oxide layer. (*a*) When both metals are normal metals, the current is proportional to the voltage, as predicted by Ohm's law. (*b*) When one metal is a normal metal and another metal is a superconductor, the current is approximately zero until the applied voltage V approaches the critical voltage $V_c = E_g/(2e)$.

Example 38-8 Superconducting Energy Gap for Mercury

Calculate the superconducting energy gap for mercury ($T_c = 4.2$ K) predicted by the BCS theory.

PICTURE The energy gap is related to the critical temperature by $E_g = 3.5\,kT_c$ (Equation 38-37).

SOLVE

1. The BCS prediction for the energy gap is $\quad E_g = 3.5kT_c$

2. Substitute $T_c = 4.2$ K:

$$E_g = 3.5kT_c$$

$$= 3.5(1.38 \times 10^{-23}\ \text{J/K})(4.2\ \text{K})\left(\frac{1\ \text{ev}}{1.6 \times 10^{-19}\ \text{J}}\right)$$

$$= \boxed{1.3 \times 10^{-3}\ \text{eV}}$$

Note that the energy gap for a typical superconductor is much smaller than the energy gap for a typical semiconductor, which is of the order of 1 eV. As the temperature is increased from $T = 0$, some of the Cooper pairs are broken. Then there are fewer pairs available for each pair to interact with, and the energy gap is reduced until at $T = T_c$ the energy gap is zero (Figure 38-31).

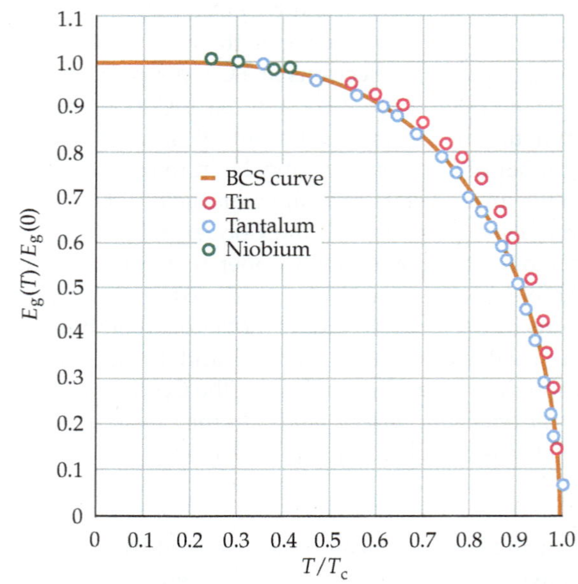

FIGURE 38-31 Ratio of the energy gap at temperature T to that at temperature $T = 0$ as a function of the relative temperature T/T_c. The solid curve is that predicted by the BCS theory.

THE JOSEPHSON EFFECT

When two superconductors are separated by a thin nonsuperconducting barrier (for example, a layer of aluminum oxide a few nanometers thick), the junction is called a **Josephson junction,** based on the prediction in 1962 by Brian Josephson that Cooper pairs could tunnel across such a junction from one superconductor to the other with no resistance. The tunneling of Cooper pairs constitutes a current, which does not require a voltage to be applied across the junction. The current depends on the difference in phase of the wave functions that describe the Cooper pairs. Let ϕ_1 be the phase constant for the wave function of a Cooper pair in one superconductor. All the Cooper pairs in a superconductor act coherently and have the same phase constant. If ϕ_2 is the phase constant for the Cooper pairs in the second superconductor, the current across the junction is given by

$$I = I_{max} \sin(\phi_2 - \phi_1) \qquad 38\text{-}38$$

where I_{max} is the maximum current, which depends on the thickness of the barrier. This result has been observed experimentally and is known as the **dc Josephson effect.**

Josephson also predicted that if a dc voltage V were applied across a Josephson junction, there would be a current that alternates with frequency f given by

$$f = \frac{2e}{h}V \qquad 38\text{-}39$$

This result, known as the **ac Josephson effect,** has been observed experimentally, and careful measurement of the frequency allows a precise determination of the ratio e/h. Because frequency can be measured very accurately, the ac Josephson effect is also used to establish precise voltage standards. The inverse effect, in which the application of an alternating voltage across a Josephson junction results in a dc current, has also been observed.

Example 38-9 Frequency of Josephson Current

Using $e = 1.602 \times 10^{-19}$ C and $h = 6.626 \times 10^{-34}$ J·s, calculate the frequency of the Josephson current if the applied voltage is 1.000 μV.

PICTURE The frequency f is related to the applied voltage V by $hf = 2eV$ (Equation 38-39).

SOLVE

Substitute the given values into Equation 38-39 to calculate f:

$$f = \frac{2e}{h}V = \frac{2(1.602 \times 10^{-19}\text{ C})}{6.626 \times 10^{-34}\text{ J}\cdot\text{s}}(1.000 \times 10^{-6}\text{ V})$$

$$= 4.835 \times 10^{8}\text{ Hz} = \boxed{483.5\text{ MHz}}$$

38-9 THE FERMI–DIRAC DISTRIBUTION

The classical Maxwell–Boltzmann distribution (Equation 17-38) gives the number dN of molecules that have energy E in the range between E and $E + dE$. The number dN is equal to the product of $g(E)\,dE$ where $g(E)$ is the **density of states** (number of energy states in the range dE) and the Boltzmann factor $e^{-E/(kT)}$, which is the probability of a state being occupied. The distribution function for free electrons in a metal is called the **Fermi–Dirac distribution.** The Fermi–Dirac distribution can be written in the same form as the Maxwell–Boltzmann distribution, where the density of states calculated from quantum theory and the Boltzmann factor is replaced by the Fermi factor. Let $n(E)\,dE$ be the number of electrons that have energies between E and $E + dE$. This number is written

$$n(E)dE = f(E)g(E)dE \qquad 38\text{-}40$$

ENERGY DISTRIBUTION FUNCTION

where $g(E)\,dE$ is the number of states that have energies between E and $E + dE$ and $f(E)$ is the probability of a state being occupied, which is the Fermi factor. The density of states in three dimensions is somewhat challenging to calculate, so we just give the result. For electrons in a metal of volume V, the density of states is

$$g(E) = \frac{8\sqrt{2}\pi m_{\text{e}}^{3/2}V}{h^3}E^{1/2} \qquad 38\text{-}41$$

DENSITY OF STATES

As in the classical Maxwell–Boltzmann distribution, the density of states is proportional to $E^{1/2}$.

At $T = 0$, the Fermi factor is given by Equation 38-24:

$$f(E) = \begin{cases} 1 & E < E_{\text{F}} \\ 0 & E > E_{\text{F}} \end{cases}$$

The integral of $n(E)\,dE$ over all energies gives the total number of electrons N. We can derive the equation

$$E_{\text{F}} = \frac{h^2}{8m_{\text{e}}}\left(\frac{3N}{\pi V}\right)^{2/3}$$

The complete Fermi-Dirac distribution function is thus

$$n(E)dE = g(E)f(E)dE = \frac{8\sqrt{2}\pi m_e^{3/2}V}{h^3}E^{1/2}\frac{1}{e^{(E-E_F)/(kT)}+1}dE \qquad 38\text{-}47$$

FERMI–DIRAC DISTRIBUTION

We can see that for those few electrons that have energies much greater than the Fermi energy, the Fermi factor approaches $1/e^{(E-E_F)/(kT)} = e^{(E_F-E)/(kT)} = e^{E_F/(kT)}e^{-E/(kT)}$, which is proportional to $e^{-E/(kT)}$. Thus, the high-energy tail of the Fermi–Dirac energy distribution decreases with increasing E as $e^{-E/(kT)}$, just like the classical Maxwell–Boltzmann energy distribution. The reason for this is that in this high-energy region there are many unoccupied energy states and few electrons, so the exclusion principle is not important. Thus, the Fermi–Dirac distribution approaches the classical Maxwell–Boltzmann distribution in the high-energy limit. This result has practical importance because it applies to the conduction electrons in semiconductors.

Example 38-10 Fermi Factor for Copper at 300 K

At what energy is the Fermi factor equal to 0.100 for copper at $T = 300$ K?

PICTURE We set $f(E) = 0.100$ in Equation 38-46, using $T = 300$ K and $E_F = 7.03$ eV from Table 38-1, and solve for E.

SOLVE

1. Solve Equation 38-46 for $e^{(E-E_F)/(kT)}$:

$$f(E) = \frac{1}{e^{(E-E_F)/(kT)}+1}$$

so

$$e^{(E-E_F)/(kT)} = \frac{1}{f(E)} - 1$$

2. Take the logarithm of both sides:

$$\frac{E-E_F}{kT} = \ln\left[\frac{1}{f(E)} - 1\right]$$

3. Solve for E. For E_F, use the value for E_F at $T = 0$ K listed in Table 38-1:

$$E = E_F + \ln\left[\frac{1}{f(E)} - 1\right]kT$$

$$= 7.03\text{ eV} + \ln\left[\frac{1}{0.100} - 1\right](8.62\times10^{-5}\text{ eV/K})(300\text{ K})$$

$$= \boxed{7.09\text{ eV}}$$

CHECK As expected, the energy is slightly above the Fermi energy when the Fermi factor is equal to 0.100.

Example 38-11 Probability of a Higher Energy State Being Occupied

Find the probability that an energy state in copper 0.100 eV above the Fermi energy is occupied at $T = 300$ K.

PICTURE The probability is the Fermi factor given in Equation 38-46, with $E_F = 7.03$ eV and $E = 7.13$ eV.

SOLVE

1. The probability of an energy state being occupied equals the Fermi factor:

$$P = f(E) = \frac{1}{e^{(E-E_F)/(kT)} + 1}$$

2. Calculate the exponent in the Fermi factor (exponents are always dimensionless):

$$\frac{E - E_F}{kT} = \frac{7.13\text{ eV} - 7.03\text{ eV}}{(8.62 \times 10^{-5}\text{ eV/K})(300\text{ K})} = 3.87$$

3. Use this result to calculate the Fermi factor:

$$f(E) = \frac{1}{e^{(E-E_F)/(kT)} + 1} = \frac{1}{e^{3.87} + 1}$$

$$= \frac{1}{48 + 1} = \boxed{0.020}$$

CHECK The probability that an energy state above the Fermi energy is occupied is less than one-half. As expected, the step 4 result is less than one-half.

TAKING IT FURTHER The probability of an electron having an energy 0.100 eV above the Fermi energy at 300 K is only about 2 percent.

Example 38-12 Probability of a Lower Energy State Being Occupied *Try It Yourself*

Find the probability that an energy state in copper 0.10 eV *below* the Fermi energy is occupied at $T = 300$ K.

PICTURE The probability is the Fermi factor given in Equation 38-46, with $E_F = 7.03$ eV and $E = 6.93$ eV.

SOLVE

Cover the column to the right and try these on your own before looking at the answers.

Steps — **Answers**

1. Write the Fermi factor:

$$f(E) = \frac{1}{e^{(E-E_F)/(kT)} + 1}$$

2. Calculate the exponent in the Fermi factor:

$$\frac{E - E_F}{kT} = \frac{6.93\text{ eV} - 7.03\text{ eV}}{(8.62 \times 10^{-5}\text{ eV/K})(300\text{ K})} = -3.87$$

3. Use your result from step 2 to calculate the Fermi factor:

$$f(E) = \frac{1}{e^{(E-E_F)/(kT)} + 1} = \frac{1}{e^{-3.87} + 1}$$

$$= \frac{1}{0.021 + 1} = \boxed{0.98}$$

CHECK As expected, the step-3 result is greater than one-half.

TAKING IT FURTHER The probability of an electron having an energy of 0.10 eV *below* the Fermi energy at 300 K is approximately 98 percent.

PRACTICE PROBLEM 38-4 What is the probability of an energy state 0.10 eV below the Fermi energy being unoccupied at 300 K?

Summary

TOPIC	RELEVANT EQUATIONS AND REMARKS	
1. The Structure of Solids	Solids are often found in crystalline form in which a small structure, which is called the unit cell, is repeated over and over. A crystal may have a face-centered-cubic, body-centered-cubic, hexagonal close-packed, or other structure depending on the type of bonding between the atoms, ions, or molecules in the crystal and on the relative sizes of the atoms.	
Potential energy	$U = -\alpha \frac{ke^2}{r} + \frac{A}{r^n}$ where r is the center-to-center separation distance between neighboring ions, α is the Madelung constant, which depends on the geometry of the crystal and is of the order of 1.8, and n is approximately 9.	38-3
2. A Microscopic Picture of Conduction		
Resistivity	$\rho = \frac{m_e v_{av}}{n_e e^2 \lambda}$ where v_{av} is the average speed of the electrons and λ is their mean free path between collisions with the lattice ions.	38-14
Mean free path	$\lambda = \frac{vt}{n_{ion}\pi r^2 vt} = \frac{1}{n_{ion}\pi r^2} = \frac{1}{n_{ion}A}$ where n_{ion} is the number of lattice ions per unit volume, r is their effective radius, and A is their effective cross-sectional area.	38-16
3. Classical Interpretation of v_{av} and λ	v_{av} is determined from the Maxwell–Boltzmann distribution, and r is the actual radius of a lattice ion. (This interpretation is not consistent with measured results.)	
4. Quantum Interpretation of v_{av} and λ	v_{av} is determined from the Fermi–Dirac distribution and is approximately constant independent of temperature. The mean free path is determined from the scattering of electron waves, which occurs only because of deviations from a perfectly ordered array. The radius r is the amplitude of vibration of the lattice ion, which is proportional to $\sqrt{T}$, so A is proportional to T.	
5. Free Electrons		
Fermi energy E_F at $T = 0$	E_F is the energy of the last filled (or half-filled) energy state.	
E_F at $T > 0$	E_F is the energy at which the probability of being occupied is $\frac{1}{2}$.	
Approximate magnitude of E_F	E_F is between 5 eV and 10 eV for most metals.	
Dependence of E_F on the number density (N/V) of free electrons	$E_F = \frac{h^2}{8m_e}\left(\frac{3N}{\pi V}\right)^{2/3}$	38-22*a*
Average energy at $T = 0$	$E_{av} = \frac{3}{5}E_F$	38-23
Fermi factor at $T = 0$	The Fermi factor $f(E)$ is the probability of a state being occupied. $f(E) = \begin{cases} 1 & E < E_F \\ 0 & E > E_F \end{cases}$	38-24
Fermi temperature	$T_F = \frac{E_F}{k}$	38-25
Fermi speed	$u_F = \sqrt{\frac{2E_F}{m_e}}$	38-26

TOPIC	RELEVANT EQUATIONS AND REMARKS
Contact potential	When two different metals are placed in contact, electrons flow from the metal with the higher Fermi energy to the metal with the lower Fermi energy until the Fermi energies of the two metals are equal. In equilibrium, there is a potential difference between the metals that is equal to the difference in the work function of the two metals divided by the electronic charge e: $$V_{\text{contact}} = \frac{\phi_1 - \phi_2}{e} \quad \text{38-27}$$
Specific heat due to conduction electrons	$$c'_{\text{V}} = \tfrac{1}{2}\pi^2 R \frac{T}{T_{\text{F}}} \quad \text{38-29}$$
6. Band Theory of Solids	When many atoms are brought together to form a solid, the individual energy levels are split into bands of allowed energies. The splitting depends on the type of bonding and the lattice separation. The highest energy band that contains electrons is called the valence band. (The lowest energy band that is not filled with electrons is called the conduction band.) In a conductor, the valence band is only partially filled, so there are many available empty energy states for excited electrons. In an insulator, the valence band is completely filled and there is a large energy gap between it and the next allowed band, the conduction band. In a semiconductor, the energy gap between the filled valence band and the empty conduction band is small; so, at ordinary temperatures, an appreciable number of electrons are thermally excited into the conduction band.
7. Semiconductors	The conductivity of a semiconductor can be greatly increased by doping. In an n-type semiconductor, the doping adds electrons at energies just below that of the conduction band. In a p-type semiconductor, holes are added at energies just above that of the valence band.
8. *Semiconductor Junctions and Devices	
**pn* Junctions	Semiconductor devices such as diodes and transistors make use of n-type semiconductors and p-type semiconductors. The two types of semiconductors are typically a single silicon crystal doped with donor impurities on one side and acceptor impurities on the other side. The region in which the semiconductor changes from a p-type semiconductor to an n-type semiconductor is called a junction. Junctions are used in diodes, solar cells, surface barrier detectors, LEDs, and transistors.
*Diodes	A diode is a single-junction device that carries current in one direction only.
*Zener diodes	A Zener diode is a diode with a very high reverse bias. It breaks down suddenly at a distinct voltage and can therefore be used as a voltage reference standard.
*Tunnel diodes	A tunnel diode is a diode that is heavily doped so that electrons tunnel through the depletion barrier. At normal operation, a small change in bias voltage results in a large change in current.
*Transistors	A transistor consists of a very thin semiconductor of one type sandwiched between two semiconductors of the opposite type. Transistors are used in amplifiers because a small variation in the base current results in a large variation in the collector current.
9. Superconductivity	In a superconductor, the resistance drops suddenly to zero below a critical temperature T_c. Superconductors with critical temperatures as high as 138 K have been discovered.
The BCS theory	Superconductivity is described by a theory of quantum mechanics called the BCS theory in which the free electrons form Cooper pairs. The energy needed to break up a Cooper pair is called the superconducting energy gap E_g. When all the electrons are paired, individual electrons cannot be scattered by a lattice ion, so the resistance is zero.
Tunneling	When a normal conductor is separated from a superconductor by a thin layer of oxide, electrons can tunnel through the energy barrier if the applied voltage across the layer is $E_g/(2e)$, where E_g is the energy needed to break up a Cooper pair. The energy gap E_g can be determined by a measurement of the tunneling current versus the applied voltage.

TOPIC	RELEVANT EQUATIONS AND REMARKS
Josephson junction	A system of two superconductors separated by a thin layer of nonconducting material is called a Josephson junction.
dc Josephson effect	A dc current is observed to tunnel through a Josephson junction even in the absence of voltage across the junction.
ac Josephson effect	When a dc voltage V is applied across a Josephson junction, an ac current is observed with a frequency $$f = \frac{2e}{h}V \quad \text{38-39}$$ Measurement of the frequency of this current allows a precise determination of the ratio e/h.
10. The Fermi–Dirac Distribution	The number of electrons with energies between E and $E + dE$ is given by $$n(E)\,dE = f(E)\,g(E)\,dE \quad \text{38-40}$$ where $g(E)$ is the density of states and $f(E)$ is the Fermi factor.
Density of states	$$g(E) = \frac{8\sqrt{2}\pi m_e^{3/2}V}{h^3}E^{1/2} \quad \text{38-41}$$
Fermi factor at $T > 0$	$$f(E) = \frac{1}{e^{(E-E_F)/(kT)} + 1} \quad \text{38-46}$$

Answers to Practice Problems

38-1	$E_F = 9.40$ eV
38-2	5.53 eV
38-3	2×10^{-13} electrons/atom
38-4	One minus the probability of the energy state being occupied. That is, $1 - 0.98 = 0.02$ or 2 percent.

Problems

In a few problems, you are given more data than you actually need; in a few other problems, you are required to supply data from your general knowledge, outside sources, or informed estimate.

Interpret as significant all digits in numerical values that have trailing zeros and no decimal points.

- • Single-concept, single-step, relatively easy
- •• Intermediate-level, may require synthesis of concepts
- ••• Challenging
- SSM Solution is in the *Student Solutions Manual*
- Consecutive problems that are shaded are paired problems.

CONCEPTUAL PROBLEMS

1 • In the classical model of conduction, the electron loses energy on average during a collision because it loses the drift velocity it had acquired since the last collision. Where does this energy appear?

2 • A metal is a good conductor because the valence energy band for electrons is (*a*) empty, (*b*) partly filled, (*c*) filled, but there is only a small gap to a higher empty band, (*d*) completely filled, (*e*) none of the above.

3 • Thomas refuses to believe that a potential difference can be created simply by bringing two different metals into contact with each other. John talks him into making a small wager and is about to win the bet. (*a*) Which two metals from Table 38-2 would demonstrate his point most effectively? (*b*) What is the value of that contact potential?

4 • (*a*) In Problem 3, which choices of different metals would make the least impressive demonstration? (*b*) What is the value of that contact potential?

5 • When a sample of pure copper is cooled from 300 K to 4 K, its resistivity decreases more than the resistivity of a sample of brass when it is cooled through the same temperature difference. Why? SSM

6 • Insulators are poor conductors of electricity because (*a*) there is a small energy gap between the filled valence band and the next higher band where electrons can exist, (*b*) there is a large energy gap between the completely filled valence band and the next higher band where electrons can exist, (*c*) the valence band has a few vacancies for electrons, (*d*) the valence band is only partly filled, (*e*) none of the above.

7 • How does the sign of the change in the resistivity of a sample of copper compare with the sign of the change in the resistivity of a sample of silicon when the temperatures of both samples increase? SSM

8 • True or false:

(*a*) Solids that are good electrical conductors are usually good heat conductors.

(*b*) At $T = 0$, an intrinsic semiconductor is an insulator.

(*c*) The Fermi energy is the average energy of an electron in a solid.
(*d*) At $T = 0$, the value of the Fermi factor can be either 1 or 0.
(*e*) Semiconductors conduct current in one direction only.
(*f*) The classical free-electron theory adequately explains the heat capacity of metals.
(*g*) The contact potential between two metals is proportional to the difference in the work functions of the two metals.

9 • Which of the following elements are most likely to act as acceptor impurities in germanium? (*a*) bromine, (*b*) gallium, (*c*) silicon, (*d*) phosphorus, (*e*) magnesium

10 • Which of the following elements are most likely to serve as donor impurities in germanium? (*a*) bromine, (*b*) gallium, (*c*) silicon, (*d*) phosphorus, (*e*) magnesium

11 • An electron hole is created when a photon is absorbed by a semiconductor. How does this hole enable the semiconductor to conduct electricity?

12 • Examine the positions of phosphorus, boron, thallium, and antimony in Table 36-1. (*a*) Which of these elements can be used to dope silicon to create an *n*-type semiconductor? (*b*) Which of these elements can be used to dope silicon to create a *p*-type semiconductor?

13 • When photons of visible light strike the *p*-type semiconductor in a *pn* junction solar cell, (*a*) only free electrons are created, (*b*) only positive holes are created, (*c*) both electrons and holes are created, (*d*) protons are created, (*e*) none of the above.

ESTIMATION AND APPROXIMATION

14 • The ratio of the resistivity of the most resistive (least conductive) material to that of the least resistive material (excluding superconductors) is approximately 10^{24}. You can develop a feeling for how remarkable this range is by considering what the ratio is of the largest values to smallest values of other material properties. Choose any three properties of materials, and using tables in this book or some other resource, calculate the ratio of the largest instance of the property to the smallest instance of that property (other than zero) and rank these in decreasing order. Can you find any other property that shows a range as large as that of electrical resistivity?

15 • A device is said to be "ohmic" if a graph of current versus applied voltage is a straight line through the origin. The resistance R of the device is the reciprocal of the slope of this line. A *pn* junction is an example of a nonohmic device, as may be seen from Figure 38-21. For nonohmic devices, it is sometimes convenient to define the *differential resistance* as the reciprocal of the slope of the I-versus-V curve. Using the curve in Figure 38-21, estimate the differential resistance of the *pn* junction at applied voltages of -20 V, $+0.2$ V, $+0.4$ V, $+0.6$ V, and $+0.8$ V.

THE STRUCTURE OF SOLIDS

16 • Calculate the center-to-center separation distance r_0 between the K^+ and the Cl^- ions in KCl. Do this by assuming that each ion occupies a cubic volume of side r_0. The molar mass of KCl is 74.55 g/mol and its density is 1.984 g/cm^3.

17 • The center-to-center separation distance between the Li^+ and Cl^- ions in LiCl is 0.257 nm. Use that value and the molar mass of LiCl (42.4 g/mol) to compute the density of LiCl.

18 • Find the value of n in Equation 38-6 that gives the measured dissociation energy of 741 kJ/mol for LiCl, which has the same structure as NaCl and for which $r_0 = 0.257$ nm.

19 •• (*a*) Use Equation 38-6 and calculate $U(r_0)$ for calcium oxide, CaO, where $r_0 = 0.208$ nm. Assume $n = 8$. (*b*) If n increases from 8 to 10, what is the fractional change in $U(r_0)$?

A MICROSCOPIC PICTURE OF CONDUCTION

20 • A measure of the density of the free electrons in a metal is the distance r_s, which is defined as the radius of the sphere whose volume equals the volume per conduction electron. (*a*) Show that $r_s = [3/(4\pi n)]^{1/3}$, where n is the free-electron number density. (*b*) Calculate r_s for copper in nanometers.

21 • (*a*) Given a mean free path $\lambda = 0.400$ nm and a mean speed $v_{av} = 1.17 \times 10^5$ m/s for the charge flow in copper at a temperature of 300 K, calculate the classical value for the resistivity ρ of copper. (*b*) The classical model suggests that the mean free path is temperature independent and that v_{av} depends on temperature. According to this model, what would ρ be at 100 K?

FREE ELECTRONS IN A SOLID

22 •• Silicon has a molar mass of 28.09 g/mol and a density of 2.41×10^3 kg/m^3. Each atom of silicon has four valence electrons and the Fermi energy of the material is 4.88 eV. (*a*) Given that the electron mean free path at room temperature is $\lambda = 27.0$ nm, estimate the resistivity. (*b*) The accepted value for the resistivity of silicon is 640 $\Omega \cdot$m (at room temperature). How does this accepted value compare to the value calculated in Part (*a*)?

23 • Calculate the number density of free electrons in (*a*) Ag ($\rho = 10.5$ g/cm^3) and (*b*) Au ($\rho = 19.3$ g/cm^3), assuming one free electron per atom, and compare your results with the values listed in Table 38-1.

24 • The density of aluminum is 2.7 g/cm^3. How many free electrons are present per aluminum atom?

25 • The density of tin is 7.3 g/cm^3. How many free electrons are present per tin atom?

26 • Calculate the Fermi temperature for (*a*) Mg, (*b*) Mn, and (*c*) Zn.

27 • What is the speed of a conduction electron whose energy is equal to the Fermi energy E_F for (*a*) Na, (*b*) Au, and (*c*) Sn?

28 • Calculate the Fermi energy for (*a*) Al, (*b*) K, and (*c*) Sn using the number densities given in Table 38-1.

29 • Find the average energy of the conduction electrons at $T = 0$ in (*a*) copper and (*b*) lithium.

30 • Calculate (*a*) the Fermi temperature and (*b*) the Fermi energy at $T = 0$ for iron.

31 •• (*a*) Assuming that each gold atom in a sample of gold metal contributes one free electron, calculate the free-electron density in gold knowing that its atomic mass is 196.97 g/mol and its density is 19.3×10^3 kg/m^3. (*b*) If the Fermi speed for gold is 1.39×10^6 m/s, what is the Fermi energy in electron volts? (*c*) By what factor is the Fermi energy higher than the kT energy at room temperature? (*d*) Explain the difference between the Fermi energy and the kT energy. SSM

32 •• The bulk modulus B of a material can be defined by $B = -V\partial P/\partial V$. (*a*) Use the monatomic ideal-gas relation $PV = \frac{2}{3}NE_{av}$, where E_{av} is the average kinetic energy, Equation 38-22 and Equation 38-23 to show that $P = \frac{2}{5}NE_F/V = CV^{-5/3}$, where C is a constant independent of V. (*b*) Show that the bulk modulus of the free electrons in a solid metal is therefore $B = \frac{5}{3}P = \frac{2}{3}NE_F/V$. (*c*) Compute the bulk modulus in newtons per square meter for the free electrons in a sample of copper and compare your result with the measured value of 140×10^9 N/m^2.

33 •• The pressure of a monatomic ideal gas is related to the average kinetic energy of the gas particles by $PV = \frac{2}{3}NE_{av}$, where n is the number of particles and E_{av} is the average kinetic energy. Use

this information to calculate the pressure of the free electrons in a sample of copper in newtons per square meter, and compare your result with atmospheric pressure, which is about 10^5 N/m^2. (*Note:* The units are most easily handled by using the conversion factors 1 N/m^2 = 1 J/m^3 and 1 eV = 1.602×10^{-19} J.) SSM

34 • Calculate the contact potential between (*a*) Ag and Cu, (*b*) Ag and Ni, and (*c*) Ca and Cu.

HEAT CAPACITY DUE TO ELECTRONS IN A METAL

35 •• Gold has a Fermi energy of 5.53 eV. Determine the molar specific heat at constant volume for gold at room temperature. SSM

QUANTUM THEORY OF ELECTRICAL CONDUCTION

36 • The resistivities and Fermi speeds of Na, Au, and Sn at $T = 273$ K are 4.2 $\mu\Omega\cdot$cm, 2.04 $\mu\Omega\cdot$cm, and 10.6 $\mu\Omega\cdot$cm, and 1.07×10^6 m/s, 1.39×10^6 m/s, and 1.89×10^6 m/s, respectively. Use those values to find the mean free paths for the conduction electrons in these elements.

37 •• The resistivity of pure copper increases by approximately $1.0 \times 10^{-8}\ \Omega\cdot$m with the addition of 1.0 percent (by number of atoms) of an impurity distributed throughout the metal. The mean free path λ depends on both the impurity and the oscillations of the lattice ions according to the equation $1/\lambda = 1/\lambda_t + 1/\lambda_i$, where λ_t is the mean free path associated with the thermal vibrations of the ions and λ_i is the mean free path associated with the impurities. (*a*) Estimate λ_i using Equation 38-14 and the data given in Table 38-1. (*b*) If r is the effective radius of an impurity lattice ion seen by an electron, the scattering cross section is πr^2. Estimate this area, using the fact that r is related to λ_i by Equation 38-16. SSM

BAND THEORY OF SOLIDS

38 • Electromagnetic radiation is incident on the surface of a semiconductor. The maximum wavelength of this light that is required if electrons are to cross the energy gap between the valence band and the conduction band is 380.0 nm. What is the energy gap, in electron volts, for the semiconductor?

39 • An electron occupies the highest energy level of the valence band in a silicon sample. What is the maximum photon wavelength that will excite the electron across the energy gap if the gap is 1.14 eV? SSM

40 • An electron occupies the highest energy level of the valence band in a germanium sample. What is the maximum photon wavelength that will excite the electron into the conduction band? In germanium, the energy gap between the valence and conduction bands is 0.74 eV.

41 • An electron occupies the highest energy level of the valence band in a diamond sample. What is the maximum photon wavelength that will excite the electron into the conduction band? In diamond, the energy gap between the valence and conduction bands is 7.0 eV.

42 •• A photon of wavelength 3.35 μm has just enough energy to raise an electron from the valence band to the conduction band in a lead sulfide sample. (*a*) Find the energy gap between the bands in lead sulfide. (*b*) Find the temperature T for which kT equals that energy gap.

SEMICONDUCTORS

43 • The donor energy levels in an n-type semiconductor are 0.0100 eV below the conduction band. Find the temperature for which $kT = 0.0100$ eV.

44 •• When a thin slab of semiconducting material is illuminated with monochromatic electromagnetic radiation, most of the radiation is transmitted through the slab if the wavelength is greater than 1.85 mm. For wavelengths less than 1.85 mm, most of the incident radiation is absorbed. Determine the energy gap of the semiconductor.

45 •• The relative binding of the extra electron in the arsenic atom that replaces an atom in silicon or germanium can be understood from a calculation of the first Bohr radius of the electron in these materials. Four of arsenic's valence electrons form covalent bonds, so the fifth electron sees a center of attraction with a charge of $+e$. This model is a modified hydrogen atom. In the Bohr model of the hydrogen atom, the electron moves in free space at a radius a_0 given by $a_0 = 4\pi\epsilon_0\hbar^2/(m_e e^2)$ (Equation 36-12). When an electron moves in a crystal, we can approximate the effect of the other atoms by replacing ϵ_0 with $\kappa\epsilon_0$ and m_e with an effective mass for the electron. For silicon, κ is 12 and the effective mass is approximately $0.2m_e$. For germanium, κ is 16 and the effective mass is approximately $0.1m_e$. Estimate the Bohr radii for the valence electron as it orbits the impurity arsenic atom in silicon and in germanium.

46 •• The ground-state energy of the hydrogen atom is given by $E_1 = -m_e e^4/(8\epsilon_0^2 h^2)$ (Equations 36-15 and 36-16 where $4\pi\epsilon_0$ is substituted for k^{-1}). Modify this equation using information in Problem 45 by replacing ϵ_0 with $\kappa\epsilon_0$ and m_e with an effective mass for the electron to estimate the binding energy of the extra electron of an impurity arsenic atom in (*a*) silicon and (*b*) germanium.

47 •• A doped n-type silicon sample has 1.00×10^{16} electrons per cubic centimeter in the conduction band and has a resistivity of $5.00 \times 10^{23}\ \Omega\cdot$m at 300 K. Find the mean free path of the electrons. Use the effective mass of $0.2m_e$ for the mass of the electrons. (See Problem 45.) Compare this mean free path with that of conduction electrons in copper at 300 K.

48 ••• In the Hall effect, the Hall coefficient R_H is the proportionality constant between the transverse electric field and the product of the applied magnetic field and the current density. That is, $E_y = R_H B_z J_x$, where the current density, the transverse electric field, and the applied magnetic field are in the $+x$, $-y$, and $+z$ directions, respectively. (The Hall effect is presented in Chapter 26.) The measured Hall coefficient of a doped silicon sample is 0.0400 V·m/(A·T) at room temperature. If all the doping impurities have contributed to the total number of charge carriers of the sample, find (*a*) the type of impurity (donor or acceptor) used to dope the sample and (*b*) the concentration of the impurities.

*SEMICONDUCTOR JUNCTIONS AND DEVICES

49 •• Simple theory for the current versus the bias voltage across a pn junction yields the equation $I = I_0(e^{eV_b/kT} - 1)$. Sketch I versus V_b for both positive and negative values of V_b using that equation.

50 • The base current in an npn transistor circuit is 25.0 mA. If 88.0 percent of the electrons entering the base from the emitter reach the collector, what is the base current?

51 •• In Figure 38-28 for the npn-transistor amplifier, suppose $R_b = 2.00$ kΩ and $R_L = 10.0$ kΩ. Suppose further that a 10.0-μA ac base current i_b generates a 0.500-mA ac collector current i_c. What is the voltage gain of the amplifier? SSM

52 •• Germanium can be used to measure the energy of incident photons. Consider a 660-keV gamma ray emitted from ^{137}Cs. (*a*) Given that the band gap in germanium is 0.72 eV, how many electron-hole pairs can be generated as this gamma ray travels through germanium? (*b*) The number N of pairs in Part (*a*) will have statistical fluctuations given by $\pm\sqrt{N}$. What then is the energy resolution of the detector in that photon energy region?

53 •• Make a sketch showing the valence and conduction band edges and Fermi energy of a *pn*-junction diode when biased (*a*) in the forward direction and (*b*) in the reverse direction.

54 •• A good silicon diode has the current-voltage characteristic given by $I = I_0(e^{eV_b/kT} - 1)$. Let $kT = 0.025$ eV (room temperature) and the saturation current $I_0 = 1.0$ nA. (*a*) Show that for small reverse-bias voltages, the resistance is 25 MΩ. *Hint: Do a Taylor-series expansion of the exponential function about* $V_b = 0$, *or use the expansion for* e^x *found in Table M-4 of the Math Tutorial.* (*b*) Find the dc resistance for a reverse bias of 0.50 V. (*c*) Find the resistance V/I for a 0.50-V forward bias. What is the current in this case? (*d*) Calculate the differential resistance dV/dI for a 0.50-V forward bias.

55 •• A long slab of silicon of thickness $T = 1.0$ mm and width $w = 1.0$ cm is placed in a magnetic field $B = 0.40$ T. The slab is in the xy plane, where the length of the slab is parallel with the x axis and the magnetic field points in the $+z$ direction. When a current of 0.20 A exists in the sample in the $+x$ direction, a potential difference of 5.0 mV develops across the width of the sample and the electric field in the sample points in the $+y$ direction. Determine the semiconductor type (n or p) and the concentration of charge carriers. (The Hall effect is presented in Chapter 26.)

THE BCS THEORY

56 • (*a*) Use Equation 38-37 to calculate the superconducting energy gap for tin, and compare your result with the measured value of 6.00×10^{-4} eV. (*b*) Use the measured value to calculate the minimum value of the wavelength of a photon that has sufficient energy to break up Cooper pairs in lead ($T_c = 3.72$ K) at $T = 0$.

57 • (*a*) Use Equation 38-37 to calculate the superconducting energy gap for lead, and compare your result with the measured value of 2.73×10^{-3} eV. (*b*) Use the measured value to calculate the minimum value of the wavelength of a photon that has sufficient energy to break up Cooper pairs in tin ($T_c = 7.19$ K) at $T = 0$. SSM

THE FERMI-DIRAC DISTRIBUTION

58 •• The number of electrons in the conduction band of an insulator or intrinsic semiconductor is governed chiefly by the Fermi factor. Because the valence band in these materials is nearly filled and the conduction band is nearly empty, the Fermi energy E_F is generally midway between the top of the valence band and the bottom of the conduction band; that is, it is at $E_g/2$, where E_g is the band gap between the two bands and the energy is measured from the top of the valence band. (*a*) In silicon, $E_g \approx 1.0$ eV. Show that in this case the Fermi factor for electrons at the bottom of the conduction band is given by $\exp(-E_g/2kT)$ and evaluate this factor. Discuss the significance of the result if there are 10^{22} valence electrons per cubic centimeter and the probability of finding an electron in the conduction band is given by the Fermi factor. (*b*) Repeat the calculation in Part (*a*) for an insulator with a band gap of 6.0 eV.

59 •• Approximately how many energy states that have energies between 2.00 eV and 2.20 eV are available to electrons in a cube of silver measuring 1.00 mm on a side?

60 •• Show that at $E = E_F$, the expression for the Fermi factor (Equation 38-24) is equal to 0.5.

61 •• (*a*) Using the equation $E_F = [h^2/(8m_e)][3N/(\pi V)]^{2/3}$ (Equation 38-22*a*), calculate the Fermi energy for silver. (*b*) Determine the average kinetic energy of a free electron and (*c*) find the Fermi speed for silver. SSM

62 •• What is the difference between the energies at which the Fermi factor is 0.9 and 0.1 at 300 K in (*a*) copper, (*b*) potassium, and (*c*) aluminum.

63 •• What is the probability that a conduction electron in silver will have a kinetic energy of 4.90 eV at $T = 300$ K? SSM

64 •• Show that $g(E) = \frac{3}{2}NE_F^{-3/2}E^{1/2}$ (Equation 38-43) follows from Equation 38-41 for $g(E)$, and from Equation 38-22*a* for E_F.

65 •• Carry out the integration $E_{av} = (1/N)\int_0^{E_F} Eg(E)dE$ to show that the average energy at $T = 0$ is $\frac{3}{5}E_F$.

66 •• The density of the electron states in a metal can be written $g(E) = AE^{1/2}$, where A is a constant and E is measured from the bottom of the conduction band. (*a*) Show that the total number of states is $\frac{2}{3}AE_F^{3/2}$. (*b*) Approximately what fraction of the conduction electrons are within kT of the Fermi energy? (*c*) Evaluate that fraction for copper at $T = 300$ K.

67 •• What is the probability that a conduction electron in silver will have a kinetic energy of 5.49 eV at $T = 300$ K?

68 •• Using the expression $g(E) = (8\sqrt{2}\pi m_e^{3/2}V/h^3)E^{1/2}$ (Equation 38-41) for the density of states, estimate the fraction of the conduction electrons in copper that can absorb energy from collisions with the vibrating lattice ions at (*a*) 77 K and (*b*) 300 K.

69 •• In an intrinsic semiconductor, the Fermi energy is about midway between the top of the valence band and the bottom of the conduction band. In germanium, the forbidden energy band has a width of 0.70 eV. Show that at room temperature the distribution function of electrons in the conduction band is given by the Maxwell–Boltzmann distribution function.

70 ••• The root-mean-square (rms) value of a variable is obtained by calculating the average value of the square of that variable and then taking the square root of the result. Use that procedure to determine the rms energy of a Fermi distribution. Express your result in terms of E_F and compare it to the average energy. Why do E_{av} and E_{rms} differ?

GENERAL PROBLEMS

71 • The density of potassium is 0.851 g/cm³. How many free electrons are there per potassium atom in a crystal of potassium?

72 • Calculate the number density of free electrons for (*a*) magnesium, which has a density of 1.74 g/cm³, and (*b*) zinc, which has a density of 7.14 g/cm³. For the calculations assume there are two free electrons per atom, and compare your results with the values listed in Table 38-1.

73 •• Estimate the fraction of free electrons in copper that are in energy states above the Fermi energy at (*a*) 300 K (about room temperature) and (*b*) 1000 K.

74 •• A certain free-electron energy state of manganese has a 10.0 percent chance of being occupied when the temperature of the manganese is $T = 1300$ K. What is the energy of the state?

75 •• The semiconducting compound CdSe is widely used for light-emitting diodes (LEDs). The energy gap in CdSe is 1.80 eV. What is the frequency of the light emitted by a CdSe LED?

76 ••• A 2.00-cm² wafer of pure silicon is irradiated with electromagnetic radiation having a wavelength of 775 nm. The intensity of the radiation is 4.00 W/m² and every photon that strikes the sample is absorbed and creates an electron-hole pair. (*a*) How many electron-hole pairs are produced in one second? (*b*) If the number of electron-hole pairs in the sample is 6.25×10^{11} in the steady state, at what rate do the electron-hole pairs recombine? (*c*) If every recombination event results in the radiation of one photon, at what rate is energy radiated by the sample?

CHAPTER 39

THE ANDROMEDA GALAXY BY MEASURING THE FREQUENCY OF THE LIGHT COMING TO US FROM DISTANT OBJECTS, WE ARE ABLE TO DETERMINE HOW FAST THE OBJECTS ARE APPROACHING TOWARD US OR RECEDING FROM US. *(NASA.)*

? Have you wondered how the frequency of the light enables us to determine the speed of recession of a distant galaxy? (See Example 39-5.)

Relativity

The theory of relativity consists of two rather different theories, the special theory and the general theory. The special theory, developed by Albert Einstein and others in 1905, describes measurements made in different inertial reference frames moving with constant velocity relative to one another. Its consequences, which can be derived with a minimum of mathematics, are applicable in a wide variety of situations encountered in physics and in engineering. On the other hand, the general theory, also developed by Einstein and others around 1916, describes accelerated reference frames and gravity. A thorough understanding of the general theory requires sophisticated mathematics, and the applications of the theory are mainly in the area of gravitation. The general theory is of great importance in cosmology, but it is rarely encountered in other areas of physics or in engineering. The general theory is applied, however, in the engineering of the Global Positioning System (GPS).*

* The satellites used in GPS contain atomic clocks.

In this chapter, we concentrate on the special theory (often referred to as special relativity*). General relativity theory (general relativity) will be discussed briefly near the end of the chapter. Special relativity is first presented in Chapter R (which precedes Chapter 11). You should consider reviewing the material in Chapter R before proceeding in this chapter.*

39-1 NEWTONIAN RELATIVITY

Newton's first law does not distinguish between a particle at rest and a particle moving with constant velocity. If there is no net external force acting, the particle will remain in its initial state, either at rest or moving with its initial velocity. A particle at rest relative to you is moving with constant velocity relative to an observer who is moving with constant velocity relative to you. How might we distinguish whether you and the particle are at rest and the second observer is moving with constant velocity, or the second observer is at rest and you and the particle are moving?

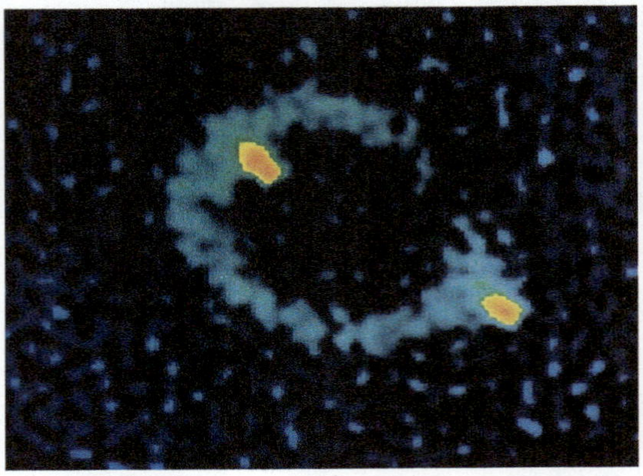

The ringlike structure of the radio source MG1131 + 0456 is thought to be due to *gravitational lensing*, first proposed by Albert Einstein in 1936, in which a source is imaged into a ring by a large, massive object in the foreground. *(NRAO/AUI.)*

Let us consider some simple experiments. Suppose we have a railway boxcar moving along a straight, flat track with a constant velocity v. (The velocity v is a signed quantity, and the sign indicates the direction of the motion of the boxcar along the track.) We note that a ball at rest in the boxcar remains at rest relative to the car. If we drop the ball, it falls straight down, relative to the boxcar, with an acceleration g due to gravity. Of course, when viewed from the reference frame of the track, the ball moves along a parabolic path because it has an initial velocity v to the right. No mechanics experiment that we can do—measuring the period of a pendulum, observing the collisions between two objects, or whatever—will tell us whether the boxcar is moving and the track is at rest or the track is moving and the boxcar is at rest. If we have a coordinate system attached to the track and another attached to the boxcar, Newton's laws hold in either system.

A set of coordinate systems at rest relative to each other is called a *reference frame*. A reference frame in which Newton's laws hold is called an *inertial reference frame*.* All reference frames moving at constant velocity relative to an inertial reference frame are also inertial reference frames. If we have two inertial reference frames moving with constant velocity relative to each other, there are no mechanics experiments that can tell us which is at rest and which is moving or if they are both moving. This result is known as the principle of **Newtonian relativity:**

> Absolute motion cannot be detected.
>
> PRINCIPLE OF NEWTONIAN RELATIVITY

This principle was well known by Galileo, Newton, and others in the seventeenth century. By the late nineteenth century, however, this view had changed. It was then generally thought that Newtonian relativity was not valid and that absolute motion could be detected in principle by a measurement of the speed of light.

ETHER AND THE SPEED OF LIGHT

We saw in Chapter 15 that the velocity of a wave depends on the properties of the medium in which the wave travels and not on the velocity of the source of the waves. For example, the velocity of sound relative to still air depends on the temperature of the air. Light and other electromagnetic waves (for example, radio

* Reference frames were first discussed in Section 3-1. Inertial reference frames were also discussed in Section 4-1.

waves and X rays) travel through a vacuum with a speed $c = 3.00 \times 10^8$ m/s that is predicted by James Clerk Maxwell's equations for electricity and magnetism. But what is this speed relative to? What is the equivalent of still air for a vacuum? A proposed medium for the propagation of light was called the *ether;* it was thought to pervade all space. The velocity of light relative to the ether was assumed to be c, as predicted by Maxwell's equations. The velocity of any object relative to the ether was considered to be the absolute velocity of the object.

Albert Michelson, first in 1881 and then again with Edward Morley in 1887, set out to measure the velocity of Earth relative to the ether by an ingenious experiment in which the velocity of light relative to Earth was compared for two light beams, one parallel to the direction of Earth's motion relative to the Sun and the other perpendicular to the direction of Earth's motion. Despite painstakingly careful measurements, they could detect no difference. The experiment has since been repeated under various conditions by a number of people, and no difference has ever been found. The absolute motion of Earth relative to the ether cannot be detected.

39-2 EINSTEIN'S POSTULATES

In 1905, at the age of 26, Albert Einstein published a paper on the electrodynamics of moving bodies.* In this paper, he postulated that absolute motion cannot be detected by any experiment. That is, there is no ether. Earth can be considered to be at rest and the velocity of light will be the same in any direction.† His theory of special relativity can be derived from two postulates. Simply stated, these postulates are as follows:

> Postulate 1: Absolute uniform motion cannot be detected.
>
> Postulate 2: The speed of light is independent of the motion of the source.
>
> EINSTEIN'S POSTULATES

Postulate 1 is merely an extension of the Newtonian principle of relativity to include all types of physical measurements (not just those measurements that are mechanical). Postulate 2 describes a common property of many waves. For example, the speed of sound waves does not depend on the motion of the sound source. The sound waves from a car horn travel through the air with the same speed, relative to the air, independent of whether the car is moving relative to the air or not. The speed of the waves depends only on the properties of the air, such as its temperature.

Although each postulate seems quite reasonable, many of the implications of the two postulates together are quite surprising and contradict what is often called common sense. For example, one important implication of these postulates is that every observer measures the same value for the speed of light independent of the relative motion of the source and the observer. Consider a light source S and two observers, R_1 at rest relative to S and R_2 moving toward S with speed v, as shown in Figure 39-1a. The speed of light measured by R_1 is $c = 3.00 \times 10^8$ m/s. What is the speed measured by R_2? The answer is *not* $c + v$. By postulate 1, Figure 39-1a is equivalent to Figure 39-1b, in which R_2 is at rest and the source S and R_1 are

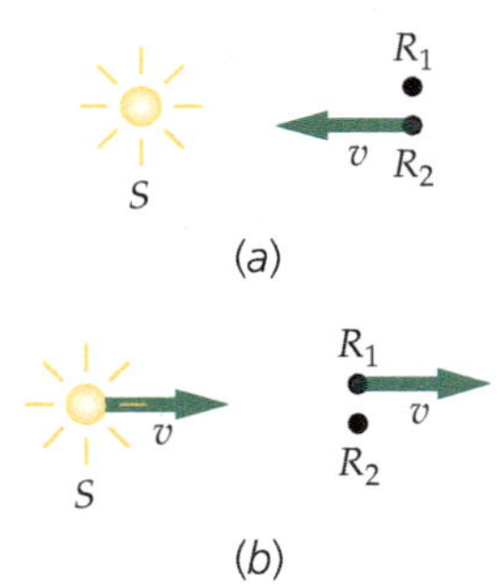

FIGURE 39-1 (a) A stationary light source S, a stationary observer R_1, and a second observer R_2 moving toward the source with speed v. (b) In the reference frame in which the observer R_2 is at rest, the light source S and observer R_1 move to the right with speed v. If absolute motion cannot be detected, the two views are equivalent. Because the speed of light does not depend on the motion of the source, observer R_2 measures the same value for that speed as observer R_1.

* *Annalen der Physik,* vol. 17, 1905, p. 841. For a translation from the original German, see W. Perrett and G. B. Jeffery (trans.), *The Principle of Relativity: A Collection of Original Memoirs on the Special and General Theory of Relativity* by H. A. Lorentz, A. Einstein, H. Minkowski, and W. Weyl, Dover, New York, 1923.

† Einstein did not set out to explain the results of the Michelson–Morley experiment. His theory arose from his considerations of the theory of electricity and magnetism and the unusual property of electromagnetic waves that they propagate in a vacuum. In his first paper, which contains the complete theory of special relativity, he made only a passing reference to the Michelson–Morley experiment, and in later years he could not recall whether he was aware of the details of the experiment before he published his theory.

moving with speed v. That is, because absolute motion cannot be detected, it is not possible to say which is really moving and which is at rest. By postulate 2, the speed of light from a moving source is independent of the motion of the source. Thus, looking at Figure 39-1*b*, we see that R_2 measures the speed of light to be c, just as R_1 does. This result is often considered as an alternative to Einstein's second postulate:

> Postulate 2 (alternate): Every observer measures the same value c for the speed of light.

This result contradicts our intuitive ideas about relative velocities. If a car moves at 50 km/h away from an observer and another car moves at 80 km/h in the same direction, the velocity of the second car relative to the first car is 30 km/h. This result is easily measured and conforms to our intuition. However, according to Einstein's postulates, if a light beam is moving in the direction of the cars, observers in both cars will measure the same speed for the light beam. Our intuitive ideas about the combination of velocities are approximations that hold only when the speeds are very small compared with the speed of light. Even in an airplane moving with the speed of sound, to measure the speed of light accurately enough to distinguish the difference between the results c and $c + v$, where v is the speed of the plane, would require a measurement with six-digit accuracy.

39-3 THE LORENTZ TRANSFORMATION

Einstein's postulates have important consequences for measuring time intervals and space intervals, as well as relative velocities. Throughout this chapter, we will be comparing measurements of the positions and times of events (such as lightning flashes) made by observers who are moving relative to each other. We will use a rectangular coordinate system xyz that has an origin O and is called the S reference frame; we will use another system $x'y'z'$ that has an origin O', is called the S' frame, and is moving with a constant velocity $\vec{v}$ relative to the S frame. Relative to the S' frame, the S frame is moving with a constant velocity $-\vec{v}$. For simplicity, we will consider the S' frame to be moving along the x axis in the $+x$ direction relative to S, where the $+x'$ direction is the same as the $+x$ direction. In each frame, we will assume that there are as many observers as are needed who have measuring devices, such as clocks and metersticks, that are identical when compared at rest (see Figure 39-2).

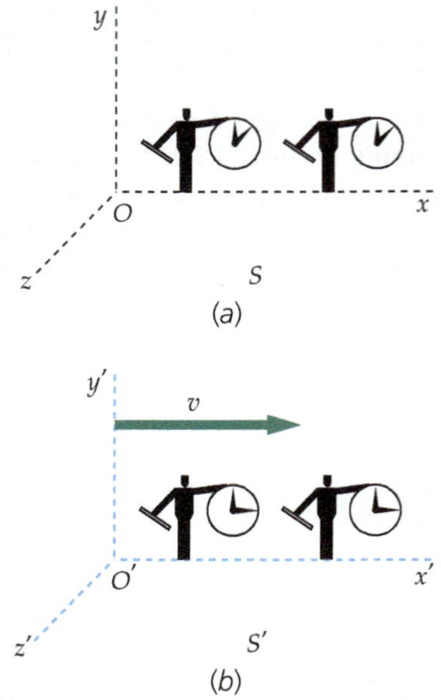

FIGURE 39-2 Coordinate reference frames S and S' moving with relative speed v. In each frame, there are observers who have metersticks and clocks that are identical when compared at rest.

We will use Einstein's postulates to find the general relation between the coordinates x, y, and z and the time t of an event as seen in reference frame S and the coordinates x', y', and z' and the time t' of the same event as seen in reference frame S', which is moving with uniform velocity relative to S. For convenience, we assume that the origins are coincident at time $t = t' = 0$. The classical relation, called the **Galilean transformation,** is

$$x = x' + vt', \quad y = y', \quad z = z', \quad t = t' \qquad \text{39-1}a$$

GALILEAN TRANSFORMATION

The inverse transformation is

$$x' = x - vt, \quad y' = y, \quad z' = z, \quad t' = t \qquad \text{39-1}b$$

These equations are consistent with experimental observations as long as v is much less than c. They lead to the familiar classical rules for velocities. If a particle has velocity $u_x = dx/dt$ in frame S, its velocity in frame S' is

$$u'_x = \frac{dx'}{dt'} = \frac{dx'}{dt} = \frac{d}{dt}(x - vt) = u_x - v \qquad \text{39-2}$$

If we differentiate this equation again, we find that the acceleration of the particle is the same in both frames:

$$a_x = \frac{du_x}{dt} = \frac{du'_x}{dt'} = a'_x$$

It should be clear that the Galilean transformation is not consistent with Einstein's postulates of special relativity. If light moves along the x axis with speed $u'_x = c$ in S', these equations imply that the speed in S' is $u_x = c + v$ rather than $u_x = c$, which is consistent with Einstein's postulates and with experiment. The classical transformation equations must therefore be modified to make them consistent with Einstein's postulates. We will give a brief outline of one method of obtaining the relativistic transformation.

We assume that the relativistic transformation equation for x is the same as the classical equation (Equation 39-1*a*) except for a constant multiplier on the right side. That is, we assume the equation is of the form

$$x = \gamma(x' + vt') \qquad 39\text{-}3$$

where γ is a constant that can depend on v and c but not on the coordinates. The inverse transformation must look the same except for the plus sign:

$$x' = \gamma(x - vt) \qquad 39\text{-}4$$

Let us consider a light pulse that starts at the origin of S at $t = 0$. Because we have assumed that the two origins are coincident at $t = t' = 0$, the pulse also starts at the origin of S' at $t' = 0$. Einstein's postulates require that the equation for the x component of the wave front of the light pulse is $x = ct$ in frame S and $x' = ct'$ in frame S'. Substituting ct for x and ct' for x' in Equation 39-3 and Equation 39-4, we obtain

$$ct = \gamma(c + v)t' \qquad 39\text{-}5$$

and

$$ct' = \gamma(c - v)t \qquad 39\text{-}6$$

We divide both sides of Equations 39-5 and 39-6 by t, and then eliminate the ratio t'/t from the two equations and determine γ. Thus,

$$\gamma = \frac{1}{\sqrt{1 - (v^2/c^2)}} \qquad 39\text{-}7$$

(Note that γ is always greater than 1, and when v is much less than c, $\gamma \approx 1$.) The relativistic transformation for x and x' is therefore given by Equation 39-3 and Equation 39-4, where γ is given by Equation 39-7. We can obtain equations for t and t' by combining Equation 39-3 with the inverse transformation given by Equation 39-4. Substituting $x = \gamma(x' + vt')$ for x in Equation 39-4, we obtain

$$x' = \gamma[\gamma(x' + vt') - vt] \qquad 39\text{-}8$$

which can be solved for t in terms of x' and t'. The complete relativistic transformation is

$$x = \gamma(x' + vt'), \quad y = y', \quad z = z' \qquad 39\text{-}9$$

$$t = \gamma\left(t' + \frac{vx'}{c^2}\right) \qquad 39\text{-}10$$

LORENTZ TRANSFORMATION

The inverse transformation is

$$x' = \gamma(x - vt), \quad y' = y, \quad z' = z \qquad 39\text{-}11$$

$$t' = \gamma\left(t - \frac{vx}{c^2}\right) \qquad 39\text{-}12$$

The transformation described by Equations 39-9 through 39-12 is called the **Lorentz transformation.** It relates the space and time coordinates x, y, z, and t of an event in frame S to the coordinates x', y', z', and t' of the same event as seen in frame S', which is moving along the x axis with speed v relative to frame S.

We will now look at some applications of the Lorentz transformation.

TIME DILATION

Consider two events, one that occurs on the x' axis at point x_0' at time t_1' in frame S' and another that occurs on the x' axis at point x_0' at time t_2' in frame S'. (Both events occur at point x_0' in S'.) We can find the times t_1 and t_2 for the events in S from Equation 39-10. We have

$$t_1 = \gamma\left(t_1' + \frac{vx_0'}{c^2}\right)$$

and

$$t_2 = \gamma\left(t_2' + \frac{vx_0'}{c^2}\right)$$

so

$$t_2 - t_1 = \gamma(t_2' - t_1')$$

The time between two events that happen at the *same place* in a reference frame is called **proper time Δt_p** between the events. In this case, the time interval $t_2' - t_1'$ measured in frame S' is proper time. The time interval Δt measured in any other reference frame is always longer than the proper time. This expansion is called **time dilation:**

$$\Delta t = \gamma\, \Delta t_p \qquad \text{39-13}$$

TIME DILATION

Example 39-1 Spatial Separation and Temporal Separation of Two Events

Two events occur at the same point x_0' at times t_1' and t_2' in frame S', which is traveling in the $+x$ direction at speed v relative to frame S. (*a*) What is the spatial separation of the events in frame S? (*b*) What is the temporal separation of the events in frame S?

PICTURE The spatial separation in S is $x_2 - x_1$, where x_2 and x_1 are the coordinates of the events in S, which are found using Equation 39-9.

SOLVE

(*a*) 1. The position x_1 in S is given by Equation 39-9 with $x_1' = x_0'$:

$$x_1 = \gamma(x_0' + vt_1')$$

2. Similarly, the position x_2 in S is given by:

$$x_2 = \gamma(x_0' + vt_2')$$

3. Subtract to find the spatial separation:

$$\Delta x = x_2 - x_1 = \gamma v(t_2' - t_1') = \boxed{\frac{v(t_2' - t_1')}{\sqrt{1 - (v^2/c^2)}}}$$

(*b*) Using the time dilation formula, relate the two time intervals. The two events occur at the same place in S', so the proper time between the two events is $\Delta t_p = t_2' - t_1'$:

$$\Delta t = t_2 - t_1 = \gamma(t_2' - t_1') = \boxed{\frac{(t_2' - t_1')}{\sqrt{1 - (v^2/c^2)}}}$$

CHECK Taking the limits of the Part-(*a*) and Part-(*b*) results as c approaches infinity gives $\Delta x = v(t_2' - t_1')$ and $\Delta t = t_2' - t_1'$, respectively. Combining these expressions gives $\Delta x = v\Delta t$. This is just the classical (nonrelativistic) equation that displacement equals velocity multiplied by time that is developed in Chapter 2 for one-dimensional motion. In addition, the equation $\Delta t = t_2' - t_1'$ is just the classical result that the length of the time between events is the same in both reference frames.

TAKING IT FURTHER Dividing the Part-(*a*) result by the Part-(*b*) result gives $\Delta x/\Delta t = v$. The spatial separation Δx of the two events in S is the distance a fixed point, such as x_0' in S', moves in S during the time interval between the events in S.

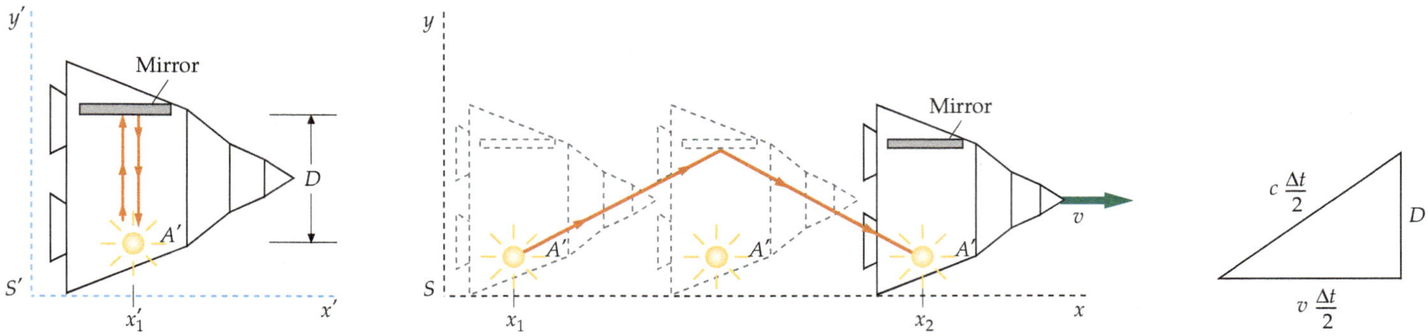

(*a*) (*b*) (*c*)

FIGURE 39-3 (*a*) Observer A' and the mirror are in a spaceship at rest in frame S'. The time it takes for the light pulse to reach the mirror and return is measured by A' to be $2D/c$. (*b*) In frame S, the spaceship is moving to the right with speed v. If the speed of light is the same in both frames, the time it takes for the light to reach the mirror and return is longer than $2D/c$ in S because the distance traveled is greater than $2D$. (*c*) A right triangle for computing the time Δt in frame S.

We can understand time dilation directly from Einstein's postulates without using the Lorentz transformation. Figure 39-3*a* shows an observer A' a distance D from a mirror. The observer and the mirror are in a spaceship that is at rest in frame S'. The observer explodes a flash gun and measures the time interval $\Delta t'$ between the original flash (Event 1) and his seeing the return flash from the mirror (Event 2). Because light travels with speed c, this time is

$$\Delta t' = \frac{2D}{c}$$

We now consider the same two events, the original flash of light and the receiving of the return flash, as observed in reference frame S, in which observer A' and the mirror are moving to the right with speed v, as shown in Figure 39-3*b*. Events 1 and 2 happen at positions x_1 and x_2, respectively, in frame S. During the time interval Δt (as measured in S) between the original flash and the return flash, observer A' and his spaceship have moved to the right a distance $v\,\Delta t$. In Figure 39-3, we can see that the path traveled by the light is longer in S than in S'. However, by Einstein's postulates, light travels with the same speed c in frame S as it does in frame S'. Because light travels farther in S at the same speed, it takes longer in S to reach the mirror and return. The time interval in S is thus longer than it is in S'. From the triangle in Figure 39-3*c*, we have

$$\left(\frac{c\,\Delta t}{2}\right)^2 = D^2 + \left(\frac{v\,\Delta t}{2}\right)^2$$

or

$$\Delta t = \frac{2D}{\sqrt{c^2 - v^2}} = \frac{2D}{c}\frac{1}{\sqrt{1 - (v^2/c^2)}}$$

Using $\Delta t' = 2D/c$, we obtain

$$\Delta t = \frac{\Delta t'}{\sqrt{1 - (v^2/c^2)}} = \gamma\,\Delta t'$$

(Equation 38-22*a*) for the Fermi energy at $T = 0$ by integrating $n(E)\,dE$ from $E = 0$ to $E = \infty$. We obtain

$$N = \int_0^{\infty} n(E)dE = \int_0^{E_F} n(E)dE + \int_{E_F}^{\infty} n(E)dE$$

$$= \int_0^{E_F} \frac{8\sqrt{2}\pi m_e^{3/2} V}{h^3} E^{1/2}\,dE + 0 = \frac{16\sqrt{2}\pi m_e^{3/2} V}{3h^3} E_F^{3/2}$$

Note that at $T = 0$, $n(E)$ is zero for $E > E_F$. Solving for E_F gives the Fermi energy at $T = 0$:

$$E_F = \frac{h^2}{8m_e}\left(\frac{3N}{\pi V}\right)^{2/3} \qquad \text{38-42}$$

which is Equation 38-22*a*. In terms of the Fermi energy, the density of states (Equation 38-41) is

$$g(E) = \frac{8\sqrt{2}\pi m_e^{3/2} V}{h^3} E^{1/2} = \frac{3}{2} N E_F^{-3/2} E^{1/2} \qquad \text{38-43}$$

DENSITY OF STATES IN TERMS OF E_F

which is obtained by solving Equation 38-42 for m_e, and then substituting for m_e in Equation 38-41. The average energy at $T = 0$ is calculated from

$$E_{av} = \frac{\int_0^{E_F} E g(E) dE}{\int_0^{E_F} g(E) dE} = \frac{1}{N}\int_0^{E_F} E g(E) dE \qquad \text{38-44}$$

where $N = \int_0^{E_F} g(E)dE$ is the total number of electrons. Substituting for $g(E)$ from Equation 38-43 and then evaluating the integral in Equation 38-44, we obtain Equation 38-23:

$$E_{av} = \tfrac{3}{5} E_F \qquad \text{38-45}$$

AVERAGE ENERGY AT $T = 0$

At $T > 0$, the Fermi factor is more complicated. It can be shown to be

$$f(E) = \frac{1}{e^{(E - E_F)/(kT)} + 1} \qquad \text{38-46}$$

FERMI FACTOR

We can see from this equation that for E greater than E_F, $e^{(E - E_F)/(kT)}$ becomes very large as T approaches zero, so at $T = 0$, the Fermi factor is zero for $E > E_F$. On the other hand, for E less than E_F, $e^{(E - E_F)/(kT)}$ approaches 0 as T approaches zero, so at $T = 0$, $f(E) = 1$ for $E < E_F$. Thus, the Fermi factor given by Equation 38-46 holds for all temperatures. Note also that for any nonzero value of T, $f(E) = \frac{1}{2}$ at $E = E_F$.

Example 39-2 How Long Is a One-Hour Nap? *Try It Yourself*

Astronauts in a spaceship traveling at $v = 0.600c$ relative to Earth sign off from space control, saying that they are going to nap for 1.00 h and then call back. How long does their nap last as measured on Earth?

PICTURE Because the astronauts go to sleep (Event 1) and wake up (Event 2) at the same place in the reference frame of the ship, the time interval for their nap of 1.00 h, as measured by a clock on the ship, is the proper time between the two events. In the reference frame of Earth, they move a considerable distance during the time between the two events. The time interval measured in Earth's frame is measured using two clocks that are stationary relative to Earth. Clock 1 is located at the position of Event 1 and measures the time of occurrence of Event 1. Clock 2 is located at the position of Event 2 and measures the time of occurrence of Event 2. The difference between the two times is longer than the proper time between the two events by the factor γ.

SOLVE

Cover the column to the right and try these on your own before looking at the answers.

Steps	Answers
1. Relate the time interval measured on Earth Δt to the proper time Δt_p (Equation 39-13).	$\Delta t = \gamma\, \Delta t_p$
2. Calculate γ for $v = 0.6c$ (Equation 39-7).	$\gamma = 1.25$
3. Substitute the value for γ to calculate the time of the nap in Earth's frame.	$\Delta t = \gamma\, \Delta t_p = \boxed{1.25\text{ h}}$

CHECK The time interval is longer in the reference frame in which the two events occur at different locations as expected.

PRACTICE PROBLEM 39-1 If the spaceship is moving at $v = 0.800c$, how long would a 1.00 h nap last as measured on Earth?

LENGTH CONTRACTION

A phenomenon closely related to time dilation is **length contraction.** The length of an object measured in the reference frame in which the object is at rest is called its **proper length** L_p. In a reference frame in which the object is moving parallel to its length, the measured length is shorter than its proper length. Consider a rod at rest in frame S' with one end at x'_2 and the other end at x'_1. The length of the rod in this frame is its proper length $L_p = x'_2 - x'_1$. Some care must be taken to find the length of the rod in frame S. In that frame, the rod is moving to the right with speed v, the speed of frame S'. The length of the rod in frame S is defined as $L = x_2 - x_1$, where x_2 is the position of one end at some time t_2, and x_1 is the position of the other end *at the same time* $t_1 = t_2$ as measured in frame S. To calculate $x_2 - x_1$ at some time t we use Equation 39-11:

$$x'_2 = \gamma(x_2 - vt_2)$$

and

$$x'_1 = \gamma(x_1 - vt_1)$$

Because $t_2 = t_1$, by subtracting the second equation from the first we obtain

$$x'_2 - x'_1 = \gamma(x_2 - x_1)$$

Solving for $x_2 - x_1$ gives

$$x_2 - x_1 = \frac{1}{\gamma}(x_2' - x_1') = (x_2' - x_1')\sqrt{1 - \frac{v^2}{c^2}}$$

or

$$L = \frac{1}{\gamma}L_p = L_p\sqrt{1 - \frac{v^2}{c^2}} \qquad 39\text{-}14$$

LENGTH CONTRACTION

Because $1/\gamma$ is less than one, it follows that the length of the rod is smaller when it is measured in a frame in which it is moving parallel to its length. Before Einstein's paper was published, Hendrik A. Lorentz and George F. FitzGerald tried to explain the null result of the Michelson–Morley experiment by assuming that distances in the direction of motion contracted by the amount given in Equation 39-14. This length contraction is now known as the **Lorentz–FitzGerald contraction.**

Example 39-3 The Length of a Moving Meterstick

A stick that has a proper length of 1.00 m moves in a direction parallel to its length with speed v relative to you. The length of the stick as measured by you is 0.914 m. What is the speed v?

PICTURE We can find v directly from Equation 39-14.

SOLVE

1. Equation 39-14 relates the lengths L and L_p and the speed v: $$L = L_p\sqrt{1 - \frac{v^2}{c^2}}$$

2. Solve for v: $$v = c\sqrt{1 - \frac{L^2}{L_p^2}} = c\sqrt{1 - \frac{(0.914\text{ m})^2}{(1.00\text{ m})^2}} = \boxed{0.406c}$$

CHECK As expected, the speed is a significant fraction of c.

An interesting example of time dilation or length contraction is the generation of muons as secondary radiation from cosmic rays. Muons decay according to the statistical law of radioactivity:

$$N(t) = N_0 e^{-t/\tau} \qquad 39\text{-}15$$

where N_0 is the number of muons at time $t = 0$, $N(t)$ is the number remaining at time t, and τ is the mean lifetime, which is approximately 2.2 μs for muons at rest. Because muons are generated (from the decay of pions) high in the atmosphere, usually several thousand meters above sea level, few muons should reach sea level. A typical muon moving with speed $0.9978c$ would travel only about 660 m in 2.2 μs. However, the lifetime of the muon measured in Earth's reference frame is increased by the factor $1/\sqrt{1 - (v^2/c^2)}$, which is 15 for this particular speed. The mean lifetime measured in Earth's reference frame is therefore 33 μs, and a muon with speed $0.9978c$ travels approximately 10 000 m during this time. From the muon's point of view, it exists for only 2.2 μs, but the atmosphere is rushing past it with a speed of $0.9978c$. The distance of 10 000 m in Earth's frame is thus contracted to only 660 m in the muon's frame, as indicated in Figure 39-4.

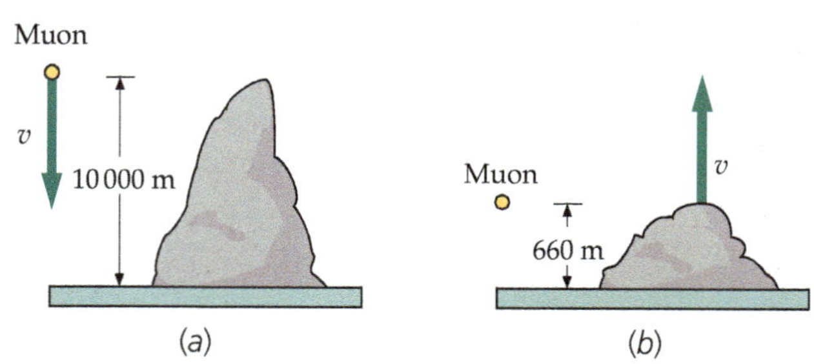

FIGURE 39-4 Although muons are created high above Earth and their mean lifetime is only about 2.2 μs when at rest, many appear at Earth's surface. (*a*) In Earth's reference frame, a typical muon moving at $0.9978c$ has a mean lifetime of 33 μs and travels 10 000 m during that time. (*b*) In the reference frame of the muon, the distance traveled by Earth is only 660 m in the muon's lifetime of 2.2 μs.

It is easy to distinguish experimentally between the classical and relativistic predictions of the observation of muons at sea level. Suppose that we observe 10^8 muons at an altitude of 10 000 m during some time interval with a muon detector. How many would we expect to observe at sea level during the same time interval? According to the nonrelativistic prediction, the time it takes for the muons to travel 10 000 m is (10 000 m)/(0.998c) ≈ 33 μs, which is 15 lifetimes. Substituting $N_0 = 1.0 \times 10^8$ and $t = 15\tau$ into Equation 39-15, we obtain

***See* Math Tutorial *for more information on the* The Exponential Function**

$$N = N_0 e^{-t/\tau} = 1.0 \times 10^8 e^{-15} = 31$$

We would thus expect all but about 31 of the original 100 million muons to decay before reaching sea level.

According to the relativistic prediction, Earth must travel only the contracted distance of 660 m in the rest frame of the muon. This trip takes only 2.2 μs = 1τ. Therefore, the number of muons expected at sea level is

$$N = N_0 e^{-t/\tau} = 1.0 \times 10^8 e^{-1} = 37 \times 10^6$$

Thus, relativity predicts that we would observe 37 million muons during the same time interval. Experiments have confirmed the relativistic prediction of 37 million muons.

THE RELATIVISTIC DOPPLER EFFECT

For light or other electromagnetic waves in a vacuum, a distinction between the motion of the source and the motion of the receiver cannot be made. Therefore, the expressions we derived in Chapter 15 for the Doppler effect cannot be correct for light. The reason it is not correct is that in Chapter 15 we assumed the time intervals in the reference frames of the source and receiver to be the same.

Consider a source moving toward a receiver with speed v, relative to the receiver. If the source emits N electromagnetic-wave crests during a time Δt_R (measured in the frame of the receiver), the first crest will travel a distance $c\,\Delta t_R$ and the source will travel a distance $v\,\Delta t_R$ measured in the frame of the receiver. The wavelength in this reference frame will be

$$\lambda' = \frac{c\,\Delta t_R - v\,\Delta t_R}{N}$$

The frequency f' observed by the receiver will therefore be

$$f' = \frac{c}{\lambda'} = \frac{c}{(c - v)}\frac{N}{\Delta t_R} = \frac{1}{1 - (v/c)}\frac{N}{\Delta t_R}$$

If the frequency of the source in the reference frame of the source is f_0, it will emit $N = f_0 \Delta t_S$ waves in the time Δt_S measured by the source. Then

$$f' = \frac{1}{1 - (v/c)}\frac{N}{\Delta t_R} = \frac{1}{1 - (v/c)}\frac{f_0\,\Delta t_S}{\Delta t_R} = \frac{f_0}{1 - (v/c)}\frac{\Delta t_S}{\Delta t_R}$$

Here Δt_S is the proper time interval (the first wave and the Nth wave are emitted at the same place in the reference frame of the source). Times Δt_S and Δt_R are related by Equation 39-13 for time dilation:

$$\Delta t_R = \gamma\,\Delta t_S = \frac{\Delta t_S}{\sqrt{1 - (v^2/c^2)}}$$

Thus, when the source and the receiver are moving toward one another we obtain

$$f' = \frac{f_0}{1 - (v/c)}\sqrt{1 - (v^2/c^2)} = \sqrt{\frac{1 + (v/c)}{1 - (v/c)}}f_0 \quad \text{approaching} \qquad 39\text{-}16a$$

This differs from our classical equation only in the time-dilation factor. It is left as a problem (Problem 25) for you to show that the same results are obtained if the calculations are done in the reference frame of the receiver.

When the source and the receiver are moving away from one another, the same analysis shows that the observed frequency is given by

$$f' = \frac{f_0}{1 + (v/c)}\sqrt{1 - (v/c^2)} = \sqrt{\frac{1 - (v/c)}{1 + (v/c)}}f_0 \quad \text{receding} \qquad 39\text{-}16b$$

An application of the relativistic Doppler effect is the **redshift** observed in the light from distant galaxies. Because the galaxies are moving away from us, the light they emit is shifted toward the longer wavelengths. (Because light that has the longest visible wavelengths appears red, this is referred to as a redshift.) The speed of the galaxies relative to us can be determined by measuring this shift.

Example 39-4 Red Light/Green Light

You are spending the day shadowing two police officers. You have just witnessed the officers pulling over a car that went through a red light. The driver claims that the red light looked green because the car was moving toward the stoplight, which shifted the wavelength of the observed light. You quickly do some calculations to see if the driver has a reasonable case.

PICTURE We can use the Doppler shift formula for approaching objects in Equation 39-16*a*. This will tell us the velocity, but we need to know the frequencies of the light. We can make good guesses for the wavelengths of red light and green light and use $c = f\lambda$ to determine the frequencies.

SOLVE

1. The observer is approaching the light source, so we use the Doppler formula (Equation 39-16*a*) for approaching sources:

$$f' = \sqrt{\frac{1 + (v/c)}{1 - (v/c)}}f_0$$

2. Substitute c/λ for f, then simplify:

$$\frac{c}{\lambda'} = \sqrt{\frac{1 + (v/c)}{1 - (v/c)}}\frac{c}{\lambda_0}$$

$$\left(\frac{\lambda_0}{\lambda'}\right)^2 = \frac{1 + (v/c)}{1 - (v/c)}$$

3. Cross multiply and solve for v/c:

$$(\lambda_0)^2\left(1 - \frac{v}{c}\right) = (\lambda')^2\left(1 + \frac{v}{c}\right)$$

$$(\lambda_0)^2 - (\lambda')^2 = \left[(\lambda_0)^2 + (\lambda')^2\right]\left(\frac{v}{c}\right)$$

$$\frac{v}{c} = \frac{(\lambda_0)^2 - (\lambda')^2}{(\lambda_0)^2 + (\lambda')^2} = \frac{1 - (\lambda'/\lambda_0)^2}{1 + (\lambda'/\lambda_0)^2}$$

4. The values for the wavelengths for the colors of the visible spectrum can be found in Table 30-1. The wavelengths for red are 625 nm or longer, and the wavelengths for green are 530 nm or shorter. Solve for the speed needed to shift the wavelength from 625 nm to 530 nm:

$$\frac{\lambda'}{\lambda_0} = \frac{530\text{ nm}}{625\text{ nm}} = 0.848$$

$$\frac{v}{c} = \frac{1 - 0.848^2}{1 + 0.848^2} = 0.163$$

$$v = 0.163c = 4.90 \times 10^7\text{ m/s} = 1.10 \times 10^8\text{ mi/h}$$

5. This speed is beyond any possible speed for a car:

The driver does not have a plausible case.

CHECK A car cannot travel at relativistic speeds, so the answer to this problem was obvious.

Example 39-5 Finding Speed from the Doppler Shift

Try It Yourself

The emission spectrum of hydrogen includes a line that has the wavelength $\lambda_0 = 656$ nm. In light reaching us from a distant galaxy, the wavelength of that spectral line is measured to be $\lambda' = 1458$ nm. Find the speed at which the distant galaxy is receding from Earth.

PICTURE Wavelength is related to frequency by $c = f\lambda$ and the received frequency is related to the unshifted frequency by the Doppler shift equation for a receding source (Equation 39-16*b*).

SOLVE

Cover the column to the right and try these on your own before looking at the answers.

Steps	Answers
1. Use Equation 39-16*b* to relate the speed v to the received frequency f' and the unshifted frequency f_0.	$f' = \sqrt{\frac{1-(v/c)}{1+(v/c)}}f_0$
2. Substitute $f' = c/\lambda'$ and $f_0 = c/\lambda_0$ and solve for v/c.	$\frac{v}{c} = \frac{1-(\lambda_0/\lambda')^2}{1+(\lambda_0/\lambda')^2} = 0.664$
	$v = \boxed{0.664c}$

CHECK As expected, the result is a significant fraction of c. This result is expected because the wavelength of the received light is large in comparison to the wavelength of the same spectral line in the reference frame of the source.

39-4 CLOCK SYNCHRONIZATION AND SIMULTANEITY

We saw in Section 39-3 that proper time is the time interval between two events that occur at the same point in some reference frame. It can therefore be measured on a single clock. (Remember, in each frame there is, in principle, a stationary clock at each point in space, and the time of an event in a given frame is measured by the clock at that point.) However, in another reference frame moving relative to the first, the same two events occur at different places, so two stationary clocks are needed in this reference frame to record the times. The time of each event is measured on a different clock, and the interval is found by subtraction of the measured times. This procedure requires that the clocks be **synchronized.** We will show in this section that

> Two clocks that are synchronized in one reference frame are typically not synchronized in any other frame moving relative to the first frame.
>
> SYNCHRONIZED CLOCKS

Here is a corollary to this result:

> Two events that are simultaneous in one reference frame typically are not simultaneous in another frame that is moving relative to the first.*
>
> SIMULTANEOUS EVENTS

* This is true unless the x coordinates of the two events are equal, where the x axis is parallel to the relative velocity of the two frames.

Comprehension of these facts usually resolves all relativity paradoxes. Unfortunately, the intuitive (and incorrect) belief that simultaneity is an absolute relation is difficult to overcome.

Suppose we have two clocks at rest, one at point A and the other at point B, where points A and B are a distance L apart in frame S. How can we synchronize the two clocks? If an observer at A looks at the clock at B and sets her clock to read the same time, the clocks will not be synchronized because of the time L/c it takes light to travel from one clock to another. To synchronize the clocks, the observer at A must set her clock ahead by the time L/c. Then she will see that the clock at B reads a time that is L/c behind the time on her clock, but she will calculate that the clocks are synchronized when she allows for the time L/c for the light to reach her. Any other observers in S (except those equidistant from the two clocks) will see the clocks reading different times, but they will also calculate that the clocks are synchronized when they correct for the time it takes the light to reach them. An equivalent method for synchronizing two clocks would be for an observer at point C, a point midway between the clocks, to send a light signal and for the observers at A and B to set their clocks to some prearranged time when they receive the signal.

We now examine the question of **simultaneity.** Suppose observers at A and B agree to explode flashguns at t_0 (having previously synchronized their clocks). The observer at C will see the light from the two flashes at the same time, and because he is equidistant from A and B, he will conclude that the flashes were simultaneous. Other observers in frame S will see the light from A or B first, depending on their location, but after correcting for the time the light takes to reach them, they also will conclude that the flashes were simultaneous. We can thus define simultaneity as follows:

Two events in a reference frame are simultaneous if light signals from the events reach an observer halfway between the events at the same time.

DEFINITION—SIMULTANEITY

To show that two events that are simultaneous in frame S are not simultaneous in another frame S' moving relative to S, we will use an example introduced by Einstein. A train is moving with speed v past a station platform. We will consider the train to be at rest in S' and the platform to be at rest in S. We have observers A', B', and C' at the front, back, and middle of the train (Figure 39-5). We now suppose that the train and platform are struck by lightning at the front and back of the train and that the lightning bolts are simultaneous in the frame of the platform S. That is, an observer C on the platform halfway between the positions A and B, where the lightning strikes, sees the light from the two strikes at the same time.

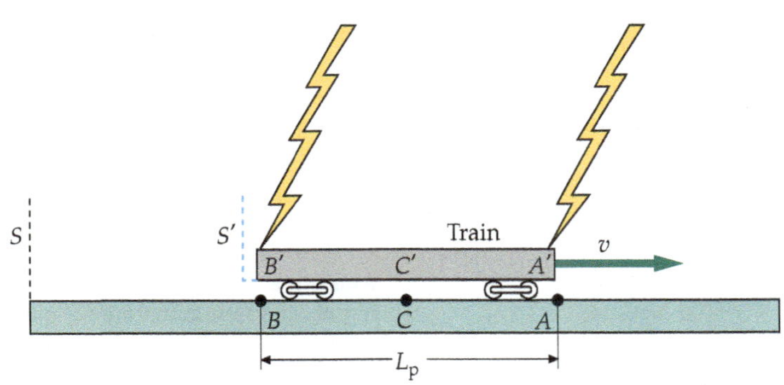

FIGURE 39-5 In frame S attached to the platform, simultaneous lightning bolts strike the ends of a train traveling with speed v. The light from the simultaneous events reaches observer C, standing on the platform midway between the events, at the same time. The distance between the bolts is $L_{\text{p platform}}$.

It is convenient to suppose that the lightning scorches both the train and platform so that the events can be easily located. Because C' is in the middle of the train, halfway between the places on the train that are scorched, the events are simultaneous in S' only if C' sees the flashes at the same time. However, the flash from the front of the train is seen by C' before the flash from the back of the train. We can understand this by considering the motion of C' as seen in frame S (Figure 39-6). By the time the light from the front flash reaches C', C' has moved some distance toward the front flash and some distance away from the back flash. Thus, the light from the back flash has not yet reached C', as indicated in the figure. Observer C' must therefore conclude that the events are not simultaneous and that the front of the train was struck before the back. Furthermore, all observers in S' on the train will agree with C' when they have corrected for the time it takes the light to reach them.

Figure 39-7 shows the events of the lightning bolts as seen in the reference frame of the train (S'). In this frame the platform is moving, so the distance between the scorch marks on the platform is contracted. The platform is shorter than it is in S, and, because the train is at rest, the train is longer than its contracted length in S. When the lightning bolt strikes the front of the train at A', the front of the train is at point A, and the back of the train has not yet reached point B. Later, when the lightning bolt strikes the back of the train at B', the back has reached point B on the platform.

The time discrepancy of two clocks that are synchronized in frame S as seen in frame S' can be found from the Lorentz transformation equations. Suppose we have clocks at points x_1 and x_2 that are synchronized in S. What are the times t_1 and t_2 on the clocks as observed from frame S' at a time t'_0?

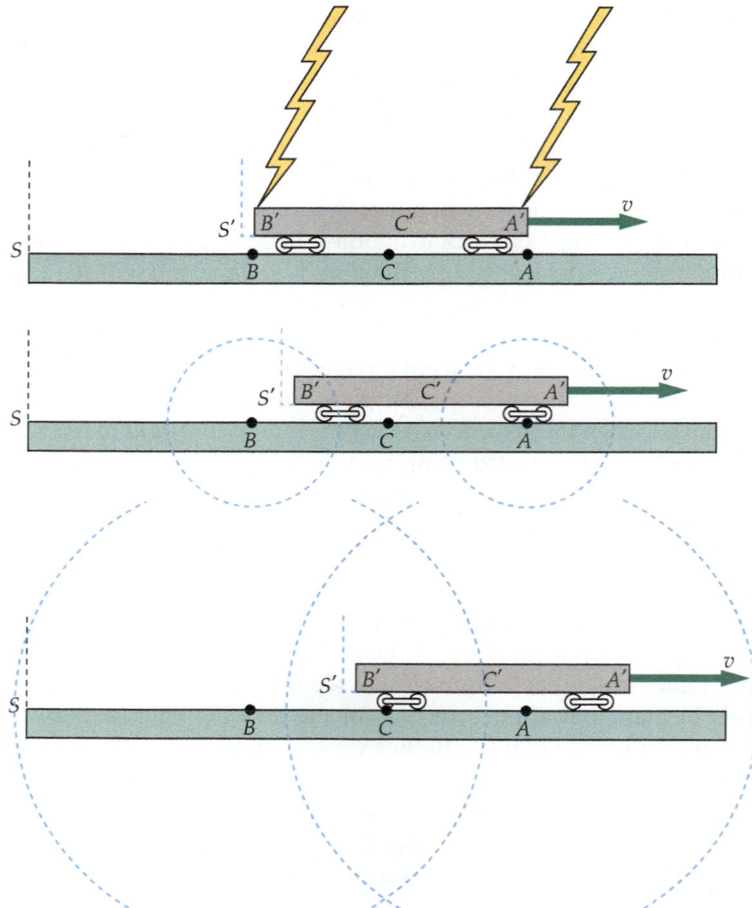

FIGURE 39-6 In frame S attached to the platform, the light from the lightning bolt at the front of the train reaches observer C', standing on the train at its midpoint, before the light from the bolt at the back of the train. Because C' is midway between the events (which occur at the front and rear of the train), the events are not simultaneous for him.

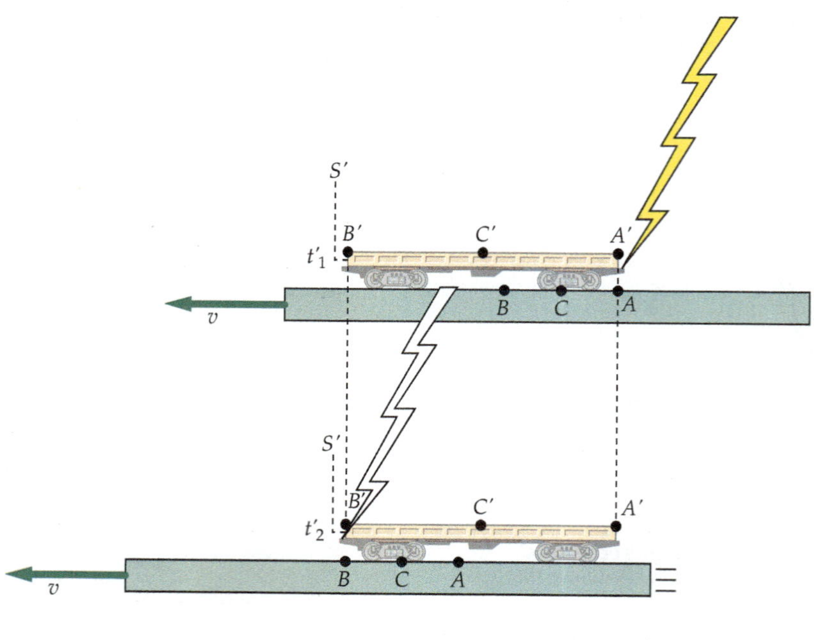

FIGURE 39-7 The lightning bolts of Figure 39-5 as seen in frame S' of the train. In this frame, the distance between A and B on the platform is less than $L_{\text{p platform}}$, and the proper length of the train $L_{\text{p train}}$ is longer than $L_{\text{p platform}}$. The first lightning bolt strikes the front of the train when A' and A are coincident. The second bolt strikes the rear of the train when B' and B are coincident.

From Equation 39-12, when $t'_1 = t'_2 = t'_0$, we have

$$t'_0 = \gamma\left(t_1 - \frac{vx_1}{c^2}\right)$$

and

$$t'_0 = \gamma\left(t_2 - \frac{vx_2}{c^2}\right)$$

Subtracting the first equation from the second, and then rearranging, gives

$$t_2 - t_1 = \frac{v}{c^2}(x_2 - x_1)$$

Note that the chasing clock (at x_2) leads the other (at x_1) by an amount that is proportional to their proper separation $L_p = x_2 - x_1$.

If two clocks are synchronized in the frame in which they are both at rest, in a frame in which they are moving along the line through both clocks, the chasing clock leads (shows a later time) by an amount

$$\Delta t_S = L_p\frac{v}{c^2} \qquad 39\text{-}17$$

where L_p is the proper distance between the clocks.

CHASING CLOCK SHOWS LATER TIME

A numerical example should help clarify time dilation, clock synchronization, and the internal consistency of these results.

Example 39-6 Synchronizing Clocks

Observer A' in a spaceship that has a flashgun and a mirror is shown in Figure 39-3 (see page 1325). Observer A' is standing next to the flashgun. The distance from the gun to the mirror is 15 light-minutes (written $15\,c\cdot\text{min}$) and the spaceship, at rest in frame S', travels with speed $v = 0.80c$ relative to a very long space platform that is at rest in frame S. The platform has two synchronized clocks, one clock at position x_1, the position of the spaceship when the observer explodes the flashgun, and the other clock at position x_2, the position of the spaceship when the light returns to the gun from the mirror. Find the time intervals between the events (exploding the flashgun and receiving the return flash from the mirror) (*a*) in the frame of the spaceship and (*b*) in the frame of the platform. Find (*c*) the distance traveled by the spaceship and (*d*) the amount by which the clocks on the platform are out of synchronization according to observers on the spaceship.

PICTURE The events occur at the same place on the spaceship, so the time between the events in frame S' is the proper time between the events.

SOLVE

(*a*) 1. In the frame of the spaceship, the light travels from the gun to the mirror and back, a total distance $D = 30\,c\cdot\text{min}$. The time required is D/c:

$$\Delta t' = \frac{D}{c} = \frac{30\,c\cdot\text{min}}{c} = \boxed{30\text{ min}}$$

2. Because the two events happen at the same place in the spaceship, the time interval is proper time:

$$\Delta t_p = \boxed{30\text{ min}}$$

(*b*) 1. In frame S, the time between the events is longer by the factor γ:

$$\Delta t = \gamma\,\Delta t_p = \gamma(30\text{ min})$$

2. Calculate γ:

$$\gamma = \frac{1}{\sqrt{1-(v^2/c^2)}} = \frac{1}{\sqrt{1-(0.80)^2}} = \frac{1}{\sqrt{0.36}} = \frac{5}{3}$$

3. Use the value of γ to calculate the time between the events as observed in frame S:

$$\Delta t = \gamma\,\Delta t_p = \tfrac{5}{3}(30\text{ min}) = \boxed{50\text{ min}}$$

(*c*) In frame *S*, the distance traveled by the spaceship is $v\,\Delta t$:

$$x_2 - x_1 = v\,\Delta t = (0.80c)(50\text{ min}) = \boxed{40\ c\cdot\text{min}}$$

(*d*) 1. The amount that the clocks on the platform are out of synchronization is related to the proper distance between the clocks L_p:

$$\Delta t_s = L_p\frac{v}{c^2}$$

2. The Part-(*c*) result is the proper distance between the clocks on the platform:

$$L_p = x_2 - x_1 = 40\ c\cdot\text{min}$$

so

$$\Delta t_s = L_p\frac{v}{c^2} = (40\ c\cdot\text{min})\frac{(0.80c)}{c^2} = \boxed{32\text{ min}}$$

TAKING IT FURTHER Observers on the platform would say that the spaceship's clock is running slow because it records a time of only 30 min between the events, whereas the time measured by observers on the platform is 50 min.

Figure 39-8 shows the situation in Example 39-6 viewed from the spaceship in S'. The platform is traveling past the ship with speed $0.8c$. There is a clock at point x_1, which coincides with the ship when the flashgun is exploded, and another at point x_2, which coincides with the ship when the return flash is received from the mirror. We assume that the clock at x_1 reads 12:00 noon at the time of the light flash. The clocks at x_1 and x_2 are synchronized in *S* but not in S'. In S', the clock at x_2, which is chasing the one at x_1, leads by 32 min; it would thus read 12:32 to an observer in S'. When the spaceship coincides with x_2, the clock there reads 12:50. The time between the events is therefore 50 min in *S*. Note that according to observers in S', this clock ticks off 50 min − 32 min = 18 min for a trip that takes 30 min in S'. Thus, observers in S' see this clock run slow by the factor 30/18 = 5/3.

Every observer in one frame sees the clocks in the other frame run slow. According to observers in *S*, who measure 50 min for the time interval, the time interval in S' (30 min) is too small, so they see the single clock in S' run too slow by the factor 5/3. According to the observers in S', the observers in *S* measure a time that is too *long* despite the fact that their clocks run too slow because the clocks in *S* are out of synchronization. The clocks tick off only 18 min, but the second clock leads the first clock by 32 min, so the time interval is 50 min.

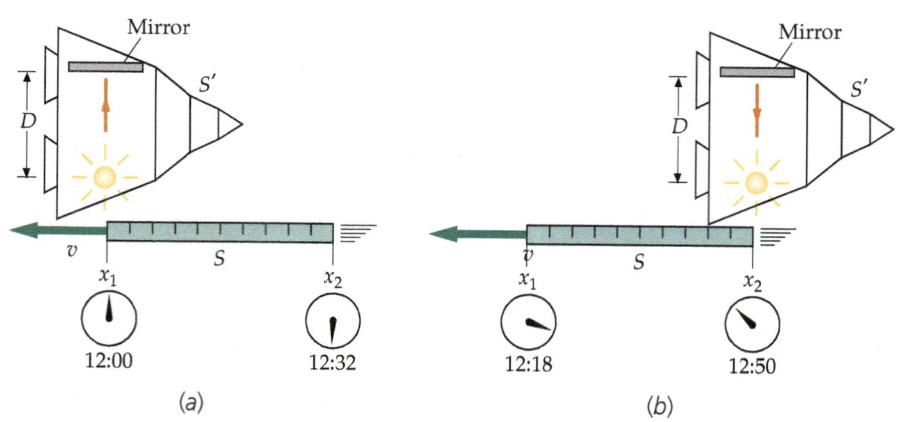

FIGURE 39-8 Clocks on a platform as observed from the spaceship's frame of reference S'. During the time $\Delta t' = 30$ min it takes for the platform to pass the spaceship, the clocks on the platform run slow and tick off (30 min)/γ = 18 min. But the clocks are unsynchronized, with the chasing clock leading by $L_p v/c^2$, which for this case is 32 min. The time it takes for the spaceship to go from x_1 to x_2, as measured on the platform, is therefore 32 min + 18 min = 50 min.

THE TWIN PARADOX

Homer and Ulysses are identical twins. Ulysses travels at high speed to a planet beyond the solar system and returns while Homer remains at home. When they are together again, which twin is older or are they the same age? The correct answer is that Homer, the twin who stays at home, is older. This problem, with variations, has been the subject of spirited debate for decades, though there are very few who disagree with the answer. The problem appears to be a paradox because of the seemingly symmetric roles played by the twins and the asymmetric result in their aging. The paradox is resolved when the asymmetry of the twins' roles is noted. The relativistic result conflicts with common sense based on our strong but incorrect belief in absolute simultaneity. We will consider a particular case with some numerical magnitudes that, though impractical, make the calculations easy.

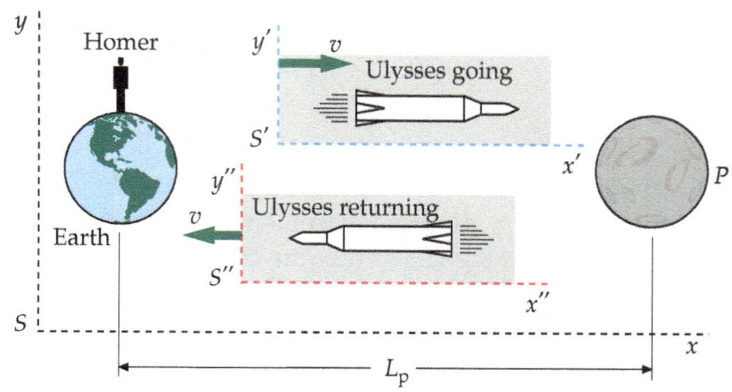

FIGURE 39-9 The twin paradox. Earth and a distant planet are fixed in frame S. Ulysses coasts in frame S' to the planet and then coasts back in frame S''. His twin Homer stays on Earth. When Ulysses returns, he is younger than his twin. The roles played by the twins are not symmetric. Homer remains at rest in one inertial reference frame, but Ulysses must go from being at rest in one inertial reference frame to another if he is to return home.

In reference frame S, Earth, planet P and Homer are at rest and Earth and planet P are a distance L_p apart (Figure 39-9). Homer is on Earth. Reference frames S' and S'' are moving with speed v toward and away from planet P, respectively. Ulysses quickly accelerates to speed v, then coasts, at rest in S', until he reaches the planet, where he quickly decelerates to a stop and is momentarily at rest in S. To return, Ulysses quickly accelerates to speed v toward Earth and then coasts, at rest in S'', until he reaches Earth, where he quickly decelerates to a stop. We can assume that the acceleration (and deceleration) times are negligible compared with the coasting times. We use the following values for illustration: $L_p = 8$ light-years $(8\,c\cdot\text{y})$ and $v = 0.8c$. Then $\sqrt{1-(v^2/c^2)} = 3/5$ and $\gamma = 5/3$.

It is easy to analyze the problem from Homer's point of view on Earth. According to Homer's clock, Ulysses coasts in S' for a time $L_p/v = 10$ y and in S'' for an equal time. Thus, Homer is 20 y older when Ulysses returns. The time interval in S' between Ulysses's leaving Earth and his arriving at the planet is shorter because it is proper time. The time it takes to reach the planet by Ulysses's clock is

$$\Delta t' = \frac{\Delta t}{\gamma} = \frac{10\text{ y}}{5/3} = 6\text{ y}$$

Because the same time is required for the return trip, Ulysses will have recorded 12 y for the round trip and will be 8 years younger than Homer upon his return.

From Ulysses's point of view, the distance between Earth and the planet is contracted and is only

$$L' = \frac{L_p}{\gamma} = \frac{8\,c\cdot\text{y}}{5/3} = 4.8\,c\cdot\text{y}$$

At $v = 0.8c$, it takes only $L'/v = 4.8c\cdot\text{y}/0.8\,c = 6$ y each way.

The real challenge in this problem is for Ulysses to understand why his twin aged 20 y during his absence. If we consider Ulysses as being at rest and Homer as moving away, Homer's clock should run slow and measure only $\frac{3}{5}(6\text{ y}) = 3.6$ y. Then why shouldn't Homer age only 7.2 y during the round trip? This, of course, is the paradox. The difficulty with the analysis from the point of view of Ulysses is that he does not remain in a single inertial reference frame. What happens while Ulysses is stopping and starting? To investigate this problem in detail, we would need to treat accelerated reference frames, a subject dealt with in the study of general relativity and beyond the scope of this book. However, we can get some insight into the problem by having the twins send regular signals to each other so that they can record the other's age continuously. If they arrange to send a signal once a year, each can determine the age of the other merely by counting the signals received. The arrival frequency of the signals will not be 1 per year because of the Doppler shift. The frequency observed will be given by Equations 39-16*a* and 39-16*b*.

Using $v/c = 0.8$ (so $v^2/c^2 = 0.64$), we have for the case in which the twins are receding from each other

$$f' = \frac{f_0}{1 + (v/c)}\sqrt{1 - (v^2/c^2)} = \frac{\sqrt{1 - 0.64}}{1 + 0.8} f_0 = \frac{1}{3} f_0$$

When they are approaching, Equation 39-16*a* gives $f' = 3f_0$.

Consider the situation first from the point of view of Ulysses. During the 6 y it takes him to reach the planet (remember that the distance is contracted in his frame), he receives signals at the rate of $\frac{1}{3}$ signal per year, and so he receives 2 signals. As soon as Ulysses turns around and starts back to Earth, he begins to receive 3 signals per year. In the 6 y it takes him to return he receives 18 signals, giving a total of 20 for the trip. He accordingly expects his twin to have aged 20 years.

We now consider the situation from Homer's point of view. He receives signals at the rate of $\frac{1}{3}$ signal per year not only for the 10 y it takes Ulysses to reach the planet but also for the time it takes for the last signal sent by Ulysses before he turns around to get back to Earth. (He cannot know that Ulysses has turned around until the signals begin reaching him with increased frequency.) Because the planet is 8 light-years away, there is an additional 8 y of receiving signals at the rate of $\frac{1}{3}$ signal per year. During the first 18 y, Homer receives 6 signals. In the final 2 y before Ulysses arrives, Homer receives 6 signals, or 3 per year. (The first signal sent after Ulysses turns around takes 8 y to reach Earth, whereas Ulysses, traveling at $0.8c$, takes 10 y to return and therefore arrives just 2 y after Homer begins to receive signals at the faster rate.) Thus, Homer expects Ulysses to have aged 12 y. In this analysis, the asymmetry of the twins' roles is apparent. When they are together again, both twins agree that the one who has been accelerated will be younger than the one who stayed home.

The predictions of the special theory of relativity concerning the twin paradox have been tested using small particles that can be accelerated to such large speeds that γ is appreciably greater than 1. Unstable particles can be accelerated and trapped in circular orbits in a magnetic field, for example, and their lifetimes can then be compared with those of identical particles at rest. In all such experiments, the accelerated particles live longer on the average than the particles at rest, as predicted. These predictions have also been confirmed by the results of an experiment in which high-precision atomic clocks were flown around the world in commercial airplanes, but the analysis of that experiment is complicated due to the necessity of including gravitational effects treated in the general theory of relativity.

39-5 THE VELOCITY TRANSFORMATION

We can find how velocities transform from one reference frame to another by differentiating the Lorentz transformation equations. Suppose a particle has velocity $u'_x = dx'/dt'$ in frame S', which is moving to the right with speed v relative to frame S. The particle's velocity in frame S is

$$u_x = \frac{dx}{dt}$$

From the Lorentz transformation equations (Equations 39-9 and 39-10), we have

$$dx = \gamma(dx' + v\,dt')$$

and

$$dt = \gamma\left(dt' + \frac{v\,dx'}{c^2}\right)$$

The velocity relative to frame S is thus

$$u_x = \frac{dx}{dt} = \frac{\gamma(dx' + v\,dt')}{\gamma\left(dt' + \dfrac{v\,dx'}{c^2}\right)} = \frac{\dfrac{dx'}{dt'} + v}{1 + \dfrac{v}{c^2}\dfrac{dx'}{dt'}} = \frac{u'_x + v}{1 + \dfrac{vu'_x}{c^2}}$$

If a particle has components of velocity along the y or z axes, we can use the same relation between dt and dt', with $dy = dy'$ and $dz = dz'$, to obtain

$$u_y = \frac{dy}{dt} = \frac{dy'}{\gamma\left(dt' + \dfrac{v\,dx'}{c^2}\right)} = \frac{\dfrac{dy'}{dt'}}{\gamma\left(1 + \dfrac{v}{c^2}\dfrac{dx'}{dt'}\right)} = \frac{u'_y}{\gamma\left(1 + \dfrac{vu'_x}{c^2}\right)}$$

and

$$u_z = \frac{u'_z}{\gamma\left(1 + \dfrac{vu'_x}{c^2}\right)}$$

The complete relativistic velocity transformation is

$$u_x = \frac{u'_x + v}{1 + \dfrac{vu'_x}{c^2}} \qquad \text{39-18}a$$

$$u_y = \frac{u'_y}{\gamma\left(1 + \dfrac{vu'_x}{c^2}\right)} \qquad \text{39-18}b$$

$$u_z = \frac{u'_z}{\gamma\left(1 + \dfrac{vu'_x}{c^2}\right)} \qquad \text{39-18}c$$

RELATIVISTIC VELOCITY TRANSFORMATION

The inverse velocity transformation equations are

$$u'_x = \frac{u_x - v}{1 - \dfrac{vu_x}{c^2}} \qquad \text{39-19}a$$

$$u'_y = \frac{u_y}{\gamma\left(1 - \dfrac{vu_x}{c^2}\right)} \qquad \text{39-19}b$$

$$u'_z = \frac{u_z}{\gamma\left(1 - \dfrac{vu_x}{c^2}\right)} \qquad \text{39-19}c$$

These equations differ from the classical and intuitive result $u_x = u'_x + v$, $u_y = u'_y$, and $u_z = u'_z$ because the denominators in the equations are not equal to 1. When v and u'_x are small compared with the speed of light c, $\gamma \approx 1$ and $vu'_x/c^2 \ll 1$. Then the relativistic and classical velocity transformation equations are the same.

Example 39-7 Relative Velocity at Nonrelativistic Speeds

A supersonic plane moves away from you, and in the $+x$ direction, at a speed of 1000 m/s (about 3 times the speed of sound) relative to you. A second plane, traveling in the same direction and ahead of the first plane, moves away from you, and away from the first plane, at a speed of 500 m/s relative to the first plane. How fast is the second plane moving relative to you?

PICTURE The speeds are so small compared with c that we expect the classical equations for combining velocities to be accurate. We show this by calculating the correction term in the denominator of Equation 39-18*a*. Let frame S be your rest frame and frame S' be the rest frame of the first plane. Then v, the velocity of S' relative to S, is 1000 m/s. The second plane has velocity $u'_x = 500$ m/s relative to S'.

SOLVE

1. Let S and S' be the reference frames of you and the first plane, respectively. Also, let u_x and u'_x be the velocities of the second plane relative to S and S', respectively. Equation 39-18*a* can be used to find u_x. The velocity of the second plane relative to you is v:

$$u_x = \frac{u'_x + v}{1 + \dfrac{vu'_x}{c^2}}$$

2. If the correction term in the denominator is negligible (compared to 1), Equation 39-18*a* gives the classical formula for combining velocities. Calculate the value of the correction term:

$$\frac{vu'_x}{c^2} = \frac{(1000)(500)}{(3.00 \times 10^8)^2} = 5.56 \times 10^{-12}$$

3. The correction term is so small that the classical and relativistic results are virtually the same:

$$u_x \approx u'_x + v$$
$$= 500 \text{ m/s} + 1000 \text{ m/s} = \boxed{1500 \text{ m/s}}$$

Example 39-8 Relative Velocity at Relativistic Speeds

Work Example 39-7 if the first plane moves with speed $v = 0.80c$ relative to you and the second plane moves with the same speed $0.80c$ relative to the first plane.

PICTURE These speeds are not small compared with c, so we need to use the relativistic expression (Equation 39-18*a*). We again assume that you are at rest in frame S and the first plane is at rest in frame S' that is moving at $v = 0.80c$ relative to you. The velocity of the second plane relative to S' is $u'_x = 0.80c$.

SOLVE

Use Equation 39-18*a* to calculate the speed of the second plane relative to you:

$$u_x = \frac{u'_x + v}{1 + \dfrac{vu'_x}{c^2}} = \frac{0.80c + 0.80c}{1 + \dfrac{(0.80c)(0.80c)}{c^2}} = \frac{1.60c}{1.64} = \boxed{0.98c}$$

CHECK As expected, the result is less than c.

The result in Example 39-8 is quite different from the classically expected result of $0.80c + 0.80c = 1.60c$. In fact, it can be shown from Equations 39-18*a*–*c* that if the speed of an object is less than c in one frame, it is less than c in all other frames moving relative to that frame with a speed less than c. (See Problem 59.) We will see in Section 39-7 that it takes an infinite amount of energy to accelerate a particle to the speed of light. The speed of light c is thus an upper, unattainable limit for the speed of a particle with mass. (There are massless particles, such as photons, that always move at the speed of light.)

Example 39-9 Relative Speed of a Photon

A photon moves along the x' axis in frame S', with speed $u'_x = c$. What is its speed in frame S?

PICTURE Use Equation 39-18*a* to calculate the speed of the photon in S.

SOLVE

The speed in S is given by Equation 39-18*a*:

$$u_x = \frac{u'_x + v}{1 + \frac{vu'_x}{c^2}} = \frac{c + v}{1 + \frac{vc}{c^2}} = \frac{c + v}{1 + \frac{v}{c}} = \frac{c + v}{\frac{1}{c}(c + v)} = \boxed{c}$$

CHECK The speed in both frames is c, independent of v. This is in accord with Einstein's postulates.

Example 39-10 Rockets Passing in Opposite Directions

Two spaceships, each 100 m long when measured at rest, travel toward each other, each with a speed of $0.85c$ relative to Earth. (*a*) What is the length of each spaceship as measured by someone at rest relative to Earth? (*b*) How fast is each spaceship traveling as measured by an observer at rest relative to the other spaceship? (*c*) What is the length of one spaceship when measured by an observer at rest relative to the other spaceship? (*d*) At time $t = 0$ on Earth, the front ends of the ships are next to each other as they just begin to pass each other. At what time on Earth are their back ends next to each other?

PICTURE (*a*) The length of each spaceship as measured on Earth is the contracted length $\sqrt{1 - (u^2/c^2)}L_p$ (Equation 39-14), where u is the speed of either spaceship relative to Earth. To solve Part (*b*), let Earth be at rest in frame S, and let the spaceship on the left (spaceship 1) be at rest in frame S', which is moving at speed $v = 0.85c$ relative to S. Then the spaceship on the right (spaceship 2) moves with velocity $u_{2x} = -0.85c$ (Figure 39-10). (*c*) The length of spaceship 2 as seen by an observer at rest relative to spaceship 1 is $\sqrt{1 - (u'^2_{2x}/c^2)}L_p$.

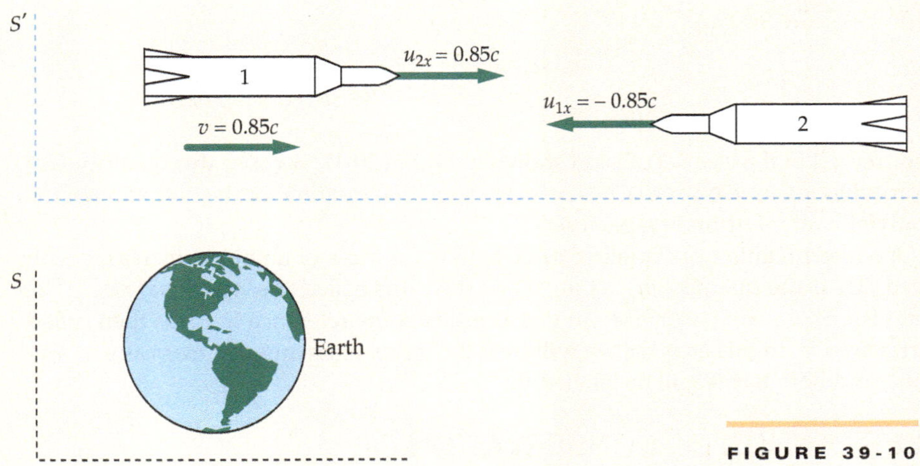

FIGURE 39-10

SOLVE

(*a*) The length of each spaceship in S, the reference frame of Earth, is the proper length divided by γ.

$$L = \frac{1}{\gamma}L_p = \sqrt{1 - \frac{|u_{2x}|^2}{c^2}}L_p = \sqrt{1 - \frac{(0.85c)^2}{c^2}}(100\text{ m}) = \boxed{53\text{ m}}$$

(*b*) Use the velocity transformation formula (Equation 39-19*a*) to find the velocity u'_{2x} of spaceship 2 as seen in frame S':

$$u'_{2x} = \frac{u_{2x} - v}{1 - \frac{vu_{2x}}{c^2}} = \frac{-0.85c - 0.85c}{1 - \frac{(0.85c)(-0.85c)}{c^2}} = \frac{-1.70c}{1.7225} = -0.987c$$

so

$$|u'_{2x}| = \boxed{0.99c}$$

(*c*) In the frame of spaceship 1, spaceship 2 is moving with speed $|u'| = 0.987c$. Use this to calculate the length of spaceship 2 as seen by an observer at rest relative to spaceship 1:

$$L = \frac{1}{\gamma}L_p = \sqrt{1 - \frac{|u'_{2x}|^2}{c^2}}L_p = \sqrt{1 - \frac{(0.987c)^2}{c^2}}(100\text{ m}) = \boxed{16\text{ m}}$$

(*d*) If the front ends of the spaceships are together at $t = 0$ on Earth, their back ends will be together after the time it takes either spaceship to move the length of the spaceship in Earth's frame:

$$t = \frac{L}{u} = \frac{53\text{ m}}{0.85c} = \frac{53\text{ m}}{(0.85)(3.00 \times 10^8\text{ m/s})} = \boxed{2.1 \times 10^{-7}\text{ s}}$$

CHECK As expected, the Part-(*c*) result is less than the Part-(*a*) result, and both results are less than the proper length of 100 m.

39-6 RELATIVISTIC MOMENTUM

We have seen in previous sections that Einstein's postulates require important modifications in our ideas of simultaneity and in our measurements of time and length. Einstein's postulates also require modifications in our concepts of mass, momentum, and energy. In classical mechanics, the momentum of a particle is defined as the product of its mass and its velocity, $m\vec{u}$, where $\vec{u}$ is the velocity. In an isolated system of particles, with no net force acting on the system, the total momentum of the system remains constant.

The reason that the total momentum of a system is important in classical mechanics is that it is conserved when there are no external forces acting on the system, as is the case in collisions. But we have just seen that $\Sigma m_i \vec{u}_i$ is conserved only in the approximation that $u \ll c$. We will define the relativistic momentum $\vec{p}$ of a particle to have the following properties:

1. In collisions, $\vec{p}$ is conserved.
2. As u/c approaches zero, $\vec{p}$ approaches $m\vec{u}$.

We will show that the quantity

$$\vec{p} = \frac{m\vec{u}}{\sqrt{1 - \dfrac{u^2}{c^2}}} \tag{39-20}$$

RELATIVISTIC MOMENTUM

is conserved in the elastic collision shown in Figure 39-11. Because this quantity also approaches $m\vec{u}$ as u/c approaches zero, we take this equation for the definition of the **relativistic momentum** of a particle.

One interpretation of Equation 39-20 is that the mass of an object increases with speed. Then the quantity $m_{\text{rel}} = m/\sqrt{1 - (u^2/c^2)}$ is called the *relativistic mass*. The relativistic mass of a particle when it is at rest in some reference frame is then called its *rest mass m*. In this chapter, we will treat the terms mass and rest mass as synonymous, and both terms will be labeled m.

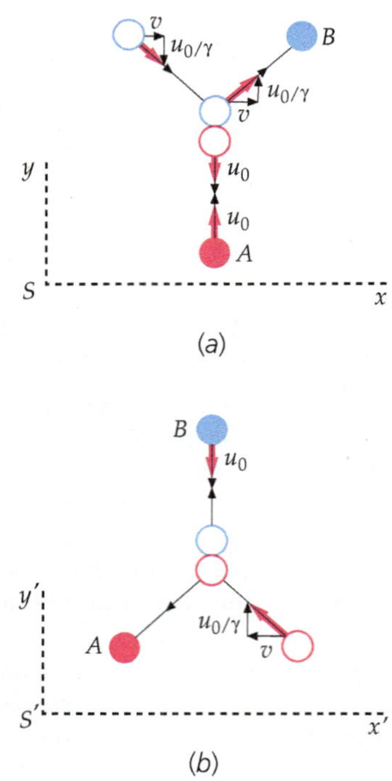

FIGURE 39-11 (*a*) Elastic collision of two identical pucks as seen in frame *S*. The vertical component of the velocity of puck *B* is u_0/γ in *S* if it is u_0 in *S′*. (*b*) The same collision as seen in *S′*. In this frame, puck *A* has a vertical component of velocity equal to u_0/γ.

ILLUSTRATION OF CONSERVATION OF THE RELATIVISTIC MOMENTUM

We consider two observers: one observer at rest in reference frame S, and the other observer at rest in frame S', which is moving to the right in the $+x$ direction with speed v relative to frame S. Each has a puck of mass m that can slide freely across a flat horizontal surface. The two pucks are identical when compared at rest. One observer launches puck A in the $+y$ direction with a speed u_0 relative to himself and the other launches puck B in the $-y$ direction with a speed u_0 relative to himself, so that each puck makes an elastic collision with the other puck, and returns to the person that launched it. Figure 39-11 shows how the collision looks in each reference frame.

We will compute the y component of the relativistic momentum of each puck in the reference frame S for the collision and show that the y component of the total relativistic momentum is zero. The speed of puck A in S is u_0, so the y component of its relativistic momentum is

$$p_{Ay} = \frac{mu_0}{\sqrt{1 - (u_0^2/c^2)}}$$

The speed of puck B in S is more complicated. Its x component is v and its y component is $-u_0/\gamma$ (Equation 39-18*b*). Thus,

$$u_B^2 = u_{Bx}^2 + u_{By}^2 = v^2 + \left[-u_0\sqrt{1 - (v^2/c^2)}\right]^2 = v^2 + u_0^2 - \frac{u_0^2 v^2}{c^2}$$

Using this result to compute $\sqrt{1-(u_B^2/c^2)}$ we obtain

$$1-\frac{u_B^2}{c^2}=1-\frac{v^2}{c^2}-\frac{u_0^2}{c^2}+\frac{u_0^2v^2}{c^4}=\left(1-\frac{v^2}{c^2}\right)\left(1-\frac{u_0^2}{c^2}\right)$$

and

$$\sqrt{1-(u_B^2/c^2)}=\sqrt{1-(v^2/c^2)}\,\sqrt{1-(u_0^2/c^2)}=\left(\frac{1}{\gamma}\right)\sqrt{1-(u_0^2/c^2)}$$

The y component of the relativistic momentum of puck B as seen in S is therefore

$$p_{By}=\frac{mu_{By}}{\sqrt{1-(u_B^2/c^2)}}=\frac{-mu_0/\gamma}{(1/\gamma)\sqrt{1-(u_0^2/c^2)}}=\frac{-mu_0}{\sqrt{1-(u_0^2/c^2)}}$$

Because $p_{By}=-p_{Ay}$, the y component of the total momentum of the two pucks is zero. If the y component of the momentum of each puck is reversed by the collision, the total momentum will remain zero and momentum will be conserved.

39-7 RELATIVISTIC ENERGY

In classical mechanics, the work done by the net force acting on a particle equals the change in the kinetic energy of the particle. In relativistic mechanics, we equate the net force to the rate of change of the relativistic momentum. The work done by the net force can then be calculated and set equal to the change in kinetic energy.

As in classical mechanics, we will define kinetic energy as the work done by the net force in accelerating a particle from rest to some final velocity u_f. Considering one dimension only, we have

$$K=\int_{u=0}^{u=u_f}F_{\text{net}}\,ds=\int_{u=0}^{u=u_f}\frac{dp}{dt}\,ds=\int_{u=0}^{u=u_f}u\,dp=\int_{u=0}^{u=u_f}u\,d\left(\frac{mu}{\sqrt{1-(u^2/c^2)}}\right) \qquad 39\text{-}21$$

where we have used $u=ds/dt$. It is left as a problem (Problem 35) for you to show that

$$d\left(\frac{mu}{\sqrt{1-(u^2/c^2)}}\right)=m\left(1-\frac{u^2}{c^2}\right)^{-3/2}du$$

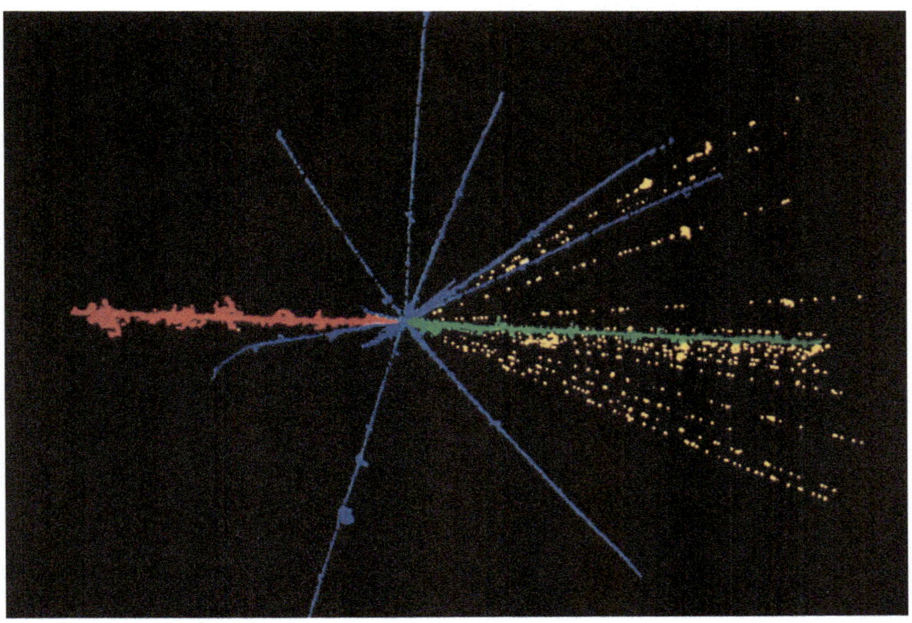

The creation of elementary particles demonstrates the conversion of kinetic energy to rest energy. In this 1950 photograph of a cosmic ray shower, a high-energy sulfur nucleus (red) collides with a nucleus in a photographic emulsion and produces a spray of particles, including a fluorine nucleus (green), other nuclear fragments (blue), and approximately 16 pions (yellow). *(© C. Powell, P. Fowler, and D. Perkins. Science Photo Library/Photo Researchers.)*

If we substitute that expression into the integrand in Equation 39-21, we obtain

$$K = \int_{u=0}^{u=u_f} u\, d\left(\frac{mu}{\sqrt{1-(u^2/c^2)}}\right) = \int_0^{u_f} m\left(1-\frac{u^2}{c^2}\right)^{-3/2} u\, du$$

$$= mc^2\left(\frac{1}{\sqrt{1-(u_f^2/c^2)}} - 1\right)$$

or

$$K = \frac{mc^2}{\sqrt{1-(u^2/c^2)}} - mc^2 \qquad \text{39-22}$$

RELATIVISTIC KINETIC ENERGY

(In this expression the final speed u_f is arbitrary, so the subscript f is not needed.)

The expression for kinetic energy consists of two terms. The first term depends on the speed of the particle. The second term, mc^2, is independent of the speed. The quantity mc^2 is called the **rest energy** E_0 of the particle. The rest energy is the product of the mass and c^2:

$$E_0 = mc^2 \qquad \text{39-23}$$

REST ENERGY

The total **relativistic energy** E is then defined to be the sum of the kinetic energy and the rest energy:

$$E = K + mc^2 = \frac{mc^2}{\sqrt{1-(u^2/c^2)}} \qquad \text{39-24}$$

RELATIVISTIC ENERGY

Thus, the work done by an unbalanced force increases the energy from the rest energy mc^2 to the final energy $mc^2/\sqrt{1-(u^2/c^2)}$. We can obtain a useful expression for the velocity of a particle by multiplying Equation 39-20 for the relativistic momentum by c^2 and comparing the result with Equation 39-24 for the relativistic energy. We have

$$pc^2 = \frac{mc^2u}{\sqrt{1-(u^2/c^2)}} = Eu$$

Dividing both sides by cE gives

$$\frac{u}{c} = \frac{pc}{E} \qquad \text{39-25}$$

Energies in atomic and nuclear physics are usually expressed in units of electron volts (eV) or mega-electron volts (MeV):

$$1\text{ eV} = 1.602 \times 10^{-19}\text{ J}$$

A convenient unit for the masses of atomic particles is eV/c^2 or MeV/c^2, which is the rest energy of the particle divided by c^2. The rest energies of some elementary particles and light nuclei are given in Table 39-1.

Table 39-1 Rest Energies of Some Elementary Particles and Light Nuclei

Particle	Symbol	Rest energy, MeV
Photon	γ	0
Electron (positron)	e or $e^-(e^+)$	0.5110
Muon	$\mu^\pm$	105.7
Pion	π^0	135.0
	$\pi^\pm$	139.6
Proton	^{1}H or p	938.272
Neutron	n	939.565
Deuteron	^{2}H or d	1875.613
Triton	^{3}H or t	2808.920
Helion	^{3}He or h	2808.391
Alpha particle	^{4}He or α	3727.379

Example 39-11 Total Energy, Kinetic Energy, and Momentum

An electron (rest energy 0.511 MeV) moves with speed $u = 0.800c$. Find (*a*) its total energy, (*b*) its kinetic energy, and (*c*) the magnitude of its momentum.

PICTURE This problem involves substituting into Equations 39-20 to 39-25.

SOLVE

(*a*) The total energy is given by Equation 39-24:

$$E = \frac{mc^2}{\sqrt{1-(u^2/c^2)}} = \frac{0.511\text{ MeV}}{\sqrt{1-0.64}} = \frac{0.511\text{ MeV}}{0.6} = \boxed{0.852\text{ MeV}}$$

(*b*) The kinetic energy is the total energy minus the rest energy:

$$K = E - mc^2 = 0.852\text{ MeV} - 0.511\text{ MeV} = \boxed{0.341\text{ MeV}}$$

(*c*) The magnitude of the momentum is found from Equation 39-20. We can simplify the momentum expression by multiplying both numerator and denominator by c^2 and using the Part-(*a*) result:

$$p = \frac{mu}{\sqrt{1-(u^2/c^2)}}$$

$$= \frac{mc^2}{\sqrt{1-(u^2/c^2)}}\frac{u}{c^2} = (0.852\text{ MeV})\frac{0.8c}{c^2} = \boxed{0.682\text{ MeV}/c}$$

CHECK The kinetic energy is less than the total energy as expected.

TAKING IT FURTHER The technique used to solve Part (*c*) (multiplying numerator and denominator by c^2) is equivalent to using Equation 39-25.

The expression for kinetic energy given by Equation 39-22 does not look much like the classical expression $\frac{1}{2}mu^2$. However, when u is much less than c, we can approximate $1/\sqrt{1-(u^2/c^2)}$ using the binomial expansion

$$(1+x)^n = 1 + nx + \frac{n(n-1)}{2}x^2 + \cdots \approx 1 + nx \qquad x \ll 1 \qquad \text{39-26}$$

Then

$$\frac{1}{\sqrt{1-(u^2/c^2)}} = \left(1 - \frac{u^2}{c^2}\right)^{-1/2} \approx 1 + \frac{1}{2}\frac{u^2}{c^2} \qquad u \ll c$$

See **Math Tutorial** *for more information on the* **Binomial Expansion**

From this result, when u is much less than c, the expression for relativistic kinetic energy becomes

$$K = mc^2\left[\frac{1}{\sqrt{1-(u^2/c^2)}} - 1\right] \approx mc^2\left[1 + \frac{1}{2}\frac{u^2}{c^2} - 1\right] = \frac{1}{2}mu^2 \qquad u \ll c$$

Thus, at low speeds, the relativistic expression is the same as the classical expression.

We note from Equation 39-24 that as the speed u approaches the speed of light c, the energy of the particle becomes very large (because $1/\sqrt{1-(u^2/c^2)}$ becomes very large). At $u = c$, the energy becomes infinite. A simple interpretation of the result is that it takes an infinite amount of energy to accelerate a particle (that has mass) to the speed of light.

In practical applications, the momentum or energy of a particle is often known rather than the speed. Equation 39-20 for the relativistic momentum and Equation 39-24 for the relativistic energy can be combined to eliminate the speed u. The result is

$$E^2 = p^2c^2 + (mc^2)^2 \qquad 39\text{-}27$$

RELATION FOR TOTAL ENERGY, MOMENTUM, AND REST ENERGY

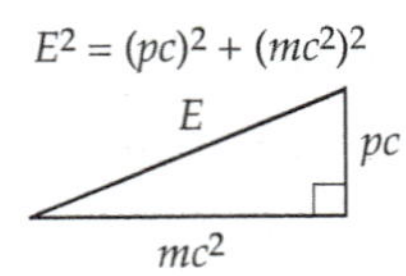

FIGURE 39-12 Right triangle to remember Equation 39-27.

This useful equation can be conveniently remembered from the right triangle shown in Figure 39-12.

PRACTICE PROBLEM 39-2

A proton (mass equal to 938 MeV/c^2) moving at speed u has a total energy of 1400 MeV. Find (*a*) $1/\sqrt{1-(u^2/c^2)}$, (*b*) the momentum of the proton, and (*c*) the speed u of the proton.

If the energy of a particle is much greater than its rest energy mc^2, the second term on the right side of Equation 39-27 can be neglected, giving the useful approximation

$$E \approx pc \qquad E \gg mc^2 \qquad 39\text{-}28$$

Equation 39-28 is an exact relation between energy and momentum for particles that do not have mass, such as photons.

MASS AND ENERGY

Einstein considered the relation $E_0 = mc^2$ (Equation 39-23) relating the energy of a particle to its mass to be the most significant result of the theory of relativity. Energy and inertia, which were formerly two distinct concepts, are related through this famous equation. As discussed in Chapter 7, the conversion of rest energy to kinetic energy with a corresponding decrease in mass is a common occurrence in radioactive decay and nuclear reactions, including nuclear fission and nuclear fusion. We illustrated this in Section 7-4 with the deuteron, whose mass is 2.22 MeV/c^2 less than the mass of its parts—a proton and a neutron. When a neutron and a proton combine to form a deuteron, 2.22 MeV of energy is released. The breaking up of a deuteron into a neutron and a proton requires 2.22 MeV of energy input. The proton and the neutron are thus bound together in a deuteron by a binding energy of 2.22 MeV. Any stable composite particle, such as a deuteron or an alpha particle (2 neutrons plus 2 protons), that is made up of other particles has a mass and rest energy that are less than the sum of the masses and rest energies of its parts. The difference in these rest energies is the binding energy of the composite particle. The binding energies of atoms and molecules are of the order of a few electron volts, which explains why there is only a negligible difference in mass between the composite particle and its parts. The binding energies of nuclei are of the order of several MeV, which explains why there is a noticeable difference in mass between the composite particle and its parts. Some very heavy nuclei, such as radium, are radioactive and decay into a less massive nucleus plus an alpha particle. In this case, the original nucleus has a rest energy greater than that of the decay particles. The excess energy appears as the kinetic energy of the decay products.

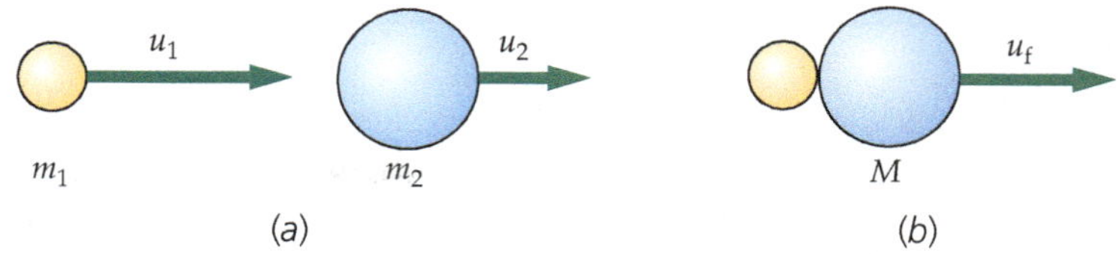

FIGURE 39-13 A perfectly inelastic collision between two particles. One particle of mass m_1 collides with another particle of mass m_2. After the collision, the particles stick together, forming a composite particle of mass M that moves with speed u_f so that relativistic momentum is conserved. Kinetic energy is lost in the process. If we assume that the total energy is conserved, the loss in kinetic energy must equal c^2 multiplied by the increase in the mass of the system.

To further illustrate the connection between mass and energy, we consider a perfectly inelastic collision of two particles. Classically, kinetic energy is lost during such a collision. Relativistically, this loss in kinetic energy shows up as an increase in rest energy of the system; that is, the total energy of the system is conserved. Consider a particle of mass m_1 moving with initial speed u_1 that collides with a particle of mass m_2 moving with initial speed u_2. The particles collide and stick together, forming a particle of mass M that moves with speed u_f, as shown in Figure 39-13. The initial total energy of particle 1 is

$$E_1 = K_1 + m_1c^2$$

where K_1 is its initial kinetic energy. Similarly the initial total energy of particle 2 is

$$E_2 = K_2 + m_2c^2$$

The total initial energy of the system is

$$E_i = E_1 + E_2 = K_1 + m_1c^2 + K_2 + m_2c^2 = K_i + M_ic^2$$

where $K_i = K_1 + K_2$ and $M_i = m_1 + m_2$ are the initial kinetic energy and initial mass of the system. The final total energy of the system is

$$E_f = K_f + M_fc^2$$

If we set the final total energy equal to the initial total energy, we obtain

$$K_f + M_fc^2 = K_i + M_ic^2$$

Rearranging gives $K_f - K_i = -(M_f - M_i)c^2$, which can be expressed

$$\Delta K + (\Delta M)c^2 = 0 \qquad \text{39-29}$$

where $\Delta M = M_f - M_i$ is the change in mass of the system.

Example 39-12 Totally Inelastic Collision

A particle of mass 2.00 MeV/c^2 and kinetic energy 3.00 MeV collides with a stationary particle of mass 4.00 MeV/c^2. After the collision, the two particles stick together. Find (*a*) the magnitude of the initial momentum of the system, (*b*) the final velocity of the two-particle system, and (*c*) the mass of the two-particle system.

PICTURE (*a*) The initial momentum of the system is the initial momentum of the incoming particle, which can be found from the total energy of the particle. (*b*) The final velocity of the system can be found from its total energy and momentum using $u/c = pc/E$ (Equation 39-25). The energy is found from conservation of energy, and the momentum from conservation of momentum. (*c*) Because the final energy and momentum are known, the final mass can be found using $E^2 = p^2c^2 + (mc^2)^2$.

SOLVE

(*a*) 1. The initial momentum of the system is the initial momentum of the incoming particle. The momentum of a particle is related to its energy and mass (Equation 39-27):

$$E_1^2 = p_1^2c^2 + (m_1c^2)^2$$

$$p_1c = \sqrt{E_1^2 - (m_1c^2)^2}$$

2. The total energy of the moving particle is the sum of its kinetic energy and its rest energy:

$$E_1 = 3.00 \text{ MeV} + 2.00 \text{ MeV} = 5.00 \text{ MeV}$$

3. Use the total energy to calculate the magnitude of the momentum:

$$p_1c = \sqrt{E_1^2 - (m_1c^2)^2} = \sqrt{(5.00 \text{ MeV})^2 - (2.00 \text{ MeV})^2} = \sqrt{21.0} \text{ MeV}$$

$$p_1 = \boxed{4.58 \text{ MeV}/c}$$

(*b*) 1. We can find the final velocity of the system from its total energy E_f and its momentum p_f using Equation 39-25:

$$\frac{u_f}{c} = \frac{p_fc}{E_f}$$

2. By the conservation of total energy, the final energy of the system equals the initial total energy of the two particles:

$$E_f = E_i = E_1 + E_2 = 5.00 \text{ MeV} + 4.00 \text{ MeV} = 9.00 \text{ MeV}$$

3. By the conservation of momentum, the final momentum of the two-particle system equals the initial momentum:

$$p_f = 4.58 \text{ MeV}/c$$

4. Calculate the velocity of the two-particle system from its total energy and momentum using $u/c = pc/E$:

$$\frac{u_f}{c} = \frac{p_fc}{E_f} = \frac{4.58 \text{ MeV}}{9.00 \text{ MeV}} = 0.509$$

$$u_f = \boxed{0.509c}$$

(*c*) We can find the mass M_f of the final two-particle system from Equation 39-27 using $pc = 4.58$ MeV and $E = 9.00$ MeV:

$$E_f^2 = (p_fc)^2 + (M_fc^2)^2$$

$$(9.00 \text{ MeV})^2 = (4.58 \text{ MeV})^2 + (M_fc^2)^2$$

$$M_f = \boxed{7.75 \text{ MeV}/c^2}$$

TAKING IT FURTHER Note that the mass of the system increased from 6.00 MeV/c^2 to 7.75 MeV/c^2. This mass increase, multiplied by c^2, equals the loss in kinetic energy of the system, as you will show in the following exercise.

PRACTICE PROBLEM 39-3 (*a*) Find the final kinetic energy of the two-particle system in Example 39-12. (*b*) Find the loss in kinetic energy, K_{loss}, in the collision. (*c*) Show that $K_{loss} = (\Delta M)c^2$, where ΔM is the change in mass of the system.

Example 39-13 Momentum and Total-Energy Conservation

A 1.00×10^6-kg rocket has 1.00×10^3 kg of fuel on board. The rocket is parked in space when it suddenly becomes necessary to accelerate. The rocket engines ignite, and the 1.00×10^3 kg of fuel are consumed. The exhaust (spent fuel) is ejected during a very short time interval at a speed of 0.500*c* relative to *S*—the inertial reference frame in which the rocket is initially at rest. (*a*) Calculate the change in the mass of the rocket–fuel system. (*b*) Calculate the final speed of the rocket u_R relative to *S*. (*c*) Again, calculate the final speed of the rocket relative to *S*, this time using classical (Newtonian) mechanics.

PICTURE The speed of the rocket and the change in the mass of the system can be calculated using conservation of momentum and conservation of energy. In reference frame *S*, the total momentum of the rocket plus fuel remains zero. After the burn, the magnitude of the momentum of the rocket equals that of the ejected fuel. Let $m_R = 1.00 \times 10^6$ kg be the mass of the rocket, not including the mass of the fuel, let $m_{Fi} = 1.00 \times 10^3$ kg be the mass of the fuel *before* the burn, and let m_{Ff} be the mass of the fuel *after* the burn. The mass of the rocket, m_R, remains fixed, but during the burn the mass of the fuel decreases. (The fuel has less chemical energy after the burn, and so has less mass as well.)

SOLVE

(*a*) 1. The magnitudes of the momentum of the rocket and the momentum of the ejected fuel are equal. For the reasons stated above, the mass of the rocket, not including the 1.00×10^3 kg of fuel, does not change during the burn:

$$p_R = p_F = p$$

$$\frac{m_R u_R}{\sqrt{1 - (u_R^2/c^2)}} = \frac{m_{Ff} u_F}{\sqrt{1 - (u_F^2/c^2)}} = p$$

$m_R = 1.00 \times 10^6$ kg, $u_F = 0.500c$, and u_R is the final speed of the rocket.

2. The total energy of the system does not change:

$$E_f = E_i$$

3. The initial energy is the rest energy of the rocket and fuel before the burn. The final energy is the energy of the rocket plus energy of the fuel. The energy of each is related to its momentum by Equation 39-27:

$$E_i = m_R c^2 + m_{Fi} c^2 = (m_R + m_{Fi})c^2$$

$$E_{Rf}^2 = p^2c^2 + (m_R c^2)^2$$

$$E_{Ff}^2 = p^2c^2 + (m_{Ff} c^2)^2$$

so

$$E_f = E_{Rf} + E_{Ff}$$

$$E_f = \sqrt{p^2c^2 + (m_R c^2)} + \sqrt{p^2c^2 + (m_{Ff} c^2)^2}$$

4. Equate the initial and final energies:

$$\sqrt{p^2c^2 + (m_R c^2)^2} + \sqrt{p^2c^2 + (m_{Ff} c^2)^2} = (m_R + m_{Fi})c^2$$

5. The step-4 result and $p = \dfrac{m_{Ff} u_F}{1 - (u_F^2/c^2)}$ (the step-1 result) constitute two simultaneous equations with unknowns p and m_{Ff}. Solving for m_{Ff} gives:

$$m_{Ff} = 866 \text{ kg}$$

so

$$m_{\text{loss}} = m_{Fi} - m_{Ff} = 1000 \text{ kg} - 866 \text{ kg} = \boxed{134 \text{ kg}}$$

(*b*) 1. To solve for u_R, we use Equation 39-25:

$$\frac{u_R}{c} = \frac{pc}{E_{Rf}}$$

2. To solve for p, we substitute the value for m_{Ff} into the Part (*a*), step-1 result:

$$p = \frac{m_{Ff} u_F}{\sqrt{1 - (u_F^2/c^2)}} = \frac{(866 \text{ kg})0.500c}{\sqrt{1 - 0.250}}$$

$$= (5.00 \times 10^2 \text{ kg})c$$

3. We use the value for p to solve for E_{Rf}:

$$E_{Rf}^2 = p^2c^2 + (m_R c^2)^2$$

$$= (5.00 \times 10^2 \text{ kg})^2c^4 + (1.00 \times 10^6 \text{ kg})^2c^4$$

$$= (1.00 \times 10^{12} \text{ kg}^2)c^4$$

so

$$E_{Rf} = (1.00 \times 10^6 \text{ kg})c^2$$

4. Using our Part (*b*), step-1 result, we solve for u_R:

$$u_R = \frac{pc^2}{E_{Rf}} = \frac{(5.00 \times 10^2 \text{ kg})c^3}{(1.00 \times 10^6 \text{ kg})c^2}$$

$$= \boxed{5.00 \times 10^{-4}c = 1.50 \times 10^{-5} \text{ m/s}}$$

(*c*) Equate the magnitude of the classical expressions for the momentum of the rocket and burned fuel and solve for u_R:

$$m_R u_R = m_F u_F$$

$$u_R = \frac{m_F}{m_R} u_F = \frac{1.00 \times 10^3 \text{ kg}}{1.00 \times 10^6 \text{ kg}} 0.500c$$

$$= 5.00 \times 10^{-4}c$$

$$= \boxed{1.50 \times 10^5 \text{ m/s}}$$

CHECK We find the result of the relativistic calculation of the final rocket speed to differ from the classical result. If carried out to five figures, the relativistic calculation gives $u_R = 4.9994 \times 10^{-4}\,c$ for the final speed of the rocket. However, the classical calculation gives $u_R = 5.0000 \times 10^{-4}\,c$. These two values differ by less than one part in 8000.

CONCEPT CHECK 39-1

If the matter being ejected were a 1.00×10^3-kg rigid block launched by a spring with one end attached to the rocket, would the mass of the block change or would the mass of the spring change?

39-8 GENERAL RELATIVITY

The generalization of the theory of relativity to noninertial reference frames by Einstein in 1916 is known as the general theory of relativity. It is much more difficult mathematically than the special theory of relativity, and there are fewer situations in which it can be tested. Nevertheless, its importance calls for a brief qualitative discussion.

The basis of the general theory of relativity is the **principle of equivalence:**

> A homogeneous gravitational field is completely equivalent to a uniformly accelerated reference frame.
>
> PRINCIPLE OF EQUIVALENCE

This principle arises in Newtonian mechanics because of the apparent identity of gravitational mass and inertial mass. In a uniform gravitational field, all objects fall with the same acceleration $\vec{g}$ independent of their masses because the gravitational force is proportional to the (gravitational) mass, whereas the acceleration varies inversely with the (inertial) mass. Consider a compartment in space undergoing a uniform acceleration $\vec{a}$, as shown in Figure 39-14*a*. No mechanics experiment can be performed inside the compartment that will distinguish whether the compartment is actually accelerating in space or is at rest (or is moving with uniform velocity) in the presence of a uniform gravitational field $\vec{g} = -\vec{a}$, as shown in Figure 39-14*b*. If objects are dropped in the compartment, they will fall to the floor with an acceleration $\vec{g} = -\vec{a}$. If people stand on a spring scale, it will read their weight of magnitude $ma = mg$.

Einstein assumed that the principle of equivalence applies to all physics and not just to mechanics. In effect, he assumed that there is no experiment of any kind that can distinguish uniformly accelerated motion from the presence of a gravitational field.

One consequence of the principle of equivalence—the deflection of a light beam in a gravitational field—was one of the first to be tested experimentally. In a region that has no gravitational field, a light beam will travel in a straight line at speed c. The principle of equivalence tells us that a region that has no gravitational field exists only in a compartment that is in free fall. Figure 39-15 shows a beam of light entering a compartment that is accelerating relative to a nearby reference frame in free fall. Successive positions of the compartment at equal time intervals are shown in Figure 39-15*a*. Because the compartment is accelerating, the distance it moves in each time interval increases with time. The path of the beam of light as observed from inside the compartment is therefore a parabola, as shown in Figure 39-15*b*.

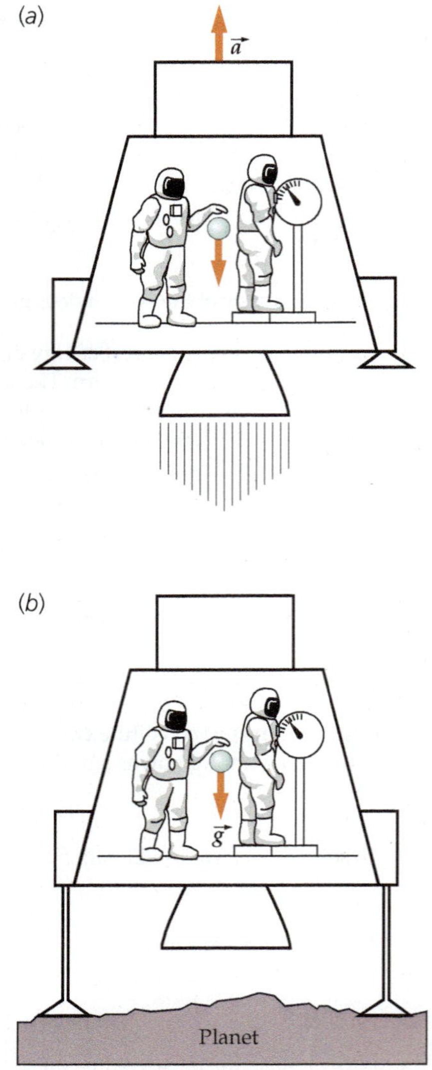

FIGURE 39-14 The results of experiments in a uniformly accelerated reference frame (*a*) cannot be distinguished from those in a uniform gravitational field (*b*) if the acceleration $\vec{a}$ and the gravitational field $\vec{g}$ have the same magnitude.

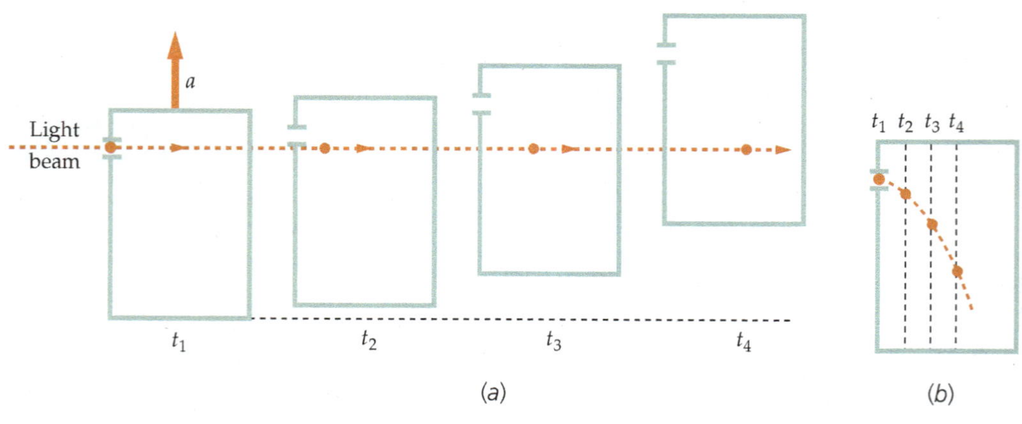

FIGURE 39-15 (*a*) A light beam moving in a straight line through a compartment that is undergoing uniform acceleration relative to a nearby reference frame in free fall. The position of the beam is shown at equally spaced times t_1, t_2, t_3, and t_4. (*b*) In the reference frame of the compartment, the light travels in a parabolic path as a ball would if it were projected horizontally. The vertical displacements are greatly exaggerated for emphasis.

But according to the principle of equivalence, there is no way to distinguish between an accelerating compartment and one moving with uniform velocity in a uniform gravitational field. We conclude, therefore, that a beam of light will accelerate in a gravitational field, just like objects that have mass. For example, near the surface of Earth, light will fall with an acceleration of 9.81 m/s^2. This is difficult to observe because of the enormous speed of light. In a distance of 3000 km, which takes light about 0.01 s to traverse, a beam of light should fall approximately 0.5 mm. Einstein pointed out that the deflection of a light beam in a gravitational field might be observed when light from a distant star passes close to the Sun, as illustrated in Figure 39-16. Because of the brightness of the Sun, this cannot ordinarily be seen. Such a deflection was first observed in 1919 during an eclipse of the Sun. This well-publicized observation brought instant worldwide fame to Einstein.

A second prediction from Einstein's theory of general relativity, which we will not discuss in detail, is the excess precession of the perihelion of the orbit of Mercury of about 0.01° per century. This effect had been known and unexplained for some time, so, in a sense, explaining it constituted an immediate success of the theory.

A third prediction of general relativity concerns the change in time intervals and frequencies of light in a gravitational field. In Chapter 11, we found that the gravitational potential energy between two masses M and m a distance r apart is

$$U = -\frac{GMm}{r}$$

where G is the universal gravitational constant, and the point of zero potential energy has been chosen to be when the separation of the masses is infinite. The potential energy per unit mass near a mass M is called the *gravitational potential* ϕ:

$$\phi = -\frac{GM}{r} \qquad 39\text{-}30$$

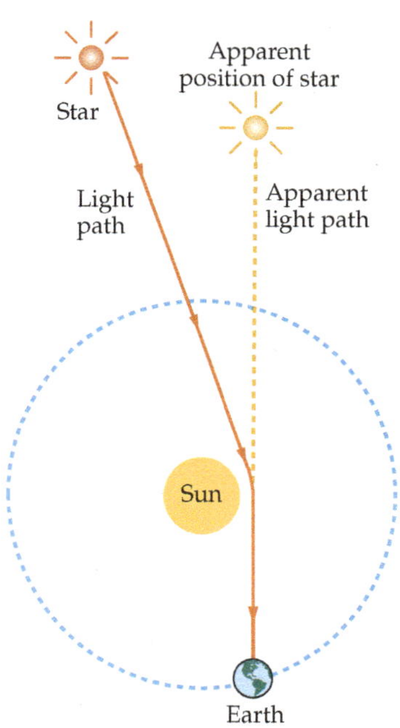

FIGURE 39-16 The deflection (greatly exaggerated) of a beam of light due to the gravitational attraction of the Sun.

The quartz sphere in the top part of the container is probably the world's most perfectly round object. It is designed to spin as a gyroscope in a satellite orbiting Earth. General relativity predicts that the rotation of Earth will cause the axis of rotation of the gyroscope to precess in a circle at a rate of approximately 1 revolution in 100 000 years. *(Michael Freeman.)*

According to the general theory of relativity, clocks run more slowly in regions of lower gravitational potential. (Because the gravitational potential is negative, as can be seen from Equation 39-30, the nearer the mass the more negative, and therefore, the lower the gravitational potential.) If Δt_1 is a time interval between two events measured by a clock where the gravitational potential is ϕ_1 and Δt_2 is the interval between the same two events as measured by a clock where the gravitational potential is ϕ_2, general relativity predicts that the fractional difference between the times will be approximately*

$$\frac{\Delta t_2 - \Delta t_1}{\Delta t} = \frac{1}{c^2}(\phi_2 - \phi_1) \qquad 39\text{-}31$$

This extremely accurate hydrogen maser clock was launched in a satellite in 1976, and its time was compared to that of an identical clock on Earth. In accordance with the prediction of general relativity, the clock on Earth, where the gravitational potential was lower, lost approximately 4.3×10^{-10} s each second compared with the clock orbiting Earth at an altitude of approximately 10 000 km. *(NASA.)*

A clock in a region of low gravitational potential will therefore run more slowly than a clock in a region of higher gravitional potential. Because a vibrating atom can be considered to be a clock, the frequency of vibration of an atom in a region of low potential, such as near the Sun, will be lower than the frequency of vibration of the same atom on Earth. This shift toward a lower frequency, and therefore a longer wavelength, is called the **gravitational redshift.**

As our final example of the predictions of general relativity, we mention **black holes,** which were first predicted by J. Robert Oppenheimer and Hartland Snyder in 1939. According to the general theory of relativity, if the density of an object such as a star is great enough, its gravitational attraction will be so great that once inside a critical radius, nothing can escape, not even light or other electromagnetic radiation. (The effect of a black hole on objects outside the critical radius is the same as that of any other mass.) A remarkable property of such an object is that nothing that happens inside it can be communicated to the outside. As sometimes occurs in physics, a simple but incorrect calculation gives the correct results for the relation between the mass and the critical radius of a black hole. In Newtonian mechanics, the speed needed for a particle to escape from the surface of a planet or a star of mass M and radius R is given by Equation 11-21:

$$v_e = \sqrt{\frac{2GM}{R}}$$

If we set the escape speed equal to the speed of light and solve for the radius, we obtain the critical radius R_S, called the **Schwarzschild radius:**

$$R_S = \frac{2GM}{c^2} \qquad 39\text{-}32$$

For an object that has a mass equal to five times that of our Sun (theoretically the minimum mass for a black hole) to be a black hole, its radius would have to be approximately 15 km. Because no radiation is emitted from a black hole and its radius is expected to be small, the detection of a black hole is not easy. The best chance of detection occurs in a binary-star system in which a black hole is a close companion to a normal star. Then both stars revolve around their center of mass and the gravitational field of the black hole will pull gas from the normal star into the black hole. However, to conserve angular momentum, the gas does not go straight into the black hole. Instead, the gas orbits around the black hole in a disk, called an accretion disk, while slowly being pulled closer to the black hole. The gas in this disk emits X rays because the temperature of the gas being pulled inward reaches several million kelvins. The mass of a black-hole candidate can often be estimated. An estimated mass of at least five solar masses, along with the emission of X rays, establishes a strong inference that the candidate is, in fact, a black hole. In addition to the black holes just described, there are supermassive black holes that exist at the centers of galaxies. At the center of the Milky Way is a supermassive black hole that has a mass of about two million solar masses.

* Because this shift is usually very small, it does not matter by which interval we divide on the left side of the equation.

Summary

TOPIC	RELEVANT EQUATIONS AND REMARKS	
1. **Einstein's Postulates**	The special theory of relativity is based on two postulates of Albert Einstein. All of the results of special relativity can be derived from these postulates. Postulate 1: Absolute uniform motion cannot be detected. Postulate 2: The speed of light is independent of the motion of the source. An important implication of these postulates is Postulate 2 (alternate): Every observer measures the same value c for the speed of light.	
2. **The Lorentz Transformation**	$x = \gamma(x' + vt'), \quad y = y', \quad z = z'$	39-9
	$t = \gamma\left(t' + \frac{vx'}{c^2}\right)$	39-10
	$\gamma = \frac{1}{\sqrt{1 - (v^2/c^2)}}$	39-7
Inverse transformation	$x' = \gamma(x - vt), \quad y' = y, \quad z' = z$	39-11
	$t' = \gamma\left(t - \frac{vx}{c^2}\right)$	39-12
3. **Time Dilation**	The time interval measured between two events that occur at the same point in space in some reference frame is called the proper time interval Δt_p between those two events. In another reference frame in which the same two events occur at different places, the time interval Δt between the events is longer by the factor γ.	
	$\Delta t = \gamma\, \Delta t_p$	39-13
4. **Length Contraction**	The length of an object measured in the reference frame in which the object is at rest is called its proper length L_p. When measured in another reference frame, the length of the object along the direction parallel to the velocity of the object is	
	$L = \frac{L_p}{\gamma}$	39-14
5. **The Relativistic Doppler Effect**	$f' = \frac{\sqrt{1 - (v^2/c^2)}}{1 - (v/c)} f_0$ approaching	39-16*a*
	$f' = \frac{\sqrt{1 - (v^2/c^2)}}{1 + (v/c)} f_0$ receding	39-16*b*
6. **Clock Synchronization and Simultaneity**	Two events that are simultaneous in one reference frame typically are not simultaneous in another frame that is moving relative to the first. If two clocks are synchronized in the frame in which they are at rest, they will be out of synchronization in another frame. In the frame in which they are moving, the chasing clock leads by an amount	
	$\Delta t_S = L_p \frac{v}{c^2}$	39-17
	where L_p is the proper distance between the clocks.	
7. **The Velocity Transformation**	$u_x = \frac{u'_x + v}{1 + (vu'_x/c^2)}$	39-18*a*
	$u_y = \frac{u'_y}{\gamma[1 + (vu'_x/c^2)]}$	39-18*b*
	$u_z = \frac{u'_z}{\gamma[1 + (vu'_x/c^2)]}$	39-18*c*

TOPIC	RELEVANT EQUATIONS AND REMARKS	
Inverse velocity transformation	$u'_x = \dfrac{u_x - v}{1 - (vu_x/c^2)}$	39-19*a*
	$u'_y = \dfrac{u_y}{\gamma[1 - (vu_x/c^2)]}$	39-19*b*
	$u'_z = \dfrac{u_z}{\gamma[1 - (vu_x/c^2)]}$	39-19*c*
8. Relativistic Momentum	$\vec{p} = \dfrac{m\vec{u}}{\sqrt{1 - (u^2/c^2)}}$	39-20
	where m is the mass of the particle.	
9. Relativistic Energy		
Kinetic energy	$K = \dfrac{mc^2}{\sqrt{1 - (u^2/c^2)}} - mc^2$	39-22
Rest energy	$E_0 = mc^2$	39-23
Total Relativistic energy	$E = K + E_0 = \dfrac{mc^2}{\sqrt{1 - (u^2/c^2)}}$	39-24
10. Useful Formulas for Speed, Energy, and Momentum	$\dfrac{u}{p} = \dfrac{pc}{E}$	39-25
	$E^2 = p^2c^2 + (mc^2)^2$	39-27
	$E \approx pc \qquad E \gg mc^2$	39-28

Answer to Concept Check

39-1 Only the rest mass of the spring would change.

Answers for Practice Problems

39-1 1.67 h

39-2 (*a*) 1.49, (*b*) $p = 1.04 \times 10^3$ MeV/c, and (*c*) $u = 0.74c$

39-3 (*a*) $K_f = E_f - M_fc^2 = 9.00\text{ MeV} - 7.75\text{ MeV} = 1.25\text{ MeV}$,

(*b*) $K_{loss} = K_i - K_f = 3.00\text{ MeV} - 1.25\text{ MeV} = 1.75\text{ MeV}$, and

(*c*) $(\Delta M)c^2 = (M_f - M_i)c^2 = 7.75\text{ MeV} - (2.00\text{ MeV} + 4.00\text{ MeV}) = 1.75\text{ MeV} = K_{loss}$

Problems

In a few problems, you are given more data than you actually need; in a few other problems, you are required to supply data from your general knowledge, outside sources, or informed estimate.

Interpret as significant all digits in numerical values that have trailing zeros and no decimal points.

- • Single-concept, single-step, relatively easy
- •• Intermediate-level, may require synthesis of concepts
- ••• Challenging
- SSM Solution is in the *Student Solutions Manual*
- Consecutive problems that are shaded are paired problems.

CONCEPTUAL PROBLEMS

1 • The approximate total energy of a particle of mass m moving at speed $u \ll c$ is (*a*) $mc^2 + \frac{1}{2}mu^2$, (*b*) $\frac{1}{2}mu^2$, (*c*) cmu, (*d*) mc^2, (*e*) $\frac{1}{2}cmu$. SSM

2 • A set of twins work in an office building. One twin works on the top floor and the other twin works in the basement. Considering general relativity, which twin will age more quickly? (*a*) They will age at the same rate. (*b*) The twin who works on the top floor will age more quickly. (*c*) The twin who works in the basement will age more quickly. (*d*) It depends on the speed of the office building. (*e*) none of the above

3 • True or false:
(*a*) The speed of light is the same in all reference frames.
(*b*) The time interval between two events is never shorter than the proper time interval between the two events.
(*c*) Absolute motion can be determined by means of length contraction.
(*d*) The light-year is a unit of distance.
(*e*) Simultaneous events must occur at the same place.
(*f*) If two events are not simultaneous in one frame, they cannot be simultaneous in any other frame.
(*g*) The mass of a system that consists of two particles tightly bound together by attractive forces is less than the sum of the masses of the individual particles when separated.

4 • An observer sees a system moving past her that consists of a mass oscillating on the end of a spring and measures the period T of the oscillations. A second observer, who is moving with the mass–spring system, also measures its period. The second observer will find a period that is (*a*) equal to T, (*b*) less than T, (*c*) greater than T, (*d*) either (*a*) or (*b*) depending on whether the system was approaching or receding from the first observer, (*e*) Not enough information is given to answer the question.

5 • The Lorentz transformation for y and z is the same as the classical result: $y = y'$ and $z = z'$. Yet the relativistic velocity transformation does not give the classical result $u_y = u'_y$ and $u_z = u'_z$. Explain why this result occurs.

ESTIMATION AND APPROXIMATION

6 •• The Sun radiates energy at the rate of approximately 4×10^{26} W. Assume that this energy is produced by a reaction whose net result is the fusion of four protons to form a single ^{4}He nucleus and the release of 25 MeV of energy that is radiated into space. Calculate the Sun's loss of mass per day.

7 •• The most distant galaxies that can be seen by the Hubble telescope are moving away from us and have a redshift parameter of about $z = 5$. [The redshift parameter z is defined as $(f - f')/f'$, where f is the frequency measured in the rest frame of the emitter and f' is the frequency measured in the rest frame of the receiver.] (*a*) What is the speed of the galaxies relative to us (expressed as a fraction of the speed of light)? (*b*) *Hubble's law* states that the recession speed is given by the expression $v = Hx$, where v is the speed of recession, x is the distance, and H, the Hubble constant, is equal to 75 km/s/Mpc, where 1 pc = 3.26 $c \cdot$y. (The abbreviation for parsec is pc.) Estimate the distance of such a galaxy from us using the information given. SSM

TIME DILATION AND LENGTH CONTRACTION

8 • The proper mean lifetime of a muon is 2.2 μs. Muons in a beam are traveling through a laboratory at 0.95c. (*a*) What is their mean lifetime as measured in the laboratory? (*b*) How far do they travel, on average, before they decay?

9 •• In the Stanford linear collider, small bundles of electrons and positrons are fired at each other. In the laboratory's frame of reference, each bundle is approximately 1.0 cm long and 10 μm in diameter. In the collision region, each particle has an energy of 50 GeV, and the electrons and the positrons are moving in opposite directions. (*a*) How long and how wide is each bundle in its own reference frame? (*b*) What must be the minimum proper length of the accelerator for a bundle to have both its ends simultaneously in the accelerator in its own reference frame? (The actual proper length of the accelerator is less than 1000 m.) (*c*) What is the length of a positron bundle in a reference frame that moves with the electron bundle?

10 • Use the binomial expansion equation

$$(1 + x)^n = 1 + nx + \frac{n(n-1)}{2}x^2 + \ldots \approx 1 + nx \qquad x \ll 1$$

to derive the following results for the case when v is much less than c.

$$(a)\ \gamma \approx 1 + \frac{1}{2}\frac{v^2}{c^2}$$

$$(b)\ \frac{1}{\gamma} \approx 1 - \frac{1}{2}\frac{v^2}{c^2}$$

$$(c)\ \gamma - 1 \approx 1 - \frac{1}{\gamma} \approx \frac{1}{2}\frac{v^2}{c^2}$$

11 •• Star A and Star B are at rest relative to Earth. Star A is 27 $c \cdot$y from Earth, and as viewed from Earth, Star B is located beyond (behind) Star A. (*a*) A spaceship is making a trip from Earth to Star A at a speed such that the trip from Earth to Star A takes 12 y according to clocks on the spaceship. At what speed, relative to Earth, must the spaceship travel? (Assume that the times for the accelerations are very short compared to the overall trip time.) (*b*) Upon reaching Star A, the spaceship speeds up and departs for Star B at a speed such that the gamma factor, γ, is twice that of Part (*a*). The trip from Star A to Star B takes 5.0 y (spaceship's time). How far, in $c \cdot$y, is Star B from Star A in the rest frame of Earth and the two stars? (*c*) Upon reaching Star B, the spaceship departs for Earth at the same speed as in Part (*b*). It takes it 10 y (spaceship's time) to return to Earth. If you were born on Earth the day the ship left Earth and you remain on Earth, how old are you on the day the ship returns to Earth?

12 • A spaceship travels to a star 35 $c \cdot$y away at a speed of 2.7×10^8 m/s. How long does the spaceship take to get to the star (*a*) as measured on Earth and (*b*) as measured by a passenger on the spaceship?

13 •• Unobtainium (Un) is an unstable particle that decays into normalium (Nr) and standardium (St) particles. (*a*) An accelerator produces a beam of Un that travels to a detector located 100 m away from the accelerator. The particles travel with a velocity of $v = 0.866c$. How long do the particles take (in the laboratory frame) to get to the detector? (*b*) By the time the particles get to the detector, half of the particles have decayed. What is the half-life of Un? (*Note:* half-life as it would be measured in a frame moving with the particles). (*c*) A new detector is going to be used, which is located 1000 m away from the accelerator. How fast should the particles be moving if half of the particles are to make it to the new detector? SSM

14 •• A clock on Spaceship A measures the time interval between two events, both of which occur at the location of the clock. You are on Spaceship B. According to your careful measurements, the time interval between the two events is 1.00 percent longer than that measured by the two clocks on Spaceship A. How fast is Spaceship A moving relative to Spaceship B. *Hint: Use one or more of the results of Problem 10.*

15 •• If a plane flies at a speed of 2000 km/h, how long must the plane fly before its clock loses 1.00 s because of time dilation? *Hint: Use one or more of the results of Problem 10.*

THE LORENTZ TRANSFORMATION, CLOCK SYNCHRONIZATION, AND SIMULTANEITY

16 •• Show that when $v \ll c$ the relativistic transformation equations for x, t, and u_x reduce to the classical transformation equations.

17 •• A spaceship of proper length $L_p = 400$ m moves past a transmitting station at a speed of $0.760c$. (The transmitting station broadcasts signals that travel at the speed of light.) A clock is attached to the nose of the spaceship and a second clock is attached to the transmitting station. The instant that the nose of the spaceship passes the transmitter, the clock attached to the transmitter and the clock attached to the nose of the spaceship are set equal to zero. The instant that the tail of the spaceship passes the transmitter a signal is sent by the transmitter that is subsequently detected by a receiver in the nose of the spaceship. (*a*) When, according to the clock attached to the nose of spaceship, is the signal sent? (*b*) When, according to the clocks attached to the nose of spaceship, is the signal received? (*c*) When, according to the clock attached to the transmitter, is the signal received by the spaceship? (*d*) According to an observer that works at the transmitting station, how far from the transmitter is the nose of the spaceship when the signal is received? SSM

18 •• In frame S, event B occurs 2.0 μs after event A, and event A occurs at the origin whereas event B occurs on the x axis at $x = 1.5$ km. How fast and in what direction must an observer be traveling along the x axis so that events A and B occur simultaneously? Is it possible for event B to precede event A for some observer?

19 •• Observers in reference frame S see an explosion located on the x axis at $x_1 = 480$ m. A second explosion occurs, 5.0 μs later, at $x_2 = 1200$ m. In reference frame S', which is moving along the x axis in the $+x$ direction at speed v, the two explosions occur at the same point in space. What is the separation in time between the two explosions as measured in S'?

20 ••• In reference frame S, events 1 and 2 are separated by a distance $D = x_2 - x_1$ and a time $T = t_2 - t_1$. (*a*) Use the Lorentz transformation to show that in frame S', which is moving along the x axis with speed v relative to S, the time separation is $t'_2 - t'_1 = \gamma(T - vD/c^2)$. (*b*) Show that the events can be simultaneous in frame S' only if D is greater than cT. (*c*) If one of the events is the *cause* of the other, the separation D must be less than cT, because D/c is the smallest time that a signal can take to travel from x_1 to x_2 in frame S. Show that if D is less than cT, t'_2 is greater than t'_1 in all reference frames. This shows that if the cause precedes the effect in one frame, it must precede it in all reference frames. (*d*) Suppose that a signal could be sent with speed $c' > c$ so that in frame S the cause precedes the effect by the time $T = D/c'$. Show that there is then a reference frame moving with speed v less than c in which the effect precedes the cause.

21 ••• A rocket that has a proper length of 700 m is moving to the right at a speed of $0.900c$. It has two clocks—one in the nose and one in the tail—that have been synchronized in the frame of the rocket. A clock on the ground and the clock in the nose of the rocket both read zero as they pass by each other. (*a*) At the instant the clock on the ground reads zero, what does the clock in the tail of the rocket read according to observers on the ground? When the clock in the tail of the rocket passes the clock on the ground, (*b*) what does the clock in the tail read according to observers on the ground, and (*c*) what does the clock in the nose read according to observers on the ground, and (*d*) what does the clock in the nose read according to observers on the rocket? (*e*) At the instant the clock in the nose of the rocket reads 1.00 h, a light signal is sent from the nose of the rocket to an observer standing by the clock on the ground. What does the clock on the ground read when the observer on the ground receives the signal? (*f*) When the observer on the ground receives the signal, he immediately sends a return signal to the nose of the rocket. What is the reading of the clock in the nose of the rocket when that signal is received at the nose of the rocket?

THE VELOCITY TRANSFORMATION AND THE RELATIVISTIC DOPPLER EFFECT

22 •• **SPREADSHEET** A spaceship, at rest in a certain reference frame S, is given a speed increase of $0.50c$ (call this increase boost 1). Relative to its new rest frame, the spaceship is given a further $0.50c$ increase 10 seconds later (as measured in its new rest frame; call this increase boost 2). This process is continued indefinitely, at 10-s intervals, as measured in the rest frame of the spaceship. (Assume that the boosts take a very short time compared to 10 s.) (*a*) Using a spreadsheet program, calculate and graph the speed of the spaceship in reference frame S as a function of the boost number for boost 1 to boost 10. (*b*) Graph the gamma factor in the same manner. (*c*) How many boosts does it take until the speed of the ship in S is greater than $0.999c$? (*d*) How far does the spaceship move between boost 1 and boost 6, as measured in reference frame S? What is the average speed of the spaceship between boost 1 and boost 6, as measured in S?

23 • Light is emitted by a sodium sample that is moving toward Earth with speed v. The wavelength of the light is 589 nm in the rest frame of the sample. The wavelength measured in the frame of Earth is 547 nm. Find v.

24 • A distant galaxy is moving away from us at a speed of 1.85×10^7 m/s. Calculate the fractional redshift $(\lambda' - \lambda_0)/\lambda_0$ that we observe the light from the galaxy to have.

25 •• Derive $f' = f_0\sqrt{1 - (v^2/c^2)}/[1 - (v/c)]$ (Equation 39-16*a*) for the frequency received by an observer moving with speed v toward a stationary source of electromagnetic waves.

26 • Show that if v is much less than c, the Doppler shift is given approximately by

$$\Delta f/f \approx \pm v/c$$

27 •• A clock is placed in a satellite that orbits Earth with an orbital period of 90 min. By what time interval will this clock differ from an identical clock on Earth after 1.0 y? (Assume that special relativity applies and neglect general relativity.) SSM

28 •• For light that is Doppler-shifted with respect to an observer, we define the redshift parameter $z = (f - f')/f'$, where f is the frequency of the light measured in the rest frame of the emitter and f' is the frequency measured in the rest frame of the receiver. If the emitter is moving directly away from the receiver, show that the relative velocity between the emitter and the receiver is $v = c(u^2 - 1)/(u^2 + 1)$, where $u = z + 1$.

29 • A light beam moves along the y' axis with speed c in frame S', which is moving in the $+x$ direction with speed v relative to frame S. (*a*) Find the x and y components of the velocity of the light beam in frame S. (*b*) Show that, according to the velocity transformation equations, the magnitude of the velocity of the light beam in S is c.

30 •• A spaceship is moving east at speed $0.90c$ relative to Earth. A second spaceship is moving west at speed $0.90c$ relative to Earth. What is the speed of one spaceship relative to the other spaceship?

31 •• A particle moves with speed $0.800c$ in the $+x''$ direction along the x'' axis of frame S'', which moves with the same speed and in the same direction along the x' axis relative to frame S'. Frame S' moves with the same speed and in the same direction along the x axis relative to frame S. (*a*) Find the speed of the particle relative to frame S'. (*b*) Find the speed of the particle relative to frame S. SSM

RELATIVISTIC MOMENTUM AND RELATIVISTIC ENERGY

32 • A proton that has a rest energy equal to 938 MeV has a total energy of 2200 MeV. (*a*) What is its speed? (*b*) What is its momentum?

33 • If the kinetic energy of a particle equals twice its rest energy, what percentage error is made by using $p = mu$ for the magnitude of its momentum?

34 •• In a certain reference frame, a particle has momentum of 6.00 MeV/c and total energy of 8.00 MeV. (*a*) Determine the mass of the particle. (*b*) What is the total energy of the particle in a reference frame in which its momentum is 4.00 MeV/c? (*c*) What is the relative speed of the two reference frames?

35 •• Show that

$$d\left(\frac{mu}{1-(u^2/c^2)}\right) = m\left(1-\frac{u^2}{c^2}\right)^{-3/2} du$$

(*Note:* This relation was used to derive the relativistically correct expression for kinetic energy (Equation 39-22).)

36 •• The K^0 particle has a mass of 497.7 MeV/c^2. It decays into a π^- and π^+, each having mass 139.6 MeV/c^2. Following the decay of a K^0, one of the pions is at rest in the laboratory. Determine the kinetic energy of the other pion after the decay and of the K^0 prior to the decay.

37 •• In reference frame S', two protons, each moving at $0.500c$, approach each other head-on. (*a*) Calculate the total kinetic energy of the two protons in frame S'. (*b*) Calculate the total kinetic energy of the protons as seen in reference frame S, which is moving with one of the protons. SSM

38 •• An antiproton $\bar{p}$ has the same mass m as a proton p. The antiproton is created during the reaction $p + p \rightarrow p + p + p + \bar{p}$. During an experiment, protons at rest in the laboratory are bombarded with protons of kinetic energy K_L, which must be great enough so that an amount of kinetic energy equal to $2mc^2$ can be converted into the rest energy of the two particles. In the frame of the laboratory, the total kinetic energy cannot be converted into rest energy because of conservation of momentum. However, in the zero-momentum reference frame in which the two initial protons are moving toward each other with equal speed u, the total kinetic energy can be converted into rest energy. (*a*) Find the speed of each proton u so that the total kinetic energy in the zero-momentum frame is $2mc^2$. (*b*) Transform to the laboratory's frame in which one proton is at rest, and find the speed u' of the other proton. (*c*) Show that the kinetic energy of the moving proton in the laboratory's frame is $K_L = 6mc^2$.

39 ••• A particle of mass 1.00 MeV/c^2 and kinetic energy 2.00 MeV collides with a stationary particle of mass 2.00 MeV/c^2. After the collision, the particles stick together. Find (*a*) the speed of the first particle before the collision, (*b*) the total energy of the first particle before the collision, (*c*) the initial total momentum of the system, (*d*) the total kinetic energy after the collision, and (*e*) the mass of the system after the collision.

GENERAL RELATIVITY

40 •• Light traveling in the direction of increasing gravitational potential undergoes a frequency redshift. Calculate the shift in wavelength if a beam of light of wavelength $\lambda = 632.8$ nm is sent up a vertical shaft of height $L = 100$ m.

41 •• Let us revisit a problem from Chapter 3: Two cannons are pointed directly toward each other, as shown in Figure 39-17. When fired, the cannonballs will follow the trajectories shown. Point P is the point where the trajectories cross each other. Ignore any effects due to air resistance. Using the principle of equivalence, show that if the cannons are fired simultaneously (in the rest frame of the cannons), the cannonballs will hit each other at point P.

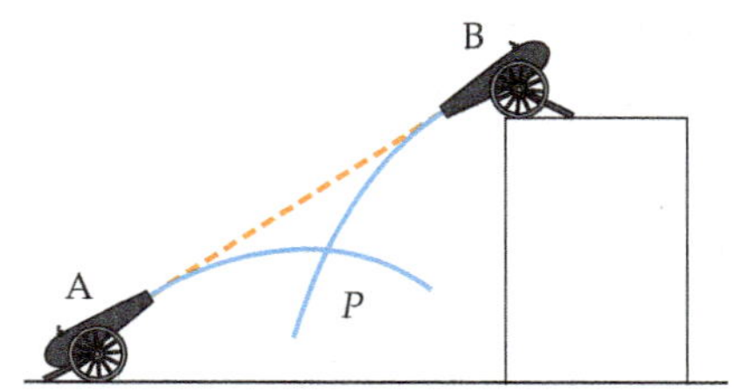

FIGURE 39-17
Problem 41

42 ••• A horizontal turntable rotates with angular speed ω. There is a clock at the center of the turntable and an identical clock mounted on the turntable a distance r from the center. In an inertial reference frame, in which the clock at the center is at rest, the clock at distance r is moving with speed $u = r\omega$. (*a*) Show that from time dilation according to special relativity, the time between ticks, Δt_0 for the clock at rest and Δt_R for the moving clock, are related by

$$\frac{\Delta t_R - \Delta t_0}{\Delta t_0} = -\frac{r^2\omega^2}{2c^2} \qquad r\omega \ll c$$

(*b*) In a reference frame rotating with the table, both clocks are at rest. Show that the clock at distance r experiences a pseudoforce $F_r = mr\omega^2$ in the rotating frame and that this is equivalent to a difference in gravitational potential between r and the origin of $\phi_r - \phi_0 = -\frac{1}{2}r^2\omega^2$. (*c*) Use the difference in gravitational potential given in Part (*b*) to show that in this frame the difference in time intervals is the same as in the inertial frame.

GENERAL PROBLEMS

43 • How fast must a muon travel so that its mean lifetime is 46 μs if its mean lifetime at rest is 2.2 μs?

44 • A distant galaxy is moving away from Earth with a speed that results in each wavelength received on Earth being shifted so that $\lambda' = 2\lambda_0$. Find the speed of the galaxy relative to Earth.

45 •• Frames S and S' are moving relative to each other along the x and x' axes (which superpose). Observers at rest in the two frames set their clocks to $t = 0$ when the two origins coincide. In frame S, event 1 occurs at $x_1 = 1.0\ c\cdot$y and $t_1 = 1.00$ y and event 2 occurs at $x_2 = 2.0\ c\cdot$y and $t_2 = 0.50$ y. The events occur simultaneously in frame S'. (*a*) Find the magnitude and direction of the velocity of S' relative to S. (*b*) At what time do both events occur as measured in S'? SSM

46 •• An interstellar spaceship travels from Earth to a star system 12 light-years away (as measured in Earth's frame). The trip takes 15 y as measured by clocks on the spaceship. (*a*) What is the speed of the spaceship relative to Earth? (*b*) When the spaceship arrives, it sends an electromagnetic signal to Earth. How long after the spaceship leaves Earth will observers on Earth receive the signal?

47 •• The neutral pion π^0 has a mass of 135.0 MeV/c^2. This particle can be created in a proton–proton collision:

$$\mathrm{p} + \mathrm{p} \rightarrow \mathrm{p} + \mathrm{p} + \pi^0$$

Determine the threshold kinetic energy for the creation of a π^0 in a collision of a moving proton and a stationary proton. (See Problem 38.)

48 •• A rocket that has a proper length of 1000 m moves away from a space station and in the $+x$ direction at $0.60c$ relative to an observer on the station. An astronaut stands at the rear of the rocket and fires a dart toward the front of the rocket at $0.80c$ relative to the rocket. How long does it take the dart to reach the front of the rocket (*a*) as measured in the frame of the rocket, (*b*) as measured in the frame of the space station, and (*c*) as measured in the frame of the dart?

49 ••• Using a simple thought experiment, Einstein showed that there is mass associated with electromagnetic radiation. Consider a box of length L and mass M resting on a frictionless surface. Attached to the left wall of the box is a light source that emits a directed pulse of radiation of energy E, which is completely absorbed at the right wall of the box. According to classical electromagnetic theory, the radiation carries momentum of magnitude $p = E/c$ (Equation 30-24). The box recoils when the pulse is emitted by the light source. (*a*) Find the recoil velocity of the box so that momentum is conserved when the light is emitted. (Because p is small and M is large, you may use classical mechanics.) (*b*) When the light is absorbed at the right wall of the box the box stops, so the total momentum of the system remains zero. If we neglect the very small velocity of the box, the time it takes for the radiation to travel across the box is $\Delta t = L/c$. Find the distance moved by the box in that time. (*c*) Show that if the center of mass of the system is to remain at the same place, the radiation must carry mass $m = E/c^2$. SSM

50 ••• Using the relativistic conservation of momentum and energy and the relation between energy and momentum for a photon $E = pc$, prove that a free electron (an electron not bound to an atomic nucleus) cannot absorb or emit a photon.

51 ••• When a moving particle that has a kinetic energy greater than the threshold kinetic energy K_{th} strikes a stationary target particle, one or more particles may be created in the inelastic collision. Show that the threshold kinetic energy of the moving particle is given by

$$K_{\text{th}} = \frac{(\Sigma m_{\text{in}} + \Sigma m_{\text{fin}})(\Sigma m_{\text{fin}} + \Sigma m_{\text{in}})c^2}{2m_{\text{target}}}$$

Here Σm_{in} is the sum of the masses of the particles prior to the collision, Σm_{fin} is the sum of the masses of the particles following the collision, and m_{target} is the mass of the target particle. Use this expression to determine the threshold kinetic energy of protons incident on a stationary proton target for the production of a proton–antiproton pair; compare your result with the result of Problem 38. SSM

52 ••• A particle of mass M decays into two identical particles, each of mass m, where $m = 0.30M$. Prior to the decay, the particle of mass M has a total energy of $4.0mc^2$ in the laboratory reference frame. The velocities of the decay products are along the direction of motion of M. Find the velocities of the decay products in the laboratory reference frame.

53 ••• A rod of proper length L_{p} makes an angle θ with the x axis in frame S. Show that the angle θ' made with the x' axis in frame S', which is moving in the $+x$ direction with speed v, is given by $\tan\theta' = \gamma\tan\theta$ and that the length of the stick in S' is $L' = L_{\text{p}}(\gamma^{-2}\cos^2\theta + \sin^2\theta)^{1/2}$.

54 ••• Show that if a particle moves at an angle θ with the x axis with speed u in frame S, it moves at an angle θ' with the x' axis in S' given by

$$\tan\theta' = \gamma^{-1}\sin\theta/[\cos\theta - (v/u)]$$

55 ••• For the special case of a particle moving with speed u along the y axis in frame S, show that its momentum and energy in frame S', a frame that is moving along the x axis with velocity v, are related to its momentum and energy in S by the transformation equations

$$p_x' = \gamma\left(p_x - \frac{vE}{c^2}\right) \qquad p_y' = p_y \qquad p_z' = p_z \qquad \frac{E'}{c} = \gamma\left(\frac{E}{c} - \frac{vp_x}{c}\right)$$

Compare these equations with the Lorentz transformation equations for x', y', z', and t'. Notice that the quantities p_x, p_y, p_z and E/c transform in the same way as do x, y, z, and ct. SSM

56 ••• The equation for the spherical wavefront of a light pulse that begins at the origin at time $t = 0$ is $x^2 + y^2 + z^2 - (ct)^2 = 0$. Frame S' moves with velocity v along the x axis. Using the Lorentz transformation, show that such a light pulse also has a spherical wavefront in frame S' by showing that $x'^2 + y'^2 + z'^2 - (ct')^2 = 0$.

57 ••• In Problem 56, you showed that the quantity $x^2 + y^2 + z^2 - (ct)^2$ has the same value (zero) in both S and S'. A quantity that has the same value in all inertial frames is called a *Lorentz invariant*. From the results of Problem 55, the quantity $p_x^2 + p_y^2 + p_z^2 - E^2/c^2$ must also be a Lorentz invariant. Show that this quantity has the value $-m^2c^2$ in both the S and S' reference frames.

58 ••• A long rod that is parallel to the x axis is released from rest. Subsequently, it is in free fall with an acceleration of magnitude g in the $-y$ direction. An observer in a rocket ship moving with speed v parallel to the x axis passes by. Using the Lorentz transformations, show that the observer on the rocket ship will measure the rod to be bent into a parabolic shape. Is the parabola concave upward or concave downward?

59 •• Show that if u_x' and v in $u_x = (u_x' + v)/[1 + (vu_x'/c^2)]$ (Equation 39-18*a*) are both positive and less than c, then u_x is positive and less than c. *Hint: Let* $u_x' = (1 - \varepsilon_1)c$ *and* $v = (1 - \varepsilon_2)c$, *where* ε_1 *and* ε_2 *are positive numbers that are less than 1.*

60 ••• In reference frame S, the acceleration of a particle is $\vec{a} = a_x\hat{i} + a_y\hat{j} + a_z\hat{k}$. Derive expressions for the acceleration components a_x', a_y', and a_z' of the particle in reference frame S' that is moving relative to S in the x direction with velocity v.

CHAPTER 40

THE DIABLO CANYON NUCLEAR POWER PLANT NEAR SAN LUIS OBISPO, CALIFORNIA. *(Tony Hertz/Alamy.)*

? How much energy is released during the fission of one gram of ^{235}U? (See Example 40-6.)

Nuclear Physics

To many chemists, the atomic nucleus is modeled as a point charge that has most of the mass of the atom. In this chapter, we will look at the nucleus from the physicist's perspective and see how the protons and neutrons that make up the nucleus have played important roles in our everyday life as well as in the history and structure of the universe.

In this chapter, we study the properties of atomic nuclei, examine radioactivity, and explore nuclear reactions. We also discuss fission and fusion. The fission of very heavy nuclei, such as uranium, is a major source of power today, while the fusion of very light nuclei is the energy source that powers the stars, including our Sun, and may hold the key to our energy needs of the future.

40-1 PROPERTIES OF NUCLEI

The nucleus of an atom has just two kinds of particles, protons and neutrons,* which have approximately the same mass (the neutron is approximately 0.2 percent more massive). The proton has a charge of $+e$, and the neutron is uncharged.

* The most prevalent hydrogen nucleus has a single proton.

The number of protons, Z, is the atomic number of the atom, which also equals the number of electrons in the atom. The number of neutrons that a nucleus has, N, is approximately equal to Z for light nuclei. For heavier nuclei, the number of neutrons is increasingly greater than Z. The total number of nucleons* $A = N + Z$ is called the **nucleon number** or **mass number** of the nucleus. A particular nuclear species is called a **nuclide.** Two or more nuclides that have the same atomic number Z but have different values for N and A are called **isotopes.** A particular nuclide is designated by its atomic symbol (for example, H for hydrogen and He for helium) and its mass number A as a superscript. The lightest element, hydrogen, has three isotopes: protium, ^{1}H, whose nucleus is just a single proton; deuterium, ^{2}H, whose nucleus is composed of one proton and one neutron; and tritium, ^{3}H, whose nucleus is composed of one proton and two neutrons. Although the mass of the deuterium atom is about twice the mass of the protium atom and the mass of the tritium atom is about three times the mass of protium, these three atoms have nearly identical chemical properties because they each have one electron. On the average, there are about three stable isotopes for each element, although some atoms have only one stable isotope while others have five or six. The most common isotope of the second lightest element, helium, is ^{4}He. The ^{4}He nucleus is also known as an α particle. Another isotope of helium is ^{3}He, and the ^{3}He nucleus is also known as helion.

Nucleons exert a strong attractive force on other nucleons. This force, called the **strong nuclear force** or the **hadronic force,** is much stronger than the electrostatic force of repulsion between the protons and is very much stronger than the gravitational forces between the nucleons. (Gravity is so comparably weak that it can always be neglected in nuclear physics.) The strong nuclear force is roughly the same between two neutrons, two protons, or a neutron and a proton. Two protons, of course, also exert a repulsive electrostatic force on each other due to their charges, which tends to weaken the attraction between them somewhat. The strong nuclear force decreases rapidly with distance, and it is negligible when two nucleons are more than a few femtometers apart.

SIZE, SHAPE, AND DENSITY

The size and shape of the nucleus can be determined by bombarding it with high-energy particles and observing the scattering. The results depend somewhat on the kind of experiment. For example, a scattering experiment using electrons measures the charge distribution of the nucleus, whereas a scattering experiment using neutrons determines the region of influence of the strong nuclear force. A wide variety of experiments suggest that most nuclei are approximately spherical, with radii given approximately by

$$R = R_0 A^{1/3} \qquad \text{40-1}$$

NUCLEAR RADIUS

where R_0 is approximately 1.2 fm. The fact that the radius of a spherical nucleus is proportional to $A^{1/3}$ implies that the volume of the nucleus is proportional to A. Because the mass of the nucleus is also approximately proportional to A, the densities of all nuclei are approximately the same. This is analogous to a drop of liquid, which also has constant density independent of its size. The **liquid-drop model** of the nucleus has proved quite successful in explaining nuclear behavior, especially the fission of heavy nuclei.

---

* The word *nucleon* refers to either a neutron or a proton that is part of a nucleus.

N AND *Z* NUMBERS

For light nuclei, the greatest stability is achieved when the numbers of protons and neutrons are approximately equal, $N \approx Z$. For heavier nuclei, instability caused by the electrostatic repulsion between the protons is minimized when there are more neutrons than protons. We can see this by looking at the N and Z numbers for the most abundant isotopes of some representative elements: for $^{16}_{8}O$, $N = 8$ and $Z = 8$; for $^{40}_{20}Ca$, $N = 20$ and $Z = 20$; for $^{56}_{26}Fe$, $N = 30$ and $Z = 26$; for $^{207}_{82}Pb$, $N = 125$ and $Z = 82$; and for $^{238}_{92}U$, $N = 146$ and $Z = 92$. (The atomic number Z has been included here as a subscript of the atomic symbol for emphasis. It is not actually needed because the atomic number is implied by the atomic symbol.)

Figure 40-1 shows a plot of N versus Z for the known stable nuclei. The curve follows the straight line $N = Z$ for small values of N and Z. We can understand this tendency for N and Z to be equal by considering the total energy of A particles in a one-dimensional box. For $A = 8$, Figure 40-2 shows the energy levels for eight neutrons and for four neutrons and four protons. Because of the exclusion principle, only two identical particles (that have opposite spins) can be in the same space state. Because protons and neutrons are not identical, we can put two each in a state, as shown in Figure 40-2*b*. Thus, the total energy for four protons and four neutrons is less than the total energy for eight neutrons (or eight protons), as shown in Figure 40-2*a*. When the Coulomb energy of repulsion, which is proportional to Z^2, is included, this result changes somewhat. For large values of A and Z, the total energy may be increased less by adding two neutrons than by adding one neutron and one proton because of the electrostatic repulsion involved in the latter case. This explains why $N > Z$ for the larger values of A (for the heavier nuclei).

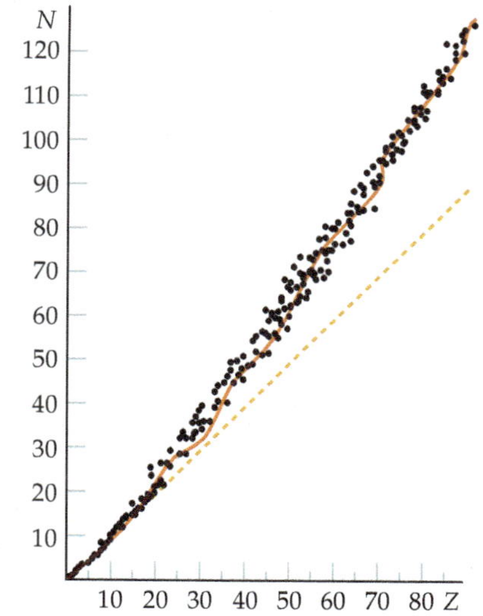

FIGURE 40-1 Plot of number of neutrons N versus number of protons Z for the stable nuclides. The dashed line is $N = Z$.

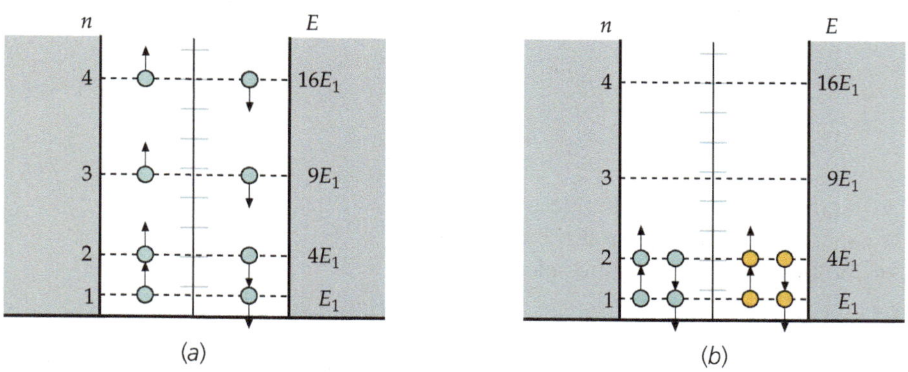

FIGURE 40-2 (*a*) Eight neutrons in a one-dimensional box. In accordance with the exclusion principle, only two neutrons (that have opposite spins) can be in a given energy level. (*b*) Four neutrons and four protons in a one-dimensional box. Because protons and neutrons are not identical particles, two of each can be in the same energy level. The total energy is much less for this case than for the case shown in Figure 40-2*a*.

PRACTICE PROBLEM 40-1

(*a*) Calculate the total energy of the eight neutrons in the one-dimensional box shown in Figure 40-2*a*. (*b*) Calculate the total energy of the four neutrons and four protons in the one-dimensional box shown in Figure 40-2*b*.

MASS AND BINDING ENERGY

The mass of a nucleus is less than the sum of the masses of its parts by E_b/c^2, where E_b is the binding energy and c is the speed of light. When two or more nucleons fuse together to form a nucleus, the total mass decreases and energy is released. Conversely, to break up a nucleus into its parts, energy is absorbed by the system and the mass of the system increases.

Atomic masses and nuclear masses are often given in unified atomic mass units (u), defined as one-twelfth the mass of a ^{12}C atom. The rest energy of one such mass unit is

$$(1\ \text{u})c^2 = 931.5\ \text{MeV} \qquad \text{40-2}$$

Consider ^{4}He, for example, which consists of two protons and two neutrons. The mass of an atom can be accurately measured in a mass spectrometer. The mass of the ^{4}He atom is 4.002 603 u and the mass of the ^{1}H atom is 1.007 825 u. These values include the masses of the electrons in the atom. The mass of the neutron is 1.008 665 u. The sum of the masses of two ^{1}H atoms and two neutrons is 2(1.007 825 u) + 2(1.008 665 u) = 4.032 980 u, which is greater than the mass of the ^{4}He atom by 0.030 377 u.* We can find the binding energy of the ^{4}He nucleus from this mass difference of 0.030 377 u by using the mass conversion factor $(1\ \text{u})c^2 = 931.5$ MeV from Equation 40-2. Then

$$(0.030\,377\ \text{u})c^2 = (0.030\,377\ \text{u})c^2 \times \frac{931.5\ \text{MeV}/c^2}{1\ \text{u}} = 28.30\ \text{MeV}$$

The total binding energy of ^{4}He is thus 28.30 MeV. In general, the binding energy of a nucleus of an atom of atomic mass M_A having Z protons and N neutrons is found by calculating the difference between the sum of the masses of the nucleons and the mass of the nucleus and then multiplying by c^2:

$$E_b = (ZM_H + Nm_n - M_A)c^2 \qquad \text{40-3}$$

TOTAL NUCLEAR BINDING ENERGY

where M_H is the mass of the ^{1}H atom and m_n is the mass of the neutron. (Note that the mass of the Z electrons in the term ZM_H is canceled by the mass of the Z electrons in the term M_A.†) The atomic masses of the neutron and of some selected isotopes are listed in Table 40-1.

Example 40-1 Binding Energy of the Last Neutron

Find the binding energy of the last neutron in a ^{4}He nucleus.

PICTURE The binding energy is energy equivalent of the mass of a ^{3}He atom plus the mass of a neutron minus the mass of a ^{4}He atom. We find the masses from Table 40-1 and multiply by c^2 to obtain the energy equivalents.

---

* Note that by using the masses of two ^{1}H atoms rather than two protons, the masses of the electrons in the atom are accounted for. We do this because it is atomic masses, not nuclear masses, that are measured directly and listed in mass tables.

† The mass associated with the binding energies of the electrons are not accounted for in this calculation.

SOLVE

1. Add the mass of the neutron to that of ^{3}He:

$$m_{^3\text{He}} + m_\text{n} = 3.016\,030\text{ u} + 1.008\,665\text{ u} = 4.024\,695\text{ u}$$

2. Subtract the mass of ^{4}He from the result:

$$\Delta m = (m_{^3\text{He}} + m_\text{n}) - m_{^4\text{He}} = 4.024\,695\text{ u} - 4.002\,603\text{ u} = 0.022\,092\text{ u}$$

3. Multiply this mass difference by c^2 and convert to MeV:

$$E_\text{b} = (\Delta m)c^2 = (0.022\,092\text{ u})c^2 \times \frac{931.5\text{ MeV}/c^2}{1\text{ u}} = \boxed{20.58\text{ MeV}}$$

CHECK As expected, the step-3 result of 20.58 MeV is less than the total binding energy of a ^{4}He nucleus. (The total binding energy of a ^{4}He nucleus is 28.30 MeV, a value that is calculated preceding Equation 40-3.)

Table 40-1 Atomic Masses of the Neutron and Selected Isotopes*

Element	Symbol	Z	Atomic mass, u
Neutron	n	0	1.008 665
Hydrogen			
Protium	^{1}H	1	1.007 825
Deuterium	^{2}H or D	1	2.014 102
Tritium	^{3}H or T	1	3.016 050
Helium	^{3}He	2	3.016 030
	^{4}He	2	4.002 603
Lithium	^{6}Li	3	6.015 125
	^{7}Li	3	7.016 004
Boron	^{10}B	5	10.012 939
Carbon	^{12}C	6	12.000 000
	^{13}C	6	13.003 354
	^{14}C	6	14.003 242
Nitrogen	^{13}N	7	13.005 738
	^{14}N	7	14.003 074
Oxygen	^{16}O	8	15.994 915
Sodium	^{23}Na	11	22.989 771
Potassium	^{39}K	19	38.963 710
Iron	^{56}Fe	26	55.939 395
Copper	^{63}Cu	29	62.929 592
Silver	^{107}Ag	47	106.905 094
Gold	^{197}Au	79	196.966 541
Lead	^{208}Pb	82	207.976 650
Polonium	^{212}Po	84	211.989 629
Radon	^{222}Rn	86	222.017 531
Radium	^{226}Ra	88	226.025 360
Uranium	^{238}U	92	238.048 608
Plutonium	^{242}Pu	94	242.058 725

* Mass values obtained at <http://physics.nist.gov/PhysRefData/Compositions/index.html>.

Figure 40-3 shows the binding energy per nucleon E_b/A versus A. The mean value is approximately 8.3 MeV. The flatness of the curve for $A > 50$ shows that E_b is approximately proportional to A. This indicates that there is saturation of nuclear forces in the nucleus as would be the case if each nucleon were attracted only to its nearest neighbors. Such a situation also leads to a constant nuclear density consistent with the measurements of the radius. If, for example, there were no saturation and each nucleon bonded to each other nucleon, there would be $A - 1$ bonds for each nucleon and a total of $A(A - 1)$ bonds altogether. The total binding energy, which is a measure of the energy needed to break all these bonds, would then be proportional to $A(A - 1)$, and E_b/A would not be approximately constant. The steep rise in the curve for low A is due to the increase in the number of nearest neighbors and therefore to the increased number of bonds per nucleon. The gradual decrease at high A is due to the Coulomb repulsion of the protons, which increases as Z^2 and decreases the binding energy. For very large A, this Coulomb repulsion is so great that a nucleus that has an A greater than approximately 300 is unstable and undergoes spontaneous fission.

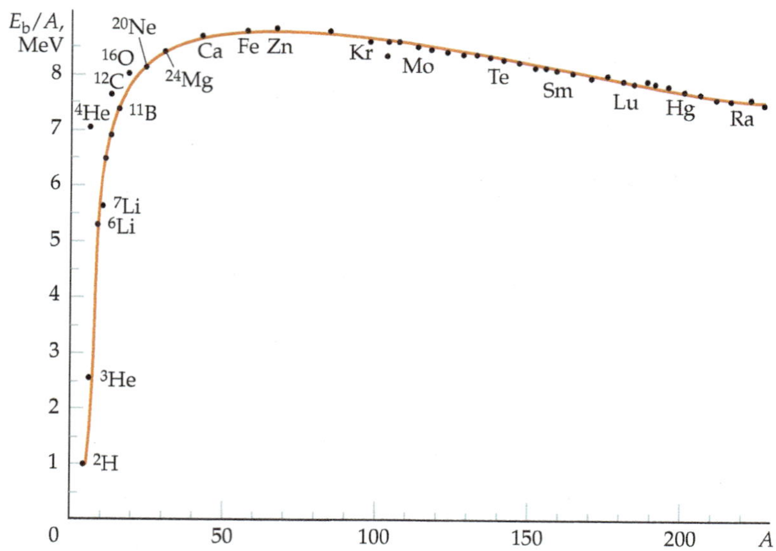

FIGURE 40-3 The binding energy per nucleon versus the nucleon number A. For nuclei that have values of A greater than 50, the curve is approximately constant, indicating that the total binding energy is approximately proportional to A.

40-2 RADIOACTIVITY

Many nuclei are radioactive; that is, they decay into other nuclei by the emission of particles such as photons, electrons, neutrons, or α particles. The terms α decay, β decay, and γ decay were used before it was discovered that α particles are ^{4}He nuclei, β particles are either electrons (β^-) or positrons* (β^+), and γ rays are photons. The rate of decay of a radioactive sample decreases exponentially with increasing time. *This exponential time dependence is characteristic of all radioactivity and indicates that radioactive decay is a statistical process.* Because each nucleus is well shielded from others by the atomic electrons, pressure and temperature changes have little or no effect on the rate of radioactive decay or other nuclear properties.

* The positron has the same mass as an electron and it has a charge of $+e$.

Let N be the number of radioactive nuclei at some time t. If the decay of an individual nucleus is a random event, we expect the number of nuclei that decay in some time interval dt to be proportional both to N and to dt. Because of these decays, the number N will decrease. The change in N between time t and time $t + dt$ is given by

$$dN = -\lambda N\, dt \tag{40-4}$$

where λ is a constant of proportionality called the **decay constant.** The rate of change of N, dN/dt, is proportional to N. This is characteristic of exponential decay. To solve Equation 40-4 for N, we first divide each side by N, thus separating the variables N and t:

$$\frac{dN}{N} = -\lambda\, dt$$

Integrating, we obtain

$$\int_{N_0}^{N'} \frac{dN}{N} = -\lambda \int_0^{t'} dt$$

or

$$\ln\frac{N'}{N_0} = -\lambda t' \tag{40-5}$$

where N' is the number of nuclei that remain at time t'. For convenience, we drop the primes from N' and t'. This introduces no ambiguity because the parameters N and t have been integrated out of the equation. Taking the exponential of each side, we obtain

$$\frac{N}{N_0} = e^{-\lambda t}$$

or

$$N = N_0 e^{-\lambda t} \tag{40-6}$$

See
Math Tutorial for more
information on
Exponential Functions

The number of radioactive decays per second is called the **decay rate** R:

$$R = -\frac{dN}{dt} = \lambda N = \lambda N_0 e^{-\lambda t} = R_0 e^{-\lambda t} \tag{40-7}$$

DECAY RATE

where

$$R_0 = \lambda N_0 \tag{40-8}$$

is the decay rate at time $t = 0$. The decay rate R is the quantity that is determined experimentally. The decay rate is also called the **activity** of the sample.

The average or **mean lifetime** τ is equal to the reciprocal of the decay constant (see Problem 40):

$$\tau = \frac{1}{\lambda} \tag{40-9}$$

The mean lifetime is analogous to the time constant in the exponential decrease in the charge on a capacitor in an *RC* circuit that we discussed in Section 25-6. After a time equal to the mean lifetime, the number of radioactive nuclei and the decay rate are each equal to $e^{-1} = 37$ percent of their original values. The **half-life** $t_{1/2}$ is defined as the time it takes for the number of nuclei and the decay rate to decrease by half. Setting $t = t_{1/2}$ and $N = N_0/2$ in Equation 40-6 gives

$$\frac{N_0}{2} = N_0 e^{-\lambda t_{1/2}} \quad \text{40-10}$$

or

$$e^{+\lambda t_{1/2}} = 2$$

Solving for $t_{1/2}$ gives

$$t_{1/2} = \frac{\ln 2}{\lambda} = (\ln 2)\tau = 0.693\tau \quad \text{40-11}$$

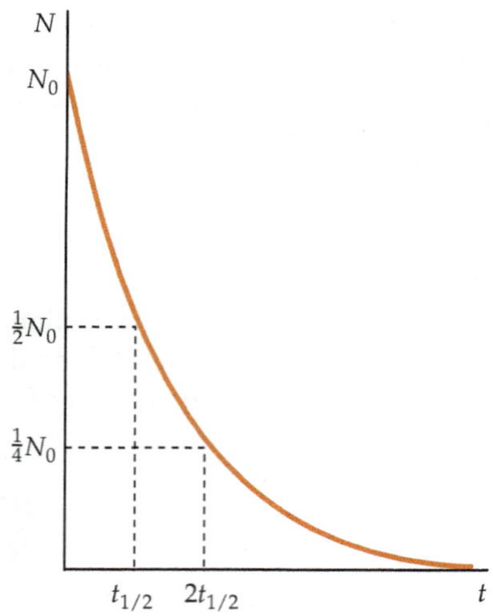

FIGURE 40-4 Exponential radioactive decay. After each half-life $t_{1/2}$, the number of nuclei remaining has decreased by one-half. The decay rate $R = \lambda N$ has the same time dependence as does N.

Figure 40-4 shows a plot of N versus t. If we multiply the numbers on the N axis by λ, this graph becomes a plot of R versus t. After each time interval of one half-life, both the number of nuclei left and the decay rate have decreased to half of their previous values. For example, if the decay rate is R_0 initially, it will be $\frac{1}{2}R_0$ after one half-life, $\left(\frac{1}{2}\right)\left(\frac{1}{2}\right)R_0$ after two half-lives, and so forth. After n half-lives, the decay rate will be

$$R = \left(\tfrac{1}{2}\right)^n R_0 \quad \text{40-12}$$

The half-lives of radioactive nuclei vary from very small times (less than 1 μs) to very large times (greater than 10^{10} y).

Example 40-2 Counting Rate for Radioactive Decay

A radioactive source has a half-life of 1.0 min. At time $t = 0$, the radioactive source is placed near a detector, and the counting rate (the number of decay particles detected per unit time) is observed to be 2000 counts/s. Find the counting rate at times $t = 1.0$ min, $t = 2.0$ min, $t = 3.0$ min, and $t = 10$ min.

PICTURE The counting rate r is proportional to the decay rate R, and the decay rate is given by $R = \left(\frac{1}{2}\right)^n R_0$ (Equation 40-12), where n is the time divided by 1.0 min.

SOLVE

1. Because the half-life is 1.0 min, the counting rate will be half as great at $t = 1.0$ min as at $t = 0$:
$$r_1 = \tfrac{1}{2}r_0 = \tfrac{1}{2}(2000 \text{ counts/s}) = \boxed{1.0 \times 10^3 \text{ counts/s at 1.0 min}}$$

2. At $t = 2.0$ min, the rate is half that at 1 min. It decreases by one-half each minute:
$$r_2 = \left(\tfrac{1}{2}\right)^2 r_0 = \tfrac{1}{4}(2000 \text{ counts/s}) = \boxed{5.0 \times 10^2 \text{ counts/s at 2.0 min}}$$
$$r_3 = \left(\tfrac{1}{2}\right)^3 r_0 = \tfrac{1}{8}(2000 \text{ counts/s}) = \boxed{2.5 \times 10^2 \text{ counts/s at 3.0 min}}$$

3. At $t = 10$ min, the rate will be $\left(\frac{1}{2}\right)^{10}$ multiplied by the initial rate:
$$r_{10} = \left(\tfrac{1}{2}\right)^{10} r_0 = \tfrac{1}{1024}(2000 \text{ counts/s}) = 1.95 \text{ counts/s} \approx \boxed{2.0 \text{ counts/s at 10 min}}$$

CHECK As expected, the counting rate decreases as the number of minutes increases.

CONCEPT CHECK 40-1

A radioactive isotope has a half-life of 10 s. You are observing a sample of this isotope. After approximately one minute of observation, there is only one atom of this isotope left in your sample. How many atoms of this isotope will be left in your sample 15 s later?

Example 40-3 Detection-Efficiency Considerations

If the detection efficiency in Example 40-2 is 20 percent, (*a*) how many radioactive nuclei are there at time $t = 0$ and (*b*) at time $t = 1.0$ min? (c) How many nuclei decay in the first minute?

PICTURE The detection efficiency depends on the probability that a radioactive decay particle will enter the detector and the probability that upon entering the detector it will produce a count. If the efficiency is 20 percent, the decay rate must be five times the counting rate.

SOLVE

(*a*) 1. The number of radioactive nuclei is related to the decay rate R and the decay constant λ:

$$R = \lambda N$$

2. The decay constant is related to the half-life:

$$\lambda = \frac{\ln 2.0}{t_{1/2}} = \frac{0.693}{1.0\ \text{min}} = 0.693\ \text{min}^{-1}$$

3. Because the detection efficiency is 20 percent, the decay rate is five times the counting rate. Calculate the initial decay rate:

$$R_0 = (5\ \text{decays/count}) \times (2000\ \text{counts/s}) = 1.0 \times 10^4\ \text{decays/s}$$

4. Substitute to calculate the initial number of radioactive nuclei N_0 at $t = 0$:

$$N_0 = \frac{R_0}{\lambda} = \frac{1.0 \times 10^4\ \text{s}^{-1}}{0.693\ \text{min}^{-1}} \times \frac{60\ \text{s}}{1\ \text{min}} = 8.66 \times 10^5 = \boxed{8.7 \times 10^5}$$

(*b*) At time $t = 1\ \text{min} = t_{1/2}$, there are half as many radioactive nuclei as at $t = 0$:

$$N_1 = \tfrac{1}{2}(8.66 \times 10^5) = 4.33 \times 10^5 = \boxed{4.3 \times 10^5}$$

(*c*) The number of nuclei that decay in the first minute is $N_0 - N_1$:

$$\Delta N = N_0 - N_1 = 8.66 \times 10^5 - 4.33 \times 10^5 = \boxed{4.3 \times 10^5}$$

CHECK The results for Parts (*b*) and (*c*) are equal, as expected. At the end of one half-life, half of the nuclei have decayed and the other half remain.

The SI unit of radioactive decay is the **becquerel** (Bq), which is defined as one decay per second:

$$1\ \text{Bq} = 1\ \text{decay/s} \tag{40-13}$$

A historical unit that applies to all types of radioactivity is the **curie** (Ci), which is defined as

$$1\ \text{Ci} = 3.7 \times 10^{10}\ \text{decays/s} = 3.7 \times 10^{10}\ \text{Bq} \tag{40-14}$$

The curie is the rate at which radiation is emitted by 1 g of radium. Because this is a very large unit, the millicurie (mCi) or microcurie (μCi) are often used.

BETA DECAY

Beta decay occurs in nuclei that have too many neutrons or too few neutrons for stability. During β decay, A remains the same while Z either increases by 1 (β^- decay) or decreases by 1 (β^+ decay).

An example of β decay is the decay of a free neutron into a proton and an electron. (The half-life of a free neutron is about 10.8 min.) The energy of β decay is 0.782 MeV, which is the difference between the rest energy of the neutron and the rest energy of the proton and an electron. More generally, during β^- decay, a nucleus of mass number A and atomic number Z decays into a nucleus, referred to as the **daughter nucleus,** of mass number A and atomic number $Z' = Z + 1$ and an electron is emitted. (The original nucleus is called the **parent.**) If the decay energy were shared by only the daughter nucleus and the emitted electron, the energy of the electron would be uniquely determined by the conservation of energy and momentum. Experiments show, however, the energies of the electrons emitted during the β^- decay of a nucleus are observed to vary from zero to the maximum energy available. A typical energy spectrum for the electrons is shown in Figure 40-5.

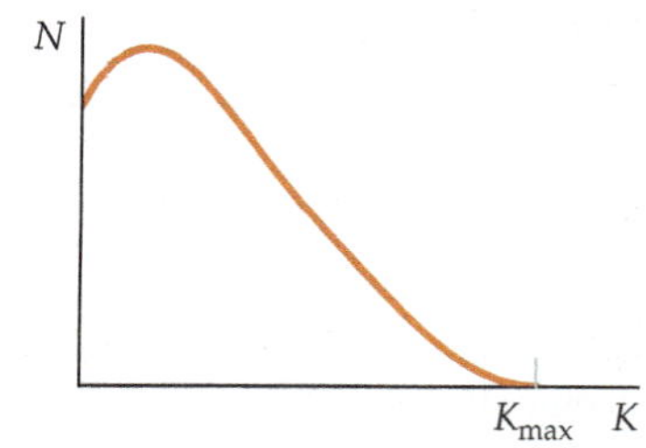

FIGURE 40-5 Number of electrons emitted during β^- decay versus kinetic energy. The fact that all the electrons do not have the same energy K_{max} suggests that another particle, one that shares the energy available for decay, is emitted.

To explain the fact that energy seemed not to be conserved during β decay, Wolfgang Pauli in 1930 suggested that a third particle, which he called the **neutrino,** is also emitted. Because the measured maximum energy of the emitted electrons is equal to the total available for the decay, the rest energy and therefore the mass of the neutrino was assumed to be zero. (It is now known that the mass of the neutrino is very small but not zero.) In 1948, measurements of the momenta of the emitted electron and the recoiling nucleus showed that the neutrino was also needed for the conservation of linear momentum during β decay. The neutrino was first observed experimentally in 1957. It is now known that there are at least three kinds of neutrinos, one (ν_e) associated with electrons, one (ν_μ) associated with muons, and one (ν_τ) associated with the tau particle, τ. Moreover, each neutrino has an antiparticle, written $\bar{\nu}_e$, $\bar{\nu}_\mu$, and $\bar{\nu}_\tau$. It is the electron antineutrino that is emitted during the decay of a neutron, which is written*

$$n \rightarrow p + e^- + \bar{\nu}_e \qquad 40\text{-}15$$

During β^+ decay, a proton changes into a neutron, and a positron (and a neutrino) is emitted. A free proton cannot decay by positron emission because of conservation of energy (the mass of the neutron and the positron is greater than the mass of the proton); however, because of binding-energy effects, a proton inside a nucleus can decay. A typical β^+ decay is

$$^{13}_{7}N \rightarrow {}^{13}_{6}C + e^+ + \nu_e \qquad 40\text{-}16$$

! Do not confuse the symbols e^- and e^+ with the symbol e. The symbols e^- and e^+ denote particles (the electron and the positron), whereas the symbol e denotes an amount of charge.

The electrons or the positrons emitted during β decay do not exist inside the nucleus. They are created during the process of decay, just as photons are created when an atom makes a transition from a higher energy state to a lower energy state.

An important example of β decay is that of ^{14}C, which is used in radioactive carbon dating:

$$^{14}C \rightarrow {}^{14}N + e^- + \bar{\nu}_e \qquad 40\text{-}17$$

The half-life for this decay is 5730 y. The radioactive isotope ^{14}C is produced in the upper atmosphere during nuclear reactions caused by cosmic rays. The chemical reactivity of a carbon atom that has a ^{14}C nucleus is the same as the chemical reactivity of a carbon atom that has a ^{12}C nucleus. For example, atoms that have these nuclei combine with oxygen to form CO_2 molecules. Because living organisms continually exchange CO_2 with the atmosphere, the ratio of ^{14}C to ^{12}C in a living organism is the same as the equilibrium ratio in the atmosphere, which is about 1.3×10^{-12}. After an organism dies, it no longer absorbs ^{14}C from the atmosphere, so the ratio of ^{14}C to ^{12}C continually decreases due to the radioactive decay of ^{14}C. The number of ^{14}C decays per minute per gram of carbon in a living organism can be calculated from the known half-life of ^{14}C and the number of ^{14}C nuclei in a gram of carbon. The result is that there are approximately 15.0 decays per minute per gram of carbon in a living organism. Using this result and the measured number of decays per minute per gram of carbon in a nonliving sample of bone, wood, or other object having carbon, we can determine the age of the sample. For example, if the measured rate were 7.5 decays per minute per gram, the age of the sample would be one half-life = 5730 years.

* This reaction is also written $n \rightarrow p + \beta^- + \bar{\nu}_e$.

Example 40-4 How Old Is the Artifact? *Context-Rich*

You have a summer job working in an archeological research lab. Your supervisor calls to tell you that they found a new bone at their current site and asks you to determine the age of the bone from a sample that she will send you. When the bone sample arrives, you take a section that contains 200 grams of carbon and you find a beta decay rate of 400 decays/min.

PICTURE There are approximately 15.0 decays per minute per gram of carbon in a living organism, and the half-life of carbon-14 is 5730 y. We need to determine the number of half-lives that have occurred since the death of the organism. We do this by using the equality $R_n = (1/2)^n R_0$ (Equation 40-12), where R_n is the current decay rate, R_0 is the initial decay rate, and n is the number of half-lives. We can determine the initial decay rate by multiplying the decay rate per gram by the mass of the carbon of the sample.

SOLVE

1. Write the decay rate after n half-lives in terms of the initial decay rate:

$$R_n = \left(\tfrac{1}{2}\right)^n R_0$$

2. Calculate the initial decay rate (the decay for 200 g of carbon when the organism died):

$$R_0 = [(15\text{ decays/min})/\text{g}](200\text{ g}) = 3000\text{ decays/min}$$

3. Substitute the values for R_0 and R_n into the step-1 equation and solve for n:

$$R_n = \left(\tfrac{1}{2}\right)^n R_0$$

$$400\text{ decays/min} = \left(\tfrac{1}{2}\right)^n 3000\text{ decays/min}$$

$$\left(\tfrac{1}{2}\right)^n = \frac{400}{3000}$$

$$2^n = \frac{3000}{400} = 7.5$$

4. We solve for n by taking the logarithm of each side:

$$n \ln 2 = \ln 7.5 \quad \Rightarrow \quad n = \frac{\ln 7.5}{\ln 2} = 2.91$$

5. The age of the bone is $nt_{1/2}$:

$$t = nt_{1/2} = 2.91(5730\text{ y}) = \boxed{1.67 \times 10^4\text{ y}}$$

CHECK If the bone were from a recently living organism, we would expect the decay rate to be a steady [(15 decays/min)/g](200 g) = 3000 decays/min. The current decay rate is given as 400 decays/min. Because 400/3000 is roughly 1/8 (actually 1/7.5), the sample must be approximately three half-lives old, which is about 3(5730 y). This is in agreement with the step-5 result of 2.91(5730 y).

PRACTICE PROBLEM 40-2 The Check of Example 40-4 states, "Because 400/3000 is roughly 1/8 (actually 1/7.5), the sample must be approximately three half-lives old. . . ." Explain why this ratio of 1/8 implies an age equal to three half-lives.

GAMMA DECAY

During γ decay, a nucleus in an excited state decays to a lower-energy state by the emission of a photon. This process is the nuclear counterpart of spontaneous emission of photons by atoms and molecules. Unlike β decay or α decay, neither the mass number A nor the atomic number Z change during γ decay. Because the spacing of the nuclear energy levels is of the order of 1 MeV (as compared with spacing of the order of 1 eV in atoms), the wavelengths of the emitted photons are of the order of 1 pm (1 pm = 10^{-12} m):

$$\lambda = \frac{hc}{E} \approx \frac{1240\text{ eV}\cdot\text{nm}}{1\text{ MeV}} = 0.00124\text{ nm} = 1.24\text{ pm}$$

The mean lifetime for γ decay is often very short. It is usually observed only because it follows either α decay or β decay. For example, if a radioactive parent nucleus decays by β decay to an excited state of the daughter nucleus, the daughter nucleus then decays to its ground state by γ emission. Direct measurements of mean lifetimes as short as approximately 10^{-11} s are possible. Measurements of mean lifetimes shorter than 10^{-11} s are difficult, but they can sometimes be made by indirect methods.

A few γ emitters have very long lifetimes, of the order of hours. The energy states that do have such long lifetimes are called **metastable states.**

ALPHA DECAY

All very heavy nuclei ($Z > 83$) are potentially unstable via α decay because the mass of the original radioactive nucleus is greater than the sum of the masses of the decay products—an α particle and the daughter nucleus. Consider the decay of ^{232}Th ($Z = 90$) into ^{228}Ra ($Z = 88$) and an α particle. This process is written as

$$^{232}\text{Th} \rightarrow {}^{228}\text{Ra} + \alpha = {}^{228}\text{Ra} + {}^{4}\text{He} \qquad 40\text{-}18$$

The mass of the ^{232}Th atom is 232.038 050 u. The mass of the daughter atom ^{228}Ra is 228.031 064 u. Adding 4.002 603 u (the mass of ^{4}He) to the mass of ^{228}Ra, we get 232.033 667 u for the total mass of the decay products. This value is less than the mass of ^{232}Th by 0.004 383 u, which multiplied by 931.5 MeV/c^2 gives 4.08 MeV/c^2 for the excess mass of ^{232}Th when compared to the total mass of the decay products. The isotope ^{232}Th is therefore potentially unstable to α decay. This decay does in fact occur in nature with the emission of an α particle of kinetic energy 4.08 MeV. (The kinetic energy of the α particle is actually somewhat less than 4.08 MeV because some of the released energy is taken up by the recoiling ^{228}Ra nucleus.)

When a nucleus emits an α particle, both N and Z decrease by 2 and A decreases by 4. The daughter of a radioactive nucleus is often itself radioactive and decays by either α decay or β decay or both. If the original nucleus has a mass number A that is 4 times an integer, the daughter nucleus and all those in the decay chain will also have mass numbers equal to 4 multiplied by an integer. Similarly, if the mass number of the original nucleus is $4n + 1$, where n is an integer, all the nuclei in the decay chain will have mass numbers given by $4n + 1$, where n decreases by one at each α decay. We can see, therefore, that there are four possible α-decay chains, depending on whether A equals $4n$, $4n + 1$, $4n + 2$, or $4n + 3$, where n is an integer. All but one of these decay chains are found on Earth. The $4n + 1$ series is not found because its longest-lived member (other than the stable end product ^{209}Bi) is ^{237}Np, which has a half-life of only 2×10^6 y. Because this period is much less than the age of Earth, this series has disappeared.

Figure 40-6 shows the thorium series, for which $A = 4n$. It begins with an α decay from ^{232}Th to ^{228}Ra. The daughter nuclide of an α decay is on the left or neutron-rich side of the stability curve (the dashed line in the figure), so it often decays by β^- decay. In the thorium series, ^{228}Ra decays by β^- decay to ^{228}Ac, which in turn decays by β^- decay to ^{228}Th. There are then four α decays to ^{212}Pb, which decays by β^- decay to ^{212}Bi. The series branches at ^{212}Bi, which decays either by α decay to ^{208}Tl or by β^- decay to ^{212}Po. The branches meet at the stable lead isotope ^{208}Pb.

The energies of α particles from natural radioactive sources range from approximately 4 MeV to 7 MeV, and the half-lives of the sources range from approximately 10^{-5} s to 10^{10} y. In general, the smaller the energy of the emitted α particle, the longer the half-life. As we discussed in Section 35-4, the enormous variation

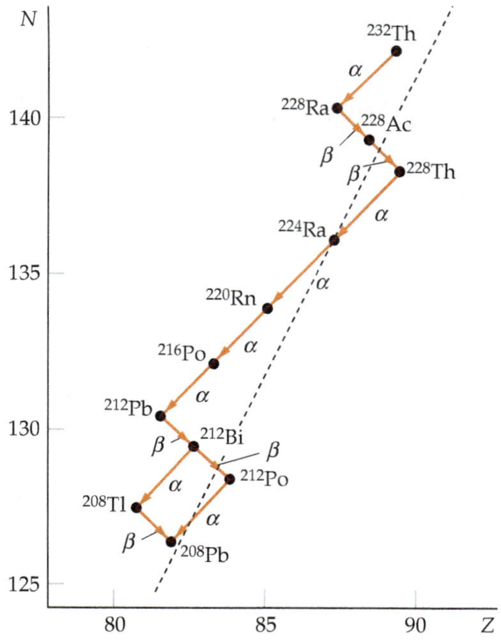

FIGURE 40-6 The thorium ($4n$) α decay series. The dashed line is the curve of stability.

in half-lives was explained by George Gamow in 1928. He considered α decay to be a process in which an α particle is first formed inside a nucleus and then tunnels through the Coulomb barrier (Figure 40-7). A slight increase in the energy of the α particle reduces the relative height $U_{max} - E$ of the barrier and also the thickness $r_1 - R$. Because the probability of penetration is so sensitive to the relative height and thickness of the barrier, a small increase in E leads to a large increase in the probability of barrier penetration and therefore to a significantly shorter lifetime. Gamow was able to derive an expression for the half-life as a function of E that is in excellent agreement with experimental results.

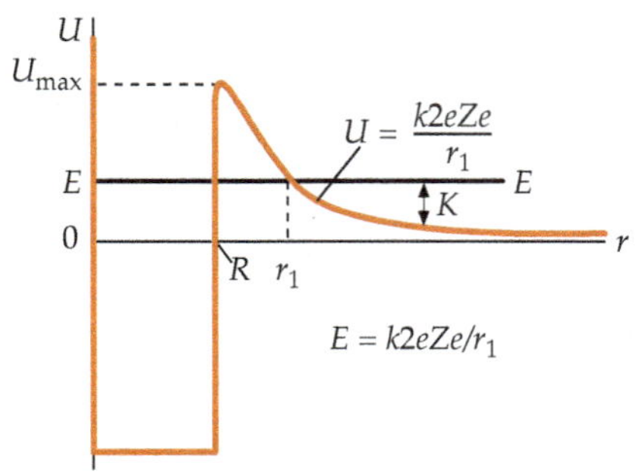

FIGURE 40-7 A model of the potential energy for an α particle and a nucleus. The strong attractive nuclear force that exists for values of r less than the nuclear radius R is indicated by the potential well. Outside the nucleus, the nuclear force is negligible, and the potential energy is given by the Coulomb potential energy function $U = +k2eZe/r$, where Ze is the nuclear charge and $2e$ is the charge of the α particle. The kinetic energy K of the α particle is equal to the energy E when the α particle is far away from the nucleus. A small increase in E reduces the relative height $U_{max} - E$ of the barrier and also reduces its thickness $r_1 - R$, leading to a much greater probability of penetration. An increase in the energy of the emitted α particles by a factor of 2 results in a reduction of the half-life by a factor of more than 10^{20}.

40-3 NUCLEAR REACTIONS

Information about nuclei is typically obtained by bombarding the nuclei with various particles and observing the results. Although the first experiments of this type were limited by the need to use naturally occurring radiation, they produced many important discoveries. In 1932, J. D. Cockcroft and E. T. S. Walton succeeded in producing the reaction

$$\mathrm{p} + {}^{7}\mathrm{Li} \rightarrow {}^{8}\mathrm{Be} \rightarrow {}^{4}\mathrm{He} + {}^{4}\mathrm{He}$$

using artificially accelerated protons. At about the same time, the Van de Graaff electrostatic generator (by R. Van de Graaff in 1931) and the first cyclotron (by E. O. Lawrence and M. S. Livingston in 1932) were built. Since then, enormous advances in the technology for accelerating and detecting particles have been made, and many nuclear reactions have been studied.

When a particle is incident on a nucleus, several different things can happen. The incident particle may be scattered, either elastically or inelastically, or the incident particle may be absorbed by the nucleus, and another particle or particles may be emitted. In inelastic scattering, the nucleus is left in an excited state and subsequently decays by emitting photons (or other particles).

The amount of energy released or absorbed during a reaction (in the center of mass reference frame) is called the **Q value** of the reaction. The Q value equals c^2 multiplied by the mass difference. When energy is released during a reaction, the reaction is said to be an **exothermic reaction.** During an exothermic reaction, the total mass of the incoming particles is greater than the total mass of the outgoing particles, and the Q value is positive. If the total mass of the incoming particles is less than that of the outgoing particles, energy is required for the reaction to take place, and the reaction is said to be an **endothermic reaction.** The Q value of an endothermic reaction is negative. In general, if Δm is the change in mass, the Q value is

$$Q = -(\Delta m)c^2 \qquad 40\text{-}19$$

Q VALUE

An endothermic reaction cannot take place below a specific threshold energy. In the laboratory reference frame in which stationary particles are bombarded by incoming particles, the threshold energy is somewhat greater than $|Q|$ because the outgoing particles must have some kinetic energy to conserve momentum.

A measure of the effective size of a nucleus for a particular nuclear reaction is the **cross section** σ. If I is the number of the incident particles per unit time per unit area (the incident intensity) and R is the number of reactions per unit time per nucleus, the cross section is

$$\sigma = \frac{R}{I} \qquad 40\text{-}20$$

The cross section σ has the dimensions of area. Because nuclear cross sections are of the order of the square of the nuclear radius, a convenient unit for them is the **barn,** which is defined as

$$1\text{ barn} = 10^{-28}\text{ m}^2 \qquad 40\text{-}21$$

The cross section for a particular reaction is a function of energy. For an endothermic reaction, it is zero for energies below the threshold energy.

Example 40-5 Exothermic or Endothermic?

Find the Q value of the reaction $\mathrm{p} + {}^{7}\mathrm{Li} \rightarrow {}^{4}\mathrm{He} + {}^{4}\mathrm{He}$ and state whether the reaction is exothermic or endothermic.

PICTURE We find the masses of the atoms from Table 40-1 and calculate the difference in the total mass of the outgoing particles and the incoming particles. The Q value is related to the change in mass Δm by $Q = -(\Delta m)c^2$. If we use the mass of protium rather than the mass of the proton, there will be four electrons on each side of the reaction, so the electron masses will cancel.

SOLVE

1. Find the mass of each atom from Table 40-1: ^{1}H 1.007 825 u; ^{7}Li 7.016 004 u; ^{4}He 4.002 603 u

2. Calculate the initial mass m_i of the incoming particles: $m_i = 1.007\,825\text{ u} + 7.016\,004\text{ u} = 8.023\,829\text{ u}$

3. Calculate the final mass m_f: $m_f = 2(4.002\,603\text{ u}) = 8.005\,206\text{ u}$

4. Calculate the change in mass:
$$\Delta m = m_f - m_i = 8.005\,206\text{ u} - 8.023\,829\text{ u} = -0.018\,623\text{ u}$$

5. Calculate the Q value:
$$Q = -(\Delta m)c^2 = (+0.018\,623\text{ u})c^2 \times \frac{931.5\text{ MeV}}{1\text{ u}} = \boxed{17.35\text{ MeV}}$$

$\boxed{Q \text{ is positive, so the reaction is exothermic.}}$

CHECK Because the initial mass is greater than the final mass, the initial energy is greater than the final energy and the reaction is exothermic, yielding 17.35 MeV.

REACTIONS WITH NEUTRONS

Nuclear reactions that involve neutrons are important for understanding nuclear reactors. The most likely reaction between a nucleus and a neutron that has an energy of more than about 1 MeV is scattering. However, even if the scattering is elastic, the neutron loses some energy to the nucleus because the nucleus recoils. If a neutron is scattered many times in a material, its energy decreases until the neutron is of the order of the energy of thermal motion kT, where k is Boltzmann's constant and T is the absolute temperature. (At ordinary room temperatures, kT is approximately 0.025 eV.) The neutron is then equally likely to gain or lose energy from a nucleus when it is elastically scattered. A neutron that has an energy of the order of kT is called a **thermal neutron.**

At low energies, a neutron is likely to be captured, producing an excited nucleus. A γ ray is then emitted from the excited nucleus. Figure 40-8 shows the neutron-capture cross section for silver as a function of the energy of the neutron. The large

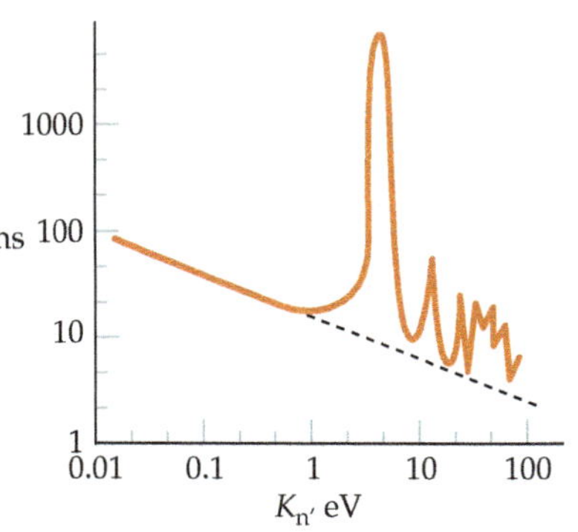

FIGURE 40-8 Neutron-capture cross section for silver as a function of the energy of the incident neutron. The straight line indicates the $1/v$ dependence of the cross section, which is proportional to the time spent by the neutron in the vicinity of the silver nucleus. Superimposed on this dependence are a large resonance and several smaller resonances.

peak in this curve is called a **resonance.** Except for the resonance, the cross section varies fairly smoothly with energy, decreasing with increasing energy roughly as $1/v$, where v is the speed of the neutron. We can understand this energy dependence as follows: Consider a neutron moving with speed v near a nucleus of diameter $2R$. The time it takes the neutron to pass the nucleus is $2R/v$. Thus, the neutron-capture cross section is proportional to the time spent by the neutron in the vicinity of the silver nucleus. The dashed line in Figure 40-8 indicates this $1/v$ dependence. At the maximum of the resonance, the value of the cross section is very large ($\sigma > 5000$ barns) compared with a value of only about 10 barns just past the resonance. Many elements show similar resonances in their neutron-capture cross sections. For example, the maximum cross section for ^{113}Cd is approximately 57 000 barns. This material is thus very useful for shielding against low-energy neutrons.

An important nuclear reaction that involves neutrons is fission, which is discussed in the next section.

40-4 FISSION AND FUSION

Figure 40-9 shows a plot of the nuclear mass difference per nucleon $(M - Zm_p - Nm_n)/A$ in units of MeV/c^2 versus A. This curve is just the negative of the binding-energy curve shown in Figure 40-3. From Figure 40-9, we can see that the values for the mass difference per nucleon for both very heavy ($A \approx 200$) and very light ($A \leq 20$) nuclides are greater than the values for nuclides of intermediate mass. Thus, energy is released when a very heavy nucleus, such as ^{235}U, breaks up into two lighter nuclei—during a process called **fission**—or when two very light nuclei, such as ^{2}H and ^{3}H, fuse together to form a nucleus of greater mass—during a process called **fusion.**

The applications of both fission and fusion to the generation of electrical power and the development of nuclear weapons have had a profound effect on our lives since the early twentieth century. The application of these reactions to the development of energy resources may have an even greater effect in the future. We will look at some of the features of fission and fusion that are important for their application in reactors to generate power.

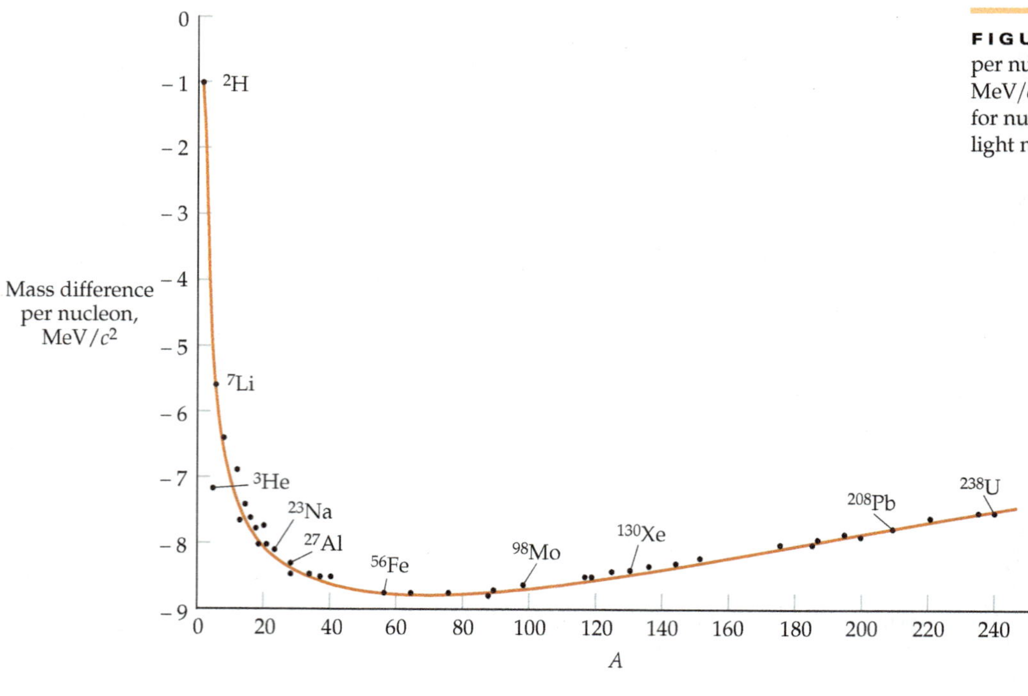

FIGURE 40-9 Plot of mass difference per nucleon $(M - Zm_p - Nm_n)/A$ in units of MeV/c^2 versus A. The mass per nucleon is less for nuclei of intermediate mass than for very light nuclei or very heavy nuclei.

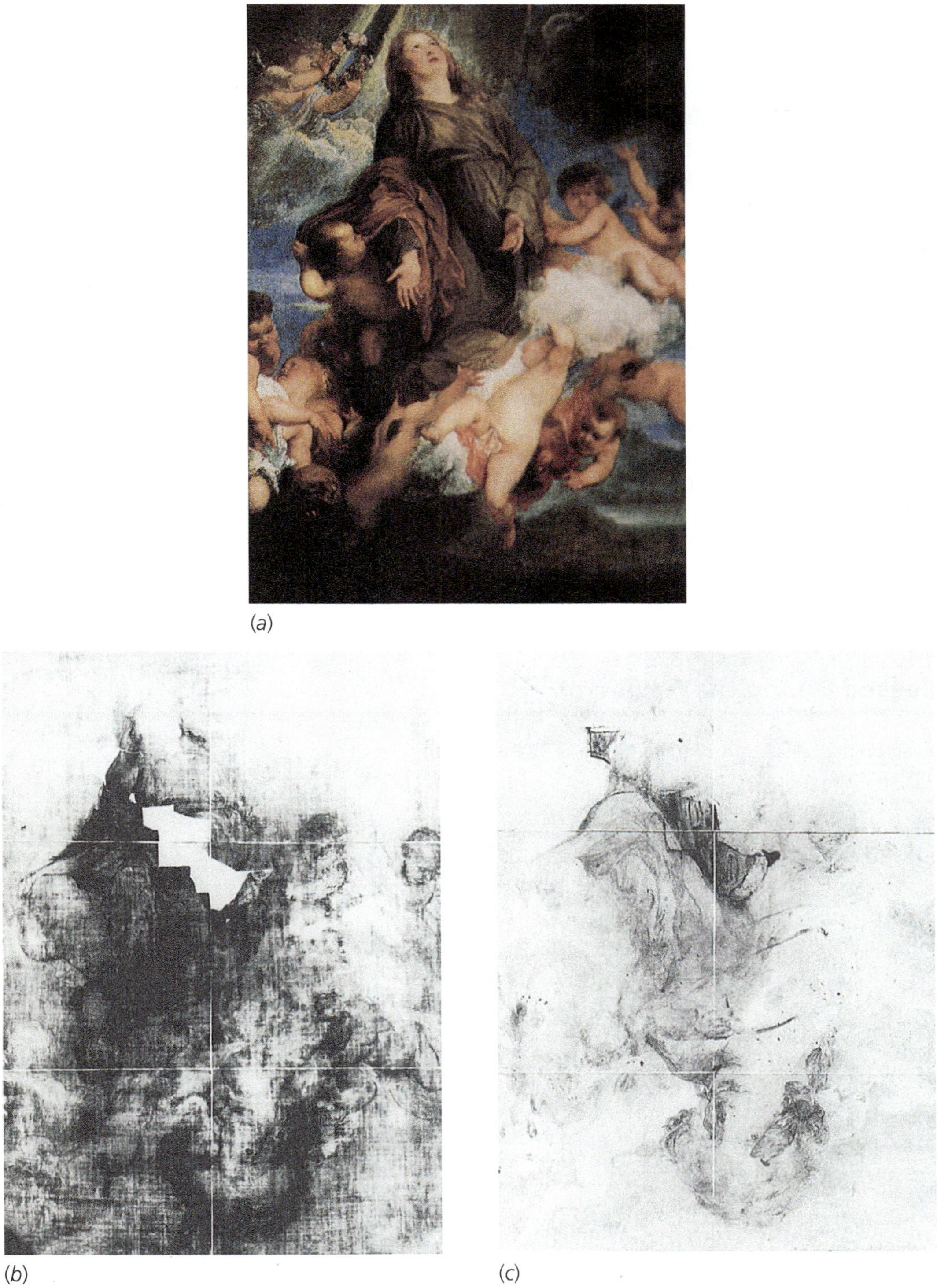

Hidden layers in paintings are analyzed by bombarding the painting with neutrons and observing the radiative emissions from nuclei that have captured a neutron. Different elements used in the painting have different half-lives. (*a*) Van Dyck's painting *Saint Rosalie Interceding for the Plague-Stricken of Palermo.* The black-and-white images in (*b*) and (*c*) were formed using a special film sensitive to electrons emitted by the radioactively decaying elements. Image (*b*), taken a few hours after the neutron irradiation, reveals the presence of manganese, found in umber, which is a dark earth pigment used for the painting's base layer. (Blank areas show where modern repairs, free of manganese, have been made.) The image in (*c*) was taken 4 days later, after the umber emissions had died away and when phosphorus, found in charcoal and boneblack, was the main radiating element. Upside down is revealed a sketch of Van Dyck himself. The self-portrait, executed in charcoal, had been overpainted by the artist. *((a) © 1991 by the Metropolitan Museum of Art. (b) and (c) Courtesy of Paintings Conservation Department, Metropolitan Museum of Art.)*

FISSION

Very heavy nuclei ($Z > 92$) are subject to spontaneous fission. They break apart into two nuclei even if the nuclei are not disturbed. We can understand this by considering the analogy of a charged liquid drop. If the drop is not too large, surface tension can overcome the repulsive forces of the charges and hold the drop together. There is, however, a certain maximum size beyond which the drop will be unstable and will spontaneously break apart. Because of spontaneous fission, an upper limit exists on the size of a nucleus and therefore on the number of elements that are possible.

Some heavy nuclei—uranium and plutonium, in particular—can be induced to fission by the capture of neutrons. During the fission of ^{235}U, for example, the uranium nucleus is excited by the capture of a neutron, causing it to split into two nuclei and emit several neutrons. The Coulomb force of repulsion drives the fission fragments apart, with the released energy eventually appearing as thermal energy. Consider, for example, the fission of a nucleus of mass number $A = 200$ into two nuclei of mass number $A = 100$. Because the rest energy for $A = 200$ is about 1 MeV per nucleon greater than that for $A = 100$, approximately 200 MeV per nucleus is released during such a fission. This is a large amount of energy. By contrast, during the chemical reaction of combustion, only about 4 eV of energy is released per molecule of oxygen consumed.

Example 40-6 Energy Released During the Fission of ^{235}U

Calculate the total energy (in kilowatt-hours) released during the fission of 1.00 g of ^{235}U, assuming that 200 MeV is released per fission.

PICTURE We need to find the number of uranium nuclei in one gram of ^{235}U, which we find using the fact that there are Avogadro's number ($N_A = 6.02 \times 10^{23}$) of nuclei in 235 grams.

SOLVE

1. The total energy is the number of nuclei multiplied by the energy per nucleus:

$$E = NE_{\text{nucleus}} = N(200\text{MeV/nucleus})$$

2. Calculate N:

$$N = \frac{6.02 \times 10^{23}\ \text{nuclei/mol}}{235\ \text{g/mol}} \times 1.00\ \text{g}$$
$$= 2.56 \times 10^{21}\ \text{nuclei}$$

3. Calculate the energy per gram in eV and convert to kW·h:

$$E = \frac{200 \times 10^{6}\ \text{eV}}{1\ \text{nucleus}} \times 2.56 \times 10^{21}\ \text{nuclei}$$
$$= 5.12 \times 10^{29}\ \text{eV} = 8.19 \times 10^{10}\ \text{J}$$
$$= 8.19 \times 10^{7}\ \text{kW}\cdot\text{s} = \boxed{2.28 \times 10^{4}\ \text{kW}\cdot\text{h}}$$

The fission of uranium was discovered in 1938 by Otto Hahn and Fritz Strassmann, who found that medium-mass elements (for example, barium and lanthanum) were produced in the bombardment of uranium with neutrons. The discovery that several neutrons were emitted during the fission process led to speculation concerning the possibility of using those neutrons to cause further

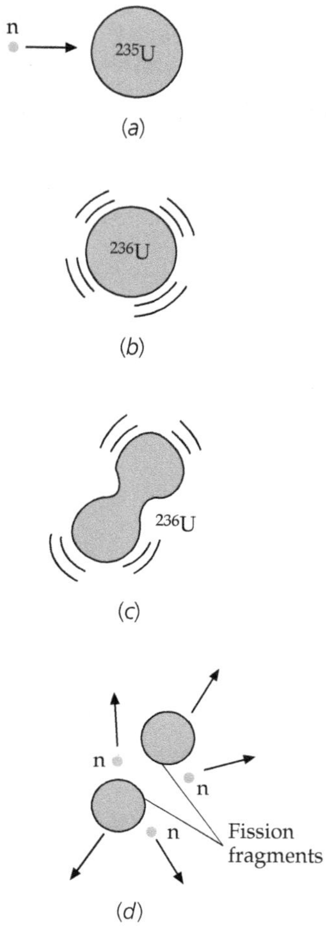

FIGURE 40-10 Schematic illustration of nuclear fission. (*a*) The absorption of a neutron by ^{235}U leads to (*b*) ^{236}U in an excited state. (*c*) The oscillation of ^{236}U has become unstable. (*d*) The nucleus splits apart into two nuclei that are less massive than the original nucleus and emits several neutrons that can produce fission in other nuclei.

fissions, thereby producing a chain reaction. When ^{235}U captures a neutron, the resulting ^{236}U nucleus emits γ rays as it de-excites to the ground state approximately 15 percent of the time and undergoes fission approximately 85 percent of the time. The fission process is somewhat analogous to the oscillation of a liquid drop, as shown in Figure 40-10. If the oscillations are violent enough, the drop splits in two. Using the liquid-drop model, Niels Bohr and John Wheeler calculated the critical energy E_c needed by the ^{236}U nucleus to undergo fission. (^{236}U is the nucleus formed momentarily by the capture of a neutron by ^{235}U.) For this nucleus, the critical energy is 5.3 MeV, which is less than the 6.4 MeV of excitation energy produced when ^{235}U captures a neutron. The capture of a neutron by ^{235}U therefore produces an excited state of the ^{236}U nucleus that has more than enough energy to break apart. On the other hand, the critical energy for fission of the ^{239}U nucleus is 5.9 MeV. The capture of a neutron by a ^{238}U nucleus produces an excitation energy of only 5.2 MeV. Therefore, when a neutron is captured by ^{238}U to form ^{239}U, the excitation energy is not great enough for fission to occur. In this case, the excited ^{239}U nucleus de-excites by γ emission and then decays to ^{239}Np by β decay, and then again to ^{239}Pu by β decay.

A fissioning nucleus can split into a pair of medium-mass nuclei, as shown in Figure 40-11. Depending on the particular reaction, 1, 2, or 3 neutrons may be emitted. The average number of neutrons emitted in the fission of ^{235}U is approximately 2.5. A typical fission reaction is

$$n + {}^{235}\text{U} \rightarrow {}^{141}\text{Ba} + {}^{92}\text{Kr} + 3n$$

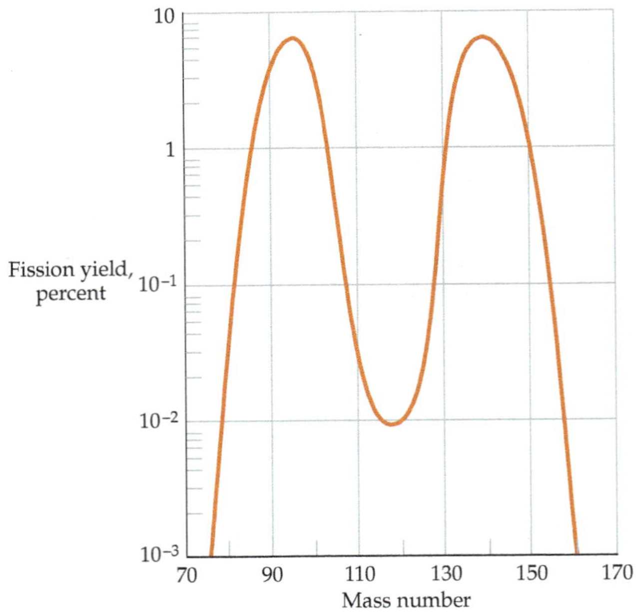

FIGURE 40-11 Distribution of the possible fission fragments of ^{235}U. The splitting of ^{235}U into two fragments of unequal mass is more likely than its splitting into fragments of equal mass.

NUCLEAR FISSION REACTORS

To sustain a chain reaction in a fission reactor, one of the neutrons (on average) that is emitted during and following* the fission of ^{235}U must be captured by another ^{235}U nucleus and cause it to fission. The **reproduction constant** k of a reactor is defined as the average number of neutrons from each fission that cause a subsequent fission. The maximum possible value of k for a uranium reactor is 2.5, but it is normally less than this for two important reasons: (1) Some of the neutrons may escape from the region containing fissionable nuclei and (2) some of the neutrons may be captured by nonfissioning nuclei in the reactor. If k is exactly 1, the reaction will be self-sustaining. If k is less than 1, the reaction will die out. If k is significantly greater than 1, the reaction rate will increase rapidly and become uncontrollable. In the design of nuclear bombs, such a runaway reaction is desired. In power reactors, the value of k must be kept very nearly equal to 1.

Because the neutrons emitted during and following fission have energies of the order of 1 MeV, whereas the chance for neutron capture leading to fission in ^{235}U is largest at small energies, the chain reaction can be sustained only if the neutrons are slowed down before they escape from the reactor. At high energies (1 MeV to 2 MeV), neutrons lose energy rapidly by inelastic scattering from ^{238}U, the principal constituent of natural uranium. (Natural uranium contains 99.3 percent ^{238}U and only 0.7 percent fissionable ^{235}U.) Once the neutron energy is below the excitation energies of the nuclei in the reactor (about 1 MeV), the main process of energy loss is by elastic scattering, in which a fast neutron collides with a nucleus at rest and transfers some of its kinetic energy to that nucleus. Such energy transfers are efficient only if the masses of the two bodies are comparable. A neutron will not transfer much energy in an elastic collision with a heavy uranium nucleus. Such a collision is like one between a marble and a billiard ball. The marble will be deflected by the much more massive billiard ball, and very little of its kinetic energy will be transferred to the billiard ball. A **moderator** consisting of material, such as water or carbon, that has light nuclei is therefore placed around the fissionable material in the core of the reactor to slow down the neutrons.

* Neutrons are sometimes emitted by the fission products. These neutrons are typically emitted a few seconds following the fission.

The inside of a nuclear power plant in Kent, England. A technician is standing on the reactor charge transfer plate, into which uranium fuel rods fit. *(© Jerry Mason/ Photo Researchers.)*

The neutrons are slowed down by elastic collisions with the nuclei of the moderator until they are in thermal equilibrium with the moderator. Because of the relatively large neutron-capture cross section of the hydrogen nucleus in water, reactors that use ordinary water as a moderator cannot easily achieve $k \approx 1$ unless they use enriched uranium, in which the ^{235}U content has been increased from 0.7 percent to between 1 percent and 4 percent. Natural uranium can be used if heavy water (D_2O) is used instead of ordinary (light) water (H_2O) as the moderator. Although heavy water is expensive, most Canadian reactors use heavy water for a moderator to avoid the cost of constructing uranium-enrichment facilities.

Figure 40-12 shows some of the features of a pressurized-water reactor commonly used in the United States to generate electricity. Fission in the core heats the water to a high temperature in the primary loop, which is closed. This water, which also serves as the moderator, is under high pressure to prevent the water from boiling.

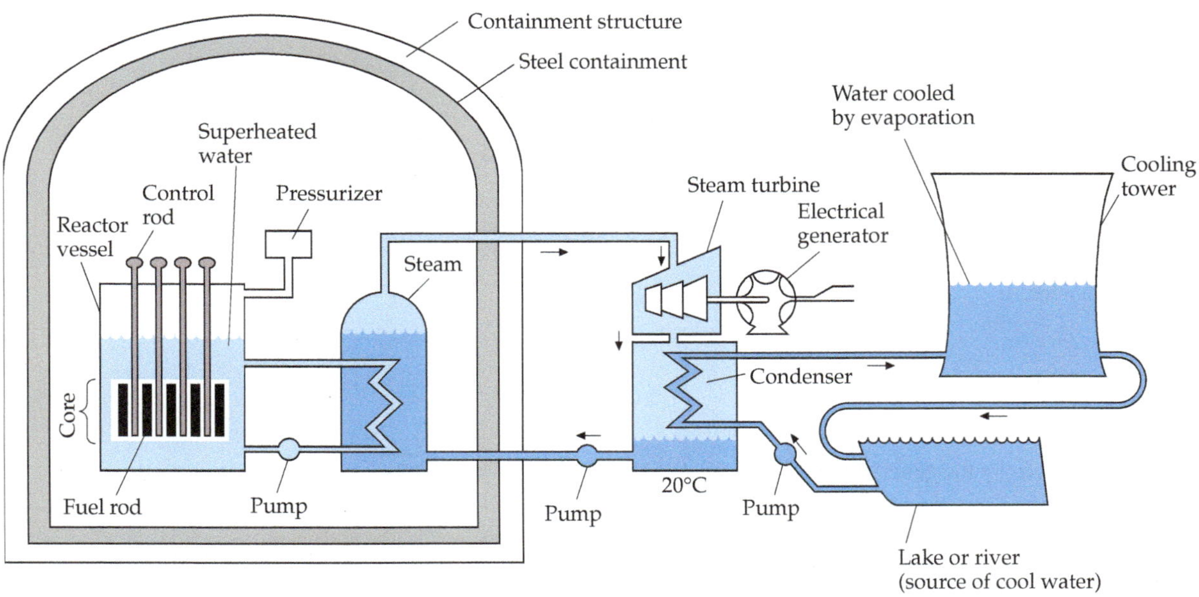

FIGURE 40-12 Simplified drawing of a pressurized-water reactor. The water in contact with the reactor core serves as both the moderator and the heat-transfer material. It is isolated from the water used to produce the steam that drives the turbines. Many features, such as the backup cooling mechanisms, are not shown here.

The hot water is pumped to a heat exchanger, where it heats the water in the secondary loop and converts the water to steam, which is then used to drive the turbines that produce electrical power. Note that the water in the secondary loop is isolated from the water in the primary loop to prevent its contamination by the radioactive nuclei in the reactor core.

The ability to control the reproduction factor k precisely is important if a power reactor is to be operated safely. Both natural negative-feedback mechanisms and mechanical methods of control are used. If k is greater than 1, the reaction rate increases and the temperature of the reactor increases. If water is used as a moderator, its density decreases with increasing temperature and the water becomes a less effective moderator. A second important control method is the use of control rods made of a material, such as cadmium, that has a very large neutron-capture cross section. To decrease the reaction rate, the control rods are inserted so that more neutrons are captured by the rods and k becomes less than 1. To increase the reaction rate, the rods are gradually withdrawn from the reactor; fewer neutrons are captured by the control rods and k becomes greater than 1.

Mechanical control of the reaction rate of a nuclear reactor using control rods is possible only because some of the neutrons emitted during the fission process are **delayed neutrons.** The time needed for a neutron to slow down from 1 MeV or 2 MeV to the thermal-energy level and then be captured is only of the order of a millisecond. If all the neutrons emitted during fission were prompt neutrons, that is, emitted immediately during the fission process, mechanical control would not be possible because the reactor would run away before the rods could be inserted farther. However, approximately 0.65 percent of the neutrons emitted are delayed by an average time of about 14 s. Those neutrons are emitted not during the fission process itself but during the decay of the fission fragments. The effect of the delayed neutrons can be seen in the following examples.

Example 40-7 Doubling Time

If the average time between fission generations (the time it takes for a neutron emitted during one fission to cause another) is $t_1 = 1$ ms $= 0.001$ s and if the average number of neutrons from each fission that cause a subsequent fission is 1.001, how long will it take for the reaction rate to double?

PICTURE The reaction rate is the number of nuclei that fission per unit time. The time to double the reaction rate is the product of the number of generations N needed to double the reaction rate and the generation time. If $k = 1.001$, the reaction rate after N generations is 1.001^N. We find the number of generations by setting 1.001^N equal to 2 and solving for N.

SOLVE

1. Set 1.001^N equal to 2 and solve for N:

$$(1.001)^N = 2$$

$$N \ln 1.001 = \ln 2$$

$$N = \frac{\ln 2}{\ln 1.001} = 693$$

2. Multiply the number of generations by the generation time:

$$t = Nt_1 = 693(0.001 \text{ s}) = \boxed{0.7 \text{ s}}$$

CHECK The step-2 result of 0.7 s is approximately 700 times the average time between generations. This many generations is plausible because the reproduction factor k is so close to 1.

TAKING IT FURTHER The doubling time of about 0.7 s is not enough time for insertion of control rods.

Example 40-8 Delayed Neutrons and Control-Rod Insertion *Try It Yourself*

Assuming that 0.65 percent of the neutrons emitted are delayed by 14 s, find the average generation time and the doubling time if $k = 1.001$.

PICTURE The doubling time is Nt_{av}, where t_{av} is the average time between generations. Since 99.35 percent of the generation times are 0.001 s and 0.65 percent are 14 s, the average generation time is 0.9935(0.001 s) + 0.0065(14 s).

SOLVE

Cover the column to the right and try these on your own before looking at the answers.

Steps	Answers
1. Compute the average generation time.	$t_{av} = 0.9935(0.001\text{ s}) + 0.0065(14\text{ s}) = 0.092\text{ s}$
2. Use your result to find the time for 693 generations.	$t = 63.8\text{ s} \approx \boxed{60\text{ s}}$

CHECK The number of delayed neutrons is approximately 0.7 percent of the total number of neutrons, but the generation time of 1 ms is approximately 0.007 percent of 14 s. Thus, an increase in the doubling time by a factor of about 100 is plausible.

TAKING IT FURTHER A doubling time of 60 s is plenty of time for mechanical insertion of control rods.

Because of the limited supply of natural uranium, the small fraction of ^{235}U in natural uranium, and the limited capacity of enrichment facilities, reactors based on the fission of ^{235}U cannot meet our energy needs for very long. A promising alternative is the **breeder reactor.** When the relatively plentiful but nonfissionable ^{238}U nucleus captures a neutron, it decays by β decay (with a half-life of 20 min) to ^{239}Np, which in turn decays by β decay (with a half-life of 2.35 days) to the fissionable nuclide ^{239}Pu. Because ^{239}Pu fissions with fast neutrons, no moderator is needed. A reactor initially fueled with a mixture of ^{238}U and ^{239}Pu will breed as much fuel as it uses or more if one or more of the neutrons emitted in the fission of ^{239}Pu is captured by ^{238}U. Practical studies indicate that a typical breeder reactor can be expected to double its fuel supply in 7 to 10 years.

There are two major safety problems inherent with breeder reactors. The fraction of delayed neutrons is only 0.3 percent for the fission of ^{239}Pu, so the time between generations is much less than that for ordinary reactors. Mechanical control is therefore much more difficult. Also, because the operating temperature of a breeder reactor is relatively high and a moderator is not desired, a heat-transfer material, such as liquid sodium metal, is used rather than water (which is the moderator as well as the heat-transfer material in an ordinary reactor). If the temperature of the reactor increases, the resulting decrease in the density of the heat-transfer material leads to positive feedback, because it will absorb fewer neutrons than before. Because of these safety considerations, breeder reactors are not yet in commercial use in the United States. There are, however, several in operation in France, Great Britain, and the former Soviet Union.

FUSION

During fusion, two light nuclei, such as deuterium (^{2}H) and tritium (^{3}H), fuse together to form a heavier nucleus. A typical fusion reaction is

$$^2\text{H} + {^3\text{H}} \rightarrow {^4\text{He}} + \text{n} + 17.6\text{ MeV}$$

The energy released in fusion depends on the particular reaction. For the ^{2}H + ^{3}H reaction, the energy released is 17.6 MeV. Although this energy is less than the energy released during a fission reaction, it is a greater amount of energy per unit mass. The energy released during this fusion reaction is (17.6 MeV)/(5 nucleons) = 3.52 MeV per nucleon. This is approximately 3.5 times as great as the 1 MeV per nucleon released in fission.

The production of power from the fusion of light nuclei holds great promise because of the relative abundance of the fuel and the absence of some of the dangers inherent in fission reactors. Unfortunately, the technology necessary to make fusion a practical source of energy has not yet been developed. We will consider the ^{2}H + ^{3}H reaction; other reactions present similar problems.

Because of the Coulomb repulsion between the ^{2}H and ^{3}H nuclei, very large kinetic energies, of the order of 1 MeV, are needed to get the nuclei close enough together for the attractive nuclear forces to become effective and to cause fusion. Such energies can be obtained in an accelerator, but because the scattering of one nucleus by the other is much more probable than fusion, the bombardment of one nucleus by another in an accelerator requires the input of more energy than is recovered. To obtain energy from fusion, the particles must be heated to a temperature great enough for the fusion reaction to occur as the result of random thermal collisions. Because a significant number of particles have kinetic energies greater than the mean kinetic energy, $\frac{3}{2}kT$, and because some particles can tunnel through the Coulomb barrier, a temperature T corresponding to $kT \approx 10$ keV is adequate to ensure that a reasonable number of fusion reactions will occur if the density of the particles is sufficiently high. The temperature corresponding to $kT = 10$ keV is of the order of 10^8 K. These temperatures occur in the interiors of stars, where such reactions are common. At these temperatures, a gas consists of positive ions and electrons and is called a **plasma.** One of the problems arising in attempts to produce controlled fusion reactions is the problem of confining the plasma long enough for the reactions to take place. In the interior of the Sun, the plasma is confined by the enormous gravitational field of the Sun. In a laboratory on Earth, confinement is a difficult problem.

The energy required to heat a plasma is proportional to the number density of its ions, n, whereas the collision rate is proportional to n^2 (the square of the number density). If τ is the confinement time, the output energy is proportional to $n^2\tau$. If the output energy is to exceed the input energy, we must have

$$C_1 n^2 \tau > C_2 n$$

where C_1 and C_2 are constants. In 1957, the British physicist J. D. Lawson evaluated these constants from estimates of the efficiencies of various hypothetical fusion reactors and derived the following relation between density and confinement time, known as **Lawson's criterion:**

$$n\tau > 10^{20}\ \text{s} \cdot \text{particles/m}^3 \qquad \text{40-22}$$

LAWSON'S CRITERION

If Lawson's criterion is met and the thermal energy of the ions is great enough ($kT \approx 10$ keV), the energy released by a fusion reactor will just equal the energy input; that is, the reactor will just break even. For the reactor to be practical, much more energy must be released.

Two schemes for achieving Lawson's criterion are currently under investigation. In one scheme, **magnetic confinement,** a magnetic field is used to confine the plasma (see Section 26-2). In the most common arrangement, first developed in the former Soviet Union and called a *tokamak*, the plasma is confined in a large toroid.

The magnetic field is a combination of the doughnut-shaped magnetic field due to the windings of the toroid and the self-field due to the current of the circulating plasma. The break-even point has almost been achieved using magnetic confinement, but we are still a long way from building a practical fusion reactor.

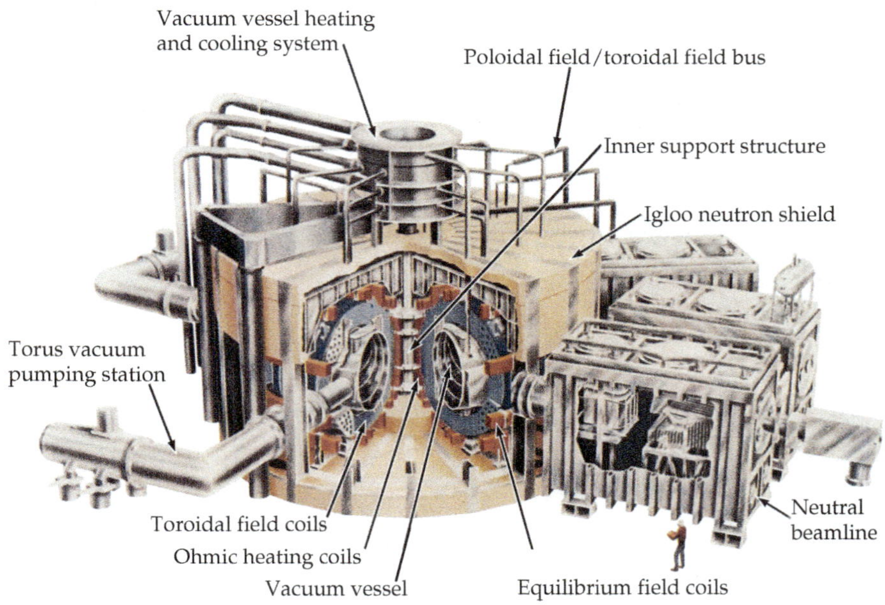

(a)

(b)

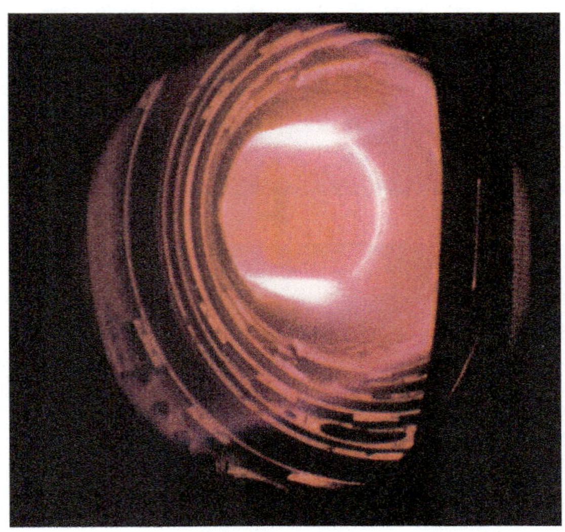

(c)

(*a*) Schematic of the Tokamak Fusion Test Reactor (TFTR). The toroidal coils, surrounding the doughnut-shaped vacuum vessel, are designed to conduct current for 3-s pulses, separated by waiting times of 5 min. Pulses peak at 73 000 A, producing a magnetic field of 5.2 T. This magnetic field is the principal means of confining the deuterium–tritium plasma that circulates within the vacuum vessel. Current for the pulses is delivered by converting the rotational energy of two 600-ton flywheels. Sets of poloidal coils, perpendicular to the toroidal coils, carry an oscillating current that generates a current through the confined plasma itself, heating it ohmically. Additional poloidal fields help stabilize the confined plasma. Between four and six neutral-beam injection systems (only one of which is shown in the schematic) are used to inject high-energy deuterium atoms into the deuterium–tritium plasma, heating beyond what could be obtained ohmically, ultimately to the point of fusion. (*b*) The TFTR itself. The diameter of the vacuum vessel is 7.7 m. (*c*) An 800-kA plasma, lasting 1.6 s, as it discharges within the vacuum vessel. *((All) Courtesy of the Princeton Plasma Physics Laboratory.)*

In a second scheme, called **inertial confinement,** a pellet of solid deuterium and tritium is bombarded from all sides by intense pulsed laser beams of energies of the order of 10^4 J lasting about 10^{-8} s. (Intense beams of ions are also used.) Computer simulation studies indicate that the pellet should be compressed to approximately 10^4 times its normal density and heated to a temperature greater than 10^8 K. This should produce approximately 10^6 J of fusion energy in 10^{-11} s, which is so brief that confinement is achieved by inertia alone.

Because the break-even point is just barely being achieved in magnetic-confinement fusion, and because the building of a fusion reactor involves many practical problems that have not yet been solved, the availability of fusion to meet our energy needs is not expected for at least several decades. However, fusion holds great promise as an energy source for the future.

(a)

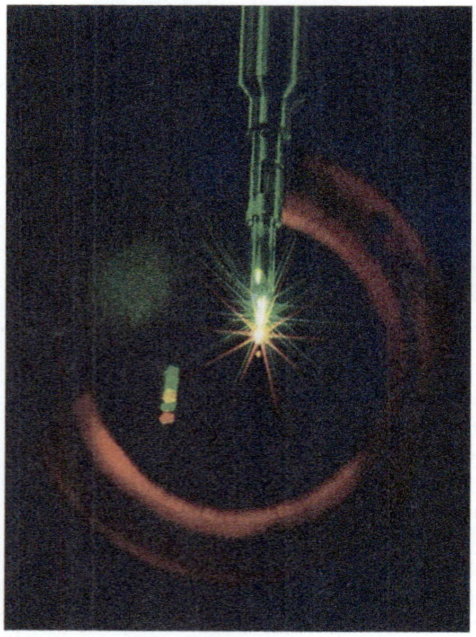

(b)

(*a*) The Nova target chamber, an aluminum sphere approximately 5 m in diameter, inside which 10 beams from the world's most powerful laser converge onto a hydrogen-containing pellet 0.5 mm in diameter. (*b*) The resulting fusion reaction is visible as a tiny star, lasting 10^{-10} s, releasing 10^{13} neutrons. *((All) Courtesy of the Lawrence Livermore National Laboratory/U.S. Department of Energy.)*

Summary

TOPIC	RELEVANT EQUATIONS AND REMARKS	
1. Properties of Nuclei	A nucleus has N neutrons, Z protons, and a mass number $A = N + Z$. For light nuclei, N and Z are approximately equal, whereas for heavy nuclei, N is greater than Z.	
Isotopes	Isotopes are two or more nuclei that have the same atomic number Z but have different values of N and A.	
Size and shape	Most nuclei are approximately spherical in shape and have a volume that is proportional to A. Because the mass is proportional to A, nuclear density is independent of A.	
Radius	$R = R_0A^{1/3} \approx (1.2\text{ fm})A^{1/3}$	40-1
Mass and binding energy	The mass of a stable nucleus is less than the sum of the masses of its nucleons. The mass difference Δm multiplied by c^2 equals the binding energy E_b of the nucleus. The binding energy is approximately proportional to the mass number A.	
2. Radioactivity	Unstable nuclei are radioactive and decay by emitting α particles (^{4}He nuclei), β particles (electrons or positrons), or γ rays (photons). All radioactivity is statistical in nature and follows an exponential decay law: $N = N_0e^{-\lambda t}$	40-6
Decay rate	$R = \lambda N = R_0e^{-\lambda t}$	40-7
Mean lifetime	$\tau = \dfrac{1}{\lambda}$	40-9
Half-life	$t_{1/2} = \tau \ln 2 = 0.693\tau$ The half-lives of α decay range from a fraction of a second to millions of years. For β decay, the half-lives range up to hours or days. For γ decay, the half-lives are usually less than a microsecond.	40-11
Decay-rate units	The number of decays per second of 1 g of radium is the curie (Ci). $1\text{ Ci} = 3.7 \times 10^{10}\text{ decays/s} = 3.7 \times 10^{10}\text{ Bq}$ $(1\text{ Bq} = 1\text{ decay/s})$	
3. Nuclear Reactions		
Q value	The Q value equals c^2 multiplied by the total mass of the incoming particles less the total mass of the outgoing particles in the center of mass reference frame. If the net mass change is Δm, the Q value is $Q = -(\Delta m)c^2$	40-19
Exothermic reaction	If total mass decreases during a reaction, Q is positive and measures the energy released.	
Endothermic reaction	If total mass increases during a reaction, Q is negative. Then $\lvert Q \rvert$ is the threshold energy for the reaction in the center of mass reference frame.	
4. Fission	Fission occurs when some heavy elements, such as ^{235}U or ^{239}Pu, capture a neutron and split apart into two nuclei. The two nuclei then fly apart because of electrostatic repulsion. A chain reaction is possible because several neutrons are emitted by a nucleus when it undergoes fission. A chain reaction can be sustained in a reactor if, on the average, one of the emitted neutrons is slowed down by scattering in the reactor and is then captured by another fissionable nucleus. Very heavy nuclei ($Z > 92$) are subject to spontaneous fission.	

TOPIC	RELEVANT EQUATIONS AND REMARKS
5. Fusion	A large amount of energy is released when two light nuclei, such as ^{2}H and ^{3}H, fuse together. Fusion takes place spontaneously inside the Sun and other stars, where the temperature is great enough (about 10^8 K) for thermal motion to bring the charged hydrogen ions close enough together to fuse. Although controlled fusion holds great promise as a future energy source, practical difficulties have thus far hindered its development.
Lawson criterion	The minimum product of particle density n and confinement time τ to get more energy out of a fusion reactor than is put in is $n\tau > 10^{20}$ s·particles/m^3.

Answer to Concept Check

41-1 The number left can be either one or zero, where zero left is more probable than one left.

Answers to Practice Problems

40-1 (*a*) $60E_1$ (*b*) $20E_1$

40-2 It is because $\frac{1}{8} = \left(\frac{1}{2}\right)^3$, so $n = 3$.

Problems

In a few problems, you are given more data than you actually need; in a few other problems, you are required to supply data from your general knowledge, outside sources, or informed estimate.

Interpret as significant all digits in numerical values that have trailing zeros and no decimal points.

- • Single-concept, single-step, relatively easy
- •• Intermediate-level, may require synthesis of concepts
- ••• Challenging
- SSM Solution is in the *Student Solutions Manual*
- Consecutive problems that are shaded are paired problems.

CONCEPTUAL PROBLEMS

1 • Isotopes of nitrogen, iron and tin have stable isotopes ^{14}N, ^{56}Fe and ^{118}Sn. Give the symbols for two other isotopes of (*a*) nitrogen, (*b*) iron, and (*c*) tin.

2 • Why is the decay chain $A = 4n + 1$ not found in nature?

3 • A decay by α emission is often followed by β decay. When this occurs, it is by β^- and not β^+ decay. Why?

4 • The half-life of ^{14}C is much less than the age of the universe, yet ^{14}C is found in nature. Why?

5 • What effect would a long-term variation in cosmic-ray activity have on the accuracy of ^{14}C dating?

6 • Why does an element that has $Z = 130$ not exist?

7 • Why is a moderator needed in an ordinary nuclear fission reactor?

8 • Explain why water is more effective than lead in slowing down fast neutrons.

9 • The stable isotope of sodium is ^{23}Na. What kind of beta decay would you expect of (*a*) ^{22}Na and (*b*) ^{24}Na?

10 • What is the advantage of a breeder reactor over an ordinary reactor? What are the disadvantages of a breeder reactor?

11 • True or false:

(*a*) In a breeder reactor, fuel can be produced as fast as it is consumed.

(*b*) The atomic nucleus is composed of protons, neutrons, and electrons.

(*c*) The mass of a ^{2}H nucleus is less than the mass of a ^{1}H nucleus plus the mass of a neutron.

(*d*) After two half-lives, all the radioactive nuclei in a given sample have decayed.

12 • Why is it that extreme changes in the temperature or the pressure of a radioactive sample have little or no effect on the radioactivity?

ESTIMATION AND APPROXIMATION

13 • We found in Chapter 25 that the ratio of the resistivity of the most insulating material to the resistivity of the least resistive material (excluding superconductors) is approximately 10^{22}. Few properties of materials show such a wide range of values. Using information in the textbook or other resources, find the ratio of largest to smallest for some nuclear properties of matter. Some examples might be the range of mass densities found in an atom, the half-life of radioactive nuclei, or the range of nuclear masses.

14 •• According to the United States Department of Energy, the U.S. population consumes approximately 10^{20} joules of energy each year. Estimate the mass (in kilograms) of (*a*) uranium that would be needed to produce this much energy using nuclear fission and (*b*) deuterium and tritium that would be needed to produce this much energy using nuclear fusion.

PROPERTIES OF NUCLEI

15 • Calculate the binding energy and the binding energy per nucleon from the masses given in Table 40-1 for (*a*) ^{12}C, (*b*) ^{56}Fe, and (*c*) ^{238}U. SSM

16 • Calculate the binding energy and the binding energy per nucleon from the masses given in Table 40-1 for (*a*) ^{6}Li, (*b*) ^{39}K, and (*c*) ^{208}Pb.

17 • Use the radius formula $R = R_0A^{1/3}$ (Equation 40-1), where $R_0 = 1.2$ fm, to compute the radii of the following nuclei: (*a*) ^{16}O, (*b*) ^{56}Fe, and (*c*) ^{197}Au.

18 • During a fission process, a ^{239}Pu nucleus splits into two nuclei whose mass number ratio is 3 to 1. Calculate the radii of the nuclei formed during the process.

19 •• The neutron, when isolated from an atomic nucleus, decays into a proton, an electron, and an antineutrino as follows: $n \rightarrow {}^1H + e^- + \bar{\nu}$. The thermal energy of a neutron is of the order of kT, where k is the Boltzmann constant. (*a*) In both joules and electron volts, calculate the energy of a thermal neutron at 25°C. (*b*) What is the speed of that thermal neutron? (*c*) A beam of monoenergetic thermal neutrons is produced at 25°C and has an intensity I. After traveling 1350 km, the beam has an intensity of $\frac{1}{2}I$. Using this information, estimate the half-life of the neutron. Express your answer in minutes. SSM

20 • Use $R = R_0A^{1/3}$ (Equation 40-1), where $R_0 = 1.2$ fm, for the radius of a spherical nucleus to calculate the density of nuclear matter. Express your answer in grams per cubic centimeter.

21 •• In 1920, 12 years before the discovery of the neutron, Ernest Rutherford argued that proton–electron pairs might exist in the confines of the nucleus in order to explain the mass number, A, being greater than the nuclear charge, Z. He also used this argument to account for the source of beta particles in radioactive decay. Rutherford's scattering experiments in 1910 showed that the nucleus had a diameter of approximately 10 fm. Using this nuclear diameter, the uncertainty principle, and that beta particles have an energy range of 0.02 MeV to 3.40 MeV, show why the hypothetical electrons cannot be confined to a region occupied by the nucleus. SSM

22 •• Consider the following fission process: ${}^{235}_{92}U + n \rightarrow {}^{95}_{37}Rb + {}^{137}_{55}Cs + 4n$. Determine the electrostatic potential energy, in MeV, of the reaction products when the surfaces of the ^{95}Rb nucleus and the ^{137}Cs nucleus are just touching immediately after being formed during the fission process.

RADIOACTIVITY

23 • Homer enters the visitors' chambers, and his Geiger beeper sounds. He shuts off the beeper, removes the device from his shoulder patch, and holds it near the only new object in the room—an orb that is to be presented as a gift from the visiting Cartesians. Pushing a button marked "monitor," Homer reads that the device is reading a counting rate of 4000 counts/s above the background counting rate. After 10 min, the counting rate has dropped to 1000 counts/s above the background rate. (*a*) What is the half-life of the source? (*b*) How high will the counting rate be (above the background counting rate) 20 min after the monitoring device was switched on?

24 • A certain source gives 2000 counts/s at time $t = 0$. Its half-life is 2.0 min. How many counts per second will it give after (*a*) 4.0 min, (*b*) 6.0 min, and (*c*) 8.0 min?

25 • The counting rate from a radioactive source is 8000 counts/s at time $t = 0$, and 10 min later the rate is 1000 counts/s. (*a*) What is the half-life? (*b*) What is the decay constant? (*c*) What is the counting rate after 20 min?

26 • The half-life of radium is 1620 y. Calculate the number of disintegrations per second of 1.00 g of radium and show that the disintegration rate is approximately 1.0 Ci.

27 • A radioactive piece of silver foil ($t_{1/2} = 2.4$ min) is placed near a Geiger counter and 1000 counts/s are observed at time $t = 0$. (*a*) What is the counting rate at $t = 2.4$ min and at $t = 4.8$ min? (*b*) If the counting efficiency is 20 percent, how many radioactive silver nuclei are there at time $t = 0$? At time $t = 2.4$ min? (*c*) At what time will the counting rate be about 30 counts/s?

28 • Use Table 40-1 to calculate the energy release, in MeV, for the α decay of (*a*) ^{226}Ra and (*b*) ^{242}Pu.

29 •• Plutonium is very toxic to the human body. Once it enters the body it collects primarily in the bones, although it also can be found in other organs. Red blood cells are synthesized within the marrow of the bones. The isotope ^{239}Pu is an alpha emitter that has a half-life of 24 360 years. Because alpha particles are an ionizing radiation, the blood-making ability of the marrow is, in time, destroyed by the presence of ^{239}Pu. In addition, many kinds of cancers will also develop in the surrounding tissues because of the ionizing effects of the alpha particles. (*a*) If a person accidentally ingested 2.0 μg of ^{239}Pu and all of it is absorbed by the bones of the person, how many alpha particles are produced per second within the body of the person? (*b*) When, in years, will the activity be 1000 alpha particles per second? SSM

30 •• Consider an alpha-emitting parent nucleus ${}^{A}_{Z}X$ initially at rest. The nucleus decays into a daughter nucleus Y and an alpha particle as follows: ${}^{A}_{Z}X \rightarrow {}^{A-4}_{Z-2}Y + {}^{4}_{2}\alpha + Q$. (*a*) Show that the alpha particle has a kinetic energy of $(A - 4)Q/A$. (*b*) Show that the kinetic energy of the recoiling daughter nucleus is given by $K_Y = 4Q/A$.

31 • The fissile material ^{239}Pu is an alpha emitter. Write the reaction that describes ^{239}Pu undergoing alpha decay. Given that ^{239}Pu, ^{235}U, and an alpha particle have respective masses of 239.052 156 u, 235.043 923 u, and 4.002 603 u, use the relations appearing in Problem 30 to calculate the kinetic energies of the alpha particle and the recoiling daughter nucleus. SSM

32 • Through a friend in the security department at the museum, Angela obtains a sample of a wooden tool handle that contains 175 g of carbon. The decay rate of the ^{14}C in the sample is 8.1 Bq. How long ago was the wood in the handle last alive?

33 • A sample of a radioactive isotope is found to have an activity of 115.0 Bq immediately after it is pulled from the reactor that formed the isotope. Its activity 2 h 15 min later is measured to be 85.2 Bq. (*a*) Calculate the decay constant and the half-life of the sample. (*b*) How many radioactive nuclei were there in the sample initially?

34 •• A 1.00-mg sample of substance that has an atomic mass of 59.934 u and emits β particles has an activity of 1.131 Ci. Find the decay constant for the substance in reciprocal seconds and find the half-life in years.

35 •• Radiation has been used for a long time in medical therapy to control the development and growth of cancer cells. Cobalt-60, a gamma emitter that emits photons that have energies of 1.17 MeV and 1.33 MeV, is used to irradiate and destroy deep-rooted cancers. Small needles made of ^{60}Co of a specified activity are encased in gold and used as body implants in tumors for time periods that are related to tumor size, tumor cell reproductive rate, and the activity of the needle. (*a*) A 1.00-μg sample of ^{60}Co, that has a half-life of

5.27 y and that is used to irradiate a small internal tumor with gamma rays, is prepared in the cyclotron of a medical center. Determine the activity of the sample in curies. (*b*) What will the activity of the sample be 1.75 y from now? SSM

36 •• (*a*) Show that if the decay rate is R_0 at time $t = 0$ and R_1 at some later time $t = t_1$, the decay constant is given by $\lambda = t_1^{-1}\ln(R_0/R_1)$ and the half-life is given by $t_{1/2} = t_1\ln(2)/\ln(R_0/R_1)$. (*b*) Use these results to find the decay constant and the half-life if the decay rate is 1200 Bq at $t = 0$ and 800 Bq at $t_1 = 60.0$ s.

37 •• A wooden casket is thought to be 18 000 years old. How much carbon would have to be recovered from the object to yield a ^{14}C counting rate of no less than 5 counts/min with a detection efficiency of 20 percent?

38 •• A sample of radioactive material is initially found to have an activity of 115.0 decays/min. After 4 d 5 h, its activity is measured to be 73.5 decays/min. (*a*) Calculate the half-life of the material. (*b*) How long (from the initial time) will it take for the sample to reach an activity of 10.0 decays/min?

39 •• The rubidium isotope ^{87}Rb is a β^- emitter that has a half-life of 4.9×10^{10} y. It decays into ^{87}Sr. This nuclear decay is used to determine the age of rocks and fossils. Rocks containing the fossils of early animals have a ratio of ^{87}Sr to ^{87}Rb of 0.0100. Assuming that there was no ^{87}Sr present when the rocks were formed, calculate the age of the fossils.

40 ••• Consider a single nucleus of a radioactive isotope that has a decay rate equal to λ. The nucleus has not decayed at $t = 0$. The probability that the nucleus will decay between time t and time $t + dt$ is equal to $\lambda e^{-\lambda t}dt$. (*a*) Show that this statement is consistent with the fact that the probability is 1 that the nucleus will decay between $t = 0$ and $t = \infty$. (*b*) Show that the expected lifetime of the nucleus is equal to $1/\lambda$. *Hint: The expected lifetime is equal to* $\int_0^\infty t\lambda e^{-\lambda t}dt$ *divided by* $\int_0^\infty \lambda e^{-\lambda t}dt$. (*c*) A sample of material contains a number of these radioactive nuclei at time $t = 0$. What is the mean lifetime of the radioactive nuclei in the sample?

NUCLEAR REACTIONS

41 • Using Table 40-1, find the Q values for the following reactions: (*a*) $^1\text{H} + {}^3\text{H} \rightarrow {}^3\text{He} + \text{n} + Q$ and (*b*) $^2\text{H} + {}^2\text{H} \rightarrow {}^3\text{He} + \text{n} + Q$.

42 • Using Table 40-1, find the Q values for the following reactions: (*a*) $^2\text{H} + {}^2\text{H} \rightarrow {}^3\text{H} + {}^1\text{H} + Q$, (*b*) $^2\text{H} + {}^3\text{He} \rightarrow {}^4\text{He} + {}^1\text{H} + Q$, and (*c*) $^6\text{Li} + \text{n} \rightarrow {}^3\text{H} + {}^4\text{He} + Q$.

43 •• (*a*) Use the values 14.003 242 u and 14.003 074 u for the atomic masses of ^{14}C and ^{14}N, respectively, to calculate the Q value (in MeV) for the β-decay reaction ${}^{14}_{6}\text{C} \rightarrow {}^{14}_{7}\text{N} + \text{e}^- + \bar{\nu}_\text{e}$. (*b*) Explain why you should not add the mass of the electron to that of atomic ^{14}N for the calculation in Part (*a*). SSM

44 •• (*a*) Use the values 13.005 738 u and 13.003 354 u for the atomic masses of ${}^{13}_{7}$N and ${}^{13}_{6}$C, respectively, to calculate the Q value (in MeV) for the β-decay reaction

$$ {}^{13}_{7}\text{N} \rightarrow {}^{13}_{6}\text{C} + \text{e}^+ + \nu_\text{e} $$

(*b*) Explain why you need to add twice the mass of an electron to the mass of ${}^{13}_{6}$C during the calculation of the Q value for the reaction in Part (*a*).

FISSION AND FUSION

45 • Assuming an average energy of 200 MeV per fission, calculate the number of fissions per second needed for a 500-MW reactor. SSM

46 • If the reproduction factor in a reactor is 1.1, find the number of generations needed for the power level to (*a*) double, (*b*) increase by a factor of 10, and (*c*) increase by a factor of 100. Find the time needed in each case if (*d*) there are no delayed neutrons, so that the time between generations is 1.0 ms, and (*e*) there are delayed neutrons that make the average time between generations 100 ms.

47 •• Consider the following fission reaction: ${}^{235}_{92}\text{U} + \text{n} \rightarrow {}^{95}_{42}\text{Mo} + {}^{139}_{57}\text{La} + 2\text{n} + Q$. The masses of the neutron, ^{235}U, ^{95}Mo, and ^{139}La are 1.008 665 u, 235.043 923 u, 94.905 842 u, and 138.906 348 u, respectively. Calculate the Q value, in MeV, for the fission reaction. SSM

48 •• In 1989, researchers claimed to have achieved fusion in an electrochemical cell at room temperature. Their now thoroughly discredited claim was that a power output of 4.00 W was produced by deuterium fusion reactions in the palladium electrode of their apparatus. The two most likely reactions are

$$ ^2\text{H} + {}^2\text{H} \rightarrow {}^3\text{He} + \text{n} + 3.27\ \text{MeV} $$

and

$$ ^2\text{H} + {}^2\text{H} \rightarrow {}^3\text{H} + {}^1\text{H} + 4.03\ \text{MeV} $$

Of the deuterium nuclei that participated in these reactions, assume half of the deuterium nuclei participated in the first reaction and the other half participated in the second reaction. How many neutrons per second would we expect to be emitted in the generation of 4.00 W of power?

49 •• A fusion reactor that uses only deuterium for fuel would have the two reactions in Problem 48 taking place in the reactor. The ^{3}H produced in the second reaction reacts immediately with another ^{2}H in the reaction

$$ ^3\text{H} + {}^2\text{H} \rightarrow {}^4\text{He} + \text{n} + 17.6\ \text{MeV} $$

The ratio of ^{2}H to ^{1}H atoms in naturally occurring hydrogen is 1.5×10^{-4}. How much energy would be produced from 4.0 L of water if all of the ^{2}H nuclei undergo fusion?

50 ••• The fusion reaction between ^{2}H and ^{3}H is

$$ ^3\text{H} + {}^2\text{H} \rightarrow {}^4\text{He} + \text{n} + 17.6\ \text{MeV} $$

Using the conservation of momentum and the given Q value, find the final energies of both the ^{4}He nucleus and the neutron, assuming the initial kinetic energy of the system is 1.00 MeV and the initial momentum of the system is zero.

51 ••• Energy is generated in the Sun and other stars by fusion. One of the fusion cycles, the proton–proton cycle, consists of the following reactions:

$$ ^1\text{H} + {}^1\text{H} \rightarrow {}^2\text{H} + \text{e}^+ + \nu_\text{e} $$

$$ ^1\text{H} + {}^2\text{H} \rightarrow {}^3\text{He} + \gamma $$

followed by

$$ ^1\text{H} + {}^3\text{He} \rightarrow {}^4\text{He} + \text{e}^+ + \nu_\text{e} $$

(*a*) Show that the net effect of these reactions is

$$ 4\,{}^1\text{H} \rightarrow {}^4\text{He} + 2\text{e}^+ + 2\nu_\text{e} + \gamma $$

(*b*) Show that 24.7 MeV is released during this cycle (not counting the additional energy of 1.02 MeV that is released when each positron meets an electron and the two annihilate). (*c*) The Sun radiates energy at the rate of approximately 4.0×10^{26} W. Assuming that this is due to the conversion of four protons into helium, γ rays, and neutrinos, which releases 26.7 MeV, what is the rate of proton consumption in the Sun? How long will the Sun last if it continues to radiate at its present level? (Assume that protons constitute about half of the total mass, 2.0×10^{30} kg, of the Sun.)

GENERAL PROBLEMS

52 • (*a*) Show that $ke^2 = 1.44$ MeV · fm, where k is the Coulomb constant and e is the magnitude of the electron charge. (*b*) Show that $hc = 1240$ MeV · fm.

53 • The counting rate from a radioactive source is 6400 counts/s. The half-life of the source is 10 s. Make a plot of the counting rate as a function of time for times up to 1 min. What is the decay constant for the source? SSM

54 • Find the energy needed to remove a neutron from (*a*) ^{4}He and (*b*) ^{7}Li.

55 • The isotope ^{14}C decays according to $^{14}\text{C} \rightarrow {}^{14}\text{N} + \text{e}^- + \bar{\nu}_\text{e}$. The atomic mass of ^{14}N is 14.003 074 u. Determine the maximum kinetic energy of the electron. (Neglect recoil of the nitrogen atom.)

56 • The density of a neutron star is the same as the density of a nucleus. If our Sun were to collapse to a neutron star, what would be the radius of that object?

57 •• Show that the ^{109}Ag nucleus is stable and does not undergo alpha decay, $^{109}_{47}\text{Ag} \rightarrow {}^4_2\text{He} + {}^{105}_{45}\text{Rh} + Q$. The mass of the ^{109}Ag nucleus is 108.904 756 u, and the products of the decay are 4.002 603 u and 104.905 250 u, respectively. SSM

58 •• Gamma rays can be used to induce photofission (fission triggered by the absorption of a photon) in nuclei. Calculate the threshold photon wavelength for the following nuclear reaction: $^2\text{H} + \gamma \rightarrow {}^1\text{H} + \text{n}$. Use Table 40-1 for the masses of the interacting particles.

59 • The relative abundance of ^{40}K (potassium 40) is 1.2×10^{-4}. The isotope ^{40}K has a molar mass of 40.0 g/mol, is radioactive, and has a half-life of 1.3×10^9 y. Potassium is an essential element of every living cell. In the human body the mass of potassium constitutes approximately 0.36 percent of the total mass. Determine the activity of this radioactive source in a student whose mass is 60 kg.

60 •• When a positron makes contact with an electron, the electron–positron pair annihilate by way of the reaction $\beta^+ + \beta^- \rightarrow 2\gamma$. Calculate the minimum total energy, in MeV, of the two photons created when a positron–electron pair annihilate.

61 •• The isotope ^{24}Na is a β emitter and has a half-life of 15 h. A saline solution containing the radioactive isotope has an activity of 600 kBq and is injected into the bloodstream of a patient. Ten hours later, the activity of 1 mL of blood from the individual yields a counting rate of 12 counts/s at a counting efficiency of 20 percent. Determine the volume of blood in the patient.

62 •• (*a*) Determine the distance of closest approach of an 8.0-MeV α particle in a head-on collision with a stationary nucleus of ^{197}Au and with a stationary nucleus of ^{10}B, neglecting the recoil of the struck nuclei. (*b*) Repeat the calculation taking into account the recoil of the struck nuclei.

63 •• Twelve nucleons are in a one-dimensional infinite square well of length $L = 3.0$ fm. (*a*) Using the approximation that the mass of a nucleon is 1.0 u, find the lowest energy of a nucleon in the well. Express your answer in MeV. What is the ground-state energy of the system of 12 nucleons in the well if (*b*) all the nucleons are neutrons so that there can be no more than 2 in each spatial state and (*c*) 6 of the nucleons are neutrons and 6 are protons so that there can be as many as 4 nucleons in each spatial state? (Neglect the energy of Coulomb repulsion of the protons.)

64 •• The helium nucleus or α particle is a very tightly bound system. Nuclei with $N = Z = 2n$, where n is an integer (for example, ^{12}C, ^{16}O, ^{20}Ne, and ^{24}Mg), can be modeled as agglomerates of α particles. (*a*) Use this model to estimate the binding energy of a pair of α particles from the atomic masses of ^{4}He and ^{16}O. Assume that the four α particles in ^{16}O form a regular tetrahedron that has one α particle at each vertex. (*b*) From the result obtained in Part (*a*) determine, on the basis of the model, the binding energy of ^{12}C and compare your result with the result obtained from the atomic mass of ^{12}C.

65 •• Nuclei of a radioactive isotope that has a decay constant of λ are produced in an accelerator at a constant rate R_p. The number of radioactive nuclei N then obeys the equation $dN/dt = R_\text{p} - \lambda N$. (*a*) If N is zero at $t = 0$, sketch N versus t for the situation. (*b*) The isotope ^{62}Cu is produced at a rate of 100 per second by placing ordinary copper (^{63}Cu) in a beam of high-energy photons. The reaction is

$$\gamma + {}^{63}\text{Cu} \rightarrow {}^{62}\text{Cu} + \text{n}$$

The isotope ^{62}Cu decays by β decay and has a half-life of 10 min. After a time long enough so that $dN/dt \approx 0$, how many ^{62}Cu nuclei are there?

66 •• The total energy consumed in the United States in 1 y is approximately 7.0×10^{19} J. How many kilograms of ^{235}U would be needed to provide this amount of energy if we assume that 200 MeV of energy is released by each fissioning uranium nucleus, that all of the uranium atoms undergo fission, and that all of the energy-conversion mechanisms used are 100 percent efficient?

67 •• (*a*) Find the wavelength of a particle in the ground state of a one-dimensional infinite square well of length $L = 2.00$ fm. (*b*) Find the momentum in units of MeV/c for a particle that has this wavelength. (*c*) Show that the total energy of an electron that has this wavelength is approximately $E \approx pc$. (*d*) What is the kinetic energy of an electron in the ground state of the well? This calculation shows that if an electron were confined in a region of space as small as a nucleus, it would have a very large kinetic energy.

68 •• If ^{12}C, ^{11}B, and ^{1}H have respective masses of 12.000 000 u, 11.009 306 u, and 1.007 825 u, determine the minimum energy, Q, in MeV, required to remove a proton from a ^{12}C nucleus.

69 ••• Assume that a neutron decays into a proton and an electron without the emission of a neutrino. The kinetic energy shared by the proton and the electron is then 0.782 MeV. In the rest frame of the neutron, the total momentum is zero, so the momentum of the proton must be equal and opposite the momentum of the electron. This determines the ratio of the kinetic energies of the two particles, but because the electron is relativistic, the exact calculation of these relative kinetic energies is somewhat challenging. (*a*) Assume that the kinetic energy of the electron is 0.782 MeV and calculate the momentum p of the electron in units of MeV/c. *Hint: Use* $E^2 = p^2c^2 + (mc^2)^2$ *(Equation 39-27)*. (*b*) Using your result from Part (*a*), calculate the kinetic energy $p^2/2m_\text{p}$ of the proton. (*c*) Because the total kinetic energy of the electron and the proton is 0.782 MeV, the calculation in Part (*b*) gives a correction to the assumption that the kinetic energy of the electron is 0.782 MeV. What percentage of 0.782 MeV is this correction? SSM

70 ••• In the laboratory reference frame, a neutron of mass m moving with speed v_L makes an elastic head-on collision with a nucleus of mass M that is at rest. (*a*) Show that the speed of the center of mass in the lab frame is $V = mv_\text{L}/(m + M)$. (*b*) What is the speed of the nucleus in the center-of-mass frame before the collision and after the collision? (*c*) What is the speed of the nucleus in the laboratory frame after the collision? (*d*) Show that the energy of the nucleus after the collision in the laboratory frame is

$$\frac{1}{2}M(2V)^2 = \frac{4mM}{(m + M)^2}\left(\frac{1}{2}mv_\text{L}^2\right)$$

(*e*) Show that the fraction of the energy lost by the neutron in the elastic collision is

$$\frac{-\Delta E}{E} = \frac{4mM}{(m + M)^2} = \frac{4(m/M)}{[1 + (m/M)]^2}$$

71 ••• (*a*) Use the result from Part (*e*) of Problem 70 to show that after N head-on collisions of a neutron with carbon nuclei at rest, the energy of the neutron is approximately $(0.714)^N E_0$, where E_0 is its original energy. (*b*) How many head-on collisions are required to reduce the energy of the neutron from 2.0 MeV to 0.020 eV, assuming stationary carbon nuclei?

72 ••• On the average, a neutron loses 63 percent of its energy in a collision with a hydrogen atom and 11 percent of its energy in a collision with a carbon atom. Calculate the number of collisions needed to reduce the energy of a neutron from 2.0 MeV to 0.020 eV if the neutron collides with (*a*) hydrogen atoms and (*b*) carbon atoms.

73 ••• Frequently, the daughter nucleus of a radioactive parent nucleus is itself radioactive. Suppose the parent nucleus, designated by P, has a decay constant λ_P, while the daughter nucleus, designated by D, has a decay constant λ_D. The number of daughter nuclei N_D are then given by the solution to the differential equation

$$dN_D/dt = \lambda_P N_P - \lambda_D N_D$$

where N_P is the number of parent nuclei. (*a*) Justify this differential equation. (*b*) Show that the solution for the equation is

$$N_D(t) = \frac{N_{P0}\lambda_P}{\lambda_D - \lambda_P}(e^{-\lambda_P t} - e^{-\lambda_D t})$$

where N_{P0} is the number of parent nuclei present at $t = 0$ when there are no daughter nuclei. (*c*) Show that the expression for N_D in Part (*b*) gives $N_D(t) > 0$ whether $\lambda_P > \lambda_D$ or $\lambda_D > \lambda_P$. (*d*) Make a plot of $N_P(t)$ and $N_D(t)$ as a function of time when $\tau_D = 3\tau_P$, where τ_D and τ_P are the mean lifetimes of the daughter and parent nuclei, respectively. SSM

74 ••• Suppose isotope A decays to isotope B and has a decay constant λ_A, and isotope B in turn decays and has a decay constant λ_B. Suppose a sample contains, at $t = 0$, only isotope A nuclei. Derive an expression for the time at which the number of isotope B nuclei will be a maximum. (See Problem 73.)

75 ••• An example of the situation discussed in Problem 73 is the radioactive isotope ^{229}Th, an α emitter that has a half-life of 7300 y. Its daughter, ^{225}Ra, is a β emitter that has a half-life of 14.8 d. In this instance, as in many instances, the half-life of the parent is much longer than the half-life of the daughter. Using the expression given in Problem 73, Part (*b*), starting with a sample of pure ^{229}Th containing N_{P0} nuclei, show that the number, N_D, of ^{225}Ra nuclei will, after several years, be given by

$$N_D = \frac{\lambda_P}{\lambda_D} N_P$$

where N_P is the number of ^{229}Th nuclei. The number of daughter nuclei are said to be in *secular equilibrium.*

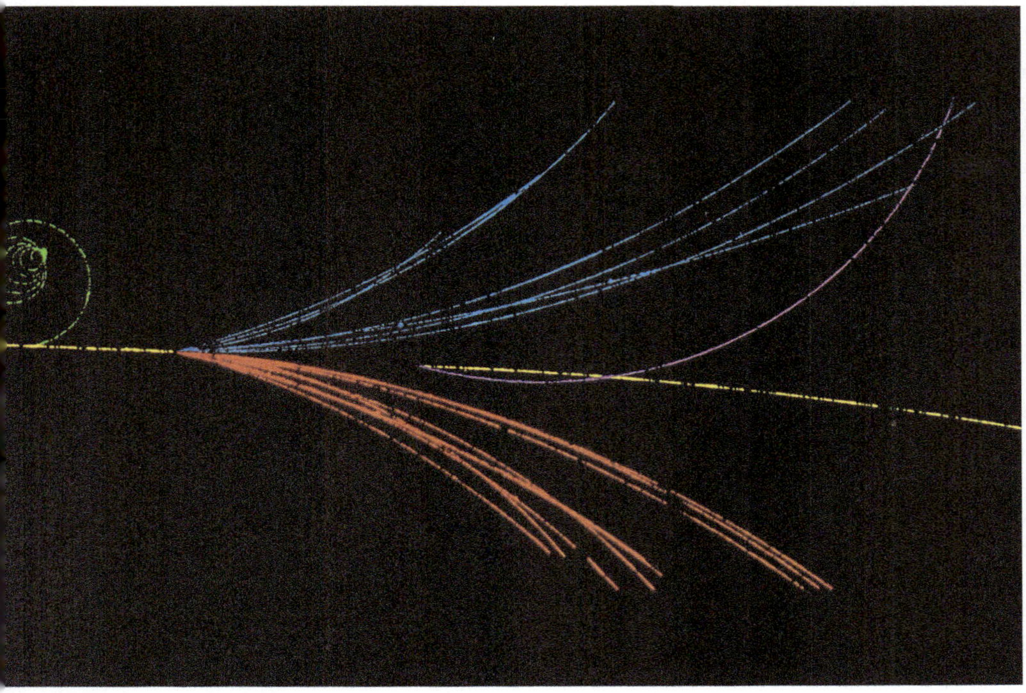

CHAPTER 41

TRACKS IN A BUBBLE CHAMBER PRODUCED BY AN INCOMING HIGH-ENERGY PROTON (YELLOW), INCIDENT FROM THE LEFT, COLLIDING WITH A PROTON AT REST. THE SMALL GREEN SPIRAL IS AN ELECTRON KNOCKED OUT OF AN ATOM. IT CURVES TO THE LEFT BECAUSE OF AN EXTERNAL MAGNETIC FIELD IN THE CHAMBER. THE COLLISION PRODUCES SEVEN NEGATIVE PARTICLES (BLUE), ALL π^-; A NEUTRAL PARTICLE Λ^0 THAT LEAVES NO TRACK; AND NINE POSITIVE PARTICLES (RED) INCLUDING SEVEN π^+, A K^+, AND A PROTON. THE Λ^0 TRAVELS IN THE ORIGINAL DIRECTION OF THE INCOMING PROTON BEFORE DECAYING INTO A PROTON (YELLOW) AND A π^- (PURPLE). *(© Lawrence Livermore Laboratory/Science Photo Library/Photo Researchers.)*

Elementary Particles and the Beginning of the Universe

? How do you determine the energy of particle interactions? (See Example 41-1.)

Items that we encounter in everyday life are made of atoms. In some sense, atoms are the building blocks of nature. However, we know that atoms are not the most fundamental constituents of matter. With the discovery of the electron by J. J. Thomson (1897), the Bohr theory of the nuclear atom (1913), and the discovery of the neutron (1932), it became clear that atoms and even nuclei have considerable structure. Indeed the once simple picture of particle physics in which there were just four "elementary" particles—the proton, neutron, electron, and photon—has become much more complex.

Since the 1950s, enormous sums of money have been spent constructing particle accelerators of greater and greater energies in hopes of finding particles predicted by various theories. At present, we know of several hundred particles that at one time or another have been considered to be elementary, and research teams at the giant accelerator laboratories around the world are searching for and finding new particles. Some of these particles have such short lifetimes (of the order of 10^{-23} s) that they can be detected only indirectly. Many particles are observed only during nuclear reactions using high-energy accelerators. In addition to the usual particle properties of mass, charge, and spin, new properties have been found and given whimsical names such as strangeness, charm, color, topness, and bottomness.

In this chapter, we will first look at the various ways of classifying the multitude of particles that have been found. We will then describe the current theory of elementary particles, called the standard model, *in which all matter in nature—from the exotic particles produced in the giant accelerator laboratories to ordinary grains of sand—is considered to be constructed from just two families of elementary particles, leptons and quarks. In the final section, we will use our knowledge of elementary particles to discuss the big bang theory which describes the origin of the universe.*

41-1 HADRONS AND LEPTONS

All the different forces observed in nature, from ordinary friction to the tremendous forces involved during supernova explosions, can be understood in terms of the four basic interactions: (1) the strong nuclear interaction (also called the hadronic interaction), (2) the electromagnetic interaction, (3) the weak (nuclear) interaction, and (4) the gravitational interaction. The four basic interactions provide a convenient structure for the classification of particles. Some particles participate in all four interactions, whereas other particles participate in only some of the interactions. For example, all particles participate in gravitational interaction, the weakest of the interactions. All particles that have electric charge participate in the electromagnetic interaction.

Particles that interact by the strong interaction are called **hadrons.** There are two kinds of hadrons: **baryons,** which have spin $\frac{1}{2}$ or $\frac{3}{2}$ or $\frac{5}{2}$, etc., and **mesons,** which have spin 0 or 1 or 2, etc. Baryons, which include nucleons, are the most massive of the elementary particles. Mesons have intermediate masses, between the mass of the electron and the mass of the proton. Particles that decay by the strong interaction have very short lifetimes, of the order of 10^{-23} s, which is about the time it takes light to travel a distance equal to the diameter of a nucleus. On the other hand, particles that decay by the weak interaction have much longer lifetimes, of the order of 10^{-10} s. Table 41-1 lists some of the properties of those hadrons that are stable against decay by the strong interaction.

Hadrons are rather complicated entities and have complex structures. If we use the term *elementary particle* to mean a point particle that has no structure and is not constructed from some more elementary entities, then hadrons are not elementary particles. It is now believed that all hadrons are composed of more fundamental entities called *quarks,* which, as far as we know, are truly elementary particles.

Particles that participate in the weak interaction but not in the strong interaction are called **leptons.** These include electrons, muons, and neutrinos, which are all less massive than the lightest hadron. The word *lepton,* meaning "light particle," was chosen to reflect the relatively small mass of the particles. However, the most recently discovered lepton, the *tau,* found by Martin Lewis Perl in 1975, has a mass of 1784 MeV/c^2, nearly twice the mass of the proton (938 MeV/c^2), so we now have a "heavy lepton." In addition, the word *muon,* short for mu-meson, is something of a misnomer. The muon is not now categorized as a meson, so it is best to refer to it

The Super-Kamiokande detector, built in Japan in 1996 as a joint Japanese–American experiment, is essentially a water tank the size of a large cathedral installed in a deep zinc mine 1 mile inside a mountain. When neutrinos pass through the tank, one of the neutrinos occasionally collides with an atom, sending blue light through the water to an array of detectors. This photograph shows the detector wall and top that have approximately 9000 photomultiplier tubes that help detect the neutrinos. Experimental results reported in June 1998 were evidence that the mass of the neutrino cannot be zero. *(ICCR (Institute for Cosmic Ray Research), The University of Tokyo.)*

Table 41-1 Hadrons That Are Stable Against Decay via the Strong Nuclear Interaction

Name	Symbol	Mass MeV/c^2	Spin, $\hbar$	Charge, e	Antiparticle	Mean Lifetime, s	Typical Decay Products*
Baryons							
Nucleon	p (proton)	938.3	$\frac{1}{2}$	+1	p^-	Infinite	
	n (neutron)	939.6	$\frac{1}{2}$	0	$\bar{n}$	930	$\mathrm{p} + \mathrm{e}^- + \bar{\nu}_e$
Lambda	Λ^0	1116	$\frac{1}{2}$	0	$\bar{\Lambda}^0$	2.5×10^{-10}	$\mathrm{p} + \pi^-$
Sigma†	Σ^+	1189	$\frac{1}{2}$	+1	$\bar{\Sigma}^-$	0.8×10^{-10}	$\mathrm{n} + \pi^+$
	Σ^0	1193	$\frac{1}{2}$	0	$\bar{\Sigma}^0$	10^{-20}	$\Lambda^0 + \gamma$
	Σ^-	1197	$\frac{1}{2}$	−1	$\bar{\Sigma}^+$	1.7×10^{-10}	$\mathrm{n} + \pi^-$
Xi	Ξ^0	1315	$\frac{1}{2}$	0	$\bar{\Xi}^0$	3.0×10^{-10}	$\Lambda^0 + \pi^0$
	Ξ^-	1321	$\frac{1}{2}$	−1	$\bar{\Xi}^+$	1.7×10^{-10}	$\Lambda^0 + \pi^-$
Omega	Ω^-	1672	$\frac{3}{2}$	−1	Ω^+	1.3×10^{-10}	$\Xi^0 + \pi^-$
Mesons							
Pion	π^+	139.6	0	+1	π^-	2.6×10^{-8}	$\mu^+ + \nu_\mu$
	π^0	135	0	0	π^0	0.8×10^{-16}	$\gamma + \gamma$
	π^-	139.6	0	−1	π^+	2.6×10^{-8}	$\mu^- + \bar{\nu}_\mu$
Kaon‡	K^+	493.7	0	+1	K^-	1.24×10^{-8}	$\pi^+ + \pi^0$
	K^0	497.7	0	0	$\bar{\mathrm{K}}^0$	0.88×10^{-10}	$\pi^+ + \pi^-$
						and	
						5.2×10^{-8}	$\pi^+ + \mathrm{e}^- + \bar{\nu}_e$
Eta	η^0	549	0	0		2×10^{-19}	$\gamma + \gamma$

* Other decay modes also occur for most particles.

† The Σ^0 is included here for completeness even though it does decay via the strong interaction.

‡ The K^0 has two distinct lifetimes, sometimes referred to as $\mathrm{K}^0_{\text{short}}$ and $\mathrm{K}^0_{\text{long}}$. All other particles have a unique lifetime.

as a muon and not as a mu-meson. As far as we know, leptons are point particles that have no structure and can be considered to be truly elementary in the sense that they are not composed of other particles.

There are six leptons. They are the electron and the electron neutrino, the muon and the muon neutrino, and the tau and the tau neutrino. (Each of the leptons has an antiparticle.) The masses of the electron, the muon, and the tau are quite different. The mass of the electron is 0.511 MeV/c^2, the mass of the muon is 106 MeV/c^2, and the mass of the tau is 1784 MeV/c^2. The standard model predicts that neutrinos, like photons, do not have mass. However, there is now strong evidence that their mass, though very small, is greater than zero. During the late 1990s, experiments using a detector in Japan called the Super-Kamiokande (Super-K) found that neutrinos emitted from the Sun arrived on Earth in much smaller numbers than the numbers that are predicted from the fusion processes in the Sun. This result can be explained if the mass of the neutrino is not zero.* In addition, a neutrino mass as small as a few eV/c^2 would have great cosmological significance. The answer to the question of whether the universe will continue to expand indefinitely or will reach a maximum size and begin to contract depends on the total mass in the universe. Thus, the answer could depend on whether the mass of the neutrino is actually zero or is merely small, because the cosmic density of each species of neutrino is ~100 per cm^3.

* The connection between the shortfall of solar-neutrino detections and the mass of the neutrino is elucidated in "On Morphing Neutrinos and Why They Must Have Mass" by Eugene Hecht, *The Physics Teacher* 41 (2003): 164–168.

(a)

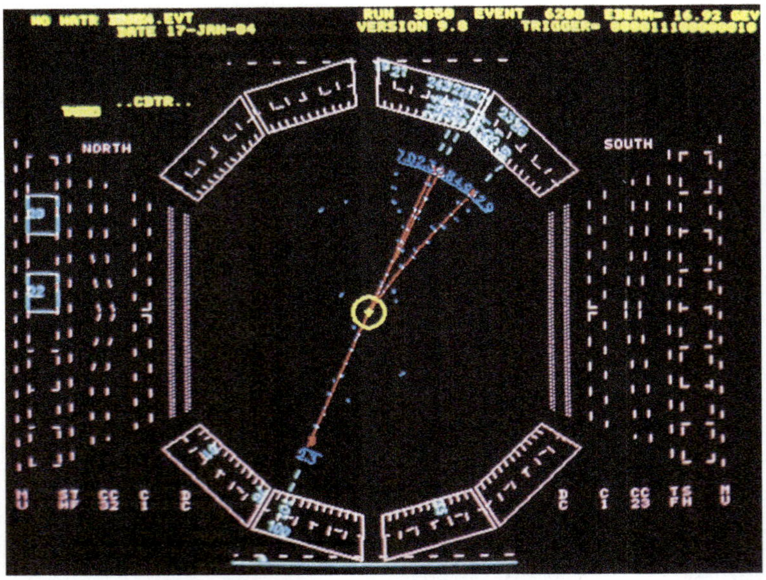

(b)

(a) A computer display of the production and decay of a τ_1 and τ_2 pair. An electron and a positron annihilate at the center marked by the yellow cross, producing a τ^+ and τ^- pair, which travel in opposite directions, but quickly decay while still inside the beam pipe (yellow circle). The τ^+ decays into two invisible neutrinos and a μ^+, which travels toward the bottom left. Its track in the drift chamber is calculated by a computer and indicated in red. It penetrates the lead–argon counters outlined in purple and is detected at the blue dot near the bottom blue line that marks the end of a muon detector. The τ^- decays into three charged pions (red tracks moving upward) plus invisible neutrinos. (b) The Mark I detector, built by a team from the Stanford Linear Accelerator Center (SLAC) and the Lawrence Berkeley Laboratory, became famous for many discoveries, including the J/ψ meson and the τ lepton. Tracks of particles are recorded by wire spark chambers wrapped in concentric cylinders around the beam pipe extending out to the ring where physicist Carl Friedberg has his right foot. Beyond this are two rings of protruding tubes, housing photomultipliers that view various scintillation counters. The rectangular magnets at the left guide the counterrotating beams that collide in the center of the detector. *((a) Science Photo Library/Photo Researchers. (b) © Lawrence Berkeley Laboratory/ Science Photo Library/Photo Researchers.)*

The observation of electron neutrinos from the supernova 1987A puts an upper limit on the masses of the neutrinos. Because the velocity of a particle that has mass depends on its energy, the arrival time of a burst of neutrinos that have mass from a supernova would be spread out in time. The fact that the electron neutrinos from the 1987 supernova all arrived at Earth within 13 s of one another results in an upper limit of about 16 eV/c^2 for their mass. Note that an upper limit does not imply that the mass is not zero. Measurements of the relative number of muon neutrinos and electron neutrinos entering the huge, underground Super-K detector suggest that at least one type of neutrino can oscillate between types (for example, between a mu neutrino and a tau neutrino). Further measurements of antineutrinos from nuclear reactors strongly show that all three types of neutrinos oscillate between types and thus have mass. Measurements made in Japan, using the *Kam*ioka Liquid Scintillator *A*nti-*N*eutrino *D*etector (KamLAND), show that oscillations from one species of neutrino to another species of neutrino can be observed over path lengths as short as 180 km (Figure 41-1).

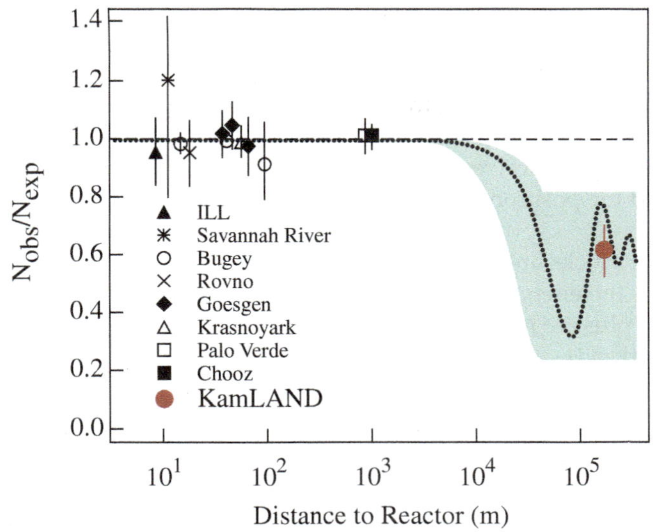

FIGURE 41-1 First evidence for antineutrino disappearance. The ratio of the number of antineutrinos observed N_{obs} to the number that one would expect to observe N_{exp} (assuming no neutrino oscillations) is plotted versus distance to the nearest antineutrino sources. The KamLAND site is 180 km from nearby antineutrino sources (nuclear reactors), while the other eight detector sites are less than 1.0 km from nearby nuclear reactors. For those eight sites, $N_{obs}/N_{exp} = 1.0$, which is what is expected assuming no neutrino oscillations. However, the KamLAND detector found $N_{obs}/N_{exp} = 0.6$. This result is strong evidence that while neutrinos do not oscillate in significant numbers while traveling over path lengths of less than 1.0 km, they do oscillate in significant numbers while traveling over path lengths only a few orders of magnitude longer than 1 km. *(© Lawrence Berkeley Laboratory/ Science Photo Library/Photo Researchers.)*

41-2 SPIN AND ANTIPARTICLES

One important characteristic of a particle is its intrinsic spin angular momentum. We have already discussed the fact that the electron has a quantum number m_s that corresponds to the z component of its intrinsic spin characterized by the quantum number $s = \frac{1}{2}$. Protons, neutrons, neutrinos, and the various other particles that also have an intrinsic spin characterized by the quantum number $s = \frac{1}{2}$ are called **spin-$\frac{1}{2}$ particles.** Particles that have spin $\frac{1}{2}$ (or $\frac{3}{2}, \frac{5}{2}$, etc.) are called fermions and obey the exclusion principle. Particles such as pions and other mesons have zero spin or integral spin ($s = 0, 1, 2$, etc.). Those particles are called bosons and do not obey the exclusion principle. That is, any number of those particles can be in the same quantum state.

Spin-$\frac{1}{2}$ particles are described by the Dirac equation, which is an extension of the Schrödinger equation that includes special relativity. One feature of Paul Dirac's theory, proposed in 1927, is the prediction of the existence of antiparticles. In special relativity, the energy of a particle is related to the mass and the momentum of the particle by $E = \pm\sqrt{p^2c^2 + m^2c^4}$ (Equation 39-27). We usually choose the positive solution and dismiss the negative-energy solution with a physical argument. However, the Dirac equation requires the existence of wave functions that correspond to the negative-energy states. Dirac got around this difficulty by postulating that all the negative-energy states were

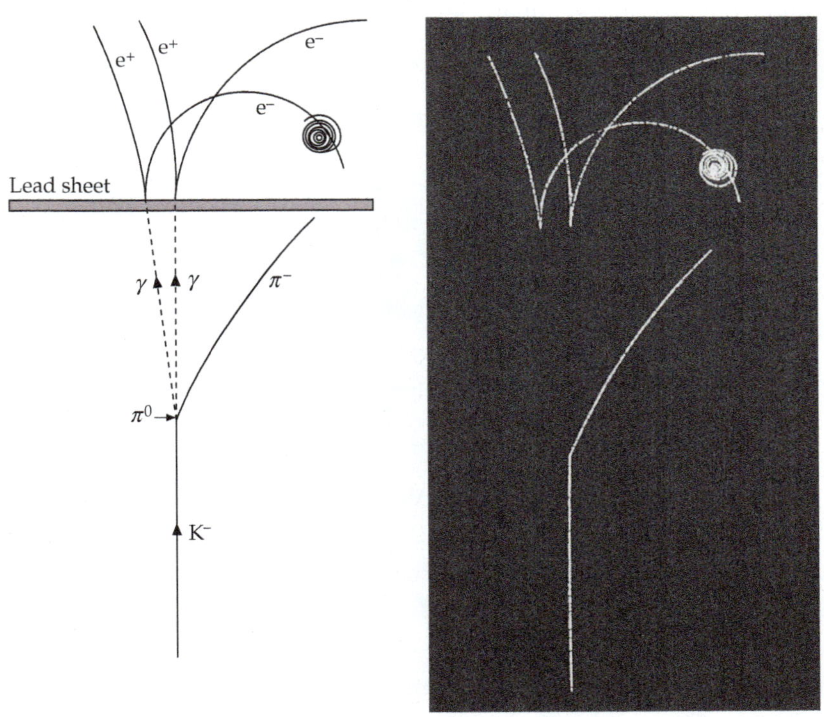

A negative kaon (K^-) enters a bubble chamber from the bottom and decays into a π^-, which moves off to the right, and a π^0, which immediately decays into two photons whose paths are indicated by the dashed lines in the drawing. Each photon interacts in the lead sheet, producing an electron–positron pair. The spiral at the right is another electron that has been knocked out of an atom in the chamber. (Other extraneous tracks have been removed from the photograph.) *(Figure 4 from "First Results from KamLAND: Evidence for Reactor Antineutrino Disappearance" by the KamLAND Collaboration,* Physical Review Letters, *Vol. 90, No. 2, December 17, 2003. Copyright © 2003 The American Physical Society. Reprinted with permission.)*

filled and would therefore not be observable. Only holes in the "infinite sea" of negative-energy states would be observed. For example, a hole in the negative sea of electron energy states would appear as a particle identical to the electron except having positive charge. When such a particle came in the vicinity of an electron the two particles would annihilate, releasing two photons having a minimum total energy of $2m_ec^2$, where m_e is the mass of the electron. This interpretation received little attention until a particle with just those properties, called the positron, was discovered in 1932 by Carl Anderson.

Antiparticles are never created alone but always in particle–antiparticle pairs. In the creation of an electron–positron pair by a photon, the energy of the photon must be at least as great as the rest energy of the electron plus the rest energy of the positron, which is $2m_ec^2 \approx 1.02$ MeV. Although the positron is stable, it has only a short-term existence in our universe because of the large supply of electrons in matter. The fate of a positron is annihilation according to the reaction

$$e^+ + e^- \rightarrow \gamma + \gamma \qquad 41\text{-}1$$

The probability of this reaction is large only if the positron and electron are moving slowly relative to one another. In the center-of-mass reference frame, the momentum of the two particles prior to annihilation is zero, so two photons moving in opposite directions are needed to conserve linear momentum.

The fact that we call electrons *particles* and positrons *antiparticles* does not imply that positrons are less fundamental than electrons. It merely reflects the nature of our universe. If our matter were made up of negative protons and positive electrons, then positive protons and negative electrons would suffer quick annihilation and would be called antiparticles.

An aerial view of the European Laboratory for Particle Physics (CERN) just outside of Geneva, Switzerland. The large circle shows the Large Electron–Positron collider (LEP) tunnel, which is 27 km in circumference. The irregular dashed line is the border between France and Switzerland. *(Richard Ehrlich.)*

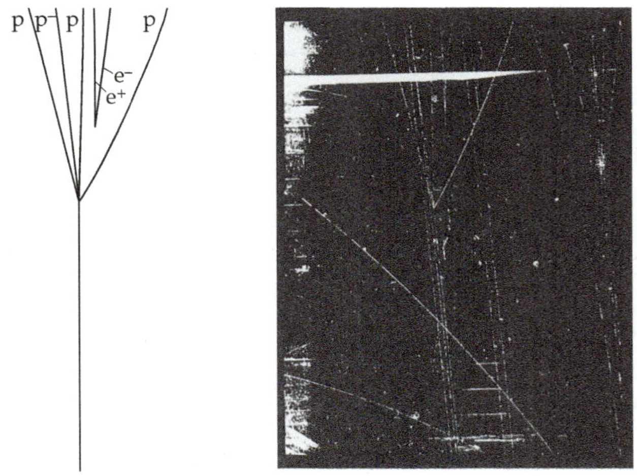

FIGURE 41-2 Bubble-chamber tracks that show the creation of a proton–antiproton pair in the collision of an incident 25-GeV proton with a stationary proton in liquid hydrogen. *(CERN.)*

The antiproton (p^-) was discovered in 1955 by Emilio Segrè and Owen Chamberlain using a beam of protons in the Bevatron at Berkeley to produce the reaction*

$$p^+ + p^+ \rightarrow p^+ + p^+ + p^+ + p^- \qquad 41\text{-}2$$

The creation of a proton–antiproton pair (Figure 41-2) requires kinetic energy of at least $2m_pc^2 = 1877$ MeV $= 1.877$ GeV in the zero-momentum reference frame in which the two protons approach each other with equal and opposite momenta. In the laboratory frame in which one of the protons is initially at rest, the kinetic energy of the incoming proton must be at least $6m_pc^2 = 5.63$ GeV (see Problem 38 of Chapter 39). This energy was not available in laboratories before the development of high-energy accelerators in the 1950s. Antiprotons annihilate with protons to produce two gamma rays in a reaction similar to the reaction in Equation 41-1.

The tunnel of the proton–antiproton collider at CERN. The same bending magnets and focusing magnets can be used for protons or antiprotons moving in opposite directions. The rectangular box in the foreground is a focusing magnet, and the next four boxes are the bending magnets. *(CERN.)*

* The antiproton is sometimes denoted by $\bar{p}$ rather than p^-. For neutral particles, such as the neutron, the bar must be used to denote the antiparticle. Thus, the antineutron is denoted by $\bar{n}$. The electron and proton are often denoted by e and p without the minus sign or plus sign superscripts.

Example 41-1 Proton–Antiproton Annihilation *Context-Rich*

You have been reading about nuclear physics and particle interactions. In particular, you have been looking at the reaction $p^+ + p^- \longrightarrow \gamma + \gamma$ (proton–antiproton annihilation). You wonder if the photons produced are visible to the human eye if the two protons are initially at rest. Are the photons visible to the human eye?

PICTURE If the photons are visible, they should have wavelengths in the visible range (400 nm to 800 nm.) Because the proton and the antiproton are at rest, conservation of momentum requires that the two photons created during their annihilation have equal and opposite momenta and therefore equal energies, frequencies, and wavelengths. Conservation of energy implies that the photons have a combined energy equal to the rest energy of the proton plus the rest energy of the antiproton (approximately 938 MeV each).

SOLVE

1. Set the total energy of the two photons, $2E_\gamma$, equal to the rest energy of the proton plus antiproton and solve for E_γ:

$$2E_\gamma = 2m_p c^2$$

so

$$E_\gamma = m_p c^2 = 938 \text{ MeV}$$

2. Set the energy of the photon equal to $hf = hc/\lambda$ and solve for the wavelength λ:

$$E_\gamma = hf = \frac{hc}{\lambda}$$

$$\lambda = \frac{hc}{E_\gamma} = \frac{1240 \text{ eV}\cdot\text{nm}}{938 \text{ MeV}}$$

$$= 1.32 \times 10^{-6} \text{ nm} = 1.32 \text{ fm}$$

3. Compare this wavelength with the wavelengths of visible light:

The photons are *not* in the visible spectrum.

CHECK In Chapter 36, we found that the energies of photons in the visible spectrum are equal to only a few electron volts. It is not a surprise to find that photons that have energies on the order of 10^9 electron volts are not in the visible spectrum.

TAKING IT FURTHER The wavelength of the photons produced by proton–antiproton annihilation is more than eight orders of magnitude less than 400 nm—the shortest wavelength in the visible spectrum.

41-3 THE CONSERVATION LAWS

One adage is "anything that can happen does." If a conceivable decay or reaction does not occur, there must be a reason. The reason is usually expressed in terms of a conservation law. The conservation of energy rules out the decay of any particle for which the total mass of the decay products would be greater than the initial mass of the particle before decay. The conservation of linear momentum requires that when an electron and a positron at rest annihilate, two photons must be emitted. Angular momentum must also be conserved during a reaction or a decay. A fourth conservation law that restricts the possible particle decays and reactions is the conservation of electric charge. The net electric charge before a decay or a reaction must equal the net charge after the decay or the reaction.

There are two additional conservation laws that are important in the reactions and the decays of elementary particles: the conservation of baryon number and the conservation of lepton number. Consider the proposed decay

$$p \rightarrow \pi^0 + e^+$$

where π is the symbol for the pion (pi-meson). This decay would conserve charge, energy, angular momentum, and linear momentum, but it does not occur. It does not conserve either lepton number or baryon number. (The proton p is a baryon, the positron e^+ is a lepton, and the π^0 is a meson.) The conservation of lepton number and baryon number implies that whenever a lepton or a baryon is created, an antiparticle of the same type is also created. We assign the **lepton number** $L = +1$ to all leptons, $L = -1$ to all antileptons, and $L = 0$ to all other particles. Similarly, the **baryon number** $B = +1$ is assigned to all baryons, $B = -1$ to all antibaryons, and $B = 0$ to all other particles. The sum of the baryon numbers and the sum of the lepton numbers cannot change during a reaction or a decay. The conservation of baryon number along with the conservation of energy implies that the least massive baryon, the proton, must be stable.

CONCEPT CHECK 41-1

Why is it that the conservation of baryon number along with the conservation of energy implies that the proton, which is the least massive baryon, must be stable?

The conservation of lepton number implies that the neutrino emitted during the β decay of the free neutron is an antineutrino:

$$n \rightarrow p^+ + e^- + \bar{\nu}_e \qquad 41\text{-}3$$

The fact that neutrinos and antineutrinos are different is illustrated by an experiment in which ^{37}Cl is bombarded with an intense antineutrino beam from the decay of reactor neutrons (reactor neutrons are fission products that are produced in nuclear reactors). If neutrinos and antineutrinos were the same, we would expect the following reaction:

$$^{37}_{17}Cl + \bar{\nu}_e \rightarrow {}^{37}_{18}Ar + e^- \qquad 41\text{-}4$$

This reaction is not observed. However, if protons are bombarded with antineutrinos, the reaction

$$p + \bar{\nu}_e \rightarrow n + e^+ \qquad 41\text{-}5$$

is observed. Note that the sum of the lepton numbers is -1 on the left side of the reaction equation in Equation 41-4 and is $+1$ on the right side of the reaction equation. But the sum of the lepton numbers is -1 on both sides of the reaction equation in Equation 41-5.

Not only are neutrinos and antineutrinos distinct particles, but the neutrinos associated with electrons are distinct from the neutrinos associated with muons. Electron-like leptons (e and ν_e), muon-like leptons (μ and ν_μ), and tau-like leptons (τ and ν_τ) are each separately conserved, so we assign separate lepton numbers L_e, L_μ, and L_τ to the particles. The leptons and their lepton numbers are listed in Table 41-2.

Table 41-2 Lepton Numbers

	L_e	L_μ	L_τ
e^-	+1	0	0
ν_e	+1	0	0
e^+	−1	0	0
$\bar{\nu}_e$	−1	0	0
μ^-	0	+1	0
ν_μ	0	+1	0
μ^+	0	−1	0
$\bar{\nu}_\mu$	0	−1	0
τ^-	0	0	+1
ν_τ	0	0	+1
τ^+	0	0	−1
$\bar{\nu}_\tau$	0	0	−1

Example 41-2 What Laws Are Being Violated? *Conceptual*

What conservation laws (if any) are violated by the following proposed decays: (*a*) $n \rightarrow p + \pi^-$, (*b*) $\Lambda^0 \rightarrow p^- + \pi^+$, and (*c*) $\mu^- \rightarrow e^- + \gamma$? ($\Lambda^0$ is the symbol for the lambda-zero particle.)

PICTURE All reactions must separately conserve energy, electric charge, baryon number, electron lepton number, muon lepton number, and tau lepton number.

SOLVE

(*a*) There are no leptons in this decay, so there is no problem with the conservation of lepton number. The net charge is zero before the decay and after the decay, so charge is conserved. Also, the baryon number is +1 both before and after the decay. However, rest energy of the proton (938.3 MeV) plus the rest energy of the pion (139.6 MeV) is greater than the rest energy of the neutron (939.6 MeV). In the rest frame of the neutron, the energy prior to the reaction (the rest energy of the neutron) is less than the total rest energy following the reaction.

This decay does not conserve energy.

(*b*) Again, there are no leptons involved, and the net charge is zero before the decay and after the decay. Also, the rest energy of the lambda-zero (1116 MeV) is greater than the rest energy of the antiproton (938.3 MeV) plus the rest energy of the pion (139.6 MeV), so in the rest frame of the lambda-zero the energy prior to the reaction (the rest energy of the lambda-zero) is greater than the total rest energy following the reaction. Energy could be conserved, with the loss in rest energy equal to the gain in kinetic energy of the decay products. There are no leptons in the reaction, so all three lepton numbers are conserved. The baryon number is +1 for the lambda particle and −1 for the antiproton and zero for the pi-meson.

This decay does not conserve baryon number.

(*c*) The μ^- has a muon lepton number (L_μ) equal to +1 and an electron lepton number (L_e) equal to 0, the e^- has $L_\mu = 0$ and $L_e = +1$, and the γ has $L_\mu = L_e = 0$.

This reaction does not conserve either muon lepton number or electron lepton number.

TAKING IT FURTHER The muon does decay by $\mu^- \rightarrow e^- + \bar{\nu}_e + \nu_\mu$, which does conserve both muon lepton numbers and electron lepton numbers.

There are some conservation laws that are not universal but apply only to certain kinds of interactions. In particular, there are quantities that are conserved during decays and reactions that occur by the strong interaction but not during decays or reactions that occur by the weak interaction. One of these quantities that is particularly important is **strangeness,** introduced by M. Gell-Mann and K. Nishijima in 1952 to explain the strange behavior of some of the heavy baryons and mesons. Consider the reaction

$$p + \pi^- \rightarrow \Lambda^0 + K^0 \tag{41-6}$$

where K is the symbol for the kaon (K-meson). The proton and the pion interact by the strong interaction. Both the Λ^0 and K^0 decay into hadrons

$$\Lambda^0 \rightarrow p + \pi^- \tag{41-7}$$

and

$$K^0 \rightarrow \pi^+ + \pi^- \tag{41-8}$$

However, the decay times for both the Λ^0 and K^0 are of the order of 10^{-10} s, which is characteristic of the weak interaction, rather than 10^{-23} s, which would be expected for the strong interaction. Other particles showing similar behavior were called **strange particles.** These particles are always produced in pairs, even when

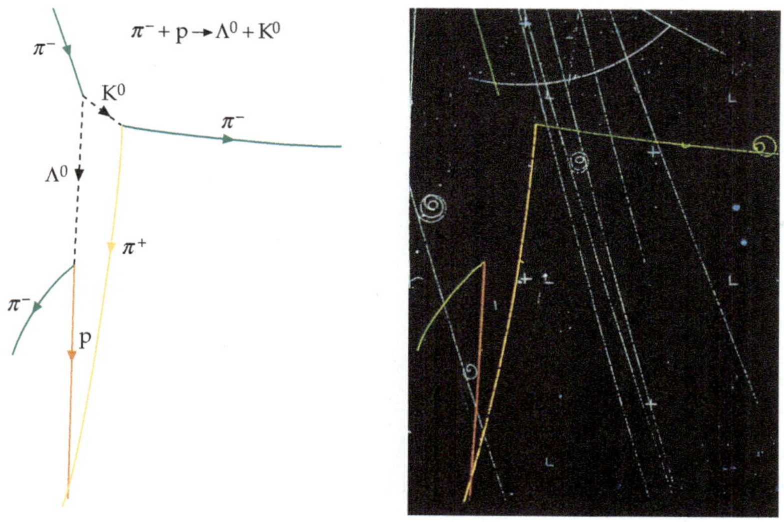

An early photograph of bubble-chamber tracks at the Lawrence Berkeley Laboratory, showing the production and the decay of two particles that have nonzero strangeness, the K^0 and the Λ^0. These neutral particles are identified by the tracks of their electrically charged decay particles. The lambda particle was named because of the similarity of the tracks of its decay particles to the uppercase Greek letter lambda (Λ). (The blue tracks are particles not involved in the reaction of Equation 41-6.) *(© Lawrence Berkeley Laboratory/Science Photo Library/Photo Researchers.)*

all other conservation laws are met. This behavior is described by assigning a new property called strangeness to the particles. During reactions and decays that occur by the strong interaction, strangeness is conserved. During reactions and decays that occur by the weak interaction, the strangeness can only change by ± 1. The strangeness of the ordinary hadrons—the nucleons and pions—was arbitrarily taken to be zero. The strangeness of the K^0 was arbitrarily chosen to be $+1$. The strangeness of the Λ^0 particle must then be -1 so that strangeness is conserved during the reaction described by Equation 41-6. The strangeness of other particles could then be assigned by looking at their various reactions and decays. During reactions and decays that occur by the weak interaction, the strangeness can change by ± 1.

Figure 41-3 shows the masses of the baryons and the mesons that are stable against decay by the strong interaction versus strangeness. We can see from this figure that the particles cluster in multiplets of one, two, or three particles of approximately equal mass, and that the strangeness of a multiplet of particles is related to the *center of charge* of the multiplet.

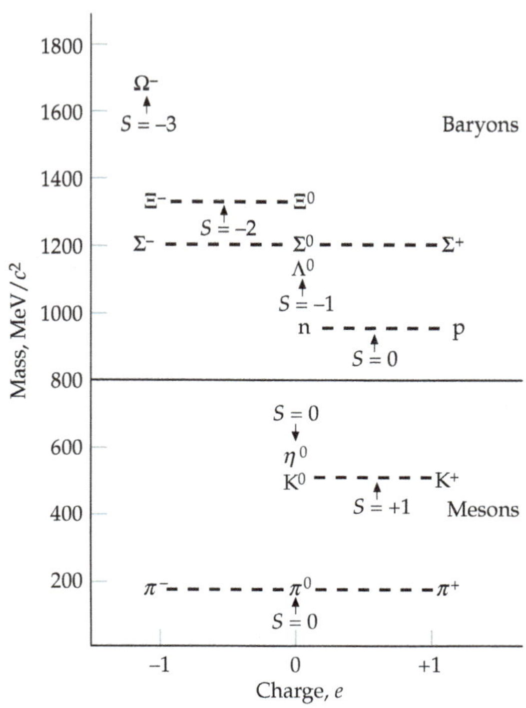

FIGURE 41-3 The strangeness of hadrons shown on a plot of mass versus charge. The strangeness of a baryon–charge multiplet is related to the number of places on the plot that the center of charge of the multiplet is displaced from that of the nucleon doublet. For each "displacement" of *e*, the strangeness changes by ± 1. For mesons, the strangeness is related to the number of places the center of charge is displaced from that of the pion triplet. Because of the unfortunate original assignment of $+1$ for the strangeness of K-mesons (kaons), all of the baryons that are stable against decay by the strong interaction have negative or zero strangeness.

Example 41-3 Strong Interaction, Weak Interaction, or No Interaction — *Conceptual*

State whether the following decays can occur by the strong interaction, by the weak interaction, or not at all: (*a*) $\Sigma^+ \rightarrow p + \pi^0$, (*b*) $\Sigma^0 \rightarrow \Lambda^0 + \gamma$, and (*c*) $\Xi^0 \rightarrow n + \pi^0$, where Σ, Λ and Ξ are the symbols for the sigma, lambda and xi particles respectively.

PICTURE We first note that the mass of each decaying particle is greater than the mass of the decay products, so none of the three reactions are in conflict with the principle of conservation of energy. In addition, no leptons are involved in any of the three decays, and charge and baryon number are both conserved during all three reactions. The decay will occur by the strong interaction if strangeness is conserved (if $\Delta S = 0$). If $\Delta S = \pm 1$, the decay will occur via the weak interaction. If $|\Delta S| > 1$, the decay will not occur.

SOLVE

(*a*) From Figure 41-3, we can see that the strangeness of the Σ^+ is -1, whereas the strangeness of both the proton and the pion is zero.

This decay is possible by the weak interaction but not by the strong interaction. It is, in fact, one of the decay modes of the Σ^+ particle which has a lifetime of the order of 10^{-10} s.

(*b*) The strangeness of both the Σ^0 and Λ^0 is -1, whereas the strangeness of the photon is zero.

This decay can proceed by the strong interaction. It is, in fact, the dominant mode of decay of the Σ^0 particle which has a lifetime of approximately 10^{-20} s.

(*c*) The strangeness of the Ξ^0 is -2, whereas the strangeness of both the neutron and the pion is zero.

Because strangeness cannot change by 2 during a decay or a reaction, this decay does not occur.

41-4 QUARKS

Leptons appear to be truly elementary particles in that they do not break down into smaller entities and they seem to have no measurable size or structure. Hadrons, on the other hand, are complex particles that have size and structure, and they decay into other hadrons. Furthermore, at the present time, there are only six known leptons, whereas there are many more hadrons. Except for the Σ^0 particle, Table 41-1 includes only hadrons that are stable against decay by the strong interaction. Hundreds of other hadrons have been discovered; their properties, such as charge, spin, mass, strangeness, and decay schemes, have been measured.

The most important advance in our understanding of elementary particles was the quark model proposed by M. Gell-Mann and G. Zweig in 1963 in which all hadrons consist of combinations of two or three truly elementary particles called **quarks.*** In the original model, quarks came in three types, called **flavors,** labeled *u*, *d*, and *s* (for *up*, *down*, and *strange*). An unusual property of quarks is that they carry fractional electron charges. The charge of the *u* quark is $+\frac{2}{3}e$ and the charge of the *d* and *s* quarks is $-\frac{1}{3}e$. Each quark has spin $\frac{1}{2}$ and a baryon number of $\frac{1}{3}$. The strangeness of the *u* and *d* quark is 0, and the strangeness of the *s* quark is -1. Each quark has an antiquark that has the opposite electric charge, baryon number, and strangeness. Baryons consist of three quarks (or three antiquarks for baryons that are antiparticles), whereas mesons consist of a quark and an antiquark, giving mesons baryon numbers $B = 0$, as required. The proton consists of the combination *uud* and the neutron consists of the combination *udd*. Baryons that have a strangeness $S = -1$ have one *s* quark. All the particles listed in Table 41-1 can be

* The name *quark* was chosen by M. Gell-Mann from a quotation from *Finnegan's Wake* by James Joyce.

constructed from these three quarks and three antiquarks.[†] The great strength of the quark model is that all the allowed combinations of three quarks or quark–antiquark pairs result in known hadrons. Strong evidence for the existence of quarks inside a nucleon is provided by high-energy scattering experiments called *deep inelastic scattering*. During these experiments, a nucleon is bombarded with electrons, muons, or neutrinos of energies from 15 GeV to 200 GeV. Analyses of particles scattered at large angles indicate that inside the nucleon are three spin-$\frac{1}{2}$ particles of sizes much smaller than that of the nucleon. These experiments are analogous to Rutherford's scattering of α particles by atoms in which the presence of a tiny nucleus in the atom was inferred from the large-angle scattering of the α particles.

In 1967, a fourth quark was proposed to explain some discrepancies between experimental determinations of certain decay rates and calculations based on the quark model. The fourth quark is labeled c for a new property called **charm.** Like strangeness, charm is conserved during strong interactions but changes by ± 1 in weak interactions. In 1975, a new heavy meson called the ψ meson was discovered that has the properties expected of a $c\bar{c}$ combination. Since then, other mesons that have combinations such as $c\bar{d}$ and $\bar{c}d$, as well as baryons having the charmed quark, have been discovered. Two more quarks labeled t and b (for *t*op and *b*ottom) were proposed in the 1970s. In 1977, a massive new meson called the ϒ (upsilon) meson or **bottomonium,** which is considered to have the quark combination $b\bar{b}$, was discovered. The top quark was observed in 1995. The properties of the six quarks are listed in Table 41-3.

The six quarks and six leptons (and their antiparticles) are thought to be the fundamental elementary particles of which all matter is composed. Table 41-4 lists the masses of the fundamental particles. In this table, the masses given for neutrinos are upper limits. The masses given for quarks are educated guesses. There is strong experimental evidence for the existence of each of these particles.

Table 41-3 Properties of Quarks and Antiquarks

Flavor	Spin	Charge	Baryon Number	Strangeness	Charm	Topness	Bottomness
Quarks							
u (up)	$\frac{1}{2}\hbar$	$+\frac{2}{3}e$	$+\frac{1}{3}$	0	0	0	0
d (down)	$\frac{1}{2}\hbar$	$-\frac{1}{3}e$	$+\frac{1}{3}$	0	0	0	0
s (strange)	$\frac{1}{2}\hbar$	$-\frac{1}{3}e$	$+\frac{1}{3}$	−1	0	0	0
c (charmed)	$\frac{1}{2}\hbar$	$+\frac{2}{3}e$	$+\frac{1}{3}$	0	+1	0	0
t (top)	$\frac{1}{2}\hbar$	$+\frac{2}{3}e$	$+\frac{1}{3}$	0	0	+1	0
b (bottom)	$\frac{1}{2}\hbar$	$-\frac{1}{3}e$	$+\frac{1}{3}$	0	0	0	+1
Antiquarks							
$\bar{u}$	$\frac{1}{2}\hbar$	$-\frac{2}{3}e$	$-\frac{1}{3}$	0	0	0	0
$\bar{d}$	$\frac{1}{2}\hbar$	$+\frac{1}{3}e$	$-\frac{1}{3}$	0	0	0	0
$\bar{s}$	$\frac{1}{2}\hbar$	$+\frac{1}{3}e$	$-\frac{1}{3}$	+1	0	0	0
$\bar{c}$	$\frac{1}{2}\hbar$	$-\frac{2}{3}e$	$-\frac{1}{3}$	0	−1	0	0
$\bar{t}$	$\frac{1}{2}\hbar$	$-\frac{2}{3}e$	$-\frac{1}{3}$	0	0	−1	0
$\bar{b}$	$\frac{1}{2}\hbar$	$+\frac{1}{3}e$	$-\frac{1}{3}$	0	0	0	−1

[†] The correct quark combinations of hadrons are not always obvious, because of the symmetry requirements on the total wave function. For example, the π^0 meson is represented by a linear combination of $u\bar{u}$ and $d\bar{d}$.

Table 41-4 Masses of Fundamental Particles

Particle	Mass
Quarks	
u (up)	336 MeV/c^2
d (down)	338 MeV/c^2
s (strange)	540 MeV/c^2
c (charmed)	1500 MeV/c^2
t (top)	174 000 MeV/c^2
b (bottom)	4500 MeV/c^2
Leptons	
e^- (electron)	0.511 MeV/c^2
ν_e (electron neutrino)	$<$ 2.2 eV/c^2
μ^- (muon)	105.659 MeV/c^2
ν_μ (muon neutrino)	$<$ 0.17 MeV/c^2
τ^- (tau)	1784 MeV/c^2
ν_τ (tau neutrino)	$<$ 28 MeV/c^2

Example 41-4 Given the Constituent Quark Species, Identify the Particle *Conceptual*

What are the properties of the particles made up of the following quarks: (*a*) $u\bar{d}$, (*b*) $\bar{u}d$, (*c*) *dds*, and (*d*) *uss*?

PICTURE Baryons are made up of three quarks, whereas mesons consist of a quark and an antiquark. We add the electric charges of the quarks to find the total charge of the hadron. We also find the strangeness of the hadron by adding the strangeness of the quarks.

SOLVE

(*a*) Because $u\bar{d}$ is a quark-antiquark combination, it has baryon number 0 and is therefore a meson. There is no strange quark here (that is, $S = 0$), so the strangeness of the meson is zero. The charge of the up quark is $+\frac{2}{3}e$ and the charge of the antidown quark is $+\frac{1}{3}e$, so the charge of the meson is $+1e$.

The quark combination $u\bar{d}$ is the π^+ meson.

(*b*) The particle $\bar{u}d$ is also a meson that has zero strangeness. Its electric charge is $-\frac{2}{3}e + \left(-\frac{1}{3}e\right) = -1e$.

The quark combination $\bar{u}d$ is the π^- meson.

(*c*) The particle *dds* is a baryon that has strangeness -1 because it has one strange quark. Its electric charge is $-\frac{1}{3}e - \frac{1}{3}e - \frac{1}{3}e = -1e$.

The quark combination *dds* is the Σ^- particle.

(*d*) The particle *uss* is a baryon that has strangeness -2. Its electric charge is $+\frac{2}{3}e - \frac{1}{3}e - \frac{1}{3}e = 0$.

The quark combination *uss* is the Ξ^0 particle.

QUARK CONFINEMENT

Despite considerable experimental effort, no isolated quark has ever been observed. It is now believed that it is impossible to obtain an isolated quark. Although the force between quarks is not known, it is believed that the potential energy of two quarks increases with increasing separation distance so that an infinite amount of energy would be needed to separate the quarks completely.

This would be true, for example, if the force of attraction between two quarks remains constant or increases with separation distance, rather than decreasing with increasing separation distance as is the case for other fundamental forces, such as the electric force between two charges, the gravitational force between two masses, and the strong nuclear force between two hadrons.

When a large amount of energy is added to a quark system, such as a nucleon, a quark–antiquark pair is created and the original quarks remain confined within the original system. Because quarks cannot be isolated, but are always bound together to form a baryon or a meson, the mass of a quark cannot be accurately known, which is why the masses listed in Table 41-4 are merely educated guesses.

41-5 FIELD PARTICLES

In addition to the six fundamental leptons and six fundamental quarks, there are other particles, called *field particles,* or *field quanta,* that are associated with the forces exerted by one elementary particle on another. In **quantum electrodynamics,** the electromagnetic field of a single charged particle is described by **virtual photons** that are continuously being emitted and reabsorbed by the particle. If we put energy into the system by accelerating the charge, some of these virtual photons are shaken off and become real, observable photons. The photon is said to mediate the electromagnetic interaction. Each of the four basic interactions can be described via mediating field particles.

The field quantum associated with the gravitational interaction, called the **graviton,** has not yet been observed. The gravitational *charge* analogous to electric charge is mass.

The weak interaction is thought to be mediated by three field quanta called **vector bosons:** W^+, W^-, and Z^0. These particles were predicted by Sheldon Glashow, Abdus Salam, and Steven Weinberg in a theory called the *electroweak theory,* which we discuss in the next section. The W and Z particles were first observed in 1983 by a group of over a hundred scientists led by Carlo Rubbia using the high-energy accelerator at CERN in Geneva, Switzerland. The masses of the $W^\pm$ particles (about 80 GeV/c^2) and the Z particle (about 91 GeV/c^2) measured during this experiment were in excellent agreement with those predicted by the electroweak theory. (The W^- particle is the antiparticle of the W^+ particle, so they must have identical masses.)

The field quanta associated with the strong force between quarks are called **gluons.** Isolated gluons have not been observed experimentally. The *charge* responsible for the strong interactions comes in three varieties, labeled *red, green,* and *blue* (analogous with the three primary colors), and the strong charge is called the **color charge.** The field theory for strong interactions, analogous to quantum electrodynamics for electromagnetic interactions, is called **quantum chromodynamics (QCD).**

Table 41-5 lists the bosons responsible for mediating the basic interactions.

Table 41-5 Bosons That Mediate the Basic Interactions

Interaction	Boson	Spin	Mass	Electric Charge
Strong	g (gluon)	1	0	0
Weak	$W^\pm$	1	80.22 GeV/c^2	$\pm 1e$
	Z^0	1	91.19 GeV/c^2	0
Electromagnetic	γ (photon)	1	0	0
Gravitational	Graviton†	2	0	0

† Not yet observed.

41-6 THE ELECTROWEAK THEORY

In the **electroweak theory,** the electromagnetic and weak interactions are considered to be two different manifestations of a more fundamental electroweak interaction. At very high energies ($\gg 100$ GeV), the electroweak interaction would be mediated by four bosons. From symmetry considerations, these would be a triplet consisting of W^+, W^0, and W^-, all of equal mass, and a singlet B^0 of some other mass. Neither the W^0 nor the B^0 would be observed directly, but one linear combination of the W^0 and the B^0 would be the Z^0 and another would be the photon. At ordinary energies, the symmetry is broken. This leads to the separation of the electromagnetic interaction mediated by the massless photon and the weak interaction mediated by the W^+, W^-, and Z^0 particles. The fact that the photon is massless and that the W and Z particles have masses of the order of 100 GeV$/c^2$ shows that the symmetry assumed in the electroweak theory does not exist at lower energies.

The symmetry-breaking mechanism is called a **Higgs field,** which requires a new boson, the **Higgs boson,** whose rest energy is expected to be of the order of 1 TeV (1 TeV $= 10^{12}$ eV). The Higgs boson has not yet been observed. Calculations show that Higgs bosons (if they exist) should be produced in head-on collisions between protons of energies of the order of a few TeV. Such energies are not presently available, but the Large Hadron Collider, a particle accelerator under construction near Geneva, Switzerland, is scheduled to come on line late in 2007. The LHC is designed to produce head-on proton–proton collisions for which each proton has an energy of 7 TeV.

41-7 THE STANDARD MODEL

The combination of the quark model, electroweak theory, and quantum chromodynamics is called the **standard model.** In this model, the fundamental particles are the leptons and quarks, each of which comes in six flavors, as shown in Table 41-4; the force carriers are the photon, the $W^\pm$ and Z particles, and the gluons (of which there are eight types). The leptons and quarks are all spin-$\frac{1}{2}$ fermions, which obey the exclusion principle, and the force carriers are integral-spin bosons, which do not obey the exclusion principle. Every interaction in nature is due to one of the four basic interactions: strong, electromagnetic, weak, and gravitational. A particle experiences one of the basic interactions if it carries a charge associated with that interaction. Electric charge is the familiar charge that we have studied previously. Weak charge, also called flavor charge, is carried by leptons and quarks. The charge associated with the strong interaction is called color charge and is carried by quarks and gluons but not by leptons. The charge associated with the gravitational force is mass. It is important to note that the photon, which mediates the electromagnetic interaction, does not carry electric charge. Similarly, the $W^\pm$ and Z particles, which mediate the weak interaction, do not carry weak charge. However, the gluons, which mediate the strong interaction, do carry color charge. This fact is related to the confinement of quarks as discussed in Section 41-4.

All matter is made up of leptons or quarks. There are no known composite particles consisting of leptons bound together by the weak force. Leptons exist only as isolated particles. Hadrons (baryons and mesons) are composite particles consisting of quarks bound together by the color charge. A result of QCD theory is that only color-neutral combinations of quarks are allowed. Three quarks of

Table 41-6 Properties of the Basic Interactions

	Gravitational	Weak	Electromagnetic	Strong	
				Fundamental	Residual
Acts on	Mass	Flavor	Electric charge	Color charge	
Particles experiencing	All	Quarks, leptons	Electrically charged	Quarks, gluons	Hadrons
Particles mediating	Graviton	$W^{\pm}, Z$	γ	Gluons	Mesons
Strength for two quarks at 10^{-18} m†	10^{-41}	0.8	1	25	(not applicable)
Strength for two protons in nucleus†	10^{-36}	10^{-7}	1	(not applicable)	20

† Strengths are relative to electromagnetic strength.

different colors can combine to form color-neutral baryons, such as the neutron and the proton. Mesons each have a quark and an antiquark and are also color-neutral. Excited states of hadrons are considered to be different particles. For example, the Δ^+ particle is an excited state of the proton. Both are made up of the *uud* quarks, but the proton is in the ground state and has spin $\frac{1}{2}$ and a rest energy of 938 MeV, whereas the Δ^+ particle is in the first excited state and has spin $\frac{3}{2}$ and a rest energy of 1232 MeV. The two u quarks can be in the same spin state in the Δ^+ without violating the exclusion principle, because they have different color. All baryons eventually decay to the lightest baryon, the proton. That the proton does not decay is consistent with the conservation of energy and conservation of baryon number.

The strong interaction has two parts, the fundamental interaction or color interaction and what is called the *residual strong interaction*. The fundamental interaction is responsible for the force exerted by one quark on another quark and is mediated by gluons. The residual strong interaction is responsible for the force between color-neutral nucleons, such as the neutron and the proton. This force is due to the residual strong interactions between the color-charged quarks that make up the nucleons and can be viewed as being mediated by the exchange of mesons. The residual strong interaction between color-neutral nucleons can be thought of as analogous to the residual electromagnetic interaction between neutral atoms that bind them together to form molecules. Table 41-6 lists some of the properties of the basic interactions.

For each particle, there is an antiparticle. A particle and its antiparticle have identical mass and spin but opposite electric charge. For leptons, the lepton numbers L_e, L_μ, and L_τ of the antiparticles are the negatives of the corresponding lepton numbers for the particles. For example, the lepton number for the electron is $L_e = +1$, and the lepton number for the positron is $L_e = -1$. For hadrons, the baryon number, strangeness, charm, topness, and bottomness are the sums of those quantities for the quarks that make up the hadron. The number of each antiparticle is the negative of the number for the corresponding particle. For example, the lambda particle Λ^0, which is made up of the *uds* quarks, has $B = 1$ and $S = -1$, whereas its antiparticle $\overline{\Lambda}^0$, which is made up of the $\overline{u}\overline{d}\overline{s}$ quarks, has $B = -1$ and $S = +1$. A particle such as the photon γ or the Z^0 particle that has zero electric charge; $B = 0$, $L = 0$, $S = 0$; and zero charm, topness, and bottomness is its own antiparticle. Note that the K^0 meson ($d\bar{s}$) has a zero value for all of these quantities except strangeness, which is +1. Its antiparticle, the $\bar{K}^0$ meson ($\bar{d}s$), has strangeness −1, which makes it distinct from the K^0. The π^+ ($u\bar{d}$) and π^- ($\bar{u}d$) are somewhat special in that they have electric charge but zero values for L, B, and S. They are antiparticles of each other, but because there is no conservation law for mesons, it is impossible to say which is the particle and which is the antiparticle. Similarly, the W^+ and W^- are antiparticles of each other.

GRAND UNIFICATION THEORIES

With the success of the electroweak theory, attempts have been made to combine the strong, electromagnetic, and weak interactions in various **grand unification theories (GUTs).** In one of these theories, leptons and quarks are considered to be two aspects of a single class of particles. Under certain conditions, a quark could change into a lepton and vice versa, even though this would appear to violate the conservation of lepton number and baryon number. One of the exciting predictions of this theory is that the proton is not stable but merely has a very long lifetime of the order of 10^{32} y. Such a long lifetime makes proton decay difficult to observe. However, projects are ongoing in which detectors monitor very large numbers of protons in search of an event indicating the decay of a proton.

41-8 THE EVOLUTION OF THE UNIVERSE

In the presently accepted model, the universe began with a singular cataclysmic event called the **big bang** and is expanding. The first evidence that the universe is expanding was the astronomer Edwin Powell Hubble's discovery of the relation between the redshifts in the spectra of galaxies and their distances from us. This relation is illustrated in Figure 41-4 for a group of spiral galaxies used by astronomers for calibrating distances. Provided that the redshift is due to the Doppler effect, the recession velocity v of a galaxy is related to its distance r from us by Hubble's law,

$$v = Hr \tag{41-9}$$

where H is the **Hubble constant.** In principle, the value of H is easy to obtain because it relies on the direct calculation of v from redshift measurements. However, astronomical distances are very challenging to measure, and they have been determined for only a fraction of the 10^{10} or so galaxies in the observable universe.

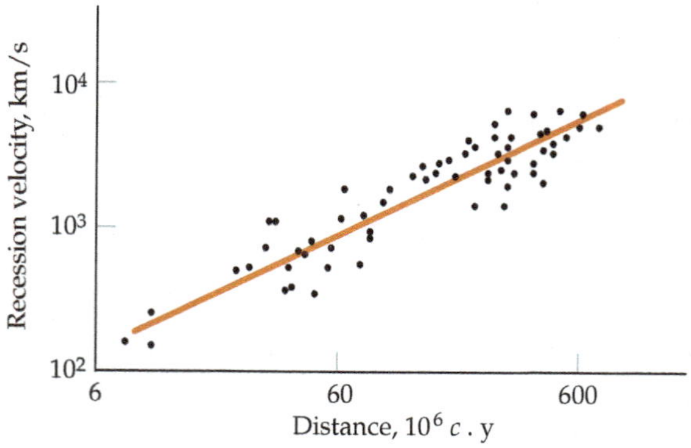

FIGURE 41-4 A plot of the recession velocities of individual galaxies versus distance.

Thus, the value of H changes as distance calibration data are refined. The currently accepted value of the Hubble constant is about

$$H = \frac{23 \text{ km/s}}{10^6 c \cdot \text{y}} \qquad 41\text{-}10$$

Hubble's law tells us that the galaxies are all rushing away from us, and those galaxies that are the farthest away are moving the fastest. However, there is no reason why our location should be special. An observer in any galaxy would make the same observations and compute the same Hubble constant. Thus, Hubble's law suggests that all of the galaxies are receding from each other at an average speed of 23 km/s per $10^6 c \cdot \text{y}$ of separation. In other words, the universe is expanding. Notice that the basic dimension of H is reciprocal time. The quantity $1/H$ is called the **Hubble age** and equals about 1.3×10^{10} y. This would correspond to the age of the universe if the gravitational pull on the receding galaxies were ignored.

Example 41-5 Using Hubble's Law

Redshift measurements of a galaxy in the constellation Virgo yield a recession velocity of 1200 km/s. How far is it to that galaxy?

PICTURE We calculate the distance from Hubble's law.

SOLVE

Use Hubble's law to find r:

$$r = \frac{v}{H} = (1200 \text{ km/s})\frac{10^6 c \cdot \text{y}}{23 \text{ km/s}} = \boxed{52 \times 10^6 c \cdot \text{y}}$$

PRACTICE PROBLEM 41-1 Show that $1/H = 1.3 \times 10^{10}$ y.

THE 2.7-K BACKGROUND RADIATION

In investigating ways of accounting for the cosmic abundance of atoms that are heavier than hydrogen atoms, cosmologists recognized that nucleosynthesis in stars could explain the abundance of atoms heavier than helium atoms but could not by itself explain the abundance of helium atoms. Helium must therefore have been formed during the big bang. To synthesize an amount of helium sufficient to account for its present abundance, the big bang would have to have occurred at an extremely high initial temperature to provide the necessary reaction rate before fusion was shut down by the decreasing density of the very rapid initial expansion. The high temperature implies a corresponding thermal (blackbody) radiation field that would cool as the expansion progressed. Theoretical analysis predicted that from the estimated time of the big bang to the present, the remnants of the radiation field should have cooled to a temperature of about 3 K, corresponding to a blackbody spectrum with peak wavelength λ_{max} in the microwave region. In 1965, the predicted cosmic background radiation was discovered by Arno Penzias and Robert Wilson at the Bell Labs. Since this landmark discovery, careful analysis has established that the temperature of the background field is 2.7281 K and has shown that it has an isotropic distribution in space.

THE BIG BANG

The singular event that initiated the expansion of the universe is thought to have been a huge explosion. The four interactions of nature (strong, electromagnetic, weak, and gravitational) initially were unified into a single interaction. Physicists have been successful in developing theoretical descriptions that unify the first three interactions, but a theory of quantum gravity, needed for the extreme densities of the single-interaction period, does not yet exist. Consequently, until the cooling universe "froze" or "condensed out" the gravitational interaction at approximately 10^{-43} s after the big bang, when the temperature was still 10^{32} K, we have no means of describing what was occurring. At this point, the average energy of the particles created would have been about 10^{19} GeV. As the universe continued to cool below 10^{32} K, the three interactions other than gravity remained unified and are described by the grand unification theories (GUTs). Quarks and leptons were indistinguishable and particle quantum numbers were not conserved. It was during this period that a slight excess of quarks over antiquarks occurred, roughly 1 part in 10^9, that ultimately resulted in the predominance of matter over antimatter that we now observe in the universe.

At 10^{-35} s, the universe had expanded sufficiently to cool to approximately 10^{27} K, at which point another phase transition occurred as the strong interaction condensed out of the GUTs group, leaving only the electromagnetic and weak interactions still unified as the **electroweak interaction.** During this period, the previously free quarks in the dense mixture of roughly equal numbers of quarks, leptons, their antiparticles, and photons began to combine into hadrons and their antiparticles, including the nucleons. By the time the universe had cooled to approximately 10^{13} K, at about $t = 10^{-6}$ s, the hadrons had mostly disappeared. This is because 10^{13} K corresponds to $kT \sim 1$ GeV, which is the minimum energy needed to create nucleons and antinucleons from the photons present by the reactions

$$\gamma \rightarrow \mathrm{p}^{+} + \mathrm{p}^{-} \qquad \text{41-11}a$$

and

$$\gamma \rightarrow \mathrm{n} + \bar{\mathrm{n}} \qquad \text{41-11}b$$

The particle–antiparticle pairs annihilated and there was no new production to replace them. Only the slight earlier excess of quarks over antiquarks led to a slight excess of protons and neutrons over their antiparticles. The annihilations resulted in photons and leptons, and after about $t = 10^{-4}$ s, those particles in roughly equal numbers dominated the universe. This was the **lepton era.** At about $t = 10$ s, the temperature had fallen to 10^{10} K ($kT \sim 1$ MeV). Further expansion and cooling dropped the average photon energy below the energy needed to form an electron–positron pair. Annihilation then removed all of the positrons as it had the antiprotons and antineutrons earlier, leaving only the small excess of electrons arising from charge conservation, and the **radiation era** began. The particles present were primarily photons and neutrinos.

Within a few more minutes, the temperature dropped sufficiently to enable fusing protons and neutrons to form nuclei that were not immediately photodisintegrated. The nuclei of deuterium, helium, and lithium were produced during this **nucleosynthesis period,** but the rapid expansion soon dropped the temperature too low for the fusion to continue and the formation of heavier elements had to await the birth of stars.

A long time later, when the temperature had dropped to about 3000 K as the universe grew to about 1/1000 of its present size, kT dropped below typical atomic ionization energies and atoms were formed. By then, the expansion had redshifted the radiation field so that the total radiation energy was about equal to the energy represented by the remaining mass. As expansion and cooling continued, the energy of the steadily redshifting radiation declined at a steady rate until, at $t = 10^{10}$ y (now), matter came to dominate the universe, with its energy density exceeding that of the 2.7-K radiation remaining from the big bang by a factor of about 1000.

Summary

TOPIC	RELEVANT EQUATIONS AND REMARKS
1. Basic Interactions	There are four basic interactions: strong, electromagnetic, weak, and gravitational.
Strong	The *charge* associated with the strong interaction is called color. Quarks and gluons have color and experience the strong interaction. Hadrons (baryons and mesons) experience a residual strong interaction resulting from the fundamental strong interaction between the quarks that make up the hadrons. Decay times by the strong interaction are typically 10^{-23} s.
Electromagnetic	All particles that have electric charge experience the force due to the electromagnetic interaction.
Weak	The *charge* associated with the weak interaction is called flavor. Quarks and leptons have flavor and experience the weak interaction. Decay times by the weak interaction are typically 10^{-10} s.
Gravitational	The *charge* associated with the gravitational interaction is called mass.
2. Fundamental Particles	There are two families of fundamental particles, leptons and quarks, each having six members. It is thought that these particles have no size and no internal structure.
Leptons	Leptons are spin-$\frac{1}{2}$ fermions: the electron e and its neutrino ν_e, the muon μ and its neutrino ν_μ, and the tau τ and its neutrino ν_τ. The electron, muon, and tau have mass, electric charge, and flavor, but not color; so they participate in the gravitational, electromagnetic, and weak interactions, but not the strong interaction. The neutrinos have flavor but no electric charge and no color. They have a very small mass.
Quarks	There are six quarks, called up u, down d, strange s, charmed c, top t, and bottom b. Each is a spin-$\frac{1}{2}$ fermion. The quarks participate in all of the basic interactions. Because they are always confined in mesons or baryons, their masses can only be estimated.
3. Hadrons	Hadrons are composite particles that are made up of quarks. There are two types of hadrons, baryons and mesons. Baryons, which include the neutron and proton, are fermions of half-integral spin consisting of three quarks. Mesons, which include pions and kaons, have zero or integral spin. Hadrons interact with each other by the residual strong interaction.
4. Field Particles	In addition to the six fundamental leptons and six fundamental quarks, there are field particles that are associated with the basic interactions. *Interaction* — *Field Particle* Gravitational — Graviton (not yet observed) Electromagnetic — Photon Weak — W^+, W^-, Z^0 Strong — Gluons
5. The Conservation Laws	Some quantities, such as energy, linear momentum, electric charge, angular momentum, baryon number, and each of the three lepton numbers, are strictly conserved during all reactions and decays. Others, such as strangeness and charm, are conserved during reactions and decays that proceed by the strong interaction but not in those that proceed by the weak interaction.
6. Particles and Antiparticles	Particles and their antiparticles have identical masses but opposite values for their other properties, such as charge, lepton number, baryon number, and strangeness. Particle–antiparticle pairs can be produced during various nuclear reactions if the energy available is greater than $2mc^2$, where m is the mass of the particle.
7. Hubble's Law	Hubble's law relates the recession velocity of a galaxy, determined from the redshift of its spectrum, to the distance of the galaxy from us: $v = Hr$ (41-9) where the Hubble constant H = 23 km/s per million light-years. From Hubble's law, we conclude that the universe is expanding and that the expansion began approximately $1/H$ years ago.

TOPIC	RELEVANT EQUATIONS AND REMARKS
8. **The Big Bang**	According to the model currently used to describe the evolution of the universe, the universe began with a big bang approximately 10^{10} years ago. The big bang model is supported by substantial experimental observations, including the isotropic, 2.7-K background blackbody radiation spectrum.

Answer to Concept Check

41-1 A proton is a baryon that has a baryon number (B) equal to 1, and all particles that are not baryons have $B = 0$. If a proton decays, conservation of baryon number implies that the decay products must contain a minimum of one baryon. In addition, conservation of energy implies that the rest mass of the decay products cannot be greater than the rest mass of the proton. Because there are no baryons that have a rest mass less than the rest mass of the proton, the proton cannot decay without either violating conservation of baryon number, conservation of energy, or both.

Problems

In a few problems, you are given more data than you actually need; in a few other problems, you are required to supply data from your general knowledge, outside sources, or informed estimate.

Interpret as significant all digits in numerical values that have trailing zeros and no decimal points.

- • Single-concept, single-step, relatively easy
- •• Intermediate-level, may require synthesis of concepts
- ••• Challenging
- SSM Solution is in the *Student Solutions Manual*
- Consecutive problems that are shaded are paired problems.

CONCEPTUAL PROBLEMS

1 • How are baryons and mesons similar? How are they different?

2 • The muon and the pion have nearly the same masses. How do the particles differ?

3 • How can you tell whether a decay proceeds by the strong interaction or the weak interaction? SSM

4 • True or false:

(*a*) All baryons are hadrons.
(*b*) All hadrons are baryons.

5 • True or false: All mesons are spin-$\frac{1}{2}$ particles.

6 • A particle that is made of exactly two quarks is (*a*) a meson, (*b*) a baryon, (*c*) a lepton, (*d*) either a meson or a baryon, but definitely not a lepton.

7 • Have any quark–antiquark combinations whose electric charge is not an integer multiplied by the fundamental charge e been observed?

8 • True or false:

(*a*) A lepton is a combination of three quarks.
(*b*) The typical times for decays by the weak interaction are orders of magnitude longer than the typical times for decays by the strong interaction.
(*c*) The muon and the pion are both mesons.

9 • True or false:

(*a*) Electrons interact with protons by the strong interaction.
(*b*) Strangeness is not conserved in reactions involving the weak interactions.
(*c*) Neutrons have zero charm.

ESTIMATION AND APPROXIMATION

10 •• Grand unification theories predict that the proton has a long but finite lifetime. Current experiments based on detecting the decay of protons in water infer that this lifetime is at least 10^{32} years. Assume 10^{32} years is, in fact, the mean lifetime of the proton. Estimate the expected time between proton decays that occur in the water of a filled Olympic-size swimming pool. An Olympic-size swimming pool is 100 m × 25 m × 2.0 m. Give your answer in days.

11 •• Table 41-6 lists some properties of the four fundamental interactions. To better understand the significance of this table, confirm the ratio of the numerical entries in the second and fourth column of the last row of the table by estimating the ratio of the electromagnetic force to the gravitational force between two protons of a nucleus.

SPIN AND ANTIPARTICLES

12 • Two pions at rest annihilate according to the reaction $\pi^+ + \pi^- \rightarrow \gamma + \gamma$. (*a*) Why must the energies of the two γ rays be equal? (*b*) Find the energy of each γ ray. (*c*) Find the wavelength of each γ ray.

13 • Find the minimum energy of the photon needed for the following pair-production reactions: (*a*) $\gamma \rightarrow \pi^+ + \pi^-$, (*b*) $\gamma \rightarrow p + p^-$, and (*c*) $\gamma \rightarrow \mu^- + \mu^+$.

THE CONSERVATION LAWS

14 • State which of the following decays or reactions violate one or more of the conservation laws, and give the law or laws violated in each case: (*a*) $p^+ \rightarrow n + e^+ + \overline{\nu}_e$, (*b*) $n \rightarrow p^+ + \pi^-$, (*c*) $e^+ + e^- \rightarrow \gamma$, (*d*) $p^+ + p^- \rightarrow \gamma + \gamma$, and (*e*) $\overline{\nu}_e + p^+ \rightarrow n + e^+$.

15 • Determine the change in strangeness in each reaction that follows, and state whether the each decay can proceed by the strong interaction, by the weak interaction, or not at all: (*a*) $\Omega^- \rightarrow \Xi^0 + \pi^-$, (*b*) $\Xi^0 \rightarrow p + \pi^- + \pi^0$, and (*c*) $\Lambda^0 \rightarrow p + \pi^-$.

16 • Determine the change in strangeness for each decay, and state whether each decay can proceed by the strong interaction, by the weak interaction, or not at all: (*a*) $\Omega^- \rightarrow \Lambda^0 + K^-$ and (*b*) $\Xi^0 \rightarrow p + \pi^-$.

17 • Determine the change in strangeness for each decay, and state whether each decay can proceed by the strong interaction, by the weak interaction, or not at all: (*a*) $\Omega^- \rightarrow \Lambda^0 + \overline{\nu}_e + e^-$ and (*b*) $\Sigma^+ \rightarrow p + \pi^0$.

18 • (*a*) Which of the following decays of the τ particle is possible?

$$\tau \rightarrow \mu^- + \overline{\nu}_\mu + \nu_\tau$$

$$\tau \rightarrow \mu^- + \nu_\mu + \overline{\nu}_\tau$$

(*b*) Explain why the other decay is not possible. (*c*) Calculate the kinetic energy of the decay products for the decay that is possible.

19 •• Using Table 41-2 and the laws of conservation of charge number, baryon number, strangeness, and spin, identify the unknown particle, symbolized by (?), in each of the following reactions: (*a*) $p + \pi^- \rightarrow \Sigma^0 + (?)$, (*b*) $p + p \rightarrow \pi^+ + n + K^+ + (?)$, and (*c*) $p + K^- \rightarrow \Xi^- + (?)$ SSM

20 •• Test the following decays for violation of the conservation of energy, electric charge, baryon number, and lepton number: (*a*) $n \rightarrow \pi^+ + \pi^- + \mu^+ + \mu^-$ and (*b*) $\pi^0 \rightarrow e^+ + e^- + \gamma$. Assume that linear momentum and angular momentum are conserved. State which conservation laws (if any) are violated in each decay.

QUARKS

21 • Find the baryon number, charge, and strangeness for the following quark combinations and identify the hadron: (*a*) *uud*, (*b*) *udd*, (*c*) *uus*, (*d*) *dds*, (*e*) *uss*, and (*f*) *dss*.

22 • Find the baryon number, charge, and strangeness for the following quark combinations: (*a*) $u\overline{d}$, (*b*) $\overline{u}d$, (*c*) $u\overline{s}$, and (*d*) $\overline{u}s$.

23 • The Δ^{++} particle is a baryon that decays by the strong interaction. Its strangeness, charm, topness, and bottomness are all zero. What combination of quarks gives a particle that has those properties?

24 • Find a possible combination of quarks that gives the correct values for electric charge, baryon number, and strangeness for (*a*) K^+ and (*b*) K^0.

25 • The D^+ meson has zero strangeness, but it has charm of +1. (*a*) What is a possible quark combination that will give the correct properties for the particle? (*b*) Repeat Part (*a*) for the D^- meson, which is the antiparticle of the D^+ meson.

26 • Find a possible combination of quarks that gives the correct values for electric charge, baryon number, and strangeness for (*a*) K^- (the K^- is the antiparticle of the K^+) and (*b*) $\overline{K}^0$.

27 •• Find a possible quark combination for the following particles: (*a*) Λ^0, (*b*) p^-, and (*c*) Σ^-. SSM

28 •• Find a possible quark combination for the following particles: (*a*) $\overline{n}$, (*b*) Ξ^0, and (*c*) Σ^+.

29 •• Find a possible quark combination for the following particles: (*a*) Ω^- and (*b*) Ξ^-.

30 •• State the properties of the particles made up of the following quarks: (*a*) *ddd*, (*b*) $u\overline{c}$, (*c*) $u\overline{b}$, and (*d*) $\overline{sss}$.

THE EVOLUTION OF THE UNIVERSE

31 • A galaxy is receding from Earth at 2.5 percent the speed of light. Estimate the distance from Earth to the galaxy. SSM

32 • Estimate the speed of a galaxy that is $12 \times 10^9\ c \cdot y$ away from us.

33 •• The Doppler frequency shift for a light from a source that is receding from a stationary receiver is given by $f' = f_0\sqrt{(1 - \beta)/(1 + \beta)}$, where $\beta = v/c$ (Equation 39-16*b*). Show that the Doppler wavelength shift for light is $\lambda' = \lambda_0\sqrt{(1 + \beta)/(1 - \beta)}$.

34 •• The red line in the spectrum of atomic hydrogen is frequently referred to as the *Hα* line, and it has a wavelength of 656.3 nm. Using Hubble's law and the Doppler equation for light from Problem 33, determine the wavelength of the *Hα* line in the spectrum emitted from galaxies at distances of (*a*) $5.00 \times 10^6\ c \cdot y$, (*b*) $5.00 \times 10^8\ c \cdot y$, and (*c*) $5.00 \times 10^9\ c \cdot y$ from Earth.

GENERAL PROBLEMS

35 • (*a*) What conditions are necessary for a particle and its antiparticle to be identical? (*b*) Find the quark combination of both the particle and the antiparticle of both the π^0 and the Ξ^0 particles. (*c*) Of the π^0 and the Ξ^0 particles, which, if any, is its own antiparticle?

36 •• The red line in the spectrum of atomic hydrogen is frequently referred to as the *Hα* line, and it has a wavelength of 656.3 nm. Light from a distant galaxy shows a redshift of the *Hα* line of hydrogen to a wavelength of 1458 nm. (*a*) What is the recessional velocity of the galaxy? (*b*) Estimate the distance to the galaxy.

37 •• (*a*) In terms of the quark model, show that the reaction $\pi^0 \rightarrow \gamma + \gamma$ does not violate any conservation laws. (*b*) Which conservation law is violated by the reaction $\pi^0 \rightarrow \gamma$? SSM

38 •• Test the following decays for violation of the conservation of energy, electric charge, baryon number, and lepton number: (*a*) $\Lambda^0 \rightarrow p + \pi^-$, (*b*) $\Sigma^- \rightarrow n + p^-$, and (*c*) $\mu^- \rightarrow e^- + \overline{\nu}_e + \nu_\mu$. Assume that linear momentum and angular momentum are conserved. State which conservation laws (if any) are violated in each decay.

39 •• Consider the following high-energy particle reaction: $p + p \rightarrow \Lambda^0 + K^0 + p + (?)$, where (?) represents an unknown particle. During this reaction, stationary protons are bombarded with a beam of high-energy protons. (*a*) Use the laws of conservation of charge number, baryon number, strangeness (Table 41-2), and spin to determine the unknown particle. (*b*) Calculate the *Q* value for the reaction. (*c*) The threshold kinetic energy K_{th} for this reaction is given by $K_{th} = -\frac{1}{2}Q(m_p + m_p + M_1 + M_2 + M_3 + M_4)/m_p$, where M_1, M_2, M_3, and M_4 are the masses of the reaction products. Find K_{th}.

40 ••• In this problem, you will calculate the difference in the time of arrival of two neutrinos of different energy from a supernova that is 170 000 light-years away. Let the energies of the neutrinos be $E_1 = 20$ MeV and $E_2 = 5$ MeV, and assume that the mass of a neutrino is 2.0 eV/c^2. Because the total energies of the neutrinos is so much greater than their rest energies, the neutrinos have speeds that are very nearly equal to c and energies that are approximately $E \approx pc$. (*a*) If t_1 and t_2 are the times that the neutrinos with speeds u_1 and u_2 take to travel a distance x, show that $\Delta t = t_2 - t_1 = x(u_1 - u_2)/u_1u_2 \approx (x\,\Delta u)/c^2$. (*b*) The speed of a neutrino of mass m and total energy E can be found from $E = mc^2/[1 - (u^2/c^2)]^{1/2}$ (Equation 39-24). Show that when $E \gg mc^2$, the speed ratio u/c is given approximately by $u/c \approx 1 - \frac{1}{2}(mc^2/E)^2$. (*c*) Use the results from Part (*a*) and Part (*b*) to calculate $u_1 - u_2$ for the energies and mass given, and calculate Δt from the result from Part (*a*) for $x = 170\,000\ c \cdot \text{y}$. (*d*) Repeat the calculation in Part (*c*) using 20 eV/c^2 for the neutrino mass.

41 ••• A Λ^0 at rest decays by the reaction $\Lambda^0 \rightarrow \text{p} + \pi^-$. (*a*) Calculate the total kinetic energy of the decay products. (*b*) Find the ratio of the kinetic energy of the pion to the kinetic energy of the proton. (*c*) Find the kinetic energies of the proton and the pion for the decay.

42 ••• A Σ^0 particle at rest decays by the reaction $\Sigma^0 \rightarrow \Lambda^0 + \gamma$. (*a*) What is the total energy (total energy includes rest energy) of the decay products? (*b*) Assuming that the kinetic energy of the Λ^0 is negligible compared with the energy of the photon, calculate the approximate momentum of the photon. (*c*) Use your result from Part (*b*) to calculate the kinetic energy of the Λ^0. (*d*) Use your result from Part (*c*) to obtain a better estimate of the momentum and the energy of the photon.

Appendix A

SI Units and Conversion Factors

Base Units*

Length	The *meter* (m) is the distance traveled by light in a vacuum in 1/299,792,458 s.
Time	The *second* (s) is the duration of 9,192,631,770 periods of the radiation corresponding to the transition between the two hyperfine levels of the ground state of the ^{133}Cs atom.
Mass	The *kilogram* (kg) is the mass of the international standard body preserved at Sèvres, France.
Mole	The *mole* (mol) is the amount of substance of a system which contains as many elementary entities as there are atoms in 0.012 kilogram of carbon 12.
Current	The *ampere* (A) is that constant current which, if maintained in two straight parallel conductors of infinite length, of negligible circular cross section, and placed 1 m apart in vacuum would produce between these conductors a force equal to 2×10^{-7} N/m of length.
Temperature	The *kelvin* (K) is 1/273.16 of the thermodynamic temperature of the triple point of water.
Luminous intensity	The *candela* (cd) is the luminous intensity in a given direction, of a source that emits monochromatic radiation of frequency 540×10^{12} hertz and that has a radiant intensity in that direction of 1/683 watt/steradian.

* These definitions are found on the Internet at http://physics.nist.gov/cuu/Units/current.html

Derived Units

Force	newton (N)	$1\ \text{N} = 1\ \text{kg}\cdot\text{m/s}^2$
Work, energy	joule (J)	$1\ \text{J} = 1\ \text{N}\cdot\text{m}$
Power	watt (W)	$1\ \text{W} = 1\ \text{J/s}$
Frequency	hertz (Hz)	$1\ \text{Hz} = \text{cy/s}$
Charge	coulomb (C)	$1\ \text{C} = 1\ \text{A}\cdot\text{s}$
Potential	volt (V)	$1\ \text{V} = 1\ \text{J/C}$
Resistance	ohm (Ω)	$1\ \Omega = 1\ \text{V/A}$
Capacitance	farad (F)	$1\ \text{F} = 1\ \text{C/V}$
Magnetic field	tesla (T)	$1\ \text{T} = 1\ \text{N/(A}\cdot\text{m)}$
Magnetic flux	weber (Wb)	$1\ \text{Wb} = 1\ \text{T}\cdot\text{m}^2$
Inductance	henry (H)	$1\ \text{H} = 1\ \text{J/A}^2$

Conversion Factors

Conversion factors are written as equations for simplicity; relations marked with an asterisk are exact.

Length

1 km = 0.6215 mi
1 mi = 1.609 km
1 m = 1.0936 yd = 3.281 ft = 39.37 in
*1 in = 2.54 cm
*1 ft = 12 in = 30.48 cm
*1 yd = 3 ft = 91.44 cm
1 lightyear = 1 $c \cdot$ y = 9.461×10^{15} m
*1 Å = 0.1 nm

Area

*1 $m^2 = 10^4\ cm^2$
1 $km^2 = 0.3861\ mi^2$ = 247.1 acres
*1 $in^2 = 6.4516\ cm^2$
1 $ft^2 = 9.29 \times 10^{-2}\ m^2$
1 $m^2 = 10.76\ ft^2$
*1 acre = 43 560 ft^2
1 mi^2 = 640 acres = 2.590 km^2

Volume

*1 $m^3 = 10^6\ cm^3$
*1 L = 1000 $cm^3 = 10^{-3}\ m^3$
1 gal = 3.785 L
1 gal = 4 qt = 8 pt = 128 oz = 231 in^3
1 $in^3 = 16.39\ cm^3$
1 ft^3 = 1728 $in.^3$ = 28.32 L
= $2.832 \times 10^4\ cm^3$

Time

*1 h = 60 min = 3.6 ks
*1 d = 24 h = 1440 min = 86.4 ks
1 y = 365.24 d = 3.156×10^7 s

Speed

*1 m/s = 3.6 km/h
1 km/h = 0.2778 m/s = 0.6215 mi/h
1 mi/h = 0.4470 m/s = 1.609 km/h
1 mi/h = 1.467 ft/s

Angle and Angular Speed

*π rad = 180°
1 rad = 57.30°
1° = 1.745×10^{-2} rad
1 rev/min = 0.1047 rad/s
1 rad/s = 9.549 rev/min

Mass

*1 kg = 1000 g
*1 tonne = 1000 kg = 1 Mg
1 u = 1.6605×10^{-27} kg
= 931.49 MeV/c^2
1 kg = 6.022×10^{26} u
1 slug = 14.59 kg
1 kg = 6.852×10^{-2} slug

Density

*1 g/cm^3 = 1000 kg/m^3 = 1 kg/L
(1 g/cm^3)g = 62.4 lb/ft^3

Force

1 N = 0.2248 lb = 10^5 dyn
*1 lb = 4.448222 N
(1 kg)g = 2.2046 lb

Pressure

*1 Pa = 1 N/m^2
*1 atm = 101.325 kPa = 1.01325 bars
1 atm = 14.7 lb/in^2 = 760 mmHg
= 29.9 inHg = 33.9 ftH_2O
1 lb/in^2 = 6.895 kPa
1 torr = 1 mmHg = 133.32 Pa
1 bar = 100 kPa

Energy

*1 kW $\cdot$ h = 3.6 MJ
*1 cal = 4.1840 J
1 ft $\cdot$ lb = 1.356 J = 1.286×10^{-3} Btu
*1 L $\cdot$ atm = 101.325 J
1 L $\cdot$ atm = 24.217 cal
1 Btu = 778 ft $\cdot$ lb = 252 cal = 1054.35 J
1 eV = 1.602×10^{-19} J
1 u $\cdot$ c^2 = 931.49 MeV
*1 erg = 10^{-7} J

Power

1 horsepower = 550 ft $\cdot$ lb/s = 745.7 W
1 Btu/h = 2.931×10^{-4} kW
1 W = 1.341×10^{-3} horsepower
= 0.7376 ft $\cdot$ lb/s

Magnetic Field

*1 T = 10^4 G

Thermal Conductivity

1 W/(m·K) = 6.938 Btu·in/(h·ft^2·F°)
1 Btu·in/(h·ft^2·F°) = 0.1441 W/(m·K)

Appendix B
Numerical Data

Terrestrial Data

Free-fall acceleration g	
Standard value (at sea level at 45° latitude)*	$9.806\,65\ \text{m/s}^2$; $32.1740\ \text{ft/s}^2$
At equator*	$9.7804\ \text{m/s}^2$
At poles*	$9.8322\ \text{m/s}^2$
Mass of Earth M_E	5.97×10^{24} kg
Radius of Earth R_E, mean	6.37×10^6 m; 3960 mi
Escape speed $\sqrt{2R_E g}$	1.12×10^4 m/s; 6.95 mi/s
Solar constant†	$1.37\ \text{kW/m}^2$
Standard temperature and pressure (STP):	
Temperature	273.15 K (0.00°C)
Pressure	101.325 kPa (1.00 atm)
Molar mass of air	28.97 g/mol
Density of air (STP), ρ_{air}	$1.217\ \text{kg/m}^3$
Speed of sound (STP)	331 m/s
Heat of fusion of H_2O (0°C, 1 atm)	333.5 kJ/kg
Heat of vaporization of H_2O (100°C, 1 atm)	2.257 MJ/kg

*Measured relative to Earth's surface.

† Average power incident normally on 1 m^2 outside Earth's atmosphere at the mean distance from Earth to the Sun.

Astronomical Data*

Earth	
Distance to moon, mean†	3.844×10^8 m; 2.389×10^5 mi
Distance to the Sun, mean†	1.496×10^{11} m; 9.30×10^7 mi; 1.00 AU
Orbital speed, mean	2.98×10^4 m/s
Moon	
Mass	7.35×10^{22} kg
Radius	1.737×10^6 m
Period	27.32 d
Acceleration of gravity at surface	$1.62\ \text{m/s}^2$
Sun	
Mass	1.99×10^{30} kg
Radius	6.96×10^8 m

* Additional solar-system data is available from NASA at <http://nssdc.gsfc.nasa.gov/planetary/planetfact.html>.

† Center to center.

Physical Constants*

Gravitational constant	G	$6.6742(10) \times 10^{-11}\ \mathrm{N \cdot m^2/kg^2}$
Speed of light	c	$2.997\,924\,58 \times 10^{8}\ \mathrm{m/s}$
Fundamental charge	e	$1.602\,176\,453(14) \times 10^{-19}\ \mathrm{C}$
Avogadro's number	N_A	$6.022\,141\,5(10) \times 10^{23}$ particles/mol
Gas constant	R	$8.314\,472(15)\ \mathrm{J/(mol \cdot K)}$
		$1.987\,2065(36)\ \mathrm{cal/(mol \cdot K)}$
		$8.205\,746(15) \times 10^{-2}\ \mathrm{L \cdot atm/(mol \cdot K)}$
Boltzmann constant	$k = R/N_A$	$1.380\,650\,5(24) \times 10^{-23}\ \mathrm{J/K}$
		$8.617\,343(15) \times 10^{-5}\ \mathrm{eV/K}$
Stefan-Boltzmann constant	$\sigma = (\pi^2/60)k^4/(\hbar^3 c^2)$	$5.670\,400(40) \times 10^{-8}\ \mathrm{W/(m^2k^4)}$
Atomic mass constant	$m_u = \frac{1}{12}m(^{12}\mathrm{C})$	$1.660\,538\,86(28) \times 10^{-27}\ \mathrm{kg} = 1\mathrm{u}$
Magnetic constant (permeability of free space)	μ_0	$4\,\pi \times 10^{-7}\ \mathrm{N/A^2}$
		$1.256\,637 \times 10^{-6}\ \mathrm{N/A^2}$
Electric constant (permittivity of free space)	$\epsilon_0 = 1/(\mu_0 C^2)$	$8.854\,187\,817 \ldots \times 10^{-12}\ \mathrm{C^2/(N \cdot m^2)}$
Coulomb constant	$k = 1/(4\pi\epsilon_0)$	$8.987\,551\,788 \ldots \times 10^{9}\ \mathrm{N \cdot m^2/C^2}$
Planck's constant	h	$6.626\,0693(11) \times 10^{-34}\ \mathrm{J \cdot s}$
		$4.135\,667\,43(35) \times 10^{-15}\ \mathrm{eV \cdot s}$
	$\hbar = h/2\pi$	$1.054\,571\,68(18) \times 10^{-34}\ \mathrm{J \cdot s}$
		$6.582\,119\,15(56) \times 10^{-16}\ \mathrm{eV \cdot s}$
Mass of electron	m_e	$9.109\,382\,6(16) \times 10^{-31}\ \mathrm{kg}$
		$0.510\,998\,918(44)\ \mathrm{MeV}/c^2$
Mass of proton	m_p	$1.672\,621\,71(29) \times 10^{-27}\ \mathrm{kg}$
		$938.272\,029(80) \times \mathrm{MeV}/c^2$
Mass of neutron	m_n	$1.674\,927\,28(29) \times 10^{-27}\ \mathrm{kg}$
		$939.565\,360(81)\ \mathrm{MeV}/c^2$
Bohr magneton	$m_B = eh/2m_e$	$9.274\,009\,49(80) \times 10^{-24}\ \mathrm{J/T}$
		$5.788\,381\,804(39) \times 10^{-5}\ \mathrm{eV/T}$
Nuclear magneton	$m_n = eh/2m_p$	$5.050\,783\,43(43) \times 10^{-27}\ \mathrm{J/T}$
		$3.152\,451\,259(21) \times 10^{-8}\ \mathrm{eV/T}$
Magnetic flux quantum	$\phi_0 = h/2e$	$2.067\,833\,72(18) \times 10^{-15}\ \mathrm{T \cdot m^2}$
Quantized Hall resistance	$R_K = h/e^2$	$2.581\,280\,7449(86) \times 10^{4}\ \Omega$
Rydberg constant	R_H	$1.097\,373\,156\,8525(73) \times 10^{7}\ \mathrm{m^{-1}}$
Josephson frequency-voltage quotient	$K_J = 2e/h$	$4.835\,978\,79(41) \times 10^{14}\ \mathrm{Hz/V}$
Compton wavelength	$\lambda_C = h/m_e c$	$2.426\,310\,238(16) \times 10^{-12}\ \mathrm{m}$

* The values for these and other constants may be found on the Internet at http://physics.nist.gov/cuu/Constants/index.html. The numbers in parentheses represent the uncertainties in the last two digits. (For example, 2.044 43(13) stands for 2.044 43 ± 0.000 13.) Values without uncertainties are exact, including those values with ellipses (such as the value of pi is exactly 3.1415. . .).

For additional data, see the following tables in the text.

Geometry and Trigonometry

$C = \pi d = 2\pi r$	definition of π
$A = \pi r^2$	area of circle
$V = \frac{4}{3}\pi r^3$	spherical volume
$A = dV/dr = 4\pi r^2$	spherical surface area
$V = A_{base}L = \pi r^2 L$	cylindrical volume
$A = \partial V/\partial r = 2\pi rL$	cylindrical surface area

$o = h\sin\theta$
$a = h\cos\theta$

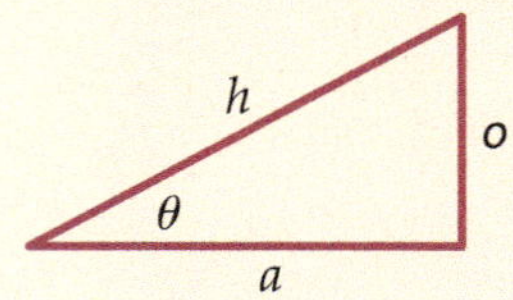

$\sin^2\theta + \cos^2\theta = 1$

$\sin(A \pm B) = \sin A\cos B \pm \cos A\sin B$

$\cos(A \pm B) = \cos A\cos B \mp \sin A\sin B$

$\sin A \pm \sin B = 2\sin[\frac{1}{2}(A \pm B)]\cos[\frac{1}{2}(A \mp B)]$

$\sin\theta \equiv y$
$\cos\theta \equiv x$
$\tan\theta \equiv \frac{y}{x}$

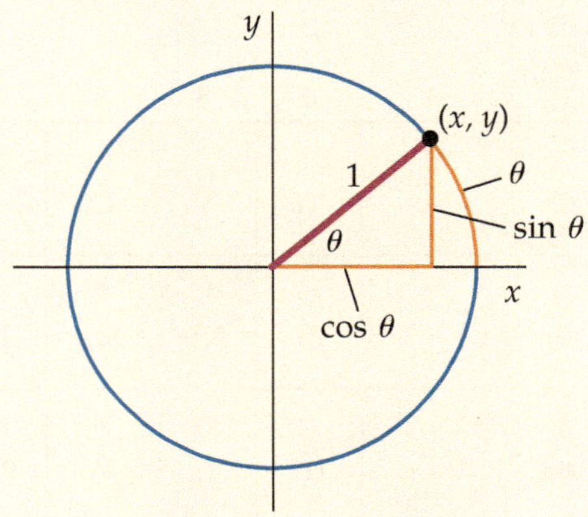

If $|\theta| \ll 1$, then
$\cos\theta \approx 1$ and $\tan\theta \approx \sin\theta \approx \theta$ (θ in radians)

Quadratic Formula

If $ax^2 + bx + c = 0$, then $x = \frac{-b \pm \sqrt{b^2 - 4ac}}{2a}$

Binomial Expansion

If $|x| < 1$, then $(1 + x)^n =$

$$1 + nx + \frac{n(n-1)}{2!}x^2 + \frac{n(n-1)(n-2)}{3!}x^3 + \ldots$$

If $|x| \ll 1$, then $(1 + x)^n \approx 1 + nx$

Differential Approximation

If $\Delta F = F(x + \Delta x) - F(x)$ and if $|\Delta x|$ is small,

then $\Delta F \approx \frac{dF}{dx}\Delta x$.

Appendix C

Periodic Table of Elements*

1	2	3	4	5	6	7	8	9	10	11	12	13	14	15	16	17	18
1 H																	2 He
3 Li	4 Be											5 B	6 C	7 N	8 O	9 F	10 Ne
11 Na	12 Mg											13 Al	14 Si	15 P	16 S	17 Cl	18 Ar
19 K	20 Ca	21 Sc	22 Ti	23 V	24 Cr	25 Mn	26 Fe	27 Co	28 Ni	29 Cu	30 Zn	31 Ga	32 Ge	33 As	34 Se	35 Br	36 Kr
37 Rb	38 Sr	39 Y	40 Zr	41 Nb	42 Mo	43 Tc	44 Ru	45 Rh	46 Pd	47 Ag	48 Cd	49 In	50 Sn	51 Sb	52 Te	53 I	54 Xe
55 Cs	56 Ba	57–71 Rare Earths	72 Hf	73 Ta	74 W	75 Re	76 Os	77 Ir	78 Pt	79 Au	80 Hg	81 Tl	82 Pb	83 Bi	84 Po	85 At	86 Rn
87 Fr	88 Ra	89–103 Actinides	104 Rf	105 Db	106 Sg	107 Bh	108 Hs	109 Mt	110 Ds	111 Rg							

Rare Earths (Lanthanides)	57 La	58 Ce	59 Pr	60 Nd	61 Pm	62 Sm	63 Eu	64 Gd	65 Tb	66 Dy	67 Ho	68 Er	69 Tm	70 Yb	71 Lu
Actinides	89 Ac	90 Th	91 Pa	92 U	93 Np	94 Pu	95 Am	96 Cm	97 Bk	98 Cf	99 Es	100 Fm	101 Md	102 No	103 Lr

*The 1–18 group designation has been recommended by the International Union of Pure and Applied Chemistry (IUPAC). Elements with atomic numbers 112, 114, and 116 have been reported but not fully authenticated as of September 2003. From http://www.iupac.org/reports/periodic_table/IUPAC_Periodic_Table-3Oct05.pdf

Atomic Numbers and Atomic Masses*

Atomic Number	Name	Symbol	Mass	Atomic Number	Name	Symbol	Mass
1	Hydrogen	H	1.00794(7)	57	Lanthanum	La	138.90547(7)
2	Helium	He	4.002602(2)	58	Cerium	Ce	140.116(1)
3	Lithium	Li	6.941(2)	59	Praseodymium	Pr	140.90765(2)
4	Beryllium	Be	9.012182(3)	60	Neodymium	Nd	144.242(3)
5	Boron	B	10.811(7)	61	Promethium	Pm	[145]
6	Carbon	C	12.0107(8)	62	Samarium	Sm	150.36(2)
7	Nitrogen	N	14.0067(2)	63	Europium	Eu	151.964(1)
8	Oxygen	O	15.9994(3)	64	Gadolinium	Gd	157.25(3)
9	Fluorine	F	18.9984032(5)	65	Terbium	Tb	158.92535(2)
10	Neon	Ne	20.1797(6)	66	Dysprosium	Dy	162.500(1)
11	Sodium	Na	22.98976928(2)	67	Holmium	Ho	164.93032(2)
12	Magnesium	Mg	24.3050(6)	68	Erbium	Er	167.259(3)
13	Aluminum	Al	26.9815386(8)	69	Thulium	Tm	168.93421(2)
14	Silicon	Si	28.0855(3)	70	Ytterbium	Yb	173.04(3)
15	Phosphorus	P	30.973762(2)	71	Lutetium	Lu	174.967(1)
16	Sulfur	S	32.065(5)	72	Hafnium	Hf	178.49(2)
17	Chlorine	Cl	35.453(2)	73	Tantalum	Ta	180.94788(2)
18	Argon	Ar	39.948(1)	74	Tungsten	W	183.84(1)
19	Potassium	K	39.0983(1)	75	Rhenium	Re	186.207(1)
20	Calcium	Ca	40.078(4)	76	Osmium	Os	190.23(3)
21	Scandium	Sc	44.955912(6)	77	Iridium	Ir	192.217(3)
22	Titanium	Ti	47.867(1)	78	Platinum	Pt	195.084(9)
23	Vanadium	V	50.9415(1)	79	Gold	Au	196.966569(4)
24	Chromium	Cr	51.9961(6)	80	Mercury	Hg	200.59(2)
25	Manganese	Mn	54.938045(5)	81	Thallium	Tl	204.3833(2)
26	Iron	Fe	55.845(2)	82	Lead	Pb	207.2(1)
27	Cobalt	Co	58.933195(5)	83	Bismuth	Bi	208.98040(1)
28	Nickel	Ni	58.6934(2)	84	Polonium	Po	[209]
29	Copper	Cu	63.546(3)	85	Astatine	At	[210]
30	Zinc	Zn	65.409(4)	86	Radon	Rn	[222]
31	Gallium	Ga	69.723(1)	87	Francium	Fr	[223]
32	Germanium	Ge	72.64(1)	88	Radium	Ra	[226]
33	Arsenic	As	74.92160(2)	89	Actinium	Ac	[227]
34	Selenium	Se	78.96(3)	90	Thorium	Th	232.03806(2)
35	Bromine	Br	79.904(1)	91	Protactinium	Pa	231.03588(2)
36	Krypton	Kr	83.798(2)	92	Uranium	U	238.02891(3)
37	Rubidium	Rb	85.4678(3)	93	Neptunium	Np	[237]
38	Strontium	Sr	87.62(1)	94	Plutonium	Pu	[244]
39	Yttrium	Y	88.90585(2)	95	Americium	Am	[243]
40	Zirconium	Zr	91.224(2)	96	Curium	Cm	[247]
41	Niobium	Nb	92.90638(2)	97	Berkelium	Bk	[247]
42	Molybdenum	Mo	95.94(2)	98	Californium	Cf	[251]
43	Technetium	Tc	[98]	99	Einsteinium	Es	[252]
44	Ruthenium	Ru	101.07(2)	100	Fermiun	Fm	[257]
45	Rhodium	Rh	102.90550(2)	101	Mendelevium	Md	[258]
46	Palladium	Pd	106.42(1)	102	Nobelium	No	[259]
47	Silver	Ag	107.8682(2)	103	Lawrencium	Lr	[262]
48	Cadmium	Cd	112.411(8)	104	Rutherfordium	Rf	[261]
49	Indium	In	114.818(3)	105	Dubnium	Db	[262]
50	Tin	Sn	118.710(7)	106	Seaborgium	Sg	[266]
51	Antimony	Sb	121.760(1)	107	Bohrium	Bh	[264]
52	Tellurium	Te	127.60(3)	108	Hassium	Hs	[277]
53	Iodine	I	126.90447(3)	109	Meitnerium	Mt	[268]
54	Xenon	Xe	131.293(6)	110	Darmstadtium	Ds	[271]
55	Cesium	Cs	132.9054519(2)	111	Roentgenium	Rg	[272]
56	Barium	Ba	137.327(7)				

* IUPAC 2005 standard atomic weights (mean relative atomic masses) as approved at the 43rd IUPAC General Assembly in Beijing, China, in August 2005 are listed with uncertainties in the last figure in parentheses. From http://www.iupac.org/reports/periodic_table/IUPAC_Periodic_Table-3Oct05.pdf

Math Tutorial

In this tutorial, we review some of the basic results of algebra, geometry, trigonometry, and calculus. In many cases, we merely state results without proof. Table M-1 lists some mathematical symbols.

Table M-1 Mathematical Symbols

Symbol	Meaning
$=$	is equal to
$\neq$	is not equal to
$\approx$	is approximately equal to
$\sim$	is of the order of
$\propto$	is proportional to
$>$	is greater than
$\geq$	is greater than or equal to
$\gg$	is much greater than
$<$	is less than
$\leq$	is less than or equal to
$\ll$	is much less than
Δx	change in x
$\lvert x\rvert$	absolute value of x
n!	$n(n-1)(n-2)\ldots1$
$\sum$	sum
lim	limit
$\Delta t \rightarrow 0$	Δt approaches zero
$\frac{dx}{dt}$	derivative of x with respect to t
$\frac{\partial x}{\partial t}$	partial derivative of x with respect to t
$\int$	integral

M-1 SIGNIFICANT DIGITS

Many numbers we work with in science are the result of measurement and are therefore known only within a degree of uncertainty. This uncertainty should be reflected in the number of digits used. For example, if you have a 1-meter-long rule with scale spacing of 1 cm, you know that you can measure the height of a box to within a fifth of a centimeter or so. Using this rule, you might find that the box height is 27.0 cm. If there is a scale with a spacing of 1 mm on your rule, you might perhaps measure the box height to be 27.03 cm. However, if there is a scale with a spacing of 1 mm on your rule, you might not be able to measure the height more accurately than 27.03 cm because the height might vary by 0.01 cm or so, depending on which part of the box you measure the height at. When you write down that the height of the box is 27.03 cm, you are stating that your best estimate of the height is 27.03 cm, but you are not claiming that it is exactly 27.030000 . . . cm high. The four digits in 27.03 cm are called **significant digits.** Your measured length, 2.703 m, has four significant digits. Significant digits are also called significant figures.

The number of significant digits in an answer to a calculation will depend on the number of significant digits in the given data. When you work with numbers that have uncertainties, you should be careful not to include more digits than the certainty of measurement warrants. *Approximate* calculations (order-of-magnitude estimates) always result in answers that have only one significant digit or none. When you multiply, divide, add, or subtract numbers, you must consider the accuracy of the results. Listed below are some rules that will help you determine the number of significant digits of your results.

1. When multiplying or dividing quantities, the number of significant digits in the final answer is no greater than that in the quantity with the fewest significant digits.
2. When adding or subtracting quantities, the number of decimal places in the answer should match that of the term with the smallest number of decimal places.
3. Exact values have an unlimited number of significant digits. For example, a value determined by counting, such as 2 tables, has no uncertainty and is an exact value. In addition, the conversion factor 0.0254000 . . . m/in is an exact value because 1.000 . . . inches is exactly equal to 0.0254000 . . . meters. (The yard is, by definition, equal to exactly 0.9144 meters, and 0.9144 divided by 36 is exactly equal to 0.0254.)
4. Sometimes zeros are significant and sometimes they are not. If a zero is before a leading nonzero digit, then the zero is not significant. For example, the number 0.00890 has three significant digits. The first three zeroes are not significant digits but are merely markers to locate the decimal point. Note that the zero after the nine is significant.
5. Zeros that are between nonzero digits are significant. For example, 5603 has four significant digits.
6. The number of significant digits in numbers with trailing zeros and no decimal point is ambiguous. For example 31000 could have as many as five significant digits or as few as two significant digits. To prevent ambiguity, you should report numbers by using scientific notation or by using a decimal point.

Example M-1 Finding the Average of Three Numbers

Find the average of 19.90, −7.524, and −11.8179.

PICTURE You will be adding 3 numbers and then dividing the result by 3. The first number has three significant digits, the second number has four, and the third number has five.

SOLVE

1. Sum the three numbers.

$$19.90 + (-7.524) + (-11.8179) = 0.55\textit{81}$$

2. If the problem only asked for the sum of the three numbers, we would round the answer to the least number of decimal places among all the numbers being added. However, we must divide this intermediate result by 3, so we use the intermediate answer with the two extra digits (italicized and red).

$$\frac{0.55\textit{81}}{3} = 0.18\textit{60333}\ldots$$

3. Only two of the digits in the intermediate answer, 0.18*60333*. . . , are significant digits, so we must round this number to get our final answer. The number 3 in the denominator is a whole number and has an unlimited number of significant digits. Thus, the final answer has the same number of significant digits as the numerator, which is 2.

The final answer is $\boxed{0.19.}$

CHECK The sum in step 1 has two significant digits following the decimal point, the same as the number being summed with the least number of significant digits after the decimal point.

PRACTICE PROBLEMS

1. $\dfrac{5.3 \text{ mol}}{22.4 \text{ mol/L}}$
2. 57.8 m/s − 26.24 m/s

M-2 EQUATIONS

An **equation** is a statement written using numbers and symbols to indicate that two quantities, written on either side of an equals sign (=), are equal. The quantity on either side of the equal sign may consist of a single term, or of a sum or difference of two or more **terms.** For example, the equation $x = 1 - (ay + b)/(cx - d)$ contains three terms, x, 1 and $(ay + b)/(cx - d)$.

You can perform the following operations on equations:

1. The same quantity can be added to or subtracted from each side of an equation.
2. Each side of an equation can be multiplied or divided by the same quantity.
3. Each side of an equation can be raised to the same power.

These operations are meant to be applied to each *side* of the equation rather than each term in the equation. (Because multiplication is distributive over addition, operation 2—and only operation 2—of the preceding operations also applies term by term.)

Caution: *Division by zero is forbidden at any stage in solving an equation; results (if any) would be invalid.*

Adding or Subtracting Equal Amounts

To find x when $x - 3 = 7$, add 3 to both sides of the equation: $(x - 3) + 3 = 7 + 3$; thus, $x = 10$.

Multiplying or Dividing by Equal Amounts

If $3x = 17$, solve for x by dividing both sides of the equation by 3; thus, $x = \frac{17}{3}$, or 5.7.

Example M-2 Simplifying Reciprocals in an Equation

Solve the following equation for x:

$$\frac{1}{x} + \frac{1}{4} = \frac{1}{3}$$

Equations containing reciprocals of unknowns occur in geometric optics and in electric circuit analysis—for example, in finding the net resistance of parallel resistors.

PICTURE In this equation, the term containing x is on the same side of the equation as a term not containing x. Furthermore, x is found in the denominator of a fraction.

SOLVE

1. Subtract $\frac{1}{4}$ from each side:
$$\frac{1}{x} = \frac{1}{3} - \frac{1}{4}$$
2. Simplify the right side of the equation by using the lowest common denominator:
$$\frac{1}{x} = \frac{1}{3} - \frac{1}{4} = \frac{4}{12} - \frac{3}{12} = \frac{4-3}{12} = \frac{1}{12} \quad \text{so} \quad \frac{1}{x} = \frac{1}{12}$$
3. Multiply both sides of the equation by $12x$ to determine the value of x:
$$12\cancel{x}\frac{1}{\cancel{x}} = \cancel{12}x\frac{1}{\cancel{12}}$$
$$\boxed{12} = x$$

CHECK Substitute 12 for x in the left side of original equation.

$$\frac{1}{x} + \frac{1}{4} = \frac{1}{12} + \frac{3}{12} = \frac{4}{12} = \frac{1}{3}$$

PRACTICE PROBLEMS Solve each of the following for x:

3. $(7.0\ \text{cm}^3)x = 18\ \text{kg} + (4.0\ \text{cm}^3)x$
4. $\frac{4}{x} + \frac{1}{3} = \frac{3}{x}$

M-3 DIRECT AND INVERSE PROPORTIONS

When we say variable quantities x and y are **directly proportional,** we mean that as x and y change, the ratio x/y is constant. To say that two quantities are proportional is to say that they are directly proportional. When we say variable quantities x and y are **inversely proportional,** we mean that as x and y change, the product xy is constant.

Relationships of direct and inverse proportion are common in physics. Objects moving at the same velocity have momenta directly proportional to their masses. The ideal-gas law ($PV = nRT$) states that pressure P is directly proportional to (absolute) temperature T, when volume V remains constant and is inversely proportional to volume, when temperature remains constant. Ohm's law ($V = IR$) states that the voltage V across a resistor is directly proportional to the electric current in the resistor when the resistance R remains constant.

CONSTANT OF PROPORTIONALITY

When two quantities are directly proportional, the two quantities are related by a *constant of proportionality.* If you are paid for working at a regular rate R in dollars per day, for example, the money m you earn is directly proportional to the time t you work; the rate R is the constant of proportionality that relates the money earned in dollars to the time worked t in days:

$$\frac{m}{t} = R \qquad \text{or} \qquad m = Rt$$

If you earn \$400 in 5 days, the value of R is \$400/(5 days) = \$80/day. To find the amount you earn in 8 days, you could perform the calculation

$$m = (\$80/\text{day})(8\ \text{days}) = \$640$$

Sometimes the constant of proportionality can be ignored in proportion problems. Because the amount you earn in 8 days is $\frac{8}{5}$ times what you earn in 5 days, this amount is

$$m = \frac{8}{5}(\$400) = \$640$$

Example M-3 Painting Cubes

You need 15.4 mL of paint to cover one side of a cube. The area of one side of the cube is 426 cm^2. What is the relation between the volume of paint needed and the area to be covered? How much paint do you need to paint one side of a cube in which the one side has an area of 503 cm^2?

PICTURE To determine the amount of paint for the side whose area is 503 cm^2, you will need to set up a proportion.

SOLVE

1. The volume V of paint needed increases in proportion to the area A to be covered.

   V and A are directly proportional.

   That is, $\frac{V}{A} = k$ or $V = kA$

   where k is the proportionality constant

2. Determine the value of the proportionality constant using the given values $V_1 = 15.4$ mL and $A_1 = 426$ cm^2:

   $$k = \frac{V_1}{A_1} = \frac{15.4\ \text{mL}}{426\ \text{cm}^2} = 0.0361\ \text{mL/cm}^2$$

3. Determine the volume of paint needed to paint a side of a cube whose area is 503 cm^2 using the proportionality constant in step 1:

   $$V_2 = kA_2 = (0.0361\ \text{mL/cm}^2)(503\ \text{cm}^2) = \boxed{18.2\ \text{mL}}$$

CHECK Our value for V_2 is greater than the value for V_1, as expected. The amount of paint needed to cover an area equal to 503 cm^2 should be greater than the amount of paint needed to cover an area of 426 cm^2 because 503 cm^2 is larger than 426 cm^2.

PRACTICE PROBLEMS

5. A cylindrical container holds 0.384 L of water when full. How much water would the container hold if its radius were doubled and its height remained unchanged?
Hint: The volume of a right circular cylinder is given by $V = \pi r^2 h$, where r is its radius and h is its height. Thus, V is directly proportional to r^2 when h remains constant.
6. For the container in Practice Problem 5, how much water would the container hold if both its height and its radius were doubled? How much water would the container hold if its radius were doubled and its height remained unchanged?
Hint: The volume V of a right circular cylinder is given by $V = \pi r^2 h$, where r is its radius and h is its height.

M-4 LINEAR EQUATIONS

A **linear equation** is an equation of the form $x + 2y - 4z = 3$. That is, an equation is linear if each term either is constant or is the product of a constant and a variable raised to the first power. Such equations are said to be linear because the plots of these equations form straight lines or planes. The equations of direct proportion between two variables are linear equations.

GRAPH OF A STRAIGHT LINE

A linear equation relating y and x can always be put into the standard form

$$y = mx + b \qquad \text{M-1}$$

where m and b are constants that may be either positive or negative. Figure M-1 shows a graph of the values of x and y that satisfy Equation M-1. The constant b, called the **y intercept,** is the value of y at $x = 0$. The constant m is the **slope** of the line, which equals the ratio of the change in y to the corresponding change in x. In the figure, we have indicated two points on the line, (x_1, y_1) and (x_2, y_2), and the changes $\Delta x = x_2 - x_1$ and $\Delta y = y_2 - y_1$. The slope m is then

$$m = \frac{y_2 - y_1}{x_2 - x_1} = \frac{\Delta y}{\Delta x}$$

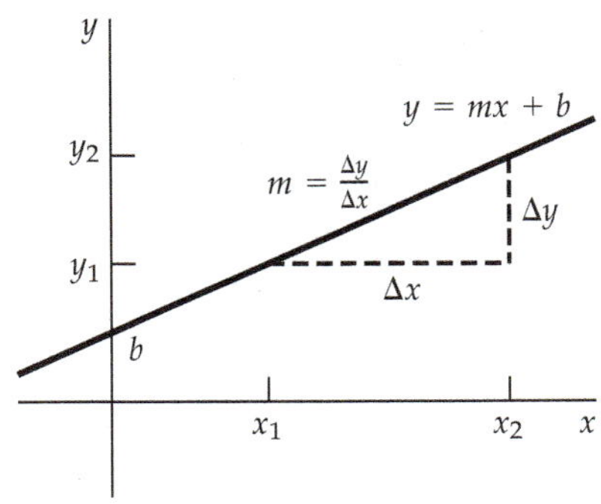

FIGURE M-1 Graph of the linear equation $y = mx + b$, where b is the y intercept and $m = \Delta y/\Delta x$ is the slope.

If x and y are both unknown in the equation $y = mx + b$, there are no unique values of x and y that are solutions to the equation. Any pair of values (x_1, y_1) on the line in Figure M-1 will satisfy the equation. If we have two equations, each with the same two unknowns x and y, the equations can be solved simultaneously for the unknowns. Example M-4 shows how simultaneous linear equations can be solved.

Example M-4 Using Two Equations to Solve for Two Unknowns

Find any and all values of x and y that simultaneously satisfy

$$3x - 2y = 8 \qquad \text{M-2}$$

and

$$y - x = 2 \qquad \text{M-3}$$

PICTURE Figure M-2 shows a graph of the two equations. At the point where the lines intersect, the values of x and y satisfy both equations. We can solve two simultaneous equations by first solving either equation for one variable in terms of the other variable and then substituting the result into the other equation.

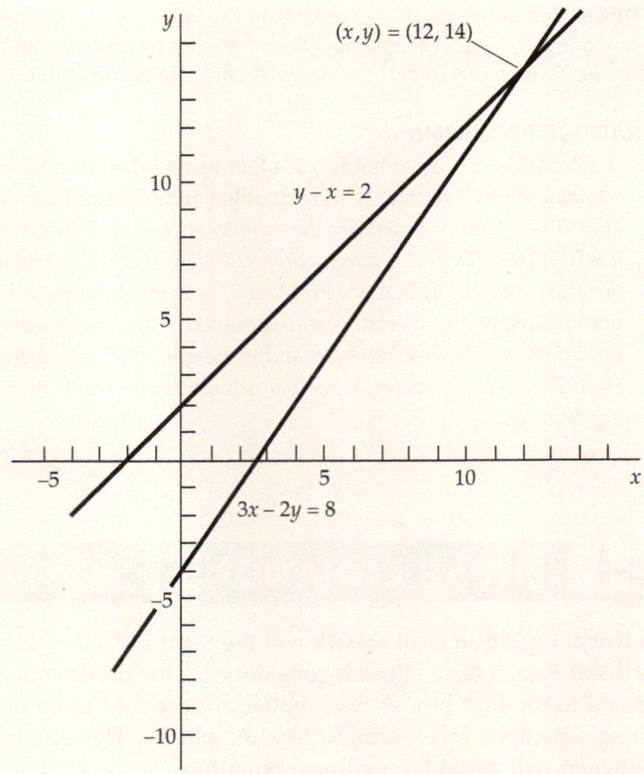

FIGURE M-2 Graph of Equations M-2 and M-3. At the point where the lines intersect, the values of x and y satisfy both equations.

SOLVE

1. Solve Equation M-3 for y: $y = x + 2$
2. Substitute this value for y into Equation M-2: $3x - 2(x + 2) = 8$
3. Simplify the equation and solve for x: $3x - 2x - 4 = 8$
   $x - 4 = 8$
   $x = \boxed{12}$
4. Use your solution for x and one of the given equations to find the value of y: $y - x = 2$, where $x = 12$
   $y - 12 = 2$
   $y = 2 + 12 = \boxed{14}$

CHECK An alternative method is to multiply one equation by a constant such that one of the unknown terms is eliminated when the equations are added or subtracted. We can multiply through Equation M-3 by 2

$$2(y - x) = 2(2)$$
$$2y - 2x = 4$$

and add the result to Equation M-2 and solve for x:

$$\cancel{2y} - 2x = 4$$
$$3x - \cancel{2y} = 8$$
$$\overline{3x - 2x = 12 \Rightarrow x = 12}$$

Substitute into Equation M-3 and solve for y:

$$y - 12 = 2 \Rightarrow y = 14$$

PRACTICE PROBLEMS

7. True or false: $xy = 4$ is a linear equation.
8. At time $t = 0.0$ s, the position of a particle moving along the x axis at a constant velocity is $x = 3.0$ m. At $t = 2.0$ s, the position is $x = 12.0$ m. Write a linear equation showing the relation of x to t.
9. Solve the following pair of simultaneous equations for x and y:

$$\frac{5}{4}x + \frac{1}{3}y = 30$$
$$y - 5x = 20$$

M-5 QUADRATIC EQUATIONS AND FACTORING

A **quadratic equation** is an equation of the form $ax^2 + bxy + cy^2 + ex + fy + g = 0$, where x and y are variables and a, b, c, e, f, and g are constants. In each term of the equation the powers of the variables are integers that sum to 2, 1, or 0. The designation *quadratic equation* usually applies to an equation of one variable that can be written in the standard form

$$ax^2 + bx + c = 0 \qquad \text{M-4}$$

where a, b, and c are constants. The quadratic equation has two solutions or **roots**—values of x for which the equation is true.

FACTORING

We can solve some quadratic equations by **factoring.** Very often terms of an equation can be grouped or organized into other terms. When we factor terms, we look for multipliers and multiplicands—which we now call **factors**—that will yield two or more new terms as a product. For example, we can find the roots of the quadratic equation $x^2 - 3x + 2 = 0$ by factoring the left side, to get $(x - 2)(x - 1) = 0$. The roots are $x = 2$ and $x = 1$.

Factoring is useful for simplifying equations and for understanding the relationships between quantities. You should be familiar with the multiplication of the factors $(ax + by)(cx + dy) = acx^2 + (ad + bc)xy + bdy^2$.

You should readily recognize some typical factorable combinations:

1. Common factor: $2ax + 3ay = a(2x + 3y)$
2. Perfect square: $x^2 - 2xy + y^2 = (x - y)^2$ (If the expression on the left side of a quadratic equation in standard form is a perfect square, the two roots will be equal.)
3. Difference of squares: $x^2 - y^2 = (x + y)(x - y)$

Also, look for factors that are prime numbers (2, 5, 7, etc.) because these factors can help you factor and simplify terms quickly. For example, the equation $98x^2 - 140 = 0$ can be simplified because 98 and 140 share the common factor 2. That is, $98x^2 - 140 = 0$ becomes $2(49x^2 - 70) = 0$, so we have $49x^2 - 70 = 0$.

This result can be further simplified because 49 and 70 share the common factor 7. Thus, $49x^2 - 70 = 0$ becomes $7(7x^2 - 10) = 0$, so we have $7x^2 - 10 = 0$.

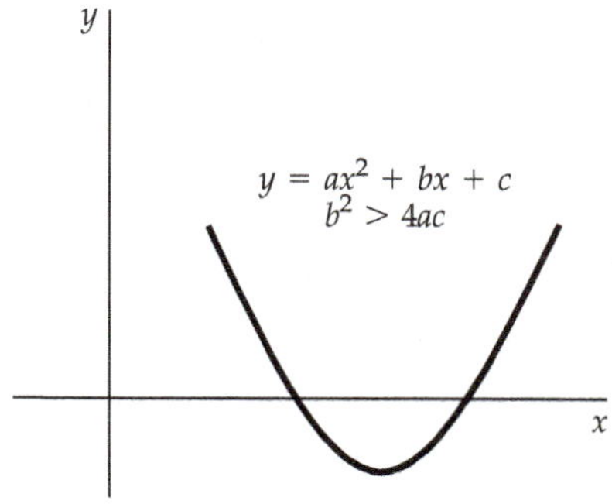

FIGURE M-3 Graph of y versus x when $y = ax^2 + bx + c$ for the case $b^2 > 4ac$. The two values of x for which $y = 0$ satisfy the quadratic equation (Equation M-4).

THE QUADRATIC FORMULA

Not all quadratic equations can be solved by factoring. However, *any* quadratic equation in the standard form $ax^2 + bx + c = 0$ can be solved by the **quadratic formula,**

$$x = \frac{-b \pm \sqrt{b^2 - 4ac}}{2a} = -\frac{b}{2a} \pm \frac{1}{2a}\sqrt{b^2 - 4ac} \qquad \text{M-5}$$

When b^2 is greater than $4ac$, there are two solutions corresponding to the + and − signs. Figure M-3 shows a graph of y versus x where $y = ax^2 + bx + c$. The curve, a **parabola,** crosses the x axis twice. (The simplest representation of a parabola in (x, y) coordinates is an equation of the form $y = ax^2 + bx + c$.) The two roots of this equation are the values for which $y = 0$; that is, they are the x *intercepts.*

When b^2 is less than $4ac$, the graph of y versus x does not intersect the x axis, as is shown in Figure M-4; there are still two roots, but they are not real numbers (see the discussion of complex numbers beginning on page M-19). When $b^2 = 4ac$, the graph of y versus x is tangent to the x axis at the point $x = -b/2a$; the two roots are each equal to $-b/2a$.

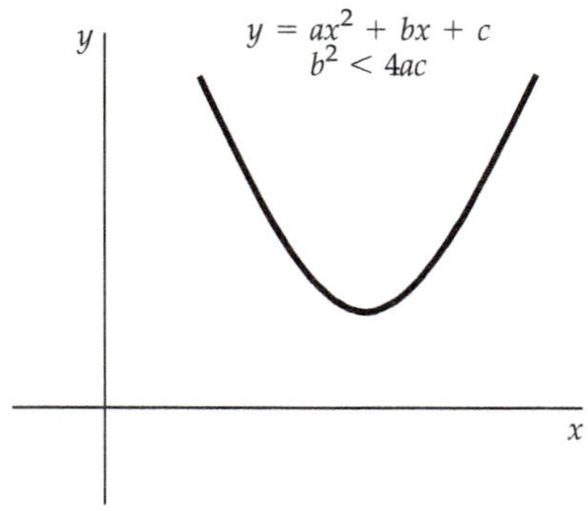

FIGURE M-4 Graph of y versus x when $y = ax^2 + bx + c$ for the case $b^2 < 4ac$. In this case, there are no real values for x for which $y = 0$.

Example M-5 Factoring a Second-Degree Polynomial

Factor the expression $6x^2 + 19xy + 10y^2$.

PICTURE We examine the coefficients of the terms to see whether the expression can be factored without resorting to more advanced methods. Remember that the multiplication $(ax + by)(cx + dy) = acx^2 + (ad + bc)xy + bdy^2$.

SOLVE

1. The coefficient of x^2 is 6 which can be factored two ways: $ac = 6$

   $3 \cdot 2 = 6$ or $6 \cdot 1 = 6$

2. The coefficient of y^2 is 10 which can also be factored two ways: $bd = 10$

   $5 \cdot 2 = 10$ or $10 \cdot 1 = 10$

3. List the possibilities for a, b, c, and d in a table. Include a column for $ad + bc$.

   If $a = 3$, then $c = 2$, and vice versa. In addition, if $a = 6$, then $c = 1$, and vice versa. For each value of a there are four values for b.

a	b	c	d	$ad + bc$
3	5	2	2	16
3	2	2	5	19
3	10	2	1	23
3	1	2	10	32
2	5	3	2	19
2	2	3	5	16
2	10	3	1	32
2	1	3	10	23
6	5	1	2	17
6	2	1	5	32
6	10	1	1	16
6	1	1	10	61
1	5	6	2	32
1	2	6	5	17
1	10	6	1	61
1	1	6	10	16

4. Find a combination such that $ad + bc = 19$. As you can see from the table there are two such combinations, and each gives the same results:

$$ad + bc = 19$$
$$3 \cdot 5 + 2 \cdot 2 = 19$$

5. Use the combination in the second row of the table to factor the expression in question:

$$6x^2 + 19xy + 10y^2 = (3x + 2y)(2x + 5y)$$

CHECK As a check, expand $(3x + 2y)(2x + 5y)$.

$$(3x + 2y)(2x + 5y) = 6x^2 + 15xy + 4xy + 10y^2 = 6x^2 + 19xy + 10y^2$$

The combination in the fifth row of the table also gives the step-4 result.

PRACTICE PROBLEMS

10. Show that the combination in the fifth row of the table also gives the step-4 result.
11. Factor $2x^2 - 4xy + 2y^2$.
12. Factor $2x^4 + 10x^3 + 12x^2$.

M-6 EXPONENTS AND LOGARITHMS

EXPONENTS

The notation x^n stands for the quantity obtained by multiplying x times itself n times. For example, $x^2 = x \cdot x$ and $x^3 = x \cdot x \cdot x$. The quantity n is called the **power,** or the **exponent,** of x (the **base**). Listed below are some rules that will help you simplify terms that have exponents.

1. When two powers of x are multiplied, the exponents are added:

$$(x^m)(x^n) = x^{m+n} \qquad \text{M-6}$$

   Example: $x^2x^3 = x^{2+3} = (x \cdot x)(x \cdot x \cdot x) = x^5$.

2. Any number (except 0) raised to the 0 power is defined to be 1:

$$x^0 = 1 \qquad \text{M-7}$$

3. Based on rule 2,

$$x^n x^{-n} = x^0 = 1$$

$$x^{-n} = \frac{1}{x^n} \qquad \text{M-8}$$

4. When two powers are divided, the exponents are subtracted:

$$\frac{x^n}{x^m} = x^n x^{-m} = x^{n-m} \qquad \text{M-9}$$

5. When a power is raised to another power, the exponents are multiplied:

$$(x^n)^m = x^{nm} \qquad \text{M-10}$$

6. When exponents are written as fractions, they represent the roots of the base. For example,

$$x^{1/2} \cdot x^{1/2} = x$$

so

$$x^{1/2} = \sqrt{x} \qquad (x > 0)$$

Example M-6 Simplifying a Quantity That Has Exponents

Simplify $\frac{x^4 x^7}{x^8}$.

PICTURE According to rule 1, when two powers of x are multiplied, the exponents are added. Rule 4 states that when two powers are divided, the exponents are subtracted.

SOLVE

1. Simplify the numerator $x^4\,x^7$ using rule 1. $\qquad x^4\,x^7 = x^{4+7} = x^{11}$

2. Simplify $\frac{x^{11}}{x^8}$ using rule 4: $\qquad \frac{x^{11}}{x^8} = x^{11}\,x^{-8} = x^{11-8} = x^3$

CHECK Use the value $x = 2$ to determine if your answer is correct.

$$\frac{2^4 2^7}{2^8} = 2^3 = 8$$

$$\frac{2^4 2^7}{2^8} = \frac{(16)(128)}{256} = \frac{2048}{256} = 8$$

PRACTICE PROBLEMS

13. $(x^{1/18})^9 =$
14. $x^6 x^0 =$

LOGARITHMS

Any positive number can be expressed as some power of any other positive number except one. If y is related to x by $y = a^x$, then the number x is said to be the **logarithm** of y to the **base** a, and the relation is written

$$x = \log_a y$$

Thus, logarithms are *exponents*, and the rules for working with logarithms correspond to similar laws for exponents. Listed below are some rules that will help you simplify terms that have logarithms.

1. If $y_1 = a^n$ and $y_2 = a^m$, then

$$y_1 y_2 = a^n a^m = a^{n+m}$$

Correspondingly,

$$\log_a y_1 y_2 = \log_a a^{n+m} = n + m = \log_a a^n + \log_a a^m = \log_a y_1 + \log_a y_2 \qquad \text{M-11}$$

It then follows that

$$\log_a y^n = n \log_a y \qquad \text{M-12}$$

2. Because $a^1 = a$ and $a^0 = 1$,

$$\log_a a = 1 \qquad \text{M-13}$$

and

$$\log_a 1 = 0 \qquad \text{M-14}$$

There are two bases in common use: logarithms to base 10 are called **common logarithms,** and logarithms to base e (where $e = 2.718\ldots$) are called **natural logarithms.**

In this text, the symbol ln is used for natural logarithms and the symbol log, without a subscript, is used for common logarithms. Thus,

$$\log_e x = \ln x \quad \text{and} \quad \log_{10} x = \log x \qquad \text{M-15}$$

and $y = \ln x$ implies

$$x = e^y \qquad \text{M-16}$$

Logarithms can be changed from one base to another. Suppose that

$$z = \log x \qquad \text{M-17}$$

Then

$$10^z = 10^{\log x} = x \qquad \text{M-18}$$

Taking the natural logarithm of both sides of Equation M-18, we obtain

$$z \ln 10 = \ln x$$

Substituting $\log x$ for z (see Equation M-17) gives

$$\ln x = (\ln 10)\log x \qquad \text{M-19}$$

Example M-7 Converting Between Common Logarithms and Natural Logarithms

The steps leading to Equation M-19 show that, in general, $\log_b x = (\log_b a)\log_a x$, and thus that conversion of logarithms from one base to another requires only multiplication by a constant. Describe the mathematical relation between the constant for converting common logarithms to natural logarithms and the constant for converting natural logarithms to common logarithms.

PICTURE We have a general mathematical formula for converting logarithms from one base to another. We look for the mathematical relation by exchanging a for b and vice versa in the formula.

SOLVE

1. You have a formula for converting logarithms from base a to base b: $\log_b x = (\log_b a)\log_a x$
2. To convert from base b to base a, exchange all a for b and vice versa: $\log_a x = (\log_a b)\log_b x$
3. Divide both sides of the equation in step 1 by $\log_a x$: $\dfrac{\log_b x}{\log_a x} = \log_b a$
4. Divide both sides of the equation in step 2 by $(\log_a b)\log_a x$: $\dfrac{1}{\log_a b} = \dfrac{\log_b x}{\log_a x}$
5. The results show that the conversion factors $\log_b a$ and $\log_a b$ are reciprocals of one other: $\dfrac{1}{\log_a b} = \log_b a$

CHECK For the value of $\log_{10} e$, your calculator will give 0.43429. For ln 10, your calculator will give 2.3026. Multiply 0.43429 by 2.3026; you will get 1.0000.

PRACTICE PROBLEMS

15. Evaluate $\log_{10}1000$.
16. Evaluate $\log_2 5$.

M-7 GEOMETRY

The properties of the most common **geometric figures**—bounded shapes in two or three dimensions whose lengths, areas, or volumes are governed by specific ratios—are a basic analytical tool in physics. For example, the characteristic ratios within triangles give us the laws of *trigonometry* (see the next section of this tutorial), which in turn give us the theory of vectors, essential in analyzing motion in two or more dimensions. Circles and spheres are essential for understanding, among other concepts, angular momentum and the probability densities of quantum mechanics.

BASIC FORMULAS IN GEOMETRY

Circle The ratio of the circumference of a circle to its diameter is a number π, which has the approximate value

$$\pi = 3.141\,592$$

The circumference C of a circle is thus related to its diameter d and its radius r by

$$C = \pi d = 2\pi r \qquad \text{circumference of circle} \qquad \text{M-20}$$

The area of a circle is (Figure M-5)

$$A = \pi r^2 \qquad \text{area of circle} \qquad \text{M-21}$$

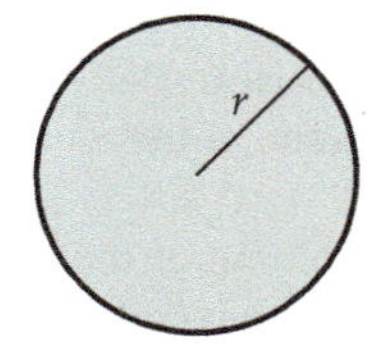

FIGURE M-5 Area of a circle.

Parallelogram The area of a parallelogram is the base b times the height h (Figure M-6):

$$A = bh$$

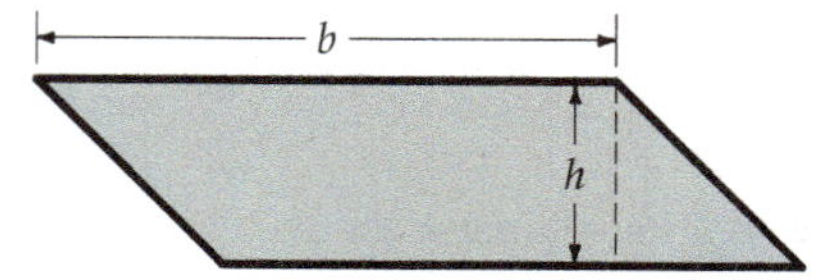

FIGURE M-6 Area of a parallelogram.

Triangle The area of a triangle is one-half the base times the height h (Figure M-7)

$$A = \frac{1}{2}bh$$

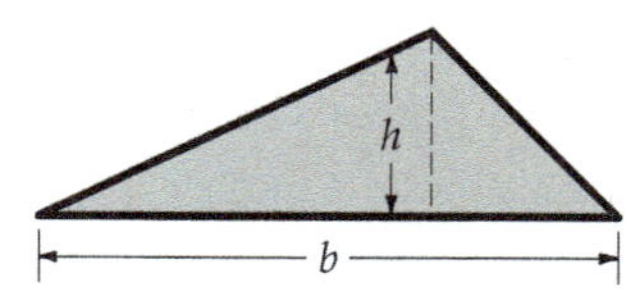

FIGURE M-7 Area of a triangle.

Sphere A sphere of radius r (Figure M-8) has a surface area given by

$$A = 4\pi r^2 \qquad \text{surface area of sphere} \qquad \text{M-22}$$

and a volume given by

$$V = \frac{4}{3}\pi r^3 \qquad \text{volume of sphere} \qquad \text{M-23}$$

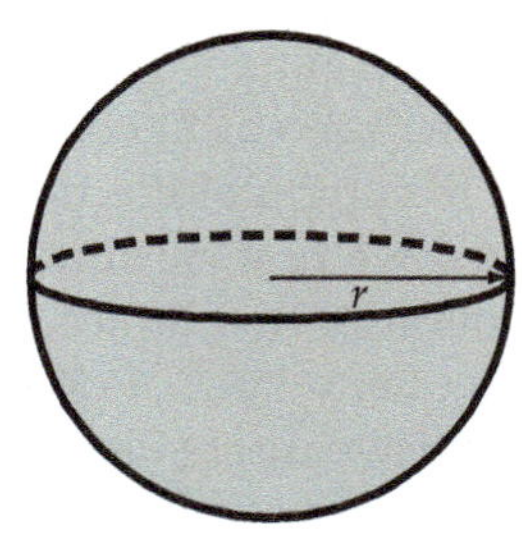

FIGURE M-8 Surface area and volume of a sphere.

Cylinder A cylinder of radius r and length L (Figure M-9) has surface area (not including the end faces) of

$$A = 2\pi rL \qquad \text{surface of cylinder} \qquad \text{M-24}$$

and volume of

$$V = \pi r^2 L \qquad \text{volume of cylinder} \qquad \text{M-25}$$

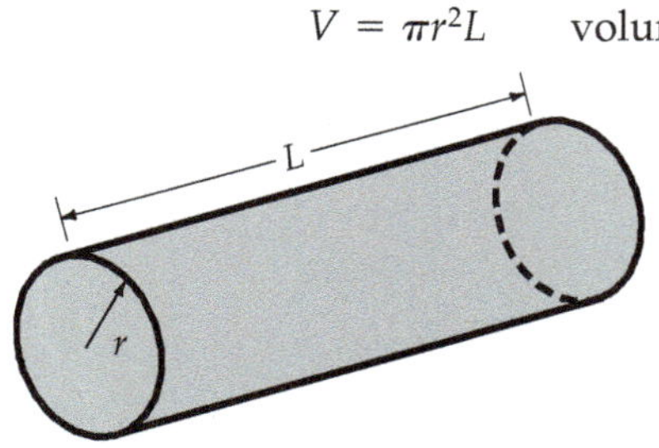

FIGURE M-9 Surface area (not including the end faces) and the volume of a cylinder.

Example M-8 Calculating the volume of a Spherical Shell

An aluminum spherical shell has an outer diameter of 40.0 cm and an inner diameter of 39.0 cm. Find the volume of the aluminum in this shell.

PICTURE The volume of the aluminum in the spherical shell is the volume that remains when we subtract the volume of the inner sphere having $d_i = 2r_i = 39.0$ cm from the volume of the outer sphere having $d_o = 2r_o = 40.0$ cm.

SOLVE

1. Subtract the volume of the sphere of radius r_i from the volume of the sphere of radius r_o: $V = \frac{4}{3}\pi r_o^3 - \frac{4}{3}\pi r_i^3 = \frac{4}{3}\pi(r_o^3 - r_i^3)$
2. Substitute 20.0 cm for r_o and 19.5 cm for r_i: $V = \frac{4}{3}\pi[(20.0\text{ cm})^3 - (19.5\text{ cm})^3] = \boxed{2.45 \times 10^3\text{ cm}^3}$

CHECK The volume of the shell is expected to be the same order of magnitude as the volume of a hollow cube with an outside edge length of 40.0 cm and an inside edge length of 39.0 cm. The volume of such a hollow cube is $(40.0\text{ cm})^3 - (39.0\text{ cm})^3 = 4.68 \times 10^3\text{ cm}^3$. The step-2 result meets the expectation that the volume of the shell is the same order of magnitude as the volume of the hollow cube.

PRACTICE PROBLEMS

17. Find the ratio between the volume V and the surface A of a sphere of radius r.
18. What is the area of a cylinder that has a radius that is 1/3 its length?

M-8 TRIGONOMETRY

Trigonometry, which gets its name from Greek roots meaning "triangle" and "measure," is the study of some important mathematical functions, called **trigonometric functions.** These functions are most simply defined as ratios of the sides of right triangles. However, these right-triangle definitions are of limited use because they are valid only for angles between zero and 90°. However, the validity of the right-triangle definitions can be extended by defining the trigonometric functions in terms of the ratio of the coordinates of points on a circle of unit radius drawn centered at the origin of the xy plane.

In physics, we first encounter trigonometric functions when we use vectors to analyze motion in two dimensions. Trigonometric functions are also essential in the analysis of any kind of periodic behavior, such as circular motion, oscillatory motion, and wave mechanics.

ANGLES AND THEIR MEASURE: DEGREES AND RADIANS

The size of an angle formed by two intersecting straight lines is known as its **measure.** The standard way of finding the measure of an angle is to place the angle so that its **vertex,** or point of intersection of the two lines that form the angle, is at the center of a circle located at the origin of a graph that has Cartesian coordinates and one of the lines extends rightward on the positive x axis. The distance traveled *counterclockwise* on the circumference from the positive x axis to reach the intersection of the circumference with the other line defines the measure of the angle. (Traveling clockwise to the second line would simply give us a negative measure; to illustrate basic concepts, we position the angle so that the smaller rotation will be in the counterclockwise direction.)

The most familiar unit for expressing the measure of an angle is the **degree,** which equals 1/360 of the full distance around the circumference of the circle. For greater precision, or for smaller angles, we either show degrees plus minutes (′)

and seconds ("), with $1' = 1°/60$ and $1'' = 1'/60 = 1°/3600$; or show degrees as an ordinary decimal number.

For scientific work, a more useful measure of an angle is the **radian** (rad). Again, place the angle with its vertex at the center of a circle and measure counterclockwise rotation around the circumference. The measure of the angle in radians is then defined as the length of the circular arc from one line to the other divided by the radius of the circle (Figure M-10). If s is the arc length and r is the radius of the circle, the angle θ measured in radians is

$$\theta = \frac{s}{r} \tag{M-26}$$

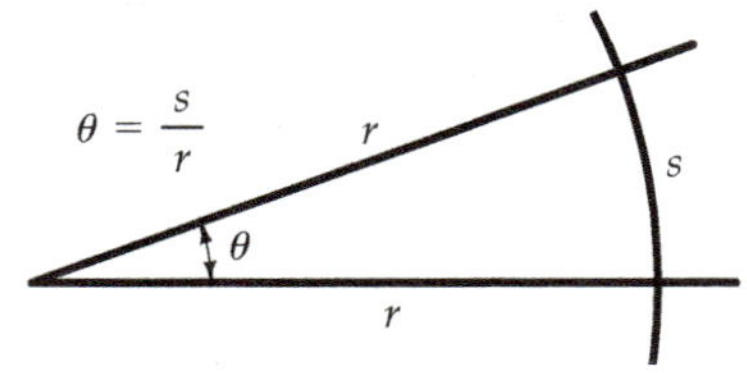

FIGURE M-10 The angle θ in radians is defined to be the ratio s/r, where s is the arc length intercepted on a circle of radius r.

Because the angle measured in radians is the ratio of two lengths, it is dimensionless. The relation between radians and degrees is

$$360° = 2\pi \text{ rad}$$

or

$$1 \text{ rad} = \frac{360°}{2\pi} = 57.3°$$

Figure M-11 shows some useful relations for angles.

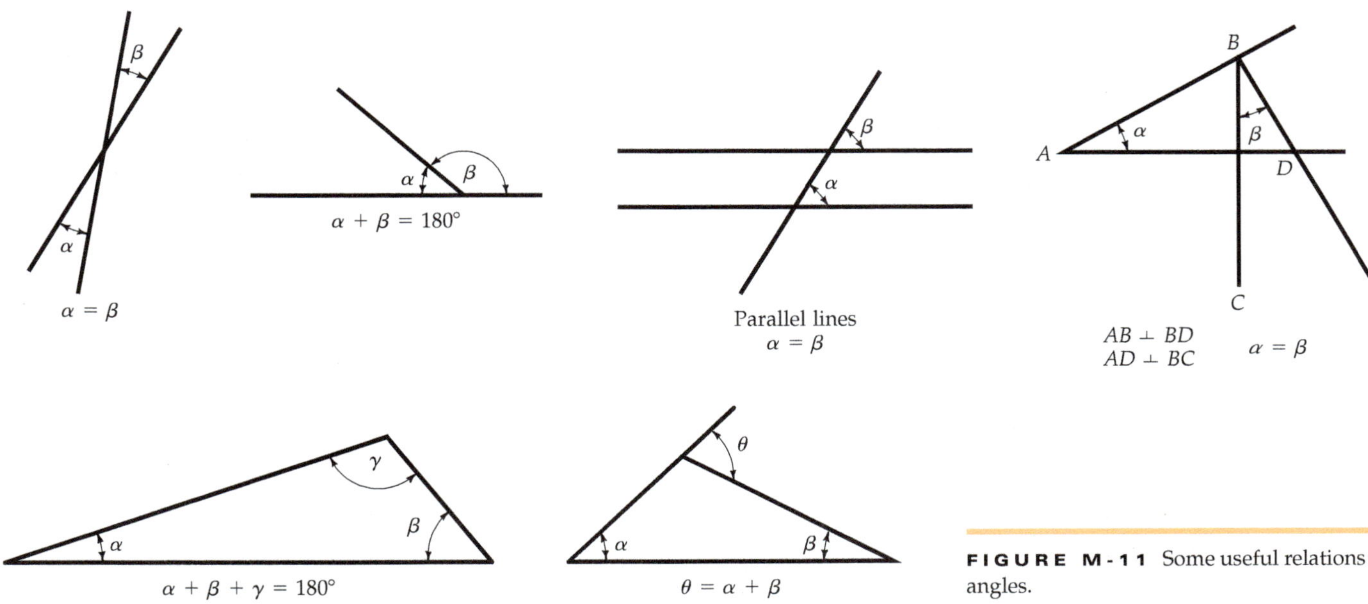

FIGURE M-11 Some useful relations for angles.

THE TRIGONOMETRIC FUNCTIONS

Figure M-12 shows a right triangle formed by drawing the line BC perpendicular to AC. The lengths of the sides are labeled a, b, and c. The right-triangle definitions of the trigonometric functions $\sin\theta$ (the **sine**), $\cos\theta$ (the **cosine**), and $\tan\theta$ (the **tangent**) for an acute angle θ are

$$\sin\theta = \frac{a}{c} = \frac{\text{Opposite side}}{\text{Hypotenuse}} \tag{M-27}$$

$$\cos\theta = \frac{b}{c} = \frac{\text{Adjacent side}}{\text{Hypotenuse}} \tag{M-28}$$

$$\tan\theta = \frac{a}{b} = \frac{\text{Opposite side}}{\text{Adjacent side}} = \frac{\sin\theta}{\cos\theta} \tag{M-29}$$

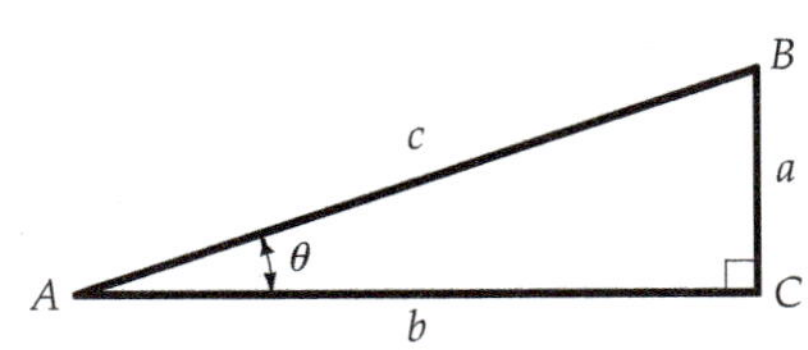

FIGURE M-12 A right triangle with sides of length a and b and a hypotenuse of length c.

(**Acute angles** are angles whose positive rotation around the circumference of a circle measures less than 90°, or $\pi/2$.) Three other trigonometric functions—the

secant (sec), the **cosecant** (csc), and the **cotangent** (cot), defined as the reciprocals of these functions—are

$$\sec\theta = \frac{c}{b} = \frac{1}{\cos\theta} \qquad \text{M-30}$$

$$\csc\theta = \frac{c}{a} = \frac{1}{\sin\theta} \qquad \text{M-31}$$

$$\cot\theta = \frac{b}{a} = \frac{1}{\tan\theta} = \frac{\cos\theta}{\sin\theta} \qquad \text{M-32}$$

The angle θ, whose sine is x, is called the arcsine of x, and is written $\sin^{-1} x$. That is, if

$$\sin\theta = x$$

then

$$\theta = \arcsin x = \sin^{-1} x \qquad \text{M-33}$$

The arcsine is the inverse of the sine. The inverse of the cosine and tangent are defined similarly. The angle whose cosine is y is the arccosine of y. That is, if

$$\cos\theta = y$$

then

$$\theta = \arccos y = \cos^{-1} y \qquad \text{M-34}$$

The angle whose tangent is z is the arctangent of z. That is, if

$$\tan\theta = z$$

then

$$\theta = \arctan z = \tan^{-1} z \qquad \text{M-35}$$

TRIGONOMETRIC IDENTITIES

We can derive several useful formulas, called **trigonometric identities,** by examining relationships between the trigonometric functions. Equations M-30 through M-32 list three of the most obvious identities, formulas expressing some trigonometric functions as reciprocals of others. Almost as easy to discern are identities derived from the **Pythagorean theorem,**

$$a^2 + b^2 = c^2 \qquad \text{M-36}$$

(Figure M-13 illustrates a graphic proof of the theorem.) Simple algebraic manipulation of Equation M-36 gives us three more identities. First, if we divide each term in Equation M-36 by c^2, we obtain

$$\frac{a^2}{c^2} + \frac{b^2}{c^2} = 1$$

or, from the definitions of $\sin\theta$ (which is a/c) and $\cos\theta$ (which is b/c)

$$\sin^2\theta + \cos^2\theta = 1 \qquad \text{M-37}$$

Similarly, we can divide each term in Equation M-36 by a^2 or b^2 and obtain

$$1 + \cot^2\theta = \csc^2\theta \qquad \text{M-38}$$

and

$$1 + \tan^2\theta = \sec^2\theta \qquad \text{M-39}$$

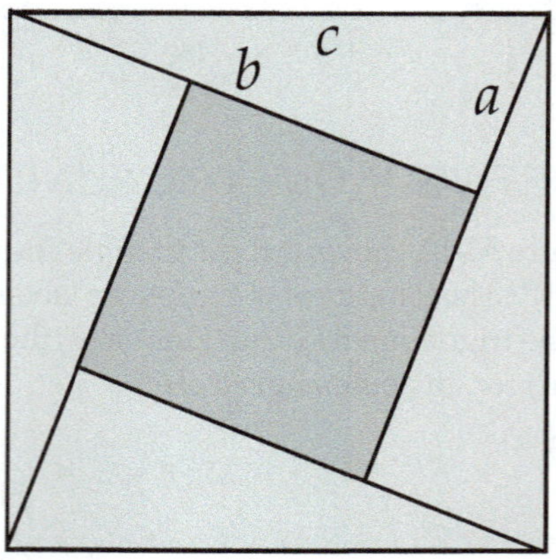

FIGURE M-13 When this figure was first published, the letters were absent and it was accompanied by the single word "Behold!" Using the drawing, establish the Pythagorean theorem ($a^2 + b^2 = c^2$).

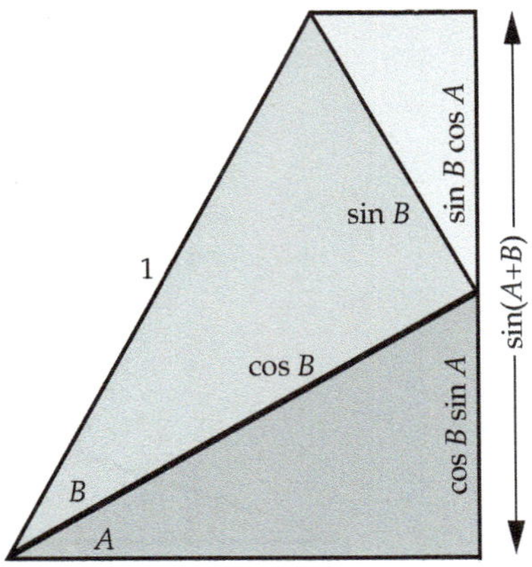

FIGURE M-14 Using this drawing, establish the identity $\sin(A + B) = \sin A \cos B + \cos A \sin B$. You can also use it to establish the identity $\cos(A + B) = \cos A \cos B - \sin A \sin B$. Try it.

Table M-2 Trigonometric Identities

$$\sin(A \pm B) = \sin A \cos B \pm \cos A \sin B$$

$$\cos(A \pm B) = \cos A \cos B \mp \sin A \sin B$$

$$\tan(A \pm B) = \frac{\tan A \pm \tan B}{1 \mp \tan A \tan B}$$

$$\sin A \pm \sin B = 2 \sin\left[\frac{1}{2}(A \pm B)\right]\cos\left[\frac{1}{2}(A \mp B)\right]$$

$$\cos A + \cos B = 2 \cos\left[\frac{1}{2}(A + B)\right]\cos\left[\frac{1}{2}(A - B)\right]$$

$$\cos A - \cos B = 2 \sin\left[\frac{1}{2}(A + B)\right]\sin\left[\frac{1}{2}(B - A)\right]$$

$$\tan A \pm \tan B = \frac{\sin(A \pm B)}{\cos A \cos B}$$

$$\sin^2\theta + \cos^2\theta = 1;\ \sec^2\theta - \tan^2\theta = 1;\ \csc^2\theta - \cot^2\theta = 1$$

$$\sin 2\theta = 2 \sin\theta \cos\theta$$

$$\cos 2\theta = \cos^2\theta - \sin^2\theta = 2\cos^2\theta - 1 = 1 - 2\sin^2\theta$$

$$\tan 2\theta = \frac{2\tan\theta}{1 - \tan^2\theta}$$

$$\sin\frac{1}{2}\theta = \pm\sqrt{\frac{1 - \cos\theta}{2}};\ \cos\frac{1}{2}\theta = \pm\sqrt{\frac{1 + \cos\theta}{2}};\ \tan\frac{1}{2}\theta = \pm\sqrt{\frac{1 - \cos\theta}{1 + \cos\theta}}$$

Table M-2 lists these last three and many more trigonometric identities. Notice that they fall into four categories: functions of sums or differences of angles, sums or differences of squared functions, functions of double angles (2θ), and functions of half angles ($\frac{1}{2}\theta$). Notice that some of the formulas contain paired alternatives, expressed with the signs $\pm$ and $\mp$; in such formulas, remember to always apply the formula with either all the "upper" or all the "lower" alternatives. Figure M-14 shows a graphic proof of the first two sum-of-angle identities.

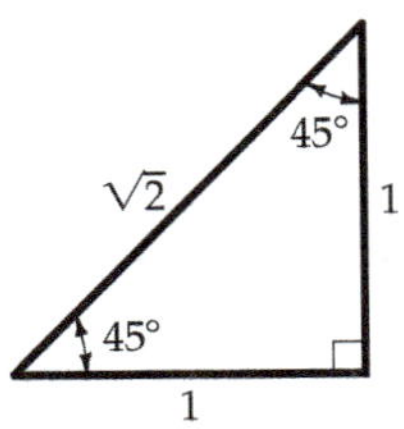

FIGURE M-15 An isosceles right triangle.

SOME IMPORTANT VALUES OF THE FUNCTIONS

Figure M-15 is a diagram of an *isosceles* right triangle (an isosceles triangle is a triangle with two equal sides), from which we can find the sine, cosine, and tangent of 45°. The two acute angles of this triangle are equal. Because the sum of the three angles in a triangle must equal 180° and the right angle is 90°, each acute angle must be 45°. For convenience, let us assume that the equal sides each have a length of 1 unit. The Pythagorean theorem gives us a value for the hypotenuse of

$$c = \sqrt{a^2 + b^2} = \sqrt{1^2 + 1^2} = \sqrt{2} \text{ units}$$

We calculate the values of the functions as follows:

$$\sin 45° = \frac{a}{c} = \frac{1}{\sqrt{2}} = 0.707 \quad \cos 45° = \frac{b}{c} = \frac{1}{\sqrt{2}} = 0.707 \quad \tan 45° = \frac{a}{b} = \frac{1}{1} = 1$$

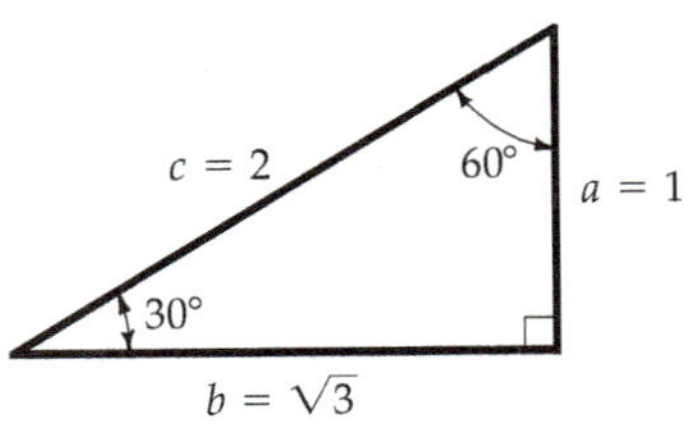

FIGURE M-16 A 30°–60° right triangle.

Another common triangle, a 30°–60° right triangle, is shown in Figure M-16. Because this particular right triangle is in effect half of an *equilateral triangle* (a 60°–60°–60° triangle, or a triangle having three equal sides and three equal angles), we can see that the sine of 30° must be exactly 0.5 (Figure M-17). The equilateral triangle must have all sides equal to c, the hypotenuse of the 30°–60° right triangle. Thus, side a is one-half the length of the hypotenuse, and so

$$\sin 30° = \frac{1}{2}$$

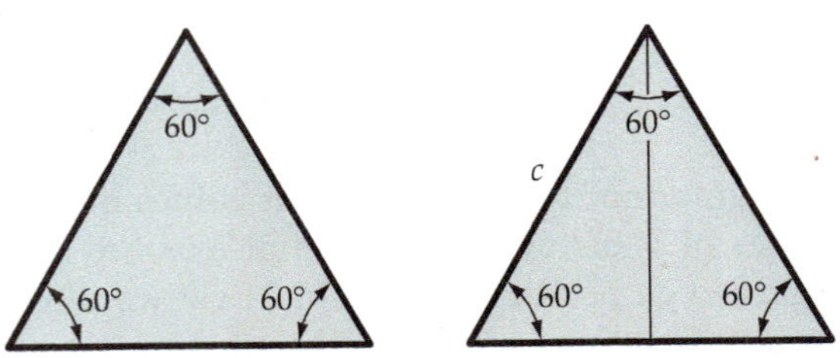

FIGURE M-17 (a) An equilateral triangle. (b) An equilateral triangle that has been bisected to form two 30°–60° right triangles.

To find the other ratios within the 30–60° right triangle, let us assign a value of 1 to the side opposite the 30° angle. Then

$$c = \frac{1}{0.5} = 2 \qquad b = \sqrt{c^2 - a^2} = \sqrt{2^2 - 1^2} = \sqrt{3}$$

$$\cos 30° = \frac{b}{c} = \frac{\sqrt{3}}{2} = 0.866 \qquad \tan 30° = \frac{a}{b} = \frac{1}{\sqrt{3}} = 0.577$$

$$\sin 60° = \frac{b}{c} = \cos 30° = 0.866 \qquad \cos 60° = \frac{a}{c} = \sin 30° = \frac{1}{2}$$

$$\tan 60° = \frac{b}{a} = \frac{\sqrt{3}}{1} = 1.732$$

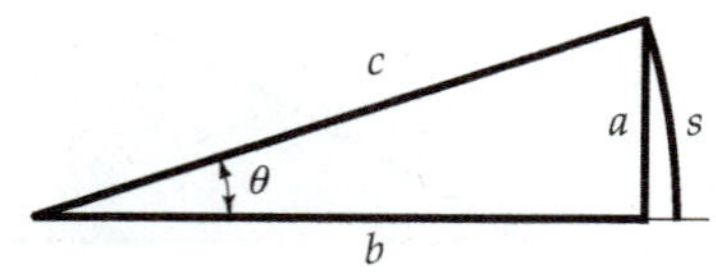

FIGURE M-18 For small angles, $\sin\theta = a/c$, $\tan\theta = a/b$, and the angle $\theta = s/c$ are all approximately equal.

SMALL-ANGLE APPROXIMATION

For small angles, the length a is nearly equal to the arc length s, as can be seen in Figure M-18. The angle $\theta = s/c$ is therefore nearly equal to $\sin\theta = a/c$:

$$\sin\theta \approx \theta \qquad \text{for small values of } \theta \qquad \text{M-40}$$

Similarly, the lengths c and b are nearly equal, so $\tan\theta = a/b$ is nearly equal to both θ and $\sin\theta$ for small values of θ:

$$\tan\theta \approx \sin\theta \approx \theta \qquad \text{for small values of } \theta \qquad \text{M-41}$$

Equations M-40 and M-41 hold only if θ is measured in radians. Because $\cos\theta = b/c$, and because these lengths are nearly equal for small values of θ, we have

$$\cos\theta \approx 1 \qquad \text{for small values of } \theta \qquad \text{M-42}$$

Figure M-19 shows graphs of θ, $\sin\theta$, and $\tan\theta$ versus θ for small values of θ. If accuracy of a few percent is needed, small-angle approximations can be used only for angles of about a quarter of a radian (or about 15°) or less. Below this value, as the angle becomes smaller, the approximation $\theta \approx \sin\theta \approx \tan\theta$ is even more accurate.

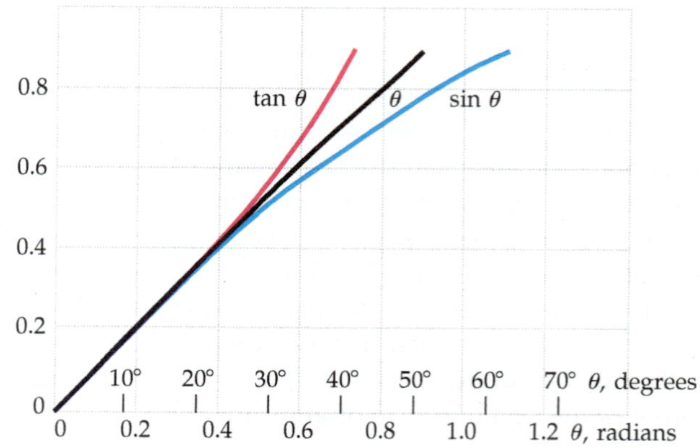

FIGURE M-19 Graphs of $\tan\theta$, θ, and $\sin\theta$ versus θ for small values of θ.

TRIGONOMETRIC FUNCTIONS AS FUNCTIONS OF REAL NUMBERS

So far we have illustrated the trigonometric functions as properties of angles. Figure M-20 shows an *obtuse* angle with its vertex at the origin and one side along the x axis. The trigonometric functions for a "general" angle such as this are defined by

$$\sin\theta = \frac{y}{c} \qquad \text{M-43}$$

$$\cos\theta = \frac{x}{c} \qquad \text{M-44}$$

$$\tan\theta = \frac{y}{x} \qquad \text{M-45}$$

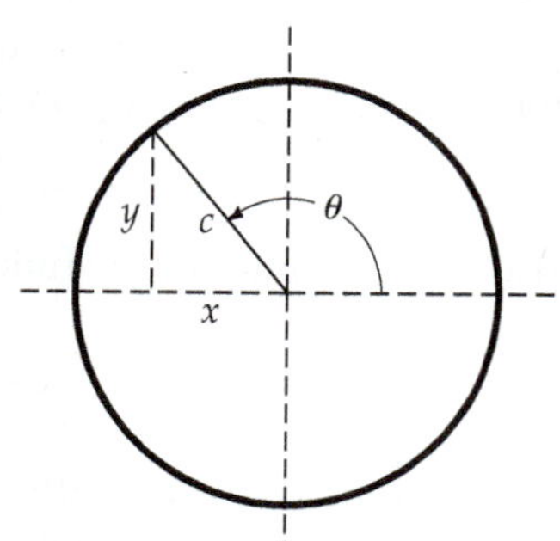

FIGURE M-20 Diagram for defining the trigonometric functions for an obtuse angle.

It is important to remember that values of x to the left of the vertical axis and values of y below the horizontal axis are negative; c in the figure is always regarded as positive. Figure M-21 shows plots of the general sine, cosine, and tangent functions versus θ. The sine function has a period of 2π rad. Thus, for any value of θ, $\sin(\theta + 2\pi) = \sin\theta$, and so forth. That is, when an angle changes by 2π rad, the function returns to its original value. The tangent function has a

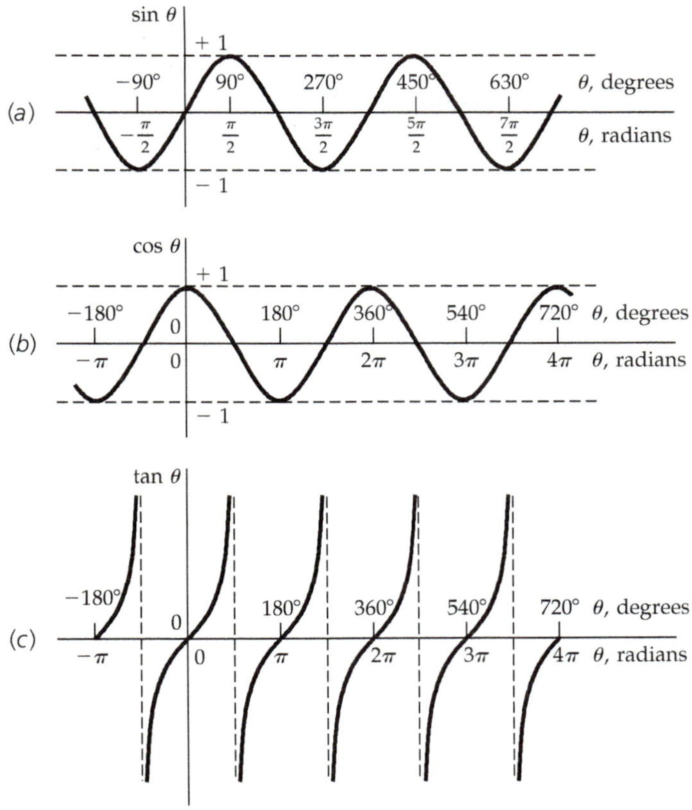

FIGURE M-21 The trigonometric functions $\sin\theta$, $\cos\theta$, and $\tan\theta$ versus θ.

period of π rad. Thus, $\tan(\theta + \pi) = \tan\theta$, and so forth. Some other useful relations are

$$\sin(\pi - \theta) = \sin\theta \qquad \text{M-46}$$

$$\cos(\pi - \theta) = -\cos\theta \qquad \text{M-47}$$

$$\sin(\tfrac{1}{2}\pi - \theta) = \cos\theta \qquad \text{M-48}$$

$$\cos(\tfrac{1}{2}\pi - \theta) = \sin\theta \qquad \text{M-49}$$

Because the radian is dimensionless, it is not hard to see from the plots in Figure M-21 that the trigonometric functions are functions of all real numbers. The functions can also be expressed as power series in θ. The series for $\sin\theta$ and $\cos\theta$ are

$$\sin\theta = \theta - \frac{\theta^3}{3!} + \frac{\theta^5}{5!} - \frac{\theta^7}{7!} + \cdots \qquad \text{M-50}$$

$$\cos\theta = 1 - \frac{\theta^2}{2!} + \frac{\theta^4}{4!} - \frac{\theta^6}{6!} + \cdots \qquad \text{M-51}$$

When θ is small, good approximations are obtained using only the first few terms in the series.

Example M-9 Cosine of a Sum

Using the suitable trigonometric identity from Table M-2, find $\cos(135° + 22°)$. Give your answer in four significant figures.

PICTURE As long as all angles are given in degrees, there is no need to convert to radians, because all operations are numerical values of the functions. Be sure, however, that your calculator is in degree mode. The suitable identity is $\cos(A \pm B) = \cos A \cos B \mp \sin A \sin B$, where the upper signs are appropriate.

SOLVE

1. Write the trigonometric identity for the cosine of a sum, with $A = 135°$ and $B = 22°$:

$$\cos(135° + 22°) = (\cos 135°)(\cos 22°) - (\sin 135°)(\sin 22°)$$

2. Using a calculator, find cos 135°, sin 135°, cos 22°, and sin 22°:

$\cos 135° = -0.7071$ $\quad\sin 135° = 0.7071$

$\cos 22° = 0.9272$ $\quad\sin 22° = 0.3746$

3. Enter the values in the formula and calculate the answer:

$$\cos(135° + 22°) = (-0.7071)(0.9272) - (0.7071)(0.3746) = -0.9205$$

CHECK The calculator shows that the $\cos(135° + 22°) = \cos(157°) = -0.9205$.

PRACTICE PROBLEMS

19. Find $\sin\theta$ and $\cos\theta$ for the right triangle shown in Figure M-12 in which $a = 4$ cm and $b = 7$ cm. What is the value for θ?
20. Find $\sin\theta$ where $\theta = 8.2°$. Is your answer consistent with the small-angle approximation?

M-9 THE BINOMIAL EXPANSION

A **binomial** is an expression consisting of two terms joined by a plus sign or a minus sign. The **binomial theorem** states that a binomial raised to a power can be written, or *expanded*, as a series of terms. If we raise the binomial $(1 + x)$ to a power n, the binomial theorem takes the form

$$(1 + x)^n = 1 + nx + \frac{n(n-1)}{2!}x^2 + \frac{n(n-1)(n-2)}{3!}x^3 + \cdots \qquad \text{M-52}$$

The series is valid for any value of n if $|x|$ is less than 1. The binomial expansion is very useful for approximating algebraic expressions, because when $|x| < 1$, the higher-order terms in the sum are small. (The order of a term is the power of x in the term. Thus, the terms explicitly shown in Equation M-52 are of order 0, 1, 2, and 3.) The series is particularly useful in situations where $|x|$ is small compared with 1; then each term is *much* smaller than the previous term and we can drop all but the first two or three terms in the expansion. If $|x|$ is much less than 1, we have

$$(1 + x)^n \approx 1 + nx, \qquad |x| \ll 1 \qquad \text{M-53}$$

The binomial expansion is used in deriving many formulas of calculus that are important in physics. A well-known use in physics of the approximation in Equation M-53 is the proof that relativistic kinetic energy reduces to the classic formula when the velocity of a particle is very small compared with the velocity of light c.

Example M-10 Using the Binomial Expansion to Find a Power of a Number

Use Equation M-53 to find an approximate value for the square root of 101.

PICTURE The number 101 readily suggests a binomial, namely, $(100 + 1)$. To approximate the answer using the binomial expansion, we must manipulate the expression to get a binomial consisting of 1 and a term less than 1.

SOLVE

1. Write $(101)^{1/2}$ to give an expression $(1 + x)^n$ in which x is much less than 1:

$$(101)^{1/2} = (100 + 1)^{1/2} = (100)^{1/2}(1 + 0.01)^{1/2} = 10(1 + 0.01)^{1/2}$$

2. Use Equation M-53 with $n = \frac{1}{2}$ and $x = 0.01$ to expand $(1 + 0.01)^{1/2}$:

$$(1 + 0.01)^{1/2} = 1 + \tfrac{1}{2}(0.01) + \frac{\frac{1}{2}\left(-\frac{1}{2}\right)}{2}(0.01)^2 + \cdots$$

3. Because $|x| \ll 1$, we expect the magnitude of terms of order 2 and higher to be significantly smaller than the magnitude of the first-order term. Approximate the binomial (1) by keeping only the zeroth and first-order terms, and (2) by keeping only the first 3 terms:

Keeping only the zeroth and first-order terms gives

$$(1 + 0.01)^{1/2} \approx 1 + \tfrac{1}{2}(0.01) = 1 + 0.005\,000\,0 = 1.005\,000\,0$$

Keeping only the zeroth, first-, and second-order terms gives

$$(1 + 0.01)^{1/2} \approx 1 + \tfrac{1}{2}(0.01) + \frac{\frac{1}{2}(-\frac{1}{2})}{2}(0.01)^2 \approx 1 + 0.005\,000\,0 - 0.000\,012\,5 = 1.004\,987\,5$$

4. Substitute these results into the equation in step 1:

Keeping only the zeroth and first-order terms gives

$$(101)^{1/2} = 10(1 + 0.01)^{1/2} \approx \boxed{10.050\,000}$$

Keeping only the zeroth, first-, and second-order terms gives

$$(101)^{1/2} = 10(1 + 0.01)^{1/2} \approx \boxed{10.049\,875}$$

CHECK We therefore expect our answer to be correct to within about 0.001%. The value of $(101)^{1/2}$, to eight figures, is 10.049 876. This differs from 10.050 000 by 0.000 124, or about one part in 10^5, and differs from 10.049 875 by about one part in 10^7.

PRACTICE PROBLEMS For the following, calculate the answer keeping the zeroth and first-order terms in the binomial series (Equation M-53), find the answer using your calculator, and show the percentage discrepancy between the two values:

21. $(1 + 0.001)^{-4}$
22. $(1 - 0.001)^{40}$

M-10 COMPLEX NUMBERS

Real numbers are all numbers, from $-\infty$ to $+\infty$, that can be *ordered*. We know that, given two real numbers, one is always equal to, greater than, or less than the other. For example, $3 > 2$, $1.4 < \sqrt{2} < 1.5$, and $3.14 < \pi < 3.15$. A number that *cannot* be ordered is $\sqrt{-1}$; we cannot measure the size of this number, and so it makes no sense to say, for example, that $3 \times \sqrt{-1}$ is greater than or less than $2 \times \sqrt{-1}$. The earliest mathematicians who dealt with numbers containing $\sqrt{-1}$ referred to these numbers as *imaginary* numbers because they could not be used to measure or count something. In mathematics the symbol i is used to represent $\sqrt{-1}$.

Equation M-5, the quadratic formula, applies to equations of the form

$$ax^2 + bx + c = 0$$

The formula shows that there are no real roots when $b^2 < 4ac$. There are, however, still two roots. Each root is a number containing two terms: a real number, and a multiple of $i = \sqrt{-1}$. The multiple of i is called an **imaginary number,** and i is called the **unit imaginary.**

A general **complex number** z can be written

$$z = a + bi \qquad \text{M-54}$$

where a and b are real numbers. The quantity a is called the real part of z, or Re(z), and the quantity b is called the imaginary part of z, or Im(z). We can represent a complex number z as a point in a plane, called the complex plane, as shown in Figure M-22, where the x axis is the **real axis** and the y axis is the **imaginary axis.** We can also use the relations $a = r\cos\theta$ and $b = r\sin\theta$ from Figure M-22 to write the complex number z in **polar coordinates** (a system in which a point is designated by the counterclockwise angle of rotation θ and the distance r in the direction of θ):

$$z = r\cos\theta + ir\sin\theta \qquad \text{M-55}$$

where $r = \sqrt{a^2 + b^2}$ is called the **magnitude** of z.

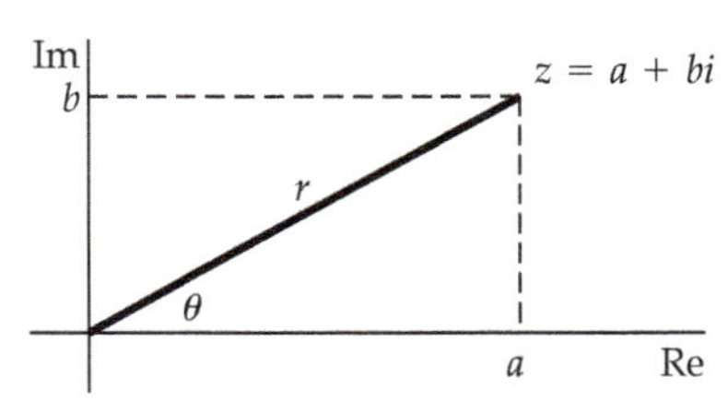

$$z = a + bi = r\cos\theta + (r\sin\theta)i = r(\cos\theta + i\sin\theta)$$

FIGURE M-22 Representation of a complex number in a plane. The real part of the complex number is plotted along the horizontal axis, and the imaginary part is plotted along the vertical axis.

When complex numbers are added or subtracted, the real and imaginary parts are added or subtracted separately:

$$z_1 + z_2 = (a_1 + ib_1) + (a_2 + ib_2) = (a_1 + a_2) + i(b_1 + b_2) \qquad \text{M-56}$$

However, when two complex numbers are multiplied, each part of one number is multiplied by each part of the other number:

$$\begin{aligned} z_1 z_2 &= (a_1 + ib_1)(a_2 + ib_2) = a_1 a_2 + i^2 b_1 b_2 + i(a_1 b_2 + a_2 b_1) \\ &= a_1 a_2 - b_1 b_2 + i(a_1 b_2 + a_2 b_1) \end{aligned} \qquad \text{M-57}$$

where we have used $i^2 = -1$.

The **complex conjugate** z^* of the complex number z is that number obtained by replacing i with $-i$ when writing z. If $z = a + ib$, then

$$z^* = (a + ib)^* = a - ib \qquad \text{M-58}$$

(When a quadratic equation has complex roots, the roots are **conjugate complex numbers,** in the form $a \pm bi$.) The product of a complex number and its complex conjugate equals the square of the magnitude of the number:

$$zz^* = (a + ib)(a - ib) = a^2 + b^2 = r^2 \qquad \text{M-59}$$

A particularly useful function of a complex number is the exponential $e^{i\theta}$. Using an expansion for e^x, we have

$$e^{i\theta} = 1 + i\theta + \frac{(i\theta)^2}{2!} + \frac{(i\theta)^3}{3!} + \frac{(i\theta)^4}{4!} + \cdots$$

Using $i^2 = -1$, $i^3 = -i$, $i^4 = +1$, and so forth, and separating the real parts from the imaginary parts, this expansion can be written

$$e^{i\theta} = \left(1 - \frac{\theta^2}{2!} + \frac{\theta^4}{4!} - \cdots\right) + i\left(\theta - \frac{\theta^3}{3!} + \cdots\right)$$

Comparing this result with Equations M-50 and M-51, we can see that

$$e^{i\theta} = \cos\theta + i\sin\theta \qquad \text{M-60}$$

Using this result, we can express a general complex number as an exponential:

$$z = a + ib = r\cos\theta + ir\sin\theta = re^{i\theta} \qquad \text{M-61}$$

If $z = x + iy$, where x and y are real variables, then z is called a **complex variable.**

COMPLEX VARIABLES IN PHYSICS

Complex variables are often used in formulas describing AC circuits: the impedance of a capacitor or an inductor includes a real part (the resistance) and an imaginary part (the reactance). (There are alternative ways, however, of analyzing AC circuits—such as rotating vectors called *phasors*—that do not require assigning imaginary values.) Complex variables are also important in the study of harmonic waves through Fourier analysis and synthesis. The time-dependent Schrödinger equation contains a complex-valued function of position and time.

Example M-11 Finding a Power of a Complex Number

Calculate $(1 + 3i)^4$ by using the binomial expansion.

PICTURE The expression is of the form $(1 + x)^n$. Because n is a positive integer, the expansion is valid for any value of x, and all terms, other than those of order n or lower must equal zero.

SOLVE

1. Write out the expansion of $(1 + 3i)^4$ to show the terms up through the fouth-order term:

$$1 + 4\cdot 3i + \frac{4(3)}{2!}(3i)^2 + \frac{4(3)(2)}{3!}(3i)^3 + \frac{4(3)(2)(1)}{4!}(3i)^4$$

2. Evaluate each term, remembering that $i^2 = -1$, $i^3 = -i$, and $i^4 = +1$: $1 + 12i - 54 - 108i + 81$

3. Show the result in the form $a + bi$: $(1 + 3i)^4 = \boxed{28 - 96i}$

CHECK We can solve the problem algebraically to show that the answer is correct. We first square $(1 + 3i)$ and then square the result, to get $(1 + 3i)^4$:

$$(1 + 3i)^2 = 1 \cdot 1 + 2 \cdot 1 \cdot 3i + (3i)^2 = 1 + 6i - 9 = -8 + 6i$$
$$(-8 + 6i)^2 = (-8)(-8) + 2(-8)(6i) + (6i)^2 = 64 - 96i - 36 = 28 - 96i$$

PRACTICE PROBLEMS Express in the form $a + bi$:

23. $e^{i\pi}$

24. $e^{i\pi/2}$

M-11 DIFFERENTIAL CALCULUS

Calculus is the branch of mathematics that allows us to deal with instantaneous rates of change of functions and variables. From the equation of a function—say, x as a function of t—we can always find x for a particular t, but with the methods of calculus you can go much further. You can know where x will have certain properties, such as a maximum or a minimum value, without having to try endless values of t. With calculus, if given the proper data, you can find, for example, the location of maximum stress on a beam, or the velocity or position of a falling object at a time t, or the energy a falling object has acquired at the time of impact. The principles of calculus are derived from examining functions at the infinitesimal level—analyzing how, say, x will change when the change in t becomes vanishingly small. We start with **differential calculus,** in which we determine the *limit* of the rate of change of x with respect to t as the change in t becomes closer and closer to zero.

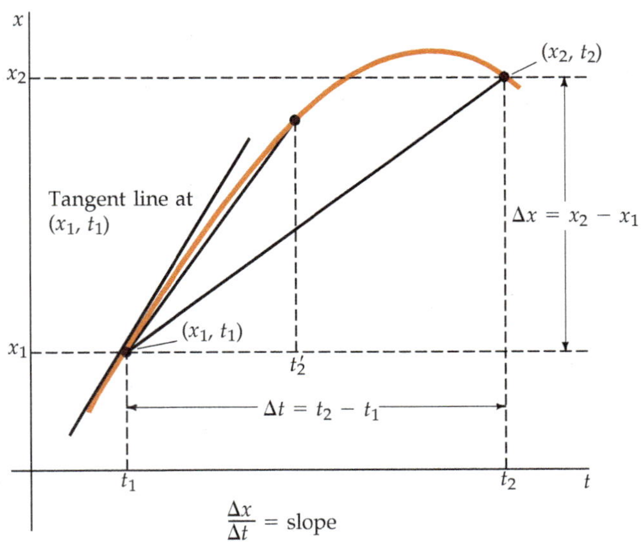

FIGURE M-23 Graph of a typical function $x(t)$. The points (x_1, t_1) and (x_2, t_2) are connected by a straight line. The slope of this line is $\Delta x/\Delta t$. As the time interval beginning at t_1 is decreased, the slope for that interval approaches the slope of the line tangent to the curve at time t_1, which is the derivative of x with respect to t.

Figure M-23 is a graph of x versus t for a typical function $x(t)$. At a particular value $t = t_1$, x has the value of x_1, as indicated. At another value t_2, x has the value x_2. The change in t, $t_2 - t_1$, is written $\Delta t = t_2 - t_1$; and the corresponding change in x is written $\Delta x = x_2 - x_1$. The ratio $\Delta x/\Delta t$ is the slope of the straight line connecting (x_1, t_1) and (x_2, t_2). If we take the limit as t_2 approaches t_1 (as Δt approaches zero) the slope of the line connecting (x_1, t_1) and (x_2, t_2) approaches the slope of the line that is tangent to the curve at the point (x_1, t_1). The slope of this tangent line is equal to the **derivative** of x with respect to t and is written dx/dt:

$$\frac{dx}{dt} = \lim_{\Delta t \to 0} \frac{\Delta x}{\Delta t} \qquad \text{M-62}$$

(When we find the derivative of a function, we say that we are **differentiating** the function; and the very small "dx" and "dt" elements are called **differentials** of x and t, respectively.) The derivative of a function of t is another function of t. If x is a constant and does not change, the graph of x versus t is a horizontal line with zero slope. The derivative of a constant is thus zero. In Figure M-24, x is not constant but is proportional to t:

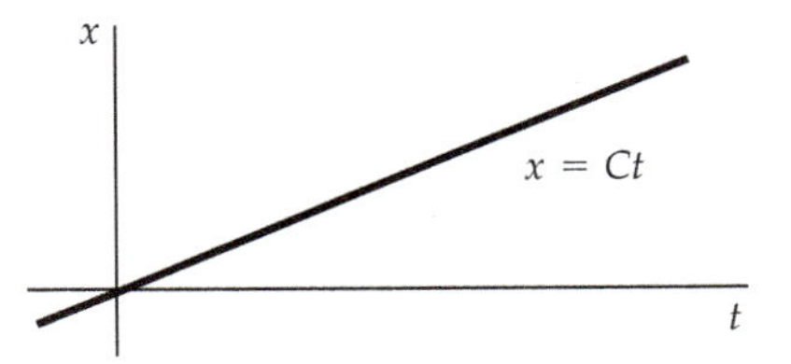

FIGURE M-24 Graph of the linear function $x = Ct$. This function has a constant slope C.

$$x = Ct$$

This function has a constant slope equal to C. Thus the derivative of Ct is C. Table M-3 lists some properties of derivatives and the derivatives of some particular functions that occur often in physics. It is followed by comments aimed at making these properties and rules clearer. More detailed discussion can be found in most calculus textbooks.

Table M-3 Properties of Derivatives and Derivatives of Particular Functions

Linearity

1. The derivative of a constant C times a function $f(t)$ equals the constant times the derivative of the function:

$$\frac{d}{dt}[Cf(t)] = C\frac{df(t)}{dt}$$

2. The derivative of a sum of functions equals the sum of the derivatives of the functions:

$$\frac{d}{dt}[f(t) + g(t)] = \frac{df(t)}{dt} + \frac{dg(t)}{dt}$$

Chain rule

3. If f is a function of x and x is in turn a function of t, the derivative of f with respect to t equals the product of the derivative of f with respect to x and the derivative of x with respect to t:

$$\frac{d}{dt}f(x(t)) = \frac{df}{dx}\frac{dx}{dt}$$

Derivative of a product

4. The derivative of a product of functions $f(t)g(t)$ equals the first function times the derivative of the second plus the second function times the derivative of the first:

$$\frac{d}{dt}[f(t)g(t)] = f(t)\frac{dg(t)}{dt} + g(t)\frac{df(t)}{dt}$$

Reciprocal derivative

5. The derivative of t with respect to x is the reciprocal of the derivative of x with respect to t, assuming that neither derivative is zero:

$$\frac{dt}{dx} = \left(\frac{dx}{dt}\right)^{-1} \quad \text{if} \quad \frac{dt}{dx} \neq 0 \quad \text{and} \quad \frac{dx}{dt} \neq 0$$

Derivatives of particular functions

6. If C is a constant, then $dC/dt = 0$.

7. $\dfrac{d(t^n)}{dt} = nt^{n-1}$ If n is constant.

8. $\dfrac{d}{dt}\sin\omega t = \omega\cos\omega t$ If ω is constant.

9. $\dfrac{d}{dt}\cos\omega t = -\omega\sin\omega t$ If ω is constant.

10. $\dfrac{d}{dt}\tan\omega t = \omega\sin^2\omega t$ If ω is constant.

11. $\dfrac{d}{dt}e^{bt} = be^{bt}$ If b is constant.

12. $\dfrac{d}{dt}\ln bt = \dfrac{1}{t}$ If b is constant.

COMMENTS ON RULES 1 THROUGH 5

Rules 1 and 2 follow from the fact that the limiting process is linear. We can understand rule 3, the chain rule, by multiplying $\Delta f/\Delta t$ by $\Delta x/\Delta x$ and noting that as Δt approaches zero, Δx also approaches zero. That is,

$$\lim_{\Delta t\to 0}\frac{\Delta f}{\Delta t} = \lim_{\Delta t\to 0}\left(\frac{\Delta f}{\Delta t}\frac{\Delta x}{\Delta x}\right) = \lim_{\Delta t\to 0}\left(\frac{\Delta f}{\Delta x}\frac{\Delta x}{\Delta t}\right) = \left(\lim_{\Delta x\to 0}\frac{\Delta f}{\Delta x}\right)\left(\lim_{\Delta t\to 0}\frac{\Delta x}{\Delta t}\right) = \frac{df}{dx}\frac{dx}{dt}$$

where we have used that the limit of the product is equal to product of the limits.

Rule 4 is not immediately apparent. The derivative of a product of functions is the limit of the ratio

$$\frac{f(t+\Delta t)g(t+\Delta t) - f(t)g(t)}{\Delta t}$$

If we add and subtract the quantity $f(t+\Delta t)g(t)$ in the numerator, we can write this ratio as

$$\frac{f(t+\Delta t)g(t+\Delta t) - f(t+\Delta t)g(t) + f(t+\Delta t)g(t) - f(t)g(t)}{\Delta t}$$
$$= f(t+\Delta t)\left[\frac{g(t+\Delta t) - g(t)}{\Delta t}\right] + g(t)\left[\frac{f(t+\Delta t) - f(t)}{\Delta t}\right]$$

As Δt approaches zero, the terms in square brackets become $dg(t)/dt$ and $df(t)/dt$, respectively, and the limit of the expression is

$$f(t)\frac{dg(t)}{dt} + g(t)\frac{df(t)}{dt}$$

Rule 5 follows directly from the definition:

$$\frac{dx}{dt} = \lim_{\Delta t\to 0}\frac{\Delta x}{\Delta t} = \lim_{\Delta x\to 0}\left(\frac{\Delta t}{\Delta x}\right)^{-1} = \left(\frac{dt}{dx}\right)^{-1}$$

COMMENTS ON RULE 7

We can obtain this important result using the binomial expansion. We have

$$f(t) = t^n$$
$$f(t+\Delta t) = (t+\Delta t)^n = t^n\left(1+\frac{\Delta t}{t}\right)^n$$
$$= t^n\left[1 + n\frac{\Delta t}{t} + \frac{n(n-1)}{2!}\left(\frac{\Delta t}{t}\right)^2 + \frac{n(n-1)(n-2)}{3!}\left(\frac{\Delta t}{t}\right)^3 + \cdots\right]$$

Then

$$f(t+\Delta t) - f(t) = t^n\left[n\frac{\Delta t}{t} + \frac{n(n-1)}{2!}\left(\frac{\Delta t}{t}\right)^2 + \cdots\right]$$

and

$$\frac{f(t+\Delta t) - f(t)}{\Delta t} = nt^{n-1} + \frac{n(n-1)}{2!}t^{n-2}\Delta t + \cdots$$

The next term omitted from the last sum is proportional to $(\Delta t)^2$, the following to $(\Delta t)^3$, and so on. Each term except the first approaches zero as Δt approaches zero. Thus

$$\frac{df}{dt} = \lim_{\Delta x\to 0}\frac{f(t+\Delta t) - f(t)}{\Delta t} = nt^{n-1}$$

COMMENTS ON RULES 8 TO 10

We first write $\sin \omega t = \sin\theta$ with $\theta = \omega t$ and use the chain rule,

$$\frac{d\sin\theta}{dt} = \frac{d\sin\theta}{d\theta}\frac{d\theta}{dt} = \omega\frac{d\sin\theta}{d\theta}$$

We then use the trigonometric formula for the sine of the sum of two angles θ and $\Delta\theta$:

$$\sin(\theta + \Delta\theta) = \sin \Delta\theta \cos \theta + \cos \Delta\theta \sin \theta$$

Because $\Delta\theta$ is to approach zero, we can use the small-angle approximations

$$\sin \Delta\theta \approx \Delta\theta \qquad \text{and} \qquad \cos \Delta\theta \approx 1$$

Then

$$\sin(\theta + \Delta\theta) \approx \Delta\theta \cos \theta + \sin \theta$$

and

$$\frac{\sin(\theta + \Delta\theta) - \sin \theta}{\Delta\theta} \approx \cos \theta$$

Similar reasoning can be applied to the cosine function to obtain rule 9.

Rule 10 is obtained by writing $\tan \theta = \sin \theta/\cos \theta$ and applying rule 4 along with rules 8 and 9:

$$\begin{aligned}\frac{d}{dt}(\tan \theta) &= \frac{d}{dt}(\sin \theta)(\cos \theta)^{-1} = \sin \theta \frac{d}{dt}(\cos \theta)^{-1} + \frac{d(\sin \theta)}{dt}(\cos \theta)^{-1} \\ &= \sin \theta(-1)(\cos \theta)^{-2}(-\sin \theta) + (\cos \theta)(\cos \theta)^{-1} \\ &= \frac{\sin^2 \theta}{\cos^2 \theta} + 1 = \tan^2 \theta + 1 = \sec^2 \theta\end{aligned}$$

To obtain rule 10, let $\theta = \omega t$ and use the chain rule.

COMMENTS ON RULE 11

Again we use the chain rule

$$\frac{de^\theta}{dt} = \frac{b\, de^\theta}{b\, dt} = b\frac{de^\theta}{d(bt)} = b\frac{de^\theta}{d\theta} \qquad \text{with} \qquad \theta = bt$$

and the series expansion for the exponential function:

$$e^{\theta + \Delta\theta} = e^\theta e^{\Delta\theta} = e^\theta\left[1 + \Delta\theta + \frac{(\Delta\theta)^2}{2!} + \frac{(\Delta\theta)^3}{3!} + \cdots\right]$$

Then

$$\frac{e^{\theta+\Delta\theta} - e^\theta}{\Delta\theta} = e^\theta + e^\theta\frac{\Delta\theta}{2!} + e^\theta\frac{(\Delta\theta)^2}{3!} + \cdots$$

As $\Delta\theta$ approaches zero, the right side of this equation approaches e^θ.

COMMENTS ON RULE 12

Let

$$y = \ln bt$$

Then

$$e^y = bt \Rightarrow t = \frac{1}{b}e^y$$

Then, using rule 11, we obtain

$$\frac{dt}{dy} = \frac{1}{b}e^y \;\therefore\; \frac{dt}{dy} = t$$

Then, using rule 5, we obtain

$$\frac{dy}{dt} = \left(\frac{dt}{dy}\right)^{-1} = \frac{1}{t}$$

SECOND- AND HIGHER-ORDER DERIVATIVES; DIMENSIONAL ANALYSIS

Once we have differentiated a function, we can differentiate the resulting derivative as long as terms remain to differentiate. A function such as $x = e^{bt}$ can be differentiated indefinitely: $dx/dt = be^{bt}$ (this function differentiates to give b^2e^{bt}, and so on).

Consider velocity and acceleration. We can define velocity as the rate of change of position of a particle, or dx/dt, and acceleration as the rate of change of velocity, or the *second* derivative of x with respect to t, written dx^2/dt^2. If a particle moves at a constant velocity, then dx/dt will equal a constant. The acceleration, however, will be zero: having constant velocity is the same as having no acceleration, and the derivative of a constant is zero. Now consider a falling object, subject to the constant acceleration of gravity: the velocity itself will be time-dependent, so the *second* derivative, dx^2/dt^2, will be a constant.

The *physical dimensions* of a derivative with respect to a variable are those that would result if the original function of the variable were divided by a value of the variable. For example, the dimension of an equation in which one term is x (for position) is that of length (L); the dimensions of the derivative of x with respect to time t are those of velocity (L/T), and the dimensions of dx^2/dt^2 are those of acceleration (L/T^2).

Example M-12 Position, Velocity, and Acceleration

Find the first and the second derivative of $x = \frac{1}{2}at^2 + bt + c$, where a, b, and c are constants. The function gives the position (in m) of a particle in one dimension, where t is the time (in s), a is acceleration (in m/s^2), b is velocity (in m/s) at a time $t = 0$, and c is the position (in m) of the particle at $t = 0$.

PICTURE Both the first and the second derivatives are sums of terms; for each differentiation we take the derivative of each term separately and add the results.

SOLVE

1. To find the first derivative, first compute the derivative of the first term:
$$\frac{d(\frac{1}{2}at^2)}{dt} = \left(\frac{1}{2}a\right)2t^1 = at$$

2. Compute the first derivative of the second and third terms:
$$\frac{d(bt)}{dt} = b, \qquad \frac{d(c)}{dt} = 0$$

3. Add these results:
$$\frac{dx}{dt} = at + b$$

4. To compute the second derivative, repeat the process for the result in step 3:
$$\frac{d^2x}{dt^2} = a + 0 = a$$

CHECK The physical dimensions show that the answer is plausible. The original function is an equation for position; all terms are in meters—the units of t^2 and t cancel the units of s^2 and s in the constants a and b, respectively. In the function for dx/dt, all terms are similarly in m/s: the constant c has differentiated to zero, and the unit for t cancels one of the units for s in the constant a. In the function for dx^2/dt^2, only the acceleration constant remains; as expected, its dimensions are L/T^2.

PRACTICE PROBLEMS

25. Find dy/dx for $y = \frac{5}{8}x^3 - 24x - \frac{5}{8}$.
26. Find dy/dt for $y = ate^{bt}$, where a and b are constants.

SOLVING DIFFERENTIAL EQUATIONS USING COMPLEX NUMBERS

A **differential equation** is an equation in which the derivatives of a function appear as variables. It is an equation in which the variables are related to each other through their derivatives. Consider an equation of the form

$$a\frac{d^2x}{dt^2} + b\frac{dx}{dt} + cx = A\cos\omega t \qquad \text{M-63}$$

that represents a physical process, such as a damped harmonic oscillator driven by a sinusoidal force, or a series *RLC* combination being driven by a sinusoidal potential drop. Although each of the parameters in Equation M-63 is a real number, the time-dependent cosine term suggests that we might find the steady-state solution to this equation by introducing complex numbers. We first construct the "parallel" equation

$$a\frac{d^2y}{dt^2} + b\frac{dy}{dt} + cy = A\sin\omega t \qquad \text{M-64}$$

Equation M-64 has no physical meaning of its own, and we have no interest in solving it. However, it is of use in solving Equation M-63. After multiplying through Equation M-64 by the unit imaginary i, we add Equation M-64 and Equation M-63 to obtain

$$\left(a\frac{d^2x}{dt^2} + ai\frac{d^2y}{dt^2}\right) + \left(b\frac{dx}{dt} + bi\frac{dy}{dt}\right) + (cx + ciy) = A\cos\omega t + Ai\sin\omega t$$

We next combine terms to get

$$a\frac{d^2(x+iy)}{dt^2} + b\frac{d(x+iy)}{dt} + c(x+iy) = A(\cos\omega t + i\sin\omega t) \qquad \text{M-65}$$

which is valid because the derivative of a sum is equal to the sum of the derivatives. We simplify our result by defining $z = x + iy$ and by using the identity $e^{i\omega t} = \cos\omega t + i\sin\omega t$. Substituting these into Equation M-65, we obtain

$$a\frac{d^2z}{dt^2} + b\frac{dz}{dt} + cz = Ae^{i\omega t} \qquad \text{M-66}$$

which we now solve for z. Once z is obtained, we can solve for x using $x = \text{Re}(z)$.

Because we are looking only for the steady-state solution for Equation M-65, we can assume its solution is of the form $x = x_0\cos(\omega t - \phi)$, where ϕ is a constant. This is equivalent to assuming that the solution to Equation M-66 is of the form $z = \eta e^{i\omega t}$, where η, pronounced eta (like beta without the b), is a constant complex number. Then $dz/dt = i\omega z$, $d^2z/dt^2 = -\omega^2 z$, and $e^{i\omega t} = z/\eta$. Substituting these into Equation M-65 gives

$$-a\omega^2 z + i\omega bz + cz = A\frac{z}{\eta}$$

Dividing both sides of this equation by z and solving for η gives

$$\eta = \frac{A}{-a\omega^2 + i\omega b + c}$$

Expressing the denominator in polar form gives

$$(-a\omega^2 + c) + i\omega b = \sqrt{(-a\omega^2 + c)^2 + \omega^2 b^2}\, e^{i\phi}$$

where $\tan\phi = \omega b/(-a\omega^2 + c)$. Thus,

$$\eta = \frac{A}{\sqrt{(-a\omega^2 + c)^2 + \omega^2 b^2}} e^{-i\phi}$$

so

$$z = \eta e^{i\omega t} = \frac{A}{\sqrt{(-a\omega^2 + c)^2 + \omega^2 b^2}} e^{i(\omega t - \phi)}$$

$$= \frac{A}{\sqrt{(-a\omega^2 + c)^2 + \omega^2 b^2}} [\cos(\omega t - \phi) + i \sin(\omega t - \phi)] \qquad \text{M-67}$$

It follows that

$$x = \text{Re}(z) = \frac{A}{\sqrt{(-a\omega^2 + c)^2 + \omega^2 b^2}} \cos(\omega t - \phi) \qquad \text{M-68}$$

THE EXPONENTIAL FUNCTION

An **exponential function** is a function of the form a^{bx}, where $a > 0$ and b are constants. The function is usually written as e^{cx}, where c is constant.

When the rate of change of a quantity is proportional to the quantity itself, the quantity increases or decreases exponentially, depending on the sign of the proportionality constant. An example of an *exponentially* decreasing function is nuclear decay. If N is the number of radioactive nuclei at some time, then the change dN in some very small time interval dt will be proportional to N and to dt:

$$dN = -\lambda N \, dt$$

where λ is the *decay constant* (not to be confused with the decay rate dN/dt, which decreases exponentially). The function N satisfying this equation is

$$N = N_0 e^{-\lambda t} \qquad \text{M-69}$$

where N_0 is the value of N at time $t = 0$. Figure M-25 shows N versus t. A characteristic of exponential decay is that N decreases by a constant factor in a given time interval. The time interval for N to decrease to half its original value is its *half-life* $t_{1/2}$. The half-life is obtained from Equation M-69 by setting $N = \frac{1}{2}N_0$ and solving for the time. This gives

$$t_{1/2} = \frac{\ln 2}{\lambda} = \frac{0.693}{\lambda} \qquad \text{M-70}$$

An example of *exponential increase* is population growth. If the number of organisms is N, the change in N after a very small time interval dt is given by

$$dN = +\lambda N \, dt$$

where λ is now the *growth constant*. The function N satisfying this equation is

$$N = N_0 e^{\lambda t} \qquad \text{M-71}$$

(Note the change of sign in the exponent.) A graph of this function is shown in Figure M-26. An exponential increase can be characterized by a doubling time T_2, which is related to λ by

$$T_2 = \frac{\ln 2}{\lambda} = \frac{0.693}{\lambda} \qquad \text{M-72}$$

Very often, we know population growth as an annual percentage increase and wish to calculate the doubling time. In this case, we find T_2 (in years) from the equation

$$T_2 = \frac{69.3}{r} \qquad \text{M-73}$$

where r is the percent per year. For example, if the population increases by 2 percent per year, the population will double every $69.3/2 \approx 35$ years. Table M-4 lists some useful relations for exponential and logarithmic functions.

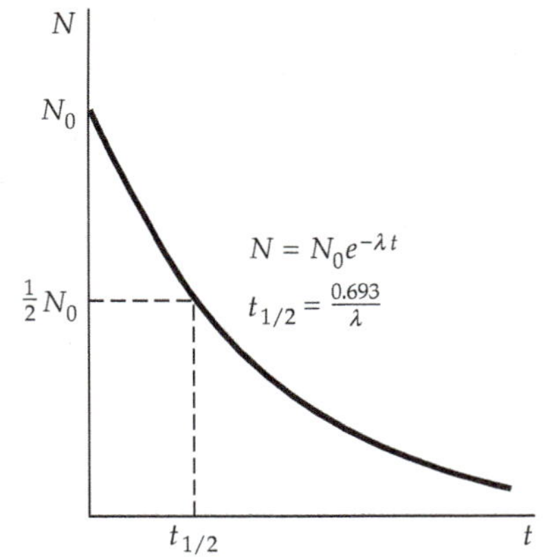

FIGURE M-25 Graph of N versus t when N decreases exponentially. The time $t_{1/2}$ is the time it takes for N to decrease by one-half.

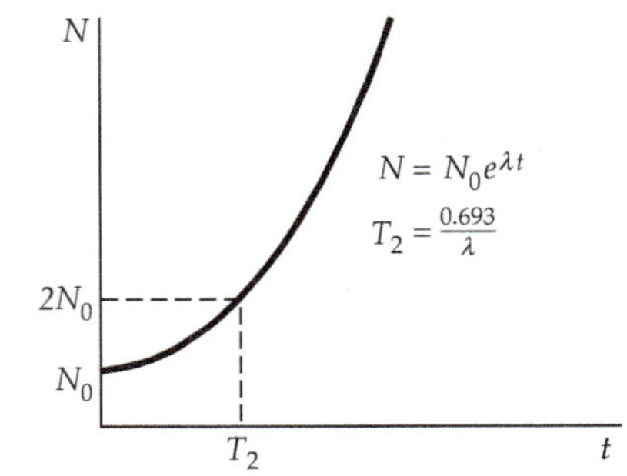

FIGURE M-26 Graph of N versus t when N increases exponentially. The time T_2 is the time it takes for N to double.

Table M-4 Exponential and Logarithmic Functions

$e = 2.718\,28$

$e^0 = 1$

If $y = e^x$, then $x = \ln y$.

$e^{\ln x} = x$

$e^x e^y = e^{(x+y)}$

$(e^x)^y = e^{xy} = (e^y)^x$

$\ln e = 1; \ln 1 = 0$

$\ln xy = \ln x + \ln y$

$\ln \frac{x}{y} = \ln x - \ln y$

$\ln e^x = x; \ln a^x = x \ln a$

$\ln x = (\ln 10) \log x$

$= 2.30\,26 \log x$

$\log x = (\log e) \ln x = 0.434\,29 \ln x$

$e^x = 1 + x + \frac{x^2}{2!} + \frac{x^3}{3!} = \ldots$

$\ln(1 + x) = x - \frac{x^2}{2} + \frac{x^3}{3} - \frac{x^4}{4}$

Example M-13 Radioactive Decay of Cobalt-60

The half-life of cobalt-60 (^{60}Co) is 5.27 y. At $t = 0$ you have a sample of ^{60}Co that has a mass equal to 1.20 mg. At what time t (in years) will 0.400 mg of the sample of ^{60}Co have decayed?

PICTURE When we derived the half-life in exponential decay, we set $N/N_0 = 1/2$. In this example, we are to find the time at which two-thirds of a sample remains, and so the ratio N/N_0 will be 0.667.

SOLVE

1. Express the ratio N/N_0 as an exponential function: $\frac{N}{N_0} = 0.667 = e^{-\lambda t}$

2. Take the reciprocal of both sides: $\frac{N_0}{N} = 1.50 = e^{\lambda t}$

3. Solve for t: $t = \frac{\ln 1.50}{\lambda} = \frac{0.405}{\lambda}$

4. The decay constant is related to the half-life by $\lambda = (\ln 2)/t_{1/2}$ (Equation M-70). Substitute $(\ln 2)/t_{1/2}$ for λ and evaluate the time: $t = \frac{\ln 1.5}{\ln 2} t_{1/2} = \frac{\ln 1.5}{\ln 2} \times 5.27\text{ y} = \boxed{3.08\text{ y}}$

CHECK It takes 5.27 y for the mass of a sample of ^{60}Co to decrease to 50 percent of its initial mass. Thus, we expect it to take less than 5.27 y for the sample to lose 33.3 percent of its mass. Our step-4 result of 3.08 y is less than 5.27 y, as expected.

PRACTICE PROBLEMS

27. The discharge time constant τ of a capacitor in an RC circuit is the time in which the capacitor discharges to e^{-1} (or 0.368) times its charge at $t = 0$. If $\tau = 1$ s for a capacitor, at what time t (in seconds) will it have discharged to 50.0% of its initial charge?

28. If the coyote population in your state is increasing at a rate of 8.0% a decade and continues increasing at the same rate indefinitely, in how many years will it reach 1.5 times its current level?

M-12 INTEGRAL CALCULUS

Integration can be considered the inverse of differentiation. If a function $f(t)$ is *integrated*, a function $F(t)$ is found for which $f(t)$ is the derivative of $F(t)$ with respect to t.

THE INTEGRAL AS AN AREA UNDER A CURVE; DIMENSIONAL ANALYSIS

The process of finding the area under a curve on the graph illustrates integration. Figure M-27 shows a function $f(t)$. The area of the shaded element is approximately $f_i\,\Delta t_i$, where f_i is evaluated anywhere in the interval Δt_i. This approximation is highly accurate if Δt_i is very small. The total area under some stretch of the curve is found by summing all the area elements it covers and taking the limit as each Δt_i approaches zero. This limit is called the **integral** of f over t and is written

$$\int f\,dt = \text{area}_i = \lim_{\Delta t_i \to 0} \sum_i f_i \Delta t_i \qquad \text{M-74}$$

The *physical dimensions* of an integral of a function $f(t)$ are found by multiplying the dimensions of the *integrand* (the function being integrated) and the dimensions of the integration variable t. For example, if the integrand is a velocity function

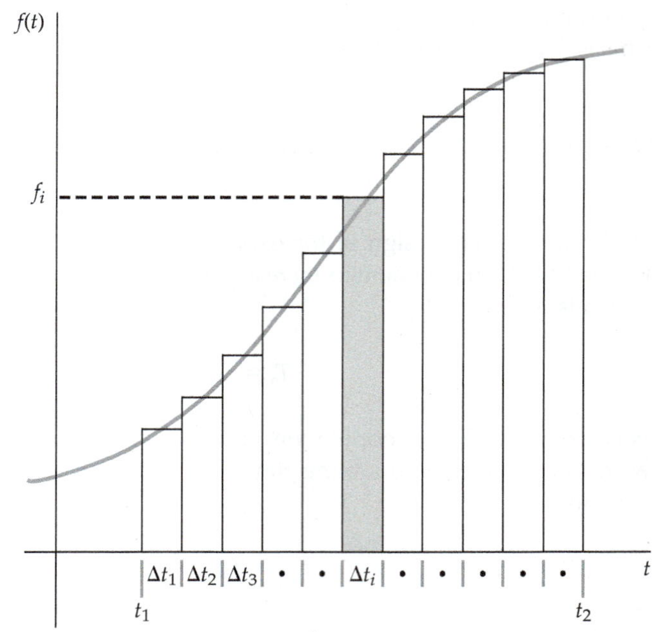

FIGURE M-27 A general function $f(t)$. The area of the shaded element is approximately $f_i\Delta t_i$, where f_i is evaluated anywhere in the interval.

$v(t)$ (dimensions L/T) and the integration variable is time t), the dimension of the integral is $L = (L/T) \times T$. That is, the dimensions of the integral are those of velocity times time.

Let

$$y = \int_{t_1}^{t} f\,dt \qquad \text{M-75}$$

The function y is the area under the f-versus-t curve from t_1 to a general value t. For a small interval Δt, the change in the area Δy is approximately $f\,\Delta t$:

$$\Delta y \approx f\,\Delta t$$

$$f \approx \frac{\Delta y}{\Delta t}$$

If we take the limit as Δt approaches 0, we can see that f is the derivative of y:

$$f = \frac{dy}{dt} \qquad \text{M-76}$$

INDEFINITE INTEGRALS AND DEFINITE INTEGRALS

When we write

$$y = \int f\,dt \qquad \text{M-77}$$

we are showing y as an **indefinite integral** of f over t. To evaluate an indefinite integral, we find the function y whose derivative is f. Because that function could contain a constant term that differentiated to zero, we include as our final term a **constant of integration** C. If we are integrating the function over a known segment—such as t_1 to t_2 in Figure M-27—we can find a **definite integral,** eliminating the unknown constant C:

$$\int_{t_1}^{t_2} f\,dt = y(t_2) - y(t_1) \qquad \text{M-78}$$

Table M-5 lists some important integration formulas. More extensive lists of integration formulas can be found in any calculus textbook or by searching for "table of integrals" on the Internet.

Table M-5 Integration Formulas†

1. $\int A\,dt = At$
2. $\int At\,dt = \frac{1}{2}At^2$
3. $\int At^n\,dt = A\frac{t^{n+1}}{n+1},\quad n \neq -1$
4. $\int At^{-1}\,dt = A\ln|t|$
5. $\int e^{bt}\,dt = \frac{1}{b}e^{bt}$
6. $\int \cos\omega t\,dt = \frac{1}{\omega}\sin\omega t$
7. $\int \sin\omega t\,dt = -\frac{1}{\omega}\cos\omega t$
8. $\int_0^\infty e^{-ax}\,dx = \frac{1}{a}$
9. $\int_0^\infty e^{-ax^2}\,dx = \frac{1}{2}\sqrt{\frac{\pi}{a}}$
10. $\int_0^\infty xe^{-ax^2}\,dx = \frac{2}{a}$
11. $\int_0^\infty x^2e^{-ax^2}\,dx = \frac{1}{4}\sqrt{\frac{\pi}{a^3}}$
12. $\int_0^\infty x^3e^{-ax^2}\,dx = \frac{4}{a^2}$
13. $\int_0^\infty x^4e^{-ax^2}\,dx = \frac{3}{8}\sqrt{\frac{\pi}{a^5}}$

† In these formulas, A, b, and ω are constants. In formulas 1 through 7, an arbitrary constant C can be added to the right side of each equation. The constant a is greater than zero.

Example M-14 Integrating Equations of Motion

A particle is moving at a constant acceleration a. Write a formula for position x at time t given that the position and velocity are x_0 and v_0 at time $t = 0$.

PICTURE Velocity v is the derivative of x with respect to time t, and acceleration is the derivative of v with respect to t. We should be able to write a function $x(t)$ by performing two integrations.

SOLVE

1. Integrate a with respect to t to find the v as a function of t. The a can be factored from the integrand because a is constant:

$$v = \int a\,dt = a\int dt$$

$$v = at + C_1$$

where C_1 represents a times the constant of integration.

2. The velocity $v = v_0$ when $t = 0$

$$v_0 = 0 + C_1 \Rightarrow C_1 = v_0$$

so $v = v_0 + at$

3. Integrate v with respect to t to find x as a function of t:

$$x = \int v\,dt = \int (v_0 + at)\,dt = \int v_0\,dt + \int at\,dt$$

$$x = v_0 \int dt + a \int t\,dt = v_0 t + \tfrac{1}{2}at^2 + C_2$$

where C_2 represents the combined constants of integration.

4. The position $x = x_0$ when $t = 0$

$$x_0 = 0 + 0 + C_2$$

so $x = x_0 + v_0 t + \tfrac{1}{2}at^2$

CHECK Differentiate the step-4 result twice to get the acceleration

$$v = \frac{dx}{dt} = \frac{d}{dt}(x_0 + v_0 t + \tfrac{1}{2}at^2) = 0 + v_0 + at$$

$$a = \frac{dv}{dt} = \frac{d}{dt}(v_0 + at) = a$$

PRACTICE PROBLEMS

29. $\int_3^6 3\,dx =$

30. $V = \int_5^8 \pi r^2\,dL =$

Answers to Practice Problems

1. 0.24 L
2. 31.6 m/s
3. 6.0 kg/cm^3
4. -3
5. 1.54 L
6. 3.07 L
7. False
8. $x = (4.5\text{ m/s})t + 3.0\text{ m}$
9. $x = 8, y = 60$
11. $2(x - y)^2$
12. $x^2(2x + 4)(x + 3)$
13. $x^{1/2}$
14. x^6
15. 3
16. ~ 2.322
17. $V/A = \tfrac{1}{3}r$
18. $A = \dfrac{2}{3}\pi L^2$
19. $\sin\theta = 0.496, \cos\theta = 0.868, \theta = 29.7°$
20. $\sin 8.2° = 0.1426, 8.2° = 0.1431$ rad
21. 0.996, 0.996 00, close to 0%
22. 0.96, 0.960 77, $\ll 1\%$
23. $-1 + 0i = -1$
24. $0 + i = i$
25. $dy/dx = \tfrac{5}{24}x^2 - 24$
26. $dy/dt = ae^{bt}(bt + 1)$
27. 0.693 s
28. 51 y
29. 9
30. $3\pi r^2$

Answers to Odd-Numbered End-of-Chapter Problems

Problem answers are calculated using $g = 9.81\ \mathrm{m/s^2}$ unless otherwise specified. Differences in the last figure can easily result from differences in rounding the input data and are not important.

Chapter 1

1 (*c*)

3 (*c*)

5 1.609×10^5 cm/mi

7 (*e*)

9 False

11

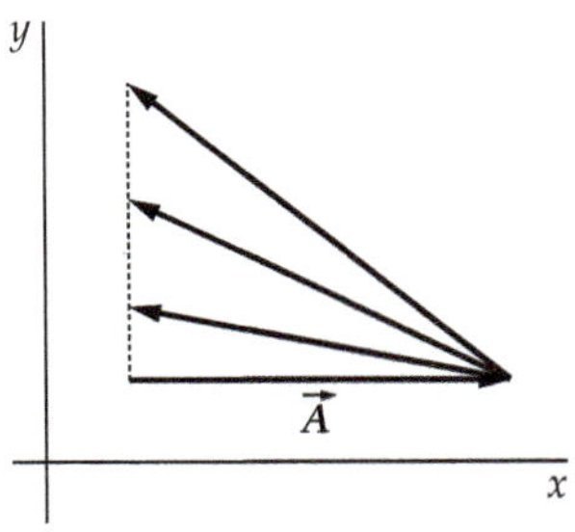

13

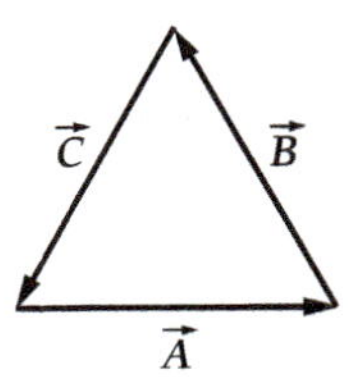

15 2.0×10^{27} molecules

17 (*a*) $\approx 3 \times 10^{10}$ diapers, (*b*) $\approx 2 \times 10^7\ \mathrm{m^3}$, (*c*) $\approx 0.8\ \mathrm{mi^2}$

19 (*a*) 50 MB, (*b*) 7×10^2 novels

21 (*a*) 40×10^{-6} W, (*b*) 4×10^{-9} s, (*c*) 3×10^6 W, (*d*) 25×10^3 m

23 (*a*) C_1 is in m; C_2 is in m/s, (*b*) C_1 is in $\mathrm{m/s^2}$, (*c*) C_1 is in $\mathrm{m/s^2}$, (*d*) C_1 is in m; C_2 is in $\mathrm{s^{-1}}$, (*e*) C_1 is in m/s; C_2 is in $\mathrm{s^{-1}}$

25 (*a*) 4.00×10^7 m, (*b*) 6.37×10^6 m, (*c*) 2.49×10^4 mi, 3.96×10^3 mi

27 210 cm

29 1.280 km

31 (*a*) 36.00 km/h·s, (*b*) 10.00 $\mathrm{m/s^2}$, (*c*) 88 ft/s, (*d*) 27 m/s

33 (*a*) 1.3×10^4 lb (*b*) 4 cases

35 (*a*) $\mathrm{m/s^2}$, (*b*) s, (*c*) m

37 $\mathrm{T^{-1}}$

39 (*a*) $\mathrm{M/T^2}$, (*b*) $\mathrm{kg \cdot m^2/s^2}$

43 $\mathrm{M/L^3}$

45 (*a*) 30 000, (*b*) 0.0062, (*c*) 0.000 004, (*d*) 217 000

47 (*a*) 1.14×10^5, (*b*) 2.25×10^{-8}, (*c*) 8.27×10^3, (*d*) 6.27×10^2

49 3.6×10^6

51 (*a*) 25.8 $\mathrm{mm^2}$, (*b*) 30.1 $\mathrm{mm^2}$

53 (*a*) $A_x = 5.0$ m, $A_y = 8.7$ m, (*b*) $v_x = -19$ m/s, $v_y = -16$ m/s, (*c*) $F_x = 35$ lb, $F_y = 20$ lb

55 You could have gone either 87 m north or 87 m south. The headings of your walk were either 60° north of east or 60° south of east, respectively.

57 (*a*) $40\hat{\imath} - 50\hat{\jmath}$, (*b*) $-51°$

59 $-0.59\hat{\imath} - 0.81\hat{\jmath}, 0.92\hat{\imath} - 0.38\hat{\jmath}, -0.51\hat{\imath} + 0.86\hat{\jmath}$

61 $\approx 3.3 \times 10^3$ mi/h, $\approx 5.3 \times 10^3$ km/h, $\approx 1.5 \times 10^3$ m/s

63 31.7 y

65 2.0×10^{23}

67 (*a*) $1.4 \times 10^{17}\ \mathrm{kg/m^3}$, (*b*) 2.2×10^2 m

69 (*a*) 4.848×10^{-6} parsec, (*b*) 3.086×10^{16} m, (*c*) 9.461×10^{15} m, (*d*) 6.324×10^4 AU, (*e*) 3.262 light-years

71 (*a*)

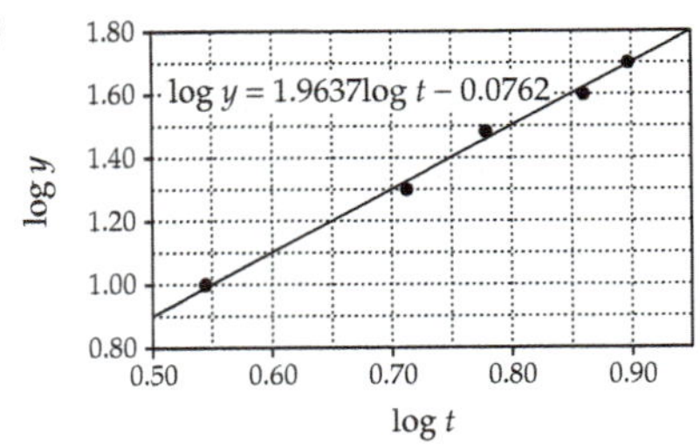

(*c*) $B = 0.84\ \mathrm{m/s^2}$, $C = 2.0$, (*d*) 1.1 s, (*e*) 1.7 $\mathrm{m/s^2}$

73 55.4×10^3 tons. The 50,000-ton claim is conservative. The actual weight is closer to 55,000 tons.

75 (*a*)

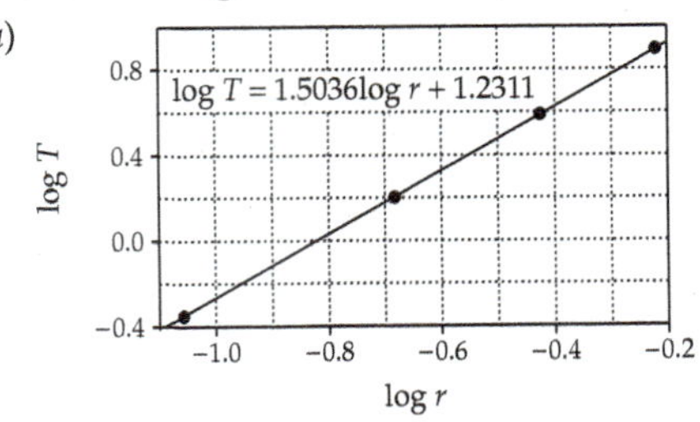

$n = 1.50$, $C = 17.0\ \mathrm{y/(Gm)^{3/2}}$, $T = [17.0\ \mathrm{y/(Gm)^{3/2}}]r^{1.50}$, (*b*) $r = 0.510$ Gm

77 (*a*) $\vec{F}_{\text{Paul}} = (35\text{ lb})\hat{\imath} + (35\text{ lb})\hat{\jmath}$ $\vec{F}_{\text{Johnny}} = (-53\text{ lb})\hat{\imath} + (-37.3\text{ lb})\hat{\jmath}$,
(*b*) $\vec{F}_{\text{Connie}} = (18\text{ lb})\hat{\imath} + (1.9\text{ lb})\hat{\jmath}$, $F_{\text{Connie}} = 18\text{ lb}$, $\theta = 6.1°$ N of E

Chapter 2

1 Zero

3 $v_{\text{av 1st half}} = 2H/T$, $v_{\text{av 2nd half}} = -2H/T$

5 (*a*) Your speed increased from zero, stayed constant for a while, and then decreased.
(*b*)

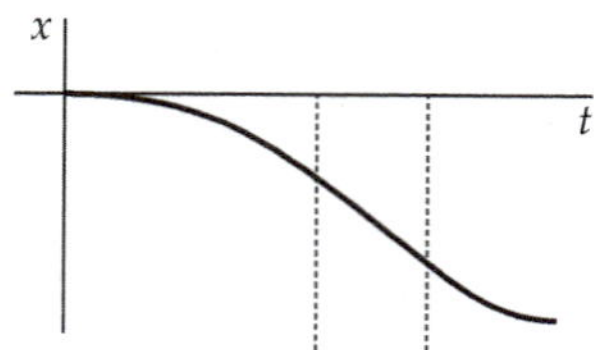

(*c*)

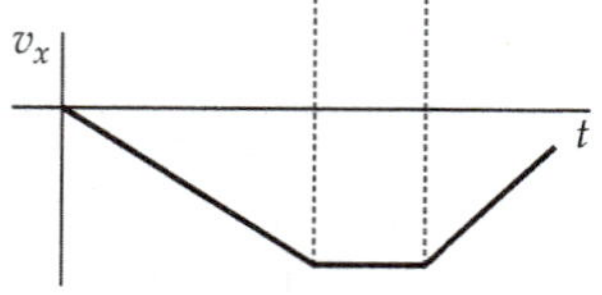

7 True

9 False. If it were true, then any time the initial and final velocities are both zero the average velocity would be zero.

11 (*a*)

13 (*a*) *b*, (*b*) *c*, (*c*) *d*, (*d*) *e*

15 (*a*) B, D, and E, (*b*) A and D, (*c*) C

17 (*a*) True, (*b*) True

19 (*a*) 0, (*b*) $-g$, (*c*) Its acceleration is greater than g in magnitude while the ball is in contact with the ceiling.

21 (*a*) False, (*b*) False, (*c*) True

23 (*a*) *c*
(*b*)

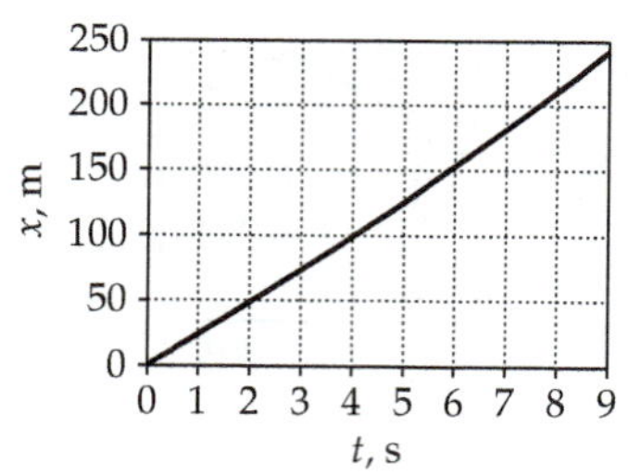

25 B is passing A.

27 (*c*)

29 (*a*) Yes, when the graphs intersect, (*b*) Yes, when the slopes of the curves have opposite signs, (*c*) Yes, when the curves have the same slope, (*d*) The two cars are farthest apart at the instant the two curves are farthest apart in the x direction.

31 $v_{\text{J}} = \frac{1}{2}v_{\text{max}}$

33 (*a*) *d*, (*b*) *b*, (*c*) None, (*d*) *c* and *d*

35 (*a*) *a*, *f*, and *i*, (*b*) *c* and *d*, (*c*) *a*, *d*, *e*, *f*, *h*, and *i*, (*d*) *b*, *c*, and *g*, (*e*) *a* and *i*, *d* and *h*, *f* and *i*

37 $-1.2 \times 10^3\text{ m/s}^2$

39 4.03 m/s^2

41 (*a*) 1.7 km ≈ 1 mi, (*b*) If the uncertainty in your time estimate is less than 1 s (±20%), the uncertainty in the distance estimate will be about 20% of 1.7 km, or approximately 300 m.

43 (*a*) 0.28 km/min, (*b*) −0.083 km/min, (*c*) 0 (*d*) 0.13 km/min

45 (*a*) 2.2 h, (*b*) $(t_{supersonic}/t_{subsonic}) \approx 0.45$

47 (*a*) 4.3 y, (*b*) 4.3×10^6 y. Because $4.3 \times 10^6\text{ y} \gg 1000\text{ y}$, Gregor does not have to pay.

49 23.5 m/s

51 (*a*) 0, (*b*) 0.3 m/s, (*c*) −2 m/s, (*d*) 1 m/s

53 $v_{\text{av}} = 122$ km/h. $v_{\text{Numerical av}} = 1.04v_{\text{av}}$. The average speed would be equal to one-third the sum of the three speeds if the three speeds were each maintained for the same length of time instead of for the same distance.

55 (*a*)

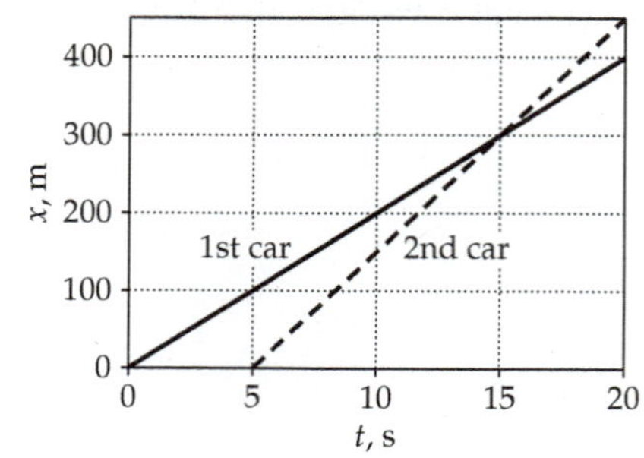

(*b*) 15 s, (*c*) 300 m, (*d*) 100 m

57 15 m/s

59 -2.0 m/s^2

61 (*a*) 2.0 m/s, (*b*) $\Delta x = (2t - 5)\Delta t + (\Delta t)^2$, (*c*) $v(t) = 2t - 5$

63 (*a*) $a_{\text{av AB}} = 3.3\text{ m/s}^2$, $a_{\text{av BC}} = 0$, $a_{\text{av CE}} = -7.5\text{ m/s}^2$, (*b*) 75 m
(*c*)

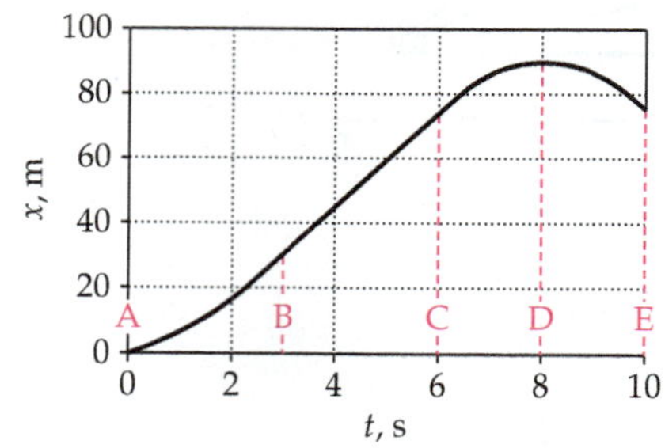

(*d*) At point D, $t = 8$ s, the graph crosses the time axis; therefore $v = 0$.

65 (*a*) 80 m/s, (*b*) 0.40 km, (*c*) 40 m/s

67 16 m/s^2

69 (*a*) 4.1 s, (*b*) 20 m, (*c*) 0.99 s and 3.1 s

71 (*a*)

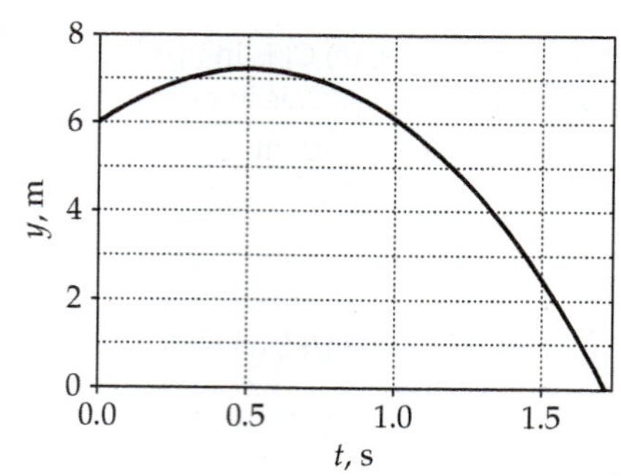

(*b*) 7.3 m, (*c*) 1.7 s, (*d*) 12 m/s

73 44 m

75 68 m/s

77 (*a*) 666 m, (*b*) 14 m/s

79 (*a*) You did not achieve your goal. To go higher, you can increase the acceleration value or the duration of the acceleration. (*b*) 138 s, (*c*) 610 m/s

81 40 cm/s, $-6.9\ \mathrm{cm/s^2}$

83 (*a*) 11 mi/h, (*b*) 0.60

85 11 m

87 28 m

89 (*a*) 2.4 m, (*b*) 1.4 s

93 (*a*) 2.1 d, (*b*) 5.8 y

95 4.8 m/s

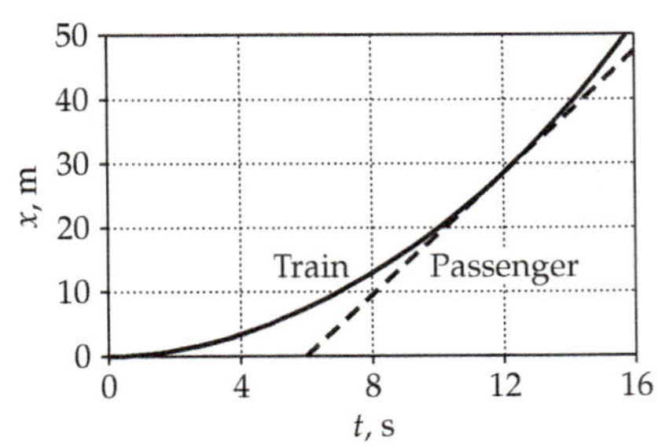

97 $h/3$

99 (*a*) 35 s, (*b*) 1.2 km

(*c*)

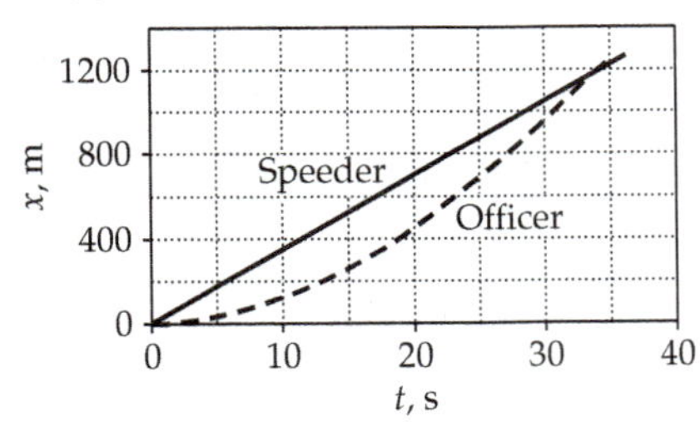

101 (*a*) $2L/3$, (*b*) $\frac{2}{3}t_{\text{fin}}$, (*c*) $\sqrt{4aL/3}$

103 (*a*)

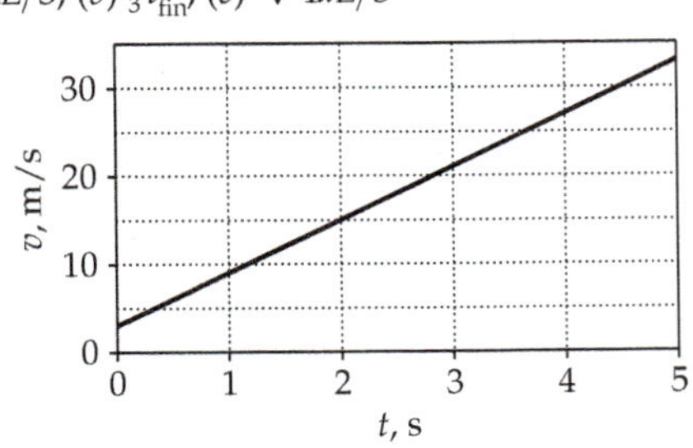

area under the curve = 90 m

(*b*) $x(t) = (3.0\ \mathrm{m/s^2})t^2 + (3.0\ \mathrm{m/s})t$, 90 m

105 $x(t) = (2.3\ \mathrm{m/s^3})t^3 - (5.0\ \mathrm{m/s})t$

107 (*a*) 0.25 m/s per box, (*b*) 0.93 m/s, 3.0 m/s, 6.0 m/s

(*c*)

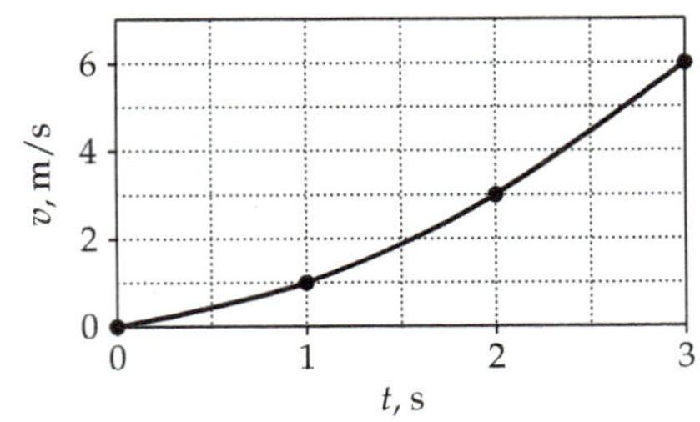

$x(3\ \mathrm{s}) = 6.5\ \mathrm{m}$

109 (*a*)

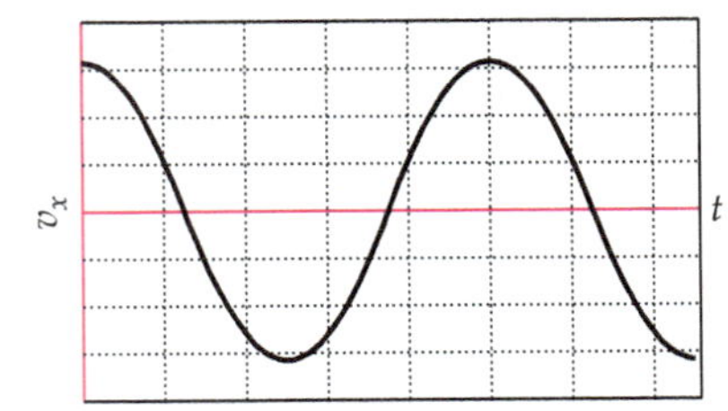

(*b*)

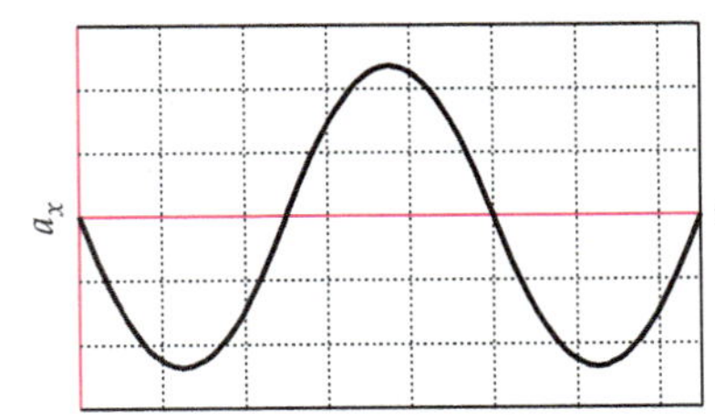

(*c*) The points at the greatest distances from the time axis correspond to turnaround points. The velocity of the body is zero at these points.
(*d*) The velocity is greatest when the slope is greatest, and vice versa. The acceleration is zero when the concavity changes sign, and the acceleration is greatest when the rate of change of the slope with respect to x is greatest.

111 (*a*) $v(t) = (0.10\ \mathrm{m/s^3})t^2 + 9.5\ \mathrm{m/s}$,
(*b*) $x(t) = \frac{1}{6}(0.20\ \mathrm{m/s^3})t^3 + (9.5\ \mathrm{m/s})t - 5.0\ \mathrm{m/s}$,
(*c*) 13 m/s, 15 m/s. v_{av} is not the same as $(v_{\text{i}} + v_{\text{f}})/2$ because the acceleration is not constant.

113 (*b*) 0.452 s, (*c*) $12.0\ \mathrm{m/s^2}$, 22.3%

115 (*a*) The maximum value of the sine function (as in $\sin \omega t$) is 1. Hence, the coefficient $B = v_{\text{max}}$.
(*b*) $a = \omega v_{\text{max}} \cos(\omega t)$. The acceleration is *not* constant.
(*c*) $|a_{\text{max}}| = \omega v_{\text{max}}$, (*d*) $x = x_0 + (v_{\text{max}}/\omega)[1 - \cos(\omega t)]$

117 (*a*) $\mathrm{s^{-1}}$, (*c*) $v_{\text{t}} = g/b$

119 (*b*) 0.762

(*c*)

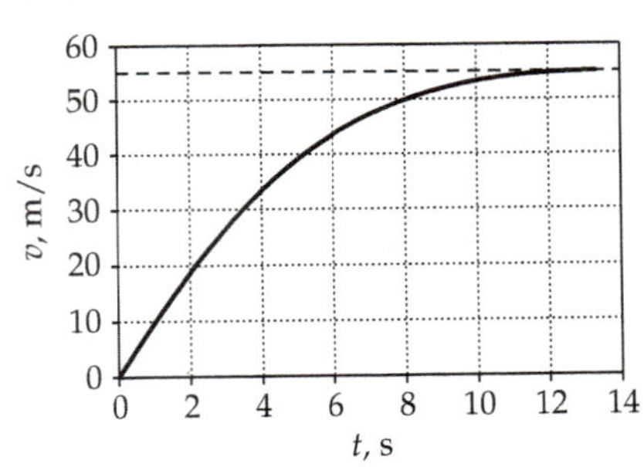

121 You should not contest your ticket.

Chapter 3

1 No. Yes.

3 Zero

5 (*e*)

7 (*c*)

9 (*a*) The velocity vector is tangent to the path.
(*b*)

11 (*a*) A car moving along a straight road while slowing down.
(*b*) A car moving along a straight road while speeding up.
(*c*) A particle moving around a circular track at constant speed.

13 (*a*)

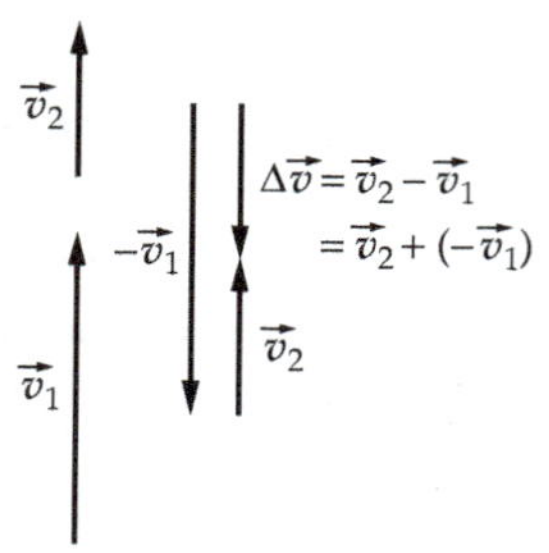

(*b*)

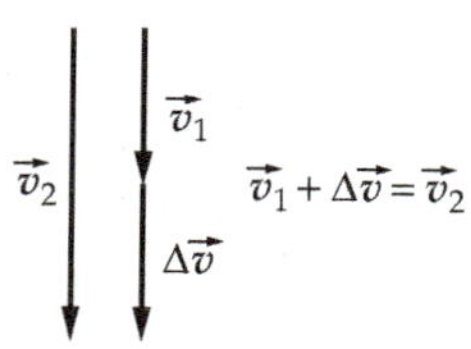

(*c*)

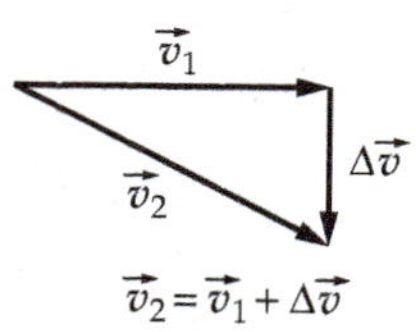

15

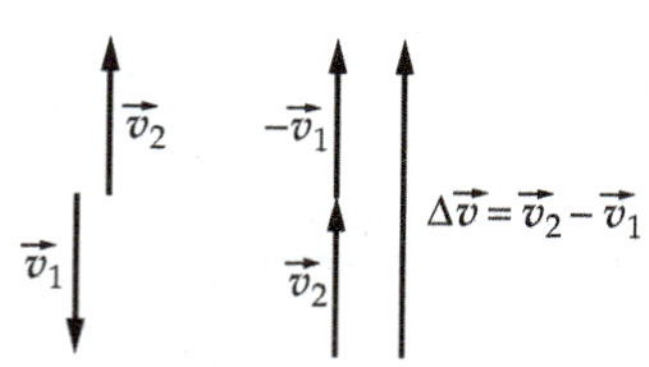

17 You must also be walking west so the rain is falling straight down relative to you.

19 (*a*) True, (*b*) True, (*c*) True

21 (*a*)

23 (*d*)

25 (*a*) False, (*b*) True, (*c*) True, (*d*) False, (*e*) True

27 Assume the $+x$ direction is east and the $+y$ direction is north. Then

(*a*)

Path	Direction of Velocity Vector
AB	North
BC	Northeast
CD	East
DE	Southeast
EF	South

(*b*)

Path	Direction of Acceleration Vector
AB	North
BC	Southeast
CD	0
DE	Southwest
EF	North

(*c*) The magnitude of the acceleration is larger on DE than on BC.

29 The droplet leaving the bottle has the same horizontal velocity as the ship. During the time the droplet is in the air, it is also moving horizontally with the same velocity as the rest of the ship. Because of this, it falls into the vessel, which has the same horizontal velocity. Because you have the same horizontal velocity as the ship does, you see things as if the ship were standing still.

31 (*a*) True, (*b*) False, (*c*) False, (*d*) True

33 (*a*)

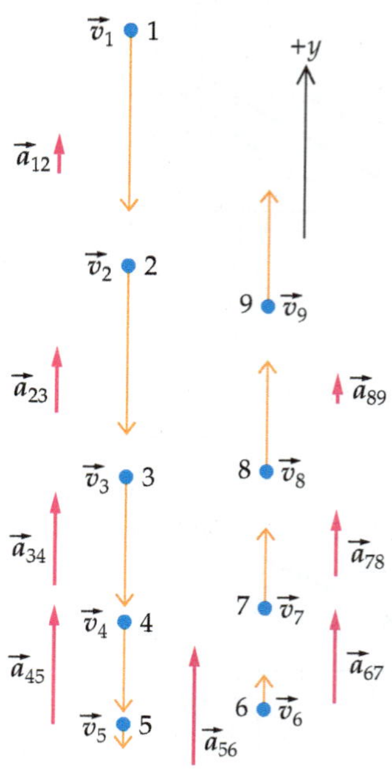

i	v_y (m/s)	Δv_y (m/s)	a_{ave} (m/s²)
1	−0.78		
2	−0.69	0.09	1.8
3	−0.55	0.14	2.8
4	−0.35	0.20	4.0
5	−0.10	0.25	5.0
6	0.15	0.25	5.0
7	0.35	0.20	4.0
8	0.49	0.14	2.8
9	0.53	0.04	0.8

(*b*) The acceleration vector always points upward, so the sign of its y component does not change. The magnitude of the acceleration vector is greatest when the bungee cord has its maximum extension.

35 $\approx 7 \times 10^4$ m/s²

37 15 m/s

39 $\Delta\vec{B} = 0, \Delta\vec{A} = -(0.25\text{ m})\hat{j} - (0.25\text{ m})\hat{i}$

41 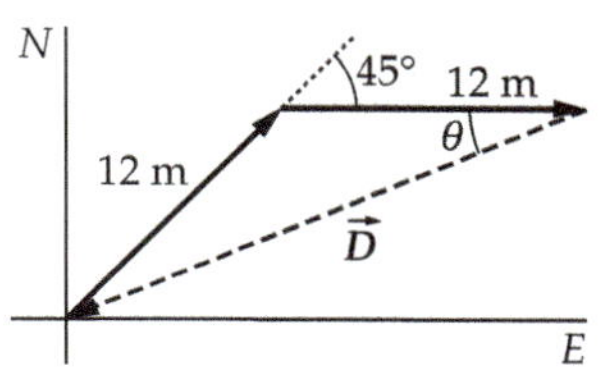

$\vec{D} \approx 22$ m @ 23° south of west

43 (*a*) $\vec{D} = (3.0\text{ m})\hat{i} + (3.0\text{ m})\hat{j} + (3.0\text{ m})\hat{k}$, (*b*) 5.2 m

45 $\vec{v}_{av} = (14\text{ km/h})\hat{i} + (-4.1\text{ km/h})\hat{j}$

47 7.2 m/s

49 (*a*) $\vec{v}_{av} = (33\text{ m/s})\hat{i} + (27\text{ m/s})\hat{j}$,
(*b*) $\vec{a}_{av} = (-3.0\text{ m/s}^2)\hat{i} + (-1.8\text{ m/s}^2)\hat{j}$

51 $\vec{v} = 30\hat{i} + (40 - 10t)\hat{j}, \vec{a} = (-10\text{ m/s}^2)\hat{j}$

53 (*a*) $\vec{v}_{av} = (20\text{ m/s})(-\hat{i} + \hat{j})$, (*b*) $\vec{a}_{av} = (-2.0\text{ m/s}^2)\hat{i}$,
(*c*) $\Delta\vec{r} = (600\text{ m})(-\hat{i} + \hat{j})$

55 (*a*) 16° west of north, (*b*) 280 km/s

57 8.5°, 2.57 h

59 You should fly your plane across the wind.

61 (*a*) $\vec{r}_{AB}(6.0\text{ s}) = (1.2 \times 10^2\text{ m})\hat{i} + (4.0\text{ m})\hat{j}$,
(*b*) $\vec{v}_{AB}(6.0\text{ s}) = (-20\text{ m/s})\hat{i} - (12\text{ m/s})\hat{j}$, (*c*) $(-2.0\text{ m/s}^2)\hat{j}$

63 (*a*) $\vec{v}_{rel} = (0.80\text{ m/s})\hat{i} - (1.2\text{ m/s})\hat{j}$, (*b*) 1.0 m/s

65 1.5×10^{-6} m/s², $1.55 \times 10^{-7}g$

67 (*a*) 463 m/s, 0.343 percent of g, (*b*) From the person toward the center of Earth, (*c*) 380 m/s, 2.76×10^{-2} m/s², (*d*) Zero

69 (*a*) 14 s, 1.8 m/s, (*b*) 0.89 m/s, 0.40 m/s²

71 (*a*) 15 cm, (*b*) The range of accelerations is 1300g to 2700g.

73 $h = (v_0^2 \sin^2\theta_0)/2g$

75 34 m/s

77 20.3 m/s, 36.1°

79 69.3°

81 (*a*) 18 m/s, (*b*) 13°

83 (*a*) 8.1 m, (*b*) 23 m/s

85 63.4° below the horizontal

87 (*a*) The kick is short. (*b*) 0.34 m under the bar, (*c*) 5.2 m

89 (*a*) 0.97 s, (*b*) 4.2 m, (*c*) 13 m/s @ 70° below the horizontal

91 (*a*) 485 km, (*b*) 1.70 km/s

93 (*a*) 194 m, (*c*) 219 m, 11 percent

95 (*b*) 80 m, (*c*) 288 m. The approximate solution is 80 m larger. (The approximation ignores higher-order terms, and they are important when the differences are not small.)

99 (*a*) 11 m/s, (*b*) 3.1 s, (*c*) $\vec{v} = (6.5\text{ m/s})\hat{i} + (-22\text{ m/s})\hat{j}$

101 (*a*) 21.5 m/s, (*b*) 3.53 s, (*c*) 19.3 m/s

103 (*a*) 7.41 m/s, (*b*) 0.756 s, (*c*) 15.9 m/s, 17.5 m/s, 25.0°

105 (*a*) 0.785 m, (*b*) 105 m

107 (*a*) 1.1 m, (*b*) 3.9 m

109 (*a*) 15 km, (*b*) 54 s

111 806 mi/h, 60.3° north of west

113 4th step

115 (*a*) $v_{min} = \dfrac{x}{\cos\theta}\sqrt{\dfrac{g}{2(x\tan\theta - h)}}$,
(*b*) $v_{min} = 26$ m/s $= 58$ mi/h,

119 (*a*) 26 m, (*b*) 7.8°

121 52.9 km, 52.8° east of north

Chapter 4

1 Yes, there are forces acting on it. They are the same as those that would act on it if it were sitting on your table at home.

3 In the limo you hold one end of the string and suspend the object from the other end. If the string remains vertical, the reference frame of the limo is an inertial reference frame. No, you cannot determine the limo's velocity

5 No. Predicting the direction of the subsequent motion requires additional information.

7 The mass of the probe is constant. However, the solar system will attract the probe with a gravitational force.

9 You and the elevator could be either descending and slowing or ascending and speeding up. In both cases your apparent weight is greater than your actual weight.

11 The most significant force in our everyday world is gravity. It literally keeps us on or near the ground. The other most relevant force is the electromagnetic force. It provides "the glue" to hold solids together and make them rigid. It is of great importance in electric circuits.

13 (*a*) Normal force, contact type, (*b*) Normal, contact, (*c*) Normal, contact, (*d*) Normal, contact, (*e*) Gravitational, action-at-a-distance
The two normal forces that the two blocks exert on each other and the two normal forces that the table and the bottom block exert on each other

15 When the plate is sitting on the floor, the normal force F_n acting upward on it is exerted by the floor and is the same size as the gravitational force F_g on the plate. Hence, the plate does not accelerate. However, to slow the plate down as it hits the floor requires that $F_n > F_g$ (or $F_n \gg F_g$, if the

floor is hard and the plate slows quickly). A large normal force exerted on delicate china can easily break it.

17 (*a*) The normal force of the block on the sprinter, in the forward direction.
(*b*) The frictional force by the ice on the puck, in the opposite direction to the velocity.
(*c*) The gravitation force by Earth on the ball, in the downward direction.
(*d*) The force by the stretched bungee cord on the jumper, in the upward direction.

19 (*a*) (2) 100 N, (*b*) Their accelerations are the same. (*c*) The directions of their acceleration are the same.

21 (*a*)

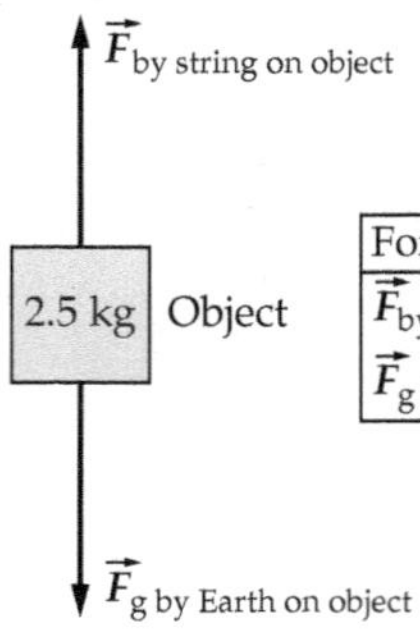

Force	Third-Law Pair
$\vec{F}_{\text{by string on object}}$	$\vec{F}_{\text{by object on string}}$
$\vec{F}_{\text{g by Earth on object}}$	$\vec{F}_{\text{g by object on Earth}}$

(*b*)

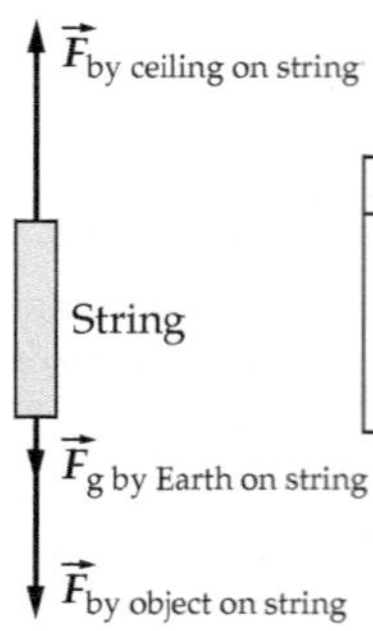

Force	Third-Law Pair
$\vec{F}_{\text{by ceiling on string}}$	$\vec{F}_{\text{by string on ceiling}}$
$\vec{F}_{\text{g by Earth on string}}$	$\vec{F}_{\text{g by string on Earth}}$
$\vec{F}_{\text{by object on string}}$	$\vec{F}_{\text{by string on object}}$

23 (*a*)

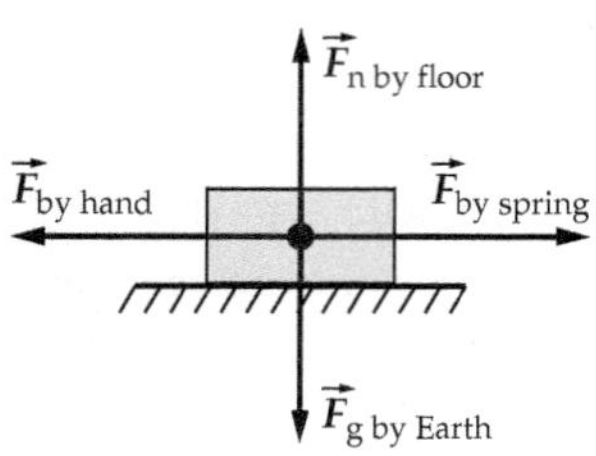

(*b*)

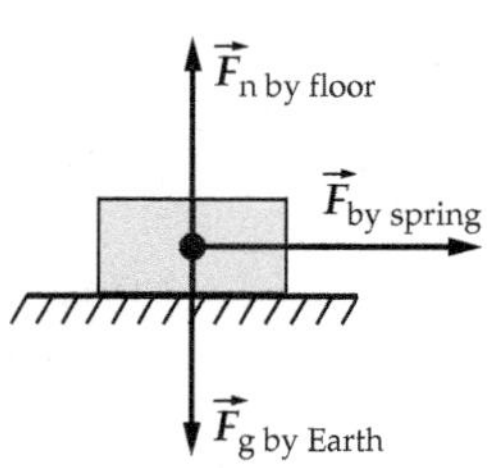

(*c*)

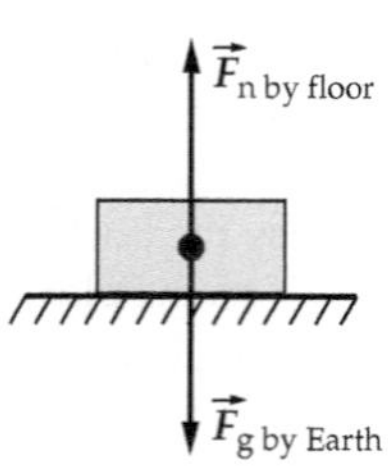

25 (*a*) m_2/m_1, (*b*) m_2/m_1, (*c*) $\Delta x = \frac{1}{2}F(m_2^{-1} - m_1^{-1})(\Delta t)^2$. Because $m_1 > m_2$, the object whose mass is m_2 is ahead.

27 3.6 kN

29 −17 kN

31 (*a*) 6.0 m/s², (*b*) 1/3, (*c*) 2.3 m/s²

33 12 kg

35 (*a*) −3.8 kN, (*b*) 3.00 cm

37 (*a*) 4.2 m/s² @ 45° from each force, (*b*) 8.4 m/s² @ 15° from $2\vec{F}$

39 (*a*) 4.0 m/s², (*b*) 2.4 m/s²

41 (*a*) $\vec{a} = (1.5\ \text{m/s}^2)\hat{\imath} + (-3.5\ \text{m/s}^2)\hat{\jmath}$,
(*b*) $\vec{v}(3.0\ \text{s}) = (4.5\ \text{m/s})\hat{\imath} + (-11\ \text{m/s})\hat{\jmath}$,
(*c*) $\vec{r}(3.0\ \text{s}) = (6.8\ \text{m})\hat{\imath} + (-16\ \text{m})\hat{\jmath}$

43 (*a*) 5.3×10^2 N, (*b*) 1.2×10^2 lb

45 (*a*) 2.45 kN, (*b*) 409 N, (*c*) 2.45 kN

47 (*a*)

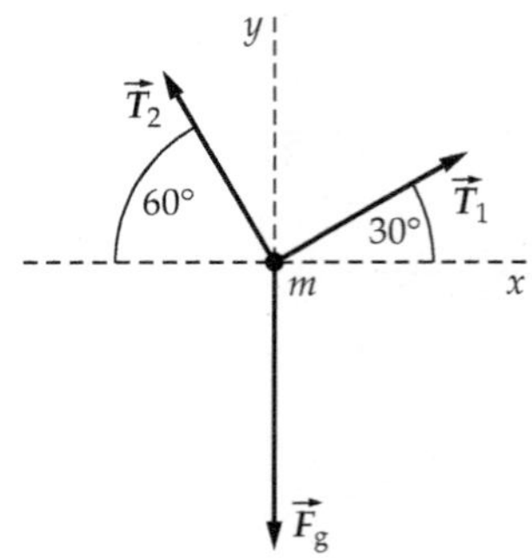

(*b*) T_2 is greater than T_1

49 (*a*) 37°, (*b*) 4.1 N, (*c*) $T_1 = 3.4$ N, $T_2 = 2.4$ N, $T_3 = 3.4$ N

51 $\vec{F}_3 = (-5.0\ \text{N})\hat{\imath} + (-26\ \text{N})\hat{\jmath}$

53 (*a*) 3.82 kN, (*b*) 4.30 kN

55 (*a*) If $T_H = 10.0$ N, then the width of the arch is 9.56 m.
(*b*) If the width of the arch is 8.00 m, then $T_H = 3.72$ N and the arch is 2.63 m high, tall enough for someone to walk through.

57 56.0 N

59 (*a*) $T = 0.42$ kN, $F_n = 0.25$ kN, (*b*) $T = mg\sin\theta$

61 0.55 kN

63 (*a*)

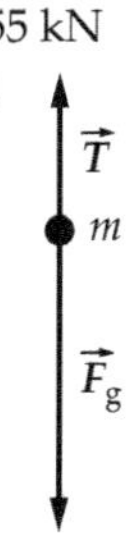

(*b*)

(*c*) No. There is no difference.

65 (*a*) 20 N, (*b*) 20 N, (*c*) 26 N, (*d*) $T_{0\rightarrow 5.0\ \text{s}} = 20$ N, $T_{5\ \text{s}\rightarrow 9\ \text{s}} = 15$ N

67 (*a*) 1.3 m/s², (*b*) $T_1 = 17$ N, $T_2 = 21$ N

69 (*a*) $a = \dfrac{F}{m_1 + m_2}m$, $F_{21} = \dfrac{Fm_1}{m_1 + m_2}$, (*b*) 0.40 m/s², 0.8 N

73 (*a*)

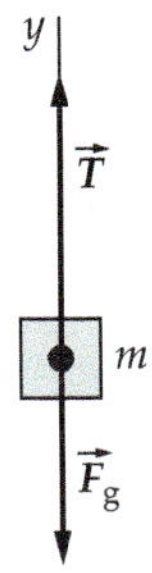

(*b*) 592 N

75 (*a*) $a = \dfrac{g(m_2 - m_1 \sin\theta)}{m_1 + m_2}$, $T = \dfrac{gm_1m_2(1 + \sin\theta)}{m_1 + m_2}$,

(*b*) 2.5 m/s^2, 37 N

77 (*a*) 1.4 m/s^2, 61 N, (*b*) $(m_1/m_2) = 1.19$

79 (*a*) 0.40 kN, (*b*) 0.37 kN

81 (*a*) 5.0 cm, (*b*) $a_{20} = 2.5\ m/s^2$, $a_5 = 4.9\ m/s^2$, $T = 25$ N

83 $m_{\text{2nd mass}} = 1.4$ kg or 1.1 kg

85 $F_{\text{on } m_2} = \left(\dfrac{m^2 + m_1^2 + m_2^2}{m + m_1 + m_2}\right)g$

87 $T_B = 305$ N, $F_{\text{mast on the deck}} = 1.55$ kN

89 (*a*) $-0.10\ km/s^2$, (*b*) 6.1 cm, (*c*) 35 ms

91 (*a*) $a = \dfrac{F}{m_1 + m_2}$, (*b*) $F_{\text{net}} = \dfrac{m_2}{m_1 + m_2}F$, (*c*) $T = \dfrac{m_1}{m_1 + m_2}F$

93 (*a*) 55.0 g, (*b*) 2.45 m/s^2, 2.03 N

95 (*a*) $T = \frac{1}{3}(F_2 + 2F_1)$, (*b*) $t_0 = (3T_0/4C)$

97 (*a*) You should throw your boot in the direction away from the closest shore. (*b*) 420 N, (*c*) 7.52 s

99 (*a*)

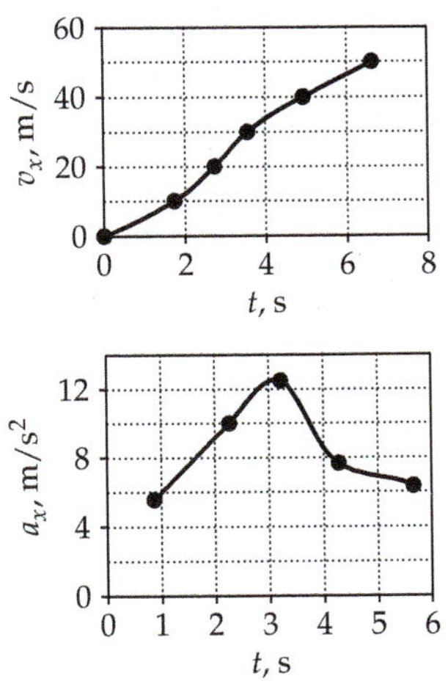

(*b*) 2.8 s to 3.6 s, (*c*) 5500 N, (*d*) 160 m

Chapter 5

1 Static and kinetic frictional forces are responsible for the accelerations. If the coefficient of static friction between the truck bed and the object is sufficiently large, then the object will not slip on the truck bed. The larger the acceleration of the truck, the larger the coefficient of static friction that is needed to prevent slipping.

3 (*d*)

5 (*c*)

7 As the spring is extended, the force exerted by the spring on the block increases. Once that force exceeds the maximum value of the force of static friction, the block will slip. As it does, it will shorten the length of the spring, decreasing the force that the spring exerts. The force of kinetic friction then slows the block to a stop, which starts the cycle over again.

9 (*a*), (*b*), and (*c*)

11 Block 1 will hit the pulley before block 2 hits the wall.

13 Air drag is proportional to the density of air and to the cross-sectional area of the object. On a warm day the air is less dense. The air is also less dense at high altitudes. Pointing his hands results in less area being presented to air drag forces and, hence, reduces them. Rounded and sleek clothing has the same effect as pointing his hands.

15 (*c*)

17 (*a*) The drag force is proportional to the area presented and some power of the speed.The drag force on the feather initially is larger because the feather presents a larger area than does the pebble. As the pebble gains speed the drag force on it increases. The drag force on the pebble eventually exceeds the drag force on the feather because the drag force on the feather cannot exceed the gravitational force on the feather.

(*b*) Terminal speed is much higher for the pebble than for the feather. The acceleration of the pebble will remain high until its speed approaches its terminal speed.

19 $a_{cm} = (m_1/(m_1 + m_2))a_1$

21 The friction of the road on the tire causes the car to slow down

23 The center of mass moves downward

25 The acceleration of the center of mass is zero.

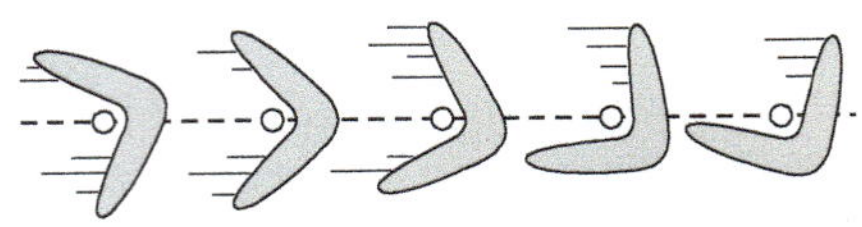

27 (*a*) M/T, kg/s, (*b*) M/L, kg/m, (*c*) ML/T^2, (*d*) 57 m/s, (*e*) 87 m/s

29 $\mu_s \approx 1.4$. This is probably not such a good idea. Tires on asphalt or concrete have a maximum coefficient of static friction of about 1.

31 (*b*)

33 (*a*) 15 N, (*b*) 12 N

35 500 N

37 (*a*) 5.9 m/s^2, (*b*) 76 m

39 (*a*) 49.1 N, (*b*) 123 N

41 (*a*) 4.6°, (*b*) 4.6°

43 0.84 m/s

45 2.4 m/s^2, 37 N

47 (*a*) 4.0 m, (*b*) 0.47

49 (*a*) 2.7 m/s^2, (*b*) 10 s

51 (*a*) 0.96 m/s², (*b*) 0.18 N

53 (*a*) The static-frictional force opposes the motion of the object, and the maximum value of the static-frictional force is proportional to the normal force F_N. The normal force is equal to the weight minus the vertical component F_y of the force F. Keeping the magnitude F constant while increasing θ from zero results in a decrease in F_y and thus a corresponding decrease in the maximum static-frictional force, f_{max}. The object will begin to move if the horizontal component F_x of the force F exceeds f_{max}. An increase in θ results in a decrease in F_x. As θ increases from 0, the decrease in F_N is larger than the decrease in F_x, so the object is more and more likely to slip. However, as θ approaches 90°, F_x approaches zero and no movement will be initiated. If F is large enough and if θ increases from 0, then at some value of θ, the block will start to move.

(*b*)

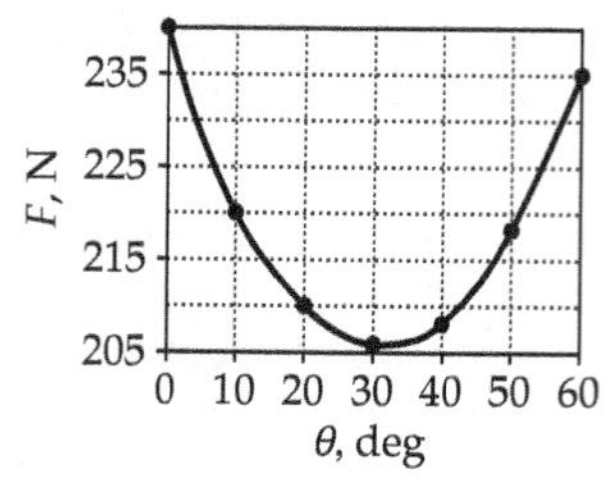

From the graph, we can see that the minimum value for F occurs when $\theta \approx 32°$.

57 (*a*) 0.24, (*b*) 1.4 m/s²

59 (*a*) 18 N, (*b*) 1.5 m/s², 2.9 N, (*c*) $a_1 = 2.0$ m/s², $a_2 = 7.8$ m/s²

61 (*a*) 5.7°, (*b*) 1.9 m/s

63 (*a*) $F_{min} = -1.6$ kg, $F_{max} = 84$ N, (*b*) $F_{min} = 5.8$ N and $F_{max} = 37$ N (The $+x$ direction is to the right.)

65 (*b*) 0.30, (*c*) 2.8 m/s

67 2.8×10^{-4} kg/m

69 $d_{n\text{ filters}} = \sqrt{n}\, d_{1\text{ filter}}$

71 25 m/s

73 (*a*) about 39 ms, (*b*) With the drag force in Problem 72, it takes about 86 times longer than it does using the centrifuge.

75 25°

77 (*a*) 1.4 m/s, (*b*) 8.5 N

79 (*a*) 8.33 m/s², upward, (*b*) 667 N, upward, (*c*) 1.45 kN, upward

81 $T_1 = [m_2(L_1 + L_2) + m_1L_1](2\pi/T)^2$, $T_1 = m_2(L_1 + L_2)(2\pi/T)^2$

83 (*a*) 53° above horizontal, 0.41 kN, (*b*) 53° below horizontal, 0.41 kN

85 (*a*) 0.40 N, (*b*) 0.644

87 52°

89 12.8 m/s

91 (*a*) 7.3 m/s, (*b*) 0.54

93 22°

95 (*a*) 7.8 kN, (*b*) −0.78 kN

97 20 km/h ≤ v ≤ 56 m/h

99 (*a*) about 60.4 m, (*b*) about 60.6 m, (*c*) about 3.3 s, (*d*) about 3.7 s, (*e*) less than

101 (0.23 cm, 0)

103 (2.0 m, 1.4 m)

105 (1.5 m, 1.4 m)

107 $(\frac{1}{4}L, \frac{1}{4}L)$

113 $\vec{v}_{cm} = (3.0\text{ m/s})\hat{i} - (1.5\text{ m/s})\hat{j}$

115 $\vec{a}_{cm} = (2.4\text{ m/s}^2)\hat{i}$

117 (*a*)

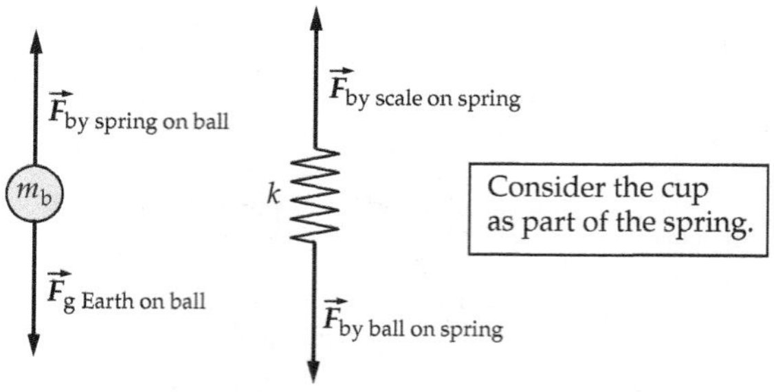

(*c*) $F_{by\ scale} = (m_{ball} + m_{platform})g$

119 (*a*)

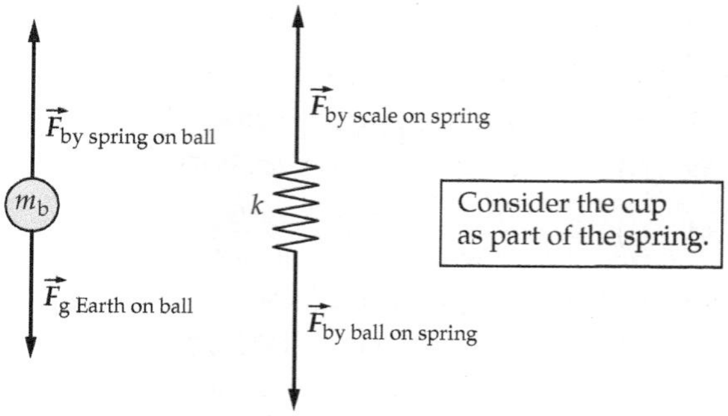

(*b*) $d' = \dfrac{m_b(g + a)}{k} > d$,

(*c*) $F_{by\ scale} = (m_b + m_p)(a + g) > F_{by\ scale\ pblm\ 117}$

121 0.51

123 1.49 kN

125 (*a*) 49 m/s², (*b*) 13 rev/min

127 (*a*) 0.19 kN, (*b*) 52 N, (*c*) 35 N, (*d*) 0.24, (*e*) 0.54 kN

129 0.43

131 (*a*)

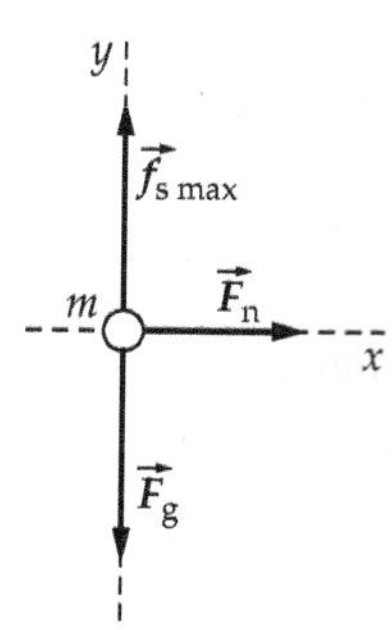

(*b*) 0.74 kN, (*c*) 20 rev/min. This result holds for all patrons, regardless of their mass.

133 Yes. Sally's claim seems to be supported by Liz's calculation.

135 4.67×10^6 m. It is inside the surface of Earth.

137 (*a*) 35 cm, (*b*) 4.7 m/s downward (*c*) 14.2 kN, 9.81 m/s^2, downward

139 (*a*) 4.91 m/s, (*b*) 1.64 m/s downward, (*c*) 1.09 m/s^2 downward

Chapter 6

1 (*a*) True, (*b*) True, (*c*) False, (*e*) True

3 (*a*) False, (*b*) True, (*c*) False, (*d*) True

5 An object moving along a curved path at constant speed has constant kinetic energy, but is accelerating (because its velocity is continually changing direction). No, because if the object is not accelerating, the net force acting on it must be zero and, consequently, its kinetic energy must be constant.

7 The work required to stretch a spring 2.0 cm is greater than the work required to stretch it 1.0 cm by a factor of 4.

9 (*d*)

11 (*a*) False, (*b*) False, (*c*) False, (*d*) False

13 (*a*) False, (*b*) False, (*c*) True, (*d*) True

15 $2\Delta t$

17 The only external force (neglecting air resistance) that does center-of-mass work on the car is the static friction force $\vec{f}_s$ exerted by the road on the tires. The positive center-of-mass work this friction force does is translated into a gain of kinetic energy.

19 (*a*) 4.5×10^{18} J, (*b*) 1%, (*c*) 1.4×10^{11} W

21 21 kJ

23 (*a*) 147 J, (*b*) 266 J

25 11 kJ, 3.5 kW

27 (*a*) 6.0 J, (*b*) 12 J, (*c*) 3.5 m/s

31 (*a*) $m(y) = 20\text{ kg} - (2.5\text{ kg/m})y$, (*b*) 0.59 kJ

33 (*a*)

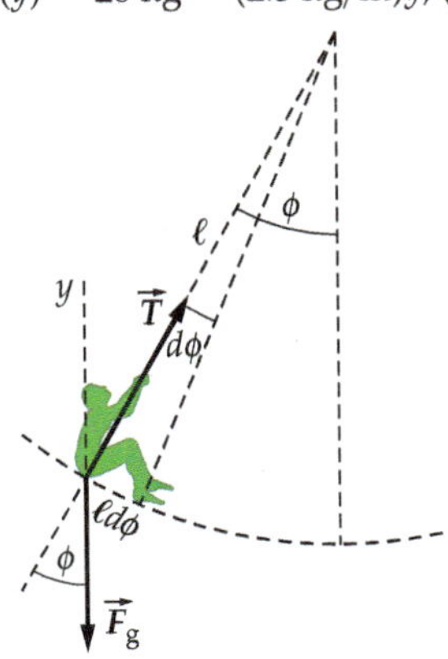

(*b*) $dW_{F_g} = -mg\ell \sin\phi \, d\phi$, (*c*) 2.5 kJ, 7.0 m/s

35 $W_{x_0 \to x} = A\left(\frac{1}{x_0} - \frac{1}{x}\right)$, $K_{x\to\infty} = \frac{A}{x_0}$, $v_{x\to\infty} = \sqrt{\frac{2A}{mx_0}}$

37 180°

39 (*a*) −24, (*b*) −10, (*c*) 0

41 (*a*) 1.0 J, (*b*) 0.21 N

43 (*b*) $6\hat{i} + 8\hat{j}$. Another vector that meets the requirements is $-6\hat{i} - 8\hat{j}$.

45 (*b*) $\vec{p} = (34\text{ kg}\cdot\text{m/s})\hat{i} + (-16\text{ kg}\cdot\text{m/s})\hat{j}$, (*c*) 0.28 kJ, 0.28 kJ

49 (*a*) $P(t) = (3.1\text{ W/s})t$, (*b*) 9.4 W

51 0.15 kW

55 (*a*) 0.381 kg/m, (*b*) 148 mi/h

57 3.2×10^5 m

59 50 kW

61 (*a*) 405 N, (*b*) 19.9 N

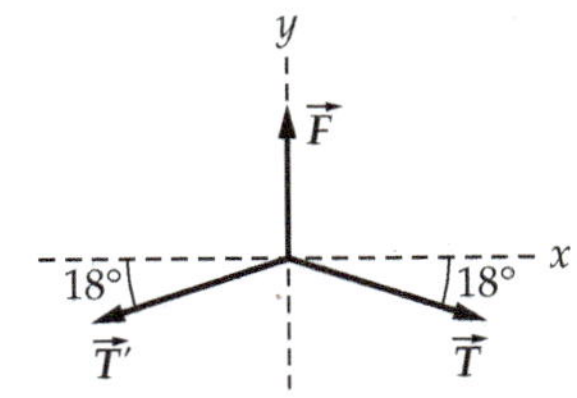

(*c*) 39.7 mJ

63 (*a*) $F(x) = mC^2x$, (*b*) $W = \frac{1}{2}mC^2x_1^2$

65 (*a*) $v = (6t^2 - 8t)$, $a = (12t - 8)$, (*b*) $P = 8mt(9t^2 - 18t + 8)$, (*c*) $W = 2mt_1^2(3t_1 - 4)^2$

67 (*a*) 208 kW, (*b*) 5.74 km

69 (*a*)

x (m)	4.0	−3.0	−2.0	−1.0	0.0	1.0	2.0	3.0	4.0
W (J)	6.0	4.0	2.0	0.5	0.0	0.5	1.5	2.5	3.0

(*b*) 28.0 J

71 (*b*) $W_{1\text{ rev cw}} = (31\text{ m})F_0$, $W_{1\text{ rev ccw}} = (-31\text{ m})F_0$

73 (*a*) $F_x = -kx\left(1 - \frac{y_0}{\sqrt{x^2 + y_0^2}}\right)$, (*c*) $v_f = \frac{L^2}{2y_0}\sqrt{\frac{k}{m}}$

Chapter 7

1 (*d*)

3 (*a*) False, (*b*) False, (*c*) False, (*d*) False

5 (*a*) True, (*b*) True, (*c*) True, (*d*) True, (*e*) False

7 (*a*) False, (*b*) False, (*c*) False, (*d*) False, (*e*) True

9 (*d*)

11 (*a*) Yes, (*b*) No, (*c*) No

13 (*a*) 25 cm, (*b*) −0.12 kJ

15 (*a*) 16 s, (*b*) 6.8 min. Not feasible that you can maintain the pace for 6.8 min.

17 1.5×10^{18} J/y, 3%

19 2.4×10^5 L/s

21 (*a*) 0.39 kJ, (*b*) 2.5 m, 4.9 m/s, (*c*) 24 J, 0.37 kJ, (*d*) 0.39 kJ, 20 m/s

23 (*a*) 10 cm, (*b*) 14 cm

25 (*a*) $F_x = C/x^2$, and $\vec{F} = F_x\hat{i}$
(*b*) If $x > 0$, $\vec{F}$ points away from the origin. If $x < 0$, $\vec{F}$ points toward the origin.
(*c*) Decrease.
(*d*) If $x > 0$, $\vec{F}$ points toward the origin. If $x < 0$, $\vec{F}$ points away from the origin.

27 $U(x) = [(-0.63\text{ kJ}\cdot\text{m})/x] + 0.30\text{ kJ}$

29 (*a*) $F_x = 4x(x+2)(x-2)$, (*b*) $x = -2$ m, $x = 0$, and $x = 2$ m, (*c*) Unstable at $x = -2$ m, stable at $x = 0$, unstable at $x = 2$ m

31 (*a*) $x = 0.0$ m and $x = 2.0$ m,
(*b*)

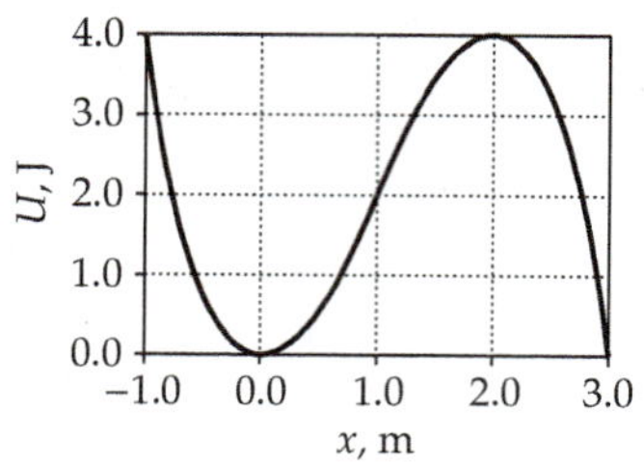

(*c*) Stable equilibrium at $x = 0$, unstable equilibrium at $x - 2.0$ m, (*d*) 2.0 m/s

33 (*a*) $U(\theta) = (m_2\ell_2 - m_1\ell_1)g\sin\theta$, (*b*) U is a minimum at $\theta = -\pi/2$, U is a maximum at $\theta = \pi/2$

35 (*a*) $U(y) = -mgy - 2Mg\left(L - \sqrt{y^2 + d^2}\right)$,
(*b*) $y = d\sqrt{m^2/4M^2 - m^2}$,
(*d*) This is a point of stable equilibrium.

37 $v = \sqrt{gL/2}$

39 (*a*) 0.858 m, (*b*) The block will retrace its path, rising to a height of 5.00 m.

41 26°

43 $U = \dfrac{[mg(\sin\theta + \mu_s\cos\theta)]^2}{2k}$

45 $6mg$

47 16.7 kN

49 $6mg$

51 (*a*) 31 m, (*b*) 34 m/s

53 (*a*) 0.15 km, (*b*) 45 m/s

55 (*a*) $K_{max} = \frac{5}{2}mgL$, (*b*) $6mg$

57 (*a*) 20°, (*b*) 6.4 m/s

59 $v_2 = L\sqrt{2(g/L)(1 - \cos\theta) + (k/m)(\sqrt{\frac{13}{4}} - 3\cos\theta - \frac{1}{2})^2}$

61 (*a*) 82 kJ, (*b*) The energy comes from the internal chemical energy in your body. (*c*) 410 kJ, (*d*) 330 kJ,

63 (*a*) 0.10 kJ, (*b*) 70 J, (*c*) 34 J, (*d*) 2.9 m/s

65 (*a*) 7.7 m/s, (*b*) 59 J, (*c*) 0.33

67 (*a*) (14 N)y, (*b*) −(14 N)y, (*c*) 2.0 m/s

69 0.87 m, 2.7 m/s

71 (*a*) 9.0×10^{13} J, (*b*) \$2.5 $\times 10^6$, (*c*) 2.8×10^4 y

73 (*a*) 3.9×10^{31} MeV, (*b*) 4.2×10^8 m/s. As expected, this result (4.2×10^8 m/s) is greater than the speed of light (and thus incorrect). Use of the nonrelativistic expression for kinetic energy is not justified.

75 1.1×10^5 reactions/s

77 0.782 MeV

79 (*a*) 1.1 kg, (*b*) 2.7×10^9 kg

81 (*a*) 6, (*b*) 0.21 eV

83 $\Delta E_{therm} = -mgv\,\Delta t\sin\theta$

85 12 m²

87 (*a*) 0.208, (*b*) 3.5 MJ

89 (*a*) $x_{1,-} = (2\mu_k mg/k) - x_i$, (*b*) $v_0 = \sqrt{(k/m)x_0^2 - 2\mu_k g x_0}$, (*c*) $\mu_k = kx_i/2mg$

91 (*a*) 11 kW, (*b*) −6.8 kW, (*c*) \$1.81, \$5.43

93 (*a*) 1.61 kJ, (*b*) 0.6 kJ, (*c*) 23 m/s

95 (*b*)

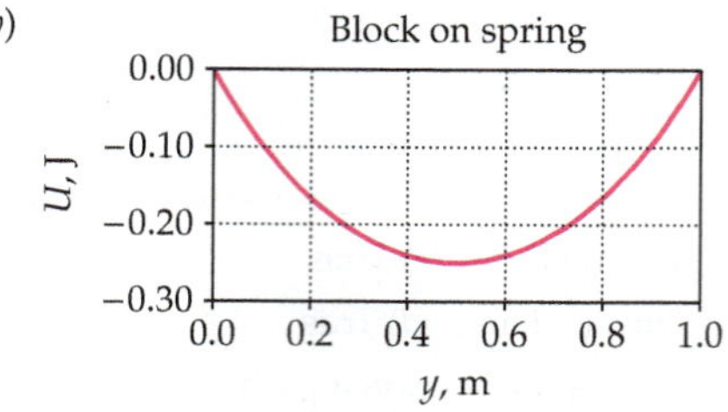

97 (*a*) 17 m, (*b*) 4.91 kN, (*c*) 4.9 m/s², (*d*) 13 kN, upward, (*e*) 5.5 kN, 64°, (*f*) 1.4 kN

99 (*a*) $F_{20} = 491$ N, $F_{30} = 981$ N, (*b*) $P_{20} = 9.8$ kW, $P_{30} = 29$ kW, (*c*) 8.8°, (*d*) 6.36 km/L

103 (*a*) $v = \sqrt{2my/(m + M)}$, (*b*) Same as in (*a*)

107 (*a*)

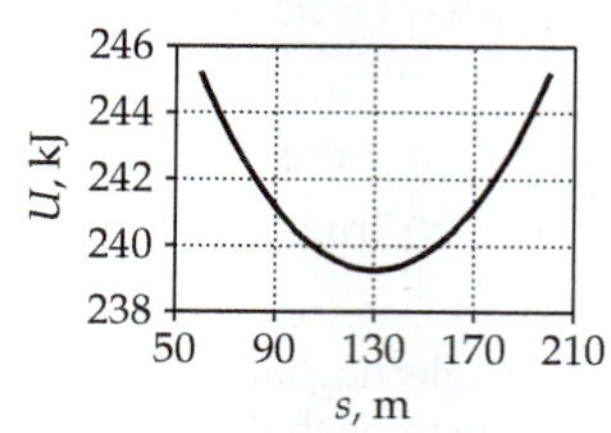

(*b*) 5.4 kJ

Chapter 8

5 The momentum of the bullet–gun system is initially zero. After firing, the bullet's momentum is directed west. Momentum conservation requires that the system's total momentum does not change, so the gun's momentum must be directed east. Kinetic energy is not conserved.

7 In a way, the rocket does need something to push upon. It pushes the exhaust in one direction, and the exhaust pushes it in the opposite direction. However, the rocket does not push against the air.

9 Think of someone pushing a box across a floor. Her push on the box is equal but opposite to the push of the box on her, but the action and reaction forces act on *different objects*. Newton's second law states that the sum of the forces acting on the box equals the rate of change of momentum of the box. This sum does not include the force of the box on her.

11 Hovering in midair while tossing objects violates the conservation of linear momentum! To throw something forward requires being pushed backward. Superheroes are not depicted as experiencing this backward motion that is predicted by conservation of linear momentum. This action does not violate conservation of energy.

13 The road. (The frictional force by the road on the tire causes the car to slow down.)

15 About 10^4

17 (*a*) False, (*b*) True, (*c*) True, (*d*) True

19 (*a*) The loss of kinetic energy is the same in both cases. (*b*) The situation in which the two objects have oppositely directed velocities.

21 (*b*)

23 The water is changing direction when it rounds the corner in the nozzle. Therefore, the nozzle must exert a force on the stream of water to change its momentum, and from Newton's third law, the water exerts an equal but opposite force on the nozzle. This requires a net force in the direction of the momentum change.

25 In the center-of-mass frame the two velocities are equal and opposite, both before and after the collision. In addition, the speed of each puck is the same before and after the collision. The direction of the velocity of each puck changes by some angle during the collision.

27 The downward force of lunar gravity and the upward thrust provided by the rocket combustion products.

29 Think of the sail facing the fan (like the sail on a square rigger might), and think of the stream of air molecules hitting the sail. Imagine that they bounce off the sail elastically—their net change in momentum is then roughly twice the change in momentum that they experienced going through the fan. Thus the change in momentum of the air is backward, so to conserve momentum of the air–fan–boat system the change in momentum of the fan–boat system will be forward.

31 (*a*) 2.34 s, (*b*) 6.7 m/s

33 5.5 m/s

35 4.0 m/s to the right

37 $\vec{v}' = 2v\hat{i} - v\hat{j}$

39 0.084

41 (*a*) 44 J, (*b*) $\vec{v}_{cm} = (1.5\text{ m/s})\hat{i}$, (*c*) $\vec{v}_{1\text{ rel}} = (3.5\text{ m/s})\hat{i}$, $\vec{v}_{2\text{ rel}} = (-3.5\text{ m/s})\hat{i}$, (*d*) 37 J

43 (*a*) 11 N · s, (*b*) 1.3 kN

45 1.81 MN · s, 0.60 MN

47 0.23 kN

49 (*a*) 1.1 N · s directed into the wall, (*b*) 0.36 kN, into the wall, (*c*) 0.48 N · s, away from the wall, (*d*) 3.8 N, away from the wall

51 (*a*) 0.20 s, (*b*) 27 ms, (*c*) Because the collision time is much shorter for the sawdust landing, the average force exerted on the vaulter by the airbag is much less than the average force the sawdust exerts on him.

53 (*a*) 20 m/s, (*b*) Twenty percent of the initial kinetic energy is transformed into thermal energy, acoustic energy, and the deformation of metal.

55 (*a*) 2.0 m/s, (*b*) The collision was not elastic.

57 $v_{\text{p f}} = -0.25$ km/s, $v_{\text{nuc f}} = 46$ m/s

59 (*a*) 5.0 m/s, (*b*) 0.25 m, (*c*) $v_{1f} = 0$, $v_{2f} = 7.0$ m/s

61 (*a*) $0.2v_0$, (*b*) $0.4v_0$

63 0.45 km/s

65

67 $h = (v^2/8g)(m_1/m_2)^2$

69 0.0529

71 (*a*) The meteorite should impact Earth along a line exactly opposite Earth's orbital velocity vector. (*b*) 2.71×10^{-15} percent, (*c*) 1.00×10^{23} kg

73 1.5×10^6 m/s

75 (*a*) $\vec{v}_1 = (312\text{ m/s})\hat{i} + (66.6\text{ m/s})\hat{j}$, (*b*) 5.6 km, (*c*) 35.8 kJ

77 0.91

79 (*a*) 20%, (*b*) 0.89

81 (*a*) 1.7 m/s, (*b*) 0.83

83 (*a*) At room temperature rubber will bounce more when it hits a stick than it will at freezing temperatures. (*b*) 3.8 cm

85 (*a*) 60°, (*b*) $v_{cf} = 2.50$ m/s, $v_8 = 4.33$ m/s

87 (*a*) $v_1 = 1.7$ m/s and $v_2 = 1.0$ m/s, (*b*) The collision was elastic.

89 5.3 m/s, 29°

91 $$K_i = K_f = \frac{p_1^2}{2}\left[\frac{m_1^2 + 6m_1m_2 + m_2^2}{m_1^2m_2 + m_1m_2^2}\right] = \frac{p'^2_1}{2}\left[\frac{m_1^2 + 6m_1m_2 + m_2^2}{m_1^2m_2 + m_1m_2^2}\right]$$

$p_1' = \pm p_1$. If $p_1' = +p_1$, the particles do not collide.

93 (*a*) $\vec{v}_{cm} = 0$, (*b*) $\vec{u}_3 = (-5.0\text{ m/s})\hat{i}$, $\vec{u}_5 = (3.0\text{ m/s})\hat{i}$, (*c*) $\vec{u}'_3 = (5.0\text{ m/s})\hat{i}$, $u'_5 = 0.75$ m/s, (*d*) $\vec{v}'_3 = (5.0\text{ m/s})\hat{i}$, $\vec{v}'_5 = (-3.0\text{ m/s})\hat{i}$, (*e*) $K_i = K_f = 60$ J

95 (*a*) 360 kN, (*b*) 120 s, (*c*) 1.72 km/s

97 (*d*) 28

99 0.19 m/s, $K_i = 31$ mJ, $K_f = 12$ mJ

101 (*a*) $\vec{p} = -(1.1 \times 10^5\text{ kg}\cdot\text{km/h})\hat{i} + (1.1 \times 10^5\text{ kg}\cdot\text{km/h})\hat{j}$, (*b*) 43 km/h. The direction of the wreckage is 46° west of north.

103 (*a*) 6.3 m/s, (*b*) 20 m

105 (*a*) The velocity of the basketball will be equal in magnitude but opposite in direction to the velocity of the baseball. (*b*) 0, (*c*) $2v$

107 (*a*) 30 km/s, (*b*) 8.1. The energy comes from an immeasurably small slowing of Saturn's orbital speed.

109 The driver was not telling the truth.

111 8.9 kg

113 (*b*) 55

115 (*a*) $$v_{2f} = \left(\frac{m_b}{m_2 + m_b}\right)\left(1 + \frac{m_1}{m_1 + m_b}\right)v,$$

$$v_{1f} = -\frac{m_2m_b(2m_1 + m_b)}{(m_1 + m_b)^2(m_2 + m_b)}v,$$

(*b*) $$\Delta K = \frac{1}{2}\frac{m_2m_b^2(2m_1 + m_b)^2}{(m_2 + m_b)^2(m_1 + m_b)^2}\left(1 + \frac{m_1m_2}{(m_1 + m_b)^2}\right)v^2.$$

This additional energy came from chemical energy in the astronaut's bodies.

117 4.53 m/s

Chapter 9

1 (*a*) The point on the rim. (*b*) Both turn through the same angle. (*c*) The point on the rim. (*d*) Both have the same angular speed. (*e*) Both have zero tangential acceleration. (*f*) Both have zero angular acceleration. (*g*) The point on the rim.

3 (*c*)

5 (*a*) Tara, (*b*) Tara, (*c*) Neither

7 By choking up, you are rotating the bat about an axis closer to the center of mass, thus reducing the bat's moment of inertia. The smaller the moment of inertia, the larger the angular acceleration (a quicker bat) for the same torque.

9 (*b*)

11 (*b*)

13 One reason is to maximize the moment arm about the line through the hinge pins for the force exerted by someone pulling or pushing on the knob.

15 (*b*)

17 (*b*)

19 (*a*)

23 12 rev

25 10%

27 Approximately 6

29 (*a*) 16 rad/s, (*b*) 47 rad, (*c*) 7.4 rev, (*d*) 4.7 m/s and 73 m/s^2

31 (*a*) 40 rad/s, (*b*) $a_t = 0.96$ m/s^2, $a_t = 0.19$ km/s^2

33 (*a*) 0.59 rad/s^2, (*b*) 4.7 rad/s

35 3.6 rad/s

37 1.0 rad/s, 9.9 rev/min

39 (*a*) 2.94 rad, (*b*) 780 d

41 60 kg·m^2

43 (*a*) 28 kg·m^2, (*b*) 32 kg·m^2

45 2.6 kg·m^2

47 (*b*) $I_{cm} = \frac{1}{12}m(a^2 + b^2)$

49 5.4×10^{-47} kg·m^2

51 1.4×10^{-2} kg·m^2

55 $I = \frac{3}{10}MR^2$

57 $I_x = 3M((H^2/5) + (R^2/20))$

59 (*a*) 1.9 N·m, (*b*) 1.2×10^2 rad/s^2, (*c*) 6.2×10^2 rad/s

61 (*a*) $a_t = g\sin\theta$, (*b*) $mgL\sin\theta$

63 (*a*) $d\tau = (2\mu_k Mg/R^2)r^2\,dr$, (*b*) $\tau = \frac{2}{3}MR\mu_k g$, (*c*) $\Delta t = (3R\omega/4\mu_k g)$

65 (*a*) 85 mJ, (*b*) 72 rev/min

67 (*a*) 19.6 kN, (*b*) 5.9 kN·m, (*c*) 0.27 rad/s, (*d*) 1.6 kW

69 (*a*) 3.6 rad/s, (*b*) 3.6 rad/s

71 3.1 m/s^2, $T_1 = 12$ N, $T_2 = 13$ N

73 30°

75 8.21 m/s

77 (*a*) $a = g/(1 + (2M/5m))$, (*b*) $T = (2mMg)/(5m + 2M)$

79 (*a*) 72 kg, (*b*) 1.4 rad/s^2, 0.29 kN, 0.75 kN

81 (*a*) $a = \dfrac{g\sin\theta}{1 + (m_1/2m_2)}$, (*b*) $T = \dfrac{m_2 g\sin\theta}{1 + (2m_2/m_1)}$, (*c*) $v = \sqrt{\dfrac{2gh}{1 + (m_1/2m_2)}}$, (*d*) $a = g$, $T = 0$, and $v = \sqrt{2gh}$

83 10 kJ

85 3.1 m/s

87 (*a*) 0.19 m/s^2, (*b*) 0.96 N

89 20°

91 (*a*) $a = \frac{2}{3}g\sin\theta$, (*b*) $f_s = \frac{1}{3}mg\sin\theta$, (*c*) $\theta_{max} = \tan^{-1}(3\mu_s)$

93 $v' = \sqrt{\frac{4}{3}}v$

95 0.22 kJ

97 (*a*) $\alpha = 2F/[R(M + 3m)]$, counterclockwise, (*b*) $a_C = F/(M + 3m)$, in direction of $\vec{F}$, (*c*) $a_{CB} = -2F/(M = 3m)$, opposite to direction of $\vec{F}$

99 (*a*) 0.40 rad/s^2, 0.20 rad/s^2, (*b*) 4.0 N, clockwise

101 (*a*) $s_1 = \frac{12}{49}(v_0^2/\mu_k g)$, $t_1 = \frac{2}{7}(v_0/\mu_k g)$, and $v_1 = \frac{5}{7}v_0$, (*b*) $\frac{5}{7}$, (*c*) $s_1 = 27$ m, $t_1 = 3.9$ s, and $v_1 = 5.7$ m/s

103 $v = (2r\omega_0/7)$

105 (*a*) 0.19 s, (*b*) 0.67 m, (*c*) 2.9 m/s

107 (*a*) $v = \frac{11}{7}v_0$, (*b*) $\Delta t = \frac{4}{7}(v_0/\mu_k g)$, (*c*) $\Delta x = \frac{36}{49}(v_0^2/\mu_k g)$

109 13 cm

111 (*a*) 7.4 m/s^2, (*b*) 15 m/s^2, (*c*) 2.4 m/s

113 (*a*) 7.8×10^2 kJ, (*b*) 90 N·m, 0.15 kN, (*c*) 1.4×10^3 rev

115 (*a*) 15 m, (*b*) 15 rad/s

117 (*a*) $\omega = \sqrt{4g/3r}$, (*b*) $F = \frac{7}{3}Mg$

119 (*a*) 32.2 rad/s, (*b*) 23° (*c*) 24 rad/s^2

121 (*a*) 14.7 m/s^2, (*b*) 66.7 cm

123 42 J

125

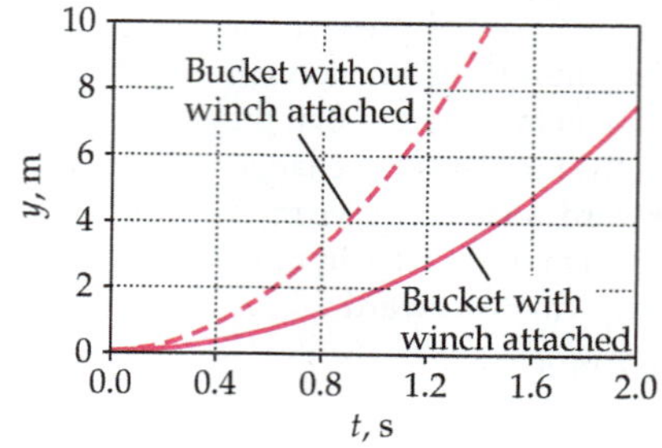

127 (*a*) 26 N, (*b*) 1.1 m/s^2, (*c*) 3.2 kg

Chapter 10

1 (*a*) True, (*b*) False, (*c*) False

3 90°

5 (*a*) L is doubled, (*b*) L is doubled

7 False. A high diver going from a tucked to a layout position.

9 (*e*)

11 The hardboiled egg is solid inside, so everything rotates with a uniform angular speed. By contrast, when you start an uncooked egg spinning, the yolk will not immediately spin with the shell, and when you stop it from spinning the yolk will continue to spin for a while.

13 (*a*)

15 (*a*) The plane tends to veer to the right. The change in angular momentum $\Delta\vec{L}_{prop}$ for the propeller is up, so the net torque $\vec{\tau}$ on the propeller is up as well. The propeller must exert an equal but opposite torque on the plane. This downward torque exerted by the prop on the plane tends to cause a downward change in the angular momentum of the plane. This means the plane tends to rotate clockwise as viewed from above.
(*b*) The plane tends to veer downward. The change in angular momentum $\Delta\vec{L}_{prop}$ for the propeller is to the right,

so the net torque $\vec{\tau}$ on the propeller is toward the right as well. The propeller must exert an equal but opposite torque on the plane. This leftward directed torque exerted by the prop on the plane tends to cause a leftward-directed change in angular momentum for the plane. This means the plane tends to rotate clockwise as viewed from the right.

17 (*a*) Your kinetic energy decreases. Increasing your moment of inertia I while conserving your angular momentum L decreases your kinetic energy $K = L^2/(2I)$.
(*b*) Extending your arms out to the side increases your moment of inertia, and decreases your angular speed. The angular momentum of the system is unchanged.

19 About 4 rev/s

21 (*a*) 33, (*b*) 33, (*c*) 8, (*d*) 14

23 (*a*) 2.4×10^{-8} kg · m²/s,
(*b*) $\ell(\ell + 1) = 5.2 \times 10^{52}$, $\ell \approx 2.3 \times 10^{26}$,
(*c*) $\Delta\ell = 2.3 \times 10^{18}$. The quantization of angular momentum is not noticed in macroscopic physics because no experiment can detect a fractional change in ℓ of 10^{-6}%.

25 (*a*) 0.331, (*b*) Because experimentally $C < 0.4$, the mass density must be greater near the center of Earth.

27 $\vec{\tau} = FR\hat{k}$

29 (*a*) $24\hat{k}$, (*b*) $-24\hat{j}$, (*c*) $-5\hat{k}$

33 $\vec{B} = 4\hat{j} + 3\hat{k}$

37 (*a*) 54 kg · m²/s, upward, (*b*) 54 kg · m²/s, downward, (*c*) 0

39 (*b*) Downward

41 (*a*) 1.3×10^{-5} kg · m²/s, away from you,
(*b*) 1.3×10^{-5} kg · m²/s, away from you,
(*c*) 1.3×10^{-5} kg · m²/s, away from you,
(*d*) 8.8×10^{-5} kg · m²/s, pointing toward you

43 (*a*) −4.9 N · m. Note that, because L decreases as the particle rotates clockwise, the angular acceleration and the net torque are both upward.
(*b*) $\omega_{\text{orbital}} = 0.48\text{ rad/s} - (0.19\text{ rad/s}^2)t$, downward.

45 (*a*) $\tau_{\text{net}} = Rg(m_2 \sin\theta - m_1)$, (*b*) $L = vR((I/R^2) + m_1 + m_2)$,
(*c*) $a = \dfrac{g(m_2 \sin\theta - m_1)}{(I/R^2) + m_1 + m_2}$

49 (*a*) 5.0 rev/s, (*b*) 0.62 kJ, (*c*) Because no external agent does work on the system, the energy comes from your internal energy.

51 10 mm/s

53 (*a*) $r_0 m v_0$, (*b*) $\frac{1}{2}mv_0^2$, (*c*) $m(v_0^2/r_0)$, $-\frac{2}{3}mv_0^2$

55 125°

57 (*a*) 3.46×10^{-47} kg · m², (*b*) 1.00 meV, 2.01 meV, 6.02 meV

59 (*a*) No, None of the allowed values of E_ℓ are equal to $3E_{0r}$.
(*b*) 2.5

61 $\vec{v}_{\text{cm}} = \dfrac{m}{M + m}\vec{v}$, $\omega = \left(\dfrac{mMd}{\frac{1}{12}ML^2(M + m) + Mmd^2}\right)v$

63 $v = \sqrt{\dfrac{(0.5M + 0.8m)(\frac{1}{3}ML^2 + 0.64mL^2)g}{0.32Lm^2}}$

65 (*a*) $v_{\text{cm}} = K/M$, (*b*) $V = 4K/M$, (*c*) $V = -2K/M$, (*d*) Yes, one point remains motionless, but only for a very brief time.

67 0.36

69 (*a*) 18 J · s, (*b*) 0.41 rad/s, (*c*) 15 s, (*d*) 0.079 J · s

71 (*a*) $\vec{L} = -(48\text{ kg} \cdot \text{m}^2/\text{s})\hat{k}$, (*b*) $\vec{\tau} = (16\text{ N} \cdot \text{m})\hat{k}$

73 (*a*) 0.24 kJ · s, (*b*) 0.31 kJ

75 (*a*) Because $\tau_{\text{net}} \neq 0$, angular momentum is not conserved.
(*b*) Because, in this frictionless environment, the net external force acting on the object is the tension force and it acts at right angles to the object's velocity, the energy of the object is conserved. (*c*) v_0

77 Yes. The solution depends only upon conservation of angular momentum of the system, so it depends only upon the initial and final moments of inertia.

81 (*a*) 0.228 rad/s, (*b*) 0.192 rad/s

83 4.47×10^{22} N · m

85 −79.9 cm

Chapter R

1 In the reference frame of the car both events occur at the same location (the location of the car). Thus, your friend's watch measures the proper time between the two events.

3 Yes. Let the initial frame of reference be frame 1. In frame 1 let L be the distance between the events, let T be the time between the events, and let the $+x$ direction be the direction of event B relative to event A. Next, calculate the value of L/T. If L/T is less than c, then consider the two events in a reference frame 2, a frame moving at speed $v = L/T$ in the $+x$ direction. In frame 2 both events occur at the same location.

5 Yes.

7 (*a*)

9 (*b*)

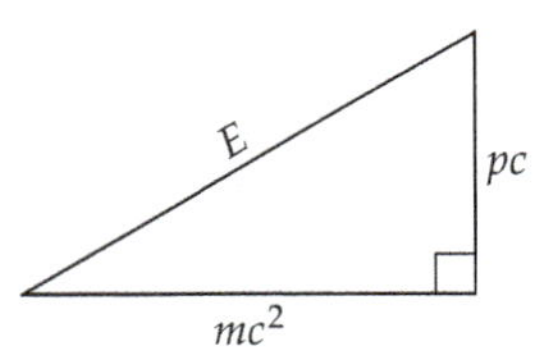

11 (*a*)

13 5.9 ns

15 (*a*) $K/E_0 = 5 \times 10^{-15}$, (*b*) $E/E_0 = 1.000\,000\,000\,000\,005\,0$,
(*c*) $K_{\text{non-rel}}/K_{\text{rel}} = 0.999\,999\,999\,999\,997\,5$

17 6.6 m

19 (*a*) 599 m

21 (*a*) 1.3×10^2 y, (*b*) 88 y

23 (*a*) 60 cm, (*b*) 2.5 ns

25 0.80c

27 (*a*) 4.50×10^{-10}%, (*b*) 142 μs less than 1 y, 142 μs

29 36 min

31 12.5 min, 12.5 min

33 60 min

35 0.400c. Event B can precede event A provided $v > 0.400c$.

37 (*a*) 11 y, (*b*) 40 y

39 (*a*) 1.01, (*b*) 1.15, (*c*) 1.67, (*d*) 7.09

41 (*a*) 0.155 mc^2, (*b*) 1.29 mc^2, (*c*) 6.09 mc^2

43 2.97 GeV

45 (*b*) 0.866c, (*c*) 0.999c

49 (*a*) 0.79%, (*b*) 69%

51 (*a*) 0.943

53 In 100 lifetimes, $d \approx 6600$ km or approximately one Earth radius. This relatively short distance should convince your classmate that the origin of the muons that are observed on Earth is within our atmosphere, and that they certainly are not from the Sun.

55 (*a*) 4.50 km/s, (*b*) 0.334 μs

57 (*a*) 0.75*c*, (*b*) 5.0 ft, (*c*) No. In your rest frame, the back end of the ladder will clear the door before the front end hits the wall of the shed, while in Ernie's rest frame, the front end will hit the wall of the shed while the back end has yet to clear the door.

Chapter 11

1 (*a*) False, (*b*) True, (*c*) True, (*d*) False

3 Earth is closest to the Sun during winter in the northern hemisphere. This is the time of fastest orbital speed. Summer would be the time for minimum orbital speed.

5 To obtain the mass M of Venus you need to measure the period T and semi-major axis a of the orbit of one of the satellites, substitute the measured values into $T^2/a^3 = 4\pi^2/(GM)$ (Kepler's third law) and solve for M.

7 (*d*)

9 (*b*)

11 (*b*)

13 You should fire the rocket in a direction to oppose the orbital motion of the satellite. As the satellite gets closer to Earth after the burn, the potential energy will decrease. However, the total mechanical energy will decrease due to the frictional drag forces transforming mechanical energy into thermal energy. The kinetic energy will increase until the satellite enters the atmosphere where the drag forces slow its motion.

15 At a point inside the sphere a distance r from its center, the gravitational field strength is directly proportional to the amount of mass within a distance r from the center, and inversely proportional to the square of the distance r from the center. The mass within a distance r from the center is proportional to the cube of r. Thus, the gravitational field strength is directly proportional to r.

17 $M_{\text{galaxy}} = 1.08 \times 10^{11}\, M_{\text{S}}$

19 About 3.0 km

21 (*a*) 6.28×10^{-4} rad/s, 2.78 h, (*b*) $L_{\text{J}} = 1.93 \times 10^{43}\ \text{kg}\cdot\text{m}^2/\text{s}$, $L_{\text{S}} = 7.85 \times 10^{42}\ \text{kg}\cdot\text{m}^2/\text{s}$, 0.70%, (*c*) $T_{\text{Sun}} = 3.64$ h, $T_{\text{Sun}} = 1.30T_{\text{max}}$

23 84.0 y

25 4.90×10^{11} m = 3.00 AU

27 (*a*) 1.6×10^{11} m = 1.1 AU, (*b*) 2.7×10^{10} m = 0.18 AU, 2.9×10^{11} m = 1.9 AU

29 (*b*) 0.73 AU, (*c*) 0.63 y

31 (*a*) 1.90×10^{27} kg

33 (*a*) 22.7 h, (*b*) 1.22×10^{9} m

35 Your weight would be ten times your weight on Earth.

37 2.27×10^4 m/s

39 (*a*) 1.4, (*b*) It is farther from the Sun than Earth. Kepler's third law $[T^2 = Cr_{\text{av}}^3]$ tells us that longer orbital periods together with larger orbital radii means slower orbital speeds, so the speed of objects orbiting the Sun decreases with distance from the Sun. The average orbital speed of Earth, given by $v = 2\pi r_{\text{ES}}/T_{\text{ES}}$, is approximately 30 km/s. Because the given maximum speed of the asteroid is only 20 km/s, the asteroid is further from the Sun.

41 (*a*) 7.37 m, (*b*) 31.9 μm

43 0.605

45 10^9 m

47 2.38 km/s

49 (*a*) 8.7 kW·h, (*b*) $500

51 6.9 km/s

53 19.4 km/s

55 13.8 km/s

57 (*a*) 7.31 h, (*b*) 1.04×10^9 J, (*c*) 8.72×10^{12} J·s

59 1.11×10^{10} J

61 $(4.0\ \text{N/kg})\hat{i}$

63 (*a*) $\vec{g} = \dfrac{Gm}{L^2}\hat{i} + \dfrac{Gm}{L^2}\hat{j}$, (*b*) $|\vec{g}| = \sqrt{2}\dfrac{Gm}{L^2}$

65 (*a*) $(-1.7 \times 10^{-11}\ \text{N/kg})\hat{i}$, (*b*) $(-8.3 \times 10^{-12}\ \text{N/kg})\hat{i}$, (*c*) 2.5 m

67 (*a*) $\frac{1}{2}CL^2$, (*b*) $\vec{g} = -\dfrac{2GM}{L^2}\left[\dfrac{L}{x_0 - L} - \ln\dfrac{x_0}{x_0 - L}\right]\hat{i}$

69 (*a*) 0, (*b*) 0, (*c*) 3.2×10^{-9} N/kg

71 $g_1 = g_2$

73 (*a*) $\dfrac{Gm(M_1 + M_2)}{9a^2}$, (*b*) $\dfrac{GmM_1}{3.61a^2}$, (*c*) 0

77 $g(x) = G\left(\dfrac{4\pi\rho_0 R^3}{3}\right)\left[\dfrac{1}{x^2} - \dfrac{1}{8(x - \frac{1}{2}R)^2}\right]$

79 $\omega = \sqrt{\dfrac{4\pi\rho_0 G}{3}}$

81 1.0 m/s

83 (*a*) $\vec{F} = -\dfrac{GMm}{d^2}\left[1 - \dfrac{d^3/4}{\{d^2 + (R^2/4)\}^{3/2}}\right]\hat{i}$

85 249 y

87 (*a*) $W = GM_{\text{E}}m\left(\dfrac{1}{r_1} - \dfrac{1}{r_2}\right)$

89 (*a*) 3.36×10^9, (*b*) 241

91 1.70 Mm

95 1.60×10^{-4}

97 $v = 1.16\sqrt{\dfrac{GM}{a}}$

99 $$g(r) = \begin{cases} 0 & r < R_1 \\ \dfrac{GM(r^3 - R_1^3)}{r^2(R_2^3 - R_1^3)} & R_1 < r \le R_2 \\ \dfrac{GM}{r^2} & R_2 < r \end{cases}$$

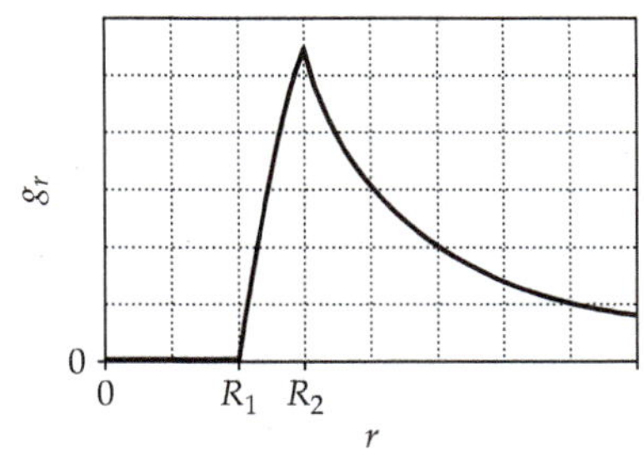

101 $g = (2G\lambda)/r$

103 (*b*) $U = -\frac{GMm_0}{L}\ln\left(\frac{x + L/2}{x - L/2}\right)$, (*c*) $F_x(x) = -\frac{GMm_0}{x^2 - (L/2)^2}$

This answer and the answer given in Example 11-8 are the same.

105 34 pN

107 (*a*) The gravitational force is greater on the lower robot, so if it were not for the cable, its acceleration would be greater than that of the upper robot and they would separate. In opposing this separation, the cable is stressed. (*b*) 2.2×10^5 m

Chapter 12

1 (*a*) False, (*b*) True, (*c*) True, (*d*) False

3 (*b*)

5 The higher the sign, the greater the torque about the horizontal axis through the lowest extremes of the posts for a given wind speed. In addition, the deeper the post hole the greater the maximum opposing torque about the same axis for a given consistency of the dirt. Thus, higher signs require deeper holes for the signposts.

7 The main reason this is done is to lower the center of gravity of the mug. The lower the center of gravity the more stable the mug is.

9 (*b*)

11 (*b*) 2.0 N/cm

13 (*b*) Taking long strides requires a large coefficient of static friction because θ is large for long strides. (*c*) If μ_s is small, that is, there is ice on the surface, θ must be small to avoid slipping.

15 84 cm

17 692 N, 2.54 kN

19 0.728 m

21 $F_1 = \frac{1}{2}Mg$, $F_2 = \frac{\sqrt{3}}{2}Mg$

23 (*a*) $\vec{F} = (30\text{ N})\hat{i} + (30\text{ N})\hat{j}$, (*b*) $\vec{F} = (35\text{ N})\hat{i} + (45\text{ N})\hat{j}$

25 (*a*) $F_n = Mg - F\sqrt{(2R - h)/h}$, (*b*) F,

(*c*) $F\sqrt{(2R - h)/h}$

27 (*a*) 6.87 N, (*b*) 1.7 N·m, (*c*) 8.3 N, 15 N

29 (*a*) 71 N, (*b*) 3.5 m, (*c*) 0.50 kN

31 (*b*) $a/3$, (*c*) $Mg/2$

33 $h = \mu_s L \tan\theta \sin\theta$

35 $\mu_s = \frac{2h}{L \tan\theta \sin\theta}$

37 59°

39 62°

41 (*a*) 42 N, (*b*) 0.14%

43 5.0°

47 (*a*) $1.4 \times 10^6\text{ N/m}^2$, (*b*) 7 mJ, (*c*) 28 mJ. There is about 4 times as much energy stored in the rubber when 0.30 kg are hung from it. That is because the stored energy increases quadratically with an increase in the mass.

49 0.69

51 Because $\text{Stress}_{\text{failing}} < \text{Stress}_{\text{cable}}$, the cable will not support the elevator.

55 1.5 kN

57 $m_1 = 0.15$ kg, $m_2 = 0.71$ kg, $m_3 = 0.36$ kg

59 1.8 kg

61 0.15

63 $\mu_s < 0.50$

65 $\mu_s = \frac{1}{2}(\cot\theta + 1)$

67 (*a*) 0.15 kN, (*b*) 3.8 m

69 $\mu_s < 0.50$

71 (*c*)

Cell	Content/Formula	Algebraic Form
B5	B4+1	$i + 1$
C5	C4+$B$1/(2*B5)	$d_i + \frac{L}{2i}$

	A	B	C	D
1	L =	0.20	m	
2		0		
3		*i*	offset	
4		1	0.100	
5		2	0.150	
6		3	0.183	
102		99	0.518	
103		100	0.519	

$d_5 = 15$ cm, $d_{10} = 26$ cm, and $d_{100} = 0.52$ cm

(*d*) No.

73 566 N

75 $F_n = 2mg$, $F = mg\frac{r}{\sqrt{R(2r - R)}}$, $F_W = mg\frac{R - r}{\sqrt{R(2r - R)}}$

Chapter 13

1 (*e*)

3 (*c*)

5 Pressure increases approximately 1 atm every 10 m of depth. To breathe requires creating a pressure of less than 1 atm in your lungs. At the surface you can do this easily, but not at a depth of 10 m.

7 (*b*)

9 False. The buoyant force on a submerged object depends on the weight of the displaced fluid, which, in turn, depends on the volume of the displaced fluid. Because the bricks have the same volume, they will displace the same volume of water.

11 Because the pressure increases with depth, the object will be compressed and its density will increase as its volume decreases. Thus, the object will sink to the bottom.

13

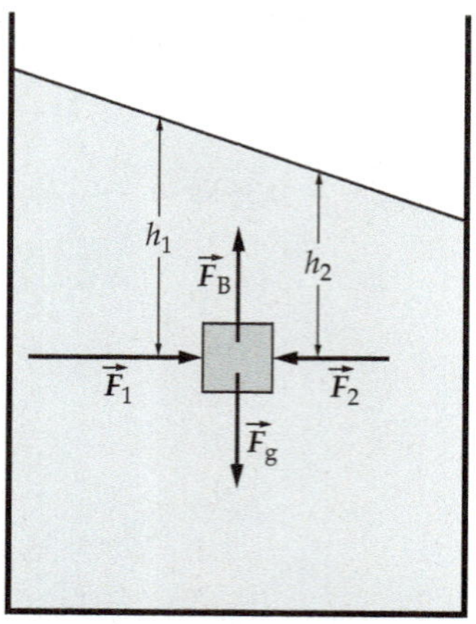

The pictorial representation shows the glass and an element of water in the middle of the glass. As is readily established by a simple demonstration, the surface of the water is not level while the glass is accelerated, showing that there is a pressure gradient (a difference in pressure) due to the differing depths ($h_1 > h_2$, and hence, $F_1 > F_2$) of water on the two sides of the element of water. This pressure gradient results in a net force on the element of water, as shown in the figure. The upward buoyant force is equal in magnitude to the downward gravitational force.

15 (*c*)

17 The mounding around entrance 1 will cause the streamlines to curve concave downward over the entrance. An upward pressure gradient produces the downward centripetal force. This means there is a lowering of the pressure at entrance 1. No such lowering occurs over entrance 2, so the pressure there is higher than the pressure at entrance 1. The air circulates in entrance 2 and out entrance 1. It has been demonstrated that enough air will circulate inside the tunnel even with the slightest breeze outside.

19 1.11 kg/m^3

21 1.0×10^2 kg

23 33.6 kg/m^3

25 0.773

27 29.8 in Hg

29 1.5 N

31 230 N

33 197 atm. Because a depth of only 2 km is required to produce a 1-percent compression, this does occur in our oceans.

35 (*a*) 15 kN, (*b*) 0.34 kg

37 45 cm

39 1.4%

41 4.36 N

43 (*a*) 11×10^3 kg/m^3, (*b*) From Table 13-1, we see that the density of the unknown material is close to that of lead.

45 800 kg/m^3, 1.11

47 250 kg/m^3

49 3.9 kg

51 3.9×10^5 N

53 2.5×10^7 kg

55 (*a*) 12.0 m/s, (*b*) 133 kPa, (*c*) The flow rates are the same.

57 (*a*) 4.6 L/min, (*b*) 7.6×10^{-2} m^2

59 144 kPa

61 0.20 kPa. Because $\Delta P \propto I_V^2$ (a plot of ΔP as a function of I_V is a parabola that opens upward), this pressure difference is the minimum pressure difference.

65 (*b*) $P_{top} = P_{at} - \rho g d$

67 (*a*) 9.28 cm/s, (*b*) 0.331 cm, (*c*) 76 cm

69 (*a*) $x = 2\sqrt{h(H - h)}$, (*c*) $x_{max} = H$

71 1.43 mm

73 90 mi/h. Because most major league pitchers can throw a fastball in the low-to-mid-90s, this abrupt decrease in drag may very well play a role in the game.

75 2.91 L/s

77 2

79 36 kg/m^3

81 0.71 kg

83 12 cm

85 One meter is a plausible diameter for such a pipe.

87 29 s

89 (*a*) 70 m^3, (*b*) 5.2 m/s^2

91 (*b*) 0.13 km^{-1}

93 39 cm^3

Chapter 14

1 (*a*) False, (*b*) True, (*c*) True

3 (*a*)

5 (*c*)

7 (*c*)

9 Neglecting the mass of the spring in your calculations results in your using a value for the mass of the oscillating system that is smaller than its actual value. Hence, your calculated value for the period will be smaller than the actual period of the system and the calculated value for the frequency, which is the reciprocal of the period, will be higher than the actual value.

Because the total energy of the oscillating system depends solely on the amplitude of its motion and on the stiffness of the spring, it is independent of the mass of the system, and so neglecting the mass of the spring would have no effect on your calculation of the system's total energy.

11 (*d*)

13 (*b*)

15 1 matches up with B, 2 matches up with D, and 3 matches up with A.

17 (*c*)

19 (*a*) True, (*b*) False, (*c*) True, (*d*) False, (*e*) True

21 (*b*)

23 (*b*)

25 About 5

27 8π

29 (*a*) 3.00 Hz, (*b*) 0.333 s, (*c*) 7.0 cm, (*d*) 0.0833 s in $-x$ direction.

31 (*a*) $x = (0.25\text{ m})\cos[(4.2\text{ s}^{-1})t]$,
(*b*) $v = -(1.0\text{ m/s})\sin[(4.2\text{ s}^{-1})t]$,
(*c*) $a = -(4.4\text{ m/s}^2)\cos[(4.2\text{ s}^{-1})t]$

33 (*a*) $x = (0.28\text{ m})\cos[(4.2\text{ s}^{-1})t - 0.45]$,
(*b*) $v = -(1.2\text{ m/s})\sin[(4.2\text{ s}^{-1})t - 0.45]$,
(*c*) $a = -(4.9\text{ m/s}^2)\cos[(4.2\text{ s}^{-1})t - 0.45]$

35 (*a*)

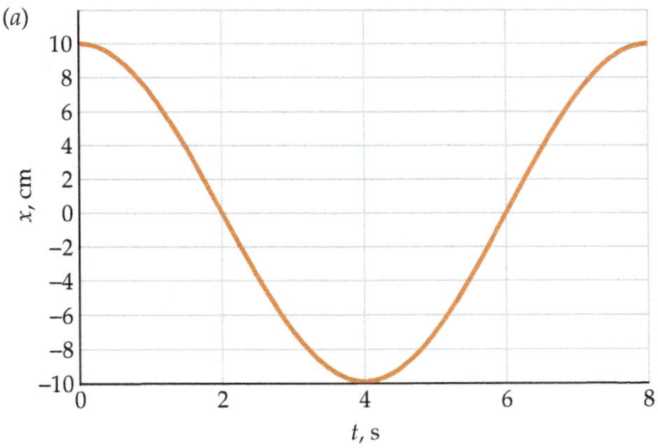

(*b*) 2.9 cm, 7.1 cm, 7.1 cm, 2.9 cm

37 (*a*) 7.9 m/s, 25 m/s^2, (*b*) 6.3 m/s, −15 m/s^2

39 (*a*) 3.1 s, 0.32 Hz, (*b*) $x = (40\text{ cm})\cos[(2.0\text{ s}^{-1})t + (\pi/2)]$

41 23 J

43 (*a*) 0.368 J, (*b*) 3.83 cm

45 1.4 kN/m

47 (*a*) 6.9 Hz, (*b*) 0.15 s, (*c*) 10 cm, (*d*) 4.3 m/s, (*e*) 1.9×10^2 m/s^2, (*f*) 36 ms, 0

49 (*a*) 0.68 kN/m, (*b*) 0.42 s, (*c*) 1.5 m/s, (*d*) 23 m/s^2

51 (*a*) 3.08 kN/m, (*b*) 4.16 Hz, (*c*) 0.240 s

53 0.262 s

55 (*a*) 1.0 Hz, (*b*) 0.50 s, (*c*) 0.29 N

57 (*a*) 6.7 cm, (*b*) 0.26 s, (*c*) Because $h < 8.0$ cm, the spring is never uncompressed. 77 cm/s

59 44 cm

61 12 s

63 11.7 s

65 $T = 2\pi\sqrt{L/(g\cos\theta)}$

67 1.1 s

69 0.50 kg · m^2

71 21.1 cm from the center of the meter stick

75 (*a*) $d = 1.64$ m, (*b*) 2.31 cm

79 (*a*) 0.31, (*b*) -3.1×10^{-2}%

81 (*a*) 1.57%, (*c*) $0.43E_0$

83 (*a*) 5.51 Pa · s, (*b*) 125

85 (*a*) 1.0 Hz, (*b*) 2 Hz, (*c*) 0.35 Hz

87 (*a*) 4.98 cm, (*b*) 14.1 rad/s, (*c*) 35.4 cm, (*d*) 1.00 rad/s

89 (*a*) 0.48 Hz, 2.1 s, (*b*) $v = -(1.2\text{ m/s})\sin\left[(3.0\text{ rad/s})t + \dfrac{\pi}{4}\right]$, (*c*) 1.2 m/s

91 The error is greater if the clock is elevated.

93 (*a*) $\mu_s = \dfrac{Ak}{(m_1 + m_2)g}$ (*b*) *A* is unchanged. *E* is unchanged. ω is reduced, and *T* is increased.

95 (*b*) 2.0 cm/s^2

101 6.44×10^{13} rad/s

103 $T = 7.78\sqrt{R/g}$

107 (*a*)

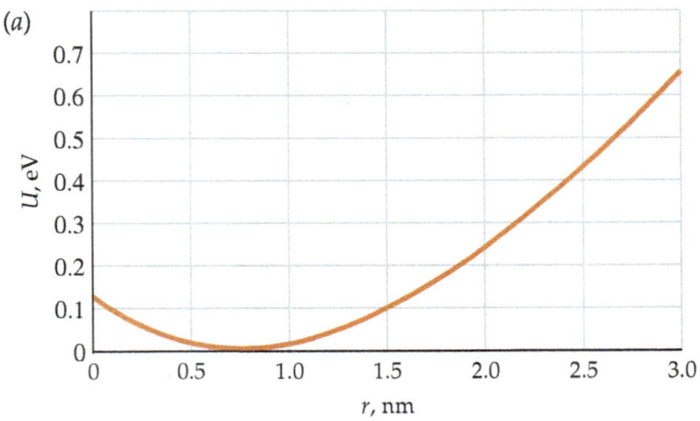

(*b*) $r = r_0$, $k = 2\beta^2 D$, (*c*) $\omega = 2\beta\sqrt{D/m}$

Chapter 15

1 The speed of a transverse wave on a uniform rope increases with increasing tension. The waves on the rope move faster as they move up because the tension increases due to the weight of the rope below.

3 (*b*)

5 The resonant (standing-wave) frequencies on a string are inversely proportional to the square root of the linear density of the string ($f = \sqrt{F_T/\mu}/\lambda$). Thus, extremely high frequencies (which might otherwise require very long strings) can be accommodated on relatively short strings if the strings are linearly denser than the high-frequency strings. High frequencies are not a problem, as they utilize short strings anyway.

7 (*c*)

9 There was only one explosion. Sound travels faster in water than air. Abel heard the sound wave in the water first, then, surfacing, heard the sound wave traveling through the air, which took longer to reach him.

11 (*b*)

13 (*a*)

15 (*b*)

17 (*a*) False, (*b*) True, (*c*) True

19 (*a*)

21 The light from the visible star will be shifted about its mean frequency periodically due to the relative approach toward and recession away from Earth as the star revolves around the common center of mass.

23

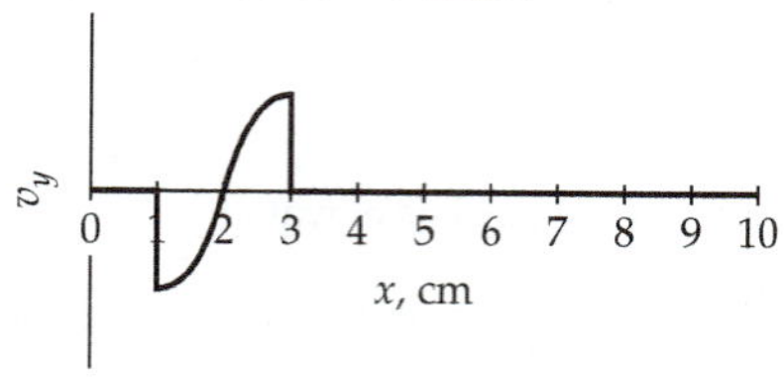

25 Path C. Because the wave speed is highest in the water, and more of path C is underwater than are paths A or B, the sound wave will spend the least time on path C.

27 11 ms

29 0.27 km/s, 21%

31 1.32 km/s

33 (*b*) 40 N, (*c*) 40.8 N, 2%

39 9.9 W

41 (*a*) 5.00 m/s in the $-x$ direction, (*b*) 10.0 cm, 50.0 Hz, 0.0200 s, (*c*) 0.314 m/s

43 (*a*) 6.8 J, (*b*) 44 W

45 (*a*) 79 mW, (*b*) The power can be increased by a factor of 100 by increasing either the frequency or the amplitude by a factor of 10, or by increasing the tension by a factor of 100. (*c*) Increasing the frequency probably would be the easiest.

47 (*a*) 0.75 Pa, (*b*) 4.00 m, (*c*) 85.8 Hz, (*d*) 343 m/s

49 (*a*) 36.4 μm, (*b*) 83.4 mPa

51 (*a*) Zero, (*b*) 3.64 μm

53 (*a*) 0.80 s, (*b*) 30 m, (*c*) 6.8 m

55 (*a*) 50.3 W, (*b*) 2.00 m, (*c*) 4.44 mW/m^2

57 (*a*) 20.0 dB, (*b*) 100 dB

59 90 dB

61 (*a*) 0.10 km, (*b*) 0.13 W

63 (*a*) 80 dB, (*b*) Eliminating the 70 dB and 73 dB sources does not reduce the intensity level significantly.

65 87.8 dB

67 57 dB

73 (*a*) 263 m/s, (*b*) 1.32 m, (*c*) 261 Hz

75 153 Hz

77 2.25×10^8 m/s

79 174 mi/h

81 (*a*) −7.78 kHz, (*b*) −4.44 kHz

83 (*a*) $f'_r = f_s(v - u_r)/(v - u_s)$

85 (*a*) 0.82 kHz, (*b*) 0.85 kHz

87 185 m, 714 Hz

89 $\lambda_{max} = (500\text{ nm})(1 + 4.36 \times 10^{-5})$ and $\lambda_{min} = (500\text{ nm})(1 - 4.36 \times 10^{-5})$

91 529 Hz, 474 Hz

93 7.99 m from the left end of the wire

95 (*a*) 55.6 N/m^2, (*b*) 3.49 W/m^2, (*c*) 0.110 W

97 77 kN

99 206 m

101 0.2 cm

103 (*a*) 10.0 m/s, (*b*) 2.00 m, (*c*) $P_{max} = 1.26 \times 10^{-4}$ kg m/s, (*d*) 3.95 mN

Chapter 16

1

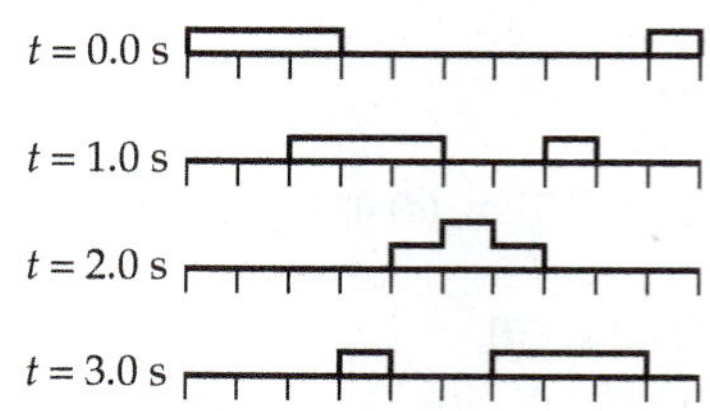

3 (*c*)

5 (*a*)

7 3.0 m

9 (*b*)

11 (*a*)

13 You could measure the lowest resonant frequency f and the length L of the pipe. Assuming the end corrections are negligible, the wavelength equals $4L$ if the pipe is stopped at one end, and is $2L$ if the pipe is open at both ends. Then use $v = f\lambda$ to find the speed of sound at the ambient temperature. Finally, use $v = \sqrt{\gamma RT/M}$ (Equation 15-5), where $\gamma = 1.4$ for a diatomic gas such as air, M is the molar mass of air, R is the universal gas constant, and T is the absolute temperature, to estimate the temperature of the air.

15 (*a*) No, (*b*) Yes

17 Standing sound waves are produced in the air columns above the water. The resonance frequency of the air columns depends on the length of the air column, which depends on how much water is in the glass.

19 The wavelength is determined mostly by the size of the resonant cavity of the mouth; the frequency of sounds he makes is equal to the wave speed divided by the wavelength. Because $v_{He} > v_{air}$ (see Equation 15-5), the resonance frequency is higher if helium is the gas in the cavity.

21 If you do not hear even one beat for the entire time the string and the tuning fork are vibrating, you can be sure that their frequencies, while not exactly the same, are very close. If the sounds of the vibrating string and the tuning fork last for 10 s, it follows that the beat frequency is less than 0.1 Hz. Hence the frequencies of the vibrating string and the tuning fork are within 0.1 Hz of each other.

23 3. The estimated frequencies agree with the observed frequencies to within 14%.

25 7.1 cm

27 (*a*) 89°, (*b*) 1.5A

29 (*a*) 0, (*b*) $2I_0$, (*c*) $4I_0$

31 (*a*) 60.0 cm, (*b*) 0.400π rad, (*c*) 24.0 m/s

33 $f_1 = 2.0$ kHz, $f_2 = 5.0$ kHz

35 (*b*)

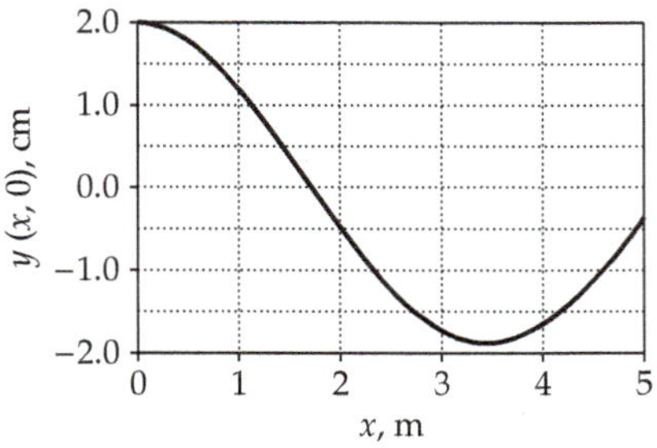

(*c*)

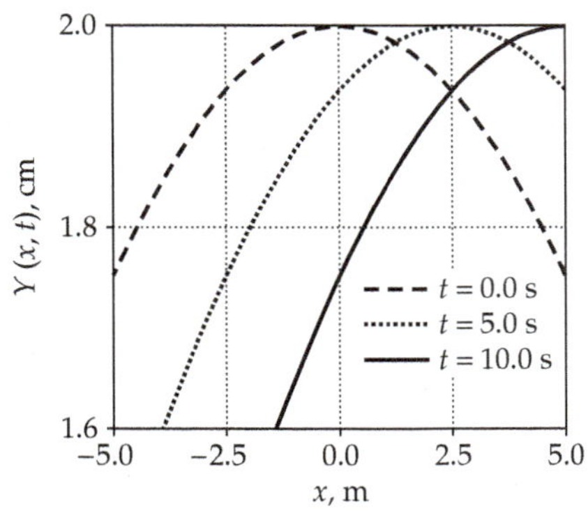

$v_{est} = 46$ cm/s, v_{est} is 92% of $v_{envelope}$

37 1.8 m, 51°

39 (*a*) 0.279 m, (*b*) 1.23 kHz, (*c*) 0.432 rad, 0.592 rad, 0.772 rad, 0.992 rad, and 1.35 rad, (*d*) 0.0698 rad

41 2.0 rad

43 (*b*) 1.2 kHz, (*c*) 15 Hz/(mi/h)

45 (*a*) 2.00 m, 2.50 Hz, (*b*) $y_3(x,t) = (4.00 \text{ mm}) \sin(\pi \text{ m}^{-1})x \cos(50.0\pi \text{ s}^{-1})t$

47 (*a*) 521 m/s, (*b*) 2.80 m, 186 Hz, (*c*) 372 Hz and 558 Hz

49 141 Hz

51 (*a*) 32.4 cm, 47.7 Hz, (*b*) 15.0 m/s, (*c*) 62.8 cm

53 (*a*) 71.5 Hz, (*b*) 5.00 kHz, (*c*) 71

55 452 Hz. Ideally, the pipe should expand so that v/L, where L is the length of the pipe, is independent of temperature.

57 (*a*) 40.0 cm, (*b*) 480 N, (*c*) You should place your finger 9.2 cm from the scroll bridge.

59 (*a*) 75 Hz, (*b*) 5th and 6th, (*c*) 2.0 m

61 (*a*) 0.574 g/m, (*b*) $\mu_E = 1.29$ g/m, $\mu_A = 2.91$ g/m, $\mu_D = 2.91$ g/m, $\mu_G = 6.57$ g/m

63 (*a*) The two sounds produce a beat because the third harmonic of the A string equals the second harmonic of the E string, and the original frequency of the E string is slightly greater than 660 Hz. (*b*) 662 Hz

65 (*a*) Because the frequency is fixed, the wavelength depends only on the tension on the string. This is true because the only parameter that can affect the wave speed on the string is the tension on the string. The tension on the string is provided by the weight hanging from its end. Given that the length of the string is fixed, only certain wavelengths can resonate on the string. Thus, because only certain wavelengths are allowed, only certain wave speeds will work. This, in turn, means that only certain tensions, and therefore weights, will work.
(*b*) Higher frequency modes on the same length of string results in shorter wavelengths. To accomplish this without changing frequency, you need to reduce the wave speed. This is accomplished by reducing the tension in the string. Because the tension is provided by the weight on the end of the string, you must reduce the weight.
(*d*) $w_1 = 19.2$ N, $w_2 = 4.80$ N, $w_3 = 2.13$ N

67 (*a*) $\Delta t = N/f_0$, (*b*) $\lambda \approx \Delta x/N$, (*c*) $k = 2\pi N/\Delta x$, (*d*) N is uncertain because the waveform dies out gradually rather than stopping abruptly at some time; hence, where the pulse starts and stops is not well defined.

69 6.74 m

71 (*a*) 1.9 cm, 3.6 m/s, (*b*) 0, 0, (*c*) −1.2 cm, −2.2 m/s, (*d*) 0, 0

73 98.0 Hz

75 (*a*) The pipe is closed at one end. (*b*) 262 Hz, (*c*) 32.7 cm

77 (*a*) $y_1(x,t) = (0.010 \text{ m})\sin[(\tfrac{1}{2}\pi \text{ m}^{-1})x - (40\pi \text{ s}^{-1})t]$,
$y_2(x,t) = (0.010 \text{ m})\sin[(\tfrac{1}{2}\pi \text{ m}^{-1})x + (40\pi \text{ s}^{-1})t]$,
(*b*) 2.00 m, (*c*) $v_y(1.0 \text{ m},t) = -(2.5 \text{ m/s})\sin[(40\pi \text{ s}^{-1})t]$,
(*d*) $a_y(1.0 \text{ m},t) = -(0.32 \text{ km/s}^2)\cos[(40\pi \text{ s}^{-1})t]$

79 $y_{res}(x,t) = (10.0 \text{ cm})\sin(kx - \omega t)$

81 (*b*) 203.4 Hz, (*c*) 203.4 Hz

83 812 Hz

85 (*a*)

Cell	Content/Formula	Algebraic Form
A6	A5+0.1	$x + \Delta x$
B4	2*B3+1	$2n + 1$
B5	(−1)^B$3*COS(B$4*$A5)/B$4*4/PI()	$\frac{4}{\pi}\sum_{n=0}^{\infty}\frac{(-1)^n \cos((2n+1)x)}{2n+1}$
C5	B5+(−1)^C$3*COS(C$4*$A5)/C$4*4/PI()	$\frac{4}{\pi}\sum_{n=0}^{\infty}\frac{(-1)^n \cos((2n+1)x)}{2n+1}$

	A	B	C	D	K	L
1						
2						
3		0	1	2	9	10
4		1	3	5	19	21
5	0.0	1.2732	0.8488	1.1035	0.9682	1.0289
6	0.1	1.2669	0.8614	1.0849	1.0134	0.9828
134	12.9	1.2030	0.9740	0.9493	0.9691	1.0146
135	13.0	1.1554	1.0422	0.8990	1.0261	0.9685

The solid curve is plotted from the data in columns A and B and is the graph of $f(x)$ for 1 term. The dashed curve is plotted from the data in columns A and F and is the graph of $f(x)$ for 5 terms. The dotted curve is plotted from the data in columns A and K and is the graph of $f(x)$ for 10 terms.

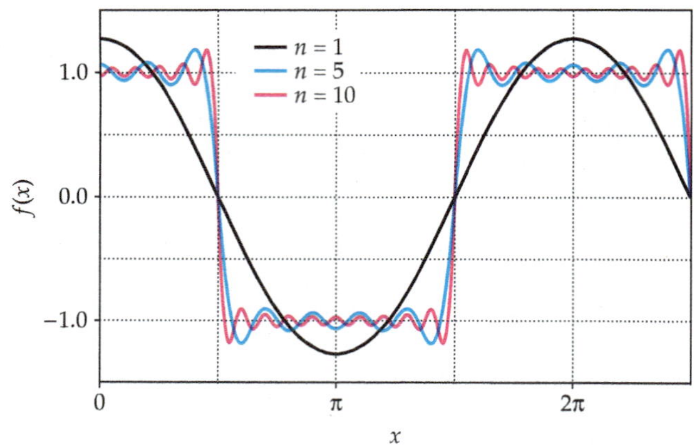

(*b*) It is equivalent to the Leibnitz formula.

87 (*b*)

Cell	Content/Formula	Algebraic Form
B1	90	L
B2	1	r
B3	340	c
B8	B7+1	$n + 1$
C7	2/$B$3*((2*(B7−1)*$B$2)^2+$B$1^2)^0.5	Δt_n

	A	B	C	D
1	L =	90	m	
2	*r* =	1	m	
3	*c* =	343	m/s	
6		*n*	*t*(*n*)	delta *t*(*n*)
7		1	0.5248	0.0001
8		2	0.5249	0.0004
206		200	2.3739	0.0114

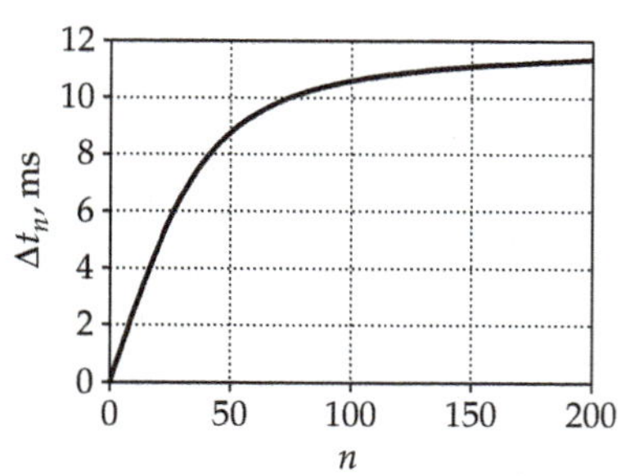

(*c*) The frequency heard at any time is $1/\Delta t_n$, so because Δt_n increases over time, the frequency of the culvert whistler decreases. $f_{\text{highest}} = 7.72$ kHz, $f_{\text{lowest}} = 85.5$ Hz

Chapter 17

1 (*a*) False, (*b*) False, (*c*) True

3 Mert's room was colder.

5 From the ideal-gas law, we have $P = nRT/V$, and V/T is the slope of the line from the origin to the point (T, V) on the graph. During the process the slope of the line from the origin to (T, V) continuously decreases, so the pressure continuously increases.

7 (*d*)

9 The average kinetic energies are equal. The ratio of their rms speeds is equal to the square root of the reciprocal of the ratio of their molecular masses.

11 False

13 It does not matter.

15 (*b*)

17 The rms speed is always somewhat greater than the speed of sound. However, it is only the component of the molecular velocities in the direction of propagation that is relevant to this issue. In addition, in a gas the mean free path is greater than the average intermolecular distance.

19 If the volume decreases the pressure increases because more molecules hit a unit of area of the walls in a given time. This happens because the number of molecules per unit volume increases as the volume decreases.

21 The average molecular speed of He gas at 300 K is about 1.4 km/s, so a significant fraction of He molecules have speeds in excess of Earth's escape velocity (11.2 km/s). Thus, they "leak" away into space. Over time, the He content of the atmosphere decreases to almost nothing.

23 About 1.2 kg/m^3

25 (*a*) 3600 K, (*b*) 230 K, (*c*) Because hydrogen is lighter than air, it rises to the top of the atmosphere. Because the temperature is high there, a greater fraction of the molecules reach escape speed. (*d*) 160 K, 10 K. Because g is less on the moon, the escape speed is lower. Thus, a larger percentage of the molecules are moving at escape speed.

27 (*a*) 1.23 km/s, (*b*) 310 m/s, (*c*) 264 m/s, (*d*) Because v_e is greater than v_{rms} for O_2, CO_2, and H_2, all three gases should be found on Jupiter.

29 (*a*) 2×10^{11} atm, (*b*) $v_{\text{rms protons}} = 5 \times 10^5$ m/s, $v_{\text{rms electrons}} = 2 \times 10^7$ m/s

31 1×10^{-4} g

33 1063°C

35 (*a*)

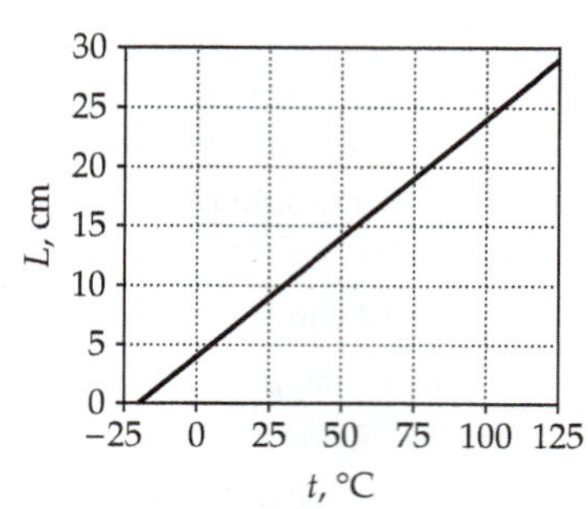

(*b*) 8.40 cm, (*c*) 107°C

37 −320°F

39 (*a*)

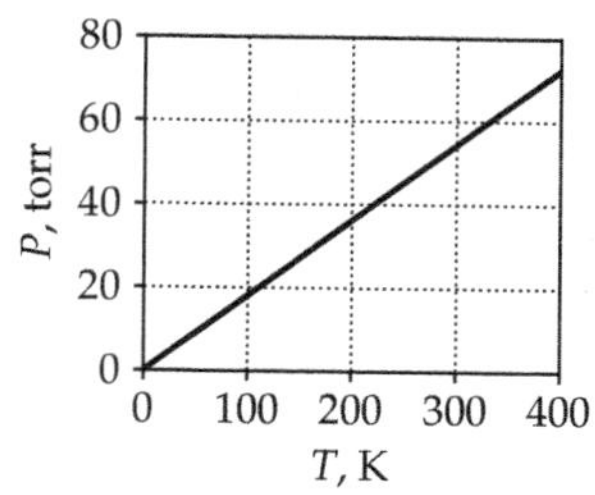

(*b*) 54.9 torr, (*c*) 3.70×10^3 K

41 $-40.0°C = -40.0°F$

43 $-183°C$, $-297°F$

45 (*a*) $R_0 = 3.91 \times 10^{-3}$ K, $B = 3.94 \times 10^3$ K, (*b*) 1.31 kΩ, (*c*) -389 Ω/K, -4.33 Ω/K, (*d*) The thermistor has greater sensitivity at lower temperatures.

47 1.79 mol, 1.08×10^{24} molecules

49 -83 *glips*

51 (*a*) 3.7×10^3 mol, (*b*) 60 mol

53 11.1 atm

55 (*a*) Air will be less dense when its water-vapor content is higher. (*b*) 18 g

57 1.1 kN

59 (*a*) 0.28 km/s, (*b*) 0.87 km/s. The rms speed of argon atoms is slightly less than one-third the rms speed of helium atoms.

61 5.0×10^5 m/s, 2.1×10^{-16} J

65 $K/\Delta U = 7.9 \times 10^4$

69 (*a*) $E_{peak} = \frac{1}{2}kT$, $E_{peak} = \frac{1}{3}E_{av}$

(*b*)

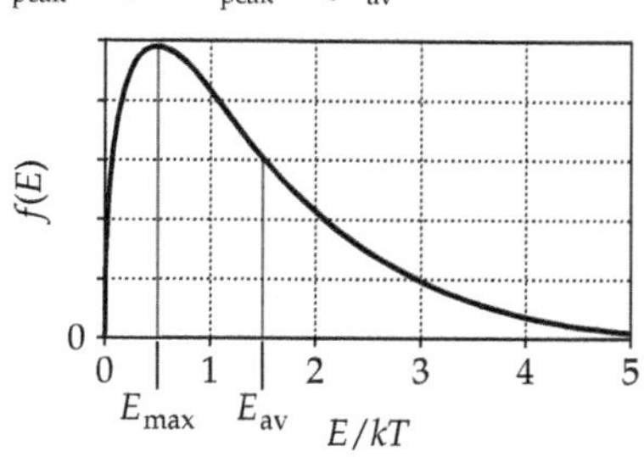

E_{max} is the most probable energy

(*c*) The graph rises from zero to the peak much more rapidly than it falls off to the right of the peak. Because the distribution is so strongly skewed to the right of the peak, the outlying molecules with relatively high energies pull the average ($3kT/2$) far to the right of the most probable value ($kT/2$).

71 (*a*) 1.2×10^2 K, (*b*) 2.4×10^2 K, (*c*) 1.4 atm

73 (*a*) To escape from the surface of a droplet of water, molecules must have enough translational kinetic energy to overcome the attractive forces from their neighbors. Therefore the molecules that escape will be those that are moving faster, leaving the slower molecules behind. The slower molecules have less kinetic energy, so the temperature of the droplet, which is proportional to the average translational kinetic energy per molecule, decreases.
(*b*) As long as the temperature is not too high, the molecules that evaporate from a surface will be only those with the most extreme speeds, at the high-energy "tail" of the Maxwell–Boltzmann distribution. Within this part of the distribution, increasing the temperature only slightly can greatly increase the percentage of molecules with speeds above a certain threshold. For example, suppose that we set an initial threshold at $E = 5kT_1$, then imagine increasing the temperature by 10%, so $T_2 = 1.1T_1$. At the threshold, the ratio of the new energy distribution to the old one is

$$\frac{F(T_2)}{F(T_1)} = \left(\frac{T_1}{T_2}\right)e^{-E/kT_2}e^{+E/kT_1} = (1.1)^{-3/2}e^{-5/1.1}e^{5} = 1.365$$

an increase of almost 37%.

75 110 mol of H_2, 55 mol of O_2

77 $4m$

79 (*a*) 142 ms, (*b*) 146 ms

81 (*a*)

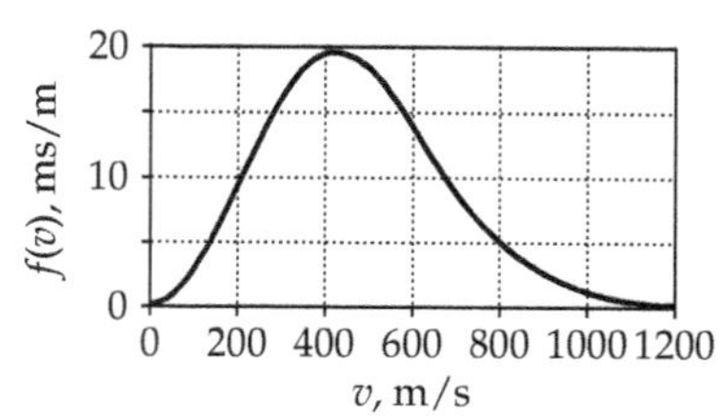

(*b*) As the temperature is increased, the graph spreads out horizontally and gets shorter vertically. More precisely, the horizontal position of the peak moves to the right in proportion to the square root of the temperature, while the height of the peak drops by the same factor, preserving the total area under the graph (which must be 1.0, the total probability of a molecule having any velocity between zero and infinity).
(*c*) A graph of $F(v)$ for nitrogen at 300 K follows. Each number in column C of the spreadsheet [shown in Part (*a*)] is approximately equal to the integral of $f(v)$ from zero up to the corresponding v value. This integral represents the probability of a molecule having a speed greater than or equal to this value of v.

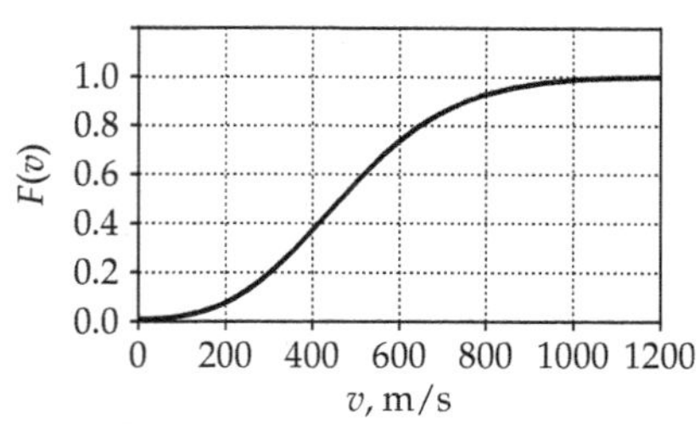

$F(v) = \int_0^v f(v')dv'$

(*d*) About 7%. Note that this value is consistent with the graph of $F(v)$ shown immediately above.
(*e*) A little under 14%.

Chapter 18

1 (*e*)

3 (*c*)

5 (*a*)

7 (*c*)

9 Yes. $\Delta E_{int} = Q_{in} + W_{on}$. If the gas does work at the same rate that it absorbs heat, its internal energy will remain constant.

11 Particles that attract each other have more potential energy the farther apart they are. In a real gas the molecules exert weak attractive forces on each other. These forces increase the internal potential energy during an expansion. An increase in potential energy means a decrease in kinetic energy, and a decrease in kinetic energy means a decrease in translational kinetic energy. Thus, there is a decrease in temperature.

13 Particles that repel each other have more potential energy the closer together they are. The repulsive forces decrease the internal potential energy during an expansion. A decrease in potential energy means an increase in kinetic energy, and an increase in kinetic energy means an increase in translational kinetic energy. Thus, there is an increase in temperature.

15 (*a*)

17 (*a*) False, (*b*) False, (*c*) True, (*d*) True, (*e*) True, (*f*) True

19 (*d*)

21 During a reversible adiabatic process, PV^{γ} is constant, where $\gamma > 1$, and during an isothermal processes, PV is constant. Thus the pressure rise during the compression is greater than the pressure drop during the expansion. The final process could be a constant volume process during which heat is absorbed from the system. A constant-volume cooling will decrease the pressure and return the gas to its original state.

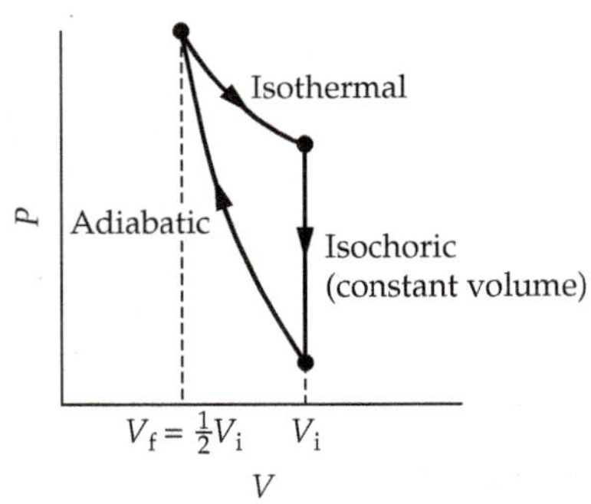

23 The temperature decreases.

25 1.6 min, an elapsed time that seems to be consistent with experience.

27 1.2×10^{-5} (or 1.2×10^{-3} percent)

29 31.3 kJ

31 48.8 g

33 12.1°C

35 4.5×10^2 kg

37 (*a*) 0°C, (*b*) 125 g

39 (*a*) 4.9°C, (*b*) No ice is left.

41 (*a*) 5.26°C, (*b*) 175 g, (*c*) No

43 618°C

45 2.20 kJ

47 54 J

49 (*a*) 6.13 W, (*b*) 38.1 min

51 (*a*) 405 J

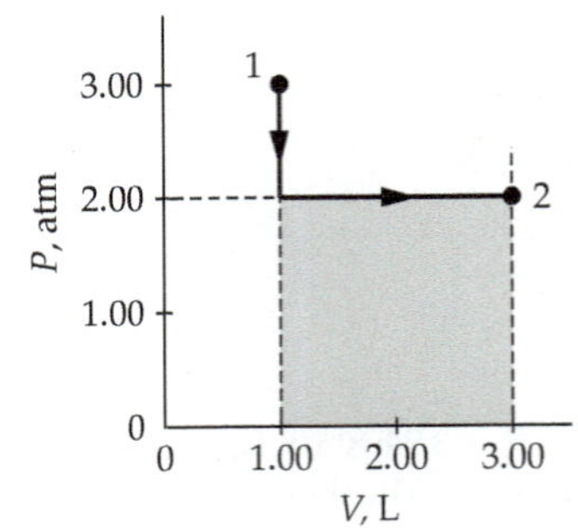

(*b*) 861 J

53 (*a*) 507 J

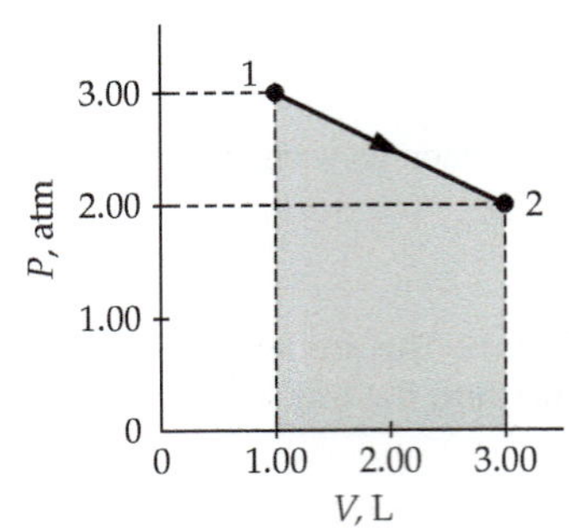

(*b*) 963 J

55 $W_{by\ gas} = \frac{3}{2}P_0V_0$

57 (*a*) 555 J, (*b*) 555 J

59 (*a*) 0.495 mol, (*b*) 3.09 kJ, (*c*) 20.8 J/mol·K, (*d*) 10.3 J/K

61 (*a*) 6.24 kJ, 6.24 kJ, (*b*) 6.24 kJ , 8.73 kJ, 2.49 kJ, (*c*) 2.49 kJ

63 59.6 L

65 $\Delta c'_p = -\frac{13}{2}R$

67 There are three translational degrees of freedom and three rotational degrees of freedom. In addition, each of the hydrogen atoms can vibrate against the oxygen atom, resulting in an additional 4 degrees of freedom (2 per atom). $C_{v\,water} = 5Nk = 5nR$

69 (*a*) 300 K, 7.80 L, 1.14 kJ, 1.14 kJ, (*b*) 5.40 L, 208 K, 0, 574 J

71 (*a*) 263 K , (*b*) 10.8 L, (*c*) 1.48 kJ, (*d*) −1.48 kJ

73 −0.14 kJ

75

Process	Q_{in} (kJ)	W_{on} (kJ)
D→A	8.98	0
A→B	13.2	−13.2
B→C	−8.98	0
C→D	−6.58	6.58

$W_{by\ gas\ total} = 6.6$ kJ

77 (*a*)

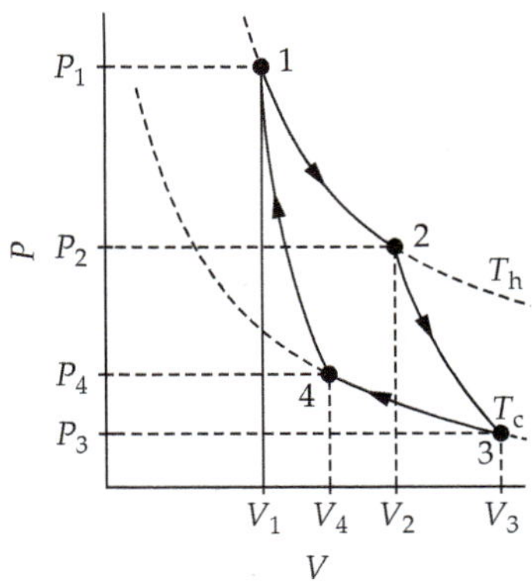

79 (*a*) 65 K, 81 K, (*b*) 1.6 kJ, (*c*) 2.2 kJ
81 (*a*) 81 K, 81 K, (*b*) 2.7 kJ, (*c*) 3.3 kJ
83 256 kcal
85 (*a*) $c(4.00 \text{ K}) = 9.20 \times 10^{-2}$ J/kg·K, (*b*) 0.0584 J/kg
87 (*a*) 2.49 kJ, (*b*) 3.20 kJ
89 171 K
91

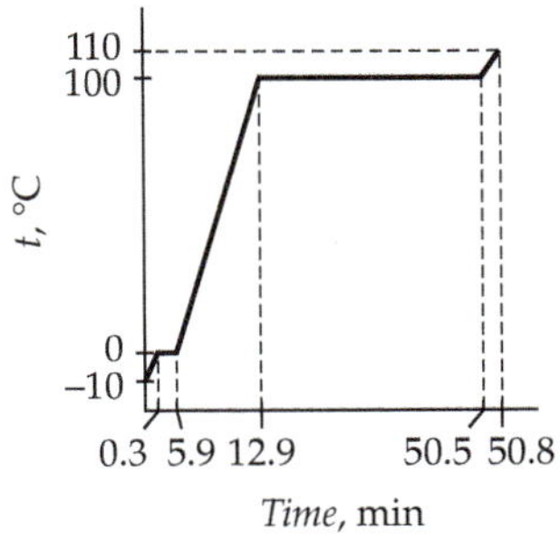

93 396 K
95 (*b*) 4.62 kJ
97 (*a*) $P_2 = \frac{1}{2}P_0$, (*b*) The gas is diatomic. (*c*) During the isothermal process the translational kinetic energy is unchanged. During the adiabatic process the translational kinetic energy increases by a factor of 1.32.

Chapter 19

1 (*c*)
3 (*a*)
5 The COP is defined so as to be a measure of the effectiveness of the device. For a refrigerator or air conditioner, the important quantity is the heat transferred from the already colder interior, Q_c. For a heat pump, the idea is to focus on the heat transferred into the warm interior of the house, Q_h.
7 Increasing the temperature of the steam increases its energy content. In addition, it increases the Carnot efficiency, and generally increases the efficiency of any heat engine.
9 A Carnot-cycle refrigerator is more efficient when the temperatures are close together because it requires less work to extract heat from an already cold interior if the temperature of the exterior is close to the temperature of the interior of the refrigerator. A Carnot-cycle heat engine is more efficient when the temperature difference is large because then more work is done by the engine for each unit of heat absorbed from the hot reservoir.
11 (*c*)
13 (*d*)
15 The cycle is that of the Otto engine (see Figure 19-3).
17

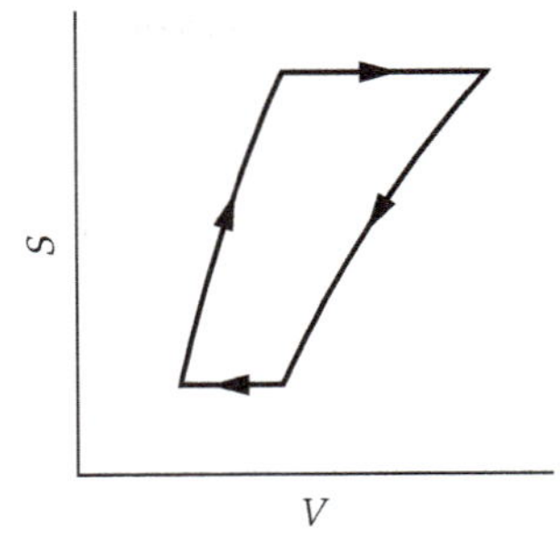

19

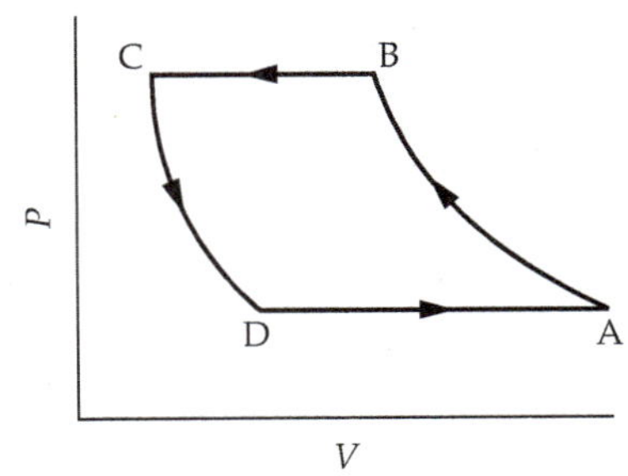

21 An increase of about 47%.
23 56%
25 (*a*) 1.7×10^{17} W, (*b*) 6.02×10^{14} J/(K·s)
27 (*a*) 500 J, (*b*) 400 J
29 (*a*) 40%, (*b*) 80 W
31 (*a*)

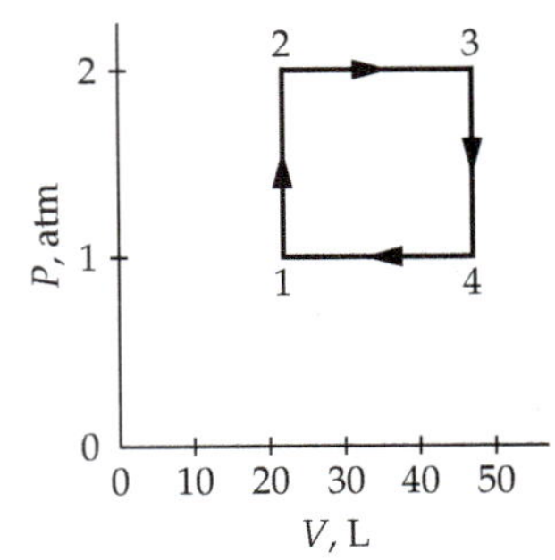

Process	W (kJ)	Q (kJ)	ΔE_{int} (kJ)
1→2	0	3.74	3.74
2→3	−4.99	12.5	7.5
3→4	0	−7.48	−7.48
4→1	2.49	−6.24	−3.75

(*b*) 15%

33 13.1%

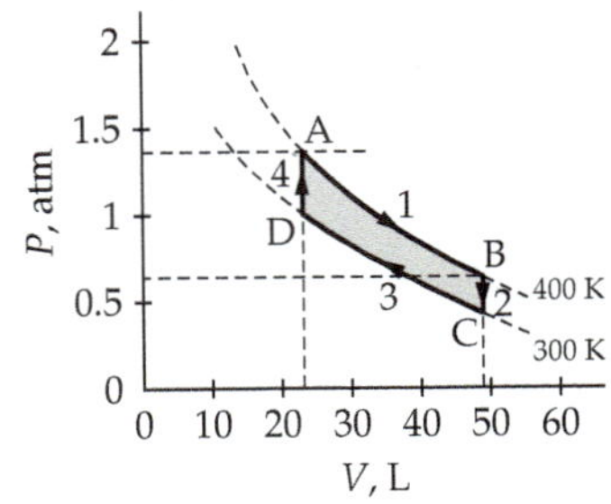

35 (*a*) $T_2 = 600$ K, $T_3 = 1800$ K, $T_4 = 600$ K, (*b*) 15%

37 (*a*) 5.16%. No contradiction. (*b*) Most warm-blooded animals survive under roughly the same conditions as humans. To make a heat engine work with appreciable efficiency, internal body temperatures would have to be maintained at an unreasonably high level.

41 (*a*) 33.3%, (*b*) 33.3 J, (*c*) 67 J, (*d*) 2.0

45 (*a*) 33%

47 (*a*) 100°C, (*b*) 3.12 kJ, $Q_{2\to3} = 0$, $Q_{3\to1} = -2.91$ kJ, (*c*) 6.7%, (*d*) 35.5%

49 (*a*) 6.3, (*b*) 3.2 kW, (*c*) 5.3 kW

51 (*a*) 0.17 MJ, (*b*) 0.12 MJ. Because the temperature difference increases when the room is warmer, the COP decreases.

53 6.05 kJ/K

55 $\Delta S_u = 2.40$ J/K and, because $\Delta S_u > 0$, the entropy of the universe increases.

57 (*a*) 0, (*b*) 267 K

59 (*a*) 244 kJ/K, (*b*) −244 kJ/K, (*c*) The entropy change of the universe is just slightly greater than zero.

61 (*a*) −117 J/K, (*b*) 138 J/K, (*c*) 20 J/K

63 (*a*) 0.42 J/K, (*b*) 125 J

65 (*a*) $W_{\text{cycle}} = 20.0$ J, (*b*) $Q_{\text{h cycle}} = 67$ J, $Q_{\text{c cycle}} = 47$ J

67 (*a*) 51%, (*b*) 0.10 MJ, (*c*) 98 kJ

69 113 W/K

71 (*a*) You should explain to him that, because the efficiency he claims for his invention (83.3%) is greater than the efficiency of a Carnot engine operating between the same two temperatures, his data are not consistent with what is known about the thermodynamics of engines. He must have made a mistake in his analysis of his data-or he is a con man looking for people to swindle.

(*b*) 135 W

73 (*a*) Process (2) is more wasteful of *available work*. (*b*) $\Delta S_1 = 1.67$ J/K, $\Delta S_2 = 0.833$ J/K

75 313 K

77 10 W

79 (*a*) 253 kPa, (*b*) 462 K, (*c*) 6.97 kJ

81 (*a*) 253 kPa, (*b*) 416 K, (*c*) 6.59 kJ

83 (*a*)

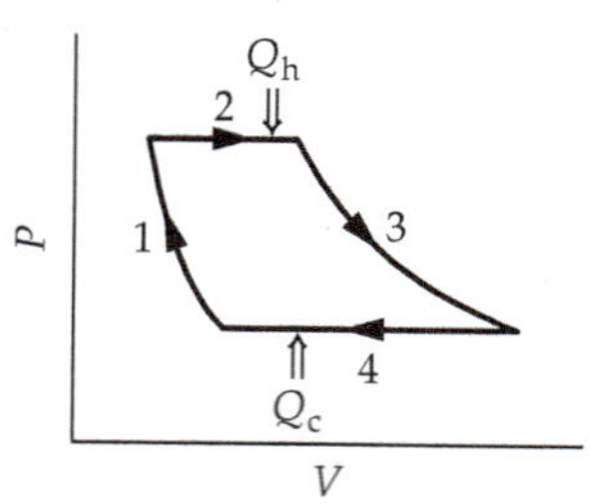

89 $T = 10^{484}$ y, $T \approx 10^{478} T_{\text{Russell}}$

Chapter 20

1 The glass bulb warms and expands first, before the mercury warms and expands.

3 Water expands greatly as it freezes. If a sealed glass bottle full of water is placed in a freezer, as the water freezes there will be no room for the expansion to take place. The bottle will be broken.

5 The strip will curl more tightly.

7 (*c*)

9 (*a*) With increasing altitude, *P* decreases; from curve OC, the temperature *T* of the liquid–gas interface decreases as the pressure decreases, so the boiling temperature decreases. Likewise, from curve OB, the melting temperature increases with increasing altitude.

(*b*) Boiling at a lower temperature means that the cooking time will have to be increased.

11 At very low pressures and temperatures, carbon dioxide can exist only as a solid or gas (or vapor above the gas). The atmosphere of Mars is 95 percent carbon dioxide. Mars, on average, is warm enough so that the atmosphere is mostly gaseous carbon dioxide. The polar regions are cold enough to enable solid carbon dioxide (dry ice) to exist, even at the low pressure.

13 (*a*)

15

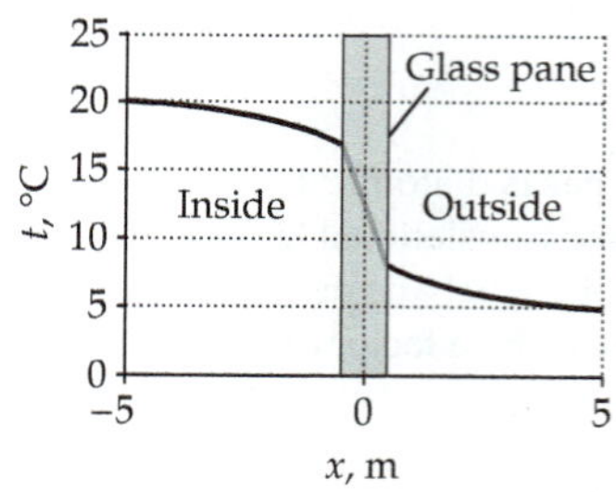

17 Your assumption was not correct and 14 mL of water overflowed.

19 17 mW/(m · K)

21 0.30 kW, 0.1 K/W

23 2.9 nm

25 8(°F · h · ft²)/Btu

27 220°C

29 $15 \times 10^{-6}\ \text{K}^{-1}$

31 $3.7 \times 10^{-12}\ \text{N/m}^2$

33 (*a*) 90°C, (*b*) 82°C, (*c*) 170 kPa

35 2.1 kBtu/h

37 (*a*) $I_{\text{Cu}} = 0.96$ kW, $I_{\text{Al}} = 0.57$ kW, (*b*) 1.53 kW, (*c*) 0.052 K/W

41 1.3 mm

43 93.5 cm^2

45 1598°C

47 2.1 km

49 5800 K

51 (*b*) The values agree to within 0.3%.

53 1.3×10^{10} kW. About 0.007 percent.

55 142 W

57 $L_2 = L_1$, $\omega_2 \approx (1 - 2\alpha\,\Delta T)\omega_1$, $E_2 = E_1(1 - 2\alpha\,\Delta T)$

59 (*a*) 0.70 cm/h, (*b*) 12 d

Chapter 21

1 The net charge on large objects is always very close to zero. Hence the most obvious force is the gravitational force.

3 (*a*) Coulomb's law is only valid for point particles. The paper bits cannot be modeled as point particles because the paper bits become polarized.

(*b*) No, the attraction does not depend on the sign of the charge on the comb. The induced charge on the paper that is closest to the comb is always opposite in sign to the charge on the comb, and thus the net force on the paper is always attractive.

5 (*a*)

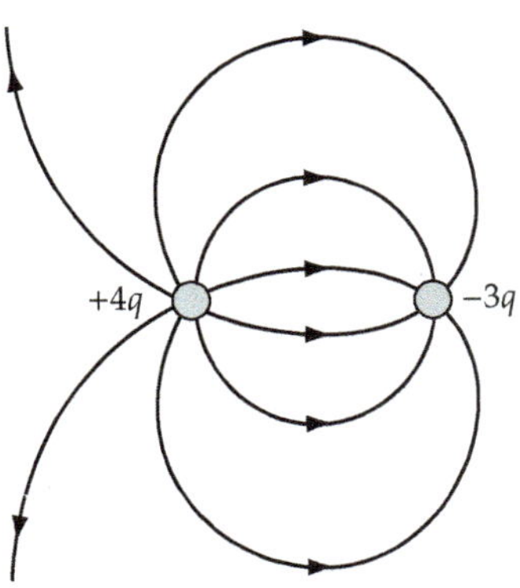

(*b*)

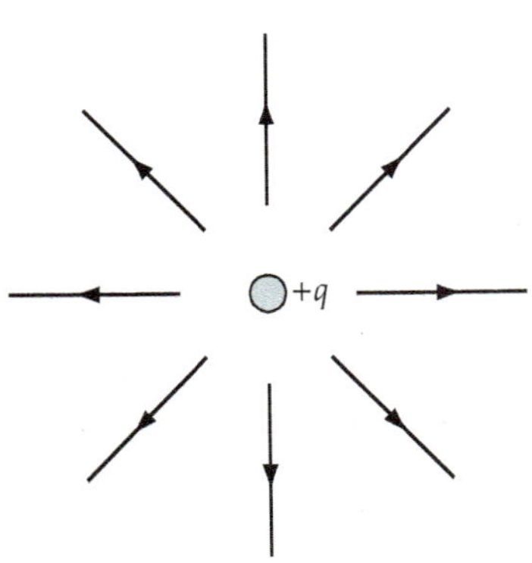

7 Assume that the rod has a negative charge. When the charged rod is brought near the aluminum foil, it induces a redistribution of charges with the side nearer the rod becoming positively charged, and so the ball of foil swings toward the rod. When it touches the rod, some of the negative charge is transferred to the foil, which, as a result, acquires a net negative charge and is now repelled by the rod.

9 (*a*) On the sphere near the positively charged rod, the induced charge is negative and near the rod. On the other sphere, the net charge is positive and on the side far from the rod. This is shown in the diagram.

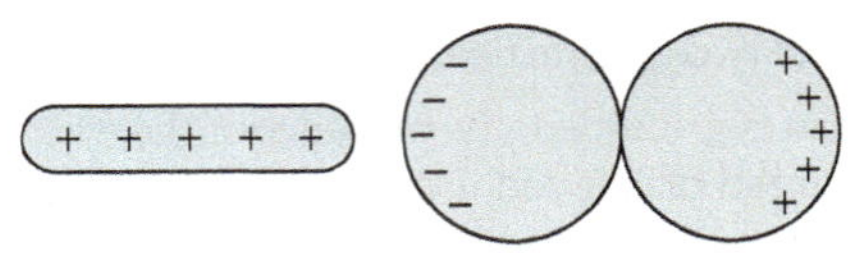

(*b*) When the spheres are separated and far apart and the rod has been removed, the induced charges are distributed uniformly over each sphere. The charge distributions are shown in the diagram.

11 (*a*) False, (*b*) True, (*c*) False, (*d*) Possibly, (*e*) False, (*f*) True

13 (*a*)

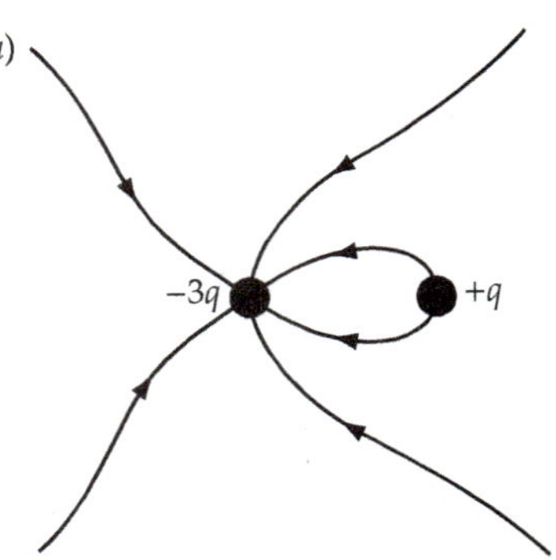

(*b*)

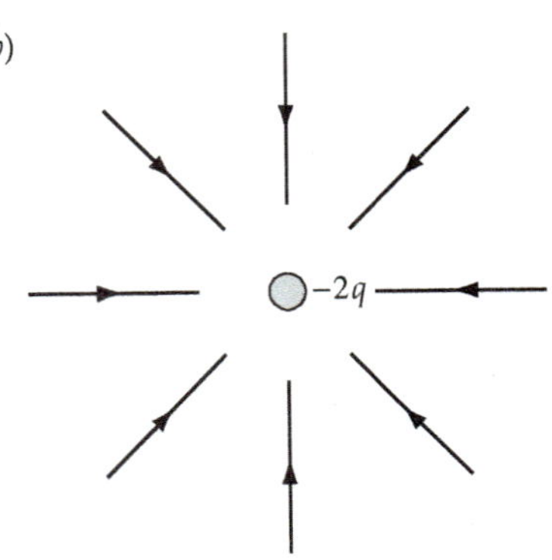

15 The dipole moment rotates back and forth in oscillatory motion. The dipole moment gains angular speed as it rotates toward the direction of the electric field, and loses angular speed as it rotates away from the direction of the electric field.

17

	1	2	3
(*a*)	down	up	up
(*b*)	up	right	left
(*c*)	down	up	up
(*d*)	down	up	up

Figure 21-23 shows the electric field due to a single dipole, where the dipole moment is directed toward the right. The electric field due two a pair of dipoles can be obtained by superposing the two electric fields.

19 Because the can is grounded, the presence of the negatively charged plastic rod induces a positive charge on it. The positive charges induced on the can are attracted, via the Coulomb interaction, to the negative charges on the plastic rod. Unlike charges attract, so the can will roll toward the rod.

21 5.0×10^{12} electrons

23 4.82×10^{7} C

25 (*a*) 2.60 h, (*b*) 2.1×10^{-13} W

27 $\vec{F}_1 = (1.5 \times 10^{-2}\text{ N})\hat{i}$

29 A distance equal to $0.41L$ from the -2.9-μC charge on the side away from the 4.0-μC charge.

31 $\vec{F}_3 = -(8.65\text{ N})\hat{j}$

33 $\vec{F}_1 = (0.90\text{ N})\hat{i} + (1.8\text{ N})\hat{j}$, $\vec{F}_2 = (-1.3\text{ N})\hat{i} - (1.2\text{ N})\hat{j}$, $\vec{F}_3 = (0.4\text{ N})\hat{i} - (0.64\text{ N})\hat{j}$

35 $\vec{F}_q = \dfrac{kqQ}{R^2}(1 + \sqrt{2})\hat{i}$

37 (*a*) $(1.0\text{ kN/C})\hat{i}$, (*b*) $(-0.36\text{ kN/C})\hat{i}$,

(*c*)

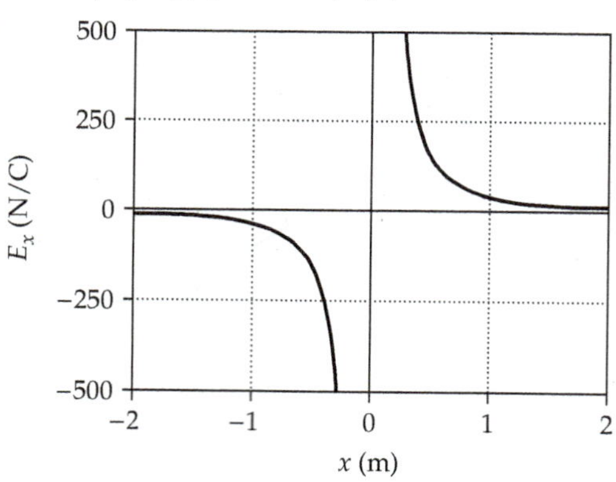

39 (*a*) $\vec{E}(0, 0) = (4.0 \times 10^{5}\text{ N/C})\hat{j}$, (*b*) $\vec{F}(0, 0) = (-1.6\text{ mN})\hat{j}$, (*c*) -40 nC

41 (*a*) 35 kN/C at 0°, (*b*) $\vec{F} = (69\ \mu\text{N})\hat{i}$

43 (*a*) 13 kN/C at 230°, (*b*) 2.1×10^{-15} N at 51°

45 (*a*) 1.9 kN/C at 230°, (*b*) 3.0×10^{-16} N at 230°

47 The charge must be placed a distance $L/\sqrt{3}$ below the midpoint of the base of the triangle, where L is the length of a side of the triangle.

49 (*a*) For a positive test charge, the equilibrium at (0, 0) is unstable for small displacements in either direction along the x axis, and stable for small displacements in either direction along the y axis.

(*b*) For a negative test charge, the equilibrium is stable at (0,0) for displacements along the x axis and unstable for displacements along the y axis.

(*c*) $q_0 = -\frac{1}{4}q$

51 (*a*) 1.76×10^{11} C/kg, (*b*) $1.76 \times 10^{13}\text{ m/s}^2$ in the direction opposite to the direction of the electric field, (*c*) 0.2 μs, (*d*) 3 mm

53 (*a*) $\vec{a} = (-5.28 \times 10^{13}\text{ m/s}^2)\hat{j}$, (*b*) 50.0 ns, (*c*) 33.4° in the $-y$ direction

55 800 μC

57 The electron strikes the lower plate 4.1 cm to the right of its initial position.

59 (*a*) 8.0×10^{-18} C·m

(*b*)

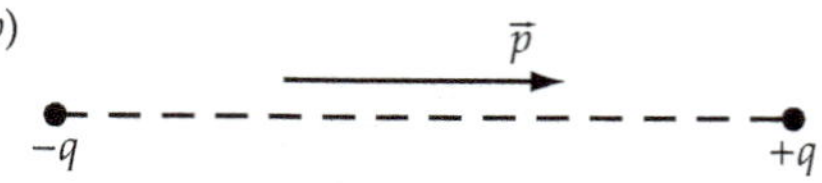

63 (*a*) 1.83×10^{6} N/C, (*b*) 1.80×10^{6} N/C. The exact and estimated values of E_p agree to within 2%. This difference is this large because the separation of the two charges of the dipole is 20% of the distance from the center of the dipole to point P.

67 (*a*) 1.8×10^{-5} C and 1.8×10^{-4} C

(*b*) -1.4×10^{-5} C and 2.1×10^{-4} C

69 (*a*) 0.225 N, downward, (*b*) 0.112 N·m, counterclockwise, (*c*) 45.8 g, (*d*) 5.00×10^{-7} C

71 (*a*) 28.0 μC and 172 μC

(*b*) 250 N

73 (*a*) $-97.2\ \mu$C, (*b*) $x = 0.0508$ m and $x = 0.169$ m

75 (*a*) 10°, (*b*) 9.9° for each

79 $v = e\sqrt{k/(2mL)}$

83 (*a*) $E_y = \dfrac{2kQy}{[y^2 + \frac{1}{4}a^2]^{3/2}}$,

(*b*) $\vec{F} = \dfrac{2kqQy}{[y^2 + \frac{1}{4}a^2]^{3/2}}\hat{j}$, where q is positive,

(*c*) $v = \sqrt{8(1 - \sqrt{2/3})}\sqrt{\dfrac{kqQ}{am}} = 1.21\sqrt{\dfrac{kqQ}{am}}$

85 4.6×10^{-14} m = 46 fm

87 (*b*) 52 μm/s

Chapter 22

1 The resultant field is directed along the dashed line, pointing away from the intersection of the two sides of the L-shaped object. This can be seen by dividing each leg of the object into 10 (or more) equal segments and then drawing the electric field on the dashed line due to the charges on each pair of segments that are equidistant from the intersection of the legs.

3 (*a*) True (assuming there are no charges inside the shell), (*b*) True, (*c*) False

5 (*a*) False, (*b*) True

7 (*a*) False, (*b*) False, (*c*) True, (*d*) False, (*e*) True

9 (*a*) radially inward, (*b*) radially outward, (*c*) radially inward

11 (*a*) radially inward, (*b*) radially inward, (*c*) The field is zero.

13 (*a*) 18 nC, (*b*) 26 N/C, (*c*) 4.4 N/C, (*d*) 2.6 mN/C, (*e*) This result is about 0.01% less than the exact value obtained in (*d*).

15 (*a*) 4.7×10^5 N/C, (*b*) 1.1×10^6 N/C, (*c*) 1.5×10^3 N/C (*d*) 1.5×10^3 N/C. This result agrees exactly, to two significant figures, with the result obtained in Part (*c*).

17 (*a*) 0.189 kQ/a^2, (*b*) 0.358 kQ/a^2, (*c*) 0.385 kQ/a^2, (*d*) 0.354 kQ/a^2, (*e*) 0.179 kQ/a^2,

(*f*)

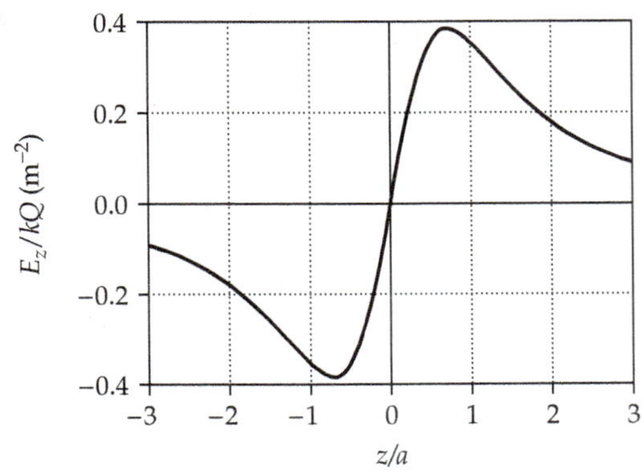

19 (*a*)

Cell	Content/Formula	Algebraic Form
B3	9.00E+09	k
B4	5.00E−10	σ
B5	0.3	r
A8	0	x_0
A9	A8+0.01	$x_0 + 0.01$
B8	2*PI()*$B$3*$B$4*(1−A8/(A8^2+$B$5^2)^2)^0.5)	$2\pi k\sigma\left(1 - \frac{x}{\sqrt{x^2 + a^2}}\right)$
C8	2*PI()*$B$3*$B$4	$2\pi k\sigma$

	A	B	C
1			
2			
3	k =	9.00E+09	N*m2/C2
4	σ =	5.00E−10	C/m2
5	a =	0.300	m
6			
7	x	E(x)	E_{sheet}
8	0.00	28.27	28.3
9	0.01	27.33	28.3
77	0.69	2.34	28.3
78	0.70	2.29	28.3

(*b*)

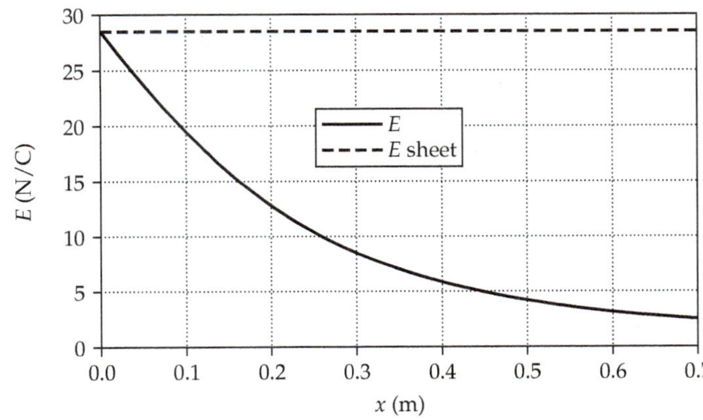

The magnitudes differ by more than 10.0 percent for $x \geq 0.0300$ m.

27 (*a*) 20.0 N·m²/C, (*b*) 17 N·m²/C

29 (*a*) 1.5 N·m²/C, 1.5 N·m²/C, (*b*) 0, (*c*) 3.0 N·m²/C, (*d*) 2.7×10^{-11} C

31 (*a*) 3.14 m², (*b*) 7.19×10^4 N/C, (*c*) 2.26×10^5 N·m²/C, (*d*) No, (*e*) 2.26×10^5 N·m²/C

33 −79.7 nC

35 $\cos\theta$

37 (*a*) $E_{r<R_1} = 0$, $\vec{E}_{R_1<r<R_1} = \frac{kq_1}{r^2}\hat{\mathbf{r}}$, $\vec{E}_{r>R_2} = \frac{k(q_1 + q_2)}{r^2}\hat{\mathbf{r}}$, (*b*) −1,

(*c*)

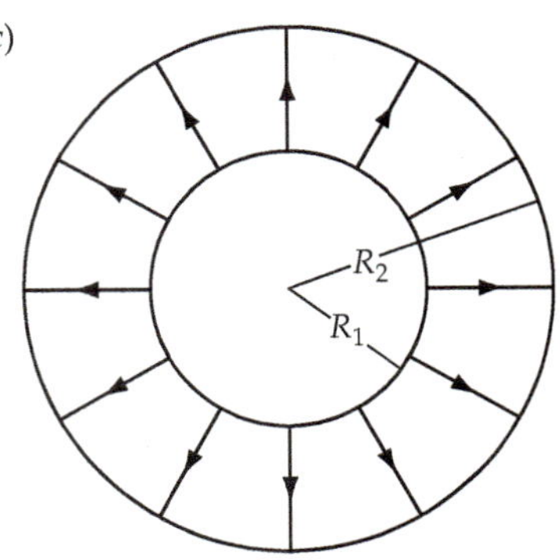

39 (*a*) 0.407 nC, (*b*) 339 N/C, (*c*) 1.00 kN/C, (*d*) 983 N/C, (*e*) 366 N/C

41 (*a*) 2.00 μC/m³, (*b*) 470 N/C

43 (*a*) $Q = 2\pi BR^2$, (*b*) $E_r = \frac{BR^2}{2\epsilon_0 r^2}$ $r > R$, $E_r = \frac{B}{2\epsilon_0}$ $r < R$

(*c*)

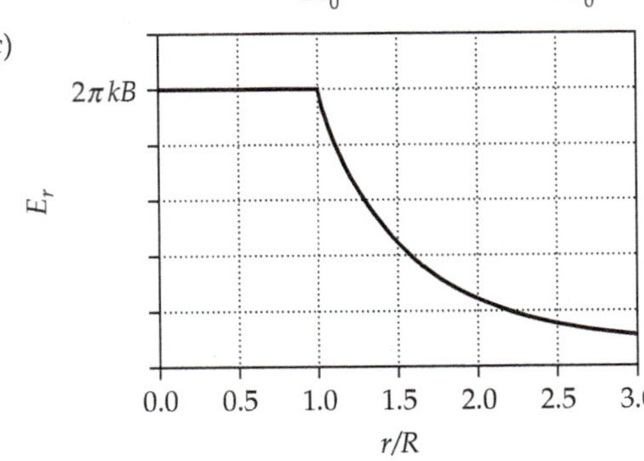

45 (a) $Q_{inside} = \frac{4\pi\rho}{3}(r^3 - R_1^3)$,

(b) $E_r = 0 \quad r < R_1, E_r = \frac{\rho}{3\epsilon_0 r^2}(R_2^3 - R_1^3) \quad R_1 < r < R_2$,

$E_r = \frac{\rho}{3\epsilon_0 r^2}(R_2^3 - R_1^3) \quad r > R_2$

47 (a) 1.41×10^6 m/s, (b) Because of its much larger mass, the impact speed of the ion will be much less than the impact speed of the electron. (The ion will impact the tube instead of the wire.)

49 (a) 679 nC, (b) 0, (c) 0, (d) 1.00 kN/C, (e) 610 N/C

51 (a) 679 nC, (b) 339 N/C, (c) 1.00 kN/C, (d) 1.00 kN/C, (e) 610 N/C

53 (a) $E_R = 0 \quad r < 1.50$ cm,

$E_R = \frac{(108\ \text{N}\cdot\text{m/C})}{R} \quad 1.50\ \text{cm} < r < 4.50\ \text{cm}$,

$E_R = 0 \quad 4.50\ \text{cm} < r < 6.50\ \text{cm}$,

$E_R = \frac{156\ \text{N}\cdot\text{m/C}}{R} \quad r > 6.50\ \text{cm}$,

(b) $\sigma_{inside} = -21.2\ \text{nC/m}^2$ and $\sigma_{outside} = 14.7\ \text{nC/m}^2$

55 (b) $E_R = \frac{b}{4\epsilon_0}R^3 \quad r < a, E_R = \frac{ba^4}{4R\epsilon_0} \quad r > a$

57 (a) 18.8 nC/m, (b) $E_R = 22.6$ kN/C $\quad R < 1.50$ cm,

$E_R = \frac{339\ \text{N}\cdot\text{m/C}}{R} \quad 1.50\ \text{cm} < R < 4.50\ \text{cm}$,

$E_R = 0 \quad 4.50\ \text{cm} < R < 6.50\ \text{cm}$,

$E_R = \frac{339\ \text{N}\cdot\text{m/C}}{R} \quad R > 6.50\ \text{cm}$

59 9.4 kN/C

61 (a) $E_r = \frac{kq}{r^2} \quad r < R_1, E_r = 0 \quad R_1 < r < R_2, E_r = \frac{kq}{r^2} \quad r > R_2$

(b)

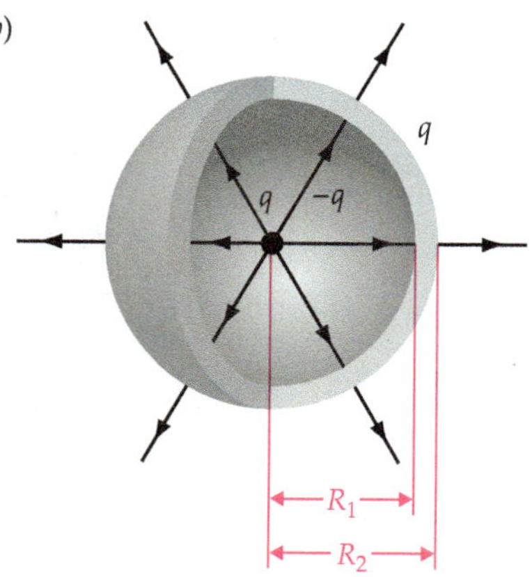

(c) $\sigma_{inner} = -\frac{q}{4\pi R_1^2}, \sigma_{outer} = \frac{q}{4\pi R_2^2}$

63 (a) $\sigma_{inner} = -0.55\ \mu\text{C/m}^2, \sigma_{outer} = 0.25\ \mu\text{C/m}^2$

(b) $E_r = (2.3 \times 10^4\ \text{N}\cdot\text{m}^2/\text{C})\frac{1}{r^2} \quad r < 60$ cm,

$E_r = 0 \quad 60\ \text{cm} < r < 90\ \text{cm}$,

$E_r = (2.3 \times 10^4\ \text{N}\cdot\text{m}^2/\text{C})\frac{1}{r^2} \quad r > 90$ cm

(c) $\sigma_{inner} = -0.55\ \mu\text{C/m}^2, \sigma_{outer} = 0.59\ \mu\text{C/m}^2$,

$E_r = (2.3 \times 10^4\ \text{N}\cdot\text{m}^2/\text{C})\frac{1}{r^2} \quad r < 60$ cm,

$E_r = 0 \quad 60\ \text{cm} < r < 90\ \text{cm}$,

$E_r = (5.4 \times 10^4\ \text{N}\cdot\text{m}^2/\text{C})\frac{1}{r^2} \quad r > 90$ cm

65 (a) $Q_{left} = 15\ \mu\text{C}$ and $Q_{right} = 65\ \mu\text{C}$,
(b) $E_{left\,x} = -68$ kN/C and $E_{right\,x} = 294$ kN/C,
$\sigma_{left} = 0.60\ \mu\text{C/m}^2$ and $\sigma_{right} = 2.60\ \mu\text{C/m}^2$

67 −115 kN/C

69 (a) $E = \frac{Q}{8\pi\epsilon_0 r^2}$, radially outward,

(b) $F = \frac{Q^2a^2}{32\pi\epsilon_0 r^4}$, radially outward,

(c) $P = \frac{Q^2}{32\pi^2\epsilon_0 r^4}$

71 (a) 3.39×10^9 N/C, toward the right, (b) 3.39×10^9 N/C, toward the right, (c) zero, (d) zero

73 (a) $\rho_0 = \frac{-e}{\pi a^3}$,

(b) $E_r(r) = \frac{ke}{r^2}\left(1 - \frac{1}{4}\left[(1 - e^{-2r/a}) - 2e^{-2r/a}\left(\frac{r}{a} + \frac{r^2}{a^2}\right)\right]\right)$

75 (a) Radially outward toward the gap, (b) $E_{center} = \frac{kQ\ell}{2\pi R^3}$

77 (a) $\vec{E} = 204$ kN/C at 56.3°,
(b) $\vec{E} = 263$ kN/C at 153°

79 (a) $v = \sqrt{\frac{2kq\lambda}{m}}$, (b) $T = \pi R\sqrt{\frac{2m}{kq\lambda}}$

81 (a) 0.997 kg, (b) 1.18 Hz

83 (b) $\vec{E}_1 = \vec{E}_2 = \frac{\rho b}{3\epsilon_0}\hat{i}$

85 $\vec{E}_1 = \left(\frac{\rho b}{3\epsilon_0} + \frac{Q}{4\pi\epsilon_0 b^2}\right)\hat{i}, \vec{E}_2 = \left(\frac{\rho b}{3\epsilon_0} - \frac{Q}{4\pi\epsilon_0 b^2}\right)\hat{i}$

87 $\frac{1}{2}R$

Chapter 23

1 The proton is moving to a region of higher potential. The proton's electrostatic potential energy is increasing.

3 The electric field is zero throughout the region.

5

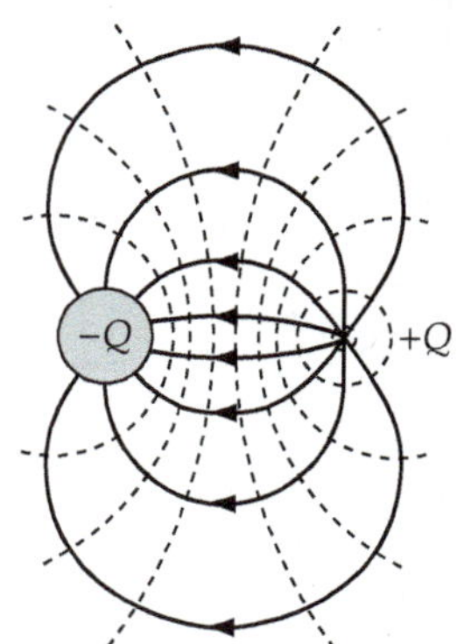

7

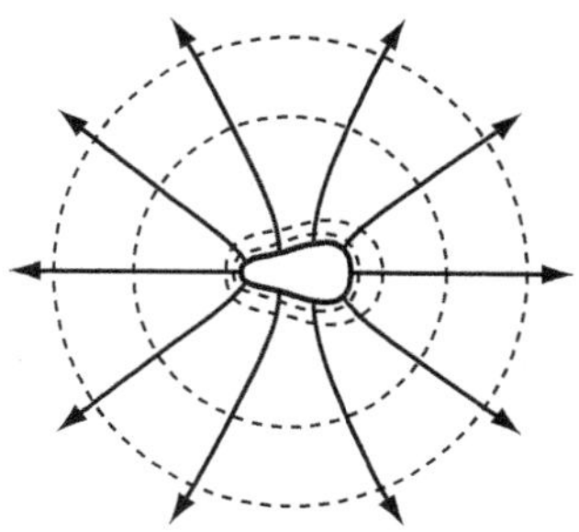

9 (*a*) 2, (*b*) 3

11 No. The local surface charge density is proportional to the normal component of the electric field, not the potential on the surface.

13 3.0×10^9 V

15 0.72 MeV

17 27 μC/m^2

19 (*a*) 4.49 kV, (*b*) 13.5 mJ

21 (*a*) -8.00 kV, (*b*) -24.0 mJ, (*c*) 24.0 mJ, (*d*) $V(x) = -(2.00 \text{ kV/m})x$

23 (*a*) 3.09×10^7 m/s, (*b*) 2.50 MV/m

25 (*a*) $r = kzZe^2/K_i$, (*b*) 46 fm, 25 fm, (*c*) No. The distance of closest approach for a 5-MeV alpha particle found above (45.5 fm) is much larger than the 7 fm radius of a gold nucleus. Hence the scattering was solely the result of the inverse-square Coulomb force.

27 (*a*) 12.9 kV, (*b*) 7.55 kV, (*c*) 4.43 kV

29 (*a*) 135 kV (*b*) 95.3 kV (*c*) Because the two field points are equidistant from all points on the circle, the answers for Parts (*a*) and (*b*) would not change.

31 (*a*) $V(x) = kq\left(\dfrac{1}{|x-a|} + \dfrac{1}{|x+a|}\right)$,

(*b*)

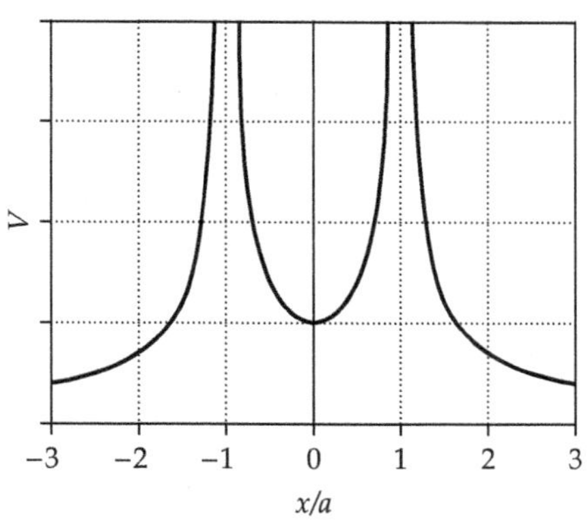

33 (*b*) at points on the z axis

35 (*a*) positive, (*b*) 25.0 kV/m

37 (*a*) +668 nC, (*b*) 3.00 kV. The plane at $x = 2.00$ m is at the higher potential.

39 (*a*) $V(x) = kq\left(\dfrac{2}{\sqrt{x^2+a^2}} + \dfrac{1}{|x-a|}\right) \quad x \neq a,$

(*b*) $E_x(x) = \dfrac{2kqx}{(x^2+a^2)^{3/2}} + \dfrac{kq}{(x-a)^2} \quad x > a,$

$E_x(x) = \dfrac{2kqx}{(x^2+a^2)^{3/2}} - \dfrac{kq}{(a-x)^2} \quad x < a$

41 (*a*) 6.02 kV, (*b*) -12.7 kV, (*c*) -42.3 kV

43 $\sim 3 \times 10^{-5}$ C/m^2

45 $V_a - V_b = \dfrac{2kq}{L}\ln\left(\dfrac{b}{a}\right)$

47 $V_a - V_b = kq\left(\dfrac{1}{a} - \dfrac{1}{b}\right)$

49

Region	$x \le 0$	$0 \le x \le a$	$x \ge a$
Part (*a*)	$\dfrac{\sigma}{\epsilon_0}x$	0	$-\dfrac{\sigma}{\epsilon_0}(x-a)$
Part (*b*)	0	$-\dfrac{\sigma}{\epsilon_0}x$	0

51 (*a*) $V(x,0) = \dfrac{kQ}{L}\ln\left(\dfrac{\sqrt{x^2+\frac{1}{4}L^2}+\frac{1}{2}L}{\sqrt{x^2+\frac{1}{4}L^2}-\frac{1}{4}L^2}\right)$

53 (*a*) $Q = \frac{1}{2}\pi\sigma_0 R^2$, (*b*) $V = \dfrac{2\pi k\sigma_0}{3R^2}\left((R^2 - 2z^2)\sqrt{z^2+R^2} + 2z^3\right)$

55 (*a*) $V(x) = \dfrac{kQ}{L}\ln\left(\dfrac{x+\frac{1}{2}L}{x-\frac{1}{2}L}\right)$

57 (*a*) $dQ = \dfrac{3Q}{R^3}r'^2\,dr'$, (*b*) $dV = \dfrac{3kQ}{R^3}r'\,dr'$, (*c*) $V = \dfrac{3kQ}{2R^3}(R^2 - r^2)$, (*d*) $dV = \left(\dfrac{3kQ}{R^3 r}\right)r'^2\,dr'$, (*e*) $V = \dfrac{kQ}{R^3}r^2$, (*f*) $V = \dfrac{kQ}{2R}\left(3 - \dfrac{r^2}{R^2}\right)$

61 (*a*) The equipotential surfaces are planes parallel to the charged planes. (*b*) The regions to either side of the two charged planes are equipotential regions, so any surface in either of these regions is an equipotential surface.

63 (*a*) 0.224 cm, closer to the wire, (*b*) 0.864 mm, (*c*) The distance between the 700 V and the 725 V equipotentials is 0.0966 mm. This closer spacing of these two equipotential surfaces was to be expected. Close to the central wire, two equipotential surfaces with the same difference in potential should be closer together to reflect the fact that the electric field strength is greater closer to the wire.

65 (*a*) 30.0 mJ, (*b*) -5.99 mJ, (*c*) -18.0 mJ

67 (*a*) 22.3 nC, (*b*) 22.3 μJ

69 $v = q\sqrt{\dfrac{6\sqrt{2}k}{ma}} = 2.91q\sqrt{\dfrac{k}{ma}}$

71 (*a*) 9.61×10^{-20} J, (*b*) 4.59×10^5 m/s, (*c*) Because $2K_{\text{i min}} > K_{\text{i escape}}$, the electron escapes from the proton with residual kinetic energy.

73 (*a*) $V(x) = \dfrac{2kq}{\sqrt{x^2+a^2}}$, (*b*) $\vec{E}(x) = \dfrac{2kqx}{(x^2+a^2)^{3/2}}\hat{i}$

75 (*a*) $V(x,y) = \dfrac{\lambda}{2\pi\epsilon_0}\ln\left(\dfrac{\sqrt{(x-a)^2+y^2}}{\sqrt{(x+a)^2+y^2}}\right)$, $V(0,y) = 0$

(*b*) $y = \pm\sqrt{21.25x - x^2 - 25}$

(*c*)

Cell	Content/Formula	Algebraic Form
A2	1.25	$\frac{1}{4}a$
A3	A2+0.05	$x + \Delta x$
B2	SQRT(21.25*A2−A2^2−25)	$y = \sqrt{21.25x - x^2 - 25}$
B4	−B2	$y = -\sqrt{21.25x - x^2 - 25}$

	A	B	C
1	x	y_{pos}	y_{neg}
2	1.25	0.00	0.00
3	1.30	0.97	–0.97
4	1.35	1.37	–1.37
5	1.40	1.67	–1.67
6	1.45	1.93	–1.93
7	1.50	2.15	–2.15
370	19.65	2.54	–2.54
371	19.70	2.35	–2.35
372	19.75	2.15	–2.15
373	19.80	1.93	–1.93
374	19.85	1.67	–1.67
375	19.90	1.37	–1.37
376	19.95	0.97	–0.97

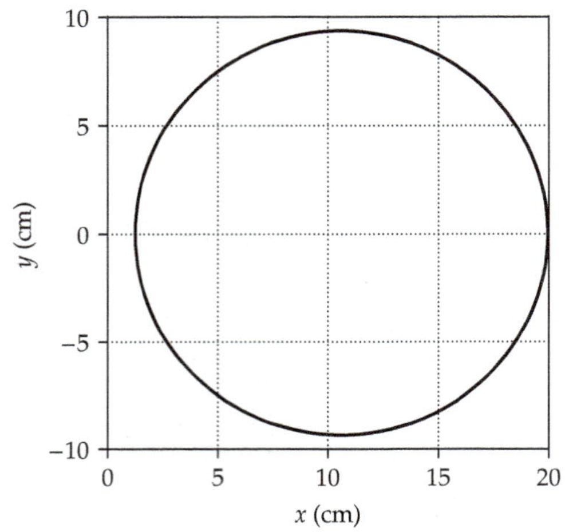

The wires are in the $y = 0$ plane.

77 (*a*) 3.56×10^8 C/m^3,

(*b*) $V(r) = \pi k a^3 \rho_0\left(\frac{1}{a} + \frac{1}{r}\right)e^{-2r/a} = ke\left(\frac{1}{a} + \frac{1}{r}\right)e^{-2r/a}$

79 (*a*) $W_{+Q\to+a} = \frac{kQ^2}{2a}$, (*b*) $W_{-Q\to0} = \frac{-2kQ^2}{a}$, (*c*) $W_{-Q\to2a} = \frac{2kQ^2}{3a}$

81 (*a*) 100 eV, (*b*) 1.38×10^5 m/s

83 $R_2 = \frac{2}{3}R_1$

85 7.1 nC

87 (*b*) $\sigma = \frac{qd}{4\pi(d^2 + r^2)^{3/2}}$

89 (*a*) $V(c) = 0, V(b) = kQ\left(\frac{1}{b} - \frac{1}{c}\right), V(a) = V(b) = kQ\left(\frac{1}{b} - \frac{1}{c}\right)$,

(*b*) $Q_b = Q, V(a) = V(c) = 0, Q_a = -Q\frac{a(c-b)}{b(c-a)}$,

$Q_c = -Q\frac{c(b-a)}{b(c-a)}, V(b) = kQ\frac{(c-b)(b-a)}{b^2(c-a)}$

93 (*a*) $R' = 0.794R$ (*b*) $\Delta E = -0.370E$

Chapter 24

1 (*c*)

3 False. The electrostatic energy density is not uniformly distributed because the magnitude of the electric field strength is not uniformly distributed.

5 1/3

7 (*a*) True, (*b*) True

9 (*a*) True, (*b*) False. Because $Q = CV$, and C increases, Q must increase. (*c*) True. $E = V/d$, where d is the plate separation. (*d*) False. $U = \frac{1}{2}QV$.

11 (*a*) $U_{\text{parallel}} = 2U_{\text{1 capacitor}}$, (*b*) $U_{\text{series}} = \frac{1}{2}U_{\text{1 capacitor}}$

13 $0.1 \text{ nF/m} \le C/L \le 0.2 \text{ nF/m}$

15 2.3 nF

17 75.0 nF

19 (*a*) 15.0 mJ, (*b*) 45.0 mJ

21 (*a*) 0.625 J, (*b*) 1.88 J

23 (*a*) 100 kV/m, (*b*) 44.3 mJ/m^3, (*c*) 88.5 μJ, (*d*) 17.7 nF, (*e*) 88.5 μJ

25 (*a*) 11 nC, (*b*) Because work has to be done to pull the plates farther apart, you would expect the energy stored in the capacitor to increase, (*c*) 0.55 μJ

27 10.00 μF

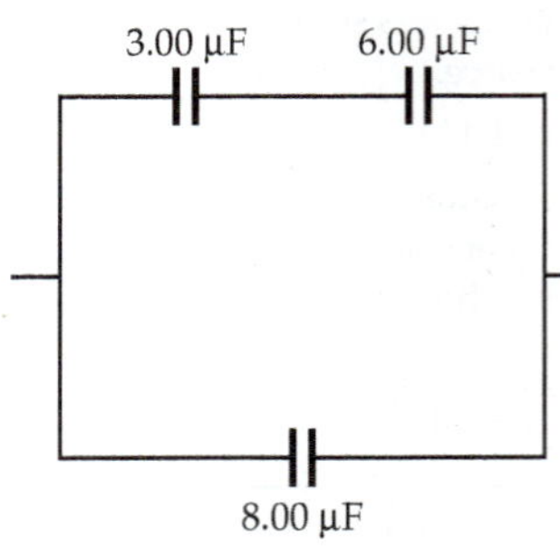

29 (*a*) 30.0 μF, (*b*) 6.00 V, (*c*) $Q_{10} = 60.0$ μC, $Q_{20} = 120$ μC, (*d*) $U_{10} = 180$ μJ, $U_{20} = 360$ μJ

31 (*a*) If their capacitance is to be a maximum, the capacitors must be connected in parallel.

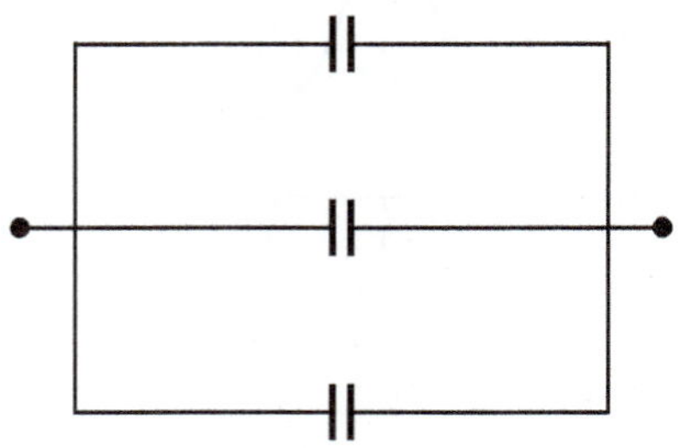

(*b*) (1) 1.67 μF

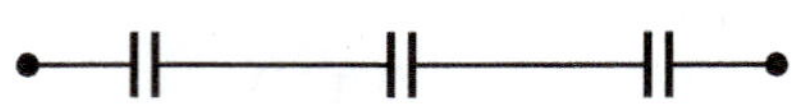

(2) 3.33 μF

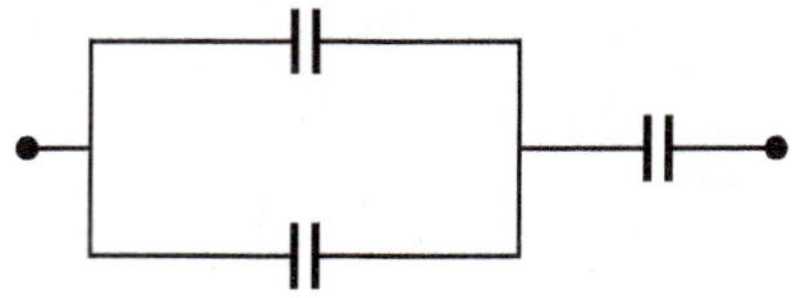

(3) 7.50 μF

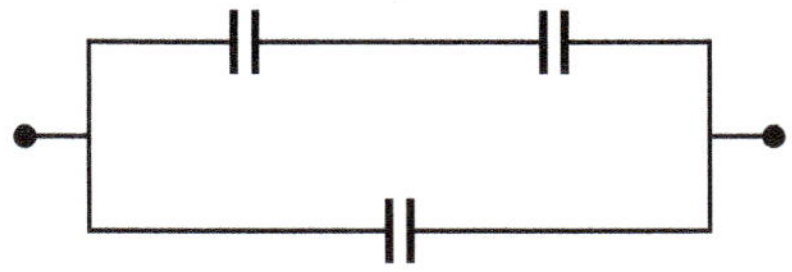

35 (a) $2C_0$, (b) $11C_0$

37 0.571 μF, 0.667 μF, 0.800 μF, 0.857 μF, 1.33 μF, 1.43 μF, 1.71 μF, 2.33 μF, 2.80 μF, 3.00 μF, 4.67 μF, 5.00 μF, 6.00 μF, 7.00 μF

39 (a) 4.80 kV, (b) 9.60 mC

41 (a) 7.9 m^2, (b) 23 V, (c) 37 μJ, (d) 0.16 J

43 (a) 1.55 pF, (b) 15.5 nC/m

45 179 pF/m

47 $\Delta C = -2\frac{P}{Y}C$

51 $R' = 2R$

53 (a) $V_{100} = V_{400} = 1.20$ kV, (b) 640 μJ

55 (a) 2.4 μF, (b) 0.4 mJ

57 (a) $V_{4.00} = V_{12.0} = 6.0$ V, (b) $U_i = 1.15$ mJ, $U_f = 0.29$ mJ

59 (a) $V_1 = V_2 = V_3 = 200$ V, (b) $Q_1 = -255\ \mu$C, $Q_2 = 145\ \mu$C, $Q_3 = 545\ \mu$C, (c) $V_1 = 127$ V, $V_2 = 36.4$ V, $V_3 = 90.9$ V

61 2.72 nF

63 (a) 50 μm, (b) 240 cm^2

65 $Q_1 = \frac{2Q}{1+\kappa}$, $Q_2 = \frac{2Q\kappa}{1+\kappa}$

67 (a) 16.7 nF, (b) 1.17 nC

69 (a) 2.1, (b) 45 cm^2, (c) 5.2 nC

71 A series combination of two of the capacitors connected in parallel with a series combination of the other two capacitors will result in total energy U_0 stored in all four capacitors.

73 2.00 μF

75 (a) $\frac{2}{3}C_0$, (b) C_0, (c) $3C_0$

77 (a) $C_{new} = \frac{\epsilon_0 A}{3d}$, (b) $V_{new} = 3V$, (c) $U_{new} = \frac{3\epsilon_0 AV^2}{2d}$, (d) $W = \frac{\epsilon_0 AV^2}{d}$

79 133 μC, 267 μC

83 (a) $U = \frac{Q^2}{2\epsilon_0 A}x$, (b) $dU = \frac{Q^2}{2\epsilon_0 A}dx$

85 (a) $U = \frac{Q^2 d}{2\epsilon_0 a[(\kappa - 1)x + a]}$, (b) $F = \frac{(\kappa - 1)Q^2 d}{2a\epsilon_0[(\kappa - 1)x + a]^2}$, (c) $F = \frac{(\kappa - 1)a\epsilon_0 V^2}{2d}$, (d) The force originates from the fringing fields around the edges of the capacitor. The effect of the force is to pull the polarized dielectric into the space between the capacitor plates.

87 (a) First show that F is inversely proportional to d for a given V_0. Because F increases as d decreases, a decrease in plate separation will unbalance the system. Hence the balance is unstable. (b) $V_0 = d_0\sqrt{\frac{2Mg}{\epsilon_0 A}}$

89 (a) $Q_1 = (200\text{ V})C_1$, $Q_2 = (200\text{ V})\kappa C_1$, (b) $U = (2.00 \times 10^4\text{ V}^2)(1 + \kappa)C_1$, (c) $U_f = (1.00 \times 10^4\text{ V}^2)C_1(1 + \kappa)^2$, (d) $V_f = 100(1 + \kappa)$ V

91 0.100 μF, 16.0 μC

93 $C = \frac{\epsilon_0 ab}{y_0}\ln(2)$

Chapter 25

1 In earlier chapters the conductors are constrained to be in electrostatic equilibrium. In this chapter, this constraint is no longer in place.

3 (c)

5 (a)

7 No, it is not necessarily true for a battery. Under normal operating conditions the current in the battery is in the direction away from the negative battery terminal and toward the positive battery terminal. That is, it opposite to the direction of the electric field.

9 (e)

11 (d)

13 You should decrease the resistance. The heat output is given by $P = V^2/R$. Because the voltage across the resistor is constant, decreasing the resistance will increase P.

15 (a)

17 (a)

19 (a) False, (b) True, (c) True

21 (b)

23 $P_2 = \frac{1}{2}P$ and $P_3 = \frac{1}{2}P$

25 1.9 kA

27 26 m

29 12 gauge

31 0.28 mm/s

33 (a) 0.21 mm/s, 0.53 mm/s, (b) 0.396

35 (a) $1.04 \times 10^8\ m^{-1}$, (b) $1.04 \times 10^{14}\ m^{-3}$, (c) 5.00 kA/$m^2$

37 0.86 s

39 (a) 33.3 Ω, (b) 0.750 A

41 1.9 V

43 63 light-years

45 1.20 Ω

47 31 mΩ

49 $R = \frac{\rho L}{\pi ab}$

51 (a) $R = \frac{\rho}{2\pi L}\ln(b/a)$, (b) 2.05 A

53 46°C

55 (a) 15.0 A, (b) 11.1 Ω, (c) 1.30 kW

57 (b) 3×10^2

59 (a) 636 K, (b) As the filament heats up, its resistance decreases. This results in more power being dissipated, further heat, higher temperature, etc. If not controlled, this thermal runaway can burn out the filament.

61 0.18 kJ

63 (*a*) 0.24 kW, (*b*) 0.23 kW, (*c*) 1.7 kJ, (*d*) 84 J

65 (*a*) 6.9 MJ, (*b*) 12.8 h

67 (*a*) 26.7 kW, (*b*) 576 kC, (*c*) 69.1 MJ, (*d*) 57.6 km, (*e*) $0.03 per km

69 $I_4 = 3.00$ A, $I_3 = 4.00$ A, $I_6 = 2.00$ A

71 (*b*) It would not affect it.

73 0.45 kΩ

75 (*a*) 6.00 Ω, (*b*) The current in both the 6.00-Ω and the 12.0-Ω resistor in the upper branch is 667 mA. The current in each 6.00-Ω resistor in the parallel combination in the lower branch is 667 mA. The current in the 6.00-Ω resistor on the right in the lower branch is 1.33 A.

77 8 pieces

79 (*a*) $R_3 = \frac{R_1^2}{R_1 + R_2}$, (*b*) 0, (*c*) $R_1 = \frac{R_3 + \sqrt{R_3^2 + 4R_2R_3}}{2}$

81 (*a*) 4.00 A, (*b*) 2.00 V, (*c*) 1.00 Ω

83 (*a*)

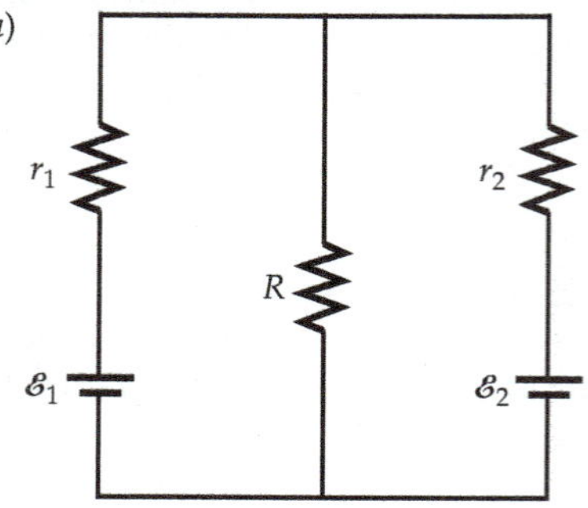

(*b*) $I_1 = -19.0$ A, $I_2 = 25.1$ A, $I_R = 6.17$ A, (*c*) Battery 2 supplies 311 W. Of the 234 W that is delivered to battery 1, 216 W goes into recharging battery 1 and 18.0 W is dissipated by the internal resistance. In addition, 76.2 W is delivered to the 2.00-Ω resistor.

85 (*a*) $I_{4\Omega} = 0.667$ A, $I_{3\Omega} = 0.889$ A, $I_{6\Omega} = 1.56$ A, (*b*) $V_{ab} = 9.33$ V, (*c*) $P_{\text{left}} = 8.00$ W, $P_{\text{right}} = 10.7$ W

87 For the series combination, the power delivered to the load is greater if $R > r$ and is greatest when $R = 2r$. If $r = R$, both arrangements provide the same power to the load. For the parallel combination, the power delivered to the load is greater if $R < r$ and is a maximum when $R = \frac{1}{2}r$.

89 $V_a - V_b = 2.40$ V

91 (*a*) 3.33 V, (*b*) 3.33 V, (*c*) 3.13 V, (*d*) 2.00 V, (*e*) 0.435 V, (*f*) $R_{\max} = 1.67$ MΩ

93 2.5 Ω

95 (*a*) 600 μC, (*b*) 0.200 A, (*c*) 3.00 ms, (*d*) 81.2 μC

97 2.18 MΩ

99 (*a*) 5.69 μC, (*b*) 1.10 μC/s, (*c*) 1.10 μA, (*d*) 6.62 μW, (*e*) 2.44 μW, (*f*) 4.19 μW

103 (*a*) 0.250 A, (*b*) 62.5 mA, (*c*) $I_2(t) = (62.5 \text{ mA})(1 - e^{-t/0.750 \text{ ms}})$

105 (*a*) 48.0 μA, (*b*) 0.866 s

107 (*a*) (1) The potential drops across R_2 and R_3 are equal, so $I_2 > I_3$. The current in R_1 equals the sum of the currents I_2 and I_3, so I_1 is greater than either I_2 or I_3. (*b*) $I_1 = 1.50$ A, $I_2 = 1.00$ A, $I_3 = 0.50$ A

109 (*a*) 43.9 Ω, (*b*) 300 Ω, (*c*) 3.8 kΩ

111 (*a*) 2.18×10^{13} s^{-1}, (*b*) 210 J/s, (*c*) 27.6 s

113 0.16 L/s

115 (*a*) 10.0 ms, (*c*) 1.00 GΩ, (*d*) 60.9 ps, (*e*) 2.89 kW

119 (*a*) $R_{\text{eq}} = \left(\frac{1 + \sqrt{5}}{2}\right)R$, (*b*) $R_{\text{eq}} = \frac{R_1 + \sqrt{R_1^2 + 4R_1R_2}}{2}$

Chapter 26

1 (*b*)

3 Because the alternating current running through the filament is changing direction every 1/60 s, the filament experiences a force that changes direction at the frequency of the current. Thus, it oscillates at 60 Hz.

5 (*a*)

9 (*a*) False, (*b*) True, (*c*) True, (*d*) True

11 According to the principle of relativity, this is equivalent to the electron moving from right to left at speed v with the magnet stationary. When the electron is directly over the magnet, the field points directly up, so there is a force directed out of the page on the electron.

13 $I \sim 2 \times 10^3$ A. You should advise him to develop some other act. A current of 2000 A would overheat the wire (which is a gross understatement).

15 (*a*) $-(3.8\ \mu\text{N})\hat{k}$, (*b*) $-(7.5\ \mu\text{N})\hat{k}$, (*c*) 0, (*d*) $(7.5\ \mu\text{N})\hat{j}$

17 0.96 N

19 $-(19\text{ fN})\hat{i} - (13\text{ fN})\hat{j} - (58\text{ fN})\hat{k}$

21 1.5 A

23 $(10\text{ T})\hat{i} + (10\text{ T})\hat{j} - (15\text{ T})\hat{k}$

27 (*a*) 87 ns, (*b*) 4.7×10^7 m/s, (*c*) 11 MeV

29 (*a*) $2v_\alpha = 2v_d = 1v_p$, (*b*) $1K_\alpha = 2K_d = 1K_p$, (*c*) $L_\alpha = 2L_d = 2L_p$

33 (*a*) 24°, 1.3×10^6 m/s, (*b*) 24°, 6.3×10^5 m/s

35 (*a*) 1.6×10^6 m/s, (*b*) 14 keV, (*c*) 7.7 eV

37 7.37 mm

39 (*a*) 63.3 cm, (*b*) 2.58 cm

41 $\Delta t_{58} = 15.7\ \mu$s, $\Delta t_{60} = 16.3\ \mu$s

43 (*a*) 21 MHz, (*b*) 46 MeV, (*c*) $f_{\text{deuterons}} = 11$ MHz, $K_{\text{deuterons}} = 23$ MeV

47 (*a*) 0.30 A·m², (*b*) 0.13 N·m

49 (*a*) 0, (*b*) 2.7×10^{-3} N·m

51 $B_{\min} = \frac{mg}{I\pi R}$

53 (*a*) $(0.84\text{ N}\cdot\text{m})\hat{k}$, (*b*) 0, (*c*) 0, (*d*) $(0.59\text{ N}\cdot\text{m})\hat{k}$

55 0.38 A·m², into the page

61 $\mu = \frac{4}{3}\pi\sigma R^4\omega$

63 (*a*) $\tau = \frac{1}{4}\pi\sigma r^4\omega B \sin\theta$, (*b*) $\Omega = \frac{\pi\sigma r^2 B}{2m}\sin\theta$

65 (*a*) 3.68×10^{-5} m/s, (*b*) 1.47 μV

67 1.0 mV

69 4

71 (*a*) 1.3 μs, (*b*) 2.4×10^6 m/s, (*c*) 0.12 MeV

75 (*a*) $B = -\frac{mg}{IL}\tan\theta$, (*b*) $\vec{a} = g\sin\theta$, up the incline

77 (*a*) $(10\text{ V/m})\hat{j}$, (*b*) The positive end has the lesser y coordinate, (*c*) 20 V

81 1.0×10^{-28} kg

Chapter 27

1 Note that, while the two far fields (the fields far from the dipoles) are the same, the two near fields (the fields near to the dipoles) are not. At the center of the electric dipole, the electric field is antiparallel to the direction of the far field above and below the dipole, and at the center of the magnetic dipole, the magnetic field is parallel to the direction of the far field above and below the dipole. It is especially important to note that while the electric field lines begin and terminate on electric charges, the magnetic field lines are continuous, i.e., they form closed loops.

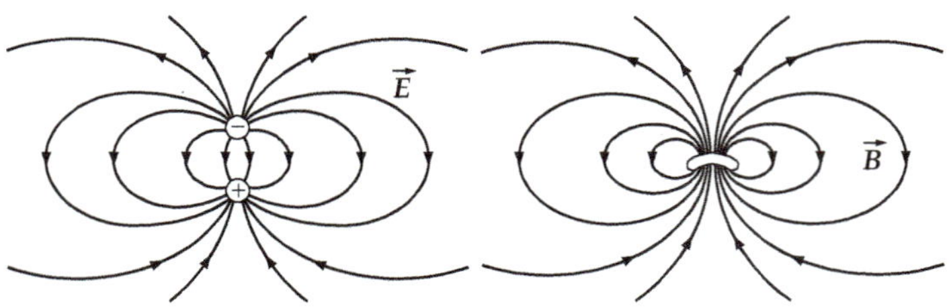

3 (*a*)

5 Both tell you about the respective fluxes through closed surfaces. In the electrical case, the flux is proportional to the net charge enclosed. In the magnetic case, the flux is always zero because there is no such thing as magnetic charge (a magnetic monopole). The source of the magnetic field is NOT the equivalent of electric charge; that is, it is NOT a thing called magnetic charge, but rather it is moving electric charges.

7 Clockwise

9 (*a*) True, (*b*) False, (*c*) True, (*d*) True

11 H_2, CO_2, and N_2 are diamagnetic ($\chi_m < 0$); O_2 is paramagnetic ($\chi_m > 0$).

13 (*a*) $\vec{B}(0, 0) = -(9.0\text{ pT})\hat{k}$, (*b*) $\vec{B}(0, 1.0\text{ m}) = -(36\text{ pT})\hat{k}$, (*c*) $\vec{B}(0, 3.0\text{ m}) = (36\text{ pT})\hat{k}$, (*d*) $\vec{B}(0, 4.0\text{ m}) = (9.0\text{ pT})\hat{k}$

15 (*a*) $\vec{B}(2.0\text{ m}, 2.0\text{ m}) = 0$,
(*b*) $\vec{B}(6.0\text{ m}, 4.0\text{ m}) = -(3.6 \times 10^{-23}\text{ T})\hat{k}$,
(*c*) $\vec{B}(3.0\text{ m}, 6.0\text{ m}) = (4.0 \times 10^{-23}\text{ T})\hat{k}$

17 $\epsilon_0\mu_0 v^2$

19 $\vec{B}(0, 3.0\text{ m}, 4.0\text{ m}) = -(9.6\text{ pT})\hat{i}$

21 (*a*) $B(0) = 54\ \mu\text{T}$, (*b*) $B(0.010\text{ m}) = 46\ \mu\text{T}$, (*c*) $B(0.020\text{ m}) = 31\ \mu\text{T}$, (*d*) $B(0.35\text{ m}) = 34\text{ nT}$

25 (*a*) $\vec{B}(-3.0\text{ cm}) = -(89\ \mu\text{T})\hat{k}$, (*b*) $\vec{B}(0) = 0$ (*c*) $\vec{B}(3.0\text{ cm}) = (89\ \mu\text{T})\hat{k}$, (*d*) $\vec{B}(9.0\text{ cm}) = -(160\ \mu\text{T})\hat{k}$

27 (*a*) $\vec{B}(-3.0\text{ cm}) = -(0.18\text{ mT})\hat{k}$, (*b*) $\vec{B}(0) = -(0.13\text{ mT})\hat{k}$, (*c*) $\vec{B}(3.0\text{ cm}) = -(0.18\text{ mT})\hat{k}$, (*d*) $\vec{B}(9.0\text{ cm}) = (0.11\text{ mT})\hat{k}$

29 (*a*) $(64\ \mu\text{T})\hat{j}$,
(*b*) $-(48\ \mu\text{T})\hat{k}$

31 (*a*) Because the currents repel, they are antiparallel.
(*b*) 39 mA

33 80 A

35 (*a*) 30 μT, down the page, (*b*) 4.5×10^{-4} N/m, toward the right

37 (*a*) 80 A, (*b*) $\vec{B}(5.0\text{ cm}, 0, 0) = -(0.24\text{ mT})\hat{j}$

39 (*a*) $\dfrac{F}{\ell} = \dfrac{3\sqrt{2}\,\mu_0 I^2}{4\pi a}$, (*b*) $\dfrac{F}{\ell} = \dfrac{\sqrt{2}\,\mu_0 I^2}{4\pi a}$

41 (*a*) 3.3 mT, (*b*) 1.6 mT

45 $B_{\text{inside}} = 0$, $B_{\text{outside}} = \dfrac{\mu_0 I}{2\pi R}$. The direction of the magnetic field is in the direction of the curled fingers of your right hand when you grab the cylinder with your right thumb in the direction of the current.

49 (*a*) $B_{R<a} = 0$, (*b*) $B_{a<R<b} = \dfrac{\mu_0 I}{2\pi R}\dfrac{R^2 - a^2}{b^2 - a^2}$, (*c*) $B_{R>b} = \dfrac{\mu_0 I}{2\pi R}$

51 (*a*) $B(1.10\text{ cm}) = 27.3\text{ mT}$, (*b*) $B(1.50\text{ cm}) = 20.0\text{ mT}$

53 (*a*) $B = B_{\text{app}} = 10.1\text{ mT}$, (*b*) $B_{\text{app}} = 10.1\text{ mT}$, $B = 1.5\text{ T}$

55 -4.0×10^{-5}

57 5.43 A/m

59

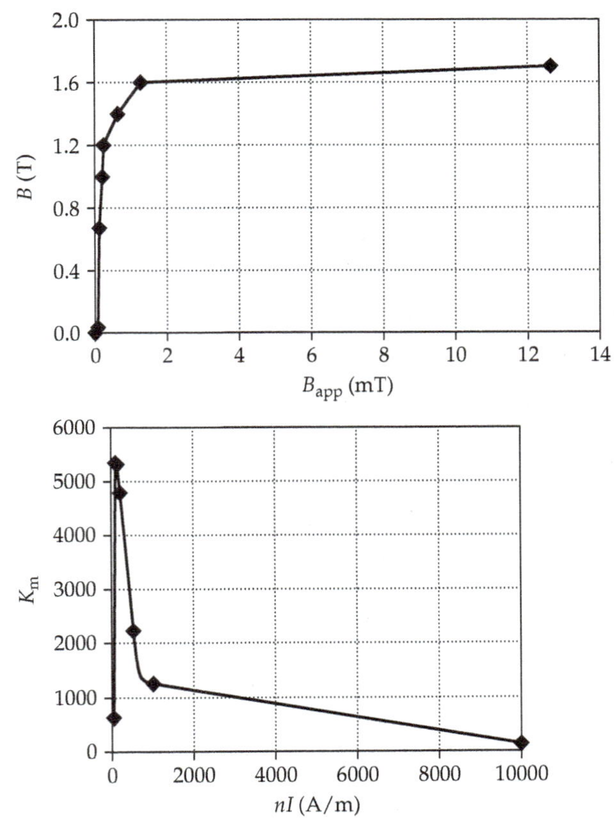

61 $1.69\mu_B$

63 (*b*) 7.46×10^{-4}

65 $B_{\text{app}} = \dfrac{\mu_0 NI}{2\pi a}$, $B = \dfrac{\mu_0 NI}{2\pi a} + \mu_0 M$

67 (*a*) 30.2 mT, (*b*) 6.96 A/m, (*c*) 30.2 mT

69 11.7, $1.48 \times 10^{-5}\text{ N/A}^2$

71 (*a*) 12.6 mT, (*b*) 1.36×10^6 A/m, (*c*) 137

73 (*a*) 1.42×10^6 A/m,
(*b*) $K_m = 90.0$, $\mu = 1.13 \times 10^{-4}\text{ T}\cdot\text{m/A}$, $\chi_m = 89$

75 (*a*) $(8.00\text{ T/m})r$, (*b*) $(3.20 \times 10^{-3}\text{ T}\cdot\text{m})\dfrac{1}{r}$, (*c*) $(8.00 \times 10^{-6}\text{ T}\cdot\text{m})\dfrac{1}{r}$, (*d*) Note that the field in the ferromagnetic region is that which would be produced in a nonmagnetic region by a current of $400I = 1600$ A. The amperian current on the inside of the surface of the ferromagnetic material must therefore be $1600\text{ A} - 40\text{ A} = 1560\text{ A}$ in the direction of I. On the outside surface there must then be an amperian current of 1560 A in the opposite direction.

77 $\vec{B}_p = \frac{\mu_0 I}{4}\left(\frac{1}{R_1} - \frac{1}{R_2}\right)$, out of the page

79 $\vec{B}_p = \frac{\mu_0}{2\pi}\frac{I}{a}(1 + \sqrt{2})$, out of the page

81 The $+x$ and $+y$ directions are up the page and to the right. (*a*) $\vec{F}_{\text{top}} = -(2.5 \times 10^{-5}\text{ N})\hat{j}$, $\vec{F}_{\text{left side}} = (1.0 \times 10^{-4}\text{ N})\hat{i}$, $\vec{F}_{\text{bottom}} = (2.5 \times 10^{-5}\text{ N})\hat{j}$, $\vec{F}_{\text{right side}} = -(0.29 \times 10^{-4}\text{ N})\hat{i}$, (*b*) $\vec{F}_{\text{net}} = (0.71 \times 10^{-4}\text{ N})\hat{i}$

83 7.1 μT, into the page

85

Cell	Formula/Content	Algebraic Form
B1	1.00E−07	$\frac{\mu_0}{4\pi}$
B2	5.00	I
B3	2.55E−03	r_0
C6	10^4*\$B\$1*2*\$B\$2*A6/\$B\$3^2	$\frac{\mu_0}{4\pi}\frac{2I}{R_0^2}R$
C17	10^4*\$B\$1*2*\$B\$2*A6/A17	$\frac{\mu_0}{4\pi}\frac{2I}{R}$

	A	B	C
1	μ/4π=	1.00E−07	N/A2
2	I=	5	A
3	R0=	2.55E−03	m
4			
5	R(m)	R(mm)	B(T)
6	0.00E+00	0.00E+00	0.00E+00
7	2.55E−04	2.55E−01	3.92E−01
105	2.52E−02	2.52+01	3.96E−01
106	2.55E−02	2.55E+01	3.92E−01

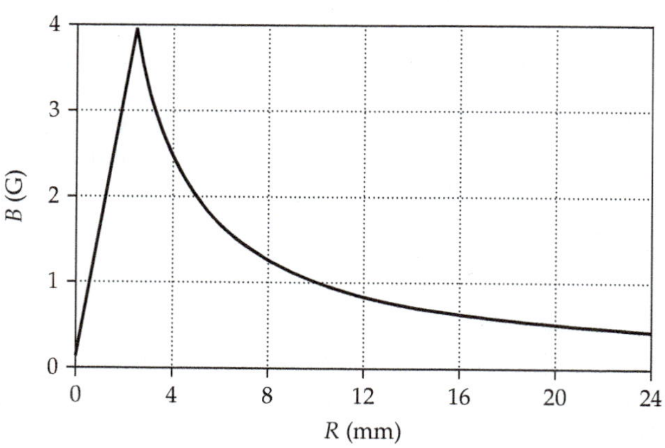

87 (*a*) $5.24 \times 10^{-2}\text{ A}\cdot\text{m}^2$, (*b*) 7.70×10^5 A/m, (*c*) 23.1 kA

89 (*a*) 15.5 GA, (*b*) Because Earth's magnetic field points down at the north pole, application of the right-hand rule indicates that the current is clockwise when viewed from above the north pole.

91 3.18 cm

93 (*a*) and (*b*) $B(5.0\text{ cm}) = B(10\text{ cm}) = 10\ \mu\text{T}$, (*c*) $B(20\text{ cm}) = 5.0\ \mu\text{T}$

95 2.24 A

97 (*c*) $B_z = \frac{1}{2}\mu_0\omega\sigma\left(\frac{R^2 + 2z^2}{\sqrt{R^2 + z^2}} - 2z\right)$

Chapter 28

1 (*a*) Orient the sheet so the normal to the sheet is both horizontal and perpendicular to the local tangent to the magnetic equator. (*b*) Orient the sheet of paper so the normal to the sheet is perpendicular to the direction of the normal described in the answer to Part (*a*).

5 (*d*)

7 The induced current is clockwise as viewed from the left. The loops repel each other.

9 (*a*) and (*b*)

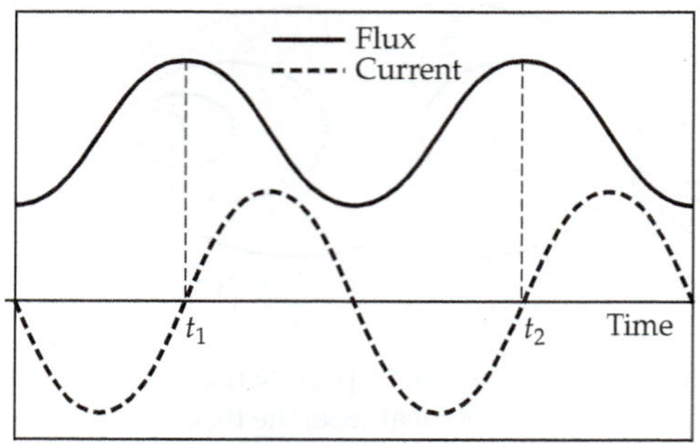

11 The magnetic field of the falling magnet sets up eddy currents in the metal tube. The eddy currents establish a magnetic field that exerts a force on the magnet opposing its motion; thus the magnet is slowed down. If the tube is made of a nonconducting material, there are no eddy currents.

13 (*c*)

15 (*a*) False, (*b*) True, (*c*) False, (*d*) True, (*e*) False

17 $u_m \approx (8 \times 10^3)u_e$

19 (*a*) 0.5 V, (*b*) 7 mV/m

21 (*a*) 0, (*b*) 14 μWb, (*c*) 0, (*d*) 12 μWb

23 $\phi_m = \pm\pi R^2 B$

25 1.68 mWb

27 (*a*) $\phi_m = \mu_0 nIN\pi R_1^2$, (*b*) $\phi_m = \mu_0 nIN\pi R_2^2$

29 $\frac{\phi_m}{L} = \frac{\mu_0 I}{4\pi}$

31 (*a*)

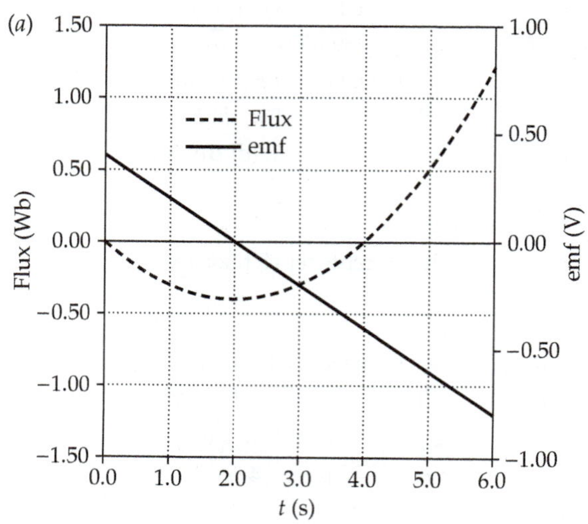

(*b*) The flux is a minimum when $t = 2.0$ s; $V(2.0\text{ s}) = 0$.

(*c*) The flux is zero when $t = 0$ and $t = 4.0$ s; $\mathcal{E}(0) = 0.40$ V and $\mathcal{E}(4.0\text{ s}) = -0.40$ V.

33 (*a*) 1.26 mC, (*b*) 12.6 mA, (*c*) 628 mV

35 79.8 μT

37 400 m/s

39 (*a*)

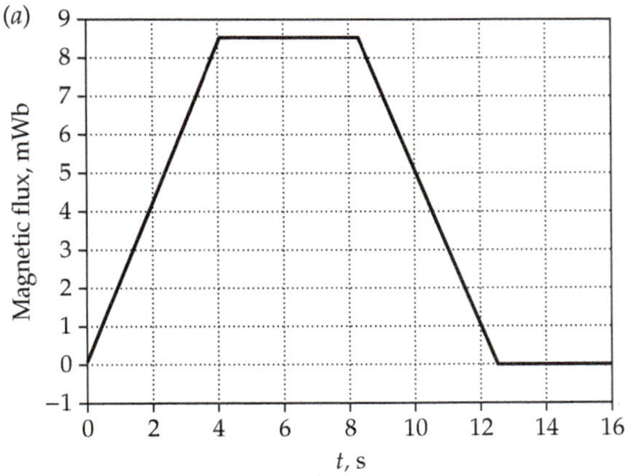

(*b*)

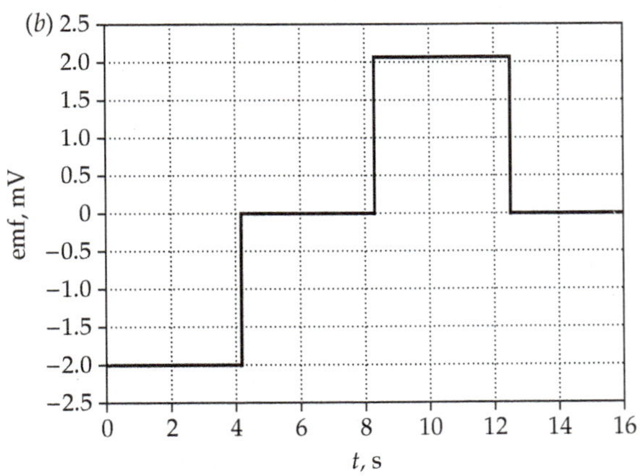

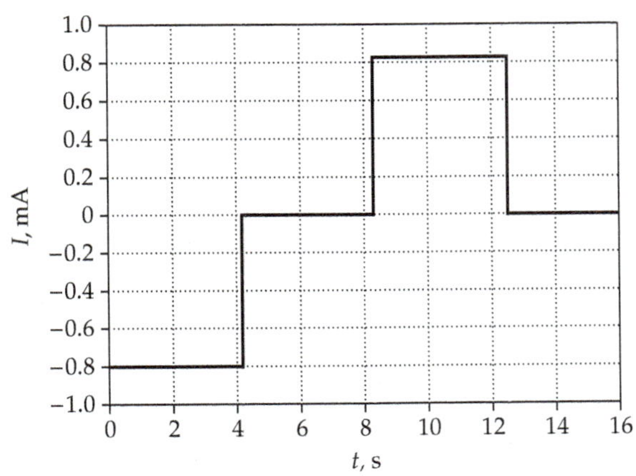

41 (*a*) $F_{\text{m}} = \dfrac{B\ell}{R}(\mathcal{E} - B\ell v)$, (*c*) 0

45 (*a*) 14 V, (*b*) 486 rev/s

47 (*a*) 24.0 Wb, (*b*) −1.60 kV

49 $L = 0$, $R = 162\ \Omega$

51 0.16 μH

57 $\dfrac{dU_m}{dx} = \dfrac{\mu_0 I^2}{16\pi}$

59 (*a*) $I = 0$, $dI/dt = 25.0$ kA/s,
(*b*) $I = 2.27$ A, $dI/dt = 20.5$ kA/s,
(*c*) $I = 7.90$ A, $dI/dt = 9.20$ kA/s,
(*d*) $I = 10.8$ A, $dI/dt = 3.38$ kA/s

61 (*a*) 44.1 W, (*b*) 40.4 W, (*c*) 3.62 W

63 (*a*) 3.00 kA/s, (*b*) 1.50 kA/s, (*c*) 80.0 mA, (*d*) 0.123 ms

65 (*a*) $I_{10\text{-}\Omega} = I_{2\text{-H}} = 1.0$ A, $I_{100\text{-}\Omega} = 0$, (*b*) $V_{2\text{-H}} = 100$ V

(*c*)

Cell	Content/Formula	Algebraic Form
B1	2.0	L
B2	100	R
B3	1	I_0
A6	0	t_0
B6	\$B\$3*EXP((−\$B\$2/\$B\$1)*A6)	$I_0 e^{-\frac{R}{L}t}$

	A	B	C
1	L=	2	H
2	R=	100	ohms
3	I_0=	1	A
4			
5	t	I(t)	V(t)
6	0.000	1.00E+00	100.00
7	0.005	7.79E-01	77.88
8	0.010	6.07E-01	60.65
9	0.015	4.72E-01	47.24
10	0.020	3.68E-01	36.79
11	0.025	2.87E-01	28.65
12	0.030	2.23E-01	22.31
32	0.130	1.50E-03	0.15
33	0.135	1.17E-03	0.12
34	0.140	9.12E-04	0.09
35	0.145	7.10E-04	0.07
36	0.150	5.53E-04	0.06

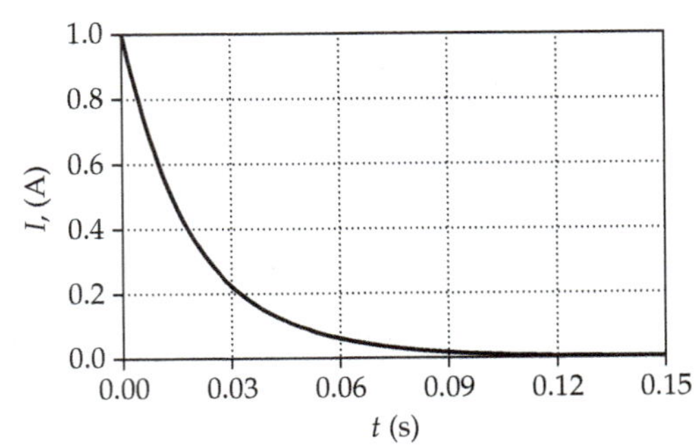

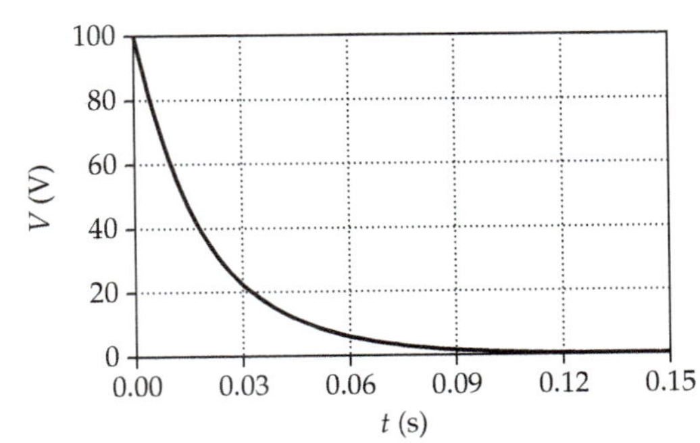

67 (*a*) 88 ms, (*b*) 35 mH

69 (*a*) 3.53 J, (*b*) 1.61 J, (*c*) 1.92 J

71 (*b*) 2.50 krad/s

75 0.28 H

77 (*a*) As the magnet passes through a loop it induces an emf because of the changing flux through the loop. This allows the coil to "sense" when the magnet is passing through it. (*b*) One cannot use a cylinder made of conductive material because eddy currents induced in it by a falling magnet would slow the magnet. (*c*) As the magnet approaches the loop the flux increases, resulting in the negative voltage signal of increasing magnitude. When the magnet is passing a loop, the flux reaches a maximum value and then decreases, so the induced emf becomes zero and then positive. The instant at which the induced emf is zero is the instant at which the magnet is at the center of the loop.

(*d*)

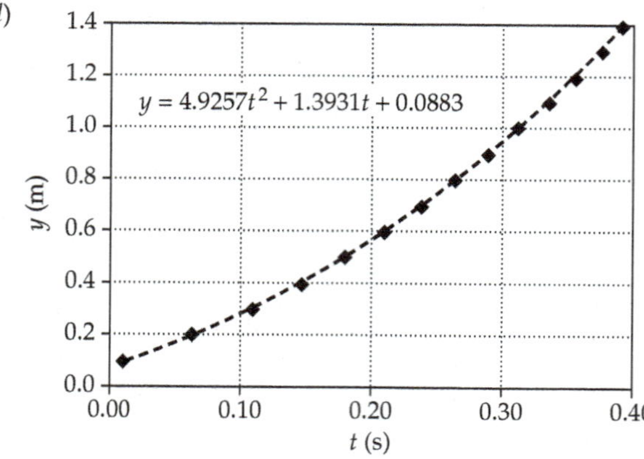

$g = 9.85\ \text{m/s}^2$

79 (*a*) $E_r = -\frac{1}{2} r\mu_0 n I_0 \omega \cos \omega t \quad r < R,$

(*b*) $E_r = -\dfrac{\mu_0 n R^2 I_0 \omega}{2r} \cos \omega t \quad r > R$

Chapter 29

1 8.33 ms

3 (*b*)

5 (*c*)

7 Yes to both questions. (*a*) While the magnitude of the charge is accumulating on either plate of the capacitor, the capacitor absorbs power from the generator. (*b*) When the magnitude of the charge is on either plate of the capacitor is decreasing, it supplies power to the generator.

9 (*a*)

11 (*a*)

13 (*a*) False, (*b*) False, (*c*) True, (*d*) True, (*e*) True, (*f*) True

15 (*a*) True, (*b*) False, (*c*) True

17 (*a*) False, (*b*) True, (*c*) True, (*d*) True, (*e*) True, (*f*) True

19 (*a*) 0.833 A, (*b*) 1.18 A, (*c*) 200 W

21 (*a*) 0.38 Ω, (*b*) 3.77 Ω, (*c*) 37.7 Ω

23 1.6 kHz

25 (*a*) 25 mA, (*b*) 18 mA

27 (*a*) 0.35 A, (*b*) 0.35 A, (*c*) $I = (0.34\ \text{A})\cos(\omega t + 0.17\ \text{rad})$

29 (*a*) 1.3 ms, (*b*) 88 mH

31 (*a*) 2.3 mJ, (*b*) 0.71 kHz, (*c*) 0.67 A

33 (*a*)

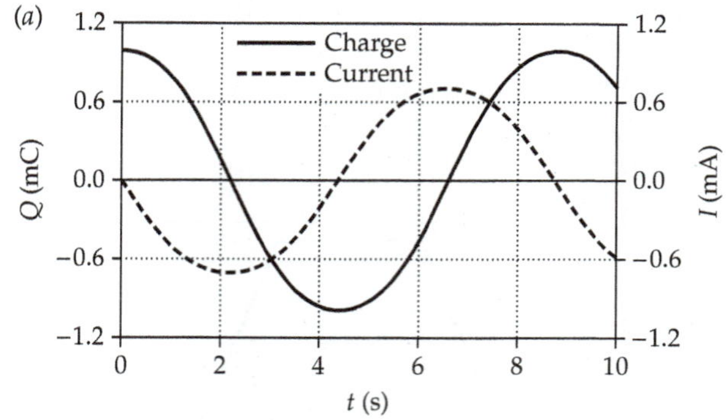

35 29.2 mH

37 (*a*) 0.33, (*b*) 27 Ω, (*c*) 0.20 H, (*d*) Because the circuit is inductive, the current lags the voltage. (*e*) 71°

39 0.397

41 (*a*) $I_{\text{rms}} = 6.23\ \text{A}$, $I_{R_L\,\text{rms}} = 2.80\ \text{A}$, $I_{L\,\text{rms}} = 5.53\ \text{A}$, (*b*) $I_{\text{rms}} = 3.28\ \text{A}$, $I_{R_L\,\text{rms}} = 2.94\ \text{A}$, $I_{L\,\text{rms}} = 1.46\ \text{A}$, (*c*) 50.2%, (*d*) 80.0%

43 60 V

45 (*a*) $\delta = \tan^{-1}\left[-\dfrac{1}{\omega RC}\right]$, (*b*) $\delta \to -90°$, (*c*) $\delta \to 0$, (*d*) For very low driving frequencies, $X_C \gg R$ and so $\vec{V}_C$ effectively lags $\vec{V}_{\text{in}}$ by 90°. For very high driving frequencies, $X_C \ll R$ and so $\vec{V}_R$ is effectively in phase with $\vec{V}_{\text{in}}$.

51 (*b*) Note that, as $\omega \to 0$, $V_L \to V_{\text{peak}}$. This makes sense physically in that, for low frequencies, X_C is large and, therefore, a larger peak voltage will appear across it than appears across it for high frequencies. Note further that, as $\omega \to \infty$, $V_L \to 0$. This makes sense physically in that, for high frequencies, X_C is small and, therefore, a smaller peak voltage will appear across it than appears across it for low frequencies.

53

Cell	Formula/Content	Algebraic Form
B1	2.00E+03	R
B2	5.00E−09	C
B3	1	$V_{\text{in peak}}$
B8	$B$3/SQRT(1+((2*PI()*A8*1000*$B$1*$B$2)^2))	$\dfrac{V_{\text{in peak}}}{\sqrt{1 + (2\pi f RC)^2}}$
C8	ATAN(2*PI()*A8*1000*$B$1*$B$2)	$\tan^{-1}(2\pi f RC)$
D8	C8*180/PI()	δ in degrees

	A	B	C	D
1	R =	1.00E+04	ohms	
2	C =	5.00E-09	F	
3	Vin peak =	1	V	
4				
5				
6	f(kHz)	V_{out}	δ(rad)	δ(deg)
7	0	1.000	0.000	0.0
8	1	0.954	0.304	17.4
56	49	0.065	1.506	86.3
57	50	0.064	1.507	86.4

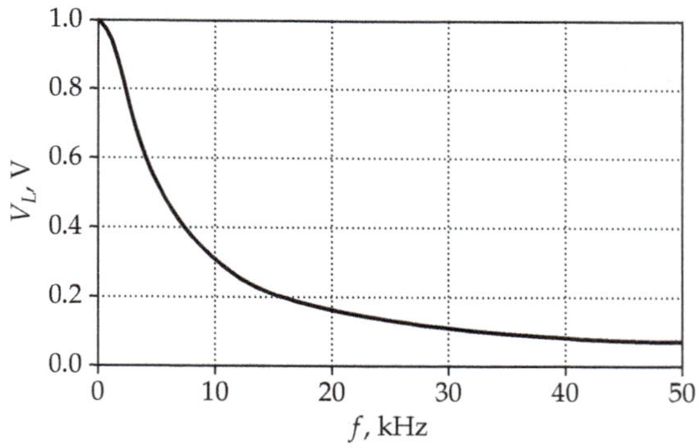

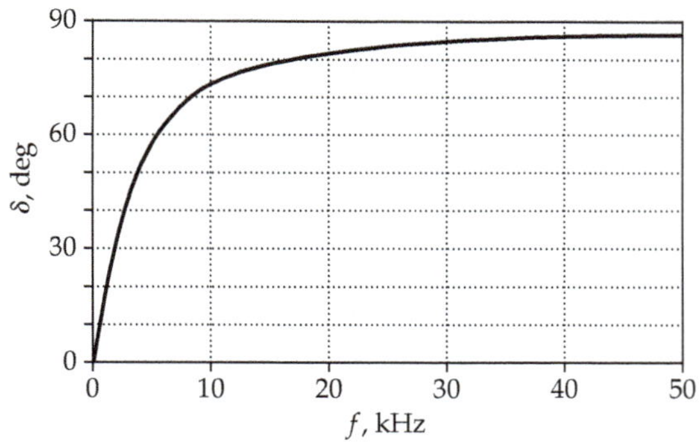

55 (b) $\Delta\omega = \dfrac{R}{L}$

57 33 μF

59 (a) $I(t) = -(19\text{ mA})\sin\left(\omega t + \dfrac{\pi}{4}\right)$, where $\omega = 1250$ rad/s, (b) 23 μF,

(c) $U_m(t) = (4.9\ \mu\text{J})\sin^2\left(\omega t + \dfrac{\pi}{4}\right)$,

$U_e(t) = (4.9\ \mu\text{J})\cos^2\left(\omega t + \dfrac{\pi}{4}\right)$, where $\omega = 1250$ rad/s, $U = 4.9\ \mu$J

61 (a) 5.4 fF, (b) $f(x) = \dfrac{70\text{ MHz}}{\sqrt{1 - (4.0\text{ m}^{-1})x}}$

65 (a) 14, (b) 80 Hz, (c) 0.27

67 (a) 10 A, (b) 53°, (c) 0.33 mF, (d) 0.13 kV

69 (a) 80 V, (b) 78 V, (c) 0.17 kV, (d) 0.11 kV, (e) 0.18 kV

71 (a)

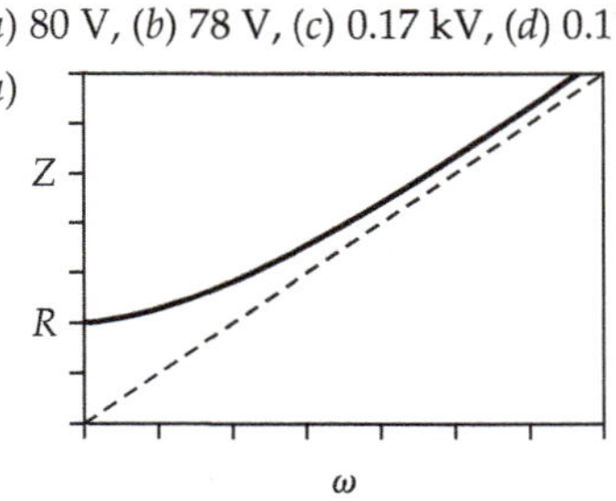

(b)

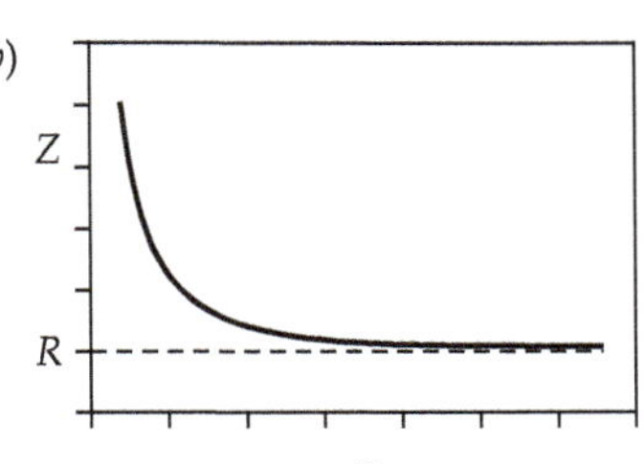

(c)

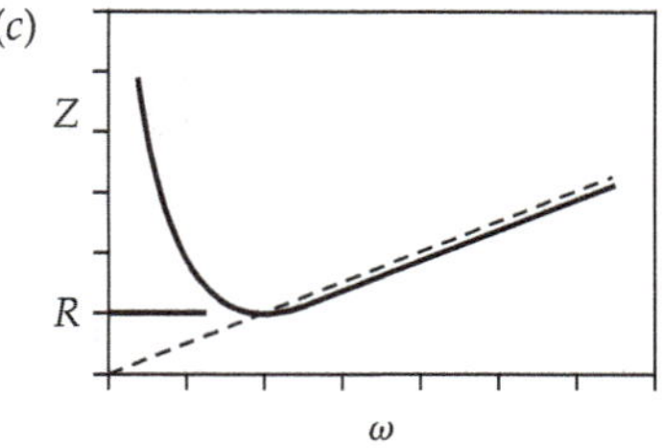

73 (a) $Z = 12\ \Omega$, (b) $R = 7.2\ \Omega$, $X = 10\ \Omega$, (c) If the current leads the emf, the reactance is capacitive.

79 (a) 1:5, (b) 50 A

81 (a) 1.5 A, (b) 19

83 3.33×10^3

85 (a) 12 V, (b) 8.5 V

87 $I_{max} = 1.06$ A, $I_{min} = -0.06$ A, $I_{av} = 0.50$ A, $I_{rms} = 0.64$ A

Chapter 30

1 (a) False, (b) True, (c) True, (d) False

3 (a) False, (b) True, (c) True, (d) True

5 (a) (1) X-rays, (2) green light, (3) red light, (b) (1) microwaves, (2) green light, (3) ultraviolet light.

7 (a) The electric dipole antenna should be oriented vertically. (b) The loop antenna and the electric dipole transmitting antenna should be in the same vertical plane.

9 (d)

11 (d)

13 2×10^{-7}

15 (a) 3.4×10^{14} V/m·s

19 (a) 10 A, (b) $\dfrac{dE}{dt} = 2.3 \times 10^{12}$ V/m·s,

(c) $\oint_C \vec{B} \cdot d\vec{\ell} = 0.79\ \mu\text{T} \cdot \text{m}$

21 580 nm, 5.17×10^{14} Hz

23 (a) 3.00×10^{18} Hz, (b) 5.45×10^{14} Hz. Consulting Table 30-1, we see that the color of light that has a wavelength of 550 nm is yellow-green. This result is consistent with those of Problem 21 and is close to the wavelength of the peak output of the Sun. Because we see naturally by reflected sunlight, this result is not surprising.

25 (a) 30°, (b) 7.1 m

27 4.13 μW/m^2

29 386 nW/m^2

31 (a) 283 V/m, (b) 943 nT, (c) 212 W/m^2, (d) 708 nPa

33 (a) 40 nN, (b) 80 nN

35 (a) 45°, (b) 5.7°

37 (a) $+x$ direction, (b) $\lambda = 0.628$ m, $f = 477$ MHz,
(c) $\vec{E}(x, t) = (194\text{ V/m})\cos[kx - \omega t]\hat{j}$,
$\vec{B}(x, t) = (647\text{ nT})\cos[kx - \omega t]\hat{k}$,
where $k = 10.0$ rad/m and $\omega = 3.00 \times 10^9$ rad/s.

39 6.10 degrees

41 (*a*) $F_{r\,\text{Earth}} = 5.83 \times 10^{8}$ N, $F_{r\,\text{Earth}} = (1.65 \times 10^{-14})F_{g\,\text{Earth}}$, (*b*) $F_{r\,\text{Mars}} = 7.18 \times 10^{7}$ N, $F_{r\,\text{Mars}} = (4.27 \times 10^{-14})F_{g,\,\text{Mars}}$, (*c*) Mars

47 2.6 mV

49 (*a*) $I = V_0\left(\frac{1}{R}\sin\omega t + \frac{\epsilon_0 \pi a^2}{d}\cos\omega t\right)$,

(*b*) $B(r) = \frac{\mu_0 V_0}{2\pi r}\left(\frac{1}{R}\sin\omega t + \omega\frac{\epsilon_0 \pi r^2}{d}\cos\omega t\right)$,

(*c*) $\delta = \tan^{-1}\left(\frac{R\omega\epsilon_0 \pi a^2}{d}\right)$

51 (*a*) $\vec{S}(x,t) = \frac{1}{\mu_0 c}[E_{10}^2\cos^2(k_1x - \omega_1 t) + 2E_{10}E_{20}\cos(k_1x - \omega_1 t)\cos(k_2x - \omega_2 t + \delta) + E_{20}^2\cos^2(k_2x - \omega_2 t + \delta)]\hat{i}$

(*b*) $\vec{S}_{\text{av}} = \frac{1}{2\mu_0 c}[E_{10}^2 + E_{20}^2]\hat{i}$

(*c*) $\vec{S}(x,t) = \frac{1}{\mu_0 c}[E_{10}^2\cos^2(k_1x - \omega_1 t) - E_{20}^2\cos^2(k_2x + \omega_2 t + \delta)]\hat{i}$,

$\vec{S}_{\text{av}} = \frac{1}{2\mu_0 c}[E_{10}^2 - E_{20}^2]\hat{i}$

53 (*a*) 9.16×10^{-15} T, (*b*) 101 mV, (*c*) 5.49 μV

55 (*a*) $\vec{E} = \frac{I\rho}{\pi a^2}\hat{i}$, where $\hat{i}$ is a unit vector in the direction of the current. (*b*) $\vec{B} = \frac{\mu_0 I}{2\pi a}\hat{\theta}$, where $\hat{\theta}$ is a unit vector perpendicular to $\hat{i}$ and tangent to the surface of the conducting cylinder.
(*c*) $\vec{S} = -\frac{I^2\rho}{2\pi^2 a^3}\hat{r}$, where $\hat{r}$ is a unit vector directed radially outward—away from the axis of the conducting cylinder.
(*d*) $\int S_n\,dA = I^2R$

57 (*a*) 574 nm (*b*) The critical radius is an upper limit and so particles smaller than that radius will be blown out.

59 3.34 mN

Chapter 31

1 (*c*)

3 (*b*)

5 The decrease in the index of refraction n of the atmosphere with altitude results in refraction of the light from the Sun, bending it toward the normal to the surfaces on constant n (that is, toward the Earth). Consequently the Sun can be seen even after it is just below the horizon.

7 The path of least time is the path through point D.

9 In resonance absorption, the molecules respond to the frequency of the light through the Einstein photon relation $E = hf$. Neither the wavelength nor the frequency of the light within the eyeball depend on the index of refraction of the medium outside the eyeball. Thus, the color appears to be the same in spite of the fact that the wavelength has changed.

11

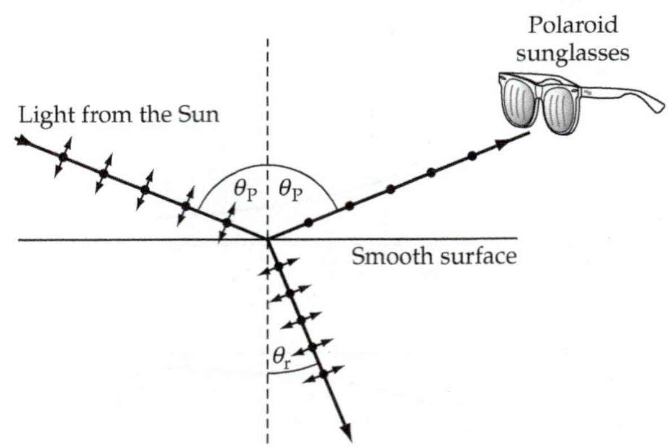

13 (*c*)

15 The population inversion between the state $E_{2\,\text{Ne}}$ and the state 1.96 eV below it (see Figure 31-51) is achieved by inelastic collisions between neon atoms and helium atoms excited to the state $E_{2\,\text{He}}$.

17 (*d*)

19 3 ps

21 (*a*) 2:00 a.m., September 1, (*b*) 2:08 a.m., September 1

23 (*a*)

25 14 ms

29

	(*a*) speed (m/s)	(*b*) wavelength (nm)	(*c*) frequency (Hz)
air	3.00×10^{8}	633	4.74×10^{14}
water	2.25×10^{8}	476	4.74×10^{14}
glass	2.00×10^{8}	422	4.74×10^{14}

31 (*a*) 50°, (*b*) 39°, (*c*) 26°

33 (*a*) 92%, (*b*) 99%

37 5.1 m wide, 2.2 m deep

39 48.8°

41 1.0×10^{2} m^2

43 1.30

45 5°

47 (*a*) 62.5°, (*b*) Yes, if $\theta \geq 41.8°$, where θ is the angle of incidence for the rays in glass that are incident on the glass-water boundary, the rays will leave the glass through the water and pass into the air.

49 1.0°

51 (*a*) 53.1°, (*b*) 56.3°

53 (*a*) $\frac{1}{8}I_0$, (*b*) $\frac{3}{32}I_0$

55 (*a*) 30°, (*b*) 1.7

57 $I_3 = \frac{1}{8}I_0\sin^2 2\omega t$

59 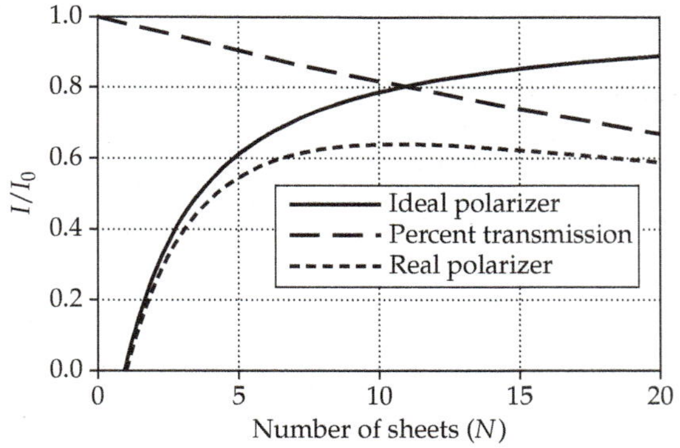

The optimum number of sheets is 11.

61 (*a*) $I_4 = 0.211I_0$, (*b*) For the single sheet between the two end sheets at $\theta = 45°$, $I_3 = 0.125I_0$. The intensity with four sheets at angles of 0°, 30°, 60° and 90° is greater than the intensity of three sheets at angles of 0°, 45° and 90° by a factor of 1.69.

63 (*a*) right circularly polarized, (*b*) $\vec{E} = E_0 \sin(kx + \omega t)\hat{j} - E_0 \cos(kx + \omega t)\hat{k}$

65 (*a*) 435 nm (*b*) 1210 nm

67 (*a*) $\lambda_{max} = 388$ nm, $\lambda_{2\to1} = 1140$ nm, $\lambda_{1\to0} = 588$ nm, (*b*) $\lambda_{1\to0} = 588$ nm, $\lambda_{3\to1} = 554$ nm

69 (*a*) 15 mJ, (*b*) 5.2×10^{16}

71 37°

73 (*a*) 36.8°, (*b*) 38.7°

75 (*a*) $x = -1.00$ m, (*b*) 26.6°, (*c*) 26.6°

77 $\theta_{p\ silicate\ flint} = 58.3°$, $\theta_{p\ borate\ flint} = 57.5°$, $\theta_{p\ quartz} = 57.0°$, $\theta_{p\ silicate\ crown} = 56.5°$

79 (*b*) $\theta_p < \theta_c$

81 (*a*) 1.33, (*b*) 37.2°, (*c*) 48.6°. Because 48.6° is also the angle of incidence at the liquid–air interface and because it is equal to the critical angle for total internal reflection at this interface, no light will emerge into the air.

83 (*c*) 1.67°

Chapter 32

1 Yes. Note that a virtual image is "seen" because the eye focuses the diverging rays to form a real image on the retina. For example, you can photograph the virtual image of yourself in a flat mirror and get a perfectly good picture.

3 (*a*) False, (*b*) False, (*c*) True, (*d*) False

5 (*a*) The mirror will produce an upright image for all object distances.

(*b*) The mirror will produce a virtual image for all object distances.

(*c*) The mirror will produce an image that is smaller than the object for all object distances.

(*d*) The mirror will never produce an enlarged image.

7 (*b*)

9 (*d*)

11 (*a*)

13 The muscles in the eye change the thickness of the lens and thereby change the focal length of the lens to accommodate objects at different distances. A camera lens, on the other hand, has a fixed focal length so that focusing is accomplished by varying the distance between the lens and the light-sensitive surface.

15 The objective lens of a microscope ordinarily produces an image that is larger than the object being viewed (see Figure 32-52), and that image is angularly magnified by the eyepiece. The objective lens of a telescope, on the other hand, ordinarily produces an image that is smaller than the object being viewed (see Figure 32-53), and that image is angularly magnified by the eyepiece. The telescope never produces a real image that is larger than the object.

17 $\frac{1}{2}R_{Earth}$

19 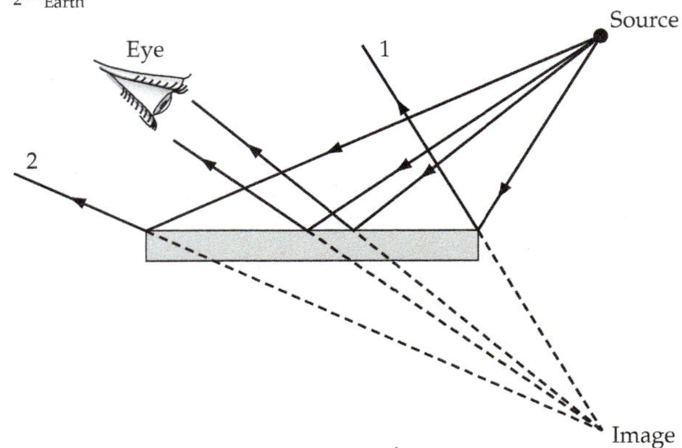

21 (*a*) 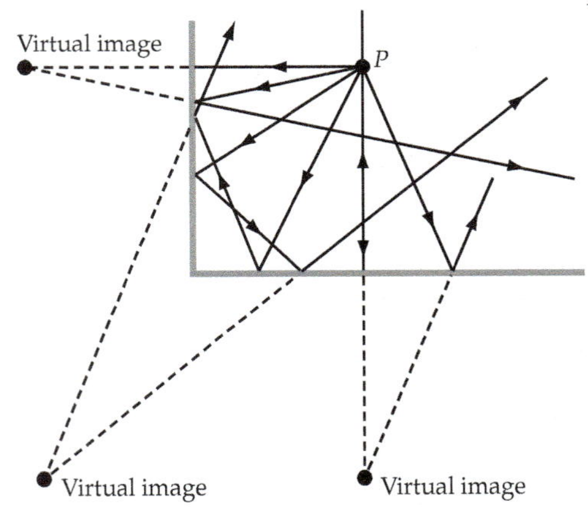

(*b*) 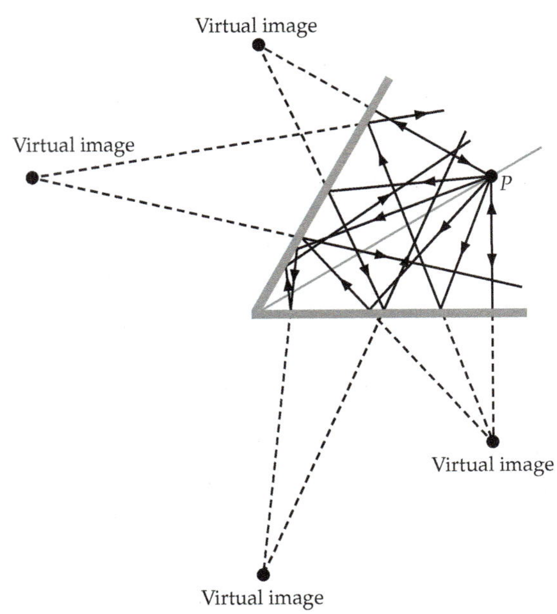

(*c*)

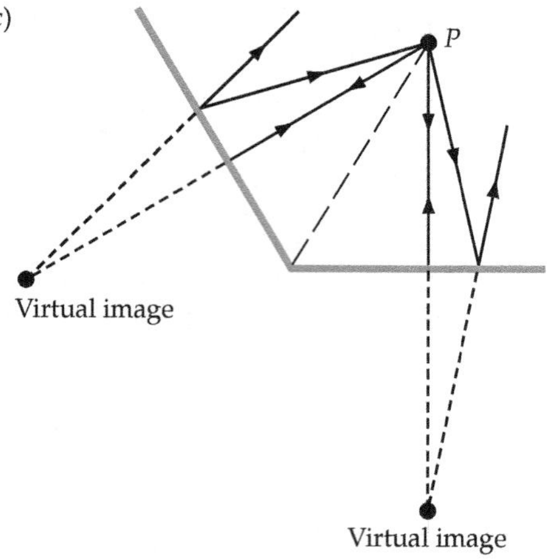

23 (*a*) For the mirror on the left, the location of the images to be 10 cm, 50 cm, 70 cm, and 110 cm behind the mirror on the left.

(*b*) For the mirror on the right, the location of the images to be 20 cm, 40 cm, 80 cm, and 100 cm behind the mirror on the right.

(*c*) The successive images are dimmer because the light travels farther to form them. The intensity falls of inversely with the square of the distance the light travels. In addition, at each reflection a small percentage of the light intensity is lost. Real mirrors are not 100% reflecting.

25 (*a*) 15 cm, 24 cm, undefined, −0.2 m,
(*b*) −0.28, −1.0, undefined, 3.0

27 (*a*) −9.9 cm, −8.0 cm, −6.0 cm, −4.8 cm,

(*b*) 0.18, 0.33, 0.50, 0.60

29 (*a*) concave, (*b*) 5.1 cm

31 The 3.7-cm-diameter image is 4.0 m in front of the mirror.

33 (*a*) −1.3 m, (*b*) convex

35 (*a*) $s' = -8.6$ cm, (*b*) 27 cm

39 (*a*) 0.6 m, (*b*) −0.8 m, (*c*) The final image is inside the rod and 0.2 m from the surface whose radius of curvature is 8.00 cm and is virtual.

41 (*a*) −30 cm, (*b*) 22 cm from the lens and on the same side of the lens as the object, (*c*) 0.27, (*d*) virtual and upright

43 (*a*) 19 cm

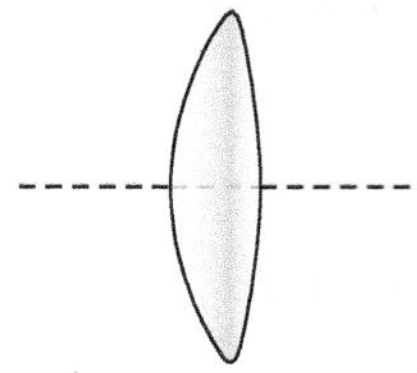

(*b*) 30 cm

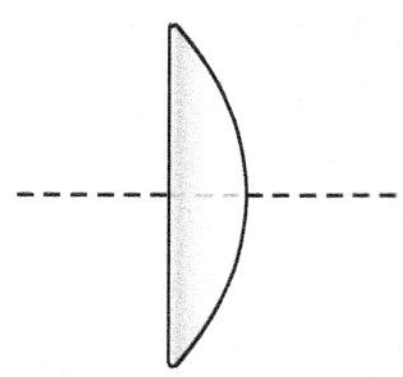

(*c*) −15 cm

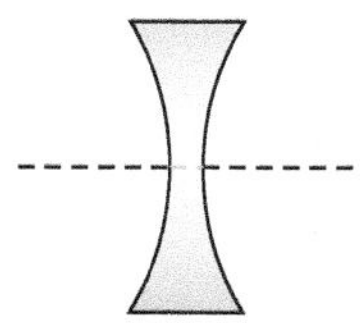

(*d*) −52 cm

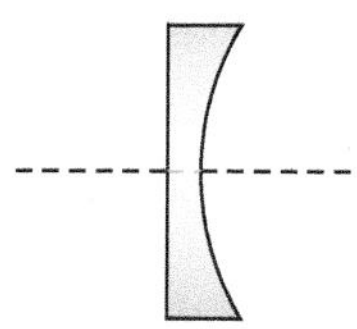

37 (*a*) $s' = -104$ cm, virtual

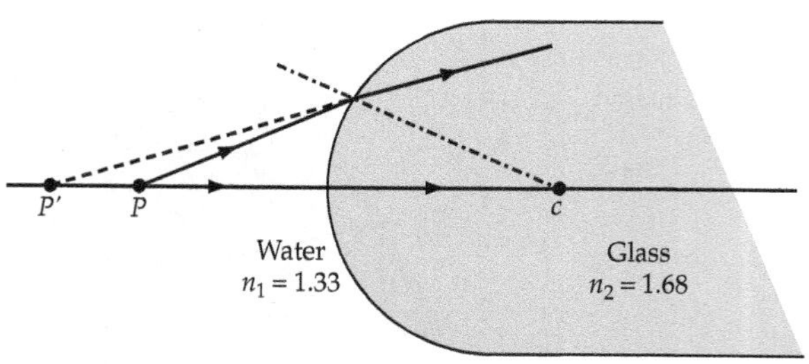

(*c*) $s' = 64$ cm, real

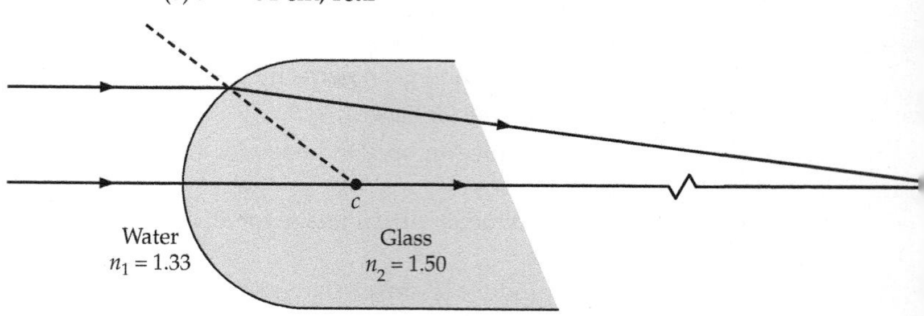

(*b*) $s' = -8.29$ cm, virtual

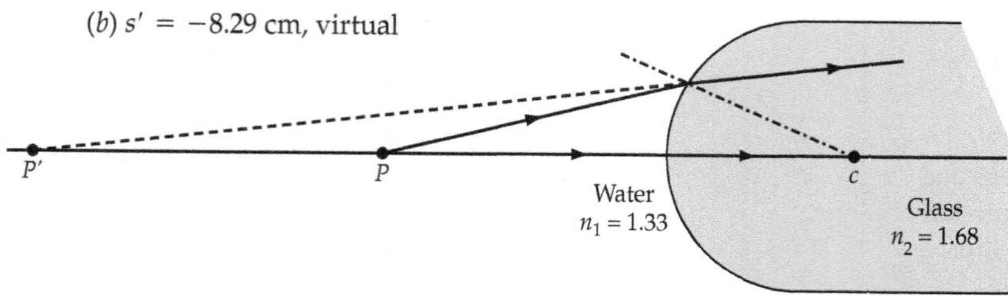

45 (*a*) $s' = 16.7$ cm, $y' = -2.00$ cm. Because $s' > 0$, the image is real and because $y'/y = -0.67$ cm, the image is inverted and diminished. These results confirm those obtained graphically. However, this ray diagram is not to scale.

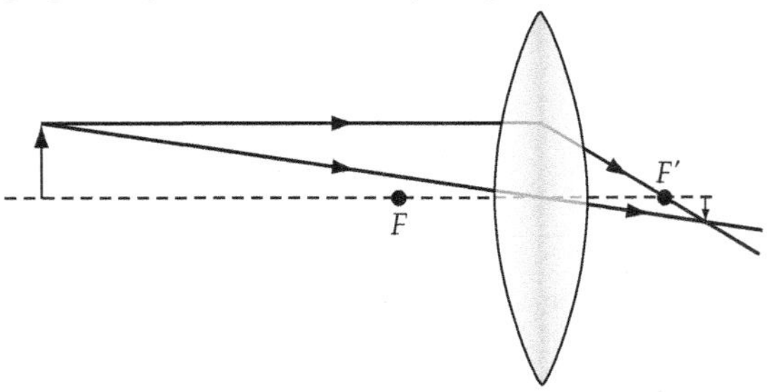

(*b*) $s' = 20.0$ cm, $y' = -3.00$ cm. Because $s' > 0$ the image is real. Because $y' = -3.00$ cm, the image is inverted and the same size as the object. These results confirm those obtained from the ray diagram.

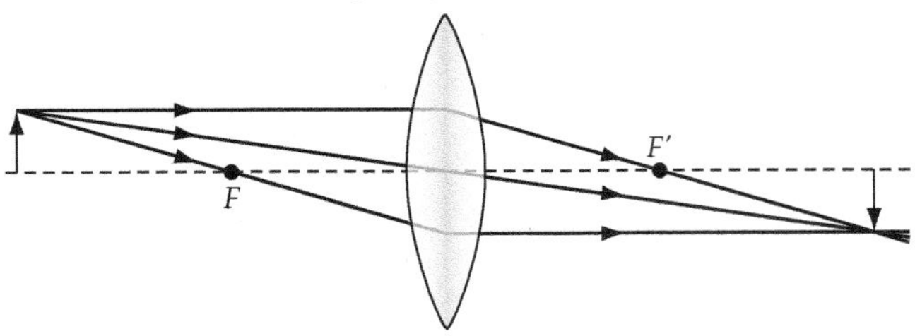

(*c*) $s' = -6.67$ cm, $y' = 1.00$ cm. Because $s' < 0$, the image is virtual. Because $y' = 1.00$ cm, the image is erect and about one-third the size of the object. These results are consistent with those obtained graphically.

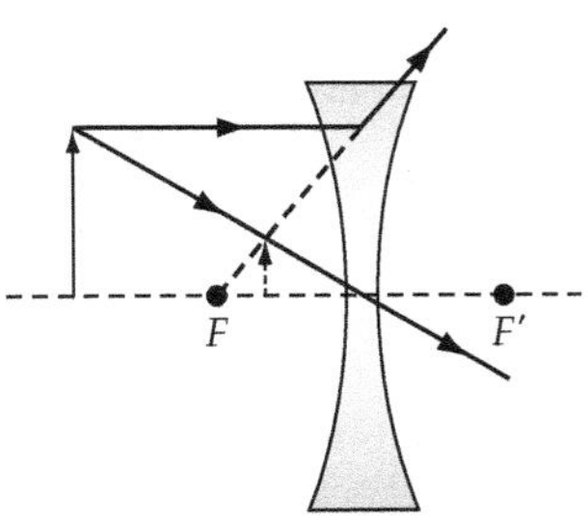

47 $r_1 = -16.2$ cm, $r_2 = \infty$,

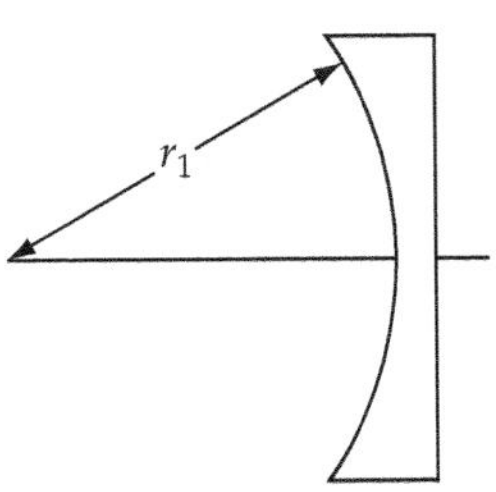

$r_1 = -32.4$ cm, $r_2 = 32.4$ cm,

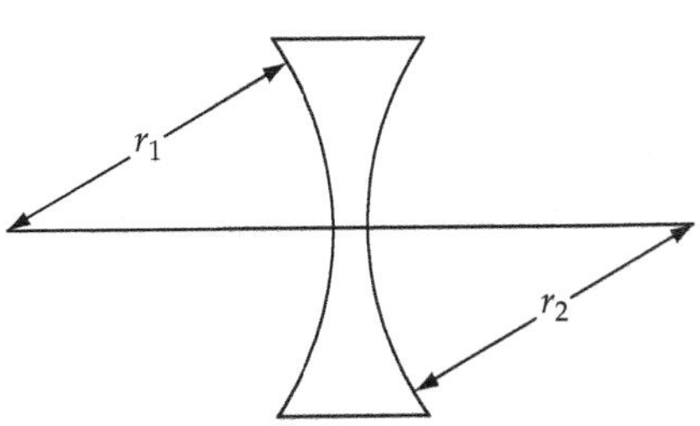

$r_1 = -5.40$ cm, $r_2 = -8.10$ cm

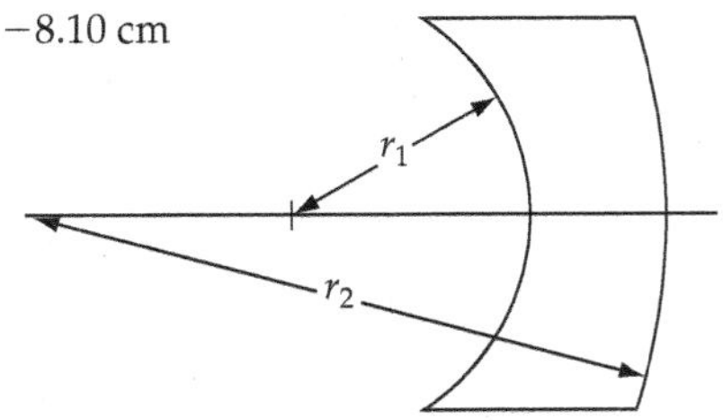

49 (*a*) The final image is 85 cm to the right of the object.

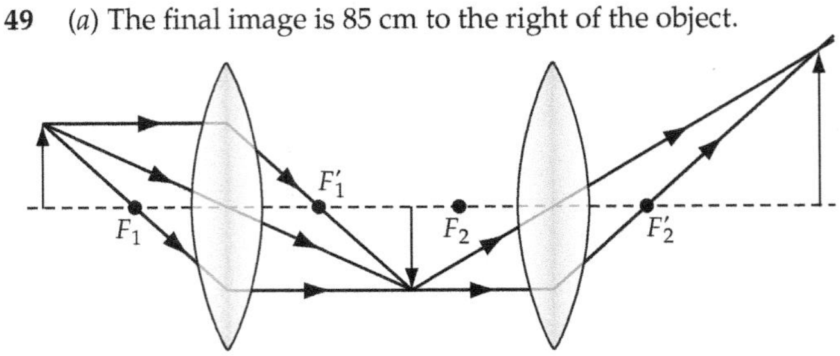

(*b*) Because $s_2' > 0$, the image is real, and because $m = m_1 m_2 = 2.0$, the image is erect and twice the size of the object.

(*c*) 2.0

51 (*b*) 3.70 m

53 (*a*) and (*b*)

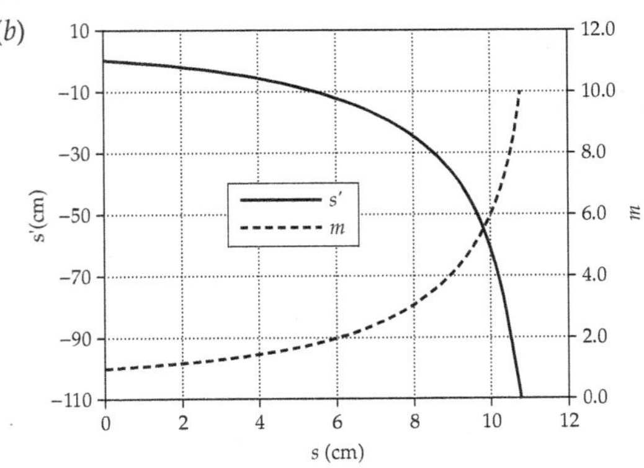

(*c*) The images are virtual and erect for this range of object distances.

(*d*) The equation for the vertical asymptote of the graph of s' versus s is $s = f$. This indicates, that as the object moves toward the second focal point, the magnitude of the image distance becomes large without limit. The equation for the vertical asymptote of the graph of m versus s is $s = f$. This indicates, that as the object moves toward the second focal point, the image becomes the large without limit. In addition, as s approaches zero s' approaches negative infinity and m approaches 1, so as the object moves toward the lens the image becomes the same size as the object and the magnitude of the image distance increases without limit.

55 (*a*) $s_2' = f_2 - 15.0$ cm, $m = m_2 = 1.00$. The final image is 20 cm from the object, virtual, erect, and the same size as the object.

(*b*)

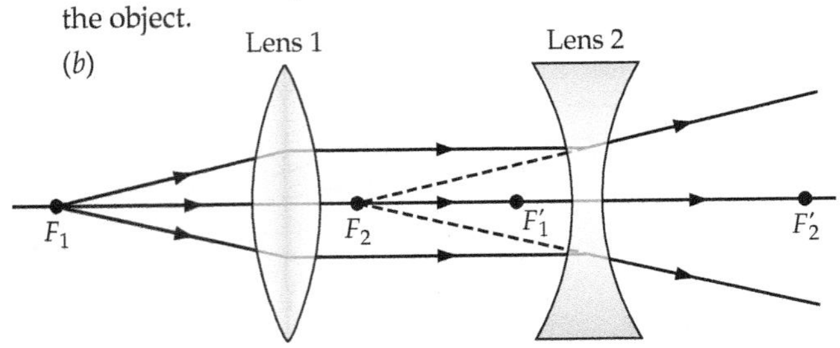

59 (*a*) The final image is 18.7 cm to the right of the second lens. (*b*) −1.53 (*c*) Because $m < 0$, the image is inverted. Because $s'_2 > 0$, the image is real.

61 Reasons for the preference for reflectors include: (1) no chromatic aberrations, (2) less expensive to shape one side of a piece of glass than to shape both sides, (3) reflectors can be more easily supported from rear instead of edges, preventing sagging and focal length changes, and (4) support from rear makes larger sizes easier to handle.

63 −1.77 mm

65 (*a*) 80.0 μrad (*b*) 1.60 mm

67 (*c*) $P_{min} = 40.0$ D, $A = 4.00$ D

69 (*c*) 6.0 D

71 0.444 D

73 3.1 D

75 5.0

77 (*a*) 3.0, (*b*) 4.0

79 (*a*) −19, (*b*) -1.9×10^2

81 −230

83 (*a*) 9.00 mm, (*b*) 0.180 mrad, (*c*) $M = -20.0$

85 (*a*) 25.0, (*b*) −1340

87 (*b*)

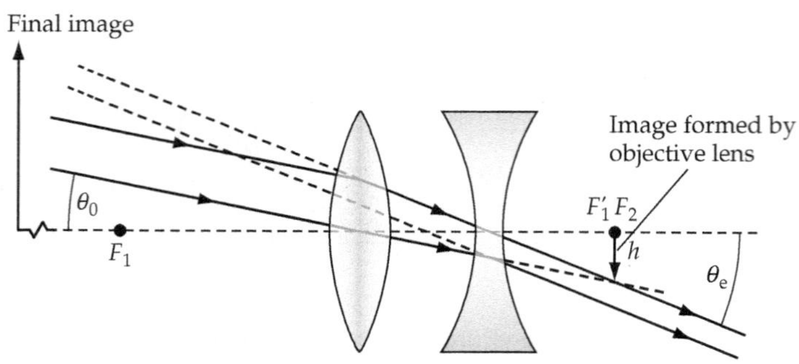

89 -6.67×10^{-3}

91 (*a*) $s = 5.0$ cm, $s' = -10$ cm

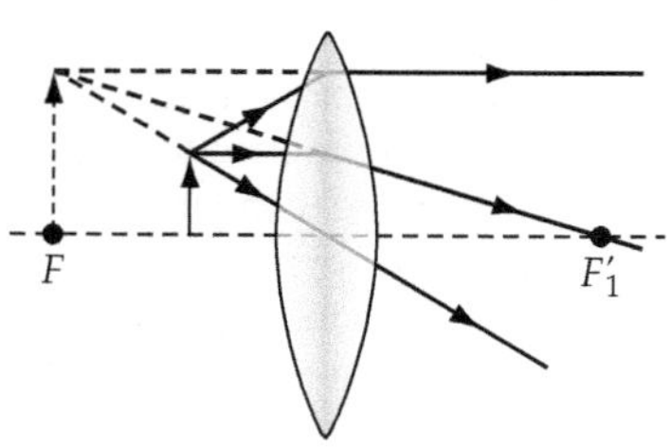

(*b*) $s = 15$ cm, $s' = 30$ cm

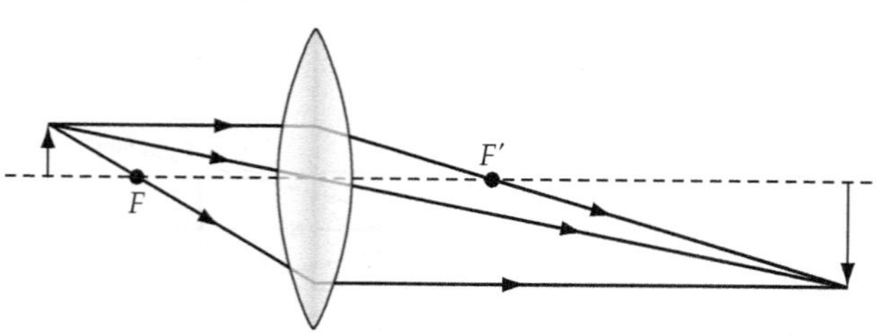

93 (*a*) The lens that has a focal length of 25 mm should be the objective. The two lenses should be separated by 210 mm. The angular magnification is −21.

(*b*)

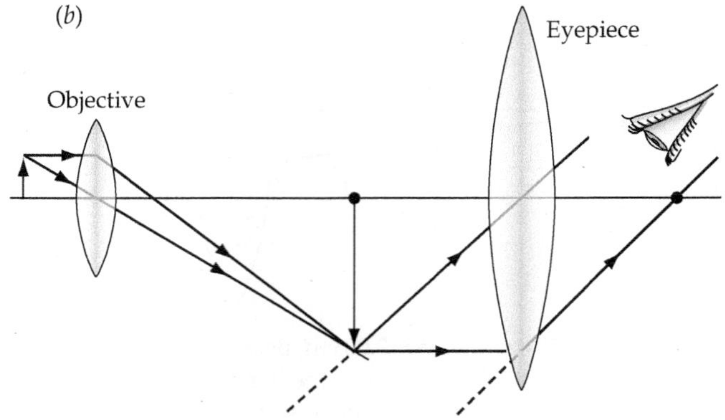

95 3.7 m

97 (*a*) 9.5 cm to the right of the second lens. (*b*) About 20% larger than the object and is inverted.

(*c*)

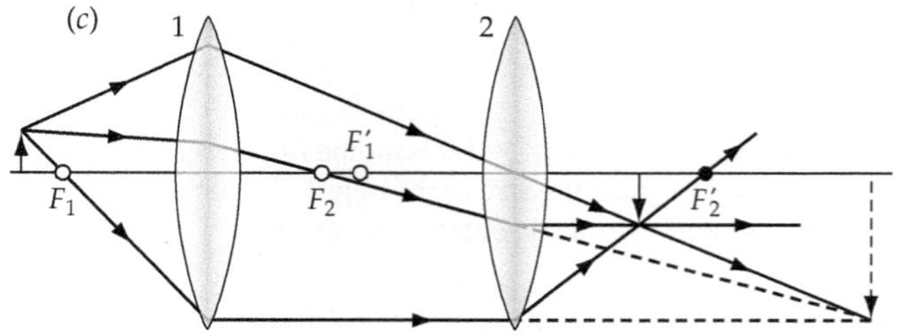

99 9.72 cm/s

101 (*a*) 18 cm from the lens, on the same side as the original object. (*b*) real and upright, (*c*) To see this image the eye must be to the left of image 4.

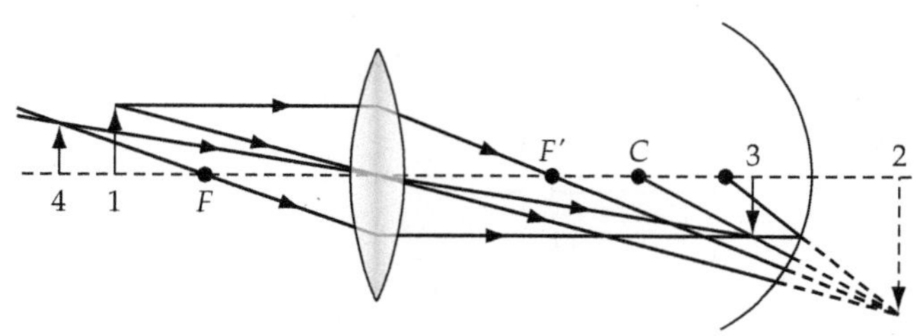

103 37 cm

105 (*a*) The final image is 13.2 cm to the left of the center of the ball. (*b*) Because s'_1 is undefined, no image is formed when the object is 20.0 cm to the left of the glass ball. (Alternatively, an image is formed an infinite distance to the left of the ball.)

Chapter 33

1 (*a*)

3 The thickness of the air space between the flat glass and the lens is approximately proportional to the square of *d*, the diameter of the ring. Consequently, the separation between adjacent rings is proportional to $1/d$.

5 Colors are observed when the light reflected from the front and back surfaces of the film interfere destructively for some wavelengths and constructively for other wavelengths. For this interference to occur, the phase difference between the light reflected from the front and back surfaces of the film must be constant. This means that twice the thickness of the film must be less than the coherence length of the light. The film is called a thin film if twice its thickness is less than the coherence length of the light.

7 (*d*)

9 (*a*)

11 (*a*)

13 (*a*)

15 (*a*) False, (*b*) True, (*c*) True, (*d*) True, (*e*) True

17 The condition for the resolution of the two sources is given by Rayleigh's criterion: $\alpha_c = 1.22\lambda/D$ (Equation 33-25), where α_c is the critical angular separation, *D* is the diameter of the aperture, and λ is the wavelength of the light illuminating (or emitted by) the objects, in this case headlights, to be resolved. Because the diameter of the pupils of your eyes are larger at night, the critical angle is smaller at night, which means that at night you can resolve the light as coming from two distinct sources when they are at a greater distance.

19 (*a*) 11 km, (*b*) 9.6 km

21 5.9 $c \cdot$y

23 $\approx$2.9 rad

25 $5.5\ \mu\text{m} < d < 5.8\ \mu\text{m}$

27 (*a*) 600 nm, (*b*) 720 nm, 514 nm, and 400 nm, (*c*) 720 nm, 514 nm, and 400 nm

29 476 nm

31 (*c*) 68, (*d*) 1.14 cm (*e*) The fringes would become more closely spaced.

33 0.535 mm, 0.926 mm

35 4.95 mm

37 (*a*) 50.0 μm, (*b*) Not with the unaided eye. The separation is too small to be observed with the naked eye. (*c*) 0.500 mm

39 625 nm and 417 nm

41 (*a*) 0.60 mrad, (*b*) 6.0 mrad, (*c*) 60 mrad

43 (*a*) 1.53 km

45 (*a*) 20.0 μm, (*b*) 9

47 8

49 $\vec{E} = 3.6\ A_0 \sin(\omega t - 0.98\ \text{rad})\hat{i}$

51 $I/I_0 = 0.0162$

53 (*b*) 6.00 mm. The width for four sources is half the width for two sources.

55 (*a*)

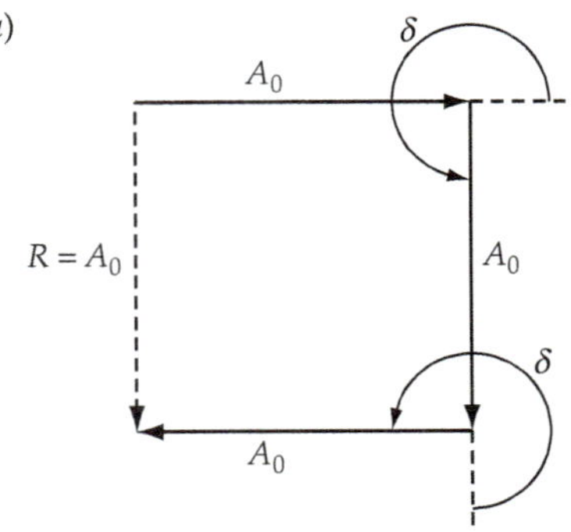

(*b*) 5.56 mW/m^2

57 (*a*) 8.54 mrad, (*b*) 6.83 cm

59 7.00 mm

61 5.00×10^9 m

63 (*a*) 86.9 mrad, 82.1 mrad, (*b*) 709 mrad, 662 mrad

65 30.0°

67 One can see the complete spectrum for only the first and second order spectra. That is, only for $m = 1$ and 2. Because $700\ \text{nm} < 2 \times 400\ \text{nm}$, there is no overlap of the second-order spectrum into the first-order spectrum; however, there is overlap of long wavelengths in the second order with short wavelengths in the third-order spectrum.

69 (*a*) 36.4 cm, 80.1 cm, (*b*) 88.4 μm, (*c*) 8000

71 3.09×10^5, $5.14 \times 10^4\ \text{cm}^{-1}$

73 (*a*) $\phi_m = \frac{1}{2}\sin^{-1}\left(m\frac{\lambda}{d}\right)$, (*b*) 32.1°

75 3.5 μm

77 3.6°, 2.5°

79 (*a*) 15.1 cm, (*b*) 3.33 m^{-1}

81 0.13 mrad

85 (*a*) 97.8 nm, (*b*) No, because 180 nm is not in the visible portion of the spectrum. (*c*) 0.273

87 12 m

Chapter 34

1 (*c*)

3 (*a*)

5 (*a*) True, (*b*) True, (*c*) True

7 (*c*)

9 According to quantum theory, the average value of many measurements of the same quantity will yield the expectation value of that quantity. However, any single measurement may differ from the expectation value.

11 2.48 pm, 2%

13 (*a*)

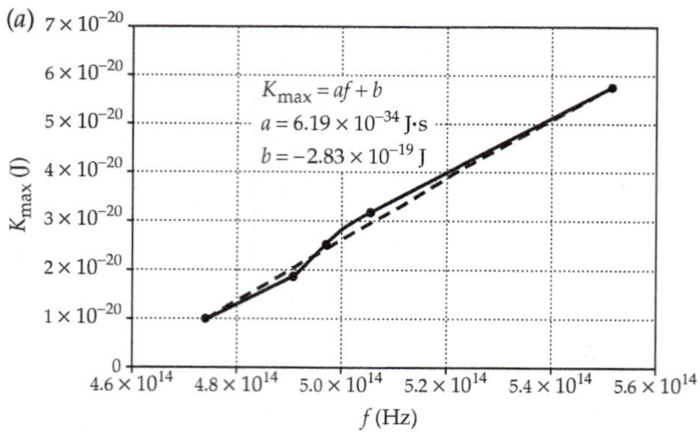

(*b*) 1.77 eV (*c*) cesium

15 (*a*) 4.14×10^{-7} eV, (*b*) 3.72×10^{-9} eV

17 (*a*) 12.4 keV, (*b*) 1.24 GeV

19 $1.95 \times 10^{16}\ \text{s}^{-1}$

21 (*a*) 4.13 eV, (*b*) 2.10 eV, (*c*) 0.78 eV, (*d*) 590 nm

23 (*a*) 653 nm, 4.58×10^{14} Hz, (*b*) 3.06 eV, (*c*) 1.64 eV

25 1.2 pm

27 0.18 nm

29 $9.32 \times 10^{-24}\ \text{kg} \cdot \text{m/s}$, $1.80 \times 10^{-23}\ \text{kg} \cdot \text{m/s}$

31 2.9 nm

33 (*a*) $p_e = 2.09 \times 10^{-22}\ \text{N} \cdot \text{s}$, $p_p = 8.97 \times 10^{-21}\ \text{N} \cdot \text{s}$,

$p_\alpha = 8.97 \times 10^{-21}\ \text{N} \cdot \text{s}$

(*b*) $\lambda_e = 3.17$ pm, $\lambda_p = 73.9$ fm, $\lambda_\alpha = 37.0$ fm

35 20.2 fm

37 0.17 nm

39 4.6 pm

41 (*a*) $E_1 = 205$ MeV, $E_2 = 818$ MeV, $E_3 = 1.84$ GeV

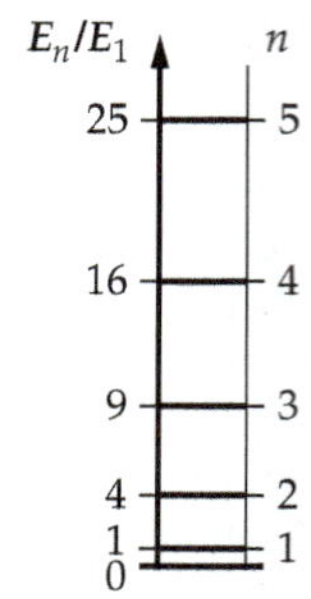

(*b*) $\lambda_{2\to1} = 2.02$ fm,

(*c*) $\lambda_{3\to2} = 1.21$ fm,

(*d*) $\lambda_{3\to1} = 0.758$ fm

43 (*a*) 0, (*b*) 1, (*c*) 0.002

45 (*a*) $L/2$, (*b*) $0.321L^2$

47 (*a*) $1/\sqrt{2}$, (*b*) 0.865

49 (*a*) 0.500, (*b*) 0.402, (*c*) 0.750

51 (*b*) For large values of *n*, the result agrees with the classical value of $L^2/3$ given in Problem 50.

53 $\langle x \rangle = 0$, $\langle x^2 \rangle = L^2\left[\frac{1}{12} - \frac{1}{2\pi^2}\right]$

55 (*a*) 3.10 eV, (*b*) 6.24×10^{16} eV, (*c*) 2.08×10^{16}

57 (*a*) 1 μm, $10^{-16}\ \text{kg} \cdot \text{m/s}$, (*b*) 2×10^{11}

59 0.2 keV

61 7×10^3 km

63 (*a*) 92 mW/m^2, (*b*) 3×10^4

67 1.3 MeV. The energy of the most energetic electron is approximately 2.5 times the rest energy of an electron.

69 1.04 eV, 554 nm

71 (*b*) 0.2% (*c*) Classically, the energy is continuous. For very large values of *n*, the energy difference between adjacent levels is infinitesimal.

73 (*a*) 6.2×10^{-4} eV/s, (*b*) 53 min

Chapter 35

1 (*a*)

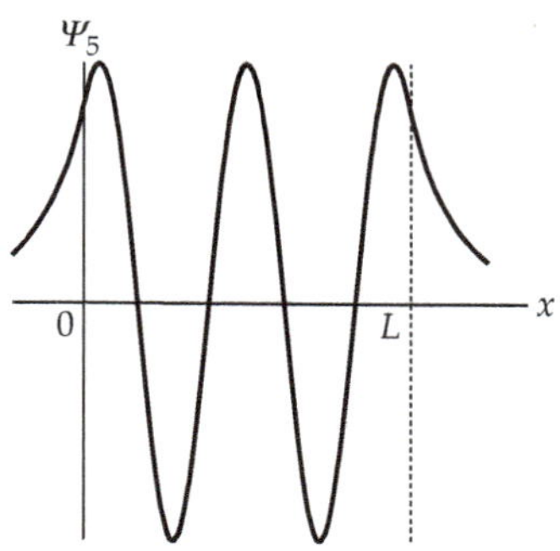

(*b*)

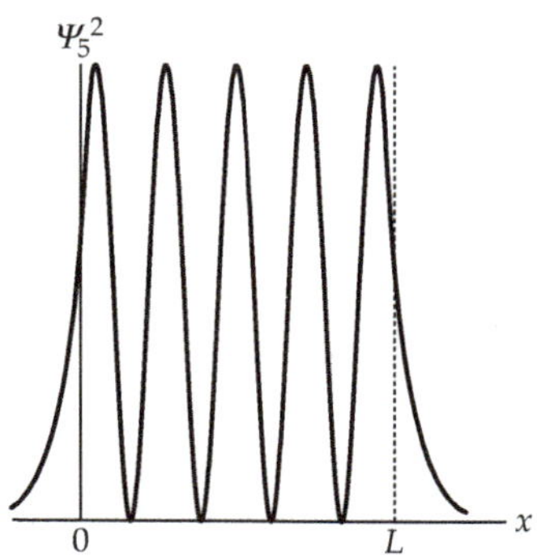

9 (*a*) 9.5 nm, (*b*) 4.1 meV

11 $\Delta x \, \Delta p = \dfrac{\hbar}{2}$

13 (*b*)

Cell	Content/Formula	Algebraic Form
A2	1.0	α
B2	(1−SQRT((A2−1)/A2))/(1+SQRT((A2−1)/A2))²	$\left(\dfrac{1-\sqrt{\dfrac{\alpha-1}{\alpha}}}{1+\sqrt{\dfrac{\alpha-1}{\alpha}}}\right)^2$
C2	1 − B2	$1-\left(\dfrac{1-\sqrt{\dfrac{\alpha-1}{\alpha}}}{1+\sqrt{\dfrac{\alpha-1}{\alpha}}}\right)^2$

	A	B	C
1	α	*R*	*T*
2	1.0	1.000	0.000
3	1.2	0.298	0.702
4	1.4	0.298	0.802
5	1.6	0.149	0.851
18	4.2	0.036	0.964
19	4.4	0.034	0.966
20	4.6	0.032	0.968
21	4.8	0.031	0.969
22	5.0	0.029	0.971

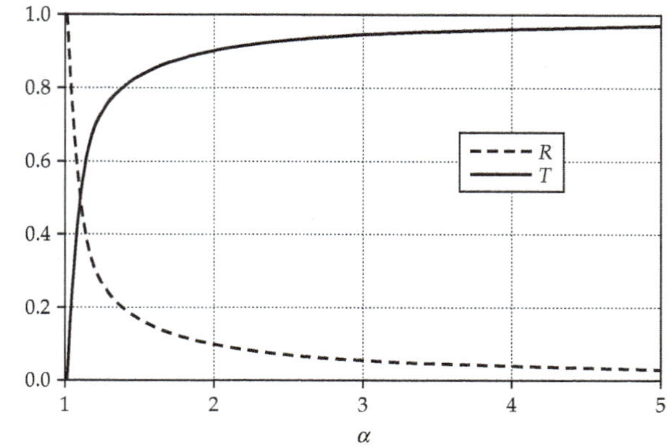

15 (*a*) 10^{-17}, (*b*) 10^{-2}

17 (*a*) $r_{1\,4.0\,\text{MeV}} = 66$ fm, $r_{1\,7.0\,\text{MeV}} = 38$ fm

(*b*) $T_{4.0\,\text{MeV}} \approx 10^{-51}$, $T_{7.0\,\text{MeV}} \approx 10^{-38}$

19 (*a*)

n_1	1	1	1	1	1	1	1	1	1	1
n_2	1	1	1	2	1	2	2	1	2	3
n_3	1	2	3	1	4	2	3	5	4	1
E	21	24	29	33	36	36	41	45	48	53

(*b*) (1, 1, 4) and (1, 2, 2)

(*c*) $\psi(1, 1, 4) = A \sin\left(\dfrac{\pi}{L_1}x\right) \sin\left(\dfrac{\pi}{2L_1}y\right) \sin\left(\dfrac{\pi}{L_1}z\right)$

21 (*a*) $\psi(x,y) = A \sin\dfrac{n\pi}{L}x \sin\dfrac{m\pi}{L}y$

(*b*) $E_{nm} = \dfrac{h^2}{8mL^2}\left(n^2 + m^2\right)$

(*c*) (1, 2) and (2, 1)

(*d*) (1, 7), (7, 1), and (5, 5)

23 $E_{110\text{ bosons}} = \dfrac{5h^2}{4mL^2}$

29 $E_0 = \dfrac{5h^2}{mL^2}, E_1 = E_2 = \dfrac{21h^2}{4mL^2}$

31 (*b*) $\langle x^2 \rangle = \dfrac{2}{L}\left(\dfrac{L^3}{24} - \dfrac{L^3}{4n^2\pi^2}\cos n\pi\right)$

35 $A_2 = \sqrt[4]{\dfrac{8m\omega_0}{h}}$

Chapter 36

1 Examination of Figure 36-4 indicates that as n increases, the spacing of adjacent energy levels decreases.

3 (*a*)

5 (*d*)

7 (*a*)

9 The energy of a bound isolated system that consists of two oppositely charged particles, such as an electron and a proton, depends only upon the principle quantum number n. For sodium, which consists of 12 charged particles, the energy of an $n = 3$ electron depends upon the degree to which the wave function of the electron penetrates the $n = 1$ and $n = 2$ electron shells. An electron in a 3s ($n = 3$, $\ell = 0$) state penetrates these shells to a greater degree than does an electron in a 3p ($n = 3$, $\ell = 1$) state, so a 3s electron has less energy (is more tightly bound) than is a 3p electron. In hydrogen, however, the wave function of an electron in the $n = 3$ shell cannot penetrate any other electron shells because no other electron shells exist. Thus, an electron in the 3s state in hydrogen has the same energy as an electron in the 3p state in hydrogen.

11 In conformity with the exclusion principle, the total number of electrons that can be accommodated in states of quantum number n is n^2 (see Problem 48). The fact that closed shells correspond to $2n^2$ electrons indicates that there is another quantum number that can have two possible values.

13 (*a*) phosphorus, (*b*) chromium

15 (*d*)

17 The optical spectrum of any atom is due to the configuration of its outer-shell electrons. Ionizing the next atom in the periodic table gives you an ion with the same number of outer-shell electrons, and almost the same nuclear charge. Hence, the spectra should be very similar.

21 (*a*) 10^5, (*b*) 10^3, (*c*) 5.08×10^4

23 (*a*) 103 nm, (*b*) 97.3 nm

25 (*a*) 1.51 eV, 821 nm
(*b*) 0.661 eV, 1880 nm, 0.967 eV, 1280 nm, 1.13 eV, 1100 nm

6→3 5→3 4→3

1100 nm 1280 nm 1880 nm

27 (*b*) $1.096850 \times 10^7\ \text{m}^{-1}$, $1.097448 \times 10^7\ \text{m}^{-1}$, R_H and $R_{H\,\text{approx}}$ agree to three significant figures.
(*c*) 0.0545%

29 (*a*) $1.49 \times 10^{-34}\ \text{J}\cdot\text{s}$, (*b*) −1, 0, +1
(*c*)

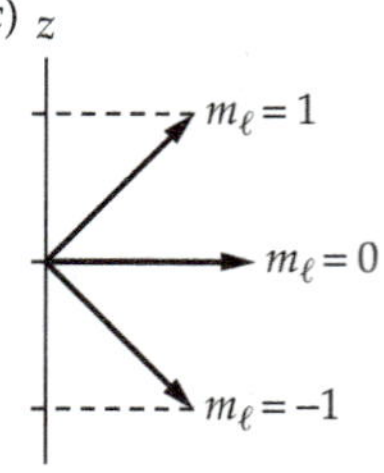

31 (*a*) 0, 1, 2, (*b*) For $\ell = 0$, $m_\ell = 0$. For $\ell = 1$, $m_\ell = -1, 0, +1$. For $\ell = 2$, $m_\ell = -2, -1, 0, +1, +2$. (*c*) 18

33 (*a*) 45.0°, (*b*) 26.6°, (*c*) 8.05°

35 (*a*) $6\hbar^2$, (*b*) $4\hbar^2$, (*c*) $2\hbar^2$

37 (*a*) 4
(*b*)

n	ℓ	m_ℓ	(n, ℓ, m_ℓ)
2	0	0	(2, 0, 0)
2	1	−1	(2, 1, −1)
2	1	0	(2, 1, 0)
2	1	1	(2, 1, 1)

39 (*a*) $\psi_{200}(a_0) = \dfrac{0.0605}{a_0^{3/2}}$, (*b*) $\left[\psi_{200}(a_0)\right]^2 = \dfrac{0.00366}{a_0^3}$
(*c*) $P(a_0) = \dfrac{0.0460}{a_0}$

41 (*a*) 9.20×10^{-4}, (*b*) 0

47 0.323

49 $\ell = 0$ or 1

51

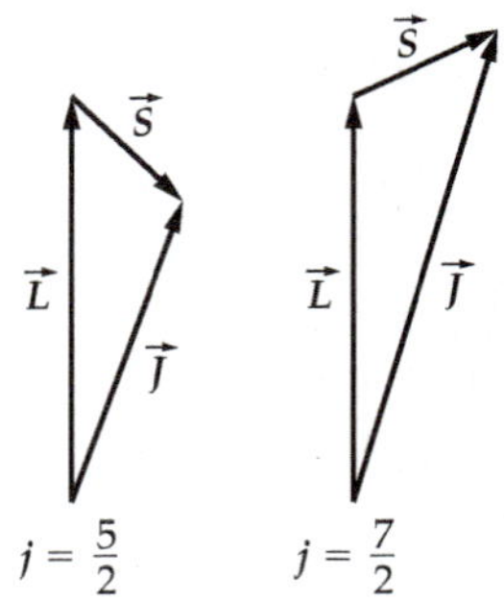

53 (*c*)

55 (*a*) $L_z = -2\hbar, -\hbar, 0, \hbar, 2\hbar$, (*b*) $L_z = -3\hbar, -2\hbar, -\hbar, 0, \hbar, 2\hbar, 3\hbar$

57 (*a*) 2s or 2p, (*b*) $1s^22s^22p^63p$, (*c*) 1s2s

59 (*a*) 0.0610 nm, 0.0578 nm, (*b*) 0.0542 nm

61 (*a*) 1.00 nm, (*b*) 0.155 nm

63 $n_i = 4$ to $n_f = 1$

65

λ, nm	n_i	n_f
164	3	2
230.6	9	3
541	7	4

67 (*a*) 1.6179 eV, 1.6106 eV, (*b*) 0.00730 eV, (*c*) 63.0 T
(*b*) No

71 (*a*) 1.06 GHz, (*b*) 28.4 cm, microwave

73 (*a*) $1.097075 \times 10^7\ \text{m}^{-1}$, (*b*) 0.179 nm

75 (*a*) $1.097074 \times 10^7\ \text{m}^{-1}$, (*b*) 0.0600 nm, (*c*) 0.238 nm

Chapter 37

1 Because the center of charge of the positive Na ion does not coincide with the center of charge for the negative Cl ion, the NaCl molecule has a permanent dipole moment. Hence, it is a polar molecule.

3 Neon occurs naturally as Ne, not Ne_2. Neon is a noble gas. Atoms of noble gases have a closed shell electron configuration.

5 The diagram would consist of a nonbonding ground state with no vibrational or rotational states for ArF (similar to the upper curve in Figure 37-4) but for ArF* there should be a bonding excited state with a definite minimum with respect to internuclear separation and several vibrational states as in the excited state curve of Figure 37-13.

7 The effective force constant from Example 37-4 is 1.85×10^3 N/m. This value is about 25% larger than the given value of the force constant of the suspension springs on a typical automobile.

9 For H_2, the concentration of negative charge between the two protons holds the protons together. In the H_2^+ ion, there is only one electron that is shared by the two positive charges such that most of the electronic charge is again between the two protons. However, in the H_2^+ ion the negative charge between the protons is not as effective as the larger negative charge between them in the H_2 molecule, and the protons should be farther apart. The experimental values support this argument. For H_2, $r_0 = 0.074$ nm, while for H_2^+, $r_0 = 0.106$ nm.

11 For more than two atoms in the molecule, there will be more than just one frequency of vibration because more relative motions are possible. In advanced mechanics, these are known as normal modes of vibration.

13 $\ell \approx 2 \times 10^{30}$, $E_{0r} \approx 5 \times 10^{-65}$ J

15 0.947 nm

17 0.44 eV

19 You should agree. The potential energy curve is shown in the following diagram. The turning points for vibrations of energy E_1 and E_2 are at the values of r where the energies equal $U(r)$. The average value of r for the vibrational levels E_1 and E_2 are labeled $r_{1\,av}$ and $r_{2\,av}$. Note that the estimate of $r_{1\,av}$ is force midway between $r_{1\,min}$ and $r_{1\,max}$. The potential is like a special spring that has a greater force constant for compressions than it has for extensions. The period of a spring-and-mass oscillator is inversely proportional to the square root of the spring constant, so our "special spring" spends more time in extension than in compression. As a result, $r_{1\,av}$ will be greater than the equilibrium radius. This argument can be extended to explain why $r_{2\,av}$ is greater than $r_{1\,av}$. It is because the "force constant" for extension, which can be estimated by taking the average slope of the potential energy curve in the region to the right of the equilibrium position, is greater for $E = E_2$ than for $E = E_1$. It is also because the "force constant" for compression is greater for $E = E_2$ than for $E = E_1$. It follows that $r_{2\,av}$ is greater than $r_{1\,av}$. Because $r_{2\,av}$ is greater than $r_{1\,av}$, it follows that as the vibrational energy of a diatomic molecule increases, the average separation of the atoms of the molecule increases and, hence, the solid expands with heating.

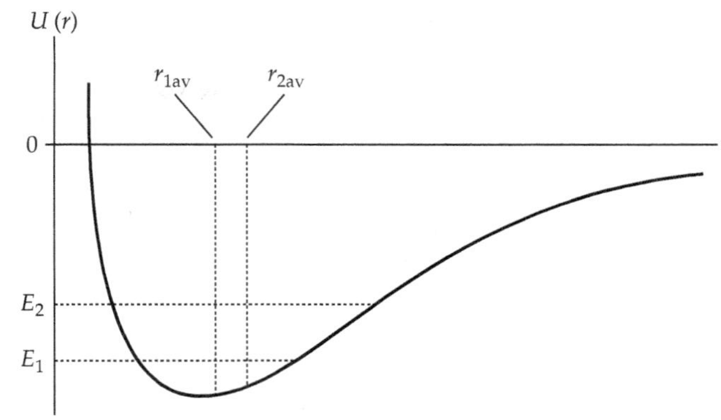

21 (a) $U_e = -6.64$ eV, (b) $E_{d\,calc} = 5.70$ eV, (c) $U_{rep} = 0.63$ eV

23 0.121 nm

25 41

27 5.6 meV

29 (a) 0.179 eV, (b) 3×10^{-47} kg·m², (c) 0.1 nm

31 (a) 1.45×10^{-46} kg·m², 0.239 meV

(b)

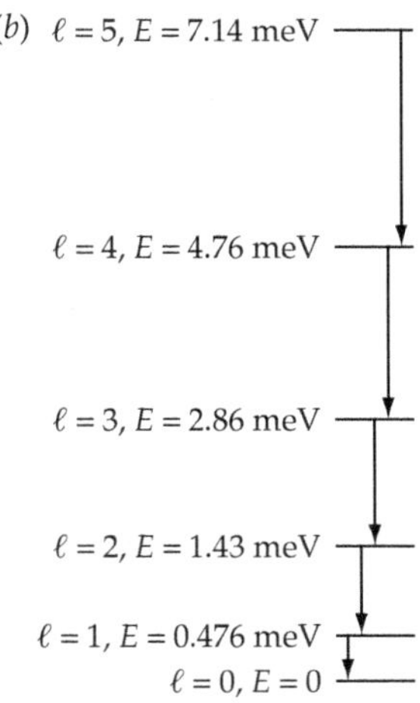

$\Delta E_{54} = 2.38$ meV, $\Delta E_{43} = 1.90$ meV, $\Delta E_{32} = 1.43$ meV, $\Delta E_{21} = 1.25$ meV, $\Delta E_{10} = 0.476$ meV

(c) $\lambda_{10} = 2600$ μm, $\lambda_{21} = 1300$ μm, $\lambda_{32} = 867$ μm, $\lambda_{43} = 650$ μm, $\lambda_{54} = 520$ μm, microwave

33 $\mu_{H^{35}Cl} = 0.972$ u, $\mu_{H^{37}Cl} = 0.974$ u,

$\dfrac{\Delta\mu}{\mu} = 0.00150$, $\Delta f/f = 0.0012$, in fair agreement (about 20% difference) with the calculated result. Note that Δf is difficult to determine precisely from Figure 37-17.

35 0.955 meV

37 1.55 kN/m

39 $r_0 = a$, $U_{min} = -U_0$, $r_0 = 0.074$ nm, $U_0 = 4.52$ eV

41 $F_x = -\dfrac{dU}{dx} \propto \dfrac{1}{x^4}$

43 (a) $\dfrac{1\text{ eV}}{\text{molecule}} = 23.0$ kcal/mol

(b) 98.2 kcal/mol

Chapter 38

1 The energy lost by the electrons in collision with the ions of the crystal lattice appears as thermal energy throughout the crystal.

3 (*a*) potassium and nickel (*b*) 3.1 V

5 The resistivity of brass at 4 K is almost entirely due to the residual resistance (the resistance due to impurities and other imperfections of the crystal lattice). In brass, the zinc ions act as impurities in copper. In pure copper, the resistivity at 4 K is due to its residual resistance. The residual resistance is very low if the copper is very pure.

7 The resistivity of copper increases with increasing temperature; the resistivity of (pure) silicon decreases with increasing temperature because the number density of charge carriers increases.

9 (*b*)

11 The excited electron is the motion of the electron in the conduction band and contributes to the current. A hole is left in the valence band allowing the positive hole to move through the band, also the motion of the hole contributes to the current.

13 (*c*)

15

V (V)	1/slope (Ω)
−20	∞
+0.2	40
+0.4	20
+0.6	10
+0.8	5

17 2.07 g/cm^3

19 (*a*) −10.6 eV, (*b*) 2.83%

21 (*a*) 0.123 μΩ · m, (*b*) 70.7 nΩ · m

23 (*a*) $n_{Ag} = 5.86 \times 10^{22}$ electrons/cm^3
(*b*) $n_{Ag} = 5.90 \times 10^{22}$ electrons/cm^3. Both these results agree with the values in Table 38-1.

25 4.0

27 (*a*) 1.07×10^6 m/s (*b*) 1.39×10^6 m/s
(*c*) 1.89×10^6 m/s

29 (*a*) 4.22 eV (*b*) 2.85 eV

31 (*a*) 5.90×10^{28} e/m^3, (*b*) 5.50 eV, (*c*) 212, (*d*) The ratio E_F/kT is equal to 212 at $T = 300$ K. The Fermi energy is the energy of the most energetic conduction electron when the crystal is at absolute zero. Because no two conduction electrons can occupy the same state, the Fermi energy is quite high compared with kT. The kT energy is the energy the average conduction electron would have when the crystal is at temperature T if the electrons did not obey the exclusion principle.

33 3.82×10^{10} N/m^2 = 3.77×10^5 atm

35 0.192 J/(mol · K)

37 (*a*) 66 nm, (*b*) 1.8×10^{-4} nm^2

39 1.09 μm

41 180 nm

43 116 K

45 $a_{B\,Si} = 3$ nm, $a_{B\,Ge} = 8$ nm

47 37.1 nm, 38.7 nm. The mean free paths agree to within about 4%.

49

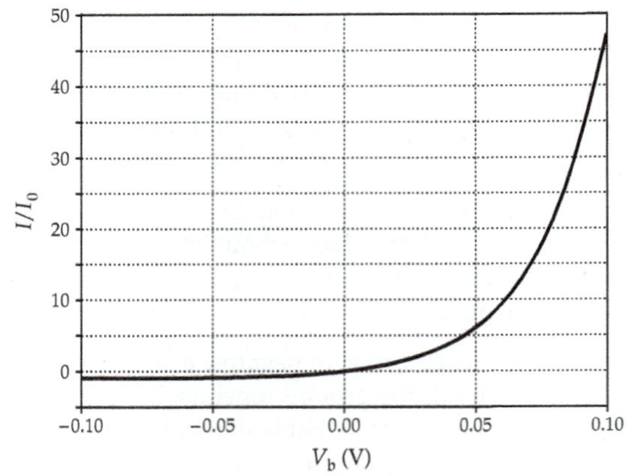

51 250

53 (*a*)

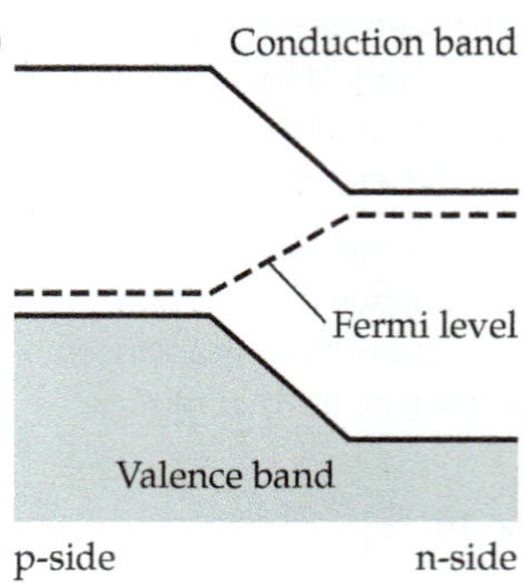

(*b*)

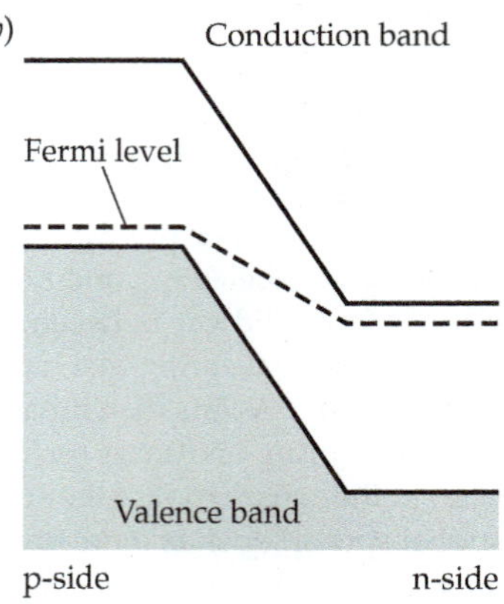

55 The charge carriers are holes and the semiconductor is p-type. 1.0×10^{23} m^{-3}

57 (*a*) 2.17 meV, $E_g \approx 0.8E_{g\,measured}$, (*b*) 0.454 mm

59 2.0×10^{18}

61 (*a*) 5.51 eV, (*b*) 3.31 eV, (*c*) 1.08×10^6 m/s

63 1

67 0.60

71 1.07

73 (*a*) 5.51×10^{-3}, (*b*) 1.84×10^{-2}

75 4.35×10^{14} Hz

Chapter 39

1 (*a*)

3 (*a*) True, (*b*) True, (*c*) False, (*d*) True, (*e*) False, (*f*) False, (*g*) True

5 Although $\Delta y = \Delta y'$, $\Delta t \neq \Delta t'$. Consequently, $\Delta y/\Delta t \neq \Delta y'/\Delta t'$.

7 (*a*) 0.946, (*b*) $1.23 \times 10^{10}\ c \cdot y$

9 (*a*) 0.98 km. The width of the beam is unchanged. (*b*) 9.6×10^7 m, (*c*) 0.10 μm

11 (*a*) 0.91 *c*, (*b*) 22 $c \cdot y$, (*c*) 101 y

13 (*a*) 0.385 μs, (*b*) 0.193 μs, (*c*) 0.998 *c*

15 1.85×10^4 y

17 (*a*) 1.76 μs, (*b*) 6.32 μs, (*c*) 3.1 μs, (*d*) 1.70 km

19 4.4 μs

21 (*a*) 2.10 μs, (*b*) 2.59 μs, (*c*) 0.49 μs, (*d*) 2.59 μs, (*e*) 4.36 h, (*f*) 18.8 h

23 2.22×10^7 m/s

27 11 ms

29 (*a*) $u_x = v$ and $u_y = \dfrac{c}{\gamma}$

31 (*a*) 0.976, (*b*) 0.997*c*

33 66.7%

37 (*a*) 290 MeV, (*b*) 629 MeV

39 (*a*) 0.943*c*, (*b*) 3.0 MeV, (*c*) 2.8 MeV/*c*, (*d*) 4.1 MeV/c^2, (*e*) 0.9 MeV

43 0.999*c*

45 (*a*) $-0.50c$, S' moves in the $-x$ direction, (*b*) 1.7 y

47 281 MeV

49 (*a*) $v = -\dfrac{E}{Mc'}$ (*b*) $d = -\dfrac{LE}{Mc^2}$

51 $K_{th} = 6m_pc^2$ in agreement with Problem 40.

Chapter 40

1 (*a*) ^{15}N, ^{16}N, (*b*) ^{54}Fe, ^{55}Fe, (*c*) ^{117}Sn, ^{119}Sn

3 Generally, decay by α emission leaves the daughter nucleus neutron rich, i.e. above the line of stability. The daughter nucleus therefore tends to decay via β^- emission which converts a nuclear neutron to a proton.

5 It would make the dating unreliable because the current concentration of ^{14}C is not equal to that at some earlier time.

7 The probability for neutron capture by the fissionable nucleus is large only for slow (thermal) neutrons. The neutrons emitted during the fission process are fast (high energy) neutrons and must be slowed to thermal neutrons before they are likely to be captured by another fissionable nucleus.

9 (*a*) β^+, (*b*) β^-

11 (*a*) True (given an unlimited supply of ^{238}U), (b) False, (c) True, (d) False

13

Material Property	Ratio (order of magnitude)
Mass density	10^{15}
Half-life	10^{15}
Nuclear masses	2

15 (*a*) $E_b = 92.2$ MeV, $E_b/A = 7.68$ MeV
(*b*) $E_b = 492$ MeV, $E_b/A = 8.79$ MeV
(*c*) $E_b = 1802$ MeV, $E_b/A = 7.57$ MeV

17 (*a*) 3.0 fm, (*b*) 4.6 fm, (*c*) 7.0 fm

19 (*a*) $E_{thermal} = 4.11 \times 10^{-21}$ J = 25.7 meV, (*b*) 2.22 km/s, (*c*) 10.1 min

23 (*a*) 5 min, (*b*) 250 Bq

25 (*a*) 200 s, (*b*) $3.5 \times 10^{-3}\ s^{-1}$, (*c*) 125 Bq

27 (*a*) 500 Bq, 250 Bq, (*b*) $N_0 = 1.0 \times 10^6$, $N_{2.4\,min} = 5.2 \times 10^5$, (*c*) 12 min

29 (*a*) $4.5 \times 10^3\ \alpha$/s, (*b*) 5.3×10^4 y

31 $^{239}_{94}\text{Pu} \rightarrow {}^{235}_{92}\text{U} + {}^{4}_{2}\alpha + Q$, $Q = 5.24$ MeV, $K_\alpha = 5.15$ MeV, $K_{^{235}\text{U}} = 89.2$ keV

33 (*a*) $\lambda = 0.133\ h^{-1}$, $t_{1/2} = 5.20$ h, (*b*) $N_0 = 3.11 \times 10^6$

35 (*a*) 1.13 mCi, (*b*) 0.898 mCi

37 About 15 g

39 7.0×10^8 y

41 (*a*) -0.764 MeV, (*b*) 3.27 MeV

43 (*a*) 0.156 MeV, (*b*) The masses given are for atoms, not nuclei, so the atomic masses are too large by the atomic number multiplied by the mass of an electron. For the given nuclear reaction, the mass of the carbon atom is too large by $6m_e$ and the mass of the nitrogen atom is too large by $7m_e$. Subtracting $6m_e$ from both sides of the reaction equation leaves an extra electron mass on the right. Not including the mass of the beta particle (electron) is mathematically equivalent to explicitly subtracting $1m_e$ from the right side of the equation.

45 $1.56 \times 10^{19}\ s^{-1}$

47 208 MeV

49 3.2×10^{10} J

51 (*c*) $3.7 \times 10^{38}\ s^{-1}$, 5.0×10^{10} y

53 $\lambda = 0.069\ s^{-1}$

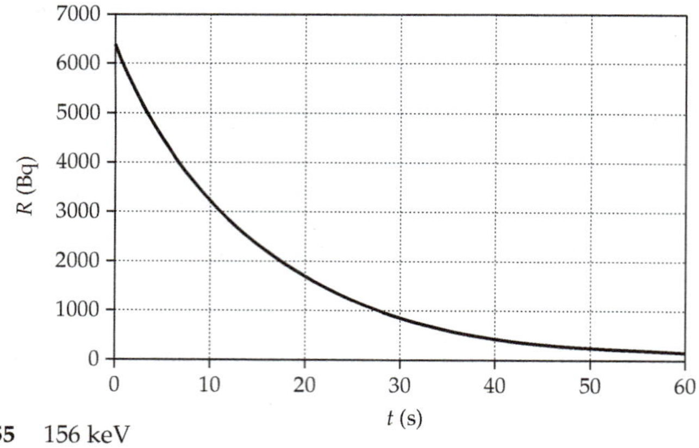

55 156 keV

59 6.7×10^3 Bq

61 6.3 L

63 (*a*) 23 MeV, (*b*) 4.2 GeV, (*c*) 1.3 GeV

65 (*a*)

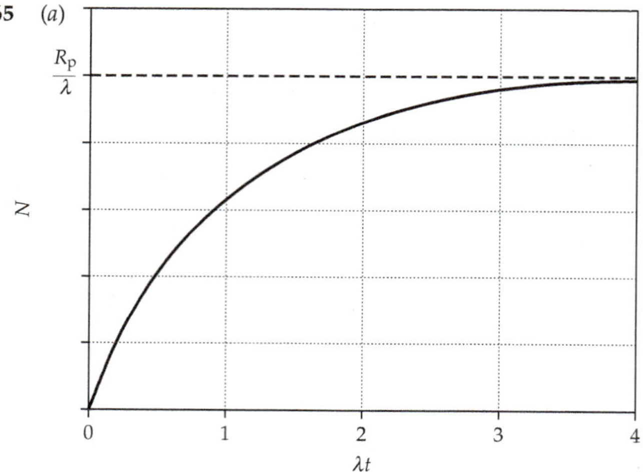

(*b*) 8.7×10^4

67 (*a*) 4.00 fm, (*b*) 310 MeV/*c*, (*d*) 310 MeV

69 (*a*) 1.188 MeV/*c*, (*b*) 752 eV, (*c*) 0.0962%

71 (*b*) 55

73 (*d*)

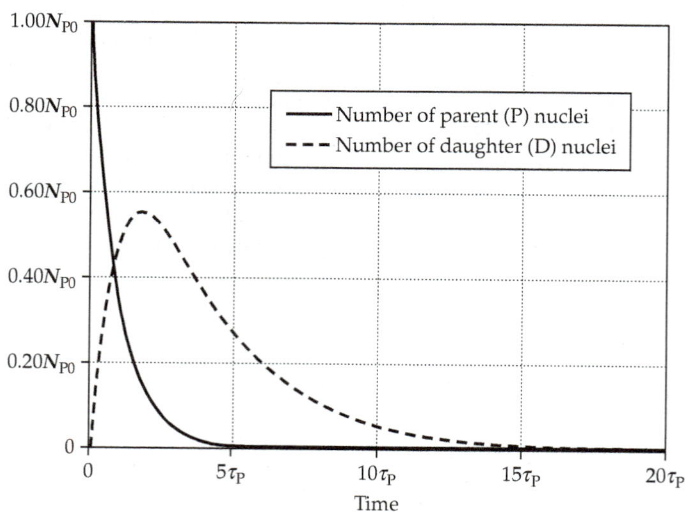

Chapter 41

1

Similarities	Differences
Baryons and mesons are hadrons, i.e., they participate in the strong interaction. Both are composed of quarks.	Baryons consist of three quarks and are fermions. Mesons consist of two quarks and are bosons. Baryons have baryon number +1 or −1. Mesons have baryon number 0.

3 A decay process involving the strong interaction has a very short lifetime (~10^{-23} s), whereas decay processes that proceed via the weak interaction have lifetimes of order 10^{-10} s.

5 False

7 No; from Table 41-3 it is evident that any quark–antiquark combination always results in an integral or zero charge.

9 (*a*) False, (*b*) True, (*c*) True

11 $\frac{F_{em}}{F_{grav}} = 1.24 \times 10^{36}$

13 (*a*) 279.2 MeV, (*b*) 1877 MeV, (*c*) 211.3 MeV

15 (*a*) Because $\Delta S = +1$, the reaction can proceed via the weak interaction, (*b*) Because $\Delta S = +2$, the reaction is not allowed, (*c*) Because $\Delta S = +1$, the reaction can proceed via the weak interaction.

17 (*a*) Because $\Delta S = +2$, the reaction is not allowed, (*b*) Because $\Delta S = +1$, the reaction can proceed via the weak interaction.

19 (*a*) K^0, (*b*) Σ^0 or Λ^0, (*c*) K^+

21

	Combination	*B*	*Q*	*S*	Hadron
(*a*)	*uud*	1	+1	0	p^+
(*b*)	*udd*	1	0	0	n
(*c*)	*uus*	1	+1	−1	Σ^+
(*d*)	*dds*	1	−1	−1	Σ^-
(*e*)	*uss*	1	0	−2	Ξ^0
(*f*)	*dss*	1	−1	−2	Ξ^-

23 From Table 41-3 we see that to satisfy the properties of charge number equal to +2 and strangeness, charm, topness, and bottomness all equal to zero, the quark combination must be *uuu*.

25 (*a*) $c\bar{d}$, (*b*) $\bar{c}d$

27 (*a*) *uds*, (*b*) $\bar{u}\bar{u}\bar{d}$, (*c*) *dds*

29 (*a*) *sss*, (*b*) *ssd*

31 $3.3 \times 10^8\ c\cdot y$

35 (*a*) Baryon number and lepton numbers are conserved quantities. A particle and its antiparticle must have baryon numbers that add to zero and lepton numbers that add to zero. Thus, for a particle and its antiparticle to be identical, its baryon number and all three of its lepton numbers must equal zero. This means it cannot be a lepton or a baryon, so it must be a meson. A particle and its antiparticle have the

complementary quark content. That is, if each quark in a particle is replaced by its antiquark, then the resulting entity is the antiparticle of the particle.
(*b*) The quark combination for the π^0 is a linear combination of $u\overline{u}$ and $d\overline{d}$ and the quark combination for the $\overline{\pi}^0$ is a linear combination of $\overline{u}u$ and $\overline{d}d$. The quark combination for the Ξ^0 is uss and that of the $\overline{\Xi}^0$ is $\overline{u}\,\overline{s}\,\overline{s}$.
(*c*) The π^0 is a meson with quark content of a linear combination of $u\overline{u}$ and $d\overline{d}$, so the π^0 is its own antiparticle.
The Ξ^0 is a baryon. As is explained in the answer to Part (*a*), a baryon cannot be its own antiparticle.

37 (*a*) The u and $\overline{u}$ annihilate, resulting in the photons.
(*b*) Two or more photons are required to conserve linear momentum.

39 (*a*) π^+, (*b*) -815 MeV, (*c*) 1.98 GeV

41 (*a*) 38 MeV, (*b*) 6.72, (*c*) 5 MeV, 33 MeV

Index

If a page number is followed by an *n*, the entry is in a footnote.

Physical Constants*

Atomic mass constant	$m_u = \frac{1}{12}m(^{12}C)$	$1\ u = 1.660\ 538\ 86(28) \times 10^{-27}$ kg
Avogadro's number	N_A	$6.022\ 1415(10) \times 10^{23}$ particles/mol
Boltzmann constant	$k = R/N_A$	$1.380\ 6505(24) \times 10^{-23}$ J/K $8.617\ 343(15) \times 10^{-5}$ eV/K
Bohr magneton	$m_B = e\hbar/(2m_e)$	$9.274\ 009\ 49(80) \times 10^{-24}$ J/T = $5.788\ 381\ 804(39) \times 10^{-5}$ eV/T
Coulomb constant	$k = 1/(4\pi\epsilon_0)$	$8.987\ 551\ 788 \ldots \times 10^9\ N \cdot m^2/C^2$
Compton wavelength	$\lambda_C = h/(m_e c)$	$2.426\ 310\ 238(16) \times 10^{-12}$ m
Fundamental charge	e	$1.602\ 176\ 53(14) \times 10^{-19}$ C
Gas constant	R	$8.314\ 472(15)$ J/(mol · K) = $1.987\ 2065(36)$ cal/(mol · K) = $8.205\ 746(15) \times 10^{-2}$ L · atm/(mol · K)
Gravitational constant	G	$6.6742(10) \times 10^{-11}\ N \cdot m^2/kg^2$
Mass of electron	m_e	$9.109\ 3826(16) \times 10^{-31}$ kg = $0.510\ 998\ 918(44)\ MeV/c^2$
Mass of proton	m_p	$1.672\ 621\ 71(29) \times 10^{-27}$ kg = $938.272\ 029(80)\ MeV/c^2$
Mass of neutron	m_n	$1.674\ 927\ 28(29) \times 10^{-27}$ kg = $939.565\ 360(81)\ MeV/c^2$
Magnetic constant (permeability of free space)	μ_0	$4\pi \times 10^{-7}\ N/A^2$
Electric constant (permittivity of free space)	ϵ_0	$= 1/(\mu_0 c^2) = 8.854\ 187\ 817 \ldots \times 10^{-12}\ C^2/(N \cdot m^2)$
Planck's constant	h	$6.626\ 0693(11) \times 10^{-34}$ J · s = $4.135\ 667\ 43(35) \times 10^{-15}$ eV · s
	$\hbar = h/(2\pi)$	$1.054\ 571\ 68(18) \times 10^{-34}$ J · s = $6.582\ 119\ 15(56) \times 10^{-16}$ eV · s
Speed of light	c	$2.997\ 924\ 58 \times 10^8$ m/s
Stefan-Boltzmann constant	σ	$5.670\ 400(40) \times 10^{-8}\ W/(m^2 \cdot K^4)$

*The values for these and other constants can be found in Appendix B as well as on the Internet at http://physics.nist.gov/cuu/Constants/index.html. The numbers in parentheses represent the uncertainties in the last two digits. (For example, 2.044 43(13) stands for 2.044 43 ± 0.000 13.) Values without uncertainties are exact. Values with ellipses are exact (like the number π = 3.1415. . .), but are not completely specified.

Derivatives and Definite Integrals

$$\frac{d}{dx}\sin ax = a\cos ax \qquad \int_0^\infty e^{-ax}\,dx = \frac{1}{a} \qquad \int_0^\infty x^2 e^{-ax^2}\,dx = \frac{1}{4}\sqrt{\frac{\pi}{a^3}}$$

$$\frac{d}{dx}\cos ax = -a\sin ax \qquad \int_0^\infty e^{-ax^2}\,dx = \frac{1}{2}\sqrt{\frac{\pi}{a}} \qquad \int_0^\infty x^3 e^{-ax^2}\,dx = \frac{4}{a^2}$$

$$\frac{d}{dx}e^{ax} = ae^{ax} \qquad \int_0^\infty xe^{-ax^2}\,dx = \frac{2}{a} \qquad \int_0^\infty x^4 e^{-ax^2}\,dx = \frac{3}{8}\sqrt{\frac{\pi}{a^5}}$$

The a in the six integrals is a positive constant.

Vector Products

$\vec{A} \cdot \vec{B} = AB\cos\theta$ $\qquad$ $\vec{A} \times \vec{B} = AB\sin\theta\,\hat{n}$ ($\hat{n}$ obtained using right-hand rule)

For additional data, see the following tables in the text.

Geometry and Trigonometry

$C = \pi d = 2\pi r$	definition of π
$A = \pi r^2$	area of circle
$V = \frac{4}{3}\pi r^3$	spherical volume
$A = \partial V/\partial r = 4\pi r^2$	spherical surface area
$V = A_{\text{base}}L = \pi r^2 L$	cylindrical volume
$A = \partial V/\partial r = 2\pi r L$	cylindrical surface area

$o = h\sin\theta$

$a = h\cos\theta$

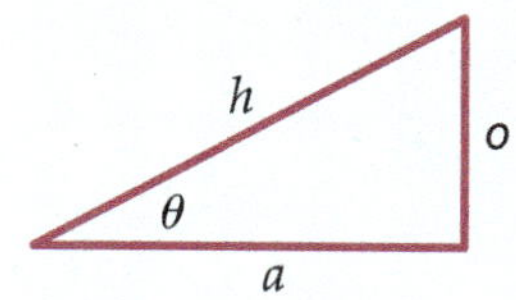

$$\sin^2\theta + \cos^2\theta = 1$$

$$\sin(A \pm B) = \sin A\cos B \pm \cos A\sin B$$

$$\cos(A \pm B) = \cos A\cos B \mp \sin A\sin B$$

$$\sin A \pm \sin B = 2\sin[\tfrac{1}{2}(A \pm B)]\cos[\tfrac{1}{2}(A \mp B)]$$

$\sin\theta \equiv y$

$\cos\theta \equiv x$

$\tan\theta \equiv \dfrac{y}{x}$

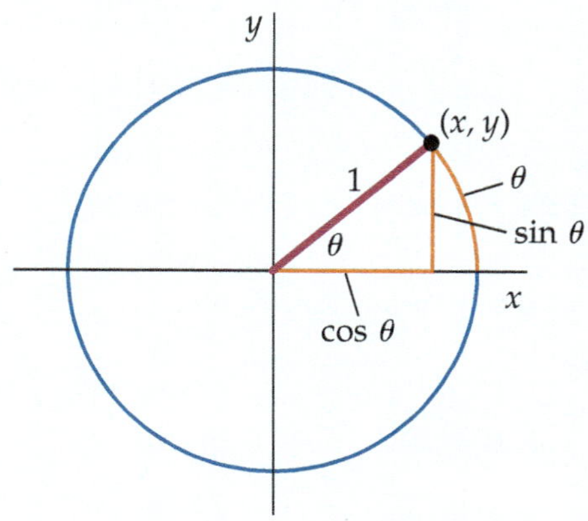

If $|\theta| << 1$, then

$\cos\theta \approx 1$ and $\tan\theta \approx \sin\theta \approx \theta$ (θ in radians)

Quadratic Formula

If $ax^2 + bx + c = 0$, then $x = \dfrac{-b \pm \sqrt{b^2 - 4ac}}{2a}$

Binomial Expansion

If $|x| < 1$, then $(1 + x)^n =$

$$1 + nx + \frac{n(n-1)}{2!}x^2 + \frac{n(n-1)(n-2)}{3!}x^3 + \ldots$$

If $|x| << 1$, then $(1 + x)^n \approx 1 + nx$

Differential Approximation

If $\Delta F = F(x + \Delta x) - F(x)$ and if $|\Delta x|$ is small,

then $\Delta F \approx \dfrac{dF}{dx}\Delta x$.